4th International Conference on New Frontiers in Physics 2015

AA004553

EPJ Web of Conferences Volume 126 (2016)

Crete, Greece
23 - 30 August 2015

Part 1 of 2

Editors:

Larissa Bravina
Yiota Foko
Sonja Kabana

ISBN: 978-1-5108-3238-1

Printed from e-media with permission by:

Curran Associates, Inc.
57 Morehouse Lane
Red Hook, NY 12571

Some format issues inherent in the e-media version may also appear in this print version.

Printed by Curran Associates, Inc. (2016)

For additional information, please contact EDP Sciences – Web of Conferences at the address below.

EDP Sciences – Web of Conferences
17, Avenue du Hoggar
Parc d'Activité de Courtabœuf
BP 112
F-91944 Les Ulis Cedex A
France

Phone: +33 (0) 1 69 18 75 75
Fax: +33 (0) 1 69 28 84 91

contact-edps@webofconferences.org

Additional copies of this publication are available from:

Curran Associates, Inc.
57 Morehouse Lane
Red Hook, NY 12571 USA
Phone: 845-758-0400
Fax: 845-758-2633
Email: curran@proceedings.com
Web: www.proceedings.com

TABLE OF CONTENTS

PART 1

PART 2

4th International Conference on New Frontiers in Physics

New Frontiers in Physics ICNFP 2015

23-30 August 2015, Crete, Greece

https://indico.cern.ch/e/icnfp2015/

International Advisory Committee

Halina Abramowicz (TAU, Tel Aviv, Israel)
Charalampos Anastasiou (ETH, Zurich, Switzerland)
Ignatios Antoniades (Ecole Polytechnique, Paris, France)
Marco Genovese (INRIM, Torino, Italy)
Rohini Godbole (IISc, Bangalore, India)
Elias Kiritsis (APC, Paris, France & UOC, Heraklion, Greece)
Shahn Majid (QML, London, UK)
Angela V. Olinto (UChicago, Illinois, USA)
Valery Rubakov (INR, Moscow, Russia)
Valerio Scarani (NUS, Singapore)
Eduard Shuryak (SBU, New York, USA)
Emmanuel Tsesmelis (CERN, Geneva, Switzerland & Univ. of Oxford, UK)
Urs Wiedemann (CERN, Geneva, Switzerland)

Main topics of the Conference

High Energy Particle Physics
Heavy Ion Physics
Superconductivity, Critical Phenomena
Quantum Physics, Quantum Entanglement
Quantum Field Theory, String Theory
Cosmology, Astrophysics, Gravity
Mathematical Physics

Local Organizing Committee

Evgeny Andronov (SPbSU, Saint Petersburg, Russia)
Panagiotis Charitos (CERN, Switzerland)
Laszlo Csernai (UiB, Bergen, Norway)
Nikolay Gulitskiy (SPbSU, Saint Petersburg, Russia)
Nikos Kallithrakas (TU, Chania, Greece)
Alisa Katanaeva (SPbSU, Saint Petersburg, Russia)
Elias Kiritsis (UoC, Heraklion, Greece & U. Paris 7, Paris, France)
Adam Kisiel (WUT, Warsaw, Poland)
Vladimir Kovalenko (SPbSU, Saint Petersburg, Russia)
Emanuela Larentzakis (OAC, Kolymbari, Greece)
Patricia Mage-Granados (CERN, Geneva, Switzerland, conference secretary)
Anton Makarov (SPbSU, Saint Petersburg, Russia)
Dmitrii Neverov (SPbSU, Saint Petersburg, Russia)
Ahmed Rebai (Complétude Ac., Nantes, France)
Andrey Seryakov (SPbSU, Saint Petersburg, Russia)
Daria Shukhobodskaia (SPbSU, Saint Petersburg, Russia)
Inna Shustina (Kiev, Ukraine)

Conference Chairs

Larissa Bravina (UIO, Oslo, Norway)
Yiota Foka (GSI, Darmstadt, Germany)
Sonia Kabana (Univ. of Nantes & Subatech, Nantes, France)

CERN EMMI GSI FAIR Subatech UNIVERSITÉ DE NANTES Saint Petersburg State University www.spbu.ru Technical University of Crete

4th International Conference on New Frontiers in Physics
23 – 30 August 2015, Kolymbari, Crete, Greece

Conference webpage: **https://indico.cern.ch/event/icnfp2015/**

International Advisory Committee

Halina Abramowicz (Tel Aviv University, Israel)
Charalampos Anastasiou (ETH Zurich, Switzerland)
Ignatios Antoniades (Ecole Polytechnique, Paris, France)
Marco Genovese (INRIM, Torino, Italy)
Rohini Godbole (Indian Institute of Science, Bangalore, India)
Elias Kiritsis (APC and University of Crete, Greece)
Shahn Majid (Queen Mary, University of London, UK)
Angela V. Olinto (University of Chicago, USA)
Valery Rubakov (INR, Moscow, Russia)
Valerio Scarani (National University of Singapore, Singapore)
Eduard Shuryak (Stony Brook University, USA)
Emmanuel Tsesmelis (CERN, Switzerland and University of Oxford, UK)
Urs Wiedemann (CERN, Switzerland)

Local Organising Committee

Evgeny Andronov (Saint Petersburg State University, Russia)
Panagiotis Charitos (CERN, Switzerland)
Laszlo Csernai (UiB, Bergen, Norway)
Nikolay Gulitskiy (Saint Petersburg State University, Russia)
Malgorzata Anna Janik (WUT, Warsaw, Poland)
Nikos Kallithrakas (TU, Chania, Greece)
Alisa Katanaeva (Saint Petersburg State University, Russia)
Elias Kiritsis (UoC, Heraklion, Greece and U. Paris 7, Paris, France)
Vladimir Kovalenko (Saint Petersburg State University, Russia)
Emanuela Larentzakis (OAC, Kolymbari, Greece)
Stefanos Leontsinis (National Technical University of Athens, Greece)
Patricia Mage-Granados (conference secretary, CERN, Switzerland)
Anton Makarov (Saint Petersburg State University, Russia)
Dmitrii Neverov (Saint Petersburg State University, Russia)
Ahmed Rebai (Completude Ac., Nantes, France)
Andrey Seryakov (Saint Petersburg State University, Russia)
Daria Shukhobodskaia (Saint Petersburg State University, Russia)
Inna Shustina (Kiev, Ukraine)

Conference Chairs

Larissa Bravina (University of Oslo, Norway)
Yiota Foka (GSI, Germany)
Sonia Kabana (University of Nantes and Subatech, France)

Organising Institutes and Sponsors

CERN - European Organization for Nuclear Research
EMMI - ExtreMe Matter Institute
GSI - Helmholtzzentrum fur Schwerionenforschung
FAIR - Facility for Antiproton and Ion Research
University of Oslo
Subatech
University of Nantes
Warsaw University of Technology
Saint Petersburg State University
Technical University of Crete
University of Bergen
University of Crete

Preface

4th International Conference on New Frontiers in Physics 2015

The conference series "New Frontiers in Physics" aims to promote scientific exchange and development of novel ideas in science with particular emphasis on interdisciplinarity. The main topics of the conference are high energy particle and nuclear physics, heavy ion physics, quantum optics and quantum entanglement, cosmology, astrophysics, gravity, mathematical physics.

The 4th International Conference on New Frontiers in Physics (ICNFP 2015) attracted about 200 participants. At the opening of the Conference Dr Konstantinos Zormpas, the General Director of the Academy which hosted the conference, welcomed the participants. During the conference 6 lectures, 81 plenary and 93 parallel talks as well as 23 posters were presented. Altogether during ICNFP 2015 there were given 203 oral or poster presentations. A special session devoted to the 50th anniversary of Hagedorn's Temperature and the Statistical Bootstrap Model has been organized on the 26th of August, with an introductory opening talk by Prof. Jan Rafelski.

A science exhibition and outreach day has been organized in the Sailing Club of Chania on the 28th August, with Opera singing by the Soprano Kalliopi Petrou and a public talk on "Black holes" ("Μικροσκοπικες Μελανες Οπες, Ενα νεο παραθυρο στο Συμπαν") by Prof. Ilias Kyritsis in Greek language in the evening. In addition, there was a Public talk in English language with title "The Old One does not throw dice, Einstein's dialog with God" by Prof. Thomas Naumann, visit to the "Park for the Preservation of Flora & Fauna", an Opera Gala (Kalliopi Petrou (soprano), Alessia Toffanin (piano)), classical music concert by Svetlana Nor (violin), Vladimir Nor (Cello) and Ruben Muradyan (piano), as well as excursions and a conference dinner with traditional Cretan Music. The "History of Crete" was presented by Emanuela Larentzakis as an after dinner talk.

We would like to thank all the conference participants as well as the members of the International Advisory Committee, the Local Organizing Committee, the Organizing Institutes and Sponsors, the Technical University of Chania, the University of Crete at Herakleion, Christina Kourkoumelis from University of Athens, the "Go-lab European Outreach project" and the Chania Sailing Club for the organization of public events and outreach, and especially the hosting Orthodox Academy of Crete for the extremely warm atmosphere of the conference. We would also like to thank Nikolay Gulitsky and Andrey Seryakov from Saint-Petersburg State University for the photos in the preface and the conference webpage and Vladimir Kovalenko from Saint-Petersburg State University and the staff of "European Physical Journal Web of Conferences" for the help with publishing the proceedings.

Larissa Bravina, Yiota Foka and Sonia Kabana,
Editors

DOI: 10.1051/epjconf/201612600001

edp sciences EPJ Web of Conferences

Statement of Peer review

In submitting conference proceedings to EPJ Web of Conferences, I certify to the Publisher that I adhere to the **Policy on Publishing Integrity** of the journal in order to safeguard good scientific practice in publishing.

1. All articles have been subjected to peer review administered by the proceedings editors.

2. Reviews have been conducted by expert referees, who have been requested to provide unbiased and constructive comments aimed, whenever possible, at improving the work.

3. Proceedings editors have taken all reasonable steps to ensure the quality of the materials they publish and their decision to accept or reject a paper for publication has been based only on the merits of the work and the relevance to the journal.

Title, date and place of the conference

International Conference on New Frontiers in Physics, 23-30.08.2015
OAC, Kolymbari, Crete, Greece

Proceedings editor(s):

Larissa Bravina, Yiota Foka, Sonja Kabana

Date and editor's signature

April, 19th, 2016

EPJ Web of Conferences **126**, 01001 (2016) DOI: 10.1051/epjconf/201612601001
ICNFP 2015

Highlights from the Telescope Array

J.N. Matthews[1,a] for the Telescope Array Collaboration

[1] *University of Utah, High Energy Astrophysics Institute, Department of Physics and Astronomy, Salt Lake City, UT 84112 USA*

Abstract. The Telescope Array measures the properties of ultra high energy cosmic ray induced extensive air showers. We do this using a variety of techniques including an array of scintillator detectors to sample the footprint of the air shower when it reaches the Earth's surface and telescopes to measure the fluorescence and Cerenkov light of the air shower. From this we determine the energy spectrum and chemical composition of the primary particles. We also search for sources of cosmic rays and anisotropy. We have found evidence of a possible source of ultra high energy cosmic rays in the northern sky. The experiment and its most recent measurements will be discussed.

1 Introduction

The Telescope Array collaboration was forged by members of the High Rsolution Fly's Eye (HiRes) and Akeno Giant Air Shower Array (AGASA) collaborations. Its purpose is to study ultra high energy cosmic rays, their spectrum, composition, and anisotropy/sources. Additional goals included understanding the differences in the observations of HiRes and AGASA as well as the transition from galactic to extra-galactic sources. The collaboration has, over time, grown to include members from the US, Japan, Russia, South Korea, and Belgium.

The Telescope Array is located in central Utah in the US at about 39.30° north latitude and 112.91° west longitude. The high energy component of the experiment consists of 38 telescopes, (9728 PMTs) at three telescope stations overlooking an array of 507 scintillator surface detectors (SD). The SD array has an area greater than 700 km^2. The main body of the Telescope Array was complete and operational as of 2008. The layout of the TA experiment is shown in Figure 1.

We use the atmosphere as a part of our calorimeter and observe the energy deposited by the primary cosmic ray and the extensive air shower it initiates. Telescopes view the longitudinal development of the air shower as it passes through the atmosphere. The three telescope stations are located on the perimeter of the SD array (on a ~30 km triangle) and view the sky over it. The northern telescope station has 14 telescopes, each with a 5.2m^2 mirror. It re-utilizes telescopes from the High Resolution Fly's Eye, HiRes-I site. The southern two stations each have 12 telescopes, with 6.8m^2 mirrors. The new telescopes at these sites use FADC electronics, while the HiRes-I telescopes are using older Sample-and-Hold electronics. The cameras in the telscopes utilize 256 PMTs for pixels, each viewing about 1° of sky. The PMTs are hexagonal and are arranged in a 16×16 hexagonal close

[a] e-mail: jnm@cosmic.utah.edu

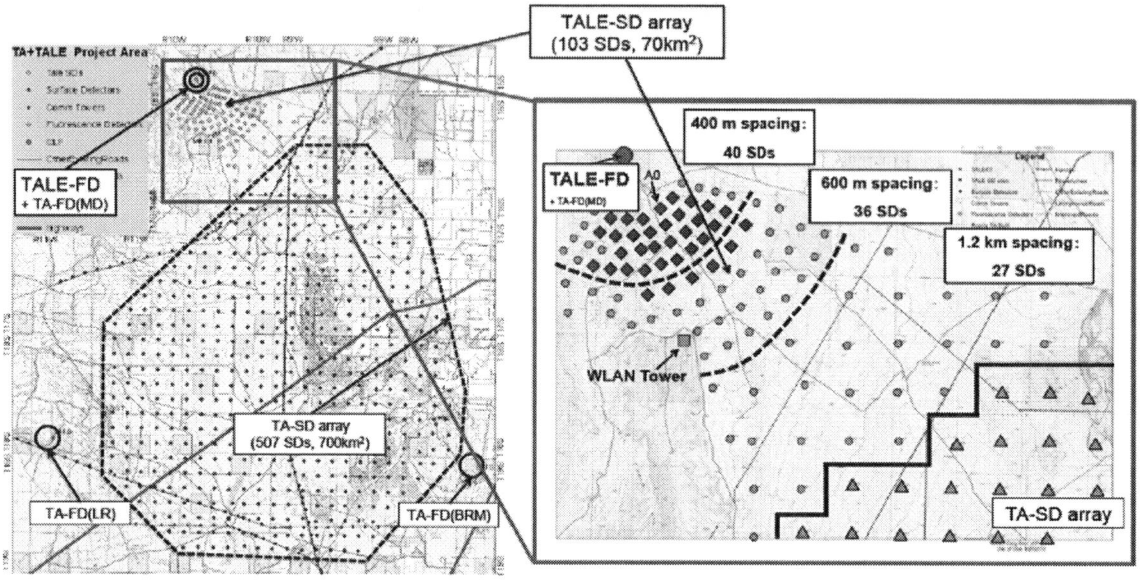

Figure 1. Layout of the Telescope Array detector site. The main array is shown on the left with a surface detector (SD) array of 507 scintillation counters surrounded by three fluorescence detector (FD) stations (MD: Middle Drum, BRM: Black Rock mesa, LR: Long Ridge) looking inward over the SD. The expanded inset on the right shows the layout of the TA Low-energy Extension (TALE). The TALE addition to Telescope Array consists of 10 new high-elevation ($31° - 59°$) fluorescence telescopes located at the MD site, and an in-fill SD array of 103 counters arranged with spacings that grows with distance from the TALE FD, in order to make optimal coverage at energies below 10^{18} eV.

pack (honey-comb) array. Each site views from 3-31° in elevation and has full azimuthal view over the SD array.

The telescopes are looking for an amount of light similar to a 50-100W UV light bulb, but 20-40 km away and moving at the speed of light. Due to the extreme sensitivity of the telescopes and cameras, the telescopes are only operated on clear moonless nights. This results in a duty cycle of about 10%.

The scintillator surface detector (SD) array, samples the density of air shower particles when they reach the Earth's surface. The 507 detectors each have two layers of half-inch (1.25 cm) thick scintillator extruded with grooves in it. Wavelengh shifting optical fibers are placed into the grooves. They gather the light, shift the wave length, and guide it to two PMTs, one for each optically separated layer. Each layer is $3m^2$ in area.

The scintillator surface detectors are deployed on a 1.2 km square grid. The are powered by solar panels and batteries for 24/7 continuous operations. They are self-calibrated by using the muon background and are read out by 2.4 GHz radio. Three radio towers divide the detectors among them. Each second the towers poll all of the detectors in their part of the array for signals, then a computer at the tower scans this for signs of a shower footprint. If a shower is found, then the tower requests that all data from the time period be sent back and recorded.

Each telescope station operates separately and independently from the SD array. The data from each can be analyzed separately or in conjunction with the other parts. Single telescope only data is known as "mono", from multiple telescopes "stereo" and telescope with SD is "hybrid". Having more information is always better since it constrains the data and reduces the uncertainties. However, one

DOI: 10.1051/epjconf/201612601001

does not always have that luxury - for example the SD data set is the largest since it operates 24/7 and the telescopes have a duty cycle of about 10%. Hence the SD ony data is used where one can tolerate the slightly larger uncertainties such as in the spectrum and anisotropy measurements. Hybrid and stereo have their own places, these are required to reduce uncertainties enough to make composition measurements.

2 Energy Spectrum

The sintillator array has the greatest number of events. Telescope Array previously published an energy spectrum from the first four years of SD operation [1]. The energy spectrum of ultra high energy cosmic rays is now updated to include data from the first seven years of data and is shown in figure 2. The Telescope Array measurements are in excellent agreement with the previous measurements by the HiRes experiment. Using the first 7 years data set, the break point for the GZK cut-off is found to be $10^{19.78\pm0.06}$ eV. Integrating an unbroken line from $10^{19.8}$ - 10^{21} eV, one would expect to find ~99 events in this range, while we actually observe 44 events. Hence, the significance of the cut-off is $\sim 6\sigma$.

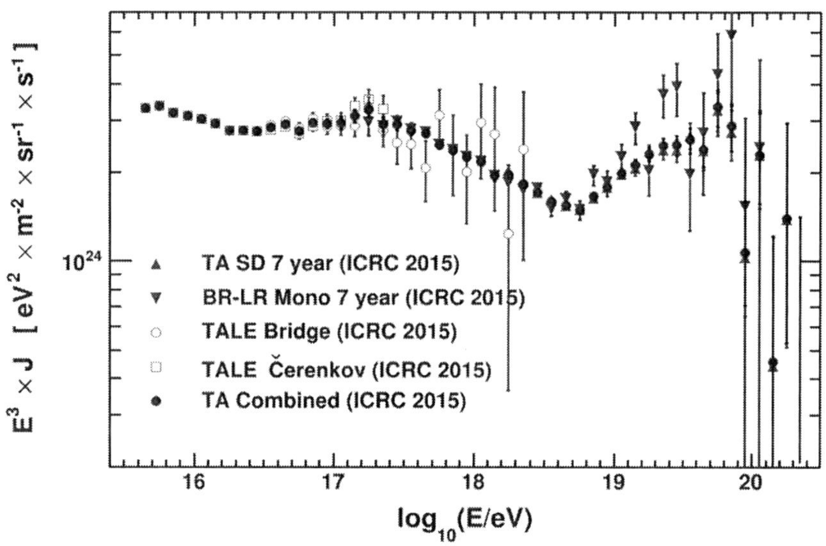

Figure 2. The combined spectrum from the Telescope Array Collaboration, including the first 7 years data from the surface detector spectrum (upright triangles), the 7 year monocular spectrum from the Black Rock and Long Ridge fluorescence detectors (inverted triangles), the TALE fluorescence spectrum (open circles), and the TALE Čerenkov spectrum (open squares). The averaged spectrum combining these four measurements is shown by the black circles.

A *dip* or *ankle* structure is also seen at $log(E/\text{eV}) = 18.70 \pm 0.02$. The simplest explanation for the combination of spectral features is that of interaction between extra-galactic cosmic ray protons and the cosmic microwave background radiation (CMBR) attenuating the flux via photo-pion production above the cut-off, and by e^+e^- pair production above the ankle[4]. The TA spectrum shows an apparent break at just above 10^{17} eV consistent with the "Second Knee" feature [5]. Another *dip* feature is seen just above the 10^{16} eV, which has been reported by other experiments [6, 7].

When the Auger is scaled up by 10% in energy (well within the energy uncertainties of both experiments, the Auger spectrum is also in good agreement with the Telescope Array and HiRes spectra at lower energies. However, for E>$10^{19.3}$ eV, the two spectra appear to diverge. The Auger spectrum has a significantly lower cut-off energy. Understanding this discrepancy is one area of joint study between the two groups.

3 Composition

We measure the composition with the Xmax technique. That is, we measure the depth of shower maximum in the atmosphere and we use monte carlo models to correlate this with chemical composition of the pimary cosmic ray. Protons have a smaller cross-section and interact deeper than heavier nucleii. The distribution also has more width to it. However, there is considerable overlap to the distributions, so one mostly looks at mean Xmax behavier and witdth behavier. These shift as a function of energy and this is called elongation.

To measure composition, one requires hybrid or stereo data in order to constrain the geometry and thereby to make a well understood measurement of Xmax, the depth of shower maximum. One advantage of stereo data is that it also provides a redundant measurement of Xmax which allows a resolution verification measurement. With hybrid, one extends the timing information in the angle vs time plots which one uses to measure the pointing direction of the incoming cosmic ray. In addition, one gets the core location of the shower footprint on the Earth's surface. These additional constraints tie down the geometry.

As of 2010, the two main measurements of composition of ultra high energy cosmic rays came from HiRes and Auger. The HiRes experiment found a composition which was very light, consistant with protons between $10^{18.2}$ and $10^{19.7}$ eV. The HiRes data is, in fact, in quite good agreement with the QGS-jet-01 proton model. The Auger experiment found that the composition was light in the lower energy range, but shifted to a heavier and heavier composition for E>$\sim 3\times10^{18}$ eV. The Auger surface detector is composed of thick water tanks. These have different sensitivities for electrons and muons. The Auger experiment has decided that the EPOS hadronic model does the best job of modeleling its data. Unfortunately, there is a fairly wide dispersion between the various hadronic models and where they place the mean Xmax of protons, helium, nitrogen, and iron.

The HiRes measurement had been made using a stereo measurement from two telescope sites about 12km apart. The Auger measurement had been done useing hybrid measurements. The Telescope Array was now in a position to make the hybrid measurements to see if this contributed to any of the difference.

At the Telescope Array, we instituted a pattern recognition cut to verify that we were seeing both the rise and the fall of the shower development profile. With these cuts, the Xmax resolution is $\sim$25 gm/cm^2 over all energies. Figure 3 shows the data-Monte Carlo comparisons for the Telescope Array hybrid Xmax data with the QGSjet II-03 model for protons and iron. In all cases, especially when one divides the set up into energy bins, the data looks much more like protons than it does iron.

Plotting the mean as a function of energy results in what is called an elongation plot. The elongation plot for the hybrid data from the Middle Drum site is shown in figure 4. As in the data-monte Carlo comparisons for the energy slices, the data looks most like the QGSjet II-03 proton Monte Carlo.

Another way to display this information is in what we call a shift-plot. We determine how much we would have to shift the Xmax distribution to get the best agreement between data and Monte Carlo. In figure 5 one can see that to get agreement with the proton Monte Carlo requires a shift of only a few grams, well within the meaurement's uncertainty. However, a shift of more than 60 grams is required

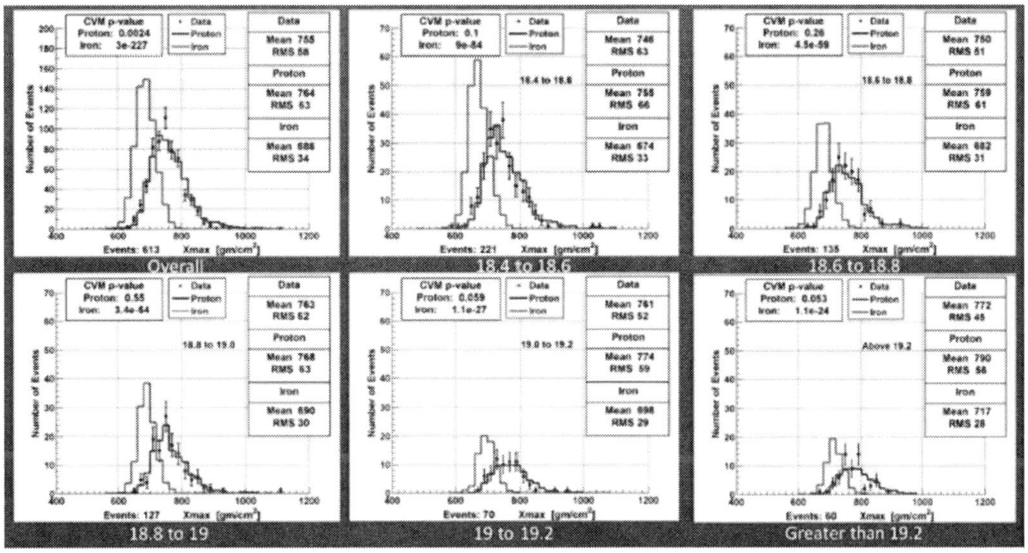

Figure 3. The comparison of the Telescope Array hybrid data with the QGSjet II-03 model. The top left histogram show the comparison of the Telescope Array data (black points) with the QGSjet II-03 model (blue for the proton model and red for the iron model) over all energies. The data is clearly more proton-like than iron-like. This is quantified by performing a Cramer von-Mises test (CvM) which yields a p-value of 0.0024 for protons, but a p-value of 10^{-227} for iron. The other five panels of the plot show the same plot, but broken up into energy bins. For all bins, including E>$10^{19.2}$ eV, the data clearly agrees much more closely with the protons than the iron. Even in that sub-set of data the CvM p-value for comparison to the proton model is 0.05, while it is 10^{-24} for iron.

for the iron Monte Carlo. In addition, the points on the plot are colored by their s-value, or probablity of agreement in the two shapes after the shifts in mean. One can see that over all energies the proton Monte Carlo has high s-values, though dropping a bit in the highest energy bin. However, over all energies, the iron Monte Carlo shows very low s-values. The data rejects iron. The same shift plot is shown for nitrogen which requires a shift of about 40 grams and also yields very low s-values over the energy range.

The two hybrid and the stereo measurements of Xmax are in good agreement with one another as well as indicating a predominately light composition. The interpretation of a light composition is fairly robust, as illustrated by the scales shown in Figure 6. To leading order, the departure of the data points from the proton line is roughly proportional to lnA where A is the nuclear mass number. Even if the scale were set to a hypothetical model that has a proton rail that is 33% deeper, the data points shown in the plot would only reach that expected for helium in this scenario. However, protons and helium are very different astrophysically. The two have much different interaction lengths. Protons with energies approaching 10^{20} eV have an interaction length of hundreds of Mpc, while helium with E>10^{19} eV has an interaction length which rapidly falls into the 10s of Mpc and lower. Simply, the helium can not reach us except from nearby sources.

4 Anisotropy

The Telescope Array collaboration has previously published that we have observed indications of a potential hotspot in the arrival direction of ultra high energy cosmic rays. [9] This publication used

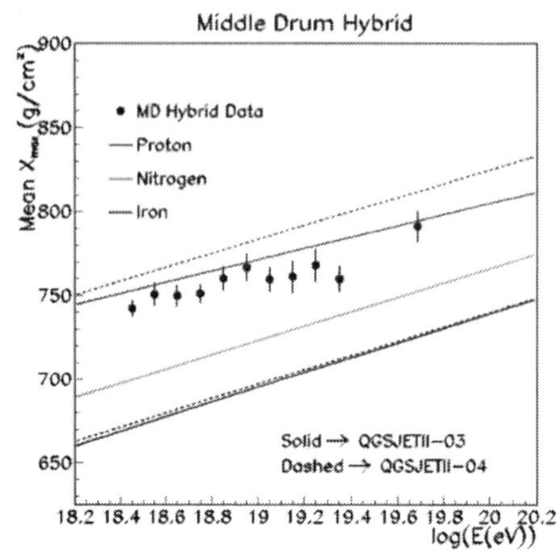

Figure 4. The Elongation Plot for the hybrid data from the Middle Drum site. The mean Xmax is plotted as a function of energy. The data is shown as the black points. The Monte Carlo results for two different hadronic models QGS-jet II-03 and QGSjet II-04 are shown for a) protons (red), b) nitrogen (green), and c) iron (blue). The data most closely resembles the QGSjet II-03 proton Monte Carlo.

the data from the first five years of SD operations. Using the Auger criteria for their published AGN correlations, we observe 72 events with E>5.7×10^{19} eV. These events are then over-sampled using 20° circles. We find that 19 of the 72 events overlap a region we call the hotspot which is centered at 146.7° in right accention and 43.2° declination. Thus, 26% of events are in about 6% of the observable area. We expect a background of 4.5 isotropic events in this area. The Li-Ma significance of this is about 5.2σ. Taking into account that this spot could appear anywhere in the sky gives us a chance probability estimate of 3.4σ.

The hotspot is about 19° off of the super-galactic plane, near Ursa Major. The Ursa Major super-cluster is extended more than ±10° from the super-galactic plane. Therefore, we can not rule out some sort of relationship between the hotspot and this super-cluster.

When we update the measurement by adding two additional years of data, there are 37 new events in the data set for a total of 109 events with E>5.7×10^{19} eV. There are 4 new events in the hotspot (3 in the 6th year and 1 in the 7th year). The expected background for this is 2.3 events. The global excess chance probabliity is 3.7×10^{-4}, which corresponds to 3.4σ the same as for the first 5 years. A K-S shows that the rate of arrival of events to the hotspot is consistant with the fluctuations expected from a Poisson distribution with a mean of 3.4 events/year. The rate of event arrivals is also inconsistant with chance excess from an isotropic distribution, a Poisson average of 0.9 events/year, at about the 2.6σ level. One interesting thing to do is to make a full-sky map combining the Telescope Array and Pierre Auger events with E>5.7×10^{19} eV. See figure 7. The plot is shown in Hammer-Aitoff projection, in Equatorial coordinates. No correction was made for the energy scale difference between the Telescope Array and Pierre Auger data sets. The thin gray line just above and left of the Telescope Array (upper) hotspot is the supergalactic plane. The pre-trial significance of this is 5.2σ (post-trial chance probablility of ~3.4σ). The Auger "warm spot" (lower) is located on the supergalactic plane at Centaurus-A and has a pretrial significance of ~3.6σ. Neither the Telescope Array, nor Pierre Auger data shows any sign of excess in the direction of Virgo.

The Auger group reported correlation of their high energy, E>5.7×10^{19} eV, events with the 472 AGNs from the Veron catalog with Z<0.018. The Telescope Array group has looked periodically for

 DOI: 10.1051/epjconf/201612601001

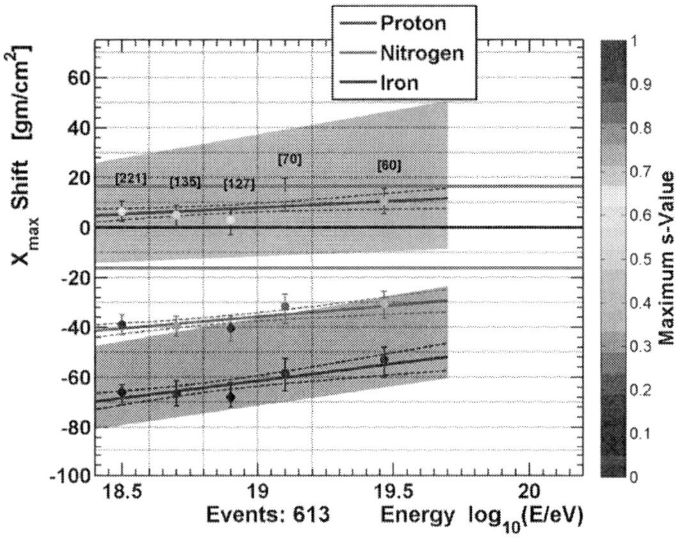

Figure 5. Shift plot showing how much one would have to shift Xmax in a given energy bin to make the mean Xmax of data align with the QGSjet II-03 Monte Carlo. Three examples are given for protons (red), iron (blue), and nitrogen (magenta). After each given shift, the goodness of fit is calulated and assigned an s-value. One can see that the proton Monte Carlo requires a small shift (well within uncertainties) and that the goodness of fit is also very good, though dropping a bit at the highest energies. The iron MC requires a shift of about 60 gm and even after that, the result has a very poor fit. The nitrogen MC does not fare much better than the iron MC. It requires a shift of about 40 gm and also has poor fit comparisons. The pink and blue bands show the spread due to using other hadronic models. Unfortunately, this spread is quite large.

this correlation, but have yet to find any significant correlation. The number of Telescope Array events correlating with AGNs is about 1σ greater than background.

We have also looked for correlation with the Large Scale Structure (LSS). This is done by comparing to the matter distribution observed by the 2MASS Galaxy redsift catalog (XSCz). The sky was divided into five intensity regions, these were then compared by a Kolmagorov-Smirnov test comparing to the expected flux distribution. For events with energies, E>10^{18} eV and E>4×10^{19} eV, the test can not distinguish between the Large Scale Structure and isotropy. However, for events E>5.7×10^{19} eV, the KS test indicated that the data is marginally incompatible with an isotropic source distribution and it is compatible with the LSS simulation.

5 The Future

The plan is to quadruple the size of the Telescope Array. We are adding 500 scintillator surface detectors to the North East and South East of the present Telescope Array site. See figure 8. The detectors will be on a grid a 2.08 km spacing and so will focus on the highest energy region. This part of the expansion has already been funded by the Japanese funding agencies. There will also be two new telescope stations which will view the sky over this expansion. A proposal was submitted in the fall of 2015 to the US NSF to fund this expansion. We hope to hear positive news in mid-2016. With this detector, we could collect the equivalent of 19 current Telescope Array SD years of data by 2020.

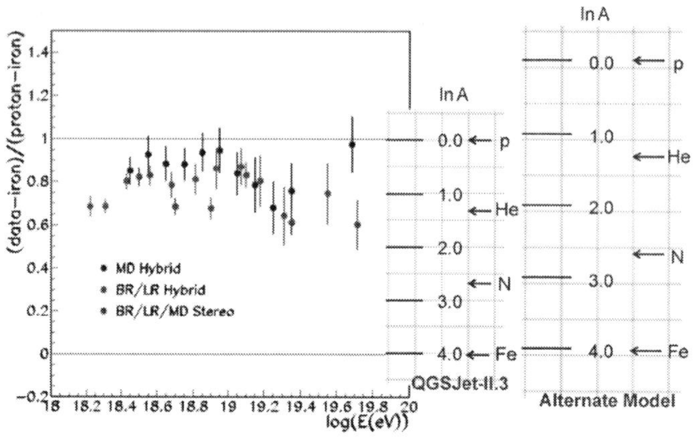

Figure 6. Summary of average X_{MAX} from Telescope Array fractional as the fractional position between the expectations of pure iron (lower lines at zero) and pure proton (upper line at 1.0) using the QGSjet-II03 hadronic model. At right is the illustrative interpretation, in terms of *lnA* of the summary plot against QGSjet-II03 and a hypothetical alternative model where the proton rail is about 33% deeper in X_{MAX}.

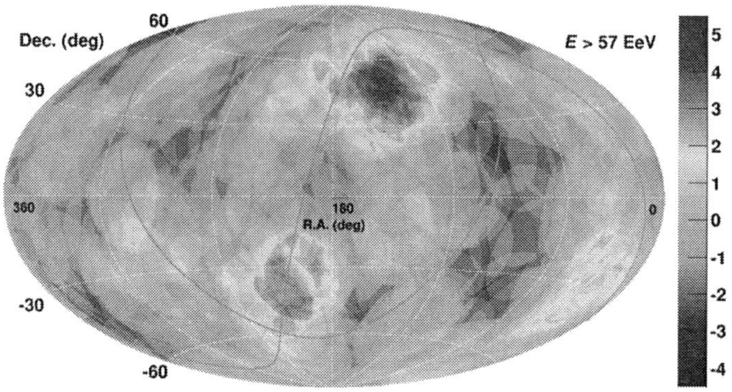

Figure 7. Full sky map combining the Telescope Array and Pierre Auger data events with E>5.7×10^{19} eV. The events have oversampling with a 20° radius circle. The Telescope Array data set includes 109 events, representing the first 7 years of data collection. The Auger data set includes 157 events, representing 10 years of data. No correction was made for the energy scale difference between the Telescope Array and Pierre Auger data sets.

With the new data from the TAx4, we will be able to quickly verify the hotspot and to begin searching for sub-structure within the hotspot. If there are really two separated sources there, we would begin to be able to resolve this.

6 Summary

The Telescope Array has measured the energy spectrum, composition, and arrival direction of ultra high energy cosmic rays in the northern hemisphere. The low energy extension, TALE is coming on line and we are pushing the energy threshold of the experiment down below 10^{16} eV. The Telescope Array and TALE have already measured the energy spectrum between 6×10^{15} eV to over 10^{20} eV with a single set of cross calibrated detectors. By looking at the composition and the energy spectrum in this range, we hope to identify the galactic to extragalactic transition. The surface array portion of

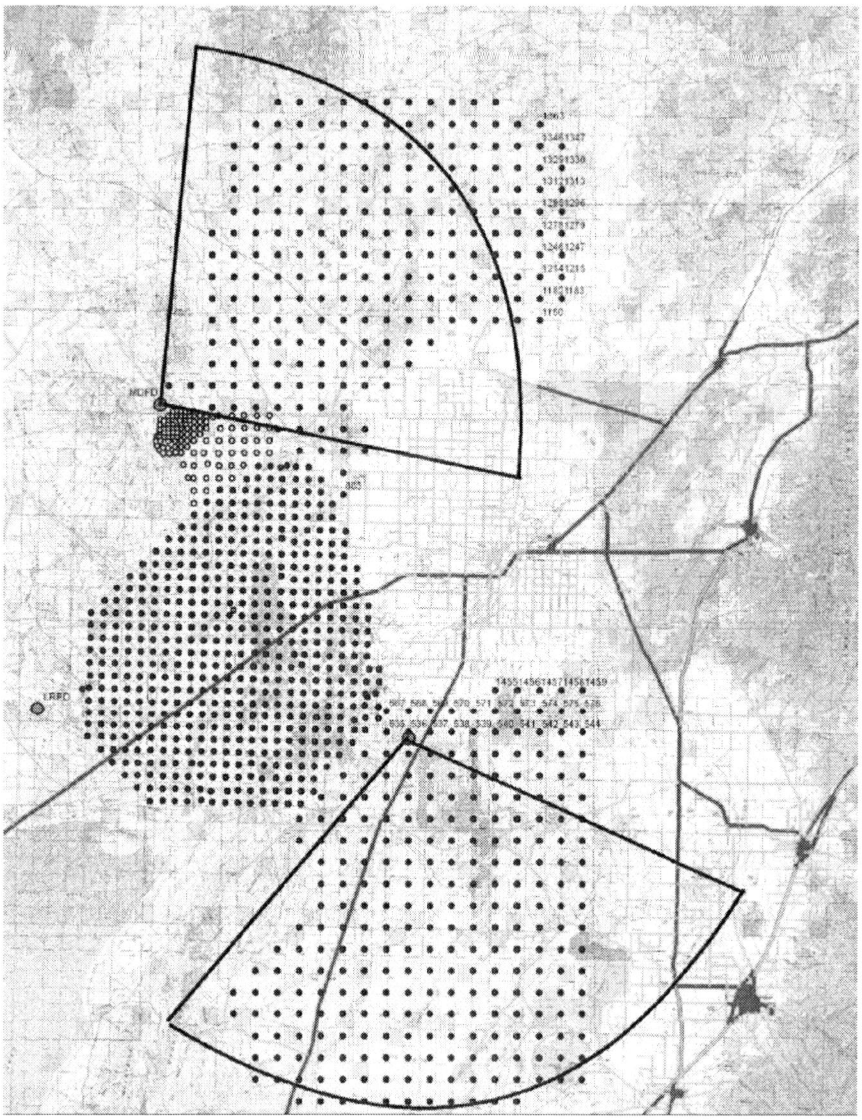

Figure 8. Map showing the Telescope Array expansion, TAx4. The main Telescope Array with 1.2 km spacing is shown as the red points. The TALE, Telescope Array Low Energy Extension is shown as yellow points. This is a graded array with 1.2, 0.6 and 0.4 km spacing. The TAx4 expansion is shown to the NE and SE of the main Telescope Array with the dark points at 2.08 km spacing. The large blue points indicate the telescope stations. New telescope shelters would be added to the northern and south-eastern telescope sites to view the sky over the TAx4 expansion.

TALE was recently funded and this will give us the capability to do hybird composition in this energy range. The composition of cosmis rays is consistant with a light composition for E>$10^{18.2}$ eV from both hybrid and stereo measurements. We have also reported on a hotspot near the direction of Ursa Major with a 3.4σ significance. We need more data to verify the source and to better to resolve it. More data is needed, but more data is coming with the expansion of TAx4.

Acknowledgements

The Telescope Array experiment is supported by the Japan Society for the Promotion of Science through Grants-in-Aids for Scientific Research on Specially Promoted Research (21000002) "Extreme Phenomena in the Universe Explored by Highest Energy Cosmic Rays" and for Scientific Research (19104006), and the Inter-University Research Program of the Institute for Cosmic Ray Research; by the U.S. National Science Foundation awards PHY-0307098, PHY-0601915, PHY-0649681, PHY-0703893, PHY-0758342, PHY-0848320, PHY-1069280, PHY-1069286, PHY-1404495 and PHY-1404502; by the National Research Foundation of Korea (2007-0093860, R32-10130, 2012R1A1A2008381, 2013004883); by the Russian Academy of Sciences, RFBR grants 11-02-01528a and 13-02-01311a (INR), IISN project No. 4.4502.13 and Belgian Science Policy under IUAP VII/37 (ULB). The foundations of Dr. Ezekiel R. and Edna Wattis Dumke, Willard L. Eccles and the George S. and Dolores Dore Eccles all helped with generous donations. The State of Utah supported the project through its Economic Development Board, and the University of Utah through the Office of the Vice President for Research. The experimental site became available through the cooperation of the Utah School and Institutional Trust Lands Administration (SITLA), U.S. Bureau of Land Management, and the U.S. Air Force. We also wish to thank the people and the officials of Millard County, Utah for their steadfast and warm support. We gratefully acknowledge the contributions from the technical staffs of our home institutions. An allocation of computer time from the Center for High Performance Computing at the University of Utah is gratefully acknowledged.

References

[1] T. Abu-Zayyad et al. (TA Collaboration), Astrophys. Journal Letters **768**, L1 (2013).
[2] K. Greisen, Phys. Rev. Lett. **16**, 222 (1966).
[3] G.T. Zatsepin and V.A. Kuz'min, JETPL **4**, 78 (1966).
[4] V. Berezinsky, A.Z. Gazizov, and S.I. Grigorieva, Physics Letters B **612**, 147 (2005).
[5] D.R. Bergman and J.W. Belz, J.Phys. **G34**, R359 (2007).
[6] W.D. Apel *et al.* (KASCADE-Grande Collaboration) Astropart. Phys. **36**, 183 (2012).
[7] M.G. Aartsen *et al.* (IceCube Collaboration) Phys. Rev. D **88**, 042004 (2013).
[8] R.U. Abbasi *et al.* (TA Collaboration), Astropart. Phys. **64**, 49 (2015).
[9] R.U. Abbasi *et al.* (TA Collaboration), Astrophysical Journal Lett. **790**, L21 (2014).
[10] T. Abu-Zayyad *et al.* (TA Collaboration), Astrophysical Journal **777**, 88 (2013).

EPJ Web of Conferences **126**, 01002 (2016) DOI: 10.1051/epjconf/201612601002
ICNFP 2015

Hamiltonian approach to QCD in Coulomb gauge: From the vacuum to finite temperatures

H. Reinhardt[1,a,b], D. Campagnari[1,c], J. Heffner[1,d], M. Quandt[1,e], and P. Vastag[1,f]

[1] *Universität Tübingen*
Institut für Theoretische Physik
Auf der Morgenstelle 14
D-72076 Tübingen

Abstract. The variational Hamiltonian approach to QCD in Coulomb gauge is reviewed and the essential results obtained in recent years are summarized. First the results for the vacuum sector are discussed, with a special emphasis on the mechansim of confinement and chiral symmetry breaking. Then the deconfinement phase transition is described by introducing temperature in the Hamiltonian approach via compactification of one spatial dimension. The effective action for the Polyakov loop is calculated and the order of the phase transition as well as the critical temperatures are obtained for the color group SU(2) and SU(3). In both cases, our predictions are in good agreement with lattice calculations.

1 Introduction

Understanding the QCD vacuum is still one of the big challenges of theoretical physics. Although substantial progress has been made during the last two decades we are far from rigorously understanding the essential features of the QCD vacuum: confinement and spontaneous breaking of chiral symmetry. Under normal conditions quarks and gluons are confined inside hadrons, and the hadrons receive most of their mass through the mechanism of spontaneous breaking of chiral symmetry. When baryonic matter is heated up or strongly compressed, hadrons melt and a plasma of quarks and gluons is formed. This deconfinement phase transition is accompanied by the restoration of chiral symmetry. We have learned a lot about this phase transition from lattice calculations at finite temperatures and vanishing baryon density, i.e. along the temperature axis of the phase diagram, see Fig. 1. However, the lattice approach fails at finite chemical potential due to the so-called sign problem (the quark determinant becomes complex). To understand the deconfinement phase transition at finite chemical potential continuum approaches are thus required, which do not face the sign problem and are able to deal with a non-zero chemical potential. I will present here such a continuum formulation, namely the

[a] *speaker*
[b] e-mail: hugo.reinhardt@uni-tuebingen.de
[c] e-mail: d.campagnari@uni-tuebingen.de
[d] e-mail: heffner@tphys.physik.uni-tuebingen.de
[e] e-mail: markus.quandt@uni-tuebingen.de
[f] e-mail: peter.vastag@uni-tuebingen.de

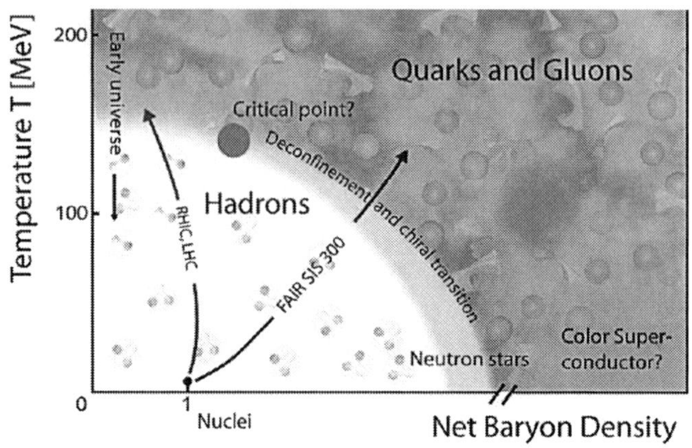

Figure 1: The phase diagram of QCD.

Hamiltonian approach. In fact, this method is currently being developed and tested at finite temperatures where it can be compared to available lattice calculations, and I will not yet present results for finite chemical potential. In the first part of my talk I will study the QCD vacuum within this approach and will then extend it to finite temperatures.

One may ask for the benefit of the Hamiltonian approach as compared to the more common functional integral approach, which is so elegant for perturbation theory and which is also the basis for the lattice calculations. From quantum mechanics we know that solving the Schrödinger equation is usually much simpler and much more efficient than calculating the functional integral directly. For example, it is quite easy to solve the Schrödinger equation for the hydrogen atom, while calculating the corresponding functional integral directly is exceedingly complicated. Also in non-perturbative continuum quantum field theory the Hamiltonian approach is in my opinion more efficient. In my talk I will use the Hamiltonian approach to QCD in Coulomb gauge first to study the QCD vacuum by means of a variational solution of the Yang–Mills Schrödinger equation. Then I will extend this approach to finite temperatures by compactifying a spatial dimension. I will first illustrate the method by means of free bosons and fermions and then apply it to QCD at finite temperatures, where I will mainly concentrate on the calculation of the effective potential of the Polyakov loop.

2 Hamiltonian approach to QCD in Coulomb gauge

The Hamiltonian approach to quantum field theory is based on the canonical quantization. We illustrate this formalism first for the Yang–Mills theory, which is defined by the classical action $S = \frac{1}{4} - \int \mathrm{d}^4x \, F^a_{\mu\nu} F^{a\mu\nu}$. Here, $F^a_{\mu\nu} = \partial_\mu A^a_\nu - \partial_\nu A^a_\mu + g f^{abc} A^b_\mu A^c_\nu$ denotes the non-Abelian field strength. For the canonical quantization, the gauge fields are considered as the "coordinates". The canonical conjugate momenta are introduced in the standard fashion by $\Pi^a_\mu(\boldsymbol{x}) = \delta S / \delta \dot{A}^{a\mu}(\boldsymbol{x})$. For the spatial components of the gauge field $A^a_i(\boldsymbol{x})$ these momenta coincide with the color electric field $\Pi^a_i(\boldsymbol{x}) = -E^a_i(\boldsymbol{x})$, while the momentum corresponding to the temporal component of the gauge field $A^a_0(\boldsymbol{x})$ vanishes. The latter causes problems in the canonical quantization which, however, can be circumvented by imposing Weyl gauge $A^a_0(\boldsymbol{x}) = 0$. One finds then the following classical Hamiltonian of Yang–Mills theory

$$H = \frac{1}{2} \int \mathrm{d}^3x \, (\boldsymbol{\Pi}^a(\boldsymbol{x}) \cdot \boldsymbol{\Pi}^a(\boldsymbol{x}) + \boldsymbol{B}^a(\boldsymbol{x}) \cdot \boldsymbol{B}^a(\boldsymbol{x})) \tag{1}$$

where $B_i^a = \frac{1}{2}\varepsilon_{ijk}F_{jk}^a$ is the color magnetic field. The theory is quantized by imposing the usual canonical commutation relations, i.e.

$$\left[A_k^a(\boldsymbol{x}), \Pi_l^b(\boldsymbol{y})\right] = \mathrm{i}\delta^{ab}\delta_{kl}\delta(\boldsymbol{x}-\boldsymbol{y}) \tag{2}$$

$$\left[A_k^a(\boldsymbol{x}), A_l^b(\boldsymbol{y})\right] = \left[\Pi_k^a(\boldsymbol{x}), \Pi_l^b(\boldsymbol{y})\right] = 0\,. \tag{3}$$

In the coordinate representation the momentum operator is therefore realized by $\Pi_k^a(\boldsymbol{x}) = \delta/(\mathrm{i}\delta A_k^a(\boldsymbol{x}))$.

Because of the Weyl gauge $A_0^a(\boldsymbol{x}) = 0$, Gauß's law is lost as an equation of motion and has to be imposed as a constraint on the wave functional

$$\hat{D}_k^{ab}(\boldsymbol{x})\Pi_k^b(\boldsymbol{x})\psi[A] = \rho_m^a(\boldsymbol{x})\psi[A]\,. \tag{4}$$

Here $\hat{D}_k^{ab}(\boldsymbol{x}) = \delta^{ab}\partial_k^x + gf^{acb}A_k^c(\boldsymbol{x})$ is the covariant derivative in the adjoint representation of the gauge group, and the operator on the left hand side, $\hat{\boldsymbol{D}}\cdot\boldsymbol{\Pi}$, is nothing but the generator of time-independent (space-dependent) gauge transformations. Furthermore, ρ_m is the color charge density of the matter field. In the absence of a matter field the right hand side of eq. (4) vanishes and the wave functional has to be invariant under time-independent gauge transformations $\psi[A^U] = \psi[A]$. With the canonical momentum replaced by the operator $\Pi_k^a(\boldsymbol{x}) = \delta/(\mathrm{i}\delta A_k^a(\boldsymbol{x}))$, eq. (1) becomes the Hamilton operator of Yang–Mills theory. With this operator one has to solve then the Schrödinger equation

$$H\psi[A] = E\psi[A] \tag{5}$$

in the Hilbert space of gauge invariant wave functionals, which has the usual scalar product known from quantum mechanics

$$\langle\phi|\ldots|\psi\rangle = \int \mathcal{D}A\,\phi^*[A]\ldots\psi[A]\,, \tag{6}$$

where $\int\mathcal{D}A$ denotes the functional integral over (time-independent) spatial components of the gauge field. One is mainly interested in the solution of the Schrödinger equation (5) for the ground state (vacuum). There have been attempts to solve the Schrödinger equation directly for the vacuum in the Hilbert space of gauge invariant wave functionals. However, so far these attempts have been successful only in 2+1 dimensions [1]. In 3+1 dimensions it seems extremly difficult to find gauge invariant representations of the wave functional except for some limited cases [2]. Therefore it is much more efficient to fix the gauge, and a convenient choice for this purpose is the Coulomb gauge $\boldsymbol{\partial}\cdot\boldsymbol{A} = 0$. This condition can be implemented in the scalar product (6) by the standard Faddeev-Popov method resulting in

$$\langle\phi|\ldots|\psi\rangle = \int \mathcal{D}A^\perp\, J(A^\perp)\phi^*[A^\perp]\ldots\psi[A^\perp]\,. \tag{7}$$

Here the integration is now over the transversal components of the gauge field $\boldsymbol{A}^\perp$ only and

$$J(A^\perp) = \mathrm{Det}(-\hat{\boldsymbol{D}}\cdot\boldsymbol{\partial}) \tag{8}$$

is the Faddeev-Popov determinant in Coulomb gauge. The Coulomb gauge fixing basically corresponds to a transition from Cartesian coordinates to curvilinear coordinates and the Faddeev-Popov determinant represents the Jacobian of this transformation.

Even if the longitudinal components of the gauge field are eliminated in Coulomb gauge the momentum operator still contains also a longitudinal component $\boldsymbol{\Pi}^{\|}$. From Gauß's law one finds for the longitudinal momentum operator when it acts on a wave functional

$$\boldsymbol{\Pi}^{\|}\,\psi[A] = -\boldsymbol{\partial}(-\hat{\boldsymbol{D}}\cdot\boldsymbol{\partial})^{-1}\rho\,\psi[A]\,, \qquad \rho^a = -f^{abc}A_k^b\Pi_k^c + \rho_m^a\,. \tag{9}$$

With this relation (9) one finds for the Yang–Mills Hamiltonian in Coulomb gauge [3]

$$H = \frac{1}{2}\int d^3x \left(J^{-1}\boldsymbol{\Pi}^a(\boldsymbol{x}) \cdot J\boldsymbol{\Pi}^a(\boldsymbol{x}) + \boldsymbol{B}^a(\boldsymbol{x}) \cdot \boldsymbol{B}^a(\boldsymbol{x})\right) + H_C$$
$$\equiv H_{YM} + H_C\,, \tag{10}$$

where

$$H_C = \frac{1}{2}\int d^3x \int d^3y\, J^{-1}\rho^a(\boldsymbol{x})J\left[(-\hat{\boldsymbol{D}} \cdot \partial)^{-1}(-\partial^2)(-\hat{\boldsymbol{D}} \cdot \partial)^{-1}\right]^{ab}(\boldsymbol{x},\boldsymbol{y})\rho^b(\boldsymbol{y}) \tag{11}$$

arises from the kinetic energy of the longitudinal components of the momentum operator and is usually referred to as the Coulomb term. Here $\rho(\boldsymbol{x})$ is the total color charge density, which includes besides the matter charge density $\rho_m(\boldsymbol{x})$ also the charge density of the gauge field $\rho_g^a(\boldsymbol{x}) = -f^{abc}A_k^b(\boldsymbol{x})\Pi_k^c(\boldsymbol{x})$, which exists obviously only in non-Abelian gauge theories. In the case of QED where the Faddeev-Popov operator becomes the Laplacian $(-\Delta)$, the term eq. (11) produces precisely the Coulomb potential between electric charge densities ρ.

The Coulomb gauge fixed Hamiltonian (10) is obviously more complicated than the original gauge invariant one (1). First, the Faddeev-Popov determinant appears in the kinetic energy of the transversal degrees of freedom [the first term in eq. (10)] and, second, the kinetic energy of the longitudinal degrees of freedom (11) is a highly non-local object. Furthermore, the Faddeev-Popov determinant appears also in the scalar product (7). This is the price one has to pay for the gauge fixing. Nevertheless it is still more convenient to work with this more complicated gauge fixed Hamiltonian than with gauge invariant wave functionals. Let us stress that by implementing Gauß's law in the gauge fixed Hamiltonian, all consequences of gauge invariance are fully taken into account.

3 Variational solution of the Yang–Mills Schrödinger equation for the vacuum

Variational calculations within the Hamiltonian approach to Yang–Mills theory in Coulomb gauge were initiated by D. Schütte in Ref. [4] who used a Gaussian trial ansatz for the vacuum wave functional. By minimizing the energy density he derived a closed set of coupled integral equations for the gluon propagator, the ghost propagator and the Coulomb potential. Later on, Ref. [5] followed the approach of Ref. [4] by using the same ansatz for the vacuum wave functional, but in addition provided the first numerical solution of the resulting integral equations. In Refs. [6, 7] we have developed in our group a variational approach to Yang–Mills theory in Coulomb gauge, which differs from previous approaches: i) in the ansatz for the vacuum wave functional, ii) in the treatment of the Faddeev-Popov determinant[1] and iii) in the renormalization. Our trial ansatz for the vacuum wave functional is given by

$$\psi[A] = \frac{1}{\sqrt{J(A)}}\exp\left[-\frac{1}{2}\int d^3x \int d^3y\, A_k^a(\boldsymbol{x})\omega(\boldsymbol{x},\boldsymbol{y})A_k^a(\boldsymbol{y})\right], \tag{12}$$

where $\omega(\boldsymbol{x},\boldsymbol{y})$ is a variational kernel, which is determined by minimizing the energy density. The ansatz (12) has the advantage that the static gluon propagator is given by the inverse of the kernel ω

$$\langle A_k^a(\boldsymbol{x})A_l^b(\boldsymbol{y})\rangle = \delta^{ab}t_{kl}(\boldsymbol{x})\frac{1}{2}\omega^{-1}(\boldsymbol{x},\boldsymbol{y}), \tag{13}$$

[1] In Refs. [4, 5] the Faddeev-Popov determinant was ignored which, however, turns out to be crucial for the infrared properties of the theory.

where $t_{kl}(\boldsymbol{x}) = \delta_{kl} - \partial_k^x \partial_l^x / \partial_x^2$ is the transversal projector. From the form of the static gluon propagator (13) it follows that the Fourier transform of $\omega(\boldsymbol{x}, \boldsymbol{y})$ represents the single-particle gluon energy. Minimizing the energy density with respect to $\omega(\boldsymbol{x}, \boldsymbol{y})$ yields the result shown in Fig. 2 (a) [8]. For large momenta the gluon energy behaves like the photon energy $\omega(p) \sim |\boldsymbol{p}|$ while at small momenta it diverges $\omega(p) \sim 1/|\boldsymbol{p}|$ implying the absence of free gluons in the infrared, which is a manifestation of confinement. The numerical calculation carried out in Ref. [9] shows that the Coulomb term (11) can be ignored in the Yang–Mills sector. Then the gluon gap equation resulting from the minimization of the energy density is given by

$$\omega^2(p) = \boldsymbol{p}^2 + \chi^2(p), \tag{14}$$

where $\chi(p)$ is the ghost loop, which is diagrammatically represented in Fig. 3. The gap equation (14) has the form of the dispersion relation of a relativistic particle except that the (effective) mass of the particle is replaced by the ghost loop $\chi(p)$. The ghost propagator is given here as the vacuum expectation value of the inverse Faddeev-Popov operator and is represented by

$$G = \langle(-\hat{\boldsymbol{D}} \cdot \boldsymbol{\partial})^{-1}\rangle = d(-\Delta)/(-\Delta), \tag{15}$$

where $d(-\Delta)$ is the so-called ghost form factor. Obviously, this form factor contains all the deviations from QED where the Faddeev-Popov operator is just given by the Laplacian. For large momenta the ghost form factor approaches unity while it diverges like $1/p$ in the infrared and thus fulfills the so-called horizon condition $d^{-1}(p = 0) = 0$ [10]. Contrary to other gauges, in Coulomb gauge the ghost form factor d has a physical meaning: Its inverse represents the dielectric function of the Yang–Mills vacuum, Ref. [11]. The horizon condition guarantees that the dielectric function vanishes in the infrared $\varepsilon(p = 0) = 0$, which implies that the Yang–Mills vacuum is a perfect color dielectricum or a so-called dual superconductor. In this way the Hamiltonian approach in Coulomb gauge relates to the dual superconductor picture of confinement introduced many years ago by Mandelstam [12] and 't Hooft [13]. The form of the dielectric function obtained here is in accord with the phenomenological bag model picture of hadrons [14]: On the one hand, inside the hadrons i.e. at small distances, the dielectric function equals unity which is the value of a normal vacuum, as in QED and also in QCD without matter, due to asymptotic freedom. On the other hand, outside the physical hadrons the infrared value $\varepsilon(p = 0) = 0$ corresponding to a confining vacuum is realized. Indeed for a medium with $\varepsilon = 0$ the electrical displacement $\boldsymbol{D} = \varepsilon\boldsymbol{E}$ vanishes implying by Gauß's law, $\boldsymbol{\partial} \cdot \boldsymbol{D} = \rho_{free}$, that no free color charges can exist.

Figure 4 compares the results of the variational calculation with the lattice results for the gluon propagator. The lattice results [15] can be nicely fitted by Gribovs' formula [16]

$$\omega(p) = \sqrt{\boldsymbol{p}^2 + M^4/\boldsymbol{p}^2} \tag{16}$$

with a mass scale of $M \simeq 880\,\text{MeV}$. The gluon energy (dashed line) obtained with the Gaussian trial wave functional agrees quite well with the lattice data in the infrared and in the ultraviolett regime but misses some strength in the mid-momentum regime. This missing strength is largely recovered when a non-Gaussian wave functional is used, Ref. [17], which includes in the exponent of the trial wave functional also cubic and quartic terms in the gauge potential. When non-Gaussian wave functionals are used, Wick's theorem is no longer valid. In Ref. [17] this problem was circumvented by exploiting generalized Dyson-Schwinger equations to express the various n-point functions occuring in the expectation values of operators like the Hamiltonian in terms of the variational kernels occuring in the exponent of the wave functional. One finds then the full curve in Fig. 4, which fits the lattice data in the mid-momentum regime much better than the Gaussian ansatz.

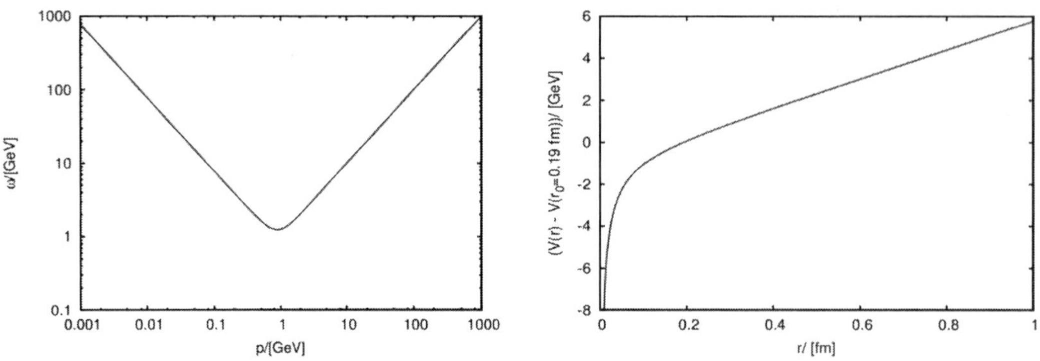

(a) (b)

Figure 2: (a) Gluon energy $\omega(p)$ following from the variation of the Yang–Mills energy density. (b) Static color charge potential obtained in the variational approach [8].

Figure 3: Diagrammatic representation of the ghost loop.

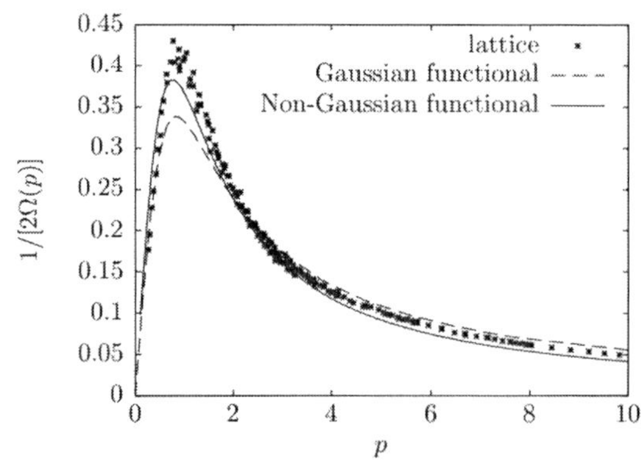

Figure 4: The gluon propagator. Comparison of the lattice result (crosses) [15] with the result of the variational calculation.

The part of the Coulomb Hamiltonian (11), which is quadratic in the charge density of the matter field ρ_m represents a two-body interaction which is induced by the Yang–Mills vacuum. The Faddeev-Popov determinant drops out from this part of the Coulomb term and its vacuum expectation value yields the static color charge potential

$$V(\boldsymbol{x},\boldsymbol{y}) = g^2\Big\langle\langle \boldsymbol{x}|(-\hat{\boldsymbol{D}}\cdot\boldsymbol{\partial})^{-1}(-\Delta)(-\hat{\boldsymbol{D}}\cdot\boldsymbol{\partial})^{-1}|\boldsymbol{y}\rangle\Big\rangle . \tag{17}$$

Figure 2 (b) shows the resulting quark-antiquark potential obtained from our variational Yang–Mills vacuum solution [8] when the vacuum expectation value is factorized resulting in

$$V(\boldsymbol{x},\boldsymbol{y}) \approx \int \mathrm{d}^3 w \int \mathrm{d}^3 z\, G(\boldsymbol{x},\boldsymbol{w}) \langle \boldsymbol{w}| - \Delta |\boldsymbol{z}\rangle\, G(\boldsymbol{z},\boldsymbol{y}), \tag{18}$$

where $G(\boldsymbol{x},\boldsymbol{y})$ is the ghost propagator (15). The obtained potential rises linearly at large distances with a coefficient given by the so-called Coulomb string tension σ_C, which on the lattice is measured to be a factor of $2\ldots3$ larger than the Wilson string tension $\sigma \approx (440\,\mathrm{MeV})^2$ [18]. At small distances the potential behaves like the Coulomb potential as expected from asymptotic freedom.

So far, we have seen signals for confinement both in the gluon propagator and in the static Coulomb potential. However, to really show that our approach yields confinement we have to calculate the vacuum expectation value of the Wilson loop

$$W[A](C) = P\exp\left[\mathrm{i}\oint \mathrm{d}\boldsymbol{x}\cdot \boldsymbol{A}(\boldsymbol{x})\right], \tag{19}$$

which is the order parameter of confinement:

$$\langle W[A](C)\rangle \sim \begin{cases} \exp(-\sigma\mathcal{A}(C)), & \text{confinement} \\ \exp(-\kappa\mathcal{P}(C)), & \text{deconfinement} \end{cases} \tag{20}$$

In the confining theory this quantity decays exponentially with the area $\mathcal{A}$ enclosed by the loop C with a coefficient referred to as Wilson string tension σ. An area law implies a linear rising potential with a slope given by σ. On the other hand in the deconfined phase the expectation value decays only with the perimeter $\mathcal{P}$ of the loop. Unfortunately the Wilson loop is difficult to calculate in a continuum theory due to path ordering. In Ref. [19] an approximate evaluation of the Wilson loop was carried out and the potential extracted from the Wilson loop is shown in Fig. 5 (a). This method works only up to intermediate distances where the small distance Coulomb behavior of the potential has not yet died out. When the perturbative part is removed, however, one does indeed observe the linear rising potential shown in Fig. 5 (b).

An order parameter (or more precisely disorder parameter) of confinement, which can be easier calculated in the continuum is the 't Hooft loop [20]. The operator $\hat{V}(C)$ corresponding to the 't Hooft loop is defined by the relation

$$\hat{V}(C_1)W(C_2) = z^{L(C_1,C_2)}W(C_2)\hat{V}(C_1) \tag{21}$$

for any two closed loops C_1 and C_2. Here $W(C_2)$ is the Wilson loop (19) and z is a non-trivial center element of the gauge group. Furthermore, $L(C_1, C_2)$ is the Gaussian linking number, which is non-zero when the two loops in the argument are non-trivially linked. The 't Hooft loop operator $\hat{V}(C)$ is dual to the Wilson loop in the sense that a perimeter law implies confinement while an area law implies deconfinement. 't Hooft gave an explicit realization of the loop operator $\hat{V}(C)$ on the lattice. In Ref. [21] a continuum representation of $\hat{V}(C)$ was derived,

$$\hat{V}(C) = \exp\left[\mathrm{i}\int_{\mathbb{R}^3} \mathcal{A}(C)\Pi\right], \tag{22}$$

where $\Pi = \delta/(\mathrm{i}\delta A)$ is the momentum operator of the gauge field and $\mathcal{A}(C)$ is the gauge potential of a so-called thin center vortex, which is a closed magnetic flux line located at the loop C. The center

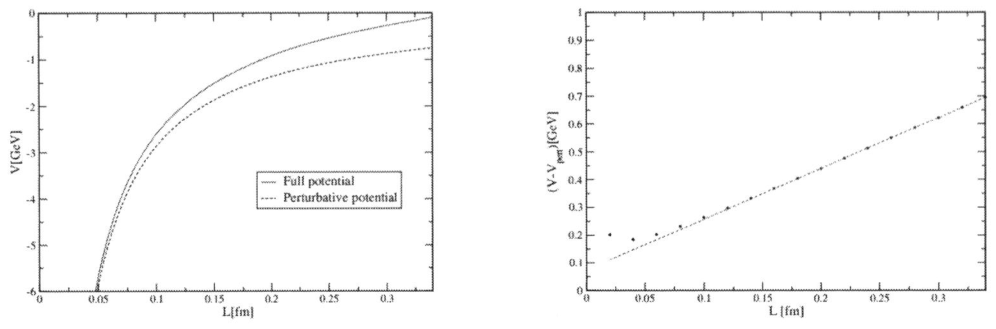

(a) (b)

Figure 5: Potential obtained from the Wilson loop.

vortex field $\mathcal{A}(C)$ is defined by its Wilson loop

$$W[\mathcal{A}(C_1)](C_2) = z^{L(C_1,C_2)}\,, \tag{23}$$

which produces a non-trivial center element z when the loop C_1 of the center vortex and the Wilson loop C_2 are non-trivially linked. Given the explicit form of the 't Hooft operator (22) and the coordinate representation of the momentum operator $\Pi = \delta/(i\delta A)$ it is not difficult to see that the 't Hooft loop operator, when acting on the wave functional, displaces the argument of the wave functional by the center vortex field

$$\hat{V}(C)\psi[A] = \psi[\mathcal{A}(C) + A]\,. \tag{24}$$

With this relation the expectation value of the 't Hooft loop operator can be easily calculated by taking the overlap of the shifted wave functional $\psi[\mathcal{A}(C) + A]$ with the unshifted one, $\psi[A]$. In Ref. [22] the 't Hooft loop was calculated with the wave functional obtained from our variational approach and indeed a perimeter law was found.

4 Variational approach to QCD in Coulomb gauge

The approach presented above for the Yang–Mills sector can be extended to full QCD. The Hamiltonian of QCD in Coulomb gauge is given by

$$H_{QCD} = H_{YM} + H_C + H_q\,. \tag{25}$$

Here H_{YM} is the Hamiltonian of the transversal gluon degrees of freedom given in eq. (10). Furthermore, H_C is the Coulomb term (11) where, however, now the charge density of the matter field is given by that of the quarks

$$\rho^a_m(\boldsymbol{x}) = \psi^\dagger(\boldsymbol{x})t^a\psi(\boldsymbol{x})\,. \tag{26}$$

Here, $\psi(\boldsymbol{x})$ is the quark field and t^a denotes the generators of the gauge group in the fundamental representation. Finally,

$$H_q = \int \mathrm{d}^3x\, \psi^\dagger(\boldsymbol{x})\left[\alpha(\boldsymbol{p} + g\boldsymbol{A}) + \beta m_0\right]\psi(\boldsymbol{x}) \tag{27}$$

DOI: 10.1051/epjconf/201612601002

is the Dirac Hamiltonian of the quarks interacting with the spatial gauge field $\boldsymbol{A}$. In the following let us concentrate on the quark sector. For the quark vacuum wave functional we have used recently the following variational ansatz [23]

$$|\phi\rangle_q = \exp\left[\int \psi_+^\dagger (s\beta + v\boldsymbol{\alpha}\cdot\boldsymbol{A} + w\beta\boldsymbol{\alpha}\cdot\boldsymbol{A})\psi_-\right]|0\rangle_q\,. \tag{28}$$

Here $\psi_\pm$ denote the positive and negative energy components of the quark Dirac spinor, while s, v, w are variational kernels to be determined by minimizing the QCD energy density. For $v = w = 0$ the ansatz reduces to a BCS type of wave functional, which was used in Refs. [24], [25], [26]. Furthermore, for $w = 0$ the ansatz was used in Refs. [27], [28]. There it was found that the inclusion of the coupling of the quarks to the transversal gluons in the vacuum wave functional substantially increases the amount of chiral symmetry breaking. With the quark wave functional (28) we have calculated the expectation value of H_{QCD} up to two loops and carried out the variation with respect to the kernel s, v, w. This results in three coupled equations, where the equations for v and w can be explicitly solved in terms of the scalar kernel s and the gluon propagator. The variational equation for the scalar kernel is usually referred to as *gap equation*. It is highly non-linear and can only be solved numerically. The extension of the ansatz for the quark wave functional with $w \neq 0$ is not just a quantitative improvement (since the variational space is increased), but it also improves the renormalization procedure, as the linear divergences in the gap equation strictly cancel. Furthermore, the surviving logarithmic divergences always come together with the square of the coupling strength and can thus be absorbed into a scale dependent renormalized effective coupling constant

$$g^2 \ln \Lambda/\mu = \tilde{g}^2(\mu)\,. \tag{29}$$

In the variational calculation of the quark sector the non-Abelian Coulomb potential shown in Fig. 2 (b) enters as input. Through the Coulomb string tension σ_C, i.e. the coefficient of the asymptotic linear slope of the potential, a scale is inherited from the gluon sector. Choosing $\sigma_C = 2\sigma$ as suggested by lattice calculations [18] a coupling of

$$\tilde{g}(\sqrt{\sigma_c}) \simeq 3.59 \tag{30}$$

is needed in order to reproduce from the variational solution the phenomenological value of the quark condensate [29]

$$\langle\bar{q}q\rangle = (-235\,\mathrm{MeV})^2\,. \tag{31}$$

In Ref. [8] the running coupling constant of Yang–Mills theory was calculated from the ghost-gluon vertex [30]. The result is shown in Fig. 6. From that figure one extracts a running coupling at the scale of the Coulomb string tension of

$$\tilde{g}(\sqrt{\sigma_C}) \simeq 3.9\,, \tag{32}$$

which is close to the value given in eq. (30) necessary to reproduce the phenomenological value of the quark condensate. Figure 7 shows the result for the variational kernels v and w. They look quite similar; however, one should notice the different scales. The kernel v is substantially larger than the kernel w. Figure 8 (a) shows the scalar form factor obtained with the coupling constant given in eq. (30) and compared to the corresponding result with the BCS-type wave function where the coupling of the quarks to the spatial gluons is neglected, i.e. eq. (28) with $v = w = 0$. The latter results in a quark condensate of $\langle\bar{q}q\rangle \simeq (-165\,\mathrm{MeV})^3$, which is substantially smaller than the phenomenological value (31). Finally, Fig. 8 (b) shows the effective quark mass obtained from the full variational calculation and the one from the BCS wave functional. Again the mass is substantially reduced when the coupling to the spatial gluons is ignored.

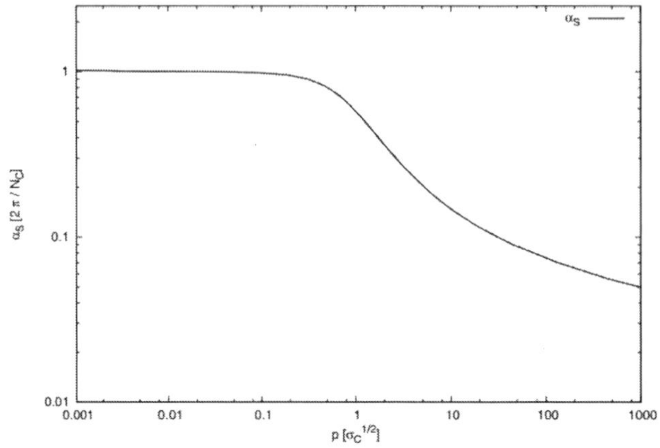

Figure 6: Running coupling constant in Ref. [8] calculated in the variational approach [6] from the ghost-gluon vertex [30].

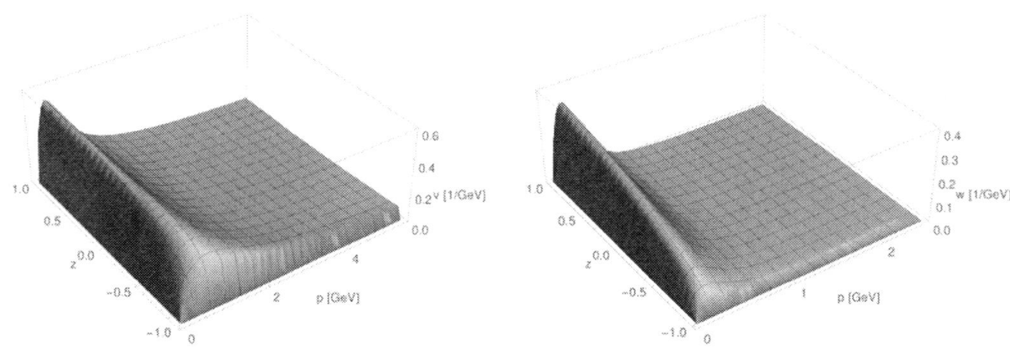

(a) (b)

Figure 7: The vector kernels (a) $v(\boldsymbol{p},\boldsymbol{q})$ and (b) $w(\boldsymbol{p},\boldsymbol{q})$ obtained from the self-consistent solution of the variational equations of the quark sector using the Gribov formula (16) as input for the gluon energy. Notice that only the section with $p = |\boldsymbol{p}| = |\boldsymbol{q}|$ and $z = \boldsymbol{p}\cdot\boldsymbol{q}/pq$ ist plotted.

5 Hamiltonian approach to finite-temperature quantum field theory by compactifying a spatial dimension

The variational Hamiltonian approach to QCD in Coulomb gauge can be straightforwardly extended to finite temperatures. For this purpose one uses a variational ansatz for the density operator of the grand canonical ensemble where the exact Hamiltonian is replaced by a single-particle Hamiltonian, with the single-particle energies determined variationally by minimizing the thermodynamic potential or, for vanishing chemical potential, the free energy. This has been done for the Yang–Mills sector in Refs. [9, 31]. Here I am going to present a more efficient approach to finite-temperature quantum field theory [33–35], which requires no ansatz for the density operator of the grand canonical ensemble.

Consider finite-temperature quantum field theory in the standard functional integral approach. Here the finite temperature is introduced by going to Euclidean space and compactifying the Euclidean time dimension by imposing periodic and antiperiodic boundary conditions for Bose and Fermi fields,

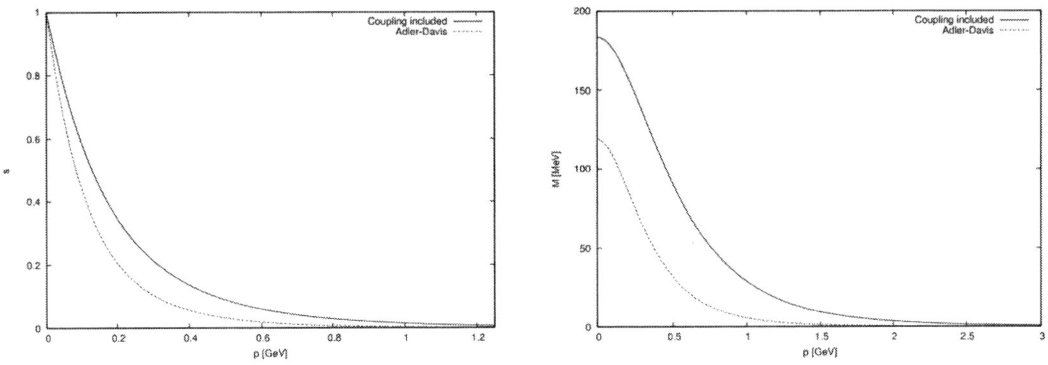

(a) (b)

Figure 8: (a) The scalar kernel $s(p)$ and (b) the effective quark mass, both obtained from the self-consistent solution of the variational equations of the quark sector using the Gribov formula (16) as input for the gluon energy.

respectively

$$A(x^0 = L/2) = A(x^0 = -L/2)$$
$$\psi(x^0 = L/2) = -\psi(x^0 = -L/2)\,. \tag{33}$$

The length of the compactified dimension L represents then the inverse temperature $T^{-1} = L$. One can now exploit the O(4) invariance of the Euclidean Lagrangian to rotate the Euclidean time axis into a space axis and, correspondingly, one space axis into the Euclidean time axis. Of course, thereby all vectorial quantities transform in the same way, i.e. we can choose the transformation:

$$x^0 \to x^3\,, \qquad A^0 \to A^3\,, \qquad \gamma^0 \to \gamma^3$$
$$x^1 \to x^0\,, \qquad A^1 \to A^0\,, \qquad \gamma^1 \to \gamma^0\,. \tag{34}$$

After this rotation we are left with the spatial periodic and antiperiodic boundary conditions

$$A(x^3 = L/2) = A(x^3 = -L/2)$$
$$\psi(x^3 = L/2) = -\psi(x^3 = -L/2)\,. \tag{35}$$

As a consequence of the O(4) rotation our spatial manifold is now $\mathbb{R}^2 \times S^1(L)$ instead of $\mathbb{R}^3$ while the temporal manifold is $\mathbb{R}$ independent of the temperature, i.e. the temperature is now encoded in one spatial dimension while time has infinite extension. We can now apply the usual canonical Hamiltonian approach to this rotated space-time manifold. As the new time axis has infinite extension $l \to \infty$, the partition function is now given by

$$Z(L) = \lim_{l\to\infty} \operatorname{Tr} \exp(-lH(L))\,, \tag{36}$$

where $H(L)$ is the usual Hamiltonian obtained after canonical quantization, however, now defined on the spatial manifold $\mathbb{R}^2 \times S^1(L)$. Taking the trace in the basis of the exact eigenstates of the Hamiltonian $H(L)$ we obtain for the partition function (36)

$$Z(L) = \lim_{l\to\infty} \sum_n \exp(-lE_n(L)) = \lim_{l\to\infty} \exp(-lE_0(L))\,. \tag{37}$$

The full partition function is now obtained from the ground state energy calculated on the spatial manifold $\mathbb{R}^2 \times S^2(L)$. Introducing the energy density $e(L)$ on $\mathbb{R}^2 \times S^1(L)$ by separating the volume Ll^2 of the spatial manifold from the energy we have

$$E_0(L) = Ll^2 e(L)\,. \tag{38}$$

It is a little exercise to show that the usual pressure P is given by

$$P = -e(L)\,, \tag{39}$$

while the usual energy density ε is obtained as

$$\varepsilon = \partial(Le(L))/\partial L - \mu\partial e(L)/\partial\mu\,. \tag{40}$$

To distinguish this quantity from the (negative) Casimir pressure $e(L)$ eq. (39), which also appears as an energy density in our formalism after the transformation eq. (34), we will denote $e(L)$ as *pseudo energy density*. Finally, after the O(4) rotation eq. (34), the finite chemical potential μ enters the single-particle Dirac Hamiltonian h in the form

$$h(\mu) = h(\mu = 0) + \mathrm{i}\mu\alpha^3\,, \tag{41}$$

where α^3 is the third Dirac matrix and $h(\mu = 0)$ the standard Dirac operator coupled to the gauge field.

To illustrate the above approach [35] let us first consider a relativistic Bose gas with dispersion relation $\omega(p) = \sqrt{p^2 + m^2}$, where we assume for simplicity a vanishing chemical potential. The usual pressure obtained from the grand canonical ensemble for such a system is given by

$$P = \frac{2}{3}\int \frac{\mathrm{d}^3 p}{(2\pi)^3}\frac{p^2}{\omega(p)} n(p)\,, \qquad n(p) = \frac{1}{\exp(\beta\omega(p)) - 1}\,, \tag{42}$$

where $n(p)$ are the finite temperature Bose occupation numbers. On the other hand for the pseudo energy density on the spatial manifold $\mathbb{R}^2 \times S^1(L)$ one finds, for the ideal Bose gas with dispersion relation $\omega(p) = \sqrt{p^2 + m^2}$ [35]

$$e(L) = \frac{1}{2}\int \frac{\mathrm{d}^2 p_\perp}{(2\pi)^2}\frac{1}{L}\sum_{n=-\infty}^{\infty}\sqrt{p_\perp^2 + p_n{}^2 + m^2}\,, \qquad p_n = \frac{2n\pi}{L}\,, \tag{43}$$

where p_n are the bosonic Matsubara frequencies. This quantity does not look at all like the negative of the pressure (42), as it should by eq. (39). In fact, as it stands it is ill defined: the integral and the sum are both divergent. To make it mathematically well defined, we first use the proper-time regularization of the square root

$$\sqrt{A} = \frac{1}{\Gamma\left(-\frac{1}{2}\right)}\lim_{\Lambda\to\infty}\left[\int\limits_{1/\Lambda^2}^{\infty} \mathrm{d}\tau\,\tau^{-\frac{3}{2}}\exp(-\tau A) - 2\Lambda + O(\Lambda^{-1})\right]. \tag{44}$$

The divergent constant appears because the limit $\Lambda \to \infty$ of the incomplete Γ-function is not smooth; it drops out when taking the difference to the zero-temperature case after eq. (46) below. With this replacement, the momentum integral in eq. (43) can be carried out in closed form. For the remaining Matsubara sum we use the Poisson resummation formula,

$$\frac{1}{2\pi}\sum_{k=-\infty}^{\infty}\exp(\mathrm{i}kx) = \sum_{n=-\infty}^{\infty}\delta(x - 2\pi n), \tag{45}$$

after which the proper-time integral can also be carried out, yielding for the pseudo energy density (43)

$$e(L) = -\frac{1}{2\pi^2} \sum_{n=-\infty}^{\infty} \left(\frac{m}{nL}\right)^2 K_2(nLm), \tag{46}$$

where $K_\nu(z)$ is the modified Bessel function. The term with $n = 0$ is divergent and represents the pseudo energy density at zero temperature, which should be eliminated from the pressure. The remaining terms $n \neq 0$ are all finite and also the remaining sum converges. This sum, however, cannot be carried out analytically for massive bosons (the same applies to the integral in the grand canonical expression (42) for the pressure). In the zero-mass limit we find from eq. (46) for the pressure (39)

$$P = \frac{\zeta(4)}{\pi^2} T^4 = \frac{\pi^2}{90} T^4, \tag{47}$$

which is Stefan-Boltzmann law, the correct result also obtained from the grand canonical ensemble. For massive bosons the evaluation of the sum in eq. (46) as well as the evaluation of the integral in eq. (42) have to be done numerically. The result is shown in Fig. 9 (a). As expected the pressure calculated from the compactified spatial dimension reproduces the result of the usual grand canonical ensemble. Figure 9 (b) shows the various contributions to the pressure. It is seen that only a few terms in the sum of eq. (46) are necessary to reproduce the result of the grand canonical ensemble to good accuracy.

In the case of the relativistic *Fermi* gas with dispersion relation $\omega(p) = \sqrt{p^2 + m^2}$ the energy density on $\mathbb{R}^2 \times S^1(L)$ is given by

$$e(L) = -2 \int \frac{d^2 p_\perp}{(2\pi)^2} \frac{1}{L} \sum_{n=-\infty}^{\infty} \sqrt{p_\perp^2 + (p_n + i\mu)^2 + m^2}, \qquad p_n = \frac{2n+1}{L}\pi, \tag{48}$$

where we have now included a non-vanishing chemical potential μ. To make this expression mathematically well-defined one has to resort again to the proper-time regularization and Poisson resummation technique sketched above. The result is

$$e(L) = \frac{2}{\pi^2} \sum_{n=0}^{\infty} \cos\left[nL\left(\frac{\pi}{L} - i\mu\right)\right] \left(\frac{m}{nL}\right)^2 K_{-2}(nLm). \tag{49}$$

Again, the term with $n = 0$ represents the zero temperature vacuum energy density, which is divergent and has to be removed. As before, this expression can only be calculated in closed form for massless particles. For the remaining sum to converge, an analytic continuation $i\mu L \to \bar{\mu} \in \mathbb{R}$ is required to carry out the sum

$$\sum_{n=1}^{\infty} (-1)^n \frac{\cos(n\bar{\mu})}{n^4} = \frac{1}{48}\left[-\frac{7}{15}\pi^2 + 2\pi^2\bar{\mu}^2 - \bar{\mu}^4\right]. \tag{50}$$

Continuing back to real chemical potentials one finds through eq. (39) for the pressure

$$P = \frac{1}{12\pi^2}\left[\frac{7}{15}\pi^4 T^4 + 2\pi^2 T^2 \mu^2 + \mu^4\right], \tag{51}$$

which is the correct result obtained also from the usual grand canonical ensemble.

In Ref. [36] the above approach was used to study Yang–Mills theory at finite temperature. For this purpose, it is merely required to repeat the variational Hamiltonian approach on the spatial manifold $\mathbb{R}^2 \times S^1(L)$. Due to the one compactified spatial dimension the 3-dimensional integral equations of the zero-temperature case are replaced by a set of 2-dimensional integral equations distinguished by different Matsubara frequencies. Below I will use this approach to calculate the effective potential of the Polyakov loop, the order parameter of confinement.

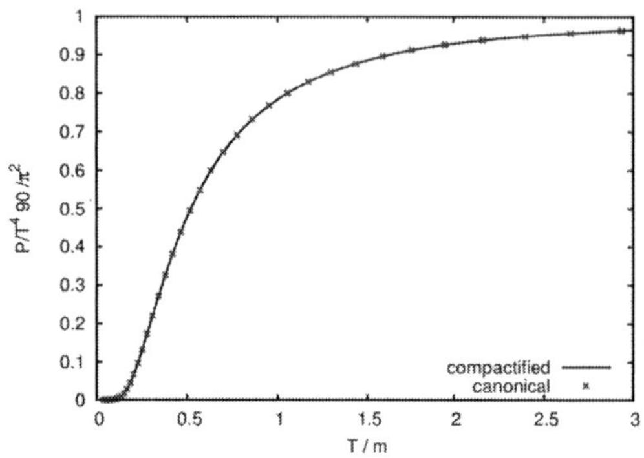

(a)

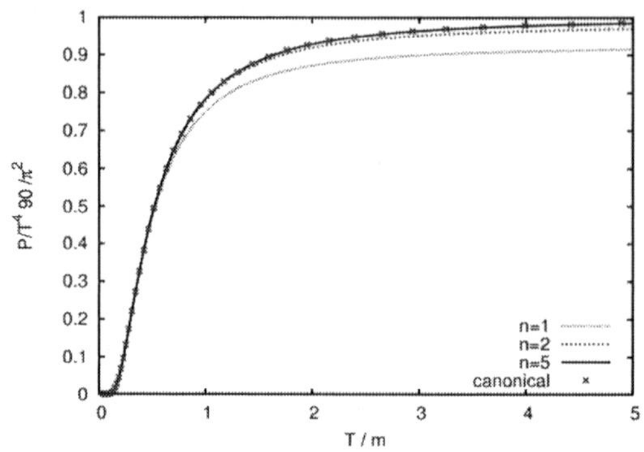

(b)

Figure 9: The pressure of a free massive Bose gas (a) calculated from eq. (46) (full curve) and from the grand canonical ensemble (42) (crosses). (b) The pressure when the summation index in eq. (46) is restricted to $|n| = 1, 2$ and 5.

5.1 The Polyakov loop

Consider SU(N) gauge theory at finite temperature, where the temperature is introduced by the usual periodic boundary condition in the temporal direction (33). Gauge transformations preserving this boundary conditions need to be periodic only up to an element z of the center $Z(N)$ of the gauge group

$$U(x^0 = L) = zU(x^0 = 0)\,, \qquad z \in Z(N)\,. \tag{52}$$

Since there are N center elements this theory has a residual global $Z(N)$ symmetry, which remains after gauge fixing. However, there are quantities which are sensitive to such a $Z(N)$ symmetry transformation. The most prominent example is the Polyakov loop

$$P[A_0](\boldsymbol{x}) = \frac{1}{d}\mathrm{tr}P\exp\left[\mathrm{i}\int_0^L \mathrm{d}x^0\, A_0(\boldsymbol{x}, x^0)\right]\,, \tag{53}$$

where $A_0 = A_0^a t^a$ is the temporal gauge field in the fundamental representation, and d is the dimension of this representation. A gauge transformation of the form (52) multiplies the Polyakov loop by the center element z

$$P[A_0^U] = zP[A_0]\,. \tag{54}$$

The expectation value of the Polyakov loop

$$\langle P[A_0](\boldsymbol{x})\rangle \sim \exp\left(-F_\infty(\boldsymbol{x})L\right) \tag{55}$$

can be shown to be related to the free energy $F_\infty(\boldsymbol{x})$ of a static color point charge located at $\boldsymbol{x}$ [37]. In a confining theory this quantity has to be infinite since there are no free color charges, while in a deconfined phase it is finite. Accordingly we find for the expectation value of the Polyakov loop

$$\langle P[A_0](\boldsymbol{x})\rangle \begin{cases} = 0 & \text{confined phase,} \\ \neq 0 & \text{deconfined phase.} \end{cases} \tag{56}$$

From eq. (54) follows that a state with vanishing expectation value of the Polyakov loop is obviously invariant with respect to the global center transformation, while in the deconfined phase the $Z(N)$ center symmetry is obviously broken. In the continuum theory the Polyakov loop can be most easily calculated in the Polyakov gauge

$$\partial_0 A_0 = 0, \qquad A_0 \text{ color diagonal.} \tag{57}$$

In this gauge one finds, for example, for the SU(2) gauge group that the Polyakov loop

$$P[A_0](\boldsymbol{x}) = \cos\left(\frac{1}{2}A_0(\boldsymbol{x})L\right) \tag{58}$$

is a unique function of the gauge field, at least in the fundamental modular region of this gauge. It can be shown, see Refs. [38], [39], that instead of the expectation value of the Polyakov loop $\langle P[A_0]\rangle$ one may alternatively use the Polyakov loop of the expectation value, $P[\langle A_0\rangle]$, or the expectation value of the temporal gauge field itself, $\langle A_0\rangle$, as order parameter of confinement. This analysis also shows that the most efficient way to obtain the Polyakov loop is to carry out a so-called background field calculation with a temporal background field $a_0(\boldsymbol{x}) = \langle A_0(\boldsymbol{x})\rangle$ chosen in the Polyakov gauge, and then calculate the effective potential $e[a_0]$ of that background field. From the minimum $\bar{a}_0$ of this potential one evaluates the Polyakov loop $P[\langle A_0\rangle] = P[\bar{a}_0]$, which can then serve as the order parameter of confinement.

Such a calculation was done a long time ago in Ref. [40], where the effective potential $e[a_0]$ was calculated in 1-loop perturbation theory. The result is shown in Fig. 10 (a). The potential is periodic due to center symmetry. The minimum of the potential occurs at the vanishing background field, which gives $P[a_0 = 0] = 1$ corresponding to the deconfined phase. This is, of course, expected due to the use of perturbation theory. Below I present the results of a non-perturbative evaluation of $e[a_0]$ in the Hamiltonian approach in Coulomb gauge.

At first sight it seems that the Polyakov loop cannot be calculated in the Hamiltonian approach due to the use of the Weyl gauge $A_0 = 0$. However, we can now use the alternative Hamiltonian approach to finite temperature introduced above, where the temperature is introduced by compactifying a spatial dimension. Here, we compactify the x_3−axis and consequently put also the background field along this axis $\boldsymbol{a} = a\boldsymbol{e}_3$. In the Hamiltonian approach the effective potential of a spatial background field $\boldsymbol{a}$ can be easily calculated by minimizing the expectation value of the Hamiltonian under the constraint $\langle \boldsymbol{A}\rangle = \boldsymbol{a}$. The resulting energy $\langle H\rangle_{\boldsymbol{a}} = L^2 l e(\boldsymbol{a})$ is then (up to the spatial volume factor) the effective potential. So the effective potential $e(\boldsymbol{a})$ is nothing but the pseudo energy density considered earlier, but now calculated in a background gauge with the contraint $\langle \boldsymbol{A}\rangle = \boldsymbol{a}$.

5.2 The effective potential of the Polyakov loop

After lengthy calculations exploiting the gluon gap equation (14) one finds for the effective potential of the Polyakov loop the following expression

$$e(a, L) = \sum_\sigma \frac{1}{L} \sum_{n=-\infty}^{\infty} \int \frac{d^2 p_\perp}{(2\pi)^2} \left(\omega(\boldsymbol{p}^\sigma) - \chi(\boldsymbol{p}^\sigma)\right) . \tag{59}$$

Here $\omega(p)$ is the gluon energy and $\chi(p)$ is the ghost loop. Furthermore, these quantities have to be taken with the momentum variable

$$\boldsymbol{p}^\sigma = \boldsymbol{p}_\perp + (p_n - \sigma \cdot a)\,\boldsymbol{e}_3 , \tag{60}$$

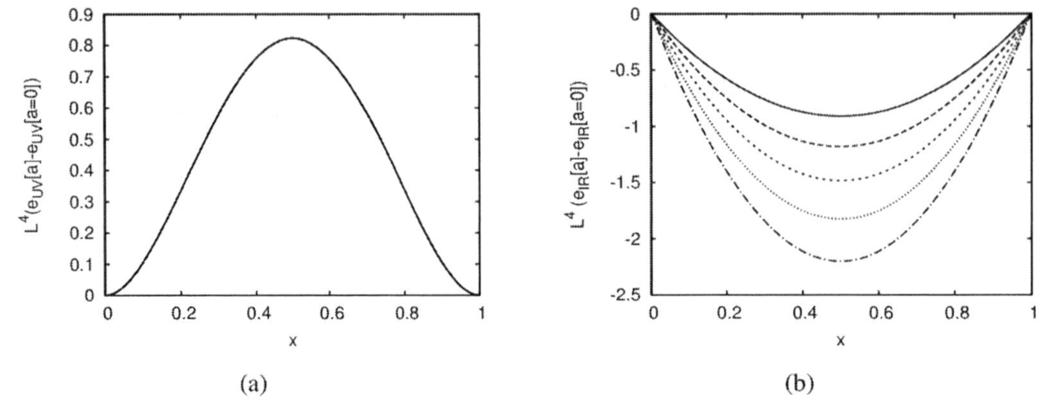

Figure 10: The effective potential of the Polyakov loop $e(a, L)$ (59) as function of the background field $x = a_3L/2\pi$. The curvature is neglected ($\chi = 0$) and the gluon energy assumed to be (a) $\omega(p) = p$ (UV-form) and (b) $\omega(p) = M^2/p$ (IR-form) respectively.

where $\boldsymbol{p}_\perp$ is the momentum corresponding to the two uncompactified space dimensions while $p_n = 2\pi n/L$ is the Matsubara frequency resulting from the compactification of the third dimension. Furthermore, $\sigma \cdot a \equiv \sigma^b a^b$ denotes the product of the color background field with the root vectors σ^b of the gauge group. Equation (59) includes also the summation over the roots σ of the gauge group. In Ref. [33, 34] the effective potential (59) was explicitly calculated using for $\omega(p)$ and $\chi(p)$ the results from the variational calculation in Coulomb gauge at zero temperature [8]. This represents certainly an approximation since, in principle, one should use the finite-temperature solutions obtained in Ref. [36].

Before I present the full results let me ignore the ghost loop $\chi(p)$ in eq. (59) and consider the ultraviolet and infrared limit of the gluon energy. If we choose the ultraviolet limit $\omega(p) = p$, we obtain from eq. (59) with $\chi(p) = 0$ precisely the Weiss potential, shown in Fig. 10 (a), which corresponds to the deconfined phase. Choosing for the gluon energy its infrared limit $\omega(p) = M^2/p$ one finds from eq. (59) with $\chi(p) = 0$ the (center symmetric) potential shown in Fig. 10 (b). From its center symmetric minimum $\bar{a} = \pi/L$ one finds a vanishing Polyakov loop $P[\bar{a}] = 0$ corresponding to the confined phase. Obviously, the deconfining phase transition results from the interplay between the confining infrared and the deconfining ultraviolet potentials. Choosing for the gluon energy the sum of the UV- and IR-parts $\omega(p) = p + M^2/p$, which can be considered as an approximation to the Gribov formula (16), one has to add the UV and IR potentials and finds a phase transition at a critical temperature $T_c = \sqrt{3}M/\pi$. With the Gribov mass $M \approx 880$ MeV this gives a critical value of $T_c \approx 485$ MeV for the color group SU(2), which is much too high as compared to the lattice value of 312 MeV [32]. One can show analytically [33, 34] that the neglect of the ghost loop $\chi(p) = 0$ shifts the critical temperature to higher values. If one uses for the gluon energy $\omega(p)$ the Gribov formula (16) and includes the ghost loop $\chi(p)$ one finds the effective potential shown in Fig. 11 (a), which shows a second order phase transition and gives a transition temperature of $T_c \approx 269$ MeV for the gauge group SU(2), which is in the right ballpark. The Polyakov loop $P[\bar{a}]$ calculated from the minimum $\bar{a}$ of the effective potential $e(a, L)$ (59) is plotted in Fig. 12(a) as function of the temperature.

The effective potential for the gauge group SU(3) can be reduced to that of the SU(2) group by noticing that the SU(3) algebra consists of three SU(2) subalgebras characterized by the three positive

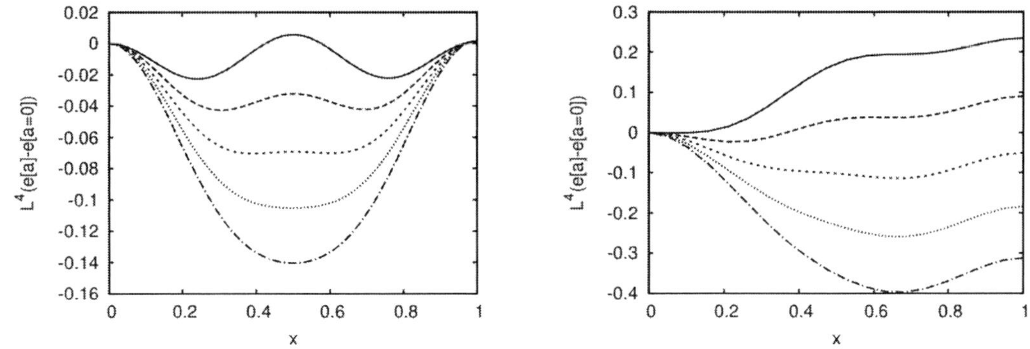

(a) (b)

Figure 11: Effective potential of the Polyakov loop as function of the background field $x = a_3 L/2\pi$ at various temperatures, for the gauge group (a) SU(2) and (b) SU(3).

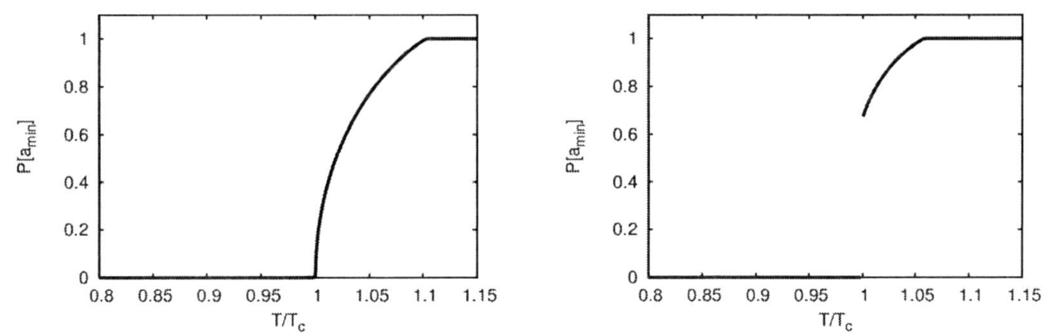

(a) (b)

Figure 12: The Polyakov loop as function of the temperature (a) for SU(2) and (b) for SU(3).

roots $\sigma = (1, 0)$, $(1/2, 1/2\sqrt{3})$, $(1/2, -1/2\sqrt{3})$. One finds

$$e_{SU(3)}(a, L) = \sum_{\sigma>0} e_{SU(2)}[\sigma](a, L) . \tag{61}$$

The resulting effective potential for SU(3) is shown in Fig. 13 as function of the components of the background field in the Cartan algebra a_3, a_8. Above and below T_c the absolute minima of the potential occur in both cases for $a_8 = 0$. Cutting the 2-dimensional potential surface at $a_8 = 0$ one finds the effective potential shown in Fig. 11 (b), which shows a first order phase transition with a critical temperature of $T_c \approx 283$ MeV. The first order nature of the SU(3) phase transition is also seen in Fig. 12 (b), where the Polyakov loop $P[\bar{a}]$ is shown as function of the temperature.

6 Conclusions

In this talk I have shown that the Hamiltonian approach to QCD in Coulomb gauge gives a decent description of the infrared properties of the theory and at the same time can be straightforwardly

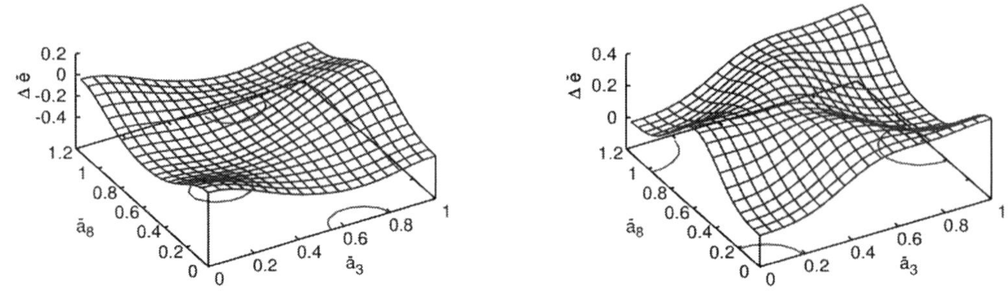

(a) (b)

Figure 13: The effective potential of the Polyakov loop for the gauge group SU(3) as function of the two Cartan components of the background field $x = a_3 L/2\pi$ and $y = a_8 L/2\pi$ for (a) $T < T_c$ and (b) $T > T_c$.

extended to finite temperatures, where it yields critical temperatures for the deconfinement phase transition in the right ballpark. Furthermore, the approach has also produced the correct order of the deconfinement phase transition, viz. first order for SU(3) and second order for SU(2). The obtained results are very encouraging and call for an extension of the present approach to non-zero chemical potentials.

References

[1] P.E. Haagensen, K. Johnson, C.S. Lam, Nucl. Phys. **B477** (1996) 273;
D. Karabali, C. Kim, V.P. Nair, Phys. Lett. **B434** (1998) 103.

[2] J.P. Greensite, Nucl. Phys. **B158** (1979) 469.

[3] N. H. Christ and T. D. Lee, Phys. Rev. D **22**, 939 (1980).

[4] D. Schutte, Phys. Rev. D **31**, 810 (1985).

[5] A. P. Szczepaniak and E. S. Swanson, Phys. Rev. D **65**, 025012 (2001).

[6] C. Feuchter and H. Reinhardt, Phys. Rev. D **70**, 105021 (2004); `arXiv:hep-th/0402106`.

[7] H. Reinhardt and C. Feuchter, Phys. Rev. D **71**, 105002 (2005)

[8] D. Epple, H. Reinhardt and W. Schleifenbaum, Phys. Rev. D **75**, 045011 (2007).

[9] J. Heffner, H. Reinhardt and D. R. Campagnari, Phys. Rev. D **85**, 125029 (2012).

[10] D. Zwanziger, Nucl. Phys. **B485** (1997) 185.

[11] H. Reinhardt, Phys. Lett. **101**, 061602 (2008).

[12] S. Mandelstam, Phys. Rep. **C23**, 245 (1976)

[13] G. 't Hooft, Phys. Scr. **25**, 133 (1982)

[14] A. Chodos, R.L. Jaffe, K. Johnson, C.B. Thorn, V.F. Weisskopf, Phys. Rev. **D9** (1974) 3471.

[15] G. Burgio, M. Quandt, H. Reinhardt, Phys. Rev. Lett. **102**, 032002 (2009).

[16] V.N. Gribov, Nucl. Phys. **B139** (1978) 1.

[17] D. R. Campagnari and H. Reinhardt, Phys. Rev. D **82**, 105021 (2010).

[18] J. Greensite and S. Olejnik, Phys. Rev. **D67** (2003) 094503;
A. Voigt, E.-M. Ilgenfritz, M. Muller-Preussker, A. Sternbeck, Phys. Rev. **D78** (2008) 014501;
G. Burgio, M. Quandt, H. Reinhardt, Phys. Rev. **D86** (2012) 045029.
[19] M. Pak and H. Reinhardt, Phys. Rev. D **80**, 125022 (2009).
[20] G. 't Hooft Nucl. Phys. **B138**, 1 (1978).
[21] H. Reinhardt, Phys. Lett. B **557**, 317 (2003).
[22] H. Reinhardt and D. Epple, Phys. Rev. D **76**, 065015 (2007).
[23] P. Vastag and H. Reinhardt, to be published
[24] J. R. Finger and J. E. Mandula, Nucl. Phys. **B199**, 168 (1982)
[25] S. L. Adler and A. C. Davis, Nucl. Phys. **B244**, 469 (1984)
[26] R. Alkofer and P. A. Amundsen, Nucl. Phys. **B306**, 305 (1988).
[27] M. Pak and H. Reinhardt, Phys. Lett. B **707**, 566 (2012).
[28] M. Pak and H. Reinhardt, arXiv:1310.1797 [hep-ph].
[29] R. Williams, C. S. Fischer and M. R. Pennington, Phys. Lett. B **645**, 167 (2007)
[30] W. Schleifenbaum, M. Leder, and H. Reinhardt, Phys. Rev. D **73**, 125019 (2006).
[31] H. Reinhardt, D. R. Campagnari and A. P. Szczepaniak, Phys. Rev. D **84**, 045006 (2011).
[32] B. Lucini, M. Teper, U. Wenger, JHEP **0401** (2004) 061.
[33] H. Reinhardt and J. Heffner, Phys. Lett. B **718**, 2 (2012).
[34] H. Reinhardt and J. Heffner, Phys. Rev. D **88**, 045024 (2013).
[35] H. Reinhardt and P. Vastag, to be published
[36] J. Heffner and H. Reinhardt, Phys. Rev. D **91**, no. 8, 085022 (2015)
[37] B. Svetitsky, Phys. Rep. **132** (1986) 1.
[38] J. Braun, H. Gies and J. M. Pawlowski, Phys. Lett. B **684**, 262 (2010).
[39] F. Marhauser and J. M. Pawlowski, `arXiv:0812.1144 [hep-ph]`.
[40] N. Weiss, Phys. Rev. D **24**, 475 (1981). D. J. Gross, R. D. Pisarski and L. G. Yaffe, Rev. Mod. Phys. **53**, 43 (1981).

EPJ Web of Conferences **126**, 02001 (2016) DOI: 10.1051/epjconf/201612602001
ICNFP 2015

Flavour physics at LHCb

B. Adeva[1,a], on behalf of the LHCb Collaboration.

[1] *University of Santiago de Compostela, Spain*

Abstract. Some selected results of the LHCb experiment, running at the LHC with pp collisions at 7 TeV and 8 TeV, are reported here, after operation with a total integrated luminosity of 3.0 fb^{-1} (Run 1). We focus on the most recent analyses on flavour physics, that include measurements of the CKM invariant phases γ and β, precision determination of the quark coupling strength V_{ub}, observation of the very rare decays $B^0_{(s)} \to \mu^+\mu^-$, search for new physics in the anomalous branching ratio of $B \to D^*\tau\bar{\nu}$, and precision angular analysis of the rare decays $B^0 \to K^{*0}\mu^+\mu^-$ and $B^0_s \to \phi\mu^+\mu^-$. Detailed comparisons are performed in all cases with the predictions of the Standard Model, and a few interesting tensions are observed.

1 The LHCb experiment

The LHCb detector [1, 2] is one of the four major detectors at the Large Hadron Collider. It is instrumented in a cone arround the proton beam axis, covering the angles between 10 and 250 mrad, where most b hadron decays produced in proton-proton collisions occur. The detector includes a high-precision tracking system with a dipole magnet, providing a measurement of momentum and impact parameter (IP), defined for charged particles as the minimum distance of a track to a primary pp interaction vertex (PV). Different types of charged particles are distinguished using information from two ring-imaging Cherenkov detectors, a calorimeter and a muon system. Simulated samples of specific signal and background decay modes of b hadrons are used at many stages throughout the analysis. These simulated events model the experimental conditions in full detail, including the pp collision, the decay of the particles, and the response of the detector [3–5].

Candidates of the different signal modes reported in this article are required to pass a trigger system [6] which reduces in real time the rate of recorded collisions from the readout clock of the LHC to approximately 4 KHz. For muon channels, a muon is typically selected with $p_T > 1.48$ GeV/c in the $\sqrt{s} = 7$ TeV collision data ($p_T > 1.76$ GeV/c in the 8 TeV data), and at least one of the final-state particles is required to have both $p_T > 0.8$ GeV/c and impact parameter larger than 100 μm with respect to all of the PVs in the event. For hadron channels, a multivariate algorithm is used for the identification of secondary vertices consistent with the decay of a b hadron [7]. In all cases, the tracks of two or more of the final-state particles are required to form a vertex that is significantly displaced from the PVs.

[a]e-mail: Bernardo.Adeva@usc.es

2 Measurement of the γ CKM phase

In the Standard Model (SM), the Flavor Changing Charged Current processes of quarks are described by a unitary complex-valued Cabibbo-Kobayashi-Maskawa (CKM) mixing matrix [8], whose elements V_{ij}, with $i = u, c, t$ and $j = d, s, b$, quantify the relative $i \leftrightarrow j$ coupling strength. This matrix originates from the misalignment between up and down type quark couplings to the Higgs boson, and being unitary, it has only one independent phase.

Every CP violation phenomenon (in the quark sector) should be related, in the SM, to this unique phase. However, up to four measurable phases can be formed from combinations of the type $V_{\alpha i}V^*_{\alpha j}V_{\beta j}V^*_{\beta i}$, and it is important to measure all of them because loop-level contributions from additional high-mass particles beyond the SM, may alter the unitarity relationships they must fulfil.

The measurement of the phase $\beta = \arg\left[-V_{cd}V^*_{cb}/(V_{td}V^*_{tb})\right]$, that provided the first evidence for CP-violation in the b-quark sector [9], is a historical example. A new competitive measurement of this parameter has been provided this year by the LHCb experiment, which is reported in the next section. Yet another sensitive test of the SM comes from measurements of the phase $\beta_s = \arg\left[-V_{ts}V^*_{tb}/(V_{cs}V^*_{cb})\right]$, only accessible from B^0_s meson decays, which is predicted to be very small in the SM. LHCb has initiated a series of measurements of this parameter using different channels, which we are not discussing here [10]. Within the precision attained so far, these measurements appear to confirm the SM prediction.

In order to disentangle the nature of new physics contributions from quantum loops, in case a significant deviation from unitarity is found, it is particularly important to have a CP-observable that only receives contributions from tree-level diagrams, and can therefore be used as a test-bench for the SM. This is precisely the case with the phase $\gamma = \arg[-V_{ud}V^*_{ub}/(V_{cd}V^*_{cb})]$ [11], the least well-measured to date of those in the unitarity triangle $V_{ud}V^*_{ub} + V_{cd}V^*_{cb} + V_{td}V^*_{tb} = 0$.

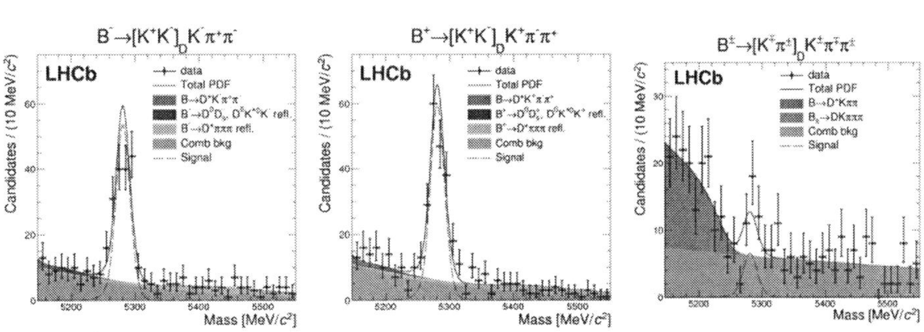

Figure 1. Mass distributions of $B^- \to DX_s^-$ candidates using GLW selections , for $B^- \to [K^+K^-]_D X_s^-$ (left), $B^- \to [K^+K^-]_D X_s^+$ (center), and the suppressed ADS mode, $B^\pm \to [K^\mp\pi^\pm]_D K^\pm\pi^\mp\pi^\mp$, sum of B^+ and B^- (right).

The phase γ can be probed by studying the interference between $b \to u$ and $b \to c$ transitions, such as the interference between $B^- \to D^0K^-$ and $B^- \to \bar{D}^0K^-$, when states accessible to both D^0 and $\bar{D}^0$ mesons are selected. A number of methods have been discussed in the literature, and are often grouped into three categories, depending on the D decay mode: (i) CP eigenstates, such as $D \to K^+K^-$ and $D \to \pi^+\pi^-$ (GLW) [12]; (ii) flavor-specific final states, such as the Cabibbo-favored and double Cabibbo suppressed $D \to K^\pm\pi^\mp$ decays (ADS) [13]; and (iii) multi-body self-conjugate final states, such as $D \to K^0_s\pi^+\pi^-$ (GGSZ) [14].

Measurements of γ have been performed from averages over several decay modes from individual experiments, and the current experimental status is $\gamma = (73^{+9}_{-10})^\circ$ by the LHCb collaboration [16],

$\gamma = (73^{+17}_{-16})^\circ$ by the BaBar collaboration [17], and $\gamma = (68^{+15}_{-14})^\circ$ by the Belle collaboration [18]. The overall precision on γ from a global fit is about 7° [19]. In order to improve the overall precision, it is important to study a wide range of final states. It has been suggested that other multi-body final states of the recoiling strange quark system could be useful [15], due to their larger branching fractions, and potentially larger interference contribution.

LHCb has performed the first ADS and GLW analyses of the decay $B^- \to DX_s^-$, where the D meson is observed through its decay to $K^\pm\pi^\mp$, K^+K^- and $\pi^+\pi^-$, and a multibody final state $X_s^- \equiv K^-\pi^+\pi^-$ is defined for the recoil system [20]. Some invariant mass spectra for the $B^- \to DX_s^-$ ADS and GLW signal modes are shown in Fig.1. The analysis uses an integrated luminosity of 3.0 fb^{-1} and includes the modes $B^- \to DX_d^-$, with lower sensitivity to γ, for normalization purposes. Significant signals are observed in the CP modes for both the favored and the suppressed B^- decays, and first evidence is seen for the ADS DCS $B^- \to D[K^+\pi^-]_D K^-\pi^+\pi^-$ decay. A fit for γ is performed, from which $\gamma = (74^{+20}_{-22})^\circ$ is found with only $B^- \to DX_s^-$ modes. Values of γ below about 25° and larger than approximately 165° are not excluded by these modes, but are excluded when other modes are considered [16]. The sensitivity to γ from this analysis makes it a promising channel for future studies.

LHCb has also explored additional multibody final states leading to possible improvement of the global precision on the γ parameter, and performed a measurement of CP observables from $B^\pm \to Dh^\pm$ decays, where D mesons are reconstructed in the ADS channel $D \to K^\mp\pi^\pm\pi^0$ and the quasi-GLW modes $D \to \pi^+\pi^-\pi^0$ and $D \to K^+K^-\pi^0$ [21]. In such cases, the interference effects that are sensitive to γ vary over the phase space of the D decay, due to the role of strongly-decaying intermediate resonances, and the integration over the phase space in general dilutes the net sensitivity. For multi body ADS/GLW modes the dilution factor can be measured with $D\bar{D}$ pairs coherently produced at the $\psi(3770)$ resonance [22]. Recent measurements of this type [23] indicate that the dilution effects in $D \to K^\mp\pi^\pm\pi^0$ and $D \to \pi^+\pi^-\pi^0$ are rather small, making these decays particularly suitable for an inclusive analysis. In particular the latter is very close to being a CP-even eigenstate, and the interference terms suffer very little dilution.

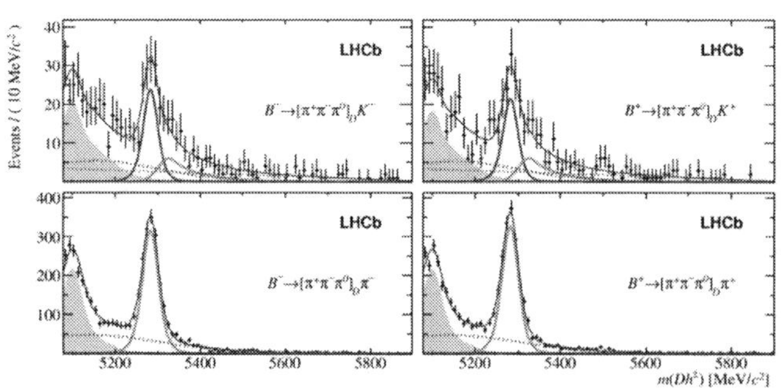

Figure 2. Invariant mass distributions of selected $B^\mp \to [\pi^+\pi^-\pi^0]_D h^\mp$ candidates, separated by B hadron charge. $B^\mp \to DK^\mp$ signal events are in the upper plots and $B^\mp \to D\pi^\mp$ in the lower plots.The red curve represents $DK^\mp$ events and the green curve represents $D\pi^\mp$ events. The grey shape indicates partially reconstructed $B^\mp$ decays and the dotted red curve indicates wrongly reconstructed D decays. The blue line represents the total PDF

As it is described in Ref. [21], twelve observables are measured by LHCb in total, including CP-asymmetries and amplitude ratios between Cabibbo favored and suppressed modes, for the above-mentioned ADS and quasi-GLW modes. Two-dimensional scans are performed for γ vs. r_B and γ vs. δ_B, where the hadronic amplitude ratio r_B and the phase difference δ_B are defined as: $A(B^- \to \bar{D}^0K^-)/A(B^- \to D^0K^-) = r_B e^{i(\delta_B-\gamma)}$. The results are compatible with the values obtained from a global analysis of other LHCb measurements sensitive to γ at tree level (also sensitive to r_B and δ_B) [16]. No evidence of CP violation is seen with the current experimental precision. First evidence is obtained for the mode $B^\mp \to [K^+K^-\pi^0]_D K^\mp$, and the channels $B^\mp \to [\pi^\mp K^\pm\pi^0]_D\pi^\mp$ $B^\mp \to [K^+K^-\pi^0]_D\pi^\mp$ have been observed for the first time. Some particular invariant mass spectra are shown in Fig.2, separated by the charge of the $B^{\mp}$candidate, as an indication of the signal significance for CP violating observables.

When analysed in the context of the underlying physics parameters, the results exhibit good consistency with other LHCb measurements, and they will be valuable in improving knowledge of γ in the unitarity triangle $V_{ud}V^*_{ub} + V_{cd}V^*_{cb} + V_{td}V^*_{tb} = 0$, when combined with results from $B^{\mp} \rightarrow DK^{\mp}$ measurements using other D decay channels.

3 New precision measurement of the β CKM phase

The violation of CP symmetry in processes involving B mesons was first observed in the "golden mode" $B^0 \rightarrow J/\psi K^0_s$ by the BaBar and Belle experiments at the asymmetric e^+e^- colliders PEP-II and KEKB. Since then, measurements of CP violation in this decay mode have reached a precision at the level of 10^{-2} [24]. LHCb has performed a new competitive measurement obtained at a hadron collider [25], where a reduced flavor tagging capability, as compared to B-factories, is compensated by a higher b hadron cross-section, with the integrated luminosity of 3.0 fb^{-1} from Run 1.

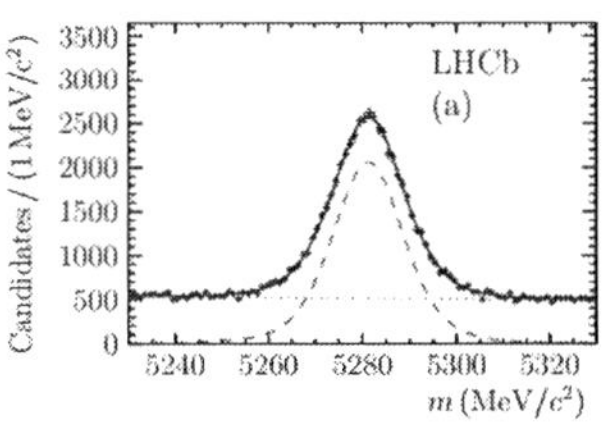

Figure 3. The distribution of the reconstructed mass of tagged $B^0 \rightarrow J/\psi K^0_s$ candidates is shown. The solid black line shows the fit projection, while the dashed (dotted) line shows the projection for the signal (background) components only.

As the $J/\psi K^0_s$ final state is common to both B^0 and $\bar{B}^0$ meson decays, the interference between the amplitudes for the direct decay and for the decay after $B^0 - \bar{B}^0$ oscillation results in a decay-time dependent CP asymmetry as follows

$$A(t) \equiv \frac{\Gamma(\bar{B}^0(t) \rightarrow J/\psi K^0_s) - \Gamma(B^0(t) \rightarrow J/\psi K^0_s)}{\Gamma(\bar{B}^0(t) \rightarrow J/\psi K^0_s) + \Gamma(B^0(t) \rightarrow J/\psi K^0_s)} = \frac{S\,\sin(\Delta m t) - C\cos(\Delta m t)}{\cosh(\frac{\Delta\Gamma t}{2}) + A_{\Delta\Gamma}\sinh(\frac{\Delta\Gamma t}{2})}$$

where $B^0(t)$ and $\bar{B}^0(t)$ indicate the flavor of the B meson at production, while t indicates the decay time. The parameters Δm and $\Delta\Gamma$ are the mass and decay width differences between the heavy and light mass eigenstates of the $B^0 - \bar{B}^0$ system, and S, C and $A_{\Delta\Gamma}$ are CP observables. As $\Delta\Gamma$ is negligible for the $B^0 - \bar{B}^0$ system, the time dependent asymmetry simplifies to $A(t) = S\,\sin(\Delta m t) - C\cos(\Delta m t)$. The $B^0 \rightarrow J/\psi K^0_s$ decay is dominated by a $\bar{b} \rightarrow c\bar{c}\bar{s}$ transition, and CP violation in the decay is expected to be negligible at the current level of experimental precision, giving $C \approx 0$, which allows to identify S with $\sin(2\beta)$.

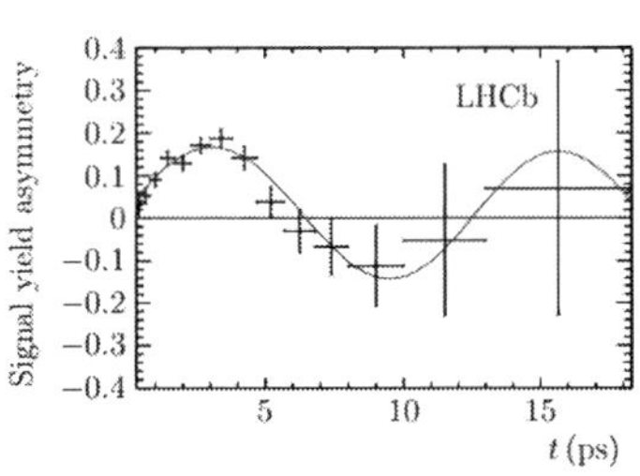

Figure 4. The time-dependent signal-yield asymmetry $(N_{\bar{B}^0} - N_{B^0})/(N_{\bar{B}^0} + N_{B^0})$ is shown. Here N_{B^0} ($N_{\bar{B}^0}$) is the number of $B^0 \rightarrow J/\psi K^0_s$ decays with a B^0 ($\bar{B}^0$) flavor tag. The solid curve is the projection of the signal PDF.

Compared to previous LHCb analysis, the effective tagging efficiency ϵ_{eff} has increased from 2.38% to 3.02%, mainly due to the inclusion of a same side pion tagger algorithm. The mass spectrum of the selected candidates is shown in Fig. 3, and the measured decay-time dependent signal-yield asymmetry is shown in Fig. 4.

The CP observables S and C are measured to be

$$\text{S} = 0.731 \pm 0.035 \text{ (stat)} \pm 0.020 \text{ (syst)}$$
$$\text{C} = \text{-}0.038 \pm 0.032 \text{ (stat)} \pm 0.005 \text{ (syst)}$$

with a statistical correlation $\rho(S, C) = 0.483$. When C is fixed to zero the measurement yields $S = \sin(2\beta) = 0.746 \pm 0.030$ $(stat)$. This result represents the most precise time-dependent CP violation measurement at a hadron collider to date. Furthermore, it has a similar precision to, and is in good agreement with, previous measurements performed at the Belle and BaBar experiments at the KEKB and PEP-II colliders [24]. This result is in excellent agreement with expectations from other measurements and improves the consistency of the CKM sector of the Standard Model. Other measurements that constraint this angle of the unitarity triangle predict $\sin(2\beta)$ as $0.771^{+0.017}_{-0.041}$ [26].

4 Precision determination of the coupling strength $|V_{ub}|$

The V_{ub} matrix element governs the most sensitive misalignment in the couplings between the up and down type quark flavors to the Higgs boson. Its apparent proportionality to the third power of λ (sine of the Cabibbo angle) remains unexplained, and it best quantifies the minimal flavor violation structure of the Standard Model. A precision measurement of the magnitude of V_{ub} is naturally achieved via the semileptonic quark-level transition $b \to u l^- \bar{\nu}_l$, which minimizes hadronic uncertainties. There are two complementary methods to perform such measurement. The simplest is to measure the branching fraction of a specific (exclusive) decay such as $\bar{B}^0 \to \pi^+ l^- \bar{\nu}$ or $B^- \to \pi^0 l^- \bar{\nu}$, where the influence of the strong interaction in the decay, encompassed by the $\bar{B}^0 \to \pi^+$ form factor, is predicted by lattice QCD (LQCD) [28] or QCD sum rules [29]. The world average from Ref. [30] is $V_{ub} = (3.28 \pm 0.29) \times 10^{-3}$, where the most precise inputs come from the Babar [31] and Belle [32] experiments. The uncertainty is dominated by the LQCD calculations, recently updated [33]. The alternative method is measure the differential decay rate in an inclusive way over all possible B meson decays containing the $b \to u l^- \bar{\nu}_l$ quark level transition. This results in $V_{ub} = (4.41^{+0.15}_{-0.17}) \times 10^{-3}$ [34], where the second uncertainty comes from theoretical calculations. The above results are summarized in Fig.5 (left).

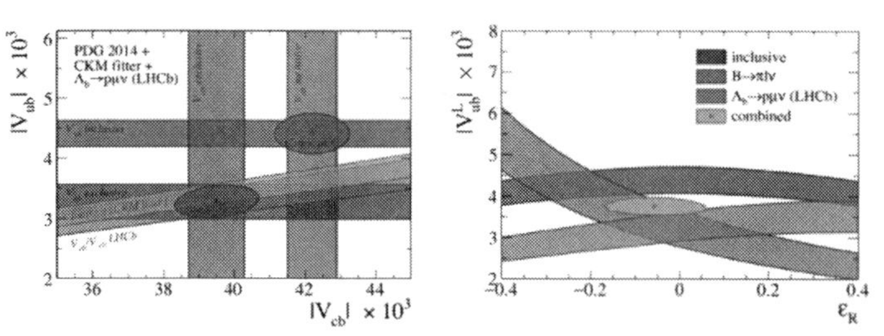

Figure 5. Summary of the exclusive and inclusive measurements indicated in the text (left) and experimental constraints on left-handed coupling, V^L_{ub} and the fractional right-handed (RH) coupling ϵ_R. While the overlap of the 68% CL bands for the inclusive and exclusive world averages suggested a RH coupling of significant magnitude, the inclusion of the LHCb measurement does not support this.

The discrepancy between the exclusive and the inclusive V_{ub} determinations has a significance of approximately 3σ and has been a long standing puzzle in flavor physics. Several explanations have been proposed, such as the presence of a right-handed $(V + A)$ W coupling [35].

LHCb has performed a measurement of the ratio of branching fractions of the Λ_b^0 into $p\mu^-\bar{\nu}_\mu$ and $\Lambda_c^+\mu^-\bar{\nu}_\mu$ final states [27]. This has been done using pp collisions from the LHC, corresponding to $2.0\, fb^{-1}$ of integrated luminosity at 8 TeV. The $b \to u$ transition $\Lambda_b^0 \to p\mu^-\bar{\nu}_\mu$ could not be considered before at B-factories, but becomes feasible at the LHC.

DOI: 10.1051/epjconf/201612602001

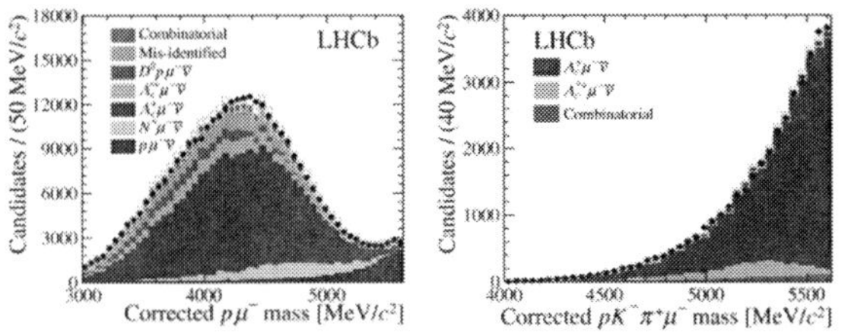

Figure 6. Fits are made for $\Lambda_b^0 \to p\mu^-\bar{\nu}$ (left) and $\Lambda_b^0 \to \Lambda_c^+(pK^-\pi^+)\mu^-\bar{\nu}$ (right) candidates. Data are represented by the black points, and the open boxes represent the statistical uncertainties from the finite size of the simulation samples used to model the mass shapes. There are no data above the nominal Λ_b^0 mass due to the removal of unphysical q^2 solutions.

To facilitate Λ_b^0 reconstruction at a hadron collider, LHCb introduces a corrected mass $M_{corr} = \sqrt{m_{h\mu}^2 + p_\perp{}^2} + p_\perp{}^2$ where h represents either the proton or the Λ_c^+ candidate, $m_{h\mu}$ is the visible mass of the $h\mu$ pair, and $p_\perp$ is the momentum of the $h\mu$ pair transverse to the Λ_b^0 flight direction.The recoil squared mass $q^2(\mu\nu)$ can be determined from the above direction, up to a two-fold ambiguity. For kinematic reasons in the experiment, it is restricted to high values, precisely where the precision on the form factors is best. Secondary vertex isolation criteria are used, and from a detailed comparison between the M_{corr} spectra shown in Fig. 6, a measurement of the ratio of branching ratios is achieved:

$$\frac{\mathcal{B}(\Lambda_b^0 \to p\mu^-\bar{\nu}_\mu)_{q^2>15GeV/c^2}}{\mathcal{B}(\Lambda_b^0 \to \Lambda_c^+\mu^-\bar{\nu}_\mu)_{q^2>7GeV/c^2}} = (1.00 \pm 0.04 \pm 0.08) \times 10^{-2}$$

using the form factor information from [36] for the restricted q^2 regions, the measurement $|V_{ub}|/|V_{cb}| = 0.083 \pm 0.004 \pm 0.004$ is obtained, where the second uncertainty arises from the uncertainty in the LQCD prediction. When the exclusive world average is used for $|V_{cb}|$ [34], the measurement is obtained

$$|V_{ub}| = (3.27 \pm 0.15 \pm 0.17 \pm 0.06) \times 10^{-3}$$

where the uncertainties indicate experimental, LQCD prediction and normalization to V_{cb}. The determination of $|V_{ub}|$ from the ratio of branching ratios depends on the size of a possible right-handed coupling [35], according to an effective Lagangian of the type

$$\mathcal{L}_{eff} = -\frac{4G_F}{\sqrt{2}} V_{ub}^L \left(\bar{u}\gamma_\mu P_L b + \epsilon_R \bar{u}\gamma_\mu P_R b\right)(\bar{\nu}\gamma^\mu P_L l) + h.c.$$

with $P_{R,L} = (1 \pm \gamma^5)/2$. The sensitivity can be appreciated in Fig. 5 (right) which shows the experimental constraints on the left-handed coupling $|V_{ub}^L|$ and the fractional right-handed coupling added to the SM, ϵ_R for different measurements. Unlike the case for the pion in $\bar{B}^0 \to \pi^+ l^- \bar{\nu}$ and $B^- \to \pi^0 l^- \bar{\nu}$ decays, the spin of the proton is non-zero, allowing an axial-vector current, which gives a different sensitivity to ϵ_R. The overlap of the bands from the previous measurements suggested a significant right-handed coupling, but the inclusion of the LHCb $|V_{ub}|$ measurement does not support that assumption.

In summary, the most precise measurement to date of $|V_{ub}|$ is reported using the exclusive decay mode $\Lambda_b^0 \to p\mu^-\bar{\nu}_\mu$. The measurement is in agreement with the exclusively measured world average [30], but disagrees with the inclusive measurement [34] at a significance level of 3.5σ. The measurement will have an important impact on the global fits to the parameters of the CKM matrix.

5 Observation of the very rare decays $B^0_{(s)} \to \mu^+\mu^-$

The decay of the B^0_s meson into dimuons is suppressed strongly, to the level $10^{-9} - 10^{-10}$. It represents a powerful probe in testing new physics effects, because the suppression originates from essential features of the heavy particle spectrum of the SM, namely: the GIM mechanism, helicity suppression and the fact that the SM contributions involve an off-diagonal element of the CKM matrix.

Since these features are of course not generally respected by generic extensions of the SM, the above decay has been searched for many years at most generations of accelerators. The Feynman diagrams are shown in Fig. 7, illustrating the suppression in the SM and the sensitivity to Supersymmetry models.

The first evidence for the $B^0_s \to \mu^+\mu^-$ decay was presented by the LHCb collaboration in 2012 [47]. The LHCb and CMS experiments have presented a joint analysis [48], in order to fully exploit the statistical power of the LHC data, and take into account the correlation between the physical quantities in common to the two analyses. The data correspond to total integrated luminosities of 25.0 fb^{-1} and 3.0 fb^{-1} for the CMS and LHCb experiments, respectively. This is equivalent to a total of approximately $10^{12} B^0_s$ and B^0 mesons produced in both experiments together, determined from the number of expected events assuming the SM branching fractions.

The branching fractions of these two decays, accounting for higher order electromagnetic and strong interaction effects, are reliably calculated in the SM. The untagged time-integrated SM predictions are $\mathcal{B}(B^0_s \to \mu^+\mu^-)_{SM} = (3.66 \pm 0.23) \times 10^{-9}$ and $\mathcal{B}(B^0 \to \mu^+\mu^-)_{SM} = (1.06 \pm 0.09) \times 10^{-10}$ [37], which use the latest lattice QCD results to compute B^0_s and B^0 meson decay constants [38].

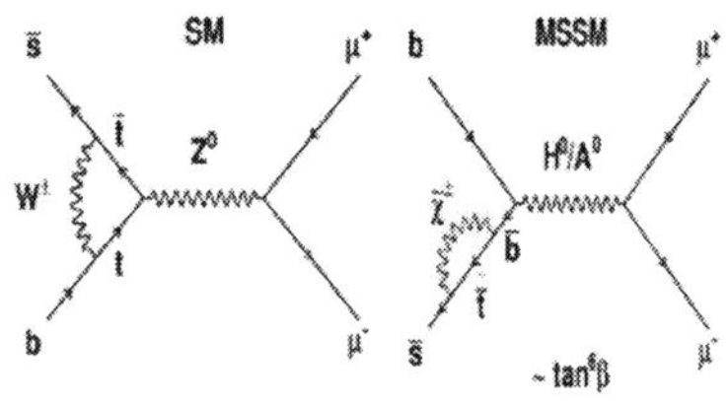

Figure 7. Some Feynman diagrams for the $B^0_s \to \mu^+\mu^-$ decay in (a) the SM and (b) Minimal Supersymmetric model.

Many theories that seek to go beyond the SM (BSM) include new phenomena and particles [39], such as in the diagram shown in Fig. 8(b), that can considerably modify the SM branching fractions. In particular, theories with additional Higgs bosons [40] predict possible enhancements of the branching fractions. A significant deviation of either of the two measurements from the SM predictions would give insight on how the SM should be extended. Alternatively, a measurement compatible with the SM could provide strong constraints on BSM theories.

The ratio of the branching fractions of the two decay modes additionally provides powerful discrimination among BSM theories [41]. In the SM it is predicted to be $\mathcal{B}(B^0 \to \mu^+\mu^-)_{SM}/\mathcal{B}(B^0_s \to \mu^+\mu^-)_{SM} = 0.0295^{+0.0028}_{-0.0025}$ [44], [45], [46]. Notably, BSM theories with the property of minimal flavor violation [42] predict the same value as the SM for this ratio.

The two experiments measure at different angular regions with respect to the LHC beams, according to their different design purposes. Their dimuon mass resolution are also different, namely ≈ 25 MeV/c^2 for LHCb, and ranging from 32-76 MeV/c^2, depending on the pseudorapidity of the two tracks for CMS. The separation between genuine $B^0_s \to \mu^+\mu^-$ decays and random combinations of two muons, most often from semi-leptonic decays of two different b hadrons, is achieved by means of boosted decision trees (BDTs) [43]. Each experiment selected the best set of discriminating variables in their respective BDT. One example is the decay length with respect to the PV. Having lifetimes of about 1.5 ps, $B^0_{(s)}$ mesons travel up to a few centimetres before they decay, with momenta between a few GeV/c and 100 GeV/c at the LHC. Candidates were categorized according to the value of the relevant BDT discriminant, to whether they were detected in CMS or LHCb, and to the kinematical configuration of both muons, in the case of CMS. 20 different categories were so defined, as described in Ref. [48].

A single dimuon mass distribution, extended over all categories, is shown in Fig. 8. Event candidates are weighted according to their values of $S/(S + B)$, where S is the expected number of B_s^0 signals and B the number of background events under the B_s^0 peak in each category.

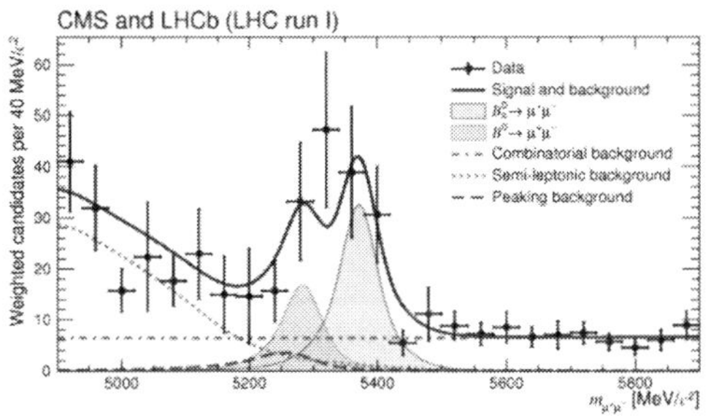

Figure 8. Weighted distribution of the dimuon invariant mass $m_{\mu^+\mu^-}$ for all categories defined within the CMS and LHCb experiments. Superimposed on the black data points are the combined fit (solid blue line) and its components: the B_s^0 (yellow shaded area) and B^0 (light blue shaded area) signal components, the combinatorial background (dashed-dotted green line); the sum of the semileptonic backgrounds (dotted salmon line); and the peaking backgrounds (dashed violet line).

Likelihood contours for $\mathcal{B}(B_s^0 \to \mu^+\mu^-)$ versus $\mathcal{B}(B^0 \to \mu^+\mu^-)$ are shown in Fig. 8. One-dimensional likelihood scans for both decay modes are displayed in the same figure. A combined fit leads to the measurements $\mathcal{B}(B_s^0 \to \mu^+\mu^-) = 2.8^{+0.7}_{-0.6} \times 10^{-9}$ and $\mathcal{B}(B^0 \to \mu^+\mu^-) = 3.9^{+1.6}_{-1.4} \times 10^{-10}$, where the uncertainties include both statistic and systematic sources. The statistical significance is computed to be 6.2σ for the $B_s^0 \to \mu^+\mu^-$ mode and 3.0σ for the $B^0 \to \mu^+\mu^-$ mode (we report the significance obtained with the FC method) .

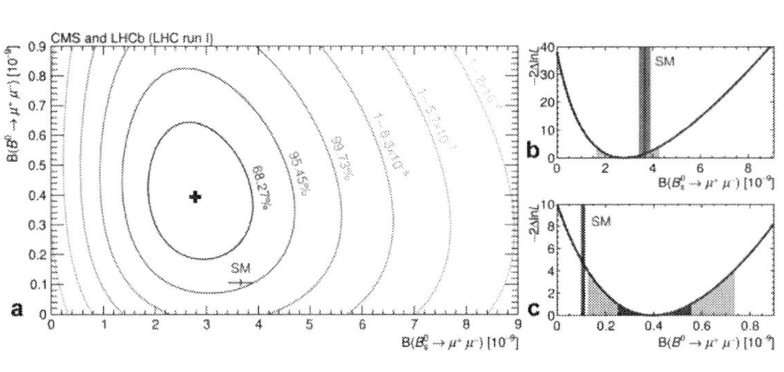

Figure 9. Likelihood contours in the $\mathcal{B}(B_s^0 \to \mu^+\mu^-)$ versus $\mathcal{B}(B^0 \to \mu^+\mu^-)$ plane. The (black) cross in (a) marks the best-fit central value. The SM expectation and its uncertainty is shown in the (red) marker. Each contour encloses a region corresponding to the reported confidence level. Variations of the test statistic -2lnL are also shown for $\mathcal{B}(B_s^0 \to \mu^+\mu^-)$ (b) and $\mathcal{B}(B^0 \to \mu^+\mu^-)$ (c). The dark and light (cyan) areas define the $\pm 1\sigma$ and $\pm 2\sigma$ confidence intervals, respectively. The SM prediction and its uncertainty is denoted with the vertical (red) band.

A fit for the ratios of the branching fractions relative to their SM predictions yields $S_{SM}^{B_s^0} = 0.76^{+0.20}_{-0.18}$ and $S_{SM}^{B^0} = 3.7^{+1.6}_{-1.4}$. The ratio of branching ratios themselves yields $\mathcal{R} = 0.14^{+0.08}_{-0.06}$ which is compatible with the SM at the 2.3σ level.

The combined analysis of the data from CMS and LHCb establishes conclusively the existence of the $B_s^0 \to \mu^+\mu^-$ decay and produces a 3σ evidence for the $B^0 \to \mu^+\mu^-$ decay. For B_s^0, this concludes a search that started more than three decades ago. A phase of even higher sensitivity and precision measurements is initiated for both decays.

6 Anomalous branching fraction of $B \to D^*\tau\bar{\nu}$

Lepton universality, which is preserved within the Standard Model, requires equality of couplings between gauge bosons and the three families of leptons. Hints for lepton nonuniversal effects in

DOI: 10.1051/epjconf/201612602001

$B^+ \to K^+e^+e^-$ and $B^+ \to K^+\mu^+\mu^-$ decays have been seen [50] , but no definitive observation of a deviation has yet been made. However, a large class of models that extend the SM contain additional interactions involving enhanced couplings to the third generation that would violate the above principle. Semileptonic decays of b hadrons to third generation leptons provide a sensitive probe for such effects. In particular, the presence of additional charged Higgs bosons, which are often required in these models, can have a significant effect on the rate of the semitauonic decay $\bar{B}^0 \to D^{*+}\tau^-\bar{\nu}_\tau$ [51].

Semitauonic decays have been observed by BaBar and Belle collaborations [52],[56]. Recently BaBar reported updated measurements [55],[56] of the ratios of branching fractions, $\mathcal{R}(D^*) \equiv \mathcal{B}(\bar{B}^0 \to D^{*+}\tau^-\bar{\nu}_\tau)/\mathcal{B}(\bar{B}^0 \to D^{*+}\mu^-\bar{\nu}_\tau)$ and $\mathcal{R}(D) \equiv \mathcal{B}(\bar{B}^0 \to D^+\tau^-\bar{\nu}_\tau)/\mathcal{B}(\bar{B}^0 \to D^+\mu^-\bar{\nu}_\tau)$, which show deviations of 2.7σ and 2.0σ, respectively, from the SM predictions [57],[58]. These ratios have been calculated with high precision, owing to the cancelation of most of the uncertainties associated with the strong interaction in the B to $D^{(*)}$ transition. Within the SM they differ because of phase-space effects due to the differing charged lepton masses.

LHCb has achieved a new measurement of $\mathcal{R}(D^*)$ using hadron collisions at the LHC with an integrated luminosity of 1.0 fb^{-1} and 2.0 fb^{-1} collected at pp center-of-mass energies of 7 TeV and 8 TeV, respectively [49]. The $\bar{B}^0 \to D^{*+}\tau^-\bar{\nu}_\tau$ decay with $\tau^- \to \mu^-\bar{\nu}_\mu\nu_\tau$ (the signal channel) and the $\bar{B}^0 \to D^{*+}\mu^-\bar{\nu}_\tau$ decay (normalization channel) produce identical visible final-state topologies; consequently both are selected by a common reconstruction procedure. The selection identifies semileptonic $\bar{B}^0$ decay candidates containing a muon candidate and a D^{*+} candidate through the decay chain $D^{*+} \to D^0(\to K^-\pi^+)\pi^+$. The selected sample contains contributions from the signal and the normalization channel, as well as several background processes from hadron collisions, which include partially reconstructed B decays and candidates from combinations of unrelated particles from different b hadron decays. The kinematic and topological properties of the various components are exploited to suppress the background contributions. The signal, the normalization component and the residual background are statistically disentangled with a multidimensional fit to the data, using template distributions derived from control samples, or from simulation validated against real data.

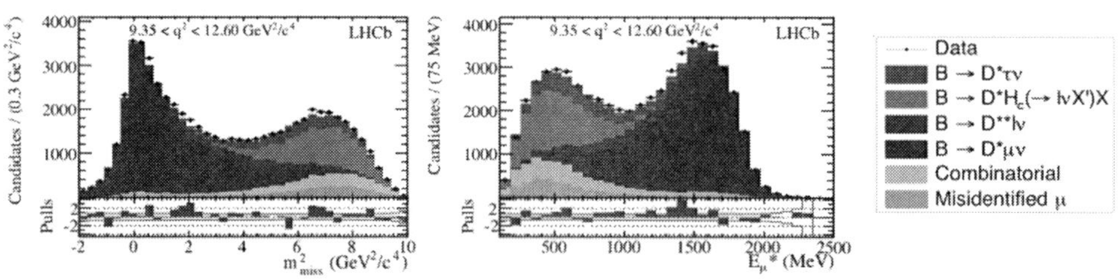

Figure 10. Distributions of m^2_{miss} (left) and E^*_μ (right) of one of the four q^2 bins of the signal data used in the fit ($9.35<q^2<12.60$ GeV^2/c^4), overlaid with projections of the fit model with all normalization and shape parameters at their best-fit values. Below each pannel differences between the data and fit are shown, normalized by the Poisson uncertainty in the data. The bands give the 1σ template uncertainties.

The separation of the signal from the normalization channel, as well as from background processes, is achieved by exploiting the distinct kinematic distributions resulting from the μ-τ mass difference and the presence of extra neutrinos from the decay $\tau^- \to \mu^-\bar{\nu}_\mu\nu_\tau$. The most discriminating kinematic variables, computed in the B rest frame, are the following quantities: the muon energy E^*_μ; the missing mass squared, defined as $m^2_{miss} = (p^\mu{}_B - p^\mu{}_D - p^\mu{}_\mu)^2$; and the squared four-momentum transfer to the lepton system, $q^2 = (p^\mu{}_B - p^\mu{}_D)^2$, where P^μ_B, P^μ_D, and P^μ_μ are the four momenta of the B meson, the D^{*+} meson and the muon. The determination of the rest-frame variables requires knowlege of the B candidate momentum vector, which is estimated from the measured parameters of the reconstructed final-state particles. The B momentum direction is determined from the unit vector to the B decay vertex from the associated PV.

The results of the fit to the signal sample are shown in Fig. 10. Values of the $\bar{B}^0 \to D^{*+}\mu^-\bar{\nu}_\mu$ form factor parameters determined from the fit agree with the current world average values. The fit finds 363000 ± 1600 $\bar{B}^0 \to D^{*+}\mu^-\bar{\nu}_\mu$ decays in the signal sample and an uncorrected ratio of yields $N(\bar{B}^0 \to D^{*+}\tau^-\bar{\nu}_\tau)/N(\bar{B}^0 \to D^{*+}\mu^-\bar{\nu}_\mu) = (4.54 \pm 0.46) \times 10^{-2}$. Accounting for the $\tau^- \to \mu^-\bar{\nu}_\mu\nu_\tau$ branching fraction and the ratio of efficiencies results in $\mathcal{R} = 0.336 \pm 0.034$, where the uncertainty includes the statistical uncertainty, the uncertainty due to form factors, and the statistical uncertainty in the kinematic distributions used in the fit. The measured value is in good agreement with previous measurements by BABAR and Belle [52],[54] and is 2.1σ greater than the SM expectation of 0.252 ± 0.003 [57].

7 Angular analysis of $b \to s\mu^+\mu^-$ decays

The decay $B^0 \to K^{*0}(\to K^+\pi^-)\mu^+\mu^-$ is a $b \to s\mu^+\mu^-$ flavor-changing neutral current (FCNC) transition that has attracted a great deal of interest at B-factories and hadron machines, in recent years. In the Standard Model (SM) the decay is forbidden at tree level and, at lowest order, only occurs via electroweak penguin and box processes. In extensions of the SM, new, heavy particles can enter in competing processes and can significantly change the branching fraction of the decay and the angular distribution of the final state particles. In Fig. 11 some representative diagrams of either case are shown.

Angular observables are of particular interest, since theoretical predictions of such observables tend to be less affected by form-factor uncertainties in the $B^0 \to K^{*0}$ transition. Hereafter K^{*0} is used to refer to the $K^{*0}(892)$.

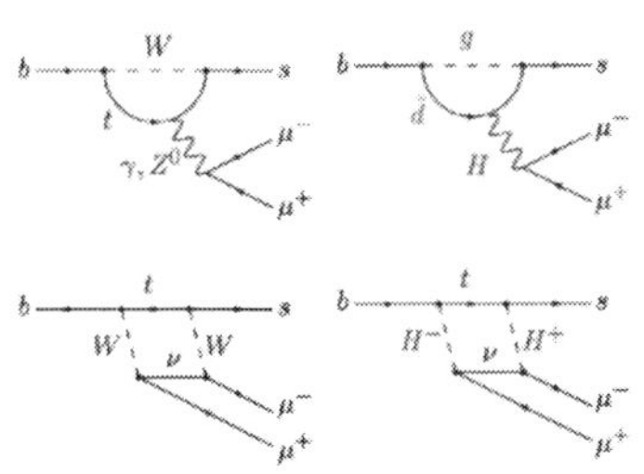

Figure 11. Some illustrating diagrams indicating the penguin loops and box contributions from the SM to $B^0 \to K^{*0}\mu^+\mu^-$ decay (left side ones), and contributions from new Higgs particles in physics beyond the SM (right side ones)

Based on 1.0 fb^{-1}, the LHCb collaboration published in 2013 an analysis of the $B^0 \to K^{*0}\mu^+\mu^-$ final state, and determined a set of angular observables that have reduced theoretical uncertainties and cannot be determined from the single angular projections [61]. The above analysis found a local deviation from the SM prediction, with a significance corresponding to 3.7σ in one observable, P_5'. Possible interpretations of this discrepancy and the consistency of all of the measurements of $b \to s$ transitions have been widely discussed in recent literature [59]. LHCb has performed an updated analysis using the full LHCb Run 1 data sample [60], corresponding to an integrated luminosity of 3.0 fb^{-1}.

The final state of the decay $B^0 \to K^{*0}\mu^+\mu^-$ can be fully described by q^2, the invariant mass of the dimuon system squared, and three decay angles $\vec{\Omega} = (\cos\theta_l, \cos\theta_K, \phi)$, where $\theta_{l,K}$ are the helicity angles in the K^{*0} and dimuon system, and ϕ describes the angle between the $\mu^+\mu^-$ plane and the $K^+\pi^-$ plane, defined in the B^0 rest frame. The differential decay rates of $\bar{B}^0 \to \bar{K}^{*0}\mu^+\mu^-$ and $B^0 \to K^{*0}\mu^+\mu^-$ are given by

$$\frac{d^4\Gamma[\bar{B}^0 \to \bar{K}^{*0}\mu^+\mu^-]}{dq^2 d\vec{\Omega}} = \frac{9}{32\pi}\sum_j I_j(q^2) f_j(\vec{\Omega}) \ , \quad \frac{d^4\bar{\Gamma}[B^0 \to K^{*0}\mu^+\mu^-]}{dq^2 d\vec{\Omega}} = \frac{9}{32\pi}\sum_j \bar{I}_j(q^2) f_j(\vec{\Omega})$$

where the terms $f_j(\vec{\Omega})$ arise from spherical harmonics and the $I_j(q^2)$ are eleven q^2 dependent angular observables. The $I_j(q^2)$ can be expressed as bilinear combinations of six complex decay amplitudes $\mathcal{A}^{L,R}_{0,\parallel,\perp}$, which correspond to different transversity states of the K^{*0} and different (left- and right-handed) chiralities of the dimuon system. Following the notation in [65], CP-averaged observables can be defined as

$$S_j = \left(I_j + \bar{I}_j\right) \Big/ \left(\frac{d\Gamma}{dq^2} + \frac{d\bar{\Gamma}}{dq^2}\right)$$

If q^2 is sufficiently large ($q^2 \gtrsim 1$ GeV^2/c^4), the muons can be considered massless and the CP-averaged observables S_1 and S_2 obey the relations $S_{1s} = 3S_{2s}$, $S_{1c} = -S_{2c}$ and $\frac{3}{4}(2S_{1s} + S_{1c}) - \frac{1}{4}(2S_{2s} + S_{2c}) = 1$ (see for example Ref. [65]). These relationships reduce the number of observables from eleven to eight. The S_{1c} observable is more commonly expressed in terms of the longitudinal polarization fraction of the K^{*0},

$$F_{\rm L} = S_{1c} = \frac{|\mathcal{A}^L_0|^2 + |\mathcal{A}^R_0|^2}{|\mathcal{A}^L_0|^2 + |\mathcal{A}^R_0|^2 + |\mathcal{A}^L_\parallel|^2 + |\mathcal{A}^R_\parallel|^2 + |\mathcal{A}^L_\perp|^2 + |\mathcal{A}^R_\perp|^2}$$

It is also conventional to replace S_{6s} by the forward-backward asymmetry of the dimuon system $A_{\rm FB}$, where $A_{\rm FB} = \frac{3}{4}S_{6s}$. Additional sets of observables, for which the leading form-factor uncertainties cancel, can be built from $F_{\rm L}$ and S_3 through S_9. Examples of such "optimized" observables include the transverse asymmetry $A_{\rm T}^{(2)}$ [63], where $A_{\rm T}^{(2)} = S_3/(1 - F_{\rm L})$ and the P' series of observables [64] with, for example, $P'_{4,5} = S_{4,5}/\sqrt{F_{\rm L}(1 - F_{\rm L})}$.

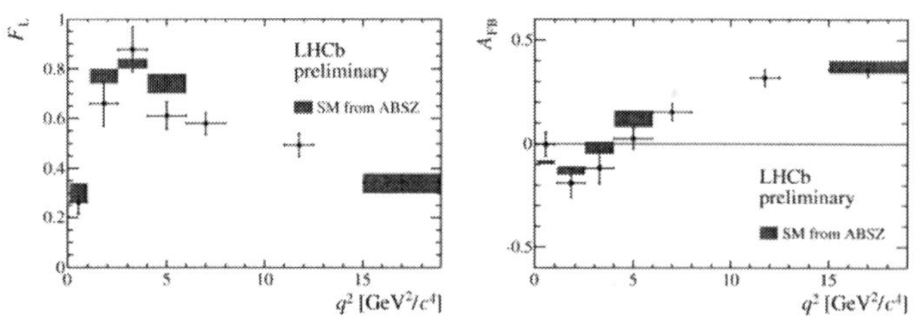

Figure 12. The CP-averaged observables $F_{\rm L}$ (left) and $A_{\rm FB}$ (right) in bins of q^2. The shaded boxes show the SM prediction taken from Ref. [71].

An unbinned maximum likelihood fit has been used to determine the CP-averaged observables $F_{\rm L}$, $A_{\rm FB}$, and S_3 through S_9. Additional observables related to the introduction of the S-wave in the $K^+\pi^-$ system are explicitely included as nuisance parameters (see Ref. [60] for more details). The data are analysed in approximately 2 GeV^2/c^4 q^2 bins. The SM predictions are based on the description in Ref. [62]. Light-cone sum rule predictions which are valid in the low-q^2 region, are combined with lattice determinations at high q^2 [66, 67]. The predictions are made in the regions $0.1 < q^2 < 6.0$ GeV^2/c^4 and $15.0 < q^2 < 19.0$ GeV^2/c^4. No predictions are included for the region close to the narrow $c\bar{c}$ resonances, the J/ψ and $\psi(2S)$, where many of the assumptions that go into the SM predictions are thought to break down.

The results on all of the observables appear largely in agreement with the SM predictions, with the exception of the observable S_5. A mild tension can also be seen in the measured $A_{\rm FB}$ distribution.

We refer the reader to Ref. [60] for the full set of measured q^2 dependent distributions, and just show in Fig. 11 the F_{L} and A_{FB} results, including the comparison with the SM predictions.

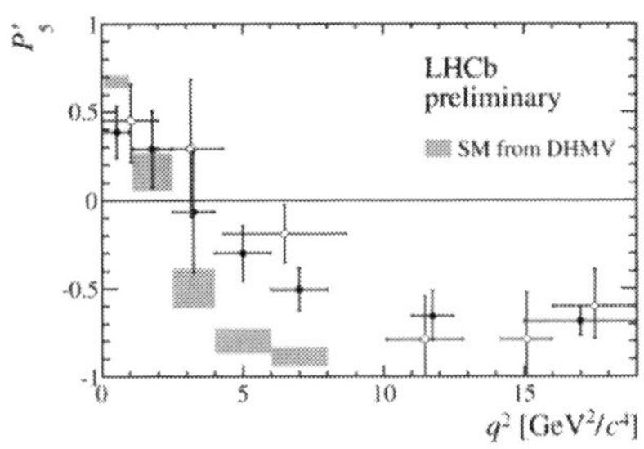

Figure 13. The observable P_5' in bins of q^2. The data of the full 3.0 fb^{-1} data sample are shown as black dots. The shaded boxes indicate the SM prediction taken from Ref. [68]. The blue open markers show the result of the 1.0 fb^{-1} analysis from Ref. [61].

The tension in S_5 can also be seen in the related distribution of $P_5' = S_5/\sqrt{F_{\mathrm{L}}(1-F_{\mathrm{L}})}$ which is shown in Fig. 13. This observable is determined by reparametrising the angular distribution accordingly. For the P_5' observable, a prediction from Ref. [68] is shown. This prediction is computed in the region $0.1 < q^2 < 8.0$ GeV^2/c^4, where a local tension with the SM prediction was seen in the earlier 1.0 fb^{-1} LHCb data analysis [61]. In the present analysis, a tension with the SM prediction (from Ref. [68]) at a level of 2.9σ is observed in each of the $4.0 < q^2 < 6.0$ GeV^2/c^4 and $6.0 < q^2 < 8.0$ GeV^2/c^4 bins. A naïve combination of these deviations, based on a χ^2 probability with two degrees of freedom and assuming the SM predictions in the two bins are uncorrelated, yields a local tension of 3.7σ. Although completely consistent with the SM predictions, the A_{FB} fit results are systematically $\lesssim 1\sigma$ below the SM prediction in the region with $1.0<q^2<6.0$ GeV^2/c^4. The zero-crossing point of A_{FB}, q_0^2, has been determined to be $(3.7^{+0.8}_{-1.1})$ GeV$^2/c^4$, where the uncertainty is purely statistical. The value of q_0^2 is in good agreement with SM predictions which are typically around 4.0 GeV$^2/c^4$, with a relative uncertainty smaller than 10% [69, 70].

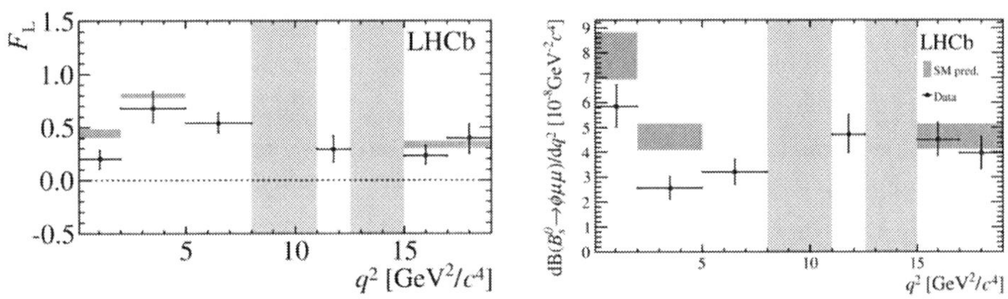

Figure 14. CP-averaged angular observable F_{L} (left) and differential branching fraction of the decay $B_s^0 \to \phi\mu^+\mu^-$ (right). Data are shown by black dots, overlaid with SM predictions [62, 71], indicated as blue shaded boxes. The vetoes excluding the charmonium resonances are indicated by grey areas.

Another $b \to s\mu^+\mu^-$ channel of interest recently studied by LHCb is the decay $B_s^0 \to \phi(\to K^+K^-)\mu^+\mu^-$ [72]. Contrary to $B^0 \to K^{*0}\mu^+\mu^-$, the above decay is self conjugate, since the final state particles provide no specific information on the $b/\bar{b}$ quark flavor. Subject to essentially the same new physics sensitivity, the $B_s^0 \to \phi\mu^+\mu^-$ channel provides good complementarity to $B^0 \to K^{*0}\mu^+\mu^-$, having an observed yield of 432 ± 24 events. With similar four-body kinematics, the full angular analysis as function of $q^2(\mu^+\mu^-)$ allows to determine four CP averages and four CP asymmetries, all of which have been determined by LHCb for the first time, using the full 3.0 fb^{-1} data sample from Run 1. Physics-wise however, the observables are different from those of the $B^0 \to K^{*0}\mu^+\mu^-$, and we refer the reader to Ref. [72] for a full description of them. In particular, there are new CP-violating asymmetries, while the S_5 and A_{FB} observables of the former channel are not accessible.

As a result of the q^2 dependent analysis of the eight angular observables, good consistency is found with the SM predictions for $B_s^0 \to \phi\mu^+\mu^-$ [72]. However, the differential branching fraction of $(2.58^{+0.38}_{-0.31} \pm 0.08 \pm 0.19) \times 10^{-8}\ \mathrm{GeV}^{-2}c^4$ for $B_s^0 \to \phi\mu^+\mu^-$ in the range $1 < q^2 < 6\ \mathrm{GeV}^2/c^4$, where precise theoretical calculations are available, is found to be more than 3σ below the SM prediction of $(4.81 \pm 0.56) \times 10^{-8}\ \mathrm{GeV}^2/c^4$ [62, 71]. In Fig. 14 the CP-averaged longitudinal polarization fraction F_{L} (left) and the differential branching fraction of $B_s^0 \to \phi\mu^+\mu^-$ (right), are shown, along with the theoretical predictions.

8 Summary

The LHCb experiment has performed a successful physics program after completion of Run 1 at the LHC, and some recent topics on flavor physics have been selected in this article. The sensitivity to $B^0_{(s)} \to \mu^+\mu^-$ has reached the level of 10^{-10} and is expected to improve with integrated luminosity. A detailed comparison with the SM predictions shows general good agreement for the selected topics, although a few interesting tensions are observed, which deserve further study. Those include hints on τ lepton non-universality in $\mathcal{R}(D^*)$, a discrepancy on the S_5 observable in $B^0 \to K^{*0}\mu^+\mu^-$, and consistently low rates of $b \to s\mu^+\mu^-$. A precision determination of V_{ub} has been achieved from a new exclusive channel, $\Lambda_b^0 \to p\mu^-\bar{\nu}_\mu$, which does not support the hypothesis of $W^\pm$ right-handed currents as a possible explanation for the still standing discrepancy between inclusive and exclusive measurements of this important parameter. The Run 2 has just started at the LHC, and we hope to learn more about the above and other flavor physics issues in the near future.

References

[1] LHCb collaboration, Alves, A.A. Jr et al., JINST 3, S088005 (2008)
[2] LHCb collaboration, Aaij et al. Int. J. Mod. Phys. A30, 1530022 (2015).
[3] T. Sjöstrand, S. Mrenna and P. Skands, Comput. Phys. Commun. 178 (2008) 852.
[4] D. J. Lange, Nucl. Inst. Meth. A462 (2001) 152.
[5] I. Belyaev et al., J. Phys. Conf. Ser. 331 (2011) 032047.
[6] Aaij et al., JINST 8, P044022 (2013).
[7] V. V. Gligorov and M. Williams, JINST 8 (2013) P02013.
[8] M. Kobayashi and T. Maskawa, Prog. Theor. Phys. 49 (1973) 652.
[9] BaBar collaboration, B. Aubert et al. Phys. Rev. Lett. 87 (2001) 091801.
Belle collaboration, K. abe et al. Phys. Rev. Lett. 87 (2001) 091802.
[10] The LHCb collaboration, R. Aaij et al., Phys. Lett. B736 (2014) 186.,
The LHCb collaboration, R. Aaij et al., Phys. Rev. D87 (2013) 112010, and
The LHCb collaboration, R. Aaij et al., Phys. Rev. Lett. 113 (2014) 211801.
[11] J. Brod and J. Zupan, JHEP 01 (2014) 051.
[12] M. Gronau and D. Wyler, Phys. Lett. B265 (1991) 172.
M. Gronau and D. London, Phys. Lett. B253 (1991) 483.
[13] D. Atwood, I. Dunietz, and A. Soni, Phys. Rev. Lett. 78 (1997) 3257.
D. Atwood, I. Dunietz, and A. Soni, Phys. Rev. D63 (2001) 036005.
[14] A. Giri, Y. Grossman, A. Soffer, and J. Zupan, Phys. Rev. D68 (2003) 054018.
[15] M. Gronau, Phys. Lett. B557 (2003) 198.

[16] Improved constraints on γ: CKM2014 update, LHCb-CONF-2014-004.
[17] BaBar collaboration, J. P. Lees et al., Phys. Rev. D87 (2013) 052015.
[18] K. Trabelsi, on behalf of the Belle collaboration at CKM2012, Cincinnati, USA, October 2012, arXiv:1301.2033.
[19] J. Charles et al. Phys. Rev. D91 (2015) 073007.
[20] The LHCb collaboration, R. Aaij et al. LHCb-PAPER-2015-020, arXiv:1505.07044.
[21] The LHCb collaboration, R. Aaij et al. Phys. Rev. D91 (2015) 112014.
[22] D. Atwood and A. Soni, Phys. Rev. D68 (2003) 033003.
[23] M. Nayak et al., Phys. Lett. B740 (2015) 1.
[24] BaBar collaboration, B. Aubert et al., Phys. Rev. D79 (2009) 072009.
Belle collaboration, I. Adachi et al., Phys. Rev. Lett. 108 (2012) 171802.
[25] The LHCb collaboration, R. Aaij et al., Phys. Rev. Lett. 115 (2015) 031601.
[26] J. Charles et al. *Current status of the Standard Model CKM fit and constraints on $\Delta F = 2$ New Physics*, arXiv:1501.05013.
[27] The LHCb collaboration, R. Aaij et al., Nature Physics DOI:10.1038/NPHYS3415.
[28] H.J. Rothe,*Lattice Gauge Theories: An introduction*, World Scientific Lecture Notes World Scientific, Singapore, 2012.
[29] C. A. Dominguez, Mod. Phys. Lett. A28 (2013) 1360002.
[30] Heavy Flavor Averaging Group, Y. Amhis et al., arXiv:1412.7515 http://www.slac.standord.edu/xorg/hfag/
[31] BaBar collaboration, J. P. Lees et al., Phys. Rev. D86 (2012) 092004.
P. del Amo Sanchez et al., Phys. Rev. D83 (2011) 032007.
[32] Belle collaboration, H. Ha et al., Phys. Rev. D83 (2011) 071101
A. Sibidanov et al., Phys. Rev. D88 (2013) 032005.
[33] RBC and UKQCD collaborations, J. M. Flynn et al., arXiv:1501.05373.
[34] Particle Data Group, K. A. Olive et al. Chin. Phys. C38 (2014) 090001.
[35] F. U. Bernlochner, Z. Ligeti, and S. Turczyk, Phys. Rev. D90 (2014) 094003 and other references quoted in [27].
[36] W. Detmold, C. Lehmer, and S. Meinel, arXiv:1503.01421.
[37] C. Bobeth et al., Phys. Rev. Lett. 112, 101801 (2014).
[38] RBC-UKQCD Collaborations, Witzel O, http://arXiv:org/abs/1311.0276 (2013).
HPQCD Collaboration, H. Na et al., Phys. Rev. D86, 034506 (2012).
Fermilab Lattice and MILC Collaborations, Bazakov et al., Phys. Rev. D85, 114506 (2012).
[39] C. Huang et al., Phys. Rev. D59, 011701 (1998).
S. Rai Choudhury and N. Gaur, Phys. Lett. B 451, 86-92 (1999).
[40] K. Babu and C. Kolda, Phys. Rev. Lett. 84, 228-231 (2000).
C. Bobeth et al., Phys. Rev. D64, 074014.
[41] A.J. Buras, Phys. Lett. B566, 115-119 (2003).
[42] D'Ambrosio et al., Nucl. Phys. B645, 155-187 (2002).
[43] Hoecker et al. Proc. Sci. Adv. Comput. Anal. Techn. Phys. Res. 040.
[44] S. Aoki et al., Eur. Phys. J. C74, 2890 (2014).
[45] Particle Data Group, J. Beringer et al., Phys. Rev. D86, 010001 (2012).
[46] Heavy Flavor Averaging Group, Y. Amis et al., arXiv:org/abs/1207.1158 (2012).
[47] LHCb Collaboration, R. Aaij et al., Phys. Rev. Lett. 110, 021801 (2013).
[48] The CMS and LHCb collaborations, Nature 522 (2015).

[49] R. Aaij et al. (LHCb collaboration), Phys. Rev. Lett. 115, 111803 (2015).
[50] R. Aaij et al. (LHCb collaboration), Phys. Rev. Lett. 113, 151601 (2014).
[51] M. Tanaka, Z. Phys. C67, 321. (1995).
[52] A. Matyja et al. (Belle collaboration), Phys. Rev. Lett. 99, 191807 (2007).
[53] B. Aubert et al. (BABAR collaboration), Phys. Rev. Lett. 100, 021801 (2008).
[54] A. Bozek et al. (Belle collaboration), Phys. Rev. D82, 072005 (2010).
[55] J. P. Lees et al. (BABAR collaboration), Phys. Rev. Lett. 109, 101802 (2012).
[56] J. P. Lees et al. (BABAR collaboration), Phys. Rev. D88, 072012 (2013).
[57] S. Fajfer, J. F. Kamenic, and I. Nisandzic, Phys. Rev. D85, 094025 (2012).
[58] J. A. Bailey et al. (Fermilab Lattice and MILC collaborations), Phys. Rev. Lett. 109, 071802 (2012).
[59] S. Descotes-Genon, J. Matias, and J. Virto, Phys. Rev. D88 (2013) 074002.
W. Altmannshofer and D. M. Straub, Eur. Phys. J. C73 (2013) 2646.
F. Beaujean, C. Bobeth, and D. Van Dyk, Eur. Phys. J. C74 (2014) 2897.
T. Hurth and F. Mahmoudi, JHEP 04 (2014) 097.
S. Jäger and J. Martin Camalich, JHEP 05 (2013) 043.
J. Lyon and R. Zwicky, arXiv:1406.0566.
W. Altmannshofer, S. Gori, M. Pospelov, and I. Yavin, Phys. Rev. D89 (2014) 095033.
A. Crivellin, G. D'Ambrosio, and J. Heeck, Phys. Rev. Lett. 114 (2015) 151801.
R. Gauld, F. Goertz, and U. Haisch, JHEP 01 (2014) 069.
F. Mahmoudi, S. Neshatpour, and J. Virto, Eur. Phys. J. C74 (2014) 2927.
A. Datta, M. Duraisamy, and D. Ghosh, Phys. Rev. D89 (2014) 071501.
[60] R. Aaij et al., The LHCb collaboration, LHCb-CONF-2015-002.
[61] R. Aaij et al, the LHCb collaboration, Phys. Rev. Lett. 111 (2013) 191801.
[62] W. Altmannshofer and D. M. Straub, Eur. Phys. J C75 (2015) 8, 382.
[63] F. Kruger and J. Matias, Phys. Rev. D71 (2005) 094009.
[64] S. Descotes-Genon, J. Matias, M. Ramon, and J. Virto, JHEP 01 (2013) 048.
[65] W. Altmannshofer et al., JHEP 01 (2009) 019 (2009).
[66] R. R. Horgan et al., Phys. Rev. D89 (2014) 094501.
[67] R. R. Horgan et al., PoS LATTICE 2014 (2015) 372.
[68] S. Descotes-Genon, L. Hofer, J. Matias, and J. Virto, JHEP 1412 (2014) 125.
[69] C. Bobeth et al. JHEP 01 (2012) 107.
[70] M. Beneke, T. Feldman, and D. Seidel, Eur. Phys. J. C41 (2005) 173.
[71] A. Bharucha, D.M. Straub, and R. Zwicky, arXiv:1503.05534.
[72] R. Aaij et al, the LHCb collaboration, JHEP 09 (2015), 179.

Observation of pentaquark resonances in $\Lambda_b \to J/\psi K^- p$ decays

B. Adeva[1,a], on behalf of the LHCb Collaboration.

[1] *University of Santiago de Compostela, Spain.*

Abstract. LHCb has recently reported the observation of exotic structures in the $J/\psi p$ channel, in $\Lambda_b \to J/\psi K^- p$ decay, which we discuss here and refer to as charmonium-pentaquark states. The data sample corresponds to an integrated luminosity of 3.0 fb^{-1} acquired with the LHCb detector from 7 and 8 TeV pp collisions. An amplitude analysis of the three-body final state reproduces the two-body mass and angular distributions. To obtain a satisfactory fit of the structures seen in the $J/\psi p$ mass spectrum, it is necessary to include two Breit-Wigner amplitudes that each describe a resonant state. Details are given on the significance, and prefered quantum numbers, of these two resonances.

1 Selection of $\Lambda_b \to J/\psi K^- p$ at LHCb

The LHCb detector [1, 2] is one of the four major detectors at the Large Hadron Collider. It is instrumented in a cone arround the proton beam axis, covering the angles between 10 and 250 mrad in the vertical plane, where most b hadron decays produced in proton-proton collisions occur. The detector includes a high-precision tracking system with a dipole magnet, providing a measurement of momentum and impact parameter (IP), defined for charged particles as the minimum distance of a track to a primary pp interaction vertex (PV). Different types of charged particles are distinguished using information from two ring-imaging Cherenkov detectors, a calorimeter and a muon system.

We discuss here data from $\Lambda_b \to J/\psi K^- p$ decay, corresponding to 1.0 fb^{-1} of integrated luminosity acquired by LHCb at $\sqrt{s} = 7$ TeV, and 2.0 fb^{-1} at $\sqrt{s} = 8$ TeV. Events are triggered by a $J/\psi \to \mu^+\mu^-$ decay, with muon transverse momentum p_T larger than 500 MeV. The dimuon system is required to form a vertex significantly displaced from the nearest pp interaction vertex, and to have an invariant mass within 120 MeV of the J/ψ mass. After applying these requirements, there is a large J/ψ signal over a small background. Only candidates with dimuon invariant mass between -48 and $+43$ MeV relative to the observed J/ψ mass peak are selected, allowing for photon radiation.

Analysis preselection requirements are imposed prior to using a gradient boosted decision tree, BDTG [3], that separates the Λ_b^0 signal from backgrounds. The p_T of individual particles must be larger than 550 MeV for muons and 250 MeV for hadrons. Each hadron must have a significantly non-zero impact parameter with respect to the pp interaction vertex, and be positively identified in the particle ID system. The K^-p system must form a vertex with $\chi^2 < 16$, as must the two muons from J/ψ decay. In addition the Λ_b^0 candidate must form a good vertex, and the vector from the primary vertex to the Λ_b^0 vertex must align with the Λ_b^0 momentum so that the cosine of the angle between them is larger than 0.999.

[a]e-mail: Bernardo.Adeva@usc.es

The BDTG technique involves a "training" procedure using sideband data background and simulated signal samples of 2×10^6 $\Lambda_b \to J/\psi K^- p$ events generated uniformly in phase space in the LHCb acceptance, using Pythia [4], with a special LHCb parameter tune [6], and the LHCb detector simulation based on Geant4 [7]. Specific backgrounds from B_s^0 and B^0 decays are vetoed, by removing combinations that when interpreted as $J/\psi K^+ \pi^-$ fall within ± 30 MeV of the B^0 mass and when interpreted as $J/\psi K^+ K^-$ fall within ± 30 MeV of the B_s^0 mass. After a relatively tight cut on the BDTG output variable, 26007 ± 166 signal candidates are selected, containing 5.4% background within ±15 MeV of the $J/\psi K^- p$ mas peak, as determined from a fit to the mass spectrum, shown in Fig. 1.

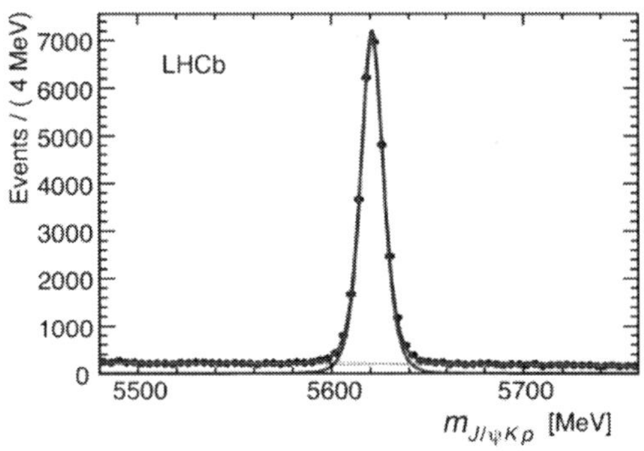

Figure 1. Invariant mass spectrum of $J/\psi K^- p$ combinations, with the total fit, signal, and background components shown as solid (blue), solid (red), and dashed lines, respectively.

2 Summary of the $\Lambda_b^0 \to J/\psi K^- p$ Dalitz plot analysis

The large yields of $\Lambda_b^0 \to J/\psi K^- p$ decays available at LHCb have been used earlier for a precise measurement of the Λ_b^0 lifetime [8]. Fig. 2 shows the Dalitz plot [9] using the $K^- p$ and $J/\psi p$ invariant masses-squared as independent variables. A distinct vertical band is observed in the $K^- p$ invariant mass distribution near 2.3 GeV^2 corresponding to the $\Lambda(1520)$ resonance. There is also a distinct horizontal band near 19.5 GeV^2. As structures are seen in both $K^- p$ and $J/\psi p$ mass distributions a full amplitude analysis is performed, using the available angular variables in addition to the mass distributions, in order to determine the resonances present. No structure is seen in the $J/\psi K^-$ invariant mass.

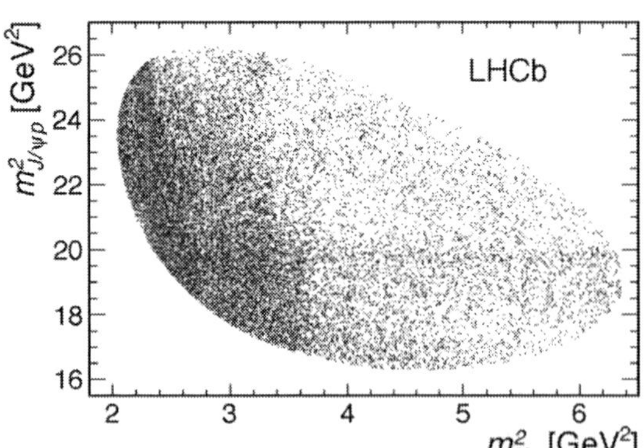

Figure 2. Invariant mass squared of $K^- p$ versus $J/\psi p$ for candidates within ±15 MeV of the Λ_b^0 mass.

The decay $\Lambda_b^0 \to J/\psi K^- p$ can proceed by the diagram shown in Fig. 3 (a), and is expected to be dominated by $\Lambda^* \to K^- p$ resonances, as are evident in the data shown in Fig. 4 (a). It could also

DOI: 10.1051/epjconf/201612602002

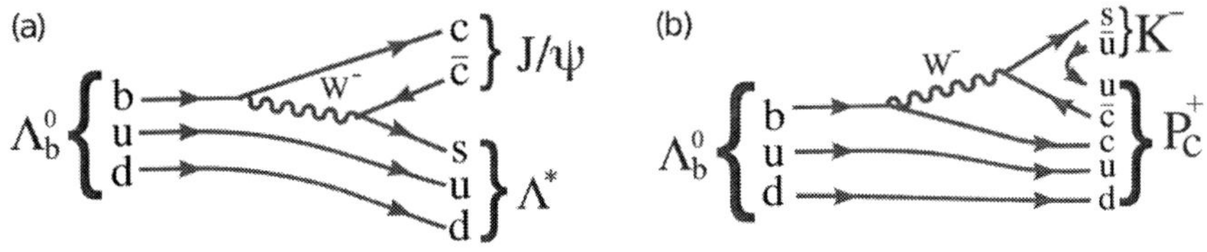

Figure 3. Feynman diagrams for the weak decay $\Lambda_b^0 \to J/\psi K^- p$ in the Standard Model (a), and for the weak decay of Λ_b^0 allowing to form a pentaquark state (b).

have exotic contributions, as indicated by the diagram in Fig. 3 (b), which could result in resonant structures in the $J/\psi p$ mass spectrum shown in Fig. 3 (b).

In practice, resonances decaying strongly into $J/\psi p$ must have a minimal quark content of $c\bar{c}uud$, and thus are charmonium pentaquarks. Such states have been labeled P_c^+, irrespective of the internal binding mechanism. In order to ascertain if the structures seen in Fig. 4 (b) are resonant in nature and not due to reflections generated by the Λ^* states, it is necessary to perform a full amplitude analysis, allowing for interference effects between both decay sequences.

The analytical structure of the helicity amplitudes used is described in Section 3, and according to it a probability density is defined (PDF) that is a function of the six independent dimensions of the $J/\psi K^- p$ phase space, namely five decay angles (denoted collectively by Ω) and the $K^- p$ invariant mass m_{Kp}. The total PDF $\mathcal{P}(m_{Kp}, \Omega \mid \vec{\omega})$ also depends on a certain number of parameters $\vec{\omega}$ that include independent helicity amplitudes and resonance constants. The sum over all candidates $-2ln\mathcal{L}(\vec{\omega}) = -2\sum_i ln\mathcal{P}$ is minimized to give the estimated values of the parameters $\vec{\omega}_{min}$, together with their covariance matrix. Such maximum likelihood fits are performed to the data, using different models of the amplitude analysis. The function $\mathcal{P}(m_{Kp}, \Omega \mid \vec{\omega})$ includes a background component with normalization fixed to be 5.4% of the total.

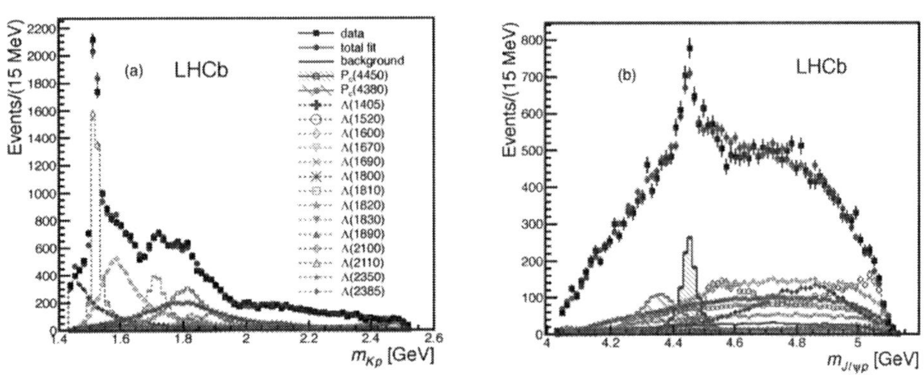

Figure 4. Fit projections for (a) m_{Kp} and (b) $m_{J/\psi p}$ for the reduced Λ^* model with two P_c^+ states. The data are shown as solid (black) squares, while the solid (red) points show the results of the fit. The solid (red) histogram shows the background distribution. The (blue) open squares with the shaded histogram represent the $P_c^+(4450)$ state, and the shaded histogram topped with (purple) filled squares represents the $P_c^+(4380)$ state. Each Λ^* component is also shown. The error bars on the points showing the fit results are due to simulation statistics.

The strategy was to first try to fit the data with a model that can describe the mass and angular ditributions including only Λ^* resonances, allowing all possible known states and decay amplitudes. We call this "extended" model. The masses and widths of the 14 Λ^* states are fixed to their PDG values, and the analysis is described in Ref. [13]. Variations of these parameters are considered in the systematic uncertainties.

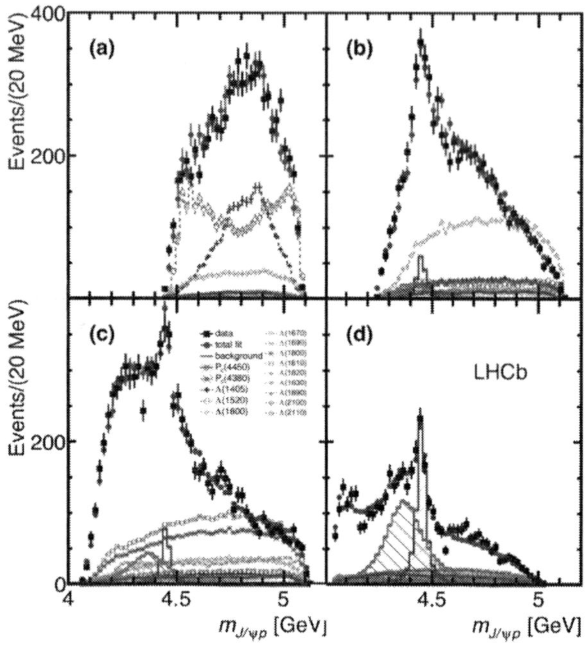

Figure 5. $m_{J/\psi p}$ in various intervals of m_{Kp} for the fit with the two P_c^+ states:
(a) $m_{Kp} < 1.55$ GeV,
(b) $1.55 < m_{Kp} < 1.70$ GeV,
(c) $1.70 < m_{Kp} < 2.00$ GeV, and
(d) $m_{Kp} > 2.00$ GeV. The data are shown as (black) squares with error bars, while the (red) circles show the results of the fit. The blue and purple histograms show the two P_c^+ states. See Fig.4 for the legend.

As this model did not give a satisfactory description of the data, one P_c^+ state was added, and when that was not sufficient, a second state was included. When determining the parameters of the P_c^+ states, a more restrictive model of the K^-p states (referred to as "reduced" model) was used that includes only the Λ^* resonances that are well motivated, and has fewer number of parameters. In both cases, the helicity couplings were expressed in terms of LS couplings, where L is the orbital angular momentum of the decay products and S is their total spin. The number of LS couplings used was also reduced when going from the extended model to the reduced model, which only kept the two lowest allowed values of L for each resonance.

Using the extended model with one resonant P_c^+ improves the fit quality, but it is still unacceptable (see Ref. [13] and supplemental material therein). Acceptable fits were only found with two P_c^+ states. The central values of the fit results are given for the reduced Λ^* model, and the differences in fitted quantities between both models are included in their systematic uncertainties.

The best fit combination finds two P_c^+ states with J^P values of $3/2^-$ and $5/2^+$, for the lower and higher mass states, respectively. The fit projections are shown in Fig. 4, and both m_{Kp} and the peaking structure in $m_{J/\psi p}$ are reproduced by the fit. The two P_c^+ states are found to have masses of $4380 \pm 8 \pm 29$ MeV and $4449.8 \pm 1.7 \pm 2.5$ MeV, with corresponding widths of $205 \pm 18 \pm 86$ MeV and $39 \pm 5 \pm 19$ MeV. The first uncertainty is statistical and the second is systematic. The mass resolution is approximately 2.5 MeV and does not affect the width determinations. The angular distributions are reasonably well reproduced. A significant Λ^* production recoiling against the J/ψ is observed with the lowest mass contributions, the $\Lambda(1405)$ and $\Lambda(1520)$ states having fit fractions of $(15 \pm 1 \pm 6)\%$ and $(19 \pm 1 \pm 4)\%$, respectively.

A comparison of the fit results with the data in m_{Kp} intervals is particularly enlightening, and it is shown in Fig. 5. As m_{Kp} moves away from the Λ^* resonances, the signal from both P_c^+ states becomes clearer. In slice (d) of Fig. 5 both states form a large part of the mass spectrum. In fact, an interference effect is observed in the data as a consequence of the fact that the two P_c^+ states have opposite parity, leading to an asymmetric distribution of the helicity angle in the P_c^+ rest frame, i.e. the angle of the proton in $J/\psi p$ rest frame with respect to the P_c^+ flight direction. This is illustrated in Fig. 6 for the entire m_{Kp} range, as described in Ref. [14] . The summed fit projections agree well with the angular distributions in the data showing that two interfering states are needed to reproduce the asymmetric distribution.

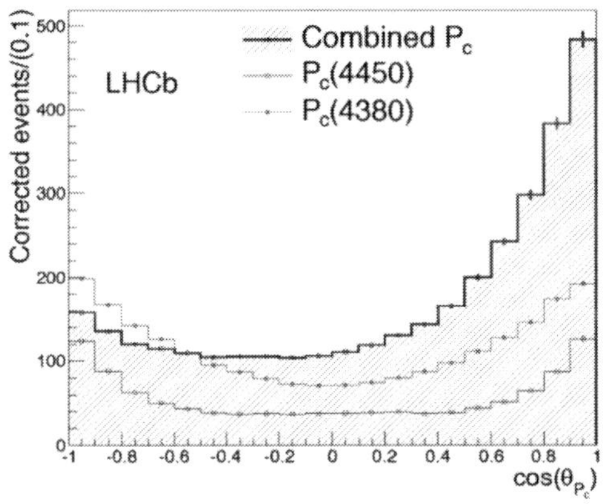

Figure 6. Distribution of the cosine of the J/ψ angle in the P_c^+ rest frame ($\theta_{P_c^+}$). The interference between opposite parity P_c^+ states is needed to explain the decay angular distribution. The fit projections are shown, indicating the combined signal (green), and the individually symmetric rates for each of the intefering amplitudes.Values of $\cos\theta_{P_c^+}$ near -1 are correlated with values of m_{Kp} near threshold, while those near +1 are correlated with higher values.

The -2ln$\mathcal{L}$ values differ by only one unit between the best fit and the parity reversed combination $(3/2^+, 5/2^-)$. Other spin-parity combinations are less likely, although the $(5/2^+, 3/2^-)$ pair changes -2ln$\mathcal{L}$ by only 2.3^2 units and therefore cannot be ruled out. All combinations $1/2^\pm$ through $7/2^\pm$ were tested, and all others are disfavored by changes of more than 5^2 in the -2ln$\mathcal{L}$ values.

Adding a single $5/2^+$ P_c^+ state to the fit with only Λ^* states reduces -2ln$\mathcal{L}$ by 14.7^2 using the extended model and adding a second lower mass $3/2^-$ P_c^+ state results in a further reduction of 11.6^2. The combined reduction of -2ln$\mathcal{L}$ by the two states together is 18.7^2. An independent calculation was performed by generating pseudoexperiments using the null hypothesis having amplitude parameters determined from the fits to the data with no or one P_c^+ state. These studies show a reduction of the significances by about 20%, giving overall significances of 9σ and 12σ, for the lower and higher mass P_c^+ states, respectively. The combined significance of two P_c^+ states is 15σ.

3 The multi-resonance model

The two interfering processes shown in Fig. 1 produce distinct decay sequences: $\Lambda_b^0 \to J/\psi\Lambda^*$, $\Lambda^* \to K^- p$ and $\Lambda_b^0 \to P_c^+ K^-$, $P_c^+ \to J/\psi p$, with $J/\psi \to \mu^+\mu^-$ in both cases. The helicity formalism is used [11], where each sequential decay $A \to BC$ contributes an amplitude term

$$\mathcal{H}^{A\to BC}_{\lambda_B,\lambda_C} D^{J_A}_{\lambda_A,\lambda_B-\lambda_C}(\phi_B,\theta_A,0)^* R_A(m_{BC}) = \mathcal{H}^{A\to BC}_{\lambda_B,\lambda_C} e^{i\lambda_A\phi_B} d^{J_A}_{\lambda_A,\lambda_A-\lambda_B}(\theta_A) R_A(m_{BC}) ,$$

where λ is the helicity (projection of the spin of the particle onto its momentum vector), and $\mathcal{H}^{A\to BC}_{\lambda_B,\lambda_C}$ are complex helicity-coupling amplitudes describing the decay dynamics. Here, θ_A and ϕ_B are the polar and azimuthal angles of B in the rest frame of A (θ_A is known as the helicity angle of A). The three angles of Wigner's D matrix are Euler angles describing the rotation of the initial coordinate system with the z axis along the helicity axis of A to the coordinate system with the z axis along the helicity axis of B. If A has a non-negligible natural width, the invariant mass distribution of the B and C daughters is described by the complex function $R_A(m_{BC})$ discussed below.

The helicity couplings can be further expressed in terms of LS couplings (B_{LS}), when the decay proceeds with orbital angular momentum L and total spin S of the A and B system. Using Clebsch-

Gordan coefficients, the helicity couplings are expressed in terms of LS couplings (B_{LS}), where L is the orbital angular momentum in the decay, and S is the total spin of A plus B.

$$\mathcal{H}^{A\to BC}_{\lambda_B,\lambda_C} = \sum_L \sum_S \sqrt{\frac{2L+1}{2J_A+1}} B_{LS} \begin{pmatrix} J_B & J_C \\ \lambda_B & -\lambda_C \end{pmatrix}\!\!\left|\begin{matrix} S \\ \lambda_B-\lambda_C \end{matrix}\right) \left(\begin{matrix} L & S \\ 0 & \lambda_B-\lambda_C \end{matrix}\right|\begin{matrix} J_A \\ \lambda_B-\lambda_C \end{matrix}\Bigg) .$$

Denoting J/ψ as ψ, the matrix element for the $\Lambda_b^0 \to J/\psi\Lambda^*$ decay sequence is

$$\mathcal{M}^{\Lambda^*}_{\lambda_{\Lambda_b^0},\lambda_p,\Delta\lambda_\mu} \equiv \sum_n \sum_{\lambda_{\Lambda^*}} \sum_{\lambda_\psi} \mathcal{H}^{\Lambda_b^0\to\Lambda_n^*\psi}_{\lambda_{\Lambda^*},\lambda_\psi} D^{\frac{1}{2}}_{\lambda_{\Lambda_b^0},\lambda_{\Lambda^*}-\lambda_\psi}(0,\theta_{\Lambda_b^0},0)^* ,$$

$$\times \mathcal{H}^{\Lambda_n^*\to Kp}_{\lambda_p,0} D^{J_{\Lambda_n^*}}_{\lambda_{\Lambda^*},\lambda_p}(\phi_K,\theta_{\Lambda^*},0)^* R_{\Lambda_n^*}(m_{Kp}) D^1_{\lambda_\psi,\Delta\lambda_\mu}(\phi_\mu,\theta_\psi,0)^* ,$$

where the x axis, in the coordinates describing the Λ_b^0 decay, is chosen to fit $\phi_{\Lambda^*} = 0$. The sum over n is due to many different Λ_n^* resonances contributing to the amplitude. Since the J/ψ decay is electromagnetic, the values of $\Delta\lambda_\mu \equiv \lambda_{\mu^+} - \lambda_{\mu^-}$ are restricted to ± 1. There are four (six) independent complex $\mathcal{H}^{\Lambda_b^0\to\Lambda_n^*\psi}_{\lambda_{\Lambda^*},\lambda_\psi}$ couplings to fit for each Λ_n^* resonance for $J_{\Lambda_n^*} = \frac{1}{2}(> \frac{1}{2})$. They can be reduced to only one (three) free B_{LS} coupling to fit if only the lowest (the lowest two) values of L are considered. The mass m_{Kp}, together with all decay angles entering Eq. 3, $\theta_{\Lambda_b^0}$, $\theta_{\Lambda_n^*}$, θ_K, θ_ψ, and θ_μ (denoted collectively as Ω), constitute the six independent dimensions of the $\Lambda_b^0 \to J\psi pK^-$ decay phase space.

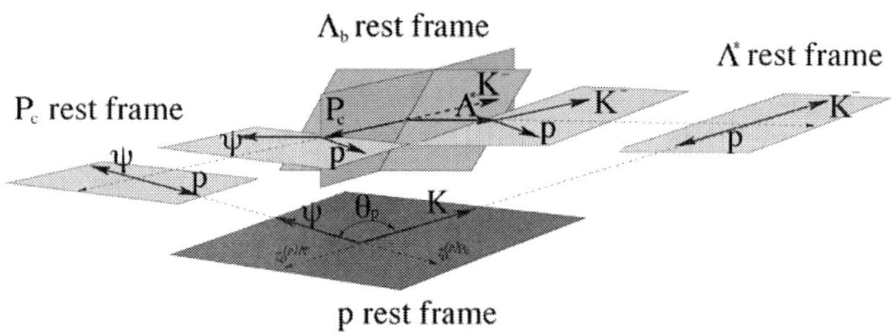

Figure 7. Definition of the decay angles in the P_c^+ decay sequence.

Similarly, the matrix element for the P_c^+ decay chain is given by

$$\mathcal{M}^{P_c}_{\lambda_{\Lambda_b^0},\lambda_p^{P_c},\Delta\lambda_\mu^{P_c}} \equiv \sum_j \sum_{\lambda_{P_c}} \sum_{\lambda_\psi^{P_c}} \mathcal{H}^{\Lambda_b^0\to P_{cj}K}_{\lambda_{P_c},0} D^{\frac{1}{2}}_{\lambda_{\Lambda_b^0},\lambda_{P_c}}(\phi_{P_c},\theta^{P_c}_{\Lambda_b^0},0)^* ,$$

$$\times \mathcal{H}^{P_{cj}\to\psi p}_{\lambda_\psi^{P_c},\lambda_p^{P_c}} D^{J_{P_{cj}}}_{\lambda_{P_c},\lambda_\psi^{P_c}-\lambda_p^{P_c}}(\phi_\psi,\theta_{P_c},0)^* R_{P_{cj}}(m_{\psi p}) D^1_{\lambda_\psi^{P_c},\Delta\lambda_\mu^{P_c}}(\phi_\mu^{P_c},\theta_\psi^{P_c},0)^* ,$$

where the angles and helicity states carry the superscript or subscript P_c to distinguish them from those defined for the Λ^* decay chain. The sum over j allows for the possibility of contributions from more than one P_c resonances. There are two (three) independent helicity couplings $\mathcal{H}^{P_{cj}\to\psi p}_{\lambda_\psi^{P_c},\lambda_p^{P_c}}$ for $J_{P_{cj}} = \frac{1}{2}(> \frac{1}{2})$, and a ratio of the two $\mathcal{H}^{\Lambda_b^0\to P_{cj}K}_{\lambda_{P_c},0}$ couplings to determine from the data.

The mass-dependent $R_{\Lambda_n^*}(m_{Kp})$ and $R_{P_{cj}}(m_{J\psi p})$ terms are given by

$$R_X(m) = B'_{L^X_{\Lambda_b^0}}(p,p_0,d)\left(\frac{p}{M_{\Lambda_b^0}}\right)^{L^X_{\Lambda_b^0}} \times BW(m|M_{0X},\Gamma_{0X}) B'_{L_X}(q,q_0,d)\left(\frac{q}{M_{0X}}\right)^{L_X} .$$

 DOI: 10.1051/epjconf/201612602002

Here p is the $X = \Lambda^*$ or P_c^+ momentum in the Λ_b^0 rest frame, and q is the momentum of either decay product of X in the X rest frame. The symbols p_0 and q_0 denote values of these quantities at the resonance peak ($m = M_{0X}$). The orbital angular momentum between the decay products of Λ_b^0 is denoted as $L^X_{\Lambda_b^0}$. Similarly L_X is the orbital angular momentum between the decay products of X. The orbital angular momentum barrier factors $p^L B_L^{'}(p, p_0, d)$ involve the Blatt-Weisskopf functions, and account for the difficulty in creating larger orbital angular momentum L, which depends on the momentum of the decay products p and on the size of the decaying particle, given by the d constant. The value $d = 3.0\ \text{GeV}^{-1} \sim 0.6$ fm was set. The relativistic Breit-Wigner amplitude is given by

$$BW(m|M_{0X}, \Gamma_{0X}) = \frac{1}{M_{0X}^2 - m^2 - iM_{0X}\Gamma(m)}, \tag{1}$$

where

$$\Gamma(m) = \Gamma_{0X}\left(\frac{q}{q_0}\right)^{2L_X+1} \frac{M_{0X}}{m} B_{L_X}^{'}(q, q_0, d)^2$$

is the mass-dependent width of the resonance. For the $\Lambda(1405)$ resonance, which peaks below the $K^- p$ threshold, a Flatté-like parametrization [12] is used, with appropriate values of the couplings for the allowed channels $\Sigma\pi$ and Kp.

To properly describe the interference between the Λ^* and P_c^+ amplitudes, the proton and muon helicity states must be rotated from the Λ^* to the P_c^+ rest frames. In the proton case, the rotation angle is θ_p (see Fig. 7), and no azimuth is involved since the Λ^* and the P_c^+ decay in the same plane. In the muon case, an azimuthal rotation about the J/ψ flight direction, by an angle α_μ, is implied in the transformation between the above reference frames, the polar angle being absent just because the μ^+ comes from J/ψ decay in both reference frames. The resulting expression is

$$|\mathcal{M}|^2 = \sum_{\lambda_{\Lambda_b^0}} \sum_{\lambda_p} \sum_{\Delta\lambda_\mu} \left| \mathcal{M}^{\Lambda^*}_{\lambda_{\Lambda_b^0},\lambda_p,\Delta\lambda_\mu} + e^{i\Delta\lambda_\mu\alpha_\mu} \sum_{\lambda_p^{P_c}} d^{\frac{1}{2}}_{\lambda_p^{P_c},\lambda_p}(\theta_p) \mathcal{M}^{P_c}_{\lambda_{\Lambda_b^0},\lambda_p^{P_c},\Delta\lambda_\mu} \right|^2 \tag{2}$$

where $\Delta\lambda_\mu = \pm 1$ accounts for both muon spin-aligned orientations. Parity conservation in Λ_b^0 production at the LHC (in absence of polarized beams) prevents Λ_b^0 longitudinal polarization, so $\lambda_{\Lambda_b^0} = +1$ and $\lambda_{\Lambda_b^0} = -1$ are equally likely, which is reflected in Eq. 2.

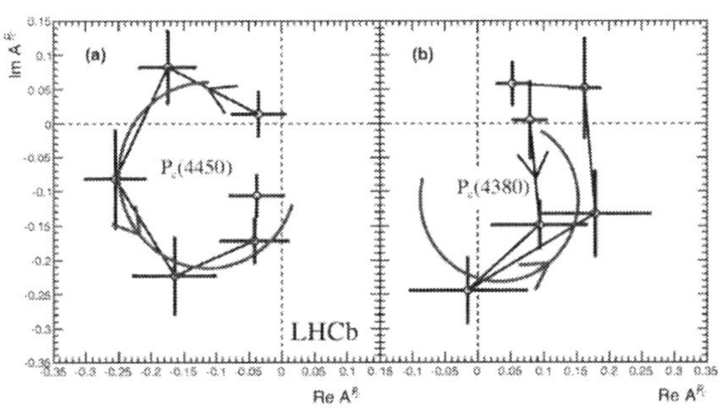

Figure 8. Fitted values of the real and imaginary parts of the amplitudes for the baseline $(3/2^-, 5/2^+)$ fit for (a) the $P_c(4450)^+$ state and (b) the $P_c(4380)^+$ state, each divided into six $m_{J/\psi p}$ bins of equal width between $-\Gamma_0$ and $+\Gamma_0$ shown in the Argand diagrams as connected points with error bars ($m_{J/\psi p}$ increases counterclockwise).The solid (red) curves are the predictions from the Briet-Wigner formula in Eq. 1 evaluated for the fitted values of the mass and width in each case.

The matrix element given by Eq. 2 is a six dimensional function of m_{Kp} and Ω and depends on the fit parameters $\vec{\omega}$, which represent independent helicity or LS couplings, and masses and widths of resonances $\mathcal{M} = \mathcal{M}(m_{Kp}, \Omega \mid \vec{\omega})$. After accounting for the selection efficiency, the probability density function (PDF) described in Section 2 is obtained.

4 Further evidence of resonant character

The resonant character of the higher mass, narrower P_c^+ state is further explored by viewing the evolution of the $m_{J/\psi p}$ dependent part of the Breit-Wigner amplitude discussed above for the $P_c(4450)^+$, in formula 1. Its magnitude and phase vary with $m_{J/\psi p}$ according to an approximately circular trajectory, as it is shown in Fig. 8 (a). The resulting Argand diagram is consistent with a rapid counterclockwise change of the $P_c(4450)^+$ phase when its magnitude reaches the maximum, a behavior characteristic of a resonance. A similar study for the wider state is also shown in Fig. 8 (b), where the fit also shows a large phase change, the amplitude values being sensitive to the details of the Λ^* model, and so this latter study is not conclusive.

5 Summary

In conclusion, a full amplitude fit has been presented to the $\Lambda_b^0 \to J/\psi K^- p$ decay. A significant Λ^* production recoiling against the J/ψ is observed, and the data cannot be satisfactorily described without including two Breit-Wigner shaped resonances in the $J/\psi p$ invariant mass distribution. The significances of the lower mass and higher mass states are 9 and 12 standard deviations, respectively. These structures cannot be accounted for by reflections from $J/\psi\Lambda^*$ resonances or other known sources. Interpreted as resonant states they must have minimal quark content $c\bar{c}uud$, and would therefore be called charmonium pentaquark states. The lighter state $P_c(4380)^+$ has a mass of $4380 \pm 8 \pm 29$ MeV and a width of $205 \pm 18 \pm 86$ MeV, while the heavier state $P_c(4450)^+$ has a mass of $4449.8 \pm 1.7 \pm 2.5$ MeV and a width of $39 \pm 5 \pm 19$ MeV. The parities of the two states are opposite with the preferred spins being $3/2$ for one state and $5/2$ for the other.

References

[1] LHCb collaboration, Alves, A.A. Jr et al., JINST 3, S088005 (2008)

[2] LHCb collaboration, Aaij et al. Int. J. Mod. Phys. A30, 1530022 (2015).

[3] A. Hoecker et al., Proc. Sci. ACAT (2007) 040.

[4] T. Sjöstrand, S. Mrenna and P. Skands, Comput. Phys. Commun. 178 (2008) 852.

[5] D. J. Lange, Nucl. Inst. Meth. A462 (2001) 152.

[6] I. Belyaev et al., J. Phys. Conf. Ser. 331 (2011) 032047.

[7] S. Agostinelli et al., Geant4 Collaboration. Nucl. Inst. Meth. Phys. Res. Sect. A 506, 250 (2003).

[8] R. Aaij et al., LHCb Collaboration, Phys. Rev. Lett. 111, 102003 (2013).

[9] R. H. Dalitz, Philos. Mag. Ser. 5 44, 1068 (1953).

[10] K. A. Olive et al. (Particle Data Group) Review of Particle Physics, Chin. Phys. C 38, 090001 (2014).

[11] S. U. Chung, CERN Report, CERN-71-08; J. D. Richman, CALT-68-1148; M. Jacob and G. C. Wick, Ann. Phys. (N.Y.) 7, 404 (1959).

[12] S. M. Flatté, Phys. Lett., 63B, 224 (1976).

[13] R. Aaij et al. (LHCb Collaboration), "Observation of $J/\psi p$ resonances consistent with pentaquark states in $\Lambda_b^0 \to J/\psi K^- p$ decays ",Phys. Rev. Lett. **115**, 072001 (2015), 12 Aug. 2015.

[14] S. Stone, Meeting of the APS Division of Particles and Fields, DPF Ann Arbor, DPF2015-457, August 2015, arXiv:1509.04051

Scale Hierarchies in Particle Physics and Cosmology

I. Antoniadis[1,2,a]

[1]*LPTHE, UMR CNRS 7589 Sorbonne Universités, UPMC Paris 6, 75005 Paris, France*
[2]*Albert Einstein Center, Institute for Theoretical Physics, Bern University, Sidlerstrasse 5, CH-3012 Bern*

Abstract. We discuss possible connections between several scales in particle physics and cosmology, such the the electroweak, inflation, dark energy and Planck scales. We then describe the phenomenology of a model of supersymmetry breaking in the presence of a tiny (tunable) positive cosmological constant. The model is coupled to the MSSM, leading to calculable soft supersymmetry breaking masses and a distinct low energy phenomenology that allows to differentiate it from other models of supersymmetry breaking and mediation mechanisms.

1 Introduction

If String Theory is a fundamental theory of Nature and not just a tool for studying systems with strongly coupled dynamics, it should be able to describe at the same time particle physics and cosmology, which are phenomena that involve very different scales from the microscopic four-dimensional (4d) quantum gravity length of 10^{-33} cm to large macroscopic distances of the size of the observable Universe $\sim 10^{28}$ cm spanned a region of about 60 orders of magnitude. In particular, besides the 4d Planck mass, there are three very different scales with very different physics corresponding to the electroweak, dark energy and inflation. These scales might be related via the scale of the underlying fundamental theory, such as string theory, or they might be independent in the sense that their origin could be based on different and independent dynamics. An example of the former constrained and more predictive possibility is provided by TeV strings with a fundamental scale at low energies due for instance to large extra dimensions transverse to a four-dimensional braneworld forming our Universe [1]. In this case, the 4d Planck mass is emergent from the fundamental string scale and inflation should also happen around the same scale [2]. We will first review this possibility, focussing on its compatibility with cosmological observations.

We will then adopt a second more conservative approach, assuming that all three scales have an independent dynamical origin. Moreover, we will assume the presence of low energy supersymmetry that allows for an elegant solution of the mass hierarchy problem, a unification of fundamental forces as indicated by low energy data and a natural dark matter candidate due to an unbroken R-parity. The assumption of independent scales implies that supersymmetry breaking should be realized in a metastable de Sitter vacuum with an infinitesimally small (tunable) cosmological constant independent of the supersymmetry breaking scale that should be in the TeV region. In a recent work [3], we studied a simple $N = 1$ supergravity model having this property and motivated by string theory.

[a]e-mail: ignatios.antoniadis@polytechnique.edu

Besides the gravity multiplet, the minimal field content consists of a chiral multiplet with a shift symmetry promoted to a gauged R-symmetry using a vector multiplet. In the string theory context, the chiral multiplet can be identified with the string dilaton (or an appropriate compactification modulus) and the shift symmetry associated to the gauge invariance of a two-index antisymmetric tensor that can be dualized to a (pseudo)scalar. The shift symmetry fixes the form of the superpotential and the gauging allows for the presence of a Fayet-Iliopoulos (FI) term, leading to a supergravity action with two independent parameters that can be tuned so that the scalar potential possesses a metastable de Sitter minimum with a tiny vacuum energy (essentially the relative strength between the F- and D-term contributions). A third parameter fixes the Vacuum Expectation Value (VEV) of the string dilaton at the desired (phenomenologically) weak coupling regime. An important consistency constraint of our model is anomaly cancellation which has been studied in [5] and implies the existence of additional charged fields under the gauged R-symmetry.

In a more recent work [6], we analyzed a small variation of this model which is manifestly anomaly free without additional charged fields and allows to couple in a straight forward way a visible sector containing the minimal supersymmetric extension of the Standard Model (MSSM) and studied the mediation of supersymmetry breaking and its phenomenological consequences. It turns out that an additional 'hidden sector' field z is needed to be added for the matter soft scalar masses to be non-tachyonic; although this field participates in the supersymmetry breaking and is similar to the so-called Polonyi field, it does not modify the main properties of the metastable de Sitter (dS) vacuum. All soft scalar masses, as well as trilinear A-terms, are generated at the tree level and are universal under the assumption that matter kinetic terms are independent of the 'Polonyi' field, since matter fields are neutral under the shift symmetry and supersymmetry breaking is driven by a combination of the $U(1)$ D-term and the dilaton and z-field F-term. Alternatively, a way to avoid the tachyonic scalar masses without adding the extra field z is to modify the matter kinetic terms by a dilaton dependent factor.

A main difference of the second analysis from the first work is that we use a field representation in which the gauged shift symmetry corresponds to an ordinary $U(1)$ and not an R-symmetry. The two representations differ by a Kähler transformation that leaves the classical supergravity action invariant. However, at the quantum level, there is a Green-Schwarz term generated that amounts an extra dilaton dependent contribution to the gauge kinetic terms needed to cancel the anomalies of the R-symmetry. This creates an apparent puzzle with the gaugino masses that vanish in the first representation but not in the latter. The resolution to the puzzle is based to the so called anomaly mediation contributions [7, 8] that explain precisely the above apparent discrepancy. It turns out that gaugino masses are generated at the quantum level and are thus suppressed compared to the scalar masses (and A-terms).

2 Effective Planck mass and the inflation scale

Low scale gravity with large extra dimensions is actually a particular case of a more general framework, where the UV cutoff is lower than the Planck scale due to the existence of a large number of particle species coupled to gravity [9]. Indeed, it was shown that the effective UV cutoff M_{UV} is given by

$$M_{UV}^2 = M_P^2/N\,, \tag{1}$$

where the counting of independent species N takes into account all particles which are not broad resonances, having a width less than their mass. For instance, in braneworld gravity with n large extra dimensions of average size R, the particle species are the Kaluza-Klein (KK) excitations of the graviton (and other possible bulk modes), whose number at a given energy scale E_* is given by

$$N \simeq R^n E_*^n\,. \tag{2}$$

Here, we work out the consequences of this scale dependence of the strength of gravity for inferring various quantities during inflation [2], which we take to be driven by a single field for economy of discussion and because the data doesn't compel us to consider otherwise [10]. As is to be expected, all dimensionless observables such as the amplitude and spectral properties of the perturbations are unaffected by the changing strength of gravity at inflationary energies. However, when one tries to *infer* an absolute energy scale for inflation, one finds that it is undetermined commensurate with (1) up to the unknown spectrum of universally coupled species between laboratory scales and the inflationary scale, the details of which we elaborate upon in the following.

According to the inflationary paradigm, the primordial perturbations observed in the CMB were created at horizon crossing during the quasi de Sitter (dS) phase of early accelerated expansion sourced by the inflaton field. Therefore all quantities that enter calculations of primordial correlation functions (which we subsequently relate to observables in the CMB) refer to quantities at the scale at which inflation occurred. We denote all quantities measured at the scale of inflation with a starred subscript. The dominant contribution to the temperature anisotropies comes from adiabatic perturbations [1] sourced by the comoving curvature perturbation $\mathcal{R}$, defined as the conformal factor of the 3-metric h_{ij} in comoving gauge:

$$h_{ij}(t,x) = a^2(t)e^{2\mathcal{R}(t,x)}\hat{h}_{ij}; \quad \hat{h}_{ij} := \exp[\gamma_{ij}] \tag{3}$$

with $\partial_i\gamma_{ij} = \gamma_{ii} = 0$ defining transverse traceless graviton perturbations. The temperature anisotropies are characterized by the dimensionless power spectrum for $\mathcal{R}$, whose amplitude is given by

$$\mathcal{P}_{\mathcal{R}} := \frac{H_*^2}{8\pi^2 M_*^2 \epsilon_*} = \mathcal{A} \times 10^{-10}, \tag{4}$$

where $\epsilon_* := -\dot{H}_*/H_*^2$, H_* being the Hubble factor during inflation. Given that $\mathcal{R}$ is conserved on super-horizon scales (in the absence of entropy perturbations), this immediately relates to the amplitude of the late time CMB anisotropies, which fixes $\mathcal{A} \sim 22.15$ [10]. The tensor anisotropies are characterized by the tensor power spectrum

$$\mathcal{P}_\gamma := 2\frac{H_*^2}{\pi^2 M_*^2}, \tag{5}$$

Taking the ratio of the above with (4), we find the tensor to scalar ratio

$$r_* := \frac{\mathcal{P}_\gamma}{\mathcal{P}_{\mathcal{R}}} = 16\epsilon_*. \tag{6}$$

Therefore any determination of r_*, either through direct measurements of the stochastic background of primordial gravitational waves or through their secondary effects on the polarization of the CMB [11–13] allows us in principle to fix the scale of inflation:

$$H_* = M_*\left(\frac{\pi^2\mathcal{A}r_*}{2\cdot 10^{10}}\right)^{1/2} := \Upsilon = 1.05\,\sqrt{r_*}\times 10^{-4}. \tag{7}$$

We see that any measurements of r_* determines the scale of inflation *up to our ignorance of the effective strength of gravity at the scale* H_*, given by $M_* \sim \frac{M_P}{\sqrt{N}}$, where N is the effective number of all universally coupled species up to the scale H_*– whether they exist in the visible sector or in any hidden sector. Note that as one lowers the scale of strong gravity, the maximum reheating temperature

[1] In what follows, we assume that all of the extra species have sufficiently suppressed couplings to the inflaton during inflation (e.g. either through derivative couplings or as Planck suppressed interactions) so that isocurvature perturbations are not significantly generated. This is trivially true for hidden sector fields.

T_i is necessarily lowered as well, since it cannot be higher than the inflation scale. Conservatively, T_i cannot be too far below the TeV scale without spoiling the standard scenarios of big bang cosmology– in particular, mechanisms for Leptogenesis and Baryogenesis which can occur no lower than the electroweak scale [14]. We note as a consistency check on the above considerations, that although additional species increase the strength of gravity, the ratio H_*^2/M_*^2 is independent of N and is fixed by observable quantities. Therefore the effects of strong gravity are evidently negligible during inflation even if M_* is much smaller than the macroscopic strength of gravity $M_{\rm pl}$. Hence inflationary dynamics, in particular the dynamics of adiabatic fluctuations remain weakly coupled independent of N and the usual computation of adiabatic correlators can be implemented [15].

In the case of extra species as KK graviton modes, the fundamental higher-dimensional gravity scale (1) with N given in (2) for $E_* = M_{UV}$ leads to the usual relation between the 4d and $(4+n)$ d Planck scales

$$M_P^2 = M_{UV}^{2+n} R^n\,. \tag{8}$$

On the other hand, during inflation N counts all KK states with mass less than the Hubble scale H_*:

$$N = (H_* R)^n\,, \tag{9}$$

and the effective gravity scale becomes

$$M_* = M_P/\sqrt{N} = M_{UV}(M_{UV}/H)^{n/2}\,, \tag{10}$$

where we used the relations (8) and (10). Equation (7) then yields:

$$H_* = M_*\Upsilon = M_{UV}(M_{UV}/H)^{n/2}\Upsilon \quad \Rightarrow \quad M_{UV}\Upsilon^{2/(n+2)}\,, \tag{11}$$

where we used eq. (10). It follows that H_* is one to three orders of magnitude below the fundamental gravity scale M_{UV} for the range $0.001 \lesssim r_* \lesssim 0.1$. The ratio H_*/M_* is of course fixed by (7). The inflation scale H_* can then be as low as the weak scale in low scale gravity models with large extra dimensions, consistently with observations.

In the following, we assume that the electroweak, inflation, gravity and dark energy scales have an independent dynamical origin and examine the corresponding conditions to the microscopic theories. More precisely, we address the question of supersymmetry breaking in dS space with an infinitesimal (tunable) cosmological constant.

3 Conventions

Throughout this paper we use the conventions of [16]. A supergravity theory is specified (up to Chern-Simons terms) by a Kähler potential $\mathcal{K}$, a superpotential W, and the gauge kinetic functions $f_{AB}(z)$. The chiral multiplets z^α, χ^α are enumerated by the index α and the indices A, B indicate the different gauge groups. Classically, a supergravity theory is invariant under Kähler tranformations, viz.

$$\begin{aligned} \mathcal{K}(z,\bar z) &\longrightarrow \mathcal{K}(z,\bar z) + J(z) + \bar J(\bar z),\\ W(z) &\longrightarrow e^{-\kappa^2 J(z)} W(z), \end{aligned} \tag{12}$$

where κ is the inverse of the reduced Planck mass, $m_p = \kappa^{-1} = 2.4 \times 10^{15}$ TeV. The gauge transformations of chiral multiplet scalars are given by holomorphic Killing vectors, i.e. $\delta z^\alpha = \theta^A k_A^\alpha(z)$, where θ^A is the gauge parameter of the gauge group A. The Kähler potential and superpotential need not be invariant under this gauge transformation, but can change by a Kähler transformation

$$\delta\mathcal{K} = \theta^A\left[r_A(z) + \bar r_A(\bar z)\right], \tag{13}$$

provided that the gauge transformation of the superpotential satisfies $\delta W = -\theta^A \kappa^2 r_A(z) W$. One then has from $\delta W = W_\alpha \delta z^\alpha$

$$W_\alpha k_A^\alpha = -\kappa^2 r_A W, \tag{14}$$

where $W_\alpha = \partial_\alpha W$ and α labels the chiral multiplets. The supergravity theory can then be described by a gauge invariant function

$$\mathcal{G} = \kappa^2 \mathcal{K} + \log(\kappa^6 W \bar{W}). \tag{15}$$

The scalar potential is given by

$$\begin{aligned} V &= V_F + V_D \\ V_F &= e^{\kappa^2 \mathcal{K}} \left(-3\kappa^2 W \bar{W} + \nabla_\alpha W g^{\alpha\bar{\beta}} \bar{\nabla}_{\bar{\beta}} \bar{W} \right) \\ V_D &= \frac{1}{2} (\mathrm{Re} f)^{-1\,AB} \mathcal{P}_A \mathcal{P}_B, \end{aligned} \tag{16}$$

where W appears with its Kähler covariant derivative

$$\nabla_\alpha W = \partial_\alpha W(z) + \kappa^2 (\partial_\alpha \mathcal{K}) W(z). \tag{17}$$

The moment maps $\mathcal{P}_A$ are given by

$$\mathcal{P}_A = i(k_A^\alpha \partial_\alpha \mathcal{K} - r_A). \tag{18}$$

In this paper we will be concerned with theories having a gauged R-symmetry, for which $r_A(z)$ is given by an imaginary constant $r_A(z) = i\kappa^{-2}\xi$. In this case, $\kappa^{-2}\xi$ is a Fayet-Iliopoulos [17] constant parameter.

4 The model

The starting point is a chiral multiplet S and a vector multiplet associated with a shift symmetry of the scalar component s of the chiral multiplet S

$$\delta s = -ic\theta, \tag{19}$$

and a string-inspired Kähler potential of the form $-p\log(s+\bar{s})$. The most general superpotential is either a constant $W = \kappa^{-3}a$ or an exponential superpotential $W = \kappa^{-3}ae^{bs}$ (where a and b are constants). A constant superpotential is (obviously) invariant under the shift symmetry, while an exponential superpotential transforms as $W \to We^{-ibc\theta}$, as in eq. (14). In this case the shift symmetry becomes a gauged R-symmetry and the scalar potential contains a Fayet-Iliopoulos term. Note however that by performing a Kähler transformation (12) with $J = \kappa^{-2}bs$, the model can be recast into a constant superpotential at the cost of introducing a linear term in the Kähler potential $\delta K = b(s+\bar{s})$. Even though in this representation, the shift symmetry is not an R-symmetry, we will still refer to it as $U(1)_R$. The most general gauge kinetic function has a constant term and a term linear in s, $f(s) = \delta + \beta s$.

To summarise,[2]

$$\begin{aligned} \mathcal{K}(s,\bar{s}) &= -p\log(s+\bar{s}) + b(s+\bar{s}), \\ W(s) &= a, \\ f(s) &= \delta + \beta s, \end{aligned} \tag{20}$$

[2] In superfields the shift symmetry (19) is given by $\delta S = -ic\Lambda$, where Λ is the superfield generalization of the gauge parameter. The gauge invariant Kähler potential is then given by $\mathcal{K}(S,\bar{S}) = -p\kappa^{-2}\log(S+\bar{S}+cV_R) + \kappa^{-2}b(S+\bar{S}+cV_R)$, where V_R is the gauge superfield of the shift symmetry.

where we have set the mass units $\kappa = 1$. The constants a and b together with the constant c in eq. (19) can be tuned to allow for an infinitesimally small cosmological constant and a TeV gravitino mass. For $b > 0$, there always exists a supersymmetric AdS (anti-de Sitter) vacuum at $\langle s + \bar{s} \rangle = b/p$, while for $b = 0$ (and $p < 3$) there is an AdS vacuum with broken supersymmetry. We therefore focus on $b < 0$. In the context of string theory, S can be identified with a compactification modulus or the universal dilaton and (for negative b) the exponential superpotential may be generated by non-perturbative effects.

The scalar potential is given by:

$$\begin{aligned} V &= V_F + V_D \\ V_F &= a^2 e^{\frac{b}{l}} l^{p-2} \left\{ \frac{1}{p}(pl - b)^2 - 3l^2 \right\} \qquad l = 1/(s + \bar{s}) \\ V_D &= c^2 \frac{l}{\beta + 2\delta l}(pl - b)^2 \end{aligned} \tag{21}$$

In the case where S is the string dilaton, V_D can be identified as the contribution of a magnetized D-brane, while V_F for $b = 0$ and $p = 2$ coincides with the tree-level dilaton potential obtained by considering string theory away its critical dimension [18]. For $p \geq 3$ the scalar potential V is positive and monotonically decreasing, while for $p < 3$, its F-term part V_F is unbounded from below when $s + \bar{s} \to 0$. On the other hand, the D-term part of the scalar potential V_D is positive and diverges when $s + \bar{s} \to 0$ and for various values for the parameters an (infinitesimally small) positive (local) minimum of the potential can be found.

If we restrict ourselves to integer p, tunability of the vacuum energy restricts $p = 2$ or $p = 1$ when $f(s) = s$, or $p = 1$ when the gauge kinetic function is constant. For $p = 2$ and $f(s) = s$, the minimization of V yields:

$$b/l = \alpha \approx -0.183268 \quad , \quad p = 2 \tag{22}$$

$$\frac{a^2}{bc^2} = A_2(\alpha) + B_2(\alpha)\frac{\Lambda}{b^3 c^2} \approx -50.6602 + O(\Lambda), \tag{23}$$

where Λ is the value of V at the minimum (i.e. the cosmological constant), α is the negative root of the polynomial $-x^5 + 7x^4 - 10x^3 - 22x^2 + 40x + 8$ compatible with (23) for $\Lambda = 0$ and $A_2(\alpha)$, $B_2(\alpha)$ are given by

$$A_2(\alpha) = 2e^{-\alpha}\frac{-4 + 4\alpha - \alpha^2}{\alpha^3 - 4\alpha^2 - 2\alpha} \quad ; \quad B_2(\alpha) = 2\frac{\alpha^2 e^{-\alpha}}{\alpha^2 - 4\alpha - 2} \tag{24}$$

It follows that by carefully tuning a and c, Λ can be made positive and arbitrarily small independently of the supersymmetry breaking scale. A plot of the scalar potential for certain values of the parameters is shown in figure 1.

At the minimum of the scalar potential, for nonzero a and $b < 0$, supersymmetry is broken by expectation values of both an F and D-term. Indeed the F-term and D-term contributions to the scalar potential are

$$\begin{aligned} V_F|_{s+\bar{s}=\frac{\alpha}{b}} &= \frac{1}{2}a^2 b^2 e^{\alpha}\left(1 - \frac{2}{\alpha}\right)^2 > 0, \\ V_D|_{s+\bar{s}=\frac{\alpha}{b}} &= \frac{b^3 c^2}{\alpha}\left(1 - \frac{2}{\alpha}\right)^2 > 0\,. \end{aligned} \tag{25}$$

The gravitino mass term is given by

$$(m_{3/2})^2 = e^{\mathcal{G}} = \frac{a^2 b^2}{\alpha^2} e^{\alpha}\,. \tag{26}$$

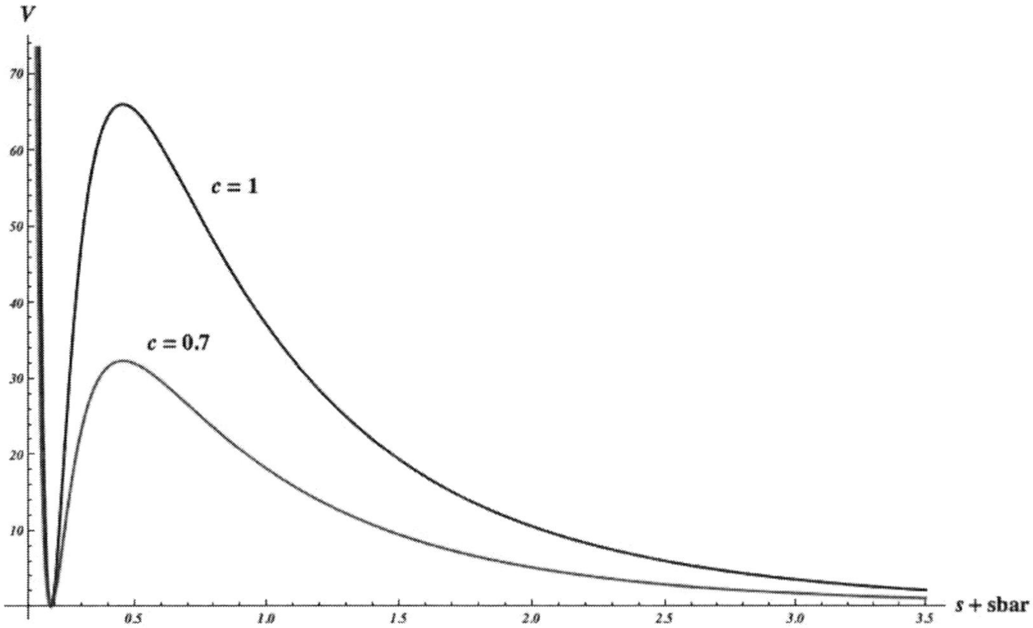

Figure 1. A plot of the scalar potential for $p = 2$, $b = -1$, $\delta = 0$, $\beta = 1$ and a given by equation (23) for $c = 1$ (black curve) and $c = 0.7$ (red curve).

Due to the Stueckelberg coupling, the imaginary part of s (the axion) gets eaten by the gauge field, which acquires a mass. On the other hand, the Goldstino, which is a linear combination of the fermion of the chiral multiplet χ and the gaugino λ gets eaten by the gravitino. As a result, the physical spectrum of the theory consists (besides the graviton) of a massive scalar, namely the dilaton, a Majorana fermion, a massive gauge field and a massive gravitino. All the masses are of the same order of magnitude as the gravitino mass, proportional to the same constant a (or c related by eq. (23) where b is fixed by eq. (22)), which is a free parameter of the model. Thus, they vanish in the same way in the supersymmetric limit $a \to 0$.

The local dS minimum is metastable since it can tunnel to the supersymmetric ground state at infinity in the s-field space (zero coupling). It turns out however that it is extremely long lived for realistic perturbative values of the gauge coupling $l \simeq 0.02$ and TeV gravitino mass and, thus, practically stable; its decay rate is [5]:

$$\Gamma \sim e^{-B} \quad \text{with} \quad B \approx 10^{300} . \tag{27}$$

5 Coupling a visible sector

The guideline to construct a realistic model keeping the properties of the toy model described above is to assume that matter fields are invariant under the shift symmetry (19) and do not participate in the supersymmetry breaking. In the simplest case of a canonical Kähler potential, MSSM-like fields ϕ can then be added as:

$$\begin{aligned} \mathcal{K} &= -\kappa^{-2}\log(s+\bar{s}) + \kappa^{-2}b(s+\bar{s}) + \sum \varphi\bar{\varphi}, \\ W &= \kappa^{-3}a + W_{MSSM}, \end{aligned} \tag{28}$$

where $W_{MSSM}(\phi)$ is the usual MSSM superpotential. The squared soft scalar masses of such a model can be shown to be positive and close to the square of the gravitino mass (TeV^2). On the other hand, for a gauge kinetic function with a linear term in s, $\beta \neq 0$ in eq. (20), the Lagrangian is not invariant under the shift symmetry

$$\delta\mathcal{L} = -\theta\frac{\beta c}{8}\,\epsilon^{\mu\nu\rho\sigma}F_{\mu\nu}F_{\rho\sigma}. \tag{29}$$

and its variation should be canceled. As explained in Ref. [5], in the 'frame' with an exponential superpotential the R-charges of the fermions in the model can give an anomalous contribution to the Lagrangian. In this case the 'Green-Schwarz' term $\text{Im}sF\tilde{F}$ can cancel quantum anomalies. However as shown in [5], with the minimal MSSM spectrum, the presence of this term requires the existence of additional fields in the theory charged under the shift symmetry.

Instead, to avoid the discussion of anomalies, we focus on models with a constant gauge kinetic function. In this case the only (integer) possibility[3] is $p = 1$. The scalar potential is given by (21) with $\beta = 0$, $\delta = p = 1$. The minimization yields to equations similar to (22), (23) and (24) with a different value of α and functions A_1 and B_1 given by:

$$\begin{aligned} b\langle s+\bar{s}\rangle &= \alpha \approx -0.233153 \\ \frac{bc^2}{a^2} &= A_1(\alpha) + B_1(\alpha)\frac{\Lambda}{a^2 b} \approx -0.359291 + O(\Lambda) \\ A_1(\alpha) &= 2e^{\alpha}\alpha\frac{3-(\alpha-1)^2}{(\alpha-1)^2}\,, \quad B_1(\alpha) = \frac{2\alpha^2}{(\alpha-1)^2}\,, \end{aligned} \tag{30}$$

where α is the negative root of $-3+(\alpha-1)^2(2-\alpha^2/2) = 0$ close to -0.23, compatible with the second constraint for $\Lambda = 0$. However, this model suffers from tachyonic soft masses when it is coupled to the MSSM, as in (28). To circumvent this problem, one can add an extra hidden sector field which contributes to (F-term) supersymmetry breaking. Alternatively, the problem of tachyonic soft masses can also be solved if one allows for a non-canonical Kähler potential in the visible sector, which gives an additional contribution to the masses through the D-term.

Let us discuss first the addition of an extra hidden sector field z (similar to the so-called Polonyi field [19]). The Kähler potential, superpotential and gauge kinetic function are given by

$$\begin{aligned} \mathcal{K} &= -\kappa^{-2}\log(s+\bar{s}) + \kappa^{-2}b(s+\bar{s}) + z\bar{z} + \sum \varphi\bar{\varphi}\,, \\ W &= \kappa^{-3}a(1+\gamma\kappa z) + W_{\text{MSSM}}(\varphi)\,, \\ f(s) &= 1\,, \quad f_A = 1/g_A^2\,, \end{aligned} \tag{31}$$

where A labels the Standard Model gauge group factors and γ is an additional constant parameter. The existence of a tunable dS vacuum with supersymmetry breaking and non-tachyonic scalar masses implies that γ must be in a narrow region:

$$0.5 \lesssim \gamma \lesssim 1.7\,. \tag{32}$$

In the above range of γ the main properties of the toy model described in the previous section remain, while $\text{Re}z$ and its F-auxiliary component acquire non vanishing VEVs. All MSSM soft scalar masses

[3] If $f(s)$ is constant, the leading contribution to V_D when $s+\bar{s} \to 0$ is proportional to $1/(s+\bar{s})^2$, while the leading contribution to V_F is proportional to $1/(s+\bar{s})^p$. It follows that $p < 2$; if $p > 2$, the potential is unbounded from below, while if $p = 2$, the potential is either positive and monotonically decreasing or unbounded from below when $s+\bar{s} \to 0$ depending on the values of the parameters.

are then equal to a universal value m_0 of the order of the gravitino mass, while the B_0 Higgs mixing parameter is also of the same order:

$$\begin{aligned} m_0^2 &= m_{3/2}^2\left[(\sigma_s+1)+\frac{(\gamma+t+\gamma t)^2}{(1+\gamma t)^2}\right], \\ A_0 &= m_{3/2}\left[(\sigma_s+3)+t\frac{(\gamma+t+\gamma t^2)}{1+\gamma t}\right], \\ B_0 &= m_{3/2}\left[(\sigma_s+2)+t\frac{(\gamma+t+\gamma t^2)}{(1+\gamma t)}\right], \end{aligned} \tag{33}$$

where $\sigma_s = -3+(\alpha-1)^2$ with α and $t \equiv \langle \mathrm{Re}\, z\rangle$ determined by the minimization conditions as functions of γ. Also, A_0 is the soft trilinear scalar coupling in the standard notation, satisfying the relation [20]

$$A_0 = B_0 + m_{3/2}\,. \tag{34}$$

On the other hand, the gaugino masses appear to vanish at tree-level since the gauge kinetic functions are constants (see (31)). However, as mentioned in Section 4, this model is classically equivalent to the theory[4]

$$\begin{aligned} \mathcal{K} &= -\kappa^{-2}\log(s+\bar{s})+z\bar{z}+\sum_{\alpha}\varphi\bar{\varphi}, \\ W &= \left(\kappa^{-3}a(1+z)+W_{\mathrm{MSSM}}(\varphi)\right)e^{bs}, \end{aligned} \tag{35}$$

obtained by applying a Kähler transformation (12) with $J = -\kappa^{-2}bs$. All classical results remain the same, such as the expressions for the scalar potential and the soft scalar masses (33), but now the shift symmetry (19) of s became a gauged R-symmetry since the superpotential transforms as $W \longrightarrow We^{-ibc\theta}$. Therefore, all fermions (including the gauginos and the gravitino) transform[5] as well under this $U(1)_R$, leading to cubic $U(1)_R^3$ and mixed $U(1)\times G_{\mathrm{MSSM}}$ anomalies. These anomalies are cancelled by a Green-Schwarz (GS) counter term that arises from a quantum correction to the gauge kinetic functions:

$$f_A(s) = 1/g_A^2 + \beta_A s \quad \text{with} \quad \beta_A = \frac{b}{8\pi^2}\left(T_{R_A}-T_{G_A}\right), \tag{36}$$

where T_G is the Dynkin index of the adjoint representation, normalized to N for $SU(N)$, and T_R is the Dynkin index associated with the representation R of dimension d_R, equal to $1/2$ for the $SU(N)$ fundamental. An implicit sum over all matter representations is understood. It follows that gaugino masses are non-vanishing in this representation, creating a puzzle on the quantum equivalence of the two classically equivalent representations. The answer to this puzzle is based on the fact that gaugino masses are present in both representations and are generated at one-loop level by an effect called Anomaly Mediation [7, 8]. Indeed, it has been argued that gaugino masses receive a one-loop contribution due to the super-Weyl-Kähler and sigma-model anomalies, given by [8]:

$$M_{1/2} = -\frac{g^2}{16\pi^2}\left[(3T_G-T_R)m_{3/2}+(T_G-T_R)\mathcal{K}_\alpha F^\alpha+2\frac{T_R}{d_R}(\log\det\mathcal{K}|_R{}'')_{,\alpha}F^\alpha\right]. \tag{37}$$

The expectation value of the auxiliary field F^α, evaluated in the Einstein frame is given by

$$F^\alpha = -e^{\kappa^2\mathcal{K}/2}g^{\alpha\bar{\beta}}\bar{\nabla}_{\bar{\beta}}\bar{W}. \tag{38}$$

[4]This statement is only true for supergravity theories with a non-vanishing superpotential where everything can be defined in terms of a gauge invariant function $G = \kappa^2\mathcal{K}+\log(\kappa^6 W\bar{W})$ [21].

[5]The chiral fermions, the gauginos and the gravitino carry a charge $bc/2$, $-bc/2$ and $-bc/2$ respectively.

Clearly, for the Kähler potential (31) or (35) the last term in eq. (37) vanishes. However, the second term survives due to the presence of Planck scale VEVs for the hidden sector fields s and z. Since the Kähler potential between the two representations differs by a linear term $b(s + \bar{s})$, the contribution of the second term in eq. (37) differs by a factor

$$\delta m_A = \frac{g_A^2}{16\pi^2}(T_G - T_R) b e^{\kappa^2 \mathcal{K}/2} g^{\alpha\bar{\beta}} \bar{\nabla}_{\bar{\beta}} \bar{W}, \tag{39}$$

which exactly coincides with the 'direct' contribution to the gaugino masses due to the field dependent gauge kinetic function (36) (taking into account a rescaling proportional to g_A^2 due to the non-canonical kinetic terms).

We conclude that even though the models (31) and (35) differ by a (classical) Kähler transformation, they generate the same gaugino masses at one-loop. While the one-loop gaugino masses for the model (31) are generated entirely by eq. (37), the gaugino masses for the model (35) after a Kähler transformation have a contribution from eq. (37) as well as from a field dependent gauge kinetic term whose presence is necessary to cancel the mixed $U(1)_R \times G$ anomalies due to the fact that the extra $U(1)$ has become an R-symmetry giving an R-charge to all fermions in the theory. Using (37), one finds:

$$M_{1/2} = -\frac{g^2}{16\pi^2} m_{3/2} \left[(3T_G - T_R) - (T_G - T_R) \left((\alpha - 1)^2 + t \frac{\gamma + t + \gamma t^2}{1 + \gamma t} \right) \right]. \tag{40}$$

For $U(1)_Y$ we have $T_G = 0$ and $T_R = 11$, for $SU(2)$ we have $T_G = 2$ and $T_R = 7$, and for $SU(3)$ we have $T_G = 3$ and $T_R = 6$, such that for the different gaugino masses this gives (in a self-explanatory notation):

$$\begin{aligned} M_1 &= 11 \frac{g_Y^2}{16\pi^2} m_{3/2} \left[1 - (\alpha - 1)^2 - \frac{t(\gamma + t + \gamma t)}{1 + \gamma t} \right], \\ M_2 &= \frac{g_2^2}{16\pi^2} m_{3/2} \left[1 - 5(\alpha - 1)^2 - 5 \frac{t(\gamma + t + \gamma t^2)}{1 + \gamma t} \right], \\ M_3 &= -3 \frac{g_3^2}{16\pi^2} m_{3/2} \left[1 + (\alpha - 1)^2 + \frac{t(\gamma + t + \gamma t^2)}{1 + \gamma t} \right]. \end{aligned} \tag{41}$$

6 Phenomenology

The results for the soft terms calculated in the previous section, evaluated for different values of the parameter γ are summarised in Table 1. For every γ, the corresponding t and α are calculated by imposing a vanishing cosmological constant at the minimum of the potential. The scalar soft masses and trilinear terms are then evaluated by eqs. (33) and the gaugino masses by eqs. (41). Note that the relation (34) is valid for all γ. We therefore do not list the parameter B_0.

In most phenomenological studies, B_0 is substituted for $\tan\beta$, the ratio between the two Higgs VEVs, as an input parameter for the renormalization group equations (RGE) that determine the low energy spectrum of the theory. Since B_0 is not a free parameter in our theory, but is fixed by eq. (34), this corresponds to a definite value of $\tan\beta$. For more details see [22] (and references therein). The corresponding $\tan\beta$ for a few particular choices for γ are listed in the last two columns of table 1 for $\mu > 0$ and $\mu < 0$ respectively. No solutions were found for $\gamma \lesssim 1.1$, for both signs of μ. The lighest supersymmetric particle (LSP) is given by the lightest neutralino and since $M_1 < M_2$ (see table 1) the lightest neutralino is mostly Bino-like, in contrast with a typical mAMSB (minimal anomaly mediation supersymmetry breaking) scenario, where the lightest neutralino is mostly Wino-like [23].

Table 1. The soft terms (in terms of $m_{3/2}$) for various values of γ. If a solution to the RGE exists, the value of $\tan\beta$ is shown in the last columns for $\mu > 0$ and $\mu < 0$ respectively.

γ	t	α	m_0	A_0	M_1	M_2	M_3	$\tan\beta(\mu > 0)$	$\tan\beta(\mu < 0)$
0.6	0.446	-0.175	0.475	1.791	0.017	0.026	0.027		
1	0.409	-0.134	0.719	1.719	0.015	0.025	0.026		
1.1	0.386	-0.120	0.772	1.701	0.015	0.024	0.026	46	29
1.4	0.390	-0.068	0.905	1.646	0.014	0.023	0.026	40	23
1.7	0.414	-0.002	0.998	1.588	0.013	0.022	0.025	36	19

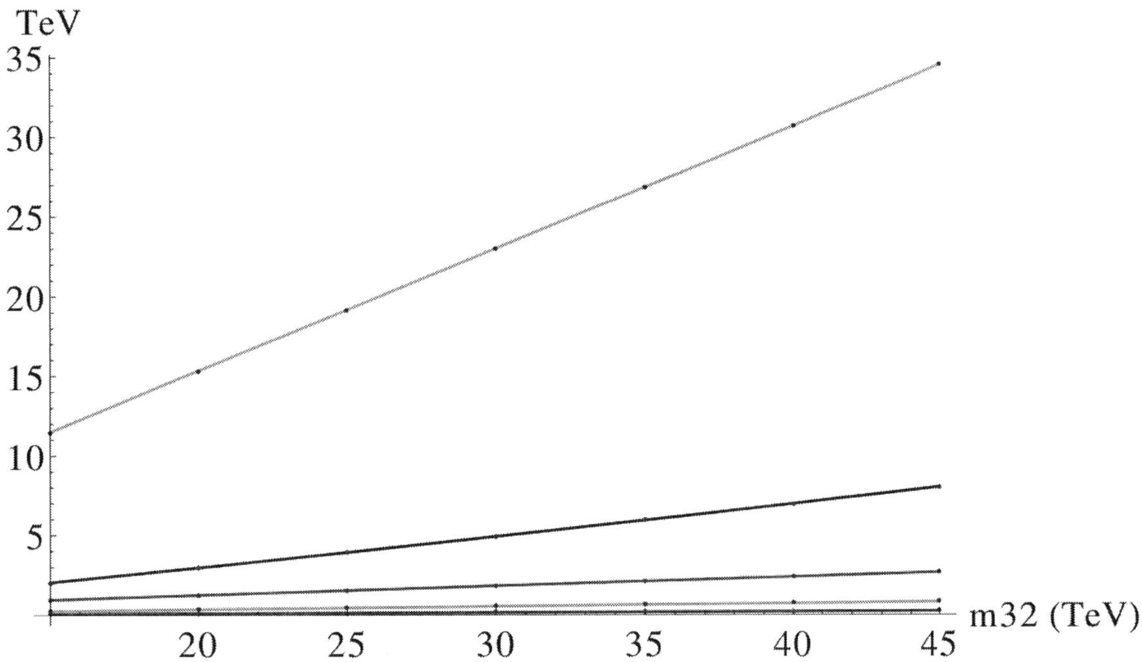

Figure 2. The masses (in TeV) of the sbottom (yellow), stop (black), gluino (red), lightest chargino (green) and lightest neutralino (blue) as a function of $m_{3/2}$ for $\gamma = 1.1$ and for $\mu > 0$. No solutions to the RGE were found when $m_{3/2} \gtrsim 45$ TeV. The lower bound corresponds to a gluino mass of 1 TeV.

To get a lower bound on the stop mass, the sparticle spectrum is plotted in Figure 2 as a function of the gravitino mass for $\gamma = 1.1$ and $\mu > 0$ (for $\mu < 0$ the bound is higher). The experimental limit on the gluino mass forces $m_{3/2} \gtrsim 15$ TeV. In this limit the stop mass can be as low as 2 TeV. To conclude, the lower end mass spectrum consists of (very) light charginos (with a lightest chargino between 250 and 800 GeV) and neutralinos, with a mostly Bino-like neutralino as LSP (80 – 230 GeV), which would distinguish this model from the mAMSB where the LSP is mostly Wino-like. These upper limits on the LSP and the lightest chargino imply that this model could in principle be excluded in the next LHC run. In order for the gluino to escape experimental bounds, the lower limit on the gravitino mass is about 15 TeV. The gluino mass is then between 1-3 TeV. This however forces the squark masses to be very high (10 – 35 TeV), with the exception of the stop mass which can be relatively light (2 – 15 TeV).

7 Non-canonical Kähler potential for the visible sector

As mentioned already in Section 4, an alternative way to avoid tachyonic soft scalar masses for the MSSM fields in the model (28), instead of adding the extra Palonyi-type field z in the hidden sector, is by introducing non-canonical kinetic terms for the MSSM fields, such as:

$$\begin{aligned} \mathcal{K} &= -\kappa^{-2}\log(s+\bar{s}) + \kappa^{-2}b(s+\bar{s}) + (s+\bar{s})^{-\nu}\sum\varphi\bar{\varphi}, \\ W &= \kappa^{-3}a + W_{MSSM}, \\ f(s) &= 1, \qquad f_A(s) = 1/g_A^2, \end{aligned} \tag{42}$$

where ν is an additional parameter of the theory, with $\nu = 1$ corresponding to the leading term in the Taylor expansion of $-\log(s+\bar{s}-\varphi\bar{\varphi})$. Since the visible sector fields appear only in the combination $\varphi\bar{\varphi}$, their VEVs vanish provided that the scalar soft masses squared are positive. Moreover, for vanishing visible sector VEVs, the scalar potential and is minimization remains the same as in eqs. (refbsalpha). Therefore, the non-canonical Kähler potential does not change the fact that the F-term contribution to the soft scalar masses squared is negative. On the other hand, the visible fields enter in the D-term scalar potential through the derivative of the Kähler potential with respect to s. Even though this has no effect on the ground state of the potential, the φ-dependence of the D-term scalar potential does result in an extra contribution to the scalar masses squared which become positive

$$\nu > -\frac{e^{\alpha}(\sigma_s+1)\alpha}{A(\alpha)(1-\alpha)} \approx 2.6 \,. \tag{43}$$

The soft MSSM scalar masses and trilinear couplings in this model are:

$$\begin{aligned} m_0^2 &= \kappa^2 a^2\left(\frac{b}{\alpha}\right)\left(e^{\alpha}(\sigma_s+1) + \nu\frac{A(\alpha)}{\alpha}(1-\alpha)\right) \\ A_0 &= m_{3/2}(s+\bar{s})^{\nu/2}(\sigma_s+3) \\ B_0 &= m_{3/2}(s+\bar{s})^{\nu/2}(\sigma_s+2) \end{aligned} \tag{44}$$

where σ_s is defined as in (33), eq. (31) has been used to relate the constants a and c, and corrections due to a small cosmological constant have been neglected. A field redefinition due to a non-canonical kinetic term $g_{\varphi\bar{\varphi}} = (s+\bar{s})^{-\nu}$ is also taken into account. The main phenomenological properties of this model are not expected to be different from the one we analyzed in section 6 with the parameter ν replacing γ. Gaugino masses are still generated at one-loop level while mSUGRA applies to the soft scalar sector. We therefore do not repeat the phenomenological analysis for this model.

References

[1] I. Antoniadis, N. Arkani-Hamed, S. Dimopoulos and G. R. Dvali, Phys. Lett. B **436** (1998) 257 [arXiv:hep-ph/9804398].
[2] I. Antoniadis and S. P. Patil, Eur. Phys. J. **C75** (2015) 182 [arXiv:1410.8845 [hep-th]].
[3] I. Antoniadis and R. Knoops, Nucl. Phys. B **886** (2014) 43 [arXiv:1403.1534 [hep-th]].
[4] F. Catino, G. Villadoro and F. Zwirner, JHEP **1201** (2012) 002 [arXiv:1110.2174 [hep-th]]; G. Villadoro and F. Zwirner, Phys. Rev. Lett. **95** (2005) 231602 [hep-th/0508167].
[5] I. Antoniadis, D. M. Ghilencea and R. Knoops, JHEP **1502** (2015) 166 [arXiv:1412.4807 [hep-th]].
[6] I. Antoniadis and R. Knoops, [arXiv:1507.06924 [hep-ph]].

[7] L. Randall and R. Sundrum, Nucl. Phys. B **557** (1999) 79 [hep-th/9810155]; G. F. Giudice, M. A. Luty, H. Murayama and R. Rattazzi, JHEP **9812** (1998) 027 [hep-ph/9810442].
[8] J. A. Bagger, T. Moroi and E. Poppitz, JHEP **0004** (2000) 009 [hep-th/9911029].
[9] G. Dvali, arXiv:0706.2050 [hep-th]; Int. J. Mod. Phys. A **25** (2010) 602 [arXiv:0806.3801 [hep-th]]; G. Dvali and M. Redi, Phys. Rev. D **77** (2008) 045027 [arXiv:0710.4344 [hep-th]]; R. Brustein, G. Dvali and G. Veneziano, JHEP **0910** (2009) 085 [arXiv:0907.5516 [hep-th]]; G. Dvali and C. Gomez, Phys. Lett. B **674** (2009) 303.
[10] P. A. R. Ade *et al.* [Planck Collaboration], Astron. Astrophys. (2014) [arXiv:1303.5076 [astro-ph.CO]]; arXiv:1303.5082 [astro-ph.CO].
[11] A. G. Polnarev, Sov. Astron. **29**, 607 (1985).
[12] M. Kamionkowski, A. Kosowsky and A. Stebbins, Phys. Rev. Lett. **78**, 2058 (1997) [astro-ph/9609132].
[13] U. Seljak and M. Zaldarriaga, Phys. Rev. Lett. **78**, 2054 (1997) [astro-ph/9609169].
[14] M. Trodden, Rev. Mod. Phys. **71**, 1463 (1999) [hep-ph/9803479]; S. Davidson, E. Nardi and Y. Nir, Phys. Rept. **466**, 105 (2008) [arXiv:0802.2962 [hep-ph]].
[15] C. Cheung, P. Creminelli, A. L. Fitzpatrick, J. Kaplan and L. Senatore, JHEP **0803**, 014 (2008) [arXiv:0709.0293 [hep-th]].
[16] D. Z. Freedman and A. Van Proeyen, Cambridge, UK: Cambridge Univ. Pr. (2012) 607 p.
[17] P. Fayet and J. Iliopoulos, Phys. Lett. B **51** (1974) 461; Phys. Lett. B **69** (1977) 489.
[18] I. Antoniadis, J. -P. Derendinger and T. Maillard, Nucl. Phys. B **808** (2009) 53 [arXiv:0804.1738 [hep-th]].
[19] J. Polonyi, Hungary Central Inst Res - KFKI-77-93 (77,REC.JUL 78) 5p.
[20] H. P. Nilles, Phys. Rept. **110** (1984) 1
[21] S. Ferrara, L. Girardello, T. Kugo and A. Van Proeyen, Nucl. Phys. B **223** (1983) 191
[22] J. R. Ellis, K. A. Olive, Y. Santoso and V. C. Spanos, Phys. Lett. B **573** (2003) 162 [hep-ph/0305212],
[23] T. Gherghetta, G. F. Giudice and J. D. Wells, Nucl. Phys. B **559** (1999) 27 [hep-ph/9904378].

EPJ Web of Conferences **126**, 02004 (2016) DOI: 10.1051/epjconf/201612602004
ICNFP 2015

Discrete Symmetries CP, T, CPT

Jose Bernabeu[1,a]

[1] *Departament of Theoretical Physics, University of Valencia & IFIC Joint Centre Univ.Valencia–CSIC*

Abstract. The role of Symmetry Breaking mechanisms to search for New Physics is of highest importance. We discuss the status and prospects of the Discrete Symmetries CP, T, CPT looking for their separate Violation in LHC experiments and meson factories.

1 Introduction

Symmetries have been an essential ingredient in the understanding of the physical laws of Nature.The assumption of the form invariance of the dynamical equations under a symmetry transformation of the physical magnitudes leads to observable consequences, with regularities, conservation laws and invariant observables that act as guiding components for the dynamics. Even more interesting, Symmetry Breaking through a definite mechanism is a source of new phenomena and new physics. In this talk I will concentrate on the Discrete Symmetries CP, T and CPT.
In Section 2 we discuss that the current level of experimental accuracy and theoretical uncertainties leaves room for additional sources of CP-Violation (CPV) beyond the Cabibbo-Kobayashi-Maskawa (CKM) mixing mechanism, as required by Baryogenesis in the Universe. We identify potential transitions in B-physics, using Flavour-Changing-Neutral-Current (FCNC) processes and CPV asymmetries, able to incorporate virtual contributions of New Physics.
Section 3 is devoted to TRV concepts and results: what is "Time Reversal" in classical and quantum mechanics, the NO-GO argument for its search with unstable particles, its By-pass using Entanglement and the Decays as Filtering Measurements for entangled meson systems in the B-Factories and in the Φ-Factory, the role of time-ordered decay channels to Flavour and CP-eigenstates for disentangling genuine separate independent asymmetries for CP, T and CPT and the 14σ observation of TRV by the BABAR experiment. Section 4 emphasizes the interest in the search of CPTV for transitions and not only in the expectation values of masses and lifetimes of elementary particles. We distinguish explicit CPT symmetry breaking mechanisms from physical scenarios in which the CPT-operator implementing the symmetry is ill-defined. For entangled systems this second alternative leads to the ω-effect, a component of the wrong symmetry in the coherent state of neutral mesons. Finally, Section 5 presents our results, conclusions and prospects.

[a]e-mail: jose.bernabeu@uv.es

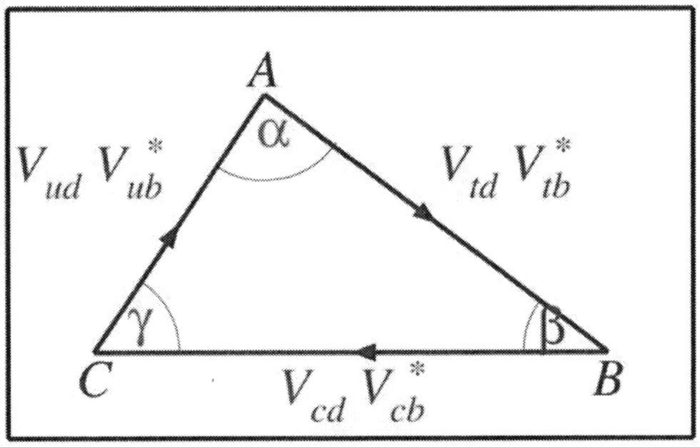

Figure 1. The bd unitarity triangle representing the CKM mixing

2 CP-Violation

The now well established CPV in the quark sector can be successfully accommodated within the Standard Model (SM) of particles and fields through the CKM quark-mixing[1]. For three families, the unitarity conditions lead to triangles like the bd-triangle of Fig.1.

Since the three sides are of comparable length the angles are sizeable and one expects large CP asymmetries in Bd decays in the SM. There are other two triangles which almost collapse to a line. This gives an intuitive understanding of why CPV is small in the leading K decays (ds triangle) and in the leading Bs decays (bs triangle). Extensive tests of the CKM mechanism using experimental data show a high degree of consistency [2]. Two B factory colliders, PEP-II at SLAC in California and KEKB at KEK in Japan, with their corresponding detectors, BABAR [3] and Belle [4], have been operating in the last decade. We now have the LHCb experiment [5] at CERN with new and complementary information about rare decays,FCNC processes and CPV. For the Bd system CPV in the interference of mixing and no-mixing amplitudes for B^0 decays is observed in decay products which are accessible by both B^0 and $\bar{B^0}$. The CPV Asymmetry can be written as

$$A_f(t) = S_f sin(\Delta mt) - C_f cos(\Delta mt) \quad (1)$$

where $\Delta\Gamma = 0$ is assumed. The most precise asymmetries are measured in the tree-dominated b$\rightarrow$ ccs transitions, such as $B^0 \rightarrow \psi$ KS, and are given by [6] S = + 0.682 ± 0.019. The penguin contributions are very small, so that one has the interpretation in the SM S = sin(2β), C=0. CPV in the interference of mixing and decay in the $B^0 \rightarrow \pi^+\pi^-$ mode is given by [6] S = −0.66± 0.06. It is interpreted in the SM as S = sin(2α).For the B^0system, the phase of the mixing amplitude is determined from the intermediate top-quark exchange in the box diagram, so that the SM interpretations of these interferences in terms of the β and α CP angles of the unitarity triangle are apparent. On the contrary, the phase γ in the unitarity triangle involves the interference of the sides for decays with charm and up quark constituents, without any relation to the mixing. The CP angle γ is thus a measure of Direct CPV. Its measurement has been undertaken by BABAR, BELLE and LHCb with the decay $B^+ \rightarrow DK^+$ and other related transitions. The extraction of γ needs a detailed analysis involving in addition the presence of strong phases associated to final-state hadronic interactions. The present average value [7] is $\gamma = (67 + -12)^o$. An ideal experiment would be one in B factories using Entanglement and detecting the pair of decays $\rightarrow \psi K^0$ and $\rightarrow \pi\pi$ at equal times without any effect of the mixing phase.

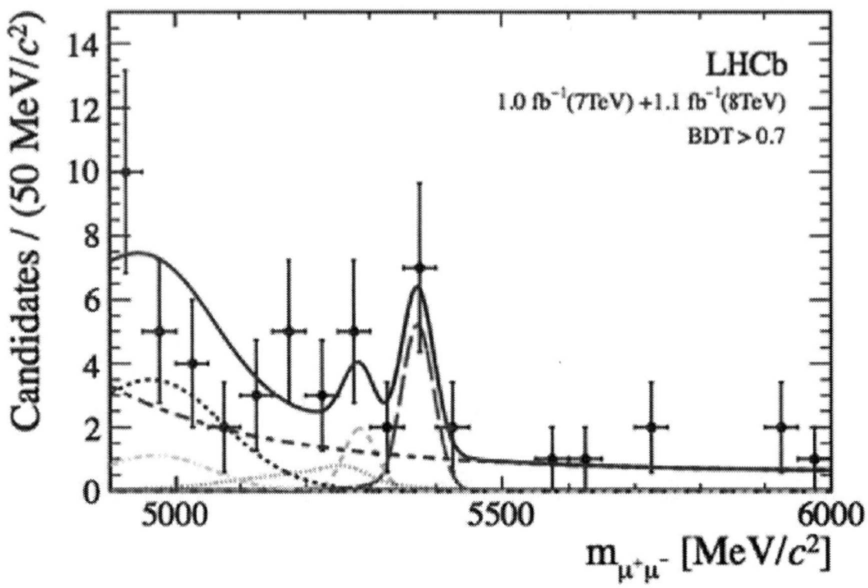

Figure 2. The invariant mass distribution of the $\mu\mu$ system showing the Bs-peak

It remains to be seen whether this gedanken experiment can become a real experiment in the upgraded SuperBELLE.
The rare decay $Bs \to \mu\mu$ has been observed by LHCb and CMS. Based on the presence of a FCNC penguin amplitude induced by Z-exchange as seen in the diagram

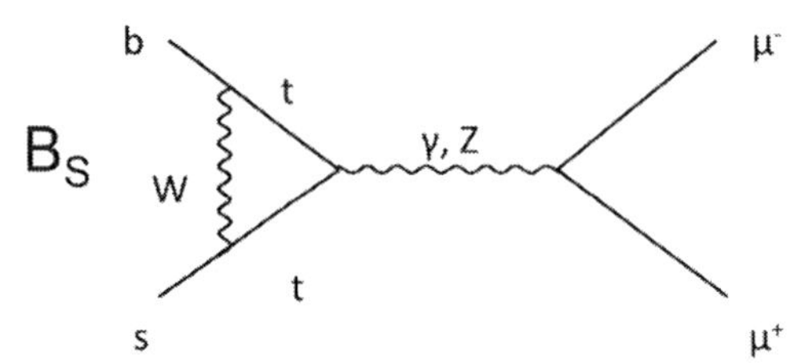

one could expect a priori New Physics virtual effects induced by Non-Decoupling of longitudinal contributions. The result [8] presented in figure 2 has represented the latest disappointment of the scientific community. The experimental Branching Ratio is

$$B(B_s^0 \to \mu^+\mu^-) = \left(3.2^{+1.4}_{-1.2}(stat)^{+0.5}_{-0.3}(syst)\right) x10^{-9} \tag{2}$$

in perfect agreement with the SM value $(3.23 \pm 0.27)x10^{-9}$. In some models, like SuperSymmetry, the result could have been orders of magnitude different from the SM value. One has to be aware, however, that in this process the FCNC penguin is projected to a pseudoscalar, so that its contraction with the leptonic vertex leads to helicity suppression and an amplitude proportional to the mass of the lepton. The FCNC b-s penguin vertex can be probed under more general conditions by searching for the processes $B \to K^* l^+ l^-$, which opens new effective current operators in scalar-vector matrix elements. In fact, the process has been observed and the analysis, using the experimental results of LHCb making use of the OPE formalism, leads to intriguing tensions with the SM expectations in some of the effective operators [9]. If this discrepancy is associated to new physics longitudinal amplitudes of the mediators, one should seriously consider the search of the process with $\nu\nu$, mediated by the Z, replacing $\mu\mu$.

3 Time Reversal

The symmetry transformation that changes a physical system with a given sense of the time evolution into another with the opposite sense is called Time-Reversal T. The T-transformation is implemented in the space of quantum states by the antiunitary operator U_T For a Hamiltonian H invariant under time reversal, [H;U_T] = 0, the time-evolution operator **U**(t;t0) transforms as

$$U_T \mathbf{U}(t, t_0) U_T^{\dagger} y = \mathbf{U}^{\dagger}(t, t_0) \tag{3}$$

The antiunitary character of U_T allows to write U_T = UK, where U is unitary ($U^{-1} = U^{\dagger}$) and K is an operator which complex conjugates all complex numbers. For the matrix elements of time-dependent transitions we have

$$\begin{aligned} \langle f|\mathbf{U}(t, t_0)|i\rangle &= \langle f|U_T^{\dagger} U_T \mathbf{U}(t, t_0) U_T^{\dagger} U_T|i\rangle \\ &= U_T f|\mathbf{U}^{\dagger}(t, t_0)|U_T i\rangle^* = \langle U_T i|\mathbf{U}(t, t_0)|U_T f\rangle \end{aligned} \tag{4}$$

where time-reversal invariance is assumed. As a consequence, the comparison between $i \rightarrow f$ and $U_T f \rightarrow U_T i$ transitions is a genuine test of this invariance.

A direct evidence for TRV would mean an experiment which, by itself, is able to establish a non-vanishing genuine TRV asymmetry independent of CPV or CPT invariance. The problem is then the filtering of definite initial and final states of the neutral meson for the Reference and T-reverse transitions, something impossible for decaying particles. The solution [10-14] arises from the quantum mechanical properties imposed by the EPR entanglement [15, 16] between the two neutral B mesons produced in the Y(4S) resonance decay. Let us suppose that the Reference transition is defined by the time-ordered decay channels l^+ first, J/Ψ K_S later, as shown in the left-hand side of figure 3. The use of Entanglement plus the Decay as a Filtering Measurement tells us that the meson transition corresponds to $\bar{B}^0 \rightarrow B_-$. In terms of meson states, the T-reverse transition is then $B_- \rightarrow \bar{B}_0$ and the question arises: Which are the time-ordered decay channels which correspond to this T-reverse transition? For definite flavour and CP eigenstates, orthogonality of $B^0, \bar{B}^0$ and B_+, B_- provides the solution: J/Ψ KL first, l^- later.

There are 8 experimentally independent Intensities for this kind of pairs of decay channels: 2 for Flavour X 2 for CP X 2 for the time-ordering of the decay channels. The four TRV independent raw symmetries measured by the BABAR experiment [17] are given in figure 4.

They correspond to the asymmetries between the Reference transitions $\bar{B}^0 \rightarrow B_-, B_+ \rightarrow B^0$, $\bar{B}^0 \rightarrow B_+, B_- \rightarrow B^0$, and their T-reverse transitions. The Δt dependence demonstrates that the TRV effect is built in the time evolution of the neutral B. The combined significance of the four measured nonvanishing asymmetries provides a conclusive 14σ TRV effect.

4 CPT invariance

The CPT theorem [19] establishes that interactions described by a Local Quantum Field Theory, with Lorentz invariance and Hermiticity, are CPT invariant. The simplest tests of CPT invariance are the equality of masses and lifetimes for a particle and its antiparticle. The most precise test comes from the mass difference between K^0 and $\bar{K}^0$ [20]. It is very important to move the tests of CPT invariance beyond the comparison of diagonal terms for particle and antiparticle by considering these

DOI: 10.1051/epjconf/201612602004

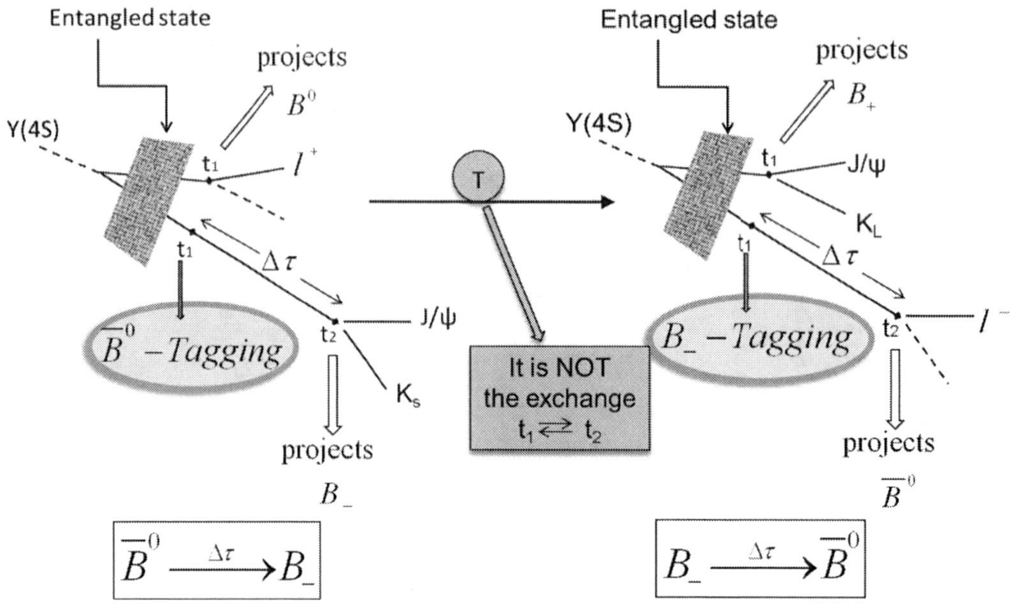

Figure 3. Basic concepts explaining the preparation and detection of initial and final meson states, for the Reference and T-reverse transitions, by means of the esperimental time-ordered decay channels to definite flavour and CP eigenstates.

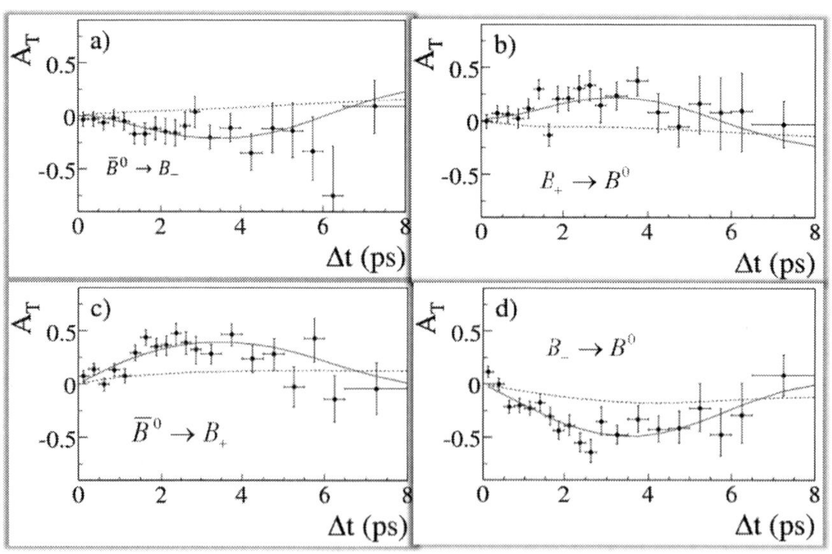

Figure 4. Experimental TRV asymmetries as function of Δt

searches for transitions. For entangled systems, the proposed tests of separate CP, T, CPT symmetries in the neutral meson systems are based on the EPR-Entanglement existing in the Meson Factories as a consequence of Particle Identity: K^0 and $\bar{K^0}$ are two states of identical particles, connected by CPT. Besides the permutation operation P for space-time properties, the strangeness charge connection is made by C, so that for bosons the indistinguishibility requirement is C P = +.
In neutral meson factories, K^0, $\bar{K^0}$ are produced by $\Phi-decay$ with J=1, S=0. This implies L=1 and C= -, so that P = -, i.e., an antisymmetric wave function. This antisymmetry is responsible for preserving $K^0\,\bar{K^0}$ terms only in the time evolution of the two-body system, including the Mixing . Similarly for orthogonal B_+B_- terms only. This correlation is perfect for tagging: Flavour-Tag, CPTag [21].
The question is [22]: What if the Particle Identity is lost? In this case, the two particle system would not satisfy the requirement C P = +. In perturbation theory, if still J=1 with C=-, this breaking leads to a mixing of the "forbidden" P = + symmetric state into the "allowed" P = - antisymmetricstate. This perturbative mixing is the ω-effect: In the time evolution of the system one finds now $\omega K^0 K^0$, ω K_+K_+ terms,..., i.e., a Demise of Tagging.
The decoherence implied by the ω-effect is best seen for equal decay channels at Δ t = 0. For the $(\pi^+\pi^-,\pi^+\pi^-)$ channel, the most prominent effect is the breaking of the Intensity I(Δ t)~ 0 for small values of Δt, a result that was a consequence of the particle identity anti-correlation: no identical states at t1=t2. The KLOE experiment has obtained the first measurement of the ω-parameter [23]:

$$\left.\begin{aligned} \mathrm{Re}(\omega) &= \left(-1.6\,^{+3.0}_{-2.1}\,{}_{\mathrm{stat}} \pm 0.4_{\mathrm{syst}}\right)\times 10^{-4} \\ \mathrm{Im}(\omega) &= \left(-1.7\,^{+3.3}_{-3.0}\,{}_{\mathrm{stat}} \pm 1.2_{\mathrm{syst}}\right)\times 10^{-4} \end{aligned}\right\} \quad |\omega| < 1.0\times 10^{-3} \text{ at } 95\% \text{ C.L.} \tag{5}$$

At least one order of magnitude improvement is expected with KLOE-2 at the upgraded DAΦNE.

5 Conclusion

Separate tests of the Discrete Symmetries CP, T, CPT are being performed in the meson factories. These studies are made possible thanks to the spectacular quantum properties of EPR entangled states. The appropriate preparation and detection of the initial and final states in meson transitions defined by Flavour-CP eigenstates decay channels are based on Entanglement and the use of the two decays as Filtering Measurements.

Acknowledgements

I would like to thank the Organising Committee of the ICNFP 2015 Conference for its superb atmosphere and the invitation extended to me. This research has been supported by MINECO and Generalitat Valenciana Projects FPA 2011-23596 and GVPROMETEO II 2013-017 and by Severo Ochoa Excellence Centre Project SEV 2014-0398.

References

[1] Cabibbo N, Phys.Rev.Lett. 10, 531 (1963)
Kobayashi M and Maskawa T, Prog.Theor.Phys. 49, 652 (1973).
[2] Antonelli M, Asner D M, Bauer D A, Becher T G, Beneke M, et al., Phys.Rept. 494, 197(2010).
[3] Aubert B et al. (BABAR Collaboration), Nucl.Instrum.Meth.A, 615(2013).
B. Aubert et al. (BABAR Collaboration), Nucl.Instrum.Meth.A 479, 1 (2002).

[4] Abashian A et al., Nucl.Instrum.Meth.A 479, 117 (2002).
[5] The LHCb Collaboration, JINST 3, S008 005(2008)
[6] Gershon T, Nir Y, in RPP, Chin. Phys. 38, Number 9 (2014).
[7] R.Aaij et al., LHCb Collaboration, arXiv:1407.6211.
[8] LHCb Collaboration, Aaij R et al., Phys. Rev. Lett. 110, 028101 (2013).
[9] Descotes-Genon S, Hofer L, Matias J, Virto J.,J.Phys.Conf.Ser. 631 (2015) 1, 012027
[10] Banuls M C and Bernabeu J, Phys.Lett.B 464, 117 (1999).
[11] Banuls M C and Bernabeu J, Nucl.Phys.B 590, 19 (2000).
[12] Wolfenstein L, Int.J.Mod.Phys.E 8, 501 (1999).
[13] Quinn H R, J.Phys.Conf.Ser. 171, 011001 (2009).
[14] Bernabeu J, Martinez-Vidal F, and Villanueva-Perez P, JHEP 1208, 064 (2012).
[15] Einstein A, Podolsky B, and Rosen N, Phys.Rev. 47, 777(1935).
[16] Reid M D, Drummond P D, Bowen W P, Cavalcanti E G, Lam P K, et al., Rev.Mod.Phys. 81,1727 (2009).
[17] BABAR Collaboration, Lees J L et al., Phys. Rev. Lett. 109(2012)211801.
[18] Carossi R et al., Phys. Lett. B 237(1990)303
[19] Lueders G, Ann. Phys. (NY) 2, 1 (1957); Pauli W, in Niels Bohr and the development of physics,Pergamon, London, 1955, p. 30;Bell J S, Proc. R. Soc. London A 231, 479 (1955)
[20] Angelopoulos A, CPLEAR collaboration, Phys. Lett. B444, 52 (1998);Abouzaid E et al, Phys. Rev. D83, 092001 (2011)
[21] Banuls M C and Bernabeu J, JHEP 9906:032(1999).
[22] Bernabeu J, Mavromatos N E and Papavassiliou J, Phys. Rev.Lett. 92:131601(2004).
[23] Di Domenico A, 0904.1976 (2009).

EPJ Web of Conferences **126**, 02005 (2016) DOI: 10.1051/epjconf/201612602005
ICNFP 2015

Proton structure and parton distribution functions from HERA

Vladimir Chekelian[1]. On behalf of the H1 and ZEUS Collaborations.
[1] *Max Planck Institute for Physics, Munich.*

Abstract. The H1 and ZEUS collaborations at the electron-proton collider HERA collected $e^{\pm}p$ scattering data corresponding to an integrated luminosity of about 1 fb^{-1}. The data were taken at proton beam energies of 920, 820, 575 and 460 GeV and an electron beam energy of 27.5 GeV, with different electric charges and longitudinal polarisation of the electron beam. Using these data inclusive neutral and charged current deep inelastic cross sections were measured over six orders of magnitude in negative four-momentum-transfer squared, Q^2, and Bjorken x. A combination of all inclusive cross sections, published by the H1 and ZEUS collaborations at HERA, was performed. Using these combined HERA data and the individual H1 and ZEUS data taken using the polarised electron beams, the proton structure functions F_2, $F_2^{\gamma Z}$, $xF_3^{\gamma Z}$ and F_L were obtained, and scaling violations, electroweak unification, and polarisation effects in the charged current process were demonstrated. The combined cross sections were used as a sole input to QCD analyses at leading, next-to-leading and next-to-next-to-leading orders, providing a new set of parton distribution functions, denoted as HERAPDF2.0. An extension of the analysis by including HERA data on charm and jet production allowed a simultaneous determination of parton distributions and the strong coupling.

1 Introduction

Deep inelastic scattering (DIS) of electrons[1] on protons at the electron-proton collider HERA has been central to the exploration of proton structure and quark–gluon dynamics as described by perturbative Quantum Chromodynamics (pQCD). HERA was operated in two phases: HERA I, from 1992 to 2000, and HERA II, from 2002 to 2007, with an electron beam energy of $E_e = 27.5$ GeV. During HERA I and II running, the H1 and ZEUS collaborations collected total integrated luminosities of approximately 500 pb^{-1} each, shared about equally between e^+p and e^-p scattering. This includes data taken with proton beam energies of $E_p = 920, 820, 575$ and 460 GeV corresponding to centre-of-mass energies of $\sqrt{s} = 318, 300, 251$ and 225 GeV. During the HERA II period, the electron beam was longitudinally polarised with left-handed and right-handed polarisations in about equal proportion. For this period, the H1 and ZEUS collaborations published both *polarised* cross sections, measured using data with the longitudinally polarised electron beams, and *unpolarised* cross sections, obtained using data with both polarisations taken together and corrected for small residual polarisation.

The combination of all H1 and ZEUS inclusive neutral (NC) and charged current (CC) *unpolarised* deep inelastic $e^{\pm}p$ cross section data was performed [1]. The correlated systematic uncertainties and

[1] The word "electron" in the paper refers to both electrons and positrons, unless otherwise stated.

global normalisations were treated such that one coherent HERA data set was obtained. The results published in [1] comprise the core of this paper.

Using these combined *unpolarised* HERA I and HERA II and *polarised* HERA II cross section data, the proton structure functions F_2, $F_2^{\gamma Z}$, $xF_3^{\gamma Z}$ and F_L were obtained, and scaling violations, electroweak unification as well as polarisation effects in CC were demonstrated.

The dependences of the combined NC and CC inclusive DIS cross sections from HERA on Bjorken x and negative four-momentum-transfer squared, Q^2, were analysed in the pQCD framework to determine a new set of quark and gluon momentum distributions in the proton, denoted as HERAPDF2.0. An inclusion of the HERA data on charm and jet production to the analysis allowed a simultaneous determination of the parton distributions and the strong coupling.

2 NC and CC cross sections and parton distributions

The neutral current $e^{\pm}p$ deep inelastic scattering cross section, after correction for QED radiative effects, can be written in the reduced form as

$$\sigma^{\pm}_{r,NC} = \frac{\mathrm{d}^2\sigma^{e^\pm p}_{\mathrm{NC}}}{\mathrm{d}x\mathrm{d}Q^2}\frac{xQ^4}{2\pi\alpha^2}\frac{1}{Y_+} = \tilde{F}_2 \mp \frac{Y_-}{Y_+}x\tilde{F}_3 - \frac{y^2}{Y_+}\tilde{F}_L\,. \tag{1}$$

Here, x is the Bjorken scaling variable, and y characterizes the inelasticity of the interaction. The helicity dependences are contained in $Y_\pm = 1 \pm (1-y)^2$. The generalised proton structure functions, $\tilde{F}_{2,3}$, occurring in Eq. 1, may be written as linear combinations of the hadronic structure functions F_2, $F_{2,3}^{\gamma Z}$, and $F_{2,3}^{Z}$, containing information on the QCD parton dynamics as well as the electroweak couplings of the quarks to the neutral vector bosons. The function F_2 is associated with pure photon exchange terms, $F_{2,3}^{\gamma Z}$ correspond to photon–Z interference and $F_{2,3}^{Z}$ correspond to pure Z exchange terms. The longitudinal structure function $\tilde{F}_L$ may be similarly decomposed. $\tilde{F}_L$ is important only at high y and is expected to be negligible at large x and Q^2.

The generalised proton structure functions depend on the electric charge of the lepton beam and on the lepton beam polarisation[2], P, on the Z-boson mass M_Z and on $\sin^2\theta_W$, with θ_W being the electroweak mixing angle.

$$\begin{aligned} \tilde{F}_2^{\pm} &= F_2 + \kappa(-v_e \mp Pa_e)F_2^{\gamma Z} + \kappa^2(v_e^2 + a_e^2 \pm 2Pv_e a_e)F_2^Z, \\ x\tilde{F}_3^{\pm} &= \kappa(-a_e \mp Pv_e)xF_3^{\gamma Z} + \kappa^2(2v_e a_e \pm P(v_e^2 + a_e^2))xF_3^Z\,. \end{aligned} \tag{2}$$

Here, $\kappa = (4\sin^2\theta_W\cos^2\theta_W)^{-1}Q^2/(Q^2+M_Z^2)$ determines the relative amount of Z to γ exchange. The quantities $v_e = -1/2+2\sin^2\theta_W$ and $a_e = -1/2$ are the vector and axial-vector couplings of the electron to the Z-boson in the Standard Model. At leading order in QCD the hadronic structure functions are related to linear combinations of sums and differences of the quark and anti-quark momentum distributions $xq(x, Q^2)$ and $x\bar{q}(x, Q^2)$ of the proton:

$$\begin{aligned} (F_2,\ F_2^{\gamma Z},\ F_2^Z) &= x\sum(e_q^2, 2e_q v_q, v_q^2 + a_q^2)(q+\bar{q}), \\ (xF_3^{\gamma Z}, xF_3^Z) &= 2x\sum(e_q a_q, v_q a_q)(q-\bar{q})\,, \end{aligned} \tag{3}$$

where v_q and a_q are the vector and axial-vector couplings of the light quarks to the Z-boson, and e_q is the charge of the quark of flavor q.

[2]The longitudinal polarisation is defined as $P = (N_R - N_L)/(N_R + N_L)$ with N_R (N_L) being the number of right-handed (left-handed) leptons in the beam.

The charged current $e^{\pm}p$ cross section of longitudinally polarised leptons and unpolarised protons can be expressed in the reduced form as

$$\tilde{\sigma}^{\pm}_{r,CC} \equiv \frac{2\pi x}{G_F^2}\left[\frac{M_W^2+Q^2}{M_W^2}\right]^2 \frac{d^2\sigma^{e^{\pm}p}_{CC}}{dxdQ^2} = \frac{1}{2}(1\pm P)\left(Y_+W_2^{\mp} \mp Y_-xW_3^{\mp} - y^2W_L^{\mp}\right), \tag{4}$$

where G_F is the Fermi constant defined using the weak boson masses M_W and M_Z. This equation shows that the cross section has a linear dependence on the longitudinal polarisation of the lepton beam. In the QPM $W_L^{\mp} \equiv 0$, and the structure functions $W_2^{\mp}$ and $xW_3^{\mp}$ are expressed as the flavor-dependent sum and difference of the quark and anti-quark momentum distributions. Only positively charged quarks contribute to W^- mediated scattering, and only negatively charged quarks couple to the exchanged W^+ boson: $W_2^- = x(U+\overline{D})$, $W_2^+ = x(\overline{U}+D)$, $xW_3^- = x(U-\overline{D})$, $xW_3^+ = x(D-\overline{U})$. Below the b-quark mass threshold, the $U, \overline{U}$ and $D, \overline{D}$ distributions represent the sums of up-type and down-type quark densities: $U = u + c$, $\overline{U} = \bar{u} + \bar{c}$, $D = d + s$, $\overline{D} = \bar{d} + \bar{s}$.

The combination of NC and CC measurements makes it possible to determine both the valence-quark distributions, $xu_v = x(U - \overline{U})$ and $xd_v = x(D - \overline{D})$, and the sea-quark distributions, $x\overline{U}$ and $x\overline{D}$. The parton distribution functions (PDFs) are determined by a fit to the x and Q^2 dependence of data using the linear DGLAP equations [2] at leading order (LO), next-to-leading order (NLO) or next-to-next-to-leading order (NNLO) in pQCD. The experimentally measured scaling violations determine the gluon distribution in the fit.

3 Combination of all inclusive NC and CC cross section data at HERA

In order to combine all inclusive *unpolarised* NC and CC cross section measurements published by H1 and ZEUS, they were translated onto common x, Q^2 grids and averaged [1]. In total 41 data sets from H1 and ZEUS entered into this combination. The references to individual data sets can be found in [1]. Separate measurements from the same data set were not translated to the same grid point and the vast majority of grid points accumulated data from both H1 and ZEUS measurements; the typical case is six measurements from six different data sets. The translation of the original measurements to the nearest grid points was performed using results of the QCD and fractal fits to the combined data; for details see [1]. The procedure applied for the combination follows very closely to the one used for the previously published combination of the HERA I data [3] and for the combination of the HERA I data and preliminary HERA II results [4].

3.1 Averaging cross sections

The averaging of the data points was performed using the HERAverager [5] tool which is based on a χ^2 minimisation method [6]. This method imposes that there is one and only one correct value for the cross section of each process at each point of the phase space. These values are estimated by optimising a vector, m^i, which is the result of the averaging for the cross sections. The χ^2 definition used takes into account the correlated and uncorrelated systematic uncertainties of the H1 and ZEUS cross-section measurements and allows for shifts of the data to accommodate the correlated uncertainties. The χ^2 function involves all measurements from all data sets and can be approximated by

$$\chi^2_{\text{tot}} \approx \chi^2_{\text{min}} + \sum_{i=1,N_M} \frac{\left[m^i - \sum_j \gamma^i_j m^i b'_j - \mu^i\right]^2}{\delta^2_{i,\text{stat}}\,\mu^i\left(m^i - \sum_j \gamma^i_j m^i b'_j\right) + \left(\delta_{i,\text{uncor}}\, m^i\right)^2} + \sum_j (b'_j)^2, \tag{5}$$

where χ^2_{min} is the minimum of χ^2_{tot}, N_M is the number of combined measurements, μ^i is the average value at point i, and γ^i_j, $\delta_{i,\text{stat}}$ and $\delta_{i,\text{uncor}}$ are its relative correlated systematic, relative statistical

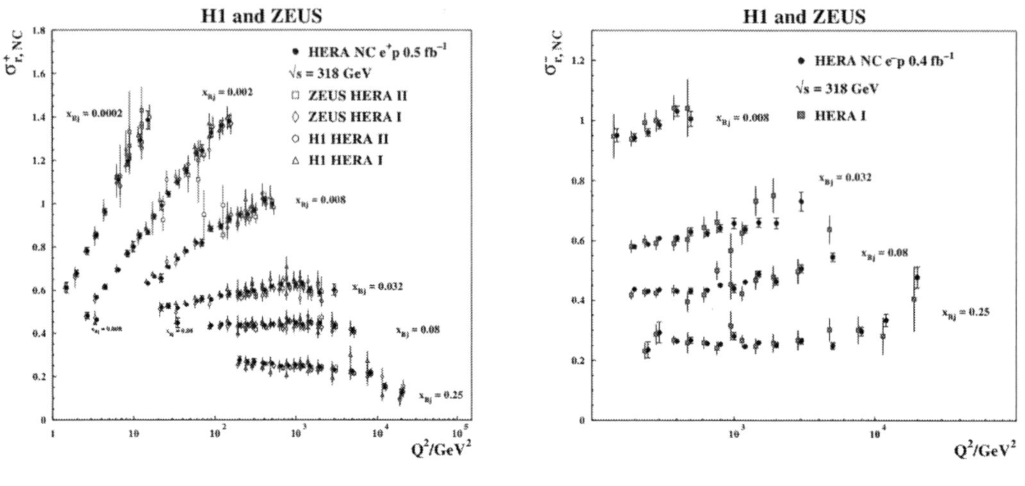

Figure 1. Left: The combined HERA data for the inclusive NC e^+p reduced cross sections as a function of Q^2 for six selected values of x, compared to the individual H1 and ZEUS data. Right: The combined HERA data for the inclusive NC e^-p reduced cross section as a function of Q^2 for four selected values of x, compared to the previously published combined results from HERA I alone.

and relative uncorrelated systematic uncertainties, respectively. The shifts b'_j, also called nuisance parameters, are related to the original correlated shifts b_j of the cross sections in units of sigma of the respective correlated systematic uncertainties through an orthogonal transformation, which is also used to determine γ^i_j [3]. The average values μ^i and systematic shifts b_j are determined analytically from a system of linear equations $\partial\chi^2_{\rm tot}/\partial m^i = 0$ and $\partial\chi^2_{\rm tot}/\partial b_j = 0$. An iterative procedure is used to determine the average of the data. It converges after two iterations.

3.2 Combined neutral and charged current cross sections at HERA

The 2927 published cross sections from the 41 data sets were combined to become 1307 combined cross-section measurements. For the resulting 1620 degrees of freedom, a $\chi^2_{\rm min} = 1687$ was obtained, indicating consistency of the input data. It was also checked that the pull distributions for the different data sets and in different corners of the phase space showed good consistency of all data.

In Fig. 1 (left) the individual H1 and ZEUS and the HERA combined reduced cross sections for NC e^+p scattering at $\sqrt{s} = 318\,\text{GeV}$ are shown as a function of Q^2 for selected values of x. The improvement due to combination is clearly visible. In Fig. 1 (right) a comparison between the combined HERA data and the combination of HERA I data alone [3] for NC e^-p scattering is shown. The integrated luminosity for e^-p scattering was very limited for the HERA I period, therefore, the improvements due to the new combination are even more substantial than for e^+p scattering.

The combined reduced cross sections for NC and CC $e^\pm p$ scattering are obtained together with their statistical, uncorrelated and correlated systematic uncertainties. At the proton beam energy of $E_p = 920\,\text{GeV}$, corresponding to the highest centre-of-mass energy of $\sqrt{s} = 318\,\text{GeV}$, they are shown in Fig. 2 for NC (left) and CC (right) as a function of Q^2 for different values of x. The total uncertainties for NC are below 1.5 % over the Q^2 range of $3 \leq Q^2 \leq 500\,\text{GeV}^2$ and below 3 % up to $Q^2 = 3000\,\text{GeV}^2$. Cross sections are provided for values of Q^2 between $Q^2 = 0.045\,\text{GeV}^2$ and $Q^2 = 50000\,\text{GeV}^2$ and for values of x between $x = 6 \times 10^{-7}$ and $x = 0.65$.

DOI: 10.1051/epjconf/201612602005

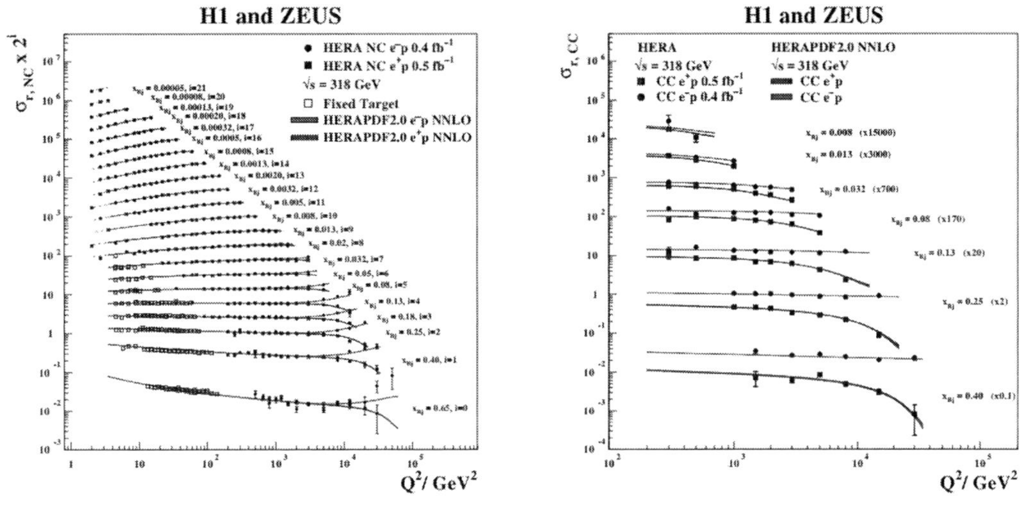

Figure 2. Left (Right): The combined HERA data for the inclusive NC (CC) e^+p and e^-p reduced cross sections at $\sqrt{s}$ = 318 GeV together with fixed-target data and the predictions of HERAPDF2.0 NNLO. The bands represent the total uncertainties on the predictions.

The combined NC and CC cross sections were integrated to obtain the single-differential cross sections $d\sigma/dQ^2$. The integration over x was performed in the region $y < 0.9$, using the shapes as predicted from HERAPDF2.0. All correlated and uncorrelated uncertainties were taken into account. The cross-sections $d\sigma/dQ^2$ are shown in Fig. 3 (left) for NC and CC e^-p and e^+p scattering together with predictions from HERAPDF2.0 NLO. Whereas the NC cross sections are three orders of magnitude larger at low $Q^2 \approx 100\,\text{GeV}^2$, where they are dominated by virtual photon exchange, the NC and CC cross sections become similar in magnitude at $Q^2 \approx 10000\,\text{GeV}^2$, i.e. at around the mass-scale squared of the electroweak bosons, demonstrating the success of electroweak unification in the Standard Model.

The total CC cross sections $\sigma_{CC}^{e^\pm p}$ measured by H1 and ZEUS in the range $Q^2 > 400$ GeV^2 and $y < 0.9$ are shown in Fig. 3 (right) as a function of the electron beam polarisation. The cross sections were obtained using the HERA II CC data with left-handed and right-handed electron beam polarisations [7, 8] and using CC data from HERA I with unpolarised electron beams [9]. The measurements are compared to expectations of the Standard Model based on the HERAPDF1.5[4] parameterisation of the parton densities of the proton. The measurements exhibit the expected linear dependence as a function of the polarisation and are consistent with the axiomatic absence of right-handed charged currents in the Standard Model.

3.3 Scaling violations of F_2 and the proton structure function $F_2^{\gamma Z}$

The dominant contribution to the NC cross section is due to $\tilde{F}_2$, see Eqs. 1-3. Therefore, the combined reduced NC cross sections in Fig. 2 (left) reveal the main properties of the proton structure function $\tilde{F}_2$. At $x \simeq 0.1$ they are independent of Q^2 demonstrating scaling of $\tilde{F}_2$ in this region. At higher and lower values of x one observes scaling violations due to emission of gluons from quarks and splitting of gluons into quark-antiquark pairs, respectively. Values of the structure function $\tilde{F}_2$ determined

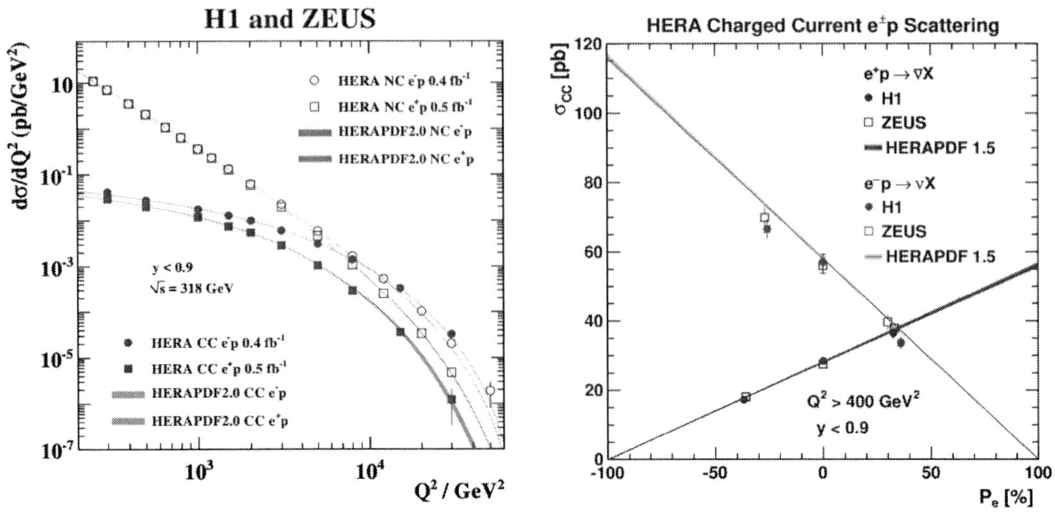

Figure 3. Left: The combined HERA NC and CC e^-p and e^+p cross sections, $d\sigma/dQ^2$, together with predictions from HERAPDF2.0 NLO. Right: The dependence of the total e^+p and e^-p CC cross sections in the region $Q^2 > 400\,\mathrm{GeV}^2$ and $y < 0.9$ on the electron beam polarisation P_e, together with the Standard Model predictions based on HERAPDF1.5.

from the reduced cross sections in the phase space, where corrections due to both $x\tilde{F}_3$ and $\tilde{F}_L$ are small, are displayed as a function of x at fixed Q^2 in Fig. 4 (left) together with HERAPDF2.0 NLO predictions. Here, the scaling violations manifest themselves by the rise towards low values of x, which is becoming steeper as Q^2 increases.

The polarisation effects in NC allow the parity violating component of the proton structure function $\tilde{F}_2$, namely $F_2^{\gamma Z}$, to be determined. For a given electron charge the difference of the left (L) and right (R) polarised NC cross sections can be written as

$$\frac{\sigma^{\pm}_{r,NC}(P^{\pm}_{L}) - \sigma^{\pm}_{r,NC}(P^{\pm}_{R})}{P^{\pm}_{L} - P^{\pm}_{R}} = \kappa\left[\mp a_e F_2^{\gamma Z} + \frac{Y_-}{Y_+} v_e x F_3^{\gamma Z} - \frac{Y_-}{Y_+}\kappa(v_e^2 + a_e^2) x F_3^{Z}\right]. \quad (6)$$

Subtracting the differences for the e^-p and e^+p interactions, the terms proportional to $xF_3^{\gamma Z}$ and xF_3^Z cancel, and $F_2^{\gamma Z}$ can be directly extracted from the cross section data. H1 performed such measurements [7] for $Q^2 \geq 200$ GeV^2 using HERA II data with the polarised electron beams. Only a weak Q^2 dependence is expected, and therefore the measurements were transformed to a common Q^2 value of 1500 GeV^2 using the H1PDF 2012 fit [7] and were averaged in each x bin. The parity violating structure function $F_2^{\gamma Z}$ measured for the first time is displayed in Fig. 4 (right).

3.4 Electron charge dependence of the NC cross section and determination of $xF_3^{\gamma Z}$

The different charges of electrons and positron beams allow the structure function $x\tilde{F}_3$ to be measured due to difference in the sign in front of the $x\tilde{F}_3$ contribution in Eq. 1. The differences in the combined e^+p and e^-p NC reduced cross sections are clearly visible at high Q^2 in Fig. 2 (left). The structure function $x\tilde{F}_3$ was extracted by subtracting the combined NC e^+p from the combined NC e^-p cross sections at HERA for $Q^2 \geq 1000\,\mathrm{GeV}^2$. The dominant contribution to $x\tilde{F}_3$ arises from the γZ interference, which allows $xF_3^{\gamma Z}$ to be determined using Eq. 2. The structure function $xF_3^{\gamma Z}$ is non-singlet

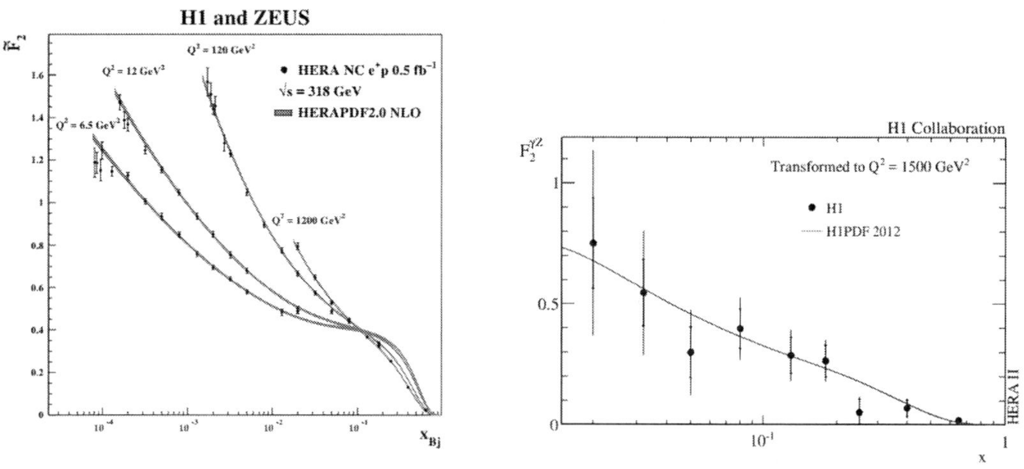

Figure 4. Left: The structure function $\tilde{F}_2$ as extracted from the HERA combined reduced NC cross sections for four values of Q^2 together with predictions using HERAPDF2.0 NLO. Right: The structure function $F_2^{\gamma Z}$ as a function of x at $Q^2 = 1500$ GeV^2 compared to predictions based on H1PDF 2012.

and shows little dependence on Q^2. The measured $xF_3^{\gamma Z}$ at different Q^2 values can thus be averaged taking into account the small Q^2 dependence. The averaged values of $xF_3^{\gamma Z}$ [1], determined for a Q^2 value of 1000 GeV^2, are shown in Fig. 5 (left). The measurement directly probes the valence-quark content of the proton and is well described by HERAPDF2.0 NLO.

3.5 Measurement of the longitudinal structure function F_L

The longitudinal structure function F_L is obtained at HERA [10, 11] in a model independent way using measurements at fixed x and Q^2 but different y. This is achieved by H1 and ZEUS using data from

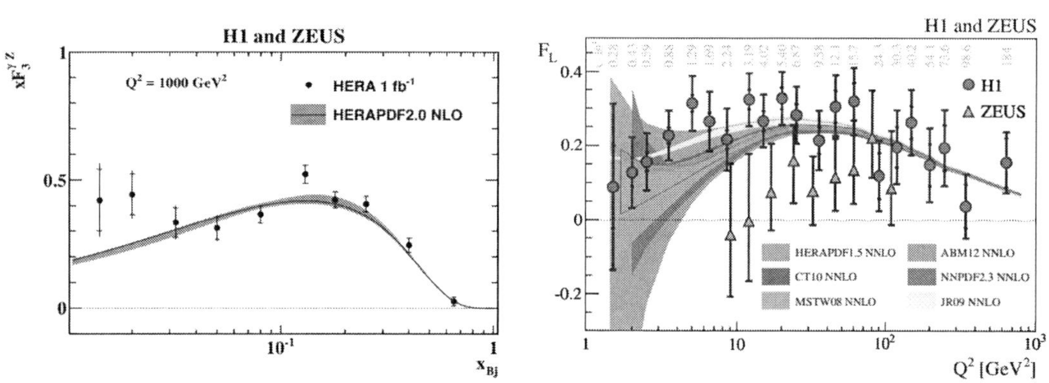

Figure 5. Left: The structure function $xF_3^{\gamma Z}$ for ten values of Q^2 together with predictions from HERAPDF2.0 NLO. Right: The H1 and ZEUS measurements of $F_L(Q^2)$. The resulting x values of the averaged F_L are given on top for each point in Q^2. The data are compared to NNLO predictions based on a selection of PDF sets as indicated.

running periods with reduced (460 GeV and 575 GeV) and nominal (920 GeV) proton beam energies. According to Eq. 1, the F_L contribution to the reduced cross section is proportional to $f(y) = y^2/[1 + (1-y)^2]$. Therefore, the F_L values can be determined as slopes of straight-line fits of the measured $\sigma_{r,NC}(x, Q^2, y)$ values as a function of the y-dependent factor $f(y)$. More sophisticated procedures to determine F_L and F_2 were applied by H1 and ZEUS to take into account correlations of data points due to systematic uncertainties. The H1 and ZEUS measurements of $F_L(x, Q^2)$ were averaged at fixed Q^2, and the resulting values of $F_L(Q^2)$ are shown in Fig. 5 (right). The H1 measurements generally lie above the ZEUS results. The differences were examined in the common Q^2 range, $9 \le Q^2 \le 110$ GeV^2. Taking into account both the correlations between the H1 and between the ZEUS measurements, a χ^2 of 11 was obtained for 8 degrees of freedom. This corresponds to a p-value of 20%, indicating consistency of the H1 and ZEUS F_L results. The data are reasonably well reproduced by theoretical predictions at NNLO. The measurements of F_L were used to determine the ratio $R = F_L/(F_2 - F_L)$. For $1.5 \le Q^2 \le 800$ GeV^2 the ratio R measured by H1 showed constant behaviour yielding the averaged value of $R = 0.23 \pm 0.04$ [10]. The corresponding ZEUS result is $R = 0.105^{+0.055}_{-0.037}$ [11] in the range $9 \le Q^2 \le 110$ GeV^2.

4 QCD analysis of the combined inclusive cross sections at HERA

The combined inclusive $e^{\pm}p$ NC and CC cross section data [1] were used as the sole input in the pQCD analysis, resulting in the PDF set called HERAPDF2.0 [1]. This analysis was performed within the HERAFitter framework [12] by fitting the predictions obtained by solving the DGLAP evolution equations at LO, NLO and NNLO in the $\overline{\text{MS}}$ scheme [13] to the combined HERA data. The DGLAP equations yield the PDFs at all scales $\mu_{\rm f}^2$ and x, if they are provided as functions of x at some starting scale. In variable-flavour-number schemes, this scale has to be below the charm-quark mass parameter, M_c, squared. It was chosen to be 1.9 GeV^2. The renormalisation and factorisation scales were taken to be Q^2. The heavy-quark coefficient functions were calculated in the general-mass variable-flavour-number scheme called RTOPT [14] for the NC structure functions. For the CC structure functions, the zero-mass approximation was used, since all HERA CC data have $Q^2 \gg M_b^2$, where M_b is the beauty-quark mass parameter in the calculation. The value of M_c and M_b were chosen after performing χ^2 scans of NLO and NNLO pQCD fits to the combined inclusive data together with the HERA combined charm data [15] or with data on beauty production from ZEUS [16] and H1 [17]. The value of the strong coupling constant was taken to be $\alpha_s(M_Z^2) = 0.118$ [18] at NLO and NNLO and $\alpha_s(M_Z^2) = 0.130$ [19] for the LO fit.

In the approach of HERAPDF, the PDFs of the proton, xf, are generically parameterised at the starting scale as $xf(x) = Ax^B(1-x)^C(1+Dx+Ex^2)$. The PDFs parameterised are the gluon distribution, xg, the valence-quark distributions, xu_v, xd_v, and the u-type and d-type anti-quark distributions, $x\bar{U}$, $x\bar{D}$. The relations $x\bar{U} = x\bar{u}$ and $x\bar{D} = x\bar{d} + x\bar{s}$ are assumed at the starting scale.

The central parameterisation is

$$
\begin{aligned}
xg(x) &= A_g x^{B_g}(1-x)^{C_g} - A'_g x^{B'_g}(1-x)^{C'_g}, && && (7)\\
xu_v(x) &= A_{u_v} x^{B_{u_v}}(1-x)^{C_{u_v}}\left(1 + E_{u_v}x^2\right), && xd_v(x) = A_{d_v} x^{B_{d_v}}(1-x)^{C_{d_v}}, &&\\
x\bar{U}(x) &= A_{\bar{U}} x^{B_{\bar{U}}}(1-x)^{C_{\bar{U}}}\left(1 + D_{\bar{U}}x\right), && x\bar{D}(x) = A_{\bar{D}} x^{B_{\bar{D}}}(1-x)^{C_{\bar{D}}}. &&
\end{aligned}
$$

In the gluon distribution, xg, an additional term of the form $A'_g x^{B'_g}(1-x)^{C'_g}$ was subtracted at NLO and NNLO to allow more flexibility for the gluon distribution. The fits without this term are marked by adding "AG" (alternative gluon) to the fit notation, e.g. HERAPDF2.0AG LO for the fit at LO.

To safely remain in the kinematic region where pQCD is expected to be applicable, only cross sections for Q^2 starting from $Q^2_{\rm min} = 3.5$ GeV^2 were used in the analysis. In the kinematic region of the fits, target-mass corrections and large-x higher-twist corrections are expected to be negligible.

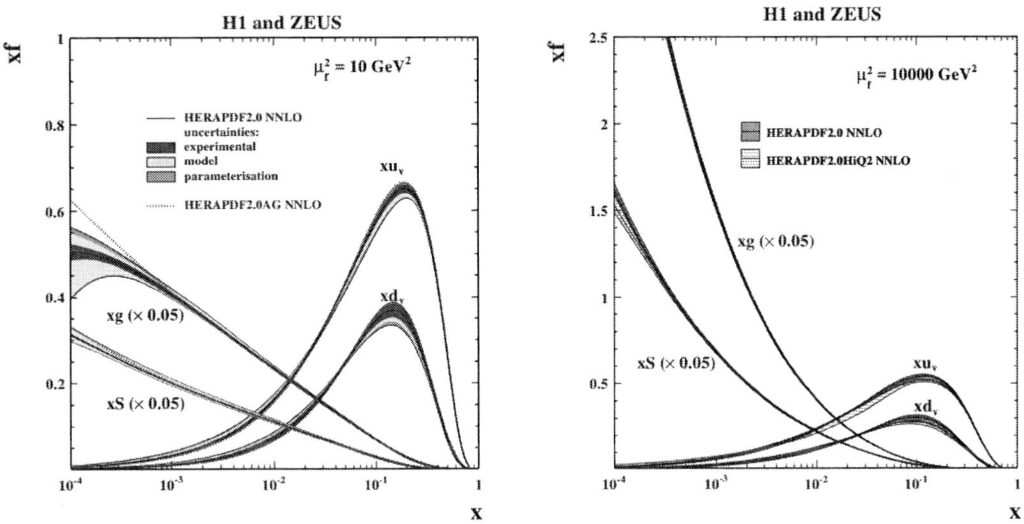

Figure 6. The parton distribution functions xu_v, xd_v, $xS = 2x(\bar{U} + \bar{D})$ and xg; the gluon and sea distributions are scaled down by a factor 20. Left: HERAPDF2.0 NNLO at $\mu_f^2 = 10\,\mathrm{GeV}^2$. The experimental, model and parameterisation uncertainties are shown. The dotted lines represent HERAPDF2.0AG NNLO with the alternative gluon parameterisation. Right: HERAPDF2.0 NNLO at $\mu_f^2 = 10000\,\mathrm{GeV}^2$ with $Q^2_{\mathrm{min}} = 3.5\,\mathrm{GeV}^2$ compared to HERAPDF2.0HiQ2 NNLO with $Q^2_{\mathrm{min}} = 10\,\mathrm{GeV}^2$. The bands represent the total uncertainties.

Experimental uncertainties were determined using the Hessian method with the criterion $\Delta\chi^2 = 1$, justified by the consistency of the input HERA data. In addition to experimental uncertainties, model and parameterisation uncertainties were also considered. The uncertainties on HERAPDF2.0 due to the choice of model settings and the form of the parameterisation were evaluated by varying the assumptions. Two kinds of parameterisation uncertainties were considered, the variation of the starting scale and the addition of terms with parameters D and E; see the generic form of parameterisation above. The only significant difference from the 14-parameter central fit came from the 15-parameter fit, for which D_{u_v} was non zero. The final parameterisation uncertainty for a given quantity is taken as the largest of the uncertainties. The total PDF uncertainty is obtained by adding in quadrature the experimental, the model and the parameterisation uncertainties. The setup used in this QCD analysis follows closely to the one established earlier for HERAPDF1.0 [3] for fitting the combined HERA I data [3].

4.1 HERAPDF2.0 and its variants

Using the standard settings discussed above, the HERAPDF2.0 sets of PDS were determined at NLO and NNLO. A summary of HERAPDF2.0 NNLO is shown in Fig. 6 (left) at the scale $\mu_f^2 = 10\,\mathrm{GeV}^2$. The experimental, model and parameterisation uncertainties are shown separately. The distribution of HERAPDF2.0AG NNLO with the alternative gluon parameterisation is shown as well. At NNLO, the gluon distribution of HERAPDF2.0 ceases to rise at low x in contrast to HERAPDF2.0AG NNLO.

It was observed that the χ^2 per degree of freedom in the fit is decreasing and getting closer to unity with increasing value of Q^2_{min}. This continues up to $Q^2_{\mathrm{min}} = 10\,\mathrm{GeV}^2$, remaining about the same for higher values of Q^2_{min}. This could be caused by non-perturbative contributions missing in the theory used in the fit. A question may arise whether including data from the kinematic region of low x and low

Q^2, i.e. below $10\,\text{GeV}^2$, in the PDF fits could introduce a bias on predictions for high x and high Q^2. To answer this, the HERAPDF2.0HiQ2 NLO and NNLO fits were performed with $Q^2_{\text{min}} = 10\,\text{GeV}^2$ and the same other settings as for HERAPDF2.0. At NLO, the gluon distributions of HERAPDF2.0 and HERAPDF2.0HiQ2 were found to be compatible within uncertainties. At NNLO, the two gluon distributions differ significantly. Using the higher Q^2_{min} at NNLO causes the gluon distribution to turn over at low x. Nevertheless, it was demonstrated that at the high scale of $\mu_f^2 = 10000\,\text{GeV}^2$, the PDFs resulting from the two fits are very similar, see Fig. 6 (right). This confirms that the value of $Q^2_{\text{min}} = 3.5\,\text{GeV}^2$ is safe for pQCD fits to HERA data and no bias was introduced for applications at higher scales like cross-section predictions for the LHC.

The PDFs of HERAPDF2.0 NNLO are directly compared in Fig. 7 to the PDFs of MMHT2014 [20], for which the same heavy-flavour scheme, i.e. RTOPT, was used. Also comparisons to the PDFs of CT10 [21], for which a heavy-flavour-scheme based on ACOT was used, and NNPDF3.0 [22], for which the FONLL scheme was used, are shown. All PDFs are compatible taking into account full uncertainties. The largest relative discrepancy ($\approx 2.5\sigma$) is found in the shape of the xu_v distribution at $x \approx 0.4$. The variants of the QCD fit, HERAPDF2.0FF3A and HERAPDF2.0FF3B, obtained using fixed-flavour-number schemes are compared in Fig. 8 to the PDFs of ABM11 FF(3N) [23] and NNPDF3.0 FF(3N) [22] with similar flavour schemes. The gluon distributions are quite similar. Some differences are observed in the xu_v and xd_v valence distributions.

An extension of the analysis by including HERA data on charm and jet production, denoted as HERAPDF2.0Jets, made a simultaneous determination of parton distributions and the strong coupling at the scale M_Z^2 possible. A full uncertainty analysis was performed including model and parameterisation uncertainties as well as additional hadronisation uncertainties on the jet data. The χ^2 and the resulting PDFs of the HERAPDF2.0Jets fit are very close to the values for the fit with fixed

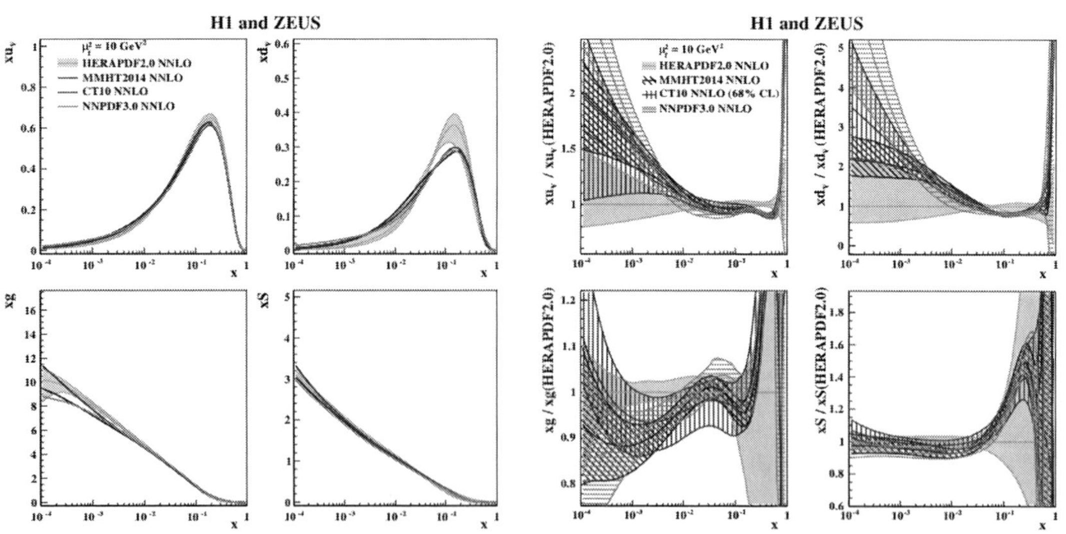

Figure 7. The parton distribution functions xu_v, xd_v, xg and $xS = 2x(\bar{U} + \bar{D})$ of HERAPDF2.0 NNLO at $\mu_f^2 = 10\,\text{GeV}^2$ compared to those of MMHT2014, CT10 and NNPDF3.0. The left panel shows the distribution with uncertainties only for HERAPDF2.0. The right panel shows the PDFs normalised to HERAPDF2.0 and with uncertainties for all PDFs.

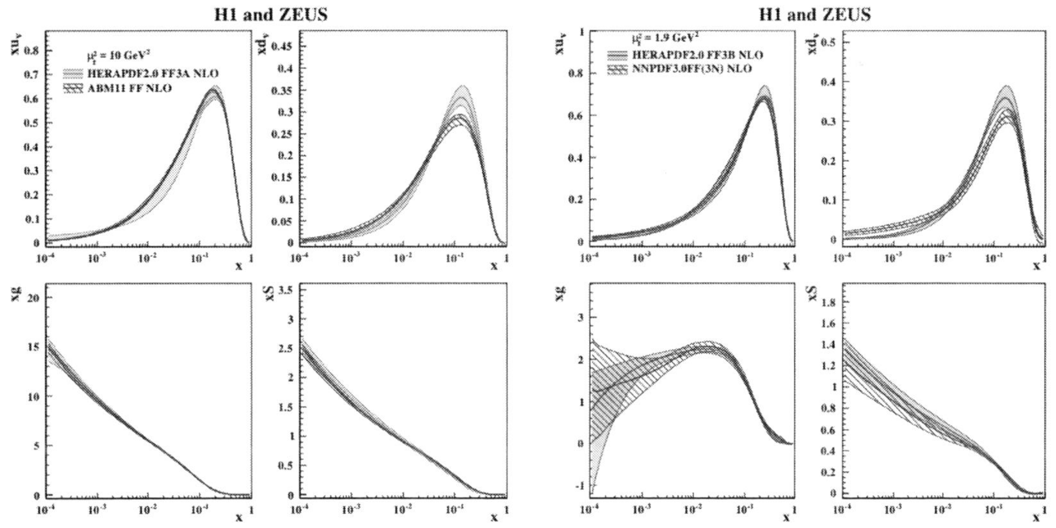

Figure 8. The parton distribution functions xu_v, xd_v, xg and $xS = 2x(\bar{U} + \bar{D})$. Left: HERAPDF2.0FF3A at $\mu_f^2 = 10\,\text{GeV}^2$ compared to ABM11 FF. Right: HERAPDF2.0FF3B at the starting scale $\mu_{f_0}^2 = 1.9\,\text{GeV}^2$ compared to NNPDF3.0FF (3N).

$\alpha_s(M_Z^2) = 0.118$, because the HERAPDF2.0Jets fit with free $\alpha_s(M_Z^2)$ yields a very similar value of

$$\alpha_s(M_Z^2) = 0.1183 \pm 0.0009(\text{exp}) \pm 0.0005(\text{model/param.}) \pm 0.0012(\text{hadron.}) \;^{+0.0037}_{-0.0030}(\text{scale}) \;.$$

The uncertainty on $\alpha_s(M_Z^2)$ due to scale uncertainties was evaluated by varying the renormalisation and factorisation scales by a factor of two, both separately and simultaneously, and taking the maximal positive and negative deviations. The result on $\alpha_s(M_Z^2)$ is compatible with the world average [18] and it is competitive with other determinations at NLO.

5 Summary and Conclusion

The H1 and ZEUS collaborations measured inclusive $e^{\pm}p$ scattering cross sections at HERA from 1994 to 2007, collecting a total integrated luminosity of about 1 fb^{-1} at four different centre-of-mass energies, with two different detectors changing and improving over time. All inclusive data were combined to create one consistent set of NC and CC cross-section measurements for unpolarised $e^{\pm}p$ scattering, spanning six orders of magnitude in both negative four-momentum-transfer squared, Q^2, and Bjorken x. Combined cross sections are provided for values of Q^2 between $Q^2 = 0.045\,\text{GeV}^2$ and $Q^2 = 50000\,\text{GeV}^2$ and values of x between $x = 6 \times 10^{-7}$ and $x = 0.65$. They are the most precise measurements ever published for ep scattering over such a large kinematic range. Using the combined HERA data and the individual data taken by H1 and ZEUS with longitudinal beam polarisation, the proton structure functions F_2, $F_2^{\gamma Z}$, $xF_3^{\gamma Z}$ and F_L were obtained, and scaling violations, electroweak unification and polarisation effects in CC were demonstrated.

The combined inclusive cross sections were used as input to a QCD analysis within the DGLAP formalism. The resulting parton distribution functions are denoted HERAPDF2.0 and are available at LO, NLO and NNLO using variable-flavour-number scheme. Two variants, HERAPDF2.0 FF3A and FF3B, using fixed-flavour number schemes, are also available at NLO. Other variants of the fit were also performed exploring an alternative, positively defined, gluon parameterisation and different Q^2_{min} values. A measurement of the strong coupling constant was performed using a perturbative QCD fit, denoted as HERAPDF2.0Jets, for which the inclusive cross sections were augmented with selected jet- and charm-production cross section data as measured by the H1 and ZEUS collaborations.

References

[1] H. Abramowicz *et al.* [H1 and ZEUS Collaborations], Eur. Phys. J. C **75**, 580 (2015), [arXiv:1506.06042].

[2] V. N. Gribov and L. N. Lipatov, Sov. J. Nucl. Phys. **15**, 438 (1972); Sov. J. Nucl. Phys. **15**, 675 (1972); L. N. Lipatov, Sov. J. Nucl. Phys. **20**, 94 (1975); Y. L. Dokshitzer, Sov. Phys. JETP **46**, 641 (1977); G. Altarelli and G. Parisi, Nucl. Phys. B **126**, 298 (1977).

[3] F. Aaron *et al.* [H1 and ZEUS Collaborations], JHEP **1001**, 109 (2010), [arXiv:0911.0884].

[4] HERAPDF1.5: URL https://www.desy.de/h1zeus/combined_results/herapdftable/ .

[5] The documentation can be found at: URL https://wiki-zeuthen.desy.de/HERAverager .

[6] F. Aaron *et al.* [H1 Collaboration], Eur. Phys. J. C **63**, 625 (2009), [arXiv:0904.0929].

[7] F. Aaron *et al.* [H1 Collaboration], JHEP **09**, 61 (2012), [arxiv:1206.7007].

[8] S. Chekanov et al. [ZEUS Collaboration], Eur. Phys. J. C **61**, 223 (2009), [arXiv:0812.4620]; H. Abramowicz et al. [ZEUS Collaboration], Eur. Phys. J. C **70**, 945 (2010), [arXiv:1008.3493].

[9] C. Adloff et al. [H1 Collaboration], Eur. Phys. J. C **13**, 609 (2000), [hep-ex/9908059]; Eur. Phys. J. C **19**, 269 (2001), [hep-ex/0012052]; Eur. Phys. J. C **30**, 1 (2003), [hep-ex/0304003];
J. Breitweg et al. [ZEUS Collaboration], Eur. Phys. J. C **12**, 411 (2000), [Erratum-ibid. C **27**, 305 (2003)], [hep-ex/9907010];
S. Chekanov et al. [ZEUS Collaboration], Phys. Lett. B **539**, 197 (2002), [Erratum-ibid. B **552**, 308 (2003)], [hep-ex/0205091]; Eur. Phys. J. C **32**, 1 (2003), [hep-ex/0307043].

[10] V. Andreev *et al.* [H1 Collaboration], Eur. Phys. J. C **74**, 2814 (2014), [arxiv:1312.4821].

[11] H. Abramowicz *et al.* [ZEUS Collaboration], Phys. Rev. D**90**, 072002 (2014), [arXiv:1404.6376].

[12] S. Alekhin *et al.* (2014), [arXiv:1410.4412]; Documentation at: URL http://www.herafitter.org .

[13] B. Fanchiotti, S. Kniehl and A. Sirlin, Phys. Rev. D **48**, 307 (1993), [hep-ph/9212285].

[14] R. S. Thorne and R. G. Roberts, Phys. Rev. D **57**, 6871 (1998), [hep-ph/9709442];
R. S. Thorne, Phys. Rev. D **73**, 054019 (2006), [hep-ph/0601245];
R. S. Thorne, Phys. Rev. D **86**, 074017 (2012), [arXiv:1201.6180].

[15] F. Aaron et al., [H1 and ZEUS Collaborations], Eur. Phys. J. C **73**, 2311 (2013), [arXiv:1211.1182].

[16] H. Abramowicz et al. [ZEUS Collaboration], JHEP **1409**, 127 (2014), [arXiv:1405.6915].

[17] F. Aaron et al. [H1 Collaboration], Eur. Phys. J. C **65**, 89 (2009), [arXiv:0907.2643].

[18] J. Beringer et al. (Particle Data Group), Phys. Rev. D **86**, 010001 (2012).

[19] J. Pumplin et al., JHEP **0207**, 012 (2002), [hep-ph/0201195].

[20] L. Harland-Lang, A. D.Martin, P. Motylinski, and R. Thorne (2014), [arXiv:1412.3989].

[21] M. Guzzi et al., Phys. Rev. D **89**, 033009 (2014), [arXiv:1302.6246].

[22] R. D. Ball et al. [NNPDF Collaboration], JHEP **1004**, 040 (2015), [arXiv:1410.8849v2].

[23] S. Alekhin et al., Phys. Rev. D **86**, 054009 (2012), [arXiv:1202.2281].

EPJ Web of Conferences **126**, 02006 (2016) DOI: 10.1051/epjconf/201612602006
ICNFP 2015

Higgs results from ATLAS

Xin Chen[1,2,a], On behalf of the ATLAS Collaboration

[1] *Department of Physics, Tsinghua University, Beijing 100084, China*
[2] *Collaborative Innovation Center of Quantum Matter, Beijing, China*

Abstract. The updated Higgs measurements in various search channels with ATLAS Run 1 data are reviewed. Both the Standard Model (SM) Higgs results, such as $H \to \gamma\gamma, ZZ, WW, \tau\tau, \mu\mu, b\bar{b}$, and Beyond Standard Model (BSM) results, such as the charged Higgs, Higgs invisible decay and tensor couplings, are summarized. Prospects for future Higgs searches are briefly discussed.

1 Introduction

The discovery of the Higgs boson [1, 2] by the ATLAS [3] and CMS experiments at the Large Hadron Collider (LHC) marks a major corner stone of the particle physics. Many new results have emerged after that, which span the aspects of its coupling strength and structure, spin/CP, mass, decay width and a broad range of BSM signatures. The most recent (up to the time of this publication) Higgs results using ATLAS 7 and 8 TeV p-p collision Run 1 data (about 25 fb^{-1} in total) are reviewed here for different Higgs decay modes. An overview of future search prospects is given at the end of the paper.

2 SM Higgs measurements

The main Higgs production modes are gluon-gluon fusion (ggH), vector-boson fusion (VBF), VH associated and $t\bar{t}H$, with their cross sections in descending order. The productions diagrams are shown in Fig. 1. However, the search is most optimized according to different Higgs decays, since their SM backgrounds are vastly different. Productions, such as VBF, have additional background rejection features (e.g., forward jets). All production and decay modes have been searched for and combined to get the most sensitive results.

2.1 $H \to \gamma\gamma$

The branching ratio of $H \to \gamma\gamma$ [4] is only 0.23% at $m_H = 125$ GeV, but it has the advantage of a small background, which mainly consists of irreducible $\gamma\gamma$ (about 77%), γ+jet and dijet events. The first process has a large theoretical uncertainty. The latter two are mainly due to π^0 faking photons, whose rate is hard to model with simulation. These backgrounds are estimated from data using mass side-bands. This channel is optimized according to different production modes. The final signal strength

[a] e-mail: xin.chen@cern.ch

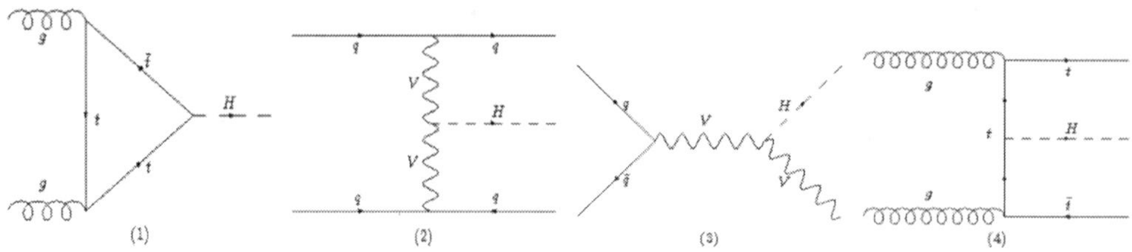

Figure 1. The Higgs production diagrams at LHC: (1) ggH, (2) VBF, (3) VH, (4) $t\bar{t}H$.

(measured rate divided by the SM prediction) based on Run 1 data is $\mu = 1.17 \pm 0.27$. Fig. 2 shows the background composition and the $\gamma\gamma$ mass spectrum.

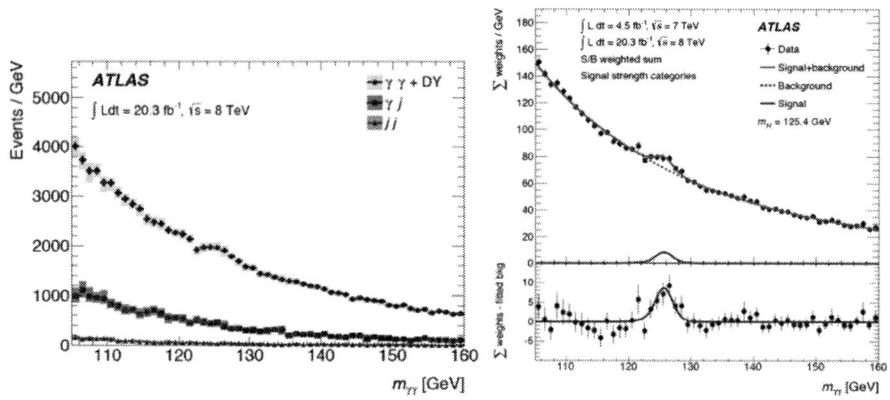

Figure 2. The background composition (left) and the $\gamma\gamma$ mass spectrum (right) in the $H \to \gamma\gamma$ channel. The mass spectrum is S/B-weighted from each category. Taken from Ref. [4].

2.2 $H \to ZZ^\star \to 4l$

Although the branching ratio of $H \to ZZ^\star \to 4l$ [5] is as small as 0.0125% at $m_H = 125$ GeV, it also has the smallest background, which mainly consists of $ZZ^\star \to 4l$ and $Zb\bar{b}$. The latter can be estimated by inverting the lepton isolation or impact parameter cuts. To maximize acceptance, loose lepton identification is applied, and the lepton p_T is as low as 6 GeV. This channel has a very rich phase space (4-body decay), and all final state particles can be reconstructed. Thus, the Matrix Element (ME) method and Boosted Decision Trees (BDT) can be used to extract the signal strength and other properties. Fig. 3 shows the 2-D m_{4l} versus BDT and the m_{4l} distributions. This channel is optimized for different production modes as well. The signal strength measured is $1.44^{+0.34}_{-0.31}(\text{stat})^{+0.21}_{-0.11}(\text{sys})$.

2.3 Higgs mass measurements

The Higgs mass measurements are only feasible in the $H \to \gamma\gamma$ and $H \to ZZ^\star \to 4l$ channels [6]. The results put a strong constraint on BSM models such as the Super Symmetry (SUSY). The $H \to \gamma\gamma$ mass is determined by the photon calibration and calo-pointing (calorimeter cell orientation and the

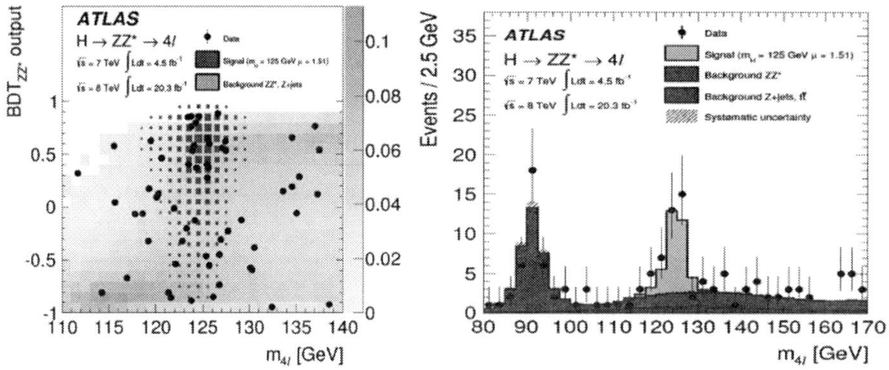

Figure 3. The m_{4l} versus BDT 2-D distribution (left) and the m_{4l} distribution (right) in the $H \rightarrow ZZ^{\star} \rightarrow 4l$ channel. Taken from Ref. [5].

primary vertex are used to measure the photon direction). The events are categorized into different photon categories (converted or unconverted, central or in the transition region) to get the best mass resolution. The final result is $125.98 \pm 0.42(\text{stat}) \pm 0.28(\text{sys})$. The $H \rightarrow ZZ^{\star} \rightarrow 4l$ mass is obtained by 2-D fit to mass and BDT distributions with a Z mass constraint. The final result is $124.51 \pm 0.52(\text{stat}) \pm 0.06(\text{sys})$. The combined mass of the two channels is $125.36 \pm 0.37(\text{stat}) \pm 0.18(\text{sys})$. It is worth to note that the two mass measurements are still compatible within 2σ level.

2.4 $H \rightarrow WW^{\star} \rightarrow 2l2\nu$

The branching ratio of $H \rightarrow WW^{\star} \rightarrow 2l2\nu$ ($l = e, \mu$) [7] is at 1% level at $m_H = 125$ GeV. However, the background is also more severe, which is dominated by top ($t\bar{t}$ and single top), Drell-Yan (for same-flavor leptons) and WW. To maximize the sensitivity, the analysis is divided into 0, 1 and 2-jet categories, as the background composition is different in different jet bins (see the N_{jet} distribution in Fig. 4). Thanks to the spin correlation between the two W bosons, the two leptons from their decay tend to be aligned with each other. This feature leads to large p_{T}^{ll}, small m_{ll} and $\Delta\Phi_{ll}$, which can help suppressing the Drell-Yan background. It is not possible to construct the Higgs mass, but a transverse mass, defined as $m_{\text{T}} = \sqrt{\left(E_{\text{T}}^{ll} + p_{\text{T}}^{\nu\nu}\right)^2 - \left|\mathbf{p}_{\text{T}}^{ll} + \mathbf{p}_{\text{T}}^{\nu\nu}\right|^2}$, can be used. The m_{T} distribution in the most sensitive $e\mu$+0-jet category (dominated by ggH) is shown in Fig. 4. The final signal strength is $1.09^{+0.23}_{-0.21}$ in this channel.

2.5 $H \rightarrow \tau\tau$

The branching ratio of $H \rightarrow \tau\tau$ [8] is 6.3% at $m_H = 125$ GeV, and it has a relatively clean background. Due to the smallness of the τ mass, the visible (e/μ or pions) and invisible (neutrinos) decay products tend to be aligned. A likelihood based algorithm, named MMC [9], is used to construct the di-τ mass. The analysis is optimized according to the leptonic or hadronic τ decay modes: $\tau_{\text{lep}}\tau_{\text{lep}}$, $\tau_{\text{lep}}\tau_{\text{had}}$, $\tau_{\text{had}}\tau_{\text{had}}$. Due to the large uncertainty from the missing transverse momentum ($E_{\text{T}}^{\text{miss}}$), the MMC mass resolution is about 20 GeV. As the Z and H mass peaks are close under this resolution, it is very important to model the $Z \rightarrow \tau\tau$ background well. This is realized by taking the $Z \rightarrow \mu\mu$ data events, and replacing the muon by simulated τ decay products. A BDT-based analysis is used to extract the signal. Fig. 5 shows the MMC distribution with all sub-channels combined.

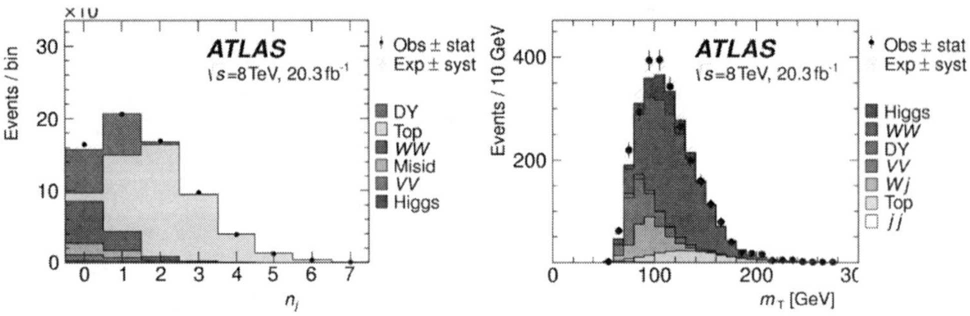

Figure 4. The jet multiplicity in the $e\mu$ category (left) and transverse mass in the $e\mu$+0-jet category (right) in the $H \to WW^\star \to 2l2\nu$ channel. Taken from Ref. [7].

The enhancement corresponds to a signal strength of $1.43^{+0.27}_{-0.26}(\text{stat})^{+0.32}_{-0.25}(\text{sys}) \pm 0.09(\text{theo})$, or a 4.5σ discovery.

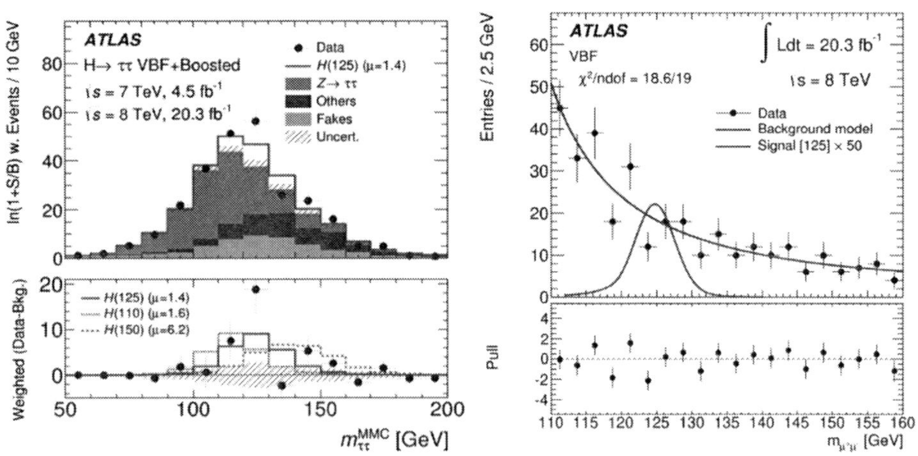

Figure 5. Left: the MMC di-τ mass weighted by weighted by $\ln(1 + S/B)$ in the $H \to \tau\tau$ channel. Right: the $\mu\mu$ mass spectrum in the $H \to \mu\mu$ channel. Taken from Ref. [8, 10].

2.6 $H \to \mu\mu$

The branching ratio of $H \to \mu\mu$ [10] is very small, and overwhelmed by the Drell-Yan $Z \to \mu\mu$ background. The analysis focuses on the VBF production, and is divided into several p_T bins of Higgs. The current observed upper limit (at 95% CL) of the signal strength is 7.0. The $\mu\mu$ mass spectrum is shown in Fig. 5.

2.7 $H \to b\bar{b}$

The $H \to b\bar{b}$ has the largest branching ratio (57.7% at m_H = 125 GeV), but suffers from a large QCD $b\bar{b}$ background. Together with the di-τ channel, they provide direct measurements of the Higgs

Yukawa couplings. The main search channels are $H \to b\bar{b}$ production associated with a vector boson (VH) [11] or a $t\bar{t}$ pair ($t\bar{t}H$) [12]. Depending on the boson (W/Z) decay modes, the former is further divided in to 0/1/2-lepton and a few boson p_{T} categories, with signal optimization in each. The major backgrounds for $V + H \to b\bar{b}$ are $W/Z + b\bar{b}/c\bar{c}$, and BDT is used to extract the signal. The $b\bar{b}$ mass with backgrounds subtracted, except for $V + Z \to b\bar{b}$, is shown in Fig. 6. The signal strength is $0.51^{+0.31}_{-0.30}(\mathrm{stat})^{+0.25}_{-0.22}(\mathrm{sys})$. In the $t\bar{t} + H \to b\bar{b}$ channel, the analysis is divided into 1- and 2-lepton modes, and several bins of different jet/b-jet numbers. Since the signal has ≥ 4 (≥ 6) jets in the 1-lepton (2-lepton) mode, and ≥ 4 b-jets, the high (low) jet/b-jet multiplicity bins are used as signal (background control) regions. Neural Network (NN) is used to extract the signal, and in the case of 1-lepton mode, ME method is also used for the NN input. The NN distribution in one of the most sensitive regions is shown in Fig. 6. The signal strength is 1.5 ± 1.1. This channel provides a direct $t\bar{t}H$ coupling measurement that otherwise can only be obtained implicitly from ggH production.

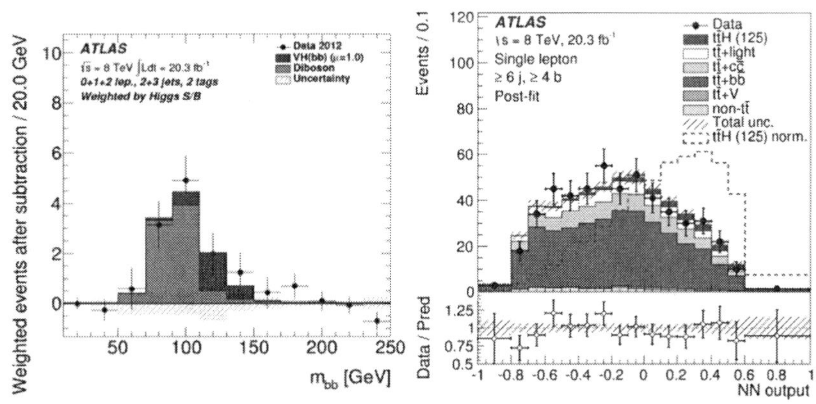

Figure 6. Left: the weighted $b\bar{b}$ mass in the $V + H \to b\bar{b}$ channel with all V decay modes combined. All backgrounds are subtracted except for $V + Z \to b\bar{b}$. Right: the NN distribution after fit in the $t\bar{t} + H \to b\bar{b}$ channel and 1-lepton + ≥ 6 jets + ≥ 4 b-jets signal region. Taken from Ref. [11, 12].

2.8 Higgs coupling combination

The statistically combined signal strength for all decay channels is $1.18^{+0.15}_{-0.14}$ [13]. Fig. 7 shows the ME level coupling strength factors for different HVV and $Hf\bar{f}$ couplings. The double minima are due to relative signs of κ_W and κ_t in the $H \to \gamma\gamma$ loop. However, the global minimum is still compatible with the SM prediction ($\kappa_V = \kappa_f = 1$) within 1σ. Fig. 7 also shows the Higgs coupling as a function of different particle mass. The fact that all the measurements lie around a straight line is a confirmation of the SM Higgs mechanism.

3 BSM Higgs searches

The BSM Higgs searches are carried out in many aspects: direct searches of new Higgs scalars such as the charged Higgs, Higgs coupling to invisible particles and the altering of its decay width, new coupling structures of Higgs coupling to bosons and fermions. The highlights of a few search results are given below.

 DOI: 10.1051/epjconf/201612602006

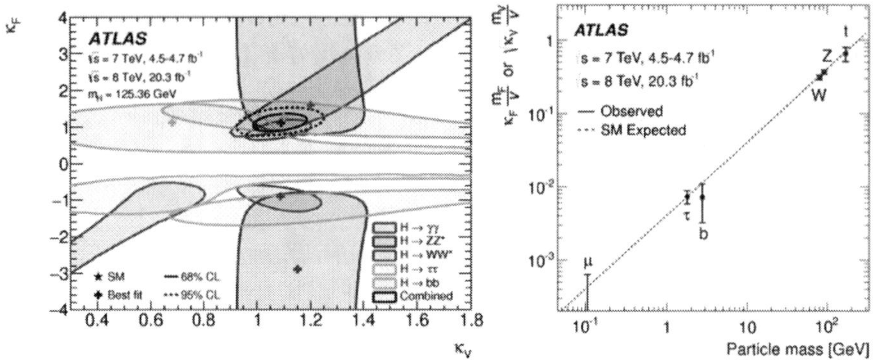

Figure 7. Left: the Higgs coupling strength for different particles with 1 and 2σ contours. All bosons and fermions share a common coupling strength factor, κ_V and κ_f, respectively, Right: the Higgs coupling as a function of different particle mass. Taken from Ref. [13].

3.1 Charged Higgs search

Charged Higgs is a benchmark of two Higgs doublet models. Its search is carried out in two phase space regions depending on whether its mass is below or above the top quark mass [14]. The main production diagrams in these two regions are shown in Fig. 8. The decay of $H^{\pm}$ searched for is $H^{\pm} \to \tau^{\pm}\nu$. In order to construct the transverse mass, all-hadronic decays of top and the hadronic τ decay are required. These events are recorded by τ+$E_{\mathrm{T}}^{\mathrm{miss}}$ triggers. At least 3 or 4 jets, one b-tagged jet and a large $E_{\mathrm{T}}^{\mathrm{miss}}$ are required. The background is divided into real or fake τ_{had} events, with the former estimated from $W \to \mu\nu$ data events (replacing the muon by simulated τ decay). Fig. 9 shows the model independent upper limits on $\mathrm{B}(t \to bH^{\pm}) \times \mathrm{B}(H^{\pm} \to \tau^{\pm}\nu)$ or $\sigma_{H^{\pm}} \times \mathrm{B}(H^{\pm} \to \tau^{\pm}\nu)$, and the exclusion regions in a bench mark point in the MSSM parameter space, $m_h^{\mathrm{mod}\pm}$, whose light Higgs mass is consistent with 125 GeV. The region with $H^{\pm}$ mass close to the top is not considered as it is phase space suppressed.

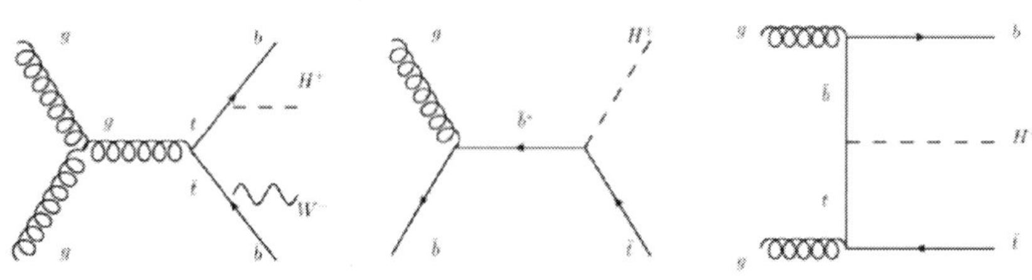

Figure 8. The production diagrams for charged Higgs in the low (left) and high (middle, right) mass regions. Taken from Ref. [14].

3.2 Off-shell Higgs

The on-shell production of diboson (WW/ZZ) can be used to probe the off-shell Higgs [15]. Unfortunately, similar to the case of double Higgs production, it receives the interference from Higgs-less box diagram diboson productions, as shown in Fig. 10. Therefore, they are considered together as the

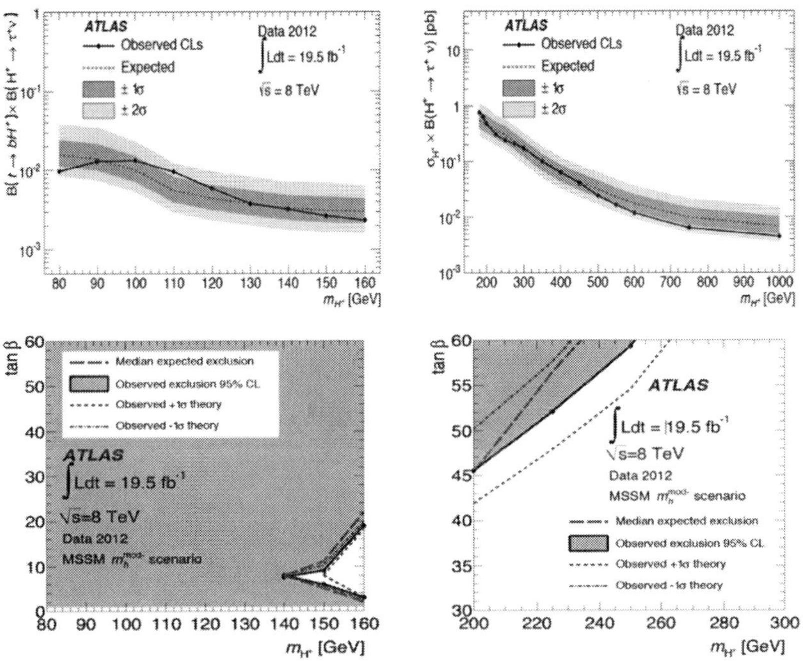

Figure 9. The 95% CL upper limit on $\mathrm{B}(t \to bH^{\pm}) \times \mathrm{B}(H^{\pm} \to \tau^{\pm}\nu)$ or $\sigma_{H^{\pm}} \times \mathrm{B}(H^{\pm} \to \tau^{\pm}\nu)$ for the low (upper left) and high (upper right) mass regions. The 95% CL exclusion regions in the $m_{H^{\pm}}$-tanβ plane in the MSSM $m_h^{\mathrm{mod}\pm}$ model in the low (bottom left) and high (bottom right) mass regions. Taken from Ref. [14].

signal. The off-shell Higgs signal strength is expressed as a function of center-of-mass scale $\hat{s}$ and the gluon and boson coupling strength factors (κ_g and κ_V):

$$\mu_{\text{off-shell}}(\hat{s}) = \frac{\sigma^{gg\to H^\star\to VV}_{\text{off-shell}}(\hat{s})}{\sigma^{gg\to H^\star\to VV}_{\text{off-shell,SM}}(\hat{s})} = \kappa^2_{g,\text{off-shell}}(\hat{s})\kappa^2_{V,\text{off-shell}}(\hat{s}). \tag{1}$$

To suppress the other diboson background, in the $ZZ \to 4l$ decay, ME method with MCFM [16] is used, while in the $ZZ \to 2l2\nu$ and $WW \to e\nu\mu\nu$ decays, the transverse mass is used as the discriminant. Unfortunately, because the cross section for box process $gg \to VV$ is not well known, the final results are expressed as a function of $R^B_{H^\star} = \kappa(gg \to VV)/\kappa(gg \to H^\star \to VV)$. The 95% CL upper limit on $\mu_{\text{off-shell}}$ is shown in Fig. 11. For $R^B_{H^\star} = 1$, the observed (expected) upper limit is 6.2 (8.1). Subject to a number of assumptions, when the off-shell results are combined with the on-shell signal strength measurement, which is defined as

$$\mu_{\text{on-shell}} = \frac{\sigma^{gg\to H\to VV}_{\text{on-shell}}}{\sigma^{gg\to H\to VV}_{\text{on-shell,SM}}} = \frac{\kappa^2_{g,\text{on-shell}}\kappa^2_{V,\text{on-shell}}}{\Gamma_H/\Gamma_H^{\text{SM}}}, \tag{2}$$

a measurement of the Higgs width can be obtained. The fitted $-2\ln L$ as a function of $\Gamma_H/\Gamma_H^{\text{SM}}$ is also given in Fig. 11 at $R^B_{H^\star} = 1$. The observed (expected) upper limit on $\Gamma_H/\Gamma_H^{\text{SM}}$ is 5.5 (8.0).

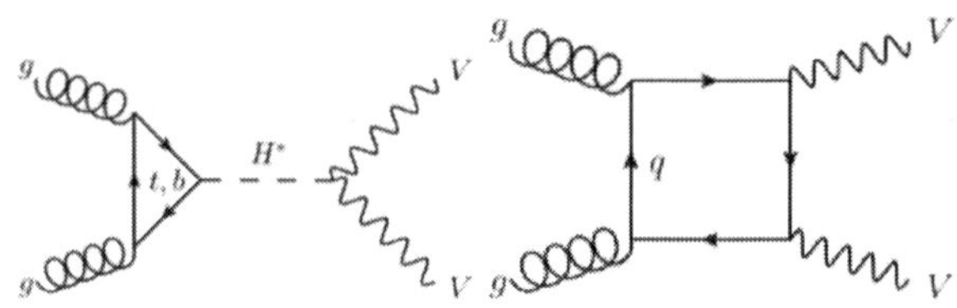

Figure 10. The on-shell production diagrams of diboson that interfere with each other. The left diagram corresponds to the off-shell Higgs production signal. Taken from Ref. [15].

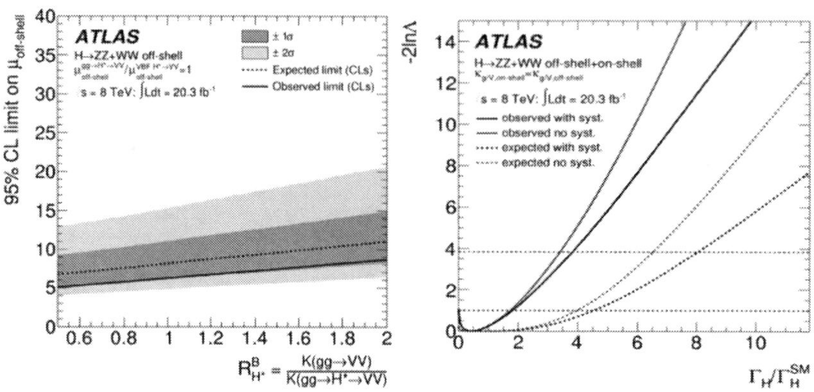

Figure 11. Left: the 95% CL limit on $\mu_{\text{off-shell}}$ as a function of $R^B_{H^*}$. Right: the $-2\ln L$ as a function of $\Gamma_H/\Gamma_H^{\text{SM}}$. Taken from Ref. [15].

3.3 $H \to$ invisible

Models like SUSY predict dark matter particles that may couple to the Higgs boson. Direct searches for $H \to$ invisible decay are carried out in the VH and VBF productions. In the VH mode, the search for $ZH \to ll$+invisible yields BR($H \to$ invisible) < 0.75 (0.62) [17] for the observed (expected) limit, and the search for $VH \to$ dijet+invisible gives BR($H \to$invisible) < 0.78 (0.86) [18] for the observed (expected). In the VBF search mode, signal events are recorded by the $E_{\text{T}}^{\text{miss}}$ trigger, and the dominant background ($Z \to \nu\nu$+jets and $W \to l\nu$+jets) is estimated with dedicated $Z \to ll$ and $W \to l\nu$ control samples. The Run 1 data gives 539 events in the signal region, while 579 events are expected. An observed (expected) limit of BR($H \to$ invisible) < 0.29 (0.35) [19] is obtained. When all three channels are combined, the observed (expected) limit is BR($H \to$ invisible) < 0.25 (0.27) [20], for which the sensitivity is dominated by the VBF channel.

3.4 Higgs tensor coupling

Using Effective Field Theory (EFT), a general HVV coupling can be expressed as (for spin-0):

$$\begin{aligned}\mathcal{L}_0 = \; & c_\alpha \kappa_{\text{SM}} \left[\tfrac{1}{2} g_{HZZ} Z_\mu Z^\mu + g_{HWW} W^+_\mu W^{-\mu}\right] X_0 + \\ & \left\{-\tfrac{1}{4}\tfrac{1}{\Lambda}\left[c_\alpha \kappa_{HZZ} Z_{\mu\nu} Z^{\mu\nu} + s_\alpha \kappa_{AZZ} \tilde{Z}^{\mu\nu}\right] - \tfrac{1}{2}\tfrac{1}{\Lambda}\left[c_\alpha \kappa_{HWW} W^+_{\mu\nu} W^{-\mu\nu} + s_\alpha \kappa_{AWW} W^+_{\mu\nu} \tilde{W}^{-\mu\nu}\right]\right\} X_0,\end{aligned} \tag{3}$$

where the first line corresponds to the SM CP-even coupling, denoted 0^+, the κ_{HVV} (κ_{AVV}) terms in the second line represent the BSM CP-even (CP-odd) couplings, denoted 0^+_h (0^-), the $\tilde{V} = \frac{1}{2}\epsilon^{\mu\nu\rho\sigma} V_{\rho\sigma}$

is the dual tensor, $c_\alpha = \cos\alpha$, $s_\alpha = \sin\alpha$, α is the mixing angle between CP-even and CP-odd components, X_0 is the spin-0 scalar, and Λ is the EFT energy scale below which the terms are valid. To probe the tensor coupling structure, two decay channels are used [21]. In the $WW \to e\nu\mu\nu$+0-jet channel, discriminants such as $m_{ll}, \Delta\phi_{ll}, p_{\mathrm{T}}^{ll}, E_{ll\nu\nu}, \Delta p_{\mathrm{T}}$ are used. In the $ZZ \to 4l$ channel, the five angles $\theta^\star, \Phi_1, \Phi, \theta_1, \theta_2$ and background rejecting BDT_{ZZ} are used. For fixed CP Higgs test, the 0^+ state is preferred than 0_h^+ and 0^- in data, as shown in Fig. 12. For the cases of mixtures of SM and BSM terms, the analysis excludes κ parameter regions far away from 0 (the SM expected value), as shown in Fig. 12 and listed in Tab. 1.

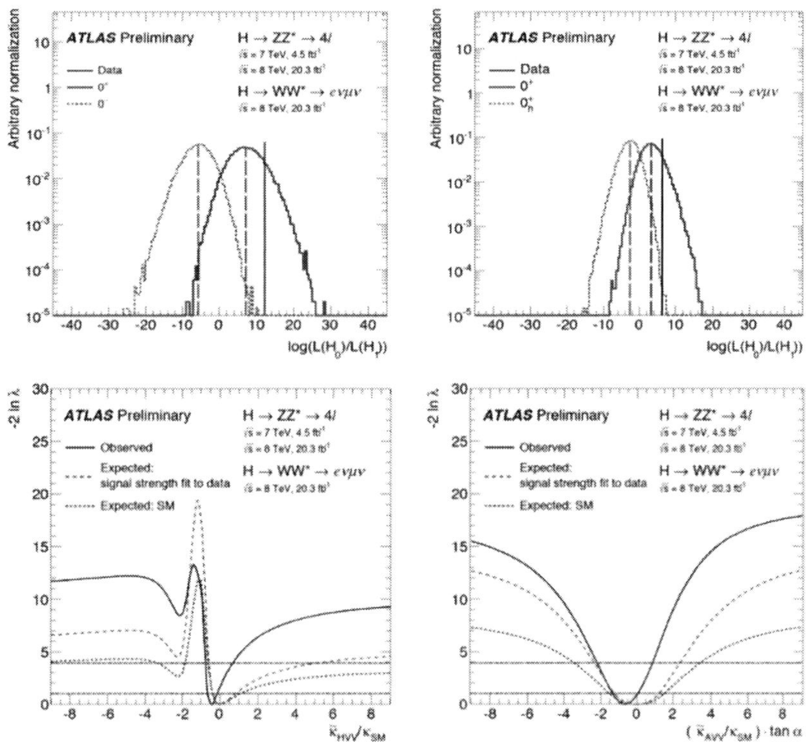

Figure 12. The likelihood ratio distributions to test the fixed Higgs CP states for 0^+ versus 0^- (top left) and 0^+ versus 0_h^+ (top right). For mixed CP states test, the $-2\ln L$ distributions are shown for SM+BSM-CP-even (bottom left) and SM+BSM-CP-odd (bottom right) cases. Taken from Ref. [21].

Table 1. The combined coupling strength ratio best fit values and 95% CL excluded regions (see Fig. 12 too), where it is defined that $\tilde{\kappa}_{AVV} = \frac{v}{4\Lambda}\kappa_{AVV}$ and $\tilde{\kappa}_{HVV} = \frac{v}{4\Lambda}\kappa_{HVV}$. Taken from Ref. [21].

coupling ratio	best fit value		95% CL exclusion region	
	expected	observed	expected	observed
$\tilde{\kappa}_{HVV}/\kappa_{\mathrm{SM}}$	0.0	-0.48	≤ -0.55 or ≥ 4.80	≤ -0.73 or ≥ 0.63
$(\tilde{\kappa}_{AVV}/\kappa_{\mathrm{SM}}) \cdot \tan\alpha$	0.0	-0.68	≤ -2.33 or ≥ 2.30	≤ -2.18 or ≥ 0.83

DOI: 10.1051/epjconf/201612602006

3.5 The Lepton Flavor Violating (LFV) Higgs coupling

Flavor Changing Neutral Current has been searched for in both the lepton and quark sectors. Due to the experimental constraint from $\mu \to e\gamma$, it is estimated that BR($H \to e\mu$) < O(10^{-8}). However, the constraint is much less stringent for the LFV decays of $H \to e\tau, \mu\tau$ (at O(10%)). The $H \to \mu\tau$ analysis [22] uses the transverse mass calculated from the $E_{\mathrm{T}}^{\mathrm{miss}}$ and the two leptons to define two signal regions:

1. SR1: $m_{\mathrm{T}}(\mu, E_{\mathrm{T}}^{\mathrm{miss}}) > 40$ GeV, $m_{\mathrm{T}}(\tau, E_{\mathrm{T}}^{\mathrm{miss}}) < 30$ GeV,
2. SR2: $m_{\mathrm{T}}(\mu, E_{\mathrm{T}}^{\mathrm{miss}}) < 40$ GeV, $m_{\mathrm{T}}(\tau, E_{\mathrm{T}}^{\mathrm{miss}}) < 60$ GeV.

The 2-D transverse mass distributions and the MMC mass in the two signal regions are shown in Fig. 13. The extracted signal rate gives BR($H \to \mu\tau$) = (0.77 ± 0.62)% based on the Run 1 data.

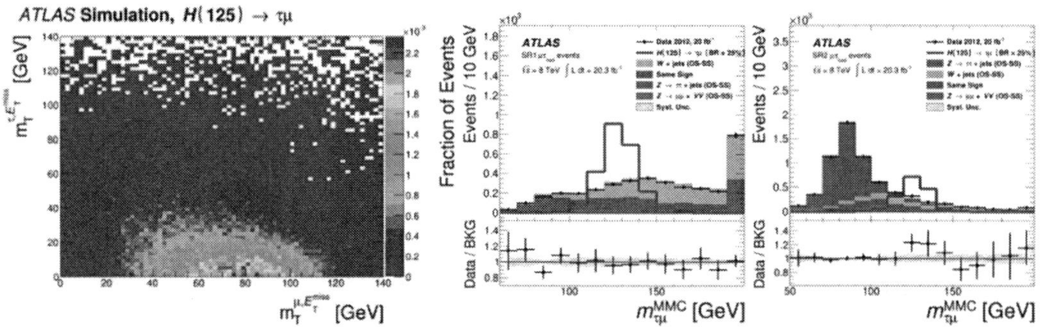

Figure 13. The transverse mass distributions in the signal (left), and the MMC distributions in SR1 (middle) and SR2 (right) signal regions. Taken from Ref. [22].

4 Higgs prospects

It is expected that during the lifetime of LHC, there will be about 3000 fb^{-1} of data collected. The current search programs will continue with more data accumulated. With 300 fb^{-1} of Run 2+3 data, the measurement precision of boson (fermion) coupling will reach a level of about 10% (20%), as shown in Fig. 14 [23]. The Higgs mass will be determined more precisely, and the width measurement may become more feasible. New Higgs physics such as the MSSM, rare or new Higgs decays may show up. With more statistics, the differential Higgs distributions, using the EFT theory, can test the Higgs properties to a much deeper level. Fig. 14 also shows the level of precision that can be achieved for Higgs couplings with particles of different masses, with 300 and 3000 fb^{-1} of data.

5 Conclusion

After the Higgs boson discovery, the main focus has shifted to its property measurements. This includes the precise mass determination (in the $\gamma\gamma$ and ZZ channels), couplings to different SM particles (bosons and fermions), CP and tensor coupling structures, total decay width, its BSM partners ($H^{\pm}$) or new decays ($H \to$ invisible). Many results are improved with advanced analysis methods since the discovery. Combination of different measurements is frequently made to test for any deviations from

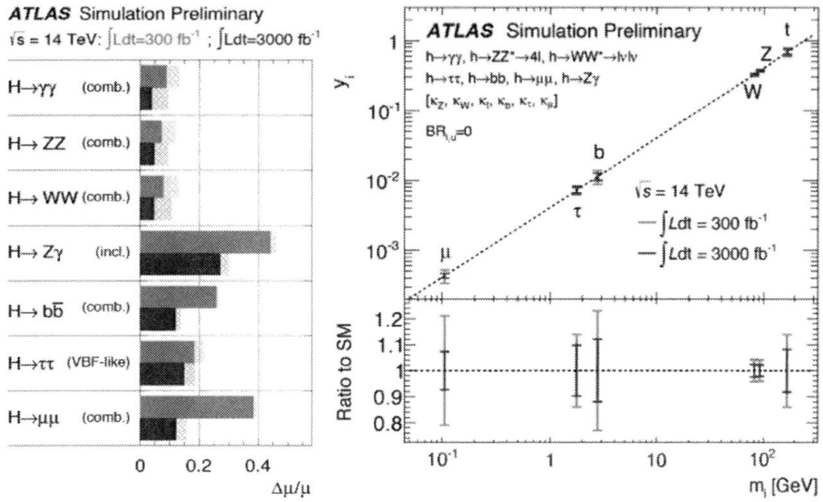

Figure 14. Left: the relative errors on the signal strength measurements in different Higgs decay modes for two luminosity scenarios (the shaded bars represent the systematic errors). Right: the Higgs coupling for particles of different masses for two luminosity scenarios. Taken from Ref. [23].

the SM prediction (so far no significant deviations observed). Our understanding of the new particle has been pushed to an unprecedented level, and will go to a much deeper level with future LHC data. Finally, it should be noted that the Higgs results summarized in this paper are not exhaustive, the analyses techniques will evolve with time, and updated results with more improvements are expected.

References

[1] ATLAS Collaboration, Phys. Lett. B 716, 1, 2012.
[2] CMS Collaboration, Phys. Lett. B 716, 30, 2012.
[3] ATLAS Collaboration, 2008 JINST 3 S08003.
[4] ATLAS Collaboration, Phys. Rev. D 90, 112015, 2014.
[5] ATLAS Collaboration, Phys. Rev. D 91, 012006, 2015.
[6] ATLAS Collaboration, Phys. Rev. D 90, 052004, 2014.
[7] ATLAS Collaboration, Phys. Rev. D 92, 012006, 2015.
[8] ATLAS Collaboration, JHEP 04, 117, 2015.
[9] A. Elagin et al., Nucl. Instrum. Meth. A 654, 481, 2011.
[10] ATLAS Collaboration, Phys. Lett. B 738, 68, 2014.
[11] ATLAS Collaboration, JHEP 01 069, 2015.
[12] ATLAS Collaboration, Eur. Phys. J. C 75:349, 2015.
[13] ATLAS Collaboration, arXiv:1507.04548.
[14] ATLAS Collaboration, JHEP 03, 088, 2015.
[15] ATLAS Collaboration, Eur. Phys. J. C 75:335, 2015
[16] J.M. Campbell et al., Phys. Rev. D 89, 05301, 2014; JHEP 04, 060, 2014.
[17] ATLAS Collaboration, Phys. Rev. Lett. 112, 201802, 2014.
[18] ATLAS Collaboration, Eur. Phys. J. C 75:337, 2015.

[19] ATLAS Collaboration, ATLAS-CONF-2015-004, https://cds.cern.ch/record/2002121.
[20] ATLAS Collaboration, arXiv:1509.00672.
[21] ATLAS Collaboration, ATLAS-CONF-2015-008, https://cds.cern.ch/record/2002414.
[22] ATLAS Collaboration, arXiv:1508.03372.
[23] ATLAS Collaboration, ATL-PHYS-PUB-2014-016, https://cds.cern.ch/record/1956710; ATL-PHYS-PUB-2014-017, https://cds.cern.ch/record/1956711.

Recent results from Daya Bay

Ming-chung Chua[1,*], on behalf of the Daya Bay Collaboration

[1]Department of Physics, The Chinese University of Hong Kong, Shatin, N. T., Hong Kong

Abstract. Utilizing powerful nuclear reactors as antineutrino sources, high mountains to provide ample shielding from cosmic rays in the vicinity, and functionally identical detectors with large target volume for near-far relative measurement, the Daya Bay Reactor Neutrino Experiment has achieved unprecedented precision in measuring the neutrino mixing angle θ_{13} and the neutrino mass squared difference $|\Delta m^2_{ee}|$. I will report the latest Daya Bay results on neutrino oscillations and light sterile neutrino search.

1 Introduction

Neutrino flavor oscillations are caused by the misalignment between neutrino flavor and mass eigenstates, parameterized by mixing angles θ_{12} and θ_{23}, and they are realized as the changing of neutrino flavors as a neutrino propagates freely. In 2012, the Daya Bay Reactor Neutrino Experiment discovered a non-zero value of $\sin^2 2\theta_{13}$ by observing the disappearance of reactor antineutrinos over kilometer distances [1], and in 2014, it reported the first measurement of the effective mass splitting $|\Delta m^2_{ee}|$ [2] by neutrino spectrum distortion [3]. In this talk, I will briefly review the Daya Bay experiment and present the latest results on neutrino oscillations and light sterile neutrino search.

2 The Daya Bay Reactor Neutrino Experiment

There are three underground experimental halls (EHs) – two near (EH1, EH2) and one far (EH3) – in the Daya Bay experiment [4], each hosting functionally identical, three-zone antineutrino detectors (ADs) – 2 in EH1 and EH2 each, 4 in EH3 - in a pool of ultrapure water segmented into the inner water shield (IWS) and outer water shield (OWS), both instrumented with photomultiplier tubes (PMTs). The water pool and Resistive Plate Chambers (RPCs) placed above it are muon veto systems, and the water buffer also serves to suppress radiation from the surrounding rocks. See Fig. 1 and 2.

* Corresponding author: mcchu@phy.cuhk.edu.hk

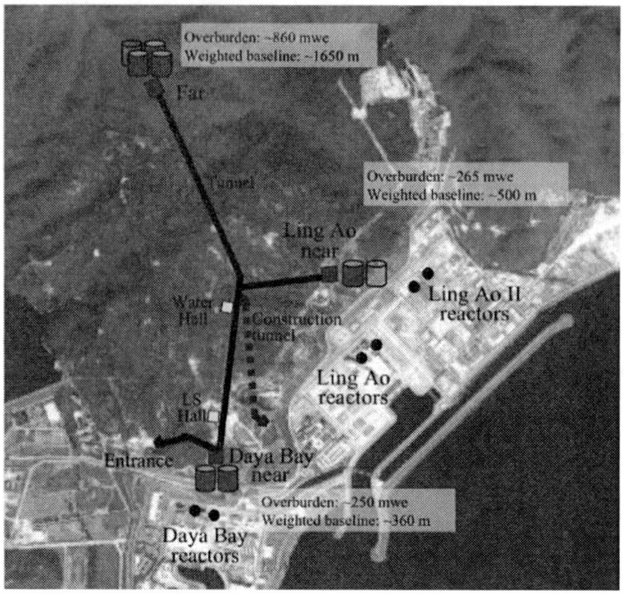

Figure 1. The layout of the Daya Bay Experiment.

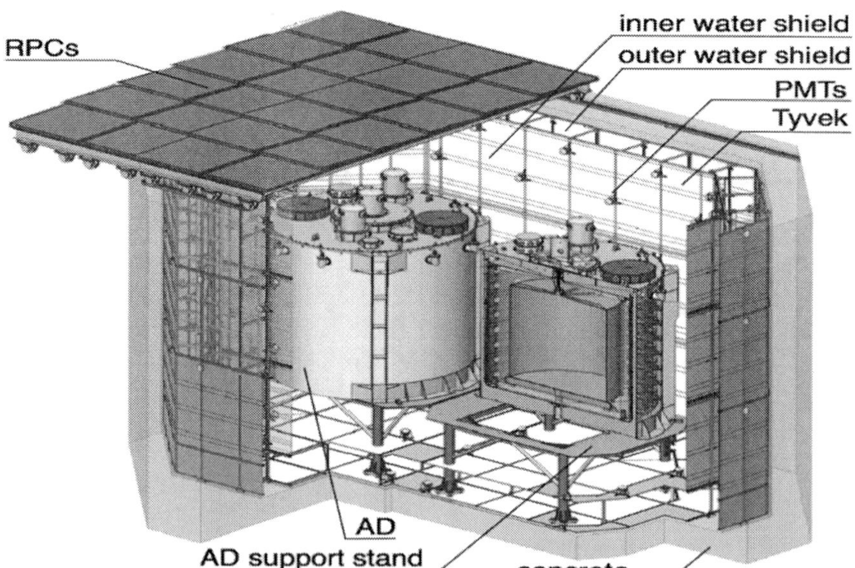

Figure 2. A water pool in each EH hosts 2 (4) ADs for measuring antineutrinos from the reactors. Adapted from [4].

Each AD contains 20 tons of gadolinium-doped liquid scintillator (LS) in the inner acrylic vessel (AV), 20 tons of undoped LS in the outer AV and 37 tons of mineral oil which buffers the 192 radially positioned 20-cm PMTs in the outermost steel tank [5]. The ADs have a light yield of ~165 photoelectrons/MeV and a reconstructed energy resolution of $\sigma_E/E \approx 8\%$ at 1 MeV.

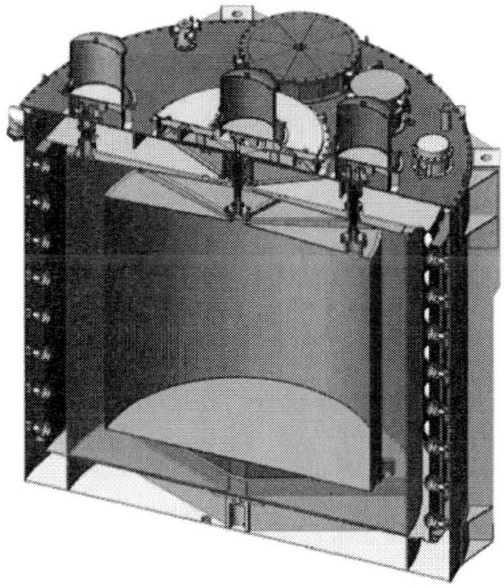

Figure 3. Each AD contains 3 zones of liquids and 192 PMTs to detect the light generated by particle interactions. Adapted from [5].

Reactor antineutrinos are detected via the inverse β-decay (IBD) reaction, $\bar{\nu}_e + p \rightarrow e^+ + n$. The energy of the prompt light emitted by the positron E_{prompt} is directly related to the neutrino energy: $E_{\bar{\nu}} = E_{\text{prompt}} + E_n + 0.78$ MeV, where E_n is the average neutron recoil energy ~ 10 keV. The neutron is captured by Gadolinium doped into the liquid scintillator in the inner acrylic vessel, with a mean capture time of ~30 μs, emitting gamma rays (totaling ~8 MeV) as a delayed signal. The requirement of prompt and delayed coincidence greatly suppresses background. The use of relative measurement by functionally identical ADs in the near and far experimental halls leads to large suppression of systematic uncertainties. Detailed discussion of the IBD selection, efficiencies, backgrounds and systematic uncertainties can be found in [4].
Fig. 4 shows the operation history of the Daya Bay experiment, number of IBD events accumulated in the near and far detectors and the measured value of $\sin^2 2\theta_{13}$ as a function of time. The experiment will continue operation until 2017, by the end of which more than 4 million IBD events will be recorded.

3 Oscillation results

The survival probability of electron antineutrinos is given by

$$P_{ee} = 1 - \sin^2 2\theta_{13}\sin^2(\Delta m_{ee}^2 L/4E) - \sin^2 2\theta_{12}\cos^4\theta_{13}\sin^2(\Delta m_{21}^2 L/4E)\,, \quad (1)$$

where E is the neutrino energy and L the distance between the neutrino source and detection.

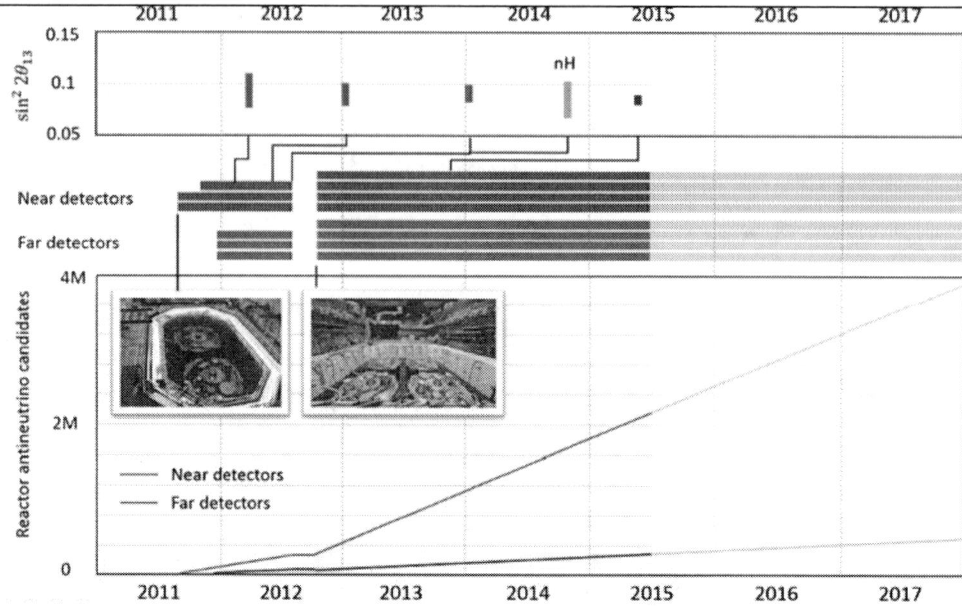

Figure 4. Operation history of the Daya Bay Experiment and the measured values of $\sin^2 2\theta_{13}$.

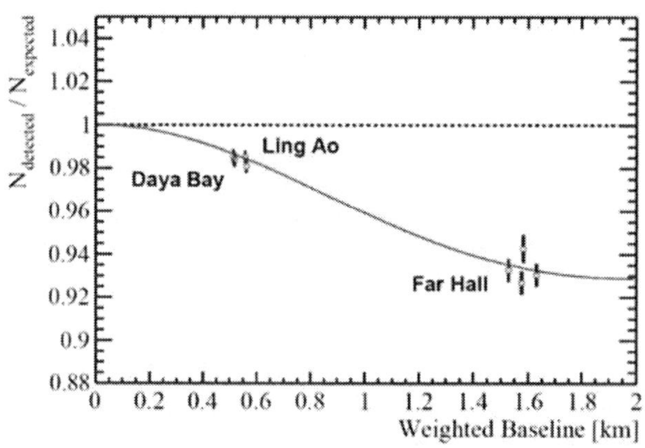

Figure 5. Measured deficit of the neutrino flux as a function of baseline, compared to the survival probability with best-fit result for $\sin^2 2\theta_{13}$. Adapted from [6].

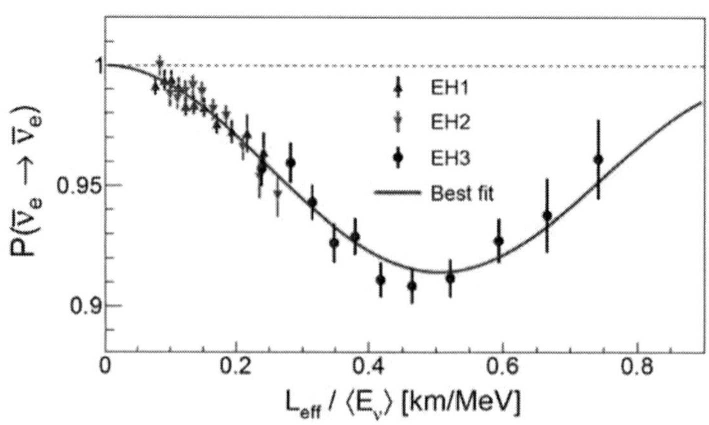

Figure 6. Distortion of neutrino energy spectrum as a function of *L*/*E*, compared to calculation of survival probability using the best-fit value of $\sin^2 2\theta_{13}$. Adapted from [6].

The neutrino mixing angle θ_{13} can therefore be measured by the deficit of the energy-integrated neutrino flux at far ADs using the near ADs as reference (the rate method), or the distortion of the neutrino energy spectrum (the rate-plus-shape method).
Figure 5 shows the rate deficit of all 8 ADs in near and far experimental halls of Daya Bay [6], whereas Figure 6 shows the spectral distortion measured at Daya Bay [6]. The results of the neutrino oscillation parameter fitting are shown in Figure 7. $\sin^2 2\theta_{13}$ is determined to be 0.084± 0.005 by Daya Bay with an unprecedentedly high precision of about 6%. An independent measurement using nH captures, which are characterized by larger target mass but more background contamination yields $\sin^2 2\theta_{13} = 0.083 \pm 0.018$ [7]. Furthermore, $|\Delta m^2_{ee}|$ is measured to be (2.42± 0.11) x10^{-3} eV^2, which agrees with results by muon neutrino disappearance experiments MINOS [8] and T2K [9], with comparable precision.

4 Search for a light sterile neutrino

A popular candidate for nonbaryonic dark matter [10] is sterile neutrinos with ~keV masses. An ev-mass-scale sterile neutrino was introduced [11] to resolve two well-known tensions between cosmic microwave background (CMB) and local measurements of the Hubble constant H_o and the amount of structures [12], as well as between CMB and lensing measurements [13]. The extra energy density of the sterile neutrinos increases the expansion rate of the universe at recombination and thus changes the acoustic scale, and their free streaming suppress small structures today. If light sterile neutrinos mix with the three active neutrinos, they could be detected via modification of the oscillation patterns of active neutrinos. The standard 3x3 neutrino mixing matrix would become a 4x4 matrix, with new mixing angles such as θ_{14} entering the survival probability of electron antineutrinos.
The Daya Bay Experiment search for a light sterile neutrino via its mixing with the active neutrinos using more than 300,000 reactor antineutrino interactions collected during the six-detector data period from December 2011 to July 2012 [14]. Using a relative measurement from its two near halls, Daya Bay is most sensitive to active-sterile neutrino mixing in the 10^{-3} $eV^2 < |\Delta m^2| < 0.3$ eV^2 range. We look for an additional spectral distortion with a frequency different from that of the atmospheric mass splitting.

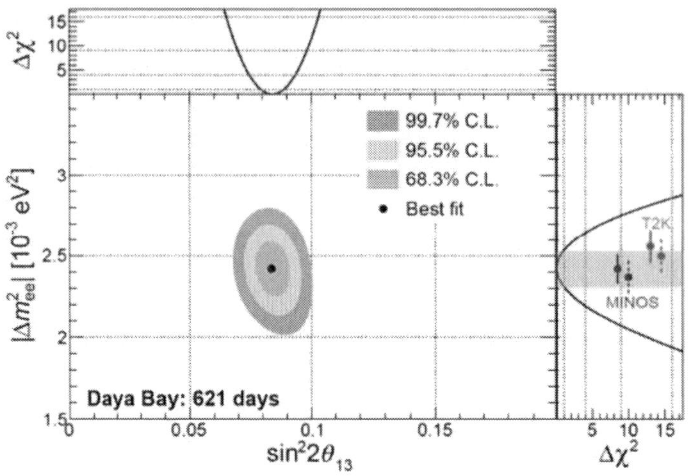

Figure 7. Rate-plus-shape results for $|\Delta m^2_{ee}|$ and $\sin^2 2\theta_{13}$. Adapted from [6].

Figure 8 shows a comparison between the Daya Bay prompt energy spectra in EH2 and EH3 and the expected ones based on EH1 spectra if a light sterile neutrino exists with $|\Delta m^2_{41}| = 4\text{x}10^{-3}$ eV^2 and $4\text{x}10^{-2}$ eV^2, assuming $\sin^2 2\theta_{14} = 0.1$. The Daya Bay data are consistent with the standard 3-flavor prediction taking into account experimental uncertainties. Exclusion of a region in the $|\Delta m^2_{41}|$ vs. $\sin^2 2\theta_{14}$ parameter space is achieved,

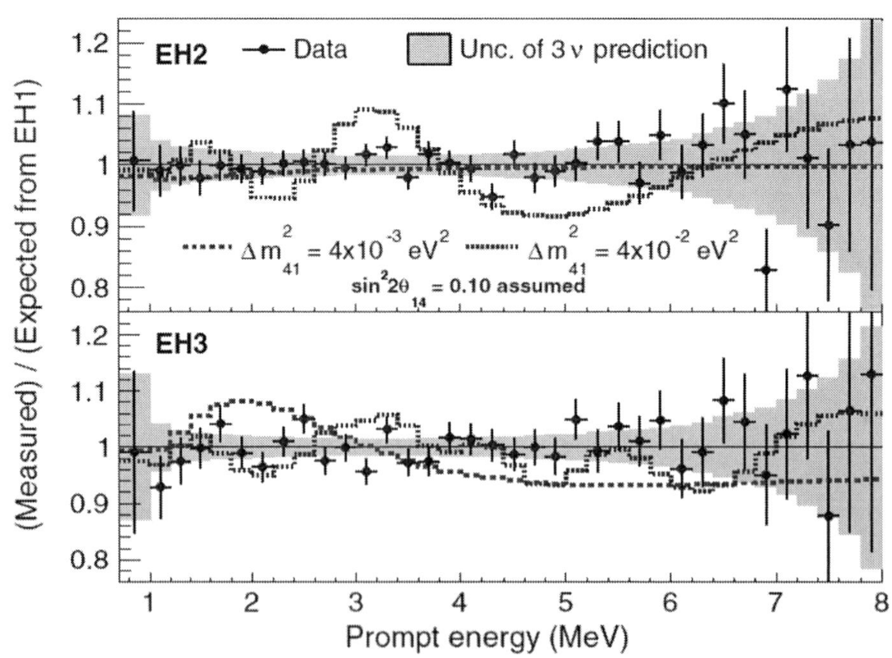

Figure 8. Measured prompt energy spectrum in EH2 and EH3 compared to expectation based on EH1 data, if a light sterile neturino exists with two values of $|\Delta m^2_{41}|$ and $\sin^2 2\theta_{14} = 0.1$. Adapted from [14].

shown in Figure 9. Daya Bay is uniquely sensitive to the 10^{-3} eV^2 < $|\Delta m^2_{41}|$ < 0.3 eV^2 range [14], which was largely unexplored previously.

5 Summary

The Daya Bay Reactor Neutrino Experiment has achieved unprecedented precision in measuring the neutrino mixing angle θ_{13} and the neutrino mass squared difference $|\Delta m^2_{ee}|$, using relative measurement by near and far detectors. Making use of several different baselines provided by the three experimental halls, the Daya Bay experiment excluded a light sterile neutrino in the 10^{-3} eV^2 < $|\Delta m^2_{41}|$ < 0.3 eV^2 range. The experiment has accumulated a large set of neutrino events, allowing for more searches for non-standard neutrino interactions and physics beyond the standard model.

Acknowledgements

The Daya Bay experiment is supported in part by the Ministry of Science and Technology of China, the United States Department of Energy, the Chinese Academy of Sciences, the National Natural Science Foundation of China, the Guangdong provincial government, the Shenzhen Municipal government, the China Guangdong Nuclear Power Group, Shanghai Laboratory for particle physics and cosmology, the Research Grants Council of the Hong Kong Special Administrative Region of China, the focused investment scheme of CUHK and University Development Fund of The University of Hong Kong, the MOE program for Research of Excellence at National Taiwan University and NSC funding support from Taiwan, the U.S. National Science Foundation, the Alfred P. Sloan Foundation, the University of Wisconsin, the Virginia Polytechnic Institute and State University, Princeton University, California Institute of Technology, University of California at Berkeley, the Ministry of Education, Youth and Sports of the Czech Republic, the Joint Institute of Nuclear Research in Dubna, Russia, and the NSFC-RFBR joint research program. We are grateful for the ongoing cooperation from the China Guangdong Nuclear Power Group and China Light & Power Company.

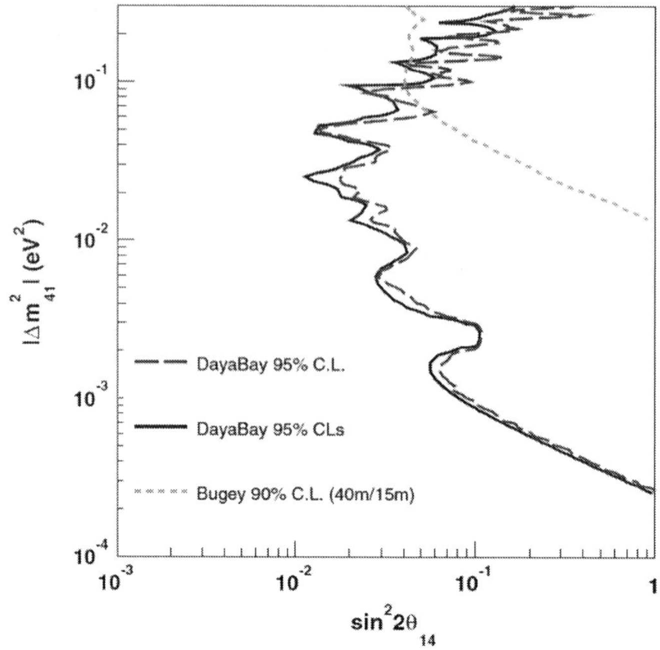

Figure 9. Exclusion plot for $|\Delta m^2_{41}|$ and $\sin^2 2\theta_{14}$. Parameter space on the right of the curves are excluded at 95% C. L. The Daya Bay result is complementary to Bugey's result and is uniquely sensitive to the 10^{-3} eV^2 < $|\Delta m^2_{41}|$ < 0.3 eV^2 range. Adapted from [14].

References

[1] F. P. An et al. (Daya Bay Collaboration), Phys. Rev. Lett. **108**, 171803 (2012); Chin. Phys. C**37**, 011001 (2013); Phys. Rev. D**90**, 071101 (2014).
[2] Δm^2_{ee} is an effective mass splitting that can be obtained by replacing $\cos^2\theta_{12}\sin^2\Delta_{31} + \sin^2\theta_{12}\sin^2\Delta_{32}$ with $\sin^2\Delta_{ee}$, where $\Delta_{ji} = 1.267\Delta m^2_{ji}(\text{eV}^2)[L(\text{m})/E(\text{MeV})]$, and $\Delta m^2_{ji} = m_j^2 - m_i^2$ is the difference between the mass-squares of the mass eigenstates ν_j and ν_i. To estimate the values of Δm^2_{31} and Δm^2_{32} from the measured value of Δm^2_{ee}, See Supplemental Material at http://link.aps.org/supplemental/10.1103/PhysRevLett.115.111802 .
[3] F. P. An et al. (Daya Bay Collaboration), Phys. Rev. Lett. **112**, 061801 (2014).
[4] F. P. An et al. (Daya Bay Collaboration), Nucl. Instrum. Methods Phys. Res., Sect. A **685**, 78 (2012).
[5] H. Band et al., JINST **8**, T11006 (2013).
[6] F P. An et al. (Daya Bay Collaboration), Phys. Rev. Lett. **115**, 111802 (2015).
[7] F P. An et al. (Daya Bay Collaboration), Phys. Rev. D **90**, 071101 (2014).
[8] P. Adamson et al. (MINOS Collaboration), Phys. Rev. Lett. **112**, 191801 (2014).
[9] K. Abe et al. (T2K Collaboration), Phys. Rev. Lett. **112**, 181801(2014).
[10] S. Dodelson and L. M. Widrow, Phys. Rev. Lett. **72**, 17(1994); A. Kusenko, Phys. Rep. **481**, 1 (2009).
[11] Mark Wyman, Douglas H. Rudd, R. Ali Vanderveld, and Wayne Hu, Phys. Rev.Lett.**112**, 051302 (2014).
[12] A. G. Riess et al., Astrophy. J. **730**, 119 (2011); A. G. Riess et al., Astrophy. J. **732**, 129 (2011).
[13] R. A. Battye and A. Moss, Phys. Rev. Lett. **112**, 051303 (2014).
[14] F P. An et al. (Daya Bay Collaboration), Phys. Rev. Lett. **113**, 141802 (2014).

EPJ Web of Conferences **126**, 02008 (2016) DOI: 10.1051/epjconf/201612602008
ICNFP 2015

Recent Borexino results and prospects for the near future

D. D'Angelo[13,a], M. Agostini[24], K. Altenmüller[22], S. Appel[22], G. Bellini[13], J. Benziger[17], D. Bick[4], G. Bonfini[12], D. Bravo[20], B. Caccianiga[13], F. Calaprice[16], A. Caminata[3], P. Cavalcante[12], A. Chepurnov[23], S. Davini[24], A. Derbin[15], L. Di Noto[3], I. Drachnev[24], A. Etenko[10], K. Fomenko[2], D. Franco[1], F. Gabriele[12], C. Galbiati[16], C. Ghiano[3], M. Giammarchi[13], M. Goeger-Neff[22], A. Goretti[16], M. Gromov[23], C. Hagner[4], E. Hungerford[25], Aldo Ianni[12], Andrea Ianni[16], K. Jedrzejczak[6], M. Kaiser[4], V. Kobychev[7], D. Korablev[2], G. Korga[12], D. Kryn[1], M. Laubenstein[12], B. Lehnert[26], E. Litvinovich[10,11], F. Lombardi[12], P. Lombardi[13], L. Ludhova[13,b], G. Lukyanchenko[10], I. Machulin[10,11], S. Manecki[20], W. Maneschg[5], S. Marcocci[24], E. Meroni[13], M. Meyer[4], L. Miramonti[13], M. Misiaszek[6,12], M. Montuschi[21], P. Mosteiro[16], V. Muratova[15], B. Neumair[22], L. Oberauer[22], M. Obolensky[1], F. Ortica[14], M. Pallavicini[3], L. Papp[22], L. Perasso[3], A. Pocar[19], G. Ranucci[13], A. Razeto[12], A. Re[13], A. Romani[14], R. Roncin[12,1], N. Rossi[12], S. Schönert[22], D. Semenov[15], H. Simgen[5], M. Skorokhvatov[10,11], O. Smirnov[2], A. Sotnikov[2], S. Sukhotin[10], Y. Suvorov[27,10], R. Tartaglia[12], G. Testera[3], J. Thurn[26], M. Toropova[10], E. Unzhakov[15], A. Vishneva[2], R.B. Vogelaar[20], F. von Feilitzsch[22], H. Wang[27], S. Weinz[28], J. Winter[28], M. Wojcik[6], M. Wurm[28], Z. Yokley[20], O. Zaimidoroga[2], S. Zavatarelli[3], K. Zuber[26], and G. Zuzel[6]

[1]*AstroParticule et Cosmologie, Université Paris Diderot, CNRS/IN2P3, CEA/IRFU, Observatoire de Paris, Sorbonne Paris Cité, 75205 Paris Cedex 13, France*
[2]*Joint Institute for Nuclear Research, 141980 Dubna, Russia*
[3]*Dipartimento di Fisica, Università degli Studi e INFN, Genova 16146, Italy*
[4]*Institut für Experimentalphysik, Universität, 22761 Hamburg, Germany*
[5]*Max-Planck-Institut für Kernphysik, 69117 Heidelberg, Germany*
[6]*M. Smoluchowski Institute of Physics, Jagiellonian University, 30059 Krakow, Poland*
[7]*Kiev Institute for Nuclear Research, 06380 Kiev, Ukraine*
[8]*Institute for Theoretical and Experimental Physics, 117218 Moscow, Russia*
[9]*Kepler Center for Astro and Particle Physics, Universität Tübingen, 72076 Tübingen, Germany.*
[10]*NRC Kurchatov Institute, 123182 Moscow, Russia*
[11]*National Research Nuclear University MEPhI (Moscow Engineering Physics Institute), 115409 Moscow, Russia.*
[12]*INFN Laboratori Nazionali del Gran Sasso, 67010 Assergi (AQ), Italy*
[13]*Dipartimento di Fisica, Università degli Studi e INFN, 20133 Milano, Italy*
[14]*Dipartimento di Chimica, Università e INFN, 06123 Perugia, Italy*
[15]*St. Petersburg Nuclear Physics Institute NRC Kurchatov Institute, 188350 Gatchina, Russia*
[16]*Physics Department, Princeton University, Princeton, NJ 08544, USA*
[17]*Chemical Engineering Department, Princeton University, Princeton, NJ 08544, USA*
[18]*Physics Department, Queen's University, Kingston ON K7L 3N6, Canada*
[19]*Amherst Center for Fundamental Interactions and Physics Department, University of Massachusetts, Amherst, MA 01003, USA*

[a]e-mail: davide.dangelo@mi.infn.it
[b]Now at RWTH Aachen University, 52074, Aachen, Germany and Forschungzentrum Jülich IKP-2, 52428 Jülich, Germany

[20] *Physics Department, Virginia Polytechnic Institute and State University, Blacksburg, VA 24061, USA*
[21] *Dipartimento di Fisica e Scienze della Terra Università degli Studi di Ferrara e INFN, Via Saragat 1-44122, Ferrara, Italy*
[22] *Physik-Department and Excellence Cluster Universe, Technische Universität München, 85748 Garching, Germany*
[23] *Lomonosov Moscow State University Skobeltsyn Institute of Nuclear Physics, 119234 Moscow, Russia*
[24] *Gran Sasso Science Institute (INFN), 67100 L'Aquila, Italy*
[25] *Department of Physics, University of Houston, Houston, TX 77204, USA*
[26] *Department of Physics, Technische Universität Dresden, 01062 Dresden, Germany*
[27] *Physics and Astronomy Department, University of California Los Angeles (UCLA), Los Angeles, California 90095, USA*
[28] *Institute of Physics and Excellence Cluster PRISMA, Johannes Gutenberg-Universität Mainz, 55099 Mainz, Germany*

Abstract. The Borexino experiment located in the Gran Sasso National Laboratory, is an organic liquid scintillator detector conceived for the real time spectroscopy of low energy solar neutrinos. The phase-I of the data taking campaign (2007 – 2010) has allowed the first independent measurements of ^{7}Be and pep solar neutrino fluxes as well as the first measurement of anti-neutrinos from the Earth. After a purification of the scintillator, Borexino is now in phase-II since 2011. Thanks to the unprecedented background levels, we have performed the first flux measurement of neutrinos from the fundamental pp reaction which powers the Sun. We review this breakthrough result and other recent results, including the latest review of our terrestrial neutrino analysis. We also discuss the upcoming measurements on middle energy solar neutrino spectral components (pep, CNO) and the new project SOX devoted to the study of sterile neutrinos via the use of a neutrino source placed in close proximity of the detector's active material.

1 Why solar neutrinos with Borexino

The Sun is an intense source of neutrinos, produced in nuclear reactions of the p-p chain and of the CNO cycle [1]. Measurements of the individual neutrino fluxes is of paramount importance for

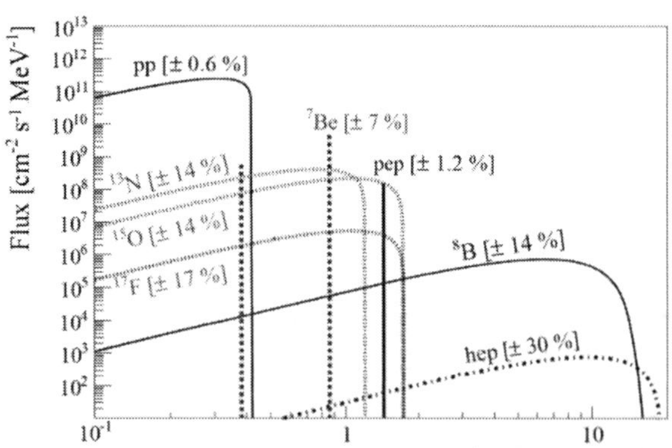

Figure 1. Solar Neutrino Spectrum.

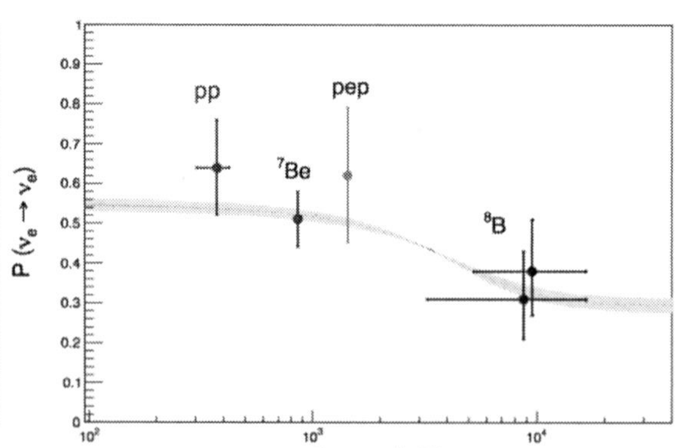

Figure 2. Survival probability of electron neutrinos as foreseen by the MSW-LMA model. Data points discussed here are also shown.

DOI: 10.1051/epjconf/201612602008

both particle physics and astrophysics. The solar neutrino spectrum can be seen in figure 1. Up to a few years ago, spectroscopical measurements were performed by water Cherenkov detectors above ~5 Mev and concerned only ^{8}B neutrinos for less then 1% of the total flux. The bulk of neutrinos at low energies were detected only with radiochemical experiments, incapable of resolving the individual components. Neutrinos are emitted in the Sun as electron-flavour neutrinos and may oscillate to a different flavour during the trajectory to the Earth. The MSW mechanism with Large Mixing Angle (LMA) foresees the survival probability for electron neutrinos on Earth after oscillation is taken into account, as shown in figure 2. Borexino was designed to perform spectroscopy of the low energy part of the solar neutrino spectrum, in particular the flux of the ^{7}Be monochromatic line at 862 keV. Borexino has largely exceeded the expected performance and broadened the physics program past the original goal.

2 The Borexino Project

Detector layout. The Borexino detector [2] is sketched in figure 3. It is located at the Gran Sasso National Laboratories (LNGS) in central Italy, at a depth of 3800 m w.e.. The active mass consists of 278 tons of pseudocumene (PC), doped with 1.5 g/l of PPO. The scintillator is contained in a thin (125 μm) nylon Inner Vessel (IV), 8.5 m in diameter. The IV is surrounded by two concentric PC buffers doped with a light quencher. The scintillator and buffers are contained in a Stainless Steel Sphere (SSS) with a diameter of 13.7 m. The SSS is enclosed in a Water Tank (WT), containing 2100 tons of ultra-pure water as an additional shield against backgrounds from the laboratory environment. The scintillation light released upon particle interaction in the pseudocumene is detected by 2212 8" PhotoMultiplier Tubes (PMTs) uniformly distributed on the inner surface of the SSS. Additional 208 8" PMTs instrument the WT and detect the Cherenkov light radiated by cosmic muons that cross the water shield.

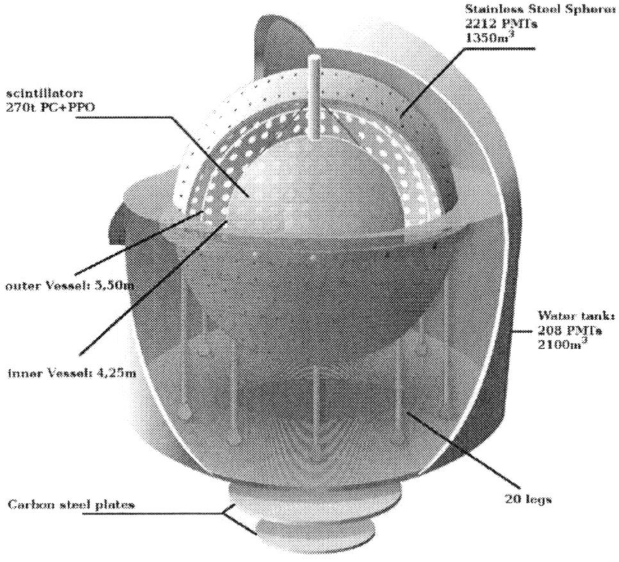

Figure 3. Sketch of the Borexino detector.

Detector performance. The Borexino scintillator has a light yield of ~10^4 photons/MeV, resulting in ~500 detected photoelectrons/MeV. The fast time response (~3 ns) of the scintillating mixture allows to reconstruct the events position by means of a time-of-flight technique with ~13 cm precision for events at the center of the detector. Depending on the analysis, the fiducial volume is defined between 75 tons and 150 tons. Neutrinos interact via elastic scattering on the electrons of the target material. The signature of ^{7}Be neutrinos is a Compton-like shoulder at 665 keV in the electron recoil spectrum. The energy resolution (1σ) at ^{7}Be energy is as low as 44 keV (or 6.6%). The inner detector trigger is defined as the coincident firing of at least 20-25 PMTs, corresponding to an energy threshold of ~40-50 keV. However the analysis threshold in phase-I was limited by the knowledge

DOI: 10.1051/epjconf/201612602008

Table 1. Background levels in the Borexino scintillator. cpd stands for "counts per day".

Isotope	Typical	Required	Before purification	After purification
^{238}U	2 10^{-5}g/g (dust)	$\leq 10^{-16}$g/g	(5.3±0.5) 10^{-18} g/g	< 0.8 10^{-19}g/g
^{232}Th	2 10^{-5}g/g (dust)	$\leq 10^{-16}$g/g	(3.8±0.8) 10^{-18} g/g	< 1.0 10^{-18}g/g
^{14}C/^{12}C	10^{-12} (cosmogenic)	$\leq 10^{-18}$	(2.69±0.06) 10^{-18}	unchanged
^{222}Rn	100 atoms/cm^3 (air)	≤10cpd/100t	1cpd/100t	unchanged
^{40}K	2 10^{-6}g/g (dust)	$\leq 10^{-18}$ g/g	≤0.4 10^{-18} g/g	unchanged
^{85}Kr	1Bq/m^3 (air)	≤1cpd/100t	(30±5) 10cpd/100t	≤5cpd/100t
^{39}Ar	17mBq/m^3 (air)	≤1cpd/100t	$\ll ^{85}$Kr	$\ll ^{85}$Kr
^{210}Po		not specified	~80 → ~5cpd/t	unchanged
^{210}Bi		not specified	~10 → ~75cpd/100t	(27±8)cpd/100t

of the ^{14}C spectral shape to ~250 keV. In phase-II we have successfully reduced our threshold to 165 keV (see Sect. 4). Pulse Shape Analysis (PSA) is performed to identify various classes of events, among which electronic noise, pile-up events, muons, α and β particles.

Borexino timeline. The phase-I of Borexino occurred between May 2007 and May 2010. Next, we proceeded to the purification of the scintillator by water extraction and sparging with a specially developed nitrogen gas, low in Kr and Ar contamination. After looping through the purification plant for six complete cycles, in November 2011 the phase-II of Borexino started and is expected to last at least until the end of 2016. The energy scale uncertainty in the range [0.2-2.0]MeV has been determined at 1.5% precision, using multiple gamma sources. The position reconstruction algorithm has been tuned using a Rn source placed in 184 positions inside the active volume. The determination of the fiducial volume is now -1.3%+0.5%. An external calibration campaign has allowed to correctly model external gamma backgrounds. The calibrations will be repeated before the end of phase-II to demonstrate stability of the detector and possibly to improve the previous results[3].

Experimental backgrounds. The background levels of the Borexino scintillator are summarised in table 1. ^{238}U and ^{232}Th contaminations are unprecedented and largely exceed the requirements of the experiment. Most other contaminants are at acceptable levels or have been reduced below required limits after purification. We observe the presence of Radon Daughters ^{210}Bi and ^{210}Po out of equilibrium. ^{210}Po is of little concern as it is an α emitter and it can be identified by PSA. ^{210}Bi contamination has been rising during the data taking of phase-I due to unclear motivations, possibly related to the movement of the scintillator. This variation was modelled and taken into account during the ^{7}Be flux modulation analysis (Sect. **??**). ^{210}B was later reduced by the purification campaign to a level which allowed the measurement of pp neutrino (Sect. 4) and will possibly allow the measurement of CNO neutrino fluxes (Sect. 7).

3 Phase-I results

^{7}Be flux. The measurement of the flux of ^{7}Be neutrinos was the primary goal of Borexino. Thanks to the unprecedented background levels, the first observation was published in summer 2007, after only 3 months of data taking. The flux was later reviewed in 2008 and 2011 reducing the error every time. The signal is extracted by a spectral fit along with other neutrino signals and background components (figure 4).

DOI: 10.1051/epjconf/201612602008

The fit is performed both using analytical spectral shapes and MonteCarlo generated curves, with or without the statistical subtraction of α events via PSA, obtaining the same result within errors. The last measured rate[4] after 741 d of live time (full phase-I) is $R_{Be} = 46.0 \pm 1.5_{stat} \pm 1.6_{syst} / d / 100 t$. For the first time the experimental error (4.8%) is lower the theoretical error (7%). The rate corresponds to a flux of $\Phi_{Be} = (3.10 \pm 0.15) \times 10^9 \, cm^{-2} \, s^{-1}$ with a survival probability $P_{ee} = 0.51 \pm 0.07$ at 862 keV.

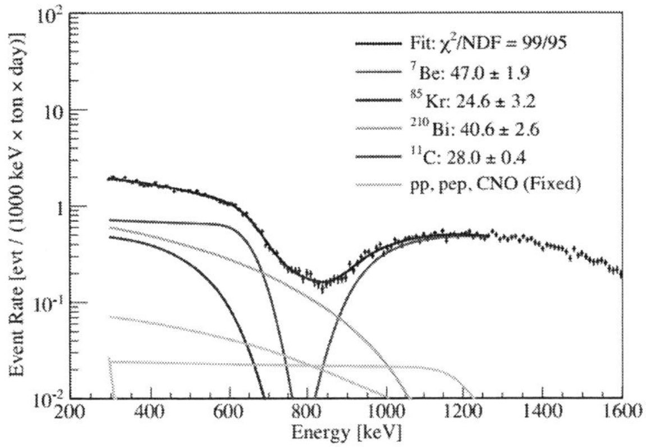

Figure 4. Fit of the energy spectrum for ^{7}Be flux analysis.

We have also carefully inspected the data set at the energy of ^{7}Be neutrino scattering looking for an eventual day-night asymmetry[5]. This was foreseen in an alternative MSW scenario, called LOW, also compatible to some extent with the solar neutrino results. The LOW scenario before Borexino could be excluded only assuming CPT invariance and using the KamLAND reactor's antineutrino results. Our data is consistent with no asymmetry: $A_{DN} = \Phi_N - \Phi_D / \Phi = 0.001 \pm 0.012_{stat} \pm 0.007_{syst}$. With this result, the MSW-LOW mechanism can be ruled out at 8.5 σ using only solar neutrino data.

^{8}B flux Borexino has measured the ^{8}B flux down to 3.0 MeV[6]. While the much larger water Cherenkov detectors can achieve better precision, Borexino holds the lowest threshold achieved so far. Lowering the threshold on ^{8}B spectroscopy represents one of the key features to inspect the transition region of the LMA scenario (see Sect. 1 and 4). The measured rate is $R_B = 0.22 \pm 0.04_{stat} \pm 0.01_{syst}$ and corresponds to a flux of $\Phi_B = (2.4 \pm 0.4_{stat} \pm 0.1_{syst}) \times 10^6 \, cm^{-2} \, s^{-1}$. The flux above 5 MeV is in good consistency with other results.

pep flux and CNO limits. As shown in figure 2, the pep neutrino energy lies at the boundary between the Vacuum and the Transition region of the MSW survival probability. Neutrinos from the pep reaction are closely related to the fundamental pp neutrinos and have their flux theoretically well constrained by this relation. In the same energy region are neutrinos from the CNO cycle reactions. CNO neutrinos are poorly constrained by the SSM and have never been detected so far. At this energy the cosmogenic background of ^{11}C is overwhelming the pep/CNO neutrino flux by about an order of magnitude. We have made the measurement[7] possible suppressing this background exploiting the three-fold coincidence between the parent muon, the ^{11}C and the neutron most often accompanying its production. By excluding the regions of the active volume close to the muon track or to the neutron capture point for a few half lives of ^{11}C (29 min), we select about 50% of the exposure where the ^{11}C background is only 10% of the total. Another powerful handle to distinguish neutrinos from is the pulse shape. The annihilation energy accompanying the ^{11}C e^+-decay may be delayed when released by ortho-positronium bound state. The rate of pep neutrinos has been extracted with a multivariate analysis based of the energy of the event, the distance from the center of the detector and a pulse shape parameter[8]. The rate is $R_{pep} = (3.1 \pm 0.6_{stat} \pm 0.3_{syst}) / d / 100 t$ corresponding to a flux of $\Phi_{pep} = (1.6 \pm 0.3) \times 10^8 \, cm^{-2} \, s^{-1}$ and a survival probability of $P^{ee}_{pep} = 0.62 \pm 0.17$ at 1.44 MeV. The flux of CNO neutrinos could not be extracted due to the spectral shape degeneracy with the ^{210}Bi background. The strongest upper limit available to date has been however obtained

 DOI: 10.1051/epjconf/201612602008

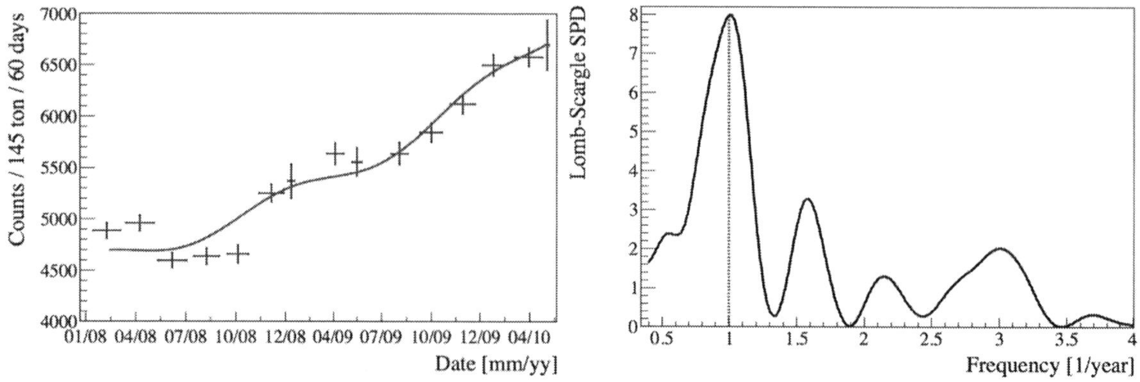

Figure 5. ^{7}Be flux modulation analysis. Left: '60 d'-binned count rate evolution. Right: '10 d'-binned Lomb-Scargle Spectral Power Density.

from this analysis. The rate of CNO neutrinos is $R < 7.1/\mathrm{d}/100\mathrm{t}$ at 95% C.L. corresponding to $\Phi_{\mathrm{CNO}} < 7.7 \times 10^8\,\mathrm{cm}^{-2}\,\mathrm{s}^{-1}$. Improving this analysis using the phase-II data to reduce errors and possibly detect CNO neutrinos is our main goal for 2016.

Annual modulation. Due to the eccentricity of the Earth's orbit, the observed solar neutrino flux is expected to undergo a yearly modulation with amplitude ±3.4% and a maximum at the beginning of January. The presence of such modulation in the ^{7}Be flux is the ultimate proof that Borexino is actually observing neutrinos from the Sun. For this analysis[8] we have defined a dynamic and enlarged 141 t FV. This has been possible thanks to the precise determination of the vessel's shape from the distribution of ^{210}Bi events deposited on the vessel's surface on a weekly basis. Nevertheless performing the spectral fit on sub-periods of the data set is not a viable method due to reduced statistics. Using the phase-I data set (850 d astr. time), we have considered the count rate in the ^{7}Be region (105-380 p.e.), which also includes background from the decay of ^{210}Bi in the scintillator. The last has been rising exponentially during the phase-I (Sect. 2). We have averaged the rate on a 60 d base (figure 5, left) and we have fitted the result with:

$$R = R_0 + R_{Bi}e^{\Lambda_{Bi}t} + \overline{R}\left[1 + 2\epsilon\cos\left(\frac{2\pi t}{T} - \phi\right)\right]$$

where R_0 accounts for a time independent background component. The rate R_{Bi} and the time constant Λ_{Bi} of ^{210}Bi background are fixed to values independently determined from a different energy interval. The period found is $T = (1.01\pm0.07)$ y and the phase, measured from Jan 1st 2008, is $\phi = (11.0\pm4.0)$ d. The average ^{7}Be rate and the eccentricity ϵ are consistent within $2\,\sigma$ with the spectral fit result and with the expected orbit eccentricity, respectively. The hypothesis of no modulation is rejected at $> 3\sigma$. An alternative approach uses Lomb-Scargle frequency analysis in 10 d binned data set. The Spectral Power Density (SPD) distribution is shown in figure 5 (right). The significance of SPD peaks has been evaluated by Monte Carlo simulations of the signal and the background. The peak at 1y with SPD=7.96 has significance largely above 3σ, while no other peak exceeds 2σ.

4 pp neutrino flux

Neutrino from the fundamental pp reaction which powers the Sun are the most elusive to detection as their energy only spans up to 420 keV, corresponding to 264 keV in the recoil electron energy spectrum. The expected rate is $(131 \pm 2)/\mathrm{d}/100\,\mathrm{t}$. Our efforts to reduce the analysis threshold and successfully fit this part of the spectrum have been twofold. First, we performed the purification campaign described in Sect. 2: ^{85}K was reduced to a level consistent with zero, ^{210}Bi by a factor ~3. Second, a large analysis effort has been done to understanding the overwhelming intrinsic background due to ^{14}C that extend up to ~220 keV. Though the Borexino scintillator was obtained from selected ancient oil batches to have a low ^{14}C content, the detector's trigger rate of $\sim 30\,\mathrm{s}^{-1}$ (~50 keV hardware threshold) is largely dominated by ^{14}C decays. Constraining the rate and correctly modelling the shape of ^{14}C was hard due to effect of the threshold. We overcame this problem studying the events that accidentally fall within the $16\,\mu$s acquisition window opened by another uncorrelated event. The ^{14}C rate was thus measured to be $(40 \pm 1)/\mathrm{s}/100\,\mathrm{t}$. ^{14}C gives a second problem: pile-up. Due to the high rate, we record events arising from a ^{14}C decay overlapped with ^{14}C, ^{210}Po, dark noise from the PMTs and virtually any other spectral species. The spectral shape for this class of events is unknown and the position reconstruction largely fails. We have therefore generated a *synthetic* pile-up sample, overlapping uncorrelated events with regular data. The obtained energy spectrum is used in the spectral fit and the rate is constrained within errors to the value obtained by this independent determination. Using data from January 2012 to May 2013 (408 d live time), we obtained the first direct detection of pp neutrinos. The fiducial volume was set to 75 t for a resulting exposure of ~85 ton·yr We performed the fit in the [165,590] keV energy range as shown in figure 6. The pp neutrino rate was determined to be:

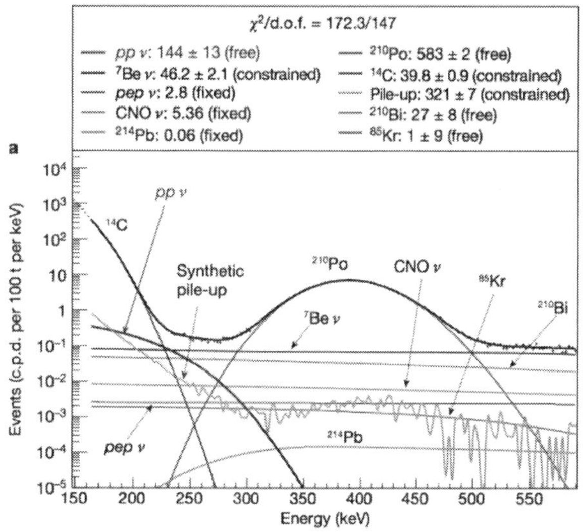

Figure 6. Fit to the energy spectrum for pp neutrino analysis.

$$R_{pp} = 144 \pm 13_{\mathrm{stat}} \pm 10_{\mathrm{sys}}/\mathrm{d}/100\,\mathrm{t},$$

rejecting the null hypothesis at 10σ. The systematic error arises from the uncertainty in the energy estimator, in the fit energy range, in the data selection, in the pile-up evaluation and in the fiducial mass. The result interpretation is threefold: (i) assuming the SSM is correct, our result indicates a survival probability of (0.64 ± 0.12) in confirmation of neutrino oscillations with MSW-LMA mechanism in the vacuum dominated region; (ii) assuming the neutrino oscillation model is correct, our result converts to an unoscillated solar neutrino flux of $(6.6 \pm 0.7) \times 10^{10}\,\mathrm{cm}^{-2}\,\mathrm{s}^{-1}$, confirming the SSM predictions on neutrinos emitted by the reaction most fundamentally bound to the energy production in the Sun; (iii) if both previous assumptions are made, the Borexino result says that the luminosity of the Sun in neutrinos matches the photon luminosity, which is a proof of the Sun stability over a period of 10^5 years. The measurement of the pp neutrino flux by Borexino was published by the Nature Journal[9].

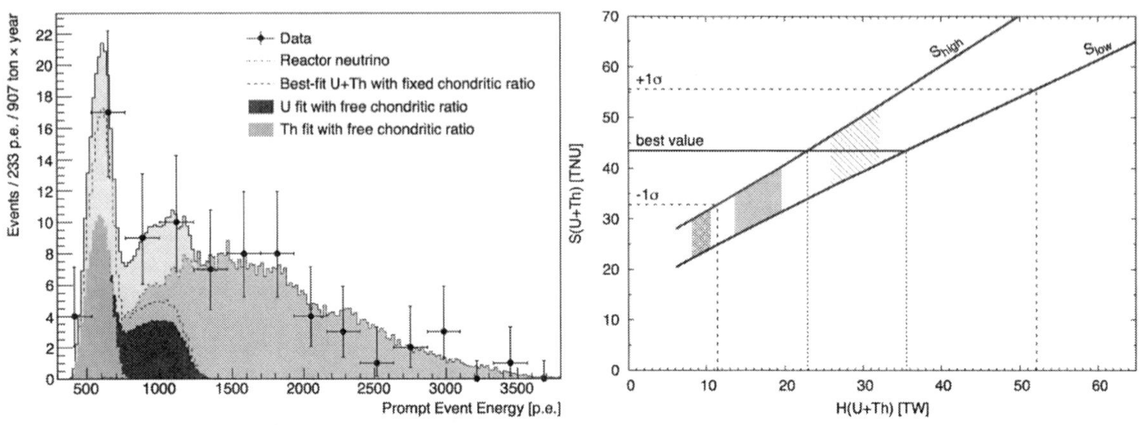

Figure 7. Left: antineutrino spectrum (prompt positron scattering). Right: Geoneutrino flux and terrestrial Heat generation foreseen by three different BSE models (coloured shaded areas). S_{high} and S_{low} are the limits where radiogenic material is concentrated at the Mantle-Crust and Core-Mantle boundary, respectively.

Survival probability after Borexino.

In figure 2, already introduced in Sect. 1, one can also see all the results of the Borexino low energy neutrino spectroscopy that we have discussed here, superimposed to the survival probability of solar electron neutrinos. The confirmation of the MSW-LMA model over a wide range of energies is striking. However, the unexplored transition region (1-3MeV) still has room for alternative models and new physics, in particular the proposed Non Standard neutrino Interactions (NSI) which foresee a different transition shape. There are two ways to test these hypotheses or confirm MSW-LMA: (i) reducing the errors on pep flux (and to a lower extent on ^{7}Be flux) and (ii) lowering the threshold on ^{8}B neutrinos to observe (or not) the expected upturn of the spectrum. The analyses we are currently performing on Borexino Phase-II data follows both approaches.

5 Geo-neutrinos

Geo-neutrinos are anti-neutrinos produced by elements in the radioactive chains of ^{238}U, ^{232}Th and by the decay of ^{40}K, with a flux of the order of $\sim 10^6$ cm^{-2} s^{-1}. These long lived elements are found in the Earth crust and mantel with unknown abundances. Earth models attribute about 19 out 44 TW of thermal heat released by the Earth to radioactive decays of these elements. Measuring Geo-neutrinos flux at different locations can help the development of Earth models and shed light on the heat prodction mechanism of the planet. Only geo-neutrinos from ^{238}U and ^{232}T chain elements can be measured by neutrino detectors, via Inverse Beta Decay with a threshold of 1.8MeV. The ratio of Th/U is supposed to be $\sim$3.9 from the analysis of chondrites, meteorites with the same composition of the Earth. After the first observation in 2010 and a revision in 2013, we present here the most recent results obtained with a statistics of 2056 d[10]. The main background component are the antineutrinos from power reactors, which we have carefully modelled from the fuel composition and up-time data provided by the power companies. Almost all other backgrounds are negligible, thanks to the tagging of the events given by the coincidence between the prompt positron scattering and the $\sim$250 μs-delayed 2.2 Mev gamma from the n capture on H. Events pairs are selected by energy cuts on both events and by their space-time correlation. The efficiency of the cuts is (0.84 ± 0.01) from simulations. We apply an en-

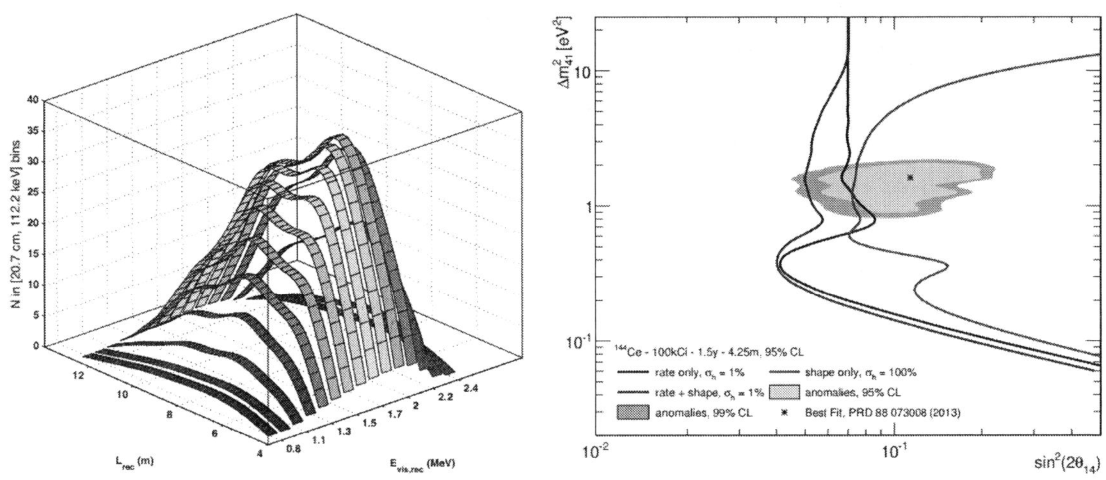

Figure 8. Oscillation wiggles in the SOX signals for a 100 kCi ^{144}Ce-^{144}Pr source (left). Sox sensitivity (right).

larged dynamic fiducial volume that selects all the scintillator volume, excluding the outermost 30 cm to the IV surface with an average mass of 225 t selected. The total exposure is (907 ± 44) ton·yr where we select 77 golden coincidences. The prompt event energy spectrum can be seen in figure 7 (left). An unbinned maximum likelihood fit returns:

$$N_{geo} = 23.7^{+6.5}_{-5.7}(\text{stat})^{+0.9}_{-0.6}(\text{syst})$$
$$N_{react} = 52.7^{+8.5}_{-7.7}(\text{stat})^{+0.7}_{-0.9}(\text{syst})$$

The non observation of geo-neutrinos is excluded at 5.9σ level. The corresponding fluxes are S_{geo} = (43.5 ± 11.1) TNU and S_{react} = (96.6 ± 15.9) TNU where 1 TNU = 1ν/ 10^{32} protons / yr. The flux of reactor antineutrinos compares well with the prediction of (87 ± 4) TNU. Freeing the Th/U ration we can resolve the two components, obtaining overall similar results and a ratio in good agreement with the chondritic ratio. We account for the local contribution of the crust N_{LOC} = (9.7 ± 1.3) TNU obtained by a a geological survey of the area around LNGS and the contribution of the Rest Of the Crust S_{ROC} = (13.7 ± 2.5) TNU as predicted by the planetary model. We can then exclude the absence of a Mantle contribution at 98% C.L. and assign $S_{Mantel} = 20.9^{+15.9}_{-10.3}$ TNU. Our results are in good agreement with the Bulk Silicate Earth model predictions as it can be seen in figure 7 (right).

6 Short distance Oscillation with boreXino (SOX)

In the past years many different experimental indications have pointed toward the existence of sterile neutrinos. Although none of them is individually strong enough to make a claim, altogether they justify a serious investigation. The hypothesised oscillation into the forth species occurs with $L/E \sim$ 1 m/MeV. In Borexino, using a neutrino source with energy of ~1 MeV, the oscillation length is significantly smaller then the detector size (~10 m) and significantly larger then the detector resolution (~12-15 cm). This would allow to see oscillation wiggles in the position distribution of events, a powerful addition to a plain disappearance analysis. The location for such a source is the ~1 m cubical pit present under the detector, which was excavated for this purpose before the detector's construction.

This non invasive deployment requires no work on the detector, it bears no risk of contamination and does not terminate the solar run of Borexino. We foresee to deploy a ^{144}Ce-^{144}Pr source in November 2016 and proceed to a 1.5 y data taking. ^{144}Ce-^{144}Pr is a β emitter of antineutrinos with energies up to 3 MeV and a decay time of ~411 d. Thanks to the neutron tagging, which makes antineutrino detection essentially background free (see Sect. 5), we can perform the measurement with a source activity of about 100-120 kCi, currently being produced at the Mayak facility (Russia). Figure 8 shows the expected signal in case of oscillation and the sensitivity of the experiment in the oscillation parameters' plane. Disappearance and wave effects will allow to clarify the matter, unambiguously proving or rejecting the hypothesis. In particular, in case of the existence of a fourth sterile neutrino with the parameters indicated, SOX[11] will surely discover the effect and measure the parameters of oscillation.

7 Conclusions and outlook

Borexino has performed low energy solar neutrino spectroscopy including ^{7}Be , pep, ^{8}B and pp neutrino fluxes. Thanks to its unprecedented background levels, which have been further reduced with a dedicated campaign of purification of the liquid scintillator, the phase-II data of Borexino currently ongoing will allow us to reduce the error on ^{7}Be , pep and ^{8}B neutrino flux measurements. Measurement of the so far undetected neutrinos from the CNO cycle reactions is of fundamental astrophysical importance in particular as the they can help resolve the solar metallicity puzzle[1]. The measurement requires an independent evaluation of the residual contamination of ^{210}Bi whose spectral shape is quasi degenerate with the CNO neutrinos. One way to do this is to infer the ^{210}Bi rate from its daughter ^{210}Po , provided equilibrium can assumed. This assumption holds if there are no convective motions in the scintillator that may bring ^{210}Po from the nylon vessel surface into the fiducial volume. We have recently installed a thermal insulation of the detector that should prevent such motion. If this will be successful we can proceed to measure the flux of CNO neutrinos, otherwise we will sensibly improve the limits we have already posed and we will consider a further purification campaign to reduce the ^{210}Bi contamination. At the end of phase-II we also expect to further improve our result on geo-neutrinos and to measure the seasonal variation effect of ^{7}Be neutrinos upon several cycles and without the background constraints of phase-I. Starting with November 2016 we will test the sterile neutrino hypothesis using a 100 kCi ^{144}Ce-^{144}Pr source placed under the detector.

References

[1] W.C. Haxton, R.G. Hamish Robertson and A. M. Serenelli, Ann. Rev. Astron. Astrophys. **51**, 21 (2013).
[2] G. Bellini et al. (Borexino coll.), *Nucl. Instrum. Methods* A **609**, 58 (2009).
[3] G. Bellini et al. (Borexino coll.), *JINST* **07**, 10018 (2012).
[4] G. Bellini et al. (Borexino coll.), *Phys. Rev. Lett.* **107**, 141302 (2011).
[5] G. Bellini et al. (Borexino coll.), *Phys. Lett.* B **707**, 22 (2012).
[6] G. Bellini et al. (Borexino coll.), *Phys. Rev.* D **82**, 033006 (2010).
[7] G. Bellini et al. (Borexino coll.), *Phys. Rev. Lett.* **108**, 051302 (2012).
[8] G. Bellini et al. (Borexino coll.), *Phys. Rev.* D **89**, 112007 (2014)
[9] G. Bellini et al. (Borexino coll.), Nature **512**, 383-386 (2014).
[10] M. Agostini et al. (Borexino coll.), *Phys. Rev.* D **92**, 031101(R) (2015).
[11] G. Bellini et al. (Borexino coll.), *JHEP* **08**, 038 (2013).

EPJ Web of Conferences **126**, 02009 (2016) DOI: 10.1051/epjconf/201612602009
ICNFP 2015

Highlights from T2K

Anna Dabrowska on behalf of the T2K collaboration[1,a]

[1] *H. Niewodniczanski Institute of Nuclear Physics PAN, Cracow, Poland*

Abstract. T2K is a long-baseline neutrino experiment designed to measure neutrino oscillations parameters. A high-intensity beam of muon neutrinos produced at the J-PARC accelerator complex is sent towards the near detector facility (the ND280 and INGRID detectors, located 280 m away from the neutrino source) and the far detector, Super-Kamiokande (295 km away). The change in the measured intensity and the composition of the neutrino beam between the near and far detectors are used to provide information on the oscillation parameters. T2K has delivered the world's best measurement of the θ_{23} angle by observing muon neutrino disappearance. It was also the first experiment to observe electron neutrino appearance (2013) with a significance of 7.3 σ, to measure the associated θ_{13} mixing angle, and to provide the first hint of a non-zero δ_{CP} phase. The first running of anti-neutrinos in the T2K experiment shows a clear dip below 1 GeV, as expected for an oscillation signal. The T2K experiment is also capable of providing information on neutrino-nucleon cross sections at energies around 1 GeV, thanks to a large amount of target material present in the near detector facility.
A summary of the recent oscillation measurements as well as selected cross section results are presented.

1 Introduction

The investigation of the neutrino properties in the past two decades conducted by numerous experiments dedicated to this subject has led to extraordinary results. The unique feature of neutrinos is that they can change or "oscillate" from one type to another as they travel over long distances. In 1998 Super-Kamiokande, as the first experiment, measured oscillations of neutrinos generated in the Earth's atmosphere [1]. Next the oscillations of neutrinos coming from the Sun were measured by the SNO experiment in 2001 [2]. The discovery of neutrino oscillations revealed that the neutrinos have finite mass, so we expect that the weak eigenstates are different from the mass eigenstates, in analogy to the quark system. The neutrino mixing is described by a 3x3 unitary matrix, called the Pontecorvo-Maki-Nakagawa-Sakata (PMNS) matrix [3]. Oscillation probabilities depend on three mixing angles (θ_{13}, θ_{12}, θ_{23}), two independent mass splittings ($|\Delta m^2|$) and one complex CP phase (δ_{CP}) which should be determined experimentally.

Since that time, various neutrino experiments (Super-Kamiokande, K2K, KamLAND, MINOS, T2K, ICARUS, OPERA, Daya Bay, RENO) have been involved in determining the above parameters, describing the phenomenon of the neutrino oscillations. Despite the fact that the values of three mixing angles and two mass splittings are now known, they need to be determined with a better precision in

[a] e-mail: Anna.Dabrowska@ifj.edu.pl

particular to check whether there is the maximal mixing in the atmospheric sector. Also, there still remain open questions such as: value of the δ_{CP} describing the CP violation in neutrino sector, and the mass hierarchy in the neutrino mass eigenstates.

2 T2K experiment

T2K is a long-baseline neutrino experiment designed to measure neutrino oscillations parameters. The main goal of T2K is to study and compare ν_μ to ν_e and $\bar{\nu}_\mu$ to $\bar{\nu}_e$ transitions in order to measure θ_{13} and explore δ_{cp}, CP violation in the lepton sector. T2K is also designed for precision measurement of ν_μ and $\bar{\nu}_\mu$ disappearance to explore θ_{23} ($\bar{\theta}_{23}$) and Δm^2_{23} ($\Delta\bar{m}^2_{23}$) parameters, i.e to test the CPT theorem or new non-standard ν interactions with matter. The cross section measurements in the near and far detectors are additional purposes of the T2K experiment.

The T2K experiment [4] is located in Japan. A high-intensity beam of muon neutrinos produced at the J-PARC accelerator complex is sent towards the near detector facility (the ND280 and INGRID detectors, located 280 m away from the neutrino source) and the far detector, Super-Kamiokande (295 km away). The change in the measured intensity and the composition of the neutrino beam between the near and far detectors are used to provide information on the oscillation parameters. The beam is produced by the conventional method. The protons accelerated to 30 GeV hit a graphite target, producing hadrons including pions and kaons. Charged hadrons are then focused by a set of three electromagnetic horns and sent to a decay tunnel where the pions and kaons decay in flight, producing neutrinos (or anti-neutrinos by the reversing current in magnetic horns).

T2K was the first long-baseline experiment using the off-axis beam technique: the far detector, similar to the near ND280 detector, is located in the direction making a 2.5 degrees angle with the beam axis. Despite some reduction in the neutrino flux passing through the far detector, a big advantage is a kinematic focusing of the off-axis beam around the energy corresponding to the maximum of the oscillation probability, which is about 600 MeV. The use of an off-axis neutrino beam, also reduces the background for ν_e appearance in the far detector by reducing the high energy tail which has a relatively large intrinsic ν_e component of the beam. Finally, the dominant interaction mode at these energies is the charged current quasi-elastic (CCQE) one, which allows for the reconstruction of the neutrino energy at the far detector. Additional significant processes are: CCQE-like multi-nucleon interaction, charged current single pion production (CCπ), neutral current single pion production (NCπ). For a precise measurement of oscillation parameters, T2K is equipped with two near detectors [4]: ND280 and INGRID. Short descriptions of the near and far detectors are included in the below subsections.

2.1 INGRID on-axis near detector

The on-axis detector INGRID [5] is used to monitor the beam rate, its direction and stability. It consists of 16 $1m \times 1m \times 1m$ cubic modules. Each module is a "sandwich" of 11 scintillator layers and 10 iron layers. They are surrounded by four veto planes. The modules are arranged as follows: seven horizontally, seven vertically, and two off-diagonally (Figure 1 left). Neutrinos are counted in the detector by reconstructing muons from neutrino charged current interactions. The profile and direction of the neutrino beam are obtained by the use of registered charged current interactions in each module.

2.2 ND280 off-axis near detector

The off-axis detector, ND280 (Figure 1 right) is used to constrain flux and cross-section systematics for oscillation analysis: it measures flux and cross section before the oscillations occur. It consists

DOI: 10.1051/epjconf/201612602009

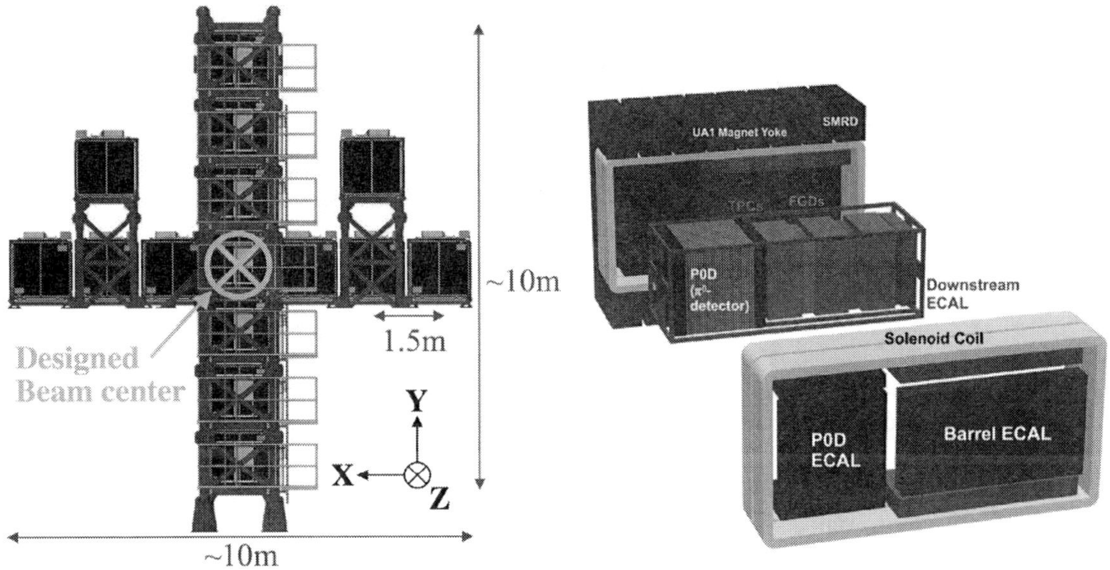

Figure 1. The INGRID on axis near detector (left) and the ND280 off axis near detector (right) [4]

of several sub-detectors: Pi-Zero Detector (P0D) and Tracker as the inner detectors, both surrounded by the electromagnetic calorimeter (ECAL) and then by the Side Muon Range Detector (SMRD). All detector components except the SMRD are placed inside a 0.2 T magnetic field produced by the recycled magnet from the UA1 experiment. The P0D subdetector, placed upstream inside the magnet is a "sandwich" of scintillator planes, lead and brass plates, and a water target. It is customized for the measurement of neutral π^0 production. Gamma rays from π^0 decay are converted to electromagnetic showers in the lead plates, and are recorded in the scintillator detectors. In the downstream of the P0D, two Fine Grained Detectors (FGDs) separated by three Time Projection Chambers (TPCs) are placed. The TPCs can measure the momentum of leptons from the curvature of the track in the magnetic field. The momentum resolution for muons is better than 10 % at 1 GeV. The FGDs consist of scintillator bars. They provide the target material for neutrino interactions and are optimized for detecting the proton recoils. By combining the TPCs and FGDs, the energy spectrum of ν_μ can be precisely measured based on CCQE (Charged Current Quasi-Elastic) neutrino interactions. The FGDs and TPCs form the so-called Tracker system, where the ν_μ and ν_e energy spectra are measured by reconstructing the lepton momentum and by separating electrons from muons using dE/dx. The other detector components, SMRD and ECAL, are installed in the area outside of the P0D and Tracker. The inner part of the detector is surrounded by an electromagnetic calorimeter which can tag escaping electrons and positrons from π^0 decays. Additionally, the outermost SMRD detector (scintillator slabs installed in the magnet yoke) is used to detect muons escaping the inner volume and to tag cosmic ray muons.

2.3 Super-Kamiokande far detector

The far detector, Super-Kamiokande(SK) [6], is a 50 kton water Cherenkov detector located 1000 m underground in the Mozumi Mine in the Japanese Alps. Its inner detector (ID), 22.5 kton of fiducial volume, is viewed by about eleven thousand 20-inch diameter PMTs. The outer detector (OD), which surrounds the ID, is also a water Cherenkov detector. It is used to veto events that enter or exit the inner detector. SK started its operation in 1996. Apart from having its own rich physics program, SK is also used as the far detector of the T2K experiment. It has an excellent ability to separate between the ν_e and ν_μ interactions, which is critical to the study of the appearance of electron neutrinos in a muon neutrino beam. It was verified that the probability of the μ/e misidentification is less than 1 % [7]. The lack of a magnetic field in the far detector makes it impossible to separate between ν and $\bar{\nu}$. The events are synchronized with the beamline with the use of a dedicated GPS system.

3 Analysis strategy

To measure the oscillation parameters, the observed number of neutrino interactions at the far detector is compared with the predicted one. The values of the oscillation parameters are then estimated using a maximum likelihood fit. The neutrino flux is predicted by simulating the hadronic interactions in the target, and the propagation and decay of the secondary particles. This simulation is tuned to the experimental results of the CERN NA61 experiment [8]. Neutrino interactions are simulated based on models with constraints from external data using the NEUT neutrino interaction generator [9]. Systematic uncertainties are incorporated in each step: flux prediction, neutrino interactions and cross sections, and the response of the detectors. Systematic uncertainties for the far detector are evaluated using atmospheric neutrino data and a π^0 control sample.

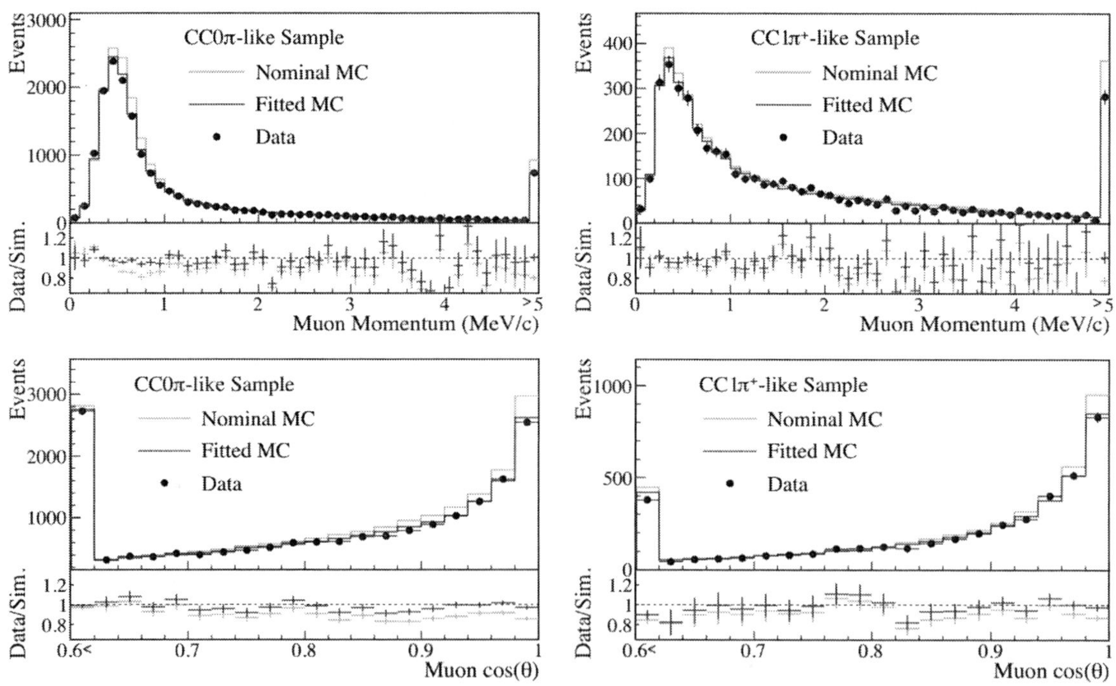

Figure 2. Muon momentum (top) and angle distribution (bottom) for the CC0π -like sample (left) and CC1π-like sample (right) selected in ND280. These are compared to the simulation, broken down into the different reaction types, with all systematic parameters set to their nominal values [10].

The off-axis near detector ND280 is used to constrain the flux uncertainties, some of the neutrino interactions and cross section uncertainties (due to the difference of target materials between the near

 DOI: 10.1051/epjconf/201612602009

and far detectors, not all of these uncertainties can be constrained using ND280). The analyses presented in this article were done based on 6.6×10^{20} protons on target for the total neutrino sample and 4.04×10^{20} protons on target for the anti-neutrino sample, respectively. In the ND280 analysis the exclusive sub-samples are selected, based on track topologies in the Tracker of the ND280 detector, to constrain the cross section components: CCQE (CC0π sub-sample), CC resonant single pion cross-section parameters (CC1π sub-sample) and CC multi pions, dominated by deep inelastic processes (CCother sub-sample) for neutrino mode [10]. Because of lower statistics in the anti-neutrino mode, the ND280 fits were done only for two sub-samples : CC1-Track and CC>1 Track to make constrains for CCQE and CCnQE (Charged Current non Quasi Elasic) interactions.

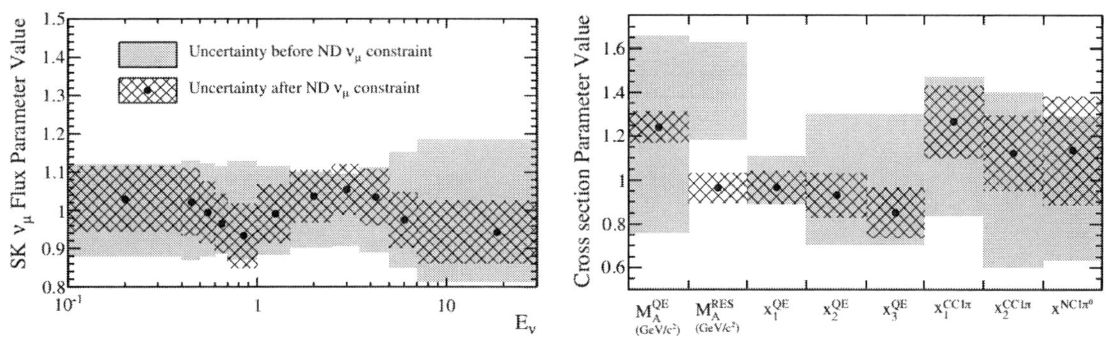

Figure 3. Prior and fitted values and uncertainties for the SK ν^{μ} flux parameters (left) and cross-section parameters (right) constrained by the near detector analysis for the oscillation analyses. Uncertainties are calculated as the square root of the diagonal elements of the relevant covariance matrix. The value of M_A^{QE} and M_A^{RES} are given in units of GeV/c^2 , and all the other parameters are multiplicative corrections [10].

For each defined sub-sample in the data and Monte Carlo, the two-dimensional plots of muon momentum and angle are prepared. Then the fit which varies the parameters related to neutrino energy spectrum and cross sections is performed to obtain the best agreement between the data and simulation. By fitting MC to the near detector data, new flux and cross-section parameters can be obtained. After the parameters adjustments, the agreements turn out to be excellent, as can be seen in Figure 2. The parameters adjusted in the ND280 analysis are also applied in the SK analysis. The uncertainties in these parameters are in general smaller than those for the prior values, as can be seen in Figure 3. The measurements done at the near detector significantly improve the experimental ability to predict the neutrino event rates and spectra at the far detector (Figure 4). The uncertainties in the predicted number of electron-like events decreases from 25 % to 6.3 %, and from 23.4 % to 7.4 % for muon-like events, respectively.

4 Oscillation analysis of joint electron neutrino appearance and muon neutrino disappearance channels

Initial T2K oscillation analyses were done separately for muon neutrino disappearance events to measure θ_{23}, $|\Delta m^2_{32}|$ and for electron neutrino appearance events to measure θ_{13} and estimate δ_{CP} parameters [11, 12]. However, to properly take into account the correlations between the estimates of all four oscillation parameters, the two sets of the above observables should be analyzed together. Four

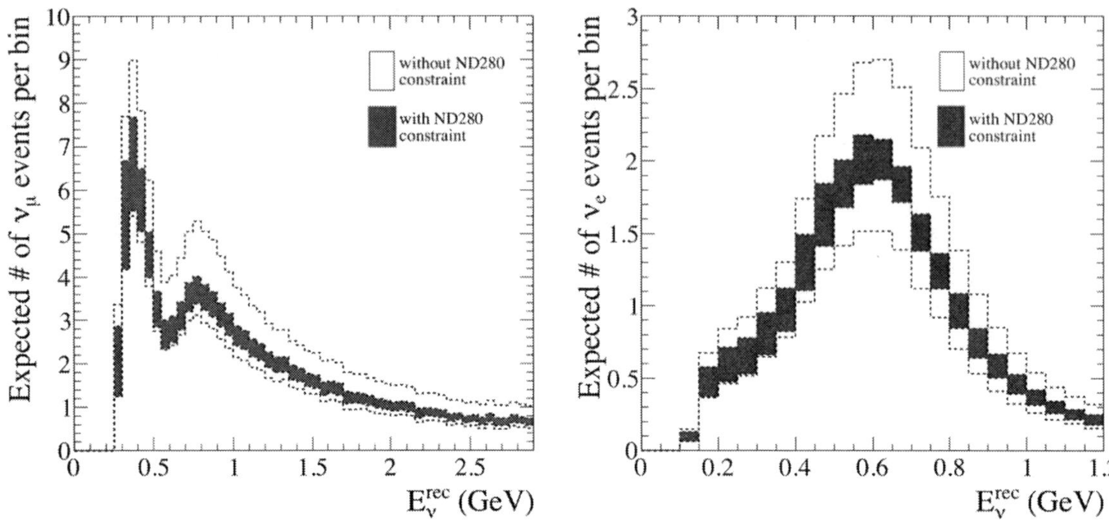

Figure 4. Total error envelopes for the reconstructed energy distributions of ν_μ CC (left) and ν_e CC (right) candidate events, using typical oscillation parameter values, with and without the ND280 constraint applied [10].

independent joint analyses were performed in T2K, and they gave consistent results. Some of them are presented below.

For the muon-like events observed at SK, fits of the rate and the reconstructed energy spectrum of the muon candidate events were done. For the electron-like events, fits of the rate and the two-dimensional distribution of the momentum and angle with the beam direction of the particle reconstructed as an electron in the event were performed (the method used in the previous T2K electron appearance analysis [12]). The point estimates of the oscillation parameters are found by minimizing the negative log-likelihood :

$$\chi^2 = -2lnL(\vec{\theta}, \vec{g}, \vec{x_s}, \vec{s})$$

where $\vec{\theta}$ represents the vector of the PMNS oscillation parameters, $\vec{g}$ is a vector containing the values of the systematic parameters constrained by the near detector, $\vec{x_s}$ are the cross-section parameters not constrained by the near detector and $\vec{s}$ are the SK detector systematic parameters [10].
The profiled $\Delta\chi^2$ of each oscillation parameter was obtained by minimizing the negative log-likelihood with respect to the systematic parameters and other three oscillation parameters using MINUIT.

Table 1. Point estimates of the oscillation parameters for the joint three-flavor oscillation frequentist analysis, for the normal hierarchy (NH) and inverted hierarchy (IH) [10].

MH	$\Delta m^2_{32}(10^{-3}eV^2/c^4)$	$\sin^2\theta_{23}$	$\sin^2\theta_{13}$	δ_{CP}	$\Delta\chi^2$
NH	2.51	0.524	0.0422	1.91	0.01
IH	2.49	0.523	0.0491	1.01	0.00

The point estimates for the oscillation parameters are summarized in Table 1. The value obtained for $\sin^2\theta_{13}$ by T2K is larger than the value found by the reactor experiments (the weighted average of the results from the three reactor experiments Daya Bay, RENO, and Double Chooz which is: $\sin^2\theta_{13reactor} = 0.0243 \pm 0.0026$ [13]). The best-fit value of $\sin^2\theta_{23}$ is consistent with maximal

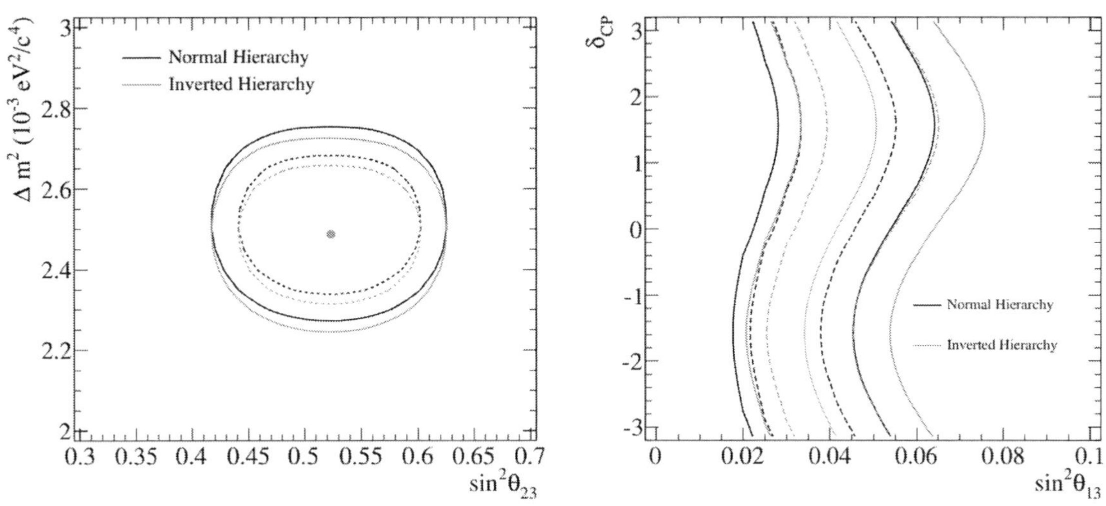

Figure 5. The 68% (dashed) and 90% (solid) C.L. regions, from the analysis without reactor data, with different mass hierarchy assumptions, using $\Delta\chi^2$ with respect to the best-fit point which comes from the inverted hierarchy obtained for the atmospheric parameters (left). The right plot shows one-dimensional confidence intervals in $\sin^2\theta_{13}$ for different values of δ_{CP} [10].

disappearance, and the difference in χ^2 between the solutions for each mass hierarchy is negligible. Figure 5 presents the 68% and 90% C.L. regions for the two mass hierarchy assumptions in the two two-dimensional oscillation parameter spaces ($\sin^2\theta_{23}$, Δm^2_{32}) and ($\sin^2\theta_{13}$, δ_{CP}), constructed using constant $\Delta\chi^2$ with respect to the inverted hierarchy best-fit point. The value of $\sin^2\theta_{23} = 0.524^{+0.057}_{-0.059}$ ($0.523^{+0.055}_{-0.065}$) in the normal (inverted) mass hierarchy scenario obtained in the T2K analysis is the world's most precise value till now.

Table 2. Point estimates of the oscillation parameters for the joint three-flavor oscillation frequentist analysis combined with the results from reactor experiments, for the normal hierarchy (NH) and inverted hierarchy (IH) [10].

MH	$\Delta m^2_{32}(10^{-3}eV^2/c^4)$	$\sin^2\theta_{23}$	$\sin^2\theta_{13}$	δ_{CP}	$\Delta\chi^2$
NH	2.51	0.527	0.0248	-1.55	0.00
IH	2.48	0.533	0.0252	-1.56	0.86

The point estimates for the oscillation parameters when the reactor measurements are included in the likelihood function are given in Table 2. The estimate for $\sin^2\theta_{13}$ is smaller than the result obtained with T2K data only, shown in Table 1. The likelihood is maximal for normal mass hierarchy and for δ_{CP} = $-\pi/2$, where the appearance probability is largest.

The confidence regions obtained in the ($\sin^2\theta_{23}$, Δm^2_{32}) parameter space are compared with the results from the Super-Kamiokande [14] and MINOS [15] experiments in Figure 6 (left). The results from T2K and MINOS used the latest value of $\sin^2\theta_{13}$ from [16] to fit this parameter whereas the result from SK has $\sin^2\theta_{13}$ fixed to the previous reactor value [13]. The distributions of δ_{CP} when fitting by marginalizing over other oscillation parameters are presented in Figure 6 (right). In this case the values of δ_{CP} around $-\pi/2$ are favored. The values of the $\Delta\chi^2$ corresponding to the

DOI: 10.1051/epjconf/201612602009

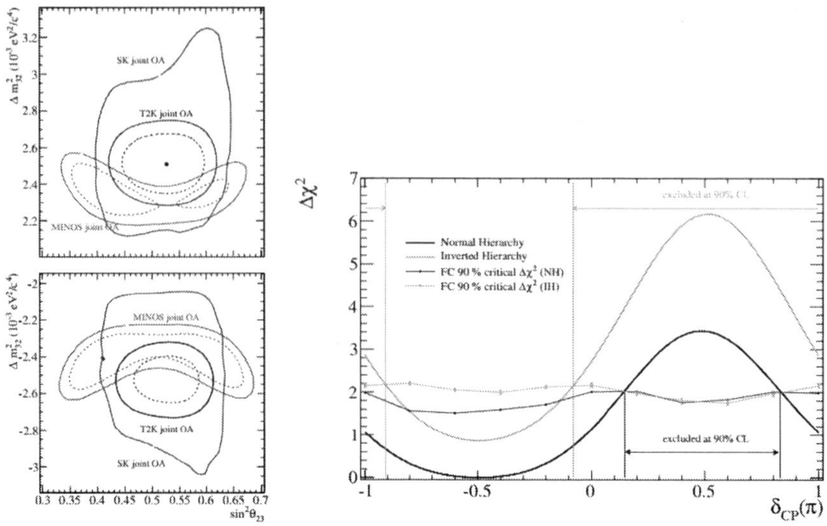

Figure 6. The 68% (dashed) and 90% (solid) C.L. regions for normal (top left) and inverted (bottom left) mass hierarchy combined with the results from reactor experiments in the ($\sin^2\theta_{23}$,Δm^2) space compared to the results from the Super- Kamiokande [14] and MINOS [15] experiments. Profiled $\Delta\chi^2$ as a function of δ_{CP} with the results of the critical $\Delta\chi^2$ values for the normal and inverted hierarchies for the joint fit with reactor constraint, with the excluded regions found overlaid (right) [10].

boundaries of 90% C.L. intervals obtained with the Feldman-Cousins approach [17] for both mass hierarchies are presented also in Figure 6 (right). The figure shows that the excluded regions for δ_{CP} at the 90% C.L are: δ_{CP} =(0.15; 0.83)π for normal hierarchy and δ_{CP} =(-0.08; 1.09)π for inverted hierarchy respectively. The normal hierarchy appears to be favored. Using a Bayesian approach, we can compare different models by looking at the posterior probabilities of different combinations of octant of $\sin^2\theta_{23}$ and mass hierarchies. The T2K data combined with the results of the reactor experiments for $\sin^2\theta_{13}$ weakly favor the normal hierarchy and the octant $\sin^2\theta_{23} > 0.5$.

5 First results from anti-neutrino running

The first preliminary result of $\bar{\nu}_\mu$ disappearance and $\bar{\nu}_e$ appearance analysis is presented here. In the anti-neutrino beam the data collected untill June 2015, which corresponds to 4.04×10^{20} POT, have been used. The applied procedure, which includes the event selection and calculation of the expected Monte Carlo events, is the same as that for neutrino analysis. From Monte Carlo simulations for a no-oscillation scenario 103.6 muon-like events are expected in the SK detector, but only 34 muon-like events were observed.

In this analysis, in order to see potential effects coming from a new physics such as CPT violation or non-standard interactions, we treat (θ_{23}, Δm^2_{32}) for neutrinos and ($\bar{\theta}_{23}$, $\Delta\bar{m}^2_{32}$) for anti-neutrinos as independent parameters and fit data with ($\bar{\theta}_{23}$, $\Delta\bar{m}^2_{32}$). Other oscillation parameters are assumed to be the same for neutrinos and anti-neutrinos and these values are fixed and taken from the ν-mode results from T2K [18] and the Particle Data Group 2014 [19]. The reconstructed anti-neutrino energy spectrum is shown in Figure 7. The evident distortion of the energy spectrum and the reduction of the observed event numbers with respect to the predicted ones can be seen.

DOI: 10.1051/epjconf/201612602009

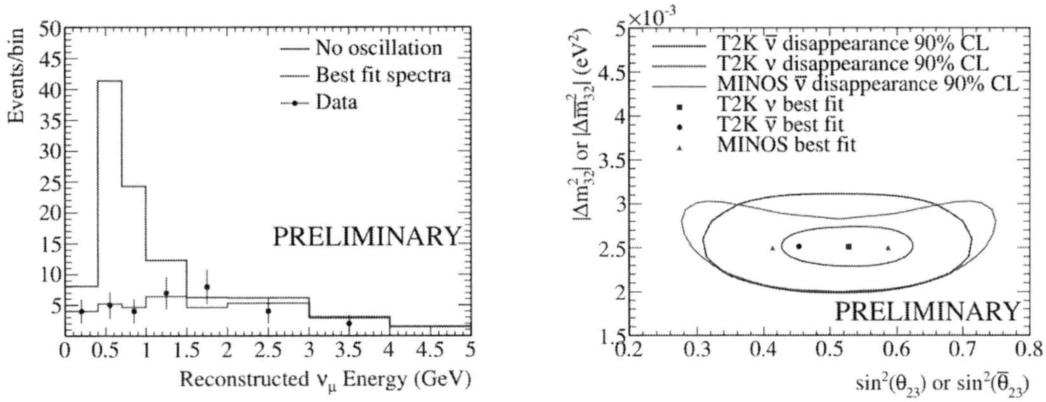

Figure 7. Distribution of reconstructed neutrino energy for 34 muon-like events observed in 4.04×10^{20} POT anti-neutrino beam data (left). Contours of ν mode and $\bar{\nu}$ mode, also for MINOS $\bar{\nu}$ results (right)

The result of the best-fit oscillation parameters for normal mass hierarchy is :

$$\Delta\bar{m}^2_{32} = 2.50^{+0.3}_{-0.2}\ 10^{-3} eV^2$$

$$\bar{\theta}_{23} = 0.46^{+0.14}_{-0.06}$$

No differences can be observed in the best-fit parameters for normal and inverted hierarchy. The results obtained here are consistent with other experiments: MINOS [20] for anti-neutrino beam and T2K neutrino beam. The comparison is shown in Figure 7 (right). The reduction of the total uncertainty on the muon anti-neutrino disappearance results is not big and passes from 14.4% without the ND280 fit to 11.6% with the use of the ND280 fit results.

In the analysis of $\bar{\nu}_e$ appearance, the selection criteria for $\bar{\nu}$ candidates in the far detector are the same as those for neutrino beam data. As a result, only three events remain as potential candidates of the $\bar{\nu}$ appearance signal. The lack of a field in the far detector makes the separation between ν and $\bar{\nu}$ impossible. The background events include: electron-like events coming from interactions in the far detector, neutrinos from oscillations: $\nu_\mu \to \nu_e$, misidentified ν_μ (or $\bar{\nu}_\mu$), and original ν_e (or $\bar{\nu}_e$) from decays of muons in the T2K beam.

Table 3. Expected number of events in SK detector with actual 4.04×10^{20} POT for normal hierarchy.

Expected events (NH)	$\delta_{CP} = -\pi/2$	$\delta_{CP} = 0$	$\delta_{CP} = +\pi/2$
Signal $\bar{\nu}_\mu \to \bar{\nu}_e$	1.961	2.636	3.288
Background $\nu_\mu \to \nu_e$	0.592	0.505	0.389
Background NC	0.349	0.349	0.349
Background other	0.826	0.826	0.826
Total	3.73	4.32	4.85

The expected event number of the background depending on δ_{CP} value and mass hierarchy is presented in Tables 3 and 4. The best-fit oscillation parameters obtained from neutrino beam analysis are assumed, and the number of expected $\bar{\nu}_e$ candidates is calculated for some δ_{CP} for normal and inverted mass hierarchy. The results, as summarized in Tables 3 and 4, changes from 3.73 to 5.45 events.

Table 4. Expected number of events in SK detector with actual 4.04×10^{20} POT for inverted hierarchy.

Expected events (IH)	$\delta_{CP} = -\pi/2$	$\delta_{CP} = 0$	$\delta_{CP} = +\pi/2$
Signal $\bar{\nu}_\mu \rightarrow \bar{\nu}_e$	2.481	3.254	3.939
Background $\nu_\mu \rightarrow \nu_e$	0.531	0.423	0.341
Background NC	0.349	0.349	0.349
Background other	0.821	0.821	0.821
Total	4.18	4.85	5.45

The collected statistics are insufficient to confirm or exclude $\bar{\nu}_e$ appearance in the T2K experiment.

6 Other analyses - cross section measurements

Before the T2K experiment started to collect data, only few measurements of the ν cross section in the range energy of the few hundreds of MeV had been performed. The T2K program of the ν cross section measurements on various nuclei is very important from the point of view of the oscillation analyses. As was shown in Section 3, the reduction of systematic uncertainties associated with neutrino interaction cross section measurement and constraints on cross section parameters allow for a more precise prediction of the number of events from neutrino interactions expected in the far detector.

Table 5. List of neutrino cross section measurements in the T2K experiment

no	Cross section	Detector	Target	Reported in
1.	ν_μCC inclusive	Tracker of ND280	CH	PRD 87 092003 (2013)
2.	ν_eCC inclusive	Tracker of ND280	CH	PRL 113 241803 (2014)
3.	ν_μNC elastic	Super-Kamiokande	Water	PRD 90 072012 (2014)
4.	ν_μCC inclusive	INGRID	CH/Fe	PRD 90 052010 (2014)
5.	ν_μCCQE	INGRID	CH	PRD 91 112002 (2015)
6.	ν_μCCQE	Tracker of ND280	CH	Accepted by PRD
7.	ν_μNC π^0	POD of ND280	CH/Water	Publication in progress
8.	ν_μCC coherent	Tracker of ND280	CH	Publication in progress
9.	ν_μCC π^+	Tracker of ND280	Water	Publication in progress
10.	ν_μCC0π^+	Tracker of ND280	CH	Publication in progress

The information about the already published or ready for publication ν cross section measurements made in both far and near detectors on Hydrocarbon, Iron, Oxygen and Carbon and for different processes is summarized in Table 5.

7 Conclusions

The T2K collaboration has presented updated results of neutrino oscillation analysis based on accumulated 6.6×10^{20} POT neutrino beam data and 4.04×10^{20} POT anti-neutrino beam data collected until June 2015.

The T2K experiment has delivered the currently most precise measurement of the angle θ_{23}, $\sin^2\theta_{23} = 0.524^{+0.057}_{-0.059}$ ($0.523^{+0.055}_{-0.065}$) in the normal (inverted) mass hierarchy scenario.

The T2K data favor larger values of $\sin^2\theta_{13}$ than the reactor experiments. The $\nu_\mu + \nu_e$ joint analysis with the reactor constraint indicates a weak preference for the normal mass hierarchy and the octant of $\sin^2\theta_{23} > 0.5$. There is a preference for values of δ_{CP} around -π/2.

In the $\bar{\nu}_\mu$ disappearance analysis, 34 $\bar{\nu}_\mu$ candidates are found where 103.6 events are expected for no oscillation. The constraints on oscillation parameters, $\Delta\bar{m}^2_{32}$ and $\bar{\theta}_{23}$, are consistent with the results from the MINOS experiment and Δm^2_{32} and θ_{23} parameters from T2K. The present $\bar{\nu}_e$ appearance analysis (three electron-like events found) is statistically limited and cannot strongly favor the appearance hypothesis in comparison with background-only hypothesis. Therefore more anti-neutrino data is needed.

It should be also added, that the cross sections of neutrino interactions measured in the near and far T2K detectors are extremely important from the point of view of the oscillation analyses presented in this paper.

Acknowledgements

This work was partially supported by the Polish National Science Centre, projects number 2011/01/M/ST2/02578 and 2014/14/M/ST2/0085

References

[1] Y.Fukuda at al., Phys. Rev. Lett. **81**, 1562–1567 (1998)
[2] Q.R.Ahmad et al., Phys. Rev. Lett. **87** (7), 071301 (2001)
[3] B.Pontecorvo, J. Exp. Theoret. Phys. **34**, 247 (1958)
Z.Maki, M.Nakagawa and S.Sakata, Prog. Theor. Phys. **28**, 870, (1962)
[4] K.Abe et al., Nucl. Instr. Meth. A **659**, 106 (2011)
[5] K.Abe et al., Nucl. Instrum. Meth. A **694**,211-223 (2012)
[6] S.Fukuda et al., Nucl. Instrum. Meth. A **501**, 418 (2003)
[7] S.Kasuga et al., Phys. Lett. B **374**,238 (1996)
[8] N.Abgrall et al., Nucl. Instrum. Meth .A **701**, 99 (2013).
[9] Y.Hayato, Acta Phys. Polon. B **40**, 2477 (2009)
[10] K.Abe et al., Phys. Rev. D **91**, 072010 (2015)
[11] K.Abe et al.,Phys. Rev. Lett. **112**, 18 181801 (2014)
[12] K.Abe et al., Phys. Rev. Lett. **112**, 061802 (2014)
[13] J.Beringer et al., Phys. Rev. D **86**, 010001 (2012)
[14] A.Himmel, AIP Conf. Proc. **1604**, 345 (2014)
[15] P.Adamson et al., Phys. Rev. Lett. **112**, 191801 (2014)
[16] J.Beringer et al., (Particle Data Group with 2013 update see: http://pdg.lbl.gov/2013/pdg2013.html) Phys. Rev.D **86**, 010001 (2012)
[17] G.J.Feldman and R. D. Cousins, Phys. Rev. D **57**, 3873 (1998)
[18] K.Abe et al., Phys. Rev. D **90**, 052010 (2014)
[19] K.A.Olive et al. Chin. Phys. C, **38**(9), 090001, (2014)
[20] P.Adamson et al. ,Phys. Rev. Lett. **108**, 191801 (2012)

EPJ Web of Conferences **126**, 02010 (2016) DOI: 10.1051/epjconf/201612602010
ICNFP 2015

Perspectives for Higgs and New Physics

Abdelhak Djouadi[1,a]

[1] *Laboratoire de Physique Théorique, CNRS & Université Paris–Sud, 91405 Orsay, France*

Abstract. The implications of the discovery of a Higgs boson at the LHC with a mass of 125 GeV are summarised in the context of the Standard Model of particle physics and in new physics scenarios beyond it, taking the example of the minimal supersymmetric Standard Model extension, the MSSM. The perspectives for Higgs and new physics searches at the next LHC upgrades as well as at future hadron and lepton colliders are then briefly summarized.

1 Introduction

The ATLAS and CMS historical discovery of a particle with a mass of 125 GeV [1] and properties that are compatible with those of a scalar Higgs boson [2, 3] has far reaching consequences not only for the Standard Model (SM) but also for new physics models beyond it. In the SM, electroweak symmetry breaking is achieved spontaneously via the Brout–Englert–Higgs mechanism [2], wherein the neutral component of an isodoublet scalar field acquires a non–zero vacuum expectation value v. This gives rise to nonzero masses for the fermions and the electroweak gauge bosons while preserving the SU(2)×U(1) gauge symmetry. One of the four degrees of freedom of the original isodoublet field, corresponds to a physical particle [3]: a scalar boson with $J^{PC} = 0^{++}$ quantum numbers under parity and charge conjugation. The couplings of the Higgs boson to the fermions and gauge bosons are related to the masses of these particles and are thus decided by the symmetry breaking mechanism. In contrast, the Higgs mass itself M_H, although expected to be in the vicinity of the weak scale $v \approx 250$ GeV, is undetermined. Let us summarise the known information on this parameter before the start of the LHC.

A direct information was the lower limit $M_H \gtrsim 114$ GeV at 95% confidence level (CL) established at LEP2 [4]. Furthermore, a global fit of the electroweak precision data to which the Higgs boson contributes, yields the value $M_H = 92^{+34}_{-26}$ GeV, corresponding to a 95% CL upper limit of $M_H \lesssim 160$ GeV [4]. From the theoretical side, the presence of this new weakly coupled degree of freedom is a crucial ingredient for a unitary electroweak theory. Indeed, the SM without the Higgs particle leads to scattering amplitudes of the W/Z bosons that grow with the square of the center of mass energy and perturbative unitarity would be lost at energies above the TeV scale. In fact, even in the presence of a Higgs boson, the W/Z bosons could interact very strongly with each other and, imposing the unitarity requirement leads to the important mass bound $M_H \lesssim 700$ GeV [5], implying that the particle is kinematically accessible at the LHC.

[a] e-mail: abdelhak.djouadi@cern.ch

Another theoretical constraint emerges from the fact that the Higgs self–coupling, $\lambda \propto M_H^2$, evolves with energy and at some stage, becomes very large and even infinite and the theory completely looses its predictability. If the energy scale up to which the couplings remains finite is of the order of M_H itself, one should have $M_H \lesssim$ 650 GeV [6]. On the other hand, for small values of λ and hence M_H, the quantum corrections tend to drive the self–coupling to negative values and completely destabilize the scalar Higgs potential to the point where the minimum is not stable anymore [6]. Requiring $\lambda \geq 0$, up to the TeV scale implies that $M_H \gtrsim 70$ GeV. If the SM is to be extended to the Planck scale $M_P \sim 10^{18}$ GeV, the requirements on λ from finiteness and positivity constrain the Higgs mass to lie in the range 130 GeV $\lesssim M_H \lesssim$ 180 GeV [6]. This narrow margin is close to the one obtained from the direct and indirect experimental constraints.

The discovery of the Higgs particle with a mass of 125 GeV, a value that makes the SM perturbative, unitary and extrapolable to the highest possible scales, is therefore a consecration of the model and crowns its past success in describing all experimental data available. In particular, the average mass value measured by the ATLAS and CMS teams, $M_H = 125.1 \pm 0.24$ GeV [7], is remarkably close to the best–fit of the precision data which should be considered as a great achievement and a triumph for the SM. In addition, a recent analysis that includes the state-of-the-art quantum corrections [8] gives for the condition of absolute stability of the electroweak vacuum, $\lambda(M_P) \geq 0$, the bound $M_H \gtrsim 129$ GeV for the present value of the top quark mass and the strong coupling constant, $m_t^{\rm exp} = 173.2 \pm 0.9$ GeV and $\alpha_s(M_Z) = 0.1184 \pm 0.0007$ [4]. Allowing for a 2σ variation of $m_t^{\rm exp}$, one obtains $M_H \geq 125.6$ GeV that is close to the measured M_H value [7]. In fact, for an unambiguous and well-defined determination of the top mass, one should rather use the total cross section for top pair production at hadron colliders which can unambiguously be defined theoretically; this mass has a larger uncertainty, $\Delta m_t \approx 3$ GeV, which allows more easily absolute stability of the SM vacuum up to M_P [9].

Nevertheless, the SM is far from being perfect in many respects. It does not explain the proliferation of fermions and the large hierarchy in their mass spectra and does not say much about the small neutrino masses. The SM does not unify in a satisfactory way the electromagnetic, weak and strong forces, as one has three different symmetry groups with three coupling constants which shortly fail to meet at a common value during their evolution with the energy scale; it also ignores the fourth force, gravitation. Furthermore, it does not contain a particle that could account for the cosmological dark matter and fails to explain the baryon asymmetry in the Universe.

However, the main problem that calls for beyond the SM is related to the special status of the Higgs boson which, contrary to fermions and gauge bosons has a mass that cannot be protected against quantum corrections. Indeed, these are quadratic in the new physics scale which serves as a cut–off and hence, tend to drive M_H to very large values, ultimately to M_P, while we need $M_H = \mathcal{O}(100\,\text{GeV})$. Thus, the SM cannot be extrapolated beyond $\mathcal{O}(1$ TeV) where some new physics should emerge. This is the reason why we expect something new to manifest itself at the LHC.

There are three avenues for the many new physics scenarios beyond the SM. There are first theories with extra space–time dimensions that emerge at the TeV scale (the cut–off is then not so high) and, second, composite models inspired from strong interactions also at the TeV scale (and thus the Higgs is not a fundamental spin–zero particle). Some versions of these scenarios do not incorporate any Higgs particle in their spectrum and are thus ruled out by the Higgs discovery. However, the option that emerges in the most natural way is Supersymmetry (SUSY) [10] as it solves most of the SM problems discussed above. In particular, SUSY protects M_H as the quadratically divergent radiative corrections from standard particles are exactly compensated by the contributions of their supersymmetric partners. These new particles should not be much heavier than 1 TeV not to spoil this compensation [11] and, thus, they should be produced at the LHC.

The Higgs discovery is very important for SUSY and, in particular, for its simplest low energy manifestation, the minimal supersymmetric SM (MSSM) that indeed predicts a light Higgs state. In the MSSM, two Higgs doublet fields H_u and H_d are required, leading to an extended Higgs consisting of five Higgs bosons, two CP–even h and H, a CP–odd A and two charged $H^\pm$ states [12]. Nevertheless, only two parameters are needed to describe the Higgs sector at tree–level: one Higgs mass, which is generally taken to be that of the pseudoscalar boson M_A, and the ratio of vacuum expectation values of the two Higgs fields, $\tan\beta = v_d/v_u$, expected to lie in the range $1 \lesssim \tan\beta \lesssim 60$. The masses of the CP–even h, H and the charged $H^\pm$ states, as well as the mixing angle α in the CP–even sector are uniquely defined in terms of these two inputs at tree-level, but this nice property is spoiled at higher orders [13]. For $M_A \gg M_Z$, one is in the so–called decoupling regime in which the h state is light and has almost exactly the SM–Higgs couplings, while the other CP–even H and the charged $H^\pm$ bosons become heavy, $M_H \approx M_{H^\pm} \approx M_A$, and decouple from the massive gauge bosons. In this regime, the MSSM Higgs sector thus looks almost exactly as the one of the SM with its unique Higgs boson.

Nevertheless, contrary to the SM Higgs boson, the lightest MSSM CP–even h mass is bounded from above and, depending on the SUSY parameters that enter the important quantum corrections, is restricted to $M_h^{\rm max} \lesssim$ 130 GeV [13] if one assumes a SUSY breaking scale that is not too high, $M_S \lesssim O$ (1 TeV), in order to avoid too much fine-tuning in the model. Hence, the requirement that the MSSM h boson coincides with the one observed at the LHC, i.e. with $M_h \approx$ 125 GeV and almost SM–like couplings as the LHC data seem to indicate, would place very strong constraints on the MSSM parameters, in particular the SUSY–breaking scale M_S. This comes in addition to the LHC limits obtained from the search of the heavier Higgs states and the superparticles.

In this talk, the implications of the discovery of the Higgs boson at the LHC and the measurement of its properties will be summarised and the prospects for the searches of new physics, in particular in the SUSY context, in the future will be summarized.

2 Implications for the Standard Model and Supersymmetry

In many respects, the Higgs particle was born under a very lucky star as the mass value of $\approx$ 125 GeV allows to produce it at the LHC in many redundant channels and to detect it in a variety of decay modes. This allows detailed studies of the Higgs properties.

2.1 Higgs production and decay

We start by summarizing the production and decay at the LHC of a light SM–like Higgs particle, which should correspond to the lightest MSSM h boson in the decoupling regime. First, for $M_H \approx$ 125 GeV, the Higgs mainly decays [14] into $b\bar{b}$ pairs but the decays into WW^* and ZZ^* final states, before allowing the gauge bosons to decay leptonically $W \to \ell\nu$ and $Z \to \ell\ell$ ($\ell = e, \mu$), are also significant. The $H \to \tau^+\tau^-$ channel (as well as the gg and $c\bar{c}$ decays that are not detectable at the LHC) is also of significance, while the clean loop induced $H \to \gamma\gamma$ mode can be easily detected albeit its small rates. The very rare $H \to Z\gamma$ and even $H \to \mu^+\mu^-$ channels should be accessible at the LHC but only with a much larger data sample.

On the other hand, many Higgs production processes have significant cross sections [15–17]. While the by far dominant gluon fusion mechanism $gg \to H$ (ggF) has extremely large rates ($\approx$20 pb at $\sqrt{s}$ = 7–8 TeV), the subleading channels, i.e. the vector boson fusion (VBF) $qq \to Hqq$ and the Higgs–strahlung (HV) $q\bar{q} \to HV$ with $V = W, Z$ mechanisms, have cross sections which should allow for a study of the Higgs particle already at $\sqrt{s} \gtrsim 7$ TeV with the $\approx$ 25 fb^{-1} data collected by each experiment. The associated process $pp \to t\bar{t}H$ (ttH) would require higher energy and luminosity.

This pattern already allows the ATLAS and CMS experiments to observe the Higgs boson in several channels and to measure some of its couplings in a reasonably accurate way. The channels that have been searched are $H \to ZZ^* \to 4\ell^\pm, H \to WW^* \to 2\ell 2\nu, H \to \gamma\gamma$ where the Higgs is mainly produced in ggF with subleading contributions from Hjj in the VBF process, $H \to \tau\tau$ where the Higgs is produced in association with one (in ggF) and two (in VBF) jets, and finally $H \to b\bar{b}$ with the Higgs produced in the HV process. One can ignore for the moment the low sensitivity $H \to \mu\mu$ and $H \to Z\gamma$ channels.

A convenient way to scrutinize the couplings of the produced H boson is to look at their deviation from the SM expectation. One then considers for a given search channel the signal strength modifier μ which for the $H \to XX$ decay mode measures the deviation compared to the SM expectation of the Higgs production cross section times decay branching fraction μ_{XX}. ATLAS and CMS have provided the signal strengths for the various final states with a luminosity of $\approx$ 5 fb^{-1} for the 2011 run at $\sqrt{s} = 7$ TeV and $\approx$ 20 fb^{-1} for the 2012 run at $\sqrt{s} = 8$ TeV. The constraints given by the two collaborations, when combined, lead to a global signal strength $\mu_{\rm tot}^{\rm ATLAS} = 1.18 \pm 0.15$ and $\mu_{\rm tot}^{\rm CMS} = 1.00 \pm 0.14$ [7]. The global value being very close to unity implies that the observed Higgs is SM–like. Hence, already with the rather limited statistics at hand, the accuracy of the ATLAS and CMS measurements is reaching the 15% level.

This is at the same time impressive and worrisome. Indeed, the main Higgs production channel is the top and bottom quark loop mediated gluon fusion mechanism and, at $\sqrt{s} = 7$ or 8 TeV, the three other mechanisms contribute at a total level below 15%. The majority of the signal events observed at LHC, in particular in the search channels $H \to \gamma\gamma, H \to ZZ^* \to 4\ell, H \to WW^* \to 2\ell 2\nu$ and to some extent $H \to \tau\tau$, thus come from the ggF mechanism which is known to be affected by large theoretical uncertainties.

Indeed, although $\sigma(gg \to H)$ is known up next–to–next–to–leading order (NNLO) in perturbative QCD (and at least at NLO for the electroweak interaction) [15, 16], there is a significant residual scale dependence which points to the possibility that still higher order contributions cannot be totally excluded. In addition, as the process is of $O(\alpha_s^2)$ at LO and is initiated by gluons, there are sizable uncertainties due to the gluon parton distribution function (PDF) and the value of the coupling α_s. A third source of theoretical uncertainties, the use of an effective field theory (EFT) approach to calculate the radiative corrections beyond NLO should also be considered [15]. In addition, large uncertainties arise when $\sigma(gg \to H)$ is broken into the jet categories $H+0j, H+1j$ and $H+2j$ [18]. In total, the combined theoretical uncertainty is estimated to be $\Delta^{\rm th} \approx \pm 15\%$ [16] and would increase to $\Delta^{\rm th} \approx \pm 20\%$ if the EFT uncertainty is also included. The a priori cleaner VBF process will be contaminated by the $gg \to H+2j$ mode making the total uncertainty in the $H+jj$ "VBF" sample also rather large [18].

Hence, the theoretical uncertainty is already at the level of the accuracy of the cross section measured by the ATLAS and CMS collaborations. Another drawback of the analyses is that they involve strong theoretical assumptions on the total Higgs width since some contributing decay channels not accessible at the LHC are assumed to be SM–like and possible invisible Higgs decays in scenarios beyond the SM do not to occur.

In Ref. [17], following earlier work [19] it has been suggested to consider the ratio $D_{XX}^{\rm p} = \sigma^{\rm p}(pp \to H \to XX)/\sigma^{\rm p}(pp \to H \to VV)$ for a specific production process p and for a given decay channel $H \to XX$ when the reference channel $H \to VV$ is used. In these ratios, the cross sections and hence, their significant theoretical uncertainties will cancel out, leaving out only the ratio of partial decay widths which are better known. The total decay width which includes contributions from channels not under control such as possible invisible Higgs decays, do not appear in the ratios $D_{XX}^{\rm p}$. Some common experimental systematical uncertainties such as the one from the luminosity measurement and the small uncertainties in the Higgs decay branching ratios also cancel out. We are thus left with only with the statistical and some (non common) systematical errors [17].

The ratios D_{XX} involve, up to kinematical factors and known radiative corrections, only the ratios $|c_X|^2/|c_V|^2$ of the Higgs reduced couplings to the particles X and V compared to the SM expectation, $c_X \equiv g_{HXX}/g^{\rm SM}_{HXX}$. For the time being, three independent ratios can be considered: $D_{\gamma\gamma}, D_{\tau\tau}$ and D_{bb}. In order to determine these ratios, the theoretical uncertainties have to be treated as a bias (and not as if they were associated with a statistical distribution) and the fit has to be performed for the two μ extremal values: $\mu_i|_{\rm exp} \pm \delta\mu_i/\mu_i|_{\rm th}$ with $\delta\mu_i/\mu_i|_{\rm th} \approx \pm 20\%$ [20].

A large number of analyses of the Higgs couplings from the LHC data have been performed and in most cases, it is assumed that the couplings of the Higgs boson to the massive W, Z gauge bosons are equal to $g_{HZZ} = g_{HWW} = c_V$ and the couplings to all fermions are also the same $g_{Hff} = c_f$. However, as for instance advocated in Ref. [21] to characterize the Higgs particle at the LHC, at least three independent H couplings should be considered, namely c_t, c_b and c_V. While the couplings to W, Z, b, τ particles are derived by considering the decays of the Higgs boson to these particles, the $Ht\bar{t}$ coupling is derived indirectly from $\sigma(gg \to H)$ and BR$(H \to \gamma\gamma)$, two processes that are generated by triangular loops involving the top quarks in the SM. One can assume, in a first approximation, that $c_c = c_t$ and $c_\tau = c_b$ and possible invisible Higgs decays are absent. In Ref. [21], a three–dimensional fit of the H couplings was performed in the space $[c_t, c_b, c_V]$, when the theory uncertainty is taken as a bias and not as a nuisance. The best-fit value for the couplings, with the $\sqrt{s}$ = 7+8 TeV ATLAS and CMS data turns out to be $c_t = 0.89$, $c_b = 1.01$ and $c_V = 1.02$, ie very close to the SM values.

2.2 Implications of the Higgs couplings measurement in the SM

The precise measurements of the Higgs couplings allow to draw several important conclusions.

i) A fourth generation fermions is excluded. Indeed, in addition to the direct LHC searches that exclude heavier quarks $m_{b'}, m_{t'} \lesssim 600$ GeV [23], strong constraints can be also obtained from the loop induced Higgs–gluon and Higgs-photon vertices in which any heavy particle coupling to the Higgs proportionally to its mass will contribute. For instance the additional 4th generation t' and b' contributions increase $\sigma(gg \to H)$ by a factor of ≈ 9 at LO but large $O(G_F m^2_{f'})$ electroweak corrections should be considered. It has been shown [23] that with a fourth family, the Higgs signal would have not been observable and the obtained Higgs results unambiguously rule out this possibility.

ii) The invisible Higgs decay width should be small. Invisible decays would affect the properties of the observed Higgs boson and could be constrained if the total decay width is determined. But for a 125 GeV Higgs, $\Gamma^{\rm tot}_H = 4$ MeV, is too small to be resolved experimentally. Nevertheless, in $pp \to VV \to 4f$, a large fraction of the Higgs cross section lies in the high–mass tail [24] allowing to to put loose constrains $\Gamma^{\rm tot}_H/\Gamma^{\rm SM}_H \approx 5$–10 [25]. The invisible Higgs decay width $\Gamma^{\rm inv}_H$ can be better constrained indirectly by a fit of the Higgs couplings and in particular with the signal strength in the $H \to ZZ$ process: $\mu_{ZZ} \propto \Gamma(H \to ZZ)/\Gamma^{\rm tot}_H$ with $\Gamma^{\rm tot}_H = \Gamma^{\rm inv}_H + \Gamma^{\rm SM}_H$; one obtains $\Gamma^{\rm inv}_H/\Gamma^{\rm SM}_H \lesssim 50\%$ at the 95% CL if the assumption $c_f = c_V = 1$ is made [20].

A more model independent approach would be to perform direct searches for missing transverse energy. These have been conducted in $pp \to HV$ with $V \to jj, \ell\ell$ and in VBF, $qq \to qq\not{E}_T$ leading to BR$_{\rm inv} \lesssim 50\%$ at 95%CL for SM–like Higgs couplings [7]. A more promising search for invisible decays is the monojet channel $gg \to Hj$ which has large rates [26]. While the most recent monojet ATLAS and CMS searches are only sensitive to BR$_{\rm inv} \sim 1$, more restrictive results can be obtained in the future.

The Higgs invisible rate and the dark matter detection rate in direct astrophysical searches are correlated in Higgs portal models and it turns out that LHC constraints are competitive [27] with those derived from direct dark matter search experiments [28].

iii) The spin–parity quantum numbers are those of a standard Higgs. One also needs to establish that the observed Higgs state is indeed a CP even scalar and hence with $J^{PC} = 0^{++}$ quantum numbers.

For the spin, the observation of the $H \to \gamma\gamma$ decay rules out the spin–1 case [29]. The Higgs parity can be probed by studying kinematical distributions in the $H \to ZZ^* \to 4\ell$ decay channel and in the VH and VBF production modes [30] and with the 25 fb^{-1} data collected so far, ATLAS and CMS found that the observed Higgs is more compatible with a 0^+ state and the 0^- possibility is excluded at the 98%CL [7]. Other useful diagnostics of the Higgs CP nature that also rely on the tensorial structure of the HVV coupling can be made in the VBF process [31]. Nevertheless, there is a caveat in the analyses relying on the HVV couplings: a CP–odd state has no tree–level VV couplings [32]. In fact, a better way to measure the Higgs parity is to study the signal strength in the $H \to VV$ channels and in Ref. [20] it was demonstrated that the observed Higgs has indeed a large CP component, $\gtrsim 50\%$ at the 95%CL. In fact, the less unambiguous way to probe the Higgs CP nature would be to look at final states in which the particle decays hadronically, e.g. $pp \to HZ \to b\bar{b}\ell\ell$ [32]. These processes are nevertheless extremely challenging even at the upgraded LHC.

2.3 Implications for Supersymmetry

We turn now to the implications of the LHC Higgs results for the MSSM Higgs sector and first make a remark on the Higgs masses and couplings, which at tree–level depend only on M_A and $\tan\beta$, when the important radiative corrections are included. In this case many parameters such as the masses of the third generation squarks $m_{\tilde{t}_i}, m_{\tilde{b}_i}$ and their trilinear couplings A_t, A_b enter M_h and M_H through quantum corrections. These are introduced by a general 2×2 matrix $\Delta\mathcal{M}^2_{ij}$ but the leading one is controlled by the top Yukawa coupling and is proportional to m_t^4, $\log M_S$ with $M_S = \sqrt{m_{\tilde{t}_1} m_{\tilde{t}_2}}$ the SUSY–breaking scale and the stop mixing parameter X_t [13]. The maximal value $M_h^{\rm max}$ is then obtained for a decoupling regime $M_A \sim \mathcal{O}(\text{TeV})$, large $\tan\beta$, large M_S that implies heavy stops and maximal mixing $X_t = \sqrt{6}M_S$ [33]. If the parameters are optimized as above, the maximal M_h value reaches the level of 130 GeV.

It was pointed out in Refs. [21, 34, 35] that when the measured value $M_h = 125$ GeV is taken into account, the MSSM Higgs sector with only the largely dominant correction discussed above, can be again described with only the two parameters $\tan\beta$ and M_A; in other words, the loop corrections are fixed by the value of M_h. This observation leads to a rather simple but accurate parametrisation of the MSSM Higgs sector, called hMSSM.

The reduced couplings of the CP–even h state (as is the case for the heavier H) depend in principle only on the angles β and α (and hence $\tan\beta$ and M_A), $c_V^0 = \sin(\beta-\alpha)$, $c_t^0 = \cos\alpha/\sin\beta$, $c_b^0 = -\sin\alpha/\cos\beta$, while the couplings of A and $H^\pm$ (as well as H in the decoupling regime) to gauge boson are zero and those to fermions depend only on β: for $\tan\beta > 1$, they are enhanced ($\propto \tan\beta$) for b, τ and suppressed ($\propto 1/\tan\beta$) for tops.

i) Implications from the Higgs mass value: In the so–called "phenomenological MSSM" (pMSSM) [37] in which the model involves only 22 free parameters, a large scan has been performed [36] using the RGE program Suspect [38] that calculates the maximal M_h value and the result confronted to the measured mass $M_h \sim 125$ GeV. For $M_S \lesssim 1$ TeV, only scenarios with X_t/M_S values close to maximal mixing $X_t/M_S \approx \sqrt{6}$ survive. The no–mixing scenario $X_t \approx 0$ is ruled out for $M_S \lesssim 3$ TeV, while the typical mixing scenario, $X_t \approx M_S$, needs large M_S and moderate to large $\tan\beta$ values. In constrained MSSM scenarios (cMSSM) such the minimal supergravity (mSUGRA) model and the gauge and anomaly mediated SUSY–breaking scenarios, GMSB and AMSB, only a few basic inputs are needed and the mixing parameter cannot take arbitrary values. A scan in these models with $M_S \lesssim 3$ TeV not to allow for too much fine-tuning [11] leads $M_h^{\rm max} \lesssim 122$ GeV in AMSB and GMSB thus disfavoring these scenarios while one has $M_h^{\rm max} = 128$ GeV in mSUGRA. In high–scale SUSY scenarios, $M_S \gg M_Z$, the radiative corrections are very large and need to be resumed [39]. For low

$\tan\beta$ values, large scales, at least $M_S \gtrsim 10^4$ GeV, are required to obtain $M_h = 125$ GeV and even higher in most cases

ii) Implications from the production rates of the observed state. Besides the corrections to the Higgs masses and couplings discussed above, there are also direct corrections to the Higgs couplings and the most ones are those affecting the $hb\bar{b}$ vertex [40] and the stop loop contributions to the $gg \to h$ production and $h \to \gamma\gamma$ decay rates [41]. A fit of the c_t, c_b and c_V couplings shows that the latter are small [20]. In turn, ignoring the direct corrections and using the input $M_h \approx 125$ GeV, one can make a fit in the plane $[\tan\beta, M_A]$. The best-fit point is $\tan\beta = 1$ and $M_A = 550$ GeV which implies a large SUSY scale, $M_S = \mathcal{O}(100)$ TeV. In all, cases one also has $M_A \gtrsim 200$–350 GeV.

iii) Implications from heavy Higgs boson searches. At high $\tan\beta$ values, the strong enhancement of the b, τ couplings makes that the $\Phi = H/A$ states decay dominantly into $\tau^+\tau^-$ and $b\bar{b}$ pairs and are mainly produced in $gg \to \Phi$ fusion with the b–loop included and associated production with b–quarks, $gg/q\bar{q} \to b\bar{b} + \Phi$ [42]. The most powerful LHC search channel is thus $pp \to gg + b\bar{b} \to \Phi \to \tau^+\tau^-$. For the charged Higgs, the dominant mode is $H^\pm \to \tau\nu$ with the $H^\pm$ light enough to be produced in top decays $t \to H^+ b \to \tau\nu b$. In the low $\tan\beta$ regime, $\tan\beta \lesssim 3$, the phenomenology of the $A, H, H^\pm$ states is richer [34]. For the production, only $gg \to \Phi$ process with the dominant t and sub-dominant b contributions provides large rates. The $H/A/H^\pm$ decay pattern is in turn rather involved. Above the $t\bar{t}$ (tb) threshold $H/A \to t\bar{t}$ and $H^+ \to t\bar{b}$ are by far dominant. Below threshold, the $H \to WW, ZZ$ decays are significant. For $2M_h \lesssim M_H \lesssim 2m_t$ ($M_A \gtrsim M_h + M_Z$), $H \to hh$ ($A \to hZ$) is the dominant $H(A)$ decay mode. But the $A \to \tau\tau$ channel is still important with rates $\gtrsim 5\%$. In the case of $H^\pm$, the channel $H^+ \to Wh$ is important for $M_{H^\pm} \lesssim 250$ GeV, similarly to the $A \to hZ$ case.

In Ref. [34] an analysis of these channels has been performed using current information given by ATLAS and CMS in the context of the SM, MSSM [43] or other scenarios. The outcome is impressive. The ATLAS and CMS $H/A \to \tau^+\tau^-$ constraint is extremely restrictive and $M_A \lesssim 250$ GeV, it excludes almost the entire intermediate and high $\tan\beta$ regimes. The constraint is less effective for a heavier A but even for $M_A \approx 400$ GeV the high $\tan\beta \gtrsim 10$ region is excluded and one is even sensitive to $M_A \approx 800$ GeV for $\tan\beta \gtrsim 50$. For $H^\pm$, almost the entire $M_{H^\pm} \lesssim 160$ GeV region is excluded by the process $t \to H^+ b$ with the decay $H^+ \to \tau\nu$. The other channels, in particular $H \to VV$ and $H/A \to t\bar{t}$, are very constraining as they cover the entire low $\tan\beta$ area that was previously excluded by the LEP2 bound up to $M_A \approx 500$ GeV. Even $A \to hZ$ and $H \to hh$ would be visible at the current LHC in small portions of the parameter space.

3 Perspectives for Higgs and New Physics

The last few years were extremely rich and exciting for particle physics. With the historical discovery of a Higgs boson by the LHC collaborations ATLAS and CMS, crowned by a Nobel prize in fall 2013, and the first probe of its basic properties, they witnessed a giant step in the unraveling of the mechanism that breaks the electroweak symmetry and generates the fundamental particle masses. They promoted the SM as the appropriate theory, up to at least the Fermi energy scale, to describe three of Nature's interactions, the electromagnetic, weak and strong forces. However, it is clear that these few years have also led to some frustration as no signal of physics beyond the SM has emerged from the LHC data. The hope of observing some signs of the new physics models that were put forward to address the hierarchy problem, that is deeply rooted in the Higgs mechanism, with Supersymmetric theories being the most attractive ones, did not materialize.

The Higgs discovery and the non–observation of new particles has nevertheless far reaching consequences for supersymmetric theories and, in particular, for their simplest low energy formulation, the MSSM. The mass of approximately 125 GeV of the observed Higgs boson implies that the scale of SUSY–breaking is rather high, at least $\mathcal{O}$(TeV). This is backed up by the limits on the masses of strongly interacting SUSY particles set by the ATLAS and CMS searches, which in most cases exceed the TeV range. This implies that if SUSY is indeed behind the stabilization of the Higgs mass against very high scales that enter via quantum corrections, it is either fine–tuned at the permille level at least or its low energy manifestation is more complicated than expected.

The production and decay rates of the observed Higgs particles, as well as its spin and parity quantum numbers, as measured by ATLAS and CMS with the $\approx$ 25 fb^{-1} data collected at $\sqrt{s}=7+8$ TeV, indicate that its couplings to fermions and gauge bosons are almost SM–like. In the context of the MSSM, this implies that we are close to the decoupling regime and this particle is the lightest h boson, while the other $H/A/H^{\pm}$ states must be heavier than approximately the Fermi scale. This last feature is also backed up by LHC direct searches of these heavier Higgs states.

This drives up to the question that is now very often asked: what to do next? The answer is, for me, obvious: we are only in the beginning of a new era. Indeed, it was expected since a long time that the probing of the electroweak symmetry breaking mechanism will be at least a two chapters story. The first one is the search and the observation of a Higgs–like particle that will confirm the scenario of the SM and most of its extensions, that is, a spontaneous symmetry breaking by a scalar field that develops a non–zero vev. This long chapter has just been closed by the ATLAS and CMS collaborations with the spectacular observation of a Higgs boson. This observation opens a second and equally important chapter: the precise determination of the Higgs profile and the unraveling of the electroweak symmetry breaking mechanism itself.

A more accurate measurement of the Higgs couplings to fermions and gauge bosons will be mandatory to establish the exact nature of the mechanism and, eventually, to pin down effects of new physics if additional ingredients beyond those of the SM are involved. This is particularly true in weakly interacting theories such as SUSY in which the quantum effects are expected to be small. These measurements could be performed at the upgraded LHC with an energy close to $\sqrt{s}=14$ TeV, in particular if a very high luminosity, a few ab^{-1}, is achieved [43, 44].

At this upgrade, besides improving the measurements performed so far, rare but important channels such as associated Higgs production with top quarks, $pp\to t\bar{t}H$, and Higgs decays into $\mu^+\mu^-$ and $Z\gamma$ states could be probed. Above all, a determination of the self–Higgs coupling could be made by searching for double Higgs production e.g. in the gluon fusion channel $gg\to HH$ [45]; this would be a first step towards the reconstruction of the scalar potential that is responsible of electroweak symmetry breaking. This measurement would be difficult at the LHC even with high–luminosity but a proton collider with an energy $\sqrt{s}=30$ to 100 TeV could do the job [44].

In a less near future, a high–energy lepton collider, which is nowadays discussed in various options (ILC, TLEP, CLIC, μ–collider) would lead to a more accurate probing of the Higgs properties [46], promoting the scalar sector to the very high–precision level of the gauge and fermion sectors achieved by the LEP and SLC colliders in the 1990s [4].

At electron-positron colliders, the process $e^+e^-\to HZ$ just looking at the recoiling Z boson allows to measure the Higgs mass, the CP parity and the absolute HZZ coupling, allowing to derive the total decay width Γ_H^{tot}. One can then measure precisely, already at $\sqrt{s}\approx 250$ GeV where $\sigma(e^+e^-\to HZ)$ is maximal, the absolute Higgs couplings to gauge bosons and light fermions from the decay branching ratios. The important couplings to top quarks and the Higgs self–couplings can measured at the 10% level in the higher-order processes $e^+e^-\to t\bar{t}H$ and $e^+e^-\to HHZ$ at energies of at least 500 GeV with a high–luminosity.

Besides the high precision study of the already observed Higgs, one should also continue to search for the heavy states that are predicted by SUSY, not only the superparticles but also the heavier Higgs bosons. The energy upgrade to $\approx$ 14 TeV (and eventually beyond) and the planed order of magnitude (or more) increase in luminosity will allow to probe much higher mass scales than presently. In fact, more generally, one should continue to search for any sign of new physics or new particles, new gauge bosons and fermions, as predicted in most of the SM extensions.

In conclusion, it is not yet time to give up on SUSY and more generally on New Physics but, rather, to work harder to be fully prepared for the more precise and larger data set that will be delivered by the upgraded LHC. It will be soon enough to "philosophize" then as the physics landscape will become more clear.

Acknowledgements
I would like the thank the organisers for their invitation to the conference and their kind hospitality. This work is supported by the ERC Advanced Grant Higgs@LHC.

References

[1] ATLAS coll., Phys. Lett. B716 (2012) 1; CMS coll., Phys. Lett. B716 (2012) 30.
[2] F. Englert and R. Brout, Phys. Rev. Lett. 13 (1964) 321; P. Higgs, Phys. Rev. Lett. 13 (1964) 508; P. Higgs, Phys. Lett. 12 (1964) 132; G. Guralnik, C. Hagen, T. Kibble, Phys. Rev. Lett. 13 (1964) 585.
[3] For a review of Higgs physics in the SM, see: A. Djouadi, Phys. Rept. 457 (2008) 1.
[4] K. Olive et al., Particle Data Group, Chin. Phys. C38 (2014) 090001.
[5] B.W. Lee, C. Quigg and H.B. Thacker, Phys. Rev. D16 (1977) 1519.
[6] N. Cabibbo et al., Nucl. Phys. B158 (1979) 295; M. Sher, Phys. Rept. 179 (1989) 273; T. Hambye and K. Riesselmann, Phys. Rev. D55 (1997) 7255; J. Ellis et al., Phys. Lett. B679 (2009) 369.
[7] Updated analyses of the Higgs sector can be found in the slides of the talks given at this conference and at the following Moriond Electroweak conference, March 2015.
[8] G. Degrassi et al., JHEP 1208 (2012) 098; F. Bezrukov et al., JHEP 1210 (2012) 140.
[9] S. Alekhin, A. Djouadi and S. Moch, Phys. Lett. B716 (2012) 214.
[10] M. Drees, R. Godbole, P. Roy, *Theory and phenomenology of sparticles*, World Sci., 2005.
[11] E. Witten, Nucl. Phys. B188 (1981) 513; R. Barbieri and G. Giudice, Nucl. Phys. B306 (1988) 63; M. Papucci, J. Ruderman and A. Weiler, JHEP 1209 (2012) 035.
[12] A. Djouadi, Phys. Rept. 459 (2008) 1; see also Eur. Phys. J. C74 (2014) 2704 for an update.
[13] S. Heinemeyer, W. Hollik and G. Weiglein, Phys. Rept. 425 (2006) 265; M. Carena and H. Haber, Prog. Part. Nucl.P hys. 50 (2003) 63; B. Allanach et al., JHEP 0409 (2004) 044.
[14] A. Djouadi, J. Kalinowski and M. Spira, Comput. Phys. Commun. 108 (1998) 56; A.Djouadi, M.Muhlleitner and M.Spira, Acta. Phys. Polon. B38 (2007) 635.
[15] J. Baglio and A. Djouadi, JHEP 1103 (2011) 055; see also: JHEP 1010 (2010) 064; J. Baglio, A. Djouadi and R. Godbole, Phys. Lett. B716 (2012) 203.
[16] S. Dittmaier et al., the LHC Higgs cross section working group, arXiv:1101.0593 [hep-ph].
[17] A. Djouadi, Eur. Phys. J. C73 (2013) 2498; A. Djouadi, J. Quevillon and R. Vega-Morales, arXiv:1509.03913 [hep-ph].
[18] S. Dittmaier et al., the LHC Higgs cross section working group: arXiv:1201.3084 [hep-ph].
[19] D. Zeppenfeld et al. Phys. Rev. D62 (2000) 013009; A. Djouadi et al., hep-ph/0002258; M. Dührssen et al., Phys. Rev. D70 (2004) 113009.
[20] A. Djouadi and G. Moreau, Eur. Phys. J. C73 (2013) 2512.
[21] A. Djouadi, L. Maiani, G. Moreau, A. Polosa, J. Quevillon, V. Riquer, Eur. Phys. J. C73 (2013) 2650; JHEP 1506 (2015) 168.
[22] A. Djouadi and A. Lenz, Phys. Lett. B715 (2012) 310; E. Kuflik, Y. Nir and T. Volansky, Phys. Rev.Lett. 110 (2013) 091801; A. Denner et al., Eur.Phys.J. C72 (2012) 1992; A. Djouadi, P. Gam-

bino, B. Kniehl, Nucl. Phys. B523 (1998) 17; A. Djouadi and P. Gambino, Phys. Rev. Lett. 73 (1994) 2528.
[23] CMS coll., Phys. Rev. D86 (2012) 112003; ATLAS coll., Phys. Rev. Lett. 109 (2012) 032001.
[24] See e.g., N. Kauer and G. Passarino, JHEP 1208 (2012) 116.
[25] ATLAS coll., ATLAS-CONF-2014-042; CMS coll., Phys. Lett. B736 (2014) 64.
[26] A. Djouadi, A. Falkowski, Y. Mambrini and J. Quevillon, Eur. Phys. J. C73 (2013) 2455.
[27] A. Djouadi, O. Lebedev, Y. Mambrini and J. Quevillon, Phys. Lett. B709 (2012) 65.
[28] M. Goodman and E. Witten, Phys. Rev. D31 (1985) 3059; M. Drees and M. Nojiri, Phys. Rev. D48 (1993) 3483; A. Djouadi and M. Drees, Phys. Lett. B484 (2000) 183.
[29] L. Landau, Dokl. Akad. Nauk Ser. Fiz. 60 (1948) 207; C. Yang, Phys. Rev. 77 (1950) 242.
[30] For a review of the CP Higgs issue, see e.g. S. Kraml (ed.) et al., hep-ph/0608079.
[31] T. Plehn, D. Rainwater and D. Zeppenfeld, Phys. Rev. Lett. 88 (2002) 051801; A. Djouadi, R. Godbole, B. Mellado, K. Mohan, Phys. Lett. B723 (2013) 307.
[32] V. Barger et al., Phys. Rev. D49 (1994) 79; B. Grzadkowski, J. Gunion and X. He, Phys. Rev. Lett. 77 (1996) 5172; J. Gunion and J. Pliszka, Phys. Lett. B444 (1998) 136; P. Bhupal Dev et al., Phys. Rev. Lett. 100 (2008) 051801; J. Ellis, V. Sanz and T. You, Phys. Lett. B726 (2013) 244.
[33] M. Carena et al., Eur. J. Phys. C26 (2003) 601; Eur. Phys. J. C73 (2013) 2552.
[34] A. Djouadi and J. Quevillon, JHEP 1310 (2013) 028.
[35] L. Maiani, A. Polosa, V. Riquer, Phys. Lett. B718 (2012) 465; ibid. B724 (2013) 274.
[36] A. Arbey et al., Phys. Lett. B708 (2012) 162; JHEP 1209 (2012) 107; Phys. Lett. B720 (2013) 153.
[37] A. Djouadi et al. (MSSM working group), hep-ph/9901246.
[38] A. Djouadi, J.L. Kneur and G. Moultaka, Comput. Phys. Commun. 176 (2007) 426; M. Muhlleitner et al., Comput. Phys. Commun. 168 (2005) 46.
[39] L. Hall and Y. Nomura, JHEP 03 (2010) 076; G. Giudice and A. Strumia, Nucl. Phys. B858 (2012) 63; N. Bernal, A. Djouadi, P. Slavich, JHEP 0707 (2007) 016.
[40] See e.g., M. Carena et al., Nucl. Phys. B577 (2000) 88.
[41] A. Djouadi et al., Phys. Lett. B435 (1998) 101; Eur. Phys. J. C1 (1998) 149; Eur. Phys. J. C1 (1998) 163; A. Arvanitaki, G. Villadoro, JHEP 02 (2012) 144; A. Delgado et al., Eur.Phys.J.C73 (2013) 2370.
[42] M. Spira et al., Phys. Lett. B264 (1991) 440; Phys. Lett. B318 (1993) 347; Nucl. Phys. B453 (1995) 17; R. Harlander and W. Kilgore, Phys. Rev. D68 (2003) 013001; R. Harlander, S. Liebler and H. Mantler, Comp. Phys. Comm. 184 (2013) 1605.
[43] ATLAS collaboration, arXiv:1307.7292; CMS collaboration, arXiv:1307.7135.
[44] S. Dawson et al., arXiv:1310.8361 [hep-ex]; N. Arkani-Hamed et al. arXiv:1511.06495 [hep-ph]; J. Baglio, A. Djouadi and J. Quevillon, arXiv:1511.07853 [hep-ph].
[45] See e.g., J. Baglio et al., JHEP 1304 (2013) 151.
[46] M. Bicer et al., arXiv:1308.6176; H. Baer et al., arXiv:1306.6352; G. Arons et al., arXiv:0709.1893; J. Aguilar-Saavedra, hep-ph/0106315; E. Accomando et al., Phys. Rept. 299 (1998) 1; A. Djouadi, Int. J. Mod. Phys. A10 (1995) 1.

EPJ Web of Conferences **126**, 02011 (2016) DOI: 10.1051/epjconf/201612602011
ICNFP 2015

Recent Results of the CMS Experiment

Tommaso Dorigo[1a] on behalf of the CMS Collaboration

[1] *INFN - Sezione di Padova*

Abstract. The CMS experiment has produced a large number of new measurements with data collected during Run 1 by the CERN Large Hadron Collider (LHC). In this report a few results in Higgs and top physics will be mentioned. After a shutdown in 2013 and 2014, the LHC restarted proton-proton collisions at the unprecedented center-of-mass energy of 13 TeV in June 2015. The data collected until August 2015 have yielded interesting events and allowed the extraction of the first results of searches at the high-energy end of mass spectra, where the effect of the higher collision energy is largest.

1 Introduction

After the 2012 discovery[1] of the Englert-Brout-Higgs boson [2–7][1], a detailed investigation of the properties of that new particle has been carried out with all the data collected in Run 1 by the Large Hadron Collider. The observable characteristics of the Higgs boson as measured by the CMS experiment currently fit perfectly to the predictions of the standard model (SM). The mining of Run 1 data in imaginative new ways has paid dividends, with the extraction of several new bits of information; in particular, differential distributions of Higgs production properties have been for the first time extracted by the analysis of CMS data. A more precise determination of the Higgs boson couplings and a study of its less frequent decay modes are foreseen with the data that has started to be collected in June 2015 at 13 TeV center-of-mass energy.

In parallel with the careful study of the Higgs boson, a number of measurements of SM processes has been systematically carried out in Run 1 data. In this report we will only mention a couple of recent results in top quark physics. The reader is referred to the public pages of the CMS experiment for a complete list of all available measurements [8].

The search for new phenomena is of course today the main focus of the CMS experiment. While the studies of 7- and 8-TeV collisions have revealed no significant departures from SM predictions, the range of possible signals of new physics hiding in the huge datasets so far collected is quite broad, and surprises are still possible. A mention of the many results produced by those searches is impossible in this brief report; a summary can be found in the reports from the parallel sessions in these proceedings.

This document is organized as follows. Section 2 provides a brief description of the experiment. In Section 3 is offered a short review of the most interesting results of the investigations of the nature of the new 125 GeV scalar particle. Section 4 describes two recent top physics results using Run 1 data. In Section 5 are highlighted a few recent results from the analysis of the 13 TeV collisions produced by the Large Hadron Collider since June 2015 . Some conclusions are offered in Section 6.

[a] e-mail: dorigo@pd.infn.it

[1] In the following, for brevity we will address the particle simply as "Higgs boson".

 DOI: 10.1051/epjconf/201612602011

2 The CMS Detector and the Datasets

CMS –an acronym for Compact Muon Solenoid– is a multi-purpose magnetic detector designed to study the multi-TeV proton-proton collisions delivered by the CERN Large Hadron Collider. The detector is located in a underground cavern at a depth of 100m in Cessy, near the border of France and Switzerland. Particles emitted in hard collisions at the center of CMS cross in succession a silicon tracker, electromagnetic and hadron calorimeters, a solenoid magnet, and muon drift chambers embedded in the solenoid iron return yoke.

2.1 Overview of the CMS Detector

The momenta of charged particles emitted in the collisions at the center of the CMS detector are measured using a 13-layer silicon pixel and strip tracker; 66 million silicon pixels of dimensions 100x150 μm are arranged in three barrel layers, and are surrounded by 9.6 million 180 μm-wide silicon strips arranged in additional concentric barrels in the central region and disks in the endcap region. In order to allow a precise measurement of charged particle momenta, the silicon tracker is immersed in the 3.8 T axial field produced by a superconducting solenoid. The tracker covers the pseudorapidity [2] range $|\eta| < 2.5$.

Surrounding the tracker are an electromagnetic calorimeter (ECAL) composed of lead tungstate crystals and a brass-scintillator hadron calorimeter (HCAL). These detectors are used to measure the energy of incident particles from the produced electromagnetic and hadronic cascades; they consist of a barrel assembly covering the central region, plus two endcaps covering the solid angle for particles emitted at lower angle with respect to the beams direction. The ECAL and HCAL cover the pseudorapidity range $|\eta| < 3.0$; at still smaller angles particles emitted in the collision encounter a steel/quartz-fiber Cherenkov forward detector (HF) which extends the calorimetric coverage to $|\eta| < 5.0$.

The outermost component of the CMS detector is the muon system, whose main component is a set of four layers of gas detectors placed within the steel return yoke. The muon system performs a high-purity identification of muon candidates and a stand-alone measurement of their momentum, and in combination with the inner tracker information it provides a high-resolution determination of muon kinematics. More detail on the CMS detector is provided elsewhere [9].

CMS collects data with a two-level trigger system. Level 1 is a hardware trigger based on custom-made electronic processors that receive as input a coarse readout of the calorimeters and muon detectors and perform a preliminary selection of the most interesting events for data analysis, with an output rate of about 100 kHz. Level 2, also called "High-Level Trigger" (HLT), uses fine-grained information from all sub-detectors in the regions of interest identified by Level 1 to produce a final decision, selecting events at a rate of about 300 Hz by means of speed-optimized software algorithms running on commercial computers.

2.2 Run 1 and Run 2

Many of the results discussed in the following sections are based on the full data samples collected in 2011 and 2012, a period referred to as Run 1. The 2011 proton-proton run of the LHC produced collisions at the center-of-mass energy of 7 TeV. In the course of seven months of data taking CMS acquired a total of 5.3 inverse femtobarns of integrated luminosity; 5.0 of these were collected with all the CMS subdetectors fully operational. In 2012 the LHC has been operating at the increased energy

[2] Pseudorapidity is defined as $\eta = -ln \tan \theta/2$, where θ is the polar angle of the particle trajectory with respect to the direction of the counter-clockwise proton beam.

of 4 TeV per beam; in total, CMS acquired a total of 21.8 inverse femtobarns of proton-proton collisions at a center-of-mass energy of 8 TeV. The LHC in 2012 ran at higher instantaneous luminosities than in 2011; this produced a high pileup of 20 to 50 simultaneous interactions per bunch crossing. CMS addressed the resulting reconstruction challenge with more sophisticated algorithms and calibration procedures, which have allowed to maintain the same physics output despite the harsher experimental conditions.

During the post-Run1 shutdown CMS prepared for Run 2 in several ways. The main challenge was to mitigate the effects of radiation on the performance of the tracker. This was mainly addressed by preparing the detector for operation at low temperatures (down to -20°C). The cooling plant and distribution were modified to prevent condensation. The central beampipe was also replaced by a narrower one in preparation for the installation, in 2016-17, of a new Pixel Tracker that will better measure the momenta and points of origin of charged particles. In the muon system, a fourth measuring station was added to each muon endcap in order to maintain discrimination between low-momentum muons and background as the LHC beam intensities increase. This was complemented by two 125-ton shielding structures at each end of detector, reducing neutron backgrounds. A better luminosity measurement is granted by a luminosity-measuring device, the Pixel Luminosity Telescope, which was installed on either side of the collision point around the beam pipe. Other improvements include the replacement of photo-detectors in the hadron calorimeter with better-performing designs and the moving of the muon readout to more accessible locations for maintenance. A first stage of a new hardware triggering system was also installed. Finally, the software and computing systems underwent a significant overhaul during the shutdown, to reduce the time needed to produce analysis datasets.

Run 2, started in June 2015, delivered proton-proton collisions to the experiments at the center-of-mass energy of 13 TeV. The data taking in 2015 was divided in a first period when proton bunches were separated by 50 ns as in Run 1, and a longer period of collisions with 25 ns spaced bunches. The shorter bunch spacing reduced the effect of pileup. A few preliminary results produced with 2015 data collected in June and July is discussed in Section 5.

3 Recent Results in Higgs Boson Physics

Due to its non-zero coupling to all massive particles, the Higgs boson can be produced through several different mechanisms in proton-proton collisions. CMS has studied all the dominant production processes: ones due to gluon-fusion diagrams, where a Higgs boson is emitted by a virtual top-quark or vector boson loop; vector-boson-fusion processes, where the Higgs is produced together with two high-rapidity hadronic jets that result from the emission of virtual W or Z bosons off the initial state quarks; and Higgs-strahlung diagrams, where the particle is radiated by a highly-off-shell W or Z boson or by a top quark.

The Higgs boson also exhibits a rich decay phenomenology. The decays to weak boson pairs ($H \to WW^{*}$, $H \to ZZ^{*}$) are possible when one of the two final-state objects is off-mass-shell: their branching ratios are comparatively small, which allows other decays to be observable. CMS has so far obtained significant signals from five different decay modes of the Higgs boson: the two mentioned above, as well as decays to b-quark pairs, τ-lepton pairs, and photon pairs. The latter, although quite rare (with a branching fraction of 2.3×10^{-3} in the SM), has in fact been crucial for the first observation of the Higgs boson. Other rare decay modes (*e.g.* $H \to Z\gamma$, $H \to \mu\mu$) will also become accessible to a direct measurement in the future.

Among all Higgs boson properties measured by CMS in Run 1, in this brief report we single out the Higgs mass. Thanks to very careful measurements exploiting both in the ZZ and the $\gamma\gamma$ decay modes,

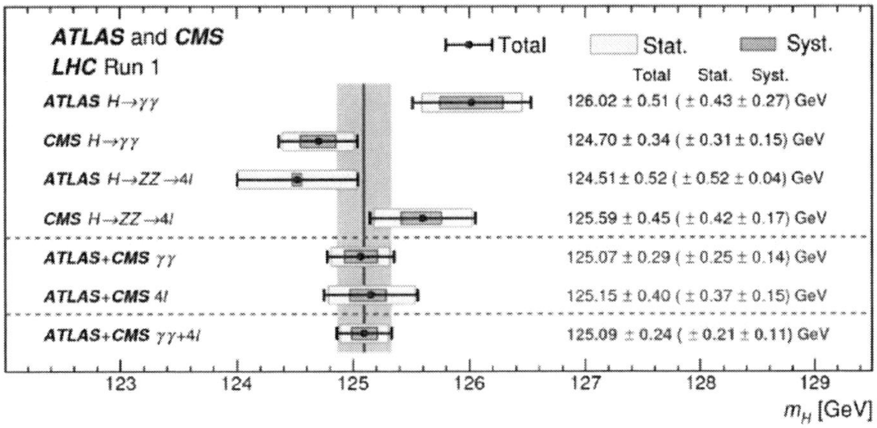

Figure 1. *Measurements of the Higgs boson mass by ATLAS and CMS in the diphoton and ZZ final states, and their combination.*

the Higgs boson mass is determined by CMS to be $M_H = 125.02 \pm 0.27 \pm 0.15$ GeV. Once combined with slightly less precise ATLAS results, the best estimate [10] is $M_H = 125.09 \pm 0.21 \pm 0.11$ GeV; in both the measurements quoted above the first uncertainty is statistical and the second includes all systematic effects. Figure 1 shows the four independent determinations by the LHC experiments and their combination results.

One recent result is worth mentioning here. Higgs boson decay candidates have been used to study differential cross sections of Higgs production in a fiducial region of the event kinematics [11]. The restriction of the measurement to a region of phase space matching the detector acceptance reduces model systematics. Cross sections have been derived in the fiducial volume at 7 and 8 TeV independently, by maximum likelihood fits to the data assuming a Higgs boson mass $M_H = 125$ GeV. The results are compared to predictions at next-to-next-to-leading order (NNLO) plus next-to-next-to-leading-logarithm (NNLL) from several generators [12–16].

Besides the cross section at 7 and 8 TeV, the considered variables for differential production cross section studies are the transverse momentum and rapidity of the four leptons in $H \to ZZ^* \to 4l$ decays, and the kinematics of the Higgs boson with respect to the leading jet. Results match nicely the precise theoretical predictions at NNLO in all considered distributions. Figure 2 (left) shows for example the measured Higgs boson cross section versus Higgs transverse momentum.

Higgs bosons have also started to be used in searches for new physics, by considering resonances that may decay to Higgs boson pairs. One example [17] is the study of models inspired by warped extra dimensions, which predict the existence of spin-0 Radions X, or spin-2 Kaluza-Klein excitations of the graviton. Due to small cross sections of the predicted phenomena, the only practical Higgs decay mode to exploit in these searches is currently the $H \to b\bar{b}$ one. Searches for resonant $X \to HH$ production and decay to $b\bar{b}b\bar{b}$ final states have been performed by CMS in the non-boosted regime ($M_X < 1$ TeV). Decays of heavy objects into HH pairs result in highly boosted b-quark jets that at detector level merge in a single observed "fat jet". These topologies can now be spotted and separated from backgrounds using jet substructure techniques. In particular, the two-subjet mass can be effectively used to distinguish H decays from generic jets. B-tagging is also used to reduce the irreducible QCD background: secondary vertices are sought in the sub-jets, or in the fat jet if the

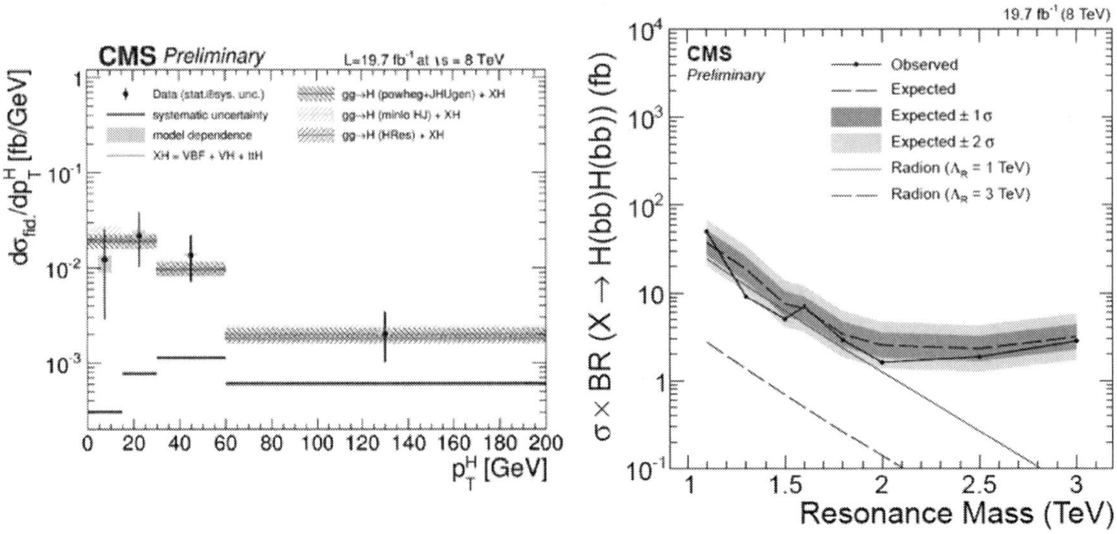

Figure 2. *Left: differential cross section of Higgs boson production as a function of Higgs transverse momentum. The data (black points with vertical bars) is compared with the predictions of several generators. The green histogram shows the predicted contribution from vector-boson-fusion and associated production processes. Right: upper limit on the product of Radion cross section times branching fraction into two Higgs bosons, with subsequent decays to bottom-quark pairs.*

sub-jets are at a radial distance $\Delta R < 0.3$ in pseudorapidity-azimuth space. B-tagged jets with mass in the 110-135 GeV range are called "H-jets". Events are retained if both jets are H-jets, and at least one of the two leading jets is b-tagged using the tight working point of the CSV algorithm (both subjets are tagged, or the fat jet is).

To further reduce backgrounds, the τ_{21} variable, describing the two-component substructure of the jet, is used to separate H-jets into a "high-purity" (HP) and a "low-purity" (LP) category. Three event categories are then created, depending on the above classification of the candidates: HP-HP, HP-LP, and LP-LP. A data-driven fit cross-checked in control samples and MC is used to extract the signal contribution and set upper limits. Figure 2 (right) shows the combined limit on Radion cross section from the combination of the search performed in the three categories.

4 Top Quark Physics

CMS produced many groundbreaking results in top quark physics with Run 1 data. Here we only mention the determination of the mass of the top quark, which is a very important parameter in the SM due to its impact in global electroweak fits and its effect on the predicted meta-stability of the Higgs boson potential. The combination of CMS measurements of the top mass [25] yields M_t = 172.44±0.13±0.47 GeV, where the first uncertainty is statistical and the second includes all systematic uncertainties. This result is the most precise to date. The CMS determination presents a small tension with the Tevatron average [24] ($M_t = 174.34 \pm 0.37 \pm 0.52$ GeV), mostly due to the high determination produced by the DZERO experiment. A summary of recent top mass determinations and combinations by CMS is given in Fig. 3 (left).

DOI: 10.1051/epjconf/201612602011

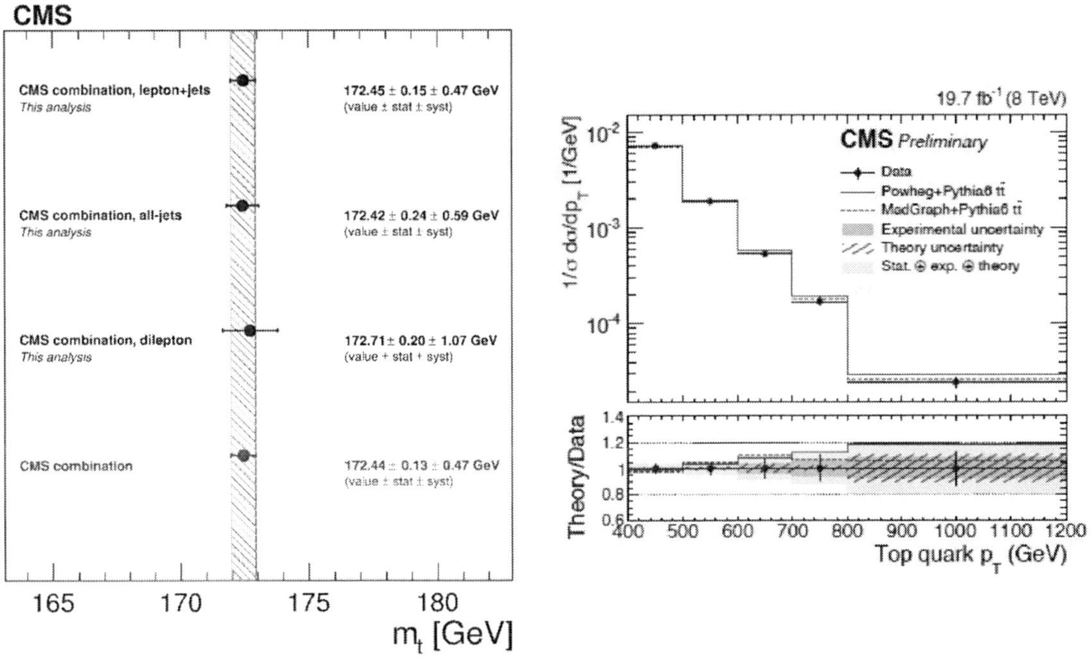

Figure 3. *Left: recent measurements of the top quark mass by the CMS experiment in the three main decay channels, and their global combination. Right: top quark p_T distribution measured by CMS with boosted objects identification techniques, compared to predictions of the Powheg generator.*

While the integrated top production cross section has been measured using all main final states and is now in the realm of precision measurements, its determination at high top-quark p_T may still offer ground for potential surprises. The high-end tail of the top quark p_T distribution has been measured by CMS [18] by searching for "boosted tops": the top quark decay products merge together as b-decay products do, creating a wide jet where substructure techniques can identify the decay and allow a reconstruction of the mass.

The boosted top technique works for $p_T > 400$ GeV, a complementary regime to the one studied with non-boosted top identification methods. A wide Cambridge-Aachen jet (CA8) is the input of the CMS hadronic top tagging algorithm in the boosted regime. This requires the presence of three subjets with all jet-pair masses above 50 GeV and a three-body-mass in the 140-250 GeV range. Opposite to the hadronic boosted top, the other top quark is required to undergo a leptonic decay. There, a regular anti-k_T (AK5) jet is used, accompanied with a lepton and neutrino signature.

A fit extracts the top pair-production signal from four independent categories of events, divided by the presence of an electron or muon in the leptonic top side and by the presence of a b-tag in the lepton-side jet. The p_T distribution of the QCD background is modeled with a data-driven technique based on a fit to the missing transverse energy distribution to obtain the normalization, plus a shape determined in a signal-poor control sample. Other backgrounds from single top and W+jets production are determined from Monte Carlo simulations. SVD unfolding is used to extract the particle-level p_T distribution of the signal component. The resulting spectrum shows a slight deficit at high p_T with respect to Powheg predictions (see Fig. 3, right).

DOI: 10.1051/epjconf/201612602011

5 A First Look at Run 2 Data

The first Run 2 data was used by CMS to verify the operational status of all detector components, checking the event reconstruction and the resolution on energy and momentum measurements using known particle signals. All the while, the highest-energy events were scrutinized carefully, as the 62.5% increase in center-of-mass energy grants large cross-section enhancements for high-mass resonances, particularly ones produced by gluon-initiated processes. Figure 4 shows an event display of an early 1-TeV dielectron event (left), and a spectacular dijet event with a jet-jet mass of 5.4 TeV (right).

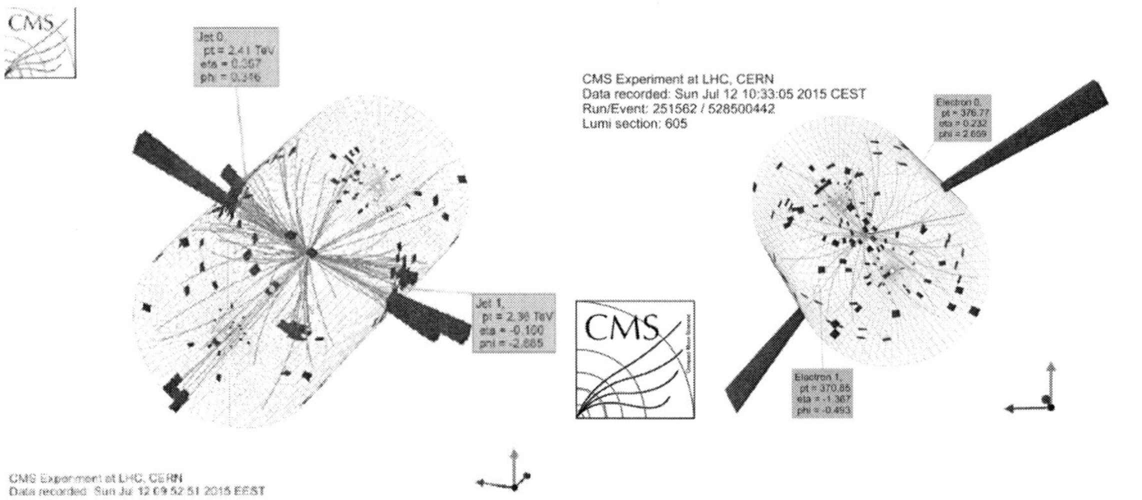

Figure 4. *Left: three-dimensional view of the energy deposits left by a high-energy dijet event collected in Run 2. The combined mass of the two leading jets is of 5.4 TeV. Right: a dielectron event with total invariant mass of 1 TeV.*

One of the first quantitative results extracted from the new data is the top pair-production cross section at 13 TeV. This measurement uses $42pb^{-1}$ of dilepton candidates in the $e\mu$ +jets final state, collected by a trigger selecting events with high-momentum electron-muon pairs. Offline, events are selected if leptons have $p_T > 20$ GeV and $|\eta| < 2.4$, and isolated from hadronic activity. Anti-k_T-clustered jets are counted if they have $p_T > 30$ GeV and $|\eta| < 2.4$. The event selection includes a requirement of at least two jets with those characteristics.

The signal is modeled with the Powheg 2.0 NLO generator interfaced with Pythia 6 for hadronization. The signal to noise ratio for events with two or more jets is already good enough that no b-tagging is necessary to further reduce backgrounds. Systematic uncertainties in the measurement include the knowledge of integrated luminosity (12%, from beam-beam scans), trigger (5%) and lepton identification (4%), and jet energy scale (3%). The $t\bar{t}$ cross section is extracted from the event counts, once backgrounds are accounted for. The result is $\sigma_{t\bar{t}} = 772 \pm 60(stat) \pm 62(syst) \pm 93(lumi)$ pb, with a total relative uncertainty of 13.5%. This measurement is consistent with the SM prediction ($\sigma_{NNLO+NNLL}(t\bar{t}) = 832^{+40}_{-46}$ pb), valid for a top quark mass of 172.5 GeV. Figure 5 compares the 13 TeV measurement with previous ones and with theory predictions as a function of the center-of-mass energy of the collisions.

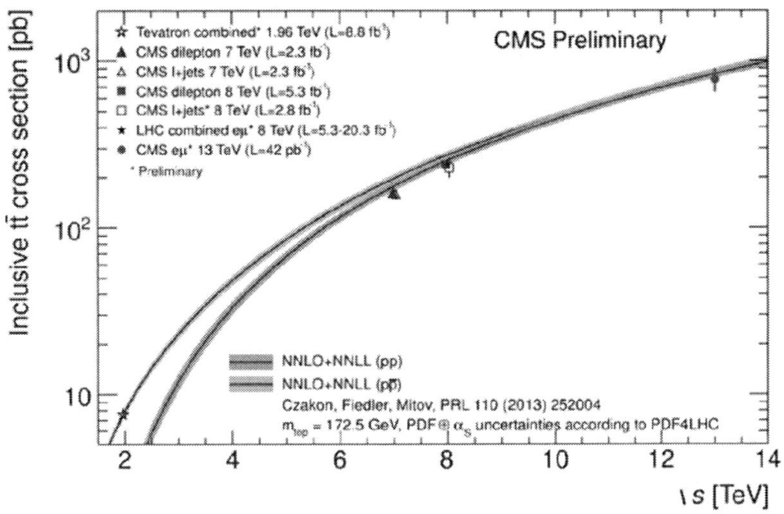

Figure 5. *Determinations of the top pair-production cross section in hadron-hadron collisions as a function of the center-of-mass energy. The recent CMS determination at 13 TeV is shown by the red marker on the right.*

6 Conclusions

The CMS experiment has exploited proton-proton collision data collected during the 2011 and 2012 runs of the Large Hadron Collider to produce a large number of groundbreaking results in precision measurements of SM observables and searches for new physics. The most recent results include differential distributions of Higgs boson production, top quark measurements at high transverse momentum, and much more. CMS also produced the most precise measurements to date of the mass of the top quark and of the Higgs boson.

The first data collected in the summer of 2015 have allowed the experiment to tune its detector components and reconstruction algorithms, and to start producing new measurements at the so far untested collision energy of 13 TeV. The measurement of top quark pair-production cross section is in excellent agreement with theoretical predictions. Events at the high-energy end of mass spectra have already started to inspire hopes that the new energy regime will provide hints of new physics.

Acknowledgements

We wish to congratulate our colleagues in the CERN accelerator departments for the excellent performance of the LHC machine. We thank the technical and administrative staff at CERN and other CMS institutes, and acknowledge support from: FMSR (Austria); FNRS and FWO (Belgium); Cap, CAPES, FAPERJ, and FAPESP (Brazil); MES (Bulgaria); CERN; CAS, Most, and NSFC (China); COLCIENCIAS (Colombia); MSES (Croatia); RPF (Cyprus); More, SF0690030s09 and ERDF (Estonia); Academy of Finland, MEC, and HIP (Finland); CEA and CNRS/IN2P3 (France); BMBF, DFG, and HGF (Germany); GSRT (Greece); OTKA and NKTH (Hungary); DAE and DST (India); IPM (Iran); SFI (Ireland); INFN (Italy); NRF and WCU (Korea); LAS (Lithuania); CINVESTAV, CONACYT, SEP, and UASLPFAI (Mexico); MSI (New Zealand); PAEC (Pakistan); MSHE and NSC (Poland); FCT (Portugal); JINR (Armenia, Belarus, Georgia, Ukraine, Uzbekistan); MON, Rosa tom, RAS and RFBR (Russia); MSTD (Serbia); MICINN and CPAN (Spain); Swiss Funding Agencies (Switzerland); NSC (Taipei); TUBITAK and TAEK (Turkey); STFC (United Kingdom); DOE and NSF (USA).

References

[1] CMS Collaboration, *Observation of a new boson at a mass of 125 GeV with the CMS experiment at the LHC*, Phys. Lett. **B716** (2012) 30-61; ATLAS Collaboration, *Observation of a new particle in the search for the standard model Higgs boson with the ATLAS detector at the LHC*, Phys. Lett. **B716** (2012) 1.

[2] F. Englert and R. Brout, *Broken symmetry and the mass of gauge vector mesons*, Phys. Rev. Lett. **13** (1964) 321-323.

[3] P. Higgs, *Broken symmetries, massless particles and gauge fields*, Phys. Lett. **12** (1964)132-133.

[4] P. Higgs, *Broken symmetries and the masses of gauge bosons*, Phys. Rev. Lett. **13** (1964) 508-509.

[5] G. Guralnik, C. Hagen, and T. Kibble, *Global conservation laws and massless particles*, Phys. Rev. Lett. **13** (1964) 585-587.

[6] P. Higgs, *Spontaneous symmetry breakdown without massless bosons*, Phys. Rev. **145** (1966) 1156-1163.

[7] T. Kibble, *Symmetry breaking in non-Abelian gauge theories*, Phys. Rev. **155** (1967) 1554-1561.

[8] See http://cms-results.web.cern.ch/cms-results/public-results/publications/.

[9] CMS Collaboration, *The CMS experiment at the CERN LHC*, J.Instrum. **3** (2008) S08004.

[10] CMS Collaboration, *Combined measurement of the Higgs boson mass in pp collisions at* $\sqrt{s} = 7$ *and* 8 *TeV with the ATLAS and CMS experiments*, Phys. Rev. Lett. **114** (2015) 191803.

[11] CMS Collaboration, *Measurement of inclusive and differential fiducial cross sections for Higgs boson production in the* $H \rightarrow 4l$ *decay channel in p-p collisions at* $\sqrt{s} = 7$ *and* $\sqrt{s} = 8$ *TeV*, CMS-PAS-HIG-14-028, arXiv:1512.08377 (2015).

[12] D. de Florian, G. Ferrera, M. Grazzini, and D. Tommasini, *Higgs boson production at the LHC: transverse momentum resummation effects in the* $H \rightarrow \gamma\gamma$, $H \rightarrow WW \rightarrow l\nu l\nu$ *and* $H \rightarrow ZZ \rightarrow 4l$ *decay modes*, J. High Energy Phys. **06** (2012) 132, doi:10.1007/JHEP06(2012)132, arXiv:1203.6321 (2012).

[13] M. Grazzini and H. Sargsyan, *Heavy-quark mass effects in Higgs boson production at the LHC*, J. High Energy Phys. **09** (2013) 129, doi:10.1007/JHEP09(2013)129, arXiv:1306.4581 (2013).

[14] S. Frixione, P. Nason, and C. Oleari, *Matching NLO QCD computations with parton shower simulations: the POWHEG method*, J. High Energy Phys. **11** (2007) 070, doi:10.1088/1126-6708/2007/11/070, arXiv:0709.2092 (2007).

[15] T. Melia, P. Nason, R. Rontsch, and G. Zanderighi, *W+W-, WZ and ZZ production in the POWHEG BOX*, J. High Energy Phys. **11** (2011) 078, doi:10.1007/JHEP11(2011)078, arXiv:1107.5051 (2011).

[16] K. Hamilton, P. Nason, and G. Zanderighi, *MINLO: multi-scale improved NLO*, J. High Energy Phys. **10** (2012) 155, doi:10.1007/JHEP10(2012)155, arXiv:1206.3572 (2012).

[17] CMS Collaboration, *Search for resonant pair production of Higgs bosons decaying to two bottom quark-antiquark pairs in proton-proton collisions at 8 TeV*, Phys. Lett. **B749** (2015) 560.

[18] CMS Collaboration, *Measurement of the differential $t\bar{t}$ production cross section for high-p_T top quarks in $e\mu$+jets final states at 8 TeV*, CMS-PAS-TOP-14-012 (2014).

[19] CMS Collaboration, *Measurement of the $t\bar{t}$ production cross section in the dilepton channel in pp collisions at $\sqrt{s} = 7$ TeV*, J. High Energy Phys. **11** (2012) 067.

[20] CMS Collaboration, *Measurement of the $t\bar{t}$ production cross section in the dilepton channel in pp collisions at $\sqrt{s} = 8$ TeV*, CMS-PAS-TOP-12-007, axXiv:1312.7582 (2013).

[21] M. Campbell and R. K. Ellis, *MCFM for the Tevatron and the LHC*, Nucl. Phys. Proc. Suppl. **205-206** (2010) 10, arXiv:1007.3492 (2010).

[22] M. Aliev et al., *HATHOR: HAdronic Top and Heavy quarks crOss section calculatoR*, Comput. Phys. Commun. **182** (2011) 1034.

[23] N. Kidonakis, *Next-to-next-to-leading soft-gluon corrections for the top quark cross section and transverse momentum distribution*, Phys. Rev. **D82** (2010) 114030.

[24] Tevatron Electroweak Working Group, *Combination of the CDF and D0 results on the mass of the top quark using up to 9.7 fb^{-1} at the Tevatron*, arXiv:1407.2682 (2014).

[25] CMS Collaboration, *Measurement of the top quark mass using proton-proton data at $\sqrt{s} = 7$ and 8 TeV*, arXiv:1509.04044, CMS-PAS-PUB-14-022 (2015), accepted for publication in Phys. Rev. D.

EPJ Web of Conferences **126**, 02012 (2016) DOI: 10.1051/epjconf/201612602012
ICNFP 2015

Big Bang, inflation, standard Physics... and the potentialities of new Physics and alternative cosmologies

Present statuts of observational and experimental Cosmology. Open questions and potentialities of alternative cosmologies

Luis Gonzalez-Mestres[1,a]

[1] *Megatrend Cosmology Laboratory, John Naisbitt University, Belgrade and Paris*
Goce Delceva 8, 11070 Novi Beograd, Serbia

Abstract. A year ago, we wrote [1] that the field of Cosmology was undergoing a positive and constructive crisis. The possible development of more direct links between the Mathematical Physics aspects of cosmological patterns and the interpretation of experimental and observational results was particularly emphasized. Controversies on inflation are not really new, but in any case inflation is not required in pre-Big Bang models and the validity of the standard Big Bang + inflation + ΛCDM pattern has not by now been demonstrated by data. *Planck* has even explicitly reported the existence of "anomalies". Remembering the far-reaching work of Yoichiro Nambu published in 1959-61, it seems legitimate to underline the need for a cross-disciplinary approach in the presence of deep, unsolved theoretical problems concerning new domains of matter properties and of the physical world. The physics of a possible preonic vacuum and the associated cosmology constitute one of these domains. If the vacuum is made of superluminal preons (superbradyons), and if standard particles are vacuum excitations, how to build a suitable theory to describe the internal structure of such a vacuum at both local and cosmic level? Experimental programs (South Pole, Atacama, AUGER, Telescope Array...) and observational ones (Planck, JEM-EUSO...) devoted to the study of cosmic microwave background radiation (CMB) and of ultra-high energy cosmic rays (UHECR) are crucial to elucidate such theoretical interrogations and guide new phenomenological developments. Together with a brief review of the observational and experimental situation, we also examine the main present theoretical and phenomenological problems and point out the role new physics and alternative cosmologies can potentially play. The need for data analyses less focused *a priori* on the standard models of Particle Physics and Cosmology is emphasized in this discussion. An example of a new approach to both fields is provided by the pre-Big Bang pattern based on a physical vacuum made of superbradyons with the spinorial space-time (SST) geometry we introduced in 1996-97. In particular, the SST automatically generates a local privileged space direction (PSD) for earch comoving observer and such a signature may have been confirmed by *Planck* data. Both superluminal preons and the existence of the PSD would have strong cosmological implications. Planck 2016 results will be particularly relevant as a step in the study of present open questions.

This paper is dedicated to the memory of Yoichiro Nambu

[a] e-mail: luis.gonzalez-mestres@cosmology.megatrend.edu.rs

1 Introduction

In [1, 2], we remind the theoretical controversies that had preceded the March 2014 BICEP2 announcement [3, 4] on the polarization of cosmic microwave background radiation (CMB). These open discrepancies had followed the 2013 *Planck* results [5–7]. Such a debate [8, 9] on the predictions of inflationary models [10, 11] involved precisely the possible generation of primordial gravitational waves leading to B-modes of the CMB, as a consequence of cosmic inflation. To date, the theoretical debate continues [12, 13] but experimental announcements have become more decoupled from theoretical controversies. The joint *Planck* – BICEP2 – Keck Array (PBKA) analysis [14, 15] found no conclusive evidence for the primordial CMB B-modes initially claimed.

Actually, as underlined in [16, 17] and later in [1, 2] and in two contributions to this Conference [18, 19], the theoretical and phenomenological situation remains uncertain and requires a long-term effort together with suitable experimental and observational programs. In particular, as pointed out in [19], if Quantum Chromodynamics played a crucial role in the transition from hadrons to quarks and gluons as fundamental objects in the 1970s, no equivalent theory exists at present for a possible similar transition from standard particles to preons with a preonic vacuum of which quarks, leptons, gauge bosons... would be excitations (the superbradyon hypothesis, formulated in 1995 [20, 21]). In spite of this difficulty, a detailed discussion on the possible origin of Quantum Mechanics from a superbradyonic vacuum with a spinorial space-time (SST) geometry is presented in [18].

The standard cosmological model (Big Bang + inflation + ΛCDM) remains far from being well-defined and well-established. In particular, it cannot completely account for recent cosmological observations and analyses by the Planck Collaboration [22–24]. 2015 Planck studies [25] do not contradict the observed asymmetry between the power spectra of two hemispheres defined by a preferred direction. Such a phenomenon agrees with a natural prediction [26–29] of the cosmic SST geometry [26, 30] that automatically generates a privileged space direction (PSD) for each comoving observer.

Standard cosmological patterns can also be challenged from a more fundamental point of view including the properties of vacuum, the space-time structure and the formation of the Universe [2, 17, 31]. A pre-Big Bang scenario can naturally make the Planck scale useless [27, 28].

As already stressed in [1, 2] and in previous work, alternative cosmologies [32, 33] must be seriously taken into account in the present situation. They can turn out to be more efficient to explain real data than the conventional cosmology based on the standard Big Bang model with inflation. Actually, cosmic inflation has no reason to have existed in a Universe involving a physical vacuum made of superbradyons [20, 21] where signals and correlations can propagate faster than light and free superluminal particles (not tachyons) may have dominated an early phase of the Universe [19, 21]. Similarly, the size of a SST Universe is expected to be much larger than that of the conventional one [1, 31]. Thus, a Universe with a SST geometry can naturally fake [28, 34] a flat standard Universe with an observed small curvature as seen with a conventional approach to cosmological modeling [35].

The present theoretical uncertainties, already discussed in [1, 2], clearly require not only a long-term effort in the theoretical domain, but also strong long-term experimental and observational programs. If CMB studies are a priority task, high-energy cosmic rays are also an important field including the ultra-high energy (UHE) domain where possible violations of the standard principles of Particle Physics and Cosmology can be searched for (see, for instance, [27, 36] and [37, 38]).

In the alternative cosmologies and patterns of Particle Physics considered in [27, 28] and in [1, 2, 18], the standard principles of Physics are local low-energy limits of a more fundamental dynamics for a sector of matter, as already postulated in [20, 21]) and dealt with in [39–41]. Together with the search for new particles at high-energy accelerators, the study of ultra-high energy cosmic rays (UHECR) can in particular provide signatures of new physics and new sectors of matter [28, 42].

In this brief review, we update the discussion of the experimental and observational situation presented in [1, 2, 28] and further study some aspects of the present theoretical problematics (see also [18, 19]). A section is devoted to memory of Yoichiro Nambu.

2 Yoichiro Nambu (1921-2015)

One half of the 2008 Nobel Prize was attributed to Yoichiro Nambu *"for the discovery of the mechanism of spontaneous broken symmetry in subatomic physics"* [43]. The press release states:

As early as 1960, Yoichiro Nambu formulated his mathematical description of spontaneous broken symmetry in elementary particle physics. Spontaneous broken symmetry conceals nature's order under an apparently jumbled surface. It has proved to be extremely useful, and Nambu's theories permeate the Standard Model of elementary particle physics. The Model unifies the smallest building blocks of all matter and three of nature's four forces in one single theory.

(end of quote)

The paper by Nambu *Quasi-Particles and Gauge Invariance in the Theory of Superconductivity* [44] was received by *The Physical Review* on July 23, 1959. In the abstract, Yoichiro Nambu starts emphasizing: *"Ideas and techniques known in quantum electrodynamics have been applied to the Bardeen-Cooper-Schrieffer theory of superconductivity"*.

Eight months later, as early as April 1960, submitted in turn to an international conference (Midwest Conference, Purdue, April 1-2, 1960) a contribution [45, 46] explicitly entitled *A 'Superconductor' Model of Elementary Particles and Its Consequences* (presented by Giovanni Jona-Lasinio in the absence of Nambu). Simultaneoulsy, another paper by Yoichiro Nambu, *Axial vector current conservation in weak interactions*, appeared in *Physical Review Letters* [47]. The article by Jeffrey Goldstone *Field theories with "Superconductor" solutions* explicitly refers to Nambu's idea [48].

Just after Goldstone's preprint, the paper by Nambu and Jona-Lasinio *Dynamical Model of Elementary Particles Based on an Analogy with Superconductivity. I* was received by *The Physical Review* on October 1960 [49]. The second article, *Dynamical Model of Elementary Particles Based on an Analogy with Superconductivity. II* [50], arrived to *The Physical Review* on May 1961. After having briefly presented the basic idea in [44, 45, 47], the authors published in these two papers a more precise illustration of the spontaneous symmetry breaking pattern Nambu had suggested.

After these major contributions, Yoichiro Nambu also played an important role in the study of the quark model and later of the string model [51, 52]. Besides his early work on dual models, a particularly important contribution (1970) is the Nambu-Goto action [53, 54]. Tetsuo Goto explicitly refers to the Copenhagen lecture by Nambu. The Nambu-Goto string is nowadays considered for cosmological applications [55] and in other fields of fundamental Physics.

Further work by Yoichiro Nambu on the dual resonance model [56] has also inspired the dual superconductor models of gluon confinement [57]. More generally, the analogies with superconductivity developed by several authors for the physics of hadrons, quark and gluons are basically updated developments of the original ideas exposed in the 1959-61 papers by Nambu and Jona-Lasinio.

2.1 From Nambu's work to preon models of vacuum?

Can the preonic vacuum be a superconductor? In a superconducting superbradyonic vacuum, free superbradyons can emerge as quasiparticles. But what about the standard particles and other similar objects? Are they similar to vorteces? Actually, in the last decades work in the field of condensed matter physics has focused on the existence and properties of solitons and similar objects in superconductors [58, 59]. Analogoust work is also carried out for superfluid systems [60]. No obvious incompatibility seems to exist between superconductivity or superfluidity and soliton-like excitations.

It is therefore not excluded that standard particles be nontrivial excitations of a superconducting or superfluid superbradyonic vacuum, adding solitons and similar objects to the basic initial picture by Nambu and Jona-Lasinio. A crucial question would then be the possible analogy with high-T_c superconductivity [61, 62]. The role of the space-time geometry must also be taken into account: a spinorial space-time can influence the properties of the superbradyonic vacuum as a possible superconductor.

Obviously, analogies with condensed matter physics and with other domains concerning matter around us can be very useful in order to develop an active reflection on the possible properties of a preonic vacuum and the way conventional particles can emerge as vacuum excitations. Work in this direction and original ideas are certainly needed, following Nambu's example.

3 Further theoretical uncertainties

The claim supported by the initial 2014 announcement of BICEP2 results [63, 64] was the possible detection, through a B-mode polarization of the cosmic microwave background radiation, of a signature of primordial gravitational waves from cosmic inflation. However, as immediately pointed out in [16, 17], the detection of primordial CMB B-modes would not have been an evidence for cosmic inflation in cosmological scenarios other than the standard Big Bang.

Inflation is not needed in pre-Big Bang patterns and, as reminded before, the cosmic SST geometry automatically generates a privileged space direction for each comoving observer in possible agreement with Planck [22] and WMAP [65] data. In the presence of this local PSD already in the early Universe, primordial CMB B-modes would have been generated through rotations around the privileged space direction and vector products by this direction. Contrary to standard calculations and conjectures involving the usual relativistic space-time, pre-Big Bang patterns incorporating a spinorial space-time geometry can naturally generate primordial vector perturbations producing CMB B-modes.

In such a pre-Big Bang scenario, the possible generation of gravitational waves from initial vector perturbations due to the local SST anisotropy must also be considered and requires a careful study. Whatever the result, the detection of primordial CMB B-modes would be a potential direct evidence for a primordial cosmic SST geometry. The question of possible preonic waves also deserves being raised. Can gravitational waves decay into preonic waves? How large is the preon critical speed?

Other pre-Big Bang models [66, 67] also lead naturally to the direct generation of primordial gravitational waves without any need for cosmic inflation.

Gravitational waves in our Universe remain nowadays a fundamental scientific interrogation. LIGO [68] reported no positive result [69] in spite of previous rumors [70]. Gravitational waves were predicted by Henri Poincaré in 1905 [71] together with his explicit formulation of special relativity. Their detection is a crucial issue for standard Physics and Cosmology.

More generally, tests of relativity and of other fundamental principles of Physics, including energy and momentum conservation, are an important research subject [27, 28]. Possible deviations from relativity at small distance scales were already considered by Albert Einstein in 1920 [72]. Even occurring at a much lower distance scale than considered by Einstein, they can play a crucial role.

At this stage, further unconventional questions must be raised. An important one is that of vacuum homogeneity for the propagation of gravitational waves, CMB, cosmic rays... The internal structure of vacuum is not really known, but usually a global homogeneity is tacitly postulated at cosmic level except for explicit effects from standard Physics and Cosmology. However, if vacuum has an internal structure more fundamental than that predicted by conventional Quantum Field Theory, its dynamics can lead to inhomogeneities at galactic, astrophysical and cosmic levels. Such inhomogeneities can in particular generate obstacles to the propagation of gravitational waves. This is just an example of possible nonstandard phenomena ignored in a conventional approach but leading potentially to important effects. Black hole physics is also expected to be significantly modified by preonic patterns.

4 BICEP, Keck Array and related experiments

With the title *The BICEP and Keck Array CMB experiments*, the public site of BICEP and Keck Array [73] presents itself as the *Official website for results from the BICEP and Keck Array series of CMB polarization experiments*. The goal of this important experimental program installed at the South Pole is thus clearly defined. We update here the description presented in [1] where more details can be found concerning the detector design and instrumentation.

The word BICEP (Background Imaging of Cosmic Extragalactic Polarization) designs a series of cryogenic experiments with adapted bolometers and sensors, deployed to the Amundsen-Scott South Pole Station. BICEP1, also called the Robinson Gravitational Wave Background Telescope [74], was the first BICEP experiment [75]. Its results [76] yielded no possible evidence for primordial CMB B-modes. An important long-term program progressively improving detectors and technologies has followed this first attempt and continues to develop.

The BICEP2 experiment [4] was installed and operated in 2010-2012 at the South Pole Station and has been the first experiment to use the antenna-coupled transition edge sensor (TES) arrays [77–80] fabricated at the Jet Propulsion Laboratory (JPL) [81].

More recent than BICEP2 and working together with it, the Keck Array [82] uses the same technology. It was completed in 2012 [83], modified in 2013-2014 [79] and further improved in 2015 [84]. Articles with Keck Array results are [15, 85, 86].

After BICEP2, BICEP3 [87] is an improved refracting telescope designed for CMB polarimetry at 95 GHz with a better rejection of background due to galactic dust. BICEP3 is presented [88] as the *"newest member of the BICEP/Keck family of inflationary probes specifically designed to measure the polarization of the cosmic microwave background (CMB) at degree-angular scales"*. Even if the detection of primordial CMB B-modes would not provide by itself an evidence for cosmic inflation [1, 2, 16, 17], the high relevance of such a program for Cosmology is obvious.

Similarly, a new (third generation) polarization sensitive receiver, SPT-3G, has been designed [89] for the South Pole Telescope with a high signal-to-noise ratio expecting to go beyond simple statistical detection. The SPT-3G focal plane is expected to start operating in 2016 [90].

POLAR [91, 92] is a long-term project to build the most sensitive cosmic microwave background polarimeter. Polar-1 will involve about 5000 bolometric detectors at 150 GHz, and will be followed by a Polar Array containing ten such telescopes.

The polarization-sensitive receiver SPTpol [97], installed on the South Pole Telescope, has measured [98] the CMB gravitational lensing potential and performed an optimal measurement of the sub-degree B-mode polarization of the CMB [99] .

In parallel, POLARBEAR [93–95], based at the ATACAMA desert, has published results [96] announcing that for the first time *"anisotropic cosmic birefringence or primordial magnetic fields have been constrained from the ground at subdegree scales"*.

The cryogenic Atacama Cosmology Telescope (ACT) [100], located at an elevation of 5190 m, is equipped with a new polarization sensitive receiver, ACTpol [101], recently optimized [102]. ACT has by now detected [103] gravitational lensing of the CMB polarization by large-scale structure, operated *"multi-wavelength detections of nine candidate gravitationally-lensed dusty star-forming galaxies (DSFGs)"* [104] and estimated the galaxy velocity dispersions and dynamical masses for 44 galaxy clusters selected via the Sunyaev-Zel'dovich effect [105].

CLASS (Cosmology Large Angular Scale Surveyor) [106–108] is also located at the Atacama Desert and will cover 70% of the sky.

SPIDER [109–112] is a balloon-borne cryogenic instrument designed to probe the possible primordial gravitational wave signal by detecting CMB B-modes at degree angular scales.

Another balloon project in this domain is PILOT (Polarized Instrument for the Long-wavelength Observations of the Tenuous ISM) [113, 114], whose goal is to characterize the polarization of the dust continuum emission in the diffuse interstellar medium. A first flight by PILOT, in September 2015, has been successful.

The above programs form a very complete and promising set to study the CMB, together with Planck and other experiments and observations.

5 The detection of cosmic neutrino background

If the cosmic microwave background radiation is given strong attention, the cosmic neutrino background (CνB) also requires careful investigation. The CMB detection projects develop technologies that will be useful in a much larger domain of Physics and Cosmology. But the cosmology and detection of the CνB deserve a detailed specific study.

The standard cosmology of the CνB has been reminded in [115] having in view the use of the Karlsruhe Tritium Neutrino (KATRIN) spectrometer [116–118] devoted to the inelastic reaction between an electron neutrino and a ^3H nucleus: $\nu \; + \; ^3H \; \rightarrow \; ^3He \; + \; e^-$.

Another experimental project using a tritium target is PTOLEMY (Princeton Tritium Observatory for Light, Early Universe Massive Neutrino Yield) [119, 120].

Neutrino capture on beta decaying nuclei had already been considered, for instance, in [121, 122].

On more general grounds, a complementary approach to the detection of cosmic neutrinos can be based on the inelastic reaction between a cosmic antineutrino and a stable nucleus, producing an electron and an unstable isotope of the next element in the periodic table.

This reaction will occur if the antineutrino energy is at least of a few MeV. Then, the emission of an electron or a photon will be a first detectable event. Subsequently, the unstable nucleus will decay generating a new signal. The delayed coincidence between the two events can provide a strong enough signature to reject background.

The reaction of cosmic antineutrinos with electron-capture decaying nuclei such as ^{163}Ho has also been considered for the detection of cosmic background neutrinos other than electron neutrinos [123, 124]. Obviously, much work remains to be done in this important field.

6 Recent Planck results

The *Planck* satellite was launched on May 2009 and took data until October 2013. The 2015 results are based on data from the entire *Planck* mission, including temperature and polarization.

The first cosmological results (2013) of the Planck Collaboration based on measurements of the temperature of the cosmic microwave background radiation and the power spectra of gravitational lensing potential were presented in [6, 125] (both revised in 2014). Constraints on inflation were discussed in [7], and have been examined again in 2015 [126].

Data on CMB anisotropy were futher analysed by *Planck* in January 2014, confirming [22] the observation of an asymmetry in power between two cosmic hemispheres defined by a preferred direction. No correction to this paper has been introduced in 2015, even it the question has been addressed [127]. Planck also reports on possible signatures of parity violation (oscillations between odd and even modes) in the hemisphere where power is larger.

Such a possible evidence for a PSD observed by *Planck* fits well with the cosmic SST prediction on the PSD [1, 2, 26–30], contrary to the Planck statement *"Although these analyses represent a step forward in building an understanding of the anomalies, a satisfactory explanation based on physically motivated models is still lacking"*. But what exactly means the expression *"physically motivated"*?.

DOI: 10.1051/epjconf/201612602012

The question of the relevance of alternative Physics and Cosmologies must clearly be raised at this stage. In any case, further measurements and analyses are required in order to elucidate the question of the possible existence of the PSD announced by *Planck*. Actually, as already discussed in [1], the *Planck* results on the possible observation of a PSD confirm the asymmetry previously found by Eriksen et al. [128], Hansen et al. [129] and Park [130] using WMAP data. A pre-Big Bang pattern with the SST geometry would naturally produce such an effect.

In March 2014, a new version of [6] equally confirmed the observation of an "anomaly" in the multipole range $20 < \ell < 40$ as compared to standard predictions of the ΛCDM pattern (Λ = cosmological constant, CDM = cold dark matter). No correction has been introduced in 2015.

6.1 Planck 2015-2016

Between January 2015 and the first days of 2016, *Planck* has posted 41 papers to *arXiv.org*, including the joint analysis with BICEP2 and the Keck Array of the results on CMB *B*-modes [15]. 28 articles are entitled *"Planck 2015 results"* [131], often repeating the 2013 subjects.

The *Planck 2015* results [131–134] are clearly more precise that those of 2013. The consistency of 2013 data is simultaneously further discussed [135]. A final release of Planck results is expected for 2016 [132, 136] and will be of particular relevance.

However, it must be noticed that Planck analyses [137] use to take as the basic reference framework the standard Big Bang + inflation pattern, together with the conventional ΛCDM model. Alternative physical and cosmological approaches become then marginal from the beginning in the interpretation of Planck results. This appears to involve some contradiction with the March 2013 statement, when *Planck* made public the first version of [22] and the ESA-*Planck* News site wrote in an article for large public [138]: *"... because precision of Planck's map is so high, it also made it possible to reveal some peculiar unexplained features that may well require new physics to be understood"*. The Planck Collaboration seemed to argue that its results systematically favor the standard cosmological pattern, except for "anomalies" potentially related to "new physics".

But if new physics can be at work, why could it not generate an alternative cosmology (f.i. involving a pre-Big Bang scenario with the superbradyonic vacuum and a cosmic SST geometry) potentially able to consistently reproduce all the *Planck* results, and more performant than the Big Bang + inflation + ΛCDM scheme? As already pointed out in [1, 2] and [27, 28], alternative cosmologies can naturally provide original solutions to important unsolved problems (see also [18, 21, 139, 140]).

In this specific case, the use of the postulates of standard Physics and standard Cosmology as the basic principles for data analysis, and simultaneously as a theory to confirm, can lead to a deformed view of the Universe, of its real content and of its history since the origin of time.

7 The Euclid and DESI projects

The Euclid project [141, 142] defines itself as *"an ESA medium class astronomy and astrophysics space mission"* devoted to the *"geometry of the dark Universe"* and aiming *"at understanding why the expansion of the Universe is accelerating and what is the nature of the source responsible for this acceleration which physicists refer to as dark energy"*.

Selected by ESA in 2011, the Euclid mission is expected to be launched in 2020. Independently, the Dark Energy Spectroscopic Instrument (DESI) [143–145] is expected to operate on Earth at the Mayall 4-Meter Telescope (Kitt Peak National Observatory, Arizona) in 2018-2022. Similar to *Euclid*, DESI plans in particular to *"probe the effects of dark energy on the expansion history"*.

Dark matter and possible dark energy are certainly essential items. But are they really leading the geometry and the evolution of the Universe?

Again, standard theoretical prejudices can influence data analysis and play a negative role if conventional cosmology is used as the basic theoretical framework.

In a cosmology built on a cosmic spinorial space-time, a remarkable feature immediately arises [1, 26, 28, 34]: already before introducing any standard matter, the expansion of the Universe follows automatically an intrinsic geometric law $H = t^{-1}$ where t is the cosmic time (age of the Universe) and H the well-known universal velocity/distance ratio dealt with by Knut Lundmark, Georges Lemaître and Edwin Hubble. The Lundmark - Lemaître - Hubble (LLH) law, [146–148] is thus obtained from the beginning as a direct consequence of the cosmic SST geometry without any further cosmological dynamics (no standard matter, yet) and with a value of H compatible with observations. Such a result illustrates again the potentialities of alternative cosmologies and deserves closer consideration, not only in the analysis of Planck data but also in view of the Euclid mission. It leads in particular to significant modifications of Friedmann-like equations [32–34]

Similarly, the question of the cosmological constant changes radically if a preonic vacuum is considered [1, 2, 28, 31, 139, 149, 150] where bosonic zero modes and standard quantum-field theoretical condensates (Higgs...) are actually not permanently condensed in vacuum. The cosmological constant problem can be solved in this way, leading to a further improvement of Friedmann-like equations.

8 Gaia

The Gaia mission [151–153] was launched in December 2013 and provides an all-sky survey of the Milky Way with a rapid analysis performed by the Gaia Science Alerts group [154] and making possible additional observations for supernovae, cataclysmic variables [155], tidal disruption of stars by massive black holes, microlensing...

Gaia defines itself, in particular, as follows [151]:

(...) Gaia provides the detailed 3D distributions and space motions of all these stars, complete to 20th magnitude. The measurement precision, reaching a few millionths of a second of arc, is unprecedented. This allows our Galaxy to be mapped, for the first time, in three dimensions. Some 10 million stars will be measured with a distance accuracy of better than 1 percent; some 100 million to better than 10 percent.

(end of quote)

The impressive set up and performances of *Gaia* will certainly lead to many results useful to several fields of space research, including Cosmology and important questions in Particle Physics.

9 Theoretical issues for Planck, Euclid and DESI

The theoretical, phenomenological, experimental and observational issues discussed in [1, 2, 18, 19] are clearly relevant for the analysis of Planck data and for the development of the Euclid mission.

Cosmological patterns usually considered as the grounds for data analysis are based on the standard space-time with four real variables and on the associated conventional relativity. The space-time felt by macroscopic "ordinary" matter and used in laboratory particle physics is extrapolated: i) to astrophysical and cosmic level, including the largest cosmological scales ; ii) to the ultimate structure of matter, assumed to be accounted for by standard quantum field theory (SQFT). Standard general relativity is also a basic ingredient providing the framework of such a cosmology. The fact that standard relativity can be just a low-energy limit for conventional particles is ignored.

Similarly, the structure of the physical vacuum is described in terms of SQFT, in spite of the fact that this description leads to the cosmological constant problem and that the usual Higgs mechanism is nothing else than condensing "elementary" vacuum excitations inside the vacuum itself without any further description of the internal structure of vacuum and of its excitations.

However, as recognized by *Planck*, the precision reached by astrophysical and cosmological measurements may allow to distinguish between the standard cosmology and alternative approaches beyond the conventional description of vacuum, particles and space-time (the Big Bang + inflation + ΛCDM pattern). In particular, a preonic cosmology with a new space-time structure like the spinorial one can lead to a new description of the beginning of the Universe [1, 2, 27, 28, 140, 150] and to related observable effects. Furthermore, the cosmic SST naturally suggests a much larger Universe than usually considered. In this SST Universe, the vacuum structure is not necessarily the same everywhere and other kinds of particles, including preons, can be the dominant free objects in some regions (see also [18]). The cosmic space curvature of the SST Universe is then naturally expected to be much smaller than the one estimated from data analyses using the ΛCDM model [31, 32].

In a general approach to the space-time geometry, the standard Lorentz metric can be just [1, 20, 28, 39, 139, 140] the naturally stable kinematics for free particles in our Universe. The speed of light can then be [20, 39] the low-energy critical speed of a family of vacuum excitations. Other kinds of particles can exist with different (larger) critical speeds and detectable signatures at accelerators, in Astrophysics and in Cosmology. Such superluminal particles would not be tachyons, and would have positive mass and energy with a speed larger than c [20]. In a SST Universe, contrary to the usual dark energy hypothesis, the observed acceleration of the expansion of the Universe can be only a remnant reflect of a fluctuation due to the early Universe dynamics and vanishing for $t \rightarrow \infty$ when the matter density tends to zero [32, 33]. Furthermore, in spite of its intrinsic positive space curvature, the SST Universe can actually describe apparent curvatures of both signs as seen by conventional matter [32, 34]. In all cases, the geometric $H\,t = 1$ law remains valid in the absence of matter.

The(1931) Big Bang hypothesis [156] was based on the recently formulated Quantum Mechanics, and assumed that the energy of the Universe had been originally concentrated in an initial quantum. At this initial stage, the notions of space and time were considered as meaningless, and Lemaître wrote *"the beginning of the world happened a little before the beginning of space and time"*. Lamaître's idea was then an impressive step forward. But more than eighty years later, our present knowledge of Particle Physics and Cosmology, together with the growing experimental and observational resources, allow to examine otherwise the actual theoretical uncertainties and experimental potentialities.

The possibility that Quantum Mechanics be no longer an exact law of Physics in the high-energy and low-distance domain must now be seriously considered [18, 27, 36]. Not only at the Planck scale, but even at a lower energy scale E_{trans} where standard physics may start being replaced by a new dominant dynamics [37, 39], and similarly above some distance scale larger than the Planck length [32, 38]. A pre-Big Bang scenario can then replace standard cosmology in such a way that the origin of time $t = 0$ makes sense and starts a real cosmological era.

Low-energy symmetries do not necessarily become more exact at very high energy [27, 42], and there is by now no experimental proof of the validity of Grand Unification. The grand unification epoch of Cosmology can therefore disappear in its standard form, and also the monopole problem. Roughly, the age of the Universe t can then be written as:

$$t = t_{PBB} + t_E \tag{1}$$

where t_{PBB} is the time during which the pre-Big Bang dynamics dominates in our side of the Universe, and t_E corresponds to the subsequent evolution. The time interval t_{PBB} can be much larger than the time scale associated to cosmic inflation in the conventional cosmology. After t_{PBB} , remnants of the pre-Big Bang era may have survived until the present time and be detectable or observable.

Strings have also been considered to build Pre-Big Bang scenarios [157, 158]. However, strings are known [159] to have an underlying composite structure. The use of strings to describe standard "elementary" particles amounts to an implicit preonic pattern [18, 19, 27, 36]. As a superbradyonic vacuum with a spinorial space-time can naturally lead to spinorial Regge trajectories of the particles generated as vacuum excitations [1, 18, 19] a spinorial string-like pattern can be imagined in an energy domain far enough from the scales characterizing the internal vacuum structure.

Obviously, this set of considerations suggests a real need for strong phenomenological flexibility when analysing Planck data or developing the Euclid and DESI projects.

9.1 Properties of new Friedmann-like equations

In a SST-based cosmology, new Friedmann-like equations [1, 32, 33] can automatically solve the flatness problem without introducing the standard dark energy and cosmological constant.

A possible way to obtain modified Friedmann equations [1, 32–34] is to consider the Einstein field equation [160, 161]:

$$R_{\mu\nu} - 1/2\, g_{\mu\nu}\, R + g_{\mu\nu}\, \Lambda = 8\pi\, G\, c^{-4}\, T_{\mu\nu} \tag{2}$$

($R_{\mu\nu}$ = Ricci curvature tensor, R = scalar curvature, $g_{\mu\nu}$ = metric tensor, $T_{\mu\nu}$ = stress-energy tensor).

In order to account for the pre-existing global cosmic curvature of the SST and the fact that, contrary to General Relativity, the SST curvature does not vanish in the absence of standard matter and of a cosmological constant, the above Einstein equation can be replaced by:

$$R_{\mu\nu} - 1/2\, g_{\mu\nu}\, R + g'_{\mu\nu}\, S + g_{\mu\nu}\, \Lambda = 8\pi\, G\, c^{-4}\, T_{\mu\nu} \tag{3}$$

where the term $g'_{\mu\nu}\, S$ accounts for the SST curvature.

In [32], the following Friedmann-like relation for the standard matter universe was considered:

$$H^2 = 8\pi\, G\, \rho/3 - k\, a^{-2}\, c^2 + t^{-2} + K + \Lambda\, c^2/3 \tag{4}$$

where ρ is the energy density associated to standard matter, c the speed of light, k the usual curvature parameter and $k\, a^{-2}\, c^2$ an extra curvature term generated by the presence of matter in the Universe. Λ stands now for a possible new version of the cosmological constant, decreasing like the matter density as the Universe expands and hence free of any cosmological constant problem. The term t^{-2} has a positive sign independent of k and dominates the large scale expansion of the Universe. K is a correction term accounting in particular for various possible effects described in [1, 32].

Similar steps allow to account for the role of the PSD, using the standard Robertson-Walker equation [162] together with (2) and (3) and modifying these equations in order to account for the local space anisotropy associated to the privileged space direction. One can then expect to suitable modify (4), basically through K and Λ even if t^{-2} remains the dominant term. The more the Universe will be large, the more the PSD will be astrophysically stable for observation.

Equation (4) will tend asymptotically to the $H\, t = 1$ law for $t \rightarrow \infty$ provided the terms other than t^{-2} vanish faster in this limit [32, 33]. We expect the matter density and all its effects in (4) to decrease like t^{-3}. This is an important difference with respect to standard cosmology, including quantum gravity, as the t^{-2} term in (4) has a deeper origin than gravity and field theory.

In the 2015 releases, the Planck Collaboration reports [134] values of 67.8 ± 0.9 km s^{-1} Mpc^{-1} for the LLH constant and of 13.799 ± 0.038 billion years for the age of the Universe. The product of the two values turns out to be around 0.956 up to less than 1.5%. Therefore, there is less than 6% left for the sum of the contributions to (4) other than t^{-2}, including those associated to dark matter and dark energy. Such a result is consistent with the previous one discussed in [32].

Thus, the purely geometric prediction of cosmic SST $H\,t = 1$ turns out to be remarkably close to the result of observational analyses. The cosmological effects of standard matter, including dark matter and dark energy, appear as a comparatively small correction that can naturally vanish in the $t \rightarrow \infty$ limit if K and Λ vary like the matter density in an expanding Universe. Actually, standard cosmology may even look rather artificial when assuming that the physical vacuum expands carrying everywhere a permanent density of harmonic-oscillator zero modes and other condensates. The cosmological constant problem is thus generated. In a preonic picture, this phenomenon can be readily avoided [1, 2, 28, 31, 139, 149, 150] assuming that the vacuum reacts generating such condensates only in the presence of standard matter and at the relevant frequencies.

For obvious reasons, the Planck analysis and the Euclid mission should not concentrate only on the standard cosmological pattern. Furthermore, possible implications of the Gödel incompleteness theorem [163, 164] for Particle Physics and Cosmology should also be studied [18, 19, 165, 166].

10 Ultra-high energy cosmic rays

Recent results from the AUGER [167] and Telescope Array [168] collaborations can be found in [169] and [170]. It clearly emerges (see, for instance, [171, 172]) that the basic dynamical mechanism at the origin of the flux suppression of ultra-high energy cosmic rays remains to be determined. As already recognized previously [173, 174], it is not yet clear if the observed fall of the UHECR spectrum corresponds to a signature of the Greisen-Zatsepin-Kuzmin (GZK) [175, 176] cutoff or to the maximum energies available at astrophysical sources. In particular, such a limitation makes difficult to interpret data [177] on UHECR traveling on moderate extragalactic distances.

Futhermore, fundamental physics at UHE including possible effects of new physics remains poorly known [2, 37, 38]. A consequent experimental and theoretical effort in this domain is clearly required. In [39, 40] and [178–180], a possible mechanism based on Lorentz symmetry violation (LSV) with a natural absolute rest frame (the vacuum rest frame, VRF) was suggested to explain a possible absence of the GZK cutoff. The basic idea was to compensate the effect of the small energy of a CMB photon with a small modification of the kinematics of the UHE particle due to LSV. Thus, the basic GZK reaction would be forbidden by the new kinematics. In the recent years, more sophisticated scenarios involving violations of fundamental principles (relativity, quantum mechanics, energy and momentum conservation...) have been considered [27, 28, 42] to describe the observed data. In such a general framework, various scenarios are possible with or without the UHECR flux suppression.

The decay of cosmic superluminal particles (not tachyons) through the Cherenkov effet in vacuum [20] was also considered since 1996 as a possible source of UHECR [30, 41, 181]. The search for more direct superbradyon signatures in cosmic-ray experiments can also be attempted. It must take into account the possible very weak interaction rate of individual superbradyons with conventional matter, as well as the violations of standard causality in experimental signatures that may result from superluminal propagation [26]. The ratio between the superbradyon critical speed c_s and the speed of light c is a basic unknown, and can be as large as the ratio between c and the speed of sound.

The question of the validity of standard low-energy symmetries at very high energy is also crucial to understand UHECR physics [27, 37, 42]. Particle propagation in vacuum at UHE must simultaneously be given special attention, considering possible vacuum inhomogeneities and unconventional interactions of UHE particles with vacuum over large distance scales.

In this respect, some orders of magnitude must be taken into account. While a present bound on the quark radius can be $\sim 10^{-16}$ cm [182], and the electron radius appears to be smaller than $\sim 10^{-18}$ cm [183], the proton radius [184] is $\sim 10^{-13}$ cm at rest and $\sim 10^{-24}$ cm at E (energy) $\sim 10^{20}$ eV. Its wavelength at $\sim 10^{20}$ eV has a similar value. Thus, the wavelength and longitudinal size of the proton at $\sim 10^{20}$ eV can even be smaller than the sizes of quarks and electrons at rest.

What can then be the interaction between the UHE proton and the internal structure of the physical vacuum including its excitations? In the case of a superbradyonic vacuum, standard relativity will not be an exact symmetry and LSV can play an important role for UHE particles. In view of the strength of standard Lorenz contraction, the shape itself of a $\sim 10^{20}$ eV proton can change tending to adapt itself to the transverse volume. The Lorentz contraction law can also change and no longer follow the relativistic formula. The proton size and structure can in any case be sensitive to the internal structure of vacuum and of its excitations. One can then imagine various unconventional effects, such as:

- An exchange of energy and momentum between the UHE proton and the vacuum internal structure. In particular, the vacuum can release a small amount of energy, allowing the proton to decay. Or conversely, take some energy and momentum from the proton and transfer them to preonic waves.

- UHECR can also be trapped in vacuum inhomogeneities at astrophysical level. For instance, if a UHE proton enters a vacuum region where the ground state energy is lower, its effective energy will be larger and a decay can occur. Subsequently, the UHE particle may not be allowed to automatically quit this vacuum region except possibly if a new interaction happens.

These are just two examples of cosmological and astrophysical unconventional mechanisms that can act on the UHECR flux in the presence of a nonstandard vacuum. Deformed quantum mechanics can produce small energy and momentum uncertainties equally leading to a UHECR flux suppression. A similar role can be played by small violations of energy and momentum conservation.

As a consequence of this kind of uncertainties, there is no real proof of the validity of models and algorithms for particle interactions at the highest energies used to analyse UHE data and study UHECR properties, even if interesting work is carried on [185]. Another open question is whether cosmic rays can be sensitive to the privileged space direction generated by the SST as it can well be the case for the CMB. Exploring possible correlations between high-energy cosmic rays data and the recent Planck observation of CMB anisotropies deserves a long-term systematic effort.

10.1 JEM-EUSO

Satellite detection of UHECR will also be an important new form of observation. The Extreme Universe Space Observatory (now JEM-EUSO [186–188]), to be hosted by the JEM module of the Japanese KIBO facility of the International Space Station (ISS) at an altitude of about 400 Km, will be the first space mission devoted to UHECR and UHE cosmic neutrinos. JEM-EUSO is planned to be launched in 2017. Using the Earth atmosphere as the target, JEM-EUSO is expected to reach unprecedented accuracy concerning the UHECR energy and direction, as well as spectrum composition.

11 On the status and potentialities of physics at accelerators

While no evidence for supersymmetry has yet been found [189], ATLAS and CMS have reported on other possible signatures of new physics. An example is provided by the (yet preliminary) diphoton signals [190–192]. Potential diboson signatures [193] also deserve close attention.

There have been a number of attempts to interpret recent LHC data using various patterns beyond the standard model. Precisely, new families of particles generated through excitations of the superbradyonic vacuum [1, 2, 18, 19, 28] appear as natural candidates for new physics at LHC. Not only with a critical speed equal to c, but also with a larger critical speed.

Possible effects of such a new fundamental dynamics on the quark-gluon plasma [194, 195], including unconventional interactions between the produced plasma and the preonic vacuum structure, must also be seriously considered. Superluminal particle production may occur in this context.

More generally, superluminal particles may be produced in LHC experiments through events usually considered as missing energy or background. The lifetime of a superluminal object initially produced with a speed larger than c is then a crucial, unknown parameter, and similarly for the interaction of the superluminal particle with the detector and its surrounding. In any case, and besides possible appropriate changes in the experimental design, new specific algorithms are required for data analysis. The direct observation and identification of a "Cherenkov" decay in vacuum [20] would be an exciting possibility, although most likely difficult in practice [196].

12 Conclusion and comments

Understanding the mathematical structure of space-time and the actual content of the physical vacuum must be a primordial goal for current and future Cosmology and Particle Physics.

As already stressed in [1], the possible existence of primordial CMB B-modes of primordial origin, would in any case not be by itself an evidence cosmic inflation. Pre-Big Bang patterns would be able to provide alternative explanations of such a signature.

Alternative particle physics and cosmologies should not be considered as marginal in data analysis. The design and realization of future experimental and observational projects should not ignore them.

The 2016 *Planck* results will obviously be very important. But a new satellite mission devoted to the CMB, with new technology allowing by further data analyses, is in any case necessary. In particular, the question of the possible existence of a privileged space direction must be clearly elucidated.

Similarly, the South Pole and Atacama programs, as well as AUGER and the Telescope Array, are crucial and must remain operating long-term projects with permanent upgrading. JEM-EUSO will complete the UHECR programs. The detection of cosmic neutrinos also requires attention.

As just discussed, CERN experiments and other high-energy programs can also play an important role in the search for new physics and the identification of basic cosmological ingredients. Possible excitations of a preonic (superbradyonic) vacuum and even free preons (superbradyons) may be produced in particle collisions or released by the physical vacuum interacting with high-energy particles.

Theory and phenomenology need in turn a profound step forward. Just as Yoichiro Nambu brought new ideas and basic techniques into superconductivity and Particle Physics, current patterns and ideas beyond standard particle physics need a more precise realization with original unconventional contributions and performing technical approaches. A reflection involving the different existing forms of matter can help to formulate a suitable theory of the physical vacuum and of its excitations.

Prospects of new physics and alternative cosmologies are also discussed in [18, 19].

References

[1] L. Gonzalez-Mestres, *BICEP2, Planck, spinorial space-time, pre-Big Bang*, invited talk at the 3rd International Conference on New Frontiers in Physics (ICNFP 2014), Kolymbari, Crete, Greece, August 23 - 30, 2014, *EPJ Web of Conferences* **95**, 03014 (2015), and references therein.

[2] L. Gonzalez-Mestres, *Tests and prospects of new physics at very high energy*, contribution the 3rd International Conference on New Frontiers in Physics, Kolymbari, Crete, Greece, August 23 - 30, 2014, *EPJ Web of Conferences* **95**, 05007 (2015), and references therein.

[3] BICEP2 Collaboration, *Detection Of B-mode Polarization at Degree Angular Scales by BICEP2*, *Physical Review Letters* **112**, 241101 (2014). Original preprint version (March 2014): arXiv:1403.3985v1.

[4] BICEP2 Collaboration, *BICEP2 II: Experiment and Three-Year Data Set*, arXiv:1403.4302.

[5] Planck mission, European Space Agency, http://sci.esa.int/planck/

[6] Planck Collaboration, *Planck 2013 results. XVI. Cosmological parameters*, arXiv:1303.5076.

[7] Planck Collaboration, *Planck 2013 results. XXII. Constraints on inflation*, arXiv:1303.5082.

[8] A. Iljjas, P.J. Steinhardt and A. Loeb, *Inflationary paradigm in trouble after Planck2013*, arXiv:1402.6980, and references therein.

[9] A. Iljjas, P.J. Steinhardt and A. Loeb, *Inflationary schism after Planck2013*, *Phys.Lett.B* **723**, 261 (2013), arXiv:1304.2785, and references therein.

[10] A.H. Guth, D.I. Kaiser and Y. Nomura, *Inflationary paradigm after Planck 2013*, arXiv: 1312.7619, and references therein.

[11] A. Linde, *Inflationary Cosmology after Planck 2013*, arXiv:1402.0526, and references therein. arXiv:1303.5082.

[12] A. Iljjas and P.J. Steinhardt, *Implications of Planck2015 for inflationary, ekpyrotic and anamorphic bouncing cosmologies*, arXiv:1512.09010

[13] A. Linde, *A brief history of the multiverse*, arXiv:1512.01203

[14] See, for instance, ESA and *Planck*, *Planck: gravitational waves remain elusive*, http://sci.esa.int/planck/55362-planck-gravitational-waves-remain-elusive/

[15] BICEP2/Keck and Planck Collaborations, *A joint analysis of BICEP2/Keck Array and Planck data*, arXiv:1502.00612.

[16] L. Gonzalez-Mestres, *CMB B-modes, spinorial space-time and Pre-Big Bang (I)*, mp_arc 14-16, and references therein.

[17] L. Gonzalez-Mestres, *CMB B-modes, spinorial space-time and Pre-Big Bang (II)*, mp_arc 14-60, and references therein.

[18] Complementary material is provided in our contribution to this conference *Spinorial space-time and the origin of Quantum Mechanics*, https://indico.cern.ch/event/344173/session/22/contribution /405

[19] See also our contribution to this conference *Spinorial space-time and Hagedorn-like temperatures*, https://indico.cern.ch/event/344173/session/22/contribution/406

[20] L. Gonzalez-Mestres, *Properties of a possible class of particles able to travel faster than light*, Proceedings of the January 1995 Moriond Workshop, Ed. Frontières, arXiv:astro-ph/9505117

[21] L. Gonzalez-Mestres, *Cosmological Implications of a Possible Class of Particles Able to Travel Faster than Light*, Proceedings of the TAUP 1995 Conference, *Nucl. Phys. Proc. Suppl.* **48** (1996), 131, arXiv:astro-ph/9601090.

[22] Planck Collaboration, *Planck 2013 results. XXIII. Isotropy and statistics of the CMB*, A&A, **571**, A23 (2014), arXiv:1303.5083 and references therein.

[23] L. Gonzalez-Mestres, *Spinorial space-time and privileged space direction (I)*, mp_arc 13-75, and references therein.

[24] See, for instance, The Kavli Foundation, *A New Baby Picture of the Universe*, http://www.kavlifoundation.org/science-spotlights/kicc-planck-universe

[25] See, for instance, the Planck Collaboration, *Planck 2015 results. XVI. Isotropy and statistics of the CMB*, arXiv:1506.07135

[26] L. Gonzalez-Mestres, *Space, Time and Superluminal Particles*, arXiv:physics/9702026.

[27] L. Gonzalez-Mestres, *Cosmic rays and tests of fundamental principles*, CRIS 2010 Proceedings, *Nucl. Phys. B, Proc. Suppl.* **212-213** (2011), 26, and references therein. The *arXiv.org* version arXiv:1011.4889 includes a relevant Post Scriptum.

[28] See, in particular, our contributions to ICNFP 2012 and ICNFP 2013 [31, 32, 37, 38].

[29] Together with the existence of a local privileged space direction for each comoving observer as predicted by the cosmic SST geometry, parity violation is a natural ingredient of the generation of such an observable effect [27, 28].

[30] L. Gonzalez-Mestres, *Physical and Cosmological Implications of a Possible Class of Particles Able to Travel Faster than Light*, contribution to the 28th International Conference on High Energy Physics, Warsaw 1996, arXiv:hep-ph/9610474, and references therein.

[31] L. Gonzalez-Mestres, *Pre-Big Bang, fundamental Physics and noncyclic cosmologies*, International Conference on New Frontiers in Physics, ICFP 2012, Kolymbari, Crete, June 10-16 2012, *EPJ Web of Conferences* **70**, 00035 (2014), and references therein. Preprint at mp_arc 13-18.

[32] L. Gonzalez-Mestres, *Pre-Big Bang, space-time structure, asymptotic Universe*, 2nd International Conference on New Frontiers in Physics, Kolymbari, Crete, Greece, August 28 - September 5, 2013, *EPJ Web of Conferences* **71**, 00063 (2014), references therein and Post Scriptum to the preprint hal-00983005.

[33] L. Gonzalez-Mestres, *Planck data, spinorial space-time and asymptotic Universe*, mp_arc 13-33, and references therein.

[34] L. Gonzalez-Mestres, *Spinorial space-time and Friedmann-like equations (I)*, mp_arc 13-80, and references therein.

[35] A recent attempt to examine the current theoretical issues of the ΛCDM model is P. Bull et al., *Beyond ΛCDM: Problems, solutions, and the road ahead*, arXiv:1512.05356 . However, the patterns we consider here seem to lie beyond the scope of this paper.

[36] L. Gonzalez-Mestres, *Preon models, relativity, quantum mechanics and cosmology (I)*, arXiv:0908.4070 , and references therein.

[37] L. Gonzalez-Mestres, *High-energy cosmic rays and tests of basic principles of Physics*, International Conference on New Frontiers in Physics, ICFP 2012, Kolymbari, Crete, June 10-16 2012, *EPJ Web of Conferences* **70**, 00047 (2014), and references therein. Preprint at mp_arc 13-19.

[38] L. Gonzalez-Mestres, *Ultra-high energy physics and standard basic principles*, 2nd International Conference on New Frontiers in Physics, Kolymbari, Crete, Greece, August 28 - September 5, 2013, *EPJ Web of Conferences* **71**, 00062 (2014), and Post Scriptum to the preprint mp_arc 14-31.

[39] L. Gonzalez-Mestres, *Vacuum Structure, Lorentz Symmetry and Superluminal Particles*, arXiv:physics/9704017.

[40] L. Gonzalez-Mestres, *Absence of Greisen-Zatsepin-Kuzmin Cutoff and Stability of Unstable Particles at Very High Energy, as a Consequence of Lorentz Symmetry Violation*, Proceedings of the 25th International Cosmic Ray Conference (held 30 July - 6 August, 1997 in Durban, South Africa), Edited by M. S. Potgieter, C. Raubenheimer, and D. J. van der Walt, Transvaal, South Africa: Potchefstroom University, 1997, Vol6, http://adsabs.harvard.edu/full/1997ICRC....6..113G and arXiv:physics/9705031 (May 1997 preprint).

[41] L. Gonzalez-Mestres, *Superluminal Particles and High-Energy Cosmic Rays*, same ICRC97 Proceedings, http://adsabs.harvard.edu/full/1997ICRC....6..109G and arXiv:physics/9705032.

[42] L. Gonzalez-Mestres, *Testing fundamental principles with high-energy cosmic rays*, HEP Europhysics Conference, Grenoble, July 2011, *PoS EPS-HEP2011* 390, and references therein.

[43] Nobel Foundation, press release *The Nobel Prize in Physics 2008. Yoichiro Nambu, Makoto Kobayashi, Toshihide Maskawa*, http://www.nobelprize.org/nobel_prizes/physics/laureates/2008/

[44] Y. Nambu, *Quasiparticles and Gauge Invariance in the Theory of Superconductivity Phys. Rev.* **117**, 648 (1960). Available, for instance, at the address http://www.mat.unimi.it/users/gaeta/SD2/risorse/nambu.pdf (Università degli Studi di Milano)

[45] Y. Nambu, *A 'Superconductor' Model of Elementary Particles and Its Consequences*, Proceedings of the Midwest Conference on Theoretical Physics, eds. F.J. Belinfante, S.G. Gartenhaus and R.W. King, Purdue University, Lafayette, Indiana, April 1-2, 1960. Online in Google Books: T. Eguchi and K. Nishijima, *Broken Symmetry: Selected Papers of Y Nambu*, World Scientific 1995.

[46] See also Y. Nambu, *Spontaneous Symmetry Breakingin Particle Physics: a Case of Cross-Fertilization*, 2008 Nobel Lecture, https://www.nobelprize.org/nobel_prizes/physics/laureates/2008/nambu_lecture.pdf.

[47] Y. Nambu, *Axial vector current conservation in weak interactions*, *Phys. Rev. Lett.* **4**, 381 (1960), available at the Princeton University address http://puhep1.princeton.edu/ kirkmcd/examples/EP/nambu_prl_4_380_60.pdf

[48] J. Goldstone, *Field theories with "Superconductor" solutions*, *Nuovo Cimento 19*, 154 (1961). CERN preprint CM-P00057225 (August 1960) http://cds.cern.ch/record/343400/files/CM-P00057225.pdf

[49] Y. Nambu and G. Jona-Lasinio, *Dynamical Model of Elementary Particles Based on an Analogy with Superconductivity. I*, *Phys. Rev.* **122**, 345 (1961). Online in Google Books: C.H. Lai, *Selected Papers on Gauge Theory of Weak and Electromagnetic Interactions*, World Scientific 1981.

[50] Y. Nambu and G. Jona-Lasinio, *Dynamical Model of Elementary Particles Based on an Analogy with Superconductivity. II*, *Phys. Rev.* **124**, 246 (1961). Available at the address http://cosmology.princeton.edu/~mcdonald/examples/EP/nambu_pr_124_246_61.pdf (Princeton University).

[51] A list of publications of Yoichiro Nambu can be found at the address http://ptps.oxfordjournals.org/content/86/iii.full.pdf+html (*Progress of Theoretical Physics Supplements*).

[52] See, for instance, P. Frampton, Pre-String Theory, arXiv:1508.06382.

[53] Y. Nambu, *Duality and hydrodynamics*, Lectures at the Copenhagen Summer Symposium (1970).

[54] T. Goto, *Relativistic quantum mechanics of one-dimensional mechanical continuum and subsidiary condition of dual resonance*, *Prog. Theor. Phys.* **46**, 1560 (1971). Available at the address http://ptp.oxfordjournals.org/content/46/5/1560.full.pdf

[55] See, for instance, A. Lazanu and E. Shellard, *Constraints on the Nambu-Goto cosmic string contribution to the CMB power spectrum in light of new temperature and polarisation data*, *JCAP* **1502**, 024 (2015), arXiv:1410.5046.

[56] Y. Nambu, *Strings, monopoles, and gauge fields*, *Phys. Rev.* D10, 4262, 1974.

[57] See, for instance, G. Ripka, *Dual superconductor models of color confinement*, Lectures delivered at the European Centre for Theoretical Studies in Nuclear Physics and Related Areas, 2002-2003, ECT, Trento, arXiv:hep-ph/0310102.

[58] See, for instance, I. Kulik, *Nonlinear waves and solitons in superconductors near critical temperature*, *Physica B + C* **107**, 169 (1981).

[59] An example of more recent work is J. Garaud, J. Carlstrom and E. Babaev, *Topological solitons in three-band superconductors with broken time reversal symmetry*, *Phys. Rev. Lett.* **107**, 197001 (2011), arXiv:1107.0995.

[60] See, for instance, M. Tylutki et al., *Dark-Bright Solitons in a Superfluid Bose-Fermi Mixture*, arXiv:1601.01471.

[61] See, for instancen S. Kivelson, D. Rokhsar and J. Sethna, *Topology of the resonating valence-bond state: Solitons and high-Tc superconductivity*, *Phys. Rev.* **B 35**, 8865 (1987)

[62] For a recent paper : A. Diens et al. *Optical excitation of Josephson plasma solitons ina cuprate superconductor*, *Nature Materials* **12**, 535 (2013), available at the address https://www.researchgate.net/publication/236076175 (*Research Gate*)

[63] BICEP2 Collaboration, *BICEP2 I: Detection Of B-mode Polarization at Degree Angular Scales*, arXiv:1403.3985v1

[64] BICEP2 Collaboration, *BICEP2 II: Experiment and Three-Year Data Set*, arXiv:1403.4302v1

[65] Wilkinson Microwave Anisotropy Probe, http://map.gsfc.nasa.gov/

[66] G. Bogdanoff, *Fluctuations quantiques de la signature de la métrique à l'échelle de Planck*, Thesis, Université de Bourgogne 1999, and related published papers.

[67] I. Bogdanoff, *Etat topologique de l'espace-temps à l'échelle 0*, Thesis, Université de Bourgogne 2002, and related published papers.

[68] LIGO Scientific Collaboration, http://www.ligo.org/

[69] The LIGO Scientific Collaboration, the VIRGO Collaboration, *An all-sky search for long-duration gravitational wave transients with LIGO*, arXiv:1511.04398.

[70] D. Castelvecchi, *Has giant LIGO experiment seen gravitational waves?*, *Nature News* 30 september 2015, http://www.nature.com/news/has-giant-ligo-experiment-seen-gravitational-waves-1.18449

[71] H. Poincaré, *Sur la dynamique de l'électron*, *Comptes rendus de l'Académie des Sciences* **140**, 1504 (1905), http://smf4.emath.fr/Publications/Gazette/2005/104/smf_gazette_104_60-70.pdf

[72] A. Einstein, *Geometrie und erfahrung*, *Preus. Akad. der Wissench., Sitzungsberichte*, part **I**, p. 123 (1921), http://www.philoscience.unibe.ch/documents/kursarchiv/WS99/Geometrie.pdf . English translation *Geometry and experience* in *Sidelights on relativity*, Methuen, London 1922, http://www.gutenberg.org/files/7333/7333-h/7333-h.htm

[73] *The BICEP and Keck Array CMB Experiments*, http://bicepkeck.org/

[74] Caltech Observational Cosmology Group, *BICEP: Robinson Gravitational Wave Background Telescope*, http://bicep.caltech.edu/public/

[75] K.W. Yoon et al., *The Robinson Gravitational Wave Background Telescope (BICEP): a bolometric large angular scale CMB polarimeter*, in *Millimeter and Submillimeter Detectors and Instrumentation for Astronomy III*, Proceedings of SPIE, Vol. 6275 (2006), arXiv:astro-ph/0606278.

[76] BICEP1 Collaboration, *Degree-Scale CMB Polarization Measurements from Three Years of BICEP1 Data*, *ApJ* **783**, 67 (2014), arXiv:1310.1422.

[77] C.L. Kuo et al., *Antenna-coupled TES bolometer arrays for CMB polarimetry*, in SPIE Proceedings Vol. 7020, Marseille, 2008, arXiv:0908.1464.

[78] A. Orlando et al., *Antenna-coupled TES Bolometer Arrays for BICEP2/Keck and SPIDER*, in SPIE Proceedings Vol. 7741, San Diego, 2010, arXiv:1009.3685.

[79] C.L. Kuo et al., *Antenna-coupled TES bolometers for the Keck Array, Spider, and Polar-1*, arXiv: 1208.1247. See also http://www.cfa.harvard.edu/CMB/bicep2/ and http://www.cfa.harvard.edu/CMB/keckarray/

[80] J.A. Brevik et al., *Initial performance of the BICEP2 antenna-coupled superconducting bolometers at the South Pole*, in SPIE Proceedings Vol. 7741, San Diego, 2010, http://authors.library.caltech.edu/22834/1/Brevik2010p12867Adaptive_Optics_Systems_Pts_1-3.pdf

[81] Jet Propulsion Laboratory, California Institute of Technology, *TES Bolometers Enable a New Probe of the Infant Universe*, http://microdevices.jpl.nasa.gov/capabilities/superconducting-devices/tes-bolometers.php

[82] C.D. Sheehy et al., *The Keck Array: a pulse tube cooled CMB polarimeter*, arXiv:1104.5516.

[83] S. Kernasovskiy et al., *Optimization and sensitivity of the Keck Array*, arXiv: 1208.0857.

[84] P. Ade et al., *Antenna-coupled TES bolometers used in BICEP2, Keck array, and SPIDER*, arXiv:1502.00619.

[85] The BICEP2, Keck Array Collaborations, *BICEP2 / Keck Array V: Measurements of B-mode Polarization at Degree Angular Scales and 150 GHz by the Keck Array*, arXiv:1502.00643.

[86] The BICEP2, Keck Array Collaborations, *BICEP2 / Keck Array VI: Improved Constraints On Cosmology and Foregrounds When Adding 95 GHz Data From Keck Array*, arXiv:1510.09217.

[87] Z.Ahmed et al., *BICEP3: a 95 GHz refracting telescope for degree-scale CMB polarization*, arXiv:1407.5928.

[88] W.L.K. Wu et al., *Initial Performance of BICEP3: A Degree Angular Scale 95 GHz Band Polarimeter*, arXiv:1601.00125.

[89] B.A. Benson et al., *A Next-Generation Cosmic Microwave Background Polarization Experiment on the South Pole Telescope*, arXiv:1407.2973.

[90] See, for instance, http://kicp.uchicago.edu/~bbenson/detectors.html (Bradford Benson, University of Chicago).

[91] POLAR CMB Polarization Experiment, http://polar-array.stanford.edu/ (Stanford University). See also Chao Lin-Kuo transparencies at the 2009 Meeting of the SLAC Users Organization.

[92] R. O'Brient et al., *Antenna-coupled TES bolometers for the Keck Array, Spider, and Polar-1*, arXiv:1208.1247.

[93] The POLARBEAR Collaboration, *A Measurement of the Cosmic Microwave Background B-Mode Polarization Power Spectrum at Sub-Degree Scales with POLARBEAR*, *Astrophysical Journal* **794**, 171 (2014), arXiv:1403.2369.

[94] Z. Kermish et al., *The POLARBEAR Experiment*, arXiv:1210.7768.

[95] D. Barron et al., *Development and characterization of the readout system for POLARBEAR-2*, arXiv:1410.7488.

[96] The POLARBEAR Collaboration, *POLARBEAR Constraints on Cosmic Birefringence and Primordial Magnetic Fields*, arXiv:1509.02461.

[97] J.E. Austermann et al., *SPTpol: an instrument for CMB polarization measurements with the South Pole Telescope*, arXiv:1210.4970.

[98] K.T. Story et al., *A Measurement of the Cosmic Microwave Background Gravitational Lensing Potential from 100 Square Degrees of SPTpol Data*, arXiv:1412.4760.

[99] R. Keisler, *Measurements of Sub-degree B-mode Polarization in the Cosmic Microwave Background from 100 Square Degrees of SPTpol Data*, arXiv:1503.02315.

[100] Atacama Cosmology Telescope (ACT), http://lambda.gsfc.nasa.gov/product/act/ (NASA) and http://www.princeton.edu/act/ (Princeton University).

[101] M.D. Niemack et al., *ACTPol: A polarization-sensitive receiver for the Atacama Cosmology Telescope*, in SPIE Proceedings Vol. 7741, San Diego, 2010, arXiv:1006.5049.

[102] R. Datta et al., *Design and Deployment of a Multichroic Polarimeter Array on the Atacama Cosmology Telescope*, arXiv:1510.07797.

[103] A. van Engelen, *The Atacama Cosmology Telescope: Lensing of CMB Temperature and Polarization Derived from Cosmic Infrared Background Cross-Correlation*, arXiv:1412.0626.

[104] T.Su et al., *On the redshift distribution and physical properties of DSFGs from ACT*, arXiv:1511.06770.

[105] C.Sifon et al., *The Atacama Cosmology Telescope: Dynamical masses for 44 SZ-selected galaxy clusters over 755 square degree*, arXiv: 1512.00910.

[106] J.W. Appel et al., *The Cosmology Large Angular Scale Surveyor (CLASS): 38 GHz detector array of bolometric polarimeters*, arXiv:1408.4789.

[107] K. Rostem et al., *Scalable background-limited polarization-sensitive detectors for mm-wave applications*, arXiv:1408.4790.

[108] D.T. Chuss et al., *Cosmology Large Angular Scale Surveyor (CLASS) Focal Plane Development*, arXiv:1511.04414.

[109] J.P. Filippini et al., *SPIDER: a balloon-borne CMB polarimeter for large angular scales*, arXiv:1106.2158.

[110] A.S. Rahlin et al., *Pre-flight integration and characterization of the SPIDER balloon-borne telescope*, arXiv:1407.2906.

[111] J. Gudmundsson et al., *The Thermal Design, Characterization, and Performance of the SPIDER Long-Duration Balloon Cryosta* arXiv:1506.06953.

[112] S. Bryan et al., *A cryogenic rotation stage with a large clear aperture for the half-wave plates in the Spider instrument*, arXiv:1510.01771.

[113] R. Misawa et al., *PILOT: a balloon-borne experiment to measure the polarized FIR emission of dust grains in the interstellar medium*, arXiv:1410.5760.

[114] The PILOT project, http://pilot.irap.omp.eu/

[115] A. Faessler, *Beta decay and the cosmic neutrino background*, 2nd International Conference on New Frontiers in Physics, Kolymbari, Crete, Greece, August 28 - September 5, 2013, *EPJ Web of Conferences*, **71**, 00044 (2014).

[116] KATRIN, https://www.katrin.kit.edu/

[117] See, for instance, R.G. Hamish Robertson, *KATRIN: an experiment to determine the neutrino mass from the beta decay of tritium*, arXiv:1307.5486.

[118] A. Faessler at al., *Search for the Cosmic Neutrino Background*, 11th International Spring Seminar on Nuclear Physics: Shell Model and Nuclear Structure: Achievements of the past two decades, Ischia May 2014, *Journal of Physics: Conference Series* **580** (2015) 012040

[119] S. Betts et al. *Development of a Relic Neutrino Detection Experiment at PTOLEMY: Princeton Tritium Observatory for Light, Early-Universe, Massive-Neutrino Yield*, arXiv:1307.4738

[120] J. Greenwald, *PPPL, Princeton launch hunt for Big Bang particles, DOE Pulse* **430**, January 12, 2015, http://web.ornl.gov/info/news/pulse/no430/story1.shtml

[121] A. Cocco, G. Mangano and M. Messina, *Probing Low Energy Neutrino Backgrounds with Neutrino Capture on Beta Decaying Nuclei, Journal of Physics: Conference Series* **110** (2008) 082014.

[122] A. Ringwald, *Prospects for the direct detection of the cosmic neutrino background*, arXiv:0901.1529

[123] A. Cocco, G. Mangano and M. Messina, *Low energy antineutrino detection using neutrino capture on electron capture decaying nuclei, Phys. Rev.* **D 79**, 053009 (2009) arXiv:0903.1217.

[124] See also Yu-Feng Li, *Detection Prospects of the Cosmic Neutrino Background, International Journal of Modern Physics* bf A 30, 1530031 2015), arXiv:1504.03966, and references therein.

[125] Planck Collaboration, *Planck 2013 results. I. Overview of products and scientific results*, arXiv:1303.5062.

[126] Planck Collaboration, *Planck 2015 results. XX. Constraints on inflation*, arXiv:1502.02114.

[127] Planck Collaboration, *Planck 2015 results. XVI. Isotropy and statistics of the CMB* arXiv: 1506.07135.

[128] H.K. Eriksen et al., *Asymmetries in the CMB anisotropy field, Astrophys.J.* **605**, 14 (2004) and Erratum *ibid.* **609**, 1198 (2004), arXiv: astro-ph/0307507.

[129] F. K. Hansen, A. J. Banday and K. M. Gorski, *Testing the cosmological principle of isotropy: local power spectrum estimates of the WMAP data, Mon.Not.Roy.Astron.Soc.* **354**, 641 (2004), arXiv:astro-ph/0404206.

[130] C.-G. Park, *Non-Gaussian Signatures in the Temperature Fluctuation Observed by the Wilkinson Microwave Anisotropy Probe*, *Mon.Not.Roy.Astron.Soc.* **349**, 313 (2004), arXiv:astro-ph/0307469.

[131] Planck mission (ESA), *Planck publications*, http://www.cosmos.esa.int/web/planck/publications

[132] Planck Collaboration, *Planck 2015 results. I. Overview of products and scientific results*, arXiv: 1502.01582.

[133] *Planck 2015 release. Explanatory supplement*, http://wiki.cosmos.esa.int/planckpla2015/

[134] Planck Collaboration, *Planck 2015 results. XIII. Cosmological parameters*, arXiv:1502.01589.

[135] Planck Collaboration, *Planck 2013 results. XXXI. Consistency of the Planck data*, arXiv: 1508.03375.

[136] The Kavli Institute for Cosmology (KICC, Cambridge, UK), *Planck*, http://www.kicc.cam.ac.uk/research/cosmic-microwave-background/planck

[137] See, for instance, the Planck Collaboration, *Planck 2015 results. XI. CMB power spectra, likelihoods, and robustness of parameters*, arXiv:1507.02704

[138] ESA-*Planck* News, *Planck reveals an almost perfect Universe*, 21 March 2013 release.

[139] L. Gonzalez-Mestres, *Lorentz symmetry violation, dark matter and dark energy*, Proceedings of the Invisible Universe International Conference (Paris 2009), *AIP Conf.Proc.* **1241** (2010),120. The *arXiv.org* version arXiv:0912.0725 contains a relevant Post Scriptum.

[140] L. Gonzalez-Mestres, *Pre-Big Bang, vacuum and noncyclic cosmologies*, Europhysics Conference on High Energy Physics, Grenoble, July 2011, *PoS EPS-HEP2011* 479, and references therein.

[141] Euclid Consortium, http://www.euclid-ec.org/

[142] R. Laureijs et al., *Euclid Definition Study Report*, arXiv:1110.3193.

[143] Dark Energy Spectroscopic Instrument (DESI), http://desi.lbl.gov/

[144] M. Levi et al. for the DESI Collaboration, *The DESI Experiment, a whitepaper for Snowmass 2013*, arXiv:1308.0847.

[145] See also K.N. Abazajian et al., *Inflation Physics from the Cosmic Microwave Background and Large Scale Structure*, Report from the "Dark Energy and CMB" working group for the American Physical Society's Division of Particles and Fields, Snowmass 2013, arXiv:1309.5381.

[146] K. Lundmark, *Curvature of Space-Time in de Sitter's World*, *MNRAS* **84** (1924), 747, http://articles.adsabs.harvard.edu/full/1924MNRAS..84..747L

[147] G. Lemaître, *Un Univers homogène de masse constante et de rayon croissant, rendant compte de la vitesse radiale des nébuleuses extra-galactiques*, *Ann. Soc. Sci. Brux.* **A 47** (1927), 49, http://articles.adsabs.harvard.edu/full/1927ASSB...47...49L

[148] E. Hubble, *A relation between distance and radial velocity among extra-galactic nebulae*, *PNAS* **15** (1929), 168, http://www.pnas.org/content/15/3/168

[149] L. Gonzalez-Mestres, *Superbradyons and some possible dark matter signatures*, arXiv:0905.4146.

[150] L. Gonzalez-Mestres, *WMAP, Planck, cosmic rays and unconventional cosmologies*, contribution to the Planck 2011 Conference, Paris, January 2011, arXiv:1110.6171.

[151] The Gaia mission, http://www.cosmos.esa.int/web/gaia/home

[152] C. Cacciani, E. Pancino and M. Bellazini, *Gaia*, to appear in *Astronomische Nachrichten*, special issue *Reconstructing the Milky Way's History: Spectroscopic Surveys, Asteroseismology and Chemodynamical Models*, Guest Editors C. Chiappini, J. Montalban, and M. Steffen, arXiv:1512.03658.

[153] L. Wyrzykowski, *First year of the Gaia Science Alerts*, XXXVII Meeting of the Polish Astronomical Society, arXiv:1601.02827.

[154] Gaia Science Alerts Working Group, https://www.ast.cam.ac.uk/ioa/wikis/gsawgwiki/

[155] R.C. Smith, *Cataclysmic Variables*, arXiv:astro-ph/0701654.

[156] G. Lemaître, *The Beginning of the World from the Point of View of Quantum Theory*, *Nature* **127**, 706 (1931).

[157] See, for instance, M. Gasperini and G. Veneziano, in *Beyond the Big Bang : Competing Scenarios for an Eternal Universe*, Ed. R. Vaas, Springer-Verlag 2007, arXiv:hep-th/0703055.

[158] M. Gasperini, *String theory and primordial cosmology*, arXiv:1402.0101.

[159] See, for instance, P. Aurenche and L. Gonzalez-Mestres, *Glueball singularity, flavor loops and the Harari-Freund picture*, *Zeitschrift für Physik C* 1, 307 (1979), CERN preprint TH.2568, and references therein.

[160] A. Einstein, *Die Feldgleichungen der Gravitation*, *Sitzungsberichte der Preussischen Akademie der Wissenschaften zu Berlin* (1915), 844.

[161] A. Einstein, *Die Grundlage der allgemeinen Relativitätstheorie*, *Annalen der Physik* **354 (7)** (1916), 769.

[162] See, for instance, M. Trodden and S.M. Carroll, *TASI lectures: Introduction to Cosmology*, arXiv: astro-ph/0401547.

[163] K. Gödel, *Über formal unentscheidbare Sätze der "Principia Mathematica" und verwandter Systeme*, *Monatshefte für Mathematik und Physik* **38**, 173 (1931), available at the addess http://www.w-k-essler.de/pdfs/goedel.pdf (Goethe Universität, Frankfurt am Main).

[164] See, for instance, Paul J. Cohen, *Set theory and the continuum hypothesis* (1966), Dover Books in Mathematics 2008.

[165] D. Castelvecchi, *Paradox at the heart of mathematics makes physics problem unanswerable*, *Nature* **528, 207 (2015)**, http://www.nature.com/nature/journal/v528/n7581/full/nature16059.htm

[166] T. Cubitt, D. Perez-Garcia and M. Wolff, *Undecidability of the spectral gap*, *Nature News*, 9 December 2015, http://www.nature.com/news/paradox-at-the-heart-of-mathematics-makes-physics-problem-unanswerable-1.18983 , and references therein.

[167] The Pierre Auger Collaboration, https://www.auger.org/

[168] The Telescope Array Collaboration, http://www.telescopearray.org/

[169] Pierre Auger Collaboration, *The Pierre Auger Observatory: Contributions to the 34th International Cosmic Ray Conference (ICRC 2015)* 1509.03732.

[170] The Telescope Array Collaboration, *The Energy Spectrum of Cosmic Rays above 1017.2 eV Measured by the Fluorescence Detectors of the Telescope Array Experiment in Seven Years*, 1511.07510.

[171] I. Valiño, *The flux of ultra-high energy cosmic rays after ten years of operation of the Pierre Auger Observatory*, in [169], p.9.

[172] R. Engel, *Upgrade of the Pierre Auger Observatory*, in [169], p. 136.

[173] The Pierre Auger Collaboration, *Hightlights from the Pierre Auger Observatory*, contribution to the ICRC 2013 Conference, arXiv:1310.4620, and references therein.

[174] The Pierre Auger Observatory, *Contributions to the 33rd International Cosmic Ray Conference (ICRC 2013)*, arXiv:1307.5059, and references therein.

[175] K. Greisen, *End of the cosmic-ray spectrum?*, *Phys. Rev. Lett.* **16**, 748 (1966). Available at the address http://physics.princeton.edu/~mcdonald/examples/EP/greisen_−prl_−16_−748_−66.pdf (Princeton University).

[176] G.T. Zatsepin and V.A. Kuz'min, *Upper Limit of the Spectrum of Cosmic Rays, JETP Letters* **4**, 78 (1966 for the original version), http://www.jetpletters.ac.ru/ps/1624/article_−24846.shtml

[177] The Telescope Array Collaboration, *Indications of Intermediate-Scale Anisotropy of Cosmic Rays with Energy Greater Than 57 EeV in the Northern S L. Gonzalez-Mestres, ky Measured with the Surface Detector of the Telescope Array Experiment*, arXiv:1404.5890.

[178] L. Gonzalez-Mestres, *Deformed Lorentz Symmetry and High-Energy Astrophysics (I)*, talk given at the ICRC 1999 Conference, Utah August 17-25, 1999, arXiv:physics/0003080.

[179] L. Gonzalez-Mestres, *Gamma and Cosmic-Ray Tests of Special Relativity*, International Symposium on High Energy Gamma-Ray Astronomy, Heidelberg, Germany, June 26-30, 2000, arXiv:astro-ph/0011181.

[180] L. Gonzalez-Mestres, *Lorentz Symmetry Violation and Acceleration in Relativistic Shocks*, arXiv:astro-ph/0011182.

[181] L. Gonzalez-Mestres, *Superluminal Matter and High-Energy Cosmic Rays*, arXiv:astro-ph/9606054, and references therein.

[182] The ZEUS Collaboration, *Search for contact interactions, large extra dimensions and finite quark radius in ep collisions at HERA, Phys.Lett.* **B 591** 23 (2004), arXiv:hep-ex/0401009.

[183] See, for instance, the ACME Collaboration, *Order of Magnitude Smaller Limit on the Electric Dipole Moment of the Electron, Science* **343**, 269 (2014) arXiv:1310.7534, and subsequent talks by Gerald Gabrielse: http://pavi14.syr.edu/Talks/Monday/Gabrielse.pdf

[184] C. Carlson, *The Proton Radius Puzzle*, arXiv:1502.05314.

[185] See, for instance, L. Cazon, for the Pierre Auger Collaboration, *Hadronic physics with the Pierre Auger Observatory*, Proceedings of CRIS2015 (Cosmic Ray International Seminar 2015), arXiv:1512.02923.

[186] Extreme Universe Space Observatory in the Japanese Experiment Module (JEM-EUSO), http://jemeuso.riken.jp

[187] A. Haungs, JTM-EUSO Collaboration, *Physics Goals and Status of JEM-EUSO and its Test Experiments*, arXiv:1504.02593.

[188] The JEM-EUSO Collaboration, *The JEM-EUSO Mission: Contributions to the ICRC 2013* arXiv:1307.7071.

[189] K. Ulmer (CMS and ATLAS collaborations), *Supersymmetry: Experimental Status*, arXiv:1601.03774.

[190] The CMS Collaboration, *CMS Physics Analysis Summary*, 18 December 2015, https://cds.cern.ch/record/2114808/files/EXO-15-004-pas.pdf

[191] The ATLAS Collaboration, ATLAS-CONF-2015-081, 15 December 2015, http://cds.cern.ch/record/2114853/files/ATLAS-CONF-2015-081.pdf

[192] The CMS Collaboration, *Search for diphoton resonances in the mass range from 150 to 850 GeV in pp collisions at* $\sqrt{s} = 8$ *TeV, Phys. Lett.* **B750**, 494 (2015), arXiv:1506.02301.

[193] The ATLAS Collaboration *Search for high-mass diboson resonances with boson-tagged jets in proton-proton collisions at* $\sqrt{s} = 8$ *TeV with the ATLAS detector, JHEP* **12**, 55 (2015), arXiv:1506.00962.

[194] A CERN collaboration devoted to the experimental study of the quark-gluon plasma using a dedicated heavy-ion detector is ALICE, http://alice-collaboration.web.cern.ch/

[195] At the Relativistic Heavy Ion Collider (RHIC, Brookhaven), the STAR Collaboration, https://www.star.bnl.gov/ , has also as primary physics task the study of the quark-gluon plasma.

[196] Such a search would concern, in particular, small-mass and massless superbradyons (to avoid a too large rest energy) as well as superluminal particles with a critical speed not too far from c.

 DOI: 10.1051/epjconf/201612602013

Precision measurement of the proton and helium flux in primary cosmic rays with the Alpha Magnetic Spectrometer on the International Space Station

M. Heil[1,a] *On behalf of the AMS collaboration*

[1] *Massachusetts Institute of Technology (MIT), Cambridge, Massachusetts 02139, USA*

Abstract. The precise measurements of the proton and helium flux in primary cosmic rays based on on data collected by the Alpha Magnetic Spectrometer during the first 30 months of operation (May 19, 2012 to November 26, 2013) onboard the International Space Station are presented. Knowledge of the rigidity dependence of the proton and helium flux is important in understanding the origin, acceleration, and propagation of cosmic rays in our galaxy. The high statistics of the measurements (300 mio. protons, 50 mio. helium) allow to study the detailed variations with rigidity of the fluxes spectral index. The spectral index of both the proton and the helium flux progressively hardens at rigidities larger than 100 GV. The rigidity dependence of the helium flux spectral index is similar to that of the proton spectral index though the magnitudes are different. Remarkably, the spectral index of the proton to helium flux ratio increases with rigidity up to 45 GV and then becomes constant; the flux ratio above 45 GV is well described by a single power law.

1 Introduction

Protons and helium nuclei in cosmic rays are believed to be mainly produced by Galactic sources such as supernova remnants. The precise proton and helium spectra in the GV - TV rigidity region give important information on the origin, acceleration, and subsequent propagation processes of cosmic rays in the Galaxy [1].

Some of the recent important measurements of the proton flux in cosmic rays are listed in Ref. [2–6]. Recent important measurements of the helium flux in cosmic rays have reported different variations of the flux with energy [4, 6–9]. Many models based on different sources, acceleration mechanisms, and diffusive propagation effects [10] were proposed to account for the hardening of the helium flux as well as for the difference of the proton and helium fluxes. The proton flux was reported in Ref. [11] and the helium flux was reported in Ref. [12] both based on data collected by the Alpha Magnetic Spectrometer (AMS) during the first 30 months (May 19, 2011 to November 26, 2013) of operation onboard the International Space Station (ISS). The Letter reporting the helium flux also reports the characteristics of the ratio of the proton flux to the helium flux.

[a]e-mail: melanie.heil@cern.ch

2 Detector

AMS is a general purpose high energy particle physics detector in space. The layout and description of the detector are presented in Ref. [13]. The key elements used in this measurement are the permanent magnet, the silicon tracker, four planes of time of flight (TOF) scintillation counters, and the array of anticoincidence counters (ACC). AMS also contains a transition radiation detector (TRD), a ring imaging Čerenkov detector (RICH), and an electromagnetic calorimeter (ECAL). The central field of the magnet [14] is 1.4 kG.

The tracker [15] has nine layers, the first (L1) at the top of the detector, the second (L2) just above the magnet, six (L3 to L8) within the bore of the magnet, and the last (L9) just above the ECAL. L2 to L8 constitute the inner tracker. Each layer contains double-sided silicon microstrip detectors which independently measure the x and y coordinates. The tracker accurately determines the trajectory of cosmic rays by multiple measurements of the coordinates with a resolution in each layer of 7-10 μm in the bending (y) direction. The inner tracker is held stable by a carbon fiber structure with negligible coefficient of thermal expansion. The stability of the inner tracker is monitored using 20 IR laser beams which penetrate layers L2 through L8 and provide sub-micron position measurements. Using cosmic rays over a 2 minute window, the position of L1 is aligned with a precision of 5 μm with respect to the inner tracker and L9 with a precision of 6 μm. Together, the tracker and the magnet measure the rigidity R of charged cosmic rays. The maximum detectable rigidity (MDR) for particles with charge $|Z| = 1$ is 2 TV over the 3 m lever arm from L1 to L9. For particles with charge $|Z| = 2$ the MDR is higher, namely 3.2 TV, as the higher signal amplitudes of $|Z| = 2$ particles provide a better coordinate resolution.

Each layer of the tracker also provides an independent measurement of the absolute value of the charge $|Z|$ of the cosmic ray. The charge resolution of the layers of the inner tracker together is $\Delta Z \simeq 0.05$ for $|Z| = 1$ particles and $\Delta Z \simeq 0.07$ for $|Z| = 2$.

Two planes of TOF counters [16] are located above L2 and two planes are located below the magnet. The overall velocity ($\beta = v/c$) resolution has been measured to be $\Delta\beta/\beta^2 = 4\%$ for $|Z| = 1$ particles and $\Delta\beta/\beta^2 = 2\%$ for $|Z| = 2$ particles. This discriminates between upward- and downward-going particles. The pulse heights of the two upper layers and of the two lower layers are combined to provide two independent measurements of the charge.

The 16 anticoincidence counters (ACC) form a cylindrical shell between the inner tracker and the magnet. Adjacent counters are combined to provide 8 readout sectors. The ACC have an efficiency of 0.999 99 to reject cosmic rays which enter the inner tracker from the side.

Before launch, at the CERN SPS, AMS was extensively calibrated with 180 and 400 GeV/c proton beams and beams of positrons, electrons, and pions from 10 to 290 GeV/c. In total, calibrations with 18 different energies and particles at 2000 positions were performed. These data allow the determination of the tracker rigidity resolution function with high precision and the verification of the absolute rigidity scale.

Monte Carlo simulated events were produced using a dedicated program developed by the collaboration based on the GEANT-4.9.6 package for protons and the GEANT-4.10.1 package for helium [17]. This program simulates electromagnetic and hadronic interactions of particles in the material of AMS and generates detector responses. The INCL++ package [18] was used to model helium-nuclear inelastic interactions below 5 GeV/nucleon and the DPMJET-II.5 package [19] was used at higher energies. The helium-nuclear elastic and quasi-elastic scattering were modeled using the measurements from [20]. As will be discussed below, we have developed a method to measure the helium-nuclear interactions with the materials in AMS and used this measurement to scale the model predictions.

The digitization of the signals is simulated precisely according to the measured characteristics of the electronics. The simulated events then undergo the same reconstruction as used for the data.

Figure 1(a) shows a comparison of the inverse rigidity for 400 GeV/c protons from the test beam and the Monte Carlo simulation. As seen, the resolution has a pronounced Gaussian core with non-Gaussian tails of ~5% and the simulation precisely reproduces the measured resolution including the tails.

3 Selection

In the first 30 months (7.96×10^7 s) AMS collected 4.1×10^{10} cosmic ray events. The effective data collection time used in this analysis are the seconds when the detector was in normal operating conditions, the AMS z axis was within 40° of the local zenith, the trigger live time exceeded 50%, and the ISS was outside of the South Atlantic Anomaly. Due to the influence of the geomagnetic field, this collection time for primary cosmic rays increases with increasing rigidity becoming constant at 6.29×10^7 s above 30 GV.

By selecting events to be downward going and to have a reconstructed track in the inner tracker with $|Z| = 1$, we obtain 1.1×10^{10} proton candidate events. By selecting $|Z| = 2$ in the inner tracker we obtain 1.7×10^9 helium candidate events. In order to have the best resolution at the highest rigidities, further selections are made by requiring the track to have hits in L1 and L9 with reconstructed charges compatible to that of the inner tracker and to satisfy additional track fitting quality criteria such as a $\chi^2/d.f. < 10$ in the bending coordinate. To further remove helium candidates which interacted within the detector, the charge measured by the upper TOF and the lower TOF is also required to be compatible with $Z = 2$,

In addition, to select only primary cosmic rays, well above the geomagnetic cutoff, the measured rigidity is required to be greater than 1.2 times the maximum geomagnetic cutoff within the AMS field of view. The cutoff was calculated by backtracing [21] particles from the top of AMS out to 50 Earth's radii using the IGRF [22] geomagnetic model. For protons the small contamination of low energy (below 2 GeV) pions produced in the upper part of the detector is removed by requiring that the mass of the selected particle is larger than 0.5 GeV/c^2. These procedures resulted in a sample of 3.0×10^8 primary cosmic rays with $Z = +1$ and 5.0×10^7 primary cosmic rays with $Z = +2$.

Since protons are the dominant component of cosmic rays, the selected sample of 3.0×10^8 events has only small contributions of other particles, mainly deuterons. The deuteron contribution decreases with rigidity; at 1 GV it is less than 2% and at 20 GV it is 0.6% [23, 24]. Deuterons were not removed. The sample also contains protons from nuclei which interact at the top of AMS (for example, in L1 or the TRD). From the measured flux [12, 25] and Monte Carlo simulation this contribution is 0.5% at 1 GV decreasing to less than 0.1% at and above 10 GV. Contamination from e^+ and e^- [26] was estimated to be less than 0.1% over the entire rigidity range. Due to the multiple independent measurements of the charge, the selected helium sample contains only a small contamination of particles which had $Z \neq 2$ at the top of AMS. Comparing the proton and helium charge distributions in the inner tracker, the proton contamination of the helium sample was measured to be less than 10^{-4} over the entire rigidity range. The sample also contains helium from other nuclei which interact at the top of AMS (for example, in L1). From the measured flux [25] and Monte Carlo simulation this contribution is below 10^{-3} for the entire rigidity range. The background contributions for protons and helium are subtracted from the fluxes and the uncertainties are accounted for in the systematic errors.

4 Analysis

The isotropic flux Φ_i for the ith rigidity bin $(R_i, R_i + \Delta R_i)$ is

$$\Phi_i = \frac{N_i}{A_i\, \epsilon_i\, T_i\, \Delta R_i} \tag{1}$$

DOI: 10.1051/epjconf/201612602013

where N_i is the number of events corrected with the rigidity resolution function (see below), A_i is the effective acceptance, ϵ_i is the trigger efficiency, and T_i is the collection time. The effective acceptance A_i was calculated using Monte Carlo samples and then corrected for small differences found between the data and Monte Carlo event selection efficiencies.

The trigger efficiency ϵ_i is measured from data with the unbiased trigger events [11, 12]. The trigger efficiency of protons ranges from 90 to 95%, while the trigger efficiency for helium was measured to range from 95 to 99.5% The inefficiencies are due to secondary δ-rays in the magnetic field entering the ACC. The Monte Carlo trigger efficiencies agree with the measured trigger efficiencies within 0.5%.

The bin-to-bin migration of events was corrected using the rigidity resolution functions obtained from Monte Carlo simulations and for protons verified with the test beam data, see for example Fig. 1(a).

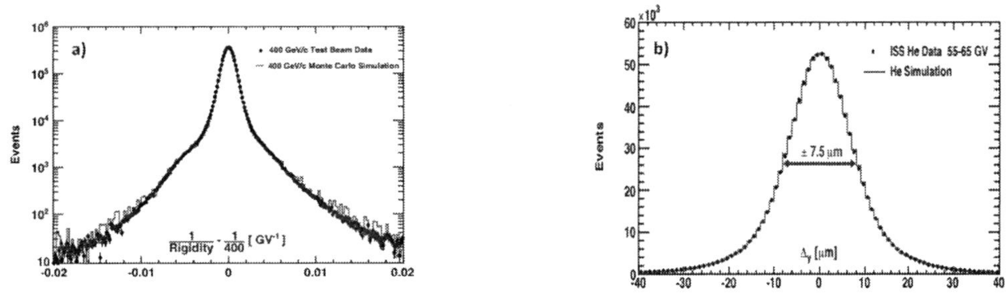

Figure 1: a) The resolution function in inverse rigidity for 400 GeV/c protons measured in the test beam compared with Monte Carlo simulated events. b) The differences of the coordinates measured in the tracker layers 3 or 5 to those obtained from the track fit excluding the tested layer for helium events for both data and Monte Carlo simulation.

Among many unfolding procedures, we selected two which are described in [11]. The validity of both were verified by our Monte Carlo simulation. The small differences between the two procedures ($< 0.5\%$) are accounted as a systematic error. We have checked the sensitivity of the results to the binning by increasing the bin width by factors of 2 and 4 as well as reducing the bin width by factors of 2 and 4. The resulting uncertainty is well within the assigned systematic errors.

Extensive studies were made of the systematic errors. The errors include the uncertainties in the trigger efficiency, the acceptance, the background contamination also accounting for helium interactions in the detector, the geomagnetic cutoff factor, the event selection, the unfolding, the rigidity resolution function, and the absolute rigidity scale. The trigger efficiency error is dominated by the statistics available from the 1% prescaled unbiased event sample [11, 12]. For protons it is negligible (less than 0.1%) below 500 GV and reaches 1.5% at 1.8 TV, for helium it is less than 0.2% below 100 GV and reaches 1% at 3 TV. The geomagnetic cutoff factor was varied from 1.0 to 1.4 and the resulting fluxes showed a systematic uncertainty of 2% at 1 GV and negligible above 2 GV.

The effective acceptance was corrected for small differences between the data and the Monte Carlo samples related to the event reconstruction and selection. The total correction for protons was found to be 5% at 1 GV decreasing below 2% above 10 GV, while the corresponding systematic uncertainty is less than 1% above 2 GV. The correction on the acceptance for helium was found to be less than 2% above 2 GV with a corresponding systematic error on the flux less than 1% below 200 GV and reaching 1.4% at 3 TV.

The detector is mostly made of carbon and aluminum. For protons the corresponding inelastic cross sections of p + C and p + Al are known to within 10% at 1 GV and 4% at 300 GV [27], and 7% at 1.8 TV from model estimations [17]. To estimate the systematic error of the proton flux due to the uncertainty in the inelastic cross sections, dedicated samples of protons were simulated with the p+C and p+Al cross sections varied by ±10%. From the analysis of these samples together with the current knowledge of the cross sections, a systematic error of 1% at 1 GV, 0.6% from 10 to 300 GV, and 0.8% at 1.8 TV was obtained. The inelastic cross sections of He+C and He+Al have only been measured below 10 GV [28]. To accurately determine the effect on the acceptance of helium interactions in the detector, we have developed a method [29] to determine the magnitude and rigidity dependence of the survival probability of helium nuclei when traversing the detector materials. For this we used a sample of primary cosmic rays collected with AMS horizontal, that is, when the ISS was oriented such that AMS was pointing within $90° \pm 10°$ of the local zenith, a total of 1.4×10^5 s. In this condition, primary helium can travel from L9 to L1 and from L1 to L9. For more details see [12]. Using the measured interaction probabilities and, in addition, above 100 GV the rigidity dependence of the cross sections from the Glauber-Gribov model [17], the systematic error on the flux due to uncertainties of helium inelastic cross sections was evaluated to be 1% below 100 GV and 2% at 3 TV.

An additional systematic error on the helium flux is due to differences in ^{3}He and ^{4}He interactions with the detector materials. This error was assigned according to the uncertainty in the helium isotopic composition. The error of 0.3% on the flux was obtained by varying the ratio of ^{3}He/^{4}He in the simulation from 0 to 0.2 [30]. The helium flux was then treated as containing only ^{4}He.

The rigidity resolution function for protons, shown in Fig. 1(a), was verified with data from both the ISS and the test beam and compared to the simulated events. For this the residuals between the hit coordinates measured in tracker layers L1 and L9 and those obtained from the track fit using the information from only the inner tracker L2 to L8 were compared between data and simulation. In order to validate the alignment of the external layers the difference between the rigidity measured using the information from L1 to L8 and from L2 to L9 was compared between data and the simulation. The resulting uncertainty on the proton MDR was estimated to be 5%. The corresponding unfolding errors on the proton flux were obtained by varying the width of the Gaussian core of the resolution function by 5% and the amplitude of the non-Gaussian tails by ~20% over the entire rigidity range and found to be 0.8% below 200 GV and 3% at 1.8 TV.

The rigidity resolution function for helium is similar to that of the protons. It was obtained from the simulations and extensively verified with the data. We list three verifications. First, the differences of the coordinates measured in the inner tracker layers to those obtained from the track fit excluding the tested layer were compared between data and simulation. This procedure directly measures the tracker coordinate accuracy of $\pm 7.5\ \mu$m for helium as shown in Fig. 1(b) for the combination of Layers 3 and 5. Second, the differences between the coordinates measured in L1 and L9 and those obtained from the track fit using the information from only the inner tracker were compared between helium events in data and simulation. Third, the RICH velocity resolution is $\Delta\beta/\beta = 8.0 \times 10^{-4}$ [31] for helium events. The rigidity resolution function up to 20 GV, including non-Gaussian tails, was obtained with data using the RICH velocity measurements only and compared with the rigidity resolution function from the simulation. The first two verifications provided the helium MDR and its uncertainty of 3.2 ± 0.16 TV. The second and third provided the uncertainty of the amplitude of the non-Gaussian tails in the helium rigidity resolution function to be 10%. The systematic error on the helium flux due to uncertainties in the rigidity resolution function was obtained by varying the width of the Gaussian core of the resolution function by 5% and the amplitude of the non-Gaussian tails by 10% over the entire rigidity range in the unfolding procedures and found to be less than 1% below 400 GV and 3.5% at 3 TV.

There are two contributions to the systematic uncertainty on the rigidity scale. The first is due to residual tracker misalignment. From the 400 GeV/c proton test beam data it was measured to be less then 1/300 TV^{-1}. For the ISS data this error was estimated by comparing the E/p ratio for electron and positron events, where E is the energy measured with the ECAL and p is the momentum measured with the tracker, see Ref. [32] for details. It was found to be 1/26 TV^{-1}, limited by the current high energy positron statistics. The second systematic error on the rigidity scale arises from the magnetic field map measurement and its temperature corrections and amounts to less than 0.5% for rigidities above 2 GV.

To ensure that the treatment of systematic errors described above is correct, we performed several additional, independent verifications. Figure 2 shows examples of the stability of the measured proton flux for different conditions (presented as the ratio to the average flux) on the left and examples of the stability of the measured helium flux for different conditions (presented as the ratio to the average flux) along with the corresponding systematic errors on the right.

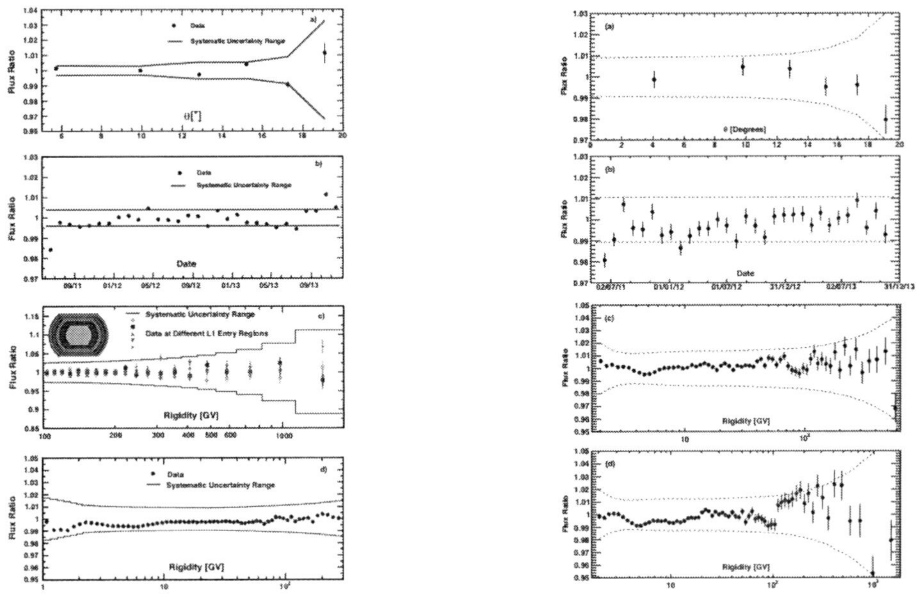

Figure 2: Verifications of the systematic errors for the proton (left) and helium (right) flux measurement. a) The variation of the flux ratio above 30 GV versus the angle θ to the AMS z axis. b) The variation of the flux ratio above 45 GV vs time. c) (left) The variation of the flux ratio vs the rigidity for different L1 entry regions (see inset). c) (right) and d) The variation of the flux ratio measured using only the inner tracker (L2 to L8) or (L1 to L8) vs the full tracker (L1 to L9). (see text for details)

Figures 2(a) show the dependence of the integral of the fluxes above 30 GV, *i.e.,* above the maximum geomagnetic cutoff, on the angle θ between the incoming particle direction and the AMS z axis; this verifies the systematic errors assigned to the acceptance. Figures 2(b) show the monthly integral fluxes above 45 GV are within the systematic errors. This verifies that the detector performance is stable over time and that the fluxes above 45 GV show no observable effect from solar modulation fluctuations for this measurement period. The variation of the proton flux due to solar modulation will be the subject of a separate publication. Figure 2(c, left) shows that the ratios of proton fluxes obtained using events which pass through different sections of L1 to the average flux is in good agreement and within the assigned systematic errors; this verifies the errors assigned to the tracker alignment.

Figures 2(c, right) and 2(d, left) show that the proton and helium fluxes obtained using the rigidity measured by only the inner tracker are in good agreement with the fluxes measured using the full lever arm. The flux ratios use the two different event samples corresponding to the inner tracker acceptance and to the L1 to L9 acceptance used for the results presented here. This verifies the systematic errors from the acceptance, the unfolding procedure, and the rigidity resolution function for two extreme and important cases. First, at the MDR of the inner tracker, where the unfolding effects and resolution functions of the inner tracker and the full lever arm are very different. Second, at low rigidities (2 to 10 GV) where the unfolding effects and the tails in the resolution functions of the inner tracker and full lever arm are also very different due to multiple and nuclear scattering. Lastly, Fig. 2(d, right) shows the good agreement between the helium flux obtained using the rigidity measured by tracker L1 to L8, MDR 1.4 TV, and the full lever arm, MDR 3.2 TV, again using different event samples, thus verifying the systematic errors on the rigidity resolution function over this extended rigidity range.

Most importantly, for both protons and helium several independent analyses were performed on the same data sample by different study groups. The results of those analyses are consistent with the ones presented here.

5 Results

The measured proton flux can be found in Ref. [11] in Table I and the helium flux is published in Ref. [12] in Table I as a function of the rigidity at the top of the AMS detector. The contributions of individual sources to the systematic errors are added in quadrature to arrive at the total systematic uncertainty of the proton an helium flux. The Monte Carlo event samples have sufficient statistics such that they do not contribute to the errors. Figure 3(a) on the left shows the proton flux as a function of rigidity with the total errors. In this and the subsequent figures, the points are placed along the abscissa at $\tilde{R}$ calculated for a flux $\propto R^{-2.7}$ [33]. Figure 3(b) on the left shows the AMS proton flux as a function of kinetic energy E_K together with the most recent results (*i.e.*, from experiments after the year 2000). Figure 3(a) on the right shows the helium flux as a function of rigidity with the total errors. Figure 3(b) on the right shows the AMS helium flux as a function of kinetic energy per nucleon E_K together with the most recent results (*i.e.*, from experiments after the year 2000).

A power law with a constant spectral index γ

$$\Phi = CR^{\gamma}, \tag{2}$$

where R is in GV and C is a normalization factor, does neither fit the proton flux nor the helium flux at the 99.9% C.L. for $R > 45\,\text{GV}$ taken solar modulation into account through the force-field approximation [34, 35]. We therefore fit the fluxes with a modified spectral index [36]

$$\Phi = C\left(\frac{R}{45\,\text{GV}}\right)^{\gamma}\left[1+\left(\frac{R}{R_0}\right)^{\Delta\gamma/s}\right]^{s} \tag{3}$$

where s quantifies the smoothness of the transition of the spectral index from γ for rigidities below the characteristic transition rigidity R_0 to $\gamma+\Delta\gamma$ for rigidities above R_0. Fitting the flux above 45 GV yields the following results for protons: $\chi^2/d.f. = 25/26$ with $\gamma = -2.849 \pm 0.002(\text{fit})^{+0.004}_{-0.003}(\text{sys})^{+0.004}_{-0.003}(\text{sol})$, $\Delta\gamma = 0.133^{+0.032}_{-0.021}(\text{fit})^{+0.046}_{-0.030}(\text{sys}) \pm 0.005(\text{sol})$, $s = 0.024^{+0.020}_{-0.013}(\text{fit})^{+0.027}_{-0.016}(\text{sys})^{+0.006}_{-0.004}(\text{sol})$, and $R_0 = 336^{+68}_{-44}(\text{fit})^{+66}_{-28}(\text{sys}) \pm 1(\text{sol})\,\text{GV}$; and these are the results for helium: $\chi^2/d.f. = 25/27$ with $\gamma = -2.780 \pm 0.005(\text{fit}) \pm 0.001(\text{sys}) \pm 0.004(\text{sol})$, $\Delta\gamma = 0.119^{+0.013}_{-0.010}(\text{fit})^{+0.033}_{-0.028}(\text{sys}) \pm 0.004(\text{sol})$, $s = 0.027^{+0.014}_{-0.010}(\text{fit})^{+0.017}_{-0.013}(\text{sys}) \pm 0.002(\text{sol})$, and $R_0 = 245^{+35}_{-31}(\text{fit})^{+33}_{-30}(\text{sys}) \pm 3(\text{sol})$ GV.

DOI: 10.1051/epjconf/201612602013

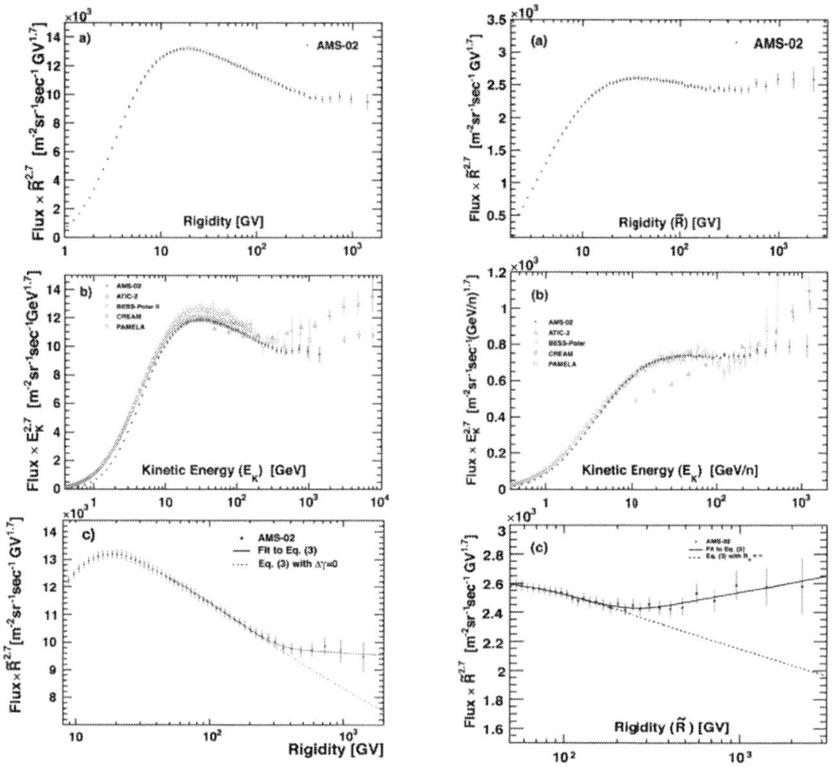

Figure 3: a) The AMS proton and helium flux multiplied by $\tilde{R}^{2.7}$ and total error as a function of rigidity. b) The fluxes as a function of kinetic energy per nucleon E_K as multiplied by $E_K^{2.7}$ compared with recent measurements [3–6, 8, 9]. For the AMS proton results $E_K \equiv \sqrt{\tilde{R}^2 + M_p^2} - M_p$ where M_p is the proton mass. For the AMS helium results $E_K \equiv (\sqrt{4\tilde{R}^2 + M^2} - M)/4$ where M is the ^{4}He mass. c) The fluxes as fitted with Eq. (3). For illustration, the fit results with $\Delta\gamma$ set to zero or $R_0 = \infty$ are also shown.

The first error quoted (fit) takes into account the statistical and uncorrelated systematic errors of the flux. The second (sys) is the error from the remaining systematic errors, namely from the rigidity resolution function and unfolding, and from the absolute rigidity scale, with their bin-to-bin correlations accounted for using the migration matrix M_{ij}. The third (sol) is the uncertainty due to the variation of the solar potential $\phi = 0.50$ to 0.62 GV [34]. The fits confirm that above 45 GV the fluxes are incompatible with a single spectral index at the 99.9% C.L. The fits are shown in Fig. 3(c) on the left for protons and 3(c) on the right for helium. For illustration, the fit results with $\Delta\gamma$ set to zero or $R_0 = \infty$ are also shown.

To obtain the detailed variation of γ with rigidity for the proton and helium flux in a model independent way, the spectral index is calculated from

$$\gamma = d[\log(\Phi)]/d[\log(R)] \tag{4}$$

over independent rigidity intervals above 8.48 GV, with a variable width to have sufficient sensitivity to determine γ. The results are presented in Figure 4(a).

As seen, the magnitude of the helium spectral index is different from that of the proton spectral index but the rigidity dependence is similar for helium and protons. In particular, the spectral indices both progressively harden with rigidity above ~100 GV.

To examine the difference between the rigidity dependence of the proton and helium fluxes, the ratio of the proton flux to the helium flux, or p/He ratio, was computed in the range where they overlap, from 1.9 GV to 1.8 TV. The statistical errors are the sum in quadrature of the relative statistical errors of the proton and helium fluxes multiplied by the p/He ratio. The systematic errors from the trigger and acceptance are likewise added in quadrature. The correlations in the systematic errors from the unfolding and the absolute rigidity scale between the proton and helium fluxes have been accounted in calculating the corresponding systematic errors of the p/He ratio. The contribution of individual sources to the systematic error are added in quadrature to arrive at the total systematic uncertainty. Figure 4(b) shows the AMS measurement with total errors, the quadratic sum of statistical and systematic errors, together with other recent measurements.

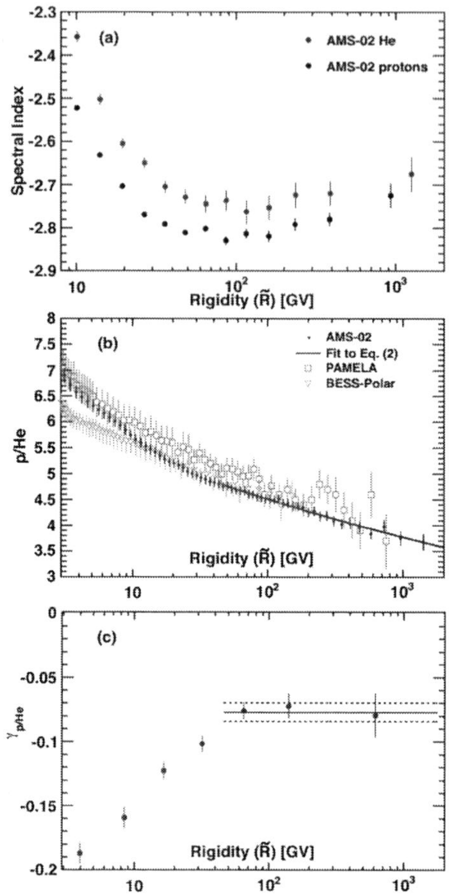

Figure 4: (a) The dependence of the helium and proton spectral indices on rigidity. (b) The p/He ratio as a function of rigidity compared with recent measurements [4, 6]. The solid blue curve indicates the fit of a single power law, Eq. (2), to the AMS data. (c) The rigidity dependence of the p/He spectral index as measured by AMS. Above 45 GV it becomes constant at $\gamma_{p/He} = -0.077 \pm 0.002(\text{fit}) \pm 0.007(\text{sys})$ as indicated by the blue lines.

Above 45 GV the p/He ratio measured by AMS is well fit with a single power law, Eq. (2), with a $\chi^2/d.f.$= 22/29 and a spectral index of $\gamma_{p/He} = -0.077 \pm 0.002(\text{fit}) \pm 0.007(\text{sys})$. The first error quoted (fit) takes into account the statistical and uncorrelated systematic errors from the flux ratio reported in [12], Table II. The second (sys) is the error from the remaining systematic errors, namely from the rigidity resolution function, unfolding and proton and helium inelastic cross sections, with proper accounting of the bin-to-bin correlations.

Figure 4(c) shows the detailed variation of $\gamma_{p/He}$ with rigidity over non-overlapping rigidity intervals. As seen, the spectral index of the ratio increases with rigidity up to about 45 GV. Above 45 GV it becomes constant.

6 Conclusion

In conclusion, precise knowledge of the proton and helium flux is important in understanding the origin, acceleration, and propagation of cosmic rays. Our precise measurements of the proton and helium flux show that the fluxes deviate from a single power law and the spectral indices progressively harden at rigidities larger than 100 GV. The magnitude of the helium spectral index is different from that of the proton spectral index, but the rigidity dependence is similar for helium and protons. The proton to helium flux ratio spectral index increases with rigidity up to about 45 GV and becomes constant for higher rigidities, i.e., the proton to helium flux ratio is well described by a single power law above 45 GV.

References

[1] I. A. Grenier, J. H. Black and A. W. Strong, Annu. Rev. Astron. Astrophys. **53**, 199 (2015); P. Blasi, Astron. Astrophys. Rev. **21**, 70 (2013); A. W. Strong, I. V. Moskalenko, and V. S. Ptuskin, Annu. Rev. Nucl. Part. Sci. **57**, 285 (2007).

[2] M. Hareyama *et al.*, J. Phys. Conf. **31**, 159 (2006); M. Boezio *et al.*, Astropart. Phys. **19**, 583 (2003); E. Diehl, D. Ellithorpe, D. Muller, S.P. Swordy, Astropart. Phys. **18**, 487 (2003); J. Alcaraz *et al.*, Phys. Lett. B **490**, 27 (2000); W. Menn *et al.*, Astrophys. J. **533**, 281 (2000); R. Bellotti *et al.*, Phys. Rev. D **60**, 052002 (1999); M. Boezio *et al.*, Astrophys. J. **518**, 457 (1999); K. Asakimori *et al.*, Astrophys. J. **502**, 278 (1998); M. Ichimura *et al.*, Phys. Rev. D **48**, 1949 (1993); I. P. Ivanenko *et al.*, in *Proceedings of the 23rd International Cosmic Ray Conference*, Calgary, 17 (1993).

[3] For the ATIC experiment see A. D. Panov *et al.*, Bull. Russian Acad. Sci. **73**, 564 (2009) extracted from D. Maurin, F. Melot, R. Taillet, Astron. Astrophys. **569**, A32 (2014).

[4] We are grateful to the BESS-Polar Collaboration for providing the BESS-Polar II proton spectrum, measured during a 25-day Antarctic balloon flight in December 2007 and January 2008. Statistical and systematic errors are combined. A publication detailing these measurements is in preparation. For a description of the BESS-Polar experiment see K. Abe *et al.*, Phys. Rev. Lett. **108**, 051102 (2012). For the BESS experiment see Y. Shikaze *et al.*, Astropart. Phys. **28**, 154 (2007); S. Haino *et al.*, Phys. Lett. B **594**, 35 (2004); T. Sanuki *et al.*, Astrophys. J. **545**, 1135 (2000).

[5] For the CREAM experiment see Y. S. Yoon *et al.*, Astrophys. J. **728**, 122 (2011).

[6] For the PAMELA experiment see O. Adriani, *et al.*, Astrophys. J. **765**, 91 (2013); O. Adriani, *et al.*, Science **332**, 69 (2011).

[7] V. A. Derbina *et al.*, Astrophys. J. **628**, L41 (2005); M. Boezio *et al.*, Astropart. Phys. **19**, 583 (2003); E. Diehl, D. Ellithorpe, D. Muller, and S.P. Swordy, Astropart. Phys. **18**, 487 (2003); J. Alcaraz *et al.*, Phys. Lett. B **494**, 193 (2000); W. Menn *et al.*, Astrophys. J. **533**, 281 (2000); R. Bellotti *et al.*, Phys. Rev. D **60**, 052002 (1999); M. Boezio *et al.*, Astrophys. J. **518**, 457 (1999); K. Asakimori *et al.*, Astrophys. J. **502**, 278 (1998); I. P. Ivanenko *et al.*, in *Proceedings of the 23rd International Cosmic Ray Conference, Calgary* (World Scientific, Singapore, 1993) p. 17.

[8] For the ATIC experiment see A. D. Panov *et al.*, Bull. Russian Acad. Sci. **73**, 564 (2009) extracted from D. Maurin, F. Melot, and R. Taillet, Astron. Astrophys. **569**, A32 (2014).

[9] For the CREAM experiment see Y. S. Yoon *et al.*, Astrophys. J. **728**, 122 (2011).

[10] See, for example, A. D. Erlykin and A. W. Wolfendale, J. Phys. G: Nucl. Part. Phys. **42**, 075201 (2015); S. Thoudam and J. R. Hörandel, Astronom. Astrophys. **567**, A33 (2014); G. Bernard, T. Delahaye, Y.-Y. Keum, W. Liu, P. Salati, and R. Taillet, Astronom. Astrophys. **555**, A48 (2013); V. S. Ptuskin, V. Zirakashvili, and E. S. Seo, Astrophys. J. **763**, 47 (2013); N. Tomassetti, Astrophys. J. Lett. **752**, L13 (2012); P. Blasi, E. Amato, and P. D. Serpico, Phys. Rev. Lett. **109**, 061101

(2012); A. E. Vladimirov, G. Jóhannesson, I. V. Moskalenko, and T. A. Porter, Astrophys. J. **752**, 68 (2012); M. A. Malkov, P. H. Diamond, and R. Z. Sagdeev, Phys. Rev. Lett. **108**, 081104 (2012); L. A. Fisk and G. Gloeckler, Astrophys. J. **744**, 127 (2012); Y. Ohira and K. Ioka, Astrophys. J. Lett. **729**, L13 (2011).

[11] M. Aguilar *et al.*, Phys. Rev. Lett. **114**, 171103 (2015);

[12] M. Aguilar *et al.*, Phys. Rev. Lett. **115**, 211101 (2015);

[13] A. Kounine, Int. J. Mod. Phys. E **21** 1230005 (2012); S. Rosier-Lees, in *Proceedings of Astroparticle Physics TEVPA/IDM*, Amsterdam, 2014 (to be published); S. C. C. Ting, Nucl. Phys. B, Proc. Suppl. **243-244**, 12 (2013); S.-C. Lee, in *Proceedings of the 20th International Conference on Supersymmetry and Unification of Fundamental Interactions (SUSY 2012)*, Beijing, 2012 (unpublished); M. Aguilar, in *Proceedings of the XL International Meeting on Fundamental Physics*, Centro de Ciencias de Benasque Pedro Pascual, 2012 (unpublished); S. Schael, in *Proceedings of the 10th Symposium on Sources and Detection of Dark Matter and Dark Energy in the Universe*, Los Angeles, 2012 (unpublished); B. Bertucci, Proc. Sci. EPS-HEP, 67 (2011); M. Incagli, AIP Conf. Proc. **1223**, 43 (2010); R. Battiston, Nucl. Instrum. Methods Phys. Res., Sect. A **588**, 227 (2008).

[14] K. Luebelsmeyer *et al.*, Nucl. Instrum. Methods Phys. Res., Sect. A **654**, 639 (2011).

[15] B. Alpat *et al.*, Nucl. Instrum. Methods Phys. Res., Sect. A **613**, 207 (2010).

[16] V. Bindi *et al.*, Nucl. Instrum. Methods Phys., Sect. A **743**, 22 (2014) and references therein.

[17] J. Allison *et al.*, IEEE Trans. Nucl. Sci. **53**, 270 (2006); S. Agostinelli *et al.*, Nucl. Instrum. Methods Phys. Res., Sect. A **506**, 250 (2003).

[18] A. Boudard, J. Cugnon, J.-C. David, S. Leray, and D. Mancusi Phys. Rev. C **87** 014606 (2013); S. Leray, D. Mancusi, P. Kaitaniemi, J. C. David, A. Boudard, B. Braunn, and J. Cugnon J. Phys.: Conf. Series **420**, 012065 (2013).

[19] J. Ranft, Phys. Rev. D **51**, 64 (1995).

[20] V. G. Ableev *et al.*, Acta Physica Polonica B **16**, 913 (1985).

[21] J. Alcaraz *et al.*, Phys. Lett. B **484**, 10 (2000); see also Ref. [24].

[22] C. C. Finlay *et al.* Geophys. J. Int. **183/3**, 1216 (2010).

[23] E. Vannuccini, in *Proceedings of the 28th International Cosmic Ray Conference*, Tsukuba, 1801 (2003).

[24] M. Aguilar *et al.*, Physics Reports **366/6**, 331 (2002).

[25] The AMS Collaboration, *Measurement of the Flux of Light Nuclei in Primary Cosmic Rays with the Alpha Magnetic Spectrometer on the International Space Station*, (to be published).

[26] M. Aguilar *et al.*, Phys. Rev. Lett. **113**, 121102 (2014).

[27] N. Abgrall *et al.* Phys. Rev. C **84**, 034604 (2011); J. R. Letaw, R. Silberberg, C. H. Tsao, Astrophys. J. Supp. **51** 271 (1983); A. S. Carroll *et al.*, Phys. Lett. B **80**, 319 (1979); S. P. Denisov, S. V. Donskov, Yu. P. Gorin, R. N. Krasnokutsky, A. I. Petrukhin, Yu. D. Prokoshkin, D. A. Stoyanova, Nucl. Phys. B **61**, 62 (1973); G. Bellettini, G. Cocconi, A. N. Diddens, E. Lillethun, G. Matthiae, J. P. Scanlon, A. M. Wetherell, Nucl. Phys. **79**, 609 (1966); T. Bowen, M. Di Corato, W. H. Moore, G. Tagliaferri, Il Nuovo Cimento **9**, 908 (1958).

[28] A. Ingemersson *et al.*, Nucl. Phys. A **676**, 3 (2000); A. Auce, R. F. Carlson, A. J. Cox, A. Ingemarsson, R. Johansson, P. U. Renberg, O. Sundberg, G. Tibell, and R. Zorro, Phys. Rev. C **50**, 871 (1994); I. Tanihata *et al.*, Phys. Lett. B **160**, 380 (1985); V. D. Aksinenko *et al.*, Nucl. Phys. A **348**, 518 (1980); E. O. Abdrahmanov *et al.*, Z. Phys. C **5**, 1 (1980); J. Jaros *et al.*, Phys. Rev. C **18**, 2273 (1978).

[29] The AMS Collaboration, *Measurement of Nuclear Interaction Cross Sections with the Alpha Magnetic Spectrometer on the International Space Station*, (to be published).

[30] The AMS Collaboration, *Measurement of the helium isotopic composition in Primary Cosmic Rays with the Alpha Magnetic Spectrometer on the International Space Station*, (to be published); O. Adriani *et al.*, Astrophys. J. **770**, 2 (2013); B. Coste, L. Derome, D. Maurin, and A. Putze, Astronom. Astrophys. **539**, A88 (2012); M. Aguilar *et al.*, Astrophys. J. **736**, 105 (2011); Z. D. Myers *et al.*, in *Proceedings of the 28th International Cosmic Ray Conference, Tsukuba* , p. 1805 ; J. Z. Wang *et al.*, Astrophys. J. **564**, 244 (2002); S. P. Ahlen *et al.*, Astrophys. J. **534**, 757 (2000); O. Reimer *et al.*, Astrophys. J. **496**, 490 (1998); Y. Hatano, Y. Fukada, T. Saito, H. Oda, and T. Yanagita, Phys. Rev. D **52**, 6219 (1995); J. J. Beatty *et al.*, Astrophys. J. **413**, 268 (1993); S. P. Jordan *et al.*, Astrophys. J. **291**, 207 (1985).

[31] F. Giovacchini, Nucl. Instrum. Methods Phys. Res., Sect. A **766**, 57 (2014).

[32] L. Accardo *et al.*, Phys. Rev. Lett. **113**, 121101 (2014); M. Aguilar *et al.*, Phys. Rev. Lett. **110**, 141102 (2013).

[33] G. D. Lafferty, T.R. Wyatt, Nucl. Instr. Methods Phys. Res., Sect. A **355**, 541 (1995). We have used Eq. (6) with $\tilde{R} \equiv x_{lw}$.

[34] I. G. Usoskin, G. A. Bazilevskaya, G. A. Kovaltsov, J. Geophys. Res. **116** A02104 (2011); K. G. McCracken, J. Beer, J. Geophys. Res. **112** A10101 (2007); I. G. Usoskin, K. Alanko-Huotari, G.A. Kovaltsov,K. Mursula, J. Geophys. Res. **110** A12108 (2005); We have also used more recent data from: http://cosmicrays.oulu.fi/phi/Phi_mon.txt.

[35] L. J. Gleeson, W. I. Axford, Astrophys. J. **154**, 1101 (1968).

[36] K. Beuermann *et al.*, Astron. Astrophys. **352**, L26 (1999).

DOI: 10.1051/epjconf/201612602014

Highlights of DAMA/LIBRA

R. Bernabei[1,2], P. Belli[2], A. d'Angelo[3,4], S. d'Angelo[1,2a], A. Di Marco[1,2], F. Montecchia[2,7], A. Incicchitti[3,4], F. Cappella[5], V. Caracciolo[5], R. Cerulli[5], C.J. Dai[6], H.L. He[6], H.H. Kuang[6], X.H. Ma[6], X.D. Sheng[6], R.G. Wang[6], and Z.P. Ye[6,8]

[1]*Dip. di Fisica, Università di Roma "Tor Vergata", I-00133 Rome, Italy*
[2]*INFN, sez. Roma "Tor Vergata", I-00133 Rome, Italy*
[3]*Dip. di Fisica, Università di Roma "La Sapienza", I-00185 Rome, Italy*
[4]*INFN, sez. Roma, I-00185 Rome, Italy*
[5]*Laboratori Nazionali del Gran Sasso, I.N.F.N., Assergi, Italy*
[6]*Key Laboratory of Particle Astrophysics, Institute of High Energy Physics, Chinese Academy of Sciences, P.O. Box 918/3, 100049 Beijing, China*
[7]*Dip. di Ingegneria Civile e Ingegneria Informatica, Università di Roma "Tor Vergata", I-00133 Rome, Italy*
[8]*University of Jing Gangshan, Jiangxi, China*

Abstract. The DAMA project develops and uses new/improved low background scintillation detectors to investigate the Dark Matter (DM) particle component(s) in the galactic halo and rare processes deep underground at the Gran Sasso National Laboratory (LNGS) of the I.N.F.N.. Here some highlights of DAMA/LIBRA (Large sodium Iodide Bulk for Rare processes) as a unique apparatus in direct DM investigation for its full sensitive mass, target material, intrinsic radio-purity, methodological approach and all the controls performed on the experimental parameters are outlined. The DAMA/LIBRA–phase1 and the former DAMA/NaI data (cumulative exposure 1.33 ton × yr, corresponding to 14 annual cycles) have reached a model-independent evidence at 9.3 σ C.L. for the presence of DM particles in the galactic halo exploiting the DM annual modulation signature with highly radio-pure NaI(Tl) target. Some of the perspectives of the presently running DAMA/LIBRA–phase2 are summarised and the powerful tools offered by a model independent strategy of DM investigation are pointed out.

1 Introduction

The DAMA/LIBRA experiment [1–15] as the former DAMA/NaI (see e.g. Refs. [8, 16, 17]), has the main aim to investigate the presence of DM particles in the galactic halo by exploiting the model-independent DM annual modulation signature (originally suggested in Refs. [18, 19]).

As a consequence of the Earth's revolution around the Sun, which is moving in the Galaxy with respect to the Local Standard of Rest towards the star Vega near the constellation of Hercules, the Earth should be crossed by a larger flux of DM particles around ≃2 June and by a smaller one around ≃2 December. In the former case the Earth orbital velocity is summed to the one of the solar system with respect to the Galaxy, while in the latter the two velocities are subtracted. Therefore the DM

[a]Deceased.

annual modulation signature has a different origin and peculiarities than the seasons on the Earth and than effects correlated with seasons.

This DM annual modulation signature is very distinctive since the effect induced by DM particles must simultaneously satisfy all the following requirements: the rate must contain a component modulated according to a cosine function (1) with one year period (2) and a phase that peaks roughly $\simeq$ 2 June (3); this modulation must only be found in a well-defined low energy range, where DM particle induced events can be present (4); it must apply only to those events in which just one detector of many actually "fires" (*single-hit* events), since the DM particle multi-interaction probability is negligible (5); the modulation amplitude in the region of maximal sensitivity must be $\simeq$ 7% for usually adopted halo distributions (6), but it can be larger (even up to $\simeq$ 30%) in case of some possible scenarios. Thus this signature is model independent, very effective and, in addition, it allows the test of a large range of DM candidates, interaction types, cross sections and of halo densities. In this case the experimental observable is not – as in other experiments – the constant part of the signal, S_0, but its modulation amplitude, S_m, as a function of energy.

This approach has several advantages; in particular, the only background of interest is the one able to mimic the signature, i.e. able to account for the whole observed modulation amplitude and to simultaneously satisfy all the many specific peculiarities of this signature. No background of this sort has been found or suggested by anyone over more than a decade. Thus, this approach does not require any identification of S_0 from the total counting rate in order to establish the presence of DM particles in the galactic halo. Therefore, the DM annual modulation signature allows one to overcome the large uncertainties associated to: (1) the many data selections/subtractions/statistical-discrimination procedures; (2) the modelling of surviving background in keV region; (3) the a priori assumption on the nature, interaction type, etc. of the DM particle(s), which are necessary in the experiments where the experimental observable is S_0. When the DM annual modulation signature is applied, S_0 can be worked out by corollary model dependent analysis to investigate on the nature of DM candidates, for each considered framework.

The full description of the DAMA/LIBRA set-up and performances during the phase1 and phase2 (presently running) and other related arguments have been discussed in details in Refs. [1–8] and references therein. Here we just point out that the sensitive part of this set-up is made of 25 highly radio-pure NaI(Tl) crystal scintillators (5-rows by 5-columns matrix) having 9.70 kg mass each one. The detectors are housed in a sealed low-radioactive copper box installed in the centre of a low-radioactive Cu/Pb/Cd-foils/polyethylene/paraffin shield. Moreover, about 1 m concrete, made from the Gran Sasso rock material, almost fully surrounds (mostly outside the barrack) this passive shield, acting as a further neutron moderator. A threefold-level sealing system prevents the detectors to be in contact with the environmental air of the underground laboratory. The radio-purity, the protocols, the procedures and the details are discussed in Refs. [1–8] and references therein.

2 DAMA/LIBRA

DAMA/LIBRA is the successor of DAMA/NaI with a higher exposed mass, higher duty cycle and increased sensitivity. Its modularity is an interesting feature to study DM and background identification. The apparatus has also the unique feature (as well as the former DAMA/NaI) that gamma calibrations are regularly performed down to the software energy threshold in the same conditions as in the production runs, without any contact of the detectors with the environment and without switching-off the electronics.

The high light yield and other response features have allowed working in a safe and reliable way down to 2 keV (DAMA/LIBRA–phase1). The light response of the detectors during phase1 typically ranged from 5.5 to 7.5 photoelectrons/keV, depending on the detector. The hardware threshold of each photomultiplier (PMT) was at single photoelectron, while a software energy threshold of 2 keV electron equivalent (hereafter keV) was used.

At the end of DAMA/LIBRA–phase1 new PMTs with higher quantum efficiency were installed, other optimisations were performed during 2011 and then data taking started as DAMA/LIBRA–phase2, allowing a decrease in the energy threshold. In addition the developments of other new electronic modules have been progressed and new-concept preamplifiers were installed. Studies on other DM features, second order effects, and several other rare processes are in progress as well as the data taking. Preliminary studies on the possibility to realise new concepts PMTs towards a possible phase3 have been started.

Among further scientific goals of this apparatus we point out here: i) investigation with high sensitivity of the DM particle component in the galactic halo by the model independent approach known as DM annual modulation signature, with highly precise determination of the modulation parameters (which carry crucial information); ii) corollary investigations on the nature of the candidate and on the many possible astrophysical, nuclear and particle physics scenarios; iii) investigations on other possible model dependent and/or model independent approaches to further investigate DM particles' features and second order effects; study of exotic scenarios (as SIMPs, neutral nuclearites, Q-balls, etc.); iv) improved search for processes of Pauli exclusion principle violation in ^{23}Na and ^{127}I; v) search for possible electric charge non-conservation (CNC) processes (electron decay into invisible channels, $e^- \rightarrow \nu_e + \gamma$, excitations of nuclear levels of ^{23}Na and ^{127}I after CNC electronic capture); vi) search for possible nucleon, di-nucleon and tri-nucleon decay into invisible channels in ^{23}Na and in ^{127}I; vii) search for solar axions by Primakoff effect in NaI(Tl); viii) search for nuclear rare decays in ^{23}Na, ^{127}I and Tl isotopes (super-dense states, cluster decay, etc.); ix) search for possible neutral particles (QED new phase) in ^{241}Am decays, etc. Most of these investigations will require further dedicated data taking and high exposure to reach competitive sensitivities. As regards the DM features, which can be suitably further exploited collecting very large exposure, see e.g. Sect. 6 of Ref. [17], Appendix of Ref. [2], Ref. [13]. In particular, that Appendix shows how the decreasing of the software energy threshold as in the present DAMA/LIBRA–phase2 also offers the unique possibility to investigate the modulation amplitude at lower energy, disentangling among many DM scenarios when a suitable exposure is collected.

3 DM annual modulation results

The total exposure of DAMA/LIBRA–phase1 is 1.04 ton × yr in seven annual cycles; when including also that of the first generation DAMA/NaI experiment it is 1.33 ton × yr, corresponding to 14 annual cycles [2–4, 8].

Figure 1 shows the time behaviour of the experimental residual rates of the *single-hit* scintillation events in the (2–6) keV energy interval for the complete DAMA/LIBRA–phase1. The residuals of the DAMA/NaI data (0.29 ton × yr) are given in Refs. [2, 16, 17]. These residual rates are calculated from the measured rate of the *single-hit* events after subtracting the constant part: $< r_{ijk} - flat_{jk} >_{jk}$. Here r_{ijk} is the rate in the considered i-th time interval for the j-th detector in the k-th energy bin, while $flat_{jk}$ is the rate of the j-th detector in the k-th energy bin averaged over the cycles. The average is made on all the detectors (j index) and on all the energy bins (k index) which constitute the considered energy interval. The weighted mean of the residuals must obviously be zero over one cycle.

A clear modulation is present in the (2–6) keV single-hit events and fulfils all the requirements of the DM annual modulation signature; in particular, no modulation is observed either above 6 keV or in the (2–6) keV multiple-hits events [2–4]. Many other analysis have been performed (Fourier analysis,

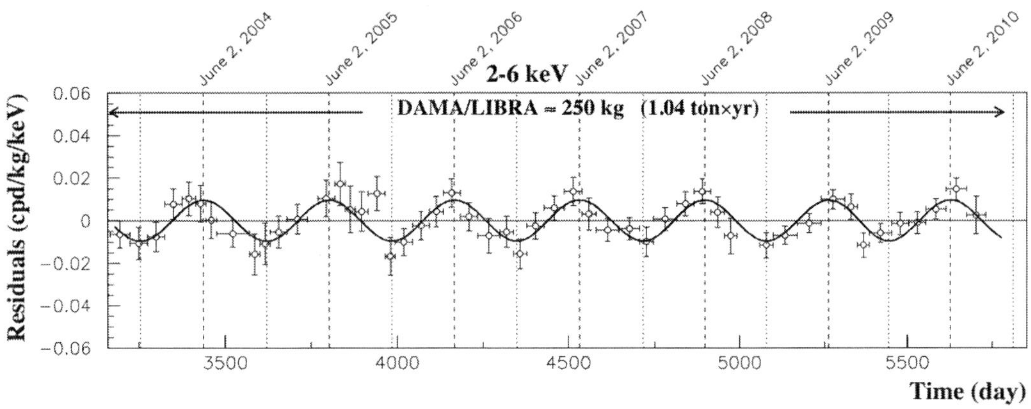

Figure 1. Experimental residual rate of the *single-hit* scintillation events measured by DAMA/LIBRA–phase1 in the (2–6) keV energy interval as a function of the time. The time scale is maintained the same of the previous DAMA papers for coherence. The data points present the experimental errors as vertical bars and the associated time bin width as horizontal bars. The superimposed curve is the cosinusoidal function behaviour $A\cos\omega(t-t_0)$ with a period $T=\frac{2\pi}{\omega}=1$ yr, a phase $t_0=152.5$ day (June 2nd) and modulation amplitude, A, equal to the central values obtained by best fit on the data points of the entire DAMA/LIBRA–phase1. The dashed vertical lines correspond to the maximum expected for the DM signal (June 2nd), while the dotted vertical lines correspond to the minimum.

maximum likelihood analysis, etc.) that further confirm the same evidence [2–4]. In general it is worth noting that this model independent signature, based on the correlation of the measured experimental rate with the Earth galactic motion, itself acts as an effective background rejection, while rejection strategies cannot safely be applied. In fact, the effect searched for (which is in most scenarios at level of few %) would be largely affected by the uncertainties associated to the rejection procedure.

The DAMA/LIBRA–phase1 data give evidence for the presence of DM particles in the galactic halo, on the basis of the exploited model independent DM annual modulation signature by using highly radio-pure NaI(Tl) target, at 7.5 σ C.L.. Including also the first generation DAMA/NaI experiment (cumulative exposure 1.33 ton $\times$ yr, corresponding to 14 annual cycles), the C.L. is 9.3 σ. In particular, with the cumulative exposure the modulation amplitude of the single-hit events in the (2–6) keV energy interval, measured in NaI(Tl) target, is (0.0112 ± 0.0012) cpd/kg/keV; the measured phase is (144 ± 7) days (corresponding to May 24 $\pm$ 7 days) and the measured period is (0.998 ± 0.002) year, values well in agreement with those expected for the DM particles.

At present status of technology DM annual modulation is the only model independent signature available in direct DM investigation that can be effectively exploited. All the many specific requirements of the signature are fulfilled by the DAMA data and no systematic or side reaction able to mimic it is available (see e.g. Refs.[2–4, 6, 7, 16, 17, 20]).

Sometimes naive statements were put forward as the fact that in nature several phenomena may show some kind of periodicity. It is worth noting that the point is whether they might mimic the annual modulation signature in DAMA/NaI and in DAMA/LIBRA, i.e. whether they might be not only quantitatively able to account for the observed modulation amplitude but also able to contemporaneously satisfy all the requirements of the DM annual modulation signature. The same is for side reactions too. This has already been deeply investigated and discussed in DAMA literature.

In particular, in Refs. [6, 7] a quantitative evaluation why the neutrons, the muons and the solar neutrinos cannot give any significant contribution to the DAMA annual modulation results and cannot mimic this signature is outlined. Table 1 summarises the safety upper limits on the contributions to the observed modulation amplitude due to the total neutron flux at LNGS, either from (α, n) reactions, from fissions and from muons' and solar-neutrinos' interactions in the rocks and in the lead around the experimental set-up; the direct contributions of muons and solar neutrinos are reported there too.

Table 1. Summary of the contributions to the total neutron flux at LNGS; the value, $\Phi^{(n)}_{0,k}$, the relative modulation amplitude, η_k, and the phase, t_k, of each component is reported. It is also reported the counting rate, $R_{0,k}$, in DAMA/LIBRA for *single-hit* events, in the (2 – 6) keV energy region induced by neutrons, muons and solar neutrinos, detailed for each component. The modulation amplitudes, A_k, are reported as well, while the last column shows the relative contribution to the annual modulation amplitude observed by DAMA/LIBRA, $S_m^{exp} \simeq 0.0112$ cpd/kg/keV [4]. For details see Ref. [7] and references therein.

	Source	$\Phi^{(n)}_{0,k}$ (neutrons cm^{-2} s^{-1})	η_k	t_k	$R_{0,k}$ (cpd/kg/keV)	$A_k = R_{0,k}\eta_k$ (cpd/kg/keV)	A_k/S_m^{exp}
SLOW neutrons	thermal n ($10^{-2} - 10^{-1}$ eV)	1.08×10^{-6}	$\simeq 0$ however $\ll 0.1$	–	$< 8 \times 10^{-6}$	$\ll 8 \times 10^{-7}$	$\ll 7 \times 10^{-5}$
	epithermal n (eV-keV)	2×10^{-6}	$\simeq 0$ however $\ll 0.1$	–	$< 3 \times 10^{-3}$	$\ll 3 \times 10^{-4}$	$\ll 0.03$
FAST neutrons	fission, $(\alpha, n) \to$ n (1-10 MeV)	$\simeq 0.9 \times 10^{-7}$	$\simeq 0$ however $\ll 0.1$	–	$< 6 \times 10^{-4}$	$\ll 6 \times 10^{-5}$	$\ll 5 \times 10^{-3}$
	$\mu \to$ n from rock (> 10 MeV)	$\simeq 3 \times 10^{-9}$	0.0129	end of June	$\ll 7 \times 10^{-4}$	$\ll 9 \times 10^{-6}$	$\ll 8 \times 10^{-4}$
	$\mu \to$ n from Pb shield (> 10 MeV)	$\simeq 6 \times 10^{-9}$	0.0129	end of June	$\ll 1.4 \times 10^{-3}$	$\ll 2 \times 10^{-5}$	$\ll 1.6 \times 10^{-3}$
	$\nu \to$ n (few MeV)	$\simeq 3 \times 10^{-10}$	0.03342*	Jan. 4th*	$\ll 7 \times 10^{-5}$	$\ll 2 \times 10^{-6}$	$\ll 2 \times 10^{-4}$
	direct μ	$\Phi^{(\mu)}_0 \simeq 20\ \mu$ m^{-2}d^{-1}	0.0129	end of June	$\simeq 10^{-7}$	$\simeq 10^{-9}$	$\simeq 10^{-7}$
	direct ν	$\Phi^{(\nu)}_0 \simeq 6 \times 10^{10}\ \nu$ cm^{-2}s^{-1}	0.03342*	Jan. 4th*	$\simeq 10^{-5}$	3×10^{-7}	3×10^{-5}

* The annual modulation of solar neutrino is due to the different Sun-Earth distance along the year; so the relative modulation amplitude is twice the eccentricity of the Earth orbit and the phase is given by the perihelion.

In any case no systematic effects or side reactions able to account for the whole observed modulation amplitude and to simultaneously satisfy all the requirements of the exploited DM signature have been found. A detailed discussion about all the related arguments can be find in Refs. [1–4, 6–8, 13, 16, 17, 20].

The long-standing annual-modulation evidence measured in DAMA experiments is model-independent, i.e. in particular independent on theoretical interpretations of the identity of DM and specifics of its interactions. It can be related to a variety of interaction mechanisms of DM particles with the detector materials and is compatible with a wide set of scenarios regarding the nature of the DM candidate and related astrophysical, nuclear and particle Physics. For example, some of the scenarios available in literature and the different parameters are discussed in Refs.[2, 8, 16] and references therein, and recently e.g. in Refs. [15, 21] . A further large literature is available on the topics; many possibilities are open.

Moreover, both negative results and positive hints are largely compatible with the DAMA model-independent DM annual modulation results in many scenarios considering the existing experimental and theoretical uncertainties (see e.g. Refs. [8, 9, 22–24]); the same holds for indirect approaches which are strongly model dependent.

4 Other signatures?

In the framework of direct DM investigations crystal scintillators represent a reliable technology to investigate model independent signatures. An apparatus made by an array of such detectors, in addition to the already mentioned advantages, allows a long term stability and effective routine calibrations down to keV energy region in the same conditions as production runs. Moreover due to the wide choice of nuclei and isotopes, they are sensitive to many candidates, interaction types and astrophysical, nuclear and particle physics scenarios. In particular in order to disentangle in the corollary investigation on the candidate particle(s) at least some of the many possible astrophysical, nuclear and particle Physics scenarios and related experimental and theoretical uncertainties, the decreasing of the software energy threshold and higher exposures are necessary.

Finally the ultimate challenge and the only effective method for such studies is the investigation of a model independent signature and of second order effects.

Many topics could be investigated:

- the peculiarities of the annual modulation phase;
- the peculiarities of the DM interaction mechanisms;
- the velocity and spatial distribution of the DM particles in the galactic halo;
- the effects induced on the DM particles distribution in the galactic halo by contributions from satellite galaxies tidal streams;
- the effects induced on the DM particles distribution in the galactic halo by the possible existence of caustics;
- the detection of possible solar wakes or gravitational focusing effect of the Sun on the DM particle;
- the investigation of possible diurnal effects and so on.

Some of these studies already started with DAMA/LIBRA–phase1([8, 13, 23] and references therein), but a further important step towards such investigations will be represented by the larger exposure collected by DAMA/LIBRA–phase2, with lower energy threshold.

4.1 Diurnal modulation

The results obtained by investigating the presence of possible diurnal variation in the low-energy single-hit scintillation events collected by DAMA/LIBRA–phase1 (1.04 ton × year exposure) have been analysed in terms of a DM second order model-independent effect due to the Earth diurnal rotation around its axis [13]. In particular, the data were analysed using the sidereal time referred to Greenwich, often called GMST.

This daily modulation of the rate on the sidereal time, expected when taking into account the contribution of the Earth rotation velocity, has several requirements as the DM annual modulation effect does. The interest in this signature is that the ratio R_{dy} of this diurnal modulation amplitude over the annual modulation amplitude is a model independent constant at given latitude; considering the LNGS latitude one has:

$$R_{dy} = \frac{S_d}{S_m} \simeq 0.016 \qquad (1)$$

Taking into account R_{dy} and the DM annual modulation effect pointed out by DAMA/LIBRA–phase1 for *single-hit* events in the low energy region, it is possible to derive the diurnal modulation amplitude expected for the same data. In particular, when considering the (2–6) keV energy

 DOI: 10.1051/epjconf/201612602014

interval, the observed annual modulation amplitude in DAMA/LIBRA–phase1 is: (0.0097 ± 0.0013) cpd/kg/keV [4] and the expected value of the diurnal modulation amplitude is $\simeq 1.5\times 10^{-4}$ cpd/kg/keV.

For completeness figure 2 shows the time and energy behaviour of the experimental residual rates of *single-hit* events both as a function of solar (*left*) and of sidereal (*right*) time, in the (2–6) keV and (6–14) keV intervals. The used time bin is 1 (either solar or sidereal) hour.

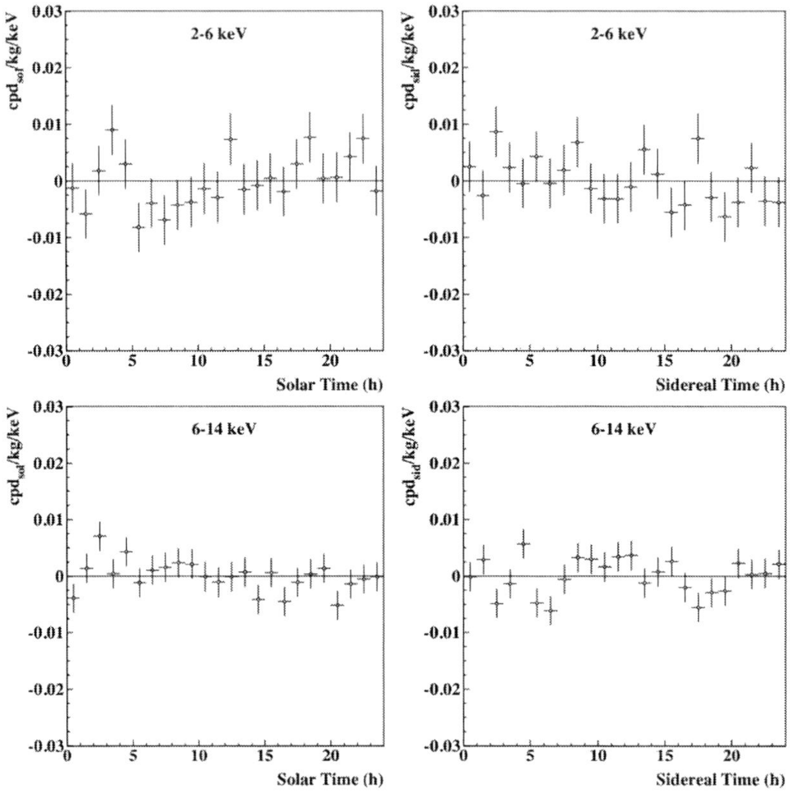

Figure 2. Experimental model-independent diurnal residual rate of the *single-hit* scintillation events, measured by DAMA/LIBRA–phase1 in the (2–6) and (6–14) keV energy intervals as a function of the hour of the solar (*left*) and sidereal (*right*) day. The experimental points present the errors as vertical bars and the associated time bin width (1 hour) as horizontal bars. The cumulative exposure is 1.04 ton × yr. See Ref. [13] for details.

The null hypothesis (absence of residual rate diurnal variation) has been tested by a χ^2 test, obtaining the results given in table 2; there the upper tail probabilities (P-values), calculated by the standard χ^2 distribution, are also reported. Thus, no diurnal variation with a significance of 95% C.L. is found at the reached level of sensitivity.

In addition to the χ^2 test, another independent statistical test has been applied: the run test [13] ; it verifies the hypothesis that the positive and negative data points are randomly distributed. The lower tail probabilities are equal to: 7% and 26% in the (2–6) and (6-14) keV energy region, respectively, for the solar case and 78% and 16% in the (2–6) and (6-14) keV energy region, respectively, for the sidereal case. Thus, in conclusion the presence of any significant diurnal variation and of time structures can be excluded at the reached level of sensitivity (see e.g. the error bars in figure 2).

Table 2. Test of absence of diurnal effect in the DAMA/LIBRA–phase1 data. The P-values, calculated by the standard χ^2 distribution, are also shown. As can be seen, the χ^2 test supports the hypothesis that the diurnal residual rates in DAMA/LIBRA–phase1 are simply fluctuating around zero.

Energy	Solar Time	Sidereal Time
2–6 keV	χ^2/d.o.f. = 25.8/24 → P = 36%	χ^2/d.o.f. = 21.2/24 → P = 63%
6–14 keV	χ^2/d.o.f. = 25.5/24 → P = 38%	χ^2/d.o.f. = 35.9/24 → P = 6%

In order to compare the experimental data with the DM diurnal effect due to the Earth rotation around its axis, the sidereal diurnal modulation amplitude of the (2–6) keV energy interval is taken into account: $A_d^{exp} = -(1.0 \pm 1.3) \times 10^{-3}$ cpd/kg/keV. Following the Feldman-Cousins procedure an upper limit can be obtained for the measured diurnal modulation amplitude: $A_d^{exp} < 1.2 \times 10^{-3}$ cpd/kg/keV (90% C.L.); thus, the DAMA/LIBRA–phase1 experimental sensitivity is not sufficient to point out this second order diurnal effect (expected amplitude $\simeq 1.5\times 10^{-4}$ cpd/kg/keV).

In conclusion, at that level of sensitivity of DAMA/LIBRA–phase1 the presence of a significant diurnal variation and of diurnal time structures in the data can be excluded for both the cases of solar and sidereal time. In particular, the sidereal diurnal modulation amplitude expected – because of the Earth diurnal motion – on the basis of the DAMA DM annual modulation results cannot be investigated at the present sensitivity; DAMA/LIBRA–phase2, presently running, with a lower software energy threshold [5] can also offer an alternative possibility to increase sensitivity to such an effect.

4.2 Daily effect on the sidereal time due to the shadow of the Earth

The results obtained in the investigation of possible diurnal effects for low-energy single-hit scintillation events of DAMA/LIBRA–phase1 (1.04 ton × year exposure) have been analysed in terms of Earth Shadow Effect, a model-dependent effect that could be expected in case of DM candidates inducing just nuclear recoils and having high cross-section with ordinary matter, which implies low DM local density in order to fulfil the DAMA/LIBRA DM annual modulation results [14].

In fact a diurnal variation of the low energy rate could be expected [25, 26] for these specific candidates, because during the sidereal day the Earth shields a given detector with a variable thickness, eclipsing the wind of DM particles. The induced effect should be a daily variation of their velocity distribution, therefore of their flux and, of course, of the signal rate measured deep underground. However, this effect is very small and would be appreciable only in case of high cross-section spin independent coupled candidates that could constitute a little fraction (ξ) in the Galactic dark halo.

The Earth's velocity in the galactic frame, $\vec{v}_e(t)$[1], defines an angle, θ, with the vector joining the centre of the Earth to the position of the laboratory. Because of the Earth's rotational motion, the θ angle varies with the diurnal sidereal time and ranges between a minimum and a maximum which depend on the laboratory position on the Earth. In particular the larger is the allowed range of θ the larger is the effect resulting. The time dependence of θ can be expressed as a function of the laboratory latitude, λ, according to $\cos\theta(t) = \cos\psi\cos\lambda\cos(\omega t + \phi_0) + \sin\psi\sin\lambda$, where $\omega = (2\pi/24^h)$ is the Earth's angular rotational velocity and t the sidereal time in hours. The angles $\psi = (48.1)^0$ and $\phi_0 = (42.4)^0$ depend on the orientation of the Earth axis in the galactic frame [27].

By the fact, the diurnal variation of the velocity distribution and of the counting rate proves undetectable for cross-sections on proton $\leq 10^{-3}$ pb, since the Earth is practically transparent to similar

[1]The Earth's orbital speed around the Sun, which is the most relevant component for the annual modulation signature among the terms contributing to $\vec{v}_e(t)$, can be neglected in the present case because the data, taken in different periods during the year, contribute approximately to the same diurnal time intervals.

particles, while high cross-section values give velocity distributions (and counting rates) significantly dependent on the considered time interval, producing a well detectable diurnal effect.

A study on diurnal variation in the rate with suitable exposure and stability could allow to investigate in some model scenarios high cross sections (σ_p) DM particle component (with small ξ) in the dark halo and decouple ξ from σ_p.

In particular, for each set of parameters [14], one can evaluate the $\xi\sigma_n$ allowed values as:

$$\xi\sigma_n = \frac{S_m^{exp}}{S'_{m,(2-4)keV}(m_{DM}, \sigma_n)}. \tag{2}$$

This corresponds, once including the experimental uncertainties on S_m^{exp}, to a band in the ξ vs σ_n plane.

Finally, for each considered set of parameters the three-dimensional allowed region in the parameter's space: ξ, σ_n, m_{DM} can be studied. In Ref. [14] results in a considered scenario and details are given.

4.3 Directionality with anisotropic scintillators

For completeness, we also mention a different approach for the direct detection of DM candidates inducing just nuclear recoils; it is based on the study of the correlation of the nuclear recoil direction with the Earth velocity [28]. This approach has relevant technical difficulties, when the detection of the short nuclear recoil track is pursued; similar activities are up to now at R&D stage.

An interesting possibility is offered by anisotropic scintillators [29, 30]. They have different light responses depending on the nucleus recoil direction with respect to the scintillator's axes. Since the response to γ/β radiation is isotropic instead, these detectors can offer the possibility to efficiently study possible anisotropy of nuclear recoils [29–32].

So directionality can be pointed out through the study of the variation in the response of anisotropic scintillation detectors during sidereal day. In fact the diurnal Earth rotation changes the impinging direction of the DM particle flux and therefore the mean direction of the recoil nuclei induced by DM particle with respect to the crystal axes.

Low background $ZnWO_4$ crystal scintillators have recently been proposed because their light output and pulse shape depend on the direction of the impinging particles with respect to the crystal axes. Both these anisotropic features can provide two independent ways to exploit the directionality approach; the features and performances of such scintillators are very promising [33] and will be explored in future.

5 Conclusions

The data of DAMA/LIBRA–phase1 have further confirmed the presence of a peculiar annual modulation of the *single-hit* events in the (2–6) keV energy region satisfying all the many requirements of the DM annual modulation signature. The cumulative exposure with NaI(Tl) target by the former DAMA/NaI and DAMA/LIBRA–phase1 is 1.33 ton × yr giving a model-independent positive evidence at 9.3 σ C.L. and a full sensitivity to many kinds of astrophysical, nuclear and particle Physics scenarios.

DAMA/LIBRA is continuously running in its new configuration (named DAMA/LIBRA–phase2) with a lower software energy threshold aiming to improve the knowledge on corollary aspects regarding the signal and on second order effects. It can play a unique role in the future, both in the investigation of DM peculiarities and in the search for rare processes. Further efforts are also in progress, in particular preliminary studies on the possibility to realise new concepts PMTs towards a possible phase3.

References

[1] R. Bernabei et al., Nucl. Instr. and Meth. A **592**, 297 (2008).
[2] R. Bernabei et al., Eur. Phys. J. C **56**, 333 (2008).
[3] R. Bernabei et al., Eur. Phys. J. C **67**, 39 (2010).
[4] R. Bernabei et al., Eur. Phys. J. C **73**, 2648 (2013).
[5] R. Bernabei et al., J. of Instr. **7**, P03009 (2012).
[6] R. Bernabei et al., Eur. Phys. J. C **72**, 2064, (2012).
[7] R. Bernabei et al., Eur. Phys. J. C **74**, 3196 (2014).
[8] R. Bernabei et al., Int. J. of Mod. Phys. A **28**,1330022 (2013) .
[9] P. Belli et al., Phys. Rev. D **84**, 055014 (2011) .
[10] R. Bernabei et al., Eur. Phys. J. C **62**, 327 (2009) .
[11] R. Bernabei et al., Eur. Phys. J. C **72**, 1920 (2012).
[12] R. Bernabei et al., Eur. Phys. J. A **49**, 64 (2013).
[13] R. Bernabei et al., Eur. Phys. J. C **74**, 2827 (2014).
[14] R. Bernabei et al., Eur. Phys. J. C **75**, 239 (2015).
[15] A. Addazi et al., Eur. Phys. J. C **75**, 400 (2015).
[16] R. Bernabei el al., La Rivista del Nuovo Cimento **26** n.1, 1-73 (2003).
[17] R. Bernabei et al., Int. J. Mod. Phys. D **13**, 2127 (2004).
[18] A.K. Drukier, K. Freese, D.N. Spergel, Phys. Rev. D **33**, 3495 (1986).
[19] K. Freese, J.A. Frieman, A. Gould, Phys. Rev. D **37**, 3388 (1988).
[20] R. Bernabei et al., Eur. Phys. J. C **18**, 283 (2000).
[21] S. Scopel et al., JCAP**07**, 041 (2015).
[22] A. Bottino, N. Fornengo, and S. Scopel, Phys. Rev. D **85**, 095013 (2012).
[23] P. Belli, A. Incicchitti and F. Cappella, Int. J. of Mod. Phys. A **29**, 1443011 (2014).
[24] R. Bernabei et al., Int. J. of Mod. Phys. A **30**, 1530053 (2015).
[25] J.I. Collar et al., Phys. Lett. B **275**, 181 (1992).
[26] J.I. Collar and F.T. Avignone III, Phys. Rev. D **47**, 5238 (1993).
[27] K.R. Lang, Astrophysical Formulae (Springer-Verlag) 1980.
[28] D.N. Spergel, Phys. Rev. D **37**, 1353 (1988).
[29] P. Belli et al., Il Nuovo Cimento C **15**, 475 (1992).
[30] R. Bernabei et al., Eur. Phys. J. C **28**, 203 (2003).
[31] N. J. C. Spooner et al., Proc. of the First International Workshop on The Identification of Dark Matter, World Scientific, Singapore, 481 (1997).
[32] Y. Shimizu et al., Nucl. Instrum. Methods A **496**, 347 (2003).
[33] F. Cappella et al., Eur. Phys. J. C **73**, 2276 (2013).

EPJ Web of Conferences **126**, 02015 (2016) DOI: 10.1051/epjconf/201612602015
ICNFP 2015

Status of the NICA project at JINR

V. Kekelidze[1,a], A. Kovalenko[1], R. Lednicky[1], V. Matveev[1], I. Meshkov[1], A. Sorin[1], and G. Trubnikov[1]

[1] *Joint Institute for Nuclear Research, Joliot-Curie, 6, Dubna, Moscow reg., Russian Federation, 141980*

Abstract. The project NICA (Nuclotron-based Ion Collider fAcility) is aimed to study hot and dense baryonic matter in heavy ion collisions in the energy range up to $\sqrt{s_{NN}}$=11 GeV, and to study nucleon spin structure in polarized proton and deuteron collisions in the energy range up to $\sqrt{s}$= 27 GeV. The heavy ion program will be performed at the Nuclotron extracted beams with the BM@N (Baryonic Matter at Nuclotron) set-up and with the MPD (MultiPurpose Detector) at the NICA collider with the average luminosity of $L = 10^{27}$ cm^{-2}s^{-1} (for $^{197}Au^{79+}$). The spin physics will be studied with the SPD (Spin Physics Detector) at the NICA collider.

1 Introduction

The NICA (Nuclotron-based Ion Collider fAcility) project is under realization at the Joint Institute for Nuclear Research (JINR) in Dubna [1–3]. The main goals of the project are study of hot and dense baryonic matter at the extreme conditions in heavy ion collisions, and investigation of nucleon spin structure in the collisions of polarized protons and deuterons. The NICA is an accelerator and experimental complex which includes collider to provide collisions of ions with atomic masses up to gold with energy range $\sqrt{s_{NN}} = 4 - 11$ GeV for Au^{79+}, and polarized protons and deuterons with energy up to $\sqrt{s}$=27 GeV for protons. This energy range will be extended to cover the lower energy region providing fixed target experiments in the beams extracted from the Nuclotron. The NICA complex comprises two collider experiments, MultiPurpose Detector (MPD) to study heavy ion collisions, and Spin Physics Detector (SPD) to study collisions of polarized particles, and one fixed target experiment – Baryonic Matter at Nuclotron (BM@N).

The physic program in heavy ion collisions is aimed to explore the QCD phase diagram in the region of hot and dense baryonic matter. The NICA energy covers the region of maximum baryonic density in phase transition. The strategy is to perform a detailed energy and system size scan with an emphasis to the production of hadrons and dileptons, event-by-event fluctuations and correlations. This program includes a search for the mixed phase of the quark and hadronic matter formed in heavy-ion collisions as a consequence of a first order phase transition. Existing data on single-particle spectra and mean multiplicities [4] suggest that this transition occurs within the NICA energy range.

The range of collision energies is promising for dilepton studies since the effect of vector meson modifications is expected to be sensitive to the baryon density, while the latter happens to reach the maximum in central Au+Au collisions at NICA [5].

[a] e-mail: kekelidze@jinr.ru

 DOI: 10.1051/epjconf/201612602015

The theory predicts strangeness enhancement in heavy-ion induced interactions (relative to elementary p+p collisions), which might be a signature for the deconfinement phase transition [6]. Taking into account small hadronic cross-sections of multi-strange hyperons, and not as strong as for other hadrons re-scattering effects of hyperons in the dense nuclear matter, nuclear objects with strangeness - hypernuclei [7] can be produced with considerably enhanced rates at the NICA energies [8]. New experimental data of NICA on (anti)hyperon and hypernuclei production will provide a valuable insight into the reaction dynamics and properties of the QCD matter.

The studies in the collider mode will be complemented by a fixed target experiment BM@N in the beams extracted from the Nuclotron at energies up to 6 GeV per nucleon. The BM@N research program on heavy-ion collisions includes [9]: investigation of reaction dynamics and nuclear EOS, study of in-medium properties of hadrons, production of (multi)-strange hyperons at the threshold and search for hypernuclei. Particle yields, ratios, transverse momentum spectra, rapidity and angular distributions, as well as fluctuations and correlations will be studied at BM@N as a function of the collision energy and centrality.

The spin physics program is under development and the corresponding detector SPD is at the design stage. The SPD experiment will provide unique opportunity to study eight Parton Distribution Functions (PDFs) in one experiment and obtain a comprehensive information on the nucleon spin structure at high statistical level with minimum systematic uncertainties. The nucleon spin structure will be studied using the Drell-Yan mechanism exploring 8 intrinsic-transverse-momentum dependent PDFs (new PDFs) at leading twist azimuthal asymmetries with different angular modulations in the hadron and spin azimuthal angles, F_h and F_s.

More detailed information on the NICA physics program is available in [10].

2 Accelerator complex

The view of the NCA complex is presented in Fig. 1.

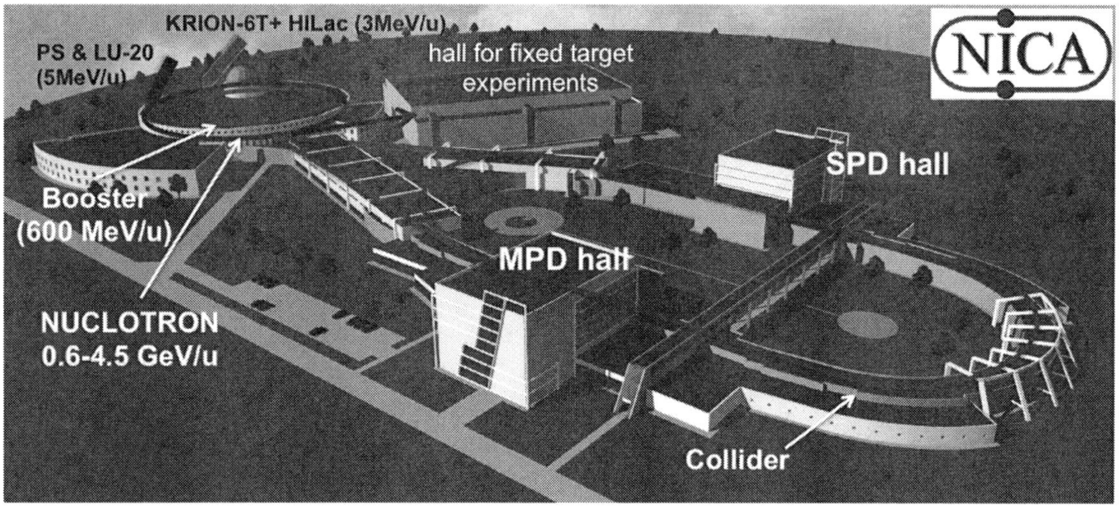

Figure 1. The NICA complex: existing facility (blue) and to be constructed one (red)

The existing facility consists of:

- injector comprising set of light ion sources including source of polarized protons and deuterons (PS), and Alvarez-type linac LU-20;
- Nuclotron – superconducting (SC) proton synchrotron with maximum magnetic rigidity of $45T \cdot m$ and the circumference of 251.52 m providing acceleration of completely stripped $^{197}Au^{79+}$ ions up to the experiment energy of 4.5 GeV/u and protons up to the energy of 12.6 GeV;
- area with fixed target experiments.

The new facility to be constructed comprises:

- new injector consisting of ESIS-type ion source that provides $^{197}Au^{31+}$ ions of the intensity of $2 \cdot 10^9$ ions per pulse of about 7 μs duration at possible repetition rate up to 50 Hz, and heavy ion linear accelerator (HILAc) consisting of RFQ and RFQ Drift Tube Linac (RFQ DTL) sections accelerating ions at A/q $\leq$8 up to the energy of 3 MeV/u at efficiency not less than 80 %;
- Booster-synchrotron housed in the Synchrophasotron yoke window and consisting of SC magnetic system providing maximum magnetic rigidity of $25T \cdot m$ at the ring circumference of 215 m, and equipped with electron cooling system that allows to provide cooling of the ion beam up to the energy of 100 MeV/u. The maximum energy of $^{197}Au^{31+}$ ions accelerated in the Booster is of 600 MeV/u; stripping foil placed in the transfer line from the Booster to the Nuclotron allows to provide the stripping efficiency at the maximum Booster energy not less than 80 %;
- transfer line transporting the particles from Nuclotron to Collider rings;
- two SC collider rings with maximum magnetic rigidity of $45T \cdot m$ and the circumference of about 503.4 m equiped with electron and stochastic cooling systems and providing gold-gold collisions at the energy range $\sqrt{s_{NN}} = 4 - 11$ GeV and p-p collisions at the energy up to $\sqrt{s} = 27$ GeV.

The NICA collider will provide average luminocity $L = 10^{27}$ cm^{-2}s^{-1} and 10^{32} cm^{-2}s^{-1} for gold-gold and proton-proton collisions, respectively. To reach such luminosity a special procedure of formation of high intensity ion bunches in the collider is developed. First, the ions accelerated in the injection accelerator chain are stored in the Collider rings with so called "barrier bucket (BB) technique" that allows to reach the required intensity of the beam. Then the beam becomes coasting after switching off the BB RF system and bunched on 22nd harmoics of the harmonic RF. When bunches are compressed sufficiently they are captured in 66th harmonics that forms short and dense bunches. Afterwards, the colliding mode is turned on by the system of the beam orbit correction.

3 MPD detector

The Multi Purpose Detector (MPD) is designed as a 4π spectrometer capable detecting of charged hadrons, electrons and photons in heavy-ion collisions at high luminosity in the energy range of the NICA collider [11]. The experimental setup (see Fig. 2) will comprise a precise 3D tracking system and a high-performance particle identification (PID) system based on the time-of-flight measurements and calorimetry. The event rate of minimum bias interactions is of about 7 kHz, and the total charge particle multiplicity exceeds 1000 in the most central Au+Au collisions at $\sqrt{s_{NN}}$=11 GeV. The detector design requires a very low material budget. The whole detector setup (Fig. 2) includes Central Detector (CD) covering pseudorapidity interval $|\eta|<2$ and two forward spectrometers (FS-A,B) for $2 < |\eta| < 3$ (optional). The cross-sectional view of the MPD Central Detector (CD) is shown in Fig. 3. The MPD CD is 9 meters long and 6.6 meters in diameter and will be constructed in two stages. At the first stage (at the end of 2019) MPD will consist of the superconducting solenoid with a flux return yoke, Time-Projection Chamber (TPC), barrel Time-Of-Flight system (TOF), Electromagnetic

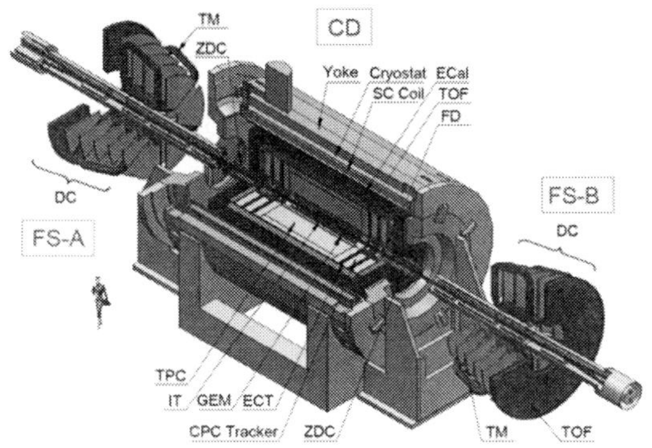

Figure 2. MPD setup.

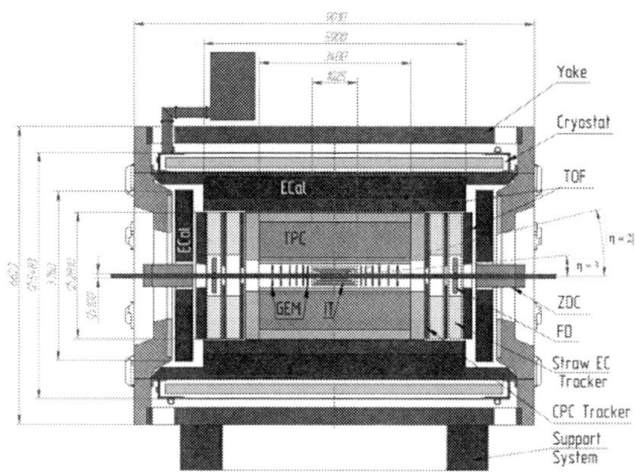

Figure 3. The MPD centlal detector cross-sectional view.

Calorimeter (ECal), Zero-Degree Calorimeter (ZDC) and Fast Forward Detector (FFD). A detailed description of the detector parts can be found in Ref. [12].

The Time-Projection Chamber is the main tracking detector of the MPD and will provide: tracking and particle ID (dE/dx measurements with a resolution better than 8%) in the pseudo rapidity range $|\eta| < 1.2$, momentum resolution for charge particles better than 3% in the transverse momentum range $0.1 < p_t < 1$ GeV/c and two-track resolution of about 1 cm. The TPC has an inner diameter of 54 cm, an outer diameter of 280 cm, and an overall length along the beam direction of 340 cm. The TPC field cage will be made of composite materials (kevlar and tedlar), with the overall thickness to be less than 5% X0. For precise tracking the relative radial magnetic and electric field components in MPD have not to exceed $\approx 5.2 \cdot 10^{-4}$. The TPC readout system at the first stage is based on the Multi-Wire Proportional Chambers (MWPC) with cathode pad readout. The gas mixture of 90% argon and 10% methane (P10) or 90%Ar and 10% CO2 is supposed to be used. In Fig. 4 a relative momentum resolution ($\Delta P/P$) for charged tracks in the MPD TPC from Monte-Carlo simulations is shown.

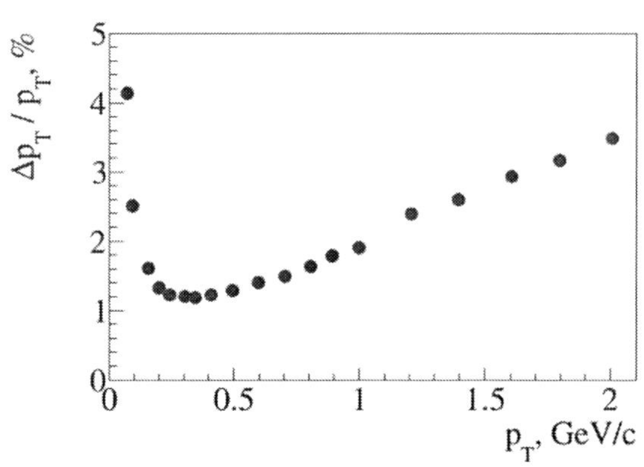

Figure 4. $\Delta P_t/P_t$ as a function of transverse momentum in MPD.

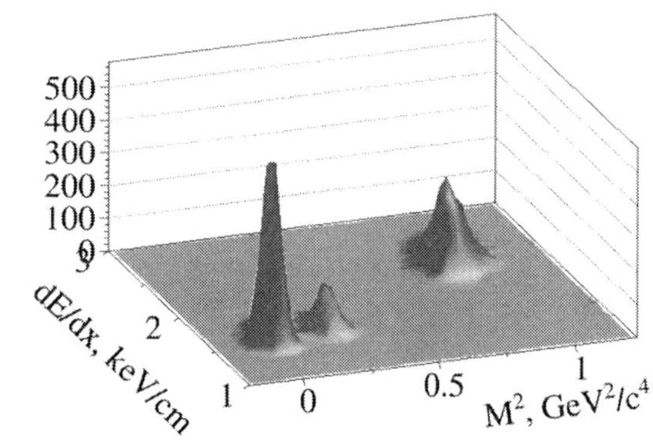

Figure 5. A typical distribution of the specific energy loss dE/dx versus mass-squared for π^+, K^+, and protons (from left to right) at p = 1 GeV/c.

Identification of charged hadrons will be achieved by the time-of-flight (TOF) measurements which are complemented by the energy loss (dE/dx) information from the TPC (see Fig. 5). The cylindrical part of the TOF MPD has a full azimuthal and $|\eta| <1.2$ coverage. The TOF outer diameter is 3.4 m and the total surface is of about 53 m^2. The basic element of the TOF system will be a Multigap Resistive Plate Chamber (MRPC). The TOF electronics will be based on the NINO specific integrated circuit [13] and TDC based on a multihit High Performance ASIC (HPTDC) chip developed at CERN [14]. The detector is segmented in azimuth into 12 sectors of 5.7 m length. Each sector consists of 4 modules and each module comprises 6 MRPCs detectors arranged in 2 layers. The overall TOF geometrical efficiency is above 95% and time resolution is better than 100 pc. It would allows to separate pions from kaons up to $p_t < 1.5$ GeV/c and identify (anti)protons up to $p_t \approx 3$ GeV/c. As can be seen from Fig. 5, the basic detector parameters, namely, dE/dx resolution of $\sigma_{dE/dx} \approx 8\%$ and TOF resolution of $\sigma_{TOF} \approx 100$ ps will provide a high degree of selectivity for hadrons at momenta below 2 GeV/c.

The primary role of the electromagnetic calorimeter is to measure the spatial position and energy of electrons and photons produced in heavy ion collisions. The electromagnetic calorimeter will be built of towers as basic elements of about 3 cm^2 cross section. Each lead-scintillator sampling tower contains 250 alternating tiles of Pb (0.275 mm) and plastic scintillator (1.5 mm). A module of 18 radiation length thickness will be $\approx$40 cm long. The cells of each tower are optically combined by 9 longitudinally penetrating wavelength shifting fibers for light collection. The light collected by these fibers will be read out by avalanche photodiodes (MAPD) units of 3x3 mm ^{2} sensitive area. The towers, mechanically grouped together, make 139 trapezium-shape modules which are combined into 48 detector sectors. The heat-producing electronics will be thus separated from the modules and mounted on the upper parts of sectors.

The FFD detector located at $\approx$75 cm on both sides of the interaction point will provide a fast trigger and timing signal for TOF. A FFD module made of 1.5 cm thick quartz Cherenkov radiator optically coupled to a multianode MCP-PMT Planacon, is capable of register high energy photons and relativistic charged particles of velocity $\beta > 0.69$ with a timing resolution better than 50 ps.

The MPD Zero Degree Calorimeter (ZDC) located in the both sides of MPD at a distance of 3.2 m from the MPD center will be used for classification of events by centrality by measuring the energy of spectators. Each ZDC semy-part consists of identical modules parallelepiped-shaped with a transverse cross-section of 10x10 cm^2 and a length of 90 cm. Each of ZDC modules consists of 60 lead-scintillator tile "sandwiches" of thickness of 16 and 4 mm for Pb and scintillator, respectively. The sampling ratio of 4÷1 will provide the compensation condition. WLS-fibers from each of 6 consecutive scintillator tiles are collected together and viewed by an avalanche photodiode at the end of each module.

4 MPD feasibility study

All the Monte-Carlo simulation and feasibility study for MPD were performed in the framework of a dedicated program tool - MPDRoot comprising interfaces to many event generators and Geant as well as detector response simulation and reconstruction algorithms.

The performance of the MPD detector for hyperon measurements was studied for a combination of Time Projection Chamber (TPC) and Time-Of-Flight system (TOF) covering the mid-rapidity region ($|\eta| < 1.3$). A material budget not exceeds 10% of X0 in the region of interest. A powerful particle ID based on ionization loss (dE/dx) measurements in the TPC and time-of-flight information from the TOF, allows one a precise trajectory reconstruction and a reliable separation of particle species.

Hyperons and cascades were reconstructed using the secondary vertex finding technique with an optimized set of topological and track quality cuts as described in [15]. In Fig. 6 are shown invariant mass distributions of (p,π^-), (Λ,π^-) and (Λ,K^-) pairs from central Au+Au collisions at center-of-mass energy $\sqrt{s} = 9$ GeV. The estimated yields of particle species for 10 weeks of data taking at the

DOI: 10.1051/epjconf/201612602015

nominal luminosity are $\approx 10^7$ and $\approx 10^6$ for Ξ^- and Ω^-, respectively. The accumulated statistics will allow one to get information on cascade production over a broad region of the QCD phase diagram and a large phase-space of the reaction.

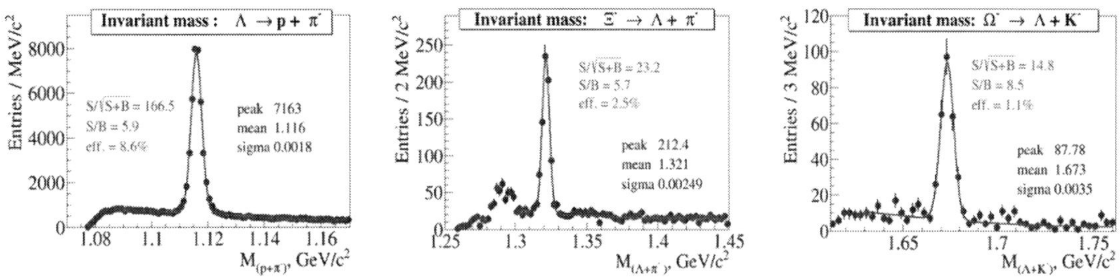

Figure 6. Invariant mass of p and π^- (left), Λ and π^- (center), Λ and K^- (right) candidates.

A detailed description of the analysis procedure and results of the MPD performance can be found in Ref. [16]. The charged hadrons and dileptons from event generators (UrQMD+Pluto) were transported through the detector setup. A realistic detector response was simulated followed by track reconstruction, TPC-TOF matching and particle identification procedures performed within the MPDRoot framework. Electron identification was achieved by using combined information about the specific energy loss dE/dx from TPC, time-of- flight from TOF and E/p information from EMC. The overall hadron rejection factor was achieved to be ≈ 3200.

The background from conversion pairs was eliminated by an extra set of topological and kinematical cuts. Fig. 7 presents a spectrum of reconstructed low-mass dielectrons after background rejection (dots), and a spectrum of true dielectrons from the event generator (line). S/B ratio in mass region $0.2 < M_{e^+e^-} < 1.2$ GeV/c^2 was estimated to be of about 10% which indicate MPD performance at the level of best published experimental results.

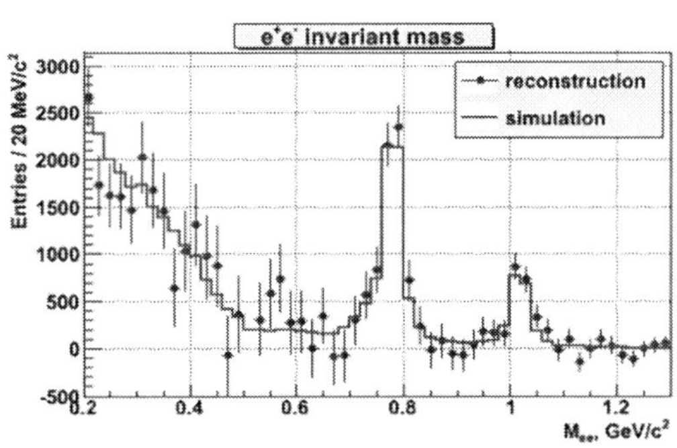

Figure 7. Invariant mass of dileptons after background subtraction.

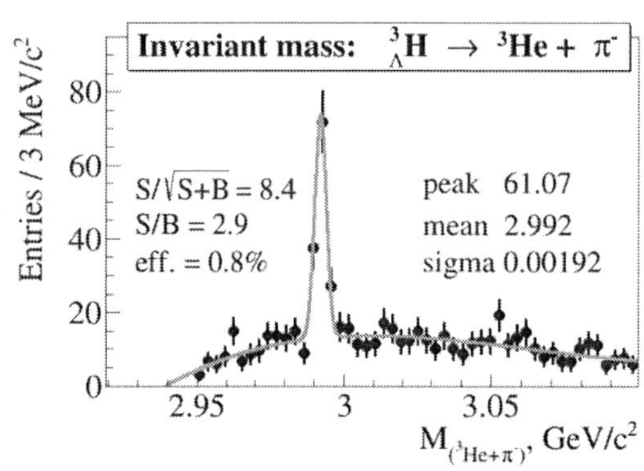

Figure 8. Invariant mass of ^{3}He and π^- candidates (MPD).

$5 \cdot 10^5$ events from the DCM-QGSM event generator corresponding ≈ 30 minutes of data taking time at NICA were used to study of MPD capability for reconstruction of hypertritons. The event reconstruction, particle ID and secondary vertex finding are described in [15]. The obtained invariant mass spectra of helium-3 and π^- candidates is shown in Fig. 8. This result demonstrates a good sensitivity of the MPD setup for hypernuclei detection. About 10^5 hypertritons will be detected in a week of data taking with a typical event rate of 7 kHz for the design NICA luminosity of $10^{27}\text{cm}^{-2}\text{c}^{-1}$. Thus, a detailed study of the production mechanism of single hypernuclei as well as an observation of double hypernuclei at NICA look feasible.

5 BM@N experiment

The BM@N) experiment will start in 2017 in a beam extracted from the Nuclotron. The Nuclotron will provide variety of beams from protons to gold ions with the kinetic energy from 1 to 6 GeV per nucleon. The ion beams of masses up to Xe will be available by the end of 2017. The gold beam with intensity $10^7/s$ in planned in 2019.

A sketch of the BM@N set-up is shown in Fig. 9. It combines high precision track measurements

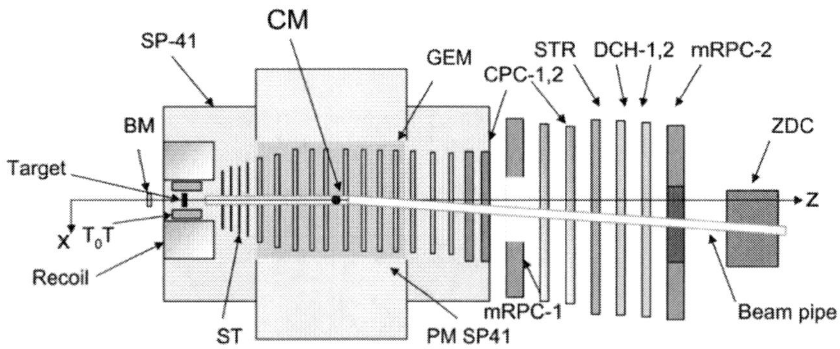

Figure 9. The BM@N experimental set-up.

with time-of-flight information for particle identification and total energy measurements for the analysis of the collision centrality. The charged track momentum and multiplicity will be measured with the set of 12 two-coordinate planes of GEM (Gaseous Electron Multipliers) detectors located downstream of the target inside the analyzing magnet complemented by drift and straw tube chambers situated outside the magnetic field. At the second stage of the BM@N experiment, at least 4 planes of two-coordinate silicon strip detectors will be installed between the GEM tracker and the target. The design parameters of the time-of-flight detectors based on multi-gap Resistive Plate Chambers (mRPC-1,2) with a strip read-out will allow to discriminate between hadrons (π, K, p) as well as light nuclei with the momentum up to a few GeV/c. The Zero Degree Calorimeter (ZDC) is designed for the analysis of the collision centrality by measuring the energy of forward going particles. The T0 detector, partially covering the backward hemisphere around the target, is planned to trigger on central heavy ion collisions and provide a start signal for the mRPC-1,2.

The detector set-up was optimized for the measurement of strange hyperons and hyper-nuclei in central Au+Au collisions. In Figures 10 and 11 are shown invariant mass of (Λ, π^-) and $(^3He, \pi^-)$ candidates, respectively, reconstructed with the GEM tracker from central Au+Au collisions at the beam kinetic energy of 4.5A GeV. The obtained results indicate a reasonable reconstruction capability for strange hyperons produced in high multiplicity central Au+Au collisions. The reconstructed signals of Ξ^-- hyperon and hyper-triton correspond to 0.9M and 2M of central collisions, respectively. The expected statistics of Ξ^--hyperons and hypertritons for a month of the BM@N operation are expected

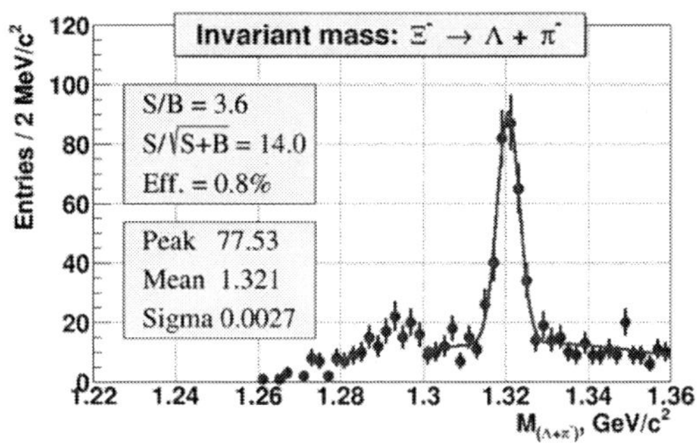

Figure 10. Invariant mass of Λ and π^-.

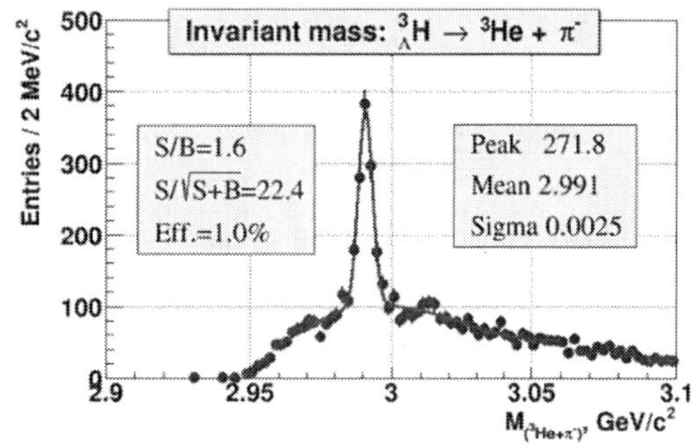

Figure 11. Invariant mass of ^{3}He and π^- candidates.

to be of 2.7M and 4M, respectively. This statistics is sufficient to perform studies of strange hyperon and hyper-nuclei production by measuring their transverse momentum spectra, rapidity and angular distributions.

6 Summary

Design and construction of the the new facility at the JINR is in progress. The NICA will provide competitive research program complementary to those of BNL, CERN and that planned at FAIR.

References

[1] G. Trubnikov, A. Kovalenko, V. Kekelidze, I. Meshkov, R. Lednicky, A. Sissakian, A. Sorin, PoS (ICHEP 2010) 523.

[2] G. Trubnikov, N. Agapov, V.Kekelidze, A. Kovalenko, V. Matveev, I.Meshkov, R. Lednicky, A. Sorin, PoS 36th International Conference of High Energy Physics (ICHEP2012), July 4-11,Melbourne, Australia.

[3] V.Kekelidze, A. Kovalenko, R. Lednicky, V. Matveev, I.Meshkov, A. Sorin, G. Trubnikov, 37th International Conference on High Energy Physics, ICHEP 2014, Valencia, Spain, 2 Jul 2014 - 9 Jul 2014 Nuclear physics B Proceedings supplements (2014).

[4] C. Alt et al, *Phys. Rev.* C **77**, 024903 (2008).

[5] J. Randrup and J. Cleymans, *Phys. Rev.* C **74**, 047901 (2006).

[6] J. Rafelski and B. Muller, *Phys. Rev. Lett.* **48**, 1066 (1982).

[7] A.K. Kermann and M.S. Weiss, *Phys. Rev.* C **8**, 408 (1973).

[8] J. Steinheimer et al., *Phys. Lett.* B **714**, 85 (2012).

[9] BMN Conceptual Design Report. http://nica.jinr.ru/files/BMN/BMN_CDR.pdf.

[10] NICA White Paper, http://nica.jinr.ru, free access

[11] K.U. Abraamyan et al., *Nucl. Instrum. Meth.* A **628**, 99 (2011).

[12] MPD Conceptual Design Report. http://nica.jinr.ru/files/CDR_MPD/MPD_CDR_en.pdf.

[13] F. Anghinolfi et al., *Nucl. Instr. Meth.* A **533**, 183-187 (2004).

[14] J. Christiansen, HPTDC V 1.3, March 2004, http://tdc.web.cern.ch/tdc/hptdc/docs/hptdc (manual, ver2.2.pdf).

[15] M. Ilieva et al. Proceedings of SQM-2015 (*to be published*).

[16] A. Zinchenko et al. Proceedings of SQM-2015 (*to be published*).

EPJ Web of Conferences **126**, 02016 (2016) DOI: 10.1051/epjconf/201612602016
ICNFP 2015

Probing antimatter gravity – The AEGIS experiment at CERN

A. Kellerbauer[1,a], S. Aghion[2,3], C. Amsler[4], A. Ariga[4], T. Ariga[4], G. Bonomi[5,6], P. Bräunig[7], J. Bremer[8], R. S. Brusa[9,10], L. Cabaret[11], M. Caccia[3,12], R. Caravita[13,14], F. Castelli[3,15], G. Cerchiari[1], K. Chlouba[15], S. Cialdi[3,15], D. Comparat[11], G. Consolati[2,3], A. Demetrio[7], L. Di Noto[13,14], M. Doser[8], A. Dudarev[8], A. Ereditato[4], C. Evans[2,3], R. Ferragut[2,3], J. Fesel[8], A. Fontana[6], S. Gerber[8], M. Giammarchi[3], A. Gligorova[17], F. Guatieri[9,10], S. Haider[8], H. Holmestad[18], T. Huse[18], E. Jordan[1], M. Kimura[4], T. Koettig[8], D. Krasnický[14], V. Lagomarsino[13,14], P. Lansonneur[19], P. Lebrun[19], S. Lehner[20], J. Liberadzka[8], C. Malbrunot[8,20], S. Mariazzi[20], V. Matveev[21,22], Z. Mazzotta[3,15], G. Nebbia[23], P. Nédélec[19], M. Oberthaler[7], N. Pacifico[17], D. Pagano[5,6], L. Penasa[9,10], V. Petráček[16], C. Pistillo[4], F. Prelz[3], M. Prevedelli[24], L. Ravelli[9,10], B. Rienäcker[8], O.M. Røhne.[18], A. Rotondi[6,25], M. Sacerdoti[3,15], H. Sandaker[18], R. Santoro[3,12], P. Scampoli[4,26], L. Smestad[8,27], F. Sorrentino[13,14], M. Špaček[16], J. Storey[4], I.M. Strojek[16], G. Testera[14], I. Tietje[8,28], E. Widmann[20], P. Yzombard[11], S. Zavatarelli[14], J. Zmeskal[20], and N. Zurlo[6,29]

(AEGIS Collaboration)

[1] *Max-Planck-Institut für Kernphysik, Saupfercheckweg 1, 69117 Heidelberg, Germany*
[2] *Politecnico of Milano, Piazza Leonardo da Vinci 32, 20133 Milano, Italy*
[3] *INFN Milano, via Celoria 16, 20133 Milano, Italy*
[4] *Laboratory for High Energy Physics, Albert Einstein Center for Fundamental Physics, University of Bern, 3012 Bern*
[5] *Department of Mechanical and Industrial Engineering, Univ. of Brescia, via Branze 38, 25123 Brescia, Italy*
[6] *INFN Pavia, via Bassi 6, 27100 Pavia, Italy*
[7] *Kirchhoff-Institut für Physik, Univ. of Heidelberg, Im Neuenheimer Feld 227, 69120 Heidelberg, Germany*
[8] *Physics Department, CERN, 1211 Geneva 23, Switzerland*
[9] *Department of Physics, University of Trento, via Sommarive 14, 38123 Povo, Trento, Italy*
[10] *TIFPA/INFN Trento, via Sommarive 14, 38123 Povo, Trento, Italy*
[11] *Laboratoire Aimé Cotton, University Paris-Sud, 91405 Orsay Cedex, France*
[12] *Department of Science, University of Insubria, via Valleggio 11, 22100 Como, Italy*
[13] *Department of Physics, University of Genova, via Dodecaneso 33, 16146 Genova, Italy*
[14] *INFN Genova, via Dodecaneso 33, 16146 Genova, Italy*
[15] *Department of Physics, University of Milano, via Celoria 16, 20133 Milano, Italy*
[16] *Czech Technical University, Prague, Brehová 7, 11519 Prague 1, Czech Republic*
[17] *Institute of Physics and Technology, University of Bergen, Allégaten 55, 5007 Bergen, Norway*
[18] *Department of Physics, University of Oslo, Sem Sælandsvei 24, 0371 Oslo, Norway*
[19] *Institute of Nuclear Physics, CNRS/IN2p3, University of Lyon 1, 69622 Villeurbanne, France*
[20] *Stefan Meyer Institute for Subatomic Physics, Austrian Academy of Sciences, Boltzmanngasse 3, 1090 Vienna, Austria*
[21] *Institute for Nuclear Research of the Russian Academy of Science, 117312 Moscow, Russia*
[22] *Joint Institute for Nuclear Research, 141980 Dubna, Russia*

[a] e-mail: a.kellerbauer@cern.ch

[23]*INFN Padova, via Marzolo 8, 35131 Padova, Italy*
[24]*University of Bologna, viale Berti Pichat 6/2, 40126 Bologna, Italy*
[25]*Department of Physics, University of Pavia, via Bassi 6, 27100 Pavia, Italy*
[26]*Department of Physics, University of Napoli Federico II, Complesso Universitario di Monte S. Angelo, 80126 Napoli, Italy*
[27]*Research Council of Norway, Drammensveien 288, 0283 Oslo, Norway*
[28]*Physics Department, Technical University Berlin, Straße des 17. Juni 135, 10623 Berlin, Germany*
[29]*Department of Civil Engineering, University of Brescia, via Branze 43, 25123 Brescia, Italy*

Abstract. The weak equivalence principle states that the motion of a body in a gravitational field is independent of its structure or composition. This postulate of general relativity has been tested to very high precision with ordinary matter, but no relevant experimental verification with antimatter has ever been carried out. The AEGIS experiment will measure the gravitational acceleration of antihydrogen to ultimately 1% precision. For this purpose, a pulsed horizontal antihydrogen beam with a velocity of several 100 m s^{-1} will be produced. Its vertical deflection due to gravity will be detected by a setup consisting of material gratings coupled with a position-sensitive detector, operating as a moiré deflectometer or an atom interferometer. The AEGIS experiment is installed at CERN's Antiproton Decelerator, currently the only facility in the world which produces copious amounts of low-energy antiprotons. The construction of the setup has been going on since 2010 and is nearing completion. A proof-of-principle experiment with antiprotons has demonstrated that the deflection of antiparticles by a few μm due to an external force can be detected. Technological and scientific development pertaining to specific challenges of the experiment, such as antihydrogen formation by positronium charge exchange or the position-sensitive detection of antihydrogen annihilations, is ongoing.

1 Introduction

The study of antimatter in the laboratory has the potential to elucidate one of the great unsolved questions in physics: Why does the observable universe appear to contain only ordinary baryonic matter and no sizable amounts of antimatter? In order to solve the baryon asymmetry riddle, several experiments at CERN's Antiproton Decelerator (AD) [1] are studying antihydrogen ($\bar{H}$), the simplest atomic antimatter system. According to the CPT theorem [2], particles and antiparticles must have exactly the same properties – up to a sign. (C stands for charge conjugation, P for parity and T for time reversal.) Thus, the comparison of different properties of hydrogen and its antimatter counterpart can shed light on the question of CPT violation. As was shown almost 50 years ago for CP violation [3], also CPT violation can result in a baryon asymmetry [4]. Until today, experimental tests have provided no indication of CPT violation.

The CPT theorem has been proven to be valid under several mild assumptions, such as locality, unitarity and Lorentz invariance. In addition, the proof requires that the physical system in question be described by a quantum field theory (QFT). Gravity is notably the only one of the four fundamental interactions which is not formulated as a QFT, but rather as a geometric property of space and time. According to the weak equivalence principle (WEP) of general relativity, the motion of a test body in a gravitational field does not depend on the body's structure and composition. In a hypothetical quantum gravity theory, a variety of exchange particles with different coupling constants and ranges is possible [5]. For instance, current results of torsion balance measurements with ordinary matter allow for the existence of additional vector (spin-1) and scalar (spin-0) exchange bosons, under the condition that their coupling constants and ranges are identical [6].

DOI: 10.1051/epjconf/201612602016

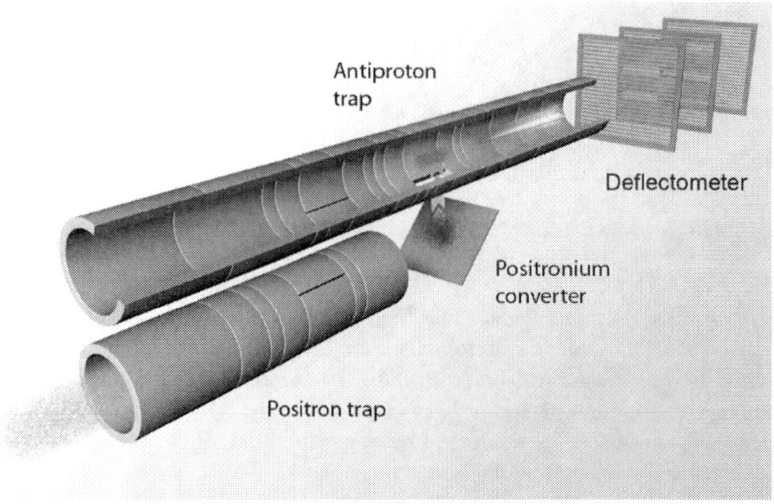

Figure 1. Three-dimensional cut-open sketch of the central part of the AEGIS apparatus, showing the low-field Penning traps, the positronium converter and the deflectometer or interferometer.

The question of an anomalous gravitational acceleration of antimatter has been the subject of considerable controversy from a theoretical point of view. Over the past decades, a number of arguments have been brought forward against it [7]. A prominent early example is Morrison's objection against anti-tensor gravity, which he showed would violate energy conservation [8]. Others are based on the observation that ordinary atoms contain virtual antiparticles due to vacuum polarization, or on the fact that certain CPT tests should be sensitive to the underlying gravitational potential. However, all of these theoretical arguments make some non-trivial assumptions, such as CPT invariance, or the somewhat arbitrary choice of an absolute gravitational potential. A comprehensive review which is relevant for a WEP test at the 1% level is given in Ref. [9]. Ultimately, a direct experimental study of antimatter gravity is required to positively resolve the issue.

While the WEP has been tested up to a relative precision of 1.8×10^{-13} with ordinary matter [10], no relevant measurement of the gravitational acceleration of antimatter has been carried out. This is because until recently only electrically charged antimatter particles – such as positrons (e^+) and antiprotons ($\bar{p}$) – were readily available, which are too sensitive to stray electromagnetic fields to permit a gravity measurement [11]. This situation changed in 2002, when copious amounts of cold $\overline{\mathrm{H}}$ were synthesized for the first time by the ATHENA experiment [12] installed at the AD. ATHENA created anti-atoms via the three-body reaction $\bar{p} + 2e^+ \longrightarrow \overline{\mathrm{H}} + e^+$. For this purpose, $\bar{p}$ and e^+ were brought into overlap in the nested electric potential of a Penning trap [13]. Today $\overline{\mathrm{H}}$ is routinely produced in large quantities by several AD experiments. Therefore, a WEP test with a neutral antimatter particle has finally come within reach.

The AEGIS experiment (Antimatter Experiment: Gravity, Interferometry, Spectroscopy) was proposed in 2007 [14, 15]. Its main goal is to measure for the first time the force acting on $\overline{\mathrm{H}}$ atoms in the gravitational field of the earth [16]. The experiment is based on the creation of a cold, bunched beam of $\overline{\mathrm{H}}$ and the measurement of its vertical displacement after a substantial horizontal flight, as shown in Fig. 1. Ultimately, the aim is to measure the acceleration g acting on the anti-atoms to 1%. The apparatus is centered around ion traps placed inside superconducting magnets for ion confinement and manipulation. AEGIS will use a novel $\overline{\mathrm{H}}$ production technique involving positronium (Ps), the bound state of e^+ and e^-. Antihydrogen will be formed via the charge exchange reaction $\mathrm{Ps}^* + \bar{p} \longrightarrow \overline{\mathrm{H}}^* + e^-$, where the asterisk denotes an excited state. In combination with an efficient cooling technique for $\bar{p}$, this scheme can create ultracold $\overline{\mathrm{H}}$ for high-precision gravity measurements.

DOI: 10.1051/epjconf/201612602016

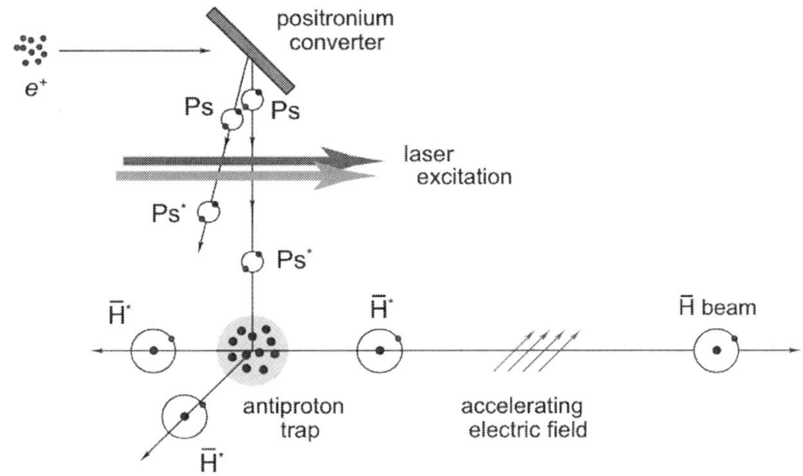

Figure 2. Scheme used in AEGIS for $\bar{\mathrm{H}}$ recombination by resonant charge exchange with positronium followed by Stark acceleration.

2 Experiment

The AEGIS experiment involves the following main experimental steps: (1) Capture of $\bar{p}$ from the AD in a Penning trap, followed by electron cooling; (2) Production of e^+ from a source of β^+-decaying radioisotope; (3) Production of Ps by implanting e^+ into a nanoporous converter material; (4) Excitation of Ps to a Rydberg state; (5) Recombination of $\bar{\mathrm{H}}$ by resonant charge exchange between Rydberg Ps and cold $\bar{p}$; (6) Formation of an $\bar{\mathrm{H}}$ beam by Stark acceleration with inhomogeneous electric fields; (7) Measurement of the vertical acceleration g in a deflectometer or interferometer. All of the aforementioned techniques have been demonstrated independently. The extraordinary challenge of AEGIS lies in combining all of them while reconciling potentially conflicting requirements, such as a high magnetic field for ion confinement and cooling and the absence of electromagnetic fields for high-precision force measurements.

2.1 Production of antihydrogen by resonant charge exchange

Antihydrogen formation by three-body recombination is a well-established method that has been demonstrated to create $\bar{\mathrm{H}}$ at high rate [17]. As mentioned above, AEGIS will make use of an alternative technique which involves Ps as precursor [18]. While it necessarily results in a lower $\bar{\mathrm{H}}$ production rate, the method has a unique advantage: A plasma of $\bar{p}$ confined in the ion trap can be pre-cooled to the temperature of the surrounding cryogenic environment or even beyond. When $\bar{\mathrm{H}}$ is formed by resonant charge exchange between Ps and $\bar{p}$, the temperature of the produced anti-atoms is close to that of the $\bar{p}$ prior to the interaction. This is because the mass of $\bar{p}$ is almost 2000 times larger than that of $e^\pm$, which therefore only impart a very small momentum on their reaction partners. Using a suitable $\bar{p}$ cooling method, such as sympathetic cooling by laser-cooled atomic [19, 20] or molecular anions [21], the technique can thus potentially create ultracold $\bar{\mathrm{H}}$.

In the AEGIS experiment, positronium will be synthesized by depositing a bunch of cold positrons into a positronium converter consisting of a nanoporous insulator or semiconductor wafer [22, 23], as shown in Fig. 2. The converter is placed a short distance away from the trap where cold $\bar{p}$ have been prepared. In the pores of the material, e^+ pick off e^- from the surface and form Ps, which thermalize with the bulk by multiple scattering. Finally, cold Ps are isotropically emitted from the surface. The conversion efficiency has been shown to be as high as 30% [24]. Before the Ps reach the $\bar{p}$ cloud where they form $\bar{\mathrm{H}}$, they are excited to a Rydberg level [25]. In this way, the cross-section of the

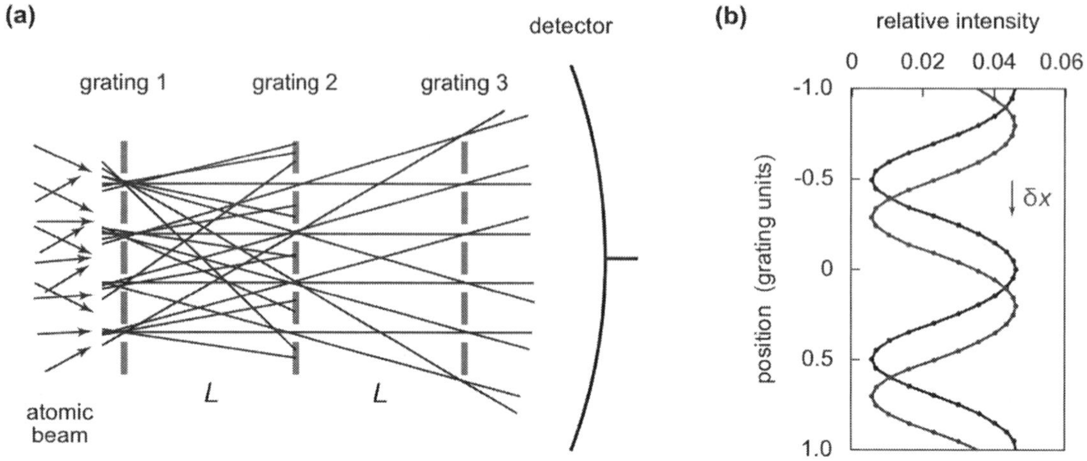

Figure 3. (a) Principle sketch of the deflectometry/interferometry technique, showing a setup with three identical gratings and a non-position-sensitive detector. (b) Monte Carlo simulation of the signal detected behind a three-grating moiré deflectometer as a function of the analysis grating's position. The red dots and solid line show the shift in the presence of a force perpendicular to the beam axis.

charge exchange reaction, which scales with the fourth power of the principal quantum number of the Ps, is significantly enhanced. Due to energy conservation, $\overline{\mathrm{H}}$ is also created in a narrow, well-defined range of Rydberg states. Finally, the $\overline{\mathrm{H}}$ cloud is formed into a beam by Stark acceleration [26].

2.2 Principle of gravity measurement

The force acting on a slow beam of massive particles can be measured with a series of slits or gratings. In the latter case, the device is called a moiré deflectometer or a matter wave interferometer, depending on the diffraction angle $\phi = \lambda_{\mathrm{dB}}/d$, where λ_{dB} is the particles' de Broglie wavelength and d is the grating period. As shown in Fig. 3(a), two material gratings spaced a distance L apart create a shadow or interference pattern a further distance L downstream. A third grating with identical grating period is placed at the location where the line pattern forms. The transmission of particles as a function of the third grating's position (in the direction of the force) is measured by an appropriate detector located beyond the device. As illustrated in Fig. 3(b), the transmission maxima yield the displacement of the beam, and hence, the force acting on it. In AEGIS, an antihydrogen beam accelerated to 500 m s^{-1} will be deflected by about 10 μm over a flight path of 1 m.

A three-grating moiré deflectometer has been used some 20 years ago to measure the acceleration due to gravity of metastable argon atoms [27]. As an alternative to a movable third grating, a position-sensitive detector can be placed in its position. In this way, the arrival positions for particles moving at different velocities can be recorded simultaneously and without any moving parts. The current AEGIS design, which is still being finalized, calls for such a setup with a sophisticated detector that can record the annihilation position of impinging anti-atoms with a vertical resolution of about 2 μm. The detector will most likely consist of a combination of complementary technologies, including a nuclear photographic emulsion and a silicon strip or pixel detector. Both types of systems have been extensively tested in the AEGIS setup and found to deliver the required position resolution and detection efficiency for an antihydrogen gravity measurement [28, 29].

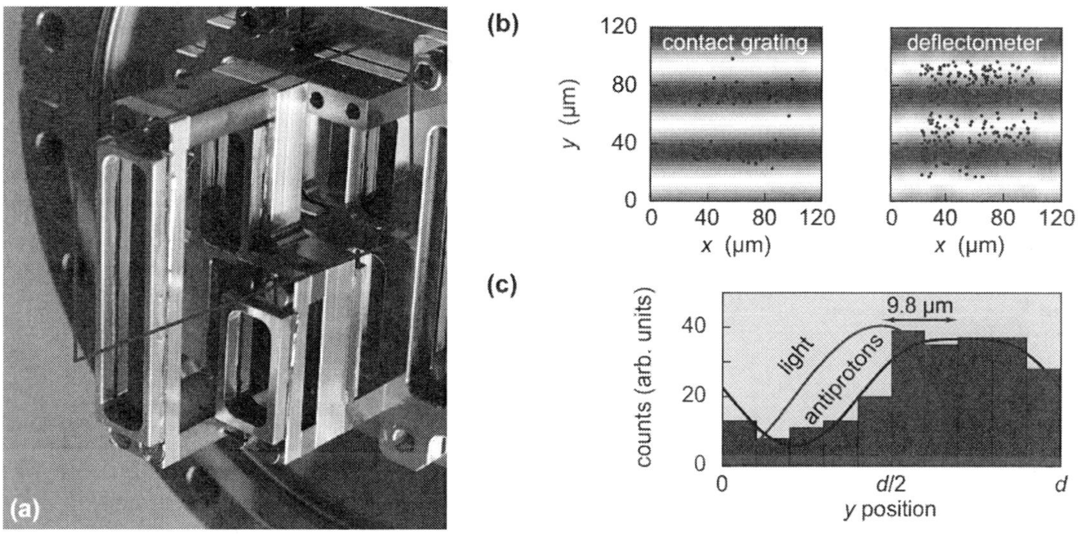

Figure 4. (a) Photograph of the miniaturized moiré deflectometer setup used in the proof-of-principle measurement with a $\bar{p}$ beam. (b) Positions of recorded $\bar{p}$ annihilation events (blue dots) and light pattern (red) behind the contact grating (left) and the deflectometer setup (right). (c) Antiproton annihilation events summed up modulo the grating period (blue bins and black line), compared to the intensity of the light pattern (red line).

3 Recent experimental results

The construction of the AEGIS apparatus began in 2010. As of 2015, all major components except for the gravimeter have been completed and installed in the experimental zone. The e^+ source and accumulator, the two superconducting magnets and all ion traps have been constructed and commissioned. Antiproton pulses from the AD have been captured and up to 4×10^5 $\bar{p}$ have been stacked in the catching trap. Storage times of close to 10 min were achieved. Similarly, several 10^8 e^+ have been confined in the accumulator for storage times of up to 20 min. A multi-layer scintillating-fiber tracker detector, to be installed around the central ion traps in order to monitor $\overline{\mathrm{H}}$ production, has been commissioned with a radioactive source [30]. The laser system for Ps excitation has been completed and commissioned [31]. In parallel, development work towards the gravimeter is ongoing. In the following, two recent significant advances towards an antimatter gravity measurement will be presented.

3.1 Deflectometry with antiprotons

In order to demonstrate the principle of a force measurement by deflectometry and to test the position resolution of the emulsion detector, we carried out a test experiment with a $\bar{p}$ beam [32]. Antiprotons are supplied directly by the AD, and hence are available in large numbers, yet they produce a star-shaped annihilation signature in the emulsion very similar to that of $\overline{\mathrm{H}}$. A miniaturized deflectometer consisting of two gratings spaced only L = 25 mm apart and with a grating period of d = 40 μm was placed in a 100-keV $\bar{p}$ beam. The setup is shown in Fig. 4(a). Annihilation events were recorded with a high-resolution photographic emulsion for 6.5 h. The reference position (without deflection) was determined by allowing part of the $\bar{p}$ beam to reach the detector via a grating placed in direct contact with it. Furthermore, both the deflectometer and the contact grating were illuminated with diffuse light, for which the material grating acted like a Talbot-Lau interferometer [33].

Since photons travel at the speed of light, their deflection during the short distance traveled is negligible. Their maxima in the plane of the detector agree with those of the antiprotons from the contact grating, as shown in the left graph of Fig. 4(b). On the other hand, antiprotons passing through the deflectometer are displaced in the positive y direction (right pane). The graph in Fig. 4(c) shows the locations of $\bar{p}$ annihilations modulo the grating period (blue bins and black line) compared to the light signal (red line). Quantitatively, a vertical deflection of $\delta y = 9.8 \pm 0.9$ (stat) $\pm$ 6.4 (syst) μm was observed, corresponding to an upward force of 530 ± 50 (stat) $\pm$ 350 (syst) aN [32]. This result is compatible with the Lorentz force due to a magnetic fringe field of about 8 G. While the force of gravity acting on $\overline{\mathrm{H}}$ atoms is many orders of magnitude smaller, their expected deflection is comparable due to the significantly lower velocity and longer flight distance in the full-scale experiment.

3.2 Positronium lifetime spectroscopy

The direct excitation of Ps to a Rydberg level is challenging due to the unavailability of a suitable laser. A scheme using $n = 2$ as intermediate state, on the other hand, has been demonstrated [34, 35]. We are pursuing an alternative path via the $n = 3$ level, which requires a lower laser power to reach saturation of the $n = 3 \longrightarrow$ Rydberg transition due to the longer lifetime of the intermediate state. The power dissipated in the cryogenic environment negatively affects both the production efficiency and the temperature of the formed $\overline{\mathrm{H}}$ and must therefore be kept to a minimum. As a prerequisite for the use of the novel excitation scheme, we studied the structure of Ps by single-shot positronium annihilation lifetime spectroscopy (SSPALS) [36]. The technique is based on monitoring the lifetime of Ps, which changes as a function of its internal quantum state. Both the excitation of Ps to a Rydberg state and two-step photodetachment were demonstrated [37].

In the setup, which is shown in Fig. 5(a), Ps was produced by directing a 3.3 keV beam of e^+ from a custom-built bunching and electrostatic transport system [38] onto a high-efficiency converter target. Both the ultraviolet (UV) laser with $\lambda_{\mathrm{UV}} \approx 204\ldots206$ nm and the infrared laser with $\lambda_{\mathrm{IR}} \approx 1064$ nm were directed onto the target region. Ps annihilations were recorded with a lead tungstate scintillator coupled to a photomultiplier placed 40 mm above the converter. In a first experiment, the IR laser was tuned to a transition from $n = 3$ to the continuum, dissociating the Ps. The free positrons were drawn away from the detector, resulting in an overall decrease of the observed annihilations, as shown by the grey line in the left pane of Fig. 5(b). The UV laser was then scanned around the $n = 1 \longrightarrow 3$ and the relative change in the number of annihilations recorded as a function of λ_{UV}. The resulting resonance (right) yields a wavelength of 205.05(2) nm [37], in good agreement with the theoretical value.

In a second measurement, the IR laser was tuned to a transition from $n = 3$ to a Rydberg state while the UV laser wavelength was kept constant on the $n = 3$ resonance. The Rydberg excitation increases the Ps lifetime because the overlap of the e^+ and e^- wavefunctions is reduced. The SSPALS data thus show a reduction of annihilations up to 300 ns after Ps production, followed by an excess in the time window 300...600 ns when Ps hit the chamber wall, as illustrated in the left pane of Fig. 5(c). The relative change in the number of annihilations at later times is shown in the right pane. Resonances of the $n = 15, 16, 17$ are clearly visible [37] and in good agreement with prior spectroscopic surveys. For higher Rydberg states, different n manifolds start to overlap in a continuum of energy levels. These measurements thus demonstrate the first laser excitation of Ps to the $n = 3$ state, enhancing the overall Ps excitation efficiency and reducing the laser energy deposited in the target region.

4 Conclusions and outlook

While the foundations of general relativity are well-tested with ordinary matter, the gravitational interaction of antimatter has never been studied experimentally. In a quantum theory of gravity, additional

DOI: 10.1051/epjconf/201612602016

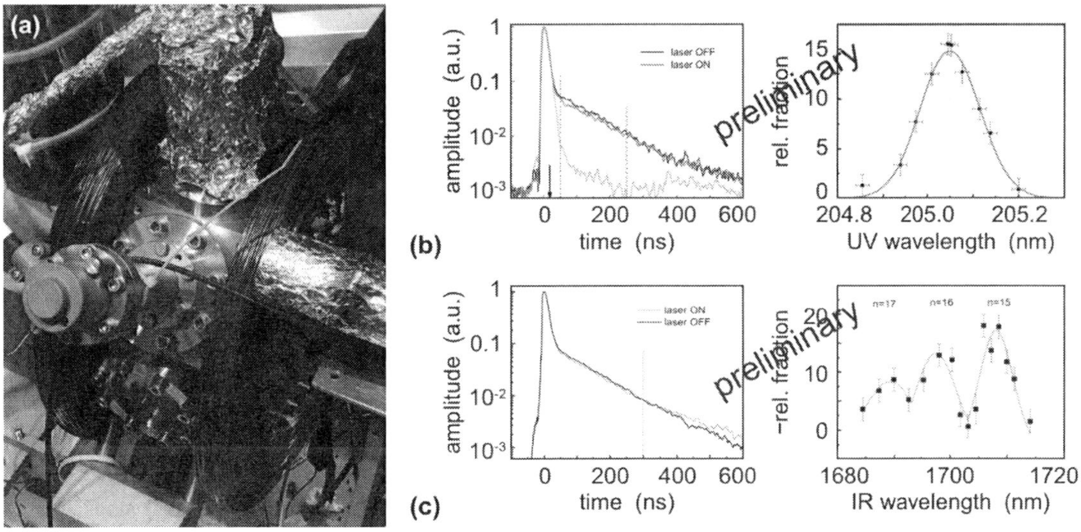

Figure 5. (a) Vacuum chamber which houses the positron annihilation lifetime spectroscopy setup. Positrons enter from the right and impinge on a central target. Lasers enter from the front; an annihilation detector is placed 40 mm above the target. (b) SSPALS data for the IR laser tuned to a transition from $n = 3$ to the continuum (left) and the resulting resonance of the $n = 1 \longrightarrow 3$ transition (right). (c) SSPALS data for the IR laser tuned to $n = 3 \longrightarrow$ Rydberg (left) and observed resonances for $n = 3 \longrightarrow 15, 16, 17$ (right) (from Ref. [37]).

exchange particles may lead to an anomalous force between matter and antimatter particles. The AEGIS experiment is gearing up to conduct the first measurement of the gravitational acceleration of an antimatter particle, by deflectometry or interferometry with a slow $\overline{\mathrm{H}}$ beam. A large part of the apparatus has been completed. Both $\bar{p}$ and e^+ have been captured and stored for extended periods in the AEGIS ion traps. The ability to detect the deflection of an antimatter particle beam by a few μm has been demonstrated. The excitation of Ps to Rydberg states via the $n = 3$ level has been achieved, paving the way for the efficient production of $\overline{\mathrm{H}}$ at low temperature. A measurement of antimatter gravity at the level of 1% is therefore within reach in the medium term.

Acknowledgements

This work was supported by the European Research Council, the European Social Fund, the Austrian Ministry for Science, Research and Economy, the German Research Foundation, the Istituto Nazionale di Fisica Nucleare, the Research Council of Norway, the Bergen Research Foundation, the Russian Ministry of Education and Science, the Russian Academy of Sciences and the John Templeton Foundation.

References

[1] J.Y. Hémery, S. Maury, Nucl. Phys. A **655**, 345c (1999), `doi:10.1016/S0375-9474(99)00223-7`

[2] W. Pauli, in *Niels Bohr and the Development of Physics*, edited by W. Pauli, L. Rosenfeld, V. Weisskopf (Pergamon Press, New York, 1955), p. 30

[3] A.D. Sakharov, JETP Lett. **5**, 24 (1967), `http://www.jetpletters.ac.ru/ps/1643/article_25089.shtml`

[4] O. Bertolami, D. Colladay, V.A. Kostelecký, R. Potting, Phys. Lett. B **395**, 178 (1997), doi:10.1016/S0370-2693(97)00062-2
[5] T. Goldman, R.J. Hughes, M.M. Nieto, in *First Workshop on Antimatter Physics at Low Energy* (Fermilab, Batavia, IL, 1986), p. 185, http://lss.fnal.gov/conf/C860410/p185.pdf
[6] T.A. Wagner, S. Schlamminger, J.H. Gundlach, E.G. Adelberger, Class. Quantum Grav. **29**, 184002 (2012), doi:10.1088/0264-9381/29/18/184002
[7] M.M. Nieto, T. Goldman, Phys. Rep. **205**, 221 (1991), doi:10.1016/0370-1573(91)90138-C
[8] P. Morrison, Am. J. Phys. **26**, 358 (1958), doi:10.1119/1.1996159
[9] M. Fischler, J. Lykken, T. Roberts, Fermilab Report FN-0822-CD-T, Fermilab, Batavia, IL (2008), http://lss.fnal.gov/archive/test-fn/0000/fermilab-fn-0822-cd-t.pdf
[10] S. Schlamminger et al., Phys. Rev. Lett. **100**, 041101 (2008), doi:10.1103/PhysRevLett.100.041101
[11] T.W. Darling, F. Rossi, G.I. Opat, G.F. Moorhead, Rev. Mod. Phys. **64**, 237 (1992), doi:10.1103/RevModPhys.64.237
[12] M. Amoretti et al. (ATHENA Collaboration), Nature **419**, 456 (2002), doi:10.1038/nature01096
[13] G. Gabrielse, S.L. Rolston, L. Haarsma, W. Kells, Phys. Lett. A **129**, 38 (1988), doi:10.1016/0375-9601(88)90470-7
[14] G. Drobychev et al. (AEGIS Proto-Collaboration), Proposal to the SPS Experiments Committee SPSC-P-334, CERN, Geneva (2007), http://cds.cern.ch/record/1037532
[15] A. Kellerbauer et al. (AEGIS Proto-Collaboration), Nucl. Instrum. Methods B **266**, 351 (2008), doi:10.1016/j.nimb.2007.12.010
[16] M. Doser et al. (AEGIS Collaboration), Class. Quantum Grav. **29**, 184009 (2012), doi:10.1088/0264-9381/29/18/184009
[17] M. Amoretti et al. (ATHENA Collaboration), Phys. Lett. B **578**, 23 (2004), doi:10.1016/j.physletb.2003.10.062
[18] C.H. Storry et al. (ATRAP Collaboration), Phys. Rev. Lett. **93**, 263401 (2004), doi:10.1103/PhysRevLett.93.263401
[19] A. Kellerbauer, J. Walz, New J. Phys. **8**, 45 (2006), doi:10.1088/1367-2630/8/3/045
[20] E. Jordan et al., Phys. Rev. Lett. **115**, 113001 (2015), doi:10.1103/PhysRevLett.115.113001
[21] P. Yzombard et al., Phys. Rev. Lett. **114**, 213001 (2015), doi:10.1103/PhysRevLett.114.213001
[22] D.W. Gidley, H.G. Peng, R.S. Vallery, Annu. Rev. Mater. Res. **36**, 49 (2006), doi:10.1146/annurev.matsci.36.111904.135144
[23] G. Consolati et al., Chem. Soc. Rev. **42**, 3821 (2013), doi:10.1039/c2cs35454c
[24] C.G. Fischer et al., Phys. Rev. B **71**, 180102 (2005), doi:10.1103/10.1103/PhysRevB.71.180102
[25] F. Castelli, I. Boscolo, S. Cialdi, M.G. Giammarchi, Phys. Rev. A **78**, 052512 (2008), doi:10.1103/PhysRevA.78.052512
[26] E. Vliegen, S.D. Hogan, H. Schmutz, F. Merkt, Phys. Rev. A **76**, 023405 (2007), doi:10.1103/PhysRevA.76.023405
[27] M.K. Oberthaler et al., Phys. Rev. A **54**, 3165 (1996), doi:10.1103/PhysRevA.54.3165

[28] S. Aghion et al. (AEGIS Collaboration), JINST **8**, P08013 (2013), doi:10.1088/1748-0221/8/08/P08013
[29] S. Aghion et al. (AEGIS Collaboration), JINST **9**, P06020 (2014), doi:10.1088/1748-0221/9/06/P06020
[30] J. Storey et al. (AEGIS Collaboration), JINST **10**, C02023 (2015), doi:10.1088/1748-0221/10/02/C02023
[31] S. Cialdi et al., Nucl. Instrum. Methods B **13**, 1527 (2011), doi:10.1016/j.nimb.2011.04.108
[32] S. Aghion et al. (AEGIS Collaboration), Nature Commun. **5**, 4538 (2014), doi:10.1038/ncomms5538
[33] E. Lau, Ann. Physik **437**, 417 (1948), doi:10.1002/andp.19484370709
[34] D.B. Cassidy, T.H. Hisakado, H.W.K. Tom, A.P. Mills, Phys. Rev. Lett. **108**, 043401 (2012), doi:10.1103/PhysRevLett.108.043401
[35] T.E. Wall et al., Phys. Rev. Lett. **114**, 173001 (2015), doi:10.1103/PhysRevLett.114.173001
[36] D.B. Cassidy, A.P. Mills, Nucl. Instrum. Methods A **580**, 1338 (2007), doi:10.1016/j.nima.2007.06.078
[37] S. Aghion et al. (AEGIS Collaboration), Phys. Rev. A (submitted 2015-09-25)
[38] S. Aghion et al. (AEGIS Collaboration), Nucl. Instrum. Methods B **362**, 86 (2015), doi:10.1016/j.nimb.2015.08.097

The Inspiring Science Education project and the resources for HEP analysis by university students

Dimitris Fassouliotis [1], Christine Kourkoumelis [1], and Stylianos Vourakis [1]

[1] *Physics Faculty and Institute of Accelerating Systems and Applications, University of Athens, Panepistimioupoli, Ilissia 15771, Greece*

Abstract. The Inspiring Science Education outreach project has been running for more than two years, creating a large number of inquiry based educational resources for high-school teachers and students. Its goal is the promotion of science education in schools though new methods built on the inquiry based education techniques, involving large consortia of European partners and implementation of large-scale pilots in schools. Recent hands-on activities, developing and testing the above mentioned innovative applications are reviewed. In general, there is a lack for educational scenaria and laboratory courses earmarked for more advanced, namely university, students. At the University of Athens for the last four years, the HYPATIA on-line event analysis tool has been used as a lab course for fourth year undergraduate physics students, majoring in HEP. Up to now, the course was limited to visual inspection of a few tens of ATLAS events. Recently the course was enriched with additional analysis exercises, which involve large samples of events. The students through a user friendly interface can analyse the samples and optimize the cut selection in order to search for new physics. The implementation of this analysis is described.

1 Introduction

The European Union (EU) has long recognized the decreasing interest of young people in science subjects and has been very supportive towards and promoting outreach projects which would bridge the gap between school science education and active research. An effective way to stimulate students' interest towards STEM subjects is to actively involve them in hands-on top level research.

The authors of this contribution have been involved in several EU outreach projects for more than a decade. Some are already finished: Learning with ATLAS@CERN (2009-2011) [1], PATHWAY to inquiry based science education [2] and Discover the COSMOS (2010-2012) [3]. Three more are on-going, Go-lab [4], Inspiring Science Education [5], which will be discussed in the next session and CREATIONS which started in October 2015.

The finished projects have already created a large number of resources addressed mainly to teachers, which are classified according to subject (keywords), age, difficulty, duration and language. The Discover the COSMOS portal [6] contains about 95,000 science education learning objects and activities connected to the science curriculum from Astronomy, Space Physics and HEP.

All projects followed the Inquiry Based pedagogical framework. Instead of just delivering lectures to the students, they were encouraged to learn by making hypothesis, perform interactive investigation to verify the hypothesis and draw conclusions by themselves, which were then discussed in the class.

The authors have been involving both high school as well as university students in activities which simulate the work of the researches at CERN. These are the subjects of the present contribution.

2 The Inspiring Science Education project and the HYPATIA demonstrator

The Inspiring Science Education (ISE) project began in April 2013 and will last until July 2016. The ISE partners design, plan and implement large-scale pilots to stimulate and evaluate innovative use of existing *eLearning* tools and digital resources for scientific disciplines and technology (STEM related subjects), enhancing science learning in 5,000 primary and secondary schools in 14 European Countries. During its first two years of operation the project has already reached 2,700 schools in Europe. In line with the ISE's inquiry learning framework, the partners have been designing the so called ISE demonstrators which are a series of success stories making effective use of a wide range of *eLearning* tools and applications. Each demonstrator is a complete lesson plan structured in several phases: Orienting and Asking Questions, Hypothesis Generation and Design, Planning and Investigation, Analysis and Interpretation and Conclusion and Evaluation. In order to evaluate their benefits and effectiveness, each phase of the demonstrator is accompanied by assessment questions (adopted according to the PISA 2012 Problem Solving Competence Framework) [7]. Currently there are 120 demonstrators. In addition to the demonstrators, ISE has harvested existing repositories with 278,000 educational resources (mainly from Open Discovery Space [8] and from Discover the COSMOS portal).

The demonstrator which is designed to provoke students' curiosity for HEP (which in most countries is absent from the national school curricula) is the HYPATIA demonstrator [9].This has been built according to the various phases of Inquiry learning referred above and shown on Fig.1.

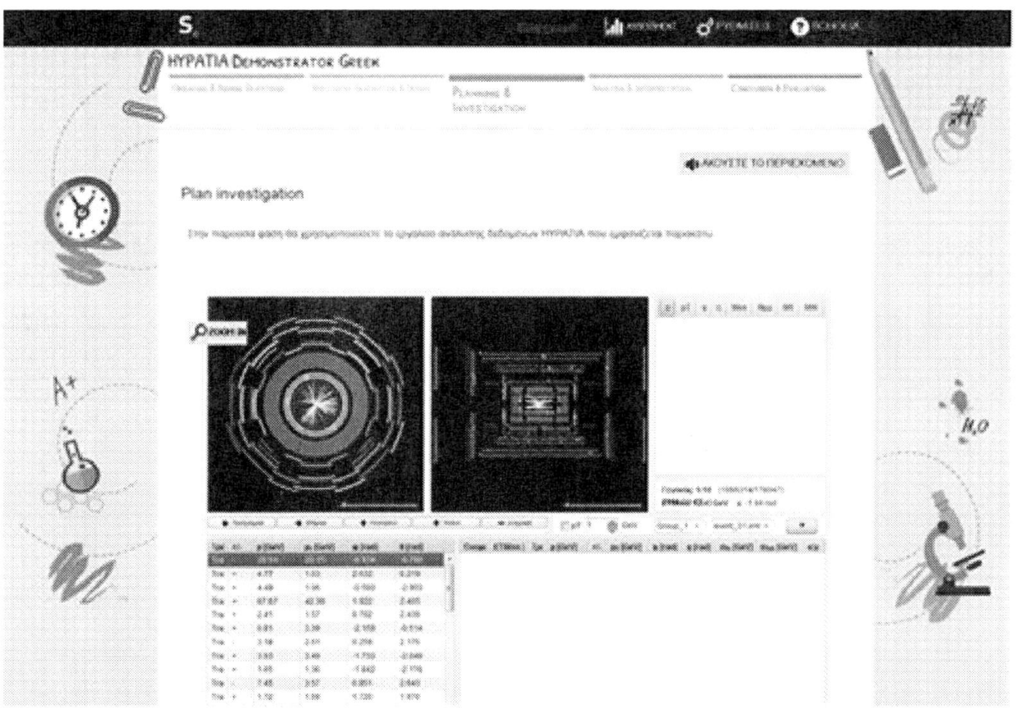

Figure 1. Screenshot of the HYPATIA demonstrator.

Special effort has been made during the design process in order to achieve the following goals:

- Engage the students in hands-on experimentation directly connected to the top-level research at CERN by analysing data almost in real-time
- The material can fit in the duration of a school lesson and is directly connected to the Big ideas of science
- The material has also been adapted to the age of students

Moreover, as previously mentioned, during the step-by-step students' involvement in the different phases of the demonstrator, PISA-like problem solving questions have been introduced, in order to measure the students' competence and gained abilities from the educational intervention.

3 Description of the "HY.P.A.T.I.A" data analysis tool

The HYPATIA demonstrator uses the embedded HYPATIA [10], [11] on-line tool. HYPATIA is an event visualization tool that uses real data from the ATLAS experiment, namely offers a graphical representation of the products of proton collisions. The students using the event display can interactively analyse the data collected almost in real-time. This allows them to take advantage of the LHC data from CERN without having to leave their classroom! The students learn how to identify particles and their interactions, and learn how the gigantic state-of-the-art ATLAS detector works. They also get a glimpse at how the data analysis that led to the discovery of the Higgs boson is performed.

The on-line version of HYPATIA has been developed since 2010 by the University of Athens. It complements the off-line version [12] which is used by the International Masterclasses [13] since 2008 by thousands of high-school students in all continents each year. The on-line version is simpler to use, yet contains all the functionalities necessary for performing the various scenaria associated with the different exercises. It includes four different parts each with an increasing level of complexity and difficulty.

The best practice for introducing HEP to 16-18 year old students is a combination of lectures, virtual tours to specific experiments and hands-on experience with HYPATIA. This combination has taken the form of the so called "mini masterclasses" which are half-day workshops, taking place locally at the interested schools. During the last five years the "mini masterclasses" were run in about 100 schools across the Greece (Fig. 2) as well as in other European sites (CERN, Ireland etc.).

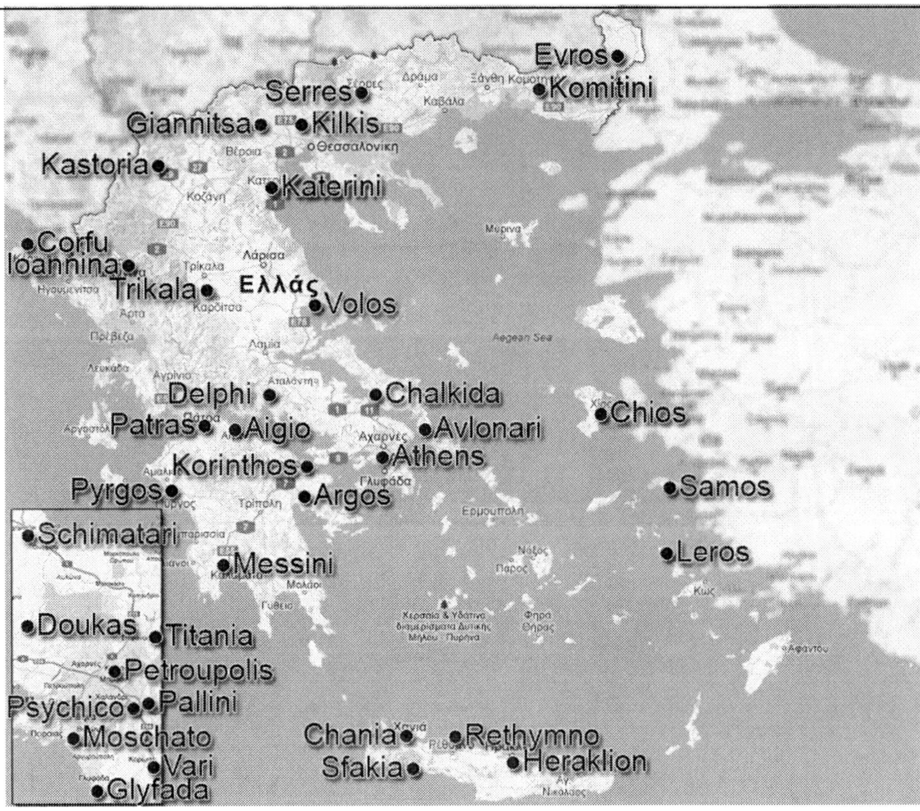

Figure 2. Map of Greece showing the schools where the "mini masterclasses" took place during the last few years.

Fig 3 shows students at the Giannitsa city (northern part of Greece) while analysing the ATLAS events during a "mini masterclass" last summer which was part of a week-long training organized by the Greek Physical Society for high performing local students.

DOI: 10.1051/epjconf/201612602017

Figure 3. Students at Giannitsa city working on the HYPATIA demonstrator.

During last semester the students' assessment has been tested in eight locations in Greece with about 250 students from 35 different schools. The preliminary results show that the students out-perform the corresponding Pisa assessment questions' scores in OECD countries.

4 The HYPATIA tool for a university laboratory

Besides the efforts to raise enthusiasm and motivate high school students, it was realized that the university laboratory exercises should include the students' involvement in most modern top research, besides the so called "classical" experiments. The students during their lab courses should perform activities which are directly connected with the most advanced research at CERN. The corresponding involvement should be at a higher level than the hands-on exercises for younger ones.

Since 2010 the HYPATIA tool is used in a lab course of the University of Athens for fourth year undergraduate physics students, majoring in HEP [14]. Each year about 50 students take part in the three hour long course performing all steps of the HYPATIA demonstrator (introductory material, become familiar with the tool, identify particles, reconstruct Z bosons in leptonic decay modes and search the Higgs boson). In addition, at the end of the lab, they have to hand-in a homework paper where they try to answer advanced questions which are related to the decays modes, widths, masses, experimental resolution and basic statistical analysis tools.

Recently an additional three hour long course has been developed which follows the above -mostly visual- inspection of events. The designed additional lab aims to permit the students to perform an in depth analysis of large samples of events. The students after visual inspection of a few tens of events, run HYPATIA in the so called "batch mode" (Fig. 4).

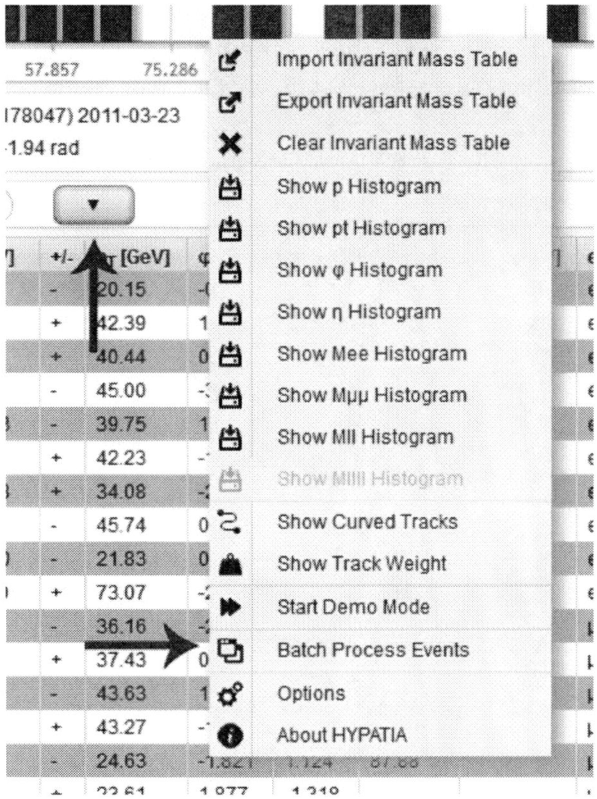

Figure 4. Drop-down HYPATIA menu for "Batch mode" running

They are able to analyse simultaneously two large samples of events and optimize the cuts for the H→4l decay searches in a way which is very similar to the one the researchers employ and which led to the Higgs boson discovery in the summer of 2012. They can analyse either two-lepton samples to look for resonances such as the J/ψ, ϒ, Z etc. or four-lepton samples to look for the Higgs boson. In the following the Higgs analysis is described.

The samples used can be either Monte-Carlo ones (signal and different backgrounds) or data[a]. The backgrounds include the $ZZ^{(*)}$ irreducible one, as well as Z + jets ($Z^{b\bar{b}}$) reducible one. The students have to realize that certain kinematical distributions of the signal are different from the corresponding ones of the background and that this is the main methodology for applying cuts in order to reduce the background and make a discovery.

They start the investigation with the Monte Carlo samples in order to optimize the cuts. The samples –in order to reduce their size- have undergone some filtering cuts: the values of the p_T of each of the four leptons should be larger than 6 GeV and the invariant mass of their two or four lepton combinations larger than 2 GeV. The variables available by the exercise where the cuts could be applied are the following:

- The momenta of all pre-selected four leptons $p_{T1,2,3,4}$
- m_{12} : the mass of the lepton pair (opposite sign, same flavour) closest in value to the Z mass
- m_{34} : the mass of the next lepton pair (opposite sign, same flavour) closest in value to the Z mass
- The largest impact[b] parameter out of the leptons in the event (divided by its error). Separately for muons ($d_{0\mu}$) and electrons (d_{0e}).

[a] Very recently the ATLAS management gave permission to use for outreach purposes 1fb^{-1} of the 2012 data which will eventually appear in the opendata portal for all LHC experiments [15]

- Distance of the lepton vertex (z_{vtx}) from the beam crossing vertex (z_0) along the z (beam) axis
- The largest isolation[c] parameter out of the leptons in the event.
- Cut on the minimum or maximum invariant mass of the four lepton combination.

The table of cuts is shown on Fig. 5 together with some pre-set values for each cut (which should be optimized during the analysis process).

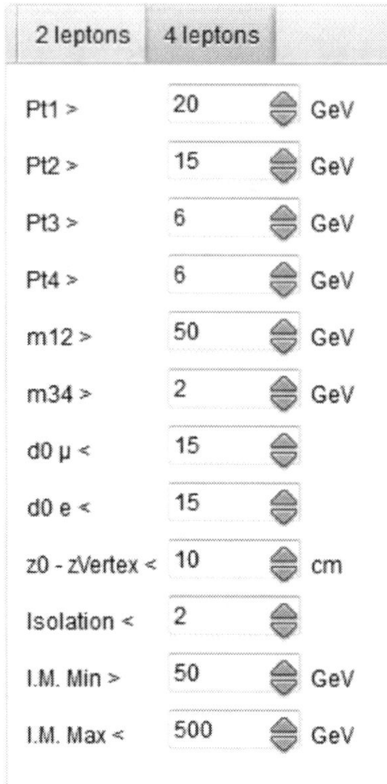

Figure 5. Tables of the variables available for optimization (together with some indicative values).

---

[b] Impact parameter is the vertical distance of the secondary lepton vertex from the interaction point

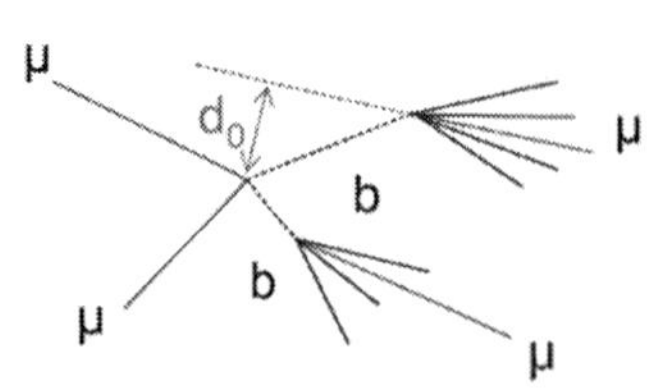

(primary vertex)

[c] The Isolation parameter is calculated by taking a cone of about 30^0 opening angle, adding all the p_Ts of tracks inside the cone (subtracting the lepton under consideration) and dividing by the p_T of the

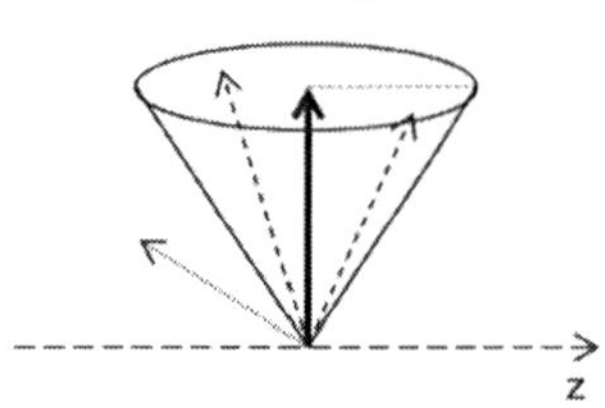

track.

The students use the so called "N-1" technique to optimize each cut. Namely they examine the distribution (histogram) of the variable under investigation having applied all the other cuts, except the cut under optimization. Its value is then decided examining the corresponding distributions for signal and background as well as the distribution of the statistical significance[d] of the signal. Such an example is given on Fig. 6. The optimal value of the cut is the point where the significance is highest. After fixing the value of the particular cut, the students proceed to the next one until they optimize all (or a subset) of them. In subsequent iterations, the students start with the new values of the cuts and re-optimize them one by one in order to arrive at a new optimal set.

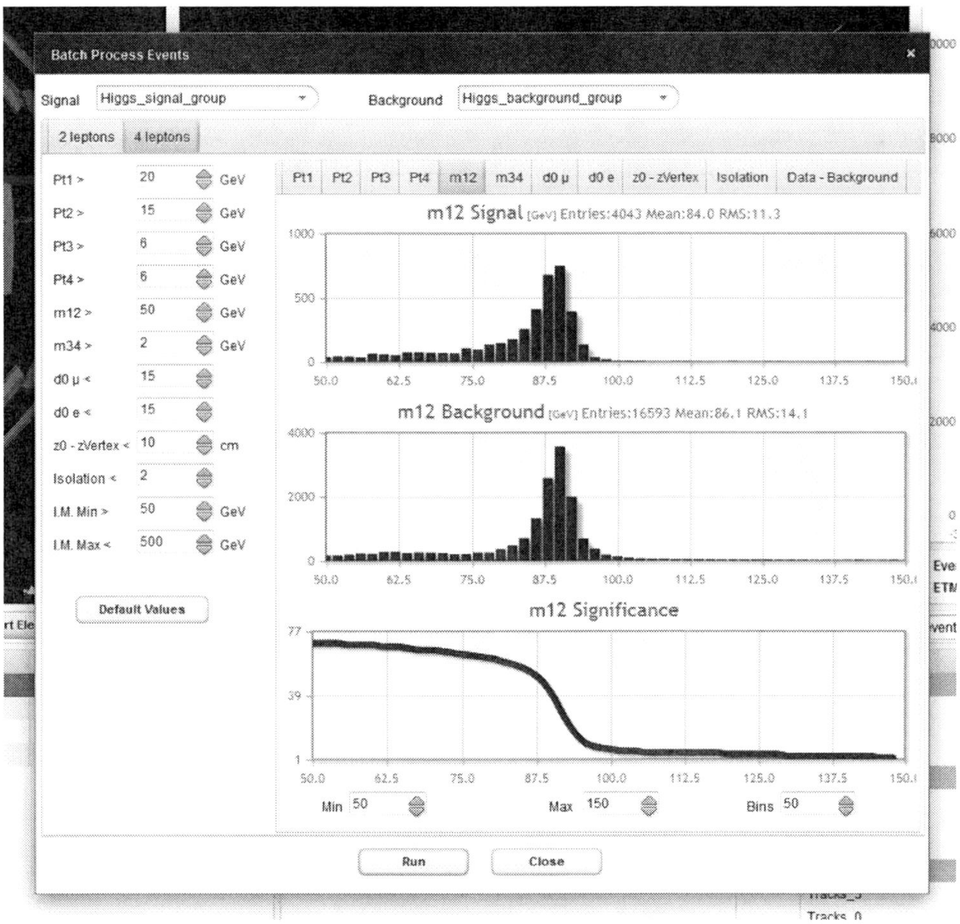

Figure 6. Histogram of a specific kinematic parameter (m_{12}). The top distribution is the distribution from the Monte-Carlo signal sample, the middle is the corresponding Monte Carlo background distribution ($ZZ^{(*)}$ plus Z+jets) and the bottom one is the statistical significance derived from the above two distributions as a function of the m_{12} value.

[d] The signal significance is defined as:

$$significance = \sqrt{2 * \left((S + B) * \ln\left(1 + \frac{S}{B}\right) - S \right)}$$

The point where the students are instructed to stop the iterations is determined by looking at a narrow region around the Higgs boson mass in the four lepton invariant mass where they should maximize, after a number of iterations, the value of the signal significance.

The present "batch mode" processing exercise can also be applied to Z'→2l Monte-Carlo sets for the Z' cut optimization. In this case the investigation is simpler and quicker because the students optimize the cuts for two leptons only, thus having fewer variables to iterte on.

The H→4l exercise is been implemented during the fall 2015 semester in the University of Athens and Birmingham advanced students' lab.

5 The Chania science fair event

In an effort to get the general public aware about CERN and HEP in general, a full day science fair was organized in the context of the ICNFP 2015 conference. The event took place on the 28th August in Chania, which is the second largest city in Crete, about 40km away from Kolymbari. It was organized for the second consecutive year and took place at the premises of the Chania Sailing club, a 16th century Venetian dockyard beautifully restored. Several conference participants took part and guided the visitors to the world of top level research. Simple experiments or Masterclasses (ATLAS and ALICE) were set up so the visitors could get hands-on experience. The simple experiments (donated by Quarknet) were "the conservation of momentum, rolling with Rutherford, probe matter, quark workbench and W signatures" [16]. They were all set up on long benches (see Fig. 7) where the public could circulate around and get involved one way or the other. At the same time, several colleagues were present to answer questions made by the public. The event lasted until late in the afternoon and there were about 150 visitors of various nationalities but mostly parents with their children.

In the evening at the same place a public lecture by Prof. E. Kiritsis of the University of Crete with title "Microscopic black holes: a new window on the universe" took place and was very well attended. In summary, the science fair had a great success and will be repeated next year for ICNFP 2016.

Figure 7. The Chania science fair set up at the Sailing Club.

6 Conclusions

Recent hands-on activities, development and testing of innovative applications towards the goal of introducing HEP to the schools and universities are reviewed. A large number of resources developed by European Outreach programs exist on the web and these have recently been enriched by applications for university students. This way many thousands of students as well as the general public can get acquainted to the top level research of CERN, and grow a real enthusiasm for it.

ACKNOWLEDGEMENT

This work was partially funded by the European Union in the context of the ISE project (Grant Agreement no. 325123) under the CIP Framework program. This document does not represent the opinion of the European Union, and the European Union is not responsible for any use that might be made of its content.

References

[1] http://www.learningwithATLAS-portal.eu/
[2] http://pathway-project.eu/
[3] http://www.discoverthecosmos.eu/
[4] http://www.go-lab-project.eu/
[5] http://inspiring-science-education.eu/
[6] http://www.portal.discoverthecosmos.eu/
[7] http://www.oecd.org/pisa/pisaproducts/46962005.pdf

[8] http://opendiscoveryspace.eu/
[9] http://portal.opendiscoveryspace.eu/edu-object/hypatia-demonstrator-english-v2-828289
[10] C. Kourkoumelis, & S. Vourakis, (2014) Hypatia-an online tool for ATLAS event visualization. *Physics Education*, **49**, 21-32
[11] http://hypatia.iasa.gr/
[12] http:/hypatia.phys.uoa.gr/
[13] http://physicsmasterclasses.org/
[14] R.M. Barnett, K.E. Johansson, C. Kourkoumelis, L. Long, J. Pequenao, C. Reimers, & P. Watkins, (2012) Learning with ATLAS experiment at CERN, *Physics Education*, **47**, 28
[15] http://opendata.cern.ch/
[16] https://quarknet.i2u2.org/data-portfolio

EPJ Web of Conferences **126**, 02018 (2016) DOI: 10.1051/epjconf/201612602018
ICNFP 2015

Thermalization and confinement in strongly coupled gauge theories

Takaaki Ishii[1,2], Elias Kiritsis[1,3], and Christopher Rosen[1,4]

[1] *Crete Center for Theoretical Physics, University of Crete, Heraklion 71003, Greece*
[2] *University of Colorado, Boulder, CO 80309, USA*
[3] *University of Paris Diderot, Sorbonne Paris Cité, APC, Paris F-75205, France*
[4] *Imperial College, London SW7 2AZ, UK*

Abstract. Quantum field theories of strongly interacting matter sometimes have a useful holographic description in terms of the variables of a gravitational theory in higher dimensions. This duality maps time dependent physics in the gauge theory to time dependent solutions of the Einstein equations in the gravity theory. In order to better understand the process by which "real world" theories such as QCD behave out of thermodynamic equilibrium, we study time dependent perturbations to states in a model of a confining, strongly coupled gauge theory via holography. Operationally, this involves solving a set of non-linear Einstein equations supplemented with specific time dependent boundary conditions. The resulting solutions allow one to comment on the timescale by which the perturbed states thermalize, as well as to quantify the properties of the final state as a function of the perturbation parameters. We comment on the influence of the dual gauge theory's confinement scale on these results, as well as the appearance of a previously anticipated universal scaling regime in the "abrupt quench" limit.

1 Introduction

The dynamical properties of strongly interacting matter are interesting both because relatively little is currently known about them, as well as the fact that they are increasingly accessible in the laboratory and large-scale experiments like the heavy ion collisions at RHIC and LHC. One standard method for calculating in strongly coupled field theories is to regularize the theory on a lattice. This method however relies on a Euclidean formulation of the field theory, and hence it is rather difficult to make contact with real-time dynamics. Finding a way around this obstacle takes special importance in the context of ongoing programs in the relativistic heavy ion collisions. At the energy scales currently accessible, the hot, dense matter produced in such collisions of heavy nuclei behaves as a strongly interacting plasma and thus novel theoretical approaches are needed to characterize its dynamical properties. One such approach is based on a holographic duality between gauge theories and string theories illustrated in figure 1. This approach was employed in [1] to model the dynamical response of a confining gauge theory at strong coupling, and that work provides the foundation for these proceedings.

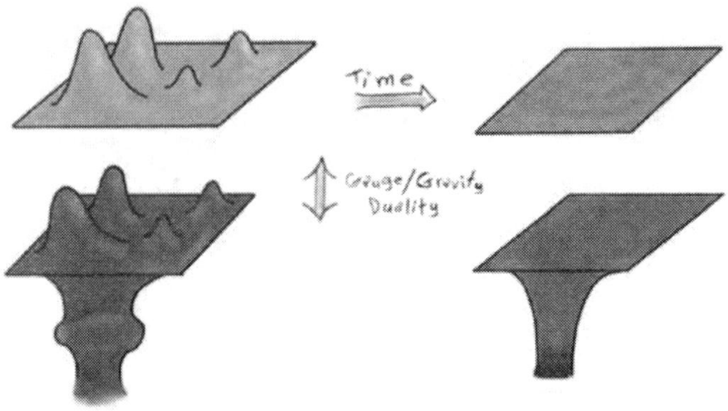

Figure 1. Cartoon depiction of holographically dual thermalization processes. On the top, a strongly coupled field theory is driven out of equilibrium by an external probe, and at late times may attain thermal equilibrium. On the bottom, the dual process is understood as a perturbation of a gravitational theory in one higher dimension, and late time thermal equilibration corresponds to black hole formation in the bulk.

2 Thermalization in Holography

Before developing the holographic model in any detail, it is useful to orient oneself with some very general expectations from the field theory perspective. Given any strongly coupled medium, a useful characterization of its properties is provided by its response to a variety of external perturbations. If the dynamics of the medium can be captured by a quantum field theory effectively described by the Lagrangian $\mathcal{L}_{\text{QFT}}$, then the application of a perturbation by a local operator O in a manner described by the source f_0 is accomplished by adding to the Lagrangian

$$\mathcal{L}_{\text{QFT}} \to \mathcal{L}_{\text{QFT}} + f_0 O. \tag{1}$$

In the special case that the source is isotropic in space and changes only in time, the probe will do work on the system as dictated by the following Ward identity:

$$\nabla^t \langle T_{tt} \rangle = \dot{f}_0 \langle O \rangle. \tag{2}$$

Here the "dot" denotes a partial derivative with respect to time and T is the stress-energy tensor.

The response of the system to the time dependent perturbation can be measured in any of a number of different ways, and these different methods generically elucidate different aspects of the system's response. A particularly natural measure is provided by the various one point functions of gauge invariant operators of the field theory. By examining the dependence of these expectation values on time, one can access both the endpoint of the dynamical evolution, as well as the route the system takes to arrive there.

One scenario that will be of particular importance in the present work arises when the late time behavior of the system is independent of time, and the various one point functions take the form

$$\langle O(t \to \infty) \rangle \sim \operatorname{tr} \rho O, \tag{3}$$

where ρ is a thermal distribution. If this is the case, then the endpoint of the evolution is a thermal state, and the equilibration process is called "thermalization".

For strongly coupled field theories with a holographic dual the thermalization process can alternatively be described in terms of weakly coupled geometric variables. These are the bulk fields of

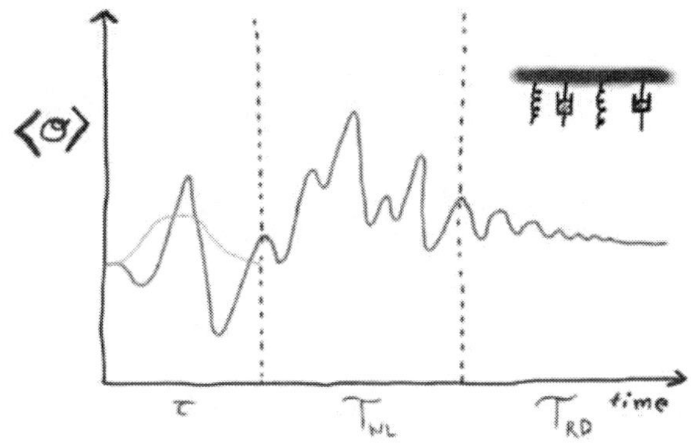

Figure 2. Typical response of a holographic system to a time-dependent perturbation. The vertical axis is the expectation value of the scalar operator determined by the behavior of the scalar near the boundary, while time is along the horizontal axis. After a time τ the system is driven out of equilibrium, and may pass through any number of non-linear regimes $\mathcal{T}_{\mathrm{NL}}$ before "ringing down" to thermal equilibrium with a characteristic time scale $\mathcal{T}_{\mathrm{RD}}$. The ring-down regime is anticipated from the well-known fact that black branes respond to small (linearized) perturbations much like a system of damped oscillators (inset).

a gravitational theory in one higher dimension. In terms of these variables, thermalization in a 3+1 dimensional field theory is understood from the dual viewpoint as horizon formation in a 4+1 dimensional theory with gravity.

The formulation of the dynamical process in terms of the dual gravity theory provides advantages both in terms of computation and intuition. In figure 2, a generic expectation for the dynamical evolution of a holographic one point function is sketched. To characterize this evolution, it is useful to imagine dividing the response into different regimes each controlled by any of an assortment of different "characteristic" time scales. For instance, if one imagines varying the source f_0 over a timescale τ, then τ is roughly the characteristic time scale governing the amount of time taken to drive the system out of equilibrium. The subsequent non-equilibrium system will evolve according to the non-linear gravitational equations of motion, and thus it could be possible that the system will pass through a highly non-linear regime, $\mathcal{T}_{\mathrm{NL}}$. This time scale could conceivably be further subdivided into any number of distinct regimes.

If a horizon forms in the bulk, it is natural to expect that a "ring-down" regime $\mathcal{T}_{\mathrm{RD}}$ defines the system's near-equilibrium behavior for the dual theory to thermalize at late times. This expectation is based on the fact that the linearized response of the final state black brane to small perturbations is controlled by the lowest lying (most real) quasi-normal mode in the spectrum at late times.

As a consequence of the preceding discussion, it is sensible to define a thermalization time as simply the sum of all the characteristic time scales defining the dynamic evolution except the time scale characterizing the application of the external source,

$$\mathcal{T}_{\mathrm{THERM}} = \mathcal{T}_{\mathrm{NL}} + \ldots + \mathcal{T}_{\mathrm{RD}}. \tag{4}$$

Thus defined, the thermalization time depends only on the out-of-equilibrium response and relaxation of the strongly coupled matter. In the majority of holographic examples of thermalization in a strongly coupled field theory found in the literature, the thermalization time appears to be dominated by the ring-down regime, i.e $\mathcal{T}_{\mathrm{THERM}} \approx \mathcal{T}_{\mathrm{RD}}$.

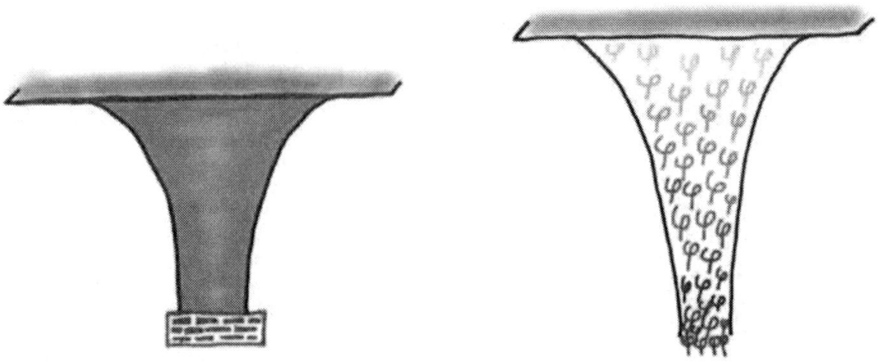

Figure 3. Examples of gravitational systems holographically dual to confining gauge theories. In the hard wall model (left) the geometry is artificially cut off at some radial position $z = z_0$, here shown as a brick wall. In bulk theories with scalar matter like the Einstein-Dilaton theories (right), it is possible to engineer a potential for the scalar which effectively shields probes from reaching arbitrarily far into the throat.

3 Confinement in Holography

One feature absent from many previous holographic investigations of thermalization is the presence of a confinement scale analogous to the Λ_{QCD} of quantum chromodynamics. It is natural to wonder if the existence of such an additional characteristic energy scale could alter the dynamical response of the theory in important ways, perhaps leading to thermalization processes in the dual field theory that lie outside the omnipresent ring-down regime.

The introduction of confinement to holographic gauge theories has a long history [2, 3], and is by now a familiar aspect of gauge/gravity duality. Confinement in any gauge theory can be identified with an "area law" behavior of the Wilson loop $\mathcal{W}$,

$$\langle \mathcal{W} \rangle \sim e^{-F_s \cdot TL} \tag{5}$$

where F_s is the QCD string tension, and TL is the area of the Wilson loop describing the separation by length L of a quark anti-quark pair propagating for time T. In a holographic theory, the Wilson loop has a dual description in terms of a certain minimal surface drooping into the bulk geometry. It can be shown [3] that to obtain the desired area law behavior of the boundary theory Wilson loop, the bulk geometry must obey certain requirements. For example, if the bulk string frame metric is written as

$$\mathrm{d}s^2 = e^{2A(z)} \left(-\mathrm{d}t^2 + \mathrm{d}\vec{x}^2 + \mathrm{d}z^2 \right) \tag{6}$$

where z is the radial coordinate orthogonal to the boundary theory directions, then the dual Wilson loop will only exhibit area law behavior if the warp factor $A(z)$ achieves a minimum at some radial coordinate z_0, and if $A(z_0)$ is finite.

Thus, roughly speaking, gravitational theories dual to confining gauge theories are "capped" in the radial direction. The mechanism for capping the geometry can be either abrupt, as in the case of the so-called hardwall model , or gentle perhaps with the aid of other bulk fields. These possibilities are illustrated and explained in more detail in figure 3.

Holographic thermalization in the hard-wall model was explored initially in [4, 5]. The primary result was that the introduction of a confinement scale indeed leads to a more complex thermalization processes. More specifically, a class of perturbations were identified that *never* thermalize [4]. This can be understood as a consequence of the following simple argument: if the perturbation to the hard-wall geometry carries sufficient energy density to produce a black brane with horizon above (closer to the boundary than) the harwall at z_0, then the perturbation is indifferent to the existence of the wall and will thermalize in a time $\mathcal{T}_{\mathrm{THERM}} \approx \mathcal{T}_{\mathrm{RD}}$. If the associated energy density is lower, the perturbation will scatter from the hard-wall and return to the boundary where it scatters once more. This process appears to repeat for as long as the numerical routines of [5] were able to run. Recently, analogous calculations were performed in the AdS soliton background [6] which is qualitatively similar to the hardwall in that there is a mass gap for black brane formation in the bulk. There too perturbations were identified which appear to scatter indefinitely, while others lead to collapse and black-brane formation.

4 Dynamical Quenches in a Confining Gauge Theory

Clearly, an important next step is to determine whether or not something similar can happen in more realistic holographic models of confining gauge theories. An example of one such class of models is the Einstein-Dilaton class, whose solutions minimize an action of the form

$$S = \frac{1}{2\kappa^2}\int d^5x\,\sqrt{-g}\left(R - \frac{4}{3}(\partial\varphi)^2 + V(\varphi)\right) - \frac{1}{\kappa^2}\int_{\partial} d^4x\,\sqrt{-\gamma}\mathcal{K} + S_{\mathrm{CT}}. \tag{7}$$

The first term in this action describes the interactions of a scalar field φ coupled to gravity, subject to the scalar potential V, while the second term is the familiar Gibbons-Hawking-York boundary term. The final term denotes a collection of bulk counterterms that must be added to the action to regularize the action on-shell, and allows for the computation of holographically renormalized correlation functions.

In a "top down" holographic model, the scalar potential V would be fully determined by the symmetries of an underlying supergravity theory. Top down models often have the advantage of providing a gravitational theory whose bulk fields are dual to explicitly identifiable operators in a known boundary gauge theory. Constructing such models with the desired dual physics is typically hard work, however, and it is often considerably more pragmatic to employ a "bottom up" approach. In a bottom up model, the scalar potential can be arbitrarily tuned to induce the desired dual physics. The freedom to tune this potential comes at the cost of rigor, as one subsequently forfeits the ability to make detailed claims about the name and operator content of the dual field theory. Nonetheless, in the present case, we will adopt a bottom up perspective and choose V such that the dual boundary gauge theory is confining at low temperatures and such that solutions with non-zero scalar correspond to relevant deformations by a dimension three scalar operator (for details, see [1]). An example of a scalar potential that realizes these features is given by

$$V(\varphi) = \frac{12(1 + a\varphi^2)^{1/4}\cosh\frac{4}{3}\varphi - b\varphi^2}{L^2} \tag{8}$$

with the parameter choice $(a, b) = (1/500, 10009/1500)$ and L the AdS scale.

The solutions to this bulk theory describe the various phases available to the dual field theory. At zero temperature, the dual gravitational solution is a horizon-less geometry with a running scalar and a fairly innocuous naked singularity in the bottom of the throat. Ideally, one would like to perturb this geometry by varying the source for the dual scalar operator in time. This would roughly correspond to driving the system out of its ground state by altering the coupling of a dimension three scalar operator. In practice, the rapidly diverging bulk scalar greatly complicates the numerical analysis and it is necessary to introduce an IR cutoff deep in the throat region of the geometry.

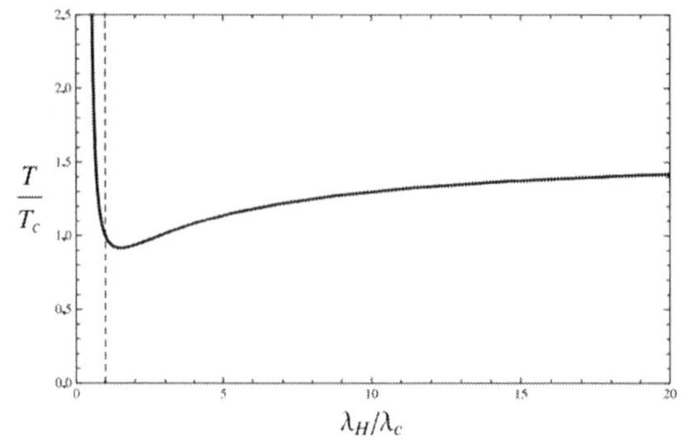

Figure 4. Plot of the temperature of the black hole solution scaled by the critical temperature as a function of the value of the scalar at the horizon, λ_H/λ_c. At $\lambda_H = \lambda_c$ there is the first order phase transition between the large black hole and the thermal gas solution which does not have a horizon, while the small black hole co-exists as a thermodynamically disfavored solution with a small horizon.

A sensible means for introducing such a cutoff can be found in another branch of solutions to the bulk action. A particularly well suited branch contains the "small black hole" solutions, which shield the singularity behind a very small horizon. These solutions provide initial state candidates in which all bulk fields are well behaved between the UV boundary and the horizon in the IR. In this way the small horizon regulates the IR, but not without introducing some drawbacks. First, the introduction of a horizon in the initial state obviously renders any information related to horizon formation beyond the reach of our computations. Second, the small black holes along this branch are never the thermodynamically preferred solutions in the dual field theory. Nonetheless, they have the advantage of retaining much of the flavor of the zero temperature solution (notably a scalar which grows rapidly in the IR) while shielding the numerical methods from the existence of the IR singularity. In figure 4 the finite temperature solutions to the model (7) are shown, parametrized by the value of the scalar $\lambda = e^{\varphi}$ at the horizon. The small black holes are the branch to the right of the minimum temperature solution.

Our computational strategy will be as follows: beginning from a small black hole initial state, we turn on a time dependent source for the boundary scalar operator of the form

$$f_0(t) = \tilde{f}_0\left(1 + \tilde{\delta}\, e^{-\frac{t^2}{2\tilde{\tau}^2}}\right), \tag{9}$$

which can be tuned by varying $\tilde{\delta}$ and $\tilde{\tau}$. This change in the scalar at the boundary propagates into the bulk, backreacting on the geometry. This dynamical process is fully described by the non-linear Einstein equations, whose causal time evolution must be solved numerically step by step in time until a final steady state solution has been obtained. At each boundary time, a standard application of the holographic dictionary allows one to read off the one-point functions of the stress-energy tensor and scalar operator in the dual gauge theory.

Evolving the solutions using the Einstein equations is difficult and requires an assortment of numerical methods tailored to the gravitational problem. While various numerical schemes have found success in the gravitational literature, we have found the characteristic formulation particularly convenient. This formulation has previously been used with success in a variety of holographic computations, starting with the important work of [7]. In this setup, the Einstein equations are recast as a nested collection of ordinary differential equations that can be solved in series when supplemented with some initial data. Independent of the choice of scheme, one must also decide which numerical techniques

 DOI: 10.1051/epjconf/201612602018

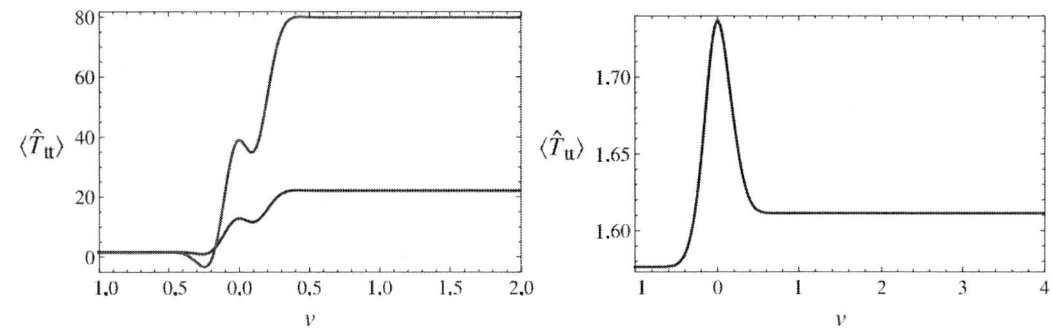

Figure 5. Time evolution of $\langle T_{tt} \rangle$ for large amplitude (left) and small amplitude (right) quenches in the small black hole initial state.

are best suited to perform the actual integration of the equations of motion. One option, which has received much attention in the numerical holography literature, are the pseudo-spectral methods. Under certain circumstances, these methods offer excellent convergence properties and are comparatively inexpensive computationally. These benefits are, unfortunately, somewhat compromised in the case that the bulk fields contain non-analytic behavior within the computational domain. Therefore, we found it best to utilize an assortment of finite difference schemes to integrate our solutions into the bulk and forward in time.

Upon successful evolution of the Einstein equations, the system's response reveals an energy density which changes markedly on the time scale of the quench $\tilde{\tau}$, followed by a rapid equilibration to its final state value. Some examples of this evolution are plotted in figure 5. By construction, this rapid change satisfies the Ward identity (2) in the presence of the time dependent source of the form (9). Indeed, both the Ward identity governing the divergence of the boundary stress energy tensor as well as the Ward identity relating the trace of this stress tensor to classical and anomalous conformal symmetry breaking terms are guaranteed to hold in any solution to the gravitational equations of motion.

To better understand the rapid approach of the perturbed system to thermal equilibrium, it is useful to look in more detail at the late time behavior of the system. In figure 6 the time dependence of the scalar one point function is plotted on a logarithmic scale. One notes immediately that after a short time (again controlled by the quench width $\tilde{\tau}$), the system oscillates with a well defined frequency ω_* while simultaneously approaching its equilibrium value exponentially with decay constant Γ. This is exactly as one would expect from a linear system controlled by an excited mode of the form $\omega_1 = \omega_* - i\Gamma$. Evidently, even in the confining model of (7-8) the thermalization time seems to be dominated by the linear regime, $\mathcal{T}_{\text{THERM}} \approx \mathcal{T}_{\text{RD}}$.

The decay width Γ is given by the imaginary part of the lowest lying scalar quasi-normal mode of the system's final state black hole. In figure 7 we quantify this by plotting Γ as a function of temperature for several thermal states of our holographic theory. As anticipated, linear scaling of Γ with temperature appears in the conformal (small scalar) limit, and deviations from this behavior are already evident at the first order phase transition where $T = T_c$. These deviations become larger as one moves further onto the small black hole branch. In this way Γ quantitatively captures the degree of the conformal symmetry breaking in the presence of the nontrivial confining scalar potential (8).

Outside of thermalization times, the holographic approach can also provide some insight into the dependence of the final state on parameters of the perturbation. In figure 8 the final state energy density is plotted as a function of quench duration with fixed (large) amplitude, and as a function of

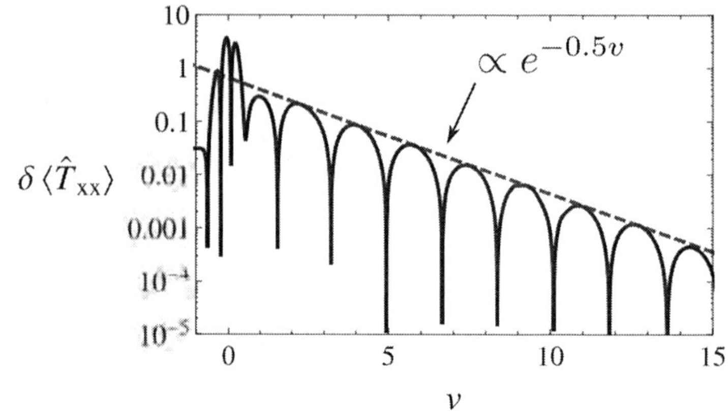

Figure 6. Generic late-time behavior of the magnitude of a one-point function's deviation from its equilibrium value. After the quench time $\sim \tilde{\tau}$ the response equilibrates much like a damped oscillator. The damped oscillations correspond to the excitations of the gravitational system's lowest lying quasi-normal mode.

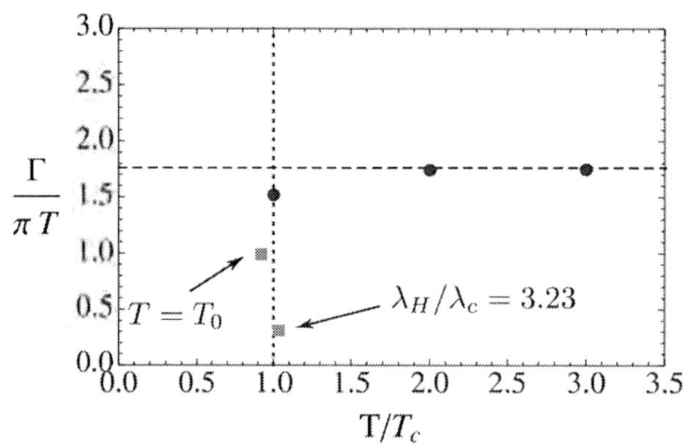

Figure 7. The temperature dependence of the decay width Γ for the lowest-lying scalar quasi-normal mode in several states of our theory. The blue circles are large black-branes whose temperature is an integer multiple of T_c. The orange squares correspond to the minimum temperature black brane at $T = T_0$ (top) and the smallest black hole we perturb in our study at $\lambda_H/\lambda_c = 3.23$ (bottom). The ratio $\Gamma/\pi T$ approaches 1.75953 (the dashed line) at high temperatures, which coincides with the expected value for perturbations of AdS_5 Schwarzschild by a dimension 3 scalar operator [8].

quench amplitude at fixed (short) duration. The most important feature shown is the appearance of simple power law scaling regimes. Combining the results of the two plots, one finds that in the limit of abrupt quenches (where the quench width is much smaller than all other dimensionful scales)

$$\langle T_{tt} \rangle_{\rm FINAL} \sim \left(\frac{\tilde{\delta}}{\tilde{\tau}}\right)^2 . \qquad (10)$$

In fact this simple scaling relation was already anticipated on very general grounds, and is a special case of the universal scaling formula in [9] for a dimension three scalar operator. That this scaling should be universal readily follows from the fact that for very fast quenches, the perturbation does not have time to propagate far from the boundary before the quench concludes. Since the near

 DOI: 10.1051/epjconf/201612602018

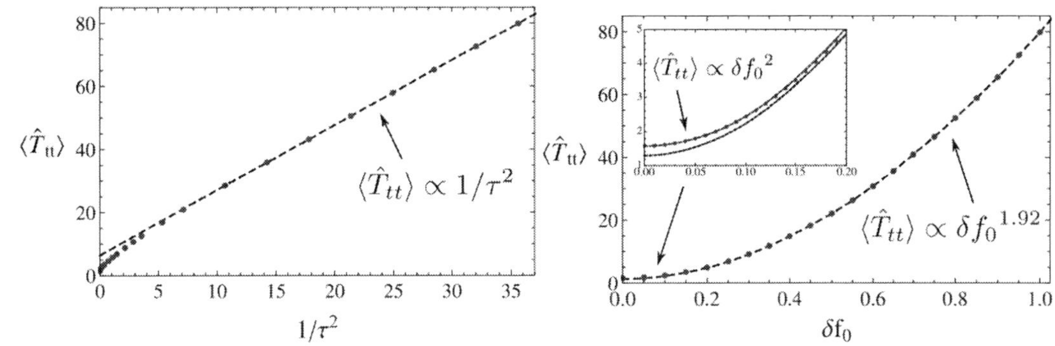

Figure 8. Final state energy density $\langle T_{tt} \rangle$ as a function of quench parameters. In the left plot the quench amplitude is held fixed, while in the right plot the quench duration is fixed.

boundary region of many holographically relevant spacetimes is asymptotically AdS, abrupt quenches are indifferent to the IR features of the geometry distinguishing different holographic spacetimes.

5 Discussion

One of the most important lessons to be extracted from our results is the domination of the thermalization time by the linear regime, even in the presence of a confinement scale in the holographic gauge theory. Indeed, in every quench that we have managed to evolve in this system, the thermalization time appears to be controlled entirely by the lowest lying quasi normal mode of the final state black brane. It is conceivable that this is due in large part to the initial state that we have chosen to perturb. As a consequence of the previously discussed IR features of very small black holes in this model, our numerical stability suffers as the size of the initial state black brane shrinks. On the other hand, it is sensible to expect that the confinement scale becomes important to the characteristics of the dynamical response when it is the dominant energy scale in the problem. Currently, the smallest black hole that we can reliably perturb has an energy density comparable to the confinement scale: $\langle T_{tt} \rangle / f_0{}^4 \sim O(1)$. Thus, it is perhaps not entirely surprising that the dynamical response of the system does not appear to pass through any regime controlled by this energy scale in our computations.

Ultimately, it would be of great interest to perturb initial states with $\langle T_{tt} \rangle / f_0{}^4 \ll 1$, or better yet the zero temperature solution itself. It is within this setting that one expects the response to the quench to be most sensitive to the diverging scalar potential in the bulk, and thus to the confinement scale of the dual boundary theory. Understanding to what extent this scalar potential is similar to the hardwall in its ability to induce scattering solutions which never collapse in the bulk is perhaps the most obvious avenue to pursue. Should such scattering solutions exist, identifying the boundary in the space of quench parameters that separates perturbations which lead to horizon formation from those which do not would permit a scaling analysis analogous to that performed in the well known case of Choptuik phenomena [10].

Last, it may prove interesting to investigate the features of other probes sensitive to the quench dynamics in this holographic theory. Some obvious contenders are various non-local probes such as Wilson loops, entanglement entropy, and two point functions of large dimension operators. These operators all share the feature that their holographic computation involves the study of worldlines or worldsheets that sag into the bulk spacetime. As the separation of boundary operators increases, these surfaces generically droop lower in the radial (holographic) direction. In this way, these probes may provide insights into the equilibration of the dual field theory at different length scales, potentially offering a more detailed understanding of the process.

References

[1] T. Ishii, E. Kiritsis, C. Rosen, JHEP **08**, 008 (2015), 1503.07766
[2] E. Witten, Adv. Theor. Math. Phys. **2**, 505 (1998), hep-th/9803131
[3] Y. Kinar, E. Schreiber, J. Sonnenschein, Nucl. Phys. **B566**, 103 (2000), hep-th/9811192
[4] B. Craps, E. Kiritsis, C. Rosen, A. Taliotis, J. Vanhoof, H.b. Zhang, JHEP **02**, 120 (2014), 1311.7560
[5] B. Craps, E.J. Lindgren, A. Taliotis, J. Vanhoof, H.b. Zhang, Phys. Rev. **D90**, 086004 (2014), 1406.1454
[6] B. Craps, E.J. Lindgren, A. Taliotis (2015), 1511.00859
[7] P.M. Chesler, L.G. Yaffe, Phys. Rev. Lett. **102**, 211601 (2009), 0812.2053
[8] A. Nunez, A.O. Starinets, Phys. Rev. **D67**, 124013 (2003), hep-th/0302026
[9] A. Buchel, R.C. Myers, A. van Niekerk, Phys. Rev. Lett. **111**, 201602 (2013), 1307.4740
[10] M.W. Choptuik, Phys. Rev. Lett. **70**, 9 (1993)

EPJ Web of Conferences **126**, 02019 (2016) DOI: 10.1051/epjconf/201612602019
ICNFP 2015

Beyond the discovery: Higgs results from CMS

Rainer Mankel[1,a], on behalf of the CMS collaboration

[1] *Deutsches Elektronen-Synchrotron (DESY), Notkestr. 85, D-22607 Hamburg, Germany*

Abstract. The observation of a Higgs boson at a mass near 125 GeV in the year 2012 has been a milestone for elementary particle physics. Since this fundamental discovery, the CMS collaboration has scrutinized the complete LHC Run I dataset in depth, and studied the properties of the observed state in full detail. This includes investigations of more elusive production and decay modes, as well as searches for first indications of an extended Higgs sector, which would represent very likely a promising gateway to new physics. This article summarizes recent Higgs results from the CMS experiment.

1 Introduction

The observation of a Higgs boson at a mass of about 125 GeV, in the following referred to as H(125), by the ATLAS and CMS experiments [1–3] has been a fundamental step forward for particle physics. Since this discovery, further Higgs research focuses on two essential questions. The first concerns the detailed properties of the new state, and whether they are precisely in accord with the expectations of the Standard Model (SM). The second question is whether there are additional Higgs bosons. Both kinds of studies have the potential to reveal physics beyond the SM. A huge number of new results have been obtained by the CMS collaboration, and selected highlights are discussed in this article.

2 Properties of the H(125) state

2.1 Bosonic decay channels

At the level of the Run I legacy results, the existence of the Higgs boson state at 125 GeV is firmly established in the di-boson decay channels, $H \to \gamma\gamma$ [4], $H \to ZZ^* \to 4\ell$ [5] and $H \to WW$ [6]. The corresponding signals are shown in Fig. 1. A recent combination of ATLAS and CMS measurements in the $\gamma\gamma$ and 4ℓ channels [7, 8] has determined the mass to be $m_H = 125.09 \pm 0.24$ GeV. The spin-parity analysis favors the $J^P = 0^+$ hypothesis, although the possibility of a mixed state with a small pseudoscalar component is not excluded yet. More detailed results concerning Higgs couplings are presented in Section 2.6.

[a] e-mail: Rainer.Mankel@desy.de

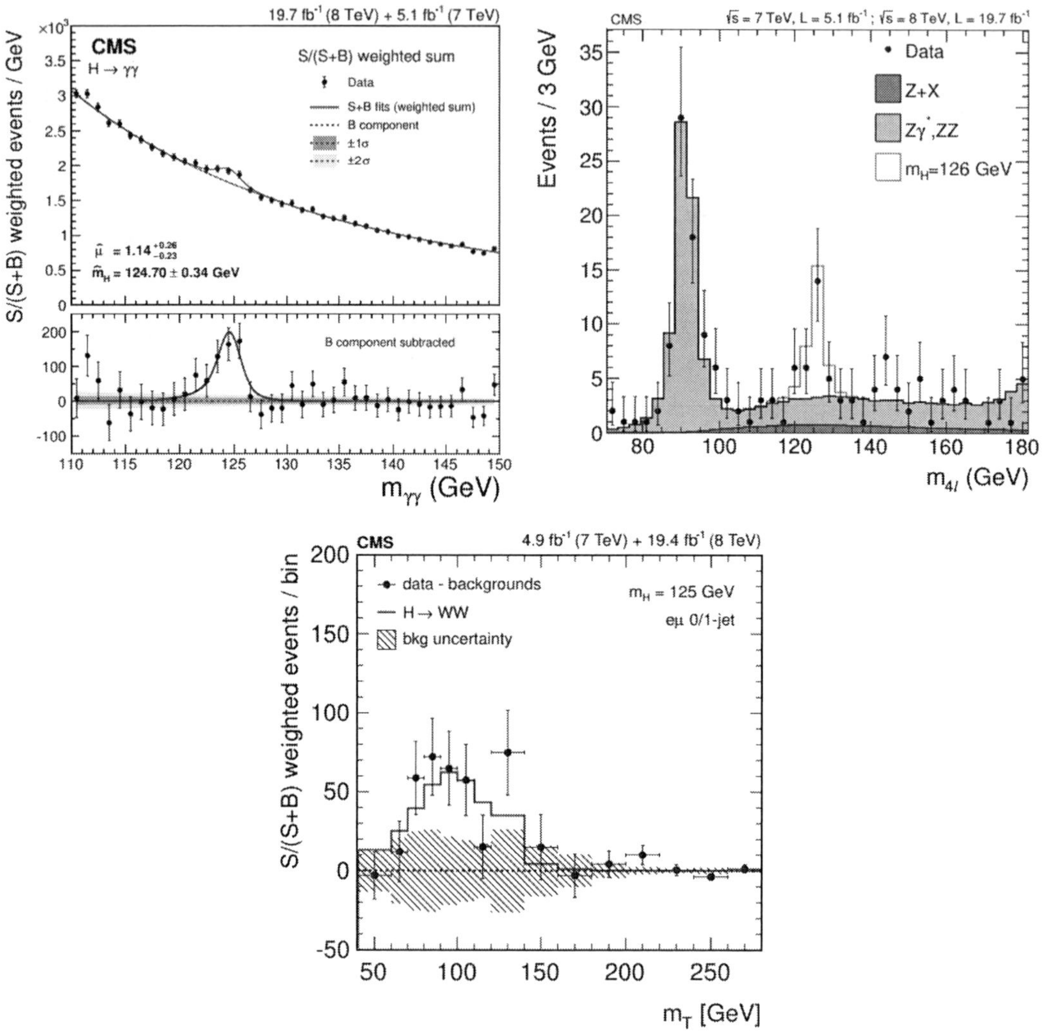

Figure 1. Higgs signals observed by CMS in the di-boson decay channels: H $\rightarrow \gamma\gamma$ (top left), H $\rightarrow$ ZZ* $\rightarrow 4\ell$ (top right) and H $\rightarrow$ WW (bottom) [4–6].

2.2 Evidence for H $\rightarrow \tau\tau$

Higgs decays into fermionic channels are far more elusive, but their study is essential to establish the nature of the Higgs boson. The leptonic mode with the highest expected branching fraction is the H$\rightarrow$ $\tau\tau$ channel. The spectrum of the reconstructed mass of τ pairs from the combined Run I data of the CMS experiment is shown in Fig. 2 (left) [9] for the most sensitive channels. While the distribution is generally in very good agreement with the expected backgrounds, an enhancement is observed which is compatible with a SM Higgs boson at a mass of 125 GeV. The excess corresponds to an observed significance of 3.2 σ (with 3.7 σ expected), and is thus already a strong indication of the Higgs boson coupling to the τ lepton as expected by the SM theory.

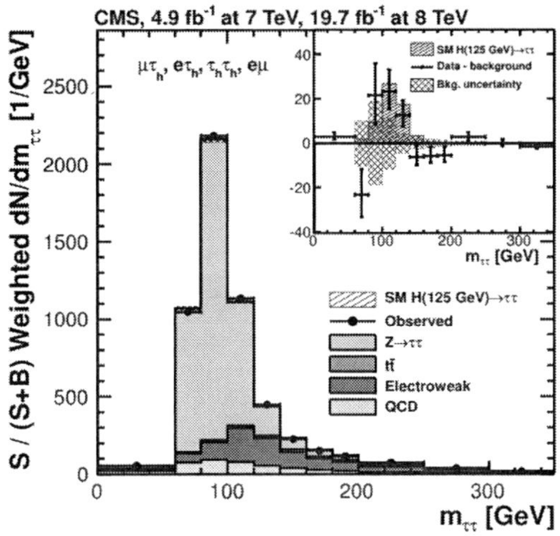

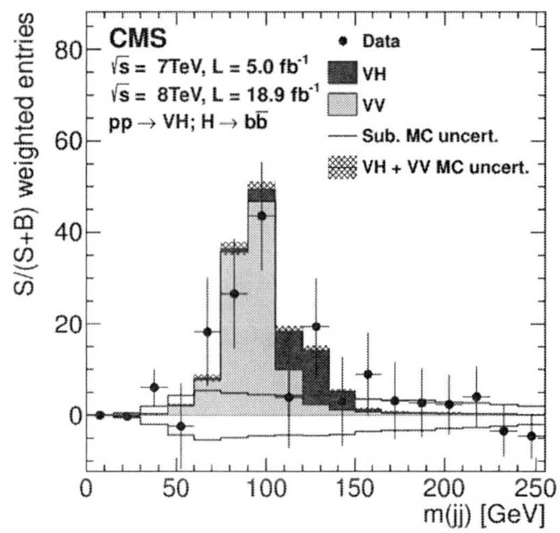

Figure 2. Left: Observed and predicted $m_{\tau\tau}$ distributions for the $\mu\tau_h$, $e\tau_h$, $\tau_h\tau_h$ and $e\mu$ channels [9]. The inset shows the difference between the observed data and expected background distributions, together with the signal distribution for a SM Higgs boson at m_H = 125 GeV. Right: Weighted b-tagged dijet invariant mass distribution, combined for all channels in the H → b$\bar{\text{b}}$ analysis [10]. All backgrounds, except dibosons, have been subtracted. The line histograms for signal and for VV backgrounds are also shown superimposed.

2.3 Search for H → b$\bar{\text{b}}$

According to the SM, the decay H → b$\bar{\text{b}}$ has the largest branching fraction, but this channel has also very difficult background conditions, which make an inclusive observation impossible with the current dataset. The situation is much more favorable for the production in association with a vector boson, which can be either a W or Z boson and is commonly referred to as "V". This leads to an improved signature with easier triggering and an improved signal-to-background ratio. The invariant mass spectrum of the pair of b-jets from this analysis [10], after background subtraction, is shown in Fig. 2 (right). A clear peak at the Z mass is visible, which is attributed to vector boson pair production, serving as a "standard candle". The invariant mass of the b-jet pair is combined with other variables discriminating between signal and background to enhance the sensitivity of the search. The multivariate analysis reveals an excess with a significance of about two standard deviations above the SM background expectation. Since the publication of [10], the modeling of the signal cross section has been improved by including the gg → ZH process [11]. The measured relative signal strength for the H → b$\bar{\text{b}}$ channel, including also the contribution from production in association with top quark pairs [14] (see Section 2.4) is σ/σ_{SM} = 0.84 ± 0.44, in good agreement with the SM. The corresponding observed significance is 2.0 σ (with 2.6 σ expected) [11]. The measurement [10] has been combined with the H → $\tau\tau$ measurement shown in Section 2.2, resulting in a joint evidence of 3.8 σ for fermionic Higgs decays [12].

Very recently, the H → b$\bar{\text{b}}$ production strength has also been analyzed in a topology expected for vector boson fusion (VBF) [13]. In this process, quarks from the initial protons emit vector bosons (W or Z), which then fuse to form the Higgs boson. No QCD color is exchanged. This leads to a very clean final state characterized by the decay products of the Higgs boson and two hadronic jets with large difference in rapidity and large invariant mass. This first search for the SM Higgs boson in a purely hadronic final state makes strong use of categorization with boosted decision trees (BDT),

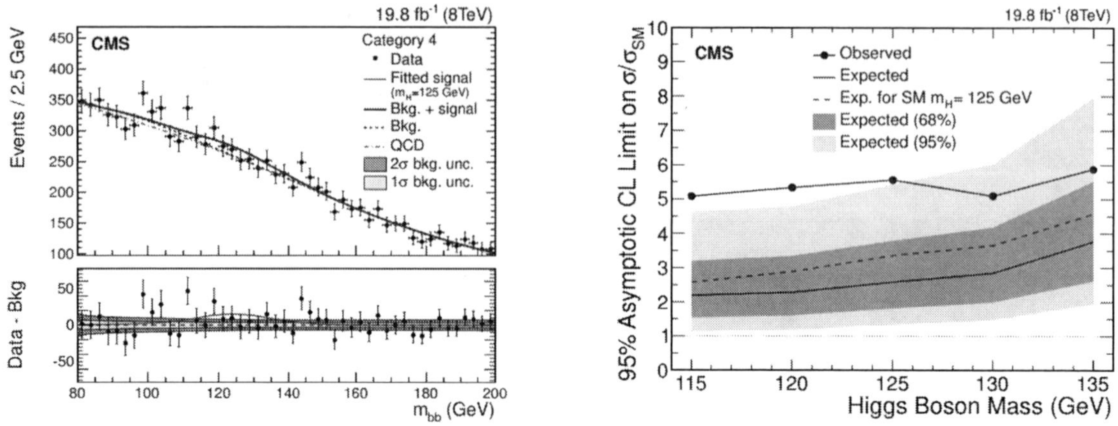

Figure 3. Left: Fit of the invariant mass of the two b-jet candidates for the Higgs boson signal (m_{H} = 125 GeV) in the VBF event categories of highest signal enrichment [13]. Data are shown by the points. The solid line is the sum of the post-fit background and signal shapes, the dashed line is the background component, and the dashed-dotted line is the QCD component alone. The bottom panel shows the background-subtracted distribution, overlaid with the fitted signal, and with the 1 σ and 2 σ background uncertainty bands. Right: Expected and observed 95% confidence level limits on the signal cross section in units of the SM expected cross section, as a function of the Higgs boson mass, including all event categories. The limits expected in the presence of a SM Higgs boson with a mass of 125 GeV are indicated by the dashed curve.

which is validated with Z($\rightarrow$ b$\bar{\text{b}}$) + jets production. The invariant mass spectrum in the most signal enriched category is shown in Fig. 3 (left); a fit indicates some enhancement near 125 GeV. A signal significance of 2.2 σ is observed, while 0.8 σ is expected. The upper limit for the cross section in Fig. 3 (right) shows a corresponding excess.

The VBF result has been combined with the BDT-based searches for H $\rightarrow$ b$\bar{\text{b}}$ in the VH and ttH (see Section 2.4) production modes [13]. The resulting combined signal strength is $\sigma/\sigma_{\mathrm{SM}}$ = 1.03 ± 0.44, and the observed signal significance is 2.6 σ (with 2.7 σ expected). In conclusion, the Run I measurements show a convincing hint for the H(125) boson coupling to b quarks in accord with the SM.

2.4 Higgs production in association with top quarks

In the SM, the Higgs boson is expected to have a particularly large Yukawa coupling to the top quark. Indirectly, this coupling is accessible via fermion loops which occur in the gluon fusion production process and the Higgs decay to photon pairs. For an unambiguous measurement, it is important to probe the top-Higgs coupling directly. This is possible through the measurement of Higgs production in association with top quark pairs. In an earlier CMS publication, the analysis combined many different topologies and decay channels, and used a BDT discriminant to enhance the signal-to-background ratio [14].

A more recent CMS analysis [15] is based on matrix elements (ME) of the signal process and the most relevant background process, pp $\rightarrow$ ttbb + X. This analysis is currently restricted to the H $\rightarrow$ b$\bar{\text{b}}$ decay channel. To each event, corresponding probability density values are assigned according to signal or background hypothesis, and the ratio, $\mathrm{P}_{\mathrm{s/b}}$, is used as a discriminant in a likelihood fit to extract the signal. This leads to an optimized separation of signal and background.

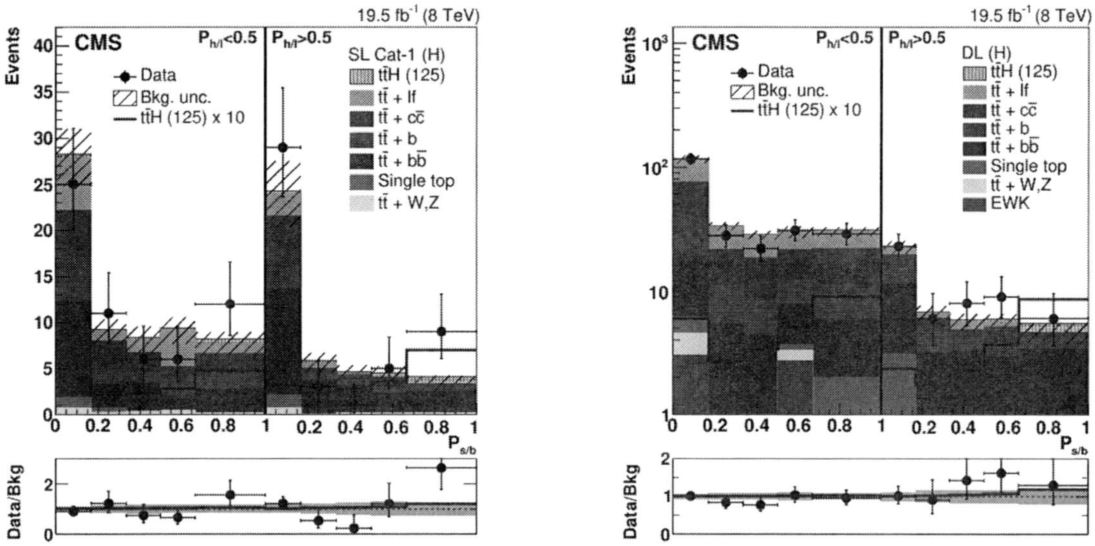

Figure 4. Distributions of the $P_{s/b}$ discriminant in the two $P_{h/l}$ bins of the ttH ME analysis in the $H \rightarrow b\bar{b}$ channel, for selected high-purity (H) categories of the single-lepton (left) and double-lepton (right) selections for data, together with the estimated background contributions. The bottom panel of each plot shows the ratio between the observed and the overall background yields. The solid blue line indicates the ratio between the signal-plus-background and the background-only distributions, enhanced by a factor of 10. The shaded and solid green bands correspond to the $\pm 1\ \sigma$ uncertainty in the background prediction after the fit.

The signal is searched for in a binned distribution of two parameters, the discriminant $P_{s/b}$ and a second discriminant $P_{h/l}$ which distinguishes between the heavy- and the light-flavor components of the $t\bar{t}$+jets background. Figure 4 shows the distribution of the data of selected high-purity categories in the single-lepton (left) and double-lepton (right) samples. The data are consistent with the background estimation; there is no evidence for a ttH signal. The results of the best fit are shown in Fig. 5 (left). The combined result corresponds to a signal strength of $\sigma/\sigma_{SM} = 1.2^{+1.6}_{-1.5}$, which is consistent with the SM. Upper limits at 95% confidence level (CL) are shown in Fig. 5 (right). In combination, the observed limit on σ/σ_{SM} is 4.2. The expected limit is 4.1 with the expected ttH signal injected, and 3.3 without. Given the fact that only the $H \rightarrow b\bar{b}$ decay channel is used, the ME method is found to be competitive with the BDT-based analysis.

2.5 Differential Higgs production cross sections

Differential cross sections are a direct test of perturbative QCD calculations in the Higgs sector. Recently, the CMS collaboration has measured such cross sections using the di-photon decay channel [16]. For example, the transverse momentum distributions of the di-photon system and the leading accompanying jet are sensitive to higher order QCD effects, while the distribution of the rapidity of the di-photon system is sensitive to the proton PDF and the production mechanism in general. All three are shown in Fig. 6, and found to be in agreement with the QCD predictions within uncertainties. In particular, there is no hint yet of extra contributions from new production processes.

DOI: 10.1051/epjconf/201612602019

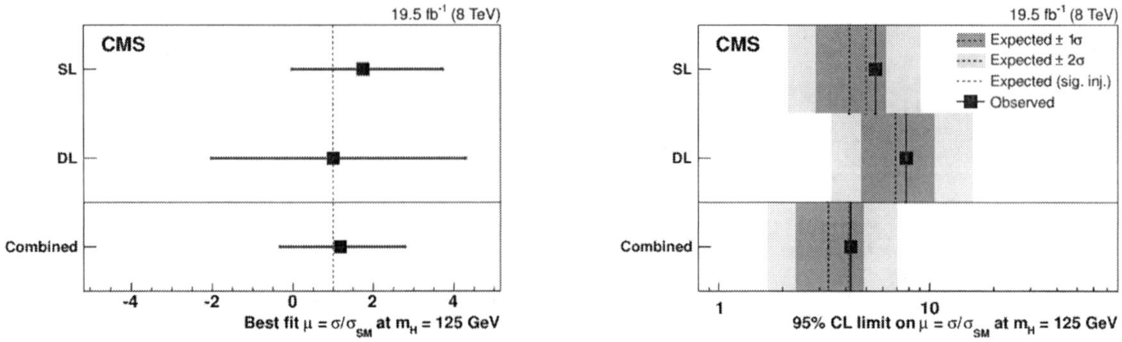

Figure 5. Left: Best-fit value of the signal strength modifier μ with its $\pm 1\ \sigma$ CL interval obtained from the individual channels and from their combination in the ttH ME analysis. Right: Observed 95% CL upper limits on μ are compared to the median expected limits under the background-only and the signal-plus-background hypotheses. The former are shown together with their $\pm 1\ \sigma$ and $\pm 2\ \sigma$ CL intervals. Results are shown separately for the individual channels and for their combination.

2.6 Higgs couplings

The measurements in the various production processes and decay channels are combined to extract information about the coupling of the Higgs boson to fundamental fermions and gauge bosons of the SM [11]. For one of the many studies, coupling modifiers for fermions and vector bosons have been introduced and profiled. The constraints from the different decay channels, expressed as 68% CL confidence regions, are shown in Fig. 7 (top left). The combined likelihood shows a clear maximum near the point $\kappa_V = \kappa_f = 1$, which corresponds to the SM expectation.

The mass dependence of the couplings can be regarded as an essential "fingerprint" of a Higgs boson. The SM predicts that the Yukawa couplings to fermions have a linear mass dependence, while the couplings to weak bosons are proportional to the square of the boson mass. For these reasons, a "reduced" weak boson coupling is computed, such that all couplings can be compared in one plot in Fig. 7 (top right). The results agree well with the linear mass dependence expected from the SM.

The same data can also be used to profile the branching ratio of the Higgs boson into invisible particles. Such a scan is shown in Fig. 7 (bottom). Also the results of the direct searches into invisible Higgs decays are included. An observed upper limit of 0.49 at 95% CL is inferred.

At the current level of measurement precision, the observed Higgs boson couples to fundamental fermions and gauge bosons as expected in the SM. On the other hand, there is still plenty of room for non-standard invisible decays. It is an important question whether the H(125) is indeed the SM Higgs boson, or just one member of an extended Higgs sector, which happens to be SM-like. Direct searches for additional Higgs bosons are essential to clarify this question, and will be discussed in the following section.

3 Higgs bosons beyond the Standard Model

Supersymmetry is a theory which naturally features extended Higgs sectors. In the minimal (MSSM) and next-to-minimal (NMSSM) supersymmetric extensions of the SM, three or five neutral Higgs bosons are predicted, respectively, as displayed in Table 1. One of these bosons must be SM-like and coincide with the observed boson at 125 GeV. In the MSSM, the H(125) state would correspond most likely to the h boson, since other assignments are already strongly disfavored by the current

DOI: 10.1051/epjconf/201612602019

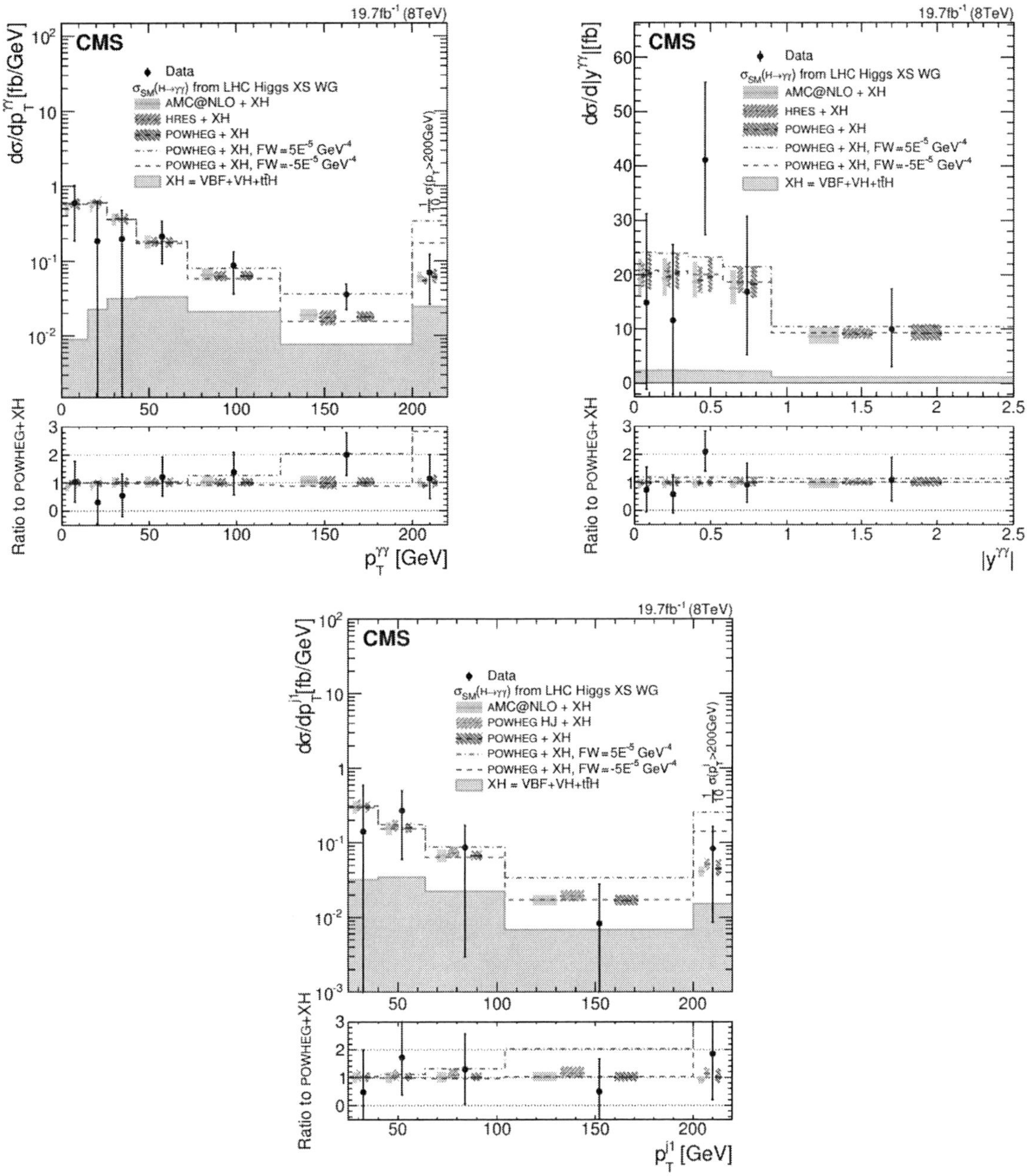

Figure 6. The $H \rightarrow \gamma\gamma$ differential cross section for inclusive events as a function of $p_T^{\gamma\gamma}$ (top left), $|y^{\gamma\gamma}|$ (top right) and p_T^{j1} (bottom) [16]. All the SM contributions are normalized to their cross section from the LHC Higgs Cross Section Working Group [17]. Theoretical uncertainties in the renormalization and factorization scales, PDF, and branching fraction are added in quadrature. The error bars on data points reflect both statistical and systematic uncertainties. The last bin of the $p_T^{\gamma\gamma}$ distribution sums the events above 200 GeV. For each graph, the bottom panel shows the ratio of data to theoretical predictions from the POWHEG generator.

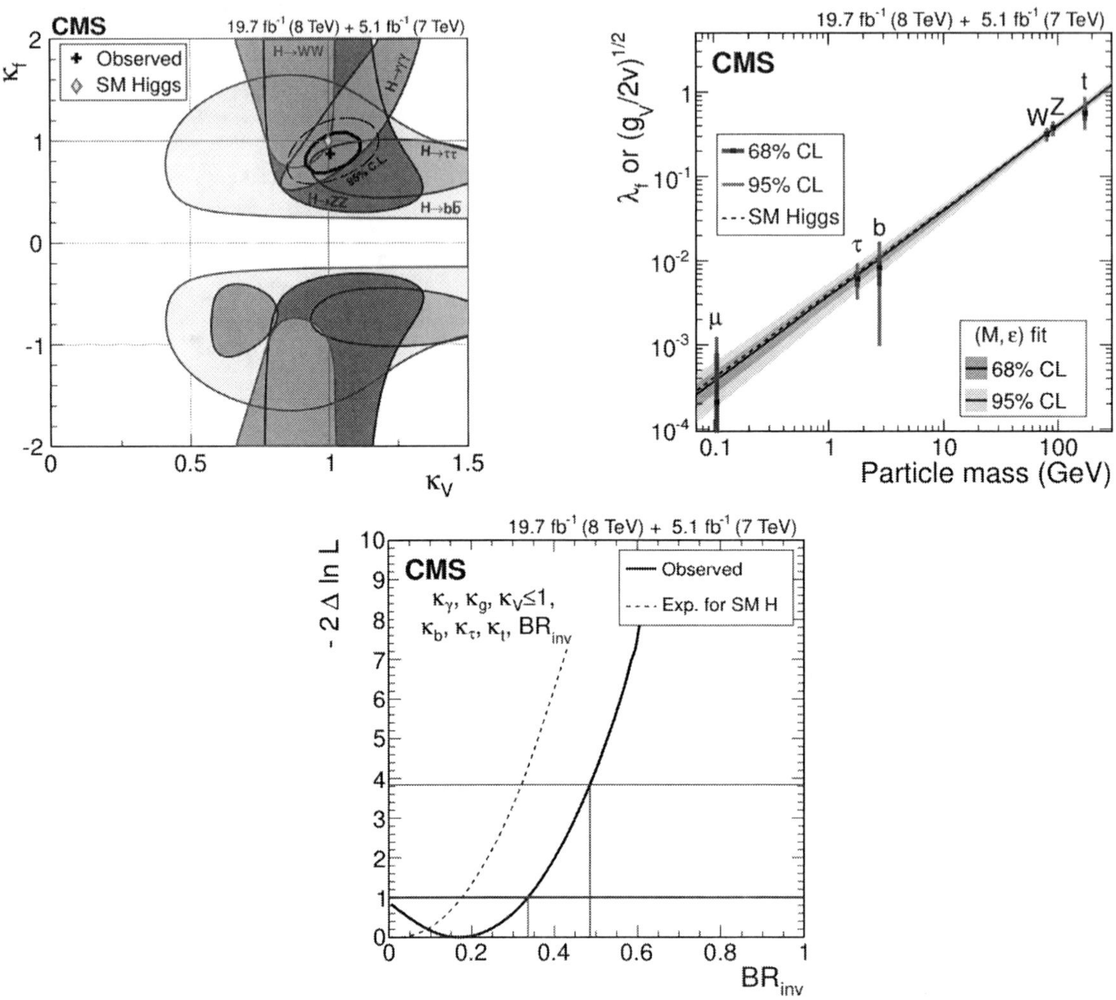

Figure 7. Top left: The 68% CL confidence regions for individual channels (coloured swaths) and for the overall combination (thick curve) for the coupling modifiers κ_V and κ_f. The cross indicates the global best-fit values. Top right: Reduced coupling constants of the Higgs boson to various fermions and bosons as a function of the particle mass. The dashed line corresponds to the SM expectation. Bottom: Result of a likelihood scan versus $BR_{BSM} = \Gamma_{BSM}/\Gamma_{tot}$. The solid curve represents the observation in data, and the dashed curve indicates the expected median result in the presence of the SM Higgs boson. The data from the searches for invisible Higgs decays are also taken into account [11].

measurements. In the NMSSM, several assignments are possible, including $H(125) \equiv h_2$, and Higgs bosons with masses below 125 GeV are thus feasible.

3.1 Search for MSSM Higgs bosons in the $b\bar{b}$ channel

With a SM-like h boson at 125 GeV, the H and A bosons are searched for at higher masses; they are expected to be degenerate, and their coupling to b quarks is enhanced by the MSSM parameter $\tan\beta$.

Table 1. Higgs sectors in the minimal (MSSM) and next-to-minimal (NMSSM) supersymmetric extensions of the SM.

Model	Structure	CP-even	CP-odd	Charged
MSSM	2 doublets	h, H	A	$H^{\pm}$
NMSSM	2 doublets + 1 singlet	h_1, h_2, h_3	a_1, a_2	$h^{\pm}$

For large $\tan\beta$ values, even at large masses the by far dominant decay mode should be the decay into b-quark pairs. The main challenge of this decay channel is the huge background rate from QCD multijet production. It is most promising to search for production in association with additional b quarks, as depicted in Fig. 8 (left), which is enhanced by a factor of $2\tan^2\beta$, including the degeneracy effect, and also leads to a better signal-to-background relation. A dedicated hadronic trigger is mandatory for this analysis.

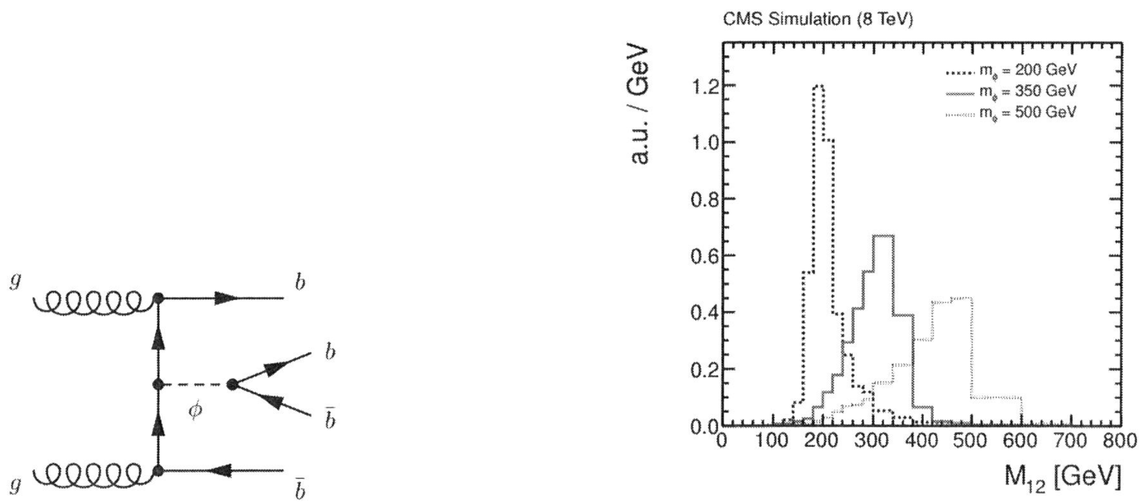

Figure 8. Left: Feynman diagram of one of the processes contributing to the MSSM $H \to b\bar{b}$ production in association with additional b quarks. Right: Distributions of the dijet invariant mass for expected signals, for Higgs boson masses of 200, 350 and 500 GeV. The vertical scale is shown in arbitrary units [18].

This CMS analysis [18] is unique at the LHC. The selection requires the three leading jets to be b-tagged, and a signal is expected to emerge as a peak in the invariant mass distribution of the two leading jets, as shown according to simulation in Fig. 8 (right). The dijet invariant mass spectrum from the triple-b-tag data sample is shown in Fig. 9 (left). The background is modeled with a set of templates reflecting the different flavor combinations, which are derived from the double-b-tag data sample, and fitted to the data. This background model is found to describe the data well, and there is no indication of a signal. The addition of signal templates to the fit does not result in significant signal strength at any of the probed Higgs masses. Upper limits on cross section time branching ratio are derived and shown in Fig. 9 (right), they agree well with the expected limits. A detailed MSSM model interpretation will be presented in Section 3.3.

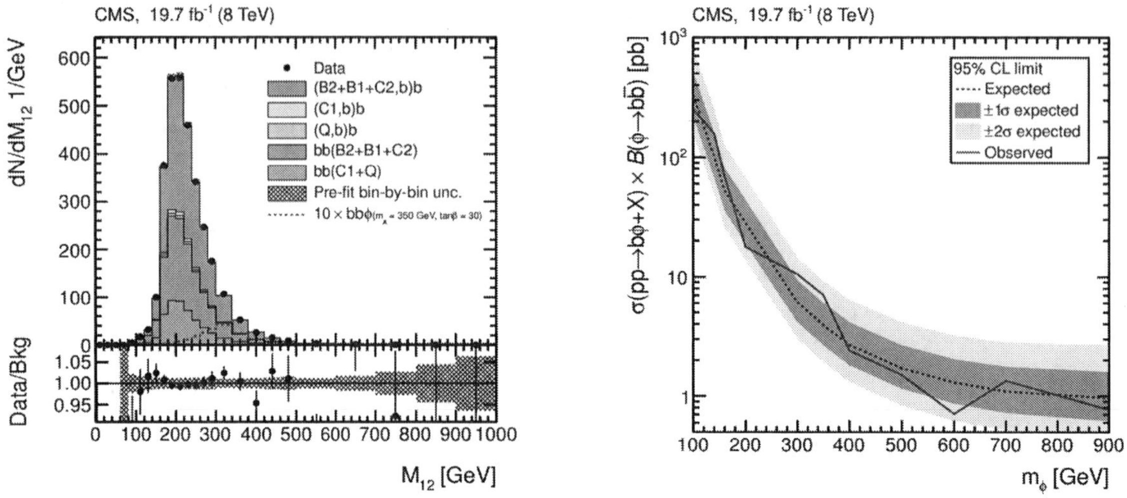

Figure 9. Left: Distribution of the dijet mass in the triple-b-tag sample in the search for MSSM Higgs bosons in the $\mathrm{b\bar{b}}$ channel, together with the corresponding distributions of the fitted background templates. The hatched area shows the total bin-by-bin background uncertainty of the templates prior to the fit. For illustration, the signal contribution expected in the $m_{\mathrm{h}}^{\mathrm{max}}$ benchmark scenario of the MSSM with $m_{\mathrm{A}} = 350$ GeV , $\tan\beta = 30$, and $\mu = +200$ GeV is overlayed, scaled by a factor 10 for better readability. In addition, the ratio of data to the background estimate is shown at the bottom. Right: Expected and observed upper limits at 95% CL on $\sigma(\mathrm{pp} \to \mathrm{b}\phi + \mathrm{X})\mathrm{BR}(\phi \to \mathrm{b\bar{b}})$, where ϕ denotes a generic neutral Higgs-like state [18].

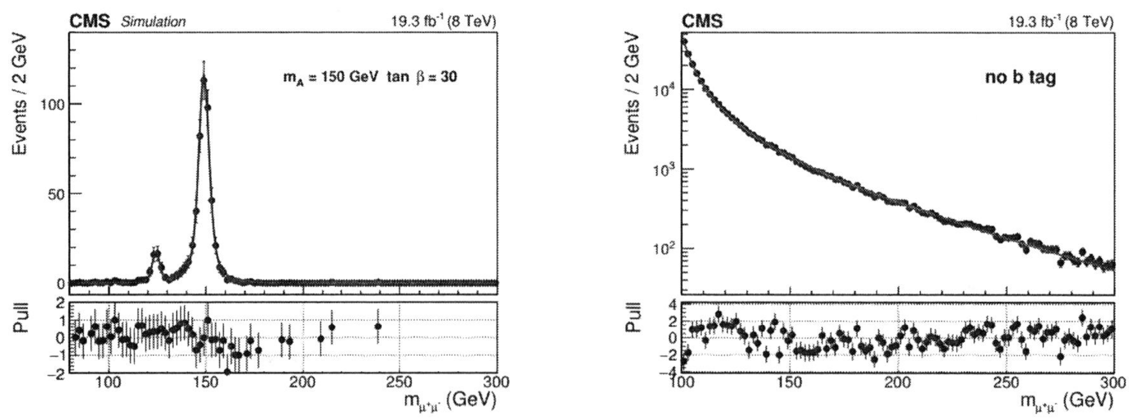

Figure 10. Invariant mass distribution of the expected signal for m_{A} =150 GeV and $\tan\beta = 30$ (left) for the MSSM Higgs boson search in the di-muon channel, and an example of the fit to the data at $\sqrt{s} = 8$ TeV including the same signal assumption (right). The distribution represents the expected number of events for an integrated luminosity of 19.3 fb^{-1}. For each plot the pull of the fit as a function of the dimuon invariant mass is shown [19].

 DOI: 10.1051/epjconf/201612602019

3.2 Search for MSSM Higgs bosons in the di-muon final state

While the branching ratio of MSSM Higgs bosons into muon pairs is expected to be small, this is compensated by the excellent mass resolution. The dominant background component originates from Drell-Yan production, and leads to a smooth background shape. The recent CMS analysis [19] distinguishes between two categories, named "b tag" and "no b tag", depending on whether the Higgs candidate is accompanied by an additional b-tagged jet or not. The signal model containing all three neutral MSSM Higgs bosons is shown for one set of parameters in Fig. 10 (left). The left of the two peaks corresponds to a h boson with SM-like properties at a mass of 125 GeV, while the right peak is the combined signal of the H and A bosons, whose cross sections are greatly enhanced due the MSSM parameter choice of $\tan\beta = 30$. The analysis proceeds with parametric signal and background models. The fit of these models to the observed di-muon mass spectrum in data is shown in Fig. 10 (right); there is no indication of an excess.

3.3 Interpretation within the MSSM

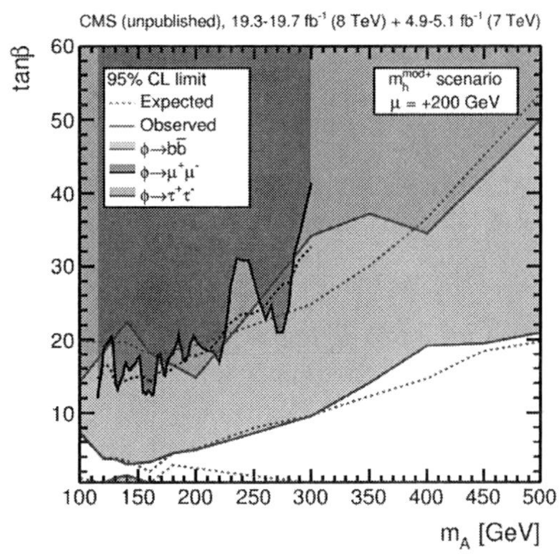

Figure 11. Comparison of the expected and observed upper limits at 95% CL for the MSSM parameter $\tan\beta$ versus m_A in the m_h^{mod+} benchmark scenario [18] obtained in different final states: $\phi \to b\bar{b}$ [18], $\phi \to \mu\mu$ [19], and $\phi \to \tau\tau$ [22]. The shaded areas mark the excluded regions.

The results from the neutral Higgs boson searches are interpreted in the m_h^{mod+} benchmark scenario [20] of the MSSM. The published CMS results in the $\phi \to \tau\tau$ channel [21] have recently been updated with an improved identification of semihadronic τ decays [22]. The results are illustrated in the MSSM parameter space for all three decay modes in Fig. 11. The most stringent MSSM limits are achieved in the $\tau\tau$ channel, while the $b\bar{b}$ and $\mu\mu$ channels are at comparable levels. At large values of m_A, a relatively wide range for $\tan\beta$ is still open according to the measurements, as can also be seen in Ref. [22].

3.4 Light NMSSM Higgs boson search

While the possibility of light Higgs bosons with SM-like couplings in the mass region below 100 GeV has already been largely excluded by measurements at LEP [23], such bosons are nevertheless feasible

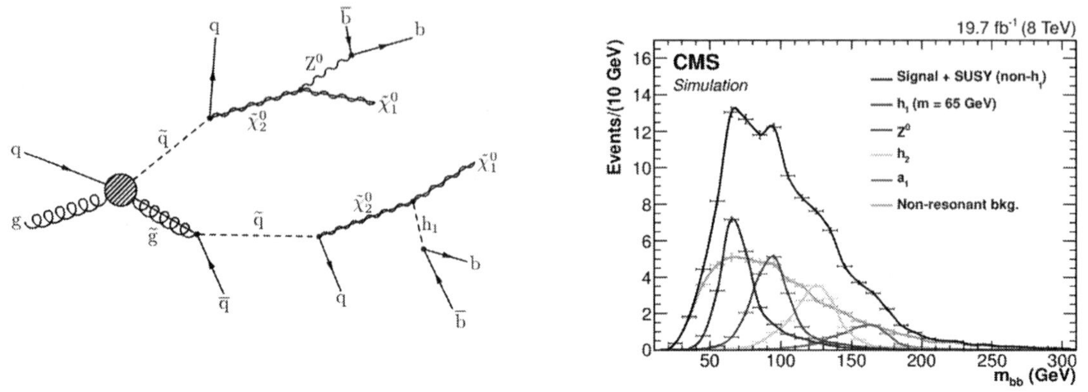

Figure 12. Left: Feynman diagram of squark-gluino production with subsequent cascade decays via neutralinos into h_1. Right: Signal + non-h_1 SUSY prediction of the modified NMSSM P4 scenario with m_{h_1} =65 GeV and normalized to the integrated luminosity of 19.7 fb^{-1} [25].

in NMSSM scenarios. In a slightly modified version of the P4 scenario [24], the h_2 state is assumed to be the H(125) boson. The lighter h_1 is assumed to have a very large singlet component, due to which it would be strongly suppressed in the standard Higgs production channels, and thus have escaped detection so far. However, it would be copiously produced in SUSY cascades, as illustrated in Fig. 12 (left). Its observation would thus imply a simultaneous discovery of supersymmetry and an extended Higgs sector.

This option has been explored for the first time at the LHC by the CMS experiment [25]. Besides the h_1, other bosons, namely Z, h_2 and a_1 should appear in the SUSY cascade as well, resulting in a complex signal mass spectrum as shown in Fig. 12 (right). The analysis searches for events with two energetic jets originating from squarks in the cascade, two b jets which might come from the Higgs decay, and large hadronic activity and missing energy. The resulting invariant b-jet pair mass spectrum measured in data is shown in Fig. 13 (left). The distribution shows no excess beyond the estimated SM backgrounds, and there is no indication for a signal. The resulting upper limit on cross section times branching ratio is shown in Fig. 13 (right). The modified NMSSM P4 scenario is excluded over the whole mass range from 30–100 GeV at $M_{\text{SUSY}} = M_{\tilde{g}} = 1$ TeV.

4 Summary

Beyond the H(125) discovery, a wealth of new CMS Higgs results have been obtained, which can be regarded as the legacy results of CMS from the LHC Run I. The properties of the observed H(125) state have already been determined to a remarkable precision. The methodology has been enhanced in many places, which is expected to fully bear fruit with the Run II data. At the current level of measurement accuracy, the properties of this state are found to be in agreement with those expected in the SM.

The quest for New Physics as it would be indicated by additional Higgs states has equally progressed. Stringent limits for the MSSM parameters have been obtained in three direct search channels, and as a result the region of low m_A is already strongly constrained, while at large values of m_A, large values of $\tan\beta$ are still possible. There is still a widely open parameter space for the Higgs sector in the NMSSM model.

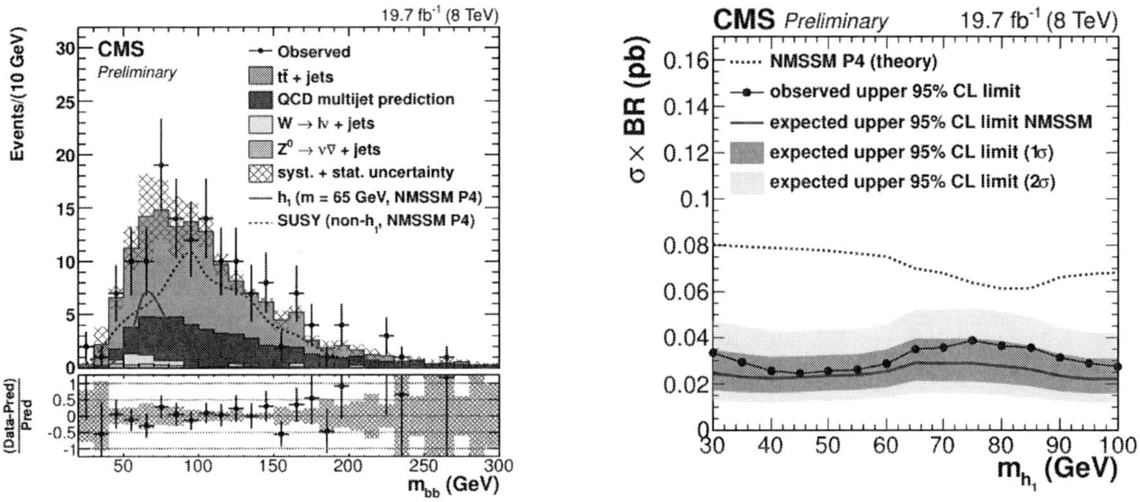

Figure 13. Left: Background-only fit result of the invariant mass of the selected b-tagged jets, where the SM background contributions are stacked, in the light NMSSM Higgs boson search. The theoretical h_1 and non-h_1 contributions are overlaid to illustrate the sensitivity of the analysis to the modified NMSSM P4 scenario expectation. Right: Upper limits for the NMSSM cross section times the branching ratio into b-quarks [25].

The Run II of the LHC at higher CM energy and large luminosity will provide a hugely extended capability for studying the properties of the H(125) boson, and searching for extensions of the Higgs sector.

References

[1] ATLAS Collaboration, "Observation of a new particle in the search for the Standard Model Higgs boson with the ATLAS detector at the LHC", Phys. Lett. B **716** (2012) 1.

[2] CMS Collaboration, "Observation of a new boson at a mass of 125 GeV with the CMS experiment at the LHC", Phys. Lett. B **716** (2012) 30.

[3] CMS Collaboration, "Observation of a new boson with mass near 125 GeV in pp collisions at $\sqrt{s}$ = 7 and 8 TeV", JHEP **06** (2013) 081.

[4] CMS Collaboration, "Observation of the diphoton decay of the Higgs boson and measurement of its properties", Eur. Phys. J. C **74** (2014) 3076.

[5] CMS Collaboration, "Measurement of the properties of a Higgs boson in the four-lepton final state", Phys. Rev. D **89** (2014) 092007.

[6] CMS Collaboration, "Measurement of Higgs boson production and properties in the WW decay channel with leptonic final states", JHEP **01** (2014) 096.

[7] P. Vanlaer, "ATLAS+CMS combined H boson mass measurement, search for offshell decays and for high mass scalars", *these proceedings*.

[8] ATLAS and CMS Collaborations, "Combined Measurement of the Higgs Boson Mass in *pp* Collisions at $\sqrt{s}$ = 7 and 8 TeV with the ATLAS and CMS Experiments," Phys. Rev. Lett. **114** (2015) 191803.

[9] CMS Collaboration, "Evidence for the 125 GeV Higgs boson decaying to a pair of τ leptons," JHEP **05** (2014) 104 .

[10] CMS Collaboration, "Search for the standard model Higgs boson produced in association with a W or a Z boson and decaying to bottom quarks", Phys. Rev. D **89** (2014) 012003.

[11] CMS Collaboration, "Precise determination of the mass of the Higgs boson and tests of compatibility of its couplings with the standard model predictions using proton collisions at 7 and 8 TeV", Eur.Phys.J. C **75** (2015) 212.

[12] CMS Collaboration, "Evidence for the direct decay of the 125 GeV Higgs boson to fermions", Nature Phys. **10** (2014) 557.

[13] CMS Collaboration, "Search for the standard model Higgs boson produced through vector boson fusion and decaying to bb", Phys. Rev. D **92** (2015) 032008.

[14] CMS Collaboration, "Search for the associated production of the Higgs boson with a top-quark pair", JHEP **09** (2014) 087.

[15] CMS Collaboration, "Search for a standard model Higgs boson produced in association with a top-quark pair and decaying to bottom quarks using a matrix element method", Eur.Phys.J. C **75** (2015) 251.

[16] CMS Collaboration, "Measurement of differential cross sections for Higgs boson production in the diphoton decay channel in pp collisions at $\sqrt{s} = 8$ TeV", arXiv:1508.07819 (2015), submitted to Eur.Phys.J. C.

[17] LHC Higgs Cross Section Working Group, "Handbook of LHC Higgs cross sections: 3. Higgs Properties", CERN Report CERN-2013-004 (2013), arXiv:1307.1347.

[18] CMS Collaboration, "Search for neutral MSSM Higgs bosons decaying into a pair of bottom quarks", JHEP **11** (2015) 071.

[19] CMS Collaboration, "Search for neutral MSSM Higgs bosons decaying to $\mu^+\mu^-$ in pp collisions at $\sqrt{s} = 7$ and 8 TeV", Phys.Lett.B **752** (2016) 221.

[20] M. Carena et al., "MSSM Higgs boson searches at the LHC: benchmark scenarios after the discovery of a Higgs-like particle", Eur.Phys.J. C **73** (2013) 2552.

[21] CMS Collaboration, "Search for neutral MSSM Higgs bosons decaying to a pair of tau leptons in pp collisions", JHEP **10** (2014) 160.

[22] CMS Collaboration, "Search for additional neutral Higgs bosons decaying to a pair of tau leptons in pp collisions at $\sqrt{s} = 7$ and 8 TeV", CMS-PAS-HIG-14-029 (2015).

[23] ALEPH, DELPHI, L3 and OPAL Collaborations, LEP Working Group for Higgs Boson Searches Collaboration, "Search for neutral MSSM Higgs bosons at LEP", Eur. Phys. J. C **47** (2006) 547.

[24] O. Stål, G. Weiglein, "Light NMSSM Higgs bosons in SUSY cascade decays at the LHC", JHEP **01** (2012) 071.

[25] CMS Collaboration, "Search for a light NMSSM Higgs boson produced in supersymmetric cascades and decaying into a b-quark pair", CMS-PAS-HIG-14-030 (2015).

EPJ Web of Conferences **126**, 02020 (2016) DOI: 10.1051/epjconf/201612602020
ICNFP 2015

Supersymmetry, Cosmological Constant and Inflation: Towards a fundamental cosmic picture via "running vacuum"

Nick E. Mavromatos[1,2,a]

[1] *Theoretical Particle Physics and Cosmology Group, Department of Physics, King's College London, Strand, London WC2R 2LS, UK*
[2] *Theory Division, Physics Department, CERN CH 1211 Geneva 23, Switzerland*

Abstract. On the occasion of a century from the proposal of General relativity by Einstein, I attempt to tackle some open issues in modern cosmology, via a toy but non-trivial model. Specifically, I would like to link together: (i) the smallness of the cosmological constant today, (ii) the evolution of the universe from an inflationary era after the big-bang till now, and (iii) local supersymmetry in the gravitational sector (supergravity) with a broken spectrum at early eras, by making use of the concept of the "running vacuum" in the context of a simple toy model of four-dimensional N=1 supergravity. The model is characterised by dynamically broken local supersymmetry, induced by the formation of gravitino condensates in the early universe. As I will argue, there is a Starobinsky-type inflationary era characterising the broken supersymmetry phase in this model, which is compatible with the current cosmological data, provided a given constraint is satisfied among some tree-level parameters of the model and the renormalised cosmological constant of the de Sitter background used in the analysis. Applying the "running vacuum" concept, then, to the effective field theory at the exit of inflation, makes a smooth connection (in cosmic time) with the radiation dominance epoch and subsequently with the current era of the Universe, characterised by a small (but dominant) cosmological-constant contribution to the cosmic energy density. In this approach, the smallness of the cosmological constant today is attributed to the failure (due to quantum gravity non-perturbative effects) of the aforementioned constraint.

1 Introduction and Summary

The year 2015 celebrates a century since the proposal by Einstein for the theory of General Relativity (GR), an elegant and quite successful (from a phenomenological point of view) classical field theory of Gravitation, generalising Newtonian gravity. GR pioneered the idea of linking the geometry of the cosmos with its dynamics, by attributing the curvature of space time to mass/energy of matter/radiation. The theory has two versions, *local* - describing the dynamics of space-time in the neighbourhood of bodies, including celestial objects, and *global*, dealing with the entire Universe as a whole, the latter being described today by the celebrated Friedmann-Lemaitre-Robertson-Walker (FLRW) model for Cosmology. The FLRW model describes quite successfully the plethora of the currently available data on Cosmology in the framework of the so-called ΛC(old)D(ark)M(atter)

[a] e-mail: Nikolaos.Mavromatos@kcl.ac.uk

(ΛCDM) model. Yet, despite these successes, there are a number of unanswered fundamental questions in both versions of GR, concerning: the quantisation of the theory, the microscopic nature of black holes and Hawking radiation, an explanation of the smallness of the (observed) cosmological constant today, a detailed microscopic theory of inflation and its graceful exit, as well as the nature of dark matter and dark energy of the Universe.

On the other side of fundamental physics, particle theory, there are also great successes but also great mysteries which still remain open. The Standard Model (SM) of Electroweak and Strong interactions, is an elegant quantum field theory that unifies three of the four forces of nature quite successfully from an experimental point of view, but leaves gravity completely out of its reach. Moreover, there can be no explanation within the SM context of the matter-over-antimatter dominance in the universe today [1]. In addition, the existence of dark matter and dark energy poses a pressing need for an extension of SM to accommodate the dark sector of the universe, *e.g.* by incorporating supersymmetry (SUSY) [2, 3], which could provide natural candidates for Dark Matter. Current searches in colliders at present, however, are not providing concrete evidence for such a framework, although it must be said that this may not be unnatural, given that the scale of supersymmetry breaking - depending on the SUSY model - may be out of reach of the current energies. Other non-trivial extensions of SM, include extra dimensional theories, as well as the whole string/brane framework, which might provide a consistent quantum framework for the unification of gravitational with the rest of the fundamental interactions in nature. In the latter framework, supersymmetry plays a crucial rôle for the stability of the vacuum in the majority of models, although there may be ways round it. Yet, no experimental evidence for the existence of strings or extra dimensions exists today.

Motivated by the lack of concrete evidence for low energy extensions of the SM, but attributing an important rôle to the supersymmetrisation of the world, not so from the stability of the vacuum view point, but rather from the perspective that supersymmetry is the maximal space-time symmetry allowed by the Coleman-Mandula theorem [4], and thus worthy of further investigation, we would like to dedicate this talk to a study of a toy model of the Universe, which could provide a non-trivial rôle of supersymmetry in ensuring inflation, graceful exit from it and subsequently connection with the FLRW cosmology until the current era, characterised by a small cosmological constant.

The inflationary paradigm [5] offers an elegant solution to the horizon and flatness problems of the standard Big Bang cosmology, whilst simultaneously seeds both the large-scale structure of the universe and temperature anisotropies of the CMB via quantum fluctuations occurring during the inflationary epoch. Thus it seems a desirable concept to keep it in any successful cosmological model for the universe evolution. The precise microscopic mechanism of inflation is however unknown at present. The current data are in agreement with a scalar field (inflaton) (or fields) with canonical kinetic term(s) slowly rolling down an almost flat potential in the context of Einstein gravity, generating in the process 50 - 60 e-folds of inflation, along with adiabatic, nearly scale invariant primordial density perturbations [6, 7]. From the best fit value of the running spectral index $n_s \sim 0.96$ for the gravitational perturbations in the slow-roll parametrisation, found by Planck [6], and the usual relations among the slow-roll inflationary parameters [7] $n_s = 1 - 6\epsilon + 2\eta\,, r = 16\epsilon$, one may deduce $r \leq 0.11$, given the non-observation of primordial gravitational wave-like (transverse and traceless) perturbations by Planck or WMAP collaborations (that is the absence of B-mode polarisations). This observational fact implies that the energy scale E_I of inflation is much smaller than the Planck scale m_P, in particular it lies in the ballpark of the scale of Grand Unification (GUT) [6, 7]: $E_I = \left(3\,H_I^2 M_{\rm Pl}^2\right)^{1/4} \simeq 2.1\times 10^{16}\times\left(\frac{r}{0.20}\right)^{1/4}$ GeV, with $M_{\rm Pl} = 2.4\times 10^{18}$ GeV the reduced Planck mass, H_I the Hubble scale during inflation and r the tensor-to-scalar perturbation ratio [7]. The upper bound on $r < 0.11$ placed by the Planck Collaboration [6] implies

$$H_I \leq 0.81\times 10^{14}\ \text{GeV} \leqslant 3.38\times 10^{-5}\, M_{\rm Pl}\,, \qquad (1)$$

which is five orders of magnitudes smaller than the reduced Planck mass. This hierarchy between inflation scale and Planck scale, already prompted researchers to seek for a natural reason for it, and already in the early days after the inflation paradigm was at play, links of the low scale of inflation with supersymmetry have been made [8].

If supersymmetry is realised in nature however, it is certainly broken, with partners masses above TeV scale, given that they have not been discovered at the large hadron collider at present. Simple realisations of global supersymmetry (SUSY) breaking, such as in the Wess-Zumino model [9], can provide, when embedded in gravitational environments, slow-roll models for inflation consistent with the Planck data [6]. Rigorous embeddings of global SUSY to local supersymmetry (SUGRA) have also been considered and explored in the literature over the years in connection with various scenarios for inflation [10], such as hybrid [11], chaotic [12], no-scale SUGRA [13], Jordan-frame-SUGRA [14] and others [15]. In the case of *some* no-scale SUGRA models, the inflationary potential can be made similar to that of the so-called Starobinsky model of inflation [16], by a specific choice of parameters. Unlike the SUGRA case, however, in the original Starobinsky model, the inflaton scalar mode represents collectively the effects of the quadratic-order scalar curvature terms (R^2) in the effective gravitational action, obtained after integrating out (conformal) matter in the early hot Universe [16]. Starobinsky-type models seem to be preferred by the Planck data [6].

In this talk I shall review a rather minimal inflationary scenario which is associated indirectly with a Starobinsky type inflation in its original sense [16]. Specifically, I will discuss the appearance of scalar inflationary modes associated with higher curvature terms of the (one-loop) effective action of simple (3+1)-dimensional $\mathcal{N} = 1$ SUGRA models without matter [17, 18], obtained by integrating out massive gravitino degrees of freedom in the broken supergavity phase. The supersymmetry breaking occurs dynamically via the formation of gravitino condensates, as a consequence of the four-gravitino interactions that characterise (any) supergravity action, via the fermionic torsion parts of the spin connection. The approach is documented in a series of previous publications [19–21] and will be reviewed briefly here. We should stress that our approach does not supersymmetrise higher curvature terms in the action to obtain Starobinksy-like effective actions, as *e.g.* in the case of [22]. For us, the higher curvature terms emerge, as we mentioned, as a result of quantum loop corrections in the broken-local-supersymmetry phase of the $\mathcal{N} = 1$ simple SUGRA action, after path-integrating out massive gravitino fields and graviton fluctuations in a de Sitter background (where such integration becomes meaningful [23]). The dynamical breaking process may be concretely realised by means of a phase transition from the supersymmetric phase where the bilinear $\langle\overline{\psi}_\mu\psi^\mu\rangle$ representing the effective scalar degree of freedom has zero vacuum expectation value, to one where $\sigma \equiv \langle\overline{\psi}_\mu\psi^\mu\rangle \neq 0$. The quantum excitations about this condensate vacuum are then identified with a gravitino condensate scalar field. Since this must be an energetically favourable process to occur, it then follows that the effective potential experienced by the gravitino condensate must be locally concave about the origin. The corresponding one-loop effective potential of the gravitino condensate scalar field, therefore, has the characteristic form of a Coleman-Weinberg double well potential, offering the possibility of hilltop-type inflation, with the condensate field playing the role of the inflaton [19, 20], which would be the simplest scenario. However, in order to guarantee a slow-roll inflation one needs unnaturally large values of the gravitino-condensate wave function renormalisation, unless transplanckian scales for supersymmetry breaking are invoked. This prompted us to discuss a second [21], rather indirect way, by means of which the gravitino condensate field is associated with inflation. This is realised via the higher (in particular quadratic-order) curvature corrections of the (quantum) effective action of the gravitino condensate field, obtained after path-integration of graviton and gravitino degrees of freedom in the massive gravitino phase. These corrections induce a Starobinsky-type inflation [16], which occurs for quite natural values of the parameters of the $\mathcal{N} = 1$ SUGRA model (actually, its

Jordan-frame variants [14]), with the inflation scale and gravitino mass in the ball park of the GUT scale.

The structure of the talk is as follows: In Section 2 we review the formalism of the dynamical breaking of SUGRA in four dimensional $\mathcal{N} = 1$ models, including superconformal extensions thereof (with broken conformal symmetry) that are necessitated for phenomenological reasons, as we shall see. In Section 3 we discuss scenarios for inflation of Starobinsky type that may occur in the massive gravitino phase, near the non-trivial minimum of the effective potential. In such scenarios, which are compatible with the Planck satellite data [6], the rôle of the inflaton field is played by the scalar mode that describes collectively the effects of scalar-curvature-squared terms that characterise the gravitational sector of the one-loop effective action in the broken SUGRA phase, after integrating out the massive gravitinos. In section 4 we apply [24] the running vacuum concept [25] in the effective field theory at the exit of the Starobinsky inflation, in order to ensure a smooth (in cosmic time) connection of that era with radiation- and matter-dominance eras of the Universe, and subsequently with the current era characterised by a small but dominant cosmological-constant type contribution to the cosmic energy density. Finally, conclusions and outlook are presented in section 5.

2 Dynamical breaking of $\mathcal{N} = 1$ $D = 4$ SUGRA models

This section is based on material presented in refs. [20]. We consider the (on-shell) action for 'minimal' $\mathcal{N} = 1$ $D = 4$ supergravity in the second order formalism [17, 18]:

$$S_{\rm SG} = \int d^4x\, e\left(\frac{1}{2\kappa^2}R\,(e) - \overline{\psi}_\mu\gamma^{\mu\nu\rho}D_\nu\psi_\rho + \mathcal{L}_{\rm torsion}\right),\ \gamma^{\mu\nu\rho} = \frac{1}{2}\left\{\gamma^\mu,\gamma^{\nu\rho}\right\},\ \gamma^{\nu\rho} = \frac{1}{2}\left[\gamma^\nu,\gamma^\rho\right], \tag{2}$$

where $\kappa^2 = 8\pi G = 1/M_{\rm Pl}^2$, in units $\hbar = c = 1$, $M_{\rm Pl}$ is the reduced Planck mass in four space-time dimensions, e is the vierbein determinant and the scalar space-time curvature $R(e)$ and gravitational covariant derivative $D_\nu\psi_\rho \equiv \partial_\nu\psi_\rho + \frac{1}{4}\omega_{\nu ab}\,(e)\,\gamma^{ab}\psi_\rho$ are defined via the torsion-free connection [1]. In the gauge condition

$$\gamma_\mu\,\psi^\mu = 0\,, \tag{3}$$

we shall work with in this article, the torsion term $\mathcal{L}_{\rm torsion}$, arising from the fermionic torsion parts of the spin connection, reads:

$$\mathcal{L}_{\rm torsion} = -\frac{1}{8}\kappa^2\left(\left(\overline{\psi}^\rho\gamma^\mu\psi^\nu\right)\left(\overline{\psi}_\rho\gamma_\mu\psi_\nu + 2\overline{\psi}_\rho\gamma_\nu\psi_\mu\right)\right). \tag{4}$$

We notice at this stage that the Fierz identities among the gravitinos, make the actual coefficients of these four-fermion terms ambiguous. This is a known ambiguity in the context of mean field theory, and only the non-perturbative physics can settle the value of the four-fermion terms [26]. Thus we may extend the definition of the torsion four-fermion interaction terms to include such ambiguities and write:

$$\mathcal{L}_{\rm torsion} = \lambda_{\rm S}\left(\overline{\psi}^\rho\psi_\rho\right)^2 + \lambda_{\rm PS}\left(\overline{\psi}^\rho\gamma^5\psi_\rho\right)^2 + \lambda_{\rm PV}\left(\overline{\psi}^\rho\gamma^5\gamma_\mu\psi_\rho\right)^2 \tag{5}$$

where the couplings $\lambda_{\rm S}$, $\lambda_{\rm PS}$ and $\lambda_{\rm PV}$ express the freedom we have to rewrite each quadrilinear in terms of the others via Fierz transformation. They satisfy the constraint: $\lambda_{\rm S} - \lambda_{\rm PS} + 4\lambda_{\rm PV} = -3\kappa^2/4$. This freedom was addressed in the second reference in [20], where we refer the reader for details.

[1] Our metric signature is $(-,+,+,+)$ and the definitions of the Ricci and Riemann curvature tensors are $R_{\mu\nu} = R^\lambda_{\ \mu\lambda\nu}$ and $R^\lambda_{\ \mu\nu\rho} = \partial_\nu\Gamma^\lambda_{\ \mu\rho} - \ldots$, respectively, with $\Gamma^\lambda_{\ \mu\nu} = \Gamma^\lambda_{\ \nu\mu}$ the torsion-free Christofel symbol, and the Ricci scalar is given by $R = g^{\mu\nu}R_{\mu\nu}$.

Following the original ideas of dynamical symmetry breaking by Nambu and Jona-Lasinio [27], we wish to linearise these four-fermion interactions via suitable auxiliary fields, *e.g.*

$$\frac{1}{4}\left(\overline{\psi}^{\rho}\psi_{\rho}\right)^{2} \sim \sigma\left(\overline{\psi}^{\rho}\psi_{\rho}\right) - \sigma^{2},$$

where the equivalence (at the level of the action) follows as a consequence of the subsequent Euler-Lagrange equation for the auxiliary scalar σ. Our task is then to look for a non-zero vacuum expectation value $\langle\sigma\rangle$ which would serve as an effective mass for the gravitino. To induce the super-Higgs effect [28] we also couple in the Goldstino, associated to global supersymmetry breaking, via [3]

$$\mathcal{L}_{\lambda} = f^{2}\det\left(\delta_{\mu\nu} + \frac{i}{2f^{2}}\overline{\lambda}\gamma_{\mu}\partial_{\nu}\lambda\right)\Bigg|_{\gamma\cdot\psi=0} = f^{2} + \dots, \tag{6}$$

where λ is the Goldstino, $\sqrt{f}$ expresses the scale of global supersymmetry breaking, and ... represents higher order terms which may be neglected in our weak-field expansion of the determinant. It is worth emphasising at this point the universality of (6); any model containing a Goldstino may be related to $\mathcal{L}_{\lambda}$ via a non-linear transformation [29], and thus the generality of our approach is preserved.

Upon the specific gauge choice (3) for the gravitino field and an appropriate redefinition, one may eliminate any presence of the Goldstino field from the final effective action describing the dynamical breaking of local supersymmetry, except the cosmological constant term f^2 in (6), which serves as a reminder of the pertinent scale of supersymmetry breaking. The non-trivial energy scale this introduces, along with the disappearance (through field redefinitions) of the Goldstino field from the physical spectrum and the concomitant development of a gravitino mass, characterises the super-Higgs effect [28]. We may then identify in the broken phase an effective action

$$S = \frac{1}{2\kappa^{2}}\int d^{4}x\, e\Big((R(e) - 2\Lambda) - \overline{\psi}_{\mu}\gamma^{\mu\nu\rho}D_{\nu}\psi_{\rho} + m_{\rm dyn}\left(\overline{\psi}_{\mu}\psi^{\mu}\right)\Big), \tag{7}$$

where Λ is the renormalised cosmological constant, to be contrasted with the (negative) tree level cosmological constant

$$\Lambda_{0} \equiv \kappa^{2}\left(\sigma^{2} - f^{2}\right), \tag{8}$$

and $m_{\rm dyn} \propto \langle\sigma\rangle$ is a dynamically generated gravitino mass, the origin of which will be explained presently. It is worth stressing at this point that Λ_0 must be negative due to the incompatibility of supergravity with de Sitter vacua; if SUGRA is broken at tree level, then of course no further dynamical breaking may take place.

For phenomenological reasons which have been outlined in detail in refs. [19, 20], and we shall discuss below, we adopt an extension of $\mathcal{N} = 1$ SUGRA which incorporates local supersymmetry in the Jordan frame, enabled by an associated dilaton superfield [14]. The scalar component φ of the latter can be either a fundamental space-time scalar mode of the gravitational multiplet, i.e. the trace of the graviton (as happens, for instance, in supergravity models that appear in the low-energy limit of string theories), or a composite scalar field constructed out of matter multiplets. In the latter case these could include the standard model fields and their superpartners that characterise the Next-to-Minimal Supersymmetric Standard Model, which can be consistently incorporated in such Jordan frame extensions of SUGRA [14, 30], leading to supersymmetric generalisations of the Higgs inflation [31]. Upon appropriate breaking of conformal symmetry, induced by specific dilaton potentials (which we do not discuss here), one may assume that the dilaton field acquires a non-trivial vacuum expectation value $\langle\varphi\rangle \neq 0$. One consequence of this is then that in the broken conformal symmetry phase, the resulting supergravity sector, upon passing (via appropriate field redefinitions) to the Einstein frame is

DOI: 10.1051/epjconf/201612602020

described by an action of the form (2), but with the coupling of the gravitino four-fermion interaction terms being replaced by

$$\tilde{\kappa} \equiv e^{-\langle\varphi\rangle}\kappa\,, \tag{9}$$

while the Einstein term in the action carries the standard gravitational coupling $1/2\kappa^2$. In this way the v.e.v. of the dilaton may be combined with the Fierz-induced ambiguities (5), so that the non-perturbative physics that determines the latter also determines the scale of the conformal symmetry breaking. As in our approach we do not discuss details of the dilaton potential, we shall consider from now on such a v.e.v., as well the coefficient λ_S of the four fravitino terms in (5) leading to the formation of scalar gravitino condensates, of interest to us here, as phenomenological. Expanding

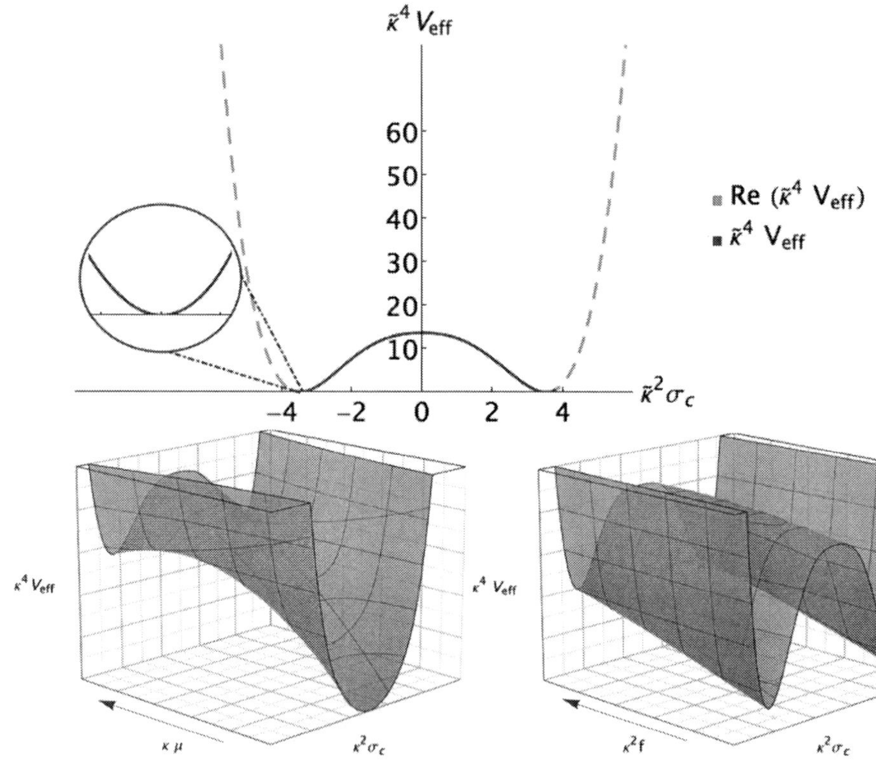

Figure 1. *Upper panel*: The one-loop effective potential (10), expressed in units of the coupling $\tilde{\kappa}$ (9). *Lower panel*: As above, but showing schematically the effect of tuning the RG scale μ and the supersymmetry breaking scale f, whilst holding, respectively, f and μ fixed. The arrows in the respective axes correspond to the direction of increasing μ and f. Pictures taken from ref. [20].

the graviton field about a de Sitter background [23] (under the assumption that it is a solution of the one-loop effective equations) with renormalised cosmological constant $\Lambda > 0$, and integrating out both bosonic and fermionic quantum fluctuations to one loop yields the following effective potential for the gravitino condensate field σ in the flat space-time limit $\Lambda \to 0$, as detailed in ref. [20],

$$V_{\rm eff} = V_B^{(0)} + V_B^{(1)} + V_F^{(1)} = -\frac{\Lambda_0}{\kappa^2} + V_B^{(1)} + V_F^{(1)}\,, \tag{10}$$

where Λ_0 is given by (8) and

$$V_B^{(1)} = \frac{45\kappa^4}{512\pi^2}\left(f^2-\sigma^2\right)^2\left(3-2\ln\left(\frac{3\kappa^2(f^2-\sigma^2)}{2\mu^2}\right)\right),$$
$$V_F^{(1)} = \left(\frac{\tilde{\kappa}}{\kappa}\right)^4\frac{\kappa^4\sigma^4}{30976\pi^2}\left(30578\ln\left(\left(\frac{\tilde{\kappa}}{\kappa}\right)^2\frac{\kappa^2\sigma^2}{3\mu^2}\right)-45867+29282\ln\left(\frac{33}{2}\right)+1296\ln\left(\frac{54}{11}\right)\right), \quad (11)$$

indicate the contributions to the effective potential from bosonic and fermionic fields respectively, and μ is an inverse renormalisation group (RG) scale, associated with a proper time short distance cut-off [20], such that the flow from the ultraviolet to the infrared limit corresponds to the direction of an increasing μ. The one-loop effective potential (10) is depicted in fig. 1 and it has the characteristic double-well shape of dynamical breaking of a symmetry. We note that as we flow from UV to IR (*i.e.* in the direction of increasing μ), we obtain the correct double-well shape required for the super-Higgs effect, and secondly that tuning f allows us to shift $V_{\rm eff}$ and thus attain the correct vacuum structure (*i.e.* non-trivial minima σ_c such that $V_{\rm eff}(\sigma_c) = 0$). Moreover, the shape of the effective potential changes, as one varies the (renormalisation) scale μ from ultraviolet to infrared values (*i.e.* flowing in the direction of increasing μ), in such a way that the broken symmetry phase (double-well shaped potential) is reached in the IR. This indicates that the dynamical generation of a gravitino mass is actually an IR phenomenon, in accordance with the rather general features of dynamical mass in field theory. In the broken phase, the mass of the gravitino condensate is then given by $m_\sigma^2 \equiv V''_{\rm eff}(\sigma_c)$, where σ_c is the minimum of $V_{\rm eff}$ and a prime denotes a functional derivative with respect to the gravitino-condensate field. As observed from (11), the bosonic contributions to the effective potential contain logarithmic terms which would contribute imaginary terms, leading to instabilities, unless

$$\sigma_c^2 < f^2 . \quad (12)$$

From (10) it is straightforward to see that this condition is equivalent to the negativity of the tree-level cosmological constant Λ_0, which is entirely sensible; if $\Lambda_0 > 0$ then SUGRA is broken at tree level (given the incompatibility of supersymmetry with de Sitter vacua) and there can be no dynamical breaking. As such, we must then tune f for a given value of μ to find self consistent minima σ_c satisfying (12), thereby ensuring a real $V_{\rm eff}$. In fact, here lies the importance of the super-Higgs effect, and thus of a non-zero positive $f^2 > \sigma_c^2 > 0$, in allowing dynamical breaking of local supersymmetry [2].

As discussed in refs. [19, 20], phenomenologically realistic situations, where one avoids trans-planckian gravitino masses, for supersymmetry breaking scales $\sqrt{f}$ at most of order of the Grand Unification (GUT) scale 10^{15-16} GeV, as expected from arguments related to the stability of the electroweak vacuum, can occur only for large $\tilde{\kappa}$ couplings, typically of order $\tilde{\kappa} \sim \left(10^3 - 10^4\right)\kappa$. Given the relation (9) this corresponds to dilaton vev of $\mathcal{O}(-10)$, where the negative sign may be familiar in the context of dilaton-influenced cosmological scenarios [33].

If we consider for concreteness the case $\tilde{\kappa} = 10^3\kappa$, which is a value dictated by the inflationary phenomenology of the model [19], we may find solutions with a vanishing one-loop effective potential at the non-trivial minima corresponding to: $\tilde{\kappa}^2\sigma_c \simeq 3.5$, $\tilde{\kappa}^2 f \simeq 3.7$, $\tilde{\kappa}\mu \simeq 4.0$, which leads to a global supersymmetry breaking scale

$$\sqrt{f} \simeq 4.7 \times 10^{15}\ \text{GeV}, \quad (13)$$

[2] In refs. [32], the importance of the super-Higgs effect was ignored, which led to the incorrect conclusion that imaginary parts exist necessarily in the one-loop effective potential, in the same class of gauges as in ref. [20] and here, and hence dynamical breaking of SUGRA was not possible. As we have seen above, such imaginary parts are absent when the condition (12) is satisfied, and thus dynamical breaking of SUGRA occurs.

and dynamical gravitino mass

$$m_{\rm dyn} \simeq 2.0 \times 10^{16}\ \text{GeV}\,. \tag{14}$$

At the non-trivial minima we find $\tilde{\kappa}^4 V_F^{(1)} \simeq -1.4$, $\tilde{\kappa}^4 V_B^{(1)} \simeq 5.9\times 10^{-13}$, with tree-level cosmological constant $\tilde{\kappa}^2 \Lambda_0 \simeq -1.4$. We thus observe that fermion contributions to the effective potential are much stronger than the corresponding bosonic contributions for the cases of large couplings $\tilde{\kappa} \gg \kappa$. These values are phenomenologically realistic, thereby pointing towards the viability (from the point of view of producing realistic results of relevance to phenomenology) of the scenarios of dynamical breaking of local supersymmetry in conformal supergravity models [3]. On the other hand, in standard SUGRA scenarios, where $\tilde{\kappa} = \kappa$, one finds, as already mentioned, transplanckian values for the dynamically generated gravitino mass [20]: $m_{\rm dyn} \simeq 2.0 \times 10^{19}$ GeV, and a global supersymmetry breaking scale $\sqrt{f} \simeq 4.7 \times 10^{18}$ GeV, far too high to make phenomenological sense.

3 Starobinsky-type inflation in the broken SUGRA phase

Starobinsky inflation is a model for obtaining a de Sitter (inflationary) cosmological solution to gravitational equations arising from a (four space-time-dimensional) action that includes higher curvature terms. Specifically, an action of the type in which the quadratic curvature corrections consist only of scalar curvature terms [16]

$$\mathcal{S} = \frac{1}{2\kappa^2} \int d^4x \sqrt{-g}\left(R + \beta R^2\right)\,, \quad \beta = \frac{8\pi}{3\mathcal{M}^2}\,, \tag{15}$$

where $\mathcal{M}$ is a constant of mass dimension one, characteristic of the model, which cannot be determined by theoretical considerations, and hence should be considered phenomenological.

The important feature of this model is that inflationary dynamics is driven purely by the gravitational sector, through the R^2 terms, and that the scale of inflation is linked to $\mathcal{M}$. From a microscopic point of view, the scalar curvature-squared terms in (15) are viewed as the result of quantum fluctuations (at one-loop level) of conformal (massless or high energy) matter fields of various spins, which have been integrated out in the relevant path integral in a curved background space-time [34].

The above considerations necessitate truncation to one-loop quantum order and to curvature-square (four-derivative) terms, which implies that there must be a region of validity for curvature invariants such that $O(R^2/M_{\rm Pl}^4) \ll 1$. This is of course a condition satisfied in phenomenologically realistic scenarios of inflation [6, 7], for which the inflationary Hubble scale H_I satisfies (1) (the reader should recall that $R = 12 H_I^2$ in the inflationary phase).

Although the inflation in this model is not driven by rolling scalar fields, nevertheless the model (15) (and for that matter, any other model where the Einstein-Hilbert space-time Lagrangian density

[3]A comment concerning SUGRA models in the Jordan frame with such large values for their frame functions is in order here. In our approach, the dilaton φ could be a genuine (dimensionless) dilation scalar field $\varphi = 2\phi$ arising in the gravitational multiplet of strings, whose low-energy limit may be identified with some form of SUGRA action. In our normalization the string coupling would be $g_s \equiv e^{\phi} = \tilde{\kappa}^{-1/2}$. In such a case, a value of $\tilde{\kappa} = e^{-\langle\varphi\rangle}\kappa = O(10^{3-4})$ would imply a large negative v.e.v. of the (four-dimensional) dilaton field of order $\langle\phi\rangle = -O(5) < 0$, and thus a weak string coupling squared $g_s = O(10^{-2})$, which may not be far from values attained in phenomenologically realistic string cosmologies [33]. On the other hand, in the Jordan-frame SUGRA models of [14], the frame function reads $\Phi \equiv e^{-\varphi} = 1 - \frac{1}{3}\left(S\overline{S} + \sum_{u,d} H_i H_i^{\dagger}\right) - \frac{1}{2}\chi\left(-H_u^0 H_d^0 + H_u^+ H_d^- + \text{h.c.}\right)$, in the notation of [30] for the various matter super fields of the next-to-minimal supersymmetric standard model that can be embedded in such supergravities. The quantity χ is a constant parameter. At energy scales much lower than GUT, it is expected that the various fields take on subplanckian values, in which case the frame function is almost one, and hence $\tilde{\kappa} \simeq \kappa$ for such models today. To ensure $\tilde{\kappa} \gg \kappa$, and thus large values of the frame function, $\Phi \gg 1$, as required in our analysis, one needs to invoke trasnplanckian values for some of the fields, $H^0_{u,d}$, and large values of χ, which may indeed characterize the inflationary phase of such theories. A similar situation occurs for the values of the higgs field (playing the role of the inflaton) in the non-supersymmetric Higgs inflation models [31].

is replaced by an arbitrary function $f(R)$ of the scalar curvature) is conformally equivalent to that of an ordinary Einstein-gravity coupled to a scalar field with a potential that drives inflation [35]. This can be seen by first linearising the R^2 terms in (15) by means of an auxiliary (Lagrange-multiplier) field $\tilde{\alpha}(x)$, and then rescaling the metric by a conformal transformation and redefining the scalar field, so that the final theory acquires canonically-normalised Einstein and scalar-field terms:

$$g_{\mu\nu} \to g^E_{\mu\nu} = (1 + 2\beta\,\tilde{\alpha}(x))\, g_{\mu\nu}\,, \quad \tilde{\alpha}\,(x) \to \rho(x) \equiv \sqrt{\frac{3}{2}}\,\ln\,(1 + 2\beta\,\tilde{\alpha}\,(x))\ . \tag{16}$$

These steps may be understood schematically via

$$\int d^4x\, \sqrt{-g}\, \left(R + \beta R^2\right) \hookrightarrow \int d^4x\, \sqrt{-g^E}\, \left(R^E + g^{E\mu\nu}\,\partial_\mu \rho\, \partial_\nu \rho - V(\rho)\right)\,, \tag{17}$$

where the arrows have the meaning that the corresponding actions appear in the appropriate path integrals, with the potential $V(\rho)$ given by:

$$V(\rho) = \frac{\left(1 - e^{-\sqrt{\frac{2}{3}}\rho}\right)^2}{4\beta} = \frac{3\mathcal{M}^2\left(1 - e^{-\sqrt{\frac{2}{3}}\rho}\right)^2}{32\,\pi}\ . \tag{18}$$

This potential is sufficiently flat for large values of ρ (compared to the Planck scale) to produce phenomenologically acceptable inflation, with the scalar field ρ playing the role of the inflaton. In fact, the Starobinsky model, with its characteristically low value of the tensor-to-scalar ratio r, provides an excellent fit to the recent Planck data on inflation [6].

Below we consider an extension [21] of the dynamically broken SUGRA analysis of the previous section to the case where the de Sitter parameter Λ is perturbatively small compared to m_P^2, but not zero, so that truncation of the series to order Λ^2 suffices. This is in the spirit of the original Starobinsky model [16], with the rôle of matter fulfilled by the now-massive gravitino field. Specifically, we are interested in the behaviour of the effective potential near the non-trivial minimum, where σ_c is a non-zero constant. It is important to notice at this point that, in contrast to the original Starobinsky model [16], where the crucial for inflation R^2 terms have been argued to arise from the *conformal anomaly* in the path integral of massless (conformal) matter in a de Sitter background, and thus their coefficient was arbitrary, in our scenario, such terms arise in the one-loop effective action of the gravitino condensate field, evaluated in a de Sitter background, after integrating out massive gravitino fields, whose mass was generated dynamically. The order of the de Sitter cosmological constant, $\Lambda > 0$ that breaks supersymmetry, and the gravitino mass are all evaluated dynamically (self-consistently) in our approach from the minimization of the effective potential. Thus, the resulting R^2 coefficient, which determines the phenomenology of the inflationary phase, is calculable [21]. Moreover, in our analysis, unlike Starobinsky's original work, we will keep the contributions from both graviton (spin-two) and gravitino quantum fluctuations. Specifically, we are interested in the behaviour of the effective potential near the non-trivial minimum, where σ_c is a non-zero constant (*cf.* fig. 2).

The one-loop effective potential, obtained by integrating out gravitons and (massive) gravitino fields in the scalar channel (after appropriate euclideanisation), may be expressed as a power series in the renormalised cosmological constant Λ [21]:

$$\Gamma \simeq S_{\rm cl} - \frac{24\pi^2}{\Lambda^2}\left(\alpha_0^F + \alpha_0^B + \left(\alpha_1^F + \alpha_1^B\right)\Lambda + \left(\alpha_2^F + \alpha_2^B\right)\Lambda^2 + \ldots\right), \tag{19}$$

where $S_{\rm cl} = -\frac{1}{2\kappa^2}\int d^4x\,\sqrt{g}\left(\widehat{R} - 2\Lambda_0\right)$ is the classical action with Λ_0 given by (8) and $\widehat{R}$ denoting the fixed S^4 background we expand around ($\widehat{R} = 4\Lambda$, Volume $= 24\pi^2/\Lambda^2$); the $\alpha_i^{B(F)}$, $i = 1, 2$ in

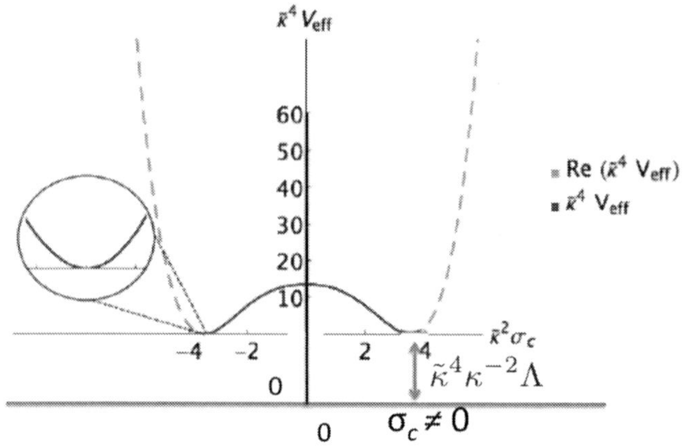

Figure 2. Generic shape of the one-loop effective potential for the gravitino condensate field σ_c in dynamically broken (conformal) Supergravity models in the presence of a non-trivial de Sitter background with cosmological constant $\Lambda > 0$ [21]. The Starobinsky inflationary phase is associated with fluctuations of the condensate and gravitational field modes near the non-trivial minimum of the potential, where the condensate $\sigma_c \neq 0$, and the potential assumes the value $\Lambda > 0$, consistent with supersymmetry breaking. The dashed green lines denote "forbidden" areas of the condensate field values, violating the condition (12), for which imaginary parts appear in the effective potential, thereby destabilising the broken symmetry phase.

(19) indicate the bosonic (B) and fermionic (F) quantum corrections at a given order ($i = 1, 2$) in Λ, respectively. The leading order term in Λ is then the effective action found in [20] in the limit $\Lambda \to 0$,

$$\begin{aligned}
\Gamma_{\Lambda\to 0} &\simeq -\frac{24\pi^2}{\Lambda^2}\left(-\frac{\Lambda_0}{\kappa^2} + \alpha_0^F + \alpha_0^B\right) \equiv \frac{24\pi^2}{\Lambda^2}\frac{\Lambda_1}{\kappa^2}, \\
\alpha_0^F &= \tilde{\kappa}^4\,\sigma_c^4\left(0.100\,\ln\left(\frac{\tilde{\kappa}^2\,\sigma_c^2}{3\mu^2}\right) + 0.126\right), \\
\alpha_0^B &= \kappa^4\,\left(f^2 - \sigma_c^2\right)^2\left(0.027 - 0.018\ln\left(\frac{3\kappa^2\left(f^2 - \sigma_c^2\right)}{2\mu^2}\right)\right).
\end{aligned} \tag{20}$$

The remaining quantum corrections , proportional to Λ and Λ^2, may be identified respectively with Einstein-Hilbert R-type and Starobinsky R^2-type terms in an effective action (21) of the form

$$\Gamma \simeq -\frac{1}{2\kappa^2}\int d^4x\,\sqrt{g}\left(\left(\widehat{R} - 2\Lambda_1\right) + \alpha_1\,\widehat{R} + \alpha_2\,\widehat{R}^2\right)\,, \tag{21}$$

where we have combined terms of order Λ^2 into curvature scalar square terms. For general backgrounds such terms would correspond to invariants of the form $\widehat{R}_{\mu\nu\rho\sigma}\,\widehat{R}^{\mu\nu\rho\sigma}$, $\widehat{R}_{\mu\nu}\,\widehat{R}^{\mu\nu}$ and $\widehat{R}^2$, which for a de Sitter background all combine to yield $\widehat{R}^2$ terms [4]. The coefficients α_1 and α_2 absorb the

[4]In general, however, the metric tensor fluctuations of such terms may differ from the ones due to the R^2 term alone. We note that, in four-dimensions, in the absence of non-trivial (*i.e.* non-constant) dilatons, one may actually deal with just one

non-polynomial (logarithmic) in Λ contributions, so that we may then identify (21) with (19) via

$$\alpha_1 = \frac{\kappa^2}{2}\left(\alpha_1^F + \alpha_1^B\right), \quad \alpha_2 = \frac{\kappa^2}{8}\left(\alpha_2^F + \alpha_2^B\right). \tag{22}$$

To identify the conditions for phenomenologically acceptable Starobinsky inflation around the non-trivial minima of the broken SUGRA phase of our model, we impose first the *cancellation* of the "classical" Einstein-Hilbert space term $\widehat{R}$ by the "cosmological constant" term Λ_1, *i.e.* that

$$\widehat{R} = 4\,\Lambda = 2\,\Lambda_1\,. \tag{23}$$

This condition should be understood as a necessary one characterising our background in order to produce phenomenologically-acceptable Starobinsky inflation in the broken SUGRA phase following the first inflationary stage, as discussed in ref. [19]. This may naturally be understood as a generalisation of the relation $\widehat{R} = 2\Lambda_1 = 0$, imposed in ref. [20] as a self-consistency condition for the dynamical generation of a gravitino mass in a Minkowski space-time, discussed in the previous section. From (23) it follows that the cosmological constant Λ satisfies the four-dimensional Einstein equations in the non-trivial minimum, and in fact coincides with the value of the one-loop effective potential of the gravitino condensate at this minimum. As we discussed in [20], this non-vanishing positive value of the effective potential is consistent with the generic features of dynamical breaking of supersymmetry [36]. In terms of the Starobinsky inflationary potential (18), the value $\Lambda > 0$ corresponds to the approximately constant value of this potential in the high φ-field regime, where Starobinsky-type inflation takes place. Thus we may set $\Lambda \sim 3\,H_I^2$, where H_I the (approximately) constant Hubble scale during inflation, which is constrained by the current data to satisfy (1).

The effective Newton's constant in (21) is then

$$\kappa_{\rm eff}^2 = \kappa^2/\alpha_1\,, \tag{24}$$

and from this, we can express the effective Starobinsky scale (15) in terms of $\kappa_{\rm eff}$ as $\beta_{\rm eff} \equiv \alpha_2/\alpha_1$. This condition thus makes a direct link between the action (19) with a Starobinsky type action (15). Comparing with (15), we can then identify the Starobinsky inflationary scale in this case as

$$\mathcal{M} = \sqrt{\frac{8\pi}{3}\frac{\alpha_1}{\alpha_2}}\,. \tag{25}$$

The coefficients α_1 and α_2 (22) have been determined in ref. [20]:

$$\begin{aligned}
\alpha_1^F &= 0.067\,\tilde{\kappa}^2\sigma_c^2 - 0.021\,\tilde{\kappa}^2\sigma_c^2\ln\left(\frac{\Lambda}{\mu^2}\right) + 0.073\,\tilde{\kappa}^2\sigma_c^2\ln\left(\frac{\tilde{\kappa}^2\sigma_c^2}{\mu^2}\right),\\
\alpha_2^F &= 0.029 + 0.014\ln\left(\frac{\tilde{\kappa}^2\sigma_c^2}{\mu^2}\right) - 0.029\ln\left(\frac{\Lambda}{\mu^2}\right),\\
\alpha_1^B &= -0.083\Lambda_0 + 0.018\,\Lambda_0\ln\left(\frac{\Lambda}{3\mu^2}\right) + 0.049\,\Lambda_0\ln\left(-\frac{3\Lambda_0}{\mu^2}\right),\\
\alpha_2^B &= 0.020 + 0.021\ln\left(\frac{\Lambda}{3\mu^2}\right) - 0.014\ln\left(-\frac{6\Lambda_0}{\mu^2}\right),
\end{aligned} \tag{26}$$

more independent invariant, $R_{\mu\nu}R^{\mu\nu}$, in addition to R^2, since the Gauss-Bonnet combination, $R_{\mu\nu\alpha\beta}R^{\mu\nu\alpha\beta} - 4\,R_{\mu\nu}R^{\mu\nu} + R^2$, is a total space-time derivative. This is not the case for Jordan-frame SUGRA, where non-trivial-dilaton factors accompany the higher-order curvature terms. Nevertheless, in the minimum of the dilaton potential, where the conformal symmetry is broken, the dilaton equals its v.e.v. which is a space-time constant and, hence, this case is equivalent to the constant-dilaton one.

where σ_c denotes the gravitino scalar condensate at the non-trivial minimum of the one-loop effective potential (*cf.* fig. 2), and $\tilde{\kappa} = e^{-\langle\varphi\rangle}\,\kappa$ is the conformally-rescaled gravitational constant in the model of ref. [14], defined previously via (9).

In ref. [21] we searched numerically for points in the parameter space such that the effective equations $\frac{\partial\Gamma}{\partial\Lambda} = 0$, $\frac{\partial\Gamma}{\partial\sigma} = 0$, are satisfied; Λ is small and positive ($0 < \Lambda < 10^{-5}M_{\rm Pl}^2$, to ensure the validity of our expansion in Λ) and $10^{-6} < \mathcal{M}/M_{\rm Pl} < 10^{-4}$, to match with known phenomenology of Starobinsky inflation [6]. For $\tilde{\kappa} = \kappa$ (*i.e.* for non-conformal supergravity), we were unable to find any solutions satisfying these constraints, which was to be expected, given the previously demonstrated non-phenomenological suitability of this simple model [20]. If we consider $\tilde{\kappa} >> \kappa$, however, for instance of order $\tilde{\kappa}/\kappa = \mathcal{O}(10^3 - 10^4)$), we find that we are able to satisfy the above constraints for a range of values [21], with gravitino mass and global SUSY breaking scale in the ball-park of GUT scale, and the Starobinsky scale of order $\mathcal{M} \sim 10^{-5}\,M_{\rm Pl}$, which leads to phenomenologically acceptable inflation in the massive gravitino phase, consistent with the Planck-satellite data [6]. In particular, typical values obtained for the parameters of our conformal SUGRA models satisfy [20, 24]

$$\Lambda \sim 3H_I^2 \sim m_{3/2}^2 \sim \tilde{\kappa}^2\sigma_c^2 \sim \kappa^2 f^2 \ll \mu^2 = 8\pi/\kappa^2\,, \sigma_c^2 \ll f^2. \tag{27}$$

Since the scale of SUSY breaking must be in the ballpark of the typical GUT scale associated to the inflation, namely $\sqrt{f} \sim 10^{16}$ GeV$\sim 10^{-2}\,M_{Pl}$, from (27) we obtain $\Lambda \sim \kappa^2 f^2 = f^2/M_{Pl}^2 \sim 10^{27}$ GeV^2. As a result, the scale of the gravitino is some two to three orders of magnitude below the GUT scale, $m_{3/2} \sim \sqrt{\Lambda} \gtrsim 10^{13}$ GeV$\sim 10^{-5}\,M_{Pl}$, which is compatible with the bound (1).

4 "Running Vacuum" and Broken SUGRA effective field theory

Exit from the inflationary phase is a complicated issue which we shall not discuss here, aside from the observation that it can be achieved by coherent oscillations of the gravitino condensate field around its minima. This is still an open issue, which may be addressed via construction of more detailed supersymmetric models, including coupling of the matter sector to gravity, which will determine the pattern of the inflaton decays.

4.1 The "Running Vacuum" Scenario and post inflationary evolution of the Universe

Nevertheless, non-trivial conclusions about the evolution of the Universe after the phase transition that describes graceful exit from inflation and a smooth connection (in the sense of an evolution in cosmic time) of this phase with the radiation-dominated and subsequent eras of the Universe till the current epoch, can be made by applying the concept of the "running vacuum" [25] (RV) immediately after the exit from inflation [24]. According to the RV hypothesis [25], the dynamical cosmic evolution ("decay") of the inflationary phase ground state ("vacuum") to the standard radiation regime can be described using a "renormalization-group-like" approach, whereby the time evolution of the vacuum energy density $\rho_\Lambda(t)$ is inherited from its dependence on a characteristic cosmic scale variable $\mu_c = \mu_c(t)$. This variable plays the role of running (mass) scale of the renormalization group (RG) approach, and a natural candidate for such scale in FLRW cosmology is the Hubble parameter $H(t)$. Therefore the proposed RG equation is [25]:

$$\frac{d\rho_\Lambda(t)}{d\ln H^2} = \frac{1}{(4\pi)^2}\sum_i \left[a_i M_i^2 H^2 + b_i H^4 + c_i\frac{H^6}{M_i^2} + \ldots\right] \tag{28}$$

The coefficients $a_i, b_i, c_i \ldots$ appearing in (28) are dimensionless and receive contributions from loop corrections of boson and fermion matter fields with different masses M_i. It must be stressed that the requirement of general covariance of the action [25] implies *only even* powers of the (cosmic-time t dependent) Hubble parameter $H(t)$ on the right-hand-side of (28).

We note that, if the evolution of the Universe is restricted to eras below the GUT scale, as is the case of the dynamically broken SUGRA discussed here, then at most the H^4 terms in (28) can contribute significantly to the $\rho_\Lambda(t)$ evolution. Then, on Integrating the RG equation (28) one obtains:

$$\rho_\Lambda(H) = \frac{\tilde{\Lambda}(H)}{\kappa^2} = \frac{3}{\kappa^2}\left(c_0 + \nu H^2 + \alpha\frac{H^4}{H_I^2}\right),$$
$$\nu = \frac{1}{48\pi^2}\sum_{i=F,B} a_i \frac{M_i^2}{M_{\rm Pl}^2}, \qquad \alpha = \frac{1}{96\pi^2}\frac{H_I^2}{M_{\rm Pl}^2}\sum_{i=F,B} b_i\,. \tag{29}$$

Here c_0 is an integration constant (with dimension +2 in natural units, *i.e.* energy squared) which can be fixed from the low energy data of the current universe, and contributes to the cosmological constant.

The RV model is based on the assumption that at any moment in cosmic time t, the vacuum is characterised by the Equation of State (EoS) of de-Sitter space time, *i.e.* $p_\Lambda(t) = -\rho_\Lambda(t) = \tilde{\Lambda}(t)/\kappa^2$, where $\tilde{\Lambda}(t)$ denotes the vacuum energy of the RV model. We stress that this EoS does not depend on whether the vacuum is dynamical or not. In contrast to other forms of dark energy, the vacuum is defined as that for which the EoS parameter ω is precisely $\omega = -1$ at any moment of the Universe's evolution.

Einstein's equations in this framework can be written as:

$$R_{\mu\nu} - \frac{1}{2}g_{\mu\nu}R = \kappa^2\,\tilde{T}_{\mu\nu}\,, \tag{30}$$

where the total stress tensor $\tilde{T}_{\mu\nu}$ is given by $\tilde{T}_{\mu\nu} \equiv T_{\mu\nu} - g_{\mu\nu}\rho_\Lambda$, with $T_{\mu\nu} = -2\partial\mathcal{L}_m/\partial g_{\mu\nu} + g_{\mu\nu}\mathcal{L}_m$ the energy-momentum tensor corresponding to the matter Lagrangian. The extra piece proportional to ρ_Λ, corresponds to the vacuum energy density associated with the presence of $\tilde{\Lambda}(t)$ (with pressure $p_\Lambda = -\rho_\Lambda$). Modeling the expanding universe in this framework as a perfect fluid with velocity 4-vector field U_μ, we have for the matter-radiation stress tensor $T_{\mu\nu} = p_m\,g_{\mu\nu} + (\rho_m + p_m)\,U_\mu U_\nu$, where ρ_m is the density of matter-radiation and $p_m = \omega_m\rho_m$ is the corresponding pressure, in which ω_m is the EoS of matter. $\tilde{T}_{\mu\nu}$ takes the same form as $T_{\mu\nu}$ but with $\rho_{\rm tot} = \rho_m + \rho_\Lambda$ and $p_{\rm tot} = p_m + p_\Lambda = p_m - \rho_\Lambda$, that is, $\tilde{T}_{\mu\nu} = (p_m - \rho_\Lambda)\,g_{\mu\nu} + (\rho_m + p_m)U_\mu U_\nu$.

The integrated form (29) of the vacuum energy density and the above considerations, provides an effective description of inflation and the subsequent stages of the FLRW Universe evolution in a smooth way [24, 25]. In particular, as discussed in [24], from Einstein's equations in the context of RV, and in particular the conservation equation (Bianchi identity) of the total stress tensor $\nabla^\mu\,\tilde{T}_{\mu\nu} = 0$, one can deduce

$$\dot{\rho}_m + 3(1+\omega_m)H\rho_m = -\dot{\rho_\Lambda}\,. \tag{31}$$

and the evolution equation of the Hubble parameter $H(t)$:

$$\dot{H} + \frac{3}{2}(1+\omega)H^2\left[1 - \nu - \frac{c_0}{H^2} - \alpha\frac{H^2}{H_I^2}\right] = 0\,, \tag{32}$$

in units of the inflation scale H_I (*cf.* (1)). Eq. (32) can be solved at various epochs [24, 25]. For instance, during inflationary era, one may neglect the term proportional to c_0 on the right-hand-side of (32), to find a constant H as a self-consistent solution during inflation, $H^2 = (1-\nu)H_I^2/\alpha$. Going away from the inflationary regime, the Hubble parameter is represented by a solution of the form

$$H(a) = \left(\frac{1-\nu}{\alpha}\right)^{1/2} \frac{H_I}{\sqrt{D\,a^{3(1-\nu)(1+\omega)}+1}}\,, \tag{33}$$

where D is a positive constant of integration and $a = a(t)$ is the scale factor of the universe. The universe will enter the standard radiation phase, with $w = 1/3$, in the case when $Da^{4(1-\nu)} \gg 1$, which is confirmed by substituting this solution (33) into (29) to obtain:

$$\rho_\Lambda(a) = \frac{\rho_I}{\alpha}\frac{1}{[1+D\,a^4]^2}\,, \tag{34}$$

where $\rho_I = 3H_I^2/\kappa^2$ is the critical density in the inflationary epoch. Then solving (31) we obtain for the radiation energy density ρ_m:

$$\rho_r(a) = \frac{\rho_I}{\alpha}\frac{D\,a^4}{[1+D\,a^4]^2}\,. \tag{35}$$

From the above expressions, it is apparent that there is no singularity in the initial state: the Universe starts at $a = 0$ with a huge vacuum energy density ρ_I/α (and zero radiation) which is progressively converted into relativistic matter. In the asymptotic radiation regime we indeed retrieve the standard behavior $\rho_r \sim a^{-4}$ with essentially negligible vacuum energy density: $\rho_\Lambda \sim a^{-8} \ll \rho_r$. Graceful exit from inflation is, therefore, naturally implemented in this formalism.

Subsequently, the universe will enter a matter dominated phase, and then the current era, in which the term c_0 in (29) becomes dominant. In this case Eq.(29) amounts to

$$\tilde{\Lambda}(H) = 3c_0 + 3\nu H_0^2 + 3\nu(H^2 - H_0^2), \tag{36}$$

which plays the rôle of the cosmological constant at the present time [24], which is positive. The reader should notice the RV corrections to the cosmic-concordance ΛCDM model.

4.2 Applying the "Running Vacuum" scenario to the Dynamically Broken SUGRA model

Let us now apply the above reasoning to the effective potential of the broken $\mathcal{N} = 1$ SUGRA model, at the exit of its inflationary phase [24]. The effective (dynamical) vacuum energy density, $\rho_\Lambda(H)$, during the inflationary phase of our SUGRA model can be extracted from the SUGRA effective action Γ (21), upon applying the constraint (23) and analytically continuing the results back to Minkowski space-time signature. In particular, the effective potential is defined as $V_{\rm eff} \equiv -\Gamma \to \int d^4x\sqrt{-g}\,\rho_\Lambda(H)$. Moreover, at the exit from inflation, any logarithmic dependence of the coefficients α_i on Λ, μ is kept fixed, and only positive integer powers of Λ are allowed to vary with cosmic time t and be set equal to $3\,H(t)^2$. This is for covariance reasons explained in [24, 25]. Doing so, we observe that the so obtained $\rho_\Lambda(H)$, remarkably, adopts precisely the generic RV structure (29) around that phase, in which the Ricci scalar is approximated by $R \simeq 12H^2$, since the Hubble parameter H remains (approximately) constant in this phase.

The imposition of the constraint (23) during the Starobinsky inflationary phase implies, as discussed previously, that the correct phenomenology is attained. So, if the constraint *was* an *exact result*, the effective vacuum energy density of the SUGRA model would correspond to the $\widehat{R^2} \to 144\,H^2$ terms in (21) with (24) playing the rôle of the effective gravitational constant,

$$\rho_\Lambda^{\rm SUGRA}(H)^{\rm exact}_{\rm constraint} = \frac{72}{\kappa_{\rm eff}^2}\frac{\alpha_2}{\alpha_1}H^4 = \frac{18}{\kappa_{\rm eff}^2}\frac{\alpha_2^F+\alpha_2^B}{\alpha_1^F+\alpha_1^B}H^4\,, \tag{37}$$

where we used (22), (26). The form (37) constitutes an admissible class of RV models (*cf.* Eq. (29)). Notice that in Eq. (37) there is no ν term. This is important, in the sense that in such a model, as a result of the effective gravitational constant (24) entering the game, which in this scenario [21] is viewed as the '*physical*' reduced Planck mass of order 10^{18} GeV, the gravitino mass and global SUSY breaking scales, (27), when expressed in terms of $\kappa_{\rm eff}$ are of order one, that is one encounters a Planck-scale gravitino. Despite this, the vanishing of ν makes the renormalization-group equation (37) a consistent one within the perturbative class of (29).

However the above construction leads to the absence of a present-era (small, positive) cosmological constant c_0. This arises from the fact that we imposed the constraint (23) exactly. It may well be that such a condition leaves (non-perturbatively, when all the higher than one-loop contributions are taken into account) a very small (constant in cosmic time) contribution $c_0 > 0$ which is preserved until the present day. Unfortunately, our one-loop construction does not allow us to explain the magnitude and the sign of this constant term, but this is equivalent to offering a solution to the cosmological constant problem, which of course our approximate one-loop analysis cannot provide. While we do not have a quantitative calculation at this point, nevertheless, the above argument provides at least an interesting qualitative explanation for it: the origin of the current cosmological term ρ_Λ^0 in the model might well be attributed to quantum (non-perturbative) effects in the SUGRA effective action, which prevent the complete cancellation (23) from being realised. The constant residue c_0 is then transferred throughout the cosmic history and pops up in our days in the form of the tiny vacuum energy contribution (36), as discussed above.

Under this assumption, the initial gravitational coupling κ, and thus the Einstein term $\frac{1}{2\kappa^2}\int d^4x\,\sqrt{-g}\,\widehat{R}$ would enter the game during the exit phase from inflation [5]. In such a case, in the exit phase, the effective vacuum energy of the SUGRA model at the inflationary phase should correspond to *both* α_1 and α_2 terms of (21), with the constraint (23) failing by a tiny amount $\tilde{c}_0 > 0$ corresponding to the present-era cosmological constant:

$$\rho_\Lambda^{\rm SUGRA}(H) = \frac{1}{\kappa^2}\left(\tilde{c}_0 + 6\alpha_1\, H^2 + 72\,\alpha_2\, H^4\right) \simeq \frac{1}{\kappa^2}\tilde{c}_0 + 1.59\,\tilde{\kappa}^2\sigma_c^2 H^2 + 12.88\, H^4\,, \tag{38}$$

where we have used the explicit form of the coefficients α_i, $i = 1, 2$ (Eqs. (22), (26)), with the scales μ and Λ *fixed* through (27).

Comparing (38) with (29) with $c_0 = \tilde{c}_0/3$, one obtains the effective values for the coefficients ν and α in the matched-SUGRA/RV model [24]:

$$\nu_{\rm eff} \simeq 0.53\,\kappa^2\tilde{\kappa}^2\sigma_c^2 \simeq O\left(\frac{m_{3/2}^2}{M_{\rm Pl}^2}\right), \qquad \alpha_{\rm eff} \simeq 4.30\, H_I^2\,\kappa^2 \simeq O\left(\frac{H_I^2}{M_{\rm Pl}^2}\right). \tag{39}$$

Thus we observe that, within the context of the pure SUGRA model, where only the gravitino plays the rôle of "matter", both coefficients are small, of typical order 10^{-9}, in accordance with their interpretation as β-function coefficients of the running vacuum energy density. In the general case where the parameters of the SUGRA model are varied from the generic values considered in (27), but within the allowed range, the values of $\nu_{\rm eff}$ and $\alpha_{\rm eff}$ can also undergo some variation and the sign of $\nu_{\rm eff}$ could change. However we stress that the sign of $\alpha_{\rm eff}$ remains always positive, which is essential for a correct description of inflation [24].

[5] The reader should bear in mind that, since during the inflationary phase the scalar degree of freedom of the Starobinsky action is slowly rolling, if there is inflation in the conformally rescaled metric (16), there is also inflation in the initial metric. The Starobinsky inflation arguments are also not affected if a small contribution to the cosmological constant, of order of the present-era one, enters the effective action (17), as this is negligible compared to the Hubble scale of inflation.

Having matched the evolution of the SUGRA model with the RV flow at the exit of the inflationary Starobinsky phase, one can apply the analysis outlined in the previous subsection 4.1, in order to smoothly connect the inflationary and the current (cosmological-constant dominated) eras of this Universe, with the interpolation of radiation and matter dominated epochs. In this approach, the smallness of the cosmological constant today is attributed to a failure of the constraint (23) due to non-perturbative quantum (supergravity) effects. This offers an interesting new insight into the cosmological constant problem, which, needless to say however, remains unsolved.

5 Conclusions

In this talk we considered a minimal inflationary scenario, by means of which a gravitino condensate in supergravity models is held responsible for breaking local supersymmetry dynamically and inducing inflation in an indirect way by means of a Starobinsky-type inflation in the massive gravitino phase. The inflaton field in this approach is associated with the scalar mode that collectively parametrizes the effects of the quadratic-curvature contributions to the one-loop quantum effective action of the gravitino condensate, after integrating out graviton and massive gravitino degrees of freedom, in a de Sitter background. The model involves parameters that assume values of a natural and phenomenologically relevant order of magnitude, specifically global supersymmetry scale and gravitino masses of the order of GUT mass scales or less. Such a scenario is a truly minimal scenario for natural inflation, in the sense that it involves two scalar primordial composite modes to achieve dynamical breaking of a gauge symmetry (supergravity) and inflation.

Moreover, the effective potential of this model has been cast in a form that allows a running vacuum scenario to be in operation, thereby implying a smooth connection of the inflationary phase to the current era, characterised by a small cosmological constant. The smallness of the latter has been attributed to non perturbative quantum (supergravity) effects that lead to the failure of the constraint (23) characterising the Starobinsky inflationary phase of the model.

From our analysis it becomes clear that, in order to ensure phenomenologically relevant supersymmetry breaking scales and gravitino masses, one needs to apply the above ideas to Jordan-frame extensions of the minimal $\mathcal{N} = 1$ $D = 4$ SUGRA model, which involve a third scalar field - the dilaton. The latter is responsible, through its appropriate potential, for the breaking of the local scale symmetry of the theory. In the context of the next to minimal supersymmetric standard model, which Jordan-frame SUGRA models can incorporate, such dilatons may be composite of appropriate matter superfields, involving Higgs (supermultiplets).

Details of the microscopic matter model are important in order to ensure the correct cosmological evolution, in particular satisfaction of the Big-Bang-Nucleosynthesis constraints. A GUT scale gravitino can be made to decay fast enough so as not to disturb the BBN, but this depends on the details of the matter sector of the theory. Such topics are left for future investigations. Nevertheless, we believe that the above simple models for dynamical breaking of supergravity and their links with Starobinsky inflation and Running Vacuum models, deserve further study, and stand a serious chance of leading to realistic phenomenological scenarios compatible with the cosmological and particle physics data.

Acknowledgements

N.E.M. would like to thank the organisers of the 4th ICNFP 2015 conference in Kolymbari (Crete, Greece) for their kind invitation to give a plenary talk. This work is supported in part by the London Centre for Terauniverse Studies (LCTS), using funding from the European Research Council via the Advanced Investigator Grant 267352 and by STFC (UK) under the research grant ST/L000326/1.

References

[1] V. A. Kuzmin, V. A. Rubakov and M. E. Shaposhnikov, Phys. Lett. B **155**, 36 (1985). doi:10.1016/0370-2693(85)91028-7; M. E. Shaposhnikov, Nucl. Phys. B **287**, 757 (1987). doi:10.1016/0550-3213(87)90127-1; M. B. Gavela, P. Hernandez, J. Orloff and O. Pene, Mod. Phys. Lett. A **9**, 795 (1994) doi:10.1142/S0217732394000629 [hep-ph/9312215]. M. B. Gavela, P. Hernandez, J. Orloff, O. Pene and C. Quimbay, Nucl. Phys. B **430**, 382 (1994) doi:10.1016/0550-3213(94)00410-2 [hep-ph/9406289].

[2] J. Wess and B. Zumino, Phys. Lett. B **49**, 52 (1974). doi:10.1016/0370-2693(74)90578-4; Nucl. Phys. B **70**, 39 (1974). doi:10.1016/0550-3213(74)90355-1;

[3] D. V. Volkov and V. P. Akulov, Phys. Lett. B **46**, 109 (1973). doi:10.1016/0370-2693(73)90490-5

[4] S. R. Coleman and J. Mandula, Phys. Rev. **159**, 1251 (1967). doi:10.1103/PhysRev.159.1251

[5] A. H. Guth, Phys. Rev. D **23**, 347 (1981). doi:10.1103/PhysRevD.23.347; A. D. Linde, Phys. Lett. B **116**, 335 (1982). doi:10.1016/0370-2693(82)90293-3; *ibid.* **108**, 389 (1982). doi:10.1016/0370-2693(82)91219-9.

[6] P. A. R. Ade *et al.* [Planck Collaboration], arXiv:1502.02114 [astro-ph.CO].

[7] J. Martin, C. Ringeval and V. Vennin, arXiv:1303.3787 [astro-ph.CO].

[8] J. R. Ellis, D. V. Nanopoulos, K. A. Olive and K. Tamvakis, Phys. Lett. B **118**, 335 (1982).

[9] D. Croon, J. Ellis and N. E. Mavromatos, Phys. Lett. B **724**, 165 (2013) [arXiv:1303.6253 [astro-ph.CO]]; J. Ellis, N. E. Mavromatos and D. J. Mulryne, JCAP **1405**, 012 (2014) [arXiv:1401.6078 [astro-ph.CO]].

[10] M. Yamaguchi, Class. Quant. Grav. **28** (2011) 103001 [arXiv:1101.2488 [astro-ph.CO]].

[11] A. D. Linde and A. Riotto, Phys. Rev. D **56** (1997) 1841 [hep-ph/9703209], E. Halyo, Phys. Lett. B **387** (1996) 43 [hep-ph/9606423].

[12] M. Kawasaki, M. Yamaguchi and T. Yanagida, Phys. Rev. Lett. **85** (2000) 3572 [hep-ph/0004243], J. Ellis, M. A. G. Garcia, D. V. Nanopoulos and K. A. Olive, arXiv:1403.7518 [hep-ph], R. Kallosh, A. Linde and A. Westphal, Phys. Rev. D **90**, no. 2, 023534 (2014) doi:10.1103/PhysRevD.90.023534 [arXiv:1405.0270 [hep-th]].

[13] J. Ellis, D. V. Nanopoulos and K. A. Olive, JCAP **1310** (2013) 009 [arXiv:1307.3537], W. Buchmuller, V. Domcke and K. Kamada, Phys. Lett. B **726** (2013) 467 [arXiv:1306.3471 [hep-th]], F. Farakos, A. Kehagias and A. Riotto, Nucl. Phys. B **876** (2013) 187 [arXiv:1307.1137], S. V. Ketov and T. Terada, JHEP **1312** (2013) 040 [arXiv:1309.7494 [hep-th]]; JHEP **1412**, 062 (2014) doi:10.1007/JHEP12(2014)062 [arXiv:1408.6524 [hep-th]]. C. Pallis, JCAP **1404** (2014) 024 [arXiv:1312.3623 [hep-ph]], S. Ferrara, R. Kallosh and A. Van Proeyen, JHEP **1311** (2013) 134 [arXiv:1309.4052 [hep-th]]; C. Pallis and N. Toumbas, arXiv:1512.05657 [hep-ph].

[14] S. Ferrara, R. Kallosh, A. Linde, A. Marrani and A. Van Proeyen, Phys. Rev. D **82**, 045003 (2010) [arXiv:1004.0712 [hep-th]]; Phys. Rev. D **83** (2011) 025008 [arXiv:1008.2942 [hep-th]].

[15] M. A. G. Garcia and K. A. Olive, JCAP **1309**, 007 (2013) doi:10.1088/1475-7516/2013/09/007 [arXiv:1306.6119 [hep-ph]]; D. Roest, M. Scalisi and I. Zavala, JCAP **1311**, 007 (2013) doi:10.1088/1475-7516/2013/11/007 [arXiv:1307.4343]; S. Ferrara, R. Kallosh, A. Linde and M. Porrati, S. Ferrara, R. Kallosh, A. Linde and M. Porrati, Phys. Rev. D **88**, no. 8, 085038 (2013) doi:10.1103/PhysRevD.88.085038 [arXiv:1307.7696 [hep-th]]; JCAP **1311**, 046 (2013) doi:10.1088/1475-7516/2013/11/046 [arXiv:1309.1085 [hep-th]].

[16] A. A. Starobinsky, Phys. Lett. B **91**, 99 (1980).

[17] D. Z. Freedman and A. Van Proeyen, *Supergravity* (Cambridge, UK: Cambridge Univ. Pr. (2012)).

[18] P. Van Nieuwenhuizen, Phys. Rept. **68** (1981) 189.

[19] J. Ellis and N. E. Mavromatos, Phys. Rev. D **88**, 085029 (2013) [arXiv:1308.1906 [hep-th]].

[20] J. Alexandre, N. Houston and N. E. Mavromatos, Phys. Rev. D **88**, 125017 (2013) [arXiv:1310.4122 [hep-th]]; Int. J. Mod. Phys. D **24**, no. 04, 1541004 (2015) doi:10.1142/S0218271815410047 [arXiv:1409.3183 [gr-qc]].

[21] J. Alexandre, N. Houston and N. E. Mavromatos, Phys. Rev. D **89**, 027703 (2014) [arXiv:1312.5197 [gr-qc]].

[22] S. V. Ketov and T. Terada, JHEP **1307**, 127 (2013) doi:10.1007/JHEP07(2013)127 [arXiv:1304.4319 [hep-th]]; K. Hamaguchi, T. Moroi and T. Terada, Phys. Lett. B **733**, 305 (2014) doi:10.1016/j.physletb.2014.05.006 [arXiv:1403.7521 [hep-ph]].

[23] E. S. Fradkin and A. A. Tseytlin, Nucl. Phys. B **234** (1984) 509.

[24] S. Basilakos, N. E. Mavromatos and J. Solà, arXiv:1505.04434 [gr-qc].

[25] I. L. Shapiro and J. Solà, Phys. Lett. **B530**, 10 (2002) [arXiv:hep-ph/0104182]; Phys. Lett. B **682**, 105 (2009) [arXiv:0910.4925; *ibid.* **B475**, 236 (2000) [hep-ph/9910462]; JHEP **0202** 006 (2002) [hep-th/0012227]; J. Solà, J. Phys. Math. Theor. **A41**, 164066 (2008) [arXiv:0710.4151]; I. L. Shapiro, J. Solà, H. Štefančić JCAP 0501 012 (2005); See also: J. Solà, J. Phys. Conf. Ser. **453** (2013) 012015 [arXiv:1306.1527]; J. Solà, and A. Gómez-Valent, Int. J. of Mod. Phys. **D24** (2015) 1541003 [e-Print: arXiv:1501.03832]; J. A. S. Lima, S. Basilakos and J. Solà, Mon. Not. Roy. Astron. Soc. **431**, 923 (2013) [arXiv:1209.2802]; S. Basilakos, J. A. S. Lima and J. Solà, Int. J. Mod. Phys. D **22**, 1342008 (2013) [arXiv:1307.6251]; *ibid.* **D23** (2014) 1442011 [arXiv:1406.2201].

[26] J. Jaeckel and C. Wetterich, Phys. Rev. D **68** (2003) 025020 [hep-ph/0207094].

[27] Y. Nambu and G. Jona-Lasinio, Phys. Rev. **122** (1961) 345.

[28] S. Deser and B. Zumino, Phys. Rev. Lett. **38** (1977) 1433.

[29] Z. Komargodski and N. Seiberg, JHEP **0909** (2009) 066 [arXiv:0907.2441 [hep-th]].

[30] M. B. Einhorn and D. R. T. Jones, JHEP **1003**, 026 (2010) [arXiv:0912.2718 [hep-ph]].

[31] F. L. Bezrukov and M. Shaposhnikov, Phys. Lett. B **659**, 703 (2008) [arXiv:0710.3755 [hep-th]]; F. Bezrukov, A. Magnin, M. Shaposhnikov and S. Sibiryakov, JHEP **1101**, 016 (2011) [arXiv:1008.5157 [hep-ph]].

[32] I. L. Buchbinder and S. D. Odintsov, Class. Quant. Grav. **6** (1989) 1955; see also S. D. Odintsov, Phys. Lett. B **213**, 7 (1988).

[33] I. Antoniadis, C. Bachas, J. R. Ellis and D. V. Nanopoulos, Phys. Lett. B **211** (1988) 393.

[34] P. C. W. Davies, S. A. Fulling, S. M. Christensen and T. S. Bunch, Annals Phys. **109**, 108 (1977); T. S. Bunch and P. C. W. Davies, Proc. Roy. Soc. Lond. A **360**, 117 (1978); N. Birrell and P. Davies, *Quantum Fields in Curved Space* (Cambridge Monogr. Math. Phys., 1982).

[35] K. S. Stelle, Gen. Rel. Grav. **9**, 353 (1978); B. Whitt, Phys. Lett. B **145**, 176 (1984).

[36] E. Witten, Nucl. Phys. B **188**, 513 (1981).

EPJ Web of Conferences **126**, 02021 (2016) DOI: 10.1051/epjconf/201612602021
ICNFP 2015

Conciliating SUSY with the Z-peaked excess

Vasiliki A. Mitsou[1,a]

[1] *Instituto de Física Corpuscular (IFIC), CSIC – Universitat de València, Parc Científic de la U.V., C/ Catedrático José Beltrán 2, E-46980 Paterna (Valencia), Spain*

Abstract. The ATLAS experiment observed an excess at the 3σ level in the channel of Z boson, jets and high missing transverse momentum in the full 2012 dataset at 8 TeV while searching for SUSY. The question arises whether the abundance and the kinematical features of this excess are compatible with the yet unconstrained supersymmetric realm, respecting at the same time the measured Higgs boson properties and dark matter density. By trying to explain this signal with SUSY we find that only relatively light gluinos together with a heavy neutralino NLSP decaying predominantly to a Z boson plus a light gravitino could reproduce the excess. We construct an explicit general gauge mediation model able to match the observed signal. More sophisticated models could also reproduce the signal, as long as it features light gluinos, or heavy particles with a strong production cross section, producing at least one Z boson in its decay chain. The implications of our findings for the Run II at LHC with the scaling on the Z peak, as well as for the direct search of gluinos and other SUSY particles, are also discussed.

1 Introduction

Supersymmetry (SUSY) [1] is an extension of the Standard Model (SM) that assigns to each SM field a superpartner with a spin differing by half a unit. SUSY solves in an elegant manner several open issues of the SM, such as the hierarchy problem, the nature of dark matter [2], and the grand unification. Although it is one of the most relevant scenarios of new physics explored at the LHC, no clear signs of SUSY have been observed so far.

Within this scenario, the ATLAS experiment [3] at the LHC [4] observed an intriguing excess of 3σ in events with e^+e^- or $\mu^+\mu^-$ pairs at the Z peak [5], accompanied by hadronic activity and missing transverse energy (MET). More precisely, the main selection criteria of the analysis are: (i) At least two same-flavour leptons with opposite electric charge with the leading (sub-leading) lepton $p_T > 25$ GeV ($p_T > 10$ GeV). Their invariant mass must be within $81 < m_{\ell\ell} < 101$ GeV; (ii) All events contain at least two jets with $p_T > 35$ GeV and $|\eta| < 2.5$, and have $E_T^{\text{miss}} > 225$ GeV and $H_T \equiv \sum_i {p_T}^{\text{jet},i} + {p_T}^{\text{lep1}} + {p_T}^{\text{lep2}} > 600$ GeV; (iii) The azimuthal angle between each of the two leading jets and E_T^{miss} higher than 0.4.

With an integrated luminosity of 20.3 fb^{-1} of pp collisions at $\sqrt{s} = 8$ TeV, the experiment observes a total of 29 pairs of electrons and muons with an invariant mass compatible with the Z-boson

[a] e-mail: vasiliki.mitsou@ific.uv.es

mass, with an expected background of 10.6 ± 3.2 events[1]. No excess over the expected background is observed outside the Z peak. The question that immediately arises is whether SUSY, or some other extension of the Standard Model (SM), can explain that excess of Z+MET events taking into account the current limits on beyond-SM (BSM) physics. A study in those terms within a SUSY framework is presented in Ref. [6] and the main conclusions are highlighted here.

As we shall see, the observed signal can only be explained if one has a large production cross section of heavy SUSY particles (gluinos or squarks) whose decay chain contains about one Z boson per parent particle. If such an explanation is indeed the answer to the observed excess, our study points out the way to confirm it in the Run II of LHC, as well as cosmological implications, in particular the particle content of dark matter in the Universe. The resulting scheme of SUSY particle mass hierarchy, including charginos and neutralinos, will be apparent.

2 Z-boson production in the MSSM

If the observed excess is confirmed, it would clearly indicate a new non-standard process producing additional Z bosons at LHC energies. Z bosons are regularly produced in the decay chains of most of the SM extensions. Still, this signal would require a significant production of Z bosons without conflicting with all other experimental searches of BSM particles. In fact, using the central value for the expected background and taking into account the BR($Z \to ee, \mu\mu$), this would imply that we have produced 273 ± 48 additional Z bosons with 20.3 fb^{-1}. Under the assumption that the Z bosons are produced in the decay chains of generic BSM particles, Y, produced in the collision, we need to produce at least $273/N(Y \to Z)$ Y particles, with $N(Y \to Z)$ the average number of Z bosons produced in the decay of a Y particle. On the other hand, the experimental cuts used in the experiment, namely $n_{\rm jets} \geq 2$, $E_{\rm T}^{\rm miss} > 225$ GeV and $H_{\rm T} > 600$ GeV, define further the characteristics of the Y particle and its decays.

We now follow a bottom-up approach attempting to identify the model within the Minimal Supersymmetric Standard Model (MSSM) framework that may yield the observed Z-boson excess[2]. We need 273 (225 at 1σ) Z bosons if we want to accommodate the observed excess. Assuming that R-parity is conserved, supersymmetric particles are produced in pairs in processes of the type $pp \to Y\bar{Y}$. Thus, the required cross section for this process would be

$$\sigma(pp \to Y\bar{Y}) = \frac{N_{\rm ev}/N(Y \to Z)}{\mathcal{L}} = \frac{137(113)/N(Y \to Z)}{20.3\ {\rm fb}^{-1}} = \frac{6.7(5.6)\ {\rm fb}}{N(Y \to Z)}, \tag{1}$$

where we take into account that two Y particles are produced in each event. So, if we obtained one Z boson for each Y-particle produced, we would need a production cross section of 6.7 ± 1.1 fb at the LHC at $\sqrt{s} = 8$ TeV. Here, we consider the production cross sections of different supersymmetric particles separately to identify the relevant Z-producing processes. However in the next section, by employing full event generation, the production of different sparticles contributes to the final Z plus jets plus MET signal.

Naively, the first option to consider in a hadron collider would be strong production of squarks or gluinos (assuming they produce Z bosons in their decays). However, current experimental searches of jets plus missing energy at LHC force the masses of these coloured particles to be high [7–9]. Nevertheless, as we will see below, in some cases we can still find cross sections of the required size.

[1] A similar analysis on Z plus $E_{\rm T}^{\rm miss}$ has been performed by CMS. However, among other differences, no cut on $H_{\rm T}$ has been applied. No deviation from SM expectations has been observed in that analysis.

[2] We assume that the acceptance of the applied selection, also taking into account the reconstruction efficiencies is ideally equal to unity. A "realistic" generation of SUSY events using detector simulation taking into account the signal acceptance and detector efficiencies follows later.

 DOI: 10.1051/epjconf/201612602021

Production cross sections of gluinos and squarks depend only on their masses and are basically independent of other MSSM parameters. In the case of gluino and squarks of the first generation, the cross section depends both on the squark and gluino masses due to t-channel contributions, but in the case of stops or sbottoms it depends only on the stop or sbottom mass. In Fig. 1 we present the production cross section of gluino pairs (left) and light-flavour squark pairs (right) calculated at NLL+NLO as a function of the gluino or squark mass.

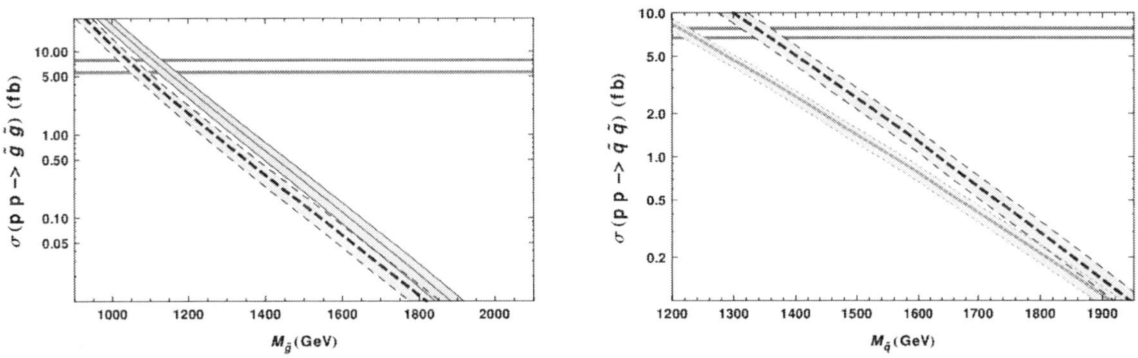

Figure 1. *Left:* Production cross section of gluino pairs as a function of the gluino mass for two fixed values of the first generation squark masses: 1000 GeV (dashed blue) and decoupled squarks (solid red). *Right:* Production cross section of squarks pairs as a function of their mass with gluino masses of 1000 GeV (dashed blue) and 2500 GeV (dotted orange). The bands correspond to $\pm 1\sigma$ theoretical uncertainty. The horizontal gray bands indicate the experimental requirements as calculated in Eq. (1). From Ref. [6].

As we can see in these figures, the required cross section is reached only for light gluino and squark masses. In the case of gluino production, the needed cross section is obtained only for $m_{\tilde{g}} \lesssim 1200$ GeV and favours heavy squark masses. In fact, these gluino masses are in the boundary of the allowed region obtained from jets plus missing-E_T searches at LHC [7–9] and would contribute significantly only if every $\tilde{g}$ produces at least a Z boson in its decay. For the production of squark pairs, present limits are $m_{\tilde{q}} \gtrsim 1400$ GeV for heavy gluinos and $m_{\tilde{q}} \gtrsim 1650$ GeV for degenerate squarks and gluinos. Under these conditions, $\sigma(pp \to \tilde{q}\tilde{q})$ is always well below the required cross section, even for $m_{\tilde{q}} \gtrsim 1400$ GeV .

Another important process is the simultaneous production of squark and gluino. However, we found that we would need both squark and gluino to be light, which is not possible if we take into account the bounds from LHC searches [7–9]. In summary, the best option seems to be gluino pair production with $m_{\tilde{g}} \lesssim 1200$ GeV with relatively heavy squarks $m_{\tilde{q}} \gtrsim 3000$ GeV , if we can get at least one Z boson in every gluino decay.

We can also consider the strong production of stop pairs, where the current bounds on stop masses are much lower, $m_{\tilde{t}} \gtrsim 650$ GeV [10, 11]. The total $\tilde{t}\tilde{t}^*$ production cross section is shown in the left panel of Fig. 2, calculated at NLL+NLO. In this case, we can see that we could reach the required cross section for $m_{\tilde{t}} \lesssim 750$ GeV which in principle could be achievable in general SUSY models (always assuming that every stop produces a Z boson in its decay). However, the cuts $H_{\mathrm{T}} > 600$ GeV and $E_{\mathrm{T}}^{\mathrm{miss}} > 225$ GeV are very restrictive. We can see this in the right panel of Fig. 2, where we show the H_{T} distribution from the decays of stop pairs with $m_{\tilde{t}_1} = 750$ GeV . As can be seen here, the H_{T} distribution peaks at $H_{\mathrm{T}} \simeq 400 - 500$ GeV as a consequence of the relatively small stop mass, and only a small fraction of the events are able to overcome the cut on H_{T} . Therefore, we must conclude that it is not possible to generate the required cross section and fulfil the requirements of the observed excess through stop production.

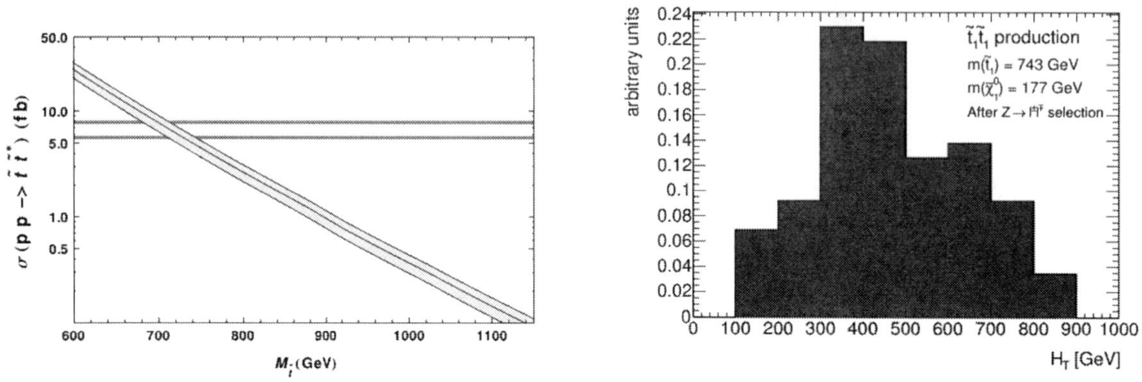

Figure 2. *Left:* Production cross section of $\tilde{t}\tilde{t}^*$ pairs as a function of the stop mass. The cross section is practically independent of other sparticle masses. *Right:* H_T distribution in arbitrary units from the decay of a pair of stops of $m_{\tilde{t}} = 743$ GeV to jets plus $E_T^{\rm miss}$ after applying the selection $Z \to \ell^+\ell^-$. From Ref. [6].

Apart from the production cross sections of coloured sparticles, we also considered weak production of charginos/neutralinos which can be large enough for light gauginos. Taking into account that the current bounds on chargino and neutralino masses are not very stringent [12, 13], electroweak production is worth exploring. In terms of gauge eigenstates the largest electroweak production cross sections are those corresponding to $\tilde{W}^0\tilde{W}^\pm$ and $\tilde{W}^+\tilde{W}^-$, corresponding to $\tilde{\chi}^0_2\tilde{\chi}^\pm_1$ and $\tilde{\chi}^+_1\tilde{\chi}^-_1$ in terms of mass eigenstates. In principle, the production cross section of charginos of $m_{\tilde{\chi}^+_1} \lesssim 350$ GeV appears to be sufficient to accommodate the required Z production if the only restriction imposed is a minimum $p_T \geq 20$ GeV for the jets. The required $E_T^{\rm miss} > 225$ GeV is easily obtained if $m_{\tilde{\chi}^0_1} \gtrsim 150$ GeV, however the high-H_T requirement proves to be too restrictive for the weak production.

Therefore we conclude that, although electroweak production could contribute efficiently to the production of additional Z bosons, these events can not overcome the experimental cuts and can not give rise to the observed signal. With this channel, we have reviewed all relevant production cross sections of different supersymmetric particles that could potentially explain the signal. The next step would be to calculate the average number of Z bosons per parent particle Y, that we use in Eq. (1).

3 Decay of sparticles to Z bosons

Z bosons are produced through the decay chains of most MSSM particles, although the multiplicity obtained per each supersymmetric particle depends on the identity of the supersymmetric particle initially produced and on the supersymmetric spectrum below its mass. The main sources of Z bosons are decays of neutralinos and charginos and also some squarks.

We can obtain Z bosons in the decays of higgsino-like neutralinos and charginos. For instance in a usual minimal Supergravity (mSUGRA) spectrum, the second neutralino will only produce Z bosons through its (relatively small) higgsino component while the two heavier neutralinos can be expected to produce a sizeable number of Z bosons. On the other hand, charginos can produce Z bosons both through the wino and from the higgsino component but only in decays of the heavier charginos, as the lightest one will only decay to a W boson and a neutralino (or lepton-slepton if $m_{\tilde{\ell}} \leq m_{\tilde{\chi}^+}$).

Let us examine the case of the heaviest electroweak neutralinos. We obtain $\mathcal{N}(\tilde{\chi}^0_2 \to Z)$ around 0.1, as expected if the higgsino content is relatively small. Then, $\mathcal{N}(\tilde{\chi}^0_3 \to Z)$ can reach at most 0.45 while the other 50% of the decays go to $\tilde{\chi}^{\pm} W^{\mp}$. Similarly, the heavy charginos produce Z bosons in their decay as can be seen in the left panel of Fig. 3. In this case $\mathcal{N}(\tilde{\chi}^+_2 \to Z)$ can be 0.3, while we have similar branching ratios to $\tilde{\chi}^0_1 W^+$ and $\tilde{\chi}^+_1 h$.

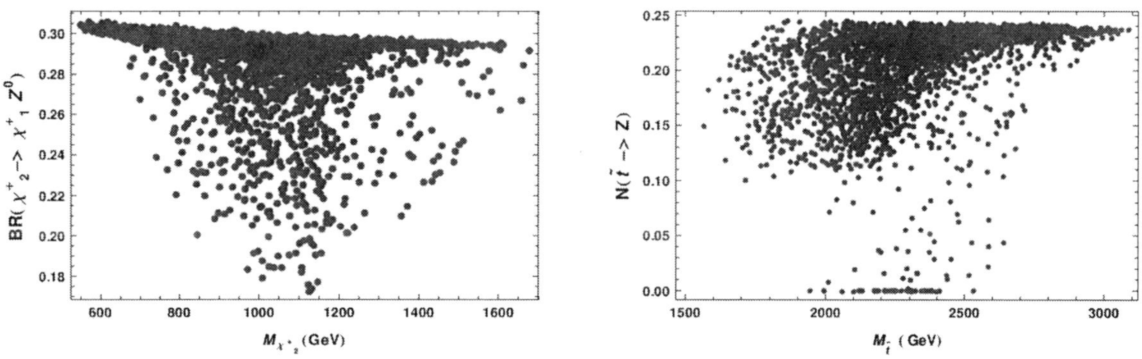

Figure 3. Average number of Z bosons, $\mathcal{N}(Y \to Z)$ in decays of $\tilde{\chi}^+_2$ (left) and $\tilde{t}$ (right) for a typical mSUGRA spectrum. From Ref. [6].

Besides chargino and neutralino decays, Z bosons couple also to sfermions through couplings that are chirality diagonal and, therefore in decays, they can only enter through chirality mixing. Although these couplings could be flavour changing, flavour mixing is bound to be small due to the stringent flavour changing neutral current constraints. Therefore we can expect a sizeable amount of Z bosons produced only through chirality mixing in third generation sfermions in decays like $\tilde{t}_2 \to \tilde{t}_1 Z$, or $\tilde{b}_2 \to \tilde{b}_1 Z$ in the large $\tan\beta$ regime.

In addition, we obtain Z bosons in the decay chains of strongly-produced sparticles. We can obtain Z bosons at different steps of the decay chain, either through the couplings of Z to sfermions or to charginos/neutralinos that we saw above. For instance, if we produce a pair of gluinos a possible decay chain would be $\tilde{g} \to \tilde{t}_2 t \to \tilde{t}_1 Z\, t \to \tilde{\chi}^+_2 b\, Z\, t \to \tilde{\chi}^+_1 Z\, b\, Z\, t \to \tilde{\chi}^0_1 W^+\, Z\, b\, Z\, t$. Therefore taking into account the corresponding branching ratios, this decay chain would contribute with $2 \times \mathrm{BR}(\tilde{g} \to \tilde{\chi}^0_1 W^+\, Z\, b\, Z\, t)$ to $\mathcal{N}(\tilde{g} \to Z)$, the number of Z bosons produced per $\tilde{g}$ produced.

As we can see in the right panel of Fig. 3, in a typical MSSM spectrum we obtain at most 0.2 Z bosons per stop (or gluino) while other squarks produce far fewer Z bosons per squark. Although the plots of Fig. 3 have been obtained from a mSUGRA spectrum, the expected number of Z bosons would be very similar in other MSSM versions, as it depends only on the spectrum below the mass of the originally produced Y particle and the content of charginos/neutralinos.

As shown in the previous section, the only possibility to explain the signal is to produce a stop or a gluino-pair as the lightest coloured sparticle, being all other squarks much heavier and only neutralinos, charginos and possibly some sleptons can be below the gluino or stop mass. Moreover, given the size of production cross sections consistent with the present searches, we need to obtain nearly one Z boson per Y particle produced. Therefore, from the expected numbers of Z bosons that we have seen in this section, we have to conclude that it is not possible to reproduce the observed signal in a MSSM with a stable (and light) neutralino.

Nevertheless, we can still consider different variations of the MSSM:

DOI: 10.1051/epjconf/201612602021

- A first possibility would be to have a light gluino below 1 TeV that can evade the bounds from jets plus missing E_T if it decays to a sufficiently heavy neutralino LSP in a sort of compressed spectrum. Under these conditions the gluino would be abundantly produced at LHC and even a small number of Z bosons per gluino could fulfil the requirements to explain the observed signal. However, this would require strongly non-universal gaugino masses and very heavy LSPs.
- A second option is to consider an MSSM where the lightest neutralino decays to a lighter gravitino plus some a Z boson. This is the case in gauge mediated MSSM [14–17] and it could be also possible in gravity mediated MSSM if the gravitino is lighter than the neutralino which then becomes the NLSP [18]. In this case, the neutralino decays to Z boson and gravitino if it is allowed by phase space and the branching ratio will depend on the lightest-neutralino composition. This is the possibility we will explore in the following.

Thus, we analyse a situation where the neutralino is the NLSP and the LSP is the gravitino. All supersymmetric particles will decay to the lightest neutralino which then decays to gravitino plus a photon, a Z boson or a Higgs. The decay width of the lightest neutralino to photon, h or Z plus gravitino [18–20] is given by

$$\begin{aligned}
\Gamma(\tilde{\chi}_1^0 \to \gamma\tilde{G}) &= \frac{|N_{11}\cos\theta_W + N_{12}\sin\theta_W|^2}{48\pi M_{Pl}^2}\frac{m_{\tilde{\chi}}^5}{m_{\tilde{G}}^2}\left[1-\frac{m_{\tilde{G}}^2}{m_{\tilde{\chi}}^2}\right]\left[1+3\frac{m_{\tilde{G}}^2}{m_{\tilde{\chi}}^2}\right], \\
\Gamma(\tilde{\chi}_1^0 \to Z\tilde{G}) &= \frac{|-N_{11}\sin\theta_W + N_{12}\cos\theta_W|^2}{48\pi M_{Pl}^2}\frac{m_{\tilde{\chi}}^5}{m_{\tilde{G}}^2}F(m_{\tilde{\chi}}, m_{\tilde{G}}, m_Z), \\
\Gamma(\tilde{\chi}_1^0 \to h\tilde{G}) &= \frac{|-N_{13}\sin\alpha + N_{14}\cos\alpha|^2}{96\pi M_{Pl}^2}\frac{m_{\tilde{\chi}}^5}{m_{\tilde{G}}^2}F(m_{\tilde{\chi}}, m_{\tilde{G}}, m_h), \qquad (2)
\end{aligned}$$

with $F(x, y, z)$ a function of the particle masses not relevant to our discussion that can be obtained from Ref. [18]. As we can see, if the lightest neutralino is bino-like, $N_{11} \simeq 1$, and the mass difference between neutralino and gravitino is larger that the Z mass, the branching ratios are $\mathrm{BR}(\tilde{\chi}_1^0 \to \tilde{G}\gamma) \simeq \cos^2\theta_W \simeq 0.8$ and $\mathrm{BR}(\tilde{\chi}_1^0 \to \tilde{G}Z) \simeq \sin^2\theta_W \simeq 0.2$. Similarly if the lightest neutralino is wino-like, the branching ratios get exchanged. From this equation we can also see that it is possible to get a very large BR to Z bosons as needed to reproduce the observed signal if $(-N_{11}\sin\theta_W + N_{12}\cos\theta_W) \simeq 1$, but this is only possible if the lightest neutralino has a very large wino component.

Although in gauge mediation models the LSP is always the gravitino, the gaugino masses in minimal models are proportional to the gauge couplings and therefore the LSP is mostly bino with small wino and higgsino components. Then, we will have to consider other extensions of the gauge mediation idea, which is the subjecy of the next section.

4 A possible explanation in General Gauge Mediation

Minimal gauge mediation predicts that all scalar and gaugino masses originate from a single scale and powers of the gauge couplings [17]. Recently a model-independent generalisation of gauge mediation was proposed under the name of General Gauge Mediation (GGM) [21, 22], where all the dependence of soft masses on the hidden sector is encoded in three real and three complex parameters obtained from a small set of current-current correlators. In these models the gaugino and sfermion masses are given by the relations

$$M_r = g_r^2 M_s \tilde{B}_r^{1/2}(0) \quad (3)$$
$$m_{\tilde{f}}^2 = g_1^2 Y_f \zeta + \sum_{r=1}^{3} g_r^2 C_2(f|r) M_s^2 \tilde{A}_r ,$$

with

$$\tilde{A}_r = -\frac{1}{16\pi^2} \int dy \left(3\tilde{C}_1^{(r)}(y) - 4\tilde{C}_{1/2}^{(r)}(y) + \tilde{C}_0^{(r)}(y)\right) , \quad (4)$$

$\tilde{B}_r^{1/2}(0)$, $C_\rho^{(r)}(y)$ (with $\rho = 0, 1/2, 1$, corresponding to scalar, fermion and vector) are associated with the current-current correlators in the hidden sector, ζ is a possible Fayet-Illiopoulos term ($\zeta = 0$ in the following), $C_2(f|r)$ the quadratic Casimirs and M_s a characteristic SUSY-breaking scale in the hidden sector.

Having six parameters, $\tilde{B}_r^{1/2}(0)$ and $\tilde{A}_r$, to fix the soft masses in the observable sector, it is clear now that we have much more freedom in GGM [23–26] and, in particular, we have

$$\frac{M_1}{g_1^2} \neq \frac{M_2}{g_2^2} \neq \frac{M_3}{g_3^2} , \quad (5)$$

as required to reproduce the observed signal at ATLAS. In particular, we need the NLSP to decay to gravitino and a Z boson with a branching ratio close to one. Fortunately, this is possible in GGM as shown in Ref. [26, 27].

In this GGM scenario, the spectrum calculator SPheno [28, 29] is used to obtain the full supersymmetric spectrum at LHC energies. We define the GGM1 parameter point with the following parameters, $M_s = 400$ TeV, $\tilde{B}_1^{1/2} = \tilde{A}_1 = 309$ TeV, $\tilde{B}_2^{1/2} = \tilde{A}_2 = 151$ TeV, $\tilde{B}_3^{1/2} = 129$ TeV, $\tilde{A}_3 = 316$ TeV and $\tan\beta = 9.8$. With these parameters we obtain the spectrum shown in Table 1.

Table 1. SUSY spectrum in the GGM1 parameter point. From Ref. [6].

Particle	$\tilde{g}$	$\tilde{\chi}_1^0$	$\tilde{\chi}_2^0$	$\tilde{\chi}_3^0$	$\tilde{\chi}_4^0$	$\tilde{\chi}_1^\pm$	$\tilde{\chi}_2^\pm$	$\tilde{G}$
Mass [GeV]	1088.0	428.4	431.34	1357.0	1360.9	429.1	1361.0	4.8×10^{-9}
Particle	$\tilde{q}_L$	$\tilde{q}_R$	$\tilde{b}_1$	$\tilde{b}_2$	$\tilde{t}_1$	$\tilde{t}_2$	$\tilde{\ell}_L$	$\tilde{\ell}_R$
Mass [GeV]	3006	2957	2876	2952	2716	2881	5863	5328
Particle	h	H	A	H^+				
Mass [GeV]	119.4	1471	1471	1473				

With respect to this spectrum, some comments are in order:

1. The two lightest neutralinos and the lightest chargino are very similar in mass, ~ 430 GeV and this allows a large neutrino mixing as required. In fact the neutralino mixing matrix is given by

$$N_{ij} \simeq \begin{pmatrix} -0.51 & 0.85 & -0.076 & 0.031 \\ 0.86 & 0.51 & -0.0024 & 0.0071 \\ -0.015 & 0.028 & 0.71 & 0.71 \\ -0.037 & 0.065 & 0.70 & -0.71 \end{pmatrix} \quad (6)$$

On the other hand the relatively large NLSP mass is needed to overcome the $E_{\mathrm{T}}^{\mathrm{miss}}$ cut.

DOI: 10.1051/epjconf/201612602021

2. The gluino is relatively light $m_{\tilde{g}} = 1088.0$ GeV which allows for a sizeable production cross section and taking into account the squark masses of order ~ 3 TeV, this mass is allowed by the latest LHC bounds.

3. The lightest Higgs mass must reproduce the observed value at LHC of $m_h \simeq 125$ GeV and in this spectrum it reaches only 119.4 GeV. This problem (typical in minimal gauge mediation models) can be solved either by increasing the stop masses taking a larger $\tilde{A}_3$ or assuming extra operators in the Higgs sector, as the dimension-5 operators proposed in Ref. [30]. Here, we assume that this problem is solved by one of these mechanisms, given it does not affect the observed phenomenology on the Z peak.

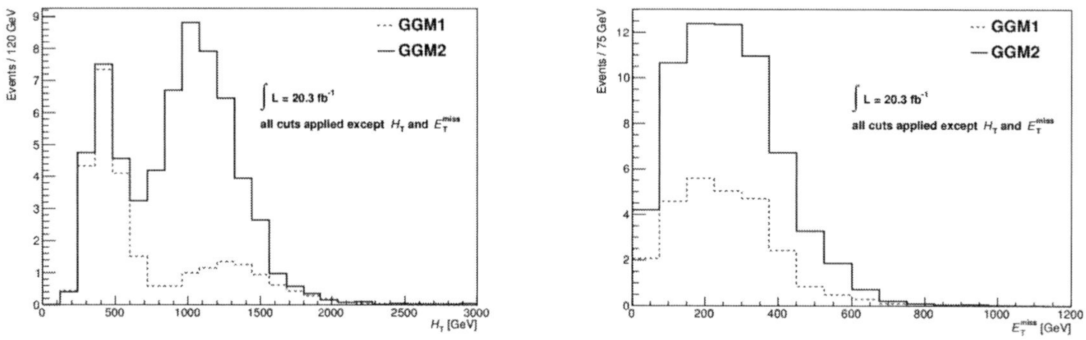

Figure 4. H_{T} (left) and $E_{\mathrm{T}}^{\mathrm{miss}}$ (right) distributions after applying the selection criteria of the Z+MET analysis [5], except for the cuts on H_{T} and $E_{\mathrm{T}}^{\mathrm{miss}}$, for the GGM1 (dashed red) and the GGM2 point (solid blue). From Ref. [6].

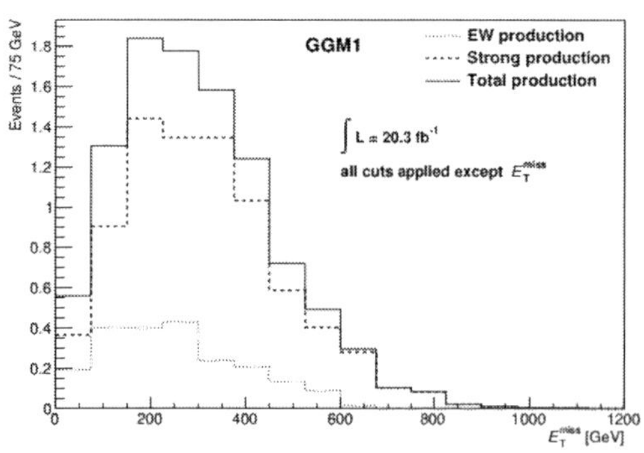

Figure 5. $E_{\mathrm{T}}^{\mathrm{miss}}$ distribution corresponding to the GGM1 point from strong production (solid blue) and electroweak production (dashed red) with the same selection as in Fig. 4, after applying the $H_{\mathrm{T}} > 600$ GeV cut. From Ref. [6].

Under these conditions, the lightest neutralino width is $\Gamma_{\tilde{\chi}_1^0} = 4.097 \times 10^{-10}$ GeV and the decay branching ratios are $\mathrm{BR}(\tilde{\chi}_1^0 \to \tilde{G}\gamma) = 1.14 \times 10^{-3}$, $\mathrm{BR}(\tilde{\chi}_1^0 \to \tilde{G}Z) = 0.997$ and $\mathrm{BR}(\tilde{\chi}_1^0 \to \tilde{G}h) = 1.35 \times 10^{-3}$. Therefore, gluinos are produced at LHC with a cross section of (8.4 ± 1.6) fb at NLL+NLO and after going through different decay chains all of them produce a Z boson plus a gravitino. In this case, the strong production represents approximately 20% of the total production cross section.

We simulate the production of supersymmetric particles at LHC at 8 TeV (LHC8) with this spectrum using Pythia8 [31] and Prospino2 [32–34] K-factors and the response of the ATLAS detector using Delphes [35]. The selection of events for this study is performed as close as possible to the search performed in ATLAS [5]. The dashed red line in Figs. 4 and 5 shows the H_T and E_T^{miss} distributions respectively for the GGM1 point after applying all the aforementioned selection cuts except for the H_T and E_T^{miss} cuts. In the H_T distribution, we can distinguish the two peaks corresponding to electroweak production at low H_T values and gluino production at higher H_T. From here, we can expect that the cut on H_T will eliminate most of the electroweak production but not the gluino production. This can be seen in Fig. 5, where the E_T^{miss} distribution is presented separately for strongly produced events (solid blue line) and for the electroweak component of the production (dashed red line) for the same selection as in Fig. 4 but after applying the H_T cut, i.e. the final selection except for the cut on E_T^{miss}. The electroweak component is significantly reduced by the H_T cut while mainly only events coming from strong production survive the cut, as expected. In fact, in this simulation of point GGM1, 99% of the gluino points and only 11% of the electroweak points have survived the H_T cut.

We see that the peak in the E_T^{miss} distribution is approximately at $m_{\tilde{\chi}_1^0}/2$, and, for $m_{\tilde{\chi}_1^0} = 425$ GeV, a reasonable fraction of the events will survive the E_T^{miss} cut at 225 GeV. In the simulation, 65% of the gluino point and 53% of the electroweak points survive this cut. However, due the relatively small production cross section, the final number of events is small. In this simulation and after applying all relevant experimental cuts, an expected signal of 6.34 ± 1.02 lepton pairs is found, to be compared with the observed excess of 19.4 ± 3.2. This number of surviving events was obtained at NLO with PYTHIA and PROSPINO2 but, unfortunately, it is still too low to explain the observations.

Trying to obtain a model able to account for the excess, we have considered a second point in our GGM scenario with a lighter gluino. The GGM2 point is obtained with $M_s = 400$ TeV, $\tilde{B}_1^{1/2} = \tilde{A}_1 = 309$ TeV, $\tilde{B}_2^{1/2} = \tilde{A}_2 = 150$ TeV, $\tilde{B}_3^{1/2} = 110$ TeV, $\tilde{A}_3 = 270$ TeV and $\tan\beta = 9.8$. With these parameters we obtain the spectrum shown in Table 2.

Table 2. SUSY spectrum for the point GGM2. From Ref. [6].

Particle	$\tilde{g}$	$\tilde{\chi}_1^0$	$\tilde{\chi}_2^0$	$\tilde{\chi}_3^0$	$\tilde{\chi}_4^0$	$\tilde{\chi}_1^\pm$	$\tilde{\chi}_2^\pm$	$\tilde{G}$
Mass (GeV)	911.4	424.9	432.7	1111.8	1117.1	425.8	1117.2	4.8×10^{-9}
Particle	$\tilde{q}_L$	$\tilde{q}_R$	$\tilde{b}_1$	$\tilde{b}_2$	$\tilde{t}_1$	$\tilde{t}_2$	$\tilde{l}_L$	$\tilde{l}_R$
Mass (GeV)	2510	2470	2400	2450	2250	2400	5890	5360
Particle	h	H	A	H^+				
Mass (GeV)	118.1	1250	1250	1253				

The neutralino mixing matrix in point GGM2 is similar to the corresponding mixing matrix in GGM1, and the $\mathrm{BR}(\tilde{\chi}_1^0 \to \tilde{G}Z) = 0.94$. However, gluino is now much lighter and the gluino-gluino cross section is now 41.6 ± 7.5 fb with PROSPINO2 at NLL+NLO, thus we can expect many more gluino pairs to be produced and a larger contribution in the final selection for this point.

The simulation for this GGM2 point is presented by the solid blue line in Fig. 4. The H_T distribution peaks at slightly lower values than in the case of GGM1, due to the slightly lower gluino mass, but it is still enough to overcome the H_T cut at 600 GeV. The E_T^{miss} distributions are similar for both GGM points, due to the very similar neutralino masses in both cases. However, in the case of GGM2 the strong production cross section is larger and much more important in relation with the electroweak production: for the GGM2 point, the strong production represents $\sim 70\%$ of the total cross section. As seen in Fig. 6, after applying the selection we obtain an expected signal of 28.0 ± 4.7 events, compatible with the excess reported by ATLAS.

We have to emphasise here that it is not difficult to obtain the observed excess for light gluino masses, and a gluino mass in between the two presented examples, $m_{\tilde{g}} \in [900, 1100]$ GeV, could

DOI: 10.1051/epjconf/201612602021

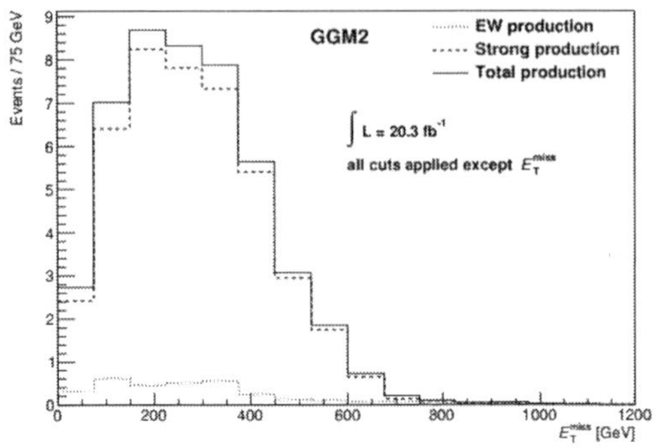

Figure 6. $E_{\mathrm{T}}^{\mathrm{miss}}$ distribution corresponding to the GGM2 point from strong production (solid blue) and electroweak production (dashed red) with the same selection as in Fig. 4, after applying the $H_{\mathrm{T}} > 600$ GeV cut. From Ref. [6].

reproduce the observed signal. However, these points may be in conflict with direct searches of jets plus $E_{\mathrm{T}}^{\mathrm{miss}}$ [7–9]. There is a tension between this excess and the bounds from gluino searches in jets plus $E_{\mathrm{T}}^{\mathrm{miss}}$. This tension will only be solved by the new data at LHC13. Therefore, we have proved that it is indeed possible to construct a supersymmetric model that accommodates the observed excess of lepton pairs on the Z peak. The simulations presented here are only a proof of existence and the final model may be very different. Nevertheless, this model will have to share the main features of the examples that we presented here.

5 Prospects for SUSY at LHC13

As we have shown in the previous sections, the excess observed in ATLAS, if due to SUSY, requires a gluino of a mass ~ 1 TeV producing nearly one Z boson per gluino in its decay. This scenario would also require relatively heavy squarks of the first generation with $m_{\tilde{q}} \gtrsim 2.5$ TeV. If this is indeed the correct explanation to the observed excess, such light gluinos would be abundantly produced at Run II in LHC together with other SUSY particles. Therefore, the results obtained at pp collisions at 13 TeV should confirm or reject this supersymmetric explanation of the ATLAS excess.

For point GGM1, the gluino pair production cross section at 8 TeV was $\sigma(pp \to \tilde{g}\tilde{g})^{\mathrm{NLL+NLO}}_{\mathrm{LHC8}} = 7.6 \pm 1.3$ fb at NLL+NLO. Similarly the production cross section at 13 TeV would be $\sigma(pp \to \tilde{g}\tilde{g})^{\mathrm{NLL+NLO}}_{\mathrm{LHC13}} = 150 \pm 16$ fb, that is, we would expect to produce 20 times more gluinos at LHC13 for the same integrated luminosity. Repeating the same exercise with point GGM2, we have a gluino pair production cross section at 8 TeV of $\sigma(pp \to \tilde{g}\tilde{g})^{\mathrm{NLL+NLO}}_{\mathrm{LHC8}} = 40 \pm 6$ fb while the production cross section at 13 TeV would be $\sigma(pp \to \tilde{g}\tilde{g})^{\mathrm{NLL+NLO}}_{\mathrm{LHC13}} = 550 \pm 70$ fb. In this point, the cross section increases by a factor ~ 15 in going from 8 to 13 TeV centre-of-mass energy. In any case, for both points this would have unambiguous signatures, both on the Z peak with a scaling of the signal found at LHC8 and in direct searches for gluinos using jets plus missing E_T in the extension of the analysis in Refs. [7–9].

Before closing, we should comment on the nature of dark matter in our scenario. As we have seen the signal seems to prefer a non-stable neutralino decaying to a very light gravitino and a Z boson. Under these conditions the neutralino mass has no relation with the dark matter abundance of the Universe and its role as dark matter component would be played by the gravitino. The gravitino mass is not bounded by the observed signal but regardless of its exact mass, unfortunately no signal of dark matter is to be expected in direct search experiments.

6 Conclusions

The ATLAS experiment has announced a 3σ excess at the Z peak consisting of 29 pairs of leptons observed to be compared with 10.6 ± 3.2 expected lepton pairs [5]. No excess outside the Z peak was observed. By trying to explain this signal with SUSY we found that only relatively light gluinos, $m_{\tilde{g}} \lesssim 1.2$ TeV, together with a heavy neutralino NLSP of $m_{\tilde{\chi}_1^0} \gtrsim 400$ GeV decaying predominantly to Z boson plus a light gravitino, such that nearly every gluino produces at least one Z boson in its decay chain, could do it.

The latter is not possible to be reproduced within minimal SUSY models, as mSUGRA, minimal gauge mediation or anomaly mediation. The requirement of a neutralino NLSP decaying to Z plus gravitino points to models of General Gauge mediation as the simplest possibility. It has been shown that a model of this class is able to reproduce the observed signal overcoming all the experimental cuts [6]. Needless to say, more sophisticated models could also reproduce the signal, however, they will always share the above mentioned features, i.e. light gluinos (or heavy particles with a strong production cross section) with an effective $\mathcal{N}(\tilde{g} \to Z) \simeq 1$.

Acknowledgments

The author acknowledges support by the Spanish Ministry of Economy and Competitiveness (MINECO) under the project FPA2012-39055-C02-01, by the Generalitat Valenciana through the project PROMETEO II/2013-017, by the Centro de Excelencia Severo Ochoa SEV-2014-0398 and by the Spanish National Research Council (CSIC) under the JAE-Doc program co-funded by the European Social Fund (ESF).

References

[1] H. P. Nilles, *Phys. Rept.* **110** (1984) 1;
S. P. Martin, *A Supersymmetry primer*, in *Perspectives on supersymmetry II* ed. G. L. Kane, World Scientific (1998) [hep-ph/9709356].

[2] For a review on dark-matter searches in colliders, see e.g.: V. A. Mitsou, *Int. J. Mod. Phys.* A **28** (2013) 1330052 [arXiv:1310.1072 [hep-ex]].

[3] G. Aad *et al.* [ATLAS Collaboration], JINST **3** (2008) S08003.

[4] L. Evans and P. Bryant, JINST **3** (2008) S08001.

[5] G. Aad *et al.* [ATLAS Collaboration], Eur. Phys. J. C **75** (2015) 318 [Erratum-ibid. **75** (2015) 463] [arXiv:1503.03290 [hep-ex]].

[6] G. Barenboim, J. Bernabeu, V. A. Mitsou, E. Romero, and O. Vives, "METing SUSY on the Z peak," arXiv:1503.04184 [hep-ph] (2015).

[7] V. Khachatryan *et al.* [CMS Collaboration], Phys. Rev. D **91** (2015) 052018 [arXiv:1502.00300 [hep-ex]].

[8] G. Aad *et al.* [ATLAS Collaboration], JHEP **1504** (2015) 116 [arXiv:1501.03555 [hep-ex]].

[9] G. Aad *et al.* [ATLAS Collaboration], JHEP **1409** (2014) 176 [arXiv:1405.7875 [hep-ex]].

[10] V. Khachatryan *et al.* [CMS Collaboration], Phys. Lett. B **736** (2014) 371 [arXiv:1405.3886 [hep-ex]].

[11] G. Aad *et al.* [ATLAS Collaboration], JHEP **1411** (2014) 118 [arXiv:1407.0583 [hep-ex]].

[12] V. Khachatryan *et al.* [CMS Collaboration], Phys. Rev. D **90** (2014) 092007 [arXiv:1409.3168 [hep-ex]].

[13] G. Aad *et al.* [ATLAS Collaboration], JHEP **1405** (2014) 071 [arXiv:1403.5294 [hep-ex]].
[14] M. Dine and A. E. Nelson, Phys. Rev. D **48** (1993) 1277 [hep-ph/9303230].
[15] M. Dine, A. E. Nelson and Y. Shirman, Phys. Rev. D **51** (1995) 1362 [hep-ph/9408384].
[16] M. Dine, A. E. Nelson, Y. Nir and Y. Shirman, Phys. Rev. D **53** (1996) 2658 [hep-ph/9507378].
[17] G. F. Giudice and R. Rattazzi, Phys. Rept. **322** (1999) 419 [hep-ph/9801271].
[18] J. L. Feng, S. Su and F. Takayama, Phys. Rev. D **70** (2004) 075019 [hep-ph/0404231].
[19] T. Moroi, "Effects of the gravitino on the inflationary universe," hep-ph/9503210 (1995).
[20] J. R. Ellis, K. A. Olive, Y. Santoso and V. C. Spanos, Phys. Lett. B **588** (2004) 7 [hep-ph/0312262].
[21] P. Meade, N. Seiberg and D. Shih, Prog. Theor. Phys. Suppl. **177** (2009) 143 [arXiv:0801.3278 [hep-ph]].
[22] M. Buican, P. Meade, N. Seiberg and D. Shih, JHEP **0903** (2009) 016 [arXiv:0812.3668 [hep-ph]].
[23] L. M. Carpenter, "Surveying the Phenomenology of General Gauge Mediation," arXiv:0812.2051 [hep-ph] (2008).
[24] A. Rajaraman, Y. Shirman, J. Smidt and F. Yu, Phys. Lett. B **678** (2009) 367 [arXiv:0903.0668 [hep-ph]].
[25] A. M. Thalapillil, JHEP **1106** (2011) 059 [arXiv:1012.4829 [hep-ph]].
[26] Y. Kats, P. Meade, M. Reece and D. Shih, JHEP **1202** (2012) 115 [arXiv:1110.6444 [hep-ph]].
[27] J. T. Ruderman and D. Shih, JHEP **1208** (2012) 159 [arXiv:1103.6083 [hep-ph]].
[28] W. Porod, Comput. Phys. Commun. **153** (2003) 275 [hep-ph/0301101].
[29] W. Porod and F. Staub, Comput. Phys. Commun. **183** (2012) 2458 [arXiv:1104.1573 [hep-ph]].
[30] M. Dine, N. Seiberg and S. Thomas, Phys. Rev. D **76** (2007) 095004 [arXiv:0707.0005 [hep-ph]].
[31] T. Sjostrand, S. Mrenna and P. Z. Skands, Comput. Phys. Commun. **178** (2008) 852 [arXiv:0710.3820 [hep-ph]].
[32] W. Beenakker, R. Hopker, M. Spira and P. M. Zerwas, Nucl. Phys. B **492** (1997) 51 [hep-ph/9610490].
[33] W. Beenakker, M. Kramer, T. Plehn, M. Spira and P. M. Zerwas, Nucl. Phys. B **515** (1998) 3 [hep-ph/9710451].
[34] W. Beenakker, M. Klasen, M. Kramer, T. Plehn, M. Spira and P. M. Zerwas, Phys. Rev. Lett. **83** (1999) 3780 [Erratum-ibid. **100** (2008) 029901] [hep-ph/9906298].
[35] J. de Favereau *et al.* [DELPHES 3 Collaboration], JHEP **1402** (2014) 057 [arXiv:1307.6346 [hep-ex]].

EPJ Web of Conferences **126**, 02022 (2016)
ICNFP 2015
DOI: 10.1051/epjconf/201612602022

Heavy flavors in nucleus-nucleus and proton-nucleus collisions

Marzia Nardi[1,a]

[1] *Istituto Nazionale di Fisica Nucleare - Sezione di Torino*
Via P. Giuria 1 - 10125 Torino (Italy)

Abstract. A multi-step setup for heavy-flavor studies in high-energy nucleus-nucleus (AA) and proton-nucleus (pA) collisions is presented. The propagation of the heavy quarks in the medium is described in a framework provided by the relativistic Langevin equation, here solved using weak-coupling transport coefficients. Successively, the heavy quarks hadronize in the medium. We compute the nuclear modification factor and the elliptic flow parameter of the final D mesons both in AA and in pA collisions and compare our results to experimental data.[b]

1 Introduction

The primary goal of the ongoing heavy-ion collision experiments at the Relativistic Heavy Ion Collider (RHIC) and at the Large Hadron Collider (LHC) is to create and study a new state of matter, the Quark-Gluon Plasma (QGP), where quarks and gluons are no longer confined, but can freely move over distances much larger than the typical hadronic size.

A nucleus-nucleus collision is a very complicated process undergoing several stages: right after the first interaction between the incident nuclei (or proton and nucleus) a dense system of partons is produced, which shortly reach a thermal equilibrium (at least locally) and form the QGP. It then expands, cools down and decades into a system of interacting hadrons. The hadronic interactions cease when the system, still expanding and colling, is too dilute. The final hadrons (or their decay products) are then detected in the experimental apparatus.

Many observables has been proposed to study the properties of the QGP. Among them, the heavy quarks (charm and beauty) play a special role, for the following reasons:

- their mass M is much larger than Λ_{QCD}: it is possible to compute cross sections and p_T spectra with perturbative techiques from the QCD theory (next-to-lead order calculations are, presently, the "status of the art");
- M in also larger than the average temperature of the medium formed in the collision: their thermal production is expected to be negligible both in the plasma and, even more, in the hadronic phase;
- they are created at the very beginning of the collision between the two incident nuclei (or proton and nucleus) and witness all the subsequent evolution;

[a] e-mail: nardi@to.infn.it

[b] Work done in collaboration with A. Beraudo, A. De Pace, M. Monteno and F. Prino (INFN, Torino)

EPJ Web of Conferences **126**, 02022 (2016) DOI: 10.1051/epjconf/201612602022
ICNFP 2015

- $M \gg gT$, with gT being the typical momentum exchange in the collision with the plasma particle: many soft scattering are needed to change significantly the momentum (and trajectory) of the heavy quark[1]; therefore heavy quarks do not thermalize with the medium in the later stage, they do not loose "memory" of their initial momentum and energy and carry precious information about the QGP phase.

In a series of papers [1–4] over the last few years we developed a complete setup (referred to as POWLANG) for the study of heavy flavour observables in high-energy nucleus-nucleus (AA) collisions, describing the initial hard production of the $Q\overline{Q}$ pairs and the corresponding parton-shower stage through the POWHEG-BOX package [5, 6] and addressing the successive evolution in the plasma through the relativistic Langevin equation. Here, following Ref. [4], we supplement our numerical tool by modeling the hadronization of the heavy quarks accounting for the presence of a surrounding medium made of light thermal partons feeling the collective flow of the local fluid cell. Moreover, we present our first (preliminary) results for proton-nucleus (pA) collisions at LHC energies.

2 Heavy flavour in proton-proton collisions

Because of their large mass, the initial production of c and b quarks is a short-distance process involving a large-momentum transfer, described by perturbative QCD (pQCD). For this purpose we rely on a standard pQCD public tool, namely POWHEG-BOX, in which the hard $Q\overline{Q}$ event is interfaced with a shower stage described by PYTHIA [7], to include the effects of Initial- and Final-State Radiation [1, 2].

Experimental data obtained in proton-proton (pp) collisions can be exploited to validate the theoretical calculations used to simulate the initial hard $Q\overline{Q}$ production. In our setup, the heavy quarks are created in pairs by the POWHEG-BOX event generator. Their momenta are not back-to-back neither along the beam-axis, due to the different Bjorken-x carried by the partons taking part in the hard event, nor in the azimuthal plane, due to the gluon radiation occurring during the hard process or the shower stage and also to the intrinsic k_T-broadening one can include in the simulation. Eventually, heavy-quark hadronization and the final decays of the D (B) mesons are simulated with PYTHIA, which is also used to describe the parton shower stage.

In Fig. 1 we show some results for pp collisions. In the left panel, the inclusive p_T distribution of D^0 mesons [3] is compared to experimental data measured by the ALICE collaboration [8]. Also shown is the FONLL systematic uncertainty band [9]. The right panel shows $D-h$ azimuthal correlations compared to preliminary ALICE data [10, 11], for two different p_T-intervals of the charmed meson. In our simulation D^0, $\bar{D}^0$ and $D^{\pm}$ are used as trigger particles and the light hadrons are limited to charged pions and kaons, protons and antiprotons, excluding the weak decays of Λ and K^0. Any D-meson is correlated with all the light hadrons (except its own decay products) created in the same event. The near-side peak takes contribution both from correlations present at the partonic level (from $Q\overline{Q}$ pairs arising from gluon splitting) and from hadrons coming from the fragmentation of the same string of the parent heavy quark. Our results include also the simulation of the Underlying Event due to Multiple Parton Interactions, performed with PYTHIA 6.4, which gives rise to the pedestal observed in Fig. 1.

The agreement between the results of POWHEG+PYTHIA event generator and the experimental data is quite good: this makes us confident that the production process is reasonably well descripted and can be used as a starting point for the simulation of AA and pA collisions.

[1]For realistic temperature $g \sim 2$, so the present condition, at least in the initial stage of the evolution, is only marginally fulfilled by charm quarks.

DOI: 10.1051/epjconf/201612602022

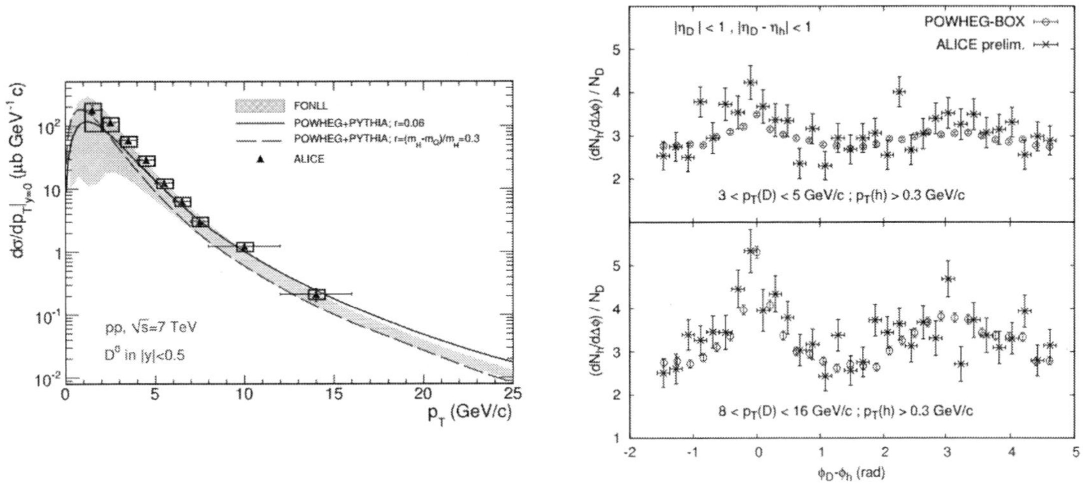

Figure 1. Results for pp collisions at $\sqrt{s_{NN}} = 7$ TeV compared to experimental data by the ALICE Collab. [8, 10, 11]. Left: p_T spectrum for D^0 mesons. Right: azimuthal $D-h$ correlations for different p_T-cuts.

3 Heavy flavour in pA and AA collisions: initial conditions

In nuclear collisions the $c\bar{c}$ or $b\bar{b}$ production given by the pQCD calculation must be modified in two ways: *i*) the nuclear parton distribution functions should be corrected for shadowing or antishadowing effects (we have adopted here the EPS09 scheme [12]); *ii*) the colliding partons acquire, on average, a larger transverse momentum during the crossing of the two nuclei (Cronin effect), which can be described by a Glauber calculation [2].

Moreover, their initial positions are distributed in the transverse plane according to the density of nucleon-nucleon collision, as computed in the (optical) Glauber model.

The heavy quarks created in nuclear collisions propagate in a strongly interacting and non-static medium, whose properties and evolution is described through hydrodynamical calculations, performed with the viscous 2+1 code of Ref. [13] (for the AA case), or with ECHO-QGP [14] (for the pA case, in 2+1 dimensions for simplicity). The assumption of longitudinal boost-invariance (implicit in the 2+1 description) restricts the validity of our calculation to a limited region around mid-rapidity.

While for the initial conditions in AA collisions one can adopt the smooth results provided by an optical-Glauber calculation, in a pA collision the event-by-event fluctuations in the initial state are extremely important, since they are the main source of anisotropic flow in the final state.

We assumed that each nucleon-nucleon collisions, occurring at the transverse coordinate $\mathbf{x}_i$ in a give initial distribution (the single nucleons being randomly located with a distribution probability given by a realistic nuclear density), deposit some entropy in the transverse plane, distributed according to a gaussian form centered around the scattering position and depending on the smearing parameter σ. For the numerical calculation we tested two values of σ : 0.2 and 0.4 fm.

The total entropy distribution thus results in:

$$s(\mathbf{x}) = \frac{K}{2\pi\sigma^2} \sum_{i=1}^{N_{coll}} \exp\left[-\frac{(\mathbf{x}-\mathbf{x}_i)^2}{2\sigma^2}\right] \tag{1}$$

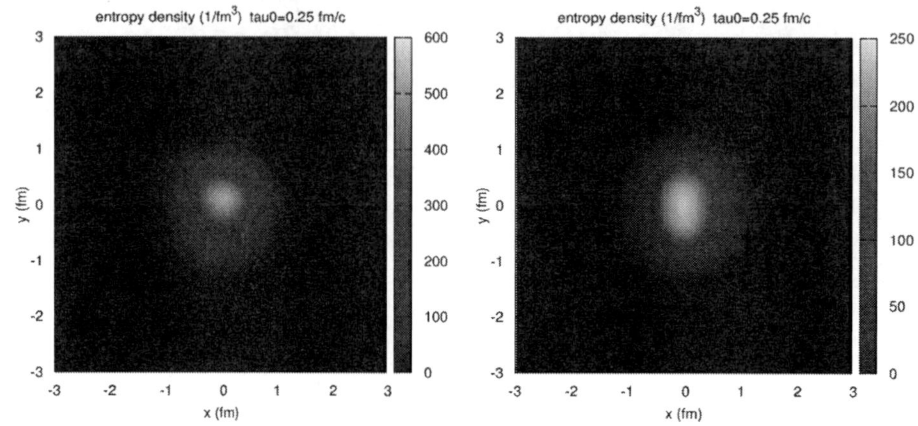

Figure 2. The initial entropy-density profile for a single p-Pb collision at $\sqrt{s_{NN}} = 5.02$ TeV (left panel) and the result of the weighted average of the 0-20% most central events (right panel).

The multiplicative constant K can be fixed, for instance, by matching the initial entropy density calculated with the Glauber Model in the optical approximation and in the Monte Carlo simulation. The initial eccentricity, which translate into a non-vanishing elliptic flow [15], can be then evaluated as (the brackets denote the average in the transverse plane, with the entropy density in Eq. 1 as a weight):

$$\epsilon_2 = \frac{\sqrt{\{y^2 - x^2\}^2 + 4\{xy\}^2}}{\{y^2 + x^2\}} \, .$$

Since a full event-by-event simulation with our hydro+transport setup would be really demanding, for the huge computing and storage resources required, we evaluate a realistic average background as follows: for a given centrality class we average over all the events of the considered percentile (based on the number of N_{part}); After rotating each of them by the event-plane angle ψ_2, we compute an average entropy-density distribution from which a realistic average eccentricity ϵ_2 can be obtained. In Fig. 2 we shown an example of a single event (left) and the average distribution (right), for a p-Pb collision at $\sqrt{s_{NN}} = 5.02$ TeV.

4 Propagation of the Heavy Quarks through the medium

The time evolution of the heavy-quark phase-space distribution $f_Q(t, x, \boldsymbol{p})$ can be described by the Boltzmann equation:

$$\frac{\mathrm{d}}{\mathrm{d}t} f_Q(t, x, \boldsymbol{p}) = C[f_Q] \, ; \tag{2}$$

$C[f_Q]$ is the Boltzmann collision integral, taking into account losses and gains due to single-particle collisions [16]:

$$C[f_Q] = \int \mathrm{d}\boldsymbol{k} \, [w(\boldsymbol{p}+\boldsymbol{k}, \boldsymbol{k}) f_Q(\boldsymbol{p}+\boldsymbol{k}) - w(\boldsymbol{p}, \boldsymbol{k}) f_Q(\boldsymbol{p})] \, ,$$

where $w(p, k)$ is the transition rate for a heavy quark changing its momentum from $\boldsymbol{p}$ to $\boldsymbol{p} - \boldsymbol{k}$.

Neglecting x- and mean field dependence of f_Q and expanding the collision integral for small momentum exchange (Landau approximation [17]) the collision integral can be expressed as:

$$C[f_Q] \simeq \int \mathrm{d}\boldsymbol{k} \left[k^i \frac{\partial}{\partial p^i} + \frac{1}{2} k^i k^j \frac{\partial^2}{\partial p^i \partial p^j} \right] [w(\boldsymbol{p}, \boldsymbol{k}) f_Q(t, \boldsymbol{p})] \; ;$$

the Boltzmann equation reduces to the Fokker-Planck equation:

$$\frac{\partial}{\partial t} f_Q(t, \boldsymbol{p}) = \frac{\partial}{\partial p^i} \left\{ A^i(\boldsymbol{p}) f_Q(t, \boldsymbol{p}) + \frac{\partial}{\partial p^j} [B^{ij}(\boldsymbol{p}) f_Q(t, \boldsymbol{p})] \right\}$$

where

$$\begin{aligned} A^i(\boldsymbol{p}) &= \int \mathrm{d}\boldsymbol{k}\, k^i w(\boldsymbol{p}, \boldsymbol{k}) = A(p) p^i \\ B^{ij}(\boldsymbol{p}) &= \frac{1}{2} \int \mathrm{d}\boldsymbol{k}\, k^i k^j w(\boldsymbol{p}, \boldsymbol{k}) = B_0(p) + (\delta^{ij} - \hat{p}^i \hat{p}^j) B_1(p) \,. \end{aligned}$$

The problem is reduced to the evaluation of three transport coefficients: a friction term $A(p)$ and two terms describing the momentum broadening $B_0(p)$ and $B_1(p)$.

The Fokker-Planck equation can be then recast into a form suitable to follow the dynamics of each individual quark: the Langevin equation

$$\frac{\Delta p^i}{\Delta t} = -\eta_D(p) p^i + \xi^i(t),$$

with the properties of the noise encoded in

$$\langle \xi^i(\boldsymbol{p}_t) \xi^j(\boldsymbol{p}_{t'}) \rangle = b^{ij}(\boldsymbol{p}_t) \frac{\delta_{tt'}}{\Delta t} \qquad b^{ij}(\boldsymbol{p}) = \kappa_{\parallel}(p) \hat{p}^i \hat{p}^j + \kappa_{\perp}(p)(\delta^{ij} - \hat{p}^i \hat{p}^j) \,.$$

The transport coefficients $\kappa_{\perp}$ and $\kappa_{\parallel}$ can be calculated within a weakly-coupled scenario: perturbative QCD and "Hard Thermal Loop" (HTL) approximation, as done in Ref. [1]. We note that the transport coefficients computed with pQCD+HTL have a strong dependence on the momentum p of the heavy quark and on the temperature T of the system.

The Langevin approach is a very convenient numerical tool and allows one to establish a link between observables and transport coefficients derived from QCD. However, it was derived starting from a soft-scattering expansion of the collision integral $C[f]$ truncated at second order (friction and diffusion terms): in Ref. [16] it was found that this approximation is excellent for b-quarks for all the evolution of the system, while for c- quarks it is valid, in a short time interval of a few fm/c after the initial collision, only for small momenta.

Having all the ingredients, we can now procede to simulate the heavy-quark propagation in the medium. This occurs through several steps, repeated iteratively until hadronization:

1. a heavy quark with momentum p is located in a position $\boldsymbol{x}$ at a given time t: the hydrodynamical simulation tells us the conditions (temperature $T(\boldsymbol{x}, t)$, energy density $\varepsilon(\boldsymbol{x}, t)$, four-velocity $u^{\mu}(\boldsymbol{x}, t)$,...) of the surrounding fluid cell;

2. we perform a Lorentz boost in the local fluid rest-frame and set the transport coefficients using the temperature and the momentum $\tilde{p}$ of the quark (in this reference system);

3. we use the Langevin equation to compute the new quark momentum $\tilde{p}'$ and its new position $\tilde{\boldsymbol{x}}'$ at the time $t + \Delta t$;

4. we perform another Lorentz boost back to the laboratory frame; we check whether the surrounding fluid cell is in the QGP phase (i.e. its temperature is higher than the deconfining temperature T_{dec}: if so, we restart from step 1, otherwise the quark hadronizes.

The time step used in our simulation is $\Delta t = 0.02$ fm/c, for the deconfining temperature T_{dec} we have tested two values: $T_{dec} = 155$ MeV (the value quoted in Lattice calculation for the QGP-hadron transition) and $T_{dec} = 170$ MeV, allowing the heavy quarks to form bound states with light quarks which may survive, in a small temperature range, in the deconfined phase [18].

5 Hadronizations of the Heavy Quarks

Finally, after the propagation through the dense medium, the heavy quarks decouple and hadronize. In the pp case, the hadronization is performed with PYTHIA, while for the nuclear collisions we introduce a new simple model to take into account the effect of the thermalized medium.

Once a heavy quark Q, during its stochastic propagation in the fireball, has reached a fluid cell below the decoupling temperature $T_{\rm dec}$, it is forced to hadronize. One extracts then a light antiquark $\overline{q}$ (up, down or strange, with relative thermal abundances dictated by the ratio $m/T_{\rm dec}$) from a thermal momentum distribution corresponding to the temperature $T_{\rm dec}$ in the local rest frame of the fluid; the local fluid four-velocity u^μ allows one to boost the momentum of $\overline{q}$ to the laboratory frame. A string is then constructed joining the endpoints given by Q and $\overline{q}$ and is then given to PYTHIA 6.4 to simulate its fragmentation into hadrons (and their final decays). In case the invariant mass of the string is not large enough to allow its decay into at least a pair of hadrons the event is resampled, extracting a new thermal parton to associate to the heavy quark.

With this hadronization scheme we can first of all provide a realistic estimate of the role of the thermal light quarks to explain peculiar features of the D meson spectra at low and moderate p_T; secondly, the complete information on all the final state particles arising from the fragmentation of the strings allows us to provide theory predictions for observables like $D-h$, $e-h$, e^+-e^-... angular correlations to be compared to existing data and possibly used as a guidance to future measurements.

Notice that further possible interactions in the hadronic phase, which might enhance the elliptic flow, are here neglected: we plan to include them in our future work.

6 Results: AA collisions

In Fig. 3 we show some POWLANG outcomes for the R_{AA} of D^0 mesons in central $(0-10\%)$ Au-Au collisions at $\sqrt{s_{NN}}=200$ GeV. HTL transport coefficients are employed and the difference between the two hadronization schemes (here taken to occur at $T_{\rm dec} = 170$ MeV, either with vacuum fragmentation or with in-medium string fragmentation, are clearly visible: in the second case the radial flow of the light thermal parton leads to the development of a bump around $p_T \sim 1.5$ GeV in qualitative agreement with the experimental data. Also shown for comparison is the result for the limiting scenario in which charmed particles reach full kinetic equilibrium with the medium [4]. Our results are compared to STAR data[19].

The ALICE Collaboration has recently released the data for the in-plane/out-of-plane nuclear modification factors of D mesons in semi-central (30- 50%) Pb-Pb collisions [20]. Experimental findings are displayed in Fig. 4 and compared to the transport predictions of the POWLANG setup: also in this case model results with in-medium fragmentation look in better agreement with the experimental data. The theoretical results are given for two different values of the deconfining temperature: we notice that the nuclear modification factor R_{AA} is not very sensitive to the value of $T_{\rm dec}$.

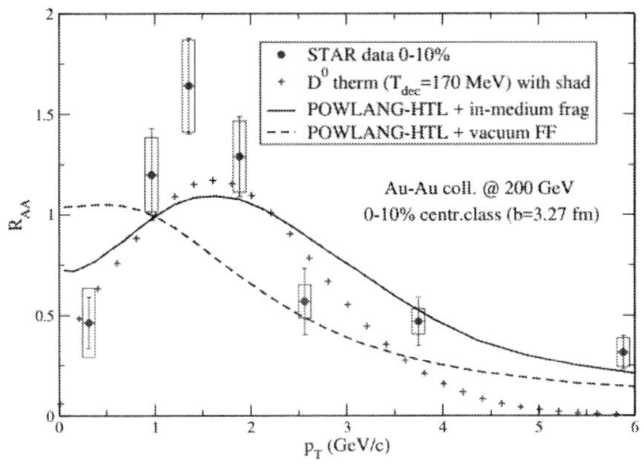

Figure 3. The R_{AA} of D^0 mesons in central Au-Au collisions at $\sqrt{s_{NN}} = 200$ GeV. POWLANG results obtained with HTL transport coefficients and a decoupling temperature $T_{\rm dec} = 170$ MeV are plotted. Also shown for comparison is the extreme case of full kinetic thermalization of D mesons. Theory curves are compared to STAR data [19]

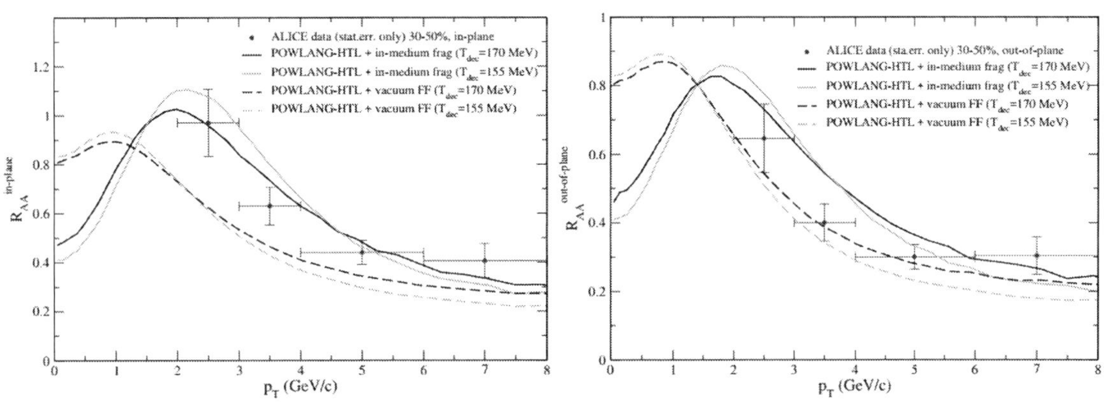

Figure 4. R_{AA} in-plane (left) and out-of-plane (right) of D mesons. ALICE data in the $30-50\%$ centrality class [20] are compared to POWLANG results obtained with different hadronization mechanisms (in-medium and vacuum fragmentation) and decoupling temperatures ($T_{\rm dec} = 170$ MeV and 155 MeV).

In Fig. 5 we address the v_2 of D mesons in semicentral (30-50%) Pb-Pb collisions at $\sqrt{s_{NN}} = 2.76$ TeV at the LHC. The effect of the new procedure for in-medium hadronization through string fragmentation is clearly visible: while POWLANG outcome with standard in-vacuum fragmentation of charm largely underpredicts the observed v_2, the additional flow acquired from the light thermal partons move the theory curves with in-medium hadronization closer to the experimental data measured by the ALICE Collaboration [21]. The value chosen for the decoupling temperature is $T_{\rm dec} = 155$ MeV; we observed that v_2 is more sensitive to $T_{\rm dec}$ than other observables, the lower value seeming to be preferred by the data: this agrees with the expectation that the elliptic flow needs more time to fully develop with respect to the quenching of the spectra. This plot also shows how a full kinetic thermalization up to large values of p_T is disfavoured by the data.

Finally we decide to address more differential observables like angular correlations between heavy flavour particles (or their decay products) and the charged hadrons produced in the same collision. In Fig. 6 we display our results for $D-h$ azimuthal correlations. All figures are obtained with weak-coupling HTL transport coefficients in the QGP phase. In general one observes a strong suppression of the away-side peak around $\Delta\phi = \pi$. Depending on the cuts imposed on the trigger particles (D-mesons) and on the associated hadrons this can be mostly due either to the energy loss (moving particles below

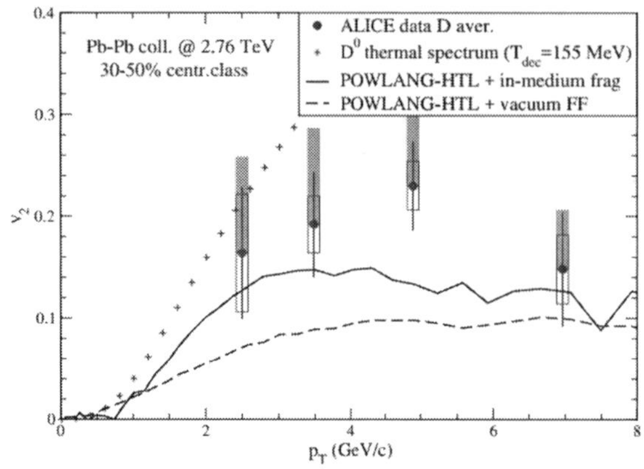

Figure 5. The v_2 of D mesons in Pb-Pb collisions at $\sqrt{s_{NN}} = 2.76$ TeV. POWLANG results (with HTL transport coefficients) with in-vacuum and in-medium HQ fragmentation at the decoupling temperature $T_{\rm dec}$ = 155 MeV compared to ALICE data [21] in the 30-50% centrality class and to the limit of kinetic thermalization.

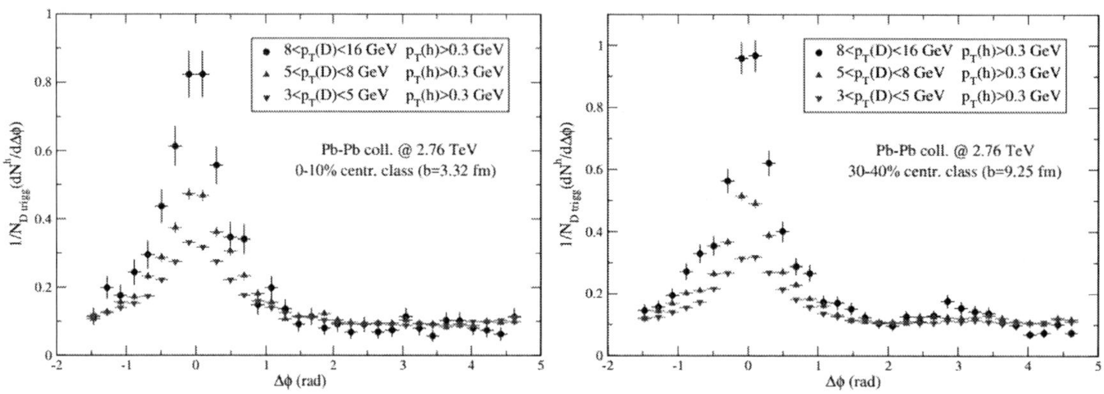

Figure 6. *D–h* correlations in Pb-Pb collisions at $\sqrt{s_{\rm NN}} = 2.76$ TeV for various p_T cuts on the trigger particle and different centrality classes: 0–10% (left) and 30–40% (right).

the p_T-cut) or to the angular decorrelation (moving particles away from $\Delta\phi = \pi$) of the parent heavy quark.

7 Results: pA collisions

For quite some time, it was commonly believed that in pA collision the formation of QGP can not occur, because the small size of the system does not allow to reach a sufficiently dense system in thermal equilibrium. Such measurements were considered just a benchmark to estimate cold nuclear-matter effects. However, recent experimental data on central p-Pb collisions at the LHC and d-Au (and ^{3}He-Au) at RHIC may suggest the formation of a medium with collective behaviour even in such small systems.

It is therefore very important to make accurate simulations of these processes in order to give a correct interpretation of current experimental data.

We show some preliminary POWLANG results for c-quarks in p-Pb collisions at LHC. Work is still in progress to extend to calculation to the bottom quarks and to simulate d-Au collisions at RHIC.

In Fig. 7 we show how the formation of a hot deconfined medium in the p-Pb collision at $\sqrt{s_{NN}}$ = 5.02 TeV affects the heavy quark and hadron spectra, by modifying their propagation and subsequent

DOI: 10.1051/epjconf/201612602022

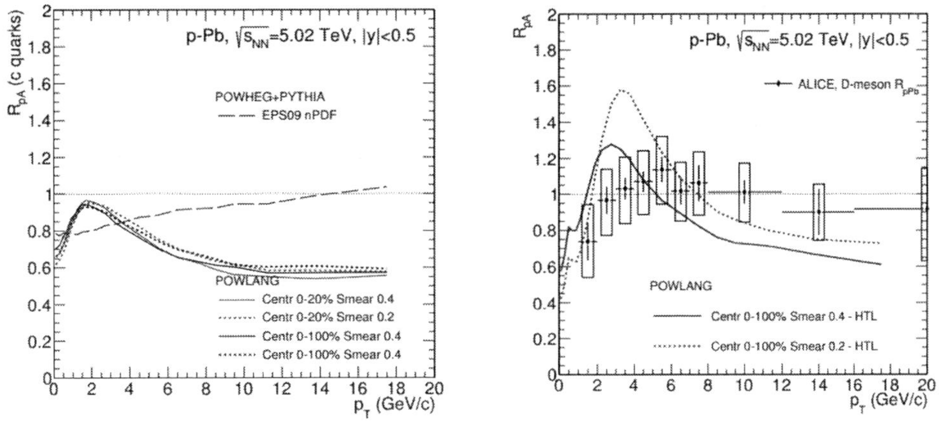

Figure 7. The nuclear modification factor of c-quarks (left) and for D mesons (right) in p-Pb collisions at $\sqrt{s_{NN}} =$ 5.02 TeV. POWLANG results (with HTL transport coefficients) with in-medium heavy-quark fragmentation, for σ = 0.2 and 0.4 fm. Experimental data from Ref. [22]

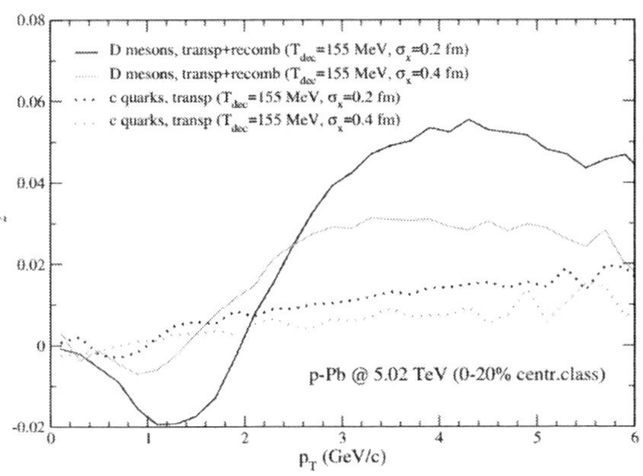

Figure 8. The v_2 of D mesons in p-Pb collisions at $\sqrt{s_{NN}} = 5.02$ TeV. POWLANG results (with HTL transport coefficients) with in-medium heavy-quark fragmentation, for σ = 0.2 and 0.4 fm. The v_2 coefficients for c-quarks alone are plotted for comparison.

hadronization. The left panel shows the nuclear modification factor for c-quarks: the blue, dashed line shows the result for the quarks produced by initial-state interaction, according to the EPS09 parton distribution functions (before the propagation in the expanding medium); the dotted and solid lines give the analogous result for the charm quarks after the propagation (both for central and minimum bias collisions, with two different choices of the smearing parameter od Eq. 1). The difference between the EPS09 curve and the other ones is a clear evidence of the effect of Langevin dynamics on the c-quark even in such a small system. In the right panel we show the result for the final D meson, with the hadronization occurring in the medium as explained in Sec. 5. The peak at $p_T = 3 \div 4$ GeV/c (more pronounced for $\sigma = 0.2$ fm) is a consequence of the radial flow of light quarks inherited by the meson. The POWLANG result is compatible, within errorbars, with the experimental data measured by the ALICE Collaboration [22].

In Fig. 8 the elliptic flow coefficient v_2 is plotted, both for c-quarks (dotted lines) and for D mesons, for a deconfining temperature of 155 MeV. Again, the contribution of the light-quark flow, inherited by the charmed meson in the hadronization process, is crucial to enhance the v_2.

8 CONCLUSIONS

The simple model to describe heavy quark hadronization in the presence of a hot deconfined medium (a Quark-Gluon Plasma) has considerably improved the agreement of the POWLANG results with the experimental data at RHIC and LHC energies. In particular, results for the nuclear modification factors and the elliptic flow of D mesons have been presented.

Recently, signatures of final-state medium effects were observed also in pA collisions: we have presented some preliminary results for p-Pb collisions at LHC energies, and we plan to continue this study in more detail in the next future.

References

[1] A. Beraudo, A. De Pace, W.M. Alberico, A. Molinari, Nucl. Phys. A **831**, 59 (2009)
[2] W.M. Alberico, A. Beraudo, A. De Pace, A. Molinari, M. Monteno, M. Nardi, F. Prino, Eur. Phys. J. C **71**, 1666 (2011)
[3] W.M. Alberico, A. Beraudo, A. De Pace, A. Molinari, M. Monteno, M. Nardi, F. Prino, M. Sitta, Eur.Phys.J. **C73**, 2481 (2013)
[4] A. Beraudo, A. De Pace, M. Monteno, M. Nardi and F. Prino, Eur.Phys.J. **C75**, 121 (2015) 3
[5] S. Frixione, P. Nason, G. Ridolfi, JHEP 0709 (2007) 126.
[6] S. Alioli, P. Nason, C. Oleari and E. Re, JHEP 1006 (2010) 043.
[7] T. Sjostrand, S. Mrenna and P.Z. Skands, JHEP 0605 (2006) 026.
[8] ALICE Collaboration (B.I. Abelev et al.), J. High Energy Phys. **1201**, 128 (2012)
[9] M. Cacciari, M. Greco, P. Nason, J. High Energy Phys. **9805**, 007 (1998); J. High Energy Phys. **0103**, 006 (2001)
[10] S. Bjelogrlić (ALICE Collab.), Nucl. Phys A **931**, 563 (2014)
[11] A. Rossi (ALICE Collab.), Nucl. Phys A **932**, 51 (2014)
[12] K.J. Eskola, H. Paukkunen, C.A. Salgado, J. High Energy Phys. **0904**, 065 (2009)
[13] P. Romatschke, U. Romatschke, Phys. Rev. Lett. **99**, 172301 (2007).
[14] ECHO-QGP project: `http://theory.fi.infn.it/echoqgp`
[15] H. Niemi, K. J. Eskola and R. Paatelainen, arXiv:1505.02677 [hep-ph].
[16] V. Greco et al., Phys. Rev. **C 90** (2014) 4, 044901
[17] B. Svetitsky Phys. Rev. **D 37**, 2484 (1988)
[18] W. M. Alberico et al. Phys. Rev. D **72** (2005) 114011.
[19] L.Adamczyk *et al.* (STAR Collab.), arXiv:1404.6185 [nucl-ex].
[20] ALICE Collaboration (B. Abelev *et al.*) 2014 *Phys. Rev.* C **90** 034904
[21] B. Abelev (ALICE Collaboration), Phys. Rev. Lett. **111** (2013) 102301.
[22] B. B. Abelev (ALICE Collaboration), Phys. Rev. Lett. **113** (2014) 23, 232301.

Highlights from Super-Kamiokande

Kimihiro OKUMURA[1,2,a]

[1] *Research Center for Cosmic Neutrinos, Institute for Cosmic Ray Research, University of Tokyo, Kashiwa, Chiba 277-8582, Japan*
[2] *Kavli Institute for the Physics and Mathematics of the Universe (WPI), The University of Tokyo Institutes for Advanced Study, University of Tokyo, Kashiwa, Chiba 277-8582, Japan*

Abstract. Recent results from Super-Kamiokande experiment are reviewed in this paper; Neutrino mass hierarchy and CP violation in the lepton sector are investigated via $\nu_\mu \rightarrow \nu_e$ oscillation of the atmospheric neutrinos. The event rate, correlation with solar activity, energy spectrum of the solar neutrinos are measured via electron elastic scattering interactions. Neutrino emission from the WIMP annihilation at the center of the Sun are searched in the GeV energy regions. New project, SK-Gd project, to enhance anti-neutrino identification capability, has been approved inside the collaboration group.

1 The Super-Kamiokande detector

Figure 1 shows the schematic view of Super-Kamiokande (SK) detector. SK is a water Cherenkov detector, which size is 41.4 m and 39.3 m in height and diameter, respectively. It is located about 1000 m underground in Kamioka mine of Gifu prefecture, Japan. The total volume is 50 kton, and the fiducial volume for analysis is 22.5 kton. The detector is separated into the inner (ID) and the outer (OD) detectors. In the inner detector, 11129 20" photomultipliers (PMTs) are placed in grid on the detector wall to take the image of Cherenkov lights, which are produced by charged particles. In the outer detector 1885 8" PMTs are placed to detect activities coming from outside such as cosmic ray muons.

SK has been running almost 20 years from April 1996. The first phase of the experiment, SK-I, began with the original configuration of 40% photo coverage. The detector underwent an implosion accident losing about half of PMTs in 2001. The rest of PMTs are re-arranged uniformly, and SK-II resumed with 20% photo coverage. After full recovery, SK-III started with 40% coverage from 2006. The frontend electronics were upgraded in SK-IV.

SK is a multi-purpose detector with rich physics. Neutrino oscillations have been studied with solar and atmospheric neutrinos. SK also plays a role of the far detector in T2K long baseline experiment, where neutrino oscillation has been studied with accelerator neutrinos. At the same time, we are searching for astronomical neutrinos, such as supernova burst and supernova relic neutrinos. Proton decay is predicted in GUT theory, which is the physics beyond standard model. WIMP annihilation signal from astronomical objects, such as the Sun, may be observable if the mass range is in the energy range from GeV to TeV.

[a] e-mail: okumura@icrr.u-tokyo.ac.jp

This paper reviews the recent results and status of SK. It goes with atmospheric neutrino (Sec. II), solar neutrino (Sec. III), solar WIMP search (Sec. IV), SK-Gd project (Sec. V), and summary (Sec. VI).

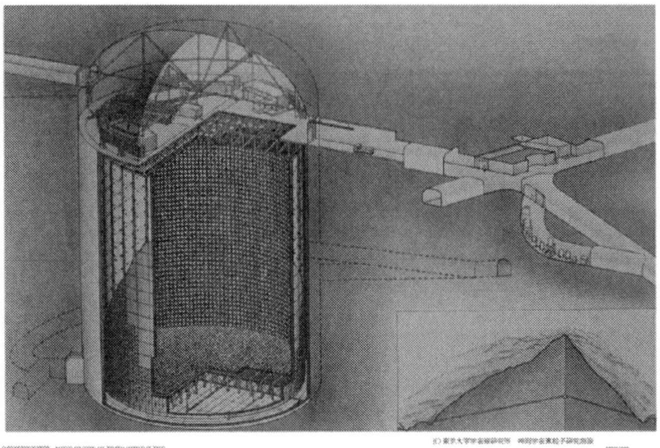

Figure 1. Schematic drawing of Super-Kamiokande detector.

Table 1. Experimental phase, observation period, and detector configuration of Super-Kamiokande

Phase	Period	# of ID PMTs (photo coverage)
SK-I	Apr. 1996 – Jul. 2001	11146 (40%)
SK-II	Oct. 2002 – Oct. 2005	5182 (20%)
SK-III	Jul. 2006 – Aug. 2008	11129 (40%)
SK-IV	Sep. 2008 – Present	11129 (40%)

2 Oscillation analysis with atmospheric neutrino

Since the existence of the neutrino mass became evident in 1998 [1], three-flavor mixing of the neutrino oscillation is widely accepted as the standard theory. The flavor eigenstate of three neutrinos (ν_α, α=e,μ,τ) are expressed by the mass eigenstate (ν_i, i=1,2,3) and PMNS [2, 3] mixing matrix as follows:

$$\begin{pmatrix} \nu_e \\ \nu_\mu \\ \nu_\tau \end{pmatrix} = \begin{pmatrix} 1 & 0 & 0 \\ 0 & \cos\theta_{23} & \sin\theta_{23} \\ 0 & -\sin\theta_{23} & \cos\theta_{23} \end{pmatrix} \begin{pmatrix} \cos\theta_{13} & 0 & \sin\theta_{13}e^{-i\delta_{CP}} \\ 0 & 1 & 0 \\ -\sin\theta_{13}e^{i\delta_{CP}} & 0 & \cos\theta_{13} \end{pmatrix} \begin{pmatrix} \cos\theta_{12} & \sin\theta_{12} & 0 \\ -\sin\theta_{12} & \cos\theta_{12} & 0 \\ 0 & 1 & 0 \end{pmatrix} \begin{pmatrix} \nu_1 \\ \nu_2 \\ \nu_3 \end{pmatrix}.$$

According to the neutrino oscillation measurements using the atmospheric [4, 5], reactor [6–9], accelerator [10–13], solar neutrinos [14, 15], the values of two mass splitting, (Δm^2_{21}, Δm^2_{32}), and three mixing angles, (θ_{12}, θ_{23}, θ_{13}), have been measured. However, still important pieces are missing, such as, CP violation phase, δ_{CP} and neutrino mass hierarchy. We have a great interest in measuring δ_{CP} since it may be the source of matter/anti-matter asymmetry universe. The mass hierarchy means the ordering of second and third mass state. The normal (inverted) hierarchy is defined as that mass of the 3rd neutrino is larger (smaller) than others, as shown in Figure 2. The octant of θ_{23}, i.e. $\theta_{23}>\pi/4$ or $<\pi/4$, is also unknown.

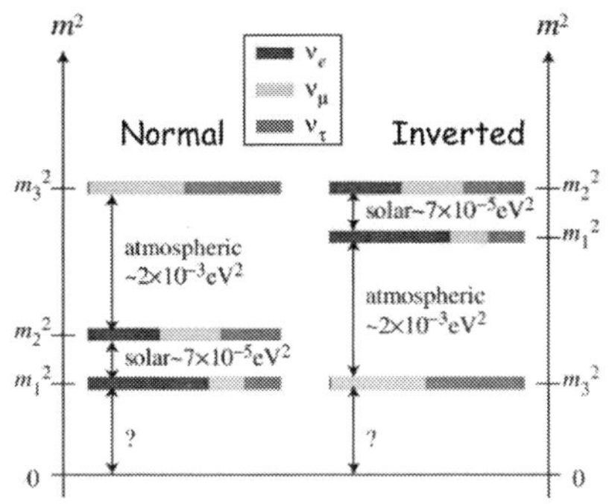

Figure 2. Scheme of normal and inverted mass hierarchy.

These parameters are possible to be probed by using atmospheric neutrinos. The atmospheric neutrinos are the decay products of secondary particles, such as pions and kaons, generated by the interactions of cosmic-ray and atmosphere. SK has been observing the atmospheric neutrino events in a wide energy range from sub-GeV up to TeV scale.

The oscillation of the upward-going electron neutrino, which travels in a long distance inside the Earth, is affected by the matter effect, and contains important information on the mass hierarchy and CP violation phase. Figure 3 shows the oscillated electron neutrino flux relative to the non-oscillated flux in the upward-going direction. A resonant-like enhancement by MSW effect is seen around multi GeV region. Interestingly the resonance occurs for either neutrino or anti-neutrino, depending on the mass hierarchy, i.e. the resonance occurs only for neutrino (anti-neutrino) oscillation with normal (inverted) hierarchy case. In addition θ_{23} octant affect the amplitude of the resonance. The absolute flux of the electron neutrinos in the range of Sub-GeV energy region is affected by δ_{CP} value.

A fit to SK data was performed with several oscillation parameter sets of δ_{CP}, θ_{23}, Δm^2_{32} in the interested regions, based on χ^2 value between data and Monte Calro (MC) expectation. This fit was also applied twice with normal and inverted hierarchy assumptions, respectively. Figure 4 shows χ^2 difference between normal and inverted mass hierarchy assumptions. The difference of minimum χ^2 values for each assumption was 3.0. This means that SK data favors normal hierarchy, but less than 2 σ significance. Also χ^2 becomes minimum in δ_{CP}~240° and $\sin^2\theta_{23}$>0.5.

We introduced T2K constraint into oscillation fit in order to improve mass hierarchy sensitivity since T2K data can reduce Δm^2_{32} and θ_{23} uncertainties. A model of T2K predictions is constructed using public available data from T2K, and χ^2 contribution from T2K data is calculated for each oscillation parameter. This is possible since the neutrino cross section and the detector efficiency are same due to the usage of the same detector between T2K and SK experiments, Another oscillation fit was performed including T2K constraint. As a result of the fit, normal hierarchy is still favored. Figure 5 shows the χ^2 difference ($\Delta\chi^2$) from best fit point for different data sample. Both SK and T2K favored around 280 degrees in δ_{CP}, but CP conservation, which corresponds to $\delta_{CP}\sim$ 0 or π, is still allowed at least 90% confidence level.

3 Observation of solar neutrino

SK is observing Boron 8 solar neutrino via electron scattering interaction in water. At present there are two main motivations to analyze solar neutrino. One motivation is to observe spectrum distortion due to the transition from vacuum oscillation below 1 MeV to matter oscillation above 5 MeV. This distortion is indicated from several solar neutrino measurements, but there is no measurement observing distorted energy spectrum in a experiment. Another motivation is day-night difference of

DOI: 10.1051/epjconf/201612602023

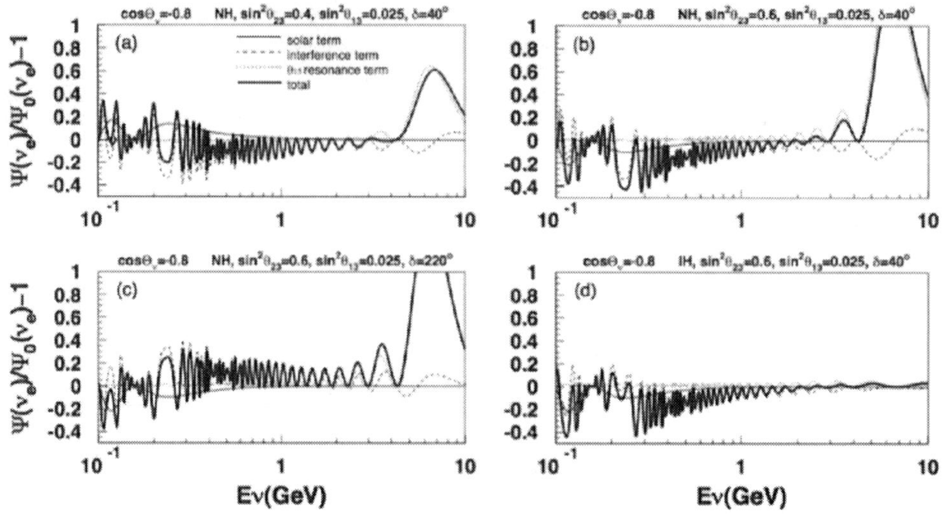

Figure 3. Oscillated ν_e flux relative to the non-oscillated flux as a function of neutrino energy for the upward-going neutrinos with zenith angle $\cos\theta$ = -0.8. Thin solid lines, dashed lines, and dotted lines correspond to the solar term, the interference term, and the θ_{23} resonance term, respectively. Thick solid lines are total fluxes. Parameters are set as ($\sin^2\theta_{12}$, $\sin^2\theta_{13}$, $\sin^2\theta_{23}$, δ_{CP}, Δm^2_{21}, Δm^2_{32}) = (0.31, 0.025, 0.6, 40°, $7.6\times10^{-5}\text{eV}^2$, $+2.4\times10^{-}5\text{eV}^2$) unless otherwise noted. The θ_{23} octant effect can be seen by comparing (a) ($\sin^2\theta_{23}$ = 0.4) and (b) ($\sin^2\theta_{23}$ = 0.6). δ_{CP} value is changed to 220° in (c) to be compared with 40° in (b). The mass hierarchy is inverted only in (d) so θ_{13} resonance (MSW) effect disappears in this plot.

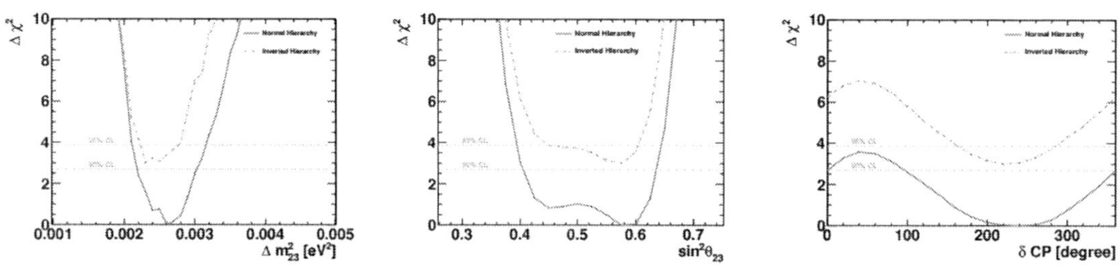

Figure 4. χ^2 difference from the global best fit point in the oscillation fits as a function of Δm^2_{23} (left), $\sin^2\theta_{23}$ (middle), δ_{CP} (right), respectively. Two lines shows the cases with normal (blue solid) and inverted (orange dashed) mass hierarchy assumptions. At each point the χ^2 is marginalized for other oscillation and nuisance parameters.

solar neutrino flux which is expected to be caused by the re-generation of Earth's matter effect when neutrino propagates the Earth.

Figure 6 shows the angular distribution of the observed events with respect to sun direction in SK-IV, and the event rate of the observed solar neutrino for each SK experiment phase. The peak due to the solar neutrino scattering is evident in the direction of the Sun. The measured event rate are consistent among the experiment phase within the uncertainties. About 77,000 events are observed from SK-I to SK-IV, and the ratio of data to unoscillated MC is calculated to be 0.4459.

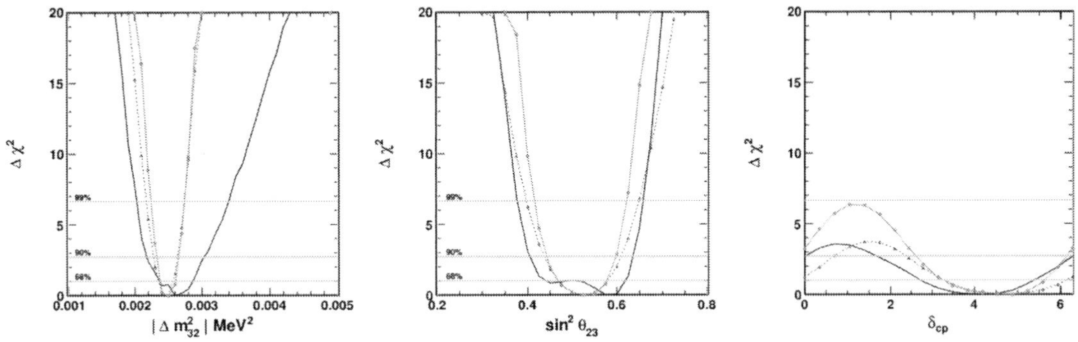

Figure 5. χ^2 difference from the best fit point in the oscillation fits using SK (solid black), T2K model (dashed purple), and SK plus T2K model (solid light blue) data sample, as a function of Δm^2_{23} (left), $\sin^2\theta_{23}$ (middle), δ_{CP} (right), respectively. Normal mass hierarchy is assumed.

The yearly event rate with the Sun spot number are shown in Figure 7 from 1996 up to the present. The observed rate is consistent with flat distribution; the p-value of the observed rate compared with the constant is 78.6%. There is no significant correlation between the solar neutrino flux and the Sun spot number,

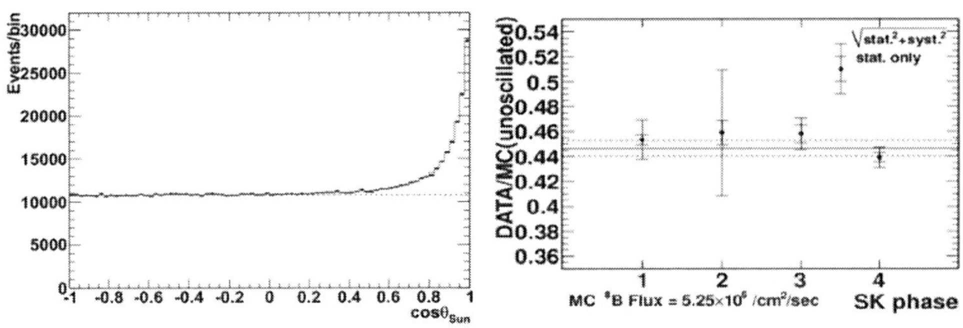

Figure 6. Angular distribution of the observed events with respect to solar direction in SK-IV (left), and the ratios of data to unoscillated Monte Carlo expectation in ^{8}B flux for each SK experiment phase. $\cos\theta_{Sun}$ represents the cosine of the inner angle between electron direction and the solar direction. The events near the peak around $\cos\theta_{Sun}=1$ corresponds to the recoiled electron events by solar neutrino interactions, and the flat components are background.

We have been trying to reduce the background events and lower the energy threshold to observe spectrum distortion since it becomes more visible in the lower energies. The energy threshold of SK-IV is 3.5 MeV in recoil electron kinetic energy, which is getting lower than previous SK phase. Figure 8 shows the combined energy spectrum for all SK periods. As a result of fit, the expected spectrum by MSW oscillation is slightly disfavored in the level of <2 σ compared to other assumptions, such as exponential quadratic functions. The de-convoluted energy spectrum to neutrino energy are also shown in Figure 8. The obtained spectrum is consistent with flat and still distortion is not observable.

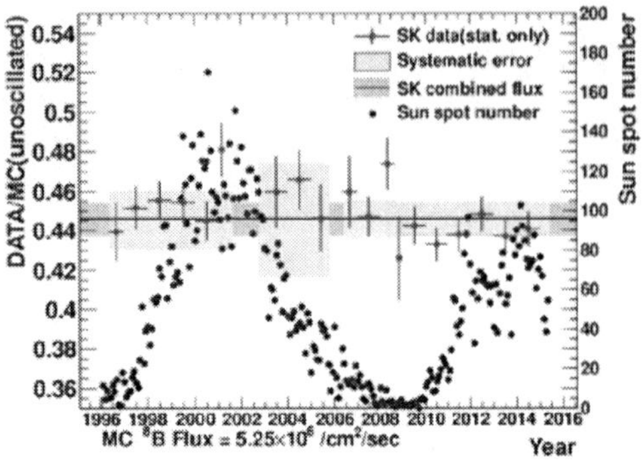

Figure 7. Ratio of data flux to unoscillated Monte Carlo expectation (red with error bar) and the Sun spot number (black dot) are shown as a function of observed date from the beginning of SK-I to the recent period of SK-IV.

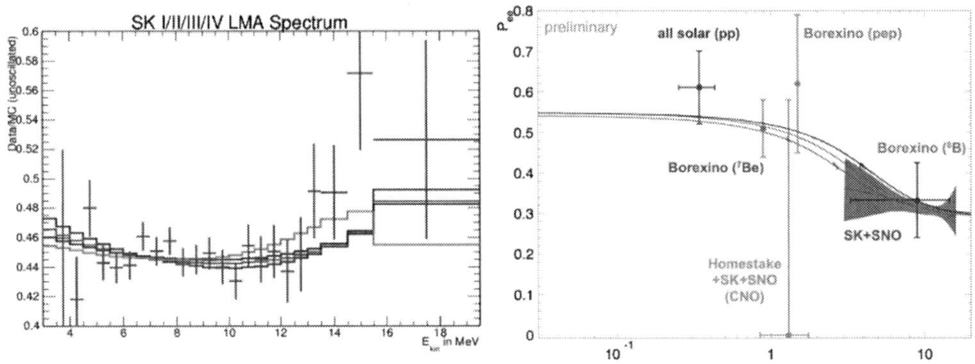

Figure 8. Combined spectrum of the recoiled electron kinematic energy from SK-I to SK-IV with the expected spectra of the various spectral assumptions (left), and the observed neutrino energy spectrum of several solar neutrino experiments (right). In the left plot the expected spectrum for the best fit oscillation parameters obtained with solar plus KamLAND (blue) and solar global (green) data are shown. The energy spectra expressed by the quadratic (black) and exponential (orange) functions are also shown. Note that the unit of the horizontal axis is the total energy of recoil electron where the detector effect such as energy resolution are not de-convoluted. The error bar represents statistical uncertainties. In the right plot the measured oscillation probability by SK plus SNO data (red) is shown with those measured with p-p solar (blue), ^{7}Be by Borexino (green), Homestake plus SK plus SNO CNO (olive), and pep by Borexino (cyan) data, as a function of neutrino energy.

4 Search for neutrino emission from solar WIMP

It is believed that Weakly Interacting Massive Particle (WIMPs) are attracted by the Sun's gravity, lose their energy by interactions, annihilate via $\chi\chi \rightarrow b\bar{b}$ / W^+W^- / $\tau^+\tau^-$ channels. Then neutrinos are expected to be produced as their decay products. SK is possible to search WIMP signal by looking for such neutrino events from the Sun. Recently some direct dark matter experiments, such as DAMA/LIBRA [16, 17], reported positive results in the low energy WIMP mass region around 10 GeV. We have improved our analysis to search for WIMP signal for such low energy region.

SK is suitable to search for these low mass WIMP signal. Figure 9 shows the angular distributions with respect to the Sun and the momentum distributions of several different energy samples. In the figure, the expected signal distributions are shown for the cases of 6 GeV/c^2 and 200 GeV/c^2 WIMP mass, respectively. The signal peak in the angular distributions will appear clearly in the corresponding energy range. These signals are expected to be distributed over the atmospheric neutrino backgrounds.

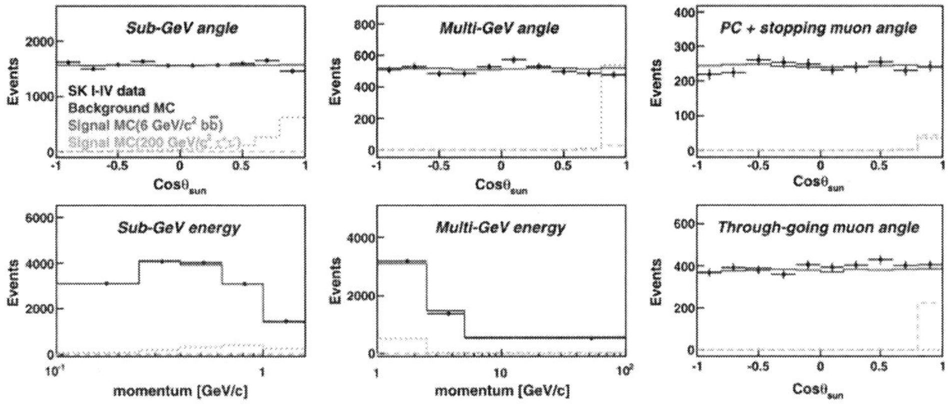

Figure 9. Angular and reconstructed momentum distributions of SK I-IV data (black crosses); atmospheric-neutrino background MC (normalized to data live time, blue solid); WIMP neutrino signal MC for the 6 GeV/c^2 $b\bar{b}$ channel (magenta dotted) / 200 GeV/c^2 $\tau^+\tau^-$ channel (cyan dashed) at 90% upper limit, magnified 30 times for visibility. Among seven sub-GeV samples, the single-ring e-like 0-decay-electron and μ-like 0,1-decay-electron samples are combined and shown in the left column. In the middle column, six multi-GeV samples are combined and shown. PC and up-μ samples are shown in the right column.

The WIMP signal was searched by the fitting method. The WIMP simulation are generated with several WIMP mass assumption and via the several annihilation modes of $\chi\chi \to b\bar{b}, W^+W^-, \tau^+\tau^-$. The angular distributions of the data are fitted with atmospheric MC (background) plus WIMP simulation data with varying their normalizations. As a result of fitting analysis, no significant excess was seen over the background distributions.

In order to have the meaningful upper limit on the WIMP cross section, the conversion factor from neutrino flux to WIMP cross sections is calculated using DarkSUSY 5.0.6 [18]. Also with several standard assumptions, which are commonly used in dark matter analysis, the upper limits on spin-dependent (SD) and spin-independent (SI) cross sections are obtained below 200 GeV/c^2 in WIMP mass as shown in Figure 10 and Figure 11. The current best limits are set in SD interaction and, some fraction of WIMP allowed regions are ruled out in SI interaction. This results has been published in [19].

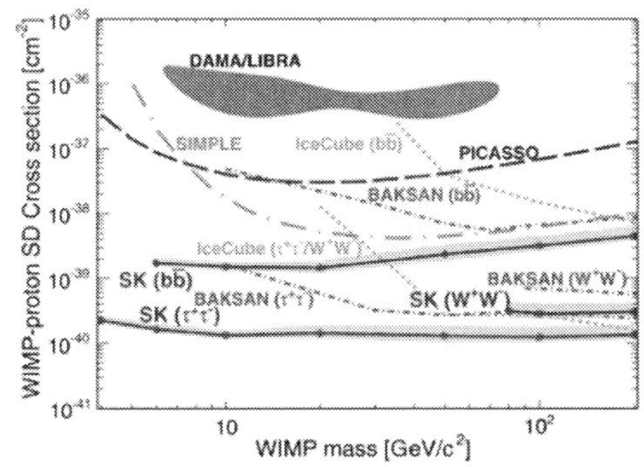

Figure 10. 90% C. L. upper limits on SD WIMP-proton cross section calculated at DarkSUSY [18] default are shown in red solid with uncertainty bands to take account uncertainties in the capture rate for the $b\bar{b}$, W^+W^- and $\tau^+\tau^-$ channels from top to the bottom. Also shown are limits from other experiments: IceCube [20] in brown dashed: (top) $b\bar{b}$ / W^+W^- or $\tau^+\tau^-$ (bottom); BAKSAN in pink dot-dashed: $b\bar{b}$ (top) / W^+W^- (middle) / $\tau^+\tau^-$ (bottom) ; PICASSO [21] (blue long-dashed); SIMPLE [22] (green long dot-dashed). The black shaded region is the 3σ C.L. signal claimed by DAMA/LIBRA [16, 17].

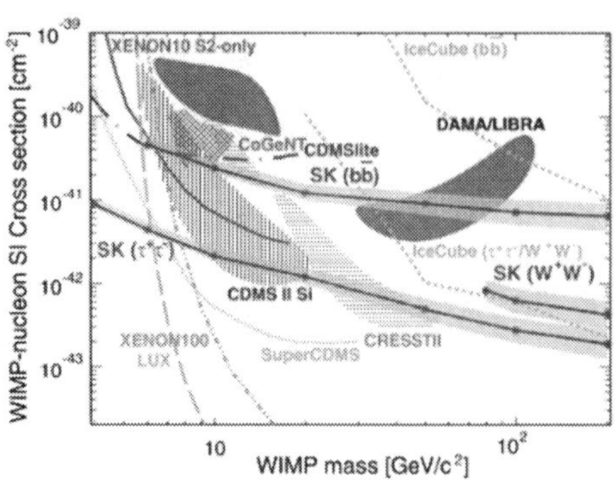

Figure 11. 90% C. L. upper limits on the SI WIMP-nucleon cross section (plotting scheme is the same as Figure 10). Also shown are event excesses or annual modulation signals reported by other experiments: DAMA/LIBRA (black shaded regions, 3σ C.L.); CoGeNT [23] (magenta diagonally cross-hatched region, 90% C.L.); CRESSTII [24] (violet horizontally-shaded regions, 2σ C.L.); CDMS II Si [25] (blue vertically-shaded region, 90% C.L.); and limits: IceCube in brown dashed: $b\bar{b}$ (top) / W^+W^- (middle) / $\tau^+\tau^-$ (bottom); SuperCDMS [26] (cyan dotted); CDMSlite [26] (blue long dot-dashed); XENON10 S2-only [27] (dark green dash triple dot); XENON100 [28] (green dash double dot); LUX [29] (orange long-dashed)

5 Super-Kamiokande Gadolinium project

SK collaboration has decided to approve new project, named as Super-Kamioande Gadolinium project (SK-Gd) in 2015 June. In the project, 0.2% concentration of Gadolinium (Gd) sulfate is added in the pure water. When the neutrons are produced via anti-neutrino interactions, they are thermalized quickly and captured into Gd with high efficiency of about 90%. The Gd atom emits multiple gammas which total energy is about 8 MeV. By detecting the prompt positron and the delayed gamma signals, the events of anti-neutrino interaction are possible to be separated from those of neutrinos.

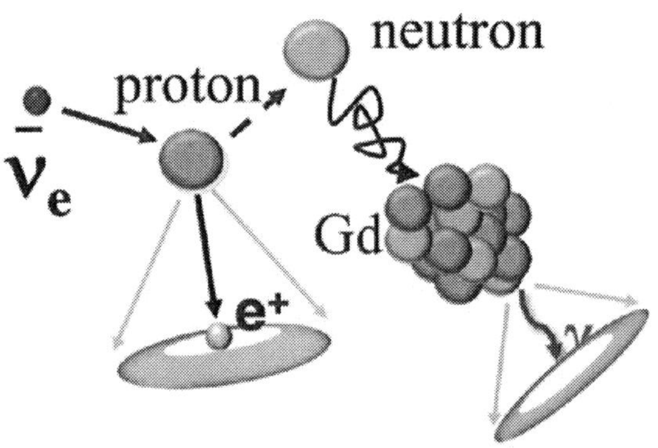

Figure 12. Schematic diagram of electron anti-neutrino reaction with Gadolinium (Gd) water. The signal is recognized by the coincidence of the prompt positron signal and the delayed gamma produced by neutron capture with Gd.

The main motivation of this project is to detect supernova relic (SNR) neutrino, which should be produced in the past supernova. It is believed that much more star formation and supernova burst occurred in the past though such star formation rate has large uncertainties from the current astronomical observations. If the amount of SNR neutrino is measured, it will be great interest in the astrophysics and the cosmology.

The attempt in detecting SNR neutrino has also been made in current SK but not detected so far in the energy window of 10-30 MeV since the data suffers from atmospheric neutrino backgrounds, such as Michel electrons produced from low energy invisible muons. Using the coincidence technique of the prompt and the delayed signals in SK-Gd, such background events should be eliminated.

In order to investigate the feasibility of the SK-Gd project, we have been tested the effect of adding Gd in water, such as the water transparency and the material effect like rust. A small test tank of 200 m^3 was build with the same detector material and photo-multipliers as used in SK to investigate the influence of Gd water to materials and the quality of Gd water. The test tank facility is shown in Figure 13. This test has been performed from October 2014 to April 2015.

The results of the water transparency and scattering measurements by using the laser injector seems good. The amount of Cherenkov light left after 15 m photon travel is estimated to 92% of pure water case at 0.2% concentration of Gadolinium (Gd) sulfate, as shown in Figure 13. Also no significant increase of the Rayleigh scattering photons, which may affect the performance of the Cherenkov light imaging, have been observed.

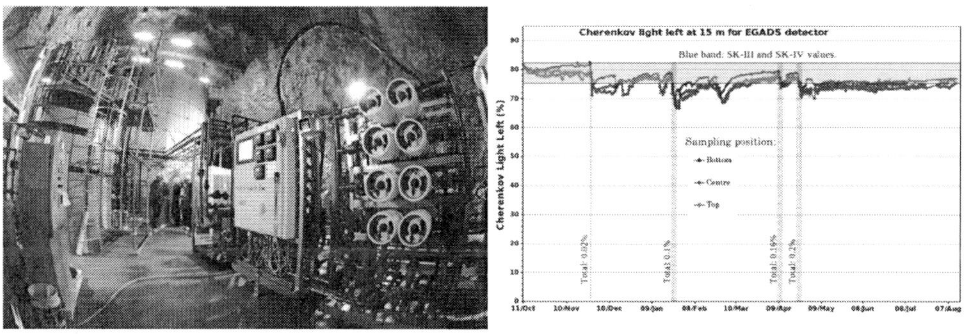

Figure 13. (Left) Photo of the EGADS tank and equipments for testing the feasibility of the Gadolinium water in SK. The 200 m^3 test tank is located at the opposite side of the experimental hall. The Gd water purification system is located in the near side. (Right) The variation of the water transparency measured in EGADS tank. The measurement started with pure water, and the gadolinium sulfate is added gradually, 0.02%, 0.1%, 0.16%, and 0.2%. The horizontal value is shown by the expected Cherenkov light left at 15 m estimated based on the water transparency measured with several wavelength of the laser injector. The difference in line color represents the the different tank location (blue:bottom, red:center, green:top). The blue shaded region corresponds to those of pure water case measured in SK-III and IV.

According to these verification studies, we, SK collaboration, decided to approve SK-Gd project in 2016 June. The actual start time will be determined by considering the T2K neutrino beam schedule.

6 Summary

Recent physics results and the status of Super-Kamiokande experiment are reviewed in this paper. The oscillation analysis using the atmospheric neutrino shows a slight preference to normal hierarchy and $\delta_{CP}{\sim}3\pi/2$, The expected distortion in the energy spectrum has not been seen from the solar neutrino observation. The stronger upper limits for SD and SI cross section are set in the mass range of 4-200 GeV/c^2 from the indirect solar WIMP search. A new project, SK-Gd, is approved and announced by the SK collaboration, which aims to detect supernova relic neutrino using neutron capture by adding Gadolinium.

References

[1] Y. Fukuda et al. (Super-Kamiokande Collaboration), Phys. Rev. Lett. **81**, 1562 (1998)
[2] Z. Maki, M. Nakagawa, S. Sakata, Prog. Theor. Phys. **28**, 870 (1962)
[3] B. Pontecorvo, Sov. Phys. JETP **26**, 984 (1968)
[4] Y. Ashie et al. (Super-Kamiokande), Phys. Rev. **D71**, 112005 (2005), `hep-ex/0501064`
[5] R. Wendell et al. (Super-Kamiokande Collaboration), Phys.Rev. **D81**, 092004 (2010), `1002.3471`

[6] S. Abe et al. (KamLAND Collaboration), Phys.Rev.Lett. **100**, 221803 (2008), 0801.4589
[7] F. An et al. (DAYA-BAY Collaboration), Phys.Rev.Lett. **108**, 171803 (2012), 1203.1669
[8] J. Ahn et al. (RENO collaboration), Phys.Rev.Lett. **108**, 191802 (2012), 1204.0626
[9] Y. Abe et al. (DOUBLE-CHOOZ Collaboration), Phys.Rev.Lett. **108**, 131801 (2012), 1112.6353
[10] M. Ahn et al. (K2K), Phys.Rev. **D74**, 072003 (2006), hep-ex/0606032
[11] P. Adamson et al. (MINOS), Phys.Rev.Lett. **106**, 181801 (2011), 1103.0340
[12] K. Abe et al. (T2K), Phys.Rev.Lett. **112**, 181801 (2014), 1403.1532
[13] K. Abe et al. (T2K), Phys.Rev.Lett. **112**, 061802 (2014), 1311.4750
[14] J. Hosaka et al. (Super-Kamiokande), Phys.Rev. **D73**, 112001 (2006), hep-ex/0508053
[15] B. Aharmim et al. (SNO), Phys.Rev. **C72**, 055502 (2005), nucl-ex/0502021
[16] R. Bernabei et al. (DAMA Collaboration), Eur.Phys.J. **C56**, 333 (2008), 0804.2741
[17] C. Savage, G. Gelmini, P. Gondolo, K. Freese, JCAP **0904**, 010 (2009), 0808.3607
[18] P. Gondolo et al., JCAP **0407**, 008 (2004), astro-ph/0406204
[19] K. Choi et al. (Super-Kamiokande), Phys. Rev. Lett. **114**, 141301 (2015), 1503.04858
[20] M. Aartsen et al. (IceCube), Phys.Rev.Lett. **110**, 131302 (2013), 1212.4097
[21] S. Archambault et al. (PICASSO Collaboration), Phys.Lett. **B711**, 153 (2012), 1202.1240
[22] M. Felizardo, T. Girard, T. Morlat, A. Fernandes, A. Ramos et al., Phys.Rev.Lett. **108**, 201302 (2012), 1106.3014
[23] C. Aalseth et al. (CoGeNT collaboration), Phys.Rev.Lett. **106**, 131301 (2011), 1002.4703
[24] G. Angloher, M. Bauer, I. Bavykina, A. Bento, C. Bucci et al., Eur.Phys.J. **C72**, 1971 (2012), 1109.0702
[25] R. Agnese et al. (CDMS Collaboration), Phys.Rev.Lett. **111**, 251301 (2013), 1304.4279
[26] R. Agnese et al. (SuperCDMS Collaboration) (2014), 1402.7137
[27] J. Angle et al. (XENON10 Collaboration), Phys.Rev.Lett. **107**, 051301 (2011), 1104.3088
[28] E. Aprile et al. (XENON100 Collaboration), Phys.Rev.Lett. **109**, 181301 (2012), 1207.5988
[29] D. Akerib et al. (LUX Collaboration) (2013), 1310.8214

The MoEDAL experiment at the LHC

Searching beyond the standard model

James L. Pinfold (for the MoEDAL Collaboration)[1,a]

[1] *University of Alberta, Physics Department, Edmonton, Alberta T6G 0V1, Canada*

Abstract. MoEDAL is a pioneering experiment designed to search for highly ionizing avatars of new physics such as magnetic monopoles or massive (pseudo-)stable charged particles. Its groundbreaking physics program defines a number of scenarios that yield potentially revolutionary insights into such foundational questions as: are there extra dimensions or new symmetries; what is the mechanism for the generation of mass; does magnetic charge exist; what is the nature of dark matter; and, how did the big-bang develop. MoEDAL's purpose is to meet such far-reaching challenges at the frontier of the field.
The innovative MoEDAL detector employs unconventional methodologies tuned to the prospect of discovery physics. The largely passive MoEDAL detector, deployed at Point 8 on the LHC ring, has a dual nature. First, it acts like a giant camera, comprised of nuclear track detectors - analyzed offline by ultra fast scanning microscopes - sensitive only to new physics. Second, it is uniquely able to trap the particle messengers of physics beyond the Standard Model for further study. MoEDAL's radiation environment is monitored by a state-of-the-art real-time TimePix pixel detector array. A new MoEDAL sub-detector to extend MoEDAL's reach to millicharged, minimally ionizing, particles (MMIPs) is under study Finally we shall describe the next step for MoEDAL called Cosmic MoEDAL, where we define a very large high altitude array to take the search for highly ionizing avatars of new physics to higher masses that are available from the cosmos.

1 Introduction

In 2010 the MoEDAL experiment [1] at the Large Hadron Collider (LHC) was unanimously approved by CERN's Research Board to start data taking in 2015. MoEDAL is a pioneering experiment designed to search for highly ionizing avatars of new physics such as magnetic monopoles or massive (pseudo-)stable charged particles. Its groundbreaking physics program defines over 30 scenarios that yield potentially revolutionary insights into such foundational questions as: are there extra dimensions or new symmetries; what is the mechanism for the generation of mass; does magnetic charge exist; what is the nature of dark matter; and, how did the big-bang develop. The LHC machine [2] brings its counter rotating beams nominally consisting of 2808 bunches of protons each with roughly 10^{11} protons per bunch. The proton beams are brought into collision at four points around the 27 km circumference LHC ring. The biggest of the LHC's seven experiments are ATLAS [3] and CMS [4], general-purpose detectors designed to investigate the largest range of physics possible. ALICE [5]

[a] e-mail: jpinfold@ualberta.ca

and LHCb [6] have detectors specialized for focussing on specific phenomena such as quark–gluon plasma formation and the investigation of the matter-antimatter asymmetry using b-hadron physics, respectively. These four detectors in huge caverns sit roughly 100m underground on the LHC ring. Prior to the approval of the MoEDAL experiment the smallest experiments on the LHC were TOTEM [7] and LHCf [8], which focus on "forward particles" – protons or heavy ions that brush past each other rather than meeting head on when the beams collide. TOTEM uses detectors positioned on either side of the CMS interaction point, while LHCf is made up of two detectors that sit along the LHC beam line, at 140 metres either side of the ATLAS collision point. MoEDAL the seventh and latest LHC experiment shares an interaction point with the LHCb experiment. It is designed to significantly expand the discovery horizon of the general-purpose LHC detectors, ATLAS and CMS in a complementary manner. The innovative MoEDAL detector employs unconventional methodologies tuned to the prospect of discovery physics. The largely passive MoEDAL detector, deployed at Point 8 on the LHC ring has a dual nature. First, it acts like a giant camera, comprised of Nuclear Track Detectors (NTDs) - analyzed offline by ultra fast scanning microscopes - sensitive only to new physics. Second, it is uniquely able to trap the particle messengers of physics beyond the Standard Model for further study. MoEDAL's radiation environment is monitored by a state-of-the-art real-time TimePix pixel detector array. In addition, we are preparing a proposal for a third active detector element capable of detecting the faintest flicker of light - single photon signals - produced by the passage of Minimally Ionizing (millicharged) Particles (mQPs) with charge as small as one thousandth that of an electron.

2 Highly ionzing particles - messengers of new physics

MoEDAL's groundbreaking physics prospectus [9] covers 34 fundamentally important Beyond the Standard Model (BSM) scenarios involving electrically and magnetically charged particles. It is designed to exploit the use of Highly Ionizing Particles (HIPs) as wide ranging avatars of new physics. One of the main objectives for MoEDAL is the state-of-the-art search for the magnetic charge carried by the monopole, just as the search for the Higgs was the prime motivation for the LHC and its general-purpose experiments.

To be detected a particle must interact with the material of the detector losing energy as they travel. Except when they are moving very near to the speed of light charged particles heavier than electrons primarily lose energy by interacting with the atoms of the material by exciting atomic electrons or ionizing atoms. The Bethe formula describes the energy loss per distance travelled of fast charged particles that are more massive than electrons such as muons, protons, pions and ions (atomic nuclei). A simplified form of the Bethe equation is:

$$-\frac{dE}{dx} = z^2\frac{KZ}{A\beta^2}\left[\frac{1}{2}ln\left(\frac{2m_e\beta^2\gamma^2}{I^2}\right)-\beta^2\right] \tag{1}$$

where, β is the speed of the particles expressed as a fraction of the speed of light (β=v/c) and $K = 4\pi N_A r_e^2 m_e c^2$. The properties of the medium are the atomic number Z, the atomic mass A and the mean excitation energy I. The constants are the electron mass, m_e, the classical electron radius, r_e and Avogardo's number N_A.

All stable or long-lived Standard Model charged particles produced in proton-proton collisions at the LHC have a single charge ($ze \rightarrow e$), are relatively light i.e. at most the mass of a proton and are moving near to the speed of light ($\beta \sim 1$). Particle are "minimum ionizing" (MIPs) if $\beta\gamma = 3 \rightarrow 4$ (or $dE/dx \sim 1 \rightarrow 2$ MeV g^{-1} cm^2).

In proton-proton collisions as the LHC electrically charged particles can only be highly ionizing if they are massive and thus moving very slowly with $\beta\gamma < 1$, and/or multiply charged. However, there are no known massive stable or long-lived particles that are multiply charged. The only way that a Standard Model particle can be highly ionizing is if it is slowing down to stop. Energy loss due to ionization by mQP is described by the Bethe-Bloch formula. Thus, all other things being equal a milli-charged particle (with q = 0.001e) will ionize 10^{-6} aless than a singly charged particle (with q=1e) and multiple detection coincidence can rule out detector related backgrounds. Thus, massive stable or pseudo-stable HIPs or mQps traversing the detector are clear messengers of new physics with no obvious Standard Model Backgrounds.

2.1 Particles with magnetic charge

A prime example of a HIP from BSM is the magnetic monopole. The ionization energy loss of a magnetically charged particle can be obtained from the Bethe formula by replacing the electric charge ze by $ng\beta$ where g is the magnetic charge and n is the number of Dirac magnetic charges and realizing that for magnetic monopoles the velocity dependence of the Lorentz force cancels the $1/\beta^2$ term in the Bethe-Bloch formula for charged particles:

$$-\frac{dE}{dx} = n(g/e)^2 \frac{KZ}{A}\left[\frac{1}{2}ln\left(\frac{2m_e\beta^2\gamma^2}{I^2}\right) - \beta^2\right] \tag{2}$$

A single Dirac (magnetic) charge is $g \approx 68.5e$. As we can see from the Bethe formula for magnetic charge given above, the energy loss of a singly charged ($n = 1$) very relativistic magnetic monopole ($\beta \approx 1$) is amazingly about 4700 times that of a proton.

The MoEDAL experiment detects HIPs using stacks of passive plastic Nuclear Track Detectors (NTDs). The HIP damages the plastic as it traverses the NTD stack. Etching the plastic in a hot sodium hydroxide solution can reveal this damage as two collinear etch-pits each with diameter ~ 5 - 10μm in each of the six layers of the stack. The NTD plastic employed by MoEDAL has a threshold of around Z/β of 5 (or 5 MIPs) where Z is the charge number of the particle and β is the velocity of the particle divided by the speed of light (v/c). A HIP messenger of new physics would leave a tell tale signature of twelve tiny etch pits (for each stack of ten NTD sheets) forming a precisely defined line pointing back to the collision. There is no known Standard Model particle that can give a trail of collinear etch pits in a MoEDAL stack.

3 Directly detecting magnetic charge

Clearly, a unique property of the magnetic monopole is that it has magnetic charge. Thus a moving magnetic monopole will be ringed by an electric field just as a moving electric charge is ringed by a magnetic field. Imagine that a magnetic monopole traverses the superconducting wire coil of a SQUID. As the monopole approaches the coil it drives an electrical current within the coil that in turn creates a changing magnetic field. This changing magnetic field creates an electric field that counters the original electric field of the moving monopole. As the wire in the coil has no resistance the overall effect is to make the total electric field vanish in the coil. Meanwhile the current continues to flow in the coil because the wire is superconducting and without resistance. One can use Faraday's Law to calculate the magnitude of the current that would be generated:

$$I = -\frac{mu_0 q_m}{L} \tag{3}$$

where, $L$is the inductance of the coil, μ_0 is the permeability of free space and q_m is the magnetic charge. Consequently, the current induced only depends on the magnetic charge and is independent of the speed, mass or direction of the magnetic monopole.

4 A description of the MoEDAL detector

The novel MoEDAL detector is deployed at IP8 on the LHC ring, depicted in figure 1, is comprised of three key subsystems. The main component of MoEDAL's NTD subsystem consists of a low threshold (LT) NTD array - comprised of an array of (400) plastic NTD stacks (25 ×x 25 cm^2) each stack made up of CR39 (3) and MAKROFOL (3) NTD sheets. This is the largest NTD array every deployed at an accelerator. The CR39 NTDs have an excellent charge resolution (≥ 0.2 of a single electric charge, based on a single measurement) and a low threshold, allowing the detection of HIPs with and ionizing power equal to or more than 5 times that of a Minimum Ionizing Particle (MIP). An illustration of the LT-NTD stack is given in figure 2 (Left).

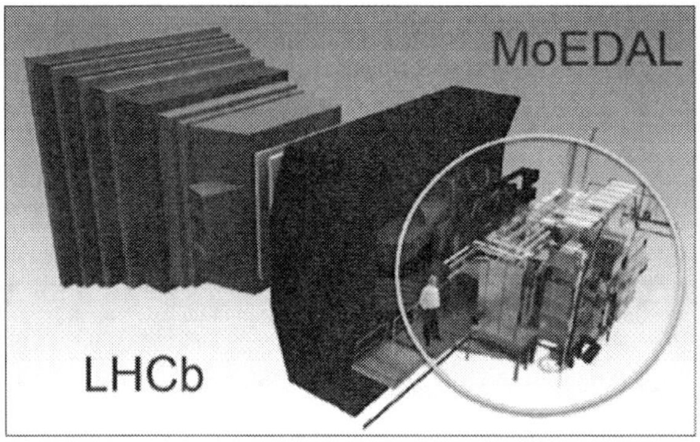

Figure 1. A view of the MoEDAL detector deployed around the LHCb's VELO detector at IP8.

The LT-NTD array has been enhanced by the Very High Charge Catcher (VHCC) subdetector – with threshold around 50 MIPs. The VHCC subdetector – comprised of two flexible low mass stacks of MAKROFOL in an aluminium foil envelope - is deployed in the forward acceptance of the LHCb experiment just after the LHCb RICH detector. The overall acceptance of the NTD system is~ 60%. The exposed NTD detectors will be etched to reveal the passage of a HIP as illustrated in figure 2 (right). The etched plastic will then be analysed at Bologna and CERN using state-of-the art ultra-fast automated optical scanning microscopes developed by the Muenster and INFN-Bologna MoEDAL groups, combined with cutting edge image analysis software. This enables low NTD thresholds to be maintained despite large beam related backgrounds. The NTD detectors will be calibrated using the heavy-ion (Fe) beam at NASA's Space Radiation Laboratory Laboratory (NSRL).

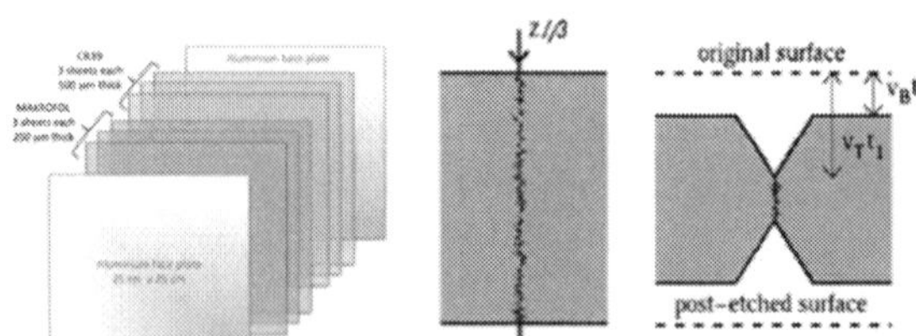

Figure 2. (Left) A MoEDAL stack comprised of CR39 and Markrofol NTD detectors. (Right) The damage zone created by the HIP that is subsequently revealed by etching in a hot NaOH solution.

The Magnetic Monopole Trapper (MMT) is the third MoEDAL sub-detector system. Its sensitive volume consists of roughly 1 tonne of pure aluminium trapping volumes deployed around the MoEDAL cavern at IP8. Aluminium is well suited as it has an anomalously large nuclear magnetic moment. After exposure the MMT trapping volumes will be monitored at the ETH SQUID facility for the presence captured monopoles. A schematic description of this facility is shown in figure 3 (left). Figure 3 (right) shows the results of an exposure of a MMT test detector at IP8 on the LHC ring. As can be seen the SQUID can detect magnetic charges greater than one tenth of a Dirac charge. Test solenoids can be used to calibrate the response of the SQUID to a trapped monopole. After the SQUID scan has been performed the trapping volumes will be sent to an underground laboratory, for example SNOLAB, to be monitored for the decays of very long lived electrically charged particles.

The use of SQUIDS to detect trapped magnetic charge has also been thoroughly tested in particle and astroparticle experiments where the search has been performed on "found" or alternate use (such as beampipes) objects on a "one-off" basis. But, MoEDAL is the first experiment ever to use purpose-made replaceable trapping volumes to capture magnetic charge. After the SQUID scan has been performed the trapping volumes will be sent to SNOLAB – 2km underground – to be monitored for the decays of very long lived electrically charged particles.

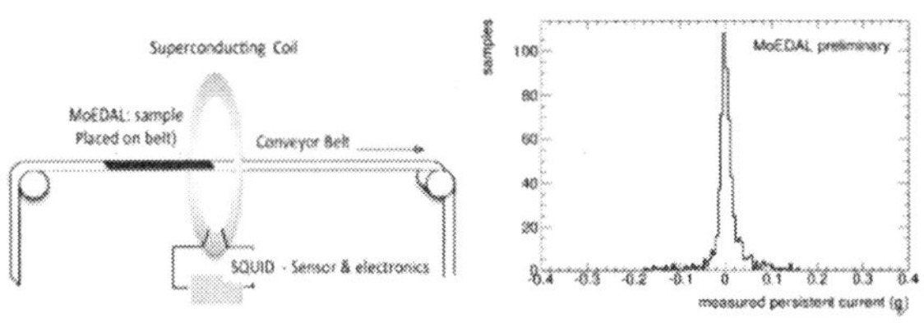

Figure 3. (Left) A schematic depiction of the use of a SQUID magnetometer to detect the presence of a trapped magnetic charge. (Right) The results of a SQUID scan of MMT elements exposed to roughly a year of LHC collisions at IP8 - as part of a MoEDAL test MMT detector deployment.

The fourth and only active sub-detector system is the TimePix pixel device array (TMPX), consisting of $\sim$ 8 devices distributed throughout the MoEDAL cavern at IP8. TMPX will be used to monitor highly ionizing beam related backgrounds. Each pixel of the innovative TimePix chip contains a preamp, a discriminator with threshold adjustment, synchronization logic and a 14-bit counter. MoEDAL uses TimePix Time-over-Threshold modes where each pixel acts as an ADC for energy measurement. Effectively, the TimePix device is a tiny electronic bubble-chamber providing a real-time 'colour' (energy dependent) movie of the background as illustrated in figure 4.

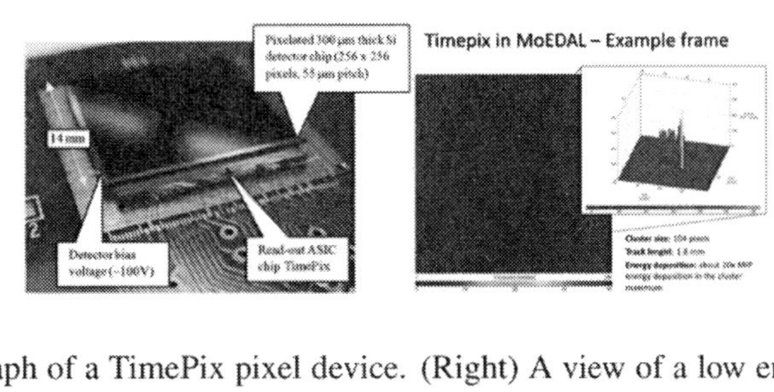

Figure 4. (Left) a photograph of a TimePix pixel device. (Right) A view of a low energy highly ionizing event in the TimePix chip's silicon sensitive volume.

DOI: 10.1051/epjconf/201612602024

4.1 A proposed minimally charged (millicharged) particle detector for MoEDAL

A key aspect of the MoEDAL experiment is that is sensitive to HIP avatars of new physics for which the standard general-purpose LHC detectors are not optimized. MoEDAL is preparing a proposal to add a new subdetector that is sensitive to mQPs. This addition is consistent with the MoEDAL collaboration's aim to extend the physics reach of the LHC in a way that is complementary to the existing capability provided by the main LHC detectors since they have no sensitivity to mQPs with charge below ~ 0.1e.

The proposed detector is similar in principle to that employed on an earlier beam dump search for MMIPs at SLAC [10]. It will employ three deep liquid scintillator layers in triple coincidence with an acceptance of 0.01

5 A Comparison of the MoEDAL detector with conventional LHC detectors

MoEDAL's exposed NTD detectors are sent for analysis to INFN-Bologna and CERN and for calibration at the NASA Space Radiation Laboratory (NSRL) heavy-ion beam facility. The exposed MMT trapping detectors are monitored for the presence of trapped magnetic charge at the ETH Zurich SQUID Facility and at SNOLAB in Canada for captured long-lived massive electrically charged particles. In this way MoEDAL's unparalleled use of passive tracking and trapping detectors has the extraordinary effect of allowing the physics to be brought to the detectors, rather than the usual situation of bringing the detector to the physics. This unique approach has the effect of adding the capabilities of the remote facilities to that of the deployed MoEDAL detector and allowing us to significantly expand the physics discovery reach of the LHC in a cost effective way.

Standard collider detectors are not optimized for the detection of HIPs such as magnetic monopoles and massive long-lived or stable charged particles. Rather, the general-purpose detectors, such as ATLAS and CMS, are designed to detect minimum-ionizing particles moving near to the speed of light. Effects arising from the particles low velocity and high-density energy deposition, such as electronics saturation, light quenching in scintillators, and adjacent hits from delta electrons, are extremely challenging to deal with. Indeed, in some cases it may be impossible to make an accurate measurement of the effective charge of the particle. For example, the resulting dead time as a result of electronics saturation may be of the order of the bunch crossing time. In addition, HIPS will be quickly absorbed within the detector mass even before they penetrate far into the inner tracking detectors.

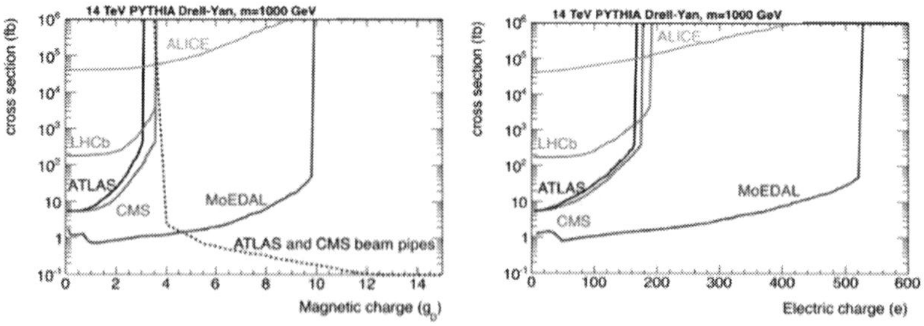

Figure 5. (The expected reach [11] of the search for direct monopole–anti-monopole Drell-Yan pair production process at the LHC (Ecm =14 TeV). Assuming the luminosity taken by LHCb/MoEDAL is 2 fb^{-1}, by ATLAS & CMS 20 fb^{-1}, and by ALICE 0.004 fb^{-1}.

On the other hand, MoEDAL does not suffer from the problems that would severely hamper detection of slow moving HIPS at the main LHC detectors. The NTD based sub-detectors have the demonstrated ability to measure accurately the ionizing power of a HIP. The MMT sub-detector, read out through a remote SQUID, can directly detect magnetic charge - no other Collider detector has this capability. MoEDAL's largely passive detection systems require no trigger, electronics, high voltage or gas supply. Its light and uniform detection technology requires minimal support structure and provides maximum cost effective coverage close to IP8 with few experimental compromises.

MoEDAL's sensitivity to highly ionizing messengers of new physics, together with its ability to make a permanent 'photographic' record of HIP BSM particles in its NTD detectors and capture such particles in its trapping detectors, is unique. This set of capabilities combined with the absence of SM backgrounds allows the definitive identification of HIP avatars of new physics with only a few candidate events detected. In an independent study [11] a comparison was made of the response of the LHC detectors to HIPs, as shown in figure 5. MoEDAL's sensitivity clearly surpasses that of the other LHC detectors.

6 MoEDAL's Physics Program, a New Light on the Terascale Frontier

MoEDAL's pioneering physics prospectus [9] covers more than 34 fundamentally important BSM scenarios allowing MoEDAL to significantly expand the LHC's discovery horizon in a complementary way. This program is illustrated in the pie chart shown in figure 6. One of the main objectives for MoEDAL is the state-of-the-art search for the magnetic charge, just as the search for the Higgs was the prime motivation for the LHC and its general-purpose experiments. For example, the work of our Theory Board has renewed interest in the LHC search for a Terascale EW Monopole [12] arising from the SM. If discovered, this would be the first topological particle to be observed – with the utmost consequences for our understanding of the Universe. But MoEDAL is designed to do much more as indicated in figure 6 and as discussed below.

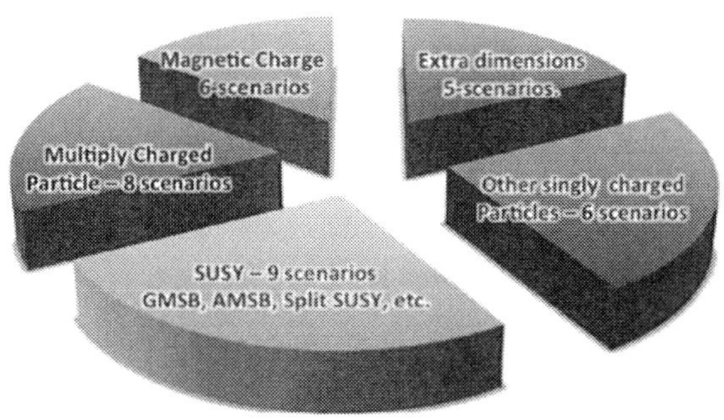

Figure 6. The MoEDAL Physics Program broken down by topic.

6.1 The hunt for magnetic charge

It was Dirac who first hypothesized that a magnetic monopole could exist within the framework of quantum mechanics [13]. He saw the monopole as the end of an infinitely thin infinitely long solenoid called a Dirac string. The annoying string does not pose a problem as long as it cannot be detected. Dirac surmised that the Dirac string would not detectable if the phase picked up by the wave function of the electron as it travels completely around the string is equal to one – in other words it has no affect. He calculated that the condition for the string not to be detected:

$$e^{iqg} = 1 \Rightarrow 2\pi n(n = 1, 2, 3..n) \tag{4}$$

Dirac was quick to realize that the above relation between electric charge (q) and magnetic charge (g) explains the quantization of electric charge, as long as at least one monopole exists. The numerical value of the magnetic charge is $g = n/2\alpha = n68.5q$ - using $1/\alpha = 4\pi/q^2$ where α is the fine structure constant ($\sim 1/137$) giving rise to the extremely highly ionizing nature of the magnetic monopole that we discussed above.

Julian Schwinger [14] generalized the quantization condition to dyons, particles that carry both electric and magnetic charge. If we consider two dyons with electric and magnetic charges (q_1, g_1) and (q2, g2) Schwinger found that each dyon is unable to detect the Dirac string of the other iff:

$$q_1 g_2 - q_2 g_1 = n \tag{5}$$

He found that the minimum charge for a megntic monopole was twice the Direc charge.

6.2 The't Hooft Polyakov monopole

In the Standard Model electroweak unification was achieved using a gauge theory developed by Sheldon Glashow, Steven Weinberg and Abdus Salam (GWS) [15]. This theory had two components called U(1) - that has the same Abelian gauge symmetry structure as Maxwell's electromagnetism and SU(2), that has a non-Abelian gauge symmetry that describes the weak interaction. These symmetries are broken at low energies by the Higgs field. In 1974 Gerard 't Hooft [16] and Alexander Polyakov [17] wrote their famous papers on a new kind of magnetic monopole that didn't have a problematic "Dirac String" with a point-like singularity at the origin. Their monopole had its origin in Georgi's and Glashow's earlier model of EW unification [18] later supplanted by GWS theory.

The Georgi-Glashow model has only one gauge symmetry, a non-Abelian symmetry known as SO(3), that is spontaneously broken by the Higgs field down to the U(1) gauge symmetry of Maxwell's equations. The quantization of charge arises naturally in such theories without the need for a magnetic monopole as was the case in Dirac's theory. However, a major implication of 't Hooft's and Polyakov's work was that any unified gauge theory like Georgi's and Glashow's inevitably possesses monopole solutions. So, magnetic monopoles that were only "nice" in Abelian theories like Maxwell's electromagnetism are now necessary in some non-Abelian theories.

The 't Hooft and Polyakov monopole arose in the situation where the Higgs field vector points away from the origin everywhere in the "hedgehog" configuration, In contrast to the Dirac monopole, which necessitates a singular point-like magnetic source at the origin, the 't Hooft-Polyakov monopole is everywhere smooth. In this case the field configuration is topologically stable, like a knot that cannot unravel. Remarkably, the Higgs-field configuration corresponding to this magnetic monopole is an example of a topological soliton, a localized and stable field excitation that behaves like an elementary particle. In the microscopic heart of the 't Hooft-Polyakov monopole the full unbroken SO(3) gauge symmetry of the unified theory reigns supreme. Far away the 't Hooft-Polyakov monopole would be indistinguishable from a Dirac monopole except that its magnetic charge is twice as much.

The roughly 100 GeV/c^2 mass of these monopoles — determined by the energy scale of the weak interaction – was ruled out by existing experimental results. But 't Hooft and Polyakov observed that the structure of the SU(5) Grand Unified Theory (GUT), which attempts to describe all known elementary particle forces except gravity, was similar in some respects to that of the Georgi-Glashow model. It was not then surprising that the SU(5) GUT model also contains a 't Hooft-Polyakov magnetic monopole solution, but with the much higher mass of 10^{16} GeV. The existence of magnetic monopoles is an unavoidable consequence of any theory that would describe all strong and electroweak forces by one GUT interaction. Unfortunately, GUT monopoles are far too massive be produced at any conceivable man-made accelerator.

6.3 The electroweak monopole

The discovery of what increasingly appears to be the Standard Model Higgs boson in July 2012 by the ATLAS and CMS [19] experiments at the Large Hadron Collider (LHC) at CERN has led to the possibility that the last piece of the Standard Model puzzle may now be in place. This discovery reinforces the electroweak theory of Glashow, Salam and Weinberg, a central pillar of the successful Standard Model. A monopole-like solution within the framework of the Standard Model would explain from first principles the quantization of the electric charge, something that the present Standard Model cannot do. In 1986 Cho and Maison published a paper suggesting that such a magnetic monopole could exist.

Cho and Maison's "Electroweak" monopole is a hybrid of the Dirac Monopole and the 't Hooft Polyakov monopole. Like the 't Hooft-Polyakov monopole its magnetic charge is twice that of the Dirac monopole. According to a recent estimate [9] the mass of the Electroweak Monopole could lie in the range 4-7 GeV/c^2, possibly low enough to be detectable at the LHC by the MoEDAL experiment. Thus, the LHC could be the first collider to produce Electroweak monopoles. Naively, one might expect the production rate would be comparable to a typical electroweak process such as WW production, above the threshold energy. If that is so then MoEDAL experiment should easily be able to detect the EW monopole, if it the threshold is within reach at the LHC. If detected the Electroweak monopole would be the first topological particle to be observed in nature.

7 Electrically charged highly ionizing partilce messengers of new physics

Electrically charged stable or pseudo-stable massive particles (SMPs), with mass greater than the proton mass do not exist within the framework of the Standard Model – they are true and certain messengers of new physics, predicted in a plethora of beyond the Standard Model scenarios. As we have seen such charged massive particles can be highly ionizing, if they have a combination of low velocity and/or multiple charge, and are thus detectable by MoEDAL. The most obvious possibility for an SMP is that one or more new states exist which carry a new conserved, or almost conserved, global quantum number. SUSY with R-parity, extra dimensions with KK-parity, and several other models fall into this category. The lightest of the new states will be stable, due to the conservation of this new 'parity', and depending on quantum numbers, mass spectra, and interaction strengths, one or more higher-lying states may also be stable or meta-stable.

In general, electrically charged stable states are on the face of it highly constrained by cosmology. Coloured particles are also strongly constrained. Consequently, most models are usually constructed to provide non-charged and colourless stable dark-matter candidates. For those models where Weakly Interacting Massive Particles (WIMPs) are the dark matter candidate, SMPs must be meta-stable states that are higher in the hierarchy of particles. Models with long-lived or stable SMPs either have a non-WIMP dark matter candidate or else do not address the problem of particulate dark matter at all. Our recent paper [9] describing MoEDAL's physics program describes nearly thirty such new physics scenarios involving electrically charged massive particles that are "detector stable" and can have a Z/β or at least around five.

7.1 Highly ionizing particles from supersymmetric scenarios

One key problem of the Standard Model is that it does not include gravity, one of the four fundamental forces. The model also fails to explain why gravity is so much weaker than the electromagnetic or nuclear forces. For example, an electrostatically charged hair comb can counteract the gravitational attraction of a whole planet on a small scrap of paper. This difference in the strength of fundamental forces is one aspect of the 'Hierarchy Problem"'.

A prime candidate for a deeper more fundamental theory that would resolve many of the important problems with the Standard Model is Supersymmetry (SUSY), where for every Standard Model particle there is a superpartner (sparticle) whose spin differs by 1/2 - for excellent reviews see, e.g. [20]. By doubling the number of particles in this way one obtains cancellations and degrees of freedom with which we can tackle such the Hierarchy Problem. Indeed, SUSY appears to be a necessary ingredient in the only available candidate for quantum theory of gravity, string theory.

If SUSY were an exact symmetry, SUSY particles would have the same masses as their SM partners. But as no SUSY particles have been observed to date we can surmise that the SUSY partners must be much heavier than the Standard Model particles and therefore SUSY must be a broken symmetry. The mass gap between Standard Model and SUSY particles cannot be much larger than $\sim$1 TeV without destroying SUSY's ability to solve the hierarchy problem. Other important benefits of SUSY are that it allows for the unification of the electromagnetic, weak and strong gauge couplings of the Standard Model – a necessary and predicts particles that are natural dark matter candidates. The Standard Model fails in both these arenas.

At present, little is known for certain about the nature of the SUSY-breaking mechanism. As a result the MoEDAL physics program explores many supersymmetric scenarios [9]. The most common SUSY framework is divided into two sectors. The usual fields of the Minimal Supersymmetric extension to the Standard Model (MSSM) i.e. quarks, leptons, gauge bosons, and two Higgs doublets, together with their supersymmetric partners, belong to the visible sector. While the second, hidden, sector contains additional fields - yet unobserved quantum superfields and the corresponding hypothetical particles, which do not interact via the Standard Model force messengers (gluons, photons, and W and Z bosons) - that lead to the breaking of Supersymmetry. The SUSY breaking, occurring at some higher energy scale in the hidden sector, is mediated to the visible sector by the exchange of weakly interacting 'messenger' particles.

As an example we consider the search for long lived stau slepton, the superpartner of the tau lepton, just one of many examples of SUSY scenarios that give rise to SMPs detectable by MoEDAL [9]. At the LHC, the signals of the supersymmetric particles depend on the nature of the lightest SUSY particle. It may be the lightest neutralino, which escapes detection and leads to missing transverse energy in the detector. Another possibility for the LSP is the gravitino ($\tilde{G}$), the superpartner of the graviton. If the LSP is the gravitino the lifetime of the NLSP slepton, decaying via $\tilde{l} \to \tilde{G}l$, is given by [21]:

$$T = \frac{1}{\Gamma(\tilde{l} \to \tilde{G}l)} = \frac{1}{48\pi M_{Pl}} \frac{m_{\tilde{l}}^5}{M_{\tilde{G}}^2} \left[1 - \frac{M_{\tilde{G}}^2}{m_{\tilde{l}}^2}\right]^4 \tag{6}$$

where, M_{Pl} is the Planck Scale, $M_{\tilde{G}}$ is the mass of the gravitino, $m_{\tilde{l}}$ is the mass of the slepton for example a stau, the superpartner of the tau. Since M_{Pl} is much larger than the electroweak scale, the NLSP lifetime is naturally long, perhaps reaching into years, as can be seen in figure 7.

MoEDAL can detect massive slowly moving ($\beta < 0.2$) very long-lived sleptons in two ways. First, via the high ionization which will leave a signal in MoEDAL's NTD detectors. Secondly, a fraction of the sleptons produced will be captured in its trapping detectors. After exposure at IP8 the

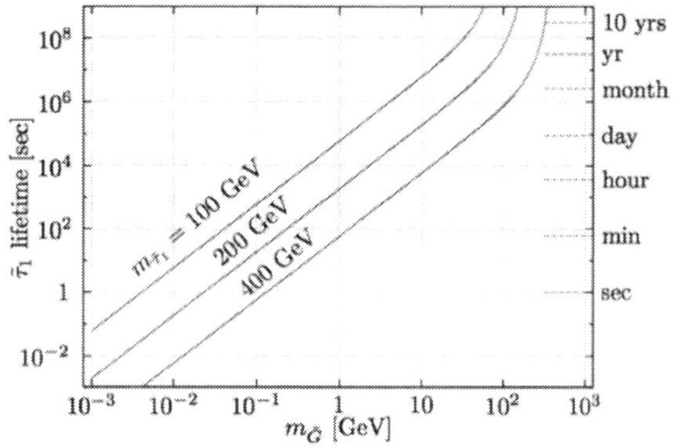

Figure 7. (The stau lifetime in a gravitino LSP scenario, for different values of the stau ($\tilde{\tau}$) mass and the gravitino ($\tilde{G}$) mass [21].

trapping volumes will be removed and monitored deep underground, where cosmic ray backgrounds are minimised, for the slow decays of any trapped sleptons. MoEDAL's sensitivity to such events complements that of the standard general-purpose LHC detectors ATLAS and CMS that are optimized for minimum ionizing relativistic particles. On the other hand, MoEDAL is optimized for HIPs.

8 Millicharged matter

Although Charged Massive Particles (ChaMPs) have been more or less ruled out as dark matter candidates, this is not easily the case for particles with tiny charge. Millicharged particles (mQPs) have been firstly proposed as a solution of the dark matter puzzle long time ago [22]. It is possible to simply add mQPs to the Standard Model for example as Standard Model neutrinos that have tiny electric charges. A small modification of the hypercharge assignments of the Standard Model particles can accommodate tiny electric charges to neutrinos (for a review see e.g. [23]).

Holdom [24] has shown that mQPs will arise naturally, and without violating charge quantization, in certain models involving what are called "shadow universes" or "mirror universes" that are hidden from our own. If there is a shadow-U(1) gauge group i.e a shadow electromagnetism, it follows there is such a thing as a shadow photon, also called a paraphoton. Holdom showed that if the photon and the paraphoton mix a particle charged under the Shadow-U(1) appears to have a very small coupling to the photon - it is a mQP.

The mQP parameter space as defined by mass and charge is highly constrained by direct searches by accelerator experiments [26][27][28][29] as well as indirect astrophysical [26],[30], [31],[32] and cosmological observations [25][30][33]. Of course, indirect observations can be evaded by adding extra degrees of freedom or by using different assumptions. In addition, the parameter space for mQPs with masses between 0.1 GeV and 100 GeV is largely unexplored by accelerator based experiments. MoEDAL's proposed mQP subdetector should be able to push into uncovered territory towards the discovery limits shown in figure 8 for mQPs with charge $\geq$ 0.001e.

9 Cosmic MoEDAL

Grand Unified Theories (GUTs) of electroweak and strong interactions predict the existence of superheavy magnetic monopoles with masses larger than 10^{16} GeV [35]. They would have been produced at the end of the GUT epoch, at a mass scale of $\sim 10^{14}$ GeV and the cosmic time of $\sim 10^{-}34$

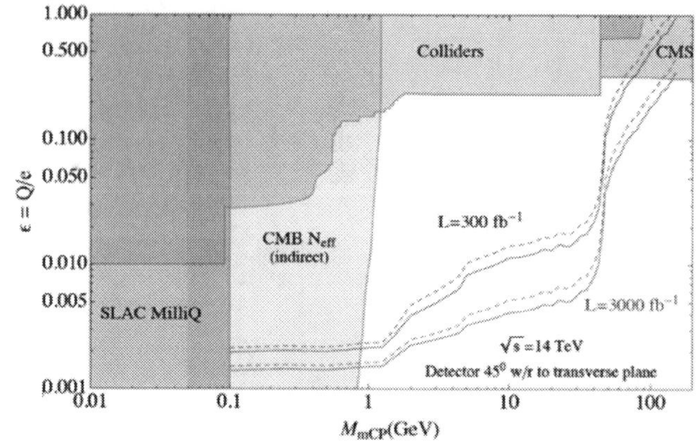

Figure 8. The existing bounds are taken from Ref. [25]. The black line shows the expected 95% C.L. exclusion (solid) and 3σ sensitivity (dashed) of the proposed experiment at centre-of-mass energy 14 TeV, assuming 300 fb^{-1} of luminosity. In blue the corresponding limits achievable in an LHC run with 3000 fb^{-1}

s. Such monopoles cannot be produced with existing accelerators, nor with any foreseen for the future. Lower mass monopoles, proposed by many authors, require a phase transition in the early universe in which a semi-simple gauge group yields a U(1) factor at a lower energy scale [35][36]. Such"Intermediate Mass Monopole" (IMMs), with masses around $10^6 \rightarrow 10^{10}$ GeV, are discussed elsewhere [36][37][38][39]. Additionally, monopoles with masses in the range that are potentially detectable at the LHC have been posited [9]. Such monopoles would be present as primary cosmic particles but also as secondaries produced in UHECR interactions in the atmosphere.

For example, an exposure at a high altitude laboratory would allow us to push the search for relativistic and non-relativistic cosmic monopoles down to relatively low masses (compared to the GUT monopole), higher magnetic charges and lower velocities [40]. The high altitude exposure could allow detection of the above mentioned particles even if they had a strong interaction cross section - as may be the case of the monopole-proton dyonic composite and the lower mass nuclearite - which could prevent them from reaching the earth surface

A large array of NTDs, of the order of 10,000 m^2 or larger, deployed for several years at high altitude would be able to push the search for HIPs in primary or secondary cosmic rays beyond the level reached by previous experiments such as SLIM high altitude array and the MACRO experiment underground. In particular, such a surface array would be sensitive to relativistic and non-relativistic magnetic monopoles with multiple magnetic charge, fluxes well under the Parker Bound, velocities down to $\beta \sim 10^{-4}$ and masses ranging from the TeV level up to the highest GUT monopole mass. The high altitude would allow for an increased possibility that particles which may have a greater than electromagnetic cross-section – as may be the case for a lower mass nuclearite or a dyonic monopole-proton bound state - can be detected. In this arena it would seem most advantageous to place an array at the highest possible altitude, for example on: Mt Chacaltaya in Bolivia (~ 5300 m); Tenerife on the Canarie Islands (~4000m); or, Mauna Kea in Hawaii (4200 m).

10 Summary and Conclusion

The MoEDAL detector employing unconventional methodologies and tuned to the prospect of discovery physics was installed at Point 8 on the LHC rings in the Winter of 2014/2015. The largely passive MoEDAL detector, has a dual nature. First, it acts like a giant camera, comprised of nuclear track detectors - analyzed offline by ultra fast scanning microscopes - sensitive only to new physics. Second, it is uniquely able to trap the particle messengers of physics beyond the Standard Model for further study. MoEDAL's radiation environment is monitored by a state-of-the-art real-time TimePix pixel detector array. The possibility of adding a new MoEDAL subdetector sensitive to mQPs is being studied.

In 2015 the LHC restarted operations at the unparalleled energy of nearly 6.5 TeV beam energy and the enormous collision rate of around a billion collisions per second. The LHC has been compared to an immense microscope allowing us to peer deeper into the Terascale quantum universe than ever before, to explore the inner structure of matter at a scale down to 10^{-19} m. The LHC will recreate the conditions that occurred when the Universe was a picosecond old providing an unprecedented laboratory for the study of the cosmology of the nascent Universe at the earliest times. In parallel, non-accelerator experiments are also exploring the Terauniverse via high-energy astrophysics.

At present it looks very much as though the Higgs boson - the last piece of the Standard Model - has been discovered. But, the LHC is only just getting started. The quest for the deeper underlying theory has now begun in earnest that brings many fundamental questions to the fore. Are there new symmetries of nature? Are there extra spatial dimensions? Is there a deeper substructure? Does magnetic charge exist? Why is gravity so weak compared to the other fundamental forces? What is the nature of dark matter? What was the physics of the earliest era of the universe? The list goes on.

The MoEDAL experiment has a huge discovery potential designed to expand the physics horizons of the LHC and provide paradigm-shifting insights into foundational questions in a complementary way to the other LHC experiments due to its enhanced sensitivity to the highly ionizing avatars of new physics such as magnetic monopoles or massive (pseudo-)stable charged particles. We are investigating the possibility to further enhance the complementary extension of the LHC's physics reach by incorporating a new sub-detector to search for minimally ionizing particles. The unprecedented but complementary nature of the MoEDAL detector and its systematics - exemplified by its ability to retain a permanent record, and even capture new particles for further study - will make MoEDAL an invaluable asset in the elucidation of any Terascale Beyond the Standard Model (BSM) scenario covered by its extensive physics repertoire.

The search for HIPs such as Q-balls, nuclearites and magnetic monopoles can be pushed to GUT mass scales, and below the Parker Bound [41], by the deployment of a 10,000 m^2 or larger COSMIC-MoEDAL array at high altitude.

References

[1] J. L. Pinfold, MoEDAL Collaboration, Technical Design Report of the MoEDAL Experiment MoEDAL Collaboration, CERN-LHCC-2009-006, MoEDAL-TDR-001 (2009).

[2] Large Hadron Collider Official Site: http://home.web.cern.ch/topics/large-hadron-collider.

[3] G. Aad, *et al.*, [ATLAS Collaboration], JINST 3, S08003 (2008).

[4] S. Chatrchyan *et al.*, [CMS Collaboration], JINST 3, S08004 (2008).

[5] K. Aamodt *et al.*, [ALICE Collaboration], JINST 3, S08002 (2008).

[6] A. A. Alves, Jr. *et al.*, [LHCb Collaboration], JINST 3, S08005 (2008).

[7] G. Anelli *et al.*, [TOTEM Collaboration], JINST 3, S08007 (2008).

[8] O. Adriani *et al.*, [LHCf Collaboration] JINST 3, S08006 (2008).

[9] B. Acharya *et al.*, [MoEDAL Collaboration], Int. J. Mod. Phys. A29, 1430050 (2014).

[10] A.A. Prinz, R. Baggs, J. Ballam, S. Ecklund, C. Fertig, J.A. Jaros, K. Kase, A. Kulikov, Willem G. J. Langeveld, R. Leonard et al., Phys. Rev. Lett. 81 1175 (1998).

[11] A. de Roeck *et al.*, Eur. Phys. J. C72, 1985 (2012)
[12] K. Kimm, J. H. Yoon and Y. M. Cho, Eur. Phys., J. C75 67 (2015).
[13] P. A. M. Dirac, Proc. Roy. Soc. Lond. A133, 60 (1931)
[14] J. S. Schwinger, Science 165 757 (1969).
[15] S. L. Glashow, Nucl. Phys. 22 no. 4 579 (1961); S. Weinberg, Phys. Rev. Lett. 19, 1264 (1967); A. Salam, Weak and electromagnetic interactions, in Elementary particle physics: relativistic groups and analyticity, N. Svartholm, ed., p. 367. Almqvist & Wiskell, Proceedings of the eighth Nobel symposium (1968); G. 't Hooft and M. Veltman, Nucl. Phys. B44, 189 (1972).
[16] G. 't Hooft, Nucl. Phys. B79, 276 (1974).
[17] A. M. Polyakov, JETP Lett. 20, 194 (1974).
[18] H. Georgi and S. Glashow, Phys. Rev. Lett. 32, 438 (1974).
[19] G. Aad *et al.*, ATLAS Collaboration, Phys. Lett. B716, 1 (2012); S. Chatrchya *et al.*, CMS Collaboration Phys. Lett. B716, 30 (2012).
[20]] S.P. Martin, hep-ph/9709356 (1997); X. Tata, hep-ph/9706307 (1997).
[21] K. Hamaguchi, M. M. Nojiri and A. de Roeck, JHEP 0703, 046 (2007).
[22] H. Goldberg and L. J. Hall, Phys. Lett. B174, 151 (1986); A. De Rujula, S. L. Glashow and U. Sarid, Nucl. Phys. B333, 173 (1990); S. Dimopoulos, D. Eichler, R. Esmailzadeh and G. D. Starkman, Phys. Rev. D41, 2388 (1990).
[23] R. Foot, G. C. Joshi, H. Lew and R. R. Volkas, Mod. Phys. Lett. A5, 95 (1990) [Erratum-ibid. 1990 A 5, 2085]; R. Foot, H. Lew and R. R. Volkas, J. Phys. G19, 361 (1993) [Erratum-ibid. 1993 G 19, 1067]; S. Davidson, S. Hannestad and G. Raffelt, JHEP 0005, 003 (2000).
[24] B. Holdom, Phys. Lett. B166, 196 (1986).
[25] H. Vogel and J. Redondo, JCAP 1402, 029 (2014).
[26] S. Davidson, S. Hannestad, and G. Raffelt, JHEP 0005, 003 (2000).
[27] A. Prinz, R. Baggs, J. Ballam, S. Ecklund, C. Fertig, *et al.*, Phys. Rev. Lett. 81 1175 (1998).
[28] A. Badertscher *et al.*, Phys.Rev. D75, 032004 (2007).
[29] S. Chatrchyan *et al.* (CMS Collaboration), Phys. Rev. D87 092008 (2013).
[30] S. Davidson, B. Campbell, and D. C. Bailey, Phys. Rev. D43, 2314 (1991).
[31] R. Mohapatra and I. Rothstein, Phys. Lett. B247, 593 (1990).
[32] S. Davidson and M. E. Peskin, Phys. Rev. D49, 2114 (1994).
[33] A. Dolgov, S. Dubovsky, G. Rubtsov, and I. Tkachev, Phys. Rev. D88, 117701 (2013).
[34] A. Haas et al., [hep-ph] arXiv:1410.6816v1 (2014).
[35] Monopoles in Quantum Field Theory, N.S. Craige, P. Goddard, W. Nahm editors, World Scientific (1982); G. Giacomelli, Rivista del Nuovo Cimento 7 (1984); Monopoles in Gauge Theories, Lecture Notes by N. Craige, G. Giacomelli, W. Nahm and Q. Shafi, World Scientific (1986).
[36] E. Huguet and P. Peter (Bound states in monopoles: sources for UHECR?) hep-ph/9901370 (1999).
[37] T. J. Weiler and T. K. Kephart, Nucl. Phys. B. Proc. Suppl. 51, 218 (1996); T.W. Kephart and T. J. Weiler, Astrop. Phys. 4, 217 (1996).
[38] C. O. Escobar and R. A. Vazquez, Astrop. Phys. 10, 197 (1999).
[39] A. De Rujula, Nucl. Phys. B435, 257 (1995).
[40]] S. Balestra *et al.*, Eur.Phys.J. C55, 57-63 (2008).
[41] E. Parker, Astrophys. J. 160, 383 (1970).

Nucleon structure studies with the COMPASS experiment at CERN

Stephane Platchkov[1,a]

[1] *IRFU/SPhN, CEA Saclay*

Abstract. The COMPASS experiment at CERN uses hadron and lepton beams for nucleon structure studies. Most of the data collected so far with a muon beam and either proton or deuteron polarised target were analysed in terms of longitudinally or transversely polarised parton distribution functions and transverse momentum-dependent distributions in the nucleon. A negative hadron beam is used to perform Drell-Yan measurements. The COMPASS large polarised target gives access to several momentum-dependent single-spin asymmetries and provides a stringent test of the fundamental QCD factorisation assumptions. With positive and negative muon beams COMPASS also studies Generalised Parton Distributions using exclusive deeply virtual Compton scattering and meson production experiments. An overview of the most recent COMPASS results is given. The expected physics outcome of the forthcoming measurements is discussed.

1 Introduction

The COMPASS fixed-target experiment at CERN [1] uses the SPS secondary and tertiary beams for nucleon structure and spectroscopy studies. The COMPASS hadron spectroscopy programme is covered elsewhere in this conference [2]. In the last few years COMPASS has made detailed studies of the nucleon structure using Deep-Inelastic Scattering (DIS) and Semi-Inclusive DIS (SIDIS). The data were collected with a positive, naturally-polarised muon beam, and either proton or deuteron longitudinally polarised target. An important share of the beam time was dedicated to measurements with a transversely polarised target. The DIS and SIDIS data are mainly used to evaluate the polarised parton distribution functions (PDF) and transverse-momentum dependent (TMD) distributions.

Since 2012 COMPASS has initiated a new physics programme, making full use of its high-acceptance experimental setup. The nucleon structure share of this programme comprises two parts. In the first part, a negative hadron beam is used to perform Drell-Yan measurements on polarised nucleon and unpolarised nuclear targets. The Sivers momentum-dependent asymmetry measured in Drell-Yan will be compared with the single-spin asymmetries already measured in SIDIS. The comparison should provide a stringent test of the fundamental QCD factorisation assumptions.

The second part of the new programme is devoted to the study of the Generalised Parton Distributions (GPD). Positive and negative muons beams scattered off an unpolarised proton target will allow measurements of both beam charge and spin sum and difference. The results will provide a determination of the transverse nucleon size as a function of the parton longitudinal momentum.

[a] e-mail: S.Platchkov@cea.fr

2 Quark contribution to the nucleon spin

Most of the information [3] about the quark helicity contribution to the spin, $\Delta\Sigma$, comes from deep-inelastic scattering (DIS) measurements of polarised leptons on polarised targets. In the recent years COMPASS has measured double-polarisation asymmetries on both proton (NH_3) and deuteron (^6LiD) targets [4, 5]. The corresponding proton, g_1^p, and deuteron, g_1^d, spin-dependent structure functions are shown in Fig. 1, together with available data from other experiments. The data are displayed as a function of the four-momentum transfer squared, Q^2, in different $x = x_{Bj}$ intervals. The COMPASS data are particularly important as they cover the low-x domain.

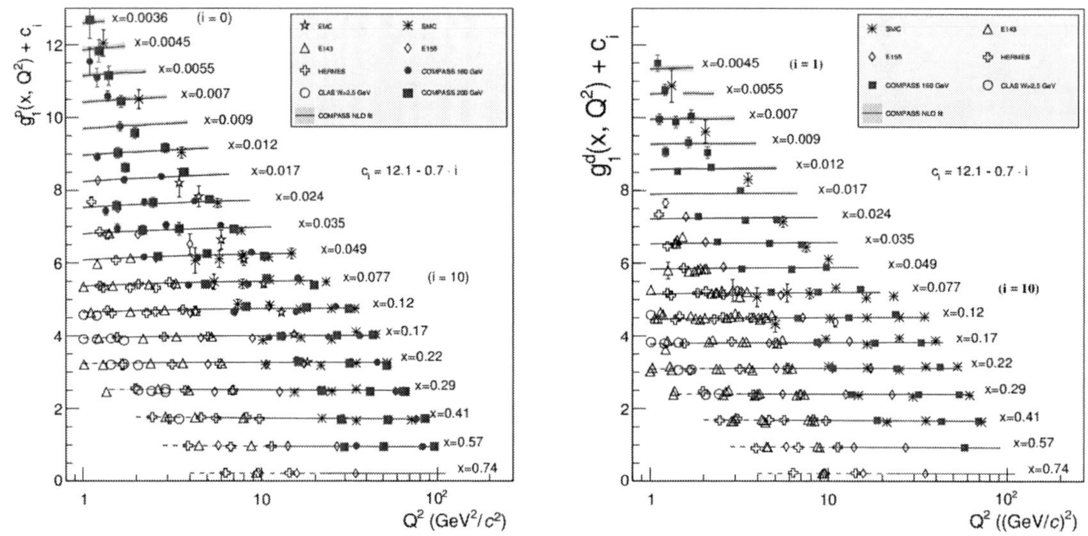

Figure 1. World data on the $g_1(x)$ structure function for the proton (left) and for the deuteron (right) a a function of Q^2, for various x values. The COMPASS data are indicated in red. The lines represent any of the solutions of the COMPASS global NLO QCD fit (see text).

The world data for the spin structure functions for proton, deuteron and neutron targets are included in a global QCD fit [5], performed at NLO. A functional form is assumed at an initial Q_0^2 scale for all singlet, triplet, and octet flavour combinations of the quark distributions, as well as for the gluon distribution. The first moments of the non-singlet distributions are fixed to the values of the axial-vector weak coupling constants measured in neutron and hyperon β-decay and assuming SU(3) flavour symmetry. The $x - Q^2$ evolution is done in the $\overline{MS}$ factorisation scheme. In order to investigate the sensitivity of the results to the functional form, the fit was performed for several sets of functional shapes. It was found that two sets of functional shapes are needed to span almost entirely the range of the possible fit results, and particularly for the less-well known gluon contributions, $\Delta g(x)$.

The sets of functional forms, corresponding to the extreme solutions for $\Delta g(x)$, are displayed in Fig. 2. According to each of the solutions, the value of the quark contribution, $\Delta\Sigma$, ranges between 0.26 and 0.34, or $\Delta\Sigma = 0.30 \pm 0.04$. On the top row are shown the quark singlet helicity $\Delta q(x)$ (the integral of which gives $\Delta\Sigma$) and the gluon helicity $\Delta g(x)$ distributions. In Fig. 2 bottom row, the quark helicities are shown separately for Δu, Δd and Δs distributions. The solid and dashed lines correspond to two extreme Δg solutions of the NLO QCD fit. The dark and light bands correspond respectively

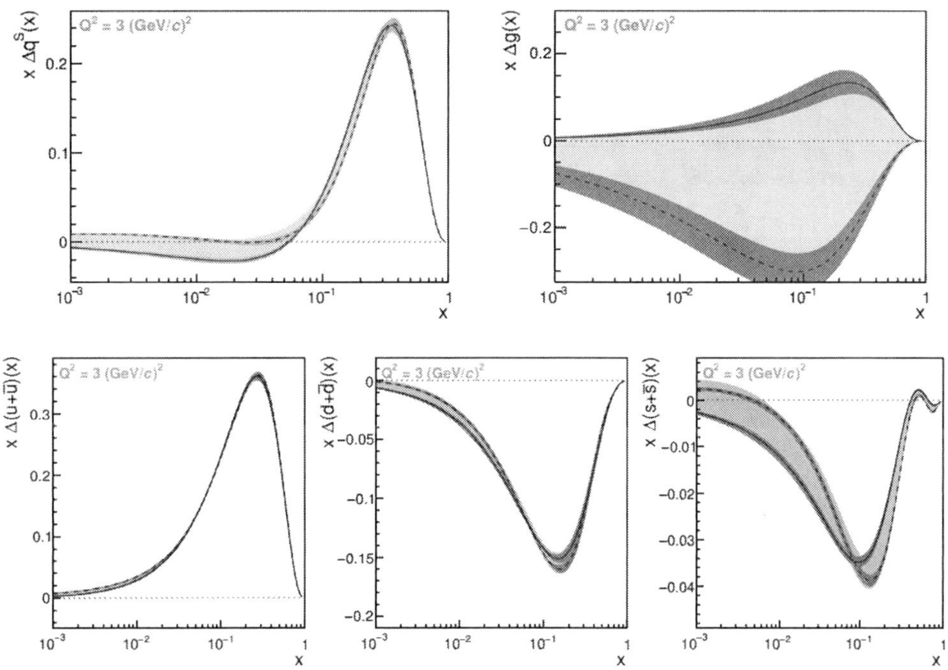

Figure 2. Top: Helicity distributions for quarks (left) and for gluons (right) at $Q^2 = 3$ (GeV/c)^2. Bottom left to right: helicity distributions for up, down, and strange quark flavour, respectively. The solid and dashed lines correspond to the two extreme solutions for the Δg distributions. The dark bands correspond to the statistical uncertainties and the light bands to the systematic uncertainties.

to the propagation of the statistical errors and the systematic uncertainties. A significant part of the systematic uncertainties is attributed to the choice of the functional form.

2.1 Gluon contribution to the nucleon spin

The range of solutions resulting from the different assumptions for the shape of the Δg(x) distribution clearly show that the inclusive data are unable to determine neither the shape, nor even the sign of Δg. Further information on Δg is inferred by considering the photon-gluon fusion (PGF) process, in which the virtual photon is absorbed by a gluon in the nucleon, giving rise to a quark-antiquark pair. There are two options for the identification of the PGF process. The first, "open charm" option, is free from physical background at LO. This fact is justified by a negligible quantity of intrinsic-charm in the nucleon [6] in the COMPASS kinematic domain. The open charm is tagged by the invariant mass reconstruction of D^0 mesons, via detection of their hadronic decay products [7]. The open charm result for $\Delta g/g$, obtained from 160 GeV/c polarised muon scattering off polarised proton and deuteron targets is shown in Fig. 3.

In the second method, called "high-p_T" method, hadron pairs with high transverse momenta p_T and $Q^2 > 1$ (GeV/c)^2 in the final state are selected [8]. Despite the large statistics available, this method requires a more complicated analysis. Indeed, several processes contribute to the measured asymmetries: the leading order virtual-photon absorption process (LP), the QCD Compton process (QCDC), and the PGF process. In order to extract $\Delta g/g$ from the high-p_T asymmetries corresponding

DOI: 10.1051/epjconf/201612602025

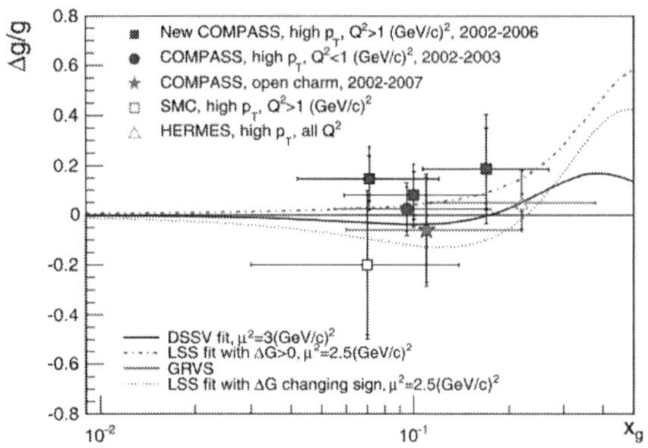

Figure 3. COMPASS results for direct $\Delta g/g$ extraction at LO using the hadron pairs method (squares) and the open charm method (star). The data points from SMC and HERMES are also displayed. The curves are NLO global fit evaluations of $\Delta g/g$ from Refs [12–14].

to the PGF process only, the contributions from the LP and QCDC processes must be evaluated and subtracted. The corresponding parton asymmetries, as well as the fractions corresponding to each physical process, are determined from a full Monte Carlo simulation of the COMPASS experiment [8]. The resulting data points are shown in Fig. 3. They agree with the open charm point from COMPASS and with the high-p_T points from SMC [9] and HERMES [10] experiments, as well as with a previous [11] evaluation by COMPASS using a quasi-real photoproduction process instead of DIS. Figure 3 also shows some of the available world data fits at NLO [12–14].

All COMPASS data on a deuteron target (2002 to 2006) were recently re-analysed, using an extension of the high-p_T method, called "all-p_T" method [15]. In this new method, the low p_T events are kept in the analysis and at least one hadron is required in the final state. The gluon polarization and leading process asymmetry (which is dominant at low p_T) are extracted simultaneously from the same data set using a Neural Network approach. The all-p_T method results in a reduction of both systematic and statistical uncertainties by more than 50%, compared to the result of Ref. [8]. The preliminary value of the gluon polarization is $\Delta g/g = 0.113 \pm 0.038 \pm 0.035$ at an average gluon momentum fraction $x_g = 0.10$ and at a scale of $\mu^2 = 3$ (GeV/c)^2. All studies indicate that the gluon polarisation is small and either slightly positive or compatible with zero in the x_g range of the data.

2.2 Polarised quark distributions

The polarised quark distributions for the light flavours are extracted from semi-inclusive DIS (SIDIS) asymmetry measurements. A total of 8 asymmetries are measured [16] at COMPASS, namely A^{π^+}, A^{π^-}, A^{K^+}, and A^{K^-} for both proton and deuteron targets. By also taking into account the two inclusive spin asymmetries, which were used for the extraction of g_1^d and g_1^p, the $\Delta u(x)$, $\Delta\bar{u}(x)$, $\Delta d(x)$, $\Delta\bar{d}(x)$, and $\Delta s(x)$ quark helicities are evaluated using a LO fit. All helicities are extracted assuming the DSS parametrisation [17] for the quark to pion and quark to kaon fragmentation functions. The results from the extractions are shown in Fig. 4.

The SIDIS results for $\Delta u(x)$ and $\Delta d(x)$ PDFs are compatible with the QCD fit results of the DIS data. However, the strange quark polarised distribution is quite different. The corresponding first moment is: $\Delta s = -0.01 \pm 0.01(stat.) \pm 0.01(syst.)$. The difference between this result and the one obtained from the inclusive asymmetries, $(\Delta s + \Delta\bar{s}) = -0.08 \pm 0.01(stat.) \pm 0.02(syst.)$, is puzzling.

2.3 Quark fragmentation functions

A possible explanation of the Δs puzzle could come from the kaon fragmentation functions (FF) used in the SIDIS analysis. Indeed, the extraction of the $\Delta s(x)$ distribution is particularly sensitive [16]

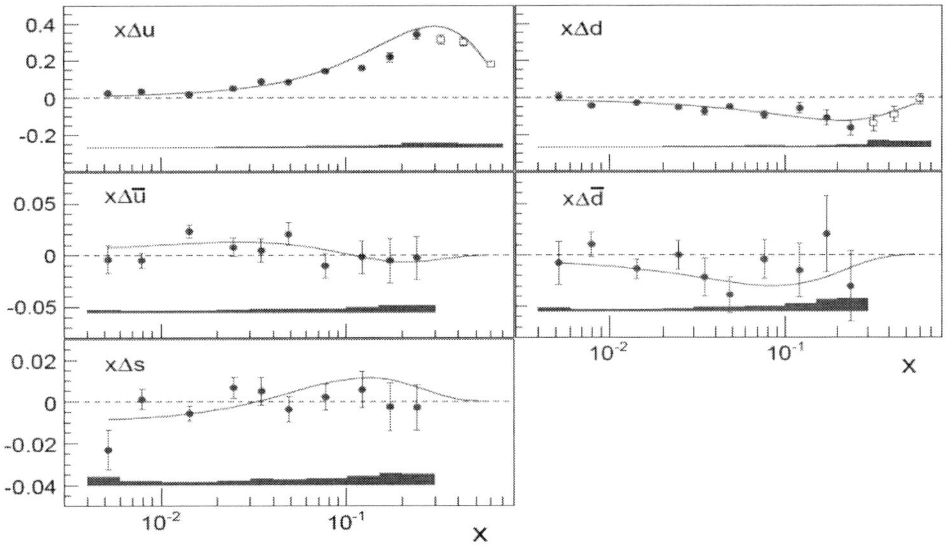

Figure 4. SIDIS results for the light quark helicities, under the assumption that at $Q^2 = 3\ (\mathrm{GeV}/c)^2$. The curves correspond to a global NLO fit taken from Ref. [13].

to the kaon FFs. The FFs are universal and describe the process of hadronisation. They can be extracted via a QCD fit to the pion and kaon multiplicities, which represents the number of produced pions and kaons per deep inelastic events. COMPASS has measured both pion and kaons multiplicities on a deuteron target. Using the π^+ and π^- multiplicities data and some additional assumptions (charged conjugation and isospin symmetry), a LO QCD fit is performed. Two pion FFs are fitted: the "favoured" FF (when the quark of the target nucleon is a valence quark of the produced hadron) and the "unfavoured" FF (when the quark of the target nucleon is a sea quark of the produced hadron). Preliminary results for the pion FFs extracted at LO [18] are shown in Fig. 5. The results are compared to the FFs used for analysing the SIDIS results. The new COMPASS FFs are at variance with one

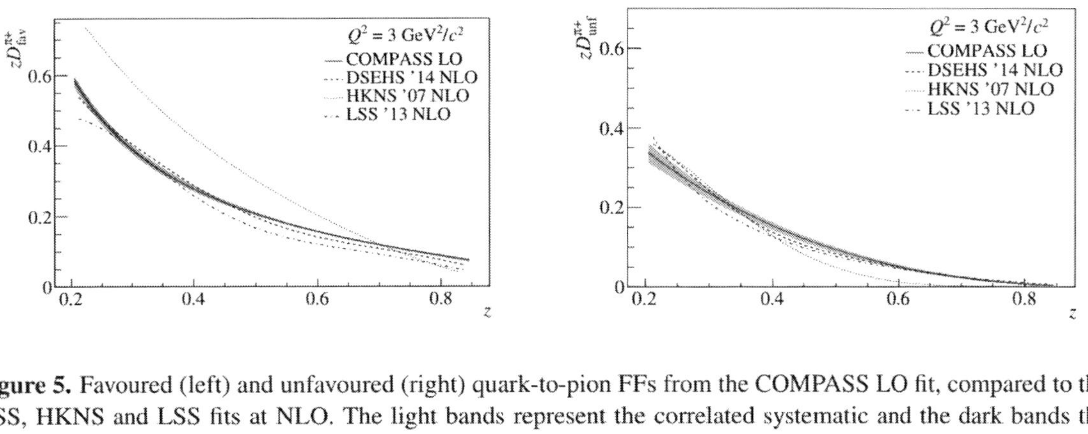

Figure 5. Favoured (left) and unfavoured (right) quark-to-pion FFs from the COMPASS LO fit, compared to the DSS, HKNS and LSS fits at NLO. The light bands represent the correlated systematic and the dark bands the total statistical uncertainties.

DOI: 10.1051/epjconf/201612602025

of the older parametrisation [19]. Recent NLO re-analyses [20, 21], which include the preliminary COMPASS pion multiplicities data, are in close agreement with the COMPASS LO fit. The observed differences between the older and new parametrisations of the pion FFs have a minor effect on the extraction of the $\Delta u(x)$ and $\Delta d(x)$ distributions. In connection with the Δs puzzle presented above, it is worth noting that the kaon FFs are much less well known than the pion FFs. COMPASS has also measured the kaon multiplicities; preliminary FF fits show large deviations from the available parametrisations.

3 The three-dimensional spin-dependent structure of the nucleon

A complete description of the partonic structure of the nucleon in QCD includes the spin and transverse momentum dependent parton distribution functions (TMD PDF). At leading twist there are eight TMD PDFs that take into account the correlations between the quark spin, the quark transverse momentum and the nucleon spin. The TMD PDFs can be accessed in SIDIS by detecting at least one of the produced hadrons. The resulting SIDIS cross section exhibits different azimuthal modulations, with different combinations of two relevant angles: ϕ_s and ϕ_h. The two angles are the azimuthal angles of the initial nucleon transverse spin vector and the produced hadron momentum, respectively, in

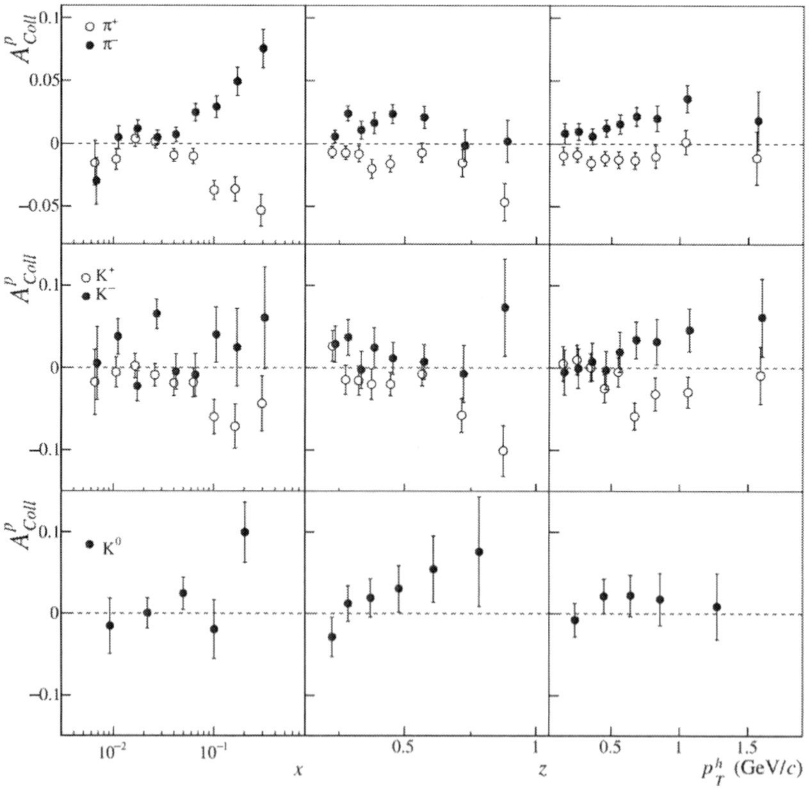

Figure 6. The Collins asymmetries for charged pions (top), charged kaons (middle) and neutral kaons (bottom) as a function of x, z and p_T.

a reference system in which the z axis is the virtual photon direction and the $x-z$ plane is the lepton plane according to Ref. [22].

Over the years COMPASS has performed extensive studies of these TMD PDFs using data for single and two hadron SIDIS off unpolarised [23], longitudinally [24] and transversely polarised protons (deuterons) [25] in a NH_3 (^{6}LiD) target. The two most thoroughly studied transverse spin asymmetries are the Collins and Sivers asymmetries.

The Collins asymmetries allow access to the transversity PDF, which is coupled to the Collins fragmentation functions. The results for the Collins asymmetries using a transversely polarised proton target [26] are shown in Fig. 6 as a function of the Bjorken variable, x, the fraction of the hadron energy relative to the virtual photon energy, z, and the transverse hadron momentum, p_T. At small x the pion asymmetries are compatible with zero, while in the valence region they show an increasing signal, which has opposite sign for positive and negative pions. The results for charged kaons, although with larger statistical uncertainties, show a similar trend. The Collins asymmetry for neutral kaons shows a positive trend with increasing z. The average asymmetry is positive but compatible with zero within the statistical uncertainty.

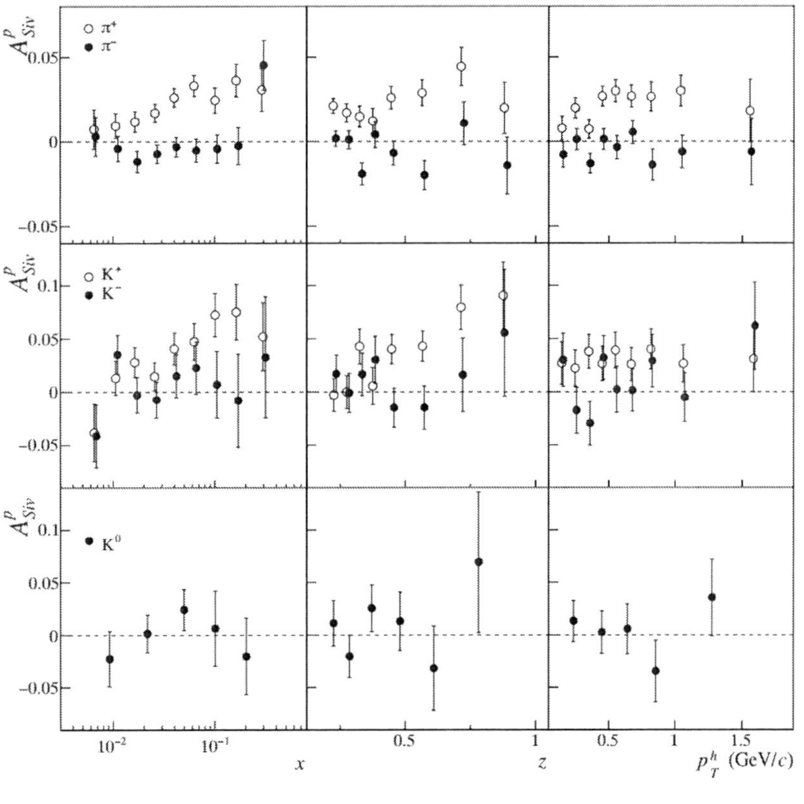

Figure 7. The Sivers asymmetries for charged pions (top), charged kaons (middle) and neutral kaons (bottom) as a function of x, z and p_T.

Figure 7 shows the COMPASS measurement of the Sivers asymmetries [26] for a transversely polarised proton target, as a function of the x, z and p_T variables. The Sivers asymmetries give access to the Sivers PDFs, which describe the correlations between quark transverse momentum and nucleon spin. The asymmetries for negative pions and kaons, as well as for neutral kaons are compatible with zero, while for positive pions and kaons there is a clear evidence for a positive signal extending over the full measured x region and increasing with z.

The new COMPASS data for the TMD PDFs provide a valuable input for multidimensional global analyses and constitute an important contribution for the understanding of the transverse momentum structure of the nucleon.

4 Generalised parton distributions

The Generalised Parton Distributions (GPD) contain additional information on the nucleon structure [27, 28]. They are accessible at COMPASS via deeply virtual Compton scattering (DVCS) and deeply virtual meson production (DVMP). At leading twist and for a spin 1/2 target, there exist four chiral-even GPDs (H, $\tilde{H}$, E, $\tilde{E}$) for each quark flavour. The first two of them describe processes that conserve nucleon helicity and the last two those that involve nucleon helicity flip. The GPDs H and E are of special interest as they are related to the total angular momentum carried by the partons in the nucleon [29]. When Fourier-transformed to impact parameter space and for the case of vanishing longitudinal momentum transfer, GPDs provide a three-dimensional description of the nucleon in a momentum-coordinate space, which is also known as "nucleon tomography" [30].

For DVCS on a proton unpolarised target in the COMPASS kinematical domain, the GPD H distribution yields the dominant contribution, while the GPD E is accessible with transverse target polarisation or with a neutron (deuteron) target. All GPDs depend on the photon virtuality Q^2, the total four-momentum transfer squared, t, between the initial and final nucleon states, and two additional

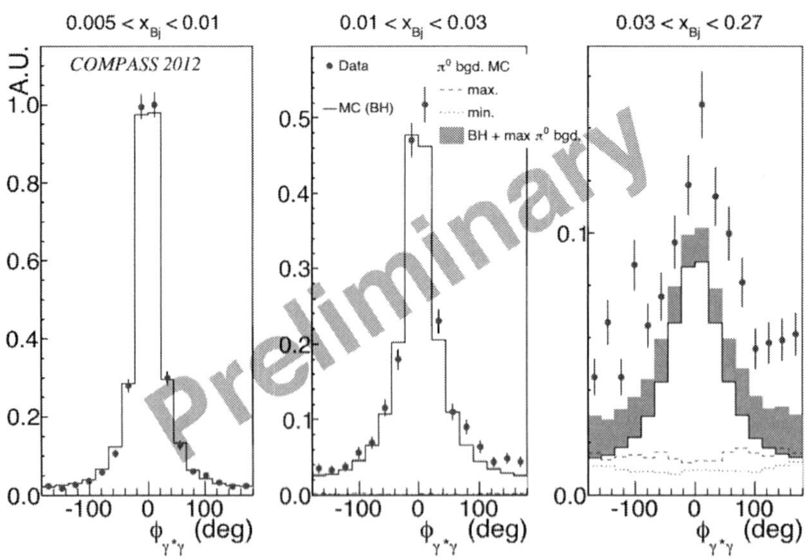

Figure 8. Exclusive single photon events obtained in the 2012 test run as a function of $\phi_{\gamma^*\gamma}$ (red dots) compared to the Bethe-Heitler contribution (black line histogram) and the π^0 contamination estimated by a Monte-Carlo simulation.

variables, x and ξ, representing the average and half the difference between the initial and final longitudinal momentum fraction of the nucleon carried by the struck parton. In DVCS and DVMP processes x is an internal variable, while the skewness ξ is related to the Bjorken variable $x_{Bj} = Q^2/2M\nu$.

The extraction of GPDs from such measurements needs experimental data on hard exclusive processes in a broad kinematic range. Up to now information in the high-energy regime at low x_{Bj} was provided by H1 and ZEUS at DESY. On the other hand, HERMES and JLAB experiments cover the low-energy domain at high x_{Bj}. The COMPASS measurements will provide a connection between these measurements by covering the region of x_{Bj} between 0.01 and 0.1, where both sea and valence quarks are important.

A competing process is the Bethe-Heitler (BH) process of elastic lepton-nucleon scattering with a hard Bremsstrahlung photon emitted by the incoming or by the outgoing lepton. It produces the same final state as DVCS so that both processes interfere at the level of amplitudes.

In the COMPASS kinematical domain the dominant process is function of the x_{Bj} variable. At low x_{Bj} almost pure BH events are recorded; these events provide a reference yield, which is used for normalisation purposes. At higher x_{Bj}, the DVCS contribution is dominant. After subtraction of the BH contribution, the data can be used for a measurement of the t-dependence of the cross section, which provides a tomographic image of the nucleon. In the intermediate x_{Bj} domain the DVCS contribution is enhanced by the BH process through their interference. The size of the interference term increases with rising x_{Bj} and provides via its characteristic $\phi_{\gamma^*\gamma}$ modulation additional possibilities to access real and imaginary part of the complex DVCS amplitude. Here $\phi_{\gamma^*\gamma}$ is the azimuthal angle between the leptonic plane (defined by the incident and scattered lepton) and the hadronic plane (defined by the virtual and real photon). The DVCS process itself is dominant for $x_{Bj} > 0.03$.

The first proof of the feasibility for DVCS was made in 2009 with a reduced experimental setup. In 2012, a four-week long pilot run has been recorded, with a mostly complete DVCS setup, which included a full-scale recoil detector and a large-size hydrogen target. The preliminary results from the pilot run are shown in Fig. 8. The data are displayed in three x_{Bj} bins as a function of $\phi_{\gamma^*\gamma}$. The data confirm the expected behaviour: in the lowest x_{Bj} bin, where the DVCS contribution is negligible, the calculated BH contribution perfectly agrees with the measurements. The π^0 contamination, when one of the decay photon mimics the high energy DVCS photon and the other photon is not detected, is estimated using a Monte-Carlo simulation.

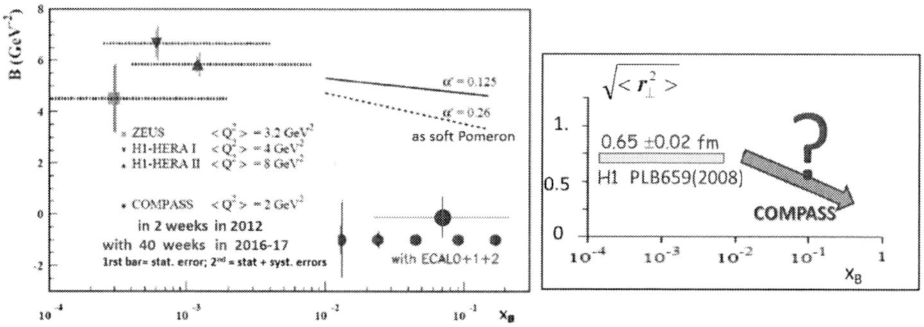

Figure 9. Projections for the measurement of the x_{Bj} dependence of the t-slope parameter $B(x_{Bj})$ of the DVCS cross section for a two years data taking, together with the expected measurement from the 2012 pilot run. The data from HERA (see text) are shown as triangles (left). Sketch of the corresponding x_{Bj} dependence of the mean proton radius (right).

The full DVCS run is planned for 2016 and 2017. The data taking will include cross sections measurements with both muon charge and polarisation states of the incoming lepton beam, $d\sigma(\mu^{+,\rightarrow})$ and $d\sigma(\mu^{-,\leftarrow})$. The real and the imaginary part of the Compton form factors (CFF) can be extracted from the sum $\mathcal{S}$ and the difference $\mathcal{D}$ of these two two cross-sections. A CFF is a sum over flavours f of convolutions of the respective GPDs with a perturbatively calculable kernel describing the hard photon-quark interaction. In the difference $\mathcal{D}$ the pure BH contribution cancels out. The analysis of the $\phi_{\gamma^*\gamma}$ dependence provides a measurements of the real part of the corresponding CFF. On the other hand, in $\mathcal{S}$ the BH contribution does not cancel and has to be subtracted. Integrating over $\phi_{\gamma^*\gamma}$ yields the t dependence of the cross section, and therefore the transverse size of the nucleon at different values of x_{Bj}.

Figure 9 left shows the t-slope parameter B as a function of x_{Bj}. In the region of low x_{Bj} this parameter has been measured in HERA by H1 [31] and ZEUS [32] to be constant. A decrease is expected at the higher x_{Bj} region covered by COMPASS. The projected COMPASS results should clearly measure this decrease. The point with a large horizontal bar represents the statistical uncertainty of the 2012 pilot run measurement. Here, the shrinkage parameter α' is related, through a simple relation, to the decrease of the nucleon size with increasing x_{Bj}. Figure 9 right summarises this information for the proton strong radius, instead of B.

5 Drell-Yan measurements

In the Drell-Yan process (DY), the annihilation of a quark from the beam with an antiquark from the target (or vice versa) produces, via a virtual time-like photon, a pair of leptons with a large invariant mass [33]. The properties of this lepton pair reflect the distributions of the partons inside the interacting hadrons. The Drell-Yan process thus provides a powerful tool for probing the parton substructure of the hadrons by allowing an extraction of the shape and the flavour dependence of the familiar PDFs in both beam and target hadrons. In specific cases and for polarised beam or target, the Drell-Yan process is also sensitive to the transverse momentum dependent (TMD) parton density distributions.

COMPASS has several unique advantages for performing Drell-Yan studies. First of all, CERN is presently the only place where beams of (positive or negatve) pions, kaons and protons beams are available. The valence antiquarks in the beam interact preferentially with the valence quarks of the target nucleons. Meson-induced Drell-Yan process is the unique way to access the quark structure of pions and kaons, as these short-lived mesons cannot be used as stable targets.

A second advantage lies in the COMPASS experimental setup. The setup, operational since 2002, was especially designed for semi-inclusive and exclusive experiments, requiring detection of several particles in the final state and therefore a large geometrical acceptance. The latter has been achieved by building two consecutive spectrometers around two large-aperture dipole magnets. As a result, the COMPASS acceptance is between 4 and 10 times larger than the acceptances of the previous Drell-Yan experiments. These factors are directly related to the final statistical accuracy achieved. Finally, COMPASS will be able to access the correlation between the nucleon's transverse spin and the transverse momentum distributions of valence quarks by making use of its large-size transversely polarised NH_3 target.

The main goal of the COMPASS studies is a measurement of the TMD Sivers asymmetry. QCD factorisation predicts that the Sivers quark distribution function, which depends on the quark transverse momentum, is not universal, i.e. when extracted from the DY process, it should have a sign oppositie to that extracted from a SIDIS reaction. The test of this fundamental prediction is considered as a crucial test of the QCD TMD factorisation.

In order to maximise the statistical accuracy, the relatively low cross-section of the Drell-Yan process is compensated by a high intensity pion beam. A maximum beam intensity achievable in

 DOI: 10.1051/epjconf/201612602025

COMPASS, close to 10^8 particles per second, is used. A thick hadron absorber is placed immediately downstream the polarised target. In its inner part, the absorber contains a core of tungsten (W) designed to stop the beam which did not interact in the polarised target. This beam plug is in itself a heavy nuclear target. A thin aluminium (Al) target is also placed half-way between the polarised target and the beam plug. The data from the Al and W targets will be used for additional unpolarised Drell-Yan studies.

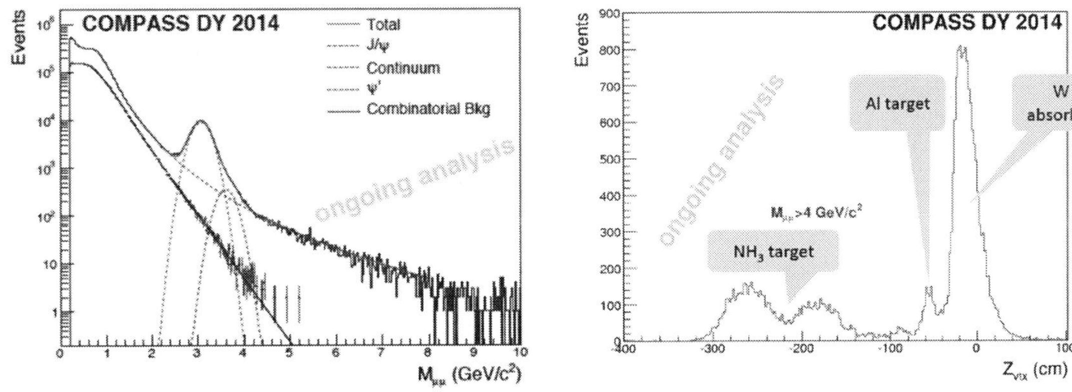

Figure 10. Dimuon mass distribution for two weeks of data taking with an unpolarized target in 2014 (left). Vertex distribution of the events from the three targets in use (right).

A full-scale DY pilot run with the nearly final COMPASS setup, but with an unpolarised target was taken in the fall of 2014. The results from the pilot run are shown in Fig. 10. The resulting mass distribution (left panel) is in good agreement with the expectations, both in terms of mass resolution (estimated by the width of the J/ψ peak) and counting rates. The reconstructed muon tracks give access to the interaction vertex, as shown in the right panel. The two NH_3 cells are well separated and the peaks from the Al and W targets are clearly visible. The COMPASS Drell-Yan measurement started in 2015. In the mass range of $4 < M_{\mu\mu} < 9$ GeV/c^2, about 800 events per day are expected. The expected statistical uncertainty for the Sivers asymmetry from the 2015 data should be of about 2%.

References

[1] COMPASS Collaboration, P. Abbon et al., Nucl. Inst. Meth. **A 577**, 455 (2007); COMPASS Collaboration, P. Abbon et al., Nucl. Inst. Meth. **A 779**, 69 (2015).
[2] F. Nerling, talk at this conference.
[3] C.A. Aidala, S.D. Bass, D. Hasch and G.K. Mallot, Rev. Mod. Phys. **85**, 655 (2013).
[4] COMPASS Collaboration, M. G. Alekseev et al., Phys. Lett. **B 690**, 466 (2010).
[5] COMPASS Collaboration, C. Adolph et al., arXiv:1503.08935.
[6] J. Alwall, PoS, DIS 2005, Madison, USA, arXiv:0508126v1.
[7] COMPASS Collaboration, C. Adolph et al., Phys. Rev. **D 87**, 052018 (2013).
[8] COMPASS Collaboration, M. G. Alekseev et al., Phys. Lett. **B 718**, 922 (2013).
[9] SMC Collaboration, B. Adeva et al., Phys. Rev. **D 70**, 012002 (2004).
[10] HERMES Collaboration, A. Airapetian et al., Journal of High Energy Physics **1008**, 130 (2010).
[11] COMPASS Collaboration, E.S. Ageev, et al., Phys. Lett. **B 633**, 25 (2006).

[12] E. Leader, A.V. Sidorov and D.B. Stamenov, Phys. Rev. **D 82**, 114018 (2010);
[13] D. de Florian, R. Sassot, M. Stratmann and W. Vogelsang, Phys. Rev. **D 80**, 034030 (2009).
[14] M. Gluck, E. Reya, M. Stratmann and W. Vogelsang, Phys. Rev. **D 63**, 094005 (2001).
[15] M. Stolarski, PoS, DIS 2014, Warsaw, Poland, 211.
[16] COMPASS Collaboration, M. G. Alekseev et al., Phys. Lett. **B 693**, 227 (2010).
[17] D. de Florian, R. Sassot and M. Stratmann Phys. Rev. **D 75**, 114010 (2007).
[18] COMPASS Collaboration, C. Adolph et al., to be published.
[19] M. Hirai, S. Kumano, T.-H. Nagai and K. Sudoh, Phys. Rev. **D 75**, 094009 (2007).
[20] E. Leader, A. Sidorov and D. Stamenov, arXiv:1312.5200.
[21] D. de Florian, et al, Phys. Rev. **D 91**, 014035 (2015).
[22] A. Bacchetta et al., Phys. Rev. **D 70**, 117504 (2004).
[23] COMPASS Collaboration, C. Adolph et al., Nucl. Phys. **B 886**, 1046 (2014).
[24] COMPASS Collaboration, M. G. Alekseev et al., Eur. Phys. J. **C 70**, 39 (2010).
[25] COMPASS Collaboration, M. G. Alekseev et al., Phys. Lett. **B 673**, 127 (2009)
[26] COMPASS Collaboration, C. Adolph et al., Phys. Lett. **B 744**, 250 (2015).
[27] D. Mueller et al., Fortsch. Phys. **42**, 101 (1994).
[28] A. V. Radyushkin, Phys. Lett. **B380**, 417 (1996).
[29] X. D. Ji, Phys. Rev. Lett. **78**, 610 (1997).
[30] M.Burkardt, Int. J. Mod. Phys. **A 18**, 173 (2003).
[31] A. Aktas et al., Eur. Phys. J. **C 44**, (2005) 1; D. Aaron et al., Phys. Lett. **B 659**, 796 (2008).
[32] ZEUS Collaboration, S. Chekanov et al., JHEP **0905**, 108 (2009).
[33] J.C. Peng and J.-W. Qiu, Prog. Part. Nucl. Phys. **76**, 43 (2014).
[34] COMPASS Collaboration, COMPASS-II Proposal, CERN–SPSC–2010–014.

EPJ Web of Conferences **126**, 02026 (2016) DOI: 10.1051/epjconf/201612602026
ICNFP 2015

ALICE overview

Sami S. Räsänen[1,2,a] on behalf of the ALICE Collaboration

[1] *University of Jyvaskyla, Department of physics, P.O. Box 35, FI-40014, University of Jyvaskyla, Finland*
[2] *Helsinki Institute of Physics, P.O.Box 64, FI-00014 University of Helsinki, Finland*

Abstract. Recent results from the ALICE experiment are presented with a particular emphasis on particle identification, the nuclear modification factor (R_{AA}) and azimuthal anisotropy (v_2). Comparison of lead-lead and proton-lead results reveals evidence of collectivity in small systems.

1 Introduction

The main goal of ultrarelativistic heavy ion collisions is to study the thermal properties of quantum chromodynamics (QCD). It is widely accepted that the thermal state of QCD, the quark-gluon plasma (QGP), is reached in high energy nucleus-nucleus collisions at BNL-RHIC and CERN-LHC. Properties of the QGP medium, such as shear viscosity to entropy ratio, existence and location of the QCD critical point, electromagnetic radiation and energy loss provide a focus for active measurements and theory development; see e.g. the Hot QCD White Paper [1]. Recently there has been a growing interest in possible collective phenomena in smaller collision systems [2], like deuteron-gold [3, 4], helium-gold [5], proton-lead [6–8] or even in proton-proton [9–11].

The ALICE experiment [12] is the dedicated heavy ion experiment at the LHC. Compared to ATLAS [13] and CMS [14], ALICE is a low luminosity experiment with more limited acceptance but equipped with excellent particle identification (PID) and tracking capabilities down to very low transverse momentum, p_T. One example of the performance of the ALICE tracking is a recent measurement of the mass difference between light nuclei and anti-nuclei [15]. The left panel of Fig. 1 shows observed counts with respect to the squared mass-over-charge ratio in the Pb – Pb collisions at $\sqrt{s} = 2.76$ GeV for selected rigidity ($p/|z|$) intervals and 2σ cut of for TPC dE/dx expectations for (anti-)deuterons and (anti-)helium-3. The mass of the light nuclei can be measured very precisely by fits to these distributions. The right panel of Fig. 1 shows the resulting mass differences and binding energy differences compared to earlier measurements. If the Charge conjugation-Parity-Time reversal (CPT) –symmetry is exact, then these differences should be zero. ALICE has now presented the most precise measurement confirming CPT invariance to hold in relative mass differences at 0.1% level, which gives stringent constraints to any effective theory where CPT is broken.

On top of the measurements of heavy ion collisions, ALICE has also a proton-proton program based upon the strengths of the experiment. One general motivation is to test and tune event generators and basic perturbative QCD predictions, which is important in its own right,but also a precise understanding of QCD background is often crucial for the electroweak measurements at the LHC.

[a] e-mail: sami.s.rasanen@jyu.fi

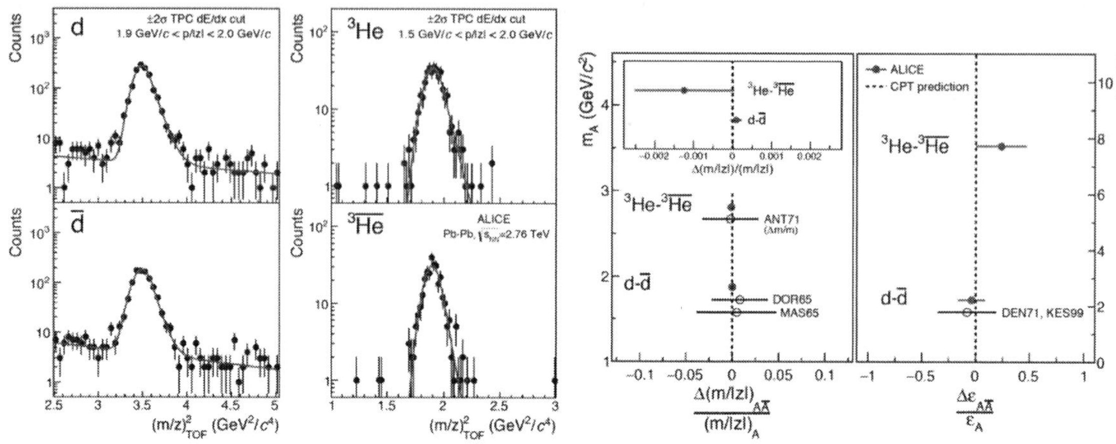

Figure 1. Left: The ALICE measurements [15] for squared mass-over-charge ratio distributions for deuterons (left) and ^{3}He (right) in selected rigidity intervals. Particle and anti-particle spectra are in the top and bottom plots, respectively. Right: Results for d-d and ^{3}He-^{3}He mass-over-charge ratio and binding energy differences compared with CPT invariance expectation (dotted lines) and other measurements.

Some recent measurements that demonstrate the PID capabilities of ALICE include identified hadrons and particle ratios at $\sqrt{s} = 7$ TeV [16] and π^0's at $\sqrt{s} = 2.76$ TeV [17]. ALICE has also measured (charged) jet spectrum at $\sqrt{s} = 7$ TeV [18] and found a good agreement with ATLAS. Here, the excellent tracking capability of ALICE was used to measure the mean number the mean number of constituents down to low-p_T jets. On top of these examples, the obvious and important part of the ALICE pp program is to measure reference spectra to the heavy ion collisions.

The rest of this paper will concentrate on summarizing recent ALICE results on the nuclear modification factor R_{AA} and elliptic flow v_2 with an emphasis on particle identification and small systems.

2 Nuclear modification factor

It has long been predicted that partons traversing hot QCD matter would lose energy and attenuate, but such a phenomena was not observed in early experiments at the CERN SPS [19]. A suppression of high transverse momentum hadron was first observed by experiments at RHIC [16, 20]. To quantify the nuclear modification, PHENIX experiment introduced the nuclear modification factor [20]

$$R_{AA}(p_T) = \frac{(\text{Yield in Pb + Pb})}{(\text{Number of collisions}) \times (\text{Yield in p + p})} = \frac{dN^{AA}/dp_T d\eta}{\langle N_{\text{bin}} \rangle \, dN^{\text{pp}}/dp_T d\eta}. \quad (1)$$

If the nucleus-nucleus collision could be described as a superposition of independent proton-proton collisions, then R_{AA} would equal unity for all p_T. At the low-p_T $R_{AA} < 1$ since the soft particle production in heavy ion collisions scales with the number of participants rather than the number of collisions [21]. At high-p_T at the SPS $R_{AA} > 1$ which is connected with Cronin enhancement [22]. At RHIC a suppression with respect to binary scaling was observed at the high-p_T.

In proton-lead collisions, the corresponding nuclear modification factor R_{pA} measures the magnitude of cold nuclear matter effects such as a modification of the nuclear parton distributions [23]. As will be seen, typically the cold nuclear effects are small at very high energy. To quantify nuclear modifications precisely, the most favourable case is when all collision systems, pp, p – Pb and Pb – Pb, are measured at the same collision energy and by the same detector. Currently at the LHC, there is a measured pp reference for the $\sqrt{s} = 2.76$ TeV lead-lead measurements, but not yet for the $\sqrt{s} = 5.02$ TeV proton-lead run. This optimal scenario will be achieved at the end of 2015 when the LHC will provide both $\sqrt{s} = 5.02$ TeV reference pp run and also Pb – Pb collisions at this same energy.

Following subsections discuss the R_{AA} of various single particle spectra. ALICE has also measured R_{AA} of jets [24]. See also the dedicated talk on jets in heavy ions collisions presented at this conference [25] .

2.1 Light flavours

The first ALICE R_{AA} measurement for charged hadrons was published in [26] and immediately made an important extension to RHIC results (not shown here). RHIC results reached up to $p_T \sim 10$ GeV/c and it was open weather the data flattens in that region or if R_{AA} starts to rise again. With the LHC data extending to 50 GeV/c (ALICE) or 100+ GeV/c (ATLAS [27], CMS [28]), it is clear that R_{AA} is again rising towards unity. This rules out a very simple phenomenological description of energy loss where every parton would loose a constant fraction of its energy, and also gives more constraints to detailed calculations.

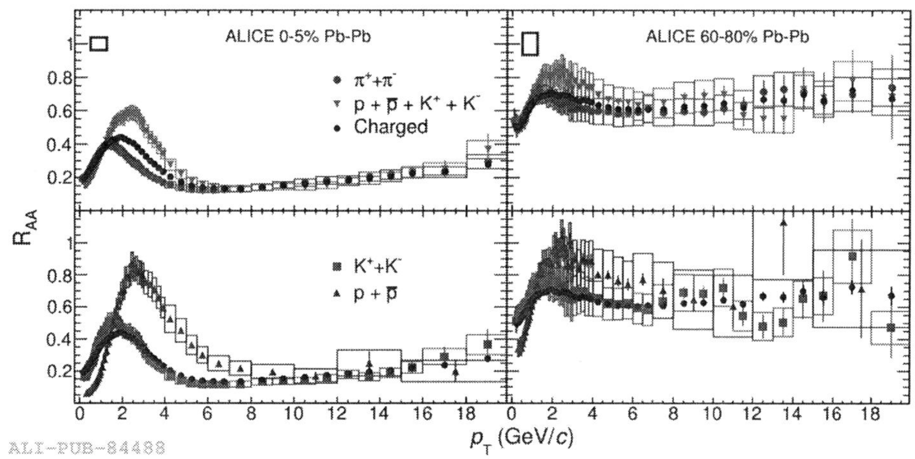

Figure 2. The nuclear modification factor R_{AA} as a function of p_T for different particle species for central (left) and peripheral (right) collisions [29].

Fig. 2 shows recent ALICE measurements of R_{AA} for identified light hadron flavours; charged pions and kaons and (anti-)protons together with all charged hadrons up to $p_T = 20$ GeV/c in central (left) and peripheral (right) collisions. In central collisions, R_{AA} is observed to be the same for all particle species for $p_T \gtrsim 10$ GeV/c suggesting that there is no direct interplay between the energy loss in the medium and the particle species composition in the hard core of the quenched jet and may disfavour models that predict significant particle species dependence also at high-p_T. The region $p_T \lesssim 2 \ldots 2.5$ GeV/c is generally associated with the thermal production which roughly scales as a number of participants. In the mid-p_T region, $3 < p_T < 10$ GeV/c, protons are clearly less suppressed than pions or kaons. Further studies are needed in order to determine whether models containing only hydrodynamics and jet quenching can describe also the intermediate p_T region or whether a different hadronization model, such as recombination [30], act in this region.

2.2 Heavy flavours

Dedicated heavy flavour (HF) theory [31] and experimental [32] talks were delivered in this conference. Here, we summarize briefly ALICE results for HF R_{AA}. One of the key motivations to study HF in nucleus-nucleus collisions have been the expectation of reduced energy loss in the medium due to so called dead cone effect [33], that arises from suppression of the forward QCD scattering amplitude due to heavy quark mass in the propagator.

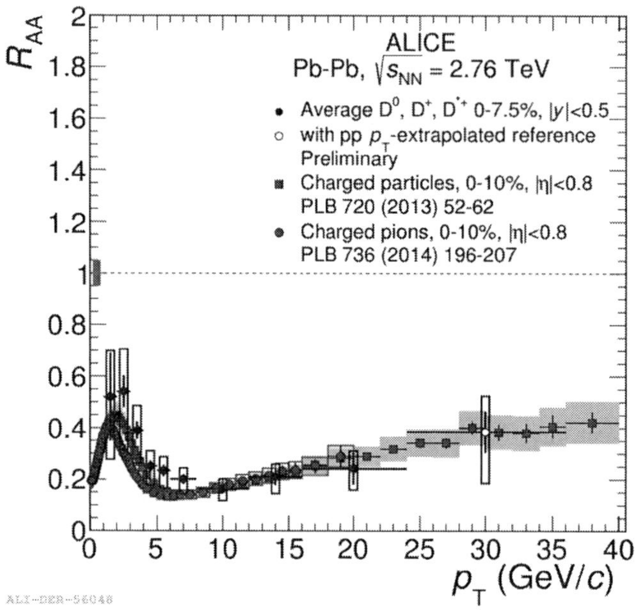

Figure 3. Average D-meson R_{AA} compared to charged hadrons and - pions.

Fig. 3 shows the average D-meson R_{AA} compared to all charged particles and charged pions. Particularly at high-p_T the results agree within one another, and also in lower p_T within the experimental uncertainties, so the dead cone effect is not obvious. Although some models can reproduce this, they tend to have difficulties in simultaneously reproducing the elliptic flow, as will be discussed later in Sec. 3.1.

Suppression of the J/ψ states in presence of QGP was predicted in 1986 [34] and it has been studied extensively at the SPS [35] and RHIC [36]. On top of the effects coming from the hot deconfined medium, there may be cold nuclear matter effects like modification of parton distribution functions or final state absorption in the nuclear medium. Both SPS and RHIC experiments found suppression beyond any cold matter effects. At the LHC, CMS has observed a sequential suppression of the higher mass bottomonium states [37] showing clear medium effects in the suppression.

An earlier measurement in ALICE [38] revealed an interesting surprise: the p_T-integrated R_{AA} of the J/ψ was less suppressed in central lead-lead collisions at the LHC as compared to central gold-gold collisions at RHIC. This is counter intuitive in a sense that naively one would expect the larger and more hot medium to give larger suppression.

Fig. 4 shows ALICE results for p_T-dependence of J/ψ nuclear modification factor R_{AA} in central lead-lead collisions measured at forward rapidity (left panel) using di-muons pairs [39] and in mid-rapidity (right panel) using di-electron pairs [40]. While in [38, 39] one compared forward LHC to mid-rapidity RHIC measurements, later measurement [40] confirmed that the observed smaller suppression at the LHC does not come from a trivial rapidity dependence. Note that the last bin in the

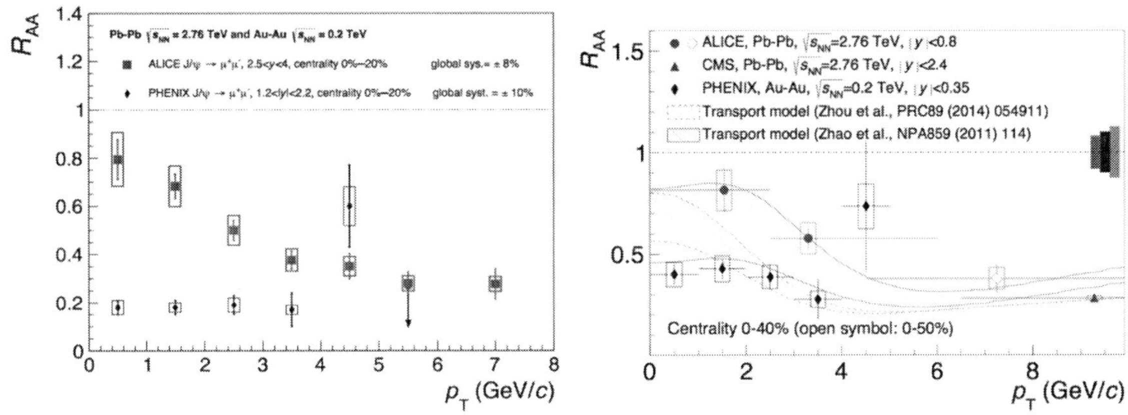

Figure 4. Nuclear modification factor R_{AA} of J/ψ in 0–20 % most central lead-lead collisions at forward rapidity [39] (left) and in 0–40 % most central collisions at mid-rapidity [40] (right).

right panel has open marker indicating that the statistics was too low to determine the mean p_T in that bin, hence the marker is placed at the bin center.

A possible explanation for smaller suppression comes from higher density of charmed quarks at the LHC that would increase the recombination probability to J/ψ. The right panel in Fig. 4 shows results from two transport models [41, 42] that can describe the rise of R_{AA} towards the small p_T. In both models the rise comes from recombination.

2.3 Proton-lead collisions

In the early RHIC era, deuteron-gold collisions were considered mainly as a control for heavy ions. It was expected that there is no medium created and the measurements would probe only cold nuclear matter effects. Originally proton-lead collisions were expected to play the same role also at the LHC. Experiments at both accelerator centers have now found that there may be collective behaviour also in small systems, as will be discussed in Sec. 3.2.

Fig. 5 shows the nuclear modification factor R_{AA} for charged hadrons [43] (left) and average D-mesons [44] (right) in minimum bias proton-lead collisions compared to central and peripheral lead-lead collisions. At high-p_T, R_{AA} is consistent with unity for both charged hadrons and average D-mesons in p − Pb collisions. On the other hand, a clear suppression is seen in central heavy ion collisions. Peripheral heavy ion results lie in between indicating smooth and gradual disappearance of the nuclear effects. The observed cold nuclear matter effects (at high-p_T) are rather modest. This was expected if the only source of cold nuclear modifications is due to the nuclear parton distribution functions, see e.g. [45]. However, saturation physics phenomenology [46] gives results that are also consistent with the data.

3 Elliptic flow

If the fireball created in the relativistic heavy ion collisions thermalizes, then there is a strong hydrodynamical pressure gradient that drives the system to collective motion towards the vacuum outside the droplet. In a very simple blast wave model, the fireball is assumed to break up instantaneously, at

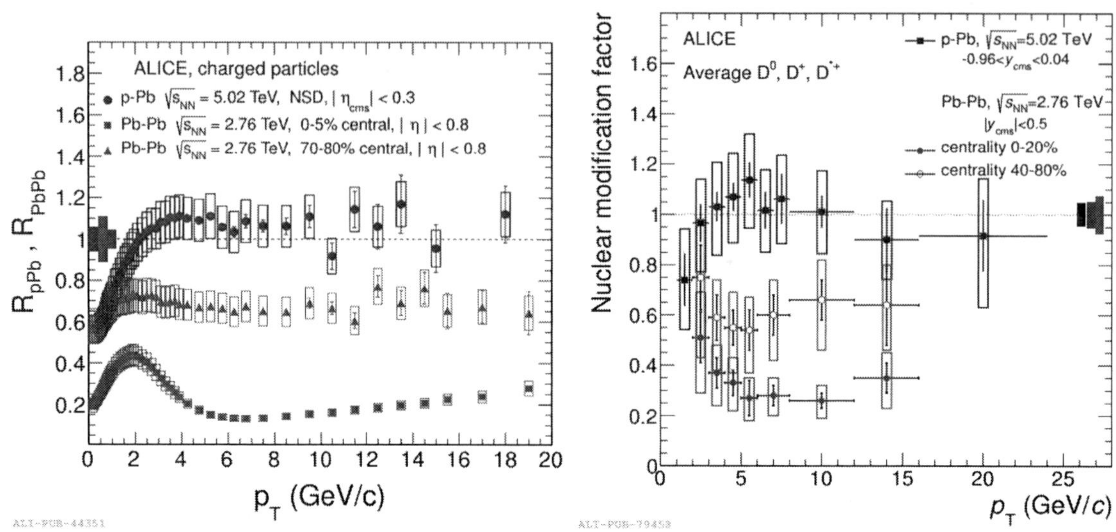

Figure 5. Nuclear modification factor R_{AA} for charged hadrons [43] (left) and average D-mesons [44] (right) in minimum bias proton-lead collisions compared to central and peripheral lead-lead collisions.

some constant proper time, parametrized collective transverse flow profile, in which case the invariant yield of thermal particles becomes [47]

$$E\frac{dN}{d^3p} \propto m_T \int_0^{R_A} dr\, r I_0\left(\frac{\beta_T \gamma_T p_T}{T_{kin}}\right) K_1\left(\frac{\gamma_T m_T}{T_{kin}}\right), \quad \text{where } \beta_T = \beta_T(r) = \beta_{max}\left(\frac{r}{R_A}\right)^{\alpha}. \tag{2}$$

The gamma-factor is related to the collective transverse flow $1/\gamma_T = \sqrt{1-\beta_T^2}$, transverse mass $m_T = \sqrt{p_T^2 + m^2}$ and R_A is the nuclear radius. The model has four parameters to be determined by a fit to the data: the kinetic freezeout temperature of the system, T_{kin}, the maximum flow velocity, β_{max}, the shape parameter for the flow profile, α, and an overall normalization related to the size (also the lifetime) of the fireball. In the case of constant radial flow $\beta_T = \beta_0 =$ (const.), i.e. shape parameter $\alpha = 0$, one finds that asymptotically at $p_T \gg \max(m, T_{kin})$

$$\frac{1}{m_T}\frac{dN}{dm_T dy} \sim \exp\left(\frac{-m_T}{T_{eff}}\right), \quad \text{where } T_{eff} \equiv T_{kin}\sqrt{\frac{1+\beta_0}{1-\beta_0}}. \tag{3}$$

This illustrates the blue shifts in spectra coming from the collective radial expansion. The increase of the apparent temperature is the dominant flow effect in heavy ion collisions.

Fig. 6 shows the blast wave parameters T_{kin} and mean of the transverse velocity $\langle\beta_T\rangle$ at the LHC (solid black) and in RHIC (dashed blue) [48]. Centrality increases when moving from left to right. Average flow velocity increases and freezeout temperature decreases with growing centrality. This suggests that more central collisions have longer lifetime and hence the flow has time to build up.

The blast wave model has several short comings. Perhaps the most important, it does not contain any dynamics and it neglects the fact that significant fraction of pions and protons originate from the decays of heavier resonances [49]. The latter could be easily taken into account by considering the resonance decay chains in the blast wave spectra but it would still only give a qualitative description.

 DOI: 10.1051/epjconf/201612602026

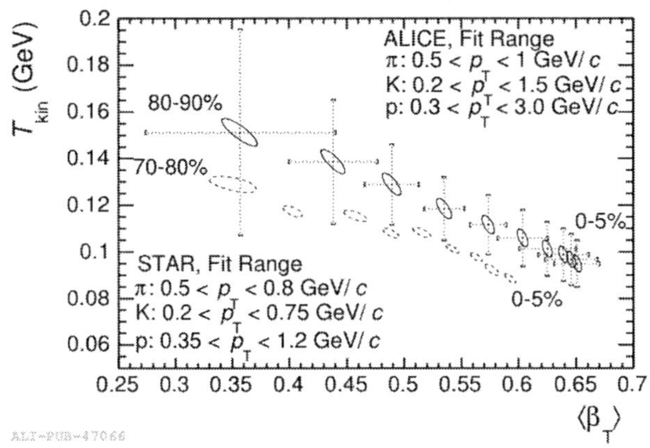

Figure 6. Blast wave fit parameters break up temperature $T_{\rm kin}$ and mean transverse velocity $\langle\beta_T\rangle$ at the LHC and in RHIC [48]. With increasing centrality, the mean velocity grows and break up temperature reduces. This can be interpreted such that in more central collisions the average lifetime of the system is longer leading to stronger flow and later break up.

Hydrodynamical modeling gives a real dynamical solution of the conservation equations and is the only known thermodynamically consistent way to treat the Equation of State of strongly interacting matter in the evolution equations. The solution of the hydrodynamical evolution equations give dynamically the local temperature, densities and collective velocity in every space-time point inside the fireball. Hence all quantitative thermodynamical characteristics should follow from detailed hydrodynamical simulations. For example, see recent results found using sophisticated theoretical frameworks like VISHNU [50], MUSIC [51] or pQCD+saturation+viscous hydrodynamics [52] in heavy ion collisions or using hydrodynamics in small systems [53].

It was realized [54] that there will be deviations of radial symmetry in the transverse flow due to anisotropic pressure gradients during the evolution. The Fourier decomposition of the angular distribution of particles with respect to the reaction plane has become the standard method with which to analyze flow anisotropies [55]. There are many experimental methods to measure those Fourier coefficients, like flow cumulants [56], but we shall not elaborate the discussion in this note.

Using Fourier decomposition, azimuthal dependence of the final hadron spectra becomes

$$E\frac{dN}{d^3p} = \frac{1}{p_{\rm T}}\frac{dN}{dp_{\rm T}dyd\phi} = \frac{1}{2\pi p_{\rm T}}\frac{dN}{dp_{\rm T}dy}\left(1+\sum_{n=1}^{\infty}2v_n(p_{\rm T})\cos(n\phi-\psi_n)\right), \tag{4}$$

where v_n are the flow harmonics and ψ_n event plane angles. The second harmonic coefficient, v_2, is called elliptic flow and it has a clear centrality dependence. Elliptic flow reflects the dominant almond-shaped geometry in the non-central collisions. The third harmonic, v_3, is called triangular flow. Triangular flow comes dominant in the ultra-central collisions and it is related to the fluctuations in the initial state of the collisions. For example, simultaneous description of the $p_{\rm T}$ and centrality dependence of the v_2 and v_3 coefficients puts constraints on the initial geometry fluctuations and the shear viscosity to entropy ratio η/s. A higher viscosity causes more rapid damping of the fluctuations (v_3) and slows the buildup of v_2. Based on these studies, it has been concluded that QGP shows features of nearly perfect liquid. [50, 51]

The hydrodynamical models [50–52] go beyond the event averages of the flow coefficients (the v_n in Eq. (4)) and study e.g. experimentally measured event-by-event fluctuations of flow [57] and correlations between event plane angles [58]. Such a global analysis provide significantly tighter constraints on QCD matter properties.

One may observe that there is no rapidity dependence in the flow coefficients in Eq. (4). Consequently, flow leads into interesting long range correlations seen in the two-particle correlations. Given the single particle distribution (4), one obtains a similar Fourier decomposition for the two particle distribution [59]

$$\frac{dN^{\rm pair}}{d(\phi_t - \phi_a)} =: \frac{dN^{\rm pair}}{d\Delta\phi} \propto 1 + \sum_{n=1}^{\infty} 2V_{n\Delta}(p_{\rm Tt}, p_{\rm Ta}) \cos(\Delta\phi), \tag{5}$$

where $\Delta\phi := \phi_t - \phi_a$ is a difference in the azimuthal angle between trigger and associated hadron that have transverse momenta $p_{\rm Tt}$ and $p_{\rm Ta}$, respectively. In the case of flow correlations only, one would have a connection to single particle flow coefficients

$$V_{n\Delta}(p_{\rm Tt}, p_{\rm Ta}) = v_n(p_{\rm Tt})v_n(p_{\rm Ta}) \tag{6}$$

and you would see the correlations extending over large rapidity gaps $\Delta\eta := \eta_t - \eta_a$. This gives an opportunity to determine flow coefficients for hadrons and also for identified flow [60]: by choosing a symmetric bins for trigger and associated particles, $p_{\rm Tmin} < p_{\rm Ta}, p_{\rm Tt} < p_{\rm Tmax}$, one finds

$$v_n^h\{2PC\} = \sqrt{V_{n\Delta}^{h-h}} \qquad v_n^i\{2PC\} = \frac{V_{n\Delta}^{h-i}}{\sqrt{V_{n\Delta}^{h-h}}}, \tag{7}$$

where $i = \pi$, K, p in case of identified flow. Above $h-h$ refers to unidentified charged hadron-hadron correlation in a symmetric $p_{\rm T}$-bin and $h-i$ to a case where one of the hadrons is identified.

So far we have related the harmonics with collectivity in the system (hydrodynamical flow). However, one should be a bit careful since particularly hydrodynamics is expected to describe the bulk of the particle production, i.e. it would be valid description at $p_{\rm T} < 1 \ldots 3$ GeV/c depending on the particle species. However, as discussed in Sec. 2, hard partons are not expected to thermalize but instead they traverse the medium losing their energy. But since the medium is expected to have a smaller geometrical extension in in-plane as compared to out-of-plane, the partons are (on the average) less suppressed at the in-plane direction, causing a positive second Fourier coefficient v_2. Hence, even though one generally speaks of "flow coefficients", the full range in $p_{\rm T}$ is not related with hydrodynamical flow.

The rest of this note concentrates on identified v_2 in Pb – Pb and p – Pb collisions.

3.1 Heavy ion collisions

In the Blast Wave model (2), radial flow breaks the $m_{\rm T}$–scaling present in the non-flowing case (i.e., if $\beta_{\rm max} = 0$ then (yield) $\sim K_1(-m_{\rm T}/T_{\rm kin})$). Heavier particles gain more momentum from the common collective flow field and this leads into mass ordering of elliptic flow at low-$p_{\rm T}$. The left panel in Fig. 7 shows recent ALICE measurement[1] [61] for elliptic flow of $\pi^\pm$, $K^\pm$, $p + \bar{p}, \Lambda + \bar{\Lambda}, \Xi + \bar{\Xi}$ and $\Omega + \bar{\Omega}$. One sees a clear mass ordering of the different hadrons and also that a sophisticated hydrodynamical simulation [63] can reproduce that qualitatively rather well, although a more detailed analysis presented in the ALICE paper shows some tensions. The study also provided evidence for constituent quark scaling [64–66] to be violated at the intermediate $p_{\rm T}$ region at the LHC as deviations up to ~ 20 % level are observed [61]

The right panel in Fig. 7 shows the elliptic flow of average D-mesons compared to charged particle v_2. The measurements [62] show that D-mesons have a positive v_2 in range $2 < p_{\rm T} < 6$ GeV/c interval with 5.7σ significance. This high-$p_{\rm T}$ range is outside the thermal region so these results, together with the D-meson R_{AA}, provide very tight constraints on the heavy quark interactions with the medium.

[1]The published paper includes also the v_2 of K^0_s and ϕ mesons, not shown in Fig. 7.

DOI: 10.1051/epjconf/201612602026

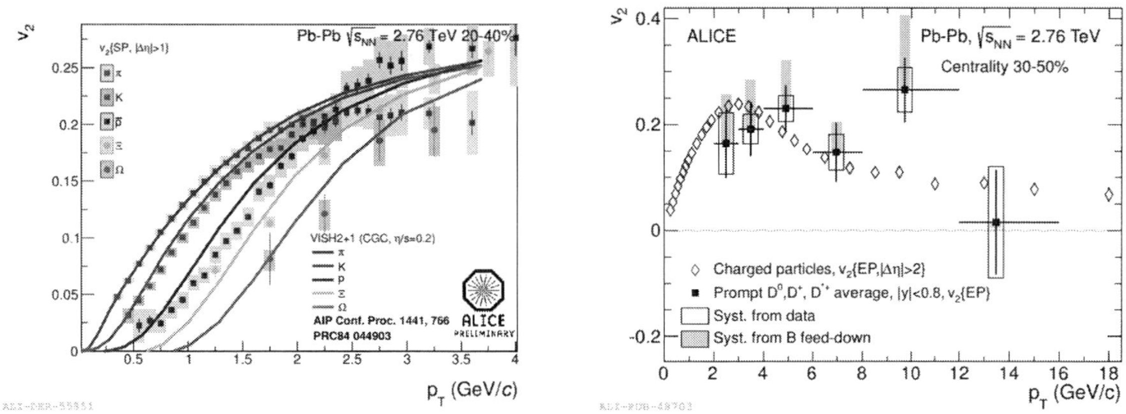

Figure 7. Left: elliptic flow v_2 for identified particles in $\sqrt{s_{\rm NN}}$ = 2.76 TeV measured by ALICE [61]. Right: D-meson and charged hadron v_2 compared [62].

3.2 Double ridge in proton-lead collisions

An interesting observation in p-Pb collisions at the LHC was the appearance of a double ridge structure in two-particle correlations [6]. Two-particle correlations are define as

$$\frac{1}{N_{\rm trig}}\frac{d^2N}{d\Delta\phi d\Delta\eta}, \tag{8}$$

where $\Delta\phi = \phi_t - \phi_a$ and $\Delta\eta = \eta_t - \eta_a$, such that trigger (t) and associated (a) transverse momenta are in the given bin, and also classify the events globally in terms of the charged particle multiplicity. Typically two-particle correlation functions have a near side peak at $(\Delta\phi, \Delta\eta) = (0, 0)$ coming from hadrons associated with the fragmentation of the same jet as the trigger particle and an away-side peak at $\Delta\phi = \pi$, which is smeared in $\Delta\eta$ due to the different momentum fractions carried by partons in the initial hard scattering. If we now assume that the hard scattering is not significantly modified in proton-lead collision, we can assume that jet correlations from the hard scattering are similar in low- and high-multiplicity p – Pb events. For the further support, we have checked that the correlation function in the low-multiplicity p – Pb collisions is observed to be very similar with the one measured in proton-proton collisions. Then we can study any residual correlations in high-multiplicity events, on top of the hard scattering, by subtracting the correlation functions measured at low-multiplicity from the correlation function measured in the high-multiplicity events. The left panel in Fig. 8 shows the resulting residual correlation function, when you perform such a subtraction [6]. One can observe a clear double ridge structure in $\Delta\phi$ that is elongated in $\Delta\eta$. Returning to the discussion of collective flow of single particles is reflected into two-particle correlations, see Eq. (5), we immediately realize that the observation very much resembles this expectation.

The right panel in Fig. 8 shows the projection of the two-dimensional subtracted correlation function (left) into $\Delta\phi$ together with Fourier decomposition including $\cos(2\Delta\phi)$ and $\cos(3\Delta\phi)$ terms resembling elliptic and triangular flow in the single particle distributions. Note, however, that you cannot use directly Eq. (7) to obtain single particle $v_{2,3}$ here because the trigger and associated bins are asymmetric. Nevertheless, we observe that the double ridge structure is dominantly given by the 2$^{\rm nd}$ harmonic and after inclusion of 3$^{\rm rd}$ harmonic the description of the data already is very good. Results from HIJING simulation, that does not have any flow effects, is also shown and there is no similar double ridge structure seen indicating that this is not a trivial long range pseudorapidity correlation.

 DOI: 10.1051/epjconf/201612602026

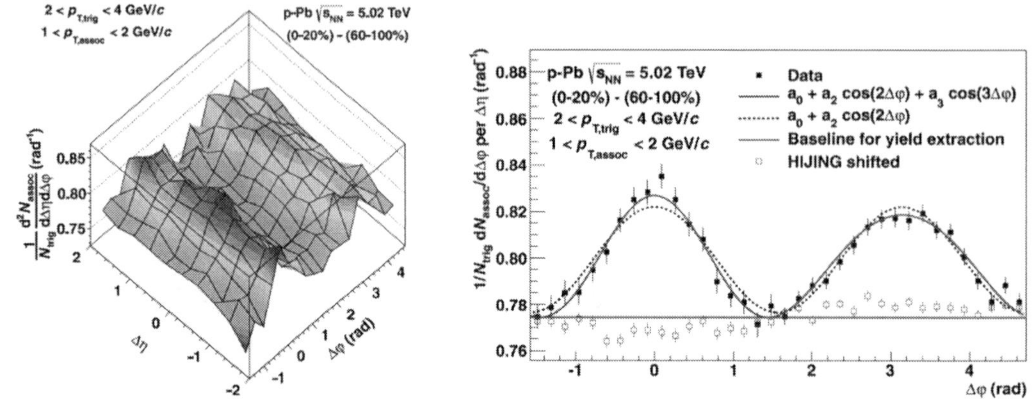

Figure 8. Left: Associated yield per trigger particle in $\Delta\phi$ and $\Delta\eta$ for pairs of charged particles with $2 < p_{\mathrm{Ttrig}} < 4$ GeV/c and $1 < p_{\mathrm{Tassoc}} < 2$ GeV/c in p – Pb collisions at $\sqrt{s_{\mathrm{NN}}} = 5.02$ TeV for the 0–20% multiplicity class,after subtraction of the associated yield obtained in the 60–100% event class. Right: as left but projected onto $\Delta\phi$ averaged over $0.8 < |\Delta\eta| < 1.8$ on the near side and $|\Delta\eta| < 1.8$ on the away side. [6]

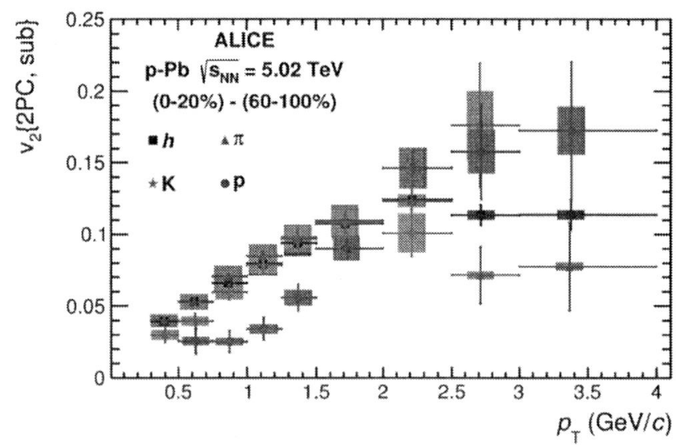

Figure 9. Fourier coefficients $v_2\{2\mathrm{PC}, \mathrm{sub}\}$ for pions, kaons and protons measured in the high-multiplicity p – Pb collisions [60].

More detailed study of the properties of the observed double ridge was presented in [60]. Fig. 9 shows the $v_2\{2\mathrm{PC}, \mathrm{sub}\}$, extracted using Eq. (7), where "sub" refers that the jet correlations are removed by subtracting the low-multiplicity correlation function from the high-multiplicity one before the projection to $\Delta\phi$. The observed v_2 values have a similar mass ordering than observed in Pb – Pb collisions, compare to left panel in Fig. 7. While not yet conclusive proof, it shows that the Fourier analysis of the double ridge leads to a mass ordering in $v_2\{2\mathrm{PC}, \mathrm{sub}\}$ that is characteristic of collective flow. The ATLAS collaboration has also measured higher harmonics using two-particle correlations [67] and the CMS collaboration has also shown that multiparticle correlations [68] give a non-zero v_2 at high-multiplicity proton-lead collisions. The latter strongly favour collective correlations.

Hydrodynamical calculations can qualitatively reproduce these observations [53]. However, one might be concerned on validity of the (viscous) hydrodynamical description due to large pressure gradients in these small system [69]. An alternative Color Glass Condensate (CGC) explanation based on two gluon scattering from the same color field gives rise to similar correlations when the target gluon densities are high (saturation) [70]. It remains to be seen if the CGC based model can also explain the observed mass ordering.

4 Summary

ALICE is the dedicated heavy ion experiment at the CERN LHC that is equipped with excellent tracking and particle identification capabilities. Recent measurements have been presented of recent ALICE results for R_{AA} and elliptic flow for identified particles in Pb – Pb and p – Pb collisions. There are interesting hints for collective behaviour in high-multiplicity p – Pb collisions. However, the effect of a Color Glass Condensate or other cold matter effects may yet provide an alternative explanation for the observed behaviour.

References

[1] Y. Akiba et al. (2015), 1502.02730
[2] I. Tserruya, These proceedings (2015)
[3] A. Adare et al. (PHENIX), Phys. Rev. Lett. **114**, 192301 (2015), 1404.7461
[4] L. Adamczyk et al. (STAR), Phys. Lett. **B747**, 265 (2015), 1502.07652
[5] A. Adare et al. (PHENIX), Phys. Rev. Lett. **115**, 142301 (2015), 1507.06273
[6] B. Abelev et al. (ALICE), Phys. Lett. **B719**, 29 (2013), 1212.2001
[7] G. Aad et al. (ATLAS), Phys. Rev. Lett. **110**, 182302 (2013), 1212.5198
[8] S. Chatrchyan et al. (CMS), Phys. Lett. **B724**, 213 (2013), 1305.0609
[9] V. Khachatryan et al. (CMS), JHEP **09**, 091 (2010), 1009.4122
[10] G. Aad et al. (ATLAS) (2015), 1509.04776
[11] V. Khachatryan et al. (CMS) (2015), 1510.03068
[12] K. Aamodt et al. (ALICE), JINST **3**, S08002 (2008)
[13] G. Aad et al. (ATLAS), JINST **3**, S08003 (2008)
[14] S. Chatrchyan et al. (CMS), JINST **3**, S08004 (2008)
[15] J. Adam et al. (ALICE), Nature Phys. (2015), 1508.03986
[16] J. Adam et al. (ALICE), Eur. Phys. J. **C75**, 226 (2015), 1504.00024
[17] B.B. Abelev et al. (ALICE), Eur. Phys. J. **C74**, 3108 (2014), 1405.3794
[18] B.B. Abelev et al. (ALICE), Phys. Rev. **D91**, 112012 (2015), 1411.4969
[19] X.N. Wang, Phys. Rev. Lett. **81**, 2655 (1998), hep-ph/9804384
[20] K. Adcox et al. (PHENIX), Phys. Rev. Lett. **88**, 022301 (2002), nucl-ex/0109003
[21] K. Aamodt et al. (ALICE), Phys. Rev. Lett. **106**, 032301 (2011), 1012.1657
[22] D. Antreasyan, J.W. Cronin, H.J. Frisch, M.J. Shochet, L. Kluberg, P.A. Piroue, R.L. Sumner, Phys. Rev. **D19**, 764 (1979)
[23] K.J. Eskola, H. Paukkunen, C.A. Salgado, JHEP **04**, 065 (2009), 0902.4154
[24] J. Adam et al. (ALICE), Phys. Lett. **B746**, 1 (2015), 1502.01689
[25] A. Shabetai, These proceedings (2015)
[26] K. Aamodt et al. (ALICE), Phys. Lett. **B696**, 30 (2013), 1012.1004
[27] G. Aad et al. (ATLAS), JHEP **09**, 050 (2015), 1504.04337
[28] S. Chatrchyan et al. (CMS), Eur. Phys. J. **C72**, 1945 (2012), 1202.2554
[29] J. Adam et al. (ALICE) (2015), 1506.07287
[30] R.J. Fries, B. Muller, C. Nonaka, S.A. Bass, Phys. Rev. **C68**, 044902 (2003), nucl-th/0306027
[31] M. Nardi, These proceedings (2015)
[32] M.J. Kweon, These proceedings (2015)
[33] Y.L. Dokshitzer, D.E. Kharzeev, Phys. Lett. **B519**, 199 (2001), hep-ph/0106202

[34] T. Matsui, H. Satz, Phys. Lett. **B178**, 416 (1986)
[35] R. Arnaldi (NA60), Nucl. Phys. **A830**, 345C (2009), 0907.5004
[36] A. Adare et al. (PHENIX), Phys. Rev. **C77**, 024912 (2008), [Erratum: Phys. Rev.C79,059901(2009)], 0903.4845
[37] S. Chatrchyan et al. (CMS), Phys. Rev. Lett. **109**, 222301 (2012), 1208.2826
[38] B. Abelev et al. (ALICE), Phys. Rev. Lett. **109**, 072301 (2012), 1202.1383
[39] B.B. Abelev et al. (ALICE), Phys. Lett. **B734**, 314 (2014), 1311.0214
[40] J. Adam et al. (ALICE), JHEP **07**, 051 (2015), 1504.07151
[41] K. Zhou, N. Xu, Z. Xu, P. Zhuang, Phys. Rev. **C89**, 054911 (2014), 1401.5845
[42] X. Zhao, R. Rapp, Nucl. Phys. **A859**, 114 (2011), 1102.2194
[43] B. Abelev et al. (ALICE), Phys. Rev. Lett. **110**, 082302 (2013), 1210.4520
[44] B.B. Abelev et al. (ALICE), Phys. Rev. Lett. **113**, 232301 (2014), 1405.3452
[45] I. Helenius, K.J. Eskola, H. Honkanen, C.A. Salgado, JHEP **07**, 073 (2012), 1205.5359
[46] P. Tribedy, R. Venugopalan, Phys. Lett. **B710**, 125 (2012), [Erratum: Phys. Lett.B718,1154(2013)], 1112.2445
[47] E. Schnedermann, J. Sollfrank, U.W. Heinz, Phys. Rev. **C48**, 2462 (1993), nucl-th/9307020
[48] B. Abelev et al. (ALICE), Phys. Rev. **C88**, 044910 (2013), 1303.0737
[49] K.J. Eskola, H. Niemi, P.V. Ruuskanen, S.S. Rasanen, Phys. Lett. **B566**, 187 (2003), hep-ph/0206230
[50] H. Song, S. Bass, U.W. Heinz, Phys. Rev. **C89**, 034919 (2014), 1311.0157
[51] C. Gale, S. Jeon, B. Schenke, P. Tribedy, R. Venugopalan, Phys. Rev. Lett. **110**, 012302 (2013), 1209.6330
[52] H. Niemi, K.J. Eskola, R. Paatelainen (2015), 1505.02677
[53] P. Bozek, W. Broniowski, G. Torrieri, Phys. Rev. Lett. **111**, 172303 (2013), 1307.5060
[54] J.Y. Ollitrault, Phys. Rev. **D46**, 229 (1992)
[55] S. Voloshin, Y. Zhang, Z. Phys. **C70**, 665 (1996), hep-ph/9407282
[56] A. Bilandzic, R. Snellings, S. Voloshin, Phys. Rev. **C83**, 044913 (2011), 1010.0233
[57] G. Aad et al. (ATLAS), JHEP **11**, 183 (2013), 1305.2942
[58] G. Aad et al. (ATLAS), Phys. Rev. **C90**, 024905 (2014), 1403.0489
[59] K. Aamodt et al. (ALICE), Phys. Lett. **B708**, 249 (2012), 1109.2501
[60] B.B. Abelev et al. (ALICE), Phys. Lett. **B726**, 164 (2013), 1307.3237
[61] B.B. Abelev et al. (ALICE), JHEP **06**, 190 (2015), 1405.4632
[62] B. Abelev et al. (ALICE), Phys. Rev. Lett. **111**, 102301 (2013), 1305.2707
[63] C. Shen, U. Heinz, P. Huovinen, H. Song, Phys. Rev. **C84**, 044903 (2011), 1105.3226
[64] D. Molnar, S.A. Voloshin, Phys. Rev. Lett. **91**, 092301 (2003), nucl-th/0302014
[65] A. Adare et al. (PHENIX), Phys. Rev. Lett. **98**, 162301 (2007), nucl-ex/0608033
[66] B.I. Abelev et al. (STAR), Phys. Rev. **C75**, 054906 (2007), nucl-ex/0701010
[67] G. Aad et al. (ATLAS), Phys. Rev. **C90**, 044906 (2014), 1409.1792
[68] V. Khachatryan et al. (CMS), Phys. Rev. Lett. **115**, 012301 (2015), 1502.05382
[69] H. Niemi, G.S. Denicol (2014), 1404.7327
[70] K. Dusling, R. Venugopalan, Phys. Rev. **D87**, 094034 (2013), 1302.7018

STAR heavy-ion highlights

Olga Rusnakova (for the STAR Collaboration)[1,a]

[1] *Czech Technical University in Prague, Faculty of Nuclear Sciences and Physical Engineering, Brehova 7, Praha 1*

Abstract. Parton energy loss, quarkonium sequential melting and particle production from electromagnetic interactions are tools to study Quark Gluon Plasma properties. The STAR detector, with large acceptance at mid-rapidity, excellent particle identification and wide transverse momentum coverage, is able to study these probes in details. Di-electron spectra form Beam Energy Scan, measurements of reconstructed jets in Au+Au collisions and quarkonium measurements in p+p and heavy ion collisions are reported in this paper.

1 Introduction

Partons with high transvere momenta (p_T), which are produced in the hard scatterings of relativistic heavy-ion collisions, interact with the surrounding hot and dense medium, Quark Gluon Plasma (QGP), via color exchange and may lose significant part of their energy. However, single partons can not be measured experimentally. As a proxy, high-$p_T$hadrons can be used to study parton energy loss. By comparing high-$p_T$hadrons in heavy-ion and p+p collisions, information about parton interactions with the QGP is obtained. The observation of single high-$p_T$hadron suppression in Au+Au collisions agrees with the parton energy loss expectations [1]. The disappearance of away-side jet peaks in dihadron correlations in Au+Au collisions, but not in d+Au collisions, further suggests such high-$p_T$hadron suppression is a hot medium effect [2]. However, the leading hadron does not carry all the information of its parent parton. Fully reconstructed jets - collimated sprays of hadrons fragmented and hadronized from a hard scattered parton allow to reconstruct the full kinematics of their parent parton. Therefore they represent a vital probe of the QGP medium. Latest STAR results on reconstructed semi-inclusive jets in Au+Au collisions will be reported in this paper. Another probe used to study the QGP properties besides jets is the quarkonium production. Quarkonia, bound states of quark and anti-quark pairs, produced in the initial hard scattering can be modified by color screening in the QGP and therefore are sensitive to the QGP temperature. In order to study possible quarkonium modification in the QGP, a reference of quarkonium production in p+p collisions is needed. Moreover, data on quarkonium production in p+p collisions constrain pQCD calculations [3, 4]. Υ (bb) suppression in Au+Au and U+U collisions will be discussed. J/ψ ($c\bar{c}$) production dependence on event multiplicity at various $p_T$at mid-rapidity in p+p collisions will be also reported. By varying the colliding energy in a wide range (7.7 - 200 GeV per nucleon-nucleon pair), RHIC has conducted a Beam Energy Scan (BES) program to explore the nuclear matter phase diagram. Results from BES - Phase I have shrunk

[a] e-mail: olga.hajkova@fjfi.cvut.cz

the region of interest for finding the critical point and the boundary of the first-order phase transition to beam energies below 20 GeV.

The STAR experiment comprises several subsystems that provide full azimuthal coverage at midrapidity ($|\eta| < 1$). A detailed description of the STAR detector can be found in [5].

2 Beam Energy Scan

In order to explore the QCD phase diagram, a Beam Energy Scan (BES) program was carried out by the STAR experiment at RHIC. A detailed description of the STAR detector can be found in [5]. The main aims of the BES program were to search for the turn-off of QGP signatures, signals of the first order phase transition and the critical point. One of the important QGP signatures is the nuclear modification factor R_{CP} - ratio of yields at most central over peripheral collisions, appropriately scaled by the corresponding number of binary collisions. For several particle species, R_{CP} values less than unity has been measured in heavy ion collisions at high energies [6].

Fig. 1 shows the R_{CP} of charged hadrons in Au+Au collisions at $\sqrt{s_{\rm NN}}$= 7.7 – 200 GeV. We observe that for $p_{\rm T} > 2$ GeV/c, the R_{CP} is less than unity at 39 GeV and then the value increases with the decreasing beam energy.

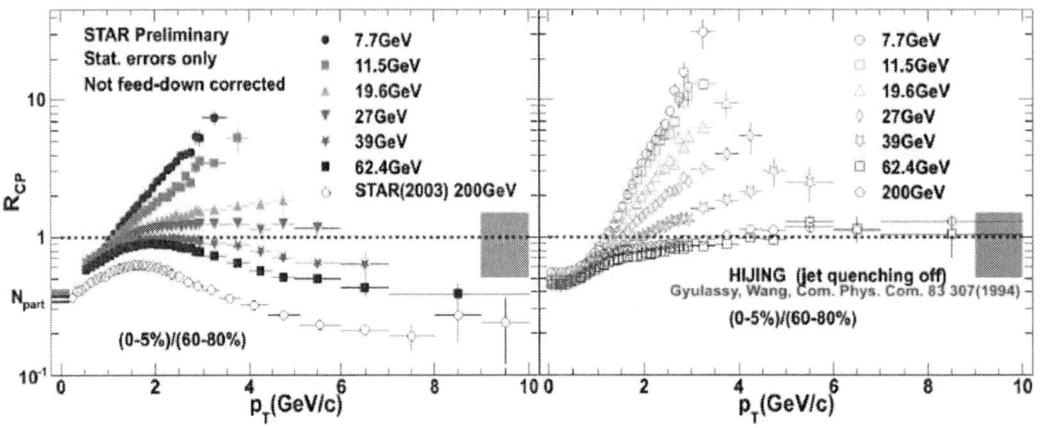

Figure 1. Left: R_{CP} for charged hadrons in Au+Au collisions at $\sqrt{s_{\rm NN}}$=7.7–39 GeV. Errors are statistical only. Grey band represents the normalization error from N_{bin}. Right: HIJING [7] theoretical prediction of R_{CP} for charged hadrons in Au+Au collisions at $\sqrt{s_{\rm NN}}$=7.7–39 GeV.

Measurements of QGP radiation would be viable to answer the question for signals of chiral symmetry restoration expected from QCD. Since the in-medium modifications of the spectral functions of short-lived resonances (such as the ρ-meson) survive the fireball evolution through the mesons' decays into $e^+ e^-$ pairs (di-electrons), low mass di-electron production (< 1 GeV/c^2) is a possible link to chiral symmetry restoration.

As can be seen on Fig. 2, excess yields in the di-electron invariant mass ($M_{e^+e^-}$) distributions at $0.2 < M_{e^+e^-} < 0.8$ GeV/c^2 over the hadronic decay electron cocktails are observed for collision energies from 19.6 to 200 GeV. A model calculation including ρ meson in-medium broadening and chiral symmetry restoration could possibly describe the di-electron spectrum down the enrgy of 20 GeV. Beam energies lower than 20 GeV however require more data for QGP phase transition search in BES - Phase II .

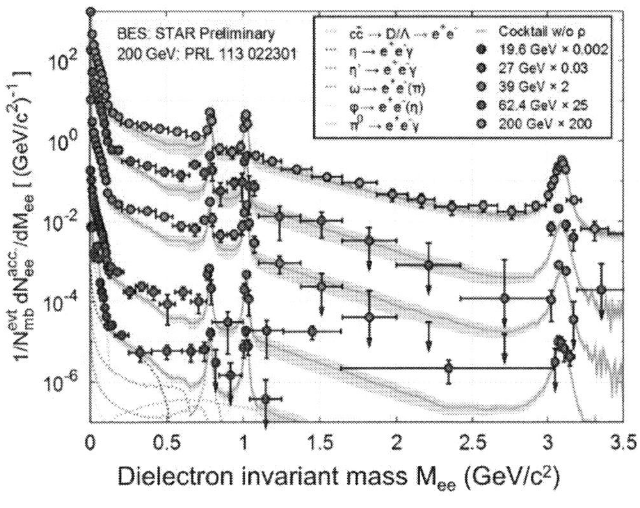

Figure 2. Measured di-electron invariant mass, M_{ee} (GeV/c), for BES Phase-I compared to cocktail simulations (grey) and model calculations [8]. The single vector meson contributions are depicted for 19 GeV. For 39 GeV, the two model contributions from hadronic gas and QGP phase are shown next to the total expected yield from cocktail and model (orange). Contributions from photon conversions are not removed from the data at 19 GeV.

The first harmonic coefficient of the Fourier expansion of the final-state momentum-space azimuthal distribution relative to the reaction plane is called directed flow v_1. Both hydrodynamic and nuclear transport models indicate that v_1 in the mid-rapidity region offers sensitivity to details of the expansion of the participant matter during the early collision stages.

Fig. 3 reports values of $v_1(y)$ slope (which is the linear term F in a cubic fit, where $v_1(y) = Fy + F_3y^3$) near mid-rapidity for anti-protons, protons and net protons and a comparison with UrQMD transport model. As can be seen, the UrQMD predictions do not describe the data at all.

The observed minimum for protons and net protons is a possible signature of a first-order phase transition between hadronic matter and a deconfined phase.

3 Semi-Inclusive Charged Jet Measurement

Jets represent a very valuable probe of the QGP medium because they preserve most of the momenta of the original hard scattered partons. However, jet reconstruction in the environment of heavy-ion collisions has to deal with an immense and highly fluctuating background, resulting in a large number of fake combinatorial jets.

In recoil-jet analysis, the background is not suppressed by a momentum cut on jet constituents. Instead, a hard event is selected by triggering on a high-$p_T$trigger hadron. An away-side jet is then reconstructed opposite in azimuth

$$|\phi_{Jet} - (\phi_{Trig} + \pi)| \leq \pi/4. \quad (1)$$

Such a jet is completely unbiased with respect to its fragmentation. The semi-inclusive recoil jet yield per trigger is an observable calculable by means of pQCD. In similar recoil jet analyses at the LHC, the background is usually reduced by subtracting corresponding recoil spectra for two different trigger hadron momentum ranges. However, in this analysis an innovative technique of mixed event is

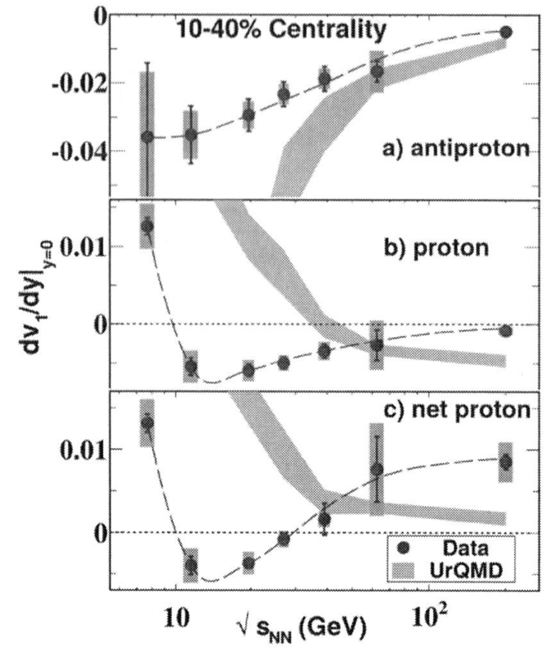

Figure 3. Directed flow slope (dv_1/dy) near mid-rapidity versus beam energy for intermediate-centrality Au+Au. Panels (a), (b) and (c) report measurement for antiprotons, protons, and net protons, respectively, along with UrQMD calculations subject to the same cuts and fit conditions. Systematic uncertainties are shown as shaded bars. Dashed curves are a smooth fit to guide the eye [9].

used to describe the combinatorial background. The mixed event is generated by randomly selecting tracks from real events in the same centrality class, event plane direction and primary vertex z position. All the tracks in the mixed event are therefore fully uncorrelated. By running the jet reconstruction algorithm on the mixed event samples with the same conditions as for real events (with the trigger hadron now chosen in a random direction) a recoil-jet distribution which describes the combinatorial background is obtained.

Analyzed data were recorded with a Minimum Bias (MB) trigger condition in Au+Au collisions at $\sqrt{s_{NN}}$=200 GeV. Jets were reconstructed from charged tracks measured by the STAR Time Projection CHamber (TPC) using the anti-k_T sequential clustering algorithm via the FASTJet software package. Jet p_T is then adjusted for background energy on an event-by-event basis according to

$$p_{T,jet}^{reco} = p_{T,jet} - \rho \times A \qquad (2)$$

where A is the jet area calculated using soft ghost particles and ρ is the background energy density calculated for each event as the median jet energy density $\frac{p^i_{T,jet}}{A_i}$ from all reconstructed jets in the event, excluding 2 (peripheral) or 3 (central events) hardest jets in the event. The background jets are reconstructed using the k_T algorithm.

The reconstructed spectra were further corrected for the background fluctuations and detector effects (tracking efficiency, momentum resolution) by the unfolding. Two methods (Bayesian unfolding,

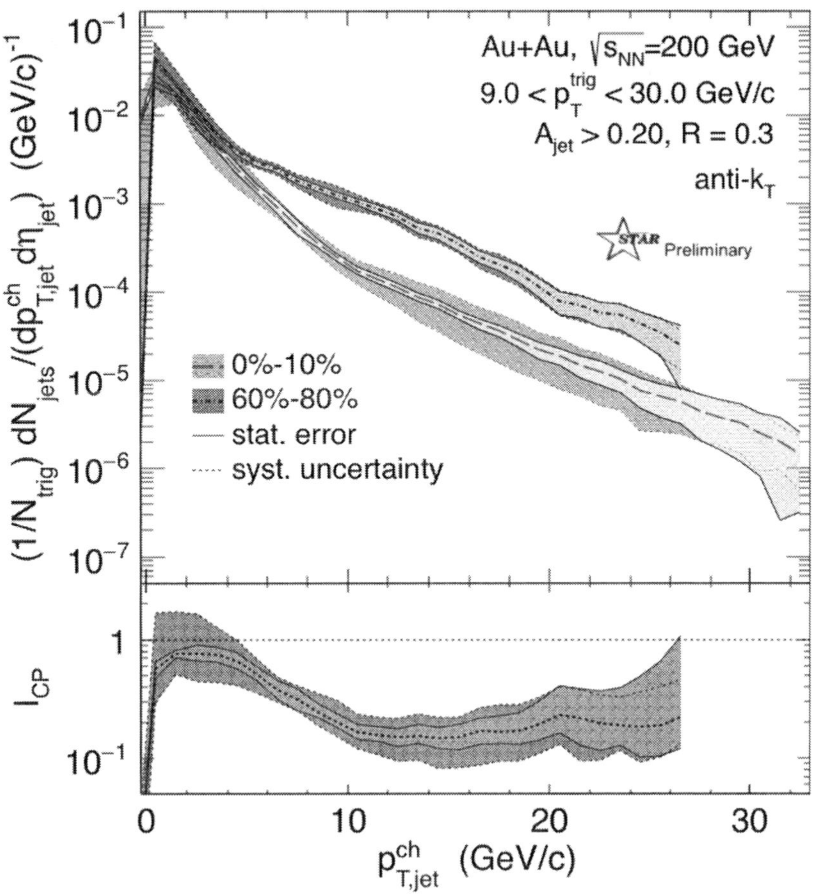

Figure 4. Upper panel: Fully corrected recoil-jet yield per trigger for central and peripheral Au+Au collisions at $\sqrt{s_{NN}}$ = 200 GeV. Bottom panel: I_{CP}, ratio of fully corrected recoil jet spectra in central (0-10%) and peripheral (60-80%) Au+Au collisions at $\sqrt{s_{NN}}$ = 200 GeV.

Singular Value Decomposition [10, 11]), several prior distributions and several values of the regularization parameter were used to check the systematic dependency of the unfolding procedure.

In-medium jet modification is measured via the ratio of the recoil yield in central and peripheral collisions (I_{CP}). Fig. 4 shows the fully corrected spectra both in peripheral and central collisions (top panel) and their ratio I_{CP} (bottom panel). For low jet momenta p_T < 5 GeV/c, I_{CP} is close to unity. For jet momenta p_T > 10 GeV/c I_{CP} drops to ~ 0.2, which indicates significant jet suppression in central heavy ion collisions compared with peripheral collisions. This is a stronger suppression than observed in a similar measurement by the ALICE collaboration in Pb+Pb collisions at $\sqrt{s_{NN}}$ = 2.76 TeV. ALICE measured $I_{CP} \approx 0.6$ for 20 < $p_{T,jet}$ < 100 GeV/c for charged jets with $R = 0.4$ [12]. However the jet p_T shift needed for the peripheral spectra to match the central spectra is similar in

DOI: 10.1051/epjconf/201612602027

both experiments, $\approx$ 8 GeV/c. Therefore, the larger suppression observed at RHIC may be due to similar out-of-cone energy transport combined with a steeper falling spectrum at RHIC energies than that at LHC energies. The different choice of trigger particle p_T and collision energies at STAR and ALICE could have an influence on the measured suppression as well.

4 Υ Measurement in Au+Au and U+U Collisions

The production of heavy quarkonia, a bound state of a heavy (c, b) quark and its antiquark, is suppressed in heavy ion collisions compared to p+p collisions due to the Debye screening of the quark potential. Charmonium ($c\bar{c}$) suppression was therefore proposed as one of the key signatures of the QGP formation [13]. Moreover, it is expected that quarkonium states with different binding energies dissociate at different temperatures, therefore their suppression pattern could serve as a QGP thermometer [14]. However in spite of the original expectations, only a weak energy dependence of J/ψ suppression has been observed over a broad range of collision energies [15, 16]. An explanation to this is that later recombination of $c\bar{c}$ pairs gives a significant contribution to the J/ψ yield which compensates for dissociation. Moreover, d+Au measurements proved that cold nuclear matter (CNM) effects also play an important role [17]. Feed-down from heavier states such as χ_c, Ψ_0 and B mesons to J/ψ also contribute to the measured yields. Bottomonia are expected to be less affected by many of these effects than charmonia, since uncorrelated recombination and co-mover absorbtion are predicted to be negligible at RHIC energies [18]. Therefore bottomonium represents a cleaner probe of the QGP matter. However, a feed-down contribution from sources as χ_b , Υ(2S) and Υ(3S) decaying to Υ(1S) is still present.

Bottomonia Υ were reconstructed in the di-electron decay channel. Fig. 5 shows the Υ nuclear modification factor, R_{AA} , at mid-rapidity as a function of number of participants, N_{part}, calculated from the Monte Carlo Glauber Model [19] together with R_{AA} measurements in Au+Au collisions at $\sqrt{s_{\rm NN}}$= 200 GeV and R_{AA} in U+U collisions at $\sqrt{s_{\rm NN}}$= 197 GeV. R_{AA} is close to unity for the collisions with a small number of participants, and significantly suppressed at larger N_{part}. U+U collisions produce higher initial energy densities than Au+Au collisions since the U-nucleus has a larger number of nucleons than an Au-nucleus. The measured U+U data confirm the Υ suppression seen in Au+Au collisions and extend the R_{AA} measurement to higher N_{part}. To test the sequential melting of the different Υ states, more statistics would be needed. Nevertheless, the emerging pattern of increasing suppression with higher N_{part} and lower binding energy is consistent with the sequential melting hypothesis.

The Au+Au Υ(1S) shows a suppression similar to that of high-p_T J/ψ mesons, which can be only explained if also cold nuclear matter effects were present [20]. In centrality integrated Au+Au collisions the (2S+3S) state yields are consistent with a complete suppression within the measurement error. Although the U+U results are consistent with Au+Au, the significance that $R_{AA} > 0$ for the Υ(2S+3S) states in U+U collisions of 0-60% centrality is 1.8 σ.

Even though comparable with each other, results form heavy ion collisions at RHIC exhibit a steeper N_{part} dependence than results from Pb+Pb collisions at 2.76 TeV at the LHC [21]. This may suggest that suppression is driven by the energy density and not strictly by the number of participant nucleons.

The model of Strickland and Bazow [22] incorporates lattice QCD results on bottomonium screening and thermal broadening, as well as the dynamical propagation of the Υ meson in the colored medium. Assuming an initial central temperature $428 < T < 444$ MeV, the scenario with a potential based on heavy quark internal energy is consistent with the observations, while the free energy based scenario is disfavoured. The model of Liu et al. [23] uses an internal-energy-based potential and

DOI: 10.1051/epjconf/201612602027

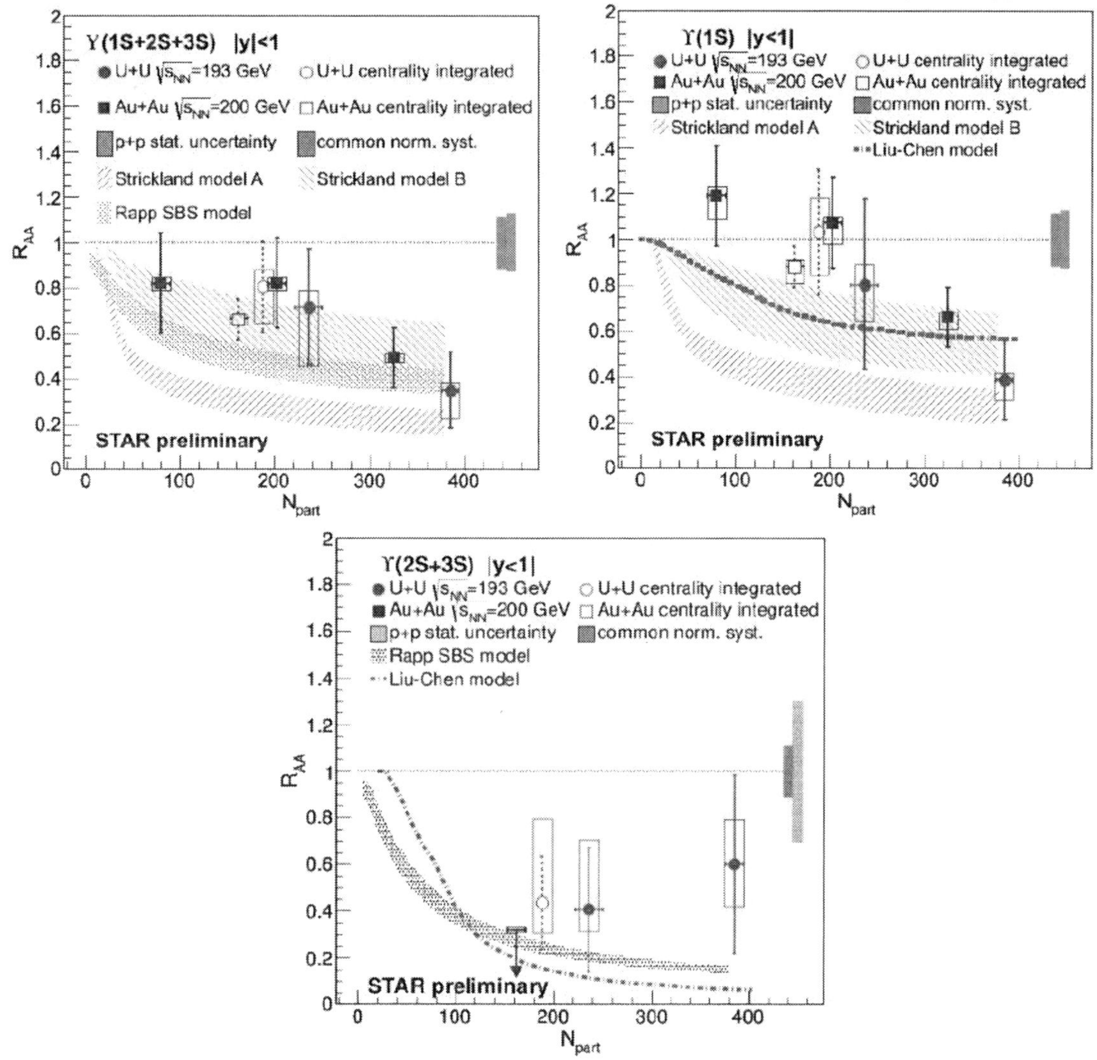

Figure 5. R_{AA} of the ϒ(1S+2S+3S) (left), ϒ(2S+3S) (bottom) and the ϒ(1S) (right) states plotted as a function of N_{part}, for two collision energies and systems: $\sqrt{s_{NN}}$= 200 GeV Au+Au (squares) and $\sqrt{s_{NN}}$= 193 GeV U+U collisions (circles), compared to different models. Open symbols represent the measurements of 0-60% centrality. The models on the bottom figure are for ϒ(2S) only.

an input temperature T =340 MeV. The model proposed by Emerick, Zhao, and Rapp [24], which includes possible CNM effects in addition, is also consistent with STAR results.

We observe a significant suppression of bottomonium states in central heavy ion collisions. While the ϒ(1S) is similarly suppressed in $\sqrt{s_{NN}}$=200 GeV Au+Au collisions to the high-p_T J/ψ mesons, there is a stronger suppression in the case of the excited states ϒ(2S+3S). This attests to the sequential melting picture in the presence of a deconfined medium. The suppression pattern seen in the case of the U+U collisions is consistent with the trend marked in Au+Au collisons and extends it towards higher N_{part} values. The indication of ϒ(1S) suppression in central Au+Au collisions is confirmed by the new measurements in U+U. There is a hint with an 1.8 σ significance that ϒ(2S+3S) is not

DOI: 10.1051/epjconf/201612602027

completely suppressed in 0-60% U+U collisions at 193 GeV. An ongoing analys of high statistics 2014 data will reduce statistical and systematical errors significantly and should provide separate R_{AA} measurements of all three Υ(1S), Υ(2S) and Υ(3S) states [25]. High statistics p+Au collisions taken in 2015 will help us gain a deeper insight to the CNM effects.

5 J/ψ Measurement in Au+Au and U+U Collisions

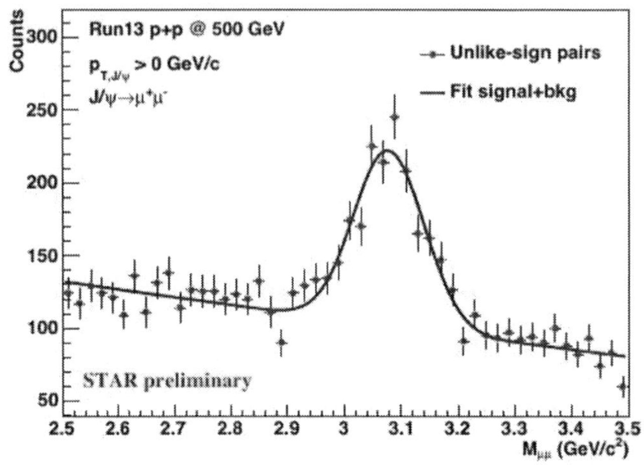

Figure 6. Invariant mass distribution of unlike-sign muon pairs (red circles) fitted with a Gaussian function describing signal and an exponential function describing background (solid line).

Measurements of J/ψ yields in p+p collisions serve as the baseline for heavy-ion collision measurements and can also constrain theoretical models of different production mechanisms [3, 4]. Recent reports from the LHC on Υ yields [26] and on D meson and J/ψ yields [27] showed enhancement of charm and bottom production in high event activity p+p collisions.

STAR measured J/ψ production from the di-electron and also in the di-muon decay channel. Fig. 6 shows the invariant mass peak of J/ψ meson reconstructed in the di-muon decay channel using the recently installed MTD detector. Fig. 7 shows relative J/ψ yield as a function of event activity in p+p collisions at $\sqrt{s}$ = 500 GeV. The event activity is characterized by the number of charged particles in pseudorapidity range $|\eta| < 0.9$ relative to that in a MB p+p event. Red points show J/ψ production from the di-muon channel with $|\eta| < 0.5$ and $p_T > 0$ GeV/c and blue points show J/ψ from the di-electron channel with $|\eta| < 1$ and $p_T > 4$ GeV/c. The black dashed line represents the linear case of a ratio of J/ψ production to the charged particle multiplicity. The low-$p_T$di-muon J/ψ yield roughly follows the dashed line while the high-$p_T$di-electron J/ψ yield rises faster than the linear trend. Enhancement of J/ψ production in the high activity events implies that hard processes dominate over soft processes in high multiplicity p+p collisions relative to low multiplicity collisions. The PYTHIA8 predictions for J/ψ p_T above 0 and 4 GeV/c are shown as the dashed lines. The shaded areas represent the statistical errors. The percolation model [28] is plotted as a solid curve. For J/ψ with $p_T > 0$ GeV/c the two models exhibit different trends in the area where the event multiplicity is 2 times higher than the MB multiplicity. The year 2015 p+p collision data will bring more statistics which will allow to discriminate the models.

DOI: 10.1051/epjconf/201612602027

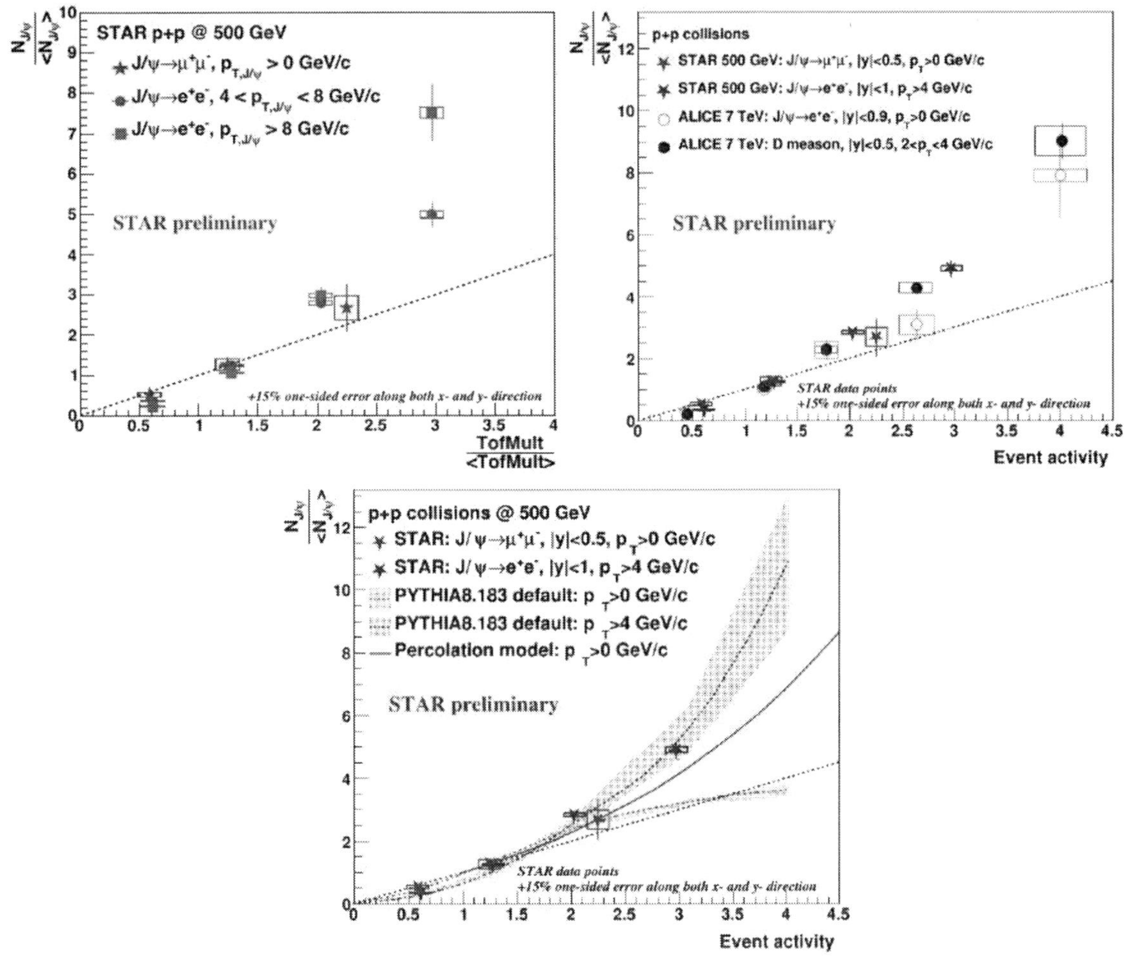

Figure 7. Dependence of relative invariant J/ψ yield on the event activity: Left: For three different J/ψ $p_T$bins. Bottom: For STAR measurement, PYTHIA8 prediction and the percolation model. Right: For STAR (500 GeV) and ALICE (7 TeV) measurements.

6 Summary and Outlook

In summary, larger jet suppression has been measured at RHIC energies than LHC energies for semi-inclusive charged jets, which however corresponds to a similar horizontal p_T shift in the jet spectrum.

Υ suppression in higher energy density U+U collisions confirms and extends the trend seen in Au+Au collisions. The precision of current measurements in the di-electron channel is however limited by the low statistics of the heavy ion and the reference samples. Ongoing analysis of the high-statistics $\sqrt{s_{NN}}$=200 GeV Au+Au dataset, taken in 2014 with the recently installed MTD detector, will strongly reduce both the statistical and systematic uncertainties by utilizing the di-muon decay channel. The event activity dependence of J/ψ production in p+p collisions is observed.

In 2014, STAR underwent two major heavy flavor detector upgrades, the MTD, to identify muons from quarkonium decay, and the HFT to precisely measure displaced secondary vertices of mesons and baryons containing heavy quarks. In 2015-2017, with STAR's enhanced capabilities in heavy flavor measurement, high multiplicity p+p and p+Au collisions will be investigated, and information on the QGP temperature and shear viscosity in Au+Au collisions at 200 GeV are also expected to come. In 2019-2020, RHIC will have a dedicated program for BES - Phase II with sufficiently high luminosity for 5-20 GeV Au+Au collisions. STAR will also carry out an inner-TPC (iTPC) upgrade, which will extend the pseudo-rapidity coverage, increase low p_T reach in tracking of particles for QGP bulk property study, and improve the dE/dx resolution of proton/kaon separations at high p_T for jet fragmentation measurements, in addition to the electron identification and acceptance for the low mass di-electron studies discussed above. The refined high-precision measurements will enable BES - Phase II to boost our understanding of the nuclear matter phase diagram.

References

[1] J. Adams et al. (STAR), Phys. Rev. Lett. **91**, 172302 (2003), nucl-ex/0305015
[2] J. Adams et al. (STAR), Phys. Rev. Lett. **91**, 072304 (2003), nucl-ex/0306024
[3] M. Butenschoen, B.A. Kniehl, Phys. Rev. **D84**, 051501 (2011), 1105.0820
[4] Y.Q. Ma, R. Venugopalan, Phys. Rev. Lett. **113**, 192301 (2014)
[5] K.H. Ackermann et al. (STAR), Nucl. Instrum. Meth. **A499**, 624 (2003)
[6] J. Adams et al. (STAR), Nucl. Phys. **A757**, 102 (2005), nucl-ex/0501009
[7] M. Gyulassy, X.N. Wang, Comput. Phys. Commun. **83**, 307 (1994), nucl-th/9502021
[8] R. Rapp, J. Wambach, Adv. Nucl. Phys. **25**, 1 (2000), hep-ph/9909229
[9] L. Adamczyk et al. (STAR), Phys. Rev. Lett. **112**, 162301 (2014), 1401.3043
[10] G. D'Agostini, ArXiv e-prints (2010), 1010.0632
[11] A. Hocker, V. Kartvelishvili, Nucl.Instrum.Meth. **A372**, 469 (1996), hep-ph/9509307
[12] J. Adam et al. (ALICE), JHEP **09**, 170 (2015), 1506.03984
[13] T. MATSUI, H. Satz, PHYSICS LETTERS B **178**, 416 (1986)
[14] A. Mócsy, P. Petreczky, Phys. Rev. Lett. **99**, 211602 (2007)
[15] B. Alessandro et al. (NA50), Eur. Phys. J. **C39**, 335 (2005), hep-ex/0412036
[16] A. Adare et al. (PHENIX), Phys. Rev. Lett. **98**, 232301 (2007), nucl-ex/0611020
[17] A. Adare et al. (PHENIX Collaboration), Phys. Rev. Lett. **107**, 142301 (2011)
[18] Z. Lin, C. Ko, Physics Letters B **503**, 104 (2001)
[19] R.J. Glauber, Nuclear Physics A **774**, 3 (2006), {QUARK} {MATTER} 2005Proceedings of the 18th International Conference on Ultra-Relativistic Nucleus–Nucleus Collisions
[20] L. Adamczyk et al. (STAR), Phys. Lett. **B735**, 127 (2014), [Erratum: Phys. Lett.B743,537(2015)], 1312.3675
[21] S. Chatrchyan et al. (CMS Collaboration), Phys. Rev. Lett. **109**, 222301 (2012)
[22] M. Strickland, D. Bazow, Nucl. Phys. **A879**, 25 (2012), 1112.2761
[23] Y. Liu, B. Chen, N. Xu, P. Zhuang, Physics Letters B **697**, 32 (2011)
[24] A. Emerick, X. Zhao, R. Rapp, Eur. Phys. J. **A48**, 72 (2012), 1111.6537
[25] C. Yang et al., Nucl. Instrum. Meth. **A762**, 1 (2014), 1402.1078
[26] S. Chatrchyan et al. (CMS), JHEP **04**, 103 (2014), 1312.6300
[27] J. Adam et al. (ALICE), JHEP **09**, 148 (2015), 1505.00664
[28] E.G. Ferreiro, C. Pajares, Phys. Rev. C **86**, 034903 (2012)

EPJ Web of Conferences **126**, 02028 (2016) DOI: 10.1051/epjconf/201612602028
ICNFP 2015

Transport phenomena in a plasma of confining gluons

Radoslaw Ryblewski[1,a]

[1] *The H. Niewodniczański Institute of Nuclear Physics, Polish Academy of Sciences, PL-31342 Kraków, Poland*

Abstract. The plasma of confining gluons resulting from the Gribov quantization is considered. In the fluid dynamical framework the non-equilibrium properties of the system are studied. In the linear response approximation the formulas for the bulk, ζ, and shear, η, viscosities of the plasma are calculated analytically. Surprisingly, the approximate scaling of the ζ/η ratio reveals the strong-coupling properties of the system under consideration.

1 Introduction

The experimental data measured in ultra-relativistic heavy-ion collisions at RHIC (Relativistic Heavy-Ion Collider) and the LHC (Large Hadron Collider) energies suggests that the new state of strongly interacting matter created in these reactions, called the quark-gluon plasma (QGP), to a good approximation behaves as a fluid with the smallest viscosity in Nature. This observation triggered enormous interest in the development of relativistic viscous [1–22], and, more recently, anisotropic [23–29] fluid dynamics.

The successful application of the fluid dynamical modelling requires, in addition to the initial conditions and the freeze-out prescription, some *a priori* knowledge about the thermodynamic as well as the transport properties of the underlying microscopic theory, namely quantum chromodynamics (QCD). While the lattice QCD simulations results on the equation of state at large and intermediate temperatures are already quite reliable [30, 31], and agree well with the results of the resummed perturbation theory methods [32–34], there is still a lot of study required to understand the dynamic properties of the system [35]. In the case of charge-free medium the two most important transport coefficients which enter the theory already at the first order in the gradient expansion are the bulk, ζ, and shear, η, viscosities.

In this work we review the main results of our recent investigations [36, 37] concerning the bulk and shear viscosities in the plasma made of confining gluons resulting from the Gribov–Zwanziger (GZ) quantization [38, 39]. We show that the GZ plasma has certain features which are characteristic for strongly-coupled systems.

2 Gribov dispersion relation

The Gribov quantization amounts to fixing the residual gauge transformations remaining after the Fadeev–Poppov procedure. In the Coulomb gauge it results in the modification of the dispersion

[a] e-mail: Radoslaw.Ryblewski@ifj.edu.pl

relation for gluons in the infrared regime [38]

$$E(\mathbf{k}) = \sqrt{\mathbf{k}^2 + \frac{\gamma_G^4}{\mathbf{k}^2}}. \tag{1}$$

In consequence the soft gluons are effectively removed from the physical spectrum, which leads to the confining properties of the system. At the same time a new scale, the so called Gribov scale γ_G, is introduced which may be found by a self-consistent solution of the gap equation [38, 39] and recently it was shown to be directly related to the (chromo)magnetic scale [40, 41]. It was Zwanziger who generalized Gribov approach to the case of finite temperatures and presented its qualitative aggreement with pure glue lattice data [39]. Recently various interesting properties of the GZ approach have been demonstrated [41–48] which makes it an interesting playground for a study of the impact of confining properties of SU(3) Yang–Mills theory on the plasma dynamics.

3 Fluid dynamical description of the Gribov–Zwanziger plasma

For bulk and shear transport coefficients it is sufficient to consider the simplest (0+1)-dimensional longitudinally boost-invariant and transversally homogeneous system, which, as a matter of fact, resembles well the early-time dynamics of ultra-relativistic heavy-ion collisions. In this case the most general form of the energy-momentum tensor is

$$T^{\mu\nu} = (\varepsilon + P_T)u^\mu u^\nu - P_T g^{\mu\nu} + (P_L - P_T)z^\mu z^\nu, \tag{2}$$

where the four-velocity is $u^\mu = (t, 0, 0, z)/\tau$, with $\tau = \sqrt{t^2 - z^2}$ being proper time, and $z^\mu = (z, 0, 0, t)/\tau$. In Eq. (2) the variables

$$\varepsilon = \int \mathrm{d}K\, E(k \cdot u)\, f(\tau, w, k_\perp), \tag{3}$$

$$P_L = \int \mathrm{d}K \frac{w^2}{\tau^2 E(k \cdot u)} \left[1 - \frac{\gamma_G^4}{(k \cdot u)^2}\right] f(\tau, w, k_\perp), \tag{4}$$

$$P_T = \int \mathrm{d}K \frac{k_\perp^2}{2\, E(k \cdot u)} \left[1 - \frac{\gamma_G^4}{(k \cdot u)^2}\right] f(\tau, w, k_\perp), \tag{5}$$

denote the energy density, transverse and longitudinal pressure, respectively [36, 37]. Here we introduced the phase space distribution function f, $\int \mathrm{d}K(\ldots) = \frac{g_0}{(2\pi)^3} \int_{-\infty}^{\infty} \frac{\mathrm{d}w}{\tau} \int \mathrm{d}^2 k_\perp (\ldots)$, $k^\mu = (|\mathbf{k}|, \mathbf{k})$ and $w = k_\| t - k^0 z$. In Eqs. (3)-(5) we employed the GZ dispersion relation (1) which has the following covariant form [36, 37]

$$E(k \cdot u) = \sqrt{(k \cdot u)^2 + \frac{\gamma_G^4}{(k \cdot u)^2}}. \tag{6}$$

The dynamical properties of the system follow from the energy and momentum conservation, $\partial_\mu T^{\mu\nu} = 0$, which reduces to the single differential equation

$$\frac{\mathrm{d}\varepsilon}{\mathrm{d}\tau} + \frac{\varepsilon + P_L}{\tau} = 0. \tag{7}$$

In Eq. (7) one can identify bulk and shear viscous pressure corrections through

$$\pi = 2(P_T - P_L)/3, \tag{8}$$

and

$$\Pi = P - P_{\text{GZ}}, \tag{9}$$

where $P = (P_L + 2P_T)/3$ and P_{GZ} is the pressure in the GZ equilibrium, that is, in the case where f has the Bose–Einstein form $f_{\text{GZ}} = \{\exp[E(k \cdot u)/T] - 1\}^{-1}$ and the temperature T follows from the Landau matching condition, $\varepsilon(T) = \varepsilon_{\text{GZ}}(T, \gamma_{\text{G}})$. Once the system is in the vicinity of the local thermal equilibrium the pressure corrections are well described by the Navier–Stokes expressions

$$\pi = \frac{4}{3}\frac{\eta}{\tau}, \qquad \Pi = -\frac{\zeta}{\tau}, \tag{10}$$

where ζ and η stand for bulk and shear viscosity, respectively.

4 Bulk and shear viscosity and ζ/η scaling

In order to obtain explicit formulas for viscosities in the GZ plasma one can use Eqs. (8)-(10) and consider small perturbations around equilibrium state. In this case one can write

$$f \approx f_{\text{GZ}} + \delta f + \cdots, \tag{11}$$

which, when used together with the Boltzmann kinetic equation in the relaxation time approximation (RTA)

$$\frac{\partial f(\tau, w, k_\perp)}{\partial \tau} = \frac{f_{\text{GZ}}(\tau, w, k_\perp) - f(\tau, w, k_\perp)}{\tau_{\text{rel}}(\tau)}, \tag{12}$$

after some algebra, leads to the following formulas

$$\zeta = \frac{g_0 \gamma_{\text{G}}^5}{3\pi^2} \frac{\tau_{\text{rel}}}{T} \int_0^\infty \mathrm{d}y \left[c_s^2 - \frac{1}{3}\frac{y^4 - 1}{y^4 + 1}\right] f_{\text{GZ}}(1 + f_{\text{GZ}}), \tag{13}$$

$$\eta = \frac{g_0 \gamma_{\text{G}}^5}{30\pi^2} \frac{\tau_{\text{rel}}}{T} \int_0^\infty \mathrm{d}y \frac{\left(y^4 - 1\right)^2}{y^4 + 1} f_{\text{GZ}}(1 + f_{\text{GZ}}), \tag{14}$$

for bulk and shear viscosity, respectively. Above $f_{\text{GZ}} = \{\exp[\gamma_{\text{G}}\sqrt{y^2 + y^{-2}}/T] - 1\}^{-1}$.

In the left panel of Fig. 1 we plot ζ and η obtained from Eqs. (13)-(14) scaled by the equilibrium entropy density, s, and the relaxation time, τ_{rel}, as functions of temperature. We observe that there is almost linear scaling of the scaled shear viscosity with temperature, similar to the ideal massless gas, $\eta/s = T/5\tau_{\text{rel}}$. We also observe a minor enhancement of bulk viscosity near the phase transition temperature $T_c = 260$ MeV. Another striking result is that, different from conventional quasiparticle approaches, at low temperatures bulk viscosity dominates over the shear viscosity.

Finally, in the right panel of Fig. 1 we show the ζ/η ratio scaled by the $1/3 - c_s^2$. We observe that at large temperatures the scaling is, to a good approximation, linear. This can be seen more explicitly when one considers the large temperature expansion of the ζ/η ratio. In this case one obtains an exact linear scaling

$$\frac{\zeta}{\eta} = \kappa_{\text{GZ}}\left(\frac{1}{3} - c_s^2\right) \qquad (T \gg \gamma_{\text{G}}), \tag{15}$$

where $\kappa_{\text{GZ}} = 5/2$. This is different from the perturbative treatments of the weakly-coupled plasmas, which predict a quadratic scaling [49–51]. Surprisingly, our result seems to be in line with the results obtained in strongly-coupled theories using gauge theory/gravity duality [52–54].

DOI: 10.1051/epjconf/201612602028

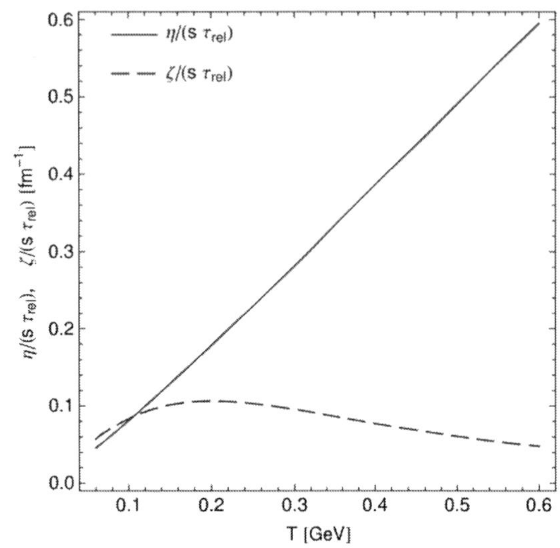

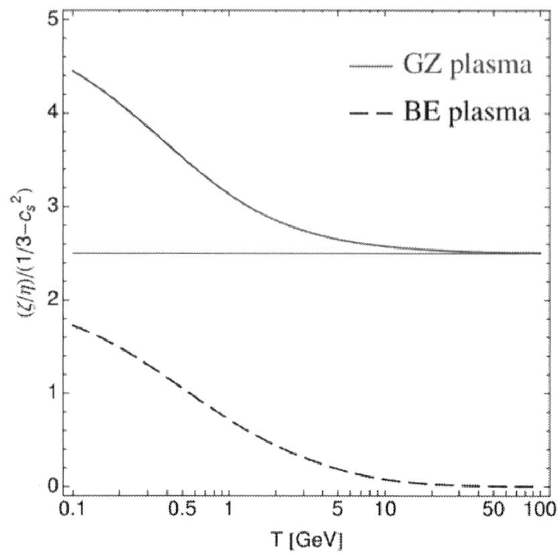

Figure 1: (Color online) Left: Analytic results on bulk (dashed blue line) and shear (solid red line) viscosity (see Eqs. (13) and (14), respectively) scaled by equilibrium entropy density and the relaxation time as functions of the temperature. Right: The bulk to shear viscosity ratio scaled by the difference $1/3 - c_s^2$ for the Gribov–Zwanziger plasma and the Bose–Einstein plasma of massive particles with the mass $m = \sqrt{2}\gamma_G$.

5 Conclusions

In this work we reviewed our recent results concerning the transport coefficients, namely shear and bulk viscosities, for a system composed of gluons resulting from the Gribov–Zwanziger quantization. The GZ procedure allows us to include the confining effects in the gluon system in a straightforwad way and, in consequence, to understand better the beviour of the SU(3) Yang-Mills plasma.

Using the fluid dynamical considerations in the small perturbation limit we found the exact analytic formulas for the transport coefficients. Based on the analysis of the obtained expressions we observe various properties of the system which are characteristic for the strongly-coupled systems. In particular, at large temperatures, we found the linear scaling of the ζ/η ratio with the difference $1/3 - c_s^2$, which is similar to the one found in strongly-coupled non-conformal gauge theory plasma. Our findings stay in line with the well established understanding of the quark-gluon plasma as the strongly-coupled fluid with small viscosity.

Acknowledgements

Author would like to thank Wojciech Florkowski, Nan Su and Konrad Tywoniuk for the fruitful collaboration. This work was supported by the Polish National Science Center grant No. DEC-2012/07/D/ST2/02125.

References

[1] A. Muronga, Phys. Rev. Lett. **88**, 062302 (2002), nucl-th/0104064
[2] U.W. Heinz, H. Song, A.K. Chaudhuri, Phys.Rev. **C73**, 034904 (2006), nucl-th/0510014

[3] R. Baier, P. Romatschke, U.A. Wiedemann, Phys. Rev. **C73**, 064903 (2006), hep-ph/0602249
[4] K. Dusling, D. Teaney, Phys. Rev. **C77**, 034905 (2008), 0710.5932
[5] M. Chojnacki, W. Florkowski, W. Broniowski, A. Kisiel, Phys. Rev. **C78**, 014905 (2008), 0712.0947
[6] M. Luzum, P. Romatschke, Phys. Rev. Lett. **103**, 262302 (2009), 0901.4588
[7] G.S. Denicol, T. Kodama, T. Koide, P. Mota, Phys. Rev. **C80**, 064901 (2009), 0903.3595
[8] P. Bozek, Phys. Rev. **C81**, 034909 (2010), 0911.2397
[9] A. El, Z. Xu, C. Greiner, Phys. Rev. **C81**, 041901 (2010), 0907.4500
[10] J. Peralta-Ramos, E. Calzetta, Phys.Rev. **C82**, 054905 (2010), 1003.1091
[11] B.H. Alver, C. Gombeaud, M. Luzum, J.Y. Ollitrault, Phys. Rev. **C82**, 034913 (2010), 1007.5469
[12] G. Denicol, T. Koide, D. Rischke, Phys.Rev.Lett. **105**, 162501 (2010), 1004.5013
[13] B. Schenke, S. Jeon, C. Gale, Phys.Rev.Lett. **106**, 042301 (2011), 1009.3244
[14] P. Bozek, Phys.Rev. **C85**, 034901 (2012), 1110.6742
[15] H. Niemi, G.S. Denicol, P. Huovinen, E. Molnár, D.H. Rischke, Phys. Rev. C **86**, 014909 (2012)
[16] P. Bozek, W. Broniowski, Phys. Rev. Lett. **109**, 062301 (2012), 1204.3580
[17] I.A. Karpenko, Yu.M. Sinyukov, K. Werner, Phys. Rev. **C87**, 024914 (2013), 1204.5351
[18] J. Peralta-Ramos, E. Calzetta, Phys.Rev. **D87**, 034003 (2013), 1212.0824
[19] P. Bozek, A. Bzdak, V. Skokov, Phys. Lett. **B728**, 662 (2014), 1309.7358
[20] G. Denicol, S. Jeon, C. Gale, Phys.Rev. **C90**, 024912 (2014), 1403.0962
[21] F.G. Gardim, J. Noronha-Hostler, M. Luzum, F. Grassi, Phys. Rev. **C91**, 034902 (2015), 1411.2574
[22] R.S. Bhalerao, A. Jaiswal, S. Pal, Phys. Rev. **C92**, 014903 (2015), 1503.03862
[23] M. Martinez, M. Strickland, Nucl. Phys. **A848**, 183 (2010), 1007.0889
[24] W. Florkowski, R. Ryblewski, Phys. Rev. **C83**, 034907 (2011), 1007.0130
[25] L. Tinti, W. Florkowski, Phys. Rev. **C89**, 034907 (2014), 1312.6614
[26] D. Bazow, U.W. Heinz, M. Strickland, Phys. Rev. **C90**, 054910 (2014), 1311.6720
[27] D. Bazow, U.W. Heinz, M. Martinez, Phys.Rev. **C91**, 064903 (2015), 1503.07443
[28] M. Nopoush, M. Strickland, R. Ryblewski, D. Bazow, U. Heinz, M. Martinez, Phys. Rev. **C92**, 044912 (2015), 1506.05278
[29] L. Tinti (2015), 1506.07164
[30] A. Bazavov et al. (HotQCD), Phys. Rev. **D90**, 094503 (2014), 1407.6387
[31] S. Borsanyi, Z. Fodor, C. Hoelbling, S.D. Katz, S. Krieg, K.K. Szabo, Phys. Lett. **B730**, 99 (2014), 1309.5258
[32] J.O. Andersen, L.E. Leganger, M. Strickland, N. Su, JHEP **08**, 053 (2011), 1103.2528
[33] S. Mogliacci, J.O. Andersen, M. Strickland, N. Su, A. Vuorinen, JHEP **12**, 055 (2013), 1307.8098
[34] N. Haque, A. Bandyopadhyay, J.O. Andersen, M.G. Mustafa, M. Strickland, N. Su, JHEP **05**, 027 (2014), 1402.6907
[35] H.B. Meyer, Eur. Phys. J. **A47**, 86 (2011), 1104.3708
[36] W. Florkowski, R. Ryblewski, N. Su, K. Tywoniuk (2015), 1504.03176
[37] W. Florkowski, R. Ryblewski, N. Su, K. Tywoniuk (2015), 1509.01242
[38] V.N. Gribov, Nucl. Phys. **B139**, 1 (1978)
[39] D. Zwanziger, Nucl. Phys. **B323**, 513 (1989)

[40] D. Zwanziger, Phys. Rev. **D76**, 125014 (2007), hep-ph/0610021
[41] K. Fukushima, N. Su, Phys.Rev. **D88**, 076008 (2013), 1304.8004
[42] R. Alkofer, L. von Smekal, Phys. Rept. **353**, 281 (2001), hep-ph/0007355
[43] M.N. Chernodub, V.I. Zakharov, Phys. Rev. Lett. **100**, 222001 (2008), hep-ph/0703167
[44] G. Burgio, M. Quandt, H. Reinhardt, Phys. Rev. Lett. **102**, 032002 (2009), 0807.3291
[45] A. Maas, Phys. Rept. **524**, 203 (2013), 1106.3942
[46] N. Su, K. Tywoniuk, Phys. Rev. Lett. **114**, 161601 (2015), 1409.3203
[47] D.E. Kharzeev, E.M. Levin, Phys. Rev. Lett. **114**, 242001 (2015), 1501.04622
[48] F.E. Canfora, D. Dudal, I.F. Justo, P. Pais, L. Rosa, D. Vercauteren, Eur. Phys. J. **C75**, 326 (2015), 1505.02287
[49] S. Weinberg, Astrophys. J. **168**, 175 (1971)
[50] S. Jeon, L.G. Yaffe, Phys. Rev. **D53**, 5799 (1996), hep-ph/9512263
[51] W. Florkowski, A. Jaiswal, E. Maksymiuk, R. Ryblewski, M. Strickland, Phys. Rev. **C91**, 054907 (2015), 1503.03226
[52] P. Benincasa, A. Buchel, A.O. Starinets, Nucl. Phys. **B733**, 160 (2006), hep-th/0507026
[53] A. Buchel, Phys. Rev. **D72**, 106002 (2005), hep-th/0509083
[54] D. Li, S. He, M. Huang, JHEP **06**, 046 (2015), 1411.5332

EPJ Web of Conferences **126**, 02029 (2016)
ICNFP 2015
DOI: 10.1051/epjconf/201612602029

Archimedes force on Casimir apparatus

V. Shevchenko[1,2,a] and E. Shevrin[1]

[1] *National Research Centre "Kurchatov Institute", ac.Kurchatova sq., 1, Moscow 123182 Russia;*

[2] *Far Eastern Federal University, Sukhanova str. 8, Vladivostok 690950 Russia*

Abstract.
The talk addresses a problem of Casimir apparatus in weak gravitational field, surrounded by a dense medium. The falling of the apparatus has to be governed by the equivalence principle, taking into account proper contributions to the weight of the apparatus from its material part and from distorted quantum fields. We discuss general ex pression for the corresponding force in terms of the effective action. By way of example we compute explicit expression for Archimedes force, acting on the Casimir apparatus of finite size, immersed into thermal bath of free scalar field. It is shown that besides universal term, proportional to the volume of the apparatus, there are non-universal quantum corrections, depending on the boundary conditions.

1 Introduction

Reflections about matter dynamics in gravitational field are among the most fruitful themes in the history of physics. Well known legends about Newton, inspired by the falling apple in his mother's garden, or about Galileo, dropping balls from the top of the Leaning Tower of Pisa are good examples. In modern times, classical tests of General Relativity such as light deflection by the Sun's gravity, weighting-the-photon experiments of Pound-Rebka type , Shapiro delay, neutron interferometry in gravitational field, and ALPHA, AEGIS and GBAR experiments at CERN, exploring falling antimatter, continue the same line of studies.

Of particular interest are quantum field theoretic physics in classical gravitational field, where Hawking radiation is the best known phenomenon. Needless to say that the problem of genuine gravitational interaction between parts of intrinsically quantum object (for example, between two entangled photons) cannot be addressed in semiclassical approach, leaving aside the fact that it is beyond our current experimental abilities. It is to be stressed that we have no direct experimental information how an elementary particle like proton gravitationally interacts with another one, and Newton gravity law has never been tested experimentally for small distances, even in 1 μm range. Therefore its naive extrapolation to distances of the order 10^{-35} meters could be plainly wrong, as various extra dimensions scenarios suggest. In other words, the "ultimate" ultraviolet fundamental

[a] e-mail: Vladimir.Shevchenko@cern.ch

scale can have nothing to do with the conventional Planck distance (calculated from long-distance asymptotic of the gravitational interaction, described by the Newton constant G).

Coming back to the case when semiclassical treatment is appropriate, the simplest example is non-relativistic motion of a test body in external weak gravitational field. The basic fact governing this type of motion is well known from school textbooks: the force acting on the body is proportional to its mass and directed along the free fall acceleration:

$$\mathbf{f} = m\mathbf{g} = \rho V \mathbf{g} \tag{1}$$

where $\rho = m/V$ is average density and V is the body's volume. Simplicity of this formula should not camouflage a highly nontrivial fact, that the force depends on the only parameter of the body - its mass (and not, for example, on its chemical composition, entropy etc). Combined with the Newton's second law of motion this fact has, of course, direct relation to the celebrated equivalence principle.

The situation gets more complex if the test body is immersed into gas or fluid. The expression (1) is to be replaced in this case by

$$\mathbf{f} = (\rho - \rho_f) V \mathbf{g} \tag{2}$$

where ρ_f is the fluid's density, and the term proportional to ρ_f is known as Archimedes force. The expression (2) has many approximations behind it and there are a few relevant small parameters. First, the independence of this force on any characteristic features of the body other than its volume is by no means trivial. It is based on smallness of a ratio of gas/fluid molecules size to that of the body (and also holes in the body's surface etc), which makes continuous medium approximation applicable.[1] Another parameter is the Planck constant $\hbar$ - the result (2) is of course purely classical and may get quantum corrections, for example, if typical quantum correlation length in the fluid is comparable with the body size. Also needless to say that (2) is valid in non-relativistic and weak gravitational field approximations. Last but not least, the expression (2) is invariant under shifts $\rho \to \rho + \text{const}$. It is "self-renormalized" in this sense and piece of vacuum (or any other medium in stationary case) with the "mass"

$$\frac{1}{c^2}\int dV \,\langle T^{00}\rangle \tag{3}$$

does not "fall" in gravitational field, because there is compensating "pressure" on this piece of exactly the same magnitude from surroundings, directed upwards.

We analyze weighting the Casimir apparatus in weak gravitational field in this talk. This problem has attracted some attention in recent years [1–8] and there used to be controversy in the literature we will mention below. We argue that the key point is physically correct definition of the weighting procedure, since there is no possibility to weight Casimir energy alone - one always measure the weight of Casimir apparatus as a whole. The weighting procedure and the results are to be universal and applicable to any Casimir apparatus, not only to two parallel plate Casimir cavity, usually taken as example. Our aim is to discuss such procedure and to apply it to concrete case of Casimir cavity in thermal bath of massless scalar field.

2 Archimedes Force

The basic ingredient is quantum field theoretical average of energy-momentum tensor $\langle T_{\mu\nu}(x)\rangle$, where average over fields is computed with the standard integration measure $\mathcal{D}\Phi$, normalized to have $\langle 1\rangle = 1$. In geometric setup used by us in this paper,[2] Casimir apparatus is encoded by some x-dependent measure deformation, $\mathcal{D}\Phi \to \mathcal{D}'\Phi$, corresponding to constraints the fields have to obey on

[1] This is just what helped Archimedes to find out the volume of King's Hiero crown in well known legend.

[2] By the word "geometric" we mean neglect of dynamical properties of the boundaries like frequency-dependence etc.

the boundary or in interior of the apparatus. For example, in classical Casimir setup of two infinite ideally conducting parallel planes interacting with electromagnetic field this deformation looks like

$$\mathcal{D}A_\rho(x) \to \mathcal{D}A_\rho(x)\,\delta(\tilde{F}^{\mu 3}(x_3 = a_1))\delta(\tilde{F}^{\mu 3}(x_3 = a_2)) \tag{4}$$

where $\tilde{F}^{\mu\nu}$ is dual field strength, a_1, a_2 - coordinates of the planes along the third axis (the axes 1 and 2 are in the planes). The form of the above expression is quite general and in many cases one can write $\mathcal{D}'\Phi = \mathcal{D}\Phi\,\Delta[\Phi]$ with some functional of the fields. One could think of Casimir plates as of "passive detectors", which measure parallel electric and normal magnetic field at the position of their location and always return zero value. This is like having quantum or classical particle in a corridor made of hard walls so that any subsequent measurement of the particle coordinate will definitely return a result inside the corridor. Uncertainty relation causes pressure on the boundaries (walls and plates in the examples above), which depends, in particular, on how hard (conductive) they are. In general words, we are interested in relations between what and how precise a detector is able to measure and its own dynamics. It is important to remember, however, that the "detectors" in the above examples are artificial and make no record of the results.

Coming back to the Casimir apparatus in weak gravitational field, writing the metric as $g_{\mu\nu} = \eta_{\mu\nu} + 2h_{\mu\nu}$, one gets for the energy shift at the leading order of semiclassical approximation [3]

$$\delta E_g = -\int d^3x\, h_{\mu\nu}(x)\langle T^{\mu\nu}\rangle(x) \tag{5}$$

In geometric setup the average $\langle T^{\mu\nu}\rangle(x)$ has two parts - the "material" one, corresponding to the objects (planes, cavities, robes, springs etc) the Casimir apparatus is made of, and the "field" part. For nonrelativistic case spatial components of the energy-momentum tensor are suppressed by inverse powers of the speed of light, $|T^{00}| \gg |T^{ij}|$ and only temporal component of the metric tensor $h_{00}(x)$ is relevant. This can be correct approximation for the material part of $\langle T^{\mu\nu}\rangle(x)$, but certainly not for its "field" part we are interested in here. Consequently it is easy to check, that various choices of the metric $h_{\mu\nu}$ lead to different answers for δE_g, even if all these choices correspond to uniform field with free fall acceleration $\mathbf{g}$. Moreover, the energy becomes orientation-dependent for some choices, in gross contradiction with the equivalence principle and scalar nature of mass. This is physically unacceptable and should be resolved.

The source of the problem was identified in [2, 3] as gauge non-invariance of (5). Indeed, (5) is invariant under weak field gauge transformation $h_{\mu\nu} \to h_{\mu\nu} + \partial_\mu\xi_\nu + \partial_\nu\xi_\mu$ only if $\partial_\mu T^{\mu\nu} = 0$. The energy-momentum tensor is covariantly conserved for the combined "material + field" system

$$\nabla_\mu T^{\mu\nu} = 0 \tag{6}$$

but not for the "field" part alone. Thus two logically possible alternative paths can be chosen: either one is to include the material part and carefully work with the full energy-momentum tensor, obeying (6), or one is to argue, that this or that choice of the metric is more physical than another choices and compute the force using the distinguished metric. Mostly the latter path was followed in the literature with the motivation for preferable role of Fermi metric choice $c^2h_{00} = \mathbf{g}\mathbf{z}\ ;\ h_{ij} = 0$ and the result for the weight of Casimir energy (in classical two plates case)

$$\mathbf{f} = \mathbf{g}\,\frac{E_C}{c^2}\,S \quad \text{where} \quad E_C = -\frac{\pi^2\hbar c}{720a^3} \tag{7}$$

and S stays for the plates area. The energy-momentum tensor of the system is given by $\langle T^{\mu\nu}\rangle = (E_C/a)\times\mathrm{diag}(1,-1,-1,3)$ between the plates and zero outside [9].

Thus, the cavity feels small upward push and Casimir energy gravitates as any other form of energy in accord with the equivalence principle. In a sense, the answer (7) could have been written without any computations, if E_C is known. However, the arguments based on physically distinguishable role of a particular parametrization of the metric are difficult if not impossible to generalize to other cases. In particular, it is not clear how to write the next order $O(h_{\mu\nu}h_{\rho\sigma})$ correction to (7). It is also important, that the result (7) by its nature should be quite general, which, however, is also not clear taking into account that the methods used for its derivation (see e.g. [2]) heavily use properties of particular two infinite parallel planes geometry. This calls for systematic derivation applicable beyond the weak field approximation and for arbitrary Casimir apparatus.

We argue here that weighting methodology suggested in [10] can be naturally adopted to the Casimir apparatus weighting problem. We consider static metric of the following form

$$ds^2 = g_{00}(x_3)c^2dt^2 + g_{33}(x_3)dx_3^2 + d\mathbf{x}_\perp^2 \tag{8}$$

with the choice of x_3-coordinate axes such that $g_{00}(x_3 = 0) = -1$ (we denote x_3 coordinate as z below for simplicity of notation). We are to weight, following the symmetry of the above choice, two large identical boxes with some identical boundary conditions for the fields on their internal boundaries (see Figure 1). We put inside the box number 2 the Casimir apparatus of much smaller size, i.e. we assume a set of conditions encoded by some functional

$$\mathcal{B}[\phi(x)] = 0 \quad \text{for} \quad x \in V \tag{9}$$

put on the fields inside the volume of the apparatus V or, in particular case, only on its boundary $S = \partial V$. There is no apparatus in the box number 1. Then, following [10] we consider a function:

$$w_k(z) = \int\limits_{S_k} d^2\mathbf{x}_\perp \sqrt{-g_{00}} \cdot g_{33}\langle T^{33}\rangle_k \tag{10}$$

The index $k = 1, 2$ is the box label and integration goes over section of the boxes at constant z (sections are assumed to be of arbitrary shape, but z-independent, i.e. geometry is cylindrical). The measures $\langle ...\rangle_k$ take into account Casimir apparatus conditions (9) in the box 2. For Minkowskii space with the metric $g_{\mu\nu} = \eta_{\mu\nu} = \text{diag}(-1, 1, 1, 1)$ the function (10) is nothing but the integrated pressure, i.e. for $z = 0$ it is the force acting on the bottom plane of the box. It is obvious that all contributions to this force from "material" parts of the boxes are identical.

The difference of the above defined integrated pressures at $z = 0$ we call, by definition, the weight of our Casimir apparatus:

$$f = w_2(0) - w_1(0) \tag{11}$$

To express this force in terms of energy-momentum tensor integrals, we are to take into account that the total energy-momentum tensor is covariantly conserved inside each box. For the metric choice (8) the equation (6) reads

$$\frac{\partial w(z)}{\partial z} = \frac{\partial \sqrt{-g_{00}}}{\partial z} \int\limits_{S_k} d^2\mathbf{x}_\perp\, g_{00}\langle T^{00}\rangle \tag{12}$$

where we have used the definition (10). Integrating (12) over the entire boxes and assuming $\lim\limits_{z\to\infty}[\langle T^{33}\rangle_2(z) - \langle T^{33}\rangle_1(z)] = 0$ (which physically corresponds to finiteness of the Casimir apparatus), we obtain the following final answer

$$f = \frac{1}{2}\int d^3\mathbf{x}\, \sqrt{-g_{00}(z)}\left(\frac{\partial g_{00}(z)}{\partial z}\right)\left[\langle T^{00}\rangle_2(z) - \langle T^{00}\rangle_1(z)\right] \tag{13}$$

We discuss this result in the next section.

3 Discussion and applications

The most important property of the expression (13) is its independence on spatial components of the metric and energy-momentum tensors. To get that the particular form (8) of the metric tensor was crucial. It is clear, in particular, that the weighting procedure we use would not be operational for the case of transverse coordinates x_1, x_2 - dependent metric tensor. On the other hand, one has no need to take $g_{33} = 1$ and nowhere we have used weak field approximation. Therefore in is legitimate to expand (13) over difference $(-1 - g_{00})$. The leading term corresponds to the standard choice $g_{00} = -1 + 2\mathbf{g}z/c^2 + O(1/c^4)$:

$$\mathbf{f}_0 = \mathbf{g}\frac{1}{c^2}\int d^3\mathbf{x}\,\left[\langle T^{00}\rangle_2(z) - \langle T^{00}\rangle_1(z)\right] \tag{14}$$

For Casimir plates discussed above equation (14) reproduces the result (7). On the other hand, in classical limit we come back to (2) taking into account that

$$\int\limits_{V_{body}} d^3\mathbf{x}\,\langle T^{00}\rangle_2(x) = mc^2 \;\; ; \quad \int\limits_{V_{body}} d^3\mathbf{x}\,\langle T^{00}\rangle_1(x) = \rho_f c^2 V_{body} \tag{15}$$

and $\langle T^{00}\rangle_2(x) = \langle T^{00}\rangle_1(x)$ for x outside the body. Needless to say that this last condition does not take place for quantum "field part" of the total energy-momentum tensor, since the body distorts fields around it and its energy is delocalized in this sense.

The important issue is UV-divergencies of (13). Since the force is physical observable, we expect they all get renormalized. The detailed picture of such renormalization can be rather tricky, as for example [1, 5] clearly show. We have to take care only about divergencies, related to the Casimir apparatus, since all other obviously cancel in (13). General intuition suggests that the former ones renormalize the nonrelativistic mass the the material objects the apparatus is made of. From this point of view, while distinction between "body" and "medium" is clear in non-relativistic and non-quantum limits, it is somehow lost in general case since the body in question - Casimir apparatus - is surrounded by the cloud of quantum fields, distorted by its presence and this distortion contributes to its total rest mass. Contrary to classical vacuum, which at least in principle can be cleaned to any desired level, allowing independent measurement of each contribution, one cannot "clean" quantum vacuum by eliminating fluctuating quantum fields out of it.

Let us also make a comment on next-to-leading corrections to (14). They come from two places: expansion of metric-dependent multiplier in (13) and expansion of energy-momentum tensor average. It is convenient to rewrite (13) as

$$f = 2\int d^3\mathbf{x}\left(\frac{\partial h_{00}(z)}{\partial z}\right)\frac{1}{\sqrt{g_{33}(z)}}\,\frac{\delta(W_2 - W_1)}{\delta g_{00}(z)} \tag{16}$$

where W_k is the corresponding effective action and the standard definition

$$\langle T^{\mu\nu}(x)\rangle = \frac{2}{\sqrt{-g}}\frac{\delta W}{\delta g_{\mu\nu}} \tag{17}$$

was used. Next-to-leading correction has the following form:

$$\begin{aligned} f = f_0 - \int d^3\mathbf{x}\left(\frac{\partial h_{00}(z)}{\partial z}\right) h_{33}(z)\left[\langle T^{00}\rangle_2(x) - \langle T^{00}\rangle_1(x)\right] + \\ +4\int d^3\mathbf{x}\left(\frac{\partial h_{00}(z)}{\partial z}\right)\int d^4x'\, h_{\alpha\beta}(x')\,\frac{\delta}{\delta g_{\alpha\beta}(x')}\,\frac{\delta(W_2 - W_1)}{\delta g_{00}(x)}\Bigg|_{h=0} \end{aligned} \tag{18}$$

We see that the dependence on spatial components of the metric appears at the next order. Another correction to the classical Archimedes force - "weight of quantum fluctuations" - corresponds to the last term. It is worth mentioning [11] that in Casimir systems with massless fields fluctuations of energy-momentum tensor components

$$\int d^4x' \, [\langle T^{\mu\nu}(x)T^{\alpha\beta}(x')\rangle - \langle T^{\mu\nu}(x)\rangle \cdot \langle T^{\alpha\beta}(x')\rangle] \tag{19}$$

are typically not small with respect to average $\langle T^{\mu\nu}(x)\rangle$, so both the second and the third terms in the right hand side of (18) are in general of the same order.

For general geometry of the Casimir body the right hand side of (13) or (14) is given by some complicated expression, and no universal dependence of the Archimedes force on the body's volume like in (2) can be expected. To get the latter universality there should be small parameter in the system, as we discussed above. An interesting example is Casimir apparatus in the thermal bath. If geometric size of the apparatus is large compared to thermal wavelength

$$r \gg \frac{\hbar c}{k_B T} \tag{20}$$

one could think of large temperature expansion as being a good approximation. Technically we can realize it using effective action and heat kernel expansion formalism in Euclidean space [12]. It is convenient to start with expression for free energy

$$\beta F_\beta = -\log \int \mathcal{D}\Phi \, e^{-S[\Phi]} \tag{21}$$

where the action for massless minimally coupled free scalar field is given by the standard expression $S = \frac{1}{2}\int_0^\beta d\tau \int d^d\mathbf{x} \, \Phi\,(-\Box)\,\Phi$ and covariant D'Alembertian is $\Box = \nabla_\mu \nabla^\mu$. In Euclidean formalism (see, e.g. [13, 14]) one considers theory in $d+1$-dimensional Euclidean space-time with the topology $\mathbb{R}^d \times \mathbb{S}^1$, where length of the latter compact dimension is denoted as β. The fields satisfy the conditions of periodicity in Euclidean time $\Phi\,(\mathbf{x}, \tau) = \Phi\,(\mathbf{x}, \tau + \beta)$. Parameter β will be associated with inverse temperature in what follows: $\beta = (k_B T)^{-1}$.

As is well known, the temperature-dependent part of one-loop Euclidean effective action can be represented in terms of the corresponding thermal heat kernel $\hat{K}^\beta(s|x, y)$:

$$\beta F_\beta = -\frac{1}{2}\int_0^\infty \frac{ds}{s} \left(\mathrm{Tr}\hat{K}^\beta(s) - \mathrm{Tr}\hat{K}(s)\right) \tag{22}$$

where $\hat{K}^\beta(s)$ is periodic in Euclidean time solution of the equation

$$\left(\frac{d}{ds} - \Box\right) \hat{K}^\beta(s|x, y) = \hat{1} \cdot \delta(s)\delta(x, y) \tag{23}$$

with s playing the role of proper time. It can be shown (see [14] and references therein) that temperature dependence of the trace of finite temperature heat kernel can be factorized as

$$\mathrm{Tr}\hat{K}^\beta(s) = \frac{\beta}{(4\pi s)^{1/2}}\, \theta_3\left(0, e^{-\frac{\beta^2}{4s}}\right) \int d^d\mathbf{x} \,\mathrm{tr}\, \hat{K}_d(s|\mathbf{x}, \mathbf{x}) \tag{24}$$

where $\theta_3(a, b)$ is Jacobi function and we denote d-dimensional zero-temperature kernel $\hat{K}^\infty(s)$ as $\hat{K}_d(s)$.

The key result [16–19] (see also review [12] and references therein) is expression for the heat kernel asymptotic expansion on the manifold $\mathcal{M}$ in powers of the proper time:

$$\int d^d\mathbf{x}\ \mathrm{tr}\ \hat{K}_d(s|\mathbf{x},\mathbf{x}) = \frac{1}{(4\pi s)^{\frac{d}{2}}}\sum_{n=0}^{\infty}\left(s^n A_n + s^{n/2}B_{n/2}\right) \tag{25}$$

where the coefficients are given by

$$A_n = \int_{\mathcal{M}} d^dx\ \sqrt{-g(x)}\, a_n(x) \quad ; \quad B_{n/2} = \int_{\partial\mathcal{M}} d^{d-1}x\ \sqrt{\gamma(x)}\, b_{n/2}(x) \tag{26}$$

for integer and half-integer powers. Here $g(x)$ and $\gamma(x)$ denote the determinants of the bulk and induced boundary metrics, respectively. The surface integrals at the boundaries $B_{n/2}$ are build of local invariants incorporating such local characteristics of the surface as its extrinsic curvature $K_{\mu\nu}$ etc.

In the problem under discussion the manifolds $\mathcal{M}$ correspond to intrinsic space inside the boxes. Let us denote as V_{box} the volume of the boxes (identical for the box 1 and the box 2), then by V we denote the volume of Casimir apparatus, located in the box 2. In the same way we denote as S_{box} the identical surface area of the boxes and by S - the surface area of Casimir apparatus. Then taking into account that $a_0 = 1$, one obtains, at the leading order, for the free energy in the box 1 (without Casimir apparatus)

$$F_\beta^{(1)} = -\frac{\pi^2}{90}\frac{1}{\beta^4}V_{box} + b\frac{\zeta(3)}{8\pi}\frac{1}{\beta^3}S_{box} + O(\beta^{-2}) \tag{27}$$

while for the box 2, excluding the apparatus:

$$F_\beta^{(2,out)} = -\frac{\pi^2}{90}\frac{1}{\beta^4}(V_{box} - V) + b\frac{\zeta(3)}{8\pi}\frac{1}{\beta^3}(S_{box} + S) + O(\beta^{-2}) \tag{28}$$

The parameter b, proportional to $b_{1/2}$ from (26), encodes boundary conditions for the field, for particular case of Dirichlet boundary conditions $b = 1$, while $b = -1$ for Neumann ones [17, 19]. In principle, one can consider the case (quasistationary for small thermal conductivity of the material the apparatus is made of) with different temperatures inside and outside the apparatus and add contribution to the total free energy in the box 2 from the internal volume of the apparatus:

$$F_\beta^{(2,in)} = -\frac{\pi^2}{90}\frac{1}{\beta_{in}^4}V + b\frac{\zeta(3)}{8\pi}\frac{1}{\beta_{in}^3}S + O(\beta^{-2}) \tag{29}$$

Then, using the relation

$$U = F_\beta - T\frac{\partial F_\beta}{\partial\beta} \tag{30}$$

for internal energy, we get for the leading and first sub-leading contributions to Archimedes force (14)

$$\mathbf{f} - m\mathbf{g} = \mathbf{g}\frac{\hbar}{c}\left(\frac{\pi^2}{30}\left(T_{in}^4 - T_{out}^4\right)\left(\frac{k_B}{\hbar c}\right)^4\cdot V - b\frac{\zeta(3)}{4\pi}\left(T_{in}^3 + T_{out}^3\right)\left(\frac{k_B}{\hbar c}\right)^3\cdot S\right) \tag{31}$$

This result is worth commenting. First, as it should be by design of our weighting procedure, all factors depending on geometry of the boxes have been cancelled. The expansion goes in parameter $(\hbar c/k_BT)(S/V)$ which is assumed to be small according to condition (20). The first term in the right hand side is nothing but the weight of thermal photon gas.[3] This term is universal and scales as

[3]This physics described by (31) should not be misinterpreted as physics of flying balloons with heated air inside, where key factor is pressure gradient dependence on temperature.

volume. The next-to-leading non-universal term depends on boundary conditions and scales as area. Due to scalar nature of the problem both inside and outside parts of the boundary contribute with the same sign and this term does not vanishes but doubles for equal inside and outside temperatures. The importance of such terms and surface-dependent effects they describe for various metrology problems like precise calibration of thermometers etc was stressed in [15].

The effects discussed above are extremely tiny. It is important, on the other hand, to remember that genuine interplay of gravity and quantum takes place not only at Planck energies, but at normal Earth-like conditions, with extreme weakness as a price to pay. We know a few examples in the history of physics then multiplicity could save the case, e.g. huge value of $N_{Avogadro}$ allows to reach limits on the lifetime of a proton far exceeding the age of the Universe, luminosity in collider physics is a key to observe strongly suppressed processes etc). In this respect, the problem to find proper (and experimentally reasonable) "multiplicity factor" for weak gravity of quantum states/energies does not look hopeless. Indeed, there are suggestions to use advanced techniques from gravitational wave search [20, 21] to nail such effects. There can surely be surprises prepared by the Lord for us here.

Acknowledgements

One of the authors (V.Sh.) thanks Yu. Gusev for discussions. The work was partly supported by the RFBR Grant 14-22-03030.

References

[1] K. A. Milton, K. V. Shajesh, S. A. Fulling and P. Parashar, Phys. Rev. D **89** 064027 (2014)
[2] K. V. Shajesh, K. A. Milton, P. Parashar and J. A. Wagner, J. Phys. A **41** 164058 (2008)
[3] S. A. Fulling, K. A. Milton, P. Parashar, A. Romeo, K. V. Shajesh and J. Wagner, Phys. Rev. D **76** 025004 (2007)
[4] K. A. Milton, P. Parashar, K. V. Shajesh and J. Wagner, J. Phys. A **40** 10935 (2007)
[5] K. A. Milton, Lect. Notes Phys. **834** 39 (2011)
[6] G. Bimonte, E. Calloni, G. Esposito, G. M. Napolitano and L. Rosa, arXiv:0904.4864 [hep-th].
[7] G. Bimonte, E. Calloni, G. Esposito and L. Rosa, Phys. Rev. D **76** 025008 (2007)
[8] G. Bimonte, E. Calloni, G. Esposito and L. Rosa, Phys. Rev. D **74** 085011 (2006); [Phys. Rev. D **75** 049904 (2007)] [Phys. Rev. D **75** 089901 (2007)] [Phys. Rev. D **77** 109903 (2008)]
[9] L. S. Brown and G. J. Maclay, Phys. Rev. **184** 1272 (1969).
[10] T.Padmanabhan, *Gravitation: Foundations and Fronties,* Cambridge University Press, 2010.
[11] V. Sopova and L. H. Ford, Phys. Rev. D **66** 045026 (2002)
[12] D. V. Vassilevich, Phys. Rept. **388** 279 (2003)
[13] Y. Gusev and A. Zelnikov, Class. Quant. Grav. **15** 13 (1998)
[14] Y. V. Gusev, Russ. J. Math. Phys. **22** 9 (2015)
[15] Y.V.Gusev, Russ. J. Math. Phys. **23** 1 (2016)
[16] B. S. DeWitt., Phys.Rev. **162** 1195 (1967).
[17] L. Dolan and R. Jackiw., Phys. Rev. **D9** 3320 (1974).
[18] J. S. Dowker and G. Kennedy., J. Phys. A: Math. Gen. **11** 895 (1978).
[19] I. G. Avramidi., Physics of Atomic Nuclei **56** 138 (1993).
[20] E. Calloni *et al.*, Phys. Rev. D **90** 022002 (2014)
[21] E. Calloni *et al.*, arXiv:1511.04269 [gr-qc].

EPJ Web of Conferences **126**, 02030 (2016) DOI: 10.1051/epjconf/201612602030
ICNFP 2015

Chaotic vortical flows and their manifestations

M. Baznat[1,a], K. Gudima[1,b], A. Sorin[1,c,d], and O. Teryaev[1,e]

[1]*Joint Institute for Nuclear Research, 141980 Dubna, Russia*

Abstract. We study vorticity and hydrodynamic helicity in semi-peripheral heavy-ion collisions using the kinetic model of Quark-Gluon Strings. The angular momentum, which is a source of P-odd observables, is preserved with a good accuracy. We observe formation of the specific toroidal structures of the vorticity field. Their existence, accompanied by the strange chemical potential, is mirrored in the polarization of hyperons of the percent order.

1 Introduction

The local violation [1] of discrete symmetries in strongly interacting QCD matter is now under intensive theoretical and experimental investigations. The Chiral Magnetic Effect (CME) uses the (C)P-violating (electro)magnetic field emerging in heavy ion collisions in order to probe the (C)P-odd effects in QCD matter.

There is an even more interesting counterpart of this effect, Chiral Vortical Effect (CVE)[2] due to coupling to P-odd medium vorticity leading to the induced electromagnetic and [3] all conserved-charge currents, in particular the baryonic ones.

The zeroth component of axial current and correspondent axial charge are related to hydrodynamical helicity

$$H \equiv \int dV(\boldsymbol{v} \cdot \boldsymbol{w}),$$

being the projection of velocity $\boldsymbol{v}$ to vorticity $\boldsymbol{w} = curl\boldsymbol{v}$. This quantity is manifesting the recently discovered [4] and confirmed [5] phenomenon of separation, i.e. mirror behavior (same magnitude but different sign) in the half-spaces separated by the reaction plane

Another important P-odd observable is the baryon polarizatiion. The mechanisms analogous to CVE (known as axial vortical effect, see [6] and Ref. therein) leads to induced axial current of strange quarks which may be converted to polarization of Λ-hyperons [3, 4, 7]. Another mechanism of this polarization is provided [8] by so-called thermal vorticity in the hydrodynamical approach.

The noncentral heavy ion collisions could naturally generate a rotation (global or local, both related to vorticity) with an angular velocity normal to the reaction plane, which is their generic qualitative feature. It is naturally to expect that angular momentum conservation plays an essential role in

[a]e-mail: baznat@theor.jinr.ru
[b]e-mail: gudima@cc.acad.md
[c]*speaker*
[d]e-mail: sorin@theor.jinr.ru
[e]e-mail: teryaev@theor.jinr.ru

DOI: 10.1051/epjconf/201612602030

the defining the quantitative properties of vortical effects. At the same time, it remains to be studied to which extent the particles carrying the main part of angular momentum participate in the collisions.

We address these problem by performing the extensive numerical simulations [4, 9]. We explore the distribution of angular momentum and find that the role of participant nucleons is relatively small, albeit noticeable. We study in some detail the structure of vorticity field and compare different approaches to polarization calculation. We found the peculiar toroidal "tire-like" structure manifesting themselves in the polarization of hyperons. Very different approaches to polarization calculation lead to surprisingly similar results.

2 Modelling velocity, vorticity and helicity in kinetic model

One of the first models designed to describe the dynamics of energetic heavy-ion collisions was the intra-nuclear cascade model developed in Dubna [10] which is based on the Monte-Carlo solution of a set of the Boltzmann-Uehling-Uhlenbeck relativistic kinetic equations with the collision terms, including cascade-cascade interactions. For particle energies below 1 GeV it is sufficient to consider only nucleons, pions and deltas. The model includes a proper description of pion and baryon dynamics for particle production and absorption processes. In the original version the nuclear potential is treated dynamically, i.e., for the initial state it is determined using the Thomas-Fermi approximation, but later on its depth is changed according to the number of knocked-out nucleons. This allows one to account for nuclear binding. The Pauli principle is implemented by introducing a Fermi distribution of nucleon momenta as well as a Pauli blocking factors for scattered nucleons.

At energies higher than about 10 GeV, the Quark-Gluon String Model (QGSM) is used to describe elementary hadron collisions [11, 12]. This model is based on the $1/N_c$ expansion of the amplitude for binary processes where N_c is the number of quark colours. Different terms of the $1/N_c$ expansion correspond to different diagrams which are classified according to their topological properties. Every diagram defines how many strings are created in a hadronic collision and which quark-antiquark or quark-diquark pairs form these strings. The relative contributions of different diagrams can be estimated within Regge theory, and all QGSM parameters for hadron-hadron collisions were fixed from the analysis of experimental data. The break-up of strings via creation of quark-antiquark and diquark-antidiquark pairs is described by the Field-Feynman method [13], using phenomenological functions for the fragmentation of quarks, antiquarks and diquarks into hadrons. The modified non-Markovian relativistic kinetic equation, having a structure close to the Boltzmann-Uehling-Uhlenbeck kinetic equation, but accounting for the finite formation time of newly created hadrons, is used for simulations of relativistic nuclear collisions. One should note that QGSM considers the two lowest SU(3) multiplets in mesonic, baryonic and antibaryonic sectors, so interactions between almost 70 hadron species are treated on the same footing. This is a great advantage of this approach which is important for the proper evaluation of the hadron abundances and characteristics of the excited residual nuclei. The energy extremes were bridged by the QGSM extension downward in the beam energy [14].

For investigation of dynamical formation of velocity $\boldsymbol{v}$ and vorticity ω ($\equiv rot\, \boldsymbol{v}$) fields in relativistic heavy ion collision the coordinate space was divided into $50 \times 50 \times 100$ cells of volume $dxdydz$ with $dx = dy = 0.6fm$, $dz = 0.6/\gamma$ fm, where γ is the gamma factor of equal velocity system of collision. In this reference system the total momentum and total energy of the produced particles were calculated in all cells for each of fixed 25 moments of time t covering the interval of 10 fm/c.

The results were averaged for about 10000 heavy ion collisions with identical initial conditions. The spectator nucleons of projectile or target ions, which at given time momentum do not undergone any individual collision, were included in evaluation of velocity. The velocity field in the given cell was defined by the following double sum over the particles in the cell and over the all simulated collisions:

$$v(x, y, z, t) = \frac{\sum_i \sum_j \boldsymbol{P}_{ij}}{\sum_i \sum_j E_{ij}} \tag{1}$$

where $\boldsymbol{P}_{ij}$ and E_{ij} are the momentum and energy of particle i in the collision j, respectively. The vorticity was calculated using the discrete partial derivatives.

We paid a special attention to the pseudoscalar characteristics of the vorticity, that is the mentioned hydrodynamical helicity $H \equiv \int dV(\boldsymbol{v} \cdot \boldsymbol{w})$ which is related to a number of interesting phenomena in hydrodynamics and plasma physics, such as the turbulent dynamo (providing possibly additional mechanism of magnetic filed generation on the later stages of heavy-ion collisions) and Lagrangian chaos. It might be compared to the analog of topological charge $Q = \int d^3x J^0(x)$ where the current $J^\mu = \epsilon^{\mu\nu\rho\gamma} u_\nu \partial_\rho u_\gamma$ (as usual, the four-velocity $u_\nu \equiv \gamma\,(1, \boldsymbol{v}\,))$ contributes to the hydrodynamical anomaly [15] and the polarization of hyperons [3, 7]. The calculation of the topological charge which is the correct relativistic generalization of the hydrodynamical helicity leads to the extra factor γ^2 in the integrand. Still as the helicity itself is a more traditional quantity, we use it for the numerical calculations.

3 Angular momentum conservation in the kinetic model

The natural source of the P-odd observables in heavy-ion collisions is the pseudovector of angular momentum. The question immediately emerges whether it is conserved in the course of evolution governed by QGSM discussed in the previous section.. To check this we calculated the angular momentum at various moments of collision taking into account both the contributions of participants and spectators. The general structure of velocity field follows the "small bang" pattern qualitatively discussed earlier and which may be quantified by the following (see fig. 1) velocity dependence allowing to extract the "little Hubble" constant.

The averaged (over mentioned 10000 events) time dependencies of angular momenta in $Au + Au$ collisions at $\sqrt{s_{NN}} = 5\ Gev$ with the impact parameter $8\ fm$ equal to the (transverse) radius of the nuclei are presented at figure (2). The fireball angular momentum and its ratio to the total angular momentum (appeared to be conserved in the applied model) are shown. Also, the behaviour of the absolute value of helicity changing in the accordance to angular momentum is shown. We found that the participants carry about 20% of angular momentum and that the total angular momentum of participants and spectators is conserved with a rather good accuracy and that helicity is changing in accordance with participants angular momentum (up to $t \sim 10\ fm/c$ and decreasing afterwards) so that at least part of angular momentum is pumped into vorticity.

4 Large-scale structures of vorticity fields

We study the qualitative structure of velocity and vorticity fields. Our observation is that while velocity field represent the "small bang" picture, vorticity field form the relative thin toroidal "tire-like" structures. They emerge for the particular orientation of reaction plane and survive after the average over various orientations

5 Hyperon polarization

We compare the two rather distinct methods of determining the hyperon polarization. The first corresponds to earlier suggested[3] and explored [4] relation to induced axial current while the second one follows the procedure [8, 16] based on thermal vorticity.

DOI: 10.1051/epjconf/201612602030

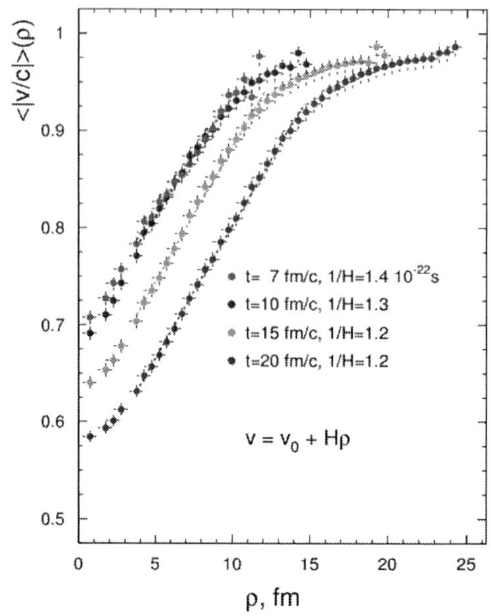

Figure 1: "Little Hubble" law

5.1 Helicity and polarization of hyperons

The polarization of hyperons is based on the discussed effect of helicity separation which is confirmed by the calculation in the hybrid UrQMD model manifests very similar behavior [17]. This is not surprising as the helicity is in fact generated at the hydrodynamical stage of the model, while the transition from kinetic to hydrodynamical stage should be performed similarly to our Eq. (1).

This effect might be qualitatively explained, if the perpendicular components of velocities (which are selected to have different signs) and the corresponding vorticities (assumed to have the same signs) provide the dominant contribution to the scalar product in the helicity definition. However, the numerical analysis [4] showed that the longitudinal components along the beam directions (z-axis) provide even larger contribution to the helicity than contributions from the transverse direction (y-axis). Note that comparable values of z and y components to helicity is due to larger $z-$components of velocity and $y-$ components of vorticity

So, such qualitative picture is oversimplified, but still provides a correct sign convention for the helicity-separation effect.

The energy dependence of helicity separation effect appears to be very weak, with the maximal value achieved around the NICA energy range.

The hydrodynamical helicity should give rise to the polarization of $\Lambda-$hyperons with the sign differing for the particles with "up" and "down" $y-$components of their momenta, so that the hyperons acquire the helicity in the course of their motion transverse to the reaction plane. As we already

DOI: 10.1051/epjconf/201612602030

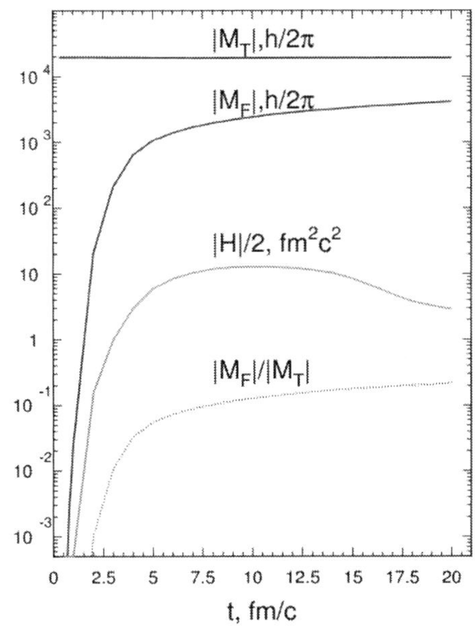

Figure 2: Angular momentum and helicity evolution

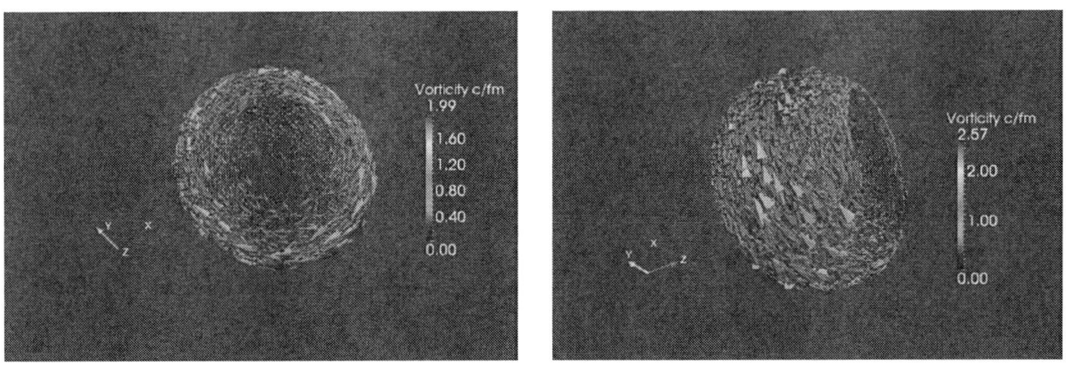

(a) (b)

Figure 3: (a) Vorticity field for the fixed orientation of reaction plane (b) Vorticity field for the orientation of reaction plane averaged in the full 2π region suggested earlier [3], the effect is pronounced at moderate (NICA) energies due to large (strange) chemical potential. the current investigation shows that, luckily, the helicity at these energies is also noticeable.

For semi-quantitative estimate of this effect one may use the average strange chiral charge produced by the zeroth component of the respective current

$$Q_5^s = \frac{N_c}{2\pi^2}\int d^3x\mu^2(x)\epsilon^{ijk}u_i\partial_j u_k = \frac{<\mu^2> N_c H}{2\pi^2},$$

where we use the mean value theorem to extract the value of the square of strange chemical potential at some point inside the integration region and get the helicity from the remaining integral. Assuming that the strange chirality is carried by the Λ hyperons whose average number in each evemt is $< N_\Lambda >$ one get the estimate for its average polarization as

$$< P_\Lambda > \sim \frac{<\mu^2> N_c H}{2\pi^2 < N_\Lambda >}$$

For numerical estimate at NICA energies, we take (see Fig. 3) $H = 30fm^2 (c = 1)$ and, as a typical values, $<\mu^2> = 900MeV^2$, $< N_\Lambda > = 15$ to get $< P_\Lambda > \sim 0.8\%$. This value is not large, but does not exclude the opportunity to measure the effect. Note that it is indirectly supporting the actual calculations of helicity as the obtained expression respects the density matrix positivity [20] limit $P_\Lambda \leq 1$. Should the helicity be much larger (resulting in the appearance of much larger value of dimensionless ratio), much larger number of hyperons production (and/or K^*−mesons) would be required to preserve the density matrix positivity. This is an example of the situation when the spin-dependent effects may be used [20] to bound the spin-averaged cross-sections from below.

Let us pass to more detailed calculations of polarization taking into account the spatial distribution of chemical potential and the kinematics of medium and produced hyperons.

$$Q_5^s = \frac{N_c}{2\pi^2}\int d^3x\mu^2(x)\gamma^2\epsilon^{ijk}u_i\partial_j u_k. \tag{2}$$

In [4] we explored the mean-value theorem, while here the spatial variation of strange chemical potential μ is taken into account. To do so, the description of kinetic distribution functions by the correspondent equilibrium equation was performed, providing the matching of kinetic and thermodynamical descriptions. As a result, the time dependence of the distribution of strange chemical potential takes the form represented at Fig. 4 The average polarization can be estimated by dividing Q_5^s (2) by the number of Λ's, assuming that the pseudovector of axial current is proportional to the pseudovector of polarization, $Q_5^s \sim < \Pi_0^{\Lambda,lab} >$. Selecting the axial charge related to the particles in the definite rapidity or transverse momentum interval, the respective dependence of polarization may be also obtained.

As the axial charge should be related to the zeroth component of hyperon polarization in laboratory frame Π_0^{lab}, the transformation to hyperon rest frame should be performed. Taking into account that polarization pseudovector should be directed along y axis (as it has to be collinear to $\boldsymbol{M}$ pseudovector), one get

$$\Pi_0^{\Lambda,lab} = \frac{\Pi_0^\Lambda p_y}{M_\Lambda}, \tag{3}$$

so that the rest frame polarization can be obtained as

$$< \Pi_0^\Lambda > = Q_5^s < \frac{M_\Lambda}{N_\Lambda p_y} > . \tag{4}$$

DOI: 10.1051/epjconf/201612602030

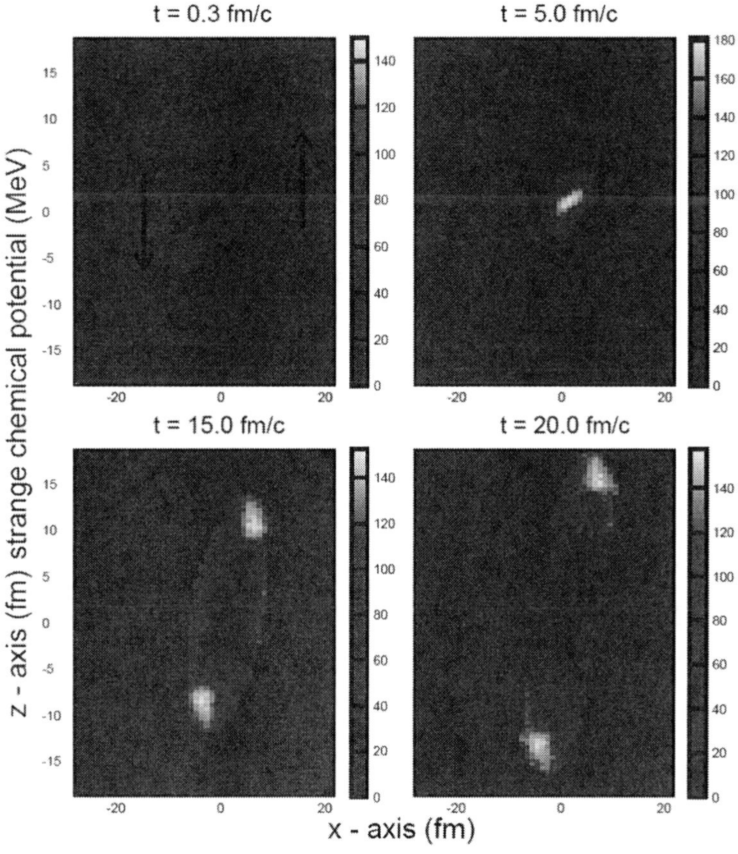

Figure 4: The time dependence of strange chemical potential.

The possible violation of positivity ($|\Pi_\Lambda| \geq 1$) or even the divergence is due to the fact, that hyperons with zero y component of the momentum should not have the zeroth component of polarization and therefore should not contribute to Q_5^s. To avoid this problem one may instead attribute the factor p_y/M to each hyperon in the denominator of (4). Nevertheless the comparison of various approaches shows the similar scale and rapidity dependence of polarization.

5.2 Polarization and thermal vorticity

Another approach to polarization is based on the so-called thermal vorticity [8]. To provide the comparison we calculated (see Fig. 6) the thermal vorticity field and respective polarization. While a

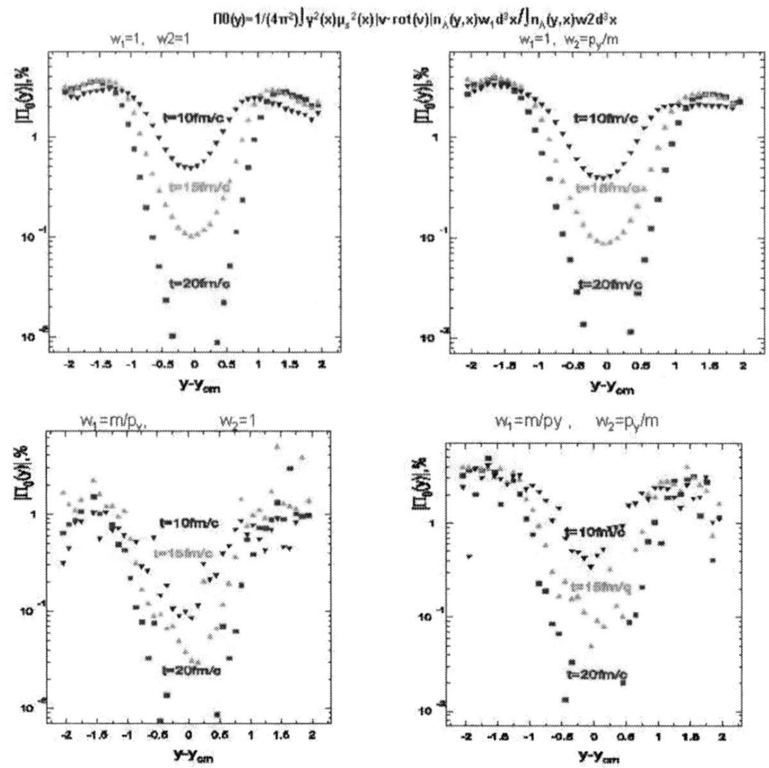

Figure 5: The rapidity dependence of polarization in helicity-based approach.

scale of the polarization in the thermal vorticity-based approach is several times larger, its rapidity dependence, surprisingly enough, appear to be similar in these rather distinct approaches.

6 Conclusions and Outlook

We investigated vorticity and hydrodynamical helicity in noncentral heavy-ion collisions in the framework of the kinetic quark-gluon string model. We have observed that the vorticity is predominantly localized in a relatively thin layer ($2 \div 3$ fm) on the boundary between the participant and spectator nucleons. This might be qualitatively understood in the spirit of the core-corona type models [18, 19].

Thus, the gradients of the velocities in the region occupied by the participants are small due to the compensation of momenta between the target and projectile particles in the c.m. frame. As the result the vorticity is substantial only in the thin transition layer between the participant (i.e., core) and the spectator (i.e., corona) regions. We found the novel effect of the helicity separation in heavy-ion collisions when it has the different signs below and above of the reaction plane. We have investigated its dependence on the type of nuclei and collision energy and observed that it is maximal in the NICA energy range. We have also calculated the degree of alignment of the velocity and vorticity which is maximal for the Beltrami flows (manifesting elegant symmetry structures [21]) whose relativistic generalization is currently under investigation [22].

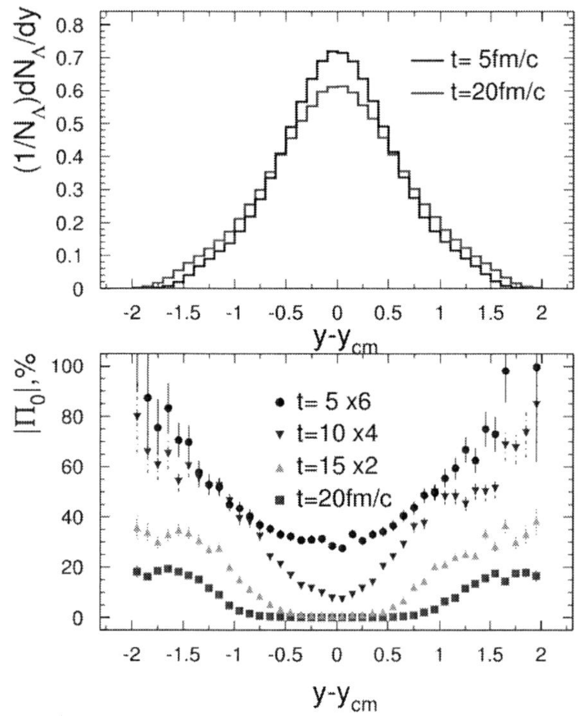

Figure 6: The rapidity dependence of the multiplicity and polarization in the thermal vorticity-based approach. The magnitudes of polarization at different times are rescaled by the identified factors.

We used the obtained values of helicity for estimates of Λ hyperons polarization in heavy-ion collisions at NICA energy range due to earlier suggested [3] mechanism. The resulting polarization is bout 1% and may be studied experimentally. Of course, its more detailed theoretical investigations are required.

In particular, the discovery of extra T^2 term [23] raises again the question why polarization was not observed at RHIC. Here one may refer to the (exponential) dilution of polarization by temperature effects (similar to what happens at much larger scale in polarized targets) although this problem certainly requires further investigation.

References

[1] K. Fukushima, D.E. Kharzeev and H.J. Warringa, Phys. Rev. D **78**, 074033 (2008)
[2] D. Kharzeev and A. Zhitnitsky, Nucl. Phys. A **797**, 67 (2007)

[3] O. Rogachevsky, A. Sorin and O. Teryaev, Phys. Rev. C **82**, 054910 (2010) [arXiv:1006.1331 [hep-ph]].

[4] M. Baznat, K. Gudima, A. Sorin and O. Teryaev, Phys. Rev. C **88**, 061901 (2013) [arXiv:1301.7003 [nucl-th]].

[5] O. Teryaev and R. Usubov, arXiv:1406.4451 [nucl-th].

[6] T. Kalaydzhyan, Phys. Rev. D **89**, no. 10, 105012 (2014) [arXiv:1403.1256 [hep-th]].

[7] J. -H. Gao, Z. -T. Liang, S. Pu, Q. Wang and X. -N. Wang, Phys. Rev. Lett. **109**, 232301 (2012)

[8] F. Becattini, L. Csernai and D. J. Wang, Phys. Rev. C **88**, no. 3, 034905 (2013) [arXiv:1304.4427 [nucl-th]].

[9] M. I. Baznat, K. K. Gudima, A. S. Sorin and O. V. Teryaev, arXiv:1507.04652 [nucl-th].

[10] V.D. Toneev, K.K. Gudima, Nucl. Phys. A **400**, 173c (1983)

[11] V.D. Toneev, N.S. Amelin, K.K. Gudima, S.Yu. Sivoklokov, Nucl. Phys. A **519**, 463c (1990).

[12] N.S. Amelin, E.F. Staubo, L.S. Csernai et al., Phys.Rev. C **44**, 1541 (1991).

[13] R.D. Field, R.P. Feynman, Nucl. Phys. B **136**, 1 (1978).

[14] N.S. Amelin, K.K. Gudima, S.Yu. Sivoklokov, V.D. Toneev, Sov. J. Nucl. Phys. **52**, 272 (1990).

[15] D.T. Son and P. Surowka, Phys. Rev. Lett. **103**, 191601 (2009) [arXiv:0906.5044 [hep-th]].

[16] L. P. Csernai, D. D. Strottman and C. .Anderlik, Phys. Rev. C **85** (2012) 054901 [arXiv:1112.4287 [nucl-th]].

[17] J. Steinheimer, private communication.

[18] J. Aichelin and K. Werner, aXiv:0810.4465[nucl-th]

[19] J. Steinheimer and M. Bleicher, arXiv:1104.3981[nucl-th]

[20] X. Artru, M. Elchikh, J. -M. Richard, J. Soffer and O. V. Teryaev, Phys. Rept. **470**, 1 (2009)

[21] P. Fre and A. S. Sorin, Phys. Part. Nucl. **46**, no. 4, 497 (2015) [arXiv:1501.04604 [math-ph]].

[22] A.S. Sorin, O.V. Teryaev, work in progress.

[23] K. Landsteiner, E. Megias, L. Melgar and F. Pena-Benitez, JHEP **1109** (2011) 121 [arXiv:1107.0368 [hep-th]].

A low-energy determination of α_s at three loops

Antonio Vairo[1,a]

[1] *Physik-Department, Technische Universität München, James-Franck-Str. 1, 85748 Garching, Germany*

Abstract. We review one of the most accurate low-energy determinations of α_s. Comparing at short distances the QCD static energy at three loops and resummation of the next-to-next-to leading logarithms with its determination in 2+1-flavor lattice QCD, we obtain $\alpha_s(1.5 \text{ GeV}) = 0.336^{+0.012}_{-0.008}$, which corresponds to $\alpha_s(M_Z) = 0.1166^{+0.0012}_{-0.0008}$. We discuss future perspectives.

1 α_s in 2015

For many years the average of the strong coupling constant, α_s, provided by the Particle Data Group (PDG) (the latest printed edition is [1]) has shown a rather stable central value and a steady decrease in the associated error, see figure 1. This satisfactory situation has however been challenged in the last years (see for example the summary tables in [2–4] and the recent collection [5]): precise determinations pointing often towards a lower value of α_s have appeared, traditionally larger and precise determinations from τ decay have become smaller and less precise [6], accurate determinations from lattice QCD have been questioned. The 2016 PDG average will reflect this changed situation and, for the first time in twenty years, the α_s world average will see an increase in the error [7].

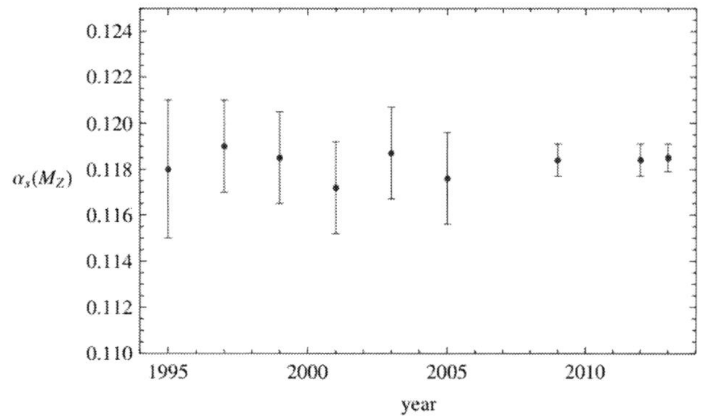

Figure 1. List of PDG averages for $\alpha_s(M_Z)$ from 1995 to 2013, source [8].

[a] e-mail: antonio.vairo@ph.tum.de

DOI: 10.1051/epjconf/201612602031

This change of scenario makes it even more important to look for accurate determinations of α_s. The extraction of α_s that follows from comparing perturbative QCD with short distance lattice computations of the QCD static energy is one of these determinations. Perturbatively this quantity is known at three loops and next-to-next-to-next-to leading logarithmic (NNNLL) accuracy. On the lattice side, the static Wilson loop is one of the most accurately known quantities that has been computed since the establishment of lattice QCD. Nowadays computations are with quarks at or very close to their physical masses so that the lattice spacing can be matched to a physical scale, eventually leading to a determination of the physical (unquenched) α_s. The strong coupling constant extracted in this way is typically at a low energy scale because the lattice cannot explore too short distances. In fact the method provides one of the most precise low-energy determinations of α_s. In this sense the determination of α_s from the QCD static energy not only provides a competitive but also a complementary determination with respect to high-energy determinations. Obviously there is also a value *per se* in having a low-energy determination of α_s, for it adds to our understanding of low-energy QCD and non-trivially constraints our modeling of it. Here we will review our latest determination of α_s from the QCD static energy [3], which improves our previous determination [9]. An earlier quenched analysis can be found in [10]. Finally, as we will discuss briefly in the outlook, although the extraction of α_s from the static Wilson loop provides already today an accurate value, there is room for improvement in several directions.

2 QCD static energy in perturbation theory

The QCD static energy, $E_0(r)$, is defined (in Minkowski spacetime) as

$$E_0(r) = \lim_{T\to\infty} \frac{i}{T} \ln \left\langle \mathrm{Tr}\,\mathrm{P}\exp\left\{ ig \oint_{r\times T} dz^\mu A_\mu \right\}\right\rangle, \tag{1}$$

where the integral is over a rectangle of spatial length r, the distance between the static quark and the static antiquark, and time length T; $\langle \dots \rangle$ stands for the path integral over the gauge fields A_μ and the light quark fields, P is the path-ordering operator of the color matrices and g is the SU(3) gauge coupling ($\alpha_s = g^2/(4\pi)$). The above definition of $E_0(r)$ is valid at any distance r. In the short range, $r\Lambda_{\rm QCD} \ll 1$ for which $\alpha_s(1/r) \ll 1$, $E_0(r)$ may be computed in perturbative QCD and expressed as a series in α_s (computed at a typical scale of order $1/r$):

$$E_0(r) = \Lambda_s - \frac{4\alpha_s}{3r}(1 + \dots), \tag{2}$$

where Λ_s is a constant that accounts for the normalization of the static energy and the dots stand for higher-order terms. The expansion of $E_0(r)$ in powers of α_s is known up to three loops. At three loops, a contribution proportional to $\ln\alpha_s$ appears for the first time. This three-loop logarithm has been computed in [11, 12]. The complete three-loop contribution can be derived from [13, 14]. The leading logarithms have been resummed to all orders in [15], providing, among others, also the four-loop contribution proportional to $\ln^2\alpha_s$. The four-loop contribution proportional to $\ln\alpha_s$ has been computed in [16]. Next-to-leading logarithms have been resummed to all orders in [17], which represents the status of the art of the QCD static energy in perturbation theory. A concise but complete summary with all relevant formulas can be found in [18]. $E_0(r)$ is, therefore, one of the best known quantities in perturbative QCD.

The appearance of $\ln\alpha_s$ terms starting from three loops signals the cancellation of contributions coming from the different energy scales $1/r$ and α_s/r:

$$\ln\alpha_s = \ln\frac{\mu}{1/r} + \ln\frac{\alpha_s/r}{\mu}, \tag{3}$$

where μ is a factorization scale. Indeed in the short range $E_0(r)$ is characterized by a hierarchy of well-separated scales, see figure 2.

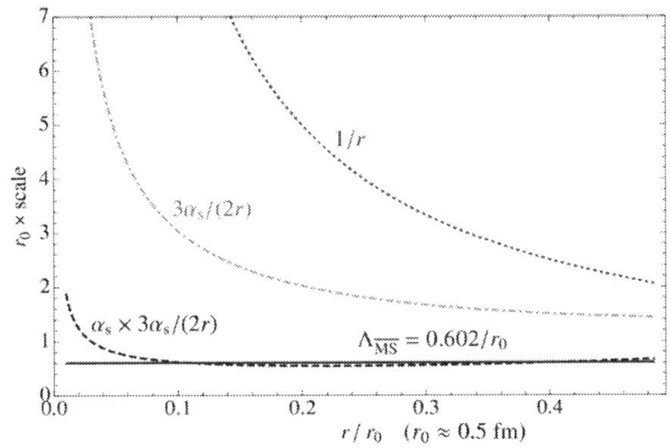

Figure 2. Energy scales for the short-range static energy in the quenched case, from [17].

The separation of contributions coming from gluons scaling like $1/r$ from contributions coming from gluons scaling like α_s/r leads to the factorization formula [11, 12]:

$$E_0(r) \;=\; \Lambda_s \;+\; V_s(r,\mu) - i\frac{V_A^2}{3}\int_0^\infty dt\, e^{-it(V_o-V_s)}\,\langle \mathrm{Tr}\,\{\mathbf{r}\cdot g\mathbf{E}(t,\mathbf{0})\,\mathbf{r}\cdot g\mathbf{E}(0,\mathbf{0})\}\rangle(\mu) + \ldots\,, \quad (4)$$

where $V_s(r,\mu) = -4\alpha_s/(3r) + \ldots$ is the color-singlet static potential, $V_o(r,\mu) = \alpha_s/(6r) + \ldots$ is the color-octet static potential, V_A a Wilson coefficient describing the chromoelectric dipole coupling, and $\mathbf{E}$ the chromoelectric field. The color-singlet static potential encodes the contributions from the scale $1/r$, while the low-energy contributions are in the term proportional to the two chromoelectric dipoles.

All the terms in (4) are well known in perturbation theory. The three-loop expression of the color-singlet static potential has been computed in [13, 14]. The Wilson coefficient V_A is 1 up to corrections of order α_s^2 [16]. The chromoelectric field correlator, $\langle \mathrm{Tr}\,\{g\mathbf{E}(t,\mathbf{0})\cdot g\mathbf{E}(0,\mathbf{0})\}\rangle$ (Wilson lines connecting the fields are understood), is known up to two loops [19]. The color-octet static potential is known up to three loops [20].

An advantage of equation (4) is that it allows for an efficient resummation of the logarithmic contributions to the static potentials and eventually to the static energy. This is achieved by evaluating the anomalous dimensions of the potentials through the computation of the ultraviolet divergences in the relevant integrals proportional to the two chromoelectric dipoles and then by solving the renormalization group equations for the static potentials [15, 17]. This is how contributions to the static energy of the type $\alpha_s^{3+n}\ln^n \alpha_s/r$ and $\alpha_s^{4+n}\ln^n \alpha_s/r$ have been computed for all n.

The perturbative expansion of V_s is affected by a renormalon ambiguity of order Λ_{QCD}. The ambiguity reflects in the poor convergence of the perturbative series.

A first method to cure the poor convergence of the perturbative series of V_s consists in subtracting a (constant) series in α_s from V_s and reabsorb it in a redefinition of the normalization constant Λ_s [21]. This is the strategy we followed in the earlier analysis [9].

A second possibility consists in considering the force

$$F(r,\nu) = \frac{d}{dr}E_0(r). \quad (5)$$

Table 1. Lattice couplings, spacings and volumes used for the extraction of α_s.

β	7.373	7.596	7.825
r_1/a	5.172(34)	6.336(56)	7.690(58)
Volume	$48^3 \times 64$	64^4	64^4

It does not depend on Λ_s and is free from the renormalon of order $\Lambda_{\rm QCD}$. In (5) we have written explicitly the dependence of F on the renormalization scale for the coupling, ν, which appears when computing F order by order in $\alpha_s(\nu)$. Once integrated upon the distance, the force gives back the static energy

$$E_0(r) = \int_{r_*}^{r} dr'\, F(r', 1/r'), \tag{6}$$

up to an irrelevant constant determined by the arbitrary distance r_*, which can be reabsorbed in the overall normalization when comparing with lattice data. Notice that the integration in (6) can be done (numerically) keeping the strong-coupling constant running at a natural scale of the order of the inverse of the distance. This scheme does not generate therefore potentially large logarithms of the type $\ln \nu r$. Such logarithms are typically generated within the first method. In the newer analysis of [3], we have adopted this second method.

3 Numerical analysis

In order to determine α_s, we match the three-loop expression of $E_0(r)$ in the short range with the determination of the same quantity in 2+1-flavor lattice QCD obtained from the tree-level improved gauge action and the Highly-Improved Staggered Quark (HISQ) action by the HotQCD collaboration [22]. The strange quark mass, m_s, has been fixed to its physical value, while the light quark masses, m_l, have been set equal to $m_s/20$. This corresponds to a pion mass of about 160 MeV in the continuum limit. Several lattice spacings and volumes have been considered, but for our determination of α_s we restricted to the three finest spacings shown in table 1. The largest gauge coupling, $\beta = 7.825$, corresponds to a lattice spacing of $a = 0.041$ fm. The lattice spacing has been fixed using the r_1 scale defined as $r^2 dE_0(r)/dr|_{r=r_1} = 1.0$. In turn, the scale r_1 could be fixed from the pion decay constant [23]: $r_1 = 0.3106 \pm 0.0017$ fm.

Because of the many lattice points available at short distance (see figure 3) we can fit α_s (or equivalently $\Lambda_{\overline{\rm MS}}$) for each lattice spacing separately and eventually average these determinations. This is different from our earlier analysis in [9], where, because of the less data points, we had to fit α_s on data points from all lattice spacings at the same time. Such a procedure introduced a sizeable extra error source due to the different lattice normalizations. This error is absent in the newer analysis in [3].

We use the following procedure: *(i)* for each lattice spacing we fit $\Lambda_{\overline{\rm MS}}$ at different orders of perturbative accuracy; the overall normalization constant that matches the lattice data with the perturbative expression of $E_0(r)$ is typically fixed on the 7th, 8th or 9th lattice point called $N_{\rm ref}$ in some of the following figures (this choice is dictated by the fact that lattice points at short distance are expected to be better described by perturbation theory, but at too short distance may be affected by lattice artefacts); *(ii)* we repeat the fits for each of the following distance ranges: $r < 0.75r_1$, $r < 0.7r_1$, $r < 0.65r_1$, $r < 0.6r_1$, $r < 0.55r_1$, $r < 0.5r_1$, and $r < 0.45r_1$, and use only the ranges where the reduced χ^2 either

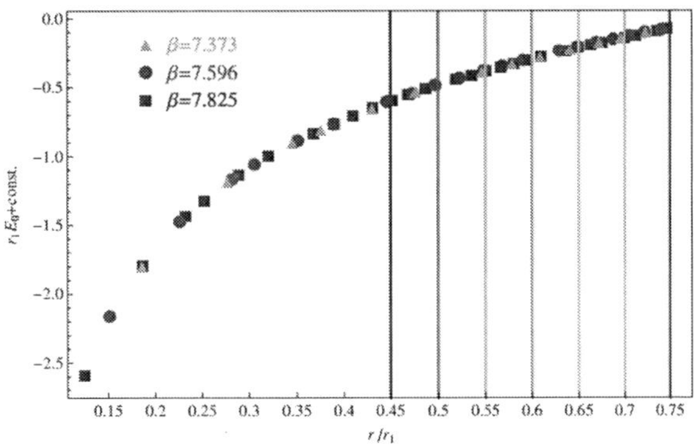

Figure 3. Data sets used in the analysis of [3]. Lattice data are from [22].

decreases or does not increase by more than one unit or is smaller than 1, when increasing the perturbative order; *(iii)* to estimate the perturbative uncertainty of the result, we repeat the fits by varying the renormalization scale in the perturbative expansion from $\nu = 1/r$ to $\nu = \sqrt{2}/r$ and $\nu = 1/(\sqrt{2}r)$, or by adding/subtracting a term $\pm 4\alpha_s^{n+2}/(3r^2)$ to the n-loop expression of the force; we take the largest uncertainty.

As expected, perturbative QCD describes better the data in the shortest distance ranges. In particular, criterion *(ii)* is fulfilled for data in the range $r < 0.6r_1$ or shorter; to be on the most conservative side, in the final result we will use only fits for $r < 0.5r_1$.

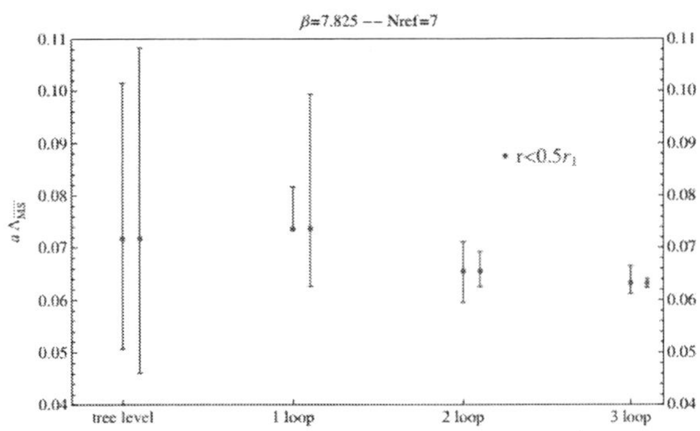

Figure 4. Results for $a\Lambda_{\overline{\rm MS}}$ from the fits to the lattice data at $\beta = 7.825$ for different orders of perturbative accuracy. The first error bar corresponds to the ν variation, whereas the second one to the addition of a generic higher-order term. The static energies have been normalized on the 7th point.

We have performed several checks and estimated several error sources. *(a)* When considering the perturbative errors of *(iii)* we notice that uncertainties due to the higher-order term $\pm 4\alpha_s^{n+2}/(3r^2)$ dominate for the static energy at tree level and one loop, while starting from two loops the dominant source of uncertainties comes from variation of ν. Uncertainties decrease in going to higher orders, and in all cases they fall inside the error bar of the lower-order determination, see figure 4. *(b)* The result is insensitive or very little sensitive to changes in the lattice spacing, in the range of short distances used (see figure 5) and in the point chosen for the normalization with the lattice data. *(c)* We have estimated and added to our determination of $\Lambda_{\overline{\rm MS}}$ a statistical error taken as the variation in

$\Lambda_{\overline{MS}}$ when one allows for the fit to be one χ^2 unit above minimum. Statistical errors are smaller than perturbative errors, but, in particular for the shortest distance ranges, not negligible. *(d)* An analysis done comparing the perturbative expression of the force (5) with the numerical derivative of the lattice data for $E_0(r)$ gives consistent results, but with larger uncertainties associated with the interpolation of the lattice points. *(e)* By repeating the fits adding a monomial term proportional to r^3 and r^2, which could be associated with gluon and quark local condensates, and also a term proportional to r, we could not find evidence for a significant non-perturbative term at short distances and the value of $\Lambda_{\overline{MS}}$ remains unchanged.

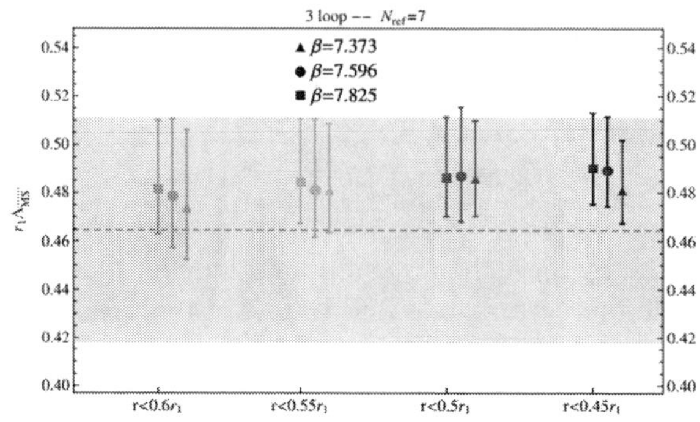

Figure 5. Results for $r_1\Lambda_{\overline{MS}}$ at three-loop accuracy for the different lattice spacings of table 1 and ranges of figure 3. We show the fit for data normalized with respect to the 7th lattice point. Similar analyses with consistent results have been done in [3] also for different normalization points. The band shows our previous result [9].

The analysis revealed that the data are not sensitive to logarithmic corrections of order $\alpha_s^{4+n}\ln^n\alpha_s/r$, moreover these corrections are much smaller than the non-logarithmic contribution of order α_s^4/r, which turns out to be numerically as large as the logarithmic corrections of order $\alpha_s^{3+n}\ln^n\alpha_s/r$, but with opposite sign. Hence we have included in our perturbative expression of the static energy only logarithmic corrections of the type $\alpha_s^{3+n}\ln^n\alpha_s/r$, whereas we have not included logarithmic corrections of the type $\alpha_s^{4+n}\ln^n\alpha_s/r$, although they are known. We chose $\mu = 1.26r_1^{-1} \sim 0.8$ GeV for the factorization scale. Variations of μ only produce small effects on the results.

As shown in figure 5, the newer analysis of [3] is consistent with the older one of [9], but with an error which is roughly half. The reduction in the error can be traced back to two main improvements. First, as already discussed in this section, the newest lattice data provide much more points in the short range so that we can fit for each lattice spacing independently and avoid normalization errors that arise when combining data from different lattices. Second, as discussed at the end of section 2, fitting with the perturbative expression obtained from (6) allows avoiding uncertainties associated with a fixed renormalization scale ν, which is a potential source of large logarithms of the type $\ln \nu r$.

4 Results

The results of the fits of $\Lambda_{\overline{MS}}$ for the different lattice spacings are summarized in table 2. The first error is the perturbative one, the second error is the one from the statistical uncertainties in the fit, and, for the last column, the third error corresponds to the one coming from the conversion from a to r_1. Errors in the right-hand side of the last column are added in quadrature. The final number for $r_1\Lambda_{\overline{MS}}$ with three flavors is obtained as a weighted average of the results for the three different lattices with

DOI: 10.1051/epjconf/201612602031

errors added linearly. Once converted in physical units, it reads

$$\Lambda_{\overline{\rm MS}} = 315^{+18}_{-12}\ \text{MeV}. \tag{7}$$

Table 2. $\Lambda_{\overline{\rm MS}}$ determinations from the three lattice spacings and averaging.

	$a\Lambda_{\overline{\rm MS}}$; $N_{\rm ref}=7$	$a\Lambda_{\overline{\rm MS}}$; $N_{\rm ref}=8$	$a\Lambda_{\overline{\rm MS}}$; $N_{\rm ref}=9$	$a\Lambda_{\overline{\rm MS}}$; range spanned	$r_1\Lambda_{\overline{\rm MS}}$; range spanned
$\beta=7.373$	$0.0957^{+0.0046}_{-0.0028}$ ±0.0017	$0.0957^{+0.0046}_{-0.0028}$ ±0.0017	$0.0957^{+0.0046}_{-0.0028}$ ±0.0017	$0.0957^{+0.0046}_{-0.0028}$ ±0.0017	$0.4949^{+0.0240}_{-0.0144}\pm0.0086\pm0.0025$ $=0.4949^{+0.0256}_{-0.0170}$
$\beta=7.596$	$0.0781^{+0.0046}_{-0.0029}$ ±0.0007	$0.0784^{+0.0043}_{-0.0027}$ ±0.0010	$0.0785^{+0.0046}_{-0.0029}$ ±0.0007	$0.0783^{+0.0048}_{-0.0031}$ ±0.0010	$0.4961^{+0.0303+0.0066}_{-0.0197-0.0061}\pm0.0044$ $=0.4961^{+0.0313}_{-0.0211}$
$\beta=7.825$	$0.0644^{+0.0032}_{-0.0019}$ ±0.0006	$0.0642^{+0.0033}_{-0.0020}$ ±0.0008	$0.0643^{+0.0032}_{-0.0020}$ ±0.0008	$0.0643^{+0.0033}_{-0.0021}$ ±0.0008	$0.4944^{+0.0256}_{-0.0159}\pm0.0065\pm0.0037$ $=0.4944^{+0.0267}_{-0.0175}$
Average					$r_1\Lambda_{\overline{\rm MS}}=0.495^{+0.028}_{-0.018}$

With this value of $\Lambda_{\overline{\rm MS}}$ we can compare the static energy as computed in perturbative QCD at three loop accuracy with resummation of the $\alpha_s^{3+n}\ln^n\alpha_s/r$ terms with lattice data. In figure 6 this comparison is done for the $\beta = 7.825$ lattice data. The comparison shows also visually (the numerical quantitative analysis has been summarized in the previous section) that perturbation theory agrees with lattice data up to about 0.2 fm.

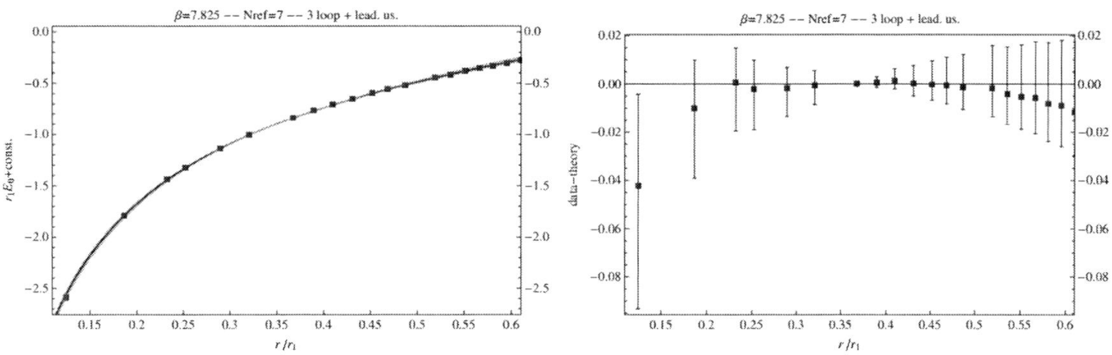

Figure 6. Left panel: comparison of the lattice data for $\beta = 7.825$ with the perturbative expression at three loops plus logarithmic corrections of the type $\alpha_s^{3+n}\ln^n\alpha_s/r$. We take $r_1\Lambda_{\overline{\rm MS}} = 0.495^{+0.028}_{-0.018}$; the grey band reflects the uncertainty in $r_1\Lambda_{\overline{\rm MS}}$. Right panel: result of subtracting the perturbative expression from the lattice data. In both panels the static energy has been normalized on the 7th lattice point.

By converting (7) to α_s at the highest energy scale we used, i.e., 5.1 GeV, and then evolving the value of α_s down to 1.5 GeV with 3 flavors, we obtain

$$\alpha_s(1.5\ \text{GeV}) = 0.336^{+0.012}_{-0.008}, \tag{8}$$

which is one of the most accurate determinations of the strong coupling constant at this low-energy scale. Finally, by evolving α_s at four loops up to the Z mass, M_Z, including the decoupling relations at the quark thresholds ($M_c = 1.6$ GeV and $M_b = 4.7$ GeV), we obtain

$$\alpha_s(M_Z) = 0.1166^{+0.0012}_{-0.0008}. \tag{9}$$

The effects of higher-order terms in the running are negligible with the current accuracies, but may become relevant when and if the precision of $\alpha_s(M_Z)$ will be reduced to the per mil level.

5 Outlook

We have computed the strong coupling constant, α_s, by fitting the static energy, as obtained in perturbation theory with three loop accuracy and with resummation of the $\alpha_s^{3+n} \ln^n \alpha_s/r$ terms, with the short-distance part of the static energy as obtained on physical lattices. The result is a very accurate determination of α_s at low energy. At the Z-mass scale the determination is competitive with the others entering the PDG average. Its central value is lower than the PDG average of 2014 [1], but closer to the expected PDG average of 2016 [7].

There are several ways in which the present determination may be improved with the potential of becoming the most precise determination of α_s. As shown in figure 4, to achieve the present accuracy the inclusion of the three-loop result is crucial but, as discussed at the end of section 3, not all of the presently available perturbative information has been used. More precise lattice data on finer lattices and with more data points at short distances could take advantage of it and improve the determination of α_s. Also, it would be important, in order to reduce possible systematic effects, to perform the same study on Wilson loops computed on different lattices with different actions.

A possible systematic effect is due to the finite lattice spacing. A continuum extrapolation would reduce this effect and allow for a precise determination of the force between static charges along the same lines developed in [24] for the quenched case. As we mentioned in section 3, a study of the force with the present lattice data, although consistent with the result obtained from the static energy, did not provide a comparable accuracy because of finite lattice spacing effects in the calculation of the slope.

Another possibility consists of computing the force directly from the lattice, i.e., not as the slope of the static energy. A matrix element that gives the force, F, between a static quark located in $\mathbf{r}$ and a static antiquark located in $\mathbf{0}$ is [25, 26]

$$F(r) = -\lim_{T\to\infty} \frac{\left\langle \mathrm{Tr}\, \mathrm{P}\, \hat{\mathbf{r}} \cdot g\mathbf{E}(t,\mathbf{r}) \exp\left\{ig \oint_{r\times T} dz^\mu\, A_\mu\right\}\right\rangle}{\left\langle \mathrm{Tr}\, \mathrm{P} \exp\left\{ig \oint_{r\times T} dz^\mu\, A_\mu\right\}\right\rangle}. \tag{10}$$

The chromoelectric field $\mathbf{E}(t,\mathbf{r})$ on the right-hand side is located at the quark line of the Wilson loop. It would be interesting to compute this matrix element and examine if, besides being an alternative way to compute the force, it may also provide a more accurate determination of α_s.

Acknowledgements

I thank Alexei Bazavov, Nora Brambilla, Xavier Garcia i Tormo, Péter Petreczky and Joan Soto for a long going collaboration on the subject. This work is supported in part by DFG and NSFC (CRC110). I also acknowledge financial support from the DFG cluster of excellence "Origin and structure of the universe" (www.universe-cluster.de).

References

[1] K. A. Olive *et al.* [Particle Data Group Collaboration], Chin. Phys. C **38** (2014) 090001.
[2] S. Moch *et al.*, arXiv:1405.4781 [hep-ph].

[3] A. Bazavov, N. Brambilla, X. Garcia i Tormo, P. Petreczky, J. Soto and A. Vairo, Phys. Rev. D **90** (2014) 7, 074038 [arXiv:1407.8437 [hep-ph]].
[4] A. H. Hoang, D. W. Kolodrubetz, V. Mateu and I. W. Stewart, Phys. Rev. D **91** (2015) 9, 094018 [arXiv:1501.04111 [hep-ph]].
[5] D. d'Enterria *et al.*, arXiv:1512.05194 [hep-ph].
[6] D. Boito, M. Golterman, K. Maltman, J. Osborne and S. Peris, Phys. Rev. D **91** (2015) 3, 034003 [arXiv:1410.3528 [hep-ph]].
[7] S. Bethke and G. Dissertori, at this conference.
[8] http://pdg.lbl.gov/2015/html/rpp_archives.html
[9] A. Bazavov, N. Brambilla, X. Garcia i Tormo, P. Petreczky, J. Soto and A. Vairo, Phys. Rev. D **86** (2012) 114031 [arXiv:1205.6155 [hep-ph]].
[10] N. Brambilla, X. Garcia i Tormo, J. Soto and A. Vairo, Phys. Rev. Lett. **105** (2010) 212001 [Phys. Rev. Lett. **108** (2012) 269903] [arXiv:1006.2066 [hep-ph]].
[11] N. Brambilla, A. Pineda, J. Soto and A. Vairo, Phys. Rev. D **60** (1999) 091502 [hep-ph/9903355].
[12] N. Brambilla, A. Pineda, J. Soto and A. Vairo, Nucl. Phys. B **566** (2000) 275 [hep-ph/9907240].
[13] C. Anzai, Y. Kiyo and Y. Sumino, Phys. Rev. Lett. **104** (2010) 112003 [arXiv:0911.4335 [hep-ph]].
[14] A. V. Smirnov, V. A. Smirnov and M. Steinhauser, Phys. Rev. Lett. **104** (2010) 112002 [arXiv:0911.4742 [hep-ph]].
[15] A. Pineda and J. Soto, Phys. Lett. B **495** (2000) 323 [hep-ph/0007197].
[16] N. Brambilla, X. Garcia i Tormo, J. Soto and A. Vairo, Phys. Lett. B **647** (2007) 185 [hep-ph/0610143].
[17] N. Brambilla, A. Vairo, X. Garcia i Tormo and J. Soto, Phys. Rev. D **80** (2009) 034016 [arXiv:0906.1390 [hep-ph]].
[18] X. Garcia i Tormo, Mod. Phys. Lett. A **28** (2013) 1330028 [arXiv:1307.2238].
[19] M. Eidemüller and M. Jamin, Phys. Lett. B **416** (1998) 415 [hep-ph/9709419].
[20] C. Anzai, M. Prausa, A. V. Smirnov, V. A. Smirnov and M. Steinhauser, Phys. Rev. D **88** (2013) 5, 054030 [arXiv:1308.1202 [hep-ph]].
[21] A. Pineda, J. Phys. G **29** (2003) 371 [hep-ph/0208031].
[22] A. Bazavov *et al.* [HotQCD Collaboration], Phys. Rev. D **90** (2014) 9, 094503 [arXiv:1407.6387 [hep-lat]].
[23] A. Bazavov *et al.* [MILC Collaboration], PoS LATTICE **2010** (2010) 074 [arXiv:1012.0868 [hep-lat]].
[24] S. Necco and R. Sommer, Nucl. Phys. B **622** (2002) 328 [hep-lat/0108008].
[25] N. Brambilla, A. Pineda, J. Soto and A. Vairo, Phys. Rev. D **63** (2001) 014023 [hep-ph/0002250].
[26] A. Pineda and A. Vairo, Phys. Rev. D **63** (2001) 054007 [Phys. Rev. D **64** (2001) 039902] [hep-ph/0009145].

Hagedorn temperature and physics of black holes

V.I. Zakharov[1,2,3,a], Thomas G. Mertens[4,5], and Henri Verschelde[5]

[1] *ITEP, B. Cheremushkinskaya 25, Moscow, 117218 Russia*
[2] *Max-Planck Institut fuer Physik, Foehringer Ring 6, Muenchen, Germany*
[3] *School of Biomedicine, Far Eastern Federal University, Sukhanova str 8, Vladivostok 690950*
[4] *Joseph Henry Laboratories, Princeton University, Princeton, NJ 08544, USA*
[5] *Ghent University, Department of Physics and Astronomy Krijgslaan, 281-S9, 9000 Gent, Belgium*

Abstract. A mini-review devoted to some implications of the Hagedorn temperature for black hole physics. The existence of a limiting temperature is a generic feature of string models. The Hagedorn temperature was introduced first in the context of hadronic physics. Nowadays, the emphasis is shifted to fundamental strings which might be a necessary ingredient to obtain a consistent theory of black holes. The point is that, in field theory, the local temperature close to the horizon could be arbitrarily high, and this observation is difficult to reconcile with the finiteness of the entropy of black holes. After preliminary remarks, we review our recent attempt to evaluate the entropy of large black holes in terms of fundamental strings. We also speculate on implications for dynamics of large-N_c gauge theories arising within holographic models.

1 Introduction.

1.1 From hadrons to black holes

This talk was presented at a session devoted to the 50th anniversary of the introduction of the Hagedorn temperature [1]. Originally, the Hagedorn temperature T_H was discussed in connection with hadronic physics. Alternatively, one can say that it referred to a mass scale of the order of the pion mass, $T_H \sim m_\pi$. We will discuss the physics of black holes, or, more precisely, the properties of the black-hole horizon. In field theory, the local temperature close to the horizon can be arbitrarily high, and this is known to be inconsistent with finiteness of the Bekenstein-Hawking entropy of black holes, see, in particular, [2][3]. The inconsistency becomes manifest at the gravitational scale, or at $T \sim M_{\text{Planck}}$. This observation serves as a motivation [4] to introduce strings at this scale. It is within this framework that we make our remarks on the explicit evaluation of the Bekenstein-Hawking entropy, see [5] and references therein. Nowadays, the literature on the subject of strings and black holes is huge. Because of the format of the talk we limit our list of references to only a few papers. General background can be found in [6].

It should be remarked that the recent firewall paradox (supposedly burning up infalling observers) has rekindled the interest in horizon physics (see [7] and subsequent work). We however are interested

[a] e-mail: vzakharov@itep.ru

in the physics as described by static observers and are hence (a priori) safely away from speculating on the experience of infalling observers.

The outline of the talk is as follows. In the Introduction we describe briefly how the notion of a limiting temperature arises within a generic string picture. In Section 2 we remind the reader of the basics of the black holes. In Section 3 we address the issue of evaluating black-hole entropy within string theory. In Section 4 we discuss briefly possible phenomenological implications within holographic models.

1.2 Hagedorn temperature

As a starting point, we can choose the assumption made by R. Hagedorn [1] that the density of hadronic states $\omega(E)$ at large energy E grows exponentially:

$$\omega(E) \sim \exp(\beta_H E) \text{ , where } \beta_H \sim m_\pi^{-1}. \tag{1}$$

Then the partition function

$$z = \int_0^\infty dE\omega(E)e^{-\beta E} \tag{2}$$

exists only as far as $\beta > \beta_H$. In the microcanonical language, there is a "limiting temperature" T_H:

$$\frac{1}{T} = \frac{\partial S}{\partial E} \tag{3}$$

and

$$T < T_H , \quad T_H \equiv \frac{1}{\beta_H} = const . \tag{4}$$

The physics behind this remarkable phenomenon is actually quite simple: if we pump energy into the system, new higher-mass states are produced rather than that the energy of already existing states is increased. The increasing of energy of existing states would mean an increasing temperature. The dominance of production of new massive states manifests the emergence of a limiting temperature.

1.3 Hagedorn temperature and strings

In view of the fact that the spectrum (1) leads to such a drastic conclusion as the existence of a limiting temperature in nature, we should probably re-examine the reasons for introducing the exponential spectrum itself.

The strongest support for the assumption (1) comes from the string model of hadrons. And, in turn, the strongest point of the string model is that it reproduces linear Regge trajectories. Indeed, the energy of a string of length L is given by:

$$E_{\text{string}} = \sigma \cdot L, \; \sigma \equiv (2\pi\alpha')^{-1} , \tag{5}$$

where σ is the string tension. For a rotating string

$$M^2 = \frac{1}{\alpha'} J, \tag{6}$$

where J is the total angular momentum. In other words, the Regge trajectories are linear. Moreover, the density of states is indeed exponential at high energy:

$$\omega(E) \sim \frac{\exp(\beta_H E)}{E^{1+D/2}} , \tag{7}$$

where $\beta_H \sim \sqrt{\alpha'}$ and D is the number of (non-compact) spatial directions. Derivations of Eqs (5), (6), (7) can be found in standard textbooks.

DOI: 10.1051/epjconf/201612602032

1.4 Limiting temperature vs phase transition

At first sight, the argumentation above looks strong enough and we could expect the existence of a limiting temperature. In fact, it was realized a long time ago that there is a viable alternative to the introduction of the Hagedorn temperature. Namely, one can argue that there is a phase transition. While at low temperatures hadrons appear to be fundamental they are in fact composite and are built up by quarks and gluons. Thus, at some critical temperature T_{cr} there is a phase transition to deconfinement. This phase transition has been observed and studied thermodynamically in great detail through lattice simulations for various non-Abelian gauge groups, including the realistic case of quantum chromodynamics.

The precise relation between T_H and T_{cr} remains somewhat obscure. Phenomenologically, it is obvious that one should have $T_H > T_{cr}$, where T_H is defined within the string model. How close T_H is to T_{cr}, remains unclear because of the uncertainties of the string models of hadrons.

Thus, we can say that the existence of T_{cr} can be traced back to the fact that it is the field theory which is fundamental, not the hadronic strings model.

As we will see next, the quantum field theory (QFT) becomes, in turn, problematic at the gravitational scale. This is revealed by considering black holes.

2 Black holes. Preliminaries

Let us first remind the reader a few well-known equations concerning black holes. The Schwarzschild geometry reads as

$$ds^2 = -\left(1 - \frac{2G_N M}{r}\right)dt^2 + \left(1 - \frac{2G_N M}{r}\right)^{-1} dr^2 + d\mathbf{x}_\perp^2, \tag{8}$$

where M is the mass of the central body, or black hole in our case and G_N is the Newton constant. Note that the G_{00} component of the metric vanishes at the horizon $r_H = 2G_N M$.

The thermodynamic entropy is proportional to the area of the black hole:

$$S_{BH} = \frac{\text{Area}}{4G_N} \tag{9}$$

and there is Hawking radiation with temperature

$$\beta_{\text{Hawking}} = 8\pi G_N M\ . \tag{10}$$

Close to the horizon, it is useful to introduce the distance ρ to the horizon, $\rho = \sqrt{8G_N M(r - 2G_N M)}$. Then for $\rho \ll 4G_N M$

$$ds^2_{\text{Rindler}} = -\frac{\rho^2}{(4G_N M)^2}dt^2 + d\rho^2 + d\mathbf{x}_\perp^2.$$

Many results apply just in this limit of the so-called Rindler space. For Euclidean time τ,

$$ds^2_{\text{Euclidean}} = \frac{\rho^2}{(4G_N M)^2}d\tau^2 + d\rho^2 + d\mathbf{x}_\perp^2,$$

which is flat space in polar coordinates. Moreover, the τ-coordinate is periodic,

$$\tau \sim \tau + \beta_{\text{Rindler}}\ , \tag{11}$$

with $\beta_{\text{Rindler}} = 8\pi G_N M \equiv \beta_{\text{Hawking}}$.

One can say that a black hole provides a "lab" to study temperatures arbitrarily high. Indeed, near the horizon the blue-shift factor is given by

$$\chi \equiv \frac{4G_N M}{\rho}\,, \tag{12}$$

where ρ is the distance to the horizon. Hence

$$\beta_{\text{local}} = \beta_{\text{Rindler}}\chi^{-1}, \;\; \beta_{\text{local}} \to 0, \text{ if } \rho \to 0. \tag{13}$$

Note that if we go beyond the Rindler approximation the overall Euclidean thermal manifold is cigar-shaped.

2.1 Black holes and limiting temperature

In quantum field theory the entropy density $s \sim T^3$ and by using (12) the total entropy stored near the horizon is estimated as

$$S \sim \int d\rho T(\rho)^3 = \frac{\text{Area}}{\epsilon^2}\,, \tag{14}$$

where ϵ is a cut off at small distances.

Not to exceed the black hole entropy (9), we need a limiting temperature (brick wall of 't Hooft [2][3]). In other words, there is a need for a modification of QFT at short distances. Strings are welcome back on the fundamental level!

In practice, to limit applicability of field theory near the horizon one introduces a so-called stretched horizon. The stretched horizon is a surface placed close to the actual horizon, in front of it, such that $G_{00} \ll 1$. For more details see [8], [9].

3 Stringy horizon

3.1 Long-string picture of L. Susskind

Consider the formation of a black hole by throwing in matter focused inside the (future) black hole. For a distant observer, the matter falls in infinitely long. As a result, the infalling matter spreads out in the transverse directions [4]. Indeed, consider the parton-model representation of the matter. Then there is diffusion of the partons in the transverse directions. And since the process takes long, the partons cover the whole area. Moreover, it is known that, say, two long strings merge into a single one, because of entropic considerations. In this way one comes to the long-string picture of L. Susskind. According to this picture near the horizon, at $\rho \sim l_s$ where l_s is the string scale, there exists a single long string [1].

Moreover, one might hope that by counting the number of states of the long string, one could reproduce the entropy (9) of the black hole:

$$S_{\text{long string}} = S_{BH} \quad (?) \tag{15}$$

Note, though, that the matching (15) is not without problems. Indeed, the density of states of a long string is exponential in its length, or mass M, see (7). On the other hand, the Bekenstein-Hawking entropy is proportional to the area of the black hole, or its mass squared, $S_{BH} \sim M^2$. To maintain (15)

[1] It is worth emphasizing that physical pictures in general relativity are observer-dependent. The infalling matter is visualized as a single long string by a distant observer.

DOI: 10.1051/epjconf/201612602032

one is forced to speculate that this apparent mismatch is removed by accounting for the self-gravitation of the long string [10].

Thus, the long-string picture has definite advantages, by resolving the ultraviolet divergence (12) through the introduction of a finite l_s and by relating the number of degrees of freedom of a black hole to the number of degrees of freedom of a long string (which is much better understood). Also, the proportionality of the entropy to the area comes out naturally because of the random walk of partons in the transverse directions.

Many questions are left open, however. In particular:

- What keeps the long string at $\rho \sim l_s$?
- How to get quantitatively $S = (\text{Area})/4G_N$?
- Qualitative picture vs fundamental strings?

These questions were addressed in [5], see also references therein.

3.2 Main results

We considered a kind of mean field approximation, when a thin shell of matter of mass δM falls into the black hole of mass M. The gravitational field of the mass M is taken into account while the self-interaction within the shell is neglected. The thin shell is described in the framework of string theory. We find that in Euclidean time the shell occupies a zero mode. This allows us to evaluate the entropy carried to the black hole by the shell. Integrating over dM reproduces the Bekenstein-Hawking entropy for black holes. Moreover, knowing the wave function of the zero mode allows us to visualize the profile of the stringy horizon at distances of order l_s from the horizon. One can summarize the results by saying that, in the approximation of the mean field, the idea of relating the entropy of (large) black holes to the stringy degrees of freedom is realized on a fully quantitative level. An important reservation is that these results hold in the case of type II superstrings and heterotic strings in Rindler space. Whereas for bosonic strings, there arises no consistent picture.

3.3 Main tool: thermal scalar

On the technical side, the results just mentioned are obtained by utilizing the construction of the so-called thermal scalar in curved space, see [11] and references therein. The theory of the thermal scalar has been developing since long, see in particular [12], [13]. Roughly speaking, the thermal scalar is a Euclidean counterpart of the Hagedorn transition. Namely, the original mechanism for the Hagedorn divergence of the partition function (2) is the production of high-mass states. A complementary view on the Hagedorn transition in Euclidean time is that it is a kind of a Higgs phase transition when the mass squared of a complex field changes its sign. This scalar field, or thermal scalar, lives in spatial dimensions only (not temporal) and its mass is given by:

$$m^2_{\text{thermal scalar}} = \frac{(\beta-\beta_H)\beta_H}{2\pi(\alpha')^2}. \tag{16}$$

Eq. (16) holds in flat space and as far as $m^2_{\text{thermal scalar}}$ is positive. What happens at $\beta < \beta_H$ is not clear a priori, for further discussion see [13].

The equivalence between the standard formulation of the Hagedorn transition and that one in terms of the thermal scalar can readily be demonstrated using the polymer, or random-walk formulation of Euclidean field theory. It is also straightforward to see that the thermal scalar corresponds to a string once wrapped around the compact Euclidean time [12]. The time dependence is fixed then by the periodicity and effectively the wave function of the thermal scalar depends only on the spatial coordinates.

To consider strings in the background field of black holes, one needs to generalize the thermal scalar to curved space, see [14], [11] and references therein.

3.4 Thermal scalar in curved space

The action for the thermal-scalar field φ is given by [11, 15]:

$$S = \int d^{D-1}x\,\sqrt{G}e^{-2\Phi} \times\Big(G_{ij}\nabla^i\varphi\nabla^j\varphi^* + \frac{1}{4\pi^2(\alpha')^2}(\beta^2 G_{00} - \beta_H^2)\varphi\varphi^*\Big), \tag{17}$$

where G_{00}, G_{ij} are the components of the metric and Φ is the dilaton field. A crucial point is that for type II strings in Rindler space, $ds^2_{\text{Rindler}} = a^2\rho^2 dt^2 + d\rho^2 + d\mathbf{x}_\perp^2$, the action (17) receives no α' corrections [11, 15]. Moreover, this is also true for heterotic strings [16] (but not for bosonic strings).

The corresponding equation for the wave functions

$$\Big(-\partial_\rho^2 - \frac{1}{\rho}\partial_\rho + \frac{1}{4\pi^2(\alpha')^2}(\beta^2 a^2\rho^2 - \beta_H^2)\Big)\varphi_n(\rho) = \lambda_n\varphi(\rho) \tag{18}$$

has solutions:

$$\varphi_n(\rho) = \exp\Big(-\frac{a\beta\rho^2}{4\pi\alpha'}\Big)L_n\Big(\frac{a\beta\rho^2}{2\pi\alpha'}\Big),\ \lambda_n = a\beta(1+2n) - 2\pi \tag{19}$$

in terms of Laguerre polynomials L_n, and $n \in \mathbb{N}$. The zero mode (n=0) at $\beta = \beta_{\text{Rindler}} = 2\pi/a$ dominates the thermal partition function.

This analysis was done for the singly wound thermal scalar state. A curiosity at this point is that states that are wound multiple times are simply absent from the spectrum. This will be important in the final section.

3.5 Picture emerging

The build-up of a black hole by throwing a thin shell of mass δM to the black hole of mass M_{initial} can be consistently described by string theory in a mean-field approximation. The shell ends up as a long string in a layer of thickness $\delta\rho \sim l_s$ near the horizon.

Moreover, the entropy of black holes is calculable without any adjustable parameters. In more detail, the density of states seen by a distant observer is

$$\omega(\delta M) \sim \frac{\exp(\beta_{\text{Hawking}}\delta M)}{\delta M},\ \text{or}\ \beta_{\text{Hawking}} = \beta_{\text{Hagedorn}}. \tag{20}$$

Integrating over δM we get the Bekenstein-Hawking entropy:

$$\delta S_{BH} = 8\pi G_N M\delta M \rightarrow S_{BH} = \frac{\text{Area}}{4G_N}. \tag{21}$$

Also, knowledge of the wave function (19) allows us to evaluate [5, 17] the profile of the energy-momentum tensor associated with the long string of mass δM:

$$\begin{aligned} \left\langle T^0_0(\mathbf{x})\right\rangle &= -2N^2\Big(\frac{2\rho^2}{\alpha'}-1\Big)e^{-\rho^2/\alpha'} ,\\ \left\langle T^\rho_\rho(\mathbf{x})\right\rangle &= 2N^2 e^{-\rho^2/\alpha'} ,\\ \left\langle T^i_j(\mathbf{x})\right\rangle &= 2N^2\delta^i_j\Big(1-\frac{\rho^2}{\alpha'}\Big)e^{-\rho^2/\alpha'} , \end{aligned} \tag{22}$$

where $i, j = 1, 2$ are transverse directions and N is a normalization factor.

Note a positive value of the radial pressure, $\left\langle T^\rho_\rho(\mathbf{x})\right\rangle$. It is this pressure that keeps the matter from collapsing onto the center. Remarkably, Eqs (22) specify a distribution of matter near the horizon, at $\rho \sim l_s$.

4 From strings to gauge theories, via holography

Amusingly enough, lessons from strings on the gravitational scale might produce a new insight into the dynamics of gauge theories. The means is holography: strings live in curved extra dimensions, while gauge theory lives on a flat boundary. Both theories are inter-related.

The most famous case of the string-gauge duality refers to $\mathcal{N} = 4$ supersymmetric Yang-Mills theory [18]. In case of non-supersymmetric Yang-Mills theories, Witten constructed a model [19] which in the far infrared limit belongs to the same universality class as large-N_c gauge theories. The model can be generalized to incorporate massless quarks [20]. From our perspective, it is crucial that the geometry in extra dimensions inherent to this model is of the same cigar-shape as we encountered while discussing large black holes above.

In more detail, the Euclidean version of the model [19] introduces two compact dimensions:

- Periodic Euclidean time $\tau \sim \tau + \beta_\tau(z)$ where the periodicity, β_τ depends on an extra coordinate z. As is common to holographic models, the z-coordinate is associated with resolution. The limit $z \to 0$ corresponds to Yang-Mills theories in the ultraviolet limit, while $z \to z_{\text{horizon}}$ corresponds to the infrared limit on the field theoretic side
- There is another periodic coordinate $\sigma \sim \sigma + \beta_\sigma(z)$. Wrapping around σ counts the topological charge associated with the corresponding stringy state

From first principles, at $T = 0$ the (τ, z) space is a cylinder and (σ, z) is cigar-shaped, $\beta_\sigma(z_{\text{horizon}}) = 0$. At the deconfining phase transition, $T = T_{cr}$ the geometries in the (τ, z) and (σ, z) coordinates are interchanged.

Now we are coming to a central point, namely how, if at all, the geometry in the extra dimensions is related to gauge theory phenomenology. One of the routes is to consider properties of so-called defects. One of the best studied examples of a defect on the field-theoretic side are instantons. On the stringy side, defects can be identified topologically, in terms of wrapping around compact directions. In the geometric language the string action is very simple,

$$(\text{action}) \sim L \cdot (\text{tension}) \ .$$

If there is a cylinder-type geometry then the wrapping number is well-defined and the probability to find a defect is suppressed by the action.

However, in case of a cigar-shaped geometry,

$$\beta_\tau(z_{\text{horizon}}) = 0 \text{ or } \beta_\sigma(z_{\text{horizon}}) = 0$$

the action for wrapped states vanishes at the horizon and the probability to find them in the vacuum state is not suppressed by their action.

Instantons are distinguished by a non-vanishing topological charge. As is mentioned above, in the geometric language, the topological charge is associated with wrapping around the σ-coordinate. The simplest geometric object which can be wrapped around the σ direction is a $D0$-brane. The corresponding defect is characterized by a topological charge determined by the wrapping number, its position in the Euclidean space and by its action, which depends on the value of its z-coordinate. Thus, the number of the parameters characterizing such $D0$ branes matches instantons of field theory, for further details and references see [21].

Let us check the gravity-gauge correspondence on the example of the instantons [21]. Consider first temperature $T = 0$ and start with the strings. Since β_σ vanishes on the horizon, $\beta_\sigma(z = z_{\text{horizon}}) = 0$, instantons are not suppressed by the action in the far infrared. This is, indeed, well known on the field theoretic side of the correspondence. At $T = T_{cr}$ the (σ, z) geometry is changed into a cylinder and instantons become suppressed according to the stringy picture. Again, this conclusion is well known in field theory and is supported by the existing phenomenology.

Now we are coming to the central point of this section, that is the manifestation of the zero mode discussed in the preceding section. Classically, the action for the defects,

$$S_{\text{defect}} = L \cdot (\text{tension})$$

vanishes for any wrapping number as far as $\beta_\sigma(z) = 0$. However, quantum-mechanically keeping a $D0$ brane at the tip of the cigar results in kinetic energy, because of the uncertainty principle. As a result, only the lowest level survives as the zero mode at the tip of the cigar (see discussion in the preceding section). The lowest level, in turn, corresponds to a single wrapping. Phenomenologically, this implies that only instantons with the topological charge $Q_{\text{top}} = \pm 1$ exist while $|Q_{\text{top}}| \geq 2$ are suppressed in the vacuum. Also, an effective infrared cut off might arise dynamically in the far infrared. These predictions which follow from holography seem to be supported by the lattice data.

Acknowledgements

This work is supported by a grant from the Russian Science Foundation (project number 16-12-10059).

References

[1] R. Hagedorn, *"Statistical thermodynamics of strong interactions at high-energies "*, Nuovo Cim. Suppl. 3 (1965) 147.

[2] G. 't Hooft, *"Dimensional reduction in quantum gravity"*, *Salamfest* , (1993) 0284.

[3] G. 't Hooft, *"On the Quantum Structure of a Black Hole"*, Nucl. Phys. B **256** (1985) 727.

[4] L. Susskind, L. Thorlacius, and J. Uglum, *"The Stretched horizon and black hole complementarity"* Phys.Rev. D48 (1993) 3743;
L. Susskind, *" Strings, black holes and Lorentz contraction"*, Phys.Rev. D49 (1994) 6606, arXiv: hep-th/9308139.

[5] T. G. Mertens, H. Verschelde, and V. I. Zakharov, *"The long string at the stretched horizon and the entropy of large non-extremal black holes"*, JHEP 1602 (2016) 041, arXiv:1505.04025 [hep-th].

[6] L. Susskind and J. Lindsay, *" An Introduction to Black Holes, Information and the String Theory Revolution: The Holographic Universe"*, (2004), World Scientific publishing company.

[7] A. Almheiri, D. Marolf, J. Polchinski and J. Sully, *"Black Holes: Complementarity or Firewalls?"*, JHEP **1302** (2013) 062 [arXiv:1207.3123 [hep-th]].

[8] T. Damour, *"Black Hole Eddy Currents"*, Phys. Rev. D18 (1978) 3598.

[9] M. Parikh and F. Wilczek, *"An Action for black hole membranes "*, Phys.Rev. D58 (1998) 064011.

[10] G. T. Horowitz and J. Polchinski, *"Selfgravitating fundamental strings"*, Phys.Rev. D57 (1998) 2557, arXiv: hep-th/9707170.

[11] T. G. Mertens, H. Verschelde, and V. I. Zakharov, *" Near-Hagedorn Thermodynamics and Random Walks: a General Formalism in Curved Backgrounds"*, JHEP 1402 (2014) 127, arXiv:1305.7443 [hep-th];
"Random Walks in Rindler Spacetime and String Theory at the Tip of the Cigar", JHEP 1403 (2014) 086, arXiv:1307.3491 [hep-th].

[12] B. Sathiapalan, *"Vortices on the String World Sheet and Constraints on Toral Compactification"*, Phys.Rev. D35 (1987) 3277;
Ya. I. Kogan, *" Vortices on the World Sheet and String's Critical Dynamics"*, JETP Lett. 45 (1987) 709.

[13] J. J. Atick and E. Witten, *"The Hagedorn Transition and the Number of Degrees of Freedom of String Theory"*, Nucl.Phys. B310 (1988) 291.

[14] M. Kruczenski and A. Lawrence, *"Random walks and the Hagedorn transition"*, JHEP 0607 (2006) 031, arXiv: hep-th/0508148.

[15] A. Giveon and N. Itzhaki, *"String theory at the tip of the cigar"*, JHEP 1309 (2013) 079, arXiv:1305.4799 [hep-th]

[16] T. G. Mertens, H. Verschelde, and V. I. Zakharov, *"Perturbative String Thermodynamics near Black Hole Horizons "*, JHEP 1506 (2015) 167, arXiv:1410.8009 [hep-th].

[17] T. G. Mertens, H. Verschelde, and V. I. Zakharov, . *"On the Relevance of the Thermal Scalar"*, JHEP 1411 (2014) 157, arXiv:1408.7012 [hep-th].

[18] J. M. Maldacena, *" The Large N limit of superconformal field theories and supergravity"*, Int. J. Theor. Phys. 38 (1999) 1113, arXiv: hep-th/9711200;
O. Aharony , S. S. Gubser, J. M. Maldacena, H. Ooguri, and Y. Oz, *" Large N field theories, string theory and gravity"*, Phys. Rept. 323 (2000) 183, arXiv: hep-th/9905111.

[19] E. Witten, *" Anti-de Sitter space, thermal phase transition, and confinement in gauge theories"*, Adv. Theor. Math. Phys. 2 (1998) 505, arXiv: hep-th/9803131.

[20] T. Sakai and Sh. Sugimoto, *" Low energy hadron physics in holographic QCD"*, Prog. Theor. Phys. 113 (2005) 843, arXiv: hep-th/0412141.

[21] O. Bergman and G. Lifschytz, *"Holographic U(1)A and string creation "*, JHEP 0704 (2007) , 043, arXiv:hep-th/0612289.

EPJ Web of Conferences **126**, 03001 (2016) DOI: 10.1051/epjconf/201612603001
ICNFP 2015

Hagedorn legacy

Johann Rafelski

Department of Physics, The University of Arizona, Tucson, AZ 85721, USA

Abstract. These remarks open the one-day session "50 years of Hagedorn's Temperature and the Statistical Bootstrap Model". These developments set the path at CERN towards the discovery of Quark-Gluon Plasma in the year 2000.

Rolf Hagedorn in his CERN office in 1977/8, Photo: Johann Rafelski

In 1964/5 Rolf Hagedorn single handiedly had set CERN on the path to the study of hot hadronic matter and quark-gluon plasma. This session of the ICFNP2015 meeting has been set aside to celebrate 50 years of Hagedorn's pivotal ideas which opened to study the high energy density matter defining our Universe in primordial times above Hagedorn temperature T_{H} ; that is, before the emergence of matter as we know it. T_{H} is the pivotal concept that gave birth to a new field of physics. Thanks to early Hagedorn ideas, we understand today the different phases of strongly interacting matter, the early Universe, and we gained a deep insight into the working of quark confinement.

Let me begin with some personal remarks: I first met 'Herr Hagedorn' in Winter 1975/76, at his Colloquium on the Statistical Bootstrap Model (SBM). After the lecture and some discussion that followed, I asked if I could visit him at CERN. I arrived as a CERN-fellow in September 1977. The two most important impressions: Herr Hagedorn was an extraordinary thorough and understanding teacher; Herr Hagedorn was available to help those who were in need. For a more comprehensive account see the special volume [1] dedicated to his life and work, see figure 1.

After the vagaries of WW II, the path of Hagedorn to CERN was direct, as was his preparation for the work and discoveries he was to accomplish. In the early 1950s Werner Heisenberg hired Rolf

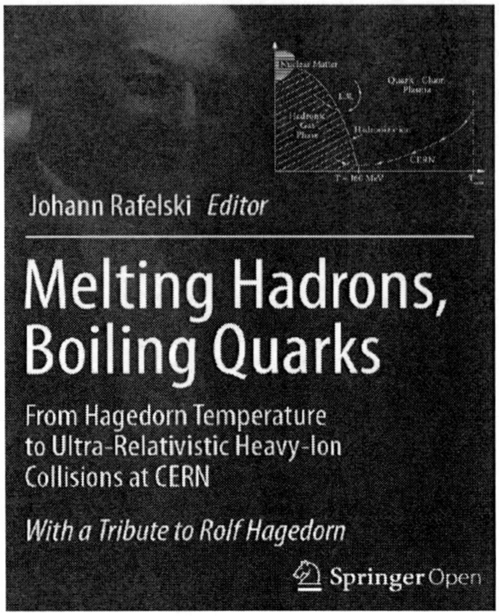

Figure 1. The tale of Hagedorn temperature and its impact on today's science is told in *Melting Hadrons, Boiling Quarks: From Hagedorn Temperature to Relativistic Heavy-Ion Collisions at CERN; With a Tribute to Rolf Hagedorn*, http://www.springer.com/de/book/9783319175447, see [1].

Hagedorn, who was trained by the preeminent thermodynamics theorist Richard Becker, to work on his cosmic emulsion-star 'hobby'. From the present day perspective this was the precursor research into hot hadronic matter. Heisenberg encouraged Hagedorn to consider CERN for continuing this work. Hagedorn consequently applied on January 27, 1954 to join the future CERN laboratory, with thermal physics and cosmic event stars as his scientific preparation. Hagedorn was hired by J.B. Adams, however, first to help build the 1st CERN accelerator, the PS. A few years on when J.B. Adams became director of CERN, he assigned Hagedorn effective January 1, 1961 to a permanent position in the CERN Theory Division, charging Hagedorn with research tasks that Hagedorn dutifully executed in the following decades: "It is intended that you devote your time partly to investigations and computing problems associated with the experimental programs of CERN, such as the use of statistical models for predicting particle production, and partly to those aspect of theoretical physics that will enable you to keep abreast of modern developments. . . ."

Hagedorn's work of 1960-64 had set the stage for his big discoveries that followed in Winter 1964/65. The pivotal period 1964/65 was a singularly critical event in the history of modern physics as several new ideas emerged, which are today merging into a unified view about our Universe:

- Hagedorn proposed the Hagedorn Temperature $T_{\rm H}$ and the statistical description of elementary particles, the statistical bootstrap model (SBM);
- Gell-Man and Zweig independently recognized that strongly interacting particles have more elementary constituents, the quarks;
- Higgs mechanism was recognized, opening the door for the standard model of particle physics;
- The cosmic microwave background was discovered helping to establish the hot big-bang theory.

Akin to quarks, the Hagedorn temperature had an uneasy birth: Hagedorn work of 1960-1964 showed that the Fermi model produces too few pions - could this mean particles are distinguishable? Hagedorn prepared a report [2] that was a direct precursor of all his coming discoveries. But he reconsidered and withdrew his preprint as new ideas were born quickly.

Within a few more weeks Hagedorn realized how particles can be made distinguishable: he introduced an exponential hadron spectrum where despite large particle abundance, no particle is the

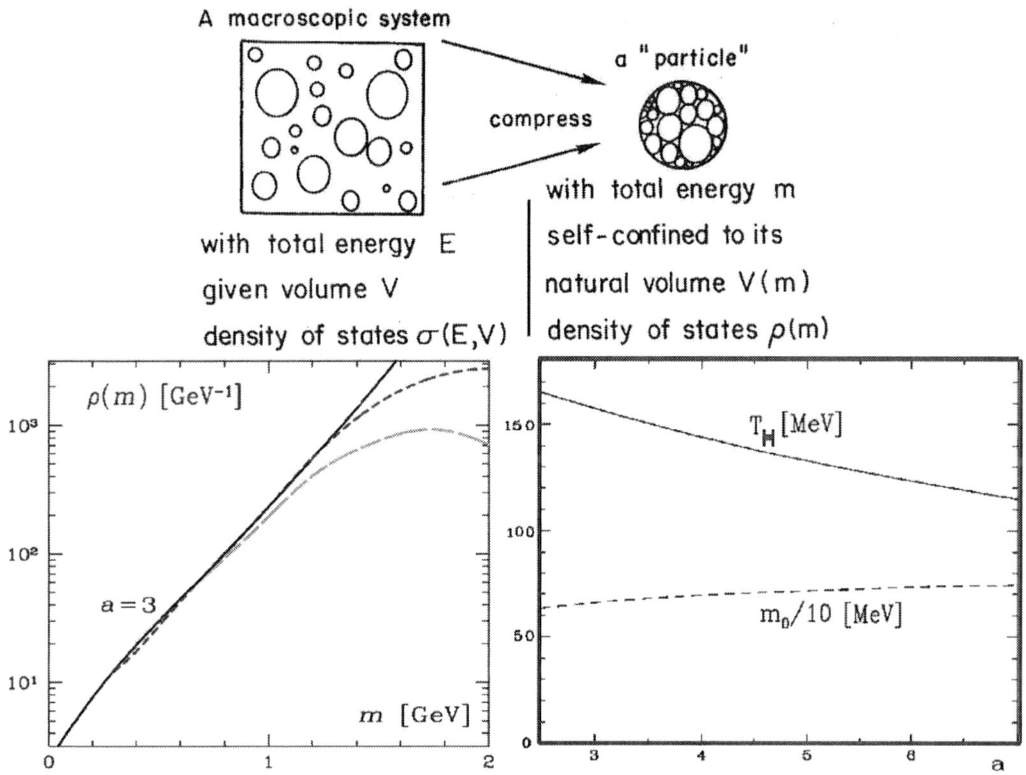

Figure 2. Top: Hagedorn's picture illustration of his model prepared for review of 1984 [4]. Bottom on left: The mass spectrum for $a = 3$, compared to two sets of experimental data: available in early days to Hagedorn (long dashed, green) and recent vintage (short-dashed, red). On right: the parameters of Eq. (1): $T_{\rm H}$ and m_0 as functions of the power index a.

same as the other. He created the bootstrap model to compute this spectrum. And, he found that the spectrum he theoretically predicted agreed with experiment. This was the Hagedorn revolution of November 1964-January 1965, published as his monumental model [3]: the SBM: Statistical Bootstrap Model.

The idea of SBM was straightforward, a volume V filled with particles to the limit is itself a new hadron. This is illustrated in figure 2, top part. This yielded the exponential mass spectrum of hadronic states

$$\rho \propto \frac{e^{m/T_{\rm H}}}{(m_0^2 + m^2)^{a/2}} , \tag{1}$$

where the exponential slope was the Hagedorn temperature $T_{\rm H}$. The index a was at first believed shown by Hagedorn to have a value $a = 2.5$.

In order to determine $T_{\rm H}$ given a limited range of experimental data available for the hadron resonance mass spectrum, see left hand, bottom, of figure 2, Hagedorn had to use a value of the parameter $a = 2.5$. A greater value $7 \geq a \geq 3$ emerged in a more realistic versions of SBM. As we see on right in figure 2 the larger is a, the smaller is the expected value of the Hagedorn temperature. For a discussion of this situation see Ref.[5]. Therefore the value of $T_{\rm H}$ which was at first reported to be at $T_{\rm H} = 164$ MeV, was recognized to be well below this upper limit. The mechanism that

DOI: 10.1051/epjconf/201612603001

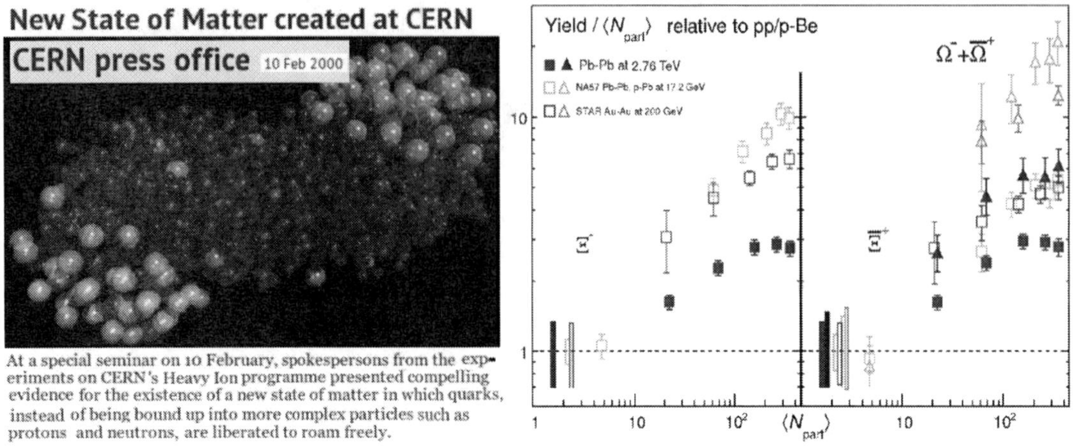

Figure 3. On left: CERN announces the discovery of quark-gluon plasma press.web.cern.ch/press-releases/2000/02/new-state-matter-created-cern; On right: the 2014 summary of strange antibaryon signature of QGP, adapted from a CERN-LHC-Alice collaboration report [6].

increased value of a and decreased value of T_{H} was the canonical conservation of baryon number, and strangeness.

The work of Hagedorn has attracted interest by many researchers; some are today here. They followed in Hagedorn's footsteps, often working at CERN. We were using statistical methods in strong interaction physics. After 1977 this effort turned into the exploration of hot hadronic matter created in nuclear collisions. This effort had created at CERN a hot spot in a new research field that was not fully appreciated everywhere at first. It was this development which made it natural that in late 1970s and early 1980s CERN became the host to a new experimental research program into the physics of relativistic heavy ion collisions. Ultimately, this led to the announcement of the discovery of quark-gluon plasma at CERN in early 2000, see figure 3 on left. Among the prominent signatures was the strange antibaryon production enhancement, the largest and predicted effect of quark deconfinement in hot hadron matter, shown on the right in figure 3, see Refs. [1, 5].

References

[1] J. Rafelski, Editor, "Melting Hadrons, Boiling Quarks - From Hagedorn Temperature to Ultra-Relativistic Heavy-Ion Collisions at CERN : With a Tribute to Rolf Hagedorn," doi:10.1007/978-3-319-17545-4 Springer Open, Heidelberg, New York (2016)

[2] R. Hagedorn, "Thermodynamics of Distinguishable Particles: A Key to High-Energy Strong Interactions?," PRINT-64-817, CERN-TH-483 doi:10.1007/978-3-319-17545-4_19

[3] R. Hagedorn, "Statistical thermodynamics of strong interactions at high-energies," Nuovo Cim. Suppl. **3** 147 (1965).

[4] R. Hagedorn, "How We Got to QCD Matter from the Hadron Side: 1984," Lect. Notes Phys. **221** 53 (1985). doi:10.1007/978-3-319-17545-4_25

[5] J. Rafelski, "Melting Hadrons, Boiling Quarks," Eur. Phys. J. A **51** 114 (2015) doi:10.1007/978-3-319-17545-4_33,10.1140/epja/i2015-15114-0

[6] B. Abelev *et al.* [ALICE Collaboration]: "Multi-strange baryon production at mid-rapidity in Pb-Pb collisions at $\sqrt{s_{NN}} = 2.76$ TeV," Phys. Lett. B **728** 216 (2014) [Erratum **734** 409 (2014)]

High temperature Bose-Einstein condensation

Viktor Begun[1,a]

[1] *Institute of Physics, Jan Kochanowski University, Kielce, Poland*

Abstract. The indications of a possible pion condensation at the LHC are summarized. The condensation is predicted by the non-equilibrium hadronization model for 2.76 TeV Pb+Pb collisions at the LHC. The model solves the proton/pion puzzle and reproduces the low p_T enhancement of the pion spectra, as well as the spectra of protons and antiprotons, charged kaons, K_S^0, $K^*(892)^0$ and $\phi(1020)$. The obtained parameters allow to estimate the amount of pion condensate on the level of 5% from the total number of pions at the LHC. The condensate is located at $p_T < 250$ MeV.

1 Introduction

Bose-Einstein condensation (BEC) was predicted right after the introduction of Bose statistics [1]. However, it took 70 years to observe BEC experimentally [2]. The main difficulty for obtaining BEC in a gas of atoms is that they form a liquid or a solid before reaching BEC. It can be avoided, if extremely low densities and temperatures are achieved. This was done only after the development of the corresponding technologies, and the leaders of the two groups that created the condensate received the Nobel Prize in 2001 [3]. It might be the reason for the widespread delusion that BEC means low temperatures and atomic gases only. However, all conventional two-quark mesons are bosons and may condense. The temperatures that are reached in high-energy collision are of the MeV scale. It corresponds to $\sim 10^{10}$ Kelvin and is far above the BEC temperatures for cold atoms. The ratio of the BEC temperature in atomic gases, $T_C(A)$, to that in the pion gas, $T_C(\pi)$, is even larger, because one should take into account a possible size of the system and the masses of the particles. One obtains 10^{12} times higher temperature for the gas of pions [4]:

$$\frac{T_C(\pi)}{T_C(A)} \simeq \frac{m_A}{m_\pi}\left(\frac{r_A}{r_\pi}\right)^2 \simeq \frac{m_A}{m_\pi}10^{10} \simeq 10^{12}\,. \tag{1}$$

Besides of much higher temperature, the properties of a BEC of mesons could be very different from the low-temperature BEC of atoms. The most obvious differences are the much smaller volumes, much higher densities, and different interaction forces involved in the formation of the high temperature BEC. The clear advantage of the pion BEC is that pion system is in the form of gas at freeze-out. The reason to consider pion BEC now is the recent data from Pb+Pb collisions at $\sqrt{s_{NN}} = 2.76$ TeV energies at Large Hadron Collider (LHC), that contain the set of puzzles.

[a]e-mail: viktor.begun@gmail.com

DOI: 10.1051/epjconf/201612603002

- The predictions of the hadron-resonance gas (HRG) were too high for ratios to pions, especially for proton to pion ratio [5].
- The best fit of the LHC data by the HRG still gives nearly three standard deviations for protons [6].
- The low-transverse-momentum pion spectra show up to 50% enhancement compared to hydrodynamic models [5].
- The temperature obtained in HRG at the LHC falls out from the freeze-out line deeply in the hadronic phase [7].

The HRG worked well for smaller collisions energies. Pions and protons are among the most abundant particles. Therefore the problems with their description look suspicious. There are several solution proposed to explain the proton to pion ratio. The hadronization and freeze-out in a chemical non-equilibrium [8]. The separate freeze-out for strange particles [9]. An incomplete list of hadrons [10], and hadronic rescattering in the final stage [11]. However, none of them is commonly accepted yet.

There are deep physical reasons for the non-equilibrium and pion condensation at the LHC. It can be due to fast expansion and overcooling of the QGP [12, 13], or due to gluon condensation in Color Glass Condensate [14], and subsequent hadronization of the low p_T gluons into low p_T pions [15]. The non-equilibrium hadronization can explain the measured particle ratios [8], and also the spectra [16, 17] very well. The obtained parameters indicate the possibility that 5% of the total number of pions can be in BEC at the LHC [18]. The calculations of the pion spectra with the assumption of BEC show that the maximal transverse momentum that the BEC can obtain due to the expansion of the fireball is $p_T < 250$ MeV [18]. It is the same region where ALICE Collaboration observes the emission of 23% of pions from a coherent source [19].

2 The non-equilibrium model

The thermodynamic motivation behind the non-equilibrium HRG (NEQ) is the following. The fast expansion of the fireball may cause it's overcooling, so that the number of quarks and anti-quarks is larger than their equilibrium value at the reached temperature. Therefore, the subsequent freeze-out may lead to the formation of mesons and baryons with the multiplicities that are higher than the corresponding equilibrium numbers at the given temperature. The phase-space distribution of the primordial particles in NEQ is similar to the usual HRG:

$$f_i = g_i \int \frac{d^3p}{(2\pi)^3} \frac{1}{\gamma_i^{-1} \exp(\sqrt{p^2+m^2}/T) \pm 1}, \tag{2}$$

where g_i is the degeneracy factor, p–momentum, m–mass, T–temperature, and

$$\gamma_i = \gamma_q^{N_q^i+N_{\bar{q}}^i} \gamma_s^{N_s^i+N_{\bar{s}}^i} \exp\left(\frac{\mu_B B_i + \mu_Q Q_i + \mu_S S_i}{T}\right). \tag{3}$$

The γ_i contains the chemical potentials and corresponding conserved charges (Q, B, S) of a particle i in equilibrium HRG (EQ), and the non-equilibrium parameters γ_q, γ_s. The N_q^i, $N_{\bar{q}}^i$ and N_s^i, $N_{\bar{s}}^i$ are the numbers of light (u, d) and strange (s) quarks and anti-quarks in the i–th hadron, correspondingly. Note the plus sign between the number of quarks and anti-quarks, in contrast to the usual conserved charges, that are defined as the difference between particles and anti-particles. One can exponentiate γ_q and γ_s and see that each particle in NEQ obtains it's own chemical potential, which can be also recalculated into a different temperature. Therefore NEQ is a particular case of the multiple-freeze out

 DOI: 10.1051/epjconf/201612603002

model, where each particle has it's own freeze-out, but the relation between them is fixed by Eq. (3). The NEQ is advocated by J. Rafelski and collaborators since a long time, see Ref. [20] for the review. It is implemented in the SHARE model [21] and allows for the HRG fit of the particle multiplicities and ratios. The THERMINATOR Monte-Carlo event generator [22] inherits the same particle and decay list and allows for further calculation of the particle spectra from the Cooper-Frye formula at the freeze-out hypersurface Σ_μ:

$$\frac{dN}{dyd^2p_T} = \int d\Sigma_\mu p^\mu f(p\cdot u), \qquad t^2 = \tau_f^2 + x^2 + y^2 + z^2, \qquad x^2 + y^2 \le r_{\max}^2\,, \tag{4}$$

where τ_f and $r_{\max}$ are the characteristic time and radius of the hypersurface. There are several hypersurfaces implemented in THERMINATOR following the paper [23]. For the LHC we use Cracow model, that assumes the Hubble-like flow of particles: $u^\mu = x^\mu/\tau_f$, because it fits the data the best. At the LHC $\mu_B \simeq \mu_Q \simeq \mu_S \simeq 0$ due to large collision energy compared to the rest mass of the colliding nuclei. Therefore, EQ model has two parameters left - temperature and volume, while NEQ has also γ_q and γ_s. The measured set of particles, that was used in [16–18, 24], includes 6 to 8 multiplicities depending on centrality. It allows to determine both the parameters in EQ and in NEQ.

There is only one additional parameter in the model that allows to describe the spectra, because the product $\pi\,\tau_f\,r_{\max}^2$ is equal to the volume per unit rapidity, while the ratio $r_{\max}/\tau_f$ determines the slopes of the spectra. The procedure is the following: first, the volume and temperature are found from the fit of all mean multiplicities, and then the $r_{\max}/\tau_f$ is found from the best fit to the spectra of pions $\pi^+ + \pi^-$ and kaons $K^+ + K^-$. This approach gives unexpectedly good results in NEQ. The fit that was made for the spectra of pions and kaons only, appears also very good for $p + \bar{p}$, K_S^0, $K^*(892)^0$ and $\phi(1020)$, see Figs. 1, 2. The temperature in EQ is much higher, therefore the amount of secondary pions is larger

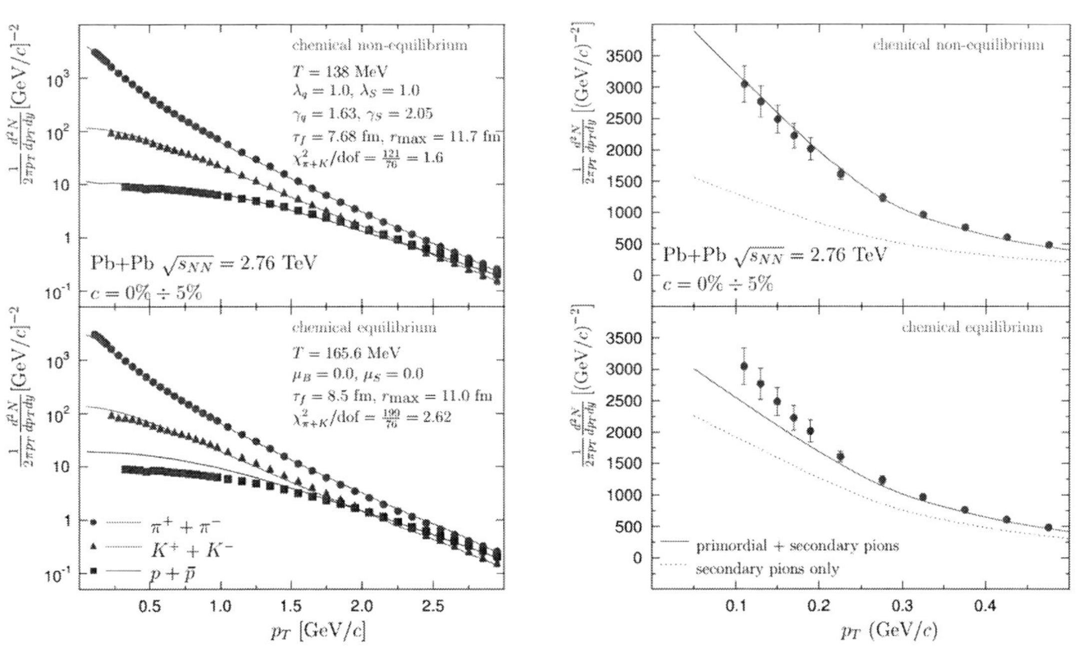

Figure 1. The spectra of pions, kaons and protons (left) in NEQ (up) and EQ (down), and the same only for pions in linear scale (right). The figures are from Ref. [17].

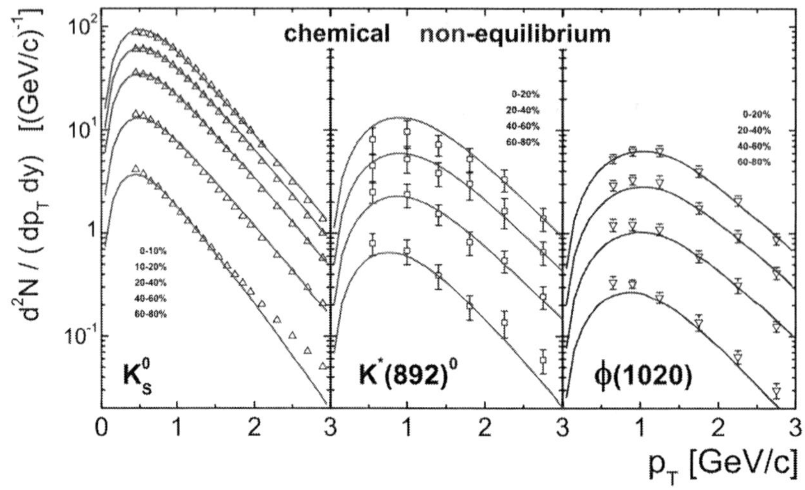

Figure 2. The spectra of strange particles in NEQ. The figure is from Ref. [16].

in NEQ. However, it is not enough to reproduce the steepness of the spectrum at $p_T < 250$ MeV, see Fig. 1 (right). The best state of the art fit of pion, kaon and proton spectra in EQ allows to describe protons taking into account rescattering mechanism [25]. Unfortunately, the authors of Ref. [25] do not provide the plot for pions in a linear scale, or the data/model ratio. However, one can see that they still have about 30% deficit of pions in the model at $p_T < 250$ MeV. The simultaneous fit of $K^*(892)^0$ and $\phi(1020)$ in NEQ is as surprising, as the applicability of the pion-kaon fit to protons. The $K^*(892)^0$ is short living, while $\phi(1020)$ is long living. They should feel a long rescattering phase differently [26]. The successful simultaneous description of protons, pions, $K^*(892)^0$ and $\phi(1020)$ in NEQ without rescattering may indicate that the rescattering phase is well parameterized by the non-equilibrium parameters (3).

The numerical value of the γ_q recalculated into the corresponding chemical potential is very close to the pion mass: $\mu_\pi = 2T \ln \gamma_q \simeq 134$ MeV $\simeq m_{\pi^0} \simeq 134.98$ MeV. It suggests that a substantial part of π^0 mesons tend to form the condensate, but this possibility is not allowed in the HRG implementation in SHARE and THERMINATOR. It can be done by explicit treatment of hadronic ground states. When chemical potential approaches the mass of a particle, $\mu \to m$, the discrete sum over the low momentum quantum levels should be taken, instead of the integral over momentum. It can be shown that in the infinite volume limit, $V \to \infty$, only the zero momentum level, $p_0 = 0$, can be taken into account [4]. Then the total mean multiplicity of a particle, $N_{\rm tot}$, receives one more term:

$$\begin{aligned} N_{\rm tot} &= \sum_j \frac{1}{\exp[(\sqrt{p_j^2+m^2}-\mu)/T]-1} \\ &\simeq \frac{1}{\exp[(m-\mu)/T]-1} + V\int_0^\infty \frac{d^3p}{(2\pi)^3}\,\frac{1}{\exp[(\sqrt{p^2+m^2}-\mu)/T]-1} \\ &= N_{\rm cond} + N_{\rm norm}\,, \end{aligned} \tag{5}$$

where $N_{\rm cond}$ is the number of particles in the condensate. When chemical potential approaches the mass, the $N_{\rm cond}$ grows infinitely and competes with arbitrarily large volume, therefore the γ_q does not grow so much. The corresponding changes were introduced into SHARE. The new fit in the NEQ

DOI: 10.1051/epjconf/201612603002

model with explicit treatment of the ground state (BEC)[1] gives rather different results, see Fig. 3. The inclusion of the ground state makes equilibrium and non-equilibrium models closer. They even

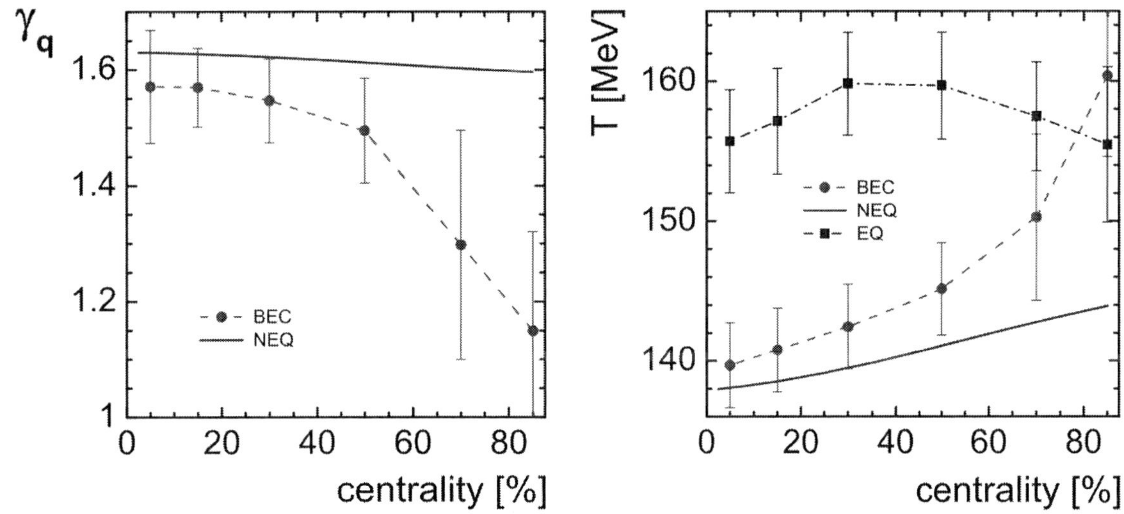

Figure 3. The non-equilibrium parameter γ_q (left) and temperature (right) as the function of centrality. The figures are from Ref. [24]

coincide at high centrality, because $\gamma_q^{\mathrm{BEC}} \simeq 1$ there, while $\gamma_q^{\mathrm{EQ}} = 1$ by definition.

Pion spectra with the condensate were calculated in Ref. [18] using the appropriately modified THERMINATOR. The condensate is at rest only in it's reference frame and is moving with the freeze-out hypersurface. The approximation (5) leads to the step at $p_T \sim 230$ MeV, because the corresponding momentum distribution is given by:

$$\frac{dN}{dyd\phi_p\, p_T dp_T} = \frac{N_{\mathrm{cond}}}{V}\frac{\tau_f^3}{m^2}\,\theta\left(r_{\max} - p_T\tau_f/m\right) . \tag{6}$$

where θ is the Heaviside step function. The inclusion of several more low lying levels would lead to finer steps that tend to a continuous line with increasing number of steps, and have the same area under the curve. The current approximation shows the maximal momentum that the condensate may obtain from the movement of the freeze-out hypersurface, which is $p_T < 250$ MeV.

The condensate rate grows with centrality in BEC, but the similar growths shows EQ model, see Fig. 4 (left). It is the finite volume effect, which meas that the ground state $N_0^{EQ} = N_{\mathrm{cond}}(\mu = 0)$ should be taken into account even in equilibrium for very peripheral, proton-proton, or proton-nucleus collisions. The gray area shows the 10% deviation from the best fit. It is well within the error bars at central, but not in peripheral collisions, see Fig. 4 (right). The combined data on multiplicities and spectra are compatible with 5% of pions in the condensate [18].

The recent PDG reviews [27] report much lower mass and width of the $f_0(500)$ resonance, or the sigma meson (σ), see Fig. 5, and Ref. [28, 29] for explanations. Lower mass results in higher multiplicity in HRG. The σ meson decays 100% into pions, therefore the inclusion of the updated σ could have added some of the missing pions and weaken the BEC signal. The Breit-Wigner distribution

[1]The same abbreviation BEC is used for the Bose-Einstein condensation and for the non-equilibrium model that allows the condensation on the ground state.

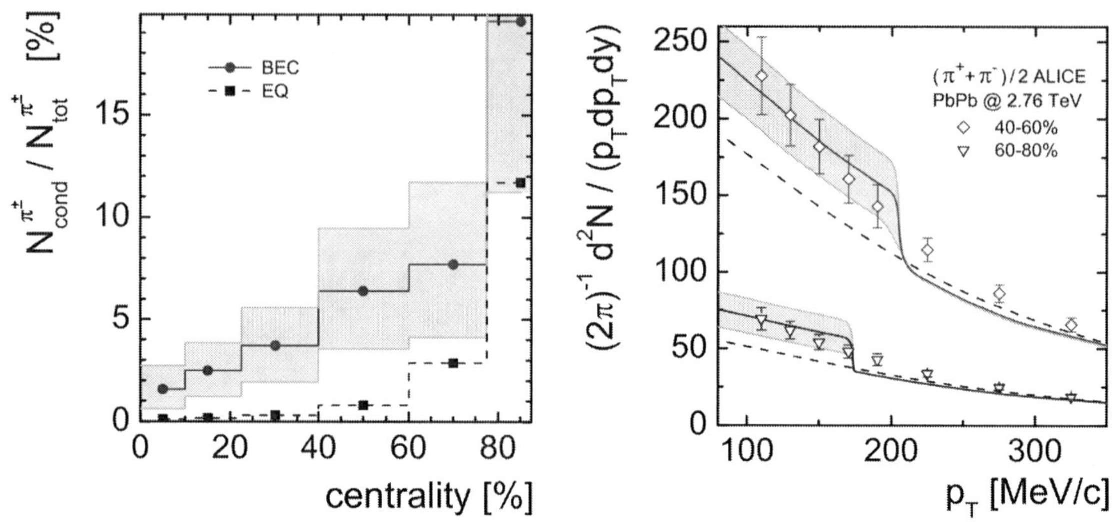

Figure 4. The pion spectra (left) and the condensate rate that was found in BEC (right) by the fit of mean multiplicities at the LHC. Solid line is for BEC, while dashed line is for EQ. The figures are from Ref. [18].

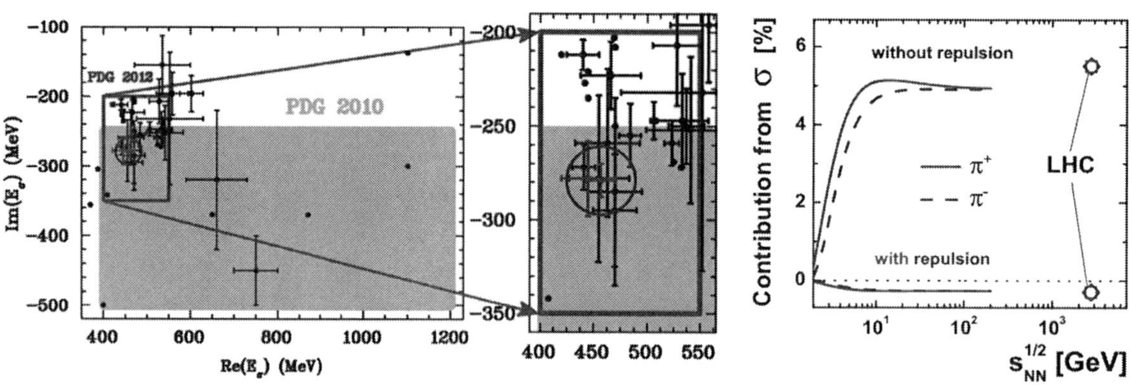

Figure 5. The change of the width and mass of the σ meson since 2010 (left), and the contribution to pion yields from the decay of σ (right). The left figure is from Ref. [30].

is not accurate enough for so wide resonances as σ. However, the width of the σ can be obtained from the derivative of the experimental $\pi\pi$ phase shift, that we take from [28]. It has attractive (0;0) and repulsive (2;0) isospin-spin channel. The attractive one is responsible for the emergence of the $f_0(500)$ resonance, however, the repulsive one cancels $f_0(500)$ until $f_0(980)$ takes over above the mass $M_{\pi\pi} \sim 0.85$ GeV, see Ref. [31]. The cancellation happens on the level of the distribution function, therefore it is present in all isospin-averaged observables. The σ implemented as a Breit-Winger pole with $M_\sigma = 484$ and $\Gamma_\sigma/2 = 255$ MeV produces up to 5% of pions, while the truth contribution is -0.3%, see Fig. 5 (right) [31]. The absence of pions from the σ decays enhances all ratios to pions, compared to the results that were obtained in models with σ. It is particularly important for the description of the K/π horn at the SPS [32, 33] and the proton-pion puzzle at the LHC.

 DOI: 10.1051/epjconf/201612603002

3 Conclusions

The non-equilibrium thermal model combined with the single freeze-out scenario explains very well the spectra of π, K, p, K_S^0, $K^*(892)^0$ and $\phi(1020)$ particles at the LHC. The introduction of the ground state decreases the non-equilibrium parameters and increases the temperature with centrality. The enhancement of the low p_T pion spectra may be interpreted as a signature of the onset of pion condensation at the LHC. The missing pions from the σ meson in HRG enhance the proton-pion puzzle at the LHC, and allow for larger amount of the condensate. Many efforts and attention of the community is required to obtain high temperature BEC and study it's properties. However, it could open a wide new field of research.

Acknowledgments
The author is thankful to W. Florkowski, M. Rybczynski, M. Gorenstein, W. Broniowski and F. Giacosa for fruitful collaboration. This work was supported by Polish National Science Center grant No. DEC-2012/06/A/ST2/00390.

References

[1] S. N. Bose, Z. Phys. **26**, 178 (1924); A. Einstein, Sitz. Ber. Preuss. Akad. Wiss. (Berlin) **1**, 3 (1925).
[2] M. H. Anderson, et al., Science **269**, 198 (1995); K. B. Davis, et al., Phys. Rev. Lett. **75**, 3969 (1995).
[3] Editor Tore Frangsmyr, *Les Prix Nobel*, (The Nobel Prizes 2001, Nobel Foundation, Stockholm, 2002) 77-108.
[4] V. V. Begun and M. I. Gorenstein, Phys. Rev. C **77**, 064903 (2008) [arXiv:0802.3349 [hep-ph]].
[5] B. Abelev *et al.* [ALICE Collaboration], Phys. Rev. C **88**, 044910 (2013) [arXiv:1303.0737 [hep-ex]].
[6] J. Stachel, A. Andronic, P. Braun-Munzinger and K. Redlich, J. Phys. Conf. Ser. **509**, 012019 (2014) doi:10.1088/1742-6596/509/1/012019 [arXiv:1311.4662 [nucl-th]].
[7] J. Cleymans, EPJ Web Conf. **95**, 03004 (2015) [arXiv:1412.7045 [hep-ph]].
[8] M. Petran, J. Letessier, V. Petracek and J. Rafelski, Phys. Rev. C **88**, no. 3, 034907 (2013) [arXiv:1303.2098 [hep-ph]].
[9] S. Chatterjee, B. Mohanty and R. Singh, Phys. Rev. C **92**, no. 2, 024917 (2015) [arXiv:1411.1718 [nucl-th]].
[10] J. Noronha-Hostler and C. Greiner, Nucl. Phys. A **931**, 1108 (2014) [arXiv:1408.0761 [nucl-th]].
[11] F. Becattini, M. Bleicher, T. Kollegger, T. Schuster, J. Steinheimer and R. Stock, Phys. Rev. Lett. **111**, 082302 (2013) [arXiv:1212.2431 [nucl-th]].
[12] E. Shuryak, arXiv:1412.8393 [hep-ph].
[13] T. Csorgo and L. P. Csernai, Phys. Lett. B **333**, 494 (1994) [hep-ph/9406365].
[14] J. P. Blaizot, F. Gelis, J. F. Liao, L. McLerran and R. Venugopalan, Nucl. Phys. A **873**, 68 (2012) [arXiv:1107.5296 [hep-ph]].
[15] F. Gelis, Nucl. Phys. A **931**, 73 (2014) [arXiv:1412.0471 [hep-ph]].
[16] V. Begun, W. Florkowski and M. Rybczynski, Phys. Rev. C **90**, no. 5, 054912 (2014) [arXiv:1405.7252 [hep-ph]].
[17] V. Begun, W. Florkowski and M. Rybczynski, Phys. Rev. C **90**, no. 1, 014906 (2014) [arXiv:1312.1487 [nucl-th]].

[18] V. Begun and W. Florkowski, Phys. Rev. C **91**, 054909 (2015) [arXiv:1503.04040 [nucl-th]].

[19] B. B. Abelev *et al.* [ALICE Collaboration], Phys. Rev. C **89**, no. 2, 024911 (2014) [arXiv:1310.7808 [nucl-ex]].

[20] J. Rafelski, Eur. Phys. J. A **51**, no. 9, 114 (2015) [arXiv:1508.03260 [nucl-th]].

[21] G. Torrieri, S. Steinke, W. Broniowski, W. Florkowski, J. Letessier and J. Rafelski, Comput. Phys. Commun. **167**, 229 (2005) [nucl-th/0404083].

[22] M. Chojnacki, A. Kisiel, W. Florkowski and W. Broniowski, Comput. Phys. Commun. **183**, 746 (2012) [arXiv:1102.0273 [nucl-th]].

[23] W. Broniowski and W. Florkowski, Phys. Rev. Lett. **87**, 272302 (2001) [nucl-th/0106050].

[24] V. Begun, EPJ Web Conf. **97**, 00003 (2015) [arXiv:1412.6532 [nucl-th]].

[25] S. Ryu, J.-F. Paquet, C. Shen, G. S. Denicol, B. Schenke, S. Jeon and C. Gale, Phys. Rev. Lett. **115**, no. 13, 132301 (2015) [arXiv:1502.01675 [nucl-th]].

[26] A. G. Knospe, C. Markert, K. Werner, J. Steinheimer and M. Bleicher, arXiv:1509.07895 [nucl-th].

[27] K. A. Olive *et al.* [Particle Data Group Collaboration], Chin. Phys. C **38**, 090001 (2014).

[28] R. Garcia-Martin, R. Kaminski, J. R. Pelaez and J. Ruiz de Elvira, Phys. Rev. Lett. **107**, 072001 (2011) [arXiv:1107.1635 [hep-ph]].

[29] J. R. Pelaez, arXiv:1510.00653 [hep-ph].

[30] R. Kaminski, Acta Phys. Polon. Supp. **8**, no. 1, 103 (2015).

[31] W. Broniowski, F. Giacosa and V. Begun, Phys. Rev. C **92**, no. 3, 034905 (2015) [arXiv:1506.01260 [nucl-th]].

[32] M. Gazdzicki and M. I. Gorenstein, Acta Phys. Polon. B **30**, 2705 (1999) [hep-ph/9803462].

[33] C. Alt *et al.* [NA49 Collaboration], Phys. Rev. C **77**, 024903 (2008) [arXiv:0710.0118 [nucl-ex]].

EPJ Web of Conferences **126**, 03003 (2016) DOI: 10.1051/epjconf/201612603003
ICNFP 2015

A possible evidence of observation of two mixed phases in nuclear collisions

K.A. Bugaev[1], A.I. Ivanytskyi[1], V.V. Sagun[1], G.M. Zinovjev[1], D.R. Oliinychenko[1,2], V.S. Trubnikov[3], and E.G. Nikonov[4]

[1] *Bogolyubov Institute for Theoretical Physics of the National Academy of Sciences of Ukraine, 03680, Kiev, Ukraine*
[2] *FIAS, Goethe University, Ruth-Moufang Str. 1, 60438 Frankfurt upon Main, Germany*
[3] *National Science Center "Kharkov Institute of Physics and Technology", 61108, Kharkov, Ukraine*
[4] *Laboratory for Information Technologies, JINR, Joliot-Curie Str. 6, Dubna 141980, Russia*

Abstract. Using an advanced version of the hadron resonance gas model we have found several remarkable irregularities at chemical freeze-out. The most prominent of them are two sets of highly correlated quasi-plateaus in the collision energy dependence of the entropy per baryon, total pion number per baryon, and thermal pion number per baryon which we found at center of mass energies 3.6-4.9 GeV and 7.6-10 GeV. The low energy set of quasi-plateaus was predicted a long time ago. On the basis of the generalized shock-adiabat model we demonstrate that the low energy correlated quasi-plateaus give evidence for the anomalous thermodynamic properties of the mixed phase at its boundary to the quark-gluon plasma. The question is whether the high energy correlated quasi-plateaus are also related to some kind of mixed phase. In order to answer this question we employ the results of a systematic meta-analysis of the quality of data description of 10 existing event generators of nucleus-nucleus collisions in the range of center of mass collision energies from 3.1 GeV to 17.3 GeV. These generators are divided into two groups: the first group includes the generators which account for the quark-gluon plasma formation during nuclear collisions, while the second group includes the generators which do not assume the quark-gluon plasma formation in such collisions. Comparing the quality of data description of more than a hundred of different data sets of strange hadrons by these two groups of generators, we find two regions of the equal quality of data description which are located at the center of mass collision energies 4.3-4.9 GeV and 10.-13.5 GeV. These two regions of equal quality of data description we interpret as regions of the hadron-quark-gluon mixed phase formation. Such a conclusion is strongly supported by the irregularities in the collision energy dependence of the experimental ratios of the Lambda hyperon number per proton and positive kaon number per Lambda hyperon. Although at the moment it is unclear, whether these regions belong to the same mixed phase or not, there are arguments that the most probable collision energy range to probe the QCD phase diagram (tri)critical endpoint is 12-14 GeV.

1 Introduction

The hadron resonance gas model (HRGM) [1] is traditionally used to extract the parameters of chemical freeze-out (CFO) from the measured hadronic yields. Its version with the multicomponent hard-core repulsion [2–4] allowed one for the first time to successfully describe the most problematic ratios K^+/π^+ with $\chi^2/dof = 3.9/14$ and Λ/π^- with $\chi^2/dof = 10.2/12$ without spoiling all other hadron yield ratios [5, 6]. Fig. 1 demonstrates the present fit quality of these traditionally problematic ratios. The achieved high quality $\chi^2/dof \simeq 0.95$ [5, 6] of data description of 111 independent hadron yield ratios measured at midrapidity in central nucleus-nucleus collisions at the center of mass energies $\sqrt{s_{NN}} = 2.7, 3.1, 3.6, 4.3, 4.9, 6.3, 7.6, 8.8, 9.2, 12.3, 17.3, 62.4, 130, 200$ GeV proves that the multi-component version of HRGM is a precise and a sensitive tool of heavy ion collision phenomenology.

Using the multicomponent version of HRGM it was possible to reveal a few novel irregularities at CFO. The most remarkable of them are two sets of highly correlated quasi-plateaus in the collision energy dependence of the entropy per baryon, total pion number per baryon, and thermal pion number per baryon which were found at the center of mass energies 3.6-4.9 GeV and 7.6-10 GeV [7] and the sharp peak of the trace anomaly found at the center of mass energy 4.9 GeV [8]. The low energy set of quasi-plateaus was predicted a long time ago [9–11] as a signal of the anomalous thermodynamic properties inside the quark-gluon-hadron mixed phase. Unfortunately, the generalized shock-adiabat model cannot be safely applied to the central nuclear collisions at $\sqrt{s_{NN}} \geq 7.6$ GeV [9]. Therefore, in order to correctly interpret the high energy quasi-plateaus here we use the results of meta-analysis [12] of the quality of data description (QDD) of 10 existing event generators of nucleus-nucleus collisions along with the thorough analysis of irregularities in the existing experimental hadron yield ratios.

The work is organized as follows. In Sect. 2 we remind the basic elements of the HRGM with multicomponent hard-core repulsion. A brief description of the meta-analysis suggested in [12] is presented in Sect. 3 along with a discussion of existing hadron multiplicity data which help to shed light on the problem of the formation of two quark-gluon-hadron mixed phases. In Sect. 4 our conclusions are formulated.

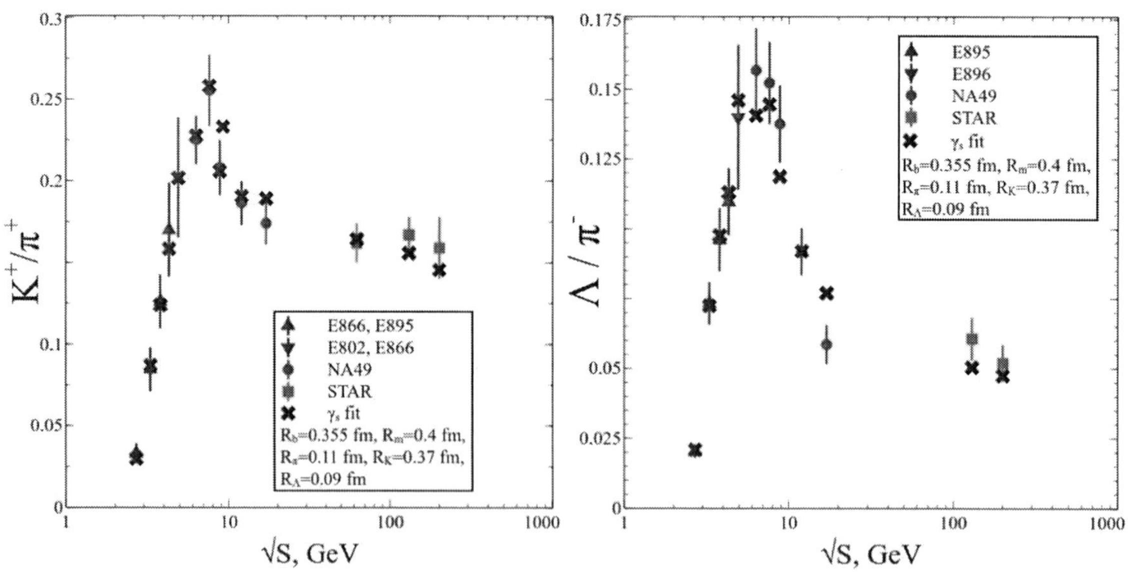

Figure 1. Collision energy dependence of K^+/π^+ and Λ/π^- hadron yield ratios which traditionally were the most problematic ones.

2 HRGM with multicomponent hard-core repulsion

The HRGM is based on the assumption of local thermal and chemical equilibrium at CFO. Hence the hadron yields produced in the collisions of large atomic nuclei can be found using the grand canonical valuables, i.e. using the temperature T, the baryonic μ_B, the strange μ_s and the isospin third projection μ_{I3} chemical potentials. As usual, the chemical potential μ_s is fixed by the condition of zero total strange charge. The possible deviation of strange charge from the full chemical equilibrium is taken into account by the parameter γ_s [13]. It changes the thermal density φ_j of hadron sort j as $\varphi_j \to \gamma_s^{S_j} \varphi_j$, where S_j is the total number of strange valence quarks and antiquarks in such a hadron.

The main difference of the present version of HRGM from the ones developed by other authors is that in our HRGM several sorts of hadrons have individual hard-core radii. Thus, it employes different hard-core radii for pions, R_π, kaons, R_K, Λ-hyperons, R_Λ, other mesons, R_m, and baryons, R_b. The best global fit of 111 independent hadronic multiplicities measured in the whole collision energy range from $\sqrt{s_{NN}} = 2.7$ GeV to $\sqrt{s_{NN}} = 200$ GeV was found for $R_b = 0.355$ fm, $R_m = 0.4$ fm, $R_\pi = 0.1$ fm, $R_K = 0.37$ fm and $R_\Lambda = 0.11$ fm with the quality $\chi^2/dof \simeq 0.95$ [5, 6]. The second virial coefficient between the hadrons of hard-core radii R_i and R_j is defined as $b_{ij} = \frac{2\pi}{3}(R_i + R_j)^3$. Taking from the thermodynamic code THERMUS [14] such characteristics of hadrons of sort i as the spin-isospin degeneracy g_i, the mass m_i and the width Γ_i one can find the set of partial pressures p_i for each hadronic component ($p = \sum_i p_i$ is total pressure) from the system

$$p_i = T\varphi_i \exp\left[\frac{\mu_i - 2\sum_j p_j b_{ji} + \sum_{jl} p_j b_{jl} p_l/p}{T}\right] . \tag{1}$$

Here $\mu_i = Q_i^B \mu_B + Q_i^{I3} \mu_{I3} + Q_i^S \mu_S$ is the full chemical potential of the i-th hadronic sort expressed via its charges $\{Q_i^A\}$ and the corresponding chemical potentials $\{\mu_A\}$ (with $A \in \{B, I3, S\}$). In the Boltzmann approximation the thermal density of i-th hadronic sort reads as

$$\varphi_i = \gamma_s^{S_i} g_i \int\limits_{M_i}^{\infty} dm\, f(m, m_i, \Gamma_i) \int \frac{k^2 dk}{2\pi^2}\, e^{-\frac{\sqrt{m^2+k^2}}{T}} , \tag{2}$$

where M_i is a threshold of its dominant decay channel and f is the normalized Breit-Wigner mass attenuation function. Thermal multiplicities $N_i^{th} = V\frac{\partial p}{\partial \mu_i}$ (V is the effective volume at CFO) should be corrected by the hadron decays after the CFO according to the branching ratios $Br_{l\to i}$. The latter define the probability of particle l to decay into a particle i. Hence the ratio of full multiplicities can be written as

$$\frac{N_i^{tot}}{N_j^{tot}} = \frac{p_i + \sum_{l\neq i} p_l Br_{l\to i}}{p_j + \sum_{l\neq j} p_l Br_{l\to j}} . \tag{3}$$

With the help of (3) we obtained the high quality fit of experimental hadron ratios measured at AGS for energies $\sqrt{s_{NN}} = 2.7, 3.1, 3.6, 4.3, 4.9$ GeV[15–24], at SPS energies $\sqrt{s_{NN}} = 6.3, 7.6, 8.8, 12.3, 17.3$ GeV measured by the NA49 Collaboration [25–33] and at RHIC energies $\sqrt{s_{NN}} = 9.2, 62.4, 130, 200$ GeV measured by the STAR Collaboration [34]. As described in [3, 4], from these data we constructed 111 independent ratios measured at 14 values of collision energies. The most important results are shown in Figs. 1 and 2. The left panel of Fig. 2 shows the highly correlated quasi-plateaus in the collision energy dependence of the entropy per baryon s/ρ_B, total pion number per baryon ρ_π^{tot}/ρ_B, and thermal pion number per baryon ρ_π^{th}/ρ_B at laboratory energies 6.9–11.6 GeV (i.e. $\sqrt{s_{NN}} = 3.6 - 4.9$ GeV) which were found in [7]. As one can see from the left panel of Fig. 2, a clear plateau is demonstrated by the thermal pion number per baryon while other quantities show quasi-plateaus.

Nevertheless, all these quasi-plateaus are important, since their strong correlation with the plateau in the thermal pion number per baryon allows one to find out their common width in the collision energy [7, 8].

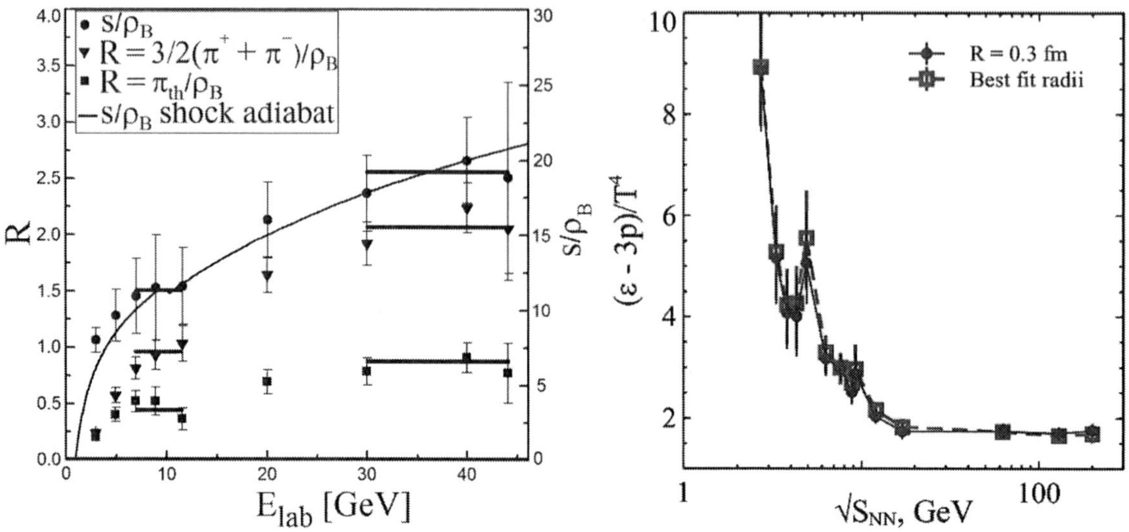

Figure 2. Left: The correlated quasi-plateaus at CFO found in [7] (see details in the text). Right: Trace anomaly as function of collision energy at CFO established in [8].

Note that these low energy quasi-plateaus were predicted about 25 years ago [9–11] as a manifestation of the anomalous thermodynamic properties of quark-gluon-hadron mixed phase. In contrast to the normal thermodynamic properties, in the medium with the anomalous ones the adiabatic compressibility of matter increases for increasing pressure. In the normal media (pure gaseous or liquid phases) there exists a repulsion between the constituents at short distances which leads to an opposite behavior of the adiabatic compressibility. Therefore, an appearance of these quasi-plateaus is a signal of the quark-gluon-hadron mixed phase formation [9–11]. Such a conclusion is strongly supported by an existence of the sharp peak of the trace anomaly $\delta = \frac{\varepsilon-3p}{T^4}$ (here ε is energy density) at $\sqrt{s_{NN}} = 4.9$ GeV [8] (see the right panel of Fig. 2). This peak is important, since in lattice QCD an inflection or a maximum point of the trace anomaly is used for a determination of the pseudo-critical temperature of the cross-over transition [35]. One may think that a sharp peak of δ at CFO is exclusively generated by the peak of baryonic density which in our HRGM also exists at $\sqrt{s_{NN}} = 4.9$ GeV. However, the real situation is more complicated. Writing the trace anomaly as

$$\delta = \frac{\varepsilon - 3p}{T^4} = \frac{Ts + \mu_B\rho_B + \mu_{I3}\rho_{I3} - 4p}{T^4} \simeq \frac{s}{T^3}\left(1 + \frac{\mu_B}{T}\frac{\rho_B}{s}\right) - 4\frac{p}{T^4}, \tag{4}$$

where in the last step the small contribution $\mu_{I3}\rho_{I3}$ related to the charge of the third isospin projection is neglected. From (4) one can easily conclude that the strong increase of δ on the collision energy interval $\sqrt{s_{NN}} = [4.3; 4.9]$ GeV is provided by a strong jump of the effective number of degrees of freedom s/T^3 on this interval [8, 36]. Note that despite the existence of a baryon density peak on this interval of collision energy, the entropy per baryon s/ρ_B is constant on it as one can see from the left panel of Fig. 2. Now it is evident that without a strong jump of the effective number of degrees of freedom s/T^3 the sharp peak of trace anomaly at CFO would not exist. At higher collision energies the trace anomaly δ decreases mainly because the ratio μ_B/T strongly decreases, while the inverse entropy per baryon decreases slowly. It is important to mention that the sharp peak of δ is seen, if the finite width of all hadronic resonances is included into the HRGM [36], while for the HRGM with a zero width of hadron resonances such a peak is washed out.

The physical origin of the trace anomaly sharp peak (and, hence, of a strong jump of the effective number of degrees of freedom s/T^3) at CFO found at $\sqrt{s_{NN}} = 4.9$ GeV is rooted in the trace anomaly peak existing at the shock adiabat [8]. The shock adiabat model reasonably well describes the hydrodynamic and thermodynamic quantities of the initial state formed in the central nucleus-nucleus collisions in the laboratory energy range 1 GeV $\leq E_{lab} \leq$ 30 GeV [8], while at higher collision energies, i.e. for $\sqrt{s_{NN}} \geq 7.6$ GeV, it can be used for qualitative estimates. In [8] it was found that the peak of δ at the shock adiabat appears at the collision energy corresponding exactly to the boundary between the quark gluon plasma (QGP) and quark-gluon-hadron mixed phase and, therefore, the trace anomaly sharp peak at CFO is a signal of QGP formation. In this respect it is interesting that in the right panel of Fig. 2 one can see a second peak of trace anomaly located at $\sqrt{s_{NN}} = 9.2$ GeV. Although the second peak of δ is less pronounced than the first one, it is also associated to the high energy set of quasi-plateaus shown in the left panel of Fig. 2 at the collision energy interval $E_{lab} = [30; 44.]$ GeV ($\sqrt{s_{NN}} = [7.6; 9.2]$ GeV). Therefore, the future experiments at RHIC, NICA and FAIR will have to find out whether the high energy peak of trace anomaly has any physical meaning.

3 Meta-analysis of quality of data description

The main objects of a meta-analysis suggested in [12] are the mean deviation squared of the quantity $A^{model,h}$ of the model M from the data $A^{data,h}$ per number of the data points n_d for a particle type h

$$\langle\chi^2/n\rangle^h_A\Big|_M = \frac{1}{n_d}\sum_{k=1}^{n_d}\left[\frac{A_k^{data,h} - A_k^{model,h}}{\delta A_k^{data,h}}\right]^2\Bigg|_M, \tag{5}$$

and its error which is defined according to the rule of indirect measurements [58] as

$$\Delta_A\langle\chi^2/n\rangle^h_A\Big|_M \equiv \left[\sum_{k=1}^{n_d}\left[\delta A_k^{data,h}\,\frac{\partial\langle\chi^2/n\rangle^h_A\big|_M}{\partial\, A_k^{data,h}}\right]^2\right]^{\frac{1}{2}} \equiv \frac{2}{\sqrt{n_d}}\sqrt{\langle\chi^2/n\rangle^h_A\Big|_M}, \tag{6}$$

where $\delta A_k^{data,h}$ is an experimental error of the experimental quantity $A_k^{data,h}$ and the summation in Eqs. (5) and (6) runs over all data point n_d at given collision energy. For a convenience the quantity defined in (5) is called the quality of data description (QDD). To get the most complete picture of the dynamics of nuclear collisions, one has to compare the available data on the transverse mass (m_T) distributions $A = \frac{1}{m_T}\frac{d^2N(m_T,y)}{dm_T dy}$, the longitudinal rapidity (y) distributions $A = \frac{dN(y)}{dy}$ and the hadronic yields (Y) measured at midrapidity $A = \frac{dN(y=0)}{dy}$ or/and the total one, i.e. measured within 4π solid angle, since right these observables are traditionally believed to be sensitive to the equation of state properties [37, 38].

The QDD of strange hadrons was found for two types of models [12]:

- **The hadron gas (HG) models** are as follows: ARC [39], RQMD2.1(2.3) [40], HSD [41–43], UrQMD1.3(2.0, 2.1, 2.3) [44], statistical hadronization model (SHM) [45] and AGSHIJET_N* [46, 47]. These models do not include the QGP formation in the process of A+A collisions.
- **The QGP models** are as follows: Quark Combination (QuarkComb) model [48], 3-fluid dynamics (3FD) model [49–52], PHSD model [53–55] and Core-Corona model [56, 57]. These generators explicitly assume the QGP formation in A+A collisions.

 DOI: 10.1051/epjconf/201612603003

A short description of these models along with the criteria of their selection can be found in the Appendix of [12].

The main idea of such a meta-analysis [12] is based on the assumption that the HG models of heavy ion collisions should provide worse description of the data above the QGP threshold energy, whereas below this threshold they should be able to better (or at least not worse) reproduce experimental data compared to the QGP models. Furthermore, it is assumed that both kinds of models should provide an equal and rather good QDD at the energy of mixed phase production. Hence, the energy of the mixed phase formation should be located below the energy at which the equal QDD is changed to the essentially worse QDD of HG models.

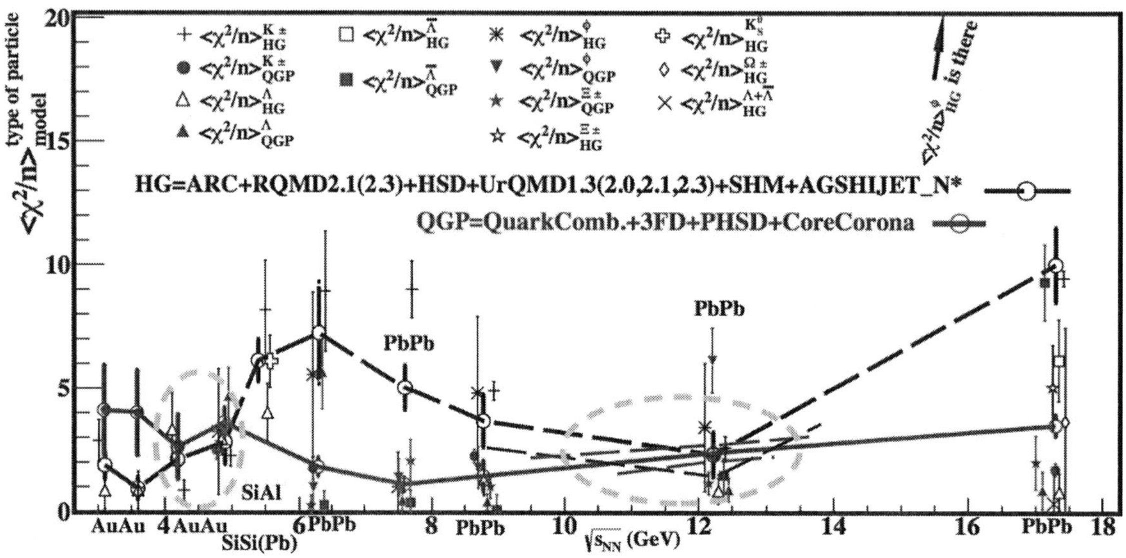

Figure 3. Comparison of $\langle\chi^2/n\rangle^{\{h\}}_{\{A\}}\Big|_{HG}$ (black symbols and dashed curve) and $\langle\chi^2/n\rangle^{\{h\}}_{\{A\}}\Big|_{\overline{QGP}}$ (red symbols and solid curve) as functions of collision energy obtained for the arithmetic averaging. The symbols of different hadrons which correspond to the same collision energy are slightly spread around the energy value for better perception. The symbols are connected by the lines to guide the eye. The numbers staying behind the short name of the model indicate the version(s) used in meta-analysis. The short dashed ovals indicate the regions of possible mixed phase formation.

Based on these assumptions the experimental data measured at the collision energies $\sqrt{s_{NN}}$ = 3.1, 3.6, 4.2, 4.9, 5.4, 6.3, 7.6, 8.8, 12.3 and 17.3 GeV were analyzed in [12]. The collision energies $\sqrt{s_{NN}} \leq 4.9$ GeV correspond to Au+Au reactions studied at AGS. At $\sqrt{s_{NN}} = 5.4$ GeV the reactions Pb+Si, Si+Si and Si+Al were also investigated at AGS, while higher collision energies correspond to Pb+Pb reactions studied at SPS. Using the definitions (5) and (6) at each collision energy the available description of the transverse mass distributions, the longitudinal rapidity distributions and the hadronic yields measured at midrapidity or/and the total one obtained by a given model was arithmetically averaged for each kind of analyzed strange hadron. The QDDs and their errors obtained for each energy and for each hadron were arithmetically averaged over the models belonging to the same type. Then the QDDs and their errors found in this way for the models belonging to the same type were arithmetically averaged for each hadron and antihadron, if available, in order to reduce the number of data for a comparison. Finally, the resulting QDDs and their errors of the same same type of model were arithmetically averaged over all hadronic species. More details can be found in [12]. The

DOI: 10.1051/epjconf/201612603003

averaged QDDs of HG models $\langle \chi^2/n \rangle^{\overline{\{h\}}}_{\{A\}}\Big|_{\overline{HG}}$ and the ones of QGP models $\langle \chi^2/n \rangle^{\overline{\{h\}}}_{\{A\}}\Big|_{\overline{QGP}}$ were found in this way together with their errors. The results are shown in Fig. 3.

From Fig. 3 one can see that the meta-analysis of work [12] leads to an independent conclusion that the mixed phase exists at the same collision energy range $\sqrt{s_{NN}} = [4.3; 4.9]$ GeV which was originally found in [7, 8]. This result is important not only to validate the entire framework of shock adiabat model used in [7, 8], but also to justify the jump of the effective number of degrees of freedom s/T^3 at CFO and a sharp peak of the trace anomaly δ at CFO as reliable signals of QGP formation. In addition the meta-analysis of QDD [12] predicts that the most probable collision energy range of the second phase transition is $\sqrt{s_{NN}} = 10 - 13.5$ GeV. Thus, the meta-analysis of QDD supports an interpretation of the high energy set of correlated quasi-plateaus as an indicator of phase transition, although it shifts this transition to slightly higher collision energies.

Unfortunately, at present it is impossible to distinguish between two possible explanations of this phenomenon. The first possible explanation is that with increasing collision energy the initial states of thermally equilibrated matter formed in nucleus-nucleus collisions move first from the hadron gas into the mixed phase, then from the mixed phase to QGP and then again they return to the same mixed phase, but at higher initial temperature and lower baryonic density. Such a scenario corresponds to the case, if QCD phase digram has a critical endpoint [12]. An alternative explanation [12] corresponds to the QCD phase diagram with the tricritical endpoint. In such a case the second phase transition is the second order phase transition of (partial) chiral symmetry restoration or a transition between quarkyonic matter and QGP [59]. It is necessary to stress, that despite the lack of a single interpretation of the second phase transition possibly existing at $\sqrt{s_{NN}} = 10 - 13.5$ GeV there are strong arguments [12] that the (tri)critical endpoint of QCD phase diagram maybe located at $\sqrt{s_{NN}} = 12 - 14$ GeV.

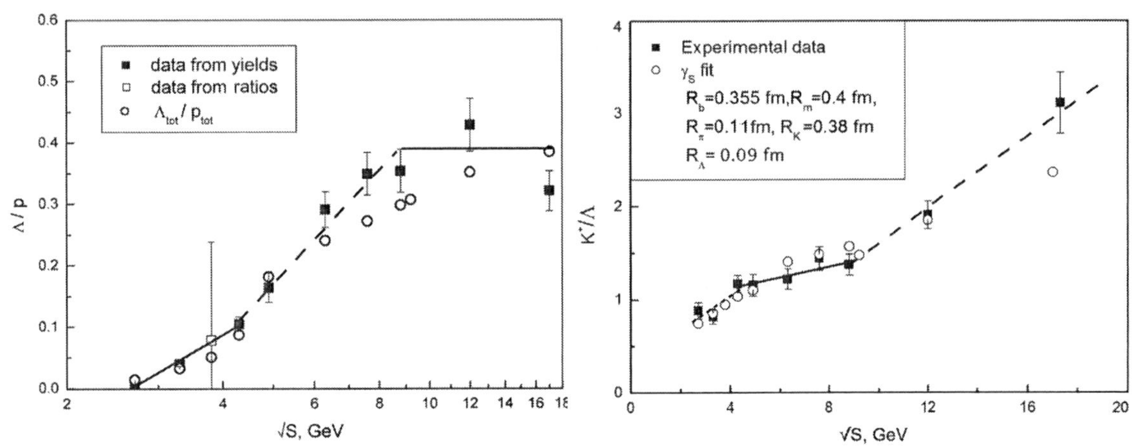

Figure 4. The collision energy dependence of the Λ/p (left) and K^+/Λ (right) ratios obtained within the present HRGM. The lines are given to guide the eye. More explanations are given in the text.

The combined conclusions obtained from inspecting two sets of correlated quasi-plateaus at CFO, two peaks of trace anomaly at CFO and the ones found out by the meta-analysis, led us a thorough analysis of the Λ/p and K^+/Λ ratios. From the left panel of Fig. 4 one can see that there are three regimes in the energy dependence of the Λ/p ratio: at $\sqrt{s_{NN}} = 4.3$ GeV the slope of this ratio clearly increases, while above $\sqrt{s_{NN}} = 8.8$ GeV it nearly saturates. The sudden increase of Λ/p slope at $\sqrt{s_{NN}} = 4.3$ GeV can be naturally explained by the idea of work [60] that the mixed phase formation can be identified by a rapid increase in the number of strange quarks per light quarks. Evidently, the Λ/p ratio is a convenient indicator because at low collision energies Λ hyperons are generated in collisions of nucleons. Moreover, such a ratio does not depend on baryonic chemical potential, since both the protons and Λ hyperons have the same baryonic charge. As it is seen from the left panel of Fig. 4, this mechanism works up to $\sqrt{s_{NN}}$ = 4.3 GeV, while an appearance of the mixed phase should lead to an increase of the number of strange quarks and antiquarks due to the annihilation of light quark-antiquark and gluon pairs. Clearly, this simple picture is well fitted into the prediction that the mixed phase can be reached at $\sqrt{s_{NN}}$ = 4.3 GeV, while QGP is formed at $\sqrt{s_{NN}} \geq 4.9$ GeV. The dramatic decrease of the slope of the experimental Λ/p ratio at $\sqrt{s_{NN}} > 8.8$ GeV which is seen in Fig. 4 can be an evidence for the second phase transformation, which we discussed above.

It is appropriate to say a few words about the experimental data of hadron yields shown in both panels of Fig. 4. For the AGS collision energies $\sqrt{s_{NN}}$ = 2.7, 3.1, and 4.3 GeV the yields of protons and kaons were, respectively, taken from Refs. [24] and [18], whereas for Λ−hyperons they were taken from Ref. [19]. Experimental yields measured at the highest AGS energy $\sqrt{s_{NN}}$ = 4.9 GeV for protons and kaon were taken from Ref. [16, 24], while for Λ they were given in Ref. [23]. The mid-rapidity yields of protons, kaons and lambdas measured at the SPS energies $\sqrt{s_{NN}}$ = 6.3, 7.6, 8.8, 12, and 17.3 GeV are provided by the NA49 collaboration in Refs. [26, 27, 29–31]. For a comparison, in Fig. 4 we also show the value with huge error bars which are found from other two ratios, Λ/π^- and p/π^-, for $\sqrt{s_{NN}}$ = 3.6 GeV.

It is interesting that the energy dependence of the K^+/Λ ratio shows the change of slopes at the same energies as the Λ/p ratio. This can be see from the right panel of Fig. 4. Note that in the dominant hadronic reactions the positive kaons and Λ hyperons are born simultaneously. Since both of these hadrons carry the strange charge, then the logic of work [60] is inapplicable to their ratio. Therefore, in contrast to an increase of the slope of the Λ/p ratio on the interval $\sqrt{s_{NN}}$ = [4.3; 8.8] GeV, the K^+/Λ ratio has a flattening of the slope on this collision energy interval. In the leading order this ratio is defined via the kaon mass m_K, the Λ mass m_Λ and two chemical potentials as $K^+/\Lambda \simeq \sqrt{\left[\frac{m_K}{m_\Lambda}\right]^3} \exp\left[\frac{m_\Lambda - m_K + 2\mu_S - \mu_B}{T}\right]$. Therefore, a small slope of this ratio at the interval $\sqrt{s_{NN}}$ = [4.3; 8.8] GeV evidences about the cancellation of energy dependencies of strange and baryonic chemical potentials, i.e. $m_\Lambda - m_K + 2\mu_S - \mu_B \simeq Const$ for $\sqrt{s_{NN}}$ = [4.3; 8.8] GeV. An increase of the K^+/Λ ratio at higher collision energies can be mainly explained by the fast decrease of the baryonic chemical potential.

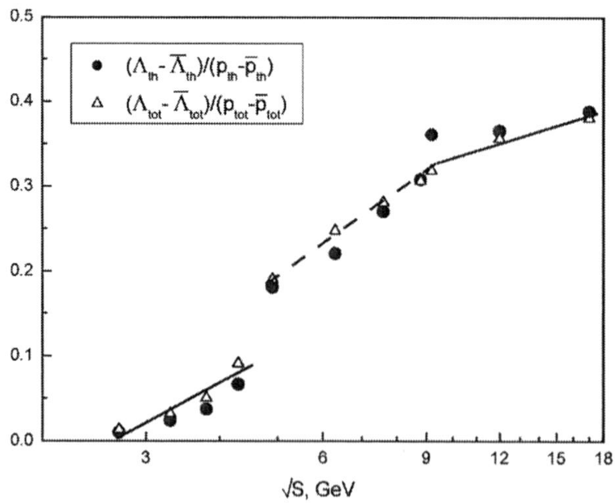

Figure 5. Most recent predictions for the collision energy dependence of the $\frac{\Delta\Lambda}{\Delta p}$ ratio. The triangles depict the ratio of total multiplicities, while the circles correspond to the ratio of thermal multiplicities. The lines are given to guide the eye.

DOI: 10.1051/epjconf/201612603003

Very recently a better description of the Λ/p ratio was achieved in [61], when more data were analyzed. Note that this result was obtained not on the expense of worsening of other hadron yield ratios. Based on this new fit of hadron yield ratios the predictions for the $\frac{\Delta\Lambda}{\Delta p} = \frac{\Lambda-\bar{\Lambda}}{p-\bar{p}}$ ratio are made. As one can see from Fig. 5 this ratio demonstrates even more dramatic changes in the collision energy dependence. Indeed, at the narrow collision energy interval $\sqrt{s_{NN}} = 4.3 - 4.9$ GeV this ratio has a strong jump, while at $\sqrt{s_{NN}} = 9.2$ GeV it shows a change of slope. Our educated guess is that the collision energy dependence of the $\frac{\Delta\Lambda}{\Delta p}$ ratio is an indicator of two phase transformations [61]. Since the observed jump of this ratio is located in the collision energy region of the mixed phase formation (i.e. with a first order phase transition), then a change of its slope at $\sqrt{s_{NN}} = 9.2$ GeV can be naturally associated with a weak first order or a second order phase transition. Note that such a hypothesis is well supported by the results of the meta-analysis [12] which is briefly summarized above.

4 Conclusions

From the discussions given in previous sections it is clear that a development of the multicomponent version of HRGM in 2012 led to a real breakthrough in our understanding of the thermodynamics at CFO. With the help of HRGM it was possible for the first time to describe the Strangeness Horn with the highest accuracy [3, 4, 6] including the topmost point. The new concept of separated CFOs for strange and non-strange hadrons with conservation laws connecting them allows one to naturally explain the appearance of apparent chemical non-equilibrium of strange particles [4]. Furthermore, a thorough analysis of the entropy per baryon and the pion number (thermal and total) per baryon at CFO led to a finding out of two sets of strongly correlated quasi-plateaus [7, 8]. Since the low energy set of quasi-plateaus was predicted in [9–11] as a signal of the quark-gluon-hadron mixed phase formation, then it was necessary to give a physical interpretation of the high energy set of quasi-plateaus.

A good hint to interpret the appearance of high energy set of quasi-plateaus is provided by the meta-analysis [12] of QDD. Since the QDD meta-analysis gave an independent evidence for the quark-gluon-hadron mixed phase formation at the narrow region of collision energy $\sqrt{s_{NN}} = 4.3 - 4.9$ GeV, then its predictions for the possible existence of another mixed phase at collision energies $\sqrt{s_{NN}} = 10 - 13.5$ GeV led us to a more thorough inspection of existing hadron multiplicity ratios. As one can see from the left panel of Fig. 4 the Λ/p ratio exhibits three different regimes in the collision energy dependence: at $\sqrt{s_{NN}} = 4.3$ GeV the slope of this ratio suddenly increases, while above $\sqrt{s_{NN}} = 8.8$ GeV it nearly saturates. As we argued above a strong increase of Λ/p slope at $\sqrt{s_{NN}} = 4.3$ GeV can be naturally explained by the idea of work [60] that the mixed phase formation can be identified by a rapid increase in the number of strange quarks per light quarks, while a dramatic decrease of the Λ/p slope at $\sqrt{s_{NN}} > 8.8$ GeV can be an evidence for the second phase transformation.

It is remarkable that the K^+/Λ ratio shows the change of slopes at the same energies as the Λ/p ratio, although the K^+/Λ ratio involves two strange particles and (in general) it strongly depends on the baryonic chemical potential. Also the trace anomaly at CFO shows two peaks at the collision energies $\sqrt{s_{NN}} = 4.9$ GeV and $\sqrt{s_{NN}} = 9.2$ GeV. The low energy trace anomaly peak can be explained within the shock adiabat model [8] as a signal of QGP formation, whereas the existence of high energy peak requires further confirmation by better experimental data. If it will be confirmed, then it will serve as a new indicator for a second phase transition. Nevertheless, already now all irregularities at CFO discussed here together with the results of the QDD meta-analysis form a coherent picture of possible observation of two mixed phases in nucleus-nucleus collisions. Therefore, the Beam Energy Scan program at RHIC has a unique chance to experimentally verify the above signals and to discover the mixed phases before the start of NICA and FAIR programs. The new observable, the $\frac{\Delta\Lambda}{\Delta p}$ ratio suggested in [61], will be very helpful for this because of its high sensitivity. However, to reach such goals the RHIC experiments should provide much smaller error bars, especially at low energies. Hence, the experiments in a fixed target mode are absolutely necessary for the success of the Beam Energy Scan program at RHIC.

Acknowledgements

The authors are thankful to D. B. Blaschke, T. Galayuk, R. A. Lacey, I. N. Mishustin, D. H. Rischke, K. Redlich, L. M. Satarov, A.V. Taranenko and K. Urbanowski for interesting discussions and valuable comments. K.A.B., A.I.I., V.V.S. and G.M.Z. acknowledge the partial support of the program "On perspective fundamental research in high-energy and nuclear physics" launched by the Section of Nuclear Physics of NAS of Ukraine. D.R.O. acknowledges a support of Deutsche Telekom Stiftung. K.A.B. is very thankful to all organizers of the ICNFP2015 for providing him with a chance to present and to discuss these results at the Conference and for a warm hospitality in OAC.

References

[1] A. Andronic, P. Braun-Munzinger and J. Stachel, Nucl. Phys. A **772**, 167 (2006) and references therein.
[2] D.R. Oliinychenko, K.A. Bugaev and A.S. Sorin, Ukr. J. Phys. **58**, 211 (2013).
[3] K. A. Bugaev, D. R. Oliinychenko, A. S. Sorin, G. M. Zinovjev, Eur. Phys. J. A **49**, 30 (2013).
[4] K. A. Bugaev *et al.*, Europhys. Lett. **104**, 22002 (2013).
[5] V. V. Sagun, Ukr. J Phys. **59**, 755 (2014).
[6] V. V. Sagun *et al.*, Ukr. J. Phys. **59**, 1043 (2014).
[7] K. A. Bugaev *et al.*, Phys. Part. Nucl. Lett. **12**, 238 (2015).
[8] K. A. Bugaev *et al.*, arXiv:1412.0718 [nucl-th] and references therein.
[9] K. A. Bugaev, M. I. Gorenstein, B. Kämpfer and V. I. Zhdanov, Phys. Rev. D **40**, 2903 (1989).
[10] K. A. Bugaev, M. I. Gorenstein and D. H. Rischke, JETP Lett. **52**, 1121 (1990).
[11] K. A. Bugaev, M. I. Gorenstein and D. H. Rischke, Phys. Lett. B **255**, 18 (1991).
[12] V. A. Kizka, V. S. Trubnikov, K. A. Bugaev and D. R. Oliinychenko, arXiv:1504.06483 [hep-ph].
[13] J. Rafelski, Phys. Lett. B **62**, 333 (1991).
[14] S. Wheaton, J. Cleymans and M. Hauer, Comput. Phys. Commun. **180**, 84 (2009).
[15] J.L. Klay *et al.*, Phys. Rev. C **68**, 054905 (2003).
[16] L. Ahle *et al.*, Phys. Lett. B **476**, 1 (2000).
[17] L. Ahle *et al.*, Phys. Lett. B **490**, 53 (2000).
[18] J.L. Klay *et al.*, Phys. Rev. Lett. **88**, 102301 (2002).
[19] C. Pinkenburg *et al.*,Nucl. Phys. A **698**, 495 (2002).
[20] J. Barrette *et al.*, Phys. Rev. C **62**, 024901 (2000).
[21] L. Ahle *et al.*, Phys. Rev. C **60**, 064901 (1999).
[22] L. Ahle *et al.*, Phys. Rev. C **58**, 3523 (1998).
[23] S. Albergo *et al.*, Phys. Rev. Lett. **88**, 062301 (2002).
[24] B. B. Back *et al.*, Phys. Rev. Lett. **87**, 1970 (2001).
[25] I. G. Bearden *et al.*, Phys. Rev. C **66**, 044907 (2002).
[26] S. V. Afanasiev *et al.*, Phys. Rev. C **66**, 054902 (2002).
[27] S. V. Afanasiev *et al.*, Phys. Rev. C **69**, 024902 (2004).
[28] C. Alt *et al.*, Phys. Lett. B **635**, 270 (2006).

[29] T. Anticic *et al.,* Phys. Rev. Lett. **93**, 022302 (2004).
[30] S. V. Afanasiev *et al.,* Phys. Lett. B **538**, 275 (2002).
[31] C. Alt *et al.,* Phys. Rev. Lett. **94**, 192301 (2005).
[32] F. Antinori *et al.,* Phys. Lett. B **595**, 65 (2004).
[33] F. Antinori *et al.,*J. Phys. G **31**, 1345 (2005) .
[34] J. Adams *et al.,* Phys. Rev. Lett. **92**, 112301 (2004).
[35] Sz. Borsanyi *et al.,* JHEP **1208**, 053 (2012).
[36] K. A. Bugaev *et al.,* Ukr. J. Phys. **60**, 181 (2015).
[37] P. Koch, B. Müller and J. Rafelski, Phys. Rep. **142**, 167 (1986) and references therein.
[38] C. M. Hung and E. V. Shuryak, Phys. Rev. C **57**, 1891 (1998) and references therein.
[39] Y. Pang, T. J. Schlagel and S.H. Kahana, Phys. Rev. Lett. **68**, 2743 (1992).
[40] H. Sorge, Phys. Rev. C **52**, 3291 (1995).
[41] W. Ehehalt and W. Cassing, Nucl. Phys. A **602**, 449 (1996).
[42] W. Cassing and E. L. Bratkovskaya, Phys. Rep. **308**, 65 (1999).
[43] J. Geiss, W. Cassing and C. Greiner, Nucl. Phys. A **644**, 107 (1998).
[44] S. A. Bass *et al.,* Prog. Part. Nucl. Phys. **41**, 225 (1998).
[45] F. Becattini, J. Manninen and M. Gazdzicki, Phys. Rev. C **73**, 044905 (2006).
[46] R. Longacre, preprint BNL-48648, C93-01-13.
[47] S. E. Eiseman *et al.,* Phys. Lett. B **297**, 44 (1992).
[48] L. X. Sun, R. Q. Wang, J. Song and F. L. Shao, Chin. Phys. C **36**, 55 (2012).
[49] Y. B. Ivanov, V. N. Russkikh and V. D. Toneev, Phys. Rev. C **73**, 044904 (2006).
[50] Yu. B. Ivanov and V. N. Russkikh, Phys. Rev. C **78**, 064902 (2008).
[51] Yu. B. Ivanov, Phys. Rev. C **87**, (2013) 064905.
[52] Yu. B. Ivanov, Phys. Rev. C **89**, 024903 (2014).
[53] W. Cassing and E. L. Bratkovskaya, Phys. Rev. C **78**, 034919 (2008).
[54] W. Cassing, Eur. Phys. J. ST **168**, 3 (2009).
[55] W. Cassing and E. L. Bratkovskaya, Nucl. Phys. A **831**, 215 (2009).
[56] P. Bozek, Acta Phys. Pol. B **36**, 3071 (2005).
[57] K. Werner, Phys. Rev. Lett. **98**, 152301 (2007).
[58] J. R. Taylor, *An introduction to error analysis* (University Science Book, Mill Valley, California, 1982).
[59] A. Andronic *et al.,* Nucl. Phys. A **837**, 65 (2010).
[60] J. Rafelski and B. Müller, Phys. Rev. Lett. **48**, 1066 (1982).
[61] K.A. Bugaev *et. al.,* arXiv:1510.03099v1 [nucl-th] (2015).

Colour particle states behaviour in the QCD vacuum

Kuvshinov V.I.[1,a] and Bagashov E.G.[1]

[1] *Joint Institute for Power and Nuclear Research - Sosny, National Academy of Sciences, 220109 Belarus, Minsk, Acad. Krasin str. 99*

Abstract. The results of an interaction of a quantum state of quark with QCD vacuum, where the latter plays a role of environment, could be treated as decoherence. This may have direct implications for the confinement of quarks phenomenon.
The general description and discussion of this process is given. Characteristics from quantum optics and information theory (purity, fidelity, von Neumann entropy) are proposed as means for numerical analysis of the process of interaction of quark colour state with stochastic vacuum.
Problems of stability of colour particles motion and order-chaos transitions are briefly discussed. It is shown that there should be a connection between the properties of QCD stochastic vacuum and Higgs boson mass and self interaction coupling constant.
The behaviour of squeezed and entangled quantum states, the interaction of colour superpositions and multiparticle states with stochastic QCD vacuum is described. It is shown that it leads to a fully mixed quantum state with equal probabilities for different colours. Decoherence rate is found to be proportional to the product of the distance between colour charges and the time during which this interaction has taken place. I.e. such an interaction seems to lead naturally to confinement of quarks.

1 Introduction

Interactions of some quantum system with the environment can be effectively represented by additional stochastic terms in the Hamiltonian of the system. In this case the averaging with respect to stochastic terms - degrees of freedom of environment - yields the density matrix of the system.

Physically the interactions with the environment result in decoherence and relaxation of quantum superpositions. Information on the initial state of the quantum system is lost after suciently large time [1–4].

Quantum decoherence is the loss of coherence or ordering of the phase angles between the components of a system in a quantum superposition. Decoherence occurs when a system interacts with its environment in a thermodynamically irreversible way. It can be viewed as the loss of information from a system due to the environment (often modeled as a heat bath). Dissipation is a decohering process by which the populations of quantum states are changed due to entanglement with a bath Relaxation usually means the return of a perturbed system into equilibrium. Each relaxation process can be characterized by a relaxation time τ.

[a]e-mail: v.kuvshinov@sosny.bas-net.by

2 Stochastic QCD vacuum

The model of QCD stochastic vacuum is one of the popular phenomenological models which explains quark confinement (Wilson loop decrease), string tensions and field congurations around static charges [5–7]. In stochastic vacuum only the second-order correlators are important and the others are negligible (they might, however, be important in coherent vacuum) [6], which means Gauss domination. This has been confirmed by lattice calculations [7, 8], although the most important evidence for this is probably Casimir scaling.

In such a situation one assumes that it is possible to calculate vacuum expectation values of gauge invariant quantities as expectation values with respect to some well-behaved stochastic gauge field. It is known that such vacuum provides confining properties, giving rise to QCD strings with constant tension at large distances.

2.1 QCD vacuum as environment

We consider QCD stochastic vacuum as the environment for colour quantum particles and average over external QCD stochastic vacuum implementations. Instead of considering nonperturbative dynamics of Yang-Mills fields one introduces external environment and average over its implementations.

As a consequence we obtain: decoherence, relaxation of quantum superpositions, information loss and confinement of colour states phenomenon. It allows us to say that white objects can be obtained as white mixtures of states described by the density matrix as a result of evolution in the QCD stochastic vacuum as environment [9–11].

3 Colour decoherence

3.1 General description

Consider propagation of heavy spinless colour particle along some fixed path γ. The amplitude is obtained by parallel transport [9–11]

$$\partial_\mu|\phi\rangle = i\hat{A}_\mu|\phi\rangle, \tag{1}$$

$$|\phi(\gamma)\rangle = \mathcal{P}\exp\left(i\int_\gamma dx^\mu \hat{A}_\mu\right)|\phi_{in}\rangle, \tag{2}$$

where $\mathcal{P}$ is the path ordering operator and $\hat{A}_\mu$ is the gauge field vector. In order to consider mixed states we introduce the colour density matrix, taking into account both colour particle and QCD stochastic vacuum (environment):

$$\hat{\rho}(loop, \gamma\bar{\gamma}) = |\phi(\gamma)\rangle\langle\phi(\gamma)|, \tag{3}$$

Here we average over all implementations of stochastic gauge field (environment degrees of freedom) - and decoherence arises due to interaction with environment.

In the model of QCD stochastic vacuum only expectation values of path ordered exponents over closed paths are defined (in order to keep the gauge invariance). Closed path corresponds to a process in which the particle-antiparticle pair is created, propagates and finally annihilates.

With the help of (1), (2), (3) we can obtain the expression for density matrix [9, 11]:

$$\hat{\rho}(loop, \gamma\bar{\gamma}) = N_c^{-1} + (|\phi_{in}\rangle\langle\phi_{in}| - N_c^{-1})W_{adj}(\gamma\bar{\gamma}). \tag{4}$$

Here N_c is the number of colours.

Colour density matrix in colour neutral stochastic vacuum can be decomposed into the pieces that transform under trivial and adjoint representations [9, 11–13]

$$\hat{\rho}_1 = N_c^{-1}\hat{I} + \rho_1^a \hat{T}_a, \tag{5}$$

and Wilson loop in fundamental representation is [6]

$$W_{fund}(loop, \gamma\bar{\gamma}) = \langle \mathrm{Tr}\, \mathcal{P} \exp\left(\int_{loop,\gamma\bar{\gamma}} i\hat{A}_\mu dx^\mu\right)\rangle. \tag{6}$$

(Non-abelian Stokes theorem).

In confinement region Wilson loop decays exponentially with the area spanned by loop, so for the rectangular loop spanning the time interval T and distance R we get:

$$\hat{\rho}(loop, \gamma\bar{\gamma}) = N_c^{-1} + (\rho_{in} - \hat{I}N_c^{-1})\exp(-\sigma_{adj}RT), \tag{7}$$

$$\hat{\rho}(L : RT \to \infty) = N_c^{-1}\hat{I}, \tag{8}$$

where $\hat{I}$ is the unit operator, $\sigma_{adj} = \sigma_{fund}G_{adj}G_{fund}^{-1}$ is the string tension in the adjoint representation, G_{adj}, G_{fund} are the eigenvalues of quadratic Casimir operators. Under Gaussian dominance string tension is

$$\sigma_{fund} = \frac{g^2}{2} l_{corr}^2 F^2, \tag{9}$$

g is coupling constant, l_{corr} correlation length in the QCD stochastic vacuum, F - average of the second cumulant of curvature tensor [7, 8].

Here all colour states are mixed with equal probabilities and all information on initial color state is lost. The stronger are the color states connected the stronger their states transform into the white mixture.

3.2 Decoherence rate, purity, von Neumann entropy

The decoherence rate of transition from pure colour states to white mixture state can be estimated on the base of purity [3]:

$$P = \mathrm{Tr}\, \hat{\rho}^2. \tag{10}$$

$$P = N_c^{-1} + (1 - N_c^{-1})\exp\left(-2\sigma_{fund}G_{adj}G_{fund}^{-1}RT\right). \tag{11}$$

When RT tends to 0, $P \to 1$, that corresponds to a pure state. When the composition RT tends to infinity, the purity tends to N_c^{-1}, which corresponds to the white mixture [11].

The rate of purity decrease is

$$T_{dec}^{-1} = -2\sigma_{fund}G_{adj}G_{fund}^{-1}R. \tag{12}$$

Left side of the equation is the inverse characteristic time of decoherence proportional to QCD string tension and distance R. It can be inferred from (4) and (11) that the stronger particle-antiparticle pair is coupled by QCD string or the larger is the distance between particle and antiparticle, the quicker the initial state tends to white mixture as a result of interaction with the QCD stochastic vacuum. Thus white states can be obtained as a result of decoherence process. The information of quark colour states is lost in hadrons due to interactions between quarks and confining non-Abelian gauge fields.

Von Neumann entropy can be used as a measure of the loss of information:

$$S = -\text{Tr}\,(\hat{\rho}\ln\hat{\rho}). \qquad (13)$$

Idempotency of the initial density matrix gives us $S = 0$ for the initial state and we will have $S = \ln N_c$ for large RT. Thus during the interaction with QCD vacuum the entropy increases and the information is being lost.

4 Colour confinement and instability of colour particle motion

The stability of quantum motion of the particles is described by quantum fidelity [14, 15]. The definition of fidelity is similar with Wilson loop definition in QCD [16]. Using the analogy between the theory of gauge fields and the theory of holonomic quantum computation [2, 16, 17], we can define the fidelity of quark (the scalar product of state vectors for perturbed and unperturbated motion) as an integral over the closed loop, with particle traveling from point x to the point y:

$$f = < \left(\langle\phi_{in}|\mathcal{P}\exp\left(\int i\hat{A}_\mu dx^\mu\right)|\phi_{in}\rangle\right) > . \qquad (14)$$

The final expression for the fidelity of the particle moving in the stochastic vacuum is

$$f = \exp\left(-\frac{1}{2}g^2 l_{corr}^2 F^2 S_\gamma\right). \qquad (15)$$

Thus fidelity for colour particle moving along the contour decays exponentially with the surface spanned over the contour, the decay rate being equal to the string tension (9). The motion becomes more and more unstable.

Sometimes fidelity is defined in another way [18–20]:

$$F(\omega,\tau) = \text{Tr}\,(\sqrt{\sqrt{\omega}\tau\sqrt{\omega}}) \qquad (16)$$

This quantity represents the square root of the probability of transition between the states described by density matrices ω and τ. If one of the states is pure, then fidelity is equal to the square root of the overlap between the states [21]. Thus sometimes its definition is squared [22]. In our case, for the final state $\hat{\rho}(loop)$ and the initial state $\hat{\rho}_{in}$ we get

$$F(\hat{\rho}_{in},\hat{\rho}(loop)) = \text{Tr}\,(\sqrt{\sqrt{\hat{\rho}_{in}}\hat{\rho}(loop)\sqrt{\hat{\rho}_{in}}}) = $$
$$= \text{Tr}\,(\hat{\rho}_{in}\sqrt{N_c^{-1} + (1-N_c^{-1})W_{adj}(\gamma)}). \qquad (17)$$

The absence of the second term under the root (when $W_{adj}(\gamma) \to 0$) returns the value of fidelity $F = N_c^{-1/2}$. We obtained some non-unit value, which means that the state has changed. And it can be seen how $W_{adj}(\gamma) \to 1$ gives us the initial state: $F = 1$.

For two random paths in Minkowski space, which are close to each other, the expression for the fidelity is similar, but now the averaging is performed with respect to all random paths which are close enough. Then the final expression is

$$f = \exp\left(-\frac{1}{2}g^2 l_{corr}\int_{\gamma_1} dx^\upsilon F_{\chi\alpha}\tilde{F}_{\upsilon\beta}\upsilon^\chi\langle\delta\chi^\alpha\delta\chi^\beta\rangle\right), \qquad (18)$$

where $\delta\chi$ is the deviation of the path γ_2 from the path γ_1, v is the four-dimensional velocity and l_{corr} is the correlation length of perturbation of the particle path expressed in units of world line length.

If unperturbed path is parallel to the time axis in Minkowski space, the particle moves randomly around some point in three dimensional space. The fidelity in this case decays exponentially with time.

Thus we have the connection between confinement and instability of colour particle motion that could be related to possible mechanisms of colour particle confinement.

5 Order-chaos transition, critical energy of and mass of Higgs boson

The increasing of instability of motion in the confinement region is also connected with existence of chaotic solutions of Yang-Mills field [9, 23], possible chaos onset [24]. Yang-Mills fields already on classical level show inherent chaotic dynamics and have chaotic solutions [23, 24]. The same is true for quarks.

It was shown that Higgs bosons and its vacuum quantum fluctuations regularize the system and lead to the emergence of order-chaos transition at some critical energy [25, 26]:

$$E_c = \frac{3\mu^4}{64\pi^2}\exp\left(1 - \frac{\lambda}{g^4}\right). \tag{19}$$

Here μ is mass of Higgs boson, λ is its self interaction coupling constant, g is coupling constant of gauge and Higgs fields. The expression is valid both for large and small distances. Important here is the value of mass of Higgs boson.

In the region of confinement there exists the point of order-chaos transition where the fidelity decreases exponentially and which is equal to string tension. This connects the properties of stochastic QCD vacuum and Higgs boson mass and self interaction coupling constant.

6 Squeezed and entangled colour states

The increasing of instability of motion in the confinement region is also connected with possible quantum entanglement as a probe of confinement and quantum squeezing of colour states [27–30]. The emergence of entagled and squeezed states in QCD becomes possible due to the four-gluon self-interaction. The three-gluon self-interaction does not lead to the effects [31–33].

The existence of quantum entanglement in Yang-Mills-Higgs theory was considered in [28] on the base of original quasiclassical formalism developed in [27]. The concept of quantum entanglement was found to be very useful as a model-independent characteristic of the structure of the ground state of quantum field theories which exhibit strong long range correlations, most notably lattice spin systems at and near the critical points and the corresponding conformal field theories [34].

Quantum entanglement was also considered as an alternative way to probe the confining properties of large-N gauge theories [29]. Quantum entanglement between the states of static quarks in the vacuum of pure Yang-Mills theory was analyzed in [30]. Hilbert space of physical states of the fields and the charges is endowed with a direct product structure by attaching an infinite Dirac string to each charge. Tracing out the gauge degrees of freedom gives the density matrix which depends on the ratio of Polyakov and Wilson loops spanned on quark world lines [35].

7 Multiparticle states in QCD vacuum

Interaction of colour superposition with stochastic QCD vacuum was described in [20]. In its essential parts the process is analogous to the one shown above in Section 3. Let us consider the cases of two and three colour particles.

7.1 Two-particle states

Consider a system of two quarks, denoted as A and B. Also consider that there is only two possible colour states for each of them: $|A\rangle$, $|B\rangle$ and $|\bar{A}\rangle$, $|\bar{B}\rangle$. Thus $N_c = 2$.

Here is the list of possible types of states, in which the system may be found (excluding some specific types e.g. as the squeezed states):

- By purity:
  - Pure;
  - Mixed.
- By separability:
  - Separable;
  - Non-separable (entangled).

Pure separable system state vector of two quarks can be presented as

$$|\phi\rangle = |a\rangle|b\rangle, \tag{20}$$

where $|a\rangle$ and $|b\rangle$ are some states of subsystems A and B. Since the system state vector can be factorized (20), we may consider the evolution of each quark independently, using the expressions given in Section 3.

Mixed separable state gives us similar picture, with the only difference being the absence of a state vector. The density matrix of the system will have the form

$$\rho_{AB} = \rho_A \rho_B, \tag{21}$$

where density matrices ρ_A and ρ_B correspond to states of A and B quarks. ρ_A and ρ_B therefore may also be considered independently with the help of the described mechanism.

We will choose pure non-separable state in the form

$$|\phi\rangle = \alpha|A\rangle|\bar{B}\rangle + \beta|\bar{A}\rangle|B\rangle. \tag{22}$$

Density matrix for such a state will have the form

$$\rho_{AB} = |\alpha|^2 |A\bar{B}\rangle\langle\bar{B}A| + \alpha\beta^* |A\bar{B}\rangle\langle B\bar{A}| + \alpha^*\beta|\bar{A}B\rangle\langle\bar{B}A| + |\beta|^2|\bar{A}B\rangle\langle B\bar{A}|. \tag{23}$$

During the evolution and the interaction with vacuum this state will take the form (8), with the exception that it should be raised to a power of $N_p = 2$, where N_p is the number of particles. It can be seen that the entanglement present in (22) vanishes, since the resulting density matrix can be presented in the form $\sum_i k_i \rho_{i1}\rho_{i2}$, where k_i is some weight factor, ρ_{i1} and ρ_{i2} - some density matrices.

Mixed entangled state is rather difficult to describe. It seems that the easiest way to formally present it in the given case is to artificially suppress (but not entirely) the non-diagonal components of the density matrix - for example by dividing them by 2. After the interaction with the stochastic vacuum, however, we will again have the state of the form (8). Before the interaction the quantum parameters of the system would be as follows:

The same parameters after the interaction with vacuum (with the state types corresponding to the initial state), also considering fidelity (where the arguments are density matrices of the initial and final states):

Table 1. Purity and entropy of different states of a two-particle system before the interaction with vacuum

State:	pure separable	mixed separable	pure entangled	mixed entangled
P (purity)	1	$\frac{1}{4} \leq$P< 1	1	$\frac{1}{4}$ <P< 1
S (entropy)	0	0<S≤ 2 ln 2	0	0<S< 2 ln 2

Table 2. General form of purity, entropy and fidelity of different states of a two-particle system after the interaction with vacuum

State:	pure separable	mixed separable	pure entangled	mixed entangled
P (purity)	$\frac{1}{4}$	$\frac{1}{4}$	$\frac{1}{4}$	$\frac{1}{4}$
S (entropy)	2 ln 2	2 ln 2	2 ln 2	2 ln 2
F (fidelity)	$\frac{1}{2}$	$\frac{1}{2}$ <F≤ 1	$\frac{1}{2}$	$\frac{1}{2}$ <F< 1

7.2 Three-particle states

Pure separable state vector can be represented as

$$|\phi\rangle = |a\rangle|b\rangle|c\rangle, \tag{24}$$

where $|a\rangle$, $|b\rangle$, $|c\rangle$ are arbitrary states of subsystems (quarks) A, B and C. As in the two-particle case, this situation is analogous to what have already been discussed: if the system's state vector can be factorized, we can discuss the evolution of quarks separately, using expressions derived in the former sections and assuming $N_c = 3$, $N_p = 3$. Of course, the number of dimensions of the density matrix would increase.

In case of mixed separable state we will have similar situation, although without the state vector. Instead of it we will use density matrix to describe the evolution of the system. The density matrix will write as:

$$\rho_{ABC} = \rho_A \rho_B \rho_C, \tag{25}$$

where the density matrices ρ_A, ρ_B, ρ_C would correspond to the states of quarks A, B, C. Each one of these states could be considered independently as well, and its interaction with the stochastic vacuum will be described by the aforementioned mechanism, beginning from the moment of the introduction of density matrix (e.g. (3)). Final state density matrix is determined by the multiplication of density matrices of the final states of the subsystems.

To illustrate the case of pure separable state let us consider the system, analogous to the hadron: here we have three quarks of different colours (and thus $N_c = 3$). We shall symbolically mark these colours by the subscripts 1, 2, 3. The three quarks themselves would be marked as A, B, C. In this case, for example, the writing B_3 would mean that the quark B is in the state that is described by the colour 3. And the writing A_1B_2 means that the system is composed of quarks A and B with colour states 1 and 2, respectively. Hadronic structure corresponds to the wave function, such as

$$|\phi\rangle = a|A_1B_2C_3\rangle + b|A_1B_3C_2\rangle + c|A_2B_1C_3\rangle + d|A_2B_3C_1\rangle + e|A_3B_1C_2\rangle + f|A_3B_2C_1\rangle, \tag{26}$$

where the complex weight factors a, b, c, d, e, f are normalized as:

$$|a|^2 + |b|^2 + |c|^2 + |d|^2 + |e|^2 + |f|^2 = 1. \tag{27}$$

This particular structure of wave function means that the quarks cannot occupy the same colour state. Wave function of antiparticle (with respect to the given hadron) would have the same structure, although instead of colour states there would be anti-colour states. So we would consider only the hadron wave function.

During the evolution and interaction with the stochastic vacuum this state will turn into a state that is described by the expression (8), which in the limit $W_{adj} \to 0$ (also taking into account $N_p = 3$ and $N_c = 3$) would give us the expression:

$$\rho_{ABC} = \text{diag}\ \frac{1}{27}. \tag{28}$$

As in the two-particle case, the mixed entangled state is impossible to represent in the generalized (analytical) form, since it can have various forms. However, it is still possible to acquire this type of state from the pure entangled state, artificially suppressing the non-diagonal terms. Here also after the interaction with vacuum under the condition $W_{adj} \to 0$ we get the state (28).

Before the interaction the parameters of the system would be as follows:

Table 3. Purity and entropy of different states of a three-particle system before the interaction with vacuum

State:	pure separable	mixed separable	pure entangled	mixed entangled
P (purity)	1	$\frac{1}{27} \leq$P< 1	1	$\frac{1}{27}$ <P< 1
S (entropy)	0	0<S$\leq 3\ln 3$	0	0<S$\leq 3\ln 3$

The same parameters after the interaction with vacuum (with the state types corresponding to the initial state), also considering fidelity (where the arguments are density matrices of the initial and final states):

Table 4. General form of purity, entropy and fidelity of different states of a three-particle system after the interaction with vacuum

State:	pure separable	mixed separable	pure entangled	mixed entangled
P (purity)	$\frac{1}{27}$	$\frac{1}{27}$	$\frac{1}{27}$	$\frac{1}{27}$
S (entropy)	$3\ln 3$	$3\ln 3$	$3\ln 3$	$3\ln 3$
F (fidelity)	$\frac{1}{\sqrt{27}}$	$\frac{1}{\sqrt{27}}$ <F$\leq$ 1	$\frac{1}{\sqrt{27}}$	$\frac{1}{\sqrt{27}}$ <F< 1

8 Conclusions

It is productive to consider QCD vacuum as an environment (in the sense of quantum optics) for colour particles. Under certain conditions such an approach leads to confinement of quarks and might serve as an explanation for it.

In the case of of stochastic (not coherent) QCD vacuum (only correlators of the second order are important) in confinement region (Wilson loop decays exponentially) we have decoherence of pure colour states into mixed white states and the onset of instability (chaoticity) of their motion. During this process quantum fidelity and purity decay exponentially (decay rate = QCD string tension).

Dynamics of Yang-Mills fields can be regularized by Higgs fields and quantum fields fluctuations. Critical point of order-chaos transition appears which corresponds to the point, where fidelity and purity drop exponentially. Quantum squeezing and entanglement accompany nonperturbative (A^4) evolution of colour particles in QCD vacuum, confinement, decoherence and instability.

For colour superposition, two and three particle state types (pure separable, mixed separable and nonsepaparable (entangled)) when $RT \to \infty$ we obtain diagonalization of density matrix, elimination of entanglement, decreasing of purity and fidelity, increasing of von Neumann entropy.

References

[1] H. Haken, P. Reineker. The Influence of Exciton Motion on Spin-Resonance. Absorption. Z. Physik. **250**, 300 (1972).

[2] P. Reineker. *Exciton Dynamics in Molecular Crystals and Aggregates*. (Springer-Verlag, Berlin, 1991).

[3] F. Haake. *Quantum Signatures of Chaos*. (Springer-Verlag, Berlin, 1991).

[4] A. Peres. *Quantum Theory: Concepts and Methods*. (Kluwer, Dordrecht, 1995).

[5] J. Ambjørn, P. Olesen. On the Formation of a Color Magnetic Quantum Liquid in QCD. Nuclear Physics B. **170**, no. 1, 60-78 (1980).

[6] Y. A. Simonov. The Confinement. Uspekhi Fizicheskih Nauk. **4** (1996) (in Russian).

[7] A. D. Giacomo, H. Dosch, V. I. Shevchenko, and Y. A. Simonov, Physics Reports **372**, no. 4, 319-368 (2002).

[8] D. S. Kuzmenko, Y. A. Simonov, et al. Vacuum, Confinement and QCD Strings in the Vacuum Correlator Method. Uspekhi Fizicheskih Nauk. **174**, no. 1 (2004) (in Russian).

[9] V. Kuvshinov, A. Kuzmin. *Gauge Fields and Theory of Deterministic Chaos* (Belorussian Science, Minsk, 2006, p. 1-268, in Russian).

[10] V. Kuvshinov, P. Buividovich. White Mixed States in QCD Stochastic Vacuum. Nonlinear Phenomena in Complex Systems. **8**, no.3, 313-316 (2005).

[11] V. Kuvshinov, P. Buividovich. Decoherence of Quark Colour States in QCD Vacuum. Acta Physica Polonica B. (Proceedings Supplement), **1**, no. 3, 579-582 (2008).

[12] A.O. Barut, R. Raczka, *Theory of Group Representations and Applications* (World Scientific, Singapore, 1986).

[13] D. Littlewood, *The Theory of Group Characters* (Clarendon Press, Oxford, 1950).

[14] A. Peres, Stability of Quantum Motion in Chaotic and Regular Systems, Phys. Rev. A30, 1610-1615 (1984).

[15] T. Prosen, M. Žnidarič, Can quantum chaos enhance the stability of quantum computation? Journal of Physics A: Mathematical and General, L681-L687 (2001).

[16] V. Kuvshinov, A. Kuzmin. Physics Letters A, **316**, 391-394 (2003).

[17] V. I. Kuvshinov, P. V. Buividovich. Fidelity and Wilson Loop for Quarks in Confinement Region. Acta Physica Polonica. **36**, no. 2, 195-200 (2005).

[18] A. Uhlmann. The "Transition Probability" in the State Space of a *-Algebra. Rep. Math. Phys., 9, 273 (1976).

[19] M. Hübner. Explicit Computation of the Bures Distance for Density Matrices, Phys. Lett. A, 163, 239-242 (1992).

[20] V. I. Kuvshinov, E. G. Bagashov. Evolution of colour superposition in the stochastic QCD vacuum. Nonlinear Phenom. Complex Syst., 16, No. 3, 242-246 (2013).

[21] J. A. Miszczak, Z. Puchala, P. Horodecki, A. Uhlmann, and K. Zyczkowski. Sub– and Super-Fidelity as Bounds for Quantum Fidelity, Quantum Information and Computation, **9**, no. 1 and 2, 0103–0130 (2009).

[22] R. Jozsa. Fidelity for Mixed Quantum States. J. Mod. Opt. **41** 2315-2323 (1994).

[23] G. K. Savvidy. Infrared Instability of the Vacuum State of Gauge Theories and Asymptotic Freedom. Physics Letters B, **71**, no. 1, 133-134 (1977).

[24] T. Kawabe et al. Phys. Rev. D. **41**, no. 6, pp. 1983-1988 (1990).

[25] V. Kuvshinov, A. Kuzmin. Nonlinear Phenomena in Complex Systems, vol. 2, no.3, 100-104 (1999).

[26] V. Kuvshinov, A. Kuzmin. J. Nonl. Math. Phys. **9**, no.4, 382-388 (2002).
[27] V. Kuvshinov, V. Marmish, V. Shaporov. Theor. Math. Phys., **139**, 477-490 (2004).
[28] V. Kuvshinov, V. Marmish. Letters in EPAN v.2, 23 (2005).
[29] L.R. Klebanov, D. Kutasov, A. Murugan. Entanglement as a Probe of Confinement. arXiv:0709.2140v2 [hep-th].
[30] T. Nashioka, T. Takayanagi. JHEP. v.01, 090 (2007).
[31] V. Kuvshinov, V. Shaporov. Acta Physica Polonica B, **30**, no. 59, (1999).
[32] V. Kuvshinov, V. Shaparau. Phys. Atom. Nucl., **65**, 309-314 (2002).
[33] V. Kuvshinov, V. Shaparau. Acta Physica Polonica B, **35**, no. 1, 443-446 (2004).
[34] P. Calabrese and J. Cardy. Int.J.Quant.Inf., **4**,429 (2006).
[35] V. I. Kuvshinov, P. V. Buividovich. Nonlinear Phenomena in Complex Systems, 13, no. 2, 149-155 (2010).

The hot Hagedorn Universe

Presented at the ICFNP2015 meeting, August 2015

Johann Rafelski[1] and Jeremiah Birrell[2]

[1] *Department of Physics, The University of Arizona, Tucson, AZ 85721, USA*
[2] *Department of Mathematics, The University of Arizona, Tucson, AZ 85721, USA*

Abstract. In the context of the half-centenary of Hagedorn temperature and the statistical bootstrap model (SBM) we present a short account of how these insights coincided with the establishment of the hot big-bang model (BBM) and helped resolve some of the early philosophical difficulties. We then turn attention to the present day context and show the dominance of strong interaction quark and gluon degrees of freedom in the early stage, helping to characterize the properties of the hot Universe. We focus attention on the current experimental insights about cosmic microwave background (CMB) temperature fluctuation, and develop a much improved understanding of the neutrino freeze-out, in this way paving the path to the opening of a direct connection of quark-gluon plasma (QGP) physics in the early Universe with the QCD-lattice, and the study of the properties of QGP formed in the laboratory.

1 The big-bang model established

Who today can remember that before 1965 the big-bang model (BBM) was challenged by those who had difficulty accepting that there is a beginning of time? And, even after the discovery of the cosmic microwave background (CMB) radiation, announced May 1965, the idea that the primordial Universe had to be hot generated an additional challenge: how could all the energy in the Universe come from an initial space-time singularity? Such conceptual difficulties were exploited by those who did not like the hot BBM model. One of us can remember the scientific disputes.

A literature search shows that Hagedorn's statistical bootstrap model (SBM) was an inadvertent solution to many conceptual difficulties. In fact, a who's who of cosmology of this period cites Hagedorn's work [1]. While it could sound presumptuous to claim that Hagedorn was the one who turned the corner in establishing the hot BBM as the standard cosmological model, at the least his contribution was very important.

The Hagedorn SBM had a built-in feature that was needed, a divergence in the energy content even for a initial singular point volume. The way this works is rather easy to explain. It is convenient to introduce the hadron mass spectrum $\rho(m)$, where

$$\rho(m)\mathrm{d}m = \text{number of hadron states in } \{m, m+\mathrm{d}m\}\ . \qquad (1)$$

In Hagedorn's SBM, an exponentially growing mass spectrum $\rho(m) \propto m^{-a}\exp(m/T_{\mathrm{H}})$, is a natural solution. Thus we have within the partition function of a hadron gas comprising many different

Table 1. Thermodynamic quantities assuming exponential form of hadron mass spectrum with pre-exponential index a: pressure P, energy density ε, energy density fluctuation $\delta\varepsilon/\varepsilon$, the heat capacity $C_V = \mathrm{d}\varepsilon/\mathrm{d}T$ and the sound velocity $v^2 = \mathrm{d}P/\mathrm{d}\varepsilon$

a	P	ε	$\delta\varepsilon/\varepsilon$	$C_V = \mathrm{d}\varepsilon/\mathrm{d}T$	$v^2 = \mathrm{d}P/\mathrm{d}\varepsilon$
5/2	$C\ln(T_0/\Delta T)$	$C/\Delta T$	C	$C/\Delta T^2$	$C\Delta T$
3	$P_0 - C\Delta T^{1/2}$	$C/\Delta T^{1/2}$	$C/\Delta T^{1/4}$	$C/\Delta T^{3/2}$	$C\Delta T$
7/2	$P_0 - C\Delta T$	ε_0	$C/\Delta T^{1/2}$	$C/\Delta T$	$C\Delta T$
4	$P_0 - C\Delta T^{3/2}$	$\varepsilon_0 - C\Delta T^{1/2}$	$C/\Delta T^{3/4}$	$C/\Delta T^{1/2}$	$C\Delta T$

particles as described by the mass spectrum, a singular contribution

$$\Delta Z(T,V) = \exp\left[V\left(\frac{T}{2\pi}\right)^{3/2}\int_{3T_\mathrm{H}}^{\infty}\rho(m)m^{3/2}\mathrm{e}^{-m/T}\mathrm{d}m\right]. \tag{2}$$

The lower limit $3T_\mathrm{H}$ on the integral dm ensures that the non-relativistic limit shown in Eq. (2) is reasonable. We can compute, for a few selected values of the power index a in $\rho(m)$ of immediate physical interest, the singular properties of the hadron gas near to $T = T_\mathrm{H} + \Delta T$ to be as shown in the table 1.

The value $a = 5/2$ which Hagedorn advocated in late 1960s shows a singular energy density and a slower logarithmically divergent pressure. This allows the Universe to originate in a singular volume condition, and yet to contain an infinite energy. This was a new type of singularity, and a revolutionary insight that impacted a few other areas of physics. We see the appearance of the term "Hagedorn temperature" in distant physics topics today.

Hagedorn was accordingly presenting his ideas also at cosmology meetings and one of the best accounts of his work on hot hadronic matter and the BBM are his lectures at the Cargese Summer school in 1971 [2], a classic of the period with very important contributions by other cosmology and astrophysics luminaries.

To close this introduction let us quote from one of Hagedorn's popular lectures of 1968: one of last paragraphs reads [3]: "At least a few theories about the beginning of the Universe assume a BBM, that is to say a creation explosion. Following previous ideas – based on traditional black body radiation – the Universe began with infinite energy density, with energy density proportional to the pressure, and infinitely high temperature. Under such extreme conditions, traditional black body radiation no longer remains, but rather the conditions are found akin to the high-energy collisions of nucleons. And then when strongly interacting matter is present the temperature cannot be infinite, but only about 10^{12} K, and the pressure is not anymore proportional to the energy density but only proportional to its logarithm. This is a different scenario of the beginning of the Universe than was previously thought."

The following is the last paragraph, equally interesting, a typical 'Hagedornian', and perhaps should be saved for the end of this lecture, but it seems that it should better be presented together with the former content: "I (Hagedorn) close with an anecdote: On the bulletin board of a German university the following could once be read among lecture announcements: *Tuesdays 9-11 AM, free for all discussion session about the structure of the Universe – only for the advanced. signed X.* We will, alas, always be beginners."

2 The particle and nuclear Universe

In the following three sub-sections we will explore key features of the early Universe from the present day perspective, connecting the field of hot hadron matter and quark gluon plasma (QGP) to the observation of the early Universe by way of the CMB. We embed our study into the Universe evolution period during the time window where the laws of nuclear science govern the result. We summarize the pivotal technical details that one can call the standard model of cosmology in the appendices: we describe the cosmology in context of the Robertson-Walker Universe, see Appendix A.1; derive key relations that characterize the Universe, specifically the thermal and free-streaming components of matter and radiation, in Appendices A.2, and A.3, respectively. All these inputs allow us to evaluate ab-initio the dynamics of the Universe. This report updates and extends earlier related efforts [5–8].

2.1 Hot Hagedorn Universe

Forwarding from the past to the present, we look again at the entries in table 1: we note that it is not necessary that the energy density is singular at $T = T_{\rm H}$; in fact should $a \geq 7/2$, we clearly see that the hadron gas singularity is dissolved. A fit to the hadron mass spectrum of the exponential shape predicts $T_{\rm H}(a = 7/2) = 151\,\text{MeV}$ and $T_{\rm H}(a = 8/2) = 144\,\text{MeV}$. This value is also confirmed by lattice gauge numerical evaluation; for details of this complicated situation we refer to the discussion in the review, Ref. [9].

It is an undisputed fact emerging from the lattice gauge numerical evaluation of strong interaction properties that a singularity at $T_{\rm H}$ is avoided. We find a smooth transformation, a position clearly taken by Hagedorn as of the early 1980s, see for example Ref. [10]. We can say that hadrons dissolve into their constituents, quarks, and that this phase transformation has been established near $T_{\rm H} = 145 + -7\,\text{MeV}$ [11, 12]. An interesting feature at the condition is that the sound velocity vanishes, first discovered in the SBM model, see last column in table 1. This has also been confirmed by lattice gauge theory. For this reason the QGP decay into hadrons is subject to non-trivial slow flow dynamics; one can argue that the fireball of hot matter sits and spits out its content that streams out freely. This feature goes well with the sudden hadronization model, but this is not our present discussion topic.

We are interested in connecting the properties of the QGP stage of the early Universe to the present day world we see around us today. To achieve this objective, we have to evolve the Universe dynamically. In figure 1 we show, as an example of such a computation, the evolution of the degrees of freedom as a function of temperature. By far the fastest and most dramatic change occurs when the two phases of strongly interacting matter meet: the Hagedorn gas phase, and the deconfined QGP, with a cross-over point at $T = T_{\rm H}$. To describe this situation in this computation we use the Wuppertal-Budapest lattice QCD equations of state [11, 12] continued to yet higher $T > 500\,\text{MeV}$ based on thermal-QCD [13]. The strong interactions phases are supplemented by ideal gases of weakly and electromagnetically interacting particles, either coupled or free-streaming (such as cold neutrinos), and we incorporate in this study all the usual elements of the ΛCDM model introduced in Appendix A.

In figure 2 we show the evolution (usual time arrow from left to right) of the Universe temperature as a function of time (left side scale) from midst of QGP at about 400 MeV down to temperature of 10 keV where the BBM nucleosynthesis (BBN) had ended. Note that there are two temperatures shown for large times: the annihilation of electrons with positrons feeds entropy only into the photon degrees of freedom and thus already free-streaming neutrinos are in comparison (dashed line) colder. The sharing of entropy from disappearing (massive) particle families results in reheating of only the coupled degrees of freedom. For reheating to differentiate the temperature, a particle must already have decoupled when annihilation of a massive degree of freedom occurs. It is a phenomenon that therefore does not occur, other than for neutrinos, for other visible (as opposed to dark) particles in

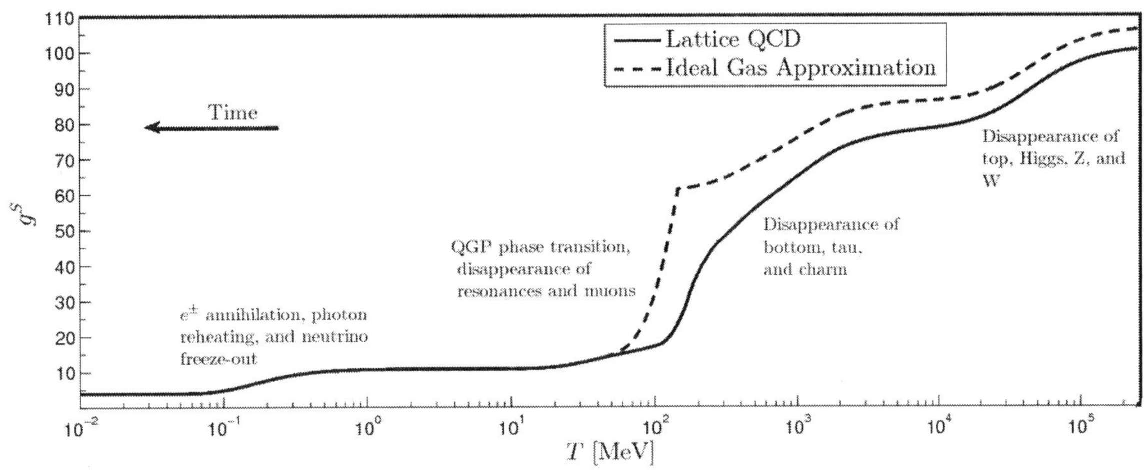

Figure 1. Active entropic degrees of freedom, g^S, in the Universe as a function of ambient temperature, T, from the electroweak transition on right, to conditions similar to present day, on left (time arrow from right to left). Note a significant modifications of the ideal gas (dashed line) model in the QGP era. The extrapolation for $T > 500\,\mathrm{MeV}$ relies on thermal-QCD approach [13].

the evolution of the Universe. However, dark particle components of dark matter will clearly be much colder and we return to discuss this situation later.

We also show the deceleration parameter q (right side scale) in figure 2. We saw earlier that $q = 1$ corresponds to a radiation dominated Universe where $P/\varepsilon = 1/3$. q decreases from unity when the Universe expansion slows down due to emergence of a massive degree of freedom. We also saw that $q = 1/2$ for a matter dominated Universe, where pressure $P \ll \varepsilon$. Note the dip around 20 μs, that is near to hadronization of QGP at T_{H}, where many very massive baryons and heavy mesons are formed and quickly disappear. Another dip appears due to the pion and muon masses becoming visible, before their particle density also disappears as the Universe cools. A clean version of the same effect is seen for $1 < t < 300\,\mathrm{s}$ when in a wide temperature range at the scale of electron mass we see the effect of the electron and positron density. What we learn is that the deceleration parameter q is a diagnostic tool for when a massive particle component in the Universe becomes partially nonrelativistic, but is in an abundance that impacts the speed of expansion.

In the analysis of the CMB, which is the present day experimental access to early the Universe, the particle content in the Universe can be studied, as the speed of expansion impacts the build-up of fluctuation structure. In this analysis one expresses the uncertainty about particle content as an uncertainty in the number of invisible neutrino degrees of freedom N_{eff}. If there is nothing unusual governing the Universe expansion, within the precision of the data the value $N_{\mathrm{eff}} \simeq 3$ will be determined.

During recent years the CMB data analysis yielded $3 \lesssim N_{\mathrm{eff}} \lesssim 4$ with the long term bias above 3.5 broken when the Planck collaboration [14] revised the fit strategy to allow a greater tension with priors, redistributing the error over other parameters of the standard model of cosmology. This is not a suitable place to discuss in any depth the question if there is, or is not, $\delta N_{\mathrm{eff}} \equiv N_{\mathrm{eff}} - 3 > 0$. Only the next generation of experiments that allow a significantly greater precision can answer this. Awaiting this and in consideration of the bias towards $\delta N_{\mathrm{eff}} \simeq 0.3$-$0.6$, we believe that the period of Universe expansion shown in figure 2 should be reexamined from the nuclear science perspective in order to understand how different nuclear science effects could have an impact on δN_{eff}.

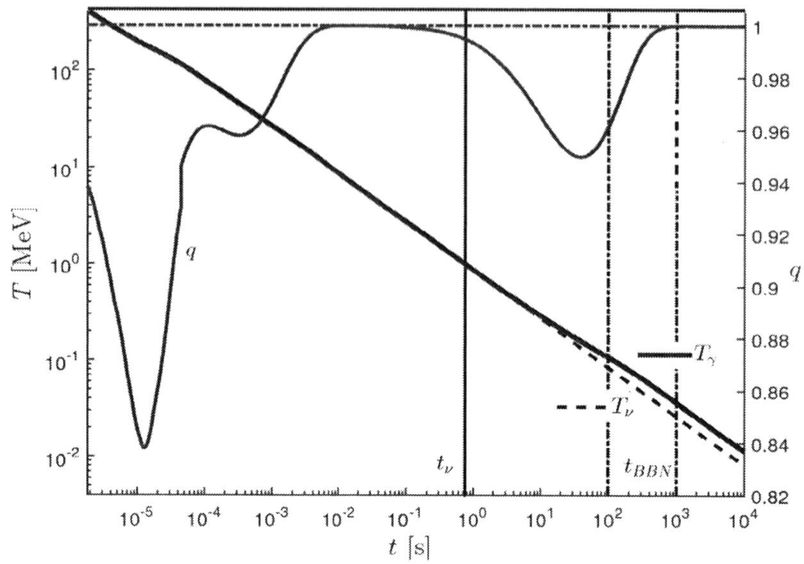

Figure 2. The temperature, left scale and deceleration parameter, right scale, as a function of the age of the Universe from 1 micro second to 10,000 seconds (10 orders of magnitude). This is the domain in the expansion of the Universe where nuclear science plays the primary role.

In the following we address two sources of $\delta N_{\rm eff}$ excess. The first originates in the generation of unknown light particles during QGP hadronization, and the other due to modification of neutrino freeze-out dynamics. Both effects can only be present if some new physics comes along. We discuss these two effects in turn, beginning with discussion of how the QGP condition can impact the interpretation of CMB temperature fluctuations.

2.2 New particles from QGP hadronization

The fact that $0 < \delta N_{\rm eff} < 1$ does not mean that there could not be one, or more, particle degrees of freedom that are the cause. As long as the particle concerned decouples within QGP we can be sure of a considerable reheating effect, given the excess QGP entropy flow from disappearing degrees of freedom into the remaining visible Universe content. This effect dilutes the free-streaming contribution of the here called 'secret' particle considered. While we invented this effect independently and study it in a model-independent way [15], two earlier and independent efforts can be found in literature, one focused on sterile neutrinos [16, 17], and the other on broken dark matter symmetries [18]. All this work assumes the presence of secret particles emerging from QGP that remain undetected today. That it must be happening somewhere in QGP is obvious considering the systematic behavior seen in figure 1 as the number of degrees of freedom drops rapidly when temperature drops towards 100 MeV.

There is a constraint from above as well, which could be a problem for Ref. [18]: if the decoupling occurs at a temperature that is too high, the contribution a particle would make to $\delta N_{\rm eff}$ could be too small to matter. We believe that the QGP hadronization point $T_{\rm H}$ is the only special characteristic point and thus this is what we consider in the following as a source condition from where a new secret particle would free-stream. Using the best lattice results to describe the domain of phase transformation of QGP we obtained [15] the contribution of unseen degrees of freedom to $\delta N_{\rm eff}$ for both fermions and bosons as shown in figure 3, as a function of the decoupling temperature $T_{d,s}$, where solid lines are for bosons. A smaller (dashed) contribution of fermions is due to their $g_s = 7/8\times$ smaller impact. Note that a sterile neutrino would count as two fermionic degrees of freedom.

The horizontal dotted lines in figure 3 show the domain $\delta N_{\rm eff} + 0.046 \in (0.36, 0.62)$ compatible with CMB data analysis. A contribution to the total $\delta N_{\rm eff}$ of 0.046 is expected by mechanisms we will

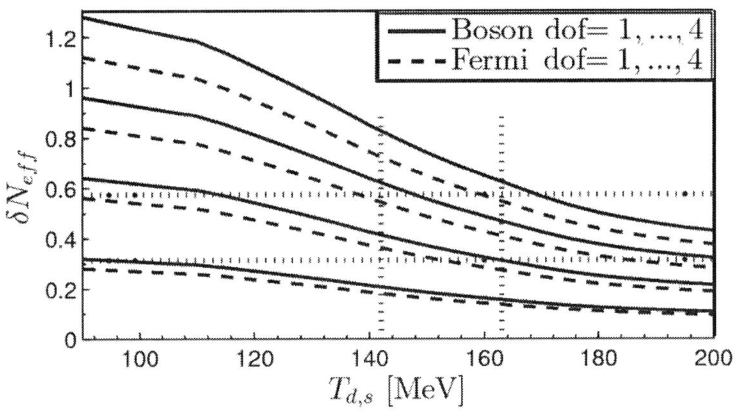

Figure 3. Solid lines bottom to top: Increase in $\delta N_{\rm eff}$ as a function of decoupling point $T_{d,s}$ due to the effect of $1, \ldots, 4$ light secret boson degrees of freedom ($g_s = 1, \ldots, 4$), and dashed lines, the same for light fermion DoF ($g_s = 7/8 \times 1, \ldots, 7/8 \times 4$). The horizontal dotted lines correspond to $\delta N_{\rm eff} + 0.046$=0.36, 0.62 . The vertical dotted lines show the reported range of QGP transformation temperatures $T_{\rm H} \in (142, 163)$ MeV.

discuss below, and thus the horizontal lines indicate the remaining contribution that the secret particles would need to make. The vertical dotted lines bracket the range of QGP transformation temperatures $T_{\rm H} \in (142, 163)$ MeV. We see that decoupling of 2-3 new particle degrees of freedom could improve the agreement between CMB data and theory of cosmological evolution. We also note that, as the decoupling temperature of the QGP increases, more secret degrees of freedom are needed to explain the presumed CMB signal. This is due to the relatively large reheating that occurs after decoupling, as we described earlier. Another feature is that for decoupling at $T = 75$ MeV we still see significant dillution due to ongoing reheating by pions and muons.

The possibility to explain data does not mean that there are secret particles that couple to (anti)quarks at hadronization of QGP, even if this hypothesis gives this consideration a sense of plausibility. However, there are already a few eternally searched for particles: axions, and, sterile neutrinos, so first we must clarify if these particles could do the job. Of the two, axions were introduced to protect parity symmetry of strong interactions and thus many firmly believe in their existence. In addition these particles are naturally related to strong interactions. To build a connection to QGP, consider if axions could couple to color field lines, which do not exist in stationary baryons or mesons. Color field lines are abundant in the deconfined QGP state. Therefore, axions could naturally decouple at $T_{\rm H}$, which would fit the above picture. And, this would have immediate laboratory consequence, with any QGP fireball possibly shining brightly; even if the axion "light" is invisible, this could be noticed due to missing energy [15] if these particles have time to chemically equilibrate in the QGP phase.

It can be safely said that strong interactions have plenty of unexplained challenges and hence the possibility of secret particle freeze-out at $T_{\rm H}$ is a viable hypothesis to be pursued in laboratory experiments searching for missing energy. The CERN experiment SHINE (NA61) – which in its name already shines secret radiation, has informally expressed interest in pursuing this opportunity. On the other end, and less laboratory verifiable, are speculations about novel symmetry breaking mechanisms specific to the early Universe. It is best here to quote Steven Weinberg's abstract [18]: "It is suggested that Goldstone bosons may be masquerading as fractional cosmic neutrinos, contributing about 0.39 to what is reported as the effective number of neutrino types in the era before recombination. The broken symmetry associated with these Goldstone bosons is further speculated to be the conservation of the particles of dark matter."

2.3 Degrees of freedom and neutrino freeze-out

The well-established effect, known to make a contribution $\delta N_{\rm eff} = 0.046$, is that the electron-positron annihilation sends entropy, albeit in a small amount, into the neutrino degrees of freedom. In fact, in figure 2 the domain of neutrino freeze-out is indicated by a vertical line and it is clear that some annihilation is already underway before neutrinos decouple and one must remember that the decoupling process is gradual and energy dependent. We have studied this effect using a novel numerical method [19].

Since there is a small $\delta N_{\rm eff}$ effect, one must ask if, and under which conditions, a larger effect could arise – which would require that in some way our extrapolation back in time to the neutrino decoupling is not entirely correct, perhaps because the conditions are so much more extreme. In fact, if we could turn up alone the mass of the electron without a change in other dimensioned natural constants that could perhaps do the trick nicely and easily as we see inspecting figure 2: an electron of several MeV would undergo pair annihilation before the computed neutrino decoupling. However, the computation of neutrino decoupling should change as well since now there are fewer electrons and positrons to scatter from. So the evaluation of what will happen when physical constants change, even one out of many, is not a trivial enterprise.

We actually needed to perform a dimensional reduction which revealed that there are two parameters that count [20]. The first one is the dimensionless Weinberg angle of electroweak interactions, which controls detail of neutrino cross sections, with the known value obtained from the mass ratio

$$\frac{M_W}{M_Z} \equiv \cos\theta_{\rm W} = 0.8815\ , \longrightarrow \ \sin^2\theta_{\rm W}\big|_{M_Z} = 0.231\ . \tag{3}$$

There is no anchor in the Standard Model of particle physics fixing $\sin^2\theta_{\rm W}$ and yet it is difficult to imagine that the ratio of masses in Eq. (3) could be modified in the early Universe. However, the value of $\sin^2\theta_{\rm W}$ varies with the running energy scale relatively strongly and thus $\sin^2\theta_{\rm W}$ could be sensitive to the temperature of the hot Universe at the low scales at which we are studying the decoupling process, without altering the elementary mass ratio that is governed by the 10^5 times higher energy scale. Hence we show the key results in figure 4 as a function of $\sin^2\theta_{\rm W}$.

The other parameter on which the freeze-out condition depends combines all dimensioned variables as follows

$$\eta \equiv M_{\rm Pl}\, m_e^3\, G_F^2\ , \qquad M_{\rm Pl}^2 \equiv \frac{1}{8\pi G_N} = 2.4354\, 10^{18}\,{\rm GeV}\ , \quad \eta_0 = 0.04421\ . \tag{4}$$

and we find that as η grows, the temperature at which the neutrino decoupling occurs drops and thus more of the annihilation entropy of electrons and positrons finds its way into neutrinos, hence increasing $N_{\rm eff}$. This is so since we can think that increasing η the Fermi coupling increases and this makes neutrino scattering on electrons stronger. It helps also that this increase could mean that the mass of the electron is increasing and thus the electron mass moves right into the neutrino decoupling scale while the increase of G_F moves the decoupling towards the electron mass. Similarly, an increase in Planck mass decreases the strength of gravity, and makes the speed of expansion slower. The Planck mass provides the scale of time in which the processes that we are looking at occur; the characteristic time constant is a small fraction of $\tau \equiv M_{\rm Pl}/m_e^2 = 6.18\,$s. Increasing $M_{\rm Pl}$ we thus give the system more time to relax and this helps to move entropy into neutrinos. All these remarks are summarized in figure 4 which presents the change in $N_{\rm eff}$ for the value $\eta/\eta_0 = 1, 2, 5, 10$ as a function of $\sin^2\theta_{\rm W}$.

The result shows that it is rather difficult to increase sufficiently the value of $N_{\rm eff}$. The difficulty of the situation is that in principle we should believe that there are only two scales, Planck $M_{\rm Pl}$ and Higgs v, and the coefficient $\eta \propto M_{\rm Pl}/v$ scales linearly with the ratio of both scales. It is hard to imagine that the Higgs scale which defines the vacuum state can be drastically modified. Similarly, it is difficult to see how there can be a change in the Planck scale allowing accommodation of a larger value of $N_{\rm eff}$, without other early Universe effects, for example, a modfied dynamics of nucleosynthesis during the

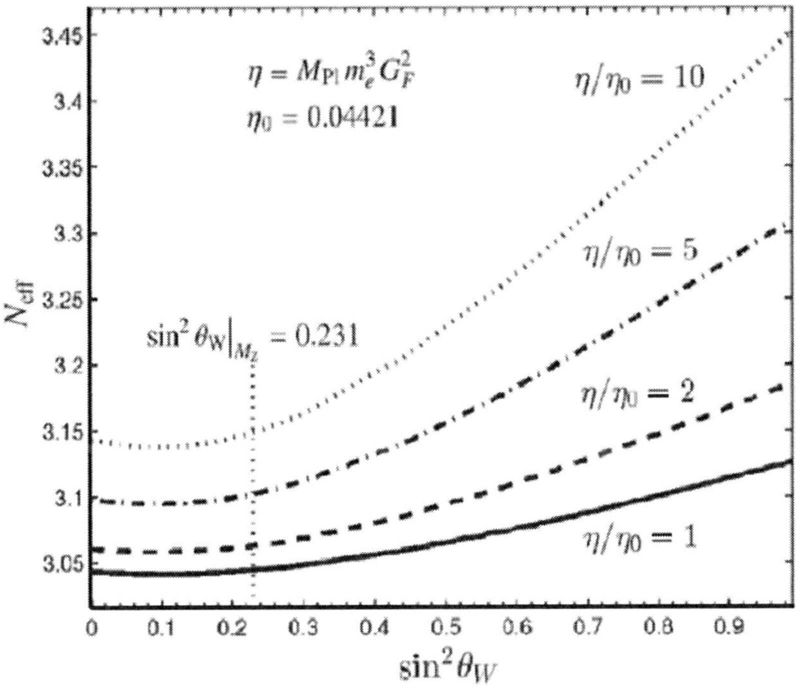

Figure 4. The dependence of $N_{\rm eff}$ on the combined parameters of the neutrino freeze-out model in the early Universe, as a function of $\sin^2\theta_{\rm W}$, with $\eta = 1, 2, 5, 10$.

immediately following BBN era. Thus, remembering that BBN works, our finding is that an allowable change in natural constants could not explain an increase in $N_{\rm eff}$.

As always, there is an 'if'. What if the mass of the electron is not really in the realm of Higgs minimal coupling? If not, it is less fixed by far and could easily change independently of other constants of nature. This would even help explain why BBN needs not worry about electron density, which as we saw in figure 2, is significant during the BBN time period. So if for example m_e could be three times greater and of the scale of the neutrino decoupling temperature, that would make the situation rather different for $N_{\rm eff}$. A reason to think in this way is that the m_e scale looks very 'electromagnetic': take any electromagnetic mass difference for any of particles around and one sees that 1.5 MeV is a typical variation. Also note that the required minimal coupling of Higgs to electron is exceptionally weak, $g = 0.5/250000 = 2\,10^{-6}$. The experimental limit on Higgs-electron coupling is 600 times above that expected by minimal coupling [21]; thus we do not really know if electron mass originates in the Higgs mechanism. Moreover, the expected small value of g renders higher order effects involving other interactions relevant, an example of this type of behavior is the fact that in an electron-positron collider Higgs would not be produced by minimal interaction, but by another higher order coupling to Higgs, see for example Ref. [22].

The conclusion of this subsection is that only if the electron is not really what Higgsologists want it to be, then there is a chance that we can, in agreement with all other early Universe considerations, create a scenario such that $N_{\rm eff}$ is modified by an increase in the electron mass present in the distant past at high temperature.

3 Conclusions

Hagedorn and his temperature were of pivotal importance in the development of the modern picture of the early Universe. Today, the experimental and theoretical insights into the working of the early

Universe and the connection of QGP and Hagedorn hadron phase offer an interesting opportunity, a fulfillment of the promise that relativistic heavy ion collisions are going to shine new light on the early Universe physics.

In the near future, experimental precision of the CMB data should improve greatly. Thus it makes good sense to characterize the evolution of the Universe from the present through to the QGP era under the most conservative assumptions about its properties, so as to create the model background against which the new data can be tested. An interesting outcome of this exercise we have obtained is an overview of the composition of the Universe across all evolution epochs and have shown here how the entropic degrees of freedom evolve.

One of the most striking outcomes is the strong dependence of the available degrees of freedom on the QCD-QGP equations of state. We believe that we are first to employ realistic lattice-QCD results in this domain and were able to present here a glimpse on our extensive study of the evolution of the Universe connecting the QGP era to the observational period after recombination.

The connection to observables involves the passage through the neutrino decoupling era, requiring a refined understanding of neutrino freeze-out, electron positron annihilation and the dependence of these processes on natural constants. Within the realm of conventional physics we have been able to solve this problem in full. We conclude that we can use these results to study the QGP phase of the early Universe, and in particular to explore it it can be a natural source of undiscovered light particles.

The interesting worry impacting this plan of action is, whether the electron is just another minimally coupled (to Higgs) elementary particle, or if its small mass signals some novel physics. If electron mass is not constant as a function of time and is also sensitive to temperature in the context of the early Universe, then our conclusions will be modified. On the other hand we gain insight into another related physics domain, the unraveling of the electron, and this is very much worth the effort.

A The ΛCDM standard model

A.1 Standard Cosmology

In order to travel back in time to the period when QGP dominated the Universe and beyond, we first need to elaborate on the relation between the expansion dynamics of the Universe and temperature. For this purpose we need some preparation in the standard cosmological (FLRW-Universe) model. We use the spacetime metric

$$ds^2 = c^2dt^2 - a^2(t)\left[\frac{dr^2}{1-kr^2} + r^2(d\theta^2 + \sin^2(\theta)d\phi^2)\right] \tag{5}$$

characterized by the scale parameter $a(t)$ of a spatially homogeneous Universe. The geometric parameter k identifies the geometry of the spacial hypersurfaces defined by comoving observers. Space is a flat-sheet for the observationally preferred value $k = 0$ [14]. In this case it can be more convenient to write the metric in rectangular coordinates

$$ds^2 = c^2dt^2 - a^2(t)\left[dx^2 + dy^2 + dz^2\right]. \tag{6}$$

We will work in units where $\hbar = 1, c = 1$.

The global Universe dynamics can be characterized by two quantities: the Hubble parameter H, a strongly time dependent quantity on cosmological time scales, and the deceleration parameter q:

$$\frac{\dot{a}}{a} \equiv H(t), \quad \frac{\ddot{a}}{a} = -qH^2, \quad q \equiv -\frac{a\ddot{a}}{\dot{a}^2}, \quad \dot{H} = -H^2(1+q). \tag{7}$$

The Einstein equations are:

$$G^{\mu\nu} = R^{\mu\nu} - \left(\frac{R}{2} + \Lambda\right) g^{\mu\nu} = 8\pi G_N T^{\mu\nu}, \quad R = g_{\mu\nu} R^{\mu\nu}. \tag{8}$$

Symmetry considerations imply that the stress energy tensor is determined by an energy density and an isotropic pressure

$$T^{\mu}_{\nu} = \mathrm{diag}(\varepsilon, -P, -P, -P). \tag{9}$$

It is common to absorb the Einstein cosmological constant Λ into the energy and pressure

$$\varepsilon_\Lambda = \frac{\Lambda}{8\pi G_N} = 25.6\,\mathrm{meV}^4, \qquad P_\Lambda = -\varepsilon_\Lambda \tag{10}$$

and we implicitly consider this done from now on. The fitted value of the dark energy is indicated

One should note that the bag constant $\mathcal{B}$ of the quark-bag model has the same behavior in regard to energy and momentum as has the Einstein cosmological parameter $\mathcal{B} \leftrightarrow \Lambda/8\pi G_N$. $\mathcal{B}$ adds positively to the energy density but negatively to the pressure, counteracting the positive particle pressure. Contrary to the initial expectation based on the quark-bag model where quark pressure is in equilibrium with the bag constant, in the dynamical Universe the appearance of such a bag term will accelerate the expansion just like today where we see an acceleration due to dark energy. The parallel meaning of $\mathcal{B}$ and $\Lambda/8\pi G_N$ relies on both quantities acting within the volume of their respective 'Universe', in the sense that $\mathcal{B}$ is strictly and only present within the volume of quark blobs – hadrons, or QGP.

Two dynamically independent equations arise using the metric Eq. (5) in Eq. (8):

$$\frac{8\pi G_N}{3}\varepsilon = \frac{\dot a^2 + k}{a^2} = H^2\left(1 + \frac{k}{\dot a^2}\right), \qquad \frac{4\pi G_N}{3}(\varepsilon + 3P) = -\frac{\ddot a}{a} = qH^2. \tag{11}$$

We can eliminate the strength of the interaction, G_N, solving both these equations for $8\pi G_N/3$, and equating the result to find a relatively simple constraint for the deceleration parameter:

$$q = \frac{1}{2}\left(1 + 3\frac{P}{\varepsilon}\right)\left(1 + \frac{k}{\dot a^2}\right). \tag{12}$$

For a spatially flat Universe, $k = 0$, note that in a matter-dominated era where $P/\varepsilon << 1$ we have $q \simeq 1/2$; for a radiative Universe where $3P = \varepsilon$ we find $q \simeq 1$; and in a dark energy Universe in which $P = -\varepsilon$ we find $q = -1$. Spatial flatness is equivalent to the assertion that the energy density of the Universe equals the critical density

$$\varepsilon = \varepsilon_{\mathrm{crit}} \equiv \frac{3H^2}{8\pi G_N}. \tag{13}$$

The CMB power spectrum is sensitive to the deceleration parameter and the presence of spatial curvature modifies q. The Planck results [14] constrain the effective curvature energy density fraction,

$$\Omega_K \equiv 1 - \varepsilon/\varepsilon_{\mathrm{crit}}, \tag{14}$$

to

$$|\Omega_K| < 0.005. \tag{15}$$

This indicates a nearly flat Universe. We will work here within an exactly spatially flat cosmological model, $k = 0$.

As must be the case for any solution of Einstein's equations, Eq. (11) implies that the energy momentum tensor of matter is divergence free:

$$T^{\mu\nu}{}_{;\nu} = 0 \Rightarrow -\frac{\dot{\varepsilon}}{\varepsilon + P} = 3\frac{\dot{a}}{a} = 3H. \tag{16}$$

A dynamical evolution equation for $\varepsilon(t)$ arises once we combine Eq. (16) with Eq. (11), eliminating H. Given an equation of state $P(\varepsilon)$, solutions of this equation describes the dynamical evolution of matter in the Universe. In practice, we evolve the system in both directions in time. On one side, we start in the present era with the energy density fractions fit by Planck [14],

$$H_0 = 67.74\text{km/s/Mpc}\ ,\quad \Omega_b = 0.05\ ,\quad \Omega_c = 0.26\ ,\quad \Omega_\Lambda = 0.69\ , \tag{17}$$

and integrate backward in time. On the other hand, we start in the QGP era with an equation of state determined by an ideal gas of SM particles, combined with a perturbative QCD equation of state for quarks and gluons, and integrate forward in time.

A.2 Matter Content

In this work, matter will be modeled by a particle distribution function $f(t, x, p)$ that, roughly speaking, gives the probability of finding a particle per unit spacial volume per unit momentum space volume at a given time. The distribution function gives the stress energy tensor, particle four-current, and entropy four-current via

$$T^{\mu\nu}(t,x) = \frac{d}{(2\pi)^3}\int p^\mu p^\nu f(t,x,p)\sqrt{|g|}\frac{d^3p}{p_0}, \tag{18}$$

$$n^\nu(t,x) = \frac{d}{(2\pi)^3}\int p^\nu f(t,x,p)\sqrt{|g|}\frac{d^3p}{p_0}, \tag{19}$$

$$s^\nu(t,x) = -\frac{d}{(2\pi)^3}\int (f\ln(f) \pm (1\mp f)\ln(1\mp f))p^\nu\sqrt{|g|}\frac{d^3p}{p_0}, \tag{20}$$

where the upper signs are for fermions, the lower for bosons, d is the degeneracy of the particle, and g is the determinant of the metric. In a flat FRW Universe, the expressions for the energy density, pressure, number density, and entropy density of a particle of mass m are

$$\varepsilon = \frac{d}{(2\pi)^3}\int f(t,x,p)E d^3p, \tag{21}$$

$$P = \frac{d}{(2\pi)^3}\int f(t,x,p)\frac{p^2}{3E}d^3p, \tag{22}$$

$$n = \frac{d}{(2\pi)^3}\int f(t,x,p)d^3p,\quad E = \sqrt{m^2+p^2}, \tag{23}$$

$$s = -\frac{d}{(2\pi)^3}\int (f\ln(f) \pm (1\mp f)\ln(1\mp f))d^3p. \tag{24}$$

A.3 Equilibrium and Free-Streaming Distribution

Free-streaming is a type of non-equilibrium distribution that is significant in cosmology. Here we outline its properties, including what distinguishes it from the equilibrium distributions. The freeze-out process, whereby a particle species stops interacting and decouples from the photon background, involves several steps that lead to the final form of the free-streaming momentum distribution. For further details, see Ref. [23].

Chemical freeze-out of a particle species occurs at the temperature, $T_{\rm ch}$, when particle number changing processes slow down and the particle abundance can no longer be maintained at an equilibrium level. Prior to the chemical freeze-out temperature, number changing processes are significant and keep the particle in chemical (and thermal) equilibrium, implying that the distribution function has the Fermi-Dirac form, obtained by maximizing entropy at fixed energy

$$f_c(t,E) = \frac{1}{\exp(E/T)+1}, \text{ for } T(t) > T_{\rm ch}. \tag{25}$$

Kinetic freeze-out occurs at the temperature, T_f, when momentum exchanging interactions no longer occur rapidly enough to maintain an equilibrium momentum distribution. When $T_f < T(t) < T_{\rm ch}$, number changing process no longer occur rapidly enough to keep the distribution in chemical equilibrium but there is still sufficient momentum exchange to keep the distribution in thermal equilibrium. The distribution function is therefore obtained by maximizing entropy, with a fixed energy, particle number, and antiparticle number separately, implying that the distribution function has the form

$$f_k(t,E) = \frac{1}{\Upsilon^{-1}\exp(E/T)+1}, \text{ for } T_f < T(t) < T_{\rm ch}. \tag{26}$$

The fugacity

$$\Upsilon(t) \equiv e^{\sigma(t)} \tag{27}$$

controls the occupancy of phase space and is necessary once $T(t) < T_{\rm ch}$ in order to conserve particle number. See Ref. [23] for a detailed discussion of its significance.

For $T(t) < T_f$ there are no longer any significant interactions that couple the particle species of interest and so they begin to free-stream through the Universe, i.e. travel on geodesics without scattering. The Einstein-Vlasov equation can be solved, see Ref. [24], to yield the free-streaming momentum distribution

$$f(t,E) = \frac{1}{\Upsilon^{-1}e^{\sqrt{p^2/T^2+m^2/T_f^2}}+1} \tag{28}$$

where the free-streaming effective temperature

$$T(t) = \frac{T_f a(t_k)}{a(t)} \tag{29}$$

is obtained by red-shifting the temperature at kinetic freeze-out.

The corresponding free-streaming energy density, pressure, and number densities are given by

$$\varepsilon = \frac{d}{2\pi^2}\int_0^\infty \frac{\left(m^2+p^2\right)^{1/2} p^2 dp}{\Upsilon^{-1}e^{\sqrt{p^2/T^2+m^2/T_f^2}}+1}, \tag{30}$$

$$P = \frac{d}{6\pi^2}\int_0^\infty \frac{\left(m^2+p^2\right)^{-1/2} p^4 dp}{\Upsilon^{-1}e^{\sqrt{p^2/T^2+m^2/T_f^2}}+1}, \tag{31}$$

$$n = \frac{d}{2\pi^2}\int_0^\infty \frac{p^2 dp}{\Upsilon^{-1}e^{\sqrt{p^2/T^2+m^2/T_f^2}}+1}, \tag{32}$$

where d is the degeneracy of the particle species. These differ from the corresponding expressions for an equilibrium distribution in Minkowski space by the replacement $m \rightarrow mT(t)/T_f$ *only* in the exponential.

The separation of the freeze-out process into these three regimes is of course only an approximation. In principle there is a smooth transition between them. However, it is a very useful approximation in cosmology. See [19, 25] for methods capable of resolving these smooth transitions.

References

[1] R. Hagedorn, "Statistical thermodynamics of strong interactions at high-energies," Nuovo Cim. Suppl. **3** 147 (1965).

[2] R. Hagedorn, *in: Proceedings, Summer School on Theoretical Physics, Cargèse Lectures in Physics: Cargese, France, September 1971* "Thermodynamics of strong interactions," Cargese Lect. Phys. **6**, 643 (1973), E. Schatzman (ed.) 731 pages, New York: Gordon and Breach (1973)

[3] R. Hagedorn, "Boiling Primordial Matter: 1968," Published as "Siedende Urmaterie" in Verhandlungen der Schweizerischen Naturforschenden Gesellschaft; 148 (1968) pp 51–63. Republished translated in Ref. [4], pp 125–138, Chapter 16, doi:10.1007/978-3-319-17545-4_16

[4] J. Rafelski, Editor, "Melting Hadrons, Boiling Quarks - From Hagedorn Temperature to Ultra-Relativistic Heavy-Ion Collisions at CERN : With a Tribute to Rolf Hagedorn," doi:10.1007/978-3-319-17545-4 Springer Open, Heidelberg, New York (2016)

[5] M. J. Fromerth and J. Rafelski, "Hadronization of the quark Universe," http://arxiv.org/abs/astro-ph/0211346 astro-ph/0211346.

[6] M. J. Fromerth, I. Kuznetsova, L. Labun, J. Letessier and J. Rafelski, "From Quark-Gluon Universe to Neutrino Decoupling," Acta Phys. Polon. B **43**, no. 12, 2261 (2012) doi:10.5506/APhysPolB.43.2261 [arXiv:1211.4297 [nucl-th]].

[7] J. Rafelski, "Connecting QGP-Heavy Ion Physics to the Early Universe," Nucl. Phys. Proc. Suppl. **243-244**, 155 (2013) doi:10.1016/j.nuclphysbps.2013.09.017 [arXiv:1306.2471 [astro-ph.CO]].

[8] J. Rafelski and J. Birrell, "Traveling Through the Universe: Back in Time to the Quark-Gluon Plasma Era," J. Phys. Conf. Ser. **509**, 012014 (2014) doi:10.1088/1742-6596/509/1/012014 [arXiv:1311.0075 [nucl-th]].

[9] J. Rafelski, "Melting Hadrons, Boiling Quarks," Eur. Phys. J. A **51** 114 (2015) doi:10.1007/978-3-319-17545-4_33,10.1140/epja/i2015-15114-0

[10] R. Hagedorn, "How We Got to QCD Matter from the Hadron Side: 1984," Lect. Notes Phys. **221** 53 (1985), reprinted in Chapter 25 of [4] doi:10.1007/978-3-319-17545-4_25

[11] S. Borsanyi, "Thermodynamics of the QCD transition from lattice," Nucl. Phys. A **904-905**, 270c (2013) doi:10.1016/j.nuclphysa.2013.01.072 [arXiv:1210.6901 [hep-lat]].

[12] S. Borsanyi, Z. Fodor, C. Hoelbling, S. D. Katz, S. Krieg and K. K. Szabo, "Full result for the QCD equation of state with 2+1 flavors," Phys. Lett. B **730**, 99 (2014) doi:10.1016/j.physletb.2014.01.007 [arXiv:1309.5258 [hep-lat]].

[13] M. Strickland, private communication.

[14] P. A. R. Ade *et al.* [Planck Collaboration], "Planck 2013 results. XVI. Cosmological parameters," Astron. Astrophys. **571**, A16 (2014) doi:10.1051/0004-6361/201321591 [arXiv:1303.5076 [astro-ph.CO]].

[15] J. Birrell and J. Rafelski, "Quark–gluon plasma as the possible source of cosmological dark radiation," Phys. Lett. B **741**, 77 (2015) doi:10.1016/j.physletb.2014.12.033 [arXiv:1404.6005 [nucl-th]].

[16] L. A. Anchordoqui and H. Goldberg, "Neutrino cosmology after WMAP 7-Year data and LHC first Z' bounds," Phys. Rev. Lett. **108** 081805 (2012) doi:10.1103/PhysRevLett.108.081805 [arXiv:1111.7264 [hep-ph]].

[17] L. A. Anchordoqui, H. Goldberg and G. Steigman, "Right-Handed Neutrinos as the Dark Radiation: Status and Forecasts for the LHC," Phys. Lett. B **718**, 1162 (2013) doi:10.1016/j.physletb.2012.12.019 [arXiv:1211.0186 [hep-ph]].

[18] S. Weinberg, "Goldstone bosons as Fractional Cosmic Neutrinos," Phys. Rev. Lett. **110**, no. 24, 241301 (2013) doi:10.1103/PhysRevLett.110.241301 [arXiv:1305.1971 [astro-ph.CO]].

[19] J. Birrell, J. Wilkening and J. Rafelski, J. Comput. Phys. **281**, 896 (2014) doi:10.1016/j.jcp.2014.10.056 [arXiv:1403.2019 [math.NA]].

[20] J. Birrell, C. T. Yang and J. Rafelski, "Relic Neutrino Freeze-out: Dependence on Natural Constants," Nucl. Phys. B **890**, 481 (2014) doi:10.1016/j.nuclphysb.2014.11.020 [arXiv:1406.1759 [nucl-th]].

[21] W. Altmannshofer, J. Brod and M. Schmaltz, "Experimental constraints on the coupling of the Higgs boson to electrons," JHEP **1505**, 125 (2015) doi:10.1007/JHEP05(2015)125 [arXiv:1503.04830 [hep-ph]].

[22] T. Han, Z. Liu, Z. Qian and J. Sayre, "Improving Higgs coupling measurements through ZZ Fusion at the ILC," Phys. Rev. D **91**, 113007 (2015) doi:10.1103/PhysRevD.91.113007 [arXiv:1504.01399 [hep-ph]].

[23] J. Birrell, C. T. Yang, P. Chen and J. Rafelski, "Relic neutrinos: Physically consistent treatment of effective number of neutrinos and neutrino mass," Phys. Rev. D **89**, 023008 (2014) doi:10.1103/PhysRevD.89.023008 [arXiv:1212.6943 [astro-ph.CO]].

[24] Y. Choquet-Bruhat, *General Relativity and the Einstein Equations*, Oxford Mathematical Monographs, OUP Oxford (2008) https://books.google.com/books?id=UjHbm5rfpi8C.

[25] G. Mangano, G. Miele, S. Pastor, T. Pinto, O. Pisanti and P. D. Serpico, "Relic neutrino decoupling including flavor oscillations," Nucl. Phys. B **729**, 221 (2005) doi:10.1016/j.nuclphysb.2005.09.041 [hep-ph/0506164].

EPJ Web of Conferences **126**, 03006 (2016) DOI: 10.1051/epjconf/201612603006
ICNFP 2015

Appearance of Hagedorn limiting temperature in microscopic model calculations

E. Zabrodin[1,2,3,4,a], L. Bravina[1,3,4], M. Bleicher[3,5], and H. Stöcker[3,6,7]

[1] *Department of Physics, University of Oslo, PB 1048 Blindern, Oslo, Norway*
[2] *Skobeltsyn Institute of Nuclear Physics, Moscow State University, RU-119991 Moscow, Russia*
[3] *Frankfurt Institute for Advanced Studies, Ruth-Moufang-Str. 1, D-60438 Frankfurt a.M., Germany*
[4] *National Research Nuclear University "MEPhI" (Moscow Engineering Physics Institute), RU-115409 Moscow, Russia*
[5] *Institute for Theoretical Physics, J.W. Goethe Universität, D-60438 Frankfurt a.M., Germany*
[6] *J.W. Goethe Universität, D-60438 Frankfurt a.M., Germany*
[7] *GSI Helmholtzzentrum für Schwerionenforschung GmbH, D-64291 Darmstadt, Germany*

Abstract. One of important consequences of Hagedorn statistical bootstrap model is the prediction of limiting temperature T_{crit} for hadron systems colloquially known as Hagedorn temperature. According to Hagedorn, this effect should be observed in hadron spectra obtained in infinite equilibrated nuclear matter rather than in relativistic heavy-ion collisions. We present results of microscopic model calculations for the infinite nuclear matter, simulated by a box with periodic boundary conditions. The limiting temperature indeed appears in the model calculations. Its origin is traced to strings and many-body decays of resonances.

1 Introduction

The statistical bootstrap model (SBM) is based on the assumption of infinite sequence of heavier and heavier resonances (or clusters), each being a constituent of a heavier resonance and, simultaneously, being composed of lighter ones. The bootstrap equation proposed by Hagedorn [1, 2] reads in a general form as

$$\rho(m)_m = \{input\ particles\} + \sum_2^{\infty} \int \ldots \int \{products\ of\ \ \delta - functions\ \ and\, \rho' s\}\,, \qquad (1)$$

where $\rho(m)$ is a density of clusters with masses between m and $m+dm$. This equation has an analytical solution [3]

$$\rho(m)_{m\to\infty} \longrightarrow \rho_0\, m^{a_H}\ \exp{(m/T_H)} \qquad (2)$$

containing two parameters: the exponent $a_H \approx -5/2$ and the limiting temperature T_H above which the hadronic matter becomes unstable. This temperature is known nowadays simply as a Hagedorn temperature, and at $T \geq T_H$ the phase transition to a new state of matter, quark-gluon plasma (QGP),

[a] e-mail: eugen.zabrodin@fys.uio.no

takes place. Further details of the SBM can be found elsewhere, see, e.g., [1, 2, 4–6] and references therein. Answering the question how the SBM is related to real collisions, Hagedorn admitted that [2] "*... All this applies to infinitely extended hadronic matter in equilibrium. Experiments unfortunately produce only microscopic lumps of such matter, which are never in equilibrium. The relation between the described model and the situation in a collision is, therefore, far from trivial. It seems, however, that at any given time equilibrium is nearly reached* **locally**, *so that the model can be applied locally and then be folded with collective motions assumed* **ad hoc** *or derived from special models.*"

The microscopic transport models [7–10], employed for calculations of *hh*, *hA* and *A*+*A* collisions at relativistic energies, utilise the colored objects, i.e. strings, in addition to the tables of known hadrons and their resonances. Strings in their turn decay into hadrons, and the string masses are not limited from the top. Can this picture be similar to that proposed by the SBM? It is very tempting, therefore, to use the microscopic model with hadrons, resonances and strings for the simulation of infinite nuclear matter. Aspects of the calculations are discussed in Sec. 2.

2 Simulation of infinite nuclear matter

A. Cubic box with periodic boundary conditions.
The standard choice for such a simulation is a cubic box with periodic boundary conditions [11–13] to ensure the energy and momentum conservation. If one particle leaves the box, another particle, fully identical to the first one, enters the box from the opposite side. The box should be neither too big nor too small, e.g. $V = 10 \times 10 \times 10 = 1000\,\mathrm{fm}^3$ [11] or $V = 5 \times 5 \times 5 = 125\,\mathrm{fm}^3$ [12]. The energy density ε, the net baryon density ρ_B, and the net strangeness density ρ_S are fixed at the initial stage. Recall, that for the infinite nuclear matter the net strangeness is zero, but for the central cell in relativistic heavy-ion collisions, which is an open system, ρ_S can be small though differ from zero [14, 15]. In case of $\rho_S = 0$ the initial configuration in the box consists of protons and neutrons uniformly distributed in the configuration space. Their momenta are then rescaled to get the required energy density. The system is ready, and one has to employ a microscopic model, such as UrQMD [7, 8] or HSD [9], to trace the evolution of the system and study the relaxation of hadron-string matter in the box to equilibrium.

First, the kinetic equilibrium should be verified. It assumes that the velocity distributions $f_i(v_j)$, $j = x, y, z$ of hadron species are close to Maxwellian and isotropic, $f_i(v_x) = f_i(v_y) = f_i(v_z)$, thus leading to isotropy of the pressure gradients. Then, the particle composition and their energy spectra should be time independent. This conditions correspond to requirements of chemical equilibrium and thermal equilibrium, respectively.

B. Features of UrQMD model.
The Ultra-relativistic Quantum Molecular Dynamics model (UrQMD) [7, 8] is a Monte-Carlo event generator designed for the description of hadron-hadron, hadron-nucleus, and nucleus-nucleus collisions in a broad energy range from hundred MeV up to several TeV. As independent degrees of freedom the model contains 55 baryon and 32 meson states, together with their antiparticles and explicit isospin-projected states, with masses up to 2.25 GeV/c. At lower energies the interaction dynamics of *hh* or $A + A$ collisions takes into account interactions between the hadrons and their excited states, resonances. At higher energies the UrQMD treats the production of new particles via formation and fragmentation of specific colored objects, strings. Strings are uniformly stretched between the quarks, diquarks and their antistates with constant string tension $\kappa \approx 1$ GeV/fm. The excited strings are fragmenting into pieces via the Schwinger mechanism of $q\bar{q}$-pair production, and the distribution of newly produced hadrons is uniform in the rapidity space.

UrQMD utilises the longitudinal mechanism of string excitation, where the string masses arise from momentum transfer. The model employs tables of the experimentally available information, such as hadron cross sections, resonance widths and decay modes. If this information is lacking, the one-boson exchange model, detailed balance considerations and isospin symmetry conditions are used. The propagation of particles is governed by Hamilton equation of motion, and hadronic cascade is used for the description of hA and $A + A$ interactions. Due to the uncertainty principle newly produced particles can interact further only after a certain formation time. Hadrons containing the valence quarks, however, can interact immediately with the reduced cross section $\sigma = \sigma_{qN}$. The Pauli principle is taken into account via the blocking of the final state, if the outgoing phase space is occupied.

C. Statistical model of ideal hadron gas.
To extract the thermodynamic characteristics, such as temperature T, baryo-chemical potential μ_B and strangeness chemical potential μ_S one has to compare the hadron energy spectra and yields, extracted from the box calculations, with those obtained in statistical model (SM) of an ideal hadron gas [16–18] with essentially the same degrees of freedom. - If the system is in thermal and chemical equilibrium, its macroscopic characteristics are fully determined by particle distribution functions

$$f(p, m_i) = \left[\exp\left(\frac{\sqrt{p^2 + m_i^2} - \mu_{\rm B} B_i - \mu_{\rm S} S_i}{T} \right) \pm 1 \right]^{-1} , \qquad (3)$$

where p, m_i and d_i is the momentum and mass of the hadron species i, respectively. Sign + stands for fermions and − for bosons. One has to know just three parameters, namely, temperature T and chemical potentials assigned to the conserved charges: baryon chemical potential $\mu_{\rm B}$ and strangeness chemical potential $\mu_{\rm S}$. Chemical potential of i-th hadron depends on baryon and strangeness content $\mu_i = B_i\mu_{\rm B} + S_i\mu_{\rm S}$. Then the expressions for particle number density n_i, energy density ε_i and pressure P read

$$n_i = \frac{g_i}{2\pi^2} \int_0^\infty p^2 f(p, m_i) dp \qquad (4)$$

$$\varepsilon_i = \frac{g_i}{2\pi^2} \int_0^\infty p^2 \sqrt{p^2 + m_i^2}\, f(p, m_i) dp \qquad (5)$$

$$P = \frac{g_i}{2\pi^2} \int_0^\infty p^2 dp \frac{p^2}{3(p^2 + m_i^2)^{1/2}} f(p, m_i). \qquad (6)$$

with g_i being the spin-isospin degeneracy factor. The entropy density s_i can be calculated either via the distribution function as

$$s_i = -\frac{g_i}{2\pi^2} \int f(p, m_i) \left[\ln f(p, m_i) - 1 \right] p^2 dp \, , \qquad (7)$$

or from the Gibbs thermodynamic identity

$$T s_i = \varepsilon_i + P_i - \mu_B \rho_{B_i} - \mu_S \rho_{S_i} \, . \qquad (8)$$

The procedure is as follows [14, 15]. The values of energy density ε, baryon density ρ_B and strangeness density ρ_S employed for the box calculations are inserted as an input to a system of the SM nonlinear equations

DOI: 10.1051/epjconf/201612603006

$$\varepsilon = \sum_i \varepsilon_i^{\rm SM}(T, \mu_{\rm B}, \mu_{\rm S}), \tag{9}$$

$$\rho_{\rm B} = \sum_i B_i \cdot n_i^{\rm SM}(T, \mu_{\rm B}, \mu_{\rm S}), \tag{10}$$

$$\rho_{\rm S} = \sum_i S_i \cdot n_i^{\rm SM}(T, \mu_{\rm B}, \mu_{\rm S}). \tag{11}$$

to determine temperature T, baryon chemical potential μ_B and strangeness chemical potential μ_S. After that all characteristics of the system in equilibrium are known and particle spectra can be compared with those obtained from microscopic model calculations.

3 Relaxation to equilibrium

Before starting the simulations one has to select energy density, net baryon density and net strangeness density in the system. As one of the possible benchmarks we choose the conditions corresponding to those in the central cell in central Pb+Pb collisions at SPS energy (E_{lab} = 160 AGeV) at t = 10 fm/c after the beginning of the collision. At this moment ε = 468 GeV/fm^3, ρ_B = 00.9 fm^{-3} and ρ_S = -0.01 fm^{-3}, and the hadron matter in the cell is already in kinetic equilibrium [15]. Figure 1 displays the difference $\sigma_x-(\sigma_y+\sigma_z)/2$ of the second moments $\sigma_j^{(2)} = \langle v_j^2\rangle - \langle v_j\rangle^2$, $j = x, y, z$ of the longitudinal and transverse velocity distributions of main hadron species.

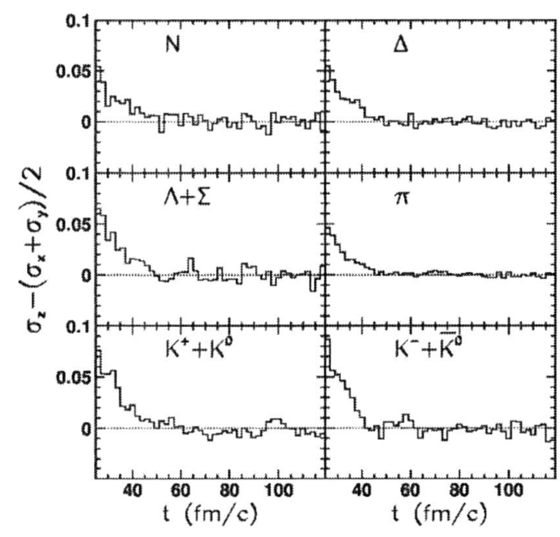

Figure 1. Anisotropy $\sigma_x - (\sigma_y + \sigma_z)/2$ of the velocity distributions of hadrons in the box with V = 125 fm^3, ε = 468 GeV/fm^3, ρ_B = 00.9 fm^{-3} and ρ_S = -0.01 fm^{-3} as a function of time.

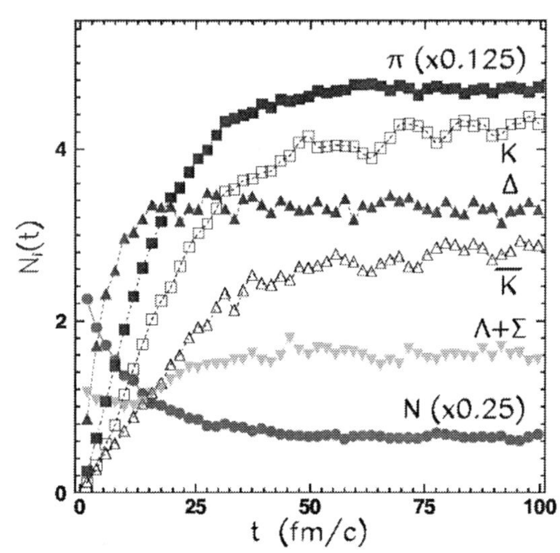

Figure 2. Time evolution of the hadron yields in the box with the same parameters as those in Fig. 1. (From [12]).

These distributions become isotropic at $t \approx 50$ fm/c. Although the yields of some hadron species are also frozen at that time, as shown in Fig. 2, for kaons and antikaons the saturation takes place later, at $t \approx 90$ fm/c only. The late relaxation of kaons is not a feature of a particular model. Similar delays for the saturation times of strange particles were observed in [13] also. Recall, that although the chemical composition of hadrons in the box is frozen, the inelastic reactions can still take place [12].

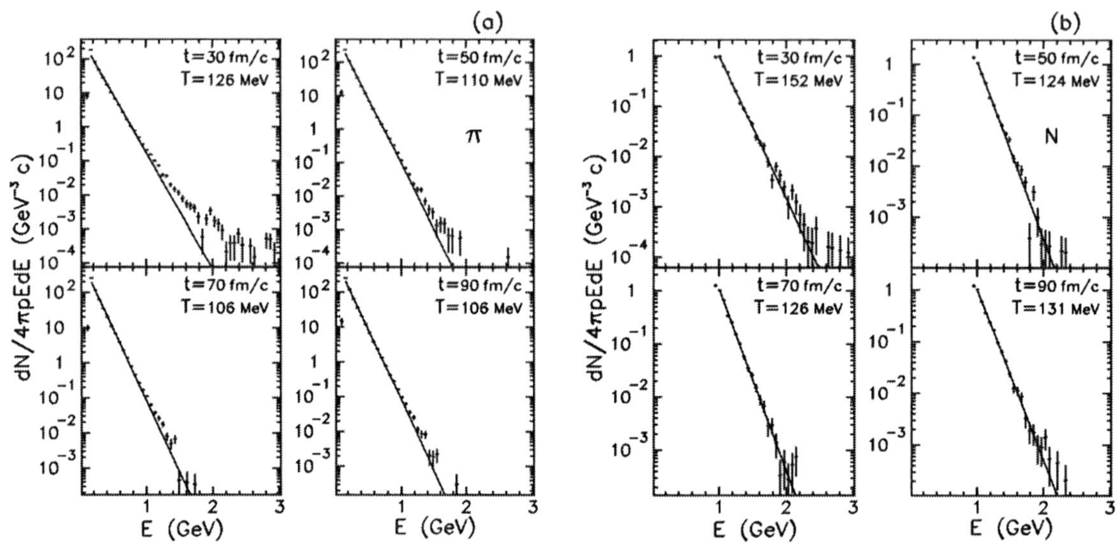

Figure 3. Energy spectra of (a) pions and (b) nucleons at 30, 50, 70 and 90 fm/c after the beginning of collisions in the box. Initial conditions and volume of the box are the same as in Fig. 1.

Finally, the conditions of thermal equilibrium should be investigated. The energy spectra of pions and nucleons at times from 30 fm/c to 90 fm/c are shown in Fig. 3(a) and Fig. 3(b), respectively. Lines indicate the results of the fit to Boltzmann-Gibbs distribution $\exp(-E_i/T + \mu_i/T)$. For pions the spectrum becomes time independent at $t \geq 70$ fm/c indicating that the thermal equilibrium is attained with the apparent temperature $T_\pi \approx 106$ MeV. Nucleons seems to be thermalized already at $t \approx 50$ fm/c with a bit higher apparent temperature $T_N \approx 127$ MeV. Still the difference between T_π and T_N is small, and the nuclear matter in the box reaches the state which is very close to thermal and chemical equilibrium.

One should not be confused here by the long times needed to reach an equilibrium state. The relaxation times to chemical and thermal equilibrium in the box, which is a closed system, are long because of the continuous production of new hadrons in energetic collisions of primary hadrons. For the realistic heavy-ion collisions equilibration process proceeds much faster. We will come to this point later in Sec. 5.

4 Appearance of limiting temperature

To decide whether or not the hadron-resonance-string matter in the box reveals the limiting temperature, one has to repeat calculations similar to those in Sec. 3 with different energy densities. In UrQMD this was done in [11] for the box with $V = 1000$ fm^3, $\rho_B = 0.16$ fm^{-3} and $\rho_S = 0$ fm^{-3}. Results of the calculations are depicted in Fig. 4. Although the maximum energy density shown in this figure is 1 GeV/fm^3, the calculations have been performed up to $\varepsilon = 5$ GeV/fm^3 [11]. The results

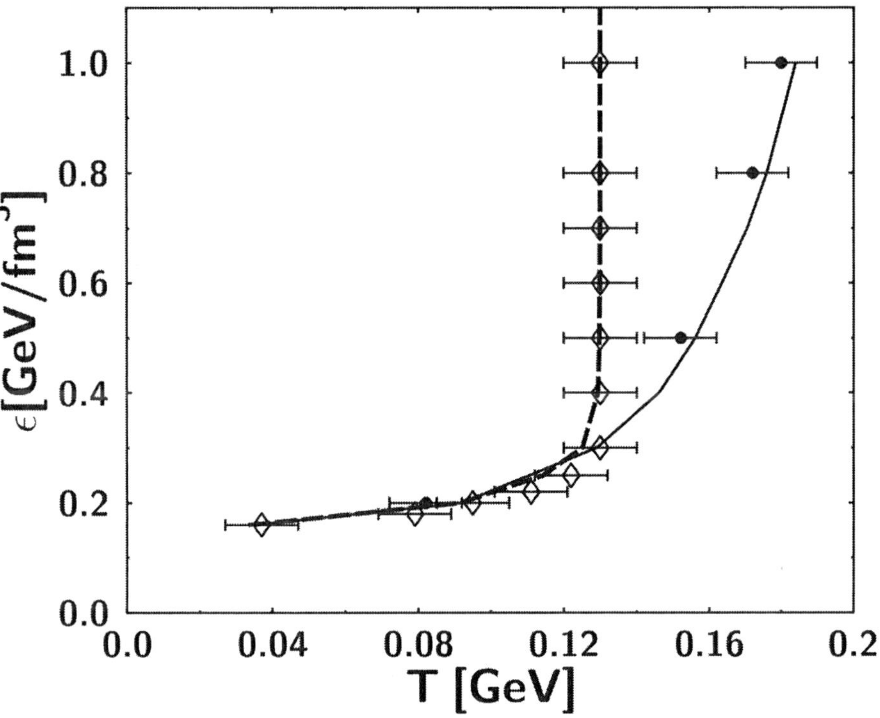

Figure 4. The energy density in the box with $V = 1000$ fm^3 versus the temperature. Open and full symbols show the UrQMD calculations with and without the strings and many-body resonance decays, respectively. Solid curve indicates the statistical model calculations, whereas dashed curve displays the SM calculations with the continuum mass spectrum. (From [11]).

remain stable, namely, at energy densities below 0.3 GeV/fm^3 the temperature rapidly rises with increasing ε, whereas at higher energy densities the temperature of hadron species in the box saturates at around $T \approx 130 \pm 15$ MeV [11]. Solid curve in Fig. 4 represents the predictions of the statistical model of ideal hadron gas with all but strings degrees of freedom. We see that in the latter case the temperature continues to increase with rising energy density. The situation can be partially improved if the string degrees of freedom are omitted in the microscopic model. However, this is not the whole story. The SM trend with infinitely rising temperature can be fully reproduced in the UrQMD calculations only if strings and (this is important to stress) many-body ($n > 2$) decays of resonances are suppressed, as shown by the full points in Fig. 4. And vise versa, can one get the limiting temperature out of the statistical model calculations? The answer is "yes", provided the continuum mass spectrum is implemented. The Hagedorn mass spectrum for the strings given by Eq.(2) was inserted in the statistical model in [11]. Six types of strings were considered, i.e. baryon strings with different strangeness content ($B = 1$; $S = 0, -1, -2, -3$), meson strings with ($B = 0$; $S = 0, 1$) and their anti-states, respectively. The upper mass limit was fixed to $M_{max} = 1$ TeV. In this case the system reveals the limiting temperature behavior, represented by the dashed curve in the same Figure.

Hadronic degrees of freedom and parameters of the string excitation play a crucial role for the evaluation of numerical value of T_{crit}. As seen in Fig. 5, in the microscopic model with shorter

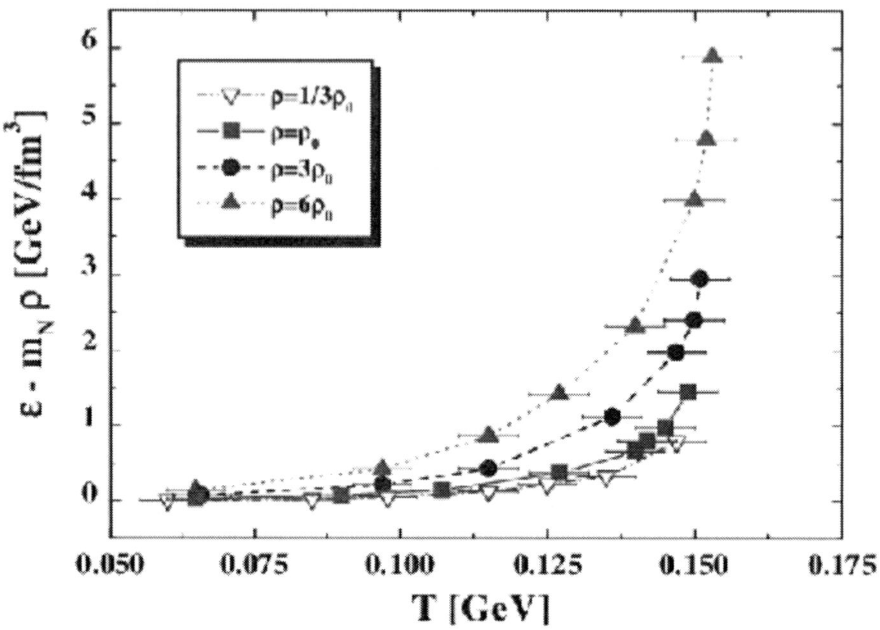

Figure 5. The same as Fig. 4 but for BUU+Lund transport model calculations [12]. The box volume is $V = 1000\ \mathrm{fm}^3$, and the baryon density varies from $1/3\rho_0$ to $6\rho_0$. (From [13]).

table of resonances and different threshold for string excitations the limiting temperature is about $T_{crit} \approx 150 \pm 5$ MeV [13]. The T_{crit}, however, is remarkably insensitive to the baryon density of the system.

The abundances of all 55 baryon states and 32 meson states, implemented in UrQMD, are presented in Fig. 6 for the case of absence of strings and suppression of many-body decays of resonances. Predictions of the statistical model are plotted onto the UrQMD results also. The agreement between the two models is fair. Two meson species, η' and f_1, are missing in the model calculations simply because these mesons are produced in the UrQMD merely via the string fragmentation. This process was blocked in the calculations. Thus we may conclude that the presence of (i) strings and (ii) many-body decays of resonances in microscopic model calculations leads to emergence of Hagedorn-like limiting temperature in particle spectra.

Last but not least, we should check another aforementioned Hagedorn statement about the possibility of local equilibrium in relativistic heavy-ion collisions and compare the results with those obtained for the infinite matter. This comparison is done in Sec. 5.

DOI: 10.1051/epjconf/201612603006

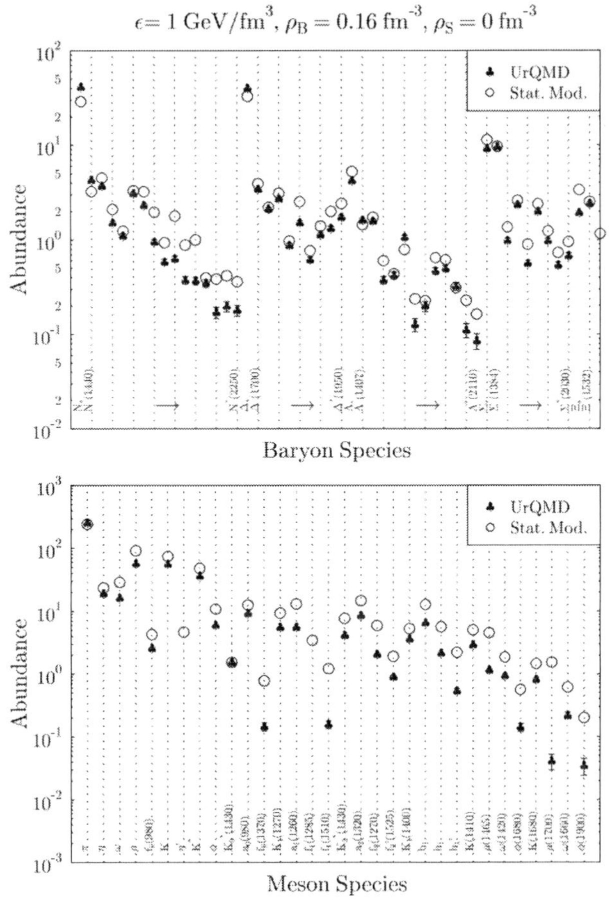

Figure 6. Yields of baryons (upper plot) and mesons (bottom plot) in UrQMD box calculations (solid symbols) and in statistical model of an ideal hadron gas (open symbols). Parameters of the box are $V = 1000$ fm^3, $\rho_B = 0.16$ fm^{-3}, $\rho_S = 0$ fm^{-3}. Strings and many-body decays of resonances are suppressed in the UrQMD calculations. (From [11]).

5 Equilibrium in the central cell and in the box

Two reactions will be discussed here, namely, central collisions of gold nuclei at AGS (E_{lab} = 10.7 AGeV) and lead nuclei at SPS (E_{lab} = 160 AGeV). Results for other energies, such as E_{lab} = 40 AGeV or $\sqrt{s_{NN}}$ = 200 GeV can be found in [19, 20]. Energy spectra of main hadron species in both reactions are displayed in Fig. 7. Here results obtained for the central cell of volume $V = 125$ fm^3 at $t = 10$ fm/c after the beginning of the collision are compared with those extracted from the box calculations with the essentially the same values of ε, ρ_B, ρ_S. The agreement between the cell and the box results is good for both energies. - Note, however, that the deviations from the SM of an ideal hadron gas become stronger with rising collision energy. - Why the hadronic matter in the cell reaches the (quasi)equilibrium much earlier compared to the matter in the box? The cell is an

DOI: 10.1051/epjconf/201612603006

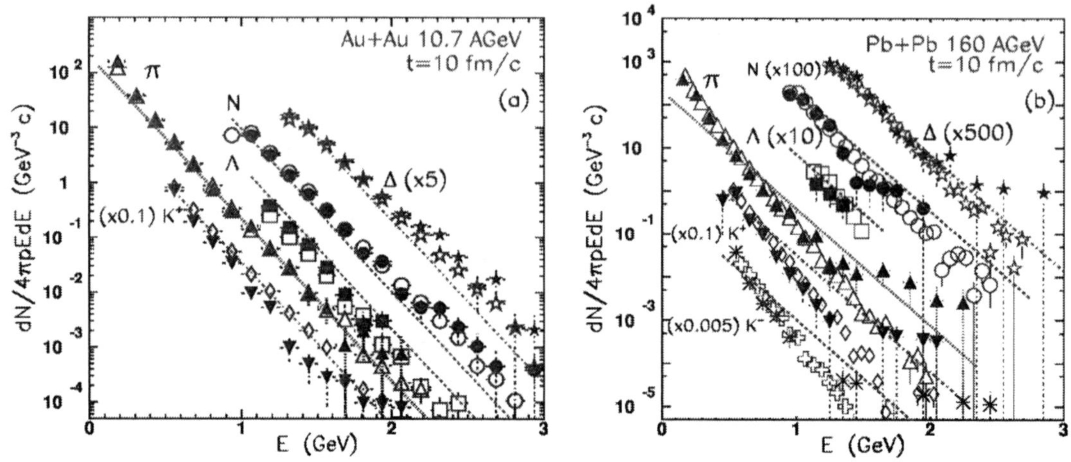

Figure 7. (a) Solid symbols: energy spectra of hadrons in the UrQMD calculations of central cell with V = 125 fm^3 at t = 10 fm/c of Au+Au collisions at E_{lab} = 10.7 AGeV. Open symbols denote the UrQMD box calculations with the same ε, ρ_B, ρ_S as in the cell. Lines are results of the fit to the SM with the parameters T = 147 MeV, μ_B = 510 MeV and μ_S = 129 MeV. (b) The same as (a) but for Pb+Pb collisions at E_{lab} = 160 AGeV.

open system, and energetic particles leave it quickly, thus resulting in effective "cooling" and faster thermalization of the rest matter in the cell compared to the box, which is a closed system where the total energy must be redistributed among all newly produced hadrons.

Still, the particle composition in the cell should be examined. We present the time evolution of hadron abundances in the cell in both reactions in Figs. 8(a) and 8(b). Here the values of ε, ρ_B, ρ_S were extracted from the cell at each moment t and used as an input for the box calculations, also plotted in these figures. Hadron multiplicities in the cell and in the box are close to each other at $t \geq 10$ fm/c. It is interesting to note that the the fraction of already "frozen" particles in the cell is quite small compared to the total abundances. Statistical model provides a good agreement for all hadrons except pions, which are underestimated. The difference becomes stronger with rising collision energy and can be attributed to lack of detailed balance in microscopic models.

Nevertheless, equilibration in the central cell sets in quite late at $t \approx 10$ fm/c compared to $\tau_0 \approx 0.5 \div 1$ fm/c used in hydrodynamic models for relativistic heavy-ion collisions. Recall, that many distributions in the cell, e.g. P vs. ε or entropy-per-baryon ratio [15, 19], indicate that quasi-equilibrium description can be applicable much earlier, already at $t \approx 3-4$ fm/c. These results are supported by the conclusions of Ref.[21], where the authors demonstrate the applicability of hydrodynamic description (under a certain conditions) to the systems being out of local equilibrium.

Thermalization rates can also be significantly increased if one directly includes the Hagedorn states, or resonances, with an exponential mass spectrum [22–24] in a hadron-resonance gas model. The simultaneous collision of several particles, most likely pions, cannot create a string, whereas it can produce a massive resonance called Hagedorn state (HS) [22]. This HS is a highly unstable state with a large decay width, therefore, the process of thermalization is speeded up extremely via the

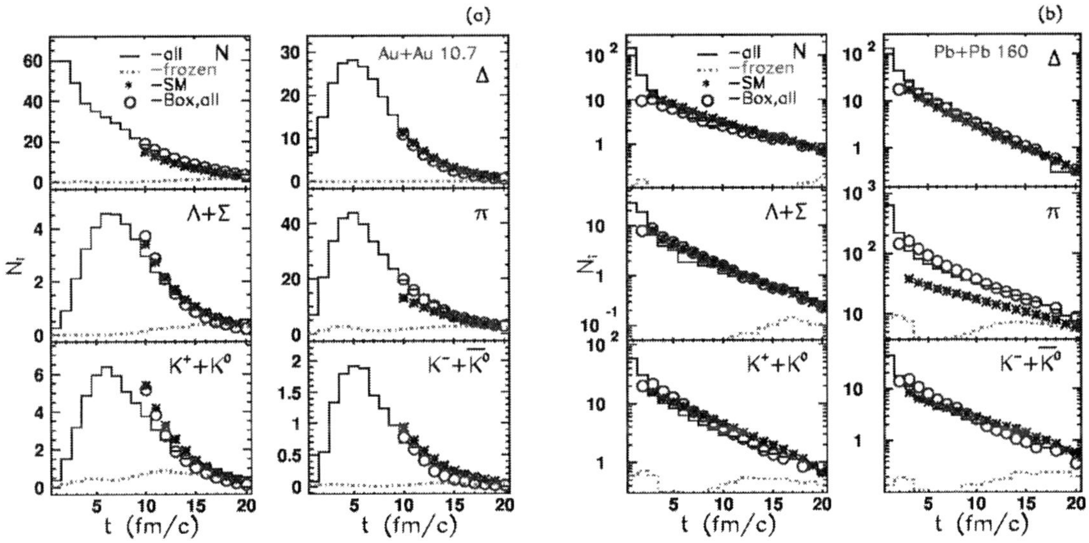

Figure 8. (a) The hadron yields in the central cell of Au+Au central collision at AGS energy (histograms) in comparison with the box calculations (open symbols) and predictions of SM (asterisks). Dash-dotted histograms show the fractions of already frozen particles in the cell. (b) The same as (a) but for Pb+Pb central collisions at SPS energy. (From [12]).

reactions like

$$n\pi \longrightarrow \{HS\} \longrightarrow n'\pi + \{\overline{K}K + \overline{B}B + \ldots\} \tag{12}$$

For instance, the times to reach chemical equilibrium for kaons and lambdas in the UrQMD box calculations are shortened to 2 fm/c and 5 fm/c, respectively [24]. Proper inclusion of the Hagedorn states to microscopic transport models deserves further investigations.

6 Conclusions

In conclusion, relaxation of hot and dense hadron-string matter to chemical and thermal equilibrium is studied. Two cases are considered. The first one is the infinite nuclear matter (closed system), represented by a box with periodic boundary conditions. The second one is a central cell (open system) in central heavy-ion collisions at ultrarelativistic energies. In the box calculations the hadron abundances saturate after a certain time, which is longer for strange particles compared to the non-strange ones. Energy spectra of different hadrons exhibit two time-independent inverse slopes corresponding to two temperatures, T_{meson} and T_{baryon}. With the rise of the energy density in the system these spectra show the existence of the Hagedorn-like limiting temperature T_{crit} = 130 ± 15 MeV (UrQMD) or T_{crit} = 150 ± 5 MeV (BUU+Lund). The absolute value of the T_{crit} depends on the table of resonances in the particular model and on the parameters of the string excitation and fragmentation. The origin of limiting temperature in the model is traced to presence of strings and many-body decays of resonances. Comparison with the statistical model of an ideal hadron gas shows that the SM is able to reproduce the UrQMD box limiting temperature behavior, provided it utilises the continuum Hagedorn-like mass spectrum of strings in addition to hadrons and resonances.

The yields and energy spectra of hadron species in the cell at $t \geq 10$ fm/c are very close to those in the box with the same values of energy density, net baryon density, and net strangeness density. The obtained results are independent on the collision energy of nuclei. The relaxation time can be significantly reduced if one allows for creation of Hagedorn states with exponential mass spectrum in the vicinity of T_{crit}.

References

[1] R. Hagedorn, Nuovo Cim. Suppl. **3**, 147 (1965)
[2] R. Hagedorn, Riv. Nuovo Cim. **6N10**, 1 (1983)
[3] W. Nahm, Nucl. Phys. **B45**, 525 (1972)
[4] M. Gorenstein, these Proceedings
[5] K. Redlich, these Proceedings
[6] J. Rafelski (Ed.), *Melting Hadrons, Boiling Quarks - From Hagedorn Temperature to Ultra-Relativistic Heavy-Ion Collisions at CERN* (Springer, Heidelberg, 2016)
[7] S.A. Bass *et al.*, Prog. Part. Nucl. Phys. **41**, 255 (1998)
[8] M. Bleicher *et al.*, J. Phys. **G25**, 1859 (1999)
[9] W. Ehehalt and W. Cassing, Nucl. Phys. **A602**, 449 (1996)
[10] N.S. Amelin, L.V. Bravina, Sov. J. Nucl. Phys. **51** (1990) 133;
L.V. Bravina *et al.*, Phys. Lett. **B354**, 196 (1995)
[11] M. Belkacem *et al.*, Phys. Rev. **C58**, 1727 (1998)
[12] L.V. Bravina *et al.*, Phys. Rev. **C62**, 064906 (2000)
[13] E.L. Bratkovskaya *et al.*, Nucl. Phys. **A675**, 661 (2000)
[14] L.V. Bravina *et al.*, Phys. Lett. **B434**, 379 (1998);
L.V. Bravina *et al.*, J. Phys. **G25**, 351 (1999)
[15] L.V. Bravina *et al.*, Phys. Rev. **C60**, 024904 (1999);
L.V. Bravina *et al.*, Nucl. Phys. **A661**, 600 (1999)
[16] J. Cleymans, K. Redlich, Phys. Rev. **C60**, 054908 (1999)
[17] A. Andronic, P. Braun-Munzinger, J. Stachel, Nucl. Phys. **A772**, 167 (2006).
[18] G. Torrieri *et al.*, Comput. Phys. Commun. **167**, 229 (2005)
[19] L.V. Bravina *et al.*, Phys. Rev. **C78**, 014907 (2008);
E.E. Zabrodin *et al.*, J. Phys. **G36**, 064065 (2009)
[20] L.V. Bravina *et al.*, Phys. Rev. **C63**, 064902 (2001);
L.V. Bravina *et al.*, Nucl. Phys. **A698**, 383c (2002);
L.V. Bravina *et al.*, J. Phys. **G27**, 421 (2001)
[21] Ph. Mota, T. Kodama, R.D. de Souza, J. Takahashi, Eur. Phys. J. **A48**, 165 (2012)
[22] C. Greiner, S. Leupold, J. Phys. **G27**, L95 (2001)
[23] J. Noronha-Hostler, C. Greiner, I. Shovkovy, Phys. Rev. Lett. **100**, 252301 (2008)
[24] M. Beitel, K. Gallmeister, C. Greiner, Phys. Rev. **C90**, 045203 (2014)

EPJ Web of Conferences **126**, 04001 (2016) DOI: 10.1051/epjconf/201612604001
ICNFP 2015

Beyond two generations

Georgios Anagnostou, on behalf of the CMS Collaboration[1,a]
[1] *Institute Of Nuclear and Particle Physics, NCSR Demokritos*

Abstract. A review is presented of searches performed by CMS experiment for new particles produced in association with or decaying to top quarks, as well as heavy top partners. The analysis presented use data collected with CMS experiment during 2012, in proton-proton collisions at a centre-of-mass energy of 8 TeV.

1 Introduction

After the discovery of a SM-like Higgs boson [1, 2], an explanation for the scale of its mass became of great importance. New particles, such as heavy top partners, could improve our understanding of the problem, through cancelations in the radiative corrections of the Higgs boson mass.

Many searches have been performed by CMS experiment [3] for new particles produced in association with or decaying to top quarks, as well as heavy top partners [4]. In this high energy regime, the top quarks become boosted and their decay products are oftenly collimated. Sophisticated algorithms have been developed for tagging top quarks and estimating the mass of their hadronic decays [5–7]. These techniques allow the reconstruction of invariant mass observables in many final states, a powerful tool to discrimimate possible new resonances from their background processes. Mass reconstruction methods are also developed for leptonic final states including one or two invisible particles [8–10].

This report summarizes many of the searches performed by the CMS for heavy top partners as well as new particles produced in association with or decaying to top quarks [8–18]: top quarks associated with possible dark matter particles (section 2) and resonances decaying to top quarks (sections 3, 4, 5). Finally, searches for heavy top partners are presented in sections 6, 7.

2 Dark matter searches

Theories that attempt to extend the Standard Model like supersymmetry or extra dimensions, predict the existence of neutral weakly interacting particles which escape detection in collider experiments. Final states with a single jet [19, 20] or photon [21, 22] associated with large missing energy have been searched by both ATLAS and CMS without any evidence for new physics. This result can be accomodated by models where the new particle converts a light quark to a top quark, predicting "monotop" candidates, that is an invisible particle in association with a top quark [23, 24]. The CMS search [11] is focused on the hadronic final state, in which top decays to a bottom quark and a W

[a]e-mail: Georgios.Anagnostou@cern.ch

 DOI: 10.1051/epjconf/201612604001

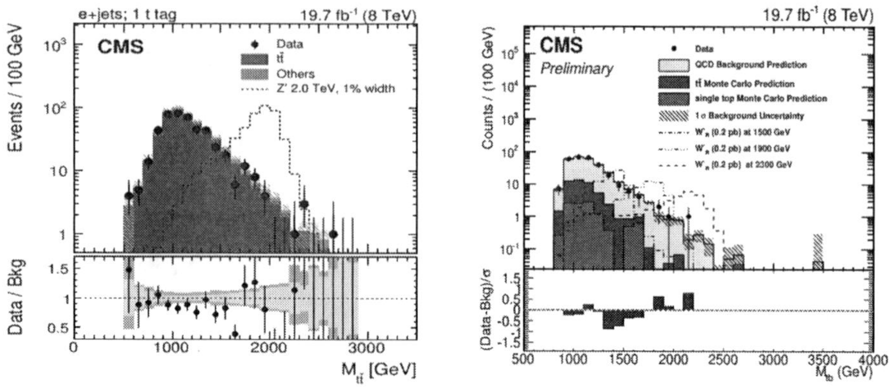

Figure 1. Invariant mass of the reconstructed $t\bar{t}$ pair in the electron+jet channel for events with 1 t-tagged jet (left) [13], and the reconstructed M_{tb} invariant mass in the all-hadronic $W' \rightarrow t\bar{b}$ search (right) [15].

boson, which further decays into a pair of quarks. Events with large missing energy and three jets are selected and one of them must be identified as originating from a b quark. The signal events as well as the number of multijet background events, are measured simulatenously using a likelihood approach. The observed lower limits for the invisible particle are 330 GeV for a scalar and 650 GeV for a vector particle respectively.

Another scenario predicts a new fermion interacting with quarks via a four-fermion contact interaction [25, 26]. The exclusion limit for the scalar case is the least stringent among all types that have been searched. In addition, the interaction coupling strength is proportional to the mass of the quark, predicting better sensitivity in final states with third-generation quarks [27]. The CMS search is focused on the scalar case, in the final state originating from the semi-leptonic top pair decay. The selection requires events with a single isolated lepton, at least three jets of which one must be identified as originating from b quark fragmentation and large missing transverse energy [12]. Further selection criteria are based on the transverse mass of the lepton and missing energy, as well as M_{T2}^{W} variable [28], in order to suppress W+jets and top pair backgrounds. The upper limits on the $pp \rightarrow t\bar{t} + X\bar{X}$ production cross section depend on the mass hypothesis of the DM particle: cross sections larger than 20 to 55 fb are excluded for DM particles ranging from 1 to 1000 GeV.

3 Resonances decaying to top pairs

A resonance decaying to top pairs is predicted by many BSM theories like leptophobic Z′ [29], extended gauge theories [30] or extensions of the Randal-Sundrum model with Kaluza-Klein (KK) excitations [31]. The CMS search for resonances decaying to top quarks [13] is performed in final states with two, one or zero leptons refered to as dilepton, lepton+jet and all hadronic channels respectively.

The dilepton final states requires two non-isolated leptons (electrons or muons) of opposite charge, at least two jets, as well as large transverse missing energy. The final observable is a mass variable constructed from the vectorial sum of the above. The lepton+jet final state selects one electron or muon, at least two jets and large missing energy. The CMS top tagging algorithm is applied to identify the fully hadronic top decays merged into a single jet, using Cambridge-Aachen jets with a distance parameter of R=0.8 (CA8 jets) [5, 6]. Events are categorized based on the lepton flavour and the number of CA8 t-tagged jets, from which no more than one is allowed to avoid overlap with the all-hadronic channel. The invariant mass of the $t\bar{t}$ system is reconstructed (Figure 1, left), by minimizing a χ^2 function based on the masses of the possible $t\bar{t}$ candidates.

The all-hadronic final state requires a di-jet topology in which each jet is consistent with the decay of a top quark. The search is performed in two different Z' mass regions: in the $M_{Z'}$ region below 1 TeV the Cambridge-Aachen (CA) jet reconstruction is applied with a distance parameter of R=1.5 and HEPTopTagger algorithm [7], whereas in the high mass region the top jets are reconstructed using an R=0.8 and CMS t-tagging algorithm [5, 6]. In order to reduce the dominant multijet production, b-tagging is applied to the subjets of top, requiring one of them to originate from the fragmentation of a b quark. The combination of the above channels for resonances decaying to top pairs gives an exclusion range up to 2.4 TeV for the narrow Z' hypothesis. For wide resonances (width 10% of its mass) the exclusion range is up to 2.9 TeV. Randall-Sundrum KK gluons decaying to top pairs are excluded for masses below 2.8 TeV.

4 Excited top quark

The proposal that the top quark might have a composite structure [32], can be directly tested by searching for an excited top quark. The CMS search [8] adopts a model in which pair produced t^* quark decays predominantly to a top quark and a gluon [33]. The analysis is targeting the lepton+jets final state ($t^*t^* \rightarrow t\bar{t}gg \rightarrow l^+\nu_l b\bar{q}q'\bar{b}gg$), requiring an isolated lepton and at least six jets, of which one must be combatible with originating from b quark fragmentation. A kinematic fit is performed to final state objects in order to reconstruct t^* candidates in each event. The fit is subject to the six mass constraints from excited top t^*, top quark and W boson decays in both brances of the pair production. The analysis estimates a lower limit for m_{t^*} of 803 GeV for the combined electron and muon channels.

A search for pair produced resonances decaying to a top quark and another parton is also performed by CMS in the dilepton final state [9]. The experimental signature constists of two isolated leptons (electrons or muons), two jets coming from the decays of b quarks and two jets originating from light flavour quarks or gluons. The interpretation is based on a spin-3/2 top quark excitation model, as in the lepton+jets final state [34]. An additional interpretation is also used, based on an R-parity violating MSSM model with the LSP bottom squark decaying into a top quark and a strange quark [35]. The invariant mass of the t^* or $\tilde{b}$ resonance can be reconstructed: given the masses of the top quark and the W boson, the $t\bar{t}$ system can be solved analytically [36, 37]. The quadric equation can be solved for the two leptons and b-tagged jets combinations to form $t\bar{t}$ candidates which can further be combined with the selected light jets to form pairs of t^* or $\tilde{b}$ candidates. The pair with the closest $M_{t,jet}$ values is chosen and the average value of the pair is used as the final reconstructed mass. The observed upper limit for t^* and $\tilde{b}$ resonances is 703 GeV and 326 GeV respectively.

5 t+b resonances

A new massive charged gauge boson W' is predicted by many extensions of the SM [38, 39] and has been searched in the lepton-neutrino, diboson and light quark final states [40–42]. The most important experimental constraints imposed by the leptonic final state, do not apply for the case of W' bosons with purely right-handed couplings and a hypothetical right-handed neutrino mass larger than a few GeV [43]. CMS has performed further searches for a $W'^+ \rightarrow t\bar{b}$ (or charge conjugate) in the lepton+jets and all-hadronic final state [14, 15].

The lepton+jets search [14] selects events with an isolated lepton (electron or muon), at least two jets of which one must originate from b-quark fragmentation. The topology $W' \rightarrow tb, t \rightarrow bW \rightarrow bl\nu$ is fully reconstructable: The z component of the neutrino momentum is calculated using the W-boson

 DOI: 10.1051/epjconf/201612604001

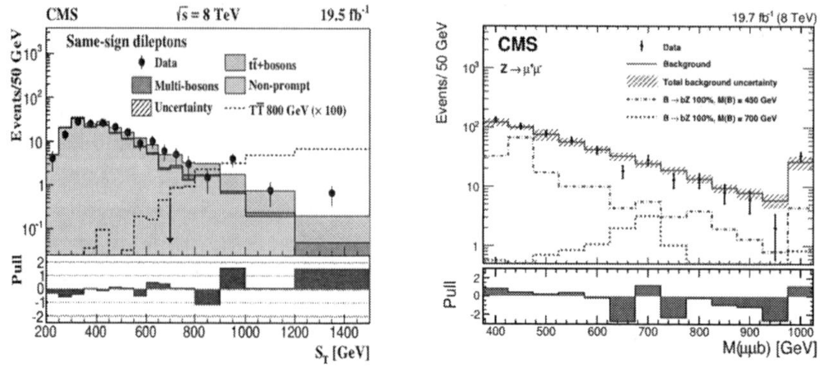

Figure 2. Observed and excpected S_T distribution (same-sign dilepton sample) in the vector-like T quark search (left) [17], and the invariant mass of reconstructed B quark candidates in the opposite-sign dimuon channel (right) [18].

mass constraint. Each W-candidate is combined with all selected jets to form top-quark candidates. The candidate with invariant mass closest to top-quark mass is chosen and is further combined with the highest -in P_T- of the remaining jets to form a W' candidate. The observed limit for purely right-handed W' bosons is 2.05 TeV. For a W' boson with purely left-handed couplings including interefence effects, the observed limit is 1.84 TeV.

In the all-hadronic $W' \to t\bar{b}$ search [15], the Cambridge-Aachen algorithm is used for jet reconstruction (CA8 jets) and the CMS top-tagger for top jet identification. The selection requires a back-to-back dijet topology composed of a t-tagged jet and a low mass b-tagged jet which further allow the reconstruction of the W' candidate mass (Figure 1, right). The production of a right-handed W' boson in the all-hadronic search is excluded for masses up to 2.02 TeV. The combination of both lepton+jets and all-hadronic channels increase the exclusion range up to 2.15 TeV.

6 Vector-like quarks

Vector-like quarks are hypothetical new quarks for which both left and right-handed fields have the same transformation properties under the electroweak gauge group [45]. They do not need to receive their masses from Yuakawa couplings to Higgs doublet, as their bare mass terms are invariant under the symmetry of the theory. So these models are constistent with existing Higgs data, in contrast with models predicting extra quarks with chiral couplings such as fourth generation quarks [46].

The CMS search for top-quark partners with an exotic charge [16] is looking for pair produced new particles $T_{5/3}$ with charge 5e/3 and decay via $T_{5/3} \to tW^+$, followed by $t \to W^+b$ (and charge conjugates). The analysis is performed in the same-sign final state: it has less $t\bar{t}$ background but also instrumental effects from charge misidentification and fake leptons which are estimated from control regions. In addition to the CMS top-tagging, a W-tagging algorithm is applied to the CA8 jets, requiring exactly two subjets and their mass to be consistent with a W boson mass [44]. The selection requires two isolated same-sign leptons, large scalar sum of the P_T of all leptons/jets and the number of CA8 subjets, leptons and anti-kt jets to be greater than six. The observed limit on the mass of $T_{5/3}$ is 800 GeV.

CMS has performed an inclusive search for a vector-like T quark with charge 2/3, which further decays to bW, tZ or tH, in single-lepton and mulitlepton final states [17]. The single lepton channel requires at least four jets and uses a BDT to further separate T-quark signal from SM background. The multilepton channel requires at least two leptons and is further divided into four mutually exclusive subchannels. The opposite-sign dilepton sample targets the bWbW final state and exploits the smallest invariant mass combination of leptons and b-jets M_{lb}^{min}, in order to suppress the $t\bar{t}$ background. A second opposite-sign dilepton channel targets events in which both leptons originate from a Z boson decay and has further selection criteria based on the increased jet activity. A same-sign dilepton sample accepts events in which at least one T quark decays to tZ or tH and uses control regions to estimate charge misidentification and non-promt lepton backgrounds (Figure 2, left). Finally, a trilepton channel also accepts events in which at least one T quark decays to tZ or tH. The non-prompt backgrounds are determined from data, as in the dilepton case. All the subchannels are using observables like H_T and S_T, in order to exploit the increased jet activity and the invisible particles included. The lower limits for a vector-like T-quark ranges between 687 and 782 GeV, depending on the possible branching ratios of the three decay modes.

CMS has also searched for a vector-like heavy B quark having an electric charge $-1/3$ [18]. The B quarks are assumed to be pair produced and decay via $B \to tW$, $B \to bZ$ or $B \to bH$. Several different mutually exclusive final states are searched in order to be sensitive to the different decay modes, based on the number of selected leptons: single leptons, dileptons with opposite or identical charges, multileptons or all-hadronic without any identified leptons. The first two final states (single lepton and same-sign dilepton) target the tW decay and use as final observable the S_T variable. The same observable is also used in the multilepton final state which is sensitive to both tW and bZ decay modes. The all-hadronic channel is designed for Higgs bosons decaying to a pair of b quarks. The Higgs boson is identified using CA8 jets that are composed of two b-tagged subjets with an invariant mass constistent with m_H. At least one CA8 jet is required to satisfy these criteria known as Higgs tagging (H-tagging). Finally, in the opposite sign lepton pair channel, the B quark candidate is fully reconstructed from its decay products ($B \to Zb, Z \to ll$), by requiring an M_{ll} close to Z boson mass and an additional b-tagged jet (Figure 2, right). The observed exclusion limit for a vector-like B quark ranges from 740 GeV to 900 GeV depending on the possible branching ratios of the different decay modes.

7 Search in the 2-dimensional mass plane

Many BSM models predict final states with two invisible particles [47–49]. The experimental searches are often based on missing energy related observables where the new physics appears in the tail of the distribution. In this case, the establishment of a possible discovery and the model constraining (e.g particle masses) are challenging problems. The suggestion to search in 2-dimensional mass space, for final states with two invisible particles is described in [50]. The reconstruction of mass peaks concentrates signal events in a small region of the two-dimensional mass space whereas background events have no reason to do the same. The reconstructed masses also give valuable information about the particle content and thus the parameters of the model.

CMS has performed a search for anything decaying like top pairs, in their dilepton final state [10]. The analysis is searching simultaneously for both a new heavy top partner T′ and a new heavy charged gauge boson W′ as predicted by the littlest higgs model [47, 48]. The selection requires two leptons, two jets and large missing energy with the same energy requirements for all reconstructed objects. The analysis is based on a two-dimensional mass reconstruction of both unknown particles in the $[M_{T'}, M_{W'}]$ plane (Figure 3). The analytic solutions [36, 37] together with constraints from the parton distribution functions (PDFs) [51–53] are used to reconstruct the masses of two unknown particles simultaneously. The region of $M_{T'}$ excluded is in the range 800-920 GeV, depending on $M_{W'}$.

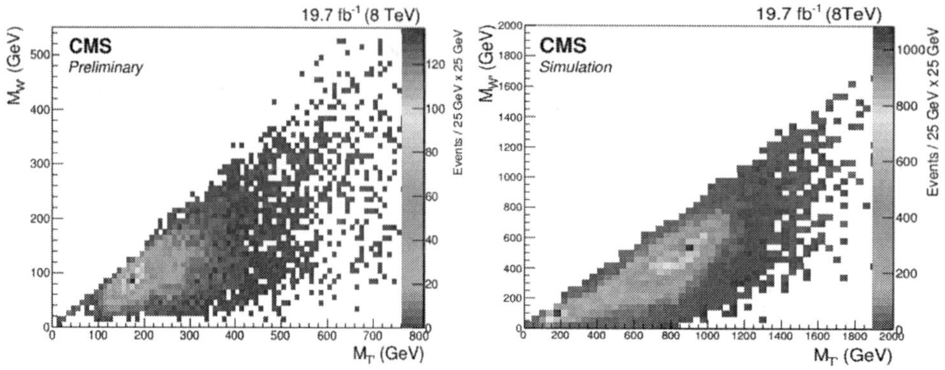

Figure 3. Mass reconstruction in two-dimensional mass space using the 2012 dataset (left) as well as simulated events for the same luminosity of both signal [$M_{T'}$ = 1000 GeV, $M_{W'}$ = 600 GeV] and background processes (right) [10].

8 Conclusions

A summary of many CMS searches with final states involving top quarks or heavy top partners is presented. Sophisticed alogorithms are developed for top tagging/mass reconstruction in final states with/without missing energy. No evidence for new physics was found for top quarks associated with possible dark matter particles, resonances decaying to top quarks as well as heavy top-partners. The analysis methods developed in Run1 will facilitate the searches for BSM physics in the new energy regime of Run2.

References

[1] ATLAS Collaboration, "Observation of a new particle in the search for the Standard Model Higgs boson with the ATLAS detector at the LHC", Phys. Lett. B **716** (2012) 1, arXiv:1207.7214.

[2] CMS Collaboration, "Observation of a new boson at a mass of 125 GeV with the CMS experiment at the LHC", Phys. Lett. B **716** (2012) 30, arXiv:1207.7235.

[3] CMS Collaboration, "The CMS experiment at the CERN LHC", JINST **3** (2008) S08004.

[4] https://twiki.cern.ch/twiki/bin/view/CMSPublic/PhysicsResultsB2G

[5] CMS Collaboration, "A Cambridge-Aachen (C-A) Based Jet Algorithm For Boosted Top-Jet Tagging", CMS Physics Analysis Summary CMS-PAS-JME-09-001, 2009.

[6] CMS Collaboration, "Boosted Top Jet Tagging at CMS", CMS Physics Analysis Summary CMS-PAS-JME-13-007, 2014.

[7] T. Plehn, M. Spannowsky, M. Takeuchi, and D. Zerwas, "Stop Reconstruction with Tagged Tops", JHEP **10** (2010) 078, arXiv:1006.2833.

[8] CMS Collaboration, "Search for pair production of excited top quarks in the lepton+jet final state", JHEP **06** (2014) 125, arXiv:1311.5357.

[9] CMS Collaboration, "Search for pair production of resonances decaying to a top quark plus a jet in final states with two leptons", CMS Physics Analysis Summary CMS-PAS-B2G-12-008, 2013.

[10] CMS Collaboration, "Search in two-dimensional mass space for $T'\bar{T}' \rightarrow W^{+\prime}bW^{-\prime}\bar{b}$ in the dilepton final state in proton-proton collisions at $\sqrt{s} = 8$ TeV", CMS Physics Analysis Summary CMS-PAS-B2G-12-025, 2015.

[11] CMS Collaboration, "Search for Monotop Signatures in proton-proton collisions at 8 Tev", Phys. Rev. Lett. **114** (2015) 101801, hep-ex:1410.1149.

[12] CMS Collaboration, "Search for the production of dark matter in association with top-quark pairs in the single-lepton final state in pp collisions at $\sqrt{s}$ = 8 TeV", JHEP **06** (2015)121, arXiv:1504.03198.

[13] CMS Collaboration, Search for resonant $t\bar{t}$ production in proton-proton collisions at *mathrm* $\sqrt{s}$ = 8 TeV", CERN-PH-EP/2015-126, arXiv:1506:03062.

[14] CMS Collaboration, "Search for $W' \to$ tb in the lepton+jets final state in pp collisions at $\sqrt{s}$ = 8 TeV", JHEP **05** (2014) 108, hep-ex:1402.2176.

[15] CMS Collaboration, "Search for $W' \to$ tb in the all-hadronic final state", CMS Physics Analysis Summary CMS-PAS-B2G-12-009, 2014.

[16] CMS Collaboration, "Search for Top-Quark Partners with Charge 5/3 in the Same-Sign Dilepton Final State", Phys. Rev. Lett. **112** (2014) 171801, hep-ex:1212:2391.

[17] CMS Collaboration, "Inclusive search for a vector-like T quark with charge 2/3 in pp collisions at $\sqrt{s}$ = 8 TeV", Phys. Lett. B **729** (2014) 149, hep-ex:1311.7667.

[18] CMS Collaboration, "Search for pair-produced vector-like B quarks in proton-proton collisions at $\sqrt{s}$ = 8 TeV", CMS Physics Analysis Summary CMS-PAS-B2G-13-006, 2015, hep-ex:1507.07129.

[19] ATLAS Collaboration, "Search for dark matter candidates and large extra dimensions in events with a jet and missing transverse momentum with the ATLAS detector", JHEP **04** (2013) 075, arXiv:1210.4491.

[20] CMS Collaboration, "Search for dark matter, extra dimensions, and unparticles in monojet events in proton-proton collisions at $\sqrt{s}$ = 8 TeV", (2014), arXiv:1408.3583.

[21] ATLAS Collaboration, "Search for dark matter candidates and large extra dimensions in events with a photon and missing transverse momentum in pp collision data at at $\sqrt{s}$ = 7 TeV with the ATLAS detector", Phys. Rev. Lett. **110**, (2013), 011802, arXiv:1209.4625.

[22] CMS Collaboration, "Search for Dark Matter and Large Extra Dimensions in pp Collisions Yielding a Photon and Missing Transverse Energy", Phys. Rev. Lett. **108**, (2012), 261803, arXiv:1204.0821.

[23] J. Andrea, B. Fuks, and F. Maltoni, "Monotops at the LHC", Phys. Rev. D **84** (2011) 074025, arXiv:1106.6199.

[24] J.-L. Agram et al., "Monotop phenomenology at the Large Hadron Collider", Phys. Rev. D **89** (2014) 014028, arXiv:1311.6478.

[25] M. Beltrán, D. Hooper, E.W. Kolb, Z.A.C. Krusberg and T.M.P. Tait, Maverick dark matter at colliders, JHEP **09** (2010) 037, arXiv:1002.4137.

[26] K. Cheung, K. Mawatari, E. Senaha, P.-Y. Tseng and T.-C. Yuan, The top window for dark matter, JHEP **10** (2010) 081, arXiv:1009.0618.

[27] T. Lin, E.W. Kolb and L.-T. Wang, Probing dark matter couplings to top and bottom quarks at the LHC, Phys. Rev. **D 88** (2013) 063510, arXiv:1303.6638.

[28] Y. Bai, H.-C. Cheng, J. Gallicchio and J. Gu, "Stop the top background of the stop search", JHEP **07** (2012) 110, arXiv:1203.4813.

[29] R. M. Harris and S. Jain, "Cross sections for leptophobic topcolor Z' decaying to top-antitop", Eur. Phys. J. C **72** (2012) 2072, arXiv:1112.4928.

[30] K. R. Lynch, S. Mrenna, M. Narain, and E. H. Simmons, "Finding Z' bosons coupled preferentially to the third family at CERN LEP and the Fermilab Tevatron", Phys. Rev. D **63** (2001) 035006, arXiv:hep-ph/0007286.

[31] K. Agashe et al., "LHC signals from warped extra dimensions", Phys. Rev. D **77** (2008) 015003, arXiv:hep-ph/0612015.

[32] H. Georgi, L. Kaplan, D. Morin and A. Schenk, "Effects of top compositeness", Phys. Rev. **D 51** (1995) 3888, hep-ph/9410307.

[33] B. Hassanain, J. March-Russell and J.G. Rosa, "On the possibility of light string resonances at the LHC and Tevatron from Randall-Sundrum throats", JHEP **07** (2009) 077, arXiv:0904.4108.

[34] C. Burges and H. J. Schnitzer, "Virtual Effects of Excited Quarks as Probes of a Possible New Hadronic Mass Scale", Nucl. Phys. B **228** (1983) 464.

[35] C. Csaki, Y. Grossman, and B. Heidenreich, "MFV SUSY: A Natural Theory for R-Parity Violation", Phys. Rev. D **85** (2012) 095009, arXiv:1111.1239.

[36] L. Sonnenschein, "Analytical solution of $t\bar{t}$ dilepton equations", Phys. Rev. D **73** (2006) 054015, arXiv:hep-ph/0603011.

[37] L. Sonnenschein, "Algebraic approach to solve t anti-t dilepton equations", Phys. Rev. D **72**, 095020, 2005, arXiv:hep-ph/0510100.

[38] M. Schmaltz, "Little Higgs review", Ann. Rev. Nucl. Part. Sci. **55** (2005) 229, hep-ph/0502182.

[39] T. Appelquist, H.-C. Cheng and B.A. Dobrescu, "Bounds on universal extra dimensions", Phys. Rev. **D 64** (2001) 035002, hep-ph/0012100.

[40] CMS collaboration, "Search for new physics in final states with a lepton and missing transverse energy in pp collisions at the LHC, Phys. Rev. **D 87** (2013) 072005, arXiv:1302.2812.

[41] CMS collaboration, "Search for exotic resonances decaying into WZ/ZZ in pp collisions at $\sqrt{s}$ =7 TeV, JHEP **02** (2013) 036, arXiv:1211.5779.

[42] CMS collaboration, "Search for narrow resonances using the dijet mass spectrum in pp collisions at $\sqrt{s}$ = 8 TeV", Phys. Rev. **D 87** (2013) 114015, arXiv:1302.4794.

[43] M. Nemevšek, F. Nesti, G. Senjanović and Y. Zhang, "First Limits on Left-Right Symmetry Scale from LHC Data", Phys. Rev. **D 83** (2011) 115014, arXiv:1103.1627.

[44] CMS Collaboration, "Identifying Hadronically Decaying Vector Bosons Merged into a Single Jet", CMS Physics Analysis Summary CMS-PAS-JME-13-006, 2013.

[45] J. A. Aguilar-Saavedra, R. Benbrik, S. Heinemeyer, and M. Perez-Victoria, "Handbook of vectorlike quarks: Mixing and single production", Phys. Rev. D **88** (2013) 094010.

[46] A. Djouadi and A. Lenz, "Sealing the fate of a fourth generation of fermions", Phys. Lett. B **715** (2012) 310.

[47] T. Han, H. E. Logan, B. McElrath, and L.-T. Wang, "Phenomenology of the little Higgs model", Phys. Rev. D **67** (2003) 095004, arXiv:hep-ph/0301040.

[48] Jürgen Reuter, Marco Tonini, Maikel de Vries, "Littlest Higgs with T-parity: Status and Prospects", JHEP **1402** (2014) 053, arXiv:1310.2918.

[49] S. P. Martin, "A Supersymmetry primer", Adv. Ser. Direct. High Energy Phys. 18 (1997) 1, arXiv:hep-ph/9709356.

[50] G. Anagnostou, "Model Independent Search in 2-Dimensional Mass Space", in 2nd International Conference on New Frontiers in Physics, EPJ Web of Conferences, Volume 71, 2014.

[51] R. H. Dalitz and G. R. Goldstein, "The decay and polarization properties of the top quark, Phys. Rev. D **45** (1992).

[52] D0 Collaboration, "Measurement of the top quark mass in final states with two leptons", Phys. Rev. D **80**, 092006 (2009).

[53] CMS Collaboration, "Measurement of the top quark mass using proton-proton data at $\sqrt{s}$ = 7 and 8 TeV", CERN-PH-EP-2015-234, arXiv:1509.04044.

EPJ Web of Conferences **126**, 04002 (2016) DOI: 10.1051/epjconf/201612604002
ICNFP 2015

Searches for the direct pair production of third generation squarks using the ATLAS detector

John Anders[1a] on behalf of the ATLAS Collaboration

[1] *The University of Liverpool*

Abstract. Third generation squarks are expected to have masses of < 1 TeV due to naturalness arguments. As such production of these squarks may be possible at the LHC. An overview of the most recent ATLAS third generation results and summary paper are presented in these proceedings. In addition to this, an overview of the outlook for such searches in Run 2 of the LHC is also presented.

1 Introduction

Supersymmetry (SUSY) is a possible extension of the Standard Model (SM) which provides solutions to questions which remain unanswered by the SM, such as the Higgs hierarchy problem, and certain SUSY scenarios also provide a candidate for dark matter [1].

The introduction of a scalar top quark ($\tilde{t}$) with mass < 1 TeV allows for cancellations of the quantum loop corrections to the Higgs mass from the top quark, thus solving the hierarchy problem. Further to this, in R-parity conserving (RPC) SUSY models the lightest SUSY particle (LSP) is stable and a dark matter candidate. In most generic searches the LSP is a neutralino ($\tilde{\chi}_1^0$) which is a superposition of the eigenstates corresponding to the Higgsinos and Bino. The main signature of such scenarios is a large amount of missing transverse momentum, E_T^{miss} due to the non-interacting LSP leaving the detector. In R-parity violating scenarios (RPV) where the LSP is not stable and can decay to SM particles, this is not the case and vastly different signatures are expected involving a high object multiplicity.

2 Overview of Recent Run 1 Analyses

A wide variety of SUSY scenarios were searched for during Run 1 of the LHC. The general analysis strategy is as follows:

Signal regions (SRs) are defined targeting a specific model and SUSY mass parameter set. Selections based on key kinematic variables are optimised such that they maximise the discovery or exclusion significance for a particular model. Control regions (CRs) are then defined to constrain the main SM background processes present in the SR. These regions are kinematically close to the SR but are designed to be orthogonal to the SR, and are enriched in the specific background process under consideration. The number of events in the CR is used to produce normalisation parameters for the

[a] e-mail: john.kenneth.anders@cern.ch

backgrounds in the SR. The normalisation parameters are validated in validation regions (VRs) which are also orthogonal to both the SRs and CRs. After validation the yields in the SRs are compared to the corrected SM prediction and if no excesses are observed, exclusion limits are placed on the models under consideration.

2.1 Search for direct charm squark pair production

A search for scharm quark pair production is presented in Ref. [2]. The analysis targets final states with two charm-tagged jets (tagged using a dedicated charm tagging algorithm with efficiency $O(20\%)$) and large E_T^{miss} to search for $\tilde{c} \rightarrow c + \tilde{\chi}_1^0$. As this analysis is a dedicated search for scharm quarks, the scharm quarks are treated as non-degenerate with respect to the other squarks. The main discriminating variable is the boost-corrected contransverse m_{CT} [3], which is used to discriminate between scharm pair production and $t\bar{t}$ pair production.

Three overlapping SRs are defined, each with increasingly tighter m_{CT} selections (the tighter the selection, the higher the $\tilde{c}$ mass targeted). The main SM processes that contribute to the SM background in the SRs are $t\bar{t}$, Z-jets and W-jets production. For each of these backgrounds a dedicated CR region is defined, which is enriched in the background under consideration.

No excesses are seen, and exclusion limits are placed in the $(\tilde{c}, \tilde{\chi}_1^0)$ mass plane, Fig. 1. The analysis extends the general exclusion for charm squarks originally set by the ATLAS 0L+2-6 jets analysis [4].

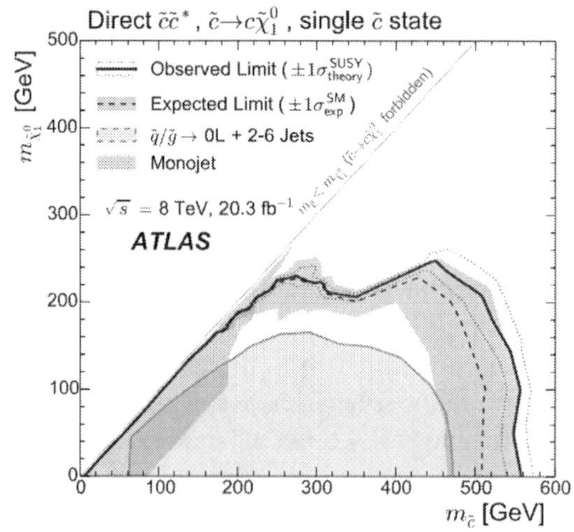

Figure 1. Exclusion limits at 95% confidence level (CL), placed in the $(\tilde{c}, \tilde{\chi}_1^0)$, mass plane (red), increasing the general exclusion placed by the ATLAS 0L+2-6 jets analysis (grey). Taken from Ref. [2].

2.2 Stop via stau

The decay processes of the SUSY particles are largely dependent upon the SUSY symmetry breaking scenario under consideration. In the case of gauge mediated symmetry breaking the LSP is the gravitino ($\tilde{G}$). Due to this, more complex decay modes are available such as the top squark decaying via an intermediate tau slepton $\tilde{\tau}$. Ref. [5] describes a search for stop pair production, where each stop decays via a stau ($\tilde{t} \rightarrow \tilde{\tau} b \nu$), with the stau subsequently decaying to a tau and a gravitino ($\tilde{\tau} \rightarrow \tau \tilde{G}$).

Three channels are investigated, differing by the decay mode for the tau pair that is under investigation, with each having dedicated SRs targeted at different regions of phase space. The main background is from $t\bar{t}$ pair production, with both real-tau, and mis-identified taus as separate components of this background. CRs are defined for both the real-tau and fake-tau $t\bar{t}$ contributions to produce separate normalisation parameters for $t\bar{t}$ pair production decaying into real taus and mis-identified taus.

No significant excesses are seen in any of the SRs, and a statistical combination of the SRs from each channel is performed to produce overall 95% CL exclusion limits in the ($\tilde{t}$,$\tilde{\tau}$) mass plane, Fig. 2.

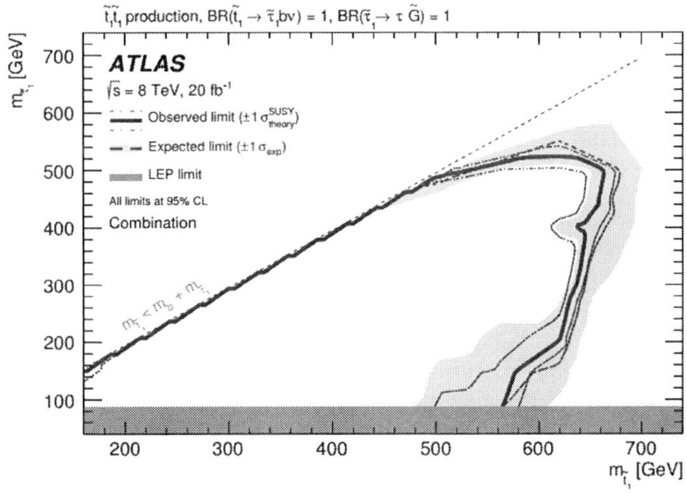

Figure 2. Exclusion limits at a 95% CL from the stop to stau analysis in the ($\tilde{t}$, $\tilde{\tau}$) mass plane, assuming the process $\tilde{t} \rightarrow \tilde{\tau} b \nu$, and subsequently $\tilde{\tau} \rightarrow \tau \tilde{G}$. Taken from Ref. [5].

2.3 RPV Stop

R-parity violating SUSY scenarios do not contain the large E_T^{miss} which is characteristic of R-parity conserving models and hence must use different methods and variables in order to discriminate signal and background processes. A search for RPV stop pair production is presented in Ref. [6], searching for stop pair production with the top squark decaying via $\tilde{t} \rightarrow cs$. Earlier experiments place limits on stop masses below 100 GeV in these scenarios [7].

The main discriminating variable in the analysis is the average jet mass of a large radius jet, where the radius (R) is dependent upon the stop mass under consideration: The main background is multi-jet production, which is estimated using a purely data driven estimate. No excesses are found, and limits are placed on the stop mass in such scenarios, with $100 < m_{\tilde{t}} < 310$ GeV excluded, Fig. 3.

3 Third Generation Summary Paper

Throughout Run 1 of the LHC many searches have been performed for third generation squarks, with limits placed on a wide variety of different scenarios. In order to fully exploit these results, the Third Generation Summary paper [8] was produced, containing interpretations of SM analyses, investigations into additional decay modes and statistical combinations of existing analyses, to combine the results and futher constrain the SUSY parameter space.

DOI: 10.1051/epjconf/201612604002

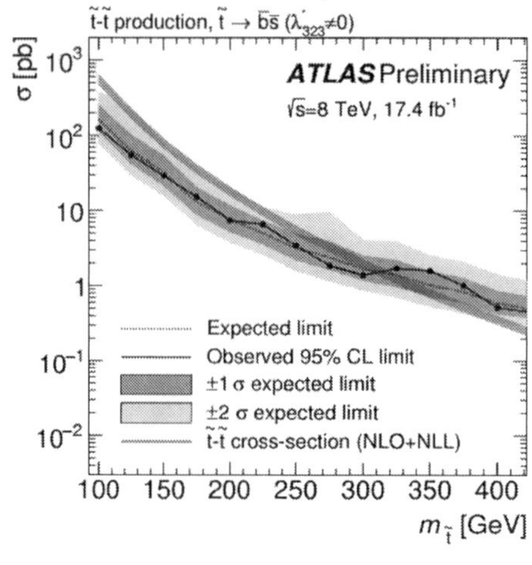

Figure 3. 95% CL exclusion Limits placed on the stop mass by the RPV stop search. The search extends the limits set by earlier experiments and excludes $m_{\tilde{t}}$ up to 310 GeV. Taken from Ref. [6].

3.1 Reinterpreting Existing Analyses

Interpretation of SM measurements into the properties of the top quark can provide limits on the top squark mass, in regions of phase space that are difficult to target with conventional analysis techniques. The reinterpretation of both the ATLAS $t\bar{t}$ cross section measurement [9] and the spin correlation measurement (the opening angle $\Delta\phi(l,l)$ between the two leptons from dileptonic $t\bar{t}$ decays) [10] place limits in the region where $m_{\tilde{t}} = m_t$, due to the difference expected in the distributions if a top squark with $m_{\tilde{t}} = m_t$ is present. The combination of the exclusion from the cross section measurement and the spin correlation measurement (Fig. 4), allow limits to be placed on light top squarks, with 150 $< m_{\tilde{t}} <$ 191 GeV excluded, with the assumption of massless neutralinos.

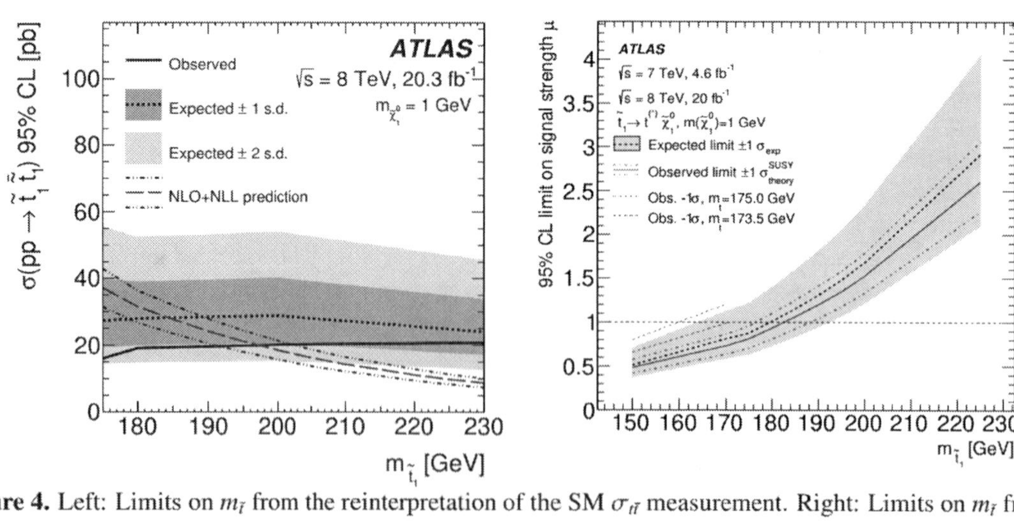

Figure 4. Left: Limits on $m_{\tilde{t}}$ from the reinterpretation of the SM $\sigma_{t\bar{t}}$ measurement. Right: Limits on $m_{\tilde{t}}$ from the reinterpretation of the $t\bar{t}$ spin correlation measurement. Taken from Ref. [8].

3.2 Investigating Additional Decay Modes

More complex decay scenarios are also considered, for example the "tbMET" and the "WW-like" analyses.

The tbMET analysis targets stop pair production where two possible decay channels are available for the stop, $\tilde{t} \rightarrow t + \tilde{\chi}_1^0$, and also decays involving charginos ($\tilde{\chi}_1^{\pm}$), $\tilde{t} \rightarrow b + \tilde{\chi}_1^{\pm}$ (with the subsequent decay $\tilde{\chi}_1^0 \rightarrow W^* + \tilde{\chi}_1^0$, as the neutralino and chargino are treated as almost mass degenerate in these scenarios). Various branching ratios of these decays are considered in the analysis, which is optimised for the 1ℓ+$2b$-jets+E_T^{Miss} final state. The main SM backgrounds are $t\bar{t}$ and W-jets production. Branching ratio dependent limits are placed in the $(m_{\tilde{t}}, m_{\tilde{\chi}_1^0})$ phase space, for the scenarios where $\Delta m(\tilde{\chi}_1^{\pm}, \tilde{\chi}_1^0)$ = 5 or 20 GeV, (Fig. 5).

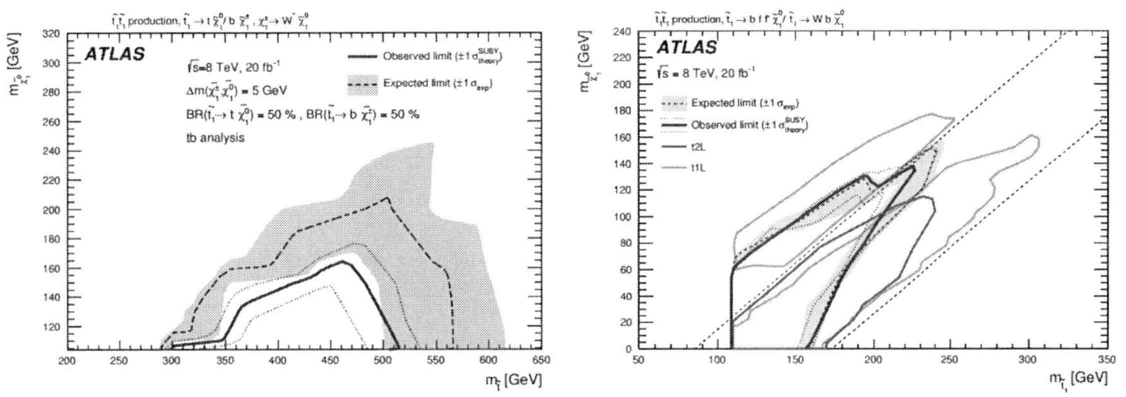

Figure 5. Left: Exclusion limits at 95% CL placed in the $m(\tilde{t}, \tilde{\chi}_1^0)$ phase space by the tbMET analysis, for a BR($\tilde{t} \rightarrow b + \tilde{\chi}_1^{\pm}$) = 50% and assuming $\Delta m(\tilde{\chi}_1^{\pm}, \tilde{\chi}_1^0)$ = 5 GeV. Right: Limits placed by the WW-like analysis which provides 95% exclusion limits in the region where the three and four body stop decay modes overlap. Taken from Ref. [8].

The WW-like analysis considers two possible mass spectra for the top squark, where either: the mass difference between the stop and neutralino being less than the W-boson mass ($m_{\tilde{t}} - m_{LSP} < m_W$) which considers the four body stop decay, $\tilde{t} \rightarrow bff\tilde{\chi}_1^0$); or a mass difference between the stop and neutralino being larger than the W mass, but smaller than the top mass ($m_W < m_{\tilde{t}} - m_{LSP} < m_t$) which considers a three body stop decay, $\tilde{t} \rightarrow bW\tilde{\chi}_1^0$. The analysis is motivated by the discrepancy between the SM expectation and the measurement of σ_{WW} reported by both ATLAS [11] and CMS [12]. The analysis is optimised for a final state consisting of two different flavour (DF, $e\mu$) opposite sign (OS) leptons, with the stransverse mass (m_{T2} [13]) as the main discriminating variable. No significant excesses were found in any of the 7 SRs, and exclusion limits are placed in the $(m_{\tilde{t}}, m_{\tilde{\chi}_1^0})$ mass plane, Fig. 5.

3.3 Limits on the Heavier Stop Eigenstate

An alternative method to investigate difficult to target regions of the stop phase space is to search for pair production of the heavier stop eigenstate ($\tilde{t}_2$), which is a convienient method to search in the region where $\Delta m(\tilde{t}, \tilde{\chi}_1^0) \approx m_t$. Such decays are searched for in three channels: $\tilde{t}_2 \rightarrow \tilde{t}_1 + Z$, $\tilde{t}_2 \rightarrow \tilde{t}_1 + h$,

$\tilde{t}_2 \rightarrow \tilde{t}_1 + \tilde{\chi}_1^0$, with many SRs defined for each. No excesses are seen in any of the channels and 95% CL exclusion limits are placed in the $m(\tilde{t}_2, \tilde{\chi}_1^0)$ phase space assuming BR = 100%, Fig. 6.

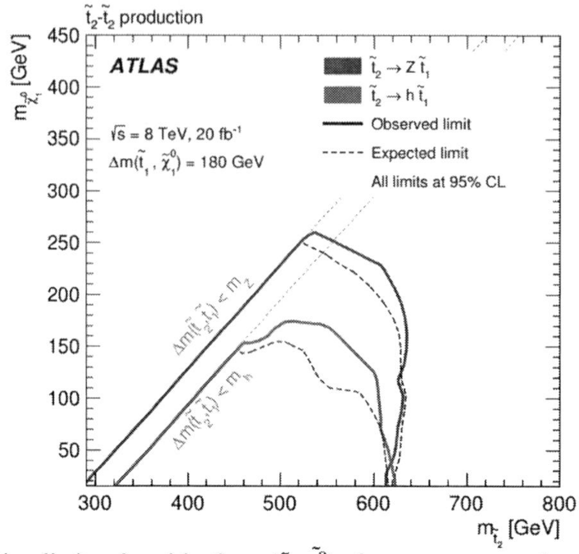

Figure 6. 95% CL exclusion limits placed in the $m(\tilde{t}_2, \tilde{\chi}_1^0)$ phase space assuming BR = 100% for the decays considered. Taken from Ref. [8].

3.4 Sbottom Limits

Sbottom decay modes are also investigated, with dedicated analyses (such as the "bbMET" search for sbottom pair production with the sbottom decaying via: $\tilde{b} \rightarrow b\tilde{\chi}_1^0$ [14]) and reinterpretations of other analyses (for example the reinterpretation the ATLAS same sign (SS) or 3ℓ strong production analysis [15] to investigate $\tilde{b} \rightarrow b\tilde{\chi}_2^0$, with $\tilde{\chi}_2^0 \rightarrow \tilde{\chi}_1^0 + h$). Limits are subsequently placed in the $m(\tilde{b}, \tilde{\chi}_1^0)$ phase space by these analyses, Fig. 7.

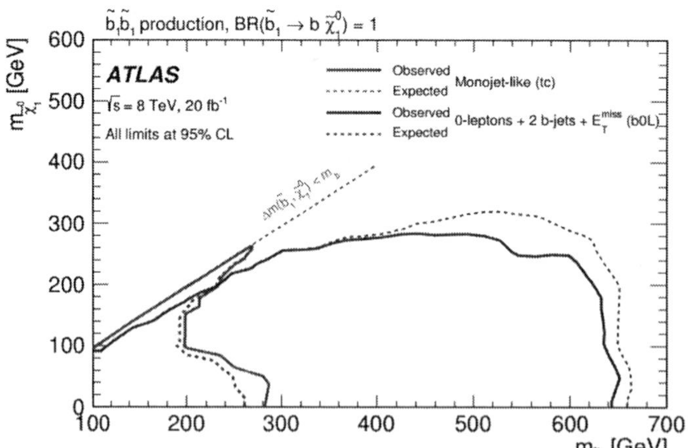

Figure 7. Exclusion limits at the 95% CL placed in the $m(\tilde{b}, \tilde{\chi}_1^0)$ phase space by the bbMET analysis (blue) and the Monojet analysis (purple). Taken from Ref. [8].

DOI: 10.1051/epjconf/201612604002

4 Outlook for LHC Run 2

Many ATLAS third generation analyses are targeting discovery with Run 2 data, one of which is the sbottom analysis. The analysis has a similar strategy to the Run 1 sbottom analysis, focusing on final states with two b-tagged jets and missing transverse energy, to search for direct sbottom pair production with the sbottom decaying via $\tilde{b} \rightarrow b\tilde{\chi}_1^0$. Preliminary studies indicate that with an integrated luminosity of 10 fb^{-1}, 3σ evidence for bottom squarks with mass up to 800 GeV may be obtained [16].

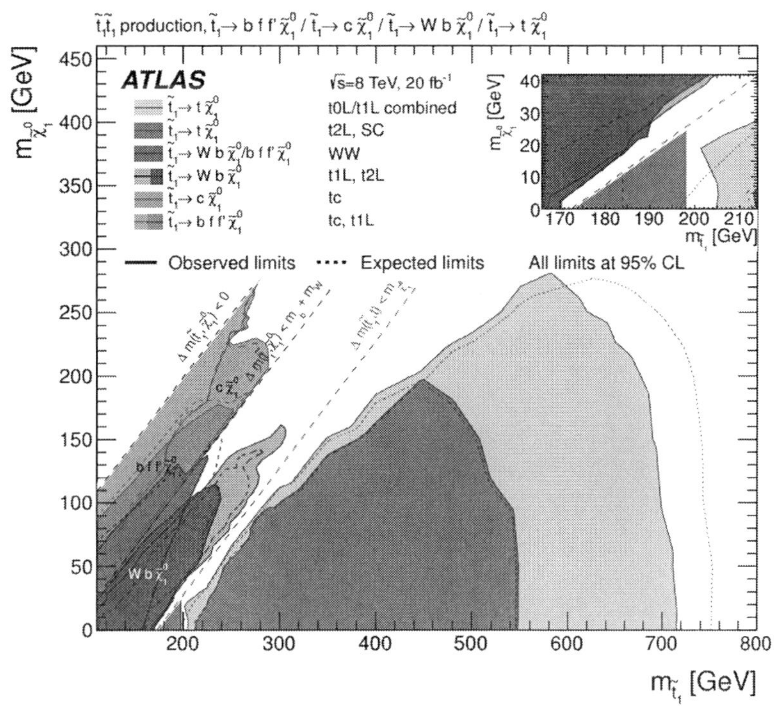

Figure 8. Exclusion limits at 95% CL in the $\tilde{t},\tilde{\chi}_1^0$ mass plane provided by the full programme of ATLAS third generation searches during LHC Run 1. Taken from Ref. [8].

5 Conclusion

ATLAS is almost complete with its dedicated program of searches for third generation squarks during Run 1 of the LHC, with limits placed on a wide variety of SUSY scenarios. Fig. 8 shows the full extent of the coverage and exclusion in the $(\tilde{t},\tilde{\chi}_1^0)$ mass plane that has been provided by Run 1 ATLAS searches, with similar coverage and exclusion also in the $(\tilde{b},\tilde{\chi}_1^0)$ mass plane.

Run 2 preparation is well underway, targeting discovery of third generation squarks with early data.

References

[1] R. Barbieri and G. Giudice, Nucl. Phys. B **306** (1988) 63–76.
[2] The ATLAS Collaboration, Phys. Rev. Lett. **144, 161801** (2015) arXiv:1501.01325 [hep-ex].
[3] D. R. Tovey, JHEP **0804** (2008) 034, arXiv:0802.2879 [hep-ph].
[4] The ATLAS Collaboration, JHEP**09** (2014) 176 arXiv:1405.7875 [hep-ex].
[5] The ATLAS Collaboration, submitted to EPJC, arXiv:1509.04976 [hep-ex].
[6] The ATLAS Collaboration, ATLAS-CONF-2015-026.
[7] CDF Collaboration, T. Aaltonen et al, Phys. Rev. Lett. **111** (2013) 031802, arXiv:1303.2699 [hep-ex].
[8] The ATLAS Collaboration, submitted to Eur. Phys. J, arXiv:1506.08616 [hep-ex].
[9] ATLAS Collaboration, JHEP **1406** (2014) 035, arXiv:1404.2500 [hep-ex].
[10] ATLAS Collaboration, Eur. Phys. J. **C74** no. 10, (2014) 3109, arXiv:1406.5375 [hep-ex].
[11] ATLAS Collaboration, Phys. Lett. **B712** (2012) 289–308, arXiv:1203.6232 [hep-ex].
[12] CMS Collaboration, Eur. Phys. J. **C73** no. 10, (2013) 2610, arXiv:1306.1126 [hep-ex].
[13] A. Barr, C. Lester, and P. Stephens, J.Phys. **G29** (2003) 2343–2363, arXiv:hep-ph/0304226 [hep-ph].
[14] The ATLAS Collaboration, JHEP **10** (2013) 189, arXiv:1308.2631 [hep-ex].
[15] The ATLAS Collaboration, JHEP **06** (2014) 035, arXiv:1404.2500 [hep-ex].
[16] The ATLAS Collaboration, ATL-PHYS-PUB-2015-005.

Recent results from NA61/SHINE

Evgeny Andronov[1,a], On behalf of the NA61/SHINE Collaboration

[1] *Saint Petersburg State University, ul. Ulyanovskaya 1, 198504, Petrodvorets, Saint Petersburg, Russia*

Abstract. The NA61/SHINE experiment aims to discover the critical point of strongly interacting matter and study the properties of the onset of deconfinement. For these goals a scan of the two dimensional phase diagram (T-μ_B) is being performed at the SPS by measurements of hadron production in proton-proton, proton-nucleus and nucleus-nucleus interactions as a function of collision energy and system size.
In this contribution intriguing results on the energy dependence of hadron spectra and yields in inelastic p+p and centrality selected Be+Be collisions will be presented. In particular, the energy dependence of the signals of deconfinement, the 'horn', 'step' and 'kink', in p+p interactions will be presented and compared with the corresponding results from central Pb+Pb collisions from NA49.

1 Introduction

NA61/SHINE [1] is a fixed target experiment at the Super Proton Synchrotron (SPS) of the European Organization for Nuclear Research (CERN). The layout of the NA61/SHINE detector is sketched in fig. 1. It consists of a large acceptance hadron spectrometer with excellent capabilities in charged particle momentum measurements and identification by a set of five Time Projection Chambers as well as Time-of-Flight detectors. The high resolution modular forward calorimeter, the Projectile Spectator Detector, measures energy flow around the beam direction, which in nucleus-nucleus reactions is primarily a measure of the number of spectator (non-interacted) nucleons and thus related to the centrality of the collision. Both primary and secondary beams are available to the experiment, allowing data taking with projectile sizes ranging from proton to lead, as well as with pions and kaons. An array of beam detectors identifies beam particles, secondary hadrons and ions as well as primary ions, and measures precisely their trajectories.

The main goal of the strong interaction programme of the experiment is to discover the Critical Point (CP) [2] of strongly interacting matter and study the properties of the onset of deconfinement (OD) [3, 4]. To achieve this goal a two-dimensional phase diagram scan - energy versus system size - is being performed by NA61/SHINE. It includes measurements of hadron production in collisions of protons and various nuclei (p+p, Be+Be, Ar+Sc, Xe+La) at a range of beam momenta (13*A* - 158*A* GeV/c). Figure 2 shows for which systems and energies data has already been collected (green), is scheduled for recording (red) or is planned (gray). Besides studying strong interactions, the experiment also performs precise hadron production reference measurements for neutrino (Fermilab and T2K) and cosmic-ray (KASCADE and Pierre Auger Observatory) physics.

[a] e-mail: evgeny.andronov@cern.ch

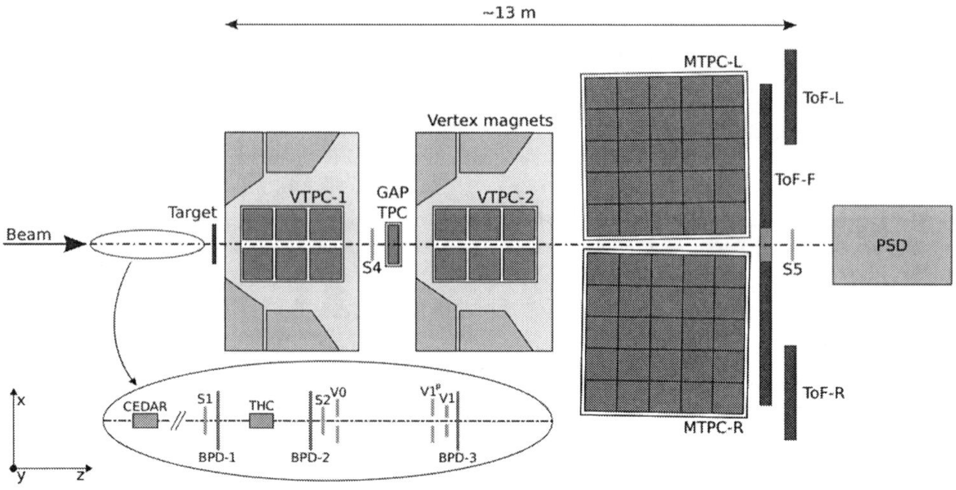

Figure 1. Schematic layout of the NA61/SHINE experiment at the CERN SPS (horizontal cut in the beam plane, not to scale).

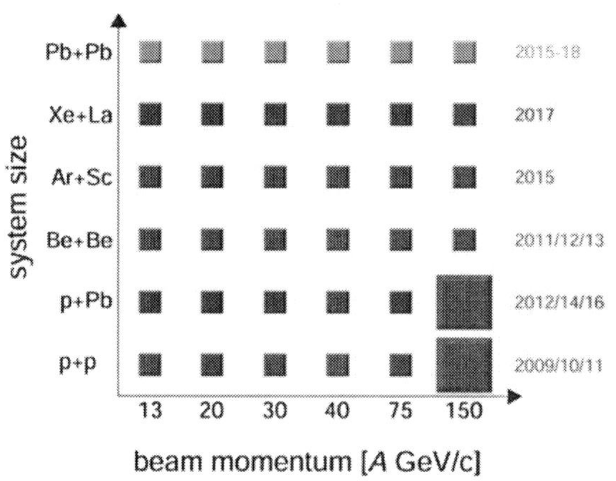

Figure 2. Data taking status of the strong interaction programme of NA61/SHINE.

2 Analysis of identified hadron spectra

The identification of particles was performed for inelastic p+p interactions and centrality selected Be+Be collisions using such techniques as the h^- method [5], dE/dx and $tof - dE/dx$ measurements. The h^- method is based on the fact that the majority of negatively charged hadrons produced in inelastic interactions at SPS energies are pions. The remaining small contamination ($\approx$ 10%) of other hadrons (K^- and $\overline{p}$) is subtracted using models of particle production (e.g. EPOS1.99 [6]). The dE/dx method is based on the particle energy loss in the TPC gas. It can be applied for π^+, π^-, K^+, K^-, p and $\overline{p}$ with total momentum (p_{tot}) in the momentum range 4 – 100 GeV/c and with transverse momentum (p_T) in the momentum range 0 – 2 GeV/c. The time-of-flight measurement together with the particle path length and momentum measured by the TPCs allow to calculate its mass m. Combined $tof - dE/dx$ analysis provides excellent separation of different hadron species close to mid-rapidity. It is performed for π^+, π^-, K^+, K^-, p and $\overline{p}$ with p_{tot} in the momentum range $2-10$ GeV/c and with p_T in the momentum range $0-2$ GeV/c. The acceptances in rapidity versus transverse momentum phase space for these methods are shown in fig. 3 for p+p interactions at 158 GeV/c.

The results presented in this paper are corrected for detector inefficiencies, feed-down from weak decays and secondary interactions, contribution from non-target interactions, as well as trigger and event selection biases. The detailed simulation of the NA61/SHINE detector response was based on GEANT3.

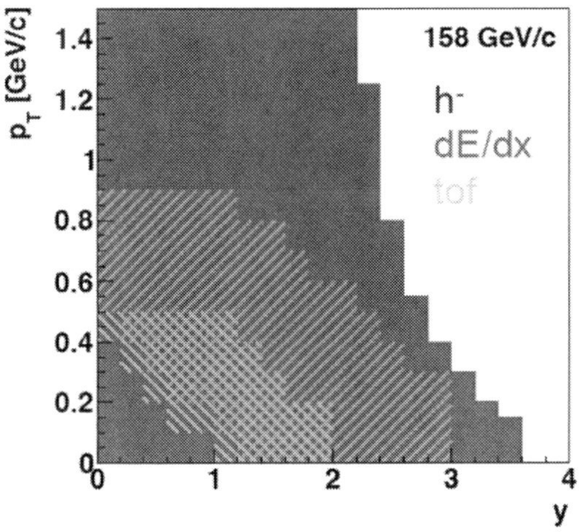

Figure 3. Acceptance for various data analysis methods for π^- from inelastic p+p interactions at 158 GeV/c. Blue, solid area: h^-; magenta shading: dE/dx; yellow shading: time-of-flight.

3 p+p results

This section presents inclusive spectra of identified hadrons produced in inelastic p+p interactions at 20, 31, 40, 80, 158 GeV/c and observables, that are usually treated as signalsof the onset of deconfinement [7].

Figure 4 (left) shows spectra of transverse mass of negatively and positively charged π, K, p and Λ [8] produced in inelastic p+p interactions at mid-rapidity. Corresponding NA49 measurements [4, 9, 10] for central Pb+Pb collisions are shown in fig. 4 (right). The data was fitted using a blast wave model parametrisation [11] $\frac{dN_i}{m_T dm_T dy} = A_i m_T K_1\left(\frac{m_T \cosh\rho}{T}\right) I_0\left(\frac{p_T \sinh\rho}{T}\right)$, where the parameter ρ is related to the transverse flow velocity β_T by $\rho = \tanh^{-1}\beta_T$. One finds that β_T is significantly smaller in p+p than Pb+Pb collisions. While the spectra are approximately exponential in p+p reactions with inverse slope parameter of the order 150 MeV as expected from the Hagedorn statistical bootstrap model [12], this exponential dependence is possibly modified in Pb+Pb interactions by the transverse flow.

The analysis of particle spectra is an important step towards the study of the system size dependence of signals of the onset of deconfinement observed in central Pb+Pb collisions ('kink', 'horn' and 'step') [4]. These signatures were predicted in the Statistical Model of the Early Stage (SMES) [13], which suggests that the phase transition line is crossed by heavy ion collisions between the top AGS energy (beam energy 11.7A GeV) and the top SPS energy (beam energy 158A GeV).

Figure 5 shows that the energy dependence of mean π multiplicity increases slower in p+p than in Pb+Pb collisions ('kink'). Hence, the two dependences cross each other at around 40A GeV/c, slightly violating the prediction of the Wounded Nucleon Model [16] $\langle\pi\rangle_{(AA)}/\langle N_W\rangle = \langle\pi\rangle_{pp}/2$, while the SMES model predicts that the creation of the QGP leads to the increase of entropy and, consequently, to the rise of the slope for the plotted dependence. The way of obtaining the mean π multiplicity in inelastic p+p interactions is described in detail in ref. [5].

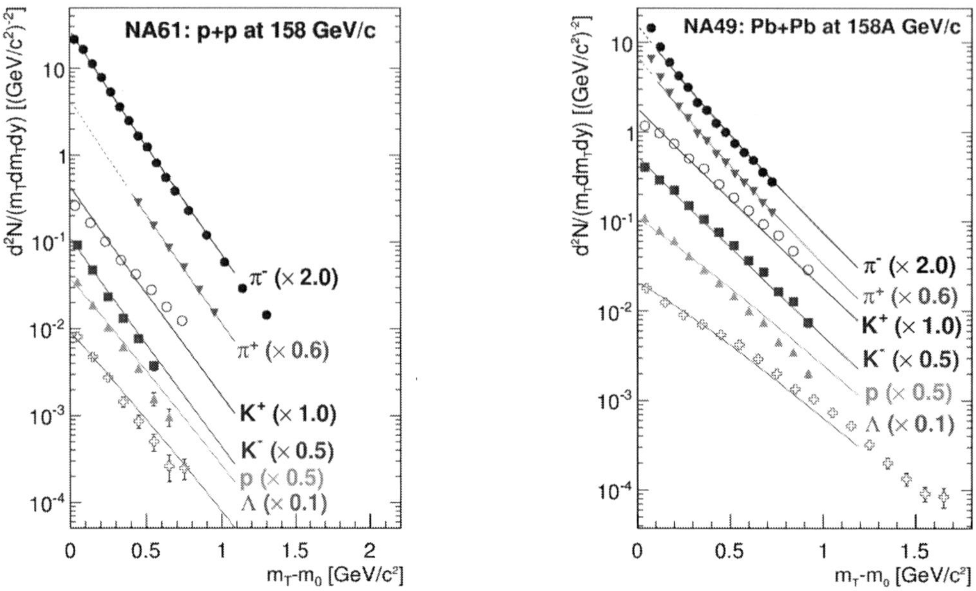

Figure 4. Transverse mass spectra at mid-rapidity. Left: NA61/SHINE inelastic p+p interactions at 158 GeV/c; right: NA49 central Pb+Pb collisions at 158*A* GeV/c. Lines show fits based on the blast wave model.

The new measurements of NA61/SHINE significantly improve the world data for the inverse-slope parameter T of m_T spectra of kaons. The m_T spectra were fitted to the exponential function $\frac{d^2n}{dp_T dy} = \frac{S\,p_T}{T^2+m_K T}\exp\left(-\frac{\sqrt{p_T^2+m_K^2}-m_K}{T}\right)$, where S is the yield integral and m_K is the K mass. Figure 6 shows the energy dependence of T for K$^+$ and K$^-$. Surprisingly the NA61/SHINE results from inelastic p+p collisions exhibit rapid changes like observed in central Pb+Pb interactions and reproduce the 'step'-like behaviour.

The energy dependence of the K$^+/\pi^+$ ratio at mid-rapidity for inelastic p+p interactions and central Pb+Pb/Au+Au collisions is presented in fig. 7 (left). Again surprisingly, the data suggests that even inelastic p+p interactions exhibit a step-like structure in the energy dependence of the K$^+/\pi^+$ ratio, which looks like a precursor of the 'horn' structure observed for Pb+Pb collisions in the same energy domain. This finding can be qualitatively explained by the modified SMES model when a phase transition is assumed in the p+p interactions and exact strangeness conservation is taken into account [29]. Furthermore, the measured ratios were compared to theoretical models, as shown in fig. 7 (right). It is evident that none of the models quantitatively describes the data, with EPOS1.99 being the most consistent.

Proton transverse momentum spectra and mean proton transverse mass $\langle m_T \rangle$, around mid-rapidity, in inelastic p+p interactions were also studied by the NA61/SHINE collaboration. The results are compared with theoretical models in fig. 8. Mean m_T of protons was calculated from the following parametrization of data: $\frac{d^2n}{dp_T dy} = \frac{S\,p_T}{T^2+m_p T}\exp\left(-\frac{\sqrt{p_T^2+m_p^2}-m_p}{T}\right)$. It exhibits a slow increase with the

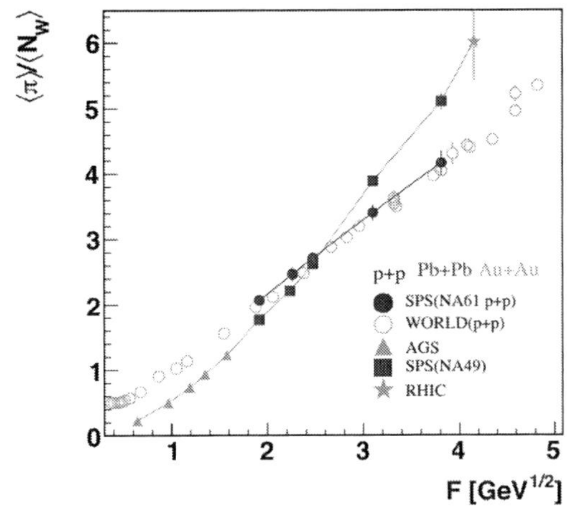

Figure 5. The ratio of mean pion multiplicity to the mean number of wounded nucleons for p+p interactions measured by the NA61/SHINE experiment (full blue circles) and other experiments (open blue circles) and central Au+Au and Pb+Pb collisions plotted as a function of collision energy $F = \left(\frac{(\sqrt{s_{NN}}-2m_N)^3}{\sqrt{s_{NN}}}\right)^{1/4}$. World data from refs. [14, 15].

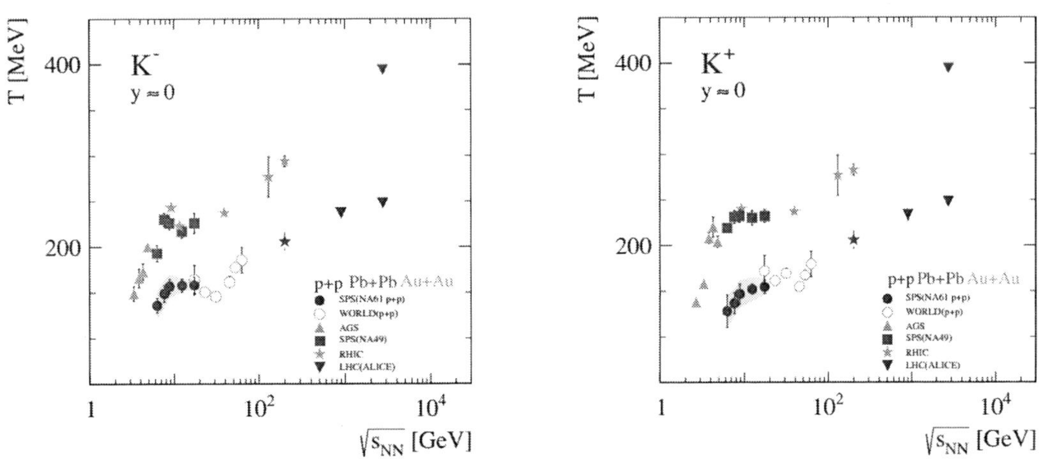

Figure 6. Energy dependence of the inverse slope parameter T of transverse mass spectra of K^- and K^+ in inelastic p+p interactions measured by the NA61/SHINE experiment (full blue circles) and other experiments (open blue circles, blue star and blue triangles) and central Au+Au and Pb+Pb interactions. World data from refs. [17–20]

collision energy in contradiction to the UrQMD or HSD models predictions, that do not reproduce this dependence neither qualitatively nor quantitatively.

DOI: 10.1051/epjconf/201612604003

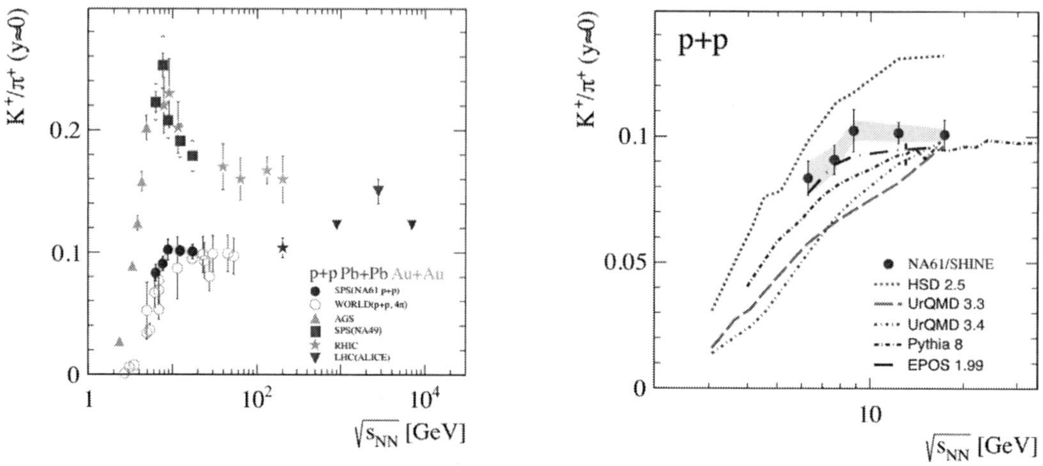

Figure 7. Energy dependence of the K^+/π^+ ratio in inelastic p+p interactions interactions measured by the NA61/SHINE experiment (full blue circles) and other experiments (open blue circles, blue star and blue triangles) compared with central Pb+Pb and Au+Au interactions (left) and theoretical model predictions (right). World data from refs. [19, 21–24], models [6, 25–28].

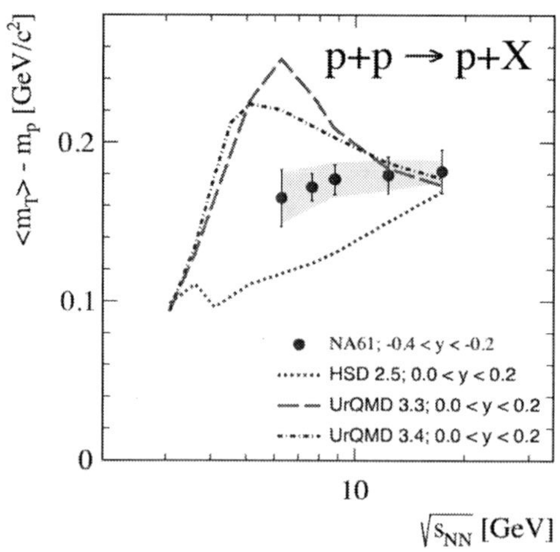

Figure 8. Comparison of energy dependence of $\langle m_T \rangle - m_p$ in inelastic p+p interactions with predictions of models [30].

4 Be+Be results

Previously, the inelastic cross section for ^{7}Be+^{9}Be was only measured at 1.45*A* GeV/c [31]. The new measurements of NA61/SHINE allowed to extend this to 13*A*, 20*A* and 30*A* GeV/c [32]. These measurements were performed using a round scintillator counter S4 placed on the beam-line behind the target, that measures the square of the charge of a particle passing through it. Inelastic interactions were selected by requiring a signal below that expected for an intact beam nucleus. Data was also taken with the target removed to be able to subtract the background caused by beam interactions with detector material.

The interaction probability in the target is given by $P_{int} = \frac{P_I - P_R}{1 - P_R}$, where P_I and P_R are the interaction probabilities when the target is inserted and removed, respectively. Using P_{int}, the cross section can be calculated from $\sigma_{inel} = \frac{1}{\rho L_{eff} N_A / A} P_{int}$, where $L_{eff} = \lambda_{abs}(1 - e^{-L/\lambda_{abs}})$ and $\lambda_{abs} = \frac{A}{\rho N_A \sigma_{inel}}$. N_A, ρ, L and A are the Avogadro constant, the target density, length and atomic number, respectively. Figure 9 shows that the new measurements are in good agreement with the previous measurement as well as the Glauber model [33] prediction.

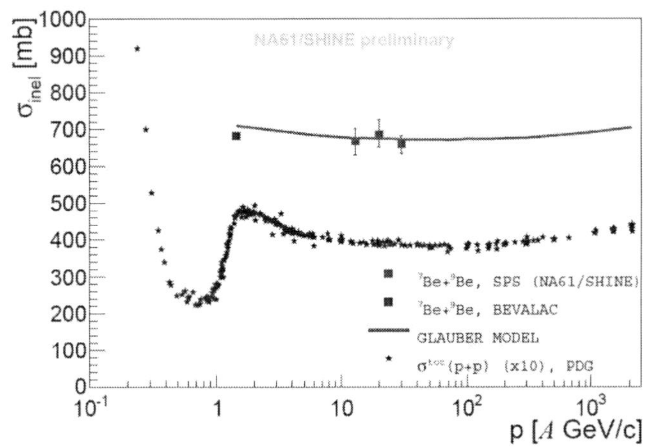

Figure 9. Inelastic cross-section of ^{7}Be+^{9}Be interactions as a function of beam momentum in comparison with the total cross-section of p+p interactions.

Preliminary results on spectra were obtained for ^{7}Be+^{9}Be interactions at beam momenta of 20A, 30A, 40A, 75A and 150A GeV/c in the centrality classes 0–5%, 5–10%, 10–15% and 15–20% [34]. The double differential spectra of π^- in rapidity and transverse momentum for ^{7}Be+^{9}Be collisions at three beam momenta and in two centrality classes are presented in fig. 10. The centrality was derived from the energy deposited in the forward calorimeter, the Projectile Spectator Detector.

To compare the shape of transverse mass spectra of negatively charged pions for ^{7}Be+^{9}Be and Pb+Pb reactions the ratio of normalized transverse mass spectra for ^{7}Be+^{9}Be/p+p and Pb+Pb/p+p was calculated. Figure 11 shows qualitatively similar behaviour for both ^{7}Be+^{9}Be (15% of the most central events) and Pb+Pb reactions (5% of the most central events for 150A GeV/c and 7.5% of the most central events for other energies).

The high-m_T regions of both ^{7}Be+^{9}Be and Pb+Pb exhibit an increase of the ratio, while for the intermediate regions a decrease is seen. This effect is stronger for central Pb+Pb collisions. Usually, this is attributed to collective flow. Also, for ^{7}Be+^{9}Be reactions the increase of the ratio at high values of m_T appears to be larger at higher beam momenta. This may suggest an increase of the magnitude of collective effects in ^{7}Be+^{9}Be collisions with increasing beam momentum. The low-m_T regions of both ^{7}Be+^{9}Be and Pb+Pb reactions exhibit an increase of the ratio. Possible explanations include isospin asymmetry of p+p data or electromagnetic effects.

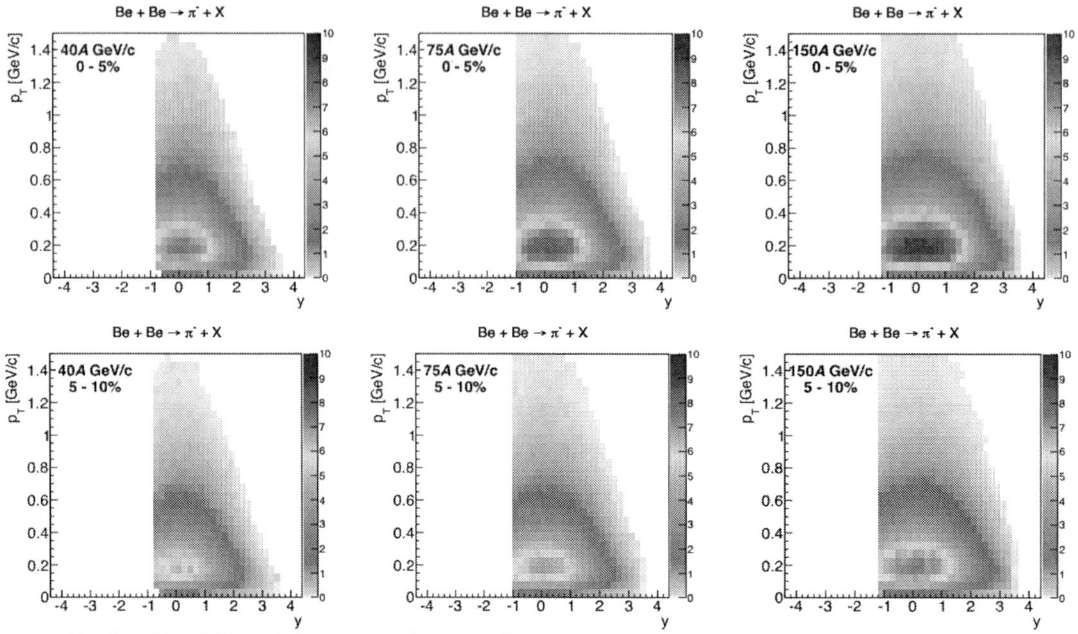

Figure 10. Double differential spectra of negatively charged pions in rapidity and transverse momentum for ^{7}Be+^{9}Be collisions at 40*A*, 75*A* and 150*A* GeV/c for 0-5% and 5-10% centrality classes.

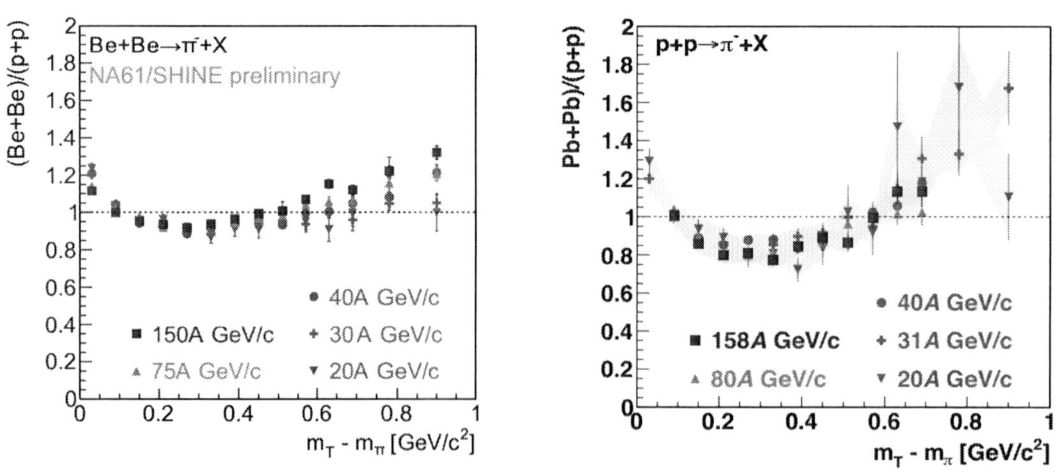

Figure 11. Ratio of normalised transverse mass spectra of π^- mesons for different system sizes. Left: (^{7}Be+^{9}Be)/(p+p) (15% of the most central events); right: (Pb+Pb)/(p+p) (5% of the most central events for 150*A* GeV/c and 7.5% of the most central events for other energies [5])

5 Conclusions

The ongoing NA61/SHINE scan program covers a wide range of beam energies and collision system sizes. The analysis of the first results from inelastic p+p and ^{7}Be+^{9}Be interactions at CERN SPS energies showed many interesting effects. In particular, the p+p data exhibited step-like structures in the energy region where the onset of deconfinement was found in central Pb+Pb collisions. Moreover, it was observed that many of the measurements could not be explained well by theoretical models. For ^{7}Be+^{9}Be reactions new results on the inelastic cross section were obtained at several energies. Transverse mass and rapidity spectra were measured in the SPS energy range for several centrality classes. An indication of a transverse flow effect was found at the highest beam momenta.

Acknowledgements

This work was supported by the Federal Agency of Education of the Ministry of Education and Science of the Russian Federation, SPbSU research grant 11.38.193.2014, the Hungarian Scientific Research Fund (grants OTKA 68506 and 71989), the János Bolyai Research Scholarship of the Hungarian Academy of Sciences, the Polish Ministry of Science and Higher Education (grants 667/N-CERN/2010/0, NN 202 48 4339 and NN 202 23 1837), the Polish National Center for Science (grants 2011/03/N/ST2/03691, 2012/04/M/ST2/00816 and 2013/11/N/ST2/03879), the Foundation for Polish Science — MPD program, co-financed by the European Union within the European Regional Development Fund, the Russian Academy of Science and the Russian Foundation for Basic Research (grants 08-02-00018, 09-02-00664 and 12-02-91503-CERN), the Ministry of Education, Culture, Sports, Science and Technology, Japan, Grant-in-Aid for Scientific Research (grants 18071005, 19034011, 19740162, 20740160 and 20039012), the German Research Foundation (grant GA 1480/2-2), the EU-funded Marie Curie Outgoing Fellowship, Grant PIOF-GA-2013-624803, the Bulgarian Nuclear Regulatory Agency and the Joint Institute for Nuclear Research, Dubna (bilateral contract No. 4418-1-15/17), Ministry of Education and Science of the Republic of Serbia (grant OI171002), Swiss Nationalfonds Foundation (grant 200020117913/1) and ETH Research Grant TH-01 07-3.

References

[1] N. Abgrall *et al.*, JINST **9**, P06005 (2014)
[2] Z. Fodor and S.D. Katz, JHEP **0404**, 050 (2004)
[3] M. Gazdzicki and M.I. Gorenstein, Acta Phys. Polon. B **30**, 2705 (1999)
[4] C. Alt *et al.*, Phys. Rev. C **77**, 024903 (2008)
[5] N. Abgrall *et al.*, Eur.Phys.J. C **74**, 2794 (2014)
[6] T. Pierog and K. Werner, Nucl.Phys.Proc.Suppl. **196**, 102-105 (2009)
[7] S. Pulawski, PoS(CPOD2014)010 (2015)
[8] A. Aduszkiewicz *et al.*, arXiv:1510.03720 [hep-ex] (2015)
[9] C. Alt *et al.*, Phys.Rev. C **73**, 044910 (2006)
[10] S. Afanasiev *et al.*, Phys.Rev. C **66**, 054902 (2002)
[11] E. Schnedermann, J. Sollfrank and U. W. Heinz, Phys.Rev. C **48**, 2462-2475 (1993)
[12] R. Hagedorn, CERN-Report 71-12 (1971)
[13] M. Gazdzicki, M. Gorenstein and P. Seyboth, Acta Phys.Polon. B **42**, 307-351 (2011)
[14] A.I. Golokhvastov, Phys. Atom. Nucl. **64**, 1841 (2001)
[15] E. Abbas *et al.*, Phys. Lett. B **726**, 610 (2013)
[16] A. Bialas, M. Bleszynski and W. Czyz, Nucl.Phys. B **111**, 461 (1976)
[17] M. Kliemant, B. Lungwitz and M. Gazdzicki, Phys. Rev. C **69**, 044903 (2004)
[18] B.I. Abelev *et al.*, Phys. Rev. C **79**, 034909 (2009)
[19] K. Aamodt *et al.*, Eur. Phys. J. C **71**, 1655 (2011)
[20] B.B. Abelev *et al.*, Phys. Lett. B **736**, 196–207 (2014)
[21] M. Gazdzicki and D. Roehrich, Z. Phys. C **65**, 215–223 (1995)

[22] M. Gazdzicki and D. Rohrich, Z. Phys. C **71**, 55–64 (1996)
[23] I. Arsene *et al.*, Phys. Rev. C **72**, 014908 (2005)
[24] B. Abelev *et al.*, Phys. Rev. Lett. **109**, 252301 (2012)
[25] W. Ehehalt and W. Cassing, Nucl.Phys. A**602**, 449-486 (1996)
[26] S. Bass *et al.*, Prog.Part.Nucl.Phys. **41**, 255-369 (1998)
[27] M. Bleicher *et al.*, J. Phys. G **25**, 1859 (1999)
[28] T. Sjostrand *et al.*, Comput.Phys.Commun. **191**, 159-177 (2015)
[29] R. V. Poberezhnyuk, M. Gazdzicki and M. I. Gorenstein, Acta Phys. Polon. B **46**, 1991 (2015), arXiv:1502.05650 [nucl-th]
[30] V.Y. Vovchenko, D. Anchishkin, and M. Gorenstein Phys. Rev. C **90** no. 2, 024916 (2014)
[31] I. Tanihata *et al.*, Phys. Rev. Lett. **55**, 2676 (1985)
[32] I. Weimer, https://edms.cern.ch/file/1308546/1
[33] W. Broniowski, M. Rybczynski and P. Bozek, Comput. Phys. Commun. **180**, 69 (2009)
[34] E. Kaptur, PoS(CPOD2014)053 (2015)

Top quark pair properties using the ATLAS detector at the LHC

Gaetano Barone[1 a]
On behalf of the ATLAS Collaboration

[1] *Brandeis University*

Abstract. These proceeding review the most recent results on the top quark measurements performed by the ATLAS detector, based upon datasets of 4.59 fb^{-1} and of 20.3 fb^{-1} of proton–proton collisions delivered by the Large Hadron Collider at the centres-of-mass energy of $\sqrt{s} = 7$ TeV and of $\sqrt{s} = 8$ TeV, respectively. The measurements of the top quark mass, the cross section of $t\bar{t}$ production in association of vector bosons (W, Z and γ) and the measurement of the charge asymmetry are detailed. The importance of evermore precise measurement of the top quark's properties is highlighted.

1 Introduction

The discovery of the top quark (t) ages, nowadays, twenty years, however, many of its properties are yet to fully constrained by the experiment. The top quark possesses a Yukawa coupling to the Higgs boson of $O(1)$ and a mass close to the scale of the Electro Weak Symmetry Braking (EWSB), on which the modern understanding of particle physics is based. Moreover, because of its large mass, thus decaying before forming bound states, and because of its almost single decay mode trough $t \to Wb$ several of its properties are directly transferred to its decay products. Undiscovered physical phenomena connected with the EWSB can manifest themselves through deviations from the predictions of the Standard Model (SM) in top quark observables. At the Large Hadron Collider (LHC), top quarks are abundantly produced tough Electro Weak (EW) interactions, so called single top, ($pp \to W^\star \to tb$) or in pairs ($t\bar{t}$) via either the strong interaction ($pp \to g \to t\bar{t}$) or EW interaction ($pp \to \gamma^\star(Z^\star) \to t\bar{t}$). Therefore, top related physics at the LHC provide an important framework for a stringent testing of the SM and for the search of new phenomena.

These proceedings focus on the measurements of the top quark properties performed by the ATLAS detector with datasets of proton–proton collisions at the centre-of-mass energies of $\sqrt{s} = 7$ TeV and 8 TeV delivered by the LHC. The most recent results of the top quark mass (m_{top}) are reviewed in Sec. 2. The measurements of $t\bar{t}$ production in association with vector bosons ($t\bar{t}V$) are discussed in Sec. 3. The review of the top quark properties is concluded by the review on the results of the charge asymmetry in Sec. 4. The concluding remarks are discussed in Sec. 5.

2 Top quark mass

The top quark mass is one of the free parameters of the SM, radiative and loop corrections relate m_{top} to the Higgs boson and to the W–boson. Therefore, the accurate experimental determination of m_{top}

[a]e-mail: gaetano.barone@cern.ch

is of great importance for stringent tests of the SM as well as for the evaluations of the stability of the EW vacuum [1].

The most recent results from the ATLAS experiment [2] determine $m_{\mathrm{top}}$with an accuracy of 1 GeV by combining the measurements in the single lepton (ℓ= electron or muon) plus jets final state and di-lepton final state. Both measurements are performed by a kinematic reconstruction of the invariant mass of the top quark decay products which allows for the extraction of the reconstructed top mass (labelled hereafter $m_{\mathrm{top}}^{\mathrm{reco}}$ for the single lepton channel and $m_{\ell b}^{\mathrm{reco}}$ for the di-lepton channel).

A parametrisation of the reconstructed top quark mass in simulation allows for a calibration of the detector response with respect to the value of m_{top} predicted in simulation. In the single lepton decay final state the response is broken down by the leading components of the systematic uncertainty which are the b-jet energy scale dependence (BJSF) and the jet energy scale dependence (JSF).

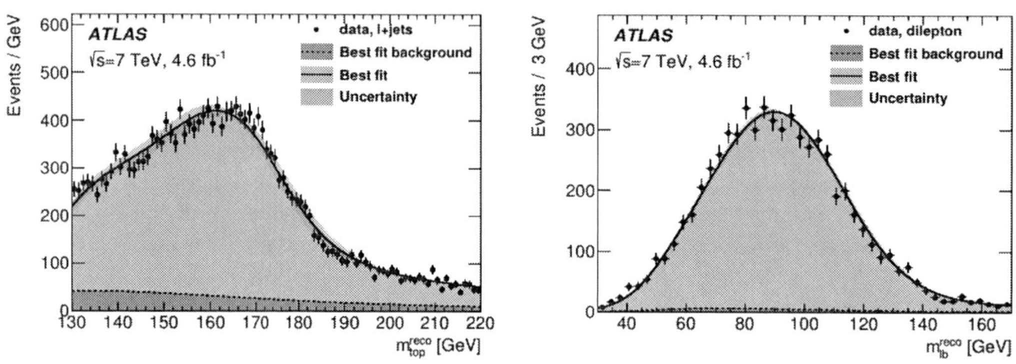

Figure 1. The fitted distributions in the data, showing $m_{\mathrm{top}}^{\mathrm{reco}}$ (left) for the single lepton (electron or muon) final state and $m_{\ell b}^{\mathrm{reco}}$ (right) for the di-lepton final state. The fitted probability density functions for the background alone and for signal-plus-background are also shown. The uncertainty bands indicate the total uncertainty on the signal-plus-background fit [2].

A template fit on data of the parametrised distributions as a function of m_{top}, see Fig. 1, allows for the extraction of the measured top quark mass. The result in the single lepton final state is

$$m_{\mathrm{top}}^{\ell+\mathrm{jets}} = 172.33 \ \pm 0.75\ (\mathrm{stat.+JSF+bJSF}) \ \pm 1.02\ (\mathrm{syst.})\ \mathrm{GeV} \tag{1}$$

where the leading systematic uncertainties are due to a residual jet energy scale and jet energy resolution mis-modelling. Respectively, in the di-lepton final state the result is:

$$m_{\mathrm{top}}^{\mathrm{dil}} = 173.79 \ \pm 0.54\ (\mathrm{stat.}) \ \pm 1.30\ (\mathrm{syst.})\ \mathrm{GeV} \tag{2}$$

with the leading systematic uncertainty arising from the b-jet energy scale mis-modelling. The combination of those two results ($m_{\mathrm{top}}^{\mathrm{comb}}$) corresponds to

$$m_{\mathrm{top}}^{\mathrm{comb}} = 172.99 \ \pm 0.48\ (\mathrm{stat.}) \ \pm 0.78\ (\mathrm{syst.})\ \mathrm{GeV} \tag{3}$$

The result of Eq. 3 is among the most precise measurements of the top quark mass to date. However, the definition of the top mass depends upon the renormalisation scheme used in simulation and it lacks a clear interpretation in a well defined theoretical picture. Several theoretical calculations in perturbative quantum chromodynamics identify the top quark mass with the pole mass $m_{\mathrm{top}}^{\mathrm{pole}}$. The pole mass can be determined experimentally from the inclusive cross section of $t\bar{t}$ pairs ($\sigma_{t\bar{t}}$). Due to the reduced sensitivity of $\sigma_{t\bar{t}}$ to m_{top} and to large uncertainties on the renormalisation and factorisation scales, the measured $m_{\mathrm{top}}^{\mathrm{pole}}$ is of a smaller precision compared to that of the measurement of Eq. 3. However, a novel technique [3] suggests that the measurement of the normalised differential cross section $\mathcal{R}(m_{\mathrm{top}}^{\mathrm{pole}}, \rho_s) = \frac{1}{\sigma_{t\bar{t}+1\mathrm{jet}}} \cdot \frac{d\sigma_{t\bar{t}+1\mathrm{jet}}}{d\rho_s}(m_{\mathrm{top}}^{\mathrm{pole}}, \rho_s)$ for $t\bar{t}$ production with at least one additional jet as a function of the invariant mass of the $t\bar{t}$+1 jet system $\rho_s = \frac{2m_0}{\sqrt{s_{t\bar{t}+1\mathrm{jet}}}}$, with m_0 an arbitrary constant, enhances the sensitivity to the top quark exploiting the phase-space dependence upon the amount of gluon radiation. The measurement of $\mathcal{R}(m_{\mathrm{top}}^{\mathrm{pole}}, \rho_s)$ in a fiducial phase-space within the detector acceptance allows for a reduction upon the theoretical uncertainties entering in the extrapolation from the measured phase-space to the one defined in simulation.

The combination of the these to factors are exploited by the ATLAS experiment and $m_{\mathrm{top}}^{\mathrm{pole}}$ is extracted from a parametrisation on the measured value of $\mathcal{R}(m_{\mathrm{top}}^{\mathrm{pole}}, \rho_s)$, see Fig. 2.

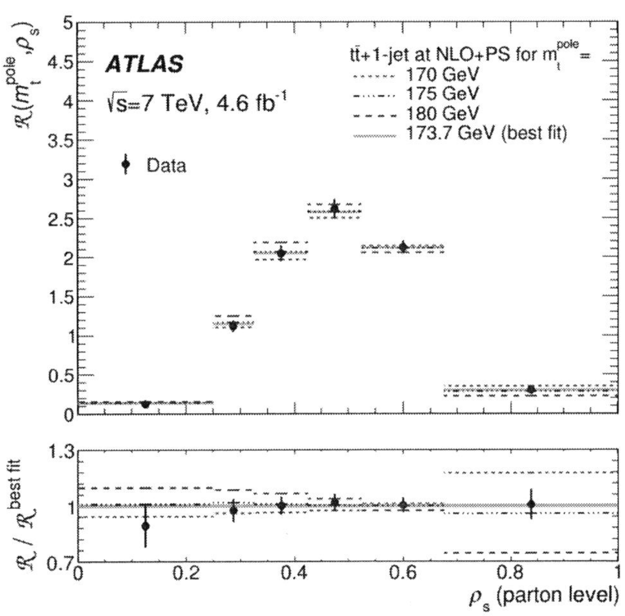

Figure 2. $\mathcal{R}(m_{\mathrm{top}}^{\mathrm{pole}}, \rho_s)$-distribution at parton level corrected for detector and hadronization effects after the background subtraction as a function of $\rho_s(m_0 = 170$ GeV). The predictions of the $\sigma_{t\bar{t}+1\mathrm{jet}}$ calculation are performed at next to leading order using three different masses ($m_{\mathrm{top}}^{\mathrm{pole}} = 170, 175$ and 180 GeV) are shown together with the result of the best fit to the data, $m_{\mathrm{top}}^{\mathrm{pole}} = 173.7 \pm 1.5$ (stat.) GeV. The black points correspond to the data. In the lower part of the figure, the ratios of the different R-distributions to the one corresponding to the best fit are shown. The shaded area indicates the statistical uncertainty [1].

The extracted value of $m_{\mathrm{top}}^{\mathrm{pole}}$ [1] corresponds to

$$m_{\mathrm{top}}^{\mathrm{pole}} = 173.7 \pm 1.5\ (\mathrm{stat.}) \pm 1.4\ (\mathrm{syst.})\ ^{+1.0}_{-0.5}\ (\mathrm{theory})\ \mathrm{GeV}, \tag{4}$$

with the measurement being limited by the statistical uncertainty and missing higher order theoretical calculations. The main systematic uncertainties arise from the jet energy scale dependence (0.9 GeV).

EPJ Web of Conferences **126**, 04004 (2016) DOI: 10.1051/epjconf/201612604004
ICNFP 2015

3 $t\bar{t}$ plus vector boson

Hypothesised new phenomena, at a higher scale of that accessible by the experiment, can be modelled by effective theories. These models are based upon anomalous couplings of vector bosons ($V = W, Z$ and photon) to the top quark that can be probed through precision measurements at lower scales. For example, models of composite top quarks [4], or with excited top quark production, followed by the radiative decay $t^\star \to t\gamma$. The measurements of the $t\bar{t}V$ cross sections as performed by the ATLAS collaboration are discussed in the following two sections. Section 3.1 focuses on the $t\bar{t}$ plus photon ($t\bar{t}\gamma$) production cross section, while Sec. 3.2 focuses on the simultaneous measurement of the $t\bar{t}$ plus W ($t\bar{t}W$) and $t\bar{t}$ plus Z ($t\bar{t}Z$) production cross sections.

3.1 Cross section of $t\bar{t}$ production association with a photon

The $t\bar{t}\gamma$ fiducial cross section ($\sigma^{\text{fid.}}_{t\bar{t}\gamma}$) is measured in the single lepton channel characterised by a high-p_{T} electron or muon, at least four high-p_{T} jets, large transverse missing momentum and an energetic photon. The final states of $t\bar{t}\gamma$ production are indistinguishable from the production of background processes which may also feature a final state energetic photon. The contribution of the background processes is estimated using techniques based on data and on well known phenomena, reducing the simulation-induced model-dependency of the result. The response for hadrons, or hadron decay products ($\pi^0 \to \gamma\gamma$) is close to that of real photon objects. Exploiting the differences in shape of the track-isolation ($p_{\text{T}}^{\text{iso}}$), defined as the scalar sum of all track momenta within a cone of radius 0.2 with respect to the photon candidate, allows for a discrimination between real photons and hadrons. The final result is extracted by a template-based likelihood fit on $p_{\text{T}}^{\text{iso}}$ from data candidates in the electron and muon channels simultaneously see Fig. 3.

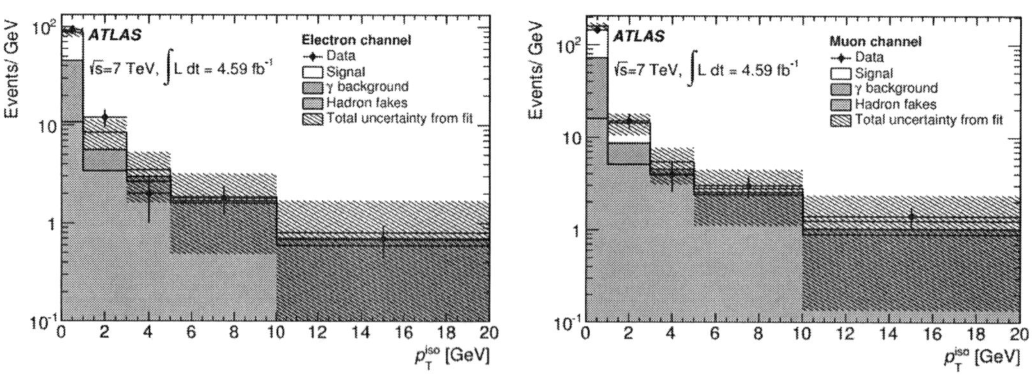

Figure 3. Results of the combined likelihood fit using the track-isolation ($p_{\text{T}}^{\text{iso}}$) distributions as the discriminating variable for the electron channel (left) and the muon channel (right). The contribution from $t\bar{t}\gamma$ events is labeled as "Signal", prompt-photon background is labeled "γ backgrounds", the contribution from hadrons misidentified as photons (as estimated by the template fit) is labeled as "Hadron fakes" [5].

The $\sigma^{\text{fid.}}_{t\bar{t}\gamma}$ is defined in a phase-space within the kinematic and geometrical detector acceptance. The definition of particles belonging to the phase-space is motivated by imposing experimentally observable selection criteria. This definition diminishes the model-dependency and augments the reproducibility of the result. The result measured cross section [5],

$$\sigma^{\text{fid.}}_{t\bar{t}\gamma} = 63 \pm 8\,(\text{stat.})\,^{17}_{-13}\,(\text{syst.}) \pm 1\,(\text{lumi.})\,\text{fb}, \tag{5}$$

is in good agreement with that of leading order calculation in the fiducial phase-space normalised to fraction of the next-to-leading order to leading order prediction [6] of 48 ± 10 fb.

3.2 Production cross section of $t\bar{t}$ paris in association with a W or Z–boson

Depending on the decays of the top quarks, W and Z–bosons the number of final state prompt isolated leptons may vary between zero and four. Only channels with two (both with same-sign and opposite-sign charge), three, and four leptons are considered in this measurement. Each channel is divided into multiple regions in data in order to enhance the sensitivity to the $t\bar{t}Z$ and $t\bar{t}W$ cross sections. The production cross sections for $t\bar{t}W$ ($\sigma_{t\bar{t}W}$) and $t\bar{t}Z$ ($\sigma_{t\bar{t}Z}$) are determined simultaneously using a binned maximum-likelihood fit over all regions and discriminant bins considered in the analysis [7]. The fit results are summarised in Fig. 4.

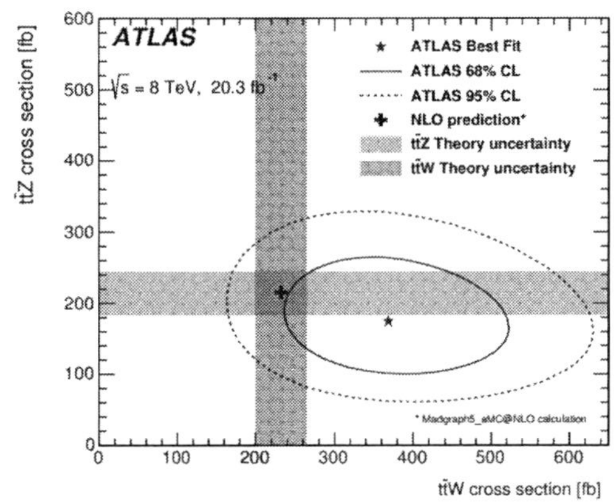

Figure 4. The result of the simultaneous fit to the $t\bar{t}W$ and $t\bar{t}Z$ cross sections along with the 68% and 95% Confidence Level (CL) uncertainty contours. The shaded areas correspond to 14% uncertainty, which includes renormalisation and factorisation scale uncertainties as well as parton density function uncertainties including α_s variations [7].

The results are [7]

$$\sigma_{t\bar{t}W} = 369\,^{86}_{-79}\ \text{(stat.)}\ \pm 44\ \text{(syst.) fb} \tag{6}$$

for the $t\bar{t}W$ production and

$$\sigma_{t\bar{t}Z} = 176\,^{52}_{-48}\ \text{(stat.)}\ \pm 24\ \text{(syst.) fb} \tag{7}$$

for the $t\bar{t}Z$ production. It can be seen in Fig. 4 that the results are in good agreement with the theoretical prediction.

4 Charge asymmetry

The difference in rapidity between top and anti-top quarks in proton–proton collisions is referred to as the charge asymmetry (A_C). At hadron colliders and the leading order contribution of A_C is zero, because of the symmetric $t\bar{t}$ production under the exchange of top and anti-top quarks. At next-to-leading order the $t\bar{t}$ paris produced from quark anti-quark annihilation, due to interferences with initial- and final-state gluon emission, produce an asymmetry in rapidity. The SM predicts values of

A_C to be of the percent level with next-to-leading order accuracy [8]. Several extensions of the SM, for example models with anomalous vector or axial-vector couplings (axigluons), predict higher values of A_C. Moreover, these models predict also differences in the spectra of A_C as a function of invariant mass of the $t\bar{t}$ system ($m_{t\bar{t}}$), of the longitudinal boost of $t\bar{t}$ ($\beta_{z,t\bar{t}}$) and of the transverse momentum of the $t\bar{t}$ system ($p_{T,t\bar{t}}$). Therefore, the measurement of A_C both inclusively and differentially is of great interest.

The analysis is carried out in the single lepton plus jets plus missing transverse momentum final state, and the event kinematics are reconstructed using a kinematic likelihood fit [10]. The reconstructed rapidity spectra are corrected for acceptance and detector resolution effects using a *Bayesian* unfolding procedure [9]. The final measured A_C is extracted from a likelihood fit, where the rapidity spectra are further discriminated with respect to the measured lepton charge.

The differential results of A_C are shown in Fig. 5 and they are compatible with the SM prediction. The inclusive $t\bar{t}$ production charge asymmetry is measured to be [9]

$$A_C = 0.009 \pm 0.005 \text{ (stat.} \oplus \text{syst.)}, \tag{8}$$

compatible with the SM prediction, $A_C = 0.0111 \pm 0.0004$ [8].

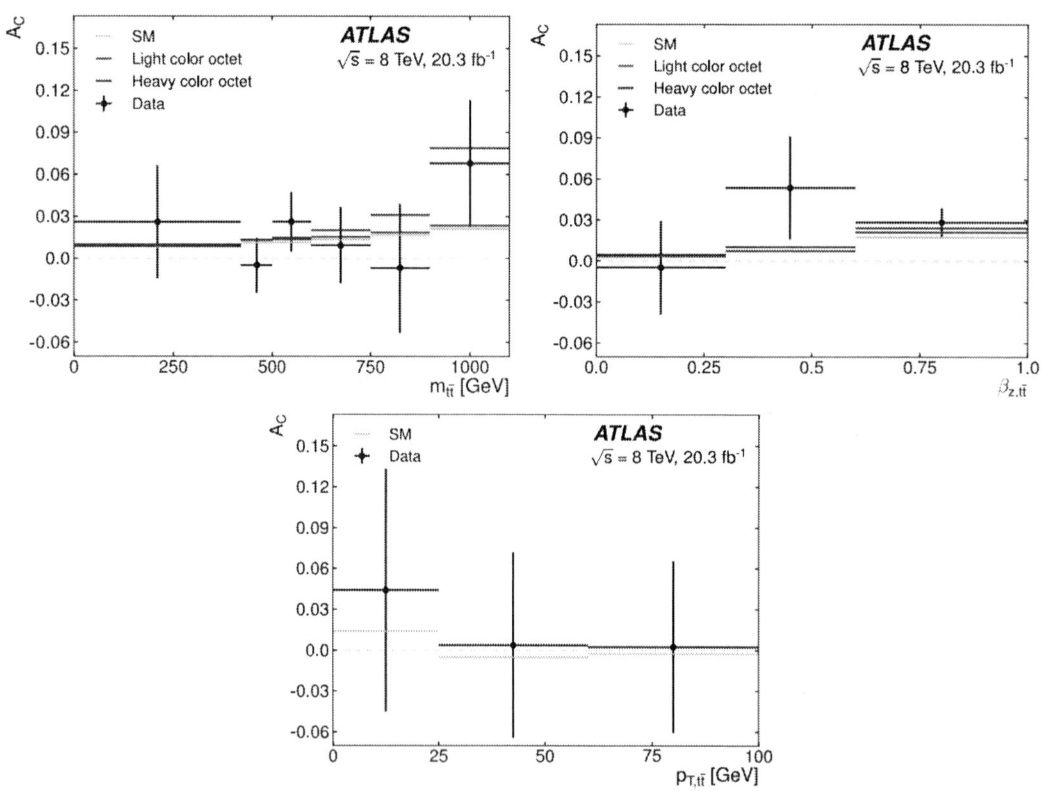

Figure 5. Measured A_C values as a function of bin-averaged $m_{t\bar{t}}$ (top left), $\beta_{z,t\bar{t}}$ (top right) and $p_{T,t\bar{t}}$ (bottom) compared with predictions for the SM [8] and for right-handed colour octets [11] with masses below the $t\bar{t}$ threshold and beyond the kinematic reach of current LHC searches. The beyond the standard model predictions are shown only for the two top plots. [9].

5 Conclusions

These proceeding reviewed some of the most recent results on the top quark measurements performed by the ATLAS detector. These results are based upon datasets of 4.59 fb^{-1} and of 20.3 fb^{-1} of proton–proton collisions delivered by the LHC at the centres-of-mass energy of $\sqrt{s} = 7$ TeV and of $\sqrt{s} = 8$ TeV, respectively. In particular, the measurements of the top quark mass, the cross section of $t\bar{t}$ production in association of vector bosons (W, Z and γ) and the measurement of the charge asymmetry were detailed.

The abundant production of $t\bar{t}$ pairs at the LHC allows for evermore precise measurements of the top quark properties leading to stringent tests of the SM and to searches for new phenomena trough deviations from the expected SM predictions.

References

[1] ATLAS Collaboration, *JHEP* **10** (2015), doi:10.1007/JHEP10(2015)121.

[2] ATLAS Collaboration, *EPJ* **C75** (2015), 7, doi:10.1140/epjc/s10052-015-3544-0.

[3] S. Alioli *et al.*, *EPJ* **C73** (2013), 2438, arXiv:1303.6415.

[4] B. Lillie J. Shu and T. M. P. Tait, *JHEP* **04**, (2008), 87.

[5] ATLAS Collaboration, *PRD* **91** (2015), 072007.

[6] K. Melnikov, A. Scharf, and M. Schulze, *PRD* **83** (2011), 074013.

[7] ATLAS Collaboration, (2015), arXiv:1509.05276.

[8] W. Bernreuther and Z.-G. Si, *PRD* **86** (2012), 034026, doi:10.1103/PhysRevD.86.034026.

[9] ATLAS Collaboration, (2015), arXiv:1509.02358.

[10] J. Erdmann *et al.*, *NIM* **A748** (2014), 18-25, doi:10.1016/j.nima.2014.02.029.

[11] J. A. Aguilar-Saavedra, *JHEP* **08** (2014), 172, doi:10.1007/JHEP08(2014)172.

Measurements of Multi-boson production, Trilinear and Quartic Gauge Couplings with the ATLAS detector

Maurice Becker[1,a], on behalf of the ATLAS Collaboration

[1] *Institute of Physics, Johannes Gutenberg University Mainz*

Abstract. The ATLAS collaboration has carried a set of measurements that provide stringent tests of the electroweak sector of Standard Model, specifically on di- and multi-boson production cross sections and on triple and quartic gauge-boson couplings. Such measurements include cross sections for WV (V=W or Z) production in the leptonic or semileptonic channels, the production of a W or Z boson in association with photons, a Z boson in the vector-boson fusion channel and two same-charge W bosons in the vector-boson scattering channel. These measurements are compared to (N)NLO predictions of the Standard Model and provide model-independent constraints on new physics, by setting limits on anomalous gauge-boson couplings.

An analysis of the inclusive four-lepton lineshape has been carried out in the mass range from 80 to 1000 GeV, where several distinct physics processes give rise to the production of 4-lepton final state, namely the single Z resonant processes, the Higgs production at 125 GeV, as well as continuum ZZ production processes with $q\bar{q}$ and gg initial states.

1 Introduction

Electroweak (EW) precision measurements at LHC provide a good way to test the expectations of the Standard Model in the EW sector. Furthermore many diboson processes play an important background role for other analyses like measurements of Higgs boson properties or searches for signatures beyond the Standard Model. Comparing expectations of EW processes with measurements allows indirect searches for new physics. Deviations from SM gauge interactions, like anomalous triple gauge couplings (aTGCs), could lead to changes in the production cross section, especially in the high boson pair mass range. One can not only search for anomalous couplings, but also the expected triple and quartic gauge coupling can be studied at proton proton collisions at the LHC like Vector Boson Fusion (VBF) or Vector Boson Scattering (VBS). Especially measuring the VBS cross section is a good way to probe EW symmetry breaking. This note presents the latest results of diboson cross section measurements using data from the ATLAS detector [1] at the LHC from 2011 and 2012 with $\sqrt{s}$ = 7 TeV and 8 TeV with an integrated luminosity of 4.7 fb^{-1} and 20.3 fb^{-1} respectively. Also VBF and VBS measurements are presented together with the first evidence of quartic gauge coupling obtained at LHC.

[a]e-mail: maurice.becker@cern.ch

2 Experimental Setup and Physics Object Definitions

Every analysis that is presented here measures the cross section of processes with at least one lepton in addition with other objects like jets, photons or missing transverse momentum ($E_{\mathrm{T}}^{\mathrm{miss}}$). How these objects are measured within the ATLAS detector shall be shortly described.
Electrons are build from tracks in the inner detector that are matched to clusters in the electromagnetic calorimeter, while muons are reconstructed by matching the inner detector tracks with the tracks of the muon spectrometer. The typical kinematic range where the leptons are measured is above a p_{T} of 15 GeV in a absolute pseudorapidity[1] region smaller than 2.47, that is covered by the tracking system.
Due to the confinement of quarks in the Standard Model quarks can not exist isolated. As a result quarks hadronize to collimated bundles of particles (jets). Jets are build by using special cluster algorithms e.g. the anti-k_t algorithm [2].
Due to the fact that neutrinos do not leave signatures in the detector it is possible to measure the size of missing transverse momentum using momentum conservation.

3 Diboson production cross section measurements

3.1 Wγ and Zγ cross section measurement

This analysis measured the integrated and differential fiducial cross sections for W (or Z) boson production in association with a high-energy photon [3], looking at leptonically decaying W or Z bosons in ATLAS data recorded 2011 at $\sqrt{s}$ = 7 TeV with 4.6 fb^{-1}. In addition to production of a W and Z boson via the s, u- and t-channels, the signal receives also contributions from W boson plus quark (gluon) when the quark (gluon) fragments to a photon.
The selection of Wγ events is done by requiring exactly one high p_{T} lepton, one high p_{T} photon, high $E_{\mathrm{T}}^{\mathrm{miss}}$ and high transverse mass of the lepton-$E_{\mathrm{T}}^{\mathrm{miss}}$ system. To select Zγ events two high p_{T} leptons from the same flavor but opposite charge are required that form an invariant mass near the Z boson mass, accompanied by a high energetic photon. In order to subtract background from events where the photon is radiated from a lepton in a bremsstrahlung process, it is required that the photon is clearly separated from the lepton. Zγ events where the Z boson decays to neutrinos are selected by only requiring an high $E_{\mathrm{T}}^{\mathrm{miss}}$ and an isolated photon.
The main background contributions are given by W+jets events where the photon results from a π^0 decay, Z events that mimic a Wγ signal and γ+jets events where a jet fakes a lepton and $E_{\mathrm{T}}^{\mathrm{miss}}$ comes from e.g. mis-measured jet energies. These backgrounds are estimated using a two-dimensional side-band method where the contribution is measured in orthogonal control regions and extrapolated to the signal region.
The results of the differential cross-section are shown in Figure 1 as a function of the E_T^{γ}. The distributions are shown at particle level where detector effects are corrected for (unfolded). The cross section is given in an inclusive region, including all additional jets that might occur in an event and in an exclusive region that adds an extra requirement that there are no additional jets in the event since the exclusive region is more sensitive to anomalous triple gauge couplings. The measured cross section is compared to different theory predictions from SHERPA [4], ALPGEN [5] and MCFM [6]. It can be seen that MCFM leads to worse results in higher E_T^{γ} regions compared to SHERPA and ALPGEN for the Wγ case due to the fact that this region is more sensitive to higher jet multiplicities that are not included in MCFM. Besides of this a good agreement between data and theory prediction can be seen and therefore also limits on anomalous triple gauge couplings were set.

[1]ATLAS uses a right-handed coordinate system where the z-axis points along the beam axis, the x-axis to the center of the LHC and the y-axis upwards. In the transverse plain cylindrical coordinates are used and the pseudorapidity is defined as $\eta = -\ln[\tan(\theta/2)]$, where θ is the polar angle.

DOI: 10.1051/epjconf/201612604005

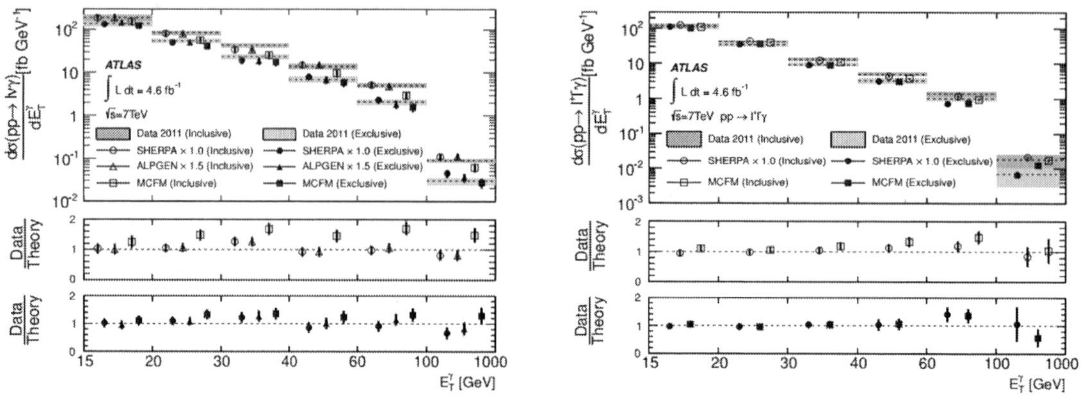

(a) (b)

Figure 1. Differential cross section of (a) $pp \to W\gamma \to l\nu\gamma$ and (b) $pp \to Z\gamma \to l^+l^-\gamma$ processes, using combined electron and muon measurements in an exclusive ($N_{jet} = 0$) and inclusive ($N_{jet} \geq 0$) region [3]. Included is besides the measured result using data theory predictions using MCFM, SHERPA and ALPGEN are shown.

3.2 Measurement of $W^{\pm}Z$ production decaying leptonically

A further process of the electroweak interaction is the production of a W and a Z boson where both decay leptonically to electron or muons. In this analysis [7] events in 13 fb^{-1} data from 2012 at $\sqrt{s} = 8$ TeV, are selected with two same flavor, opposite charge leptons within the Z mass window ($|m_{ll} - 91.1876| < 10$ GeV) and one additional lepton that forms together with the missing transverse momentum a W boson. One background that passes the selection criteria comes from events with three real electrons, amounting to 8% of the selected events (e.g. ZZ events where one lepton escapes the detector acceptance and thus faking $E_{\mathrm{T}}^{\mathrm{miss}}$ which is estimated via Monte Carlo simulation). Another background comes from processes where one lepton is faked (e.g. jet faking an electron) which is estimated using a data driven method and amounts about 17% of the selected events.
The integrated cross section, measured to $20.3^{+0.8}_{-0.7}(stat.)^{+1.2}_{-1.1}(syst.)^{+0.7}_{-0.6}(lumi.)$ pb, is in good agreement with the theory prediction of MCFM of 20.3 ± 0.8 pb.

3.3 Measurement of W^+W^- production decaying leptonically

The $W^+W^- \to l^+\nu l^-\nu$ cross section is measured by selecting events with two leptons with opposite charge and large missing transverse momentum [8]. Since top pair production ($t\bar{t}$) would lead to the same signature it is required that there is no jet with $p_{\mathrm{T}} > 25$ GeV in an event. With this cut a large fraction of the $t\bar{t}$ background is rejected as it is shown in Figure 2(a). To reject Z+jets events, where a miscalibrated jet could fake $E_{\mathrm{T}}^{\mathrm{miss}}$ and the charge of the leptons is misidentified, it is required that the dilepton invariant mass is outside of the Z mass window. The remaining $t\bar{t}$ and Z+jets background is estimated using data driven techniques. The measured total cross section is $71.4^{+1.2}_{-1.2}(stat.)^{+5.0}_{-4.4}(syst.)^{+2.2}_{-2.1}(lumi.)$ pb and shown together with a theory prediction in Figure 2(b). It can be seen that the combined cross section differs by about 2.1 σ from the SM prediction. Possible explanations for this difference could be missing NNLO corrections in QCD [9] and/or jet-veto resummation [10].

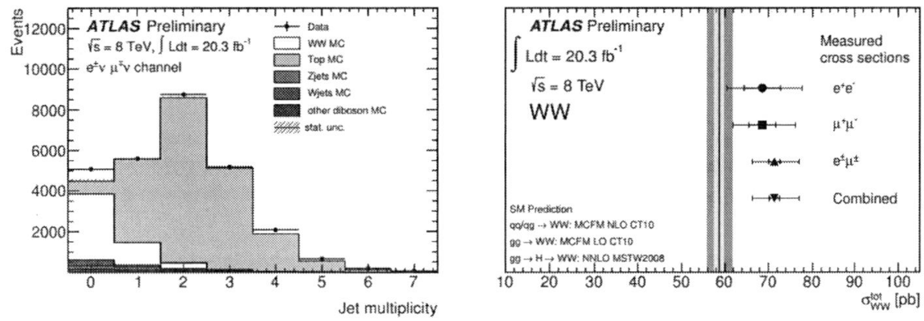

(a) (b)

Figure 2. (a): Jet multiplicity for W^+W^- events in the $e^{\pm}\nu\mu^{\mp}\nu$ channel without applying a jet veto [8]. (b): Measured and predicted W^+W^- cross section for three different channels together with their combination [8]. The green and yellow bands represent the PDF and the total theory uncertainties.

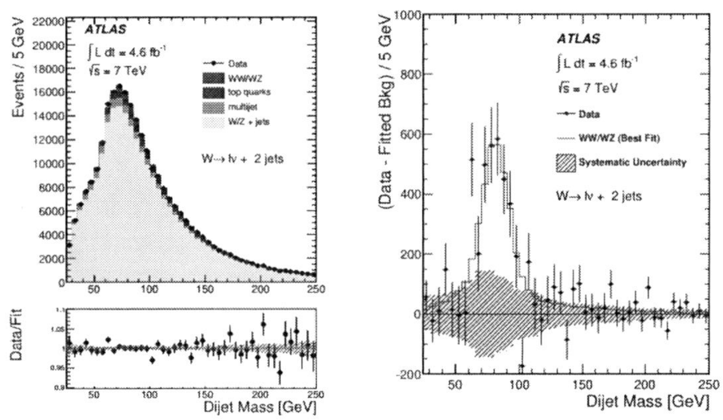

(a) (b)

Figure 3. (a): Distributions of the dijet invariant mass for the sum of the electron and muon channels after the likelihood fit in semileptonic decaying WZ events. (b): Distribution of the background-subtracted data for the sum of the electron and muon channels in semileptonic decaying WZ events [11].

3.4 Measurement of WW/WZ production decaying semileptonic

An analysis of the WZ/WZ production where one of the bosons decays hadronically has also been performed using the 2011 data of ATLAS [11]. The selection requires high $E_{\mathrm{T}}^{\mathrm{miss}}$, a high p_{T} lepton forming together high transverse mass. Events are selected where there is one leptonically decaying W boson and hadronically decaying Z or W boson. Due to the jet energy resolution it is not possible to distinguish between the bosons. It is also not possible to distinguish this signature from background where there is only one W or Z boson with additional jets from higher order QCD processes as it can be seen in Figure 3(a). In order to extract the signal, the background is fitted with a likelihood approach where the shape is taken from simulation. The result after subtracting the background can be seen in Figure 3(b). The measured cross section is 68 ± 7*(stat.)* ± 19*(syst.)* pb and in good agreement with the theory expectation of 61.1 ± 2.2 pb from MC@NLO [12]. The large systematic uncertainty on the measured result is dominated by the uncertainties of the jet energy scale and modeling of the background fit.

 DOI: 10.1051/epjconf/201612604005

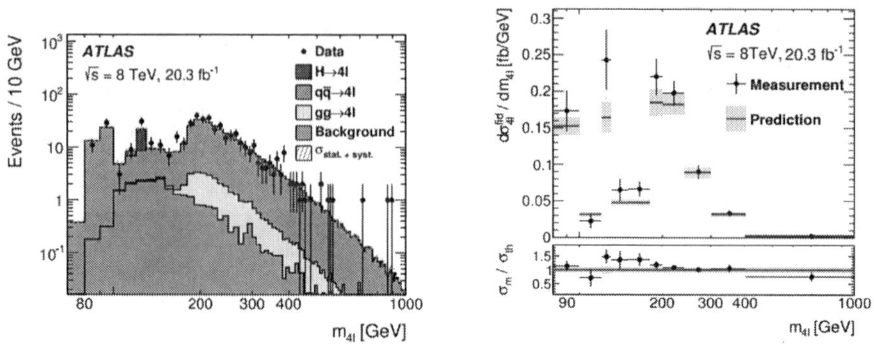

(a) (b)

Figure 4. (a): Data and MC prediction comparison for selected events as a function of the invariant mass m_{4l} in selected events with four charged leptons. (b): The measured differential cross-section distributions (the black points) of m_{4l} from events with four charged leptons unfolded into the fiducial volume, and compared to theory predictions (red histogram) [13].

3.5 Measurement of four lepton production

There are different ways in pp collisions to produce 4 leptons in the final state. The investigation of this signature is a great way to test the electroweak predictions over a large kinematic range [13]. Events with four charged leptons come from resonant Z and Higgs decays but also from the non-resonant ZZ continuum which are produced by $q\bar{q}$ and gg initial states. The selection requires 4 high p_{T} leptons that form pairs with same flavor but opposite charge in an event. This signature is very clean and the background is very small and arises from events where a lepton is "faked". To get an estimate of this background a data driven technique is used. The resulting invariant mass of the four lepton system can be seen in Figure 4(a) and shows that the different processes contribute in different kinematic regions. Figure 4(b) shows the resulting differential cross section as a function of the invariant mass of the four lepton system in a range from 80 GeV to 1 TeV in comparison with the theory prediction which is overall in good agreement with data. Since the measurement is also sensitive to high m_{4l} it is possible to estimate the gluon contribution above 180 GeV which is only known at LO QCD precision. In this region a fit to data is performed that takes all non gg contribution and background into account. The cross section taking the integrated number of events from the fit is divided by the LO theory prediction. The result is $\mu_{gg} = 2.4 \pm 1.0(stat.) \pm 0.5(syst.) \pm 0.8(theory)$ which shows that this measurement can be used to constrain the theory prediction of the gg to four lepton process.

4 Vector Boson Fusion (VBF) and Vector Boson Scattering (VBS)

4.1 Measurement of electroweak production of Z boson in association with two jets

The production of a Z boson with two additional jets is mainly dominated by the Drell Yan process where additional jets arise from strong interaction (e.g. gluon radiation) but there is also the possibility to produce a Z boson + two jets via the electroweak vector boson fusion. Since there is large interference between these two processes it is not possible to completely separate those. The main goal of the analysis is on the one hand to provide the inclusive and differential cross section for Zjj events that provide constraints on the theoretical modeling of QCD-initiated processes and on the other hand to measure the electroweak Zjj production cross section [14]. Therefore the selection of

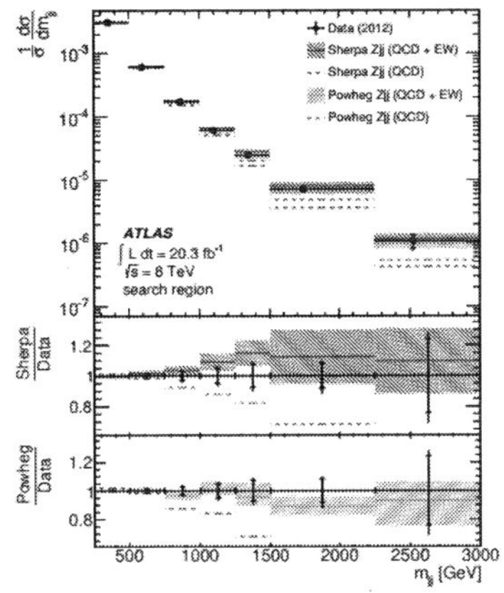

(a)

Figure 5. Unfolded normalized differential cross section distribution as a function of dijet invariant mass of the electro weak production of a Z boson with two jets [14].

events is divided in different regions that have different sensitivities to the electroweak process. The base line is just the requirement of a pair of high p_T leptons within in the Z mass range and moderate cuts on the jet p_T. Due to the special kinematics of VBF processes the two jets tend to scatter into very forward and backward regions, which leads to large invariant masses of the jet jet system due to large rapidity separation between them. In addition no further hadronic activity is expected between the jets. Thus cuts on m_{jj} and the numbers of additional jets between the forward-backward jets leads to larger sensitivity for electroweak VBF. A result of the differential cross-section in the most sensitive region is shown in Figure 5 where it is visible that a description of the spectrum is not possible with only Zjj QCD processes but predictions by POWHEG and SHERPA are in good agreement when adding the EW processes. The inclusive cross section of electroweak Zjj is found to be $54.7 \pm 4.6(stat.)^{+9.8}_{-10.4}(syst.) \pm 1.5(lumi.)$ fb, which is in good agreement with the theory prediction. Therefore limits on anomalous triple gauge couplings were set.

4.2 Measurement of $W^{\pm}W^{\pm}jj$ production

A key process to check the nature of electroweak symmetry breaking is the scattering of two massive vector bosons (VBS). Therefore an analysis was performed that shows first evidence of VBS contributions in a $W^{\pm}W^{\pm}jj$ end state [15]. The required final state consists of two W bosons and two jets and is dominated by QCD contributions. For VBF the jets tend to have large p_T and a large rapidity gap so that selecting events with high m_{jj} or large rapidity differences of the jets enhances the VBF contribution. In order to suppress background from Drell Yan and $t\bar{t}$ production it is required that the leptons have same charge. The cross section is measured in two different kinematic regions: One more inclusive, containing also QCD contributions, and one with an additional cut on the rapidity difference, $|\Delta y_{jj}| > 2.4$ (VBS region). For the VBS region the significance of the combined signal and background over the background only hypothesis is 3.6 standard deviations. The resulting cross section, measured in the inclusive and the VBS region, can be seen in Figure 6 and is found to be in good agreement with theory prediction within the given uncertainties that are mainly dominated by statistical uncertainties. Therefore limits are set on anomalous quartic gauge couplings.

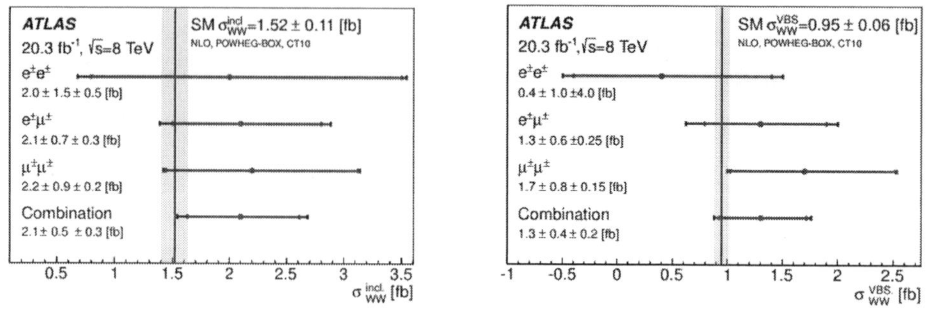

(a) (b)

Figure 6. The measured cross sections of $W^{\pm}W^{\pm}jj$ production are shown for the inclusive region (a) and the VBS region (b) compared to the predicted Standard Model cross section [15]. The inner red error band represents the statistical error, and the blue band represents the total error on each measurement. The uncertainties on the cross sections quoted are statistical and systematic respectively.

4.3 Evidence of $W\gamma\gamma$ production

Due to the non-Abelian structure of the electroweak sector of the Standard Model also quartic gauge couplings are allowed. Deviations from this SM expectation would be a hint for physics beyond the Standard Model. Therefore an analysis was performed that looks for $W\gamma\gamma$ production in 2012 data of ATLAS with $\sqrt{s} = 8$ TeV [16]. $W\gamma\gamma$ candidates are formed by requiring a high p_T lepton, large E_T^{miss} and two isolated photons. Contributions to this event topology can arise from W production with photons emitted in the initial or final state and from triple or quartic gauge coupling processes. The cross section of the process is measured in an inclusive region where no restriction on the recoil system is done and an exclusive region with the restriction that there is no additional jet with p_T >30 GeV in the event which is more sensitive to quartic gauge couplings. The main background contribution originates from events where jets are mis-identified as leptons, which is estimated using a data driven approach. The invariant mass system of the two photons, shown in Figure 7, shows good agreement between data and expectation when taking the $W\gamma\gamma$ signal contribution into account. The combined significance of signal plus background over background only hypothesis is above 3σ. The total inclusive cross section is measured to $6.1^{1.1}_{-1.0}(stat.) \pm 1.2(syst.) \pm 0.2(lumi.)$ fb which is in agreement with the theory prediction. Since no significant differences between measurement and expectation are found, limits where set on anomalous quartic gauge couplings.

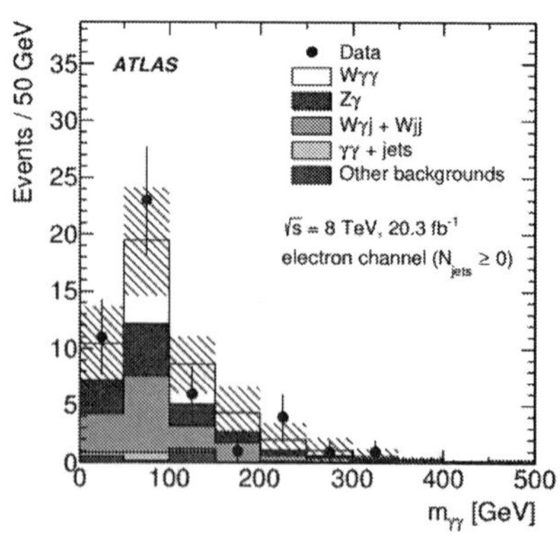

Figure 7. (a): Diphoton invariant mass distribution of selected $W\gamma\gamma$ events in the electron channel. The expected signal based on the SHERPA prediction is shown. The hashed areas show the total systematic and statistical uncertainty on the background estimate [16].

 DOI: 10.1051/epjconf/201612604005

5 Conclusion and Outlook

Different diboson cross section measurements have been presented. The production cross section of WZ and WW decaying leptonically and semi-leptonically were compared to theory predictions at $\sqrt{s}$ = 8 TeV and no deviations were found to the Standard Model expectation. Also a differential cross section measurement of $W\gamma$ and $Z\gamma$ production using data at $\sqrt{s}$ = 7 TeV is in agreement with theory predictions and could be used for further theoretical constraints. A differential cross section measurement of the invariant mass spectrum of four charged leptons was presented, which can be used as input for theoretical modeling.

A measurement of the Zjj production cross section at $\sqrt{s}$ = 8 TeV was shown, which includes contributions from vector boson fusion processes, and is in good agreement with the Standard Model expectation. An additional measurement looking at processes with $W^{\pm}W^{\pm}jj$ in the final state showed the first evidence of electroweak vector boson scattering in this channel. First evidence of electroweak $W\gamma\gamma$ production via quartic gauge coupling was observed which is in good agreement with the Standard Model expectation.

All presented measurements are summarized in Figure 8. It can be seen that the LHC gives a great opportunity to measure electroweak processes over very large range of production cross section. Since there is no deviation to the SM found, limits are set on anomalous triple or quartic gauge couplings.

ATLAS data taking restarted in 2015 with $\sqrt{s}$ = 13 TeV. The higher center of mass energy results in higher cross section for the electroweak processes and once enough data is taken, there are also new channels where the electroweak sector of the Standard Model can be tested.

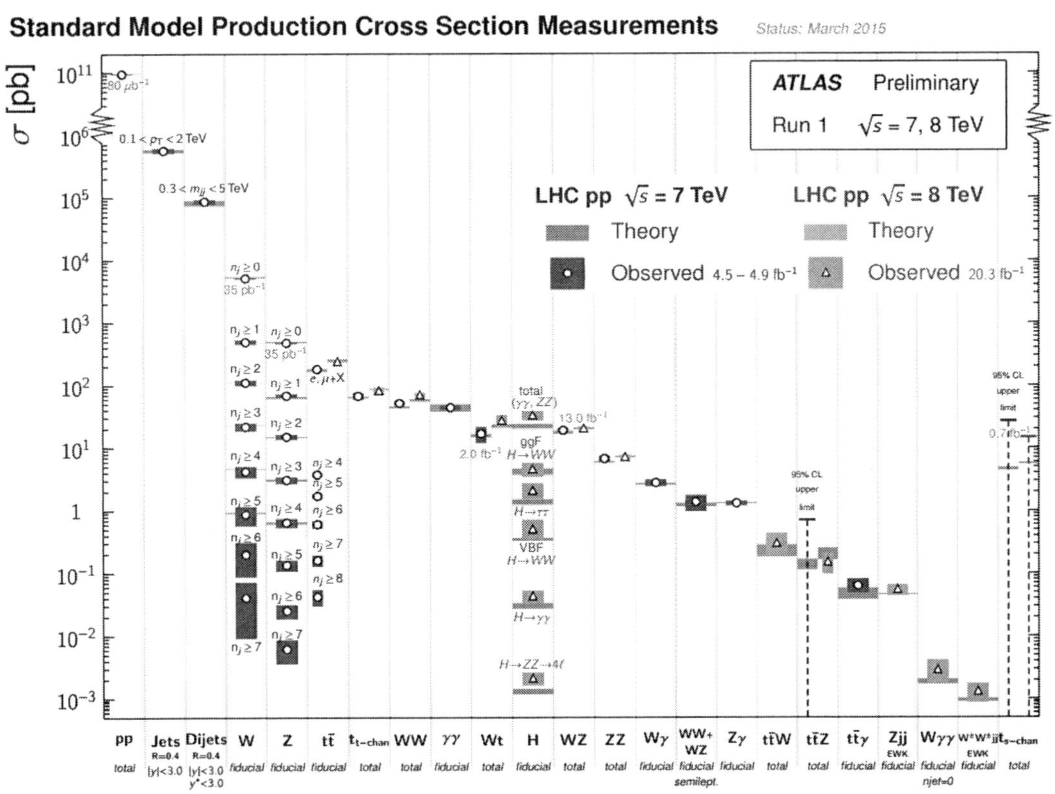

Figure 8. Summary of several Standard Model total and fiducial production cross section measurements, corrected for leptonic branching fractions, compared to the corresponding theoretical expectations. All theoretical expectations were calculated at NLO or higher. [17].

References

[1] ATLAS Collaboration, JINST **3**, S08003 (2008)
[2] M. Cacciari, G.P. Salam, G. Soyez, JHEP **04**, 063 (2008)
[3] ATLAS Collaboration, Phys. Rev. D **87**, 112003 (2013)
[4] T. Gleisberg et al., JHEP **02**, 056 (2004)
[5] M.L. Mangano et al., JHEP **07**, 001 (2003)
[6] J.M. Campbell, R.K. Ellis, Nucl. Phys. Proc. Suppl. **205-206**, 10 (2010)
[7] ATLAS Collaboration, ATLAS-CONF-2013-021, http://cdsweb.cern.ch/record/1525557 (2013)
[8] ATLAS Collaboration, ATLAS-CONF-2014-033, http://cdsweb.cern.ch/record/1728248 (2014)
[9] T. Gehrmann et al., Phys. Rev. Lett. **113**, 212001 (2014)
[10] P. Jaiswal, T. Okui, Phys. Rev. **D90**, 073009 (2014)
[11] ATLAS Collaboration, JHEP **2015**, 49 (2015)
[12] S. Frixione, B.R. Webber, JHEP **2002**, 029 (2002)
[13] ATLAS Collaboration, arXiv:1509.07844 (2015)
[14] ATLAS Collaboration, JHEP **04**, 031 (2014)
[15] ATLAS Collaboration, Phys. Rev. Lett. **113**, 141803 (2014)
[16] ATLAS Collaboration, Phys. Rev. Lett. **115**, 031802 (2015)
[17] https://atlas.web.cern.ch/Atlas/GROUPS/PHYSICS/CombinedSummaryPlots/SM/ (2015)

Anisotropic flow fluctuations in Pb+Pb collisions at LHC

L.V Bravina[2,3,4,a], E.S. Fotina[1], V.L. Korotkikh[1], I.P. Lokhtin[1], L.V. Malinina[1,5], E.N. Nazarova[1], S.V. Petrushanko[1], A.M. Snigirev[1], and E.E. Zabrodin[1,2,3,4]

[1] *Skobeltsyn Institute of Nuclear Physics, Moscow State University, RU-119991 Moscow, Russia*
[2] *Department of Physics, University of Oslo, PB 1048 Blindern, Oslo, Norway*
[3] *Frankfurt Institute for Advanced Studies, Ruth-Moufang-Str. 1, 60438 Frankfurt a.M., Germany*
[4] *National Research Nuclear University "MEPhI" (Moscow Engineering Physics Institute), RU-115409 Moscow, Russia*
[5] *Joint Institute for Nuclear Researches, Dubna, Russia*

Abstract. Fluctuations of anisotropic flow in lead-lead collisions at LHC energies arising in HYDJET++ model are studied. It is shown that intrinsic fluctuations of the flow which appear mainly because of the fluctuations of particle multiplicity, momenta and coordinates are insufficient to match the measured experimental data, provided the eccentricity of the freeze-out hypersurface is fixed at any given impact parameter b. However, when the variations of the eccentricity in HYDJET++ are taken into account, the agreement between the model results and the data is drastically improved. Both model calculations and the data are filtered through the unfolding procedure. This procedure eliminates the non-flow fluctuations to a higher degree, thus indicating a dynamical origin of the flow fluctuations in HYDJET++ event generator.

1 Introduction

Anisotropic flow of secondary particles in ultra-relativistic heavy-ion collisions is very sensitive to creation of even a small amount of quark-gluon plasma (QGP). This topic is a quickly developing branch of modern high energy physics. The present status of the theory and the experiment can be found, e.g., in topical reviews [1–7] and references therein. To analyse the flow the particle distribution in azimuthal plane is decomposed into the Fourier series [8, 9]

$$\frac{dN}{d\phi} \propto 1 + 2\sum_{n=1}^{\infty} v_n(p_T)\cos\left[n(\phi - \Psi_n)\right] \quad (1)$$

$$v_n = \langle \cos\left[(n(\phi - \Psi_n)\right]\rangle , \quad (2)$$

with p_T, ϕ and Ψ_n being the particle transverse momentum, the particle azimuthal angle and the azimuth of the participant plane of n-th order, respectively. The Fourier harmonics v_n are called directed flow (v_1), elliptic flow (v_2), triangular flow (v_3), quadrangular flow (v_4) and so forth. As a collective effect, the flow is nicely described by hydrodynamic models [10] (see, however, [11]), although the p_T-distributions of the flow harmonics are usually traced to transverse momenta $p_T \simeq 2$ GeV/c, and hard

[a]e-mail: larissa.bravina@fys.uio.no

processes are not considered. The HYDJET++ model [12–14], in contrast, combines parametrised hydrodynamic treatment of soft processes with the description of hard jets traversing hot and dense partonic medium. The interplay between soft hydro-like processes and jets within the HYDJET++ was studied in a number of papers [15–23]. The model predicted correctly violation of the number-of-constituent-quark (NCQ) scaling of v_2 at LHC, whereas the scaling certainly holds for hadrons produced in soft processes [15, 16], rise of the high-p_T tail of $v_4(p_T)/v_2^2(p_T)$ ratio [17, 18], the important role of elliptic and triangular flow non-linear contributions to higher order harmonics [14, 19–21] and to the long-range angular di-hadron correlations leading to the formation of characteristic "ridge" structures [22]. Our recent study [23] was devoted to event-by-event (EbyE) fluctuations which alter the resulting anisotropic flow. To what extent these fluctuations are taken into account in the HYDJET++? - The answer to this question is given below. Section 2 presents basic principles of the anisotropic flow generation in the model.

2 Generation of anisotropic flow in HYDJET++

Detailed description of the event generator HYDJET++, which merges thermalized hydro-like state with separated chemical and thermal freeze-outs [24, 25] with in-medium fragmentation of hard partons [26], can be found elsewhere [12]. The fine tuning of free model parameters was done in [13] to match the bulk observables in lead-lead collisions at LHC. Implementation of triangular flow in addition to the elliptic one provided the possibility to describe the basic features of higher flow harmonics up to hexagonal flow [14, 21]. This is done in a straightforward manner. Namely, the elliptic spatial deformation of the freeze-out hypersurface depends on the spatial anisotropy parameter $\varepsilon(b)$, where b is the impact parameter, as [25]

$$R_{ell}(b,\varphi) = R_{fo}(b)\sqrt{\frac{1-\varepsilon^2(b)}{1+\varepsilon(b)\cos 2\varphi}} \qquad (3)$$

Here parameter $R_{fo}(b) = R_0\sqrt{1-\varepsilon(b)}$ determines the scale of the transverse size of the fireball at freeze-out and φ is the azimuthal angle between the transverse velocity of the fluid element and the impact parameter axis. In contrast to transverse isotropic parametrisation, the angle φ in HYDJET++ does not coincide with the azimuthal angle ϕ but is linked to it via the nonzero flow anisotropy parameter $\delta(b)$

$$\tan\varphi = \sqrt{\frac{1-\delta(b)}{1+\delta(b)}}\tan\phi \qquad (4)$$

Both parameters $\varepsilon(b)$ and $\delta(b)$ are proportional to the initial spatial anisotropy $\varepsilon_0 = b/(2R_A)$.

To extend the model to triangular flow, triangular deformations should be included [14]

$$R_{triang}(b,\varphi) = R_{ell}(b,\varphi)\{1+\varepsilon_3(b)\cos[3(\varphi-\Psi_3)]\}\,. \qquad (5)$$

Since no correlations between the reaction planes of elliptic and triangular flows are observed, i.e. $v_2(\Psi_3) = v_3(\Psi_2) \equiv 0$, the azimuth Ψ_3 is randomly generated, while Ψ_2 is always set to zero in model calculations. A new free parameter $\varepsilon_3(b)$ entering Eq. (5) controls the degree of triangularity of the system. It is important to stress that, although the three parameters $\varepsilon(b)$, $\delta(b)$ and $\varepsilon_3(b)$ were up to recent times fixed in the model, e.g., $\varepsilon_3(b_1) = \varepsilon_3(b_2)$ if $b_1 = b_2$, the flow harmonics $v_n(p_T)$ must be obtained only after averaging over a sample of generated events due to event-by-event fluctuations. The flow fluctuations in HYDJET++ arise because of (i) fluctuations in particle multiplicity, (ii) fluctuations in particle coordinates and directly linked to it (iii) fluctuations in particle momenta (via the correlations with radial flow), (iv) decays of resonances, and (v) production of (mini)jets. These fluctuations are investigated in Sec. 3.

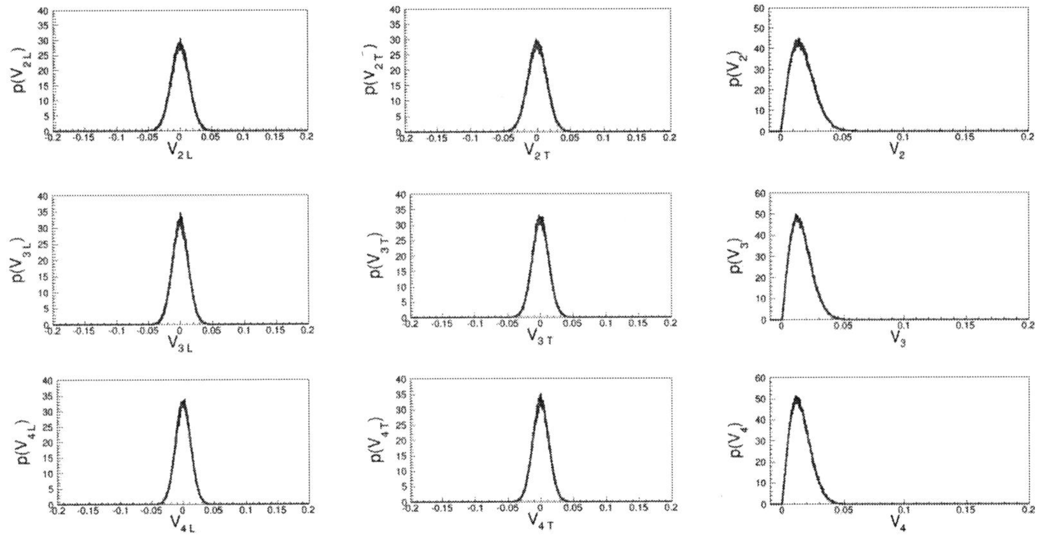

Figure 1. The probability density distributions for V_{nL} (left column), V_{nT} (middle column), and for the modulus of the flow vector $V_n = |\vec{V}_n|$, $n = 2, 3, 4$ (right column) for central Pb+Pb collisions in HYDJET++ with $b = 0$. Dashed red curves in the plots indicate the fit of simulated HYDJET++ points to Eq. (6) (left), Eq. (7) (middle) and Eq. (8) (right), respectively.

3 Event-by-event flow in HYDJET++

Let us determine the flow vector $\vec{V}_n$ and its modulus $V_n = \vec{V}_n$, $n = 2, 3, 4$ in a single event. - From here we use capital letters to distinguish the components of EbyE-determined flow vector $\vec{V}_n$ from the flow harmonics v_n, obtained after averaging of $\cos[n(\varphi - \Psi_n)]$ over all particles in all events. - Since we know the direction of the "true" vector $\vec{V}_n$ for a given event with the impact parameter b in model calculations, it is possible to decompose the obtained flow vector into longitudinal part $\vec{V}_{nL}$, which is parallel to the direction of the "true" vector, and the transverse part $\vec{V}_{nT}$ perpendicular to it.

Firstly, one has to study the intrinsic fluctuations in HYDJET++ arising in a system when the anisotropic flow is absent. This situation corresponds to the case of perfect central collisions with zero impact parameter. The all three parameters are zero, $\varepsilon(0) = \delta(0) = \varepsilon_3(0) = 0$, and the probability distributions for both EbyE flow components, $\vec{V}_{nL}$ and $\vec{V}_{nT}$, and the modulus V_n should be represented by Gaussian functions [27, 28]

$$P(\vec{V}_{nL}) = \frac{1}{2\pi\sigma_n^2(0)} \exp\left[-\frac{\vec{V}_{nL}^2}{2\sigma_n^2(0)}\right], \qquad (6)$$

$$P(\vec{V}_{nT}) = \frac{1}{2\pi\sigma_n^2(0)} \exp\left[-\frac{\vec{V}_{nT}^2}{2\sigma_n^2(0)}\right], \qquad (7)$$

$$P(V_n) = \frac{V_n}{\sigma_n^2(0)} \exp\left[-\frac{V_n^2}{2\sigma_n^2(0)}\right]. \qquad (8)$$

DOI: 10.1051/epjconf/201612604006

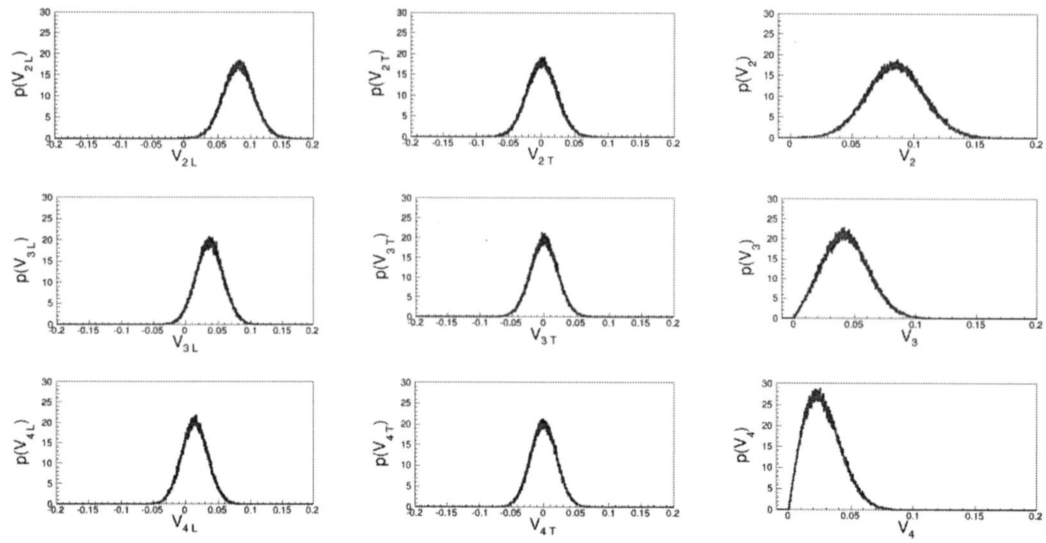

Figure 2. The same as Fig. 1 but for 20-25% centrality collisions. Dashed red curves in the plots indicate the fit of simulated HYDJET++ points to Eq. (9) (left), Eq. (10) (middle) and Eq. (11) (right), respectively.

The results of model calculations fitted to Eqs.(6)-(8) are displayed in Fig. 1. One may see that the simulated distributions and the fit are on the top of each other. Both longitudinal and transverse distributions of the flow vector are symmetric with respect to zero, meaning that the event-integrated flow is essentially zero. Note, that for any n-th component of the flow vector all Gaussians depend on a single parameter σ_n. For instance, the modulus mean is $\langle V_n \rangle = \sqrt{\frac{\pi}{2}}\,\sigma_n$, whereas the width of the $P(V_n)$ distribution reads $\sigma_{V_n} = \sqrt{2 - \frac{\pi}{2}}\,\sigma_n$. As seen in Fig. 1, the parameter σ_n is almost the same for all three components, $\sigma_2 \approx \sigma_3 \approx \sigma_4 = 0.013$. Although σ_n depends on a number of model parameters, it is especially sensitive to the mean multiplicity, which controls the centrality variation of σ_n.

The case of collisions with non-zero impact parameters is a bit more complex. For these collisions the probability distributions given by Eqs.(6)-(8) should be modified accordingly [27, 28]

$$P(V_{nL}) = \frac{1}{\sqrt{2\pi\sigma_n^2(b)}} \exp\left[-\frac{(V_{nL} - \langle V_{nL} \rangle)^2}{2\sigma_n^2(b)}\right], \tag{9}$$

$$P(V_{nT}) = \frac{1}{\sqrt{2\pi\sigma_n^2(b)}} \exp\left[-\frac{(V_{nT})^2}{2\sigma_n^2(b)}\right], \tag{10}$$

$$P(V_n) = \frac{V_n}{\sigma_n^2(b)} \exp\left[-\frac{V_n^2 + \langle V_{nL} \rangle^2}{2\sigma_n^2(b)}\right] I_0\left(\frac{V_n \langle V_{nL} \rangle}{\sigma_n^2(b)}\right), \tag{11}$$

where I_0 is the modified Bessel function of the first kind. The probability densities of the components and modulus of the flow vector, calculated in lead-lead collisions with centrality 20-25%, are shown in Fig. 2. Results of the fit to Eqs.(9)-(11) are plotted onto the generated distributions as well. Several features should be mentioned here. The width σ_{V_n} and the average modulus $\langle V_n \rangle$ now depend on both

DOI: 10.1051/epjconf/201612604006

$\sigma_n(b)$ and $\langle V_{nL}\rangle$. Also note that $\langle V_n\rangle \neq \langle V_{nL}\rangle$ although $\langle V_{nL}\rangle = v_n(b)$, $n = 2,3$, because the values of free parameters $\varepsilon(b)$, $\varepsilon_3(b)$, $\delta(b)$, responsible for emergence of spatial and momentum anisotropy in the model, are tuned to provide the correct description of elliptic and triangular flow at given impact parameter b. Then, the EbyE flow distributions at non-zero impact parameter, shown in Fig. 2, are broader compared to that at $b = 0$, presented in Fig. 1, but the widths of the longitudinal and transverse parts are still almost equal, $\sigma_2 \simeq \sigma_3 \simeq \sigma_4 = 0.02$. The values of V_{nL} at which $P(V_{nL})$ reach their maxima become smaller with rising harmonic number n thus reflecting the fact that $v_4 < v_3 < v_2$ at this centrality.

We demonstrated that event-by-event flow fluctuations arise in the model even when the parameters, responsible for the creation of spatial anisotropy, are fixed. The next step is the comparison of intrinsic EbyE fluctuations in HYDJET++ with the experimentally measured ones. This non-trivial problem is considered in Sec. 4.

4 Unfolding procedure and comparison with data

The EbyE distributions of harmonics of anisotropic flow in lead-lead collisions at LHC were studied, e.g., by ATLAS Collaboration in [29]. However, the results were obtained after the application of the so-called *unfolding* procedure in order to extract the "true" value of the flow vector and get rid of the non-flow effects caused by the finite event multiplicities, jet fragmentation and decays of resonances. The procedure is cumbersome, so often people simply rescale their predictions [30, 31] to make a comparison with the data. Analysis below shows that such a simplistic approach is not always justified. Let us consider implementation of the unfolding procedure in details.

As an input, one selects the spectra of charged particles with $p_T \geq 0.5$ GeV/c and $|\eta| < 2.5$, corresponding to ATLAS transverse momentum and pseudorapidity cuts. Angular distribution of particles is modified as

$$\frac{dN}{d\varphi} \propto 1 + 2\sum_{n=1}^{\infty} V_n^{\rm obs}\cos\left[n(\varphi - \Psi_n^{\rm obs})\right] = 1 + 2\sum_{n=1}^{\infty}\left(V_{n,x}^{\rm obs}\cos n\varphi + V_{n,y}^{\rm obs}\sin n\varphi\right) , \qquad (12)$$

with $V_n^{\rm obs}$ being the magnitude of the observed per-particle flow vector, whereas $\Psi_n^{\rm obs}$ represents the azimuth of the event plane.

Then, the single-particle event-by-event distributions are constructed

$$\begin{aligned} V_n^{\rm obs} &= \sqrt{(V_{n,x}^{\rm obs})^2 + (V_{n,y}^{\rm obs})^2} , \\ V_{n,x}^{\rm obs} &= V_n^{\rm obs}\cos n\Psi_n^{\rm obs} = \langle\cos n\varphi\rangle , \\ V_{n,y}^{\rm obs} &= V_n^{\rm obs}\sin n\Psi_n^{\rm obs} = \langle\sin n\varphi\rangle . \end{aligned} \qquad (13)$$

Note, that the final "unfolded" distribution is insensitive to the method used for the extraction of the magnitude of the observed per-particle vector $V_n^{\rm obs}$ [29]. The two sub-events (2SE) method subdivides the event sample further into two sub-groups containing charged particles emitted in (i) forward and (ii) backward hemisphere in the c.m. system. The difference between the EbyE flow vectors of the two sub-events is fitted to the Gaussian with the width $\delta_{\rm 2SE} = 2\delta$, which enters the response function [29]

$$P(V_n^{\rm obs}|V_n) \propto V_n^{\rm obs}\exp\left[-\frac{(V_n^{\rm obs})^2 + V_n^2}{2\delta^2}\right] I_0\left(\frac{V_n^{\rm obs}V_n}{\delta^2}\right) . \qquad (14)$$

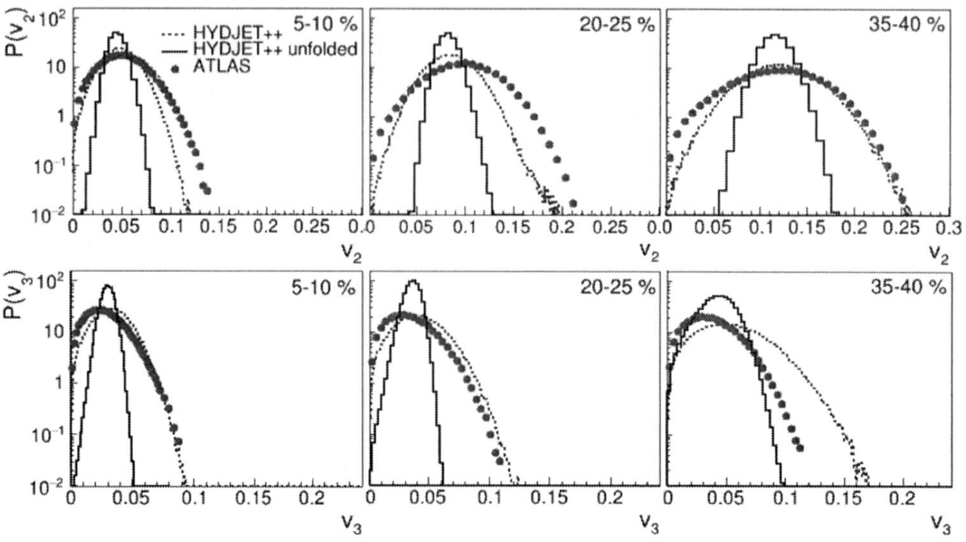

Figure 3. The probability density distributions of elliptic flow V_2 (upper row) and triangular flow V_3 (bottom row) in three centrality intervals: $5-10\%$ (left), $20-25\%$ (middle) and $35-40\%$ (right). Dashed and solid histograms present the results for simulated HYDJET++ events before and after the unfolding procedure, respectively. The full circles are the ATLAS data from [29].

The obtained response function is used then as an input to find the Bayesian unfolding matrix via the iteration procedure [32]. The contributions arising mostly due to non-flow processes and effects of finite particle multiplicity can be subtracted by the EbyE unfolding procedure in order to get the dynamical flow fluctuations only. How does it work?

Figure 3 displays the probability density distributions of elliptic and triangular EbyE flows obtained in three centrality intervals: $\sigma/\sigma_{geo} = 5 - 10\%$, $20 - 25\%$, and $35 - 40\%$ in the version of HYDJET++ with unsmeared eccentricity parameters. At first glance, the model results shown by dashed curves are not far from the ATLAS data. However, the experimental results are obtained by means of the unfolding procedure. It is unfolding that makes the HYDJET++ distributions much narrow, see solid histograms in Fig. 3, thus meaning that the dynamical fluctuations in the model are not strong enough to match the data.

The situation is changed dramatically if we allow for variations of $\varepsilon(b)$ and $\varepsilon_3(b)$. Now the values of both parameters are not fixed anymore but rather smeared normally around the unsmeared values with the widths proportional to these values. The proportionality coefficients are tuned to fit the data at a single arbitrary centrality, say $10 - 15\%$ or $20 - 25\%$, and the found values are used then for all other centralities.

Results obtained in HYDJET++ with the smeared anisotropy parameters are shown in Fig. 4. Now the initial $P(V_n)$ distributions are broader than the unfolded ones only in the areas of relatively high flow values. The agreement of unfolded spectra with the data is very good. Since unfolding suppresses strongly the non-flow fluctuations, both Fig. 3 and Fig. 4 indicate the dynamical origin of the flow fluctuations in HYDJET++.

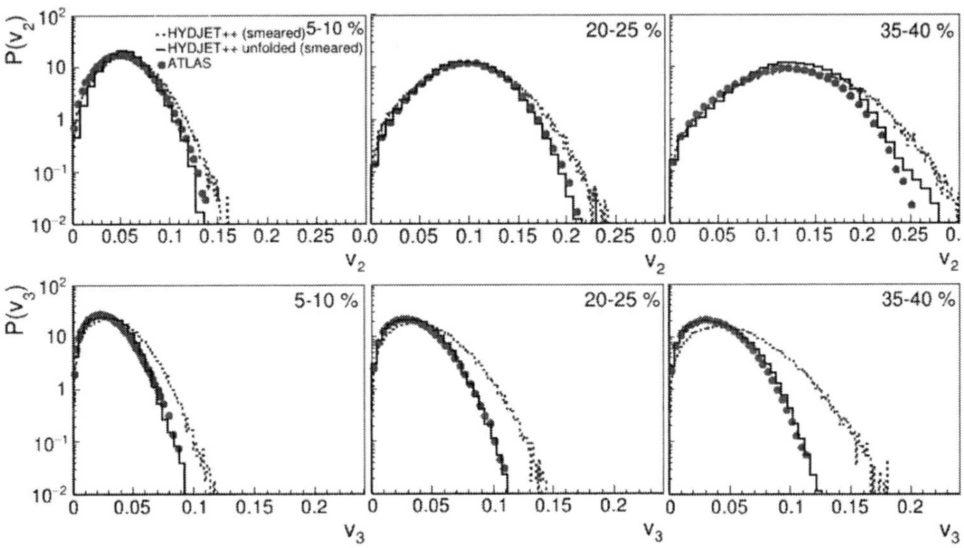

Figure 4. The same as Fig. 3 but for the model calculations with additional smearing of the anisotropy parameters. The full circles are the ATLAS data from [29].

5 Conclusions

We analyse the event-by-event fluctuations of elliptic and triangular flows in lead-lead collisions at $\sqrt{s_{NN}}$ = 2.76 TeV within the HYDJET++ model. To compare the model results with the data the unfolding procedure was employed. Unfolding eliminates to a great extent the non-flow contributions caused by, e.g. decays of resonances, finite event multiplicities and jet fragmentation. These effects cannot be taken away by the rescaling of the distributions. Therefore, for correct comparison of "apples" to "apples" one has to follow the procedure implemented for reconstruction of the experimental data.

The dynamical origin of the flow fluctuations in the model is traced to the correlations between the momenta and coordinates of final particles and the velocities of hadronic fluid elements. Gaussian smearing of the model parameters $\varepsilon(b)$ and $\varepsilon_3(b)$, responsible for spatial asymmetry of the freeze-out hypersurface, has proven to be absolutely necessary for good quantitative description of elliptic and triangular flow fluctuations. These fluctuations can be further linked to the initial-state fluctuations, provided the response of the system is linear.

Acknowledgments

Fruitful discussions with K. Bugaev, G. Eyyubova and K. Redlich are gratefully acknowledged. This work was supported in parts by the Department of Physics, UiO, and by Russian Scientific Fund under grant No.14-12-00110 in a part of computer simulation of anisotropic flow fluctuations in Pb+Pb collisions and the event-by-event unfolding analysis.

References

[1] L. P. Csernai and H. Stöcker, J. Phys. G **42**, 124001 (2014).
[2] H. G. Ritter and R. Stock, J. Phys. G **42**, 124002 (2014).
[3] J. Jia, J. Phys. G **42**, 124003 (2014).
[4] G. S. Denicol, J. Phys. G **42**, 124004 (2014).
[5] H. Petersen, J. Phys. G **42**, 124005 (2014).
[6] Z. Tang, Nu Xu, K. Zhou, and P. Zhuang, J. Phys. G **42**, 124006 (2014).
[7] R. Snellings, J. Phys. G **42**, 124007 (2014).
[8] S. A. Voloshin and Y. Zhang, Z. Phys. C **70**, 665 (1996).
[9] A. M. Poskanzer and S. A. Voloshin, Phys. Rev. C **58**, 1671 (1998).
[10] S. A. Voloshin, A. M. Poskanzer, and R. Snellings, in *Relativistic Heavy Ion Physics*, Landolt-Börnstein Database Vol. 23, edited by R. Stock (Springer, Berlin, 2010), p.5–54.
[11] E. E. Zabrodin, C. Fuchs, L. V. Bravina, and Amand Faessler, Phys. Lett. B **508**, 184 (2001).
[12] I. P. Lokhtin *et al.*, Comput. Phys. Commun. **180**, 779 (2009).
[13] I. P. Lokhtin *et al.*, Eur. Phys. J. C **72**, 2045 (2012).
[14] L. V. Bravina *et al.*, Eur. Phys. J. C **74**, 2807 (2014).
[15] G. Eyyubova *et al.*, Phys. Rev. C **80**, 064907 (2009).
[16] E. Zabrodin *et al.*, J. Phys. G. **37**, 094060 (2010).
[17] E. Zabrodin, G. Eyyubova, L. Malinina, and L. Bravina, Acta Phys. Polon. B **5**, 349 (2012).
[18] L. Bravina, B. H. Brusheim Johansson, G. Eyyubova, and E. Zabrodin, Phys. Rev. C **87**, 034901 (2013).
[19] E. Zabrodin *et al.*, EPJ Web Confer. **71**, 00142 (2014).
[20] E. Zabrodin *et al.*, EPJ Web Confer. **95**, 03039 (2015).
[21] L. V. Bravina *et al.*, Phys. Rev. C **89**, 024909 (2014).
[22] G. Eyyubova *et al.*, Phys. Rev. C **91**, 064907 (2015).
[23] L. V. Bravina *et al.*, Eur. Phys. J. C **75**, 588 (2015) (arXiv:1509.02692 [hep-ph]).
[24] N. S. Amelin *et al.*, Phys. Rev. C **74**, 064901 (2006).
[25] N. S. Amelin *et al.*, Phys. Rev. C **77**, 014903 (2008).
[26] I. P. Lokhtin and A. M. Snigirev, Eur. Phys. J. C **46**, 211 (2006).
[27] S. A. Voloshin, A. M. Poskanzer, A. Tang, and G. Wang, Phys. Lett. B **659**, 537 (2008).
[28] R. S. Bhalerao and J.-Y. Ollitrault, Phys. Lett. B **641**, 260 (2006).
[29] G. Aad *et al.* (ATLAS Collaboration), JHEP **11**, 183 (2013).
[30] H. Niemi, K. J. Eskola, and R. Paatelainen, arXiv:1505.02677 [hep-ph].
[31] M. Rybczynski and W. Broniowski, arXiv:1510.08242 [nucl-th].
[32] T. Adye, arXiv:1105.1160 [physics.data-an]

Investigation of particle coherence in Pb+Pb collisions at LHC

Bengt Henrik Brusheim Johansson[1,a]

[1] *Department of Physics, P.b. 1048 Blindern, 0316 Oslo, Norway*

Abstract. The observed particle distributions in Pb+Pb collisions at LHC are investigated. Monte Carlo simulations of the azimuthal anisotropy i.e. flow, and transverse momentum spectra provide insights in e.g. hadron genesis and in medio scattering. Simulations are made using the HYDJET++ model, which is based on parameterization of soft processes and generation of hard physics. The second and third event planes are implemented at this stage, thus generating the second and third order flows which are believed to provide the main part of the observed azimuthal distribution. Here, the ellipticity driven elliptic, and fluctuation driven triangular flow are used as differentiated observables. The elliptic and triangular flow are simulated in agreement with experimental data for $p_T < 3.5$ GeV and centralities $0 - 50\%$. In connection to the simulation of azimuthal anisotropy, the particle transverse momentum spectra are simulated with accuracy, which provides a high degree of reproduction of geometric features. In extension to these simulations, the dependences on fragmentation processes are investigated through the second and third order flows, which are displaying sensitivity to particle coherence. In the simulations, particle coherence is thus seen to manifest itself through the azimuthal anisotropy.

1 Introduction

In processes involving factorization of abstracted observables; *in praegressus* mechanics may be studied by extensive modelling of the evolutionary processes; or of selected parts of it. In the high energy laboratories, the heavy ion collisions are believed to produce exotic phases which consist of partially factorized matter. Experiments at the Relativistic Heavy Ion Collider (RHIC) have rendered evidence of a hot and dense matter produced in the early stages. This phase is hypothesized as a perfect liquid [1–4]; and is the common view of today. The postulated state then undergoes phase transitions when evolving, hadronizing; and later interacting with the particle detectors, thus producing the observed distributions [5, 6]. The (strongly) interacting matter thus displays collectivity in form of e.g. azimuthal anisotropy [7–10]. The fluid postulate suggests a (partially) hydrodynamic description. This description provides a "flow" due to pressure gradients (eccentricity) in the produced fireball (1)

[a] e-mail: bhjohan@uio.no

$$\epsilon_n = -e^{-i\phi_n}\frac{\langle r^n e^{i\phi}\rangle}{\langle r^n\rangle}. \tag{1}$$

Here, the initial state event plane is denoted by ϕ_n and the transverse radius is denoted "r". The eccentricities are believed to evolve from the initial state while undergoing significant *in medio* modulations; thus rendering in an anisotropic momentum distribution; which is described in terms of flow terms (2) in the general momentum distribution (3). Thus; the azimuthal anisotropy provides a main observable in the investigations of *in medio* processes such as jet production and modulation. The mentioned flow-terms corresponds by order n to the terms in (1); and are thus projected as

$$v_n = \langle \cos n(\phi - \Psi_n)\rangle. \tag{2}$$

Here, ϕ and Ψ_n are the observed particle azimuth and the nth event plane angle. The flow terms are projected out of the Fourier expansion [11, 12] of the observed momentum particle distribution as

$$E\frac{d^3N}{d^3p} = \frac{1}{\pi}\frac{d^2N}{dp_T^2 dy}\left[1 + 2\sum_{n=1}^{\infty} v_n \cos n\,(\phi - \Psi_n)\right]. \tag{3}$$

Here; the total momenta is p and the transverse momenta and rapidity are denoted by p_T and y. The energy is denoted by E. The flow displays characteristics depending on the order n; i.e. the second order flow depends on the main geometric mode of the initial conditions; i.e. the ellipticity, $n = 2$ in (1). The third order mode; triangularity, is necessarily dependent on fluctuations in the initial energy distribution in order to produce the non-zero flow; due to the odd geometry in (1). Higher order flow, i.e. higher order terms in (3); here represented by the flows of order 4-6, are also expected to display a differentiated behaviour, i.e. sensitivity to fluctuations, thus being of interest. The higher flow-harmonics may be viewed upon as complex objects; thus possibly factorizable. Factorizations of the higher order flow holds information which can be extracted through different projections of the simulations. In this paper; the higher order harmonics [13, 14] are to be examined in the second, third; and fifth order planes [15]; thus providing a base for the simulations of higher order flows. These higher order flows are then related to factorizations of the lower order flows in order to investigate processes through their coherence characteristics. Also; the particle spectra has to be included in the investigations. Thus, the elliptic and the triangular flow are to be implemented in the model in order to simulate a fireball with the main geometric modes; ellipticity and fluctuations in terms of triangularity. The yield is then simulated on the freeze-out hyper surfaces in order to simulate the azimuth and spectra of the heavy ion collisions.

2 Method

In the model used for present paper (HYDJET++); the hydro dynamical processes are parameterized on freeze-out hyper surfaces of the expanding fireball [16, 17]; and forms in superposition with hard particles the simulated distributions. In present model [18, 19], the expanding fireball is modulated by *in medio* collisions [20–22] and radiation [23, 24]. Also, shadowing effects [25] are accounted for in the model. In order to simulate the azimuthal anisotropy; the fireball transverse radius "R"

$$R\,(b,\phi) = R_f\,(b)\,\frac{\sqrt{1-\epsilon^2(b)}}{\sqrt{1+\epsilon(b)\cos 2\phi}}, \tag{4}$$

is modulated in terms of azimuth ϕ and the impact-parameter dependent model-eccentricity $\epsilon(b)$; with R_f as the fireball RMS radius. Also, the linear transverse rapidity profile; and thus the particle velocity are modulated in terms of a second model parameter $\delta\,(b)$. The modulation of the transverse velocity u^T is

$$u^x = \sqrt{1+\delta\,(b)}\sinh\tilde{\rho_u}\cos\phi, \quad u^y = \sqrt{1-\delta\,(b)}\sinh\tilde{\rho_u}\sin\phi, \tag{5}$$

where the transverse rapidity is further modulated by event plane angles, e.g. $\tilde{\rho_u} = \tilde{\rho_u}(\Psi_3)$. In the model, the generation of the third order flow, v_3, is carried out by introducing a model triangularity, ϵ_3. The third event plane is generated uniformly in the model; only weakly correlated with the reaction plane. At this moment; the model does not propagate the initial conditions through an evolution iteration. Thus; the event-by-event fluctuations are quite artificial in the simulations.

3 Results

The momentum azimuthal anisotropy is investigated by projecting the flow terms in (3). These terms are further projected in terms of particle production mode as an attempt to separate different types of coherency. The transverse momentum spectra are simulated for centralities and displayed for reference to abundances and particle production modes. The

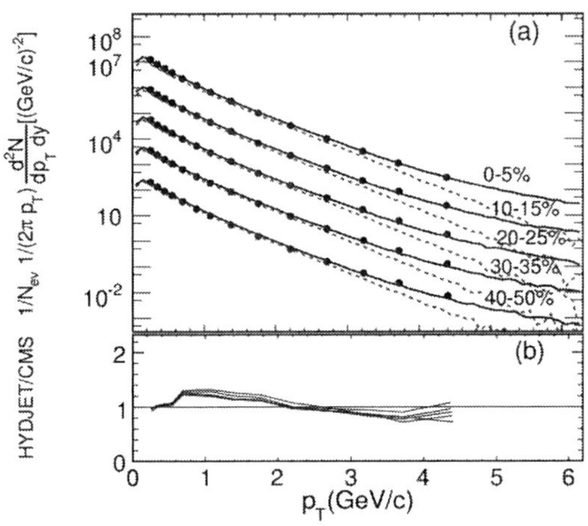

Figure 1. Transverse momentum spectra for unidentified particles (a) (——) and hydro dynamical particles (- - - -). Particle spectra are simulated for: $\sigma/\sigma_{geo} = 0-5\%$; $\sigma/\sigma_{geo} = 10-15\%$; $\sigma/\sigma_{geo} = 20-25\%$; $\sigma/\sigma_{geo} = 30-35\%$; and $\sigma/\sigma_{geo} = 40-50\%$. Experimental data [6] are denoted by ●. Comparison between experimental data and simulations in the lower pane (b).

particle spectra displays differentiation in transverse momentum in Fig.1a: the lower transverse momentum regime is to a high degree produced in hydrodynamic processes. Due to this mode, the spectra displays a thermal-exponential nature for the $p_T \leq 2$ GeV/c regime. For the higher transverse momentum regime; the defining particle production mode is fragmentation, which produces a power-law spectrum, differentiated from the thermal part of the spectra. The spectra displays a high degree of accuracy when compared to CMS data [6], as seen in the bottom pane of Fig.1b.

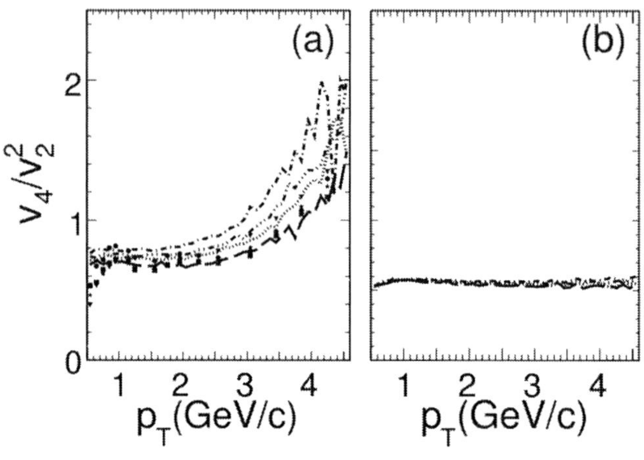

Figure 2. Ratio v_4/v_2^2 for unidentified- (a); and hydro dynamical particles (b). Simulations are made for: $\sigma/\sigma_{geo} = 10-20\%$ (— — —); $\sigma/\sigma_{geo} = 20-30\%$ (· · · · · ·); $\sigma/\sigma_{geo} = 30-40\%$ (— · · —); and $\sigma/\sigma_{geo} = 40-50\%$ (— · —). ALICE [26] data are denoted by: $\sigma/\sigma_{geo} = 10-20\%$ (●); $\sigma/\sigma_{geo} = 20-30\%$ (■); $\sigma/\sigma_{geo} = 30-40\%$ (★); and $\sigma/\sigma_{geo} = 40-50\%$ (▼)

The azimuthal anisotropy is in present paper displayed as higher order flows compared to the corresponding factorization in terms of lower order flows. The v_4 flow is simulated in the model Ψ_2 plane; and is compared to the factorization v_2v_2. The ratio is also compared to experimental ALICE data [26] (Fig.2a); due to the available ALICE $v_4(\Psi_2)$ projection. The agreement with experimental data is fair; although deviations are seen for centralities. The experimental data seem to scale for centralities to a greater extent than the simulations with an ATLAS setting might indicate. This feature is likely due to the absence of a model Ψ_4 plane. The Ψ_4 plane might then provide a higher degree of scaling for the simulated ratio (Fig.2b); due to higer correlation to jets. The fourth order Ψ_2 flow displays scaling; largely depending on the common event plane. The ratio is seen to be weakly dependent on centrality due to elevated fluctuation levels. It is also seen in the simulations that the ratio acts as an observer of jet-production (Fig.2) [27]. Further; a fifth order model event plane is established in order to investigate the pentagonal (v_5) flow compared to its v_2v_3 factorization, and also the effect of the event plane mixing. The inclusive v_5/v_3v_2 ratio displays in Fig.3a; a fair level of agreement with experimental data [5]. The greatest prediction power of the model seems to be located at the mid- and low ($p_T \leq 3$ GeV/c) regime. The experimental ratio displays high levels of fluctuations; thus difficult to analyse at this stage. Some deviations are also seen for centralities; in particular for the central $10-20\%$ hits. The hydro dynamical projection in Fig.3b of the ratio displays a weak or non-existent scaling for transverse momentum and centralities due to coherency from the elliptic plane as seen from the ratio hierarchy. The planes are thus seen to correlate weakly. Jets are seen to elevate the ratio; even for the lower transverse momentum regime due to the fluctuation dependence of the ratio.

In connection to the simulation of the v_5/v_3v_2 plane; the effects of event plane mixing is studied through the $v_5(\Psi_2, \Psi_3)/v_3v_2$ ratio. For the fifth order, mixed event plane; flow is also simulated using combinations of the second and third event plane as of Fig.4a. The simulations display a reverse centrality hierarchy compared to the "plain" fifth order ratio. The elliptic plane provides coherency which counteracts the fluctuation dependent behaviour of the ratio. Here; the hydrodynamic projection, displayed in Fig.4b displays scaling both in transverse momentum and centrality in contrast to the case of non-event plane mixing. This is due to wave number coherence scaling: similar to the case of Fig.2b. It is now seen that the jets provides the ratio hierarchy. In order to move the investigation of correlators forward; the sixth order factorization v_3^2/v_2^3 is investigated for inclusive particles and also as a hydro dynamical projection. The flow displays weak scaling and dependence of transverse momentum and a strong centrality dependence due to the fluctuation/coherence nature of the ratio; similar to the case of Fig.5. The agreement with experimental data is high (Fig.5a). In particular for the more peripheral "tuned" hits. The hydro dynamical projection; displayed in Fig.5b displays a strong ellipticity dependence due to the differentiation in coherency. Jet production is thus seen to provide for an approximate transverse momentum wave-number scaling.

4 Discussion and Conclusion

The simulations of the higher order flows display a high degree of reproduction. The second, third; and fifth order flows are fully implemented in the model thus exhibiting a high degree

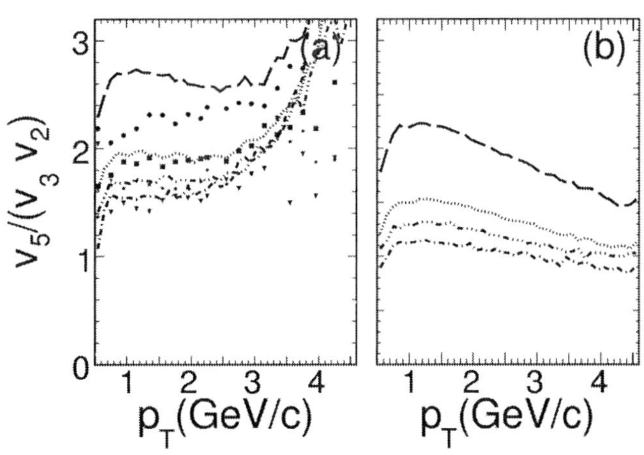

Figure 3. Ratio $\frac{v_5}{v_2 v_3}$ for unidentified- (a); and hydro dynamical particles (b). Simulations are made for: $\sigma/\sigma_{geo} = 10 - 20\%$ (— — —); $\sigma/\sigma_{geo} = 20 - 30\%$ (······); $\sigma/\sigma_{geo} = 30 - 40\%$ (— · · —); and $\sigma/\sigma_{geo} = 40 - 50\%$ (— · —). Experimental data are included for comparison [5]. The data are denoted by: $\sigma/\sigma_{geo} = 10 - 20\%$ (●); $\sigma/\sigma_{geo} = 20 - 30\%$ (■); $\sigma/\sigma_{geo} = 30 - 40\%$ (★); and $\sigma/\sigma_{geo} = 40 - 50\%$ (▼)

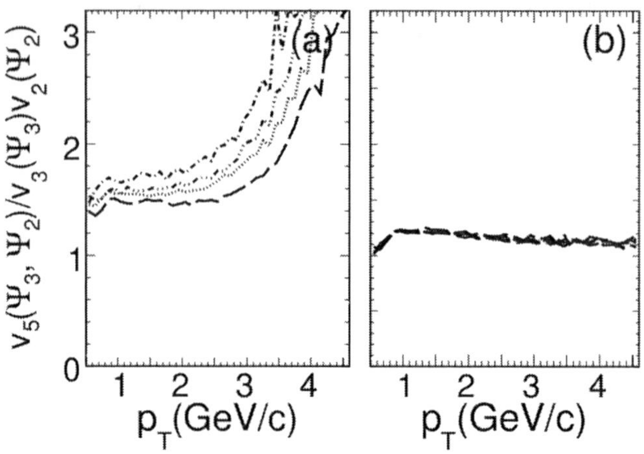

Figure 4. Ratio $\frac{v_5(\Psi_2,\Psi_3)}{v_2 v_3}$ for unidentified- (a); and hydro dynamical particles (b). Simulations are made for: $\sigma/\sigma_{geo} = 10 - 20\%$ (— — —); $\sigma/\sigma_{geo} = 20 - 30\%$ (· · · · · ·); $\sigma/\sigma_{geo} = 30 - 40\%$ (— · · —); and $\sigma/\sigma_{geo} = 40 - 50\%$ (— · —). Data are denoted by: $\sigma/\sigma_{geo} = 10 - 20\%$ (●); $\sigma/\sigma_{geo} = 20 - 30\%$ (■); $\sigma/\sigma_{geo} = 30 - 40\%$ (★); and $\sigma/\sigma_{geo} = 40 - 50\%$ (▼)

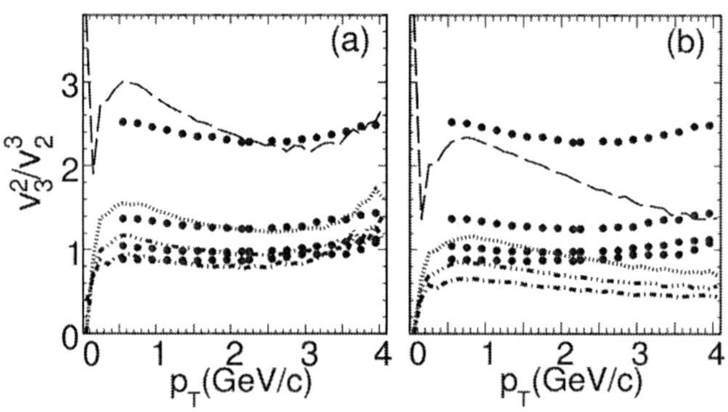

Figure 5. Ratio v_3^2/v_2^3 for unidentified- (a); and hydro dynamical particles (b). Experimental data for inclusive particles are included for comparison [5] and are denoted by ● . Simulations are made for: $\sigma/\sigma_{geo} = 10 - 20\%$ (— — —); $\sigma/\sigma_{geo} = 20 - 30\%$ (· · · · · ·); $\sigma/\sigma_{geo} = 30 - 40\%$ (— · · —); and $\sigma/\sigma_{geo} = 40 - 50\%$ (— · —).

of accuracy. The simulations of these particular flows are thus believed to represent a high degree of robustness and reliability. The flow ratios are displaying two fundamental types of behaviour. First; the mixed event plane type; which displays hydro dynamical factorization. The hydro dynamical flow is here seen as factorized due to the common event plane and the absence of jet fluctuations, thus nullifying the differentiation in coherency. This common factorization seems to represent the geometric limit. Fluctuations are thus seen to form the mixed event plane centrality hierarchy alone in the model.

Second; the non-mixed event plane ratios displays a dependence on transverse momentum and also centrality due to the skewness in wavenumbers and geometric mode (coherence). The centrality hierarchy is here seen as strongly coherence dependent in contrast to the case of event plane mixing. *In conclusion*; the pure and event plane mixed flow displays differences in the level of factorization in the model. Thus, fluctuations are the only provider of flow hierarchy for the mixed event plane flow. Pure flow depends on initial geometry; thus displaying the relations between the event planes.

Acknowledgements

The author wish to thank his supervisors L. Bravina, E. E. Zabrodin; and the associated research group. The author also wish to thank the ICNFP2015 organizing committee; and the people of Crete for an amazing conference and an awe-inspiring visit.

References

[1] I. Arsene et al., Nucl. Phys. A **757**, 1 (2005)
[2] B.B. Back et al., Nucl. Phys. A **757**, 28 (2005)
[3] J. Adams et al., Nucl. Phys. A **757**, 102 (2005)
[4] K. Adcox et al., Nucl. Phys. A **757**, 184 (2005)
[5] G. Aad et al., Phys. Rev. C **86**, 014907 (2012)
[6] S. Chatrchyan et al., Phys. Rev. C **87**, 014902 (2013)
[7] U. Heinz, R. Snellings, Annu. Rev. Nucl. Part. Sci. **63**, 123 (2013)
[8] S.A. Voloshin, Nucl. Phys. A **904-905**, 90c (2013)
[9] H.G. Ritter, R. Stock, J. Phys. G: Nucl. Part. Phys. **41**, 124002 (2014)
[10] R. Snellings, J. Phys. G: Nucl. Part. Phys. **41**, 124007 (2014)
[11] S. Voloshin, Y. Zhang, Z. Phys. C **70**, 665 (1996)
[12] A.M. Poskanzer, S.A. Voloshin, Phys. Rev. C **58**, 1671 (1998)
[13] J. Velkovska, J. Phys. G: Nucl. Part. Phys. **38**, 124011 (2011)
[14] J. Jia, J. Phys. G: Nucl. Part. Phys. **38**, 124012 (2011)
[15] L. Bravina, B.H. Brusheim Johansson, E.E. Zabrodin, G. Eyyubova, V.L. Korotkikh, I.P. Lokhtin, L.V. Malinina, S.V. Petrushanko, A.M. Snigirev, Phys. Rev. C **89**, 024909 (2014)
[16] N.S. Amelin, R. Lednicky, I.P. Lokhtin, L.V. Malinina, A.M. Snigirev, I.A. Karpenko, Y.M. Sinyukov, I. Arsene, L. Bravina, Phys. Rev. C **74**, 064901 (2006)
[17] N.S. Amelin, R. Lednicky, T.A. Pocheptsov, I.P. Lokhtin, L.V. Malinina, A.M. Snigirev, I.A. Karpenko, Y.M. Sinyukov, Phys. Rev. C **77**, 014903 (2008)
[18] I.P. Lokhtin, A.M. Snigirev, Eur. Phys. J. C **45**, 211 (2006)
[19] I.P. Lokhtin, L.V. Malinina, S.V. Petrushanko, A.M. Snigirev, I. Arsene, K. Tywoniuk, Comp. Phys. Comm. **180**, 779 (2009)
[20] J.D. Bjorken, FERMILAB-PUB-82/59-THY (1982)
[21] E. Braaten, M. Thoma, Phys. Rev. D **44**, 1298 (1991)
[22] I.P. Lokhtin, A.M. Snigirev, Eur. Phys. J. C **16**, 527 (2000)

[23] R. Baier, Y.L. Dokshitzer, A.H. Mueller, D. Shiff, Phys. Rev. C **60**, 064902 (1999)
[24] R. Baier, Y.L. Dokshitzer, A.H. Mueller, D. Shiff, Phys. Rev. C **64**, 057902 (2001)
[25] K. Tywoniuk, I. Arsene, L. Bravina, A.B. Kaidalov, E. Zabrodin, Phys. Lett. B **657**, 170 (2007)
[26] B. Abelev et al., Phys. Lett. B **719**, 18 (2013)
[27] L. Bravina, B.H. Brusheim Johansson, G. Eyyubova, E.E. Zabrodin, Phys. Rev. C **87**, 034901 (2013)

EPJ Web of Conferences **126**, 04008 (2016) DOI: 10.1051/epjconf/201612604008
ICNFP 2015

Double unification of particles with fields and electricity with gravity in non-empty space of continuous complex energies

Igor E. Bulyzhenkov[1,2,3,a]

[1] *Moscow Institute of Physics and Technology, 9 Institutskiy per., Dolgoprudny, Moscow reg., 141700, Russia*
[2] *National Science and Technology University MISiS, 4 Leninsky pros., Moscow, 119049, Russia*
[3] *Lebedev Physics Institute RAS, 53 Lininsky pros., Moscow, 119991, Russia*

Abstract. Non-empty space reading of Maxwell equations as local energy identities explains why a Coulomb field is carried rigidly by electrons in experiments. The analytical solution of the Poisson equation defines the sharp radial shape of charged elementary densities which are proportional to continuous densities of electric self-energy. Both Coulomb field and radial charge densities are free from energy divergences. Non-empty space of electrically charged mass-energy can be described by complex analytical densities resulting in real values for volume mass integrals and in imaginary values for volume charge integrals. Imaginary electric charges in the Newton gravitational law comply with real Coulomb forces. Unification of forces through complex charges rids them of radiation self-acceleration. Strong gravitational fields repeal probe bodies that might explain the accelerated expansion of the dense Metagalaxy. Outward and inward spherical waves form the standing wave process within the radial carrier of complex energy.

1 Introduction

Recent measurements of dynamical electric fields associated with moving charges "support the idea of a Coulomb field carried *rigidly* by the electron beam" [1]. This experimental result seems natural only in non-empty space physics, where charged radial fields represent distributions of continuous elementary matter in the very radial structure of its Coulomb/Newton fields. Non-empty space (inferred by Plato and Aristotle as a continuous material plenum beyond observations) can be better justified through extended electric charges in the laboratory, rather than through extended masses in observed weak gravitation. Since the days of the ancient Greeks, non-empty space for physical reality had been repeatedly claimed by many thinkers, including Einstein: "A coherent field theory requires that all elements be continuous... And from this requirement arises the fact that the material particle has no place as a basic concept in a field theory. Thus, even apart from the fact that it does not include gravitation, Maxwell's theory cannot be considered as a complete theory" (translation [2]). In this paper we try to comment quantitatively on this Einstein's vision of a coherent theory of pure classical fields which could indeed strike out the particle as an independent concept.

The postulated point-particle paradigm results not only in the mathematical divergence of the Coulomb energy, but also in the physical inconsistence of the microscopic electron theory. Any

[a] e-mail: bulyzhenkov.ie@mipt.ru

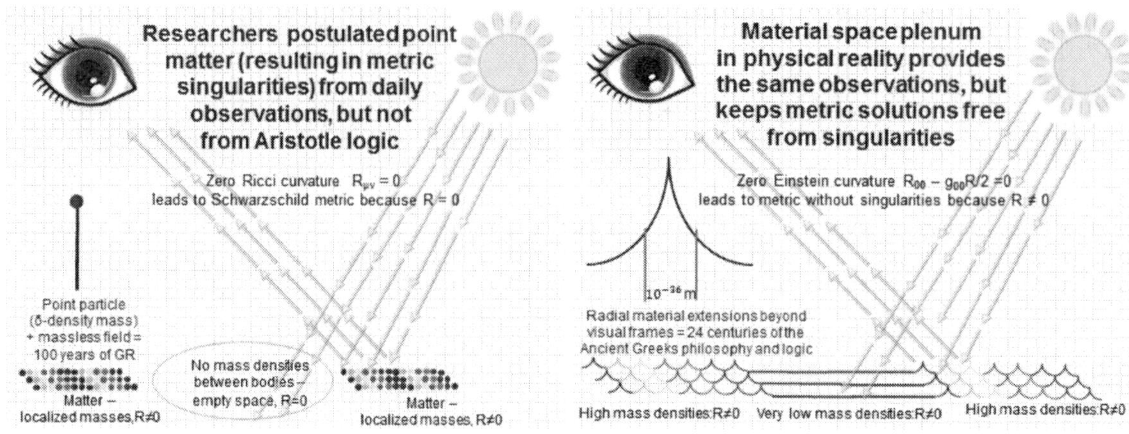

Figure 1. Empty space and point particles vs non-empty space physics of radial energy carries

point source in the microscopic Maxwell-Lorentz equations may be considered as "an attempt which we have called intellectually unsatisfying" according to De Broglie [3]. Einstein also criticized his 1915 field equation for the point gravitational source: "it resembles a building with one wing built of resplendent marble and the other built of cheap wood" (translation [2]). The Dirac delta-operator formalism for the point charge seems a provisional modeling of physical reality until local analytical charge-field relations can be finally proposed for Maxwell's electrodynamics and Einstein's gravitation. Recall that modern researchers postulate point matter (resulting in metric singularities) from daily observations, but not from analytical mathematics or logic. The latter had lead Aristotle (after Plato's aether and material forms) to the conclusion that space is to be a continuous material plenum without empty regions (material bodied cannot be placed to such empty regions without logical inconsistencies). Non-dual physical reality of non-empty space with extremely low and extremely high mass-energy densities provides the same observations in practice as the dual (massive particle + massless field) physical model, but keeps metric solutions free from singularities (see comments on the Figure 1). Indeed, the scalar Ricci curvature $R = g^{\mu\nu}R_{\mu\nu}$ is proportional to a scalar mass density in non-empty 3-space and, therefore, $R \neq 0$ in all space-time points. From here, static metric in non-dual reality of inhomogeneous mass-energy densities is to be defined by a zero Einstein curvature $R^o_o - R/2 = 0$, rather than by a zero Ricci curvature $R^o_o = 0$ for the conventional dual model in question.

As to non-empty space for the electromagnetic domain, a continuously distributed elementary charge was reasonably inferred by Mie in order to derive properties of charges from properties of fields (and potentials) and to overcome the energy divergence flaw in the Coulomb field center. Regretfully, in 1912-1913 and later the 'Theory der Matter' [4] had not found the appropriate (logarithmic) post-Coulomb potential for strong fields. And the promising non-empty space concept had not been justified before the mathematical era of quantum non-locality. The formal probabilities for the delta-operator 'dice' in the same empty space arena postponed the search of rigorous analytical solutions for Mie's nonlocal matter. In 1984 Schwinger intuitively proposed [5] to extend the point classical electron by the exponential radial distribution like in the Yukawa field potential. However, the annoying problem of the unphysical point source and the Coulomb energy divergence are still unresolved satisfactorily within the trustful mathematical formalism of classical fields. These permanent challenges of contemporary theoretical and mathematical physics may be considered as a motivation for our reinforcement of Mie and Schwinger nonlocal (astro)particle by the continuous radial density over the infinite world volume or the entire Universe.

2 Radial densities of imaginary electric charges

Let us revisit classical electromagnetic equations regarding coherent solutions for a distributed particle and its field. By following Mie, we assume that it is possible to relate analytically electric, $\vec{d}(x)$, and magnetic, $\vec{b}(x)$, field intensities in the Maxwell-Lorentz equations [6],

$$\begin{cases} div\ \vec{d}(\mathbf{x},t) = 4\pi\rho(\mathbf{x},t) \\ div\ \vec{b}(\mathbf{x},t) = 0 \\ c\ curl\ \vec{b}(\mathbf{x},t) = 4\pi\rho(\mathbf{x},t)\vec{v} + \partial_t\vec{d}(\mathbf{x},t) \\ c\ curl\ \vec{d}(\mathbf{x},t) = -\partial_t\vec{b}(\mathbf{x},t), \end{cases} \tag{1}$$

to the local charge, $\rho(\mathbf{x},t)$, and current, $\rho(\mathbf{x},t)\vec{v}$, densities of the extended electron. Contrary to Lorentz, the electron's charge distribution was not postulated by Mie within a microscopic spatial volume (electron's fields were assigned by Lorentz to charge-free regions or to the supposed empty space around charges). The mathematical equation $div\ \vec{d}(\mathbf{x},t) = 4\pi\rho(\mathbf{x},t)$ can be rigorously resolved under the non-empty space assumption by admitting that the elementary charge and its field coexist together in all space points $\mathbf{x}$ of the infinite Universe. In other words, we tend to maintain the idea of Mie (and Einstein) that the elementary electric (and gravitational) charge is to be integrated into its spatial field structure with instantaneous local relations between scalar functions $\rho(\mathbf{x},t)$ and $\vec{d}(\mathbf{x},t)\cdot\vec{d}(\mathbf{x},t)$. Then locally paired, *rigid* motion of Coulomb radial fields and electron's radial shapes can maintain the precision experiment [1] as a direct justification of the non-empty space paradigm for physical reality.

The mass density $m_o n(r)$ of the distributed radial electron with the analytical density $n(r)$ possesses the same active mass-energy density as the passive mass-energy of electron's gravitational field (due to the Principle of Equivalence). Similarly, the charge density $\rho(r) = qn(r)$ of the same electron should possess a self-energy density which is to be equal to the Coulomb field energy density. Therefore, the passive charge density of the extended particle is to be proportional to the active field energy density

$$\rho(\mathbf{x},t) = [\pm\vec{d}(\mathbf{x},t)]^2/4\pi\varphi_o \tag{2}$$

in the rest frame of reference. Such physical relations between Coulomb's field intensity $\vec{d}(\mathbf{x},t)$ and particle's density $\rho(\mathbf{x},t)$ can be expected [7] among correct solutions to Maxwell's equations if the Einstein Principle of Equivalence works for electricity. Moreover, the electric charge q, with self-energy $q\varphi_o$, and its density ρ, with self-energy density $\rho\varphi_o$, has the energy nature under the universal potential $\varphi_o = const$ like a mass has the energy nature due to $E = mc^2$. Maxwell-Lorentz's equations can be equally discussed as for electric currents with the electric density ρ as well as for energy flows with the electric energy density $\rho\varphi_o$. In general, electric self-energy $E_q = q\varphi_o$ of the charge q should be added to the mechanical self-energy $E_m = mc^2$ of a charged body.

One can say that EM energy flows with their natural conservations/dissipations are more fundamental for unification of electrodynamics and thermodynamics than electric currents with the well defined conservation of charges but ill-defined dissipation notions. Anyway, classical electrodynamics can fluently employ the charge self-energy (justified so far only in quantum electrodynamics) in Mie's non-empty space with material fields. Recall, that these fields can represent continuous particle-charges. The delta-operator description of localized charges, say on separated capacitor's plates, can also introduce the electrostatic energy of such point charges in addition to the equal amount of their Coulomb field energy (separated spatially from such point charges). Contrary to diverging self-energies of point particles, self-energy of the spatially distributed charge can be finite and can be described through analytical densities with the clear physical meaning.

The local equality of the charge energy density, $\rho\varphi_o$, and the Coulomb field energy density, $\vec{d}^2/4\pi$, assumes finite electrostatic energy integrals of an elementary continuous carrier of electricity,

$$\int \frac{\vec{d}^2(\mathbf{x},t)}{4\pi} dv = \varphi_o \int \rho(\mathbf{x},t) dv \equiv \varphi_o q \neq \infty. \tag{3}$$

This divergence-free energy distributions for a continuous elementary carrier 'at rest' ($\vec{v} = \vec{b} = 0$) complies with $div\, \vec{d}(r) = 4\pi\rho(r)$ under exact radial solutions for field and charge densities,

$$\begin{cases} \vec{d}(r) = q\hat{\mathbf{r}}/r(r+r_o), \\ \rho(r) = qr_o/4\pi r^2(r+r_o)^2, \end{cases} \tag{4}$$

where the particle's density $n(r) = r_o/4\pi r^2(r+r_o)^2$ in non-empty space replicates the Dirac operator density $\delta(r)$ under the empty space approximation of reality. The exact solutions (4) match the finite balance of electric energy (3) and rids Maxwell electrodynamics of the Coulomb energy divergence. Notice that the extended classical particle in (4) is distributed over macro and mega scales (as well as over Planck scales and the quantum microworld) that denies the localized particle.

The field flux of the radial charge distribution $\rho(r)$ depends on a selected radius R for a Gaussian sphere,

$$4\pi R^2 \hat{\mathbf{r}}\vec{d}(R) \equiv q(R) = \int_0^R \rho(r) 4\pi r^2 dr = \int_0^R \frac{q(\infty) r_o dr}{(r+r_o)^2} = q(\infty)\frac{R}{R+r_o}, \tag{5}$$

where the complex parameter r_o defines the half-charge scale $|r_o|$ for the infinite spatial (astro)distribution (4) of the total charge $q \equiv q(\infty)$. In other words, any non-zero value $r_o \neq 0$ unavoidably results in the global astrodistribution of the elementary charge over the entire Universe. Therefore, the non-point Maxwell electron (4) cannot be localized in principle within the microscopic (and even macroscopic) volume assumed in the Lorentz model. The classical charge density is rigidly bound with the electric field energy density, $\rho \propto \vec{d}^2$. The equivalent charge and its field counter-flows of imaginary electric energies fill non-empty world space continuously together with collinear (residual) co-flows of real mass-energies within the elementary radial carrier [7]. The exact mathematical solution (4) to the classical field equations (1) admits the global material overlap of all elementary charges and rejects the empty space simplification of reality through separated fields and particles.

One can use imagine numbers for the 'negative' elementary charge of the electron, $\int \rho(\mathbf{x},t)dv \equiv q \equiv ie \Rightarrow -ie_o$. The charge self-potential $\varphi_o \equiv E_q/q = q/r_o = c^2/\sqrt{G}$ we defined via the light speed limit c and the Cavendish constant G. The similar charge to spatial scale ratio, $\sqrt{G}m/r_m = c^2/\sqrt{G}$, arises in gravitation of continuous masses [7], where $mc^2 \equiv \sqrt{G}m\varphi_o$ is the body relativistic energy. We keep the fundamental potential $\varphi_o \equiv c^2/\sqrt{G} = 3.48 \times 10^{27} StatV = 1.04 \times 10^{27} V$ for self-energies of both mechanical and electric charges,

$$E = (\sqrt{G}m + q)\varphi_o = mc^2 + ieG^{-1/2}c^2, \tag{6}$$

in order to extend the Einstein mechanical formula on electricity.

Despite the electron possesses in (6) only imaginary electric energy ($-i \cdot 1{,}04 \times 10^{24} KeV$ of the imaginary charge $q = -ie_o$) next to 511 KeV of the real mechanical energy, paired interactions of imaginary electric charges correspond to real Coulomb forces and to real interaction energies.

Contrary to an electric charge defined in real numbers, like the inertial mass, the imaginary electric charge exhibits a correct direction of the radiation self-force, which is proportional to $q^2 = -e_o^2 < 0$. Ultimately, unified densities of gravito-mechanic and electric charges can be described in complex functions. Real integrals over spatial densities provide masses of energy carriers, imaginary integrals - electric charges.

3 Logarithmic strong field potential for complex energy carriers

The strong-field potential $W(\mathbf{x}) = -\varphi_o ln[1 + \sqrt{G}M/|\mathbf{x}|\varphi_o]$ of the gravitational charge $\sqrt{G}M$ could be introduced [7] for interaction with other gravitational charges $\sqrt{G}m$. This radial potential for static gravitational fields matches zero Einstein curvature, $G_o^o \equiv R_o^o - R/2 = 0$, and the flat-space subinterval, $(g_{oi}g_{oj}g_{oo}^{-1} - g_{ij})dx^i dx^j = \delta_{ij}dx^i dx^j$, in the relativistic four-interval:

$$ds^2 = \frac{c^2 dt^2}{(1 + \sqrt{G}M/r\varphi_o)^2} - dr^2 - r^2 d\theta^2 - r^2 sin\theta^2 d\phi^2 \tag{7}$$

The mechanical self-energy $\varphi_o \sqrt{G}m = mc^2$ of the charge $\sqrt{G}m$ arises to compensate the integral self-action of these charge densities $\rho_m(\mathbf{x}) = \nabla^2 W(\mathbf{x})/4\pi = mr_o/4\pi\mathbf{x}^2(r_o + |\mathbf{x}|)^2$ in its negative interaction potential $W(\mathbf{x})$,

$$mc^2 \equiv \varphi_o \sqrt{G}m \equiv \int \varphi_o \rho_m(\mathbf{x}) d^3x = -\int W(\mathbf{x})\rho_m(\mathbf{x}) d^3x \text{ or}$$
$$\int d^3x[\varphi_o + W(\mathbf{x})]\nabla^2 W(\mathbf{x})/4\pi = 0 \tag{8}$$

This universal compensation of Einstein's mechanical self-energy mc^2 by the self-interaction of mass densities in negative gravitational potentials takes place due to the integral mathematical equality $\int_o^\infty dx[ln(1 + x^{-1})]/(1 + x)^2 \equiv \int_o^\infty dx/(1 + x)^2 = 1$.

The gravity-type Poisson equation, $\nabla^2 W = 4\pi\rho \Rightarrow \varphi_o^{-1}(\nabla W)^2$, reads the imaginary (astro)electron as a non-linear field composition with respect to the radial field intensity $d(r)$ or the electron's interaction potential $W(r)$, with $d(r) = -\partial_r W(r)$ and

$$\frac{\partial_r[r^2 \partial_r W(r)]}{4\pi r^2} = \frac{[\partial_r W(r)]^2}{4\pi\varphi_o} = \frac{qr_o}{4\pi r^2(r + r_o)^2}. \tag{9}$$

Such a s non-linear equation reveals the imaginary post-Coulomb potential with the negative, gravitational sign for the continuous radial carrier of electricity,

$$W(r) = -\frac{q}{r_o} ln\left(1 + \frac{r_o}{r}\right) \equiv -\varphi_o ln\left(1 + \frac{q}{\varphi_o r}\right). \tag{10}$$

This logarithmic potential reproduces the regular Coulomb law on measurable scales with an opposite, Newtonian sign $(-q/r_o)ln[(r + r_o)/r] \approx (-q/r)$, when $|r_o| \ll r$. The unified Newton-type potential (10) leads to real Coulomb forces, $q_1[-\nabla(-q_2 r_o^{-1})ln(1 + r_o r^{-1})] = ie_1(-ie_2)\hat{\mathbf{r}}/r(r + r_o) \approx e_1 e_2 \hat{\mathbf{r}}/r^2$, with mutual repulsion of like electric charges e_1 and e_2, and attraction of unlike ones. The radiation self-force, $f_{rad} = 2q^2\dddot{\mathbf{v}}/3c^3 = 2(-ie_o)^2\ddot{\mathbf{v}}/3c^3 = -2e_o^2\ddot{\mathbf{v}}/3c^3$, takes the physical, damping direction for the imaginary charge, contrary to a mathematical self-acceleration of the real electric charge.

The gravitational sign in the Poisson equation for imaginary electric energies suggests to inverse positive and negative imaginary densities of electric charges, $\rho \rightarrow -\rho$, in the Maxwell-Lorentz equations (1), which are irrelevant to the conventional signs of electrons and protons. The electrostatic solution for charged imaginary densities in (1) with the gravitational type direction of the radial Coulomb field, $\vec{D}(\mathbf{x}) = -Q\hat{\mathbf{x}}/|\mathbf{x}|(|\mathbf{x}| + Q\varphi_o^{-1})$, facilitates the unification of gravity and electricity on the basis of one complex energy charge $Q\varphi_o \equiv (q_m + iq_e)c^2/\sqrt{G} \Rightarrow (\sqrt{G}m + ie)c^2/\sqrt{G}$. The Maxwell-Lorentz equations (1) for the massless elementary charge can be extended on complex energy flows of the radial carrier with the elementary charge density $\rho(\mathbf{r}, t) \equiv (\sqrt{G}m + ie)n(\mathbf{x}, t) = -div\,\vec{D}(\mathbf{x}, t)/4\pi$,

$$\begin{cases} div\,\vec{B}(\mathbf{x}, t)\varphi_o = 0 \\ c\ curl\,\vec{B}(\mathbf{x}, t)\varphi_o = \vec{v} div\,\vec{D}(\mathbf{x}, t)\varphi_o + \partial_t\vec{D}(\mathbf{x}, t)\varphi_o \\ c\ curl\,\vec{D}(\mathbf{x}, t)\varphi_o = -\partial_t\vec{B}(\mathbf{x}, t)\varphi_o. \end{cases} \tag{11}$$

One may indeed remove from the Maxwell-Lorentz system of equations for the electron its density $\rho(\mathbf{r}, t)$, which is the function of fields in non-dual physics rather than an independent entity in the empty space paradigm.

The unified gravi-electric potential of the static system of complex charges $q_k = \sqrt{G}m_k + ie_k \equiv \varphi_o z_k$ is given by the analytic function,

$$W(\mathbf{x}) = -\frac{c^2}{\sqrt{G}} ln\left(1 + \frac{z_1}{|\mathbf{x}-\mathbf{a}_1|} + \frac{z_2}{|\mathbf{x}-\mathbf{a}_2|} + ... + \frac{z_n}{|\mathbf{x}-\mathbf{a}_n|}\right), \tag{12}$$

which is responsible for the Equivalence Principle of passive and active energy densities in each point of the non-empty space plenum,

$$\epsilon_p(\mathbf{x}) \equiv \frac{[-\nabla W(\mathbf{x})]^2}{4\pi} = \frac{\varphi_o \nabla^2 W(\mathbf{x})}{4\pi} \equiv \epsilon_a(\mathbf{x}). \tag{13}$$

The volume energy integral of overlapping material spaces with paired interactions demonstrates the system energy conservation due to interference exchanges $2ab$ in join densities $(a + b)^2$,

$$\mathcal{E}_{sys} \equiv \int \epsilon_p(\mathbf{x}) c^2 d^3x = \frac{\varphi_o^2}{4\pi} \int \left(\frac{\frac{(\mathbf{x}-\mathbf{a}_1)z_1}{|\mathbf{x}-\mathbf{a}_1|^3} + \frac{(\mathbf{x}-\mathbf{a}_2)z_2}{|\mathbf{x}-\mathbf{a}_2|^3} + ... + \frac{(\mathbf{x}-\mathbf{a}_n)z_n}{|\mathbf{x}-\mathbf{a}_n|^3}}{1 + \frac{z_1}{|\mathbf{x}-\mathbf{a}_1|} + \frac{z_2}{|\mathbf{x}-\mathbf{a}_2|} + ... + \frac{z_n}{|\mathbf{x}-\mathbf{a}_n|}} \right)^2 d^3x \equiv \varphi_o^2 \sum_{k=1}^{n} z_k \equiv \sum_{k=1}^{n} \mathcal{E}_k, \tag{14}$$

and the system energy compensation by self-interaction of charged densities in the potential (12),

$$\frac{\varphi_o}{4\pi} \int \left(\frac{\frac{(\mathbf{x}-\mathbf{a}_1)z_1}{|\mathbf{x}-\mathbf{a}_1|^3} + \frac{(\mathbf{x}-\mathbf{a}_2)z_2}{|\mathbf{x}-\mathbf{a}_2|^3} + ... + \frac{(\mathbf{x}-\mathbf{a}_n)z_n}{|\mathbf{x}-\mathbf{a}_n|^3}}{1 + \frac{z_1}{|\mathbf{x}-\mathbf{a}_1|} + \frac{z_2}{|\mathbf{x}-\mathbf{a}_2|} + ... + \frac{z_n}{|\mathbf{x}-\mathbf{a}_n|}} \right)^2 \left[1 - ln\left(1 + \frac{z_1}{|\mathbf{x}-\mathbf{a}_1|} + \frac{z_2}{|\mathbf{x}-\mathbf{a}_2|} + ... + \frac{z_n}{|\mathbf{x}-\mathbf{a}_n|}\right)\right] d^3x \equiv 0. \tag{15}$$

4 Gravitation repulsion by dense mass-energy fields

For a pure mechanical system, for example, the volume energy (14) of nonempty space at $z_k \Rightarrow Gm_k/c^2 \equiv r_k$ can be represented analytically at $r_k + r_l \ll |\mathbf{a}_k - \mathbf{a}_l|$ as a sum of GR metric energy,

$$\mathcal{E}_{GR} \approx \int \frac{\varphi_o^2 r_1^2 d^3x}{4\pi|\mathbf{x}|^4 (1 + \frac{r_1}{|\mathbf{x}|} + \frac{r_2}{|\mathbf{a}_1-\mathbf{a}_2|} + ... + \frac{r_n}{|\mathbf{a}_1-\mathbf{a}_n|})^2} + \int \frac{\varphi_o^2 r_2^2 d^3x}{4\pi|\mathbf{x}|^4 (1 + \frac{r_1}{|\mathbf{a}_2-\mathbf{a}_1|} + \frac{r_2}{|\mathbf{x}|} + ... + \frac{r_n}{|\mathbf{a}_2-\mathbf{a}_n|})^2} \tag{16}$$

$$+ ... + \int \frac{\varphi_o^2 r_n^2 d^3x}{4\pi|\mathbf{x}|^4 (1 + \frac{r_1}{|\mathbf{a}_n-\mathbf{a}_1|} + \frac{r_2}{|\mathbf{a}_n-\mathbf{a}_2|} + ... + \frac{r_n}{|\mathbf{x}|})^2} = \sum_{i=1}^{n} \mathcal{E}_i \sqrt{g_{oo}^{\neq i}(\mathbf{a}_i)} \approx \sum_{i=1}^{n} \left(m_i c^2 - m_i \sum_{k \neq i}^{n} \frac{Gm_k}{R_{ik}} \right) > 0,$$

with $E_2 \equiv m_2 c^2 \sqrt{g_{oo}^{\neq 2}(\mathbf{a}_2)} \equiv m_2 c^2 / \left(1 + \frac{r_1}{|\mathbf{a}_2-\mathbf{a}_1|} + \frac{r_3}{|\mathbf{a}_2-\mathbf{a}_3|} + ... + \frac{r_n}{|\mathbf{a}_2-\mathbf{a}_n|}\right)$, for instance, and the interference (dark) part of the system energy,

$$E_{dark} = \frac{\varphi_o^2}{4\pi} \sum_{i=1}^{n} \sum_{k \neq i}^{n} \int_0^{2\pi} d\varphi \int_0^{\infty} r^2 dr \int_0^{\pi} \frac{r_i r_k \left(r^2 - R_{ik} r cos\theta\right) sin\theta d\theta}{r^3 (R_{ik}^2 + r^2 - 2R_{ik} r cos\theta)^{3/2}} \approx \sum_{i=1}^{n} \left(m_i \sum_{k \neq i}^{n} \frac{Gm_k}{R_{ik}} \right) > 0. \tag{17}$$

Any mechanical system tends to an equipartition distribution of energy between available degrees of freedom that results from (16)-(17) to Newtonian gravitational attraction (16) in week fields with low dark energy (17), but to gravitational repulsion in dense energies with an excess of the dark component. Indeed, the radial fall speed $dr/dt = \pm c\sqrt{g_{oo}(1-g_{oo})}$ can be found from the GR interval equation $(dr/ds)^2 + (rd\varphi/ds)^2 - E_m^2/m^2 g_{oo} = -1$ at $E_m/m \equiv c^2\sqrt{g_{oo}}/\sqrt{1-v^2c^{-2}} = const \Rightarrow 1$ and $d\varphi/ds = 0$, $ds = \sqrt{g_{oo}}cdt\sqrt{1-v^2c^{-2}}, v^2 = (dr/\sqrt{g_{oo}}dt)^2$. Then, coordinate time acceleration d^2r/dt^2 in the flat space metric (7) with $g_{oo}(r) = r^2/(r+r_o)^2$,

$$d^2\mathbf{r}/dt^2 = -c^2 r_o \mathbf{r}(r^2 - 2r_o r - r_o^2)/(r+r_o)^5, \tag{18}$$

universally describes both the Newton attraction $-r_o c^2\mathbf{r}/r^3$ for $r_o \ll r$ and the strong field GR repulsion $+\mathbf{r}c^2/r_o^2$ for $r \ll r_o \equiv GM/c^2$.

The radial dimension of the Metagalaxy is less than its gravitational scale $R_o = GM_{Meta}/c^2$ as is widely accepted. According to (18), this dense Metagalaxy should repeal its material elements behind $R_o(1+\sqrt{2})$. The strong-field limit, when $r \ll r_o$ and $dr/dt \approx cr/R_o \Rightarrow rH_o$, corresponds to the measured Hubble velocities of expanding galaxies at $R_o \Rightarrow c/H_o = 1,3\times 10^{26}m$ that computes $M_{Meta} = R_o c^2/G = 1.8\times 10^{53}kg$ very close to direct observations. Acceleration of the Universe expansion in the dense mass-energy limit, $d^2r/dt^2 = c^2 r/R_o^2 \Rightarrow rH_o^2$, is proportional to the distance r that redefine the Hubble velocity as $V_H = rf(t) = rH_o(1+H_o t)$.

5 Maxwell equations are local energy identities in non-dual physics

By pairing locally energy densities of the radial field and the continuous particle, one also unifies in (10) gravi-mechanical and electric self-energies $q\varphi_o \Rightarrow Q\varphi_o = (\sqrt{G}m + ie)c^2/\sqrt{G}$. Such a double unification (particle with field and gravity with electricity that was predicted many years ago [2]) assumes the application of the complex charge $q \Rightarrow Q = (\sqrt{G}m + ie)$ in the logarithmic potential (10) for non-empty space. The sphere of the microscopic radius $|r_o| = |Q/\varphi_o|$ contains exactly half of the complex electron charge Q. The other half of the elementary astrocharge is distributed over micro, macro, and mega scales in the Universe, which is already known [8] as the non-local material system.

The electron's density scale $|r_o| \approx e_o\sqrt{G}/c^2 = 1,38\times 10^{-34}cm$ is even less than the Planck's length, $l_p \equiv \sqrt{\hbar c}(\sqrt{G}/c^2) = 1,62\times 10^{-33}cm$, and surely is much smaller of the current limit $10^{-17}cm$ for space inhomogeneity measurements. The ratio of classical and quantum scales, $|r_o|^2/l_p^2 \equiv \alpha \approx 1/137$, can shed some light on the physical origin of the Sommerfeld (fine-structure) constant. The very small spatial scale $10^{-34}cm$ for main densities of the radial electron explains the formal success of the δ-operator modeling of the nonlocal energy carrier by the point matter approximation. The formal interaction of such point-like concentrations of energy with external weak distributions of energy is conceptually incorrect for extended charges due to their global overlap in the entire Universe.

One can verify quantitatively that the complex astrocharge distribution, $\rho(r) = Qn(r) = (\sqrt{G}m + ie)n(r)$ with $r_o = Q\sqrt{G}/c^2$, generates the potential (10) under the regular integral rule for classical gravitation and electrodynamics,

$$\begin{aligned}\frac{(-Q)}{r_o}ln\left(1+\frac{r_o}{r}\right) &= \int\frac{(-Q)n(r')dv'}{|\vec{r}-\vec{r}'|} = -\int_o^\infty\int_{-\pi/2}^{\pi/2}\int_o^{2\pi}\frac{d\phi'\,sin\theta'\,d\theta'\,r'^2dr'}{\sqrt{r^2+r'^2-2rr'cos\theta'}}\frac{Qr_o}{4\pi r'^2(r'+r_o)^2}\\ &= -\int_o^\infty\frac{dr'Qr_o}{(r'+r_o)^2}\left(\frac{|r'+r|-|r'-r|}{2rr'}\right) = -\int_r^\infty\frac{Qdr'}{r_o}\left(\frac{1}{r'}-\frac{1}{r'+r_o}\right) = \int_r^\infty\hat{\mathbf{r}}'\vec{D}(r')dr' = W(r).\end{aligned} \tag{19}$$

Notice that the potential $W(r)$ coincides with the work associated with the replacement of a unit probe (positive) charge from the point r to ∞ against the radial field $\hat{\mathbf{r}}'\vec{D}(r') = D(r) = -\partial_r W(r)$. The integration over r' within $0 \leq r' \leq r$ vanishes identically in (19) in agreement with the physical meaning of potentials for a probe body.

Like the first pair of Maxwell field identities, $\nabla_\lambda F_{\mu\nu} + \nabla_\mu F_{\nu\lambda} + \nabla_\nu F_{\lambda\mu} \equiv 0$, the second pair of Maxwell current equations can also be considered (in the non-empty space paradigm) as identities for balanced momentums $\rho\varphi_o u^\mu/c$ of (yin-yang) paired energy densities,

$$\begin{cases} [\nabla_\lambda F_{\mu\nu}(x) + \nabla_\mu F_{\nu\lambda}(x) + \nabla_\nu F_{\lambda\mu}(x)]\varphi_o/4\pi c \equiv 0 \\ [\delta^\mu_\lambda - u^\mu(x)u_\lambda(x)]\nabla_\nu F^{\nu\lambda}(x)\varphi_o/4\pi c \equiv 0. \end{cases} \tag{20}$$

Here we used the following identical relations for classical fields:

$$\begin{cases} \nabla_\nu F^{\nu\mu}(x) \equiv 4\pi j^\mu(x)/c \equiv 4\pi\rho(x)u^\mu(x) \\ u_\mu(x)\nabla_\nu F^{\nu\mu}(x) \equiv 4\pi\rho(x) \\ \nabla_\nu F^{\nu\mu}(x) \equiv u^\mu(x)u_\lambda(x)\nabla_\nu F^{\nu\lambda}(x). \end{cases} \tag{21}$$

The energy-momentum balance of paired 4-vector flows in the second identity (20) clams the absence of curl energy currents along directions normal to the 4 velocity u^μ of field-energy densities, where $u_\mu u^\mu = 1$. These is no much sense to introduce particles at all in the non-dual scheme for field-energy identities (20). In other words, material reality is nothing but inhomogeneous energy flows through all spatial points of the Universe.

Once one identified electric currents with field-energy flows in the Maxwell-Lorentz theory for extended electrons, the variation techniques for non-dual material fields should be different from textbook variation procedures of the empty space physics. Indeed, the current Maxwell equations were derived in the dual approach from the summary action of fields and particles varied and fixed independently, like $\delta(A_\mu j^\mu) = j^\mu\delta A_\mu$. In non-dual physics, variations of self-energy of material flows in their potentials $\delta(A_\mu j^\mu)$ are equal to the sum $j^\mu\delta A_\mu + A_\mu\delta j^\mu$, with the double variational result

$$\delta(cA_\mu\nabla_\nu F^{\nu\mu}/4\pi) = (c/4\pi)[\delta A_\mu\nabla_\nu F^{\nu\mu} + A^\mu\nabla^\nu(\partial_\mu\delta A_\nu - \partial_\nu\delta A_\mu)] \Rightarrow (c/2\pi)\nabla_\nu F^{\nu\mu}\delta A_\mu = 2j^\mu\delta A_\mu. \tag{22}$$

Therefore, the Lagrangian for non-dual charged energies should posses double electromagnetic densities $F_{\mu\nu}F^{\mu\nu}$ in the non-empty space action,

$$S_{n-em} = -\frac{1}{c}\int d\Omega\left[\frac{1}{c}A_\mu j^\mu + \frac{1}{8\pi}F_{\mu\nu}F^{\mu\nu}\right], \tag{23}$$

compared to the dual (field + matter) approach. Then the variations of (23) with respect to four field potentials A_μ result in the Maxwell relation $4\pi j^\mu/c = \nabla_\nu F^{\nu\mu}$ or energy flow identities in (20) which is the Lagrange equation of motion. It is worth to note, that the non-empty space action (23) vanished on the Lagrange variation trajectories:

$$\begin{aligned} S_{n-em} &= -\frac{1}{c}\int d\Omega\left[\frac{1}{c}A_\mu j^\mu + \frac{1}{8\pi}F_{\mu\nu}F^{\mu\nu}\right] = -\int\frac{d\Omega}{8\pi c}\left[2A_\mu\nabla_\nu F^{\nu\mu} + (\partial_\mu A_\nu - \partial_\nu A_\mu)F^{\mu\nu}\right] \\ &= -\int\frac{d\Omega}{4\pi c}A_\mu\nabla_\nu(F^{\nu\mu} - F^{\mu\nu}) - \oint\frac{dS_\mu}{4\pi c}A_\nu F^{\mu\nu} = 0 + 0 = 0. \end{aligned} \tag{24}$$

In other words, self interactions of extended charges in electrodynamics balance their electric self-energies like gravitational self-interactions in (8) balance mechanical self-energies mc^2 of elementary mass-energy carriers. This is an expected result that motion along Lagrange trajectories, like (20), nullifies the action integral. Recall that the conventional current+field electromagnetic action in dual physics with $F_{\mu\nu}F^{\mu\nu}/16\pi$ instead of $F_{\mu\nu}F^{\mu\nu}/8\pi$ does not match such a path nullification in the Lagrange variation formalism.

6 Inward and outward longitudinal waves within the continuous electron

Sphere radial currents or moving electric densities under strict spherical symmetry do not generate magnetic fields in Maxwell's electrodynamics as is well known. Such magnetic-field-free currents may equally have inward and outward directions with respect to a center of the radial electron. Time-varying currents of charged densities should be balanced by time-varying Maxwell displacement currents. In other words, the static radial shape of the elementary charge (associated with its static Coulomb field) may be accompanied by radial wave modulations with longitudinal electric induction.

Periodical inward, $(-\hat{\mathbf{r}}V_l)\rho_{in}(r,t)$, and outward, $\hat{\mathbf{r}}V_l\rho_{out}(r,t)$, currents, where $\hat{\mathbf{r}} \equiv \vec{r}/r$, are balanced by radial displacement currents $\partial_t d_{in}(r,t)\hat{\mathbf{r}}$ and $\partial_t d_{out}(r,t)\hat{\mathbf{r}}$, respectively, with $div[d_{in}(r,t)\hat{\mathbf{r}}] = 4\pi\rho_{in}(r,t)$ and $div[d_{out}(r,t)\hat{\mathbf{r}}] = 4\pi\rho_{out}(r,t)$:

$$\left\{V_l\partial_r r^2[d_{out}(r,t) - d_{in}(r,t)] + r^2\partial_t[d_{out}(r,t) + d_{in}(r,t)]\right\}\hat{\mathbf{r}}/r^2 = 0. \tag{25}$$

This classical equation for balanced radial currents with the spherical symmetry has wave solutions along both outward and inward directions,

$$\begin{cases}\vec{d}_{out}(r,t) = \hat{\mathbf{r}}qa_{out}cos(\omega t - kr + \varphi_{out})/r^2\\ \vec{d}_{in}(r,t) = -\hat{\mathbf{r}}qa_{in}cos(\omega t + kr + \varphi_{in})/r^2\end{cases} \tag{26}$$

These wave modulations with $k = \omega/V_l$ are responsible for dynamical reshaping of the continuous elementary charge q in external fields. Such longitudinal waves within the elementary charged continuum keep its energy, while transverse electromagnetic waves are responsible for energy exchanges between different carriers of energy. In general, the dimensionless wave amplitudes a_{out} and a_{in} may have different complex values related to power of longitudinal waves within the extended electron. However, there are no internal power flows within the elementary continuum in equilibrium. In this case, superposition of outward and inward waves (26) can exist as a standing wave,

$$\vec{d}_{stan}(r,t) = \frac{qa\hat{\mathbf{r}}}{r^2}[cos(\omega t - kr) - cos(\omega t + kr)] = \frac{2qa\hat{\mathbf{r}}}{r^2}sinkr \cdot sin\omega t, \tag{27}$$

with its time-averaged energy integral over the elementary charge volume,

$$E_{a,\omega} = \int_o^\infty \frac{(2qa)^2 sin^2kr < sin^2\omega t >_t}{4\pi r^4}4\pi r^2 dr = 2q^2a^2\int_o^\infty dr\frac{sin^2kr}{r^2} = \pi a^2q^2\frac{\omega}{V_l}. \tag{28}$$

A set of internal standing waves (27) with different frequencies ω and amplitudes a may play in Maxwell's electrodynamics the same role as the foam of virtual photons around to the quantum electron in wave mechanics. De Broglie was first who assigned in 1923 the internal wave process to each elementary particle with inertia. And later he specified this wave process for his unique physical interpretation of quantum phenomena [10]. We tend to support the internal wave process in therms of radial wave modulations (26)-(28) which shed some "classical' light on the celebrated double-slit experiment. In fact, not only quantum mechanics but also Maxwell's theory of classical electric fields and electric energy flows may comment on the wave motion of continuous electrons. One day classical and quantum electrodynamics will lead to similar wave physics for the elementary charge.

7 Conclusions

Unified electric currents and mass flows of the elementary continuous carrier are proportional to electric energy and mas-energy flows, respectively, of this carrier. The radial Maxwell-Lorentz electron

(1)-(4) tends to replicate (astro)distributions of the quantum electron in the De Broglie - Bohm formulation of wave mechanics [9]. In particular, De Broglie periodical oscillations, $\psi = \psi_o expi(\omega t - \mathbf{kx})$, of elementary particle's matter everywhere in the Universe may be associated in the developed classical approach to internal wave modulations (26) - (27) within the continuous electron.

Each radial electron possesses a fine, but finite, density in all spatial points. Information free elastic reshaping of these elementary densities might occur with internal longitudinal waves without energy losses/gains, while free transverse waves and other dissipative signals (with information and mutual energy exchanges between overlapping microcosms of elementary energies) obey the speed restriction of Special Relativity regarding centers of radial carriers.

The experimental result [1] for the rigid motion of Coulomb's fields confirms the non-empty paradigm for overlapping material bodies. In non-dual physical reality, nebular dense fronts of infinitely extended masses (observed as the planets) are moving within, not outside, the r^{-4} mass-energy distribution of the radial Sun. Such a global overlap of locally undivided continuous carriers of mass-energy in joint non-empty space explains not only the electromagnetic laboratory experiment [1], but also the gravitational Laplace problem of the Solar system stability.

Outward spherical waves were considered in the conventional empty space physics as retarded waves which have already left their point source. Inward spherical waves were called non-physical waves by the dual model, because these (advanced) waves have not yet met their point source in question. The non-empty space paradigm for radial material sources provides equal physical rights to outward and inward waves in reality of non-local bodies. In this way, intercepted energy flows along advanced directions toward moving centers of cosmic bodies, as was claimed by astrophysicists N. Kozyrev [11], may be considered as a justification of the non-empty space in reality. Similarly, inward radial signals might be expected under acceleration of electric charges when their radial structures attain external energy. It appears that an interesting development of experiments [1] might be their reiteration with proton beams at the Large Hadron Collider. In general, wave energy flows along a computed advanced axis toward moving centers of radial energies may ultimately justify the non-empty space paradigm for non-dual reality of the nonlocal world.

References

[1] R. de Sangro, G. Finocchiaro, P. Patteri, M. Piccolo, G. Pizzella, *Measuring propagation speed of Coulomb fields*, Eur. Phys. J. C **75**, 137 (2015)

[2] M.-A. Tonnelat, *The principles of electromagnetic theory and relativity* (Riedel Publishing Co., Dordrecht, 1966)

[3] L. de Broglie, *New perspectives in physics* (Basic Books, New York, 1962)

[4] G. Mie, Ann. der Physik **37**, 511 (1912); *ibid* **39**, 1 (1912); *ibid* **40**, 1 (1913)

[5] J. Schwinger, in *Quantum, Space and Time - The Quest Continues* (Ed. A.O. Barut, A. Van der Merwe, and J.-P. Vigier, Cambridge University Press, 1984) p.620

[6] H.A. Lorentz, *The Theory of Electrons* (Leipzig, 1916)

[7] I.E. Bulyzhenkov, Int. Jour. of Theor. Phys. **47**, 1261 (2008); *Pure field electrodynamics of continuous complex charges: 4th year course tutorial* (MIPT, Moscow, 2015) ISBN 978-5-7417-0554-4

[8] A. Aspect, P. Grangier, G. Roger, Phys. Rev. Let. **47**, 460 (1981), *ibid* **49**, 91 (1982)

[9] D. Bohm, Phys. Rev. **85**, 166 (1952); D. Bohm and B.J. Hiley, *The Undivided Universe*, Routledge, London, 1993

[10] L. de Broglie, *Reinterpretation of Wave Mechanics* (Gauthier-Villars, Paris, 1971)

[11] N.A. Kozyrev, *Selected Works* (St-Petersburg State University, St-Petersburg, 1991, in Russian)

AUTHOR INDEX

AUTHOR INDEX

AUTHOR INDEX

AUTHOR INDEX

EDP Sciences – Web of Conferences
17, Avenue du Hoggar
Parc d'Activité de Courtabouf
BP 112
F-91944 Les Ulis Cedex A
France

ISBN 978-1-5108-3238-1

9 781510 832381

CURRAN ASSOCIATES INC.
proceedings.com

4th International Conference on New Frontiers in Physics 2015

EPJ Web of Conferences Volume 126 (2016)

Crete, Greece
23 - 30 August 2015

Part 2 of 2

Editors:

Larissa Bravina
Yiota Foko
Sonja Kabana

ISBN: 978-1-5108-3238-1

Printed from e-media with permission by:

Curran Associates, Inc.
57 Morehouse Lane
Red Hook, NY 12571

Some format issues inherent in the e-media version may also appear in this print version.

Printed by Curran Associates, Inc. (2016)

For additional information, please contact EDP Sciences – Web of Conferences
at the address below.

EDP Sciences – Web of Conferences
17, Avenue du Hoggar
Parc d'Activité de Courtabœuf
BP 112
F-91944 Les Ulis Cedex A
France

Phone: +33 (0) 1 69 18 75 75
Fax: +33 (0) 1 69 28 84 91

contact-edps@webofconferences.org

Additional copies of this publication are available from:

Curran Associates, Inc.
57 Morehouse Lane
Red Hook, NY 12571 USA
Phone: 845-758-0400
Fax: 845-758-2633
Email: curran@proceedings.com
Web: www.proceedings.com

TABLE OF CONTENTS

PART 1

PART 2

EPJ Web of Conferences **126**, 04009 (2016) DOI: 10.1051/epjconf/201612604009
ICNFP 2015

Precision measurements of Standard Model parameters and Review of Drell-Yan and vector boson plus jets measurements with the ATLAS detector

Noemi Calace[1,a], on behalf of the ATLAS Collaboration

[1] *Département de physique nucléaire et corpusculaire, Section de Physique, Université de Genève, 24 Quai Ernest-Ansermet, Genève, Switzerland*

Abstract. The inclusive productions of the W and the on- or off-shell $Z/\gamma*$ bosons are standard candles at hadron colliders, while the productions of light and heavy-flavour jets in association with a W or a Z boson are important processes to study quantum-chromo-dynamics (QCD) in multi-scale environments. The measurements of their production cross-sections integrated and differential in several variables have been measured at 7 and 8 TeV centre-of-mass energies and are compared to high-order QCD calculations and Monte Carlo simulations. These measurements have an impact on our knowledge of the parton densities of the proton, and test soft resummation effects and hard emissions for small and large momentum transfers and in multi-scale processes. Precision measurements of fundamental Standard Model parameters in Drell-Yan final states are also performed to describe the angular distributions of the decay lepton.
Run-1 studies carried out by the ATLAS Collaboration are re-viewed and first LHC Run-2 results are included.

1 Introduction

The ATLAS Collaboration [1] has carried out multiple measurements of vector boson production both inclusively and with associated light or heavy flavour jets at the Large Hadron Collider (LHC). Thanks to their large production cross section, events containing a W or a Z boson are produced abundantly at the LHC. These events provide the ideal test to probe high-order quantum-chromo-dynamics (QCD), measure electro-weak parameters and constrain parton distribution functions (PDFs). In the last years, these measurements have reached high experimental precision allowing detailed comparisons with theoretical predictions at next-to-next-to-leading-order (NNLO) QCD. Moreover, the dynamical effects of the strong interaction can be probed by measuring QCD radiation in the different kinematic ranges: the high and low vector boson p_T regimes can probe high-order QCD corrections and resummation techniques, respectively.
The studies of W or Z boson events where accompanying jets are explicitly reconstructed, either inclusively in jet flavour or exclusively as heavy-flavour jets, provide precision tests of high-order QCD calculations and can also be used to constrain PDFs. Furthermore, such processes constitute an important background to Standard Model (SM) as well as new physics searches.

[a]e-mail: noemi.calace@cern.ch

The measurements presented here are based on the data collected by the ATLAS detector during the LHC Run-1 proton-proton collision running at 7 TeV and 8 TeV centre-of-mass energy, including also the first LHC Run-2 results at 13 TeV centre-of-mass energy.

2 Forward-backward asymmetry of electron and muon pair-production

The vector and axial-vector couplings in the neutral current annihilation process $q\bar{q} \to Z/\gamma^* \to l^+l^-$ lead to a forward-backward asymmetry $A_{\rm FB}$ in the polar angle distribution of the final state lepton l^- with respect to the quark direction in the rest frame of the dilepton system. The ATLAS experiment has measured the forward-backward asymmetry in electron and muon pairs from Z/γ^* boson decays and extracted the weak mixing angle [2] based on the full set of proton-proton (pp) collision data collected in 2011 at the LHC at a centre-of-mass energy of $\sqrt{s} = 7$ TeV, corresponding to an integrated luminosity of 4.8 fb^{-1}.

The data are analysed over a range of dilepton invariant masses from 66 GeV to 1000 GeV in the central-central electron (where both electrons are reconstructed in the central region of the detector $|\eta| < 2.47$) and muon channels, and up to 250 GeV in the central-forward electron channel. The latter includes events where one electron is reconstructed in the forward pseudorapidity range ($2.5 < |\eta| < 4.9$). The forward-backward asymmetry is measured separately for the three channels as a function of the dilepton invariant mass and unfolded for detector effects and final-state radiation. An additional unfolding procedure is used to further correct the observed values of $A_{\rm FB}$ to remove dilution effects, which occur when the wrong choice is made for the direction of the quark. The resulting $A_{\rm FB}^{\rm cor}$ values are found to be in agreement with the corresponding Standard Model predictions. The final $A_{\rm FB}^{\rm cor}$ values as a function of the dilepton invariant mass for the central-central electron, central-forward electron and muon channels are shown in Figure 1.

Several SM parameters can be extracted from the dependence of the $A_{\rm FB}$ values on the invariant dilepton mass. One of these is the electroweak mixing angle $\sin^2\theta_{\rm eff}^{\rm lept}$, also related to the effective leptonic weak mixing angle $\sin^2\theta_{\rm eff}^{\rm lept}$. The value of the $A_{\rm FB}$ at the peak of the Z/γ^* resonance, $A_{\rm FB}^{0,l}$, can be also written as a function of the asymmetry parameters A_l and A_q,

$$A_{\rm FB}^{0,l} = \frac{3}{4}A_l A_q,$$

with l (q) denoting the leptons (quarks) in the final (initial) state and with the parameters A_l and A_q directly related to the electroweak vector and axial-vector couplings.

In the analysis the effective weak mixing angle ($\sin^2\theta_{\rm eff}^{\rm lept}$) and the muon asymmetry parameter A_μ are extracted from the forward-backward asymmetry measurements $A_{\rm FB}$.

For the effective weak mixing angle, the results from the electron and muon final states are combined:

$$\sin^2\theta_{\rm eff}^{\rm lept} = 0.2308 \pm 0.0005\,(\text{stat.}) \pm 0.0006\,(\text{syst.}) \pm 0.0009\,(\text{PDF}) = 0.2308 \pm 0.0012\,(\text{tot.}).$$

The dominant uncertainty comes from knowledge of the PDFs.

The result from the muon channel, when converted to the asymmetry parameter A_μ, yields

$$A_\mu = 0.153 \pm 0.007\,(\text{stat.}) \pm 0.006\,(\text{syst.}) \pm 0.007\,(\text{PDF}) = 0.153 \pm 0.012\,(\text{tot.}),$$

where the uncertainties are propagated from the uncertainties on $\sin^2\theta_{\rm eff}^{\rm lept}$. The measurement is in good agreement with the best previous ones.

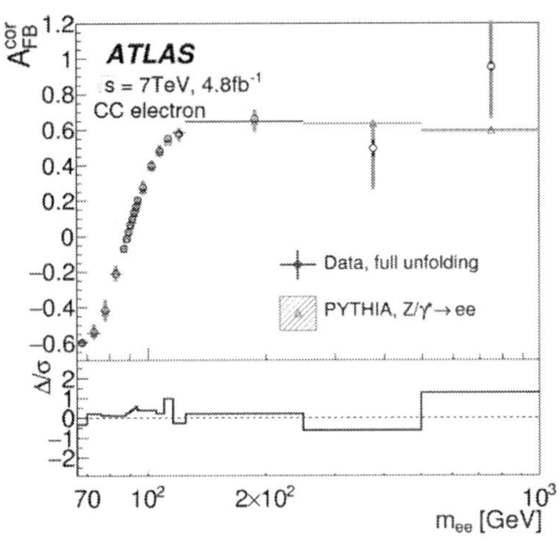

(a)

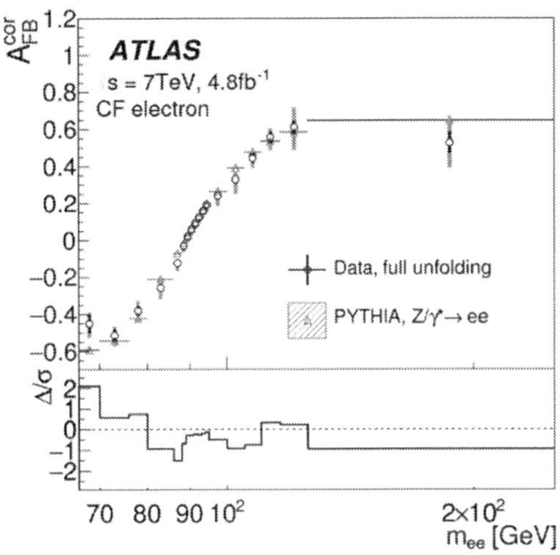

(b)

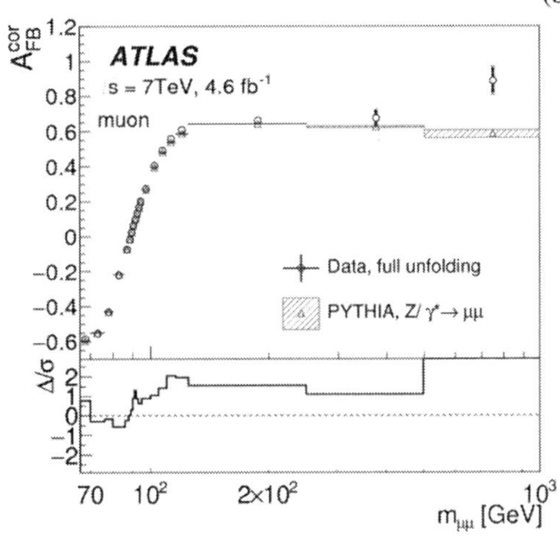

(c)

Figure 1: Forward-backward asymmetry ($A_{\mathrm{FB}}^{\mathrm{cor}}$) values as a function of the dilepton invariant mass for the (a) central-central, electron, (b) central-forward electron and (c) muon channels. Leptons are unfolded to Born level to account for mass bin migration and dilution effects are corrected. The measurement is extrapolated to the full phase space, and the results are compared to truth-level MC information. For the data, the black inner error bars represent the statistical component and the lighter outer error bars the total error (statistical and systematic added in quadrature). The boxed shaded regions for the MC expectations represent only the statistical uncertainty; theoretical uncertainties are included in the data systematic uncertainties. The lower panel of each plot shows the pull value, Δ/σ, for each mass bin, where Δ is the difference between data and simulation and σ is the sum in quadrature of the data and simulation uncertainties. Taken from Ref. [2].

3 Z/γ^* transverse momentum and ϕ_η^*

The measurements of boosted Z-boson production, i.e. with non-zero p_{T}, can be used to constrain theoretical models of perturbative QCD (pQCD). The low Z-p_{T} region can probe resummation techniques whereas the high Z-p_{T} regime probes higher-order QCD corrections. The measurement is an important input to constrain the parton shower parameters of MC generators and provide a better understanding of the Higgs boson production. A correct modelling of the vector boson p_{T} spectrum is crucial for many physics analyses, e.g. the W mass measurement. Using pp collision data at a centre-of-mass energy of $\sqrt{s}$ = 7 TeV at the LHC, the ATLAS collaboration performed the measurements of the Z/γ^* boson transverse momentum spectrum, in the $Z/\gamma^* \rightarrow e^+e^-$ and $Z/\gamma^* \rightarrow \mu^+\mu^-$ channels. Normalised differential cross sections as a function of the Z/γ^* boson transverse momentum are measured for transverse momenta up to 800 GeV. The measurement in performed inclusively up to Z/γ^* rapidities of 2.4, as well as in three rapidity bins [3]. The cross-section measurements are then compared to pQCD and resummed predictions.

Figure 2a shows the ratio of the measured cross section as a function of the boson transverse momentum with predictions from Fewz [4–6] and Dynnlo [7, 8]. The large discrepancy at low p_{T} is due to the not inclusion of resummation effects. In Figure 2b, the measurement is compared to ResBos [9, 10] which is in agreement within the estimated uncertainties at low p_{T}, as it includes resummation effects. However, the precision of direct measurements of the Z/γ^* spectrum at low p_{T} at the LHC is limited by the experimental resolution and systematic uncertainties: the angular correlations of the lepton pair, ϕ^*_η, can probe the low p_{T} domain providing a better experimental resolution and a smaller sensitivity to experimental systematic uncertainties [11]. As it is defined [12], ϕ^*_η depends exclusively on the directions of the two lepton tracks, which are better measured than their momenta.

The $Z/\gamma^* \rightarrow e^+e^-$ and $Z/\gamma^* \rightarrow \mu^+\mu^-$ decays produced in pp collisions at a centre-of-mass energy of $\sqrt{s}$ = 7 TeV are used for this measurement. The data were collected with the ATLAS detector at the LHC and correspond to an integrated luminosity of 4.6 fb^{-1}. Normalised differential cross sections as a function of ϕ^*_η are measured separately for electron and muon decay channels. These channels are then combined for improved accuracy. The results are compared to QCD calculations and to predictions from different Monte Carlo event generators.

The best descriptions of the measured ϕ^*_η spectrum are provided by Sherpa and Powheg+Pythia8 MC event generators (Figure 3). For ϕ^*_η values above 0.1, predictions from Sherpa are able to reproduce the data to within ~ 2%. The low ϕ^*_η part of the spectrum is, however, described less accurately than by ResBos.

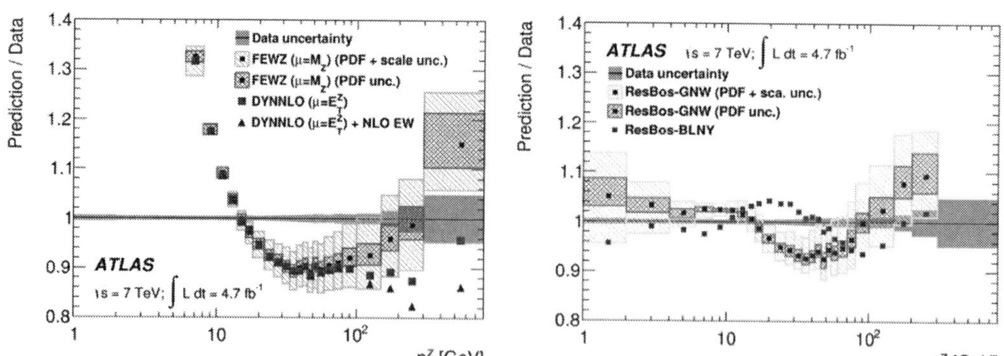

(a) (b)

Figure 2: Comparison of the Z boson p_{T} distributions predicted by different computations: Fewz and Dynnlo (a) and ResBos (b) with the Born-level combined measurement. Taken from Ref. [3].

DOI: 10.1051/epjconf/201612604009

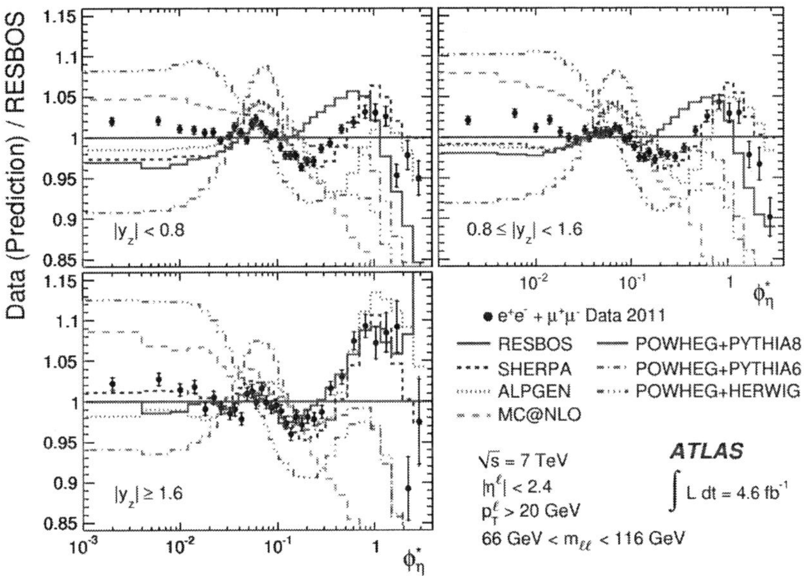

Figure 3: The ratio of the combined normalised differential cross section $1/\sigma^{\text{fid}} \cdot d\sigma^{\text{fid}}/d\phi^*_\eta$ to the ResBos predictions as a function of ϕ^*_η in three ranges of $|y_Z|$. The inner and outer error bars on the data points represent the statistical and total uncertainties, respectively. The uncertainty due to quantum-electrodynamic first-state radiation is included in the total uncertainties. The measurements are also compared to predictions from different MC event generators. Taken from Ref. [11].

4 W + jets measurement

W boson plus jets production has been measured in proton–proton collisions at $\sqrt{s} = 7$ TeV with the ATLAS experiment [13]. Final states with up to seven jets are measured, with comparisons to next-to-leading-order (NLO) QCD predictions for up to five jets. With an integrated luminosity of 4.6 fb^{-1}, the data set allowed an exploration of a large kinematic range, including jet production up to a transverse momentum of 1 TeV in a large rapidity acceptance region $|y| < 4.4$.
At these high transverse momenta higher-order electroweak effects can become as important as those from pQCD corrections. The cross section has been measured inclusively and differentially in several observables. Figure 4 shows the cross section for the production of W + jets as a function of the inclusive jet multiplicity. The measured cross section is compared to a number of theoretical predictions at both leading-order (LO) and NLO in pQCD. The leading-order predictions are using Alpgen [14], Sherpa [15, 16] and HEJ [17, 18]. The NLO predictions are obtained using MEPS@NLO [15, 16] and BlackHat+Sherpa [19–21]. All agree within the assigned experimental uncertainties. Alpgen and Sherpa show some opposite trend towards higher jet multiplicities.

5 R_{jets} measurement

The R_{jets} mesurement, defined as the ratio $R_{\text{jets}} = \sigma(W + \text{jets})/\sigma(Z + \text{jets})$, directly probes the difference between the kinematic distributions of the jet system recoiling against the W or Z bosons. In comparison to separate W + jets and Z + jets cross section measurements, the R_{jets} measurement is a more precise test of pQCD, since some experimental uncertainties and effects from non-perturbative

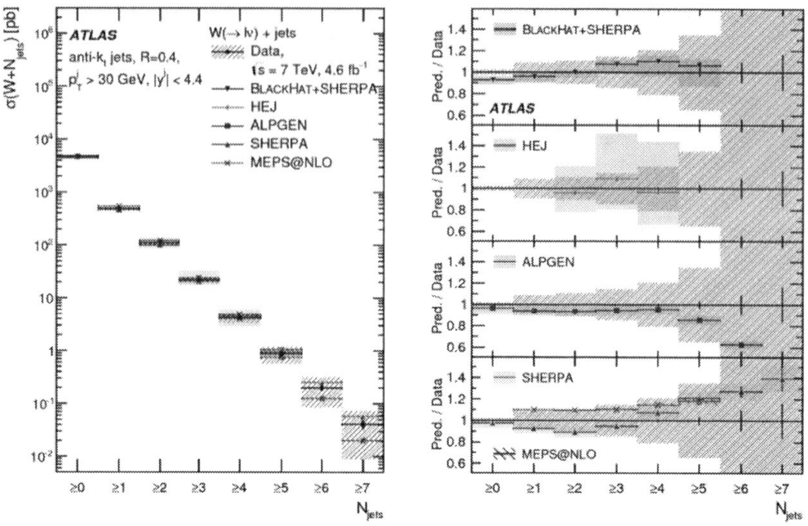

Figure 4: Cross section for the production of W + jets as a function of the inclusive jet multiplicity. For the data, the statistical uncertainties are shown by the vertical bars, and the combined statistical and systematic uncertainties are shown by the black-hashed regions. The data are compared to predictions from BlackHat+Sherpa, HEJ, Alpgen, SHERPA and MEPS@NLO. The left-hand plot shows the differential cross sections and the right-hand plot shows the ratios of the predictions to the data. The theoretical uncertainties on the predictions are described in the text. Taken from Ref. [13].

processes, such as hadronization and multi-parton interactions, are greatly reduced in the ratio. This allows precise comparisons with state-of-the-art Monte Carlo simulations and NLO pQCD calculations to be made. 7 TeV pp collisions are used to perform this measurement [22].
The R_{jets} measurement is done for the electron and muon decay channels of the W and Z bosons for jets with transverse momentum $p_{\text{T}} > 30$ GeV and rapidity $|y| < 4.4$. The measurements of the electron and muon channels are performed in slightly different phase spaces and combined in a common phase space. An extensive set of exclusive and inclusive differential measurements have been measured and compared to a variety of theoretical predictions.
Figure 5 shows R_{jets} as a function of the inclusive and exclusive jet multiplicity (N_{jets}) up to four jets. The ratio is compared to a prediction calculated with BlackHat+Sherpa and the prediction of the Alpgen and Sherpa generators. Small deviations can be seen at larger N_{jets} for the comparison with Sherpa.

6 Run-2 results

After a two year long shut down, ATLAS started to take data from pp collision data at a centre-of-mass energy of $\sqrt{s} = 13$ TeV. The study of leptonic W and Z boson decays is ideal for validating the early detector performance, such as lepton reconstruction, and the analysis software chain.

6.1 W and Z boson production cross sections

W and Z boson production cross sections have been measured by the ATLAS experiment with pp collisions corresponding to an integrated luminosity of approximately 85 pb^{-1}, collected at $\sqrt{s} = 13$

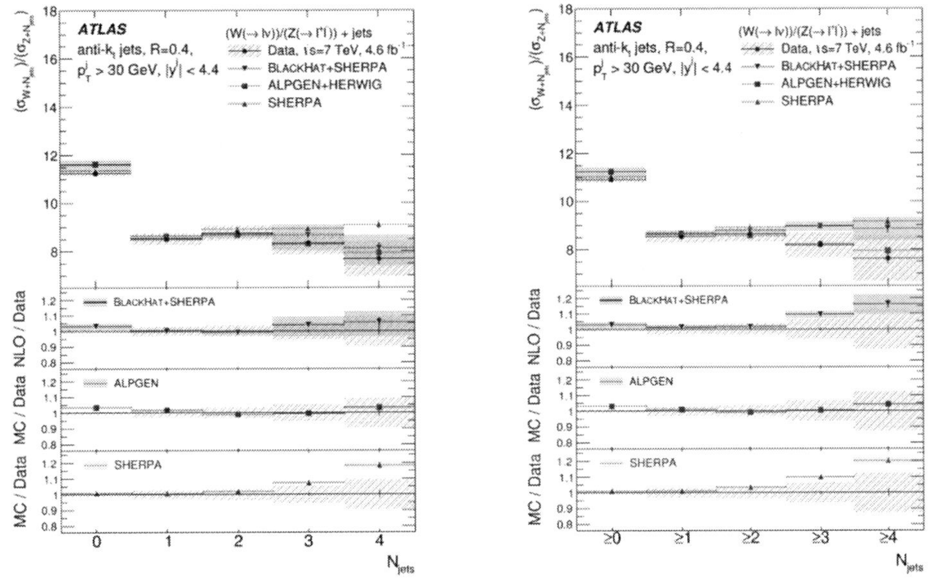

Figure 5: The ratio of W + jets and Z + jets production cross sections, R_{jets} , as a function of exclusive jet multiplicity, N_{jets} , (a) and inclusive jet multiplicity (b). Ratios of the BLACKHAT+SHERPA NLO calculation and the ALPGEN and SHERPA generators to the data are shown in the lower panels. Vertical error bars show the respective statistical uncertainties. The hatched error band shows statistical and systematic uncertainties added in quadrature for the data. The solid error bands show the statistical uncertainties for the ALPGEN and SHERPA predictions, and the combined statistical and theoretical uncertainties for the BLACKHAT+SHERPA prediction. Taken from Ref. [22].

TeV [23]. In particular, the total inclusive production cross sections for the $W \rightarrow e\nu$, $W \rightarrow \mu\nu$, $Z \rightarrow e^+e^-$, $Z \rightarrow \mu^+\mu^-$ processes, the charge-specific cross sections for W^+ and W^- boson production and the cross-section ratios of W^+ to W^- production and of W to Z production have been measured. Figure 6a shows the dilepton mass distribution after a $Z \rightarrow e^+e^-$ selection. In Figure 6b the W boson transverse mass for the $W \rightarrow \mu\nu$ channel selection is shown. The data for both selections are in good agreement with the expectation. The systematic uncertainties shown are obtained using very early studies as well as extrapolation from the Run-1 uncertainties.
The cross section results in the electron and muon channel are then combined taking into account correlated systematic uncertainties. The combined fiducial cross sections are compared in Figure 7 to the predictions which are calculated using different PDF sets. The measurements agree well with the predictions. The measurement experimental precision is already comparable to PDF uncertainties; however, the experimental precision is diluted by the uncertainty of the preliminary luminosity calibration.
The ratios of the combined fiducial cross sections for W^+ to W^- boson production and for W to Z boson production are compared to theory predictions in Figures 8a and 8b. For the ratio $R_{W^+/W^-} = \sigma^{fid}_{W^+}/\sigma^{\text{fid}}_{W^-}$, there is a significant scatter for different PDF predictions and the accuracy of the experimental result is comparable to the spread among them. For the ratio $R_{W/Z} = \sigma^{\text{fid}}_{W}/\sigma^{\text{fid}}_{Z}$, the predictions agree within quoted uncertainties and the measurement is consistent with all of them.

DOI: 10.1051/epjconf/201612604009

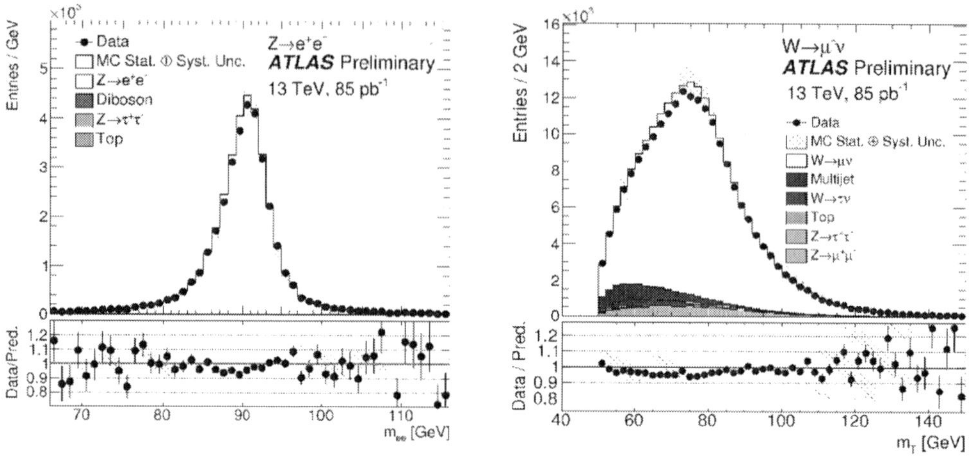

(a) (b)

Figure 6: Dilepton mass distribution after the $Z \to e^+e^-$ selection (a) and transverse mass distribution, calculated from the lepton and the missing transverse energy E_T^{miss} from $W \to \mu\nu$ selection (b). The expected contributions from all backgrounds are estimated with Monte Carlo simulations, except for the multi-jet background in the W boson selection which is estimated with a data-driven method. Systematic uncertainties for the signal and background distributions are combined in the shaded band, the statistical uncertainty is shown on the data points. The luminosity uncertainty is not included. Taken from Ref. [23].

6.2 Z + jets measurement

Using data corresponding to an integrated luminosity of 85 pb^{-1} at 13 TeV, preliminary measurements of the cross section for the production of a Z boson in association with jets have been performed by the ATLAS collaboration [24].
The Z boson is identified using its decays to electron and muon pairs ($Z \to e^+e^-$, $Z \to \mu^+\mu^-$) produced in association with one or more jets in the kinematic range of $p_T^{\text{jet}} > 30$ GeV and $|y^{\text{jet}}| < 2.5$. In order to improve the precision of the measurement, the results in the $Z \to e^+e^-$, $Z \to \mu^+\mu^-$ channels are combined, taking into account correlated systematic uncertainties.
The fiducial production cross sections for Z bosons plus inclusive one to four jet multiplicities have been measured and ratios of the production cross section for successive jet multiplicities have also been derived.
As shown in Figure 9, the measured fiducial cross sections for the various jet multiplicities and the ratios of successive jet multiplicities have also been compared to predictions from Sherpa and from MadGraph. The predictions show a reasonable agreement with the observed cross sections and cross section ratios within the uncertainties.

7 Conclusions

The analyses presented here cover a range of precision measurements of W and Z production at the ATLAS experiment in proton-proton collisions ranging from a center of mass energy from 7 TeV to 13 TeV. ATLAS has also carried out a wide range of measurements of vector boson with accompanying

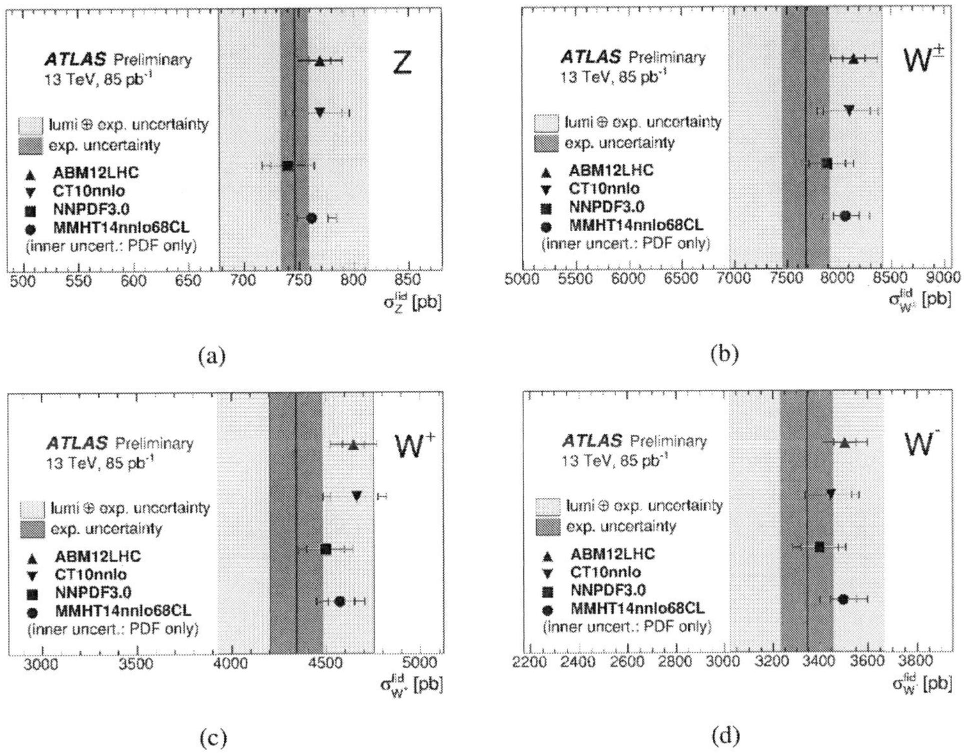

Figure 7: Predictions for the fiducial cross sections $\sigma_{Z}^{\rm fid}$, $\sigma_{W^{\pm}}^{\rm fid}$, $\sigma_{W^{+}}^{\rm fid}$ and $\sigma_{W^{-}}^{\rm fid}$ for the four PDFs CT10nnlo, NNPDF3.0, MMHT14nnlo68cl, and ABM12LHC compared to the measured fiducial cross sections (red line). The green (cyan) band corresponds to the experimental uncertainty without (with) the luminosity uncertainty. The inner error bar of the predictions represents the PDF uncertainty while the outer error bar includes the sum of all other systematics. Taken from Ref. [23].

jet production to test and constrain QCD calculations.
All 7 and 8 TeV precision analyses show overall a good agreement with the theory predictions. First studies of the 13 TeV data show a good understanding of the early ATLAS detector performance and are agreeing with the theoretical expectation.

References

[1] ATLAS Collaboration, JINST **3**, S08003 (2008)
[2] ATLAS Collaboration, JHEP **09**, 049 (2015)
[3] ATLAS Collaboration, JHEP **09**, 145 (2015)
[4] K. Melnikov and F. Petriello, Phys. Rev. **D74** 114017 (2006)
[5] R. Gavin, Y. Li, F. Petriello, and S. Quackenbush, Comput. Phys. Commun. **182**, 2388-2403 (2011)
[6] Y. Li and F. Petriello, Phys. Rev. **D86**, 094034 (2012)
[7] S. Catani, L. Cieri, G. Ferrera, D. de Florian, and M. Grazzini, Phys. Rev. Lett. **103**, 082001 (2009)

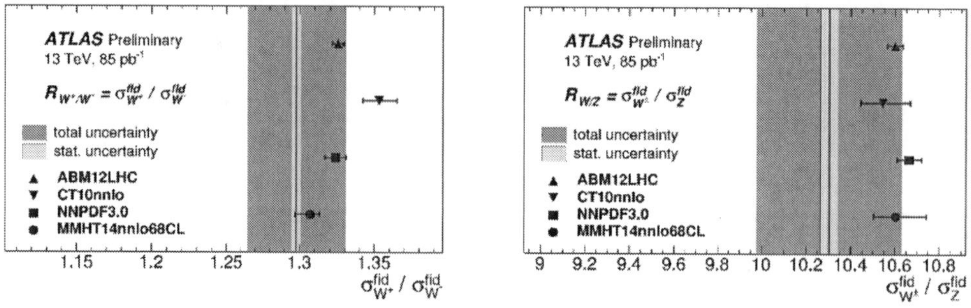

(a) (b)

Figure 8: Ratio of W^+ to W^--boson (a) and Ratio of $W^\pm$ to Z-boson (b) production fiducial cross sections (red line) compared to predictions based on different PDF sets. The inner shaded band corresponds to the statistical uncertainty while the outer band shows statistical and systematic uncertainties added in quadrature. The theory predictions are given with the corresponding PDF uncertainties shown as error bars. Scale uncertainties are not included in the error bars of the predictions. Taken from Ref. [23].

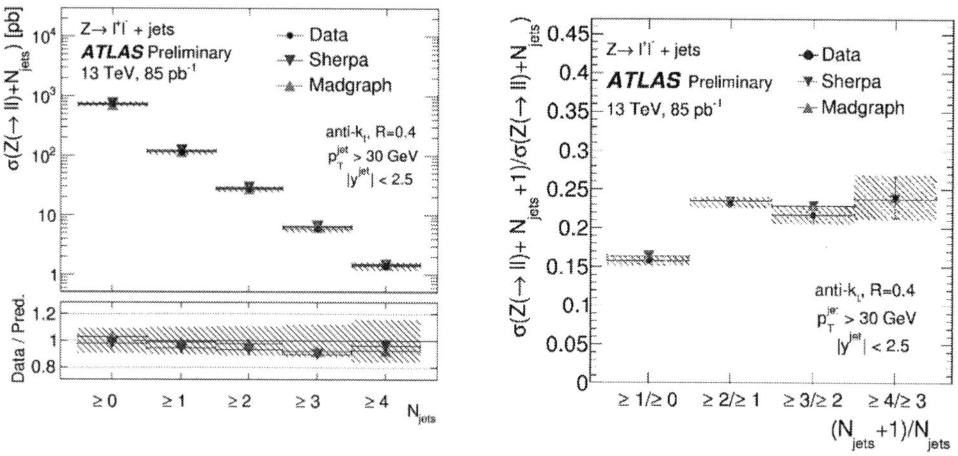

(a) (b)

Figure 9: Combined $Z \to l^+l^-$ plus $\geq N_{\rm jets}$ fiducial cross sections (a), and cross section ratios for successive jet multiplicities. Results are reported at the Born level and compared to predictions from Sherpa and MadGraph. The hatched error band corresponds to the total uncertainty of the results: systematic, statistical, and luminosity uncertainties for the fiducial cross section results (a), systematic and statistic uncertainties for the ratio results (b). Taken from Ref. [24].

[8] S. Catani and M. Grazzini, Phys. Rev. Lett. **98**, 222002 (2007)
[9] C. Balazs and C. P. Yuan, Phys. Rev. **D56**, 5558-5583 (1997)
[10] M. Guzzi, P. M. Nadolsky, and B. Wang, Phys. Rev. **D90** no. 1, 014030 (2014)

[11] ATLAS Collaboration, Phys. Lett. **B720**, 32-51 (2013)
[12] A. Banfi, et al., Eur. Phys. J. **C71**, 1600 (2011)
[13] ATLAS Collaboration, Eur. Phys. J. **C75** no. 2, 82 (2015)
[14] M. L. Mangano, M. Moretti, F. Piccinini, R. Pittau, and A. D. Polosa, JHEP **07**, 001 (2003)
[15] T. Gleisberg, S. Hoeche, F. Krauss, M. Schonherr, S. Schumann, F. Siegert, and J. Winter, JHEP **02**, 007 (2009)
[16] S. Hoeche, F. Krauss, M. Schonherr, and F. Siegert, JHEP **04**, 027 (2013)
[17] J. R. Andersen, T. Hapola, and J. M. Smillie, JHEP **09**, 047 (2012)
[18] J. R. Andersen and J. M. Smillie, JHEP **01**, 039 (2010)
[19] Z. Bern, L. J. Dixon, F. Febres Cordero, S. Hoeche, H. Ita, D. A. Kosower, D. Maitre, K. J. Ozeren, Phys. Rev. **D88** no. 1, 014025 (2013)
[20] C. F. Berger, Z. Bern, L. J. Dixon, F. Febres Cordero, D. Forde, T. Gleisberg, H. Ita, D. A. Kosower, and D. Maitre, Phys. Rev. **D80**, 074036 (2009)
[21] C. F. Berger, Z. Bern, L. J. Dixon, F. Febres Cordero, D. Forde, T. Gleisberg, H. Ita, D. A. Kosower, and D. Maitre, Phys. Rev. Lett. **106**, 092001 (2011)
[22] ATLAS Collaboration, JHEP **07**, 032 (2013)
[23] ATLAS Collaboration, ATLAS-CONF-2015-039, (2015)
[24] ATLAS Collaboration, ATLAS-CONF-2015-041, (2015)

Review of ϕ_3 measurements at Belle

Daniel Červenkov[1,a] on behalf of the Belle collaboration

[1] *Faculty of Mathematics and Physics, Charles University in Prague, Czech Republic*

Abstract. One of the main goals of the Belle experiment (and the upcoming Belle II experiment) is to measure parameters of electro-weak CP violation (CPV). A useful way to parametrize CPV is through the use of unitarity triangles. A brief introduction to unitarity triangle measurements of CP violation is presented, followed by a summary of recent results from the Belle experiment concerning measurements of the least well constrained angle of the standard unitarity triangle — ϕ_3/γ.

1 Unitarity Triangle

Charge-parity (CP) symmetry in weak interactions is broken by a single irreducible complex phase in the Cabibbo-Kobayashi-Maskawa (CKM) quark mixing matrix [1], [2]. This matrix is embedded in the charged current Lagrangian

$$\mathcal{L}_{CC} = \frac{g}{2\sqrt{2}}(\bar{u}\ \bar{c}\ \bar{t})\gamma^{\mu}(1-\gamma_5)\begin{pmatrix} V_{ud} & V_{us} & V_{ub} \\ V_{cd} & V_{cs} & V_{cb} \\ V_{td} & V_{ts} & V_{tb} \end{pmatrix}\begin{pmatrix} d \\ s \\ b \end{pmatrix} W^{+}_{\mu} + \text{h.c.}\,. \tag{1}$$

It is useful to introduce unitarity triangles in order to parametrize the CKM matrix. The matrix is unitary, which can be written down in the following way,

$$\begin{pmatrix} V^*_{ud} & V^*_{cd} & V^*_{td} \\ V^*_{us} & V^*_{cs} & V^*_{ts} \\ \mathbf{V^*_{ub}} & \mathbf{V^*_{cb}} & \mathbf{V^*_{tb}} \end{pmatrix}\begin{pmatrix} \mathbf{V_{ud}} & V_{us} & V_{ub} \\ \mathbf{V_{cd}} & V_{cs} & V_{cb} \\ \mathbf{V_{td}} & V_{ts} & V_{tb} \end{pmatrix} = \begin{pmatrix} 1 & 0 & 0 \\ 0 & 1 & 0 \\ \mathbf{0} & 0 & 1 \end{pmatrix}. \tag{2}$$

Eq. 2 is equivalent to 9 complex scalar equations. Let us focus on the ones that are equal to 0. Out of these, only 3 are independent. We can write down, e.g., the equation comprised of the terms marked in bold.

$$V_{ud}V^*_{ub} + V_{cd}V^*_{cb} + V_{td}V^*_{tb} = 0\,. \tag{3}$$

This is a sum of 3 complex numbers that is equal to 0 — any such equation represents a triangle in the complex plane (however, the triangle might be degenerate — have no area).

The triangle we picked is the standard choice, as it has the benefit of its angles being of the same order. This in turn means, they can be measured independently with reasonable relative errors. Furthermore, the triangle is usually rotated so one of its sides lies on the real axis and it is normalized to a unit length, as can be seen in Fig. 1.

[a] e-mail: cervenkov@ipnp.mff.cuni.cz

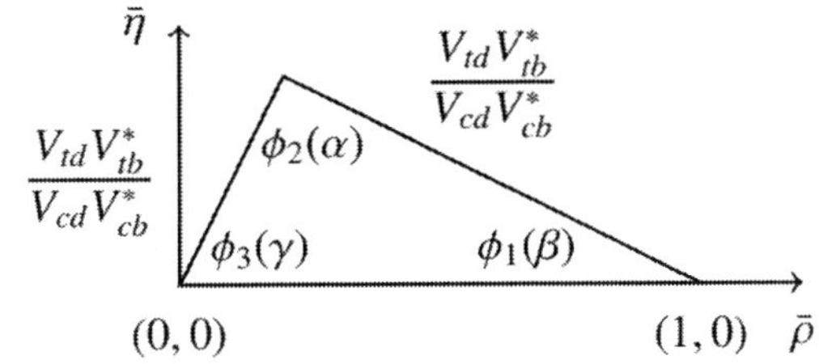

Figure 1. The standard unitarity triangle (Belle angle naming convention first, BaBar convention in parentheses).

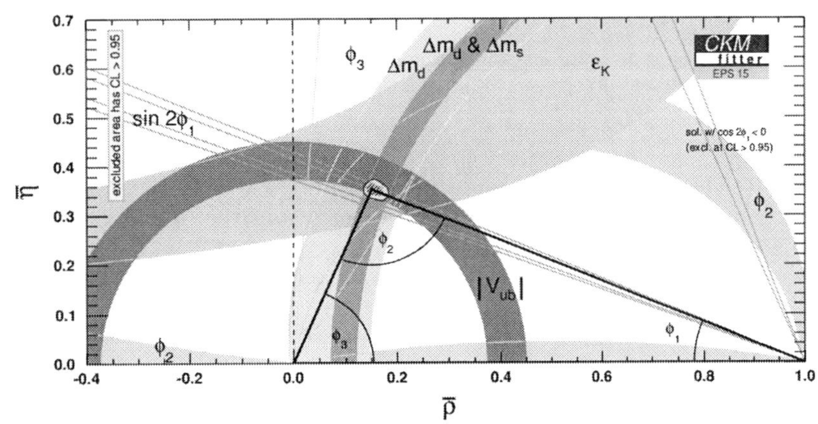

Figure 2. CKMfitter group's global unitarity triangle fit.

The fact, that there is a non-trivial, non-degenerate unitarity triangle implies CP violation. Moreover, careful measurement of the triangle can reveal beyond Standard Model (BSM) physics. Telltale BSM signs would be if the triangle is actually not a triangle (its sides don't close), or if there is a discrepancy between measurements from tree-dominated processes and loop-dominated processes.

The various CKM measurements are combined into a single plot by the CKMfitter group [3], shown in Fig. 2. It's also worth noting that ϕ_3 is the least well constrained angle of the unitarity triangle.

2 Methods for Measuring ϕ_3

The most powerful and theoretically clean way of measuring ϕ_3 is based on the interference between $b \to \bar{u}cs$ and $b \to u\bar{c}s$ tree level amplitudes [4]. What we mean by saying theoretically clean is that there is no penguin graph contribution to these processes and consequently no theoretical uncertainty connected to the hadronic parameters, as they can be obtained from experiment. It is worth noting that ϕ_3 is the only parameter of CP violation that can be measured solely from tree-level processes. However, ϕ_3 measurement precision is limited by small branching fractions of the involved processes.

An example of the mentioned type of processes is $B^\pm \to D^{(*)}K^\pm$ followed by $D \to f$, and $B^\pm \to \bar{D}^{(*)}K^\pm$ followed by $\bar{D} \to f$, where f is a *common final state*. Interference between these two paths gives rise to direct CP violation. Diagrams of the relevant decays are depicted in Fig. 3.

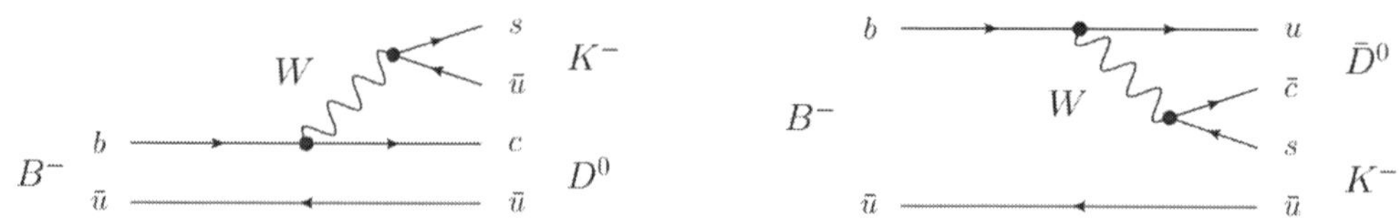

Figure 3. Diagrams of the two tree-level $B^- \to DK^-$ decays.

Most analyses neglect the effects of neutral D meson mixing and CP violation as these are expected to be small [5]. However, D mixing and CP violation can be included at no cost to the uncertainty by using measured values from other studies [6].

2.1 GLW Method

The method was first proposed by Gronau, London and Wyler [4], [7]. Their idea was that a B meson can decay weakly to a state with a D^0 or $\bar{D}^0$. But if we reconstruct the neutral D meson from a final state that is a CP-eigenstate, we actually select a superposition $(D^0 \pm \bar{D}^0)/\sqrt{2}$. We label these states D_{CP^+} and D_{CP^-} for the CP-even and CP-odd states respectively. We can then construct 4 observables that encode the CP violation parameters,

$$\mathcal{R}_{CP^\pm} = 2\frac{\Gamma(B^- \to D_{CP^\pm}K^-) + \Gamma(B^+ \to D_{CP^\pm}K^+)}{\Gamma(B^- \to D_{fav}K^-) + \Gamma(B^+ \to D_{fav}K^+)} = 1 + r_B^2 \pm 2r_B\cos(\delta_B)\cos(\phi_3), \tag{4}$$

$$\mathcal{A}_{CP^\pm} = \frac{\Gamma(B^- \to D_{CP^\pm}K^-) - \Gamma(B^+ \to D_{CP^\pm}K^+)}{\Gamma(B^- \to D_{CP^\pm}K^-) + \Gamma(B^+ \to D_{CP^\pm}K^+)} = \pm r_B\sin(\delta_B)\sin(\phi_3)/R_{CP^\pm}, \tag{5}$$

where D_{fav} signifies that the D meson is reconstructed in a favored hadronic mode such as $D^0 \to K^-\pi^+$, r_B is the magnitude of the ratio of $B \to \bar{D}^0K^-$ and $B \to D^0K^-$ amplitudes and δ_B is the strong phase difference between them. $\mathcal{A}_{CP^\pm}$ different from 0 means CP is violated in these processes.

Notice that the system is not over-constrained by the 4 observables even though there are 3 parametes, as $\mathcal{R}_{CP^+}\mathcal{A}_{CP^+} = -\mathcal{R}_{CP^-}\mathcal{A}_{CP^-}$.

2.2 ADS Method

Atwood, Dunietz and Soni [8] realized that CKM-suppressed decays can also be used to measure ϕ_3. Let us consider the process $B^- \to [K^+\pi^-]_DK^-$, where the parentheses represent a final state that was produced from an intermediate D resonance. The full final state can be reached in two ways — either CKM-favored $B^- \to D^0K^-$ followed by CKM-suppressed $D^0 \to K^+\pi^-$, or CKM-suppressed $B^- \to \bar{D}^0K^-$ followed by CKM-favoured $D^0 \to K^-\pi^+$.

In contrast to the GLW method, when the D meson decays to a non-CP-eigenstate, one has to factor in the magnitude of the ratio of the suppressed and favored D decays r_D as well as their relative strong phase δ_D, much like we did for the B decay. Fortunately, these hadronic parameters can be obtained from mixing measurements [9].

The ADS observables are similar to the GLW ones, however there are only 2 of them as they are charge-averaged,

$$\mathcal{R}_{ADS} = \frac{\Gamma(B^- \to [K^+\pi^-]_D K^-) + \Gamma(B^+ \to [K^-\pi^+]_D K^+)}{\Gamma(B^- \to [K^-\pi^+]_D K^-) + \Gamma(B^+ \to [K^+\pi^-]_D K^+)} = r_B^2 + r_D^2 + 2r_B r_D \cos(\delta_B + \delta_D)\cos(\phi_3). \quad (6)$$

$$\mathcal{A}_{ADS} = \frac{\Gamma(B^- \to [K^+\pi^-]_D K^-) - \Gamma(B^+ \to [K^-\pi^+]_D K^+)}{\Gamma(B^- \to [K^+\pi^-]_D K^-) + \Gamma(B^+ \to [K^-\pi^+]_D K^+)} = 2r_B r_D \sin(\delta_B + \delta_D)\sin(\phi_3)/\mathcal{R}_{ADS}. \quad (7)$$

2.3 Dalitz Plot Method (GGSZ)

The Dalitz plot analysis method for ϕ_3 measurements was proposed by Giri, Grossman, Soffer and Zupan [10] and independently by Bondar [11]. The idea behind it is to use (usually self-conjugate) multi-body final states accessible to both D^0 and $\bar{D}^0$ mesons. One then measures relative phases and magnitudes of their amplitudes for D mesons coming from B decays such as $B^\pm \to DK^\pm$.

Let us consider a $B^\pm \to [K_S^0\pi^+\pi^-]_D K^\pm$ process. Its amplitude can be written down as

$$A_{B^+}(s_+, s_-) = \bar{A}_D(s_+, s_-) + r_B e^{i(\delta_B+\phi_3)} A_D(s_+, s_-), \quad (8)$$

$$A_{B^-}(s_+, s_-) = A_D(s_+, s_-) + r_B e^{i(\delta_B+\phi_3)} \bar{A}_D(s_+, s_-), \quad (9)$$

where we have introduced the Dalitz variables $s_\pm = m^2_{K_S^0\pi^\pm}$ and A_D is the amplitude of the D decay. The strong phase $\delta_B \equiv \delta_B(s_+, s_-)$ has to have a large variation over the Dalitz plot; if it were constant, there would be no ϕ_3 sensitivity.

Two approaches are possible — model independent, binned analysis (as proposed in the original paper) and model dependent, unbinned analysis. The former divides the Dalitz plot into bins across which there is a small strong phase variation. The events in one bin are then treated equally. Extra input in the form of strong phase measurements from charm factories are required for this method [12].

The second method employs a certain model of the strong phase distribution function across the Dalitz phase-space. While introducing an obvious model uncertainty, it has a higher statistical power, which can be very desirable in low yield analyses.

2.4 Continuum Suppression

A vital procedure in these measurements is called continuum suppression. It distinguishes B meson decay events from $e^+e^- \to q\bar{q}$ type events, where q stands for u, d, s or c type quarks. The name *continuum* comes from the fact, that these events hadn't gone through an intermediate $\Upsilon(4S)$ resonance, in contrast to the B meson decay events.

These backgrounds can be quite prominent, but have very different event shapes (solid angle distribution of decay particles); see Fig. 4. This allows us to exploit a variety of observables such as thrust, sphericity, Fox-Wolfram moments, Δz and flavor tag to discriminate against these backgrounds. Usually neural networks or boosted decision trees are used to combine many of the listed variables into a single one with high discriminating power.

3 Measurements

All the listed analyses use the entire 711 fb^{-1} Belle data sample containing $772\times10^6 B\bar{B}$ events. Listed uncertainties are statistical first, systematic second.

DOI: 10.1051/epjconf/201612604010

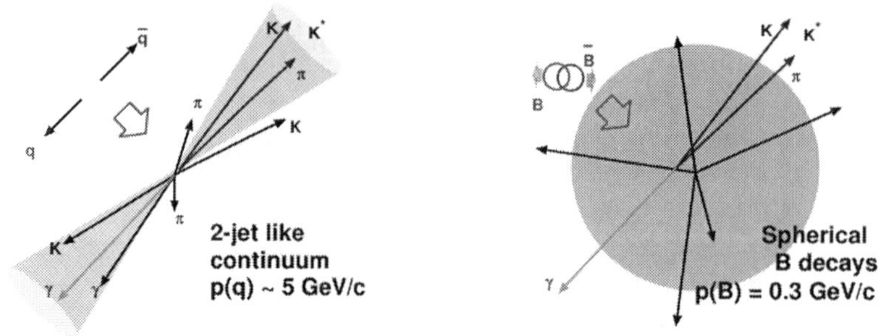

Figure 4. Event shapes for continuum (left) and B decay events (right) [13].

3.1 $B^- \to D^{*0}K^-, D^{*0} \to D^0\pi^0, D^0 \to K_S^0\pi^0, K_S^0\eta$ (GLW)

It is of note, that this measurement is a B-factory exclusive — it would be very difficult to study this channel on hadron machines, because of the multiple neutral final-state particles.

The analysis was carried out in both $D^{*0} \to D^0\pi^0$ and $D^{*0} \to D^0\gamma$ modes. Combining them with CP-even D^0 final-states $\pi^+\pi^-$ and K^+K^-, yields the following results [14],

$$R_{CP+} = +1.19 \pm 0.13 \pm 0.03, \tag{10}$$
$$R_{CP-} = +1.03 \pm 0.13 \pm 0.03, \tag{11}$$
$$A_{CP+} = -0.14 \pm 0.10 \pm 0.01, \tag{12}$$
$$A_{CP-} = +0.22 \pm 0.11 \pm 0.01. \tag{13}$$

The observables show clear hints of CP asymmetry.

3.2 $B^- \to D^0K^-, D^0 \to K^+\pi^-$ (ADS)

The authors updated this study with the full Belle dataset, adding 20% more data to the previous study [15]. They also incorporated several improvements such as a 2D fit for signal yield extraction and a neural network for background suppression. This resulted in a first observation of the suppressed mode, with a 4.1σ significance (including systematic uncertainties) [16],

$$R_{DK} = (1.63^{+0.44+0.07}_{-0.41-0.13}) \times 10^{-2}, \tag{14}$$
$$A_{DK} = -0.39^{+0.26+0.04}_{-0.28-0.13}. \tag{15}$$

3.3 $B^0 \to D^0K^{*0}, K^{*0} \to K^+\pi^-, D^0 \to K^-\pi^+$ (ADS)

A major advantage of this analysis is that it uses a self-tagging channel, because of the K^* decay. This in turn boosts efficiency. While the study uses the ADS method, the fact that K^* has a natural width larger than the experimental resolution, leads to some complications which require a modified definition of the ADS observables. Further details can be found in [17]. The obtained result is [18]

$$R_{DK^{*0}} = (4.1^{+5.6+2.8}_{-5.0-1.8}) \times 10^{-2}. \tag{16}$$

As the $R_{DK^{*0}}$ value is not significant, an upper limit was established

$$R_{DK^{*0}} < 0.16 \quad (95\%\ \text{C.L.}). \tag{17}$$

3.4 $B^- \to D^0 K^-, D^0 \to K^+\pi^-\pi^0$ (ADS)

The fact that there is a 3-body final-state in this ADS analysis means that the strong phase difference between the interfering processes can vary over the phase-space. This can "dilute" direct CP violation effects. To quantify this dilution, correlated $D\bar{D}$ production is used [19], [20]. Fortunately, the effect is small in this channel and allows for a CP violation measurement [21]:

$$R_{DK} = (1.98 \pm 0.62 \pm 0.24) \times 10^{-2},$$
$$A_{DK} = 0.41 \pm 0.30 \pm 0.05.$$

This is a first evidence for the suppressed decay, with a 3.2σ significance.

3.5 $B^- \to D^0 K^-, D^0 \to K^0_S\pi^+\pi^-$ (Dalitz)

This analysis is the first model-independent measurement of ϕ_3 in the relevant channel. The authors report a result of [22]

$$\phi_3 = (77^{+15.1}_{-14.9} \pm 4.1 \pm 4.3)^\circ, \tag{18}$$

where the third uncertainty comes from the precision of the strong-phase parameters obtained by CLEO, which are an external input to this analysis.

This uncertainty is comparable to the model uncertainties from the latest Belle and BaBar measurements: $3^\circ - 9^\circ$. For future experiments, the model uncertainty is expected to dominate as there will be more statistics and, possibly, better systematics control. On the other hand, the precision of the strong-phase parameters measurement will improve as BES-III results supersede CLEO's.

3.6 $B^0 \to D^0 K^{*0}, K^{*0} \to K^+\pi^-, D^0 \to K^0_S\pi^+\pi^-$ (Dalitz)

This analysis also uses the model-independent approach and resulted in establishing an upper limit on the suppressed vs. favored ratio $r_S < 0.87$ at the 68% confidence level [23]. This value is closely related to ϕ_3 sensitivity because the statistical uncertainty of ϕ_3 measurements is proportional to $1/r_S$.

The authors also report the "raw" observables $x_\pm$, $y_\pm$ which are defined as

$$x_\pm = r_S \cos(\delta_S \pm \phi_3),$$
$$y_\pm = r_S \sin(\delta_S \pm \phi_3)$$

and are measured to be

$$x_- = +0.4^{+1.0+0.0}_{-0.6-0.1} \pm 0.0,$$
$$y_- = -0.6^{+0.8+0.1}_{-1.0-0.0} \pm 0.1,$$
$$x_+ = +0.1^{+0.7+0.0}_{-0.4-0.1} \pm 0.1,$$
$$y_+ = +0.3^{+0.5+0.0}_{-0.8-0.1} \pm 0.1,$$

where the third uncertainty again comes from the precision of the strong-phase parameters, as in Sec. 3.5.

These observables have the benefit, that they can be readily merged with similar results from other studies for a combined measurement.

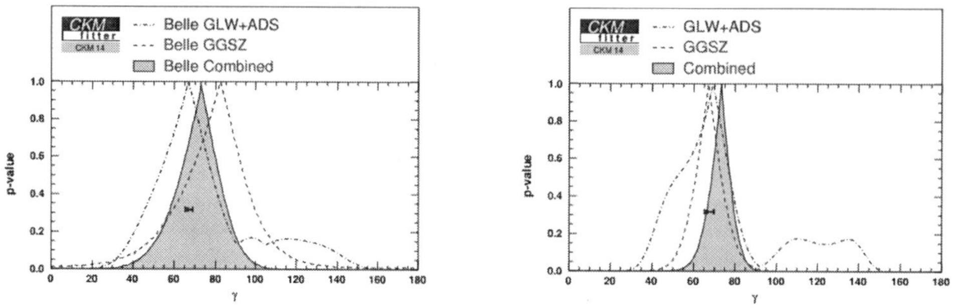

Figure 5. ϕ_3 combination fit; Belle (left) and world (right) [3].

4 Conclusion

Precision of the measurements of the ϕ_3 angle of the unitarity triangle is currently dominated by statistical uncertainty. The current Belle and world combination ϕ_3 fits can be seen in Fig. 5.

With the new generation of flavor experiments, Belle II and the LHCb upgrade, along with inputs from charm threshold provided by BES-III, the world combination of the least well-known parameter of the unitarity triangle — ϕ_3, could reach sub-degree precision. It is important we continue constraining the unitarity triangle as New Physics *could* very well be hiding there.

Acknowledgement

The author would like to thank the Grant Agency of the Charles University in Prague (GAUK), which supported this work via the grant #1370313.

References

[1] N. Cabibbo, Phys. Rev. Lett. **10**, 531 (1963)
[2] M. Kobayashi, T. Maskawa, Progr. Theoret. Phys. **49**, 652 (1973)
[3] J. Charles et al. (CKMfitter Group), Eur. Phys. J. **C41**, 1 (2005), `hep-ph/0406184`
[4] M. Gronau, D. London, Phys. Lett. **B253**, 483 (1991)
[5] A.F. Falk, Y. Grossman, Z. Ligeti, Y. Nir, A.A. Petrov, Phys. Rev. **D69**, 114021 (2004), `hep-ph/0402204`
[6] M. Rama, Phys. Rev. **D89**, 014021 (2014), `1307.4384`
[7] M. Gronau, D. Wyler, Phys. Lett. **B265**, 172 (1991)
[8] D. Atwood, I. Dunietz, A. Soni, Phys. Rev. D **63**, 036005 (2001)
[9] K.A. Olive et al. (Particle Data Group), Chin. Phys. **C38**, 090001 (2014)
[10] A. Giri, Y. Grossman, A. Soffer, J. Zupan, Phys. Rev. D **68**, 054018 (2003)
[11] A. Bondar, unpublished, Belle Dalitz analysis meeting, (2002)
[12] J. Libby et al. (CLEO Collaboration), Phys. Rev. D **82**, 112006 (2010)
[13] M. Nakao, unpublished, Belle Analysis School, (2009)
[14] K. Trabelsi, Y. Horii et al. (Belle Collaboration), unpublished, CKM Preliminary, (2012)
[15] Y. Horii, K. Trabelsi, H. Yamamoto et al. (Belle Collaboration), Phys. Rev. D **78**, 071901 (2008)

[16] Y. Horii, K. Trabelsi, H. Yamamoto et al. (Belle Collaboration), Phys. Rev. Lett. **106**, 231803 (2011), 1103.5951
[17] B. Aubert et al. (BaBar Collaboration), Phys. Rev. D **80**, 031102 (2009)
[18] K. Negishi, Y. Horii, Y. Onuki, T. Sanuki, H. Yamamoto et al. (Belle Collaboration), Phys. Rev. **D86**, 011101 (2012), 1205.0422
[19] N. Lowrey et al. (CLEO Collaboration), Phys. Rev. **D80**, 031105 (2009), 0903.4853
[20] J. Libby et al., Phys. Lett. **B731**, 197 (2014), 1401.1904
[21] M. Nayak, J. Libby, K. Trabelsi et al. (Belle Collaboration), Phys. Rev. **D88**, 091104 (2013), 1310.1741
[22] H. Aihara et al. (Belle Collaboration), Phys. Rev. **D85**, 112014 (2012), 1204.6561
[23] A. Abdesselam et al. (Belle Collaboration) (2015), 1502.07550

EPJ Web of Conferences **126**, 04011 (2016) DOI: 10.1051/epjconf/201612604011
ICNFP 2015

Single top-quark production cross-section and properties using the ATLAS detector at the LHC

Andrew M. Chegwidden[1,a], On behalf of the ATLAS Collaboration

[1] *Department of Physics and Astronomy, Michigan State University*

Abstract. ATLAS measurements of single top-quark processes are summarised. Measurements using data recorded at the LHC proton-proton collider at centre-of-mass energies of $\sqrt{s}$=7 TeV and 8 TeV are presented. The measurements are performed using the semi-leptonic decay mode of the top quark. Production cross-sections for the t-channel and Wt modes of single top-quark production and the extraction of the CKM matrix element $|V_{tb}|$ are shown. All measurements are compared to theoretical calculations. In addition, the s-channel production mode is explored along with limits on exotic production modes.

1 Introduction

Studying single top-quark processes allows for both a further understanding of the Standard Model (SM) as well as a probe for beyond the Standard Model (BSM) theories. Single top-quark processes also allow for unique tests of the SM. The ratio of the parton distribution functions (PDF) for the up and down quarks and the CKM matrix element $|V_{tb}|$ are easily extracted from measurements of single top-quark production cross-sections. Many BSM searches utilise single top-quark final states as their signature or as a background. These proceedings present measurements of single top-quark processes in proton-proton (pp) collisions at centre-of-mass energies of $\sqrt{s}$=7 TeV and 8 TeV using data recorded by the ATLAS detector [1] at the LHC [2].

2 Measurements

The production of single top quarks at the LHC occurs through three relevant subprocesses: t-channel production where a space-like W boson is exchanged, associated production of a top-quark and an on-shell W boson (Wt-channel), and s-channel production which is mediated by an off-shell W boson.

[a] e-mail: andrew.chegwidden@cern.ch

DOI: 10.1051/epjconf/201612604011

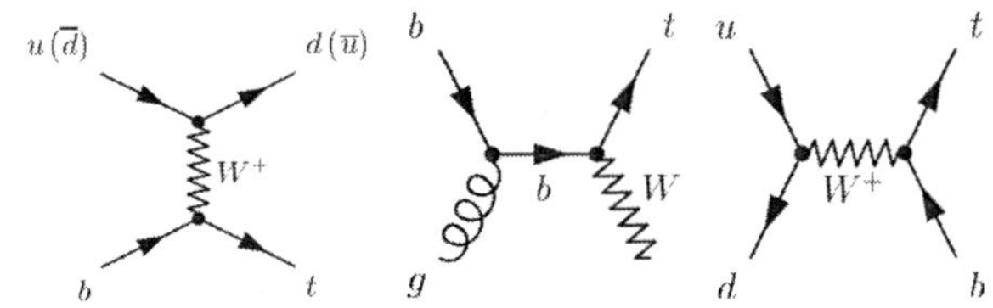

(a) (b) (c)

Figure 1. Feynman diagrams of single top-quark production processes: (a) t-channel production, (b) associated Wt production, and (c) s-channel production.

The Feynman diagrams for these processes are shown in Figure 1.

2.1 Measurement of the t-channel cross-section

The t-channel production mode is the dominant single top production mode at the LHC. The t-channel cross-section is predicted at next-to-leading order with next-to-next-to-leading logarithm gluon resummation. They are calculated to be $\sigma_t = 64.6.8^{+2.6}_{-1.7}$ pb and $\sigma_t = 87.8^{+3.4}_{-1.9}$ pb for $\sqrt{s}$=7 TeV and $\sqrt{s}$=8 TeV, respectively [3]. The analysis uses the semi-leptonic decay of the W boson such that the final state consists of two high-p_{T} jets (one forward light-quark from the initial hard scatter and one central b-tagged jet from the W decay), one isolated high p_{T} lepton, and large $E_{\mathrm{T}}^{\mathrm{miss}}$. This selection is applied to reconstructed events and is also used to define a fiducial region based on particle-level objects. Signal discrimination from background events is enhanced through the use of a neural-network (NN) based on 14 final state variables. The two most discriminating variables are the pseudorapidity[1] of the untagged jet $|\eta(j)|$ and the reconstructed mass of the top-quark $m(\ell\nu b)$.

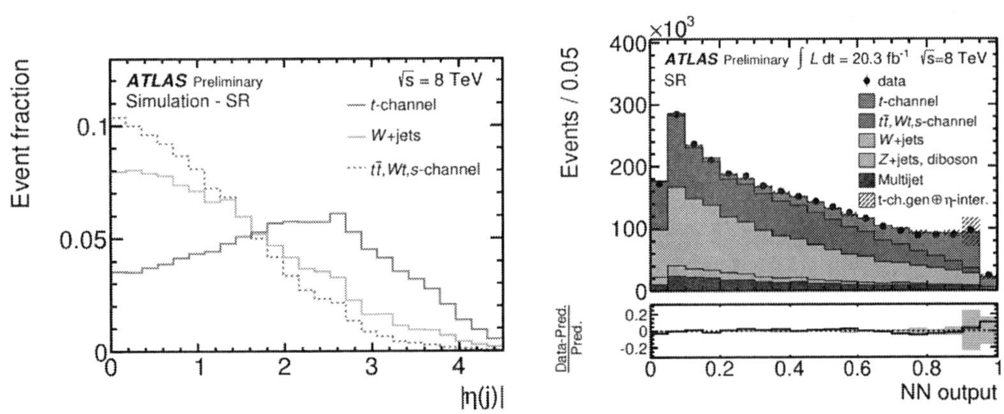

(a) (b)

Figure 2. Distributions of the (a) most discriminating variable, the η of the untagged jet and of (b) the NN output-distribution for the combined electron and muon channel are shown. The shaded band includes the t-channel generator uncertainty and the jet energy scale η-intercalibration uncertainty. The Signal Region (SR) corresponds to events with exactly one lepton, missing transverse momentum, and exactly one b-tagged jet [4].

[1] ATLAS uses a right-handed coordinate system with its origin at the nominal interaction point (IP) in the centre of the detector and the z-axis along the beam pipe. The x-axis points from the IP to the centre of the LHC ring, and the y-axis points upward. Cylindrical coordinates (r, ϕ) are used in the transverse plane, ϕ being the azimuthal angle around the beam pipe. The pseudorapidity is defined in terms of the polar angle θ as $\eta = -\ln\tan(\theta/2)$.

DOI: 10.1051/epjconf/201612604011

The $|\eta(j)|$ distribution and the output of the NN can bee seen in Figures 2(a) and 2(b), respectively [4]. The estimated number of t-channel single top-quark events is obtained from a binned maximum-likelihood fit of the NN output and is used to calculate a fiducial cross-section. The fiducial cross-section is calculated using the inclusive cross-section for each MC generator and the corresponding selection efficiency of the particle-level selection. A fiducial cross-section of $\sigma_{\text{fid}} = 3.37 \pm 0.05(\text{stat}) \pm 0.48(\text{syst})$ pb is obtained [4].

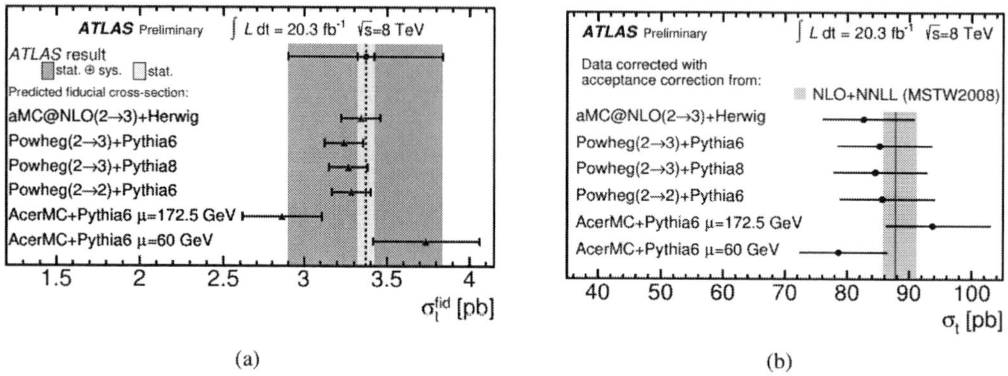

Figure 3. (a) The t-channel fiducial cross-section predictions for different NLO MC generators together with the measured value and (b) the total t-channel cross-section measurements obtained from the fiducial cross-section using the acceptance of different MC generators [4].

Figure 3(a) shows the extrapolation of the fiducial cross-section to the full phase space using the acceptance from various MC generator models while Figure 3(b) shows the extrapolated inclusive cross-section [4]. Using the acceptance of the fiducial volume from the aMC@NLO + Herwig event generator [5], an inclusive t-channel cross-section of $\sigma_t = 82.6 \pm 1.2(\text{stat}) \pm 11.4(\text{syst})$ pb is obtained [4].

2.2 Measurement of the Wt-channel cross-section

The associated production of a top quark and a W boson is the second-largest production mode of single top events at the LHC. The Wt-channel cross-section is predicted at next-to-leading order with next-to-next-to-leading logarithm gluon resummation. They are calculated to be $\sigma_{Wt} = 15.74^{+1.17}_{-1.21}$ pb and $\sigma_{Wt} = 22.37 \pm 1.52$ pb for $\sqrt{s}$=7 TeV and $\sqrt{s}$=8 TeV, respectively [6]. The analysis uses events with a final state consisting of one electron, one oppositely charged muon, $E_{\text{T}}^{\text{miss}}$, and at least one b-tagged central jet. Signal discrimination from background events is enhanced through the use of a Boosted Decision Tree (BDT) classifier.

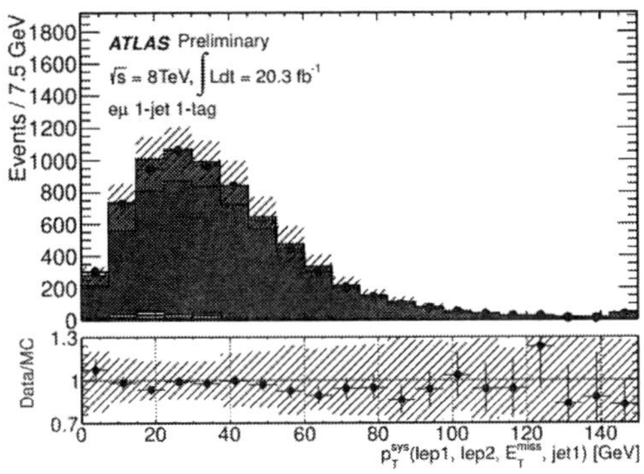

(a)

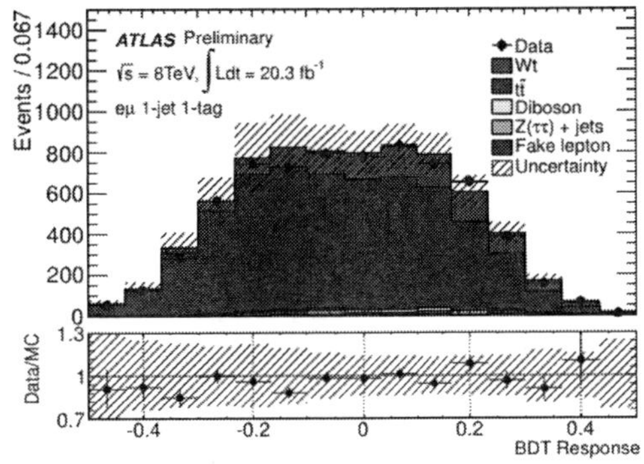

(b)

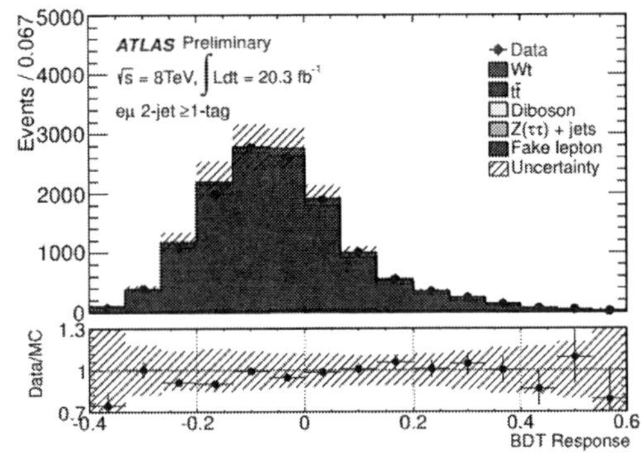

(c)

Figure 4. (a) Distribution of the the p_T of the vectorial sum of the leptons, E_T^{miss}, and the leading jet. BDT classifier distributions for (b) 1-jet and (c) 2-jet events. The uncertainty band is the quadratic sum of all systematic uncertainties [7].

Of the 19 variables used in the analysis, the one with the highest discriminating power is shown in Figure 4(a). Classifiers are trained for both 1-jet and 2-jet events. The outputs for the 1-jet and 2-jet classifiers are shown in Figures 4(b) and 4(c), respectively. A maximum-likelihood fit to the BDT classifier distributions is performed to extract the signal cross-section. The result of the fit is a Wt-channel cross-section measurement of $\sigma_{Wt} = 27.2 \pm 2.8(\text{stat}) \pm 5.4(\text{syst})$ pb [7].

2.3 CKM matrix element $|Vtb|$

Single top-quark production in the t- and Wt-channels proceeds via the W-t-b vertex. As such, the cross-section for the production modes is proportional to $|V_{tb}|^2$. The SM predicts $|V_{tb}|$ to be close to 1.0 and its value is sensitive to new physics such as additional quark generations. Measurements of $|V_{tb}|$ are independent of the assumption of the number of quark generations as well as the unitarity of the CKM matrix. The only assumptions made are that $|V_{tb}| \gg |V_{td}|$, $|V_{ts}|$, meaning $\mathcal{B}(t \to bW) = 1$, and that the W-t-b vertex is a SM-like weak coupling. The observed value of $|V_{tb}|^2$ is calculated by dividing the measured cross-section by the theoretical prediction,

$$|V_{tb} \cdot f|^2 = \frac{\sigma_{\text{exp}}}{\sigma_{\text{theory}}} \tag{1}$$

and with a fixed top-quark mass of 172.5 GeV. In Equation (1), f is a coupling which is allowed to be bigger than the SM value of 1. The values of $|V_{tb} \cdot f|$ extracted from the t- and Wt-channel cross-sections are $|V_{tb}| = 0.97^{+0.09}_{-0.10}$ [4] and $|V_{tb}| = 1.10 \pm 0.12$ [7], respectively. Lower limits on the value of $|V_{tb}|$ at the 95% confidence level using the t- and Wt-channel production cross-sections are $|V_{tb}| > 0.78$ [4] and $|V_{tb}| > 0.72$ [7], respectively.

3 SM and BSM searches

Many BSM searches have been conducted at the LHC involving events having final states similar to that of single top-quark processes. While the s-channel production mode has been observed at the Tevatron, it has proven difficult to observe at the LHC. Searches involving additional heavy gauge bosons and dark matter candidates have also been conducted.

3.1 Search for SM s-channel production

Single top-quark production in the s-channel was first observed at the Tevatron in proton-antiproton ($p\bar{p}$) collisions by the CDF and D0 collaborations [8]. Measuring the s-channel production mode at the LHC is difficult as the signal-to-background ratio is much smaller in pp collisions than in $p\bar{p}$ collisions due to the quark-antiquark initial state. The s-channel cross-section is predicted at next-to-leading order with next-to-next-to-leading logarithm gluon resummation. They are calculated to be $\sigma_s = 4.63^{+0.20}_{-0.18}$ pb and $\sigma_s = 5.61 \pm 0.22$ pb for $\sqrt{s}$=7 TeV and $\sqrt{s}$=8 TeV, respectively [9]. The analysis selects events with a single lepton, exactly two b-tagged jets, and large $E_{\mathrm{T}}^{\mathrm{miss}}$. To discriminate between signal and background events, a BDT event classifier is utilised. Nineteen final state variables are used as input to the BDT such that the expected signal significance is maximised. The variable with the highest discriminating power is the difference in azimuthal angle between the leading b-tagged jet and the top-quark candidate reconstructed with the sub-leading b-tagged jet.

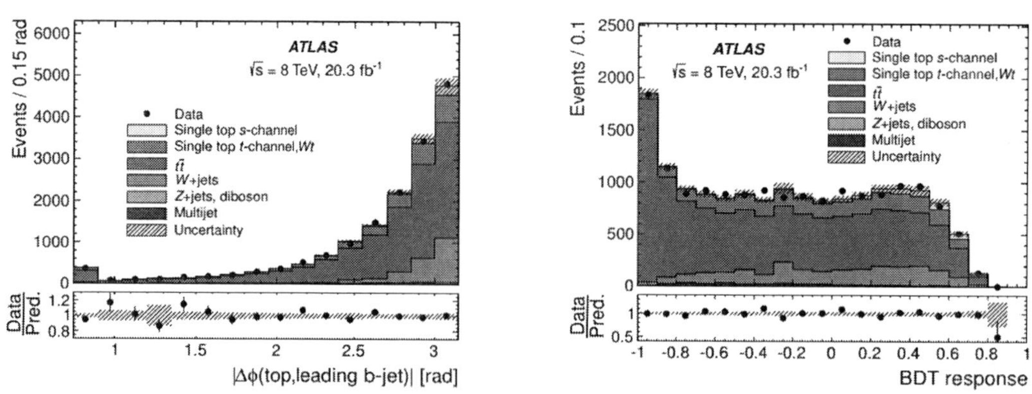

(a) (b)

Figure 5. Distributions of (a) the most discriminating variable used in BDT classifier, the $|\Delta\phi|$ between the leading b-tagged jet and the top-quark candidate reconstructed with the sub-leading b-tagged jet and (b) observed and predicted values of the BDT response. The uncertainty band represent the normalisation uncertainty of all processes after the fit and added in quadrature with Monte Carlo statistical uncertainty [10].

Distributions of the most discriminating variable and of the BDT output are shown in Figures 5(a) and 5(b), respectively. The signal is extracted in a binned maximum-likelihood fit of the BDT output distribution. The fit gives a cross-section of $\sigma_s = 5.0 \pm 4.3$ pb, which is consistent with the SM. The observed upper limit on the s-channel cross-section is 14.6 pb at the 95% CL [10].

3.2 Search for additional heavy gauge bosons, W'

The single top-quark final state can be used to search for additional charged heavy gauge bosons. The analysis performs a search in the $W' \to t\bar{b} \to \ell\nu b\bar{b}$ decay mode. This process is similar to the SM s-channel production mode except it is mediated by the W', a heavier version of the W boson. The analysis utilises a BDT classifier in order to discriminate signal from background. Four BDTs are trained for two-jet and three-jet multiplicity events and W' boson signals with left-handed and right-handed couplings to the top quark. Ten (eleven) variables are used as inputs in the BDT in 2-jet (3-jet) searches. Of these variables, $m_{t\bar{b}}$ and $p_{\mathrm{T}}(t)$ have the highest discriminating power.

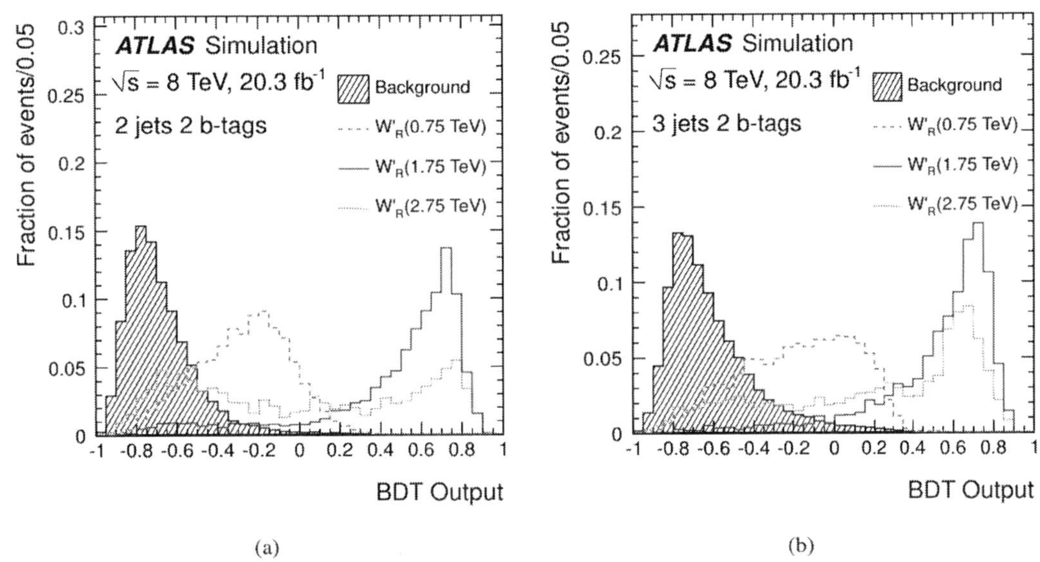

Figure 6. Distributions of the BDT output for three different mass hypotheses of the W'_R boson in (a) the 2-jet and (b) the 3-jet multiplicity signal regions. Electron and muon channels are combined and all distributions are normalised to unity [11].

The BDT outputs for 2-jet and 3-jet multiplicity events are shown in Figures 6(a) and 6(b) respectively. As no excess in data is observed, exclusion limits are set. For a left-handed (right-handed) W' boson, masses below 1.70 (1.92) TeV are excluded at the 95% CL [11].

3.3 Monotop

Many BSM theories predict the production of events with large missing transverse momentum in association with a single reconstructed object. Such searches are colloquially known as Mono-X searches and have taken place at the LHC where the single object is taken to be a photon, a heavy quark, or a W or Z boson. The monotop analysis follows the spirit of these searches where the final state consists of a singly-produced top quark in association with a significant amount of missing transverse momentum. This missing transverse momentum could correspond to one or several undetected neutral particles.

These particles can either be stable or weakly interacting and can be interpreted as dark matter candidates. The analysis considers two main production modes. The first is the resonant production of a +2/3 charged spin-0 boson, S, decaying into a right-handed top-quark and a neutral colour-singlet spin-1/2 fermion, f_{met}. The second is the non-resonant production of a neutral, color singlet, spin-1 boson, v_{met}, in association with a right-handed top-quark.

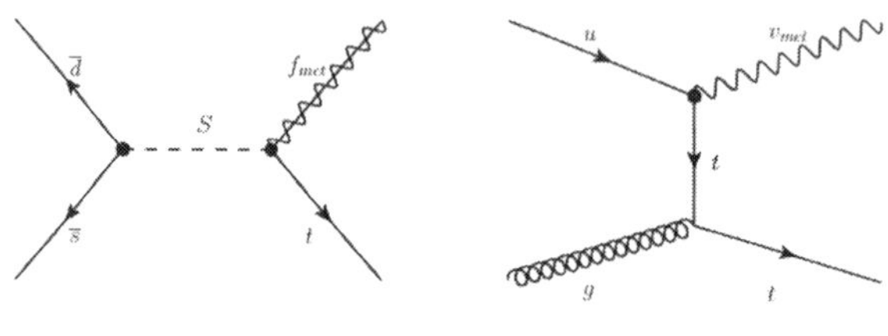

(a) (b)

Figure 7. Feynman diagrams of leading-order processes leading to monotop events through (a) the production of a colored scalar resonance S decaying into a top-quark and a spin-1/2 fermion f_{met} in the resonant model and (b) the t-channel non-resonant production of a top quark in association with a spin-1 boson v_{met} in the non-resonant model [12].

The Feynman diagrams for these two production modes are shown in Figure 7. The analysis utilises a cut-based approach for the discrimination of the signal from the background. The monotop signal is prominent in regions of phase space characterised by high transverse mass calculated with the lepton and missing transverse momentum, $m_{\text{T}}(\ell, E_{\text{T}}^{\text{miss}})$, as well as small values of $|\Delta\phi(\ell, b)|$.

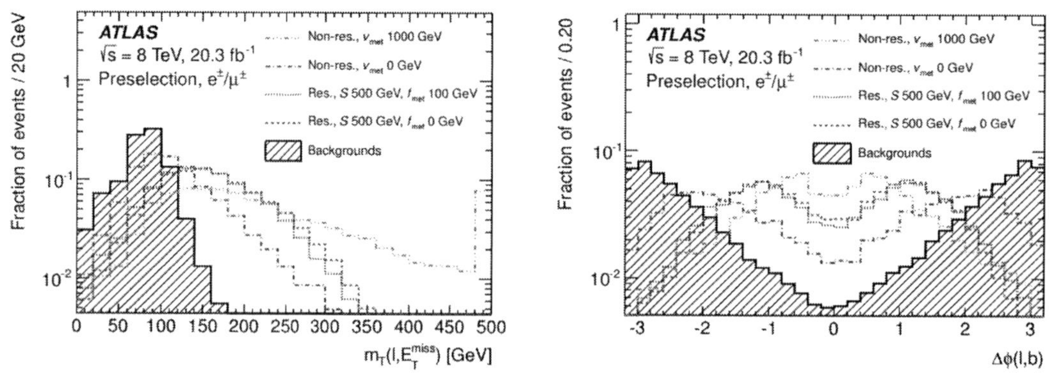

(a) (b)

Figure 8. Distributions normalised to unity of (a) $m_{\text{T}}(\ell, E_{\text{T}}^{\text{miss}})$ and of (b) $|\Delta\phi(\ell, b)|$. The expected distributions for the resonant model with $m(S) = 500$ GeV are shown for the $m(f_{\text{met}}) = 0$ GeV and $m(f_{\text{met}}) = 100$ GeV hypotheses, as well as for the non-resonant model for the $m(v_{\text{met}}) = 0$ GeV and $m(v_{\text{met}}) = 1000$ GeV hypotheses. All distributions are compared to the expected distribution for the backgrounds [12].

Figure 8 shows the distribution of the two variables used in the analysis to discriminate signal from background. As no excess is observed in data, 95% CL upper limits on the signal production cross-sections are set.

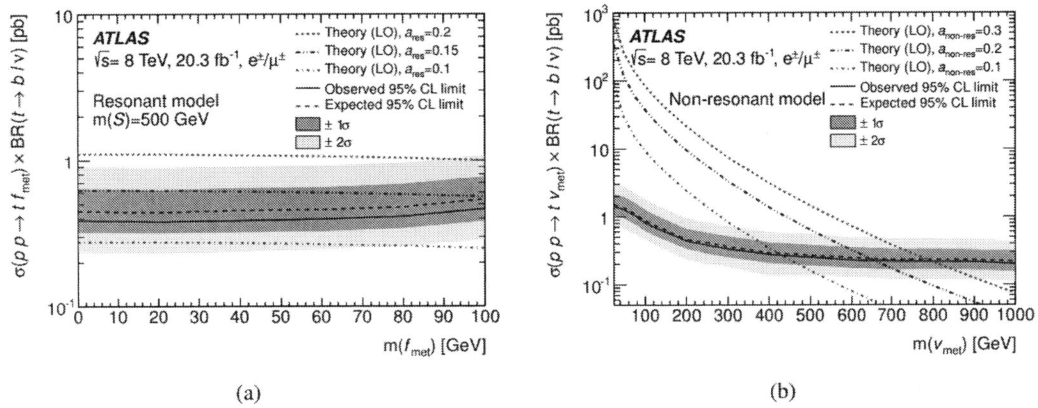

Figure 9. Observed and expected limits on the cross-section times branching ratio for the (a) resonant model with $m(S) = 500$ GeV and (b) non-resonant model, as a function of $m(f_{\rm met})$ and $m(v_{\rm met})$, respectively [12].

Figure 9 shows the limit at the 95% CL on the production cross-section times branching ratio as a function of the invisible state, for each of the two signal models. In the case of resonant production, cross-sections are excluded across the entire mass range. In the non-resonant production mode, cross-sections up to $m(v_{\rm met}) = 657$ GeV are excluded [12].

4 Summary

Single top-quark measurements and properties were presented along with current limits on exotics production modes and an upper limit on the SM s-channel production cross-section. Single top-quark events continue to serve as a test of the SM as well as a probe for BSM theories. Many new results should be forthcoming as data are collected from $\sqrt{s}$=13 TeV collisions.

Acknowledgments

The author wishes to thank the organisers for the invitation and gratefully acknowledges the support of the National Science Foundation (grant PHY-1068318).

References

[1] ATLAS Collaboration, JINST **3** (2008) S08003.
[2] L. Evans and P. Bryant (editors), JINST **3** (2008) S08001.
[3] N. Kidonakis,Phys.Rev.D **83** (2011) 091503.
[4] ATLAS Collaboration, ATLAS-CONF-2014-007, http://cds.cern.ch/record/1668960 .
[5] G. Corcella et al., JHEP 0101 (2001) 010.
[6] N. Kidonakis,Phys.Rev.D **82** (2010) 054018.
[7] ATLAS Collaboration, ATLAS-CONF-2013-100, http://cds.cern.ch/record/1600799 .
[8] CDF and D0 Collaborations, T. Aaltonen et al, Phys.Rev.Lett. **112** (2014) 231803.
[9] N. Kidonakis, Phys.Rev.D **81** (2010) 054028.
[10] ATLAS Collaboration, Phys.Lett.B **740** (2014) 118.
[11] ATLAS Collaboration, Phys.Lett.B **743** (2015) 233-255.
[12] ATLAS Collaboration, EPJC **75** (2015) 79.

DOI: 10.1051/epjconf/201612604012

New tests of variability of the speed of light.

Mariusz P. Dąbrowski[1,2,3,a], Vincenzo Salzano[1], Adam Balcerzak[1], and Ruth Lazkoz[4]

[1] *Institute of Physics, University of Szczecin, Wielkopolska 15, 70-451 Szczecin, Poland*
[2] *National Centre for Nuclear Research, Andrzeja Sołtana 7, 05-400 Otwock, Poland,*
[3] *Copernicus Center for Interdisciplinary Studies, Sławkowska 17, 31-016 Kraków, Poland*
[4] *Fisika Teorikoaren eta Zientziaren Historia Saila, Zientzia eta Teknologia Fakultatea, Euskal Herriko Unibertsitatea, 644 Posta Kutxatila, 48080 Bilbao, Spain*

Abstract. We present basic ideas of the varying speed of light cosmology, its formulation, benefits and problems. We relate it to the theories of varying fine structure constants and discuss some new tests (redshift drift and angular diameter distance maximum) which may allow measuring timely and spatial change of the speed of light by using the future missions such as Euclid, SKA (Square Kilometer Array) or others.

1 Introduction - main frameworks of varying constants theories

In 1937 Paul Dirac [1] made interesting remarks about the relations between atomic and cosmological quantities bearing in mind that the gravitational constant is proportional to the Hubble parameter $G \propto H(t) = (da/dt)/a$ and concluding that the former must evolve in time – $G(t) \propto 1/t$, and the scale factor $a(t) \propto t^{1/3}$. The conclusion was to explain that gravity is "weak" compared to electromagnetism since the universe is "old" i.e. $F_e/F_p \propto (e^2/m_e m_p)t \propto t$, where e is the charge, m_e electron and m_p proton mass. First fully quantitative framework of varying constant theory was proposed by Brans and Dicke [2] (scalar-tensor gravity) in which the gravitational constant G was associated with an average gravitational potential (scalar field) ϕ surrounding a given particle: $< \phi >= GM/(c/H_0) \propto 1/G = 1.35 \times 10^{28} g/cm$. In this approach the scalar field gives the strength of gravity $G = 1/16\pi\Phi$ and, as it emerged later, the action which reads as

$$S = \int d^4x \sqrt{-g}\left(\Phi R - \frac{\omega}{\Phi}\partial_\mu\Phi\partial^\mu\Phi + \Lambda + L_m\right) \tag{1}$$

in fact, relates also to the low-energy-effective superstring theory for $\omega = -1$ where the string coupling constant $g_s = \exp(\phi/2)$ changes in time, and ϕ is the dilaton field related to Brans-Dicke field $\Phi = \exp(-\phi)$ [3].

2 Benefits and problems of varying c theories

Though attempts were performed already by Einstein [4], then by Dicke [5], Petit [6], and Moffat [7], the most popular approach was found by Albrecht and Magueijo [8] who introduced a scalar field

[a] e-mail: mpdabfz@wmf.univ.szczecin.pl

$c^4 = \psi(x^\mu)$, $\mu = 0, 1, 2, 3$, with the action (R - Ricci scalar, Λ - the cosmological constant, L_m - matter, L_ψ - scalar field)

$$S = \int d^4x\sqrt{-g}\left[\frac{\psi(R+2\Lambda)}{16\pi G} + L_m + L_\psi\right]. \tag{2}$$

This model breaks Lorentz invariance (relativity principle and light principle), so that there is a preferred frame (called cosmological or CMB frame) in which the field is minimally coupled to gravity. The Riemann tensor is computed in such a frame for a constant $c = \psi^{1/4}$ and no additional derivative terms of the type $\partial_\mu\psi$ appear in this frame (though they do in other frames). Einstein equations remain the same form, except that c now varies.

The varying c theories can be related to varying fine structure constant α (or charge $e = e_0\epsilon(x^\mu)$) theories [9, 10]

$$S = \int d^4x\sqrt{-g}\left(R - \frac{\omega}{2}\partial_\mu\psi\partial^\mu\psi - \frac{1}{4}f_{\mu\nu}f^{\mu\nu}e^{-2\psi} + L_m\right) \tag{3}$$

with $\psi = \ln\epsilon$ and $f_{\mu\nu} = \epsilon F_{\mu\nu}$ is the electromagnetic tensor. This is due to the definition of the fine structure constant

$$\alpha(x^\mu) = \frac{e^2}{\hbar c(x^\mu)}. \tag{4}$$

Assuming linear expansion of the field ψ, $e^\psi = 1 - 8\pi G\zeta(\psi - \psi_0) = 1 - \Delta\alpha/\alpha$ with the constraint on the local equivalence principle violence $|\, \zeta \,| \leq 10^{-3}$, we have the relation to dark energy [25, 26]:

$$w + 1 = \frac{(8\pi G\frac{d\psi}{d\ln a})^2}{\Omega_\psi}, \tag{5}$$

where w is the barotropic index, Ω_ψ is the dimensionless density parameter of the ψ field. The field equations for Friedmann universes based on the action (3) are [11]

$$\frac{\dot{a}^2}{a^2} = \frac{8\pi G}{3}\left(\varrho_r + \varrho_\psi\right) - \frac{kc^2}{a^2}, \tag{6}$$

$$\frac{\ddot{a}}{a} = -\frac{8\pi G}{3}\left(\varrho_r + 2\varrho_\psi\right), \tag{7}$$

$$\ddot{\psi} + 3\frac{\dot{a}}{a}\dot{\psi} = 0, \tag{8}$$

where $\varrho_r \propto a^{-4}$ stands for the density of radiation while

$$\varrho_\psi = \frac{p_\psi}{c^2} = \frac{\sigma}{2}\dot{\psi}^2 \tag{9}$$

stands for the density of the scalar field ψ (standard with $\sigma = +1$, and phantom with $\sigma = -1$) and

$$\alpha = \alpha_0 e^{2\psi}. \tag{10}$$

Applying the simplest method, one derives Einstein-Friedmann equations generalized to varying speed of light (VSL) theories and varying gravitational constant G theories as (ϱ - mass density; $\varepsilon = \varrho c^2(t)$ - energy density in $Jm^{-3} = Nm^{-2} = kgm^{-1}s^{-2}$)

$$\varrho(t) = \frac{3}{8\pi G(t)}\left(\frac{\dot{a}^2}{a^2} + \frac{kc^2(t)}{a^2}\right), \tag{11}$$

$$p(t) = -\frac{c^2(t)}{8\pi G(t)}\left(2\frac{\ddot{a}}{a} + \frac{\dot{a}^2}{a^2} + \frac{kc^2(t)}{a^2}\right), \tag{12}$$

DOI: 10.1051/epjconf/201612604012

Table 1. The set of singularities for Friedmann geometry [15–17]

Type	Name	t sing.	a(t_s)	$\varrho(t_s)$	p(t_s)	$\dot{p}(t_s)$ etc.	w(t_s)
0	Big-Bang (BB)	0	0	∞	∞	∞	finite
I	Big-Rip (BR)	t_s	∞	∞	∞	∞	finite
I_l	Little-Rip (LR)	∞	∞	∞	∞	∞	finite
I_p	Pseudo-Rip (PR)	∞	∞	finite	finite	finite	finite
II	Sudden Future (SFS)	t_s	a_s	ϱ_s	∞	∞	finite
II_g	Gen. Sudden Future (GSFS)	t_s	a_s	ϱ_s	p_s	∞	finite
III	Finite Scale Factor (FSFS)	t_s	a_s	∞	∞	∞	finite
IV	Big-Separation (BS)	t_s	a_s	0	0	∞	∞
V	w-singularity (w)	t_s	a_s	0	0	0	∞

and the generalized conservation law is obtained from (11) and (12) as

$$\dot{\varrho}(t) + 3\frac{\dot{a}}{a}\left(\varrho(t) + \frac{p(t)}{c^2(t)}\right) = -\varrho(t)\frac{\dot{G}(t)}{G(t)} + 3\frac{kc(t)\dot{c}(t)}{4\pi G a^2} . \tag{13}$$

2.1 Benefits: solution to the horizon, flatness, and singularity problems

VSL theories solve basic problems of standard cosmology such as the flatness problem and the horizon problem. The first one can be coped with, when one inserts (13) into the Friedmann equation (11) to get

$$\frac{\dot{a}^2}{a^2} = \frac{8\pi G_0 C}{3}a^{-3(w+1)} + \frac{kc_0^2 a^{2n-2}(2n-1)}{2n+3w+1} , \tag{14}$$

and the density term (with an ansatz for the variability of $c = c_0 a^n$, with n = const, and C =const.) will dominate the curvature term at large scale factor if

$$2 \geq 2n + 3(w+1). \tag{15}$$

The second one is solved bearing in mind that for large scale factor the solution is $a(t) = t^{2/3(w+1)}$ and the proper distance to the horizon reads as

$$d_H = c(t)t = c_0 a^n(t)t = c_0 a_0^n t^{(3w+3+2n)/3(w+1)} \tag{16}$$

so that the scale factor grows faster than d_H under the same condition as in (15).

Varying constants can also remove ("regularize") or change the nature of singularities within the framework of Friedmann geometry [12]. In Table 1 we see the properties of these singularities. It shows how the enumerated quantities behave at the singularity $t = t_s$: the scale factor a, the mass density ρ, the pressure p, the pressure derivatives, and the barotropic index w.

Interesting remarks related to regularizing singularities are as follows:

- In order to regularize an SFS or an FSF singularity by varying $c(t)$, the light should slow and eventually stop propagating at a singularity. This is in analogy to loop quantum cosmology (LQC), where in the anti-newtonian limit $c = c_0\sqrt{1-\varrho/\varrho_c} \longrightarrow 0$ for $\varrho \longrightarrow \varrho_c$ with ϱ_c being the critical density [13]. The low-energy limit $\varrho \ll \varrho_0$ gives the standard value $c \rightarrow c_0$.)

- To regularize an SFS or an FSF by varying gravitational constant $G(t)$ - the strength of gravity has to become infinite at an initial (curvature) singularity. Effectively, a new singularity - of strong coupling for a physical field such as $G \propto 1/\Phi$ appears. Such problems were already dealt with in superstring and brane cosmology where both the curvature singularity and a strong coupling singularity show up [3].

2.2 Problems: derivation of the field equations from a proper action

As it was already mentioned, the equations (11)-(13) have just been obtained in a special frame - the one in which c is a constant and does not lead to any extra boundary terms (apart from standard ones). Einstein equations were simply generalized:

$$G_{\mu\nu} - g_{\mu\nu}\Lambda = \frac{8\pi G}{\psi} T_{\mu\nu}, \tag{17}$$

while the action (2) varied in a standard way leads to different field equations

$$G_{\mu\nu} - g_{\mu\nu}\Lambda = \frac{8\pi G}{\psi} T_{\mu\nu} - \frac{1}{\psi}\psi_{;\nu;\mu} + \frac{1}{\psi}\Box\psi. \tag{18}$$

The application of Bianchi identity to (17) gives a conservation equation with dynamical ψ

$$T^{\mu\nu}{}_{;\mu} = -T^{\mu\nu}\psi_{;\mu} \tag{19}$$

If ψ was supposed to be a dynamical matter field, then one could get the evolution equation using the Lagrangian

$$L_\psi = -\frac{\omega}{16\pi G\psi}\dot{\psi}^2, \tag{20}$$

but working only in a preferred frame and with ψ not coupled to $\sqrt{-g}$. The best formulation was recently proposed by Moffat [18].

2.3 Benefits of varying α cosmology

Since one does not brake Lorentz invariance in varying fine structure constant α theories, then there are no such problems in these models - the standard variational principle applies and the dynamical equation for the scalar field is given.

According to the definition, any variability of c (or e, $\hbar$) is related to the variability of α:

$$\frac{\Delta\alpha}{\alpha} = -\frac{\Delta c}{c}. \tag{21}$$

The best constraint on $\Delta\alpha$ which comes from 2 billion years ago is from Oklo natural nuclear reactor and reads as $\Delta\alpha/\alpha = (0.15 \pm 1.05) \cdot 10^{-7}$ at z = 0.14. There are other constraints e.g. from VLT/UVES (Very Large Telescope/Ultraviole Echelle Survey) quasars: $\Delta\alpha/\alpha = (0.15 \pm 0.43) \cdot 10^{-5}$ at $1.59 < z < 2.92$, and from SDSS (Sloan Digital Sky Survey) quasars: $\Delta\alpha/\alpha = (1.2 \pm 0.7) \cdot 10^{-4}$ at $0.16 < z < 0.8$.

2.4 α-dipole

According to [19] there is anisotropy in the variability of the fine structure constant in the sky (α-dipole at $R.A.17.4 \pm 0.9h$, $\delta = -58 \pm 9$ as measured independently by Keck Telescope ($\Delta\alpha < 0$) and VLT. Some specific measurements of α are listed below (in parts per million; UVES - Ultraviolet and Visual Echelle Telescope, HARPS - High Accuracy Radial velocity Planet Searcher, LP - Large Program measurement):

Table 2. Measurements of α

Object	z	$\Delta\alpha/\alpha$	Spectrograph	Ref.
HE0515−4414	1.15	−0.1 ± 1.8	UVES	Molaro et al. (2008) [20]
HE0515−4414	1.15	0.5 ± 2.4	HARPS/UVES	Chand et al. (2006) [21]
HE0001−2340	1.58	−1.5 ± 2.6	UVES	Agafonowa et al. (2011) [22]
HE2217−2818	1.69	1.3 ± 2.6	UVES–LP	Molaro et al. (2013) [23]
Q1101−264	1.84	5.7 ± 2.7	UVES	Molaro et al. (2008) [20]

2.5 Strongest bound – atomic clock Rosenband bound at $z = 0$

Rosenband [24] measurement gives the following bound at $z = 0$ (present)

$$\left(\frac{\dot{\alpha}}{\alpha}\right)_0 = (-1.6 \pm 2.3) \times 10^{-17}\text{yr}^{-1}\,, \tag{22}$$

which can be transformed onto the bound for the scalar field coupling ξ:

$$\left|\frac{\dot{\alpha}}{\alpha}\right|_0 = |\xi| H_0 \sqrt{3\Omega_{\Phi 0} \mid 1 + w_{\Phi 0} \mid}\,, \tag{23}$$

and translates for $H_0 = (67.4 \pm 1.4)\,\text{km.s}^{-1}\text{Mpc}^{-1}$ Planck value) into the conservative (3σ) bound

$$|\xi| \sqrt{3\Omega_{\Phi 0} \mid 1 + w_{\Phi 0} \mid} < 10^{-6}. \tag{24}$$

3 Redshift drift test of varying c models

Redshift drift measurement [27] is to collect data from two light cones separated by 10-20 years to look for a change in redshift of a source as a function of time.

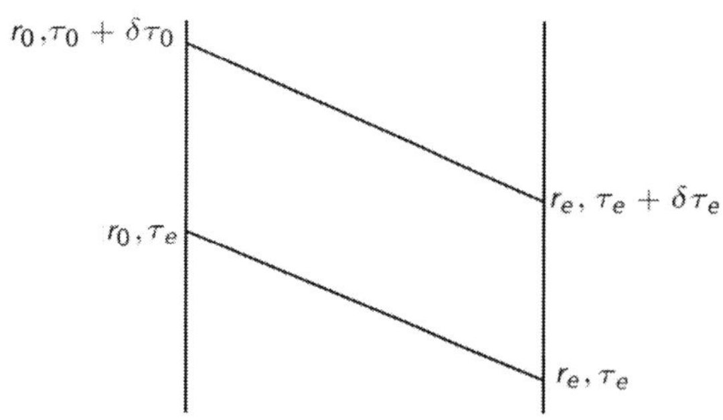

Figure 1. The idea of redshift drift measurement.

There is a relation between the times of emission of light by the source t_e and $t_e + \Delta t_e$ and the times of their observation at t_o and $t_o + \Delta t_o$ which in VSL theory generalizes into [28]

$$\int_{t_e}^{t_o} \frac{c(t)dt}{a(t)} = \int_{t_e+\Delta t_e}^{t_o+\Delta t_o} \frac{c(t)dt}{a(t)}\,, \tag{25}$$

and for small Δt_e and Δt_o transforms into

$$\frac{c(t_e)\Delta t_e}{a(t_e)} = \frac{c(t_0)\Delta t_o}{a(t_o)}\,. \tag{26}$$

The definition of redshift in VSL theories remains the same as in standard Einstein relativity i.e. $1 + z = a(t_0)/a(t_e)$. Using (26) we have

$$\frac{\Delta z}{\Delta t_0} = \frac{\Delta z}{\Delta t_0}(z, n) = H_0(1 + z) - H(z)(1 + z)^n \ . \tag{27}$$

In the limit $n \to 0$ the formula (27) reduces to the one of the standard Friedmann universe [27]. Bearing in mind the definitions of dimensionless density parameters Ω_i, and assuming flat universe one has

$$H^2(z) = H_0^2 \left[\Omega_{m0}(1 + z)^3 + \Omega_\Lambda\right] \tag{28}$$

and so (27) gives

$$\frac{\Delta z}{\Delta t_0} = H_0 \left[1 + z - \sqrt{\Omega_{m0}(1 + z)^{3+2n} + \Omega_\Lambda(1 + z)^{2n}}\right] \tag{29}$$

which can further be rewritten to define a new redshift function

$$\tilde{H}(z) \equiv (1 + z)^n H(z) = H_0 \sqrt{\sum_{i=1}^{i=k} \Omega_{wi}(1 + z)^{3(w_{eff}+1)}} \ , \tag{30}$$

where $w_{eff} = w_i + \frac{2}{3}n$. The redshift drift can be measured by future telescopes such as E-ELT (European-Extremely Large Telescope), TMT (Thirty Meter Telescope), GMT (Giant Magellan Telescope) as well as gravitational wave detectors (DECIGO) DECi-Hertz Interferometer Gravitational Wave Observatory and BBO (Big Bang Observer).

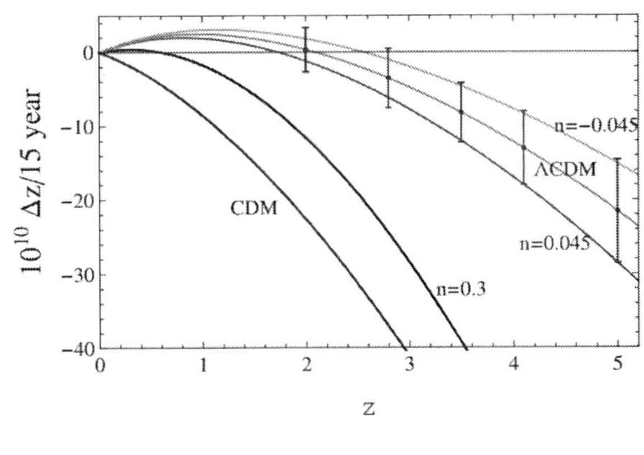

Figure 2. The redshift drift effect for 15 year period of observations for various values of the varying speed of light parameter n. The error bars are taken from [29] and presumably show that for $|n| < 0.045$ one cannot distinguish between VSL models and ΛCDM models.

This relation is presented in Fig. 2 from which one cas easily see that for small values of the parameter n (small variation of c) the dark energy can be mimicked while for large values of n there is a clear distinction between dark energy which can be detected.

4 Measuring c by future galaxy surveys

Speed of light c appears in many observational quantities. Among them in the angular diameter distance [33]

$$D_A = \frac{D_L}{(1 + z)^2} = \frac{a_0}{1 + z} \int_{t_1}^{t_2} \frac{c(t)dt}{a(t)}, \tag{31}$$

where D_L is the luminosity distance, a_0 present value of the scale factor (normalized to $a_0 = 1$ later), and we have taken the spatial curvature $k = 0$ (otherwise there would be sin or sinh in front of the integral). Using the definition of redshift and the dimensionless parameters Ω_i we have

$$D_A = \frac{1}{1+z} \int_0^z \frac{c(z)dz}{H(z)}, \tag{32}$$

where

$$H(z) = \sqrt{\Omega_{r0}(1+z)^4 + \Omega_{m0}(1+z)^3 + \Omega_\Lambda}. \tag{33}$$

4.1 Angular diameter distance maximum

Due to the expansion of the universe, there is a maximum of the distance at

$$D_A(z_m) = \frac{c(z_m)}{H(z_m)}, \tag{34}$$

which can be obtained by simple differentiating (32) with respect to z:

$$\frac{\partial D_A}{\partial z} = -\frac{1}{(1+z)^2} \int_0^z \frac{c(z)dz}{H(z)} + \frac{1}{1+z} \frac{c(z)}{H(z)} = 0. \tag{35}$$

In a flat $k = 0$ cold dark matter (CDM) model, there is a maximum at $z_m = 1.25$ and $D_A \approx 1230$ Mpc. For the standard ΛCDM model of our interest the maximum is at $1.4 < z_m < 1.8$. The product of D_A and H gives **exactly** the speed of light c at maximum (the curves intersect at z_m):

$$D_A(z_m)H(z_m) = c_0 \equiv 299792.458 \ \ \mathrm{kms}^{-1} \tag{36}$$

if we believe it is constant (defined officially by Bureau International des Poids et Mesures (BIPM) [30] and a relative error is claimed to be 10^{-9} [31]).

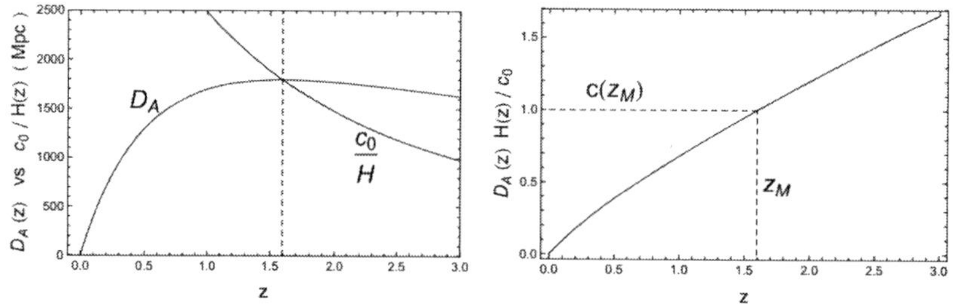

Figure 3. D_A and $H(z)$ crossing plots at z_m.

Measuring z_m is problematic if one uses D_A only (this is because of a large plateau around z_m which makes it difficult to avoid errors from small sample of data – besides, one has binned data, observational errors, and intrinsic dispersion). However, one can appeal to an independent measurement of $c_0/H(z)$ which is the radial (line-of-sight) mode of the baryon acoustic oscillations (BAO) surveys for which $D_A(z)$ is the tangential mode [32]. In other words, we have both tangential and horizontal modes as

$$y_t = \frac{D_A}{r_s}, \qquad y_r = \frac{c}{Hr_s}, \tag{37}$$

where

$$r_s = \int_{z_{dec}}^{\infty} \frac{cc_s(z)dz}{H(z)} \tag{38}$$

is the sound horizon size at decoupling and c_s the speed of sound. The measurements of BAO are from BOSS DR11 CMASS [34]

$$\frac{D_V}{r_s(z_d)} = 13.85 \pm 0.17 \quad \text{at} \quad \bar{z} = 0.57, \tag{39}$$

where the volume-averaged distance is

$$D_V = \left[(1+z)^2 cz \frac{D_A^2}{H}\right]^{\frac{1}{3}}, \tag{40}$$

and from BOSS DR11 LOWZ [35]

$$D_V = (1264 \pm 25)\left(\frac{r_s(z_d)}{r_{s,fid}(z_d)}\right) \quad \text{at} \quad \bar{z} = 0.32. \tag{41}$$

4.2 The method to measure c.

In Ref. [33] a new method to measure c based on the relation (35) was proposed, and it is composed of the following steps:

1. Independently measuring $D_A(z)$ and $H(z)$.
2. Calculating z_m.
3. Getting the product $D_A(z_m)H(z_m) = c(z_m)$.
4. Calculating $\Delta c = c(z_m) - c_0$, first assuming that $c(z_m)$ may not be equal to c_0.
5. Determining possible level of variability/constancy of c.

For this sake the background ΛCDM model with an ansatz [36]

$$c(a) \propto c_0 \left(1 + \frac{a}{a_c}\right)^n \tag{42}$$

is taken into account, where a_c is the scale factor at the transition epoch from some $c(a) \neq c_0$ (at early times) to $c(a) \to c_0$ (at late times to now). Three scenarios are considered [33]:
1) standard case $c = c_0$;
2) $a_c = 0.005$, $n = -0.01 \to \Delta c/c \approx 1\%$ at $z \propto 1.5$;
3) $a_c = 0.005$, $n = -0.001 \to \Delta c/c \approx 0.1\%$ at $z \propto 1.5$. After using 10^3 Euclid project [37] mock data simulations [38], one obtains the following results:
1) $z_m = 1.592^{+0.043}_{-0.039}$ (fiducial model input $z_m = 1.596$) and $c/c_0 = 1 \pm 0.009$;
2) $z_m = 1.528^{+0.038}_{-0.036}$ (fiducial $z_m = 1.532$) and $c(z_m)/c_0 = 1.00925 \pm 0.00831$;
and

$$< c(z_m)/c_0 - 1\sigma_{c(z_m)/c_0} >= 1.00094^{+0.00014}_{-0.00033}, \tag{43}$$

so that *a possible detection by Euclid of* 1% *variation at* 1σ*-level in future will be possible.*
3) $z_m = 1.584^{+0.042}_{-0.039}$ (fiducial $z_m = 1.589$) and $c(z_m)/c_0 = 1.00095 \pm 0.00852$
and

$$< c(z_m)/c_0 - 1\sigma_{c(z_m)/c_0} >= 0.99243^{+0.00016}_{-0.00013}, \tag{44}$$

so that a detection by Euclid of 0.1% variation at 1σ-level will *not* be possible.

4.3 Other surveys and perspectives

In fact, Euclid will have $1/10$ of the error bars of the current missions like WiggleZ Dark Energy Survey (e.g. [39]). Other missions which will be competitive to Euclid and useful for our task will be: Dark Energy Spectroscopic Instrument (DESI) [40]; Square Kilometer Array (SKA) [41]; Wide-Field Infrared Survey Telescope (WFIRST) [42] (especially having largest sensitivity at potential z_m region i.e. $1.5 < z < 1.6$).

5 Conclusions

Varying speed of light c (and related to them varying fine structure constant α) theories which attract more interest among physicists have both advantages as well as problems. The advantages of them is that they solve the flatness, horizon problems, and in some special cases, the singularity problem. However, their violation of Lorentz invariance leads to a choice of a preferred frame and a drop of standard variational principle. On the other hand, α-varying theories have better formulation and due to the definition of α, they can be related to varying-c theories.

In this paper we have proposed some new tests to check variability of c in future telescope/space missions. The first was the redshift drift test which gives clear prediction for redshift drift effect which can potentially be measured by future telescopes like E-ELT, TMT, GMT, DECIGO/BBO. The second was to use baryon acoustic oscillations test and the Hubble function test to independently measure the radial D_A and tangential mode c/H of the volume distance D_V at the angular diameter distance maximum z_m.

Putting this last method in simple terms we have considered a "cosmic" measurement of the speed of light c with D_A giving the dimension of length playing the role of a "cosmic ruler" and $1/H$ giving the dimension of time playing the role of a "cosmic clock"/"chronometer" i.e.

$$c = \frac{D_A}{\left(\frac{1}{H}\right)}. \tag{45}$$

We have checked that 1% variability of c can be tested at 1σ level by Euclid mission. It is likely that such variability will also be possible to test by SKA and WFIRST.

Acknowledgements
The research of V.S., M.P.D., and A.B. was supported by the Polish National Science Center Grant DEC-2012/06/A/ST2/00395.

References

[1] P.A.M. Dirac, Nature **139**, 323 (1937); Proc. Roy. Soc. A**165**, 189 (1938).
[2] C.H. Brans, R.H. Dicke, Phys. Rev. **124**, 925 (1961).
[3] J. Polchinski, *String Theory* (Cambridge University Press, Cambridge, 1998).
[4] A. Einstein, Jahrbuch für Radioaktivität und Elektronik **4**, 411 (1907).
[5] R.H. Dicke, Rev. Mod. Phys. **29**, 363 (1957).
[6] J.-P. Petit, Mod. Phys. Lett. A **3**, 1527 (1988).
[7] J. Moffat, Int. J. Mod. Phys. D **2**, 351 (1993).
[8] A. Albrecht, J. Magueijo, Phys. Rev. D**59**, 043516 (1999).
[9] J.K. Webb et al. Phys. Rev. Lett. **87**, 091301 (2001).
[10] J.D. Barrow, H. Sandvik, J. Mageuijo, Phys. Rev. D**65**, 063504 (2002).

[11] J.D. Barrow, D. Kimberly, J. Magueijo, Class. Quantum Grav. **21**, 4289 (2004).
[12] M.P. Dąbrowski, K. Marosek, Journ. Cosmol. Astrop. Phys. **02**, 012 (2013).
[13] T. Cailleteau, J. Mielczarek, A. Burrau, and J. Grain, Class. Quantum Grav. **29**, 095010 (2012).
[14] M.P. Dąbrowski, K. Marosek, A. Balcerzak, Mem. Soc. Astron. Ital. **85**, 44-49 (2014).
[15] S. Nojiri, S.D. Odintsov and S. Tsujikawa, Phys. Rev. D **71**,063004 (2005).
[16] M.P. Dąbrowski, T. Denkiewicz, AIP Conference Proceedings **1241**, 561 (2010).
[17] M.P. Dąbrowski, Mathematical Structures of the Universe, M. Eckstein, M. Heller, S.J. Szybka (eds.), (Copernicus Center Press, Kraków), 99 (2014).
[18] J.W. Moffat, arXiv: 1501.01872.
[19] J.K. Webb et al., Phys. Rev. Lett. **107**, 191101 (2011).
[20] P. Molaro, D. Reimers, I. I. Agafonova, S. A. Levshakov, Eur. Phys. JST **163**, 173 (2008).
[21] H. Chand, R. Srianad, P. Petitjean, B. Aracil, R. Quast and D. Reimers, Astron. Astroph. **451**, 45 (2006).
[22] I. I. Agafonova, P. Molaro and S. A. Levshakov, Astron. Astroph. **529**, A**28** (2011).
[23] P. Molaro et al., Astron. Astroph. **555**, A68 (2013).
[24] T. Rosenband et al., Science **319**, 1808 (2008).
[25] T. Denkiewicz, M.P. Dąbrowski, C.J.A.P. Martins, P. Vielzeuf - Phys. Rev. D **89**, 083514 (2014).
[26] M.P. Dąbrowski, T. Denkiewicz, C.J.A.P. Martins, P. Vielzeuf - Phys. Rev. D **89**, 123512 (2014).
[27] A. Sandage, Astrophys. J. **136**, 319 (1962); A. Loeb, Astrophys. J. **499**, L11 (1998).
[28] A. Balcerzak, M.P. Dąbrowski, PLB **728**, 15 (2014).
[29] C. Quercellini et al., Phys. Rep. **521**, 95 (2012).
[30] www.bipm.org
[31] K.M. Evenson, Phys. Rev. Lett. **29**, 1346 (1972).
[32] S. Nesseris, L. Perivolaropoulos, Phys.Rev. D**73**, 103511 (2006).
[33] V. Salzano, M.P. Dąbrowski, R. Lazkoz, Phys. Rev. Lett. **114**, 101304 (2015).
[34] L. Samushia et al., Mon. Not. R. Astron. Soc. **439**, 3504 (2014).
[35] R. Tojeiro et al., arXiv: 1401.1768.
[36] J. Magueijo, Rep. Prog. Phys. **66**, 2025 (2003).
[37] L. Laureijs et al. 0912.0914 (Euclid Collaboration).
[38] A. Font-Ribeira et al., Journ. Cosmol. Astropart. Phys. **05**, 023 (2014).
[39] B.D. Sherwin, arXiv: 1207.4543.
[40] M. Levi et al., Astroph. Journ. **803**, 21 (2015).
[41] Ph. Bull et al., 1405.1452)
[42] D. Spergel et al., arXiv: 1305.5425.

EPJ Web of Conferences **126**, 04013 (2016) DOI: 10.1051/epjconf/201612604013
ICNFP 2015

Inclusive searches for squarks and gluinos with the ATLAS detector

Ingrid Deigaard[1,a] on behalf of the ATLAS Collaboration

[1] *Nikhef National Institute for Subatomic Physics, Amsterdam, The Netherlands*

Abstract. Despite the absence of experimental evidence, weak scale supersymmetry remains one of the best motivated and studied Standard Model extensions. This talk summarises recent ATLAS results on inclusive searches for supersymmetric squarks and gluinos, including third generation squarks produced in the decay of gluinos. The searches involved final states containing jets, including those tagged as originating from b-quark decays, missing transverse momentum with and without light leptons, taus or photons. Sensitivity projections for the data that will be collected in 2015 are also presented.

1 Introduction

Supersymmetry (SUSY) [1] is one of the most studied theoretical extensions of the Standard Model (SM). SUSY expands the symmetry group of the SM to include a symmetry between bosons and fermions leading to a doubling of the particle content of the SM. To ensure that the proton is stable in SUSY, R-parity is introduced. This leaves the lightest supersymmetric particle (LSP), normally either the gravitino, $\tilde{G}$, or the lightest neutralino, $\tilde{\chi}_1^0$, stable providing a candidate for Dark Matter.

These proceedings will focus on the searches for the first and second generation of scalar quarks (squarks, $\tilde{q}$) and the fermionic partner of the gluon (gluino, $\tilde{g}$). Of the production cross section of SUSY particles, the cross section for squark and gluino production are the largest. Many decays of squarks and gluinos are considered in simplified models - from direct decays ($\tilde{q} \rightarrow q\tilde{\chi}_1^0$) to more complex decays going through several steps ($\tilde{g} \rightarrow qq\tilde{\chi}_1^\pm \rightarrow qq\nu_\tau\tilde{\tau} \rightarrow qq\nu_\tau\tau\tilde{\chi}_1^0$), where branching fractions are set to 100%. More complex theories, such as MSUGRA, are studied as well. The various decays of squarks and gluinos in the simplified models lead to a big variety of final states all of which will have large missing transverse momentum, E_T^{miss}, originating from the only weakly interacting LSP escaping the detector unnoticed.

A selection of searches for squarks and gluinos performed with the 20 fb^{-1} data collected by the ATLAS experiment [2] with $\sqrt{s}$ = 8 TeV are presented. Unfortunately, no excesses have been observed therefore the results will be presented as limits on various SUSY scenarios. Furthermore, the prospects of the SUSY searches for squarks and gluinos with data collected in 2015 at $\sqrt{s}$ = 13 TeV is presented.

[a] e-mail: ingrid.deigaard@cern.com

2 Selection of recent searches for squarks and gluinos at $\sqrt{s} = 8$ TeV

Searches for SUSY are based on a good understanding of the SM background. SUSY searches are normally performed in places of phase space where the predicted SM cross section is low, i.e. regions with many high-p_T objects and high E_T^{miss}.

There are three methods for estimating the SM background: fully data driven, partial data driven, and purely from simulation. The major irreducible backgrounds ($W+$ jets, $Z+$ jets, and top production) are estimated using a partial data driven technique. In this method the simulated background is normalised in control regions where the yield of the background under study is enhanced. The control regions are designed to be kinematically close and orthogonal to regions where the SUSY signal is enhanced. Through transfer factors, the background estimation is calculated in the signal enhanced regions.

All background estimates derived from control regions are checked by calculating transfer factors to other regions (validation regions) that are close to the signal enhanced regions as well. The background estimation is validated by comparing the number of expected events in the validation regions with observation.

2.1 Searches for direct squark decays

In the simplified model of direct squark decays, the first two generations of squarks are assumed to be light. The squarks will be produced in pairs and each squark will decay directly to a quark and a $\tilde{\chi}_1^0$ (the LSP) with a 100% branching fraction. The masses of the squark and the LSP are varied independently. There are three analyses in ATLAS that are sensitive to this model: the monojet analysis, the 0-lepton 2-6 jets analysis, and the razor analysis.

The monojet analysis [3, 4] selects events with one high-p_T jet, high E_T^{miss}, and no leptons (e/μ). The variables used to define the signal regions are the p_T of the leading jet and E_T^{miss}. This analysis targets especially a compressed mass-spectrum where the mass difference between the squark and the LSP is small ($\lesssim 50$ GeV).

The 0-lepton, 2-6 jets analysis [4, 5] selects events with 2-6 jets inclusively, high E_T^{miss}, and no leptons (e/μ). The variables used to discriminate between SUSY signal and SM background are the effective mass, $m_{\mathrm{eff}} = E_T^{miss} + \sum_{i=1}^{n_{\mathrm{jets}}} |p_T^i(\mathrm{jet})|$, and $E_T^{miss}/m_{\mathrm{eff}}$. The analysis defines 17 signal regions with various number of jets, and cuts on the discriminating variables targeting many different SUSY models.

The razor analysis [4] selects at least two jets, high E_T^{miss}, and no leptons. Two signal regions are defined using a set of variables called the razor variables [6]. The razor variables are designed to discriminate between events with particles coming from a SUSY decay and events coming from SM processes.

The combined exclusion limit of direct squark decays based on the best expected exclusion of the three analyses is seen in Fig. 1. It is seen that the 0-lepton, 2-6 jets analysis, and the razor analysis have similar sensitivity reach, whereas the monojet analysis is strongest in the area of small mass difference between the squarks and the LSP. For a LSP with a mass of 100 GeV, squarks with masses up to 850 GeV are excluded.

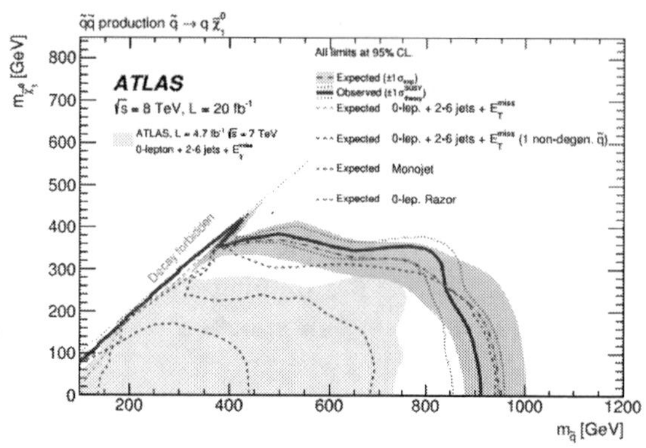

Figure 1. Exclusion limits in the mass plane of the LSP and first- and second-generation squarks assuming direct squark decays. The solid and the dashed red lines show respectively the combined observed and expected 95% CL exclusion limits. Expected limits from the individual analyses are shown as well. A previous result from ATLAS using 7 TeV proton-proton collisions is represented by the grey area. A limit obtained with the 0-lepton, 2-6 jets analysis in a scenario with only one light-flavour squark produced is shown in green. Taken from Ref. [4].

2.2 Searches for one-step squark and gluino decays

The simplified SUSY models of one-step decays of squarks and gluinos considered here assume pair production of either squarks or gluinos and an immediate decay through a chargino ($\tilde{\chi}_1^{\pm}$), $\tilde{q} \to q\tilde{\chi}_1^{\pm} \to qW^{\pm}\tilde{\chi}_1^0$ and $\tilde{g} \to qq\tilde{\chi}_1^{\pm} \to qqW^{\pm}\tilde{\chi}_1^0$. The masses of the squarks/gluinos and $\tilde{\chi}_1^0$ are kept as free parameters while the mass of $\tilde{\chi}_1^{\pm}$ is fixed to $x = (m_{\tilde{\chi}_1^{\pm}} - m_{\tilde{\chi}_1^0})/(m_{\tilde{q}/\tilde{g}} - m_{\tilde{\chi}_1^0}) = 1/2$.

There are two analyses which show similar sensitivity to squark and gluino production with one-step decays: the 0-lepton, 2-6 jets analysis, and the 1-lepton analysis [4, 7]. The 1-lepton analysis selects events with 1 soft or hard lepton (e/μ), high $E_{\mathrm{T}}^{\mathrm{miss}}$, and more than two high-p_{T}jets. The 0-lepton and the 1-lepton analyses are orthogonal as their signal and control regions don't overlap between the two analyses thus they can be statistically combined to improve their collected sensitivity. The combination define a common set of systematics that can be correlated between the two analyses [4].

As seen in Fig. 2, the exclusion limit for the statistical combination of the 0- and 1-lepton analyses is as strong or stronger than each of the individual analyses at any mass point. For a massless LSP, the exclusion limits of squark and gluino masses reach 810 GeV and 1275 GeV, respectively.

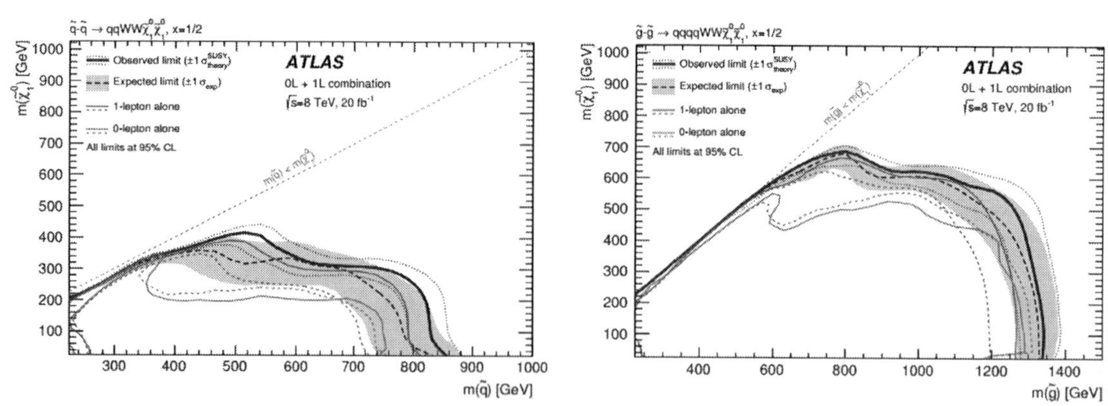

Figure 2. Observed and expected exclusion limits for simplified models of squark-pair (left) and gluino-pair (right) production with one-step decays via $\tilde{\chi}_1^{\pm}$ into a W-boson and $\tilde{\chi}_1^0$. Squark (left)/gluino (right) and neutralino masses in the area below the observed limit are excluded at 95% CL. The individual limits from the 0-lepton and the 1-lepton analyses are overlaid in green and magenta, respectively. Taken from Ref. [4].

2.3 Searches for gluinos decaying through third generation squarks

Next, a simplified model of gluino pair production is considered with subsequent decay through an off-shell scalar top quark, $\tilde{t}_1$, as $\tilde{g} \rightarrow t\tilde{t}_1$. The stop quark is assumed to be the lightest squark, but heavier than the gluino, thus the stop originating from the gluino decay is produced off-shell and decays immediately with a 100% branching fraction as $\tilde{t}_1 \rightarrow t\tilde{\chi}_1^0$. If $m_{\tilde{g}} < 2m_t + m_{\tilde{\chi}_1^0}$, one if the top quarks will be produced off-shell leading to a five-body decay, $\tilde{g} \rightarrow tbW\tilde{\chi}_1^0$. The decay of a gluino through a stop quark leads to final states with b-jets, decay products from W-bosons, and $E_{\mathrm{T}}^{\mathrm{miss}}$. There are two analyses that are sensitive to this SUSY model: the same-sign/3-leptons (SS/3L) analysis, and the 0-1 lepton and 3 b-jets analysis.

The SS/3L analysis [4, 8] selects events with either two leptons (e/μ) of identical electrical charge or 3 leptons together with up to 3 jets originating from b-quarks, and high $E_{\mathrm{T}}^{\mathrm{miss}}$. Five signal regions are defined using various jet and b-jet multiplicity. The effective mass, m_{eff}, is used as a discriminating variable.

The 0-1 lepton, 3 bjets analysis [4, 9] selects events without of with one lepton, 4-7 jets of which at least three are b-jets, and high $E_{\mathrm{T}}^{\mathrm{miss}}$. Nine signal regions are defined based on lepton multiplicity and jet multiplicity. Three variables are used to discriminate signal from background: m_{eff}, $m_{\mathrm{eff}}/\sum_{i=1}^{4}|p_{\mathrm{T}}^{i}(\mathrm{jet})|$, and the transverse mass, $m_{\mathrm{T}} = \sqrt{2p_{\mathrm{T}}^{\ell}E_{\mathrm{T}}^{\mathrm{miss}}(1-\cos\Delta\phi(\ell, E_{\mathrm{T}}^{\mathrm{miss}}))}$.

The limits in the $(m_{\tilde{g}}, m_{\tilde{\chi}_1^0})$-plane of the SS/3L, and the 0-1 lepton, 3 b-jets analyses are shown together with the combined expected and observed exclusion limits in Fig. 3. The Off-shell/On-shell diagonal separates the region where the gluino decays to two on-shell top quarks (below the line) from the region where the gluino decays into an on-shell and an off-shell top quark (above the line). For a low $m_{\tilde{\chi}_1^0}$, gluino masses up to 1320 GeV have been excluded.

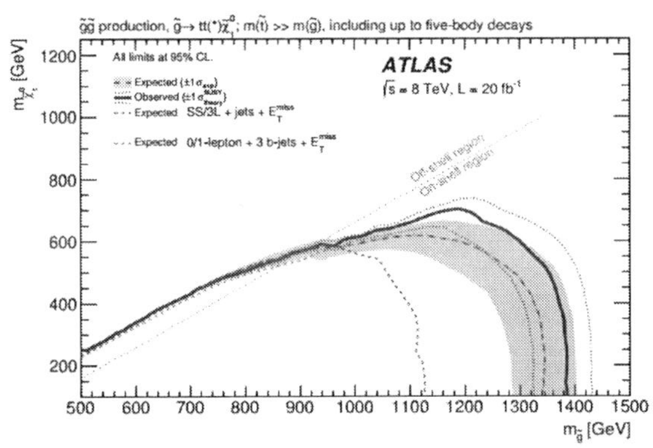

Figure 3. Exclusion limits for the simplified models of gluino decaying through off-shell scalar top quarks. In the region below the grey dashed line labelled "On-shell region", the gluinos decay to two real top quarks, whereas in the "Off-shell region", the gluino decay involves an off-shell top quark. The solid red line and the dashed red line show respectively the combined observed and expected 95% CL exclusion limits. Expected limits from the SS/3L, and the 0-1 lepton, 3 b-jets analyses are shown in purple and cyan, respectively. Taken from Ref. [4].

2.4 Searches for gauge-mediated SUSY breaking models

In general gauge-mediated SUSY breaking models (GGM), the lightest SUSY particle is the gravitino, $\tilde{G}$. The final state depends on the nature for the second lightest particle, the NLSP ($\tilde{\chi}_1^0$). In these proceedings we consider a GGM SUSY model with higgsino-bino mixing in the NLSP. We consider gluino pair production with a one-step decay, $\tilde{g} \rightarrow qq\tilde{\chi}_1^0 \rightarrow qq(\gamma/h)\tilde{G}$, where the higgs, h, is assumed to decay directly to a pair of b-quarks. This leads to final states with photons, jets, b-jets, and $E_{\mathrm{T}}^{\mathrm{miss}}$.

The analysis sensitive to this GGM scenario, is selecting events with at least one photon, 2-4 jets, 1 or 2 b-jets, $E_{\mathrm{T}}^{\mathrm{miss}}$, and no leptons [10]. Two signal regions are defined for this SUSY model using the invariant mass of the two b-jets, and the transverse mass of the photon and the $E_{\mathrm{T}}^{\mathrm{miss}}$.

The expected and observed limits on the GGM scenario with higgsino-bino mixing in the NLSP, is shown in Fig. 4. For NLSP masses below approximately 450 GeV, the direct production of gaugino becomes dominant thus making the analysis insensitive to the gluino mass.

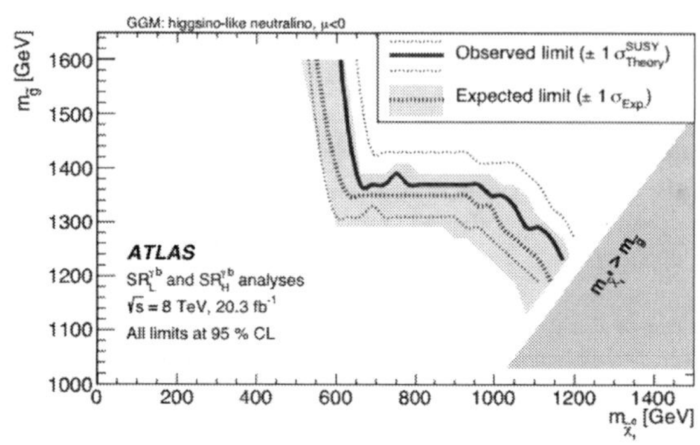

Figure 4. Exclusion limits in the gluino-NLSP mass plane, for the higgsino-bino GGM model. The solid red line and the dashed red line show the combined observed and expected 95% CL exclusion limits, respectively. Taken from Ref. [10].

2.5 Searches for CMSSM/mSUGRA

The mSUGRA/CMSSM model is given by five parameters: a universal scalar mass (m_0), a universal gaugino mass ($m_{1/2}$) , a universal trilinear scalar coupling (A_0), all defined at the grand unification scale, $\tan\beta$, and the sign of the higgsino mass parameter (μ). The masses of all the SUSY particles increase with increasing $m_{1/2}$. The masses of the squarks and the sleptons will depend on m_0 as well. In the model shown here, the values $\tan\beta = 30$, $A_0 = -2m_0$ and $\mu > 0$ are chosen, such that the lightest scalar Higgs boson mass is approximately 125 GeV in a large fraction of the (m_0, $m_{1/2}$) parameter space.

The observed and expected limits of all analysis providing limits in this CMSSM/mSUGRA model are shown in Fig. 5 along with the best combined expected and observed limits. It is seen that at low m_0, the combination of the 0- and 1-lepton analyses is the strongest, whereas at higher m_0 the 0-1 lepton, 3 b-jets analysis is the strongest. [4]

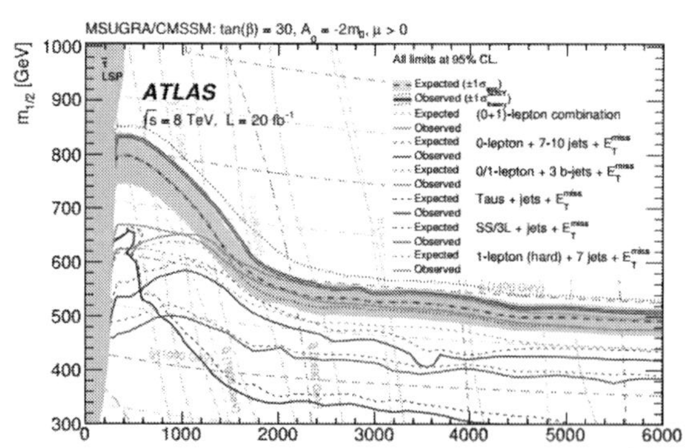

Figure 5. Exclusion limits in the (m_0, $m_{1/2}$) plane for the mSUGRA/CMSSM model. The solid red line and the dashed red line show the combined observed and expected 95% CL exclusion limits, respectively. Expected (dashed lines) and observed (solid lines) limits from the individual analyses are also shown. Taken from Ref. [4].

3 Prospects for searches for squarks and gluinos at $\sqrt{s} = 13$ TeV

With an increase in the collision energy of the LHC from $\sqrt{s} = 8$ TeV to $\sqrt{s} = 13$ TeV, the production cross section for heavy particles, such as squarks and gluinos, will increase more than the cross section of the SM background. This provides a better opportunity to discover SUSY particles if they exist.

The prospects of searching for SUSY at $\sqrt{s}$ = 13 TeV have been studied for two simplified SUSY models both assuming gluino pair production, but with different decays. One of the models assume a direct decay of the gluino, $\tilde{g} \to qq\tilde{\chi}_1^0$, with a massless LSP, $m_{\tilde{\chi}_1^0} = 0$ GeV, similar to the squark model described in Sec. 2.1. The other model studied assumes a one-step decay of the gluino, $\tilde{g} \to qq\tilde{\chi}_1^{\pm} \to qqW^{\pm}\tilde{\chi}_1^0$, assuming $m_{\tilde{\chi}_1^0} = 25$ GeV and that the mass difference between the gluino and the chargino is $x = (m_{\tilde{\chi}_1^{\pm}} - m\tilde{\chi}_1^0)/(m_{\tilde{g}} - m_{\tilde{\chi}_1^0}) = 1/2$, similar to the model described in Sec. 2.2.

Two analyses are studying their sensitivity to these models at $\sqrt{s}$ = 13 TeV; the 0 lepton, 2-6 jets analysis is studying the SUSY model containing the direct decay of the gluino, and the 1 lepton analysis is studying its sensitivity to the model with the one-step decay of the gluino [11]. Both analyses are relying on the corresponding analyses performed at $\sqrt{s}$ = 8 TeV, described in Ref. [5, 7]. The analyses are based on simulations of the Standard Model background with an assumption on the systematic uncertainty on the background of 20% for the 0 lepton analysis and 25% for the 1 lepton analysis. Four integrated luminosity scenarios are studied, 1 fb^{-1}, 2 fb^{-1}, 5 fb^{-1}, and 10 fb^{-1}, with corresponding optimisations of the signal regions for each luminosity.

The sensitivity for simplified SUSY models with direct gluino decays and one-step gluino decays are shown in Fig. 6 left and right, respectively, with the gluino masses already excluded by the $\sqrt{s}$ = 8 TeV analyses shaded in grey. The sensitivity for observation of SUSY particles at $\sqrt{s}$ = 13 TeV is expected to exceed that of full $\sqrt{s}$ = 8 TeV data sample at an integrated luminosity of 5 fb^{-1} for directly decaying gluinos and 2 fb^{-1} for gluinos decaying through charginos.

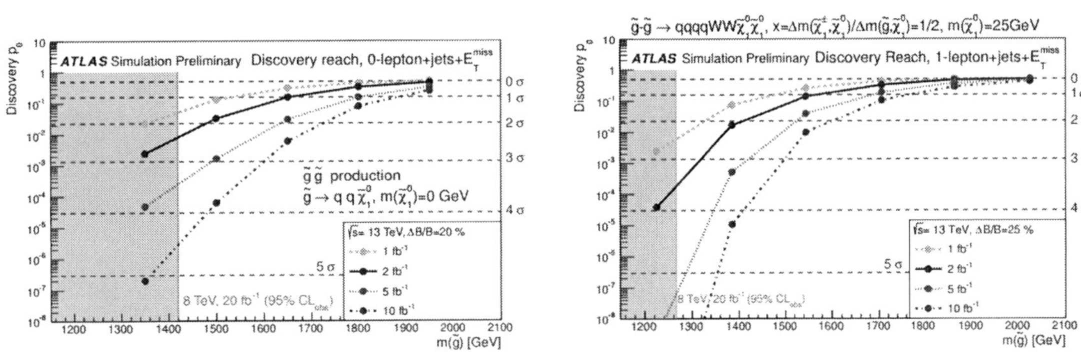

Figure 6. Discovery p_0 values as a function of the gluino mass in the gluino pair production and direct to two quarks and $\tilde{\chi}_1^0$ (left), and decay via $\tilde{\chi}_1^{\pm}$ and two quarks to $\tilde{\chi}_1^0$ and a W-boson. The neutralino mass is set to $m_{\tilde{\chi}_1^0} = 0$ GeV (left) and $m_{\tilde{\chi}_1^0} = 25$ GeV (left). Different integrated luminosities were assumed in the optimisation: 1, 2, 5 or 10 fb-1. The total uncertainty on the background is assumed to be 20 % (left) and 25 % (right). Taken from Ref. [11].

4 Conclusion

A wide range of searches for squarks and gluinos have been performed at $\sqrt{s}$ = 8 TeV with the ATLAS detector. No excesses have been observed above the Standard Model expectation. A selection of searches for squarks and gluinos have been presented in Sec. 2.

With the increase in collision energy of the LHC to $\sqrt{s}$ = 13 TeV, a new era for SUSY searches can begin, as much higher masses of squarks and gluinos can be probed due the vast increase in their production cross section. Thus, already with an integrated luminosity of 2 fb^{-1}, the sensitivity for observation of gluino pair production will exceed the current limits.

References

[1] S. P. Martin, [arXiv:hep-ph/9709356], and references therein.
[2] ATLAS Collaboration, JINST **3** (2008) S08003.
[3] ATLAS Collaboration, Phys. Rev. D **90** (2014) 052008.
[4] ATLAS Collaboration, Journal of High Energy Physics **10** (2015) 054.
[5] ATLAS Collaboration, Journal of High Energy Physics **09** (2014) 176.
[6] C. Rogan, arXiv:hep-ph/1006.2727 (2010).
[7] ATLAS Collaboration, Journal of High Energy Physics **04** (2015) 116.
[8] ATLAS Collaboration, Journal of High Energy Physics **06** (2014) 035.
[9] ATLAS Collaboration, Journal of High Energy Physics **10** (2014) 024.
[10] ATLAS Collaboration, Phys. Rev. D **92** (2015) 072001.
[11] ATLAS Collaboration, ATL-PHYS-PUB-2015-005, https://twiki.cern.ch/twiki/bin/view/AtlasPublic/PUBnotes.

LHCf experiment: forward physics at LHC for cosmic rays study

M. Del Prete[1,2,a], O. Adriani[1,2], E. Berti[1,2], L. Bonechi[1,2], M. Bongi[1,2], G. Castellini[3], R. D'Alessandro[1,2], M. Haguenauer[4], Y. Itow[5,6], K. Kasahara[7], K. Kawade[5], Y. Makino[5], K. Masuda[5], E. Matsubayashi[5], H. Menjo[8], G. Mitsuka[2,5], Y. Muraki[5], Y. Okuno[5,9], P. Papini[1], A-L. Perrot[10], S. Ricciarini[1,2], T. Sako[5], N. Sakurai[6], Y. Sugiura[5], T. Suzuki[7], T. Tamura[11], A. Tiberio[1,2], S. Torii[6], A. Tricomi[12,13], W.C. Turner[14], and Q.D. Zhou[5]

[1] *INFN section of Florence, Italy*
[2] *University of Florence, Italy*
[3] *IFAC-CNR, Italy*
[4] *Ecole-Polytechnique, France*
[5] *Solar-Terrestrial Environment Laboratory, Nagoya University, Japan*
[6] *Kobayashi-Maskawa Institute for the Origin of Particles and the Universe, Nagoya University,Japan*
[7] *RISE, Waseda University, Japan*
[8] *Graduate school of Science, Nagoya University, Japan*
[9] *Nagasaki Institute Applied Science, Japan*
[10] *CERN, Switzerland*
[11] *Kanagawa University, Japan*
[12] *INFN Section of Catania, Italy*
[13] *University of Catania, Italy*
[14] *LBNL, Berkeley, USA*

Abstract. The LHCf experiment, optimized for the study of forward physics at LHC, completes its main physics program in this year 2015, with the proton-proton collisions at the energy of 13 TeV. LHCf gives important results on the study of neutral particles at extreme pseudo-rapidity, both for proton-proton and for proton-ion interactions. These results are an important reference for tuning the models of the hadronic interaction currently used for the simulation of the atmospheric showers induced by very high energy cosmic rays. The results of this analysis and the future perspective are presented in this paper.

1 Introduction

LHC-forward experiment (LHCf) measures neutral particles production in a very forward region in proton-proton and proton-ion collisions at the Large Hadron Collider. The main purpose of LHCf is to improve hadronic interaction models used by Monte Carlo (MC) simulations of cosmic ray experiments. Highest energy cosmic rays can only be detected from secondary particles which are produced by the interaction of the primary particles with nuclei of the atmosphere, the so-called air showers. It is possible to reconstruct the kinematic parameters and the type of primary particle

[a]e-mail: marina.delprete@fi.infn.it

DOI: 10.1051/epjconf/201612604014

studying the development of air showers. The energy flow of secondary particles is concentrated in the forward direction. Measurement of particle production at lager pseudo-rapidity, small angles, are very important. Moreover in the forward region soft QCD interactions dominates this MC simulations of air showers are based on phenomenological model, therefore inputs from experimental data are crucial. LHC accelerator gives the possibility to study a wide range of collision energies, from 0.9 TeV to 13 TeV in the center of mass frame, which corresponds to an energy range in the laboratory frame from 10^{14} eV to 10^{17} eV, this energy range covers the "knee" region of cosmic rays spectrum, which occurs at 10^{15} eV.

For these reasons LHCf can give more information to improve hadronic interaction models used in MC simulations to reproduce the development of air showers.

2 The LHCf experiment

LHCf is composed of two independent detectors, called Arm1 and Arm2 located 140 meters away from ATLAS interaction point, IP1 [1]. Arm1 is placed in the side of IP8 direction, toward LHCb experiment, and Arm2 is in the opposite direction toward IP2 and ALICE experiment. Detectors are placed inside Target Neutral Absorber (TAN), where the beam pipe turns into two separates tubes. The LHCf detectors occupy the first 30 cm of the TAN slots, facing the interaction point, thus minimizing the material between them and IP1. This is the best condition allowed at the LHC for measuring fluxes and energy spectra of high energy forward neutral particles produced by the interacting beams. Since charged particles are deviated by the D1 dipole magnet (which bends colliding beams into the two separate beam pipes), only neutral particles, mainly photons and neutrons, reach the detector.

Each detector is made of two sampling and imaging calorimeters called towers. This is required to minimize the overlap of signals due to the different simultaneous showers when detecting two gamma system decay. The longitudinal dimension of towers is of 44 radiation lengths, which correspond to 1.6 nuclear interaction lengths. The transverse cross sections are different for the two detectors. Arm1 has dimension $20x20\,mm^2$ for the small tower and for the large tower $40x40\,mm^2$. The Arm2 transverse cross sections are $25x25\,mm^2$ for the small tower and $32x32\,mm^2$ for the large tower.

Each tower is composed of 16 tungsten layers and 16 plastic scintillator layers to measure energy and also contains 4 position sensitive layers that are different for the two detector [2]. Arm1 uses scintillating fiber (SciFi) to measure position of the neutral particles in the towers, in Arm2 the position is measured by silicon microstrip detectors.

The different geometries of the two detectors have been carefully studied to maximize the accessible kinematic region and exploit the different characteristics of the position sensitive detectors. The smaller tower of each detector is placed on the beam center and covers the pseudo-rapidity range $\eta >$ 9.6, while the larger tower covers the pseudo-rapidity range 8.4 $<\eta <$ 9.4. Energy resolution obtained with these detectors configuration is better than 5 % for photons and of about 40 % for neutrons.

3 Results from LHCf

Until now, data were taken in three different runs of LHC in 2010, 2013 and 2015. In 2010 LHC produced proton-proton collisions at $\sqrt{s} = 900\,GeV$ and at $\sqrt{s} = 7\,TeV$; in 2013 proton-lead collisions at $\sqrt{s_{NN}} = 5.02\,TeV$ and proton-proton collisions at $\sqrt{s} = 2.76\,TeV$ were produced. In 2015 a special low luminosity proton-proton run at $\sqrt{s} = 13\,TeV$ was dedicated to LHCf. Main results by the LHCf experiment are published in papers [3–6]. In this section we will present results for the inclusive neutron spectra measured in p-p collisions at $\sqrt{s} = 7\,TeV$ in section 3.1, for the p_T vs the pseudo rapidity π_0 spectra in p-Pb collisions at $\sqrt{s_{NN}} = 5.02\,TeV$ in section 3.2 and preliminary analysis from the data taken in proton-proton collisions at $\sqrt{s} = 13\,TeV$ in section 3.3.

DOI: 10.1051/epjconf/201612604014

3.1 Neutron spectra in p-p collisions at $\sqrt{s} = 7\,TeV$

The data used in this analysis were taken in 2010 in low luminosity conditions, that are optimal for LHCf experiment due to the reduced event pile-up effect. Integrated luminosities were $0.68\,nb^{-1}$ and $0.53\,nb^{-1}$ for Arm1 and Arm2 respectively, with a typical luminosity of $6.4 \cdot 10^{28}\,cm^{-2}s^{-1}$, derived from LHCf front counters [7]. Performance of detectors for hadron measurements have been carefully studied by Monte Carlo simulations. The data analysis method for neutron energy spectra closely follows that used for the photon component, which is described in [8].

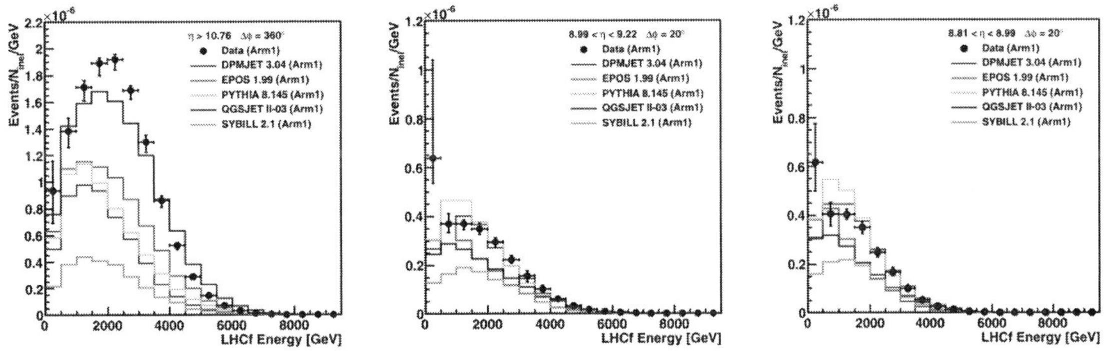

Figure 1. Preliminary neutron spectra measured by LHCf Arm1 in the $\sqrt{s} = 7\,TeV$ p-p collisions in three different rapidity regions. η is > 10.75, $8.99 < \eta < 9.22$ and $8.81 < \eta < 8.99$, in the left, central and right plots respectively.

Particle identification has been carried out by looking at the longitudinal shower development, using two dimensional cuts in the L20% and L90% plane, where $L20\%$ and $L90\%$ are the longitudinal depths containing 20% and 90% of the total deposited energy, respectively. A two dimensional cut on parameter $L2D = L90\% - 1/4 \cdot L20\%$ is then applied. Impact position of particle is reconstructed by the transverse distribution of energy deposit in position sensitive layer. Since the energy resolution

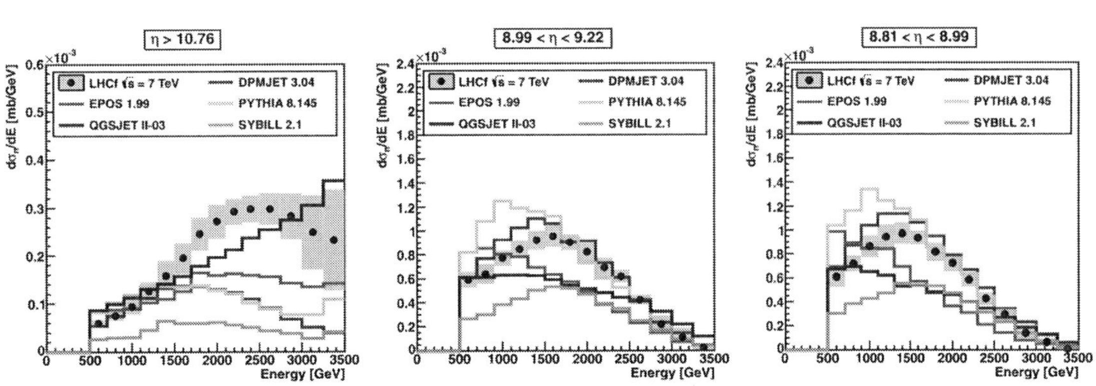

Figure 2. Preliminary neutron spectra measured by LHCf in the $\sqrt{s} = 7\,TeV$ p-p collisions in three different rapidity regions, compared with the models predictions.

for neutrons is of about 40%, true energy spectra are smeared by detector response and it is necessary to apply leakage corrections to the energy. Smeared energy spectra of neutrons measured by Arm1 and Arm2 are compared with MC simulation predictions of several models are shown in figure 1 in three pseudo-rapidity ranges. Unfolding technique is therefore needed to estimate true energy spectra, we use a multidimensional-spectra unfolding Bayesian method relying on energy and P_T measured variables [9]. Preliminary results after unfolding are shown in figure 2. A pronounced high energy neutron peak in the very forward region ($\eta > 10.76$), that is predicted only by the QGSJET-II model, is evident from these results. The presence of this peak is a clear indication of small inelasticity in the very forward direction of the p-p collision. We therefore expect that the baryons produced in the first interaction of a primary cosmic ray with the atmosphere carry on a large fraction of the primary energy, leading to deeply penetrating VHECR induced atmospheric showers.

3.2 π_0 spectra in p-Pb collision at $\sqrt{s_{NN}} = 5.02\,TeV$

During 2013 p-Pb run at the nucleon-nucleon center-of-mass energy of $\sqrt{s_{NN}} = 5.02\,TeV$, Arm2 was installed on p-remnant side for almost all the time, but it also operated for some hours on the Pb-remnant side. The spectra measured in the p-remnant side provide new information on the effect of

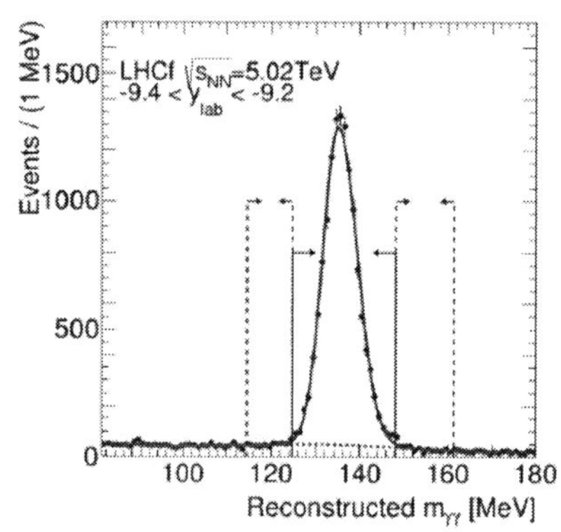

Figure 3. Two-photon invariant mass distributions in rapidity range $-9.4 < y_{LAB} < -9.2$. Peak is fitted with an asymmetric Gaussian (blue solid curve), background is fitted with a polynomial (blue dashed curve). Signal and background windows are indicated by solid and dashed black lines respectively.

the dense nuclear matter in the forward particle production, which is particularly important for HECR Physics. Infact heavy nuclei and light nuclei can be involved in the the primary interactions of cosmic rays in the Earth atmosphere [10].

The invariant mass spectrum of the photons pairs for events where each calorimeter tower is hit by one photon, in figure 3, demonstrates the good performance of the detector in the reconstruction of a clean π_0 sample. A clean peak corresponding to the neutral pion events is prominent over a small background due to pairs of uncorrelated photons. Once the π_0 events are identified, transverse momentum and rapidity are evaluated, neutral pion P_T spectra is measured in different rapidity ranges, and the results have been compared to the predictions of different hadronic interaction models. According to preliminary simulations, approximately half of the forward produced π_0 comes from UltraPeripheral Collisions (UPC) of the proton in the electromagnetic field produced by the Pb nucleus. To obtain pure QCD spectra the effect of the UPC is subtracted. The Weizsacher-Williams approximation is used to estimate the energy distribution of the virtual photons emitted by the fully ionized Pb nucleus,

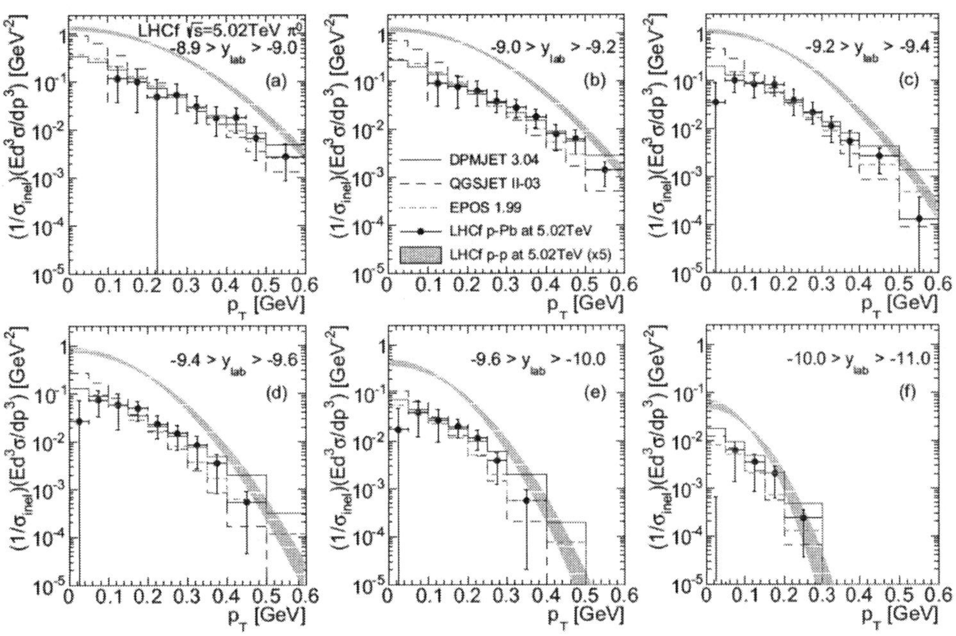

Figure 4. Transverse momentum distribution of neutral pions in several rapidity regions. Coloured histograms show results from Monte Carlo simulations with different hadronic interaction models, black points show experimental data in p-Pb collisions at 5.02 TeV, gray shaded areas represent the extrapolation at 5.02 TeV from data with p-p collisions (multiplied by a factor 5).

and finally the electromagnetic interactions between the virtual photons and the proton is simulated with the SOPHIA model [11].

Neutral pion P_T spectra were measured in different rapidity ranges, and the results were compared to the predictions of different hadronic interaction MC models. Moreover spectra are compared with the spectra in p-p collisions at 5.02 TeV. Th p-p spectra were estimated by interpolating the LHCf results from the p-p runs at 2.76 TeV and 7 TeV taken in the 2013 and 2010 respectively. In figure 4 are shown the results. The better data agreement is with the EPOS and DPMJET simulations and the P_T spectra in p-Pb collisions are harder than in p-p collisions.

Figure 5 shows the measured R_{pPb} compared with the predictions by different models. LHCf data show a strong suppression for neutral pion production for rapidity values $y_{lab} > 8.9$. R_{pPb} varies from 0.1 at $P_T \sim 0.1\ GeV$ to 0.3 at $P_T \sim 0.6\, GeV$. Hadronic interaction models predict also small values of R_{pPb}, showing an overall agreement with the LHCf data within the experimental uncertainty

3.3 Preliminary analysis at $\sqrt{s} = 13\,TeV$ p-p run

In June 2015 there was a special low luminosity p-p run at $\sqrt{s} = 13\,TeV$ was dedicated for LHCf data measurement. The luminosity was $10^{29}\,cm^{-2}s^{-1}$, ~ 5 orders of magnitudes below the nominal one of LHC. In preparation of this data taken both the detectors were upgraded. Due to the higher radiation damage expected, all plastic scintillators were replaced by GSO scintillators, with larger radiation hardness. Also scintillating fiber in Arm1 were replaced by GSO bars and in Arm2 silicon detectors position were changed to better catch showers in calorimeters and a new microstrip bonding scheme

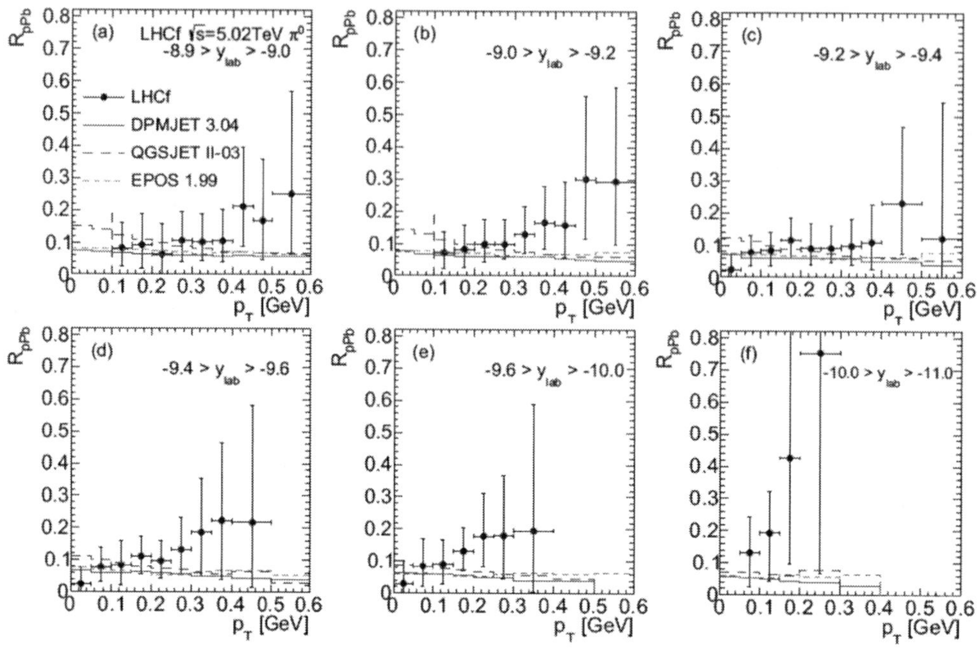

Figure 5. Nuclear modification factor for neutral pions in p-Pb collisions at 5.02 TeV. Coloured histograms show results from Monte Carlo simulations with different hadronic interaction models, experimental data are shown as black points. Nuclear modification factor grows with energy as expected.

has been done to reduce the signal amplitude and avoid saturation of electronics with higher energy signals [12]. From 8th to12^{nd} of June LHCf takes successfully more than 32 hours of physics data . This operation was in collaboration with the ATLAS experiment, the trigger signals of LHCf were sent to ATLAS that recorded their data accordingly. The analyses of common events will permit to classify the events based on the nature of the processes, diffractive dissociation and non-diffractive. Selecting only diffractive events using ATLAS informations will improve the study of soft QCD interactions.

The high statistic and the good performance of detectors have permitted to detect clearly during the operation, with a preliminary analysis, the mesons π_0 and η, in the invariant mass of the two gammas (figure 6).

4 Conclusion and future prospects

LHCf experiment has successfully completed the first proposed program. LHCf performed measurements on forward production of neutral particles in proton-proton and proton-lead collisions at LHC accelerator. These measurements are necessary to calibrate hadronic interaction models used in cosmic rays physics to understand the development of atmospheric showers. We measured photons, neutral pions and neutrons for p-p interactions at $\sqrt{s}$ = 900 GeV and 2.76, 7, 13 TeV and p-Pb interactions at $\sqrt{s}$ = 5.02 TeV.

The analysis of hadronic component for p-p 7 TeV run and the π_0 analysis for p-Pb run are completed. The analysis of data taken in p-p run at 13 TeV is ongoing as the the analysis in common with Atlas collaboration on data taking in 2013 is ongoing.

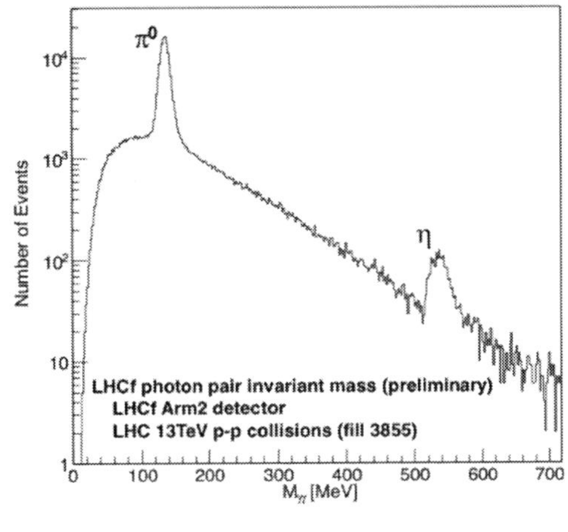

Figure 6. Photon pair invariant mass obtained with the data of the fill 3855 in Arm2. The mass of π_0 and η are clearly reconstructed at $\sim 140\, MeV$ and $\sim 550\, MeV$ respectively.

The future prospects of the experiment are to complete the data taking with a p-Pb run at the top LHC energy in the 2016 or 2017. Another opportunity will given moving Arm1 detector to RICH, and LHCf will be RICHf, and take data in p-p collisions at $510\, TeV$ in the 2017.

References

[1] O. Adriani et al., JINST **3**, S08006 (2008)
[2] O. Adriani et al., JINST **5**, P01012 (2010)
[3] O. Adriani et al., Physics Letters B **715**, 298-303 (2012)
[4] O. Adriani et al., Physics Letters B **703**, 128-134 (2011)
[5] O. Adriani et al., Physical Review D **86**, 092001 (2012)
[6] O. Adriani et al., Physical Review C **89**, 065209 (2014)
[7] K. Taki et al., JINST **7**, T01003 (2012)
[8] K. Kawade et al., JINST **9**, P03016 (2014)
[9] G. D' Agostini, Nucl. Instrum. Meth. A **362**, 487 (1995)
[10] O. Adriani, et al., CERN-LHCC-2011-015, LHCC-I-021 (2011)
[11] A. Mucke et al., Computer Physics Communications **124**, 290 (1999)
[12] Y. Makino et al. PoS TIPP2014 **028**, (2014)

EPJ Web of Conferences **126**, 04015 (2016) DOI: 10.1051/epjconf/201612604015
ICNFP 2015

SHiP: a new multipurpose beam-dump experiment at the SPS

H.B. Dijkstra[1,a]

[1] *CERN, Geneva, Switzerland*

Abstract. SHiP is an experiment to look for very weakly interacting particles at a new to be constructed beam-dump facility at the CERN SPS. The SHiP Technical Proposal has been submitted to the CERN SPS Committee in April 2015. The 400 GeV/c proton beam extracted from the SPS will be dumped on a heavy target with the aim of integrating 2×10^{20} proton on target in five years. A detector located downstream of the target, based on a long vacuum tank followed by a spectrometer and particle identification detectors, will allow probing a variety of models with light long-lived exotic particles and masses below a few GeV/c^2. The main focus will be the physics of the so-called Hidden Portals, i.e. search for Dark Photons, Light scalars and pseudo-scalars, and Heavy Neutral Leptons (HNL). The sensitivity to HNL will allow for the first time to probe, in the mass range between the kaon and the charm meson mass, a coupling range for which Baryogenesis and active neutrino masses could also be explained. Integrated in SHiP is an Emulsion Cloud Chamber, already used in the OPERA experiment, which will allow to study active neutrino cross-sections and angular distributions. In particular SHiP can distinguish between ν_τ and $\bar{\nu}_\tau$, and their deep inelastic scattering cross sections will be measured with statistics three orders of magnitude larger than currently available.

1 Introduction

With the discovery of the Higgs boson[1][2] all the constituents of the Standard Model (SM) have now been observed. The data indicates that the SM provides a phenomenological description of electroweak and strong interactions at the Fermi scale, including the Higgs sector of electroweak symmetry breaking. Nature has chosen masses for the Higgs boson and the top quark such that the lifetime of the SM vacuum greatly exceeds the lifetime of the Universe[3]. In addition, the Landau pole in the Higgs self-interaction potential, which indicates when the theory would become unphysical, suggests this will be well above the Planck scale[4]. Hence, the SM could be an effective, weakly-coupled field theory all the way up to the Planck scale. However, experimental observations like neutrino oscillations, the existence of dark matter (DM) and the baryon asymmetry of the Universe (BAU) prove that the SM is an incomplete description of nature. While this necessitates the extension of the theory, neither experiment nor theory provide guidance on the mass scale of physics beyond the SM (BSM), nor on the coupling strength of any new particles to the SM particles.

Over the next decades the Fermi-mass scale and even beyond will be comprehensively explored either directly by ATLAS and CMS at the LHC and possibly at a FCC facility, or indirectly assuming generic couplings at experiments like LHCb, Belle2 and NA62. Hidden particles, which interact very

[a]e-mail: Hans.Dijkstra@cern.ch

weakly with the SM particles, are predicted in many theoretical models capable of explaining the known shortcomings of the SM. A large part of their accessible parameter space remains hitherto unexplored.

The goal of the proposed Search for Hidden Particles (SHiP) experiment is to extend the search for BSM physics by searching directly for these very weakly interacting particles. Models with no new physics between the Fermi and Planck scales try to extend the SM using the smallest possible set of fields and renormalizable interactions. For example this "*Minimality principle*" motivates νMSM[5][6] which attempts to explain the pattern of neutrino masses, DM and the observed BAU by introducing three, new, right-handed Majorana Heavy Neutral Leptons (HNL). The lightest of these, N_1, provides the DM candidate, while $N_{2,3}$ are responsible for the baryon asymmetry. Through the seesaw mechanism[7] these HNLs also allow the pattern of neutrino masses and oscillations to be explained. This paper will concentrate on the search for these HNL particles. The full physics programme of SHiP is elaborated in the SHiP Physics Proposal[8].

The next section will present the experiments with the best sensitivity to date. The best limits are obtained for hidden particles with couplings to active neutrinos, which can be produced in semi-leptonic decays of strange or charmed particles. A new general purpose fixed target facility at the CERN SPS accelerator would provide an unique combination of intensity and energy capable of producing the large yields required, combined with the most favourable experimental conditions. With SPS performance equivalent to what was achieved for the CNGS[9] programme, the charm yield with a 400 GeV/c beam-dump will provide almost two orders of magnitude more charm than what can be obtained at the high luminosity LHC. Section 3 describes all aspects of the required new beam line, its dump and the SPS operational scenario. The SHiP facility does not interfere with approved SPS experiments.

The detectors for the hidden particle search are described in section 4, which includes a discussion of backgrounds and gives the expected sensitivity reach for HNLs.

Active neutrinos can be considered as the previous generation of hidden particles, and they are produced copiously at the SHiP facility. Hence, in addition to a hidden particles spectrometer, the SHiP experiment contains an emulsion based detector which will allow studying the properties of ν_τ and $\bar{\nu}_\tau$ independently. Section 5 will show that SHiP will increase the ν_τ statistics by almost three orders of magnitude compared to all previous experiments combined.

2 Past Experiments

Previous experiments have made essential contributions to constraining the parameter space for HNLs. The most significant limits below the charm mass [1] have been obtained in the fixed target experiments PS191, CHARM and NuTeV, shown in Figure 1.

The pioneer experiment, PS191, was specifically designed to search for HNLs at the CERN PS. No signal candidates survived rigorous visual tests at the final stage of the event selection leading to the limits on their coupling strength U_μ^2 to ν_μ shown in Figure 1. With $\sqrt{s}$ = 6 GeV/c^2 no charm is produced, and hence the results of PS191 are limited to masses below 450 MeV/c^2. The high energies available at the CERN SPS and the FNAL accelerators allowed extending the searches for HNLs to higher masses using the HNL production via mixing to active neutrinos from charmed decays. CHARM searched for HNL decays in the $e^+e^-\nu$, $\mu^+\mu^-\nu$ and $e\mu\nu$ final states. No signal candidate was found resulting in limits on U_e^2 and U_μ^2, as shown in Figure 1 for U_μ^2. NuTeV searched for events with a muon and a charged track originating from a common decay vertex in their helium volume.

[1]Searches in B and Z^0 decays are in general sensitive to larger masses, but they only cover orders of magnitude larger couplings.

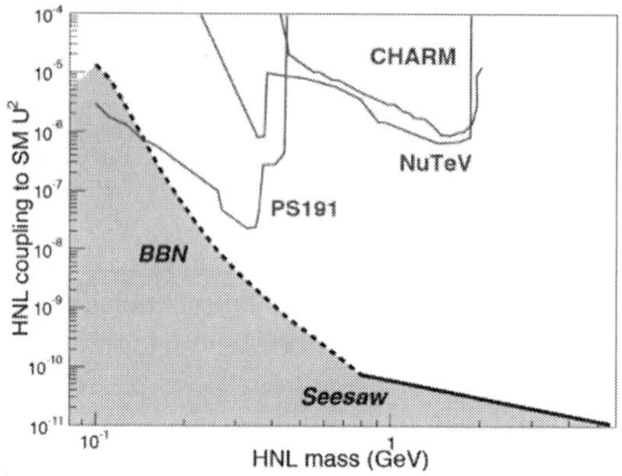

Figure 1. Sensitivity contours for the HNL coupling to active neutrino, $U^2 = U_e^2 + U_\mu^2 + U_\tau^2$ as function of the HNL mass assuming $U_e^2 : U_\mu^2 : U_\tau^2 = 1 : 16 : 3.8$. Experiments are discussed in the text.

No signal events passed the selection criteria which is consistent with the expected background of 0.57±0.15 events originating mainly from active neutrino interactions within the decay volume and in its vicinity. The corresponding NuTeV limit on U_μ^2 is also shown in Figure 1.

Table 1 lists the relevant parameters of these three experiments, in comparison with those planned for SHiP. SHiP can obtain about four orders of magnitude more yield for the same coupling by a combination of increasing the number of protons on target (p.o.t.), moving closer to the target, and increasing the fiducial decay volume.

Table 1. Comparison of the experimental conditions of HNL search experiments.

Experiment	PS191	NuTev	CHARM	SHiP
Reference	[10]	[11]	[?]	[13]
Proton energy (GeV/c)	19.2	800	400	400
p.o.t. ($\times 10^{19}$)	0.86	0.25	0.24	20
Decay volume (m^3)	360	1100	315	1780
Decay volume pressure (bar)	1, He	1, He	1, Air	10^{-6}
Distance to target (m)	128	1400	480	120
Angle to beam (mrad)	40	0	10	0

3 The SHiP Facility at the SPS

The SHiP experiment requires a total of 2×10^{20} p.o.t. over a period of about five years. The CERN management set up a task force of the CERN accelerator sector and the occupational health and safety and environmental protection unit to study an implementation of a 400 GeV/c proton beam for SHiP with minimal modifications to the SPS complex, and maximum use of existing transfer lines. Figure 2 shows the location of the proposed facility in the SPS accelerator complex. The location is ideally suited since it allows a full integration on CERN land with almost no impact on the existing facilities.

The SHiP facility shares the TT20 transfer line with the other North Area facilities, allowing switching on a cycle-by-cycle basis with other North Area beam destinations. Contrary to the beam extraction

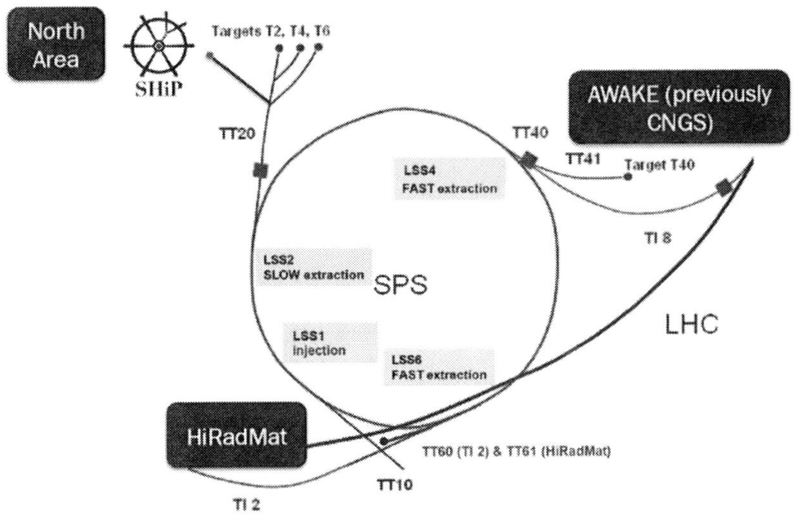

Figure 2. Overview of the SPS accelerator complex. The SHiP facility is located in the North Area and shares the TT20 transfer line with the fixed target programmes.

used for the CNGS programme, SHiP requires a spill of about one second to reduce background from muon pile-up. Taking into account that 20 % of the SPS time will be used for injecting protons into the LHC, and the number of protons required for the other targets in the North Area, leads to a SPS cycle with 4×10^{13} p.o.t. per 1 s spill, and 10^6 spills per year.

In order to maximize the production of heavy mesons and minimize the production of neutrinos and muons, the target must be designed with a material of the shortest possible nuclear interaction length and with dimensions to contain the proton shower. Consequently, the SHiP production target is one of the most challenging aspects of the facility due to the very high average beam power (up to 350 kW) deposited on the target. The peak power during the spill amounts to 2.56 MW. For this reason, the design of the target relies on the energy dilution produced by the large beam spot and the beam sweep produced at the beginning of the SHiP beam line. The required performance is achieved with a longitudinally segmented hybrid target consisting of blocks of four interaction lengths of titanium-zirconium doped molybdenum (TZM) alloy (58 cm) in the core of the shower followed by six interaction lengths of pure tungsten (58 cm). The target is embedded in a massive cast iron bunker (440 m^3), which acts as additional hadron absorber.

4 Hidden Particle Spectrometer

HNLs couple to the standard model active neutrinos, albeit with a very small coupling strength, and can hitherto replace neutrinos in any semi-leptonic decay if kinematically allowed. A 400 GeV/c proton on the Mo target produces $\sim$ 0.008 D-hadrons. HNLs produced from a semileptonic charm decay have a relatively large opening angle to the beam direction, hence the spectrometer should be placed as close to the target as possible. The small coupling results in a long lifetime, typically $c\tau = 50$ km, requiring a long decay volume. The spectrometer has been optimised to detect the decay of HNLs to $l^{\pm}l^{\mp}\nu$ and $l^{\pm}h^{\mp}$, with $h = \pi,\ \rho$. The protons that impinge on the target produce $\sim 10^{10}$ muons/s. To maximize the sensitivity of the experiment, this muon flux has to be reduced to less than $\sim 10^{5}$/s to avoid mimicking the signal due to μ-interactions, or pile-up of muons.

Figure 3 shows an overview of the spectrometer, with a total length of ~130 m. On the left is the target complex followed by the hadron absorber. This is followed by a 48 m long complex of warm magnets to sweep the muon flux out of the acceptance of the spectrometer. After this muon shield, the first 10 m is used for the ν_{τ} detector, and this is followed by a large vacuum vessel with a magnetic spectrometer.

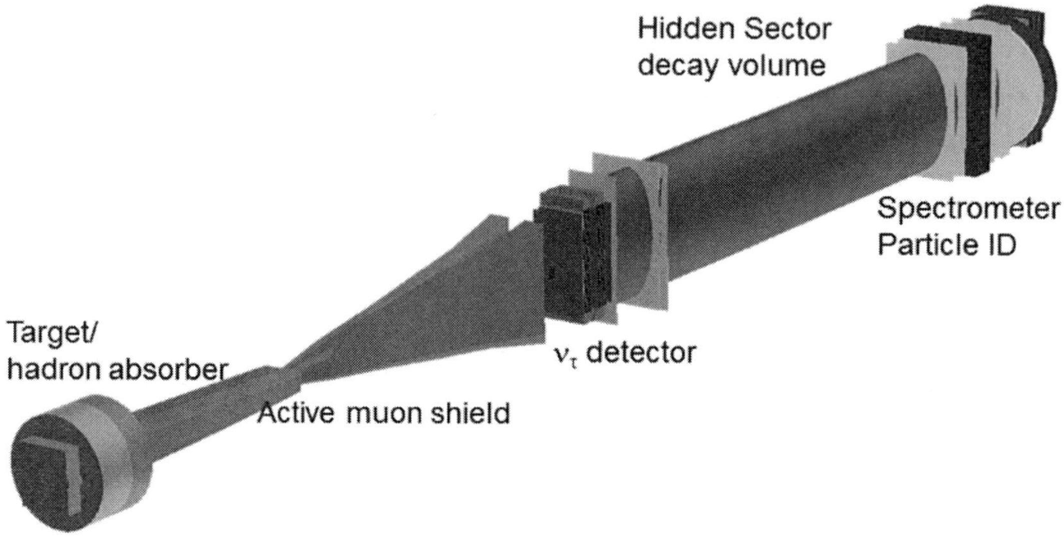

Figure 3. Overview of the SHiP spectrometer.

Large momentum muons are predominately produced in decay of η, ρ and ϕ-mesons. Their rate up to 350 GeV/c is prohibitively large, hence the muon shield needs a total $\int B_y dl \approx 85$ Tm to bend these particles out of the acceptance of the spectrometer. Figure 4 shows on the left side the trajectory of muons with an initial momentum of 350 GeV/c through 48 m of warm Fe-magnets with a field of 1.8 T. Large momentum muons with an angle which is not too large in the opposite direction from the magnetic bending direction are swept out far enough to miss the contour of the HNL decay-vessel. However, as is shown in the right panel of Figure 4, the return field bends back lower momentum muons back towards the detector. To prevent this, the first section of field $0 < z < 19$ m is used to separate μ^{+} and μ^{-} to either side of the beam direction. This separation creates enough room to place the return field for the subsequent magnets $19 < z < 48$ m in the central region, where there are no charged particles. Hence, particles bend back into the acceptance of the spectrometer will encounter the field again and will be swept back out. With this configuration of magnets, the total number of muons above 3 GeV/c in the spectrometer is less than 10 kHz.

The spectrometer to detect the HNL decay products consists of a $\sim$ 60 m long evacuated (10^{-6} bar) vessel. At the end of the vessel a 0.65 Tm spectrometer magnet surrounded by four straw stations allows to determine the momentum of the decay products with $\Delta p/p < 1$ %. The vessel is followed by an electromagnetic calorimeter to provide e, γ and π identification, and a hadron calorimeter to provide π/μ discrimination especially at low momenta. Larger momentum muons are identified with the muon

DOI: 10.1051/epjconf/201612604015

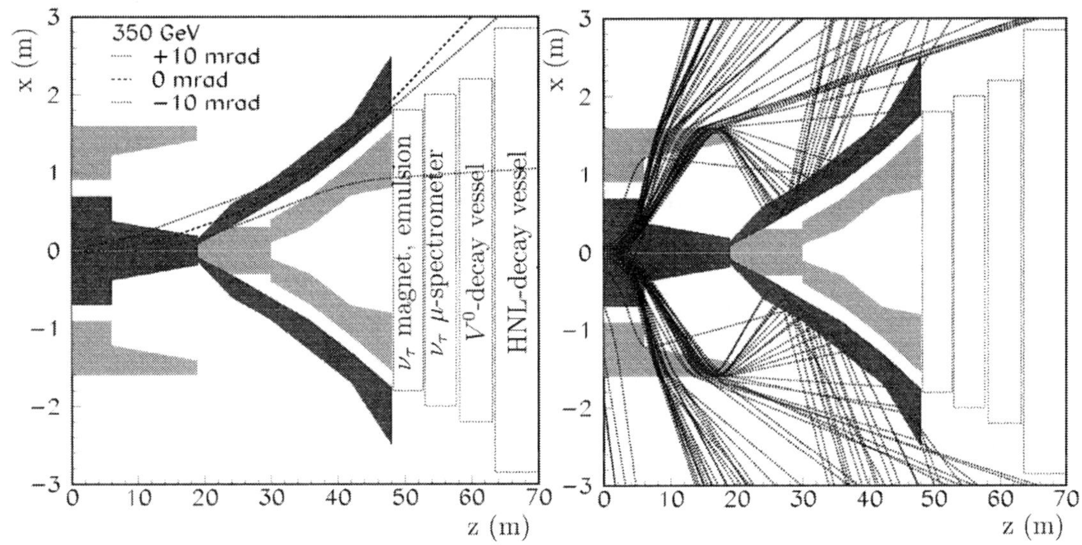

Figure 4. A horizontal cut of the active muon shield. The blue and green show the regions of field and return-field respectively, both are oriented perpendicular to this view. Also sketched are the outlines of the various detector elements. On the left the trajectory of three 350 GeV/c muons with a range of initial angles. On the right the shield is overlayed with a selection of low momentum muon trajectories.

spectrometer downstream of the calorimeters. While the spectrometer has moderate requirements, the real challenge is to tag all the possible background sources, such as:

- Neutrino- and muon-induced backgrounds: neutrinos and muons coming from hadron decays produced in the proton interactions can interact inelastically with the material surrounding the decay volume. These interactions can generate particles that enter the decay volume and mimic signal events.
- Random combination of tracks in the fiducial volume from muons, or other charged particles from interactions in the proximity of the detector, which enter the decay volume and together fake decay vertices which resemble signal events.
- Cosmic muons entering the decay volume can create background in the same way as the previous two classes.

To tag these backgrounds, the vessel is surrounded by:

- The entrance window is covered by a large array of scintillators detecting with high efficiency any charged particle which enters the vessel.
- 5 m after the entrance window, inside the vessel, a thin straw chamber detects in addition the decay products of V^0-particles which have entered the vessel.
- The entire vessel is surrounded by liquid scintillator read by ~ 1700 photo-detectors.
- Just after the exit window of the vessel, a large array of scintillators form a TOF detector with a resolution of ~ 100 ps to reduce the random combination of tracks.

SHiP has simulated all backgrounds with effective statistics larger than expected in 2×10^{20} p.o.t., and has not found any background events which could mimic a HNL decay. For the sensitivity contours 0.1 background events will be assumed.

The HNL yield depends on the hierarchy of the active neutrino masses and on the relative strength of the HNL couplings to the three SM flavours U_e^2, U_μ^2, U_τ^2. Figure 5 shows two scenarios which conform to existing theoretical studies[16], $U_e^2 : U_\mu^2 : U_\tau^2 \sim 52 : 1 : 1$ and inverted hierarchy of active neutrino masses, and $U_e^2 : U_\mu^2 : U_\tau^2 \sim 1 : 16 : 3.8$ and normal hierarchy of active neutrino masses.

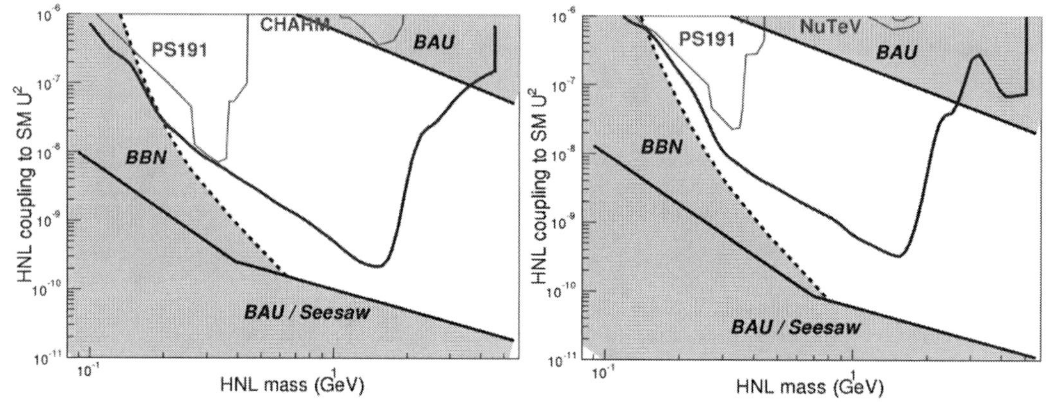

Figure 5. Sensitivity regions in the parameter space of νMSM, for two scenarios where U_e^2(left) and U_μ^2(right) dominate. The blue curves can be interpreted as 3σ evidence if two events are observed in SHiP. In the region below BAU, the observed Baryogenesis can be generated via oscillations of the two heaviest HNLs. The regions below the "seesaw" and "BBN" lines are excluded by neutrino oscillations experiments and Big Bang Nucleosynthesis respectively.

5 Active ν Physics

The tau neutrino detector shown in Figure 3 is located immediately downstream of the muon shield. It consists of a neutrino emulsion target in a magnetic field of 1 T, followed by a Muon Magnetic Spectrometer (MMS). The emulsion target is made of Opera-type modules which employ the Emulsion Cloud Chamber (ECC) technology. The charges of the hadrons produced in the neutrino interactions are identified using a tracker consisting of active planes which interleave the ECC modules. Muons are identified by the MMS which consists of a warm iron dipole magnet instrumented with active layers based on the Opera Resistive Plate Chambers and the Opera Drift Tube Tracker.

With this addition, the SHIP detector will have the unique potential to explore the physics of tau neutrinos. Whereas DONUT at Fermilab observed nine tau neutrino candidates[14] and OPERA at LNGS five candidates from oscillations[15], the SHiP experiment will collect ~ 3000 reconstructed tau neutrino interactions during the above mentioned data-taking period. Both the ν_τ and $\bar{\nu}_\tau$ cross sections will be measured independently for the first time! SHiP is unique in its capability to accumulate large statistics of all the three active neutrino and anti-neutrino flavours. The ν_e cross-section will be studied at high energies, and since its production is dominated by charm decays, this measurement will also provide the normalization for the search for hidden particles. In addition the strange-quark content of the nucleon will be measured by means of the charmed hadron production in anti-neutrino interactions and the collection of a few million muon neutrinos will contribute to the studies of structure functions and to the measurement of the Weinberg angle.

6 Conclusion

SHiP intends to search for very weakly interacting particles in a hitherto unexplored region of parameter space. The covered region is particularly promising, since if such particles are found, the νMSM model[5][6] can link these particles to active neutrino mixing and Baryogenesis. In addition SHiP will measure ν_τ and $\bar{\nu}_\tau$ cross sections independently and with unprecedented statistics.

The discovery of a very weakly interacting hidden sector would lead to a dramatic breakthrough in our fundamental understanding of particle physics. In addition, it would open up a new field of experimental particle physics, which would benefit from accelerators capable of exceeding the already unique performance of the SPS.

References

[1] G. Aad et al., Phys.Lett. **B716**, 1–29, (2012).
[2] S. Chatrchyan et al., Phys.Lett. **B716**, 30–61, (2012).
[3] D. Buttazzo, G. Degrassi, P. P. Giardino, G. F. Giudice, F. Sala, et al., JHEP. **1312**, 089, (2013), and references herein.
[4] J. Ellis, J. Espinosa, G. Giudice, A. Hoecker, and A. Riotto, Phys.Lett. **B679**, 369–375, (2009)
[5] T. Asaka, S. Blanchet, and M. Shaposhnikov, Phys.Lett. **B631**, 151–156, (2005).
[6] T. Asaka and M. Shaposhnikov, Phys.Lett. **B620**, 17–26, (2005).
[7] Proceedings, International Conference, SEESAW25, Paris, France, June 10-11, 2004, World Scientific, 2005, and references therein.
[8] S. Alekhin et al., CERN-SPSC-2015-017 (2015).
[9] D.Autiero et al., Nucl.Phys.B **189**, 263-270, (2009).
[10] G. Bernardi, G. Carugno, J. Chauveau, F. Dicarlo, M. Dris, et al., Phys.Lett. **B203**, 332, (1988).
[11] A. Vaitaitis et al., Phys. Rev. Lett. **83**, 4943–4946, (1999).
[12] F. Bergsma et al., Phys. Lett. **B166**, 473, (1986).
[13] M. Anelli et al., CERN-SPSC-2015-016 and CERN-SPSC-2015-040, 2015.
[14] K. Kodama et al., Phys.Rev. **D78**, 052002, (2008).
[15] N. Agafonova et al., PTEP. **2014(10)**, 101C01, (2014).
[16] D. Gorbunov and M. Shaposhnikov, JHEP. **0710**, 015, (2007).

Analytical Formulae linking Quark Confinement and Chiral Symmetry Breaking *

Takahiro M. Doi[1,a], Krzysztof Redlich[2,3,4], Chihiro Sasaki[2,5], and Hideo Suganuma[1]

[1] *Department of Physics, Kyoto University, Kyoto 606-8502, Japan*
[2] *Institute of Theoretical Physics, University of Wroclaw, PL-50204 Wroclaw, Poland*
[3] *Extreme Matter Institute EMMI, GSI, Planckstr. 1, D-64291 Darmstadt, Germany*
[4] *Department of Physics, Duke University, Durham, North Carolina 27708, USA*
[5] *Frankfurt Institute for Advanced Studies, D-60438 Frankfurt am Main, Germany*

Abstract. Dirac spectrum representations of the Polyakov loop fluctuations are derived on the temporally odd-number lattice, where the temporal length is odd with the periodic boundary condition. We investigate the Polyakov loop fluctuations based on these analytical relations. It is semi-analytically and numerically found that the low-lying Dirac eigenmodes have little contribution to the Polyakov loop fluctuations, which are sensitive probe for the quark deconfinement. Our results suggest no direct one-to-one corresponding between quark confinement and chiral symmetry breaking in QCD.

1 Introduction

It is one of the most important problems in particle and nuclear physics to understand the nonperturbative properties of QCD, such as confinement and chiral symmetry breaking, including these relation. By numerous efforts, these properties have been partially understood.

The Polyakov loop is an order parameter of quark confinement in the quenched QCD [1]. Moreover, recently it is pointed out that the Polyakov loop fluctuations are very sensitive probes for the quark deconfinement even in full QCD with light quarks [2].

The low-lying eigenmodes of the Dirac operator are important for chiral symmetry breaking, for example known as Banks-Casher relation [3]. At low temperature, the low-lying Dirac modes exist and have dominant contribution to chiral condensate while at high temperature these modes disappear and chiral symmetry is restored.

In the presence of light dynamical quarks, both quark deconfinement and chiral restoration are not phase transition but crossover and take place in the same temperature region [4–7]. This observation seems evidence that confinement and chiral symmetry breaking are strongly correlated in QCD. However, there are some contrary observations that chiral restoration does not immediately mean the quark deconfinement. For example, "hadrons" can be observed as bound states after removal of low-lying

*This proceeding of ICNFP 2015 has the same contents as that of the conference "Lattice 2015" because of very large overlap between our presentations on these two conferences. The proceeding of the lattice-conference will be published as Proc. Sci. (Lattice 2015), 315 (2015).

[a]e-mail: doi@ruby.scphys.kyoto-u.ac.jp

Dirac modes [8]. Also, it is shown that low-lying Dirac modes are not important for confinement properties such as the Polyakov loop and the linear confining potential of a quark-antiquark system [9].

In this paper, we derive some analytical relations to express the Polyakov loop and its fluctuations by Dirac eigenmodes. Then we semi-analytically and numerically show that low-lying Dirac modes have negligible contribution to the Polyakov loop fluctuations based on these analytical relations. This talk is mainly based on our recent work [10].

2 Polyakov loop fluctuations and Dirac mode on a lattice

In this section, we review the Polyakov loop fluctuations [2] and Dirac-mode expansion [9] as the basics of our work. In this paper, we consider the SU(N_c) lattice QCD formalism on a square lattice with spacing a. We denote each site as $s = (s_1, s_2, s_3, s_4)$ with $s_\mu = 1, 2, \cdots, N_\mu$ and a link variable as $U_\mu(s) = e^{iagA_\mu(s)}$ with a gauge field, $A_\mu(s) \in su(N_c)$, and the gauge coupling g. We use spatially symmetric lattice, i.e., $N_1 = N_2 = N_3 \equiv N_\sigma$ and $N_4 \equiv N_\tau$, with $N_\sigma \geq N_\tau$. We define all the γ-matrices to be hermite as $\gamma_\mu^\dagger = \gamma_\mu$.

2.1 Polyakov loop fluctuations

For each gauge configuration, the Polyakov loop L and the Z_3 rotated Polyakov loop $\tilde{L}$ are defined as

$$L \equiv \frac{1}{N_c V} \sum_s \mathrm{tr}_c\{\prod_{i=0}^{N_\tau - 1} U_4(s + i\hat{4})\}, \qquad \tilde{L} \equiv L e^{2\pi ki/3} \quad (k = 0, \pm 1), \tag{1}$$

where $\hat{\mu}$ is the unit vector in direction of μ in the lattice unit, $V = N_\sigma^3 N_\tau$ is the volume of 4-dimensional lattice, $k = 0$ is taken in the confinement phase, and k in the deconfinement phase is chosen such that the transformed Polyakov loop $\tilde{L}$ lies in its real sector. Using the Z_3 rotated Polyakov loop $\tilde{L}$, three Polyakov loop susceptibilities are introduced as

$$T^3 \chi_A = \frac{N_\sigma^3}{N_\tau^3}[\langle |L|^2 \rangle - \langle |L| \rangle^2], \tag{2}$$

$$T^3 \chi_L = \frac{N_\sigma^3}{N_\tau^3}[\langle (L_L)^2 \rangle - \langle L_L \rangle^2], \tag{3}$$

$$T^3 \chi_T = \frac{N_\sigma^3}{N_\tau^3}[\langle (L_T)^2 \rangle - \langle L_T \rangle^2], \tag{4}$$

where $L_L \equiv \mathrm{Re}(\tilde{L})$ and $L_T \equiv \mathrm{Im}(\tilde{L})$, and $\langle x \rangle$ denotes an average over all gauge configurations. Also, their ratios are introduced as,

$$R_A \equiv \frac{\chi_A}{\chi_L}, \qquad R_T \equiv \frac{\chi_T}{\chi_L}. \tag{5}$$

By finite-temperature lattice QCD calculation, it is found that the ratio R_A is a very sensitive probe for the quark deconfinement even in the presence of light dynamical quarks [2].

2.2 Operator formalism and Dirac mode on a lattice

The link-variable operator $\hat{U}_{\pm\mu}$ is defined by the matrix element

$$\langle s|\hat{U}_{\pm\mu}|s'\rangle = U_{\pm\mu}(s)\delta_{s\pm\hat{\mu},s'}. \tag{6}$$

At the temporal boundary, we impose the temporal antiperiodicity to the link-variable operator as

$$\langle \mathbf{s}, N_t|\hat{U}_4|\mathbf{s}, 1\rangle = -U_4(\mathbf{s}, N_t) \tag{7}$$

to impose the temporal antiperiodicity to the Dirac operator for finite-temperature formalism. Using the link-variable operator, the Polyakov loop is expressed as

$$L = -\frac{1}{N_c V}\mathrm{Tr}_c\{\hat{U}_4^{N_\tau}\} = \frac{1}{N_c V}\sum_s \mathrm{tr}_c\{\prod_{n=0}^{N_t-1} U_4(s+n\hat{t})\}, \tag{8}$$

where Tr_c denotes the functional trace, $\mathrm{Tr}_c \equiv \sum_s \mathrm{tr}_c$, and tr_c is taken over color index.

Unlike the Polyakov loop, a functional trace of a product of the link-variable operators corresponding to the non-closed path vanishes:

$$\mathrm{Tr}_c(\hat{U}_{\mu_1}\hat{U}_{\mu_2}\cdots\hat{U}_{\mu_{N_P}}) = \mathrm{tr}_c\sum_s U_{\mu_1}(s)\cdots U_{\mu_{N_P}}(s+\sum_{k=1}^{N_P-1}\hat{\mu}_k)\langle s+\sum_{k=1}^{N_P}\hat{\mu}_k|s\rangle = 0, \tag{9}$$

with $\sum_{k=1}^{N_P}\hat{\mu}_k \neq 0$ for any non-closed path with its length N_P. This notable property of the link-variable operator is satisfied due to the definition of the link-variable operator and easily understood by Elitzur' s theorem [11] that a vacuum expectation value of a gauge-variant operator is zero.

The Dirac operator $\hat{\not{D}}$ on the lattice is defined by

$$\hat{\not{D}} = \gamma_\mu\hat{D}_\mu = \frac{1}{2a}\sum_{\mu=1}^{4}\gamma_\mu(\hat{U}_\mu - \hat{U}_{-\mu}). \tag{10}$$

The eigenvalue equation of the Dirac operator can be expressed as

$$\hat{\not{D}}|n\rangle = i\lambda_n|n\rangle, \quad \lambda_n \in \mathbf{R} \tag{11}$$

because of the anti-hermiticity of the Dirac operator. These Dirac eigenstates $|n\rangle$ have the completeness of $\sum_n |n\rangle\langle n| = 1$. Due to $\{\hat{\not{D}}, \gamma_5\} = 0$, the chiral partner $\gamma_5|n\rangle$ is also an eigenstate with the eigenvalue $(-i\lambda_n)$. Known as the Banks-Casher relation [3], the low-lying Dirac modes have the dominant contribution to the chiral condensate $\langle\bar{\psi}\psi\rangle$ and thus these modes are essential modes for chiral symmetry breaking.

3 Dirac spectrum representation of Polyakov loop fluctuations

In this section, we shortly show the derivation of the Dirac spectrum representations of the Polyakov loop fluctuations and then discuss the relation between confinement and chiral symmetry breaking in QCD. The detailed derivation is shown in Ref. [10, 12, 13].

We consider a temporally odd-number lattice with odd-number temporal size N_τ. On such a lattice, we introduce a functional trace I defined as

$$I = \mathrm{Tr}_{c,\gamma}(\hat{U}_4\hat{\not{D}}^{N_\tau-1}), \tag{12}$$

where $\mathrm{Tr}_{c,\gamma} \equiv \sum_s \mathrm{tr}_c \mathrm{tr}_\gamma$, and tr_γ is taken over spinor indexes. Substituting the definition of the Dirac operator (10), the functional trace I is expressed as a sum of products of odd-number link-variable operators because N_τ is odd number. Note that most of the terms in the expansion of I exactly vanish because one cannot make a closed loop by using odd-number link-variable operators on a square lattice. Thus there is only contribution from the closed path due to the temporal periodicity, that is the Polyakov loop L. In this way, the functional trace I can be expressed as

$$I = \frac{12V}{(2a)^{N_\tau - 1}} L. \tag{13}$$

On the other hand, using the completeness of the Dirac mode, the functional trace is expressed as

$$I = \sum_n \langle n|\hat{U}_4 \hat{\not{D}}^{N_\tau - 1}|n\rangle = i^{N_\tau - 1} \sum_n \lambda_n^{N_\tau - 1} \langle n|\hat{U}_4|n\rangle. \tag{14}$$

Therefore, we derive the Dirac spectrum representation of the Polyakov loop:

$$L = \frac{(2ai)^{N_\tau - 1}}{12V} \sum_n \lambda_n^{N_\tau - 1} \langle n|\hat{U}_4|n\rangle. \tag{15}$$

Since the identity (15) is exactly satisfied for each gauge-configuration, the Z_3 rotated Polyakov loop $\tilde{L}$ can be also expressed by the Dirac modes as

$$\tilde{L} = \frac{(2ai)^{N_\tau - 1}}{12V} \sum_n \lambda_n^{N_\tau - 1} e^{2\pi k i/3} \langle n|\hat{U}_4|n\rangle, \tag{16}$$

where k is defined in Eq. (1). Correspondingly, we can express $L_L \equiv \mathrm{Re}(\tilde{L})$, $L_T \equiv \mathrm{Im}(\tilde{L})$, and $|L|$ by the Dirac modes. From these expressions and Eqs. (2)-(5), one can derive analytical relations connecting the Polyakov loop fluctuations and the Dirac modes. In particular, we focus on the Dirac spectrum representation of R_A:

$$R_A = \frac{\left\langle \left|\sum_n \lambda_n^{N_\tau - 1} \langle n|\hat{U}_4|n\rangle\right|^2 \right\rangle - \left\langle \left|\sum_n \lambda_n^{N_\tau - 1} \langle n|\hat{U}_4|n\rangle\right| \right\rangle^2}{\left\langle \left(\sum_n \lambda_n^{N_\tau - 1} \mathrm{Re}\left(e^{2\pi k i/3} \langle n|\hat{U}_4|n\rangle\right)\right)^2 \right\rangle - \left\langle \sum_n \lambda_n^{N_\tau - 1} \mathrm{Re}\left(e^{2\pi k i/3} \langle n|\hat{U}_4|n\rangle\right) \right\rangle^2}. \tag{17}$$

Note that the series of analytical relations is exactly satisfied in both full QCD and quenched QCD.

As mentioned above, the ratio R_A is a sensitive probe for the quark deconfinement and the Dirac modes are strongly related to chiral symmtery breaking. Since Eq. (17) is an analytical relation connecting the ratio R_A and the Dirac modes, we can extract from it the information of the relation between confinement and chiral symmetry breaking in QCD. The damping factors $\lambda_n^{N_\tau - 1}$ in sums over all the Dirac modes appearing in Eq. (17) play important roles. Because these damping factors are negligibly small with small eigenvalues $|\lambda_n| \simeq 0$, the contribution from the low-lying Dirac modes to the ratio R_A is strongly suppressed while these modes are responsible for saturating the chiral condensate. In other words, important modes for chiral symmetry breaking are not important for chiral symmetry breaking. This analytical discussion is consistent with other numerical studies of the Dirac mode expansion [9], which show that confinement properties such as the Polyakov loop and the confining linear force defined from the Wilson loop are almost unchanged after removal of the low-lying Dirac modes.

Not only the damping factor $\lambda_n^{N_\tau - 1}$, but also the Dirac-mode matrix element $\langle n|\hat{U}_4|n\rangle$ should be taken into account. If the matrix element increases stronger than the damping factor as $\sim 1/\lambda_n^{N_\tau - 1}$ in the small-$|\lambda|$ region, the above analytical observation is not correct. However, the matrix element $\langle n|\hat{U}_4|n\rangle$ does not change our analytical expectation qualitatively. The detailed numerical results of the nontrivial behavior of the matrix element $\langle n|\hat{U}_4|n\rangle$ are shown in Ref. [13].

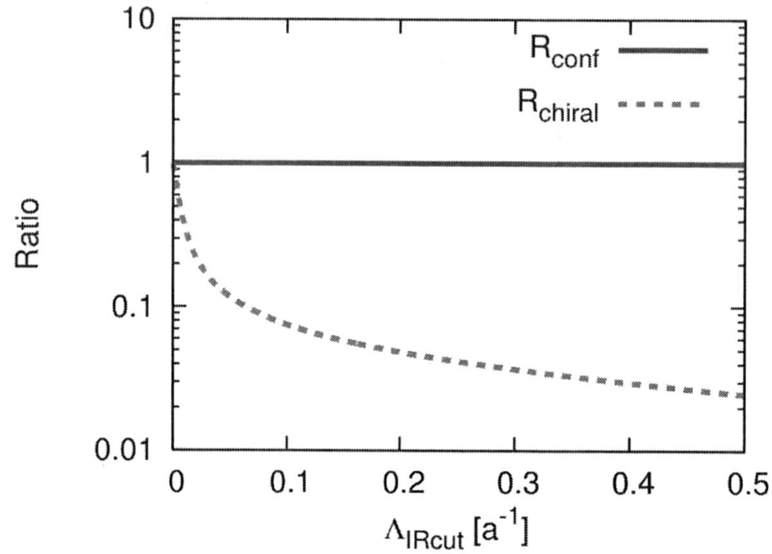

Figure 1. The numerical results for the $R_{\rm chiral}$ and $R_{\rm conf}$ plotted against the infrared cutoff Λ in lattice units at $\beta = 5.6$ on $10^3 \times 5$ lattice taken from Ref. [10]. The quark mass of $m = 5$ MeV is taken for calculation of the ratio $R_{\rm chiral}$.

4 Numerical analysis

In this section, we show the numerical results based on the analytical relation (17) to quantitatively confirm the above analytical discussion that the low-lying Dirac modes have negligible contribution to the ratio R_A.

Since the Dirac eigenvalue λ_n and the Dirac-mode matrix element $\langle n|\hat{U}_4|n\rangle$ are gauge-invariant and can be calculated by solving the Dirac eigenvalue equation Eq. (11), we can calculate the contribution from each Dirac mode. We introduce the infrared cutoff Λ and define the Λ-dependent Polyakov loops,

$$|L|_\Lambda = \frac{(2a)^{N_\tau - 1}}{12V} \left| \sum_{|\lambda_n|>\Lambda} \lambda_n^{N_\tau - 1} \langle n|\hat{U}_4|n\rangle \right|, \tag{18}$$

for the modulus, and

$$(L_L)_\Lambda = C_\tau \sum_{|\lambda_n|>\Lambda} \lambda_n^{N_\tau-1} \mathrm{Re}\left(e^{2\pi k i/3} \langle n|\hat{U}_4|n\rangle\right), \tag{19}$$

$$(L_T)_\Lambda = C_\tau \sum_{|\lambda_n|>\Lambda} \lambda_n^{N_\tau-1} \mathrm{Im}\left(e^{2\pi k i/3} \langle n|\hat{U}_4|n\rangle\right). \tag{20}$$

for the real and the imaginary part, respectively, with $C_\tau = (2ai)^{N_\tau-1}/12V$. Then, the Λ-dependent Polyakov loop susceptibilities, $(\chi_A)_\Lambda$, $(\chi_L)_\Lambda$, $(\chi_T)_\Lambda$, are defined using the corresponding Polyakov loops, $|L|_\Lambda$, $(L_L)_\Lambda$ or $(L_T)_\Lambda$, and their ratios are also defined as

$$(R_A)_\Lambda = \frac{(\chi_A)_\Lambda}{(\chi_L)_\Lambda}, \qquad (R_T)_\Lambda = \frac{(\chi_T)_\Lambda}{(\chi_L)_\Lambda}. \tag{21}$$

For comparison, we introduce the cutoff-dependent chiral condensate $\langle\bar{\psi}\psi\rangle_\Lambda$ as

$$\langle\bar{\psi}\psi\rangle_\Lambda = -\frac{1}{V}\sum_{|\lambda_n|\geq\Lambda}\frac{2m}{\lambda_n^2+m^2}, \tag{22}$$

where m is the current quark mass. In order to see the contribution from the low-lying Dirac modes to the ratio R_A and chiral condensate, we introduce two quantities $R_{\rm conf}$ and $R_{\rm chiral}$ as

$$R_{\rm conf} = \frac{(R_A)_\Lambda}{R_A}, \qquad R_{\rm chiral} = \frac{\langle\bar{\psi}\psi\rangle_\Lambda}{\langle\bar{\psi}\psi\rangle}. \tag{23}$$

The ratios $R_{\rm conf}$ and $R_{\rm chiral}$ are calculated in the SU(3) lattice QCD at the quenched level. In our calculation, the standard plaquette action is used on $10^3 \times 5$ lattice with two different $\beta \equiv \frac{2N_c}{g^2}$: $\beta = 5.6$ for the confinement phase and $\beta = 6.0$ for the deconfinement phase. For each β, 20 gauge-configurations are taken every 500 sweeps after the thermalization of 5000 sweeps.

In Fig. 1, the numerical results for $R_{\rm conf}$ and $R_{\rm chiral}$ in the confinement phase are shown with various values of the infrared cutoff Λ. We take the quark mass $m = 5$ MeV for the chiral condensate. From Fig. 1, one can confirm that the quantity $R_{\rm chiral}$ is largely reduced by removal of the low-lying Dirac modes and then the low-lying Dirac modes are important for chiral symmetry breaking. Nevertheless, the quantity $R_{\rm conf}$, that is the ratio R_A, is almost unchanged by removing the low-lying Dirac modes even with large cutoff $\Lambda \simeq 0.5$ GeV. The results in the deconfinement phase show the similar behavior with the confinement phase. In this way, we have numerically confirmed that the low-lying Dirac modes have little contribution to the ratio R_A.

5 Summary

We have derived the Dirac spectrum representations of the Polyakov loop and its fluctuations on the temporally odd-number lattice. Based on the analytical relations, it is semi-analytically and numerically shown that the low-lying Dirac modes have negligible contribution to the Polyakov loop fluctuations, which are good probes for the quark deconfinement although these modes have dominant contribution to the chiral condensate. That is, the important modes for chiral symmetry breaking are not important for confinement. Our results suggest no direct one-to-one corresponding between confinement and chiral symmetry breaking in QCD.

Acknowledgements

T.M.D. is supported by Grant-in-Aid for JSPS Fellows (Grant No.15J02108), and H.S. is supported by Grant No.15K05076 from Japan Society for the Promotion of Science. K.R. acknowledges partial support of the U.S. Department of Energy under Grant No. DE-FG02-05ER41367, K.R. and C.S. are partly supported by the Polish Science Foundation (NCN) under Maestro Grant No. DEC-2013/10/A/ST2/00106. The lattice QCD calculations were performed on NEC-SX8R and NEC-SX9 at Osaka University.

References

[1] H.J. Rothe, “Lattice Gauge Theories", (World Scientific, 2012), and references therein.

[2] P.M. Lo, B. Friman, O. Kaczmarek, K. Redlich and C. Sasaki, Phys. Rev. D**88**, 014506 (2013); Phys. Rev. D**88**, 074502 (2013).

[3] T. Banks and A. Casher, Nucl. Phys. B**169**, 103 (1980).
[4] F. Karsch, Lect. Notes Phys. **583**, 209 (2002), and references therein.
[5] F. R. Brown, F. P. Butler, H. Chen, N. H. Christ, Z. Dong, W. Schaffer, Leo I. Unger and Alessandro Vaccarino, Phys. Rev. Lett. **65**, 2491 (1990).
[6] Y. Aoki, G. Endrodi, Z. Fodor, S. D. Katz, and K. K. Szabo, Nature (London) **443**, 675 (2006).
[7] A. Bazavov *et al.*, Phys. Rev. D**85**, 054503 (2012).
[8] C.B. Lang and M. Schrock, Phys. Rev. D**84**, 087704 (2011); L.Ya. Glozman, C.B. Lang, and M. Schrock, Phys. Rev. D**86**, 014507 (2012).
[9] S. Gongyo, T. Iritani and H. Suganuma, Phys. Rev. D**86**, 034510 (2012); T. Iritani and H. Suganuma, Prog. Theor. Exp. Phys., **3**, 033B03 (2014).
[10] T. M. Doi, K. Redlich, C. Sasaki and H. Suganuma, Phys. Rev. D**92**, 094004 (2015).
[11] S. Elitzur, Phys. Rev. D**12**, 3978 (1975).
[12] H. Suganuma, T. M. Doi, T. Iritani, arXiv:1404.6494 [hep-lat]; Proc. Sci. (Lattice 2013), 374 (2013); Proc. Sci. (QCD-TNT-III), 042 (2014); EPJ Web of Conf. **71**, 00129 (2014); Proc. Sci. (Hadron 2013), 121 (2014).
[13] T. M. Doi, H. Suganuma, T. Iritani, Phys. Rev. D**90**, 094505 (2014); Proc. Sci. (Lattice 2013), 375 (2013); Proc. Sci. (Hadron 2013), 122 (2014).

EPJ Web of Conferences **126**, 04017 (2016) DOI: 10.1051/epjconf/201612604017
ICNFP 2015

Decay of charmonium states into a scalar and a pseudoscalar glueball

Walaa I. Eshraim[1,a]

[1] *Frankfurt Institute for Advanced Studies, Goethe University, Ruth-Moufang-Str. 1, D–60438 Frankfurt am Main, Germany*

Abstract. In the framework of a chiral symmetric model, we expand a $U(4)_R \times U(4)_L$ symmetric linear sigma model with (axial-)vector mesons by including a dilaton field, a scalar glueball, and the pseudoscalar glueball. We compute the decay width of the scalar charmonium state $\chi_{C0}(IP)$ into a predominantly scalar glueball $f_0(1710)$. We calculate the decay width of the pseudoscalar charmonium states $\eta_C(IS)$ into a predominantly scalar glueball $f_0(1710)$ as well as into a pseudoscalar glueball with a mass of 2.6 GeV (as predicted by Lattice-QCD simulations) and with a mass of 2.37 GeV (corresponding to the mass of the resonance $X(2370)$). This study is interesting for the upcoming PANDA experiment at the FAIR facility and BESIII experiment. Moreover, we obtain the mixing angle between a pseudoscalar glueball, with a mass of 2.6 GeV, and the charmonium state η_C.

1 Introduction

Charmonium is a bound state of a charm and an anti-charm quark. It is one of the simplest bound states of Quantum Chromodynamic (QCD), the strong interaction theory of quarks and gluons, like positronium in Quantum Electrodynamic (QED). The first charmonium state has been discovered (J/ψ) with quantum number $J^{PC} = 1^{--}$ in November 1974 at BNL [1] and SLAC [2]. Since that time great scientific experimental [3] and theoretical process [4] have been achieved for charmonium spectroscopy. All charmonium states have been observed below the open-charm threshold [5]. The study of the charmonium spectroscopy is very neccessary for understanding hadronic dynamics as the hydrogen atom [6] and the strong interaction physics.

The colourless bound states of gluons are called glueballs. There is an active experimental and theoretical research program, searching for states that are (predominantly) glueballs as seen in Refs.[7]. These efforts are important for understanding glueball properties which gives insight into the non-perturbative behavior of QCD. By Lattice QCD [8], the glueball spectrum has been predicted where the third lightest glueball is a pseudoscalar state ($J^{PC} = 0^{-+}$), denoted as $\widetilde{G}$, with a mass of 2.6 GeV. The mass of a pseudoscalar glueball $\widetilde{G}$ could be also lower than the lattice-QCD prediction and might have already been observed in the BESIII experiment. At that experiment the pseudoscalar resonances have been investigated in J/ψ decays [9]. Particularly, the resonance $X(2370)$ is a good candidate because it has been clearly observed in the $\pi^+\pi^-\eta'$ channel and is

[a]e-mail: weshraim@th.physik.uni-frankfurt.de

quite narrow ($\sim$ 80 MeV). There is a good candidate for the scalar glueball which is the resonance $f_0(1710)$, because its mass is very close to lattice-QCD predictions, and because it is produced in the gluon-rich decay of the J/ψ, as seen in Refs. [10–12]. Moreover, in Ref. [13], the three physical resonances $f_0(1370)$, $f_0(1500)$, and $f_0(1710)$ are assigned predominantly to the nonstrange meson $\sigma_N \equiv (u\bar{u} + d\bar{d})/\sqrt{2}$, the hidden-strange meson $\sigma_S \equiv s\bar{s}$, and the scalar glueball $G \equiv gg$, respectively, as a result of solving the mixing between the three bare fields (σ_N, σ_S, G).

The extended Linear Sigma Model (eLSM), includes (axial-)vector and (pseudo)scalar charmed mesons, which was introduced in Refs. [14, 15] as a result of the extension from $N_f = 3$ to $N_f = 4$ case. The symmetry properties of the eLSM have been discussed in Ref. [16]. The eLSM has four charmonium states, which are the ground-state (axial-)vector $J/\psi(1S)$ and $\chi_{c1}(1P)$ with quantum numbers $J^{PC} = 1^{-+}$ and $J^{PC} = 1^{++}$ as well as (pseudo-)scalar ground states $\eta_c(1S)$ and $\chi_{c0}(1P)$ with quantum numbers $J^{PC} = 0^{-+}$ and $J^{PC} = 0^{++}$, respectively, [14–16]. In the present work, we include the scalar glueball to be a dynamical field in the study of charmonum states, whereas it was a frozen field the $N_f = 4$ study of masses of open and hidden charmed mesons as well as of the decay of open charmed mesons [14, 15]. Including the scalar glueball in the eLSM is important to guarantee dilation invariance of the model. Furthermore, the pseudoscalar glueball has first been included into the eLSM via the interaction of the pseudoscalar glueball and (pseudo)scalar mesons in the case of $N_f = 3$ as seen in Ref.[17].

In this work, we study the decay properties of the charmonium state χ_{C0} into a scalar glueball and scalar-isoscalar states as well as of the charmonium state η_C into a scalar glueball, a pseudoscalar glueball, and a scalar-isoscalar states in the eLSM . We compute also the mixing angle between charmonium state η_C and a pseudoscalar glueball.

2 A $U(4)_R \times U(4)_L$ Linear Sigma Model with Glueballs

The dilaton field, a scalar glueball G, and a pseudoscalar glueball $\tilde{G}$ are included in the extended Linear Sigma model (eLSM) with (pseudo)scalar and (axial-)vector mesons as seen in Refs. [17–19]. The corresponding Lagrangian describing only the interaction of a scalar glueball and a pseudoscalar glueball with mesons in the eLSM and relevant for the decay of charmonium states into glueballs reads

$$\mathcal{L} = \mathcal{L}_{dil}(G) - m_0^2\left(\frac{G}{G_0}\right)^2 \mathrm{Tr}(\Phi^\dagger\Phi) + \mathrm{Tr}\left\{\left[\left(\frac{G}{G_0}\right)^2 \frac{m_1^2}{2} + \Delta\right]\left[(L^\mu)^2 + (R^\mu)^2\right]\right\}$$
$$+ c(\det\Phi - \det\Phi^\dagger)^2 + ic_{\tilde{G}\Phi}\tilde{G}(\det\Phi - \det\Phi^\dagger) + \ldots, \qquad (1)$$

where

$$\mathcal{L}_{dil}(G) = \frac{1}{2}(\partial_\mu G)^2 - \frac{1}{4}\frac{m_G^2}{\Lambda_G^2}\left[G^4 \ln\left(\frac{G}{\Lambda_G}\right) - \frac{G^4}{4}\right], \qquad (2)$$

is a dilation Lagrangian which describes the scalar dilaton filed that is represented by a scalar glueball $G \equiv |gg>$ with quantum number $J^{PC} = 0^{++}$, and emulates the trace anomaly of pure Yang-Mills QCD [19, 20]. The dimensionful parameter $\Lambda_G \sim N_C\,\Lambda_{QCD}$ sets the energy scale of low-energy QCD and is identical to the minimum of the dilaton potential G_0, $G_0 = \Lambda_G$ [18]. Lattice QCD gives the mass of a scalar glueball m_G of about $(1.5 - 1.7)$ GeV [21]. As a result of Ref. [13], the resonance $f_0(1710)$ is predominantly a scalar glueball with a 86 percent admixture. At the classical level of the Yang-Mills sector of QCD, the dilatation symmetry, $x^\mu \to \lambda^{-1}x^\mu$, is realized and explicitly broken due to the logarithmic term of the dilaton field potential [22]. This breaking leads to the divergence of the corresponding current [18]:

$$\partial_\mu J^\mu_{dil} = T^\mu_{dil,\mu} = -\frac{1}{4}m_G^2\Lambda_G^2. \tag{3}$$

Note that we include the scalar glueball in the eLSM to combine the dilatation invariance with the meson mass terms. A further breaking of dilatation and chiral symmetry are represented by the chiral anomaly term $c(det\Phi - det\Phi^\dagger)^2$ with a dimensionful constant c. The quark-mass term $\mathrm{Tr}\left[\Delta(L^{\mu 2} + R^{\mu 2})\right]$ contributes to the masses of the (axial-)vector mesons and also breaks the dilatation and chiral symmetry due to nonzero current quark masses [14], with Δ defined as

$$\Delta = \begin{pmatrix} 0 & 0 & 0 & 0 \\ 0 & 0 & 0 & 0 \\ 0 & 0 & \delta_S & 0 \\ 0 & 0 & 0 & \delta_C \end{pmatrix}, \tag{4}$$

where $\delta_S \sim m_S^2$ and $\delta_C \sim m_C^2$. The left-handed L^μ and right-handed R^μ represent the (axial-)vector multiples, see Ref.[14] for the case of $N_f = 4$. Furthermore, the last field included in the eLSM is the pseudoscalar glueball $\widetilde{G} \equiv |gg\rangle$, with quantum numbers $J^{pc} = 0^{-+}$, via $ic_{\tilde{G}\Phi}\tilde{G}(\det\Phi - \det\Phi^\dagger)$. This term couples the pseudoscalar glueball with (pseudo)scalar mesons. where $c_{\tilde{G}\phi}$ is a dimensionless coupling constant. This term has been successfully studied in the decay of a pseudoscalar glueball into scalar and pseudoscalar mesons in the case of $N_f = 3$ as seen in Ref.[17]. In eLSM Lagrangian (1) the dots refer to further chirally invariant terms, which are irrelevant for the present work, see Refs.[14, 15].

In this framework, we use the multiplets of scalar and pseudoscalar mesons, Φ, in the case of $N_f = 4$, as seen in Ref.[15] as

$$\Phi = (S^a + iP^a)t^a = \frac{1}{\sqrt{2}}\begin{pmatrix} \frac{(\sigma_N+a_0^0)+i(\eta_N+\pi^0)}{\sqrt{2}} & a_0^+ + i\pi^+ & K_0^{*+} + iK^+ & D_0^{*0} + iD^0 \\ a_0^- + i\pi^- & \frac{(\sigma_N-a_0^0)+i(\eta_N-\pi^0)}{\sqrt{2}} & K_0^{*0} + iK^0 & D_0^{*-} + iD^- \\ K_0^{*-} + iK^- & \overline{K}_0^{*0} + i\overline{K}^0 & \sigma_S + i\eta_S & D_{S0}^{*-} + iD_S^- \\ \overline{D}_0^{*0} + i\overline{D}^0 & D_0^{*+} + iD^+ & D_{S0}^{*+} + iD_S^+ & \chi_{C0} + i\eta_C \end{pmatrix}, \tag{5}$$

where S and P refer to scalar and pseudoscalar mesons, respectively, whereas t^a are the generators of the group $U(N_f)$. The multiplet Φ transforms as $\Phi \to U_L\Phi U_R^\dagger$ under $U_L(4)\times U_R(4)$ chiral transformations, while $U_{L(R)} = e^{-i\theta^a_{L(R)}t^a}$ is an element of $U(4)_{R(L)}$, under parity which $\Phi(t, \overrightarrow{x}) \to \Phi^\dagger(t, -\overrightarrow{x})$, and under charge conjugate $\Phi \to \Phi^\dagger$. The determinant of Φ is invariant under $SU(4)_L \times SU(4)_R$, but not under $U(1)_A$ because $\det\Phi \to \det U_A\Phi U_A = e^{-i\theta_A^0\sqrt{2N_f}}\det\Phi \neq \det\Phi$. The pseudoscalar glueball $\widetilde{G}$ is also chirally invariant, which is invariant under parity ($\widetilde{G} \to -\widetilde{G}$), but not invariant under the axial $U_A(1)$ transformation.

In the Lagrangian (1), all states are assigned as physical resonances to light quark-antiquark states with mass $\lesssim$ 2 GeV [17] and heavy quark-antiquark states [14]. In the present work, we need to deal only with the scalar and pseudoscalar mesons. For the scalar sector S, the isotriplet field $\overrightarrow{a_0}$ and kaonic field K^{*0} are assigned to the physical states $a_0(1450)$ and $K_0^*(1430)$, respectively. In the scalar-isoscalar sector, the identification of the scalar glueball G is still uncertain, the two most likely candidates are $f_0(1500)$ and $f_0(1710)$ and/or admixtures of them. The bare nonstrange field

$\sigma_N \equiv \left|\bar{u}u + \bar{d}d\right\rangle / \sqrt{2}$, the bare strange field $\sigma_S \equiv |\bar{s}s\rangle$ and the scalar glueball (G) are assigned from the following mixing matrix which is constructed in Ref. [13]:

$$\begin{pmatrix} f_0(1370) \\ f_0(1500) \\ f_0(1710) \end{pmatrix} = \begin{pmatrix} 0.94 & -0.17 & 0.29 \\ 0.21 & 0.97 & -0.12 \\ -0.26 & 0.18 & 0.95 \end{pmatrix} \begin{pmatrix} \sigma_N \equiv (\bar{u}u + \bar{d}d)/\sqrt{2} \\ \sigma_S \equiv \bar{s}s \\ G \equiv gg \end{pmatrix}. \tag{6}$$

As a conclusion of Ref.[13], we obtain that the resonances $f_0(1370)$ and $f_0(1500)$ are predominantly $\left|\bar{u}u + \bar{d}d\right\rangle / \sqrt{2}$ and $|\bar{s}s\rangle$, respectively. However the resonance $f_0(1710)$ is predominantly a scalar glueball G. In the scalar charm sector, open charmed meson D_0^*, strange-cahrm meson D_{S0}^* and scalar-isoscalar state χ_{C0} are assigned to the resonances $D_0^*(2400)$, $D_{S0}^*(2317)$ and the graund charm-anticharm resonance $\chi_{C0}(IP)$, respectively. Now let us turn to the pseudoscalar sector P, the light quark-antiquark present the pion isotriplet $\vec{\pi}$ and the kaon isodoublet K [19], the isoscalar fields nonstrange $\eta_N \equiv |\bar{u}u + \bar{d}d\rangle/\sqrt{2}$ and strange $\eta_S \equiv \bar{s}s\rangle$ which are mixed in the physical fields η and η' as

$$\eta = \eta_N \cos\varphi + \eta_S \sin\varphi,\ \eta' = -\eta_N \sin\varphi + \eta_S \cos\varphi, \tag{7}$$

where the mixing angle is $\varphi \simeq 44.6°$ [19]. In the charm sector, there are the well-established D resonance, the open strange-charmed resonance D_S, and charmonium ground state $\eta_C(IS)$ [14].

From the eLSM features, spontaneous symmetry breaking occurs when $m_0^2 < 0$, and consequently the scalar-isoscalar fields as well as the glueball field G condense [19, 23]. One has to shift the scalar-isoscalar fields by their vacuum expectation values to implement this breaking [14] as

$$\begin{aligned} G &\to G + G_0\,, & \sigma_N &\to \sigma_N + \phi_N\,, \\ \sigma_S &\to \sigma_S + \phi_S\,, & \chi_{C0} &\to \chi_{C0} + \phi_C\,. \end{aligned} \tag{8}$$

where G_0 is the gluon condensate equal to $\Lambda \approx 3.3$ GeV [13]. ϕ_N, ϕ_S, and ϕ_C represent the chiral-nonstrange, strange, and charm quark-antiquark condensates [14, 15], respectively.

3 Results

In the case of $N_f = 3$ most of the parameters are fixed, in the low-energy study [19]. We are left with only three new parameters related to the charm sector in the case of $N_f = 4$, δ_C, ϕ_c, ε_C, which are determined by a fit, with χ^2/d.o.f $\simeq 1$, including masses of open and hidden charmed mesons (see details in Ref. [14]).

In the present work, we have to change the value of the parameter c, which is the coefficient of the axial anomaly term, to fit the results of the decay widths of χ_{c0} and η_C. Therefore, for the determination of c, we use the decay widths of χ_{c0} into $\eta\eta$ and $\eta'\eta'$, which are [24].

$$\Gamma^{exp}_{\chi_{c0}\to\eta\eta} = (0.031 \pm 0.0039)\text{MeV} \ \text{and}\ \Gamma^{exp}_{\chi_{c0}\to\eta'\eta'} = (0.02 \pm 0.0035)\text{MeV}\,.$$

We then perform a fit by minimizing the χ^2-function,

$$\chi^2(c) \equiv \left(\frac{\Gamma^{th}_{\chi_{c0}\to\eta\eta}(c) - \Gamma^{exp}_{\chi_{c0}\to\eta\eta}}{\xi\Gamma^{exp}_{\chi_{c0}\to\eta\eta}}\right)^2 + \left(\frac{\Gamma^{th}_{\chi_{c0}\to\eta'\eta'}(c) - \Gamma^{exp}_{\chi_{c0}\to\eta'\eta'}}{\xi\Gamma^{exp}_{\chi_{c0}\to\eta'\eta'}}\right)^2, \tag{9}$$

which gives $c = 7.178 \times 10^{-10}$ MeV^{-4} with χ^2/d.o.f $= 0.18$ where $\xi = 1$. The coupling constant $c_{\tilde{G}\Phi}$ describes the coupling of the pseudoscalar glueball $\tilde{G}$ and (pseudo)scalar mesons. By comparing the interaction Lagrangians of the pseudoscalar glueball with (pseudo)scalar mesons in the two cases $N_f = 3$ and $N_f = 4$ [16], we get

$$c_{\widetilde{G}\Phi} = \frac{\sqrt{2}\, c_{\widetilde{G}\Phi(N_f=3)}}{\phi_C} = 0.036. \tag{10}$$

The three-body decay of the charmonium state χ_{C0} into a scalar glueball and (pseudo)scalar mesons are reported in Table 1.

Table 1. The partial decay widths of χ_{c0}.

Decay Channel	theoretical result [MeV]	Experimental result [MeV]
$\Gamma_{\chi_{c0}\to f_0(1370)\eta\eta}$	4.10^{-4}	-
$\Gamma_{\chi_{c0}\to f_0(1500)\eta\eta}$	3.10^{-3}	-
$\Gamma_{\chi_{c0}\to f_0(1370)\eta'\eta'}$	27.10^{-4}	-
$\Gamma_{\chi_{c0}\to f_0(1370)\eta\eta'}$	89.10^{-6}	-
$\Gamma_{\chi_{c0}\to f_0(1500)\eta\eta'}$	11.10^{-3}	-
$\Gamma_{\chi_{c0}\to f_0(1710)\eta\eta}$	8.10^{-5}	-
$\Gamma_{\chi_{c0}\to f_0(1710)\eta\eta'}$	3.10^{-5}	-

The decay widths of the charmonium state η_C into a scalar glueball and (pseudo)scalar mesons are presented in Table 2.

Table 2. The partial decay widths of η_c.

Decay Channel	theoretical result [MeV]	Experimental result [MeV]
$\Gamma_{\eta_c\to f_0(1370)\eta}$	0.00018	-
$\Gamma_{\eta_c\to f_0(1500)\eta}$	0.006	-
$\Gamma_{\eta_c\to f_0(1710)\eta}$	0.000032	-
$\Gamma_{\eta_c\to f_0(1370)\eta'}$	0.027	-
$\Gamma_{\eta_c\to f_0(1500)\eta'}$	0.024	-
$\Gamma_{\eta_c\to f_0(1710)\eta'}$	0.0006	-

There are no experimental data for these decays firm quoted in the PDG with which we can compare.

The interaction Lagrangian $[ic_{\widetilde{G}\Phi}\widetilde{G}(\det\Phi - \det\Phi^\dagger)]$ contains only one decay process which describes the decay of the pseudoscalar charmonium η_c into a pseudoscalar glueball $\widetilde{G}$ through the channel $\eta_C \to \widetilde{G}\pi\pi$. By using the corresponding decay width for the three-body case [16], the decay width of the pseudoscalar charmonium state η_c into a pseudoscalar glueball with a mass of 2.6 GeV (as predicted by lattice QCD in the quenched approximation [21]) is

$$\Gamma_{\eta_C\to\pi\pi\widetilde{G}(2600)} = 0.124 \text{ MeV}, \tag{11}$$

and for a mass of the charmonium state η_c which is about of 2.37 GeV (corresponding to the mass of the resonance $X(2370)$ measured in the BESIII experiment [9])

$$\Gamma_{\eta_C\to\pi\pi\widetilde{G}(2370)} = 0.16 \text{ MeV}\,. \tag{12}$$

These results could be tested in the PANDA experiment at the upcoming FAIR facility.

The mixing between the charmonium state χ_{c0} and the scalar glueball G is neglected because it is expected to be small.

3.1 Mixing of a pseudoscalar glueball $\widetilde{G}$ and a charmonium state η_C

The full interaction Lagrangian of the mixing between the pseudoscalar glueball $\widetilde{G}$ and η_C has the form

$$\mathcal{L}_{\widetilde{G},\eta_C} = \frac{1}{2}(\partial_\mu \widetilde{G})^2 + \frac{1}{2}(\partial_\mu \eta_C)^2 - \frac{1}{2}m_{\widetilde{G}}^2 \widetilde{G}^2 - \frac{1}{2}m_{\eta_c}^2 \eta_c^2 + Z_{\widetilde{G}\eta_C} \widetilde{G}\,\eta_C \tag{13}$$

where

$$Z_{\widetilde{G}\eta_C} = \frac{-1}{4} c_{\widetilde{G}\Phi} Z_{\eta_C} \phi_N^2 \phi_S . \tag{14}$$

The physical fileds η_C and $\widetilde{G}$ can be obtained through an SO(2) rotation

$$\begin{pmatrix} \widetilde{G}' \\ \eta_C' \end{pmatrix} = \begin{pmatrix} cos\phi & sin\phi \\ -sin\phi & cos\phi \end{pmatrix} = \begin{pmatrix} \widetilde{G} \\ \eta_C \end{pmatrix}, \tag{15}$$

with

$$m_{\eta_C'}^2 = m_{\widetilde{G}}^2 sin^2\phi + m_{\eta_C}^2 cos^2\phi + Z_{\widetilde{G}\eta_C} sin(2\phi), \tag{16}$$

$$m_{\widetilde{G}'}^2 = m_{\widetilde{G}}^2 cos^2\phi + m_{\eta_C}^2 sin^2\phi - Z_{\widetilde{G}\eta_C} sin(2\phi), \tag{17}$$

where the mixing angle ϕ reads

$$\phi = \frac{1}{2} \arctan\left[\frac{-c_{\widetilde{G}\Phi} Z_{\eta_c} \phi_N^2 \phi_S}{2(m_{\eta_C}^2 - m_{\widetilde{G}}^2)}\right]. \tag{18}$$

$c_{\widetilde{G}\Phi}$ is a dimensionless coupling constant between $\widetilde{G}\Phi$ determined by Eq.(10). We then obtain the mixing angle of the pseudoscalar glueball $\widetilde{G}$ and the pseudoscalar charm-anticharm meson η_c to be $-1°$, for a mass of the pseudoscalar glueball of 2.6 GeV, as predicted by lattice-QCD simulations [21].

4 Conclusion

In the present work we have represented a chirally invariant linear sigma model with (axial-)vector mesons in the four-flavour case, $N_f = 4$, by including a dilaton field and a scalar glueball field, and describing the interaction of the pseudoscalar glueball with (pseudo-)scalar mesons. We have calculated the decay widths of the hidden-charmed meson χ_{c0} into two and three scalar-isoscalar states as well as into a scalar glueball G (Tables 1), which is predominantly $f_0(1710)$. We have also computed the decay widths of the pseudoscalar charmonium state η_C into scalar-isoscalar states (Table 2) and into a pseudoscalar glueball $\widetilde{G}$, through the channel $\eta_C \to \pi\pi\widetilde{G}$. The latter is obtained from the interaction term of the pseudoscalar glueball. In addition, we have evaluated the mixing angle between the pseudoscalar glueball and η_c, which is very small and equal to $-1°$. The results presented in this work are interesting for the upcoming PANDA experiment at the FAIR facility as well as for the BESIII experiment.

Further applications of the described approach are to calculate the decay widths of hidden charmonium states into light mesons [25].

Acknowledgments

The author thanks D. H. Rischke for cooperation, and S. Schramm for useful discussion. Financial support from Female program for HIC for FAIR is acknowledged.

References

[1] J. J. Aubert *et.al.*, Phys. Rev. Lett. **33**, 1404 (1974).

[2] J. E. Augustin *et.al.*, Phys. Rev. Lett. **33**, 1406 (1974).

[3] H. B. Li, Nucl. Phys. Proc. Suppl. **233**, 185 (2012) doi:10.1016/j.nuclphysbps.2012.12.075 [arXiv:1209.3059 [hep-ex]]; A. Anastassov *et al.* [CLEO Collaboration], Phys. Rev. D **65**, 032003 (2002) doi:10.1103/PhysRevD.65.032003 [hep-ex/0108043]; P. Lees *et al.* (BaBar collaboration), Phys. Rev. D **81**, 052010 (2010); S.-K. Choi *et al.* (Belle collaboration), Phys. Rev. Lett. **89**, 102001 (2002); P. Lees *et al.* (BaBar collaboration), Phys. Rev. D **81**, 052010 (2010).

[4] M. Neubert, Phys. Rept. **245**, 259 (1994) doi:10.1016/0370-1573(94)90091-4 [hep-ph/9306320]; L. -P. Sun, H. Han and K. -T. Chao, arXiv:1404.4042 [hep-ph]; M. S. Khan, DESY-THESIS-2014-003; N. Brambilla *et al.*, Eur. Phys. J. **C71** (2011) 1534, [arXiv:1010.5827]

[5] E. Eichten, K. Gottfried, T. Kinoshita, K. D. Lance, and T. M. Yan, Phys. Rev. D 17, 3090 (1978); Phys. Rev. D 21, 203 (1980); K.A. Olive et al. (Particle Data Group); Chin. Phys. C 38, 090001 (2014).

[6] V. A. Novikov *et. al.*, Phys. Rep. **41C**, 1 (1978).

[7] F. E. Close, Rept. Prog. Phys. **51**, 833 (1988); S. Godfrey and J. Napolitano, Rev. Mod. Phys. **71**, 1411 (1999) [arXiv:hep-ph/9811410]; C. Amsler and N. A. Tornqvist, Phys. Rept. **389**, 61 (2004); E. Klempt and A. Zaitsev, Phys. Rept. **454**, 1 (2007) [arXiv:0708.4016 [hep-ph]].

[8] M. Loan, X. Q. Luo and Z. H. Luo, Int. J. Mod. Phys. A **21**, 2905 (2006) [arXiv:hep-lat/0503038]; E. B. Gregory, A. C. Irving, C. C. McNeile, S. Miller and Z. Sroczynski, PoS **LAT2005**, 027 (2006) [arXiv:hep-lat/0510066]; Y. Chen *et al.*, Phys. Rev. D **73**, 014516 (2006) [arXiv:hep-lat/0510074].

[9] M. Ablikim *et al.* (BES Collaboration), Phys. Rev. Lett. **95**, 262001 (2005); N. Kochelev and D. P. Min, Phys. Lett. B **633**, 283 (2006); M. Ablikim *et al.* (BES III Collaboration), Phys. Rev. Lett. 106.072002 (2011).

[10] C. Amsler and F. E. Close, Phys. Rev. D **53** (1996) 295 [arXiv:hep-ph/9507326]. W. J. Lee and D. Weingarten, Phys. Rev. D **61**, 014015 (2000). [arXiv:hep-lat/9910008]; F. E. Close and A. Kirk, Eur. Phys. J. C **21**, 531 (2001). [arXiv:hep-ph/0103173]. H. Y. Cheng, C. K. Chua and K. F. Liu, Phys. Rev. D **74** (2006) 094005 [arXiv:hep-ph/0607206]. V. Mathieu, N. Kochelev and V. Vento, Int. J. Mod. Phys. E **18** (2009) 1 [arXiv:0810.4453 [hep-ph]].

[11] H. Y. Cheng, C. K. Chua and K. F. Liu, Phys. Rev. D **74** (2006) 094005 [arXiv:hep-ph/0607206].

[12] P. Chatzis, A. Faessler, T. Gutsche and V. E. Lyubovitskij, Phys. Rev. D **84** (2011) 034027 [arXiv:1105.1676 [hep-ph]]; T. Gutsche, Prog. Part. Nucl. Phys. **67** (2012) 380.

[13] S. Janowski, F. Giacosa and D. H. Rischke, Phys. Rev. D **90**, no. 11, 114005 (2014) [arXiv:1408.4921 [hep-ph]].

[14] W. I. Eshraim, F. Giacosa and D. H. Rischke, Eur. Phys. J. A **51**, no. 9, 112 (2015) [arXiv:1405.5861 [hep-ph]].

[15] W. I. Eshraim, PoS QCD **-TNT-III**, 049 (2013) [arXiv:1401.3260 [hep-ph]]; W. I. Eshraim and F. Giacosa, EPJ Web Conf. **81**, 05009 (2014) [arXiv:1409.5082 [hep-ph]]; W. I. Eshraim, EPJ Web Conf. **95**, 04018 (2015) [arXiv:1411.2218 [hep-ph]]; W. I. Eshraim, J. Phys. Conf. Ser. **599**, no. 1, 012009 (2015) [arXiv:1411.4749 [hep-ph]].

[16] W. I. Eshraim, arXiv:1509.09117 [hep-ph].

[17] W. I. Eshraim, S. Janowski, F. Giacosa and D. H. Rischke, Phys. Rev. D **87**, 054036 (2013) [arXiv:1208.6474 [hep-ph]]; W. I. Eshraim, S. Janowski, A. Peters, K. Neuschwander and F. Giacosa, Acta Phys. Polon. Supp. **5**, 1101 (2012) [arXiv:1209.3976 [hep-ph]]; W. I. Eshraim and S. Janowski, PoS ConfinementX **118**, (2012) [arXiv:1301.3345 [hep-ph]]; W. I. Eshraim and S. Janowski, J. Phys. Conf. Ser. **426**, 012018 (2013) [arXiv:1211.7323 [hep-ph]].

[18] S. Janowski, D. Parganlija, F. Giacosa and D. H. Rischke, Phys. Rev. D **84**, 054007 (2011) doi:10.1103/PhysRevD.84.054007 [arXiv:1103.3238 [hep-ph]].

[19] D. Parganlija, P. Kovacs, G. Wolf, F. Giacosa and D. H. Rischke, Phys. Rev. D **87**, 014011 (2013) [arXiv:1208.0585 [hep-ph]].

[20] C. Rosenzweig, A. Salomone and J. Schechter, Phys. Rev. D 24, 2545 (1981); A. Salomone, J. Schechter and T. Tudron, Phys. Rev. D 23, 1143 (1981); C. Rosenzweig, A. Salomone and J. Schechter, Nucl. Phys. B 206, 12 (1982) [Erratum-ibid. B 207, 546 (1982)]; A. A. Migdal and M. A. Shifman, Phys. Lett. B 114, 445 (1982); H. Gomm and J. Schechter, Phys. Lett. B 158, 449 (1985); R. Gomm, P. Jain, R. Johnson and J. Schechter, Phys. Rev. D 33, 801 (1986).

[21] C. J. Morningstar and M. J. Peardon, Phys. Rev. D **60** (1999) 034509 [hep-lat/9901004]; M. Loan, X. Q. Luo and Z. H. Luo, Int. J. Mod. Phys. A **21**, 2905 (2006) [arXiv:hep-lat/0503038]; E. B. Gregory, A. C. Irving, C. C. McNeile, S. Miller and Z. Sroczynski, PoS **LAT2005**, 027 (2006) [arXiv:hep-lat/0510066]; Y. Chen *et al.*, Phys. Rev. D **73**, 014516 (2006) [arXiv:hep-lat/0510074].

[22] A. Salomone, J. Schechter and T. Tudron, Phys. Rev. D **23**, 1143 (1981); H. Gomm and J. Schechter, Phys. Lett. B **158**, 449 (1985); A. A. Migdal and M. A. Shifman, Phys. Lett. B **114**, 445 (1982).

[23] D. Parganlija, F. Giacosa and D. H. Rischke, Phys. Rev. D **82**, 054024 (2010) [arXiv:1003.4934 [hep-ph]].

[24] J. Beringer *et al.* (Particle Data Group), Phys. Rev. D **86**, 010001 (2012).

[25] W. I. Eshraim and D. H. Rischke, in preparation.

EPJ Web of Conferences **126**, 04018 (2016) DOI: 10.1051/epjconf/201612604018
ICNFP 2015

Production of non-photonic electrons in U+U collisions at $\sqrt{s_{NN}} = 193$ GeV at the STAR experiment

Katarina Gajdosova on behalf of the STAR Collaboration[1,a]

[1] *Czech Technical University in Prague,*
Faculty of Nuclear Sciences and Physical Engineering,
Břehová 7, 115 19 Prague 1, Czech Republic

Abstract. Heavy quarks are suitable probes to study the properties of Quark-Gluon Plasma (QGP), a strongly interacting medium, which can be created in ultrarelativistic heavy-ion collisions at the Relativistic Heavy Ion Collider (RHIC). Non-photonic electrons (NPE) that originate from semileptonic decays of D and B mesons can serve as a good proxy for heavy flavor quarks. Nuclear modification factor R_{AA} of NPE is measured in heavy-ion collisions, which is sensitive to the effect of QGP on heavy quarks.
Measurements of NPE R_{AA} in Au+Au collisions at $\sqrt{s_{NN}} = 200$ GeV reveal that NPE production is strongly suppressed. In year 2012 STAR collected data from U+U collisions at $\sqrt{s_{NN}} = 193$ GeV. In most central collisions higher energy density can be achieved in comparison to collisions of gold nuclei. In this proceedings the preliminary results on NPE in 0-5% most central U+U collisions at a transverse momentum range $1.2 < p_T < 6$ GeV/c are presented. The nuclear modification factor in U+U collisions is compared to that in Au+Au collisions and theoretical models.

1 Introduction

A new state of QCD matter is created during ultrarelativistic heavy-ion collisions at RHIC. One of the suitable probes to study the hot and dense medium, Quark-Gluon Plasma, are heavy quarks charm (c) and bottom (b). Due to their large masses, heavy quarks can be created only in the scatterings with large momenutm transfers which happen during the early stages of heavy-ion collisions before creation of QGP. Thus, their initial production is not affected by this medium, while the final distribution of particles composed of these heavy quarks is modified by the interaction between heavy quarks and QGP and other possible nuclear matter effects.

To quanitfy the effects of nuclear matter vs. the production in p+p collisions we commonly use the nuclear modification factor, R_{AA}. R_{AA} is defined as the ratio of particle production in heavy-ion collisions to particle production in proton-proton collisions scaled by the mean number of binary collisions obtained from Glauber model [1]:

$$R_{AA} = \frac{1}{\langle N_{bin} \rangle} \frac{dN^2_{AA}/dydp_T}{dN^2_{pp}/dydp_T}. \quad (1)$$

[a] e-mail: katka.gajdosova19@gmail.com

If $R_{AA} = 1$, a heavy-ion collision is just a superposition of multiple proton-proton collisions. A so-called suppression is observed when $R_{AA} < 1$, which means that at given p_T and rapidity there are less particles produced in heavy-ion collisions compared to proton-proton collisions due to possible energy loss effects from QGP on heavy quarks.

Heavy quarks have been studied via open heavy flavor measurements. There are two popular approaches to these measurements: direct reconstruction of D mesons via hadronic decay channels, and indirect reconstruction of electrons/muons from semileptonic decays of charm and bottom hadrons. Non-photonic electrons from semileptonic decays of D and B mesons are a good proxy for heavy flavor quarks. Therefore, they are a suitable observable for the study of heavy flavor energy loss mechanisms in QGP. Semileptonic decay channel gained interest due to lower background compared to direct reconstruction in heavy-ion collision environment with thousands of hadrons per event, and the possibility of triggering for high-p_T electrons in electromagnetic calorimeter. The D meson production was measured via the hadronic decay channels in p+p collisions at $\sqrt{s}$ = 200 [2] and 500 GeV [3], also in Au+Au collisions at $\sqrt{s_{NN}}$ = 200 GeV [4] and U+U collisions at $\sqrt{s_{NN}}$ = 193 GeV [3]. The semileptonic decay channels of open heavy flavor have been measured in p+p collisions at $\sqrt{s}$ = 200 GeV [5], [6], [7]. Studies of NPE in p+p collisions present an important test of perturbative QCD calculations and are also essential for NPE studies in heavy-ion collisions where they serve as a baseline. The theoretical predictions of FONLL are able to describe the production of NPE and D mesons in p+p collisions well within uncertainties. Modification of NPE spectra in Cu+Cu and Au+Au collisions has been investigated at $\sqrt{s_{NN}}$ = 200 GeV [8], [9], [10], [11], [12]. Eventually, it is important to understand the Cold Nuclear Matter effects and be able to distinguish them from QGP effects. Thus, NPE were also studied in d+Au collisions at $\sqrt{s_{NN}}$ = 200 GeV [13], [14]. System size dependence of NPE suppression was described in details in [15]. All above mentioned measurements show strong suppression which indicates that heavy quarks lose energy when traversing the QGP.

Uranium nuclei have more nucleons than nuclei of gold. Therefore, by colliding uranium nuclei one can achieve higher energy density than in Au+Au collisions. In Fig. 1 (left) the ratio of energy density of orientation-averaged U+U to Au+Au collisions is plotted as a function of centrality. It is possible to have up to 20% more energy density in uranium nuclei collisions [17]. Therefore, study of NPE production in central U+U collisions with highest energy density achieved at RHIC can reveal whether NPE production will be more suppressed in U+U collisions in comparison to Au+Au collisions at the same centrality class.

2 Detector layout

The Solenoidal Tracker at RHIC (STAR) [16] covers 2π in azimuth and two units of pseudorapidity around mid-rapidity and is wrapped inside a solenoidal magnet, which has a field strength of 0.5 T.

The STAR detector is composed of various subdetectors, each of them fulfilling different task in the particle detection. The detector is shown in Fig. 1 (right).

The main detector of the STAR is the Time Projection Chamber (TPC), which is a gas detector designed for tracking and particle identification using their ionization energy loss in the gas. The Time of Flight (ToF) detector is surrounding the TPC and is able to measure the velocity of particles which improves electron identification at low p_T. The energy of electrons is obtained from the Barrel Electromagnetic Calorimeter (BEMC), which is outside of the ToF. This detector is important in the electron identification at high transverse momentum p_T.

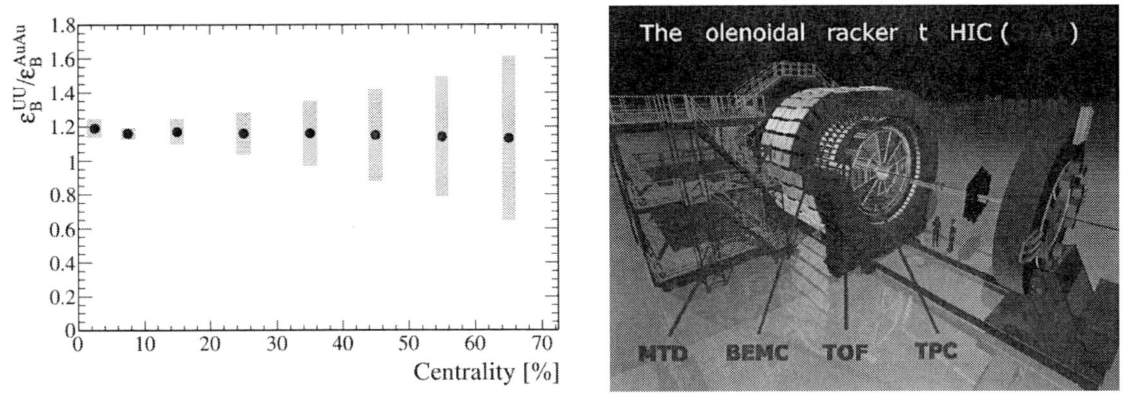

Figure 1: Left: Ratio of energy densities in U+U and Au+Au collisions as a function of centrality. Taken from Ref. [17]. Right: View of the STAR detector.

3 Analysis procedure

The data sample used for the analysis presented in this proceedings comes from 0-5% most central U+U collisions at $\sqrt{s_{NN}}$ = 193 GeV taken during the year 2012 (run12) at the STAR experiment. The total number of events was 40 M. In this analysis the TPC and BEMC detectors were used for particle identification, tracking and extraction of energy information. The centrality was selected based on ToF and Zero Degree Calorimeter (ZDC) detectors.

As was already mentioned above, non-photonic electrons originate in semileptonic decays of open heavy flavor mesons. It is not possible to reconstruct the invariant mass of $D(B)$ meson due to missing neutrino from the decay, so we measure a continuous spectrum. This measurement is accompanied by large background caused by photonic electrons, which are always created in pairs e^+e^- and mainly come from photon conversions $\gamma \rightarrow e^+ + e^-$ and Dalitz decays of η and π^0. This background has to be subtraced from the electron sample. Thus, the non-photonic electron yield is obtained according the following formula

$$N_{npe} = N_{inclusive}\epsilon_{purity} - N_{photonic}/\epsilon_{photonic}, \quad (2)$$

where N_{inc} is the inclusive electron candidate yield, ϵ_{purity} is the purity of the inclusive electron sample, $N_{photonic}$ is the photonic electron yield and $\epsilon_{photonic}$ is the photonic electron reconstruction efficiency. First, the inclusive electron sample is obtained using selection criteria from TPC and BEMC detectors and corrected with the purity for hadron contamination. Then, the photonic electrons are subtracted. We cannot detect all photonic electrons pairs. Therefore, the photonic electron yield needs to be corrected for the photonic electron reconstruction efficiency.

The purity of inclusive electron sample is calculated from the normalized electron energy loss distribution $n\sigma_e$ in TPC which is fitted with multi-Gaussian function at various p_T bins. Figure 2 (left) shows $n\sigma_e$ distribution for the p_T bin 2.0 < p_T < 3.0 GeV/c. Due to large remaining hadron contamination, the multi-Gaussian function is used. The purity is then defined as the ratio of integral under electron Gaussian to the multi-Gaussian within the $n\sigma_e$ cut. The inclusive electron sample in this analysis is rather clean with ϵ_{purity} almost equal unity due to tight cut on $n\sigma_e$, constant over transverse momentum range: -0.5 < $n\sigma_e$ < 2.5 (also plotted in Fig. 2 (left) with dark green lines)

In Fig. 2 (right) the invariant mass of electron pairs with trigger electrons at 2.0 < p_T < 3.0 GeV/c is shown. Electrons are paired with each other and grouped together according to their charges.

DOI: 10.1051/epjconf/201612604018

The unlike-sign distribution (blue color) represents all e^+e^- pairs while like-sign distribution (red color) characterises the combinatorial background which is subtracted from the unlike-sign pairs. The resulting distribution - unlike-like sign (black color) - contains the real photonic electron pairs. As can be seen in Fig. 2 (right): only at low m_{ee} there are entries reflecting the low invariant mass of γ or π^0, η.

The photonic electron reconstruction efficiency is calculated with Monte Carlo (MC) simulation. First, particle tracks are generated which originate at a collision vertex. These simulated tracks are embedded into a real collision and propagated through the detector using GEANT, simulating the detector response for each particle. The simulation data are then reconstructed using the same reconstruction chain used in real data reconstruction. Final photonic electron reconstruction efficiency varies from 30% at low p_T to 50% at high p_T.

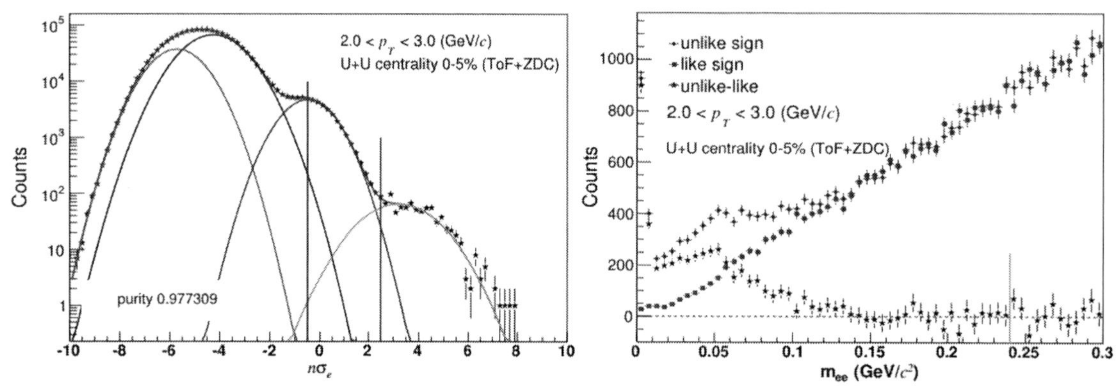

Figure 2: Left: Distribution of standard ionization energy loss in the TPC. Each particle contribution is denoted in different colors. The blue line represents π, the pink line $K + p$, the red line are electrons, and finally the green line are the so called merged pions which wrongly reconstruct two tracks as one. Right: Invariant mass of electron pairs.

3.1 Invariant yield of non-photonic electrons

The invariant yield of non-photonic electrons is calculated as follows:

$$\frac{1}{2\pi p_T}\frac{d^2N}{dp_T d\eta} = \frac{1}{2}\frac{1}{2\pi p_T}\frac{1}{\Delta p_T}\frac{1}{\Delta\eta}\frac{N_{npe}}{N_{events}}\frac{1}{\epsilon_{bemc}\epsilon_{n\sigma_e}\epsilon_{trk}}, \tag{3}$$

where N_{events} is total number of events of data sample used for this analysis, and ϵ_{bemc}, $\epsilon_{n\sigma_e}$ and ϵ_{rec} are the different efficiency components in the analysis. ϵ_{emc} is the efficiency of electron identification cuts using BEMC detector which goes from 30% at low p_T up to 70% at high p_T. $\epsilon_{n\sigma_e}$ is the efficiency of electron identification using energy loss in TPC and is about 50% through the whole p_T range used in this analysis. Finally, ϵ_{trk} is the efficiency of single track reconstruction in TPC. Monte Carlo simulations were used to calculate this efficiency too and the result is about 40%.

Invariant yield of non-photonic electrons is shown in Fig. 3 in a transverse momentum range $1.2 < p_T < 6.0$ GeV/c. The background contribution from J/ψ decays was estimated and subtracted from the NPE yield. Blue lines represent statistical uncertainties and blue boxes show systematical

 DOI: 10.1051/epjconf/201612604018

uncertainties. The main sources of systematic uncertainties are estimation of photonic electron reconstruction efficiency, purity estimation, and calculation of BEMC efficiency ϵ_{bemc}, $\epsilon_{n\sigma_e}$ and single track reconstruction efficiency ϵ_{trk}. The largest source is from BEMC efficiency ϵ_{bemc}, while the lowest one is from ϵ_{trk}.

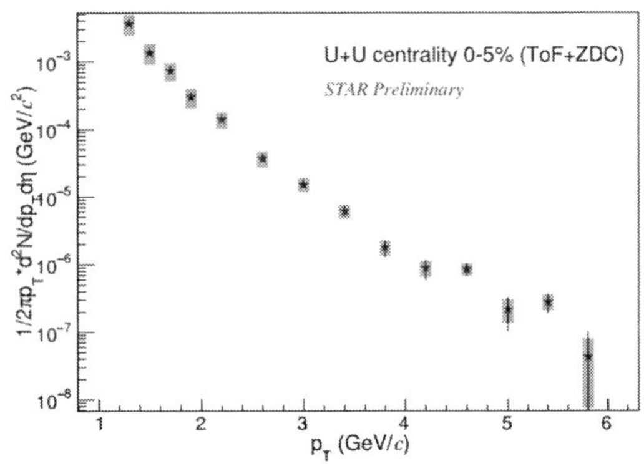

Figure 3: Invariant yield of non-photonic electrons as a function of p_T in 0-5% most central U+U collisions at $\sqrt{s_{NN}}$ = 193 GeV.

4 Nuclear modification factor

Invariant yield of NPE in U+U collisions was scaled by the mean number of binary collisions N_{bin} = 1156 ± 163 for 0-5% most central events and divided by the NPE yield in p+p collisions taken during the year 2012 at the STAR experiment [18]. The nuclear modification factor of NPE in 0-5% most central U+U collisions in 1.2 < p_T < 6.0 is plotted in Fig. 4 (left). It is consistent with 1 for p_T < 2.5 GeV/c and reveals strong suppression for p_T > 2.5 GeV/c. Yield of NPE in Au+Au collisions [18] was obtained in the same way as the one in U+U collisions and the same p+p reference was used. Nuclear modification factor of NPE in U+U collisions is consistent within errors but there is indication of lower R_{AA} than in Au+Au collisions. R_{AA} of NPE in U+U collisions is compared to theoretical model including CNM effects (Cronin effect and initial state energy loss), QGP effects such as jet quenching, and new attenuation mechanism for open heavy flavor mesons, called collisional dissociation, originating from short formation times of D and B mesons. The dashed green line represents the case with lower initial state energy loss but large Cronin effect and the dot-dashed magenta line the case with higher initial state energy loss and small Cronin effect [19], [20], [21].

In addition, nuclear modification factor of NPE as a function of number of participants is plotted in Fig. 4 (right). R_{AA} of NPE in U+U and Au+Au collisions are plotted with blue and red closed symbols, respectively. The statistical uncertainties are represented by the vertical lines and systematical uncertainties are shown by the filled boxes. R_{AA} in U+U collisions extends the trend in Au+Au collisions and the overall suppression is similar within uncertainties to the one of D meson [4] and light hadrons [22].

5 Summary

In summary, first preliminary results of non-photonic electrons in 0-5% most central U+U collisions at the STAR experiment were shown in this proceedings. The nuclear modification factor of NPE

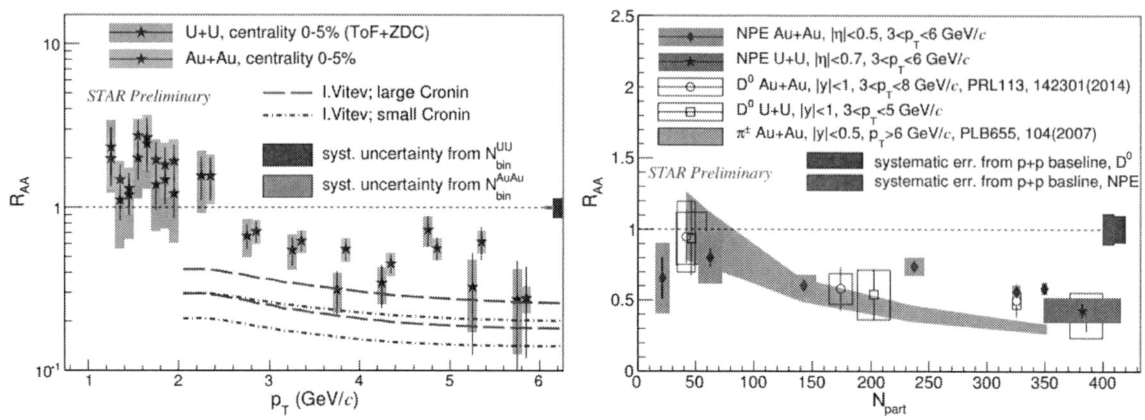

Figure 4: Left: Nuclear modification factor of NPE versus p_T in U+U collisions compared to AuAu collisions and model [19], [20], [21]. Right: Nuclear modification factor of NPE in U+U and Au+Au collisions, D mesons in U+U and Au+Au collisions [4] and $\pi^{\pm}$ in Au+Au collisions [22] versus number of participants.

in U+U collisions was shown in transverse momentum range $1.2 < p_T < 6.0$ GeV/c and was also compared to R_{AA} in Au+Au collisions and model including CNM effects, QGP effects and collisional dissociation. R_{AA} of NPE in U+U collisions is similar within uncertainties, but there is indication of lower R_{AA} than that in Au+Au collisions at the same centrality class. Also, R_{AA} of NPE versus mean number of participants was shown. The nuclear modification factor of NPE in U+U collisions extends the trend of that in Au+Au collisions. R_{AA} of NPE was also compared to the one of D mesons measured via hadronic decay channel and R_{AA} of light hadrons. The magnitude of suppression is similar within uncertainties.

Acknowledgements

This work was supported by Grant Agency of the Czech Technical University in Prague, grant No. SGS13/215/OHK4/35/14 and by Grant Agency of the Czech Republic, grant No. 13-20841S.

References

[1] M. L. Miller *et al.*, Ann. Rev. Nucl. Part. Sci. **57**, 205 (2007).
[2] L. Adamczyk *et al.* (STAR Collaboration), Phys. Rev. D **86**, 072013 (2012).
[3] Z. Ye for the STAR Collaboration, Quark Matter proceedings, Nucl. Phys. A **931**, 520 (2014).
[4] L. Adamczyk *et al.* (STAR Collaboration), Phys. Rev. Lett. **113**, 142301 (2014).
[5] H. Agakishiev *et al.* (STAR Collaboration), Phys. Rev. D **83**, 052006 (2011).
[6] A. Adare *et al.* (PHENIX Collaboration), Phys. Rev. Lett. **97**, 252002 (2006).
[7] A. Adare *et al.* (PHENIX Collaboration), Phys. Rev. Lett. **96**, 032001 (2006).
[8] A. Adare *et al.* (PHENIX Collaboration), Phys. Rev. C **86**, 024909 (2012).
[9] B. I. Abelev *et al.* (STAR Collaboration), Phys. Rev. Lett. **98**, 19301 (2007); Erratu m-ibid. **106**, 159902 (2011).

[10] A. Adare *et al.* (PHENIX Collaboration), Phys. Rev. Lett. **98**, 172301 (2007).
[11] A. Adare *et al.* (PHENIX Collaboraton), Phys. Rev. C **84**, 044905 (2010).
[12] A. Adare *et al.* (PHENIX Collaboration), arXiv:1509.04662 [nucl-ex] (2015).
[13] A. Adare *et al.* (PHENIX Collaboration), Phys. Rev. Lett. **109**, 242301 (2012).
[14] A. Adare *et al.* (PHENIX Collaboration), Phys. Rev. Lett. **112**, 252301 (2014).
[15] A. Adare *et al.* (PHENIX Collaboration), Phys. Rev. C **90**, 034903 (2014).
[16] K.H. Ackermann *et al.*, Nucl. Instrum. Meth. **A 499**, 624 (2003).
[17] D. Kikola, G. Odyniec, R. Vogt, Phys. Rev. C **84**, 054907 (2011).
[18] X. Bai for the STAR Collaboration, poster contribution at Quark Matter 2015.
[19] R. Sharma, I. Vitev, B.-W. Zhang, Phys. Rev. C **80**, 054902 (2009).
[20] R. Sharma, I. Vitev, Phys. Rev. C **87**, 044905 (2013).
[21] Z.-B. Kang *et al.*, Phys. Rev. Lett. **114**, 092002 (2015).
[22] B. I. Abelev *et al.* (STAR Collaboration), Phys. Lett. B **655**, 104 (2007).

EPJ Web of Conferences **126**, 04019 (2016) DOI: 10.1051/epjconf/201612604019
ICNFP 2015

Stochastic Navier–Stokes Equation with Colored Noise: Renormalization Group Analysis

N. V. Antonov[1,a], N. M. Gulitskiy[1,b], and A. V. Malyshev[1,2,c]

[1] *Department of Theoretical Physics, Faculty of Physics, Saint Petersburg State University, Ulyanovskaja 1, Saint Petersburg–Petrodvorez, 198504 Russia*
[2] *State Scientific Centre "Institute for Theoretical and Experimental Physics" of National Research Centre "Kurchatov Institute", B. Cheremushkinskaya 25, Moscow, 117218 Russia*

Abstract. In this work we study the fully developed turbulence described by the stochastic Navier–Stokes equation with finite correlation time of random force. Inertial-range asymptotic behavior is studied in one-loop approximation and by means of the field theoretic renormalization group. The inertial-range behavior of the model is described by limiting case of vanishing correlation time that corresponds to the nontrivial fixed point of the RG equation. Another fixed point is a saddle type point, i.e., it is infrared attractive only in one of two possible directions. The existence and stability of fixed points depends on the relation between the exponents in the energy spectrum $\mathcal{E} \propto k^{1-y}$ and the dispersion law $\omega \propto k^{2-\eta}$.

1 Introduction and description of the model

One of the possible ways to consider the fully developed turbulence within the framework of some microscopic model is to study the stochastic Navier–Stokes equation with random external force [1]. It has the form

$$\partial_t v_i + (v_l \partial_l) v_i + \partial_i \wp = \nu_0 \partial^2 v_i + \phi_i, \tag{1}$$

where $v_i(x)$ is a transverse (owing to the incompressibility) velocity field, $x \equiv \{t, \mathbf{x}\}$, $\partial_t \equiv \partial/\partial t$, $\partial_i \equiv \partial/\partial x_i$, ν_0 is molecular kinematic viscosity, $\partial^2 = \partial_i \partial_i$ is the Laplace operator, and $\wp = -\partial^{-2}(\partial_i v_l)(\partial_l v_i)$ is pressure. The simplest way is to assume that random force ϕ_i is decorrelated in space and time [2]. In this case, we preserve the Galilean symmetry of the system, therefore this approach is very interesting from the physical point of view. On the other hand, it is intriguing to consider such a model with colored noise. At the microscopical level such correlations of the random force ϕ_i arise if the noise is itself the result of removing "faster" degrees of freedom. In Fourier space these correlations can be represented in the form [3]

$$\left\langle \phi_i(\omega, \mathbf{k})\, \phi_j(\omega', \mathbf{k}') \right\rangle \propto \delta(\omega + \omega')\delta(\mathbf{k} + \mathbf{k}') P_{ij}(\mathbf{k}) D_\phi(\omega, k), \tag{2}$$

[a] e-mail: n.antonov@spbu.ru
[b] e-mail: n.gulitskiy@spbu.ru
[c] e-mail: a.v.malyshev@spbu.ru

where $P_{ij}(\mathbf{k}) = \delta_{ij} - k_i k_j / k^2$ is the transverse projector. We choose the function $D_\phi(\omega, k)$ in the form [4]

$$D_\phi(\omega, k) = \widetilde{D}_0 \frac{k^{8-d-y-2\eta}}{\omega^2 + v_0^2 u_0^2 k^{4-2\eta}}; \tag{3}$$

here $k \equiv |\mathbf{k}|$ is the wave number, $D_0(k) = \widetilde{D}_0 k^{8-d-y-2\eta}$ is the zero-time (power-law) correlation function of the extracted degrees of freedom, a new parameter u_0 is needed for the reason of dimensionality, and $\widetilde{D}_0 > 0$ is an amplitude factor. From (3) it follows that the energy spectrum of the velocity in the inertial range has the form $\mathcal{E} \propto k^{1-y}$, while the correlation time at the momentum k scales as $k^{-2+\eta}$.

This means that the random force ϕ_i is simulated by the following statistical ensemble: it is assumed to be Gaussian, homogeneous, with the zero mean and the following correlation function:

$$\left\langle \phi_i(t, \mathbf{x}) \phi_j(t', \mathbf{x}') \right\rangle = \widetilde{D}_0 \int \frac{d\omega}{2\pi} \int_{k>m} \frac{d\mathbf{k}}{(2\pi)^d} P_{ij}(\mathbf{k}) \frac{k^{8-d-(y+2\eta)}}{\omega^2 + v_0^2 u_0^2 k^{4-2\eta}} e^{i\mathbf{k}\cdot(\mathbf{x}-\mathbf{x}') - i\omega(t-t')}. \tag{4}$$

Here d is an arbitrary (for generality) dimension of $\mathbf{x}$ space, and $1/m$ is an integral turbulence scale, related to stirring. The function (4) involves two independent exponents y and η, which in the renormalization group (RG) approach play the role of two formal expansion parameters. Such ensemble was employed in some models, studied in [4–8]. It was shown that, depending on the values of the exponents y and η, the model reveals various types of inertial-range scaling regimes with nontrivial anomalous exponents, which were explicitly derived to the first [4] and second [5] orders of double expansion in y and η.

Depending on the parameter u_0, the function (4) demonstrates two interesting limiting cases: if $u_0 \to 0$, the situation corresponds to the independent of time ("frozen") velocity field, the case $u_0 \to \infty$ corresponds to the rapid-change (zero-time correlated) model. The relations

$$\widetilde{D}_0 / v_0^5 u_0^2 = \tilde{g}_0 \equiv \Lambda^{y+2\eta} \tag{5}$$

define the coupling constant $\tilde{g}_0$, which plays the role of the expansion parameter in the ordinary perturbation theory, and the characteristic ultraviolet (UV) momentum scale Λ.

Such ensemble with arbitrary function $D_\phi(\omega, k) \propto f\left(Wk^2/\omega^3\right)$ instead of (3) was used in [9] (here W is dissipation rate of energy). The first order term of the expansion of viscosity in the turbulent Reynolds number was obtained.

A very powerful method to study the various statistical models of turbulence is provided by the field theoretic renormalization group; see the monographs [10, 11] and references therein. In a number of papers this approach was applied to the case of passive vector (magnetic) fields advected by a turbulent flow with some prescribed properties: large-scale anisotropy, helicity, compressibility, finite correlation time, non-Gaussianity, a more general form of nonlinearity, etc.; see [12–16] and references therein.

2 Field theoretic formulation of the model

The stochastic problem (1) and (4) is equivalent to the field theoretic model of the set of two fields $\Phi \equiv \{v', v\}$ with the De Dominicis–Janssen action functional

$$S(\Phi) = \frac{1}{2} v_i' D_v v_k' + v_k' \left[-\partial_t - (v_i \partial_i) + \nu_0 \partial^2 \right] v_k. \tag{6}$$

Here D_v is the correlator (4), the needed integrations over $x = (t, \mathbf{x})$ and summations over the vector indices are implied [10, 11]. The field theoretic formulation means that statistical averages of random quantities in the original stochastic problem coincide with functional averages with weight $\exp \mathcal{S}(\Phi)$ with the action (6).

The model (6) corresponds to a standard Feynman diagrammatic technique with the triple vertex $-v'_k(v_i\partial_i)v_k$ and two bare propagators. In the frequency-momentum representation, the triple vertex corresponds to the expression

$$V_{ijk} = \frac{1}{2} i \left(\delta_{ik}\, k_j^{v'} + \delta_{ij}\, k_k^{v'} \right), \tag{7}$$

where $k^{v'}$ is the momentum of the field v'; in the diagrams it is represented by the point, in which three lines connect with each other. The two propagators are determined by the quadratic (free) part of the action functional and are represented in the diagrams as slashed straight (the slashed end corresponds to the field v') and straight (the end without a slash corresponds to the field v) lines, respectively. In the frequency-momentum representation they have forms

$$\left\langle v_i v'_j \right\rangle_0 = \frac{P_{ij}(\mathbf{k})}{-i\omega + \nu_0 k^2}; \tag{8}$$

$$\left\langle v_i v_j \right\rangle_0 = \widetilde{D}_0 \frac{P_{ij}(\mathbf{k})}{\omega^2 + \nu_0^2 k^4} \frac{k^{8-d-(y+2\eta)}}{\omega^2 + \nu_0^2 u_0^2 k^{4-2\eta}}. \tag{9}$$

From the analysis of canonical dimensions it follows that for any $d > 2$ superficial divergences can be present only in the 1-irreducible functions of two types. The first example is the function $\langle v'_\alpha v_\beta \rangle_{1-\mathrm{ir}}$, for which the formal index of divergence is $d_\Gamma = 2$. Another possibility is the function $\langle v'_\alpha v_\beta v_\gamma \rangle_{1-\mathrm{ir}}$ with $d_\Gamma = 1$.

3 Feynman diagrammatic technique

Consider the generating functional of the 1-irreducible Green's functions:

$$\Gamma(\Phi) = \mathcal{S}(\Phi) + \widetilde{\Gamma}(\Phi), \tag{10}$$

where for the functional arguments we have used the same symbols $\Phi = \{\boldsymbol{v}, \boldsymbol{v}'\}$ as for the corresponding random fields; $\mathcal{S}(\Phi)$ is the action functional (6) and $\widetilde{\Gamma}(\Phi)$ is the sum of all the 1-irreducible diagrams with loops [11]. This means that

$$\Gamma_2^{\alpha\beta} = \left[i\omega - \nu_0 p^2 \right] P_{\alpha\beta}(\mathbf{p}) + \Sigma_{\alpha\beta}, \tag{11}$$

where $\mathbf{p}$ is an external momentum, $P_{\alpha\beta}(\mathbf{p})$ is the transverse projector and $\Sigma_{\alpha\beta}$ is the "self-energy operator," diagrammatic representation of which is given in Fig. 1.

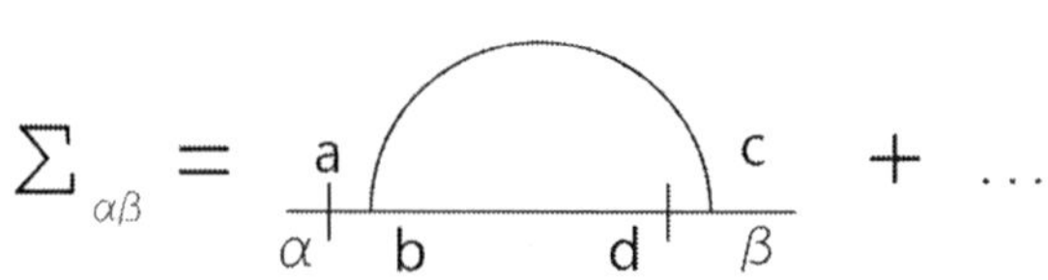

Figure 1. The one-loop approximation of the 1-irreducible response function $\langle v'_\alpha v_\beta \rangle_{1-\mathrm{ir}}$

Scalarization of the expression (11) produces

$$\Gamma_2 = i\omega - \nu_0 \mathbf{p}^2 + \Sigma, \tag{12}$$

and after calculations for the one-loop approximation of the scalar quantity Σ one obtains

$$\Sigma = -\frac{1}{16}p^2 \times \tilde{g}_0 \nu_0 \frac{u_0^3 d(d-1) + 3u_0^2 d(d-1) + 2u_0(d^2 - d + 2)}{d(d+2)(u_0+1)^3} C_d \frac{m^{-y}}{y}, \tag{13}$$

where $C_d \equiv S_d/(2\pi)^d$. Moreover, we use the minimal subtraction (MS) renormalization scheme, in which all the anomalous dimensions γ are independent of the regularizers such as y and η, and we may choose them arbitrary with the only restriction – our diagrams have to remain UV finite; see [5] for a detailed discussion. The most convenient way which we adopt in what follows is to put $\eta = 0$.

The one-loop expansion of the second divergent function $\langle v'_\alpha v_\beta v_\gamma \rangle_{1\text{-ir}}$ has the form

$$\langle v'_\alpha v_\beta v_\gamma \rangle_{1\text{-ir}} = V_{\alpha\beta\gamma} + (\Delta_1 + \Delta_2 + \Delta_3), \tag{14}$$

where $V_{\alpha\beta\gamma}$ is the vertex (7), and diagrams Δ_1, Δ_2, and Δ_3 are presented in Figs. 2a–2c.

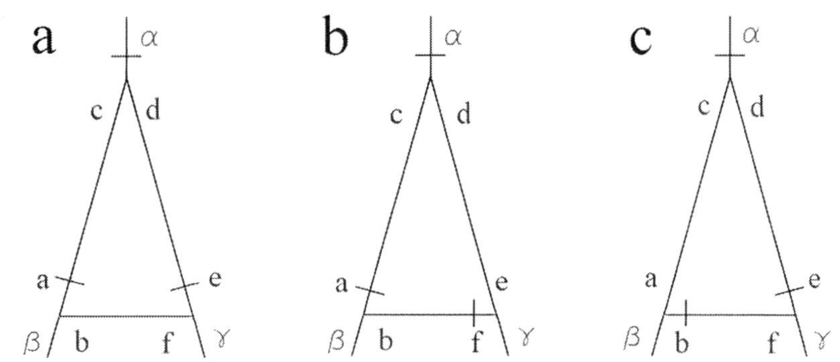

Figure 2. The one-loop approximation of the 1-irreducible function $\langle v'_\alpha v_\beta v_\gamma \rangle_{1\text{-ir}}$

Integrations over the momenta in the diagrams depicted in Figs. 1–2 are performed via averaging over the angles:

$$\int d\mathbf{k} f(\mathbf{k}) = S_d \int_m^\infty dk\, k^{d-1} \langle f(\mathbf{k}) \rangle, \tag{15}$$

where $\langle \cdots \rangle$ is the averaging over the unit sphere in the d-dimensional space, S_d is its surface area, and $k = |\mathbf{k}|$.

For the divergent part of diagram 2a, which is proportional to an external momentum $\mathbf{p}$, we have

$$\Delta_1 = \frac{i}{4}\tilde{g}_0 \frac{u_0(u_0+2)}{4(u_0+1)^2} \frac{(d+1)p_\alpha \delta_{\beta\gamma} - p_\beta \delta_{\alpha\gamma} - p_\gamma \delta_{\alpha\beta}}{d(d+2)} C_d \frac{m^{-y}}{y}. \tag{16}$$

Two remaining diagrams give

$$\Delta_2 = \Delta_3 = -\frac{i}{4}\tilde{g}_0 \frac{u_0(u_0^2 + 3u_0 + 4)}{8(u_0+1)^3} \frac{(d+1)p_\alpha \delta_{\beta\gamma} - p_\beta \delta_{\alpha\gamma} - p_\gamma \delta_{\alpha\beta}}{d(d+2)} C_d \frac{m^{-y}}{y}. \tag{17}$$

Using the transversality condition $\partial_i v_i = 0$ together with expression (7) and moving the derivative in the vertex from the field $\mathbf{v}'$ to the field $\mathbf{v}$ (using integration by parts) one can conclude that the term proportional to p_α gives no contribution, therefore the sum of the three triangle diagrams gives

$$\Delta_1 + \Delta_2 + \Delta_3 = \frac{i}{8}\tilde{g}_0 \frac{u_0}{(u_0+1)^3} \frac{p_\beta \delta_{\alpha\gamma} + p_\gamma \delta_{\alpha\beta}}{d(d+2)} C_d \frac{m^{-y}}{y}. \tag{18}$$

4 Renormalization of the model and fixed points

The model (6) is multiplicatively renormalizable with two independent renormalization constants Z_1 and Z_2; the renormalized action functional has the form

$$\mathcal{S}_R(\Phi) = \frac{1}{2} v'_i D_v v'_k + v'_k \left[-\partial_t - Z_1 (v_i \partial_i) + Z_2 \nu_0 \partial^2 \right] v_k. \tag{19}$$

These renormalization constants can be found from the requirement of UV finiteness of the expressions (12) and (14). Taking into account expressions (13) and (18) and introducing new coupling $g = \tilde{g} C_d / 4d(d+2)$ for the anomalous dimensions γ_1 and γ_2 one obtains

$$\gamma_1 = g \frac{u}{(u+1)^3}; \tag{20}$$

$$\gamma_2 = g \frac{u^3 d(d-1) + 3u^2 d(d-1) + 2u(d^2-d+2)}{4(u+1)^3}. \tag{21}$$

Dimensionality considerations together with expressions (9) and (19) give

$$g_0 = g\mu^y Z_g, \quad u_0 = u\mu^\eta Z_u; \quad \gamma_g = 2\gamma_1 - 3\gamma_2, \quad \gamma_u = -\gamma_2, \tag{22}$$

where μ is "reference mass" (additional free parameter of the renormalized theory) in the MS renormalization scheme. One of the basic RG statements is that the asymptotic behavior of the model is governed by the fixed points $\{g^*, u^*\}$, defined by the relations $\beta_g = 0$, $\beta_u = 0$. From the expressions (22) it follows that

$$\beta_g = g(-y - 2\gamma_1 + 3\gamma_2); \quad \beta_u = u(-\eta + \gamma_2). \tag{23}$$

The type of a fixed point (IR/UV attractive or a saddle point), i.e., the character of the RG flow in the vicinity of the point, is determined by the matrix $\Omega_{ik} = \partial\beta_i/\partial g_k$, where β_i is a full set of β-functions and g_k is a full set of couplings. For an IR attractive fixed point, the matrix Ω has to be positive, i.e., the real parts of all its eigenvalues are positive.

From the analysis of β functions it follows that if $\frac{1}{3} < \alpha < \frac{1}{3} + \frac{4/3}{3d(d-1)+2}$ the system possesses fixed point $\{g^*, u^*\}$ with coordinates

$$u^* = \frac{-3 + \sqrt{1 - \frac{16(\alpha-1)}{d(d-1)(3\alpha-1)}}}{2}; \tag{24}$$

$$g^* = \frac{3\alpha - 1}{2} y \frac{(u^*+1)^3}{u^*}, \tag{25}$$

where $\alpha = \eta/y$; see Fig. 3.

Moreover, it turns out that one of the two eigenvalues of the matrix Ω for this point is less than zero, thus, this fixed point is saddle one, i.e., it is IR attractive only in one of the two possible directions.

Another interesting case to be considered is $u^* \to \infty$. From (4) it follows that this case corresponds to the rapid-change model; this means that in this situation one should obtain the well-known fixed point of the model with zero-time correlations [2]. This is indeed so: taking into account that

$$\gamma_2 = \tilde{g} \frac{d(d-1)}{4} \quad \text{at} \quad u^* \to \infty, \tag{26}$$

one can obtain that in this case

$$\hat{g}^* = y \frac{4(d+2)}{3(d-1)}, \tag{27}$$

where $\hat{g} = \tilde{g} C_d / 4$. This fixed point is IR attractive if $y > 0$, $\eta > y/3$.

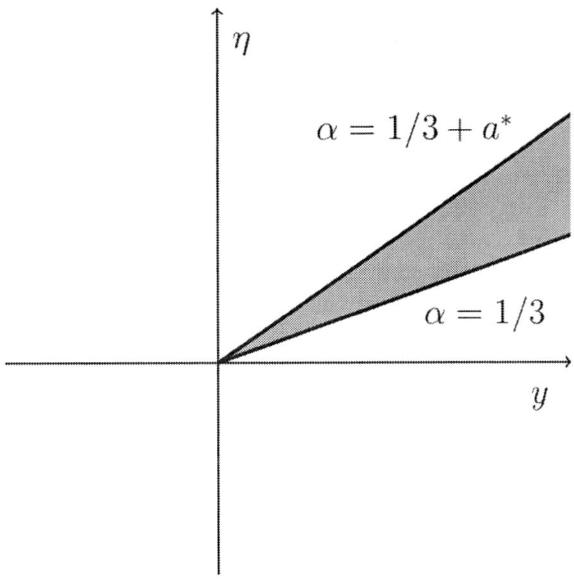

Figure 3. Domain of the existence of the fixed point in the model (6); $a^* = \frac{4/3}{3d(d-1)+2}$

5 Critical dimensions

In the leading order of IR asymptotic behavior, Green's functions satisfy the RG equation with the substitution $g \to g^*$ and $u \to u^*$. This feature, together with canonical scale invariance, gives us critical dimensions of the fields in the model, which, in fact, govern the asymptotic behaviour of arbitrary correlation functions.

If $u^* \to \infty$ one obtains

$$\Delta_v = 1 - y/3, \quad \Delta_{v'} = d - 1 + y/3, \tag{28}$$

which is in agreement with [2].

The saddle fixed point (24)–(25) gives

$$\Delta_v = 1 + \frac{\eta - y}{2}, \quad \Delta_{v'} = d - 1 + \frac{\eta - y}{2}. \tag{29}$$

6 Conclusion

We applied the field theoretic renormalization group to the analysis of the Navier–Stokes equation with colored random force, i.e., to the model with arbitrary *finite* correlation time of the velocity field. The critical dimensions of the fields are calculated. As it should be for the case of infinite fixed point, they coincide with the zero-time rapid-change model considered earlier in [2].

One loop approximation provides that, depending on the two exponents y and η that describe the energy spectrum $\mathcal{E} \propto k^{1-y}$ and the dispersion law $\omega \sim k^{2-\eta}$ of the velocity field, the possible *nontrivial* types of the IR behavior appear to reduce to the only one limiting case: the rapid-change type behavior, realized for $y > 0$, $\eta > y/3$. Another fixed point, realized for $\frac{1}{3} < \alpha < \frac{1}{3} + \frac{4/3}{3d(d-1)+2}$, where $\alpha = y/\eta$, is a saddle type point. The fact that the only IR attractive fixed point corresponds to vanishing correlation time means, in particular, that the Galilean symmetry, violated by the colored random force, is automatically restored in the IR limit. This gives a quantitative illustration of the general concept that the symmetries of the Navier–Stokes equation, broken spontaneously and by initial or boundary conditions, are restored in the statistical sense for the fully developed turbulence; see the discussion in [1].

Acknowledgements

The authors are indebted to L. Ts. Adzhemyan, Michal Hnatič, and Juha Honkonen for discussions.

The work was supported by the Saint Petersburg State University within the Research Grant No. 11.38.185.2014 and by the Russian Foundation for Basic Research within the Project No. 16-32-00086. A.V.M. acknowledges the support from the FAIR-Russia Research Center. N.M.G. and A.V.M. were also supported by the Dmitry Zimin's "Dynasty" foundation.

References

[1] U. Frisch, *Turbulence: The Legacy of A.N. Kolmogorov* (Cambridge Univ. Pr., Cambridge, 1995)
[2] L.Ts. Adzhemyan, A.N. Vasil'ev, Yu.M. Pis'mak, Theor. Math. Phys., **57**(2), 1131 (1983)
[3] A.S. Monin and A.M. Yaglom, *Statistical fluid mechanics*, vol. 2 (MIT Press, Cambridge, 1975)
[4] N.V. Antonov, Physica D, **144**:3-4, 370 (2000);
N.V. Antonov, Phys. Rev. E **60**, 6691 (1999).
[5] L.Ts. Adzhemyan, N.V. Antonov, J. Honkonen, Phys. Rev. E **66**, 036313 (2002)
[6] N.V. Antonov and A.V. Malyshev, J. Stat. Phys., **146/1**, 33 (2012)
[7] N.V. Antonov and N.M. Gulitskiy, Phys. Rev. E **92**, 043018 (2015)
[8] N.V. Antonov and N.M. Gulitskiy, Theor. Math. Phys., **176**(1), 851 (2013)
[9] D. Carati, Phys. Fluids A **2**, 1854 (1990);
D. Carati and K. Chriaa, Phys. Fluids A **5**, 3023 (1993)
[10] J. Zinn-Justin, *Quantum Field Theory and Critical Phenomena* (Clarendon, Oxford, 1989)
[11] A. N. Vasil'ev, *The field theoretic renormalization group in critical behavior theory and stochastic dynamics* (Chapman & Hall/CRC, Boca Raton, 2004)
[12] N.V. Antonov, J. Phys. A: Math. Gen. **39**, 7825 (2006).
[13] N.V. Antonov and N.M. Gulitskiy, Phys. Rev. E **91**, 013002 (2015)
[14] N.V. Antonov and N.M. Gulitskiy, Lecture Notes in Comp. Science, **7125/2012**, 128 (2012);
N.V. Antonov and N.M. Gulitskiy, Phys. Rev. E **85**, 065301(R) (2012);
N.V. Antonov and N.M. Gulitskiy, Phys. Rev. E **87**, 039902(E) (2013)
[15] E. Jurčišinova and M. Jurčišin, J. Phys. A: Math. Theor., **45**, 485501 (2012);
E. Jurčišinova and M. Jurčišin, Phys. Rev. E **88**, 011004(R) (2013);
E. Jurčišinova and M. Jurčišin, Phys. Rev. E **91**, 063009 (2015)
[16] E. Jurčišinova, M. Jurčišin, and Remecký, Phys. Rev. E **88**, 011002 (2013);
E. Jurčišinova and M. Jurčišin, Phys. Part. Nucl. 44, 360 (2013);
E. Jurčišinova, M. Jurčišin, and P. Zalom, Phys. Rev. E **89**, 043023 (2014)

EPJ Web of Conferences **126**, 04020 (2016) DOI: 10.1051/epjconf/201612604020
ICNFP 2015

Forward and small-x physics at CMS

Paolo Gunnellini[1 a]
on behalf of the CMS Collaboration

[1] *Deutsches Elektronen-Synchrotron (DESY) Notkestrasse 85, 22607, Hamburg (Germany)*

Abstract. A review of experimental measurements from the Compact Muon Solenoid Experiment at the LHC Run 1 is presented. A particular focus is given to jet results in the central and forward regions, and to measurements sensitive to the behaviour of small-x gluons inside the proton and to double parton scattering.

1 Introduction

The study of the dynamics of hadron jets in proton-proton collisions yields important information about underlying partonic processes. The production of jets with large transverse momenta p_T is thought to be originated from scatterings of partons which can be described by the theory of strong interactions. The cross section of such a proton-proton collision process is factorized in a convolution of the cross section for a partonic subprocess and the parton distribution functions (PDF). The QCD semi-hard interactions experienced by quarks and gluons are treated analytically through the so-called parton shower calculations. Different approaches to perform the parton evolution have been developed, depending on which kinematic region of the phase space is probed and several evolution equations are available for the calculations. At large splitting scales Q^2 and not too large longitudinal momentum fraction x, the Dokshitzer-Gribov-Lipatov-Altarelli-Parisi (DGLAP) equations [5] are good tools to calculate the partonic evolution with great success in reproducing a wide range of measurements. In the regime of low-x, though, the approximations used by the DGLAP approach break down and the formalism needs to be improved by the Balitsky-Fadin-Kuraev-Lipatov (BFKL) evolution scheme. Furthermore, at high factorization scales, the gluon distributions become large at low values of longitudinal momentum fractions of the colliding protons. In this regime, on the one hand parton recombination, i.e. saturation (see e.g. [1]), may occur. On the other hand, the probability to have more than one partonic interaction becomes non-negligible, leading to the production of pairs of physics objects at high transverse momentum via double parton scattering (DPS).

2 Jet measurements

At the Large Hadron Collider (LHC), various jet measurements, such the inclusive [2, 3] or the multijet [4] cross section for high p_T jets, have been measured at a collision energy $\sqrt{s}$ = 7 TeV by the CMS experiment and are in good agreement with predictions obtained at next-to-leading order (NLO)

[a]e-mail: paolo.gunnellini@desy.de

quantum chromodynamics (QCD). This good description of data represents a nice success of the current formalism used for calculating partonic interactions in pp collisions. Many attempts have been tried in multijet measurements to reach regions of the phase space where BFKL effects might appear but so far, no stringent evidence has been observed yet. In particular, by looking at more sophisticated scenarios than inclusive jets, e.g. by measuring jets at high pseudorapidities or by requiring correlations among selected jets, one has the possibility to be sensitive to lower and lower x values. Some of these scenarios have been measured by the CMS experiment to investigate different evolution models and are presented in the following.

The single-jet cross section has been measured [6] for the two leading, i.e. with highest p_T, jets in inclusive dijet events containing at least one forward (in the pseudorapidity range $3.2 < |\eta| < 4.7$) and one central jet (defined by the region $|\eta| < 2.8$). The data are based on 3.14 pb^{-1} of integrated luminosity at $\sqrt{s} = 7$ TeV. Jets are reconstructed using the anti-k_T algorithm (R = 0.5) in the p_T range 35-150 GeV. The single-jet cross section is shown in Figure 1, as a function of the p_T of the forward jet. The total systematic uncertainties are of the order of 20-30%, dominated by the absolute jet energy scale. The measurement, compared to perturbative QCD calculations, as implemented in the parton-shower event generators PYTHIA [7, 8] and HERWIG++ [9] at leading order (LO) or POWHEG+PYTHIA [10] at NLO, is found to be in agreement within the current experimental and theoretical uncertainties. Predictions from Monte Carlo (MC) event generators implementing different kinds of parton evolution, such the combined DGLAP+BFKL resummation model of CASCADE [11], are also found to be in good agreement with the measured dijet cross section.

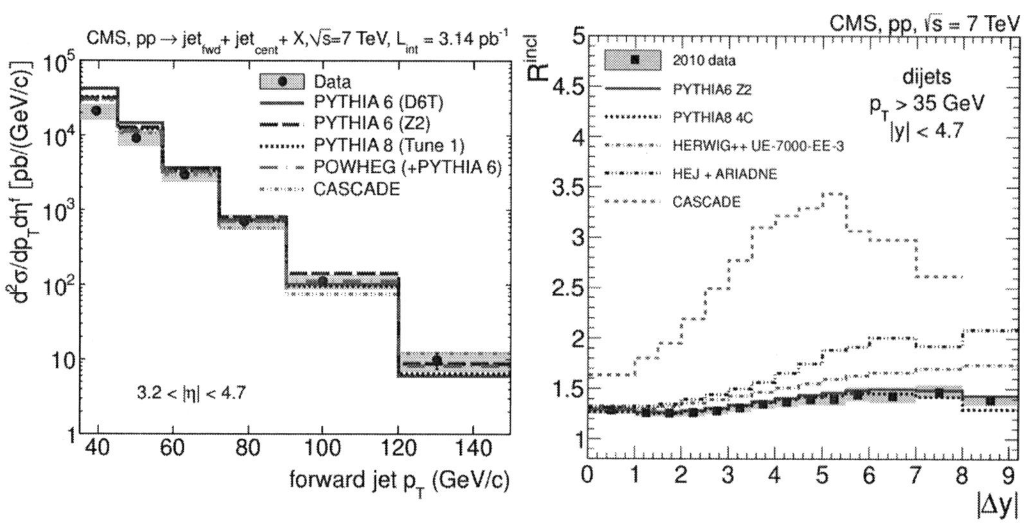

Figure 1. Differential cross sections as a function of the forward jet p_T for dijet events with at least one central jet and one forward jet, compared to predictions from several models (left). Ratios of the inclusive to exclusive dijet cross sections as a function of the rapidity separation Δy between the two jets compared to the predictions of the DGLAP-based MC generators PYTHIA6, PYTHIA8 and HERWIG++, as well as of CASCADE and HEJ+ARIADNE which incorporate elements of the BFKL approach (right).

A production of at least two jets is also studied in pp collisions at $\sqrt{s} = 7$ TeV as function of the rapidity separation Δy between the jets [12], for jets with $p_T > 35$ GeV and $|y| < 4.7$. Two classes of events are considered:

- inclusive events, with at least one pair of jets;
- exclusive events, with exactly one pair of jets.

The ratio R^{incl} of the cross section of all pairwise combinations of jets from the inclusive sample to the exclusive dijet cross section is shown on Figure 1(right) as a function of Δy. This observable is expected to be sensitive to effects beyond collinear factorization, as the radiation probability increases with $|\Delta|$y. The predictions of the MC event generators PYTHIA6 and PYTHIA 8 agree with the measurements, while predictions from the HERWIG++ event generator exhibit a more pronounced rise with $|\Delta y|$. The BFKL-motivated generators CASCADE and HEJ+ARIADNE [13] predict for these ratios a significantly stronger rise than observed.

Finally, for a further investigation of possible BFKL effects, correlations in azimuthal angle of the most forward and most backward in rapidity jets, (the so-called Mueller-Navelet jets) are measured with data collected in pp collisions with the CMS detector at the LHC at $\sqrt{s} = 7$ TeV. The azimutal-angle distributions between Mueller-Navelet jets are measured for jets with $p_T > 35$ GeV and $|y| < 4.7$. The results are presented as a function of the difference in azimuthal angle between the two selected jets in different ranges of rapidity separation Δy, reaching a Δy up to 9.4 in Figure 3. The results are compared to predictions of various Monte-Carlo event generators and analytical predictions based on the DGLAP and BFKL parton evolution schemes.

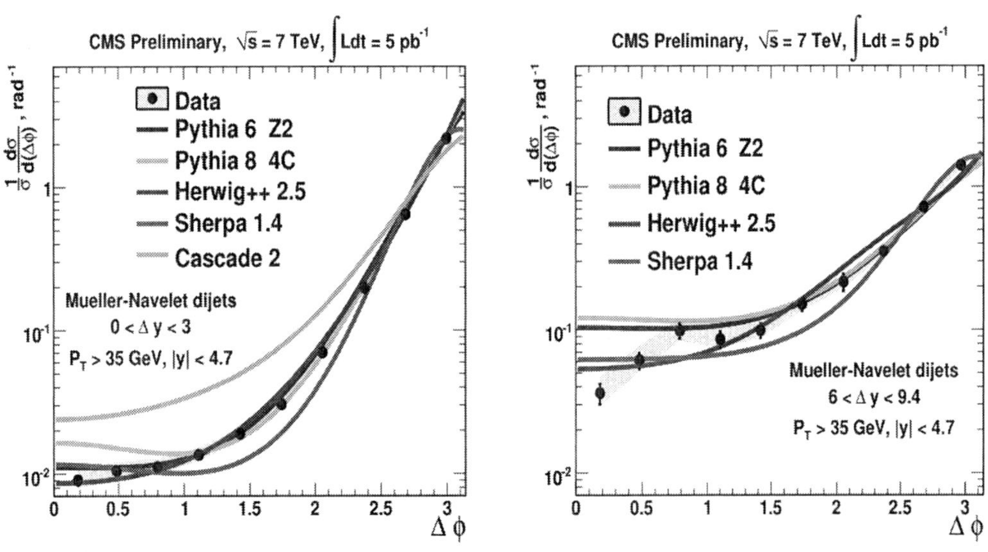

Figure 2. Distribution of azimuthal angle difference, $\Delta\phi$, between MN jets in rapidity interval $\Delta y < 3.0$ (left) and $6 < \Delta y < 9.4$ and its comparison to the predictions from LL based MC generators PYTHIA 6 tune Z2, PYTHIA 8 tune 4C, HERWIG++ 2.5, SHERPA 1.4 and LL BFKL motivated MC generator CASCADE.

From these measurements, no clear evidence of BFKL effects or of measurements deviating from DGLAP evolution scheme is obtained. MC event generators which are based on the DGLAP formalism are able to reproduce well a wide range of jet measurements. On the one hand, this might be due to the fact that the sensitivity to sufficiently small x values is not yet achieved reached at $\sqrt{s}$ = 7 TeV; if this is the case, measurements at a higher collision energy during LHC RunII will scan a still unexplored phase space which might reveal BFKL signals. On the other hand, BFKL effects might be hidden in the measured phase space due to many background contributions. It is indeed very difficult to select and isolate a "pure" gluonic ladder in a regime where the BFKL scheme should rule the parton evolution. At high p_T, the DGLAP approach is expected to dominate, while for decreasing p_T, increasing activity from MPI contributes to the final state. Independently on the p_T threshold, quark ladders following a DGLAP-based evolution add a major background and make the search of BFKL effects very complicated. New strategies for tackling these issues are under discussion and are summarized in the following:

- measurements of more exclusive scenarios, e.g. rapidity gap in dijet inclusive or exclusive ratios [14];
- measurements of more inclusive scenarios selecting corners of the phase space, e.g. jets in the forward region, in final states with a high jet multiplicity or associated to heavy-flavour production;

By pursuing these ideas, the new LHC phase will surely be an exciting and vivid period for searching BFKL effects.

3 Small-x gluon studies

A large set of measurements has been also performed by the CMS experiment for investigation of the dynamics of gluons at small longitudinal fractional momenta x. The integrated yields of events with a leading charged particle or a leading charged-particle jet with p_T above a given threshold have been measured in pp collisions at $\sqrt{s} = 8$ TeV in a data sample corresponding to an integrated luminosity of 45 μb^{-1} [15]. The yields are found to be relatively flat in the p_T region around 1 GeV (where the fixed-order perturbative parton-parton cross section diverges in the absence of any mechanism that saturates or unitarises the perturbative QCD scattering), followed by a steep decrease for $p_T >$ 10 GeV. The flattening behaviour observed at very low p_T is reproduced by models which apply a cutoff of the partonic cross section at values between 3 and 5 GeV, with the best description provided by the EPOS event generator [16]. Further comparisons of this measurement with various other phenomenological predictions of hadronic interaction models will help to improve the description of the transition between the perturbative and non-perturbative QCD regimes, which is believed to be dominated by the effects of parton density saturation and MPI.

A study at generator level, aiming for an investigation of the small-x gluon behaviour, has shown an interesting feature. Even though demonstrated only at a qualitative level, an increase of the gluon density at small-x values tends to be preferred for a consistent description of measurements at central and forward pseudorapidities. In particular, it has been observed that the description of charged-particle multiplicity at high pseudorapidities [17] provided by tunes using the NNPDF2.3LO PDF [18], e.g. Monash [19] and CUETP8M1 [17], is better than tunes using other PDF sets. This is due to the different small-x behaviour between NNPDF2.3LO and the other PDF. In fact, at low parton scales Q^2, the NNPDF2.3LO PDF has a gluon distribution at small x values that is larger than in HERAPDF1.5LO [20]. This contributes to more particles (and more energy) in the forward region. An investigation performed within the CMS collaboration confirmed this conclusion. By taking the HERAPDF1.5LO PDF set as a default, the low-x gluon distribution has been artificially increased according to a function (see Figure 1(left)) with two free parameters, x_0 and p_1.

The x_0 parameter sets the initial point of the x gluon increase, while the steepness of the rise is ruled by the p_1 parameter. The nominal gluon distribution in HERAPDF1.5LO and various distributions obtained by applying different modifications are shown in Figure 4(left). A tune based on a parton distribution function using $x_0 = 2{\cdot}10^{-5}$ and $p_1 = 5$, resulted to improve the description of charged-particle multiplicity consistently at both central and forward pseudorapidities (see Figure 4(right)) with respect to the tune CUETP8S1 [17] based on the nominal HERAPDF1.5LO PDF

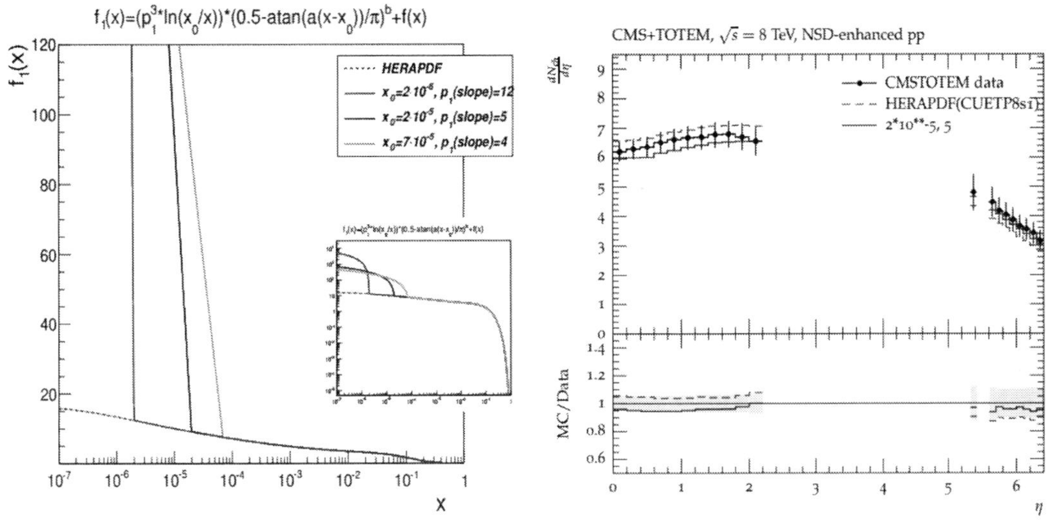

Figure 3. Gluon distribution as a function of the longitudinal fractional momentum x for the nominal HERA-PDF1.5LO PDF set and for the modified sets with different values of x_0 and (left). Comparison of charged-particle multiplicity as a function of η for the CUETP8S1-HERAPDF tune and a tune performed with modified low-x gluon behaviour (right).

set. Note that the range of modified x values in the considered gluon distributions are below the range of the HERA deep inelastic scattering data. Future measurements sensitive to this low-x regime are strictly needed to confirm the results of this qualitative study and to further constrain the behaviour of the gluon density.

4 Double parton scattering measurements

A wide range of measurements sensitive to DPS contributions has been also carried out within the CMS physics program for different physics channels. The investigated channels are characterized by a high number of physics objects in the final state, e.g. jets. Measurements related to the W+two-jet [21], four-jet [22], γ+three-jet [23] and two-b+two-light-jet [24] final states have been released by the CMS experiment. The current strategy for DPS-extraction analyses is based on the association of the selected objects in two pairs and on the measurement of differential cross sections as a function of correlation observables defined by the kinematical topology of the final state. The two pairs can be produced by two independent scatterings in case of a DPS event and the correlation observables are expected to be able to disentangle this occurrence. Studies at generator level have shown that variables which are most sensitive to a DPS contribution are the ones which consider the relative topology of all selected objects, not only of the objects of one of the two pairs [25]. One of such observables is $\Delta\mathrm{S}$, which corresponds to the azimuthal angle between the two selected pairs and is defined as:

$$\Delta\mathrm{S} = \arccos\left(\frac{\vec{p}_{\mathrm{T}}(\mathrm{pair}_1)\cdot\vec{p}_{\mathrm{T}}(\mathrm{pair}_2)}{|\vec{p}_{\mathrm{T}}(\mathrm{pair}_1)|\cdot|\vec{p}_{\mathrm{T}}(\mathrm{pair}_2)|}\right) \tag{1}$$

For the W+two-jet channel, pair_1 and pair_2 are, respectively, the reconstructed W and the dijet system, while in the four-jet (two-b+two-light-jet) final state, they correspond to the hard (bottom) and soft (light) dijet systems. In Figure 4, the differential cross section normalized to the number of selected events as a function of the ΔS observable measured by CMS is shown for various final states.

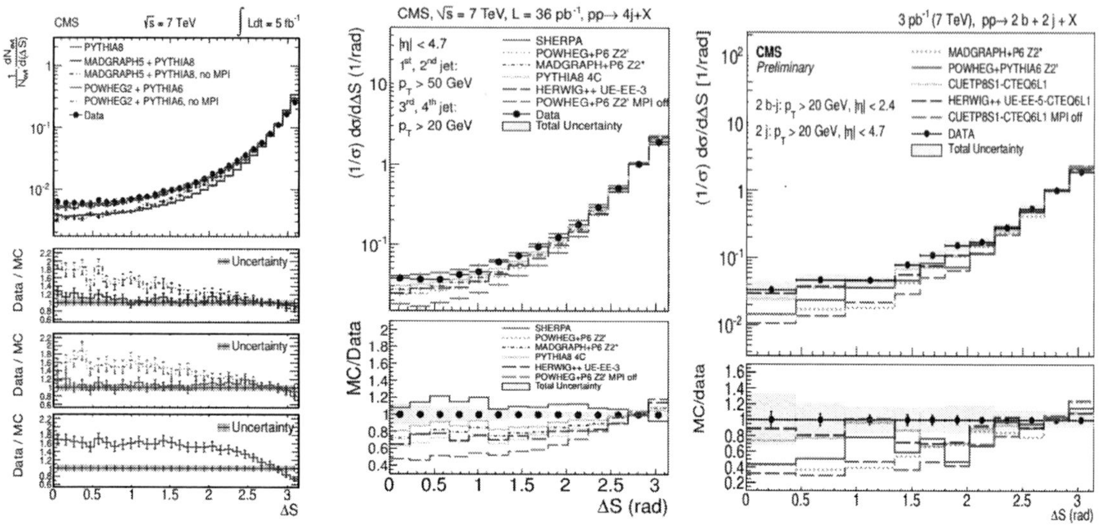

Figure 4. CMS data on the ΔS observable for various measured final states: W+two-jet (left), four-jet (center), γ+three-jet (right) channels.

In each of them, the measured cross sections have a similar shape with a falling distribution from correlated configurations at high values, down to uncorrelated jet topologies at low values. The measured cross sections have been also compared to predictions obtained with various MC event generators, implementing different matrix elements (ME) interfaced with parton shower (PS) and underlying event (UE) simulation. The PYTHIA 8 and HERWIG++ event generators simulate Leading Order (LO) 2→2 processes, SHERPA [26] and MADGRAPH [27] generate LO diagrams with higher number of partons in the final state, while POWHEG produces Next-to-Leading-Order (NLO) processes, with real and virtual corrections included in the ME. Predictions without the contribution of MPI are also considered in the comparison. All predictions include a simulation of parton-shower and underlying-event effects generated according to updated tunes.

In general, for all considered channels, predictions of multileg or NLO event generators are able to reproduce reasonably well the measured distributions. However, some room for improvement is available for the simulation of hard MPI contribution in the current models, in order to achieve a better description of the cross section at low values of ΔS. Predictions from event generators without the simulation of MPI do not reproduce the measured points, mainly in the region at low values of ΔS, where a DPS contribution is expected. This comparison clearly indicates the need for MPI in the current models to achieve a correct description of the DPS-sensitive observables. A quantitative extraction of the DPS contribution necessarily relies on the measurement of correlation observables and can be carried out through the so-called template [21] or the fitting [17] methods.

5 Summary and conclusions

The CMS experiment is very active in the research about many aspects concerning QCD and the understanding of the partonic behaviour in proton-proton collisions. A large set of jet measurements over different regions of phase space shows that current predictions based on the DGLAP evolution scheme are able to reproduce the data with a very good level of agreement. Investigations on observables sensitive to small-x gluon dynamics did not give clear evidence of BFKL effects, yet. Various ideas and alternative approaches to tackle them in future measurements are under discussion.

The measurement of the integrated jet cross section performed with the CMS experiment as a function of the leading charged-particle and jet can be used as a probe of the transition from the perturbative to the non-perturbative region. The distributions, steeply falling at large transverse momenta and flattening out at p_T values between 1 and 10 GeV, show effects of parton density saturation which are reproduced by models applying a cutoff of the partonic cross section at around 3–5 GeV. However, qualitative studies at generator level show that a better description of measurement at central and forward rapidities is achieved in case an increase at small-x values is introduced for the gluon density.

Measurements sensitive to hard multiple interactions are available in many different physics channels. Observables investigating correlations in azimuthal angle between the objects selected in the considered final states can not be described by current models without implementing a simulation of multiple-parton interactions. These measurements can be used in the future to study the dependence of the double parton scattering contribution on the partonic initial state in terms of x and flavour, and on the scale of the process.

References

[1] A. H. Mueller, Nucl. Phys. B **335** (1990) 115
[2] CMS Collaboration [CMS Collaboration], section at sqrt(s) = 8 TeV," CMS-PAS-SMP-14-001
[3] S. Chatrchyan *et al.* [CMS Collaboration], Phys. Rev. Lett. **107** (2011) 132001
[4] P. Kokkas [CMS Collaboration], PoS EPS **-HEP2011** (2011) 277
[5] G. Altarelli and G. Parisi, Nucl. Phys. B **126** (1977) 298
[6] S. Chatrchyan *et al.* [CMS Collaboration], JHEP **1206** (2012) 036
[7] T. Sjostrand, S. Mrenna and P. Z. Skands, JHEP **0605** (2006) 026
[8] T. Sjöstrand, S. Mrenna and P. Z. Skands, Comput. Phys. Commun. **178** (2008) 852
[9] S. Gieseke, D. Grellscheid, K. Hamilton, A. Papaefstathiou, S. Platzer, P. Richardson, C. A. Rohr and P. Ruzicka *et al.*, arXiv:1102.1672 [hep-ph]
[10] J. M. Campbell, R. K. Ellis, P. Nason and G. Zanderighi, JHEP **1308** (2013) 005
[11] H. Jung *et al.*, Eur. Phys. J. C **70** (2010) 1237
[12] S. Chatrchyan *et al.* [CMS Collaboration], Eur. Phys. J. C **72** (2012) 2216
[13] J. R. Andersen and J. M. Smillie, JHEP **1106** (2011) 010
[14] C. Royon, AIP Conf. Proc. **1654** (2015) 070002
[15] V. Khachatryan *et al.* [CMS Collaboration], arXiv:1507.00233 [hep-ex]
[16] T. Pierog, I. Karpenko, J. M. Katzy, E. Yatsenko and K. Werner, Phys. Rev. C **92** (2015) 3, 034906
[17] CMS Collaboration [CMS Collaboration], CMS-PAS-GEN-14-001
[18] R. D. Ball *et al.* [NNPDF Collaboration], Nucl. Phys. B **877** (2013) 290
[19] P. Skands, S. Carrazza and J. Rojo, Eur. Phys. J. C **74** (2014) 8, 3024
[20] A. M. Cooper-Sarkar, PoS DIS **2014** (2014) 032
[21] S. Chatrchyan *et al.* [CMS Collaboration], JHEP **1403** (2014) 032
[22] S. Chatrchyan *et al.* [CMS Collaboration], Phys. Rev. D **89** (2014) 9, 092010
[23] CMS Collaboration [CMS Collaboration], CMS-PAS-FSQ-12-017
[24] CMS Collaboration [CMS Collaboration], CMS-PAS-FSQ-13-010

[25] P. Gunnellini, DESY-THESIS-2015-010

[26] T. Gleisberg, S. Hoeche, F. Krauss, M. Schonherr, S. Schumann, F. Siegert and J. Winter, JHEP **0902** (2009) 007

[27] J. Alwall et al, JHEP 06 (2011) 128

ATLAS Soft (and some harder) QCD: Selected Results

RWL Jones[1] for the ATLAS Collaboration

[1] *Department of Physics, Lancaster University, Lancaster LA1 4YB, UK*

Abstract. Recent results on soft and hard QCD production are presented. Results derive from the 900 GeV and 7, 8 and 13 TeV runs.

1 Introduction

ATLAS excels at high-p_T physics, but also provides a window into important softer QCD processes. These have intrinsic interest but also provide the understanding that underpins the searches for new physics. ATLAS [1] now has data from 900GeV up to 13TeV in the same detector, allowing scale evolution to be probed.

The most relevant trigger and detector systems for the results presented here are the minimum bias trigger scintillators (MBTS), which have been refurbished for the 13 TeV running (Run 2); and the inner detector, with good p_T resolution, and excellent lifetime resolution; an Insertable B-Layer (IBL) [2] was added for Run 2, dramatically improving the lifetime resolution of the detector (from 100 to 50 fs) and enhancing the pattern recognition. The MBTS is made up of 2 cm thick disks of scintillator 3.6 m from the interaction region, giving a pseudorapidity coverage $2.07 < |\eta| < 3.86$. It has a high efficiency that is untiform to within 1%.

The studies presented consider various classes of pp interaction, and a glossary is needed. If both protons remain intact and there is no additional activity, the interaction is elastic. If one proton breaks up, it is single dissociation. If both break up, it is double dissociation. If the protons remain intact, but there is activity arising from pomerons or gluons exchanged between the two protons, there is central diffraction; and if there is the same activity accompanied by both protons breaking up, there is non-diffractive dissociation.

2 Bose-Einstein Correlations

This studies the correlations in phase space between two identical bosons arising from the symmetry of their wave functions. The enforced symmetry enhances likelihood of two particles being close in phase space, and allows one to probe the source of the bosons in size and shape. By investigating how the enhancement of close particle pairs depends on particle multiplicity and transverse momentum probes the production mechanism. The closeness of phase space is measured using the virtuality of particle pair, $Q^2 = -\left(p_1^2 - p_2^2\right)$, where p_i is the particle 4-momentum. The strength of the effect is measured using a correlation function:

$$C_2(Q) = N^{\text{like-sign}}(Q)/N^{\text{ref}}(Q) = c_0\,[1 + \Omega\,(\lambda, QR)]\,(1 + \epsilon Q)\,,$$

DOI: 10.1051/epjconf/201612604021

where C_0 is a normalisation factor, ϵ accounts for long-range effects, R is the effective radius of the source and λ is the strength of the effect, being 0 for a coherent source and 1 for a chaotic source. $N^{\text{like-sign}}(Q)$ is the number of like-sign particle pairs in a region of Q and the reference sample $N^{\text{ref}}(Q)$ is the expected pair distribution without a Bose-Einstein effect. The reference sample can be formed using unlike sign particle pairs, particles from opposite hemispheres, or by mixing particles from different collisions. To remove the effect of resonances etc, the study uses the ratio

$$R_2(Q) = C_2(Q)/C_2^{\text{Monte Carlo}}(Q),$$

where the Monte Carlo lacks Bose Einstein correlations. A current ATLAS study [4] sees a clear Bose-Einstein signal, and extends studies to very high charged particle multiplicity events (with > 108 charged particles reconstructed); this study uses oppositely charged particles (which are overwhelmingly pions). The fits work best assuming an exponential source, but results are also produced assuming a Gaussian source to compare with earlier measurements. Looking at the fitted source strength and radius as a function of multiplicity, both are seen to be essentially independent of centre of mass energy. However, the source radius growth saturates at high multiplicity as seen in figure 1, which is the first time that this has been observed. The behaviour is consistent with pomeron-based models.

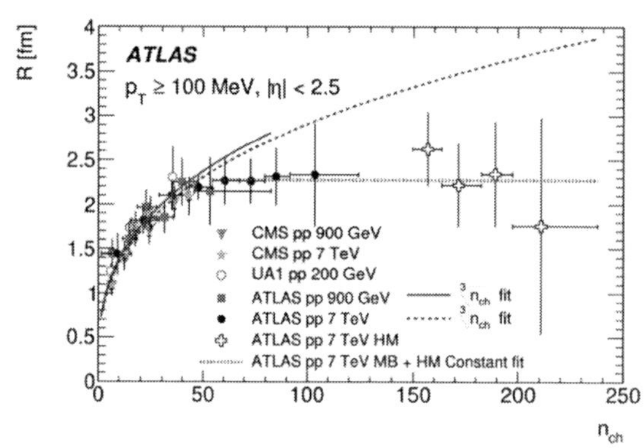

Figure 1. The source radius plotted as a function of event multiplicity [4].

The strength and radius can also be studied as a function of $k_T = |\overline{p}_{T,1} + \overline{p}_{T,2}|/2$. This again shows no centre-of-mass energy dependence within the uncertainties, but when broken down by multiplicity region, the source radius decreases with increasing k_T, but increases with increasing multiplicity.

3 Two Photon Scattering

The LHC can be an effective two-photon collider. The predictions for the expected behaviour use the Equivalent Photon Approximation with absorbtive corrections for finite proton size. For the experimental study, the channel $\gamma\gamma \to ll$ is chosen [5]. There there is an elastic signal but dissociative backgrounds. The process $\gamma\gamma \to (W \to l\nu)(W \to l\nu)$ is the other important background. The signal selection uses the distance of dilepton from other activity (see figure 2) and the dilepton mass to discriminate from the background.

The experimental results using the 7 TeV dataset improve on the precision of an earlier CMS result. Both in the electron and muon modes, the results agree with that earlier result, and with equivalent photon approximation predictions after absorptive corrections for finite proton size.

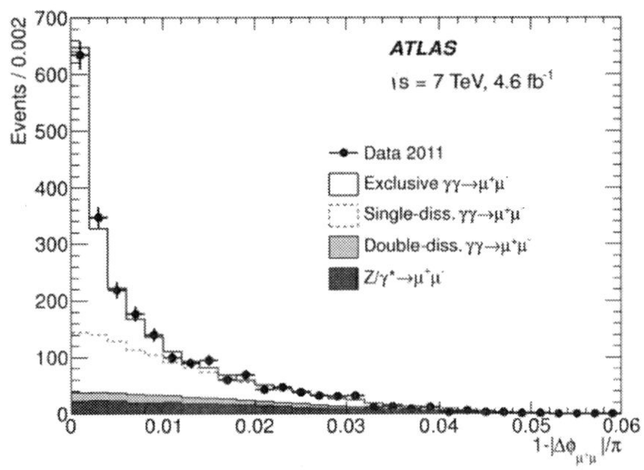

Figure 2. The acoplanarity in candidate $\gamma\gamma \rightarrow \mu^+\mu^-$ events [5].

4 Recent results on minimum bias events

Unlike the harder collisions described above, soft particle production cannot be calculated from first principles. Instead, running models tuned to the data must be used. The tuning uses a wide range of measurements. Track multiplicities are now available up to 13 TeV in ATLAS. The track-based analysis [6] uses dedicated runs with low pile-up of collisions in each readout of the detector, and events with only one reconstructed vertex are selected; the 13 TeV sample had 10 million events, corresponding to an integrated luminoisty of $169\mu b^{-1}$. It also employs a special version of the track reconstruction that reconstructs tracks down to 100 MeV momentum.

The differential cross-sections with respect to pseudorapidty η and p_T have been measured and compared to various models. Most models have a similar shape in η but differ in normalisation. Herwig is the exception, and gives a poor agreement with the data; it is notable for having been tuned to the underlying event data. However, most models give a good agreement over 10 orders of magnitude in p_T, especially the EPOS model.

The charged particle multiplicity (n_{ch}) distribution (within the p_T and η acceptances) have also been measured, and at low n_{ch} no Monte Carlo model reproduces the data well. It is notable the models that lack colour reconnection (such as QGSJET) fail to reproduce the scaling with n_{ch} seen in the data, see for example figure 3.

The mean charged particle multiplicity $< n_{ch} >$ evolution with centre-of-mass energy $\sqrt{s}$ can also now be studied over a large energy range in a single experiment, and is well modelled except by HERWIG++.

5 Recent results on the underlying event

The underlying event is made up from all particles in the collision except those from the hard scatter. Its $\sqrt{s}$ dependence has also been examined with the new 13 TeV data, and allows one to test the multi-parton interaction modelling in various models. The technique used is to define a region sensitive to the underlying event in the plane perpendicular to the beam; this region is transverse to the direction of an object representing the hard scatter. The object used to define the hard scatter direction in the 7 TeV data is either a Z [7] or the leading jet [8]; for the 13 TeV data it is the leading track [9]. Either the mean charged multiplicity or the mean scalar sum of particle p_T in bins of pseudorapidity and angle to the leading object can be studied. These can be studied as a function of the p_T of the leading

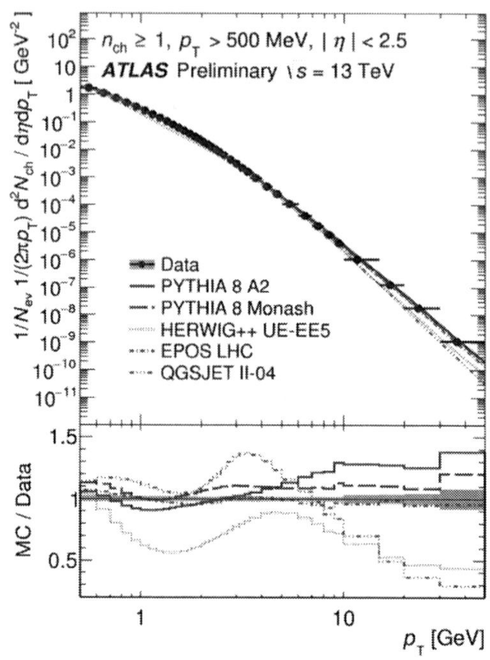

Figure 3. The multiplicity in minimum bias events as a function of pseudorapidity and track p_T [6].

object. The data are broadly compatible between hard object definitions, particularly at high-p_T of the leading object. The Z and jet measures agree particularly well at high p_T, while at low p_T the selections for the Z introduce biases.

For the 13 TeV data, the uncorrected mean scalar sum of particle p_T and n_{ch} have been studied as a function of angle to the leading object. PYTHIA tuned to the minimum bias data agrees well for a leading object $p_T > 1$ GeV, while the best agreement is with Monte Carlos tuned to lower energy underlying event data for leading $P_T > 5$ GeV. The mean charged multiplicity and scalar sum of particle p_T has also been studied in the transverse and 'toward' (i.e. around the leading object) regions. It is hard for models to simultaneously describe the results for these two quantities, and (in contrast to the minimum bias case) EPOS gives a poor description of the data.

6 Inelastic cross-section measurements

The total inelastic cross-section can be determined using two distinct methods. The first uses the optical theorem to relate the elastic events measured in the ALFA Roman pot detectors to the inelastic cross section [10]. The other, new, result counts the inelastic events triggering the MBTS directly with a large rapidity gap [11]. The mass of the largest of the two jets triggering the MBTS is calculated; the coverage of the MBTS imposes a fiducial requirement that this is greater than 13 GeV. Again, low pile-up data is used. The fraction of event with only hits on one side of the MBTS is used to constrain the diffractive contribution. The fiducial cross-section is found to be 65.2 ± 0.8(exp.) ± 5.9(lum.)mb, where the luminosity uncertainty dominates. The models tend to predict a slightly higher cross-section but largely agree within the uncertainties.

7 Jet results

Many results on jets, usually using the anti-kT algorithm with a cone size $R = 0.4$ are now available for the 13 TeV results [12]. A cross-section for jets with $|y| < 0.5$ has been measured and found to be in good agreement with Next to Leading Order calculations for various parton distribution functions. Energy-energy calculations have also been used to obtain a determination of the strong coupling [13] that is in good agreement with measurements from many other processes as shown in figure 4. Run to the Z pole, the result is $\alpha_s(m_Z) = 0.1173 \pm 0.0010(\text{exp.})^{+0.0063}_{-0.0020}(\text{scale}) \pm 0.0017(\text{PDF}) \pm 0.0002(\text{NPC})$, where the last term describes the uncertainties from non-perturbative corrections.

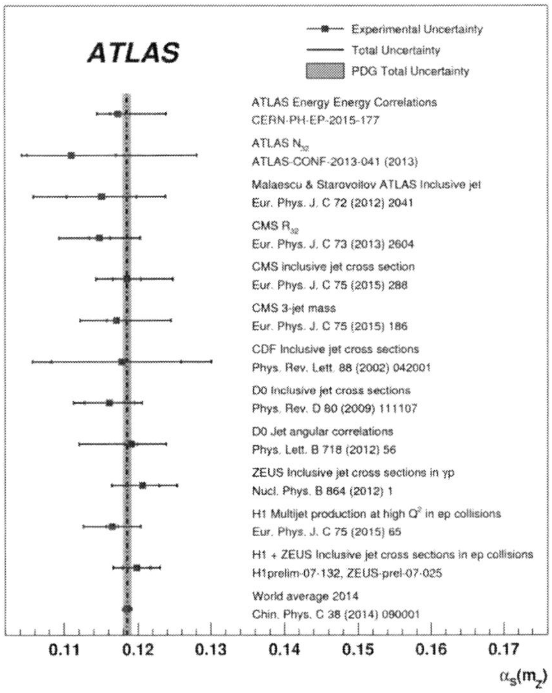

Figure 4. The strong coupling determined from energy-energy correlations in ATLAS compared to determinations in other processes; the plot is from the additional material provided for reference [13], and can be found at https://atlas.web.cern.ch/Atlas/GROUPS/PHYSICS/PAPERS/STDM-2014-03/#auxstuff

References

[1] ATLAS Collaboration, JINST **3**, S08003 (2008)
[2] ATLAS IBL Collaboration, CERN-LHCC-2010-013 (2010) https://cds.cern.ch/record/1291633; ATLAS Collaboration, CERN-LHCC-2010-013 (2013) https://cds.cern.ch/record/1291633
[3] ATLAS TDAQ Collaboration, JINST **9**, C10020 (2014)
[4] ATLAS Collaboration, Eur. Phys. J. **C75**, 466 (2015)
[5] ATLAS Collaboration, Phys. Lett. **B749**, 242 (2015)
[6] ATLAS Collaboration, CERN-EP-2016-014 (2015) arXiv:1602.01633 [hep-ex]
[7] ATLAS Collaboration, Eur. Phys. J. **C74**, 3195 (2014)

DOI: 10.1051/epjconf/201612604021

[8] ATLAS Collaboration, Eur. Phys. J. **C74**, 2965 (2014)
[9] ATLAS Collaboration, ATL-PHYS-PUB-2015-019 (2015) https://cds.cern.ch/record/2037684
[10] ATLAS Collaboration, Nucl. Phys. **B**, 486 (2014)
[11] ATLAS Collaboration, ATLAS-CONF-2015-038 (2015) https://cds.cern.ch/record/2045064
[12] ATLAS Collaboration, ATLAS-CONF-2015-034 (2015) https://cds.cern.ch/record/2038145
[13] ATLAS Collaboration, Phys. Lett. **B750** 427 (2015)

EPJ Web of Conferences **126**, 04022 (2016) DOI: 10.1051/epjconf/201612604022
ICNFP 2015

Recent results from ATLAS on B Physics and Quarkonia

RWL Jones[1] for the ATLAS Collaboration

[1] *Department of Physics, Lancaster University, Lancaster LA1 4YB, UK*

Abstract. Recent results from the extensive programme of heavy flavour and onia studies in ATLAS are presented. These benefit from the very high integrated luminosity in the first running period at the LHC at 7 and 8 TeV, and some are now extended to include information from the latest 13 TeV running.

1 Introduction

Heavy flavour physics in ATLAS [1] has both intrinsic interest as laboratory for the standard model and as a potential window into new physics. ATLAS has the advantage of high integrated luminosities and good muon triggers, and is particularly well suited for multi-muon final states. It also has excellent charged particle tracking and lifetime resolution, which has been further improved in 2015 by the addition of the insertable B-layer (IBL) [2], reducing from 100 fs to 50 fs. The momentum resolution is less than 10% up to 1 TeV. The typical heavy flavour primary trigger require either two muons of at least 4 GeV momentum transverse to the beam (p_T) or else p_T thresholds of 4 and 6 GeV when rates must be reduced. Prescales are applied when necessary.

2 Onium Production

The foundation of many of the heavy flavour studies in ATLAS is onium reconstruction, and particularly the J/ψ. ATLAS has extended its studies of J/ψ and $\psi(2S)$ production from its studies at 7 and 8 TeV centre of mass energies to include the new 13 TeV data [3].

2.1 J/ψ and $\psi(2S)$ production

Both prompt and non-prompt production is studied. The two onia have individual strengths; the $\psi(2S)$ has no significant feed-down from higher states, and allows the $J^{PC} = 1^{--}$ states to be studied. The J/ψ, which is more plentiful, has contributions from 1^{--} and J^{++} in comparable amounts.

The studies use 4 GeV thresholds for both muons, and corrections are made for trigger and reconstruction efficiencies. The acceptance depends on the spin alignment; current measurements support the central assumption of no alignment. The prompt and non-prompt components are distinguished by constructing the pseudo-proper decay time using the transverse displacement of the the decay vertex from the interaction point and the p_T of the onium. The prompt production is compared with non-relativistic QCD predictions (NRQCD, see for example [4] and [5]), and good agreement is found across the wide range of p_T, with no dependence on rapidity, y. The non-prompt production is assumed to be dominated by onia from B-decays, and are compared to fixed order, next to leading logarithm (FONLL) predictions. Here, there is evidence that the $\psi(2S)$ predictions tend to overestimate the differential cross sections. Thus far, double differential cross-sections are available from the 13 TeV data [6]. The results now span centre of mass energies from 2.76 to 13 TeV within one experiment and with self-consistent analyses. Strikingly, there is no significant change in the the fraction of non-prompt production with either rapidity or centre of mass energy, as illustrated in figure 1 and 2.

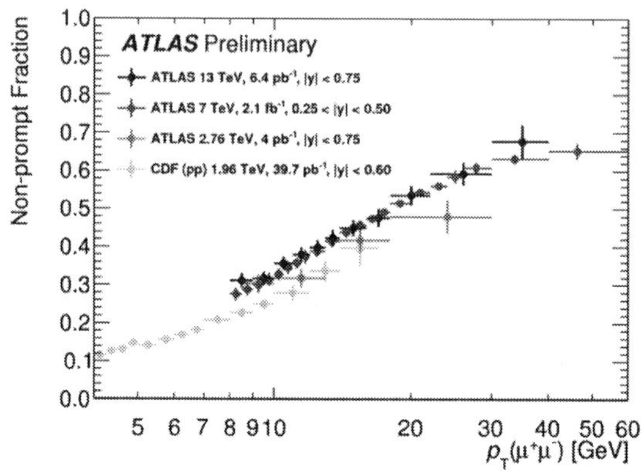

Figure 1. The fraction of prompt J/ψ production as a function of p_T for different centre of mass energies observed in ATLAS pp collisions and in CDF $p\bar{p}$ collisions [6].

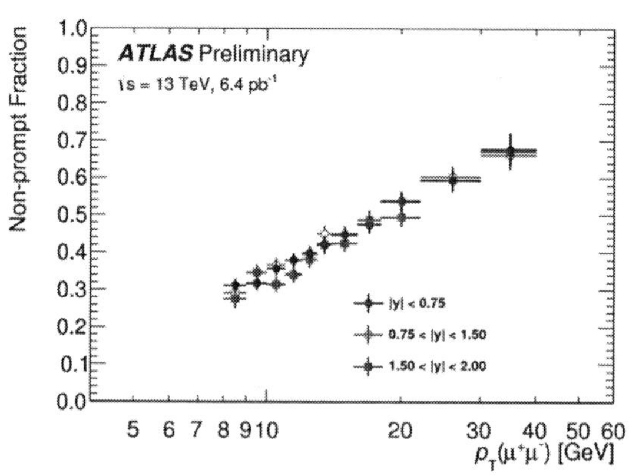

Figure 2. The fraction of prompt J/ψ production as a function of p_T for different rapidity y intervals as observed in the 13 TeV data [6].

2.2 Excited B_c production

With the first observation of the $\chi_b(3P)$ in 2012 [7], ATLAS showed its potential for new particle discoveries. The ATLAS search has continued with the observation of a state consistent with an excited B_c state [8]. This was observed both in the 7 TeV and 8 TeV data. The $B_c^{\pm}$ was reconstructed from its decay to $J/\psi(\mu\mu)\pi^{\pm}$. Opposite charged particles (where the particles are inconsistent with being muons and consistent with originating from the same interaction as the $B_c^{\pm}$ are added in pairs to the $B_c^{\pm}$, and the mass difference spectrum $Q = m(B_c^{\pm}\pi^+\pi^-) - m(B_c^{\pm} - 2 \cdot m(\pi^{\pm})$ is constructed. A new structure is observed in both samples, consistent with predictions for a $B_c^{\pm}(2S)$ meson. The total significance for the new particle is 5.2σ when look-elsewhere effects are taken into account.

2.3 Search for hidden beauty in $\Upsilon(1S)\pi\pi$

Searches have also been made for hidden beauty states in the $\Upsilon(1S)\pi\pi$ decay mode (where the $\Upsilon(2S)$ and $\Upsilon(3S)$ regions are excluded) [9]. This has resulted in the most sensitive X_b production search in the mass range $m(\Upsilon(1S)\pi\pi) > 10.1$ GeV. This state would be an analogue to the $X(3872)$ narrow resonance observed in the charmonium sector, which is still an unresolved mystery. This has revealed no evidence for the $\Upsilon(1^3D_J)$, $\Upsilon(10860)$ and $\Upsilon(11020)$ states, or in general for a beauty analogue to the $X(3872)$.

2.4 Charmonia in association with Vector Bosons

The high available statistics combined with the tools for studying charmonia allow for the study of associated production between intermediate vector bosons and charmonia. These allow for test of QCD predictions at the perturbative/non-perturbative boundary. Anomalous rates provide evidence for processes beyond the standard model from charged Higgs decays, light scalars etc. For the prompt associated production, the colour octet and colour singlet contributions are uncertain and can be probed. Similarly, the contributions from single parton scattering and double parton scattering (where the two objects come from separate processes) are not well understood, and can be disentangled. The latter can be an important uncertainty for other studies, such as the measurement of the Higgs couplings.

ATLAS has measured the associated production of a $W^{\pm}$ and a prompt J/ψ at 7 TeV [10]. It is found that the single parton scattering dominates at low p_T of the J/ψ, although considerable double parton scattering is evident. The NLO colour singlet predictions are well below the observed rates, while the colour octet has a considerably smaller colour octet contribution from non-relativistic QCD (NRQCD). The combined theory predictions fall short of the observations.

ATLAS has also measured the associated production of a Z and a prompt J/ψ at 8 TeV [11]. Both prompt and non-prompt J/ψ have been studied, with unambiguous observations in both modes, and both the azimuthal angle between the Z and the J/ψ, $\Delta\phi$, and the differential cross-section as a function of the p_T of the J/ψ have been studied. The smallest bin of $\delta\phi$ for the prompt case allows the maximum allowed rate for double parton scattering to be set. Double parton scattering makes up $29 \pm 9\%$ of the prompt production but only $8 \pm 2\%$ of the non-prompt. The p_T spectra again exceed the combined colour octet and colour singlet predictions. The associated prompt production represents $(63 \pm 13 \pm 5 \pm 10) \times 10^{-7}$ of the inclusive Z production, and the non-prompt represents $(102 \pm 15 \pm 5 \pm 3) \times 10^{-7}$ of the inclusive Z production both in the range $8.5 < p_T(J/\psi) < 100$ GeV and $|y(J/\psi)| < 2.1$.

3 B physics measurements

ATLAS has a large and established B-physics programme, with many recent additions making use of the full run 1 statistics. The recent results involve investigations of new decay channels, the decay properties and the investigation of potential CP violation and new physics using specific channels.

3.1 Decays of the $B^{\pm}$

ATLAS has produced recent results on the decay of the B_c^+ (and its charge conjugate, implied), in the decay mode $B_c^+ \to J/\psi D_s^+$ and $B_c^+ \to J/\psi D_s^{*+}$ [12]. The D_s is reconstructed in its decay to $\phi(KK)\pi^+$. The equivalent D_s^{*+} is not explicitly reconstructed using the soft π^0 or photon from the decay, but rather using a cascade vertex fit, using pointing, mass and vertex constraints. A two-dimensional fit is then performed to the fitted mass and helicity angle to measure the branching ratios in each channel and the relative transverse polarisation

$$\Gamma_{\pm\pm}(B_c^+ \to J/\psi D_s^{*+})/\Gamma(B_c^+ \to J/\psi D_s^{*+}) = 0.38 \pm 0.23\,(\text{stat.}) \pm 0.07\,(\text{syst.})\,;$$

this is in good agreement with LHCb and previous measurements. The relative branching ratio for $B_c^+ \to J/\psi D_s^+$ and $B_c^+ \to J/\psi D_s^{*+}$ relative to $B_c^+ \to J/\psi \pi^+$ are higher than in pre-LHC measurements, but for the $B_c^+ \to J/\psi D_s^+$ the LHCb measurement agrees with ATLAS; no LHCb result exists as yet for the other case.

3.2 Observation of $\Lambda_B^{\to}\psi(2S)\Lambda$

ATLAS has made the first observation of the decay mode $\Lambda_B^{\to}\psi(2S)\Lambda$ using its 8 TeV data [13]. This was measured in the kinematic range $p_T(\Lambda_b^0) > 10$ GeV and $|\eta((\Lambda_b^0)| < 2.1$. The following relative branching fraction has been measured:

$$\frac{\Gamma(\Lambda_B^0 \to \psi(2S)\Lambda)}{\Gamma(\Lambda_B^0 \to J/\psi\Lambda)} = 0.501 \pm 0.033\,(\text{stat.}) \pm 0.016\,(\text{syst.}) \pm 0.011\,(\mathcal{B})\,;$$

here, $\mathcal{B}$ is the uncertainty from the branching fraction for each of the onium states to muons.

3.3 New measurements using the decay $B_s \to J/\psi\phi$

ATLAS has presented a new determination of the CP-violating phase, ϕ_s, and the B_s meson width difference using the decays $B_s \to J/\psi\phi$ [14]. This analysis uses both the 7 TeV and 8 TeV datasets, which have then been combined. The CP violating phase is potentially sensitive to new physics contributions, while the width difference (measuring the lifetime difference between the CP even and odd eigenstates) is less sensitive to new physics but constrains the phenomenological models. As the decay mode chosen has a vector-vector final state, a time dependent flavour tagged analysis allows the two eigenstates to be resolved, characterised by three decay angles in the transversity basis. The analysis has the advantage of high statistics and relatively easy triggering, and is free of lifetime cuts. In this case, the term charge tagging refers to a determination if the initial quark leading to B_s decay was a quark or an antiquark. Opposite side tagging is used, which means the initial quark is inferred from the decay of the other associated B-hadron created in the event. The data is calibrated from the data using the self-tagging decay $B^{\pm} \to J/\psi K^{\pm}$.

The final analysis uses an unbinned maximum likelihood fit to nine physics parameters, and the three of most physics interest are shown in table 1. Both ϕ_s and $\Delta\Gamma_s$ are consistent with other measurements and with the standard model prediction, as illustrated for the combined result in figure 3. With the greatly improved lifetime resolution in the new run, these results can be greatly improved, limited only by the available trigger bandwidth for the foreseeable future.

Table 1. The three most relevant physics parameters of the nine fitted in the decay using the unbinned maxiumum likelihood fit to the 7 and 8 TeV data [14]; the results are from the paper as submitted for publication and so a minor modification from those presented at the conference.

Parameter	Value	Stat.	Syst.	units
ϕ_s	-0.098	0.084	0.040	rad
$\Delta\Gamma_s$	0.083	0.011	0.007	ps^{-1}
Γ_s	0.677	0.003	0.003	ps^{-1}

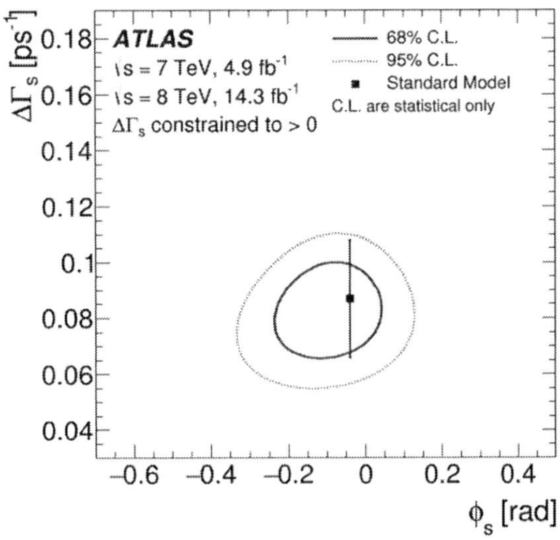

Figure 3. The combined 7 and 8 TeV results for ϕ_s and $\Delta\Gamma$ along with the Standard Model expectation [14].

References

[1] ATLAS Collaboration, JINST **3**, S08003 (2008)
[2] ATLAS Collaboration, CERN-LHCC-2012-009 (2012) https://cds.cern.ch/record/1451888
[3] ATLAS Collaboration, **CERN-PH-EP-2015-292** (2015) arXiv:1512.03657 [hep-ex]
[4] G.T. Bodwin, E. Braaten and G. P. Lepage, Phys. Rev. **D51**, 1125 (1995)
[5] M. Butenschoen and B. A. Kniehl, Nucl. Phys. Proc. Suppl. **222-224**, 151 (2012)
[6] ATLAS Collaboration, **ATLAS-CONF-2015-030** (2015) https://cds.cern.ch/record/2037967
[7] ATLAS Collaboration, Phys. Rev. Lett. **108**, 152001 (2012)
[8] ATLAS Collaboration, Phys. Rev. Lett. **113**, 212004 (2014)
[9] ATLAS Collaboration, Phys. Lett. **B740**, 199 (2015)
[10] ATLAS Collaboration, JHEP **04**, 172 (2014)
[11] ATLAS Collaboration, Eur. Phys. J. **C75**, 229 (2015)
[12] ATLAS Collaboration, Eur. Phys. J. C **76** 1 (2016)
[13] ATLAS Collaboration, Phys. Lett. B **751**, 63 (2015)
[14] The ATLAS Collaboration, CERN-PH-EP-2015-166 (2015) arXiv:1601.03297 [hep-ex]

EPJ Web of Conferences **126**, 04023 (2016) DOI: 10.1051/epjconf/201612604023
ICNFP 2015

Results from the Double Chooz experiment

Michiru Kaneda[1,a] on behalf of the Double Chooz Collaboration

[1] *Tokyo Institute of Technology,2-12-1 Oookayama, Meguro-ku, Tokyo, Japan*

Abstract. Recent results from the Double Chooz experiment on the neutrino mixing angle θ_{13} are presented. Two detectors are located at distances of 400 m and 1050 m from the reactor cores of the Chooz nuclear power plant, to measure the original neutrino flux from the reactor cores and the disappearance of neutrinos, respectively. The Far Detector has taken data since 2011 while the Near Detector started the data taking in 2014. The latest far detector only result with gadolinium capture events is $\sin^2 2\theta_{13} = 0.090^{+0.032}_{-0.029}$. Studies using hydrogen capture events also have been improved and the combined result of gadolinium and hydrogen capture events is obtained as $\sin^2 2\theta_{13} = 0.088 \pm 0.033$.

1 Introduction

In the standard three-flavor framework, neutrino oscillations are described by the three mixing angles, three mass squared differences and a CP-violating phase. While other parameters except for CP-violating phase have been measured, the mixing angle θ_{13} remained unknown until recently and reactor experiments [1–4] have proved it as non-zero value. In the reactor neutrino experiments, a disappearance of electron antineutrino, $\bar{\nu}_e$, is observed. The survival probability of $\bar{\nu}_e$ is described as:

$$P_{\bar{\nu}_e \to \bar{\nu}_e} = 1 - \sin^2 2\theta_{13} \sin^2 \left(\frac{1.27 \Delta m^2_{13} L}{E_{\bar{\nu}_e}} \right), \tag{1}$$

where L is a distance and $E_{\bar{\nu}_e}$ is energy of $\bar{\nu}_e$. The mass squared difference of Δm^2_{13} is measured by the MINOS experiment as $\Delta m^2_{13} = 2.44^{+0.90}_{-0.10}$ [5]. The reactor neutrino has energy of O(MeV), and the first maximum oscillation point is about 1 km.

$\bar{\nu}_e$ is detected by using the inverse β-decay (IBD) process: $\bar{\nu}_e + p \to e^+ + n$. The positron annihilates with an electron and observed as the prompt signal with an energy deposit related to the incoming neutrino energy as: $E \simeq E_\nu - 0.8\text{MeV}$. The neutron is captured on either Gd or H and observed as the delayed signal. Gd capture occurs after a mean time of 30 μs and emits 8 MeV signal, while H capture occurs after a mean time of 200 μs and emits 2.2 MeV signal. A background is strongly suppressed by the coincidence of these prompt and delayed signals.

2 The Double Chooz experiment

The Double Chooz experiment has two detectors in the Chooz nuclear power plant in France, which has two reactor cores producing 4.25 GW_{th} each. The Near Detector (ND) and the Far Detector (FD) are located at $\sim$ 400 m and $\sim$ 1050 m from the cores, respectively.

[a]e-mail: kaneda@hep.phys.titech.ac.jp

These two detectors were made based on the identical design. The detector consists of the Inner Detector (ID), the Inner Veto (IV) and the Outer Veto (OV). The innermost of the ID is the neutrino target (NT) filled with 10 m^3 Gd-loaded liquid scintillator. The NT is surrounded by a 55 cm thick Gd-free liquid scintillator, called γ-catcher (GC), to capture γ-rays escaping from the ND. The outside of the GC is the Buffer region filled with mineral oil and the 390 photomultiplier tubes (PMTs) placed there. The IV is a 50 cm thick liquid scintillator layer with 78 PMTs, surrounding the Buffer region. A stainless steel vessel separates the IV and the Buffer region. The OV is the detector of plastic scintillator strips covering the top of the Double Chooz detector, to identify cosmic muon background events.

Three types of backgrounds are accounted for the analysis: cosmogenetic isotopes, such as ^{9}Li, emitting a neutron with their β decay; correlated signals due to proton recoil from spallation fast neutrons and stopping muons producing a Michel-electron; and accidental background. All these events are estimated from the data and subtracted from the final IBD candidates.

The FD has worked since 2011 while the ND started to take data in 2014 and all results on θ_{13} in this presentation used only the FD data.

3 Gadolinium Analysis Result

The neutrino candidates are selected from samples from which events within 1 ms after a muon event and light noise events are rejected. The prompt and the delayed candidates are selected with energy of 0.5MeV $< E_{\text{prompt}} <$ 20MeV and 4MeV $< E_{\text{delayed}} <$ 10MeV, respectively. The coincidence selection is applied by using differences of timings (ΔT) and the positions (ΔR) of the prompt and the delayed signals as: 0.5μs $< \Delta T <$ 150μs and $\Delta R <$ 100cm. To select isolated events, the prompt events that have other events within 200 μs before and 600 μs after are rejected. For the data that has the OV information, no OV hit is required. The veto using the IV information is deployed to suppress fast neutron and accidental events. For stopping muon events, the vertex reconstruction goodness method (F_v veto) is used. Cosmogenetic events are rejected by ^{9}Li likelihood method. The details of the selection can be found in [1].

The data of 460.67 live days is used and 17351 IBD candidates are selected. Two different analysis methods are attempted for the oscillation analysis: reactor rate modulation analysis and rate + shape analysis.

For the reactor rate modulation (RRM) analysis, the data is divided into different reactor configurations: two reactor cores on, one of the reactor cores is off, and both reactor cores are off. By comparing observed and expected IBD event rates at different configurations, θ_{13} and the background rate is simultaneously extracted from the linear correlation between the observed and the expected events.

Figure 1 shows the correlation of the expected and observed event rates with the best-fit prediction in the left, and the contour plot of allowed region on (θ_{13}, B) plane in the right. The best fit result of the RRM is $\sin^2 2\theta_{13} = 0.090^{+0.034}_{-0.035}$.

The rate + shape analysis uses the prompt energy spectrum. The data is divided into 40 energy bins and compared to Monte Carlo simulation. The shape of cosmogenetic background is estimated from ^{9}Li vetoed events. The correlated background is assumed to be flat with energy and extrapolated from the higher energy region. The accidental background is estimated from the off-time window events.

Figure 2 shows the prompt energy spectrum and the ratio of the data to the non-oscillation prediction. The best fit value is $\sin^2 2\theta_{13} = 0.090^{+0.032}_{-0.029}$.

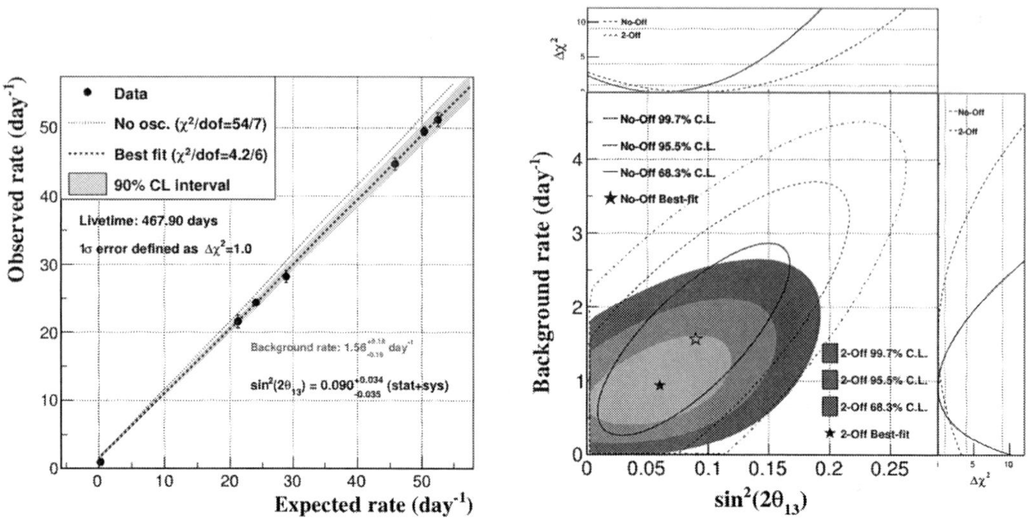

Figure 1. RRM fit including the reactor-off data for Gd analysis. The left plot shows (θ_{13}, B) fit, superimposed to the null oscillation hypothesis assuming the background estimates. The right plot shows 68.3, 95.5 and 99.7 % C.L. allowed regions on (θ_{13}, B) plane.

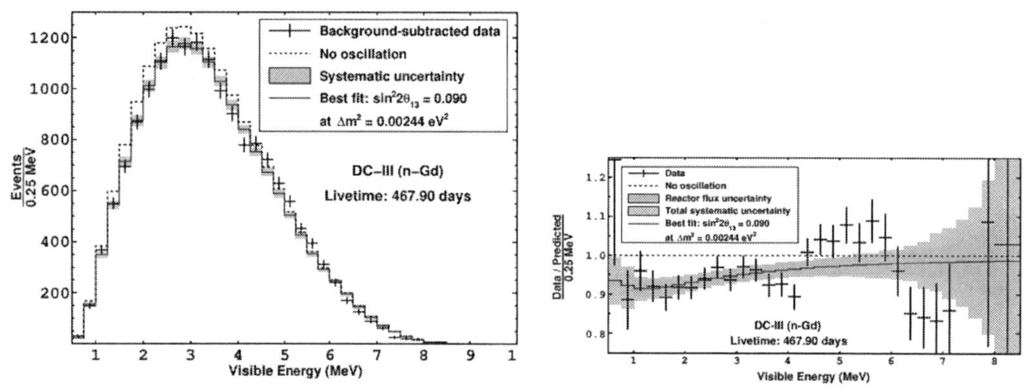

Figure 2. Prompt energy spectrum of IBD candidate of Gd analysis. In the left plot, black points shows data, blue dashed line shows the prediction of non-oscillation prediction, and red line shows the best fit. Background components of accidental, cosmogenetic (^{9}Li + ^{8}He) and fast neutron and stopping muons are also shown. The right plot shows the ratio of the data to the non-oscillation prediction.

4 Hydrogen Analysis Result

H events have more statistics than Gd events, while more accidental background events appear because the delayed energy is lower and ΔT of prompt and delayed events is longer.

The basic event selections are similar with the Gd analysis. The prompt and the delayed candidates are selected with energy of 1.0MeV $< E_{\text{prompt}} <$ 20MeV and 1.3MeV $< E_{\text{delayed}} <$ 3MeV, respectively. The correlation selections are 0.5μs $< \Delta T <$ 800μs and $\Delta R <$ 120cm. For isolation cut, no event is required 800 μs before and 900 μs after prompt signal. OV-veto, IV-veto and F_v veto are also applied as Gd analysis.

To improve the background rejection, an artificial neutral network (ANN) method has been developed with three parameters of delayed energy, ΔR and ΔT. ANN method reduces $\sim$ 95 % of backgrounds compared to previous cut-based analysis [6].

In addition, flash-ADC information for each PMT is used for multiplicity pulse shape (MPS) veto. In the fast neutron events, additional prompt recoils can happen and it makes additional pulses. Rejecting events with such an additional pulses results 25 % reduction of the fast neutron events.

Further details of H analysis are described in [2].

For the H analysis, the data of 452.72 live days is used and 31835 IBD candidates are selected. The best fit value is $\sin^2 2\theta_{13} = 0.095^{+0.038}_{-0.039}$ for the RRM analysis, and $\sin^2 2\theta_{13} = 0.124^{+0.030}_{-0.039}$ for the rate + shape analysis.

The combined result of Gd and H events is also obtained for the RRM analysis. Figure 3 shows the correlation plot and the counter plot for Gd and H combined analysis.

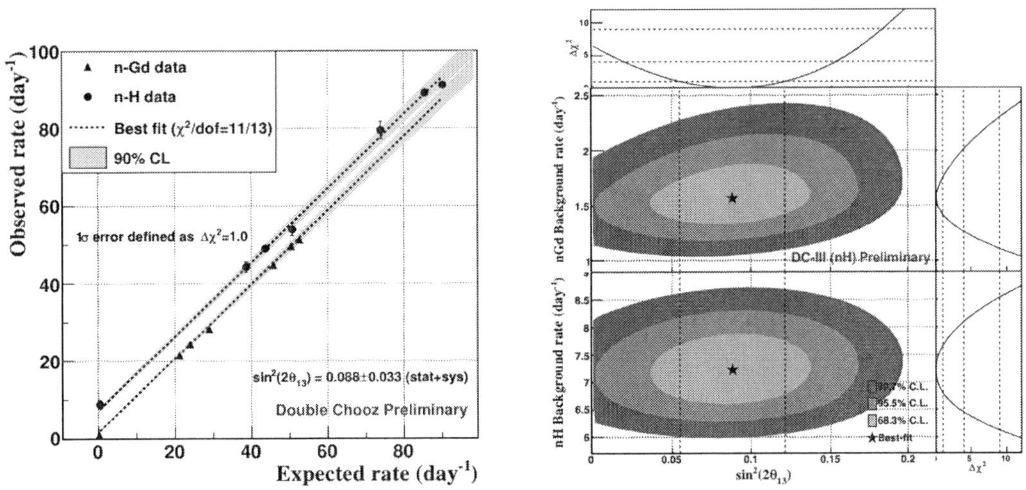

Figure 3. Gd and H combined RRM RRM fit including the reactor-off data. The left plot shows (θ_{13}, B) fit, superimposed to the null oscillation hypothesis assuming the background estimates. The right plot shows 68.3, 95.5 and 99.7 % C.L. allowed regions on (θ_{13}, B) plane.

The best value is obtained as $\sin^2 2\theta_{13} = 0.088 \pm 0.030$.

5 Spectrum distortion

A spectrum distortion is found above 4 MeV of the prompt energy as shown in Figure 2. There is an excess around 5 MeV in observed IBD rate, while it was confirmed as insignificant effect on θ_{13} results because the statistics and the oscillation effect are much more in the lower energy region. And a strong correlation with the reactor power is reported in [1].

6 Conclusion

Double Chooz published θ_{13} measurement results for Gd and H analysis. In each analysis, the RRM analysis and the rate + shape analysis are attempted to obtain θ_{13} value. Gd analysis shows

$\sin^2 2\theta_{13} = 0.090^{+0.034}_{-0.035}$ and $\sin^2 2\theta_{13} = 0.090^{+0.032}_{-0.029}$ for the RRM analysis and the rate + shape analysis, respectively. Gd and H combined result is obtained as $\sin^2 2\theta_{13} = 0.088 \pm 0.030$.

The ND data taking started in 2014. The current large uncertainty on the neutrino flux will be reduced by the FD data and the uncertainty on $\sin^2 2\theta_{13}$ will be reduced as 10 % level in a few years.

References

[1] Y. Abe *et. al.*, JHEP **1410**, 086 (2014).
[2] Y. Abe *et. al.*, arXiv:1510.08937 (2015).
[3] F.P. An *et. al.*, *Phys. Rev. Lett.* **115**, 11 (2015).
[4] J.K. Ahn *et. al.*, *Phys. Rev. Lett.* **108**, 191802 (2012).
[5] P. Adamson *et. al.*, *Phys. Rev. Lett.* **112**, 191801 (2014).
[6] Y. Abe *et. al.*, *Phys. Lett.* B. **723**, 66-70 (2013).

EPJ Web of Conferences **126**, 04024 (2016) DOI: 10.1051/epjconf/201612604024
ICNFP 2015

Oblique corrections in the Dine-Fischler-Srednicki axion model

Alisa Katanaeva[1,a] and Domènec Espriu[2]

[1] *V. A. Fock Department of Theoretical Physics, Saint-Petersburg State University, 1 ul. Ulyanovskaya, 198504, Russia*
[2] *Departament d'Estructura i Constituents de la Matèria, Institut de Ciències del Cosmos (ICCUB), Universitat de Barcelona, Martí Franquès 1, 08028 Barcelona, Spain*

Abstract. In the Minimal Standard Model (MSM) there is no degree of freedom for dark matter. There are several extensions of the MSM introducing a new particle - an invisible axion, which can be regarded as a trustworthy candidate at least for a part of the dark matter component. However, as it is extremely weakly coupled, it cannot be directly measured at the LHC. We propose to explore the electroweak sector indirectly by considering a particular model that includes the axion and derive consequences that could be experimentally tested.
We discuss the Dine-Fischler-Srednicki (DFS) model, which extends the two-Higgs doublet model with an additional Peccei-Quinn symmetry and leads to a physically acceptable axion. The non-linear parametrization of the DFS model is exploited in the generic case where all scalars except the lightest Higgs and the axion have masses at or beyond the TeV scale. We compute the oblique corrections and use their values from the electroweak experimental fits to put constraints on the mass spectrum of the DFS model.

1 Introduction

In this paper we reexamine the Dine-Fischler-Srednicki-Zhitnitsky (DFSZ) model. It is the two-Higgs doublet model (2HDM) containing an additional singlet, endowed with a Peccei-Quinn (PQ) symmetry. Introduction of the PQ symmetry in the Standard Model (SM) leads to the solution of the strong CP problem but induces the necessity of two Higgs doublets and the presence of an axion. The last becomes physically acceptable in the DFSZ model. From one point of view, the model includes a lot of new physics coming from the 2HDM. From another, an invisible axion is a possible candidate at least for a part of the dark matter. Both these reasons, as well as the interplay between the 2HDM content and the axion, make this model interesting to study.

The discovery of a Higgs-like particle with $m_h \sim 126$ GeV and the development of experiments, now probing the predictions of the standard electroweak theory with sufficient accuracy, impose the constraints on any potential new physics that might exist at higher energies. Nevertheless, there is still a room for a wide variety of the 2HDM scenarios. In the same time the introduction of an axion restricts the number of possibilities, making the phenomenological consequences of the DFSZ model more rigorous.

[a] e-mail: alice.katanaeva@gmail.com

2 Fields and symmetries

The DFSZ model originated as a solution to several problems. Firstly, Peccei and Quinn [1] proposed to introduce an additional global $U(1)_{PQ}$ symmetry to the SM to rotate away the CP violating term. Then, the introduction of two Higgs doublets is necessary to make the SM invariant under the $U(1)_{PQ}$ transformation. The last thing is to make a new Goldstone boson of the necessarily broken PQ symmetry - an axion - invisible, meaning it should have a very light mass and be very weakly coupled to ordinary matter. It wasn't achieved in the original PQ model. However, later Dine, Fischler and Srednicki [2] and Zhitnitsky [3] independently proposed a generalization of the Peccei-Quinn scheme with a harmless axion. Addition of a complex scalar field gives an axion the right properties.

So, we work with two Higgs doublets and a complex scalar singlet:

$$\phi_1 = \begin{pmatrix} \alpha_+ \\ \alpha_0 \end{pmatrix}; \quad \phi_2 = \begin{pmatrix} \beta_+ \\ \beta_0 \end{pmatrix}; \quad \phi. \tag{1}$$

The vevs of the fields are

$$\langle\phi_1\rangle = \begin{pmatrix} 0 \\ v_1 \end{pmatrix}; \quad \langle\phi_2\rangle = \begin{pmatrix} 0 \\ v_2 \end{pmatrix}; \quad \langle\phi\rangle = v_\phi. \tag{2}$$

We can define the usual electroweak vacuum expectation value as $v^2 = (v_1^2 + v_2^2)/2 = 246$ GeV and the well-known 2HDM parameter $\tan\beta = v_2/v_1$. The DFSZ model includes a 2HDM of type II: ϕ_1 couples only to right-handed charge 2/3 quarks, ϕ_2 couples only to the right-handed charge $-1/3$ quarks and to right-handed charged leptons. The Yukawa terms have the structure:

$$\mathcal{L}_Y = G_1 \bar{q}_L \tilde{\phi}_1 u_R + G_2 \bar{q}_L \phi_2 d_R + G_3 \bar{l}_L \phi_2 e_R + h.c., \tag{3}$$

and similarly for other quarks and leptons. Here $\tilde{\phi}_i = i\tau_2\phi_i^*$.

For the potential we choose the one respecting CP, $SU(2)\times U(1)$ and $U(1)_{PQ}$ symmetries:

$$V(\phi,\phi_1,\phi_2) = \lambda_\phi(\phi^*\phi - V_\phi^2)^2 + \lambda_1(\phi_1^\dagger\phi_1 - V_1^2)^2 + \lambda_2(\phi_2^\dagger\phi_2 - V_2^2)^2 + \lambda_3(\phi_1^\dagger\phi_1 - V_1^2 + \phi_2^\dagger\phi_2 - V_2^2)^2 +$$
$$+ \lambda_4\left[(\phi_1^\dagger\phi_1)(\phi_2^\dagger\phi_2) - (\phi_1^\dagger\phi_2)(\phi_2^\dagger\phi_1)\right] + (a\phi_1^\dagger\phi_1 + b\phi_2^\dagger\phi_2)\phi^*\phi + c(\phi_1^\dagger\phi_2\phi^2 + \phi_2^\dagger\phi_1\phi^{*2}). \tag{4}$$

Furthermore, we consider an additional possible symmetry - the custodial symmetry. It is an approximate symmetry in the SM keeping the equality $\rho \equiv \frac{M_W^2}{M_Z^2\cos^2\theta_W} \simeq 1$ even when $g' \neq 0$, with θ_W being the Weinberg angle and M_W, M_Z – the electroweak gauge boson masses.

To make more evident such global symmetries of the potential of the DFSZ model one can introduce a matrix notation [4]. Firstly, we construct the 2×2 matrices from the fields of Higgs doublets:

$$\Phi_{12} = (\tilde{\phi}_1\ \phi_2) = \begin{pmatrix} \alpha_0^* & \beta_+ \\ -\alpha_- & \beta_0 \end{pmatrix}, \quad \Phi_{21} = (\tilde{\phi}_2\ \phi_1) = \begin{pmatrix} \beta_0^* & \alpha_+ \\ -\beta_- & \alpha_0 \end{pmatrix} = \tau_2\Phi_{12}^*\tau_2. \tag{5}$$

Secondly, we consider the following combinations:

$$I = \Phi_{12}^\dagger\Phi_{12} = \begin{pmatrix} \phi_1^\dagger\phi_1 & \tilde{\phi}_1^\dagger\phi_2 \\ -\phi_1^\dagger\tilde{\phi}_2 & \phi_2^\dagger\phi_2 \end{pmatrix}, \quad J = \Phi_{12}^\dagger\Phi_{21} = \begin{pmatrix} \phi_2^\dagger\phi_1 & 0 \\ 0 & \phi_2^\dagger\phi_1 \end{pmatrix}. \tag{6}$$

Their vevs are

$$\langle I\rangle = \begin{pmatrix} v_1^2 & 0 \\ 0 & v_2^2 \end{pmatrix}, \quad \langle J\rangle = v_1 v_2. \tag{7}$$

This vacuum is not invariant under the full group $SU(2)_L \times SU(2)_R$. However, if $v_1 = v_2$, then $\langle I \rangle$ is proportional to the 2×2 identity matrix and the vacuum preserves a group $SU(2)_V$ (the V stands for "vectorial"), corresponding to identical matrices, i.e. $L = R$. This remaining group preserved by the vacuum is the custodial-symmetry group.

The last thing is to define the constant matrix W,

$$W = (V_1^2 + V_2^2)\frac{\mathbf{I}}{2} + (V_1^2 - V_2^2)\frac{\tau_3}{2} = \begin{pmatrix} V_1^2 & 0 \\ 0 & V_2^2 \end{pmatrix}, \tag{8}$$

and then, the potential can be written as:

$$V(\phi, I, J) = \lambda_\phi(\phi^*\phi - V_\phi^2)^2 + \frac{\lambda_1}{4}\{\mathrm{Tr}\,[(I-W)(1+\tau_3)]\}^2 + + \frac{\lambda_2}{4}\{\mathrm{Tr}\,[(I-W)(1-\tau_3)]\}^2 + \tag{9}$$
$$+ \lambda_3\,[\mathrm{Tr}(I-W)]^2 + \frac{\lambda_4}{4}\mathrm{Tr}\left[I^2 - (I\tau_3)^2\right] + \frac{1}{2}\mathrm{Tr}\,[(a+b)I + (a-b)I\tau_3]\,\phi^*\phi + \frac{c}{2}\mathrm{Tr}(J\phi^2 + J^\dagger\phi^{*2}).$$

A custodial global $SU(2)_L \times SU(2)_R$ transformation acts on our fields as

$$\Phi_{ij} \to L\Phi_{ij}R^\dagger, \quad I \to RIR^\dagger, \quad J \to J. \tag{10}$$

If $SU(2)_L \times SU(2)_R$ is to be a symmetry, the parameters of the potential have to be set according to the custodial relations: $\lambda_1 = \lambda + \lambda_B$, $\lambda_2 = \lambda - \lambda_B$, $\lambda_4 = 2\lambda + \lambda_{4B}$, $V_1^2 = V^2 + V_B^2$, $V_2^2 = V^2 - V_B^2$, $a + a_B = b$. All the "B" parameters vanish in the limit of custodial symmetry. In total, there are 11 parameters of which 7 are custodial preserving and 4 are custodial breaking.

3 Effective potential

In this section we integrate out the heavy scalars in order to build a low-energy effective theory at the TeV scale with an axion and a light Higgs.

We decompose the matrix-valued Φ_{12} field in the following form $\Phi_{12} = \mathcal{U}\mathcal{M}_{12}$. $\mathcal{U} = \exp\left(i\frac{\vec{G}\cdot\vec{\tau}}{v}\right)$ is a 2×2 matrix containing the three Goldstone bosons G_i associated to the breaking of $SU(2)_L \times U(1)_Y$ to $U(1)_{em}$. Note that the matrices I and J entering the DFSZ potential are actually independent of $\mathcal{U}$. The effective potential then does depend only on the degrees of freedom contained in $\mathcal{M}_{12}$.

There is also the singlet field in the scalar potential, it can be parametrized as: $\phi = \rho + iG_\phi$. The phase of ϕ does not drop from the potential automatically because of the c term in (4). G_ϕ mixes with the usual 0^- scalar from the 2HDM. To have a well-defined massless state we need to find a suitable phase both in $\mathcal{M}_{12}$ and in ϕ that drops from the potential.

It is straightforward to define the phases from the requirement that the kinetic terms are diagonal and exhibit the canonical normalization. Then, the non-linear parametrization of Φ_{12} reads as[1]

$$\Phi_{12} = \mathcal{U}M_{12}U_a, \tag{11}$$

with a unitary matrix containing the axion being $U_a = \exp\left(i\frac{2a_\phi}{\sqrt{v_\phi^2 + v^2 s_{2\beta}^2}}\begin{pmatrix} s_\beta^2 & 0 \\ 0 & c_\beta^2 \end{pmatrix}\right)$ and the heavy scalars collected in the matrix

$$M_{12} = \sqrt{2}\begin{pmatrix} (v+H)c_\beta - \left(S - \frac{iv_\phi}{\sqrt{v_\phi^2 + v^2 s_{2\beta}^2}}\tilde{A}_0\right)s_\beta & \sqrt{2}H_+ c_\beta \\ \sqrt{2}H_- s_\beta & (v+H)s_\beta + \left(S + \frac{iv_\phi}{\sqrt{v_\phi^2 + v^2 s_{2\beta}^2}}\tilde{A}_0\right)c_\beta \end{pmatrix}. \tag{12}$$

[1]Here we introduce the short-hand notation $s_{m\beta}^n \equiv \sin^n(m\beta)$ and $c_{m\beta}^n \equiv \cos^n(m\beta)$.

The field redefinition we use is as follows:

$$v + H = \frac{c_\beta}{\sqrt{2}}\Re[\alpha_0] + \frac{s_\beta}{\sqrt{2}}\Re[\beta_0], \quad S = -\frac{s_\beta}{\sqrt{2}}\Re[\alpha_0] + \frac{c_\beta}{\sqrt{2}}\Re[\beta_0], \quad H_\pm = \frac{c_\beta\beta_\pm - s_\beta\alpha_\pm}{2}. \tag{13}$$

The singlet field is non-linearly parametrized as

$$\phi = \left(v_\phi + \rho - i\frac{v s_{2\beta}}{\sqrt{v_\phi^2 + v^2 s_{2\beta}^2}}\tilde{A}_0\right)\exp\left(i\frac{a_\phi}{\sqrt{v_\phi^2 + v^2 s_{2\beta}^2}}\right). \tag{14}$$

The fields of H, S and ρ are not mass eigenstates but they have vanishing vevs. The mass eigenstates are defined through the rotation matrix R:

$$H = \sum_{i=1}^{3} R_{Hi}h_i, \quad S = \sum_{i=1}^{3} R_{Si}h_i, \quad \rho = \sum_{i=1}^{3} R_{\rho i}h_i. \tag{15}$$

H, S are also called interaction eigenstates. For instance, H has the same coupling to the gauge fields as the SM Higgs boson, but the Higgs mass $m_h \simeq 126$ GeV is attributed to the lightest of h_i states.

The construction of the effective Lagrangian for the DFSZ model goes with inclusion of additional light particles explicitly as dynamical states:

$$\mathcal{L} = \frac{v^2}{4}\left(1 + 2g_1\frac{h}{v} + g_2\frac{h^2}{v^2} + ...\right)\mathrm{Tr}\,\mathcal{D}_\mu\mathcal{U}^\dagger\mathcal{D}^\mu\mathcal{U} + \left(\frac{v_\phi^2}{v_\phi^2 + v^2\sin^2 2\beta}\right)\partial_\mu a_\phi\partial^\mu a_\phi + \frac{1}{2}\partial_\mu h\partial^\mu h -$$
$$- V(h) + \sum_{i=0}^{13} a_i\left(\frac{h}{v}\right)O_i + \mathcal{L}_{ren}, \tag{16}$$

where the O_i is a set of local gauge invariant operators [4],

$$\mathcal{D}_\mu\mathcal{U} = D_\mu\mathcal{U} + \mathcal{U}(\partial_\mu U_a)U_a^\dagger, \quad V(h) = \frac{m_h^2}{2}h^2 - d_3(\lambda v)h^3 - d_4\frac{\lambda}{4}h^4, \tag{17}$$

$$\mathcal{L}_{ren} = \frac{c_1}{v^4}(\partial_\mu h\partial^\mu h)^2 + \frac{c_2}{v^2}\mathrm{Tr}\,\mathcal{D}_\mu\mathcal{U}^\dagger\mathcal{D}^\mu\mathcal{U} + \frac{c_3}{v^2}(\partial_\mu h\partial^\nu h)\,\mathrm{Tr}\,\mathcal{D}^\mu\mathcal{U}^\dagger\mathcal{D}_\nu\mathcal{U}. \tag{18}$$

The terms in $\mathcal{L}_{ren}$ are required for renormalizability [5] at the one-loop level and play no role in the discussion. The couplings a_i are functions of h/v and can be regularly expanded. Their constant parts $a_i(0)$ are related to the electroweak precision parameters.

4 Mass eigenstates

Having defined the fields we proceed with the description of the mass spectrum of the model (following [6]). We have two doublets and a singlet, so a total of $4 + 4 + 2 = 10$ spin-zero particles. Three particles are eaten by the gauge bosons and 7 scalars fields are left in the spectrum: two charged Higgs bosons, two 0^- states and three neutral 0^+ states.

The charged Higgs bosons have a mass $m_{H_\pm}^2 = 8\left(\lambda_4 v^2 + \frac{c v_\phi^2}{s_{2\beta}}\right)$. In the 0^- sector A_0 and G_ϕ fields mix forming the massless state, the axion $a_\phi = \frac{v s_{2\beta}A_0 + v_\phi G_\phi}{\sqrt{v_\phi^2 + v^2 s_{2\beta}^2}}$, and the massive state $\tilde{A}_0 = \frac{v_\phi A_0 - v s_{2\beta}G_\phi}{\sqrt{v_\phi^2 + v^2 s_{2\beta}^2}}$, with $m_{\tilde{A}_0}^2 = 8c\left(\frac{v_\phi^2}{s_{2\beta}} + v^2 s_{2\beta}\right)$.

 DOI: 10.1051/epjconf/201612604024

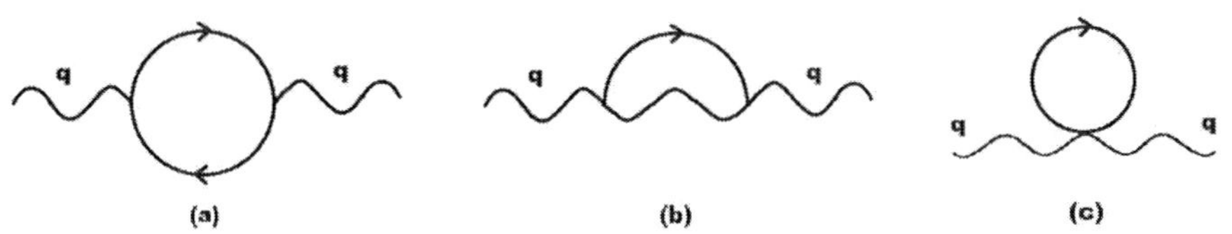

Figure 1. Types of Feynman diagrams occurring in the calculation of the vacuum polarizations.

As was mentioned before, there is mixing in the 0^+ sector. We mark the corresponding 0^+ mass eigenstates as h_i. The mass matrix can be diagonalized in the limit of large v_ϕ, which is astrophysically constrained to be at least of order 10^7 GeV. In the paper [6] an expansion in v/v_ϕ is carried out to the second order to get the masses of h_i. It is argued that the nominal expansion in powers of v/v_ϕ is applicable in a number of cases.

The cases of special interest are the ones with parameters a, b or c of order $O(v^2/v_\phi^2)$ but $c \gg \lambda_i v^2/v_\phi^2$ or $c \sim \lambda_i v^2/v_\phi^2$. For instance, in the last case ρ is a mass eigenstate with mass $m_{h_3}^2 = 4\lambda_\phi v_\phi^2$. The two remaining masses are

$$m_{h_1,h_2}^2 = 8v^2\left(K \mp \sqrt{K^2 - L}\right), \tag{19}$$

where $K = 2\left(\lambda_1 c_\beta^2 + \lambda_2 s_\beta^2 + \lambda_3\right) + \frac{v_\phi^2 c}{2v^2 s_{2\beta}}$, $L = 8\left[(\lambda_1\lambda_2 + \lambda_1\lambda_3 + \lambda_2\lambda_3)\, s_{2\beta}^2 + \frac{v_\phi^2 c}{2v^2 s_{2\beta}}\left(\lambda_1 c_\beta^4 + \lambda_2 s_\beta^4 + \lambda_3\right)\right]$.

These cases can provide much richer phenomenology (with h_2, charged Higgs and $\tilde{A}_0$ at the weak scale) than others not discussed here. One can notice that the presence of these light states requires that some couplings are rather small which my seem odd or fine-tuned. For a discussion on the 'naturalness' of this possibility see [7].

5 Oblique corrections

5.1 Definitions and experimental measurements

If the new physics scale is significantly higher than the electroweak scale, new physics effects from virtual particles in loops are expected to contribute predominantly through vacuum polarization corrections to the electroweak precision observables.

One can parametrize possible departures from the SM with the so-called oblique parameters. In our calculations we use the parameters ϵ_1, ϵ_2, and ϵ_3 defined in [8] as follows:

$$\epsilon_1 \equiv \frac{1}{M_W^2}\left[A^{33}(0) - A^{11}(0)\right], \quad \epsilon_2 = F^{11}(M_W^2) - F^{33}(M_W^2), \quad \epsilon_3 = \frac{c}{s}F^{30}(M_Z^2). \tag{20}$$

A^{ij} and F^{ij} are the coefficients in the vacuum-polarization tensors

$$\Pi_{\mu\nu}^{ij}(q) = -ig_{\mu\nu}\left[A^{ij}(0) + q^2 F^{ij}(q^2)\right] + q_\mu q_\nu \text{ terms}, \tag{21}$$

where ij may be either WW, W_3W_3 or W_3B and q_μ is the four-momentum of the gauge boson. In the effective theory ϵ_1, ϵ_2, and ϵ_3 receive one-loop contributions from the leading $O(p^2)$ term and the tree-level contributions from the $a_i(0)$:

$$\epsilon_1 = 2a_0(0) + \dots, \quad \epsilon_2 = -g^2 a_8(0) + \dots, \quad \epsilon_3 = -g^2 a_1(0) + \dots, \tag{22}$$

 DOI: 10.1051/epjconf/201612604024

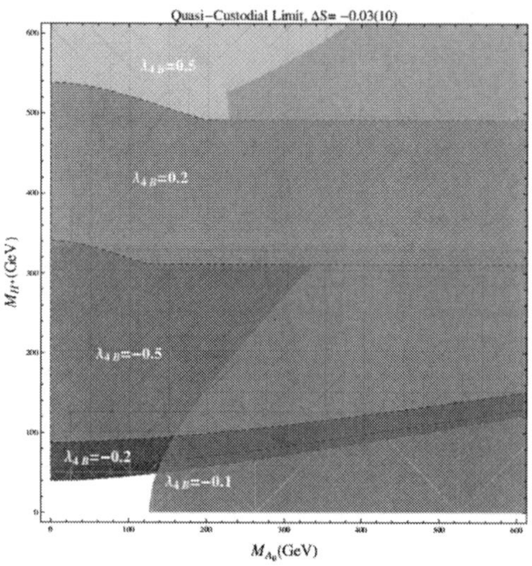

Figure 2. Exclusion plot imposed by the constraint from ΔS on the $\tilde{A}_0$ and $H_\pm$ in the 'quasi-custodial' limit for different values of the symmetry breaking parameter λ_{4B}. Grey regions are excluded by stability.

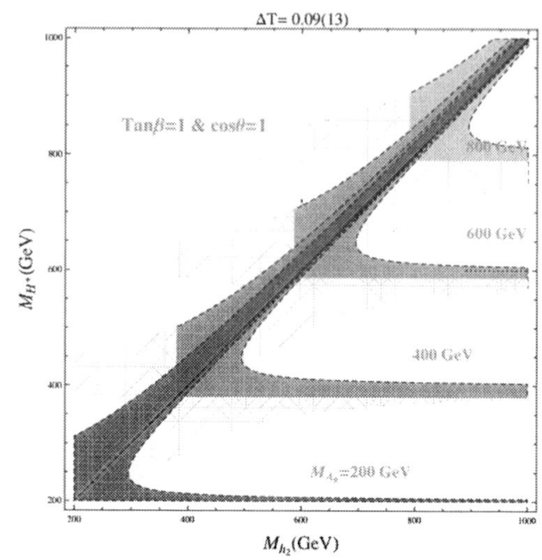

Figure 3. Exclusion plot imposed by the constraint from ΔT on the h_2 and $H_\pm$. The successive horizontal bands correspond to different values of m_{A_0}. The stability bounds are implemented.

where the ellipses symbolize the one-loop $\frac{v^2}{4}\,\mathrm{Tr}\, D_\mu \mathcal{U}^\dagger D^\mu \mathcal{U}$ contributions.

The Feynman diagrams which contribute to the vacuum polarization in the DFSZ model are depicted in Fig. 1. The wavy lines correspond to gauge bosons and the usual ones to various scalar fields. Considering the gauge invariant kinetic term $\mathcal{L}_{\rm kin} = \frac{1}{2}(\partial_\mu\phi)^*\partial^\mu\phi + \frac{1}{4}\,\mathrm{Tr}\left[(D_\mu\Phi_{12}^\dagger)D^\mu\Phi_{12}\right]$, one can get all types of interactions and consequently all contributions to ϵ_i.

Experimental constraints on the oblique parameters are obtained from the global fits of the electroweak sector of the SM. For a long time, such fits have been used to exploit measurements of electroweak precision observables at lepton colliders (LEP, SLC), together with measurements at hadron colliders (Tevatron, LHC), and accurate theoretical predictions at multi-loop level, to constrain free parameters of the SM, such as the Higgs and top masses. Today, all fundamental SM parameters entering these fits are experimentally determined, including information on the Higgs couplings, and the global fits are used as powerful tools to assess the validity of the theory and to constrain scenarios for new physics.

While performing global fits([9],[10]), it is more common to work with another set of parameters S, T and U determined firstly in [11]. The connection between S, T and U, defined relative to the SM ($\Delta T = T - T^{SM}$, *etc.*), and ϵ_i is as follows:

$$\Delta T = \frac{\epsilon_1 - \epsilon_1^{SM}}{\alpha}, \quad \Delta U = -\frac{4s_W^2(\epsilon_2 - \epsilon_2^{SM})}{\alpha}, \quad \Delta S = \frac{4s_W^2(\epsilon_3 - \epsilon_3^{SM})}{\alpha}. \tag{23}$$

Here, $\alpha = e^2/(4\pi) = g^2 s_W^2/(4\pi)$ is the fine-structure constant, $s_W = \sin\theta_W$ is the sine of the weak mixing angle θ_W.

DOI: 10.1051/epjconf/201612604024

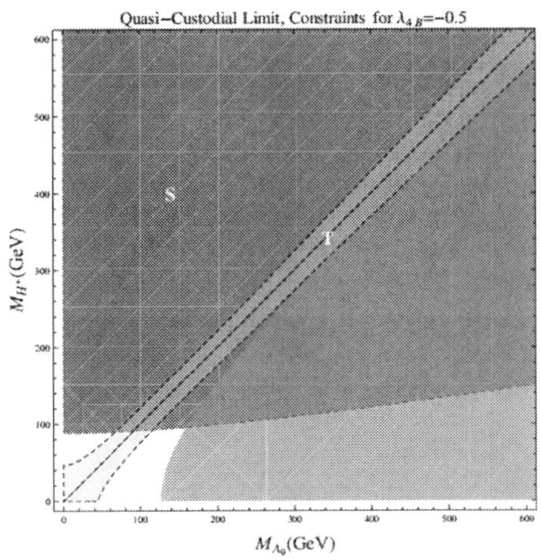

Figure 4. Restrictions both from ΔS and ΔT in the 'quasi-custodial' limit for $\lambda_{4B} = -0.5$. Grey regions are excluded by stability.

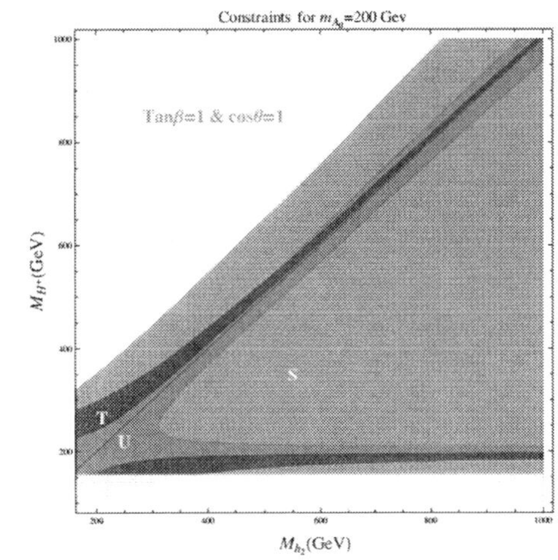

Figure 5. Restrictions from ΔS, ΔT and $\Delta U \simeq 0$ on the mass spectrum. The stability bounds are implemented.

The experimental determinations of ΔS, ΔT and ΔU are:

$$\text{from the Gfitter group [9]: } \Delta S = 0.05 \pm 0.11, \quad \Delta T = 0.09 \pm 0.13, \quad \Delta U = 0.01 \pm 0.11; \tag{24}$$

$$\text{from the PD paper [10]: } \Delta S = -0.03 \pm 0.10, \quad \Delta T = 0.01 \pm 0.12, \quad \Delta U = 0.05 \pm 0.10. \tag{25}$$

ΔU is considered to be very small in most new physics models and therefore often set to zero in the global fits. This changes the experimental limits on ΔS, ΔT a bit due to correlations between the parameters. The relevant constraints in the $\Delta U = 0$ scenario are [9]:

$$\Delta S|_{\Delta U=0} = 0.06 \pm 0.09, \quad \Delta T|_{\Delta U=0} = 0.1 \pm 0.07. \tag{26}$$

5.2 Spectrum implications

With both theoretical expressions for the oblique parameters and experimental bounds on them, we can explore the allowed range of masses. As we are basically interested in the possibility of obtaining a lightish spectrum, we discuss the case where the c parameter scales as v^2/v_ϕ^2.

We will assume two types of settings. First, a 'quasi-custodial' one, which means that the custodial symmetry is broken only via the coupling $\lambda_{4B} = \lambda_4 - 2\lambda$ being non-zero. The rotation matrix of 0^+ states is equal to unity. The mass spectrum gets an additional restriction: $m_{H_\pm}^2 = m_{h_2}^2 + 8v^2\lambda_{4B}$. In Fig. 2 we provide an exclusion plot from the constraints on ΔS as an example. Notice, that the negative values of λ_{4B} allow much lighter spectra than the positive ones.

Then, the general setting, when the three masses $m_{\tilde{A}_0}$, $m_{H_\pm}$ and m_{h_2} are unrelated, except for the eventual lack of stability of the potential. The rotation matrix can be different form the identity. However, experimentally the parameter $\cos\theta$, appearing on the diagonal of the rotation matrix, is known [6] to be very close to one. We present a range of masses allowed by the constraints on ΔT in Fig. 3.

Furthermore, we depict constraints from several oblique parameters simultaneously. For the particular choice of λ_{4B} in the 'quasi-custodial' setting we have the exclusion plot both from ΔS and ΔT in Fig. 4. ΔT significantly narrows the allowed region, while ΔU gives nothing new and is not present in the plot. For $\lambda_{4B} = 0.2$ the possible range would start at about 1 TeV range, while for negative λ_{4B} it goes down to 100 GeV. If we completely give up the custodial symmetry, ΔT determines the most severe restrictions on the spectrum generally. However, $\Delta U \simeq 0$ scenario brings even more restrictions, see Fig. 5. As ΔU gets closer to exact zero its region shrinks to two lines, which overlap with ΔT region only asymptotically at infinity.

6 Conclusions

The nature of electroweak symmetry breaking keeps being an important issue in particle physics today. The Standard Model of particle physics contains a mechanism for electroweak symmetry breaking, and the discovery of the Higgs boson at the LHC proves its consistency. However, the Minimal Standard Model still has a several well-known problems. One of them being the absence of the degree of freedom for dark matter.

The Dine-Fischler-Srednicki-Zhitnitsky model is a model with the electroweak symmetry breaking pattern, similar to the SM. The model contains an invisible axion which is an interesting candidate for dark matter. Being extremely weakly coupled, the axion cannot be directly detected at the LHC. Hence, we make an investigation what indirect consequences of the axion presence can be seen experimentally.

We consider the constraints from electroweak precision parameters, expressed in terms of the oblique parameters, to get the restrictions on the model spectrum. The large scale, appearing in the DFSZ model to make the axion nearly invisible, seems to generate a very heavy and inaccessible spectrum of the new physics. However, we discuss cases in which a rather light spectrum appears, with even a possibility to be tested at the LHC.

References

[1] R. D. Peccei and H. R. Quinn, Phys. Rev. Lett. **38** (1977) 1440; Phys. Rev. D16 (1977) 1791.

[2] M. Dine, W. Fischler and M. Srednicki, Phys. Lett. B **104**, 199 (1981).

[3] A.R. Zhitnitsky, Sov. J. Nucl. Phys. **31**, 260 (1980) [Yad. Fiz. **31**, 497 (1980)].

[4] P. Ciafaloni and D. Espriu, Phys. Rev. D **56**, 1752 (1997) [hep-ph/9612383].

[5] R.L. Delgado, A. Dobado and F. J. Llanes-Estrada, JHEP **1402**, 121 (2014); D. Espriu, F. Mescia and B. Yencho, Phys. Rev. D **88**, 055002 (2013) [arXiv:1307.2400 [hep-ph]].

[6] D. Espriu, F. Mescia and A. Renau, arXiv:1503.0295 [hep-ph].

[7] R. R. Volkas, A. J. Davies and G. C. Joshi, Phys. Lett. B **215**, 133 (1988); R. Foot, A. Kobakhidze, K. L. McDonald and R. R. Volkas, Phys. Rev. D **89**, no. 11, 115018 (2014) [arXiv:1310.0223 [hep-ph]].

[8] G. Altarelli and R. Barbieri, Phys. Lett. B **253**, 161 (1990).

[9] M. Baak, J. Cuth, J. Hallec, A. Hoecker, R. Kogler, K. Monig, M. Schott, J. Stelzer, Eur. Phys. J. C **74**, 3046 (2014) [arXiv:1407.3792v1 [hep-ph]].

[10] K.A. Olive et al. (Particle Data Group), Chin. Phys. C **38**, 090001 (2014).

[11] M. Peskin and T. Takeuchi, Phys. Rev. Lett. **65**, 1964 (1990).

 DOI: 10.1051/epjconf/201612604025

Magnetic monopole search with the MoEDAL test trapping detector

Akshay Katre (On behalf of the MoEDAL Collaboration)[1,a]

[1] *Particle physics department, University of Geneva, 1211 Geneva 4, Switzerland*

Abstract. MoEDAL is designed to search for monopoles produced in high-energy Large Hadron Collider (LHC) collisions, based on two complementary techniques: nuclear-track detectors for high-ionisation signatures and other highly ionising avatars of new physics, and trapping volumes for direct magnetic charge measurements with a superconducting magnetometer. The MoEDAL test trapping detector array deployed in 2012, consisting of over 600 aluminium samples, was analysed and found to be consistent with zero trapped magnetic charge. Stopping acceptances are obtained from a simulation of monopole propagation in matter for a range of charges and masses, allowing to set model-independent and model-dependent limits on monopole production cross sections. Multiples of the fundamental Dirac magnetic charge are probed for the first time at the LHC.

1 Introduction

Magnetic monopoles were first postulated by Dirac in 1931, who showed that with their existence, electric charge quantisation could be explained as a natural consequence of angular momentum quantisation [1]. Their introduction would also add symmetry to Maxwell's equations of electromagnetism. 't Hooft and Polyakov, in 1974, independently demonstrated that a Grand Unified Theory (GUT) scheme possesses a monopole solution when a U(1) subgroup of electromagnetism that is embedded into a larger gauge group is spontaneously broken by the Higgs mechanism [2, 3]. Monopole solutions have been proposed to arise within the electroweak theory itself [4], which relies on spontaneous gauge symmetry breaking. The Cho-Maison electroweak monopole would have a mass of the order of several TeV [5, 6] and is possibly within the range of the Large Hadron Collider (LHC). With the abundance of monopole theories and no definite estimate on its mass, the search for free magnetic charges in nature is compelling.

Dirac proposed that the monopoles should carry a magnetic charge g equal to the multiple of a fundamental unit of magnetic charge referred to as the Dirac charge g_D: $g = n.g_D$, where g_D is equivalent to 68.5 times the charge of an electron. The minimum value of the quantisation number n varies between different theories. According to Dirac, $n = 1, 2, 3..$, while according to Schwinger, as well as, Cho and Maison, $n = 2, 4, 6..$ [4, 7], and $n = 3, 6, 9..$ or $n = 6, 12, 18..$ if one considers the elementary charge to be carried by the down quark. As a result of the high value of the Dirac magnetic charge, a high velocity monopole is expected to suffer energy losses in matter over ≈4700 times higher than a muon [8–10]. Thus, in general monopoles in nature will manifest themselves as

[a]e-mail: akshay.k@cern.ch

DOI: 10.1051/epjconf/201612604025

highly ionising particles. A free magnetic charge will also induce a persistent current when passed through a superconducting loop. Searches for monopoles using these two signatures (high ionisation and magnetic induction) were performed in cosmic rays and in matter, and at accelerators each time a new energy regime was reached [11].

At colliders, three kinds of techniques are commonly used to search for magnetic monopoles: (1) General-purpose detectors with high ionisation energy loss detection capabilities (e.g. OPAL at LEP [12] and CDF at the Tevatron [13]); (2) dedicated deployment of nuclear-track detectors [14] around the interaction points (e.g. at LEP [15, 16] and at the Tevatron [17]); and (3) the induction technique applied to accelerator and detector material in which monopoles have stopped and remained trapped (e.g. at HERA [18] and at the Tevatron [19, 20]). These searches have excluded the presence of monopoles with charges equal to or above the Dirac charge and masses up to 400 GeV. Masses higher by one order of magnitude can be probed at the LHC. For complementary searches, the LHC programme should include all three of these techniques [21]. The general-purpose ATLAS detector has performed searches for monopoles at 7 and 8 TeV centre of mass energies [22, 23]. The MoEDAL experiment near the LHCb interaction point uses a combination of in-flight detection with nuclear-track detectors and re-usable aluminium trapping detectors [24].

2 Experimental setup

The Monopole & Exotics Detector at the LHC (MoEDAL) experiment is a dedicated experiment designed to enhance, in a complementary way, the physics reach of the LHC. The primary objectives of MoEDAL is to directly search for the Magnetic Monopole or Dyon and other highly ionising Stable (or meta- stable) Massive Particles (SMPs) at the LHC [25].

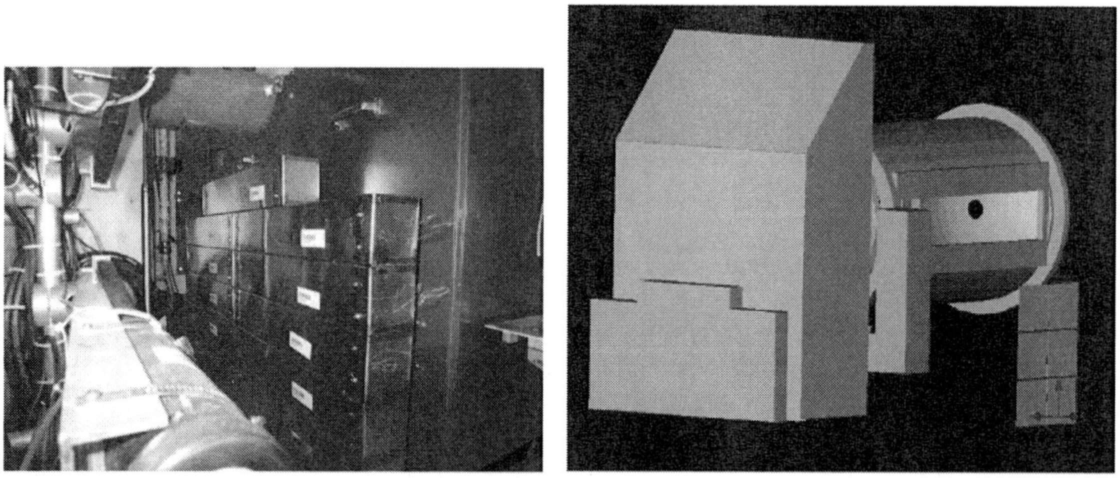

Figure 1. Left: Photograph of the 2012 MMT test array placed downstream of the LHCb stacked under the beam pipe 1.8 m away from the interaction point along the back of the VELO vacuum vessel. Right: Simulation of the LHCb cavern with the MMT array [26, 27].

Deployed around the LHCb experiment's Vertex Locator (VELO) cavern, it is a unique and largely passive LHC detector that comprises four sub-detector systems. The largest sub-detector systems are arrays of two nuclear-track detectors which are optimised to probe a range of particle charges. The

DOI: 10.1051/epjconf/201612604025

TimePix pixel devices form a real-time radiation monitoring system devoted to monitoring of highly-ionising backgrounds in the MoEDAL cavern. The Magnetic Monopole Trapper (MMT) is the other passive sub-detector system that is capable of capturing long-lived charged particles and measure directly magnetic charge properties of particles.

2.1 MMT sub-detector system

The MoEDAL MMT test detector system is an aluminium volume placed in the vicinity of the LHCb interaction point. Aluminium is a good choice for the trapping volume material for three important reasons. First, the anomalously large magnetic moment of aluminium nucleus means that it will strongly bind a trapped monopole. Second, aluminium does not present a problem with respect to activation. Lastly, aluminium allows a cost effective approach to the construction of the MMT detector. In September 2012, an initial MMT test array comprising 11 boxes each containing 18 rods of 60 cm length and 2.5 cm diameter was installed and exposed to 0.75 fb^{-1} of 8 TeV proton-proton collisions. The boxes were placed downstream of the LHCb detector, 1.5 m away from the interaction point, immediately in front of the LHCb VELO vacuum vessel, covering 1.3% of the total solid angle (Fig. 1). After the completion of the run, the rods are cut into samples of 20 cm length (except for samples

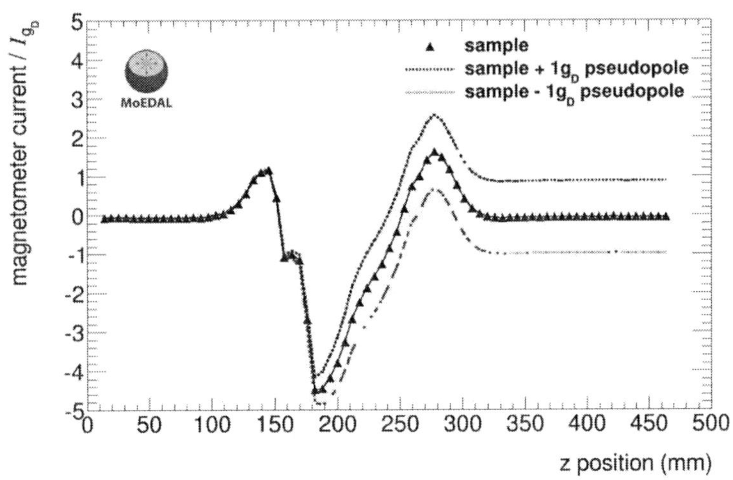

Figure 2. The plot shows the response of the magnetometer current at 76 different positions for a single sample as it traverses through the magnetometer. The dashed blue and red lines indicate the expected response from the magnetometer if a magnetic monopole of charge +1 and -1 g_D is trapped in the sample.

from box 11, which were cut into a mix of 10, 15, 20 and 30 cm samples), which total to 606 samples that are later passed through the magnetometer for measurements.

A DC-SQUID rock magnetometer based at the Laboratory of Natural Magnetism (ETH-Zurich) was used. The magnetometer is capable of detecting monopole charges much less and much greater than the Dirac charge as was demonstrated in previous studies [28]. The calibration was performed with a convolution method applied to a dipole sample, and cross-checked using thin solenoids which mimic a monopole of well known magnetic charge [28]. The response of the magnetometer was found to be linear and charge symmetric which allowed to express the measured currents in units of Dirac magnetic charge. Fig. 2 shows the magnetometer current response as one sample traverses through the magnetometer. The dashed lines show the expected response from the sample if a magnetic monopole of charge $\pm g_D$.

Each of the samples was passed through the magnetometer at least once over a 7-day measurement campaign in September 2013 [1]. On average, every tenth measurement was performed with an empty

[1]The samples from Box 11 were measured earlier in May 2013

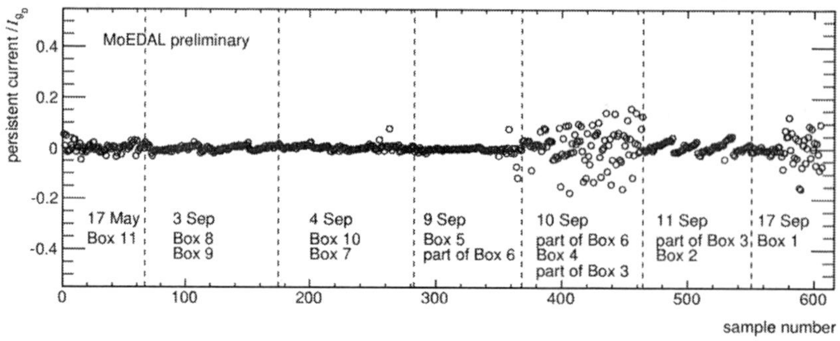

Figure 3. The persistent current (in terms of Dirac charge) measured for 606 samples from the MMT boxes during the September 2013 measurement campaign [29].

sample holder for an offset subtraction. The relevant value is the persistent current, defined as the difference between the currents measured after and before passage of the sample through the sensing coil, and then subtracting the difference obtained with a nearby empty holder measurement (Fig. 3).

Whenever the persistent current differs from zero by more than one fourth of the Dirac charge, the sample is considered as a candidate and measured again several times. A sample that contained a trapped monopole would consistently yield the same value for repeated measurements, while deviations in the first measurement correspond to instrumental effects (Fig. 4). The factors that affect the magnetometer response could be due to flux jumps occurring when the slew rate is large, noise currents in the SQUID feedback loop, small variations in the length of the sample holder from one run to the next and accumulation of condensed water and ice in the magnetometer tube near the cold sensing region.

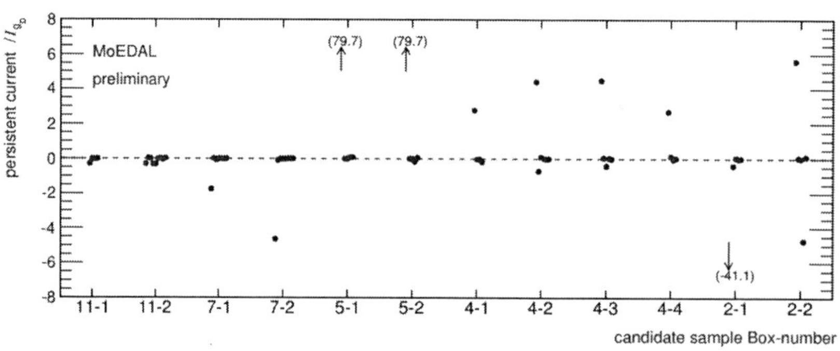

Figure 4. Results of multiple persistent current measurements (in units of Dirac charge) for the samples that yielded large ($|g|>0.25g_D$) values for the first measurement [29].

3 Monopole simulation

For model-dependent and model-independent interpretations of the results, magnetic monopoles were simulated using Drell-Yan (DY) and Single-monopole production [29].

The leading-order DY process produces a pair of heavy monopoles from initial pp state by quark-antiquark annihilation into a virtual photon using MADGRAPH5 Monte-Carlo event generator. The monopole coupling to the Z boson is set to zero. PYTHIA is used for the initial-state radiation and the hadronisation and the underlying event. The kinematic distributions for the spin-1/2 monopoles are shown in Fig. 5.

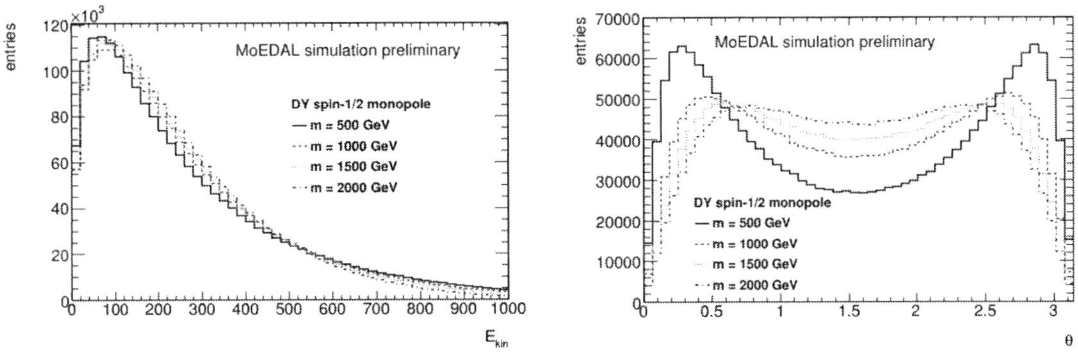

Figure 5. The kinematic distributions of spin-1/2 pair-produced monopoles considered for model-dependent results [29].

The Single-monopole samples are generated with a flat kinetic energy distribution ranging from 0 to 10000 GeV and a flat θ and ϕ distribution which encompasses the angular distribution of the trapping detector, i.e., 2.4 rad $< \theta <$ 3.0 rad and -2.7 rad$< \phi <$ -0.5 rad.

The acceptance of the trapping detector is defined on an event-by-event basis as the probability that the monopole stops inside one of the aluminium rods[2]. It is determined by propagating monopoles into the geometry model. The acceptance is defined in a unique way such that it only depends on the geometry of the detector and not on the production model. Since the collisions are symmetric with respect to the azimuthal angle ϕ, only two kinematic variables are needed to define the acceptance in a model-independent manner. The two variables chosen here are the longitudinal kinetic energy E_z^{kin} and the polar angle θ, after restricting the denominator of the acceptance definition to the range -2.7 rad $< \phi <$ -0.5 rad (encompassing the extent of the trapping detector). Thus for Single-monopole Monte-Carlo samples, the acceptance is mapped for all mass and charge combinations as a function of E_z^{kin} and θ (with -2.7 $< \theta <$ -0.5 rad), as show in Fig. 6 for monopoles with m = 1000 GeV. These two-dimensional histograms contain all the information needed to obtain the acceptance in any given pair-production model to a good approximation.

The DY acceptances can be computed in two ways. The first method requires full simulated pair-produced monopoles to be tracked through the geometry of the MoEDAL detector in order to compute the acceptances. The more computationally efficient way is to use the Single-monopole efficiency maps (described above) folded with the DY pair-production kinematics for both spin-1/2 and spin-0 monopoles distributions.

4 Systematics

The uncertainty in the amount of material in the geometry description of MoEDAL is the dominant source of systematic error. The VELO region within the acceptance of the LHCb detector is designed

[2]In the case of pair-produced DY monopoles, at least one of the two monopoles is required to stop inside the aluminium rods.

 DOI: 10.1051/epjconf/201612604025

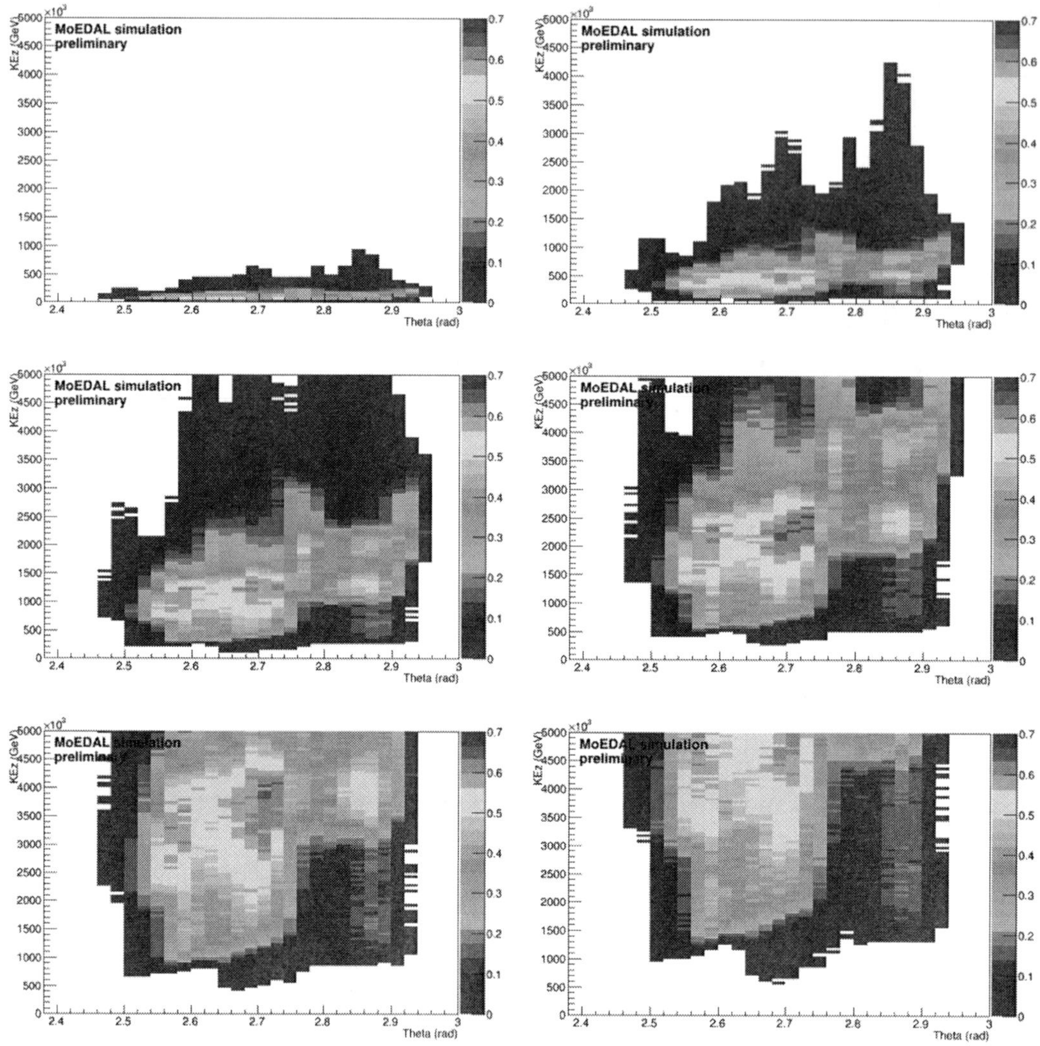

Figure 6. The acceptance of the MoEDAL test array in the kinetic energy and θ parameter space for monopole charges from 1.0 - 6.0 gD [29].

to minimise the amount of material obstructing the passage of measured particles. This region has been well modelled by GEANT4. However, the MoEDAL test trapping detector considered here lies just upstream of the VELO vacuum vessel in a more complex region containing: cables, manifolds, and piping, where detailed drawings are not always available. In order to account for possible imperfect understanding of the material in this arena two geometry models were constructed to conservatively describe the minimum and maximum amount of material. These models are based on a detailed physical examination of the material in this region. The uncertainty on the material description has a direct consequence on the acceptance maps, that affects both model-independent and model-dependent results. At the time of writing these proceedings, the determination of these systematic uncertainties is ongoing and will be documented in future publications [29].

5 Results and conclusions

MoEDAL is designed for passive detection of magnetic monopoles, both in-flight (with the track-etch technique) and trapped (with the induction technique). This is the first time at a collider experiment with a dedicated scalable and recyclable monopole trapping array has been deployed. Under the assumption of monopole capture by aluminium nuclei, there were no magnetic monopoles candidates found in 606 samples of the MMT test array exposed to 0.75 fb^{-1} of 8 TeV proton-proton collisions.

Despite its small acceptance and modest received luminosity MoEDAL's test MMT detector is capable of probing ranges of charge, mass and energy that were not accessed by other LHC experiments. Additionally, this technique can yield results very quickly and would allow for and unambiguous background-free assessment of a signal. A new, larger trapping detector array with a sensitive mass of around 1 tonne was deployed in 2014 in the LHCb cavern, allowing to perform a search in 13 TeV collisions in the near future.

References

[1] P.Dirac, *Quantised Singularities in the Electromagnetic Field*, Proc. Roy. Soc. **A 133** (1931) 60.

[2] G. 't Hooft, *Magnetic Monopoles in Unified Gauge Theories*, Nucl. Phys. B **79** (1974) 276.

[3] A. Polyakov, *Particle Specturm in the Quantum Field Theory*, JETP Lett **20** (1974) 194.

[4] Y. Cho and D. Maison, *Monopole configuration in Weinberg-Salam model*, Phys. Lett. B **391** (1997) 360.

[5] T. Kirkman and C. Zachos, *Asymptotic Analysis of the Monopole Structure*, Phys. Rev. D **24** (1981) 999.

[6] Y. M. Cho, K. Kim and J.H. Yoon, *Finite Energy Electroweak Dyon*, Eur. Phys. J. C **75**, no. 2, 67 (2015) [arXiv:1305.1699 [hep-ph]].

[7] J. Schwinger, *Magnetic charge and the charge quantization condition*, Phys. Rev. D **12** (1975) 3105.

[8] S. Ahlen, *Stopping-power formula for magnetic monopoles*, Phys. Rev. D **17** (1978) 229.

[9] S. Ahlen, *Theoretical and experimental aspects of the energy loss of relativistic heavily ionizing particles*, Rev. Mod. Phys. **52** (1980) 121.

[10] S. Ahlen and K. Kinoshita, *Calculation of the stopping power of very-low-velocity magnetic monopoles*, Phys. Rev. D **26** (1982) 2347.

[11] M. Fairbairn, A. Kraan, D. Milstead, T. Sjostrand, P. Skands, and T. Sloan, *Stable massive particles at colliders*, Phys. Rept. **438** (2007) 1, arXiv:0611040 [hep-ph].

[12] OPAL Collaboration, *Search for Dirac magnetic monopoles in e^+e^- collisions with the OPAL detector at LEP2*, Phys. Lett. B **663** (2008) 37, arXiv:0707.0404 [hep-ex].

[13] CDF Collaboration, *Direct search for Dirac magnetic monopoles in pp collisions at $\sqrt{s} = 1.96$ TeV*, Phys. Rev. Lett. **96** (2006) 201801, arXiv:0509015 [hep-ex].

[14] J. Pinfold, *Searching for the magnetic monopole and other highly ionizing particles at accelerators using nuclear track detectors*, Radiat. Meas. **44** (2009) 834.

[15] K. Kinoshita, R. Du, G. Giacomelli, L. Patrizii, F. Predieri, P. Serra, M. Spurio, and J. Pinfold, *Search for highly ionizing particles in e^+e^- annihilations at sqrt(s) = 91.1 GeV*, Phys. Rev. D **46** (1992) 881.

[16] J. Pinfold, R. Du, K. Kinoshita, B. Lorazo, M. Regimbald, and B. Price, *A Search for highly ionizing particles produced at the OPAL intersection point at LEP*, Phys. Lett. B **316** (1993) 407.

[17] M. Bertani, G. Giacomelli, M. Mondardini, B. Pal, L. Patrizii, F. Predieri, P. Serra-Lugaresi, G. Sini, M. Spurio, V. Togo, and S. Zucchelli, *Search for Magnetic Monopoles at the Tevatron Collider*, Europhys. Lett. **12** (1990) 613.

[18] H1 Collaboration, *A direct search for stable magnetic monopoles produced in positron-proton collisions at HERA*, Eur. Phys. J. C **41** (2005) 133, arXiv:0501039 [hep-ex].

[19] G. Kalbfleisch, K. Milton, M. Strauss, L. Gamberg, E. Smith, and W. Luo, *Improved Experimental Limits on the Production of Magnetic Monopoles*, Phys. Rev. Lett. **85** (2000) 5292, arXiv:0005005 [hep-ex].

[20] G. Kalbfleisch, W. Luo, K. Milton, E. Smith, and M. Strauss, *Limits on production of magnetic monopoles utilizing samples from the D0 and CDF detectors at the Tevatron*, Phys. Rev. D **69** (2004) 052002, arXiv:0306045 [hep-ex].

[21] A. De Roeck, A. Katre, P. Mermod, D. Milstead, and T. Sloan, *Sensitivity of LHC experiments to exotic highly ionising particles*, Eur. Phys. J. C **72** (2012) 1985, arXiv:1112.2999 [hep-ph].

[22] ATLAS Collaboration, *Search for magnetic monopoles in sqrt(s) = 7 TeV pp collisions with the ATLAS detector*, Phys. Rev. Lett. **109** (2012) 261803, arXiv:1207.6411 [hep-ex].

[23] ATLAS Collaboration, *Search for magnetic monopoles and stable particles with high electric charges in 8 TeV pp collisions with the ATLAS detector*, CERN-PH-EP-2015-174, arxiv:1509.08059.

[24] MoEDAL Collaboration, *Technical Design Report of the Moedal Experiment*, CERN-LHCC-2009-006 ; MOEDAL-TDR-001 (2009).

[25] B. Acharya *et al.* [MoEDAL Collaboration], *The Physics Programme Of The MoEDAL Experiment At The LHC*, Int. J. Mod. Phys. A **29**, 1430050 (2014) [arXiv:1405.7662 [hep-ph]].

[26] S. Agostinelli, *et al.*, *GEANT4: A Simulation toolkit*, Nucl.Instrum.Meth. **A506** (2003) 250–303. doi:10.1016/S0168-9002(03)01368-8.

[27] Geant4 Collaboration, http://geant4.cern.ch/support/

[28] A. De Roeck, H.-P. Hächler, A. M. Hirt, M. Dam Joergensen, A. Katre, P. Mermod, D. Milstead, and T. Sloan, *Development of a magnetometer-based search strategy for stopped monopoles at the Large Hadron Collider*, Eur. Phys. J. C **72** (2012) 2212, arXiv:1206.6793 [physics.ins-det].

[29] B. Acharya et al. [MoEDAL Collaboration], *Search for magnetic monopoles with the MoEDAL trapping detector in 8 TeV proton-proton collisions at the LHC*, to be submitted (2015).

[30] K. Milton, *Theoretical and experimental status of magnetic monopoles*, Rep. Prog. Phys. 69 (2006) **1637**, arXiv:0602040 [hep-ex].

EPJ Web of Conferences **126**, 04026 (2016) DOI: 10.1051/epjconf/201612604026
ICNFP 2015

Interacting relativistic quantum dynamics for multi-time wave functions

Matthias Lienert[1,a]

[1] *Mathematical Institute, Ludwig-Maximilians-University, Theresienstr. 39, 80333 Munich, Germany*

Abstract. In this paper, we report on recent progress about a rigorous and manifestly covariant interacting model for two Dirac particles in 1+1 dimensions [9, 10]. It is formulated using the multi-time formalism of Dirac, Tomonaga and Schwinger. The mechanism of interaction is a relativistic generalization of contact interactions, and it is achieved going beyond the usual functional-analytic Hamiltonian method.

1 Introduction

The multi-time formalism has the virtue to provide a manifestly covariant representation of the quantum state in the Schrödinger picture. The basic idea, suggested early on by the Nobel laureates Dirac [1], Tomonaga [2] and Schwinger [3] is to regard the usual *single-time wave function* of the N-particle Schrödinger equation,

$$\varphi : \mathbb{R}^{3N} \times \mathbb{R} \longrightarrow \mathcal{S}, \quad (\mathbf{x}_1, ..., \mathbf{x}_N, t) \longmapsto \varphi(\mathbf{x}_1, ..., \mathbf{x}_N, t), \tag{1}$$

where $\mathcal{S}$ is a spin space, as the special case of a more general object,

$$\psi : \Omega \subset \mathbb{R}^{4N} \longrightarrow \mathcal{S}, \quad (x_1, ..., x_N) \longmapsto \psi(x_1, ..., x_N), \tag{2}$$

the *multi-time wave function* ($x_k = (t_k, \mathbf{x}_k)$), namely the special case of equal times, i.e.

$$\varphi(\mathbf{x}_1, ..., \mathbf{x}_N, t) = \psi(t, \mathbf{x}_1, ..., t, \mathbf{x}_N). \tag{3}$$

In (2), Ω denotes a suitable domain. Most naturally, it is given by the set of *space-like configurations* [4–6], as a collection of N points which are space-like related is the most natural relativistic generalization of "N space-points at a time".

While the transformation properties of φ under Lorentz transformations $\Lambda \in \mathcal{L}_+^\uparrow$ cannot easily be defined, the ones of ψ are straightforwardly given by:

$$\psi'(x_1, ..., x_N) = S[\Lambda] \otimes \cdots \otimes S[\Lambda]\psi(\Lambda^{-1}x_1, ..., \Lambda^{-1}x_N), \tag{4}$$

where the matrices $S[\Lambda]$ form a spinor representation of the Lorentz group.

[a] e-mail: lienert@math.lmu.de

The multi-time wave function is assumed to attain physical meaning by a generalized Born rule, i.e. the postulate that a suitable quadratic expression in $\psi(x_1, ..., x_N)$ yields the probability density to find particle 1 at $x_1 \in \Sigma$, particle 2 at $x_2 \in \Sigma$ and so on, for any space-like hypersurface Σ [2, 4].

As evolution equations, one usually considers N simultaneous Schrödinger equations of Hamiltonian form, each of which expresses the time evolution in one of the t_k's:

$$i\frac{\partial}{\partial t_k}\psi = H_k\psi, \quad k = 1, ..., N, \tag{5}$$

where the H_k are differential operators on an appropriate function space. The familiar Schrödinger equation for φ naturally follows from (5) using (3) and the chain rule.

However, in contrast to the usual Schrödinger equation, (5) is not always consistent but possesses a joint solution ψ for every initial $\psi_0 = \psi(t_1 = 0, ..., t_N = 0)$ if and only if [7]:

$$[i\partial_{t_j} - H_j, i\partial_{t_k} - H_k] = 0 \;\; \forall j, k. \tag{6}$$

In fact, this *consistency condition* is fairly restrictive. It was for example recently shown in [7] that if

$$H_k = H_k^{\text{Dirac}} + V_k(x_1, ..., x_N), \tag{7}$$

where H_k^{Dirac} is the free Dirac Hamiltonian acting on the coordinates and spin index of the k-th particle and $V_k(x_1, ..., x_N)$ is a potential acting on the spin index of the k-th particle, then the multi-time evolution equations are gauge-equivalent to the case of purely external potentials $V_k \equiv V_k(x_k)$.

Therefore, the question of alternative mechanisms of interaction arises. Instead of going to quantum field theory (QFT) and trying to face the ultraviolet divergence problem there (see [5, 8] for multi-time formulations of QFTs), we explore a different possibility to construct a rigorous model here: *relativistic contact interactions*.

The main idea is that boundary conditions may lead to interaction effects even though the formal operators H_k in (5) are the free ones. This clearly avoids the use of potentials and therefore the before-mentioned no-go theorem. However, remains a non-trivial task to implement the idea, as Hamiltonian methods are not suitable to treat multi-time equations on a bounded domain Ω such as the space-like configurations. We shall present an alternative way here. The main virtues of the considered model are its rigorous, interacting and manifestly covariant character as well as its relative simplicity when compared with other rigorous interacting models. (It is constructed so as to be explicitly solvable.)

2 Definition of the model

Consider two mass-less spin-$\frac{1}{2}$ particles in 1+1 dimensions. As mentioned before, the natural domain is given by the set of space-like configurations, i.e.

$$\Omega := \{(t_1, z_1, t_2, z_2) \in \mathbb{R}^2 \times \mathbb{R}^2 : (t_1 - t_2)^2 - (z_1 - z_2)^2 < 0\}. \tag{8}$$

The multi-time wave function is a map

$$\psi : \Omega \longrightarrow \mathbb{C}^2 \otimes \mathbb{C}^2 \tag{9}$$

and we denote its spin components by $\psi_i(t_1, z_1, t_2, z_2)$, $i = 1, 2, 3, 4$.
Note that Ω is the disjoint union of the two open sets Ω_1, defined by $z_1 < z_2$, and Ω_2, defined by $z_1 > z_2$. Because of this feature, one may choose to consider each Ω_i separately, which we do in the

following, focusing on Ω_1.
As evolution equations, we prescribe the free mass-less Dirac equations for each particle, i.e.

$$i\gamma_k^\mu \partial_{k,\mu}\psi(x_1,x_2) = 0, \quad k=1,2 \tag{10}$$

in covariant notation, or

$$\begin{aligned} i\frac{\partial}{\partial t_1}\psi(t_1,z_1,t_2,z_2) &= -i\,\sigma_3\otimes 1_2\frac{\partial}{\partial z_1}\psi(t_1,z_1,t_2,z_2),\\ i\frac{\partial}{\partial t_2}\psi(t_1,z_1,t_2,z_2) &= -i\,1_2\otimes\sigma_3\frac{\partial}{\partial z_2}\psi(t_1,z_1,t_2,z_2). \end{aligned} \tag{11}$$

when solved for the time derivatives, as in eq. (5). Here, 1_2 stands for the 2×2 unit matrix,

$$\gamma^0 = \sigma_1 = \begin{pmatrix} 0 & 1\\ 1 & 0\end{pmatrix}, \quad \gamma^1 = \sigma_1\sigma_3 = \begin{pmatrix} 0 & -1\\ 1 & 0\end{pmatrix}, \tag{12}$$

and σ_i, $i=1,2,3$ denote the Pauli matrices.
Initial data are given on an equal-time hyperplane, e.g. for $t_1=t_2=0$:

$$\psi_i(0,z_1,0,z_2) = g_i(z_1,z_2), \quad z_1<z_2 \;\;(\text{on } \Omega_1), \tag{13}$$

where g_i are given C^1-functions.
Finally, boundary conditions are prescribed (as limits) on the set of *coincidence points*

$$C := \{(t_1,z_1,t_2,z_2)\in\mathbb{R}^2\times\mathbb{R}^2 : t_1=t_2,\ z_1=z_2\}. \tag{14}$$

In [6, 9] it was shown that a both physically preferred and mathematically admissible class of boundary conditions is given by (using the basis in spin space corresponding to (11), (12)):

$$\lim_{\varepsilon\to 0}\psi_2(t,z-\varepsilon,t,z+\varepsilon) = e^{-i\theta}\lim_{\varepsilon\to 0}\psi_3(t,z-\varepsilon,t,z+\varepsilon), \quad t,z\in\mathbb{R}. \tag{15}$$

for any $\theta\in[0,2\pi)$. The limiting process is well-defined, as we aim at C^1-solutions (in both space and time). We shall explain the special features of this class of boundary conditions in the following.

3 Results

3.1 Probability conservation

In order for a quantum-mechanical equation to have statistical meaning, probability conservation has to be ensured. The system (10) possesses the *conserved tensor current*

$$j^{\mu\nu}(x_1,x_2) = \overline{\psi}(x_1,x_2)\gamma_1^\mu\gamma_2^\nu\psi(x_1,x_2) \tag{16}$$

where $\overline{\psi} := \psi^\dagger\gamma_1^0\gamma_2^0$. "Conserved" means $\partial_{1,\mu}j^{\mu\nu} = \partial_{2,\nu}j^{\mu\nu} = 0$. Furthermore, $j^{\mu\nu}$ is positive-definite as $j^{00} = \psi^\dagger\psi \geq 0$. Therefore, we can express probability conservation on a general space-like hypersurface Σ (and in the domain Ω_1) as [9]:

$$\int_{(\Sigma\times\Sigma)\cap\Omega_1} d\sigma_{1,\mu}\, d\sigma_{2,\nu}\, j^{\mu\nu}(x_1,x_2) = 1, \quad \text{independently of } \Sigma. \tag{17}$$

By treating $d\sigma_{1,\mu}\, d\sigma_{2,\nu}\, j^{\mu\nu}(x_1,x_2)$ as an $2d$-form ($d=1$ here), constructing a closed surface including $(\Sigma_i\times\Sigma_i)\cap\Omega_1$ for two different space-like hypersurfaces Σ_i, $i=1,2$ and using Stokes' theorem in its differential geometric form, one can show [9]:

Theorem 3.1 *Probability conservation on* Ω_1 *in the sense of* (17) *is ensured by the boundary conditions* (15).

3.2 Lorentz invariance

The main goal of the multi-time formulation is manifest covariance. Apart from the boundary conditions, the Lorentz invariance of the model is already clear from its definition. In order to check the covariance of the boundary conditions, we explicitly determine the matrices $S[\Lambda]$ in (4).
According to the representation theory of Dirac spinors, we have:

$$S[\Lambda] = \exp\left(\tfrac{\beta}{4}[\gamma^0,\gamma^1]\right), \quad \beta \in \mathbb{R}. \tag{18}$$

Using the representation (12), we explicitly obtain:

$$S[\Lambda] = \begin{pmatrix} \cosh\beta + \sinh\beta & 0 \\ 0 & \cosh\beta - \sinh\beta \end{pmatrix} \tag{19}$$

and it follows:

$$S[\Lambda] \otimes S[\Lambda] = \begin{pmatrix} (\cosh\beta + \sinh\beta)^2 & & & \\ & 1 & & \\ & & 1 & \\ & & & (\cosh\beta - \sinh\beta)^2 \end{pmatrix}. \tag{20}$$

Therefore, under a Lorentz transformation Λ, the components ψ_2, ψ_3 transform as:

$$\psi_i(x_1, x_2) \stackrel{\Lambda}{\longmapsto} \psi_i(\Lambda^{-1}x_1, \Lambda^{-1}x_2) \ \text{ for } i = 2, 3. \tag{21}$$

Because in addition the boundary conditions are prescribed on the Lorentz invariant set C, we obtain:

Theorem 3.2 *The model, as defined by* (8), (10), (13) *and* (15), *is Lorentz invariant.*

3.3 Existence and uniqueness of solutions

The previous two subsections have established some of the most important physical properties of the model. Of course, also mathematical sense has to be given to it. This requires to establish the existence and uniqueness of solutions of the initial boundary value problem (11), (13) and (15).

In order to understand the meaning of the multi-time evolution equations (11) in more detail, we write them out in matrix-vector form:

$$i\frac{\partial}{\partial t_1}\begin{pmatrix} \psi_1 \\ \psi_2 \\ \psi_3 \\ \psi_4 \end{pmatrix} = -i\begin{pmatrix} 1 & & & \\ & 1 & & \\ & & -1 & \\ & & & -1 \end{pmatrix}\frac{\partial}{\partial z_1}\begin{pmatrix} \psi_1 \\ \psi_2 \\ \psi_3 \\ \psi_4 \end{pmatrix},$$

$$i\frac{\partial}{\partial t_2}\begin{pmatrix} \psi_1 \\ \psi_2 \\ \psi_3 \\ \psi_4 \end{pmatrix} = -i\begin{pmatrix} 1 & & & \\ & -1 & & \\ & & 1 & \\ & & & -1 \end{pmatrix}\frac{\partial}{\partial z_2}\begin{pmatrix} \psi_1 \\ \psi_2 \\ \psi_3 \\ \psi_4 \end{pmatrix}. \tag{22}$$

For example, for ψ_1 we have:

$$\left(\frac{\partial}{\partial t_1} + \frac{\partial}{\partial z_1}\right)\psi_1 = 0, \quad \left(\frac{\partial}{\partial t_2} + \frac{\partial}{\partial z_2}\right)\psi_1 = 0 \quad \Rightarrow \quad \psi_1(t_1, z_1, t_2, z_2) = f_1(z_1 - t_1, z_2 - t_2) \tag{23}$$

where f_1 is a C^1-function.

We see that each component of ψ is constant along certain two-dimensional surfaces in four-dimensional space, e.g. defined by $z_1 - t_1 = c_1, z_2 - t_2 = c_2$ for ψ_1 and constants $c_1, c_2 \in \mathbb{R}$. In analogy to the method of characteristics in PDE theory, we call these surfaces the *multi-time characteristics* of ψ. They allows for a powerful method to study the existence and uniqueness theory of our model [9]. To see whether ψ_i is uniquely determined at a point $p \in \Omega_1$, one considers the multi-time characteristic which includes p. Then one calculates whether it intersects either the boundary set C or the initial data surface $t_1 = t_2 = 0$ in a single point q. If for all i and all p there exists a path connecting p with the respective q while staying in the multi-time characteristic intersected with the domain Ω_1, then there exists a unique solution of the initial boundary value problem. At the boundary, it has to be made clear, in addition, that for every t exactly one side of (15) is already uniquely determined by initial conditions.

By this method, it is even possible to determine the explicit solution [6, lemma 3.6.1]:

Theorem 3.3 *The unique solution of the initial boundary value problem* (11), (13) *and* (15) *on* Ω_1, *where the initial data are supposed to satisfy the boundary conditions, is given by:*

$$\begin{aligned}
\psi_1(t_1, z_1, t_2, z_2) &= g_1(z_1 - t_1, z_2 - t_2), \\
\psi_2(t_1, z_1, t_2, z_2) &= \begin{cases} g_2(z_1 - t_1, z_2 + t_2) & \text{for } z_1 - t_1 < z_2 + t_2 \\ e^{-i\theta} g_3(z_2 + t_2, z_1 - t_1) & \text{for } z_1 - t_1 \geq z_2 + t_2 \end{cases} \\
\psi_3(t_1, z_1, t_2, z_2) &= \begin{cases} g_3(z_1 + t_1, z_2 - t_2) & \text{for } z_1 + t_1 < z_2 - t_2 \\ e^{i\theta} g_2(z_2 - t_2, z_1 + t_1) & \text{for } z_1 + t_1 \geq z_2 - t_2 \end{cases} \\
\psi_4(t_1, z_1, t_2, z_2) &= g_4(z_1 + t_1, z_2 + t_2).
\end{aligned} \tag{24}$$

3.4 Interaction

The main point about the model is to achieve a mechanism of interaction by boundary conditions. Of course, the question of how to decide whether interaction is present arises. In non-relativistic quantum mechanics, one simply defines interactions by the presence of an interaction potential. Of course, such a definition is limited to a particular mathematical framework and is not general enough here. Instead, we try to capture the essence of interaction as follows: *A model is said to be interacting if it generates entanglement, i.e. if there are initial product wave functions* $\psi(0, z_1, 0, z_2) = \phi(z_1) \otimes \chi(z_2)$ *which cannot be written as product wave functions for later times* t_1, t_2.

With the aid of the explicit solution formula, this criterion is easy to check. Using

$$\begin{aligned}
g_1(z_1, z_2) &= \phi_1(z_1)\chi_1(z_2), \quad g_2(z_1, z_2) = \phi_1(z_1)\chi_2(z_2), \\
g_3(z_1, z_2) &= \phi_2(z_1)\chi_1(z_2), \quad g_4(z_1, z_2) = \phi_2(z_1)\chi_2(z_2)
\end{aligned} \tag{25}$$

in (24), one indeed finds, as a consequence of the case differentiation for ψ_2, ψ_3:

Theorem 3.4 *The model defined by* (11), (13) *and* (15) *on* Ω_1 *is interacting.*

By studying the behavior of localized wave packets in more detail (compare [9]), one can see that the interaction corresponds to a scattering process with range zero, where the two particle bounce off each other when they meet. In addition, the scattering is associated with a spin flip and relative phase shift of the components ψ_2, ψ_3 corresponding to $|\uparrow\downarrow\rangle \leftrightarrow e^{-i\theta}|\downarrow\uparrow\rangle$.

4 Summary and outlook

Indeed, a rigorous interacting relativistic model could be constructed, conforming to the strictest requirements of relativistic invariance, as embodied by the multi-time formalism. Its relative simplicity when compared with other models of this kind makes it particularly attractive as a pedagogical example. It is, furthermore, a counterexample to the claim in the "folklore" of relativistic quantum mechanics that the creation and annihilation of particles were necessary form a purely mathematical point of view to achieve relativistic interactions.

The developed methods, such as the formulation of probability conservation for N particles on space-like hypersurfaces as well as the generalized method of characteristics, may be of interest for both scientists working in relativistic quantum mechanics as well as mathematical physics.

Besides this, one may ask whether or how the model can be generalized to more realistic situations. As shown in [10], the extension to $N \geq 3$ particles is straightforward. The formulation of a similar model for higher dimensions, however, does not seem feasible, as the set of coincidence points then is too low-dimensional to have impact on the time evolution [9].

It will be an interesting future task to develop further generalizations with respect to the massive case as well as to a variable particle number. The assumption of zero masses has so far been a technical one whose main advantage is to render the model explicitly solvable. However, besides the strategy for the existence proof, nothing essential is expected to change for non-zero masses, since the boundary conditions are extracted from the tensor current $j^{\mu\nu}$ which stays the same in the massive case. With regard to the variable particle number case, the idea would be to choose the particle-position representation of QFT (see [5]) such that $\psi = (\psi^{(0)}, \psi^{(1)}, \psi^{(2)}, ...)$ where $\psi^{(n)}$ is an n-particle wave function. When compared to the N-particle case, additional possibilities for the boundary conditions appear, namely by relating $\psi^{(n)}$ with $\psi^{(m)}$ for $n \neq m$. This corresponds to the recently introduced idea of "interior boundary conditions" which was used in [11, 12] to avoid ultraviolet divergencies for a simple non-relativistic QFT. Studying the variable particle number case for our model may help to extend this idea to a relativistic setting.

References

[1] P. A. M. Dirac. *Proc. R. Soc. Lond. A*, **136**:453–464, 1932.

[2] S. Tomonaga. In J. Schwinger, editor, *Selected Papers on Quantum Electrodynamics*, pages 156–168. Dover, 1958.

[3] J. Schwinger. *Phys. Rev.*, **74**(2162):1439–1461, 1948.

[4] F. Bloch. *Phys. Z. d. Sowjetunion*, **5**:301–315, 1934.

[5] S. Petrat and R. Tumulka. *Ann. Phys.*, **345**:17–54, 2014.

[6] M. Lienert. PhD thesis, 2015. Mathematical Institute, Ludwig-Maximilians-University, Munich, Germany. https://edoc.ub.uni-muenchen.de/18705/.

[7] S. Petrat and R. Tumulka. *J. Math. Phys.*, **55**(032302), 2014.

[8] S. Petrat and R. Tumulka. *J. Phys. A*, **47**(11):112001, 2014.

[9] M. Lienert. *J. Math. Phys.*, **56**(4):042301, 2015.

[10] M. Lienert and L. Nickel. *J. Phys. A*, **48**(32):325301, 2015. arXiv:1502.00917.

[11] S. Teufel and R. Tumulka. Preprint: arXiv:1505.04847 (2015)

[12] S. Teufel and R. Tumulka. To appear in F. Finster, J. Kleiner, C. Röken, and J. Tolksdorf (editors), *Quantum Mathematical Physics – A Bridge between Mathematics and Physics*, Basel: Birkhäuser (2015). Preprint: arXiv:1506.00497.

DOI: 10.1051/epjconf/201612604027

Polarizability of pseudoscalar mesons from the lattice calculations

E.V. Luschevskaya[1,2,3,a], O.E. Soloveva[1,2], and O.V. Teryaev[4,2]

[1]*State Scientific Center of the Russian Federation - Institute for Theoretical and Experimental Physics, 117218 Moscow, Russia*
[2]*National Research Nuclear University "MEPhI" (Moscow Engineering Physics Institute), Kashirskoe highway 31, 115409 Moscow, Russia*
[3]*School of Biomedicine, Far Eastern Federal University, 690950 Vladivostok, Russia*
[4]*Joint Institute for Nuclear Research, Dubna, 141980, Russia*

Abstract. We explore the energy dependence of neutral and charge pions of the background constant abelian magnetic field in SU(3) lattice gauge theory without dynamical quarks. The energy of neutral pseudoscalar meson diminishes with the field, while the energy of charged one increases according with the theoretical expectations. Based on these data we estimate the magnetic polarizabilities of π^0 and $\pi^{\pm}$ mesons for various quark masses.

1 Introduction

Quantum Chromodynamics in a strong magnetic field presents an interesting topic for research. Magnetic fields of hadronic scale could exist in the Early Universe [1], in cosmic objects like magnetars and can be created in terrestrial laboratories (RHIC, LHC, FAIR, NICA) [2]. Background magnetic field also makes possible to calculate the magnetic polarizabilities of hadrons. To get the dipole magnetic polarizability and hyperpolarizability we measure the energy of a meson as a function of the uniform abelian field. The magnetic polarizabilities are important physical characteristics and describe the distibution of quark currents inside the hadron in the external field. It characterizes its internal structure. The magnetic field is also interesting quantity in connection with COMPASS [3] and JLab experiments. Comparison with the predictions of the experiments and chiral perturbation theory [4, 5] presents interest for fundamental science.

2 Details of calculations

We mention the details of calculations only breafly, because it is presented in our previous works [6, 7]. For the generation of $SU(3)$ gauge configurations we use the tadpole improved Lüscher-Weisz action. We solve Dirac equation numerically

$$D\psi_k = i\lambda_k\psi_k, \quad D = \gamma^{\mu}(\partial_{\mu} - iA_{\mu}) \tag{1}$$

[a]e-mail: luschevskaya@itep.ru

and find eigenfunctions ψ_k and eigenvalues λ_k for a test quark in the external gauge field A_μ. We find eigenstates of Dirac operator to calculate the correlators. From the correlators we obtain ground state energies. For the calculation of the fermion spectrum we use the Neuberger overlap operator [8]. Fermion fields obey periodical boundary conditions in space and antiperiodic boundary conditions in time. The background gauge field A_μ is a superposition of non-abelian $SU(3)$ gluon field and $U(1)$ abelian uniform magnetic field. We introduce the magnetic field in the symmetric gauge

$$A_{\mu ij} \to A_{\mu ij} + A^B_\mu \delta_{ij}, \; A^B_\mu(x) = \frac{B}{2}(x_1\delta_{\mu,2} - x_2\delta_{\mu,1}). \tag{2}$$

Magnetic field is directed along z axis and its value is quantized

$$qB = \frac{2\pi k}{(aL)^2}, \; k \in \mathbb{Z}, \tag{3}$$

where $q = -1/3\,e$ is the elementary quark charge.

We consider two types of quarks (u and d) which are degenerate in mass. Our simulations have been carried out on symmetrical lattices with the lattice volumes 16^4, 18^4, 20^4, lattice spacings 0.095 fm, 0.115 fm, 0.125 fm and various bare quark masses in the interval [0.007, 0.06]. We use 200-250 statistical independent gauge field configurations for the each lattice data set.

We calculate the following observables in the coordinate space and background gauge field A

$$\langle \psi^\dagger(x) O_1 \psi(x) \psi^\dagger(y) O_2 \psi(y) \rangle_A, \tag{4}$$

where $O_1, O_2 = \gamma_5, \gamma_\mu$ are Dirac gamma matrices, $\mu, \nu = 1,..,4$ are Lorentz indices, $x = (\mathbf{n}a, n_t a)$ and $y = (\mathbf{n}'a, n'_t a)$ are the lattice coordinates. The spatial lattice coordinate $\mathbf{n}, \mathbf{n}' \in \Lambda_3 = \{(n_1, n_2, n_3) | n_i = 0, 1, ..., N-1\}$, n_t, n'_t are the numbers of lattice sites in the time direction. For the observables (4) the following equation is fulfilled

$$\langle \bar\psi O_1 \psi \bar\psi O_2 \psi \rangle_A = -\mathrm{Tr}\,[O_1 D^{-1}(x,y) O_2 D^{-1}(y,x)] + \mathrm{Tr}\,[O_1 D^{-1}(x,x)]\mathrm{Tr}\,[O_2 D^{-1}(y,y)]. \tag{5}$$

where $D^{-1}(x,y)$ is the Dirac propagator.

We perform Fourier transformation of (5) numerically and choose $\langle \mathbf{p} \rangle = 0$ as we are interested in the meson ground state energy. To obtain the masses we expand the correlation function to the exponential series

$$\tilde{C}(n_t) = \langle \psi^\dagger(\mathbf{0}, n_t) O_1 \psi(\mathbf{0}, n_t) \psi^\dagger(\mathbf{0}, 0) O_2 \psi(\mathbf{0}, 0) \rangle_A = \sum_k \langle 0|O_1|k\rangle\langle k|O_2^\dagger|0\rangle e^{-n_t a E_k}, \tag{6}$$

For a large n_t the main contribution to the correlator (6) comes from the ground state and due to the periodic boundary conditions has the following form

$$\tilde{C}_{fit}(n_t) = A_0 e^{-n_t a E_0} + A_0 e^{-(N_T - n_t) a E_0} = 2A_0 e^{-N_T a E_0/2} \cosh((\frac{N_T}{2} - n_t) a E_0), \tag{7}$$

where A_0 is a constant, E_0 is the ground state energy, a is the lattice spacing.

3 Results

3.1 Magnetic dipole polarizability of π^0

To calculate the magnetic polarizability of mesons we consider sufficiently large magnetic fields. The magnetic field of hadronic scale allows to measure the response of the internal structure to this field.

DOI: 10.1051/epjconf/201612604027

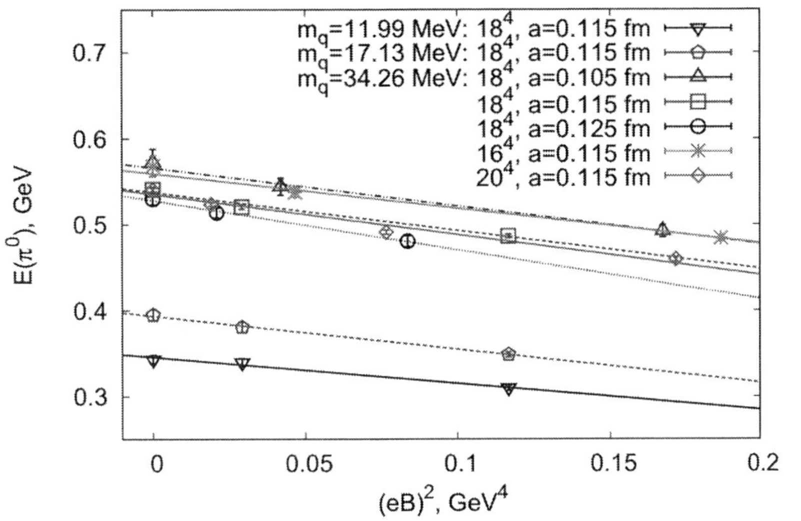

Figure 1. The energy of π^0 ground state as a function of the magnetic field value squared for various lattice volumes, spacings and the quark masses. Points correspond to the lattice data, lines are the lattice data fits obtained using the function (8).

For the neutral pion this is manifested in the deviation of the particle energy from the constant value, for the charged particle we observe the deviation of pion energy from Landau level.

We calculate the energy of neutral pion from the correlation function $C^{PSPS} = (\langle\bar{\psi}_d(\vec{0}, n_t)\gamma_5\psi_d(\vec{0}, n_t)\bar{\psi}_d(\vec{0}, 0)\gamma_5\psi_d(\vec{0}, 0)\rangle + \langle\bar{\psi}_u(\vec{0}, n_t)\gamma_5\psi_u(\vec{0}, n_t)\bar{\psi}_u(\vec{0}, 0)\gamma_5\psi_u(\vec{0}, 0)\rangle)/\sqrt{2}$. In Fig.1 we show the energy of π^0 ground state versus the field value squared for the lattice volumes 16^4, 18^4, 20^4, lattice spacings 0.095 fm, 0.105 fm, 0.115 fm, 0.125 fm and quark masses 11.99 MeV, 17.13 MeV, 34 MeV.

The each set of lattice data is fitted at $(eB)^2 \in [0, 0.3\ \text{GeV}^4]$ using the function

$$E = E(B = 0) - 2\pi\beta_m(eB)^2, \tag{8}$$

where $E(B = 0)$ and β_m are the fit parameters. At the considered range of fields the energy dependence is linear on the magnetic field and we calculate the value of β_m from its slope.

The magnetic polarizability is shown in Fig.2 for the lattice volume 18^4 as a function of the quark mass, the β_m values are also depicted for the lattice volumes 16^4, 20^4 and quark mass 34.26 MeV.

The value of β_m diminishes with the quark mass and extrapolation to the chiral limit gives the number $(3.2\pm0.4)\cdot10^{-4}\ \text{fm}^3$, $\chi^2/d.o.f = 0.715$ for the lattice volume 18^4 and lattice spacing 0.115 fm.

The values of magnetic polarizability for various quarks masses are a bit higher than assessments obtained in our previous work [6], where for the neutral particles we considered very simple model with only one type of quarks d. Correlation functions for d and u quarks behave differently in the external magnetic field because u and d quarks possess different absolute value of electric charges, in this work we take this into account. Further improvement of the accuracy in β_m determination might be carried out due to studing the effects of finite lattice volume and lattice spacing, consideration of weaker magnetic fields.

The value of magnetic dipole polarizability obtained within the framework of chiral perturbation theory is equal to $\beta_m = (1.5 \pm 0.3) \cdot 10^{-4}\ \text{fm}^3$ [4].

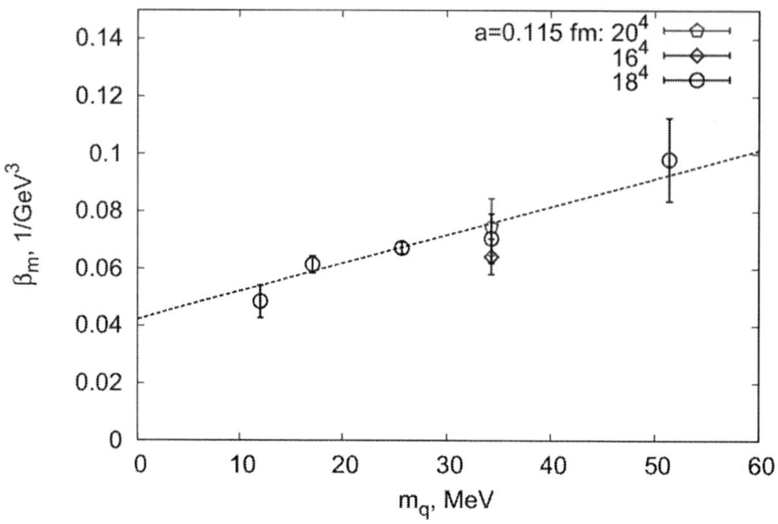

Figure 2. The dipole magnetic polarizability of neutral pion for the lattice volumes 16^4, 18^4, 20^4 and lattice spacing 0.115 fm depending on the quark mass.

3.2 Magnetic hyperpolarizability of π^0

At larger magnetic fields the terms with higher degrees on the magnetic field give contribution to the π^0 energy. In Fig.3 we show the energy of neutral pion for magnetic fields in the interval from 0 to 1.7 GeV^4 with the lattice data fits described by the formula

$$E = E(B = 0) - 2\pi\beta_m(eB)^2 - 2\pi\beta_m^h(eB)^4 - k(eB)^6, \tag{9}$$

where the mass of pion $E(B = 0)$, the dipole magnetic polarizability β_m, the magnetic hyperpolarizability β_m^h and k are fit parameters. The term proportional to B^3 is parity forbidden and π_0 can't decay to three (magnetic) photons.

The decay $\pi^0 \rightarrow 4\gamma_M$ is allowed, so the term $\sim (eB)^4$ appears in formula 9. The values β_m^h are negative for the all sets of data represented in Fig.3. For example for the lattice volume 18^4, quark mass 34.26 MeV we get the value $\beta_m^h = (-7 \pm 2) \cdot 10^{-7}$ fm^7 at lattice spacing $a = 0.095$ fm and $\beta_m^h = (-6.7 \pm 1.5) \cdot 10^{-7}$ fm^7 at $a = 0.115$ fm. For the lowest quark mass 17.13 MeV we obtain $\beta_m^h = (-8 \pm 2) \cdot 10^{-7}$ fm^7, $\chi^2/d.o.f. = 1.433$ using for fitting the formula (9).

3.3 Magnetic dipole polarizability of $\pi^\pm$

The energy levels of free charged (non-pointlike) particle in a constant abelian magnetic field parallel to z axis are described by the formula

$$E^2 = p_z^2 + (2n + 1)|qH| - gs_z qH + E^2(H = 0) - 2\pi\beta_m H^2, \quad H = eB. \tag{10}$$

We consider the particle momentum $p_z = 0$, ground state $n = 0$, meson charge $q = \pm 1$, spin $s_z = 0$ and $g = 0$. $E^2(H = 0)$ is the pion energy at zero magnetic field (the mass). So in our case

$$E^2 = |qH| + E^2(H = 0) - 2\pi\beta_m H^2, \quad H = eB. \tag{11}$$

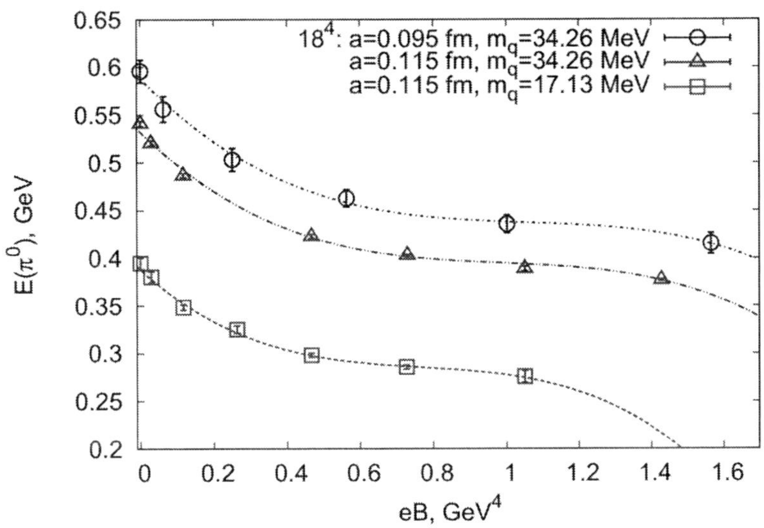

Figure 3. The energy of π^0 ground state as a function of the field value for the lattice volumes 16^4, 18^4, lattice spacings $a = 0.095$ fm, 0.105 fm, 0.115 fm and the quark mass equal to 34.26 MeV. Curves correspond to the fits of lattice data by the function (9).

We calculate the energy of charged pion from the correlation function $C^{PSPS} = \langle \bar{\psi}_u(\vec{0}, n_t)\gamma_5\psi_u(\vec{0}, n_t)\bar{\psi}_d(\vec{0}, 0)\gamma_5\psi_d(\vec{0}, 0)\rangle$. Its energy is depicted in Fig.4 for the lattice volumes 18^4, 20^4, lattice spacings 0.095 fm, 0.115 fm, 0.125 fm and quark masses 17.13 MeV, 34.26 MeV. In Fig.4 we also show the lattice data fits obtained using the formula (11).

The value of magnetic dipole polarizability has positive value for the all sets of data considered and its absolute value increases with the quark mass. For example fitting the lattice data at $eB \in [0; 2.5]$ for the volume 18^4, spacing 0.115 fm gives the $\beta_m = (2.8 \pm 0.2) \cdot 10^{-4}$ fm^3, $\chi^2/d.o.f = 4.48$ for the quark mass 17.13 MeV and the value $\beta_m = (1.3 \pm 0.2) \cdot 10^{-4}$ fm^3, $\chi^2/d.o.f = 3.68$ for the quark mass 34.26 MeV. Chiral perturbation theory predicts the value of magnetic dipole polarizability equal $\beta_m = -2.76 \cdot 10^{-4}$ fm^3 [5].

The values of magnetic polarizability obtained for the same lattice quark mass but for different lattice volumes and lattice spacings agree between each other within the errors. The magnetic polarizability depends strongly on the quark mass and chiral extrapolation will be done in the following work.

4 Conclusions

We calculate the ground state energies of neutral and charged pions depending on the value of the external abelian magnetic field for various lattice volumes, spacings and quark mass. From energy dependence we calculate the magnetic dipole polarizability for and magnetic hyperpolarizability of π^0. In the chiral limit the magnetic dipole polarizability is equal to $(3.2 \pm 0.4) \cdot 10^{-4}$ fm^3 for the lattice volume 18^4 and lattice spacing 0.115 fm, that is close to the prediction od ChPT. The further improvement in accuracy might be done in the following work. We didn't observe strong dependence of the β_m value from the lattice volume and lattice spacing. However it strongly depends on the lattice quark mass.

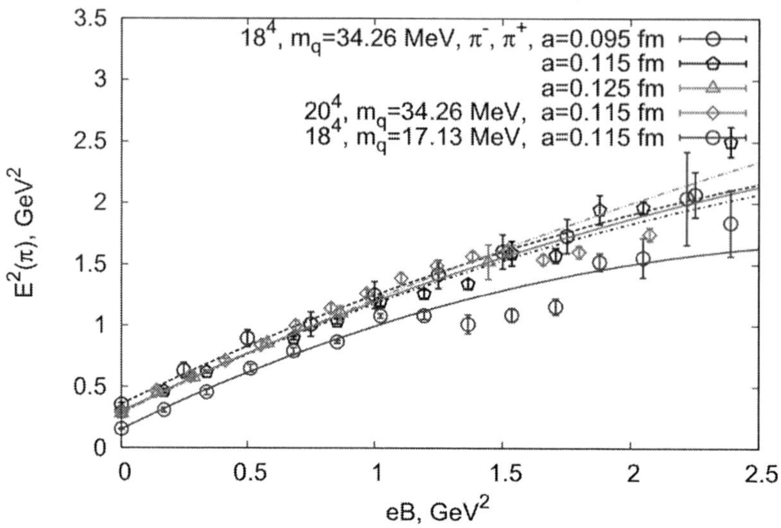

Figure 4. The energy squared of π^{+-} ground state versus the field value squared for the lattice volumes 18^4, 20^4, various lattice spacings and the quark masses 17.13 MeV and 34.26 MeV. Fits of lattice data are performed using the function (11)

We also find the contribution of hyperpolarizability β_m^h to the neutral pion energy. This is very tiny effect and we try to essess its value: for the all sets $\beta_m^h < 0$, for the lowest quark mass $\beta_m^h = (-8 \pm 2) \cdot 10^{-7}$ fm^7.

We calculate the magnetic dipole polarizability of charged pion for several lattice spacings, two lattice volumes and two bare lattice quark mass 34.26 MeV and 17.13 MeV. For the $m_q = 34.26$ MeV and $m_q = 17.13$ MeV we obtain $\beta_m = (1.3 \pm 0.2)10^{-4}$ fm^3 and $\beta_m = (2.8 \pm 0.2)10^{-4}$ fm^3 respectively. The dipole magnetic polarizability depends on the lattice quark mass and chiral extrapolation will be done soon. The sign of $\beta_m(\pi^\pm)$ doesn't agree with the prediction of ChPT [5].

Acknowledgements

The authors are grateful to FAIR-ITEP supercomputer center where these numerical calculations were performed. This work is completely supported by a grant from the Russian Science Foundation (project number 16-12-10059).

References

[1] D. Grasso and H.R. Rubinstein, Phys. Rept. **348**, 163 (2001), arXiv: astro-ph/0009061.
[2] V. Skokov, A. Illarionov and V. Toneev, *Int. J. Mod. Phys.* A **24**, 5925 (2009), arXiv: 0907.1396[nucl-th].
[3] C. Adolph et al. (COMPASS Collaboration), Phys. Rev. Lett. **114**, 062002 (2015).
[4] J. Gasser, M.A. Ivanov and M.E. Sainio, Nuclear Physics B **728**, 31 (2005).
[5] A. Aleksejevs and S. Barkanova, Nucl. Phys. Proc. Suppl. 245, **17** (2013), arXiv:1309.3313 [hep-ph].

[6] E.V. Luschevskaya, O.E. Solovjeva, O.E. Kochetkov and O.V. Teryaev, Nuclear Physics B **898**, 627 (2015).
[7] E.V. Luschevskaya, O.E. Kochetkov, O.V. Teryaev and O.E. Solovjeva, JETP Letters **101**, Issue 10, 674 (2015).
[8] H. Neuberger, Phys. Lett. B **417**, 141 (1998), arXiv: hep-lat/9707022.

Searches for New Physics with Lepton Flavours and Multi-lepton Final States in ATLAS

Romain Madar[1,a] on behalf of the ATLAS collaboration

[1] *Université Clermont Auvergne, Université Blaise Pascal, CNRS/IN2P3, Laboratoire de Physique Corpusculaire, BP 10448, F-63000 CLERMONT-FERRAND, France*

Abstract. Lepton flavours and multi-lepton signatures are powerful probes to search for new physics. Also, the lepton sector of the Standard Model has to be extended to describe the observed neutrino oscillations. ATLAS experiment searches for lepton flavour violation, new gauge boson allowing to unify quark and lepton fields, heavy charged and neutral leptons as well as for new phenomena in multiphoton final states are summarized in this document. The most recent results will be discussed and compared with existing results.

1 Introduction

The leptonic sector of the Standard Model (SM) is a promising area to search for new phenomena beyond the SM (BSM). In a bottom-up approach, the neutrino oscillations is probably one of the best established and mesured BSM observation. This phenomena suggests that neutrinos are massive but the mechanism which induces neutrino masses remains an open question. Another consequence of the neutrino oscillation is the violation of lepton flavour in the neutral lepton sector. Searching for lepton flavor violation (LFV) in the charged sector becomes then a natural approach to better identify a new physics beyond what is currently known. Also, the LFV might be also connected to one of the most fundamental open question: the matter-antimatter asymmetry observed in our Universe [1]. In a top-down approach, the leptonic sector is involved in Grand Unified Theories (GUT) which predict new particles. One of the key ideas is to consider that quarks and leptons are two components of a unique, larger, mathematical object. This can be realised using larger groups of symmetry which lead to new gauge bosons coupled to leptons but also to quarks. In all the SM extensions addressing these issues, new states involving leptons are predicted and could be produced at the LHC.

2 Lepton Flavour Violation in Z Boson Decay

The LFV is observed in the neutral lepton sector via neutrino oscillations phenomenon. It can be searched at the LHC in the charged lepton sector using the $Z \to e\mu$ decay signature. In the SM, the level of LFV in the charged lepton sector is highly suppressed. Typically, the SM predicts $\mathcal{BR}(\mu \to e\gamma) \sim 10^{-54}$ [2], which is well below any experimental sensitivity. An observation

[a] e-mail: romain.madar@cern.ch

of LVF in the charged sector would lead to a clear signature for BSM physics. Experimental upper limits on $\mathcal{BR}(Z \to e\mu)$ were obtained at LEP ($< 1.7\ 10^{-6}$) but also indirectly using $\mu \to 3e$ constraints reintrepreted in term of $Z^* \to e\mu$ decay ($< 10^{-12}$). The ATLAS result [3] is based on an analysis of collisions where an isolated electron and muon are reconstructed, with an invariant mass compatible with the Z-boson mass. The signal would then appear as a mass-localised excess of events. The SM backgrounds are mainly reducible. The $WW \to e\nu_e\ \mu\nu_\mu$ production will produce a large missing transverse momentum ($E_{\mathrm{T}}^{\mathrm{miss}}$) due to escaping neutrinos coming from W-boson decay, as shown in Figure 1. The top quark pair production can be suppressed disgarding collision with a large number of jets. The only irreducible background is the $Z \to \tau\tau$ process where each tau lepton decays into a muon and a electron respectively. No excess of events is found on top of the expected SM background and an upper limit at 95% confidence level (CL) on the number of signal events is obtained based on an unbinned fit of data to the (e, μ) invariant mass distribution as shown in Figure 1. This upper limit is then converted into a limit on the branching ratio $\mathcal{BR}(Z \to e\mu)$. The obtained constraint reaches $7.5\ 10^{-7}$ and is twice better than LEP direct limit.

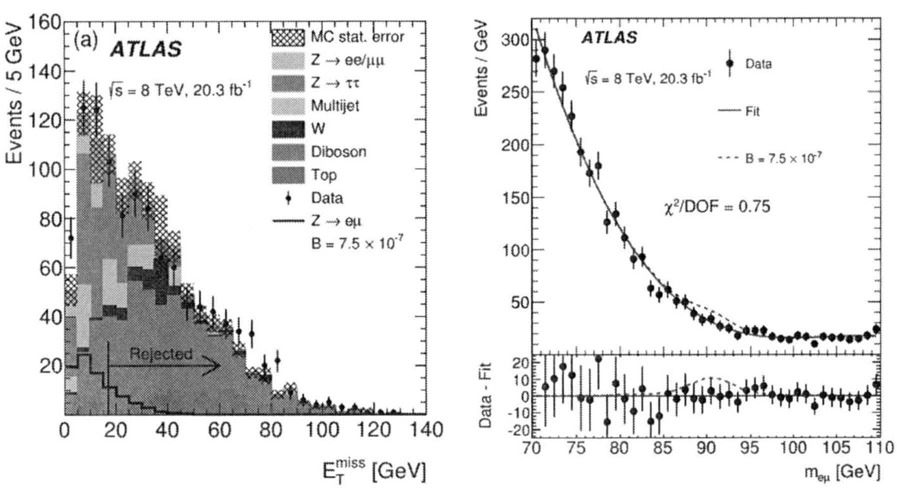

Figure 1. Missing transverse energy distribution (left) and invariant mass distribution of the (e, μ) system (right). Taken from Ref. [3].

Being a BSM phenomenon, it is natural to expect that LFV migh also occur in new neutral resonance sectors. ATLAS obtained constraints on this possible LFV signature [4] based on two benchmarks scenarios: a supersymmetric extension of the SM where the new resonance is a scalar ($\tilde{\nu}_\tau$) and an additional gauge sector with a new vectorial field (Z'). The strategy is the same as in [3] except that all the three possibles final states are analysed namely $e\mu$, $e\tau$, $\mu\tau$ where tau lepton decay in hadrons. Final states with hadronically decaying tau leptons lead to two specific experimental challenges. First, the background due to parton-initiated jets faking hadronic tau leptons can be large and is particularly difficult to accurately model using simulated collisions. The prediction of this background has to be extracted from real data in relevant signal-free control regions. Second, the reconstuction of the mass of the resonance cannot be fully performed because of the neutrinos coming from tau leptons decay. To cope with this issue, neutrino kinematics can be deduced from the missing transverse momentum and assuming they are emitted in the same direction as visible tau decay products - this assumption being valid for high masse resonances.

DOI: 10.1051/epjconf/201612604028

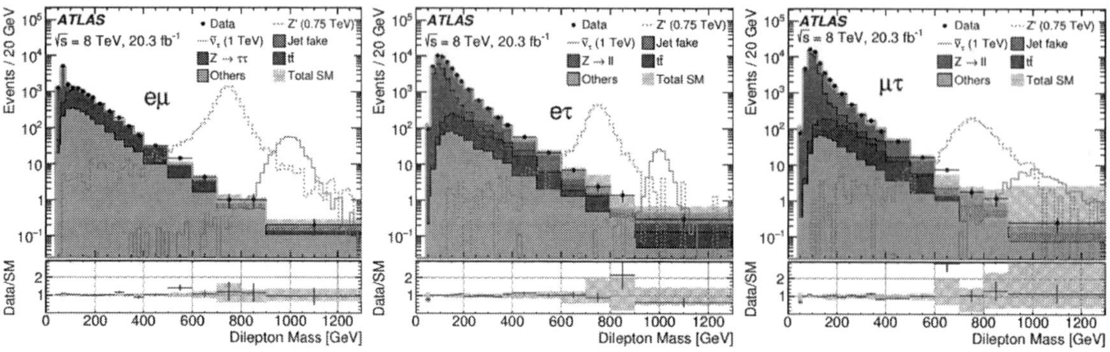

Figure 2. Dilepton invariant mass distribution for the $e\mu$ (left), $e\tau$ (middle) and $\mu\tau$ (right) final state. The expected background, observed data and expected signals (for the two considered benchmarks) are shown. Taken from Ref. [4].

Figure 2 shows the invariant mass of the (ℓ, ℓ') system for the three considered final states. The large contamination from parton-initiated jets is clearly visible for the $e\tau$ and $\mu\tau$ channels. Since no excess is observed in data, upper limits on the signal cross-section are extracted for the three different channels and the two benchmark signals. Thes excluded cross-section can be interpreted in excluded mass for the new resonances using the prediction of the two particular models. The typical excluded mass scale goes from 1.7 to 2.5 TeV. The CMS experiment has similar results [5] although a proper comparison is difficult due to significant differences in the theoretical benchmarks.

3 Leptoquarks Searches

The leptoquarks (LQ) are colored gauge bosons predicted by many GUTs attempting to unify lepton and quark fields in a same group representation. The parameter space of these theories is quite large but can be highly reduced by exisiting constraints on flavour changing neutral currents (FCNC) and on the proton lifetime. After taking these constraints into account, the phenomenology of these SM extensions includes three generations of leptoquarks corresponding to the three generation of fermions (LQ_i, $i = \{1, 2, 3\}$) - all other structure being strongly constrained by FCNC. These bosons can decay into either a charged lepton and quark or into a neutrino and a quark - all other couplings being strongly constrained by the proton lifetime. Experimentally, the relevant decay mode is the visible one $LQ_i \to \ell_i q_i$ and its branching ratio is written β.

At the LHC, leptoquarks can be pair or single produced. The search performed by ATLAS experiment [6] focuses on pair-produced leptoquarks of first and second generation[1] in $eejj$ and $\mu\mu jj$ final states respectively, the main SM background being Z+jets production. The search strategy is based on the specific topology of leptoquarks pair production. First, the high mass of these new gauge bosons lead particles with high transverse momentum (p_{T}) in the final state. This property helps to discriminate the signal from SM backgrounds with the distribution of S_{T}, defined as the scalar p_{T} sum of the two leading leptons and two leading jets, as shown in Figure 3. Second, the mass of the produced leptoquarks can be partially reconstructed using (ℓ, j) pairing combinatorics - each (ℓ, j) system coming from LQ decay having the same invariant mass.

[1]The third generation can be probed using bottom quark and top quark final states but the strategy is slightly different and this analysis is not detailed in this document.

 DOI: 10.1051/epjconf/201612604028

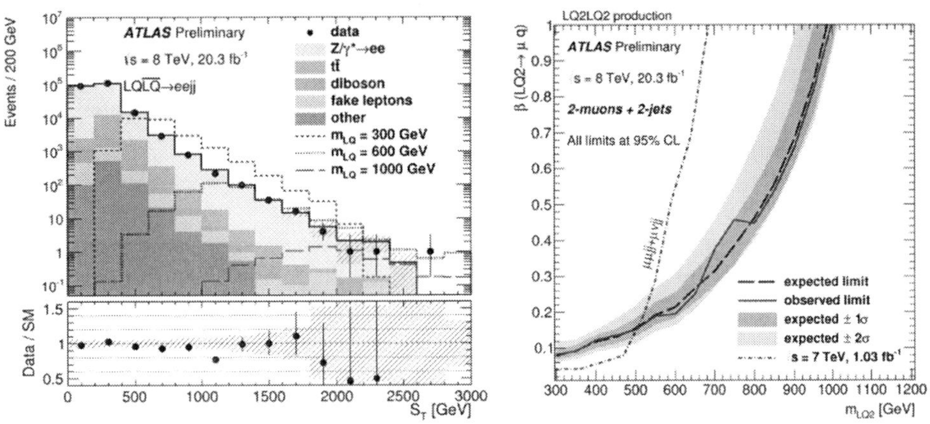

Figure 3. Distribution of scalar p_T sum for the $eejj$ final state (left) and excluded signal at 95% confidence level in the (β, m) parameter space for the $\mu\mu jj$ final state (right). Taken from Ref. [6].

Since no excess of data is observed, LQ must be heavy or weakly coupled to SM particles if they are realised in Nature. An upper limit at 95% C.L. on the cross-section of LQ pair production is obtained based on cut-and-count analysis in several signal regions tuned separately for each mass hypothesis. This limit can be interpreted in term of constraints in the model parameter space, namely the leptoquark mass (m_{LQ}) and the branching ratio of LQ decay into ℓj. Figure 3 shows the obtained constraint in the (m_{LQ},β) plane for the second generation leptoquarks. Overall, the typical excluded leptoquark mass scale is 1 TeV. CMS experiment has similar results [7] although a search for single LQ production only were performed in the electron channel.

4 Heavy Lepton Searches

Neutrino masses, offering the simplest explanation of flavour oscillations, are not described in the SM and can also be solved by BSM dynamics which predict new heavy leptonic states. ATLAS experiment performed searches for three types of heavy leptons exploiting several final states [8, 10, 11].

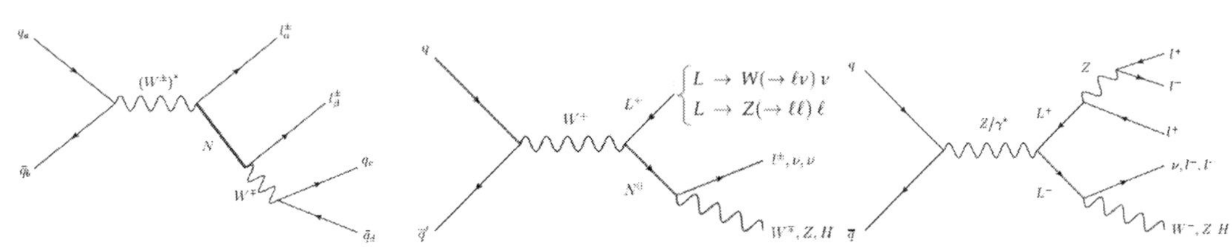

Figure 4. Distribution of scalar p_T sum for the $eejj$ final state (left) and excluded signal at 95% confidence level in the (β, m) parameter space for the $\mu\mu jj$ final state (right).

The first type is a generic heavy Majorana neutrino N appearing in two different classes of model: type-I seesaw mechanism allowing generating neutrino masses and left-right symmetric models (where left-handed and right-handed fermions interact in the same way) having additionnal gauge bosons (W_R', Z'). The second type of heavy lepton consists in a weak-triple lepton (L^+, N^0, L^-) involved in type-III seesaw mechanism, and the last type is a heavy lepton having only vectorial couplings (vector-like lepton) which are predicted by many BSM theories like composite Higgs boson or wraped extra-dimensions. Figure 4 shows examples of feynman diagrams for the different signals process which could occurs at the LHC.

ATLAS experiment focused on heavy Majorana neutrino search using two leptons having the same electric charge [8], as showns in first diagram of Figure 4. The particular signal topology allows reducing the SM background, dominated by charge mis-reconstruction and mis-identification of leptons. In the type-I seesaw signal, the hardonic decay of the W boson makes the di-jet mass peaking at m_W. In the case of the left-right symmetric model, the heavy Z' is produced in a s-channel and follows the decay chain

$$pp \to Z' \to NN \text{ and } N \to W_R^{\pm\prime}(\to jj)\,\ell^{\pm}$$

The Z' mass can be then reconstructed from the invariant mass of all objects in the final state. Since no excess is observed in data, upper limits on cross-section can be obtained and as well as a reinterpretation in term of model parameters. Figure 5 shows the region of the left-right symmetric model parameters which are excluded by ATLAS experiment. In summary, Majorana neutrino is excluded up to 300 GeV for a given coupling to lepton $|V_{\ell N}|^2 = 10^{-1}$ (the excluded region in the mass-coupling plane is also derived). CMS experiment obtained better constraint but analyzed the muon channel only [9].

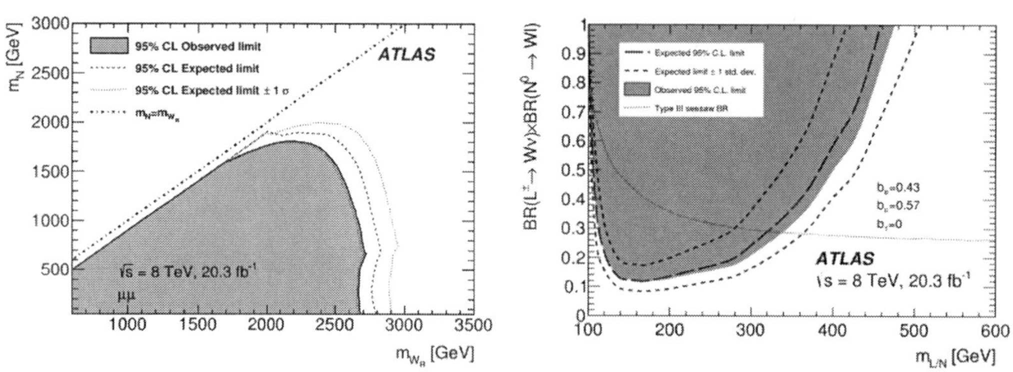

Figure 5. Excluded region at 95% C.L. for the left-right symmetric model (left) and type-III seesaw mechanism (right). Taken from Ref. [8].

The models based on a triplet of heavy leptons with $L \to W\nu$ have two type of experimental signatures, as shown in the second diagram of Figure 4. One contains two same-sign leptons with two jets and the other has two oppsite-sign leptons with tow jets, coming from the W boson decay in each case. Event selection is optimized for each of the six categories to increase the signal over background ratio. Since no excess is observed in data, an upper limit were obtained and reinterpreted in constraints of term of model parameters. Figure 5 shows the excluded regions the ($m_{N/L}$, $\mathcal{BR}_{L\to W\nu} \times \mathcal{BR}_{N\to W\ell}$) plane, where $m_{N/L}$ is the mass of the triplet. The typical excluded mass scale is reaches 350 GeV for this type of heavy leptons.

It is possible to extend the search to the case where $L \to Z\ell$ by considering several multi-lepton final states [11]. Models based on vector-like lepton can also lead to the same kind of experimental

 DOI: 10.1051/epjconf/201612604028

signature. In this case, there are 6 signal regions can be analyzed to probe the presence of signal. Since no excess is observed, upper limits are derived and typical excluded mass scales reach 400 GeV for type-III seesaw leptons and 160 GeV for vector-like leptons.

5 Multi-photons Final States Searches

The search for new physics in multi-photon final states is also connected to rare Z decays and more generally to new dynamics in the lepton sector. The lepton flavor violation in Z decay was discussed in section 2, but this is not the only way Z boson properties can be affected by new physics. Indeed, a Z decay into three photons would be clear signature of new physics as it is highly suppressed in the SM ($\mathcal{BR}(Z \to 3\gamma) \sim 10^{-10}$). In addition, many BSM models predict multi-photon final state such as nMSSM predicting a pseudo-scalar boson a coupled to Higgs bosons and photons or additional gauge sector extensions predicting Z' coupled to a and photon fields. Figure 6 shows the feynman diagrams corresponding to the signal processes of these SM extensions at the LHC, which are searched for by ATLAS experiment [12].

Figure 6. Feynman diagrams corresponding to the signal in the nMSSM (left) and in case of new gauge sector (right).

In this kind of final state, the main background comes from parton-initiated jets that are wrongly reconstructed as photons. Being challenging to accurately model using simulation, the prediction of this SM background is obtained using real data in signal-free control regions. The search strategies are different for $Z \to 3\gamma$ search and for new resonances search. The first one is based on count of events with three well isolated photons having an invariant mass compatible with the Z-boson mass. For resonnances searches, a localised excess of event is searched for in each photon pair invariant mass spectrum, as shown in Figure 7. There is no evidence for new physics in both cases and upper limits are then derived. The constraint on $\mathcal{BR}(Z \to 3\gamma)$ reaches the level of 2.2 10^{-6}, which is five times better than LEP limit. Concerning the a and Z' resonance searches, the excluded production cross-section is of the order of the 1 fb for a-boson mass between 10 and 240 GeV and a Z'-boson mass between 100 and 500 GeV, as shown in Figure 7. These are the first constraints on this type of models.

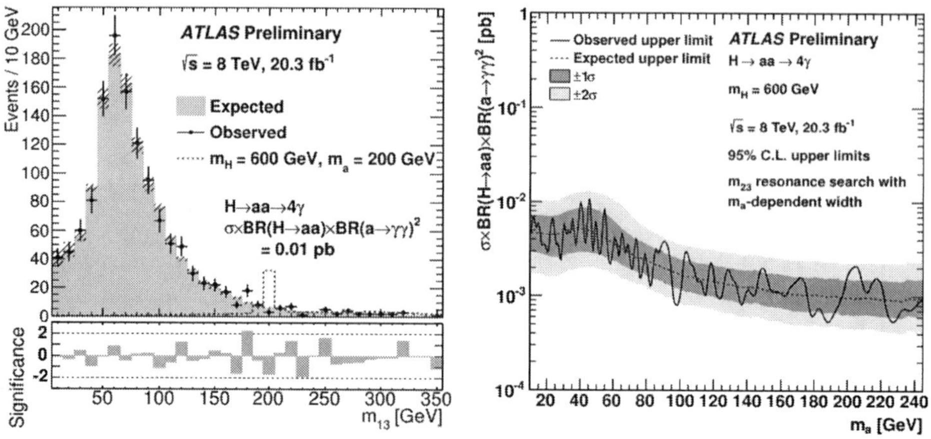

Figure 7. Distribution of invariant mass of the leading and third-leading photon system (left) and upper limit on cross-section production for nMSSM signal (right). Taken from Ref. [12].

6 Summary

The lepton sector of the SM is very promising to search for new physics given the observed neutrino oscillations, and could possibly address the matter-antimatter asymmetry problem. Search for lepton flavor violation in the SM were pushed twice below existing LEP limits and LFV beyond the SM sector is now excluded to typical mass scale of 2.5 TeV. The interesting approach of unifying quarks and leptons predicting new gauge bosons is also now excluded up to a typical mass scale of 1 TeV. Heavy leptons that could explain neutrino oscillations are now constrained to be typically heaviest than 300 GeV. Finally, multiphoton final states allow constraining rare Z-boson decay in three photon to the level of 10^{-6} which is five times better than LEP limit. It also provides the first constraints on particular BSM signals leading to multiphoton final states.

References

[1] M. Fukugita and T. Yanagida, Phys. Lett. B. **174**, 45 (1986)
[2] S. M. Bilenky, S.T. Petcov, B. Pontecorvo, Phys. Lett. B **67** 309 (1977)
[3] ATLAS Collaboration, Phys. Rev. D. **90**, 072010 (2014)
[4] ATLAS Collaboration, Phys. Rev. Lett. **115**, 031801 (2015)
[5] CMS Collaboration, PAS EXO-13-002 (2015)
[6] ATLAS Collaboration, arXiv:1508.04735 [hep-ex] (2015), submitted to EPJC
[7] CMS Collaboration, arXiv:1509.03750 [hep-ex] (2015), submitted to PRD
[8] ATLAS Collaboration, JHEP **07** 162 (2015)
[9] CMS Collaboration, Phys. Lett. B **748**, 144-166 (2015)
[10] ATLAS Collaboration, JHEP **09** 108 (2015)
[11] ATLAS Collaboration, Phys. Rev. D. **92** 032001 (2015)
[12] ATLAS Collaboration, arXiv:1509.05051 [hep-ex] (2015), submitted to EPJC

EPJ Web of Conferences **126**, 04029 (2016) DOI: 10.1051/epjconf/201612604029
ICNFP 2015

Aspects of Meson Condensation

Andrea Mammarella[1,a]

[1] *Laboratori Nazionali del Gran Sasso*

Abstract. In this work pion and kaon condensation in the framework of chiral perturbation theory is studied. I consider a system at vanishing temperature with nonzero isospin chemical potential and strangeness chemical potential; meson masses and mixing in the normal phase, the pion condensation phase and the kaon condensation phase are described. There are differences with previous works, but the results presented here are supported by both theory group analysis and by direct calculations. Some pion decay channels in the normal and the pion condensation phases are studied, finding a nonmonotonic behavior of the Γ decay as a function of μ_I.

1 Introduction

The properties of strongly interacting matter in an isospin and/or strangeness rich medium are relevant in a wide range of phenomena including the astrophysics of compact stars and heavy-ion collisions. It is known that depending on the value of the isospin chemical potential, μ_I, and on the value of the strangeness chemical potential, μ_S, three different phase can be realized: the normal phase, the pion condensed (πc) phase and the kaon condensed (Kc) phase [1–3]. The realization of a mesonic condensate can drastically change the low energy properties of matter, including the mass spectrum and the lifetime of mesons.

Previous analyses of the meson condensed phases by QCD-like theories were developed in [4, 5]. Pion condensation in two-flavor quark matter was studied in [2, 6] and in three-flavor quark matter in [3]. In particular, the phase diagram as a function of μ_I and μ_S was presented in [3]. Finite temperature effects in $SU(2)_L \times SU(2)_R$ chiral perturbation theory (χPT) have been studied in [7–10]. One remarkable property of quark matter with nonvanishing isospin chemical potential is that it is characterized by a real measure, thus the lattice realization can be performed with standard numerical algorithms [11, 12]. The πc phase and the Kc phase have been studied by NJL models in [13–15] and by random matrix models in [16]. All these models find results in qualitative and quantitative agreement, and in particular, the phase diagram of matter has been firmly established. However, regarding the low energy mass spectrum in three-flavor quark matter, we found that it was only studied in [3]. Our results are in disagreement with those of [3], the most relevant difference is in the mixing between mesonic states. Regarding the pion decay, previous works focused on density and temperature effects in standard decay channels [17–19], but not all the decay channels have been

[a] e-mail: andrea.mammarella@lngs.infn.it

considered.

In this article I briefly review how to include chemical potentials in χPT [20–24], then I describe some phenomenological properties related to this inclusion, like the existence of the different phases already listed, the meson masses and mixing. I will also show how the inclusion of chemical potentials affects pion decay channels [25].

The paper is organized as follows. In Sec. 2 I describe the model and how it predicts different phases. In Sec. 3 I show how group theory tools can be used to calculate the mass eigenstates in the condensed phases and I list and discuss the results obtained. In Sec.4 I discuss the impact of chemical potentials on charged pion decays. In Sec. 5 I summarize the results.

2 Model

2.1 Lagrangian and definitions

In this section I briefly review the model that I am going to use in the following. It is the one described in [3]. The general $O(p^2)$ Lorentz invariant Lagrangian density describing the pseudoscalar mesons can be written as

$$\mathcal{L} = \frac{F_0^2}{4}\mathrm{Tr}(D_\nu\Sigma D^\nu\Sigma^\dagger) + \frac{F_0^2}{4}\mathrm{Tr}(X\Sigma^\dagger + \Sigma X^\dagger)\,, \tag{1}$$

where Σ corresponds to the meson fields, $X = 2B_0(s + ip)$ describes scalar and pseudoscalar external fields and the covariant derivative is defined as

$$D_\mu\Sigma = \partial_\mu\Sigma - \frac{i}{2}[v_\mu, \Sigma] - \frac{i}{2}\{a_\mu, \Sigma\}\,, \tag{2}$$

with v_μ and a_μ the external vectorial and axial currents, respectively. The Lagrangian has two free parameters F_0 and B_0, related to the pion decay and to the quark-antiquark condensate, respectively, see for example [20–24].

The Lagrangian density is invariant under $SU(N_f)_L \times SU(N_f)_R$ provided the meson field transform as

$$\Sigma \to R\Sigma L^\dagger\,, \tag{3}$$

and the chiral symmetry breaking corresponds to the spontaneous global symmetry breaking $SU(N_f)_L \times SU(N_f)_R \to SU(N_f)_{L+R}$. In standard χPT, the mass eigenstates are charge eigenstates as well. Thus mesons are particles with a well defined mass and charge. The presence of a medium can change this picture. In particular, if the vacuum carries an electric charge, then the mass eigenstates will not typically be charge eigenstates. The presence of a medium can be taken into account by considering appropriate external currents in Eq. (1).

At vanishing temperature the vacuum is determined by maximizing the Lagrangian density with respect to the external currents. The pseudoscalar mesons are then described as oscillations around the vacuum. We use the same nonlinear representation of [3] corresponding to

$$\Sigma = u\bar{\Sigma}u \qquad \text{with} \qquad u = e^{iT\cdot\phi/2}\,, \tag{4}$$

where T_a are the $SU(N_f)$ generators and $\bar{\Sigma}$ is a generic $SU(N_f)$ matrix to be determined by maximizing the static Lagrangian. The reasoning behind the above expression is that under $SU(N_f)_L \times SU(N_f)_R$ mesons can be identified as the fluctuations of the vacuum as in Eq. (3) with $\theta_a^R = -\theta_a^L = \phi_a$.

In the following we will assume that $a_\mu = 0$, $p = 0$, $X = 2GM$, where M is the $N_f \times N_f$ diagonal quark mass matrix and G is a constant, that with these conventions is equal to B_0. Moreover, we will assume that

$$v^\nu = -2eQA^\nu - 2\mu\delta^{\nu 0} = -\frac{2}{3}(\mu_B - \mu_S)I\delta^{\nu 0} - \tilde{A}_3^\nu \lambda_3 - \tilde{A}_8^\nu \lambda_8\,, \tag{5}$$

meaning that the vectorial current consists of the electromagnetic field and a quark chemical potential, with μ a $SU(N_f) \times SU(N_f)$ matrix in flavor space. Its explicit expression is:

$$\mu = \text{diag}\,(\mu_u, \mu_d, \mu_s) = \text{diag}\left(\frac{1}{3}\mu_B + \frac{1}{2}\mu_I, \frac{1}{3}\mu_B - \frac{1}{2}\mu_I, \frac{1}{3}\mu_B - \mu_S\right) = \frac{\mu_B - \mu_S}{3} I + \frac{\mu_I}{2}\lambda_3 + \frac{\mu_S}{\sqrt{3}}\lambda_8\,, \tag{6}$$

It is important to remark that this model only holds for $|\mu_B| \lesssim 940$ MeV, $|\mu_I| \lesssim 770$ MeV and $|\mu_S| \lesssim 550$ MeV [3].

2.2 Groud state and different phases

To find the ground state we have to substitute (4) in the Lagrangian (1). It is not necessary to use a complete SU(3) parametrization for $\bar{\Sigma}$, but is sufficient:

$$\bar{\Sigma} = \begin{pmatrix} 1 & 0 & 0 \\ 0 & \cos\beta & -\sin\beta \\ 0 & \sin\beta & \cos\beta \end{pmatrix} \begin{pmatrix} \cos\alpha & \sin\alpha & 0 \\ -\sin\alpha & \cos\alpha & 0 \\ 0 & 0 & 1 \end{pmatrix} \begin{pmatrix} 1 & 0 & 0 \\ 0 & \cos\beta & \sin\beta \\ 0 & -\sin\beta & \cos\beta \end{pmatrix}, \tag{7}$$

because $\bar{\Sigma}$ has to be orthogonal to the chemical potential in the SU(3) generator space [25].
Substituting (7) in (1) and maximizing, we find three different vacua, that implies that there are three ground states and therefore three different phases [3]:

- Normal phase:

$$\mu_I < m_\pi\,, \tag{8}$$

$$\mu_S < m_K - \frac{1}{2}\mu_I\,, \tag{9}$$

characterized by

$$\alpha_N = 0, \quad \beta_N \in (0, \pi), \quad \bar{\Sigma}_N = \text{diag}(1, 1, 1)\,. \tag{10}$$

- Pion condensation phase:

$$\mu_I > m_\pi\,, \tag{11}$$

$$\mu_S < \frac{-m_\pi^2 + \sqrt{(m_\pi^2 - \mu_I^2)^2 + 4m_K^2\mu_I^2}}{2\mu_I}\,, \tag{12}$$

characterized by

$$\cos\alpha_\pi = \left(\frac{m_\pi}{\mu_I}\right)^2, \quad \beta_\pi = 0\,, \tag{13}$$

$$\bar{\Sigma}_\pi = \begin{pmatrix} \cos\alpha_\pi & \sin\alpha_\pi & 0 \\ -\sin\alpha_\pi & \cos\alpha_\pi & 0 \\ 0 & 0 & 1 \end{pmatrix} = \frac{1 + 2\cos\alpha_\pi}{3} I + i\lambda_2 \sin\alpha_\pi + \frac{\cos\alpha_\pi - 1}{\sqrt{3}}\lambda_8\,. \tag{14}$$

- Kaon condensation phase:

$$\mu_S > m_K - \frac{1}{2}\mu_I, \tag{15}$$

$$\mu_S > \frac{-m_\pi^2 + \sqrt{(m_\pi^2 - \mu_I^2)^2 + 4m_K^2\mu_I^2}}{2\mu_I}, \tag{16}$$

characterized by

$$\cos\alpha_K = \left(\frac{m_K}{\frac{1}{2}\mu_I + \mu_S}\right)^2, \beta_K = \pi/2, \tag{17}$$

$$\bar{\Sigma}_K = \begin{pmatrix} \cos\alpha & 0 & \sin\alpha \\ 0 & 1 & 0 \\ -\sin\alpha & 0 & \cos\alpha \end{pmatrix} = \tag{18}$$

$$\frac{1 + 2\cos\alpha_K}{3} I + \frac{\cos\alpha_K - 1}{2\sqrt{3}}\left(\sqrt{3}\lambda_3 - \lambda_8\right) + i\lambda_5 \sin\alpha_K .$$

Note that the kaon condensation can only happen for

$$\mu_S > \bar{\mu}_S = m_K - \frac{m_\pi}{2}. \tag{19}$$

3 Meson Masses and Mixing

3.1 Mixing

As we have seen in Sec. 2.2 the ground state in the condensed phases is not diagonal and thus has SU(3) charges that cause symmetry breaking. It is useful to study the breaking pattern to learn something about the possible meson mixing in the condensed phases.

The starting Lagrangian has an $SU(3)_L \times SU(3)_R$ simmetry, broken to $SU(3)_V$ by the quark masses. The introduction of chemical potentials via the external current (5) further reduce this symmetry to $U(1)_{L+R} \times U(1)_{L+R}$, meaning that the λ_3 and λ_8 terms in (5) break isospin and hypercharge conservation. When the system enters one of the condensed phases the vacuum acquires a charge and thus there is no symmetry left.

In all these cases we can use the quantum numbers of the $SU(2)$ subgroups of $SU(3)$ to label the states and to find out the ones that can mix. In figure 1 these quantum numbers are represented. We need only two of them, because they are not independent. In table 1 are shown the states that can mix and the related quantum numbers.

Unfortunately, this is not sufficient to determine if π_0 and η can mix, because they do not have a well defined U and V spin, so we have to study deeply how the ground state affect them. Let us first consider the normal phase. In the normal phase there is no operator that can induce the mixing of the mesonic states, thus the mesonic states remain unchanged but the Q_3 and Q_8 charges will induce Zeeman-like mass splittings.

In any of the condensed phases, there is an additional charge that is spontaneously induced, and the corresponding operator will lead to mixing.

Let us first focus on isospin (or T-spin). We have to consider two cases. Suppose that the vacuum has a charge that commutes with T^2, as in the πc phase, say the charge corresponding to $T_2 = i(T_- - T_+)$, see Eq. (4). The $T_\pm$ operators can induce mixing among the charged pions and among the kaons.

Table 1. Mixing mesons with the corresponding T-spin and U-spin quantum numbers. These quantum numbers label the $SU(3)$ subspace spanned by the corresponding mesonic states. The π_0 and the η do not appear because they are not U-spin eigenstates.

Mixing states	(T, U)
π_+, π_-	$(1, 1/2)$
K_+, K_-	$(1/2, 1/2)$
$K_0, \bar{K}_0$	$(1/2, 1)$

On the other hand, T-spin conservation does not allow the $|\pi_0\rangle = |T = 1, T_3 = 0\rangle$ to mix with the $|\eta\rangle = |T = 0, T_3 = 0\rangle$.

Now suppose instead that the vacuum has a charge that does not commute with T^2 as in the Kc phase, see Eq. (18). Any operator that does not commute with isospin will commute with U-spin or with V-spin. In the Kc phase $Q_5|0\rangle \neq 0$, then the vacuum is not invariant under this charge. However, since $[T_5, U] = 0$ it follows that U-spin is conserved. The lowering and raising operators inducing the mixing will be $U_\pm$. Regarding the π_0 and the η, in this case we have that $|U = 1, U_3 = 0\rangle$ and $|U = 0, U_3 = 0\rangle$ do not mix. Since $|U = 1, U_3 = 0\rangle = \frac{|\pi_0\rangle + \sqrt{3}|\eta\rangle}{2}$ and $|U = 0, U_3 = 0\rangle = \frac{\sqrt{3}|\pi_0\rangle - |\eta\rangle}{2}$, these will be the mass eigenstates.

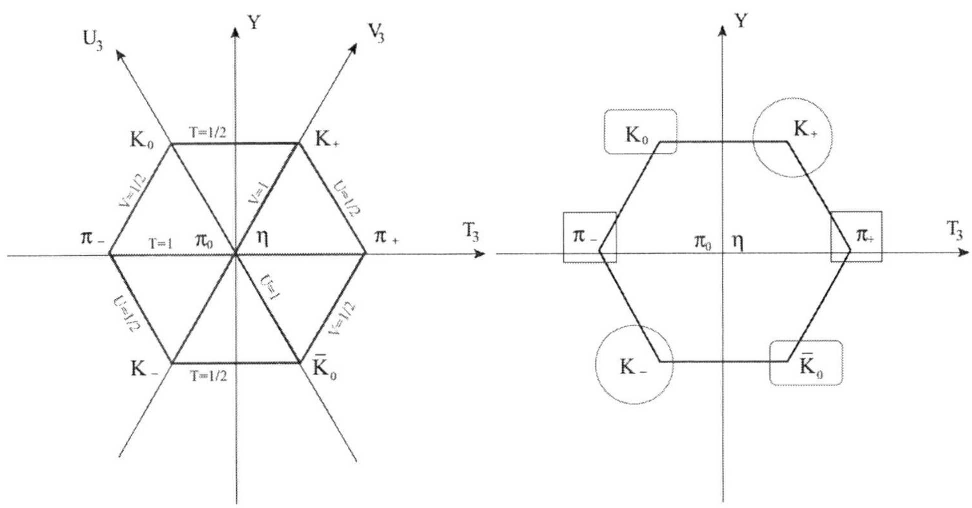

Figure 1. (color online) Weight diagrams of the mesonic octet. In the top panel we have reported the axes corresponding to the third component of isospin (or T-spin), U-spin, V-spin and hypercharge, Y. Any mesonic state belongs to a multiplet of T-spin, U-spin and V-spin as indicated in the diagram. In the bottom panel we have indicated the mesonic states having the same T-spin, U-spin and V-spin quantum numbers. Only the states marked with the same symbol can mix. The π_0 and the η are not simultaneous T-spin, U-spin and V-spin eigenstates; their mixing depends on the spontaneously induced charge of the vacuum.

DOI: 10.1051/epjconf/201612604029

3.2 Masses

The mass eigenstates are found diagonalizing the Lagrangian in the different phases. They present the mixing predicted by the group theory analysis of the previous subsection. Remember the definitions:

$$m_\pi^2 = 2Gm/F_0^2 \,, \tag{20}$$

$$m_K^2 = G(m + m_s)/F_0^2 \,, \tag{21}$$

$$m_{\eta^0}^2 = 2G(m + 2m_s)/3F_0^2 = (4m_K^2 - m_\pi^2)/3 \,. \tag{22}$$

Meson masses have been calculated in [25]. They are, in the normal phase:

$$m_{\pi^0} = m_\pi \,, \tag{23}$$

$$m_{\pi^\pm} = m_\pi \mp \mu_I \,, \tag{24}$$

$$m_{\eta^0} = \sqrt{(4m_K^2 - m_\pi^2)/3} \,, \tag{25}$$

$$m_{K^\pm} = m_K \mp \frac{1}{2}\mu_I \mp \mu_S \,, \tag{26}$$

$$m_{K^0/\bar{K}^0} = m_K \pm \frac{1}{2}\mu_I \mp \mu_S \,. \tag{27}$$

In the condensed phases, as said, the mass eigenstates are mixture of normal phase mesons, so I will list them with a ~. They have been calculated in [25]. In the pion condensation phase, their masses are:

$$m_{\tilde{\pi}_0} = \mu_I \tag{28}$$

$$m_{\tilde{\pi}_+} = 0 \tag{29}$$

$$m_{\tilde{\pi}_-} = \frac{\sqrt{3m_\pi^4 + \mu_I^4}}{\mu_I} \tag{30}$$

$$m_{\tilde{\eta}}^2 = m_\eta^2 + \frac{1}{3}m_\pi^2\left(\frac{m_\pi^2 - \mu_I^2}{\mu_I^2}\right) \tag{31}$$

$$m_{\tilde{K}^-/\tilde{K}^+} = \pm\frac{1}{2}\left(\frac{m_\pi^2}{\mu_I} + 2\mu_S \pm \sqrt{\left(\frac{m_\pi^2}{\mu_I} + 2\mu_S\right)^2 + 4M_4^2}\right) \tag{32}$$

$$m_{\tilde{K}^0/\tilde{\bar{K}}^0} = \pm\frac{1}{2}\left(\frac{m_\pi^2}{\mu_I} - 2\mu_S \pm \sqrt{\left(\frac{m_\pi^2}{\mu_I} - 2\mu_S\right)^2 + 4M_6^2}\right) . \tag{33}$$

with:

$$s_{45} = \mu_I \cos\alpha + 2\mu_S \,, \tag{34}$$

$$M_4^2 = M_5^2 = m_k^2 + \frac{1}{4}\mu_I^2 - \mu_S^2 - \frac{1}{2}(\mu_I^2 + 2\mu_I\mu_S)\cos\alpha_\pi \,, \tag{35}$$

$$s_{67} = \mu_I \cos\alpha_\pi - 2\mu_S \,, \tag{36}$$

$$M_6^2 = M_7^2 = m_k^2 + \frac{\mu_I^2}{4} - \mu_S^2 - \frac{\mu_I^2 - 2\mu_I\mu_S}{2}\cos\alpha_\pi \,. \tag{37}$$

In the kaon condensation phase, their masses are:

$$m_{\tilde{\pi}^\pm} = \mp\frac{1}{2}\left(\frac{\mu_I}{2}(3+\cos\alpha_K) + \mu_S(\cos\alpha_K - 1) \mp \sqrt{\left(\frac{\mu_I}{2}(3+\cos\alpha_K) + \mu_S(\cos\alpha_K - 1)\right)^2 + 4M_1'^2}\right), \tag{38}$$

$$m_{\tilde{K}^+} = 0, \tag{39}$$

$$m_{\tilde{K}^-} = \sqrt{M_5'^2 + u_{45}^2}, \tag{40}$$

$$m_{\tilde{K}^0/\tilde{\bar{K}}^0} = \pm\frac{1}{2}(u_{67} \pm \sqrt{u_{67}^2 + 4M_6'^2}), \tag{41}$$

$$m_{\tilde{\pi}^0/\tilde{\eta}} = \sqrt{\frac{M_3'^2 + M_8'^2 \mp \sqrt{4u_{38}^2 + (M_3'^2 - M_8'^2)^2}}{2}}. \tag{42}$$

with:

$$u_{12} = \frac{1}{2}((3+\cos\alpha_K)\mu_I + 2(\cos\alpha_K - 1)\mu_S), \tag{43}$$

$$M_1'^2 = M_2'^2 = m_\pi^2 - \frac{1}{2}\mu_I^2(1+\cos\alpha_K) - 4\mu_I\mu_S(\cos\alpha_K - 1), \tag{44}$$

$$M_5'^2 = m_k^2\cos\alpha_K - \frac{1}{4}(\mu_I + 2\mu_S)^2\cos(2\alpha_K), \tag{45}$$

$$u_{45} = \frac{1}{4}\left(\frac{1+3\cos\alpha_K)}{\cos\alpha_K}\right)(\mu_I + 2\mu_S), \tag{46}$$

$$M_6'^2 = M_7'^2 = m_k^2 + \frac{\mu_I - 2\mu_S}{4}((\cos\alpha(\mu_I + 2\mu_S) - 2\mu_I), \tag{47}$$

$$u_{67} = \frac{1}{2}((-3+\cos\alpha)\mu_I + 2(1+\cos\alpha)\mu_S), \tag{48}$$

$$M_3'^2 = \frac{1}{24}\left[\cos\alpha_K\left(2m_k^2 + 9m_\pi^2 + 6\frac{G}{F_0^2}m_s\right) + m_\pi^2(16 - 6\cos^2\alpha_K)\right], \tag{49}$$

$$M_8'^2 = \frac{1}{8\cos\alpha_K}(2m_k^2 + (1+2\cos\alpha_K)m_\pi^2 + 6\frac{G}{F_0^2}m_s\cos\alpha_K), \tag{50}$$

$$u_{38} = \frac{1}{8\sqrt{3}}(-2m_k^2 + (3+2\cos\alpha_K)m_\pi^2 - 6\frac{G}{F_0^2}m_s). \tag{51}$$

Plot of these masses as a function of μ_I for fixed values of μ_S are shown in figure 2.

4 Pion decays

It has been shown how the introduction of chemical potentials can change meson masses and mixing. Here I will describe how it affects the charged pions decay. The processes are:

$$\pi^+ \longrightarrow \ell^+\nu_\ell, \tag{52}$$

$$\pi^- \longrightarrow \ell^-\bar{\nu}_\ell, \tag{53}$$

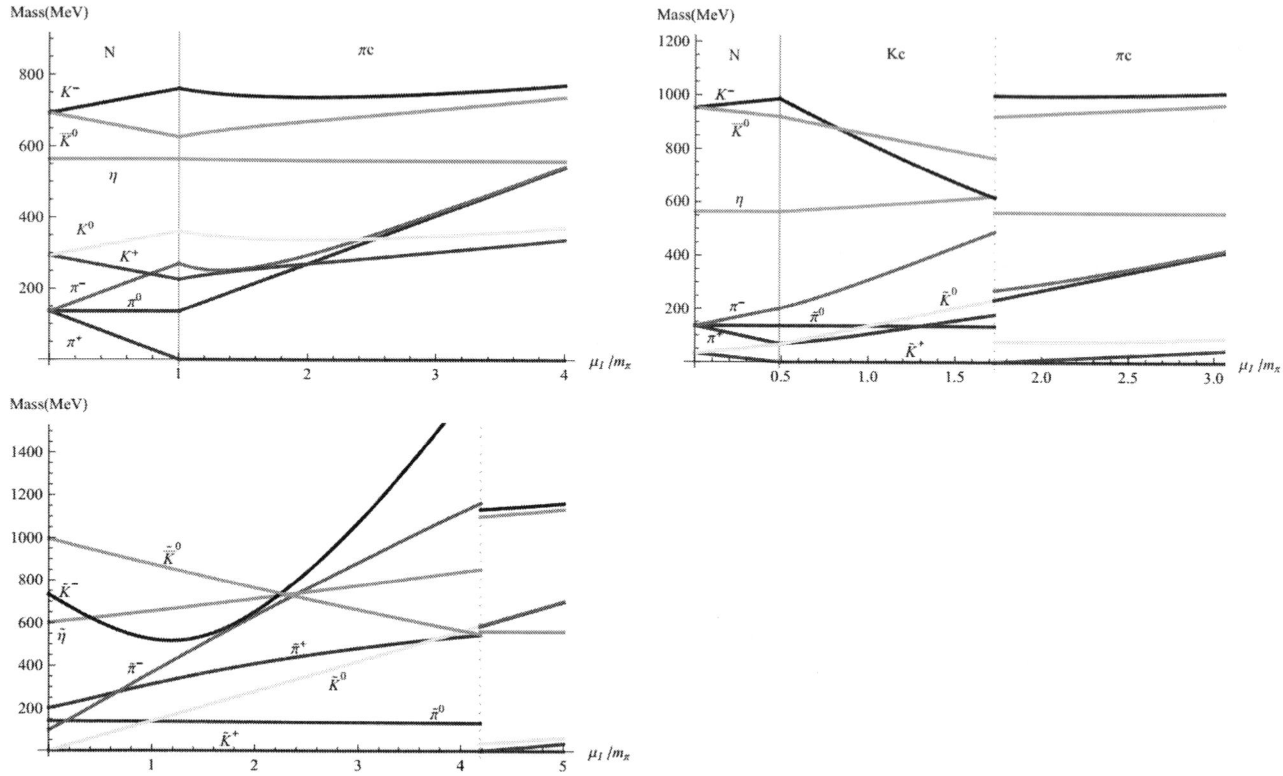

Figure 2. (color online) Mass spectrum of the pseudoscalar mesonic octet. Top panel, results obtained for $\mu_S = 200$ MeV. The vertical solid line represents the second order phase transition between the normal phase and the pion condensation phase. In this case the strange quark chemical potential is below the threshold value for kaon condensation, 425 MeV, thus the kaon condensed phase does not take place for any value of μ_I. Middle panel, results obtained for $\mu_S = 460$ MeV. The vertical solid line represents the second order phase transition from the normal phase to the kaon condensation phase. The dashed line corresponds to the first order phase transition between the kaon condensed phase to the pion condensed phase. Bottom panel, results obtained for $\mu_S = 550$ MeV, corresponding to the largest value of μ_S.

that in the Standard Model (SM) have the decay width:

$$\Gamma^0_{\pi\to\ell\nu_\ell} = \frac{G_F^2 F_0^2 V_{ud}^2 m_\ell^2 m_\pi}{4\pi}\left(1 - \frac{m_\ell^2}{m_\pi^2}\right)^2, \tag{54}$$

where G_F is the Fermi constant, V_{ud} is the ud CKM matrix element, m_ℓ and m_π are the lepton and pion masses.
For the following calculation I will assume $\mu_S = 0$ for simplicity.
In the normal phase chemical potentials do not mix states, so the only change is that we have to replace m_π with $m_{\pi_\pm}$ from (24).
In the pion condensation phase the $\tilde{\pi}_+$ is massless, so it does not decay, while the $\tilde{\pi}_-$ is a mixture of

 DOI: 10.1051/epjconf/201612604029

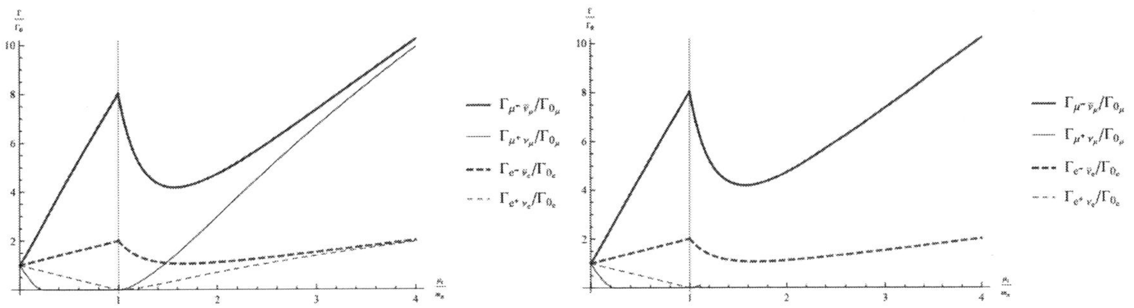

Figure 3. (color online) Normalized leptonic decay rates of charged pions in normal phase and in the condensed phase normalized to the value in vacuum. The phase transition between the normal phase and the pion condesed phase corresponds to the solid vertical line. Top, results obtained for vanishing leptonic chemical potential. Bottom, results obtained assuming weak equilibriub. In this case the decay in positively charged leptons is Pauli blocked. In both plots the thick solid line represents $\Gamma_{\mu^-\bar{\nu}_\mu}/\Gamma_{0_\mu}$, the thin solid line represents $\Gamma_{\mu^+\nu_\mu}/\Gamma_{0_\mu}$, the thick dashed line represents $\Gamma_{e^-\bar{\nu}_e}/\Gamma_{0_e}$ and the thin dashed line represents $\Gamma_{e^+\nu_e}/\Gamma_{0_e}$

$\pi_\pm$ so it can decay in both $\ell^+\nu_\ell$ and $\ell^-\bar{\nu}_\ell$. The related decay widths are:

$$\frac{\Gamma_{\tilde{\pi}_-\to\ell^+\nu_\ell}}{\Gamma^0_{\pi\to\ell\nu_\ell}} = \frac{|U^*_{21}\cos\alpha + iU^*_{22}|^2}{2}\frac{m_{\tilde{\pi}^-}}{m_\pi}\left(\frac{1-m_\ell^2/m_{\tilde{\pi}^-}^2}{1-m_\ell^2/m_\pi^2}\right)^2, \tag{55}$$

$$\frac{\Gamma_{\tilde{\pi}_-\to\ell^-\bar{\nu}_\ell}}{\Gamma^0_{\pi\to\ell\nu_\ell}} = \frac{|U^*_{21}\cos\alpha - iU^*_{22}|^2}{2}\frac{m_{\tilde{\pi}^-}}{m_\pi}\left(\frac{1-m_\ell^2/m_{\tilde{\pi}^-}^2}{1-m_\ell^2/m_\pi^2}\right)^2, \tag{56}$$

where:

$$\begin{pmatrix}\tilde{\pi}_+\\ \tilde{\pi}_-\end{pmatrix} = \begin{pmatrix}U_{11} & U_{12}\\ U_{21} & U_{22}\end{pmatrix}\begin{pmatrix}\pi_+\\ \pi_-\end{pmatrix}, \tag{57}$$

The plot of these decay widths are shown in figure 3.

5 Conclusion

I have shown how meson physics in presence of chemical potentials can be described using Chiral Perturbation Theory. This lead to three different phases, a normal phase, a pion condensation phase and a kaon condensation phase. The condensed phases have a charge and thus can generate mixing among mesons.

In this context I have illustrated how the mixing is influenced by model symmetries and that groups theory constraints the mixing possibilities. These results, obtained by group theory alone, is expected to hold in any theory describing meson states. Then I have listed the masses of mesons in the three phases, calculated using the Lagrangian (1). These masses are in perfect agreement with the group theory reasoning.

I have also described how the charged pion decay is influenced by the chemical potentials, showing that their inclusion in the model leads to a significant asymmetry between the decay in $\ell^+\nu_\ell$ and the decay in $\ell^-\bar{\nu}_\ell$.

These results can be applied for example in the physics of compact stars, the study of the cosmic ray, the study of some nuclear decays.

Acknowledgements

This work is supported by Progetto Speciale Multiasse "Sistema Sapere e Crescita" PO FSE Abruzzo 2007 - 2013.

References

[1] A. B. Migdal, E. Saperstein, M. Troitsky, and D. Voskresensky, *Phys.Rept.*, **192**, pp. 179–437, 1990.
[2] D.T. Son and M. A. Stephanov, *Phys.Rev.Lett.*, **86**, pp. 592–595, 2001.
[3] J. Kogut and D. Toublan, *Phys.Rev.*, **D64**, p. 034007, 2001.
[4] J. Kogut, M. A. Stephanov, and D. Toublan, *Phys.Lett.*, **B464**, pp. 183–191, 1999.
[5] J. Kogut, M. A. Stephanov, D. Toublan, J. Verbaarschot, and A. Zhitnitsky, *Nucl.Phys.*, **B582**, pp. 477–513, 2000.
[6] D. Son and M. A. Stephanov, *Phys.Atom.Nucl.*, **64**, pp. 834–842, 2001.
[7] M. Loewe and C. Villavicencio, *Phys.Rev.*,**D67**, p. 074034, 2003.
[8] M. Loewe and C. Villavicencio, *Phys.Rev.*, **D70**, p. 074005, 2004.
[9] L.-y. He, M. Jin, and P.-f. Zhuang, *Phys.Rev.*,**D71**, p. 116001, 2005.
[10] T. Xia and P. Zhuang, arXiv:1411.6713v1 [hep-ph], 2014.
[11] M. G. Alford, A. Kapustin, and F. Wilczek, *Phys.Rev.*, **D59**, p. 054502, 1999.
[12] J. Kogut and D. Sinclair, *Phys.Rev.*,**D66**, p. 034505, 2002.
[13] D. Toublan and J. Kogut, *Phys.Lett.*, **B564**, pp. 212–216, 2003.
[14] A. Barducci, R. Casalbuoni, G. Pettini, and L. Ravagli, *Phys.Rev.*,**D69**, p. 096004, 2004.
[15] A. Barducci, R. Casalbuoni, G. Pettini, and L. Ravagli, *Phys.Rev.*, **D71**, p. 016011, 2005.
[16] B. Klein, D. Toublan, and J. Verbaarschot, *Phys.Rev.*,**D72**, p. 015007, 2005.
[17] A. Barducci, R. Casalbuoni, S. De Curtis, R. Gatto, and G. Pettini, *Phys.Rev.*, **D42**, pp. 1757–1763, 1990.
[18] C. Dominguez, M. Loewe, and J. Rojas, *Phys.Lett.*, **B320**, pp. 377–380, 1994.
[19] M. Loewe and C. Villavicencio, arXiv:1107.3859 [hep-ph], 2011.
[20] J. Gasser and H. Leutwyler, *Annals Phys.*, **158**, p. 142, 1984.
[21] H. Leutwyler, *Annals Phys.*, **235**, pp. 165–203, 1994.
[22] G. Ecker, *Prog.Part.Nucl.Phys.*, **35**, pp. 1–80, 1995.
[23] S. Scherer, *Adv.Nucl.Phys.*, **27**, p. 277, 2003.
[24] S. Scherer and M. R. Schindler, arXiv:hep-ph/0505265, 2005.
[25] A. Mammarella and M. Mannarelli, arXiv:1507.02934 [hep-ph].

EPJ Web of Conferences **126**, 04030 (2016) DOI: 10.1051/epjconf/201612604030
ICNFP 2015

Recent observation of D^{**} and D_s^{**} at Belle

Dmitry Matvienko[1,a]
on behalf of the Belle Collaboration

[1] *Budker Insitute of Nuclear Physics, SB RAS, Novosibirsk, 630090 and Novosibirsk State University, Novosibirsk, 630090*

Abstract. We report on the first observation of $\bar{B}^0 \to D_1(2430)^0\omega$, $\bar{B}^0 \to D_1(2420)^0\omega$ and $\bar{B}^0 \to D_2^*(2460)^0\omega$ decays and determination of their rates as well as the fractions of longitudinal polarizations of the ω for these decays. We also discuss improved measurements of the $B \to DD_{s0}^*(2317)$ branching fractions with a search for isospin partners of the $D_{s0}^*(2317)^+$ state, $\mathbf{z}^0$ and $\mathbf{z}^{++}$. These measurements use a data sample containing 772 million $B\bar{B}$ events collected at the $\Upsilon(4S)$ resonance with the Belle detector at the KEKB asymmetric-energy e^+e^- collider.

1 Introduction

Since 2004, significant success in observation of orbitally and radially excited heavy-light mesons has been achieved in several experiments, such as Belle [1], *BABAR* [2], LHCb [3] etc.

The relativistic quark model [4] predicts four physical states with a P-wave orbital excitation in the charm and charm-strange meson families. They have spin-parities $J^P = 0^+$ $(j = 1/2)$, 1^+ $(j = 1/2)$, 1^+ $(j = 3/2)$ and 2^+ $(j = 3/2)$, where $j = s \pm 1$ and s is the spin of the light quark in the meson. The states with $j = 1/2$ are expected to be broad (hundreds of MeV$/c^2$), while the states with $j = 3/2$ are expected to be narrow (tens of MeV$/c^2$ or smaller). All these states have been discovered [5]. They are $D_0^*(2400)$, $D_1(2430)$, $D_1(2420)$ and $D_2^*(2460)$ (D^{**} states) in the charm sector as well as $D_{s0}^*(2317)$, $D_{s1}(2460)$, $D_{s1}(2536)$ and $D_{s2}(2573)$ (D_s^{**} states) in the charm-strange sector. The observed masses and widths for the D^{**} states are in reasonable agreement with the quark model predictions. However, this is not the case for the D_s^{**} states. The conventional $c\bar{s}$ hypothesis contradicts to experimentally measured masses, widths and relative branching fractions of the P-wave excitations of the D_s meson.

Here, we discuss two recent Belle measurements of the neutral B decays to the $D^{**}\omega$ final state [6], where D^{**} states are reconstructed in the $D^*\pi$ mode, and the neutral and charged B decays to the $\bar{D}D_{s0}^{*+}(2317)$ final state [7]. The Belle detector [8] at the KEKB asymmetric-energy e^+e^- (3.5 on 8 GeV) collider [9] is used to study B mesons produced in the reaction $e^+e^- \to \Upsilon(4S) \to B\bar{B}$. The $B\bar{B}$ integrated luminosity of 711 fb^{-1} contains $(772 \pm 11) \times 10^6$ $B\bar{B}$ pairs. The studies presented here, $\bar{B}^0 \to D^{**}\omega$ and $B \to \bar{D}D_{s0}^{*+}(2317)$, use the complete Belle data set.

[a] e-mail: d.v.matvienko@inp.nsk.su

2 D^{**} production in semileptonic and hadronic B decays

Study of D^{**} production in semileptonic B decays is of interest for several reasons. Firstly, $\bar{B} \to D^{**} l^- \bar{\nu}_l$ decays constitute a significant fraction of B semileptonic decays and may help to explain the origin of the $\sim 3\sigma$ discrepancy between exclusive and inclusive determinations of the Cabibbo-Kobayashi-Maskawa matrix element $|V_{cb}|$. It is therefore interesting to learn more about the composition of the final hadronic state X_c in the inclusive $\bar{B} \to X_c l^- \bar{\nu}_l$ decay. Moreover, the measured decay properties for $\bar{B} \to D^{**} l^- \bar{\nu}_l$ can be compared with QCD sum rules [10] and predictions from heavy quark effective theory (HQET) [11]. It has been found that these properties for narrow $j = 1/2$ D^{**} states are in serious conflict with the theoretical expectations. Semileptonic B decay rates to the D^{**} states depending on the invariant four-velocity transfer w are estimated as:

$$\frac{d\Gamma(B \to D^{**}_{j=1/2} l\nu)}{dw} \propto G_F^2 |V_{cb}|^2 K_{1/2}(w) |\tau_{1/2}(w)|^2,$$
$$\frac{d\Gamma(B \to D^{**}_{j=3/2} l\nu)}{dw} \propto G_F^2 |V_{cb}|^2 K_{3/2}(w) |\tau_{3/2}(w)|^2,$$

where K_j are known kinematical factors with hierarchy of $K_{1/2}(w) < K_{3/2}(w)$, and $\tau_j(w)$ are Isgur-Wise functions with predicted $\tau_{1/2}(w) << \tau_{3/2}(w)$ behavior [11]. Thus, HQET and QCD sum rules predict the suppression of $j = 1/2$ states relative to the $j = 3/2$. However, experimental measurements demonstrate comparable rates for $j = 1/2$ and $j = 3/2$. Herewith, the observed rates for $j = 3/2$ states are in good agreement with predictions. This is a so called "1/2 semileptonic puzzle". There are different attempts to solve this problem [12, 13] but, at this moment, the full understanding is absent. To elucidate this problem, it is useful to study non-leptonic B decays to D^{**}.

Non-leptonic B decays to D^{**} states and a light hadronic system (to be specific we will discuss a pion here) can be divided into three classes. Class I $\bar{B}^0 \to D^{**+}\pi^-$ decays describe the color-favored pion emission process, class II $\bar{B}^0 \to D^{**0}\pi^0$ decays correspond to the color-suppressed D^{**} emission process and class III $B^- \to D^{**0}\pi^-$ decays comprise these two emission processes. Experimental results for non-leptonic relative decay rates are shown in Table 1 for class I and class III decays. Class II $\bar{B}^0 \to D^{**0}\pi^0$ decays have not been observed so far. All entries corresponding to the class

Table 1. Observed branching fraction (BF) products for $B \to D^{**}\pi$ class I and class III decays.

Class	BF product	Value
	$\mathcal{B}(\bar{B}^0 \to D_0^*(2400)^+\pi^-) \times \mathcal{B}(D_0^*(2400)^+ \to D^0\pi^+)$	$(6.0 \pm 3.0) \times 10^{-5}$
	$\mathcal{B}(\bar{B}^0 \to D_1(2430)^+\pi^-) \times \mathcal{B}(D_1(2430)^+ \to D^{*0}\pi^+)$	$< 7 \times 10^{-5}$
Class I	$\mathcal{B}(\bar{B}^0 \to D_1(2420)^+\pi^-) \times \mathcal{B}(D_1(2420)^+ \to D^{*0}\pi^+)$	$(3.7 \pm 0.9) \times 10^{-4}$
	$\mathcal{B}(\bar{B}^0 \to D_2^*(2460)^+\pi^-) \times \mathcal{B}(D_2^*(2460)^+ \to D^0\pi^+)$	$(2.2 \pm 0.4) \times 10^{-4}$
	$\mathcal{B}(\bar{B}^0 \to D_2^*(2460)^+\pi^-) \times \mathcal{B}(D_2^*(2460)^+ \to D^{*0}\pi^+)$	$(2.5 \pm 0.6) \times 10^{-4}$
	$\mathcal{B}(\bar{B}^- \to D_0^*(2400)^0\pi^-) \times \mathcal{B}(D_0^*(2400)^0 \to D^-\pi^+)$	$(6.4 \pm 1.4) \times 10^{-4}$
	$\mathcal{B}(\bar{B}^- \to D_1(2430)^0\pi^-) \times \mathcal{B}(D_1(2430)^0 \to D^{*-}\pi^+)$	$(5.0 \pm 1.3) \times 10^{-4}$
Class III	$\mathcal{B}(\bar{B}^- \to D_1(2420)^0\pi^-) \times \mathcal{B}(D_1(2420)^0 \to D^{*-}\pi^+)$	$(6.8 \pm 1.5) \times 10^{-4}$
	$\mathcal{B}(\bar{B}^- \to D_2^*(2460)^0\pi^-) \times \mathcal{B}(D_2^*(2460)^0 \to D^-\pi^+)$	$(3.5 \pm 0.4) \times 10^{-4}$
	$\mathcal{B}(\bar{B}^- \to D_2^*(2460)^0\pi^-) \times \mathcal{B}(D_2^*(2460)^0 \to D^{*-}\pi^+)$	$(2.2 \pm 1.1) \times 10^{-4}$

III decays in Table 1 are of the same order, whereas class I decays to the broad $j = 1/2$ states are one order of magnitude smaller than decays to the narrow $j = 3/2$ states. Moreover, class I and class III decays to the $j = 3/2$ states are similar in size. This hierarchy of the decay rates is theoretically understood. Using the factorisation and predictions for $\tau_j(w)$ functions in heavy quark limit together with finite corrections to HQET one obtains striking agreement between theory and experiment for class I decays. Class III results can be also explained because there is an extra D^{**} emission diagram here, which is expected to be suppressed for $j = 3/2$ states, but not for $j = 1/2$ due to heavy quark symmetry (HQS). However, there are no observed class II decays with the D^{**} emission process only. Therefore, it is interesting to confirm this prediction from experimental measurements. The recent Belle study of the $\bar{B}^0 \to D^{**0}\omega$ decays solves this problem.

2.1 $B \to D^{**}\omega$ analysis strategy

Observation of the class II hadronic decays allows one to answer the following questions: what are the $j = 1/2$ and $j = 3/2$ rates in class II decays and what is the tensor rate in these decays? It is obvious that the tensor state $D_2^*(2460)^0$ cannot be generated by the V-A current and should be absent in the factorisation approximation. Moreover, it should be suppressed by HQS. Nevertheless, soft collinear effective theory (SCET) [14] predicts nontrivial production of this state as well as the equality of brancnhing fractions and strong phases in the decays $\bar{B}^0 \to D_2^*(2460)^0 M$ and $\bar{B}^0 \to D_1(2420)^0 M$, where $M = \pi, \rho, K$ or K^* with longitudinal polarization.

We have chosen to analyse class II $\bar{B}^0 \to D^{**0}\omega$ decays for several reasons. Firstly, lower fractions of $q\bar{q}$ continuum and combinatorial backgrounds are expected in this mode relative to the $\bar{B}^0 \to D^{**0}\pi^0$ decays. Further, we can measure the polarisations and partial-wave fractions of the D^{**} states. Finally, we can perform the coherent study of the ρ-meson-like state in the $\omega\pi$ final state.

All B candidates in this analysis are identified using two kinematic variables: the energy difference $\Delta E = \sum_i \sqrt{|\mathbf{p}_i^*|^2 c^2 + m_i^2 c^4} - E^*_{\rm beam}$ and the beam-constrained mass $M_{\rm bc} = \sqrt{E^{*2}_{\rm beam}/c^4 - |\sum_i \mathbf{p}_i^*|^2/c^2}$, where the summation is over all particles forming the B candidate, $\mathbf{p}_i^*$ and m_i are their three-momenta and masses, respectively, and $E^*_{\rm beam}$ is the beam energy. All quantities are defined in the e^+e^- c.m. frame. These variables allow one to select the signal sample with the small background fraction. We use a tight cut on $M_{\rm bc}$ corresponding to about ± 2 times the mass resolution and loose cuts on ΔE and invariant mass $M(\pi^+\pi^-\pi^0)$ of the ω.

Figure 1 shows the distribution of the selected events in the $\Delta E - M(\pi^+\pi^-\pi^0)$ plane, where we define four regions to distinguish between signal and background:the signal region I and sideband regions II, III and IV. The background component in the signal region I is described by the functions, which are determined in the control sideband regions II, III and IV in Fig. 1.

To distinguish the contributions for narrow and broad D^{**} states in the signal region, coherent amplitude analysis is performed. This analysis minimises the likelihood function in the six-dimensional kinematic phase space of the decay. The likelihood function is constructed from the background and signal probability density functions (PDF).

In total, six variables are required to fully describe $D^*\omega\pi$ kinematics. We define two sets of kinematic variables, corresponding to the $\omega\pi$ and D^{**} production. Each of the sets includes the invariant mass squared of the intermediate state ($\omega\pi$ or D^{**}) and five angular variables. The angles describing $\omega\pi$ production are defined in Fig. 2. Similar variables are also defined for the D^{**} contribution. All the D^{**} states are well distinguished by their angular distributions.

The signal PDF is equal to the decay matrix element squared, where a six-dimensional amplitude is written as a sum of the Breit-Wigner factors with angular parts. These factors describe the production and decay of the intermediate resonances, ρ-meson-like and D^{**} states [15]. The model parameters in the matrix element and observables are determined from the unbinned likelihood fit in the signal region.

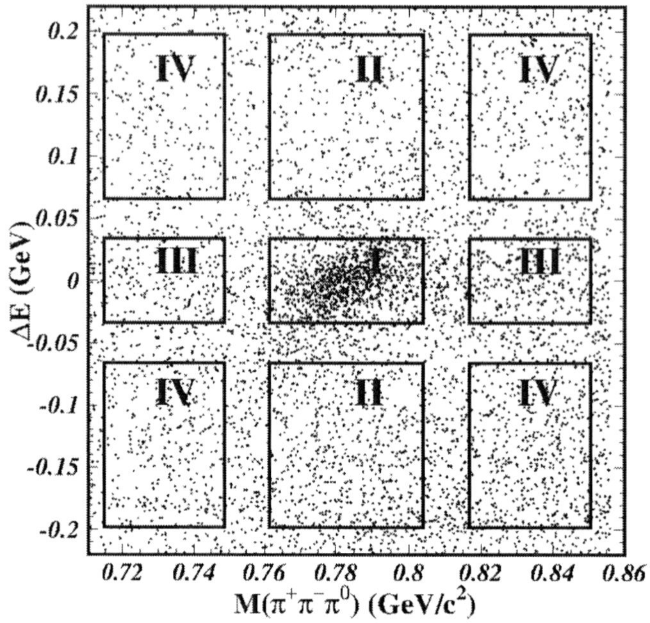

Figure 1. Distribution of ΔE versus $M(\pi^+\pi^-\pi^0)$ for the selected $\bar{B}^0 \to D^{*+}\omega\pi^-$ candidates. The signal region (I) and sideband regions (II, III and IV) are shown.

2.2 D^{**} polarizations and branching fraction products

The distributions of the $D^*\pi$ and $\omega\pi$ invariant masses squared for the $D^{*+}\omega\pi^-$ candidates in the signal $\Delta E - M(\pi^+\pi^-\pi^0)$ region I are shown in Figs. 3 (a) and (b). These distributions correspond to kinematic regions of phase space enriched (Fig. 2 (a)) and depleted (Fig. 2 (b)) with D^{**} mesons. The results of the fit with the considered signal model are superimposed. The plots demonstrate a reasonable description of the data by the fit. The fit results require large signals for the $\rho(770)$ and $\rho(1450)$ contributions with branching fraction products of $(1.48 \pm 0.27\,(\text{stat.})\,^{+0.15}_{-0.09}\,(\text{syst.})\,^{+0.21}_{-0.56}\,(\text{model})) \times 10^{-3}$ and $(1.07\,^{+0.15}_{-0.31}\,(\text{stat.})\,^{+0.06}_{-0.13}\,(\text{syst.})\,^{+0.40}_{-0.02}\,(\text{model})) \times 10^{-3}$, respectively. The combined first-class current branching fraction product $(1.90 \pm 0.11\,(\text{stat.})\,^{+0.11}_{-0.13}\,(\text{syst.})\,^{+0.02}_{-0.06}\,(\text{model})) \times 10^{-3}$ is well measured and has small model uncertainties.

Color-suppressed decays saturated by the $\bar{B}^0 \to D_1(2430)^0\omega$ decays with branching fraction product of $(2.5 \pm 0.4\,(\text{stat.})\,^{+0.7}_{-0.2}\,(\text{syst.})\,^{+0.4}_{-0.1}\,(\text{model})) \times 10^{-4}$ and $\bar{B}^0 \to D_1(2420)^0\omega$ with a significantly smaller branching fraction product of $(0.7 \pm 0.2\,(\text{stat.})\,^{+0.1}_{-0.0}\,(\text{syst.}) \pm 0.1\,(\text{model})) \times 10^{-4}$ are observed in the studied mode. The fit results also require a non-trivial fraction of $(0.4 \pm 0.1\,(\text{stat.})\,^{+0.0}_{-0.1}\,(\text{syst.}) \pm 0.1\,(\text{model})) \times 10^{-4}$ for the non-factorizable $\bar{B}^0 \to D_2(2460)^0\omega$ decay.

The individual partial-wave fractions for D^{**} production have been measured with the comparable probabilities of about 30%. The longitudinal polarizations for D^{**}-production have been also measured. The results, $(63.0 \pm 9.1(\text{stat.}) \pm 4.6(\text{syst.})^{+4.6}_{-3.9}(\text{model.}))\%$ for the $D_1(2430)^0$ production and $(67.1 \pm 11.7(\text{stat.})^{+0.0}_{-4.2}(\text{syst.})^{+2.3}_{-2.8}(\text{model.}))\%$ for the $D_1(2420)^0$ production, have large errors, but imply non-trivial non-factorizable QCD effects in the color-suppressed channel [16].

 DOI: 10.1051/epjconf/201612604030

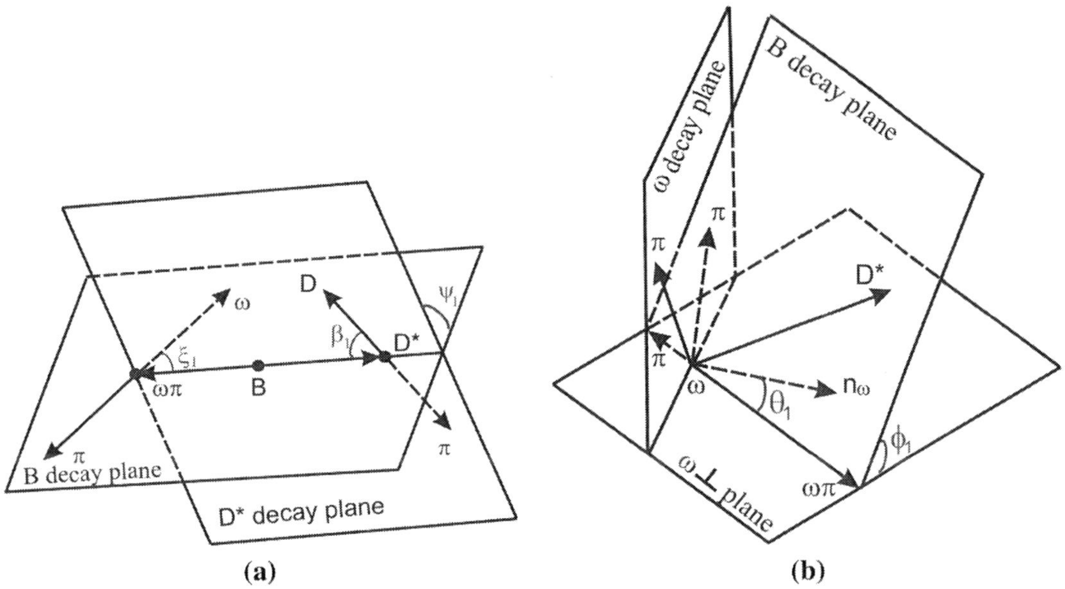

Figure 2. Kinematics of a $\bar{B}^0 \to D^{*+}\omega\pi^-$ decay mediated by an $\omega\pi^-$ intermediate resonance. The diagram in (a) defines two polar angles ξ_1 and β_1 and one azimuthal angle ψ_1. The diagram in (b) defines one polar angle θ_1 and one azimuthal angle ϕ_1. The direction n_ω in (b) corresponds to the vector normal to the ω decay plane.

3 Is the $D^*_{s0}(2317)$ molecule or tetraquark?

The $D^{*+}_{s0}(2317)$ state has been first observed by BaBar [17] and confirmed by CLEO [18] in 2003. Its mass has been measured below the DK threshold despite the potential model prediction based on the $J^P = 0^+$ P-wave $c\bar{s}$ hypothesis. Moreover, CLEO placed a stringent 90% confidence level upper limit on the relative partial width of the radiative $D^{*+}_{s0}(2317) \to D^{*+}_s\gamma$ and strong $D^{*+}_{s0}(2317) \to D^+_\pi\pi^0$ decays, which is 0.059.

The observation of a sub-threshold mass has led to speculations about the nature of its quark bound state. The most popular hypotheses are DK molecule [19] and tetraquark [20] models. Both these models predict the mass of the $D^{*+}_{s0}(2317)$ below the DK threshold. However, they have significantly different estimations of the radiative and strong widths of studied state.

If it is assumed that the $D^{*+}_{s0}(2317)$ is a candidate for DK molecule with isospin quantum number I of $I = 0$, we derive the following estimations for radiative $\Gamma_{\rm rad}$ and strong $\Gamma_{\rm str}$ widths: $\Gamma_{\rm rad} \sim 120$ KeV and $\Gamma_{\rm str} \sim 10$ KeV, where their ratio is near the experimental upper limit. According to heavy quark symmetry, the molecular hypothesis predicts also BK and B^*K molecules. Their radiative decays to $B^{(*)}\gamma$ final states are the best chance for discovery of these states.

It is difficult to satisfy the upper limit of 0.059 in the frame of the tetraquark hypothesis with $I = 0$, while the assignment $I = 1$ changes the situation [20]. In other words, if $D^{*+}_{s0}(2317)$ is assigned to the isotriplet scalar four-quark meson, the CLEO upper limit can be easily satisfied because $\Gamma_{\rm rad} \sim 0.8$ keV and $\Gamma_{\rm str} \sim (4-5)$ MeV in this model. The isotriplet tetraquark model predicts the existence of isospin partners for the $D^{*+}_{s0}(2317)$ state, $\mathbf{z}^0$ and $\mathbf{z}^{++}$. The recent Belle study of the $B^+ \to \bar{D}^0 D^{*+}_{s0}(2317)$ and $B^0 \to D^- D^{*+}_{s0}(2317)$ decays tests this model.

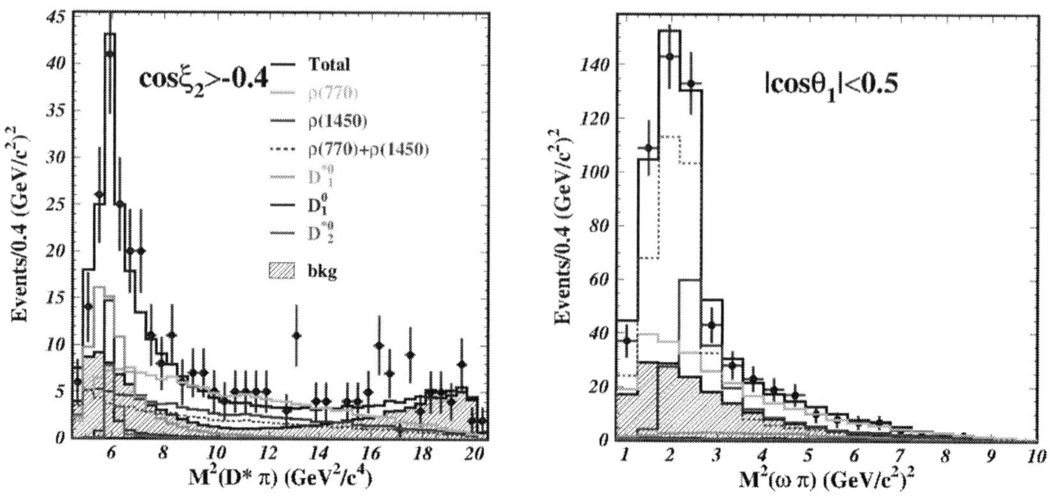

(a) **(b)**

Figure 3. Distributions of the $D^*\pi$ and $\omega\pi$ invariant masses squared for selected $D^{*+}\omega\pi^-$ candidates corresponding to the D^{**} (a) enriched and (b) depleted kinematic regions of phase space.

3.1 Measurement of $B \to \bar{D}D_{s0}^{*+}(2317)$

All B candidates are also identified by the ΔE and $M_{\rm bc}$ variables as in $\bar{B}^0 \to D^{**0}\omega$ decays but with a sufficiently broad window $M_{\rm bc} > 5.20\,{\rm GeV}/c^2$. The $D_{s0}^{*+}(2317)$ candidates are reconstructed in the $D_s^+\pi^0$ mode and D_s^+ mesons are identified via their $\pi^+K^+K^-$ decay mode. Candidates for D^- mesons in the neutral B decays are formed via the $K^+\pi^-\pi^-$ decay mode and D^0 mesons in the charged B decays via the $K^-\pi^+$ and $K^+\pi^+\pi^-\pi^-$ decay modes.

We select events in the three-dimensional region of $M_{\rm bc}$, ΔE and invariant mass $M(D_s^+\pi^0)$ variables and perform the unbinned likelihood fit in this space. We form the signal region in these dimensions and use here the signal PDFs in the fit. The outside region is described by the signal and background PDFs in the appropriate variables. The background PDF includes two components to describe combinatorial and peaking backgrounds. The latter peaks in ΔE and $M_{\rm bc}$ (but not $M(D_s^+\pi^0)$ mostly from three-body $B \to \bar{D}\pi^0 D_s^+$ decays. Figure 4 shows the results of the fit for the neutral B decays in the signal regions of the two variables not being plotted. The measured branching fraction product is $\mathcal{B}(B^0 \to D^-D_{s0}^*(2317)^+) \times \mathcal{B}(D_{s0}^*(2317)^+ \to D_s^+\pi^0) = (1.02^{+0.13}_{-0.12} \pm 0.11) \times 10^{-3}$. Similar fit results are obtained for the charged B decays. The measured branching fraction product is $\mathcal{B}(B^+ \to \bar{D}^0 D_{s0}^*(2317)^+) \times \mathcal{B}(D_{s0}^*(2317)^+ \to D_s^+\pi^0) = (0.80^{+0.13}_{-0.12} \pm 0.12) \times 10^{-3}$. All these results agree well with the existing PDG world-average values and significantly improve their precision [5].

3.2 Search for isospin partners of the $D_{s0}^{*+}(2317)$

The isospin partners of the $D_{s0}^{*+}(2317)$, $\mathbf{z}^0$ and $\mathbf{z}^{++}$, have been searched for in the $B^0 \to \bar{D}^0 D_s^+\pi^-$ and $B^+ \to D^-D_s^+\pi^+$ decay channels. Since the $\mathbf{z}^0$ and $\mathbf{z}^{++}$ are hypothetical, their masses are expected to lie within a $\pm 10\,{\rm MeV}/c^2$ mass region of the $D_{s0}^{*+}(2317)$ mass. To be certain that we cover the whole reasonable mass region, we scan for $\mathbf{z}^0$ and $\mathbf{z}^{++}$ signals in 13 adjacent mass bins, covering a ± 32.5 interval around the measured $D_{s0}^{*+}(2317)$ mass, which is 2317.8 MeV.

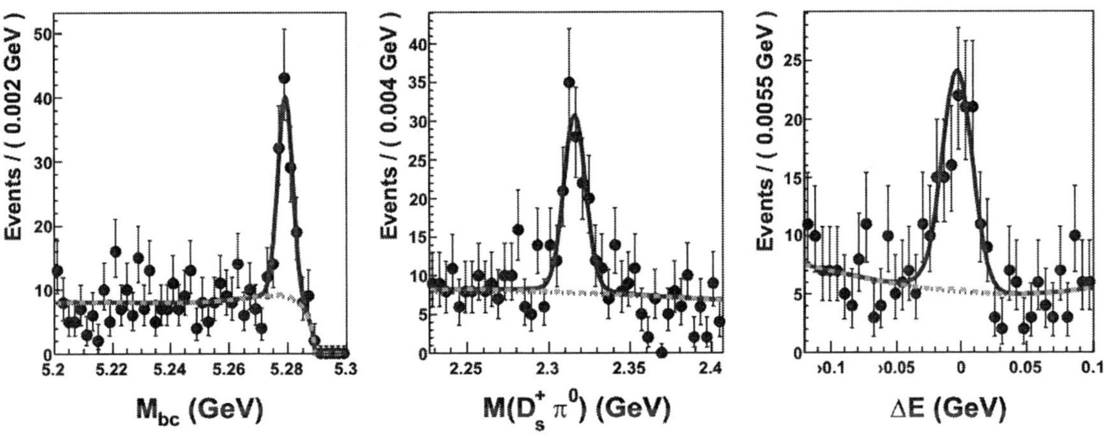

Figure 4. The M_{bc} (left), $M(D_s^+\pi^0)$ (center) and ΔE (right) distributions for projections of the $B^0 \to D^- D_{s0}^{*+}$ candidates that are in the signal regions of the two variables are not being plotted. The red short-dashed curve is the fitted background; the green long-dashed curve describes the peaking background and the solid blue curve includes the signal.

A three-dimensional fit to the data, similar to the measurement of $B \to \bar{D}D_{s0}^{*+}(2317)$, is performed in each studied mass bin. The possible peaking backgrounds are studied using a Monte Carlo sample of generic B meson decays without $\mathbf{z}^0$ and $\mathbf{z}^{++}$ signals. None of the fits returns a positive $\mathbf{z}^0$ or $\mathbf{z}^{++}$ signal with a statistical significance of more than 1.3σ. Assuming the masses of the $\mathbf{z}^0$ and $\mathbf{z}^{++}$ are the same as the $D_{s0}^{*+}(2317)$ we obtain the following 90% confidence level upper limits on the product branching fractions: $\mathcal{B}(B^0 \to \bar{D}^0\mathbf{z}^0) \times \mathcal{B}(\mathbf{z}^0 \to D_s^+\pi^-) < 0.64 \times 10^{-4}$ and $\mathcal{B}(B^+ \to D^-\mathbf{z}^{++}) \times \mathcal{B}(\mathbf{z}^{++} \to D_s^+\pi^+) < 0.27 \times 10^{-4}$. The upper limits on the product branching fractions in other mass bins are close to the values obtained in the central bin. All these upper limits are significantly smaller than the tetraquark predictions with $I = 1$.

4 Conclusion

The recent Belle measurement of $\bar{B}^0 \to D^{*+}\omega\pi^-$ showed the relative dominance of the broad $D_1(2430)^0$ production relative to the narrow $D_1(2420)^0$ in agreement with HQET studies. The significant $\bar{B}^0 \to D_2(2460)^0\omega$ decay has been also observed. Multidimensional coherent amplitude analysis allows one to measure non-factorizable D^{**} polarisations and their partial-wave rates. Moreover, it allows one to perform the first consistent study of the $\rho(770)$ and $\rho(1450)$ states in B meson decays.

New Belle measurements of $B \to \bar{D}D_{s0}^{*+}(2317)$ decays are consistent with the previous experimental results but they have higher precision. The upper limits for the $\mathbf{z}^0$ and $\mathbf{z}^{++}$ decay rates have been obtained. The results are more than one magnitude smaller than the $D_{s0}^{*+}(2317)$-tetraquark predictions with isospin $I = 1$.

Acknowledgements

This work is supported by the Ministry of Education and Science of the Russian Federation and the Russian Foundation for Basic Research under Grant No. 14-02-31527.

References

[1] K. Abe *et al.* (Belle Collaboration), Phys. Rev. D **69**, 112002 (2004); A. Kuzmin *et al.* (Belle Collaboration), Phys. Rev. D **76**, 012006 (2007); D. Liventsev *et al.* (Belle Collaboration), Phys. Rev. D **77**, 091503 (2008).

[2] B. Aubert *et al.* (BaBar Collaboration), Phys. Rev. Lett. **101**, 261802 (2008); B. Aubert *et al.* (BaBar Collaboration), Phys. Rev. D **79**, 112004 (2009).

[3] R. Aaji *et al.* (LHCb Collaboration), Phys. Rev. D **87**, 092001 (2013); R. Aaji *et al.* (LHCb Collaboration), Phys. Rev. D **92**, 032002 (2015).

[4] S. Godfrey and K. Moats, arXiv:1510.08305 and references therein.

[5] K.A. Olive *et al.* (Particle Data Group), Chin. Phys. C **38**, 090001 (2014).

[6] D. Matvienko *et al.* (Belle Collaboration), Phys. Rev. D **92**, 012013 (2015).

[7] S.-K. Choi *et al.* (Belle Collaboration), Phys. Rev. D **91**, 092011 (2015).

[8] A. Abashian *et al.* (Belle Collaboration), Nucl. Instrum Methods Phys. Res., Sect. A **479**, 117 (2002); also see detector section in J. Brodzicka *et al.*, Prog. Theor. Exp. Phys. **2012**, 04D001 (2012).

[9] S. Kurokawa and E. Kikutani, Nucl. Instrum. Methods Phys. Res., Sect. A **499**, 1 (2003), and other papers included in this volume; T. Abe *et al.*, Prog. Theor. Exp. Phys. **2013**, 03A001 (2013) and following articles up to 03A011.

[10] N. Uraltsev, Phys. Lett. B **501**, 86 (2001).

[11] M. Neubert, Phys. Rep. **245**, 259 (1994).

[12] R. Clein, T. Mannel, F. Shahriaran and D. van Dyk, Phys. Rev. D **91**, 094034 (2015).

[13] B. Blossier, arXiv:1411.3563.

[14] S. Mantry, Phys. Rev. D **70**, 114006 (2004).

[15] D.V. Matvienko, A.S. Kuzmin, and S.I Eidelman, J. High Energy Phys. 09 (2011) 129.

[16] H.Y. Cheng, C.K. Chua, and A. Soni, Phys. Rev. D **71**, 014030 (2005).

[17] B. Aubert *et al.* (BaBar Collaboration), Phys. Rev. Lett. **90**, 242001 (2003).

[18] D. Besson *et al.* (CLEO Collaboration), Phys. Rev. D **68**, 032002 (2003).

[19] T. Barnes, F.E. Close and H.J. Lipkin, Phys. Rev. D **68**, 054006 (2003); A.P. Szczepaniak, Phys. Lett. B **567**, 23 (2003).

[20] A.Hayashigaki and K.Terasaki, Prog. Theor. Phys. **114**, 1191 (2005).

EPJ Web of Conferences **126**, 04031 (2016) DOI: 10.1051/epjconf/201612604031
ICNFP 2015

Search for sterile neutrino mixing in the $\nu_\mu \rightarrow \nu_\tau$ appearance channel with the OPERA detector

N. Mauri[1,a] on behalf of the OPERA Collaboration

[1] *INFN - Sezione di Bologna, Viale Berti Pichat 6/2, I-40127 Bologna, Italy*

Abstract. The OPERA experiment has observed muon neutrino to tau neutrino oscillations in the atmospheric sector in appearance mode. Five ν_τ candidate events have been detected, a number consistent with the expectation from the "standard" 3ν framework. Based on this result new limits on the mixing parameters of a massive sterile neutrino have been set. The analysis is performed in the 3+1 neutrino model.

1 Introduction

The discovery of neutrino oscillations established the fact that neutrinos have a non-zero mass, pointing to physics beyond the Standard Model. Since the first observation of the disappearance of atmospheric muon neutrinos in 1998, many experiments contributed to draw a coherent picture of the new "standard" 3ν framework. The oscillations are formalized by the PMNS unitary matrix U_{PMNS} through which three flavour eigenstates are mixed with three mass eigenstates [1, 2]. After less than 20 years, the three mixing angles $\theta_{12}, \theta_{13}, \theta_{23}$ parameterized in U_{PMNS} are measured as well as the two squared mass differences $\Delta m^2_{21}, \Delta m^2_{32}$ determining the oscillation frequencies [3].

However, a certain number of so-called neutrino "anomalies" show tensions with this well established 3ν framework [4]. Different kinds of experiments, both in appearance and disappearance [5–8], point towards a new parameter region $\Delta m^2 \sim 1$ eV^2. Since the number of flavour-active neutrinos is bound to 3 by the coupling with the Z^0, the new squared mass difference could be accounted for by the existence of a sterile neutrino. A sterile neutrino is a neutral lepton which does not couple to $W^\pm/Z^0$ bosons. It is not an exotic particle, e.g. right-handed neutrinos are singlets under $SU(2)_L$ gauge. Sterile neutrinos can mix with the active ones and the mass scale of additional states has no strong a priori. The experimental "anomalies" hint to the existence of light $O(1)$ eV steriles.

The OPERA experiment was designed to observe $\nu_\mu \rightarrow \nu_\tau$ oscillations in appearance mode, in order to unambiguously prove the $\nu_\mu \rightarrow \nu_\tau$ transition at the atmospheric scale in the 3ν framework. OPERA recently assessed the discovery of ν_τ appearance in the CNGS ν_μ beam with a significance of 5.1σ [9]. The number of ν_τ events detected by OPERA is compatible with the expectations from the standard 3ν framework. We can therefore set limits on the oscillations induced by a sterile neutrino, here formalized by the 3 + 1 neutrino model.

[a] e-mail: nicoletta.mauri@bo.infn.it

2 The 3+1 neutrino model

The new 3+1 model should include the well established 3ν framework. The additional sterile eigenstate can be represented through an extended 4×4 mixing matrix U which slightly perturbs the inner 3×3 U_{PMNS} and extends the mixing to a 4th neutrino. In this 3+1 neutrino model, the first three mass eigenstates are mostly composed by flavour-active states with the well known mixing, while the forth mass eigenstate is almost exclusively composed by the sterile state. The 4th mass term is supposed to be heavier than the other three ones, since the experimental anomalies suggest a $\Delta m^2 \sim \mathrm{eV}^2$ and the cosmological limits on the sum of neutrino masses favour a positive Δm^2_{41} [10]. In the 3+1 model, defining $\Delta_{ij} = 1.27\,\Delta m^2_{ij}\,L/E$ (where Δm^2_{ij} is expressed in eV^2, L in km and E in GeV), the probability for $\nu_\mu \to \nu_\tau$ transitions in given by

$$\begin{aligned} P_{\nu_\mu\to\nu_\tau}(E) &= 4|U_{\mu3}|^2|U_{\tau3}|^2\sin^2\Delta_{31} + 4|U_{\mu4}|^2|U_{\tau4}|^2\sin^2\Delta_{41} \\ &+ 2\mathcal{R}[U_{\mu3}U^*_{\tau3}U^*_{\mu4}U_{\tau4}]\sin 2\Delta_{31}\sin 2\Delta_{41} \\ &- 4\mathcal{I}[U_{\mu3}U^*_{\tau3}U^*_{\mu4}U_{\tau4}]\sin^2\Delta_{31}\sin 2\Delta_{41} \\ &+ 8\mathcal{R}[U_{\mu3}U^*_{\tau3}U^*_{\mu4}U_{\tau4}]\sin^2\Delta_{31}\sin^2\Delta_{41} \\ &+ 4\mathcal{I}[U_{\mu3}U^*_{\tau3}U^*_{\mu4}U_{\tau4}]\sin 2\Delta_{31}\sin^2\Delta_{41} \end{aligned} \quad (1)$$

In the above formula, the solar-driven term is neglected due to the CNGS baseline and neutrino energy ($\Delta_{21} \approx 4 \times 10^{-3}$). In the first row of Eq. 1, the first term corresponds to the standard oscillation probability for ν_τ appearance. The second term corresponds to the oscillation purely induced by the mixing with the sterile state, with an effective mixing parameter $\sin 2\theta_{\mu\tau} = 2|U_{\mu4}||U_{\tau4}|$. The remaining four terms correspond to the interference between active and sterile neutrinos, and depending on the extra terms in the extended U matrix they can lead to an enhancement or a suppression of the expected ν_τ. These interference terms depend on the possible CP–violating phase $\phi_{\mu\tau} = Arg(U^*_{\mu4}U_{\tau4}U_{\mu3}U^*_{\tau3})$ and on the mass hierarchy of standard neutrinos (Normal Hierarchy, NH, $\Delta m^2_{31} > 0$, or Inverted Hierarchy, IH, $\Delta m^2_{31} < 0$).

The number of ν_τ events observed by OPERA is compared to the number of events expected from the 3+1 model as a function of the new mixing angles and squared mass difference. In Sect. 4 these numbers are used to constrain the parameters appearing in Eq. 1.

3 The OPERA experiment

OPERA is a long baseline experiment located in the underground Gran Sasso laboratory and exposed to the CNGS ν_μ beam from 2008 to 2012. The experiment was designed to perform a direct observation of ν_τ charged-current interactions on event-by-event basis, with a signal to background ratio of $O(10)$.

The topological observation of the short-lived τ decay requires an exceptional granularity integrated in a large mass. This is achieved in OPERA using Emulsion Cloud Chambers (ECC) and real-time electronic detectors with a modular structure [11]. The ECC basic unit is a "brick" composed of 57 nuclear emulsion films, providing the necessary micrometric accuracy, interleaved with 56 lead plates, providing the necessary mass. The target is a hybrid modular structure in which the bricks are arranged in vertical "walls", transverse to the beam direction, alternated with planes of plastic scintillator strips. The detector is made of two identical Super Modules, each consisting of a target section and a muon spectrometer. The latter is a dipolar iron magnet instrumented with Resistive Plate Chambers (RPC) and high precision drift tube detectors (PT). The electronic detectors trigger the read-out and allow identifying the brick containing the neutrino interaction. If a muon

DOI: 10.1051/epjconf/201612604031

track crosses the spectrometers, the RPCs and PTs are used to measure the muon charge and momentum [12]. The neutrino interaction vertex and its full topology are reconstructed in the brick with a sub-μm spatial resolution.

3.1 $\nu_\mu \to \nu_\tau$ search

In five years of data taking, from 2008 to 2012, OPERA integrated a CNGS beam luminosity corresponding to 17.97×10^{19} protons on target (p.o.t.), the 80% of the nominal design value. A total of 19505 CNGS events have been registered correspondingly. Only interactions inside the OPERA target volume are analysed for oscillation studies. These internal events are further classified by an automatic algorithm into 0μ and 1μ through the identification of a muon track [12] or the total number of fired scintillator or RPC planes.

The data set used for the $\nu_\mu \to \nu_\tau$ search is composed by internal 0μ events and 1μ events with a muon momentum below 15 GeV/c, for a total of 5408 fully analysed events in the ECCs. In this sample, after the whole scanning and decay search procedure [13], 5 ν_τ candidate events are observed. The total expected background is 0.25 ± 0.05 events. The background fluctuation probability is evaluated with two test statistics, one based on the Fisher's method and the other on the profile likelihood ratio. The background-only hypothesis is excluded with a 5.1σ significance by both methods [9].

The expected signal from the standard 3ν model, assuming $\Delta m^2_{32} = 2.44 \times 10^{-3}$ eV^2 [3] and a maximal mixing, considering all the τ decay channels, amounts to 2.64 ± 0.53 events, compatible with the observed ν_τ candidates [9].

4 Constraints on the mixing with a sterile neutrino

In this paper we interpret the OPERA results in the context of the 3+1 neutrino model, deriving limits on the oscillations induced by a massive sterile neutrino. Limits on the sterile neutrino mixing based on the observation of 4 ν_τ candidate events have been recently published by the OPERA Collaboration [14]. Here we report on the preliminary updated analysis considering the current number of 5 ν_τ events [15, 16].

The results of this analysis are given in terms of the effective mixing parameter $\sin 2\theta_{\mu\tau} = 2|U_{\mu 4}||U_{\tau 4}|$ (see Sect. 2), which is the leading mixing term at short baseline experiments, allowing a direct comparison with their results.

The number μ of expected ν_τ events in the 3+1 model is evaluated as:

$$\mu = A \int \phi_{\nu_\mu}(E)\, P_{\mu\tau}(E)\, \sigma_{\nu_\tau}(E)\, \epsilon_{\nu_\tau}(E)\, \mathrm{d}E + n_b \tag{2}$$

where A is a normalization factor taking into account the delivered p.o.t. and the target mass, ϕ_{ν_μ} is the CNGS neutrino flux, $P_{\mu\tau}$ is the oscillation probability given by Eq. 1, σ_{ν_τ} is the ν_τ CC cross section, ϵ_{ν_τ} is the ν_τ identification efficiency, and n_b is the number of expected background events.

A gaussian prior is assumed for the atmospheric squared mass difference Δm^2_{31} with parameters determined from global fits, i.e. $(2.47 \pm 0.06) \times 10^{-3}$ eV^2 for normal hierarchy and $(-2.34 \pm 0.06) \times 10^{-3}$ eV^2 for inverted hierarchy [3].

In order to derive limits on the sterile neutrino mixing, the test statistics $-2 \ln \lambda_p$ is used, where λ_p is the profile likelihood ratio. In this counting analysis, the likelihood is given by a poissonian depending on the number of expected events μ (Eq. 2) and on the number of observed events $n_{\nu_\tau} = 5$. Two independent implementations of the method give similar results. Details on the statistical treatment of both analyses can be found in [15, 16].

 DOI: 10.1051/epjconf/201612604031

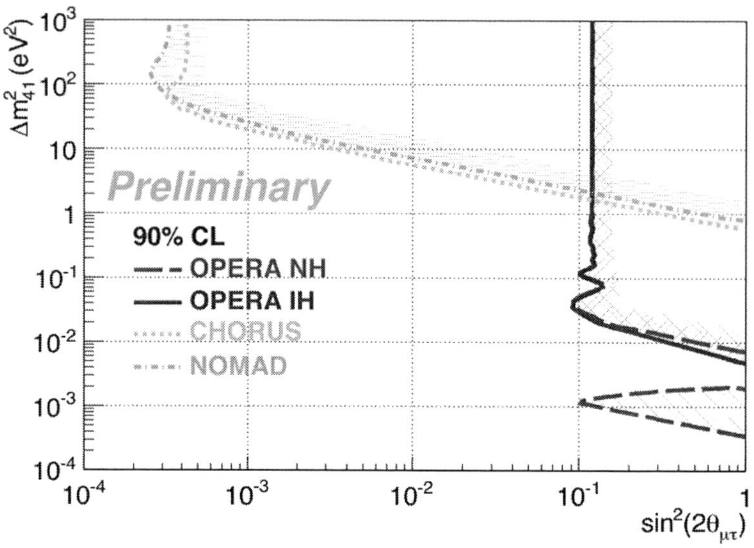

Figure 1. Limits on the sterile neutrino mixing derived from the OPERA data. Preliminary 90% C.L. exclusion limits in the Δm^2_{41} versus $\sin^2 2\theta_{\mu\tau}$ parameter space for normal (red) and inverted (blue) hierarchy. The CHORUS and NOMAD 90% C.L. excluded regions are also shown.

The limits on the sterile neutrino parameters are derived minimizing $-2\ \ln \lambda_p$ and are shown in Fig. 1. The 90% C.L. exclusion regions in the $(\Delta m^2_{41}, \sin^2 2\theta_{\mu\tau})$ parameter space are obtained for both normal (NH) and inverted (IH) mass hierarchies. The OPERA results in this parameter space can be directly compared to the limits given by the CHORUS and NOMAD short baseline $\nu_\mu \to \nu_\tau$ searches [17, 18].

The OPERA sensitivity extends the limits on Δm^2_{41} to 10^{-2} eV^2 for a relatively large mixing, i.e. $\sin^2 2\theta_{\mu\tau} > 0.5$, for both hierarchies. In case of normal hierarchy, a small region at $\Delta m^2_{41} \sim 10^{-3}$ eV^2 is also excluded, since the number of expected ν_τ events in this model is lower than the number expected from the 3ν framework. The suppression is due to the interference terms in Eq. 1. As said in Sect. 2, cosmological limits on the sum of neutrino masses favour $\Delta m^2_{41} > 0$, which is the assumption of the analysis here presented. For negative values of Δm^2_{41} the exclusion plots are similar but with NH and IH regions exchanged.

At large values of Δm^2_{41} (> 1 eV^2), there is almost no correlation between Δm^2_{41} and the effective mixing parameter, and the dependence of the oscillation probability $P_{\nu_\mu \to \nu_\tau}$ on Δm^2_{41} vanishes. In this context, which is in fact the phenomenological 3+1 model, the limits are expressed in the $(\phi_{\mu\tau}, \sin^2 2\theta_{\mu\tau})$ space. In Fig. 2 the 90% C.L. exclusion regions are shown for NH and IH hierarchies. Marginalizing over the phase $\phi_{\mu\tau}$, values of $\sin^2 2\theta_{\mu\tau} > 0.119$ are excluded at 90% C.L. Given the definition of the effective mixing parameter $\sin 2\theta_{\mu\tau} = 2|U_{\mu 4}||U_{\tau 4}|$, the upper limit $\sin^2 2\theta_{\mu\tau} < 0.119$ can be translated into an exclusion curve in the $(|U_{\mu 4}|^2, |U_{\tau 4}|^2)$ space, as shown in Fig. 3 together with the unitarity bound $|U_{\mu 4}|^2 + |U_{\tau 4}|^2 \leq 1$.

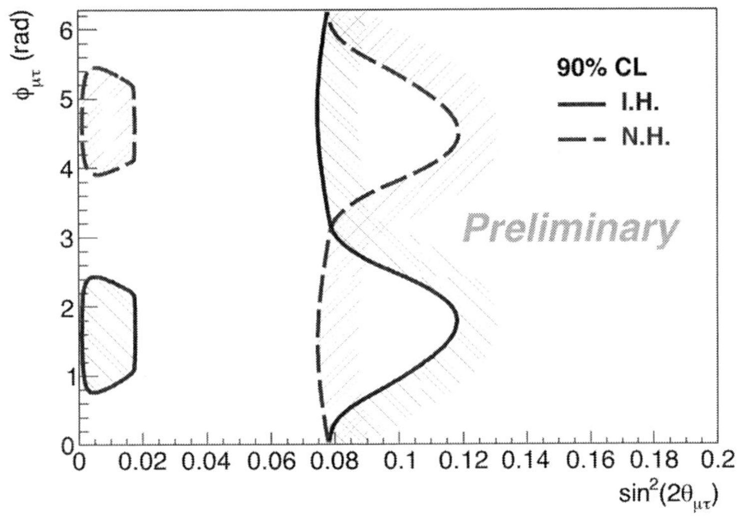

Figure 2. Preliminary 90% C.L. exclusion limits in the $\phi_{\mu\tau}$ versus $\sin^2 2\theta_{\mu\tau}$ parameter space for normal (red) and inverted (blue) hierarchy, assuming $\Delta m^2_{41} > 1$ eV^2.

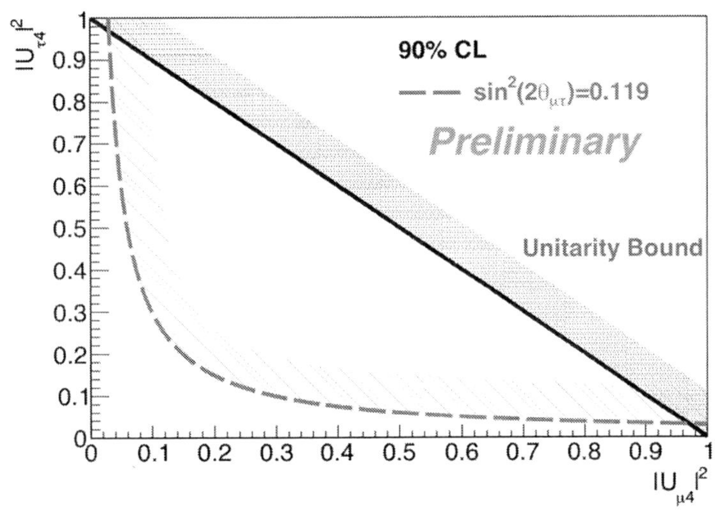

Figure 3. Preliminary 90% C.L. exclusion limits in the $|U_{\tau4}|^2$ versus $|U_{\mu4}|^2$ parameter space (cyan line), assuming the upper limit $\sin^2 2\theta_{\mu\tau} < 0.119$ for $\Delta m^2_{41} > 1$ eV^2. The unitarity bound is also shown (black line).

5 Conclusions

The OPERA experiment has observed $\nu_\mu \to \nu_\tau$ oscillations in the atmospheric sector in appearance mode detecting 5 ν_τ event candidates with an expected background of 0.25 events. The 5.1σ discovery of ν_τ appearance is compatible with the standar 3ν oscillation framework.

In this paper, OPERA results have been used to derive limits on the oscillations induced by a sterile neutrino, obtaining 90% C.L. exclusion regions for three different parameter spaces. The effective mixing parameter of the 3 + 1 neutrino model, for $\Delta m^2_{41} > 1$ eV^2, is constrained at 90% C.L. at $\sin^2 2\theta_{\mu\tau} < 0.119$. Based on this upper limit, constraints on $|U_{\tau4}|^2$ and $|U_{\mu4}|^2$ have also been set.

The OPERA sterile mixing search in appearance mode provides complementary information with respect to disappearance experiments, and the long baseline search further complements the short baseline one, providing stringent limits in the effective mixing angle.

References

[1] B. Pontecorvo, Zh. Eksp. Teor. Fiz. **33** 549 (1957); B. Pontecorvo, Zh. Eksp. Teor. Fiz. **34** 247 (1958); B. Pontecorvo, Sov. Phys. JETP **26** 984 (1968)

[2] Z. Maki, M. Nakagawa and S. Sakata, Prog. Theor. Phys. **28** 870 (1962)

[3] K. A. Olive *et al.*, [Particle Data Group], Chin. Phys. C **38** 090001 (2014)

[4] K. N. Abazajian *et al.*, "Light sterile neutrinos: a White Paper", arXiv:1204.5379.

[5] A. Aguilar *et al.* (LSND Collaboration), Phys. Rev. D **64**, 112007 (2001)

[6] A. A. Aguilar-Arevalo *et al.* (MiniBooNE Collaboration) Phys. Rev. Lett. **110**, 161801 (2013)

[7] G. Mention *et al.*, Phys. Rev. D **83**, 073006 (2011)

[8] C. Giunti and M. Laveder, Phys. Rev. C **83**, 065504 (2011)

[9] N. Agafonova *et al.* (OPERA Collaboration), Phys. Rev. Lett. **115**, 121802 (2015)

[10] P. A. R. Ade *et al.* (Planck Collaboration), submitted to A&A (2015), arXiv:1502.01589

[11] R. Acquafredda *et al.* (OPERA Collaboration), JINST **4**, P04018 (2009)

[12] N. Agafonova *et al.* (OPERA Collaboration), New J. Phys. **13**, 053051 (2011)

[13] N. Agafonova *et al.* (OPERA Collaboration), Eur. Phys. J. C **74**, 2986 (2014)

[14] N. Agafonova *et al.* (OPERA Collaboration), JHEP **06**, 069 (2015)

[15] S. Dusini, G. Sirri and M. Tenti, OPERA Public Note 181 (2015)

[16] A. Longhin, A. Paoloni and F. Pupilli, OPERA Public Note 180 (2015)

[17] CHORUS Collaboration, Nucl. Phys. B **793**, 326 (2008)

[18] NOMAD Collaboration, Nucl. Phys. B **601**, 3 (2001)

EPJ Web of Conferences **126**, 04032 (2016) DOI: 10.1051/epjconf/201612604032
ICNFP 2015

Anomalous transport in second order hydrodynamics *

Eugenio Megías[1,a] and Manuel Valle[2]

[1] *Max-Planck-Institut für Physik (Werner-Heisenberg-Institut), Föhringer Ring 6, D-80805, Munich, Germany*
[2] *Departamento de Física Teórica, Universidad del País Vasco UPV/EHU, Apartado 644, 48080 Bilbao, Spain*

Abstract. We study the non-dissipative transport effects appearing at second order in the hydrodynamic expansion for a non-interacting gas of chiral fermions by using the partition function formalism. We discuss some features of the corresponding constitutive relations, derive the explicit expressions for the conductivities and compare with existing results in the literature.

1 Introduction

Hydrodynamics is an effective description of out-of-equilibrium systems in which it is assumed local thermodynamical equilibrium. The hydrodynamical systems should obey the conservation laws of the energy-momentum tensor and spin one currents, and these quantities are written in terms of fluid variables in the so-called constitutive relations. Some of the transport phenomena are related to dissipative effects, as they lead to entropy production: examples are the shear viscosity η and bulk viscosity ζ [1]. However new phenomena on the hydrodynamics induced by quantum anomalies have recently received much attention and interest. In presence of anomalies the currents are no longer conserved, and this has important effects in the constitutive relations. Two relevant phenomena appear at first order in the hydrodynamic expansion as a consequence of chiral anomalies: the *chiral magnetic effect*, which is responsible for the generation of an electric current induced by a magnetic field [2], and the *chiral vortical effect*, in which the electric current is induced by a vortex [3]. It is believed that these phenomena can produce observable effects in heavy ion physics [4], as well as in condensed matter systems [5]. These effects are non-dissipative, and the associated conductivities are almost completely fixed by imposing the requirement of zero entropy production. At second order a plethora of dissipative and non-dissipative conductivities have been studied, see e.g. [6] and references therein.

Some methods to compute the transport coefficients from a microscopic theory, either dissipative or non-dissipative, include kinetic theory [7, 8], Kubo formulae [9], diagrammatic methods [10] and fluid/gravity correspondence [11]. Recently it has been proposed a new formalism to obtain the non-dissipative part of the anomalous constitutive relations, and it is based on the existence of an equilibrium partition function in a stationary background. It has been observed that the equations of hydrodynamics are significantly constrained by the requirement of consistency with the partition function [12, 13], and these constraints seem to overlap with the ones obtained from the existence of an

*Talk given by E. Megías at the 4th International Conference on New Frontiers in Physics (ICNFP 2015), 23-30 August 2015, Kolymbari, Crete, Greece.

[a] e-mail: emegias@mppmu.mpg.de

entropy current with non-negative divergence [14]. In this work we study, within this formalism, the non-dissipative constitutive relations up to second order in the hydrodynamic expansion for an ideal gas of chiral fermions. We will be able to get explicit results for some of the transport coefficients.

2 Hydrodynamics of relativistic fluids

Hydrodynamics is based on the assumption that the scales of variation of the observables are much longer than any microphysical scale in the system, and the result can be organized in a gradient expansion, also called hydrodynamic expansion [1]. The constitutive relations for the energy-momentum tensor and charged currents write generally in the form

$$\langle T^{\mu\nu}\rangle = (\varepsilon + \mathcal{P})u^\mu u^\nu + \mathcal{P}g^{\mu\nu} + \langle T^{\mu\nu}\rangle_{\text{diss \& anom}}\,, \tag{1}$$

$$\langle J^{\mu}\rangle = \rho u^\mu + \langle J^{\mu}\rangle_{\text{diss \& anom}}\,, \tag{2}$$

where ε is the energy density, $\mathcal{P}$ the pressure, ρ the charge density and u^μ the local fluid velocity. In addition to the equilibrium contributions, there are extra terms which lead to dissipative and anomalous effects. Within the Landau frame [1] and in presence of external electromagnetic fields, these terms write up to first order in derivatives as [2]

$$\langle T^{\mu\nu}\rangle_{\text{diss \& anom}} = -\eta P^{\mu\alpha}P^{\nu\beta}\left(\nabla_\alpha u_\beta + \nabla_\beta u_\alpha - \frac{2}{3}g_{\alpha\beta}\nabla^\lambda u_\lambda\right) - \zeta P^{\mu\nu}\nabla^\alpha u_\alpha + \cdots\,, \tag{3}$$

$$\langle J^{\mu}\rangle_{\text{diss \& anom}} = -\sigma T P^{\mu\nu}\nabla_\nu\left(\frac{\mu}{T}\right) + \sigma E^\mu + \sigma^{\mathcal{B}}B^\mu + \sigma^{\mathcal{V}}\omega^\mu + \cdots\,, \tag{4}$$

where $P^{\mu\nu} = G^{\mu\nu} + u^\mu u^\nu$, and the electric, magnetic fields and vorticity are defined as $E^\mu = F^{\mu\nu}u_\nu$, $B^\mu = \frac{1}{2}\epsilon^{\mu\nu\rho\lambda}u_\nu F_{\rho\lambda}$ and $\omega^\mu = \epsilon^{\mu\nu\rho\lambda}u_\nu\nabla_\rho u_\lambda$ respectively. The coefficients appearing in Eqs. (3)-(4) are the shear η and bulk η viscosities, the electric σ, chiral magnetic $\sigma^{\mathcal{B}}$ and chiral vortical $\sigma^{\mathcal{V}}$ conductivities respectively. At this point it is worth analyzing the parity $\mathcal{P}$ and time reversal $\mathcal{T}$ properties of these coefficients. The spatial component of the charged current, J^i, is $\mathcal{P}$-odd and $\mathcal{T}$-odd, while the magnetic field and vorticity are $\mathcal{P}$-even and $\mathcal{T}$-odd. Then one concludes from Eq. (4) that $\sigma^{\mathcal{B}}$ and $\sigma^{\mathcal{V}}$ are $\mathcal{P}$-odd and $\mathcal{T}$-even. On the other hand, the second law of thermodynamics states the increase of entropy with time, i.e.

$$\frac{\partial}{\partial t}s > 0\,, \tag{5}$$

and this means that only $\mathcal{T}$-odd contributions can lead to entropy production of the system. This is not the case of the chiral conductivities, so that these terms should be related to non-dissipative transport. A similar analysis for the shear, bulk and electric conductivities implies that these transport coefficients are $\mathcal{P}$-even and $\mathcal{T}$-odd, and they are associated with dissipative transport phenomena. In Sections 3 and 4 we present the partition function formalism, which is suitable to compute $\mathcal{T}$-even conductivities. We use this formalism to obtain the constitutive relations up to second order in the hydrodynamic expansion in Section 5.

3 Equilibrium partition function formalism to hydrodynamics

We present in this section the main ingredients of the equilibrium partition function formalism introduced in Refs. [12, 13] (see also e.g. Refs [14–17]). Let us consider a relativistic invariant Quantum

[1]This frame is defined as the one in which the viscous terms are transverse, $u_\mu\langle T^{\mu\nu}\rangle_{\text{diss \& anom}} = 0 = u_\mu\langle J^\mu\rangle_{\text{diss \& anom}}$.
[2]In this work we will consider (3 + 1) space-time dimensions.

Field Theory with a time independent $U(1)$ gauge connection on the manifold

$$ds^2 = G_{\mu\nu}dx^\mu dx^\nu = -e^{2\sigma(x)}(dt + a_i(x)dx^i)^2 + g_{ij}(x)dx^i dx^j\,, \tag{6}$$
$$\mathcal{A} = A_0(x)dx^0 + \mathcal{A}_i(x)dx^i\,. \tag{7}$$

The fields σ, a_i, g_{ij}, A_0 and $\mathcal{A}_i$ are smooth functions of the spatial coordinates x. As usual, the partition function of the system writes

$$Z = \mathrm{Tr}\, e^{-\frac{H-\mu_0 Q}{T_0}}\,, \tag{8}$$

where H is the Hamiltonian of the theory, Q is the charge associated to the gauge connection, while T_0 and μ_0 are the temperature and chemical potential at equilibrium. An obvious question is: what is the dependence of Z on the fields σ, a_i, g_{ij}, A_0 and $\mathcal{A}_i$? This has important implications for the hydrodynamics of the system, as we will see later. To answer this, the standard procedure is to build the most general partition function of the system consistent with the allowed symmetries, in particular: i) 3-dim diffeomorphism invariance; ii) Kaluza-Klein invariance, i.e. $t \to t+\phi(x)\,,\ x \to x$; and iii) $U(1)$ time-independent gauge invariance (up to an anomaly). It is convenient to introduce the combination $A_i \equiv \mathcal{A}_i - A_0 a_i$, which is invariant under the Kaluza-Klein transformation.

From the partition function we can compute the energy-momentum tensor and $U(1)$ charged current by performing the appropriate t-independent variations, i.e.

$$\delta \log Z = \frac{1}{T_0}\int d^3x \sqrt{g_3}\left(-\frac{1}{2}T_{\mu\nu}\delta g^{\mu\nu} + J^\mu \delta\mathcal{A}_\mu\right), \tag{9}$$

where $g_3 = \det(g_{ij})$. In particular, for a general partition function of the form $\log Z = \mathcal{W}(e^\sigma, A_0, a_i, A_i, g^{ij}, T_0, \mu_0)$, one gets

$$\langle J^i\rangle = \frac{T_0}{\sqrt{-G}}\frac{\delta\mathcal{W}}{\delta A_i}\,, \qquad \langle J_0\rangle = -\frac{T_0 e^{2\sigma}}{\sqrt{-G}}\frac{\delta\mathcal{W}}{\delta A_0}\,, \tag{10}$$
$$\langle T_0^{\ i}\rangle = \frac{T_0}{\sqrt{-G}}\left(\frac{\delta\mathcal{W}}{\delta a_i} - A_0\frac{\delta\mathcal{W}}{\delta A_i}\right), \qquad \langle T_{00}\rangle = -\frac{T_0 e^{2\sigma}}{\sqrt{-G}}\frac{\delta\mathcal{W}}{\delta\sigma}\,, \tag{11}$$

and a similar expression for $\langle T^{ij}\rangle$. This illustrates the fact that $\mathcal{W}$ plays the role of a generating functional for the hydrodynamic constitutive relations, and one expects that its form matches order by other with the derivative expansion in hydrodynamics. In the rest of this section we present the most important properties of this partition function up to second order in derivatives.

3.1 Equilibrium partition function at zeroth order

We study first the zeroth order in derivatives. The most general partition function at this order, which is consistent with the symmetries mentioned above, reads [12]

$$\mathcal{W}_0 = \int d^3x \sqrt{g_3}\frac{e^\sigma}{T_0}P(e^{-\sigma}T_0, e^{-\sigma}A_0)\,, \tag{12}$$

where P is an arbitrary function of its arguments. From Eq. (12) and after applying the variational formulae (10)-(11), one gets

$$\langle J^0\rangle = e^{-\sigma}\partial_b P\,, \qquad \langle J^i\rangle = 0\,, \tag{13}$$
$$\langle T^{ij}\rangle = P g^{ij}\,, \qquad \langle T_{00}\rangle = e^{2\sigma}(P - a\partial_a P - b\partial_b P)\,, \qquad \langle T_0^{\ i}\rangle = 0\,, \tag{14}$$

where for convenience we have defined the variables $a \equiv e^{-\sigma} T_0$ and $b \equiv e^{-\sigma} A_0$. This result should correspond to the equilibrium contribution of the hydrodynamic constitutive relations. Finally, after a comparison with Eqs. (1)-(2) one gets

$$\mathcal{P} = P \,, \qquad \varepsilon = -P + a\partial_a P + b\partial_b P \,, \qquad \rho = \partial_b P \,, \qquad u^\mu = e^{-\sigma}(1,0,0,0) \,. \tag{15}$$

Note that P is identified with the pressure of the system, and the thermodynamic variables ε, $\mathcal{P}$ and ρ are determined in terms of this single *master* function. Thermodynamical consistency requires to identify the local value of the temperature, T, and chemical potential, μ, with a and b respectively.

3.2 Equilibrium partition function at higher orders

Let us discuss now the properties of the partition function at higher derivative orders. The most general partition function at first order in the derivative expansion was presented in [12], and it reads

$$\mathcal{W}_1 = \int d^3x \sqrt{g_3} \left[\alpha_1(T,\nu)\epsilon^{ijk} A_i F_{jk} + T_0 \alpha_2(T,\nu)\epsilon^{ijk} A_i f_{jk} + T_0^2 \alpha_3(T,\nu)\epsilon^{ijk} a_i f_{jk} \right] , \tag{16}$$

where

$$T = e^{-\sigma} T_0 \,, \qquad \nu = \frac{A_0}{T_0} \,, \tag{17}$$

while $F_{ij} = \partial_i A_j - \partial_j A_i$ and $f_{ij} = \partial_i a_j - \partial_j a_i$. Eq. (16) contains only $\mathcal{P}$-odd contributions, as one cannot build $\mathcal{P}$-even contributions to the partition function at first order in derivatives. Following the method of Section 4, a computation of the partition function for an ideal gas of Weyl fermions leads to the following result [17]

$$\alpha_1(T,\nu) = -\frac{C}{6}\nu \,, \quad \alpha_2(T,\nu) = -\frac{1}{2}\left(\frac{C}{6}\nu^2 - C_2\right) , \quad \alpha_3(T,\nu) = 0 \,, \tag{18}$$

where

$$C = \frac{1}{4\pi^2} \,, \qquad C_2 = \frac{1}{24} \,, \tag{19}$$

are constants related to the axial anomaly [3, 18] and gauge-gravitational anomaly [19] respectively. After a suitable computation of the constitutive relations as it will be explained in Section 5, one gets the well known expressions for the chiral conductivities in the Landau frame [3]

$$\sigma^{\mathcal{B}} = C\mu - \frac{\rho}{\varepsilon + \mathcal{P}}\left(\frac{1}{2}C\mu^2 + C_2 T^2\right) , \qquad \sigma^{\mathcal{V}} = \frac{1}{2}C\mu^2 + C_2 T^2 - \frac{\rho}{\varepsilon + \mathcal{P}}\left(\frac{1}{3}C\mu^2 + 2C_2 T^2\right)\mu \,, \tag{20}$$

where we have used that $\nu = \mu/T$. These results have been obtained in a wide variety of methods, see e.g. [2, 3, 9, 18–25].

Finally, the most general partition function at second order in derivatives is built from seven scalar and two pseudo-scalar quantities as follows [14, 16]

$$\begin{aligned} \mathcal{W}_2 = \int d^3x \sqrt{g_3} \Big[& M_1 g^{ij} \partial_i T \partial_j T + M_2 g^{ij} \partial_i \nu \partial_j \nu + M_3 g^{ij} \partial_i \nu \partial_j T \\ & + T_0^2 M_4 f_{ij} f^{ij} + M_5 F_{ij} F^{ij} + T_0 M_6 f_{ij} F^{ij} + M_7 R \\ & + N_1 \epsilon^{ijk} \partial_i A_0 f_{jk} + T_0^{-1} N_2 \epsilon^{ijk} \partial_i A_0 F_{jk} \Big] , \end{aligned} \tag{21}$$

where R is the Ricci scalar in 3 dim, with $M_i = M_i(T,\nu)$ and $N_i = N_i(T,\nu)$. We keep for the moment both $\mathcal{P}$-odd and $\mathcal{P}$-even contributions.

[3] We have considered a fluid with an anomalous charge $U(1)$. The extension to $U_V(1) \times U_A(1)$ would require the introduction of both vector and axial gauge connections.

4 Second order partition function

We present in this section the procedure to compute the second order partition function of Eq. (21) for a theory of free massless Dirac fermions. We refer to Ref. [17] for full details in the computation.

4.1 Free theory of Dirac fermions

We will derive the partition function by using Pauli-Villars (PV) regularization, and this demands the consideration of the massive theory for the vacuum contribution. The action of the theory is

$$S = \int d^4x \sqrt{-G}\mathcal{L}, \qquad \text{where} \qquad \mathcal{L} = -i\bar{\Psi}\underline{\gamma}^\mu \nabla_\mu \Psi + im\bar{\Psi}\Psi . \tag{22}$$

The space-time dependent Dirac matrices satisfy $\{\underline{\gamma}^\mu(x), \underline{\gamma}^\nu(x)\} = 2G^{\mu\nu}(x)$, and they are related to the Minkowski matrices by $\underline{\gamma}^\mu(x) = e^\mu_a(x)\gamma^a$, where $e^\mu_a(x)$ is the vierbein, $\{\gamma^a, \gamma^b\} = 2\eta^{ab}$ and $\eta^{ab} = \mathrm{diag}(-1, 1, 1, 1)$. The $U(1)$ current and energy-momentum tensor write

$$J^\mu = -\bar{\Psi}\underline{\gamma}^\mu\Psi , \qquad T_{\mu\nu} = \frac{i}{4}\bar{\Psi}\left[\underline{\gamma}_\mu \overrightarrow{\nabla}_\nu - \overleftarrow{\nabla}_\nu \underline{\gamma}_\mu + (\mu \leftrightarrow \nu)\right]\Psi , \tag{23}$$

where it has been assumed that the spinor field $\Psi = \begin{pmatrix} \psi_L \\ \psi_R \end{pmatrix}$ satisfies the Dirac equation. Using the explicit form of the background Eqs. (6)-(7) one has

$$J_0 = -e^{-\sigma}\psi^\dagger\psi , \qquad J^i = -\psi^\dagger\sigma_i\psi , \tag{24}$$

$$T_{00} = \frac{i}{2}e^\sigma\left(\psi^\dagger\partial_t\psi - \partial_t\psi^\dagger\psi\right) + e^\sigma A_0\psi^\dagger\psi - \frac{1}{4}e^{3\sigma}\epsilon^{ijk}\partial_j a_k\psi^\dagger\sigma_i\psi , \tag{25}$$

where σ_i are the Pauli matrices, and ψ is the two-component Weyl fermion ψ_L. The same expressions are obtained for ψ_R, but with opposite sign in J^i.

4.2 Thermal Green's function

The expectation values of J_μ and $T^{\mu\nu}$ at equilibrium may be computed from the thermal Green's function, defined as

$$\langle T\psi(-i\tau, \boldsymbol{x})\psi^\dagger(0, \boldsymbol{x}')\rangle_{T_0} = T_0\sum_n e^{-i\omega_n\tau}\mathcal{G}(\boldsymbol{x}, \boldsymbol{x}', \omega_n) , \qquad \omega_n = 2\pi T_0\left(n + \frac{1}{2}\right) , \tag{26}$$

where T denotes time ordering. The explicit calculation of the Green's function is the most computationally demanding step in the derivation. After rotating to imaginary time $t \to -i\tau$, the Green's function satisfies

$$-\sqrt{-G}\,\gamma^0\underline{\gamma}^0(i\omega_n - \mathcal{H})\mathcal{G}(\boldsymbol{x}, \boldsymbol{x}', \omega_n) = \delta^{(3)}(\boldsymbol{x} - \boldsymbol{x}') , \tag{27}$$

where $\mathcal{H}$ is the Hamiltonian

$$\mathcal{H} = -i\left(\frac{1}{4}{\omega_0}^{ab}\gamma_{ab} - iA_0\right) - \frac{i}{g^{00}}\underline{\gamma}^0\left(\underline{\gamma}^k\nabla_k - m\right) , \tag{28}$$

with ${\omega_\mu}^{ab}$ the spin connection and $\gamma_{ab} = \frac{1}{2}[\gamma_a, \gamma_b]$. Then by expanding Eq. (27) in derivatives, the Green's function can be computed recursively order by order in a derivative expansion, i.e. $\mathcal{G} =$

$\mathcal{G}_0 + \mathcal{G}_1 + \mathcal{G}_2 + \cdots$, where $\mathcal{G}_\ell$ is the contribution at order ℓ. Note that to get $\mathcal{G}_\ell$ one needs to know the Green's function at all the orders lower than ℓ. The complete expressions for $\mathcal{G}$ up to second order are very lengthy and will not be presented here. Finally, from Eqs. (24)-(26) one obtains the precise form of the current and energy-momentum tensor up to ℓ-th order in the derivative expansion

$$\langle J_0\rangle_\ell = T_0 \sum_n \left[-e^{\sigma}\mathrm{tr}\,\mathcal{G}_\ell(\boldsymbol{x},\boldsymbol{x},\omega_n)\right] , \qquad \langle J^i\rangle_\ell = -T_0 \sum_n \mathrm{tr}\left[\sigma_i\,\mathcal{G}_\ell(\boldsymbol{x},\boldsymbol{x},\omega_n)\right] , \tag{29}$$

$$\langle T_{00}\rangle_\ell = T_0 \sum_n \left[e^{\sigma}(i\omega_n + A_0)\,\mathrm{tr}\,\mathcal{G}_\ell(\boldsymbol{x},\boldsymbol{x},\omega_n) - \frac{1}{4}e^{3\sigma}\epsilon^{ijk}\partial_j a_k\,\mathrm{tr}\left[\sigma_i\,\mathcal{G}_{\ell-1}(\boldsymbol{x},\boldsymbol{x},\omega_n)\right]\right] . \tag{30}$$

Note that $\langle T_{00}\rangle_\ell$ receives a contribution $\propto \mathrm{rot}\,\boldsymbol{a}\cdot\langle \boldsymbol{J}\rangle_{\ell-1}$.

4.3 Charge density at second order

To get $\mathcal{W}_2$ it is enough to compute $\langle J_0\rangle_2$ and $\langle T_{00}\rangle_2$ including only bilinear terms $\sim \partial_i X \partial_j Y$. The evaluation of Eq. (29) produces the following renormalized expression

$$\begin{aligned}\langle J_0\rangle_2 = &\frac{1}{24\pi^2}\left(-\nabla^i A_0 \nabla_i \sigma + \frac{1}{2}e^{2\sigma} f_{ij}F^{ij} + \frac{1}{2}A_0 e^{2\sigma} f_{ij}f^{ij}\right)\left(\ln\frac{e^{2\sigma}\bar{M}^2}{T_0^2} + Q(\nu)\right)\\ &+ \frac{1}{48\pi^2}\left(\nabla^i A_0 \nabla_i A_0 + \frac{e^{2\sigma}}{2}A_0^2 f_{ij}f^{ij} + \frac{e^{2\sigma}}{2}F_{ij}F^{ij} + e^{2\sigma}A_0 f_{ij}F^{ij}\right)\frac{1}{T_0}Q'(\nu)\\ &- \frac{1}{24\pi^2}A_0\nabla^i\sigma\nabla_i\sigma + \frac{1}{8\pi^2}\nabla^i A_0\nabla_i\sigma + \frac{1}{96\pi^2}e^{2\sigma} f_{ij}F^{ij} + \frac{1}{32\pi^2}e^{2\sigma}A_0 f_{ij}f^{ij} + \frac{A_0}{48\pi^2}R ,\end{aligned} \tag{31}$$

where $Q(\nu)$ is the analytic continuation of the series $Q(\nu) = -2\sum_{n=1}^{\infty}(-1)^n \cosh(n\nu)\log(n^2)$. A similar expression is obtained for $\langle T_{00}\rangle_2$. To derive this result we have regularized the vacuum contribution in a gauge invariant way by using the PV regularization procedure. We have defined the rescaled PV mass as $\bar{M} = 2^{-3/2}e^{\gamma_E}M$. The distinction between vacuum and finite temperature and chemical potential contributions in the expectation values can be obtained, for instance, by considering the Poisson summation formula in the Matsubara sums.

4.4 Partition function at second order

By using the variational formulae (10)-(11) with Eq. (21), and after a comparison with the expressions of $\langle J_0\rangle_2$ and $\langle T_{00}\rangle_2$, one gets a system of 14 equations and 7 functions of two arguments. After consistently solving these equations we get the following result

$$N_1(T,\nu) = 0 = N_2(T,\nu) , \qquad M_1(T,\nu) = -\frac{1}{144}\frac{1}{T} - \frac{1}{48\pi^2}\frac{\nu^2}{T} , \tag{32}$$

$$M_2(T,\nu) = \frac{1}{48\pi^2}T\left(\ln\frac{\bar{M}^2}{T^2} + Q(\nu) - 1\right) , \qquad M_3(T,\nu) = -\frac{1}{12\pi^2}\nu , \tag{33}$$

$$M_4(T,\nu) = -\frac{1}{96\pi^2}\frac{\nu^2}{T}\left(\ln\frac{\bar{M}^2}{T^2} + Q(\nu) + 3 - 6\pi^2 C\right) + \frac{1}{288}\frac{1}{T} - \frac{C_2}{8T} + \frac{1}{384\pi^2}\frac{1}{T^3}M^2\ln 2 , \tag{34}$$

$$M_5(T,\nu) = -\frac{1}{96\pi^2}\frac{1}{T}\left(\ln\frac{\bar{M}^2}{T^2} + Q(\nu)\right) , \qquad M_6(T,\nu) = -\frac{1}{48\pi^2}\frac{\nu}{T}\left(\ln\frac{\bar{M}^2}{T^2} + Q(\nu) + 2 - 6\pi^2 C\right) , \tag{35}$$

$$M_7(T,\nu) = -\frac{1}{288}T - \frac{1}{96\pi^2}T\nu^2 + \frac{1}{96\pi^2}\frac{1}{T}M^2\ln 2 , \tag{36}$$

where the constants C and C_2 are given by Eq. (19). This result corresponds to a free theory of one left Weyl fermion. The coefficients obtained with one free Dirac fermion are twice these expressions. Note that these terms depend on the renormalization scale. The combination of terms proportional to M^2 in M_4 and M_7 is a pure renormalization effect, and they can be renormalized by adding a counterterm proportional to the Ricci scalar $\tilde{R}$ of the 3 + 1 dimensional metric,

$$\mathcal{W}_2^{\rm ct} = -\frac{M^2 \ln 2}{96\pi^2} \int d^4x \sqrt{-G}\,\tilde{R}\,. \tag{37}$$

The renormalized partition function is $\mathcal{W}_2^{\rm ren} = \mathcal{W}_2 + \mathcal{W}_2^{\rm ct}$ and the counterterm exactly cancels the M^2 terms in $\mathcal{W}_2$, so that the renormalized coefficients $M_{4,7}^{\rm ren}$ are the same as $M_{4,7}$ after removing these contributions. On the other hand, this theory only violates conformal invariance because renormalization effects, which lead to a logarithmic dependence $\sim \ln\frac{\bar{M}}{T}$. The anomalous partition function reads

$$\begin{aligned}\mathcal{W}_{\rm anom} &= \frac{1}{24\pi^2}\int d^3x\sqrt{g_3}\frac{1}{T}\ln\frac{\bar{M}}{T}\times\left(e^{-2\sigma}g^{ij}\partial_iA_0\partial_jA_0 - \frac{1}{2}A_0^2 f_{ij}f^{ij} - \frac{1}{2}F_{ij}F^{ij} - A_0 f_{ij}F^{ij}\right)\\ &= -\frac{1}{48\pi^2}\int d^4x\sqrt{-G}\ln\frac{\bar{M}}{T}\mathcal{F}_{\mu\nu}\mathcal{F}^{\mu\nu}\,,\end{aligned} \tag{38}$$

where $\mathcal{F}_{\mu\nu}$ is the four dimensional field strength of the gauge field $\mathcal{A}_\mu$. This is in agreement with the form of the local covariant action for the trace anomaly [26, 27].

5 Non-dissipative constitutive relations

In this section we determine partially the non-dissipative part of the second order constitutive relations in terms of the functions $M_i(T,\nu)$. The outline of the procedure may be sketched by

$$\langle O_\ell\rangle_{\rm eq} = \delta(O_{\rm perfect\,fluid} + O_1 + \ldots + O_{\ell-1}) + O_\ell\,, \tag{39}$$

where $O_\ell \equiv T^{\mu\nu}_{(\ell)}$, $J^\mu_{(\ell)}$, and $\delta(O_{\rm perfect\,fluid} + \cdots)$ is a correction of order ℓ due to all changes proportional to derivatives of the background evaluated in the constitutive relations of lower orders, see e.g. Ref. [16]. As an example $\delta O_{\rm perfect\,fluid}$ receives corrections of the fluid velocity $\propto \delta u_{(1)}$ where $u = u_{(0)} + \delta u_{(1)} + \cdots$. The most general non-dissipative form of the constitutive relations in the Landau frame at second order can be expressed as

$$\begin{aligned}T_{(2)\mu\nu} &= T\left(\kappa_1\tilde{R}_{\langle\mu\nu\rangle} + \kappa_2 u^\alpha u^\beta\tilde{R}_{\langle\mu\alpha\nu\rangle\beta} + \kappa_3\nabla_{\langle\mu}\nabla_{\nu\rangle}\nu + \lambda_3\,\omega_{\langle\mu\alpha}\omega^\alpha{}_{\nu\rangle} + \lambda_4\,\mathfrak{a}_{\langle\mu}\mathfrak{a}_{\nu\rangle}\right) + \cdots\\ J_{(2)\mu} &= \upsilon_1 P_{\mu\alpha}u_\nu\tilde{R}^{\nu\alpha} + \upsilon_2 P_{\mu\alpha}\nabla_\nu\mathcal{F}^{\nu\alpha} + \cdots\,,\end{aligned} \tag{40}$$

where $\mathfrak{a}_\mu = u^\alpha\nabla_\alpha u_\mu$, the vorticity tensor is $\omega_{\mu\nu} \equiv \frac{1}{2}P_\mu^\alpha P_\nu^\beta\left(\nabla_\alpha u_\beta - \nabla_\beta u_\alpha\right)$ and $X_{\langle\mu\nu\rangle}$ stands for the traceless and symmetric projection transverse to u^μ, see e.g. [6, 12, 16]. The goal is to determine the coefficients κ_i, λ_i and υ_i by comparison with the partition function $\mathcal{W}_2$. After using the variational formulae Eqs. (10)-(11) with Eq. (21), we arrive at the following general result

$$\kappa_1 = -2M_7^{\rm ren}\,, \qquad \kappa_2 = -2M_7^{\rm ren} - 2T\frac{\partial M_7^{\rm ren}}{\partial T}\,, \qquad \kappa_3 = 2\frac{\partial M_7^{\rm ren}}{\partial\nu}\,, \tag{41}$$

$$\lambda_3|_{\mu=0} = 16T^2M_4^{\rm ren} - 6M_7^{\rm ren} - 2TM_7^{\rm ren\,\prime}(T)\,, \tag{42}$$

$$\lambda_4|_{\mu=0} = -2T^2M_1 + 4TM_7^{\rm ren\,\prime}(T) + 2T^2M_7^{\rm ren\,\prime\prime}(T)\,, \tag{43}$$

$$\upsilon_1 = 4T^2(2\nu M_5 - M_6) - \frac{8\rho}{\varepsilon+P}T^3\left(M_4^{\rm ren} + \nu^2M_5 - \nu M_6\right)\,, \tag{44}$$

$$\upsilon_2 = -4TM_5 + \frac{2\rho}{\varepsilon+P}T^2(2\nu M_5 - M_6)\,. \tag{45}$$

The coefficients λ_3 and λ_4 multiply terms that are bilinear in derivatives, and these are more difficult to obtain. However they can be directly inferred at zero chemical potential from the results derived in Section 5 of Ref. [12]. In the particular case of the free field theory of Weyl fermions, the explicit result of these transport coefficients are

$$\kappa_1 = \frac{T}{144} + \frac{1}{48\pi^2}\frac{\mu^2}{T}, \qquad \kappa_2 = 2\kappa_1, \qquad \kappa_3 = -\frac{\mu}{24\pi^2}, \qquad \lambda_3|_{\mu=0} = 0 = \lambda_4|_{\mu=0}, \tag{46}$$

$$\upsilon_1 = -\frac{1}{2}\left(C - \frac{1}{3\pi^2}\right)\mu + \frac{\rho}{\varepsilon + P}\left[\frac{1}{2}\left(C - \frac{1}{6\pi^2}\right)\mu^2 + \left(C_2 - \frac{1}{36}\right)T^2\right], \tag{47}$$

$$\upsilon_2 = \frac{1}{24\pi^2}\left(\ln\frac{\bar{M}^2}{T^2} + Q\left(\frac{\mu}{T}\right)\right) - \frac{\rho}{\varepsilon + P}\frac{1}{4}\left(C - \frac{1}{3\pi^2}\right)\mu. \tag{48}$$

The only second order coefficient which shows sensitivity to the renormalization scale is υ_2. Notice also that there is no mixture between the conformal and chiral anomaly contributions in the constitutive relations.

Let us compare these transport coefficients with some existing results in the literature. The values of $\kappa_{1,2}$ and λ_3 have been computed in Ref. [28] at zero chemical potential. Our values for κ_1 and κ_2 are in agreement with this reference after setting $\mu = 0$, but the vanishing value of $\lambda_3|_{\mu=0}$ is in contrast with the result obtained in this work, where they find $\lambda_3^{\text{Moore,Sohrabi}} = -T^2/24$ for a Weyl fermion. This difference is related to the second term in the rhs of Eq. (30), which seems to be not included in the diagrammatic computation of these authors. Note however that this coefficient was previously computed in Ref. [8] from kinetic theory by some of the same authors, and a vanishing value was obtained. Finally the coefficients κ_3 and υ_2 were explicitly computed within a holographic model in 5 dim in Refs. [18, 20, 24]. A comparison between the weak coupling results presented above and the explicit expressions from the holographic model, leads to

$$[\kappa_3]_{\text{weak coupling}} \propto [\kappa_3]_{\text{strong coupling}}, \qquad [\upsilon_2]_{\text{weak coupling}} \sim c(T) + \frac{5}{112\pi^4}\frac{\mu^2}{T^2} \propto [\upsilon_2]_{\text{strong coupling}}, \tag{49}$$

in the regime $\mu \ll T$, where $c(T)$ has a logarithmic dependence on T in the free fermion theory, while it is a constant in the holographic result. These coefficients receive contributions not induced by chiral anomalies, and so we cannot expect that the results from both approaches agree. However, we do confirm agreement in the parametric dependence in μ and T.

6 Conclusions and discussion

In this work we have studied second order transport effects induced by external electromagnetic fields, vortices and curvature, in a relativistic fluid for a free theory of chiral fermions in (3+1) dim. We have addressed the computation by using the equilibrium partition function formalism, which can only account for non-dissipative effects, i.e. transport coefficients multiplying quantities that survive in equilibrium. Examples of such contributions at first order in the hydrodynamic expansion are the chiral magnetic and vortical conductivities, which are $\mathcal{P}$-odd and $\mathcal{T}$-even. The situation is slightly different at second order, as we find that the parity violating part of the partition function vanishes, and the nonzero non-dissipative coefficients we have obtained are $\mathcal{P}$-even and $\mathcal{T}$-even. An explanation for this is that the only possible pseudo-scalars appearing at second order are $\mathcal{T}$-odd, and so if the underlying Hamiltonian is invariant under $\mathcal{T}$, this part of the partition function should vanish [17].

We have shown that the renormalization effects of the conformal anomaly mix with the chiral anomaly in some terms of the partition function, however this mixture does not appear in the constitutive relations. We examined the non-dissipative constitutive relations in the Landau frame, and derived the dependence with temperature and chemical potential of five transport coefficients: $\kappa_{1,2,3}$ and $\upsilon_{1,2}$. We also obtained the zero chemical potential value for two additional coefficients $\lambda_{3,4}$. Some of them: $\kappa_{1,2,3}$, υ_2 and λ_3; were previously computed in the literature by using different methods, and our results are consistent except for the latter. Finally let us mention that, as it has been pointed out in Refs. [12, 17], the vanishing value of λ_4 at zero chemical potential is required by conformal invariance, while this only requires that $\lambda_3 = \text{cons} \cdot T$. The vanishing value of λ_3 is due to a further cancellation.

Acknowledgements

This work has been supported by Plan Nacional de Altas Energías grant FPA2012-34456, Spanish Consolider-Ingenio 2010 Programme CPAN (CSD2007-00042) and by the Basque Government under grant IT559-10. The research of E.M. is supported by the European Union under a Marie Curie Intra-European Fellowship (FP7-PEOPLE-2013-IEF) with project number PIEF-GA-2013-623006.

References

[1] P. Kovtun, J. Phys. **A45**, 473001 (2012)
[2] K. Fukushima, D.E. Kharzeev, H.J. Warringa, Phys. Rev. **D78**, 074033 (2008)
[3] D.T. Son, P. Surowka, Phys. Rev. Lett. **103**, 191601 (2009)
[4] B. Keren-Zur, Y. Oz, JHEP **06**, 006 (2010)
[5] G. Basar, D.E. Kharzeev, H.U. Yee, Phys. Rev. **B89**, 035142 (2014)
[6] D.E. Kharzeev, H.U. Yee, Phys. Rev. **D84**, 045025 (2011)
[7] P.B. Arnold, G.D. Moore, L.G. Yaffe, JHEP **0011**, 001 (2000)
[8] M.A. York, G.D. Moore, Phys. Rev. **D79**, 054011 (2009)
[9] K. Landsteiner, E. Megias, F. Pena-Benitez, Lect. Notes Phys. **871**, 433 (2013)
[10] J.L. Manes, M. Valle, JHEP **01**, 008 (2013)
[11] S. Bhattacharyya, V.E. Hubeny, S. Minwalla, M. Rangamani, JHEP **0802**, 045 (2008)
[12] N. Banerjee, J. Bhattacharya, S. Bhattacharyya, S. Jain, S. Minwalla et al., JHEP **1209**, 046 (2012)
[13] K. Jensen, M. Kaminski, P. Kovtun, R. Meyer, A. Ritz et al., Phys. Rev. Lett. **109**, 101601 (2012)
[14] S. Bhattacharyya, JHEP **07**, 139 (2014)
[15] J.L. Mañes, M. Valle, JHEP **1311**, 178 (2013)
[16] S. Bhattacharyya, J.R. David, S. Thakur, JHEP **1401**, 010 (2014)
[17] E. Megias, M. Valle, JHEP **11**, 005 (2014)
[18] J. Erdmenger, M. Haack, M. Kaminski, A. Yarom, JHEP **0901**, 055 (2009)
[19] K. Landsteiner, E. Megias, F. Pena-Benitez, Phys. Rev. Lett. **107**, 021601 (2011)
[20] N. Banerjee, J. Bhattacharya, S. Bhattacharyya, S. Dutta, R. Loganayagam et al., JHEP **1101**, 094 (2011)
[21] K. Landsteiner, E. Megias, L. Melgar, F. Pena-Benitez, JHEP **1109**, 121 (2011)
[22] K. Jensen, R. Loganayagam, A. Yarom, JHEP **1302**, 088 (2013)
[23] E. Megias, K. Landsteiner, F. Pena-Benitez, Acta Phys. Polon. Supp. **6**, 45 (2013)
[24] E. Megias, F. Pena-Benitez, JHEP **1305**, 115 (2013)
[25] K. Landsteiner, E. Megias, F. Pena-Benitez, Phys. Rev. **D90**, 065026 (2014)
[26] M. Giannotti, E. Mottola, Phys. Rev. **D79**, 045014 (2009)
[27] C. Eling, Y. Oz, S. Theisen, S. Yankielowicz, JHEP **1305**, 037 (2013)
[28] G.D. Moore, K.A. Sohrabi, JHEP **11**, 148 (2012)

EPJ Web of Conferences **126**, 04033 (2016) DOI: 10.1051/epjconf/201612604033
ICNFP 2015

Highlights from the COMPASS experiment at CERN

Hadron spectroscopy and excitations

Frank Nerling[1,a], on behalf of the COMPASS Collaboration

[1] *Universität Mainz, Institut für Kernphysik, Germany*

Abstract. The COMPASS experiment at the CERN-SPS studies the spectrum and the structure of hadrons by scattering high energy hadrons and polarised muons off various fixed targets. Recent results for the hadron programme comprise highlights from different topics. A selective overview is given and, among others, the following results are discussed. The precise determination of the pion polarisability, a long standing puzzle that has been solved now, is presented as well as measurements of radiative widths. The observation of a new narrow axial-vector state, the $a_1(1420)$, as well as deeper insights into the exotic 1^{-+}-wave, which is under study since decades by several experiments, are discussed and further, the search for the charmonium-like exotic $Z_c(3900)$ state in the COMPASS data is covered.

1 Introduction

The COMPASS fixed-target experiment [1] at CERN-SPS is a facility to study quantum chromodynamics (QCD). It is dedicated to study the perturbative and non-perturbative regime of QCD, and probe the structure and dynamics of hadrons. Using a muon beam, COMPASS investigates nucleon structure by measuring helicity, transversity and general parton distribution functions — the recent highlights from the COMPASS muon programme are summarised in [2]. Hadron excitations and spectroscopy, including the search for exotic states, are investigated using hadron beams on a liquid hydrogen (proton) and different nuclear targets. A selection of recent results of the COMPASS hadron programme are discussed and summarised here.

The two-stage spectrometer (Fig. 1) is equipped with electromagnetic and hadronic calorimeters, providing detection of charged and neutral final state particles with a homogeneous acceptance over a wide kinematic range, especially covering a large range in momentum transfer. At lowest reduced four-momentum transfers between the beam particle and the target of $t' < 10^{-3}$ $(\mathrm{GeV}/c)^2$, Primakoff reactions are studied, from which the pion polarisability has been extracted at high experimental precision, Chiral perturbation theory (ChPT) is tested and the radiative decay widths of the $a_2(1320)$ and the $\pi_2(1670)$ have been measured in pion-photon reactions. At low four-momentum transfers of $0.1\ (\mathrm{GeV}/c)^2 < t' < 1\ (\mathrm{GeV}/c)^2$, the mass spectrum of hadrons is investigated, including searches for (spin-) exotic mesons.

[a] e-mail: nerling@cern.ch

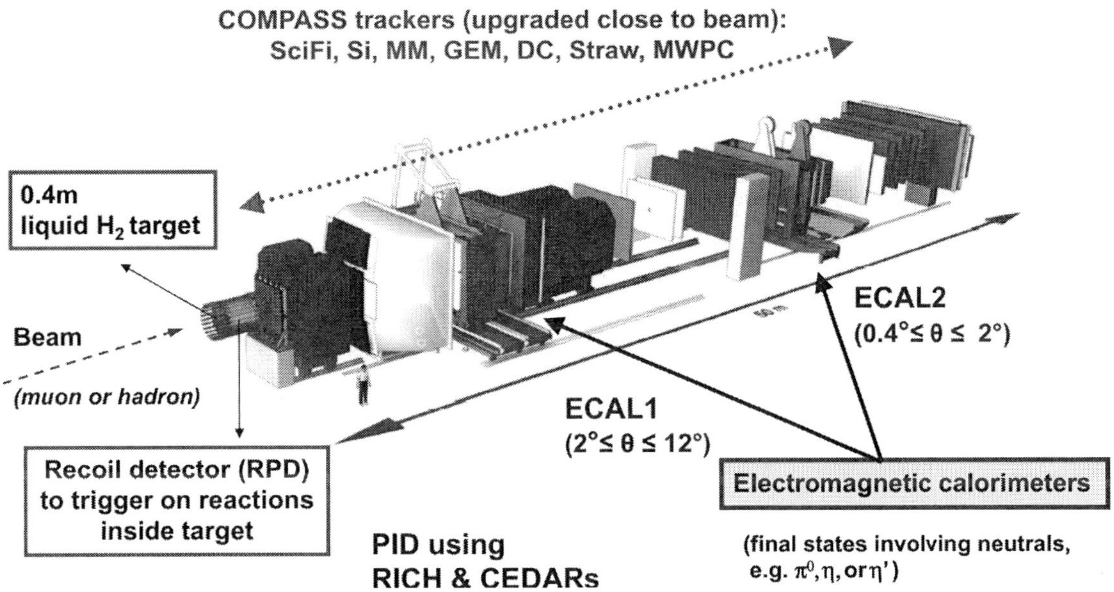

Figure 1. Schematic drawing of the COMPASS two-stage spectrometer as used for the hadron runs 2008/09.

2 Results on hadron excitations

For the understanding of QCD and the low-momentum expansion ChPT, in which the pions are identified as the Goldstone bosons as a result of the spontaneous chiral symmetry breaking, the pion properties are a crucial input. Whereas pion-pion scattering has extensively been studied and successfully described within ChPT, experimental results of pion-photon reactions in terms of the most basic process provided still a puzzle: Previous measurements of the pion polarisability, the leading structure-dependent term of Compton scattering, provided resultant values significantly larger as expected from theory, *cf.* Fig. 2. Apart from the pion polarisability in Primakoff reactions (Sec. 2.1), also pion-photon interactions with more than one pion in the final state are studied (Secs. 2.2 and 2.3).

2.1 Measurement of pion polarisability

The rigidity of an extended object against deformations by an external electric or magnetic field are described by the so-called electric and magnetic polarisabilities α and β, respectively. As the pion is not a point-like particle, it is sensitive to deformations by such external fields, and this tiny effect, called pion polarisability, is under the assumption of $\alpha_\pi = -\beta_\pi$ predicted in Chiral Perturbation Theory (ChPT) to be $5.7 \times 10^{-4}\ \text{fm}^3$ [3].

Polarisabilities are usually measured in Compton scattering experiments. Even though pions can hardly be used as a fixed-target, they can be scattered off the Coulomb potential of a heavy nucleus, like a Nickel (Ni) target. In the corresponding $\pi^-\gamma \to \pi^-\gamma$ scattering process, the polarisabilities manifest as a deviation from the Born cross-section of a point-like particle. The measurement of the cross-section of this process allows to extract the pion polarisabilities via the deviation from the assumption for a point-like particle.

Such a measurement is experimentally demanding and the systematics need precisely to be controlled. An important advantage of the COMPASS experiment is that we can perform this measurement with the pion and with the muon beam (quickly switchable within about two hours), in which the latter acts as a control measurement of the point-like muon.

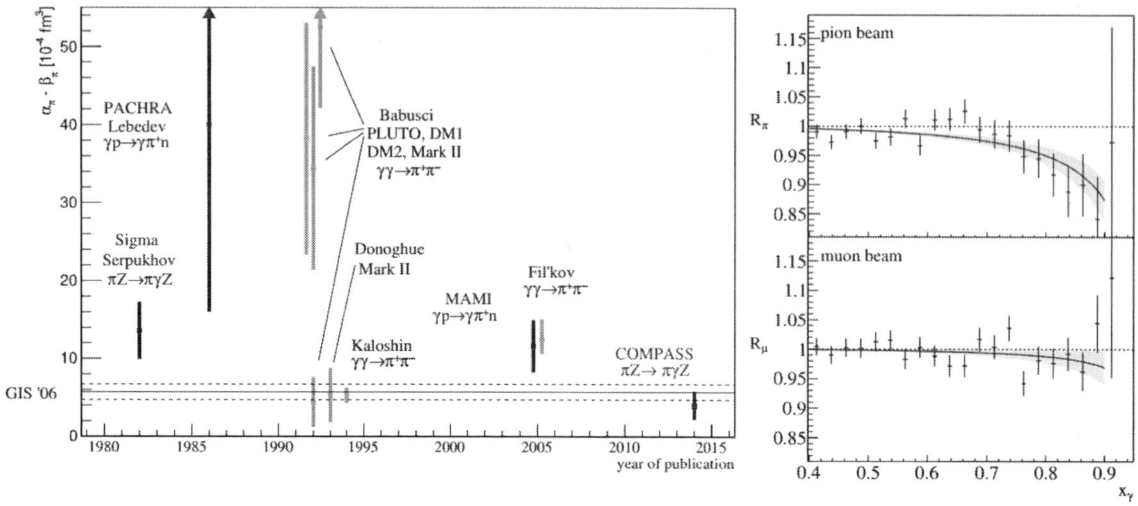

Figure 2. *Left*: Compilation of experimental measurements of the pion polarisability $\alpha_\pi - \beta_\pi$ including the new COMPASS result and the theoretical prediction from ChPT [3], plot taken from [4]. *Right:* Ratio R of the measured cross-section over the simulated one for a point-like object for the pion beam data *(top)* and for the control measurement with the muon beam data *(bottom)* [5].

The COMPASS data of scattering a pion beam off a Ni target comprises the Compton part of the cross-section [5]. The $\pi^- Ni \to \pi^- \gamma Ni$ events are selected by requiring one negatively charged track from the vertex inside the Ni target, a high-energetic shower in the electromagnetic calorimeters for the final state photon, whereas the photon exchange is ensured by a cut on low photon virtualities Q^2, energy conservation is applied to select exclusive reactions, *i.e.* $\Delta E = E_{beam} - E_{\pi'} - E_\gamma \approx 0$, neglecting the (tiny) nuclear recoil energy. In order to stay in the kinematic region of stable trigger and muon identification efficiencies, the ratio of the final state photon and the initial beam energy $x_\gamma = E_\gamma / E_{beam}$ is required to be between 0.4 and 0.9. Under the assumption of $\alpha_\pi = -\beta_\pi$ that has to be made due to the limited statistics at lower x_γ (*i.e.* the phase space region sensitive to $\alpha_\pi + \beta_\pi$), the polarisation should manifest in terms of a decreasing ratio R of the measured cross-section over the simulated one for a point-like particle with increasing x_γ. This behaviour has exactly been measured for the pion beam data, *i.e.* for the extended object of the pion, as displayed in Fig. 2 (right, top). To extract the polarisability α_π from the measured ratio R, the photon energy spectrum is examined according to the following relation, which is a simplified relation of the Primakoff cross-section in one-photon exchange approximation:

$$R = \frac{\sigma(x_\gamma)}{\sigma_{\alpha_\pi=0}(x_\gamma)} = 1 - \frac{3}{2} \cdot \frac{m_\pi{}^3}{\alpha} \cdot \frac{x_\gamma{}^2}{1 - x_\gamma} \alpha_\pi .$$

For the muon beam data (Fig. 2, right, bottom), the size of the fake polarisability of $(0.5 \pm 0.5_{stat}) \times 10^{-4}$ fm^3 is within the statistical uncertainty compatible with zero. This value is taken as an estimate of the systematic error due to apparative imperfections, not described by the MC simulation. For the pion beam data (Fig. 2, right, top), a pion polarisability of $\alpha_\pi = (2.0 \pm 0.6_{stat}) \times 10^{-4}$ fm^3 is determined from the fit. This pion polarisability value measured by COMPASS is in tension with previous experiments and in agreement with the prediction from ChPT (Fig. 2, left).

DOI: 10.1051/epjconf/201612604033

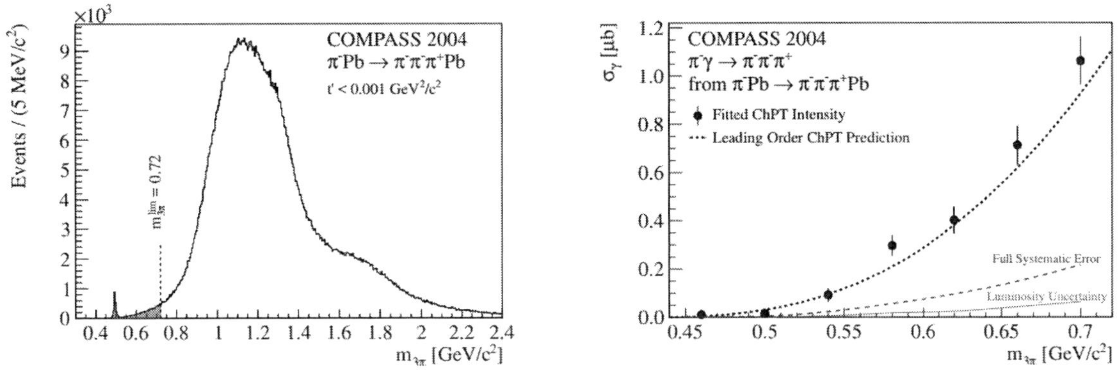

Figure 3. *Left:* Spectrum of the 3π mass from $\pi^- Pb \rightarrow \pi^-\pi^+\pi^- Pb$ in the Primakoff region of four-momentum transfer $t' < 10^{-3}\,\mathrm{GeV}^2/c^2$. The region of interest in view of ChPT up to $m_{3\pi} \approx 0.72\,\mathrm{GeV}/c^2$ is indicated. At $m_{K^-} \approx 0.49\,\mathrm{GeV}/c^2$, a peak from the admixture of incident beam kaons decaying into $(3\pi)^-$ is visible that are used for flux normalisation. *Right:* First measurement of the photo-production cross-section σ_γ in this mass range, found in agreement with the prediction from leading order ChPT [6].

2.2 Measurement of chiral dynamics

From the same data, further processes on chiral dynamics are accessible, like *e.g.* the chiral anomaly in $\pi^-\gamma \rightarrow \pi^-\gamma$, the corresponding analysis is still under way, the same holds for the neutral pion case of two-pion production $\pi^-\gamma \rightarrow \pi^-\pi^0\pi^0$ at low energy. Based on the 2004 pilot run data (Primakoff kinematics), the two-pion production for the charged case $\pi^-\gamma \rightarrow \pi^-\pi^+\pi^-$ has been analysed [6].

Figure 3 (left) shows the mass spectrum of the outgoing three-pion final state. Of particular interest in view of chiral dynamics is the low mass tail up to three pion masses, as indicated in the figure. Performing a partial-wave analysis (PWA) including amplitudes from ChPT calculations substituting the isobaric partial waves in this mass range, after flux normalisation on the observed kaon decays (using the kaon admixture in the beam), the absolute cross-section is, within the experimental uncertainty of 20 %, found in agreement with the leading order ChPT prediction.

In the 3π mass spectrum (Fig. 3, left), structures from the well-known resonances $a_1(1260)$ and $a_2(1320)$ are visible at higher masses — the corresponding radiative couplings have been measured as discussed in the next section (Sec. 2.3).

2.3 Measurement of radiative width

Electromagnetic transitions can be measured via radiative decays of a resonance $X \rightarrow \pi\gamma$. The direct measurement of $\pi\gamma$ is experimentally difficult. Alternatively, the inverse process of scattering a pion off a Coulomb field producing an intermediate resonance X, *i.e.* the Primakoff reaction, can be used. The Coulomb potential of a heavy nucleus acts as a (quasi-real) photon source, and the Primakoff production cross-section of a resonance X is proportional to the width of the radiative decay of that resonance: $\sigma_{\mathrm{Primakoff}}(X) \sim \Gamma_0(X \rightarrow \pi\gamma)$. Due to the exchange particle being a quasi-real photon, Primakoff produced resonances have predominantly spin projection $M{=}1$. Even though states with spin projection $M{=}1$ can also be produced diffractively, the cross-section here is proportional to $t'^{|M|}e^{-bt'}$, so that the diffractive production is highly suppressed at very small values of t'.

Based on the negative pion beam data on a thin lead (Pb) target, the COMPASS Collaboration studied this reaction and measured the radiative widths of the $a_2(1320)$ and the $\pi_2(1670)$ [7]. Performing a partial-wave analysis, the data is decomposed into spin-parity states, and thus also the $M{=}1$ states are identified. To extract the radiative widths of the $a_2(1320)$ and the $\pi_2(1670)$,

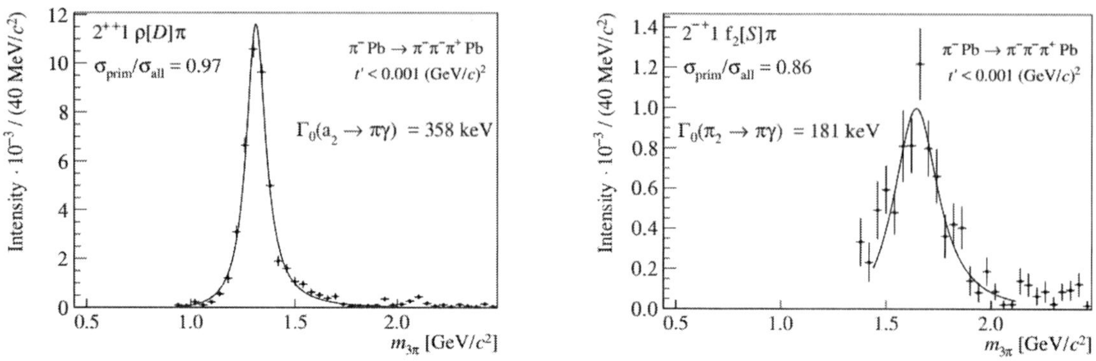

Figure 4. Results of the measured radiative widths of the $a_2(1320)$ *(left)* and the $\pi_2(1670)$ *(right)* (reprinted with kind permission of EPJ from [7], copyright Società Italiana di Fisica / Springer-Verlag 2014). As compared to previous measurements, the COMPASS result for the $a_2(1320)$ is the most precise one, whereas for the $\pi_2(1670)$, the COMPASS value is the first measurement at all.

the two partial waves $J^{PC}M^{\epsilon}\,(isobar)\,\pi = 2^{++}1^{\epsilon}\rho(770)\pi$ D-wave and $2^{-+}1^{\epsilon}f_2(1270)\pi$ S-wave have been considered, respectively. For the events analysed, namely at $t'<10^{-3}\,\text{GeV}^2/\text{c}^2$, the contribution of Primakoff production is 97 % and 86 % of the intensities found in the $2^{++}1^{\epsilon}\rho(770)\pi\,D$ and $2^{-+}1^{\epsilon}f_2(1270)\pi$ S-waves, respectively. The intensities for the two different reflectivities $\epsilon = \pm 1$ have to be summed up incoherently as the production plane at such small t' is not measured sufficiently precise due to limited experimental resolution. That there are further possible decay channels (than $\rho(770)\pi$ for the $a_2(1320)$ and $f_2(1270)\pi$ for the $\pi_2(1670)$) has been taken into account by consideration of the relevant branching fractions. The fits to the data are shown in Fig. 4, from which a width of $\Gamma_0(a_2(1320) \to \pi\gamma) = (358\pm6\pm42)$ keV and $\Gamma_0(\pi_2(1670) \to \pi\gamma) = (181\pm11\pm27)$ keV have been measured for the $a_2(1320)$ and the $\pi_2(1670)$, respectively. Both results are consistent with predictions based on the VMD model [8]. Compared to previous measurements, the COMPASS result for $\Gamma_0(a_2(1320) \to \pi\gamma)$ is the most precise one, whereas for $\Gamma_0(\pi_2(1670) \to \pi\gamma)$ the COMPASS result is the first measurement at all.

3 Results on hadron spectroscopy

There are many states that have been observed so far in the light meson sector, whereas quite some of the observations still need confirmation. Comparing the experimental situation to the predictions based on the naive quark model, quite some agreement is found, however, not all the predictions are perfectly consistent with experiment: Some of the predicted states have not yet been observed, also there are observed states that are not expected, see *e.g.* [9]. Given the relatively large widths of many of these states ($\sim$100 MeV/c^2) together with the high level density of the mesons with (u,d,s) constituent quarks, there are quite some overlaps and mixing, complicating the analyses.

Of particular interest are exotic mesons, not fitting the constituent quark model. QCD allows for and predicts such states, like glue-balls, hybrids or tetraquarks according to various models. Mesons of same spin-parity quantum numbers mix, and the unambiguous observation is thus difficult. So-called spin-exotic states with J^{PC} quantum numbers not accessible by a simple $q\bar{q}$ configuration are on the other hand very promising to search for as they do not mix with ordinary mesons. The experimental observation of such spin-exotic states would be a fundamental confirmation of QCD, for a recent overview, see *e.g.* [10].

The COMPASS data give access to all decay channels spin-exotic hybrid candidates have been reported in so far by different experiments, like $\rho(770)\pi$, $\eta(')\pi$, $f_1(1285)\pi$ and $b_1(1235)\pi$. The lightest

DOI: 10.1051/epjconf/201612604033

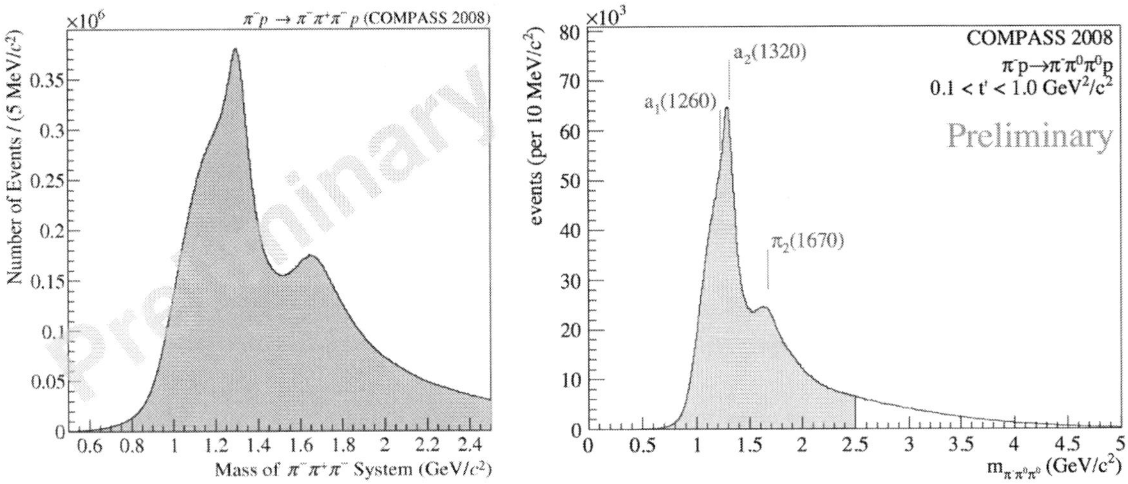

Figure 5. Invariant mass spectrum of the exclusively produced, outgoing $(3\pi)^-$ system for the charged *(left)* and the neutral *(right)* decay modes [14, 15].

hybrid candidate with exotic $J^{PC} = 1^{-+}$, the famous $\pi_1(1600)$, has been searched for and is studied in two different decay channels simultaneously, namely in $\pi^-\pi^+\pi^-$ and $\pi^-\pi^0\pi^0$ final states [11–16] produced in diffractive pion dissociation $\pi^- p \to (3\pi)^- p'$. Since the reconstruction depends on different parts of the detector, the observation of a state in both channels provides an independent confirmation within the same experiment, the possibility of cross-checks and control of systematics.

The invariant mass of the diffractively produced $(3\pi)^-$ system is shown for the charged and neutral $\rho\pi$ decay modes in Fig. 5 (left) and (right), respectively. In the range of four-momentum transfer $0.1\,(\mathrm{GeV}/c)^2 < t' < 1.0\,(\mathrm{GeV}/c)^2$ used for PWA, about 50 million and 3.5 million exclusive $(3\pi)^-$ events have been reconstructed from the 2008 data (of negative pion beam impinging on a proton target) in the mass range of 0.5 and 2.5 GeV/c^2 for the charged and the neutral mode, respectively. By the interaction with the target, the beam pion is excited to an intermediate resonance X^- that subsequently decays into a di-pion resonance, the so-called isobar decaying into two pions, and a bachelor pion with the relative orbital angular momentum L between them. For the results presented here, the $\rho(770)$, $f_0(980)$, $f_2(1270)$, $f_0(1500)$, $\rho_3(1690)$ and a broad $(\pi\pi)_s$ component have been used. The set of partial waves contains 87 partial waves up to $L = 6$ and an incoherent wave, the so-called "flat wave", having an isotropic angular distribution, representing events of three uncorrelated pions.

Given the observed dependence on t' and the large statistics, the PWA method has been extended with respect to the scheme of a two-step PWA as applied previously, for a complete description including all details, see [17]. The data has been divided into 11 and 8 bins of t' such that about equal statistics is contained in each bin, and (as previously) into 20 MeV/c^2 and 40 MeV/c^2 wide $m_{3\pi}$ bins for the charged and the neutral mode data, respectively. The first step analysis, the so-called mass-independent PWA, is carried out independently for each mass bin and independently for the different ranges of t'. This maximum likelihood (rank-1) fit of the model to the data extracts the production amplitudes, taking into account the detector acceptance. Finally, the analysis is completed by the second step analysis, the so-called mass-dependent fit. In this χ^2 fit of Breit-Wigner amplitudes to a subset of the spin density matrix, resonance parameters are extracted, for which not only the intensities but also the interferences are taken into account. Also it takes into account the observed t' dependencies by performing a simultaneous optimisation of the resonant parameters in all t' regions.

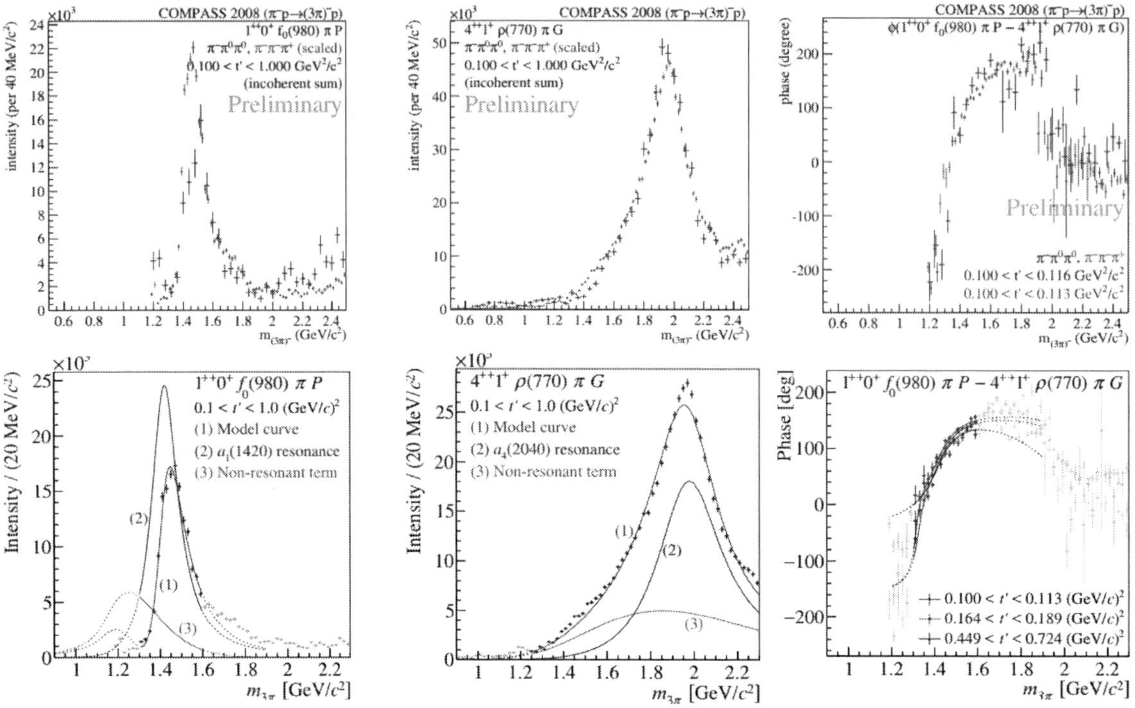

Figure 6. *Top:* First step PWA result, charged versus neutral decay mode [14]: The fitted intensity for the $1^{++}0^{+}\,f_0(980)\,\pi$ P-wave shows a narrow structure at about $1.4\,\mathrm{GeV}/c^2$, it is similarly observed for both, neutral and charged $(3\pi)^-$ mode *(left)*. In the resultant intensity for the $4^{++}1^{+}\,\rho(770)\,\pi$ G-wave, the known $a_4(2040)$ is similarly observed *(centre)*. The relative phase of the two shows a clean rapid phase motion in the signal region. *Bottom:* Second step PWA result, charged decay mode [18]: Complete fit result in terms of the resultant $1^{++}0^{+}\,f_0(980)\,\pi\,P$ intensity (incoherent sum over t' ranges) *(left)*, the first step result is given by the black data points and the resonance model fit overlaid in red consists of the BW describing the narrow structure, that we call the $a_1(1420)$ (blue curve), and the non-resonant background contribution (green curve). The corresponding result is shown for the $a_4(2040)$ observed in the $4^{++}1^{+}\,\rho(770)\,\pi$ G-wave *(centre)*. The relative phase of the $a_1(1420)$ against the $a_4(2040)$ *(right)* is shown for three different t' ranges, consistently showing a clean rapid phase rotation in the signal region.

3.1 Observation of a new axial-vector state $a_1(1420)$

Figure 6 (top/left) shows the first step PWA result for one out of the 87 partial waves, namely the $1^{++}0^{+}\,f_0(980)\pi$ P-wave, in terms of the fitted intensity for both decay modes. Indeed, the incoherent sum over the different t' ranges is shown here, and the charged is normalised to the neutral mode data using the integral. Good agreement on the fit result is found for the two channels. For both decay modes, the $1^{++}0^{+}\,f_0(980)\pi$ P-wave intensity, containing only a very small fraction of the total observed intensity of about 0.25 % [18], shows a clear signal slightly above $1.4\,\mathrm{GeV}/c^2$. Rapid phase rotations with respect to known resonances, like *e.g.* the $a_4(2040)$ (Fig. 6, top/centre), are observed in the signal region (Fig. 6, top/right), as it is expected for a resonance. A phase does not depend on the amount of amplitude, and the phase motion is observed independent of t'.

In the second analysis step, a Breit-Wigner resonance model with a minimal set of three waves is used, namely the $2^{++}1^{+}\,\rho(770)\pi\,D$, $4^{++}1^{+}\,\rho(770)\pi\,G$ and $1^{++}0^{+}\,f_0(980)\pi\,P$ waves. For the charged mode data, the completed PWA result is show for the $1^{++}0^{+}\,f_0(980)\pi$ P-wave in Fig. 6 (bottom/left) and for the $4^{++}0^{+}\,\rho(770)\pi$ G-wave in Fig. 6 (bottom/centre), the phase difference between them is given in Fig. 6 (bottom/right). From this fit, we obtain a mass of $m = 1414^{+15}_{-13}\,\mathrm{MeV}/c^2$ and a width

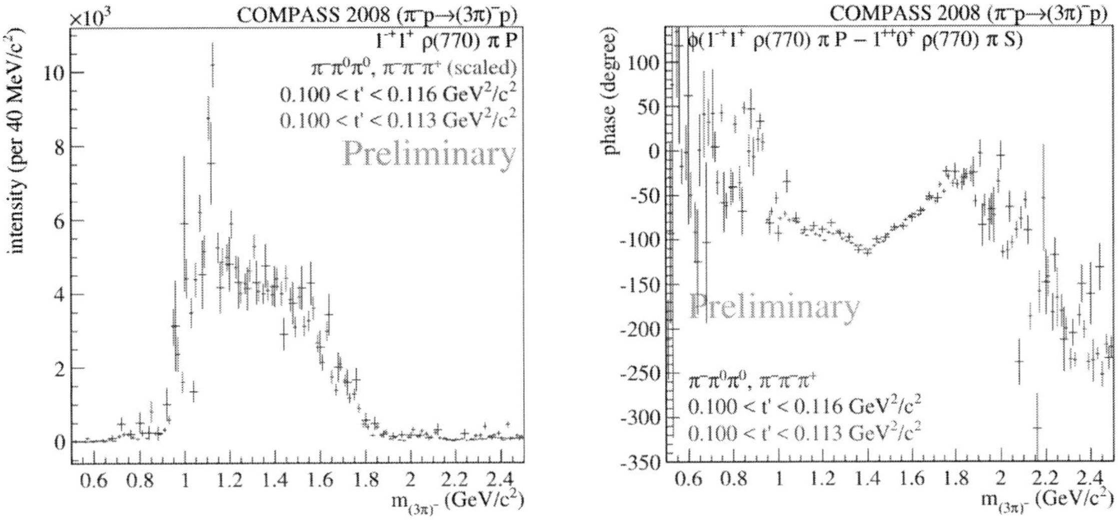

Figure 7. First step PWA result for the exotic $1^{-+}1^{+}\,\rho(770)\pi$ P-wave in $(3\pi)^{-}$ final states, charged versus neutral mode data. The resultant fitted intensities are found in good agreement for the two channels *(left)*. Shown are the results for the lowest range of four-momentum transfer t', where the signal on top of the broad bump at about 1.6 GeVc^2 has the lowest evidence as compared to the higher t' regions, in which the signal to background ratio improves with increasing t'. The relative phases with respect to the $1^{++}0^{+}\,\rho(770)\pi$ S-wave, in which the well-known $a_1(1260)$ resonance is observed, is displayed *(right)*, showing a clean phase motion in the range of about 1.4 and 1.8 GeV/c^2 consistently for the two decay modes.

of $\Gamma = 153^{+8}_{-23}$ MeV/c^2. The interpretation of this new isovector state is still unclear. The properties of the $a_1(1420)$ suggest it to be the isospin partner of the $f_1(1420)$, especially supported by the strong coupling to the $f_0(980)$ (and not to ρ) that is interpretable as a $K\bar{K}$ molecule. The $a_1(1420)$ and the $f_1(1420)$ may possibly be the first observed isospin partners for a $K\bar{K}\pi$ moleculartype excitation, which suggests further studies of the $K\bar{K}\pi$ final state. A first, very promising look at the $f_1(1420)$ observed decaying to $K\bar{K}\pi$ in the (same) COMPASS 2008 data set has already been provided [19], including first very preliminary PWA confirming (due to spin assignment) the state in the $K\bar{K}\pi$ invariant mass indeed to be the $f_1(1420)$ and not the $\eta(1295)$.

3.2 Status of the search for the spin-exotic $\pi_1(1600)$ resonance

In the 2004 pilot run data (π^- beam, Pb target), a significant J^{PC} spin-exotic signal has been observed at $1660\pm10^{+0}_{-64}$ MeV/c^2. It shows a clean phase motion against well-known resonances and is consistent with the disputed $\pi_1(1600)$ [20]. The high statistics of the 2008 proton target data allows the search for exotic states in different decay modes in the same experiment [11]. Employing the same PWA method as in [20], the results obtained for the $(1^{-+})1^{+}\rho^{-}\pi^{0}$ and $(1^{-+})1^{+}\rho^{0}\pi^{-}$ intensity and relative phase are similar and consistent with the previous observations [12, 13]. Apart of the established resonances $a_1(1260)$, $a_2(1320)$, $\pi_2(1670)$, also $\pi(1800)$ and $a_4(2040)$, an exotic signal in the 1^{-+} wave at around 1.6 GeV/c^2 has been observed, that shows a clean phase motion with respect to well-known resonances. These results are consistently obtained for both $\rho\pi$ decay modes, neutral and charged. The extension to an additional t' binning and performing the PWA independently also in different ranges of t' [17] allows for disentangling resonant from non-resonant particle production (*e.g.* dynamically produced components caused by the Deck effect [21]).

DOI: 10.1051/epjconf/201612604033

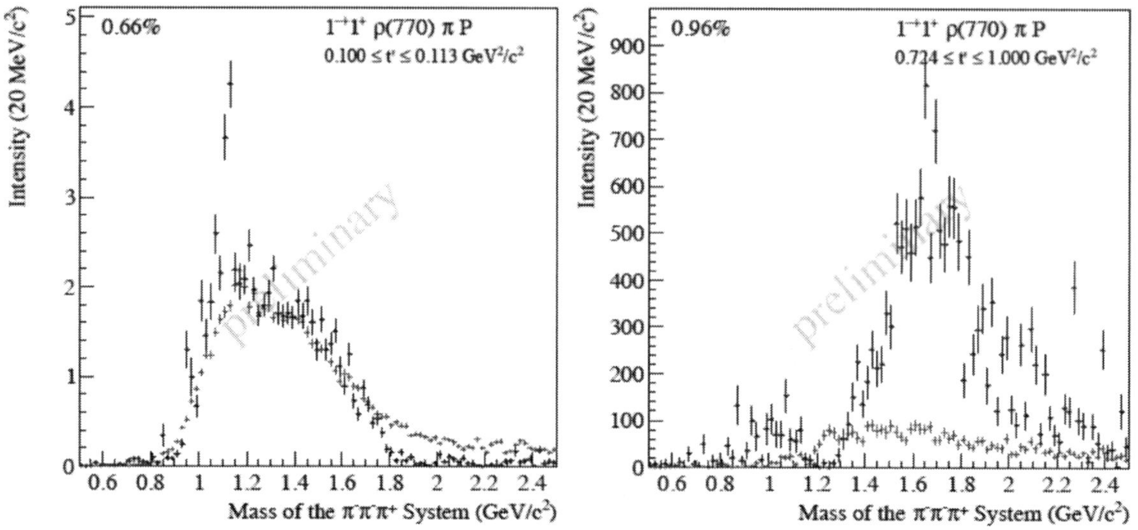

Figure 8. First step PWA result for the intensity of the exotic $1^{-+}1^{+}\,\rho(770)\,\pi$ P-wave for the charged decay mode (blue) overlaid with a projection of a simulated Deck-effect in this wave (green), shown for the range of lowest *(left)* and highest *(right)* values of t' [16].

A dependency of the exotic $1^{-+}1^{+}\rho(770)\pi\,P$ wave intensity on the squared four-momentum t' is observed. Whereas a significant signal is observed at around $1.6\,\text{GeV}/c^2$ on top of a moderate background at larger values of t', the signal to background ratio decreases with decreasing t'. The fitted intensity of the exotic $1^{-+}1^{+}\rho(770)\pi\,P$ wave is shown for the range of smallest t' for the charged and neutral decay modes in Fig. 7 (left). Apart from the good agreement between the two channels, we observe mainly a broad bump. Even though there is hardly evidence for a signal at about $1.6\,\text{GeV}/c^2$ in this plot, *i.e.* for this t' range of $0.100 < t' < 0.113\,\text{GeV}/c^2$, the corresponding relative phase with respect to the $1^{++}0^{+}\rho(770)\pi$ S-wave, in which the well-known $a_1(1260)$ is observed, a clean phase motion is consistently observed for the two different decay modes in the range of about 1.4 and $1.8\,\text{GeV}/c^2$. This is exactly the mass range where a significant exotic 1^{-+} signal is observed for ranges of larger t' (see e.g. Fig. 8, right). The relative phase of two resonances is independent of the given absolute amplitudes, and a consistent phase motion is observed in all bins of t'.

The observed dependency of the signal to background ratio on t' could be caused by the interference of a resonance with a large non-resonant background contribution, possibly introduced and to be explained by the Deck effect [21]. In order to further investigate this explanation, pseudo data were generated according to a dedicated model [22]. This data has been analysed as the real data, including the decomposition into partial waves, in order to study the Deck-amplitude contribution to the different partial waves included in the PWA. Figure 8 shows the promising outcome of the study for the exotic 1^{-+} partial wave, exemplary for the ranges of smallest and largest t' for the charged decay mode. The Deck amplitude intensities are normalised to the total intensity (summed over all mass and t' bins) in the given wave. A large part of the exotic 1^{-+} wave intensity at small values of t' can be described as a contribution by the non-resonant Deck process (Fig. 8, left), whereas the contribution at large t' is almost negligible according to this ansatz.

For completeness, further relevant results and ongoing analyses that are not dicussed in detail here are the PWA results of diffractive $(\eta\pi)^-$ and $(\eta'\pi)^-$ production [23] and preliminary PWA results for central production (*i.e.* via double pomeron exchange) of $\pi^+\pi^-$ and K^+K^- from the 2009 COMPASS proton beam data [24].

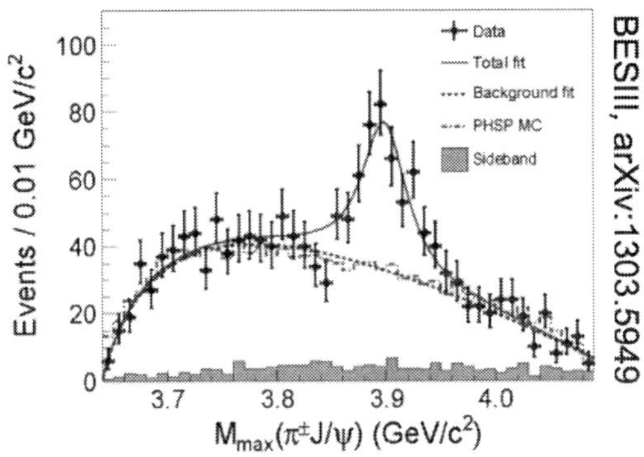

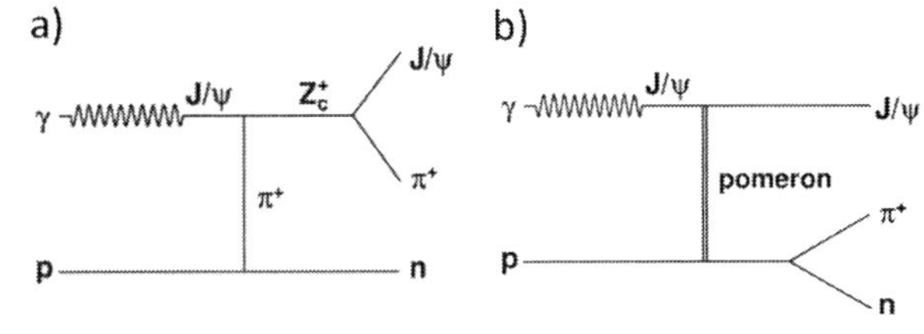

Figure 9. *Left:* The charmonium-like exotic state $Z_c(3900)$ decaying to $J/\psi\pi^{\pm}$ as observed by BESIII [26]. *Right:* In the COMPASS muon data, the $Z_c(3900)$ should be produced via photoproduction according to the process given under a), the main background originates from pomeron exchange reactions, b).

3.3 Search for the manifestly exotic $Z_c(3900)$ state

Even though the COMPASS muon beam data is mainly dedicated to study (spin-dependent) parton distribution functions, a hadron spectroscopy result has been obtained based on the full set of COMPASS data collected with a muon beam between 2002 and 2011, namely an upper limit on the photoproduction of the $Z_c(3900)$ [25]. The charged charmonium-like exotic state $Z_c(3900)$ was observed decaying to $J/\psi\pi^{\pm}$ by the BESIII [26] Collaboration and confirmed by Belle [27] in 2013, see Fig. 9 (left) for the observation plot, later also the neutral partner was confirmed [28].

As the photon may act as a J/ψ according to the VDM, the $Z_c(3900)$ should be produced via interaction of the incoming photon with the virtual pion from the nucleon target according to the diagram in Fig. 9 (right, a), and the cross-section for this process should be sizable [29].

The reconstructed exclusive production channel is $\mu^+N \rightarrow \mu^+Z_c^{\pm}(3900)N \rightarrow \mu^+J/\psi\pi^{\pm}N$, with J/ψ decaying to $\mu^+\mu^-$, means the final state to detect is $\mu^+\mu^+\mu^-\pi^{\pm}N$. The event selection of the exclusively produced $\mu^+J/\psi\pi^{\pm}$ is straight forward: A vertex is reconstrcuted having exactly three outgoing muons and a pion, to identify the J/ψ, a mass cut on the reconstructed $\mu^+\mu^-$ invariant mass is applied as indicated in Fig. 10 (left), energy balance is requested and, in order to suppress background from pomeron exchange reactions (Fig. 10, right, b), a momentum cut $p_{\pi^{\pm}} > 2$ GeV/c for the pions is applied. In addition, in order to make use of the measurement of the cross-section $\sigma(\gamma N \rightarrow J/\psi N)$ provided by the NA14 experiment [30] for absolute normalisation, the exclusive μ^+J/ψ sample is selected as well using the same selection criteria. The ratio of acceptances for both samples equals in first approximation about the acceptance for the additional pion that we know to be $0.5 \pm 0.1_{syst}$, averaged over all setup and target configuartions.

The resultant COMPASS plot for the search for the $Z_c(3900)$ is given in Fig.10 (right), where the reconstructed $J/\psi\pi$ invariant mass is plotted and the expectation region for a signal of the $Z_c(3900)$ is indicated. No signal of exclusive photoproduction of the $Z_c^{\pm}(3900)$ state and the decay into $J/\psi\ \pi^{\pm}$ has been found. Thus, an upper limit was determined on the product of the cross-section and the relative $Z_c^{\pm}(3900) \rightarrow J/\psi\pi^{\pm}$ decay probability normalised to the cross-section of incoherent exclusive

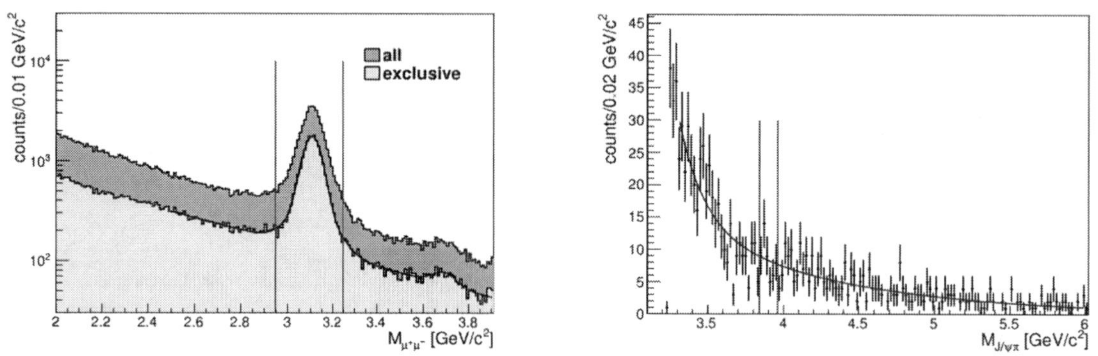

Figure 10. *Left:* Reconstructed $\mu^+\mu^-$ invariant mass for all events and for the exclusive sample selected, the mass cut applied to identify the J/ψ is indicated. *Right:* Final plot of the reconstructed $J/\psi\pi$ invariant mass distribution, where the mass range of the expected $Z_c(3900)$ signal is indicated — no signal is observed [25].

photo-production of J/ψ mesons. We obtain the following result [25]

$$\frac{BR(Z_c^{\pm}(3900) \to J/\psi\,\pi^{\pm}) \times \sigma_{\gamma\,N\to Z_c^{\pm}(3900)\,N}}{\sigma_{\gamma\,N\to J/\psi\,N}}\bigg|_{\langle\sqrt{s_{\gamma N}}\rangle=13.8\ GeV} < 3.7\times10^{-3}. \tag{1}$$

Assuming $\sigma_{\gamma\,N\to J/\psi\,N} = 14.0\pm1.6_{\text{stat.}}\pm2.5_{\text{syst.}}$ nb as measured by the NA14 Collaboration for $\sqrt{s}_{\gamma\,N} = 13.7\ GeV$ [30], the result can be presented as

$$BR(Z_c^{\pm}(3900) \to J/\psi\,\pi^{\pm}) \times \sigma_{\gamma\,N\to Z_c^{\pm}(3900)\,N}\bigg|_{\langle\sqrt{s_{\gamma\,N}}\rangle=13.8\ GeV} < 52\text{ pb}. \tag{2}$$

This result has been obtained within the framework of the Z_c production mechanism proposed in Ref. [29]. Unless the assumptions made therein are wrong, a conclusion is that the regarded decay channel $Z_c^{\pm}(3900) \to J/\psi\,\pi^{\pm}$ can not be the dominant one. This result is an important input for the clarification of the nature of this charmonium-like $Z_c^{\pm}(3900)$ state that due to the charge and the quark content of $c\bar{c}$ is manifestly an exotic one.

3.4 Summary and outlook

The COMPASS experiment at CERN/SPS is well suited for high precision measurements in the fields of hadron excitations and spectroscopy, completing together with the hadron structure studies the broad and rich COMPASS hadron physics programme.

Chiral perturbation theory has been confirmed by the COMPASS measurement of the pion polarisability $(\alpha_\pi - \beta_\pi)$ at high precision, and chiral dynamics has been observed (again in agreement with ChPT) in the process $\pi\gamma \to 3\pi$ at low relative momenta. Moreover, radiative couplings of meson resonances have been measured at unprecedented precision.

A possible new isovector resonance with $J^{PC} = 1^{++}$, a mass of about $1420\,\text{MeV}/c^2$ and a width of about 150 MeV c^2 has been observed in $f_0(980)\pi P$ decay mode. It shows a rapid phase motion with respect to established resonances. The extension to a t'-resolved partial-wave analysis provides deeper understanding of the exotic signal that we observe at around $1.6\,\text{GeV}/c^2$ in the $1^{-+}1^{+}\,\rho(770)\pi\,P$ wave, showing resonant behaviour. The intensity observed at small momentum transfers t' seems to a large extent to be attributed to contributions from the Deck effect, whereas merely a negligible Deck contribution seems to play a role at larger values of t'. Deeper insights on the disputed $\pi_1(1600)$ hybrid candidate are not only gained via the two $(3\pi)^-$ decay channels but also in the $\eta\pi$ and $\eta'\pi$ channels.

As an outlook, the 2012 Primakoff data set with considerably larger statistics will further improve the pion polarisability measurement. Thanks to the extended x_γ range, also a measurement of $(\alpha_\pi + \beta_\pi)$ will be feasible, also a first measurement of the kaon polarisability is foreseen. For spectroscopy, the analysis of $(5\pi)^-$, $(\eta\eta\pi)^-$ and also $(f_1\pi)^-$ will allow for access towards higher masses, and complete the list of channels for spin-exotic search.

References

[1] COMPASS Collaboration, P. Abbon *et al.*, Nucl. Instrum. Meth. **A577** (2007) 455-518, Nucl. Instrum. Meth. **A779** (2015) 69-115; arXiv:1410.1797 [phys./ins-det].
[2] S. Platchkov, for the collaboration, these proceedings (2015).
[3] J. Gasser, M.A. Ivanov, and M.E. Sainio, Nucl. Phys. B **745**, 84 (2006).
[4] T. Nagel, PhD thesis, TU München (2012), CERN-THESIS-2012-138.
[5] COMPASS Collaboration, C. Adolph *et al.*, Phys. Rev. Lett. **114**, 062002 (2015).
[6] COMPASS Collaboration, C. Adolph *et al.*, Phys. Rev. Lett. **108**, 192001 (2012).
[7] COMPASS Collaboration, C. Adolph *et al.*, European Physical Journal A **50**, 79 (2014).
[8] J.L. Rosner, Phys. Rev. D **23**, 1127 (1981).
[9] J. Beringer *et al.*, (Particle Data Group), Phys. Rev. D **86** (2012) 010001.
[10] C.A. Meyer, Y. Van Haarlem, Phys. Rev. C **82**, 025208 (2010).
[11] F. Nerling, for the collaboration, AIP Conf. Proc. **1257** (2010) 286; arXiv:1007.2951[hep-ex].
[12] F. Haas, for the collaboration, Conf. Proc., Hadron2011, Munich (2011); arXiv:1109.1789v2[hep-ex].
[13] F. Nerling, for the collaboration, Conf. Proc. MESON2012, EPJ Web Conf. 37, 09025 (2012); arXive:1401.4943[hep-ex].
[14] S. Uhl, for the collaboration, Conf. Proc. Hadron2013, PoS, 087 (2013); arXive:1401.4943v2[hep-ex].
[15] F. Haas, PhD thesis, TU München, January 2014; CERN-THESIS-2013-277.
[16] S. Uhl, for the collaboration, Conf. Proc. PANIC 2014, DESY-PROC-2014-04, 169 (2014).
[17] COMPASS Collaboration, C. Adolph *et al.*, subm. to Phys. Rev. D, (2015); arXiv:1509.00992[hep-ex].
[18] COMPASS Collaboration, C. Adolph *et al.*, Phys. Rev. Lett. **115**, 082001 (2015).
[19] J. Bernhard and F. Nerling, for the collaboration, Conf. Proc. Hadron2011, Munich (2011); arXiv:1109.0219[hep-ex].
[20] COMPASS collaboration, M. Alekseev *et al.*, Phys. Rev. Lett, **104**, 241803 (2010).
[21] R.T. Deck, Phys. Rev. Lett. **13**, 169 (1964).
[22] C. Daum *et al.*, Nucl. Phys. B **182**, 269 (1981).
[23] COMPASS Collaboration, C. Adolph *et al.*, Phys. Lett. B **740**, 303 (2015).
[24] A. Austregesilo, Conf. Proc. Hadron2013, PoS, 102 (2013).
[25] COMPASS Collaboration, C. Adolph *et al.*, Phys. Lett. **B742**, 330 (2015).
[26] BESIII Collaboration, M. Ablikim *et al.*, Phys. Rev. Lett. **110**, 252001 (2013).
[27] Belle Collaboration, Z.Q. Liu *et al.*, Phys. Rev. Lett. **111**, 019901 (2013).
[28] BESIII Collaboration, M. Ablikim *et al.*, Phys. Rev. Lett. **113**, 212002 (2014).
[29] Q.-Y. Lin *et al.*, Phys. Rev. D **88**, 114009 (2013); arXiv:1304.0121[hep-ex].
[30] NA14 Collaboration, R. Barate *et al.*, Z. Phys. C **33**, 505, (1987).

Open-charm production measurements with ALICE at the LHC

P. Pagano[1,2,a] on behalf of the ALICE Collaboration

[1] *Salerno University - Italy*
[2] *I.N.F.N. - Gruppo Collegato di Salerno*

Abstract. The LHC heavy-ion physics program aims at investigating the properties of strongly-interacting matter under extreme conditions of temperature and energy density where the formation of the Quark-Gluon Plasma (QGP) is expected. Heavy-flavour hadrons, containing charm and beauty quarks, are considered efficient probes to investigate the properties of the QGP produced in heavy-ion collisions. Heavy quarks are produced in hard partonic scattering processes in the initial stage of hadronic collisions and propagate through the hot and dense medium created in the collision losing energy interacting with the medium via radiative and collisional processes. The high precision tracking, good vertexing capabilities and excellent particle identification offered by the ALICE experiment allow us to measure particles containing heavy quarks in a wide transverse-momentum range in pp, p-Pb and Pb-Pb collisions. A review of the main results on prompt D-mesons production, reconstructed via their hadronic decays at mid-rapidity, in pp collisions at $\sqrt{s} = 7$ TeV, p-Pb collisions at $\sqrt{s_{NN}} = 5.02$ TeV and Pb-Pb collisions at $\sqrt{s_{NN}} = 2.76$ TeV will be shown.

1 Introduction

The ALICE experiment at the LHC aims to investigate the properties of the high energy-density and strongly-interacting matter created in heavy-ion collisions at ultrarelativistic energies. Charm and beauty quarks are an excellent probe to study it. Due to their large masses, they are produced predominantly in hard parton scattering processes at the early stages of the collisions. Therefore, they pass through all the phases of the system evolution losing energy in their interactions with other medium constituents. The flavour is conserved in these interactions so heavy quarks can be considered sensitive probes of the medium properties. Open heavy-flavour hadron production is expected to be sensitive to the energy density of the system through the mechanism of in-medium energy loss of heavy quarks [1–3]. Quantum Chromodynamics (QCD) theoretical calculations predict a dependence of the energy loss on the mass of the parton traversing the medium, according to the dead cone effect [4], which suppresses the induced gluon radiation at small angles for heavy partons. Furthermore, the energy loss through the medium depends on the parton colour charge and path length. This results in an expected hierarchy of the parton energy loss, with beauty quarks losing less energy than charm quarks, and charm quarks losing less energy than light quarks and gluons. One of the observables that are sensitive to the interaction of hard partons with the medium is the nuclear modification factor, R_{AA},

[a] e-mail: paola.pagano@sa.infn.it

defined as the ratio of the particle yield measured in Pb–Pb collisions and the yield in pp collisions scaled by the number of binary collisions. R_{AA} is expected to be equal to one for heavy flavours in the absence of medium effects. The expected hierarchy of the energy loss described above can be probed comparing the R_{AA} of different particle species, respectively, B, D mesons and light-hadrons [5]. Further knowledge of the properties of the medium created in heavy-ion collisions can be gained from the study of the azimuthal anisotropy of open heavy flavours: the initial spatial asymmetry is transformed into an asymmetry in momentum via hydrodynamic expansion of the medium. This is quantified in terms of the second harmonic coefficient v_2 of the Fourier expansion of the D-meson azimuthal distribution. The elliptic flow parameter v_2 brings information on the medium transport properties, in particular on the question whether heavy quarks take part in the collective expansion of the medium and on the path-length dependence of energy loss.

The interpretation of particle production measurements in Pb-Pb collisions requires detailed studies and understanding of their production in pp and p-Pb collisions. pp collisions provide the essential reference for the nuclear modification factor and a sensitive test of perturbative QCD models. In addition, with pp collisions it is possible to gain information on the role of Multiple-Parton Interactions (MPI), i.e. several hard partonic interactions occurring in a single collision between two nucleons. Measurements of open heavy-flavour in p–Pb collisions allow to disentangle cold nuclear matter (CNM) effects (such as transverse momentum broadening, gluon saturation at low Bjorken-x [6], modification of the nuclear parton distribution functions [7, 8], energy loss [9]) from final-state effects that occur in a deconfined medium.

2 ALICE detector layout and open heavy-flavour reconstruction

The ALICE detector [10] is well suited for heavy-flavour measurements at the LHC in a wide momentum and rapidity range. The detectors in the central barrel used for this analysis are: the Inner Tracking System (ITS), the Time Projection Chamber (TPC) and the Time-of-Flight detector (TOF). The ITS comprises six layers of silicon pixel, strip and drift detectors and is used for the tracking of charged particles and the reconstruction of the primary and secondary vertex. Track reconstruction in the central barrel ($|\ \eta\ |< 9$) starts from the TPC, which also provides particle identification (PID) by measuring the specific energy loss $\mathrm{d}E/\mathrm{d}x$ of charged particles passing throught it. Further PID information is given by the TOF via the time stamp of the charged particle hits on the detector with respect to the time of the interaction. The cooperation of these high performing detectors allows the excellent performance needed for heavy-flavour analyses. Moreover the V0 scintillator detectors at forward and backward rapidities provide minimum bias trigger and are used to determine the centrality of Pb-Pb collisions.

Prompt D mesons are fully reconstructed through their hadronic decay channels: $D^0 \rightarrow K^-\pi^+$, $D^+ \rightarrow K^-\pi^+\pi^+$, $D^{*+} \rightarrow D^0\pi^+$, $D_s^+ \rightarrow \phi\pi^+ \rightarrow K^+K^-\pi^+$ and their charge conjugates. The reconstruction is based on the selection of their decay secondary vertex, displaced by few hundred μm from the primary vertex and topological selections, combined with particle identification criteria, applied to reduce the high combinatorial background.

The signal yield is obtained by fitting the invariant mass distributions of the candidates, while the contribution from beauty feed-down is estimated based on FONLL pQCD calculations and subtracted in order to evaluate the results for prompt D meson yield [11].

The D-meson yield is measured as a function of the charged-particle multiplicity in pp and p-Pb collisions. The D meson yields and the charged-particle multiplicity are both normalized to the multiplicity-integrated values. In the case of p-Pb collisions, the nuclear modification factor R_{pPb} is defined as the D-meson cross section in p-Pb collisions divided by the measured cross section in pp collisions scaled by the lead mass number.

The p_T-differential cross section measurements of prompt D meson have been performed in pp collisions at $\sqrt{s} = 2.76$ TeV [12] and $\sqrt{s} = 7$ TeV [13] and the results are well described by perturbative QCD. Because of the limited statistics of the $\sqrt{s}= 2.76$ TeV sample, a $\sqrt{s}$ extrapolation is applied to define the pp reference needed for the R_{AA} calculation.

The prompt D-meson elliptic-flow parameter v_2 is also reported as a function of p_T in semi-peripheral collisions (30-50% centrality class).

3 Results

The p_T-differential cross sections for D^0, D^+, D^{*+} and D_s^+ at $\sqrt{s} = 7$ TeV are shown in Fig. 1, together with FONLL and GM-VFNS model calculations. Within uncertainties, theoretical predictions and measurements are in agreement. Nevertheless, it can be noted that the measurements tend to be higher than the central value of the FONLL prediction; for GM-VFNS, instead, the data lie on the lower side of the predictions.

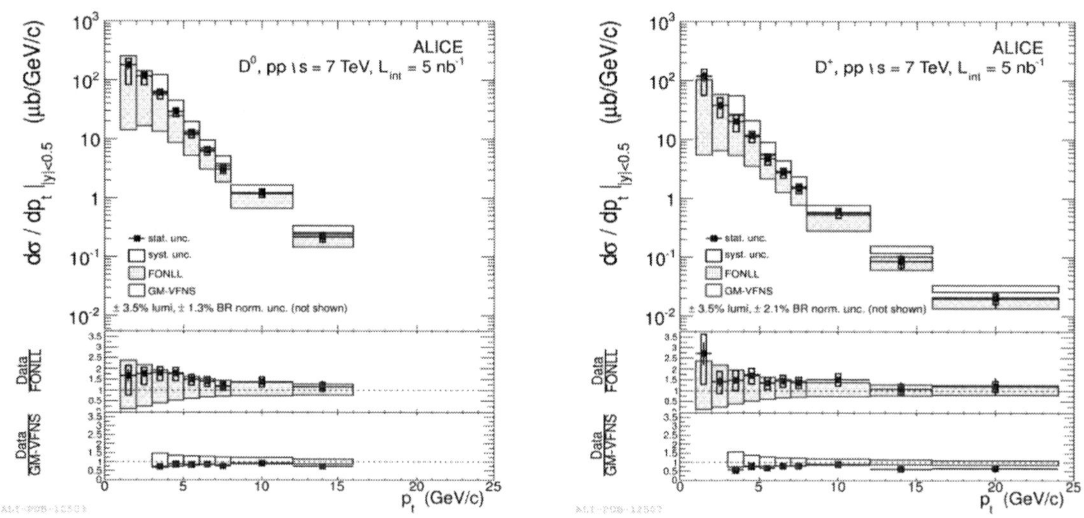

Figure 1. p_T-differential inclusive cross section for prompt D^0 and D^+ [12] mesons in pp collisions at $\sqrt{s} = 7$ TeV compared with FONLL [7] and GM-VFNS [15] theoretical predictions. The symbols are positioned horizontally at the centre of each p_T interval.

In the left panel of Fig. 2 the average of D^0, D^+ and D^{*+} self-normalized yield as a function of relative charged- particle multiplicity is shown for pp and p-Pb collisions in the transverse-momentum range $2 < p_T < 4$ GeV/c. The result for pp collisions indicates that D-meson yields increase with the charged-particle multiplicity by a factor of about 15 in the range between 0.5 and 6 times $<dN_{ch}/d\eta>$. The increase of the heavy-flavour self-normalized yields with the charged-particle multiplicity is qualitatively consistent with the theoretical calculations including the contribution from MPI to heavy-flavor production at LHC energies [16]. A similar behaviour was observed also for p-Pb collisions. However, in this case, high-multiplicity events also come from the higher number of binary nucleon-nucleon collisions.

In the right panel of Fig. 2 shows the average R_{pPb} and R_{AA} of prompt D^0, D^+ and D^{*+} mesons as a function of p_T [18, 19]. The R_{pPb} of D mesons is consistent with unity for $p_T > 2$ GeV/c, and is described by calculations that include CNM effects [20–24].

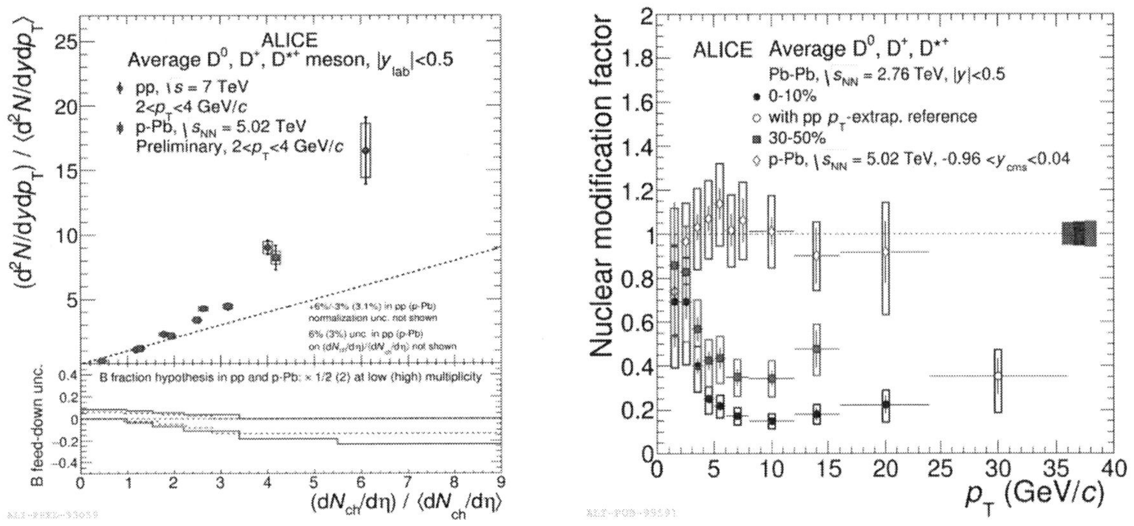

Figure 2. Left: Average D-meson yields as a function of charged-particle multiplicity in pp at $\sqrt{s}$ = 7 TeV [17] and p-Pb at $\sqrt{s_{NN}}$ = 5.02 TeV for 2 $<p_T<$ 4 GeV/c. D mesons are reconstructed in $|y_{lab}|$<0.5. The results are presented in the top panels with their statistical (vertical bars) and systematic (boxes) uncertainties, except for the feed-down fraction uncertainty that is drawn separately in the bottom panels. Right: Prompt D-meson nuclear modification factor (average of D^0, D^+ and D^{*+}) as a function of p_T in Pb-Pb collisions at $\sqrt{s_{NN}}$ = 2.76 TeV in the 0-10% and 30-50% centrality classes [19] superimposed with the same measurement in p-Pb collisions at $\sqrt{s_{NN}}$ = 5.02 TeV [18].

The nuclear modification factor in Pb–Pb collisions, R_{AA}, shows a significant suppression for p_T > 3 GeV/c in both the 0–10% and 30–50% centrality classes; in particular a larger suppression is observed for the most central collisions, as expected. As the R_{pPb} remains consistent with unity over the entire measured p_T range, it is concluded that the suppression in Pb–Pb collisions can be attributed to final-state effects in the medium rather than CNM effects.

In the left panel of Fig. 3 the D-meson R_{AA}, that it is the average of D^0, D^+ and D^{*+}, as a function of the centrality of Pb-Pb collisions, expressed as the average number of nucleons that participate in the collision N_{part}, is shown for 8 < p_T < 16 GeV/c [25]. The centrality dependence of D-meson R_{AA} was studied in the range between 0 and 80%. In the same figure the D-meson R_{AA} is compared to the non-prompt J/Ψ R_{AA} coming from beauty-hadron decays measured by the CMS Collaboration [26]. The chosen p_T ranges imply that both D and B mesons probed by these measurement have average p_T of about 10 GeV/c in a slightly different y range. A suppression that increases towards central collisions was observed for both D mesons and non-prompt J/Ψ R_{AA}. The measurements show a D-meson R_{AA} lower than the non-prompt J/Ψ R_{AA} in central collisions, suggesting a larger energy loss for charm than for beauty quarks, as expected from theoretical calculations that predict a mass-dependent energy loss. The predictions from a model [27], including mass-dependent energy loss for charm and beauty quarks, are superimposed to the measurements: the D-meson and non-prompt J/Ψ R_{AA} can be well described by the this calculation.

Figure 3 (right) shows the comparison of the D-meson R_{AA} with that of charged hadrons and pions as a function of p_T. The measurements show a similar suppression from mid to high p_T but this is still consistent with models including parton charge and mass dependence of energy loss.

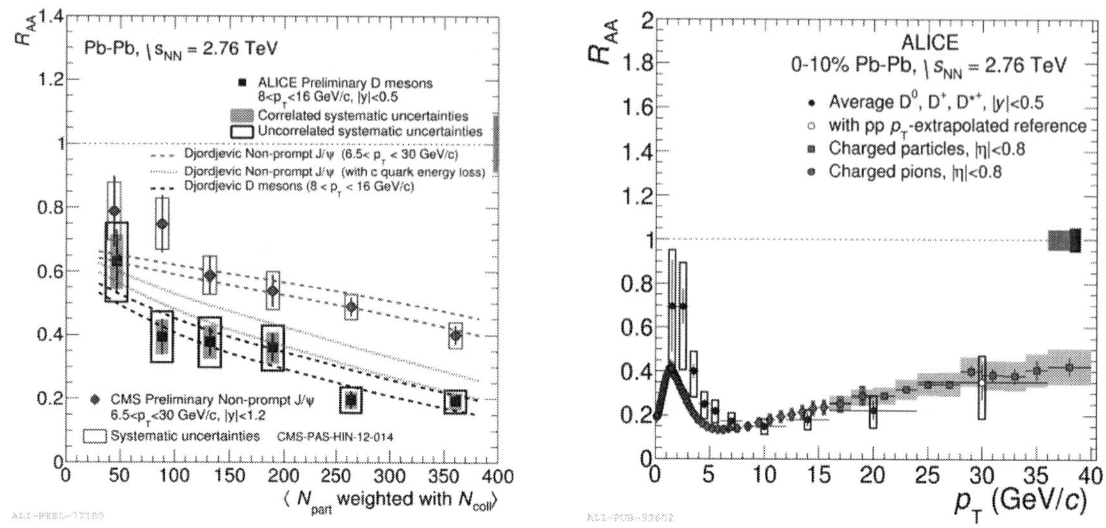

Figure 3. Left: Average D^0, D^+ and D^{*+} R_{AA} vs N_{part} for the 8 $<p_T<$ 16 GeV/c (black points) [25]. Non-prompt J/ψ R_{AA} vs N_{part} measured by CMS for $|y|<1.2$, 6.5 $<p_T<$ 30 GeV/c (blue points) [26]. The data points are compared to Djordjevic models for D mesons (dashed black line), non-prompt J/ψ (dashed blue line) and also considering the charm mass for the non-prompt J/ψ (dotted blue line). Right: Average of prompt D^0, D^+ and D^{*+} R_{AA} in the centrality class 0-10% compared to pions and charged particles.

The average v_2 of the three D-meson species measured in 30-50 % central Pb-Pb collisions, shown in the right panel of Fig. 4, is larger than zero with a 5.7 σ significance in the interval 2 $<p_T<$ 6 GeV/c. This result indicate that at low p_T charm quarks participate in the collective motion of the system. A postive v_2 is also observed for $p_T>$ 6 GeV/c but for a more qualitative study of the path-length dependence of parton energy loss via v_2 measurements more statistics is needed.

In both panels of Fig. 4 [36] the ALICE measurements are compared to various theoretical calculations including in-medium parton energy loss [28–35]. Most models describe qualitatively well the prompt D-meson R_{AA} as a function of p_T, in particular the models including both collisional and radiative energy loss. The low p_T v_2 result is best described by models that include the collective expansion of the medium and a mechanism of hadronisation via recombination of charm quarks with light quarks. Nevertheless, the simultaneous description of the large suppression of D mesons in central collisions and their anisotropy in non-central collisions is challenging for theoretical models.

4 Conclusions

The ALICE experiment at the LHC has an excellent performance for heavy-flavor measurements, thanks to tracking, vertexing and PID. Results on the D-meson production measurements from ALICE were reported in pp, p-Pb and Pb-Pb collisions. The p_T-differential cross sections for prompt D mesons in pp collisions are described within uncertainties by theoretical predictions based on pQCD calculations. The D-meson self-normalized yields exhibit an increasing trend with increasing charged-particle multiplicity in pp collisions, suggesting that MPIs affect the hard momentum scale relevant for heavy-flavour production. A similar increase is observed in p–Pb collisions, but in this case a higher number of binary nucleon-nucleon collisions also contributes to high-multiplicity events. The R_{pPb} of prompt D mesons in p-Pb collisions is consistent with unity within uncertainties, providing

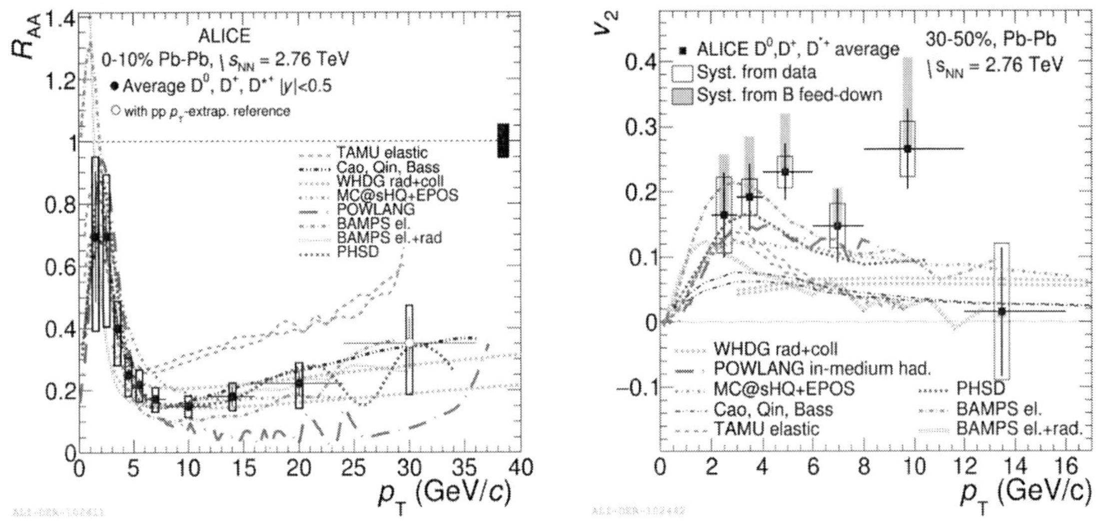

Figure 4. Left: Model comparisons for average D meson R_{AA} in the 0–20% centrality class. Right: Model comparisons for average D meson v_2 in the 30–50% centrality class. In both panels the seven model calculations are: WHDG rad+coll [28], POWLANG [29], Cao, Qin, Bass [30], MC@sHQ+EPOS [31], BAMPS [32], TAMU elastic [33], UrQMD [34, 35]. The models WHDG rad+coll, POWLANG, TAMU elastic and UrQMD are shown by two lines that represent their uncertainty.

no indication for large CNM effects at high p_T. In the most centra Pb-Pb collisions, a strong suppression of heavy-flavour yields is observed at intermediate and high p_T with respect to binary-scaled pp collisions. As the influence of CNM effects is small in the measured p_T range, these results in Pb–Pb collisions can be interpreted as a final-state effect due to in-medium parton energy loss, consistent with the expected parton-mass dependence. The v_2 measured in Pb–Pb semi-central collisions is larger than zero at low p_T, suggesting that heavy quarks participate in the collective motion of the system.

References

[1] M. Gyulassy and M. Plumer, Phys. Lett. **B243** 432 (1990).
[2] R. Baier, Y. L. Dokshitzer, A. H. Mueller, S. Peigne and D. Schiff, Nucl. Phys. **B484** 265 (1997).
[3] M. H. Thoma and M. Gyulassy, Nucl. Phys. **B351** 491 (1991).
[4] Y. L. Dokshitzer, D. E. Kharzeev, Phys. Lett. **B519** (2001) 199.
[5] A. Adil, I.Vitev, Phys. Lett. B **649** (2007)139.
[6] H. Fujii, K. Watanabe, Nucl. Phys. A **915** (2013) 1.
[7] K. Eskola, H. Paukkunen and C. Salgado, JHEP **04** (2009) 065.
[8] D. de Florian and R. Sassot, Phys.Rev. D **69** (2004) 074028.
[9] B. B. Abelev et al., ALICE Collaboration, Int. J. Mod. Phys. A **29** (2014) 1430044.
[10] B. B. Abelev et al., ALICE Collaboration, Int. J. Mod. Phys. A **29** (2014) 1430044.
[11] M. Cacciari et al., CERN-PHTH 2011 227.
[12] B. Abelev et al., ALICE Collaboration, JHEP **07** (2012) 191.
[13] B. Abelev et al., ALICE Collaboration, JHEP **01** (2012) 128.

[14] B. Abelev et al., ALICE Collaboration, Phys. Lett. B **718** (2012) 279.
[15] G. Aad et al., ATLAS Collaboration, Eur.Phys. J. C **72** (2012).
[16] K. Aamodt et al., ALICE Collaboration, JINST **3** S08002 (2008).
[17] B. Abelev et al., ALICE Collaboration, JHEP 09 (2012) 148.
[18] B. Abelev et al., ALICE Collaboration, Phys. Rev. Lett. **113** (23) (2014) 232301.
[19] B. Abelev et al., ALICE Collaboration, Preprint (2015). arXiv:1509.06888.
[20] R. Sharma, I. Vitev, B.-W. Zhang, Phys. Rev. C **80** (2009) 054902.
[21] K. J. Eskola, H. Paukkunen, C. A. Salgado, JHEP **04** (2009) 065.
[22] M. L. Mangano, P. Nason, G. Ridolfi, Nucl. Phys. B **373** (1992) 295–345.
[23] D. Stump, et al., JHEP **10** (2003) 046.
[24] H. Fujii, K. Watanabe, Nucl. Phys. A **920** (2013) 78–93.
[25] B. Abelev et al., ALICE Collaboration, Preprint (2015). arXiv:1506.06604.
[26] V. Khachatryan et al., CMS Collaboration, CMS-PAS-HIIN-12-014.
[27] Djordjevic, Phys. Lett. B **734** (2014) 286-289.
[28] W.A. Horowitz, M. Gyulassy, Nucl Phys. A **872** (2011)265.
[29] M.Monteno et al., Jour. Phys. G **38** (2011) 124144.
[30] S. Cao, G. Qin, S.A. Bass, Phys.Rev. C **88** (2013) 044907.
[31] E. Retinskaya, M. Luzum, J. Ollitrault, Phys.Rev. C **89** (2014) 014902.
[32] J. Uphoffa, O. Fochlera, Z. Xub, C. Greinera, Phys.Lett. B **717** (2012) 430.
[33] He, Min et al., Phys.Rev. B **735** (2014) 445.
[34] T. Lang, H. van Hees, J. Steinheimer, M. Bleicher, Phys. Rev. C **93** 014901 (2016).
[35] T. Lang et al, J. Phys. Conf. Ser. **426** 012032 (2013).
[36] B. Abelev et al., ALICE Collaboration, Phys. Rev. C 90 (2014)034904.

EPJ Web of Conferences **126**, 04035 (2016)
ICNFP 2015
DOI: 10.1051/epjconf/201612604035

Search for the dark photon in π^0 decays

Roberto Piandani[1,a]

[1] *INFN Pisa*

Abstract. A sample of 1.69×10^7 fully reconstructed $\pi^0 \to \gamma e^+ e^-$ in the kinematic range $m_{ee} > 10MeV/c^2$ with a negligible background contamination collected by the NA48/2 experiment at CERN in 2003-04 is analysed to search for the dark photon (A') via the decay chain $\pi^0 \to \gamma A'$, $A' \to e^+ e$. No signal is observed, and preliminary exclusion limits on space of dark photon mass $m_{A'}$ and mixing parameter ϵ^2 are reported.

1 Introduction

High intensity kaon experiments provide opportunities for precision studies of π^0 decay physics due to the fact that kaons represent a source of tagged neutral pion decays, mainly via their $K^\pm \to \pi^\pm \pi^0$, $K^\pm \to e^\pm \pi^0 \nu$ and $K^\pm \to \mu^\pm \pi^0 \nu$ decays. One of them is the NA48/2 experiment at the CERN SPS, which collected a large sample of charged kaon ($K^\pm$) decays in flight in 2003–04 corresponding to $\sim 2 \times 10^{11}$ $K^\pm$ decays in the fiducial decay volume [1].
This large sample of π^0 mesons produced and decaying in vacuum collected by NA48/2 allows for a high sensitivity search for the dark photon (A'), a hypothetical gauge boson appearing in hidden sector new physics models with an extra U(1) gauge symmetry. In these models the interaction of the dark photon with the visible sector proceed through kinetic mixing with the Standard Model (SM) hypercharge [2]. Such scenarios with GeV-scale dark matter provide possible explanations to the observed rise in the cosmic-ray positron fraction with energy and the muon gyromagnetic ratio ($g - 2$) measurement [3]. From the experimental point of view, the dark photon is characterized by two a priori unknown parameters, the mixing parameter ϵ and the mass $m_{A'}$. The A' production and the subsequent decay can be obtained from the following decay chain:

$$K^\pm \to \pi^\pm \pi^0,\ \pi^0 \to \gamma A',\ A' \to e^+ e^-$$

with three charged particles and a photon in the final state. The expected branching fraction of the π^0 decay is [4]

$$B(\pi^0 \to \gamma A') = 2\epsilon^2 \left(1 - \frac{m_{A'}^2}{m_{\pi^0}^2}\right)^3 \times B(\pi^0 \to \gamma\gamma) \quad (1)$$

which is kinematically suppressed as $m_{A'}$ approaches m_{π^0}. In the mass range $2m_e \ll m_{A'} < m_{\pi^0}$ accessible in this analysis, the dark photon is below threshold for all decays into SM fermions except

[a] for the NA48/2 Collaboration: Cambridge, CERN, Dubna, Chicago, Edinburgh, Ferrara, Firenze, Mainz, Northwestern, Perugia, Pisa, Saclay, Siegen, Torino, Wien.
e-mail: roberto.piandani@cern.ch

$A' \rightarrow e^+e^-$, while the allowed loop-induced decays ($A' \rightarrow 3\gamma, A' \rightarrow \nu\bar{\nu}$) are strongly suppressed. Therefore, assuming that the dark photon decay only in SM particles, $B(A' \rightarrow e^+e^-) \simeq 1$. The expected total dark photon decay width is [5]

$$\Gamma(A' \rightarrow e^+e^-) = \frac{1}{3}\alpha\epsilon^2 m_{A'}\sqrt{1-\frac{4m_e^2}{m_{A'}^2}}\left(1+\frac{2m_e^2}{m_{A'}^2}\right)$$

It follows that, for $2m_e \ll m_{A'} < m_{\pi^0}$, the dark photon mean proper lifetime $\tau_{A'}$ satisfies the relation

$$c\tau_{A'} \simeq 0.8\mu m \times \left(\frac{10^{-6}}{\epsilon^2}\right) \times \left(\frac{100MeV}{m_{A'}}\right)$$

The NA48/2 analysis is performed assuming that the dark photon decays at the production point (*prompt decay*), which is valid for sufficiently large values of the mass ($m_{A'} > 10MeV/c^2$) and the mixing parameter ($\epsilon^2 > 5 \times 10^{-7}$). In this condition the dark photon signature is identical to the Dalitz decay $\pi^0 \rightarrow \gamma e^+e^-$, which therefore represents an irreducible background and determines the sensitivity. The largest π^0_D sample is obtained through the reconstruction of $K^\pm \rightarrow \pi^\pm\pi^0$ and $K^\pm \rightarrow \mu^\pm\pi^0\nu$ decays (denoted $K_{2\pi}$ and $K_{\mu3}$). In Addition, the $K^\pm \rightarrow \pi^\pm\pi^0\pi^0$ decay (denoted $K_{3\pi}$) is considered as a background in the $K_{\mu3}$ sample.

2 The NA48/2 experiment

The NA48/2 beam line has been designed to deliver simultaneous narrow momentum band K^+ and K^- beams following a common beam axis derived from the primary 400 GeV/c protons extracted from the CERN SPS. Secondary beams with central momenta of (60 ± 3) GeV/c (r.m.s.) were used. The beam kaons decayed in a fiducial decay volume contained in a 114 m long cylindrical vacuum tank. The momenta of charged decay products were measured in a magnetic spectrometer, housed in a tank filled with helium placed after the decay volume. The spectrometer comprised four drift chambers (DCHs), two upstream and two downstream of a dipole magnet which provided a horizontal transverse momentum kick of 120 MeV/c to singly-charged particles. Each DCH was composed of eight planes of sense wires. A plastic scintillator hodoscope (HOD) producing fast trigger signals and providing precise time measurements of charged particles was placed after the spectrometer. Further downstream was a liquid krypton electromagnetic calorimeter (LKr), an almost homogeneous ionization chamber with an active volume of 7 m^3 of liquid krypton, $27X_0$ deep, segmented transversally into 13248 projective 2×2 cm^2 cells and with no longitudinal segmentation. The LKr information is used for photon measurements and charged particle identification. An iron/scintillator hadronic calorimeter and muon detectors were located further downstream. A dedicated two-level trigger was in operation to collect three-track decays with an efficiency of about 98%. A detailed description of the detector can be found in Ref. [6].

3 Event selection

The full NA48/2 data sample is used for the analysis. The $K_{2\pi}$ and $K_{\mu3}$ with the following $\pi^0 \rightarrow \gamma e^+e^-$ event selection requires a three-track vertex reconstructed in the fiducial decay region formed of a pion or muon candidate track and two opposite-sign electron ($e^\pm$) candidate tracks. Charged particle identification is based on the ratio of energy deposition in the LKr calorimeter (E) to the momentum measured by the spectrometer (p). Pions and muons are kinematically constrained to the momentum range above 5 GeV/c, while the momentum spectra of electrons originating from π^0 decays are soft,

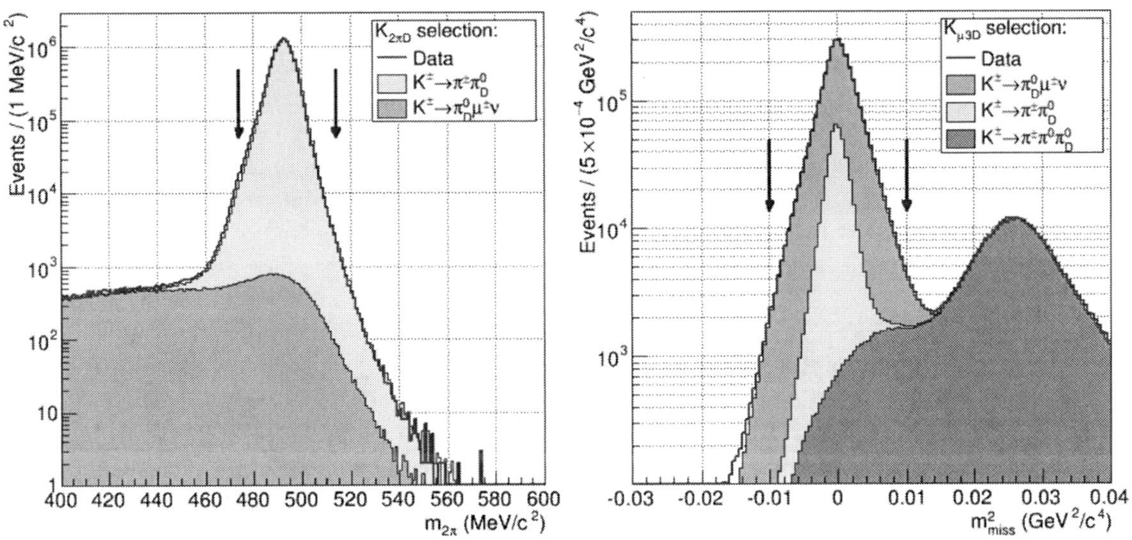

Figure 1. Data and Monte Carlo reconstructed $\pi^{\pm}\pi^{0}_{D}$ invariant mass for the event passing the $K_{2\pi}$ selection (left). Data and Monte carlo reconstructed invariant mass for the event passing the $K_{\mu 3}$ selection (right). The arrows represent the selection cuts.

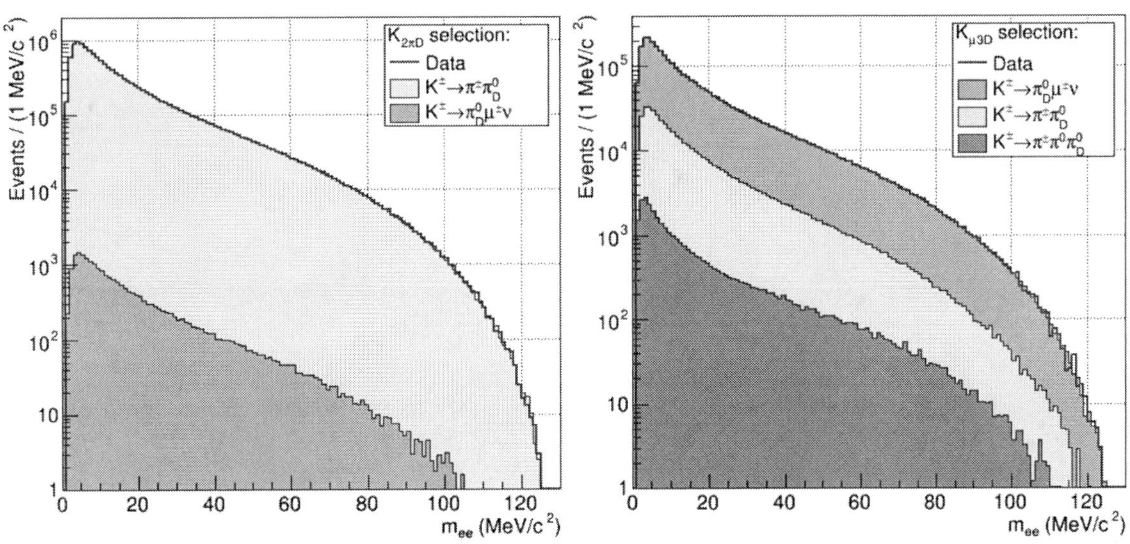

Figure 2. Data and Monte Carlo reconstructed m_{ee} invariant mass for the event passing the $K_{2\pi}$ selection (left). Data and Monte carlo reconstructed m_{ee} invariant mass for the event passing the $K_{\mu 3}$ selection (right).

peaking at 3 GeV/c. Therefore,$p > 5\ GeV/c$ and $E/p < 0.85$ ($E/p < 0.4$) are required for the pion (muon) candidate, while $p > 2.75\ GeV/c$ and $(E/p)_{min} < E/p < 1.15$, where $(E/p)_{min} = 0.80$ for $p < 5\ GeV/c$ and $(E/p)_{min} = 0.85$ otherwise, are required for the electron candidates. Furthermore, a single insolated LKr energy deposition cluster is required and considered as the photon candidate. The reconstructed invariant mass of the $e^+e^-\gamma$ system is required to be compatible with the π^0 mass, $|m_{e^+e^-\gamma} - m_{\pi^0}| < 8\ Mev/c^2$ (this interval corresponds to ±5 times the resolution on $m_{e^+e^-\gamma}$).
For the $K_{2\pi}$ selection, the reconstructed invariant mass of the $\pi^\pm e^+e^-\gamma$ (figure 1) system should be compatible with the $K^\pm$ mass, $474\ MeV/c^2 < m_{\pi e^+e^-\gamma} < 514\ MeV/c^2$. For the $K_{\mu3}$ selection, the squared missing mass $m^2_{miss} = (P_K - P_\mu - P_{\pi^0})^2$ (figure 1) should be compatible with tne neutrino mass ($|m^2_{miss}| < 0.01\ GeV^2/c^4$), where P_μ and P_{π^0} are the reconstructed $\mu^\pm$ and π^0 four-momenta, and P_K is the nominal kaon four-momentum.
At the end of the two selections a sample of 1.38×10^7 (0.31×10^7) fully reconstructed π^0_D decay candidates comining from $K_{2\pi}$ ($K_{\mu3}$) with a negligible background is selected. Correcting the observed number of candidates for acceptance and trigger efficiency, the total number of $K^\pm$ decays in the 98 m long fiducial decay region for the analysed data sample is found to be $N_K = (1.57 \pm 0.05) \times 10^7$, where the quoted error is dominated by the external uncertainty on the π^0_D decay branching fraction. The reconstructed e^+e^- invariant mass spectra are displayed in figure 2. In addition to the above individual $K_{2\pi}$ and $K_{\mu3}$ selections, a *joint* dark photon selection is also considered: an event passes the joint selection if it passes either the $K_{2\pi}$ or the $K_{\mu3}$ selection. The acceptance of the joint selection for any process is equal to the sum of acceptances of the two mutually exclusive individual selections.

4 The π^0 Dalitz simulation

The π^0_D decay is simulated using the following lowest-order differential decay rate

$$\frac{d^2\Gamma}{dxdy} = \Gamma_0 \frac{\alpha}{\pi} |F(x)|^2 \frac{(1-x)^3}{4x}\left(1 + y^2 + \frac{r^2}{x}\right)$$

with x and y, two kinematic variables, defined as

$$x = \frac{(q_1 + q_2)^2}{m^2_\pi} = \frac{m^2_{ee}}{m^2_{\pi^0}} \quad y = \frac{2p(q_1 - q_2)}{m_{\pi^0}(1-x)}$$

$r = 2m_e/m_{\pi^0}$, q_1 , q_2 and p are the four-momenta of the electrons ($e^\pm$) and the neutral pion, Γ_0 is the rate of the $\pi^0 \to \gamma\gamma$ decay, and F(x) is the pion transition form factor (TFF).
Radiative corrections to the differential rate are implemented following the classical approach of Mikaelian and Smith [7]: the differential decay rate is modified using the following formula

$$\frac{d^2\Gamma^{rad}}{dxdy} = \delta(x,y)\frac{d^2\Gamma}{dxdy}$$

which does not account for the emission of inner bremsstrahlung photons.
The TFF is conventionally parameterized as $F(x) = 1 + ax$. The transition form factor slope parameter a is expected from vector meson dominance models to be $a \simeq 0.03$, and detailed theoretical calculations based on dispersion theory yield $a = 0.0307 \pm 0.0006$ [8]. Experimentally, the PDG average $a = 0.032 \pm 0.004$ [9] is determined mainly from a $e^+e^- \to e^+e^-\pi^0$ rate measurement in the space-like region by the CELLO experiment [10]. The precision on the used radiative corrections to the π^0_D decay is limited: in particular, the missing correction to the measured TFF slope due to two-photon exchange is estimated to be $\Delta a = +0.005$ [11]. Therefore the background description cannot benefit

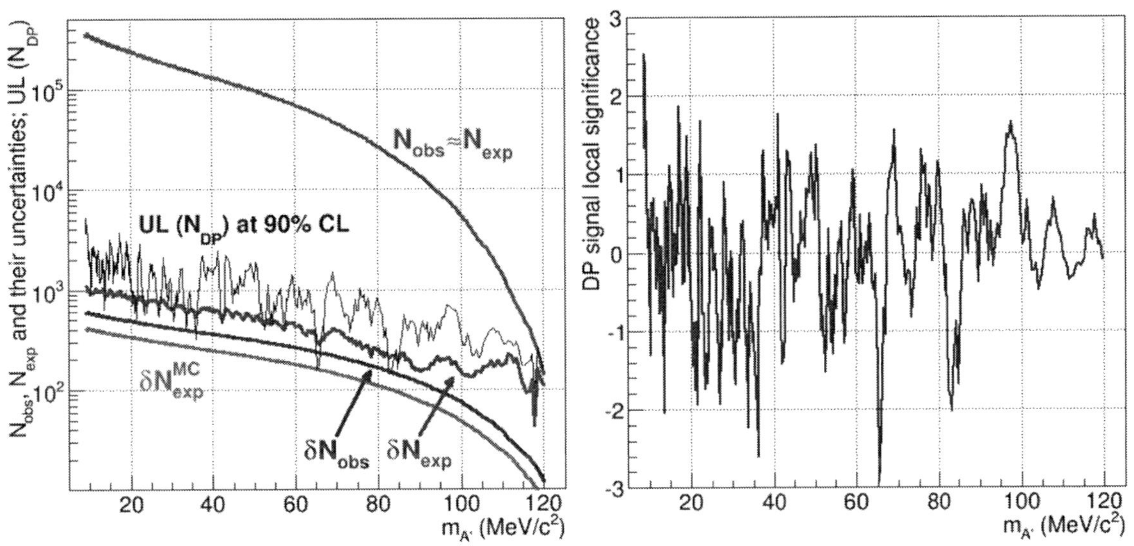

Figure 3. Numbers of observed data events and expected π_D^0 background events passing the joint selection (indistinguishable in a logarithmic scale), estimated uncertainties and obtained upper limits at 90% CL on the numbers of dark photon candidates (N_{DP}) for each mass value $m_{A'}$ (left). Estimated local significance of the dark photon signal for each A' mass value (right).

from the precise inputs on the TFF slope [8, 9], and an "effective" TFF slope obtained from a fit to the data m_{ee} spectrum itself is used to obtain a satisfactory background description (as quantified by a χ^2 test) over the range $m_{ee} > 8\ MeV/c^2$ used for the dark photon search. The low m_{ee} region is not considered for the search as the acceptance computation is less robust due to the steeply falling geometrical acceptance at low m_{ee} and decreasing electron identification efficiency at low momentum.

5 Search for the dark photon

A scan for a dark photon signal in the mass range $9\ MeV/c^2 \leq m_{A'} < 120\ MeV/c^2$ is performed. The lower extent of the considered mass range is determined by the limited precision of MC simulation of background at low mass, while at the upper limit of the mass range the signal acceptance drops to zero. The mass step of the scan and the width of the dark photon signal mass window around the assumed mass are determined by the resolution δm_{ee} on the e^+e^- invariant mass. The resolution on m_{ee} as a function of m_{ee} evaluated with MC simulation is parameterized as $\sigma_m(m_{ee}) = 0.067\ MeV/c^2 + 0.0105 \cdot m_{ee}$, and varies from $0.16\ MeV/c^2$ to $1.33 MeV/c^2$ over the mass range of the scan. The mass step of the scan is set to be $\sigma_m/2$, while the signal region mass window for each dark photon mass hypothesis is defined as $\pm 1.5\sigma_m$ around the assumed mass. The mass window width has been optimised with MC simulations to obtain the highest sensitivity to the dark photon signal, determined by a trade-off between π_D^0 background fluctuation and signal acceptance.
For each considered dark photon mass value, the number of observed data events N_{obs} passing the joint selection is compared to the expected number of background events N_{exp}. The latter is evaluated from MC simulations, corrected for the trigger efficiency measured from control data samples passing the joint selection. The numbers of observed and expected events for each mass value and their estimated uncertainties δN_{obs} and δN_{exp} are shown in figure 3. The uncertainty $\delta N_{obs} = \sqrt{N_{obs}}$ is statistical, while the uncertainty δN_{exp} has contributions from the limited size of the generated MC samples and

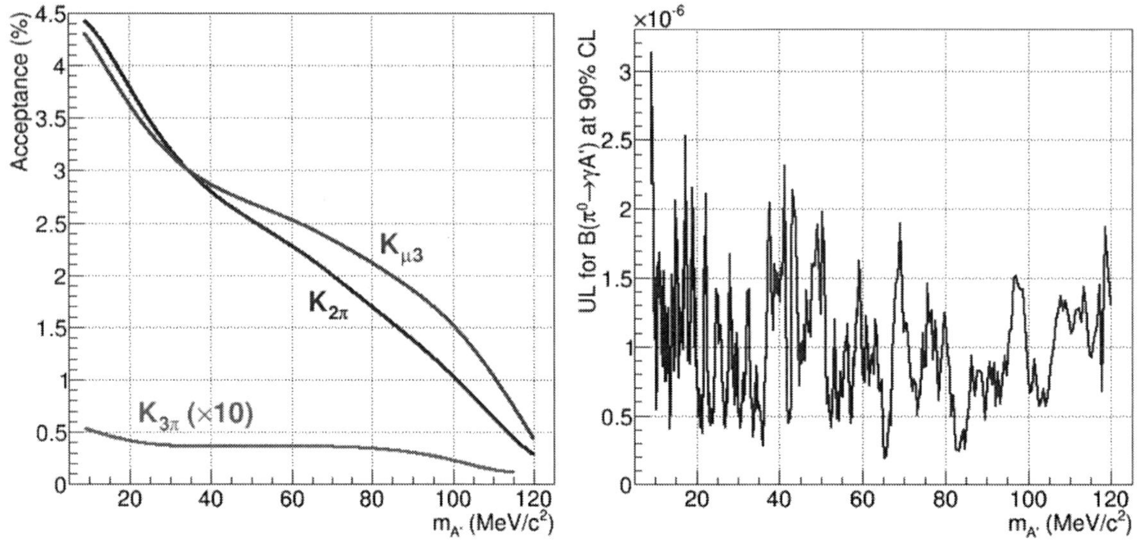

Figure 4. Acceptances of the joint dark photon selection for $K_{2\pi}$, $K_{\mu 3}$ and $K_{3\pi}$ depending on the assumed DP mass, evaluated with MC simulations. The $K_{3\pi}$ acceptance is scaled by a factor of 10 for visibility (left). Obtained upper limits on $B(\pi^0 \to A'\gamma)$ at 90% CL for each A' mass value (right).

the statistical errors on the trigger efficiencies measured in the dark photon signal region. The local statistical significance of the dark photon signal for each mass value estimated as

$$Z = (N_{obs} - N_{exp}) / \sqrt{(\delta N_{obs})^2 + (\delta N_{exp})^2}$$

Confidence intervals at 90% CL for the number of $A' \to e^+e^-$ decay candidates (N_{DP}) in each mass hypothesis are computed from N_{obs} , N_{exp} and δN_{exp} using the Rolke–López method [12] assuming Poissonian (Gaussian) errors on the numbers of observed (expected) events.

Upper limits at 90% CL on the branching fraction $B(\pi^0 \to A'\gamma)$ for each dark photon mass value with the assumption $B(A' \to e^+e^-) = 1$ (which is a good approximation for $m_{A'} < 2m_\mu$ if A' decays to SM fermions only) are computed using the relation

$$B(\pi^0 \to A'\gamma) = \frac{N_{DP}}{N_K e_1 e_2 [B(K_{2\pi}) A_{DP}(K_{2\pi}) + B(K_{\mu 3}) A_{DP}(K_{\mu 3}) + 2B(K_{3\pi}) A_{DP}(K_{3\pi})]}$$

where $A_{DP}(K_{2\pi})$, $A_{DP}(K_{\mu 3})$ and $A_{DP}(K_{3\pi})$ are the acceptances of the joint dark photon selection for $K_{2\pi}$, $K_{\mu 3}$ and $K_{3\pi}$ decays, respectively, followed by the prompt $\pi^0 \to A'\gamma$, $A' \to e^+e^-$ decay chain. The trigger efficiencies e_1 ane e_2 are taken into account neglecting their variations over the m_{ee} invariant mass, variations measured to be at the level of a few permille. Event distributions in the angle between e^+ momentum in the e^+e^- rest frame and the e^+e^- momentum in the π^0 rest frame are identical for the decay chain involving the dark photon ($\pi^0 \to A'\gamma$, $A' \to e^+e^-$) and the π^0_D decay, up to a negligible effect of the radiative corrections that should not be applied in the former case (found to influence the acceptance at the level below 1% in relative terms). Therefore dark photon acceptances are evaluated using the MC samples produced for background description, and no dedicated MC productions are required. The resulting upper limits on $B(\pi^0 \to A'\gamma)$ and signal acceptances are shown in figure 4. The upper limits are $O(10^{-6})$ and do not exhibit a strong dependence on the A' mass.

 DOI: 10.1051/epjconf/201612604035

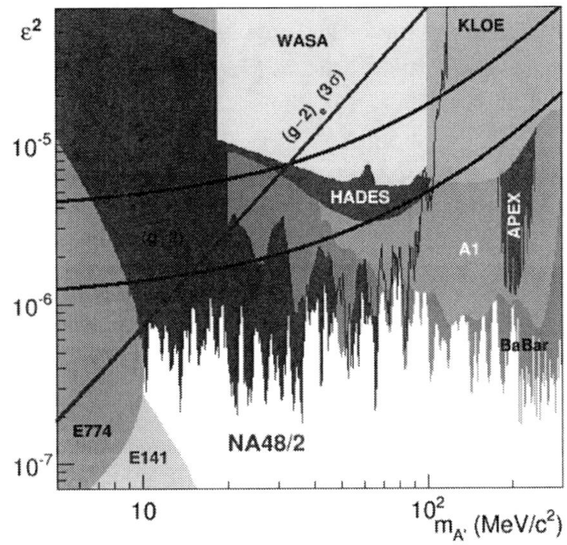

Figure 5. The NA48/2 preliminary upper limits at 90% CL on the mixing parameter ϵ^2 versus the A' mass, compared to the other published exclusion limits from meson decay, beam dump and e^+e^- collider experiments. Also shown are the band for the muon $(g-2)$ and the region excluded by the electron $g-2$ measurement.

Upper limits at 90% CL on the mixing parameter ϵ^2 for each dark photon mass value calculated from the $B(\pi^0 \to A'\gamma)$ upper limits using equation 1 are shown in figure 5, together with the constraints from the SLAC E141 and FNAL E774 [13], KLOE [14], WASA [15], HADES [16], A1 [17], APEX [18] and BaBar [19] experiments. Also shown is the band in the $(m_{A'}, \epsilon^2)$ plane where the discrepancy between the measured and calculated muon $(g-2)$ values falls into the $\pm 2\sigma$ range due to the dark photon contribution, as well as the region excluded by the electron $(g-2)$ measurement [3, 20].
The obtained upper limits on ϵ^2 represent an improvement over the existing data in the dark photon mass range $9-70$ MeV/c^2 , and exclude the whole favoured region by muon $g-2$ [21]. The most stringent limits (2×10^{-7}) are achieved at $m_{A'} \sim 10$ MeV/c^2. The sensitivity of the prompt A' decay search is limited by the irreducible π^0_D background. The achievable upper limit on ϵ^2 scales as the inverse square root of the integrated beam flux, which means that the possible improvements to be made with this technique using larger future $K^\pm$ samples are modest.

References

[1] J. R. Batley *et al.*, Eur. Phys. J. **C52**, 875 (2007).
[2] B. Holdom, Phys. Lett. **B166**, 196 (1986).
[3] M. Pospelov, Phys. Rev. **D80**, 095002 (2009).
[4] B. Batell, M. Pospelov and A. Ritz, Phys. Rev. **D80**, 095024 (2009).
[5] B. Batell, M. Pospelov and A. Ritz, Phys. Rev. **D79**, 115019 (2009).
[6] V. Fanti *et al.*, (NA48 Collaboration), Nucl. Instrum. Meth. A **574** 433 (2007).
[7] K. O. Mikaelian and J. Smith, Phys. Rev. **D5**, 1763 (1972).
[8] M. Hoferichter *et al.*, Eur. Phys. J. **C74**, 3180 (2014).
[9] K. A. Olive *et al.* (Particle Data Group), Chin. Phys. **C38**, 090001 (2015).
[10] H. J. Behrend *et al.*, Z. Phys. **C49**, 401 (1991).

[11] K. Kampf, M. Knecht and J. Novotný, Eur. Phys. J. **C46**, 191 (2006).
[12] W. A. Rolke and A. M. López, Nucl. Instrum. Meth. A **458**, 745 (2001).
[13] S. Andreas, C. Niebuhr and A. Ringwald, Phys. Rev. **D86**, 095019 (2012).
[14] D. Babusci *et al.*, (KLOE-2 collaboration), Phys. Lett. **B720**, 111 (2013).
[15] P. Adlarson *et al.*, (WASA-at-COSY collaboration), Phys. Lett. **B726**, 187 (2013).
[16] G. Agakishev *et al.*, (HADES collaboration), Phys. Lett. **B731**, 265 (2014).
[17] H. Merkel *et al.*, (A1 collaboration), Phys. Rev. Lett. **112**, 221802 (2014).
[18] S. Abrahamyan *et al.*, (APEX collaboration), Phys. Rev. Lett. **107**, 191804 (2011).
[19] J. P. Lees *et al.*, (BaBar collaboration), Phys. Rev. Lett. **113**, 201801 (2014).
[20] H. Davoudiasl, H. S. Lee and W. J. Marciano, Phys. Rev. **D89**, 095006 (2014).
[21] R. J. Batley *et al.*, Phys. Lett. **B746**, 178 (2015).

The NA62 experiment at CERN

Mauro Piccini[1,a]

[1] *INFN - Sezione di Perugia*

Abstract. The rare decays $K \to \pi\nu\bar{\nu}$ are excellent processes to make tests of new physics at the highest scale complementary to LHC thanks to their theoretically cleanness. The NA62 experiment at CERN SPS aims to collect of the order of 100 events in two years of data taking for the decay $K^+ \to \pi^+\nu\bar{\nu}$, keeping the background at the level of 10%. Part of the experimental apparatus has been commissioned during a technical run in 2012. The diverse and innovative experimental techniques will be explained and some preliminary results obtained during the 2014 pilot run will be reviewed.

1 Introduction

NA62 is the last generation kaon experiment at CERN SPS aiming to study the decay $K^+ \to \pi^+\nu\bar{\nu}$. The goal of the experiment is to measure the decay branching ratio ($O(10^{-10})$) with 10% accuracy, collecting about 100 events in two years of data taking and assuming a 10% signal acceptance. The proton beam extracted from the SPS in the north area at CERN fulfills such demanding request in

[a]On behalf of the NA62 Collaboration: G. Aglieri Rinella, R. Aliberti, F. Ambrosino, B. Angelucci, A. Antonelli, G. Anzivino, R. Arcidiacono, I. Azhinenko, S. Balev, M. Barbanera, J. Bendotti, A. Biagioni, L. Bician, C. Biino, A. Bizzeti, T. Blazek, A. Blik, B. Bloch-Devaux, V. Bolotov, V. Bonaiuto, M. Bragadireanu, D. Britton, G. Britvich, M.B. Brunetti, D. Bryman, F. Bucci, F. Butin, E. Capitolo, C. Capoccia, T. Capussela, V. Carassiti, N. Cartiglia, A. Cassese, A. Catinaccio, A. Cecchetti, A. Ceccucci, P. Cenci, V. Cerny, C. Cerri, B. Checcucci, O. Chikilev, R. Ciaranfi, G. Collazuol, A. Conovaloff, P. Cooke, P. Cooper, G. Corradi, E. Cortina Gil, F. Costantini, F. Cotorobai, A. Cotta Ramusino, D. Coward, G. D'Agostini, J. Dainton, P. Dalpiaz, H. Danielsson, J. Degrange, N. De Simone, D. Di Filippo, L. Di Lella, S. Di Lorenzo, N. Dixon, N. Doble, B. Dobrich, V. Doria, V. Duk, V. Elsha, J. Engelfried, T. Enik, N. Estrada, V. Falaleev, R. Fantechi, V. Fascianelli, L. Federici, S. Fedotov, M. Fiorini, J. Fry, J. Fu, A. Fucci, L. Fulton, S. Gallorini, S. Galeotti, E. Gamberini, L. Gatignon, G. Georgiev, A. Gianoli, M. Giorgi, S. Giudici, L. Glonti, A. Goncalves Martins, F. Gonnella, E. Goudzovski, R. Guida, E. Gushchin, F. Hahn, B. Hallgren, H. Heath, F. Herman, O. Hutanu, D. Hutchcroft, L. Iacobuzio, E. Iacopini, E. Imbergamo, O. Jamet, P. Jarron, K. Kampf, J. Kaplon, V. Karjavin, V. Kekelidze, S. Kholodenko, G. Khoriauli, A. Khotyantsev, A. Khudyakov, Yu. Kiryushin, A. Kleimenova, K. Kleinknecht, A. Kluge, A. Korotkova, M. Koval,. V Kozhuharov, M. Krivda, Y. Kudenko, J. Kunze, G. Lamanna, C. Lazzeroni, R. Lenci, M. Lenti, E. Leonardi, P. Lichard, R. Lietava, L. Litov, D. Lomidze, A. Lonardo, N. Lurkin, D. Madigozhin, G. Maire, A. Makarov, C. Mandeiro, I. Mannelli, G. Mannocchi, A. Mapelli, F. Marchetto, R. Marchevski, S. Martellotti, P. Massarotti, K. Massri, P. Matak, E. Maurice, E. Menichetti, E. Minucci, M. Mirra, M. Misheva, N. Molokanova, J. Morant, M. Morel, M. Moulson, S. Movchan, D. Munday, M. Napolitano, I. Neri, F. Newson, A. Norton, M. Noy, G. Nuessle, T. Numao, V. Obraztsov, A. Ostankov, S. Padolski, R. Page, V. Palladino, A. Pardons, C. Parkinson, E. Pedreschi, M. Pepe, F. Perez Gomez, M. Perrin-Terrin, L. Peruzzo, P. Petrov, F. Petrucci, R. Piandani, M. Piccini, D. Pietreanu, J. Pinzino, M. Pivanti, I. Polenkevich, L. Pontisso, I. Popov, Yu. Potrebenikov, D. Protopopescu, F. Raffaelli, M. Raggi, P. Riedler, A. Romano, P. Rubin, G. Ruggiero, V. Russo, V. Ryjov, A. Salamon, G. Salina, V. Samsonov, C. Santoni, E. Santovetti, G. Saracino, F. Sargeni, S. Schifano, V. Semenov, A. Sergi, M. Serra, A. Sher, S. Shkarovskiy, D. Soldi, A. Sotnikov, V. Sougonyaev, M. Sozzi, T. Spadaro, F. Spinella, R. Staley, M. Statera, P. Sutcliffe, N. Szilasi, D. Tagnani, M. Valdata-Nappi, P. Valente, M. Vasile, T. Vassilieva, B. Velghe, M. Veltri, S. Venditti, R. Volpe, M. Vormstein, H. Wahl, R. Wanke, P. Wertelaers, A. Winhart, R. Winston, E. Worcester, B. Wrona, O. Yushchenko, M. Zamkovsky, A. Zinchenko.

terms of intensity. The NA62 detector must be able to reject background events from decay channels with branching ratios up to 10 orders of magnitude higher than the signal and with similar experimental signature, such as $K^+ \to \mu^+\nu$. To this purpose, good kinematic rejection, particle identification (PID) and veto systems are required. The required resolution on the measurement of kinematic parameters (momenta, missing mass) is obtained tracking both the decaying kaon with a silicon pixel detector (GTK) and the parent charged particles with a straw spectrometer. The key elements of the PID in NA62 are: a differential Cherenkov detector (KTAG) to identify kaons in the charged beam, a Ring Imaging Cherenkov detector (RICH) to identify π^+ and μ^+ in the momentum range between 15 and 35 GeV/c and three scintillators detector acting as hadron calorimeters and muon counters at the end of the detector region. Veto systems working at different angles with respect to the beam direction, especially to suppress background coming from kaon decays with photons in the final state, complete the experimental setup: Large Photon Vetos (LAV) made by lead glass ad large angle, a Liquid Krypton calorimeter (LKr) at intermediate angles and two shashlik detectors at small angles. The detectors have to stand a rate of about 10 MHz of events, high performances read-out and trigger systems have been developed for this purpose. High resolution on time measurement is also required in order to reduce background produced by event overlaps. In the following sections, after a short theoretical introduction, a review of the detectors and of the results obtained during the 2014 pilot run will be presented.

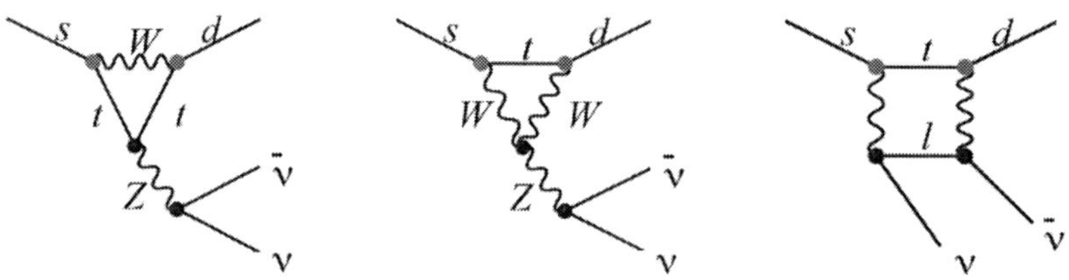

Figure 1. Z-penguin diagrams and box diagram contributing to the process $K \to \pi\nu\bar{\nu}$.

2 Theoretical framework and experimental status of the art

The decay $K \to \nu\bar{\nu}$ are flavor-changing neutral current (FCNC) processes that probe the $s \to dnu\bar{\nu}$ transition via the Z-penguin and the box diagram shown in Figure 1. They are highly GIM suppressed and their Standard Model (SM) rates are very small. For the reasons listed below the SM calculation for their branching ratios (BRs) is particularly clean [2]:

- The loop amplitudes are dominated by the top-quark contributions. The neutral decay violates CP; its amplitude involves the top-quark contribution only. Small corrections to the amplitudes from the lighter quarks come into play for the charged channel.
- The hadronic matrix element for these decays can be factorized and obtained from the precise experimental measurement of the $K \to \pi e\nu$ (Ke3) rate.
- There are no long-distance contributions from processes with intermediate photons.

In the SM, BR($K_L \to \pi^0\nu\bar{\nu}$) = $(2.43\pm0.39\pm0.06)\times10^{-11}$ and BR($K^+ \to \pi^+\nu\bar{\nu}$) = $(7.81\pm0.75\pm0.29)\times 10^{-11}$ [3]. The first uncertainties listed for each BR derive from the input parameters. The second and smaller uncertainties listed demonstrate the size of the intrinsic theoretical uncertainties. Giving the corrections from lighter-quark contributions, these are slightly larger for the charged channel. Because the SM rates are small and predicted very precisely, the BRs for these decays are sensitive probes for new physics. In evaluating the rates for the FCNC kaon decays, the different terms of the operator product expansion are differently sensitive to modifications from a given new-physics scenario. If BR($K_L \to \pi^0 \nu\bar{\nu}$) and BR($K^+ \to \pi^+ \nu\bar{\nu}$) are ultimately both measured, and one or both BRs is found to differ from its SM value, it may be possible to characterize the physical mechanism involved [3], e.g., a mechanism with minimal flavor violation [4], manifestations of supersymmetry [5], a fourth generation of fermions [6], Higgs compositeness as in the littlest Higgs model [7], or an extra-dimensional mechanism such as in the Randall-Sundrum model [8]. The most precise experimental result for the BR($K^+ \to \pi^+ \nu\bar{\nu}$) has been obtained by the dedicated experiments E787 and E949 at the Brookhaven National Laboratory [9] and is based on a total of 7 events using a decay-at-rest technique; the combined measurement is BR($K^+ \to \pi^+ \nu\bar{\nu}$) = $(17.3 \pm 11.5) \times 10^{-11}$. For the neutral channel only an upper limit for the BR has been measured [10], BR($K^0 \to \pi^0 \nu\bar{\nu}$) < 2.6×10^{-8} at 90% confidence level.

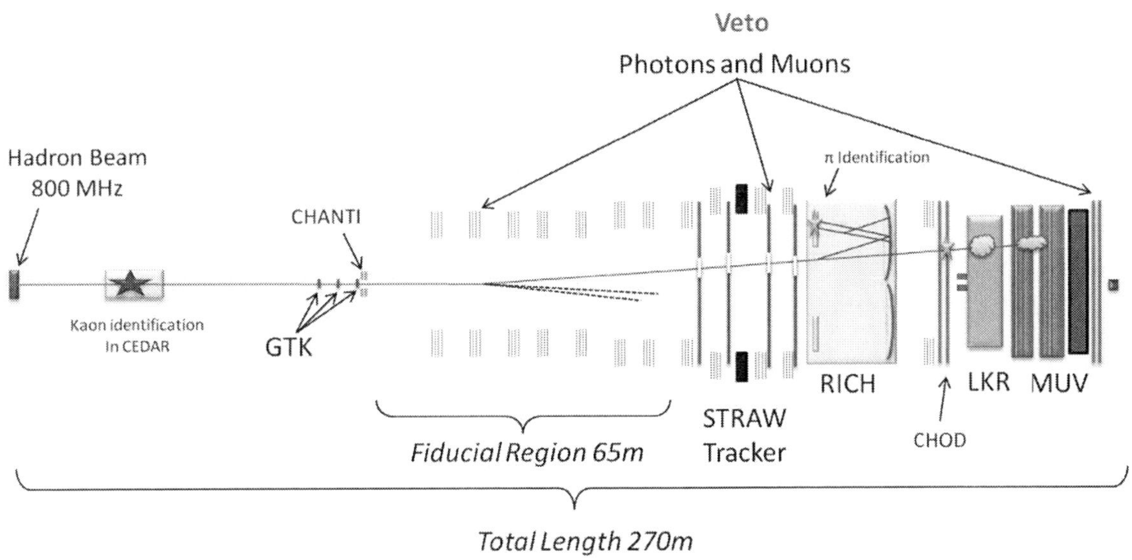

Figure 2. Schematic view of the NA62 experiment showing the main sub-detectors (not to scale). The main elements for the det ection of the K^+ decay products are located along a 170 m long region starting about ~ 100 m downstream of the beryllium target. Useful K^+ decays will be detected within a 65 m long fiducial decay region.

3 The NA62 detector

NA62 uses a decay-in-flight technique, differently from the kaon decay at rest approach which was at the basis of the previous experiments.

The presence of two undetectable neutrinos in the final state reduces the signal signature to one high momentum charged track with nothing else, which has to be discriminated against background coming from all other kaon decays. The high momentum of the incoming beam (75 GeV/c) improves the background rejection and sets the longitudinal scale of the experiment. To achieve the required background suppression different principles have to be combined and the resulting requirements are outlined here:

- High intensity and good timing: a high intensity kaon beam is essential in order to reach sensitivity to a branching ratio of $\sim 10^{-10}$. The incoming secondary beam from the SPS provides a particle rate of 750 MHz, containing about 6% of kaons delivering roughly $45 \cdot 10^6$ decays in the fiducial region per spill. Precise timing (in the range 100 - 150 ps) of the K^+ and the π^+ allows precise matching of the particles in the decay. The time resolution is essential to keep wrong associations below 1%.
- Low-mass tracking: track position and momenta have to be measured with high accuracy in low mass detectors. They are essential because inelastic scattering of beam particles in the detector material can produced an isolated π^+ appearing like a signal event and hence contributing to the background.
- Hermetic vetoing for photons and muons: the kinematic rejection must be accompanied by direct vetoing for photons (in particular for the $K^+ \to \pi^+\pi^0$ background) requiring a typical inefficiency of 10^{-4} for high energetic photons. About two third of K^+ decays contain muons in the final state, therefore a muon veto system is mandatory both in the trigger and off-line.
- Particle ID: several detectors must complement the event information with direct evidence on the particle species.

The experimental setup, shown in Figure 2, consists of a 100 m long beam line to select the appropriate secondary beam component produced by protons from the SPS CERN accelerator, followed by a 80 m long evacuated volume which defines the decay region. Along the beam line, different detectors are distributed as listed below.

3.1 Differential Cherenkov counter (KTAG)

The KTAG identifies the K^+ component in the beam with respect to the other beam particles by employing an upgraded differential Cherenkov counter (CEDAR). The CEDAR filled with nitrogen gas is placed in the incoming beam to positively identify the kaon component in the high rate environment. It is designed to identify particles of a specific mass by making the detector blind to the Cherenkov light produced by particles of different mass. The KTAG upgrade of NA62 acts in the Cherenkov light detection stage: to cope with the expected 45 MHz kaon rate, 384 photomultipliers (PMTs) grouped in 8 light boxes are placed behind the 8 annular slits. A preliminary analysis of 2014 data shows a kaon-identification efficiency better than 95% and a time resolution below 100 ps.

3.2 Beam tracker (GTK)

The GigaTracker (GTK) comprises three stations measuring time, direction and momentum of the beam particles before entering the decay region. The three stations are placed along the beam line just before the fiducial region. Each GTK station is a hybrid silicon micro-pixel detector with a total number of 18000 $300 \times 300\ \mu m^2$ pixels grouped in 10 read-out chips. The expected performances on the track measurement are a 0.4% resolution for the momentum, 16 mrad for the direction and 200 ps for the timing. One chip per station has been commissioned during the 2014 run.

3.3 Guard-Ring detector (CHANTI)

This detector is a guard-ring counter following the last GTK station to veto beam particles which scatter inelastically in the material of the third GTK station. It consists of six stations of x-y plastic scintillator bars surrounding the beam, covering an angle between 0.034 and 1.38 rad. The expected efficiency in vetoing signal-like events is 99%. Preliminary 2014 data analysis shows a time resolution of the order of 1 ns.

3.4 Pion spectrometer (STRAW)

The STRAW Tracker measures the coordinates and momentum of secondary charged particles originating from the decay region. To minimise multiple scattering the chambers are built of ultra-light material and are installed inside the vacuum tank. Two straw chambers are located on each side of a large aperture dipole magnet, providing a vertical B-field of 0.36 T. Each chamber is equipped with 1792 straw tubes which are arranged in four "views" providing measurements of four coordinates x, y, u, v. The requirements for the detector performances are a relative momentum resolution of 1%, a spatial resolution of 130 μm per coordinate and a very low track reconstruction inefficiency. The detector has been fully commissioned during the 2014 run.

3.5 Pion Timing (CHOD)

The CHOD is a Charged-particle HODoscope (inherited from NA48) to detect tracks with precise measurements of the arrival time and impact point, able to provide a fast signal used to drive the trigger. It is composed of two planes made of 64 plastic scintillators, one with vertical and one with horizontal slabs. The CHOD can provide the timing of charged decay products with a resolution of about 220 ps.

3.6 Ring image Cherenkov detector (RICH)

The RICH detector is placed downstream of the last straw chamber. It consists of a 17.5 m long cylindrical vessel with a diameter varying from 3.4 to 4 m filled with neon at atmospheric pressure and room temperature. An aluminum beam pipe spanning the length of the RICH keeps non-decayed beam particles in vacuum. The Cherenkov light is reflected by a mosaic of 18 hexagonal and 2 semi-hexagonal (close to the beam pipe) mirrors with 17 m focal length, made of 2.5 cm thick glass. They are placed at the downstream end of the vessel and they cover a total surface of about 3 m diameter. The reflected light is collected by 2×960 photomultiplier tubes (PMTs) with 18 mm diameter size and suitable quantum efficiency and timing performances. The PMTs are assembled in a compact hexagonal packing into two aluminum disks placed at the entrance window. Winston cones carved into the disks and covered with aluminized mylar are used to convey the light onto the active PMT area, thus increasing light collection. Quartz windows are used to separate the PMTs from the neon in order to avoid electrical discharges at the working voltage (900 V). The mirror segments are individually hanged on a light aluminum honeycomb structure and moved by means of two piezo-motors each, in order to align the light toward the two PMTs disks. The RICH allows an additional rejection of the muon background. The requested μ suppression factor is 10^{-2} between 15 and 35 GeV/c (the momentum range of interest), and a time resolution of 70 ps has been already measured during the 2014 run.

3.7 Large-Angle Photon Vetoes (LAV)

The LAV system provides full coverage for decay photons with polar angles from 8.5 to 50 mrad, with an inefficiency of $10^{-3} \div 10^{-4}$ on photons down to 150 MeV. It consists of 12 stations, whose diameter increases with distance from the target, containing 4 or 5 layers of azimuthally staggered lead glass blocks, for a total of 2500 channels. The time resolution observed in 2014 run is about 1 ns.

3.8 Small-Angle Photon Vetoes (IRC and SAC)

A Small Angle Calorimeter (SAC) and an Intermediate Ring Calorimeter (IRC) give photon coverage below 1 mrad. Both detectors are made of consecutive lead and plastic scintillator plates. The inefficiency for 1 GeV photons is below 10^{-4}. A time resolution of 3 ns was measured in the 2014 run.

3.9 Quasi-homogeneous liquid kripton calorimeter (LKr)

The LKr Calorimeter (inherited from NA48) measures precisely the electro-magnetic energy with the possibility to discriminate between positrons and pions using their shower properties. It covers intermediate angle between 1 and 8.5 mrad. It is divided in 13248 cells and the electromagnetic showers are fully detected through ionisation of low energy charged particles. The reconstructed shower time resolution is 500 ps, and the space resolution of the order of 1 mm; the total inefficiency for 10 GeV photons is below 10^{-5}. The new readout has been validated during the 2014 run.

3.10 Hadron and muon detectors (MUV)

NA62 uses a fast scintillator (MUV3) for direct muon vetoing and 2 sampling hadron calorimeters (MUV1+2) which measure the deposited hadron energy in the event, distinguishing hadron showers from muons. MUV1 and MUV2 modules are iron-scintillator sandwich calorimeters with scintillator strips alternately oriented in the horizontal and vertical directions. This system supplements and provides redundancy with respect to the RICH in the detection and rejection of muons. Only one of these was installed in 2014. The MUV3 is a vertical array of fast scintillator tiles placed after an iron wall and is used to detect non-showering muons. Each tile is read by 2 PMTs to avoid the effect of Cherenkov photons on time resolution. Data from the 2014 run show a time resolution below 500 ps.

3.11 Trigger and data acquisition

The experiment makes use of an integrated trigger and data acquisition system with three trigger levels. The lowest level, level 0, is implemented directly in the digital readout card for each detector subsystem. The detector hits are resolved into quantities that are used in trigger logic to decide which events will be read out for level 1. Level 0 will process about 10 MHz of detector hits; about 1 MHz of events will be sent to level 1. The level 1 trigger is implemented in software. It is the first asynchronous trigger level and will reduce the rate of events seen by level 2 by an order of magnitude applying specific requests on the data collected by each detector. The level 2 trigger is implemented in the event builder running on the acquisition PC farm; it is the first trigger level which makes use of the full event information. The O(100 kHz) of events input to level 2 are reduced to a few kHz of events ultimately written to disk.

4 Analysis strategy

NA62 detectors are designed to measure the K^+ and the secondary particles from kaon decays occurring in the decay volume. The signature of the signal is one track in the final state matched with one K^+ track in the beam. The measurable quantities are $\vec{p}_{K^+}$ (the kaon momentum-vector), $\vec{p}_{\pi^+}$ (the pion momentum-vector) and the angle $\theta_{\pi K}$ (angle between the pion and the kaon).

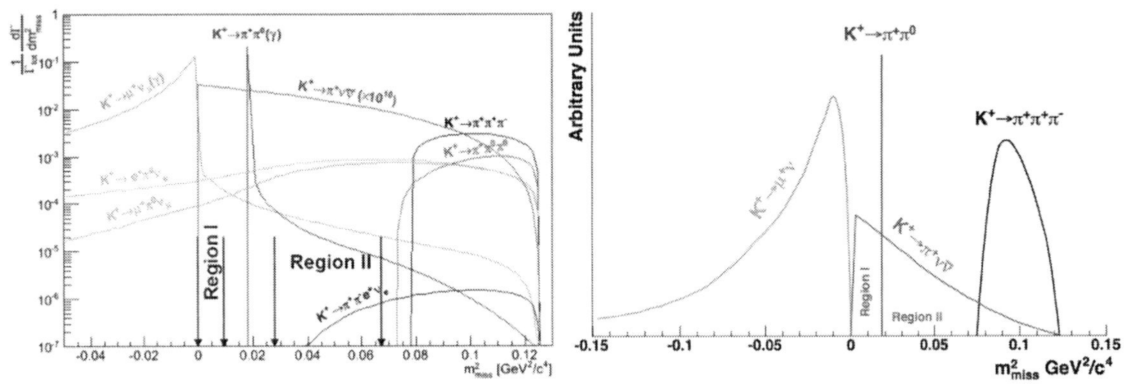

Figure 3. Left: Distribution of m^2_{miss} for the main kaon decays and the signal, with the hypothesis that the charged decay product is a pion. The two signal regions, $[0, 0.01]GeV^2/c^4$ and $[0.026, 0.068]GeV^2/c^4$, are indicated. Right: Distribution of m^2_{miss} for the signal and the main background sources after kinematic cuts.

The main kinematic variable to discriminate the signal from the overwhelming background is the squared missing mass m^2_{miss}. It is defined for the events, when only one charged track from Kaon decay is detected assuming the track belongs to a charged pion:

$$m^2_{miss} = (p_{K^+} - p_{\pi^+}). \qquad (1)$$

Figure 3 (left) shows the m^2_{miss} distribution in log scale for the main decay channels and the $K^+ \to \pi^+ \nu\bar{\nu}$ decay (red curve). The signal decay differential width is magnified by 10^{10}. Two signal regions (I and II), on each side of the $K^+ \to \pi^+\pi^0$ peak, are chosen, where more than 90% of the main K^+ decays do not contribute. The kinematics background suppression is of the order of 10^5. The further suppression of background at the order of 10^7 is done by means of the detector related cuts (timing, particle identification, muon and photon veto).

Table 1. Expected signal and background from K^+ decays estimated from NA62 sensitivity studies.

Decay	event/year
$K^+ \to \pi^+ \nu\bar{\nu}$	45
$K^+ \to \pi^+\pi^0$	5
$K^+ \to \pi^+\pi^+\pi^-$	1
$K^+ \to \pi^+\pi^- e^+\nu_e$	< 1
$K^+ \to \mu^+\nu_\mu\gamma$	1.5
other rare decays	0.5
Total background	< 10

Table 1 shows the expected number of signal and background events per year of data taking estimated from MC simulation. Given this expectation, NA62 will be able to reach a 10% precision in a measurement of $BR(K^+ \to \pi^+\nu\bar{\nu})$. The main background remains the $K^+ \to \pi^+\pi^0(\gamma)$ decay, where two photons from π^0 decay are not detected and the signal event topology is mimicked. To suppress this background very good photon veto ability is required. The aimed π^0 suppression factor is 10^8. The second important background comes from the $K^+ \rightarrow \mu^+\nu_\mu(\gamma)$ decay. The designed overall suppression factor for muons is of the order of 10^5.

In the analysis, the π^+ momentum will be required to be less than 35 GeV/c. In this way, the momentum of the π^0 amounts to at least 40 GeV/c. Such a large energy deposit can hardly be missed in the calorimeters. In addition only events with the K^+ vertex reconstructed in the fiducial volume are selected. The m^2_{miss} distribution for signal and the three main background sources after the kinematic cuts is shown in Figure 3 (right).

5 2014 pilot run

Data recorded during the pilot run in 2014 are very successfully used to understand the performance of the NA62 sub-detectors. However, only a small part of it was recorded with stable run conditions and is useful for preliminary physics studies. The beam intensity was 5% of the nominal one and two triggers have been used to collect a minimum bias sample and a $\pi\nu\bar{\nu}$-like without photon rejection sample. Only about 1% of the available data have been studied. The experiment was at a preliminary stage: data were not yet readout from GTK and the analysis used the mean nominal direction and momentum of the kaon beam; LAV information was not yet used for photon veto; only preliminary detector time alignment and energy calibration were used.

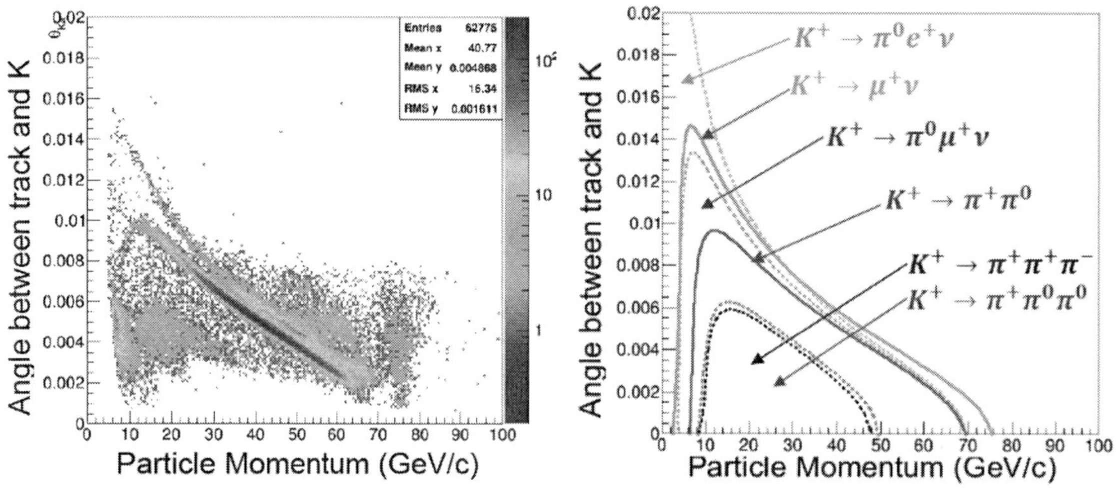

Figure 4. Angle between the reconstructed track and the nominal kaon direction as a function of particle momentum [GeV /c] in 2014 data (left) and as expected from pure kinematics (right).

The distribution of the angle between the kaon and the secondary track as a function of the track momentum is shown in Figure 4 (left), for events with only 1 track in the spectrometer and after requiring the presence of a kaon in the KTAG. A reasonable agreement can be observed if compared with the expectations for the main kaon decays (right): the suppression of the $K^+ \rightarrow \mu^+\nu$ component is due to the muon rejection at trigger level. The intersection between the track and the nominal beam direction is used to reconstruct the vertex and suppresses the background from kaon interactions.

After requiring the vertex position to be reconstructed in the fiducial decay region, and selecting the track momentum between 15 and 35 GeV/c, the squared missing mass distribution shown in Figure 5 is obtained. The $K^+ \rightarrow \mu^+\nu$ background is well suppressed with the help of RICH and muon veto

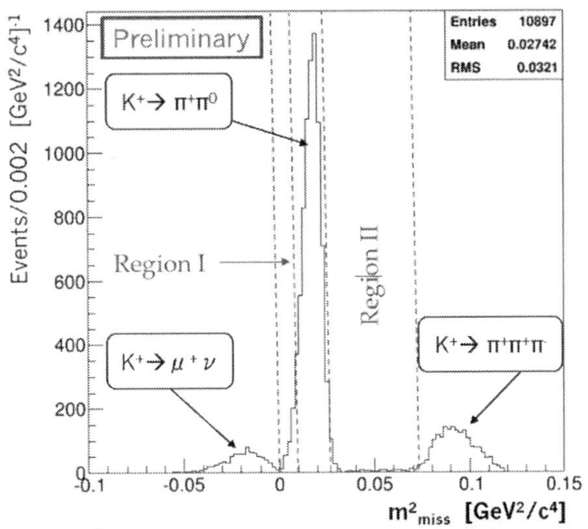

Figure 5. the m^2_{miss} distribution after the signal event selection. Two regions for the final selection of signal events are indicated by the vertical red-dashed lines. The kaon decays contributing to the distribution are indicated.

system. The muon background suppression is expected to be improved during the 2015 physics run with the help of MUV1. The three-track background, mainly coming from the $K^+ \to \pi^+\pi^+\pi^-$ decay is also well suppressed and does not contribute to the signal selected region. The main background contribution comes from the $K^+ \to \pi^+\pi^0$ decay. This particular background dominates in the signal region after the final selection, especially in region I. LAV information is not yet exploited both at trigger level and analysis level, preventing an additional suppression of $K^+ \to \pi^+\pi^0$ while the use of mean beam momentum and direction, instead of event by event GTK information, worsens the m^2_{miss} resolution.

References

[1] na62.web.cern.ch/na62/Documents/TechnicalDesign.html, 2010.
[2] J. Brod, M. Gorbahn and E. Stamou, Phys. Rev. D 83, 034030 (2011).
[3] D. M. Straub, arXiv:1012.3893 [hep-ph].
[4] T. Hurth, G. Isidori, J. F. Kamenik and F. Mescia, Nucl. Phys. B 808, 326 (2009).
[5] G. Isidori, F. Mescia, P. Paradisi, C. Smith and S. Trine, JHEP 0608, 064 (2006).
[6] A. J. Buras, B. Duling, T. Feldmann, T. Heidsieck and C. Promberger, JHEP 1009, 104 (2010).
[7] M. Blanke, A. J. Buras, B. Duling, S. Recksiegel and C. Tarantino, Acta Phys. Polon. B 41, 657 (2010) [arXiv:0906.5454 [hep-ph]].
[8] M. Blanke, A. J. Buras, B. Duling, K. Gemmler and S. Gori, JHEP 0903, 108 (2009) [arXiv:0812.3803 [hep-ph]].
[9] E949 COLLABORATION; ARTAMONOV, A.V. ET AL. Phys. Rev. Lett. 101 191802, 2008.
[10] E391a Collaboration (J.K. Ahn et al.) Phys.Rev. D81 (2010) 072004.

DOI: 10.1051/epjconf/201612604037

Search for the Standard Model Higgs boson produced in association with top quarks with the ATLAS detector

Yang Qin[1,a] On behalf of the ATLAS Collaboration

[1] *University of Manchester*

Abstract. After the discovery of a new particle consistent with the Higgs boson in 2012, the direct observation and measurement of the coupling of this particle to top quarks has become increasingly important. In this talk, a review of the latest ATLAS results on the search for the Higgs boson produced in association with top quarks, $t\bar{t}H$, is presented. In particular, three analyses aiming at different $t\bar{t}H$ final states are reviewed. These include a search with the Higgs boson decaying to $b\bar{b}$, a search in multileptonic final states and a search with $H \to \gamma\gamma$. The results from these analyses are combined, giving a $t\bar{t}H$ signal strength of 1.8 ± 0.8 and excluding a $t\bar{t}H$ signal 3.2 larger than the SM expectation at 95% confidence level for a SM Higgs boson with a mass of 125.36 GeV.

1 Introduction

The discovery of a new particle in consistency with the Standard Model (SM) [1–3] Higgs boson [4–7] was reported by the ATLAS [8] and the CMS [9] collaborations in July 2012. Ever since the discovery, the properties of the scalar boson have been extensively studied. One of the most important aspects is to measure the coupling of the Higgs boson to fermions, i.e. the Yukawa coupling. As the top quark is the heaviest fermion, it gives the strongest Yukawa coupling with a coupling constant at the order of one. The measurement of the top-Higgs Yukawa coupling, y_t, is of particular importance for both understanding the SM electroweak symmetry breaking and testing Beyond Standard Model (BSM) theories. With a value close to unity, y_t is also expected to provide insight into the scale of new physics [10].

The $t\bar{t}H$ process, shown in Figure 1(a) and 1(b), provides a direct probe to y_t, while other processes are only sensitive through loop effects, for example the gluon-gluon fusion shown in Figure 1(c). However, the small production rate of the $t\bar{t}H$ process makes the relevant search very challenging.

Another process that is sensitive to the top Yukawa coupling is the Higgs production associated with a single top quark, tH [1]. As an example, the Feynman diagrams for the t-channel tH production are shown in Figure 2, where the Higgs boson can be emitted from either the top quark or the W boson. The SM predicts an almost perfect destructive interference between the two processes shown in Figure 2(a) and Figure 2(b), resulting in a suppressed total tH cross-section. The tH production therefore is sensitive not only to the magnitude of y_t, but also to the relative sign of y_t and g_{HWW}, which represents the coupling between the Higgs boson and the W boson. In BSM theories, y_t and g_{HWW} can have an anomalous correlation, leading to a boosted inclusive tH production cross-section.

[a]e-mail: yang.qin@cern.ch

[1]For simplicity, tH also refers to $\bar{t}H$ throughout these proceedings.

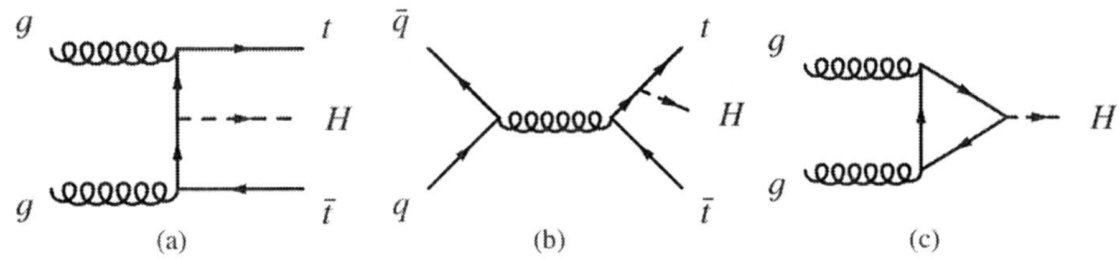

Figure 1. The Feynman diagrams of (a,b) tree level $t\bar{t}H$ processes demonstrating a direct coupling of the top quark and the Higgs boson and (c) the Higgs production in gluon-gluon fusion with a top-Higgs coupling at the loop level.

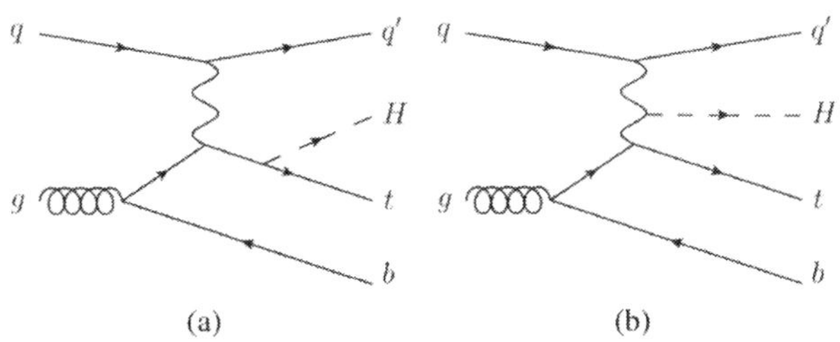

Figure 2. The Feynman diagrams of the t-channel tH production, where the Higgs boson can be emitted from either (a) the top quark or (b) the W boson [11].

The $t\bar{t}H$ process produces complex final states, since both the $t\bar{t}$ pair and the Higgs boson can undergo multiple possible decays, which in combination provides a variety of final states. In these proceedings, three analyses recently performed by the ATLAS experiment probing different $t\bar{t}H$ final states are reviewed. All three analyses discussed in these proceedings were designed to be orthogonal, and contributed to the combined Higgs boson property measurement described in Ref. [12].

2 Search for $t\bar{t}H$ with $H \to b\bar{b}$ at $\sqrt{s}$ = 8 TeV

The dominant decay mode for a Higgs boson with a mass around 125 GeV is $H \to b\bar{b}$, with a branching ratio of approximately 58%. However, it usually suffers from a large background associated with heavy flavour jets, especially the $t\bar{t}$+$b\bar{b}$ background which has the same final states as the $t\bar{t}H$ signal. In this analysis, two channels based on the decay of the $t\bar{t}$ pair are included, the single lepton channel and the dilepton channel. In the single lepton channel one of the top quarks decays leptonically, $t \to Wb \to l\nu b$, while the other decays hadronically, $t \to Wb \to q\bar{q}'b$ [2]. In the dilepton channel, both the top and anti-top quark decay leptonically. The lepton in both cases exclusively refers to an electron or muon given the complexity introduced by hadronically decaying taus.

Since this analysis is heavily dominated by background, the selected events are categorised into different regions defined by the number of jets (j) and number of b-tagged jets (b) in order to improve the search sensitivity. In the single lepton channel, nine independent regions are considered: (4j, 2b), (4j, 3b), (4j, 4b), (5j, 2b), (5j, 3b), (5j, ≥4b), (6j, 2b), (6j, 3b) and (6j, ≥4b). In the dilepton channel, six regions are considered: (2j, 2b), (3j, 2b), (3j, 3b), (≥4j, 2b), (≥4j, 3b) and (≥4j, ≥4b). As is shown in Figure 3, the signal-to-background ratios S/B and the sensitivities, represented by $S/\sqrt{B}$, are very low in all regions, where S and B refer to the expected signal for a Higgs boson with m_H = 125 GeV and the total background respectively. The regions with $S/B > 1\%$ and $S/\sqrt{B} > 0.3$ are referred to as "signal-rich regions", The remaining regions are referred to as "signal-depleted regions", and are used to constrain systematic uncertainties on the background processes.

[2]Charge conjugate states are implied here.

 DOI: 10.1051/epjconf/201612604037

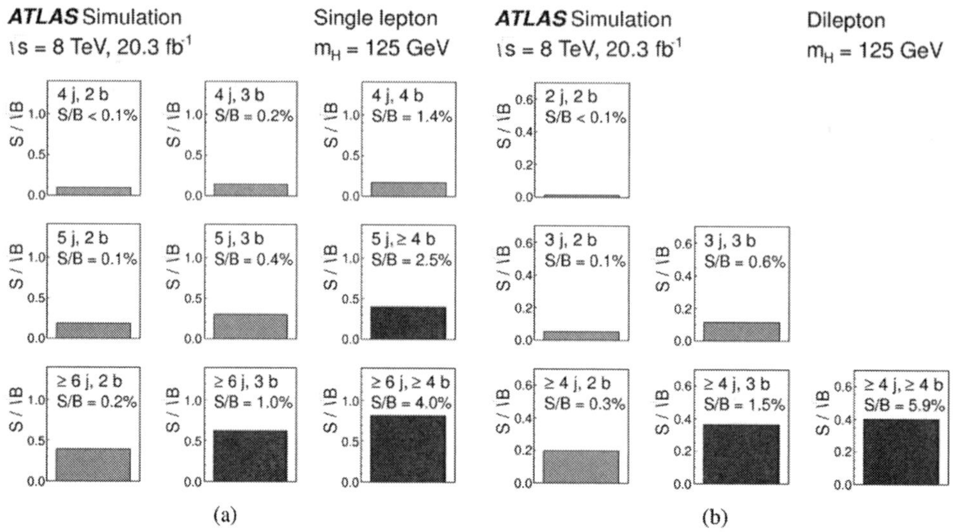

Figure 3. The S/B and $S/\sqrt{B}$ in each region considered in (a) the single lepton channel and (b) the dilepton channel. The "signal-rich regions" are shown in red, and the "signal-depleted regions" in blue [13].

The dominant background in this analysis is $t\bar{t}$+jets, which has been decomposed into $t\bar{t}$+light, $t\bar{t}+c\bar{c}$ and $t\bar{t}+b\bar{b}$. In the signal-rich regions the largest component is the $t\bar{t}+b\bar{b}$ background. Therefore the understanding of this background is vital to this analysis. The modelling of the $t\bar{t}+b\bar{b}$ background is done using an inclusive $t\bar{t}$ sample generated with Powheg+Pythia6 [14–17]. However, since Powheg+Pythia only models $t\bar{t}+b\bar{b}$ with parton shower, it is then reweighted to a fully matched next-to-leading order (NLO) prediction with massive b quarks [18], which generates $t\bar{t}+b\bar{b}$ processes at NLO following the four-flavour scheme using the Sherpa+OpenLoops framework [19, 20].

A simultaneous profiled likelihood fit is performed across all regions. This allows the impact of the systematic uncertainties on the search sensitivity to be reduced by exploiting the constraining power provided by the high statistics in the background-dominated regions. Given the extremely small signal expectation, the analysis uses Neural Networks (NN) to exploit the separation between signal and background in each signal-rich region. Various categories of variables are considered in the construction of the NN, including object kinematics, global event variables, event shape variables and object pair properties. In addition to the kinematic variables, two variables calculated by the Matrix Element Method (MEM) [21] are used to construct the NN in the single lepton channel, which carry significant separation power into the NN discriminants. These two variables are the Neyman-Pearson likelihood ratio and the logarithm of the summed signal likelihoods. A maximum of twelve (ten) variables are used to build the NN in the single lepton (dilepton) channel. An additional NN is built in the (3j, 3b) region in the dilepton channel despite of its low $S/\sqrt{B}$, because its relatively high S/B ratio can further improve the sensitivity. Furthermore, the (5j, 3b) region in the single lepton channel has a significant amount of both $t\bar{t}$+light and $t\bar{t}$+HF (HF for "heavy flavour") background, where $t\bar{t}$+HF refers inclusively to $t\bar{t}+c\bar{c}$ and $t\bar{t}+b\bar{b}$. A dedicated NN is built in this region to separate the $t\bar{t}$+light and $t\bar{t}$+HF background. This provides strong constraints on the normalisation of the $t\bar{t}$+HF background. In other background-dominated regions where the sensitivity is low, $H_{\mathrm{T}}^{\mathrm{had}}$ and H_{T} are used as the discriminant variables respectively in the single lepton channel and the dilepton channel, where H_T^{had} is defined as the scalar sum of the jet p_T [3], and H_T is the scalar sum of the jet and lepton p_T.

The post-fit NN distribution in the most sensitive region in the two channels are shown in Figure 4. A reasonable agreement between data and the prediction as well as the separation power provided by the NN between the total background and the $t\bar{t}H$ signal are presented in both channels. The results of the statistical analysis are summarised in Figure 5. The best-fit signal strength is μ (m_H = 125 GeV) = 1.5 ± 1.1, shown in Figure 5(a). The observed (expected) significance of the signal is 1.4 (1.1) standard deviations, corresponding to an observed (expected) p-value of 8 (15) %. Figure 5(b) shows the observed upper limits on the $t\bar{t}H$ signal strength and those expected with and without assuming a SM Higgs boson with m_H = 125 GeV. Using the CL$_s$ method [22–24], an observed (expected) 95% confidence level (CL) upper limit on the signal strength of 3.4 (2.2) is obtained. The expected upper limit on the signal strength with the presence of a SM Higgs boson is found to be 3.1.

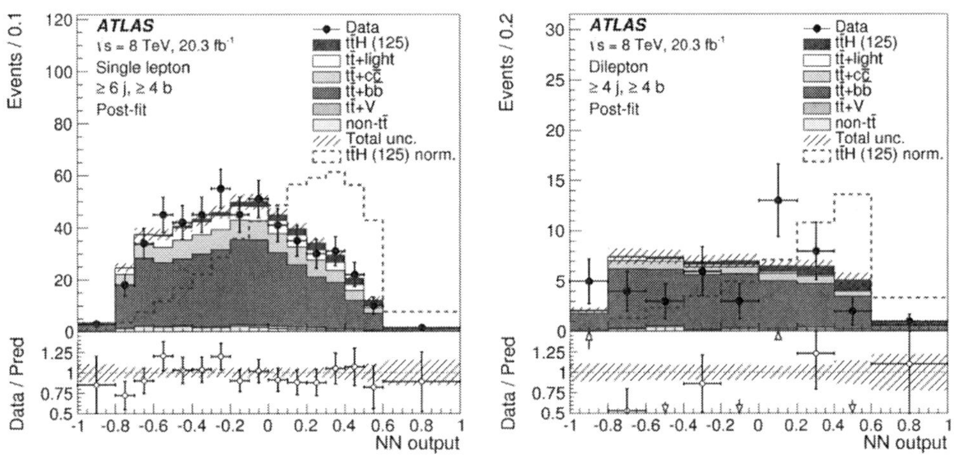

(a) (b)

Figure 4. The post-fit NN distribution in (a) the (6j, ≥4b) region in the single lepton channel and (b) the (≥4j, ≥4b) region in the dilepton channel [13].

3 Search for $t\bar{t}H$ with multilepton final states at $\sqrt{s}$ = 8 TeV

This analysis is primarily sensitive to the $H \to WW^*$, $H \to ZZ^*$ and $H \to \tau\tau$ decays, with one or two leptons from the decay of the $t\bar{t}$ pair. A total of five multileptonic final states are considered, with different sensitivities on the three Higgs decay modes. These include the final states with two same-charge-sign (SS) light leptons (e or μ) with no additional hadronically decaying τ lepton (denoted as τ_{had}), three light leptons, two SS light leptons with one τ_{had}, four light leptons and one lepton with two τ_{had} candidates. The five final states with their corresponding fraction of expected $t\bar{t}H$ signal yields from different Higgs decay modes are listed in Table 1.

[3] ATLAS uses a right-handed coordinate system with its origin at the nominal interaction point (IP) in the centre of the detector and the z-axis coinciding with the axis of the beam pipe. The x-axis points from the IP to the centre of the LHC ring, and the y-axis points upward. Cylindrical coordinates (r, ϕ) are used in the transverse plane, ϕ being the azimuthal angle around the beam pipe. The pseudorapidity is defined in terms of the polar angle θ as $\eta = -\ln[\tan(\theta/2)]$. Transverse momentum and energy are defined as $p_T = p\sin\theta$ and $E_T = E\sin\theta$, respectively.

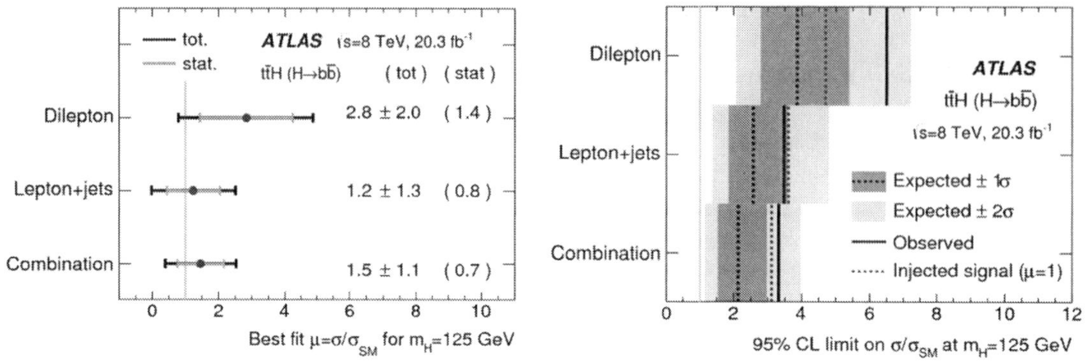

(a) (b)

Figure 5. The (a) fitted values of the $t\bar{t}H$ signal strength and their uncertainties and (b) 95% CL upper limits on $\sigma(t\bar{t}H)$ relative to the SM prediction, $\sigma/\sigma_{\mathrm{SM}}$, for the individual channels and their combination [13].

Table 1. Fraction of expected $t\bar{t}H$ signal from different Higgs decay modes in each of the five final states included in the analysis [25].

	Higgs boson decay mode			
Category	WW^*	$\tau\tau$	ZZ^*	Other
$2\ell 0\tau_{\mathrm{had}}$	80%	15%	3%	2%
3ℓ	74%	15%	7%	4%
$2\ell 1\tau_{\mathrm{had}}$	35%	62%	2%	1%
4ℓ	69%	14%	14%	4%
$1\ell 2\tau_{\mathrm{had}}$	4%	93%	0%	3%

The observed and expected number of jets spectrum in each signal region of the five categories are shown Figure 6. In general, a reasonable agreement between data and prediction is presented, whereas in $2\ell 1\tau_{\mathrm{had}}$ (6(c)) and 4ℓ (6(d)) categories only one event is observed. As can be seen from the histograms, the leading background is from vector boson production in association with top quark pairs, $t\bar{t}V$, which includes $t\bar{t}Z$ and $t\bar{t}W$. Another dominant background is $t\bar{t}$ with non-prompt leptons, including light leptons from the decay of heavy flavour hadrons and, in the categories with τ_{had}, misidentified τ_{had}.

Figure 7 shows the best-fit values of the $t\bar{t}H$ signal strength for a Higgs with $m_H = 125$ GeV extracted from the individual categories and their combination. The best-fit signal strength given by the combination is $\mu = 2.1^{+1.4}_{-1.2}$. The observed (expected) significance of signal under the background only hypothesis ($\mu = 0$) is 1.8 (0.9) standard deviations, corresponding to a p-value of 3.7 (18) %. This compares to an observed significance of 0.9 standard deviations assuming the presence of the SM $t\bar{t}H$ signal, corresponding to a p-value of 18%. Using the CL$_s$ method, the observed (expected) 95% CL upper limit on the signal strength is found to be 4.7 (2.4).

4 Search for $t\bar{t}H$ with $H \to \gamma\gamma$ and constraints on y_t at 7 TeV and 8 TeV

The search in the $H \to \gamma\gamma$ decay channel is a particular challenge, as this channel has an extremely small branching ratio. For a Higgs with a mass around 125 GeV, the branching ratio is BR($H \to \gamma\gamma$) = 2.28×10^{-3}. But the search in this channel benefits from the good diphoton invariant mass ($m_{\gamma\gamma}$)

DOI: 10.1051/epjconf/201612604037

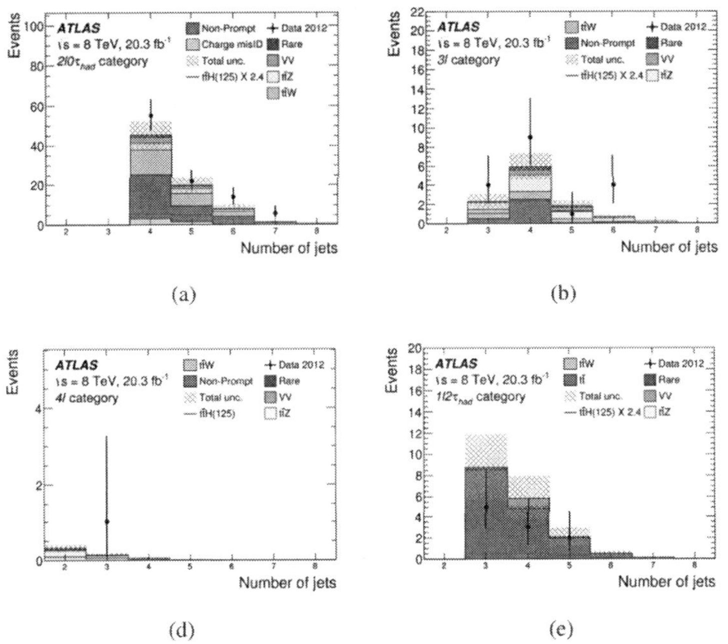

Figure 6. The observed and expected number of jets spectrum in each signal region [25].

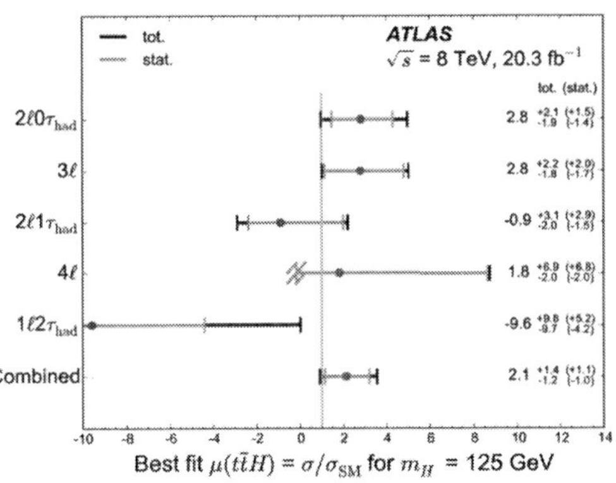

Figure 7. Best-fit values of the signal strength in each of the fitted categories. the truncated lower uncertainty for the $4l$ category is due to the negative expected total yield given by a non-physical signal strength of $\mu < -0.17$ [25].

resolution, which is estimated to be approximately 1.5 GeV. Additionally, the diphoton background has a continuously decreasing $m_{\gamma\gamma}$ spectrum, which can be constrained by sideband data. Hence, there is no need to rely on the theoretical assumptions on the modelling of the background.

Apart from the search for the $t\bar{t}H$ production, this analysis also provides constraints on the value of y_t. A scale factor, κ_t, is introduced to describe the relation between y_t and its expected value in the SM: $y_t = \kappa_t y_t^{\mathrm{SM}}$. The variation of κ_t affects not only the $t\bar{t}H$ and tH production rate, but also the $H \to \gamma\gamma$ branching ratio and other Higgs boson production processes [26]. Figure 8 illustrates the dependence of the $t\bar{t}H$ and the tH production cross section and the BR($H \to \gamma\gamma$) on the value of κ_t. The value of $\kappa_t = 1$ represents the SM. The tH production has a minimum around the SM expectation given the destructive interference between y_t and g_{HWW}. For $\kappa_t = 0$, the $t\bar{t}H$ production is turned off, and the contribution from y_t to the tH production and BR($H \to \gamma\gamma$) vanishes, resulting in an increase in their values due to the removal of the destructive interference. For $\kappa_t < 0$, a further increase can be seen, because the interference between the contribution of top quarks and W bosons becomes constructive.

The minimum of BR($H \to \gamma\gamma$) appears at $\kappa_t = +4.7$ due to the cancellation between the contribution from top quarks and W bosons to the loop-induced $H \to \gamma\gamma$ decay.

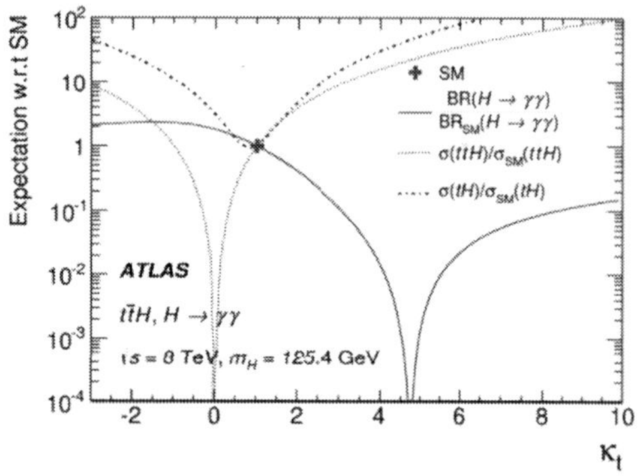

Figure 8. The $t\bar{t}H$ and tH production cross section as well as the $H \to \gamma\gamma$ branching ratio divided by their SM expectations as a function of κ_t [11].

This analysis was preformed on both 7 and 8 TeV data. The selected events in this analysis are categorised based on the decay of the $t\bar{t}$ pair: the leptonic category has at least one lepton; the hadronic category has no lepton in the final state. The analysis searches for a local event excess in the $m_{\gamma\gamma}$ window of (120, 130) GeV. An unbinned signal-plus-background likelihood fit is performed in the range of (105, 160) GeV. Figure 9 shows the fitted distribution compared to the data at 8 TeV for both the leptonic and the hadronic category. In the $m_{\gamma\gamma}$ range of (120, 130) GeV, five candidate events are observed in total. The expected signal yield is 1.3 events, while the total background is estimated to be $4.6^{+1.3}_{-0.9}$ events.

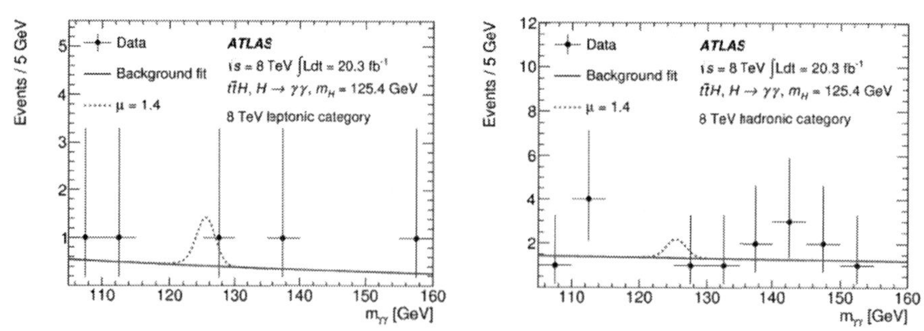

(a) (b)

Figure 9. The $m_{\gamma\gamma}$ distribution for (a) the leptonic channel and (b) the hadronic channel for data at 8 TeV. An unbinned signal-plus-background likelihood fit to the full spectrum is used to estimate the continuum background (dashed line) and the SM Higgs boson production with $m_H = 125.4$ GeV [11].

The best-fit signal strength for $t\bar{t}H$ is $\mu_{t\bar{t}H}$ ($m_H = 125.4$ GeV) $= 1.3^{+2.5}_{-1.7}$(stat.)$^{+0.8}_{-0.4}$(syst.), assuming SM values for all other $H \to \gamma\gamma$ contribution. The results interpreted as 95% CL limits for a Higgs boson with $m_H = 125.4$ GeV are summarised in Figure 10. The observed and expected upper limits on the $t\bar{t}H$ signal strength times BR($H \to \gamma\gamma$) are shown in Figure 10(a). In combination the observed (expected) limit is 6.7 (4.9). The limits as a function of κ_t are shown in Figure 10(b), where the continuum background plus the SM Higgs boson production ($\kappa_t = +1$) is used as the alternative hypothesis. The shape of the distribution is a result of the combined dependence of the $t\bar{t}H$ and tH

production cross section and BR($H \to \gamma\gamma$), as is shown in Figure 8. The observed (expected) lower and upper limits on κ_t are -1.3 and +8.0 (-1.2 and +7.8).

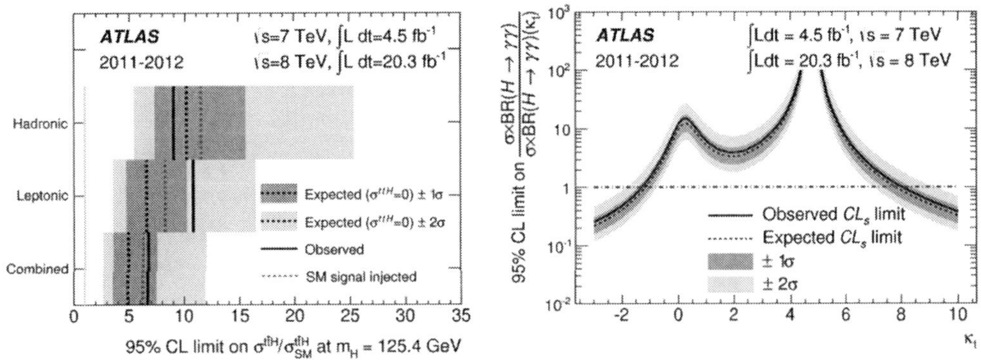

(a) (b)

Figure 10. The observed and expected 95% CL limits on (a) the $t\bar{t}H$ signal strength times BR($H \to \gamma\gamma$) and (b) κ_t (using $\kappa_t = +1$ as the alternative hypothesis) for a Higgs boson with $m_H = 125.4$ GeV [11].

5 Summary

The ATLAS Collaboration has performed searches for the SM Higgs boson produced in association with top quarks in various final states and have exploited constraints on the top Yukawa coupling, y_t. All results so far are consistent with the SM prediction. A combined measurement of the Higgs boson production, decay and coupling has been performed, taking into account the results from the three analyses reviewed in these proceedings, with minor modifications to the analyses to have a consistent SM Higgs boson mass of 125.36 GeV and treatment against the tH processes. The combined $t\bar{t}H$ signal strength is measured to be 1.8 ± 0.8. A $t\bar{t}H$ signal 3.2 larger than the expectation of the SM prediction has been excluded. In LHC Run 2 starting from June 2015, a much higher $t\bar{t}H$ production rate is expected given the increased luminosity and the collision energy of 13 TeV. Moreover, the relative increase of the $t\bar{t}H$ production cross section with respect to that of the background $t\bar{t}$ process is expected to be 3 times larger, which significantly improves the search sensitivity. Therefore, the observation and measurement of top Yukawa coupling in Run 2 is highly promising.

References

[1] S. L. Glashow, Nucl. Phys. **22** (1961) 579.
[2] S. Weinberg, Phys. Rev. Lett. **19** (1967) 1264–1266.
[3] A. Salam, Proc. of the 8th Nobel Symposium (1969) 367.
[4] F. Englert and R. Brout, Phys. Rev. Lett. **13** (1964) 321.
[5] P. W. Higgs, Phys. Rev. Lett. **13** (1964) 508.
[6] P. W. Higgs, Phys. Lett. **12** (1964) 132.
[7] G. Guralnik, C. Hagen, and T. Kibble, Phys. Rev. Lett. **13** (1964) 585.
[8] ATLAS Collaboration, Phys. Lett. **B716** (2012) 1, `arXiv:1207.7214 [hep-ex]`.

[9] CMS Collaboration, Phys. Lett. **B716** (2012) 30, arXiv:1207.7235 [hep-ex].
[10] F. Bezrukov and M. Shaposhnikov, arXiv:1411.1923 [hep-ph].
[11] ATLAS Collaboration, Phys. Lett. **B740** (2015) 222–242, arXiv:1409.3122 [hep-ex].
[12] ATLAS Collaboration, submitted to Eur. Phys. J. C (2015), arXiv:1507.04548 [hep-ex].
[13] ATLAS Collaboration, Eur. Phys. J. **C75** no. 7, (2015) 349, arXiv:1503.05066 [hep-ex].
[14] P. Nason, JHEP **11** (2004) 040, arXiv:hep-ph/0409146.
[15] S. Frixione, P. Nason, and C. Oleari, arXiv:0709.2092 [hep-ph].
[16] S. Alioli, P. Nason, C. Oleari, and E. Re, JHEP **06** (2010) 040, arXiv:1002.2581 [hep-ph].
[17] T. Sjöstrand, S. Mrenna, and P. Skands, JHEP **05** (2006) 026, arXiv:hep-ph/0603175.
[18] F. Cascioli, P. Maierhöfer, N. Moretti, S. Pozzorini, and F. Siegert, Phys. Lett. **B734** (2014) 210–214, arXiv:1309.5912 [hep-ph].
[19] T. Gleisberg et al., JHEP **0902** (2009) 007, arXiv:0811.4622 [hep-ph].
[20] F. Cascioli, P. Maierhöfer, and S. Pozzorini, Phys. Rev. Lett. **108** (2012) 111601, arXiv:1111.5206 [hep-ph].
[21] K. Kondo, J.Phys.Soc.Jap. **57** (1988) 4126–4140.
[22] T. Junk, Nucl. Instr. Meth. A **434** (1999) 435, arXiv:hep-ex/9902006.
[23] A. L. Read, J. Phys. G **28** (2002) 2693.
[24] G. Cowan, K. Cranmer, E. Gross, and O. Vitells, Eur. Phys. J. C **71** (2011) 1554, arXiv:1007.1727 [physics.data-an].
[25] ATLAS Collaboration, Phys. Lett. **B749** (2015) 519–541, arXiv:1506.05988 [hep-ex].
[26] LHC Higgs Cross Section Working Group Collaboration, J. R. Andersen et al., arXiv:1307.1347 [hep-ph].

EPJ Web of Conferences **126**, 04038 (2016) DOI: 10.1051/epjconf/201612604038
ICNFP 2015

Measurements of the properties of the Higgs boson using the ATLAS detector

Michaela Queitsch-Maitland[1,a], On behalf of the ATLAS Collaboration.

[1] *Department of Physics & Astronomy, University of Manchester, Oxford Road, Manchester, M13 9PL, UK.*

Abstract. Measurements of the properties of the Higgs boson with the ATLAS detector are presented. The spin, parity and tensor couplings of the Higgs boson in the $H \to \gamma\gamma$, $H \to ZZ^* \to 4\ell$ and $H \to WW^* \to e\nu\mu\nu$ channels are investigated. The results are based on pp collision data collected by the ATLAS detector corresponding to an integrated luminosity of 4.7 fb^{-1} at centre-of-mass energy $\sqrt{s} = 7$ TeV, and 20.3 fb^{-1} at $\sqrt{s} = 8$ TeV. Measurements of the fiducial and differential cross sections for Higgs boson production in the $H \to \gamma\gamma$ and $H \to ZZ^* \to 4\ell$ are also presented using 20.3 fb^{-1} of $\sqrt{s} = 8$ TeV data. The cross section measurements in the $H \to \gamma\gamma$ channel are then used to set limits on anomalous Higgs boson interactions in an effective Lagrangian. All results are consistent with expectations for a Standard Model Higgs boson.

1 Introduction

The discovery of a Higgs boson [1, 2] in 2012 by the ATLAS [3] and CMS [4] Collaborations at the Large Hadron Collider (LHC) marked an exciting new era in experimental particle physics. With more data, emphasis has now shifted to detailed analysis of the properties of the new particle. These measurements are crucial to determine whether this new particle is indeed the Higgs boson as predicted by the Standard Model (SM), or whether it is a more exotic Higgs boson described by a theory beyond the Standard Model (BSM).

This note presents a selection of measurements and their interpretations that probe different properties of Higgs boson production. The compatibility of the data with different spin, parity and tensor coupling scenarios of the Higgs boson are tested in the $H \to \gamma\gamma$, $H \to ZZ^* \to 4\ell$ and $H \to WW^* \to e\nu\mu\nu$ decay channels. Measurements of fiducial and differential cross section for Higgs boson production in the $H \to \gamma\gamma$, $H \to ZZ^* \to 4\ell$ decay channels are also presented. Lastly an interpretation of the measured differential cross sections in the $H \to \gamma\gamma$ channel, using an effective Lagrangian to probe anomalous Higgs boson couplings, is shown. The results are based on pp collision data collected by the ATLAS detector, and correspond to an integrated luminosity of 4.7 fb^{-1} at centre-of-mass energy $\sqrt{s} = 7$ TeV, and 20.3 fb^{-1} at $\sqrt{s} = 8$ TeV.

2 Spin and parity interpretations

In the SM the Higgs boson is a CP-even scalar particle, $J^{CP} = 0^{++}$. Theories beyond the SM may introduce Higgs bosons with different spin-CP structure, and may include CP-mixing. The *Higgs char-*

[a] e-mail: michaela.queitsch-maitland@cern.ch

acterisation framework [5], which uses an *effective field theory* (EFT) approach, is used to confront the data with different spin-parity scenarios. The SM hypothesis $J^P = 0^+$ is compared to alternative models including a pseudo-scalar boson ($J^P = 0^-$), a BSM scalar boson ($J^P = 0^+_h$), and a graviton-like tensor model ($J^P = 2^+$) with universal ($\kappa_q = \kappa_g$) and non-universal ($\kappa_q \neq \kappa_g$) couplings.

The measurements [6, 7] perform fits of discriminant observables sensitive to the spin and parity of the signal. For $H \to \gamma\gamma$ the data are divided into categories based on $|\cos\theta^*|$, the production angle of the two photons in the Collins-Soper reference frame [8], and the p_T of the diphoton system. The $H \to ZZ^* \to 4\ell$ channel uses a matrix element based discriminant. For the $H \to WW^* \to e\nu\mu\nu$ decay channel a Boosted Decision Tree (BDT) for each model is trained with observables based on the lepton kinematics and missing transverse energy.

The test statistic, $\tilde{q}$, that is used to distinguish between two spin-parity hypotheses is based on a ratio of profile likelihoods evaluated under either the SM or alternative spin-parity hypothesis. All considered non-SM spin hypotheses, shown in Figure 1, are excluded at more than 99.9% CL_s.

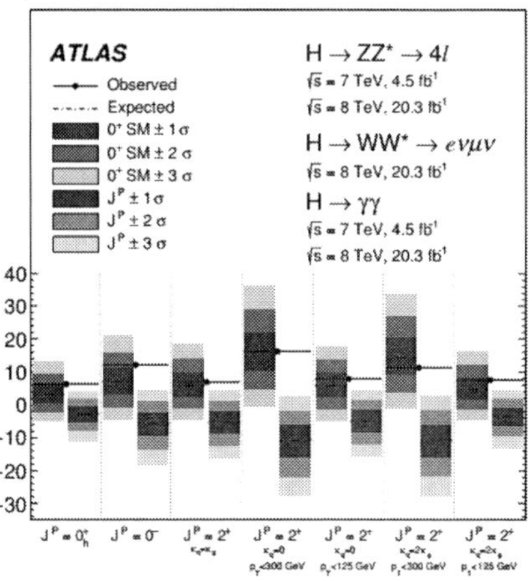

Figure 1. Distributions of the test statistic for the SM Higgs boson (blue) and alternative J^P hypothesis (red). The observed values are indicated by the black points. These fits are performed combining the $H \to \gamma\gamma$, $H \to ZZ^* \to 4\ell$ and $H \to WW^* \to e\nu\mu\nu$ decay channels. See Auxiliary Material of [7].

In addition to fixed spin and parity tests, which probe only one CP eigenstate, the tensor structure of the HVV interaction was also investigated. This is done assuming a spin-0 hypothesis and testing a mixture of SM + BSM CP-even and SM + BSM CP-odd independently. Limits are set on the parameters κ_{HVV}/κ_{SM} and $(\kappa_{AVV}/\kappa_{SM})\tan\alpha$. The likelihood scans of these parameters are shown in Figure 2. The best fit values for the parameters are within the SM expectation.

3 Fiducial and differential cross sections

The measurement of fiducial and differential cross sections for Higgs boson production offer an alternative way of probing the properties of the Higgs boson. Cross sections allow a wide range of physics to be investigated, such as the theoretical modelling of gluon-fusion production, spin-CP, and associated jet activity. Fiducial cross sections are defined within a *fiducial volume*, which is specified by physics object definitions and kinematic selection requirements. The fiducial cross section for a fiducial region, i, is given by

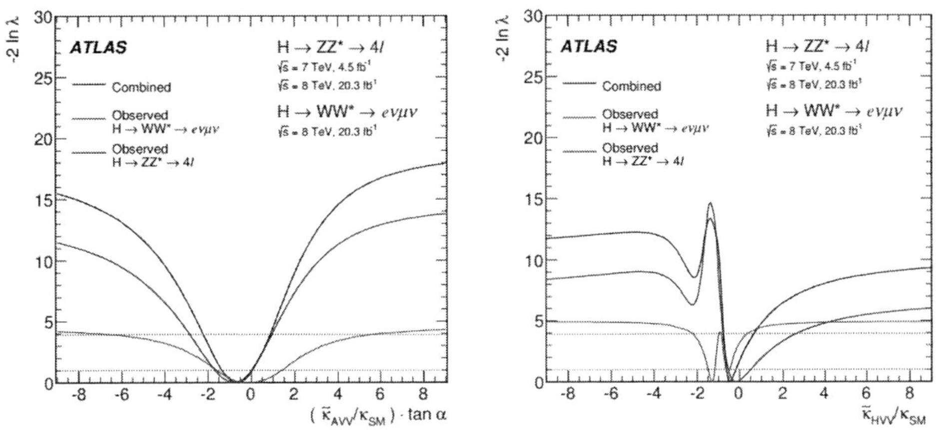

Figure 2. Observed distribution for the test statistic for the $H \to ZZ^* \to 4\ell$ and $H \to WW^* \to e\nu\mu\nu$ analyses and their combination, shown as a function of $\kappa_{\mathrm{HVV}}/\kappa_{\mathrm{SM}}$ and $(\kappa_{\mathrm{AVV}}/\kappa_{\mathrm{SM}})\tan\alpha$. The 68% and 95% CL exclusion regions are indicated by the corresponding horizontal lines [7].

$$\sigma_i^{fid} = \frac{\nu_i^{sig}}{c_i \int L dt} \tag{1}$$

where c_i is a correction factor, which is determined using simulated Higgs boson event samples and accounts for the difference in yield between particle level and detector level. These correct for inefficiency and resolution effects of the detector. Therefore measurements of these fiducial cross sections can be interpreted in a wide variety of theoretical models at the particle level, and the theoretical uncertainties are mostly factorised from the measurement. The particle level selection criteria are chosen to be as close to the selection applied to the data as possible to minimise the extrapolation of phase space and to reduce uncertainties.

The measured fiducial cross sections for the $H \to \gamma\gamma$ [9] and $H \to ZZ^* \to 4\ell$ [10] decay channels are

$$\sigma_{\gamma\gamma}^{fid} = 43.2 \pm 9.4(\text{stat.})^{+3.2}_{-2.9}(\text{syst.}) \pm 1.2(\text{lumi.})\ \text{fb}$$
$$\sigma_{\text{LHC-XS}}^{fid} = 30.5 \pm 3.3\ \text{fb}$$

$$\sigma_{4\ell}^{fid} = 2.11^{+0.53}_{-0.47}(\text{stat.}) \pm 0.08(\text{syst.})\ \text{fb}$$
$$\sigma_{\text{LHC-XS}}^{fid} = 1.3 \pm 0.13\ \text{fb}$$

where $\sigma_{\text{LHC-XS}}^{fid}$ are the current best estimate of the predicted SM cross sections, in the respective fiducial regions and for a Higgs boson mass of 125.4 GeV. Details of the definitions of the fiducial regions can be found in [9, 10].

The differential cross sections are measured for many observables that are sensitive to various properties of the Higgs boson. For example, the p_{T} of the Higgs boson and the jet multiplicity, N_{jets}, shown in Figure 3 are sensitive to the theoretical modelling of gluon fusion as well as additional jet activity produced in association with the Higgs boson.

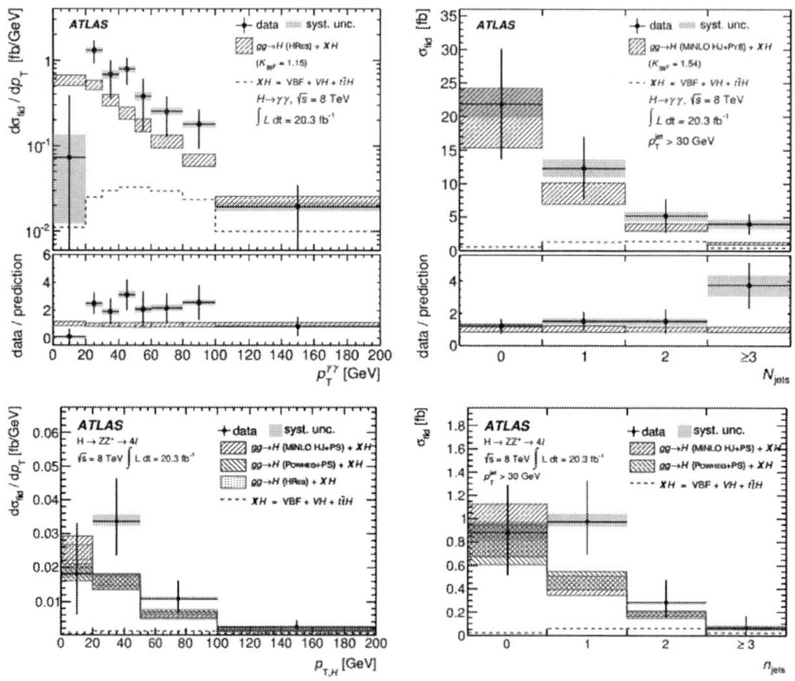

Figure 3. Measured differential cross sections as a function of the Higgs boson p_T (left) and N_{jets} (right) for the $H \rightarrow \gamma\gamma$ (top) and $H \rightarrow ZZ^* \rightarrow 4\ell$ (bottom) channels. The black points represent the data, where the total uncertainty is given by the vertical black bar and the systematic uncertainty by the grey band. The SM predictions are shown in coloured hatched boxes, with the width of the band representing the theoretical uncertainty [9, 10].

Inclusive cross sections,

$$\sigma_i^{incl} = \frac{\sigma_i^{fid}}{\alpha_i \mathcal{B}} = \frac{\nu_i^{sig}}{\alpha_i \mathcal{B} c_i \int L dt} \tag{2}$$

which are corrected for the acceptance of the fiducial phase space, α_i, and the branching fraction, $\mathcal{B}$ are also extracted. In this way data in the $H \rightarrow \gamma\gamma$ and $H \rightarrow ZZ^* \rightarrow 4\ell$ are combined to reduce statistical uncertainties through a likelihood fit. The resulting total cross section and prediction [11] are

$$\sigma_{data}^{pp\rightarrow H} = 33.0 \pm 5.3(\text{stat.}) \pm 1.6(\text{syst.})\ \text{pb}$$
$$\sigma_{\text{LHC-XS}}^{pp\rightarrow H} = 22.16 \pm 1.99(\text{scale}) \pm 2.00(\text{pdf})\ \text{pb}.$$

The normalised differential cross sections for the Higgs boson p_T and the leading jet p_T are shown in Figure 4. All cross section measurements are consistent with the expectations for a SM Higgs boson.

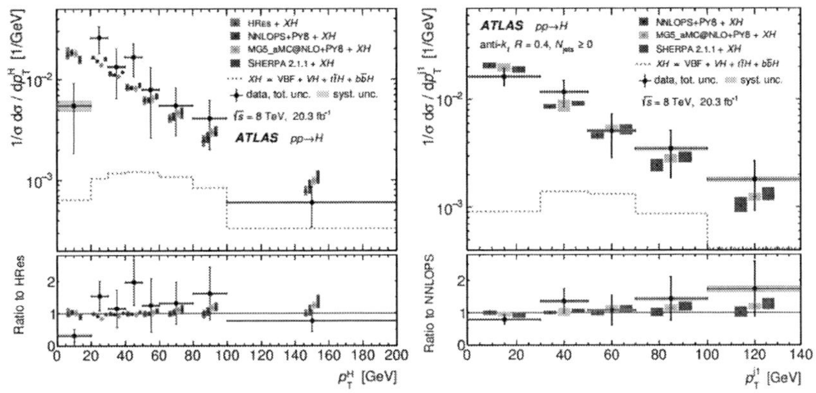

Figure 4. Normalised differential cross section for Higgs p_T (left) and leading jet p_T (right) combining data in the $H \to \gamma\gamma$ and $H \to ZZ^* \to 4\ell$ channels. The data are compared to different state-of-the-art theoretical predictions [11].

4 Interpretations of cross section measurements using an effective Lagrangian

The cross section measurements in the $H \to \gamma\gamma$ channel discussed in Section 3 were also used to set limits on *Wilson coefficients* for the *strongly interacting light Higgs* (SILPH) EFT formalism [12]. The measurement uses five of the differential distributions in [9], where the statistical correlations between bins of observables in data are evaluated using pseudo-experiments [13]. An example for these correlations are shown in Figure 5 for diphoton p_T and N_{jets}.

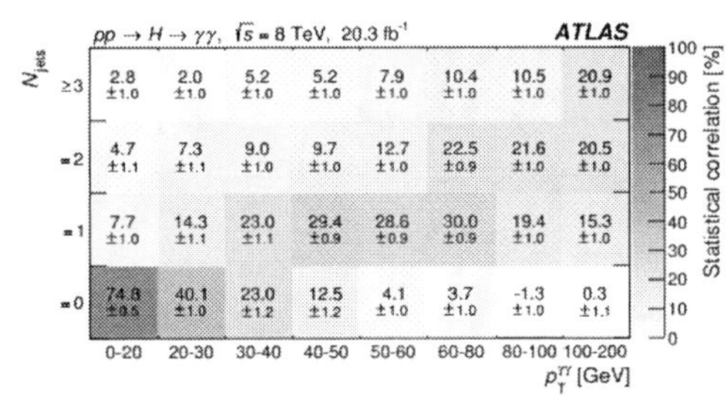

Figure 5. Statistical correlations between individual bins in diphoton p_T and N_{jets} evaluated using a bootstrapping procedure. The errors represent the statistical precision of the pseudo-experiments and the data [13].

The Lagrangian is described by

$$\mathcal{L} = \bar{c}_\gamma O_\gamma + \bar{c}_g O_g + \bar{c}_{\text{HW}} O_{\text{HW}} + \bar{c}_{\text{HB}} O_{\text{HB}}$$
$$+ \tilde{c}_\gamma O_\gamma + \tilde{c}_g O_g + \tilde{c}_{\text{HW}} O_{\text{HW}} + \tilde{c}_{\text{HB}} O_{\text{HB}} \quad (3)$$

where $\bar{c}_i$ and $\tilde{c}_i$ are Wilson coefficients that specify the strength of new CP-even and CP-odd interactions respectively. The O_i are dimension six operators described in [12]. The terms with γ, g, and HW or HB affect the coupling of the Higgs to photons, gluons, and weak vector bosons respectively.

 DOI: 10.1051/epjconf/201612604038

Limits are set on the Wilson coefficients by constructing a χ^2 function with respect to data and each prediction. Scans of one or two Wilson coefficients are carried out at a time. The 68% and 95% confidence level (CL) regions for example scans are shown in Figure 6. No significant deviations from the SM predictions are observed, and limits are set on the Wilson coefficients.

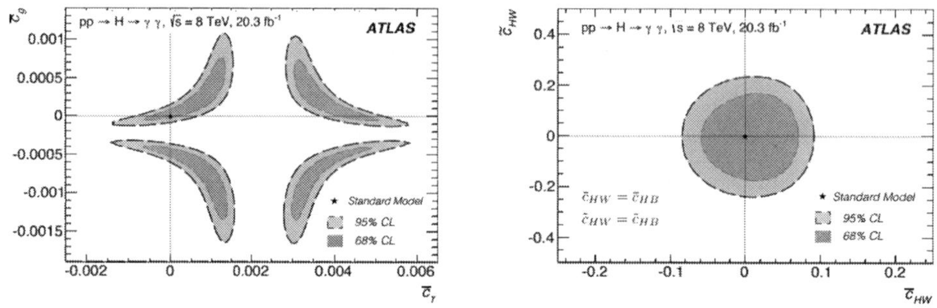

Figure 6. The 68% (dark) and 95% (light) confidence regions for fits to the $\bar{c}_\gamma$ and $\tilde{c}_g$ (left), and $\bar{c}_{\mathrm{HW}}$ and $\tilde{c}_{\mathrm{HW}}$ (right) Wilson coefficients. All other coefficients are set to zero, except for the fits to $\bar{c}_{\mathrm{HW}}$ and $\tilde{c}_{\mathrm{HW}}$ (right), where $\bar{c}_{\mathrm{HW}} = \bar{c}_{\mathrm{HB}}$ and $\tilde{c}_{\mathrm{HW}} = \tilde{c}_{\mathrm{HB}}$. The marker indicates the SM value [13].

5 Summary and Conclusion

Measurements probing the properties of the Higgs boson at the LHC are reported, based on pp collision data collected by the ATLAS detector corresponding to an integrated luminosity of 4.7 fb^{-1} at centre-of-mass energy $\sqrt{s} = 7$ TeV, and 20.3 fb^{-1} at $\sqrt{s} = 8$ TeV. The compatibility of the data with different spin, parity and tensor coupling scenarios of the Higgs boson are tested in the $H \to \gamma\gamma$, $H \to ZZ^* \to 4\ell$ and $H \to WW^* \to e\nu\mu\nu$ decay channels. Fiducial and differential cross sections for Higgs boson production in the $H \to \gamma\gamma$, $H \to ZZ^* \to 4\ell$ decay channels are presented. In the $H \to \gamma\gamma$ channel a selection of the measured differential cross sections are used to set limits on anomalous Higgs boson couplings using an effective Lagrangian. No significant deviations from the SM predictions are observed in any of the measurements discussed.

References

[1] ATLAS Collaboration, Phys. Lett. **B716**, 1-29 (2012).
[2] CMS Collaboration, Phys. Lett. **B716**, 30-61 (2012).
[3] ATLAS Collaboration, JINST **3** S08003 (2008).
[4] CMS Collaboration, JINST **3** S08004 (2008).
[5] P. Artoisenet, et al., JHEP **11**, 043 (2013).
[6] ATLAS Collaboration, Eur. Phys. J. **C75**, 231 (2015).
[7] ATLAS Collaboration, Eur. Phys. J. **C75**, 476 (2015).
[8] J. Collins and D. Soper, Phys. Rev. D **16**, 2219 (1977).
[9] ATLAS Collaboration, JHEP **09**, 112 (2014).
[10] ATLAS Collaboration, Phys. Lett. **B738**, 234-253 (2014).

[11] ATLAS Collaboration, Phys. Rev. Lett. **115**, 091801 (2015).
[12] R. Contino, et al., JHEP **1307**, 035 (2013).
[13] ATLAS Collaboration, Phys. Lett. **B753**, 69-85 (2016).

EPJ Web of Conferences **126**, 04039 (2016) DOI: 10.1051/epjconf/201612604039
ICNFP 2015

Absence of the Gribov ambiguity in a special algebraic gauge

Haresh Raval[a]

Department of Physics, Indian Institute of Technology Bombay, Mumbai-400076, India

Résumé. The Gribov ambiguity exists in various gauges except algebraic gauges. However in general, algebraic gauges are not Lorentz invariant, which is their fundamental flaw. Here we discuss a quadratic gauge fixing, which is Lorentz invariant. We show that nontrivial copies can not occur in this gauge. We then provide an example of spherically symmetric gauge field configuration and prove that with a proper boundary condition on the configuration, this gauge removes the ambiguity on a compact manifold $\mathbb{S}^3$.

1 Introduction

Defining the path integral in gauge theories has a major issue of infinite redundant functional integrations. The fact that the Yang-Mills action is invariant under the gauge transformation is the cause of the issue. The issue is addressed by invoking a gauge condition such as the Landau gauge $\partial^\mu A_\mu = f$. However, it is shown in ref. [1] that even after the Landau gauge fixing, there still exist a finite number of equivalent configurations, which contribute to the measure of the path integral. This implies that the Landau gauge does not uniquely choose a configuration, the problem known as the Gribov ambiguity. We need only inequivalent configurations in the measure in order to properly quantize the theory. The inequivalent configurations can be extracted out by restricting the space of integration to the fundamental modular region C^0, where the faddeev-poppov operator has positive eigenvalues [1]. However, the region C^0 still contains Gribov copies [1]. The restriction on the space of integration is achieved by adding suitable terms to the effective action S_{eff} resulting from the Landau gauge fixing [2, 3]. This modified action is known as Gribov-Zwanziger action. The same ambiguity is shown to exist in all the covariant gauges [4].

An essential reason why some gauges have the ambiguity is the differential operator involved in the gauge. Algebraic gauges are ambiguity free since they do not have a differential operator, but they have one disadvantage. In general, they violate the Lorentz invariance, which is a basic requirement for any theory. Whereas, the gauge under consideration in this paper is Lorentz invariant. Alternative formulations addressing the Gribov ambiguity are suggested in ref. [5, 6]. The former ref. particularly is an approach using Lorentz invariant algebraic gauge conditions.

The contents of this paper are arranged as follows : in the next section, we discuss a particular quadratic gauge and its consequences at infra-red scale. In sec. 3, we show that non trivial configurations can not occur in the present gauge. In sec. 4, we examine a case of the spherically symmetric gauge configuration. We prove that when a proper boundary condition on the gauge configuration at ∞

[a] e-mail: haresh@phy.iitb.ac.in

is taken into account, the quadratic gauge uniquely chooses the configuration on a compact manifold $\mathbb{S}^3$.

2 A quadratic gauge and Effective Lagrangian

There have been studies using quadratic gauges in several contexts. A few of the references are [7–12]. Here we consider a particular quadratic gauge introduced in the ref. [13] in the context of infra-red phenomena in QCD.

$$H^a[A^\mu(x)] = A^a_\mu(x)A^{\mu a}(x) = f^a(x); \quad \text{for each } a \tag{1}$$

where $f^a(x)$ is an arbitrary function of x. This gauge condition results in the effective Lagrangian of the form [13]

$$\begin{aligned}\mathcal{L}_{eff} &= \mathcal{L}_{YM} + \mathcal{L}_{GF} + \mathcal{L}_{ghost} \\ &= -\frac{1}{4}F^a_{\mu\nu}F^{\mu\nu a} - \frac{1}{2\zeta}(A^a_\mu A^{\mu a})^2 - \overline{c^a}A^{\mu a}(D_\mu c)^a\end{aligned} \tag{2}$$

where the first term is Yang-Mills Lagrangian with $F^a_{\mu\nu}(x) = \partial_\mu A^a_\nu(x) - \partial_\nu A^a_\mu(x) - gf^{abc}A^b_\mu(x)A^c_\nu(x)$, second and third terms are gauge fixing and ghost Lagrangian respectively and $(D_\mu c)^a = \partial_\mu c^a - gf^{abc}A^b_\mu c^c$.

The ghost Lagrangian contains a term $gf^{abc}\overline{c^a}c^c A^{\mu a}A^b_\mu$. In the ghost condensed phase as shown in ref. [13], the term gives two strong signatures of the color confinement [13] 1. Abelian dominance and 2. A pole of the off-diagonal gluon propagator is on imaginary p^2 axis. The propagator not having poles on a real p^2 axis is a sufficient condition for the confinement [14].

In this paper, we aim to address another crucial non-perturbative phenomenon, the Gribov ambiguity. Thus, the purpose of this work with the results described above is also to justify the motivation and suitability of a choice of the quadratic gauge for the non-perturbative problems in QCD.

3 Absence of Gribov copies in the quadratic gauge

We now prove that non trivial Gribov copies do not exist in the present gauge due to its non linear nature. Under an infinitesimal gauge transformation $U = 1 + \alpha$, the quadratic gauge for the gauge transformed field is given by

$$\begin{aligned}\tilde{A}^a_\mu\tilde{A}^{\mu a} &= (A^a_\mu + (D_\mu\alpha)^a)(A^{\mu a} + (D^\mu\alpha)^a) \\ &= A^a_\mu A^{\mu a} + 2A^{\mu a}(D_\mu\alpha)^a + (D_\mu\alpha)^a(D^\mu\alpha)^a\end{aligned} \tag{3}$$

Hence, the gauge equivalence $\tilde{A}^a_\mu\tilde{A}^{\mu a} = A^a_\mu A^{\mu a}$ yields

$$2A^{\mu a}(D_\mu\alpha)^a + (D_\mu\alpha)^a(D^\mu\alpha)^a = 0 \tag{4}$$

Therefore, $(D_\mu\alpha)^a = 0 \Rightarrow \tilde{A}^a_\mu = A^a_\mu$ or $(D_\mu\alpha)^a = -2A^a_\mu \Rightarrow \tilde{A}^a_\mu = -A^a_\mu$
So in neither case any new configuration is present. For the finite transformation also the same result is true. A finite transformation of the gauge field is given by

$$\tilde{A}_\mu = A_\mu + i(D_\mu U)U^{-1} \tag{5}$$

where $D_\mu U = \partial_\mu U + i[A_\mu, U]$. Now, it is easy to see that solving a gauge equivalence $\tilde{A}^a_\mu \tilde{A}^{\mu a} = A^a_\mu A^{\mu a}$ for each a, amounts to solving a gauge equivalence $\tilde{A}^\mu \tilde{A}_\mu = A_\mu A^\mu$. Hence, we solve a latter equivalence which gives the following eq.

$$iA_\mu (D^\mu U)U^{-1} + i(D^\mu U)U^{-1}A_\mu + i^2 (D^\mu U)U^{-1}(D_\mu U)U^{-1} = 0$$

$$\{A_\mu + \frac{i}{2}(D_\mu U)U^{-1}, i(D^\mu U)U^{-1}\} = 0 \tag{6}$$

Therefore, $i(D_\mu U)U^{-1} = 0 \Rightarrow \tilde{A}_\mu = A_\mu$ or $i(D_\mu U)U^{-1} = -2A_\mu \Rightarrow \tilde{A}_\mu = -A_\mu$
Multiplying with the generator T^a and then taking a trace on both sides in each case would yield a configuration which we would have got by solving a former gauge equivalence. Thus, we conclude that the gauge is Gribov ambiguity free.

4 Spherically symmetric gauge field configuration and the quadratic gauge

Here we demonstrate that the quadratic gauge uniquely picks up a spherically symmetric configuration on a compact manifold $\mathbb{S}^3$, when a proper boundary condition on the field is required to be satisfied. Compactification of a euclidean space $\mathbb{R}^N$ to a compact manifold $\mathbb{S}^N$ is achieved by the condition $U(\infty) = I$ [4]. Since the space in this example is $\mathbb{R}^3$, the condition would compactify it to $\mathbb{S}^3$. We begin by adopting a parameterization for a vector potential shown in ref. [1]

$$A_i = f_1(r)\frac{\partial \hat{n}}{\partial x_i} + f_2(r)\hat{n}\frac{\partial \hat{n}}{\partial x_i} + f_3(r)\hat{n}n_i, \qquad i = 1, 2, 3 \tag{7}$$

Where $n_i = \frac{x_i}{r}$, $r = \sqrt{\Sigma x_i^2}$, $\hat{n} = in_j\sigma_j$ σ_j are Pauli matrices , $\hat{n}^2 = -1$. For simplicity we choose $A_0 = 0$. Now, the spherically symmetric operator is given by

$$U = \exp\left(\frac{\alpha(r)}{2}\hat{n}\right) = \cos\left(\frac{\alpha(r)}{2}\right) + \hat{n}\sin\left(\frac{\alpha(r)}{2}\right) \tag{8}$$

Therefore, the compactification condition $U(\infty) = I$ implies $\alpha(\infty) = 4\pi n$; n is an integer. The gauge transformation $A_\mu \longrightarrow \tilde{A}_\mu = UA_\mu U^{-1} + U^{-1}(\partial_\mu U)$ results in transformations of f_1, f_2 and f_3 as follows

$$\tilde{f}_1 = f_1\cos\alpha + (f_2 + \frac{1}{2})\sin\alpha$$

$$\tilde{f}_2 + \frac{1}{2} = -f_1\sin\alpha + (f_2 + \frac{1}{2})\cos\alpha$$

$$\tilde{f}_3 = f_3 + \frac{1}{2}\dot{\alpha} \tag{9}$$

where overdot indicates differentiation with respect to r. Now, a th component of A_i can be derived using following formula

$$A_i^a = \frac{1}{2}Tr(A_i\sigma_a) \tag{10}$$

after some algebra, we get

$$A_1^1 = i[f_1(\frac{1}{r} - \frac{x_1^2}{r^3}) + f_3\frac{x_1^2}{r^2}] \tag{11a}$$

$$A_2^1 = i[-f_1\frac{x_1x_2}{r^3} + f_2\frac{x_3}{r^2} + f_3\frac{x_1x_2}{r^2}] \tag{11b}$$

$$A_3^1 = i[-f_1\frac{x_1x_3}{r^3} - f_2\frac{x_2}{r^2} + f_3\frac{x_1x_3}{r^2}] \tag{11c}$$

$$A_1^2 = i[-f_1\frac{x_1x_2}{r^3} - f_2\frac{x_3}{r^2} + f_3\frac{x_1x_2}{r^2}] \tag{11d}$$

$$A_2^2 = i[f_1(\frac{1}{r} - \frac{x_2^2}{r^3}) + f_3\frac{x_2^2}{r^2}] \tag{11e}$$

$$A_3^2 = i[-f_1\frac{x_2x_3}{r^3} + f_2\frac{x_1}{r^2} + f_3\frac{x_2x_3}{r^2}] \tag{11f}$$

$$A_1^3 = i[-f_1\frac{x_1x_3}{r^3} + f_2\frac{x_2}{r^2} + f_3\frac{x_1x_3}{r^2}] \tag{11g}$$

$$A_2^3 = i[-f_1\frac{x_2x_3}{r^3} - f_2\frac{x_1}{r^2} + f_3\frac{x_2x_3}{r^2}] \tag{11h}$$

$$A_3^3 = i[f_1(\frac{1}{r} - \frac{x_3^2}{r^3}) + f_3\frac{x_3^2}{r^2}] \tag{11i}$$

We now impose a boundary condition on A_k^j s. We require that

$$A_k^j \longrightarrow 0 \text{ as } \frac{1}{r}, \text{ as } r \longrightarrow \infty \tag{12}$$

From Eq.s (11), it is clear that this condition is achievable and the boundary condition on f_1, f_2 and f_3 can be easily interpreted, which is as following

$$f_1, f_2 \longrightarrow const. \text{ as } r \longrightarrow \infty \text{ and } f_3 \longrightarrow 0 \text{ as fast as } \frac{1}{r} \text{ as } r \longrightarrow \infty \tag{13}$$

Here we make one note. We want to address the ambiguity on $\mathbb{S}^3$, therefore a boundary condition on f_3 needs to be little stronger (faster than $\frac{1}{r}$ as $r \to \infty$) because of the eq. for copies (24) that we shall come across later in the section. Hence, we consider a stronger condition on f_3 only. We will use these boundary conditions to prove our claim. We first evaluate a condition

$$A_i^a A^{ia} = A_1^a A^{1a} + A_2^a A^{2a} + A_3^a A^{3a}; \text{ for each } a \tag{14}$$

For example taking $a = 1$, the gauge above takes the form

$$\begin{aligned} A_i^1 A^{i1} &= (A_1^1)^2 + (A_2^1)^2 + (A_3^1)^2 \\ &= -\Big[\frac{f_1^2}{r^2}(1 - \frac{x_1^2}{r^2}) + \frac{f_2^2}{r^2}(1 - \frac{x_1^2}{r^2}) + f_3^2\frac{x_1^2}{r^2}\Big] \end{aligned} \tag{15}$$

In spherical polar coordinates, the condition can be written as

$$= -\frac{1}{r^2}(f_1^2 + f_2^2) + \sin^2\theta\cos^2\phi\Big(\frac{1}{r^2}(f_1^2 + f_2^2) - f_3^2\Big) \tag{16}$$

Hence,

$$\tilde{A}_i^1 \tilde{A}^{i1} = -\frac{1}{r^2}(\tilde{f}_1^2 + \tilde{f}_2^2) + \sin^2\theta\cos^2\phi\Big(\frac{1}{r^2}(\tilde{f}_1^2 + \tilde{f}_2^2) - \tilde{f}_3^2\Big) \tag{17}$$

For $\tilde{A}_i^1 \tilde{A}^{i1} = A_i^1 A^{i1}$ to happen

$$\frac{1}{r^2}[(\tilde{f}_1^2 + \tilde{f}_2^2) - (f_1^2 + f_2^2)] + \sin^2\theta\cos^2\phi[\tilde{f}_3^2 - f_3^2 - \frac{1}{r^2}((\tilde{f}_1^2 + \tilde{f}_2^2) - (f_1^2 + f_2^2))] = 0 \tag{18}$$

DOI: 10.1051/epjconf/201612604039

Since α is a function of r only, the first term and the coefficient of $\sin^2\theta\cos^2\phi$ in a second term of Eq. (18) must individually vanish, giving us two different copy equations

$$\tilde{f}_1^2 + \tilde{f}_2^2 = f_1^2 + f_2^2 \;\Rightarrow\; f_2 + \frac{1}{2} = -f_1 \cot\frac{\alpha}{2} \tag{19}$$

$$\tilde{f}_3^2 = f_3^2 \;\Rightarrow\; f_3\dot{\alpha} + \frac{1}{4}\dot{\alpha}^2 = 0 \;\Rightarrow\; \quad \dot{\alpha} = 0 \text{ or } \dot{\alpha} = -4f_3 \tag{20}$$

For a copy to exist, Eq. (19) has to be satisfied, with a parameter α in it satisfying either of two eq.s in Eq. (20). There are two choices to make since f_1 and f_2 are arbitrary functions.

1. $f_2 + \frac{1}{2} \neq -f_1 \cot\frac{\alpha}{2}$
2. $f_2 + \frac{1}{2} = -f_1 \cot\frac{\alpha}{2}$.

If

$$f_2 + \frac{1}{2} \neq -f_1 \cot\frac{\alpha}{2} \tag{21}$$

then it is clear that no copy exists for this choice.
However, if

$$f_2 + \frac{1}{2} = -f_1 \cot\frac{\alpha}{2} \tag{22}$$

then we encounter two copies corresponding to eq.s $\dot{\alpha} = 0$ and $\dot{\alpha} = -4f_3$. They are obtained by putting Eq. (22) in the transformation (9)

$$\tilde{f}_1 = -f_1$$
$$\tilde{f}_2 = f_2$$

Therefore for $\dot{\alpha} = 0$ (putting $\dot{\alpha} = 0$ back in transformation (9)), we obtain

$$\tilde{f}_1 = -f_1, \qquad \tilde{f}_2 = f_2, \qquad \tilde{f}_3 = f_3$$

which yields a copy

$$\begin{aligned} \tilde{A}_j^j &= A_j^j - 2if_1\Big(\frac{1}{r} - \frac{x_j^2}{r^3}\Big) \\ \tilde{A}_k^j &= A_k^j + 2if_1\frac{x_j x_k}{r^3} \end{aligned} \tag{23}$$

However, on a compact manifold $\mathbb{S}^3$ this copy no longer exists. Because $\dot{\alpha} = 0 \Rightarrow \alpha = const.$ everywhere including infinity. Setting $\alpha(r) = \alpha(\infty) = 4\pi n$, for which the copy Eq. (22) implies $f_2 = \infty$ everywhere giving a copy which is also ∞. We want finite copies of A_k^j which is well behaved and finite at finite distances, which is not possible for $\dot{\alpha} = 0$ on $\mathbb{S}^3$. Therefore, the Eq. (22) is not valid on $\mathbb{S}^3$, thus the copy vanishes on it.

Now, we are left with only one copy which corresponds to

$$\dot{\alpha} = -4f_3 \Rightarrow \alpha = -4\int f_3\, dr + const. \tag{24}$$

Putting $\dot{\alpha} = -4f_3$ back in transformation (9), we get

$$\tilde{f}_1 = -f_1, \qquad \tilde{f}_2 = f_2, \qquad \tilde{f}_3 = -f_3$$

which yields a copy

$$\begin{aligned} \tilde{A}^j_j &= -A^j_j \\ \tilde{A}^j_k &= -A^k_j \end{aligned} \tag{25}$$

It can also be removed on $\mathbb{S}^3$. We recall boundary conditions (13). Since $f_3 \longrightarrow 0$ faster than $\frac{1}{r}$ as $r \longrightarrow \infty$, Eq. (24) $\Rightarrow \alpha(\infty) = const.$. As for the previous copy, we set $\alpha(\infty) = 4\pi n$ for which Eq. (22) implies $f_2 \longrightarrow \infty$ as $r \longrightarrow \infty$. Hence it is clear that on $\mathbb{S}^3$, Eq. (22) is an obstruction for the boundary condition on f_2 (Eq. (13)) to be satisfied therefore not valid. Therefore this copy does not exist on $\mathbb{S}^3$.

The result is true for stronger boundary conditions such as $\frac{1}{r^2}$ and e^{-r}. Similarly, it can be shown that the condition for other two components, $\tilde{A}^2_i\tilde{A}^{i2} = A^2_i A^{i2}$ and $\tilde{A}^3_i\tilde{A}^{i3} = A^3_i A^{i3}$, produce same two equations for copy.

Whereas for coulomb gauge, we have [2]

$$\frac{\partial A_i}{\partial x_i} = \hat{n}(\dot{f}_3 + \frac{2}{r}f_3 - \frac{2}{r^2}f_1) \tag{26}$$

Because Pauli matrices σ_a are unit vectors in 2×2 matrix space, the condition

$$\frac{\partial \tilde{A}^a_i}{\partial x_i} = \frac{\partial A^a_i}{\partial x_i} \tag{27}$$

for all three components yields the equation

$$\ddot{\alpha} + \frac{2}{r}\dot{\alpha} - \frac{4}{r^2}\Big((f_2 + \frac{1}{2})\sin\alpha + f_1\cos\alpha\Big) = 0 \tag{28}$$

This equation is known to be solvable and therefore the ambiguity exists.

5 conclusion

We discussed a particular quadratic gauge, which is a Lorentz invariant algebraic gauge and showed how the non linearity does not allow the ambiguity to exist. Later we worked out an example of spherically symmetric configuration in the quadratic gauge and proved that the configuration with a proper boundary condition does not have any copy on $\mathbb{S}^3$. Thus, through the work in this paper with the results of ref. [13], the choice of the quadratic gauge for infra-red phenomena is also justified.

Acknowledgment

Haresh is sincerely thankful to Professor Urjit A. Yajnik for useful comments on the subject.

References

[1] V. N. Gribov, *Nucl. Phys.* **B139** 1 (1978)
[2] D. Zwanziger *Nucl. Phys.* **B323** 513-544 (1989)
[3] D. Zwanziger, *Nucl. Phys.* **B321** 591-604 (1989)
[4] I. Singer, *Comm. Math. Phys.* **60** 7 (1978)
[5] A. A. Slavnov, *JHEP* **08** 047 (2008)
[6] J. Serreau and M. Tissier, *Phys. Lett. B* **712** 97-103 (2012)

[7] Ken-ichi SHIZUYA, *Nucl. Phys.*, **B109** 397-420 (1976)
[8] S. Weinberg, *Phys. Lett. B* **91** 51-55 1980
[9] Ashok Das, *Pramana* **16** 409-416 (1981)
[10] Ashok Das, M.A. NAMAZIE, *Phys. Lett. B* **99** 463-466 (1981)
[11] S. D. Joglekar, Phys. Rev. D **10** 4095 (1974)
[12] S. D. Joglekar and B. P. Mandal, Phys. Rev. D **51**, 1919 (1995)
[13] Haresh Raval and Urjit A. Yajnik, Phys. Rev. D **91**, 085028 (2015)
[14] C. D. Roberts, A. G. Williams, and G. Krein, Int. J. Mod.Phys. **A 07**, 5607 (1992)

EPJ Web of Conferences **126**, 04040 (2016) DOI: 10.1051/epjconf/201612604040
ICNFP 2015

Searches for electroweak production of supersymmetric gauginos and sleptons with the ATLAS detector

Huan Ren - on behalf of ATLAS collaboration[1,a]

[1] *Institute of High Energy Physiscs, Beijing, China*

Abstract. This paper shows the summary of the results from some recent research by ATLAS SUSY electroweak group(EWK). The sub-topics are including direct stau production, same-sign chargino pair production via vector boson fusion(VBF) and compressed spectra in direct production.

1 Introduction

Since the discovery of Higgs boson at LHC, the most essential part of standard model(SM) was confirmed. However, we still have some remaining issues after this: hierachy problem, dark matter, origin of gravity and no gauge unification at higher scale. Supersymmetry(SUSY) has become a very appealing extension of SM to answer these questions. Currently, we have set limits of most SUSY models probe masses up to $900GeV$(squarks) and $1.4TeV$(gluinos).

The cross-sections of different SUSY procedure are shown on Fig. 1. The EWK sector provides a promising approach for new physics among all SUSY searches. The hadronic activity is quite low in a certain EWK process despite the low production cross-section. For most of the EWK models, we have 1-4 leptons in the final state, as well as large missing transverse energy(Etmiss) and 0-2 jets(or b-jets).

2 Search for direct stau production in events with at least two hadronic taus and missing transverse momentum using multivariate analysis technique

This part is an update of JHEP 1410(2014)96.

2.1 direct stau production

The direct stau production(see Fig. 2) have the final states: two opposite-sign hadronically decaying taus, no jet, and large E^T_{miss}. Limited sensitivity was observed using cut-and-count analysis due to the low cross-section of direct stau process. Multivariate analysis(MVA) is used to improve this result. We defined one boosted decision tree(BDT) signal region(SR) for this intension, as shown on Fig. 3.

[a]e-mail: huan.ren@cern.ch

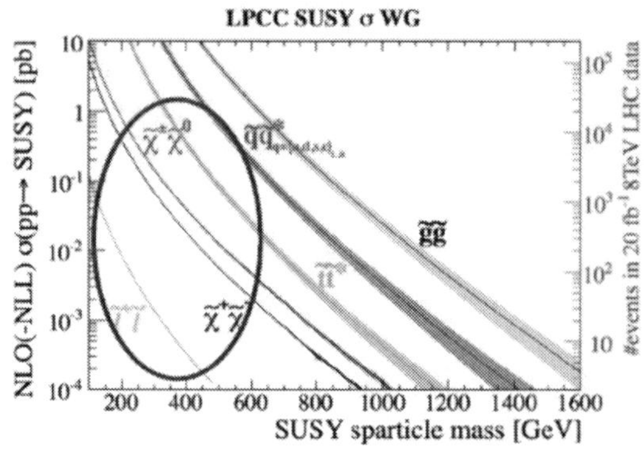

Figure 1. SUSY cross-sections for different production against SUSY particle mass.

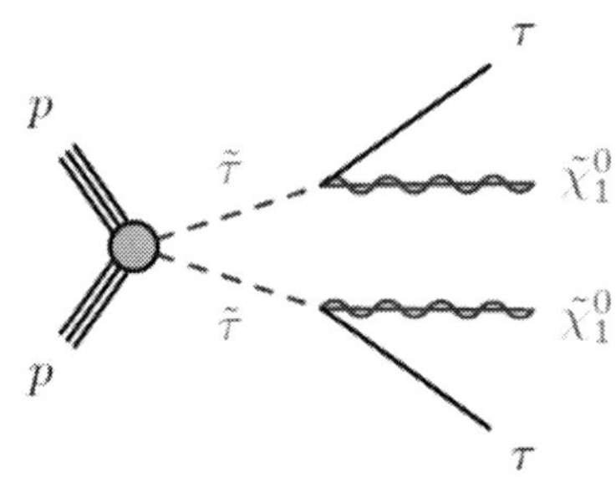

Figure 2. Feynman diagram of the pair production of charged staus decaying into 2 taus in the final state.

2.2 background

The main background of this study includes:

- W+jets(1real + 1fake lepton). This background is estimated by a normalization to data in dedicated WCR.
- multi-jet. We apply an "ABCD" estimation to this background source, using tau-ID and m_{T2} as the two uncorrelated variables, shown in Fig. 4.
- Other sub-dominant background. This is estimated by Monte Carlo(MC) simulation.

2.3 results

The results of this direct stau update was shown in Fig. 5. Significant improvement is observed in the low neutralino mass region using MVA compared to cut-and-count analysis.

DOI: 10.1051/epjconf/201612604040

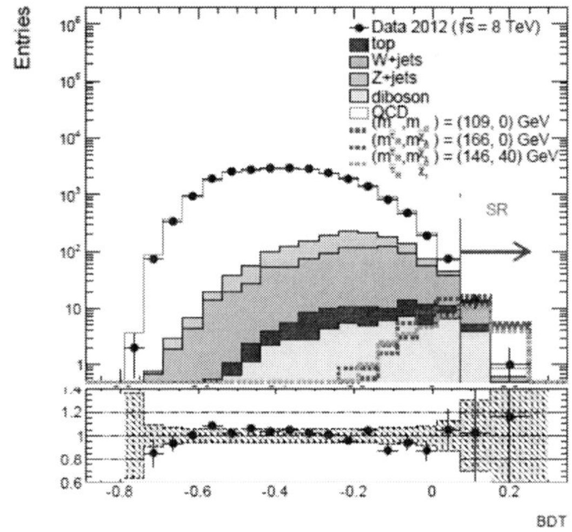

Figure 3. Distribution of BDT response and the SR definition.

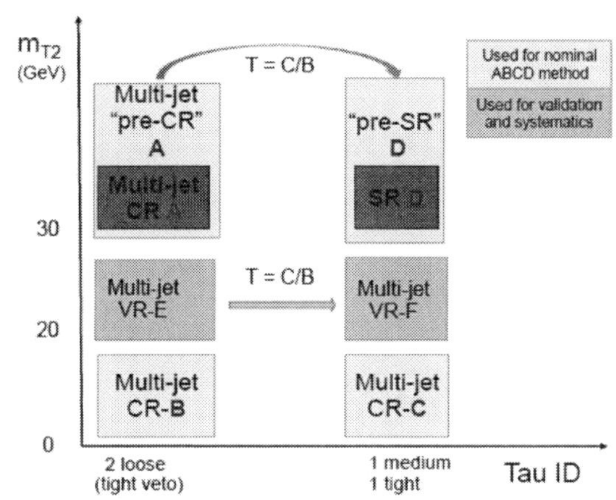

Figure 4. Introduction of "ABCD" estimation. Two uncorrelated variables are used on each axis, defining 3 CRs with enough statistics and 1 SR. The extrapolation from the CRs provides a good estimation to the event yields in SR.

3 Search for supersymmetry in compressed scenarios with two and three leptons and missing transverse momentum in the final state

3.1 Same-sign C1C1

The same-sign(SS) C1C1(two chargino-one mediate process) production(see Fig. 6) have a final state with: two SS light leptons, ≥2jet and large E^T_{miss}.

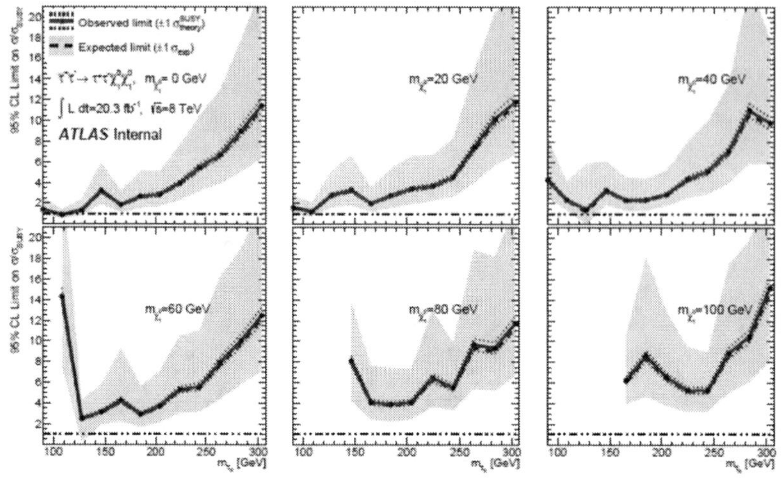

Figure 5. Upper limits on the signal strength for production of left and right-handed stau pairs for different neutralino masses.

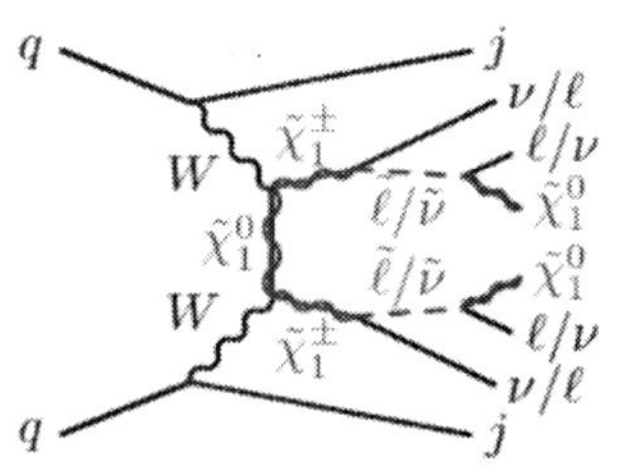

Figure 6. Feynman diagram for VBF production of C1C1 with two SS leptons in the final states.

3.2 background

The SR of this analysis includes the following main background:

- "prompt" leptons(diboson, H), estimated from MC simulation directly.
- non-prompt(fake) leptons(W+jets, $t\bar{t}$) - Fake Factor Method was adopted for the estimation.
- Charge-misID leptons - measured from dedicated control region(CR).

The disctributions of E^T_{miss} and Mjj are shown on Fig. 7.

3.3 results

The fitting results of of this topic was shown on Fig.8. The best observed upper limit is found for $\widetilde{\chi}^0_1$ mass of 120GeV and m($\widetilde{\chi}^{\pm}_1$)-m($\widetilde{\chi}^0_1$)=25GeV. Slightly stronger sensitivity is also observed for higher $\widetilde{\chi}^{\pm}_1$ masses.

4 Search for supersymmetry in compressed scenarios

The compressed scenarios are some SUSY scenarios with small mass difference between sparticles and neutralinos, leading to low-momentum leptons in final states. Generally, low mass-splitting, high

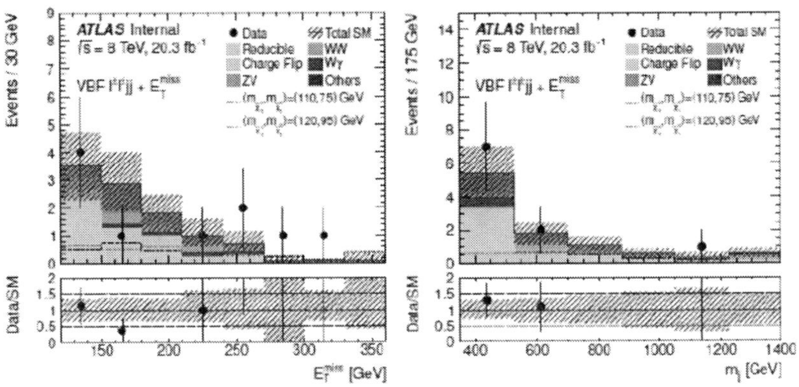

Figure 7. Distribution of the E_{miss}^{T} and di-jet invariant mass in SR.

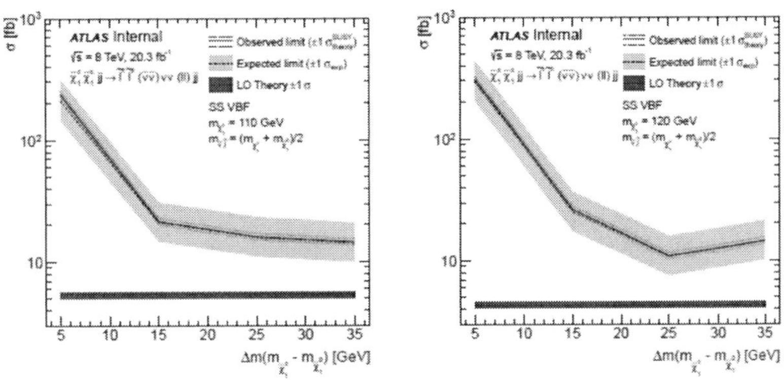

Figure 8. 95% CL exclusion limits on the cross-section for VBF production of $\widetilde{\chi}_1^{\pm}\widetilde{\chi}_1^{\pm}$. The limits have been set with respect to the mass difference between the $\widetilde{\chi}_1^{\pm}$ and $\widetilde{\chi}_1^{0}$.

soft decay products and SM-like behaviour leads to many difficulties for this analysis. The results of re-optimization of three sub-topics are shown:

- two opposite-sign(OS) light leptons.
- two same-sign(SS) light lrptons.
- three light leptons.

4.1 Two OS light leptons: Event selection and background estimation

The experimental signature of this study is: two OS light leptons and E_{miss}^{T} shown in Fig. 9. Two SRs are defined for this analysis, requiring a high pT initial-state-radiation(ISR) jets, which are sensitive to small and moderate mass splittings. Discriminate variabls "Super razor" are also used in the SR definition, shown on Fig. 10. The main background in the SRs are:

- Irreducibel background(WW, top, ZV), estimated by normalized MC simulation in dedicated CRs.
- Reducibel background(all fake sources) - Matrix Method adopted for this background source.

- Others(Higgs, Z+jets) - measured by MC simulation directly.

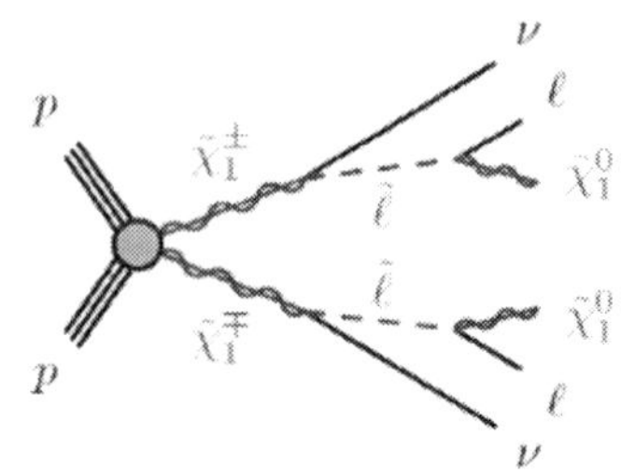

Figure 9. Feynman diagram for compressed scenario with two OS leptons in the final states.

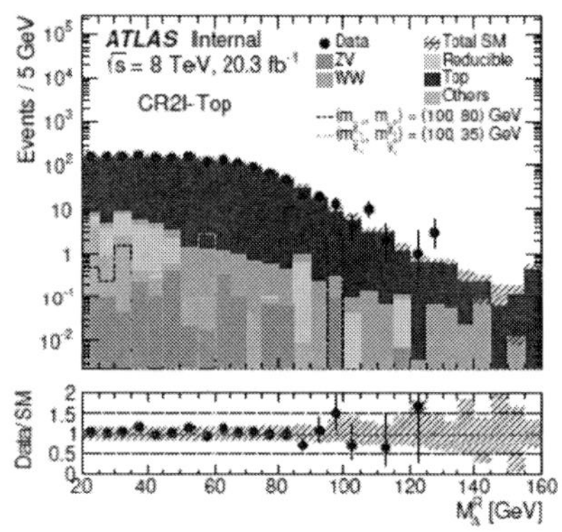

(a) M_{Δ}^{R} in the top CR

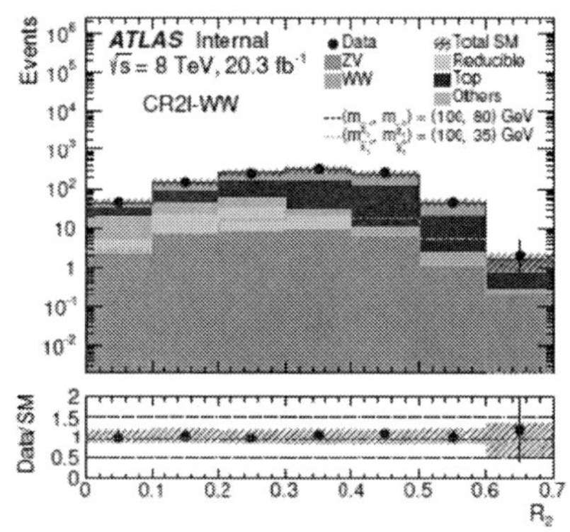

(c) R_2 in the WW CR

Figure 10. Distributions of the "super razor" varibales in SRs.

4.2 Two SS light leptons: Event selection and background estimation

The experimental signature of this scenario is: two SS light leptons and E_{miss}^{T} shown in Fig. 11. 8 BDTs are trained to cover four different mass splittings for m($\tilde{\chi}_1^{\pm}$)-m($\tilde{\chi}_1^0$) = 20, 35, 60, 100GeV. For each mass splitting, one region requires an ISR jet, and the other applies a jet-veto, shown on Fig. 12.

The SM background in the SRs are:

- "prompt" leptons(diboson, triboson, $t\bar{t}$V, tZ, H), estimated from MC simulation.
- non-prompt(fake) leptons(W+jets, $t\bar{t}$) - Matrix Method for the estimation, except Wy by MC simulation.
- Charge-misID leptons - charge-misID rate measured from control region(CR).

4.3 Three light leptons: Event selection and background estimation

The final state of this scenario is: exactly three light leptons(1 same-flavor OS pair) and E_{miss}^{T} shown in Fig. 13.

DOI: 10.1051/epjconf/201612604040

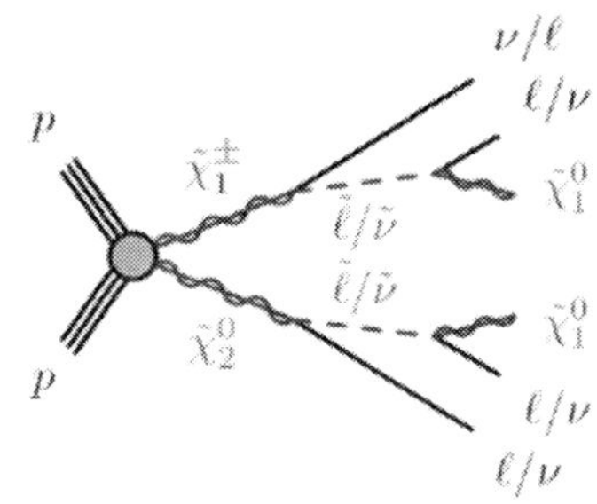

Figure 11. Feynman diagram for compressed scenario with two SS leptons in the final states.

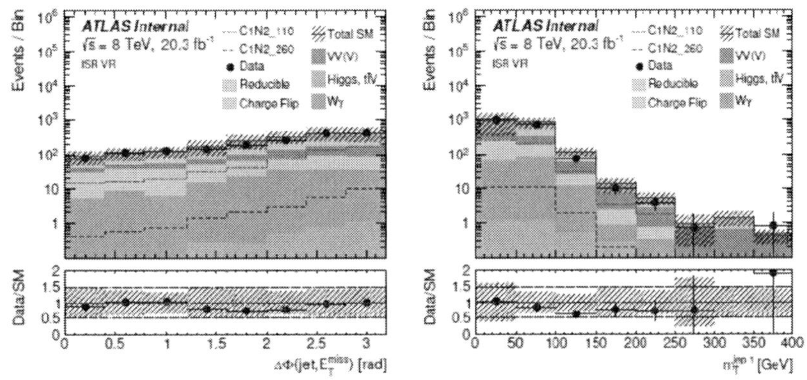

(a) $\Delta\phi(\text{jet}, E_T^{\text{miss}})$ in ISR VR

(b) m_T^{lep1} in ISR VR

Figure 12. Distributions of $\Delta\phi$ and m_T^{lep1} in SRs.

Two SRs are defined with low pT leptons to target mass splittings of 4-15 and 15-25GeV, while another two SRs which request a jet with pT>50GeV to target ISR events for both splitting regions.

The MET distribution in the SRs are shown on Fig. 14, with the following SM background:

- irreducible(prompt) leptons, including diboson and triboson, estimated from MC simulation.
- reducible(fake) leptons, including Vjets, WW, top and $t\bar{t}$, used Matrix Method for the estimation.

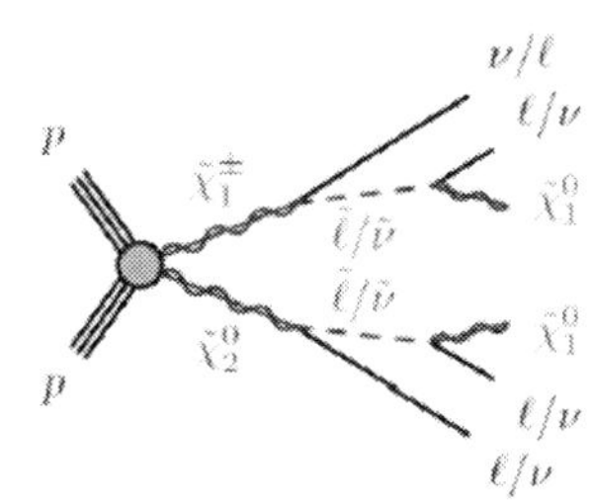

Figure 13. Feynman diagram for compressed scenario with 3 leptons in the final states.

DOI: 10.1051/epjconf/201612604040

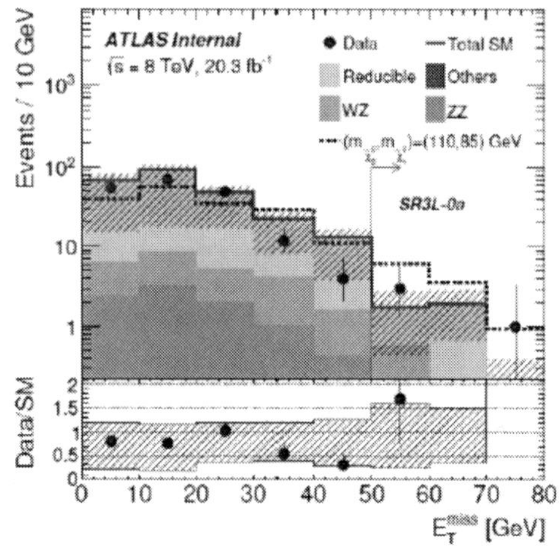

(a) E_T^{miss} in SR3ℓ-0a

Figure 14. Distributions of E^{T}_{miss} in SR.

4.4 results

The fitting results of all the three scenarios are shown on Fig. 15 and Fig. 16. For the OS2l analysis, re-optimized analysis nicely complements the already published one in the region of low mass splittings close to the diagonal.

For the SS2l/3l analysis, re-optimized analysis nicely complements the already published one in the region of low mass splittings close to the diagonal. The combination of the new analyses give an improved sensitivity to compressed scenarios up to $\widetilde{\chi}_1^{\pm}$ masses of 250 GeV.

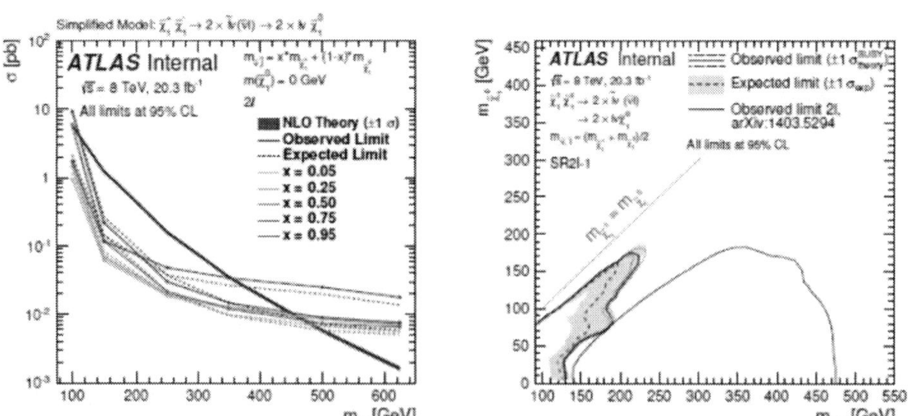

(a) Variable $\tilde{\ell}_L$ mass

(b) m($\tilde{\ell}_L$)=0.5×m($\tilde{\chi}_1^{\pm}$)+0.5×m($\tilde{\chi}_1^0$)

Figure 15. 95% CL exclusion limits on the cross-section for compressed scenario with 2 OS leptons in the final states.

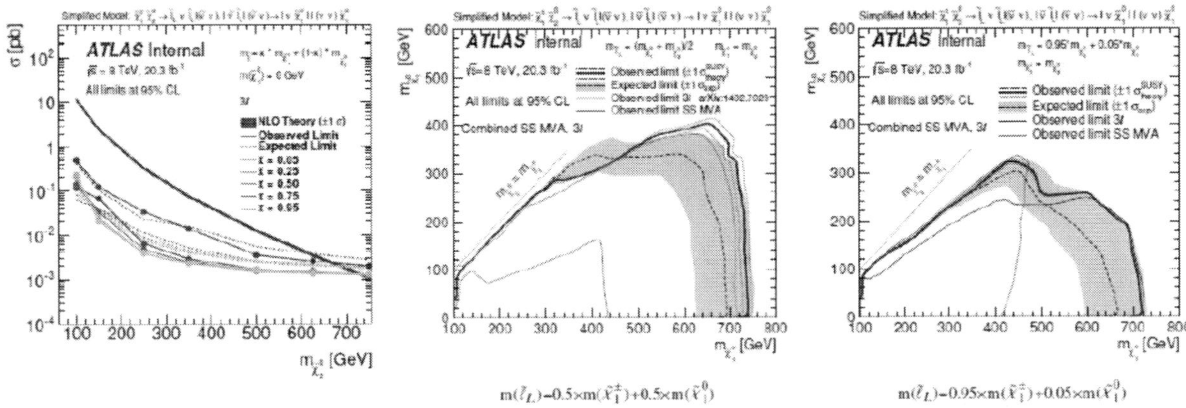

Figure 16. 95% CL exclusion limits on the cross-section for compressed scenario with 2 SS leptons or 3 leptons in the final states.

5 Conclusions

Various searches in EWK SUSY sector at ATLAS:
https://twiki.cern.ch/twiki/bin/view/AtlasPublic/SupersymmetryPublicResults

Newly updated results for EWK SUSY searching are presented, shown on Fig.17. No significant excess observed beyond SM expectation. Higher sensitivity and expanding exclusion/discovery contour is expected with 13/14TeV LHC RUN2 data.

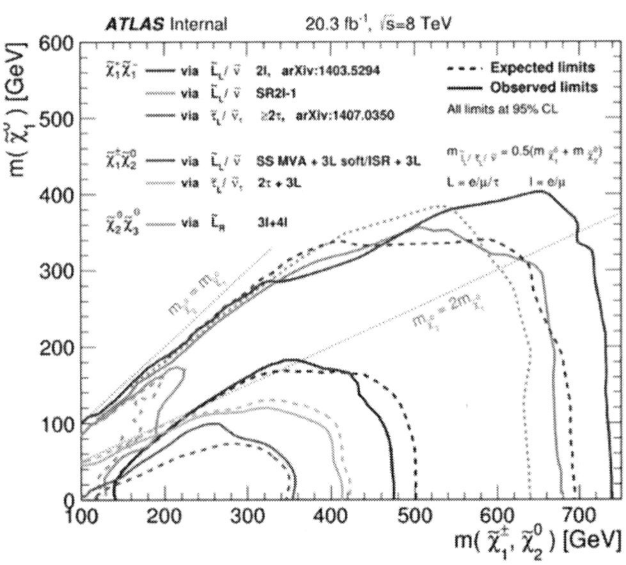

Figure 17. 95% CL exclusion limits on the cross-section for compressed scenario with 2 SS leptons or 3 leptons in the final states.

References

[1] ATLAS collaboration, Searches for electroweak production of supersymmetric gauginos and sleptons with the ATLAS detector, JHEP 1410(2014)96

[2] ATLAS collaboration, EWK summary paper, ATLAS-SUSY-2014-05

EPJ Web of Conferences **126**, 04041 (2016) DOI: 10.1051/epjconf/201612604041
ICNFP 2015

Reconstruction and identification of tau leptons in CMS

Francesco Romeo[1,a] on behalf of the CMS Collaboration

[1] *IHEP of Beijing, 19B Yuquan Lu, Shijingshan District, Beijing, 100049, China*

Abstract. Tau leptons constitute an important experimental signature for analyses at the CERN LHC related to Higgs boson, Standard Model, and beyond the Standard Model measurements. We describe the algorithm used by the CMS experiment to reconstruct and identify decays of tau leptons into hadrons and a neutrino during Run 1 of the CERN LHC. The performance of the algorithm is studied in proton-proton collisions recorded at a center-of-mass energy of 8 TeV, corresponding to an integrated luminosity of 19.7 fb^{-1}. The algorithm achieves an identification efficiency of typically 50-60%, with misidentification rates for quark and gluon jets, electrons and muons that vary between per mille and percent levels.

1 Introduction

Tau leptons constitute an important experimental signature for analyses at the CERN LHC related to Higgs boson, Standard Model, and beyond the Standard Model measurements [1–6]. These analyses benefit from the high reconstruction and identification performances of the CMS tau algorithm [7], whose the description is the main subject of this proceeding.

The tau lepton has a mass of 1.777 GeV and a lifetime of 290.6 fs ($c\tau \approx 87\mu m$) [8]. The branching fractions for the main tau decay modes are given in Table 1.

Table 1. Main decay channels of the tau lepton. The symbol h denotes a charged meson (pion or kaon).

Decay channel	Resonance	BR (%)
$\tau^- \to e^-\bar{\nu}_e\nu_\tau$		17.8
$\tau^- \to \mu^-\bar{\nu}_\mu\nu_\tau$		17.4
$\tau^- \to h^-\nu_\tau$		11.5
$\tau^- \to h^-\pi^0\nu_\tau$	ρ (770)	26.0
$\tau^- \to h^-\pi^0\pi^0\nu_\tau$	a_1 (1260)	9.5
$\tau^- \to h^-h^+h^-\nu_\tau$	a_1 (1260)	9.8
$\tau^- \to h^-h^+h^-\pi^0\nu_\tau$		4.8
Others		3.2
All modes containing hadrons		64.8

It decays into hadrons in about two-third of the cases, typically into either one or three charged mesons (π or, less frequently, K) and up to two neutral pions. The π^0 decays instantaneously in $\gamma\gamma$. The remaining one-third of the case the τ lepton decays into an electron or muon and two neutrinos. The electrons and muons originating from tau decays are reconstructed and identified using standard

[a] e-mail: fromeo@cern.ch

CMS algorithms [9, 10], while the algorithm for the tau reconstruction and identification presented in this paper focuses on "hadronic" tau decays, denoted by τ_h.

The main challenge in identifying τ_h decays is related to quark and gluon jets that can mimic τ_h decays and have a cross section for QCD jet production that exceeds by many orders of magnitude the rate at which tau leptons are produced at the LHC. To mitigate the misidentification of quark and gluon jets as τ_h, we exploit some of the tau properties: small track multiplicity, collimation and isolation of the decay products, non-negligible lifetime. The misidentification of electrons or muons as τ_h represents another possible source of background and specific identification criteria have been developed in order to reduce this possibility.

2 The CMS detector

The Compact Muon Solenoid (CMS) [11] is a general-purpose detector which operates at LHC. The natural coordinate frame used to describe the CMS detector geometry is a right-handed cartesian system with its origin at the nominal interaction point, the x-axis pointing radially inward-towards the centre of the LHC, the y-axis pointing up (perpendicular to the LHC plane), and the z-axis along the anticlockwise-beam direction. The polar angle θ is measured from the positive z-axis and the azimuthal angle ϕ is measured in the transverse (x, y) plane. The pseudorapidity, η, is defined as $-ln(tan\frac{\theta}{2})$, the radius $r = \sqrt{x^2 + y^2}$. The CMS subdetectors, radially installed inside-out, are: tracker, electromagnetic calorimeter (ECAL), hadronic calorimeter (HCAL), muon systems. A detailed description of the CMS detector can be found in [11].

3 The hadronic tau reconstruction and identification

In this section we describe the hadrons-plus-strips (HPS) algorithm used by CMS to tag τ_h. The algorithm follows two main stages: the reconstruction step, in which charged and neutral particles are combined to look for specific τ_h decays; and the identification step, in which some discriminators are defined to separate τ_h decays from quark and gluon jets, electrons, and muons.

3.1 The hadronic tau reconstruction

CMS uses a particle-flow (PF) algorithm [12, 13] to provide a global event description. The information available in all CMS subdetectors is employed to reconstruct and identify individual particles: muons, electrons, photons, charged and neutral hadrons. These particles are then used to reconstruct higher-level objects like isolation variables for leptons or the vector imbalance in transverse momentum in the event (whose module will be referred to as E_T^{miss}). Particles reconstructed by the PF algorithm are also used as input to reconstruct jets, using the anti-k_t algorithm [14] with a distance parameter R = 0.5. Jets are the seed for the HPS algorithm.

In order to reconstruct the tau decay modes, the HPS algorithm considers charged particles and neutral pions. The charged particles are the constituents of the seeding jet which have a minimum transverse momentum, p_T, of 0.5 GeV. The neutral pions, given the high probability for photons from $\pi^0 \to \gamma\gamma$ to convert to e^+e^- pairs, are accounted for by clustering the photon and electron constituents of $p_T > 0.5$ GeV of the jet that seeds the tau reconstruction into $\eta - \phi$ "strips". The strip size is 0.05×0.20, enlarged in ϕ-direction to account for the bending of e^+e^- pairs produced by photon conversions in the 3.8 T magnetic field of CMS. Only strips with $p_T > 2.5$ are considered.

For each jet, charged hadrons and strips, from which the name of the algorithm hadrons-plus-strips originates, are combined to reproduce the tau decays of Table 1, except for the decay mode $h^-h^+h^-\pi^0$

that has a branching fraction of 4.8% and suffers huge contamination by jets. Mass window cuts are applied to test the compatibility of each hypothesis with the signatures expected for the different hadronic tau decay modes given in Table 1:

- $h^{\pm}h^{\mp}h^{\pm}$**:** Combination of three charged particles that sum to unit charge and have a mass $0.8 < m_{\tau} < 1.5$ GeV.
- $h^{\pm}\pi^{0}\pi^{0}$**:** Combination of a single charged particle with two strips with $0.4 < m_{\tau} < 1.2 \cdot \sqrt{p_T[\text{GeV}]/100}$ GeV. The size of the mass window is enlarged for τ_h candidates of high p_T to account for resolution effects. For hadronic tau candidates of $p_T < 100$ GeV (> 1111 GeV) the upper limit of the mass window is set to 1.2 GeV (4.0 GeV).
- $h^{\pm}\pi^{0}$**:** Combination of one charged particle plus one strip with $0.3 < m_{\tau_h} < 1.3 \cdot \sqrt{p_T[\text{GeV}]/100}$ GeV. For τ_h candidates of $p_T < 100$ GeV (> 1044 GeV) the upper limit of the mass window is set to 1.3 GeV (4.2 GeV).
- $h^{\pm}$**:** A single charged particle with no strips.

The four-momentum of each τ_h candidate hypothesis (p_T, η, ϕ, and mass) is given by the momentum sum of charged particles plus strips that are included in the respective decay mode combination. The two decay modes $h^{\pm}\pi^{0}$ and $h^{\pm}\pi^{0}\pi^{0}$ are analysed together and referred to as $h^{\pm}\pi^{0}s$. The distribution of tau decay modes reconstructed in $Z/\gamma^{*} \to \tau\tau$ events and of the reconstructed mass of the τ_h candidates in those events are shown in Fig. 1.

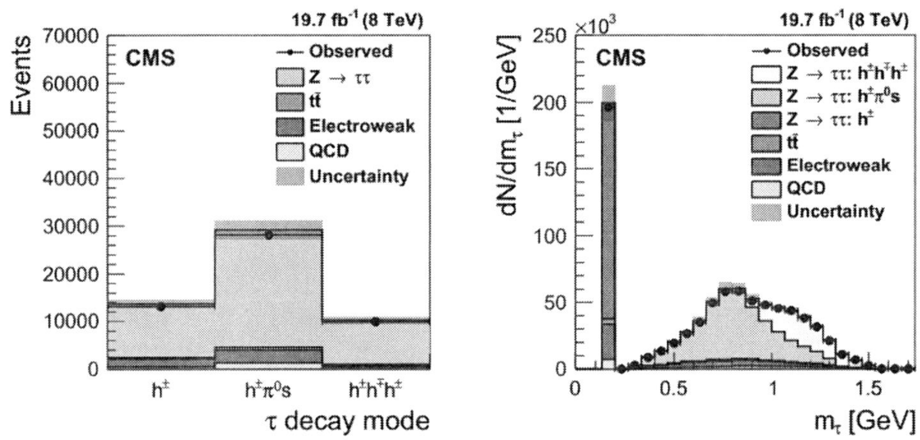

Figure 1. Distributions of the reconstructed tau decay mode (left) and of τ_h candidate mass (right) in $Z/\gamma^{*} \to \tau\tau$ events selected in data compared to the Monte-Carlo expectation.

3.2 The hadronic tau identification

Reconstructed τ_h candidates are required to satisfy some identification criteria in order to mitigate the background related to quark and gluon jets or due to electrons and muons.

The HPS algorithm has two isolation discriminators: cut-based and MVA-based (where MVA stands for Multivariate Analysis).

Cut-based: The isolation of τ_h candidates is computed by summing the transverse momenta of charged particles of $p_T > 0.5$ GeV plus photons of $E_T > 0.5$ GeV, reconstructed by the PF algorithm, within an isolation cone of size $\Delta R = 0.5$, centered on the τ_h direction. The effect of pileup is reduced by requiring the tracks associated to charged hadrons considered in the isolation p_T sum to be compatible with the production vertex of the τ_h candidate within a distance of 0.2 cm in direction of the beamline and 0.03 cm in the transverse plane. Charged hadrons used to built the τ_h candidate are excluded from the isolation p_T-sum, as are photons used to built any of the strips. The effect of pileup on the photon isolation is compensated on statistical basis, by applying so-called $\Delta\beta$ corrections:

$$I_\tau = \sum p_T^{charged}(d_Z < 0.2\text{cm}) + \max\left(p_T^\gamma - \Delta\beta, 0\right), \quad \Delta\beta = 0.46 \cdot \sum p_T^{charged}(d_Z > 0.2\text{cm}) \quad (1)$$

Two sets of cut-based tau isolation discriminators are computed, by requiring that at least 8 (HPS combined isolation 8-hit) and 3 (HPS combined isolation 3-hit) hits in the pixel plus silicon strip tracking detectors are associated to each track. Loose, medium and tight working-points are defined by requiring the isolation p_T-sum defined by Eq. 1 not to exceed thresholds of 2.0, 1.0 and 0.8 GeV respectively.

MVA-based: A boosted decision tree discriminator (BDT) [15] is used to discriminate hadronic tau decays ("signal") from quark and gluon jets ("background"). The variables used as inputs to the BDT are:

- The charged particle isolation p_T-sum $\sum p_T^{charged}(d_Z < 0.2\text{cm})$ and the neutral particle isolation p_T-sum $\sum p_T^\gamma$ defined in Eq. 1 as separate inputs.
- The reconstructed tau decay mode.
- The transverse impact parameter, d_0, of the highest p_T track of the τ_h candidate and its significance, d_0/σ_{d_0}.
- The τ_h flight distance (for the $h^\pm h^\mp h^\pm$ decay mode)

The position of the primary event vertex is refitted after excluding the tracks associated to the τ_h candidate. The inputs are augmented by the p_T and η of the hadronic tau candidate and by the $\Delta\beta$ correction defined by Eq. 1. The purpose of the p_T and η variables is to parametrise possible dependencies of the other input variables on transverse momentum and pseudo-rapidity. The $\Delta\beta$ correction parametrises the dependency on pileup for the neutral particle isolation p_T-sum.

The BDT is trained on event samples produced using the Monte-Carlo (MC) simulation. Samples of $Z/\gamma^* \to \tau\tau$, Higgs $\to \tau\tau$, $Z' \to \tau\tau$, and $W' \to \tau\nu_\tau$ events, covering the range 20-2000 GeV in τ_h candidate p_T, are used for the "signal" category. Reconstructed τ_h candidates are required to match hadronic tau decays within $\Delta R < 0.3$ on generator level. W+jets and QCD multi-jet events are used for the "background" category. The τ_h candidates that match leptons originating from the W decays are excluded from the training. Half of the available event statistics is used for the training, the other half for evaluating the MVA performance and conducting overtraining checks.

Different working-points, corresponding to different τ_h identification efficiencies and jet $\to \tau_h$ misidentification rates, are defined by varying the cut on the MVA output.

The performance of the various isolation discriminators described so far (3-hits, 8 hits, MVA with and without lifetime information) are compared in Fig. 2, in terms of efficiency versus fake-rate (probability of misidentification). The efficiency is computed separately in simulated $Z/\gamma^* \to \tau\tau$ (left) and $Z' \to \tau\tau$ (right). The fake-rate is measured in simulated samples of multijet events. The inclusion of tau lifetime information in the MVA discriminator reduces the jet $\to \tau_h$ fake-rate by about a factor two, for the same tau identification efficiency, compared to the cut based tau isolation discriminator.

Electrons and muons may be reconstructed as τ_h in the h decay mode or, in case of electrons that radiate a bremsstrahlung photon which subsequently converts, in the $h + \pi^0$ decay mode. Specific discriminators have been developed to mitigate this contamination.

An MVA-based discriminator is used to separate hadronic taus from electrons. The discriminator is built using the energy deposits in the ECAL and HCAL calorimiters as inputs. Track-related variables as the p_T, η, or the number of hits in the tracker detector are also used.

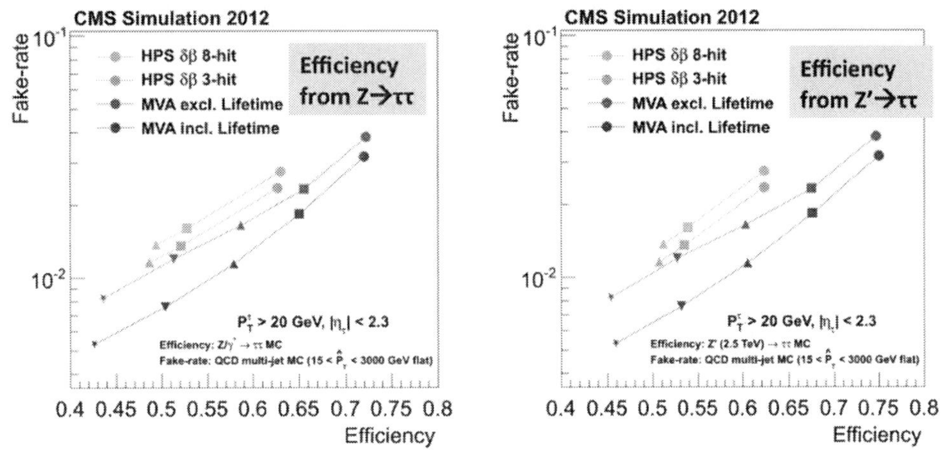

Figure 2. τ_h identification efficiency versus fake-rate for HPS cut-based isolation with 3-hits (green) and 8-hits (blue) and MVA-based isolation with (red) and without (pink) tau lifetime information. For a given discriminator, the markers correspond to the different working points.

The discriminant against muons vetoes τ_h candidates when signals in the muon system are found near the τ_h direction or when the sum of the energies in the ECAL and HCAL corresponds to < 0.2 of the momentum of the leading track of the τ_h candidate.

4 The hadronic tau performance with the LHC Run 1 at $\sqrt{s}$ = 8 TeV

The performance of the τ_h algorithm is studied in proton-proton collisions recorded at a center-of-mass energy of 8 TeV, corresponding to an integrated luminosity of 19.7 fb^{-1}. Different samples of events are used to check the τ_h reconstruction and identification in data and to validate the rates with which quark and gluon jets, electrons, and muons get misidentified as τ_h decays.

4.1 Measurement of the hadronic tau identification efficiency

The τ_h identification efficiency measurement is performed on a sample of high-purity $Z/\gamma^* \to \tau\tau$ in the $\mu\tau_h$ final state. The events are selected by pairing an isolated muon with $p_T > 25$ GeV and the highest p_T jet with at least 20 GeV, $|\eta| < 2.3$ that is separated from the muon by $\Delta R(\sqrt{(\Delta\phi)^2 + (\Delta\eta)^2}) > 0.5$. The muon and the p_T leading track of the jet are required to have opposite charge and to be compatible with originating from the same vertex. The transverse mass of the muon and E_T^{miss}, $m_T(\mu, E_T^{miss}) = \sqrt{2p_T^\mu E_T^{miss}(1 - cos\Delta\phi)}$, lower than 40 GeV and the topological cut $p_\zeta - 1.85p_\zeta^{vis} > -15$, described in [7], are further demanded. A veto on the presence of b quark jets and extra leptons is asked. No τ_h identification criteria are directly applied during the event selection.

The τ_h identification efficiency is obtained through a simultaneous fit of the number of $Z/\gamma^* \to \tau\tau$ events, N_{pass}^τ and N_{fail}^τ, with hadronic taus passing ("pass" region) and failing ("fail" region) the τ_h identification discriminant. The τ_h identification efficiency is taken as the parameter of interest μ in the fit. The number of $Z/\gamma^* \to \tau\tau$ events in the pass and fail regions as a function of μ is given by $N_{pass}^\tau = \mu N^{Z/\gamma^* \to \tau\tau}$ and $N_{fail}^\tau = (1 - \mu)N^{Z/\gamma^* \to \tau\tau}$ respectively. The multiplicity of tracks within a cone of size $\Delta R < 0.5$ centered on the τ_h direction is used to perform the fit. For more details on the fitting procedure, the background contaminations, and the systematic uncertainties estimations we turn to the main article [7].

Figure 3 shows the results of the efficiency measurement as a function of p_T for the cut-based (left) and MVA (right) isolation discriminators for different working points, considering both data (solid symbols) and MC simulation (open symbols). The MVA-based isolation performs better, reaching an efficiency of up to 70%. All data-to-MC ratios are compatible with unity within the estimated uncertainties of about 5%.

The efficiency to reconstruct and identify τ_h of higher p_T in more dense hadronic environments is measured using $t\bar{t}$ events. A detailed description of this measurement is reported in [7].

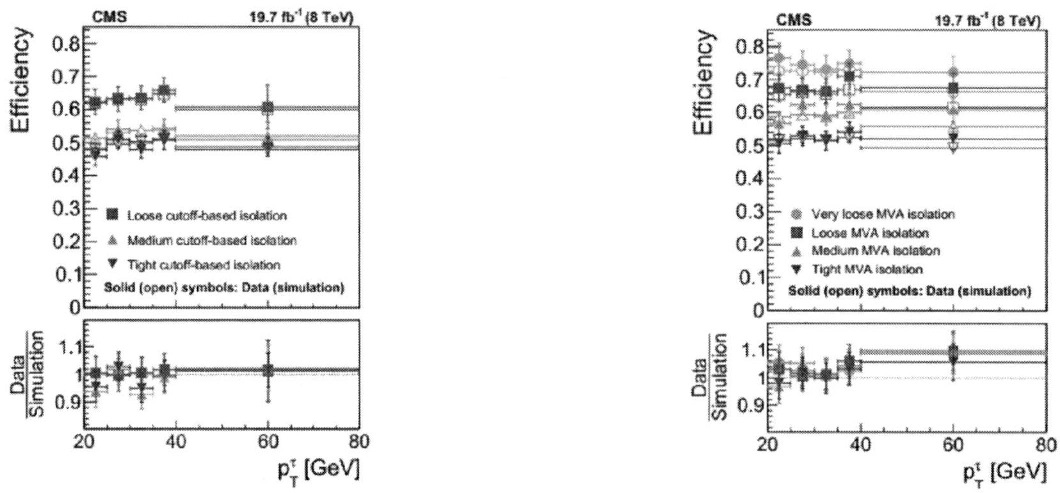

Figure 3. τ_h identification efficiency measured in $Z/\gamma^* \to \tau\tau \to \mu\tau_h$ events as function of p_T for the cut-based (left) and MVA-based (right) tau isolation discriminants, for data and MC expectation.

4.2 Measurement of the hadronic tau energy scale

The energy scale for τ_h (referred to as τES) is defined as the average reconstructed τ_h energy relative to the generator level energy of the visible τ decay products. MC-to-data τES corrections are determined in $Z/\gamma^* \to \tau\tau$ events selected in the $\mu\tau_h$ channel as described in Sec. 4.1, with the additional requirement that the selected 6jet passes the medium working point of the MVA-based τ_h isolation.

The corrections are obtained by fitting the reconstructed mass of the τ_h candidate, m_{τ_h}, and the mass of muon and τ_h candidates, m_{vis}. The m_{τ_h} and m_{vis} templates are computed by changing the τ_h four-momentum as a function of τES, and recomputing m_{τ_h} and m_{vis} after each such change.

The τES is measured separately for the reconstructed τ_h decay modes and as a function of p_T. The variable m_{τ_h} is seen to be the more sensitive observable compared to m_{vis}, as indicated by smaller uncertainties. τES correction is always found to be lower than 2%. There is no indication of a dependence of the measured τES corrections on τ_h p_T.

4.3 Measurement of the rate of jets, electrons, and muons misidentified as hadronic tau

The rate for quark and gluon jets to be misidentified as τ_h decays is measured in W+jets and multijet events. W+jets events are selected requiring the presence of a muon of $p_T > 25$ GeV, $|\eta| < 2.1$, the $m_T(\mu, E_T^{miss}) > 50$ GeV, and at least one jet of $p_T > 20$ GeV, $|\eta| < 2.3$ separated from the muon by $\Delta R > 0.5$. Multijet events are selected demanding the presence of two jets with $p_T > 20$ GeV, $|\eta| < 2.3$. The misidentification rates measured in W+jets and in multijet events are shown in Fig. 4.

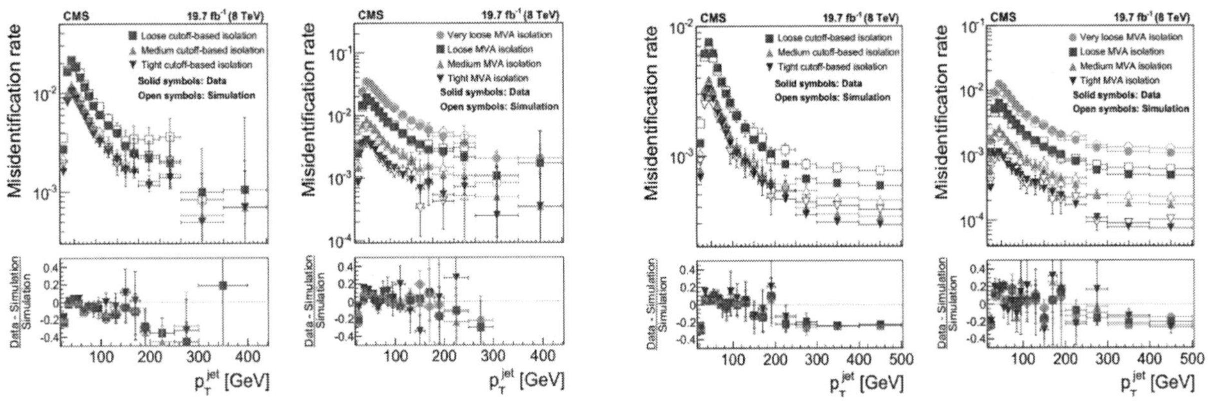

Figure 4. Fraction of times as a function of jet p_T in which quark and gluon jets in W+jets (left) and multijet (right) events pass the cutoff-based and MVA-based τ_h isolation discriminant.

In general, the misidentification rates are higher in W+jets than in multijet events. The difference is due to the higher fraction of quark jets in W+jets events that typically have a lower particle multiplicity and are more collimated than gluon jets. Overall, the jet $\to \tau_h$ misidentification rates vary between low percentages and about 10^{-4}, decreasing as a function of jet p_T, as the particle multiplicity increases for higher energy jets. MVA-based isolation, which includes τ lifetime information, performs better than cut-based isolation.

Notable differences are observed in the comparison between data and expectations. The rates for jet $\to \tau_h$ measured in data exceed the MC expectation at low p_T while the rates measured at high p_T fall short of the simulation. The magnitude of the change on data/MC is $\approx 20\%$.

The fraction of times in which electrons or muons pass the τ_h identification criteria of the dedicated discriminants against electrons or muons, are measured through the tag-and-probe technique using $Z/\gamma^* \to ee$ and $Z/\gamma^* \to \mu\mu$ events. The measured $e \to \tau_h$ ($\mu \to \tau_h$) fake-rate is on the level of per-mille for a τ_h efficiency of 90% (98%). The rates measured in data exceed the MC prediction by up to a factor 1.7 for electrons. The full details of these measurements are reported in [7].

5 First results with hadronic taus with the LHC Run 2 at $\sqrt{s}$ = 13 TeV

Preliminary results related to τ_h has been obtained by analysing proton-proton collisions recorded at a center-of-mass energy of 13 TeV, corresponding to an integrated luminosity of 5.6 pb^{-1}. $Z/\gamma^* \to \tau\tau \to \ell\tau_h$, $\ell = e, \mu$, have been selected by requiring the presence of an isolated electron (muon) with $p_T > 20$ ($p_T > 18$) GeV, $|\eta| < 2.1$ and an isolated τ_h with $p_T > 20$ GeV, $|\eta| < 2.3$ passing the identification criteria against electrons and muons that have an opposite charge to the other lepton. $m_T(\mu, E_T^{miss}) < 40$ GeV is also required.

Figure 5 shows the visible mass distribution (left) and fully reconstructed mass using the SVFit [16] algorithm (right). The simulation is normalised to the total integrated luminosity for all contributions except from the QCD background that is estimated by same sign data. A good agreement is observed overall.

6 Conclusions

The algorithm used by the CMS experiment for the reconstruction and identification of hadronic τ decays in Run 1 data from the LHC have been presented, and their performance validated with proton-proton collision data recorded at $\sqrt{s} = 8$ TeV, corresponding to an integrated luminosity of 19.7 fb^{-1}.

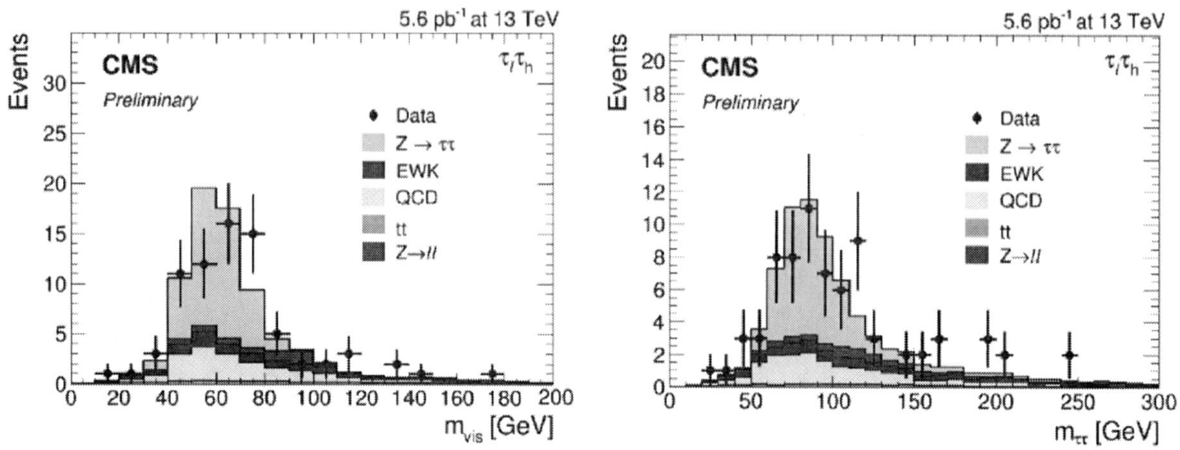

Figure 5. The visible mass distribution (left) and fully reconstructed mass using the SVFit algorithm (right) obtained by analysing proton-proton collisions recorded at a center-of-mass energy of 13 TeV, corresponding to an integrated luminosity of 5.6 pb^{-1}.

The algorithm achieves a τ_h identification efficiency of typically 50-60%, and misidentification rates for quark and gluon jets, electrons, and muons that vary between the per mille and per cent level. The τ_h identification efficiency measured in the data agrees with the MC expectation within the uncertainty of the measurement of about 5%. The measured jet $\rightarrow \tau_h$ misidentification rates are about 20% higher than predicted for low-p_T jets and 20% lower for high-p_T jets. The measured misidentification rate for electrons exceeds the MC expectation by up to a factor 1.7.

Preliminary results related to τ_h obtained by analysing $Z/\gamma^* \rightarrow \tau\tau \rightarrow \ell\tau_h$, $\ell = e, \mu$, events from proton-proton collisions recorded at $\sqrt{s} = 13$ TeV and corresponding to an integrated luminosity of 5.6 pb^{-1} show good τ_h performances in view of the LHC Run 2.

References

[1] CMS Collaboration, JHEP **05**, 104 (2014)
[2] CMS Collaboration, JHEP **10**, 160 (2014)
[3] CMS Collaboration, Phys. Lett. **B 739**, 23 (2014)
[4] CMS Collaboration, JHEP **08**, 117 (2011)
[5] CMS Collaboration, Phys. Rev. Lett. **110**, 081801 (2013)
[6] CMS Collaboration, Phys. Lett. **B 716**, 82 (2012)
[7] CMS Collaboration, Submitted to JINST, arXiv:1510.07488 (2015)
[8] Particle Data Group, Chin.Phys. **C38**, 090001 (2014)
[9] CMS Collaboration, JINST **10**, P06005 (2015)
[10] CMS Collaboration, JINST **7**, P10002 (2012)
[11] CMS Collaboration, JINST **3**, S08004 (2008)
[12] CMS Collaboration, CMS-PAS-PFT-10-001, (2009)
[13] CMS Collaboration, CMS-PAS-PFT-10-002, (2010)
[14] Cacciari et al., JHEP **04**, 063 (2008)
[15] Hoecker et al., PoS **ACAT**, 040 (2007)
[16] L. Bianchini et al., J.Phys.Conf.Ser. **513**, 022035 (2014)

Perfect Abelian dominance of confinement in mesons and baryons in SU(3) lattice QCD

Naoyuki Sakumichi[1,a] and Hideo Suganuma[2]

[1] *Theoretical Research Division, Nishina Center, RIKEN, Wako, Saitama 351-0198, Japan*
[2] *Department of Physics, Kyoto University, Kitashirakawaoiwake, Sakyo, Kyoto 606-8502, Japan*

Abstract. For a long time, the quark confinement mechanism has been one of the most difficult problems in theoretical physics. In particular, there is no clear correspondence between the confinement and non-Abelian nature of QCD. We study the static interquark potential and its Abelian projection in both mesons and baryons in the maximally Abelian (MA) gauge in SU(3) quenched lattice QCD. Remarkably, we find that the quark confining force in QCD can be perfectly described only with Abelian variables in the MA gauge, which we call "perfect Abelian dominance" of the quark confinement.

1 Introduction

The mechanism of quark confinement is an important long-standing problem in quantum chromodynamics (QCD). To explain the mechanism, Nambu, 't Hooft and Mandelstam [1–3] proposed the dual-superconductor theory in the 1970s. As shown in figure 1, in this theory, the QCD vacuum is assumed to be an electromagnetic dual version of a superconductor, and one-dimensional squeezing of the interquark color-electric flux is caused by the dual version of the Meissner effect, similar to the formation of the Abrikosov vortex in type-II superconductor. The one-dimensional squeezing yields a linear interquark potential at long distance and explain confinement. However, there are two large gaps between the dual superconductor and the QCD vacuum: (i) the dual superconductor is governed by an Abelian U(1) gauge theory, while QCD is a non-Abelian SU(3) gauge theory; (ii) the dual superconductor requires the condensation of color-magnetic monopoles, while QCD does not have such monopoles as elementary degrees of freedom. To fill in the above two gaps, 't Hooft [4, 5] proposed a mathematical procedure of Abelian projection to reduce QCD to an Abelian gauge theory including monopole degrees of freedom. In particular, as a special Abelian projection, the maximally Abelian (MA) projection can extract infrared-relevant Abelian degrees of freedom from QCD in lattice calculations, which is called "Abelian dominance". Figure 2 shows the procedure of the MA projection: (i) the diagonal part of gluons is maximized by minimizing the off-diagonal part under the gauge transformation, which is called MA gauge fixing [6–11]; (ii) SU(3) QCD is projected onto a $U(1)^2$ Abelian gauge theory by dropping the off-diagonal part of gluons. Many lattice QCD studies have shown that the MA-projected Abelian theory well reproduces interquark potential at long distances, i.e, the string tension σ (the infrared linear slope of the potential) is reproduced by the Abelian-projected one $\sigma^{\rm Abel}$ in both SU(2) [6–10] and SU(3) [12, 13] color QCD.

[a]e-mail: sakumichi.naoyuki@ocha.ac.jp (Present affiliation is Ochanomizu University.)

In this study, we investigate the Abelian dominance of quark confinement of the quark-antiquark ($Q\bar{Q}$) and three-quark (3Q) potentials in SU(3) QCD at the quenched level. Remarkably, we found perfect Abelian dominance of the string tensions in $Q\bar{Q}$ [14] and 3Q [15] systems, i.e., $\sigma^{\rm Abel} \simeq \sigma$.[1] To obtain $\sigma^{\rm Abel} \simeq \sigma$, it is necessary to use large-volume lattices of more than about 2 fm [14, 15]. These observations of $\sigma^{\rm Abel} \simeq \sigma$ indicate that the Abelianization of QCD can be realized without loss of the quark-confining force via the MA projection.

This paper is organized as follows. In Sect. 2, we describe the numerical methods to calculate the MA-projected Abelian theory and the $Q\bar{Q}$ and 3Q potentials. In Sect. 3, we show the numerical results and perfect Abelian dominance of confinement in $Q\bar{Q}$ and 3Q systems. In Sect. 4, we summarize the main results of this study. This paper is based on our previous publications [14, 15], in which the details of the formulations and results are described.

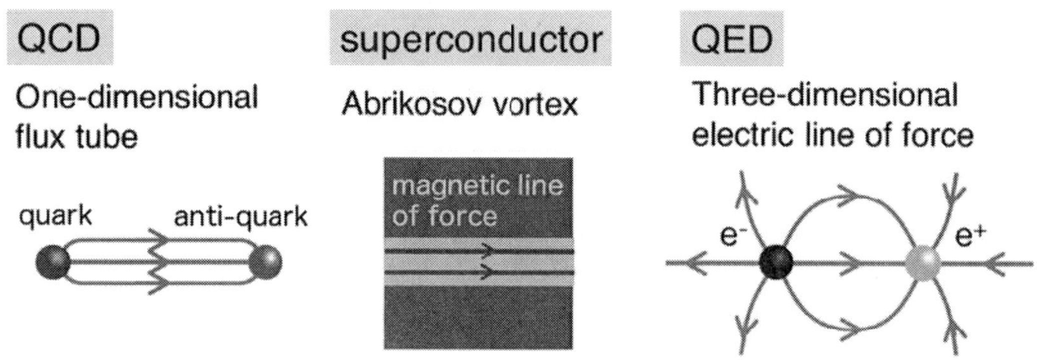

Figure 1. In superconductors, magnetic flux is repelled due to Cooper-pair condensation, and is squeezed into a one-dimensional tube like the Abrikosov vortex. In the dual-superconductor picture, the QCD vacuum is regarded as an electromagnetic dual version of the superconductor: the interquark color-electric flux is squeezed into a one-dimensional form due to magnetic-monopole condensation.

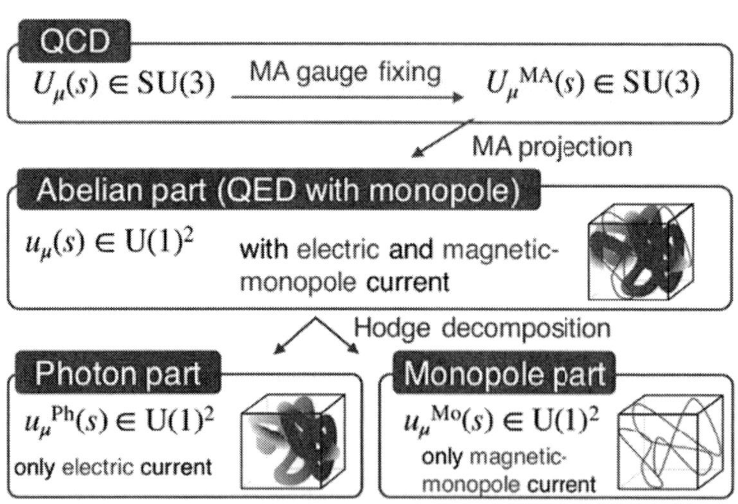

Figure 2. Schematic figure of Abelianization of QCD and the dual-superconductor scenario of confinement. In the MA gauge, QCD becomes Abelian-like, and the monopole current topologically appears. The Abelian part $u_\mu(s)$ is factorized into the regular (photon) part $u_\mu^{\rm Ph}(s)$ and the singular (monopole) part $u_\mu^{\rm Mo}(s)$ with respect to the Hodge decomposition as $u_\mu(s) = u_\mu^{\rm Ph}(s)u_\mu^{\rm Mo}(s)$. The monopole part has confinement, chiral symmetry breaking, and instantons, while the photon part does not have all of them [9, 17–19].

[1]Perfect Abelian dominance was found also in SU(2) quenched QCD [16]. $\sigma^{\rm Abel}/\sigma = 1.02(2)$ for $\beta = 2.5$ on 24^4 lattice, when the physical volume is $La \simeq 2.0$ fm.

2 Numerical methods: the MA projection and the QQ and 3Q potentials.

In this study, we perform SU(3) quenched lattice QCD simulations using the standard plaquette action [20]. We investigate $12^3 32$, $16^3 32$, $20^3 32$, and 32^4 lattices at $\beta \equiv 6/g^2 = 5.8$–6.4, with the gauge coupling g. Here, we identify 32 as the temporal sizes and the others (12^3, 16^3, 20^3, 32^3) as the spatial ones. We summarize the simulation condition and related quantities in table 1. We sample the gauge configuration every 500 sweeps, after a thermalization of 20000 sweeps.

The procedure of MA gauge fixing and MA projection in the lattice QCD formalism is shown in figure 2. The link variable

$$U_\mu(s) = \exp(iagA_\mu(s)) \in \mathrm{SU}(3) \tag{1}$$

describes the gauge field, with the lattice spacing a and the gluon field $A_\mu(s)$. To perform the SU(3) MA gauge fixing, we maximize

$$R_{\mathrm{MA}}[U_\mu(s)] \equiv \sum_s \sum_{\mu=1}^{4} \mathrm{tr}\left(U_\mu^\dagger(s)\vec{H}U_\mu(s)\vec{H}\right) = \frac{1}{2}\sum_s \sum_{\mu=1}^{4}\left(\sum_{i=1}^{3}|U_\mu(s)_{ii}|^2 - 1\right) \tag{2}$$

under the SU(3) gauge transformation. Here, $\vec{H} = (T_3, T_8)$ denotes the generators of the Cartan subalgebra. Let $U_\mu^{\mathrm{MA}}(s)$ be the link variables in the MA gauge. The Abelian part of the link variables

$$u_\mu(s) = \exp\left(i\theta_\mu^3(s)T_3 + i\theta_\mu^8(s)T_8\right) \in \mathrm{U}(1)_3 \times \mathrm{U}(1)_8$$

are extracted by maximizing the norm

$$R_{\mathrm{Abel}} \equiv \frac{1}{3}\mathrm{Re\,tr}\left(U_\mu^{\mathrm{MA}}(s)u_\mu^\dagger(s)\right) \in \left[-\frac{1}{2}, 1\right]. \tag{3}$$

In this study, we numerically maximize R_{MA} in Eq. (2) for each configuration until it converges. Here, we use the overrelaxation method [12] for the maximization algorithm to improve convergence. We remark that there remains the residual $\mathrm{U}(1)^2$ gauge symmetry in the MA gauge. In fact, R_{MA} in Eq. (2) is invariant under the $\mathrm{U}(1)^2$ gauge transformation. Thus, the Abelian link variables $u_\mu(s)$ behave as gauge variables in $\mathrm{U}(1)^2$ lattice gauge theory, which is similar to the compact QED. As mentioned above, the MA-projected Abelian theory has not only the electric current, but also the magnetic-monopole current.

Table 1. The simulation conditions (β, the lattice size L^3L_t, and the gauge configuration number N_{con}), the related quantities (lattice spacing a and the physical spatial size La), and the results of the string tension ratio $\sigma^{\mathrm{Abel}}/\sigma$, $\sigma_{3\mathrm{Q}}^{\mathrm{Abel}}/\sigma_{3\mathrm{Q}}$. The lattice spacing a is determined so as to reproduce the string tension of $\sigma = 0.89$ GeV/fm in the $\mathrm{Q\bar{Q}}$ potential. The label of (ET) means the fit analysis only with the lattice data of equilateral-triangle 3Q configurations. The values in parentheses denote the standard error.

β	L^3L_t	N_{con}	a [fm]	La [fm]	$\sigma^{\mathrm{Abel}}/\sigma$	$\sigma_{3\mathrm{Q}}^{\mathrm{Abel}}/\sigma_{3\mathrm{Q}}$	$\sigma_{3\mathrm{Q}}^{\mathrm{Abel}}/\sigma_{3\mathrm{Q}}$(ET)
6.4	32^4	200	0.0582(2)	1.86(1)	1.015(09)		
6.0	32^4	200	0.1022(5)	3.27(1)	1.009(10)		
5.8	$16^3 32$	2000	0.148(1)	2.37(3)	0.99(3)	0.97(1)	1.01(3)
6.0	$20^3 32$	1000	0.1022(5)	2.05(1)	0.97(1)	0.95(1)	0.97(3)
6.0	$16^3 32$	600	0.102(1)	1.64(1)	0.94(1)		
6.0	$12^3 32$	400	0.104(1)	1.25(4)	0.94(3)		
6.2	$16^3 32$	400	0.075(1)	1.20(1)	0.95(2)		

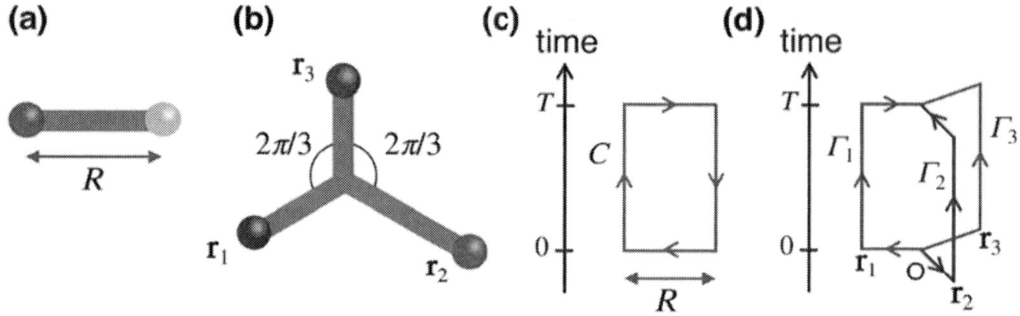

Figure 3. (a) The flux-tube configuration of the mesonic $Q\bar{Q}$ system. (b) The flux-tube configuration of the baryonic 3Q system with the minimal value of the total flux-tube length. There appears a physical junction linking the three flux tubes at the Fermat point. (c) The trajectory of the $(Q\bar{Q})$ Wilson loop W. The quark and anti-quark are generated at $t = 0$, are spatially fixed in $\mathbb{R}^3$ for $0 < t < T$, and are annihilated at $t = T$. (d) The trajectory of the 3Q Wilson loop W_{3Q}. The three quarks are generated at $t = 0$, are spatially fixed in $\mathbb{R}^3$ for $0 < t < T$, and are annihilated at $t = T$.

As shown in figure 3, we calculate the $Q\bar{Q}$ and 3Q potentials [21, 22]

$$V(r) = -\lim_{t\to\infty} \frac{1}{t} \ln \left\langle W\left[U_\mu(s)\right]\right\rangle, \tag{4}$$

$$V_{3Q}(r) = -\lim_{t\to\infty} \frac{1}{t} \ln \left\langle W_{3Q}\left[U_\mu(s)\right]\right\rangle \tag{5}$$

from the $Q\bar{Q}$ Wilson loop $W\left[U_\mu(s)\right] \equiv \mathrm{tr}\left(\prod_C U_\mu(s)\right)$ and the 3Q Wilson loop

$$W_{3Q}\left[U_\mu(s)\right] \equiv \frac{1}{3!} \sum_{a,b,c} \sum_{a'b'c'} \epsilon_{abc}\epsilon_{a'b'c'} X_1^{aa'} X_2^{bb'} X_3^{cc'}, \tag{6}$$

respectively. Here, $\langle \cdots \rangle$ is the statistical average over the gauge configurations. In the $Q\bar{Q}$ and 3Q Wilson loops, $\prod_C U_\mu(s)$ and $X_k \equiv \prod_{\Gamma_k} U_\mu(s)$ are the path-ordered product of the link variables along the path denoted by C and Γ_k in figure 3-(c) and (d), respectively. The $Q\bar{Q}$ and 3Q Wilson loops represent that the gauge-invariant $Q\bar{Q}$ or 3Q state is generated at $t = 0$ and is annihilated at $t = T$ with the quarks spatially fixed in $\mathbb{R}^3$ for $0 < t < T$.

We also calculate the MA projection of the $Q\bar{Q}$ and 3Q potentials

$$V^{\mathrm{Abel}}(r) = -\lim_{t\to\infty} \frac{1}{t} \ln \left\langle W\left[u_\mu(s)\right]\right\rangle, \tag{7}$$

$$V_{3Q}^{\mathrm{Abel}}(r) = -\lim_{t\to\infty} \frac{1}{t} \ln \left\langle W_{3Q}\left[u_\mu(s)\right]\right\rangle \tag{8}$$

from the Abelian Wilson loop in the MA gauge $W\left[u_\mu(s)\right]$ and the 3Q one $W_{3Q}\left[u_\mu(s)\right]$, respectively. They are invariant under the residual Abelian gauge transformation.

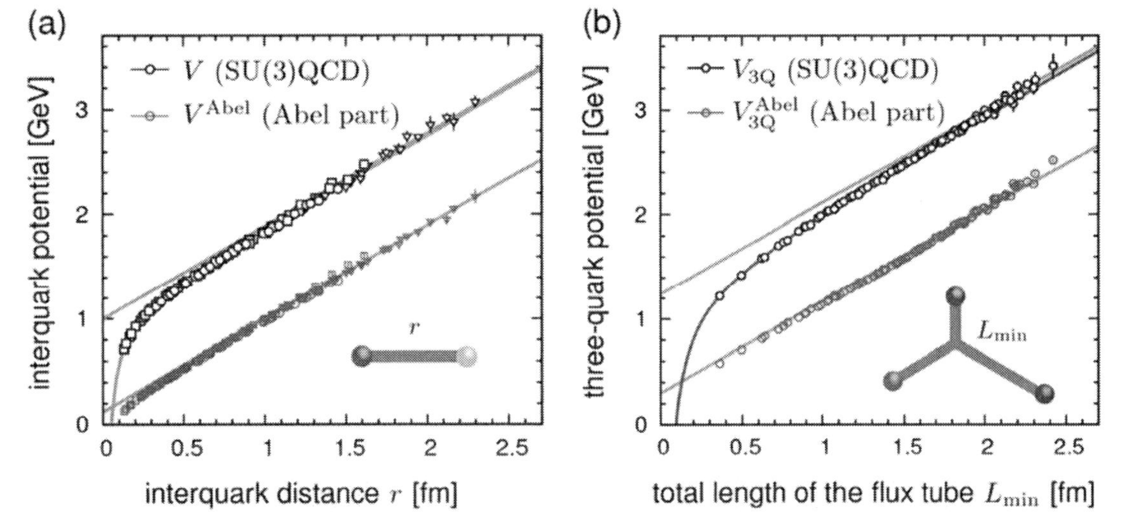

Figure 4. SU(3) quenched lattice QCD results of MA projection of (a) $Q\bar{Q}$ and (b) 3Q potentials taken from Refs. [14] and [15], respectively. In each panel, the black and blue labels denote the original SU(3) potential and the Abelian part, respectively. The circles, triangles, and squares denote the results from $16^3 32$ lattice at $\beta = 5.8$, 32^4 lattices at 6.0 and 6.4, respectively. The gray curves are obtained by the best fit with Eqs. (9) and (11) (Y ansatz). The slopes of the *parallel* orange lines for the panels (a) and (b) are σ and σ_{3Q}, respectively. The error bars in each panel denote the statistical errors estimated with the jackknife method.

3 Perfect Abelian dominance of confinement in mesons and baryons

In this section, we explain lattice QCD results of the original $Q\bar{Q}$ and 3Q potentials $V(r)$ and V_{3Q}, and their Abelian part, $V_{\rm Abel}(r)$ and $V_{3Q}^{\rm Abel}$. Figure 4 (a) shows the $Q\bar{Q}$ potential on the 32^4 lattice at $\beta = 6.4, 6.0$ and on the $16^3 32$ lattice at $\beta = 5.8$. Here, we investigate the on- and off-axis interquark directions as $(1,0,0), (1,1,0), (1,1,1), (2,1,0), (2,1,1), (2,2,1)$. Figure 4 (b) shows the 3Q potential on the $16^3 32$ lattice at $\beta = 5.8$.

First, we explain that the original $Q\bar{Q}$ and 3Q potentials are well reproduced by

$$V(r) = \sigma r - \frac{A}{r} + C, \tag{9}$$

$$V_{3Q}(\mathbf{r}_1, \mathbf{r}_2, \mathbf{r}_3) = \sigma_{3Q} L_{\rm min} - \sum_{i<j} \frac{A_{3Q}}{|\mathbf{r}_i - \mathbf{r}_j|} + C_{3Q}, \tag{10}$$

respectively. The form (10) is called the Y ansatz [21, 22]. Here, $\mathbf{r}_1, \mathbf{r}_2$, and $\mathbf{r}_3$ are the positions of the three quarks, and $L_{\rm min}$ is the minimum flux-tube length connecting the three quarks as shown in figure 3 (b). In fact, figure 4 (a) shows that the best-fit of Eq. (9) (gray curve) well reproduces the original $Q\bar{Q}$ potential. The 3Q potential is rather complicated, because it generally depends on the relative position of the three quarks. However, as shown in figure 4 (b), we can see that all the lattice data of V_{3Q} (at $\beta = 5.8$) are approximately described by a single-valued function of the total length of the flux tube, $L_{\rm min}$. The main reason is that the three-body confinement term $\sigma_{3Q} L_{\rm min}$ is relevant in the Y ansatz (10) except for short distances. When the 3Q system forms an equilateral triangle, one

DOI: 10.1051/epjconf/201612604042

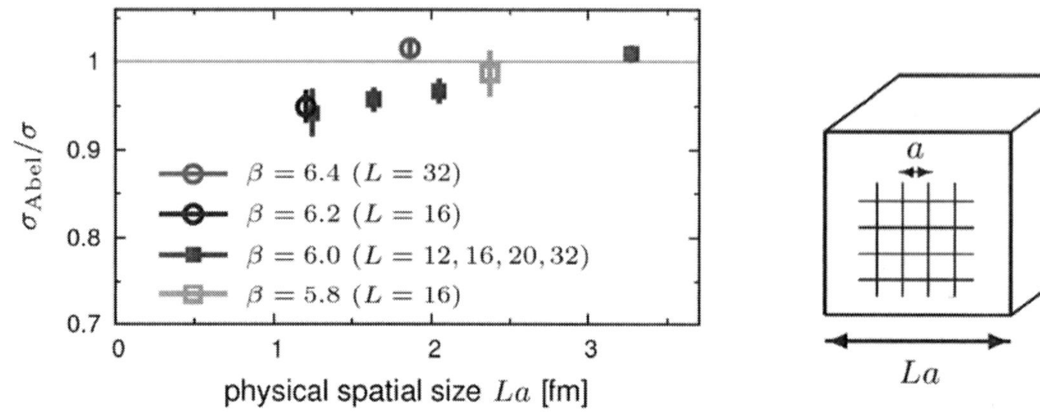

Figure 5. Physical spatial-size dependence of $\sigma_{\rm Abel}/\sigma$ taken from Ref. [15]. Here, σ and $\sigma_{\rm Abel}$ are the string tensions of the $Q\bar{Q}$ potential for SU(3) QCD and the Abelian part, respectively. Perfect Abelian dominance ($\sigma_{\rm Abel}/\sigma \simeq 1$) seems to be realized when the spatial size La is larger than about 2 fm.

finds $L_{\rm min} = \sqrt{3}\,|\mathbf{r}_i - \mathbf{r}_j|$ for any $i \neq j$, and the Y ansatz (10) becomes

$$V_{3Q}(L_{\rm min}) = \sigma_{3Q} L_{\rm min} - 3\sqrt{3}\,\frac{A_{3Q}}{L_{\rm min}} + C_{3Q}. \tag{11}$$

In figure 4 (b), we show that the best-fit of Eq. (11) (gray curve) well reproduces all the lattice data of original 3Q potential, other than the equilateral-triangle 3Q systems. These functional forms (9) and (10) indicate the flux-tube picture on the confinement mechanism as shown in figure 3 (a) and (b); that is, valence quarks are linked by the color flux tube as a quasi-one-dimensional object [20, 23]. The strength of quark confinement is controlled by the string tension of the flux tube, σ or σ_{3Q}, i.e., the infrared linear slopes of the confinement terms in both potentials.

Second, we explain "perfect Abelian dominance" of confinement in mesonic $Q\bar{Q}$ potential [14]. In figure 4 (a), the Coulomb-plus-linear ansatz (9) well reproduce all the lattice data $V(r)$. For a larger interquark distance r than 1 fm, the linear quark-confining potential $\sigma r + C$ describes $V(r)$ [upper straight line in figure 4 (a)]. The Abelian part $V^{\rm Abel}(r)$ (blue labels) has a significant agreement with $\sigma r + C'$ [lower straight line in figure 4 (a)] at large distances, which illuminates "perfect Abelian dominance" of confinement in the $Q\bar{Q}$ potential. It is likely that a large physical volume approximately larger than $(2\ {\rm fm})^3$ is necessary for the perfect Abelian dominance of the string tension, as shown in table 1 and figure 5.

Third, we explain "perfect Abelian dominance" of confinement in baryonic 3Q potential [15]. We note that the perfect Abelian dominance of the $Q\bar{Q}$ potential does not necessarily mean that of the 3Q potential, because one cannot superpose solutions in QCD even at the classical level. Indeed, a 3Q system cannot be described by the superposition of the interaction between two quarks, as is suggested from the functional form (10) of the 3Q potential [21, 22]. We find, however, "perfect Abelian dominance" of confinement in baryonic 3Q potential with an accuracy within a few percent deviation as described below. In figure 4 (b), the Y ansatz (11) well reproduces all the lattice data $V_{3Q}(r)$. When the total flux-tube length $L_{\rm min}$ is larger than 1 fm, V_{3Q} is described by the linear 3Q-confining potential $\sigma_{3Q} L_{\rm min} + C_{3Q}$ [upper straight line in figure 4 (b)]. Remarkably, the Abelian part $V^{\rm Abel}(r)$ has a significant agreement with $\sigma_{3Q} L_{\rm min} + C'_{3Q}$ [lower straight line in figure 4 (b)] at large distances, other than the equilateral-triangle 3Q systems, which is the plausible evidence for perfect

Abelian dominance of quark confinement in baryons. The ratios of the string tensions, $\sigma_{3Q}^{\rm Abel}/\sigma_{3Q}$ are summarized in table 1.

4 Summary

We have studied the MA projection of quark confinement in the mesonic quark-antiquark and baryonic three-quark potentials in the SU(3) QCD with several spacings and volume lattices. We have found perfect Abelian dominance of the string tension with an accuracy within a few percent deviation, $\sigma \simeq \sigma^{\rm Abel} \simeq \sigma_{3Q} \simeq \sigma_{3Q}^{\rm Abel}$, in $Q\bar{Q}$ and 3Q potentials simultaneously. In spite of the non-Abelian nature of QCD, quark confinement is entirely kept in the Abelian sector of QCD in the MA gauge. In other words, Abelianization of QCD can be realized without the loss of the quark-confining force via the MA projection. Such Abelianization scheme of QCD is expected to partially reduce the difficulty stemming from non-Abelian nature of QCD, and is meaningful to understand the quark confinement mechanism.

Acknowledgments

The authors thank Hideaki Iida and Toru T. Takahashi. N. S. is supported by a Grant-in-Aid for JSPS Fellows and by JSPS KAKENHI Grant 250588, 15K17725. H. S. is supported by the Grant for Scientific Research [No.15K05076] from the Ministry of Education, Science and Technology of Japan. The lattice calculations were partially performed on NEC-SX8R at Osaka University. This work was partially supported by RIKEN iTHES Project.

References

[1] Y. Nambu, Phys. Rev. D **10**, 4262 (1974).
[2] G. 't Hooft, in *High Energy Physics*, (Editorice Compositori, Bologna, 1975).
[3] S. Mandelstam, Phys. Rep. **23**, 245 (1976).
[4] G. 't Hooft, Nucl. Phys. **B190**, 455 (1981).
[5] Z. F. Ezawa and A. Iwazaki, Phys. Rev. D **25**, 2681 (1982).
[6] A. S. Kronfeld, G. Schierholz, and U.-J. Wiese, Nucl. Phys. **B293** 461 (1987).
[7] A. S. Kronfeld, M. L. Laursen, G. Schierholz, and U.-J. Wiese, Phys. Lett. B **198**, 516 (1987).
[8] T. Suzuki and I. Yotsuyanagi, Phys. Rev. D **42**, 4257 (1990).
[9] J. D. Stack, S. D. Neiman, and R. J. Wensley, Phys. Rev. D **50**, 3399 (1994).
[10] K. Amemiya and H. Suganuma, Phys. Rev. D **60**, 114509 (1999).
[11] M. A. L. Capri, D. Fiorentini, and S. P. Sorella, Phys. Rev. D **91**, 125004 (2015).
[12] J. D. Stack, W. W. Tucker, and R. J. Wensley. Nucl. Phys. **B639**, 203 (2002).
[13] V. G. Bornyakov *et al.* (DIK Collaboration), Phys. Rev. D **70**, 074511 (2004); **70**, 054506 (2004).
[14] N. Sakumichi and H. Suganuma, Phys. Rev. D **90**, 111501 (2014).
[15] N. Sakumichi and H. Suganuma, Phys. Rev. D **92**, 034511 (2015).
[16] S. Kato, K.-I. Kondo, and A. Shibata, Phys. Rev. D **91**, 034506 (2015).
[17] O. Miyamura, Phys. Lett. B **353**, 91 (1995).
[18] R. M. Woloshyn, Phys. Rev. D **51**, 6411 (1995).
[19] H. Suganuma, A. Tanaka, S. Sasaki, and O. Miyamura, Nucl. Phys. **B** (Proc. Suppl.) **47**, 302 (1996).

[20] H. J. Rothe, *Lattice Gauge Theories*, 4th ed. (World Scientific, Singapore, 2012), and references therein.
[21] T. T. Takahashi, H. Matsufuru, Y. Nemoto, and H. Suganuma, Phys. Rev. Lett. **86**, 18 (2001).
[22] T. T. Takahashi, H. Suganuma, Y. Nemoto, and H. Matsufuru, Phys. Rev. D **65**, 114509 (2002).
[23] H. Ichie, V. Bornyakov, T. Streuer, and G. Schierholz, Nucl. Phys. **A721**, C899 (2003).

EPJ Web of Conferences **126**, 04043 (2016) DOI: 10.1051/epjconf/201612604043
ICNFP 2015

The reconstruction of jets, missing E_{T} and boosted heavy particles with ATLAS in Run 2

Claudio Santoni, on behalf of the ATLAS Collaboration[1,a]

[1] *UBP IN2P3/CNRS Clermont-Ferrand France*

Abstract. The reconstruction of jets, missing E_{T} and boosted heavy particles decaying hadronically has proved to be of extreme importance in Run 1 of the LHC, and has great potential to uncover new physics with Run 2 data. The excellent ATLAS detector capabilities, in particular its high-resolution longitudinally segmented calorimeter and inner detector, have enabled the development of complex clustering and calibration algorithms for the reconstruction of such quantities. In this talk the performance of the tools determined using Run 2 data are presented.

1 Introduction

The ATLAS experiment [1] at the LHC is a multi-purpose particle detector with a forward-backward symmetric cylindrical geometry and a near 4π coverage in solid angle. Studies of the first data collected by ATLAS at center of mass energy $\sqrt{s}$=13 TeV in June-July 2015 allowed to test the performance of the detector. Experimental measurement were compared with the simulated ones. Some results are presented in this talk. Section 2 gives a brief description of the ATLAS detector. The reconstruction and calibration of the jets [2] are discussed in Section 3. Section 4 covers the details of the determination of large distance parameter jet substructure and jet tagging [3]. Finally, Section 5 presents the performance of the missing transverse energy reconstruction [4]. Conclusions are drawn is Section 6.

2 The ATLAS detector

The ATLAS detector [1] consists of an inner tracking detector (ID) residing in a 2 T axial magnetic field provided by a superconducting solenoid, surrounded by electromagnetic and hadronic calorimetry, and a muon spectrometer. The ID covers the pseudo-rapidity range $|\eta| < 2.5$ and consists of a silicon pixel detector, a silicon micro-strip detector (SCT) and a transition radiation tracker (TRT). During the first LHC long shutdown, a new tracking layer, known as the Insertable B-Layer (IBL) [5], was added close to the beam pipe. The high-granularity lead/liquid-argon (LAr) sampling electromagnetic calorimeter covers the region $|\eta| < 3.2$. The iron-scintillator tile calorimeter provides hadronic coverage in the central pseudo-rapidity range $|\eta| < 1.7$. LAr technology is also used for the hadronic calorimeters in the end-cap region $1.5 < |\eta| < 3.2$ and for both electromagnetic and hadronic measurements in the forward region, to $|\eta| < 4.9$. The muon spectrometer (MS) surrounds the calorimeters. It

[a] e-mail: claudio.santoni@cern.ch

DOI: 10.1051/epjconf/201612604043

consists of three large superconducting air-core toroid magnets, precision tracking chambers providing accurate muon tracking out to $|\eta| = 2.7$ and additional detectors for triggering in the region $|\eta| < 2.4$. A two-level trigger system is used to select events. There is a low-level hardware trigger which reduces the incoming data rate, and a high-level software trigger which selects interesting final state events.

3 Inputs to jet reconstruction and the properties of the jets

The production of energetic jets of particles in p-p collisions is one of the most numerous processes observed at LHC. The measurements allow to check perturbative QCD predictions and to search new physics. The performance of the jet reconstruction was checked using early 2015 data corresponding to an integrated luminosity of approximately 6.4 pb^{-1} [2]. In the analysis the same selection criteria were applied to both the MC simulation and to the data. Jets with transverse momentum $p_{\mathrm{T}}^{\mathrm{jet}} > 25$ GeV were considered.

3.1 Jet reconstruction

The reconstruction of jets requires [6]: a) the finding and calibration of group of calorimeter cells topologically connected (topo-clusters) [7], b) jet clustering of topo-clusters (anti-k_t algorithm [8] with R = 0.4) and correction for pile-up, c) MC calibration of the jet vs. pseudo-rapidity and d) residual calibration using in situ measurements.

3.2 Jet reconstruction performance

In general the MC simulation describes the data distributions to within about 10%. The experimental and simulated $p_{\mathrm{T}}^{\mathrm{jet}}$ distributions are compared in Fig. 1 (left). In the case of the $\eta_{\mathrm{det}}^{\mathrm{jet}}$ distribution (Figure 1 (right)) the big disagreement between data and MC in the gap region ($1.2 < |\eta| < 1.8$) is due to the miss-calibration of the special scintillation counters. Large differences are observed also in the forward regions ($3.1 < |\eta| < 4.9$). The jet azimuth angle distribution is well modeled [2].

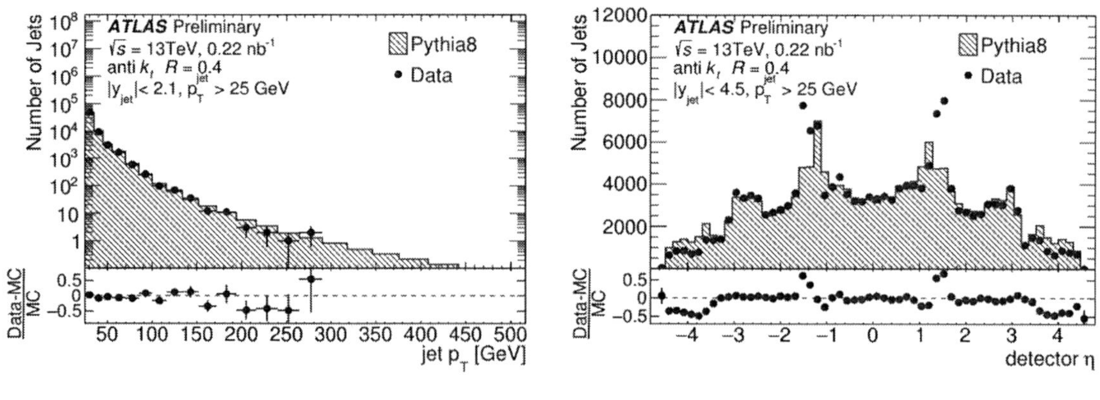

Figure 1. $p_{\mathrm{T}}^{\mathrm{jet}}$ distributions (left). $\eta_{\mathrm{det}}^{\mathrm{jet}}$ distributions (right). The relative differences data MC are shown in the plots [2].

Figure 2 (left) shows the average cluster multiplicity vs. $p_{\mathrm{T}}^{\mathrm{jet}}$. The differences between data and MC are < 20%. As shown in Fig. 2 (right) the average of the fraction of the energy carried by the leading cluster $\langle f_{\mathrm{clus}} \rangle$ vs. $p_{\mathrm{T}}^{\mathrm{jet}}$ is well modeled in MC. About 25% of the jet momentum is contained within

 DOI: 10.1051/epjconf/201612604043

a single cluster. The multiplicity and the kinematic distributions of the charged particles associated to jets are also well described in MC [2].

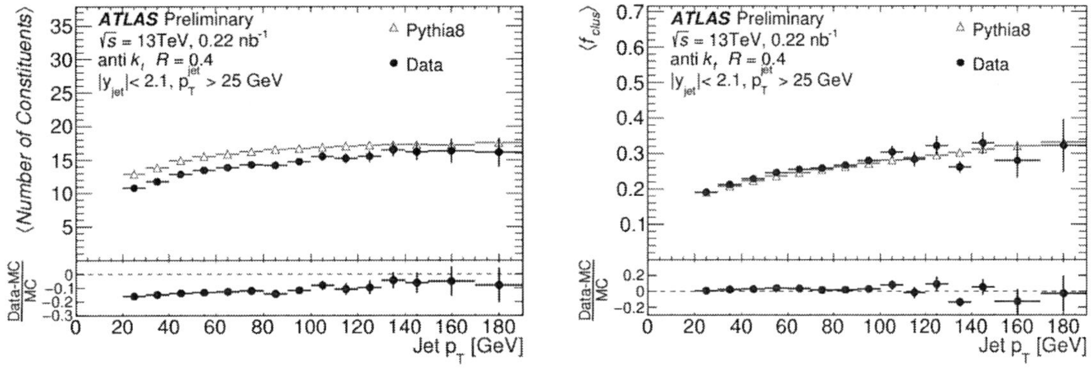

Figure 2. Average cluster multiplicity vs. p_T^{jet} (left). Average of the fraction of the energy carried by the leading cluster vs. p_T^{jet} (right). The relative differences data MC are shown in the plots [2].

The comparison with MC shows reasonable understanding of the longitudinal jet energy profile including the leakage in the muon system. As an example, Fig. 3 (left) shows the energy deposited in the first radial module of the hadronic calorimeter TileCal as a function of p_T^{jet}. The transversal profile of the jet clusters and of the jet associated tracks are generally well modeled. The jets clusters average width vs. p_T^{jet} is shown in Fig 3 (right).

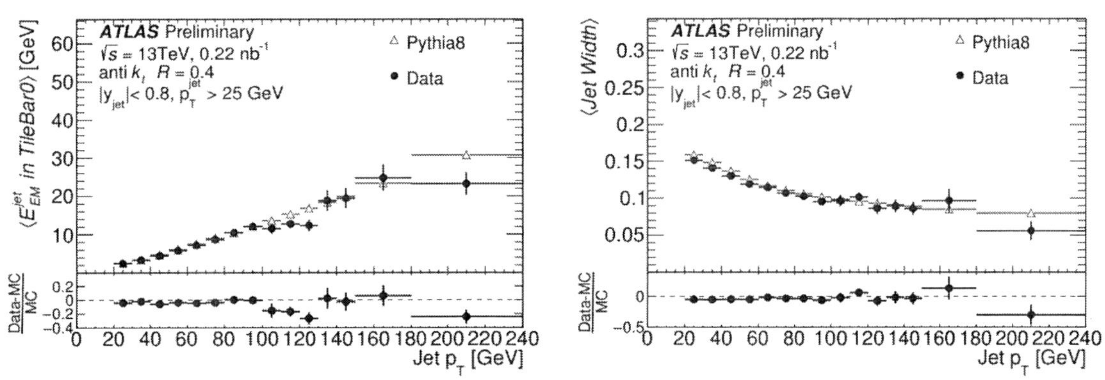

Figure 3. (left) Energy deposited in the first radial module (TileBar0) of the hadronic calorimeter TileCal vs. p_T^{jet}. (right) Jets width vs. p_T^{jet}. The relative differences data MC are shown in the plots [2].

4 Jet substructure performance

Hadronic decays of Lorentz-boosted massive particles are gaining prominence both as useful probes of physics beyond the Standard Model (see for example [9]) and in precision measurements of Standard Model observables. When a boosted heavy particle of mass m and transverse momentum p_T decays, the average angular separation between the decay products scales with m/p_{T}. If this angle is small enough, the decay products are reconstructed as a single jet, and jet substructure techniques become a powerful toolkit for discriminating heavy particle decays from continuum multi-jet processes. The performance studies presented in the talk was focused on jets with $p_{\mathrm{T}}^{\mathrm{jet}}$>200 GeV. The sample of data analyzed corresponds to an integrated luminosity of about 50 pb^{-1} [3].

4.1 Reconstruction of large radius Jets

In ATLAS three different large-radius jet algorithms are considered where the inputs, combination ordering, stopping criterion, and grooming approach vary [4]. One of the three methods is discussed in this talk. Trimmed Jets are built from LCW [6] topological clusters with the anti-k_t algorithm [8] using R = 1.0. They are then trimmed [10] using $R_{\mathrm{sub}} = 0.2$ k_t sub-jets, removing those whose p_{T} fraction is less than $f_{\mathrm{cut}} = 5\%$ of the jet p_{T}. The trimmed jet $p_{\mathrm{T}}^{\mathrm{jet}}$ is further calibrated to account for the residual detector response [11]. The diagram in Fig. 4 depicts the jet trimming procedure.

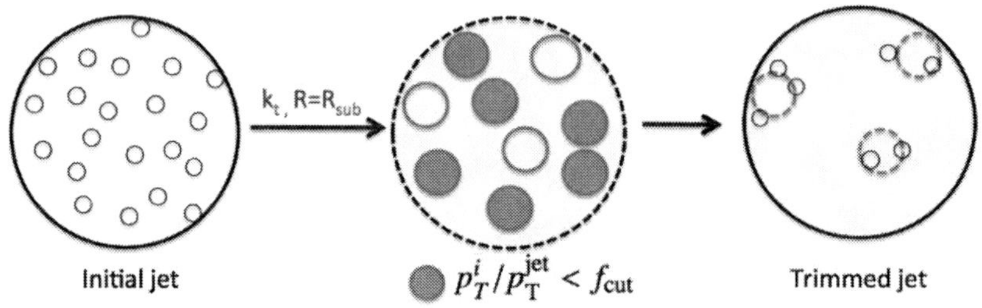

Figure 4. Diagram depicting the jet trimming procedure [11].

4.2 Performance of the reconstruction of large radius Jets

Figure 5 (left) shows the large-jet constituent multiplicity distribution after trimming. There are on average fewer clusters in data than in MC, but the level of disagreement is < 15%. The distribution of the number of trimmed jets sub-jets is shown in Fig. 5 (right). The agreement is good for small sub-jet multiplicities, it degrades for values > 3.

Figure 6 (left) shows the transverse momentum distribution of the trimmed jets. The mass of a jet is given by the difference between the squared sums of the energy E_i and momenta p_i of the constituents. The averages jet mass as a function of $p_{\mathrm{T}}^{\mathrm{jet}}$ obtained using experimental and simulated data are shown in Fig. 6 (right). The shape is modeled in MC within 5-10%.

The mass of the jets discussed in the previous subsection can be used for discriminating heavy particle decays from continuum multi-jet processes. To this aim additional substructure quantities were studied in ATLAS [3]. In this talk the N-subjettiness [12, 13] variables, denoted τ_{N}, were discussed. They attempt to determine the degree to which a given large-radius jet is composed of N or fewer sub-jets. The ratio of the resulting variables is useful in discriminating N from (N-1)-body structures within jets. For instance, two body decay boson jets have smaller values of $\tau_{21} = \tau_2/\tau_1$ than QCD jets. The τ_{21} N-subjettiness ratios distributions obtained using the early 2015 data are reported in Fig. 7. Typical differences of about 10-15% are observed.

DOI: 10.1051/epjconf/201612604043

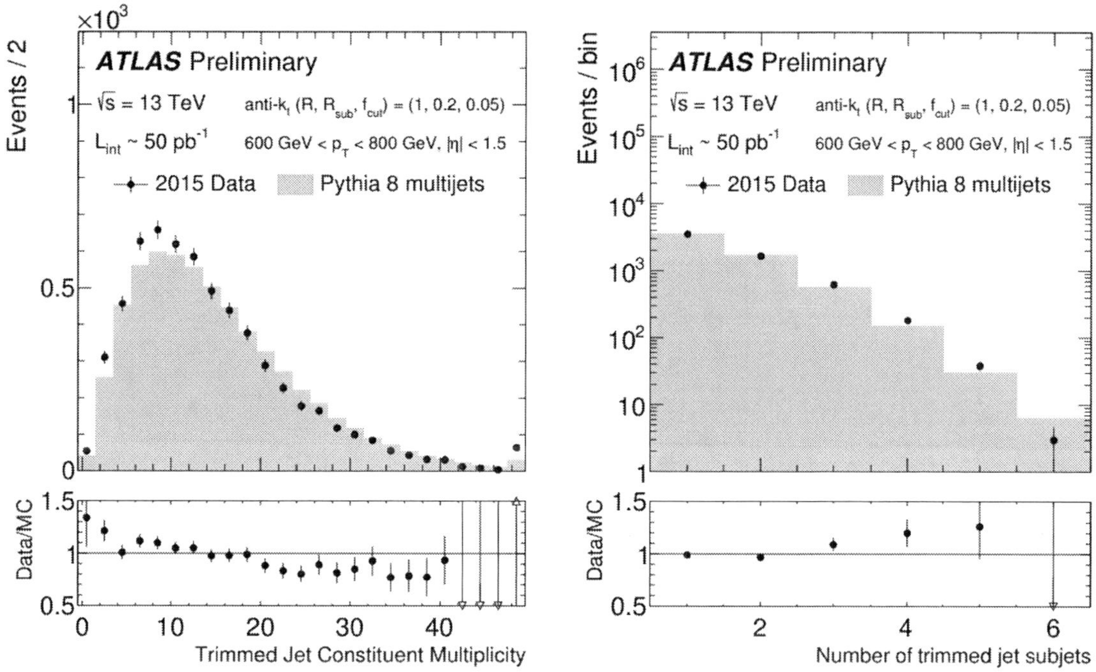

Figure 5. (left) Trimmed jet constituent multiplicity. (right) Distribution of the number of the trimmed jets. The ratios data MC are shown in the plots [3].

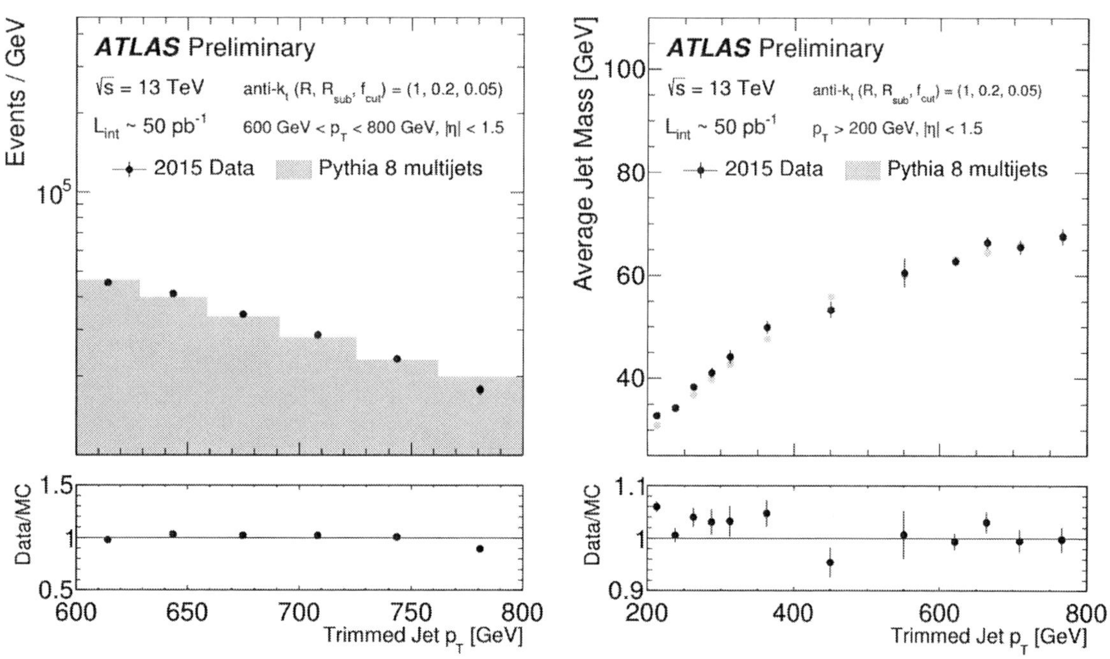

Figure 6. (left) Trimmed jets $p_{\mathrm{T}}^{\mathrm{jet}}$ distribution. (right) Average jet mass as a function of the jet transverse momentum $p_{\mathrm{T}}^{\mathrm{jet}}$. The ratios data MC are shown in the plots [3].

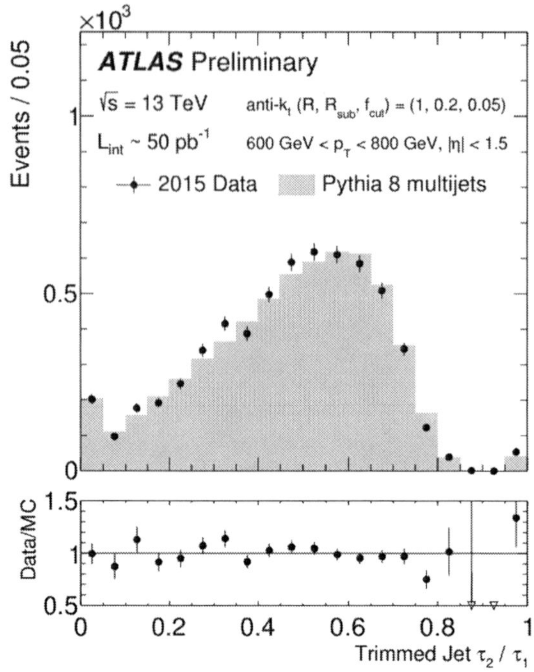

Figure 7. τ_{12} N-subjettiness ratio distributions. The ratios data MC are shown in the plot [3].

5 Missing transverse momentum reconstruction

The missing transverse momentum reconstruction is indicative of weakly interacting stable particles in the final state. The results presented in the talk were obtained analyzing $Z \to \mu^+\mu^-$ decay events. The analyzed sample corresponds to an integrated luminosity of approximately 6.4 pb^{-1} [4].

5.1 Missing E_T reconstruction

Using the Track-based Soft Term method [4], the missing transverse momentum components are obtained using the measured $x(y)$ component of the momentum of all the reconstructed physics objects like jets $E^{\mathrm{jet}}_{x(y)}$, electrons $E^{e}_{x(y)}$, photons $E^{\gamma}_{x(y)}$, taus $E^{\tau}_{x(y)}$ and muons $E^{\mu}_{x(y)}$. The soft term, $E^{\mathrm{soft}}_{x(y)}$, in Eq.(1) does not include contributions from neutral particles and corresponds to the sum of the components of the momentum of all tracks associated with the primary vertex but not matching to any physics object:

$$E^{\mathrm{miss}}_{x(y)} = -(E^{\mathrm{jet}}_{x(y)} + E^{e}_{x(y)} + E^{\gamma}_{x(y)} + E^{\tau}_{x(y)} + E^{\mu}_{x(y)} + E^{\mathrm{soft}}_{x(y)}) \quad (1)$$

One obtains:

$$E^{\mathrm{miss}}_{\mathrm{T}} = \sqrt{E^{\mathrm{miss}}_{x} + E^{\mathrm{miss}}_{y}} \quad (2)$$

The reconstructed missing energy is relatively insensitive to pile-up effects but does not include contributions from neutral particles.

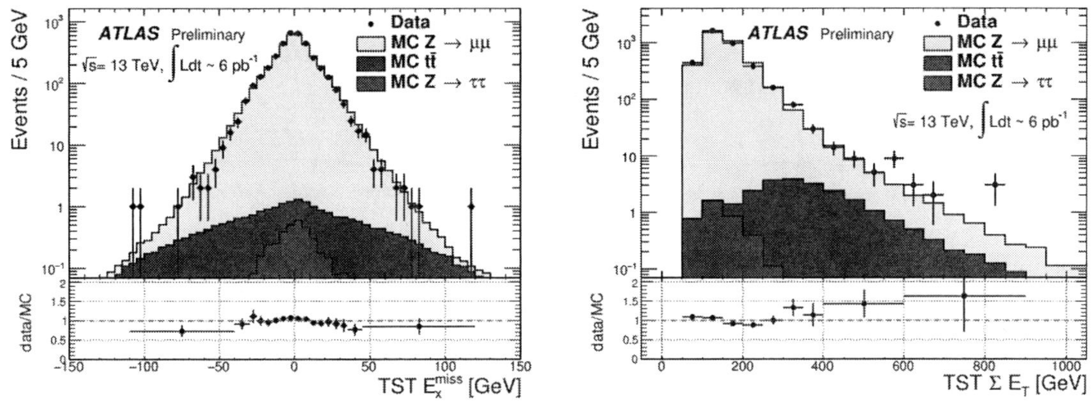

Figure 8. Comparison of the experimental and simulated distributions of E_x^{miss} (left) and $E_{\text{T}}^{\text{miss}}$ (right). The ratios data MC are shown in the plots [4].

5.2 Missing E_{T} reconstruction performance

The agreement in the bulk of the experimental and simulated distributions is within 20%. As an example the E_x^{miss} and the $E_{\text{T}}^{\text{miss}}$ distributions are shown in Fig. 8.

In $Z \to \mu^+\mu^-$ events neutrinos are produced only through heavy flavour meson decays, so this channel has very little genuine missing energy and the RMS of the $E_{\text{T}}^{\text{miss}}$ distributions is indicative of the measurement resolution. Figure 9 (left) shows the resolution of the $E_{x(y)}^{\text{miss}}$ distributions as a function of the number of primary vertices. A bin is required to have a minimum of 200 events to be considered for the resolution curve. Good agreement is found between experimental and simulated data results.

In $Z \to \mu^+\mu^-$ events the mean value of $E_{\text{T}}^{\text{miss}}$ projected onto the unit vector A_Z parallel to momentum of the Z particle, p_{T}^Z, is a measure of the $E_{\text{T}}^{\text{miss}}$ scale. The negative bias of about 5 GeV observed

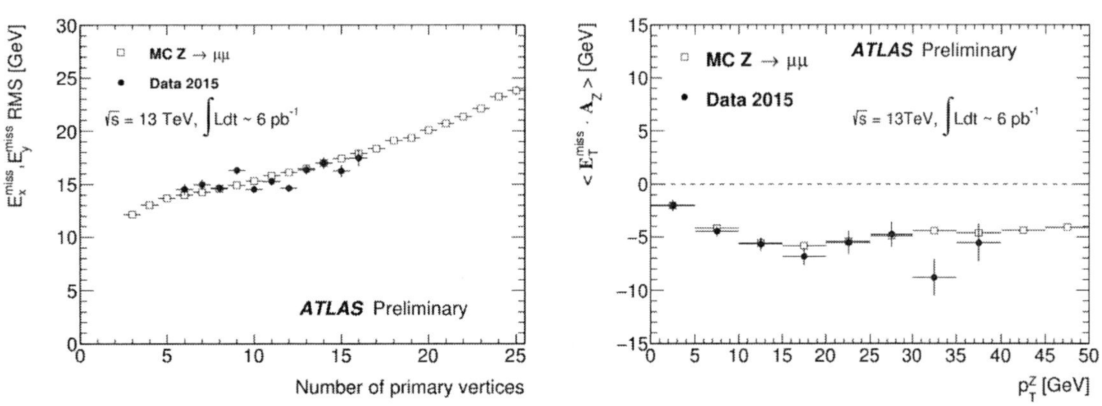

Figure 9. (left) Resolution of the $E_{x(y)}^{\text{miss}}$ determinations as a function of the number of primary vertices. (right) Mean value of the projection of $E_{\text{T}}^{\text{miss}}$ along the Z particle transverse momentum, p_{T}^Z, as a function p_{T}^Z. A bin is required to have a minimum of 200 events to be considered for the plots [4].

in data and MC (see Fig. 9 (right)) indicates an underestimation of the $E_{\mathrm{T}}^{\mathrm{miss}}$ determination due to a) soft neutral particles and b) limited acceptance of the Inner Detector (ID). A bin is required to have a minimum of 200 events to be considered for the scale plot.

6 Conclusions

ATLAS performance concerning the reconstruction of jet reconstruction, the determination of the structures of large radius jets and the missing transverse energy determination has been studied and compared to MC results using first 13 TeV collision data collected in 2015. Reasonable agreement between data and MC simulation is observed.

References

[1] ATLAS Collaboration, The ATLAS Experiment at the CERN Large Hadron Collider, JINST 3 (2008) S08003.

[2] ATLAS Collaboration, Properties of Jets and Inputs to Jet Reconstruction and Calibration with the ATLAS Detector Using Proton-Proton Collisions at $\sqrt{s}$=13 TeV, ATL-PHYS-PUB-2015-036.

[3] ATLAS Collaboration, Performance of jet substructure techniques in early $\sqrt{s}$=13 TeV *pp* collisions with the ATLAS detector, ATLAS-CONF-2015-035.

[4] ATLAS Collaboration, Performance of missing transverse momentum reconstruction for the ATLAS detector in the first proton-proton collisions at $\sqrt{s}$= 13 TeV, ATL-PHYS-PUB-2015-027.

[5] M Capeans et al., ATLAS Insertable B-Layer Technical Design Report, CERN-LHCC-2010-013. ATLAS-TDR-19 (2010), url: https://cdsweb.cern.ch/record/1291633.

[6] ATLAS Collaboration, Jet energy measurement with the ATLAS detector in proton-proton collisions at $\sqrt{s}$=7 TeV, Eur. Phys. J. C 73 (2013) 2304, arXiv: 1112.6426 [hep-ex].

[7] W. Lampl et al., Calorimeter clustering algorithms: description and performance, ATL-LARG-PUB-2008-002 (2008), url: https://cdsweb.cern.ch/record/1099735.

[8] M. Cacciari, G. P. Salam and G. Soyez, The anti-k_t jet clustering algorithm, JHEP 0804 (2008) 063, arXiv: 0802.1189 [hep-ph].

[9] ATLAS Collaboration, Search for high-mass diboson resonances with boson-tagged jets in proton-proton collisions at $\sqrt{s}$= 8 TeV with the ATLAS detector, submitted to JHEP, arXiv:1506.00962 [hep-ex].

[10] D. Krohn, J. Thaler and L.-T. Wang, Jet Trimming, JHEP 1002 (2010) 084, arXiv: 0912.1342 [hep-ph].

[11] ATLAS Collaboration, Performance of jet substructure techniques for large-R jets in proton-proton collisions at $\sqrt{s}$=7 TeV using the ATLAS detector, JHEP 1309 (2013) 076, arXiv: 1306.4945 [hep-ex].

[12] J. Thaler and K. Van Tilburg, Identifying Boosted Objects with N-subjettiness, JHEP 1103 (2011) 015, arXiv: 1011.2268 [hep-ph].

[13] J. Thaler and K. Van Tilburg, Maximizing Boosted Top Identification by Minimizing N-subjettiness, JHEP 1202 (2012) 093, arXiv: 1108.2701 [hep-ph].

EPJ Web of Conferences **126**, 04044 (2016) DOI: 10.1051/epjconf/201612604044
ICNFP 2015

Correlations and fluctuations in p+p and Be+Be at the SPS energies from NA61/SHINE

Andrey Seryakov[1,a], for the NA61/SHINE Collaboration

[1] *Saint Petersburg State University, ul. Ulyanovskaya 1, 198504, Saint Petersburg, Russia*

Abstract. Preliminary results of three different analyses are presented for the lightest collision systems studied in the energy-system size scan program of NA61/SHINE: two-particle correlations in p+p collisions at 20 – 158 GeV/c beam momentum, pseudorapidity correlations in Be+Be collisions at 150A GeV/c and transverse momentum and multiplicity fluctuations of charged particles in p+p (20 – 158 GeV/c) and Be+Be (19A – 150A GeV/c) collisions at the CERN SPS.

1 Introduction

The main goal of the NA61/SHINE fixed target experiment is to uncover properties of the onset of deconfinement by systematic study of collision energy and nuclear mass dependence of hadron production in the region of the threshold for quark-gluon plasma creation. A second aim is to find evidence of a critical point of strongly interaction matter. In this contribution a selected set of preliminary results on fluctuations and correlations from NA61/SHINE [1] is presented. They include:

- Two-particle correlations in p+p collisions at the SPS energy. Their study is widely used in current experiments (RHIC, LHC), as a tool to estimate the influence on multiplicity correlations of different particle sources in heavy ion collisions (see some NA61/SHINE results presented earlier in Refs. [2, 3]).
- Correlations in pseudorapidity in Be+Be collisions at 150A GeV/c. Their study is an effective method for probing the initial conditions for the formation of the QGP [4, 5].
- Multiplicity and transverse momentum fluctuations in p+p [6] and Be+Be collisions at the SPS energy. This analysis is based on measurements of strongly intensive quantities [7], which are preferable to the scaled variance ω, an intensive measure. Intensive quantities are independent of the system size, strongly intensive quantities are in addition independent of the unavoidable system volume fluctuations. The expected signal of the critical point is a non-monotonic dependence of fluctuation measures [8] on the energy and/or the size of the collision system.

1.1 Two-particle correlations

These correlations are defined for two different particles from the same event, that have a difference in pseudo-rapidity $\Delta\eta = \eta_1 - \eta_2$ and azimuthal angle $\Delta\phi = \phi_1 - \phi_2$. The correlation function is calculated

[a] e-mail: seryakov@yahoo.com, andrey.seryakov@spbu.ru

as:

$$C(\Delta\eta,\Delta\phi) = \frac{N^{pairs}_{mixed}}{N^{pairs}_{data}}\frac{S(\Delta\eta,\Delta\phi)}{M(\Delta\eta,\Delta\phi)}; S(\Delta\eta,\Delta\phi) = \frac{d^2N^{signal}}{d\Delta\eta d\Delta\phi}; M(\Delta\eta,\Delta\phi) = \frac{d^2N^{mixed}}{d\Delta\eta d\Delta\phi}; \tag{1}$$

where $S(\Delta\eta,\Delta\phi)$ and $M(\Delta\eta,\Delta\phi)$ are the distributions for pairs from data and mixed events, respectively. The $\Delta\phi$ range is folded. Pseudorapidity is calculated in the center of mass system with the pion mass assumed for all charged particles, allowing to compare our results with the RHIC and LHC results.

1.2 Pseudorapidity correlations

Correlations in pseudorapidity were studied between observables B and F in different pseudo-rapidity windows. The correlation coefficient is calculated as:

$$b_{rel}[B,F] = \frac{<F>}{<B>}\frac{<BF> - <B><F>}{<F^2> - <F>^2}, \tag{2}$$

where $< B >$ and $< F >$ are the event averages in the «backward» and «forward» pseudorapidity windows respectively. In this work we present three types of pseudorapidity correlations:

- Backward multiplicity N_B and forward multiplicity N_F: $b_{rel}[N_B, N_F]$.
- Backward mean transverse momentum $Pt_B = \frac{\sum_{i=1}^{N_B} pt_{B_i}}{N_B}$ and N_F: $b_{rel}[Pt_B, N_F]$.
- Backward and forward mean transverse momenta Pt_B and Pt_F: $b_{rel}[Pt_B, Pt_F]$.

1.3 Multiplicity and transverse momentum fluctuations

The strongly intensive quantities Δ and Σ proposed in Refs. [9] are defined as:

$$\Delta[A,B] = \frac{1}{C_\Delta}(<B>\omega[A] - <A>\omega[B]), \tag{3}$$

$$\Sigma[A,B] = \frac{1}{C_\Sigma}(<B>\omega[A] + <A>\omega[B] - 2(<AB> - <A><B>)), \tag{4}$$

where A and B are extensive event variables and $\omega[A] = \frac{<A^2> - <A>^2}{<A>}$ is the scaled variance. For the study of multiplicity-transverse momentum fluctuations we use the scalar sum of transverse momenta $A = Pt = \sum_{i=1}^{N} pt_i$ and the track multiplicity $B = N$ with normalization factors $C_\Delta = C_\Sigma =< N > \omega[pt]$.

2 Analysis procedure

All results were corrected for biases due to trigger, off-line selection, track and event selection etc. by using events obtained from the EPOS 1.99 model event generator and comparing results before and after GEANT detector simulation and reconstruction. Electrons and positrons were removed by a dE/dx cut. Possible effects from hard scattering were reduced by a cut on particle transverse momentum $pt < 1{,}5$ GeV/c. All results were calculated inside the NA61/SHINE acceptance, the influence of which will be shown for the example of the EPOS 1.99 model in the next chapter. Only statistical errors are shown for pseudo-rapidity correlations. More information on the two-particle correlation and fluctuation analysis procedure for p+p collisions can be found in Ref. [6] and Ref. [2] respectively. The centrality of Be+Be collisions was determined by using information from the Projectile Spectator Detector [1].

DOI: 10.1051/epjconf/201612604044

3 Two-particle correlations in p+p collisions

The correlation function $C(\Delta\eta, \Delta\phi)$ for p+p collisions for pair combinations of various charges at 158 GeV/c and for different energies is shown in figures 1 and 2 respectively. There are several notable structures: the first is a maximum at $(\Delta\eta, \Delta\phi) = (0, \pi)$, likely a result of resonance decays and momentum conservation. The maximum is stronger for unlike-sign pairs and significantly weaker for same charge pairs. The second is an enhancement at $(\Delta\eta, \Delta\phi) = (0, 0)$, probably due to Coulomb interactions (unlike-sign pairs) and quantum statistics (same charge pairs). The maximum decreases and the enhancement increases with growing collision energy.

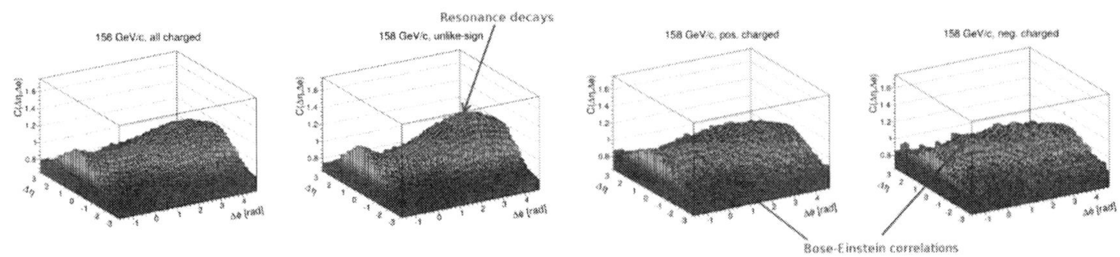

Figure 1. Results on $\Delta\eta\Delta\phi$ correlations for inelastic p+p interactions at 158 GeV/c. From left to right: results for all, unlike-sign, positive and negative charged particles. The maximum at $(\Delta\eta, \Delta\phi) = (0, \pi)$ is probably a result of resonance decays and momentum conservation. The enhancement at $(\Delta\eta, \Delta\phi) = (0, 0)$ is likely due to Coulomb interactions and quantum statistics.

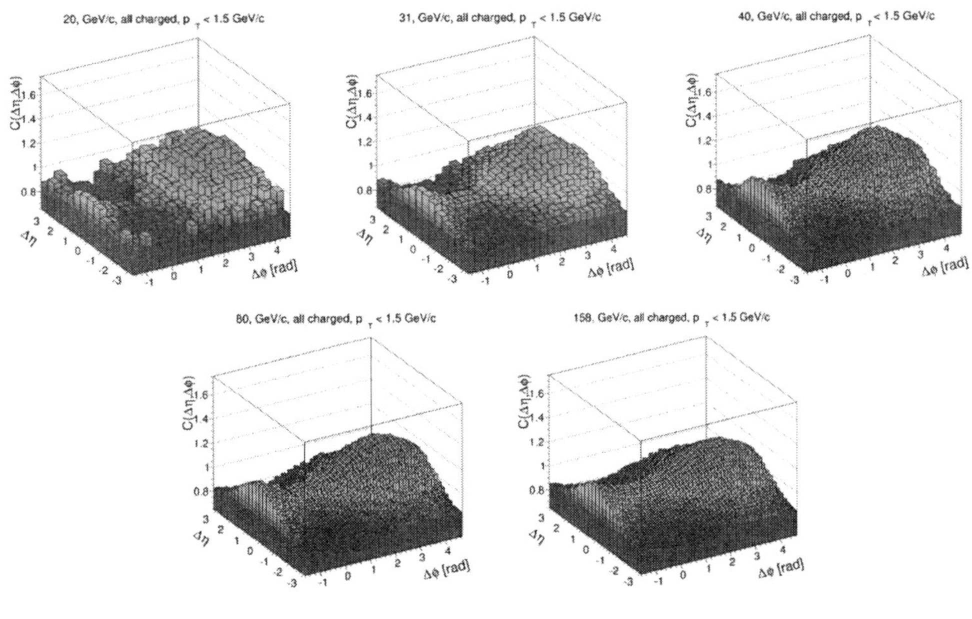

Figure 2. Energy dependence of $\Delta\eta\Delta\phi$ correlations for inelastic p+p interactions.

Results of calculations with the EPOS 1.99 and UrQMD models are also in figure 3 for 20 and 158 GeV beam momentum. A comparison with the data (left column) shows that the EPOS model predictions are consistent with the measurements whereas the UrQMD calculations are not.

DOI: 10.1051/epjconf/201612604044

The model results in the NA61/SHINE acceptance and 4π acceptance are very similar (compare figure 4 center and right).

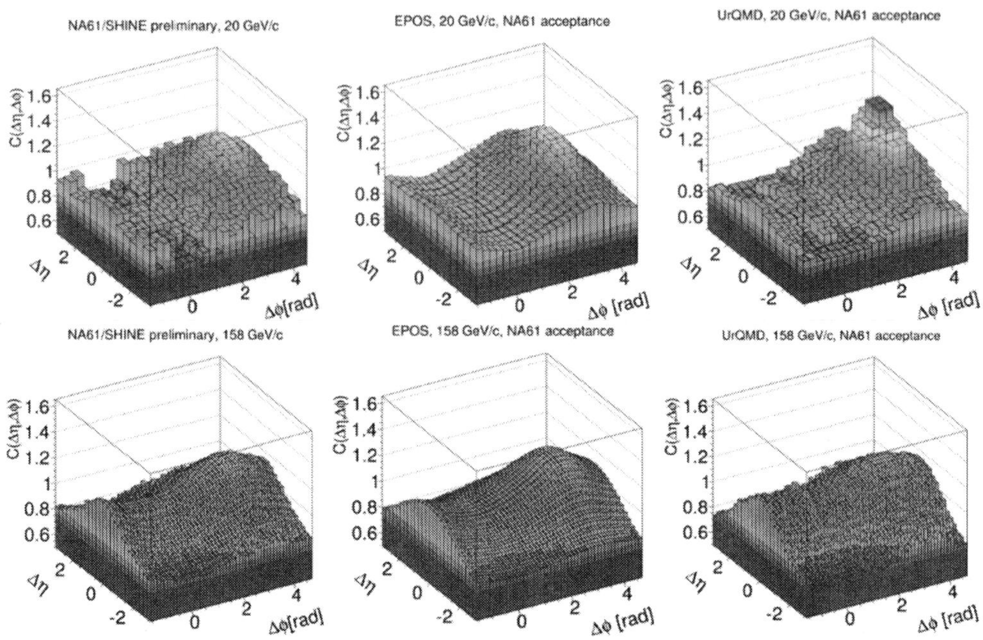

Figure 3. $\Delta\eta\Delta\phi$ correlations for all charged particles in inelastic p+p interactions at 20 (upper row) and 158 (bottom row) GeV/c. NA61/SHINE (left column), EPOS 1.99 inside the NA61/SHINE acceptance (middle column) and UrQMD in the NA61/SHINE acceptance (right column).

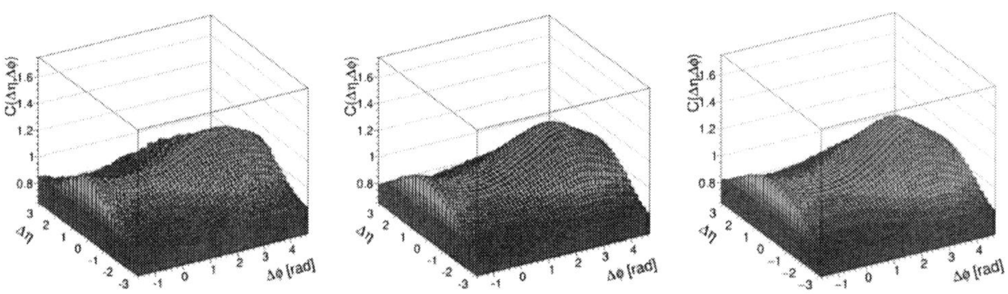

Figure 4. $\Delta\eta\Delta\phi$ correlations for all charged particles inside the NA61/SHINE acceptance in inelastic p+p interactions at 158 GeV/c. NA61/SHINE (left), EPOS 1.99 inside the NA61/SHINE acceptance (middle) and EPOS 1.99 in full 4π acceptance (right).

4 Pseudorapidity correlations in Be+Be 150A GeV/c for the 0-20% most central collisions

In this work two sets of pseudorapidity windows *B* and *F* were used: short-range (figure 5 left) and with fixed forward window (figure 5 right). A strong dependence of the correlations on the window

DOI: 10.1051/epjconf/201612604044

position is observed. EPOS 1.99 describes the data only qualitatively. Significant multiplicity–mean transverse momentum long-range correlations were observed (figure 10) that might be a signal of collective effects.

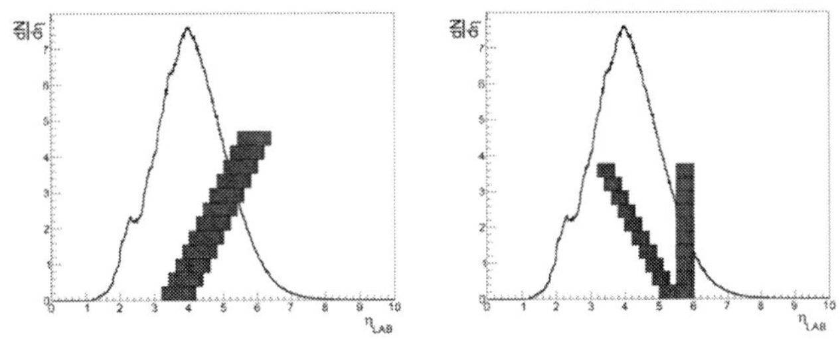

Figure 5. Illustration of connected (left) and disconnected (right) windows configurations superimposed on the inclusive pseudorapidity distribution inside the NA61/SHINE acceptance.

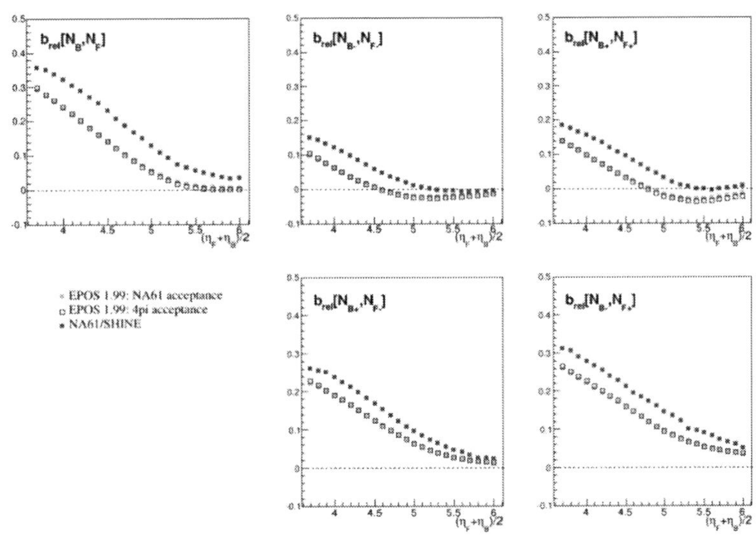

Figure 6. Correlation coefficient of multiplicities for different charge combinations as a function of pseudorapidity of the B and F windows connection point (see figure 5 left) for the 0-20% most central Be7+Be9 collisions at 150A GeV/c. EPOS 1.99 model predictions inside the NA61/SHINE acceptance (grey squares) and in 4π-acceptance (open squares) are shown for comparison.

DOI: 10.1051/epjconf/201612604044

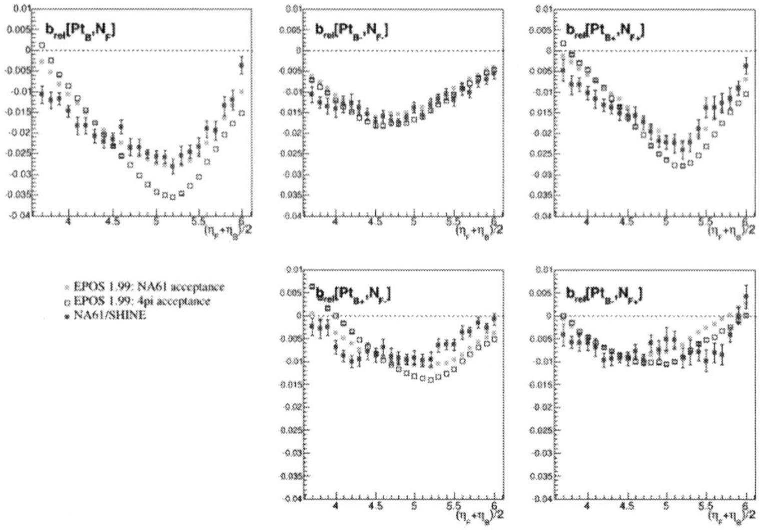

Figure 7. Correlation coefficient of event mean transverse momenta and multiplicity for different charge combinations as a function of pseudorapidity of the B and F windows connection point (see figure 5 left) for the 0-20% most central Be7+Be9 collisions at 150A GeV/c. EPOS 1.99 model predictions inside the NA61/SHINE acceptance (grey squares) and in 4π-acceptance (open squares) are shown for comparison.

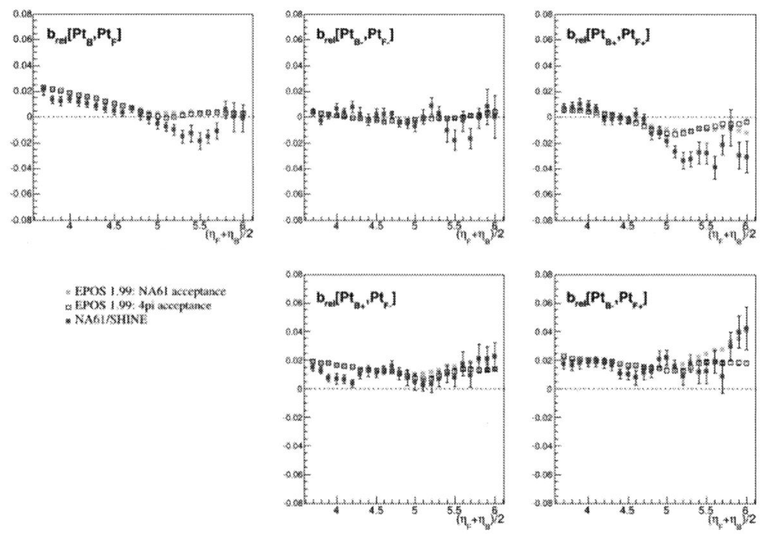

Figure 8. Correlation coefficient of event mean transverse momenta for different charge combinations as a function of pseudorapidity of the B and F windows connection point (see figure 5 left) for the 0-20% most central Be7+Be9 collisions at 150A GeV/c. EPOS 1.99 model predictions inside the NA61/SHINE acceptance (grey squares) and in 4π-acceptance (open squares) are shown for comparison.

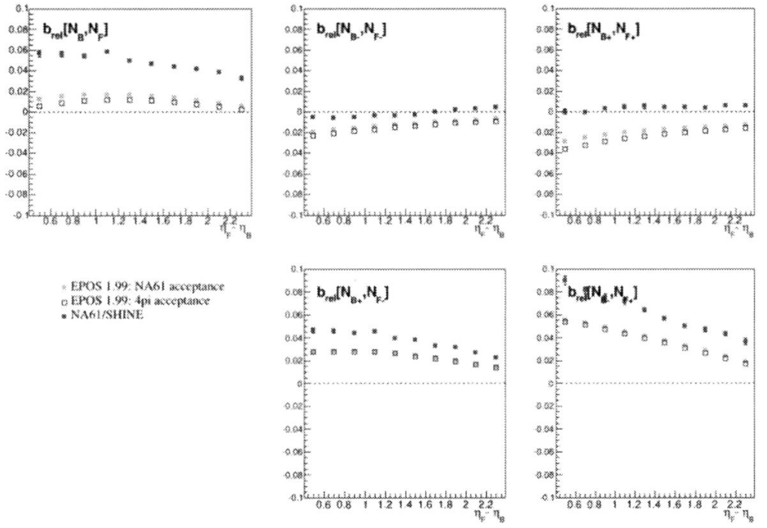

Figure 9. Correlation coefficient of multiplicities for different charge combinations as a function of pseudorapidity distance between B and F windows (see figure 5 right) for the 0-20% most central Be7+Be9 collisions at 150A GeV/c. EPOS 1.99 model predictions inside the NA61/SHINE acceptance (grey squares) and in 4π-acceptance (open squares) are shown for comparison.

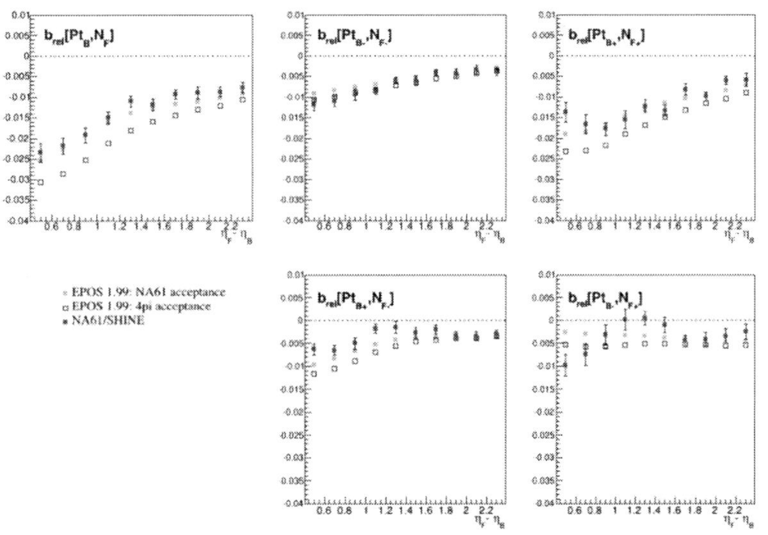

Figure 10. Correlation coefficient of event mean transverse momenta and multiplicity for different charge combinations as a function of pseudorapidity distance between B and F windows (see figure 5 right) for the 0-20% most central Be7+Be9 collisions at 150A GeV/c. EPOS 1.99 model predictions inside the NA61/SHINE acceptance (grey squares) and in 4π-acceptance (open squares) are shown for comparison.

DOI: 10.1051/epjconf/201612604044

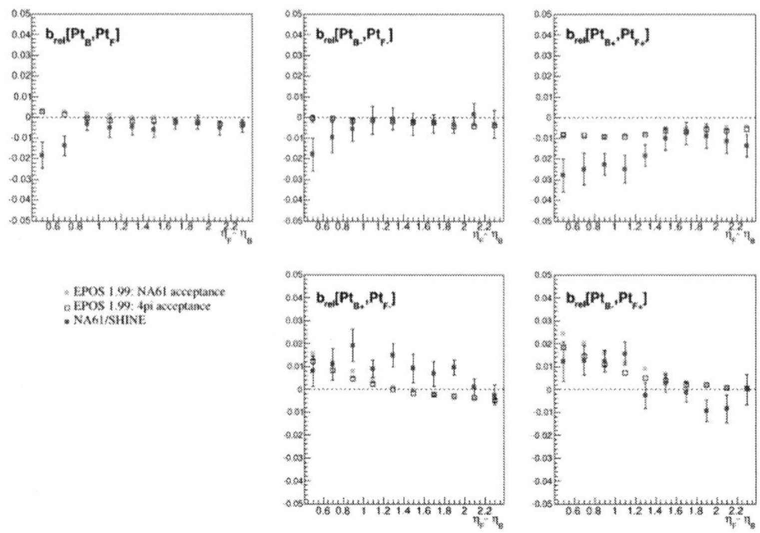

Figure 11. Correlation coefficient of event mean transverse momenta for different charge combinations as a function of pseudorapidity distance between B and F windows (see figure 5 right) for the 0-20% most central Be7+Be9 collisions at 150A GeV/c. EPOS 1.99 model predictions inside the NA61/SHINE acceptance (grey squares) and in 4π-acceptance (open squares) are shown for comparison.

5 Strongly intensive measures of transverse momentum and multiplicity fluctuations in p+p and Be+Be collisions

Measurements of the strongly intensive quantities $\Delta[Pt,N]$ and $\Sigma[Pt,N]$ are presented in figure 12 as a function of the center of mass energy and the number of wounded nucleons [8]. The dependence is flat, therefore these quantities does not show any evidence of the critical point in p+p and Be+Be collisions at the SPS energies.

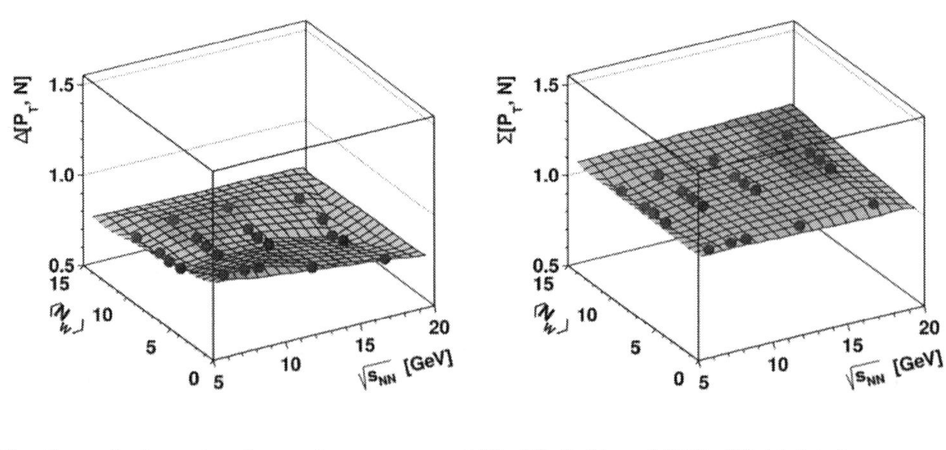

Figure 12. Strongly intensive fluctuation measures $\Delta[Pt,N]$ (left) and $\Sigma[Pt,N]$ (right) for p+p and Be+Be (preliminary results) collisions at several center of mass energies and collision centralities (defined via the mean number of participating nucleons N_w). The flat behaviour means no evidence of the critical point in this region.

6 Conclusions

The 2-particle correlations as function of pseudorapidity and azimuthal angle difference in proton-proton reactions at the SPS energies show structures coming mainly from resonance decays, conservation laws, quantum statistics and Coulomb interactions. Significant multiplicity–mean transverse momentum long-range correlations were observed in Be+Be collisions at 150A GeV/c (figure 10) that can be considered as a signal of collective effects. Strongly intensive fluctuation measures $\Delta[Pt, N]$ and $\Sigma[Pt, N]$ do not show any evidence of the critical point in p+p and Be+Be collisions at the SPS energies.

Acknowledgments

This work was supported by the Federal Agency of Education of the Ministry of Education and Science of the Russian Federation and SPbSU research grant 11.38.193.2014.

References

[1] N.Abgrall et al. (NA61 Coll.), JINST **9**, P06005 (2014).
[2] B.Maksiak for the NA61/SHINE Collaboration, Proceedings of Science PoS (CPOD2014) 055.
[3] A.Seryakov for the NA61/SHINE Collaboration, arXiv:1510.00847 [nucl-ex], J. Phys.: Conf. Ser. **668**, 012077 (2016).
[4] M.Braun and C.Pajares, Phys.Lett.B **287**, 154 (1992).
[5] L.McLerran, Nucl.Phys.A **699**, 73c (2002)
[6] A.Aduszkiewicz et al. (NA61 Coll.), arXiv:1510.00163v2 [hep-ex], submitted to Eur.Phys.J. C (2016).
[7] M.Gorenstein and M.Gazdzicki, Phys.Rev.C **84**, 014904 (2011).
[8] M.Gazdzicki and P.Seyboth, arXiv:1506.08141 [nucl-ex] (2015).
[9] M.Gazdzicki, M.Gorenstein, and M.Mackowiak-Pawlowska, Phys. Rev. C **88**, 024907 (2013).

 DOI: 10.1051/epjconf/201612604045

Precision X-ray spectroscopy of kaonic atoms as a probe of low-energy kaon-nucleus interaction

H. Shi[1,a], M. Bazzi[1], G. Beer[2], G. Bellotti[3,4], C. Berucci[5,1], A. M. Bragadireanu[1,6], D. Bosnar[7], M. Cargnelli[5], C. Curceanu[1], A. D. Butt[3,4], A. d'Uffizi[1], C. Fiorini[3,4], F. Ghio[8], C. Guaraldo[1], R.S. Hayano[9], M. Iliescu[1], T. Ishiwatari[5], M. Iwasaki[10], P. Levi Sandri[1], J. Marton[5], S. Okada[10], D. Pietreanu[1,6], K. Piscicchia[1,11], A. Romero Vidal[12], E. Sbardella[1], A. Scordo[1], D.L. Sirghi[1,6], F. Sirghi[1,6], H. Tatsuno[13,14], O. Vazquez Doce[15], E. Widmann[5], and J. Zmeskal[5]
(SIDDHARTA Collaboration)

[1] *INFN, Laboratori Nazionali di Frascati, Via E. Fermi 40, I-00044 Frascati(Roma), Italy,*
[2] *Department of Physics and Astronomy, University of Victoria, P.O. Box 1700 STN CNC, Victoria BC V8W 2Y2, Canada,*
[3] *Politecnico di Milano, Dipartimento di Elettronica, Informazione e Bioingegneria, Milano, Italy,*
[4] *INFN Sezione di Milano, Via Celoria 16 - 20133 Milano, Italy,*
[5] *Stefan-Meyer-Institut für Subatomare Physik, Boltzmanngasse 3, 1090 Wien, Austria,*
[6] *IFIN-HH, Institutul National pentru Fizica si Inginerie Nucleara Horia Hulubbei, Reactorului 30, Magurele, Romania,*
[7] *Department of Physics, Facaulty of Science, University of Zagreb, Bijenicka 32, HR-10000 Zagreb, Croatia,*
[8] *INFN Sezione di Roma I and Instituto Superiore di Sanita, I-00161 Roma, Italy,*
[9] *Department of Physics, School of Science, The University of Tokyo, Bunkyo-ku Hongo 7-3-1, Tokyo, Japan,*
[10] *RIKEN, Institute of Physical and Chemical Research, 2-1 Hirosawa, Wako, Saitama 251-0198, Japan,*
[11] *Museo Storico della Fisica e Centro Studi e Ricerche "Enrico Fermi", Piazza del Viminale 1-00184Roma, Italy,*
[12] *Universidade de Santiago de Compostela, Casas Reais 8, 15782 Santiago de Compostela, Spain,*
[13] *National Institute of Standards and Technology (NIST), Boulder, CO, 80303, USA,*
[14] *High Energy Accelerator Research Organization (KEK), Tsukuba, 305-0801, Japan,*
[15] *Excellence Cluster Universe, Technische Universität München, Boltzmannstraße 2, D-85748 Garching, Germany,*

Abstract. In the exotic atoms where one atomic $1s$ electron is replaced by a K^-, the strong interaction between the K^- and the nucleus introduces an energy shift and broadening of the low-lying kaonic atomic levels which are determined by only the electromagnetic interaction. By performing X-ray spectroscopy for Z=1,2 kaonic atoms, the SIDDHARTA experiment determined with high precision the shift and width for the $1s$ state of K^-p and the $2p$ state of kaonic helium-3 and kaonic helium-4. These results provided unique information of the kaon-nucleus interaction in the low energy limit.

[a] e-mail: hexishi@lnf.infn.it

1 Introduction

Precision X-ray spectroscopy of the Z=1,2 kaonic atoms holds the key to understand the low-energy interaction between the kaon and the nucleus. In kaonic hydrogen which is the simplest kaonic atom, the strong interaction induced shift and width of the $1s$ state can be deduced from the K-series X-rays, and they are related to the s-wave K^-p scattering length a_{K^-p} through the Deser-Trueman formula [1]. This scattering length consists of two isospin dependent components of a_0 (I=0) and a_1 (I=1), which are among the most fundamental parameters for the low-energy $\bar{K}N$ interaction, and which can only be derived experimentally from X-ray spectroscopy. To disentangle these two components, precise measurement of the kaonic deuterium $1s$ shift and width is necessary, further taking into account higher order contributions from the three-body interaction [2]. Experimentally this measurement is extremely difficult mainly due to the small X-ray yield and the broad width of the $1s$ level. Until now an unambiguous detection of the kaonic deuterium X-rays has not been achieved and the measurement remains to be the major challenge of upcoming experiments.

For the Z=2 kaonic atoms, there had been a long-standing discrepancy between the theoretical and the measured strong-interaction induced shift of the $2p$ level of kaonic helium-4. This "kaonic helium puzzle" was solved by the KEK E570 experiment which showed the $2p$ level shift was 2 ± 2(stat) ± 2(syst) eV [3], refuting the previous average shift of about - 40 eV from three early experiments [4–6]. More essentially, the precise measurement of the kaonic helium $2p$ level shift is a test ground for possible deeply-bound kaonic nuclear states predicted by Akaishi and Yamazaki [7, 8]. A series of electronvolt precision measurement of the shift and the width of the $2p$ level for both kaonic helium-3 and kaonic helium-4 atoms can effectively narrow down the allowed range of the nuclear potential of kaon in the search of deeply-bound states.

The measurements we have performed in the SIDDHARTA experiment will be introduced in detail next.

2 The SIDDHARTA experiment

2.1 Experimental method

The experiment was performed at the DAΦNE e^-e^+ collider in Frascati, Italy. The beam energies are tuned to produce ϕ (1020) resonance at rest in the lab frame, and then with a 49% branching ratio the ϕ decays into a pair of K^+K^-. As the kaon detectors, we used two plastic scintillators placed above and below the interaction point as illustrated in Fig. 1. The kaons are distinguished from the minimum ionizing particles using the time of flight information at the kaon detectors. And then from the coincidence of the two scinillators we defined the kaon trigger, which marks the timing of the incident kaons.

A fraction of the negatively charged kaons that made the kaon trigger will stop inside the volume of the gaseous target placed about 20 cm above the interaction point to form kaonic atoms. With the time correlation between the subsequent X-rays detected by the SDDs and the kaon trigger, one can distinguish the kaon origin X-rays from the background coming from the processes that are asynchronous to the kaon pair production.

With a total active area of 144 cm^2, the SDDs cover about 10% of the solid angle around the target cell. The energy calibration of the SDDs were done every few hours by shining Ti and Cu foils with an X-ray tube. Using the K_α lines of Ti (4.5 keV) and Cu (8.0 keV), we determined the scale of the energy spectra, and the energy resolution at 6 keV was stable at about 150 eV (FWHM) throughout our measurement. At the operating temperature of 140 K, the SDDs showed a timing resolution of 700 FWHM, effectively rejected the asynchronous background by four orders of magnitude. The

DOI: 10.1051/epjconf/201612604045

configuration and the performance of the detectors are discussed in detail in previous publications of the SIDDHARTA experiment [9, 10].

During the data taking in 2009, we accumulated data with gaseous targets of hydrogen (1.3 g/l), deuterium (2.50 g/l), helium-3 (0.96 g/l), and helium-4 (1.65 g/l and 2.15 g/l).

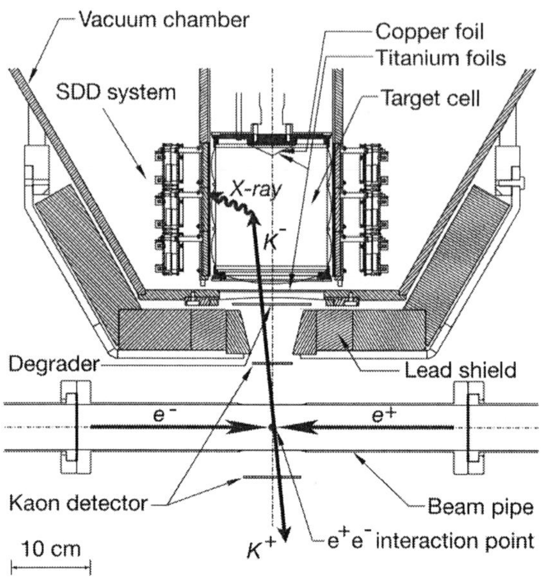

Figure 1. A schematic cutaway view of the SIDDHARTA setup at the DAΦNE interaction point. The charged kaon pairs are identified with two plastic scintillators, and the K^- induced X-rays detected by the SDDs are identified from the time correlation to the kaon pair events.

2.2 Kaonic hydrogen and deuterium

The main objective of the SIDDHARTA experiment is the precise measurement of the kaonic hydrogen K-series X-rays, thus most of the beam time was dedicated to the hydrogen target measurement to a total intergrated luminosity of 340 pb^{-1}. A total amount of 100 pb^{-1} integrated luminosity was dedicated to the first exploratory kaonic deuterium measurement.

In the final analysis to derive the $1s$ shift ϵ_{1s} and width Γ_{1s} of kaonic hydrogen, the kaonic X-ray spectra from the two data sets were fitted simultaneously. As the yield of the kaonic deuterium K-series X-rays is more than one order of magnitude smaller than that of the kaonic hydrogen, the spectrum from the deuterium target measurement helped determining the background X-rays from other kaonic atoms when the K^- stopped inside the target cell made of Kapton. This combined analysis contributed to reduce the systematic errors in the kaonic hydrogen results, which achieved the best precision up to date for the $1s$ shift ϵ_{1s} = -283 ± 36 (stat.) ± 6 (syst.) eV and for the width Γ_{1s} = 541 ± 89 (stat.) ± 22 (syst.) eV [10]. The results solved the discrepancy between two recent kaonic hydrogen X-ray experiments KpX in KEK [11] and DEAR at DAΦNE [12], in that the two results do not overlap with each other within error bars. The improved precision presents more stringent constrains to the theoretical study of the low-energy QCD near the K^-p threshold.

From the deuterium target spectrum which showed no significant amount of signal of kaonic deuterium X-rays, we evaluated an upper limit for the yield of the kaonic deuterium K_α X-ray as

$Y(K_\alpha) < 0.0039$ (C.L. 90%) [13]. It presents an important reference to plan for the future precision measurement of kaonic deuterium X-rays.

2.3 Kaonic helium-3 and helium-4

The strong interaction induced $2p$ level shift and width are measured for both kaonic helium-3 and helium-4 atoms, and it is the first time that kaonic helium-3 X-rays are measured. The results of the shift as listed in 1 are consistent with the results of E570 [3], thus a zero-compatible shift of the $2p$ level from experiment is established, which is in agreement with the theoretical estimations in [6, 15]. We have not found abnormally large widths which can directly support the estimations made in conjunction with the prediction of a deeply-bound kaon states[7, 8].

Table 1. Results on the energy shifts (ΔE_{2p}) and widths (Γ_{2p}) of the kaonic helium-3 and kaonic helium-4 $2p$ states [9, 14].

Target	ΔE_{2p} [eV]	Γ_{2p} [eV]
helium-4	+5 ± 3 (stat.) ± 4 (syst.)	14 ± 8 (stat.) ± 5 (syst.)
helium-3	-2 ± 2 (stat.) ± 4 (syst.)	6 ± 6 (stat.) ± 7 (syst.)

3 Future perspectives

Following the method successfully developed in the SIDDHARTA experiment, the extended SIDDHARTA-2 collaboration is in preparation of a series of upgrades and modifications of the apparaturs, aiming at the first precision measurement of kaonic deuterium X-rays [16].

Acknowledgements

We thank C. Capoccia, G. Corradi, B. Dulach, and D. Tagnani from LNF-INFN; and H. Schneider, L. Stohwasser, and D. Stükler from Stefan-Meyer-Institut, for their fundamental contribution in designing and building the SIDDHARTA setup. We thank as well the DAΦNE staff for the excellent working conditions and permanent support. Part of this work was supported by the European Community-Research Infrastructure Integrating Activity "Study of Strongly Interacting Matter" (HadronPhysics2, Grant Agreement No. 227431, and HadronPhysics3 (HP3) Contract No. 283286) under the Seventh Framework Programme of EU; HadronPhysics I3 FP6 European Community program, Contract No. RII3-CT-2004- 506078; Austrian Science Fund (FWF) (P24756-N20); Austrian Federal Ministry of Science and Research BMBWK 650962/0001 VI/2/2009; Croatian Science Foundation under Project No. 1680; Romanian National Authority for Scientific Research, Contract No. 2-CeX 06-11-11/2006; and the Grant-in-Aid for Specially Promoted Research (20002003), MEXT, Japan.

References

[1] S. Deser, *et al.*, Phys. Rev., **96**, 774–776, (1954).
[2] U. -G. Meissner, *et al.*, Eur. Phys. J. C**35**, 349–357(2004).
[3] S. Okada, *et al.*, Phys. Lett. B**653**,387–391 (2007).
[4] C. Wiegand and R. Pehl, Phys. Rev. Lett. **27**, 1410–1412(1971).
[5] C. J. Batty, *et al.*, Nucl. Phys. A**326**, 455–462(1979).

[6] S. Baird, *et al.*, Nucl. Phys. A**392**, 297–310(1983).
[7] Y. Akaishi, T. Yamazaki, Phys. Rev. C**55**, 044005 (2002).
[8] Y. Akaishi, in: Proc. Inter. Conf. on Exotic Atoms EXA05 (2005), http://dx.doi.org/10.1553/exa05s45.
[9] M. Bazzi, *et al.*, Phys. Lett. B**681**, 310–314(2009); M. Bazzi, *et al.*, Phys. Lett. B**697**, 199–202(2011).
[10] M. Bazzi, *et al.*, Phys. Lett. B**704**, 113–117(2011).
[11] M. Iwasaki, *et al.*, Phys. Rev. Lett. **78**, 3067–3069(1997);
T. M. Ito, *et al.*, Phys. Rev. C**58**, 2366–2382(1998).
[12] G. Beer, *et al.*, Phys. Rev. Lett. **94**, 212302(2005).
[13] M. Bazzi, *et al.*, Nucl. Phys. A**907**, 19–77(2013).
[14] M. Bazzi, *et al.*, Phys. Lett. B**714**, 40–43(2012).
[15] C. J. Batty, Nucl. Phys. A**508**, 89c(1990).
[16] SIDDHARTA-2 Collaboration, Proposal at Laboratori Nazionali di Frascati, INFN, "The upgrade of the SIDDHARTA apparaturs for an enriched scientific case", (2010).

EPJ Web of Conferences **126**, 04046 (2016) DOI: 10.1051/epjconf/201612604046
ICNFP 2015

The magnetic polarizabilities and g-factor of the neutral and charged ρ mesons in a strong magnetic field on the lattice

O.E. Solovjeva[1,2,a], E.V. Luschevskaya[1,2,4], and O.V. Teryaev[3,2]

[1] *State Scientific Center of the Russian Federation - Institute for Theoretical and Experimental Physics , Bolshaya Cheremushkinskaya 25, 117218 Moscow, Russia*
[2] *National Research Nuclear University "MEPhI", Kashirskoe highway 31, 115409 Moscow, Russia*
[3] *Joint Institute for Nuclear Research, 141980 Dubna, Russia*
[4] *School of Biomedicine, Far Eastern Federal University, 690950 Vladivostok, Russia*

Abstract. Study of the ground state energies of neutral and charged vector meson was performed in $SU(3)$ lattice gauge theory in the strong abelian magnetic field. Our method allows us to distinguish the ground state energy of mesons for various spin projections on the axis of the external magnetic field . We are observe the splitting the ground state energy of ρ mesons depending on spin projections: for neutral ρ^0 meson with zero spin projection on the axis of the field ground state energy decreases, while the energies with non-zero spins increase with the field value. For charged $\rho^{\pm}$ mesons we observe the agreement between Landau level picture and our data at magnetic field less then 0.6GeV^2. The magnetic polarizabilities of vector mesons were found for the lattice volume 18^4 fitting the ground state energies. We found the g-factor of $\rho^{\pm}$ in the chiral limit. This value is compatible with experimental determination and theoretical value from relativistic quark model.

1 Introduction

Magnetic polarizabilities and g-factors are fundamental characteristics of particles. This parameters are interesting for understanding the internal structure of hadrons and effects which may appear in electromagnetic field. Conception of polarizability was first used for the nuclear matter in the analysis of the scattering of low energy gamma quanta by atomic nuclei by Migdal [1]. For hadrons the notion of polarizability was discussed in papers [2]-[4]. New notion of polarizability assumes process of the Compton scattering on particle. The electric and magnetic polarizabilities determine the response of hadron to two-photon interactions.
In this work we will discuss the behaviour of the ground state energy of ρ mesons in the abelean strong magnetic field from the $SU(3)$ lattice gauge theory. The g-factor and magnetic polarizabilities are parameters which determine the behaviour of the ground state energy.

[a] e-mail: oesolovjeva@gmail.com

2 Details of calculations

The detailed description of the technical details our calculations are presented in [8],[9]. We generate 200−300 $SU(3)$ statistically independent lattice gauge configurations for lattice volumes 16^4, 18^4 and lattice spacings $a = 0.105$ fm, 0.115 fm and 0.125 fm. The $U(1)$ external magnetic field is included only into the Dirac operator which is used for the calculation of eigenfunctions ψ_k and eigenvectors λ_k of a test quark in a background gauge field A_μ.

This field is a sum of non-abelian $SU(3)$ gluonic field and $U(1)$ abelian constant magnetic field.

$$A_{\mu ij} \longrightarrow A_{\mu ij} + A^B_\mu \delta_{ij}, \tag{1}$$

where

$$A^B_\mu(x) = \frac{B}{2}(x_1\delta_{\mu,2} - x_2\delta_{\mu,1}). \tag{2}$$

The twisted boundary conditions are imposed because it is necessary to take into account periodic boundary conditions for fermions. [5].

The value of the magnetic field in finite volume is quantized

$$qB = \frac{2\pi k}{(aL)^2}, \quad k \in \mathbb{Z}, \tag{3}$$

where $q = -1/3\,e$. It follows to a minimal value of magnetic field $(eB)^{1/2} = 372\ Mev$ for lattice volume 20^4 and lattice spacing $a = 0.115$ fm.

One should note here that our parameters of lattice for the following values of magnetic fields corresponding to the lattice spacing cut-off: 2.5GeV^2 for $a = 0.125$ fm, 2.9GeV^2 for lattice spacing $a = 0.115$ fm and $eB \sim 3.5$ GeV^2 for $a = 0.105$ fm etc.
To determine the ground state energy is necessary to calculate the correlation function interpolation operators in Euclidean space :

$$\langle \psi^\dagger(x) O_1 \psi(x) \psi^\dagger(y) O_2 \psi(y) \rangle_A, \tag{4}$$

where interpolation operator is $\psi^\dagger(x)O_1\psi(x)$, analogue creation operator in quantum electrodynamics, which creates a particle with certain quantum numbers : for ρ meson $O_1, O_2 = \gamma_\mu, \gamma_\nu$, $\mu, \nu = 1, .., 4$ are Lorentz indices. This correlation function were calculated separately for every value of external magnetic field. For an isovector states we can express the correlation function through the Dirac propagator:

$$\langle \bar\psi O_1 \psi \bar\psi O_2 \psi \rangle_A = -\mathrm{Tr}\,[O_1 D^{-1}(x,y) O_2 D^{-1}(y,x)] \tag{5}$$

We calculated the Dirac propagator using its eigenvectors and eigenvalues

$$D^{-1}(x,y) = \sum_{k<M} \frac{\psi_k(x)\psi_k^\dagger(y)}{i\lambda_k + m}. \tag{6}$$

where $M = 50$ is the number of the lowest eigenmodes.

On the other hand one can make 3-dimentional Fourier transformation of correlation function.For the ground energy state we consider zero momentum $\mathbf{p} = 0$, because in this case $E_0 = m_0$ in zero magnetic field. The correlation function is expand to exponential series.

$$\tilde C(n_t) = \langle \psi^\dagger(\mathbf{0}, n_t) O_1 \psi(\mathbf{0}, n_t) \psi^\dagger(\mathbf{0}, 0) O_2 \psi(\mathbf{0}, 0) \rangle_A = \sum_k \langle 0|O_1|k\rangle\langle k|O_2^\dagger|0\rangle e^{-n_t a E_k} \tag{7}$$

where a is the lattice spacing, n_t is the number of nodes in the time direction, E_k is the energy of the state with quantum number k. If we consider large n_t the main contribution in (7) gives ground state. After taking into account the periodic boundary conditions on the lattice the main contribution to the ground state has the following form

$$\tilde{C}_{fit}(n_t) = A_0 e^{-n_t a E_0} + A_0 e^{-(N_T - n_t) a E_0} = 2A_0 e^{-N_T a E_0/2} \cosh((\frac{N_T}{2} - n_t) a E_0) \quad (8)$$

where A_0 is a constant, E_0 is the energy of the ground state. Using this formula we fit our data of correlators (4) and find the ground state energy as a fit parameter. In order to minimize the errors and exclude the contribution of excited states we take various values of n_t from the interval $5 \leq n_t \leq N_T - 5$. We also use a smeared gaussian source and point sink for our calculations.
The correlation functions for various spatial directions are given by the following relations

$$C_{ij}^{VV} = \langle \bar{\psi}(\mathbf{0}, n_t) \gamma_i \psi(\mathbf{0}, n_t) \bar{\psi}(\mathbf{0}, 0) \gamma_j \psi(\mathbf{0}, 0) \rangle, \quad (9)$$

where i,j=1,2,3 for x,y,z components respectively. The form of the density matrix for vector particle with spin $s = 1$ gives the formulas for energies of meson with various spin projections on the axis of the external magnetic field.

For the $s_z = 0$ one can obtain the energy of the ground state from the C_{zz}^{VV} correlator. The combinations of correlators

$$C^{VV}(s_z = \pm 1) = C_{xx}^{VV} + C_{yy}^{VV} \pm i(C_{xy}^{VV} - C_{yx}^{VV}). \quad (10)$$

gives the energies of mesons with $s_z = +1$ and $s_z = -1$.

3 Results

3.1 Neutral ρ meson

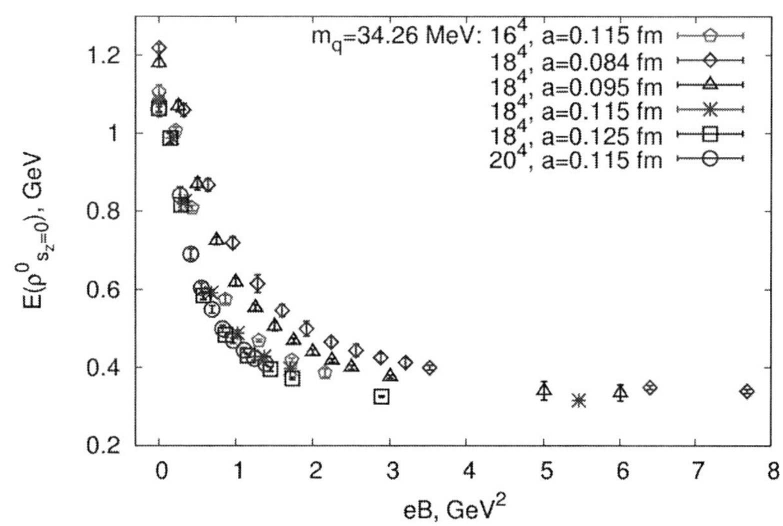

Figure 1. The ground state energy of neutral ρ^0 meson with $s_z = 0$

In the case of neutral ρ^0 meson, we observe a quadratic dependence on the magnetic field. Term with magnetic polarizability gives the contribution to the energy of ρ meson.

We describe the energy of the neutral ρ^0 meson by the following formula:

$$E = E(B = 0) - 2\pi\beta_m(eB)^2, \tag{11}$$

where we use "natural" units $\hbar = c = 1$, but $e^2 = 1/137$ in Gaussian units.
We calculate the energy for every values of magnetic field and find $E(B = 0)$ and β_m from the fit.

In Fig. 1 we show the ground state energy of neutral ρ^0 meson with $s_z = 0$ as a function of the magnetic field for various lattice spacings and volumes. We had a fit of the interval $(eB)^2 \in [0, 0.3\ \text{GeV}^4]$ by the quadratic function (11). This interval allow doesn't correct for a term $\sim (eB)^4$ in the result. From this figure one can see that the ground state energy of ρ^0 meson with $s_z = 0$ decreases with the value of magnetic field. The magnetic polarizability was found as a parameter of the fit by formula 11: $\beta_m^{s_z=0}(\rho^0) = (0.47 \pm 0.03)\ \text{GeV}^{-3}$

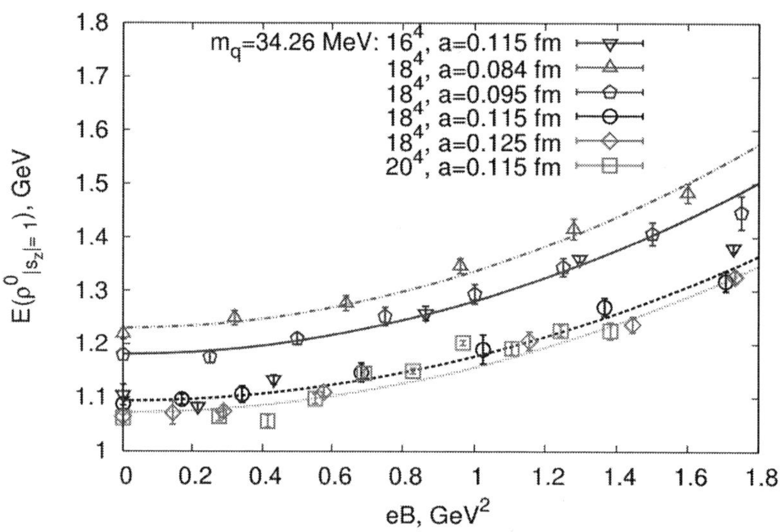

Figure 2. The ground state energy of neutral ρ^0 meson with $|s_z| = 1$

Fig. 2 presents the energy of ρ^0 with nonzero spin projections on the axis of the external magnetic field. This ground state energy increases with the value of magnetic field for the all sets of lattice data. It is worth noting that the ground state energy ρ^0 mesons with $s = -1$ and $s_z = +1$ are equal this follows from the vanishing terms iC^{VV}_{xy} and iC^{VV}_{yx} in (10) for neutral particles. We found the magnetic polarizability and observe the dependence of the results on the lattice spacing so we make an extrapolation to the continuum limit. After extrapolation the polarizability value is $\beta_m^{|s|=1}(\rho^0) = (-0.0235 \pm 0.0023)\ \text{GeV}^{-3}$ for the lattice volume 18^4 and bare quark mass $m_q = 34\ Mev$. One can see that the magnetic polarizability at $s_z = 0$ has another sign from the case of nonzero spin. This phenomena to be the result of the anisotropy created by the magnetic field.

3.2 Charged $\rho^{\pm}$ meson

Well-known formula of Landau level describing the energy levels of free charged pointlike particle in a background magnetic field it is not suitable for strong magnetic field.

$$E^2 = |qB| - gs_z qB + m^2(B = 0), \tag{12}$$

here g-factor characterizes magnetic properties of the particle, q is the charge of the particle, $m^2(B = 0)$ is the particle mass at $B = 0$.

We have to take into account the magnetic polarizabilty when particle is not pointlike. In the relativistic case the energy levels has the following form

$$E^2 = |qB| - gs_z qB + m^2(B = 0) - 4\pi m\beta(qB)^2, \tag{13}$$

where β is the magnetic polarizability

We consider ρ^- mesons while ρ^+ corresponds to the reversal of the direction of magnetic field. From the Fig. 3 one can see the energy of the charged vector ρ meson with $s_z = 0$. Our data obey

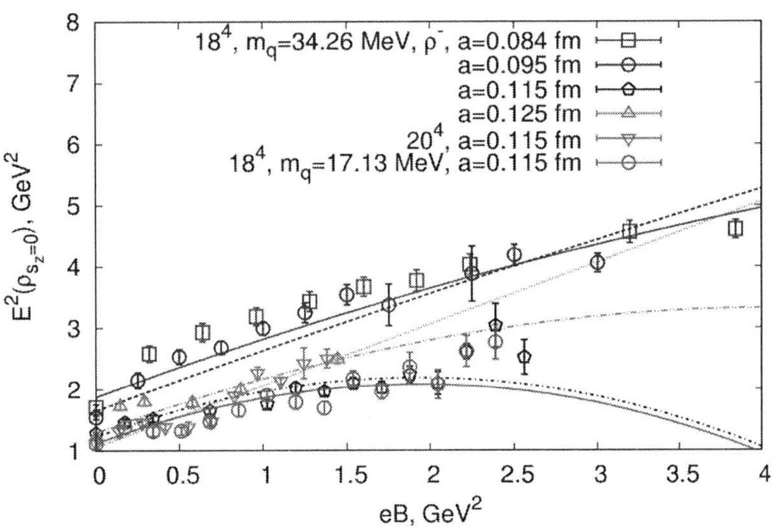

Figure 3. The ground state energy of charged $\rho^{\pm}$ meson with $s_z = 0$

formula (12) at $eB \in [0, 0.6\ \mathrm{GeV}^2]$. We fitting our data by the formula (13) for varios lattice volume and lattice spacing . We represent the energies of the ρ^- with spin projections $s_z = -1$ and $s_z = +1$ in Figs. 4 and 5 correspondingly. The energy of the ρ^- ground state with $s_z = -1$ decreases with the field value and the data agree with the fit Landau levels at $eB \in [0, 0.6\ \mathrm{GeV}^2]$. It is noticeable that we didn't observe the appearance of the tachyonic mode. If we consider strong magnetic field then we have to take into account magnetic dipole polarizabilty, megnetic hyperpolarizability and so on. It is follows, that energy does not equal zero.

From the data obtained one can estimate of g-factor of the $\rho^{\pm}$ meson.It is possible at small magnetic when our data described by the Landau level. We evaluate value of g-factor from the following relation:

DOI: 10.1051/epjconf/201612604046

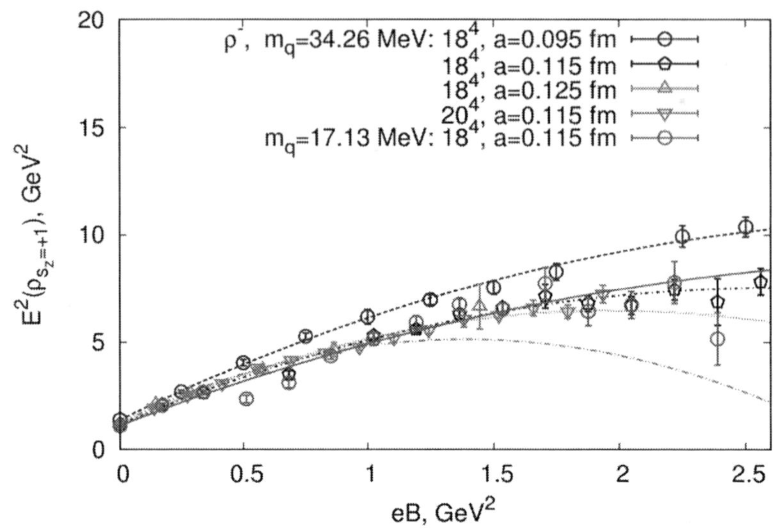

Figure 4. The ground state energy of charged $\rho^{\pm}$ meson with $s_z = +1$

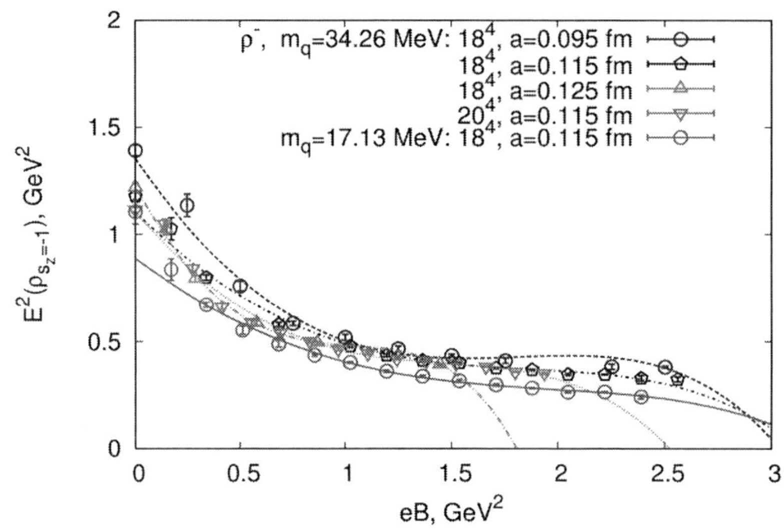

Figure 5. The ground state energy of charged $\rho^{\pm}$ meson with $s_z = -1$

$$g = \frac{E^2(s = +1) - E^2(s = -1)}{2(eB)}, \tag{14}$$

To clarify the value we made extrapolation to $m_q = 0$ and obtain $g = 2.4 \pm 0.2$ for the lattice volume 18^4 and lattice spacing $a = 0.115$ fm. It is clear that this value is compatible with recent experimental determination [6]. From theoretical prediction[7] $g \approx 2.37$, this value close to our result. In following work we plan to increase accuracy of the g-factor of the $\rho^{\pm}$ meson.

4 Conclusions

In this work we present the study of the ground state energies of neutral and charged vector meson in the strong abelian magnetic field from the $SU(3)$ lattice gauge theory. The energy of ρ^0 meson with zero spin projection $s_z = 0$ on the axis of the external magnetic field decreases, while the energies with non-zero spins $s_z = -1$ and $+1$ increase with the value of the magnetic field. Magnetic polarizability were calculated for neutral ρ^0 meson. It should be noted that magnetic polarizabilities with spin projections directed along magnetic field have different sign unlike the case another spin projections. This is due to the anisotropy of the magnetic field created
In the case of charged $\rho^{\pm}$ mesons we observe the agreement between Landau level picture and our data at small magnetic fields . The magnetic polarizabilities of neutral vector mesons were found for the lattice volumes 18^4 fitting the ground state energies. The g-factor of $\rho^{\pm}$ mesons is found from its the ground state energies with various spins. After chiral extrapolation we obtain the value $g = 2.4 \pm 0.2$ for the lattice volume 18^4 and lattice spacing $a = 0.115$ fm.

Acknowledgements

The authors are grateful to FAIR-ITEP supercomputer center where these numerical calculations were performed. This work is completely supported by a grant from the Russian Science Foundation (project number 16-12-10059).

References

[1] A.B. Migdal, J.Phys. USSR **8**, 331 (1944)
[2] A. Klein, Phys. Rev. **99**, 998 (1955)
[3] A.M. Baldin, Nucl. Phys.**18**, 310 (1960)
[4] V.A. Petrun'kin, JETP **13**, 804 (1961)
[5] M.H. Al-Hashimi and U.J. Wiese, Annals Phys. **324**, 343 (2009), arXiv: 0807.0630 [quant-ph].
[6] D. G. Gudino and G. T. Sanchez (2015), arXiv: 1305.6345 [hep-ph]
[7] A. M. Badalian, Yu. A. Simonov Phys. Rev. D **87**, 074012 (2013)
[8] E.V. Luschevskaya, O.E. Solovjeva, O.E. Kochetkov and O.V. Teryaev, Nucl. Phys. B **898**, 627 (2015).
[9] E.V. Luschevskaya, O.E. Kochetkov, O.V. Teryaev and O.E. Solovjeva,JETP Letters **101**, Issue 10, 674 (2015).

EPJ Web of Conferences **126**, 04047 (2016) DOI: 10.1051/epjconf/201612604047
ICNFP 2015

Very-high-energy astrophysical neutrinos with IceCube

Ignacio Taboada[1,a] for the IceCube Collaboration

[1] *School of Physics and Center for Relativistic Astrophysics. Georgia Institute of Technology. Atlanta, USA.*

Abstract. IceCube is a $\gtrsim$TeV neutrino observatory operating at the South Pole. IceCube has observed a flux of neutrinos of astrophysical origin, with energies beyond 2 PeV. However the sources of these neutrinos have not yet been identified. A summary of various IceCube observations is presented. The results discussed were obtained through several different analysis methods, which have varying sensitivity to the different neutrino flavors. A discussion of the spectral fit obtained for the various event selections is included, as well as the constraints on the astrophysical neutrino flavor flux ratio. Several attempts by IceCube to identify the sources of these neutrinos are described. These include studies correlating neutrino events with catalogs of sources as well as self-correlations among IceCube's neutrinos. The observations of astrophysical neutrinos are limited by statistics. So an upgrade of IceCube, including a larger detector and a surface veto is planned. This upgrade is briefly discussed.

1 Introduction

Almost three years ago, IceCube reported the discovery of an astrophysical flux of neutrinos in the 60 TeV to 1 PeV energy range [1]. The observed distribution of neutrinos is consistent with isotropy over the entire (4π sr) sky and there are no obvious individual astrophysical sources responsible for this flux. The discovery study, initially conducted with 2 years of IceCube data has been updated to include another 2 years of data [2, 3]. With additional data, the highest energy observed has increased to ~2 PeV. The existence of an astrophysical neutrino flux has been verified by several subsequent analyses e.g. Refs. [4, 5].

It is a reasonable assumption that cosmic ray sources are also sources of IceCube's neutrinos. It is however possible to explain the neutrino observation assuming that, e.g. dark matter decays into the observed flux. Cosmic rays are charged particles detected at Earth almost isotropically. Cosmic rays are observed in energies from 10^9 eV to 10^{20} eV. Below $\sim 10^{14}$ eV, balloons and satellites observe cosmic rays directly and measure their composition (i.e. the flux proportion of protons, helium, ..., iron). The cosmic ray spectrum is often described by a power law in energy, although there's observational evidence of a large number of features in the spectrum. Direct measurements allow for precise knowledge of the spectra. At higher energies, the steeply falling cosmic ray flux requires the use of ground based detectors. Ground detectors measure the cosmic ray spectra and composition indirectly, via the observation of air showers, which results in less precise measurements. At the highest energies, $> 10^{18}$ eV, the Pierre Auger observatory and Telescope Array/HiRes, disagree on the the composition,

[a] e-mail: itaboada@gatech.edu

with the former preferring heavier components (iron-like) and the latter preferring lighter components (proton-like).

There isn't certainty yet on the origin of the cosmic rays. Magnetic confinement of cosmic rays in our Galaxy is expected below the so called cosmic-ray *knee* (4×10^{15} eV), and it is presumed, in this energy range that sources are galactic. The name knee is derived from a softening of the cosmic ray spectrum from an index of ~ -2.7 below the knee to ~ -3 above the knee. At energies above the *ankle* (4×10^{18} eV) sources are expected to be extra-galactic. It is unclear whether between the knee and the ankle, sources are galactic or extragalactic. For galactic cosmic rays there is evidence that the sources are young, less than a few hundred years old, supernova remnants (SNRs). Motivation for SNRs being the source of galactic cosmic rays is that the mechanical power of all ejecta by supernovae in our galaxy, exceeds the power needed to sustain the cosmic ray population by two orders of magnitude. Shocks in supernova remnants and particle acceleration, such as first order Fermi acceleration, explain the transfer of kinetic energy of the supernova ejecta to cosmic rays.

Sources of cosmic rays are also expected to be sources of γ-rays and neutrinos. This is because a fraction of the accelerated cosmic rays should interact with material/photon fields at the source or near the source. This interaction results in neutral and charged pions that decay into γ-rays and neutrinos respectively. Observations of various galactic sources, e.g. Tycho's supernova, SN1572 [6] are consistent with the expectations of cosmic ray production. The young SNR hypothesis is also consistent with the observation of star burst galaxies with γ-rays [7, 8]. However, it is possible to produce GeV to TeV photons without cosmic ray acceleration and therefore observing γ-rays does not uniquely identify a source as a cosmic ray generator. Furthermore, given a galactic cosmic ray source with emission up to the knee, γ-rays up to 100 TeV are expected. These observations have not been unequivocally made yet, although there are several candidates, e.g. HESS-J1641-463[9].

Neutrino observations are generally assumed to be uniquely tied to cosmic rays. However as already explained, a diffuse source may have a different explanation. A *point-source* of neutrinos, however, would be a smoking gun for the observation of a cosmic ray source. IceCube's main science objective, although not the only one, is to detect astrophysical neutrinos and find the sources of cosmic rays.

These proceedings summarize observations by IceCube of the very-high-energy neutrinos of astrophysical origin.

2 IceCube

IceCube is a very-high-energy neutrino telescope in operation at the South Pole. Neutrinos are detected by observing secondary particles produced in neutrino-matter interactions. These secondaries produce Cherenkov radiation as they travel in very transparent Antarctic ice. IceCube observes the blue Cherenkov light with a 3-dimensional array of optical sensors (photo-multiplier tubes). The sensors are arranged in the bottom kilometer of vertical strings that are 2.5 km long. To compensate for the very low neutrino-matter interaction cross-section, the instrumented volume is gigantic: approximately one cubic kilometer for IceCube. A total of 5,160 optical sensors in 86 strings comprise IceCube. There is no sharp energy threshold for neutrino detection in IceCube, with the minimum energy instead, dependent on the selection criteria of each particular analysis. In most studies it is approximately 1 TeV. A section of IceCube is instrumented with higher density of optical sensors, allowing for a lower threshold. This subdetector is known as DeepCore. Another subdetector of IceCube is IceTop, an air shower array on the surface which is sensitive to cosmic rays of several PeV and higher. Figure 1 shows IceCube and its subcomponents.

IceCube construction began in 2005 and finished in December 2010. The modular design of IceCube allowed for initial partial operation. In these proceedings the moniker IC-XX represents the

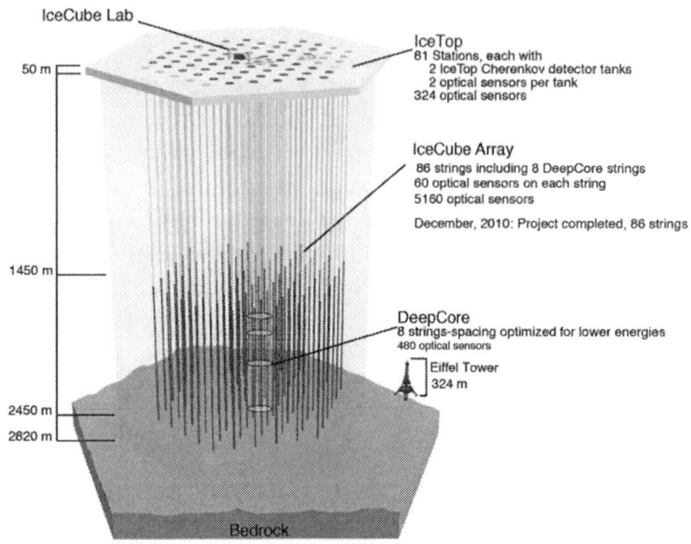

Figure 1. The IceCube neutrino observatory. A total of 86 strings with optical sensors are deployed between 1,450 and 2,450 m of depth. A subset of strings have been installed closer in horizontal distance and with a higher vertical density of sensors. This subdetector, DeepCore, is located at the center bottom of the rest of IceCube, so that the main array can be used as a muon veto. The surface array, IceTop, consists of 81 pair of ice tanks. Tanks have 2 photomultipliers each, to be sensitive to ground level particles from air showers. Each pair of tanks is located above a string. The detector is located ~1 km away from the geographical south pole.

number of strings, XX, in a given operational configuration. IceCube operates with extremely high duty cycle, over 99%. And ~99% of the optical sensors are operational.

There are two main channels for neutrino detection by IceCube. *Tracks* are made by muons, which are the product of charged current ν_μ interactions [1] with rock or ice near IceCube. A small fraction of the muon channel also arises from ν_τ charged current interactions in which the resulting τ decays into a muon (17.4% branching ratio). The track channel significantly increases the effective size of the detector since, at TeV energies and higher the range of muons is in the kilometer scale. The long lever arm of tracks in IceCube allows for good directional reconstruction, from $\sim 1^\circ$ at 1 TeV to better than $\sim 0.4^\circ$ above hundreds of TeV. Because neutrino interaction vertices are often outside of the detector, the track channel does not measure the neutrino energy directly.

For the *shower or cascade* channel neutrinos of all flavors are detected via several mechanisms including $\bar{\nu}_e + e^-$ interactions 6.3 PeV, i.e. the Glashow resonance; ν_e charged current interactions and neutral current interactions for all flavors. The majority, of ν_τ charged current interactions result in cascade channel events. Particle cascades are initiated when the initial neutrino interaction produces particles which subsequently interact through processes such as bremsstrahlung, e-pair production and hadronic interactions resulting in a geometric increase in the number of particles. This increase in particles repeats until a critical energy at which the radiative energy losses are no longer the most significant energy loss mechanism for particles propagating in the ice. A typical shower in ice will have a total length of ~10 m which is point like compared to the dimensions of IceCube. The lack of lever arm implies that angular resolution is only due to directionality in Cherenkov radiation. This information is partially washed out by optical scattering in the ice. At PeV energies, cascades can be reconstructed to $\sim 10^\circ$ and at tens of TeV to $\sim 30^\circ$. On the other hand, cascades are typically detected inside of the detector and all the visible energy is well measured. Furthermore, for most (the exception is the usually sub-dominant neutral current interaction) neutrino interaction channels, the visible energy is well correlated with the neutrino energy. At energies beyond ~500 TeV ν_τ interactions can have many complex topologies. However this subject will not be explored further in these proceedings.

[1]IceCube is unable to distinguish ν from $\bar{\nu}$

The bulk of measurements, 3 kHz, by IceCube is of atmospheric muons, which are produced in cosmic ray air showers. Atmospheric neutrinos are also easily observable with IceCube, at a rate of ~300 per day with the track channel (exclusively ν_μ) and lower with atmospheric neutrino induced cascades (a mixture of ν_e and ν_μ). The spectrum of atmospheric muons and neutrinos is softer than that of their parent cosmic rays by approximately one power of energy. Because cosmic rays sources are expected to produce neutrinos with very hard spectra, $dN/dE \propto E^{-2}$, energy is one of the possible characteristics that aids the identification of astrophysical neutrinos.

3 Discovery of a very-high-energy astrophysical neutrino flux

Initial evidence for an astrophysical flux was made with the observation of two cascade-like events of PeV energy in a search for ultra-high-energy neutrinos through the Berezinsky-Zatsepin effect, i.e. neutrinos from product of the highest energy cosmic rays, $\sim 10^{19}$ eV, and the cosmic microwave background. These two events [10] did not match the ultra-high-energy neutrino expectation but were too high-energy for the expectations of atmospheric neutrinos.

A follow-up study was designed that focused on events in which the neutrino vertex is contained in a volume slightly less than half that of IceCube. The outer layer of the detector, top and sides, is used to tag down- and through-going events - the signature of atmospheric muons and thus cosmic-ray air showers. This veto, along with an energy proxy cut (total amount of light observed by the optical sensors) can be used to reduce the background of down-going muons. The energy proxy cut is approximately equivalent to 30 TeV. Atmospheric neutrinos are also a background. However down-going atmospheric neutrinos are vetoed quite efficiently because the air showers that produce them would also produce muons that would be found by the veto.

This method was initially applied to two years of data, one with IC-79 and the first year of IC-86 and was the basis for the initial discovery claim [1]. This study has been repeated for a third [2] and fourth year [3] of data.

The following numbers correspond to the "4-year study", of 1347 days livetime starting on May 2010. The atmospheric background from tagged events is 12.6±5.1 events. The atmospheric neutrino background is $9.0^{+8.0}_{-2.2}$. A total of 54 events were found in this study. However posterior examination shows that two are background: one shows coincident activity in IceTop, indicating a simultanous air shower, and the other event corresponds to two almost simultanous but uncorrelated muons, with the first muon *disarming* the veto and the second one faking a signal. The atmospheric origin of the observations is rejected at the 6.5σ level. Of the 52 remaining "good" events, 39 are showers and 13 are tracks - which is consistent with expectations. A sky map with the 52 "good" events is shown in figure 2.

Using all 54 events, the best fit flux for astrophysical neutrinos is

$$E^2 \frac{dN}{dE} = (2.2 \pm 0.7) \times 10^{-8} (E/100 \text{ TeV})^{-0.58 \pm 0.25} \text{ GeV}^{-1} \text{ s}^{-1} \text{ sr}^{-1} \text{ cm}^{-2}. \quad (1)$$

And this flux is valid in the range 27 TeV to 2 PeV. It is also possible to fit the spectrum well including a cutoff, in which case a harder spectral index of -2.31 is found and a cutoff of 2.7 PeV [3].

Southern hemisphere events (down-going) are distributed isotropically, but there is a deficit of northern hemisphere events. This is easily explained by Earth absorption of neutrinos, which is relevant for energies above ~ 100 TeV and increases with energy. The flux at Earth's surface is consistent with isotropy.

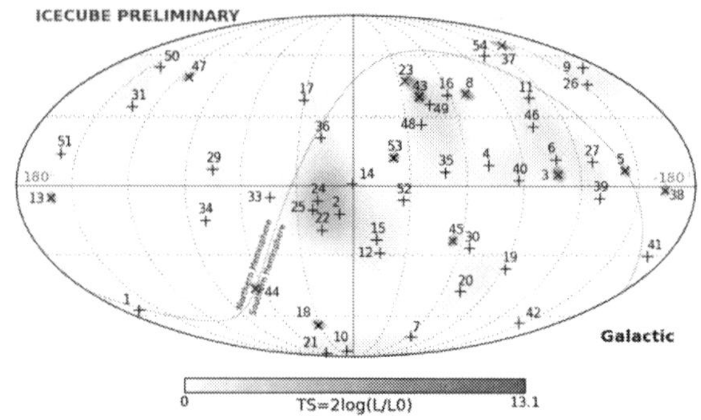

Figure 2. Directions of the 52 candidate astrophysical neutrinos in galactic coordinates. Crosses indicate shower-like events and "X"s show tracks. The color scale is the significance of event excess in a given direction, which depends on the angular uncertainty for each event. Two events are omitted as described in the text. The dashed line shows the northern and southern sky. There is no evidence for a point source in this data set or a correlation with the galactic plane.

4 Confirmation with muon neutrinos

The *traditional* neutrino astronomy channel is the track channel in which Earth acts as a filter by blocking up-going (nothern sky) atmospheric muons and only atmospheric ν_μ remain as a significant background. This background can be further reduced via energy or energy-proxies. The most recent study by IceCube is for "6 years" including IC-40, IC-59, IC-79 ($\approx$1 year each) and 3 years of IC-86 [3]. The astrophysical flux is also identified in this study with a best fit spectrum:

$$E^2 \frac{dN}{dE} = 0.78^{0.29}_{-0.25} \times 10^{-8} (E/100 \text{ TeV})^{-0.6\pm0.13} \text{ GeV}^{-1} \text{ s}^{-1} \text{ sr}^{-1} \text{ cm}^{-2} \quad (2)$$

which is valid for the energy range 160 TeV to 10 PeV. The atmospheric only hypothesis is rejected at the 6.0σ level. Figure 3 shows the energy spectrum of events as found by this study. The flux is also consistent with the expectation of isotropically distributed sources - although in this case the sensitivity of the analysis is limited to declinations between -5 to 90°.

Other studies have been conducted by IceCube that are also sensitive to the astrophysical flux [5, 11]. In particular, the starting event method used for discovery has been used with an energy-dependent veto region that allows the extension of observations to as low as 10 TeV. In this case the spectrum seems to be well described by a power law with index -2.49. There is an apparent discrepancy in spectrum as measured with starting events and with tracks. It should be noted that the sensitivity in energy range of both methods overlap but are not identical. The limited statistics in number of astrophysical events is such that inconsistent power law fits can describe the data of each method well, even if the true flux has a more complicated description than a power law. It is possible that two or more populations of sources produce the observed flux. However the statistics necessary to test this hypothesis are beyond the reach of IceCube.

5 The search for sources of the very-high-energy neutrinos

A very large number of explanations for the origin of IceCube's astrophysical flux have been offered in the literature. Consistency with isotropy for the observations obtained with both methods described

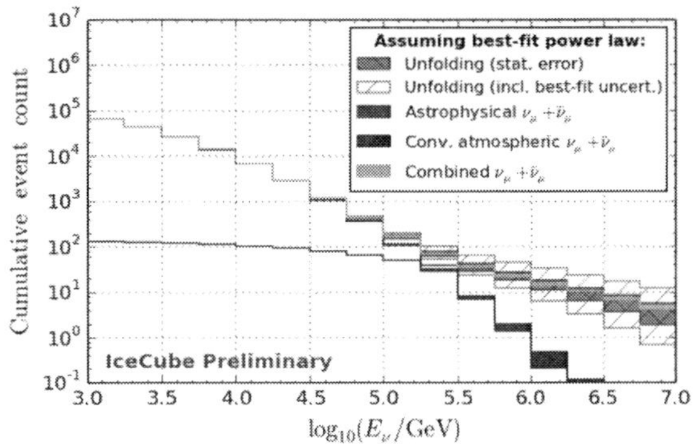

Figure 3. Energy spectrum of up-going ν_μ event candidates. The red line corresponds to the best fit astrophysical flux assuming a power law in energy. Atmospheric neutrino background is shown in blue.

indicates a dominant extragalactic origin. The data in both cases has been studied for correlation with the galactic plane. No significant correlation is found in any of the analysis. For the starting event method used for discovery, the best correlation is for $\pm 2.5°$ above and below the galactic plane with p-value of 7%. This indicates that the contribution from objects in the galactic plane is not dominant.

Searches for self-correlation in the starting event sample fail to find one or more point sources of neutrinos. With an event selection similar to that used for the track study described in section 4, a large collection of ν_μ events has been accumulated of a period of 6 years. A self-correlation study on directions with this data set also fails to find point sources. This latter study is so sensitive that if the astrophysical flux were due to a handful of sources, they would have already been detected as point sources. This lack of observations in point sources places a limit on the number of (neutrino) visible sources to >100 over the entire sky (4π sr) [2].

IceCube has searched, with null results, for neutrinos in correlation with over 1000 gamma ray bursts (GRBs) reported by various satellites. For GRBs, it is assumed that neutrinos are produced simultanously with the keV-MeV photons, the *prompt* phase. IceCube studies conclude that during prompt phase, at most 1-3% of the observed astrophysical flux is due to GRBs [14]. Even if the correlation time is allowed to expand, GRBs cannot produce more than 12% of the spectrum for times of ± 20 hours around GRBs.

Active galactic nuclei are another prominent candidate for cosmic ray sources [15]. A study searching for correlation with a stacked list of blazars detected by Fermi-LAT, suggests that less than 30% of the observed flux originates from these. This limit however is quite model dependent given the strong time variability of blazars. Finally a study on the diffuse extragalactic background as observed by Fermi-LAT, rules out star-burst galaxies as the major component of the neutrino astrophysical flux [16].

So the most popular candidates, GRBs, AGNs and star-burst galaxies are constrained or ruled out. The question of the origin of the astrophysical neutrinos remains open.

It is interesting to study the flavor flux ratio of neutrinos. In the most naive interpretation, neutrinos are expected to be produced at cosmic ray sources with a flavor flux ratio of 1:2:0 ($f_{\nu_e} : f_{\nu_\mu} : f_{\nu_\tau}$). And vacuum oscillations over astronomical baselines are such that a ratio of 1:1:1 is expected at Earth. This expectation can be modified, e.g. assuming strong magnetic fields at the sources that cool down some of the particles (muons) that produce neutrinos leading to a ratio of 0.4:0.6:0.6 at Earth [12]. Both the shower and the track channels are sensitive to mixtures of the three flavors. IceCube has conducted a study [13] in which various observations of the astrophysical flux are studied jointly. It is found that IceCube's data is consistent with all expected astrophysical flavor flux ratios. This can be seen in figure 4.

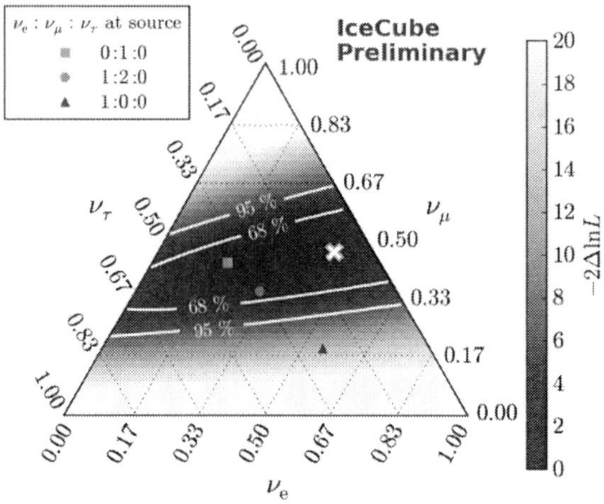

Figure 4. Flavor triangle plot, all three flavor components are free to float in a global fit of IceCube observations, but assumed to all follow the same energy spectrum. In this case spectral index, flux normalization and cutoff are fitted. The standard astrophysical assumption 1:2:0 at the source. Pure flavor composition, of any type, are ruled out.

6 Conclusions and Outlook

IceCube continues to collect that with very high uptime, stability and quality. The astrophysical flux has been observed with various methods with a combined sensitivity to 3 orders of magnitude in energy. The observed spectrum appears consistent with a dominant extragalactic origin, but the most well known candidates, GRBs, AGNs and star burst galaxies have been constrained or ruled out.

IceCube is proposing the construction of *IceCube Gen2*, which would consist of a larger underice component as well as a large area surface detector to act as a very wide angle veto.

References

[1] Aarsten M.G. et al. (IceCube collaboration), Science **342**, 6161 (2013)
[2] Aartsen M.G. et al. (IceCube collaboration), Phys. Rev. Lett. 113, 101101 (2014)
[3] Aarsten M.G. et al. (IceCube collaboration), Proc. of the 34th Int. Cosmic Ray Conf. arXiv:1510.05223, The Hague (2015).
[4] Aartsen M.G. et al. (IceCube collaboration), Phys. Rev. Lett. **115**, 081102 (2015)
[5] Aartsen et al. M.G. (IceCube collaboration) Phys. Rev. **D91**, 022001 (2015)
[6] Acciari V.A. et al. (VERITAS collaboration), ApJ **730**, L20 (2011)
[7] Acciari V.A. et al. (VERITAS collaboration), Nature **462**, 770 (2009)
[8] Abdo A.A. (Fermi collaboration), ApJ **709**, L152 (2010)
[9] Abramowski A. et al. (HESS collaboration) ApJ **794**, L1 (2014)

DOI: 10.1051/epjconf/201612604047

[10] Aarsten M.G. et al. (IceCube collaboration) Phys. Rev. Lett. **111**, 021103 (2013).
[11] Aarsten et al. M.G. (IceCube collaboration), ApJ **809**, 98 (2015)
[12] Kashti T. Waxman E., Phys. Rev. Lett. **95**, 181101 (2005)
[13] Aartsen M.G. et al. (IceCube collaboration), Phys. Rev. Lett. **114**, 171102 (2015).
[14] Aartsen M.G. et al. (IceCube collaboration), ApJ **805**, L5 (2015)
[15] Padovani P., Petropoulou M., Giommi P. and Resconi E. Submitted to MNRAS. arXiv:1506.09135
[16] Bechtol K., Ahlers M., Di Mauro M, Ajello M and Vandenbroucke J. Submitted to Phys. Rev. Lett. arXiv:1511.00688

EPJ Web of Conferences **126**, 04048 (2016) DOI: 10.1051/epjconf/201612604048
ICNFP 2015

What could the LHC teach us on the structure of space-time?

George Triantaphyllou[1, a]

[1] *National Technical University of Athens*
58 SINA Str., GR 106 72 ATHENS, Greece

Abstract. Collision energies of proton beams now available at the LHC increase the probability of discovering the inner works of the Brout-Englert-Higgs (BEH) mechanism within the foreseeable future. Nevertheless, they are still several orders of magnitude below the scale where a possible non-trivial structure of space-time would be detectable. Apart from remaining completely silent on the issue of the fundamental nature of elementary particles and the space in which they propagate, one may try to speculate on this matter by carefully extrapolating existing scientific methods and knowledge to Planck energies. In this talk, an effort is made to logically link some potential discoveries at the LHC with specific space-time structures. Since such links are inevitably weak due to the huge energy hierarchy between the electro-weak and the Planck scales, our goal does not exceed a mere presentation of naturalness and self-consistency arguments in favor of some of the possible outcomes, placing particular emphasis on the scenario of the mirror world.

1 Introduction

After several decades of efforts both in the experimental and in the theoretical front, high-energy physicists are awaiting for fresh collision data from the upgraded LHC which could potentially shed more light on the origin of electro-weak symmetry breaking. The energies of the proton beams increase the probability of discovering new physics beyond the Standard Model (SM) which might lie behind the BEH mechanism. Even though no significant deviations from the Glashow-Weinberg-Salam model containing just an elementary Higgs field have been detected so far, it is well known that the mass of such a field is unstable against quantum corrections. If nothing new is introduced to stabilize its mass, not only would it acquire a Planck-scale mass, but it would also render all other known particles much heavier than they are found to be. This instability reminds us that when a theory attempts to introduce an arbitrary hierarchy between energy scales with no special reason, nature has a way, of quantum mechanical origin, to eliminate this unnaturalness. There have been several proposals over the years trying to explain the stabilization of the electro-weak scale to such small values compared to the scale where gravity effects might set in, altering probably in a fundamental way the inner structure of space-time. Some of them, apart from introducing new fields or phenomena at energies in principle accessible at the LHC, allow us to speculate also on what general form such alterations might take.

[a] e-mail: gtriantaphyllou@aya.yale.edu

In the following, we choose to discuss some of these alternatives, one more extensively than the others since it has not been studied as extensively as the others, which could potentially hide behind the BEH mechanism, and which have already started being under scrutiny by LHC data. The first introduces large extra dimensions, the second introduces supersymmetric particles, and the third mirror particles, all expected near the TeV scale which is relevant to the BEH mechanism. This choice is obviously not exhaustive, since it is always possible that Nature keeps something else in store for us, but it is mainly guided by the characteristic differences of their implications on the structure of space-time. The huge hierarchy between the electro-weak and the Planck energy scales render the ground on which such speculations reside, and especially attempts to establish a relevant "one-to-one" correspondence, shaky. Nevertheless, the form which the stabilizing mechanism will be hopefully found to take is one of our last chances for several decades to come, at least regarding accelerator experiments, to constitute a basis of a guess concerning the true nature of elementary particles, their interactions and space-time. This is also due to the fact that the proposals discussed here are compatible with the "desert hypothesis", introducing new Physics around the scale of 1 TeV and then nothing else until scales within 1-3 orders of magnitude away from the Planck scale. This situation highlights the critical conjuncture of present-day high-energy Physics in an effort to discover the fundamental principles of Nature. Therefore, although the final results should be taken *cum grano salis*, the importance of the questions involved render the relevant theoretical speculations valuable.

2 Phenomenological inductive approach

2.1 Potential implications

One very interesting alternative which could stabilize the electro-weak scale is to assume that there are extra spacial dimensions compactified on a radius relevant to the BEH mechanism [1]. This eliminates by "brute force" the hierarchy between the Planck scale and the electro-weak scale by a radical alteration of the structure of space-time at scales which could be in principle accessible at the LHC, or the colliders next to it, by the production of mini black-holes and other phenomena [2]. Even though models with large extra dimensions are usually confronted with problems related to the electron magnetic moment which is measured experimentally very precisely, they offer a typical example of a way by which the BEH mechanism could signal a very important shift of our understanding of the nature of space-time. Recent experimental bounds on such models are found in [3].

Another alternative which has been particularly popular during the last decades is to assume that the known elementary particles have "supersymmetric partners", i.e. fields with the same gauge charges but spins differing by one half unit from the spins of the original particles. The existence of such hypothetical particles leads to the cancellation of quantum corrections which destabilize the mass of elementary Higgs scalars, offering thus an effective solution to the hierarchy problem. In this scenario, the required stabilization is due to a new space-time symmetry called "supersymmetry" which accounts for the existence of these new "partners". These new elementary particles can be studied perturbatively since they are weakly coupled at LHC energies. Moreover, they may lead to a unification of the gauge-coupling strengths at much higher energies [4], which would not only imply that proton decay should be detectable in future experiments but also that a unifying theory can incorporate all this rich structure of fields in a symmetric way. In this respect, the new space-time symmetry proposed, apart from solving the hierarchy problem, is an important ingredient in attempts unifying gravitational interactions with the other known gauge interactions in the context of string theory [5].

According to this theory, elementary particles at Planck energies can be considered as strings propagating perturbatively on a continuous space-time background. Moreover, extra space dimensions are

expected to influence directly all relevant physical considerations, allowing membrane solutions with dimensionality higher than the one of strings [6]. From such a view point, it is apparent that a supersymmetric solution to the hierarchy problem might also signal a radical shift of our understanding of the fundamental nature of particles and space-time. Even thought the "stringy" nature of particles is not expected to be detectable at the LHC, unless some of the extra dimensions are large, discovering supersymmetric particles would offer a mild indication that the unification program leading to string theory is in the right direction. Although it would be far from being a proof of the "stringy" nature of elementary particles and of extra dimensions at Planck energies, it would at least fulfill the minimal requirement of self-consistency, which, due to the importance and difficulty of the problems at hand, is not a trivial achievement. Constraints on relevant models can be found in [7].
Last but not least, we come to the third alternative, which will be mostly discussed here, according to which known fermions have mirror "partners" usually called "katoptrons". Contrary to considering fermions having zero-spin "partners", katoptron theory introduces new fermion fields with opposite chirality from the known fermions. Apart from carrying the same known gauge charges, katoptrons interact with each other via a new gauge interaction described by an SU(3) mirror-family symmetry. This interaction becomes naturally strong at around 1 TeV and breaks the electro-weak symmetry dynamically. In this scenario, the BEH mechanism is an effective description of a strongly-coupled theory similar to QCD, where instead of elementary Higgs fields or free mirror fermions, one expects several bosonic bound states, i.e. spin-0 and spin-1 mirror mesons which might show up in the ATLAS and CMS experiments at the LHC. Stabilization of the electro-weak scale is achieved by the compositeness of the Higgs fields, which can be traced back to a gauge symmetry.
Mirror fermions were first proposed in the Lee-Yang *locus classicus* on parity-symmetry violation [8] and were later used in [9] in order to break the electro-weak symmetry dynamically via mirror-fermion condensates. This was proposed in a spirit similar to [10], which in its turn was inspired by the fermion condensates appearing in BCS superconductivity [11]. Using a new gauge interaction felt only by katoptrons and becoming strong around 1 TeV assuming unification of all gauge-coupling strengths around the Planck energy was first achieved in [12]-[13], in continuation of a long tradition starting from Pati-Salam models, freeing the theory from Landau poles and completing it group-theoretically [14]. Similarly to supersymmetric theories, providing a unifying framework not only leads to a prediction of proton decay, but also enables unification of all interactions including gravity, albeit in a context distinct from string theory. In fact, considering elementary particles as vacancies within an emergent dynamical discrete space-time leads to models incorporating mirror fermions [13]. Therefore, detection of mirror mesons or other katoptron effects at the LHC, far from being a direct probe of Planck-scale Physics, could be considered as a possible indication of, or even more conservatively just as consistent with, some kind of space-time discreteness. Before returning to this important issue more extensively in the next section, we briefly review some of the possible relevant signals accessible at the LHC.

2.2 Mirror mesons and their decays

A classical study of hadron-collider phenomenology of models based on strongly-interacting fermions, which initially went under the name of "technicolor", is [15]. The relevant program here consists of studying similar production and decay mechanisms of mesons formed as bound states of the hypothetical fermions introduced, since confinement is not expected to allow katoptrons to propagate freely in space. Therefore, the focus of LHC experiments trying to test such theories should be on excesses of di-boson and di-fermion production originating from new intermediate states analogous to QCD hadrons having either zero spin, i.e. mirror scalar and pseudo-scalar mesons like pions, or spin-1, i.e. mirror vector mesons like rhos. Non-perturbative dynamics do not allow a detailed quantitative study of the relevant cross-sections as in weakly-coupled theories. Nevertheless, it is still possible to make some general qualitative remarks on what should be generally expected.

We first note that SM quantum numbers for mirror fermions give rise to a mirror chiral symmetry $SU(8)_L \times SU(8)_R$ subsequently broken by the strong coupling of the mirror family group SU(3) down to a diagonal vector SU(8) flavor symmetry corresponding to four isospin doublets, i.e. one color-singlet and one color-triplet, analogous to Gell-Mann's flavor symmetry in QCD. This leads to a mirror meson spectrum corresponding to the broken generators of the axial SU(8) symmetry, thus including, apart from the resonances σ^K, η^K and ω^K, with the superscript K a reminder of their katoptron content, analogous to the scalar σ, pseudo-scalar η and vector ω mesons in QCD, sixty-three (pseudo-scalar) mirror pions π^K, along with sixty-three (spin-1) mirror ρs, ρ^K having the same quantum numbers except from spin, their number being equal to the dimension of the adjoint representation of the flavor group. Furthermore, one can not exclude the existence of the corresponding parity-even and heavier excited states of these mesons.

In this scenario, the σ^K meson has quantum numbers equal to the SM Higgs particle, while three mirror pions, $\pi^{K\,2\,0}$ and $\pi^{K\,2\,\pm}$, are "eaten" by the electro-weak gauge bosons, becoming thus the longitudinal components of $W^{\pm}$ and Z^0. This leaves us with five remaining color-singlet pseudo-scalars, i.e. three electrically neutral along with one positively and one negatively charged, from the initially eight color-singlet mirror pseudo-scalar mesons, in addition to eight color-singlet mirror vector mesons:

$$\pi^{K\,1\,0},\ \pi^{K\,1\,\pm},\ \pi^{K\,1\,0\prime}, \eta^K\ (\text{spin}-0)\ \text{ and }\ \rho^{K\,1,2\,0},\ \rho^{K\,1,2\,\pm},\ \rho^{K\,0\prime}\ \text{and}\ \omega^K\ \ (\text{spin}-1). \tag{1}$$

Regarding the rest of the mirror mesons, eight of each kind are color-triplets denoted by

$$\pi_3^{K\,1,2,2\prime,5},\ \bar{\pi}_3^{K\,1,2,2\prime,5}\ (\text{spin}-0)\ \text{ and }\ \rho_3^{K\,1,2,2\prime,5},\ \bar{\rho}_3^{K\,1,2,2\prime,5}\ (\text{spin}-1), \tag{2}$$

all of them fractionally charged (either $-\frac{1}{3}$ for $\pi_3^{K\,1}$, $\rho_3^{K\,1}$, or $\frac{2}{3}$ for $\pi_3^{K\,2,2\prime}$, $\rho_3^{K\,2,2\prime}$, or $\frac{5}{3}$ for $\pi_3^{K\,5}$, $\rho_3^{K\,5}$) and usually called "lepto-quarks", and four of each kind are color-octets, denoted by

$$\pi_8^{K\,0},\ \pi_8^{K\,\pm},\ \pi_8^{K\,0\prime}\quad (\text{spin}-0)\ \text{ and }\ \rho_8^{K\,0},\ \rho_8^{K\,\pm},\ \rho_8^{K\,0\prime}\quad (\text{spin}-1) \tag{3}$$

respectively [15]. Note that isospin-singlet mesons are denoted either by σ^K, η^K, ω^K or by primed symbols above, the rest being members of isospin triplets. Another obvious fact to bear in mind is that mirror mesons with equal charge and color may mix with each other, like $\pi^{K\,1\,0}$ with $\pi^{K\,1\,0\prime}$, $\pi_3^{K\,2}$ with $\pi_3^{K\,2\prime}$, $\pi_8^{K\,0}$ with $\pi_8^{K\,0\prime}$, and similarly for their vector-meson counterparts.

The masses of these mesons are not easy to estimate, and large-N arguments of [15] are questionable due to the fact that the katoptron family group becomes strongly coupled and breaks at energies $\Lambda_K \sim$ 1 TeV from SU(3) down to SU(2). Therefore, one expects a doubling of the meson mass spectrum due to the hierarchy, denoted by r, of the corresponding energy scales of mirror-fermion chiral symmetry breaking, or even its tripling according to the extent by which the breaking of the remaining SU(2) family symmetry affects the corresponding mirror-meson masses. Estimating this hierarchy yields a factor r of around

$$r \sim \exp\Big(C_2(SU(3)) - C_2(SU(2))\Big) \sim 5.75, \tag{4}$$

with $C_2(g)$ the quadratic Casimir invariant of a Lie algebra g [16], noting that $r{\sim}5$ is easier reconcilable with the weak scale and with present experimental bounds. Therefore, the two lighter mirror generations are expected to have masses of around Λ_K/r=200 GeV. In the following, the lighter mirror mesons, denoted by π_A^K and ρ_A^K, omitting numerical superscripts and color subscripts, correspond to katoptrons with masses M_A~200-300 GeV, bearing in mind that they might be further split into two distinct subgroups according to the mass of their mirror-fermion content. Similarly, the heavier mirror mesons are denoted by π_B^K and ρ_B^K, corresponding to katoptrons masses M_B~Λ_K~1-1.5 TeV.

Allowing a slight digression, note that the mass hierarchy between third-generation SM quarks and the other fermion generations can be traced back to the hierarchy of mirror-fermion chiral-symmetry-breaking scales [16]. This happens because the breaking of the mirror-family symmetry allows the quantum-mechanical mixing of mirror fermions with the known fermion fields, providing thus a mass "feed-down" mechanism which is free not only of flavor-changing-neutral currents, but also of too-light pseudo-Goldstone bosons and of a too-large $\Delta\rho$ parameter [17], all problems usually confronting extended-technicolor (ETC) models. Furthermore, it might provide for deviations from QCD-like strong dynamics accommodating a small S parameter, possibly related to a reduced or even inverted mass hierarchy between vector and axial-vector mirror mesons. In addition, the generalized mass matrix mixing mirror- with ordinary fermions provides an origin not only for the Cabibbo-Kobayashi-Maskawa (CKM) matrix but also for small neutrino masses via a see-saw mechanism [12]. Moreover, this framework allows the introduction of complex phases violating the weak CP symmetry potentially leading to the observed baryon asymmetry of the universe. Last, the simultaneous existence of two groups of light and heavy mesons is not only a by-product of the breaking of the mirror-family group, but it is also required for self-consistency, since mesons only as light as the recently discovered Higgs-like scalar would be hardly able to reproduce the electro-weak scale correctly [16].

To return to the issue at hand, order-of-magnitude estimates based on QCD for mirror mesons would give masses of around $2M_A$~400-600 GeV for the lighter η_A^K, ρ_A^K and ω_A^K and $2M_B$~2-3 TeV for the heavier η_B^K, ρ_B^K and ω_B^K. On the contrary, mirror pions should be relatively light if considered as Nambu-Goldstone bosons of the broken mirror chiral symmetry. However, since the gauged mirror family symmetry is consecutively broken at around 1 TeV and 170 GeV, mirror-pion and σ^K masses are also expected to be grouped either around $M_A/2$~100-150 GeV (σ_A^K, π_A^K) or around $M_B/2$~0.5-0.75 TeV (σ_B^K, π_B^K). Note that colored mirror pions receive additional contributions to their masses due to QCD, which would be on the order of $\sqrt{\alpha_s}M_{\rho^K}\sim$ 200-300 GeV were the katoptron-family group to remain unbroken and have a single scale M_{ρ^K} for ρ^K masses, with α_s the value of the QCD coupling at the mirror pion mass. The breaking of the family symmetry implies however that these contributions might lie in the range of 0.1-0.8 TeV depending on the mass-spectrum group to which the colored meson belongs. In this picture, the meson σ_A^K corresponds to the "Higgs-like" particle detected at the LHC, in analogy to QCD where the lowest-lying resonance heavier than the three pseudo-scalar pions, the analogues of which are here "eaten" by $W^\pm$ and Z^0, is the scalar meson σ. The fact that the mass of the scalar particle detected is lower than double the masses not only of the electro-weak gauge bosons but of the top-quark as well might partially explain its relatively small width compared to the one of the sigma meson in QCD, which mainly decays into two pions.

In any case, the dynamics involved are so complicated that one should consider the mass ranges above with a fair amount of flexibility, which should be reflected in the relevant computer simulations and in the LHC data analysis. In fact, due to the breaking of the mirror family symmetry, mirror fermions might not have enough time to hadronize, either decaying before meson formation or forming very broad resonances. Assuming that mirror mesons do form, the discussion above implies a potentially very rich phenomenology which should show up in particle excesses in several processes [18]. Given precise experimental constraints, one should aim at characteristic signals which not only are more probable to be distinguished from background but also which can be distinguished from similar signals predicted by alternative theories. First of all, mirror meson phenomenology seems similar to what is expected in ETC scenaria following [19]. Therefore, only by focusing on the distinctive features of katoptron theory may one be able to favor one model over the other. Since the katoptron mass feed-down mechanism is based on quantum mixing instead of ETC interactions, the relevant effects

on the heavier fermions like the top quark should be more pronounced. In particular, the CKM matrix element $|V_{tb}|$ is expected to deviate from unity by about 5%. In addition, deviations from the SM left- and right-handed weak couplings of top quark should show up in conveniently-chosen asymmetries. Last, at much higher energies where katoptrons are no longer interacting strongly with each-other, or in the case where katoptrons do not hadronize, a forward-backward asymmetry of mirror fermions opposite to the one expected from SM fermions, or similar deviations from standard left-right weak-charge assignment, should be detectable. Given that such measurements require precision that can be easier achieved by a high-energy (3-4 TeV [18]) electron collider like CLIC and ILC or a muon collider rather than a hadronic collider, one should seek other methods of signal differentiation according to each model. One important difference for instance is the breaking of the katoptron-family gauge group, which, apart from rendering katoptrons unstable and unlikely to be connected to Dark Matter (DM) particles, leads to the lifting of the degeneracy of the mirror meson spectrum, something not expected in usual technicolor models. This leads to some relatively heavier pseudo-scalar mirror-meson masses (belonging to the "B" group) and some relatively lighter vector mirror-meson masses (belonging to the "A" group) than in typical technicolor models. This implies the prediction of many more resonances, but it changes the luminosities needed to detect their decays over background, since cross-sections depend not only on the new meson masses but also on the size of the new -unbroken-strong gauge group [15]. Taking the scalar meson σ_A^K as an example, its decay to two photons via new heavy-fermion triangle Feynman diagrams depends quadratically on the number of "colors" of the new strong interaction. Therefore, lack of observation of large deviations of the two-photon Higgs decay width from its SM value is consistent with the breaking of the mirror family group, possibly indicating even a tripling of the mirror-meson mass spectrum. We discuss in the following some processes of potential interest for the LHC assuming just a doubling of the mass spectrum. According to similar analyses [15][18][20], promising candidates detectable over background with the least integrated luminosity are the decays of the neutral color-octet pseudo-scalar mirror pions, $\pi_{8\ A,B}^{K\ 0\prime}$ produced mainly by gluon fusion but also quark-anti-quark annihilation, to a top-anti-top quark pair.
The special interest in this decay is due to the fact that mirror-pion decay widths $\Gamma(\pi^K \longrightarrow \bar{f}_i f_j)$ stemming from Yukawa interactions depend quadratically on the final-fermion masses $m_{i,j}$:

$$\Gamma(\pi_{8\ A,B}^K \longrightarrow \bar{f}_i f_j) \sim \Gamma(\pi_{3\ A,B}^K \longrightarrow \bar{f}_i f_j) \sim \frac{G_F p(m_i^2 + m_j^2)}{16\pi} \sim p\Big((m_i/2.1\ \text{TeV})^2 + (m_j/2.1\ \text{TeV})^2\Big), \quad (5)$$

where G_F is the Fermi constant and p is the fermion momentum in the mirror-pion rest frame. Although factors depending on the size of the new gauge group lead to a relative enhancement of the production mechanism in usual technicolor models, in katoptron models their effect is questionable due to the breaking of the SU(3) mirror-family group [18]. The main background to this process comes from QCD, which can be significantly reduced via stringent rapidity cuts since, contrary to this background, mirror-pion decay products are expected to be spherically distributed. Furthermore, the lighter mirror pion $\pi_{8\ A}^{K\ 0\prime}$ should be easier detectable due to the fall of cross sections with energy.
Nevertheless, there are two problems apart from the large QCD background rendering the observability of this process questionable. First, the pion mass should preferably be larger than twice the mass of the top quark, i.e. larger than about 350 GeV. This exceeds our previous estimates for the masses of the lighter mirror pions, even though it cannot be completely ruled out beforehand. Second, the breaking of the mirror-family symmetry leads to mixing of ordinary fermions with their mirror partners favoring processes which respect fermion generations [16]. Therefore, "group B" mirror scalar and pseudo-scalar mesons are expected to mainly couple to "group B" mirror vector mesons and third-generation ordinary fermions, while "group A" mirror scalars and pseudo-scalars are expected to mainly couple to "group A" mirror vector mesons and first- and second-generation ordinary fermions. Consequently,

the decay of $\pi_{8\,A}^{K\,0\prime}$ to a bottom-anti-bottom quark pair, even when kinematically allowed, is expected to be suppressed due to the small mixing between first- and second-generation katoptrons with third-generation SM fermions. Processes of main interest are thus "generation-respecting", allowing obvious extension of this argument to the aforementioned possible tripling of the mirror-meson mass spectrum. Therefore, heavier color-octet mirror pions, expected to have masses well over 1 TeV, i.e. large enough for on-shell production of top quarks, are expected to couple to third-generation fermions, leading to:

$$gg, \bar{q}q \longrightarrow \pi_{8\,B}^{K\,0\prime} \longrightarrow \bar{t}t \text{ or } \bar{b}b. \tag{6}$$

However, since cross-sections decrease with energy quite fast, the signal needed to exceed the relevant SM background requires much higher integrated luminosities than the ones estimated for lighter mirror mesons, unless $\pi_{8\,B}^{K\,0\prime}$ finally turns out to be much lighter, and this process' advantage over alternative ones becomes questionable. Other processes of interest are the decays of heavy neutral color-singlet mirror spin-0 mesons to a $\tau\bar{\tau}$ or a diphoton pair: $gg \longrightarrow \sigma_B^K$ or $\pi_B^{K\,1\,0\prime} \longrightarrow \bar{\tau}\tau$ or $\gamma\gamma$, enjoying the advantage of being QCD-background free. Nevertheless, since the cross-sections in the difermion case depend on the masses of the final fermions and in the diboson case correspond to a higher-loop process, b-quark, τ-lepton and $\gamma\gamma$ final states have a disadvantage when compared to $\bar{t}t$ pairs *ceteris paribus*.

A different promising process is related to the pair production of color-octet mirror pions π_8^K via $\bar{q}q$ annihilation and gluon fusion, which can be further enhanced by intermediate mirror vector mesons. Take as an example the following generation-respecting processes:

$$\begin{aligned} gg &\longrightarrow \text{direct}, \ g, \ \rho_{8\,A}^{K\,0\prime} \longrightarrow \pi_{8\,A}^{K\,+}\,\pi_{8\,A}^{K\,-} \longrightarrow \bar{s}c + \bar{c}s \\ gg &\longrightarrow \text{direct}, \ g, \ \rho_{8\,B}^{K\,0\prime} \longrightarrow \pi_{8\,B}^{K\,+}\,\pi_{8\,B}^{K\,-} \longrightarrow \bar{b}t + \bar{t}b, \end{aligned} \tag{7}$$

where "direct" denotes either t- and u-channels or absence of any s-channel intermediate state, and symbols on either side of "commas" refer to processes interfering with each other. The quarks produced are generally not collinear with their anti-particles, leading to signals which may in principle be distinguished from QCD background via appropriate cuts. Note however that it might be difficult to distinguish the signals discussed so far from similar processes originating from elementary scalars predicted by supersymmetric theories [15]. This regards especially the σ_A^K and π_A^K mirror mesons, since π_B^K mirror mesons have masses generally too heavy for supersymmetry, and mirror vector mesons have spin incompatible with the fields introduced by usual supersymmetric models. Another process of interest is the decay of charged mirror vector mesons, produced by interactions between up- and down- quarks q_i and q_j, to an electro-weak gauge boson pair, in analogy to [15]:

$$\bar{q}_i q_j \longrightarrow \rho_{A,B}^{K\,2\,\pm} \longrightarrow W^{\pm} + Z^0. \tag{8}$$

Recent event excesses at around 2 TeV at the LHC [21] might be related to a process involving $\rho_B^{K\,2\,0,\pm}$ or to heavier excitations of these resonances. Nonetheless, models with extra massive vector gauge bosons mimic the mirror-meson effects of this process. Therefore, such enhancements should be considered in conjunction with other effects in order to draw safer conclusions on the theory behind them.

The problem of signal ambiguity is not so acute in the case of a set of quite interesting processes regarding the pair production of lepto-quarks π_3^K. Lepto-quarks can be pair-produced via quark-anti-quark annihilation and gluon fusion, a process which can be similarly further enhanced by intermediate mirror vector mesons. Since lepto-quarks are expected to decay predominantly into one heavier quark and one heavier lepton, they are expected to lead to characteristic signatures at the LHC. For instance, generation-respecting processes like

$$\begin{array}{lll} gg & \longrightarrow & \text{direct}, \; g, \; \rho^{K\,0\prime}_{8\;A} \longrightarrow \pi^{K\,5}_{3\;A}\,\bar{\pi}^{K\,5}_{3\;A} \longrightarrow \bar{\mu}c + \bar{c}\mu \;\text{ or }\; \bar{e}u + \bar{u}e \\ gg & \longrightarrow & \text{direct}, \; g, \; \rho^{K\,0\prime}_{8\;B} \longrightarrow \pi^{K\,5}_{3\;B}\,\bar{\pi}^{K\,5}_{3\;B} \longrightarrow \bar{\tau}t + \bar{t}\tau \\ gg & \longrightarrow & \text{direct}, \; g, \; \rho^{K\,0\prime}_{8\;B} \longrightarrow \pi^{K\,2}_{3\;B}\,\bar{\pi}^{K\,2}_{3\;B} \longrightarrow \bar{\nu}_\tau t + \bar{t}\nu_\tau \;\text{ or }\; \bar{\tau}b + \bar{b}\tau, \end{array} \tag{9}$$

where the quarks and the leptons are generally expected to be non-collinear with their anti-particles, may lead to signals which can be in principle distinguished from background via appropriate directional cuts. Furthermore, note that the quadratic dependence of the lepto-quark decay width on the final-fermion masses renders it so small for all SM fermions except for the top quark, that present experimental lower bounds on lepto-quark masses can be improved considerably [22].

In any case, the number of detected mirror mesons and the luminosities required to have clear signals based on excess production of heavy fermion flavors depend crucially, apart from the widths of the new resonances, on the experimental resolution in invariant mass and direction of the reconstructed decay channels. Furthermore, of interest might be rarer processes involving ρ^K_3 vector mesons, general π^K_B or ρ^K_B decays down to π^K_A meson pairs, and heavier mirror hadrons possibly existing which are singlets under the mirror-family group and involve three katoptrons (like mirror protons p^K and mirror neutrons n^K), or even four and five katoptrons (analogous to QCD tetra- and penta-quarks). Moreover, one should study processes possibly violating lepton universality like leptonic B-meson decays due to the preferential coupling of katoptrons to heavier fermions. Note that the list of processes presented here is not exhaustive, nature being able to reveal its secrets in high-energy experiments in ways which are not *a priori* considered as the most probable. In any case, computer simulations of the processes above would give us a better picture of what to expect at the RUN II of the LHC.

3 Analytical deductive approach

3.1 Space-time structure

Supposing that signals like the above do appear, what would be the implications on our understanding of space-time? Taking an opposite direction from [13], we start here from first principles and via a top-down approach end up with a picture consistent with potential experimental findings. Although going from higher to lower energies might appear in certain aspects as non-rigorous and speculative, it facilitates focusing on various open issues to be resolved later. In the case of large extra dimensions and string theory, one assumes a continuous space-time background on which various fields propagate. The size and symmetry of these dimensions is left to be determined, while the internal symmetries are chosen in a self-consistent, *ad hoc* manner. On the other hand, katoptron theory introduces the "optimal connectivity" principle, with the fundamental degrees of freedom a multiset of identical, distinct elementary entities ("nodes") connected to each other optimally. Connections between members of this multiset, considered as identical Platonic copies of a single Parmenidean archetypical entity suited for an axiomatic formulation, lead to the emergence of space-time and elementary-particle symmetries. Each link connecting two nodes is associated with λ (or J in [13]) corresponding to a negative energy connected to energy-minimization arguments and proportional to the cosmological constant. The system grows by gradually connecting more nodes forming a complex network as a discrete space-time where points are replaced by nodes. The number of links of each node is not specified *a priori* but is determined by the dynamics. The system's self-organized criticality leads to optimal configurations, which are usually laminated lattices.

Considering only a finite number of connections for each node complies with Pythagorean parsimony and is much more, not less, conservative than assuming that each space-time point has an uncountably

infinite number of nearest neighbors, as implied by space-time continuum. Elementary particles appear here as finitely-extended (in agreement with Aristotle's arguments) topological defects coming in finite species (analogous to Empedocles' and Plato's finite classification of elements, contrary to Democritus' infinite species of "atoms" of various colors and shapes), like lattice vacancies within a discrete, dynamical space-time forming a dendrite, a "world crystal" [23]. This approach, reminiscent of Plotinus views on matter which, due to its variability, cannot correspond directly to a truly fundamental essence, introduces a duality between space-time and matter which may be also regarded as an intriguing instance of a *yin-yang* dialectic connection. This picture might also address non-locality issues in quantum mechanics, since one or a few missing connections between defects with a common origin spatially receding from each other, which in a dual picture leads to quantum entanglement, explains the long-range correlations needed to interpret phenomena like the EPR paradox.
We study qualitatively this network via the q=1 Potts Hamiltonian $H_P = -\lambda \sum_{<i,j>} \delta(s_i, s_j)$ generally used to analyze disorder-order phenomena like percolation, spin-glass transitions, and even cognition modeling, the sum over the nodes i, j being over nearest neighbors, $s_{i,j} = 0$ or 1, $\lambda > 0$ the coupling strength and $\delta(s_i, s_j)$=1 when s_i=s_j=1, being zero otherwise. This system of nodes modeling the "pre-bib-bang" world where usual interactions and space-time have not yet emerged lies initially in a highly-probable state having large temperature T and entropy, without need for fine-tuned cosmological initial conditions. Below the critical temperature T_c=$\lambda/(k_B \ln 2)$ however, the system forms configurations based on sphere packings maximizing the Newton (kissing) number and leading to optimal information propagation. For d=7, this yields the E_7 root lattice formed by the unit-norm imaginary Cayley octonions, and for d=8, the root lattice of E_8 (both exceptional symmetries used in particle Physics since [24]), which is even and self-dual, a property crucial for consistent quantization. The E_8 lattice is more stable than all lower-d and known higher-d configurations at least up to the optimal Leech lattice in d=24, while known optimal sphere packings for 8<d<16 are non-lattice, and for d>8 they do not correspond to roots of a Lie-group. Moreover, the asymptotic behavior of the potential $\mathcal{V}(d)$=$-\lambda\phi(d)K(d)$ for large d, with $K(d)$ the maximum possible (instead of using the maximum known in $\mathcal{V}_{opt}$[13]) kissing number of a d-dimensional sphere packing and $\phi(d)$ its density, after using the Kabatianskii-Levenshtein bounds [25] for $\phi(d) \xrightarrow{d\to\infty} 2^{\alpha_\phi d}$ and $K(d) \xrightarrow{d\to\infty} 2^{\alpha_K d}$, is

$$\mathcal{V}(d, \lambda = 1) \xrightarrow{d\to\infty} -2^{\alpha d} \to 0^-, \tag{10}$$

where $\alpha \equiv \alpha_\phi + \alpha_K = -0.599 + 0.401 \sim -0.2$, implying stability of the system only for $d \lesssim d_{max} \sim 100$ (a very rough estimate, a challenge for future computations being to possibly exclude potential minima other than the actually known $\mathcal{V}_{actual}(d=\{{8 \atop 24}\}, \lambda=1) \sim \{{-60 \atop -394}\}$, with $\mathcal{V}_{actual}(d=48, \lambda=1) \sim -1$ for instance not proven to be optimal for d=48) and preference for discreteness. A higher-d analogue of spin-glass phase transitions explains thus the emergence of a lattice L_G based on the roots of $G \equiv E_8 \times E_8'$ at the beginning of our Universe [13], as a "liquid-to-solid" (freezing) , "glass-to-crystal" transition. Information and coding theory employ L_G also, suggesting the use of a fuzzy version of node multisets as a quantum-computer model of the Universe reminiscent of the Ulam-von Neumann cellular automata. Furthermore, the complexity of this lattice indicates that attempts to consider particles as extended objects or to discretize space need several new dimensions.
Cosmological implications of the above include [13]: (a) Bursts of γ-rays originating from transitions of metastable "world-crystal" regions to stabler ones; (b) structure formation and intergalactic voids considered as regions with distinct lattice topologies emerging from initial quantum inhomogeneities; (c) creation of (barred) spiral galaxies as an indirect sign of space-time discreteness, recalling that detection of spirals in solid-state Physics is an indication of crystallization, galactic bars seen as topological defects between "helicoidal dislocations"; (d) Dark Matter interpreted as vacancies within regions with different structure than L_G, either of glass, amorphous (non-lattice) type, or like the Leech

lattice or with just one E_8, without katoptrons or a BEH mechanism. In the latter case one would have DM "baryons" clamped in heavy dark objects analogous to neutron stars, to which usual baryogenesis arguments do not apply. Visible galaxies should be then considered as a kind of entropic droplets, crystal bubbles within a glass "mosaic" of DM (possibly amorphous) regions, similarly to quartz (silicon crystal) islands within silicon-glass. Estimating the ratio R of crystal-to-glass-type volumes using Maxwell-Boltzmann statistics yields $R=1-\exp(-\Delta F/(k_B T_{eff}))$, ΔF measuring the free-energy difference between states and T_{eff} a suitable temperature. A future analogue of Bragg crystallography might test the validity of this picture; (e) order-of-magnitude estimate of the total number of elementary particles (recall that related rudimentary scientific estimates date back to Archimedes' heliocentric model in "*Psammites*"), seen as defects with total entropy S_p corresponding to an enthalpy cost H, as a function of the entropy S_B of the Universe determined by its size, i.e. $S_p=S_B\exp(-H/(k_B T_{eff}))$; (f) a solution to the black-hole (BH) information paradox, since here a BH is literally a "hole" within the space-time fabric gradually shrinking by Hawking radiation. Reconstruction of this fabric needs a precise number and structure of links between nodes, indicating that the information of each defect (particle) participating in the formation of a BH is stored in its surface topology, which in parallel connects the BH entropy with its area.

3.2 Emergence of time, space, gravity and gauge symmetry

We now replace s_i with Grassmann variables Ψ_i coupled via $\mathcal{E}_{ij}$ reflecting node connectivity in discretized spinor-gravity[26] inspired by Akama's spinor pre-geometry [27], assumed to belong to the $q=1$ Potts-model universality class. The non-compressibility (i.e.non-commutativity) of fermions maintains the lattice's discreteness, explaining the Heisenberg-uncertainty and Pauli-exclusion principles. Embedding the complexified L_G in complex Euclidean space suitable for longer wavelengths (lower momenta p) endows it with an isometry $G(\mathbb{C})$, yielding an action, starting from $S_{lat}=\sum_{<i,j>}\mathcal{E}_{ij}\bar{\Psi}_i\Psi_j$, over a compact toric Kähler manifold $T^d_{G(\mathbb{C})}\equiv\mathbb{C}^d/L_{G(\mathbb{C})}$ possessing a complex Lie-group structure:

$$S_f(\Psi)=\int_{T^d_{G(\mathbb{C})}} d^dx\,\det\left(\frac{i}{2}\bar{\Psi}\gamma^m\partial_\mu\Psi+\text{h.c.}\right)\equiv\int_{T^d_{G(\mathbb{C})}} d^dx\,\det(\tilde{E}^m_\mu)=\int_{T^d_{G(\mathbb{C})}} d^dx\,\det(E^m_\mu)\sum_{N=0}^{\infty}\frac{\left[\sum_{M=1}^{\infty}\frac{(-1)^{M+1}\mathrm{tr}\left(((E^{-1}\tilde{E})^n_\nu-\delta^n_\nu)^M\right)}{M}\right]^N}{N!} \quad (11)$$

using Jacobi's formula, with $d=16$ complex dimensions, h.c. standing for "hermitian conjugate", γ^m appropriate Dirac matrices, E^m_μ the elements of a suitable matrix E, lower-case Greek and Latin indices correspond to space-time and to the "internal" Lorentz symmetry, $\tilde{N}\equiv MN\leq d$ due to the Cayley-Hamilton theorem, and integration over the torus $T^d_{G(\mathbb{C})}$ implies a UV cut-off $M_{\text{Pl}}\sim L^{-1}_{\text{Planck}}$ due to a minimal distance between nodes. The variables $\bar{\Psi}_i$, Ψ_j correspond in a dual sense to lattice vacancies and consist of two Weyl spinors of opposite chirality following the Poincaré-Hopf index theorem, transforming like **(248,1)** (SM fermions) and **(1,248)** (katoptrons) under $E_8\times E'_8$. The conceptual advantage of the above consists in transferring the original non-commutativity of space-time nodes to the fields propagating in it so as to perform integrations and regularization preserving the symmetries. Defining a generating functional $W(J)\equiv\ln\left(\int\mathcal{D}\Psi\exp(\tilde{S}_J-S_f)\right)=S_J-\Gamma_{eff}(E)$ with $\left\{\begin{smallmatrix}\tilde{S}_J\\ S_J\end{smallmatrix}\right\}=\int d^dx J^\mu_m\left\{\begin{smallmatrix}\tilde{E}^m_\mu\\ E^m_\mu\end{smallmatrix}\right\}$ for a source J^μ_m, and $\Gamma_{eff}(E)$ the effective action after a Legendre transformation, we choose the matrix $E^m_\mu=\frac{\delta W(J)}{\delta J^\mu_m}\equiv<\tilde{E}^m_\mu>\sim\delta^m_\mu M_{\text{Pl}}$ for $\mu, m=0,...,3$ (0 otherwise), with elements vacuum-expectation values ("vevs") of composite bosons in a mean-field approximation. Further symmetry breaking of $G\rightarrow H\equiv E_7\times SL(2,\mathbb{C})\times E'_7\times SL(2,\mathbb{C})'$ via vevs of antisymmetric bi-fermion operators transforming like

248×248→248$_{\mathbf{a}}$ under each of the 2 E_8s yields a 4 real-d action $S_{eff}(E^m_\mu, \psi, A_\mu, \phi_{\rm infl})$ with Einstein-Hilbert and Yang-Mills terms, leading to the usual least-action-principle Euler-Lagrange equations in a lower order Taylor expansion for fluctuations $\delta E^m_\mu \ll E^m_\mu$ and for $\tilde{E}^m_\mu - E^m_\mu \equiv \bar{\psi}\gamma^m\partial_\mu\psi \ll E^m_\mu$:

$$S_{eff} \sim \int_\epsilon d^4x \sqrt{\det\left(-g_{\mu\nu}(E)\right)}\Big(kR(E) + k'\mathrm{tr}\left([D_\mu(A), D_\nu(A)]^2\right) + \bar{\psi}\gamma^\mu\left(D_\mu(A) + \delta_\mu(E)\right)\psi + \mathcal{L}(\phi_{\rm infl}) + ...\Big), \quad (12)$$

with ϵ a $L_{\rm Planck}$-regularization from $L_{G(\mathbb{C})}$, k, k' constants, $\gamma^\mu = \gamma^m E^\mu_m$, the δE^m_μ-induced Ricci scalar $R(E)$ (omitting small terms consistent with present gravitational tests) and derivative correction $\delta_\mu(E) \equiv (E^\nu_m \partial_\mu(E^m_\nu) + E^m_\mu \partial_\nu(E^\nu_m))/2$, and $D_\mu(A)$ the covariant derivative of the $E_7 \times E'_7$ Kaluza-Klein compactification gauge fields A_μ. The metric is $g_{\mu\nu}(E) = E^m_\mu(x)E^n_\nu(x)\eta_{mn}$, with $\eta_{mn} = \mathrm{diag}(-1, 1, 1, 1)$ Minkowski's space-time as the $SL(2, \mathbb{C})$ subgroups of the two E_8s break to their diagonal subgroup $SO(1,3)_D$ due to M_{Pl}-scale vevs of $2\tilde{N}$-fermion operators $(O(p/M_{\rm Pl})^{\tilde{N}}$ after Fourier transformation) for $\tilde{N}=2$ in Eq.(11) coupling the left- and right-handed sectors and transforming like **(1,3,1,3)** under H. At slightly lower energies, after the self-breaking only of the SM-fermion $SU(3)$ family group due to a parity-violating L_G asymmetry, similar operators transforming like **(24,24)** under $\tilde{H} \equiv SU(5) \times SU(5)' \subset E_7 \times E'_7 \subset H$ lead to $\tilde{H} \to SU(5)_D$, obviating the need for both outer automorphisms in [13]. Moreover, $\mathcal{L}(\phi_{\rm infl})$ is a Lagrangian for inflatons as compactified gauge-field components, including scalar kinetic terms, a "sombrero" potential [13] and the cosmological constant, generating entropy and a time arrow, diluting the 't Hooft-Polyakov monopole primordial abundance and explaining the Universe's 4-d expansion, isotropy, homogeneity and flatness, which in parallel reinforces the validity of our approximation.

The action above leads to quantized gravitational equations $\frac{\delta\Gamma_{eff}(E)}{dE^\mu_m} = J^m_\mu$, with the metric and the gauge fields induced by composite-boson operators. Further symmetry breaking takes place starting from a unified critical coupling at $M_{\rm Pl}$, estimated via Schwinger-Dyson equations as $\alpha(\mathrm{M_{Pl}}) \sim 1/C_2(E_8) \sim 0.03$, down to the SM having the known fermion-family structure with a non-perturbative BEH mechanism based on katoptrons at $\Lambda_K \sim M_{\rm Pl}\exp\left(-1.23C_2(E_8)\right) \sim 1$ TeV [13] in a way similar to QCD asymptotic freedom [28], followed by a corresponding mirror mesonic spectrum.

4 Conclusions

Our purpose has been to link potential discoveries at the LHC with theories having implications on cosmology and on our understanding of space-time. Theories accommodating forthcoming experimental data might be found to be consistent, if not indicative, either of an extended nature of elementary particles or of a certain discreteness of space-time which should not be treated as background but as emergent, dynamical and non-perturbative in a sense dual to elementary particles. Regarding the mirror world, we sketched the conceptual path leading from the optimal connectivity principle to strongly-interacting mirror fermions and to the experimental detection of their mesonic bound states. Distinguishing the relevant signals from alternative models should focus our attention on distinctive processes related to particles such as lepto-quarks and perform relevant computer simulations.

Acknowledgements

Particular thanks are due to T. Papadopoulou for her support and encouragement, and to the organizers of an excellent ICFP2015 in the OAC (Kolymbari). This research has been co-financed in part by the E. U. (European Social Fund-ESF) and Greek national funds through the Operational Program "Education and Lifelong Learning" of the National Strategic Reference Framework (NSRF) - Research Funding Program: THALES. Investing in knowledge society through the European Social Fund.

References

[1] N. Arkani-Hamed, S. Dimopoulos, G. Dvali, *Phys. Lett.* **B429** (1998) 263; I. Antoniadis *et. al.*, *Phys. Lett.* **B436** (1998) 257; E. G. Floratos, G. K. Leontaris, *Phys. Lett.* **B465** (1999) 95.
[2] S. Giddings, S. Thomas, *Phys. Rev.* **D65** (2002) 056010; T. Rizzo, *Phys. Rev.* **D60** (1999) 075001.
[3] ATLAS coll., *Phys. Rev.* **D87** (2013) 1, 015010; CMS coll., arXiv:1408.3583 [hep-ex].
[4] S. Dimopoulos, H. Georgi, *Nucl. Phys.* **B193** (1981) 150.
[5] M. Green, J. Schwarz, E. Witten, Superstring theory, *Vol. 1,2* (2012), Cambridge University Press.
[6] M. Axenides, E. G. Floratos, L. Perivolaropoulos, *Phys. Rev.* **D66** (2002) 085006.
[7] ATLAS coll., arXiv:1508.06608 [hep-ex]; CMS coll., *JHEP* **04** (2015) 124.
[8] T. D. Lee, C. N. Yang, *Phys. Rev.* **104** (1956) 254.
[9] F. Wilczek, A. Zee, *Phys. Rev.* **D25** (1982) 553.
[10] S. Weinberg, *Phys. Rev.* **D13** (1976) 974; L. Susskind, *Phys. Rev.* **D20** (1979) 2619.
[11] J. Bardeen, L. N. Cooper, J. R. Schrieffer, *Phys. Rev.* **106** (1962) 162.
[12] G. Triantaphyllou, *Eur. Phys. J.* **C10** (1999) 703.
[13] G. Triantaphyllou, *El. J. Th. Phys.* **10** (2013) 135; *ibid., Eur. Phys. J. Web Conf.* **70** (2014) 00051.
[14] J. C. Pati, A. Salam, *Phys. Rev.* **D8** (1973) 1240; H. Georgi, S. L. Glashow, *Phys. Rev. Lett.* **32** (1974) 438.
[15] E. Eichten, I. Hinchliffe, K. Lane, C. Quigg, *Rev. Mod. Phys.* **56** (1984) 579.
[16] G. Triantaphyllou, *Mod. Phys. Lett.* **A16** (2001) 53.
[17] G. Triantaphyllou, *J. Phys.* **G26** (2000) 99.
[18] G. Triantaphyllou, *Int. J. Mod. Phys.* **A15** (2000) 265.
[19] E. Fahri, L. Susskind, *Phys. Rep.* **74** (1981) 277.
[20] T. Appelquist, G. Triantaphyllou, *Phys. Rev. Lett.* **69** (1992) 2750.
[21] ATLAS coll., arXiv:1506.00962 [hep-ex]; CMS Collaboration, *JHEP* **1408** (2014) 173.
[22] K.A. Olive et al. (Particle Data Group), *Chin. Phys.* **C38** (2014) 090001.
[23] P. Jizba, H. Kleinert, F. Scardigli, *Phys. Rev.* **D81** (2010) 084030.
[24] P. Sikivie, F. Gürsey, *Phys. Rev.* **D16** (1977) 816; I. Bars, M. Gunaydin, *Phys. Rev. Lett.* **45** (1980) 859; N. S. Baaklini, *Phys. Lett.* **B91** (1980) 376.
[25] G. A. Kabatianskii, V. I. Levenshtein, *Probl. Inform. Transm.* **14**, 1 (1978) 1.
[26] A. Hebecker, C. Wetterich, *Phys. Lett.* **B574** (2003) 269.
[27] K. Akama, *Prog. Th. Ph.* **60** (1978) 1900; D. Amati, G. Veneziano, *Nucl. Phys.* **B204** (1982) 451.
[28] D. J. Gross, F. Wilczek, *Ph.Rev.Lett.* **30** (1973), 1343; H. D. Politzer, *Ph.Rev.Lett.* **30** (1973), 1346.

EPJ Web of Conferences **126**, 04049 (2016) DOI: 10.1051/epjconf/201612604049
ICNFP 2015

Beyond-the-Standard Model Higgs Physics using the ATLAS Experiment

Loan Truong[1,2,3a] on behalf of the ATLAS Collaboration

[1] *INFN Gruppo Collegato di Udine, Sezione di Trieste, Udine, Italy*
[2] *ICTP, Trieste, Italy*
[3] *SISSA, Trieste, Italy*

Abstract. Recent searches for the Higgs boson in the context of beyond the Standard Model performed by the ATLAS experiment are presented: high mass Higgs boson searches, lepton flavour violating Higgs decay, NMSSM, constraint from the search for three photons. The interpretation based on the measurements of Higgs couplings is shown, along with the constraint on the Higgs boson invisible decays. The search for invisible decays of a Higgs boson produced in association with a Z boson was performed with both $\sqrt{s} = 7$ and 8 TeV while the rest were performed using the $\sqrt{s} = 8$ TeV data of proton-proton collisions collected by the ATLAS experiment. No significant excess of data over the predicted background is observed and limits are placed in certain quantities depending on the searches.

1 Introduction

The ATLAS experiment at the LHC discovered the Higgs boson in July 2012. So far, there have been many measurements on its couplings and mass [1–5], the results are consistent with the Standard Model (SM) predictions. Nevertheless, these measurements have not yet excluded a large range of other extensions of the SM.

Exploring whether there are additional Higgs bosons or exotic decay of the SM Higgs boson can give us direct evidences about physics beyond the Standard Model (BSM). Those include the two Higgs doublet models (2HDMs) [6–9], next-to-minimal superymmetric SM (NMSSM) [10, 11], composite Higgs models [12, 13], which favour the existence of other Higgs bosons in the high-mass regime, as well as lepton flavour violating (LFV) Higgs decay, three photons Higgs decay. There are strong evidences of dark matter from astrophysical observations which could be explained by the existence of weakly interacting massive particles (WIMPs, see Ref. [14] and the references therein). The observed Higgs boson might decay to dark matter or other stable or long-lived particles which do not interact significantly with a detector [15–19] leading to Higgs boson invisible decay.

The report is organized as follows: Sec. 2 describes the high mass-Higgs boson searches, Sec. 3 describes the LFV Higgs decay search, Sec. 4 describes the search for Higgs bosons

[a] e-mail: lotruong@cern.ch

in the context of NMSSM, Sec. 5 describes the search for a Higgs boson decaying to three photons or more, Sec. 6 summarizes some constraints on BSM models using Higgs boson couplings and mass measurements, Sec. 7 describes the combination of the searches for invisible decays of the Higgs boson. Finally concluding remarks are in Sec. 8. Details about the ATLAS detector can be found in Ref. [20].

2 High masses searches

Two searches for a high-mass Higgs boson H were performed in the $H \to WW \to \ell\nu\ell\nu, \ell\nu qq$ [21] and $H \to ZZ \to \ell\ell\ell\ell, \ell\ell\nu\nu, \ell\ell qq, \nu\nu qq$ [22] channels. No significant deviations were observed. The results were interpreted separately for VBF, ggF production modes as upper limits at 95% CL (see Fig. 1 and Tab. 1) on $\sigma_H \times BR(H \to WW/ZZ)$ for heavy Higgs boson which has a narrow width.

Table 1: Upper limits on the production cross section of the high mass-Higgs boson in different weak vector boson decay modes [21, 22].

Limits on	m_H [GeV]	Upper Limit [fb] for H production mode	
		ggF	VBF
$\sigma_H \times BR(H \to WW)$	1500	21	6
$\sigma_H \times BR(H \to ZZ)$	195÷950	530÷8	310÷9

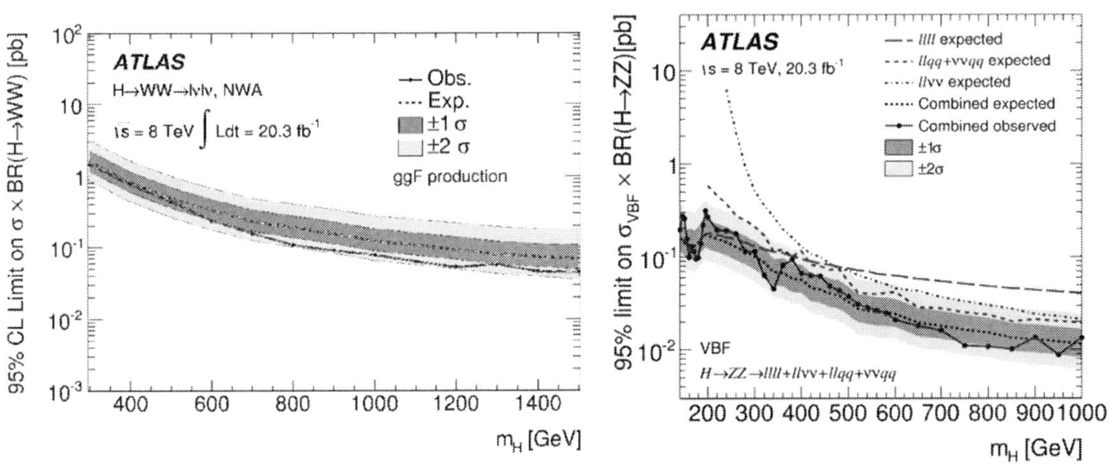

(a) $H \to WW \to \ell\nu\ell\nu$ (b) $H \to ZZ \to \ell\ell\ell\ell, \ell\ell\nu\nu, \ell\ell qq, \nu\nu qq$

Figure 1: Limits set on the production cross section of the high mass-Higgs boson in (a) WW [21], (b) ZZ [22] resonance searches.

3 Lepton flavour violating decays

The LFV Higgs decays ($H \to \tau\mu, \tau e, \mu e$) arise at tree level based on assumed flavour violating Yukawa interactions [23]. A search for LFV Higgs decays was done in the channel $H \to \mu\tau_{had}$

 DOI: 10.1051/epjconf/201612604049

in hadronic τ decays [24]. A fit to the reconstructed mass distribution in data was performed (see Fig. 2) resulting in the best fit of $BR(H \to \mu\tau) = 0.77 \pm 0.62\%$, upper limit on $BR(H \to \mu\tau)$ at 95% CL: 1.85%(1.24%) obs(exp).

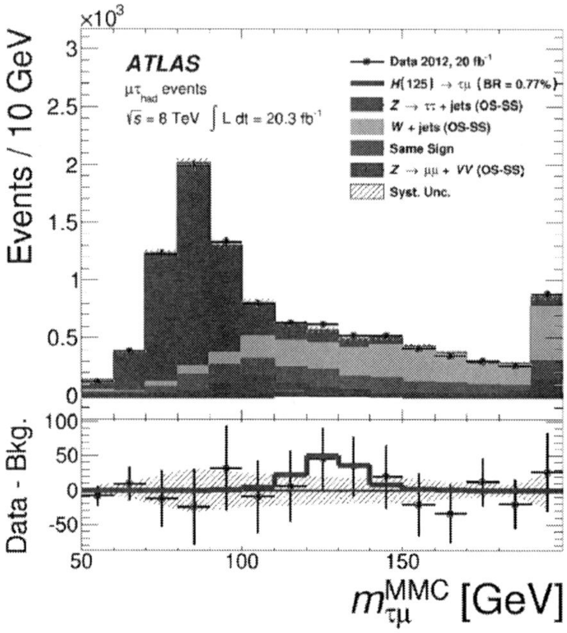

Figure 2. The reconstructed mass of the system of observed muon, τ (hadronic decay products) & $E_{\mathrm{T}}^{\mathrm{miss}}$ objects [24] by means of the Missing Mass Calculator distribution.

4 NMSSM

The NMSSM contains an additional pseudoscalar Higgs boson (a), generally assumed to have a mass lower than the observed Higgs boson (h) since its mass is protected by a Peccei–Quinn symmetry. A search for the decay to a pair of the lightest neutral pseudoscalar Higgs a of either the 125 GeV Higgs (h) or a second CP-even Higgs (H) was performed [25], where one a boson decays to 2 μ and the other decays to 2 τ. As a result, the most stringent upper limit on the branching ratio of the h boson decaying to the non-SM particles was set at 3.5% for $m_a = 3.75$ GeV (see Fig. 3).

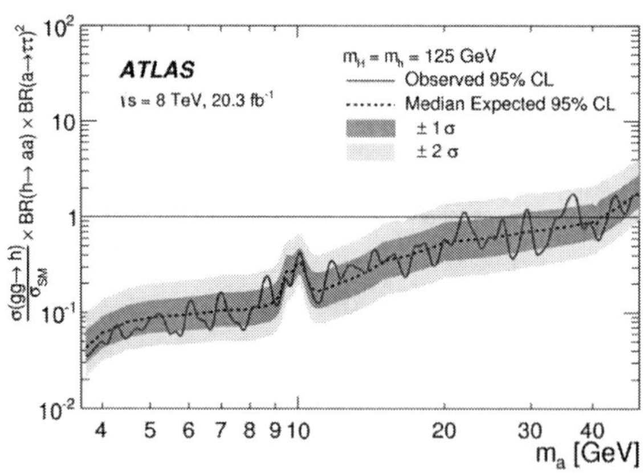

Figure 3. Limit for $\sigma_{ggH_{SM}} \times BR$ of h SM decays to aa $\times BR$ of a decays to $\tau\tau$ vs m_a [25].

5 Search for 3 photons

A search for events with at least 3 γ was done [26]. The model-independent interpretations are the first of their kind. One of the interpretations was performed for a SM Higgs boson (h) decaying to four photons via a pair of intermediate pseudoscalar particles (a). Limits on the cross section times $BR(h \to aa) \times BR(a \to \gamma\gamma)^2$ was set to be $< 10^{-3}\sigma_{SM}$ for 10 GeV $< m_a < 62$ GeV (see Fig. 4).

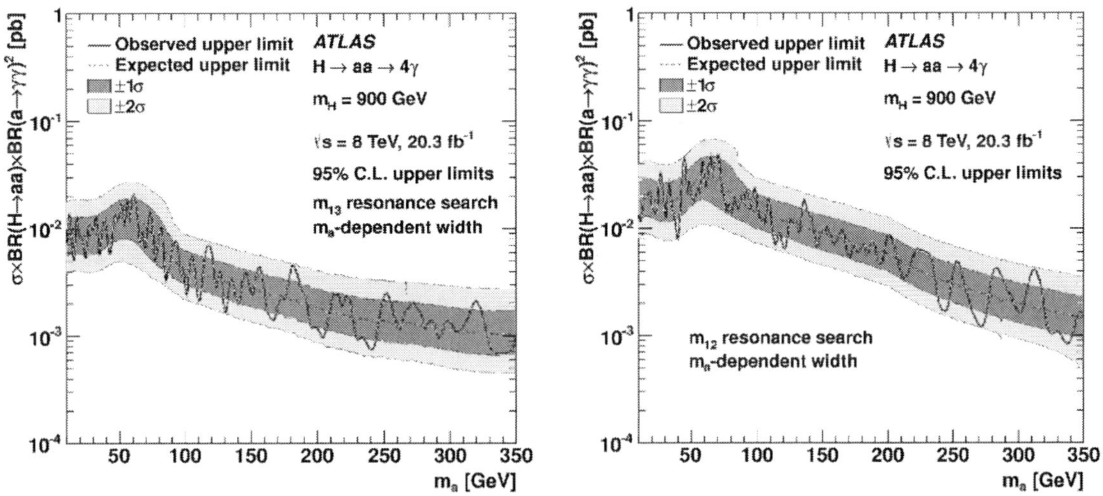

Figure 4: Left: For SM Higgs $h \to aa \to \gamma\gamma\gamma\gamma$: $\sigma \times BR(h \to aa) \times BR(a \to \gamma\gamma)^2 < 10^{-3} \times \sigma_{SM}$ for 10 GeV $< m_a <$ 62 GeV. Right: for heavy Higgs boson-like scalar: $\sigma_H \times BR(H \to aa) \times BR(a \to \gamma\gamma)^2 <$ 0.02 to 0.001 pb (depending upon m_H,m_a) [26].

6 Beyond the Standard Model constraints via Higgs boson couplings

BSM constraints [27] were performed using the measured production and decay rates of the Higgs boson ($\gamma\gamma$, ZZ, WW, $Z\gamma$, bb, $\tau\tau$, & $\mu\mu$; $t\bar{t}h$ with $h \to \gamma\gamma$, $b\bar{b}$ & multileptons). The constraints include the probe on the scaling of the couplings with mass, setting limits on parameters in extensions of the SM such as composite Higgs boson, 2HDMs. They are described in details below.

6.1 Mass scaling of couplings

In this analysis, the observed rates in different channels were used in a fit to determine how the Higgs boson couplings to other particles scale with the masses of those particles [28]. Each coupling (with fermion f and boson V) was scaled in terms of "vacuum expectation value (vev)" M and mass scaling parameter ϵ (in the SM vev $v \approx 246$ GeV and $\epsilon \to 0$) as $\kappa_f = v\,\frac{m_f^{\epsilon}}{M^{1+\epsilon}}$, $\kappa_V = v\frac{m_V^{2\epsilon}}{M^{1+2\epsilon}}$, where m is the mass of the particle. As a result of the fit, the best fit value is consistent within one standard deviation with the SM (see Fig. 5), observed values for ϵ and M are shown in Tab. 2.

DOI: 10.1051/epjconf/201612604049

Table 2: Observed and expected measurements of the mass scaling parameter ϵ and the vev parameter M [27].

Parameter	Obs.	Exp.
ϵ	0.018 ± 0.039	0.000 ± 0.042
M	224^{+14}_{-12} GeV	246^{+19}_{-16} GeV

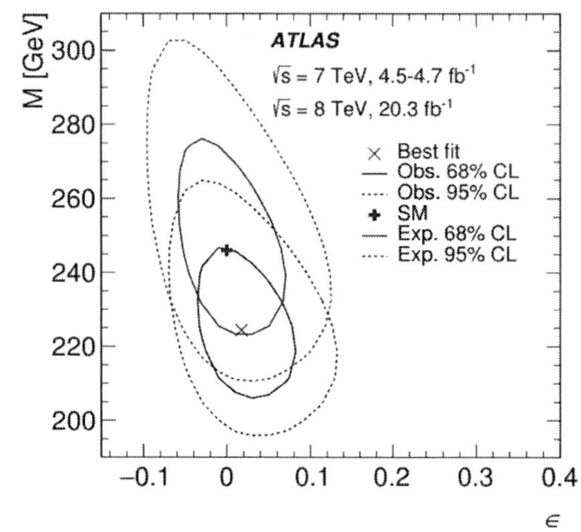

Figure 5. 2-D confidence intervals as a function of the mass scaling factor ϵ and the vev parameter M [27].

6.2 Minimal composite Higgs model

Minimal Composite Higgs Models (MCHM) [12, 13] provides a possible explanation for the scalar naturalness problem. It suggests that the Higgs boson is a composite, pseudo-Nambu–Goldstone boson rather than an elementary particle. Higgs couplings are modified as functions of compositeness scale-f: $\xi = v^2/f^2$. There are different MCHM models with different modifications of the Higgs coupling such as MCHM4 [12] with $\sqrt{\kappa} = \kappa_V = \kappa_F = 1 - \xi$ and MCHM5 [13] with $\kappa_V = 1 - \xi, \kappa_F = \frac{1-2\xi}{\sqrt{1-\xi}}$. The SM is recovered in the limit $\xi \to 0$, namely $f \to \infty$. A fit was performed in the parameter space of $[\kappa_F, \kappa_V]$ (see Fig. 6). As a result, limits at 95% CL are obtained in the order of obs (exp), for MCHM4: f > 710 (510) GeV, MCHM5: f > 780 (600) GeV.

6.3 Two Higgs Doublet Models

2HDMs models introduce two complex SU(2) doublets of scalar fields and there are 4 types of them (as summarized in Tab. 3). They predict five Higgs bosons: two neutral CP-even bosons h and H, one neutral CP-odd boson A, and two charged bosons $H^{\pm}$. When considering the CP-conserving case, there are 6 sensitive parameters in the analysis: 4 masses $m_h, m_H, m_{H^\pm}, m_A$ & 2 mixing angles α, β. Where $\tan\beta = v_1/v_2$ is the ratio of the two Higgs doublets' vevs which satisfy $v_1^2 + v_2^2 = v^2 \approx (246\ \text{GeV})^2$, α: mixing angle between h & H. Assumptions for interpretations are that the 125 GeV boson is the light Higgs, no radiative corrections from BSM for the production of Higgs boson, only SM decays. Gauge invariance requires the

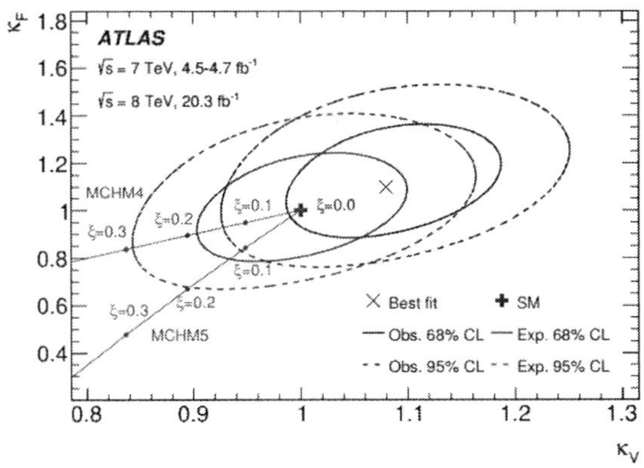

Figure 6. 2-D likelihood contours in the (κ_V, κ_F) coupling scale factor plane [27].

Table 3: Couplings of the light Higgs boson h to weak vector bosons (κ_V), up-type quarks (κ_u), down-type quarks (κ_d), and charged leptons (κ_ℓ), expressed as ratios to the corresponding SM predictions in 2HDMs of various types.

Coupling scale factor	Type I (fermiophobic)	Type II (MSSM-like)	Lepton-specific	Flipped
κ_V	$\sin(\beta-\alpha)$			
κ_u	$\cos(\alpha)/\sin(\beta)$			
κ_d	$\cos(\alpha)/\sin(\beta)$	$-\sin(\alpha)/\cos(\beta)$	$\cos(\alpha)/\sin(\beta)$	$-\sin(\alpha)/\cos(\beta)$
κ_ℓ	$\cos(\alpha)/\sin(\beta)$	$-\sin(\alpha)/\cos(\beta)$	$-\sin(\alpha)/\cos(\beta)$	$\cos(\alpha)/\sin(\beta)$

couplings of the two neutral, CP-even Higgs bosons to vector bosons relative to their SM values to be:

$$g_{hVV}^{2HDM}/g_{hVV}^{SM} = \sin(\beta-\alpha)$$
$$g_{HVV}^{2HDM}/g_{HVV}^{SM} = \cos(\beta-\alpha) \tag{1}$$

with the convention: $\sin(\beta-\alpha) \geq 0$. SM-like alignment limit is retrieved at $\cos(\beta-\alpha) = 0$.

Figure 7 shows the regions of the $[\cos(\beta-\alpha), \tan\beta]$ plane that are excluded at 95% CL for each of the four types of 2HDMs, overlaid with the exclusion limits expected for the SM Higgs sector. A physical boundary of $\kappa_V \leq 1$ is there in all four 2HDM types and restricts the profile likelihood ratio. The data are consistent with the alignment limit within approximately one standard deviation or better in each of the model.

7 Higgs invisible decays

Limits on the branching ratio of Higgs boson invisible decay [27] are set and subsequently used to constrain on a Higgs-portal dark matter model [29].

7.1 The combination of Higgs invisible decay channels

Direct searches for the SM Higgs boson h decaying into invisible particles used the following strategies: selecting events with large missing transverse momentum (E_T^{miss}), using particles

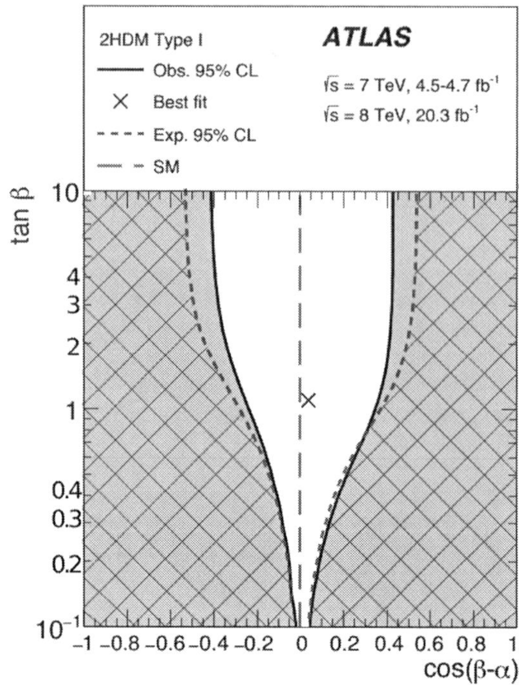

(a) Type I

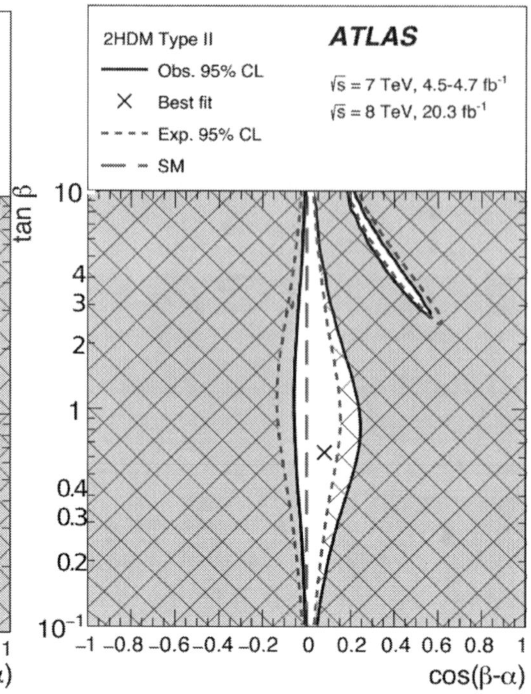

(b) Type II

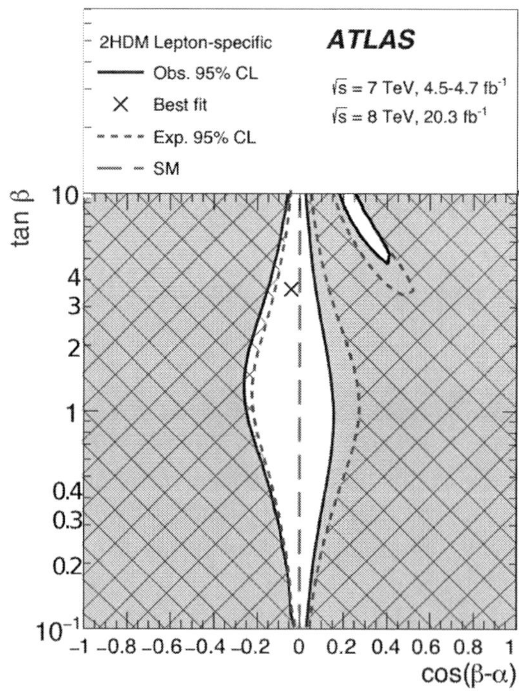

(c) Lepton-specific

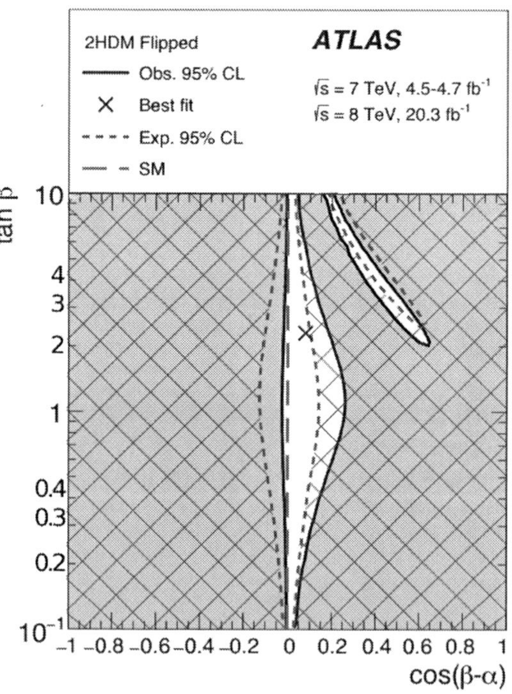

(d) Flipped

Figure 7: Regions of the $(\cos(\beta-\alpha),\ \tan\beta)$ plane of four types of 2HDMs excluded by fits to the measured rates of Higgs boson production and decays. The likelihood contours where $-2\ln\Lambda = 6.0$, corresponding approximately to the 95% CL (2 std. dev.), are indicated for both the data and the expectation assuming the SM Higgs sector. The cross in each plot marks the observed best-fit value. The light shaded and hashed regions indicate the observed and expected exclusions, respectively. The plots are shown for $\sin(\beta-\alpha) \geq 0$ without loss of generality [27].

produced associated with the Higgs boson to tag the Higgs, assuming productions (and acceptance) as in the SM where $BR(h \to ZZ \to 4\nu) = 1.2 \times 10^{-3}$. The SM branching ratio is very little so any observation would be an indication of BSM physics. They include the analyses: Z($\to$ ll) h $\to$ inv ($E_{\mathrm{T}}^{\mathrm{miss}}$) [30], W/Z($\to$ jj) h $\to$ inv ($E_{\mathrm{T}}^{\mathrm{miss}}$) [31], VBF ($\to$ jj) h $\to$ inv ($E_{\mathrm{T}}^{\mathrm{miss}}$) [32], where the first analysis contains the 7 TeV data.

Table 4 shows upper limits on the BR($h \to$ inv) at 95% CL and other CL for different analyses and their statistical combination. The combination of the invisible channels only resulted in BR($h \to inv$) < 0.25 obs (0.27 exp).

Table 4: Summary of upper bounds on BR($h \to$ inv) [27].

Results	Observed	-2σ	-1σ	Expected	$+1\sigma$	$+2\sigma$
VBF h	0.28	0.16	0.21	0.31	0.41	0.56
$Z(\to \ell\ell)h$	0.75	0.33	0.45	0.62	0.86	1.19
$V(\to jj)h$	0.78	0.46	0.62	0.86	1.19	1.60
Combined Results	0.25	0.13	0.18	0.27	0.35	0.47

Combined result was then combined statistically with the measured visible decay rates of the Higgs boson. A physical boundary of $BR_{inv} > 0$ was required. The most general result uses independent coupling parameters κ_W, κ_Z, κ_t, κ_b, κ_τ, κ_μ, κ_g, κ_γ, $\kappa_{Z\gamma}$ and BR$_{inv}$. Upper limit of 0.23(0.24) obs(exp.) were set at 95% CL. The likelihood scan for that statistical fit is shown in Fig. 8.

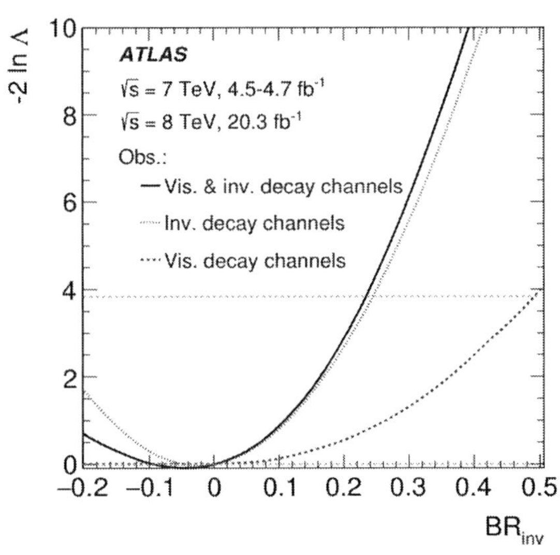

Figure 8. Likelihood scans of the Higgs' invisible branching ratio [27].

7.2 Higgs Portal Interpretation

The 90%CL upper limit of 0.22 (0.23) obs (exp) on $BR(H \to inv)$ was used to constrain on a Higgs-portal dark matter scenario [29], where the Higgs boson is the only mediator for interactions between WIMPs and other SM particles. It is sensitive for WIMP's mass $< m_h/2$. Limits are set on the cross section between WIMPs and nucleons as a function of the

WIMP mass, as shown in Fig. 9. The form factor (Higgs-nucleon coupling) f_N is taken to be $0.33^{+0.30}_{-0.07}$ [29].

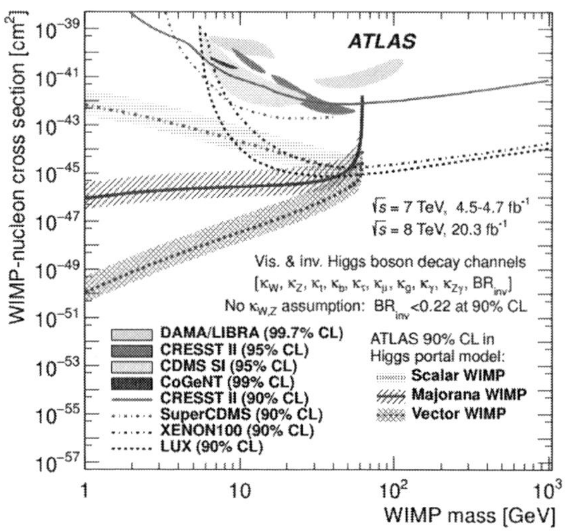

Figure 9. Upper limit at 90% CL on the WIMP-nucleon scattering cross section, ATLAS's compared with the direct searches [27].

8 Conclusion

The ATLAS collaboration has performed many searches for BSM Higgs bosons using the 7 TeV and 8 TeV datasets: high mass Higgs search, LFV Higgs decay, Higgs decays to pseudoscalar particles in the context of NMSSM, Higgs decays to four photons via pseudoscalar particles, Higgs decay into invisible particles. No significant excess for BSM Higgs physics has been observed yet. Precise measurements of the Higgs boson couplings allow to constrain new phenomena: mass scaling (ϵ : 0.018 ± 0.039, M : $224 + 14 - 12$ GeV), Minimal Composite Higgs models ($f > 710(780)$ GeV MCHM4(5)), Two Higgs Doublet Models (Alignment limit within 1σ), Higgs to invisible decays ($BR_{inv} < 0.23$) and subsequent constraint on Higgs-portal dark matter. Much better results with the new data from the $\sqrt{s} = 13$ TeV run at the LHC are expected.

References

[1] ATLAS Collaboration, Eur. Phys. J. **C75**, 476 (2015), 1506.05669
[2] CMS Collaboration (CMS), Phys. Rev. **D92**, 012004 (2015), 1411.3441
[3] ATLAS and CMS Collaborations, Phys. Rev. Lett. **114**, 191803 (2015), 1503.07589
[4] ATLAS Collaboration (2015), 1507.04548
[5] CMS Collaboration, Eur. Phys. J. **C75**, 212 (2015), 1412.8662
[6] LHC Higgs cross section working group, S. Dittmaier, C. Mariotti, G. Passarino, S. Heinemeyer, R. Tanaka (Eds.) (2013), 1307.1347
[7] T. Lee, Phys. Rev. D. **8**, 1226 (1973)
[8] J.F. Gunion, H.E. Haber, Phys. Rev. D. **67**, 075019 (2003), hep-ph/0207010
[9] G. Branco et al., Phys. Rept. **516**, 1 (2012), 1106.0034
[10] U. Ellwanger, J.F. Gunion, C. Hugonie (2001), pp. 178–188, hep-ph/0111179

[11] D. Curtin et al., Phys. Rev. **D90**, 075004 (2014), 1312.4992
[12] K. Agashe, R. Contino, A. Pomarol, Nucl. Phys. **B 719**, 165 (2005), hep-ph/0412089
[13] R. Contino, L. Da Rold, A. Pomarol, Phys. Rev. D **75**, 055014 (2007)
[14] D.C. et al., Astrophys. J. **648**, L109 (2006), astro-ph/0608407
[15] I. Antoniadis, M. Tuckmantel, F. Zwirner, Nucl.Phys. **B707**, 215 (2005), hep-ph/0410165
[16] N. Arkani-Hamed, S. Dimopoulos, G. Dvali, J. March-Russell, Phys.Rev. **D65**, 024032 (2002), hep-ph/9811448
[17] A. Datta, K. Huitu, J. Laamanen, B. Mukhopadhyaya, Phys.Rev. **D70**, 075003 (2004), hep-ph/0404056
[18] S. Kanemura, S. Matsumoto, T. Nabeshima, N. Okada, Phys.Rev. **D82**, 055026 (2010), 1005.5651
[19] A. Djouadi, A. Falkowski, Y. Mambrini, J. Quevillon, Eur.Phys.J. **C73**, 2455 (2013), 1205.3169
[20] ATLAS Collaboration, JINST **3**, S08003 (2008)
[21] ATLAS Collaboration (2015), 1509.00389
[22] ATLAS Collaboration (2015), 1507.05930
[23] R. Harnik, J. Kopp, J. Zupan, JHEP **03**, 026 (2013), 1209.1397
[24] ATLAS Collaboration (2015), 1508.03372
[25] ATLAS Collaboration, Phys. Rev. **D92**, 052002 (2015), 1505.01609
[26] ATLAS Collaboration (2015), 1509.05051
[27] ATLAS Collaboration (2015), 1509.00672
[28] J. Ellis, T. You, JHEP **06**, 103 (2013), 1303.3879
[29] A. Djouadi, O. Lebedev, Y. Mambrini, J. Quevillon, Phys.Lett. **B709**, 65 (2012), 1112.3299
[30] ATLAS Collaboration (ATLAS), Phys. Rev. Lett. **112**, 201802 (2014), 1402.3244
[31] ATLAS Collaboration (ATLAS), Eur. Phys. J. **C75**, 337 (2015), 1504.04324
[32] ATLAS Collaboration (2015), 1508.07869

EPJ Web of Conferences **126**, 04050 (2016) DOI: 10.1051/epjconf/201612604050
ICNFP 2015

A new approach to analytic, non-perturbative, gauge-invariant QCD renormalization is described, with applications to high energy elastic pp-scattering.

H. M. Fried[1], P. H. Tsang[1,a], Y. Gabellini[2], T. Grandou[2], and Y-M. Sheu[1,2]

[1] *Brown University, Providence, RI 02912, USA*
[2] *Institut Non Linéaire de Nice, Université Nice Sophia Antipolis, France*

Abstract. A new non-perturbative, gauge-invariant model QCD renormalization is applied to high energy elastic pp-scattering. The differential cross-section deduced from this model displays a diffraction dip that resembles those of experiments. Comparison with ISR and LHC data is currently underway.

1 Introduction

We start with the Schwinger Generating Functional, GF, for QCD, with gluon operators in an Arbitrary, relativistic, gauge. The GF is rearranged in terms of a "Reciprocity Relation" under a "Gaussian Linkage Operation". The GF now depends upon two functionals of A,

$$\mathfrak{Z}_{QCD}[j,\bar{\eta},\eta] = \mathfrak{N} e^{-\frac{i}{2}\int \frac{\delta}{\delta A}\cdot D_c^{(0)}\cdot \frac{\delta}{\delta A}} \cdot e^{-\frac{i}{4}\int \mathbf{F}^2 + \frac{i}{2}\int \mathbf{A}\cdot(-\partial^2)\cdot \mathbf{A}} \cdot e^{i\int \bar{\eta}\cdot \mathbf{G_c}[\mathbf{A}]\cdot\eta + \mathbf{L}[\mathbf{A}]}\Big|_{A=\int \mathbf{D_c^{(0)}}\cdot \mathbf{j}} \tag{1}$$

where the quark line, $\mathbf{G_c}[\mathbf{x},\mathbf{y}|\mathbf{A}] = [\mathbf{m} + \gamma\cdot(\delta - \mathbf{igA}\tau)]^{-1}$ and virtual quark loop, $\mathbf{L}[\mathbf{A}] = \ln[\mathbf{1} - \mathbf{i}\gamma\mathbf{A}\tau_\mathbf{c}[\mathbf{0}]]$.

The GF can now can be rearranged into gauge-invariant form. This was overlooked for decades [1].

Now combine with Fradkin expressions for the quark line, $\mathbf{G_c}[\mathbf{A}]$, and quark loop, $\mathbf{L}[\mathbf{A}]$. Efim S. Fradkin gave expressions for $\mathbf{G_c}[\mathbf{A}]$ and $\mathbf{L}[\mathbf{A}]$ in gaussian form [4]. These are exact.

The $\mathbf{F^2}$ can be rewritten using Halpern's half century old expression [2],

$$e^{-\frac{i}{4}\int \mathbf{F}^2} = N\int d[\chi] e^{\frac{i}{4}\int \chi^2 + \frac{i}{2}\int F\cdot\chi}, \tag{2}$$

where $\chi^a_{\mu\nu} = -\chi^a_{\nu\mu}$.

With the $exp[-\frac{i}{4}\int \mathbf{F}^2]$ in the GF in Gaussian form, under χ fields, the relevant Gaussian Functional operations can be performed exactly. This corresponds to the summation of all Feynman graphs of gluons exchanged between quarks.

Then the explicit cancellation of all the gauge-dependent gluon propagators is obtained [3].

[a] e-mail: peter_tsang@brown.edu

2 Explicit Gauge Invariance

A rearrangement can now be made to formally insure gauge-invariance, even though the GF still apparently contains gauge-dependent gluon propagators.

$$\mathfrak{Z}_{QCD}[j,\bar{\eta},\eta] = \mathfrak{N} e^{-\frac{1}{2}\int \frac{\delta}{\delta A}\cdot D_c^{(0)}\cdot\frac{\delta}{\delta A}} \cdot e^{-\frac{i}{4}\int \mathbf{F}^2 + \frac{i}{2}\int \mathbf{A}\cdot(-\partial^2)\cdot\mathbf{A}} \cdot e^{i\int \bar{\eta}\cdot\mathbf{G_c[A]}\cdot\eta + \mathbf{L[A]}}\Big|_{A=\int \mathbf{D_c^{(0)}}\cdot\mathbf{j}} \tag{3}$$

Gives 2n-point functions:

$$= \mathfrak{N}\int d[\chi] e^{\frac{i}{4}\int\chi^2} e^{\mathfrak{D}_A^{(0)}} e^{\frac{i}{2}\int \chi\cdot\mathbf{F} + \frac{i}{2}\int \mathbf{A}\cdot(\mathbf{D_c^{(0)}})^{-1}\cdot\mathbf{A}} G_c(1|gA)G_c(2|gA)e^{L[A]}|_{A=0} \tag{4}$$

Then,

$$e^{\mathfrak{D}_a}F_1[A] = exp[\frac{i}{2}\int \bar{Q}\cdot D_c^{(0)}\cdot(1-\bar{K}\cdot D_c^{(0)})^{-1}\cdot\bar{Q} - \frac{1}{2}tr\ln(1-D_c\cdot\tilde{K})]$$
$$\cdot\ exp[\frac{1}{2}\int A\cdot\bar{K}\cdot(1-D_c^{(0)}\cdot\bar{K})^{-1}\cdot A + i\int\bar{Q}\cdot(q-\bar{K}\cdot D_c^{(0)})^{-1}] \tag{5}$$

where

$$D_c^{(0)}\cdot(1-\bar{K}\cdot D_c^{(0)})^{-1} = D_c^{(0)}\cdot[1-(\hat{K}+(D_c^{(0)})^{-1})\cdot D_c^{(0)}]^{-1} = -(\tilde{K}_{\mu\nu}^{ab} + gf^{abc}\chi_{\mu\nu}^{c})^{-1} = -\hat{\mathbf{K}}^{-1} \tag{6}$$

$$e^{\mathfrak{D}_A}F_1[A]F_2[A] = exp[-\frac{i}{2}\int\bar{Q}\cdot\hat{\mathbf{K}}^{-1}\cdot\bar{Q} + \frac{1}{2}tr\ln\hat{\mathbf{K}} + \frac{1}{2}tr\ln(-D_c^{(0)})]$$
$$\cdot\ exp[\frac{i}{2}\int\frac{\delta}{\delta A'}\cdot\mathbf{D_c^{(0)}}\cdot\frac{\delta}{\delta\mathbf{A}'}]\cdot\mathbf{exp}[\frac{\mathbf{i}}{\mathbf{2}}\int\frac{\delta}{\delta\mathbf{A}'}\cdot\hat{\mathbf{K}}^{-1}\cdot\frac{\delta}{\delta\mathbf{A}'} - \int\bar{\mathbf{Q}}\cdot\hat{\mathbf{K}}^{-1}\cdot\frac{\delta}{\delta\mathbf{A}'}]$$
$$\cdot\ (e^{\mathfrak{D}_A}F_2[A']) \tag{7}$$

$$e^{\mathfrak{D}_A}F_1[A]F_2[A] = \mathfrak{N}exp[-\frac{i}{2}\int\bar{Q}\cdot\hat{\mathbf{K}}^{-1}\cdot\bar{Q} + \frac{1}{2}tr\ln\hat{\mathbf{K}}]\cdot exp[\frac{i}{2}\int\frac{\delta}{\delta A}\cdot\hat{\mathbf{K}}^{-1}\cdot\frac{\delta}{\delta} - \int\bar{Q}\cdot\hat{\mathbf{K}}^{-1}\cdot\frac{\delta}{\delta A}]\cdot exp[L[A]] \tag{8}$$

As one sees in above equation, all the explicit gauge dependent propagators cancels. This is gauge invariant by means of gauge-independence. It deserves to be emphasized that Gauge Independence is the strongest form of Gauge Invariance. Feynman had long hoped for this for QED.

The $-\hat{\mathbf{K}}^{-1}$ above, also written as $(f\cdot\chi)^{-1}$ represents infinite gluon exchanges summed. This term is the *Gluon Bundle*, GB, exchanged between two quarks as shown in figure 1.

Figure 1. A *Gluon Bundle, GB, is the term* $(f\cdot\chi)^{-1}$, *representing the exchange of all gluons summed.*

All the gaussian linkage operations can then be carried through exactly, corresponding to the summation of all gluons changed between any pair of quark (and/or anti-quark) lines, and including the cubic and quartic gluon interactions. See figure 2.

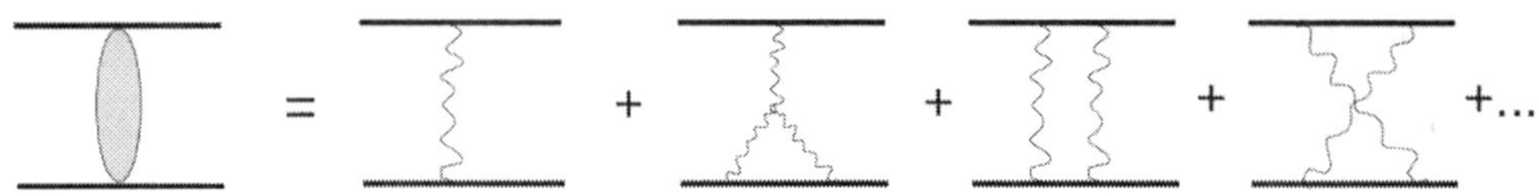

Figure 2. A *Gluon Bundle, GB, representing the exchange of all gluons summed.*

The result is explicit cancellation of all gauge-dependent gluon propagators, with resulting GF exhibiting Manifest Gauge Independence. One also finds a new, exact property of non-perturbative, gauge-invariant QCD, where the space-time coordinates of both ends of a GF are equal, modulo small uncertainties in their transverse coordinates. There is an Effective Locality between interacting quarks, and, changes all the remaining functional integrals from Schwinger into ordinary integrals. One can actually complete these integrals [3]

The summation of all gluon exchanges, a Gluon Bundle, GB, is a Gaussian functional derivative on quarks. figure 2.

An important point is that quarks are never observed individually, and, thus, cannot have fixed coordinates. The correct coordinates for quarks include transverse quark fluctuations. We believe we know how to do this, the work is still underway for putting quark fluctuations from first principles. What we have done here is to introduce phenomenlogical transverse fluctuation amplitudes for every quark-gluon vertex, replacing the usual gluon-quark current interaction at the same space-time point $\int d^4x\bar{\psi}(x)\gamma_\mu A^a_\mu(x)\tau_a\psi(x)$ by $\int d^2x'_\perp \int d^4x a(x_\perp - x'_\perp)\bar{\psi}(x')\gamma_\mu\tau_a A^a_\mu(x)\psi(x')$, with $a(x_\perp - x'_\perp)$ real and symmetric, and $x'_\mu = (x'_\perp, x_L, x_0)$.

The probability of finding two quarks separated by a transverse (or impact parameter) distance is then $\varphi(b) = \int \frac{d^2q}{(2\pi)^2} e^{iqb}|\tilde{a}(q)|^2$.

We chose a deformed gaussian $\varphi(b) = \varphi(0)e^{(\mu b)^{2+\xi}}$ with deformation parameter ξ real and small. A straight forward calculation yields, for small ξ, $V(r) \approx \xi\mu(\mu r)^{1+\xi}$. Perhaps it is important to emphasize again that all asymptotic quark states are hadronic bound states of quarks; and for such a bound state we can specify longitudinal and time coordinates, but not transverse coordinates since they are always fluctuating. The conventional "static quark" approximation used in model binding potential calculations in all non-perturbative amplitudes are plagued with divergences. All non-perturbative amplitudes are plagued with absurdities without taking such "transverse imprecision" into account.

Substituting our potential into a Schrodinger binding equation, using the "quantic" approximation then yields $\mu \sim m_\pi$, with $\xi \approx 0.1$. This is sensible since the maximum fluctuations should be less than m_π^{-1}.

Our results encompasses two different lattice calculations, $V \sim r$ and $V \sim r\ln(r)$. All lattice and other model calculations of $q - \bar{q}$ binding correspond to an amplitude containing only one or two Casimir $SU(3)$ invariants, C_2 or C_3, whereas our amplitude contains both [6]. We used the well-known half century old Eikonal function relation with potential. The minimum bound state energy for the pion shows that most of the pion's mass comes from the gluons forming the GB and relatively little from the quark masses.

3 Nuclear physics from QCD

In the same light, nucleon binding is examined. Here is the first (to our knowledge) example of nucleon binding, for a model deuteron, from basic QCD. We have performed a qualitative model, without electrical charge and nucleon spins which can always be added in, to describe the essence of Nuclear Physics. Assuming an average quark for ease of calculation, an attractive potential is obtained [7]. Quark binding takes place for $r_{ij} \approx m_\pi^{-1}$, but for nucleon binding that takes place at larger distances, extraction and regularization of the logarithmic UV divergence loop will contribute two essential features. 1) The loop stretches, so distances larger than m_π^{-1} can easily enter. 2) It provides a crucial change of sign for the effective n-n binding potential, figure 3, [5]. We expect and hope that nuclear physicists will employ such effective potentials to discuss heavy nuclei.

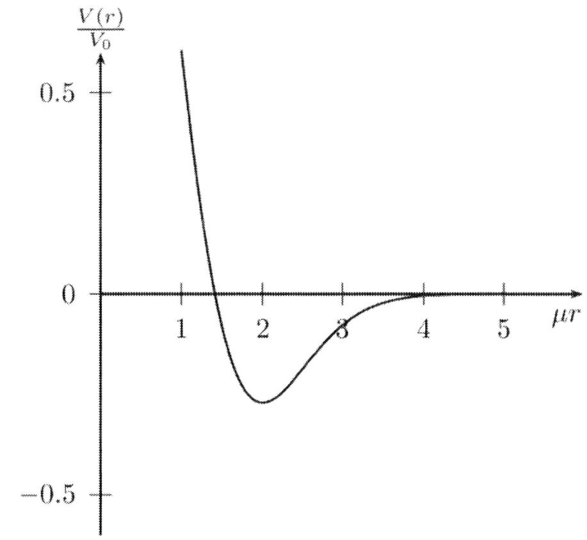

Fig. 3. Nucleon potential with $V_0 = c(\frac{g_R^2}{4\pi})\mu$.

Figure 3. Nucleon Nucleon binding is mediated by exchange of Gluon Bundles with quark loop that is able to stretch to distances greater than that of a pion.

All the basic radiative correction structure of non-perturbative QCD comes from interacting closed-quark-loops with GB's. A single dressed quark has an amplitude proportional to

$$N\int d[\chi]e^{i\frac{\chi^2}{4}}(det(gf\cdot\chi)^{-\frac{1}{2}})e^{\hat{\mathfrak{D}}_A}\cdot G_c[A]e^{L[A]}|_{A\to 0}, \tag{9}$$

While two scattering quarks are described by

$$N\int d[\chi]e^{i\frac{s\chi^2}{4}}(det(gf\cdot\chi)^{-\frac{1}{2}})e^{\hat{D}_A}G_c^{(1)}[A]G_c^{(2)}[A]e^{L[A]}|_{A\to 0}, \tag{10}$$

where $\hat{\mathfrak{D}}_A = \frac{i}{2}\int\frac{\partial}{\partial A}(gf\cdot\chi)^{-1}\frac{\partial}{\partial A}$.

Every GB exchanged is represented by the linkage operator connecting the two $G_c[A]$'s to each other, and the $G_c[A]$'s to $L[A]$. Explicit calculation shows that all self-energy graphs vanish either by asymmetry of the $(f\cdot\chi)^{-1}$ color and indices or by explicit loop integration. See figure 4 Non-perturbative QCD turns out to be far simpler than QED.

4 QCD renormalization

The radiative corrections of QCD enter when there is momentum transfer between one quark and another quark, where momentum transfer passes through intemediate GBs and/or closed quark loops.

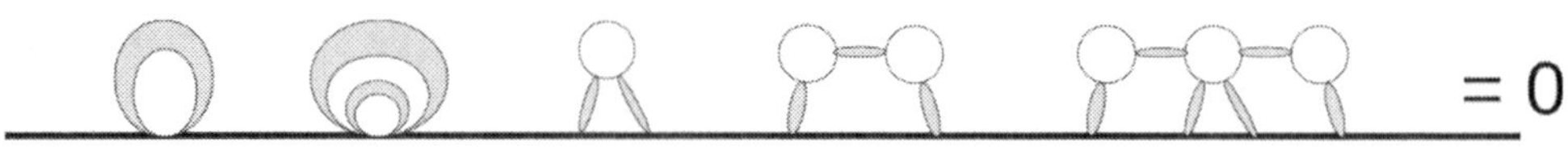

Figure 4. All self energy graphs with Gluon Bundles are zero, 0.

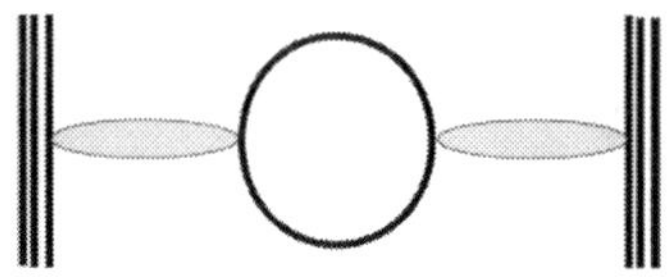

Figure 5. Nucleon Nucleon binding is mediated by exchange of Gluon Bundles with quark loop that is able to stretch to distances greater than that of a pion.

We use an exact functional cluster expansion described in Chapter 2.5 of [8]. In our particular choice of renormalization, we choose $\delta^2\ell = \kappa$, where δ represent point where GB connects to a quark loop, *ell* has the expected UV log divergence and κ is a finite positive constant, figure 6

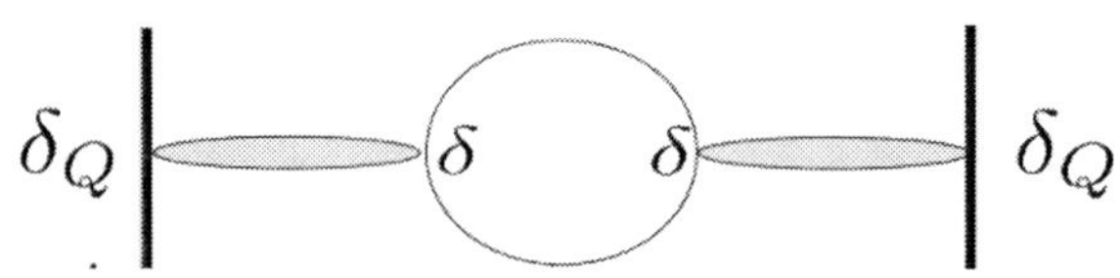

Figure 6. We chose a renormalization scheme where two connections of Gluon Bundles, δ, multplied by a quark loop with logarithmic divergence, ℓ is set to finite quantity κ to be determined by experiments.

With this particular choice if renormalization, GB chain graphs are non-zero. All other closed loops entering into the functional cluster expansion vanish. These GB chain graphs form a geometric series which can be summed, and is everywhere finite. See figure 7.

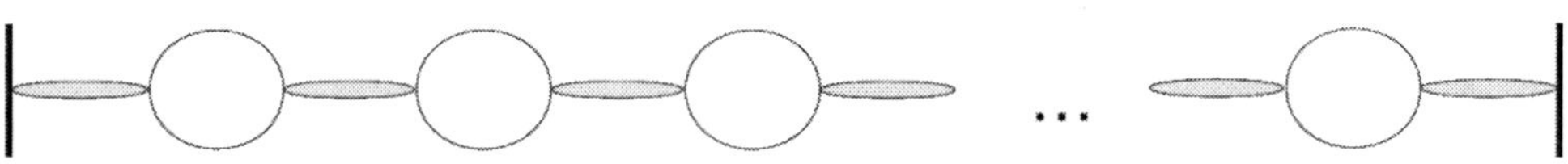

Figure 7. We set $\delta^2 \cdot \ell = \kappa$, where κ is finite and to be determined by experiments. This simplifies all posible loop connections with Gluon Bundles to only the straight chains. This is the first attempt and, thus far, compares well with experiments.

This puts us in a position where we can compare with high energy experiment events.

5 Comparison with High Energy elastic pp-scattering experiments

In high energy hadron scattering, pp-scattering in particular, there has always been a 'diffraction dip' that was difficult to explain and certainly not from first principles. There are form factors, or other methods attempted. For us, we calculate the differential cross section from, intuitively trying a GB exchanged between two hadrons (3 quarks that are not breaking up) and a GB-virtual quark loop-GB configuration. Just to get an intuitive understanding. Low and behold, we have a diffraction dip as seen in experiments. It comes from quark-GB-quark and quark-GB-virtualquarkloop-GB-quark, figure 8, [10].

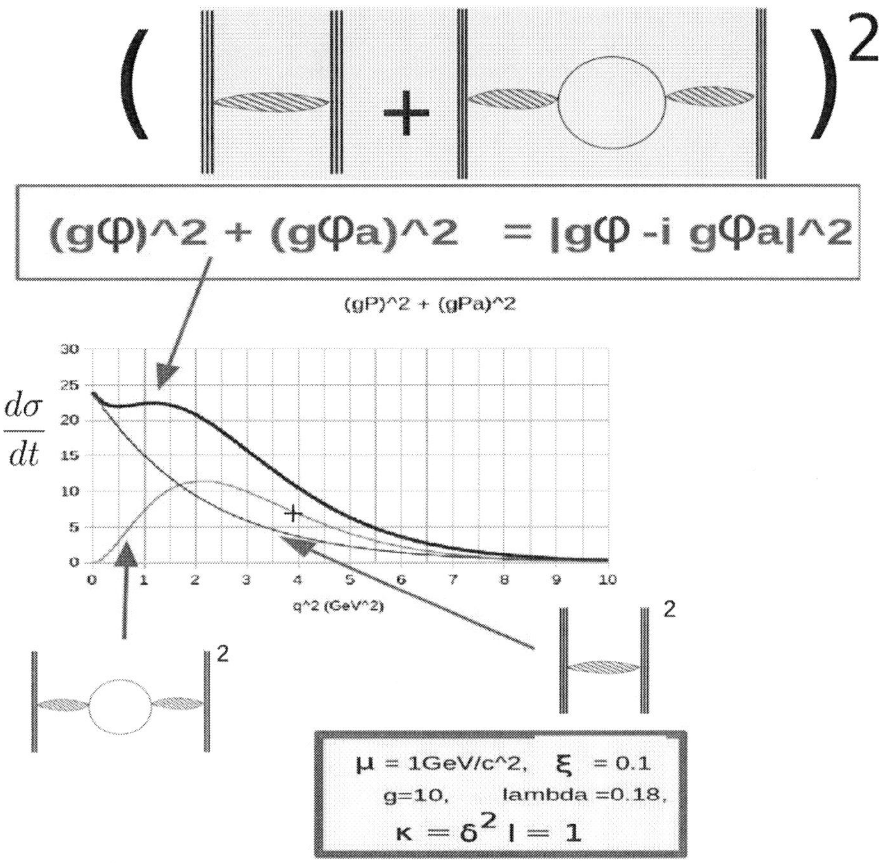

(H.M.Fried, Y.Gabellini, T.Grandou, Y-Ming Sheu, PHT. Annals of Physics 2015)

Figure 8. Elastic differential cross-section is calculated from exchange of Gluon Bundles between hadrons and exchange of Gluon Bundles between hadrons with additional quark loop. We are using one loop as first calculation. Infinitely many loop chains is already calculated and will be put in in work to appear.

The contribution from purely GB's exchanged between two nucleons provide an amplitude with exponential fall off, while the one-loop-term provides a rising function, figure 8, that, when both combined explains the diffraction dip. The exact form that includes infinite sum of all chain loops are currently underway. We expect favorable comparison with the experimental measurements [12].

The δ_q in our amplitude that connects to physical quark lines have units of time, giving us inverse eenergy relation. Raising $1/E$ to the first power provided too strong an energy dependence for positions of the dips. With δ_q proportional to $(1/m)(m/E)^p$, the power p can be deduced from data. See figure 9.

We have 3 parameters, coupling g, $\lambda\kappa$ and p that will be determined from experiment. More detailed analysis is currently underway and, of course, to data at 7 TeV and above.

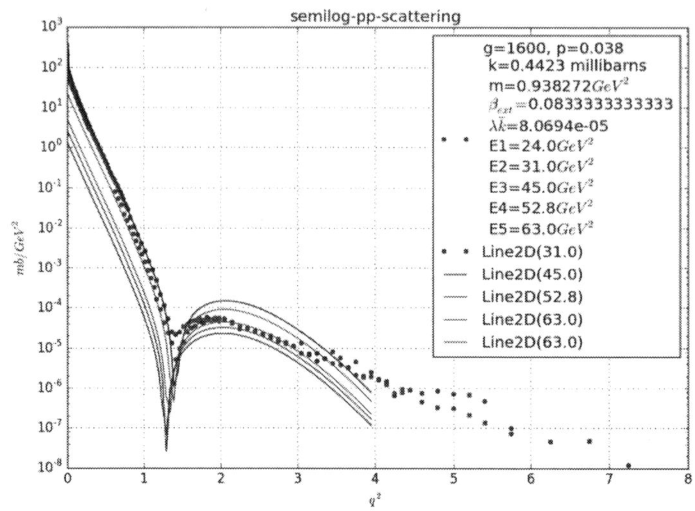

Figure 9. Early comparisons of Gluon Bundle exchanges and the one-loop-term amplitude compares well with Intersecting Storage Ring data of elastic pp-scattering. Data points are in small circular points. Our calculations are solid lines. There is the expected movement of the dip to smaller q^2 as energy, $\sqrt{s}$ is increased. [12]

Acknowledgements

This publication was made possible through the support of a Grant from the Julian Schwinger Foundation. The opinions expressed in this publication are those of the authors and do not necessarily reflect the views of the Julian Schwinger Foundation. We especially wish to thank Mario Gattobigio for his many kind and informative conversations relevant to the Nuclear Physics aspects of our work. It is also a pleasure to thank Mark Restollan, of the American University of Paris, for his kind assistance in arranging sites for our collaborative research when in Paris.

References

[1] H. M. Fried, Y. Gabellini, T. Grandou, and Y.-M. Sheu, *Eur. Phys. J. C* **65**, 395-411 (2010).

[2] M. B. Halpern, *Phys. Rev. D* **16** (1977) 1798, and H. Reinhardt, K. Langfeld, L. von Smekal, *Phys. Lett. B.* **300**, 111-117 (1993).

[3] H. M. Fried, T. Grandou, and Y.-M. Sheu, *Ann. Phys.* **327**, 2666-2690 (2012).

[4] E. S. Fradkin, *Dokl. Akad. Nauk SSSR* **98**, (1954) 47; and E. S. Fradkin, *Nucl. Phys.* **76**, 588-624 (1966).

[5] H. M. Fried, Y. Gabellini, T. Grandou, and Y.-M. Sheu, *Ann. Phys.* **338**, 107-122 (2013).

[6] T. Grandou, *Eur. Phys. Lett.* **107**, 11001 (2014); arXiv:1402.7273 [hep-th].

[7] H. M. Fried, T. Grandou, and Y.-M. Sheu, *Ann. Phys.* **344**, 78-96 (2014).

[8] H. M. Fried, *Basics of Functional Methods and Eikonal Models* (Editions Frontiéres, Gif-sur-Yvette Cedex, France, 1990).

[9] H. M. Fried, *Modern Phys. Letts.* **A 28**, 1230045 (2013).

[10] H. M. Fried, P.H.Tsang, Y. Gabellini, T. Grandou, and Y.-M. Sheu, *Ann. Phys.* **359**, 1-19 (2015).

[11] H. M. Fried and Y. Gabellini, *Ann. Phys.* **327**, 1645-1667 (2012).

[12] U. Amaldi and K.R. Schubert, Nucl. Phys. **B166**, 301 (1980). and M. Ambrosio *et al.*, (CERN-Naples-Pisa-Stony Brook Collaboration), Phys. Lett. **115B**, 495 (1982). and N.A. Amos *et al.*, Nucl. Phys. **B262**, 689 (1985). and A. Breakstone *et al.* (Ames-Bologna-CERN-Dortmund-Heidelberg-Warsaw Collaboration), Nucl. Phys. **B248**, 253 (1984).

[13] The TOTEM Collaboration *et al., Eur. Phys. Lett.* **95**, 41001 (2011). and The TOTEM Collaboration *et al., Eur. Phys. Lett.* **101**, 21002 (2013).

[14] Y.-M. Sheu, *Finite-Temperature QED: General Theory and Bloch-Nordsieck Estimates of Fermion Damping in a Hot Medium*, PhD Thesis, Brown University, May 2008.

EPJ Web of Conferences **126**, 04051 (2016) DOI: 10.1051/epjconf/201612604051
ICNFP 2015

New Quantum Effect: Emission of Cosmic X– or γ–rays by Moving Unstable Particles at Late Times

K. Urbanowski[1,a]

[1] *University of Zielona Góra, Institute of Physics, ul. Prof. Z. Szafrana 4a, 65–516 Zielona Góra, Poland.*

Abstract. A quantum effect induced by the late time properties of decaying states is discussed and its possible observational consequences are analyzed. It is shown that charged unstable particles as well as neutral unstable particles with non–zero magnetic moment which were able to survive sufficiently long may emit electromagnetic radiation. The nonclassical behavior of unstable particles at late times when deviations of the decay law from the exponential form begin to dominate is a source of the mechanism responsible for this effect. Analyzing the transition times region between exponential and non-exponential form of the survival amplitude it is found that at this time region the instantaneous energy of the unstable particle can take very large values, much larger than the energy of this state at times from the exponential time region. Results obtained for the model considered suggest that this new purely quantum mechanical effect may be responsible for causing unstable particles produced by astrophysical sources and moving with relativistic velocities to emit electromagnetic–, X– or γ–rays at some time intervals from the transition time regions.

1 Introduction

Theoretical predictions of deviations from the exponential form of the decay law at suitably long times resulted in increasing interest in studies of long time properties of unstable states. The predicted effect is almost negligible small and therefore very difficult to its confirmations in laboratory experiments. On the other hand, numbers of created unstable particles during some astrophysical processes are so large that some of them can survive up to times t at which the survival probability depending on t transforms from the exponential form into the inverse power–like form. Here we show at this time region a new quantum effect can be observed: A very rapid fluctuations of the instantaneous energy of unstable particles can take place. These fluctuations of the instantaneous energy should manifest itself as fluctuations of the velocity of the particle. Based on results presented in [1] we present in this talk some new results and we show that this effect may cause unstable particles to emit electromagnetic radiation of a very wide spectrum: from radio frequencies through X–rays to γ–rays of very high energies and that the astrophysical processes are the place where this effect should be observed.

[a]e-mail: K.Urbanowski@if.uz.zgora.pl;k.a.urbanowski@gmail.com

2 On unstable states

Studying properties of unstable states $|\phi\rangle \in \mathcal{H}$ (where $\mathcal{H}$ is the Hilbert space of states of the considered system) one usually starts from an analysis of their survival probability. The survival probability (the decay law), $\mathcal{P}_\phi(t)$ of an unstable state $|\phi\rangle$ decaying in vacuum is given by the following relation

$$\mathcal{P}_\phi(t) = |a(t)|^2, \tag{1}$$

where $a(t)$ is the probability amplitude of finding the system at the time t in the initial state $|\phi\rangle$ prepared at time $t_0 = 0$,

$$a(t) = \langle\phi|\phi(t)\rangle. \tag{2}$$

and $|\phi(t)\rangle$ is the solution of the Schrödinger equation for the initial condition $|\phi(0)\rangle = |\phi\rangle$:

$$i\hbar\frac{\partial}{\partial t}|\phi(t)\rangle = H|\phi(t)\rangle, \tag{3}$$

where H denotes the total self–adjoint Hamiltonian for the system considered.

From the literature it is known that the amplitude $a(t)$, and thus the decay law $\mathcal{P}_\phi(t)$ of the unstable state $|\phi\rangle$, are completely determined by the density of the energy distribution $\omega(\mathcal{E})$ for the system in this state [2]

$$a(t) = \int_{Spec.(H)} \omega(\mathcal{E})\, e^{-\frac{i}{\hbar}\mathcal{E}t}\, d\mathcal{E}. \tag{4}$$

where $\omega(\mathcal{E}) \geq 0$ and $a(0) = 1$.

Khalfin in [3] assuming that the spectrum of H must be bounded from below, ($Spec.(H) = [E_{min}, \infty)$ and $E_{min} > -\infty$), and using the Paley–Wiener Theorem [4] proved that in the case of unstable states there must be $|a(t)| \geq A \exp[-b\, t^q]$, for $|t| \to \infty$. Here $A > 0$, $b > 0$ and $0 < q < 1$. Therefore the decay law $\mathcal{P}_\phi(t)$ of unstable states decaying in the vacuum, (1), can not be described by an exponential function of time t if time t is suitably long, $t \to \infty$, and that for these lengths of time $\mathcal{P}_\phi(t)$ tends to zero as $t \to \infty$ more slowly than any exponential function of t. This effect was confirmed in experiment described in the Rothe paper [5], where the experimental evidence of deviations from the exponential decay law at long times was reported. In the light of the result of Rothe's group the following problem seems to be important: If (and how) deviations from the exponential decay law at long times affect the energy of the unstable state and its decay rate at this time region.

Analysis of late times properties of unstable states becomes easier if to express $a(t)$ in the following form

$$a(t) = a_{exp}(t) + a_{lt}(t), \tag{5}$$

where $a_{exp}(t)$ is the exponential part of $a(t)$, that is $a_{exp}(t) = N \exp[-i\frac{t}{\hbar}(E_\phi^0 - \frac{i}{2}\Gamma_\phi^0)]$, ($E_\phi^0$ is the energy of the system in the state $|\phi\rangle$ measured at the canonical decay times, i.e. when $\mathcal{P}_\phi(t)$ has the exponential form, Γ_ϕ^0 is the decay width, N is the normalization constant), and $a_{lt}(t)$ is the late time non–exponential part of $a(t)$. From the literature it is known that the characteristic feature of survival probabilities $\mathcal{P}_\phi(t)$ is the presence of sharp and frequent fluctuations at the transition times region, when contributions from $|a_{exp}(t)|^2$ and $|a_{lt}(t)|^2$ into $\mathcal{P}_\phi(t)$ are comparable and that the amplitude $a_{lt}(t)$ and thus the probability $\mathcal{P}_\phi(t)$ exhibits inverse power–law behavior at the late time region for times t much later than the crossover time T. The crossover time T can be found by solving the following equation,

$$|a_{exp}(t)|^2 = |a_{lt}(t)|^2. \tag{6}$$

In general $T \gg \tau_\phi$, where $\tau_\phi = \hbar/\Gamma_\phi^0$ is the lifetime of ϕ. Formulae for T depend on the model considered (i.e. on $\omega(E)$) in general). A typical form of the decay curves, that is the form of the non–decay probability $\mathcal{P}_\phi(t)$ as a function of time t is presented in Figs (1), (2). The results of calculations presented in these Figures were obtained for the Breit–Wigner energy distribution function, $\omega(E) \equiv \omega_{BW}(E)$, where

$$\omega_{BW}(E) \stackrel{\text{def}}{=} \frac{N}{2\pi}\Theta(E - E_{min})\frac{\Gamma_\phi^0}{(E - E_\phi^0)^2 + (\frac{\Gamma_\phi^0}{2})^2}, \tag{7}$$

and $\Theta(E)$ is the unit step function. The cross–over time T is a function of the parameter

$$s_0 \stackrel{\text{def}}{=} \frac{E_\phi^0 - E_{min}}{\Gamma_\phi^0}. \tag{8}$$

The larger s_0 the larger T (see Figs (1), (2)).

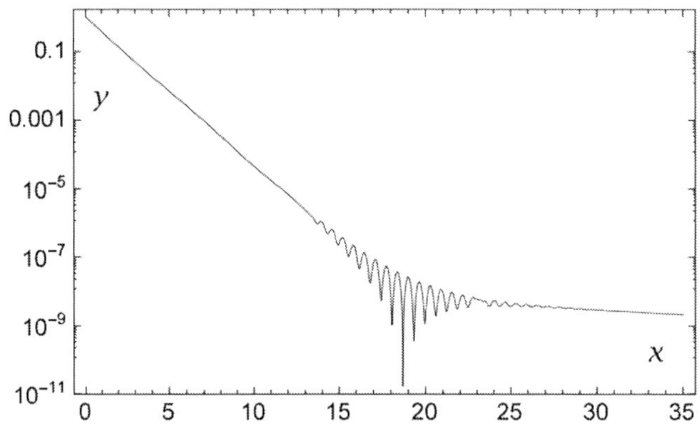

Figure 1. The general, typical form of the decay curve $\mathcal{P}_\phi(t)$: The case $s_0 = 10$. Axes: $y = \mathcal{P}_\phi(t)$ — the logarithmic scale; $x = t/\tau_\phi$ — the time is measured in lifetimes.

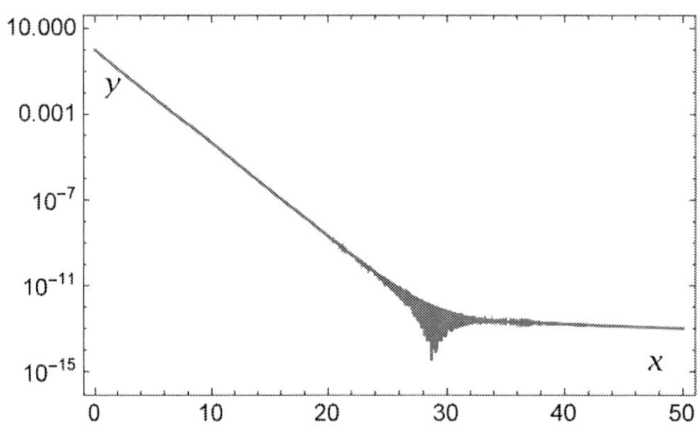

Figure 2. The same as in Fig (1) for the case $s_0 = 100$.

Fluctuations of $\mathcal{P}_\phi(t)$ at the transition times region, $t \sim T$, which one can see in Figs (1), (2), can be explained as follows: Using decomposition (5) one finds that

$$\mathcal{P}_\phi(t) = |a_{exp}(t)|^2 + 2\Re\,[a_{exp}(t)\,a_{lt}^*(t)] + |a_{lt}(t)|^2. \tag{9}$$

At times $t \sim T$ the dominating term in this decomposition of $\mathcal{P}_\phi(t)$ is the interference term $2\Re\,[a_{exp}(t)\,a_{lt}^*(t)]$ which is the source of the large fluctuations of $\mathcal{P}_\phi(t)$ mentioned.

3 Energy of unstable states

We need know properties of the energy of unstable states at late times. In the general case the energy and the decay rate of the system in the state $|\phi\rangle$ under considerations, (to be more precise the instantaneous energy and the instantaneous decay rate), can be found using the following relations (for details see [6–9]).

$$\mathcal{E}_\phi \equiv \mathcal{E}_\phi(t) = \Re\,(h_\phi(t)), \quad \gamma_\phi \equiv \gamma_\phi(t) = -\,2\,\Im\,(h_\phi(t)), \tag{10}$$

where $\Re\,(z)$ and $\Im\,(z)$ denote the real and imaginary parts of z respectively and $h_\phi(t)$ denotes the "effective Hamiltonian" for the one–dimensional subspace of states $\mathcal{H}_{\|}$ spanned by the normalized vector $|\phi\rangle$,

$$h_\phi(t) \stackrel{\text{def}}{=} i\hbar\,\frac{\partial a(t)}{\partial t}\,\frac{1}{a(t)}, \tag{11}$$

or equivalently

$$h_\phi(t) \equiv \frac{\langle\phi|H|\phi(t)\rangle}{\langle\phi|\phi(t)\rangle}. \tag{12}$$

We have $\mathcal{E}_\phi(t) \simeq E^0_\phi$ and $\varGamma_\phi(t) \simeq \varGamma^0_\phi$ at the canonical decay times. At asymptotically late times there is [1, 6, 8, 10]

$$\mathcal{E}_\phi(t) \simeq E_{min} + \frac{c_2}{t^2} + \frac{c_4}{t^4}\ldots, \quad (\text{for } t \gg T), \tag{13}$$

$$\varGamma_\phi(t) \simeq \frac{c_1}{t} + \frac{c_3}{t^3} + \ldots \quad (\text{for } t \gg T), \tag{14}$$

where $c_i = c_i^*$, $i = 1, 2, \ldots$, ($c_1 > 0$ and the sign of c_i for $i \geq 2$ depends on the model considered), so $\lim_{t\to\infty} \mathcal{E}_\phi(t) = E_{min}$ and $\lim_{t\to\infty} \varGamma_\phi(t) = 0$. Results (13) and (14) are rigorous.

The basic physical factor forcing the amplitude $a(t)$ to exhibit inverse power law behavior at $t \gg T$ is the boundedness from below of $\sigma(H)$. This means that if this condition is satisfied and $\int_{-\infty}^{+\infty} \omega(E)\,dE \equiv \int_{E_{min}}^{+\infty} \omega(E)\,dE < \infty$, then all properties of $a(t)$, including the form of the time–dependence at times $t \gg T$, are the mathematical consequence of them both. The same applies by (11) to the properties of $h_\phi(t)$ and concerns the asymptotic form of $h_\phi(t)$ and thus of $\mathcal{E}_\phi(t)$ and $\varGamma_\phi(t)$ at $t \gg T$.

Starting from relations (2), (4), (11) and assuming the form of $\omega(E)$ and performing all necessary calculations numerically one can find $\mathcal{E}_\phi(t)$ for all times t including times $t \gg T$. A typical behavior of the instantaneous energy $\mathcal{E}_\phi(t)$ at the transition time region is presented in Figs (3), (4) and (5). In these Figures the function $\kappa(t)$ is defined as follows

$$\kappa(t) = \frac{\mathcal{E}_\phi(t) - E_{min}}{E^0_\phi - E_{min}}, \tag{15}$$

and all calculations were performed for the Breit–Wigner energy distribution function (7).

One can see in Figs (1), (2) that the characteristic feature decay curves is the presence of sharp and frequent oscillations at the transition times region. This means that derivatives of the amplitude $a(t)$ may reach extremely large values for some times from the transition time region and the modulus of these derivatives is much larger than the modulus of $a(t)$, which is very small for these times. This explains why in this time region the real and imaginary parts of $h_\phi(t) \equiv \mathcal{E}_\phi(t) - \frac{i}{2}\,\gamma_\phi(t)$, which can be expressed by the relation (11), ie. by a large derivative of $a(t)$ divided by a very small $a(t)$, reach values much larger than the energy E^0_ϕ of the the unstable state measured at times for which the decay curve has the exponential form.

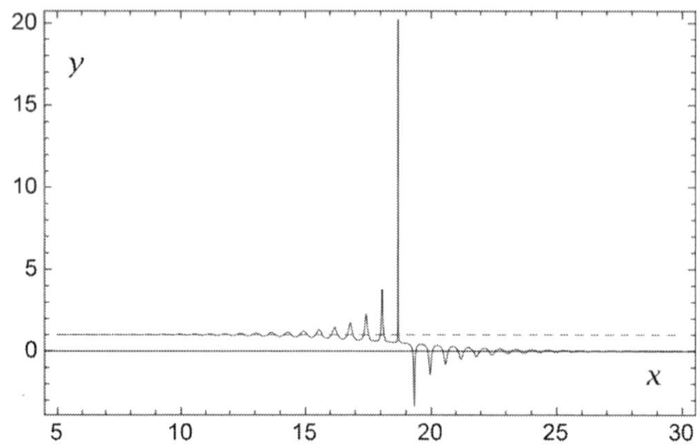

Figure 3. The instantaneous energy $\mathcal{E}_\phi(t)$ in the transitions time region: The case $s_0 = 10$. Axes: $y = \kappa(t)$, $x = t/\tau_\phi$ — the time is measured in lifetimes. The dashed line denotes the case $\kappa(t) = 1$, that is $\mathcal{E}_\phi(t) = E^0_\phi$.

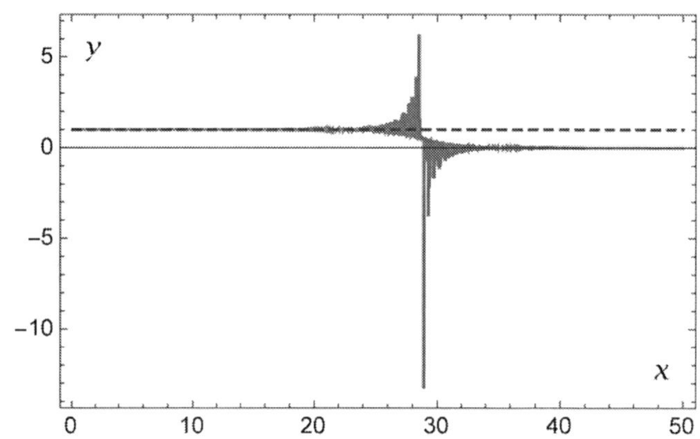

Figure 4. The same as in Fig (3) for the case $s_0 = 100$.

4 Some observable effects

Note that from the point of view of a frame of reference in which the time evolution of the unstable system was calculated the Rothe experiment as well as the picture presented in Figs (1), (2) and (3), (4), (5) refer to the rest coordinate system of the unstable system considered.

Properties of the ratio $\kappa(t)$ taking place for some time intervals mean that the instantaneous energy $\mathcal{E}_\phi(t)$ at these time intervals differs from the energy E^0_ϕ measured at times from the exponential decay time region (i.e. at the canonical decay times). The relation (12) explains why such an effect can occur. Indeed, if to rewrite the numerator of the righthand side of (12) as follows,

$$\langle\phi|H|\phi(t)\rangle \equiv \langle\phi|H|\phi\rangle\, a(t) + \langle\phi|H|\phi(t)\rangle_\perp, \tag{16}$$

where $|\phi(t)\rangle_\perp = Q|\phi(t)\rangle$, $Q = \mathbb{I} - P$ is the projector onto the subspace od decay products, $P = |\phi\rangle\langle\phi|$ and $\langle\phi|\phi(t)\rangle_\perp = 0$, then one can see that there is a permanent contribution of decay products described by $|\phi(t)\rangle_\perp$ to the energy of the unstable state considered. The intensity of this contribution depends

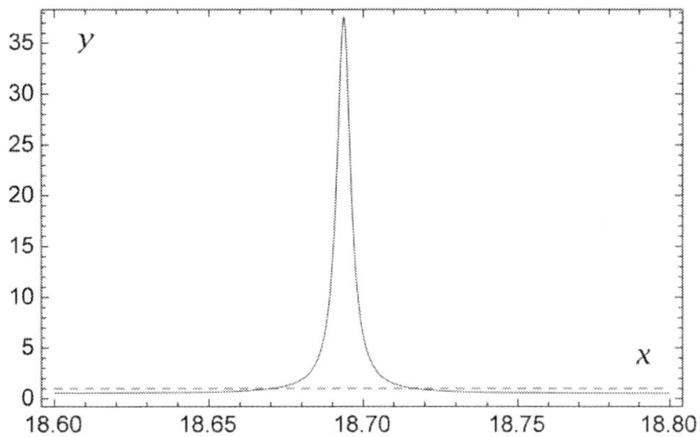

Figure 5. The highest local maximum of the instantaneous energy $\mathcal{E}_\phi(t)$ presented in Fig. (3). Axes: $y = \kappa(t)$, $x = t/\tau_\phi$ — the time is measured in lifetimes. The dashed line corresponds to the case $\kappa(t) = 1$, or $\mathcal{E}_\phi(t) = E_\phi^0$.

on time t. This contribution into the instantaneous energy is practically constant in time at canonical decay times whereas at the transition times, when $t \sim T$, it is fluctuating function of time and the amplitude of these fluctuations may be significant. What is more relations (12) and (16) allow one to proof that in the case of unstable states $\Re\,[h_\phi(t)] \neq const$. Using these relations one obtains that

$$h_\phi(t) = \langle\phi|H|\phi\rangle + \frac{\langle\phi|H|\phi(t)\rangle_\perp}{a(t)}. \tag{17}$$

From this relation one can see that $h_\phi(0) = \langle\phi|H|\phi\rangle$ if the matrix elements $\langle\phi|H|\phi\rangle$ exists. It is because $|\phi(t = 0)\rangle_\perp = 0$ and $a(t = 0) = 1$. Now if to assume that for $0 \le t_1 \neq t_2$ there is $\Re\,[h_\phi(0)] = \Re\,[h_\phi(t_1)] = \Re\,[h_\phi(t_2)] = const$ then one immediately conclude that $\Re\,[h_\phi(t)] = \langle\phi|H|\phi\rangle$ for any $t \ge 0$. Unfortunately such an observation contradicts implications of (17): From this relation it follows that $\Re\,[\frac{\langle\phi|H|\phi(t)\rangle_\perp}{a(t)}] \neq 0$ for $t > 0$ and thus $\Re\,[h_\phi(t > 0)] \neq \langle\phi|H|\phi\rangle \equiv \Re\,[h_\phi(0)]$ which shows that $\Re\,[h_\phi(t)] \equiv \mathcal{E}_\phi(t)$ can not be constant in time.

Astronomical observations provide information that there are sources of unstable particles in space that emit them with relativistic or ultra–relativistic velocities in relation to an external observer so many of these particles move in space with ultra high energies. Thus one meets the following problem: What effects can be observed by an external observer when the unstable particle, say ϕ, which survived up to the transition times region, $t \sim T$, or longer is moving with a relativistic velocity in relation to this observer. The distance d from the source reached by this particle is of order $d \sim d_T$, where $d_T = v^\phi \cdot T'$ and $T' = \gamma T$, $\gamma \equiv \gamma(v^\phi) = (\sqrt{1-\beta^2})^{-1}$, $\beta = v^\phi/c$, v^ϕ is the velocity of the particle ϕ. (For simplicity we assume that there is a frame of reference common for the source and observer both and that they do not move with respect to this frame of reference). So, in the case of moving particles created in astrophysical processes one should consider the effect shown in Figs (3), (4) together with the fact that the particle gains extremely huge energy, W^ϕ, which should be conserved.

Let us consider the case when the unstable particle under considerations has a constant momentum $\vec{p}$, $|\vec{p}| = p > 0$. Next, let us assume for simplicity that $\vec{v} = (v_1, 0, 0) \equiv (v^\phi, 0, 0)$ then there is $\vec{p} = (p, 0, 0)$. Now, let Λ_p be the Lorentz transformation from the reference frame O, where the momentum of unstable particle considered is zero, $\vec{p} = 0$, into the frame O' where the momentum of this particle is $\vec{p} \equiv (p, 0, 0) \neq 0$ or, equivalently, where its velocity $\vec{v}^\phi$ equals $\vec{v}^\phi = \vec{v}_p^\phi \equiv \vec{p}/m\gamma$, and let

E be the energy of the particle in the reference frame O and E' be the energy of the moving particle and having the momentum $\vec{p} \neq 0$. In this case the corresponding 4–vectors are: $\wp = (E/c, 0, 0, 0) \in O$ and $\wp' = (E'/c, p, 0, 0) = \Lambda_p \wp \in O'$. There is $\wp' \cdot \wp' \equiv (\Lambda_p \wp) \cdot (\Lambda_p \wp) = \wp \cdot \wp$ in Minkowski space, which is an effect of the Lorentz invariance. (Here the dot "$\cdot$" denotes the scalar product in Minkowski space). Hence, in our case: $\wp' \cdot \wp' \equiv (E'/c)^2 - p^2 = (E/c)^2$ because $\wp \cdot \wp \equiv (E/c)^2$ and thus $(E')^2 = c^2 p^2 + E^2$. There is $p \equiv \frac{E}{c} \gamma \beta$ in the case of a free particle. Therefore $c^2 p^2 + E^2 \equiv E^2 \gamma^2 \beta^2 + E^2 = E^2(1 + \beta^2 \gamma^2) \equiv E^2 \gamma^2$. The last relation means that the energy, E', of moving particles equals

$$E' = \sqrt{E^2 + c^2 p^2} \equiv E \gamma \stackrel{\text{def}}{=} W^{\phi}. \tag{18}$$

So in the case of the instantaneous energy, $\mathcal{E}_{\phi}(t)$ considered in previous Sections we can write that

$$W^{\phi} \equiv \mathcal{E}_{\phi}(t_k)\, \gamma_k = W^{\phi}(t_k), \tag{19}$$

at the instant t_k (where $\gamma_k = \gamma(t_k)$.

We have $\mathcal{E}_{\phi}(t) \equiv E_{\phi}^0$ at canonical decay times. Thus, $W^{\phi} \equiv E_{\phi}^0 \gamma$, at these times (where $\gamma = \gamma(t \sim \tau_{\phi}))$ and at times $t \gg \tau_{\phi}$, $t \sim T$ we have $\mathcal{E}_{\phi}(t) \neq E_{\phi}^0$. Now denoting by W_1^{ϕ} the energy of the moving particle at the canonical decay times, by W_2^{ϕ} the energy of the particle at transition times and by W_3^{ϕ} the energy at asymptotically late times (here simply $W_i^{\phi} = W^{\phi}(t_i)$, $i = 1, 2, 3$) and taking into account the principle of conservation of energy, we can conclude that

$$W_1^{\phi} = W_2^{\phi} = W_3^{\phi}. \tag{20}$$

The most useful relation follows from (20) for $t_1 \ll T'$ and $t_2 \sim T'$, (where $T' = \gamma T$): $W_1^{\phi} \equiv E_{\phi}^0 \gamma = W_2^{\phi}$, or the equivalent one,

$$E_{\phi}^0 \gamma_1 \equiv \mathcal{E}_{\phi}(t_2)\, \gamma(t_2). \tag{21}$$

It is convenient to rewrite the last relation as follows,

$$\gamma_1 = \frac{\mathcal{E}_{\phi}(t_2)}{E_{\phi}^0} \gamma(t_2). \tag{22}$$

The ratio $\frac{\mathcal{E}_{\phi}(t_2)}{E_{\phi}^0}$ can be extracted from Figs (3), (4), (5):

$$\frac{\mathcal{E}_{\phi}(t)}{E_{\phi}^0} = 1 + (\kappa(t) - 1) \frac{E_{\phi}^0 - E_{min}}{E_{\phi}^0}. \tag{23}$$

Thus if $\kappa(t) \neq 1$ then $\gamma_2 \neq \gamma_1$, which means that $v_2^{\phi} \neq v_1^{\phi}$ or that $a_{average} = \frac{v_2^{\phi} - v_1^{\phi}}{t_2 - t_1} \neq 0$. So, the moving charged unstable particle ϕ has to emit electromagnetic radiation at the transition time region. The same concerns neutral unstable particles with non–zero magnetic moment. The energy of the electromagnetic radiation emitted by such a charged particle in a unit of time can be found using the Larmor formula:

$$P = \frac{1}{6\pi\epsilon_0} \frac{q^2 \dot{v^{\phi}}^2}{c^2} \gamma^6, \tag{24}$$

where P is total radiation power, q is the electric charge, ϵ_0 – permittivity for free space. There is an analogy in classical physics of the effect described above: A conservation of the angular momentum and a pirouette like effect.

5 Some estimations

From the results presented in Figs (3), (5) one can find coordinates of, e.g., the highest maximum: $(x_{mx}, y_{mx}) = (18.69, 37.68)$. They allow one to find that $\kappa(t_{mx}) = y_{mx} = 37.68$. Coordinates of points of the intersection of this maximum with the straight line $y = 1$ are equal: $(x_1, y_1) = (18.67, 1.0)$ and $(x_2, y_2) = (18.72, 1.0)$. Thus $\Delta x = x_2 - x_1 = 0.05$. (Here $x = t/\tau_\phi$), and this Δx allows one to find $\Delta t = t_{mx} - t_1$. Let us consider now μ meson as an example. In such a case: $E^0_\mu - E_{min} = m_{\mu^\pm} - (m_e + m_{\bar{\nu}_e} + m_{\nu_\mu}) \simeq 105$ [MeV], ($\tau_\mu = 2,198 \times 10^{-6}$[s] and the crossover time equals: $T^\mu \simeq 165\tau_\mu = 0,37 \times 10^{-3}$[s], [1]). Using these data one finds the relation of $\gamma(t_1)$ with $\gamma(t_{mx})$ and using this relation one can extract Δv^μ. Next using such obtained Δv^μ and $\Delta t = t_{mx} - t_1$ extracted from Fig (5) one can use the Larmor formula (24) and to calculate the total radiation power P. The result is $P \sim 0.84$ [keV/s].

Astrophysical and cosmological processes in which extremely huge numbers of unstable particles are created seem to be a possibility for the above discussed effect to become manifest. The fact is that the probability $\mathcal{P}_\phi(t) = |a(t)|^2$ that an unstable particle ϕ survives up to time $t \sim T$ is extremely small. According to estimations of the luminosity of some γ–rays sources the energy emitted by these sources can even reach a value of order 10^{52} [erg/s], and it is only a part of the total energy produced there. If to consider a source emitting energy 10^{50} [erg/s] then, eg., an emission of $\mathcal{N}_0 \simeq 6.25 \times 10^{47}$ [1/s] particles of energy 10^{18} [eV] is energetically allowed. The same source can emit $\mathcal{N}_0 \simeq 6.25 \times 10^{56}$ [1/s] particles of energy 10^9 [eV] and so on.

Within the model considered in this talk the cross–over time T is given by the following approximate relation (valid for $E^0_\phi/\varGamma^0_\phi \gg 1$),

$$\varGamma^0_\phi T \equiv \frac{T}{\tau_\phi} \sim 2 \ln \Big[2\pi\Big(\frac{E^0_\phi - E_{min}}{\varGamma^0_\phi}\Big)^2\Big], \tag{25}$$

(see, eg. [6, 8, 11]). As it was mentioned, e.g. in [12], a typical value of the ratio $(E^0_\phi - E_{min})/\varGamma^0_\phi$ is $(E^0_\phi - E_{min})/\varGamma^0_\phi \geq O(10^3 - 10^6)$. Taking $(E^0_\phi - E_{min})/\varGamma^0_\phi = 10^6$ one obtains from (25) that $\mathcal{N}_\phi(T) \equiv \mathcal{P}_\phi(T)\, \mathcal{N}_0 \sim 2.53 \times 10^{-26}\, \mathcal{N}_0$. So there are $\mathcal{N}_\phi(T) \sim 14 \times 10^{21}$ particles per second of energy $W^\phi = 10^{18}$ [eV] or $\mathcal{N}_\phi(T) \sim 14 \times 10^{30}$ particles of energy $W^\phi = 10^9$ [eV] in the case of the considered example and T calculated using (25).

6 Conclusions

Analyzing properties of evolving in time unstable states we found that their instantaneous energy, $\mathcal{E}_\phi(t)$, can not be constant in time and fluctuates with the increasing amplitude. The amplitude of these fluctuations is negligible small at canonical decay time and it grows up to the transition time region, where it becomes significant. In the case of moving unstable particles these fluctuations at the transition time region, $t \sim T$, (where T is the crossover time — see (6) and (25)), cause fluctuations of the velocity of these particles at this time region, which forces the charged particles to emit electromagnetic radiation. This observation is valid also in the case of moving neutral unstable particles with a non–zero magnetic moment: They have to emit electromagnetic radiation in the transition time region too. This means that ultra relativistic unstable charged particles or unstable particles with non–zero magnetic moment can emit X– or γ–rays in this time region. A chance to observe this effect in the Earth laboratories is rather very small. On the other hand astrophysical sources are able to create such numbers $\mathcal{N}_0$ of unstable particles that sufficiently large number $\mathcal{N}_\phi(T) \gg 1$ of them has to survive up to times T, and therefore they have to emit electromagnetic radiation at this time region. Hence there is a chance to observe this effect analyzing the data of astronomical observations. The expected spectrum of this radiation can be very wide: From radio frequencies up to γ–rays.

Acknowledgments: The work was supported by the NCN grant No DEC-2013/09/B/ST2/03455.

References

[1] K. Urbanowski and K. Raczyńska, Phys. Letters, **B 731**, 236, (2014) (*Preprint* 1303.6975 [astro–ph.HE])
[2] S. Krylov and V. A. Fock, Zh. Eksp. Teor. Fiz., **17**, 93, (1947).
[3] L. A. Khalfin, Zh. Eksp. Teor. Fiz., **33**, 1371, (1957); [Sov. Phys. – JETP, **6**, 1053, (1958)].
[4] R. E. A. C. Paley and N. Wiener, *Fourier transforms in the complex domain*, (New York: American Mathematical Society, 1934)
[5] C. Rothe, S. I. Hintschich and A. P. Monkman, Phys. Rev. Lett., **96** 163601, (2006).
[6] K. Urbanowski, Cent. Eur. J. Phys., **7** 696, (2009).
[7] K. Urbanowski, International Journal of Modern Physics, **A 6** 1051, (1991).
[8] K. Urbanowski, European Physical Journal, **D 54** 25, (2009).
[9] F. Giraldi, Eur. Phys. J., **D 69**: 5 (2015).
[10] K. Urbanowski, Physical Review Letters, **107** 209001, (2011).
[11] K. Urbanowski, European Physical journal **C 58**, 151, (2008).
[12] L. M. Krauss and J. Dent, Phys. Rev. Lett., **100**, 171301, (2008).

EPJ Web of Conferences **126**, 04052 (2016) DOI: 10.1051/epjconf/201612604052
ICNFP 2015

Identified charged hadron production in pp and Pb–Pb collisions with ALICE at the LHC

Maria Vasileiou[1,a] on behalf of the ALICE Collaboration
[1]*University of Athens, Faculty of Physics*

Abstract. Nuclear matter under extreme conditions can be investigated in ultra-relativistic heavy-ion collisions. The measurement of transverse momentum distributions and yields of identified particles is a fundamental step in understanding collective and thermal properties of the matter produced in such collisions. The ALICE Experiment results on identified charged hadron production are presented for pp collisions at $\sqrt{s}$ = 0.9, 2.76 and 7 TeV and for Pb–Pb collisions at $\sqrt{s_{\rm NN}}$ = 2.76 TeV. Spectral shapes, production yields and nuclear modification factors are shown and compared to previous experiments and Monte Carlo predictions. The spectral shapes in Pb–Pb collisions indicate a strong increase of the radial flow velocity with respect to RHIC energies, which in hydrodynamic models is expected as a consequence of the increasing particle density. The observed suppression of high transverse momentum particles in central Pb–Pb collisions provides evidence for strong parton energy loss in the hot and dense medium.

1 Introduction

The spectra of identified particles provide a tool to study the dynamics of the quark-gluon plasma created in Pb–Pb collisions and particle production mechanisms in different momentum regions. Studying the low $p_{\rm T}$ spectra ($p_{\rm T}$ < 2 GeV/c), information about the production of soft particles from the thermalized medium can be obtained. Measurements in the high $p_{\rm T}$ region ($p_{\rm T}$ > 8 GeV/c) is important to study the hard production in the perturbative regime and jet quenching phenomena. The intermediate $p_{\rm T}$ range provides information of the so-called baryon anomaly phenomena first observed at RHIC, that is the enhancement of the baryon over meson ratio as a function of $p_{\rm T}$ with increasing centrality of the collision.

2 ALICE detector

ALICE (A Large Ion Collider Experiment) [1] is a general-purpose detector for heavy-ion studies at the CERN LHC (Large Hadron Collider). It has unique particle identification (PID) capabilities among the LHC experiments, allowing for measurements of particles in a wide range in $p_{\rm T}$. The ALICE experiment consists of a combination of central-barrel detectors and several forward detector systems. The central barrel, located inside a solenoidal magnet (B = 0.5 T), covers the mid-rapidity region $|\eta|$ < 0.9 over the

[a] Corresponding author: mvasili@phys.uoa.gr

full azimuthal angle. It includes a six-layer high-resolution Inner Tracking System (ITS), a large-volume Time Projection Chamber (TPC), electron and charged-hadron identification detectors which exploit Transition Radiation (TRD) and Time Of Flight (TOF) techniques, respectively. Small area detectors for high p_T particle-identification (HMPID), photon and neutral-meson measurement (PHOS), photon and electron ID and jet reconstruction (EMCal) complement the central barrel. The forward-rapidity system includes a single-arm muon spectrometer covering the pseudorapidity range $-4.0 \leq \eta \leq -2.5$ and several smaller detectors for triggering, multiplicity measurements and centrality determination.

3 Results

The results presented here are obtained from a data sample collected during the LHC Pb–Pb run at $\sqrt{s_{NN}} = 2.76$ TeV in 2010 and 2011, the p–Pb run at $\sqrt{s_{NN}} = 5.02$ TeV in the beginning of 2013 and the pp run at $\sqrt{s} = 2.76$ TeV in 2011.

3.1 Identified particle spectra and ratios

The spectra for summed charge states in central (0–5%) collisions are compared to hydrodynamical models and previous results in Au-Au collisions at $\sqrt{s_{NN}} = 200$ GeV in Fig.1 [2]. A dramatic change in spectral shapes from RHIC to LHC energies is observed, with the protons in particular showing a flatter distribution.

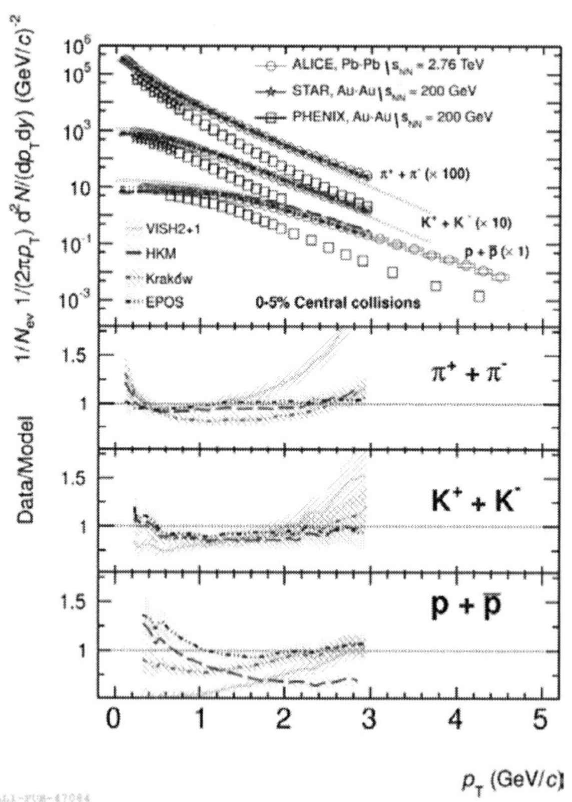

Figure 1. (Color online) Spectra of particles for summed charge states in the centrality bin 0–5% compared to hydrodynamical models and results from RHIC at $\sqrt{s_{NN}} = 200$ GeV. Systematic uncertainties plotted (boxes); statistical uncertainties are smaller than the symbol size.

The comparison to models based on hydrodynamics shows that VISH 2+1 [3] describes the pion and kaon spectra for $p_T < 1.5$ GeV/c qualitatively well, but fails in the description of protons, which can be attributed to a lack of an explicit description of the hadronic phase. HKM [4] inserts a hadronic cascade after the hydrodynamic phase, which further transports them until final decoupling. This improves the description of the protons for $p_T < 1.0$ GeV/c. The Kraków [5] model provides a better description of the spectra than the other two models. This model uses temperature (T_{ch}) fluctuations, consequence of non-equilibrium due to the bulk viscosity. The EPOS [6] model uses breakup of flux tubes which either contribute to the bulk or escape the medium as jets. This model describes qualitatively well the data in a broad p_T range. A comparison between RHIC and LHC energies based on the values of $<\beta_T>$ and T_{kin} from the combined blast wave fits [7] is shown in Fig.2[2]. For central collisions, about 10% stronger radial flow is observed at the LHC with respect to RHIC. The value of $<\beta_T>$ extracted from the fit increases with centrality, while T_{kin} decreases, similar to what was observed at lower energies . This is interpreted as a possible indication of a more rapid expansion with increasing centrality [7].

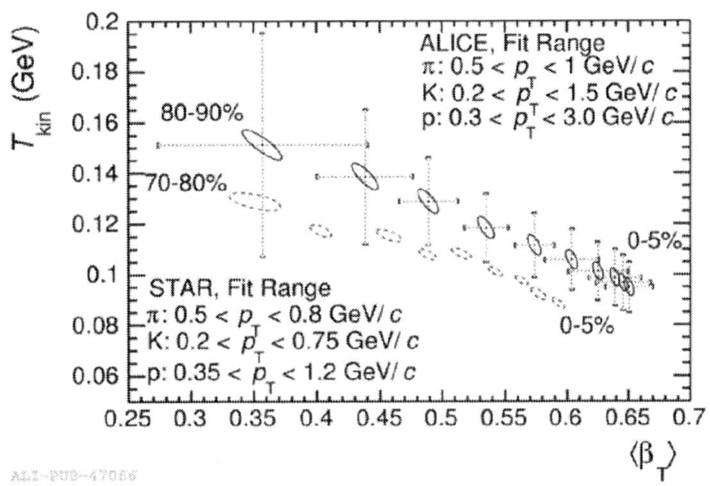

Figure 2. (Color online) Results of blast-wave fits compared to similar fits at RHIC energies. The uncertainty contours include the effect of the bin-by-bin systematic uncertainties and the dashed error bars represent the full systematic uncertainty. The STAR contours include only statistical uncertainties.

Fig.3 [8] shows the invariant yields measured in Pb–Pb collisions compared to those in pp collisions scaled by the number of binary collisions, N_{coll} [9], obtained for the measured pp cross section [10]. For peripheral Pb–Pb collisions the shapes of the invariant yields are similar to those observed in pp collisions. For central Pb–Pb collisions, the spectra exhibit a reduction in the production of high-p_T particles with respect to the reference which is characteristic of jet quenching.Fig.4 [11] shows the ratio of proton to pion spectra and kaon to pion spectra in Pb–Pb and pp collisions. For $1.5 < p_T < 5$ GeV/c the Pb–Pb ratios are strongly enhanced with respect to the pp measurement. This enhancement seems to be consistent with a common velocity boost [11] (radial flow) which leads to a mass-dependent modification of the p_T spectra. The kaon to pion ratio exhibits a bump at $p_T \sim 3$ GeV/c, which is qualitatively described in models including coalescence as a mechanism for parton hadronization in the low p_T region. The Kraków and EPOS [12,13] models, based on a hydrodynamically expanding medium, are in better agreement with the data than the Fries [14] model, which assumes recombination as the dominant hadronization mechanism, although more recent recombination calculations also show good agreement with the measured proton to pion ratio [15]. In central collisions, the ϕ - meson spectrum (not shown, see [16]) is similar in shape to the proton spectrum, supporting the dominance of radial flow as the ϕ is a meson with a mass close to the proton mass. At high momenta (> 10 GeV/c)

the particle ratios in pp and Pb–Pb collisions are equal, indicating vacuum-like hadronization through fragmentation.

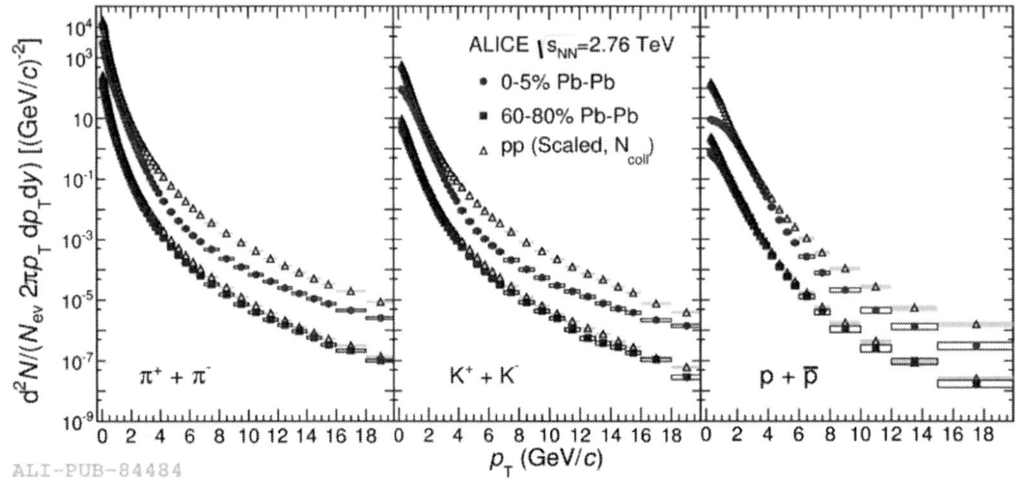

Figure 3. (Color online) Solid markers represent the invariant yields of identified particles in central and peripheral Pb–Pb collisions. Open points show the pp reference yields scaled by the average number of nucleon-nucleon collisions for 0-5% (upper) and 60-80% (lower). The statistical and systematic uncertainties are shown as vertical error bars and boxes, respectively.

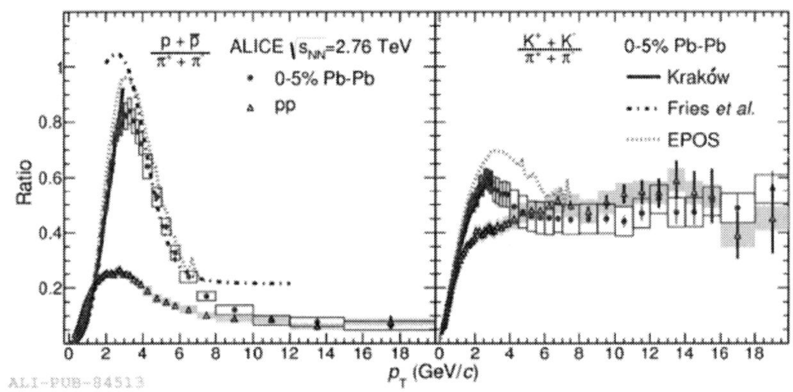

Figure 4. (Color online) Particle ratios as a function of p_T measured in pp and the most central, 0–5%, Pb–Pb collisions. Statistical and systematic uncertainties are displayed as vertical error bars and boxes, respectively. The theoretical predictions refer to Pb–Pb collisions.

3.2 Nuclear modification factor

Fig.5 [11] shows the nuclear modification factor R_{AA} as a function of p_T,defined as the ratio of the Pb–Pb spectra to the N_{coll} scaled pp spectra shown in Fig.3.The R_{AA} for the sum of kaons and protons is included as it allows a quantitative comparison to the R_{AA} of pions. For $p_T < 10$ GeV/c protons appear to be less suppressed than kaons and pions, consistent with the particle ratios shown in Fig.4. At larger p_T (> 10 GeV/c) all particle species are equally suppressed; so despite the strong energy loss observed in the most central heavy-ion collisions, the particle composition and ratios at high p_T are similar to those in vacuum. In Fig. 6 [17] the measurement of the nuclear modification factor in p–Pb collisions

R_{pPb} is compared to that in central (0-5% centrality) and peripheral (70-80%) Pb–Pb collisions R_{PbPb}. Proton-lead collisions provide a control experiment to clearly establish whether the initial state of the colliding nuclei plays a role in the observed high-p_T hadron production in Pb–Pb collisions. R_{pPb} is observed to be consistent with unity for transverse momenta higher than about 2 GeV/c. This important measurement demonstrates that the strong suppression observed in central Pb–Pb collisions at the LHC is not due to an initial-state effect, but it is a final state effect related to the hot matter created in high-energy heavy-ion collisions.

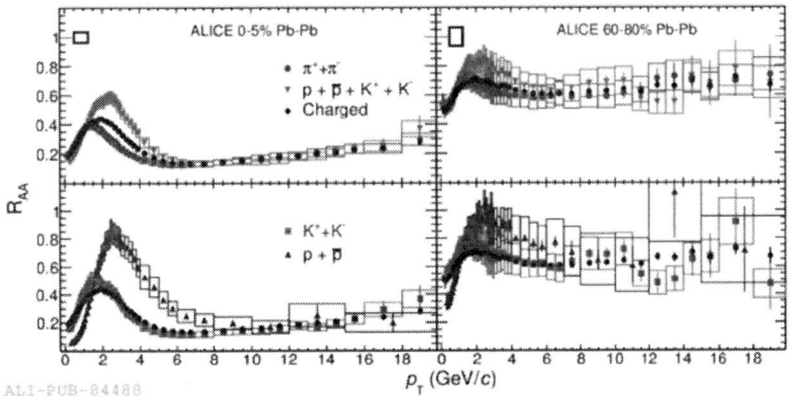

Figure 5. (Color online) The nuclear modification factor R_{AA} as a function of p_T for different particle species. Results for 0–5% (left) and 60–80% (right) collision centralities are shown. Statistical and systematic uncertainties are plotted as vertical error bars and boxes around the points, respectively. The total normalization uncertainty (pp and Pb–Pb) is indicated by the black boxes in the top panels [19].

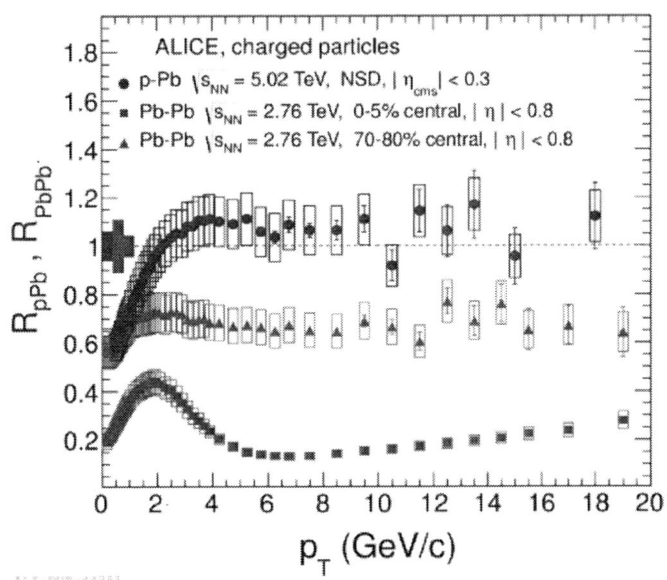

Figure 6. (Color online) The nuclear modification factor of charged particles as a function of transverse momentum in p–Pb collisions at $\sqrt{s_{NN}} = 5.02$ TeV compared to measurements in central (0-5%) and peripheral (70-80%) Pb–Pb collisions at $\sqrt{s_{NN}} = 2.76$ TeV.

In Fig. 7 [18] we compare the measurement of the nuclear modification factor for inclusive primary charged-particle ($h\pm$) production in p–Pb collisions to that in central (0–5 % centrality) Pb–Pb collisions measured by ALICE [19] and CMS [20]. The ALICE p–Pb data show no sign of nuclear matter modification of hadron production at high p_T and are fully consistent with the observation of binary collision scaling in Pb–Pb of observables which are not affected by hot QCD matter (direct photons [21] and vector bosons [22, 23]).

4 Conclusions

The spectra of identified charged particles measured by ALICE in pp and Pb–Pb collisions have been presented. Central Pb–Pb collisions spectral shapes indicate a strong increase of the radial flow velocity with $\sqrt{s_{NN}}$, which in hydrodynamic models is expected as a consequence of the increasing particle density. Strong high-p_T production suppression is measured in central Pb–Pb collisions. No suppression is observed at high-p_T in minimum bias p–Pb collisions. So, the strong suppression of high-p_T hadrons observed in central Pb–Pb collisions is not due to an initial state effect but may instead be a signature of the hot matter created in Pb–Pb collisions.

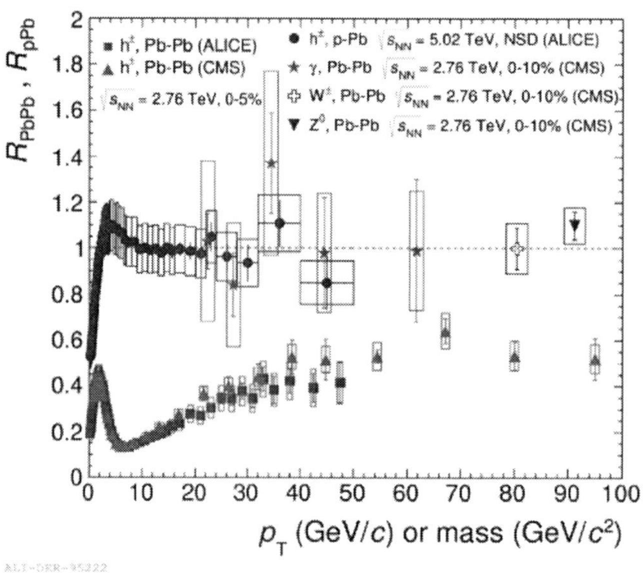

Figure 7. (Color online) Transverse momentum dependence of the nuclear modification factor R_{pPb} of charged particles ($h\pm$) measured in minimum-bias (NSD) p–Pb collisions at $\sqrt{s_{NN}} = 5.02$ TeV in comparison to data on the nuclear modification factor R_{pPb} in central Pb–Pb collisions at $\sqrt{s_{NN}} = 2.76$ TeV. The Pb–Pb data are for charged particle [19, 20], direct photon [21], Z_0 [22] and $W_\pm$ [23] production. All data are for midrapidity.

References

1. B. Abelev et al., Int. J. Mod. Phys. A **29,**143004 B.
2. Abelev et al., Phys. Rev. C **88,** 044910 (2013)
3. C. Shen, U. Heinz, P. Huovinen, and H. Song, Phys. Rev. C **84**, 044903 (2011)
4. I. Karpenko, Y. Sinyukov, and K. Werner, Phys. Rev. C **87**, 024914 (2013)
5. P. Bozek, Phys. Rev. C **85**, 034901 (2012)
6. K.Werner, I. Karpenko, M. Bleicher, T. Pierog, and S. Porteboeuf-Houssais, Phys. Rev. C **85**, 064907 (2012)
7. J. Adams et al., Nucl. Phys. A **757**, 102 (2005)
8. B. Abelev et al., Phys. Lett. B **736,** 196 (2014)
9. B. Abelev et al., Phys. Rev. C **88,** 044909 (2013)
10. B. Abelev et al., Eur. Phys. J. C **73,** 2456 (2013)
11. B. Abelev et al., Phys. Lett. B **736**, 196 (2014)

12. P. Bozek, I. Wyskiel-Piekarska, Phys. Rev. C **85**, 064915 (2012)
13. K. Werner, I. Karpenko, M. Bleicher, T. Pierog, S. Porteboeuf-Houssais, Phys. Rev. C **85,** 064907 (2012)
14. R. J. Fries, B. Muller, C. Nonaka, and S. A. Bass, Phys. Rev. Lett. **90**, 202303 (2003)
15. V. Minissale, F. Scardina and V. Greco, arXiv:1502.06213v2 [nucl-th] (2015)
16. B. Abelev et al., Phys. Rev. C **91**, 024609 (2015)
17. B. Abelev at al., Phys. Rev. Lett. **110**, 082302 (2013)
18. B. Abelev et al., Eur. Phys. J. C **74**:3054 (2014)
19. B. Abelev et al., Phys. Lett. B **720**, 52 (2013)
20. S. Chatrchyan et al., Eur. Phys. J. C **72**, 1945 (2012)
21. S. Chatrchyan et al., Phys. Lett. B **710**, 256 (2012)
22. S. Chatrchyan et al., Phys. Rev. Lett. **106**, 212301 (2011)
23. S. Chatrchyan et al., Phys. Lett. B **715**, 66 (2012)

EPJ Web of Conferences **126**, 04053 (2016) DOI: 10.1051/epjconf/201612604053
ICNFP 2015

Search for charged lepton flavour violation with the MEG and MEG II experiments

Marco Venturini[1,2,a], on behalf of the MEG collaboration

[1] *Scuola Normale Superiore - Piazza dei Cavalieri 7, 56126 Pisa, Italy*
[2] *INFN, Sezione di Pisa - Largo B. Pontecorvo 3, 56127 Pisa, Italy*

Abstract.
In the next decade Charged Lepton Flavour Violation (CLFV) experiments will play a key role in the search for new physics. Given that flavour oscillations occur among neutrinos, unmeasurably small amplitudes are expected for CLFV processes, while in many new physics scenarios such processes are strongly enhanced. A new generation of experiments looking for CLFV muon decays will be performed at the Paul Scherrer Institut (MEG II, Mu3e), at Fermilab (Mu2e) and at J-Parc (DeeMee, COMET), with sensitivities improved by one or two order of magnitudes with respect to current limits. The MEG experiment has performed the most recent improvement in the search for the CLFV decay $\mu^+ \to e^+\gamma$. With the analysis of half of the collected statistics, a new upper limit on the branching ratio has been set $\mathcal{BR}(\mu \to e\gamma) < 5.7 \times 10^{-13}$ at 90% CL, and the final result is soon to be released. A substantial improvement of MEG results requires an improvement of detector performances, in order to reject the background contributions which limit the signal sensitivity. This will be carried out by a short-term upgrade of the apparatus, MEG II. The major modifications of the experimental apparatus consist in the replacement of the current positron tracking system with a new high transparency and high granularity spectrometer and the substitution and rearrangement of some of the photosensors in the liquid-xenon photon detector. MEG II will start taking data in 2016, with a sensitivity on the $\mu^+ \to e^+\gamma$ decay of about 5×10^{-14}. Therefore as MEG performed the last improvement in the search for CLFV, MEG II will start a new era probing unexplored regions of new physics scenarios.

1 Introduction

The large number of experimental evidences for neutrino oscillations supports that lepton flavour is violated in the neutral sector and thus nonvanishing amplitudes are expected at loop level for flavour changing processes among charged leptons. However charged lepton flavour violation (CLFV) is very small in the Standard Model (SM): for instance the $\mu \to e\gamma$ decay is strongly suppressed by the fourth power of the the ratio between the neutrino and the W boson masses [1]. With an expected branching ratio of about 10^{-54}, this makes Standard-Model CLFV non detectable. The $\mu \to e\gamma$ process is strictly related to two other CLFV processes: $\mu \to eee$ and coherent muon-to-electron conversion in muonic atoms. For both processes the SM prediction [2] is below 10^{-50}. Nevertheless in many new physics

[a] e-mail: marco.venturini@pi.infn.it

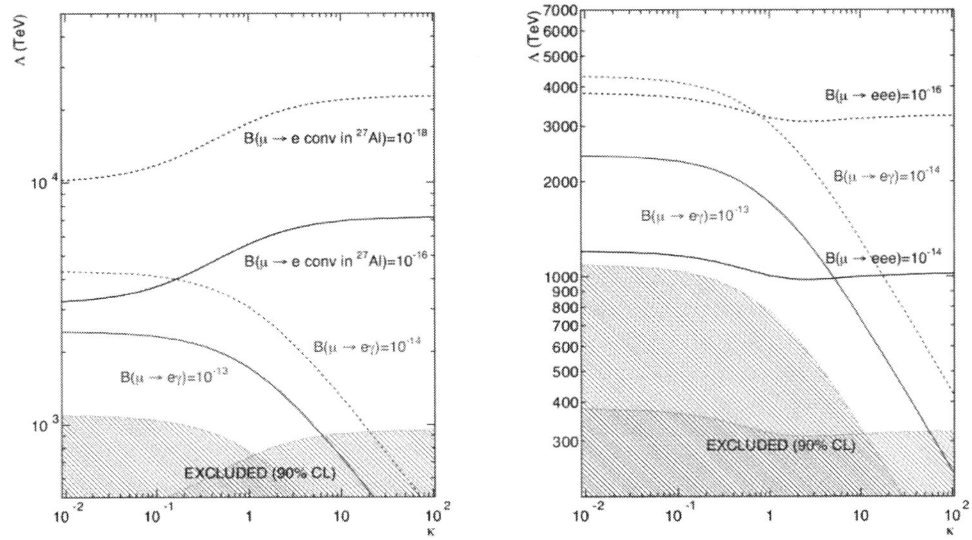

Figure 1. Sensitivities in $\mu \to e\gamma$, $\mu \to eee$ and $\mu - e$ conversion plotted in the (Λ, κ) plane [3].

scenarios much larger amplitudes are foreseen. For instance in supersymmetric models, the large mixings occurring among superparticles enhance the $\mathcal{B}(\mu \to e\gamma)$ up to $10^{-14} \div 10^{-12}$. To evaluate the deep interconnection among muon CLFV channels we can use the following effective Lagrangian [3]

$$\mathcal{L} = \frac{m_\mu}{(\kappa+1)\Lambda^2}\bar{\mu}_R\sigma_{\alpha\beta}e_L F^{\alpha\beta} + \frac{\kappa}{(\kappa+1)\Lambda^2}\bar{\mu}_R\gamma_\alpha e_L \bar{f}\gamma^\alpha f \; + \; h.c. \tag{1}$$

where κ is an adimensional parameter and Λ represents the energy scale of CLFV physics; m_μ is the muon mass, $F^{\alpha\beta}$ is the photon filed strength and f represents a general fermion that is an electron in $\mu \to eee$ or a quark in $\mu - e$ conversion. We recognise a dipole term, which powers $\mu \to e\gamma$ directly and $\mu \to eee$ and $\mu - e$ conversion at order α, and a four-fermion contact term which mediates only the non-radiative processes. Muon $g-2$ anomaly can be taken into account by adding an additional term, analogous to the first in equation (1) with $(\bar{\mu}_R, e_L)$ replaced with $(\bar{\mu}_R, \mu_L)$. Figure 1 shows the parameters space that can be explored for a given sensitivity in the three CLFV channels. Note that SM extensions do not populate the plane evenly, so different models enhance the processes in different ways.

The current upper limits on such processes are [4–6]

$$\begin{aligned} \mathcal{B}(\mu \to e\gamma) &< 5.7 \times 10^{-13}\,(90\%\,\text{C.L.}) && \text{from MEG} \\ \mathcal{B}(\mu \to eee) &< 1.0 \times 10^{-12}\,(90\%\,\text{C.L.}) && \text{from SINDRUM} \\ \mathcal{CR}(\mu\text{Au} \to e\text{Au}) &< 7 \times 10^{-13}\,(90\%\,\text{C.L.}) && \text{from SINDRUM II} \end{aligned}$$

where for muon conversion the measured quantity is the ratio to nuclear capture of muons $\mathcal{CR}$. It is clear that the search of all three CLFV processes is important in order to understand the mechanisms beyond lepton flavour violation. The MEG experiment searches for $\mu \to e\gamma$ at the Paul Scherrer Institut (PSI) in Switzerland, while other experiments in the USA, Europe and Japan are planned to search for $\mu \to eee$ (Mu3e at PSI) and $\mu - e$ conversion (Mu2e at Fermilab, Comet at J-Park).

2 The MEG experiment

2.1 The $\mu \to e\gamma$ decay

The $\mu^+ \to e^+\gamma$ process is a two-body decay: in the reference frame where the μ^+ is at rest the positron and the photon are emitted back-to-back, with the same energy ($\sim m_\mu/2$) and simultaneously. Two major backgrounds can mimic the signal signature: radiative muon decays (RMD) in which neutrinos carry little energy (physics background) and random coincidences of uncorrelated positrons and photons (accidental background). In RMDs, positron and photon observables have continuous distributions whose limits are the values expected for the signal. The number of observed events is given by the integral of the RMD differential branching ratios around the signal region of the phase space for a width given by detector resolutions. However in MEG (and in general in the most recent experiments) the RMD accounts only for the 10% of the background events. The major contribution comes from accidental detection of a positron and a photon originated from uncorrelated processes. Positrons are mostly produced via Michel decay, while photons with an energy of about $\sim m_\mu/2$ can come from RMDs, positron annihilation-in-flight and *bremsstrahlung*.

2.2 The MEG detector

The MEG apparatus, sketched in Figure 2, is tailored to take advantage of the well-defined kinematics of the $\mu \to e\gamma$ process [7]. The πE5 muon beam line delivers in MEG $3 \times 10^7\ \mu^+$/s with an average momentum of 28 MeV/c. The beam is stopped in a 205 μm polyethylene target: the emerging positrons are tracked by a magnetic spectrometer composed of a set of drift chamber modules and timing counters, immersed in a non-uniform magnetic field provided by the COBRA superconducting magnet. Photons energy, time and conversion point are measured by a liquid xenon detector.

The non-uniformity of the COBRA magnetic field reduces the dependence of the positron bending radius on the emission angle (COnstant Bending RAdius) and strongly reduces the occupancy of the drift chambers by sweeping away particles with low longitudinal momentum. The drift chamber system is composed of 16 trapezoidal modules, placed in a half circle with radial orientation. Each module contains two independent detector planes, made of an array of alternating field and sense

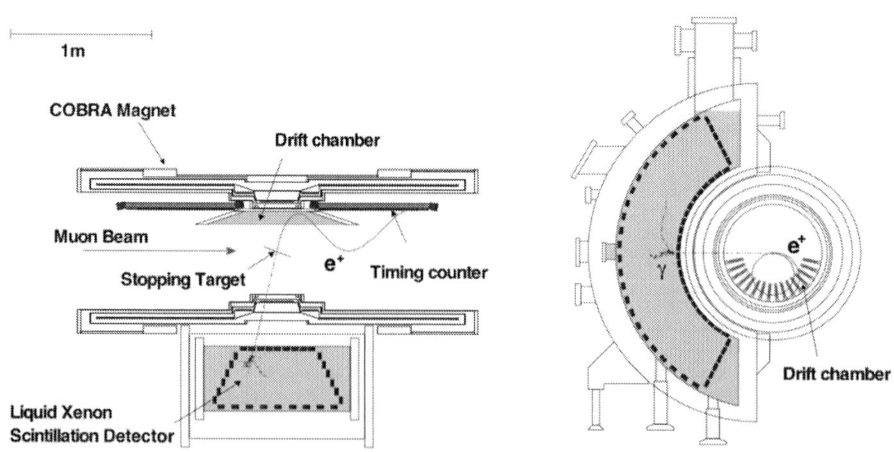

Figure 2. Schematic view of the MEG experiment with a signal event displayed.

Table 1. Measured resolutions for MEG and expected for MEG II. $(\hat{u}, \hat{v}, \hat{w})$ is the reference system in which the calorimeter is a parallelepiped with $\hat{u} \parallel \hat{z}$, $\hat{v} \parallel \hat{\varphi}$ and $\hat{w} \parallel -\hat{r}$.

Variable	MEG	MEG II
$\sigma E_\gamma(\%)$	2	1.0
γ position (mm)	$5(u,v)$, $6(w)$	$2.4(u)$, $2.2(v)$, $3.1(w)$
σP_e (keV)	306	130
e^+ angle (mrad)	$8.4(\varphi_e)$, $9.4(\vartheta_e)$	$5.3(\varphi_e)$, $4.8(\vartheta_e)$
$\sigma t_{e\gamma}$ (ps)	122	84
e^+ efficiency (%)	40	88
γ efficiency (%)	63	69
trigger efficiency(%)	99	99

wires enclosed by two cathode foils each, equipped with Vernier pads[1] for the determination of the position of the hit along the wire. The modules are filled with a 50:50 helium-ethane gas mixture, while the remaining volume inside COBRA is filled with pure helium. The single-hit resolution, *i.e.* the accuracy in the determination of the impact parameter on a wire, is about 200 μm. The resolution on the positron angle is determined from data of events where the positron makes two turns in the drift chamber system. The obtained values are $\sigma_{\vartheta_e} \sim 9.4$ mrad and $\sigma_{\varphi_e} \sim 8.4$ mrad. The energy resolution of the drift chamber is 330 keV, measured by fitting the measured Michel spectrum.

Two identical hodoscopes (timing counters) are placed at the two sides of the drift chambers along the beam axis for the determination of positron time. Each hodoscope consists of a layer of scintillating fibres, read out by avalanche photodiodes, placed on the top of an array of 15 scintillating bars, which are coupled to fine-mesh fast photomultipliers. Bars and fibres are oriented perpendicularly to each other, with the bars being parallel to the beam axis. The timing resolution of the detector is about 70 ps, determined through tracks hitting on multiple bars.

The photon detector is a homogeneous calorimeter with 900 litres of liquid xenon viewed by 846 vacuum-ultra-violet sensitive photomultipliers, and placed in a cryostat that keeps xenon temperature stable at about 165 K. The small radiation length ($X_0 = 2.7$ cm) and the high light–yield (about 40 scintillation photons per keV of deposited energy) confer to liquid xenon high detection efficiency, small leakage for electromagnetic showers and good energy resolution. In addition, all the three main mechanisms involved in scintillation have time constants below 50 ns. Such a fast response results in reduced pile–up probability and in a superior timing resolution (better than 100 ps). Energy and timing resolution of the calorimeter are extracted from the calibrations. Energy resolution depends on the position of γ-ray conversion: it is about 2% for conversions within 3 cm from the inner face and 1.6% for deeper conversions. The acceptance for photons emerging from the target is about 11 %, and this sets the acceptance of the apparatus.

For the measurement and the monitor of the apparatus performance a rich set of calibrations has been developed. They involve both single detectors (energy scale, resolution variations...) and their intercalibration. They are performed with not only the muon beam but also with a positron beam, cosmic rays and gamma lines resulting from nuclear interactions induced by protons from a Cockcroft-Walton accelerator. A special calibration is performed with charge exchange reactions $\pi^- p \to \pi^0(\gamma\gamma)n$ from a π^- beam (obtained changing the settings of the $\pi E5$ beam line) stopped in a liquid hydrogen target which temporarily replaces the polyethylene one. Photons from π^0 decays are used to calibrate

[1] each cathode foil has a periodic double-wedge structure with the two half-foils (Vernier pads) having independent readout, so that the measurement of the asymmetry in charge collection on the two pads conveys the information on the z-coordinate. In the two cathode foils enclosing a plane of the detector, the patterns are displaced by a quarter-period.

DOI: 10.1051/epjconf/201612604053

the response of the xenon detector at an energy (55 MeV) close to that of signal photons. The overall resolutions of the MEG detector on the kinematic variables are reported in Table 1.

In MEG the trigger is implemented in a cascaded custom made VME boards, performing a basic event reconstruction with signals from xenon detector and timing counters, digitized by a 100-MHz ADC. Signals from the detectors are digitized by a waveform digitizer chip, the Domino Ring Sampler 4 (DRS4) [8]. This is necessary to exploit the detectors timing resolutions and to reduce the pile up of events.

2.3 The analysis

For the measurement of the branching ratio of the process a likelihood analysis is performed [4]. The distribution of events is studied in terms of 5-dimensional vectors $\vec{x} = (E_\gamma, E_e, t_{e\gamma}, \vartheta_{e\gamma}, \varphi_{e\gamma})$. The analysis follows a blinding procedure: events falling in a blinding window containing the signal region (48 MeV $< E_\gamma <$ 58 MeV and $|t_{e\gamma}| < 1$ ns) are written in separate files which are not opened until the analysis procedure is completely defined.

The first step of the analysis is the determination of the Probability Density Functions (PDFs) for signal $S(\vec{x})$, radiative backgrounds $R(\vec{x})$ and accidental background $B(\vec{x})$, using both calibration data and distributions in the sidebands, which are the regions of the $(t_{e\gamma}, E_\gamma)$ plane next to the blinding box. Once the PDFs are determined, the number of signal, RMD and accidental events ($N_{\rm sig}$, $N_{\rm RMD}$, $N_{\rm ACC}$) is extracted by maximizing the following likelihood function:

$$\mathcal{L}(N_{\rm sig}, N_{\rm RMD}, N_{\rm ACC}) = \frac{e^{-N}}{N_{\rm obs}!} \prod_{i=1}^{N_{\rm obs}} \left[N_{\rm sig}\, S(\vec{x}_i) + N_{\rm RMD}\, R(\vec{x}_i) + NathrmACC\; B(\vec{x}_i)\right] \times$$
$$\times \exp\left[-\frac{(N_{\rm RMD} - \langle N_{\rm RMD}\rangle)^2}{2\sigma^2_{\rm RMD}}\right] \exp\left[-\frac{(N_{\rm ACC} - \langle N_{\rm ACC}\rangle)^2}{2\sigma^2_{\rm ACC}}\right]$$

where $N_{\rm obs}$ is the number of events detected in the signal window and $N = N_{\rm sig} + N_{\rm RMD} + N_{\rm ACC}$. Estimates of the number of background events in the signal region $\langle N_{\rm RMD,ACC}\rangle$ with corresponding variances $\sigma^2_{\rm RMD,ACC}$ are obtained from the sidebands. This additional measurement is included in the analysis by means of the two Gaussian terms in the likelihood function. Figure 3 shows the likelihood (blue solid line) fit projected on the five kinematic variables. Individual best fits for signal (green), RMD (red) and accidental background (magenta) are reported.

For the determination of the branching ratio, the number of measured events is normalized to the numbers of measured Michel positrons and RMDs events. The combination of these two independent methods leads to a 4% uncertainty in the branching ratio estimate. The computation of the confidence interval is based on frequentist method with a profile likelihood-ratio ordering, where the number of RMD events and the number of accidental events are treated as nuisance parameters.

With the analysis of data collected in the period 2009-2011, an upper limit on the branching ratio of 5.7×10^{-13} was obtained at 90% of confidence level (CL). The analysis of the full set of data collected in the period 2009-2013, with a double statistics, is close to completion, with a sensitivity of 4×10^{-13} on the branching ratio at 90% CL. However a substantial improvement of MEG results requires an improvement of detector performances, in order to reject the background contributions which limit the signal sensitivity. This will be carried out by a short-term upgrade of the apparatus, MEG II.

DOI: 10.1051/epjconf/201612604053

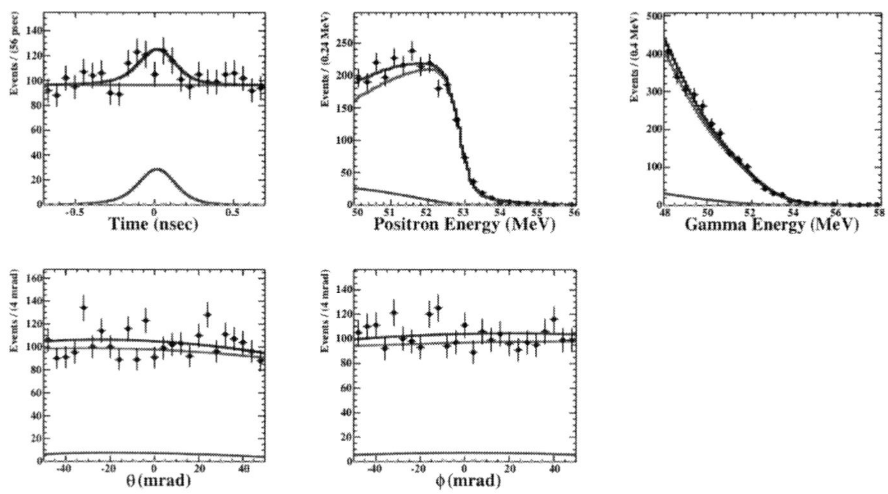

Figure 3. Likelihood fit result for 2009–2011 combined dataset. Data (black points) are shown with the projections of the 5D fit (blue line) on each of the kinematic variables. Individual best fits for signal (green), RMD (red) and accidental background (magenta) are also shown.

3 The MEG upgrade

The major modifications of the MEG apparatus [9] are shown in Figure 4: the drift chamber and the timing counters are replaced while the calorimeter undergoes substitutions and rearrangements of the photosensors.

The new positron spectrometer consists of a drift chamber and two pixelated timing counters. In the new configuration positrons traverse less material along their path, and the capabilities of matching the information from the two detectors are improved. The new tracker is a single-volume cylindrical drift chamber, composed of ten layers of approximately squared drift cells with side ranging from 6 mm (in the innermost layer) to 9 mm (in the outermost one) at alternating stereo angles θ_s ranging from $\pm 6°$ to $\pm 8°$. High transparency is achieved using an ultra-low mass gas mixture with helium and isobutane 85:15. This is fundamental for keeping the Coulomb multiple scattering at a minimum for the low-momentum positrons from μ^+ decays. Gold-plated tungsten wires with 20 μm diameter are used as anodes, while cathode and guard wires are 40-50 μm silver-plated aluminium wires. The resulting radiation length per track turn is about $1.6 \times 10^{-3} X_0$. Due to the large number of hits, a rough determination of each hit's longitudinal coordinate conveys precious information for pattern recognition strategies. For this sake, the readout of wires on both ends will permit to exploit charge division and time propagation difference. The high transparency of the chamber has the drawback of a poor ionisation statistics, which results in a bias and a worsening of resolution in the estimate of the impact parameter in the drift cells. For a clean measurement of the single-hit resolution several drift chamber prototypes were tested in a cosmic ray facility set up at INFN Sezione di Pisa [10]. The measured single-hit resolution is $110\,\mu$m, averaged over a large range of angles and of impact parameters. The longitudinal coordinate of hits will be determined by exploiting the stereo angle, with a resolution $\sigma_z = \sigma_r / \sin\theta_s \approx 1$ mm, about a factor 100 better than the $\sim$ 10 cm resolution of the z reconstructed with charge division and relative time.

The new pixelated timing counters consist of 2×256 scintillator tiles (120×50×5 mm^3). The pixels are read out at two sides by 6+6 Silicon PhotoMultipliers (SiPMs) mounted in series connection on a

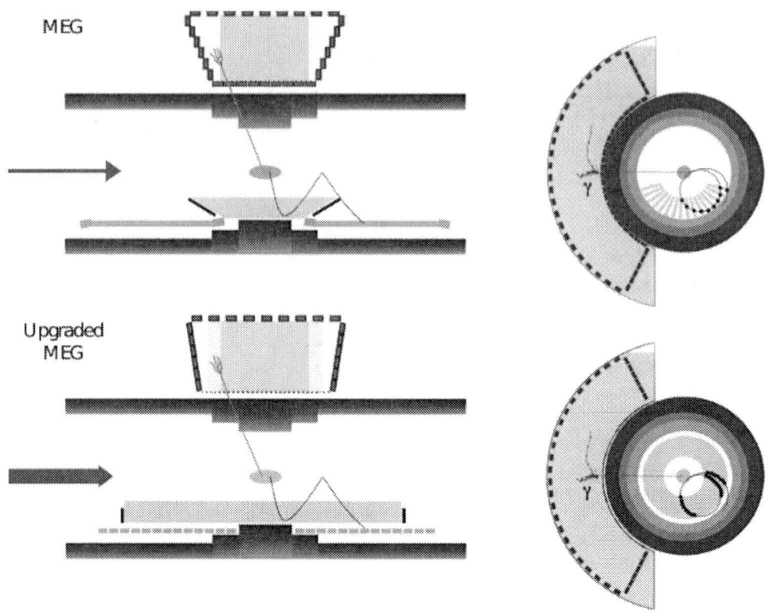

Figure 4. Improvements of the MEG detector. The upgraded apparatus consists of a new tracker with higher resolution and granularity and an improved matching with a new pixelated timing counter. The xenon calorimeter gains a larger acceptance and a better scintillation light collection. See full text for more details.

Printed Circuit Board acting as a frame: this solution keeps the material budget along the positron path to a minimum and maximizes the detector coverage. The segmentation of the timing counters brings an intrinsic potential in improving the timing resolution, coming from the possibility of averaging the positron hit time over the multiple hit pixels ($\sigma \sim 1/\sqrt{N}$). The average number of tiles hit by a positron is $\simeq 8$: from a single-pixel time resolution of $\simeq 90$ ps we reduce to ~ 35 ps, nearly two times better than previous timing counter.

The main issue of MEG xenon calorimeter is the dependence of resolutions on the depth of the γ-conversion. This is due to the granularity of the 2" PMTs (with an area of 81 cm^2) on the front face distributed in a mesh of 6.2 cm side. In MEG II the PMTs in the entrance face will be replaced by smaller photosensors, 12×12 mm^2 SiPMs. Also the imaging power is thus greatly increased and pile up events can be better identified. The layout of the lateral faces will be modified as well in order to avoid shadow areas, which result in a reduced acceptance. The new structure is visible in Figure 4, where the wider acceptance region is highlighted.

In Table 1 the resolutions expected for MEG II are reported. Better detector resolutions will permit a higher beam rate, about 7×10^7 μ/s. In three years of data taking MEG II expected sensitivity on the branching ratio of $\mu \to e\gamma$ is about 5×10^{-14}.

4 Conclusion

The final MEG result is soon to be released, with a sensitivity of about 4×10^{-13} on the branching ratio. At the same time MEG II is performing an engineering run with a drift chamber mock up and a sector of the timing counter. The mock up drift chamber has the same structure of the final one, but only one layer of anode wires.

The full apparatus is thus expected to be on beam in Autumn 2016. In three years of data taking the expected sensitivity is 5×10^{-14}, about ten times better than the current limit set by MEG.

References

[1] S. M. Bilenky, S. Petcov, and B. Pontecorvo, *Phys.Lett.*, vol. B67, p. 309, 1977.

[2] W. J. Marciano, T. Mori, and J. M. Roney, *Ann. Rev. Nucl. Part. Sci.*, vol. 58, pp. 315–341, 2008.

[3] A. de Gouvêa, P. Vogel, *Progress in Particle and Nuclear Physics*, vol. 71, pp. 75-92, 2013.

[4] J. Adam and et al., *Phys. Rev. Lett.*, vol. 110, p. 201801, May 2013.

[5] U. Bellgardt *et al.*, *Nucl. Phys.*, vol. B299, p. 1, 1988.

[6] W. H. Bertl *et al.*, *Eur. Phys. J.*, vol. C47, pp. 337–346, 2006.

[7] J. Adam *et al.*, *Eur.Phys.J.*, vol. C73, p. 2365, 2013.

[8] S. Ritt, R. Dinapoli, and U. Hartmann, *Nucl. Instrum. Meth.*, vol. A623, pp. 486–488, 2010.

[9] A. M. Baldini and et al., *ArXiv e-prints, 1301.7225*, Jan. 2013.

[10] L. Galli, E. Baracchini, S. Bettarini, F. Bosi, E. Cavallaro, S. Dussoni, M. Minuti, F. Morsani, D. Nicolo, G. Signorelli, F. Tenchini, M. Venturini, and J. Walsh, *Nuclear Science, IEEE Transactions on*, vol. 62, no. 1, pp. 395–402, 2015.

EPJ Web of Conferences **126**, 04054 (2016) DOI: 10.1051/epjconf/201612604054
ICNFP 2015

Exotica in CMS

Claudia-Elisabeth Wulz[1,a], on behalf of the CMS Collaboration

[1] *Institute of High Energy Physics of the Austrian Academy of Sciences, Nikolsdorfergasse 18, A-1050 Vienna, Austria*

Abstract. Selected results on exotica searches with the CMS detector are presented. The main topics are dark matter, boosted objects, long-lived particles and classic narrow resonance searches. Most of the analyses were performed with data recorded at at centre-of-mass energy of 8 TeV, but first results obtained at 13 TeV are also shown.

1 Introduction

At the Large Hadron Collider (LHC) a Higgs boson compatible with the minimal standard model has been discovered. Extensions of this model predict additional Higgs bosons and a wealth of other exotic particles. Searches for new phenomena, however, have been inconclusive so far. The increase of the LHC's centre-of-mass energy from 8 to 13 TeV has opened up new areas of phase space to find or to exclude new particles. In this report we will concentrate on a few captivating recent topics in the field of exotica, including dark matter, boosted objects, long-lived particles and classic narrow resonance searches. Results from proton-proton data recorded with the CMS experiment [1] are presented. An overview of relevant current beyond-standard-model physics results is available on dedicated web pages [2], in particular under "Exotica" and "Beyond 2 Generations".

2 Dark matter searches

One of the biggest puzzles today is the nature of dark matter. Beside axions, weakly interacting massive particles (WIMP) are well-motivated candidates. A large range of experiments is currently searching for them in space, underground, or at the surface of the earth. Direct detection is based on the scattering of WIMP dark matter off atomic nuclei, whereas indirect detection makes use of secondary radiation emitted in their pair annihilation. At the LHC one can also search for dark matter particles emerging in collisions of protons. The methods are based on the detection of particles produced in association with the dark matter. They are are complementary to the ones used in direct or indirect searches and are particularly well suited to low-mass WIMPs and spin-dependent couplings. Two scenarios for producing dark matter particles (DM, χ) are possible (Fig. 1). The first one assumes a contact interaction, in the framework of an effective field theory. In the second, a mediator, light enough to be produced at the LHC, is exchanged. Since dark matter is known to interact only weakly with standard model particles it can only be detected indirectly, through an imbalance of the total momentum of all particles reconstructed in the plane transverse to the LHC beams. Its magnitude is often

[a] e-mail: claudia.wulz@cern.ch

Figure 1. Feynman diagrams for the pair production of dark matter particles for the case of a contact interaction (left) and the exchange of a mediator (right).

denoted as missing transverse energy (E_T^{miss}). We focus on the search for single objects (mono-jet, mono-Z, mono-photon) as well as dijets, accompanied by E_T^{miss}.

In the mono-jet search [3] the event selection requires either $E_T^{miss} > 120$ GeV calculated from calorimeter information, or a jet reconstructed with the particle-flow technique [4, 5] with transverse momentum $p_T > 80$ GeV, and $E_T^{miss} > 105$ GeV, where E_T^{miss} is also reconstructed with the particle-flow algorithm, but excludes muons. As signal events typically contain jets from initial-state radiation, a second jet is allowed only if its distance in azimuth (φ) from the jet with the highest p_T is within 2.5 radians, thus suppressing the dijet QCD background. Events with more than two jets with $p_T >$ 30 GeV and pseudorapidity $|\eta| < 4.5$ are discarded, thereby significantly reducing backgrounds from $t\bar{t}$ and multijet events. The dominant backgrounds come from $Z(\rightarrow \nu\nu)$ + jets and $W(\rightarrow \ell\nu)$ + jets events, and are estimated from data samples of $Z(\rightarrow \mu\mu)$ and $W(\rightarrow \mu\nu)$ events, respectively. We studied scalar, vector, axial-vector and tensor interactions for a Dirac DM particle. Fig. 2 (left) shows CMS upper limits for the DM-nucleon cross sections for spin-independent couplings as a function of the mass M_χ, derived from the mono-jet search. Curves from other experiments are overlaid, as well as an area with a possible signal seen by CDMS [6]. Should an effective theory not be valid at LHC energies, we considered the case of an s-channel mediator with vector interactions. The resonant enhancement in the production cross section, once the mass of the mediator (M) is within the kinematic range and can be produced on-shell, can be clearly seen in Fig. 2 (centre), which shows results for two dark matter masses and different mediator widths Γ.

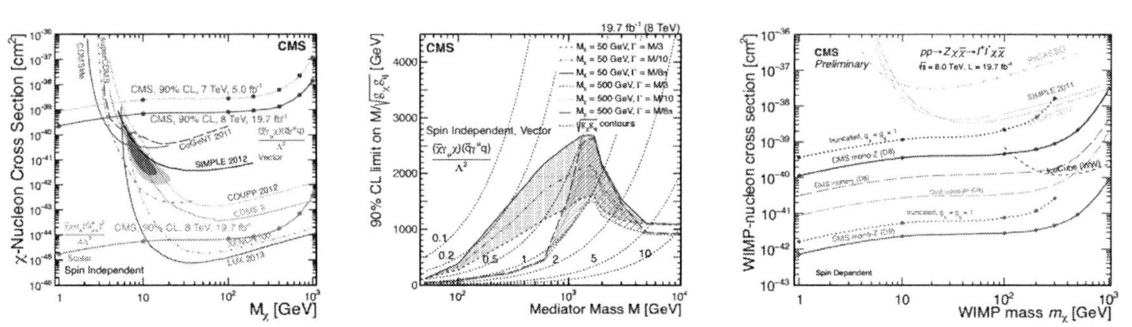

Figure 2. Upper limits on the χ-nucleon cross section, at 90% confidence level, plotted against the dark matter particle mass and compared with previous results, for spin-independent interactions and the mono-jet channel (left), and for spin-dependent interactions and the mono-Z channel (right); Observed limits on the mediator mass divided by roots of DM and quark couplings, M / $\sqrt{g_\chi g_q}$, obtained from the mono-jet analysis (centre).

Other searches for DM have been performed in the mono-Z [7] and mono-photon [8] channels. Fig. 2 (right) shows CMS limits for spin-dependent interactions obtained in these channels, together with mono-jet results and those of other experiments. D8 and D9 denote specific axial-vector and tensor couplings, respectively.

An innovative DM search using the razor technique in the dijet channel has also been performed, which in particular sheds light on the range of validity of effective field theories [9]. Razor variables, computed from the jet transverse momenta and E_T^{miss}, quantify the balance of the jet momenta. They have been developed in particular to suppress backgrounds from QCD. It has been shown that these variables are also sensitive to direct DM production [10]. The event selection is looser than that used in the mono-jet searches. However, a comparable sensitivity is achieved, as can be seen from Fig. 3, which shows spin-independent (left) and spin-dependent (right) upper limits on the DM-nucleon scattering cross section obtained with the dijet razor analysis, together with results from other experiments.

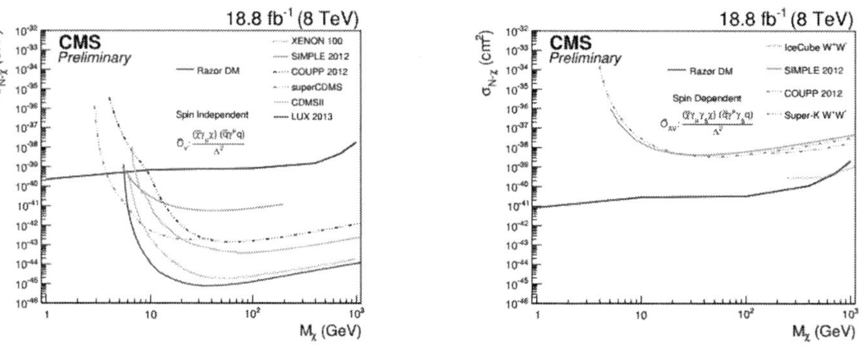

Figure 3. Upper limits at 90% confidence level on the DM-nucleon scattering cross section as a function of the DM mass in the case of spin-independent vector- (left) and spin-dependent axial-vector currents (right).

As mentioned before, DM production below a cut-off energy scale Λ can be described as a contact interaction between two quarks and two DM particles. In the case of s-channel production through a heavy mediator, Λ is identified with M/g_{eff}, where $g_{\text{eff}} = \sqrt{g_\chi g_q}$ is an effective coupling, determined by the couplings of the mediator to quark and DM fields. Lower limits on the cut-off scale are shown in Fig. 4. Following studies presented in Refs. [11–13], we use the variable R_Λ to define the range of validity of an effective field theory. R_Λ was computed as a function of the effective coupling g_{eff} in the range $0 < g_{\text{eff}} < 4\pi$. The contours corresponding to $R_\Lambda = 80\%$ for different values of g_{eff} are also outlined in Fig. 4. For values of $g_{\text{eff}} \gtrsim 2$, the limits set by the analysis lie above the $R_\Lambda = 80\%$ contours.

3 Long-lived particles

Long-lived particles are predicted in many beyond-standard-model theories such as in gauge- or anomaly-mediated supersymmetry (SUSY) breaking scenarios, in R-parity violating SUSY and split SUSY models, or in hidden valley scenarios implying a dark sector. Signatures of long-lived particles could be quite unusual. They could imply displaced objects, disappearing or kinked tracks, and delayed objects. Displaced objects would manifest themselves through their origin from a vertex displaced by O(10) mm from the primary vertex. Disappearing or kinked tracks would appear after a flight distance of O(100) mm. Delayed objects refer to very long-lived or stable particles, which

DOI: 10.1051/epjconf/201612604054

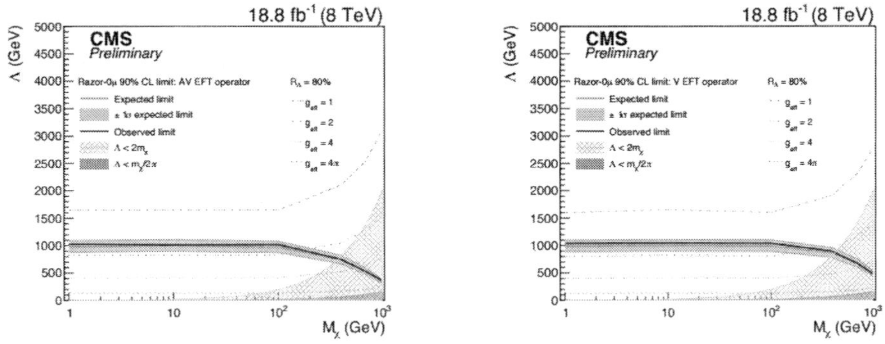

Figure 4. Lower limits at 90% CL on the cutoff scale Λ as a function of the DM mass in the case of axial-vector (left), and vector currents (right). The validity of the effective field theory for different values of the effective coupling g_{eff} is quantified by R_Λ = 80% contours.

penetrate the detector and have decay lengths in excess of O(1) m. Other possible signatures, such as lepton jets, are not discussed in this report. We focus on neutral particles decaying to photons [14] or muons [15], and on a reinterpretation of results on heavy charged particles [16].

In models with gauge-mediated SUSY breaking (GMSB) the long-lived lightest neutralino ($\tilde{\chi}^0_1$) decays to a gravitino ($\tilde{G}$) and a photon (γ), represented in Fig. 5 by two possible diagrams of squark and gluino pair-production processes that result in a diphoton final state.

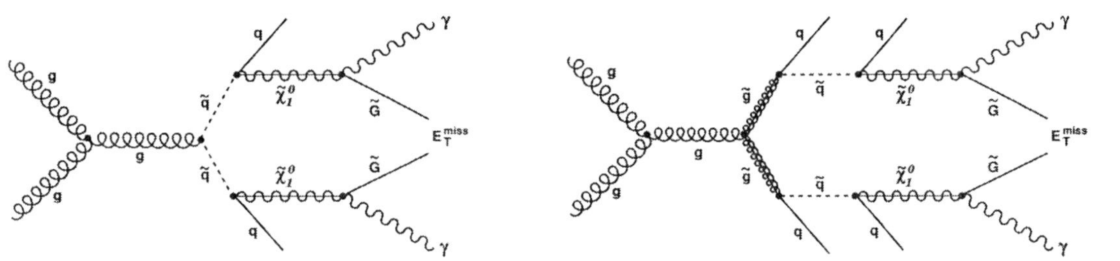

Figure 5. Feynman diagrams for the squark (left) and gluino (right) pair production processes of long-lived lightest neutralinos decaying to gravitinos and photons, resulting in diphoton final states.

The event selection for this decay channel required two photons with at least one converted to an e^+e^- pair, two or more jets, and $E_T{}^{\text{miss}}$. The latter will exhibit a long tail if it originates from gravitons, whereas it will quickly fall off for standard model and instrumental backgrounds. Requiring at least one conversion allows reconstructing the photon direction from the momentum of the electron/positron track pair. By extrapolating the photon direction back to the beam axis the impact parameter of the photon $|d_{xy}|$ can be measured as shown in Fig. 6 (left). A scenario with neutralino decay lengths $c\tau$ between 0.4 and 100 cm has been considered. We obtain an exclusion region as a function of both the neutralino mass and its mean lifetime in the context of the SPS8 [17] GMSB scenario, shown in Fig. 6 (right). The mass of the lightest neutralino is restricted to values $m(\tilde{\chi}^0_1)$ > 250 GeV, for a mean neutralino decay length of 10 cm. It can also be seen that the mean decay length must be greater than 11 cm, for $m(\tilde{\chi}^0_1)$ = 160 GeV. We have obtained roughly a factor of ten improvement with respect to the 7 TeV results [18].

DOI: 10.1051/epjconf/201612604054

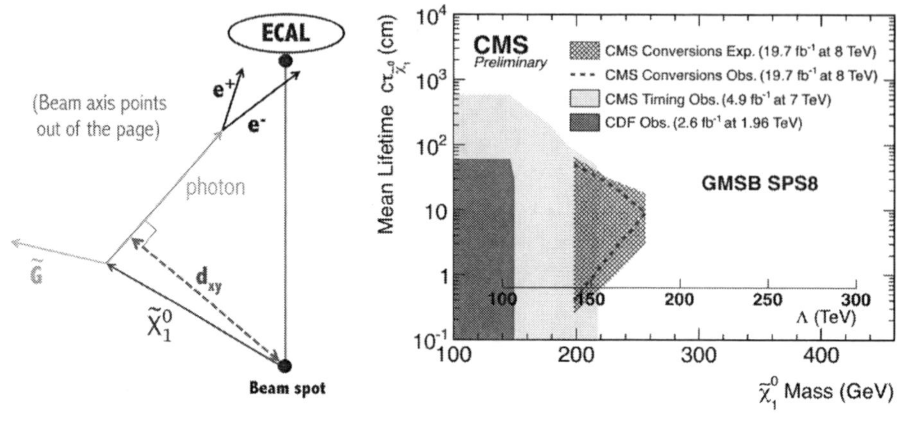

Figure 6. $\tilde{\chi}_1^0$ decay to γ + $\tilde{G}$ in the plane transverse to the LHC beams, with the photon converting to an e^+e^- pair and the subsequent reconstruction of the transverse impact parameter in the electromagnetic calorimeter (left); Exclusion plot in the plane defined by the $\tilde{\chi}_1^0$ mass and its mean lifetime in the context of the SPS8 GMSB scenario. The scale Λ is also shown.

Two models have been considered in a study of long-lived particles decaying to muons [15], with a topology requiring two muons originating from a displaced vertex, detected in the muon chambers only and not in the inner tracker of CMS. The muon reconstruction and selection efficiencies evaluated by Monte Carlo simulations have been cross-checked with cosmics data. Limits have been derived for a non-standard-model Higgs boson H decaying to two long-lived spin-0 bosons X), and for a squark pair, where each squark decays to a quark and a long-lived neutralino, which in turn decays to two muons and a neutrino in an R-parity-violating scenario. The study is orthogonal to a previous one that used only the tracker [19]. It improves sensitivity to particles that are particularly long-lived. The two analyses have been combined. The example plots of Fig. 7 show limits on the cross sections multiplied by the branching fractions into dimuons for the long-lived particles in the two models, with various combinations of masses for X bosons and squark/neutralino mass ratios.

A hidden-valley benchmark model has been assumed in a search for displaced jets [20]. Decays of particles from the hidden valley occur via a hidden-sector mediator as they do not couple directly to standard-model particles. A long-lived, spinless, neutral exotic particle X^0 which decays to $q\bar{q}$ is assumed. In this model, the X^0 is pair-produced in the decay of a non-standard-model Higgs boson, i.e. $H^0 \to 2X^0$, $X^0 \to q\bar{q}$, where the Higgs boson is produced through gluon-gluon fusion. This model predicts up to two displaced dijet vertices within the volume of the CMS tracker per event. The topology requirement was therefore two jets originating from the same displaced vertex. An event with this topology is shown in Fig.8 (left). The yellow cones represent the jet pair. The displaced vertex, with a transverse distance from the primary vertex of 5 cm, is defined by the five black tracks belonging to Jet 0 and one black track belonging to Jet 1. Fig.8 (right) illustrates expected and observed upper limits for the cross section times branching fraction for $H^0 \to 2X^0$ with $X^0 \to q\bar{q}$, at 95% confidence level, for a Higgs boson mass of 1000 GeV and two X^0 boson masses, 150 and 350 GeV, as a function of decay length.

Examples of so-called heavy stable charged particles (HSCP) are long-lived charginos, sleptons, or R-hadrons, composite colourless states made of a squark or gluino and light quarks or gluons. Search strategies exploit characteristic signatures in the detector, according to the nature of the HSCP. The energy available at the LHC is such that particles with masses greater than 100 GeV and lifetimes

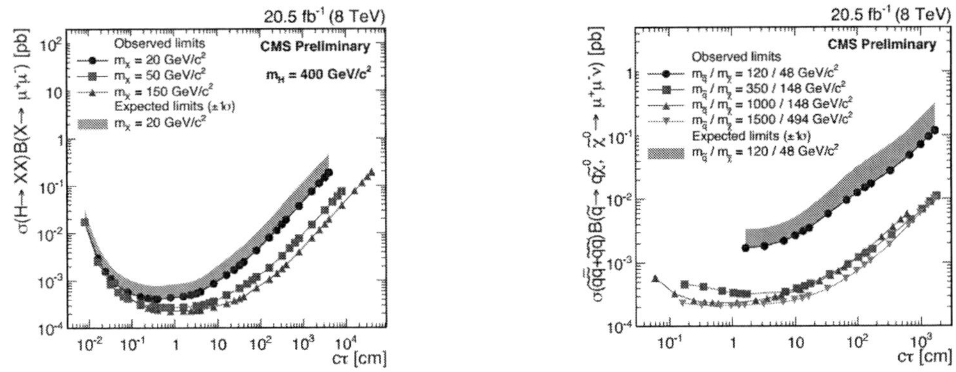

Figure 7. 95% confidence level upper limits on $\sigma(\mathrm{H} \to \mathrm{XX})\, \mathcal{B}(\mathrm{X} \to \mu^+\mu^-)$ for $\mathrm{m_H} = 400$ GeV with various X mass points (left). The shaded band shows the $\pm 1\sigma$ range of variation of the expected limits for the case of a 20 GeV X boson mass. Corresponding bands for the other X boson masses show a similar level of agreement and are omitted for clarity; 95% CL upper limits on $\sigma(\tilde{q}\bar{\tilde{q}} + \tilde{q}\tilde{q})\, \mathcal{B}(\tilde{q} \to q\tilde{\chi}^0, \tilde{\chi}^0 \to \mu\mu\nu)$ as a function of the neutralino lifetime. Shaded bands show the $\pm 1\sigma$ range of variation of the expected limits for the case of a 120 GeV squark and a 48 GeV neutralino mass. Corresponding bands for the other squark and neutralino masses show a similar level of agreement and are omitted for clarity.

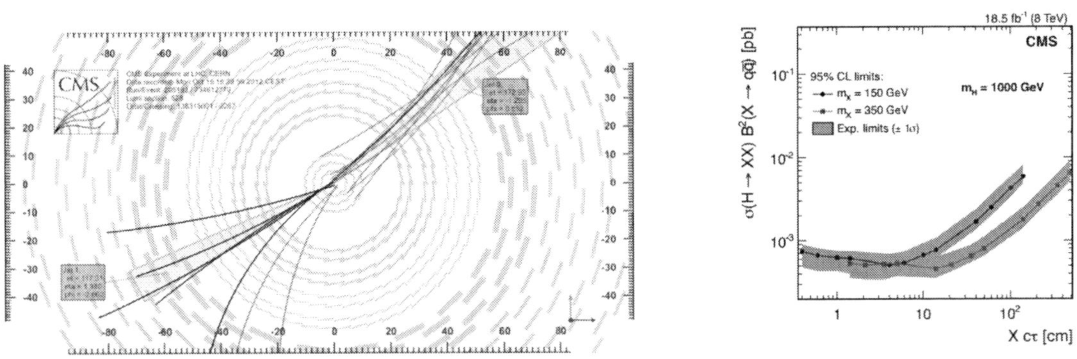

Figure 8. Candidate event with displaced jets (left); Expected and observed upper limits for the cross section times branching fraction for $\mathrm{H}^0 \to 2\mathrm{X}^0$, $\mathrm{X}^0 \to q\bar{q}$ (right).

greater than O(1) ns could be observed as high-momentum tracks with anomalously large energy losses through ionization (dE/dx). These particles could also be highly penetrating such that the fraction reaching the muon chambers would be sizable. The muon system could therefore be used to help with identification and with the measurement of the long time-of-flight of the particles. Nuclear interactions may also lead to charge exchange, such that the HSCP becomes neutral and can therefore no longer be detected in parts of the tracker or the muon chambers.

We concentrate in this report on a reinterpretation of previous results on long-lived chargino production [21], in the context of the phenomenological minimal supersymmetric standard model (pMSSM) and an anomaly-mediated SUSY breaking model (AMSB) [16]. In the pMSSM, the parameter sub-space for particle masses up to 3 TeV and charginos ($\tilde{\chi}^{\pm}$) with a mean proper decay length greater than 50 cm was considered. Fig. 9 (left) shows the fraction of parameter points excluded as a function of the chargino lifetime. The fraction of excluded model points with a chargino lifetime longer than 1000 ns (10 ns) is 100.0% (95.9%). Although these values depend on the random point sampling in the 19-dimensional pMSSM parameter space, it is remarkable that a high fraction of the points predicting long-lived charginos are excluded. The results represent the first constraints on the pMSSM obtained at the LHC.

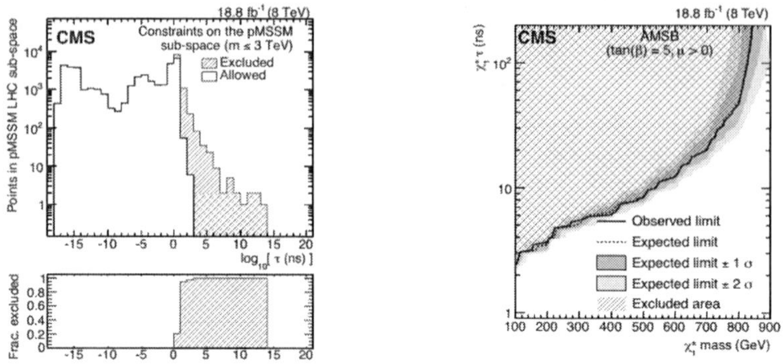

Figure 9. Number of pMSSM parameter points excluded or allowed as a function of the chargino lifetime (left); Observed and expected excluded and allowed regions in the chargino mass and lifetime parameter space in the context of the AMSB model with $\tan\beta = 5$ and $\mu > 0$ (right).

In the AMSB model, the small mass difference between the lightest chargino $\tilde{\chi}_1^{\pm}$and the lightest neutralino $\tilde{\chi}_1^0$ can lead to a long chargino lifetime, expected to be of the order of a nanosecond or longer. It is determined by the mass splitting between the two particles. Charginos with lifetimes $\gtrsim$ 100 ns (3 ns) and masses up to about 800 GeV (100 GeV) are excluded at 95% confidence level, as can be seen from Fig. 9 (right).

4 Boosted objects

Particles with a high Lorentz boost have been recognized as tools for discovery already in 1994 [22]. Only recently, however, have the reconstruction techniques become sophisticated enough to exploit them. If a boosted particle disintegrates, the decay products will be emitted close together in space. Jets, in particular, may overlap to a point where they merge into a single "fat" jet. The internal structure of these jets is a key signature to identify boosted objects among the abundant jet production at the LHC. Many searches use a variety of recently proposed substructure observables and techniques such as jet grooming to remove noise and pile-up, or pruning to remove soft, large-angle particles, as well as tagging of b and t quarks, or W and Z in jets. These new analysis methods led to improvements in sensitivity for high-mass objects up to a factor of about ten.

In the following we describe searches for boosted objects X decaying to WH, ZH or VV, where V denotes a vector boson W or Z. We have performed the first search for a resonance decaying to WH, with $W \to \ell\nu$ and $H \to b\bar{b}$ [23]. The search strategy is close to the one for high-mass WW resonances, additional b tagging requirements. The main backgrounds come from W + jets, WW/ZZ and $t\bar{t}$. A narrow resonance W′ is assumed. Fig. 10 (left) shows expected and observed upper limits on the product of the W′ production cross section and the branching fraction of $W' \to WH \to \mu\nu b\bar{b}$ at 95% confidence level. The cross section for the production of a W′ in the Little Higgs model [24] and a recently proposed model of a Heavy Vector Triplet [25] HVT, in a specific incarnation called scenario B, multiplied by its branching fraction for the relevant process, is overlaid. The number of events as a function of the reconstructed WH mass for the electron channel and the associated backgrounds are depicted in Fig. 10 (centre). As can also been seen in the right plot of the figure, which shows the p-values for the combined electron/muon channels as a function of the W′ mass, a slight but still

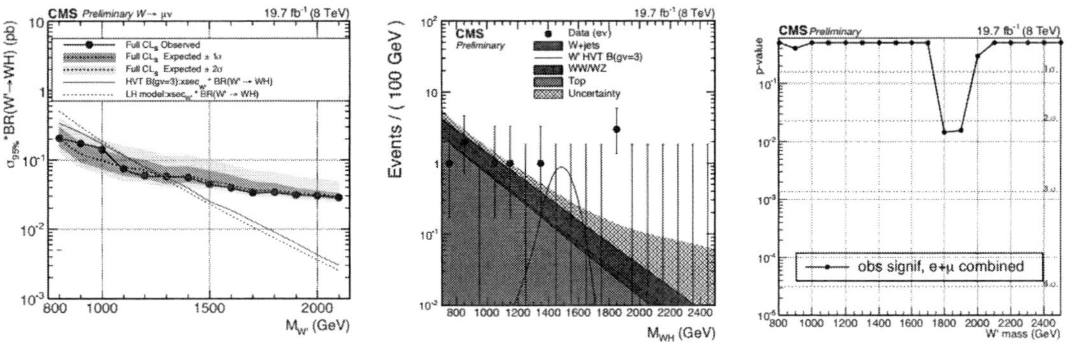

Figure 10. Observed and expected 95% confidence level upper limits on the product of the W′ production cross section and the branching fraction of W→ WH for the muon channel (left); Distributions in M_{WH} for data and expected backgrounds for the electron channel (centre); Local significance of the combined electron and muon data probing a narrow WH resonance (right).

insignificant excess around 1900 GeV is visible. Overall, in the context of the Little Higgs model, we set a lower limit on the W′ mass of 1.4 TeV. In the HVT model with scenario B we set a lower limit on the W′ mass of 1.5 TeV when combining the electron and muon channels.

Another novel analysis using τ pairs in the boosted regime for the process X → ZH → qq$\tau\tau$ was performed [26]. Six τ decay channels were considered, the purely leptonic ($\tau_e\tau_e$, $\tau_e\tau_\mu$, $\tau_\mu\tau_\mu$), the semi-leptonic ($\tau_e\tau_h$, $\tau_\mu\tau_h$) and the all-hadronic ($\tau_h\tau_h$), where the subscripts denote the characterising decay products of the τ. The main backgrounds differ according to the decay categories. For the leptonic ones they are Z/γ + jets, for the semi-leptonic $t\bar{t}$ and W + jets, and for the hadronic one they come from QCD. The on-line event selection required a single jet reconstructed by the anti-k_T algorithm with a distance parameter of 0.5 and $p_T > 320$ GeV, or $H_T > 650$ GeV, where H_T is the scalar sum of the transverse energy of all the jets in the event. Fig. 11 (left) shows the ZH mass distribution together with background estimates in the $\tau_\mu\tau_h$ category. A signal for a Z′ with mass $M_{Z'} = 1.5$ TeV is superimposed. From a combination of all possible decay modes of the τ leptons, production cross

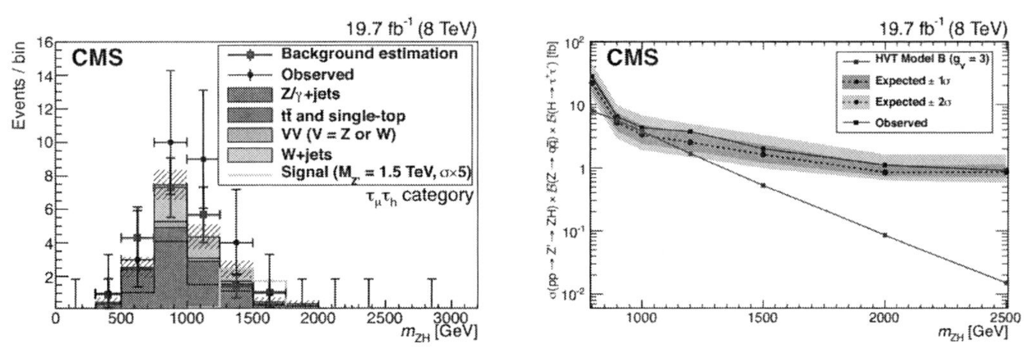

Figure 11. Observed ZH mass distribution for the $\tau_\mu\tau_h$ category along with the corresponding Monte Carlo expectations for signal and background, as well as a background estimation derived from data in red (left); Expected and observed upper limits on the Z′ cross section times branching fraction for Z′ → ZH for the six analysis categories combined (right).

 DOI: 10.1051/epjconf/201612604054

sections in a range between 0.9 and 27.8 fb, depending on the resonance mass in the range 0.8 to 2.5 TeV, are excluded at a 95% confidence level, as can be seen from Fig. 11 (right).

Fig. 12 is a summary plot of several searches for narrow spin-1 resonances decaying to a pair of bosons VV or VH. The curves represent the comparison of the results obtained in different final states and based on data corresponding to an integrated luminosity of 19.7 fb^{-1} recorded in proton-proton collisions at $\sqrt{s}$ = 8 TeV. The black dashed curve represents the theoretical prediction for the sum of the production cross sections of the three spin-1 resonances degenerate in mass in the HVT model scenario B. In certain channels there appears to be a slight excess around 1.8 to 1.9 TeV resonance mass.

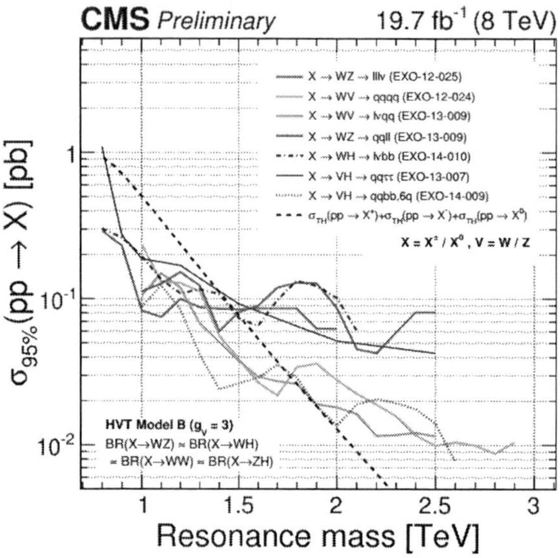

Figure 12. Observed 95% confidence level upper limits on the production cross sections of neutral (X^0) and charged ($X^\pm$) narrow spin-1 resonances decaying into a pair of bosons VV or VH. The black dashed curve represents the theoretical prediction for the sum of the production cross sections of the three spin-1 resonances degenerate in mass in the HVT model scenario B.

5 Narrow resonances

CMS has also performed classic narrow-resonance searches. Interpretations are possible for many exotica scenarios. A first glimpse of $\sqrt{s}$ = 13 TeV data is shown in the dijet spectrum in Fig.13 (left) [27]. The background is estimated from data and parameterized by a smooth function indicated by the red curve. A QCD Monte Carlo prediction, which agrees well with the data, is superimposed in blue. The highest-energy jet pair recorded so far has an invariant mass of 5.4 TeV.

With the new data at $\sqrt{s}$ = 13 TeV CMS is already sensitive beyond the $\sqrt{s}$ = 8 TeV reach for resonances with masses larger than about 5 TeV, as can be seen from Fig.13 (right), which shows upper limits on cross section times branching fraction times acceptance for dijet resonances of the type gluon-gluon, quark-gluon, and quark-quark, compared to theoretical predictions for string resonances, scalar diquarks, excited quarks, axigluons, colorons, color octet scalars S8, new gauge bosons W′ and Z′, and Randall-Sundrum gravitons. The mass limit on string resonances at 5.1 TeV in the quark-gluon channel already exceeds the previous limit at 5.0 GeV [28].

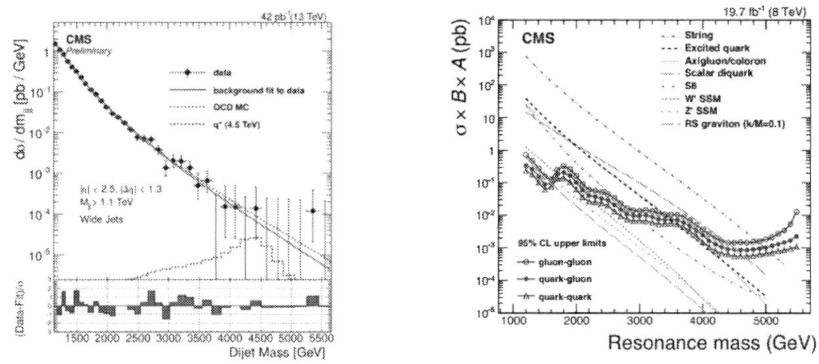

Figure 13. Dijet spectrum for $\sqrt{s}$ = 13 TeV data (left); The observed 95% confidence level upper limits on $\sigma \times \mathcal{B} \times A$ for dijet resonances of the type gluon-gluon, quark-gluon, and quark-quark, compared to theoretical predictions for several kinds of exotic particles (right).

References

[1] CMS Collaboration, JINST **3**, S08004 (2008)
[2] http://cms-results.web.cern.ch/cms-results/public-results/publications/
[3] CMS Collaboration, EPJC **75**, 235 (2015)
[4] CMS Collaboration, CMS-PAS PFT-09-001 (2009)
[5] CMS Collaboration, CMS-PAS PFT-10-001 (2010)
[6] CDMS Collaboration, PRL **111**, 251301 (2013)
[7] CMS Collaboration, CMS-PAS EXO-12-054 (2015)
[8] CMS Collaboration, CMS-PAS EXO-12-047 (2014)
[9] CMS Collaboration, CMS-PAS EXO-14-004 (2015)
[10] P. J. Fox, R. Harnik, R. Primulando, and C.-T. Yu, PRD **86**, 015010 (2012)
[11] G. Busoni, A. De Simone, E. Morgante, and A. Riotto, PLB **728**, 412 (2014)
[12] G. Busoni et al., JCAP **06**, 060 (2014)
[13] G. Busoni et al., JCAP **09**, 022 (2014)
[14] CMS Collaboration, CMS-PAS EXO-14-017 (2015)
[15] CMS Collaboration, CMS-PAS EXO-14-012 (2015)
[16] CMS Collaboration, EPJC **75**, 325 (2015)
[17] C. Allanach et al., EPJC **25**, 113 (2002)
[18] CMS Collaboration, JHEP **1211**, 172 (2012)
[19] CMS Collaboration,PRD **91**, 052012 (2015)
[20] CMS Collaboration, PRD **91**, 012007 (2015)
[21] CMS Collaboration,JHEP **07**, 122 (2013)
[22] M. Seymour, Z. Phys. C **62**,127 (1994)
[23] CMS Collaboration, CMS-PAS EXO-14-010 (2015)
[24] T. Han, H. E. Logan, B. McElrath, and L.-T. Wang, PRD **67**, 095004 (2003)
[25] D. Pappadopulo, A. Thamm, R. Torre, and A.Wulzer, JHEP **09**, 060 (2014)
[26] CMS Collaboration, PLB **748**, 255 (2015)
[27] CMS Collaboration, CMS-PAS EXO-15-001 (2015)
[28] CMS Collaboaration, PRD **91**, 052009 (2015)

The universal wave function interpretation of string theory

Dr. Zhi Gang Sha[1], and Rulin Xiu[2a]

[1] *Institute of Soul Healing and Enlightenment, 30 Wertheim Court, Unit 27D, Richmond Hill, Ontario L4B 1B9, Canada*
[2] *Hawaii Theoretical Physics Research Center, 16-266 E. Kipimana St, Keaau, HI 96749, USA*

Abstract. In this work, we will show that a deeper understanding of space-time provided by both quantum physics and general relativity can lead to a new way to understand string theory. This new way of understanding and applying string theory, the universal wave function interpretation of string theory (UWFIST), may yield to a more powerful string theory and testable prediction. We will show how to derive UWFIST and what new result we can obtain from UWFIST. We will demonstrate that UWFIST indicates that the observed space-time and all phenomena are the projections from the world-sheet hologram. UWFIST provides the possible source for dark energy and dark matter and the explanation about why the dark energy and dark matter is beyond the detection of our current detector. We will show that UWFIST may also yield correct prediction of the cosmological constant to be of the order 10-121 in the unit of Planck scale. It may also help us understand and derive the energy source for inflation and the flatness of our observed 4-dimensional universe. UWFIST may also make other testable predictions that may be detected by interferometers.
We conclude that UWFIST has the potential to make string theory a more powerful physics theory that can yield testable predictions. It is worth further investigation by more physicists.

Quantum physics has been proven to be the fundamental physics for all the elementary particles and three fundamental forces, electromagnetic, weak and strong forces. General Relativity is the proven theory for gravity. Many ways are proposed to incorporate gravity and general relativity into quantum physics. String theory is so far the only mathematically consistent theory for quantum gravity (Ref 1,2).

How to understand and apply String Theory has been a long debated subject. The currently accepted interpretation is that string theory studies the quantum dynamics of a string. The vibration of string creates particles and interactions. The scale of the string is set to be Planck Scale. The finite size of the string length helps result in a consistent quantum gravity and gauge theory. The achievement of string theory cannot be over estimated. This theory has the potential to predict the dimension of space-time, the particle and interaction spectrum, to unify all the forces and fundamental particles, and more. However, with all these great potentials and promises, string theory has not yet made a single testable prediction. Something is still missing in our understanding of string theory.

In this work, we will show a different way to derive string theory. We will show that the deeper understanding of space-time can lead to a new way to understand string theory. This new way of

[a] E-mail: rulin@htprc.org

understanding and applying string theory may yield to a more powerful string theory and testable prediction.

In this paper, we will first derive the uncertainty relation between space-time. Then we use space-time uncertainty relation to derive a string action. We will show that this provides a new way to understand string theory. Here we call it the universal wave function interpretation of string theory (UWFIST). We will discuss the possible testable predictions we can obtain from UWFIST. We will demonstrate that UWFIST indicates the possible source of dark energy and dark matter and provides explanation about why the dark energy and dark matter is beyond the detection of our current detectors. It also has the potential to give us the right estimate about cosmological constant as well as explain the energy source for the inflation. It may also unriddle why our observed 4-dimensional space-time is flat in the most part.

Derivation of Uncertainty Relation Between Space and Time

Quantum physics has completely changed our views about space-time in a way even more dramatic than the way Einstein has changed our view about space-time. According to quantum physics, how accurate we can measure time and space depends on the amount of energy and momentum we use. It takes energy and momentum to measure time and space. More specifically, according to quantum physics, to measure the time duration $\Delta\tau$, it takes the energy $\Delta E \sim \hbar/\Delta\tau$. Similarly, to measure space size $\Delta\sigma$, it takes the momentum $\Delta P \sim \hbar/\Delta\sigma$. Here $\hbar$ is the reduced Planck Constant, $\hbar = h/2\pi$. And h is the Planck Constant. It is equal to 6.626×10^{-34} joul • second.

If one takes gravity into consideration, when one measures time interval $\Delta\tau$, the energy ΔE used for time measurement will curve the space-time according to general relativity. It will create a black hole with the horizon on the order of $G\Delta E/c^4$. Here G is the gravitational constant with the value approximately 6.674×10^{-11} N $\cdot$ m^2/kg^2 and c is the speed of light with the value approximately 3×10^8 meter/second. When $\Delta\sigma$ is smaller than $G\Delta E/c^4$, no information can come out. Therefore, the measurable causal region is:

$$\Delta\sigma \geq G\Delta E/c^4.$$

This indicates an uncertainty relation between space $\Delta\sigma$ and time $\Delta\tau$:

$$\Delta\sigma\,\Delta\tau \geq l_p\, t_p \quad (1)$$

Here l_p is the Planck length, $l_p = (\hbar G/c^3)^{1/2} = 1.616 \times 10^{-35}$ meter. And t_p is the Planck time, $t_p = (\hbar G/c^5)^{1/2} = 5.39 \times 10^{-44}$ second.

The requirement (1) is the condition for the causal region. We call the relation (1) an innate uncertainty relation between space and time. It comes about when we take both quantum effect and gravity into account.

A new way to derive and understand string theory

The space-time uncertainty relation (1) indicates that space and time are quantized. With the space-time uncertainty relation (1), there are two ways to quantize space and time in quantum mechanics.

In Quantum Physics, the uncertainty relations between time τ and energy H

$$\Delta H \Delta \tau \geq \hbar$$

indicates the non-commutation relationship between time and H

$$[H, \tau] = i\hbar.$$

The uncertainty relationship between space σ and momentum P

$$\Delta P \Delta \sigma \geq \hbar$$

indicates the non-commutation relationship between σ and P

$$[P, \sigma] = i\hbar.$$

Using the same argument, we can see that the uncertainty relation (1) indicates the non-commutation relation between space and time:

$$[\sigma, \tau] = i\, l_p t_p \tag{2}$$

The first way to quantize space and time is to consider space and time as operators and they do not commute with each other $[\sigma,\tau] = i\, l_p t_p$. This will lead to non-commutative space-time geometry. Many people have been exploring and studying physics with non-commutative space time geometry hoping to find a consistent theory for quantum gravity.

Here we are going to explore the second way to quantize space and time. In quantum physics, the uncertainty relation $\Delta H \Delta \tau \geq \hbar$ and the non-commutation relationship between time and H, $[H, \tau] = i\hbar$ indicates an action A_t in the form

$$A_t = i \int_0^t H\, d\tau/\hbar.$$

This action A_t will evolve wave function Ψ from Ψ(0) at the time 0 to Ψ(t) at the time t in the form:

$$\Psi(t) = exp(i \int_0^t H\, d\tau/\hbar)\ \Psi(0).$$

In the same way, the uncertainty relation $\Delta P \Delta \sigma \geq \hbar$ and the non-commutation relation between space coordinate σ and momentum P, $[P, \sigma] = i\hbar$, indicates an action A_x in the form

$$A_x = i \int_0^x P\, d\sigma/\hbar.$$

This action A_x will evolve wave function Ψ from Ψ(0) at the space 0 to Ψ(x) at the space x in the form:

$$\Psi(x) = exp(i \int_0^x P\, d\sigma/\hbar)\ \Psi(0).$$

Using the same quantization procedure, we propose that the space time uncertainty relation (1) and non-commutation relation (2) indicates an action A_s in the form:

$$A_s = \int d\tau d\sigma / l_p t_p \tag{3}$$

This action A_s evolves the wave function from $\psi(0,0)$ at the space 0 and time 0 to $\psi(L,T)$ at the space L and time T in the following way:

$$\psi(L,T) = exp(i \int_0^T d\tau \int_0^L d\sigma / l_p t_p)\ \psi(0,0) \qquad (4)$$

It is easy to notice that action (3) is a string action (Ref 1,2). This demonstrates from another angle that when we take gravity into consideration, it is necessary to use a "string" action (3) instead of "particle" action $A = \int d\tau$ to obtain a consistent theory.

We will call the string theory expressed in the formula (4) the universal wave function interpretation of string theory. This universal wave function interpretation of string theory deviates from the usual interpretation of string theory in two ways. Firstly, the world-sheet space and time integration of the action (3) is over the causal region. i.e.

$$A_s = \int_0^T d\tau \int_0^L d\sigma / l_p^2$$

Here T and L are the age and length of the causal horizon. In the usual interpretation of string theory, σ is the string coordinate. L is taken to be the length of the string. It is set to be Planck scale. In the formula (4), σ and τ represent proper space and time. T and L are now variables. They change as universe evolves or as space and time pass by.

The second difference between the usual string theory and the string theory expressed in the equation (4) is that the universal wave function Ψ(L,T) are introduced. Here Ψ(0,0) and Ψ(L,T) represents the universal wave function at the initial proper space and time and at the proper space L and time T.

We call Ψ(L,T) the universal wave function for a good reason. For example, if we extend the string action to the cases of superstring and heterotic string with background fields created in the proper space-time, the universal wave function Ψ can include all fundamental particles, all gauge interactions, gravity, and more. It also includes the large structures of our universe up to the scale of horizon. In other words, the universal wave function could be the wave function of our universe.

Except for these two major conceptual differences, the usual string theory calculation can still be applied here. We will call this new way to understand and apply string theory as Universal Wave Function Interpretation of String Theory (UWFIST). In the following, we will discuss what UWFIST can tell us about our universe.

Our observed space-time and target space-time

In UWFIST, the action (3) is in the proper space-time. The proper space and time (σ, τ) form the world sheet. The normal observed space-time is expressed by the normal space-time coordinate $X^\mu(\sigma,\tau)$, which is a projection from the proper space-time to the normal space-time. Because $X^\mu(\sigma,\tau)$ defines an embedding:

World-sheet -> Target Space,

the observed space-time is the target space.

With the introduction of the target space-time coordinate $X^{\mu}(\sigma,\tau)$, the equation (4) becomes:

$$A_s = (1/l_p t_p) \int d\tau d\sigma\, g^{1/2} g^{ab}\, \partial a X\mu \partial b X^{\mu} \quad (5)$$

Here $g^{\alpha\beta}$ is the metric tensor on the world-sheet formed by proper space and time (τ, σ) and g = - det $g^{\alpha\beta}$.

As shown in string theory (Ref 1,2), the general form of action (5) in the presence of massless background field is in the form:

$$A_s = [i\alpha \int_0^T d\tau \int_0^L d\sigma\, g^{1/2} (g^{\alpha\beta} G^{\mu\nu} \partial_\alpha X_\mu \partial_\beta X_\nu + \varepsilon^{\alpha\beta} B^{\mu\nu} \partial_\alpha X_\mu \partial_\beta X_\nu + 1/4\alpha\, \Phi\, R)] \quad (6)$$

Here $G^{\mu\nu}(\tau, \sigma)$, $B^{\mu\nu}(\tau, \sigma)$, $\Phi(\tau, \sigma)$ are metric tensor, anti-symmetric tensor, and scalar background fields of the target space. The target space-time coordinate $X^{\mu}(\tau, \sigma)$ is a vector projection from the proper space-time to the target space. The target space background fields $G^{\mu\nu}(\tau, \sigma)$, $B^{\mu\nu}(\tau, \sigma)$, and $\Phi(\tau, \sigma)$ are various tensor projections from proper space and time.

Utilizing the Feynman Path Integral definition for the evolution of wave function as explained in Richard Feynman's work (Ref 3), the universal wave function in the presence of background fields now becomes:

$$\Psi(X^{\mu}(L,T), G^{\mu\nu}(L,T), B^{\mu\nu}(L,T), \Phi(L,T)) =$$

$$\int DX^{\mu}\, Dg^{\mu\nu}\, DB^{\mu\nu}\, D\Phi\, \exp[i\alpha \int_0^T d\tau \int_0^L d\sigma\, g^{1/2} (g^{\alpha\beta} G^{\mu\nu} \partial_\alpha X_\mu \partial_\beta X_\nu + \varepsilon^{\alpha\beta} B^{\mu\nu} \partial_\alpha X_\mu \partial_\beta X_\nu + 1/4\alpha\, \Phi\, R)] \quad (7)$$

We can extend $\Psi(X^{\mu}(L,T), G^{\mu\nu}(L,T), B^{\mu\nu}(L,T), \Phi(L,T))$ to include other massive vibrations. We can also extend the above string action to the cases of superstring and heterotic string to include all fundamental particles, all gauge interactions, gravity, and more. This universal wave function tells us the probability for our universe to be at space-time coordinate $X^{\mu}(L,T)$ and background fields $G^{\mu\nu}(L,T)$, $B^{\mu\nu}(L,T)$, $\Phi(L,T)$. We can do a Fourier transformation to transform the wave function $\Psi(X^{\mu}(L,T), G^{\mu\nu}(L,T), B^{\mu\nu}(L,T), \Phi(L,T))$ which are expressed in the target space represented by $X^{\mu}(L,T)$, $G^{\mu\nu}(L,T)$, $B^{\mu\nu}(L,T)$, $\Phi(L,T))$ to the wave function $\Psi(\omega^{\mu},\kappa^{\mu\nu},\ldots)$ expressed in the vibrational space expressed by the various vibrations $(\omega^{\mu},\kappa^{\mu\nu},\ldots)$. The wave function $\Psi(\omega^{\mu}(L,T),\kappa^{\mu\nu}(L,T),\ldots)$ expresses what kind of vibrations in our universe and how much they exist at proper time T and horizon length L in our universe.

The Existence of Dark Energy and Dark Matter

A natural indication and consequence of UWFIST is the existence of dark energy and dark matter.

The space-time coordinate $X^{\mu}(\tau, \sigma)$ is composed of vibrations in the form $\exp[i\pi n(\sigma+c\tau)/L]$ and $\exp[i\pi n(\sigma-c\tau)/L]$. Here n is an integer. Unlike the normal string theory, in which L is the Planck length, now L is causal horizon distance. The largest wavelength is on the order of the length of the causal horizon. These vibrations are very fine. They are very dark in the sense that it is very difficult to detect them. In fact, to detect the vibration with the wavelength L, it takes the time L/c. If L is the horizon distance, this means that it takes the time of the age of the visible universe or a detector as large as the whole visible universe to detect these vibrations. The existence of these vibrations may explain the existence of dark energy.

There also exist vibrations with small masses, which can be as small as $h/(Tc^2)$. These are extremely small masses. Our current detectors cannot detect these small massive particles. The existence of these vibrations with extremely small masses may explain the existence of dark matter.

We conclude that UWFIST indicates the existence of dark energy and dark matter. It also explains why the dark energy and dark matter is beyond the detection of our current detector.

Hologram projection and equation of motion

The proper space σ and time τ form the world-sheet. The world-sheet has the Weyl symmetry, the local rescaling invariance of the world-sheet metric. The action (6) is invariant under the local rescaling of the world-sheet metric:

$$g'^{\alpha\beta}(\tau,\sigma) = \exp(2\omega(\tau,\sigma))\, g^{\alpha\beta}(\tau,\sigma),$$
$$X'\mu(\tau,\sigma) = X\mu(\tau,\sigma)\,,$$
$$G'^{\mu\nu}(\tau,\sigma) = G^{\mu\nu}(\tau,\sigma),$$
$$B'^{\mu\nu}(\tau,\sigma) = B^{\mu\nu}(\tau,\sigma),$$
$$\Phi'(\tau,\sigma) = \Phi(\tau,\sigma),$$

for arbitrary $\omega(\tau, \sigma)$.

Weyl invariance of world-sheet indicates that if we expand or contract the world-sheet, the world-sheet will look the same. In other words, the world-sheet is a hologram. Our observed space-time $X^\mu(\tau, \sigma)$, particles, gauge and gravity interactions are the projection from this hologram. In other words, the observed universe is a projection from the world-sheet hologram.

The fact that our observed space-time is a projection from a hologram has great consequences. We believe many result can be obtained from this fact. Here we will just point out a few of such results.

One of the results we can derive from the fact that our observed universe is a projection from the world-sheet hologram is the classical equation of motion regarding the background field $G^{\mu\nu}$, $B^{\mu\nu}$, Φ, and more expanded as the functions of X^μ (Ref 1, 2). The detailed derivation is presented in the reference (1, 2). In the nutshell, it is obtained by calculating the higher order corrections to the action (6). The Weyl invariance is automatically preserved at the first order. However, the higher order corrections threaten to violate it. Requiring the preservation of Weyl Invariance at the higher orders leads to the classical equation of motion for the background fields $G^{\mu\nu}$, $B^{\mu\nu}$, and Φ. These classical equations of motion include Einstein's general relativity. It is the natural extension of general relativity because they naturally include gauge interactions, different particles, and other background fields and matters.

It is interesting to see that our universal wave function interpretation of string theory includes the characteristics of both quantum physics and Einstein's general relativity. Both the wave function and equation of motion come out of it naturally. It unifies quantum physics and Einstein's general relativity in such a fascinating way.

World-Sheet Contains All The Information of Our Universe

In the universal wave function interpretation of string theory, as we have discussed above, the world-sheet is the hologram from which our observed universe is projected. Because our observed universe is the projection from the world-sheet, the world sheet contains all the information about our universe. In the following, we will demonstrate this explicitly.

There are demonstrations in black hole physics and strong evidence that the area of any surface limits the information content of adjacent space-time regions at 1.4×10^{69} bits per square meter. This is often called Holographic principle (Ref 4,5,6,7,8,9). If we consider T and L as the time length and distance of the horizon of our universe, from the Holographic principle, the maximum amount of information in our universe should be proportional to TL.

Now we will demonstrate that the observable information created on the world-sheet (T, L) is also proportional to TL. Firstly, we will show that the space-time uncertainty relation indicates that there exist only finite observable states on the world sheet.

The smallest energy and momentum it takes to measure time duration $\Delta\tau$ and space length $\Delta\sigma$ is on the order of: $\Delta E\Delta\tau \sim \hbar$, $\Delta E \sim \hbar/\Delta\tau$, and $\Delta p\Delta\sigma \sim \hbar$, $\Delta p \sim \hbar/\Delta\sigma$, and $\Delta\sigma\,\Delta\tau \geq l_p\, t_p$, thus

$$\Delta E\Delta p \sim \hbar^2/(\Delta\sigma\Delta\tau) \leq c\hbar^2/(l_p^2).$$

Since $\Delta E = \hbar\omega$, $\Delta p = \hbar k$, ω and k relate to the frequency and wavelength of the vibrations used in the measurement, we obtain:

$$\omega k \leq c/l_p^2$$

Since $\omega = n/T$, $n = 1,2, \ldots$, and $k = m/L$, $m = 1,2, \ldots$, we have:

$$\omega k = nm/(TL) \leq c/(l_p^2),$$

$$nm \leq cTL/(l_p^2)$$

Therefore there are maximum $TLc/\,l_p^2$ number of energy states. Since each energy state has 2 vibrational states, corresponding to $e^{iwt\pm ikx}$ states, the number of the total possible states is $2\text{^}(cTL/\,l_p^2)$. The amount of information is:

$$Information = log_2(2\text{^}(cTL/\,l_p^2)) = (cTL/\,l_p^2)$$

We can see that the observable information on the world-sheet is also proportional to TL in Planck Units.

We conclude that world-sheet contains all the information of our universe.

Calculation of Vacuum Energy and Its Implication For Inflation

Now we are going to calculate the vacuum energy in our observed space-time. Here we define the vacuum energy to be the lowest energy state of the universal wave function. To calculate the vacuum energy, we need to calculate the wave function in formula (7). Here we will only make a simple estimate of the vacuum energy. We will discuss the detailed calculation in our future work.

Now let's assume that the energy is conserved in the process of the projection from world-sheet to our target space. Under this assumption, by calculating the vacuum energy of the world-sheet, we can obtain the amount of vacuum energy of our observed universe. In the following, we will calculate the vacuum energy of the world sheet. From that, we deduce the vacuum energy density of our observed universe and the cosmological constant. We will show that this calculation can lead to the prediction of cosmological constant consistent with the current experimental data.

In the above, we have shown that there is finite number of possible energy states in the causal horizon. Here we assume supersymmetry is broken at the space and time scale T and L. If the supersymmetry is preserved, the total vacuum energy remains zero because the contributions from the bosonic and fermionic part cancel out.

We further assume that on the world-sheet, the energy at the state (n,m) with $\omega_n = n/T$ and $k_m = m/L$ is in the form of a harmonic oscillator:

$$E_{n,m} = (n + ½)\, h/T.$$

The lowest energy of vacuum fluctuation is $E_{0n,m} = h/2T$.

The total lowest vacuum energy is:

$$E_{vac} = h/2T \; x \; TL/\, l_p^{\,2} = hL/(2\, l_p^{\,2}). \tag{8}$$

It is proportional to the length of the causal horizon.

If we assume the total vacuum energy of the world-sheet is the same as the total vacuum energy of target space, then we can use the total vacuum energy (8) to derive the vacuum energy density and cosmological constant of our universe. The lowest vacuum energy density in the four dimensional space-time target space is then:

$$\rho_{vac} = E_{vac}/(4\pi\, L^3/3) = \rho_p\, l_p^{\,2}/(2L^2). \tag{9}$$

Here ρ_p is the Planck energy density,

$$\rho_p = E_p/(4\pi\, l_p^{\,3}/3),$$

and E_p is the Planck energy $E_p = \hbar/t_p$, t_p and l_p are the Planck time and Planck length $l_p = c\, t_p = (\hbar G/c^3)^{1/2}$. The equation (9) tells us that the vacuum energy density observed in three space dimensions is inversely proportional to the square of the length of horizon L.

From the result (8), we can predict that if the observed space is one dimension, the energy density is a constant $h/(2\, l_p^{\,2})$. In 2-dimensional observed space, the energy density is inversely proportional to the length of horizon L. In n-dimensional observed space, the energy density is proportional to L^{1-n}.

Our result (9) indicates the vacuum energy as the energy source for inflation as well as for the continuous expansion of our universe. It shows us that if our universe starts at the Planck scale l_p at this initial state, the vacuum energy density is half of the Planck energy density $\rho_{vac} = \rho_p/2$. This large vacuum energy density provides the energy source for inflation, the rapid initial expansion. As universe expands, the vacuum energy density decreases, but is non-zero. Therefore our universe will continue to expand but its expansion will slow down.

Now let us assume that L is the Hubble radius with L = c/H. Here c is the speed of light and H is the Hubble constant. Hubble constant is the unit of measurement used to describe the expansion of the universe. It has also been used to estimate the age of our universe. With this assumption, we get:

$$\rho_{vac} = E_{vac}/(4\pi\, L^3/3) = 3h/(8\,\pi\, cL^2 l_p^{\,2}) = 3c^2/(8\pi G L^2) = 3H_0^{\,2}/(8\pi G) = \rho_c \tag{10}$$

Since the relation (10) is general for all L and T, it shows that the lowest vacuum energy density is equal to the critical energy density for 4-dimensional space-time and remains to be so as universe expands. This may provide a possible way to explain why our universe's energy density is so close to critical energy density. It indicates our universe may start and remain to be naturally flat for the most part in the 4-dimensional observed space-time.

The result (10) can explain why our observed 4-dimensional space-time is at the critical density at the vacuum state. It also indicates that for the observed space-time other than 4 dimensions of space-time, the vacuum energy density is not at the critical energy density. Could this be the reason why we can only observe 4-dimensional space-time?

The recent Wilkinson Microwave Anisotropy Probe (WMAP) measurements have led to the conclusion of a flat universe, i.e. our universe's energy density is of the value of critical energy density with only a 0.5% margin of error [10,11,12,13,14,15].

Inflation is so far the mostly accepted way to explain the flatness of the universe and to solve the fine-tuning problems with the initial conditions of the universe. Inflation is the proposed rapid exponential expansion of the early universe by a factor of at least 10^{78} in volume, driven by a negative-pressure vacuum energy density [16]. Inflation hypothesis is proposed to account for why our universe appears flat, homogeneous, and isotropic and to explain the origin of the large-scale structure of the cosmos. The hypothetical particle or field thought to be responsible for inflation is called inflaton. However, the detailed particle physics mechanism responsible for inflation or inflaton remains unknown. Furthermore, the abundance of "bad inflation" in the quantum theory and the realization that inflation is eternal has been challenging the real predictive power of inflation theory [17,18, 19, 20].

UWFIST provides a simple explanation and calculable prediction from the first principle about why our universe experiences fast inflation at the early stage, continue to expand, and at the same time remains flat as seen from the 4-dimensional space-time point of view. It further predicts that the observed space-time with dimensions other than four is not flat.

Calculation of Cosmological Constant

The non-zero cosmological constant was first indicated by the observations presented in 1998 of distance–red shift relation for Type Ia supernovae that the expansion of the universe is accelerating [10,11]. When combined with measurements of the cosmic microwave background radiation these data provided the first direct evidence that cosmological constant Λ is non-zero, with $\Lambda \sim 1.7 \times 10^{-121}$ in Planck units. This result has been supported by more recent WMAP measurements independently. Although there are other possible causes of an accelerating universe, the cosmological constant appears to be the simplest solution.

There are many efforts to derive the value of the cosmological constant and to understand why the cosmological constant Λ is of the magnitude of 10^{-121} in Planck units. One approach is to assume parallel universes. One uses anthropic selection and prior probability distribution for Λ over the multiverse to find the most probable value that allows galaxies to form [21,22,23]. J.D. Barrow and D.J Shaw and other people make the cosmological constant into a field and try to determine the cosmological constant from the equation of motion [24].

Next we will estimate the value of the cosmological constant using the vacuum energy density we have calculated (9). We know that:

$$l_p^2/2L^2 \sim t_p^2/2t_u^2 \sim 10^{-122},$$

here t_u is the age of the universe. Since the cosmological constant $\Lambda_c = 8\pi\, \rho_{vac}$, we obtain:

DOI: 10.1051/epjconf/201612604055

$$\Lambda_c = 4\pi \rho_p \ t_p^2/t_u^2 \sim 10^{-121} \rho_p.$$

It is quite amazing that in UWFIST, using the age of the universe and the fundamental constants ħ and G we are able to derive the value of cosmological constant to the same order of magnitude as the observed value.

Testable String Theory

It is also important to point out that UWFIST may be tested by experiment. From the string action in the presence of the background fields (5), we can see that space-time coordinates $X_\mu((\tau, \sigma)$ and gravity $G^{\mu\nu}(\tau, \sigma)$ are quantum entangled. Further we can see that because the commutation relationship: $[\sigma,\tau] = i\ l_p t_p$ and the observed target space coordinate $X^\mu(\tau, \sigma)$ is the function of σ and τ, there exists a similar commutation relationship for $X^\mu(\tau, \sigma)$ at different directions.

Craig J. Hogan has pointed out that this Planckian scale commutation relationship may be detected by interferometers (Ref 25). This may provide a way to experimentally test UWFIST. We hope future research studies can further explore this.

Multiverse

It is interesting to notice that our wave function interpretation of String Theory supports the multiverse hypothesis (Ref 26, 27). The calculation of the universal wave function in the presence of background fields using equation (5) gives the expression of the possible universes.

Multiverse has been hypothesized in modern theoretical physics to solve many unsolved problems in cosmology. However, the mathematical expression of the multiverse is usually obtained in an arbitrary way. The structure of the multiverse, the nature of each universe within it and the relationships among the various constituent universes, depend on the specific multiverse hypothesis considered. In wave function interpretation of string theory, the possible multiverse can be calculated mathematically from string theory using the equation (5).

Conclusion

In this paper, we propose a new way, the universal wave function interpretation of string theory (UWFIST), to understand and apply string theory. From the derived space-time uncertainty relation, we show that string theory can be understood as calculating the evolution of universal wave function by the string action created by the proper space-time. Our observed universe is the projection from world-sheet hologram to target space. We demonstrate that this new understanding of string theory can lead to testable prediction from string theory. It can provide the calculation of dark energy and dark matter and understanding about why dark energy and dark matter are beyond our current detection. It may lead to the prediction of cosmological constant consistent with the experimental result. It may also show us the possible energy source for the inflation and expansion of our universe. It may also explain why our universe remains flat in 4-dimension space-time. UWFIST deserves further investigation by more astrophysicists and cosmologists.

UWFIST also makes string theory a more powerful physics theory. It may enable string theory to make testable predictions. It is worth further investigation by more string theorists.

Acknowledgements

We want to thank professor Mary K Gaillard, Dr. Ervin Laszlo, and many others for their help, support and encouragement with this project. I want to thank Rasha Blais and Alan Chuck for editing this paper.

References

1. Michael Green, John H. Schwarz and Edward Witten *Superstring theory*. Cambridge University Press. *Vol. 1: Introduction*. ISBN 0-521-35752-7. *Vol. 2: Loop amplitudes, anomalies and phenomenology*. ISBN 0-521-35753-5 (1987).
2. Polchinski, Joseph *String theory*. Cambridge University Press. *Vol. 1: An Introduction to the Bosonic String*. ISBN 0-521-63303-6.*Vol. 2: Superstring Theory and Beyond*. ISBN 0-521-63304-4 (1998).
3. Feynman, Richard P.; Hibbs, Albert. *Quantum Mechanics and Path Integrals*. McGraw Hill. ISBN 0-07-020650-3 (1965).
4. Bekenstein, Jacob D. "Universal upper bound on the entropy-to energy ratio for bounded systems". *Physical Review D* **23** (215): 287–298 (January 1981).
5. 't Hooft, Gerard *Dimensional Reduction in Quantum Gravity*. pp. 10026. arXiv:gr-qc/9310026. Bibcode 1993gr.qc....10026T (1993).
6. Don N. Page. Information in Black Hole Radiation, hep-th/9306083, Phys. 36 6377-6396 (1995)
7. Susskind, Leonard. "The World as a Hologram". *Journal of Mathematical Physics* **36** (11): 6377–6396. arXiv:hep-th/9409089. Bibcode 1995JMP....36.6377S. doi:10.1063/1.531249 (1995)
8. 20. Majumdar, Parthasarathi. "Black Hole Entropy and Quantum Gravity". *ArXiv: General Relativity and Quantum Cosmology* **73**: 147. arXiv:gr-qc/9807045. Bibcode 1999InJPB..73..147M (1998).
9. 21. Bousso, Raphael. "The holographic principle". *Reviews of Modern Physics* **74** (3): 825–874. arXiv:hep-th/0203101. Bibcode 2002RvMP...74..825B. doi:10.1103/RevModPhys.74.825 (2002).
10. Joanne C. Baker, Keith Grainge, M.P. Hobson, Michael E. Jones, R. Kneissl, A.N. Lasenby, C.M.M. O'Sullivan, Guy Pooley, G. Rocha, Richard Saunders, P.F. Scott, E.M. Waldram. "Observational Evidence from Supernovae for an Accelerating Universe and a Cosmological Constant". *The Astronomical Journal* **116** (3): 1009–1038. arXiv:astro-ph/9805201. Bibcode 1998AJ....116.1009R. doi:10.1086/300499 (September 1998).
11. S. Perlmutter, G. Aldering, G. Goldhaber, R.A. Knop, P. Nugent, P.G. Castro, S. Deustua, S. Fabbro, A. Goobar, D.E. Groom, I. M. Hook, A.G. Kim, M.Y. Kim, J.C. Lee, N.J. Nunes, R. Pain, C.R. Pennypacker, R. Quimby, C. Lidman, R.S. Ellis, M. Irwin, R.G. McMahon, P. Ruiz-Lapuente, N. Walton, B. Schaefer, B.J. Boyle, A.V. Filippenko, T. Matheson, A.S. Fruchter, N. Panagia, H.J.M. Newberg, W.J. Couch (The Supernova Cosmology Project). "Measurements of Omega and Lambda from 42 High-Redshift Supernovae". *The Astrophysical Journal* **517** (2): 565–586. arXiv:astro-ph/9812133. Bibcode 1999ApJ...517..565P. doi:10.1086/307221 (June 1999).
12. Bennett, C.; et al. "The Microwave Anisotropy Probe (MAP) Mission". *Astrophysical Journal* **583** (1): 1–23. arXiv:astro-ph/0301158. Bibcode 2003ApJ...583....1B. doi:10.1086/345346 (2003a).

13. Bennett, C.; et al. "First-Year Wilkinson Microwave Anisotropy Probe (WMAP) Observations: Foreground Emission". *Astrophysical Journal Supplement* **148** (1): 97–117. arXiv:astro-ph/0302208. Bibcode 2003ApJS..148...97B. doi:10.1086/377252 (2003b).
14. Hinshaw, G.; et al. "Three-Year Wilkinson Microwave Anisotropy Probe (WMAP1) Observations: Temperature Analysis". *Astrophysical Journal Supplement* **170** (2): 288–334. arXiv:astro-ph/0603451. Bibcode 2007ApJS..170..288H. doi:10.1086/513698 (2007).
15. Hinshaw, G. et al. (WMAP Collaboration). "Five-Year Wilkinson Microwave Anisotropy Probe Observations: Data Processing, Sky Maps, and Basic Results". *The Astrophysical Journal Supplement* **180** (2): 225–245. arXiv:astro-ph/id=0803.0732. Bibcode 2009ApJS..180..225H. doi:10.1088/0067-0049/180/2/225. http://map.gsfc.nasa.gov/universe/uni_shape.html (Feb 2009).
16. Guth, Alan "The Inflationary Universe". Basic Books (1998).
17. Steinhardt, Paul J. and Neil Turok "Endless Universe: Beyond the Big Bang." Doubleday (2007).
18. *Physical Review Letters* **59** (22): 2607–2610. Bibcode 1987PhRvL..59.2607W. doi:10.1103/PhysRevLett.59.2607. PMID 10035596.
19. Penrose, Roger. *The Road to Reality: A Complete Guide to the Laws of the Universe.* London: Vintage Books, p. 755 (2004).
20. Penrose, Roger "Difficulties with Inflationary Cosmology". *Annals of the New York Academy of Sciences* **271**: 249–264. Bibcode:1989NYASA.571..249P. doi:10.1111/j.1749-6632.1989.tb50513 (1989).
21. Dicke, R. H. "Dirac's Cosmology and Mach's Principle". *Nature* **192** (4801): 440–441. Bibcode 1961Natur.192..440D. doi:10.1038/192440a0 (1961).
22. Weinberg, S. "Anthropic bound on the cosmological constant" (1987).
23. Steinhardt, Paul J. "The inflation debate: Is the theory at the heart of modern cosmology deeply flawed?" (*Scientific American*, April; pp. 18-25) (2011).
24. Barrow, John D. and Shaw, Douglas J. "The Value of the Cosmological Constant". General Relativity and Gravitation 43, 2555-2560 (2011). arXiv:1105.3105 [gr-qc] (May 2011).
25. Craig J. Hogan *Interferometers as Probes of Planckian Quantum Geometry.* arXiv:1002.4880 (2012)
26. Steinhardt, Paul J. "The inflation debate: Is the theory at the heart of modern cosmology deeply flawed?" (*Scientific American*, April; pp. 18-25) (2011).
27. Barrow, John D. and Shaw, Douglas J. "The Value of the Cosmological Constant". General Relativity and Gravitation 43, 2555-2560. arXiv:1105.3105 [gr-qc] (2011).

EPJ Web of Conferences **126**, 04056 (2016) DOI: 10.1051/epjconf/201612604056
ICNFP 2015

ICARUS T600: physics results and future activities

Andrea Zani[1,2,a], On behalf of the ICARUS Collaboration

[1] *INFN, Sezione di Pavia, via Agostino Bassi 6, 27100 Pavia, Italy*
[2] *CERN, route de Meyrin 385, 1217 Meyrin, Switzerland*

Abstract. The ICARUS T600 detector has proven the effectiveness of Liquid Argon TPC technology with a successful three-year long run at the INFN Gran Sasso National Laboratories (LNGS).
ICARUS T600 LNGS data strongly contributed to the global effort in searching for neutrino oscillations mediated by sterile states. A definitive clarification of the sterile neutrino puzzle will come from the new multi-station, Short Baseline Neutrino (SBN) experiment at the Booster Neutrino Beam (BNB) at Fermilab, where a refurbished T600 will act as Far Detector.
Presently the T600 detector is located at CERN, undergoing a major overhauling which will allow its participation in the 5 years program of the Fermilab upcoming SBN project. Moreover, ICARUS T600 is also foreseen to collect data with the NUMI Beam to be later exploited in the wider picture of the DUNE Long-Baseline project, recently undertaken by the United States community.
This contribution will report the present physics results obtained by the ICARUS Collaboration, as well as describe the overhauling activities and the physics program for the detector in its future deployment at Fermilab.

1 The ICARUS T600 detector

The ICARUS T600 detector is the largest operational Liquid Argon (LAr) Time Projection Chamber (TPC) built so far [1, 2]. It was deployed and successfully operated in the Hall B of the underground Gran Sasso Laboratory (LNGS), Italy, for more than three years, exposed to both the CERN-Neutrino-to-Gran-Sasso (CNGS) beam, and cosmic rays. The CNGS facility provided an almost pure muon-neutrino beam peaked in the $10 \leq E_\nu \leq 30$ GeV range, with a $\nu_e/\bar{\nu}_e$ contamination of less than 1%. From October 2010 to December 2012, the detector collected neutrinos corresponding to a total of 8.6×10^{19} 400-GeV Protons-On-Target (POT) with a recording efficiency exceeding 93%.

The T600 detector is made of two side-by-side cryostats housing identical half-modules: each of them contains two TPCs sharing a common cathode. The half-modules internal dimensions are $3.6 \times 3.9 \times 19.6$ m^3; they are filled with a total of 760 tons of ultrapure Liquid Argon. Each TPC has maximal drift length of 1.5 m, over which a constant electric field $E_D = 500$ V/cm is present. Charged particles interacting in the TPC active volume produce both scintillation light and ionization electrons: the latter are drifted by the field to the

[a] e-mail: andrea.zani@cern.ch

anode, which is made by three parallel wire planes facing the drift volume, 3 mm apart and with a 3 mm pitch. Each wire plane is oriented at a different angle (0°, +60°, −60°) with respect to the horizontal direction, for a total of about 54000 instrumented wires. Seventy-four 8" photomultipliers (PMTs) in total, positioned behind the wires, collect the produced scintillations light, thus yielding information about the track arrival time. In order to collect the UV light generated in liquid Argon (at λ = 128 nm), the PMT windows are coated with Tetra-phenyl Butadiene, TPB, a wavelength shifter re-emitting photons at around λ = 430 nm.

The measured drift time, combined with the electron drift velocity (v_D = 1.59 mm/μs at $E = E_D$), provides the event coordinate in the drift direction; the composition of the three views from the TPC wires instead yields the track position in the anode plane. Matching these pieces of information allows obtaining a full 3D reconstruction of the tracks, with a precision of few mm^3.

The electronic chain is designed to allow for continuous read-out, digitization (10-bit FADC) and independent waveform recording for each wire signal of the four TPCs; 400 ns non-synchronous sampling is applied on the waveforms. The trigger relies on detection of the UV scintillation light produced by the primary track, in coincidence with the CERN-SPS proton extraction time for the CNGS beam.

Two main items may be highlighted, which assured ICARUS T600 excellent technical and physical results: the capability to achieve unprecedented levels of liquid argon purity, and LAr unique features, which allow for highly-efficient e/γ separation. Continuous gas (2.5 m^3/h) and liquid (100 m^3/d) recirculation, as well as standard commercial Hydrosorb/Oxysorb™ filters, were used to reduce the concentration of electronegative impurities (O_2, N_2, H_2O): this has to be kept far below the value of 0.1 ppb - O_2 equivalent, to ensure ionization charges travel through the entire drift volume. Such system allowed obtaining, during the LNGS run, an average electron life-time > 7 ms (max charge attenuation along the drift: 12%), and a maximum value of 16 ms (with a new liquid pump), which translates into an impurity concentration lower than 20 ppt O_2 equivalent, as shown in figure 1 [3]. Such a result represents a fundamental step towards the construction of a new generation of detectors, with drift paths many meters long.

The precise spatial and calorimetric reconstruction achievable in the T600, along with the possibility to measure local energy depositions (dE/dx) wire by wire, allows the recognition of single- and double-ionising particles events: it is thus possible to efficiently perform e/γ separation, and full reconstruction of π^0's. Such feature ensures therefore negligible contamination of π^0's from NC (Neutral Current) and $\nu_\mu CC$ in selected ν_e-event samples, which is fundamental for the searches of non-standard $\nu_\mu \to \nu_e$ oscillations.

2 Sterile neutrino searches

World experimental observations of neutrino oscillations have so far established a consistent paradigm consisting of three neutrino flavors (ν_e, ν_μ, ν_τ) resulting from the mixing of three mass eigenstates (ν_1, ν_2, ν_3) with relatively small mass differences: $\Delta m^2_{31} \simeq 2.4 \times 10^{-3}$ eV^2 and $\Delta m^2_{21} \simeq 7.5 \times 10^{-5}$ eV^2 [4]. Nonetheless, more recent experiments have reported several anomalies which, if confirmed, could hint at the presence of additional non weakly interacting, hence *sterile*, neutrino flavours with larger mass-squared differences participating in the mixing. On the other side, anyway, Big Bang cosmology and the data from the Planck satellite [5] still favour models with at most one further ν-flavour, its mass being < 0.4 eV.

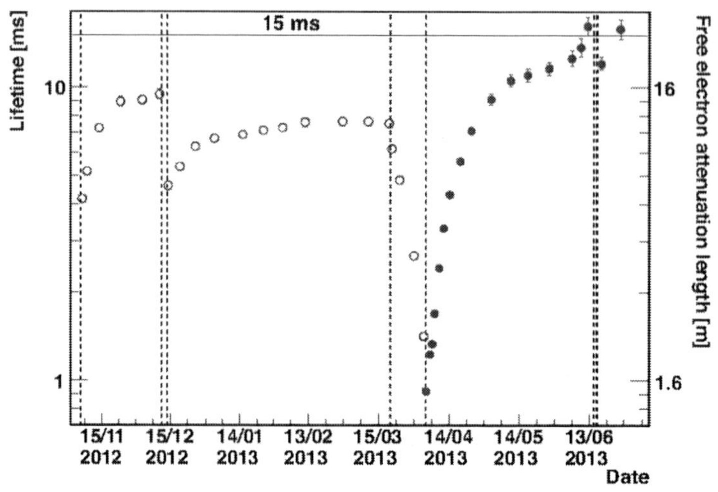

Figure 1. Electron life-time τ_e in the T600 East half-module, as a function of time [3]. Stops of recirculation due to pump maintenance are evident from the decrease in purity. The installation of a new faster pump from April 2013 translates in a much larger purification speed (cfr. slope of data points).

Two distinct classes of neutrino anomalies have been reported so far: first, the apparent disappearance signal in low energy $\bar{\nu}_e$ from nuclear reactors, beyond the expected θ_{13} effect [6], and from Mega-Curie radioactive ν_e-sources in the Gallium experiments [7, 8] originally designed to detect solar neutrinos. On the other hand, there is evidence for an $\nu_e/\bar{\nu}_e$ excess in experiments using muon beams from particle accelerators [9, 10] (the *LSND and MiniBooNE anomalies*).

All these data may be interpreted by postulating the existence of at least one additional sterile neutrino state with mass $\lesssim 1$ eV^2. However, an issue with such explanation is due to some tension between appearance and disappearance data, in particular since no evidence of ν_μ disappearance is observed in the $\Delta m^2_{41} \sim 1$ eV^2 region.

2.1 Search for LNSD-like effect with ICARUS and the CNGS beam

The ICARUS T600 detector has conducted a search for a LSND-like, ν_e-appearance effect during its run at LNGS, with the CNGS beam, characterised by $L/E_\nu \sim 36.5$ m/MeV ($L \simeq 732$ km). This is quite far from the value ~ 1 of LSND, nonetheless a search was performed, on a sample of 2650 ν-events from the CNGS beam (7.93×10^{19} POT). The expected background takes into account intrinsic ν_e contamination, θ_{13}-driven oscillations and decays of τ's from $\nu_\mu \rightarrow \nu_\tau$ oscillations: weighting for the identification efficiency, $8.4 \pm 1.1\,\nu_e$ events can be predicted in the sample. The actual number of observed events [11, 12] is *seven*, which yields no exotic oscillation evidence (see f.i. fig 2).The following limits on exotic oscillations probability can be therefore derived [11, 12], as:

$$P(\nu_\mu \rightarrow \nu_e) \leq 3.85 \times 10^{-3} \text{ (90\% C.L.)} \qquad P(\nu_\mu \rightarrow \nu_e) \leq 7.60 \times 10^{-3} \text{ (99\% C.L.)} \tag{1}$$

The corresponding exclusion plot is reported in figure 3. Such result is corroborated by the OPERA experiment [13]. Taking into account these limits and all other world results, there still remains a region in the parameter space, centered around $\Delta m^2_{41} \sim 0.5$ eV^2 and $\sin^2 2\theta_{14} \sim 0.005$, where overall agreement holds between the described negative results and the positive signals from LSND and MiniBooNE.

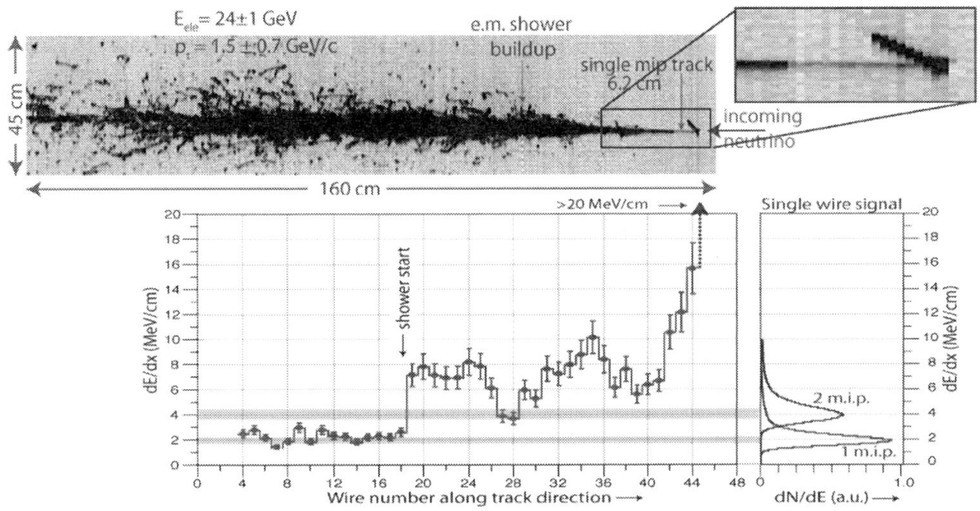

Figure 2. Example of $\nu_e CC$ interaction from the CNGS beam The single-ionizing particle behaviour is clearly visible before the development of the shower, both in the TPC Collection view and in the dE/dx analysis below.

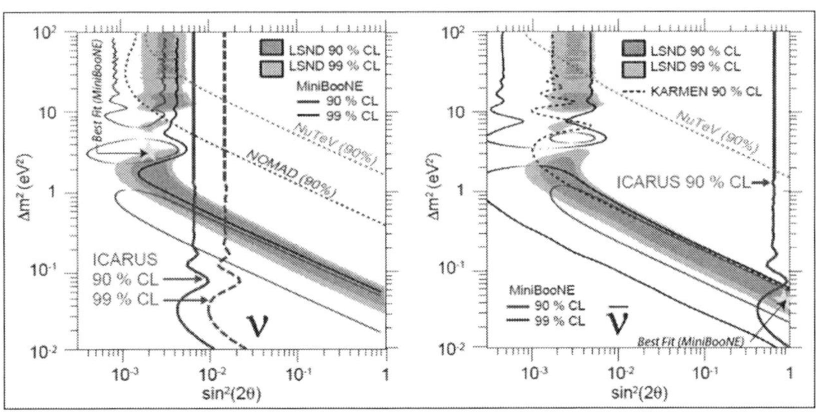

Figure 3. ICARUS limits on the search for non-standard $\nu_\mu \rightarrow \nu_e$ oscillations with CNGS beam, compared with previous results. OPERA analysis aligns with ICARUS findings.

2.2 The Short Baseline Neutrino (SBN) program at FNAL

In order to definitely solve the sterile neutrino puzzle, a whole new short baseline program is being developed, involving diffrerent Collaborations and a global effort. The Short Baseline Neutrino program [15], based at the Fermilab National Accelerator Laboratory (FNAL), foresees the deployment on the Booster Neutrino Beam (BNB) of multiple LAr-TPCs, namely SBND (82 t active volume), MicroBooNE (89 t) and ICARUS T600 (476 t), at a distance of 150, 470 and 600 m respectively. Such configuration will allow simultaneous observations of ν-interactions at different distances, thus yielding separate identification of the Δm^2_{14} and $\sin^2(2\theta_{14})$ parameters. The detectors will operate at $L/E_\nu \sim 1$ and collect millions of events, thus being able to identify effects at the percent level. The expectation is to have perfectly superimposed neutrino spectra at the different stations, in case of oscillation absence: on the contrary any difference may imply a new physics signal. Furthermore the possibility to intervene on beam optics, flipping between neutrino and anti-neutrino modes, assures the chance to disentangle appearance/disappearance effects.

A significant effort has been produced jointly by the three Collaboration to evaluate with MonteCarlo simulations the sensitivity of the program to appearance/disappearance signals, also considering one new channel (at least for this type of detector), i.e. ν_μ disappearance, connected to ν_e appearance via the relation:

$$\sin^2\left(2\theta_{\mu e}\right) \leq \frac{1}{4}\sin^2\left(2\theta_{\mu x}\right)\sin^2\left(2\theta_{ex}\right) \tag{2}$$

The sensitivity evaluation has been made by considering a joint exposure of three years (6.6×10^{20} POT from the BNB), or six years for MicroBooNE, and this should allow covering the LSND 99% CL region at the 5σ level, at the least for the ν_e channel (see fig. 4). On the other hand, the sensitivity in the ν_μ disappearance channel at 600 m distance would be limited to the very-low energy bins (0.2-0.4 GeV) for small mass differences ($\Delta m^2_{14} < 0.5$ eV^2): for this reason, it is being considered to move, at a later stage, one of the two T600 half-modules further away from the beam target, at 1500 m distance, to enhance the effect. However such implementation is still at the level of a consideration and a decision is still to be made.

2.2.1 *Cosmic background on the surface and mitigations*

The US program will have to face a rather new challenge for LAr-TPCs: the operation at shallow depths. All detectors will be deployed at few meters below ground level, covered by a three-meter concrete shielding. Nonetheless still several cosmic rays will cross the LAr active volumes during data taking: in particular for the T600 $\sim$ 12 muon tracks per half-module are expected in the 1 ms drift readout window, as estimated during the 2001 test-run performed at sea-level in Pavia [1]. This situation is completely different with respect to Gran Sasso, where each trigger contained only one event, either from beam or cosmic background: therefore new solutions must be devised to tag cosmic events, and to reconstruct position in space and timing of tracks, in order to associate them (or not) to the beam trigger.

The most serious background source is due to photons associated to cosmic muons: they can mimic $\nu_e CC$ interactions by producing electrons via Compton effect or pair production, and are most dangerous especially when they enter the active volume while their associated μ does not (as it may cross the non-active LAr region)

Some solutions are already planned for background mitigation and are being implemented:

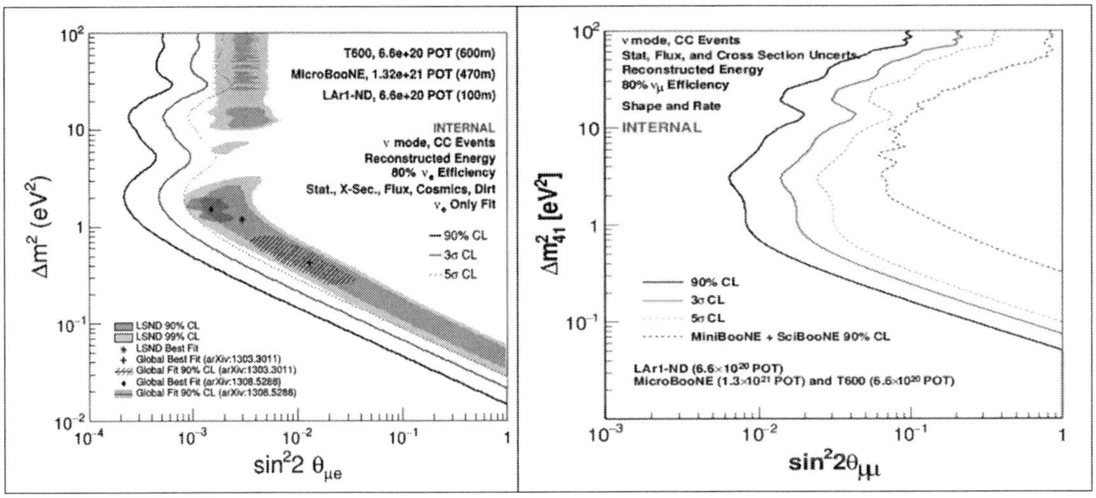

Figure 4. Expected sensitivity for the SBN program in the ν_e-appearance (left) and ν_μ-disappearance channels (right), for three years of data taking and 6.6×10^{20} POT from the Booster Neutrino Beam.

- software-wise, a first cut on γ's associated to muons can be made with geometrical selections directly on the TPC images: f.i., by defining a minimum distance between the γ and the μ (a cylinder around the muon track) which corresponds to associated particles.
- The achievement of 1 ns timing accuracy for light collection with the internal PMT system will allow exploiting the bunched structure of the BNB, which is characterized by 2 ns spills every 19 ns. For this purpose, the light collection system is being significantly upgraded, as it will be further detailed in section 3.
- The implementation of a so-called *Cosmic Ray Tagger* (CRT), i.e. an active layer surrounding as much as possible the TPC and yielding time information of external tracks. The combination of timing from both the CRT and internal PMT system will allow recognising cosmic tracks, and, with the mentioned timing accuracy, distinguish them (that enter the LAr volume) from long muons in $\nu_\mu CC$ beam events (going outward, possible source of auto-veto). Presently the CRT design foresees two-layer scintillator panels, for x-y position reconstruction, with embedded optical fibers and read by Silicon PhotoMultipliers (SiPMs). The panels will be installed and integrated in the outer supporting structure (warm vessel) of the T600.

It is important to note that the CRT, as well as all the software tools to be implemented for neutrino events reconstruction and selection, are being actually developed jointly by the three Collaborations, as they are common to the whole SBN program.

3 Overhauling activities at CERN

Before heading to FNAL, the T600 detector has to undergo a significant overhauling procedure [15]. This will allow introducing technological developments, such as new state-of-the-art, better performing components, while at the same time maintaining the already achieved performance. This activity is taking place at CERN, and it is regulated by a Memorandum of Understanding between CERN and INFN. Completion of the refurbishment is scheduled for the end of 2016, so that detector installation at FNAL can start in 2017.

The overhauling activity at CERN covers important systems of the T600, in particular:

- flattening of the existing cathode panels, to get improved planarity;
- new cold vessels made of Aluminum extruded profiles, and new purely passive insulation;
- refurbished cryogenic plant, organised in skids for independent testing/shipping;
- upgraded light collection system;
- faster warm electronic and read-out chain.

The last two items, objects of significant R&D, are further detailed in the following.

As already mentioned in section 2.2.1, light collection in the new SBN program will be fundamental, both for trigger and for tagging of cosmic background tracks. Therefore, a significant upgrade is planned for the PMT system: new 8" Hamamatsu photomultipliers (PMTs) will be installed, 90 per TPC, yielding larger Quantum Efficiency (QE), around 7%, and larger photocathodic coverage, of the order of 5%, with respect to the past [1]. The chosen 10-dynode, R5912 Hamamatsu PMTs are characterised by high linearity and nanosencond-high time resolution, and will be equipped with new voltage dividers and shielding grid, to avoid induced spurious signals on the TPC wires, as seen during the LNGS run.

The high time resolution of these tubes allows exploiting the bunch structure of the BNB beam, effectively selecting events from the spill; on the other hand the improved coverage and QE leads to high spatial resolution, which in turn allows localization and selection of events in the active volume. A full simulation of the new light collection system has been developed, to optimize the lay-out and number of PMTs: 90 samples per TPC provide longitudinal resolution < 0.5 m; furthermore, by setting a 10 phe threshold on the detected fast light component, it is possible to reduce the residual error on event localization to better than ± 20 cm, with wrongly reconstructed events limited to the detector boundaries. The same simulation was later developed for the chosen configuration, with the implementation of a neural network used to identify events and separate straight tracks (mainly cosmic muons) from electromagnetic showers, possibly due to $\nu_e CC$ beam events. The network was tested with MC events and it was possible to obtain very high efficiency, with mis-identification at the level of few percent.

The ICARUS electronic chain is being remade from scratch, with the idea to maintain the same structure, but with newer, higher-performance components: the architecture is based on analogue low noise warm front-end amplifiers, multiplexed 10-bit 2.5 MHz ADCs (resulting in 400 ns non-synchronous sampling) and digital VME modules for local storage, data compression and trigger information. With this configuration it was achieved, during the LNGS run, a signal-to-noise ratio better than 10 and a ~ 0.7 mm single-point resolution. While this structure is still very well suited for operation with even larger TPCs, it is important to insert newer components, and improve some aspects:

- introduction of serial ADCs, one per channel, to allow for synchronous sampling;
- adoption of modern serial bus architecture, with optical links for Gbit/s transmission rate;
- new compact design, with the digital part residing in a single FPGA, and the resulting boards housed in newly-designed mini-crates directly mounted on the flanges on top of the detector (significant decrease of volume).

New electronic boards containing all the mentioned improvements are being developed by the Padova group in close collaboration with CAEN industry. At the same time, the ICARUS group at CERN has tested for a few months a different solution, i.e. cold front-end boards developed by BNL and already in use by MicroBooNE. The results were quite satisfying, with the achievement of a signal-to-noise ratio ~ 20 on both Induction and Collection wire planes, but the solution was eventually dropped, due to short time available for installation.

4 Conclusions

The ICARUS T600 detector has operated smoothly for three years in the Gran Sasso underground laboratories, establishing the success of Liquid Argon TPC technology for large volumes, and paving the way for multikton-scale detectors. During the LNGS run a search was performed for non-standard $\nu_\mu \to \nu_e$ oscillations driven by supposed sterile neutrino states, but no evidence of oscillation was actually observed.

The ICARUS Collaboration is now embarking in the SBN program at FNAL, specifically dedicated to sterile neutrino searches. The program brings together three experiments (SBND, MicroBooNE and ICARUS), on the Booster Neutrino Beam, to cover in 5 years the $(\Delta m^2_{14}, \sin^2 2\theta_{14})$ parameter space and finally solve the puzzle. At the same time the ICARUS detector, also positioned off-axis on the NuMi beam, will collect $\nu_e CC$ events, to provide initial data for the future Long Baseline FNAL experiment, LBNF/DUNE (former LBNE) [17].

Before heading to the US, the T600 was moved to CERN, where it is undergoing a major technological overhaul, that will prepare the detector for its deployment at FNAL.

References

[1] Amerio S. *et al.* (ICARUS Collab.), Nucl. Instr. and Meth. A **527**, 329-410 (2004)
[2] Rubbia C. *et al.* (ICARUS Collab.), JINST **6**, P07011 (2011)
[3] Antonello M. *et al.* (ICARUS Collab.), JINST **9**, P12006 (2014)
[4] Olive K.A. *et al.* (Particle Data Group), Chin. Phys. C, **38**, 090001 (2014)
[5] Planck Collaboration, A&A, 571, A16 (2014)
[6] Mention G. *et al.*, Phys. Rev. D **83**, 073006 (2011) and reference therein
[7] Abdurashitov J.N. *et al.*, Phys. Rev. C **80**, 015807 (2009)
[8] Kaether F., Hampel W., Heusser G., Kiko J. & Kirsten T., Phys. Lett. B **685**, 47 (2010)
[9] Aguilar A. *et al.*, Phys. Rev. D **64**, 112007 (2001)
[10] Aguilar-Arevalo A. A. *et al.* (MiniBooNE Collab.), Phys. Rev. Lett. **110**, 161801 (2013)
[11] Antonello M. *et al.* (ICARUS Collab.), Eur. Phys. J. **C73**, 2345 (2013)
[12] Antonello M. *et al.* (ICARUS Collab.), Eur. Phys. J. **C73**, 2599 (2013)
[13] Agafonova N. *et al.* (OPERA Collab.), JHEP **1307**, 004 (2013)
[14] Armbruster B. *et al.* (KARMEN Collaboration), Phys. Rev. D **65**, 112001 (2002)
[15] SBND, MicroBooNE and ICARUS Collaborations, SBN-doc-269-v5, arXiv:1503.01520v1 [physics.ins-det] (2015)
[16] http://www.gtt.fr
[17] http://www.dunescience.org/ ; Adams C. *et al.* (LBNE Collaboration), FERMILAB-PUB-14-022, arXiv:1307.7335v3 [hep-ex] (2013)

EPJ Web of Conferences **126**, 04057 (2016) DOI: 10.1051/epjconf/201612604057
ICNFP 2015

Phase structure of Gross-Neveu model with compactification in the presence of external magnetic field

R.N. Zhokhov[1,a], P. B. Kolmakov[3], V.Ch. Zhukovsky[3], and K.G. Klimenko[1,2]

[1] *Institute for High Energy Physics, 142281, Protvino, Moscow Region, Russia*
[2] *University "Dubna" (Protvino branch), 142281, Protvino, Moscow Region, Russia*
[3] *Faculty of Physics, Moscow State University, 119991, Moscow, Russia*

Abstract.
In the paper there have been studied Gross-Neveu model in (2+1)-dimensional space-time with one compactified dimension in presence of external magnetic field at finite temperature. Magnetic field is directed along the uncompactified dimension that is along the axis of the cylinder on which the system lives. Chiral symmetry breaking and corresponding phase structure of the model is investigated in the leading order of $\frac{1}{N}$ expansion.

1 Introduction

Gross-Neveu model in the background of curved space-time and in space-time of nontrivial topology was considered in a range of papers ([3–5, 9, 10]). One simple example of such spaces with nontrivial topology is $R^d \times S^1 \times \cdots \times S^1$ space-time. Such space-times are used in models with compactified extra dimensions.

By studying models in the simplest space-time with nontrivial topology you hope to make generalization and predict properties of the models in spaces with more complicated structure .

Let us consider a very simple example, this is very important for the considered model. Let us show that in case of space-time $M^d \times S^1$ constant vector potential along compactified dimension can not be gauged away. Assume that

$$A_\mu = 0 \text{ for } \mu = 1, \ldots, d, \quad A_{d+1} = const.$$

Gauge transformations have the form :

$$\psi \longrightarrow e^{ie\alpha}\psi, \tag{1}$$

$$A_\mu \longrightarrow A_\mu - \partial_\mu \alpha.$$

Due to the fact that boundary condition is imposed on ψ $\psi(x_1, ..., x_d, x_{d+1} + L) = \psi(x_1, ..., x_d, x_{d+1})$, α satisfies the following condition :

$$\alpha(x_1, ..., x_d, x_{d+1} + L) = \alpha(x_1, ..., x_d, x_{d+1}) + 2\pi e n, \tag{2}$$

[a] e-mail: rzn95@mail.ru

where L- length of compactified dimension.

This condition is satisfied if $\alpha(x) = 2\pi n \frac{x}{L}$.

Transformation for A_{d+1} has the form

$$A_{d+1} \rightarrow A_{d+1} - \frac{2\pi e n}{L}. \tag{3}$$

Hence in general A_{d+1} can not be gauged away.

In this paper we consider Gross-Neveu model in $R^2 \times S^1$ space-time in presence of external magnetic field directed along uncompactified dimension at finite temperature and chemical potential.

2 Model and its thermodynamic potential

Consider Gross-Neveu model in $R^2 \times S^1$ space time (one spatial dimension is compactified), it means that our system lives on the cylinder (there are two spatial dimensions, one uncompactified and one compactified).

Let us think that the cylinder is embeded in (3+1)-dimensional Minkowski space-time, in which there is homogeneous magnetic field directed along the axis of the cylinder.

On the surface of the cylinder vector potential has only one constant component along compactified dimension. As was shown in the introduction it can not be gauged away, that is there exists non zero magnetic flux through the cylinder (along the axis of the cylinder). Moreover we assume that magnetic field interacts with the spin of the fermions, which is introduced as an additional index of the spinor field ψ. Note, that the spin introduced in such a way does not have any relation to Lorentz group.

Lagrangian that describes this model has the following form:

$$\mathcal{L} = \bar{\psi}^s i\gamma^\mu D_\mu \psi^s + \frac{G}{2N}(\bar{\psi}^s\psi^s)^2 + \mu\bar{\psi}^s\gamma^0\psi^s + \frac{\sigma_s}{2} g\mu_B B\bar{\psi}^s\gamma^0\psi^s, \tag{4}$$

γ-matrices have the following form:

$$\gamma^0 = \sigma^3, \gamma^1 = i\sigma^1, \gamma^2 = i\sigma^2. \tag{5}$$

In this paper we will work in irreducible representation. In such models one often uses reducible representation in which you can define γ_5 matrix. Generalization to that case is obvious. Flavour index of ψ is implied that runs from 1 to N.

Spin index s has two components 1 and 2.

Interaction of the magnetic field B with the spin is introduced by the last term in the Lagrangian where $\sigma_{1,2} = \pm 1$, μ_B - Bohr magneton and B - magnetic field.

μ - chemical potential, G - coupling constant, covariant derivative $D_\mu : D_\mu = \partial_\mu - ieA_\mu$.

Vector potential A_μ has the following components: $A_0 = 0$, $A_1 = 0$, $A_2 = const$. Since one dimension is compactified we have to impose a boundary condition. For ψ the boundary condition:

$$\psi(x^0, x^1, x^2 + L) = e^{2\pi i\alpha}\psi(x^0, x^1, x^2). \tag{6}$$

The model has $U(N)$ flavour symmetry.

Moreover the Lagrangian is invariant with respect to discrete chiral transformation which has the form:

$$\begin{aligned} &\psi'_L(x_0, x_1, x_2) = \pm\psi_L(x_0, x_1, -x_2), \bar{\psi}'_L(x_0, x_1, x_2) = \pm\bar{\psi}_L(x_0, x_1, -x_2), \\ &\bar{\psi}'_R(x_0, x_1, x_2) = \mp\bar{\psi}_R(x_0, x_1-, x_2), \bar{\psi}'_R(x_0, x_1, x_2) = \mp\bar{\psi}_R(x_0, x_1, -x_2), \end{aligned} \tag{7}$$

where $\psi_{R,L}(x) = \frac{1 \pm i\gamma^2}{2}\psi(x)$, $\bar{\psi}_{R,L}(x) = \bar{\psi}(x)\frac{1 \mp i\gamma^2}{2}$.

We can work in reducible representation and define chiral symmetry using γ_5 matrix but as we pointed out before this changes nothing only there will be additional degrees of freedom and by slightly changing the definition of the parameter g we have doubled thermodynamic potential. It is obvious that it's imposable to define chiral symmetry using γ_5 matrix in irreducible representation for there is no γ_5 matrix. One can see that the term with the chemical potential has the same form as the term describing interaction of magnetic field with the spin thus one can combine them and write the Lagrangian in the following form:

$$\mathcal{L} = \bar{\psi}^s i\gamma^0 \partial_0 \psi^s + \bar{\psi}^s i\gamma^1 \partial_1 \psi^s + \bar{\psi}^s i\gamma^2 (\partial_2 - ieA_2)\psi^s + \frac{G}{2N}(\bar{\psi}^s \psi^s)^2 + \mu_s \bar{\psi}^s \gamma^0 \psi^s, \quad (8)$$

where we introduced two chemical potentials corresponding to particles with spin projection plus one half and minus one half $\mu_{\uparrow\downarrow} = \mu \pm \delta\mu$, $\delta\mu = \frac{1}{2} g\mu_B B$.

Let us introduce auxiliary scalar field $\sigma(x)$

$$\mathcal{L} = \bar{\psi}^s i\gamma^0 \partial_0 \psi^s + \bar{\psi}^s i\gamma^1 \partial_1 \psi^s + \bar{\psi}^s i\gamma^2 (\partial_2 - ieA_2)\psi^s - \bar{\psi}^s \sigma \psi^s - \frac{N}{2G}\sigma^2 + \mu_s \bar{\psi}^s \gamma^0 \psi^s. \quad (9)$$

Lagrangian (9) is equivalent to the Lagrangian (8), one can make sure of this using an equation of motion for the scalar field σ, which has the following form

$$\sigma = -\frac{N}{G}\bar{\psi}^s \psi^s.$$

Let us take the functional integral over the fermion field and obtain the expression for the effective action. As a result we obtain for effective action in the leading order of in $\frac{1}{N}$-expansion.

$$S_{\text{eff}} = -i \ln \int D[\bar{\psi}(x)] D[\psi(x)] e^{i \int d^3x (\bar{\psi}(i\gamma^0(\partial_0 + \mu_s) + i\gamma^1 \partial_1 + i\gamma^2(\partial_2 - ieA_2) - \sigma)\psi - \frac{N}{2G}\sigma^2)} = \quad (10)$$

$$= -\int d^3x \frac{N}{2G}\sigma^2 - i \ln \det \left(i\gamma^0(\partial_0 + \mu_s) + i\gamma^1 \partial_1 + i\gamma^2(\partial_2 - ieA_2) - \sigma \right). \quad (11)$$

Let us introduce thermodynamic potential that is defined by $V_{\text{eff}} = -\frac{S_{\text{eff}}}{\int d^3x}$.

Calculating determinant over spinor indices and using the formula $\ln\det = \operatorname{tr}\ln$, we obtain

$$V_{\text{eff}} = \frac{N}{2G}\sigma^2 + i\frac{N}{L}\sum_{s=1,2}\sum_{n=-\infty}^{\infty}\int \frac{d^2p}{(2\pi)^2} \ln\left((p_0 + \mu_s)^2 - p_1^2 - (\frac{2\pi}{L}(n+\alpha) - eA_2)^2 - \sigma^2 \right) \quad (12)$$

It is convenient to introduce Aharonov-Bohm phase ϕ,: $\phi = \frac{eLA_2}{2\pi} = \frac{e\pi L^2 B}{8\pi}$.

In the term $(\frac{2\pi}{L}(n+\alpha) - eA_2) = \frac{2\pi}{L}(n + \alpha - \frac{eLA_2}{2\pi}) = \frac{2\pi}{L}(n + \alpha - \phi)$ we denote $\alpha - \phi$ as α, for both the phase in the boundary condition and Aharonov-Bohm phase contribute in α.

Performing Wick rotation $p_0 \to -ip_0$, we obtain

$$V_{\text{eff}} = \frac{N}{2G}\sigma^2 - \frac{N}{L}\sum_{s=1,2}\sum_{n=-\infty}^{\infty}\int \frac{d^2p}{(2\pi)^2} \ln\left((p_0 - i\mu_s)^2 + p_1^2 + (\frac{2\pi}{L})^2(n+\alpha)^2 + \sigma^2 \right). \quad (13)$$

In order to consider our theory at finite temperature we should make use of the following prescription:

$$\int \frac{dp_0}{2\pi} f(p_0) \to \frac{1}{\beta}\sum_{l=-\infty}^{\infty} f(p_{0l}), \text{ where } p_{0l} = \frac{2\pi}{\beta}(l + \frac{1}{2}), \frac{1}{\beta} = T. \quad (14)$$

We obtain for the thermodynamic potential the following expression:

$$V_{\text{eff}} = \frac{N}{2G}\sigma^2 - \frac{N}{\beta L}\sum_s \sum_{l=-\infty}^{\infty}\sum_{n=-\infty}^{\infty}\int \frac{dp_1}{2\pi}\ln((\frac{2\pi}{\beta}(l+\frac{1}{2}) - i\mu_s)^2 + (\frac{2\pi}{L})^2(n+\alpha)^2 + p_1^2 + \sigma^2). \quad (15)$$

3 Calculation of the thermodynamic potential

From now on we assume for simplicity that $\mu = 0$, for in this case $\delta\mu$ plays the same role as μ. If we want to obtain expression for both nonzero μ and $\delta\mu$, we need to divide the contribution by chemical potential placing before the sum $\frac{1}{2}$, then in the first part instead of chemical potential we should write $\mu + \delta\mu$ and in the second part $\mu - \delta\mu$.

Also we rewrite ln in a more convenient form making use of the fact that one can omit terms that are independent of σ.

$$V_{\text{eff}} = \frac{N}{2G}\sigma^2 - \frac{2N}{\beta L}\sum_{l=-\infty}^{\infty}\sum_{n=-\infty}^{\infty}\int \frac{dp_1}{2\pi}\ln\left(1 + \frac{\sigma^2}{(\frac{2\pi}{\beta}(l+\frac{1}{2}) - i\delta\mu)^2 + (\frac{2\pi}{L})^2(n+\alpha)^2 + p_1^2}\right). \quad (16)$$

We can take the sum over l using the following formula:

$$\sum_{n=-\infty}^{\infty}\ln\left(1 + \frac{b^2}{a^2 + (n+\alpha)^2}\right) = \quad (17)$$

$$= \int d\tau \ln(1 + \frac{b^2}{a^2+\tau^2}) + \ln\frac{1 - 2\cos(2\pi\alpha)e^{-2\pi\sqrt{a^2+b^2}} + e^{-4\pi\sqrt{a^2+b^2}}}{1 - 2\cos(2\pi\alpha)e^{-2\pi\sqrt{a^2}} + e^{-4\pi\sqrt{a^2}}}.$$

As a result we get the following expression for the thermodynamic potential:

$$V_{\text{eff}} = V_0 + V_L + V_{\mu T},$$

where

$$V_0 = \frac{N}{2G}\sigma^2 - 2N\int \frac{d^3p}{(2\pi)^3}\ln(p^2+\sigma^2)$$

- thermodynamic potential of Gross-Neveu model without any external conditions,

$$V_L = -\frac{2N}{L}\int \frac{d^2p}{(2\pi)^2}\ln\frac{1 - 2\cos(2\pi\alpha)\mathrm{e}^{-\mathrm{L}\sqrt{p^2+\sigma^2}} + \mathrm{e}^{-2\mathrm{L}\sqrt{p^2+\sigma^2}}}{1 - 2\cos(2\pi\alpha)\mathrm{e}^{-\mathrm{L}\sqrt{p^2}} + \mathrm{e}^{-2\mathrm{L}\sqrt{p^2}}}$$

- contribution to the thermodynamic potential due to compactification,

$$V_{\mu T} = -\frac{2N}{\beta L}\sum_{n=-\infty}^{\infty}\int \frac{dp_1}{2\pi}\left(\ln(1 + e^{-\beta\sqrt{p_1^2+\sigma_n^2}+\beta\delta\mu}) + \ln(1 + e^{-\beta\sqrt{p_1^2+\sigma_n^2}-\beta\delta\mu})\right), \quad (18)$$

where $\sigma_n = \sqrt{(\frac{2\pi}{L})^2(n+\alpha)^2 + \sigma^2}$ - contribution to the thermodynamic potential due to temperature and chemical potential.

The expression (18) in the case of nonzero μ can be rewritten in the form

$$V_{\mu T} = -\frac{N}{2\beta L}\sum_{n=-\infty}^{\infty}\int \frac{dp_1}{2\pi}\left(\ln(1 + e^{-\beta E_\uparrow^+}) + \ln(1 + e^{-\beta E_\downarrow^+}) + \ln(1 + e^{-\beta E_\uparrow^-}) + \ln(1 + e^{-\beta E_\downarrow^-})\right),$$

where $E_s^{\pm} = E_p \pm \mu_s$ and $E_p = \sqrt{(\frac{2\pi}{L})^2(n+\alpha)^2 + p_1^2 + \sigma^2}$.
For V_0 we have

$$V_0(\sigma) = \frac{N}{3\pi}\sigma^3 + \frac{N}{2g}\sigma^2. \tag{19}$$

Rewrite the expression for V_L with use of special functions.

$$V_L = \frac{N}{\pi L^3}Li_3(e^{-L\sigma+2\pi i\alpha}) + \frac{N}{\pi L^3}Li_3(e^{-L\sigma-2\pi i\alpha}) + \frac{N\sigma}{\pi L^2}Li_2(e^{-L\sigma+2\pi i\alpha}) + \tag{20}$$

$$+\frac{N\sigma}{\pi L^2}Li_2(e^{-L\sigma-2\pi i\alpha}) - \frac{N}{\pi L^3}Li_3(e^{2\pi i\alpha}) - \frac{N}{\pi L^3}Li_3(e^{-2\pi i\alpha}).$$

where $Li_\nu(z) = \sum_{n=1}^{\infty} \frac{z^n}{n^\nu}$.
Derivative of V_{eff} with respect to σ has the following form:

$$\frac{dV_{\text{eff}}}{d\sigma} = \frac{N}{\pi}\sigma\left((\sigma - \sigma_0) + \frac{1}{L}\ln(1 - 2\cos(2\pi\alpha)e^{-L\sigma} + e^{-2L\sigma})\right) + \tag{21}$$

$$+\frac{2N\sigma}{L}\sum_{n=-\infty}^{\infty}\int\frac{dp_1}{2\pi}\frac{1}{1+e^{(\beta\sqrt{p_1^2+\sigma_n^2}-\beta\delta\mu)}}\frac{1}{\sqrt{p_1^2+\sigma_n^2}} + \frac{2N\sigma}{L}\sum_{n=-\infty}^{\infty}\int\frac{dp_1}{2\pi}\frac{1}{1+e^{(\beta\sqrt{p_1^2+\sigma_n^2}+\beta\delta\mu)}}\frac{1}{\sqrt{p_1^2+\sigma_n^2}}.$$

One can see that $\sigma = 0$ is a solution of stationary equation. Besides if $\sigma \to \infty$ then $V_{\text{eff}} \to \infty$, so if at a point $\sigma = 0$ the thermodynamic potential has its maximum then minimum is is located at $\sigma \neq 0$. if at a point $\sigma = 0$ the thermodynamic potential has its minimum then in the absence of the other minimums this minimum is the global one. If there are other minimums then one has to compare values of the thermodynamic potential at these minimums and find the global one. It turns out that in the case of zero chemical potential $\mu = 0$ without of Zeeman interaction $\delta\mu = 0$ there are no other minimums and in order to find minimum of the thermodynamic potential it is enough to calculate only second derivative of the thermodynamic potential at the point $\sigma = 0$.

Second derivative of the thermodynamic potential at the point $\sigma = 0$ is

$$\left.\frac{d^2V_{\text{eff}}}{d\sigma^2}\right|_{\sigma=0} = \frac{N}{\pi}\left(-\sigma_0 + \frac{1}{L}\ln(2 - 2\cos(2\pi\alpha))\right) + \tag{22}$$

$$+\frac{2N}{L}\sum_{n=-\infty}^{\infty}\int\frac{dp_1}{2\pi}\frac{1}{1+e^{(\beta\sqrt{p_1^2+(\sigma_n^0)^2}-\beta\delta\mu)}}\frac{1}{\sqrt{p_1^2+(\sigma_n^0)^2}} + \frac{2N}{L}\sum_{n=-\infty}^{\infty}\int\frac{dp_1}{2\pi}\frac{1}{1+e^{(\beta\sqrt{p_1^2+(\sigma_n^0)^2}+\beta\delta\mu)}}\frac{1}{\sqrt{p_1^2+(\sigma_n^0)^2}},$$

where $\sigma_n^0 = \sigma_n |_{\sigma=0}$.

In the case if α tends to integer, $\alpha = n + \varepsilon$ when $\varepsilon \to 0$, the first term has the following asymptotic $\frac{N}{\pi L}\ln(\varepsilon^2)$.

Studying the behavior of other terms one can find that divergent part of second derivative of second derivative of the thermodynamic potential at the point $\sigma = 0$ equals 0:

$$\left.\frac{d^2V_{\text{eff}}}{d\sigma^2}\right|_{\sigma=0}^{div} = \frac{N}{\pi L}\ln(\varepsilon^2) - \frac{N}{\pi L}\ln(\varepsilon^2) = 0.$$

Let us consider the case $T = 0$, i.e. $\beta \to \infty$
Second derivative of the thermodynamic potential at the point $\sigma = 0$ in this case has the form:

$$\left.\frac{d^2V_{\text{eff}}}{d\sigma^2}\right|_{\sigma=0} = \frac{N}{\pi}\left(-\sigma_0 + \frac{1}{L}\ln(2-2\cos(2\pi\alpha))\right) + \frac{2N}{\pi L}\sum_{n=-\infty}^{\infty}\theta(\mu_n - 1)\ln\left(\mu_n + \sqrt{\mu_n^2 - 1}\right), \quad (23)$$

where $\mu_n = \frac{\delta\mu}{\frac{2\pi}{L}|n+\alpha|}$

Let us consider the case without Zeeman interaction at zero temperature $T = 0, \delta\mu = 0$. In this case

$$\frac{dV_{\text{eff}}}{d\sigma} = \frac{N}{\pi}\sigma\left(\sigma - \sigma_0 + \frac{1}{L}\ln(1 - 2\cos(2\pi\alpha)e^{-L\sigma} + e^{-2L\sigma})\right) = 0. \quad (24)$$

In this case one can find an explicit expression for σ:

$$\sigma = \frac{1}{L}\text{arcch}\left(\frac{e^{L\sigma_0} + 2\cos(2\pi\alpha)}{2}\right). \quad (25)$$

This expression has been obtained in the following papers [3],[4],[5] .

In the case $B = 0$ if $\frac{1}{6} < \alpha < \frac{5}{6}$ there exist $L < L_c$, when the symmetry of the model is restored, where L_c - critical radius of compactification:

$$L_c = \frac{1}{\sigma_0}\ln(2 - 2\cos(2\pi\alpha)). \quad (26)$$

In he case of other α symmetry is broken for all L. It means that if $\alpha = 0$ the symmetry is broken for all L, in case of $\alpha = \frac{1}{2}$: $L_c = \frac{2\ln 2}{\sigma_0}$

Let us rewrite the expression for integral $\int \frac{dp_1}{2\pi}\ln\left(1 + \mathrm{e}^{-\beta\sqrt{p_1^2+\sigma_n}\pm\beta\delta\mu}\right)$, that stand in the expression of thermodynamic potential using special functions . If $\sigma_n > \delta\mu$.

Make use of the formula that can be found in ([7]),

$$\int_a^{\infty} x^{n+1}(x^2 - a^2)^{\beta-1}e^{-px}dx = (-1)^n\frac{(2a)^{\beta+\frac{1}{2}}}{2\sqrt{\pi}}\Gamma(\beta)\frac{\partial^n}{\partial p^n}[p^{\frac{1}{2}-\beta}K_{\beta+\frac{1}{2}}(ap)]. \quad (27)$$

We obtain that

$$\int_{-\infty}^{\infty}\frac{dp_1}{2\pi}\ln\left(1 + e^{-\beta\sqrt{p_1^2+\sigma_n^2}\pm\beta\delta\mu}\right) = \frac{1}{\pi}\sum_{m=1}^{\infty}(-1)^{m-1}\frac{e^{\pm\beta\delta\mu m}}{m}\sigma_n K_1(\beta\sigma_n m). \quad (28)$$

As a result we get the following expression for $V_{\mu T}$

$$V_{\mu T} = -\frac{2N}{\pi\beta L}\sum_{\sigma_n<\delta\mu}\int\frac{dp_1}{2\pi}\ln\left(1 + \mathrm{e}^{-\beta\sqrt{p_1^2+\sigma_n^2}+\beta\delta\mu}\right) + \quad (29)$$

$$+\frac{2N}{\pi\beta L}\sum_{\sigma_n>\delta\mu}\sum_{m=1}^{\infty}(-1)^m\frac{\sigma_n}{m}K_1(\beta\sigma_n m)e^{\beta\delta\mu m} + \frac{2N}{\pi\beta L}\sum_{n=-\infty}^{\infty}\sum_{m=1}^{\infty}(-1)^m\frac{\sigma_n}{m}K_1(\beta\sigma_n m)e^{-\beta\delta\mu m}.$$

4 Effective potential for which $\left.\frac{d^2V_{\rm eff}}{d\sigma^2}\right|_{\sigma=0}$ is explicitly finite for all values of α

In this section we will get the expression for thermodynamic potential that has explicitly finite form of its second derivative at the point $\sigma = 0$ for all values of α.

Let us go back to the expression (15). Now take the sum over n first . Making use of the formula (17), we get

$$V_{\rm eff} = V_0 + V_{\beta\mu 0} + V_{L\beta\mu}, \tag{30}$$

$$V_{\beta\mu 0} = -\frac{2N}{\beta}\int \frac{d^2p}{(2\pi)^2}\ln\left(\frac{1 + 2ch\,(\beta\delta\mu))\,e^{-\beta\sqrt{p^2+\sigma^2}} + e^{-2\beta\sqrt{p^2+\sigma^2}}}{1 + 2ch\,(\beta\delta\mu))\,e^{-\beta\sqrt{p^2}} + e^{-2\beta\sqrt{p^2}}}\right), \tag{31}$$

$$V_{L\beta\mu} = -\frac{2N}{\beta L}\sum_{n=-\infty}^{\infty}\int_{-\infty}^{\infty}\frac{dp_1}{2\pi}\ln\left(1 - 2cos\,(2\pi\alpha)\,e^{-L\sqrt{p_1^2+\sigma_l^2}} + \mathrm{e}^{-2\mathrm{L}\sqrt{p_1^2+\sigma_l^2}}\right), \tag{32}$$

where $\sigma_l = \sqrt{(\frac{2\pi}{\beta}(l+\frac{1}{2}) - i\delta\mu)^2 + \sigma^2}$.

After not complicated calculation the expression for thermodynamic potential can be brought to the form

$$V_{\rm eff}(\sigma) = \frac{N}{\pi}\left(\frac{\sigma^3}{3} - \frac{\sigma^2\sigma_0}{2}\right) + \sum_{\pm}\frac{N}{\pi\beta^3}Li_3(-e^{-\beta\sigma\pm\beta\delta\mu}) + \sum_{\pm}\frac{N\sigma}{\pi\beta^2}Li_2(-e^{-\beta\sigma\pm\beta\delta\mu}) + \tag{33}$$

$$+\frac{4N}{\beta L}\sum_{n=1}^{\infty}\sum_{l=-\infty}^{\infty}\frac{\sigma_l}{n}K_1(L\sigma_l n)\cos(2\pi\alpha n).$$

The second and the third terms in the thermodynamic potential can be written in that form only if $\sigma > \delta\mu$.

Derivative of the thermodynamic potential is

$$\frac{dV_{\rm eff}}{d\sigma} = \frac{N}{\pi}\sigma\left((\sigma-\sigma_0) + \frac{1}{\beta}\ln(1 + 2\cosh(\beta\delta\mu)e^{-\beta\sigma} + e^{-2\beta\sigma})\right) - \tag{34}$$

$$-\frac{4N\sigma}{\beta}\sum_{l=-\infty}^{\infty}\sum_{n=1}^{\infty}K_0(L\sigma_l n)\cos(2\pi\alpha n).$$

Second derivative of the thermodynamic potential at the point $\sigma = 0$ is

$$\left.\frac{d^2V_{\rm eff}}{d\sigma^2}\right|_{\sigma=0} = \frac{N}{\pi}\left(-\sigma_0 + \frac{1}{\beta}\ln(2 + 2\cosh(\beta\delta\mu))\right) - \tag{35}$$

$$-\frac{4N}{\beta}\sum_{l=0}^{\infty}\sum_{n=1}^{\infty}\left(K_0(L\sigma_l^{+0}n) + K_0(L\sigma_l^{-0}n)\right)\cos(2\pi\alpha n),$$

where

$$\sigma_l^{\pm} = \sqrt{(\frac{2\pi}{\beta}(l+\frac{1}{2}) \pm i\delta\mu)^2 + \sigma^2},\quad \sigma_l^{\pm 0} = \sigma_l^{\pm}\,|_{\sigma=0}\,. \tag{36}$$

Consider the limit $L \to \infty$ (this case has been considered in [6]). One can see that in the expression for first derivative of the thermodynamic potential (34) the last term tends to zero as $O(\frac{1}{L}e^{-L})$. As a result we get

$$\frac{dV_{\rm eff}}{d\sigma} = \frac{N}{\pi}\sigma\left((\sigma-\sigma_0) + \frac{1}{\beta}\ln(1 + 2\,\mathrm{ch}(\beta\delta\mu)e^{-\beta\sigma} + e^{-2\beta\sigma})\right).$$

One can easily solve this for σ

$$\sigma = \frac{1}{\beta}\text{arcch}\left(\frac{e^{\beta\sigma_0} - 2\,\text{ch}(\beta\delta\mu)}{2}\right).$$

In the case $\delta\mu < 1$ ($\delta\mu$ is made dimensionless by σ_0, i.e. $\delta\mu < \sigma_0$) there exists β_c, such that if $\beta < \beta_c$ the symmetry is restored.

β_c can be found from the following equation:

$$e^{\beta\sigma_0} = 2 + e^{\beta\delta\mu} + e^{-\beta\delta\mu}.$$

One can see that with increasing value of $\delta\mu$ the value of β_c is increasing and tends to infinity if $\delta\mu \to 1$. Starting from $\delta\mu = 1$ critical temperature β_c does not exist, that is the symmetry is always restored for any values of β.

5 Numeric investigation and phase portrait

One can investigate the phase structure of the model and draw a full phase portrait of it only by numerical analysis of thermodynamic potential that is finding its global minimum point.

Let us consider the case of zero chemical potential without Zeeman interaction. In that case external magnetic field influence the system only via Aharonov-Bohm phase.

First consider the case for periodic boundary condition $\alpha = 0$.

If $B = 0$ (Fig. 1) one can see that by decreasing β (that is increasing the temperature T) the symmetry is restored. By increasing length of compactified dimension L the curve tends to a constant value $\beta = \beta_c, \beta_c = 2\ln 2$ and this value is reached already at L equals 3.

Let us assume that external magnetic field is nonzero. By increasing B one can see that there appears a bump (Fig.2), that is in corresponding region L the symmetry is restored for $\beta > \beta_c$

At rather large B (approximately $220 - 226$) the bump goes to infinity, one can see this at (Fig.3)

By further increasing B there appears one more bumps (Fig. 3, 4), etc.

One can observe that at $B = 2073$ (Fig. 5) two bumps go to infinity (though there are other bumps at the figure that do not go to infinity). This means that for infinite values of β at some values of L symmetry is restored, then it breaks down and then restores again by decreasing L.

At large values of B (Fig. 6) there are many bumps that go to infinity. One can see that all bumps going to infinity do not go beyond values of $L > L_c$ as in case of ($T = 0$). It means that magnetic field virtually do not influence the curve of phase transition of the system for $L > L_c$.

Let us now consider the case of antiperiodic boundary condition $\alpha = \frac{1}{2}$.

In tis case at zero magnetic field phase portrait of the system is symmetric under interchange of β and L, a it should be according to (15) if $B = 0, \alpha = \frac{1}{2}$.one can see that the curve if $L \to \infty$ tends to β_c, and if $\beta \to \infty$ the curve tends to L_c (Fig.8).

If the magnetic field is nonzero $B \neq 0$, then if one increase B in order to restore chiral symmetry one has to decrease length of compactified dimension L to lower value than in case of $B = 0$ (Fig. 7, 8). For large enough values of the magnetic field and large values of β ($\beta > \beta_c$) there appears regions of L, in which first the symmetry is restored, then by decreasing L it breaks down and then is restored again, as it were in the case of periodic boundary condition. Values of L, at which the symmetry is restored once for all, decrease by increasing values of magnetic field, and number of regions in which the symmetry alternatively restores and breaks down becomes larger (Fig. 9, 10).

Comparing the phase portrait for zero magnetic field (Fig. 7) and for nonzero one (Fig.. 9, 10), one can see that magnetic field could break the symmetry in the region $L < L_c$.

At Fig. 1–6 curve of phase transition between phase $\sigma = 0$(left) and $\sigma \neq 0$ (right) is depicted for periodic boundary condition ($\alpha = 0$) without taking into account Zeeman interaction ($\delta\mu = 0$) for different values of magnetic field B.

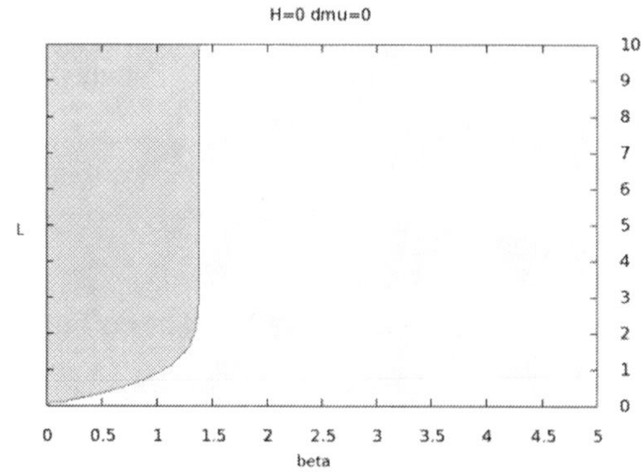

Figure 1. $\alpha = 0, B = 0$.

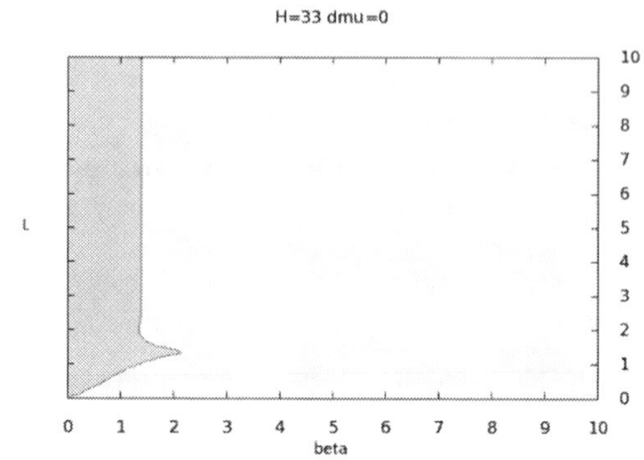

Figure 2. $\alpha = 0, B = 207$.

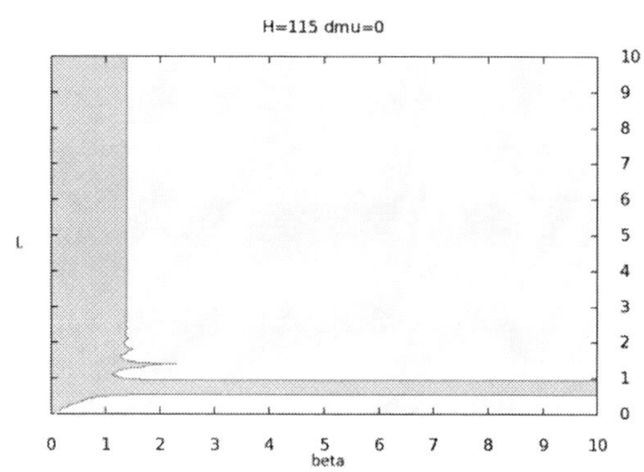

Figure 3. $\alpha = 0, B = 722$.

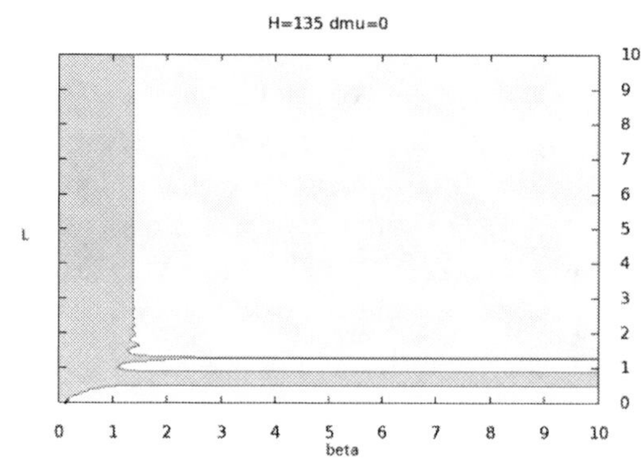

Figure 4. $\alpha = 0, B = 848$.

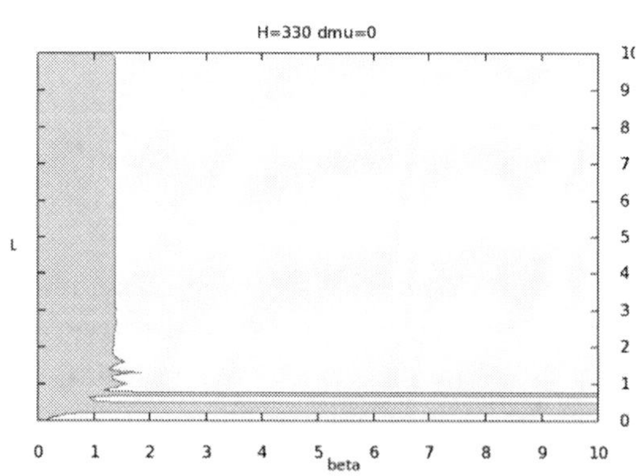

Figure 5. $\alpha = 0, B = 2073$.

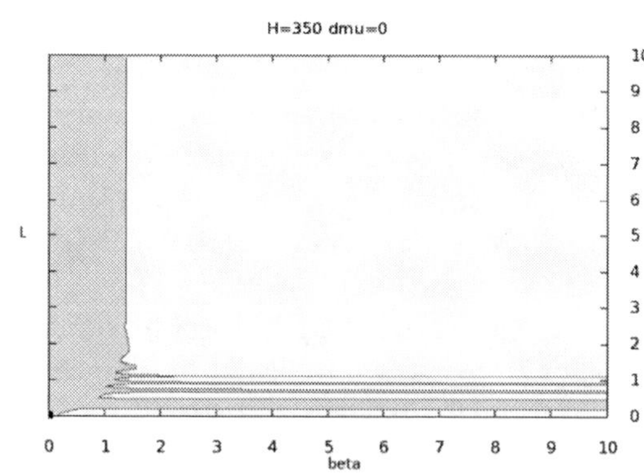

Figure 6. $\alpha = 0, B = 2198$.

At Fig. 7-10 curve of phase transition between phase $\sigma = 0$(left) and $\sigma \neq 0$ (right) is depicted for antiperiodic boundary condition ($\alpha = \frac{1}{2}$) without taking into account Zeeman interaction ($\delta\mu = 0$) for different values of magnetic field B.

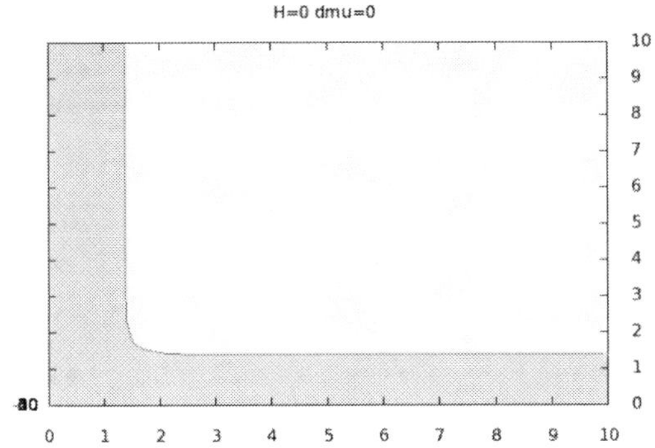

Figure 7. $\alpha = \frac{1}{2}, B = 0$.

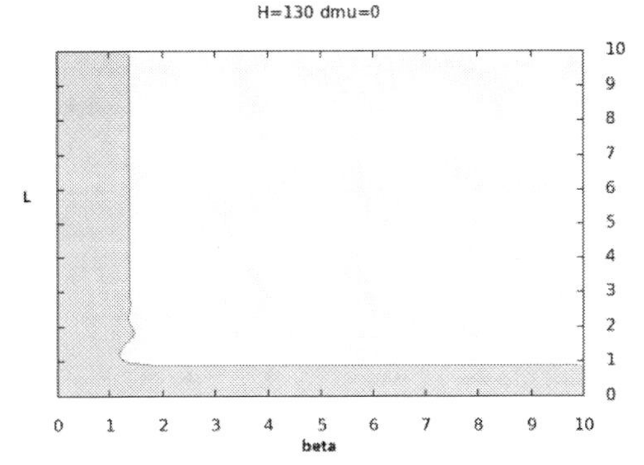

Figure 8. $\alpha = \frac{1}{2}, B = 816$.

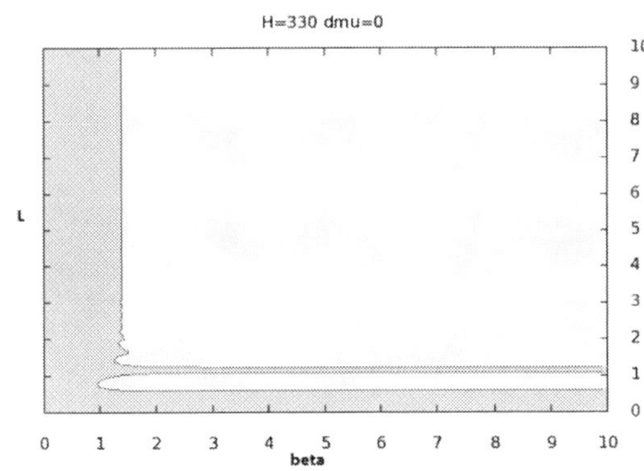

Figure 9. $\alpha = \frac{1}{2}, B = 2072$.

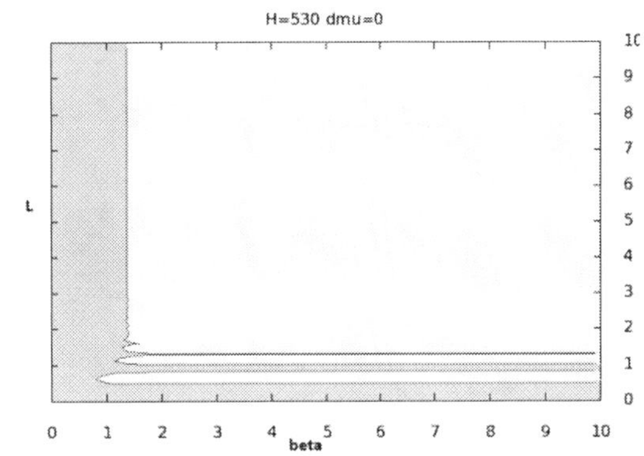

Figure 10. $\alpha = \frac{1}{2}, B = 3328$.

6 Calculation of using method of ζ-regularization.

In this section we will obtain an expression for thermodynamic potential in case of $R^1 \times S^1 \times S^1$ space-time (the second S^1 implies that the model is considered at finite temperature), for arbitrary boundary conditions on the spinor field ψ, that is for arbitrary α.We consider that there is external magnetic field B, but we will not take into account Zeeman interaction $\delta\mu = 0$, and consider case of zero chemical potential $\mu = 0$. We will make use of a method called ζ-regularization . This method was used in paper [10] in order to investigate the influence of compactification and nonzero temperature on chiral symmetry breaking in two dimensional Gross-Neveu model in case of $\alpha = 0, \alpha = \frac{1}{2}$. Moreover in this paper there was formulated this method for $R^d \times S^1 \times S^1$ space time with arbitrary d.

Here this method is generalized for arbitrary α and equation for the phase portrait is obtained.

First we will obtain an asymptotic of thermodynamic potential at a point $\sigma = 0$ up to terms quadratic over σ, that will allow us to get second derivative of the thermodynamic potential at the point $\sigma = 0$ and determine if the point $\sigma = 0$ is local minimum or maximum. In case of $\delta\mu = 0$

thermodynamic potential always has only one minimum hence having the asymptotic at point $\sigma = 0$ allows us to determine if minimum of thermodynamic potential locates at point $\sigma = 0$ or at some point $\sigma \neq 0$ that is if the chiral symmetry is broken or not. Then using method of ζ-regularization we will obtain an expression for thermodynamic potential in two form. This expressions coincides with those obtained using cut-off regularization.

First we will obtain asymptotic of thermodynamic potential at a point $\sigma = 0$. Let us go back to the expression (15). For simplicity assume that $\mu = 0$

$$V_{\text{eff}} = \frac{N}{2G}\sigma^2 - \frac{N}{\beta L}\sum_s\sum_l\sum_n\int\frac{dp_1}{2\pi}\ln\left(\left(\frac{2\pi}{\beta}\left(l+\frac{1}{2}\right) - i\delta\mu\right)^2 + \left(\frac{2\pi}{L}\right)^2(n+\alpha)^2 + p_1^2 + \sigma^2\right). \quad (37)$$

Using the following relation $\ln a = -(a^{-s})'|_{s=0}$, we get

$$V_{\text{eff}} = \frac{N}{2G}\sigma^2 + \Omega'(s)|_{s=0}, \quad (38)$$

where

$$\Omega(s) = \frac{2N}{\beta L}\sum_l\sum_n\int\frac{dp_1}{2\pi}\left(\frac{(\frac{2\pi}{\beta}(l+\frac{1}{2}) - i\delta\mu)^2 + (\frac{2\pi}{L})^2(n+\alpha)^2 + p_1^2 + \sigma^2}{\mu^2}\right)^{-s} \quad (39)$$

and μ-dimensionful parameter with the dimension of mass.

Make use of the formula

$$a^{-s} = \frac{1}{\Gamma(s)}\int_0^\infty dt\, t^{s-1}e^{-at}$$

and if $\left(\frac{\pi}{\beta}\right)^2 + \sigma^2 > \delta\mu^2$ we obtain

$$\Omega(s) = \frac{N}{\beta L}\sum_{l=-\infty}^{\infty}\sum_{n=-\infty}^{\infty}\int_{-\infty}^{\infty}\frac{dp_1}{2\pi}\frac{1}{\Gamma(s)}\int_0^\infty dt\, t^{s-1}e^{-((\frac{2\pi}{\beta}(l+\frac{1}{2})-i\delta\mu)^2+(\frac{2\pi}{L})^2(n+\alpha)^2+p_1^2+\sigma^2)\frac{t}{\mu^2}}.$$

Calculating the Gaussian integral $\int_{-\infty}^{\infty}\frac{dp_1}{2\pi}e^{-p_1^2 t} = \frac{1}{2\sqrt{\pi t}}$, we get

$$\Omega(s) = \frac{N}{\beta L}\mu^{2s}\sum_{l=-\infty}^{\infty}\sum_{n=-\infty}^{\infty}\frac{1}{2\sqrt{\pi}}\frac{1}{\Gamma(s)}\int_0^\infty dt\, t^{s-\frac{3}{2}}e^{-(\frac{2\pi}{\beta})^2((l+\frac{1}{2})-\frac{i\beta\delta\mu}{2\pi})^2 t}e^{-(\frac{2\pi}{L})^2(n+\alpha)^2 t}e^{-\sigma^2 t}.$$

Let us use Poisson resummation formula

$$\sum_{n=-\infty}^{\infty}e^{-a(n+b)^2} = \sqrt{\frac{\pi}{a}}\sum_{n=-\infty}^{\infty}e^{-\frac{\pi^2 n^2}{a}}\cos(2\pi bn), \quad (40)$$

and transform the sum of exponents into the form

$$\sum_{n=-\infty}^{\infty}e^{-(\frac{2\pi}{L})^2(n+\alpha)^2 t} = \frac{L}{2\sqrt{\pi t}}\left(1 + 2\sum_{n=1}^{\infty}e^{-\frac{L^2 n^2}{4t}}\cos(2\pi\alpha n)\right)$$

and

$$\sum_{l=-\infty}^{\infty}e^{-(\frac{2\pi}{\beta})^2(l+\frac{1}{2}-\frac{i\beta\delta\mu}{2\pi})^2 t} = \frac{\beta}{2\sqrt{\pi t}}\left(1 + 2\sum_{l=1}^{\infty}e^{-\frac{\beta^2 l^2}{4t}}(-1)^l\cosh(\beta\delta\mu l)\right).$$

As a result for $\Omega(s)$ we obtain

$$\Omega(s) = \frac{N}{\sqrt{\pi}\beta L}\frac{1}{\Gamma(s)}\mu^{2s}\int_0^\infty dt\, t^{s-\frac{3}{2}}\times \tag{41}$$

$$\times\frac{L}{2\sqrt{\pi t}}\left(1+2\sum_{n=1}^{\infty}e^{-\frac{L^2n^2}{4t}}\cos(2\pi\alpha n)\right)\frac{\beta}{2\sqrt{\pi t}}\left(1+2\sum_{l=1}^{\infty}e^{-\frac{\beta^2l^2}{4t}}(-1)^l\cosh(\beta\delta\mu l)\right)e^{-\sigma^2 t}.$$

Generalization to the case of $\delta\mu \neq 0$ is not achieved and here we assume $\delta\mu = 0$ and get:

$$\Omega(s) = \frac{N}{4\pi^{\frac{3}{2}}}\frac{1}{\Gamma(s)}\mu^{2s}\int_0^\infty dt\, t^{s-\frac{5}{2}}\left(1+2\sum_{n=1}^{\infty}e^{-\frac{L^2n^2}{4t}}\cos(2\pi\alpha n)\right)\left(1+2\sum_{l=1}^{\infty}e^{-\frac{\beta^2l^2}{4t}}(-1)^l\right)e^{-\sigma^2 t}. \tag{42}$$

There is the following relation, representation of exponent

$$e^{-A} = \frac{1}{2\pi i}\int_{Re(z)=c>0} dz\Gamma(z)A^{-z},$$

where one has to integrate over a contour $Re(z) = c > 0$. From now on wee assume that integration is over contour $Re(z) = c > \frac{3}{2}$.

Using this formula we could represent exponents as contour integrals and after not complicated transformations we obtain

$$\Omega(s) = \frac{N\sigma^3}{4\pi^{\frac{3}{2}}}\left(\frac{\mu}{\sigma}\right)^{2s}\frac{\Gamma(s-\frac{3}{2})}{\Gamma(s)}+ \tag{43}$$

$$+\frac{N\sigma^3}{4\pi^{\frac{3}{2}}}\frac{1}{\Gamma(s)}\left(\frac{\mu}{\sigma}\right)^{2s}\frac{1}{2\pi i}\int dz\Gamma(z)\Gamma\left(z+s-\frac{3}{2}\right)\sum_{(l,n)\neq(0,0)}\frac{(-1)^l\cos(2\pi\alpha n)}{(\beta^2l^2+L^2n^2)^z}\left(\frac{\sigma}{2}\right)^{-2z}.$$

In order to get V_{eff}, one has to take a derivative with respect to s at a point $s = 0$. In both terms nonzero contribution is provided from deriving of Γ-function in denominator. We get a function

$$\frac{d}{ds}\left(\frac{1}{\Gamma(s)}\right)\bigg|_{s=0} = -\frac{\Gamma'(s)}{\Gamma^2(s)}\bigg|_{s=0} = -\frac{\psi(s)}{\Gamma(s)}\bigg|_{s=0} = 1$$

$$\psi(s) = \frac{\Gamma'(s)}{\Gamma(s)}.$$

ψ has poles of 1-st order at points $s = -1, -2, -3, \ldots$.

$Res_{s=0}(\psi(s)) = -1$, for $\psi(z+1) = \psi(z) + \frac{1}{z}$.

As a result the first term in (43) together with the first term in (38) give the following contribution to thermodynamic potential:

$$\frac{N}{2G_r}\sigma^2 + \frac{N\sigma^3}{3\pi},$$

where - G_r -renormalized coupling constant, subscript r we inserted after ζ-regularization of expression for thermodynamic potential. In reality it remains literally the same and equals the coupling constant G.

Using the following notation $\frac{1}{G_r} = -\frac{\sigma_0}{\pi}$, one can see that the expression for thermodynamic potential in the case $L = \infty, \beta = \infty, \mu = 0, \delta\mu = 0, B = 0$ coincides with (19), when cut-off regularization was used.

Let us introduce a function Z

$$Z(a,b,\alpha,s) = \sum_{(l,n)\neq(0,0)} \frac{\cos(2\pi\alpha n)}{(a^2l^2+b^2n^2)^s},$$

which in case of $\alpha = 0$ is reduced to Epshtein Z-function

$$Z(a,b,s) = \sum_{(n,m)\neq(0,0)} \frac{1}{(a^2n^2+b^2m^2)^s}.$$

Let us write the relation which helps to bring the sum in the expression for $V_{\rm eff}$, to Z-function.

$$\sum_{(n,m)\neq(0,0)} \frac{(-1)^n}{(a^2n^2+b^2m^2)^s} = 2Z(2a,b,s) - Z(a,b,s), \tag{44}$$

$$\sum_{(n,m)\neq(0,0)} \frac{(-1)^{n+m}}{(a^2n^2+b^2m^2)^s} = Z(a,b,s) - 2Z(2a,b,s) - 2Z(a,2b,s) + 4Z(2a,2b,s), \tag{45}$$

$$\sum_{(l,n)\neq(0,0)} \frac{(-1)^l\cos(2\pi\alpha n)}{(a^2l^2+b^2n^2)^s} = 2Z(2a,b,\alpha,s) - Z(a,b,\alpha,s). \tag{46}$$

An integral in the expression (43) for $V_{\rm eff}$ in the case of $\alpha = 0$ with the help of formula (44), and in the case of $\alpha = \frac{1}{2}$ with the help of formula (45), can be brought down to the integral of the form

$$I = \frac{N\sigma^3}{4\pi^{\frac{3}{2}}}\frac{1}{2\pi i}\int dz\Gamma(z)\Gamma\left(z-\frac{3}{2}\right)Z(\beta,L,z)\left(\frac{\sigma}{2}\right)^{-2z}.$$

It can be shown that a function $\Gamma(z)Z(\beta,L,z)\Gamma\left(z-\frac{3}{2}\right)$ has poles at points $z = \frac{3}{2} - n, n = 0,1...$ due to $\Gamma\left(z-\frac{3}{2}\right)$ and at points $z = 0, z = 1$ due to $\Gamma(z)Z(\beta,L,z)$.

The pole at $z = \frac{3}{2}$ gives a result independent of σ, and can be omitted. The pole at $z = 1$ gives a term proportional to σ. The pole at $z = \frac{1}{2}$ gives a term proportional to σ^2. Other poles gives higher orders of σ.

Using the relation $\Gamma(-\frac{1}{2}) = -2\sqrt{\pi}$ one can get

$$Z(\beta,L,\frac{1}{2}) = \frac{8}{\beta}\sum_{m=1}^{\infty}\sum_{n=1}^{\infty} K_1(2\pi\frac{L}{\beta}nm)\cos(2\pi\alpha m).$$

As a result

$$I = -\frac{N}{2}\frac{1}{\beta L}\sigma - \frac{2N\sigma^2}{\pi\beta}\sum_{m=1}^{\infty}\sum_{n=1}^{\infty} K_1(2\pi\frac{L}{\beta}nm)\cos(2\pi\alpha m) + O(\sigma^3).$$

The first term gives no contribution to thermodynamic potential (it can be seen from (44), (45)). As a result for $V_{\rm eff}$ in the case of $\alpha = 0$ we get

$$V_{\rm eff} = \frac{N}{\pi}(\frac{\sigma^3}{3} - \frac{\sigma^2\sigma_0}{2}) + \frac{2N\sigma^2}{\pi\beta}\sum_{m=1}^{\infty}\sum_{n=1}^{\infty}(K_1(2\pi\frac{L}{\beta}nm) - K_1(\pi\frac{L}{\beta}nm)) + O(\sigma^3). \tag{47}$$

In the case of $\alpha = \frac{1}{2}$ $V_{\rm eff}$ has the form

$$V_{\rm eff} = \frac{N}{\pi}\left(\frac{\sigma^3}{3} - \frac{\sigma^2\sigma_0}{2}\right) + \frac{2N\sigma^2}{\pi\beta}\sum_{m=1}^{\infty}\sum_{n=1}^{\infty}(2K_1(4\pi\frac{L}{\beta}nm) - 2K_1(2\pi\frac{L}{\beta}nm)+ \tag{48}$$

$$+K_1(\pi\frac{L}{\beta}nm) - K_1(2\pi\frac{L}{\beta}nm)) + O(\sigma^3).$$

For arbitrary value of α we get for thermodynamic potential

$$V_{\rm eff} = \frac{N}{\pi}\left(\frac{\sigma^3}{3} - \frac{\sigma^2\sigma_0}{2}\right) + \frac{2N\sigma^2}{\pi\beta}\sum_{m=1}^{\infty}\sum_{n=1}^{\infty}(K_1(2\pi\frac{L}{\beta}nm) - K_1(\pi\frac{L}{\beta}nm))\cos(2\pi\alpha m) + O(\sigma^3). \tag{49}$$

Now let us obtain expressions for thermodynamic potential in case of $\nu \neq 0$, using method of ζ-regularization. In order to do that let us go back to the expression (42)

$$\tilde{\Omega}(s) = \frac{N}{4\pi^{\frac{3}{2}}}\frac{\mu^{2s}}{\Gamma(s)}\int_0^\infty dt\, t^{s-\frac{5}{2}}e^{-\sigma^2 t} + \frac{N}{2\pi^{\frac{3}{2}}}\frac{\mu^{2s}}{\Gamma(s)}\int_0^\infty dt\, t^{s-\frac{5}{2}}\sum_{n=1}^{\infty}e^{-\frac{L^2n^2}{4t}-\sigma^2 t}\cos(2\pi\alpha n)+$$

$$+\frac{N}{\sqrt{\pi}\beta L}\frac{1}{\Gamma(s)}\mu^{2s}\int_0^\infty dt\, t^{s-\frac{3}{2}}\times \tag{50}$$

$$\times\frac{L}{2\sqrt{\pi t}}\left(1 + 2\sum_{n=1}^{\infty}e^{-\frac{L^2n^2}{4t}}\cos(2\pi\alpha n)\right)\frac{\beta}{2\sqrt{\pi t}}2\sum_{l=1}^{\infty}e^{-\frac{\beta^2l^2}{4t}}(-1)^l\,{\rm ch}(\beta\nu l)e^{-\sigma^2 t}.$$

In the third term of (50) we can use Poisson resummation formula backwards we get for thermodynamic potential

$$V_{\rm eff} = \frac{N}{\pi}\left(\frac{\sigma^3}{3} - \frac{\sigma^2\sigma_0}{2}\right) + \frac{N}{2\pi^{\frac{3}{2}}}\int_0^\infty dt\, t^{-\frac{5}{2}}\sum_{n=1}^{\infty}e^{-\frac{L^2n^2}{4t}-\sigma^2 t}\cos(2\pi\alpha n)+ \tag{51}$$

$$+\frac{N}{\pi L}\int_0^\infty dt\, t^{-2}\sum_{n=-\infty}^{\infty}\sum_{l=1}^{\infty}e^{-\frac{\beta^2l^2}{4t}-\sigma_n^2 t}(-1)^l\,{\rm ch}(\beta\nu l).$$

Second term in the expression for $V_{\rm eff}$ we can transform to

$$\frac{2N}{\pi L^3}\sum_{n=1}^{\infty}\frac{e^{-\sigma Ln}}{n^3}\cos(2\pi\alpha n) + \frac{2N\sigma}{\pi L^2}\sum_{n=1}^{\infty}\frac{e^{-\sigma Ln}}{n^2}\cos(2\pi\alpha n).$$

The third term in the expression for $V_{\rm eff}$ can be transformed to

$$\frac{2N}{\pi L\beta}\sum_{\sigma_n>\nu}\sum_{l=1}^{\infty}\frac{\sigma_n}{l}K_1(\beta\sigma_n l)(-1)^l e^{\beta\nu l} + \frac{2N}{\pi L\beta}\sum_{n=-\infty}^{\infty}\sum_{l=1}^{\infty}\frac{\sigma_n}{l}K_1(\beta\sigma_n l)(-1)^l e^{-\beta\nu l}+$$

$$+\frac{N}{2\pi L}\int_0^\infty dt\, t^{-2}\sum_{\sigma_n<\nu}\sum_{l=1}^{\infty}e^{-\frac{\beta^2l^2}{4t}-\sigma_n^2 t}(-1)^l e^{\beta\nu l}.$$

One can see that for $\sigma > \nu$ expression for thermodynamic potential coincides with the expression (29).

Now let us obtain an expression for thermodynamic potential in another form. Again go back to the expression (41), now apply Poisson resummation formula backwards to sum over l.

For thermodynamic potential we get

$$V_{\text{eff}} = \frac{N}{\pi}\left(\frac{\sigma^3}{3} - \frac{\sigma^2\sigma_0}{2}\right) + \frac{N}{2\pi^{\frac{3}{2}}}\int_0^{\infty} dt\, t^{-\frac{5}{2}} \sum_{l=1}^{\infty} e^{-\frac{\beta^2 l^2}{4t} - \sigma^2 t}(-1)^l \operatorname{ch}(\beta v l) + \tag{52}$$

$$+\frac{N}{\pi L}\sum_{l=-\infty}^{\infty}\sum_{n=1}^{\infty}\int_0^{\infty} dt\, t^{-2} e^{-\frac{L^2 n^2}{4t} - \sigma_l^2 t}\cos(2\pi\alpha n).$$

For $\sigma > v$ the second term can be transformed to

$$\frac{N}{\pi\beta^3}\left(Li_3\left(-e^{-\beta l(\sigma+v)}\right) + Li_3\left(-e^{-\beta l(\sigma-v)}\right) + \beta\sigma(Li_2\left(-e^{-\beta l(\sigma+v)}\right) + Li_2\left(-e^{-\beta l(\sigma-v)}\right)\right).$$

The third one can be transformed to

$$\frac{4N\sigma}{\pi\beta L}\sum_{l=-\infty}^{\infty}\sum_{n=1}^{\infty}\frac{\sigma_l}{n}K_1(L\sigma_l n)\cos(2\pi\alpha n).$$

One can see that this expression coincides with the expression (33).

7 Conclusions and some numerical estimation

There have been obtained expression for thermodynamic potential in two forms each is useful in different situations.

In case of periodic ($\alpha = 0$), if there is nonzero magnetic flux then it can restore the symmetry at zero temperature. However, the symmetry can be restored only for not very large radius $L < L_c$.

In the case of antiperiodic boundary condition ($\alpha = \frac{1}{2}$) phase portrait of the system is symmetric in the plane of (L,β), that is compactification restores the symmetry of the model as increasing of temperature (β is analogous to L). In some situation in this case magnetic field can break the symmetry.

For all boundary condition when magnetic fields are large there is a region of L, where the symmetry can be restored and broken down several times by not so large changes of L.

Using method of ζ-regularization we obtained the expression for the thermodynamic potential and its behavior at zero, which can be used in determining if zero is local minimum or local maximum of the thermodynamic potential.

Let us do some numerical estimation.

One can see that magnetic field always stands in the expression $\frac{\Phi}{\Phi_0}$, where Φ- magnetic flux, $\Phi_0 = \frac{2\pi}{e}$,

$\Phi = LA_2 = \frac{L^2 B}{4\pi}$. Magnetic flux contribute if at least $\frac{\Phi}{\Phi_0} \sim 10^{-2}$. Assume that $L \sim 10^{-7}$cm.

Let us estimate magnetic field that correspond to such magnetic flux.

$$\frac{\Phi}{\Phi_0} = \frac{eL^2 B}{8\pi^2} = \frac{1}{8\pi^2}\left(\frac{L}{\lambda_c}\right)^2 \frac{B}{B_c}, \tag{53}$$

where $B_c = \frac{m^2}{e} \sim 10^{13}$ G, $\lambda_c = \frac{1}{m}$ -Compton wavelength. Restoring $\hbar, c$, we write $\lambda_c = \frac{\hbar}{mc} \sim 10^{-13}$m. Instead of speed of light c if we consider application to graphene we should take Fermi velocity $v_F = \frac{c}{300}$,and in the following estimation we should perform transformation $L \to v_F L$. Estimate for the field is $B \sim 10^3$ G=0.1 T.

DOI: 10.1051/epjconf/201612604057

References

[1] D.J. Gross and A. Neveu, Phys. Rev. D **10**, 3235 (1974).

[2] A.S. Vshivtsev, B.V. Magnitsky, V.C. Zhukovsky and K.G. Klimenko, Phys. Part. Nucl. **29**, 523 (1998) [Fiz. Elem. Chast. Atom. Yadra **29**, 1259 (1998)].

[3] D. Y. Song, Phys.Rev D v.48, n 8, 1993

[4] A.V. Gamayun and E.V. Gorbar, arXiv:hep-ph 0411157v2, 2004

[5] I. V. Krive and S. A. Naftulin, Nuclear Physics B364 (1991) 541-550, 1991

[6] H. Caldas and R.O. Ramos, Phys. Rev. B **80**, 115428 (2009).

[7] Integrals and series
A.P. Prudnikov , J.A. Bruchkov, O.I. Marichev, "Nauka", 1981

[8] D. Ebert and K. G. Klimenko, Phys.Rev. D 82, 025018 2010

[9] D. Ebert, K.G. Klimenko, Phys. Rev. **D80**, 125013 (2009).

[10] A.S.Vshivtsev, A.G.Kisunko, K.G. Klimenko, D.V. Peregudov, IHEP 96-58

EPJ Web of Conferences **126**, 05001 (2016) DOI: 10.1051/epjconf/201612605001
ICNFP 2015

Confining potential of Y-string on the lattice at finite T

Ahmed S. Bakry [a], Xurong Chen, and Peng-Ming Zhang

Institute of Modern Physics, Chinese Academy of Sciences, Lanzhou 730000, and State Key Laboratory of Theoretical Physics, Institute of Theoretical Physics, Beijing 100190, China.

Abstract. The potential due to a system of three static quark ($3Q$) is studied using SU(3) lattice QCD at finite temperature with Polyakov loops operators. We focused our analysis on the large distance properties of the $3Q$ potential and found a good fit behavior to the Y-string model formula. In addition to the linearly confining term proportional to the minimal length of the Y-string, we observed that the subleading logarithmic term, which is proportional to Dedekind η function and accounts for the Y-string's quantum fluctuations, is necessary to reproduce the quark anti-quark $Q\bar{Q}$ string tension of the corresponding mesonic system at finite temperature.

1 Introduction

The three-quark confining potential is a fundamental quantity in elementary particle physics and is essential for understanding baryon structure [1]. It remains, however, an insurmountable task to put an analytic form to the binding forces between quarks in the non-perturbative region of QCD from first principles.

Perturbative QCD provides a good description to the short-distance aspects of the $3Q$ potential as a two-body (Coulombic) one-gluon-exchange (OGE) interaction potential. More recently, it has been shown [2] that the breakdown of the two body force in the short-range happens at two loops when the first genuine three-body force appears. Discussions concerning the intermediate and large distances, however, are usually carried out either on phenomenological bases [3–6], making use of the strong-coupling expansion [7, 8] arguments or with the lattice calculations of the three-quark Wilson loop [9].

Lattice simulations have shown that the $Q\bar{Q}$ potential is linearly rising [10]. In a string phenomenology, the linear rise is consistent with the formation of a stringlike flux tube linking the color sources. The quantum fluctuations of the string produce the well-known Coulomb-like sub-leading corrections to the $Q\bar{Q}$ potential, namely, the Lüscher term. Simulations of many confining gauge models [11–17] have verified the existence of this term.

In the baryon, the fit analysis of the lattice data would suggest two types of parametrization [9] depending on the interquark distances, i.e., the Δ and Y-type potential. The Δ-potential describes a sum of two-body forces and is proportional to the perimeter of the $3Q$ triangle with a string tension half that of the corresponding $Q\bar{Q}$ system. On the other hand, the Y-potential approximates a three-body force proportional to the minimal length of the three strings with the same string tension as

[a]e-mail: abakry@impcas.ac.cn

the corresponding $Q\bar{Q}$ system. In this picture, the three body force is conceived as a result of the formation of a Y-shaped string linking the three-static quarks. This realization is consistent with the strong coupling approximation and with the dual superconducting picture of QCD [7, 18, 19].

However, many subtleties in the lattice simulations have made the exact distances at which each behavior takes over a subject of debate [3, 4, 9, 20–22]. In addition, unlike the simpler $Q\bar{Q}$ case, the sub-leading corrections arising from the quantum fluctuations of the Y-string linking the $3Q$ quark system yet has not been verified in SU(3) gauge theory. These stringy fluctuations are expected to be more enhanced at temperatures very close to the deconfinement point.

Usually in lattice simulations the potential energy is extracted from trial states that maximally overlap with the ground state using Wilson loop operators [9, 22–24]. Otherwise, the potential may be extracted with the use of Polyakov loop operators. These thermal loops address the free-energy of the system [25] and are suitable operators to study lattice gauge theories coupled to heatbath.

At zero temperature, the numerical evaluation of the $3Q$ potential with the use of Polyakov loop operators in 4D SU(3) Yang-Mills theory is known to be computationally challenging. At finite temperature, on the other hand, the loop operators assume a short extent in the time direction. By the employment of adequate levels of UV filtering [26] such that the physical measurements of the $3Q$ potential are intact, the evaluation of the two-point correlation function to extract the potential and even the three-point function corresponding to the energy density can be viable [27, 28] at finite temperature. The simulations for the energy density in the $3Q$ system, for example, have revealed a filled Δ-shape profile even at large distances.

A more profound analysis of these filled-Δ-shaped profiles revealed an underlying system of three-convoluted Y-shaped-Gaussian [28]. The energy density profile is peculiar [28] and has not been observed at zero temperature with the employment of Wilson's loop operator [29–31]. Nevertheless, the characteristics of the associated potential of the static $3Q$ system at finite temperature is not yet fully understood.

In this study, we discuss the $3Q$ potential derived from the Y-string model and investigate concisely how it compares to the corresponding lattice data of the SU(3) pure Yang-Mills theory at finite temperature. This may delineate the fine properties of the energy density and potential of the $3Q$ system even at low and zero temperatures. Our goal is to find out if the Y-law derived from the effective bosonic string model [6] provides a satisfactory description of the lattice data and reproduces the corresponding $Q\bar{Q}$ string tension. To scrutinize further the nature of the $3Q$ potential data, we draw a comparison with other fit ansatzes such as the Δ-type potential and the so-called bare Y-ansatz with the string's sub-leading corrections suppressed.

The contents are organized as follows: In Section (II), we discuss the Y-string model at finite temperatures. In section (III) we describe the $3Q$ potential measurements. The fit analysis of the lattice data is depicted in section (IV). The conclusion is provided in the last section.

2 String's phenomenology of the $3Q$ potential

A pure Yang-Mills vacuum can admit stable string-like objects which confine static color charges and give rise to the linearly rising potential. In a mesonic string, the quantum fluctuations entail the Lüscher subleading correction [25, 32] to the linearly rising potential. This has been found in consistency with very precise lattice measurements of the the $Q\bar{Q}$ potential for color source separation commencing from distance $R = 0.4$ fm [25].

At high temperature, the thermal behavior of the free-string manifests at source separation distance scales larger than what one would expect normally in the zero temperature regime $R \approx 0.8$ fm [26]. The inclusion of higher-order string's self-interactions improves the match between the lattice data and the string model [33].

In a static baryon, the Y-string system is expected [6] to be a stable configuration in the IR region of the theory. This system accounts for three strings originating from a node to the static three quarks. A theoretical development discussing the effects of the quantum fluctuations of the Y-string on the $3Q$ potential has been reported [6]. The calculations of the Casimir energy in this model have indicated a geometrical Lüscher-like sub-leading term for the Y-($3Q$) potential [5]. The Lüscher-like term is a gauge-group-independent and depends only on the geometry of the three static charges.

The simulation of 3-Potts model showed a $3Q$ potential [5] consistent with the Lüscher-like correction γ/L_Y to the linearly rising Y-ansatz, and hence the applicability of the Y-baryonic string picture at large distances.

The Y-string may be described by the Nambu-Goto type string action with Dirichlet boundary condition at the static color sources. In this formalism, the action is proportional to the total area of the three blade world sheet system swept by the fluctuating world lines of the strings, in addition to a boundary term which accounts for the junction fluctuations. This Y-string system [6] has a mixed Dirichlet-Neumann condition [34] and the string's partition function $Z_i(\phi)$ in D dimensions [6] is given by

$$Z_i(\phi) = e^{-\frac{\sigma}{2}\int|\partial\xi_{\mathrm{min},i}|^2}|\det(-\Delta_{\Theta_i})|^{-(D-2)/2}\,, \tag{1}$$

where $\xi_{\mathrm{min},i}$ is the minimal-area solution for a given junction configuration $\phi(t)$, and Δ_{Θ_i} denotes the Laplacian acting on the domain (blade) Θ_i.

Jahn and De Forcrand [6] calculated the Casimir energy for the baryonic potential V_{3Q}. This was done by evaluating the determinant of the Laplacian in Eq. (1) by conformally mapping the resulting domains to rectangles. This results in an expression for the determinant in terms of the Dedekind η function and a Gaussian function of the junction fluctuations, $\mathbf{e_i}\cdot\phi_{\mathbf{w_n}}$, which in 4D is given by

$$\det(-\Delta_{\Theta_i}) = \eta^2\left(\frac{iL_T}{2L_i}\right)\exp\left(-\frac{1}{12\pi}\sum_{w_n} w_n^3\coth(w_nL_i)|\mathbf{e_i}\cdot\phi_{\mathbf{w_n}}|^2\right),$$

with $w_n = 2\pi n/L_T$, where $L_T = 1/T$ is the time extent of the blade, and L_i are the lengths of the strings. The sum over all eigen-energies would result in a Lüscher-like correction to the V_{3Q} potential at zero temperature [6]. The baryonic potential V_{3Q} then reads

$$V_{3Q}(L_i) = \sigma L_Y + V_{\parallel} + 2V_{\perp} + O(L_i^{-2})\,,$$

with

$$V_{\parallel}(L_i) = -\sum_i\frac{1}{2L_T}\eta\left(\frac{iL_T}{2L_i}\right) + \sum_{w=0}\frac{1}{L_T}\ln\left[\frac{1}{3}\sum_{i<j}\coth(wL_i)\coth(wL_j)\right],$$

for the in-plane component and

$$V_{\perp}(L_i) = -\sum_i\frac{1}{2L_T}\eta\left(\frac{iL_T}{2L_i}\right) + \sum_{w=0}\frac{1}{L_T}\ln\left[\frac{1}{3}\sum_i\coth(wL_i)\right], \tag{2}$$

is the potential component due to the perpendicular fluctuations such that $L_Y \equiv L_1 + L_2 + L_3$ is the minimal of the sum of the lengths of the three strings. The corresponding mesonic limit would read

$$V_{\perp} = -\frac{1}{2L_T}\eta\left(\frac{iL_T}{2L_1}\right) - \frac{1}{2L_T}\eta\left(\frac{iL_T}{2L_2}\right) + \sum_{w=0}\frac{1}{L_T}\ln\left[\frac{1}{2}\coth(wL_1) + \coth(wL_2)\right]. \quad (3)$$

The quark anti-quark ($Q\bar{Q}$) potential is then

$$V_{Q\bar{Q}} = \sigma(L_1 + L_2) + \frac{1}{L_T}\ln\left[\eta\left(\frac{iL_T}{2(L_1 + L_2)}\right)\right], \quad (4)$$

which is in agreement with the mesonic string potential [33]. Expressing the sum in Eq. (2) in terms of Dedekind η functions, the potential in the $3Q$ channel would read

$$V_{3Q} = \sigma L_Y - \frac{\gamma}{L_T}\ln\left[\eta\left(\frac{iL_T}{2L_Y}\right)\right], \quad (5)$$

where γ is a geometrical factor that can be evaluated numerically by solving Eq. (2).

3 The lattice QCD measurement for the 3Q potential

A transfer matrix interpretation to the Polyakov loops correlator allow to obtain the $3Q$ static potential V_{3Q} with the center symmetry preserving operator

$$\begin{aligned}\langle \mathcal{P}_{3Q}\rangle &= \langle \mathcal{P}(\vec{r}_1)\mathcal{P}(\vec{r}_2)\mathcal{P}(\vec{r}_3)\rangle, \\ &= \exp(-V_{3Q}(\vec{r}_1, \vec{r}_2, \vec{r}_3; T)), \end{aligned} \quad (6)$$

where the Polyakov loop is given by

$$P(\vec{r}) = \frac{1}{3}\mathrm{Tr}\left[\prod_{n_{t=1}}^{N_t} U_{\mu=4}(\vec{r}, n_t)\right]. \quad (7)$$

In the context of the Polyakov loop method we have contributions of temperature-dependent effects to the free energy of a system of three static charges coupled to a heatbath. In the following we describe the lattice arrangement and parameters used to extract the above correlator Eq. (6) at finite temperature.

The gauge configurations were generated using the standard Wilson gauge action. The two lattices employed in this investigation are of a typical spatial size of $3.6^3\mathrm{fm}^3$. Performing the simulations on large enough lattice sizes would be beneficial to gain high statistics in a gauge-independent manner and also minimizing the mirror effects and correlations across the boundaries as a by-product [31, 35].

The SU(3) gluonic gauge configurations have been generated employing a pseudo-heatbath algorithm [36, 37] updating the corresponding three SU(2) subgroup elements [38]. Each update step consists of one heatbath and 5 micro-canonical reflections. We chose to perform our analysis with lattices as fine as $a = 0.1$ fm by adopting a coupling of value $\beta = 6.00$, with temporal extents of $N_t = 8$, and $N_t = 10$ slices, which correspond to temperatures $T/T_c \simeq 0.9$, and $T/T_c \simeq 0.8$, respectively.

We perform a set of measurements $n_{\mathrm{sub}} = 20$ separated by 70 sweeps of updates. Each set of measurements is taken following 2000 updating sweeps. These sub-measurements are binned together in evaluating Eq. (6). The total measurements taken on 500 bins. In this investigation, we have taken 10,000 measurements at each temperature. The measurements are taken on hierarchically generated configurations.

An ultraviolet (UV) filtering step [39] precedes our measurements of the three Polyakov loop correlator Eq. (6). We have previously shown that with adequate levels of 4D smearing the effects on the $3Q$ potential can be kept intact. We have shown that this is consistent with a mean square diameter of the Brownian motion smaller than any interquark separation.

4 Numerical results and discussions

For the study of the $3Q$ potential V_{3Q}, we are interested in the large-distance behavior which is directly related to the properties of the confining force rather than the short-distance behavior [9] that is well described by the two-body Coulomb-type potential.

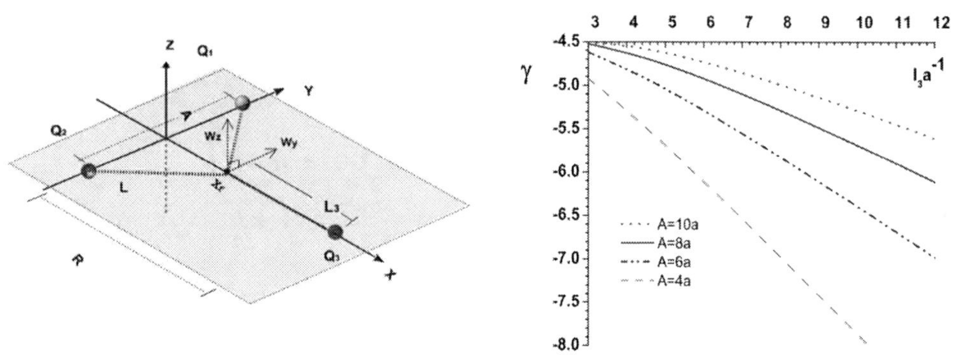

Figure 1. (Left) Schematic diagram showing the configuration of the Y string relative to the quark source positions. The junction's locus is fixed at Fermat point x_f. (Right) The geometrical factor γ of the log term in Eq. (5). The solid lines correspond to γ as a function of the third string length L_3 for three isosceles triangles with bases of length $A = 0.6$, $A = 0.8$ fm and $A = 1.0$ fm. The solid lines correspond to γ as a function of the third string length L_3a^{-1}.

Table 1. Lists the returned values of the $\chi^2_{\text{dof}}(x)$ for fits of the lattice data to the string model formula Eq. (5), the fits are for isosceles triangle quark configurations of base width $A = 6\,a$ and $A = 8\,a$ at two temperatures.

Fit Range	R=5-12		R=6-12		R=7-12	
Fit Parameters	χ^2	C	χ^2	C	χ^2	C
$T/T_c = 0.8$						
$A = 6a$	15.06	-4.17	7.61	-4.17	1.93	-4.16
$A = 8a$	3.99	-3.82	2.1	-3.83	1.38	-3.84
$T/T_c = 0.9$						
$A = 6a$	46.0	-3.05	18.6	-3.04	5.50	-3.03
$A = 8a$	2.3	-2.77	2.27	-2.77	2.26	-2.77

We evaluate the correlator Eq. (6) to extract the heavy quarks potential for a planar $3Q$ arrangement corresponding to isosceles triangles of base width A and height R as illustrated in Fig. 1. Figure 1 also shows the numerical values of the geometrical factor γ as a function of triangle length L_3 for isosceles configurations $L_1 = L_2$ evaluated by solving Eq. (2) for the Y-string potential.

The measured numerical values of the potential are reported in Fig. 2 and Fig. 3 for isosceles $3Q$ triangles $3Q$ corresponding to bases of width, $A = 0.6$ and 0.8 fm, at the two considered temperatures $T/T_c = 0.8$ and $T/T_c = 0.9$. The numerical values of the potential have been normalized to their values at isosceles height $R = 1.2$ fm.

The potential data for each base are extracted in accord to Eq. (5) by varying the height of the triangle, R, for each fixed base, as shown in Fig. 1. Our consideration of different isosceles $3Q$ configuration is intended to enable a systematic identification of the distance scales at which the string model holds.

Formula Eq. (5) sums up the contributions to the $3Q$ potential resulting from both the in-plane and the perpendicular fluctuations as discussed above. Before fitting this string potential to the lattice data, the Y-string's configuration has to be fixed to its minimal value, i.e., with the node's position at Fermat point of the $3Q$ triangular configuration. The position of the Fermat point of the planar isosceles arrangement is given by $R = \dfrac{A}{2\sqrt{3}}$.

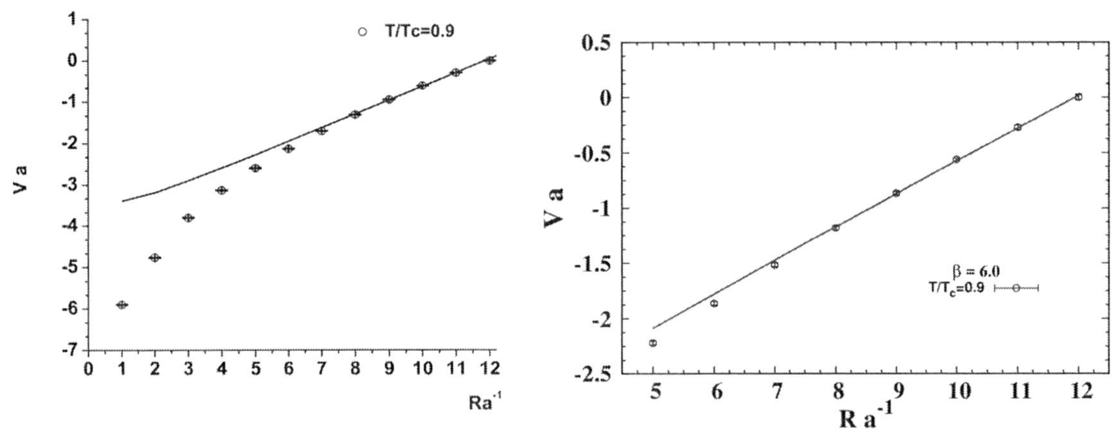

Figure 2. (Left) The lattice data of the $Q\bar{Q}$ potential at $T/T_c = 0.9$. (Right) the corresponding $3Q$ potential of a planar $3Q$ isosceles triangle with a base length $A = 0.6$ fm. The solid lines correspond to the best fits to the string model Eq. (4) and Eq. (5), respectively.

Table I summarizes the returned values of χ^2 from the resultant fits of the $3Q$ potential at both temperatures. The rescaling parameter C is the only fit parameter used.

In general, the fits show strong dependency on the fit range with the inclusion of the points at small Q_3 sources separations. Figure 2, for example, depicts the potential data corresponding to an isosceles $3Q$ triangle of the smallest width considered $A = 0.6$ fm and the corresponding fits to Eq. 5. The points at small isosceles height $R = 0.5$ fm and $R = 0.6$ fm deviate from the Y-string model curve. These deviations at relatively small heights and bases can be conceived as relevant to self-interactions of the two strings connecting the two quarks at the base, Q_1 and Q_2, in addition to the interaction of the node with the third quark Q_3. That is, higher order effects are expected to be more pronounced at higher temperatures.

The string length in the case of the Y-string linking any two quark sources is greater than the mesonic string of a corresponding $Q\bar{Q}$ pair. A fit of the potential to the mesonic string potential Eq. (4) is shown in the right graph of Fig.2. The best fits are returned if only the last four points are included, i.e., from quark separation distances $R = 0.8$ fm to $R = 1.2$ fm. The value returned for the string tension $\sigma_{Q\bar{Q}}a^{-2} = 0.032$ is the one used as input for the string tension in formula Eq. 5 for the baryonic Y-string potential.

With the increase of the width of the base of the isosceles to $A = 0.8$ fm the values of χ^2 in Table I reduce even for small isosceles heights of $R = 0.5$ fm and $R = 0.6$ fm. A wider base of the isosceles triangle would increase the length of the string linking any two quarks $L_1 + L_3$. This approaches the length at which the corresponding free mesonic string commence to match the lattice data. One can also add to this the observation that the self-interactions of the adjacent strings in the Y-string configuration are expected to be negligible with the increase of the source separation at the base.

We list in Table II values of the χ^2 returned from fits of the $3Q$ potential to a Δ-ansatz [4, 9] given by the form

$$V_{3Q}(\vec{r}_1, \vec{r}_2, \vec{r}_3) = -\frac{1}{2}A_{Q\bar{Q}} \sum_{i<j} \frac{1}{|\vec{r}_i - \vec{r}_j|} + \frac{1}{2}\sigma_{Q\bar{Q}} \sum_{i<j} |\vec{r}_i - \vec{r}_j|. \tag{8}$$

The constant $A_{Q\bar{Q}}$ determines the strength of the OGE Coulombic term derived from perturbative QCD (see Ref. [4, 9]). The value of the string tension is taken to the half that of the mesonic string. To appreciate the effects of the junction fluctuations we consider also the fits to a Y-ansatz which in addition to the perturbative Coulombic term includes only the linearly confining term corresponding to the classical configuration of the string $L_{Y_{min}}$ with the sub-leading corrections of Eq. (5) suppressed, i.e.,

$$V_{3Q}(\vec{r}_1, \vec{r}_2, \vec{r}_3) = -\frac{1}{2}A_{Q\bar{Q}} \sum_{i<j} \frac{1}{|\vec{r}_i - \vec{r}_j|} + \sigma_{QQ}L_Y. \tag{9}$$

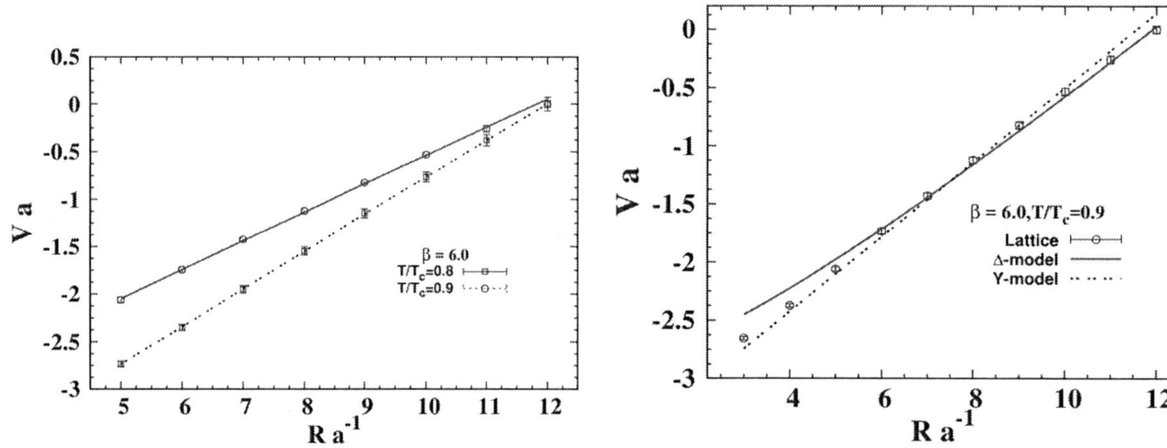

Figure 3. The $3Q$ potential versus the height length of isosceles triangular configuration of the planar $3Q$ with bases of length $A = 0.8$ fm. (Left) The lines correspond to the $3Q$ potential according to the best fits to the string picture formula of Eq. (5). (Right) The lines correspond to fits to Eq. (8) and Eq. (9) for fit range $R \in [5, 12]$.

Let us consider the highest temperature $T/T_c = 0.9$ and focus our comparison on the triangular configuration of the base width $A = 0.8$ fm. The values in Table II indicate higher χ^2 for the Δ-shape

Table 2. The returned $\chi^2_{\text{dof}}(x)$ for fits of the lattice data to $3Q$ isosceles of width $A = 0.8$ fm at $T/T_c = 0.9$. The fits compare Eq. (8) for the Δ-ansatz, Eq. (5) for the Y-string model and Eq. (9) for a bare Y-ansatz.

Fit Range	R=5-8	R=8-12	R=5-12
Fit Parameters	χ^2	χ^2	χ^2
Δ-ansatz	27.43	9.63	40.5
Y-string model	1.25	6.85	8.09
Y-bare ansatz	2.36	40.84	89.71

parametrization if the fit range includes all the considered points $R \in [5, 12]$. For points corresponding only to larger heights of the isosceles $R \in [8, 12]$, we interestingly observe that the bare Y-model, accounting merely to the classical configuration, returns high values of χ^2 compared to the Y-string model Eq. 5. At large distances and temperatures [28] the Y-string assumes a higher energy content and broadening profile. This can be read as well from the value of the χ^2 corresponding to the Δ-parametrization Eq. 8 if the same fit range is considered. The returned χ^2 values are relatively smaller and closer to the corresponding values of the Y-string model Eq. (5). A consistent physical realization of these observations is that the fluctuations of the node within the static baryon broaden largely enough to give similar effects to a hypothetical Δ-string circumventing the $3Q$ triangle.

The Y-string model Eq. (5) provides a good description of the lattice numerical data as indicated in Fig. 3 and Tables I-II. Since the model's formula contains a term encoding the effects of the fluctuations of the baryonic node, we could understand by drawing a comparison with the fit behavior of the other ansatzes, Eq. (8) and Eq. (9), to what extent the string fluctuations are relevant at the short and long distance scales.

5 Conclusion

The static $3Q$ potential has been studied in lattice QCD at finite temperature using Polyakov loops as a quark source operator. We discussed the potential energy of a Y-string at finite temperature with a fit analysis to the lattice data. The Y-string model provides a suitable description of the numerical data for fit ranges commencing from distances greater than 0.5 fm. The Y-string formula includes a three-body linear confinement term $\sigma_{3Q} L_Y$ in addition to a sub-leading logarithmic term due to the string's Gaussian fluctuations. We found the sub-leading corrections are necessary to reproduce the mesonic string tension measured with the corresponding free string model formula [40].

At finite temperature, the baryonic gluonic fields are always of a filled Δ-type [27, 28] with characteristics that have been found to be consistent with the Y-string model describing a system of fluctuating strings [28]. We found here that the data corresponding to the $3Q$ potential match the Y-string model as well. These two results could be of significant interest to the study of QCD strings and the gluonic fields in general and may be promoted into a novel picture for the properties of the confining force in the baryon even at low and zero temperatures.

Acknowledgment

This work has been funded by the Chinese Academy of Sciences President's International Fellowship Initiative, Grant No. 2015PM062, NSFC Grants (No. 11035006, No. 11175215, No. 11175220), and the Hundred Talents Program of the Chinese Academy of Sciences (Y101020BR0).

References

[1] F.J. Llanes-Estrada, O.I. Pavlova, R. Williams, Eur. Phys. J. **C72**, 2019 (2012), 1111.7087
[2] N. Brambilla, J. Ghiglieri, A. Vairo, Phys.Rev. **D81**, 054031 (2010), 0911.3541
[3] O. Andreev (2015), 1511.03484
[4] T.T. Takahashi, H. Suganuma, Y. Nemoto, H. Matsufuru, Phys. Rev. **D65**, 114509 (2002)
[5] P. de Forcrand, O. Jahn, Nucl. Phys. **A755**, 475 (2005), hep-ph/0502039
[6] O. Jahn, P.D. Forcrand, Nucl. Phys. B - Proc. Suppl. **129-130**, 700 (2004), lattice 2003
[7] J. Kogut, L. Susskind, Phys. Rev. D **11**, 395 (1975)
[8] T.Z. Nakano, K. Miura, A. Ohnishi, Prog. Theor. Phys. **123**, 825 (2010), 0911.3453
[9] T.T. Takahashi, H. Matsufuru, Y. Nemoto, H. Suganuma, Phys. Rev. Lett. **86**, 18 (2001)
[10] M. Creutz, Phys. Rev. **D21**, 2308 (1980)
[11] M. Luscher, P. Weisz, JHEP0207 **049** (2002)
[12] K.J. Juge, J. Kuti, C. Morningstar, Phys. Rev. Lett. **90**, 161601 (2003), hep-lat/0207004
[13] G.S. Bali, C. Schlichter, K. Schilling, Phys. Rev. D **51**, 5165 (1995)
[14] P. Pennanen, A.M. Green, C. Michael, Phys. Rev. **D56**, 3903 (1997), hep-lat/9705033
[15] M. Caselle, F. Gliozzi, U. Magnea, S. Vinti, Nucl. Phys. **B460**, 397 (1996), hep-lat/9510019
[16] F. Gliozzi, M. Pepe, U.J. Wiese, Phys.Rev.Lett. **104**, 232001 (2010), 1002.4888
[17] F. Gliozzi, M. Pepe, U.J. Wiese (2010), 1010.1373
[18] S. Capstick, N. Isgur, Phys. Rev. D **34**, 2809 (1986)
[19] N. Brambilla, G.M. Prosperi, A. Vairo, Phys. Lett. B **362**, 113 (1995)
[20] R. Sommer, J. Wosiek, Phys. Lett. B **149**, 497 (1984)
[21] H. Thacker, E. Eichten, J. Sexton, Nucl. Phys. B - Proc. Suppl. **4**, 234 (1988)
[22] C. Alexandrou, P. de Forcrand, O. Jahn, Nucl. Phys. Proc. Suppl. **119**, 667 (2003)
[23] C. Alexandrou, P. De Forcrand, A. Tsapalis, Phys. Rev. **D65**, 054503 (2002), hep-lat/0107006
[24] A.S. Bakry, D.B. Leinweber, A.G. Williams, Annals Phys. **326**, 2165 (2011), 1102.3477
[25] M. Luscher, K. Symanzik, P. Weisz, Nucl. Phys. **B173**, 365 (1980)
[26] A.S. Bakry, D.B. Leinweber, A.G. Williams, Phys.Rev. **D85**, 034504 (2012), 1011.1380
[27] A.S. Bakry, D.B. Leinweber, A.G. Williams, Phys. Rev. **D91**, 094512 (2015), 1107.0150
[28] A.S. Bakry, X. Chen, P.M. Zhang, Phys. Rev. **D91**, 114506 (2015), 1412.3568
[29] F. Okiharu, R. Woloshyn, Nucl.Phys.Proc.Suppl. **129**, 745 (2004), hep-lat/0310007
[30] Bissey et al., Nucl.Phys.Proc.Suppl. **141**, 22 (2005), hep-lat/0501004
[31] F. Bissey et al., Phys. Rev. D **76**, 114512 (2007), hep-lat/0606016
[32] M. Luscher, P. Weisz, JHEP **07**, 049 (2002), hep-lat/0207003
[33] M. Caselle, M. Hasenbusch, M. Panero, P. Provero, Nuclear Physics B **119**, 499 (2003)
[34] G. 't Hooft (2004), hep-th/0408148
[35] G.S. Bali, K. Schilling, C. Schlichter, Phys. Rev. **D51**, 5165 (1995), hep-lat/9409005
[36] K. Fabricius, O. Haan, Phys. Lett **B143**, 459 (1984)
[37] A.D. Kennedy, B.J. Pendleton, Phys. Lett **B156**, 393 (1985)
[38] N. Cabibbo, E. Marinari, Phys. Lett. **B119**, 387 (1982)
[39] A.S. Bakry et al., Phys. Rev. D **82**, 094503 (2010,hep-lat/1004.0782)
[40] N. Cardoso, P. Bicudo, Phys. Rev. D **85**, 077501 (2012)

Opto-box: Optical modules and mini-crate for ATLAS pixel and IBL detectors

David Bertsche[1,a], on behalf of the ATLAS Collaboration
[1]*University of Oklahoma, United States*

Abstract. The opto-box is a custom mini-crate for housing optical modules which process and transfer optoelectronic data. Many novel solutions were developed for the custom design and manufacturing. The system tightly integrates electrical, mechanical, and thermal functionality into a small package of size 35x10x8 cm^3. Special attention was given to ensure proper shielding, grounding, cooling, high reliability, and environmental tolerance. The custom modules, which incorporate Application Specific Integrated Circuits, were developed through a cycle of rigorous testing and redesign. In total, fourteen opto-boxes have been installed and loaded with modules on the ATLAS detector. They are currently in operation as part of the LHC run 2 data read-out chain. This conference proceeding is in support of the poster presented at the International Conference on New Frontiers in Physics (ICNFP) 2015 [1].

1 Introduction

ATLAS is a general-purpose particle detector located at the Large Hadron Collider (LHC) at CERN [2]. The Pixel and IBL detectors are the innermost systems which provide crucial tracking and vertexing functions. These detectors produce electronic signals which must be converted to optical signals for read-out. The opto-box is a custom mini-crate for housing the optoelectronic signal transceiver modules (opto-boards) which process and transfer this optoelectronic data.

During the first LHC long shut down (LS1) the opto-boards were moved to a new location outside of the Inner Detector. This decision was motivated by past pixel module failures in order to increase ease of access to the system during future data taking runs. This system was required to fit into the available space, so the opto-box was designed and constructed for this purpose. New high reliability opto-boards (including Application Specific Integrated Circuits, ASICs) were also developed and produced. In addition, this system was implemented to support the new Insertable B-Layer (IBL) of pixel modules also installed during LS1. The full system requirements and specifications are available in the engineering documentation [3].

2 Design

Each opto-box is of size 35x10x8 cm^3, having 30 wire bundles, 72 fibre cables, and two cooling pipes connected. For ease of installation and maintenance, much consideration was given to cable management and physical integration with the already existing detector components.

The opto-box is required to house, cool, and power the opto-boards, one of which is shown in Figure 1. An important consideration in the design was the integration of electrical, mechanical, and thermal systems. Several novel solutions were developed to meet these specific requirements.

[a]e-mail: david.bertsche@cern.ch

Figure 1: Opto-boards are optoelectronic signal transceiver modules that convert signals between electrical and optical form by using vertical-cavity surface-emitting-lasers (VCSELs). Optical fibers carry signals between the opto-boards and the read-out electronics.

The opto-box was designed as a micro-crate, with a slot and spring mechanism that allows insertion of the opto-boards from one side. A detailed mechanical design drawing of the opto-box is shown in Figure 2, where all the components shown provide mechanical support. In addition, some components also provide other functions. For example, the motherboard Printed Circuit Board (PCB) on the bottom distributes power and control signals to the opto-boards and the cooling rail at the top carries away excess heat generated by the boards.

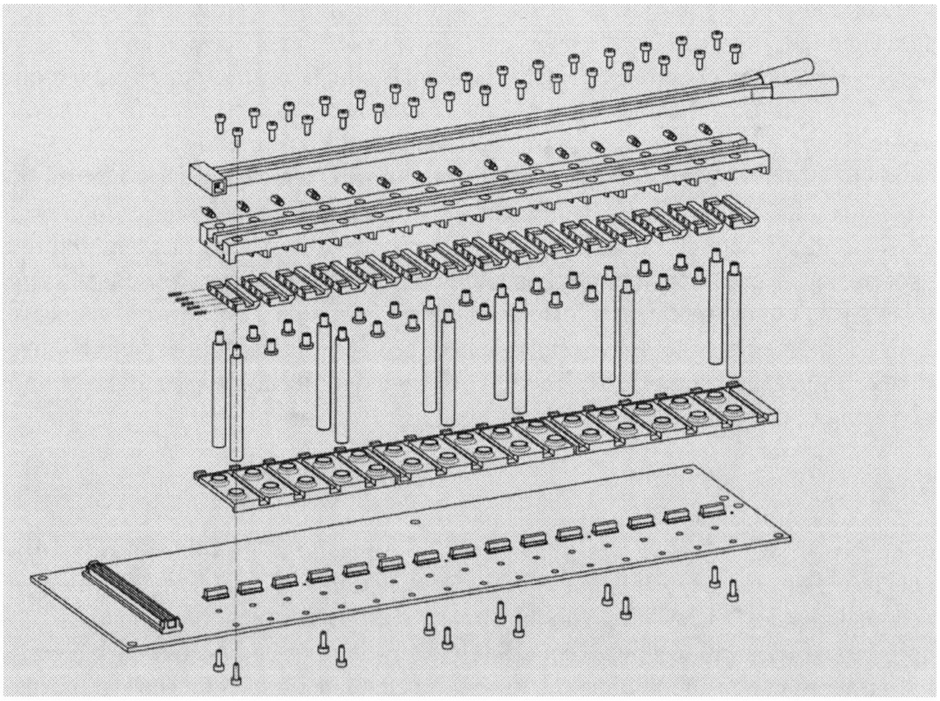

Figure 2: Computer Aided Design (CAD) image of the opto-box frame.

Cooling is provided by means of thermal contact with an aluminium cooling rail embedded with copper pipes through which C_6F_{14} fluid is chilled to 16°C is circulated by the ATLAS monophase cooling system. This fluorocarbon is chosen because of its inert and radiation-hard properties. The expected maximum power usage per opto-board is 1685 mW and 2385 mW for Pixel and IBL boards respectively.

New pixel electrical readout (ER) bundle connectors were developed, as shown in Figure 3. They contain an integrated PCB and micro-connectors which enables the connection of the ER bundle, the opto-board and the opto-box using only one connector.

Figure 3: Computer Aided Design (CAD) image of the electrical readout cables used to attach each of the opto-boards to the detector. 24 Electrical Readout (ER) bundles carry electrical signals between the opto-box and the Pixel and IBL Detectors. 24 connector boards connect the ER bundles, the motherboard and the opto-boards.

Figure 4 shows a CAD drawing of the entire opto-box design, complete with all components. In total each opto-box integrates two PCBs (each with 300 conductive traces across 8 layers), four types of connectors, and 2284 electrical connections. Two versions of the opto-box were designed, one with 15 slots for the IBL detector, and one with 24 slots for the Pixel detector. In fact, two mirror image versions of the Pixel opto-box were also designed. This allows the Pixel opto-boxes to be installed back-to-back in pairs. There are two IBL and twelve Pixel opto-boxes installed on the ATLAS Detector, for a total of fourteen.

3 Validation

Reliability was built into the opto-box system. Electromagnetic shielding continuity is provided by using shielded connectors, PCB edge plating and PCB copper pours. High quality parts were used in the construction, including custom machined metal structures, IPC class 3 PCBs and aerospace grade connectors.

Many tests were performed at all stages of the fabrication and installation process. Electrical continuity of each PCB was verified at the PCB fabrication facility prior to delivery to CERN. The PCB boards were assembled with the connectors at the CERN PCB assembly facility. The assembled boards were again tested with a custom made test apparatus. Two of the PCB boards were found to be defective and had to be reassembled.

A custom made thermal test apparatus was used to ensure that no heat load is dissipated into surrounding detector systems. A prototype opto-box was loaded with dummy boards that used resistors to simulate heat loads. Temperatures were measured at various points using thermistors and the data

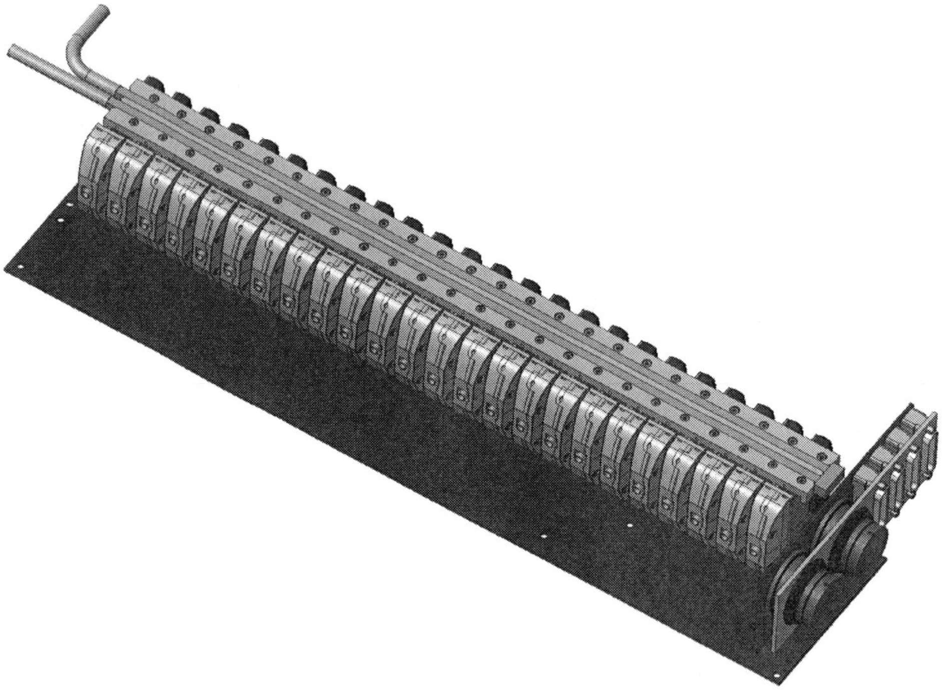

Figure 4: Computer Aided Design (CAD) image of the opto-box design for the Pixel Detector. The opto-boxes for the IBL Detector have fewer opto-board slots, but otherwise identical. The cooling pipe carries away heat using C_6F_{14} fluid at 16°C. The motherboard distributes power and control signals to the opto-boards. The end board transfers power and control signals from six cables into the motherboard.

was read out using a custom-written MATLAB program. This test verified that the cooling system was adequate.

The opto-board design was iterated until it passed rigorous testing at humidity and temperature levels far above expected operating conditions. The cooling pipes were pressure tested before and after installation to ensure that there were no leaks.

After assembly and installation all the components of each opto-box were again tested and validated for use. PCB coupons have been reserved for accelerated age studies, which will be used to forestall future failures.

4 Installation

The opto-boxes are installed on each end of barrel section of the LAr electromagnetic calorimeter, shown in Figure 5. They are still deep inside ATLAS, just outside the Inner Detector system, and can only be accessed when the end caps are removed. But this is still more accessible than their previous location and increases serviceability. Figure 6 shows one pair of the opto-boxes as installed in the ATLAS Detector.

Custom tooling was developed for installing in the opto-boards to avoid damage and to gain enough access within limited space. In addition, since the opto-boxes are installed side-by-side,

 DOI: 10.1051/epjconf/201612605002

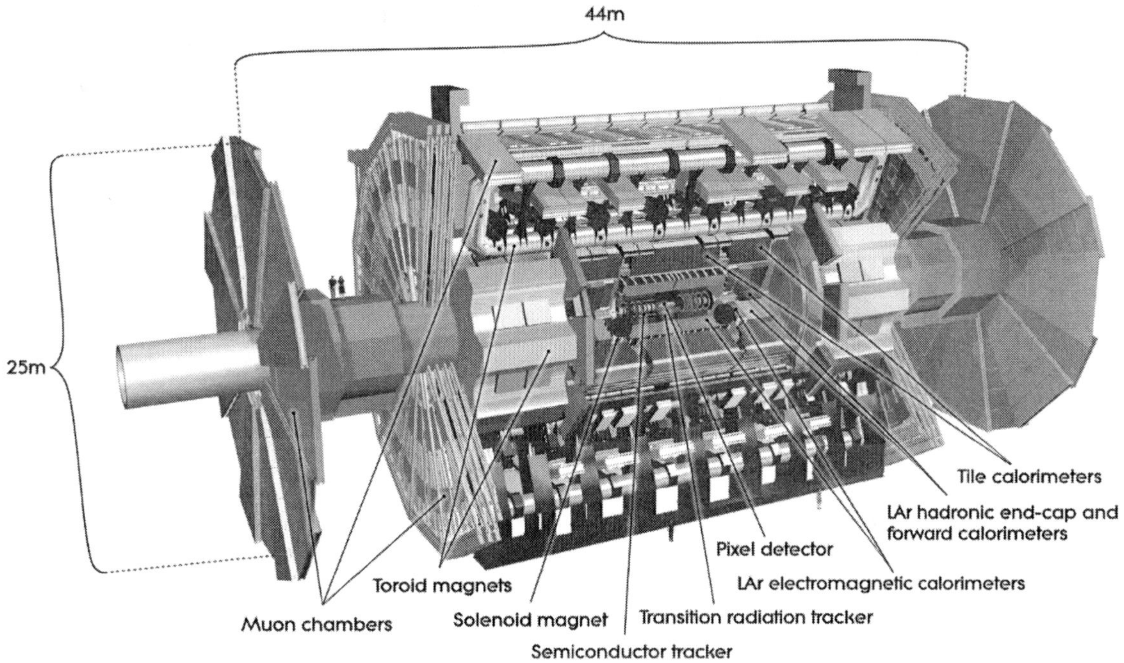

Figure 5: The opto-boxes are installed symmetrically on both sides of the ATLAS Detector. Their approximate locations are indicated by the red dots. They are installed on each end of barrel section of the LAr electromagnetic calorimeter, which is colored yellow in this figure.

mirror image Pixel Detector versions were made in order to facilitate easier access for the delicate opto-boards.

5 Operation

The opto-boxes and opto-boards began use in 2015 as ATLAS started to take run 2 data and are part of the opto-electronic readout chain. Figures 7 and 8 show that the cooling system and power systems are functional and operating within the expected parameters.

Figure 6: The white circle shows one Pixel opto-box pair installation location on the ATLAS Detector when it was still open during LS1.

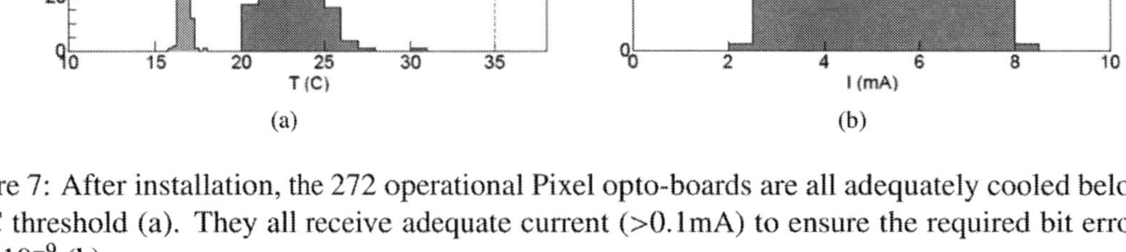

Figure 7: After installation, the 272 operational Pixel opto-boards are all adequately cooled below the 35°C threshold (a). They all receive adequate current (>0.1mA) to ensure the required bit error rate of < 10^{-9} (b).

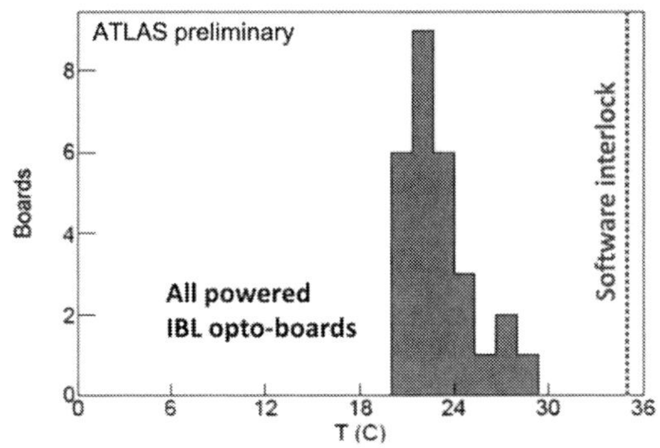

(a)

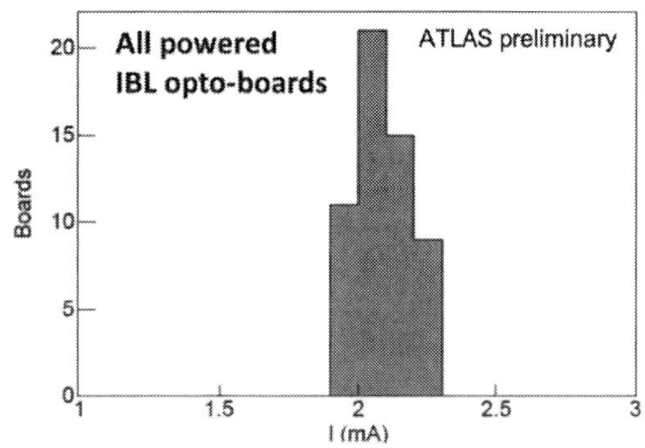

(b)

Figure 8: After installation, the 28 operational IBL opto-boards are all adequately cooled below the 35°C threshold (a). They all receive adequate current (>0.1mA) to ensure the required bit error rate of $< 10^{-9}$.

6 Summary

The opto-boxes provide ATLAS Pixel and IBL Detectors with reliable, tightly integrated, and serviceable mini-crates and modules for the optoelectronic data transfer system. They are currently operational as an integral part of data collection for LHC run 2.

References

1. "Opto-Box", Poster, ATL-INDET-SLIDE-2015-461, https://cdsweb.cern.ch/record/2045532

2. "The ATLAS Experiment at the CERN Large Hadron Collider", ATLAS Collaboration, 2008 JINST 3 S08003

3. "Engineering specification for the Pixel services after the nSQPs installation", ATLAS Project Document, ATL-IP-ES-0148, https://edms.cern.ch/document/1101163/4

EPJ Web of Conferences **126**, 05003 (2016) DOI: 10.1051/epjconf/201612605003
ICNFP 2015

First indication of LPM effect in LHCf, an LHC experiment

M. Del Prete[1,2,a] on behalf of LHCf collaboration

[1] *INFN section of Florence, Italy*
[2] *University of Florence, Italy*

Abstract. The Large Hadron Collider forward (LHCf) experiment is dedicated to the measurement of very forward neutral particle production in the high energy hadron-hadron collisions at LHC. The aim of the experiment is to improve the cosmic ray air shower development models and its setup gives an important opportunity to directly measure the Landau Pomeranchunk Migdal (LPM) effect in heavy absorber. This work presents the analysis for LPM effect on data taken in 2010 and 2013 at $\sqrt{s}$ = 7 TeV and $\sqrt{s_{NN}}$ = 5.02 TeV respectively. We study the interactions of gamma, mainly produced by π_0 decay, in one of the calorimeter pairs of LHCf. This is composed by 16 Tungsten layers as absorbers and 16 plastic scintillators for energy measurements (Arm2). We use three parameters to describe the mean shower profile with respect the photon mean energy. The results are compared with Epics MonteCarlo simulation with LPM active and inactive models, shows a first evidence of LPM effect.

1 Introduction

LHCf, LHC-forward experiment, measures neutral particles production in a very forward region in proton-proton and proton-ion collisions at the Large Hadron Collider. LHCf has the aim to improve hadronic interaction models of Monte Carlo simulations used in cosmic rays. Highest energy cosmic rays can only be detected from secondary particles which are produced by the interaction of the primary particle with nuclei of the atmosphere, the air showers. It is possible to reconstruct the kinematic parameters and the type of primary particle studying the development of air showers. Since the energy flow of secondary particles is concentrated in the forward direction, the measurement of particle production at small angles is very important.

LHC accelerator gives the possibility to study collision at very high energies, which corresponds to an energy range in the laboratory frame from 10^{14} eV to 10^{17} eV. This energy range covers the "knee" region of cosmic rays spectrum, which occurs at 10^{15} eV. The capability to measure the photon shower development in a range of energies up to $\sim 10^{13}$ eV and the specific setup of the experiment give a unique opportunity to study the Landau Pomeranchunk Migdal effect in heavy absorbers.

In the 1953 Lev Landau and Isaak Pomeranchuk noticed that at high energies and large densities the bremsstrahlung and pair production does not follow the Bethe-Heitler formula [1, 2] In the 1956 Arkady Migdal elaborated the proper quantum mechanic treatment for the cross sections. The final cross sections of the bremsstrahlung and pair production are suppressed since the momentum

[a]e-mail: marina.delprete@fi.infn.it

transferred, p_T, between the nucleus and the electron is small and the interaction occurs over distance comparable to the mean free path of incoming particle and the coherence among scattering center is lost.

2 The LHCf experiment

LHCf is composed of two independent detectors, called Arm1 and Arm2 located 140 meters away from ATLAS interaction point, IP1 [3, 4]. Arm1 is placed in the side of IP8 direction, toward LHCb experiment, and Arm2 in the opposite direction toward IP2 and ALICE experiment. Detectors are placed inside the Target Neutral Absorber, where the beam pipe turns into two separates tubes. The LHCf detectors occupy the first 30 cm of the TAN slots, facing the interaction point and minimizing the material between them and IP1. This is the best condition for measuring fluxes and energy spectra of high energy forward neutral particles produced by the interacting beams. Each detector is made of two sampling and imaging calorimeters called towers. The longitudinal dimension of towers is of 44 radiation lengths, which correspond to 1.6 nuclear interaction lengths. The transverse cross sections are for Arm1 $20x20$ mm^2 $40x40$ mm^2 for the small and for the large tower respectively, for Arm2 transverse cross sections are $25x25$ mm^2 for the small tower and $32x32$ mm^2 for the large tower.

Each tower is composed of 16 tungsten layers and 16 plastic scintillator layers to measure energy and also contains 4 position sensitive layers that are different for the two detector [5–8]. Arm1 uses scintillating fiber to measure position of the neutral particles in the towers, in Arm2 the position is measured by silicon microstrip detectors.

The pseudo-rapidity range is $\eta > 9.6$ for the small tower and $8.4 < \eta < 9.4$ for larger tower. Energy resolution obtained with these detectors configuration is better than 5 % for photons and of about 40 % for neutrons.

3 LPM effect and LHCf

LHCf measures photons of very high energy and at small angles. The fine longitudinal segmentation of its calorimeters is well suited for the measure of the shower development. For these reasons is the optimal tool to detect LPM effect in the bremsstrahlung and pair production processes. LPM effect is characterized by the material dependent component E_{LPM}, computed as [9]:

$$E_{LPM} = \frac{m^2c^3}{4\pi\hbar}\alpha X_0 \tag{1}$$

This definition of E_{LPM} is the most used but theoretical calculations differs of numerical factors depending of the models used [10–12]. LHCf has tungsten absorbers with $Z = 74$ and radiation lengths $X_0 = 3.5$ mm, so we have from equation (1) $E_{LPM} = 2.7$ TeV.

We test the capability of LHCf to detect LPM effect by comparing MC simulations to the shower development measured in the small tower, the one at smaller angle and thus with larger statistic. We use EPICS montecarlo to simulate mono-energetic 1000 photons impinging the small tower with a energy $E_\gamma = 2.48$ TeV and with and without LPM effect. We select photons $2X_0$ away from the edge of the calorimeter. After the cuts the events studied are 382, the results are in figure 1 where is evident that the shower development is delayed and elongated when the LPM effect is active in the simulation.

3.1 Data and MCs used

In this analysis we study the photons produced in p-Pb interactions at $\sqrt{s} = 5.02$ TeV, data taken in 2013, and in p-p at 7 TeV in data taken in 2010. We noticed that in p-p data the agreement data-

DOI: 10.1051/epjconf/201612605003

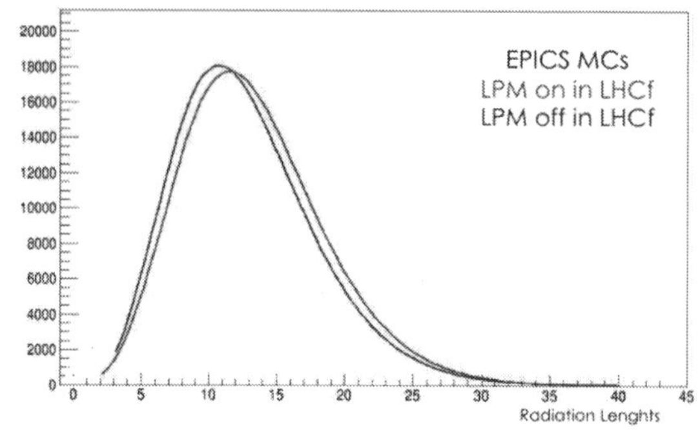

Figure 1. Simulation of photons showers, of energy ~ 2.48 TeV. The figure shows the average longitudinal development of the photon shower in the small tower, where LPM as turned on (red) and off (blue).

MCs is not as good as in p-Pb data. This problem is being investigated. In this paper the differenc is attribuited to systematic effects.

The montecarlo simulations are done with EPICS to simulate 3000 mono-energetic photons impinging the small tower with different mean energies and with LPM effect turned on and off. The energies simulated are from 200 GeV to 3200 GeV in steps of 40 GeV.

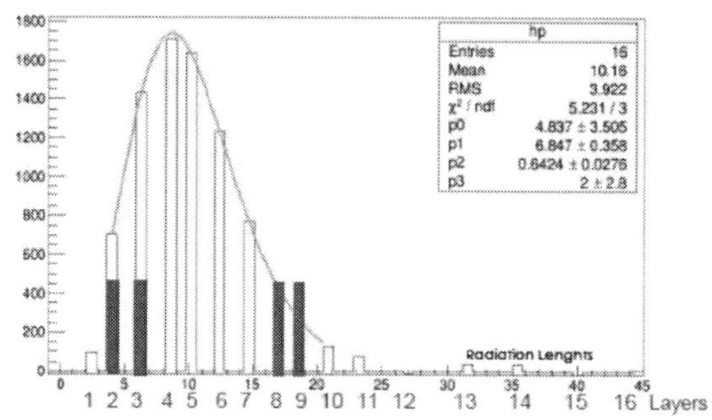

Figure 2. Figure shows the shower profile of one event from LHCf data. The fit (see text) is green, in red the scintillators interested in this analysis. The 16 entries are the scintillators in ARM2, in the panel are shown the fit parameters. The fit is done from layer 2 to layer 10 to better identify the maximum position.

We have to consider that up now the MC software have not been checked for LPM effect at high energy where LHCf works. All the tests have been done comparing the test beam data at fixed target with the simulation for energies below ~ 400 GeV. Therefore we do not expect that EPICS prediction to be in perfect agreement with the data of much larger energies, as those collected by LHCf [13].

3.2 Data analysis

In this work we select, for data and MCs, a fiducial region of the Small Tower of ARM2 $2X_0$ ($X_0 = 3.5$ mm) away from the lateral edge to avoid leakage problems. The events studied are in the energies range of 50 GeV $< E_\gamma <$ 3100 GeV to have stable data and statistically significance.

The parameters of the shower shape profile are measured using the scintillators signal in the layers of ARM2. In this way we can detect the delay and elongation due to the LPM effect. For each events

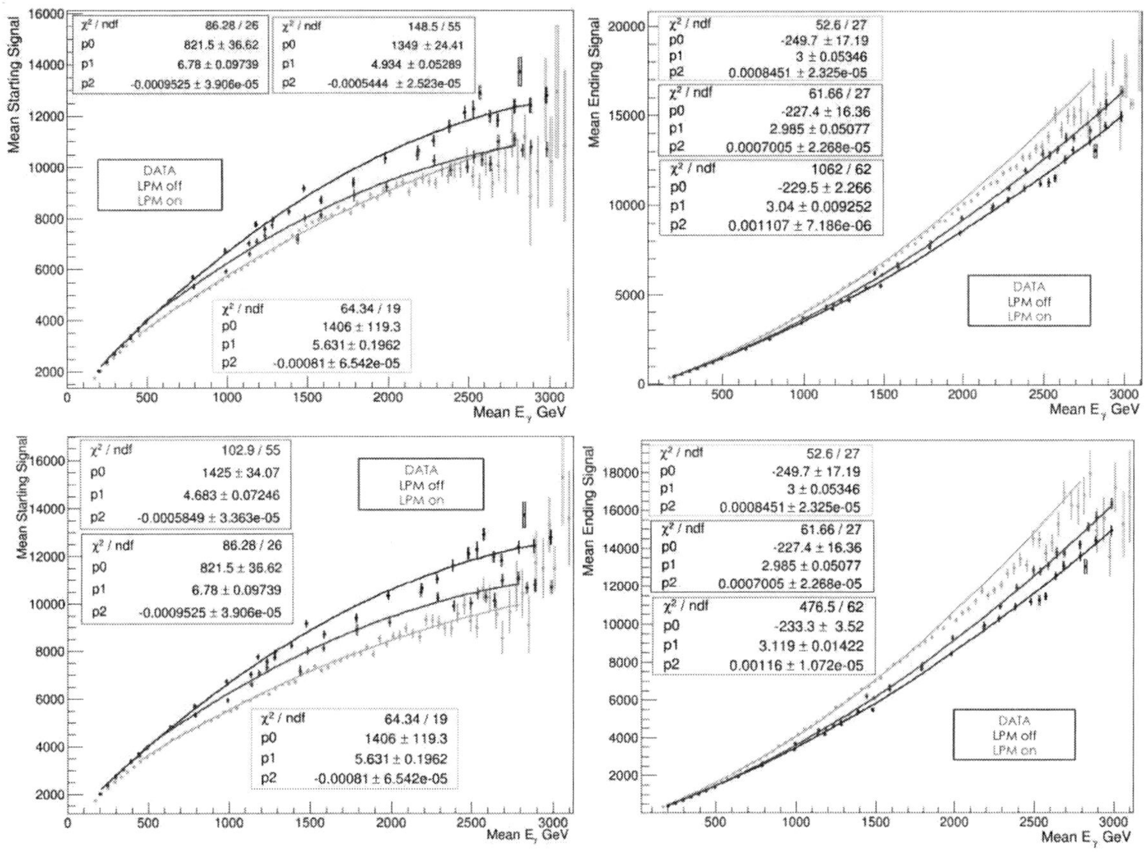

Figure 3. Figure shows the parameters S_{st} and S_{end} for the p-Pb and p-p data in the first and second raws respectively. We have a good agreement between the data and LPM on simulations. The curves are fitted with a second order polynomial function.

and energy intervals of ±20 GeV we compute three parameters and their mean. We define a '*starting point*', S_{st}, a '*ending point*', S_{end} (see figure 2):

$$S_{st} = Signal_{layer2} + Signal_{layer3} \qquad S_{end} = Signal_{layer8} + Signal_{layer9}$$

These parameters are sum of the respective scintillator signals. The longitudinal position of the shower maximum, X_{max}, and its amplitude , Y_{max}, are defined as the maximum of the fit function, the Longo Landau function [15]:

$$f(x) = p_0 \cdot p_2 \cdot (x + p3)^{p_1} \cdot p_2^{p_1} \cdot \exp(-p_2 \cdot (x + p_3)) \tag{2}$$

In this function the variable x is the longitudinal coordinate, p_0 is a normalization factor, p_1 and p_2 are parameters that describe the development of the shower by Gamma function and p_3 s the translation parameter that define the starting point of the shower.

The LPM effect is detected when the shower is delayed and elongated with respect to the simulations with LPM switched off. In this case expect that the parameter S_{st} will be smaller than for the LPM off simulations [14]. For the same reason the parameters S_{end} and X_{max} will be larger for LPM on than for LPM off.

DOI: 10.1051/epjconf/201612605003

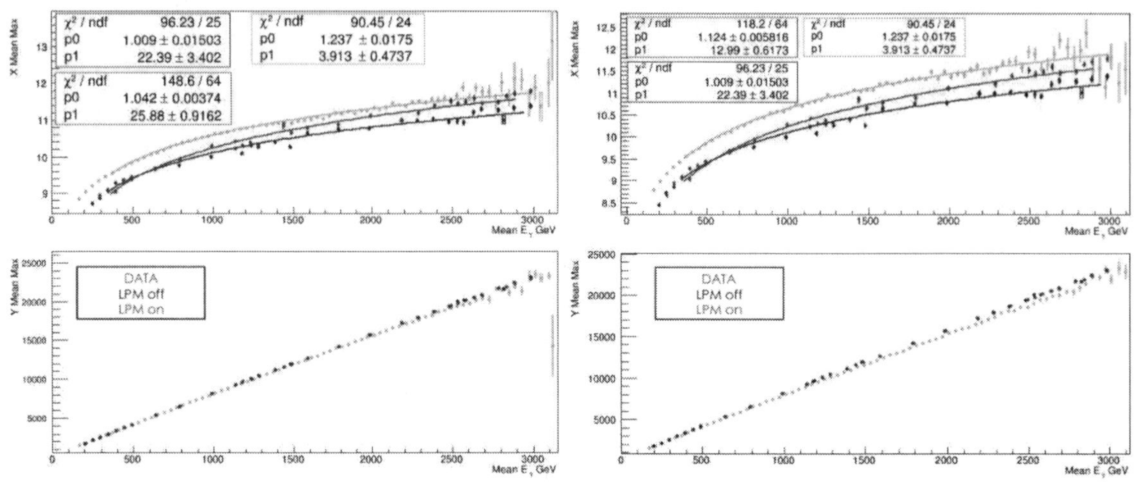

Figure 4. Figure shows the parameters X_{max}, in the first raw, and Y_{max}, in the second raw. On the left column shows p-Pb run and on the right the p-p run. The curves of X_{max} are fitted with the logaritminc function $p_0 \cdot ln(p_1 \cdot E)$.

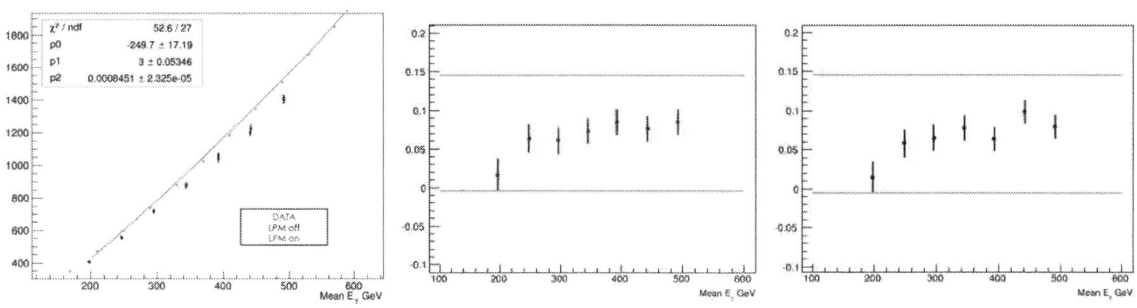

Figure 5. Figure shows the values of the parameters S_{end} in the region $E_\gamma < 600$ GeV, where EPICS makes equal predictions (first plot). We compare the MCs with the data values obtained by fit interpolation. $DS_{end}\%$ for LPM on and off using the data from the p-Pb run are shown in the second and third plots.

The first step is to compare the described parameters which are obtained by EPICS simulations with and without LPM effect to what obtained in LHCf experiment data. In figures 3 and 4 are shown the parameters S_{st}, S_{end} and X_{max} for the p-Pb and p-p runs. In particular figure 3 shows the results of more direct parameters S_{st} and S_{end} for the p-Pb run in the first and p-p run in the second raw respectively. We have for both the parameters a better, but not perfect, agreement between the LHCf data and the simulations with LPM turned on. Figure 4 shows the parameters X_{max} and Y_{max}, on the left for the p-Pb and on the right for p-p runs. In this part of the analysis we have not so clear evidence of the difference between the simulations for LPM on and off and a poor agreement between the data and the LPM on curves. For this elaborated analysis based on the fit with empirical function we need to understand better the fit function, moreover this parameter could be sensitive to LPM effect at higher energy [12].

All the parameters studied have a little disagreement between the data and the simulations for LPM on and off. Work is in progress to understand and correct for this effect. In order to understand quantitatively the amount of agreement, we consider the low energy region $E_\gamma < 600$ GeV where EPICS predictions are the same for LPM turned on and off, but differ from data.

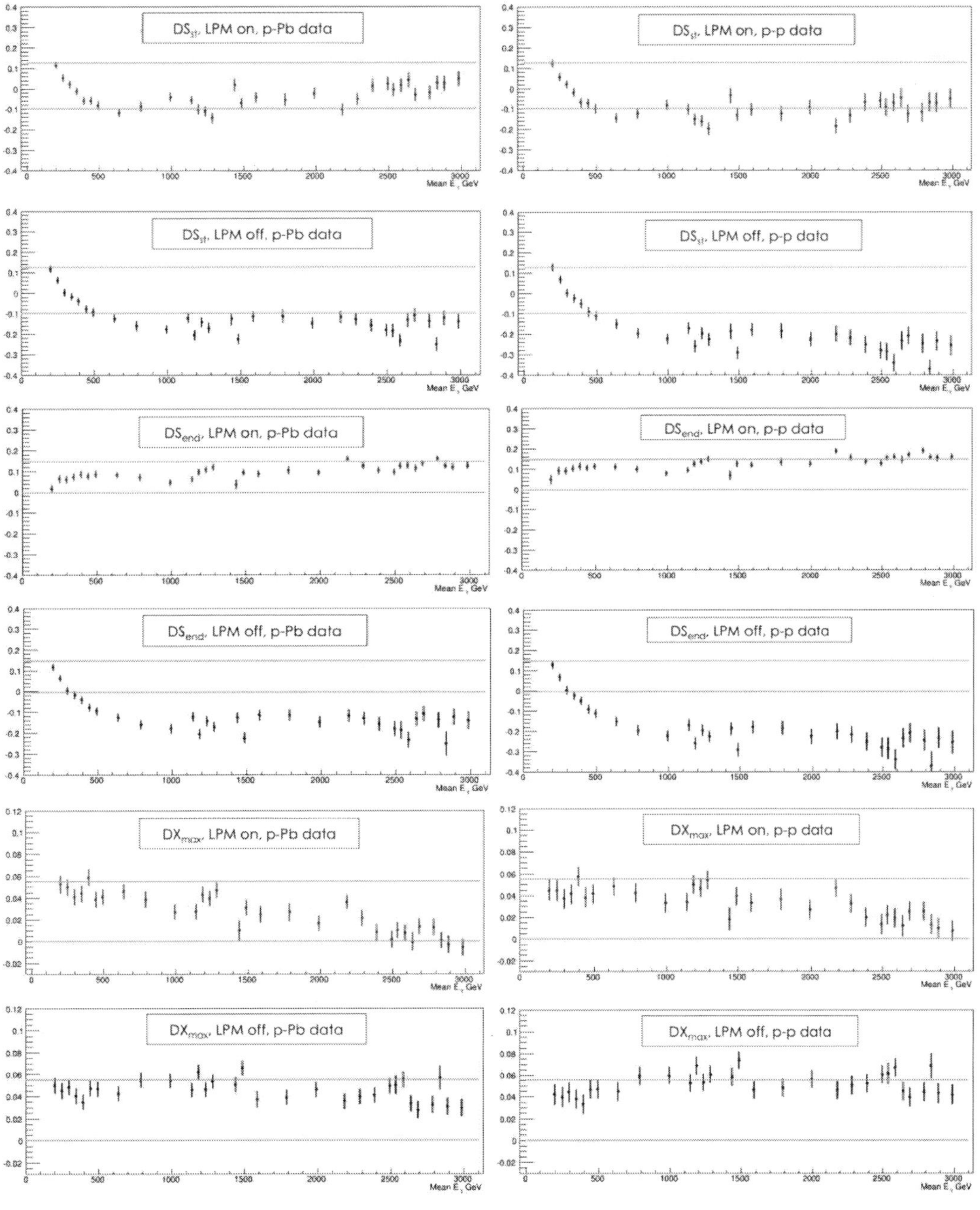

Figure 6. Figure shows $DS_{st}\%$, $DS_{end}\%$ and $DX_{max}\%$ and the 'compatibility region', described by the two horizontal yellow lines. In red the parameters computed with the data and the Montecarlo for LPM on, in blue with the data and simulations for LPM off. The left column shows the results for the p-Pb data, on the right for the p-p data. In the first and second raws are $DS_{st}\%$ values, the third and fourth $DS_{end}\%$ values and in fifth and sixth the values of $DX_{max}\%$.

In this energy region we compute the percentage difference between data and MC and define a maximum acceptable variation of the parameters, the 'compatibility region'. In this energy region we compare the data to EPICS expectations by computing the percentage differences data-MC for S_{st}, S_{end} and X_{max}:

$$DS_{st}\% = \frac{S_{st}^{data} - S_{st}^{MC}}{S_{st}^{data}} \quad DS_{end}\% = \frac{S_{end}^{data} - S_{end}^{MC}}{S_{end}^{data}} \quad DX_{max}\% = \frac{X_{max}^{data} - X_{max}^{MC}}{X_{max}^{data}}$$

Figure 5 shows the results for the p-Pb data and EPICS simulations for $E_\gamma < 600$ GeV. The left plot shows S_{end} (data and MC), the second and third plots $DS_{end}\%$ computed from EPICS with LPM turned on and off. $DS_{end}\%$ depends on E_γ, but in the same way for LPM on and off. We take the range of values of $DS_{end}\%$ as the amount of disagreement data-MC for the parameter S_{end}.

For the time being we take this range of $DS_{end}\%$ values as 'compatibility region', a measure of the confidence for MC predictions on this energies range. The same analysis was done for the parameters $DS_{st}\%$ and $DX_{max}\%$. The definition of compatibility region for $DS_{st}\%$, $DS_{end}\%$ and $DX_{max}\%$are $(-0.1, 0.125)$, $(-0.005, 0.145)$ and $(0, 0.055)$ respectively. We obtain similar values for the data collected in p-p collisions. We will assume that, at higher energies, the 'compatibility region' still describes the uncertainties in the comparison data-MC.

A summary of this analysis is presented in figure 6, The $DS_{st}\%$ as a function of E_γ is presented in the first and second raws (at the left p-Pb at the left p-p data). Data are compared to EPICS predictions with LPM turned on (first raw) and turned off (second raw). Similarly $DS_{end}\%$ and $DX_{max}\%$ follow.

Conclusions are not the same for all the three parameters that we have studied, probably due to different sensitivity to the LPM effect. The clearest results comes from $DS_{st}\%$ and $DS_{end}\%$ that measure the shower delay. Both parameters cluster in the 'compatibility region' when the data are compared to MC LPM turned on. The comparison of the data to EPICS LPM turned off shows the parameters stay outside the region. This supports the description of data with LPM effect. $DX_{max}\%$ provides more ambiguous results and the parameter does not either of the EPICS modes. A possible explanation is that the fit with empirical function (2) is too sensitive to experimental uncertainties. More work on this issue is in progress.

4 Conclusion

From this first study we can say that the statistics S_{st} and S_{end} are reasonable well described by Epics software with LPM effect, for energies, $E_\gamma < 3.1$ TeV where the LPM effect is very small. More elaborated analysis based on fit needs more work, X_{max} is not in clear agreement with simulations of LPM on or off. In any case to have stronger evidence we need more MCs for comparison and data at higher energies.

Data taken at 13 TeV, in June of 2015 will have more statistics in the optimal energies range and will confirm this evidence, there will more useful the study of X_{max}, Y_{max} and the definition of a size of shower profile, for example the size of the shower at $\frac{2}{3}$ of Y_{max}, to avoid some possible problem of the softwares in the simulation of the amplitude of the LPM effect at high energy.

References

[1] A. B. Migdal, Phys. Rev., **103**,1811-1820 (1956).

[2] H. Bethe et W. Heitler, Proc. Royal Society of London, **146**, 83-112 (1934).

[3] O. Adriani et al., JINST **3**, S08006 (2008).

DOI: 10.1051/epjconf/201612605003

[4] O. Adriani et al., JINST **5**, P01012 (2010).
[5] O. Adriani et al., Physics Letters B **715**, 298-303 (2012).
[6] O. Adriani et al., Physics Letters B **703**, 128-134 (2011).
[7] O. Adriani et al., Physical Review D **86**, 092001 (2012).
[8] O. Adriani et al., Physical Review C **89**, 065209 (2014).
[9] Landau, L. D., and I. J. Pomeranchuk, Dokl. Akad. Nauk SSSR **92**, 535 (1953).
[10] S. Klein, Rev. Mod. Phys., **71**, 5 (1999).
[11] Landau, L. D., and I. J. Pomeranchuk, Dokl. Akad. Nauk SSSR **92**, 735 (1953).
[12] T. Stanev et al., Phys. Rev. D, **25**, 1291-1304 (1982).
[13] K.Yanagisawa, K.Yoshida, Y.Komori, T.Kobayashi and J.Nishimura, 0732, 33RD ICRC (2013)
[14] K.Yoshida, Y.Komori, K.Yanagisawa, T.Kobayashi, Y.Sato and J.Nishimura, 0733, 33RD ICRC (2013)
[15] E. Longo and I. Sestili, NIM **128**, 283 (1975).

EPJ Web of Conferences **126**, 05004 (2016) DOI: 10.1051/epjconf/201612605004
ICNFP 2015

Performance of the ALICE secondary vertex b-tagging algorithm

G. Eyyubova[1,2,a] and L. Kramarik[1,b] on behalf of the ALICE collaboration

[1] *FNSPE, Czech Technical University in Prague*
[2] *SINP MSU, Russia*

Abstract. The identification of jets originating from beauty quarks in heavy-ion collisions is important to study the properties of the hot and dense matter produced in such collisions. A variety of algorithms for b-jet tagging was elaborated at the LHC experiments. They rely on the properties of B hadrons, i.e. their long lifetime, large mass and large multiplicity of decay products. In this work, the b-tagging algorithm based on displaced secondary-vertex topologies is described. We present Monte Carlo based performance studies of the algorithm for charged jets reconstructed with the ALICE tracking system in p-Pb collisions at $\sqrt{s_{\rm NN}} = 5.02$ TeV. The tagging efficiency, rejection rate and the correction of the smearing effects of non-ideal detector response are presented.

1 Introduction

Jet reconstruction provides access to the kinematics of partons produced in hard scatterings in the initial stage of heavy-ion collisions and that later suffer energy loss in the medium via gluon radiation and elastic collisions. Reconstructed jets are therefore a powerful tool to study the properties of the quark-gluon plasma (QGP) created in heavy-ion collisions at high energy. The parton energy loss in the QGP depends on the colour charge and mass of the parton. Due to the larger colour charge factor for gluons, energy loss via gluon radiation should be larger for gluon jets than quark jets. For heavy quarks it is expected that radiative energy loss is suppressed due to coherence effects [1, 2]. This should lead to a smaller modification of the transverse momentum spectra for particles containing heavy quarks than for inclusive charged hadrons. On the other hand, for heavy quarks in the QGP, collisional energy loss may be important at low transverse momentum $p_{\rm T}$. The jet quenching phenomenon and the properties of produced matter can be further understood by measuring beauty jets at low jet transverse momentum in comparison with that of light-flavour jets [3].

The ALICE detector [4] is capable to discriminate jets originating from b quarks with different tagging algorithms. Here, we study the performance of the b-jet tagging algorithm based on displaced secondary-vertex topologies for p-Pb collisions at $\sqrt{s_{\rm NN}} = 5.02$ TeV. The p-Pb collisions serve as a test for cold nuclear effects which are important in measurements of jet quenching in Pb-Pb collisions. The study is based on Monte Carlo (MC) simulations.

[a]e-mail: Gyulnara.Eyyubova@fjfi.cvut.cz
[b]e-mail: Lukas.Kramarik@cern.ch

2 Analysis details

In this analysis we use MC with simulated responses of ALICE detector by GEANT3 [5] for p-Pb collisions at $\sqrt{s_{\rm NN}} = 5.02$ TeV. A MC simulation for pp collisions at $\sqrt{s} = 5.02$ TeV with a Lorentz boost to get the same rapidity shift as in p-Pb collisions is used to study detector response for jets without underlying event. PYTHIA6, with tune Perugia 2011 [6], is employed for MC pp simulations. The MC for p-Pb collisions is represented by a cocktail of event generators PYTHIA6 (Perugia 2011) + HIJING [7] which correspond to hard physics and p-Pb background, respectively.

Schematically, the analysis of b jets can be represented by the following analysis steps: track selection suitable for b-tagging, jet reconstruction from these tracks and underlying background subtraction under the jet area, tagging with a particular algorithm and, finally, corrections of jet spectrum. Corrections of the raw tagged spectrum $\frac{{\rm d}N^{\rm tag}}{{\rm d}p_{\rm T}}$ have to be made for detector effects (unfolding) and for tagging efficiency and purity. In case of p-Pb collisions, unfolding also includes a correction for fluctuations of the underlying background under the jet area.

The efficiency of b-tagging, $\epsilon_{\rm b}$, and the purity, $f_{\rm b}$, are defined as the fraction of "true" b jets after tagging w.r.t total number of "true" b jets before tagging and w.r.t. total number of tagged jets, respectively. In order to estimate the efficiency and the purity, we use the "true" underlying flavour of a jet, which can be identified in MC by different labelling methods. The performance of the tagging algorithm is investigated via mistagging rate, which is the efficiency of mistakenly tagging as beauty a jet originating from a c-quark or a light-flavour parton. Mistagging rate studies are needed to extract sample with high purity of b jets. High purity allows to suppress the contamination of light-flavour and charm jets, which is important since the measured fraction of b jets to inclusive jets is at the level of 2-4% (as measured by CMS experiment), both in pp collisions at $\sqrt{s} = 7$ TeV [8] and in p-Pb collisions at $\sqrt{s_{\rm NN}} = 5.02$ TeV [9]. Below, we discuss the analysis steps in more details.

2.1 Track selection

In the MC simulations, charged tracks are reconstructed with simulated responses of Inner Tracking System (ITS) and Time Projection chamber (TPC) in a pseudorapidity region $|\eta| < 0.9$ and transverse momentum $0.15 < p_{\rm T} < 100$ GeV/c. The track selection is optimized for the reconstruction of beauty jets. The main track cuts are the minimum number of points in the TPC (70 out of 159) and the maximum distance of closest approach to primary vertex (in xy-plane and in z direction). Tracks are also required to have at least one point in the Silicon Pixel Detector (SPD). Because of inefficient regions in the SPD, the azimuthal distribution of these high-quality tracks is not completely uniform. This anisotropy is compensated by considering in addition tracks without reconstructed track points in the SPD. Track fit for these tracks is not constrained to the primary vertex. For the tracks used for secondary vertex reconstruction in jets, the point in SPD is required.

2.2 Jet reconstruction and background estimation

For the estimation of the background density and fluctuations in jet reconstruction, we use a similar approach as it was applied by ALICE for charged jet measurements in p-Pb collisions [10]. Namely, the anti-$k_{\rm T}$ algorithm from the FastJet package [11] with resolution parameter of $R = 0.4$ is used to reconstruct charged jets. We use the CMS method [12], which is suited for more sparse environment in p-Pb collisions w.r.t Pb-Pb collisions, for the calculation of the background density, ρ. The background fluctuations are calculated with a random cone approach as:

$$\delta p_{\rm T} = \sum_i p_{{\rm T},i} - A_{cone} \cdot \rho,$$

where the sum is over the track p_{T} in a cone and A_{cone} is the cone area. For more details see [10]. Only jets with jet axis in the pseudorapidity interval $|\eta_{\mathrm{jet}}| < 0.5$ are considered, which ensures that the jet is fully located in the TPC acceptance. In this study, the background fluctuations are measured with HIJING p-Pb events with particle transport through the ALICE detector. The background fluctuations under the jet area are found to be slightly smaller in the HIJING simulation than in the p-Pb data [10].

2.3 Correction for detector effects and background fluctuations

The jet momentum distribution is distorted by the detector mainly due to tracking inefficiency and momentum resolution. In order to correct for detector effects the detector simulation with GEANT is performed. Reconstructed jets at particle level without detector effects and jets at the detector level after particle transport through the ALICE detector are matched and based on it the detector response matrix is built. From the inverse response matrix and the momentum distribution at detector level one finds the true jet momentum distribution. The unfolding procedure has to be carrefully scrutinized since small perturbations in the measured distribution lead to large fluctuations in the solution. Several regularization methods exist. The Singular Value Decomposition (SVD) approach [13] is used in this performance study.

After subtracting the constant background in each event in order to correct the jet p_{T}, one has to keep in mind that the background is not necessary constant, but may differ for different jets in a given event. This is corrected statistically (not event by event) via an unfolding technique with a background fluctuation matrix $f(\delta p_{\mathrm{T}})$, where the background fluctuations δp_{T} are described in Sec. 2.2. The actual unfolding is done with a matrix which is a product of two matrices: detector response matrix and background fluctuation matrix.

2.4 Secondary vertex tagging algorithm

Due to the long life-time of B hadrons ($\approx$ 500 μm), in most cases their decay vertex is displaced from the primary vertex of a collision. This algorithm reconstructs secondary vertices (SV) in a jet and uses their properties to discriminate b jets among lighter flavour jets. The tracks participating in SV reconstruction are tracks belonging to a jet with additional requirements: the point in the SPD (as mentioned above) and $p_{\mathrm{T,track}} > 1$ GeV/c. Only secondary vertices made of three tracks are considered. All combinations of three tracks, satisfying the requirement, are used to build secondary vertices in a jet. The quality of vertex reconstruction is characterized by the dispersion of the tracks in the vertex $\sigma_{\mathrm{vtx}} = \sqrt{d_1^2 + d_2^2 + d_3^2}$, where $d_{1,2,3}$ are the distances of the three tracks from SV. The sign of the SV flight distance is defined w.r.t. the jet direction, so that the signed length is $L = |\vec{L}|\mathrm{sign}(\vec{L}\cdot p_{jet})$, where $\vec{L}$ is the SV position. The cuts on the most displaced SV found in the jet, namely SV dispersion, σ_{vtx}, and the significance of flight distance in a transverse plane, $L_{xy}/\sigma_{L_{xy}}$, are used for b-jet tagging in this analysis.

2.5 MC labelling

To assign the "true" flavour of reconstructed jets in the MC simulations, the MC event history is used. The labelling procedure is not unambiguous and is not strictly identical for different MC generators (it depends on parton shower implementation). Here, we treat a jet as coming from a beauty quark, if a B hadron was found in a cone $\Delta R < 0.7$ around jet axis. If no beauty hadron is present, but charm hadron was found in a cone $\Delta R < 0.7$, the jet is labelled as c-jet. All other jets are considered as light-flavour jets.

3 Performance of SV b-tagging algorithm

The discrimination power of the signed flight distance significance can be judged by the distribution of this variable for jets of different flavours, which is shown in Fig.1. Secondary vertices are searched in jets with $p_{\mathrm{T,jet}} > 20$ GeV/c and the most displaced vertex in a jet is considered. The larger the cut value of $L_{xy}/\sigma_{L_{xy}}$, the more light and charm jets are rejected compared to beauty jets. As mentioned above, the tagging procedure uses cuts on $L_{xy}/\sigma_{L_{xy}}$ and σ_{vtx} of SV. Different values of cuts yield different algorithm performance. For a particular algorithm an operating point is defined based on inclusive b-tagging efficiency and mistagging rate.

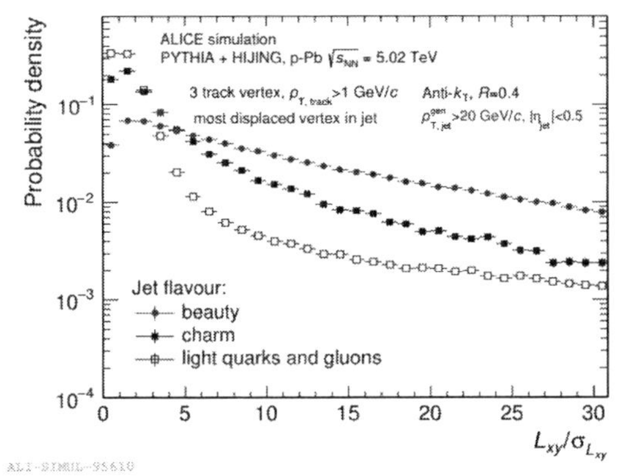

Figure 1. Probability distribution of the signed flight distance significance of the most displaced secondary vertex, found in charged jets with $p_{\mathrm{T}} > 20$ GeV/c in p-Pb simulations at $\sqrt{s_{\mathrm{NN}}} = 5.02$ TeV. The different jet flavours are assigned as described in Sec. 2.5.

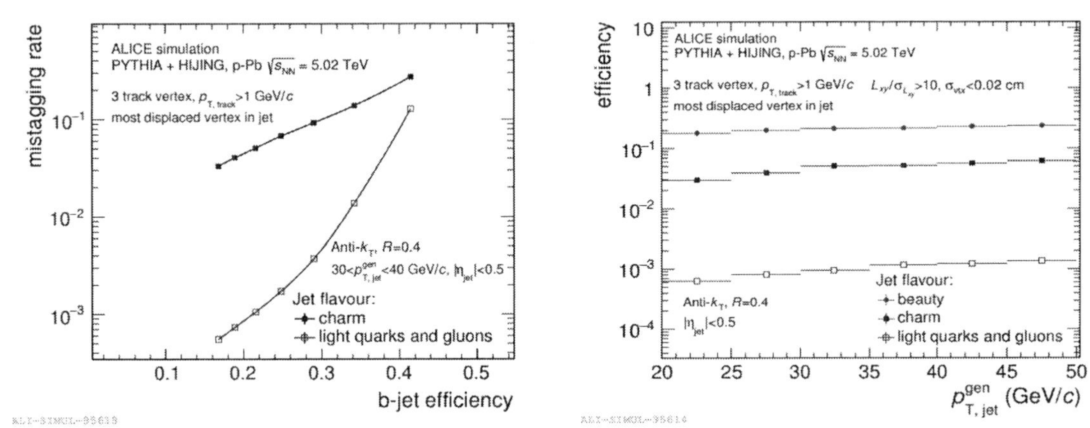

Figure 2. Left: Mistagging rate vs the b-tagging efficiency of SV tagging algorithm for different operating points for p_{T} range $30 < p_{\mathrm{T,jet}} < 40$ GeV/c. Right: The $p^{\mathrm{gen}}_{\mathrm{T,jet}}$ dependence of the b-tagging efficiency and mistagging rate for $L_{xy}/\sigma_{L_{xy}} > 10$ and $\sigma_{\mathrm{vtx}} < 0.02$ cm.

Figure 2, left, shows the mistagging rate vs the b-tagging efficiency in jet p_{T} range $30 < p_{\mathrm{T,jet}} < 40$ GeV/c for different operating points, which are obtained by varying the cuts on $L_{xy}/\sigma_{L_{xy}}$ (from 2 to 14), while the cut $\sigma_{\mathrm{vtx}} < 0.02$ cm is kept fixed. The performance may depend on $p_{\mathrm{T,jet}}$. In the considered region, $30 < p_{\mathrm{T,jet}} < 40$ GeV/c, the mistagging and tagging efficiencies are almost flat. Looser cuts result in larger statistics and higher tagging efficiency, but also higher mistagging rate, and therefore reduce purity of sample. The operation point is chosen in a way that the mistagging efficiency for light-flavour jets would be about two orders of magnitude lower than the tagging efficiency. This comes from the CMS measurements on the ratio of b jets to inclusive jets, in which the b-jet fraction was found to be at the level of 2-4% in pp collisions at $\sqrt{s} = 7$ TeV [8]. The tagging and mistagging efficiencies $\epsilon(p_{T,jet}^{gen})$ at particle level for the chosen operating point with $L_{xy}/\sigma_{L_{xy}} > 10$ and $\sigma_{\rm vtx} < 0.02$ cm are reported in Fig. 2, right. The b-tagging efficiency is at the level 0.2. The efficiency to tag light-flavour jets is two orders of magnitude lower and the efficiency to tag charm jet is about 3 to 5 times lower than the b-tagging efficiency.

The correction for detector effects is tested with two detector response matrices: the detector response for b-jets and the one for inclusive jets. Figure 3 shows the ratio of two unfolded b-jets $p_{\rm T}$-spectra. In both cases the unfolding procedure yields similar result. This ensures that beauty and inclusive jets have similar detector response matrices and therefore the detectors response for inclusive jets can be used to unfold the tagged b-jet spectrum.

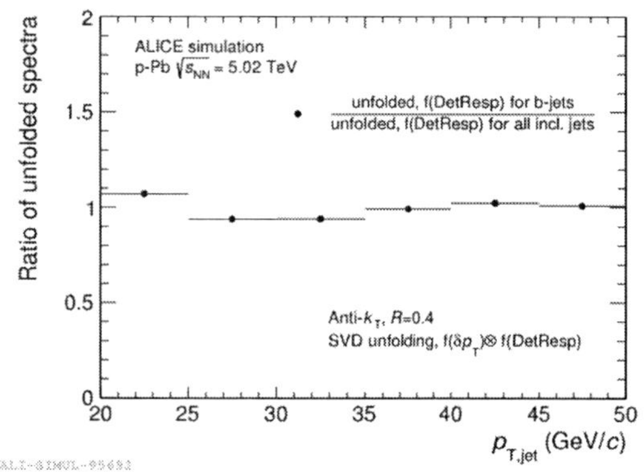

Figure 3. Comparison of SVD unfolding of b-jet spectrum with two matrices: detector matrix for inclusive jets and for b jets. Both matrices are combined with background fluctuation matrix from MC. The ratio of the two unfolded results is shown.

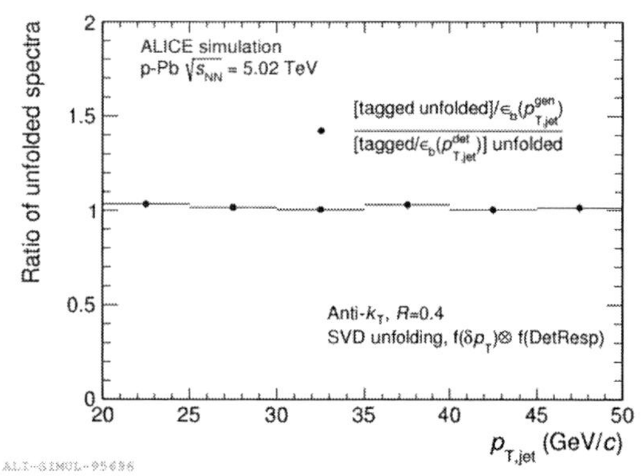

Figure 4. Comparison of two sequences of corrections: SVD unfolding and correction for tagging performance and vice versa. The ratio of the two corrected spectra is shown.

The different sequences of corrections are further studied in order to estimate the impact of background fluctuations on the stability of results. We perform the two scenarios of corrections:

1. The measured tagged b-jet spectrum is first unfolded for detector response and background fluctuations, then corrected for the b-tagging efficiency as a function of $p_{\rm T}^{\rm gen}$ (at particle level).

2. The measured tagged b-jet spectrum is first corrected for the b-tagging efficiency as a function of $p_{\rm T}^{\rm det}$ and then unfolded for detector response and background fluctuations.

The comparison of these two scenarios is shown in Fig. 4 as the ratio between the two corrected spectra. The two spectra are in agreement with each other, suggesting that both scenarios can be applied in p-Pb collisions.

4 Summary

The performance of the b-jet tagging algorithm based on displaced secondary vertices was studied with MC simulation of p-Pb events for ALICE detector. The presented results were carried out in the jet p_T range $20 < p_{T,jet} < 50$ GeV/c.

For the selected operating point the tagging efficiency is at the level 20% while the mistagging efficiency for light-flavour jets is two orders of magnitude lower, which should result in high purity of the algorithm. The rejection of tracks with V0 topology is expected to enhance the algorithm performance by reducing light-flavour contamination and is currently under investigation.

Corrections for detector response and background fluctuations were studied. It was found that the b-jet spectrum can be corrected with a detector response matrix for inclusive jets. The order of corrections (tagging efficiency vs unfolding) gives compatible results.

The tagging purity itself is not discussed here. Studies to estimate it via MC and data-driven approaches are ongoing.

Acknowledgment

This work was supported by the European social fund within the framework of realizing the project "Support of inter-sectoral mobility and quality enhancement of research teams at Czech Technical University in Prague", CZ.1.07/2.3.00/30.0034 and by Grant Agency of the Czech Technical University in Prague, grant No. SGS13/215/OHK4/35/14.

References

[1] Y. L. Dokshitzer, D. E. Kharzeev, Phys. Lett. **B519**, 199 (2001).
[2] N. Armesto, C. A. Salgado, U. A. Wiedemann, Phys. Rev. **D69**, 114003 (2004).
[3] J. Huang, Z. Kang, I. Vitev, Phys.Lett. **B726**, 251 (2013).
[4] ALICE Collaboration, B. Abelev et al., Int. J. Mod. Phys. **A29**, 1430044, (2014).
[5] R. Brun, F. Carminati, CERN Program Library Long Writeup W5013, (1993).
[6] T. Sjoestrand, S. Mrenna, P. Skands, JHEP **05**, 026 (2006);
P. Z. Skands, Phys. Rev. **D82**, 074018, (2010).
[7] X.N. Wang, M. Gyulassy, Phys.Rev. **D44**, 3501, (1991).
[8] S. Chatrchyan *et al.* [CMS Collaboration], JHEP 04 (2012) 084.
[9] V. Khachatryan *et al.* [CMS Collaboration], arXiv:1510.03373 (2015).
[10] J. Adam *et al.* [ALICE Collaboration], Phys. Lett. **B749**, 68 (2015).
[11] M. Cacciari, G. P. Salam, Phys. Lett. **B641**, 57, (2006).
[12] S. Chatrchyan *et al.* [CMS Collaboration], JHEP **1208**, 130, (2012).
[13] A. Hocker, V. Kartvelishvili, Nucl. Instrum. Meth. **A372**, 469, (1996).

EPJ Web of Conferences **126**, 05005 (2016) DOI: 10.1051/epjconf/201612605005
ICNFP 2015

Spinorial Regge trajectories and Hagedorn-like temperatures

Spinorial space-time and preons as an alternative to strings

Luis Gonzalez-Mestres[1,a]

[1] *Megatrend Cosmology Laboratory, John Naisbitt University, Belgrade and Paris*
Goce Delceva 8, 11070 Novi Beograd, Serbia

Abstract. The development of the statistical bootstrap model for hadrons, quarks and nuclear matter occurred during the 1960s and the 1970s in a period of exceptional theoretical creativity. And if the transition from hadrons to quarks and gluons as fundamental particles was then operated, a transition from standard particles to preons and from the standard space-time to a spinorial one may now be necessary, including related pre-Big Bang scenarios. We present here a brief historical analysis of the scientific problematic of the 1960s in Particle Physics and of its evolution until the end of the 1970s, including cosmological issues. Particular attention is devoted to the exceptional role of Rolf Hagedorn and to the progress of the statistical boostrap model until the experimental search for the quark-gluon plasma started being considered. In parallel, we simultaneously expose recent results and ideas concerning Particle Physics and in Cosmology, an discuss current open questions. Assuming preons to be constituents of the physical vacuum and the standard particles excitations of this vacuum (the superbradyon hypothesis we introduced in 1995), together with a spinorial space-time (SST), a new kind of Regge trajectories is expected to arise where the angular momentum spacing will be of 1/2 instead of 1. Standard particles can lie on such Regge trajectories inside associated internal symmetry multiplets, and the preonic vacuum structure can generate a new approach to Quantum Field Theory. As superbradyons are superluminal preons, some of the vacuum excitations can have critical speeds larger than the speed of light c, but the cosmological evolution selects by itself the particles with the smallest critical speed (the speed of light). In the new Particle Physics and Cosmology emerging from the pattern thus developed, Hagedorn-like temperatures will naturally be present. As new space, time, momentum and energy scales are expected to be generated by the preonic vacuum dynamics, the Planck scale does not necessarily make sense in the new scenario. It also turns out that two potential evidences for a superbradyonic vacuum with a SST geometry exist already: i) the recent results on quantum entanglement at large distances favoring superluminal propagation of signals and correlations ; ii) the anisotropy of the cosmic microwave background radiation between two hemispheres observed by the Planck Collaboration, in agreement with the predictions of cosmic SST automatically generating a privileged space direction for each comoving observer. Simultaneously to the discussion of the large number of open questions, we comment on the required experimental and observational programs.

This paper is dedicated to the memory of Rolf Hagedorn

[a] e-mail: luis.gonzalez-mestres@cosmology.megatrend.edu.rs ; personal e-mail: lgmsci@yahoo.fr

 DOI: 10.1051/epjconf/201612605005

1 Introduction

January 1965, 50 years ago, is the date of the historical CERN preprint by Rolf Hagedorn on the statistical thermodynamics of strong interactions at high energies [1]. It was a long and detailed paper, published the same year in the *Supplemento al Nuovo Cimento* [2]. In this early work, a value of 158 MeV for the hadronic highest temperature was already used to fit data. In a CERN preprint of October 1964 [3, 4], Hagedorn had used the title *Thermodynamics of Distinguishable Particles: A Key to High-Energy Strong Interactions?* and referred to a hadron highest temperature close to the pion mass. The abstract of [1] starts saying:

In this statistical-thermodynamical approach to strong interactions at high energies it is assumed that higher and higher resonances of strongly interacting particles occur and take part in the thermodynamics as if they were particles.

(end of quote)

It was simultaneously more and more clear that hadrons are not elementary particles. In 1963-64, the quark model had also been formulated [5, 6] by Murray Gell-Mann [7], George Zweig [8, 9] and André Petermann [10]. Although the questions raised with such a hypothesis were more fundamental, this new approach to the internal structure of hadrons did not invalidate the grounds of the Hagedorn thermodynamics even if an upgrade was necessary in the 1970s.

Later (1966 and 1967), together with Johannes Ranft, Hagedorn produced two other preprints entitled similarly [11, 12] and published in 1968 in the *Supplemento al Nuovo Cimento* [13, 14]. The three papers on the statistical thermodynamics of strong interactions at high energies, together with a september 1967 preprint by Hagedorn entitled *On the hadronic mass spectrum* [15, 16] were resumed by Rolf Hagedorn in a November 1967 CERN preprint [17] that was published in *Il Nuovo Cimento A* [18]. In December 1967, a new preprint version of [11] also appeared [19].

Simultaneously to the statistical bootstrap work of Hagedorn, attempts to relate direct-channel resonances to crossed-channel Regge poles culminated in July 1968 with the Veneziano work [20, 21] introducing the dual models. A few months before, the Harari-Freund picture had been formulated [22, 23] associating through the duality relation Regge poles to direct-channel resonances and the Pomeron to the background. Explicit N-particle dual amplitudes started being built in 1969 and an intermediate particle spectrum described by an infinite set of harmonic oscillators was found (see, for instance, [24–26] and on Nambu's work [27]). The string model developed then quickly and led, in particular, to a "parton view" of dual amplitudes [28, 29] as early as 1970. The authors of [28] state *"that the sum of a large number of "fishnet" Feynman graphs of very high order can be approximated by generalized Veneziano amplitudes"*. Similarly, the abstract of [29] explicitly refers to a "parton picture" of hadrons to claim that *"the limit of the sums of very high order, unrenormalized Feynman diagrams lead to dual N point functions"*. Thus, the underlying composite nature of the internal structure of dual amplitudes (strings) was clearly pointed out three years before the formulation of Quantum Chromodynamics by David Gross, Frank Wilczek and David Politzer [30, 31].

According to Gabriele Veneziano [32], Rolf Hagedorn emphasized in 1994, thirty years after formulating his statistical boostrap theory, that "the dual resonance model or string theory gives a microscopic explanation" for the hadron spectrum he had found using his own approach. After reporting this statement by Hagedorn, Veneziano develops a chapter entitled *A Surprise That Should Not Have Been One* and refers to an article by John Ellis [33] entitled *Hagedorn's Reincarnation in String Theory*. Two recent papers [34, 35] illustrate the current diversity of approaches and applications concerning string theory with Hagedorn thermodynamics

It remains to be seen how a Hagedorn-like thermodynamics can apply to alternative physics involving a new space-time geometry, new ultimate constituents of matter and possibly new Regge-like trajectories. We attempt here to discuss such questions and to suggest new ideas on the subject [36].

2 Hadrons, strings, bootstrap, quarks and gluons in the 1970's

I arrived to CERN, as a fellow, in January 1978. My paper *Structure of the cylinder term in the topological expansion*, with Patrick Aurenche, was then a preprint to appear in *Physical Review* [37]. The goal of the paper was to solve a long-term phenomenological controversy concerning the Pomeron and the f-trajectory and to definitely refute the arguments claiming that the Pomeron trajectory was actually a "promoted f" associated to what was called "f extinction". Such claims had been standing for years, aiming to contradict the Harari-Freund picture as well as the idea that the Pomeron could actually be associated to the "cylinder term" (the box diagram with the "double twist" of the quark lines in the crossed channel) of the dual field theory (see also [38, 39]).

As early as 1970, it had been found [40] that the non-planar box diagram of dual field theory (DFT) yielded a Pomeron-like Regge trajectory with a slope equal to half the initial Regge slope. In spite of its obvious advantages, interpreting this singularity as the Pomeron was rejected by many phenomenologists who stated that this identification would lead to practical inconsistencies and could not reproduce a multiperipheral pattern. In [37], followed by [38, 39]), it was shown that such claims were wrong and that the Pomeron obtained from DFT calculations carries automatically, in its internal structure (the properties of the intermediate states when interpreted in terms of particles and of internal constituents), a consistent multiperipheral description of "soft" hadronic physics.

The practical sense of the concept of bootstrap was thus tacitly debated: should one require to get the laws of Physics from a minimal phenomenological information? Fashionable phenomenology of the 1970's had often been reduced to applying multiperipheral "unitarity corrections" to the exchange of the f trajectory through the production of "clusters" without overlapping in the rapidity space. The DFT Pomeron was thus ignored or rejected, in spite of the fact that dual field theory was the natural consequence of the factorization of dual amplitudes leading to a hadron spectrum consistent with Hagedorn thermodynamics. In [37], taking the two-particle s-channel discontinuity of the one-loop double-twisted graph of DFT, it was shown that the dual field theoretical diagram associated to the Pomeron naturally leads to a multiperipheral pattern involving two kinds of regions in the rapidity space: those with a single jet dominated by a standard Regge exchange (the f, basically) and the regions with a double jet associated to the dual Pomeron exchange.

The analysis was pursued in [38], discussing in particular the role of flavour-loop corrections. In [39], a detailed analysis of the transverse momentum of the intermediate states confirmed the consistency of the dual Pomeron with multiperipheral phenomenology. In the two-jet region, the square of the total transverse momentum of each jet was shown to be proportional to the rapidity length, as expected from an uncorrelated distribution in the rapidity variable. Two-jet models for the Pomeron had been first considered in 1972-1973 by Huan Lee [41] and Gabriele Veneziano [42], but the DFT internal structure of two-jet states remained to be studied explicitly as in [37–39].

Simultaneously to our 1977-79 work, phenomenological studies became less favorable to the Pomeron - f identity. In the 1980's the Pomeron + f pattern was commonly used. In the late 1970's, the phenomenology of Quantum Chromodynamics had also started developing [43, 44].

Furthermore, as early as June 1975, a preon model for quarks and leptons was suggested by Jogesh Pati and Abdus Salam [45, 46]. When the rishon model was published in 1979 [47, 48], I had been considering the same pattern since some time but I had hesitated to publish it. After the papers by Haim Harari and Michael Shupe appeared, I proposed a more sophisticated model involving preonic internal symmetries like SU(2) and a confinement SO(3) [49]. 15 years later [50, 51], I suggested an alternative approach where preons are not constituents of quarks and leptons but of the physical vacuum (the superbradyon approach, where superbradyons are nontachyonic superluminal preons). The standard "elementary" particles are then excitations of the preonic vacuum and can be compared to phonons or solitons in condensed matter [36, 52]. See also [53, 54] and [55, 56].

 DOI: 10.1051/epjconf/201612605005

2.1 The quark-gluon plasma

What about the quark-gluon plasma, that constitutes today a direct experimental evidence for Hagedorn thermodynamics and may even be used to search for superbradyons [27, 36]? In 1975, after Quantum Chromodynamics with asymptotic freedom had been formulated, Nicola Cabibbo and Giorgio Parisi [57] pointed out that the Hagedorn temperature was not necessarily a limiting temperature but could also be associated to a second order phase transition, in particular deconfining quarks.

The new ingredient in the work by Cabibbo and Parisi was to deal with hadrons as extended objects and not just point-like as in the original Hagedorn papers. In January 1985, Rolf Hagedorn wrote [58] on this subject concerning his work of the 1960s:

The missing luck (or missing cleverness) was mainly due to the assumption of pointlike particles. With one exception [POM51] all statistical models made that same mistake at that time. It turned out that for point particles the singularity was such that, when approaching the transition temperature T_0, the energy density would diverge and thus the other phase would be out of reach. Hence T_0 was interpreted as the ultimate temperature.

(en of quote)

The reference [POM51] corresponds to a 1951 article by Isaak Pomeranchuk [59]. Reviews and contributions on the quark-gluon plasma and the experimental studies of this now established phenomenon have been presented at this conference by several authors.

In a preonic picture of vacuum [27, 36, 53, 54, 56], and assuming the standard particles to be vacuum excitations, an equivalent of the updated Hagedorn thermodynamics must in principle be considered at a scale where quarks, gluons, leptons and the photon can no longer be viewed as point-like objects. It may even happen that several such scales occur, especially in the presence of spinorial Regge trajectories and other new sets of particles generated by the dynamics of the preonic vacuum.

3 Spinorial space-time, superbradyonic vacuum and new particles

The spinorial space-time (SST) and the superbradyonic vacuum are discussed in detail in [36], focusing on their possible implications for the origin and actual properties of Quantum Mechanics as already considered in [53, 54] and in [56]. The cosmic SST is centered at the origin of the Universe $\xi = 0$ (vanishing cosmic space and time) where ξ stands for the cosmic space-time position described by a SU(2) spinor in which the four standard real coordinates have been replaced by two complex ones. The space associated to ξ_0 corresponds then to the cosmic hypersphere made of the points whose cosmic spinorial position has the same modulus (the associated cosmic time) as that of ξ_0.

Such a cosmic SST can be complemented, in a suitable approximation and at small enough cosmic distances [36], by a local SST where the SU(2) group acts on the spinorial distance $\xi - \xi_0$. The local world incorporating a preonic (superbradyonic) vacuum with a space-time geometry described by the local SST, where the spin-1/2 particles are actual representations of the local SU(2) space rotation group, can naturally generate spinorial Regge trajectories with an angular momentum spacing of 1/2 instead of 1. The sets of particles lying on these new trajectories can include, for instance, new kinds of matter resulting from specific excitations of the superbradyonic vacuum. The "elementary" particles of the standard model can lie on a spinorial Regge trajectory of this kind with internal symmetries added to the space-time SU(2) and generated as a low-energy limit of preon dynamics.

With such a local SST, a spin-1/2 particle with position at ξ_0 would be a spinorial wave function "centered" at ξ_0 and taking significant values at small enough values of $| \xi - \xi_0 |$ (the modulus of the spinorial distance). The existence of "elementary" spin-3/2 or spin-5/2 particles may thus be naturally generated. The standard Poincaré group is no longer an exact symmetry but a local low-energy limit, and an alternative to supersymmetry (approximate and broken) can emerge [36, 55, 60].

3.1 Search for new particles generated by the preonic vacuum

The search for new particles generated by a local SST geometry together with a preonic vacuum structure is obviously an important and urgent task, already for current CERN experiments. For instance, recent announcements by CMS [61] and ATLAS [62] mention to have observed in both experiments a signal in the diphoton invariant mass spectrum around 750 GeV. Even if the statistical significance remains weak, such a signal is already the subject of an increasing number of theoretical papers traying to explain its origin through all kinds of "classical" and "exotic" phenomena. But actually, such a particle, if it really exists, can be one of the low-energy objects just described generated in a preonic vacuum with a spinorial space-time or another unconventional space-time. If the standard particles actually belong to the same family of preonic vacuum excitations, the couplings for the decay of the possibly detected particles will not necessarily be radically different from the conventional ones.

Actually, as pointed out in [36], a possible direct signature already exists for the superbradyonic vacuum: the long-distance entanglement recently observed in [63, 64], clearly violating Bell inequalities [65, 66]. A strongly superluminal signal and correlation propagation in vacuum can be at the origin of such an entanglement effect observed at a distance of 1.3 kilometers.

Thus, assuming preons to be the constituents of the physical vacuum (the superbradyon hypothesis [50, 51]) instead of "quark like" constituents of the standard particles as initially postulated by authors of the 1970's articles, opens the way to a large set of phenomenological potentialities that must be seriously explored [27]). If the standard "elementary" particles are actually low-energy excitations of a superbradyonic vacuum with a spinorial space-time geometry, and if other excitations of this vacuum exist (the recently observed LHC dipole signal can be an example), new superluminal particles (not tachyons) with a critical speed not radically different from c can be not too weakly coupled to standard particles, and a new Hagedorn-like thermodynamics can hold for such a global dynamical structure.

3.2 Some open questions

At this stage, a series of crucial questions arises. In particular: : i) can superbradyons exist as free particles? ; ii) can they exist as confined in a "plasma" resulting from collisions between standard particles? ; iii) what are the space, time, momentum, energy... scales at which standard particles start existing in the superbradyonic vacuum? ; iv) does the Planck scale still make any sense? ; v) what is (are) the critical speed(s) of superbradyons, inside vacuum and possibly as free particles? ; vi) what are the thermodynamics and the internal temperature of the physical vacuum? ; vii) what can be the physical laws governing superbradyon interactions? ... Obvioulsy, a long-term work is required in experimental particle physics and cosmological observations, including new accelerators, satellite programs and Earth-based detection of ultra-high energy cosmic rays (UHECR) [53, 56].

Also, as discussed in [36], the possibility of a contradiction at very small space-time scales between the local SST (or a similar unconventional space-time) and the extrapolation of the macroscopic cosmic one must be taken into account, as well as its possible consequences for particle propagation and the generation of Quantum Mechanics (see also [54, 67] and [68, 69]. In such a scenario, the relevant space-time scale corresponds to the spinorial size of the particle internal structure, and slightly above this scale. Again, the conventional Planck scale [70] can be totally bypassed by such a fundamental process and in any case a long-term experimental and observational effort will be required to find signatures of the relevant space, time, momentum and energy scales for the generation of standard particles. The question of a Hagedorn-like themodynamics sensitive to these scales must also be raised like, more generally, that of the internal vacuum and particle structure.

Another important question just tacitly raised is: do all the particles generated as excitations of the surperbradyonic vacuum have the same critical speed? This can be natural for all particles lying on the same multiplets and spinorial Regge trajectories as the conventional particles, but less obvious for other vacuum excitations that can possibly have critical speeds larger than c. The spinorial size of such particles would in principle be different from that of the standard particles. Similarly, the vacuum structure in terms of superbradyons ("plasma"?, "condensed matter"?) requires a serious reflexion.

4 Cosmology with SST and a superbradyonic vacuum

Applications to Astrophysics of the statistical thermodynamics of hadrons were already suggested by Rolf Hagedorn as early as April 1969, leading to a September 1969 preprint published in 1970 [71]. Hagedorn thermodynamics is now a permanent ingredient of string theory and Cosmology.

An alternative to strings and standard Cosmology is provided by preon models and pre-Big Bang [53, 56]. In this kind of patterns, Hagedorn-like temperatures are also expected to be present.

In all cases, assuming superluminal free particles can exist, simple phase space considerations can explain that they have gradually disappeared at early stages of the history of our Universe [53, 56]. For a given energy, the momentum phase space is smaller when the critical speed is larger. The production of such particles becomes therefore more difficult. Simultaneously, these particles decay into other particles with a smaller critical speed. This kind of considerations would apply to superbradyons if they can be materialized as free particles, as well as to vacuum excitations with a critical speed larger than c. Thus, the actual cosmological evolution is finally dominated by the particles with the lowest critical speed (the standard particles and possible similar new ones). Some superluminal particles can remain, forming a sea with speeds close to c, and be part of the dark matter [55, 72].

Pre-Big Bang with a superbradyonic vacuum eliminates any need for an inflationary period and, simultaneously, replaces the Planck scale by new scales associated to the dynamics introduced [53, 56]. The phases of the history of the Universe generated by this dynamics can naturally involve a Hagedorn-like thermodynamics. Similarly, the SST Universe can be hyperspherical but appear to us as flat because of its size beyond direct observation [53, 56].

Actually, just as long-distance quantum entanglement provides a potential evidence for the superbradyonic vacuum, a specific prediction of the cosmic SST geometry seems to have been confirmed by observation. A 2013 Planck result [73] found the existence of a preferred space direction with a clear difference in power spectrum between the two hemispheres. Such an announcement seems to have survived to 2015 analyses and corrections [74, 75]. This 2013 Planck result corresponds to a natural prediction of the SST already found in [76] and stressed in [55, 56]. The local privileged space direction (PSD) for each comoving observer is a direct consequence of the cosmic spinorial coordinates and corresponds to multiplying the cosmic spinor by a complex phase (see also [53, 56]).

Thus, the SST + superbradyonic vacuum scenario may have already been confirmed by experimental and observational data. Further work in this direction is obviously required.

Again, important questions remain open and require further study [27]. What can be the dynamics behind the cosmic spinorial space-time, in particular around $\xi = 0$ when the superbradyonic vacuum starts developping ? And if standard particles are soliton-like solutions of the vacuum equations, what is the situation concerning plane waves? How does the vacuum interact with conventional matter, including the quark-gluon plasma? Can there be exchanges of energy and momentum? Can the quark-gluon plasma emit superluminal particles? What is the difference between the vacuum configuration in regions of space with almost no standard matter and in the presence of important matter densities?

The study of such questions will in particular involve thermodynamical issues and considerations, new theoretical efforts, new experiments and observations beyond the standard Casimir effect [77], further work on quantum field theory and on the vacuum structure [53, 56]... Preonic approaches appear particularly well suited to settle problems related to the quantum-field-theoretical structure of vacuum and can yield natural, powerful solutions (see also [36] and references therein).

5 Conclusion and comments

Resulting from an original, long-term work made in the 1960's, the statistical thermodynamics of strong interactions has become a basic component of Particle Physics and Cosmology.

After the initial Hagedorn work and the 1975 paper by Cabibbo and Parisi [57], Rolf Hagedorn, Istvan Montvay and Johann Rafelski (December 1978) [78] and subsequently Johann Rafelski and Rolf Hagedorn (April 1979) [79] presented an updated version of the statistical boostrap model studying the thermodynamics of hot nuclear matter. Later, Johann Rafelski suggested to study the properties of nuclear matter through the annihilation of antiprotons with momenta of 0.5-1.5 GeV/c at LEAR (Low Energy Antiproton Ring) colliding with heavy nuclei [80]. A subsequent paper by Johann Rafelski, Hans-Thomas Elze and Rolf Hagedorn [81] focused on hot hadronic matter and hot quark-gluon plasma. Thus, Hagedorn was already involved in the search for the quark-gluon plasma at CERN.

If the theoretical transition from hadrons to quarks was a fundamental one in the 1960s and 1970s, the same can happen at present with a transition from standard particles to preons if the basic idea is correct. But contrary to the Quantum Chromodynamics of the 1970s, there is by now no precise theory of preon dynamics. The superbradyon hypothesis helps to clarify the situation by setting preons to be (naturally superluminal) constituents of the physical vacuum and not "quark-like" constituents of the standard particles as initially assumed in the 1970s. But exploring the internal vacuum dynamics will certainly be much more difficult than building a theory of quarks with a suitable phenomenology.

In [36], a possible fundamental origin of Quantum Mechanics has been dealt with based on the spinorial space-time with a dynamical preonic (superbradyonic) vacuum. Experimental tests of Quantum Mechanics, including at the highest cosmic-ray energies and searching for signatures of a deformation, can help to understand vacuum and (possibly) superbradyon dynamics.

As previously emphasized, recent results on long-distance quantum entanglement and the privileged space direction recently observed by Planck are a strong potential evidence for a superbradyonic vacuum with a SST geometry. New theoretical studies, as well as new generations of experiments and observations, must pay particular attention to these two features of data. Actually, a global long-term experimental an observational effort will be required to really understand the properties of the preonic vacuum and superbradyons, assuming that these hypotheses correspond to the physical reality.

Spinorial Regge trajectories and other kinds of families of excitations of the superbradyonic vacuum with SST, including the standard particles and members of associated multiplets and trajectories, deserve an active exploration at accelerators without forgetting the possible production of superluminal objects. Such an alternative to standard strings does not appear to be less justified.

References

[1] R. Hagedorn: *Statistical thermodynamics of strong interactions at high-energies*, CERN preprint CM-P0057114, http://cds.cern.ch/record/346206/files/CM-P0057114.pdf

[2] R. Hagedorn: *Statistical thermodynamics of strong interactions at high-energies*, *Nuovo Cim. Suppl.* **3**, 147 (1965). Also published in *Quark-Gluon Plasma: Theoretical Foundations*, edited by J.I. Kapusta, B. Müller and J. Rafelski, North-Holland (Elsevier) 2003.

[3] R. Hagedorn, *Thermodynamics of Distinguishable Particles: A Key to High-Energy Strong Interactions?*, CERN preprint CM-P0056976, 12 October 1964, http://cds.cern.ch/record/929730/files/CM-P0056976.pdf

[4] The main work of Rolf Hagedorn is explained and reproduced in the book *Melting Hadrons, Boiling Quarks* (Springer), edited by Johann Rafelski. Freely available at the address http://www.springer.com/la/book/9783319175447

[5] V. Petrov, *Half a Century with Quarks*, arXiv:1412.8681.

[6] G. Zweig, *Concrete Quarks - The Beginning of the End*, talk given at the 2nd International Conference on New Frontiers in Physics, Kolymbari, Crete, Greece, August 28 - September 5, 2013, *EPJ Web of Conferences* **71**, 00146 (2014).

[7] M. Gell-Mann,*A schematic model of baryons and mesons*, *Phys. Lett.* **8**, 214 (1964). Available at the address http://hep.caltech.edu/gm/images/quarks.pdf (Caltech).

[8] G. Zweig, *An SU_3 model for strong interaction symmetry and its breaking*, CERN TH preprint 17 January 1964, http://cds.cern.ch/record/352337/files/CERN-TH-401.pdf

[9] G. Zweig, *An SU_3 model for strong interaction symmetry and its breaking. II*, CERN TH preprint 21 February 1964, http://cds.cern.ch/record/570209/files/CERN-TH-412.pdf

[10] A. Petermann, *Propriétés de l'étrangeté et une formule de masse pour les mésons vectoriels*, *Nucl. Phys.* **63**, 349 (1965), received 30 December 1963.

[11] R. Hagedorn and J. Ranft, *Statistical thermodynamics of strong interactions at high-energies - II (Momentum spectra of particles produced in high energy pp collisions)*, CERN preprint CM-P0057348, 30 September 1966, http://cds.cern.ch/record/933372/files/ CM-P0057348.pdf

[12] R. Hagedorn and J. Ranft, *Statistical thermodynamics of strong interactions at high-energies - III. Heavy pair (quark) production rates*, CERN preprint CM-P0057552, 3 April 1967, http://cds.cern.ch/record/936485/files/CM-P0057552.pdf

[13] R. Hagedorn and J. Ranft, *Statistical thermodynamics of strong interactions at high-energies. 2. Momentum spectra of particles produced in high energy pp collisions, Nuovo Cim. Suppl.* **6**, 169 (1968).

[14] R. Hagedorn and J. Ranft, *Statistical thermodynamics of strong interactions at high energies. 3. Heavy-pair (quark) production rates, Nuovo Cim. Suppl.* **6**, 311 (1968).

[15] R. Hagedorn, *On the hadronic mass spectrum*, CERN preprint CM-P0057745, 12 september 1967, http://inspirehep.net/record/1384659/files/CM-P00057745.pdf

[16] R. Hagedorn, *On the hadronic mass spectrum*, *Nuovo Cimento* **A52**, 1336 (1967). Available in open access in the book *Melting Hadrons, Boiling Quarks*.

[17] R. Hagedorn, *Hadronic matter near the boiling point*, CERN preprint CM-P0057767, 27 november 1967, http://cds.cern.ch/record/347805/files/CM-P0057767.pdf

[18] R. Hagedorn, *Hadronic matter near the boiling point*, *Nouvo Cimento* **A56**, 1027 (1968).

[19] R. Hagedorn and J. Ranft, *Statistical thermodynamics of strong interactions at high-energies - II. Momentum spectra of particles produced in high energy pp collisions*, CERN preprint CM-P0057765, December 1967, http://cds.cern.ch/record/938671/files/CM-P0057765.pdf

[20] G. Veneziano, *Construction of a crossing symmetric, Regge behaved amplitude for linearly rising trajectories*, CERN preprint CM-P0057839, 19 July 1968,)http://cds.cern.ch/record/390478/files/CM-P0057839.pdf

[21] G. Veneziano, *Construction of a crossing symmetric, Regge behaved amplitude for linearly rising trajectories*, *Nuovo Cimento A57*, 190 (September 1968).

[22] H. Harari, *Pomeranchuk Trajectory and its relation to low-energy scattering amplitudes*, *Phys. Rev. Lett.* **20**, 1395 (June 1968). Received 26 February 1968. Available at the address http://www.weizmann.ac.il/particle/harari//files/PhysRevLett.20.1395.pdf (Weizmann Institute)

[23] P. Freund, *Finite-Energy Sum Rules and Bootstraps*, *Phys. Rev. Lett.* **20**, 235 (January 1968), received 8 December 1967.

[24] S. Fubini and G. Veneziano, *Level Structure of Dual Resonance Models*, *Nuovo Cimento 64A*, 811 (1969).

[25] K. Bardakci and S. Mandelstam, *Analytic Solution of the Linear-Trajectory Bootstrap, Phys. Rev. 184*, 1640 (1969).

[26] M. Virasoro, *Subsidiary Conditions and Ghosts in Dual Resonance Models, Phys.Rev.* **D1**, 2933 (1970).

[27] See also our talk *Big Bang, inflation, standard Physics... and the potentialities of new Physics and alternative cosmologies*, https://indico.cern.ch/event/344173/session/4/contribution/277 (this Conference).

[28] B. Sakita and M. A. Virasoro, *Dynamical Model of Dual Amplitudes Phys. Rev.Lett. 24*, 1146 (1970).

[29] H.B. Nielsen and P. Olesen, *A Parton View on Dual Amplitudes, Phys. Lett.* **32B**,203 (1970). CERN preprint CM-P00058602.pdf (May 1970), http://cds.cern.ch/record/350954/files/CM-P00058602.pdf

[30] D. Gross and F. Wilczek, *Ultraviolet Behavior of Non-Abelian Gauge Theories, Phys. Rev. Lett.* **30**, 1343 (1973), http://www.aps.org/about/pressreleases/upload/Ultraviolet_Behavior_of_Non_Abelian_Gauge_Theory.pdf

[31] H.Politzer, *Reliable Perturbative Results for Strong Interactions? Phys. Rev. Lett.* **30**, 1346 (1973), http://wwwphy.princeton.edu/~kirkmcd/examples/EP/politzer_prl_30_1346_73.pdf

[32] G. Veneziano, *The Hagedorn Spectrum and the Dual Resonance Model: An Old Love Affair*, event *Hagedorn's Legacy*, CERN, 13 November 2015, https://indico.cern.ch/event/446994/. Published in *Melting Hadrons, Boiling Quarks*.

[33] J. Ellis, *Hagedorn's Reincarnation in String Theory*, 27 Jun 1994 Meeting, Divonne-les-Bains, France. Published in *Hot Hadronic Matter*, edited by Jean Letessier, Hans H. Gutbrod and Johann Rafelski, NATO ASI Series, Volume **346** (Springer, 1995).

[34] C. Thorn, *String Bits at Finite Temperature and the Hagedorn Phase*, arXiv:1507.03036, and references therein.

[35] T. Mertens (Thesis), *Hagedorn String Thermodynamics in Curved Spacetimes and near Black Hole Horizons*, arXiv:1506.07798, and references therein.

[36] Complementary material is provided in our contribution to this conference *Spinorial space-time and the origin of Quantum Mechanics*, https://indico.cern.ch/event/344173/session/22/contribution /405

[37] P. Aurenche and L. Gonzalez-Mestres, *Phys. Rev.* **D 18**, 2995 (1978). Received 29 August 1977, revised manuscript received 17 January 1978. Preprint LPTHE 77/33.

[38] P. Aurenche and L. Gonzalez-Mestres, *Glueball singularity, flavour loops and the Harari-Freund picture, Zeitschrift für Physik* **C1**, 307 (1979) https://cds.cern.ch/record/133138/files/CM-P00061765.pdf

[39] L. Gonzalez-Mestres and P. Aurenche, *Glueball-Pomeron identity and the transverse structure of two-jet events, Zeitschrift für Physik* **C2**, 339 (1979), http://cds.cern.ch/record/133286/files/197904147.pdf

[40] G. Frye and L. Susskind, *Non-planar dual symmetric loop graphs and the pomeron, Phys. Lett.* **B31**, 589 (1970).

[41] Huan Lee, *How to Generate the Pomeranchukon from the Background in a Dual Multiperipheral Model, Phys. Rev. Lett.* **30**, 719 (1973).

[42] G. Veneziano, *Origin and intercept of the Pomeranchuk singularity, Physics Letters* **B43**, 413 (1973).

[43] K. Shizuya and S.-H.H. Tye, *QCD Phenomenology of Gluon Jets, Phys. Rev.* **D20**, 1101 (1979) and Erratum, *Phys. Rev.* **D21**, 313 (1980). Fermilab preprint FERMILAB-Pub-79/16-THY,

http://lss.fnal.gov/archive/1979/pub/Pub-79-016-T.pdf

[44] G.P. Lepage and S. Brodsky, *Exclusive Processes in Perturbative Quantum Chromodynamics*, *Phys. Rev.* **D22**, 2157 (1980), and references therein. Preprint SLAC-PUB-2478 (March 1980), http://inspirehep.net/record/153236/files/slac-pub-2478.pdf

[45] J. Pati and A. Salam, *Quarks, leptons and pre-quarks*, ICTP (Trieste) report IC1975106, available by a search at the ICTP archive : http://cat.ictp.it/F/?func=find-b&local_base=WEBALL

[46] See also J. Pati, A. Salam and J. Strathdee, *Are Quarks Composite?*, *Phys. Lett.* **B59**, 265 (1975). Preprint available by a search at the ICTP archive.

[47] H. Harari, *A schematic model of quarks and leptons*, *Phys. Lett.* **B86**, 83 (1979). Available at the address http://www.weizmann.ac.il/particle/harari//files/schematic_model_of_quarks_and_leptons.pdf (Weizmann Institute).

[48] M. Shupe, *A composite model of leptons and quarks*, *Phys. Lett.* **B86**, 87 (1979).

[49] L. Gonzalez-Mestres, *Quarklike composite model of quarks and leptons*, *Phys. Rev.* **D23**, 2055 (1981). Received 28 March 1980.

[50] L. Gonzalez-Mestres, *Properties of a possible class of particles able to travel faster than light*, Proceedings of the January 1995 Moriond Workshop, Ed. Frontières, arXiv:astro-ph/9505117.

[51] L. Gonzalez-Mestres, *Cosmological Implications of a Possible Class of Particles Able to Travel Faster than Light*, Proceedings of the TAUP 1995 Conference, *Nucl. Phys. Proc. Suppl.* **48** (1996), 131, arXiv:astro-ph/9601090.

[52] L. Gonzalez-Mestres, *Vacuum Structure, Lorentz Symmetry and Superluminal Particles*, arXiv:physics/9704017

[53] L. Gonzalez-Mestres, *BICEP2, Planck, spinorial space-time, pre-Big Bang*, invited talk at the 3rd International Conference on New Frontiers in Physics (ICNFP 2014), Kolymbari, Crete, Greece, August 23 - 30, 2014, *EPJ Web of Conferences* **95**, 03014 (2015), and references therein.

[54] L. Gonzalez-Mestres, *Tests and prospects of new physics at very high energy*, contribution the 3rd International Conference on New Frontiers in Physics, Kolymbari, Crete, Greece, August 23 - 30, 2014, *EPJ Web of Conferences* **95**, 05007 (2015), and references therein.

[55] L. Gonzalez-Mestres, *Cosmic rays and tests of fundamental principles*, CRIS 2010 Proceedings, *Nucl. Phys. B, Proc. Suppl.* **212-213** (2011), 26, and references therein. The *arXiv.org* version arXiv:1011.4889 includes a relevant Post Scriptum.

[56] See also our contributions to the ICNFP 2012 and ICNFP 2013 Conferences.

[57] N. Cabibbo and G. Parisi, *Exponential hadronic spectrum and quark liberation*, *Phys. Lett.* **B59**, 67 (1975).

[58] R. Hagedorn, *Miscellaneous elementary remarks about the phase transition from a hadron gas to a gluon plasma*, CERN preprint TH.4100/85, January 1985, http://cds.cern.ch/record/158166/files/198504017.pdf

[59] I. Pomeranchuk, *On the theory of production of many particles in a single act*, *Doklady Akademii Nauk sssr* **78**, 889 (1951).

[60] L. Gonzalez-Mestres *Pre-Big Bang, fundamental Physics and noncyclic cosmologies*, presented at the International Conference on New Frontiers in Physics, ICFP 2012, Kolymbari, Crete, June 10-16 2012, *EPJ Web of Conferences* **70**, 00035 (2014), and references therein. Preprint version at mp_arc 13-18.

[61] The CMS Collaboration, *CMS Physics Analysis Summary*, 18 December 2015, https://cds.cern.ch/record/2114808/files/EXO-15-004-pas.pdf

[62] The ATLAS Collaboration, ATLAS-CONF-2015-081, 15 December 2015, http://cds.cern.ch/record/2114853/files/ATLAS-CONF-2015-081.pdf

 DOI: 10.1051/epjconf/201612605005

[63] B. Hensen et al., *Experimental loophole-free violation of a Bell inequality using entangled electron spins separated by 1.3 km*, *Nature* **526**, 682 (2015), arXiv:1508.05949.

[64] M. Giustina et al., *A significant-loophole-free test of Bell's theorem with entangled photons*, arXiv:1511.03190.

[65] J.S. Bell, *Speakable and Unspeakable in Quantum Mechanics: Collected Papers on Quantum Philosophy*, Cambridge University Press, 2004.

[66] J. Clauser, M. Horne, A. Shimony and R. Holt, *Proposed experiment to test local hidden-variable theories*, *Physical Revew Letters* **23**, 880 (1969). Available at http://users.unimi.it/aqm/wp-content/uploads/CHSH.pdf

[67] L. Gonzalez-Mestres, *Quantum Mechanics and the Spinorial Space-Time*, mp_arc 15-86.

[68] L. Gonzalez-Mestres, *Quantum Mechanics, preonic vacuum and space-time contradiction*, mp_arc 15-90.

[69] L. Gonzalez-Mestres, *Quantum Mechanics, space-time, preons and entanglement*, mp_arc 15-92

[70] The standard Planck scale is actually very old: M. Planck, *Über irreversible Strahlungsvorgänge,*, *S.B. Preuss. Akad. Wiss.* (1889), 440 and *Annalen der Physik* **1**, 69 (1900). Available at the address http://vincent.rossetto.free.fr/physique/ref/planck1900.pdf

[71] R. Hagedorn, *Thermodynamics of strong interactions at high-energy and its consequences for astrophysics*, *Astron. Astrophys.* **5**, 184 (1970), CERN preprint CM-P00058250, http://cds.cern.ch/record/350701/files/CM-P00058250.pdf

[72] L. Gonzalez-Mestres, *Lorentz symmetry violation, dark matter and dark energy*, contribution to the Invisible Universe International Conference, Paris June 29 - July 3, 2009. The *arXiv.org* version arXiv:0912.0725 contains a Post Scriptum and an additional note.

[73] The Planck Collaboration, *Planck 2013 results. XXIII. Isotropy and statistics of the CMB*, *Astron.Astrophys. 571*, A23 (2014), arXiv:1303.5083.

[74] The Planck Collaboration, *Planck 2015 results. XVI. Isotropy and statistics of the CMB*, arXiv:1506.07135.

[75] See, for instance, E. Martinez-Gonzalez on the behalf of the Planck Collaboration, transparencies presented at the Fundamental Cosmology meeting, Santander June 17th-19th, 2015.

[76] L. Gonzalez-Mestres, *Space, Time and Superluminal Particles*, arXiv:physics/9702026.

[77] For a recent review on the Casimir effet, see for instance G. L. Klimchitskaya and V. M. Mostepanenko, *Casimir and van der Waals forces: Advances and problems*, arXiv:1507.02393.

[78] R. Hagedorn, I. Montvay and J. Rafelski, *Thermodynamics of hot nuclear matter from the statistical bootstrap model*, Workshop on Hadronic Matter at Extreme Energy Density "Ettore Majorana", Erice, 13 - 21 October 1978, p.49. Available at the address http://www.physics.arizona.edu/ rafelski/Books/78EriceHagRaf.pdf (University of Arizona).

[79] J. Rafelski and R. Hagedorn, *Thermodynamics of hot nuclear matter in the statistical bootstrap model*, 17th International Winter Meeting on Nuclear Physics, Bormio, Italy, 22 - 27 Jan 1979, p.531. CERN preprint CM-P0005555 (April 1979), http://cds.cern.ch/record/134674/files/CM-P00055555.pdf

[80] J. Rafelski, $\bar{p}$ *annihilation on heavy nuclei*, *Phys. Lett.* **B91**, 281 (1980). CERN preprint TH.260, available at http://ccdb5fs.kek.jp/cgi-bin/img/allpdf?197911276

[81] J. Rafelski, H.-Th.Elze and R. Hagedorn, *Hot hadronic and quark matter in $\bar{p}$ annihilation on nuclei*, 5th European Symposium on Nucleon-antinucleon Interactions, Bressanone, Italy, 23 - 28 Jun 1980, p. 357, CERN preprint CM-P00061794, https://cds.cern.ch/record/982871/files/CM-P00061794.pdf

EPJ Web of Conferences **126**, 05006 (2016) DOI: 10.1051/epjconf/201612605006
ICNFP 2015

Spinorial space-time and the origin of Quantum Mechanics

The dynamical role of the physical vacuum

Luis Gonzalez-Mestres[1,a]

[1] *Megatrend Cosmology Laboratory, John Naisbitt University, Belgrade and Paris*
Goce Delceva 8, 11070 Novi Beograd, Serbia

Abstract. Is Quantum Mechanics really and ultimate principle of Physics described by a set of intrinsic exact laws? Are standard particles the ultimate constituents of matter? The two questions appear to be closely related, as a preonic structure of the physical vacuum would have an influence on the properties of quantum particles. Although the first preon models were just « quark-like » and assumed preons to be direct constituents of the conventional « elementary » particles, we suggested in 1995 that preons could instead be constituents of the physical vacuum (the superbradyon hypothesis). Standard particles would then be excitations of the preonic vacuum and have substantially different properties from those of preons themselves (critical speed...). The standard laws of Particle Physics would be approximate expressions generated from basic preon dynamics. In parallel, the mathematical properties of space-time structures such as the spinoral space-time (SST) we introduced in 1996-97 can have strong implications for Quantum Mechanics and even be its real origin. We complete here our recent discussion of the subject by pointing out that: i) Quantum Mechanics corresponds to a natural set of properties of vacuum excitations in the presence of a SST geometry ; ii) the recently observed entanglement at long distances would be a logical property if preons are superluminal (superbradyons), so that superluminal signals and correlations can propagate in vacuum ; iii) in a specific description, the function of space-time associated to the extended internal structure of a spin-1/2 particle at very small distances may be incompatible with a continuous motion at space and time scales where the internal structure of vacuum can be felt. In the dynamics associated to iii), and using the SST approach to space-time, a contradiction can appear between macroscopic and microscopic space-times due to an overlap in the time variable directly related to the fact that a spinorial function takes nonzero values simultaneously in a whole time interval. Then, continuous motion can be precluded at very small space-time scales. If discrete motion is required at such scales, the situation will possibly be close to that generating the Feynman path integral. More generally, Quantum Mechanics can naturally emerge from the spinorial space-time and from other unconventional space-time structures in a fundamental preon dynamics governing the properties of vacuum. In such scenarios, the application of Gödel - Cohen mathematics to quantum-mechanical calculations can possibly yield substantially different results from those recently obtained using the standard quantum approach without any preonic underlying structure. This is also a crucial open question for Quantum Mechanics and Particle Physics.

This paper is dedicated to the memory of Bernard d'Espagnat

[a] e-mail: luis.gonzalez-mestres@cosmology.megatrend.edu.rs ; personal e-mail: lgmsci@yahoo.fr

1 Introduction

What are the actual origin and meaning of Quantum Mechanics? Is it an ultimate fundamental property of Physics, or a dynamical feature of standard matter generated by a more basic scenario related to a deeper set of laws of Nature? In the present situation, the answer to this question is far from obvious, even if Quantum Mechanics is in practice dealt with as an ultimate principle of Physics [1].

The origin and the ultimate nature of Quantum Mechanics stay by now unknown, even if a link between wave-particle duality and entropic uncertainty is strongly suggested [2, 3] and applications of entropic uncertainty relations are currently given particular attention [4, 5].

Following our previous work, a reflexion was developed in [6, 7] on the possible origin of Quantum Mechanics from a preonic (superbradyonic?) vacuum with a specific (spinorial) space-time geometry. New ideas submitted to this Conference [8, 9] introduce explicit patterns illustrating the possible grounds of such a hypothesis. The constructions considered are specific examples of a more general approach. Some of the consequences of these patterns and of the new kind of theory considered are analyzed in [10] and further in what follows.

Describing and developing the suggested new approach, we present here explicit examples of attempts to explain and enforce Quantum Mechanics starting from a more fundamental scenario. The explicit models illustrating the basic idea use a spinorial space-time (SST) together with a pre-existing vacuum dynamics that can be associated to a preonic picture. Quantum Mechanics appears as a natural property for standard particles in a general superbradyonic scenario with SST.

It furthermore turns out, in such an explicit picture, that Quantum Mechanics can even be a direct consequence of the original time uncertainty induced by the SST when conventional particles are described as spinorial extended objects. In this kind of situation, the basic properties of the SST appear to be crucially different from those of a standard Euclidean or relativistic space-time where there is no obvious contradiction between the macroscopic space-time and the space-time felt by ultimate matter at very small scales as a natural extrapolation relies both descriptions.

In a possible version of the SST picture for the internal structure and motion of standard particles [8, 9], the natural time spread of a spinorial extended function used to describe the content of a conventional particle can naturally generate an overlap between structures of this kind "centered" at different times. Such an overlap can in turn lead to an overall structure for particle propagation incompatible with the initial dynamics that produces an isolated single particle solution to the basic vacuum equations for a given "central" time. This incompatibility would prevent continuous motion of standard particles and lead instead to a Feynman-like discrete path. Standard Quantum Mechanics would then be a large-scale limit of such a Feynman-like path [11, 12].

Quantum Mechanics can be in this case the dynamical expression of a contradiction between two space-time structures if the macroscopic space-time incorporates a curved space and a time origin of the Universe, and if simultaneously an ultimate structure of matter exists beyond standard particles at very small scales with an unconventional local space-time. The SST (cosmic and local) can provide a simple example of this space-time contradiction.

Similarly, the observed quantum entanglement at large distances would be a natural property of a superbradyonic vacuum where superluminal signals and correlations can in principle propagate, contrary to the standard space-time with the speed of light as the critical speed. Globally, a preonic scenario with a superbradyonic vacuum seems to lead to more natural properties for standard particles than the conventional postulates of Particle Physics [6, 7].

Recently, the Gödel incompleteness theorem [13, 14] has been claimed [15, 16] to lead to unsolvable calculations in quantum physics for conventional particles. Standard Quantum Mechanics has been used to reach such a conclusion. However, the situation can possibly be different in our preonic approach if the Gödel - Cohen incompleteness actually applies to the vacuum structure and dynamics and to the excitations of vacuum as described in terms of vacuum dynamics.

The search for signatures of deformations of Quantum Mechanics at ultra-high energy (possible expressions of a preonic dynamics) is a relevant basic phenomenology [17, 18]. The nature and origin of Quantum Mechanics are also crucial for cosmological patterns [19].

2 The spinorial space-time (SST)

The spinorial space-time (SST) we introduced in 1996-97 [20, 21] can possibly be the natural frame to describe a world where fermions exist. Its implications turn out to be highly nontrivial for Particle Physics and Cosmology, at small and large distances [22].

From a mathematical point of view, the SST is a SU(2) space-time with two complex coordinates replacing the four standard real ones. Its properties far from conventional, with possible cosmological implications coming directly from this mathematical structure, have been dealt with in several subsequent papers including in particular [17, 23], [24, 25] and [7].

SU(2) is just the universal covering group of the standard space rotations usually described by SO(3). But this simple transition from SO(3) to SU(2), allowing spinors and fermions to be actual representations of the group associated to space rotations, turns out to have important and previously unexpected consequences for both Cosmology and Particle Physics.

In particular, the SST illustrates how nontrivial space-time can be as compared to the usual euclidean and relativistic pictures. An example is provided by Quantum Mechanics may, that turn out to be a consequence of the properties of such a nontrivial space-time.

2.1 SST and cosmic coordinates

If ξ is a SU(2) spinor describing the cosmic SST coordinates (two complex variables replacing the standard four real ones) of a point of our space-time, it is possible to associate to ξ a positive SU(2) scalar $\mid \xi \mid$ such that $\mid \xi \mid^2 = \xi^\dagger \xi$ (the dagger stands for hermitic conjugate). A new invariant with respect to space SU(2) transformations is thus obtained.

A possible definition of cosmic time (in principle equivalent to the age of the Universe) can then be $t = \mid \xi \mid$. Associated to this definition of time is a space given by the S^3 hypersphere $\mid \xi \mid = t$ with an additional spinorial structure that would not exist in a S^3 space hypersphere obtained from a relativistic approach to Cosmology. Other definitions of the cosmic time t in terms of $\mid \xi \mid$ in the SST (f.i. $t = \mid \xi \mid^2$) are possible, but they lead to similar cosmological results as long as a single-valued function of $\mid \xi \mid$ is used to define the cosmic time.

With these definitions, the origin of cosmic time naturally associated to the beginning of the Universe is given by the point $\xi = 0$ where the initial space is contracted to a single point.

One then gets an expanding universe where cosmological comoving frames are described by straight lines crossing the time origin $\xi = 0$ and are transformed into eachother by the elements of the cosmic SU(2). Thus, the SST geometry provides in a natural way a local privileged rest frame for each comoving observer, compatible with existing cosmological observations.

Such an approach clearly suggests potential limitations of general relativity and standard cosmology. Rather than an intrinsic fundamental property of space and time, conventional relativity can be expected to be a low-energy symmetry of standard matter similar to the effective Lorentz-like symmetry of the kinematics of low-momentum phonons or solitons in a condensed medium [26, 27] where the speed of sound or the maximum speed of solitons plays the role of the critical speed. The speed of light would then be the critical speed of a family of vacuum excitations (the standard particles) without being directly related to any intrinsic space-time geometry [27, 28].

2.2 SST transformations and coordinates

Then, taking $t = \mid \xi \mid$, and ξ_0 to be the observer position on the $\mid \xi \mid = t_0$ cosmic hypersphere, space translations inside this hypersphere (the space at $t = \mid \xi_0 \mid$) and simultaneously on all the constant-time hyperspheres of the SST are described by SU(2) transformations acting on the cosmic spinor space, i.e. $\xi = U \xi_0$ with:

$$U = exp\,(i/2\ t_0^{-1}\ \vec{\sigma}.\vec{\mathbf{x}}) \equiv U(\vec{\mathbf{x}}) \tag{1}$$

$\vec{\sigma}$ being the vector formed by the usual Pauli matrices and the vector $\vec{\mathbf{x}}$ the spatial position (in time units, at the present stage) of ξ with respect to ξ_0 at constant time t_0.

As already shown in [20, 21], an attempt to associate to the cosmic spinor ξ a space-like position vector with real cosmic space coordinates defined by $\vec{\mathbf{x}_c} = \xi^\dagger \vec{\sigma} \xi$ does not actually generate such spatial coordinates. One gets instead $\mid \xi \mid^2$ times a unit vector defining a local privileged space direction (PSD) "parallel" (in the SST) to the cosmic spinor ξ. In other words, the direction of $\xi^\dagger \vec{\sigma} \xi$ corresponds to the set of points whose SST space-time position is equal to ξ times a complex phase. This definition turns out to be invariant under the cosmic SU(2) transformations [24, 25] and generates a natural local anisotropy able to produce observable signatures at the cosmological level.

To define the standard real space coordinates in the SST, a space origin ξ_0 at the cosmic time $t_0 = \mid \xi_0 \mid$ is necessary, as in (1). Such coordinates correspond to a local description of the S^3 hypersphere.

Space rotations with respect to a fixed point ξ_0 are then given by SU(2) transformations acting on the spatial position vector $\vec{\mathbf{x}}$ defined by (1). A standard spatial rotation around ξ_0 is now given by a SU(2) element $U(\vec{\mathbf{y}})$ turning $U(\vec{\mathbf{x}})$ into $U(\vec{\mathbf{y}})\ U(\vec{\mathbf{x}})\ U(\vec{\mathbf{y}})^\dagger$. The vector $\vec{\mathbf{y}}$, related to $U(\vec{\mathbf{y}})$ in a similar way to (1), provides the rotation axis and angle. If a spin-1/2 particle is present at the position $\vec{\mathbf{x}}$ with an associated spinor ξ_p describing its internal structure, then ξ_p transforms into $\xi'_p = U(\vec{\mathbf{y}})\ \xi_p$.

3 Some properties of a SST Universe

Nontrivial properties of the cosmic spinorial space-time turn out to be close to cosmological observations if the Universe is assumed to be described by a global SST geometry.

In particular, three basic cosmological phenomena are automatically generated by the SST in a purely geometric way [17, 23] and without any explicit presence of conventional matter:

i) The standard Lundmark-Lemaître-Hubble (LLH) [22] relation between relative velocities and distances at cosmic scale, with a ratio H (velocity/distance) equal to the inverse of the age of the Universe ($H = t^{-1}$) and quite compatible with observational data.

ii) The privileged space direction (PSD) for each comoving observer, leading (with parity violation) to a local asymmetry between two cosmic hemispheres as possibly observed by *Planck* [19, 29].

iii) Furthermore, in the direct SST formulation, space translations form a (non-abelian) compact group, contrary to standard space-time geometry.

More details, including a study of the cosmological implications of these properties, are given in [17, 23], [24, 25] and [6, 7, 19].

3.1 SST at small distances

Similarly to the cosmic SST with the associated SU(2) structure, a local spinorial space-time with its own SU(2) group can be potentially associated to each point ξ_0 in the SST, taking ξ_0 as the local (preonic) space-time origin in order to describe physics at small distances.

If such a local SST description makes sense, the internal properties of the standard "elementary" particles can possibly be described by structure functions taking values in this local SST. Similarly, such a local SST can be relevant to describe the internal structure of the vacuum itself.

A spin-1/2 particle wave or structure function with position at ξ_0 would then be a spinorial wave function "centered" at ξ_0 and taking significant values around this space-time position. However, $| \xi_0 |$ is far from being the only value of time concerned, as will be discussed in what follows.

Such a small-distance local SST is just an illustrative example of the nontrivial space-time structures that can manifest themselves when trying to describe the internal structure of conventional "elementary" particles and the local properties of the physical vacuum where they are generated.

Assuming that the (extended) internal structure of a standard "elementary" particle belongs to a representation of the local space-time SU(2), the allowed spin (spinorial angular momentum) values would be multiples of 1/2, including 0, 1/2, 1, 3/2 and 2 but also possibly higher spins contrary to standard assumptions. Regge-like trajectories spaced by 1/2 in angular momentum cannot be excluded in this approach [6, 7, 19, 32]. They can play an important role in Particle Physics and in Cosmology.

In particular, all particles of standard physics can thus be generated as vacuum excitations belonging to representations of the local space-time SU(2), and the existence of "elementary" spin-3/2 particles may even be natural in such a pattern. As the Poincaré group is no longer an exact symmetry, an alternative to the supersymmetry approach to space-time and internal symmetries can emerge as a new (approximate and broken) symmetry escaping standard theorems and no-go constraints [17, 23].

The search for new particles belonging to spinorial (with 1/2 spin spacing) Regge trajectories or to similar sets generated by the SST-based dynamics is therefore an important subject and should in particular be performed in current CERN experiments. Similarly for new families of particles associated to vacuum excitations, including possible superluminal ones and free superbradyons.

4 SST and quantum mechanics

What can then be the origin of the complex wave functions used in standard Quantum Mechanics (SQM)? The SST geometry can possibly play a significant role, if the excitations of the physical vacuum associated to standard particles correspond to (extended) functions of the internal spinorial complex coordinates inside this vacuum. Such functions of the vacuum structure, as well as the scalar products of these complex coordinates, will naturally be complex quantities and *a priori* compatible with a quantum-mechanical description of conventional matter.

In the SST, the four standard real space-time coordinates are replaced by two complex ones and spinors replace the usual four-vectors. Then, a function Ψ describing the internal structure of a scalar particle can naturally be complex and, for instance, proportional to the hermitic scalar product of two SST spinors ξ_1 and ξ_2 : $\Psi \sim \xi_2^\dagger \xi_1$ where ξ_1 and ξ_2 can be generated by preonic constituents of the scalar particle. The additivity of this kind of spinors can be at the origin of the additivity of the quantum wave functions of standard particles.

Thus, the usual complex wave properties of quantum particles can be a direct consequence of the more fundamental SST geometry holding already inside the preonic vacuum.

4.1 Fron SST to quantum objects

As pointed out in [6, 7], one can also hope that quantization be generated as a natural property in a SST-based preonic description of standard particles. The most basic "elementary" particles of the standard model, the fermions like quarks and leptons, can in principle be formed as the simplest particle-like vacuum excitations if the vacuum structure and dynamics incorporate a spinorial space-time geometry. The SST is just an example of the potentialities of alternative space-time geometries.

What can then be the internal structure of "elementary" fermions? A classical SST internal wave function for a spin-1/2 particle around a space-time origin ξ_0 can be, for instance, of the form:

$$\Psi_{sp} (\xi) = F(| \xi - \xi_0 |^2) (\xi - \xi_0) \quad (2)$$

together with:

$$\Psi^{*}_{sp}\,(\xi) = F(|\,\xi - \xi_0\,|^2)\,(\xi - \xi_0)^{*} \qquad (3)$$

where * stands for complex conjugate (opposite spin direction). The real function F contains a suitable cutoff in $|\,\xi - \xi_0\,|^2$ to ensure a finite particle size.

It then turns out that the wave functions Ψ_{sp} and Ψ^{*}_{sp} explicitly violate standard local causality, as they take nonzero values for past and future values of cosmic time around $|\,\xi_0\,|$. They replace distances on the S^3 space-like hypersphere associated to ξ_0 by direct spinorial distances on the SST. Thus, F tacitly defines a space-time distance scale λ_{SST} below which causality does no longer hold in the conventional sense used for the standard space-time. The time-dependence of $F(|\,\xi - \xi_0\,|^2)$ naturally introduces an intrinsic time uncertainty for the particles so described.

It seems reasonable to conjecture that λ_{SST} corresponds to the space-time scale below which conventional physics does no longer hold. Its definition from (2) and (3) has preceded the introduction of any characteristic speed relating space and time units. At this stage, only time units are actually present [21, 23]. As discussed in [7], we expect a relevant velocity scale to emerge from the basic dynamical process that generates such a spin-1/2 particle. Similarly, the propagation of the particle will produce a measurable critical speed. Distance scales will thus arise together with the generation and the propagation of the particle.

In such a scenario, vacuum excitations possibly described by complex wave functions spreading over larger (space-time) distance scales than the λ_{SST} scale can be associated to combinations of plane waves, leading to practical definitions of h, energy and momentum. At this stage, some uncertainty relations can be generated from the intrinsic SST time uncertainty just analysed.

A crucial question remains to be dealt with: that of the quantization of vacuum excitations. In other words, why are there particles instead of just waves? The SST appears also to be able to solve this enigma. If the spinorial wave function of a spin-1/2 excitation of vacuum describes a deformation of the vacuum structure, it seems reasonable that its normalization be determined by the requirement of preserving the validity of the same (preonic) fundamental equations that generate the ground-state configuration of the physical vacuum. Just as the fundamental vacuum structure would be a unique solution of these equations, the same can happen for the lowest-energy local spinorial excitation of each relevant degree of freedom. The Pauli exclusion principle for spin-1/2 particles can also be generated in this way, if the vacuum structure does not allow for several identical spin-1/2 deformations at once in the same space-time region at very small scales.

Thus, Quantum Mechanics can have in any case a simple physical interpretation as a result of vacuum dynamics rather than as an unexplained, independent fundamental principle.

It must also be noticed that if vacuum excitations were not quantized, decays would in principle be possible and one could expect none of such excitations to be stable. Thus, quantization appears to be required in order to allow for a Universe with stable standard particles.

4.2 How the SST geometry can possibly forbid continuous motion and generate the Feynman path integral

An explicit, alternative, way to generate Quantum Mechanics in a SST scenario can be through the contradiction between a local SST and the cosmic, macroscopic one.

A continous motion of a spin-1/2 particle described by a spinorial function of the type $\Psi_{sp}\,(\xi)$ "centered" around a time-dependent ξ_0 would generate a significant time overlap in a region of the order of the size λ_{SST} covered by Ψ_{sp}. But the fact that a vacuum excitation defined by $\Psi_{sp}\,(\xi)$ be a solution of the equations of vacuum dynamics does not imply that this would be the case for the superimposed structure generated by continuous motion.

Ψ_{sp} (ξ) can be an acceptable excitation of vacuum dynamics, but not necessarily its continous motion in a time interval of the order of the particle spinorial size. If the continous motion of the spin-1/2 structure is not acceptable by the vacuum dynamics, only a discrete motion will be possible with structures of the Ψ_{sp} type separated by large enough space-time intervals.

If the only solution for spin-1/2 excitations compatible with vacuum dynamics can be a discrete distribution with Ψ_{sp}'s separated by space-time distances larger than the particle size, several discrete trajectories can in principle be possible for a given particle. This would generate a situation potentially close to the grounds of the Feynman path integral. The limit of such a dynamics when space-time distances become much larger than λ_{SST} can indeed be similar to the Feynman approach to Quantum Mechanics and provide its fundamental origin.

Given the nature and the implications of the space-time contradiction, we expect Fermi statistics to naturally hold for such spin-1/2 quantum particles.

Similarly, the Bose statistics for integer spin particles will in this case be naturally generated if they are made of two spin-1/2 structures and their internal size is at least larger than 2 λ_{SST} . More sophisticated compositions for particles of all spins can obviously be imagined, provided the above description or a similar one applies for the "most elementary" conventional particles.

At scale-time distances much larger than λ_{SST} but much smaller than the cosmic time t, a local SST (the same just described or a more macroscopic version) centered around a comoving observer placed at ξ_0 can still allow for a sufficiently exact SU(2) symmetry acting on a spinor $\xi - \xi_0$. Local conservation laws can be defined in this way with the local SU(2) accounting in particular for space rotations, and time invariance still leading to a suitable energy conservation law. Space translations involving low enough distances can be phenomenologically described by SU(2) rotations around a point placed on the same straight line with respect to the point $\xi = 0$ of the cosmic SST and at a distance to ξ_0 much larger than the space distance considered. But this requires the knowledge of the comic space-time origin. Strictly speaking, a cosmic rotation around $\xi = 0$ is required to really describe space translations at a finite cosmic time together with the associated privileged space direction. A purely local definition of translations would "lose the memory" of the geometric PSD.

5 Superdradyons, vacuum, matter and space-time

In [30] and [17], we remind and analyse this statement by Abdus Salam in his December 1979 Nobel lecture [31] concerning the original preon model:

"Einstein knew that nature was not economical of structures: only of principles of fundamental applicability. The question we must ask ourselves is this: have we yet discovered such principles in our quest for elementarity, to justify having fields with such large numbers of components as elementary.

Recall that quarks carry at least three charges (colour, flavour and a family number). Should one not, by now, entertain the notions of quarks (and possibly of leptons) as being composites of some more basic entities (PRE-QUARKS or PREENS), which each carry but one basic charge."

(Salam cites here several articles)

These initial preon models were "quark-like", and attempted to better describe the variety of quantum numbers of the "elementary" particles by assuming quarks and leptons to be made of "more elementary" constituents. But the question whether the ultimate constituents of matter should really obey the same laws of Physics (standard relativity, quantum mechanics...) as quarks and leptons was only raised in [26, 28] and in our subsequent papers.

Similarly, the preons initially considered by Salam and other authors were direct constituents of the standard particles. The possibility that they actually be constituents of the physical vacuum, and the standard particles excitations of this vacuum, was an original suggestion of [26, 28] followed by [20, 21], [27] and subsequent work on the subject [32].

The approach presented here is therefore radically different from the initial preon models, and the deformation of Quantum Mechanics at high energy a logical consequence of the way this law of standard paricles is generated. Similarly, we expect Quantum Gravity to be deformed at high energy and no longer make sense beyond the space and time scales where Quantum Mechanics begins to exist. The effects of the preonic vacuum on renormalization at high energy should also be taken into account and can in particular solve the cosmological constant problem [17, 22].

As already emphasized in [26, 28] and in subsequent papers such as [20, 27], if the standard "elementary" particles are actually excitations of a more fundamental preonic vacuum structure, we do not expect the preon critical speed to be equal to the speed of light c . Just as the speed of light is much larger than that of sound in condensed matter, the critical speed of preons is expected to be much larger than c (the superbradyon hypothesis [26], superbradyons are not tachyons). This can be a crucial point to understand quantum mechanics and its phenomenological implications.

5.1 Superbradyons faster than light

Assuming superbradyons can also exist as free particles in our Universe with a critical speed c_s much larger than the speed of light c, it was already pointed out in [21] that:

"A superluminal particle moving at speed $\vec{\mathbf{v}}$ with respect to the vacuum rest frame, and emitted by an astrophysical object, can reach an observer, moving with laboratory speed $\vec{\mathbf{V}}$ with respect to the same frame, at a time (as measured by the observer) previous to the emission time. Such a phenomenon will happen if $\vec{\mathbf{v}}.\vec{\mathbf{V}} > c^2$, and the emitted particle will be seen to evolve backward in time (but it evolves forward in time in the vacuum rest frame)."

(end of quote)

This conclusion was based on a simple calculation. In the rest frame of a standard particle moving with speed $\vec{\mathbf{V}}$ with respect to the vacuum rest frame (VRF), if $\vec{\mathbf{v}}$ is the speed of the superluminal particle with respect to the VRF and assuming for simplicity $\vec{\mathbf{v}}$ to be parallel to $\vec{\mathbf{V}}$, the speed $\vec{\mathbf{v}}'$ of the superluminal particle in the rest frame of the standard particle (the laboratory frame) is given by:

$$\vec{\mathbf{v}}' = (\vec{\mathbf{v}}_i - \vec{\mathbf{V}})\,(1 - \vec{\mathbf{v}}_i.\vec{\mathbf{V}}\,c^{-2})^{-1} \tag{4}$$

leading to a singularity at $\vec{\mathbf{v}}_i.\vec{\mathbf{V}} = c^2$ which corresponds to a change in the arrow of time.

At $v > c^2\,V^{-1}$, a superluminal particle moving forward in time in the vacuum rest frame appears as moving backward in time to an observer made of ordinary matter and moving with speed $\vec{\mathbf{V}}$ in the same frame. It must be noticed, however, that the interaction between the physical vacuum and an important concentration of matter (f.i. inside a galaxy) can modify local properties of such vacuum. Thus, it is not excluded that the local VRF be changed and that, actually, our rest frame be close to the local rest frame of vacuum leading to a small value of $\vec{\mathbf{V}}$ and to a much larger value of $v > c^2\,V^{-1}$ than initially expected. Dynamical estimates and tests of the local VRF are un important subject.

In any case, it clearly follows from this explicit example that the mathematical connection between macroscopic and preonic time is far from trivial and can involve significant contradictions.

In general, we do not expect to find remnant superbradyons traveling around us at a speed larger than c because of their "Cherenkov" decay in vacuum [17, 26] spontaneously emitting standard particles since an early stage of the Universe formation. As superbradyons are constituents of vacuum, a confinement mechanism can be at work and it may happen that not all of them be allowed to exist as free particles. Free superbradyons, if they can exist, are expected to interact weakly with standard matter. They may have kept a speed close to c or lost energy in further collisions with "ordinary" particles. They can be a significant part of the cosmic and galactic dark matter.

High-energy astrophysical events can involve superbradyon emission at speeds initially larger than c and potentially able to reach the Earth with a superluminal velocity.

6 Preonic vacuum, space-time and Quantum Field Theory

A preonic description of vacuum and standard particles can also play an important role in Quantum Field Theory at large energy and momentum scales. It can in particular generate natural solutions to standard problems in Particle Physics and Cosmology, including the energy dependence of boson fields and the cosmological constant problem [17, 22].

Standard Quantum Field Theory (SQFT) develops a purely phenomenological approach to vacuum structure, where the only relevant information concerns boson condensates and the bosonic zero modes. Nothing is known in SQFT about a possible underlying preonic structure that can in particular account for such condensates and zero modes, modify their properties and lead to a new high-energy description of boson fields with a new energy dependence of the parameters of zero modes naturally solving the cosmological constant problem [17, 23]. Taking into account the internal vacuum structure, SQFT would undergo increasing modifications as the distance scale decreases and the effects of the underlying preonic dynamics become more and more able to influence dynamics and calculations. Renormalization may thus become an easier task for currently unsolved problems.

Similarly, fermion quantum fields would become operators acting on the preonic vacuum structure and such that the square of any fermionic operator vanishes identically.

With a spinorial space-time structure, the standard harmonic expansion of the quantum fields would be replaced by its equivalent on the S^3 space hypersphere. The SU(2) operators acting on the SST and on the matter properties would then generate the relevant observables and the equivalent of momentum conservation corresponding to the invariance under SST translations (cosmic rotations around the origin $\xi = 0$). An "almost energy conservation" would be associated to the "almost time invariance" (in this case, the "almost" would be of geometric origin) of the laws of Physics.

6.1 Preons and Quantum Gravity

As the preonic structure generates new space, time, momentum and energy scales, the Planck scale is expected to no longer really make sense in patterns incorporating preon dynamics [17, 22, 23].

In particular, preon dynamics can produce observable effects at scales closer to us than the Planck scale. Ultra-high energy cosmic-ray (UHECR) experiments are particularly important to search for possible signatures of such new Physics [17, 22].

The theory of Quantum Gravity would be strongly modified [6, 7] by a preonic scenario invalidating the standard role of the Planck scale and dynamically generating Quantum Mechanics.

6.2 Some open questions

A particularly important question is that of the vacuum rest frame previously discussed, and already considered in [26, 28] when formulating the superbradyon hypothesis. The natural assumption is to identify the VRF with the reference frame of the cosmic microwave background radiation that can in turn be identified with the associated comoving frame of the cosmic SST.

But strictly speaking, this astrophysical determination of the VRF is based only on specific standard matter measurements and it is not certain that it would correspond exactly and in all cases to the local internal rest frame of a preonic vacuum.

The precise fundamental space-time structures of the physical vacuum and of our Universe remain essential open questions. The internal space-time structure of the preonic vacuum is expected to incorporate some crucial aspects of the SST, so as to be able to generate and describe spin-1/2 standard particles. But apart from this requirement, a more general view of space and time is in principle allowed and can lead to a birth of Quantum Mechanics similar to that previously considered, including the possible effect of a space-time contradiction between preon dynamics and the macroscopic world.

DOI: 10.1051/epjconf/201612605006

7 Some experimental and cosmological considerations

A this stage, an important question arises: how to test experimentally the relevant new physics associated to the scenarios just described? This turns out to be a difficult task.

The possible tests of standard basic principles with ultra-high energy cosmic-ray (UHECR) data [17, 27] have already been discussed in previous papers [7, 18], including possible deformations of Quantum Mechanics [30]. New generations of UHECR experiments will certainly be required.

But the experimental and instrumental achievements of the recent period can also make feasible some attempts of direct searches for free superbradyons.

Thirty years ago, Goodman and Witten suggested [33] dark matter detection through nucleus recoil, the maximum recoil energy being $\epsilon = 4\ E\ m/M$ where E and m are the kinetic energy and the mass of the incoming particle, and M the mass of the target nucleus. Detectors developed for this goal include superheated superconducting granules [34, 35], crystal scintillators [36] presently used by DAMA/LIBRA [37] or the luminescent bolometer [38, 39] currently used by CRESST [40, 41].

For a dark matter superbradyon with speed c [17], the same formula used by Goodman and Witten would yield $\epsilon = 8\ E^2\ (Mc^2)^{-1}$ for an incoming momentum $p = 2\ E\ c^{-1}$. But the superbradyon-nucleus elastic cross-section can in practice be small even if we expect p to be in most cases compatible with coherent scattering. As the superbradyon rest energy will be much larger than its kinetic energy, the most interesting signature would possibly be the explosion of the target nucleus or of surrounding material due to a large energy transfer in an inelastic collision. Building large specific detectors devoted to such inelastic collisions between superbradyons and standard matter deserves a serious study. Experiments like AUGER or the Telescope Array can be sensitive to such events generated at the atmosphere or near a detector, even if possible background must be considered in detail.

Possible superbradyon production at high-energy accelerators should also carefully studied and searched for, looking specially for atypical events with a large missing energy without a significant missing momentum. The rest energy $m\ c_s^2$ where m is a possibly small superbradyon mass, or even the superbradyon kinetic energy if its speed is large enough, would be much larger than $p\ c$ where p is the momentum of the produced superbradyon. A spontaneous decay of the particle may complete such a signature if its speed is larger than c. The very large number of accelerator events can compensate a very weak superbradyon cross-section. In particular, proton-antiproton and electron-positron collisions at the highest possible energies, as well as nucleus-nucleus collisions, should be made available for such searches. Superluminal excitations of the preonic vacuum should also be searched for.

Detecting a high-speed superbradyon (with velocity $v \gg c$) emitted by an astrophysical explosion or a similar event would be a unique scientific opportunity. But the probability of such a performance seems difficult to estimate. Cosmology can possibly provide an additional signature for primordial superbradyons and other ultimate constituents of vacuum and matter, with a preonic era replacing the Big Bang + inflation scenario [6, 28]. The space-time scale at which the preonic vacuum generates standard "elementary" particles can be a new fundamental scale larger than the Planck scale and leading to a new phase of the history of the Universe.

8 Superluminal preons, Bell inequalities, Gödel indetermination...

In [7], the following question was explicitly raised about the origin of Quantum Mechanics:

"The subject of the origin of Quantum Mechanics is a very fundamental one and, by now, looks particularly mysterious. It therefore seems necessary to elucidate if it can be naturally understood in terms of a preonic (superbradyonic?) underlying vacuum structure and in the framework of a specific space-time geometry (the SST?)."

(end of quote)

Actually, as discussed in Section 4, Quantum Mechanics seems indeed to be a possible natural property of standard matter in a vacuum with a spinorial space-time geometry.

Furthermore, the recently observed large-distance quantum entanglement [42, 43] violating Bell inequalities [44] can then be naturally produced in a superbradyonic vacuum [10, 26].

Simultaneously, a difficulty expected from a logical point of view has been raised for standard Quantum Mechanics in [15, 16] and already in [45, 46] (see also [47, 48]) : if Quantum Mechanics is to be considered as based on a closed, consistent system of axioms, it cannot be complete and must lead to explicit incompleteness effects. The authors provide the example of an attempt to calculate the gap between the lowest energy level of electrons in a material and the next level, using an infinite lattice. Although a finite result is obtained with a finite lattice, no infinite lattice limit seems to exist.

8.1 Superbradyons and entanglement

A superbradyonic vacuum would obviously violate a basic principle leading to the Bell inequalities [44]: the requirement that any theory of Physics be "local" in the sense that no influence or information should propagate faster than light. Precisely, the authors of [42, 43] report a clear violation of the CHSH (Clauser et al.) - Bell inequality [49].

If preons are superluminal, faster-than-light propagation of signals and correlations in vacuum will necessarily be allowed. The speed of light will just not be a fundamental parameter of space, time and ultimate superbradyonic constituents inside the vacuum. As previously explained, the superbradyon critical speed c_s is expected to be much larger than c just as c is much larger than the speed of sound.

Therefore, a superbradyonic vacuum would be naturally consistent with the data announced in [42, 43]. If $c_s \simeq 10^6\ c$ (10^6 is the ratio between c and the speed of sound), the time scale associated to a distance of 1.3 Km would be $\simeq 4.10^{-12}$ seconds.

It clearly appears that a superbradyonic vacuum can not only generate Quantum Mechanics but naturally produce the observed entanglement.

8.2 Gödel indetermination and preonic vacuum

In [46], the authors explicitly claim that "for any consistent, recursive axiomatisation of mathematics, there exist specific Hamiltonians for which the presence or absence of a spectral gap is independent of the axioms". But what can be the situation if the version of Quantum Mechanics used for quantum calculations is not a consistent system of axioms but an approximation emerging from the preonic vacuum dynamics where particles are vacuum excitations and Quantum Mechanics is generated?

And if Quantum Mechanics is actually generated in a superbradyonic vacuum with a more fundamental dynamics, is it possible to obtain a deformed, approximate version of the quantum theory such that all relevant calculations of Quantum Physics escape the consequences of Gödel incompleteness?

Further work in this direction is obviously required, searching in particular for a pattern where incompleteness would possibly apply to the preonic vacuum dynamics but quantum calculations would escape its implications, at least with a reasonable approximate version of Quantum Mechanics.

8.3 Preons and deformed quantum mechanics

As previously considered, if Quantum Mechanics is generated from a preonic vacuum of from a space-time contradiction between the preonic vacuum and the macroscopic world, a deformed version of Quantum Mechanics is expected hold at very high energy. The usual quantum-mechanical equations of standard Physics will then be a low-energy limit.

Possible deformations of Quantum Mechanics have been considered, in particular, in [17, 18] and in [7, 30]. Cosmology can also provide relevant signatures [6, 7, 19].

 DOI: 10.1051/epjconf/201612605006

9 Conclusion and comments

A possible fundamental origin of Quantum Mechanics has been considered and naturally generated from the spinorial space-time with a dynamical preonic vacuum. In one of the versions dealt with, the time dispersion of the spinorial extended objects plays a crucial role to prevent a continuous motion of vacuum spinorial excitations and produce a Feynman-like path integral scenario.

A crucial question concerning the mathematical space-time structure of the preonic vacuum is that of the critical speed of preons. In connection with this basic interrogation and with recent data on entanglement, the direct experimental search for superluminal particles becomes an important subject.

Exploring unconventional space-time geometries other than SST, compatible with a similar description of the preonic vacuum and leading to similar results, is obviously an urgent task.

Entanglement is by itself a potential evidence for a superbradyonic vacuum with a SST geometry. Possible deformations of quantum mechanics at ultra-high energy, as well as cosmological data on the early Universe, can also provide tests of the pattern suggested. And can some suitable version of deformed Quantum Mechanics escape incompleteness in a preonic pattern?

The question of the origin of Quantum Mechanics clearly requires further work.

References

[1] See, for instance, T. Hollowood, *Copenhaguen Quantum Mechanics*, arXiv:1511.01069, and references therein.

[2] P.J. Coles, J. Kaniewski and S. Wehner, *Equivalence of wave-particle duality to entropic uncertainty, Nature Communications* **5**, 5814 (2014), arXiv:1403.4687.

[3] S. Wehner and A. Winter, *Entropic uncertainty relations - A survey, New Journal of Physics* - Special Issue on Quantum Information and Many-Body Theory, **12**, 025009 (2010), arXiv:0907.3704.

[4] M. Berta, S. Wehner and M. Wide, *Entropic uncertainty and measurement reversibility*, arXiv:1511.00267.

[5] P.J. Coles et al., *Entropic Uncertainty Relations and their Applications*, arXiv:1511. 04857.

[6] L. Gonzalez-Mestres, *BICEP2, Planck, spinorial space-time, pre-Big Bang*, invited talk at the 3rd International Conference on New Frontiers in Physics (ICNFP 2014), Kolymbari, Crete, Greece, August 23 - 30, 2014, *EPJ Web of Conferences* **95**, 03014 (2015), and references therein.

[7] L. Gonzalez-Mestres, *Tests and prospects of new physics at very high energy*, contribution the 3rd International Conference on New Frontiers in Physics, Kolymbari, Crete, Greece, August 23 - 30, 2014, *EPJ Web of Conferences* **95**, 05007 (2015), and references therein.

[8] L. Gonzalez-Mestres, *Quantum Mechanics and the Spinorial Space-Time*, mp_arc 15-86.

[9] L. Gonzalez-Mestres, *Quantum Mechanics, preonic vacuum and space-time contradiction*, mp_arc 15-90.

[10] L. Gonzalez-Mestres, *Quantum Mechanics, space-time, preons and entanglement*, mp_arc 15-92

[11] An original paper by R.P. Feynman is *Space-Time Approach to Non-Relativistic Quantum Mechanics, Reviews of Modern Physics* **20**, 367. Available at the address (Universitat de Barcelona) http://hermes.ffn.ub.es/luisnavarro/Nuevo_maletin/Feynman_Approach_1948.pdf

[12] An edition by D. Styer (Dover Publications, 2010) of the book by R.P. Feynman and A.R. Hibbs *Quantum Mechanics and Path Integrals* (McGraw-Hill, 1965) is presented at the address http://www.oberlin.edu/physics/dstyer/FeynmanHibbs/ (Oberlin College of Arts and Sciences).

[13] K. Gödel, *Über formal unentscheidbare Sätze der "Principia Mathematica" und verwandter Systeme, Monatshefte für Mathematik und Physik* **38**, 173 (1931), available at the addess http://www.w-k-essler.de/pdfs/goedel.pdf (Goethe Universität, Frankfurt am Main).

[14] See, for instance, Paul J. Cohen, *Set theory and the continuum hypothesis* (1966), Dover Books in Mathematics 2008.

[15] D. Castelvecchi, *Paradox at the heart of mathematics makes physics problem unanswerable*, *Nature* **528, 207 (2015)**, http://www.nature.com/nature/journal/v528/n7581/full/nature16059.htm

[16] T. Cubitt, D. Perez-Garcia and M. Wolff, *Undecidability of the spectral gap*, *Nature News*, 9 December 2015, http://www.nature.com/news/paradox-at-the-heart-of-mathematics-makes-physics-problem-unanswerable-1.18983 , and references therein.

[17] L. Gonzalez-Mestres, *Cosmic rays and tests of fundamental principles*, CRIS 2010 Proceedings, *Nucl. Phys. B, Proc. Suppl.* **212-213** (2011), 26, and references therein. The *arXiv.org* version arXiv:1011.4889 includes a relevant Post Scriptum.

[18] L. Gonzalez-Mestres, *High-energy cosmic rays and tests of basic principles of Physics*, presented at the International Conference on New Frontiers in Physics, ICFP 2012, Kolymbari, Crete, June 10-16 2012, *EPJ Web of Conferences* **70**, 00047 (2014), and references therein. Preprint version at mp_arc 13-19.

[19] See also our talk *Big Bang, inflation, standard Physics... and the potentialities of new Physics and alternative cosmologies*, https://indico.cern.ch/event/344173/session/4/contribution/277 (this Conference).

[20] L. Gonzalez-Mestres, *Physical and Cosmological Implications of a Possible Class of Particles Able to Travel Faster than Light*, contribution to the 28th International Conference on High Energy Physics, Warsaw 1996, arXiv:hep-ph/9610474, and references therein.

[21] L. Gonzalez-Mestres, *Space, Time and Superluminal Particles*, arXiv:physics/9702026.

[22] For a reminder, see also our contributions to the ICNFP 2012 and ICNFP 2013 Conferences.

[23] L. Gonzalez-Mestres *Pre-Big Bang, fundamental Physics and noncyclic cosmologies*, International Conference on New Frontiers in Physics, ICFP 2012, Kolymbari, Crete, June 10-16 2012, *EPJ Web of Conferences* **70**, 00035 (2014), and references therein. Preprint at mp_arc 13-18.

[24] L. Gonzalez-Mestres, *Spinorial space-time and privileged space direction (I)*, mp_arc 13-75, and references therein.

[25] L. Gonzalez-Mestres, *Spinorial space-time and Friedmann-like equations (I)*, mp_arc 13-80, and references therein.

[26] L. Gonzalez-Mestres, *Properties of a possible class of particles able to travel faster than light*, Proceedings of the January 1995 Moriond Workshop, Ed. Frontières, arXiv:astro-ph/9505117.

[27] L. Gonzalez-Mestres, *Vacuum Structure, Lorentz Symmetry and Superluminal Particles*, arXiv:physics/9704017

[28] L. Gonzalez-Mestres, *Cosmological Implications of a Possible Class of Particles Able to Travel Faster than Light*, Proceedings of the TAUP 1995 Conference, *Nucl. Phys. Proc. Suppl.* **48** (1996), 131, arXiv:astro-ph/9601090.

[29] See, for instance, The Planck Collaboration, *Planck 2013 results. XXXI. Consistency of the Planck data*, arXiv:1508.03375, and references therein.

[30] L. Gonzalez-Mestres, *Preon models, relativity, quantum mechanics and cosmology (I)*, arXiv:0908.4070.

[31] Abdus Salam, *Gauge Unification of Fundamental Forces*, Nobel lecture, 8 December 1979, http://www.nobelprize.org/nobel_prizes/physics/laureates/1979/salam-lecture.pdf

[32] See also our contribution to this conference *Spinorial space-time and Hagedorn-like temperatures*, https://indico.cern.ch/event/344173/session/22/contribution/406

[33] M. Goodman and E.Witten, *Detectability of certain dark matter candidates*, *Physical Review* D 31, 3059 (1985).

[34] See, for instance, L Gonzalez-Mestres and D. Perret-Gallix, *Recent results and prospects on superheated superconducting particle detectors*, Proceedings of the Workshop on Superconductive Particle Detectors, Villa Gualino October 1987, preprint version at http://www.iaea.org/inis/collection/NCLCollectionStore/_Public/20/057/20057526.pdf

[35] L. Gonzalez-Mestres, *Physics prospects with superheated superconducting granules detectors: advantages and limitations, possible improvements and alternatives*, Proceedings of the Moriond Workshop, January 1990. Preprint available at http://ccdb5fs.kek.jp/cgi-bin/img/allpdf?200032735

[36] See, for instance, L. Gonzalez-Mestres and D. Perret-Gallix, *Recent results on detector developments for low-energy neutrinos and dark matter: SSG and DSC devices*, Proceedings of the III ESO-CERN Symposium, Bologna, May 1988. Preprint version at http://ccdb5fs.kek.jp/cgi-bin/img/allpdf?198809030

[37] The DAMA Project, http://people.roma2.infn.it/ dama/

[38] L. Gonzalez-Mestres and D. Perret-Gallix, *Detection of low energy solar neutrinos and galactic dark matter with crystal scintillators*, Proceedings of the International Conference on Advanced Technology and Particle Physics, Villa Olmo (Como), June 1988. Preprint available at http://ccdb5fs.kek.jp/cgi-bin/img/allpdf?198809032

[39] L. Gonzalez-Mestres and D. Perret-Gallix, *Cryogenic detectors: status and prospects*, minis-rapporteur talk at the XXIV International Conference on High-Energy Physics, München August 1988. Preprint at http://www.iaea.org/inis/collection/NCL
CollectionStore/_Public/21/000/21000903.pdf

[40] The CRESST Experiment, http://www.cresst.de/

[41] The CRESST Experiment, *The CRESST Dark Matter Search - Status and Perspectives*, arXiv:1509.09124.

[42] B. Hensen et al., *Experimental loophole-free violation of a Bell inequality using entangled electron spins separated by 1.3 km*, *Nature* **526**, 682 (2015), arXiv:1508.05949.

[43] M. Giustina et al., *A significant-loophole-free test of Bell's theorem with entangled photons*, arXiv:1511.03190.

[44] J.S. Bell, *Speakable and Unspeakable in Quantum Mechanics: Collected Papers on Quantum Philosophy*, Cambridge University Press, 2004.

[45] T. Cubitt, D. Perez-Garcia and M. Wolff, *Undecidability of the Spectral Gap (short version)*, arXiv:1502.04135.

[46] T. Cubitt, D. Perez-Garcia and M. Wolff, *Undecidability of the Spectral Gap (full version)*, arXiv:1502.04573.

[47] J. Bausch et al., *Size-Driven Quantum Phase Transitions*, arXiv: 1512.05687.

[48] G. De las Cuevas et al., *Fundamental limitations in the purifications of tensor networks*, arXiv1512.05709.

[49] J. Clauser, M. Horne, A. Shimony and R. Holt, *Proposed experiment to test local hidden-variable theories*, *Physical Revew Letters* **23**, 880 (1969). Available at http://users.unimi.it/aqm/wp-content/uploads/CHSH.pdf

[50] L. Gonzalez-Mestres, *Pre-Big Bang, space-time structure, asymptotic Universe*, talk given at the 2nd International Conference on New Frontiers in Physics, Kolymbari, Crete, Greece, August 28 - September 5, 2013, *EPJ Web of Conferences* **71**, 00063 (2014). See also the Post Scriptum to the preprint version, hal-00983005.

[51] L. Gonzalez-Mestres, *Ultra-high energy physics and standard basic principles*, 2nd International Conference on New Frontiers in Physics, Kolymbari, Crete, Greece, August 28 - September 5, 2013, *EPJ Web of Conferences* **71**, 00062 (2014). See also the Post Scriptum at mp_arc 14-31.

Performance Studies of Micromegas Chambers for the New Small Wheel Upgrade Project

S. Leontsinis[1,2,a] and K. Ntekas[1,2,b]
on behalf of the ATLAS Muon Collaboration

[1] *National Technical University of Athens*
[2] *Brookhaven National Laboratory*

Abstract. The ATLAS collaboration has chosen the Micromegas technology along with the small-strip Thin Gap Chambers for the upgrade of the inner muon station in the high-rapidity region, the so called New Small Wheel upgrade project. It will employ eight layers of Micromegas and eight layers of small-strip Thin Gap Chambers per wheel. The New Small Wheel project requires fully efficient Micromegas chambers, able to cope with the maximum expected rate of $15\,\mathrm{kHz/cm^2}$ featuring single plane spatial resolution better than $100\,\mu\mathrm{m}$. The Micromegas detectors will cover a total active area of $\sim 1200\,\mathrm{m}^2$ and will be operated in a moderate magnetic field ($\leq 0.3\,\mathrm{T}$). Moreover, together with their precise tracking capability the New Small Wheel Micromegas chambers will contribute to the ATLAS Level-1 trigger system. Several studies have been performed on small ($10\times10\,\mathrm{cm}^2$) and medium ($1\times0.5\,\mathrm{m}^2$) size prototypes using medium ($1-5\,\mathrm{GeV/c}$) and high momentum ($120 - 150\,\mathrm{GeV/c}$) hadron beams at CERN. A brief overview of the results obtained is presented.

1 Introduction

The upgrade of the Large Hadron Collider (LHC) at CERN foresees a luminosity increase by a factor 5. To cope with the corresponding rate increase, the ATLAS detector [1] needs to be upgraded. The upgrade will proceed in two steps: Phase-I in the LHC shutdown 2018/19 [2] and Phase-II in 2023-25 [3]. For the ATLAS Muon Spectrometer the largest Phase-I upgrade concerns the replacement of the two first muon stations, called Small Wheels (SW), in the high-rapidity region with a New Small Wheel system (NSW) [4]. The NSW is expected to cope with the highest rates expected after Phase-II and enhance the performance of the forward muon system by adding trigger functionality to the first muon station.

The NSW employs eight layers of Micromegas detectors (MM) and eight layers of small-strip Thin Gap Chambers (sTGC) [5]. Both technologies will provide tracking and triggering information, resulting into a fully redundant system. Each of the two wheels will consist of sixteen sectors divided into eight large and eight small sectors according to their size. A graphical representation of one

[a] e-mail: Stefanos.Leontsinis@cern.ch
[b] e-mail: Konstantinos.Ntekas@cern.ch

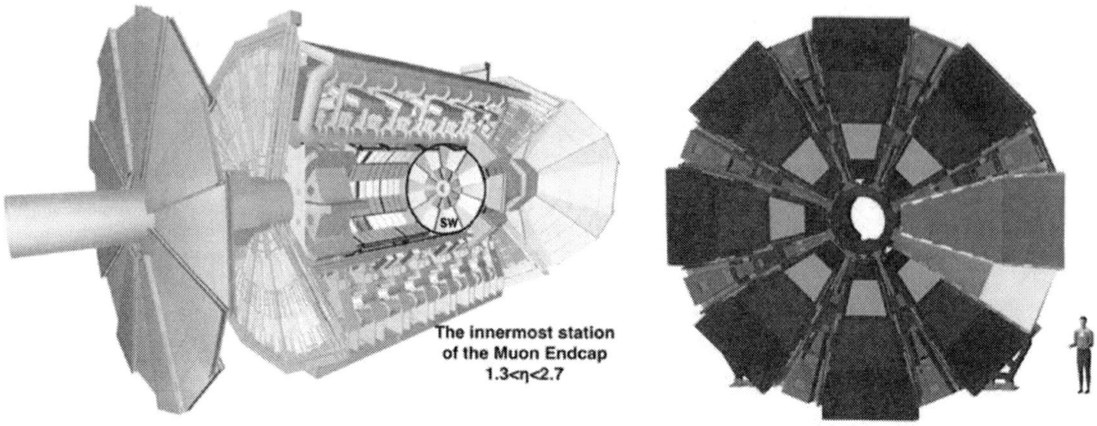

Figure 1. Left: The muon system of the ATLAS experiment. The SW region corresponds to the innermost muon endcap region. Right: Graphical representation of an assembled NSW. Two micromegas sectors, one small and one large, are also shown. Actually the MM will not be visible sitting in between the 2 sTGC quadruplets as shown in Fig. 2.

NSW is shown in Fig. 1 (right). The two technologies will be combined in four wedges per sector following the sTGC-MM-MM-sTGC configuration as shown in Fig. 2. A wedge is a quadruplet combining four layers of the same detector technology.

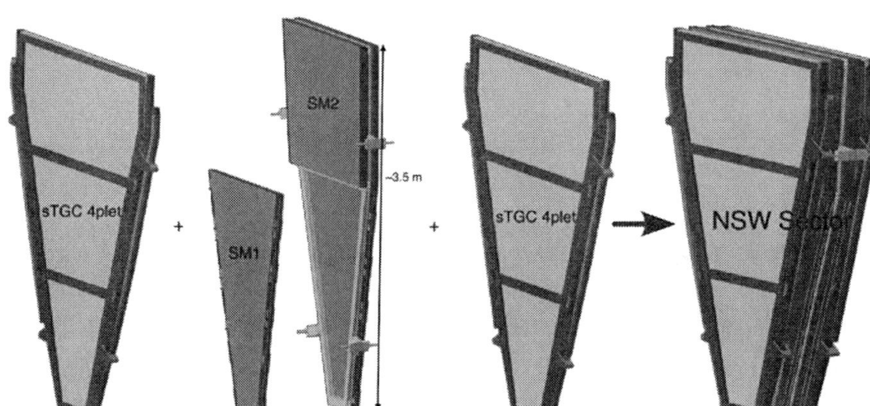

Figure 2. Combining the four MM wedges (quadruplets) into a NSW small sector. The MM detectors will be placed in between the two sTGC quadruplets. The two different types of small MM modules (SM1 and SM2) along with the size of the MM sector in the radial direction are shown.

The main characteristics of the resistive MM detector technology along with performance studies of small chambers in test beams are presented in the following sections. The results demonstrate the excellent performance of the MM chambers that fulfil the requirements of the NSW project in terms of hit reconstruction accuracy and efficiency. A brief description of the reconstruction techniques that have been developed to take full advantage of the fine intrinsic detector spatial resolution is also given followed by detailed investigation of the small efficiency losses that are measured and are attributed to the internal structure of the detector.

2 The resistive strip MM detector

The MM technology was developed in the middle 1990's introducing a micropattern gaseous detector with two asymmetric regions [6]. Standard MM detectors consist of a planar (drift) electrode, a a few millimetres thick gas gap acting as conversion and drift region, and a thin metallic mesh placed at $\sim 100 - 150\,\mu$m distance from the readout electrode, creating the amplification region. The HV potentials are chosen such that the electric field in the drift region is a few hundred V/cm, and $40 - 50\,$kV/cm in the amplification region achieving gas gain values of the order of 10^4 for the commonly used Ar+7%CO_2 gas mixture. The large multiplication factor is essential for the detection of minimum ionising particles with high efficiency.

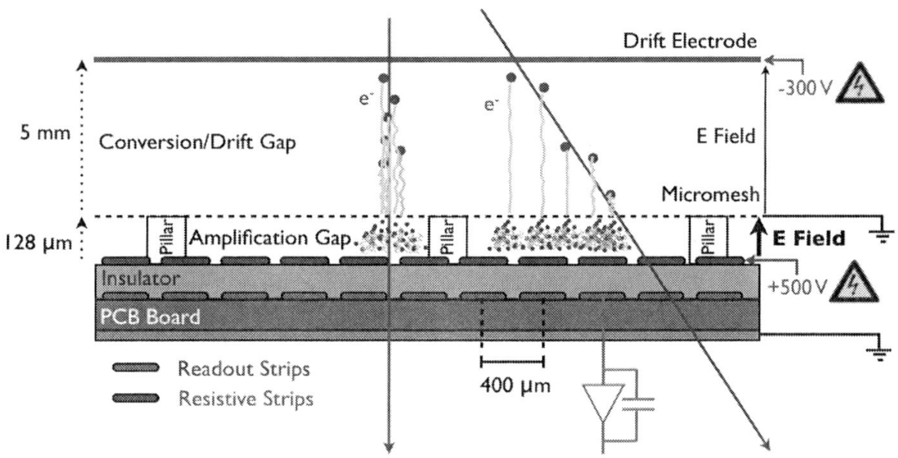

Figure 3. Cross-section of the resistive MM internal structure. A layer of resistive strips matching the read-out pattern makes the detector insensitive to discharges coming from massive ionisations.

Charged particles traversing the drift space ionize the gas molecules and the electrons, liberated by the ionization process, drift towards the mesh (in tens of nanoseconds). With an electric field in the amplification region $\sim 50 - 100$ times stronger than the drift field, the mesh is transparent to more than 95 % of the electrons. The electron avalanche takes place in the thin amplification region in about a nanosecond, resulting in a fast pulse on the readout strip [7]. The ions that are produced during the avalanche process move towards the mesh with velocities about 200 times smaller than the electrons producing also a signal, though smaller, on the readout strips. The drifting of the electrons and the avalanche formation in the amplification region, for perpendicular and inclined tracks, are illustrated in Fig. 3.

In the case of highly ionizing particles the Raether limit [8] may be reached and discharges (sparks) start to occur. Sparks may damage the detector structure or the readout electronics and lead to large dead times as a result of high voltage breakdown. By adding a layer of resistive strips on top of a thin

DOI: 10.1051/epjconf/201612605007

insulator directly above the readout electrode, as shown in Fig. 3, the MM becomes spark-insensitive. The readout electrode is no longer directly exposed to the charge created in the amplification region. The signal is induced on the resistive strips and then capacitively coupled to the readout strips. By adding the resistive protection some fraction of the signal amplitude is lost but the chamber can be operated at higher gas gain because sparking is reduced by about three orders of magnitude and the effect is only locally constrained in a small region of the detector [9]. Hence, resistive MM chambers can be efficiently operated in a high-rate environment as was demonstrated by testing small resistive prototypes inside the ATLAS experiment with particle rates up to $80\,\mathrm{kHz/cm^2}$ [10, 11].

3 Performance studies

3.1 Test beam activities

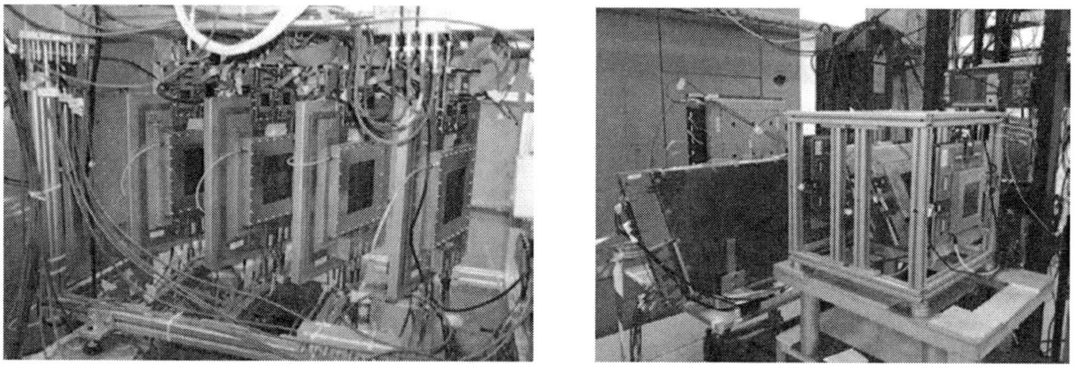

Figure 4. Photos from test beam periods at CERN (SPS H6 and PS T9 respectively). Left: A hodoscope of 8 small ($10 \times 10\,\mathrm{cm^2}$) resistive MM were exposed to a high momentum ($120 - 150\,\mathrm{GeV/c}$) π^+ beam during the summer of 2012. Right: The same hodoscope along with medium size prototypes were tested in a medium momentum ($5 - 10\,\mathrm{GeV/c}$) p beam during the summer of 2014.

In order to demonstrate the excellent performance of the MM technology and also to optimise the design and operational parameters of the detector several resistive MM chambers have been tested with medium/high energy hadron beams. Photographs from two of the experimental set-ups are shown in Fig. 4. For the results that are presented here the front-end electronics were based on the APV25 ASIC (128 channels) [12], which will not be employed in the NSW, readout through the Scalable Readout System (SRS) developed by the RD51 collaboration [13]. The front-end ASIC providing the trigger and tracking primitives for both the MM and sTGC of the NSW will be the VMM [14] currently being produced in its second prototype version. Extensive studies of the first version of the VMM using small resistive MM chambers are reported in [15, 16].

3.2 Hit & track reconstruction techniques

One of the most attractive characteristics of the MM technology is its excellent spatial resolution due to the fine readout segmentation. Moreover, because of the asymmetric field configuration and the large multiplication in the amplification region, the chambers are sensitive to a single primary ionisation electron. Using the time evolution of the integrated charge per channel, sampled every 25 ns by the APV25 chip, the arrival time of each electron is measured along with the total charge, allowing the detector to work as TPC allowing for precise reconstruction of track segments with a single detector plane. Using the recorded charge and time information, two different hit reconstruction techniques have been developed:

- **(Charge) Centroid :** Average of the strip position weighted by the strip charge.
- **μTPC** : 2D track reconstruction transforming the arrival time into distance. The hit is defined as the interpolation of the track in the middle of the drift gap. The track angle is also measured using the slope of the reconstructed segment.

The charge centroid method reconstructs hits with high accuracy when the particle traverses the chamber plane perpendicularly. In this case, the charge is shared among a few strips allowing for a very precise charge interpolation. The spatial resolution obtained with this method is well below 100 μm for perpendicular tracks. The μTPC method does not provide accurate results in this case as the pulse in each strip is the aggregation of pulses induced by more than one primary ionisation cluster generated at different times. When the particle crosses the detector under an angle the signal induced in each strip is most probably coming from one primary cluster with several strips fired from a single track. In this occasion, the μTPC method provides a very accurate measurement of the particle hit position on the detector. In the case of inclined tracks, owing to the distribution of the primary ionisation clusters along several readout strips, the strip signal amplitude becomes sensitive to the primary cluster charge fluctuations. As a consequence, the accuracy of the charge centroid method deteriorates with increasing track inclination angles and it cannot be used for non-perpendicular tracks.

The hit positions reconstructed with the centroid and μTPC methods can also be integrated in a combined hit per gas gap. The combination algorithm weights the contribution of each of the two hits (centroid and μTPC) with the size of the track footprint [17]. In the case of large clusters the μTPC method contributes more while in clusters with a size smaller than four strips the combined hit is primarily defined by the centroid method. The combined point provides a single plane hit reconstruction with a resolution better than 100 μm independently of the track incident angle [18].

3.3 Refinement of the μTPC method

The performance of the μTPC track reconstruction method is of great importance for the MM community and also for the ATLAS NSW upgrade project where the particle track angles expected are between 8 and 30°. One of the issues that were studied was the refinement of the clustering algorithm to properly treat events with more than one track, or events with noise and/or delta rays ($\sim$ 10% of the events). A pattern recognition technique, using the Hough transform, has been developed which filters out the real tracks increasing the efficiency of the event selection and improving the accuracy of the reconstructed track by removing outlying strips [16].

Moreover, due to the readout/resistive elements geometry, there is a non-negligible capacitance between the strips causing the ionisation charge to be shared among the neighbouring strips. The effect has been extensively studied in simulation concluding that $\sim$ 10% of the charge induced on a strip is capacitively coupled to its neighbours[1]. This may bias the reconstructed track by artificially increasing the cluster size in cases where the strips on the edges of the cluster have signal coming only from the coupling with their neighbours. A filtering algorithm has been developed identifying

[1]The effect depends on the strip pitch and width. For this study a geometry with strips 300 μm wide separated by a gap of 100 μm was used.

such hits and discarding them from the track reconstruction procedure.

Another parameter that should be treated carefully is the assignment of the hit position with respect to the strip pitch. In the offline reconstruction each hit position is usually assigned in the middle of the strip pitch. In fact, this is correct only for the middle strips of the cluster where the hit position is uniformly distributed along the strip pitch. For the edges of the cluster it is more probable for the hit position to be towards the strip pitch edge neighbouring to the cluster strips. The assignment of the hit in the middle of the strip pitch for the cluster edges results in biasing the reconstructed track towards larger inclination values, thus a method has been developed tuning the hit position on the cluster edges based on their charge [19].

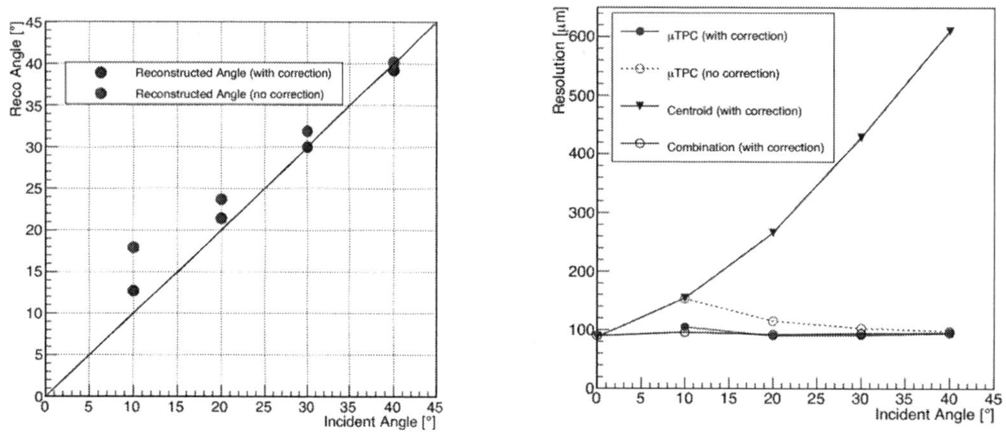

Figure 5. Left: The reconstructed incident track angle, peak value of the angular distribution, versus the true inclination angle before (red points) and after the correction (blue points). Right: The spatial resolution along the different incident angles measured before and after the correction with the μTPC (red circles) and the centroid (blue triangles) methods. The resolution measured when combining the two methods, after the correction, is also shown (black circles) [20] .

The refined μTPC reconstruction algorithm was applied in data taken during July 2012 test beam at CERN where small size ($10 \times 10\,\text{cm}^2$) chambers were exposed to a $120\,\text{GeV/c}\ \pi^+$ beam and results with and without the new recipe are compared. The improvement in the accuracy of the μTPC method, using the techniques described above, is not only evident in the better agreement of the measured incident angle with the true one shown in Fig. 5 (left), but also in the measurement of the spatial resolution which is displayed in Fig. 5 (right). The impact of the correction is more significant for smaller incident angle values where a 30% improvement in the measured spatial resolution is achieved. The results for the combined point are also shown with the black line on the right plot of Fig. 5. The spatial resolution of the combined point has been measured to be better than 100 μm for inclination angles $0° - 40°$.

3.4 Efficiency studies

In addition to the precise hit position reconstruction the NSW MM detectors should be fully efficient even at the highest rates of $15\,\text{kHz/cm}^2$ that are expected after the LHC upgrade. In this direction the efficiency of the small MM chambers has been studied in test beams in order to fully characterize and understand possible contributions to the detector inefficiency. Results for chambers with different characteristics and with different track inclination angles are presented.

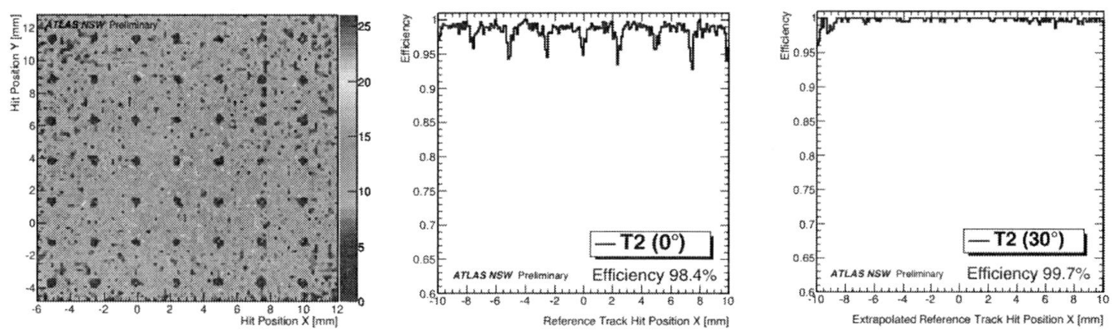

Figure 6. Left: The 2-D distribution of the hit positions measured in a MM with two-dimensional readout reveals the inefficient areas of the chamber corresponding to the pillars that define the width of the amplification region. Middle: Using a reference track the efficiency is plotted as a function of the extrapolated reference hit position on the chamber plane. A geometrical inefficiency of 1.6%, owing to the pillars, is measured along a 2 cm region for perpendicular tracks. Right: In the case of tracks inclined by 30° the efficiency becomes insensitive to the pillars ($\sim$ 100%) as several strips are fired by a single charged particle [20].

By using a MM with two-dimensional readout the profile of the beam can be directly obtained by reconstructing the centroid hit in the two coordinates. The resulting distribution obtained for perpendicular tracks is shown in Fig. 6 (left). Only a small rectangular area of the detector is shown (18×18 mm^2) corresponding to the central area of the beam with the highest statistics. The histogram population increases by going from the blue coloured areas to the green ones. Several inefficient spots can be distinguished on the two dimensional maps corresponding to the cylindrical pillar structure appearing every 2.5 mm.

By comparing the hits reconstructed in a single chamber with the reference chambers of the experimental set-up a measurement of the efficiency can also be performed. The measured efficiency is plotted as a function of the reference track hit position for a resistive MM chamber with a single readout layer in Fig. 6 (middle, right). In the middle, the case where the chamber plane is positioned vertically with respect to the beam axis is shown. Local efficiency drops, by 5%, owing to the pillars appear every 2.5 mm. The total measured efficiency is rather high with only $\sim$ 1.6% geometrical inefficiency attributed to the pillars. If however the chamber is inclined by 30° with respect to the beam the pillars are no longer visible as shown in Fig. 6 (right). The larger track footprint for inclined tracks makes the efficiency measurement insensitive pillars and a perfect efficiency is reached.

4 Conclusions

After an intensive R&D phase, which begun in 2007, the MM technology has evolved significantly over the last years and it was finally approved as one of the technologies that will be used in the NSW upgrade project of the ATLAS experiment. This is the first application of a Micropattern Gaseous

Detector (MPGD) technology in a very large scale experiment and in a high rate environment. A series of performance studies is ongoing in order to optimise the detector performance and operating conditions. Being the primary precision tracker of the NSW efforts are mainly focussed in the development of the μTPC reconstruction method allowing for a precise segment reconstruction using a single detector gap. Moreover, a single plane hit reconstruction accuracy better than 100 μm independently of the track angle has been achieved by combining different reconstruction techniques. A very high efficiency has been measured for the MM chambers tested in test beams at CERN. The 1.6% geometrical inefficiency coming from the pillar structure supporting the mesh is evident only for the perpendicular tracks. For larger inclination angles, like the track angles expected in the NSW ($8^\circ - 30^\circ$), the efficiency remains unaffected by the pillars reaching $\sim$ 100% values.

Acknowledgements

This research has been co-financed by the European Union (European Social Fund - ESF) and Greek national funds through the Operational Program "Education and Lifelong Learning" of the National Strategic Reference Framework (NSRF) - Research Funding Program: **THALES**. Investing in knowledge society through the European Social Fund.

References

[1] ATLAS Collaboration, JINST 3 S08003 (2008).
[2] ATLAS Collaboration, CERN-LHCC-2011-012, LHCC-I-020 (2011).
[3] ATLAS Collaboration, CERN-LHCC-2012-022, LHCC-I-023 (2012).
[4] ATLAS Collaboration, CERN-LHCC-2013-006, ATLAS-TDR-020 (2013).
[5] V. Smakhtin *et al.*, Nucl. Instr. Meth. A 598 196-200 (2009).
[6] Y. Giomataris *et al.* , Nucl. Instr. Meth. A376, 29-35, (1996).
[7] I. Giomataris *et al.* , Nucl.Instrum.Meth. A560, 405-408, (2006).
[8] H. Raether, Z. Phys., 112 (1939), p. 464
[9] T. Alexopoulos *et al.*, Nucl. Instr. Meth. A 640 110-118 (2011).
[10] T. Alexopoulos *et al.*, ATL-MUON-PUB-2013-002 (2013).
[11] Y. Kataoka *et al.*, JINST 9 C03016 (2014).
[12] M.J. French *et al.*, Nucl. Instr. Meth. A466, 359-365, (2001).
[13] S. Martoiu *et al.*, Nucl. Sc. Symp. and Med. Imag. Conf., IEEE, pp.2036,2038 (2011).
[14] G. De Geronimo *et al.*, IEEE Transactions on Nuclear Science, vol. 60, no. 3, (2013).
[15] G. Iakovidis, CERN-THESIS-2014-148 (2014).
[16] T. Alexopoulos *et al.*, ATL-UPGRADE-PUB-2014-001 (2014)
[17] C. Bini, JINST 9, C02032 (2014)
[18] S. Leontsinis, CERN-THESIS-2015-077 (2015).
[19] K Ntekas, ATL-MUON-PROC-2014-011 (2014)
[20] ATLAS NSW Public Results : https://twiki.cern.ch/twiki/bin/view/AtlasPublic/NSWPublicResults

EPJ Web of Conferences **126**, 05008 (2016)
ICNFP 2015 DOI: 10.1051/epjconf/201612605008

Femtoscopy with unlike-sign kaons at STAR in 200 GeV Au+Au collisions

Jindřich Lidrych[1,a] (for the STAR Collaboration)

[1] *Faculty of Nuclear Sciences and Physical Engineering*
Czech Technical University in Prague
Břehová 7
115 19 Prague 1

Abstract.
In the collisions of heavy ions the nuclear matter can undergo a phase transition from hadrons to a state of deconfined quarks and gluons called the Quak-Gluon Plasma. Femtoscopic measurements of two-particle correlations at small relative momenta reveal information about the space-time characteristics of the system at the moment of particle emission. The correlations result from quantum statistics, final-state Coulomb interactions, and the strong final-state interactions between the emitted particles.
It has been predicted that correlations due to the strong final-state interactions in a system where a narrow resonance is present will be sensitive, in the region of the resonance, to the source size and momentum-space correlations. Such a measurement can provide complementary information to the measurements at very low relative momenta. This paper presents the preliminary results of a STAR analysis of unlike-sign kaon femtoscopic correlations in Au+Au collisions at $\sqrt{s_{NN}}$=200 GeV, including the region of $\phi(1020)$ resonance. The experimental results are compared to a theoretical prediction that includes the treatment of resonance formation due to the final-state interactions.

1 Introduction

In 1960 Goldhaber and collaborators observed in proton-antiproton annihilations an excess of pairs of identical pions produced at small relative momenta [1]. These observed correlations, as experimenters correctly asserted, came as a result of quantum statistics. Based on this observation the theoretical background of femtoscopy was developed by G. I. Kopylov and M. I. Podgoretsky in the 1970s [2]. Since then the femtoscopic measurements of two-particle correlations at low relative momenta became a standard tool for extracting the space-time extents of particle emitting sources. Nowadays femtoscopic studies include identical particles, as well as non-identical interacting particles [3–6].

The approach proposed by Lednicky [7] extends the femtoscopic formalisms to higher relative momenta between the two emitted particles in a system where the final-state interactions (FSI) contain a narrow, near-threshold, resonance. It is predicted that the correlation function will be more sensitive in the region of the resonance, where the strength of the correlation should change with the source size,

[a]e-mail: jlidrych@gmail.com

r as $\sim r^{-3}$ in comparison with measurements at the very low relative momenta, where the correlation function depends on r^{-2} or r^{-1}. In addition, these measurements will be statistically advantageous, since the particle spectra fall rapidly at low relative momenta.

Pairs of unlike-sign kaons are ideally suited for such femtoscopic analysis as they contain the narrow $\phi(1020)$ resonance. The $\phi(1020)$ resonance is characterized by the decay width $\Gamma = 4.3$ MeV and the decay momentum in the rest frame $k^* = 126$ MeV/c. The use of kaons is also advantageous due to the fact that the emission source function is less affected by weak decays of resonances. From previous STAR source imaging analysis [4], the kaon source function is known to be well-described by a Gaussian form.

2 Data analysis

The data used for this analysis were collected in Au+Au collisions at $\sqrt{s_{\rm NN}} = 200$ GeV by the Solenoidal Tracker at RHIC (STAR) in 2011. The STAR [8] is a multi-purpose detector, which excels in tracking and identification of charged particles at mid-rapidity with full coverage in azimuthal angle. The most important subdetectors for this analysis are the Time Projection Chamber (TPC) [9] and the Time of Flight (ToF) [10]. The TPC records charged particle tracks and measures their momentum p and identifies them via specific ionization loss dE/dx. For kaon selection, the particles were required to have $|n\sigma_K| < 3$, where $n\sigma_K$ is the distance from the expected mean $<dE/dx>$ expressed in terms of standard deviation units σ_K. The ToF measures particle velocity β which is used to calculate particle mass m according to the relation

$$\frac{1}{\beta} = \sqrt{1 + m^2/p^2}, \tag{1}$$

where the momentum p is measured by the TPC. Due to a good time resolution, which is less than 100 ps, the ToF is able to separate charged kaons from other hadrons up to $p \sim 1.55$ GeV/c, as shown in Figure 1. Due to this fact, only primary tracks at mid-rapidity $|\eta| < 1$ with momentum $p \in [0.15, 1.55]$

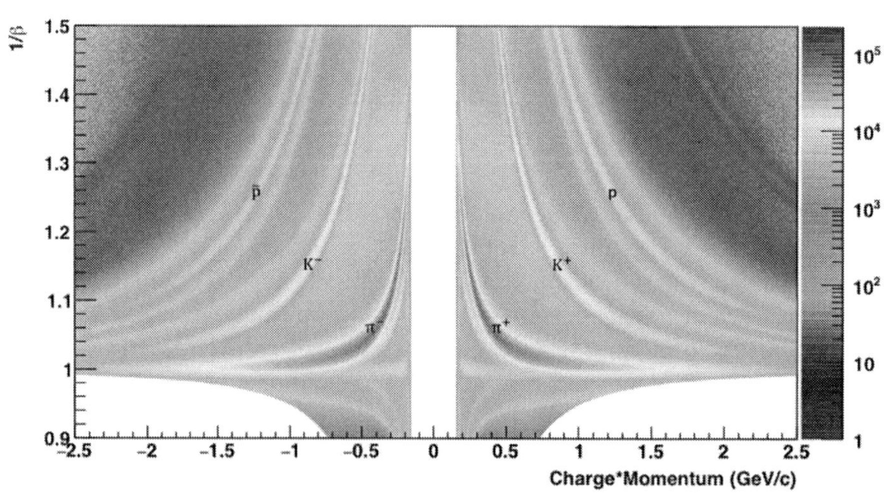

Figure 1. The relative velocity$1/\beta$ (from ToF) of charged particle as a function of momentum p (from TPC).

GeV/c, which have signal from the ToF and satisfy cut criteria on mass: $0.21 < m^2 < 0.28$ GeV$^2/c^4$ were used here.

3 Construction of correlation function

Experimentally, the two-particle correlation function $CF(q_{inv})$ is constructed as a ratio of the correlated two-particle distribution from the same event, $N_{same}(q_{inv})$, and the uncorrelated two-particle distribution from mixed events, $N_{mixed}(q_{inv})$:

$$CF\,(q_{inv}) = \frac{N_{same}(q_{inv})}{N_{mixed}(q_{inv})} = \frac{real\ pairs}{mixed\ pairs}, \tag{2}$$

where $q_{inv}^2 = -(p_1^\mu - p_2^\mu)^2$. In the pair rest frame $q_{inv} = 2k^*$. The technique of event mixing is used to obtain the uncorrelated two-particle distribution. The events are mixed within sub-classes with similar values of the primary vertex position along the beam direction and the multiplicity.

4 Unlike-sign 1D correlation functions

In this analysis, the correlation functions are constructed for 5 centralities and 4 different bins of transverse pair momentum $k_T = (\vec{p_1} + \vec{p_2})_T/2$, where $\vec{p_1}$ and $\vec{p_2}$ are the momenta of the first particle and the second particle, respectively. In Figure 2 there are the STAR preliminary results of K^+K^- correlation functions from Au+Au collisions at $\sqrt{s_{NN}} = 200$ GeV. While at the low q_{inv}, the attractive Coulomb interaction and strong interaction in s-wave can be observed, in the region of $q_{inv} \sim 0.25$GeV/c the strong interaction in p-wave via $\phi(1020)$ resonance in FSI is present. As can be seen, the correlation function is sensitive to the source size. In particular, a strong dependence on collision centrality and on the pair k_T is observed in the resonance region.

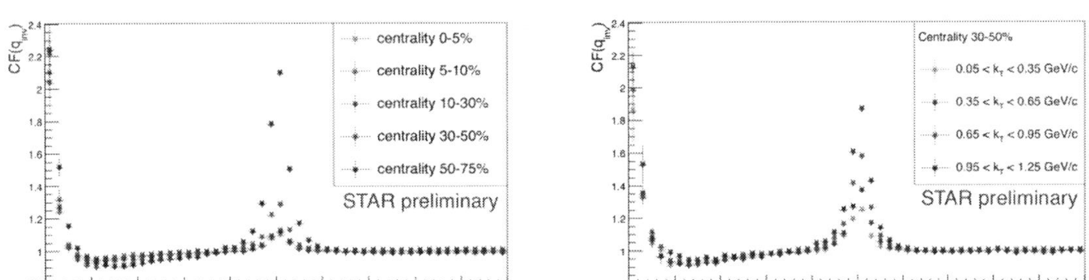

Figure 2. Left: Centrality dependence of one dimensional unlike-sign correlation function from Au+Au collisions at $\sqrt{s_{NN}} = 200$ GeV. Right: k_T dependence of one-dimensional unlike-sign correlation function for centrality 30-50% from Au+Au collisions at $\sqrt{s_{NN}} = 200$ GeV.

Regarding the centrality dependence, where the correlation functions are integrated over all k_T, the height of the ϕ peak decreases significantly with centrality. Similarly a strong dependence is observed as a function of k_T, as shown on right panel of Figure 2 for centrality $30 - 50\%$, mid-peripheral events. The k_T dependence of femtoscopic correlations reflects the dynamics of the system. The observed increase in the correlation strength with the k_T qualitatively agrees with the effects expected from a system undergoing a transverse expansion where pairs with the larger transverse momentum are emitted from a smaller effective source than the pairs with the smaller k_T [11].

For comparison to the theoretical prediction of [7], additional physical effects have to be taken into account, since they decrease the strength of the measured correlation function. These effects are contained in the so-called correlation strength, the λ parameter. In this analysis, the λ parameter is obtained from fitting an experimental correlation function of like-sign kaons. The effects of momentum resolution, feed-down and residual correlations have not been studied yet.

5 Like-sign 1D correlation functions and fitting

The one-dimensional like-sign correlation functions were constructed by the same method as the unlike-sign correlation functions introduced in the previous sections. Figure 3 shows STAR preliminary results of K^+K^+ and K^-K^- correlation functions from Au+Au collisions at $\sqrt{s_{NN}} = 200$ GeV. As the pairs consist of two identical particles with the same charge, the repulsive Coulomb interaction and Bose-Einstein statistics can be observed at low q_{inv}.

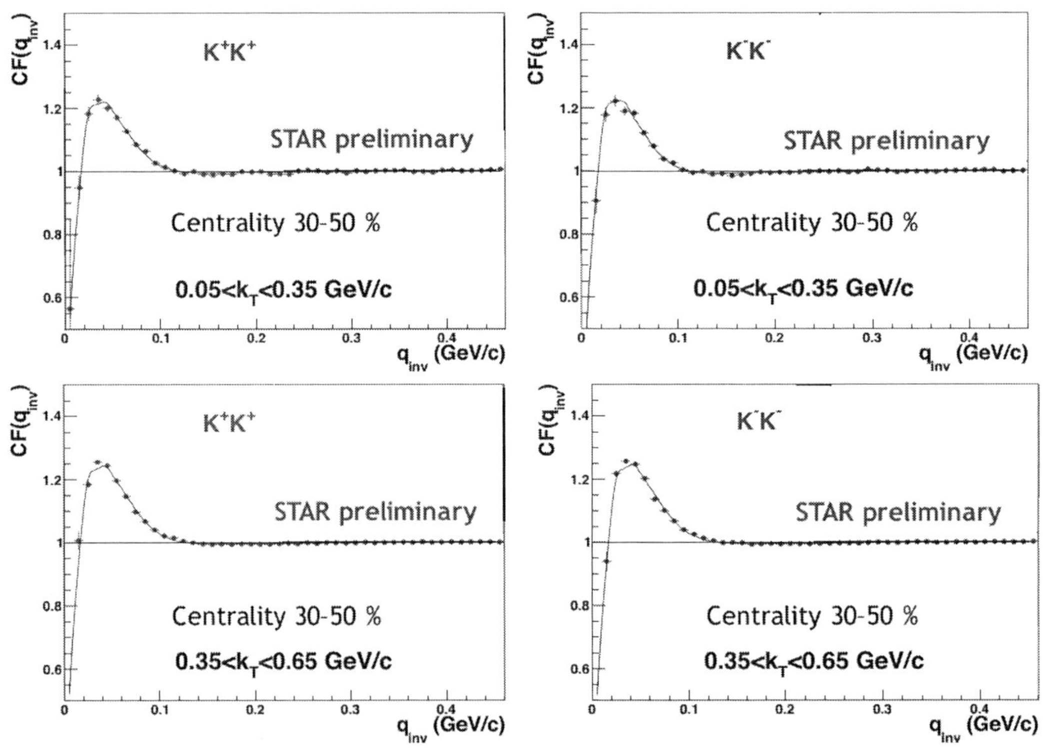

Figure 3. One dimensional K^+K^+ and K^-K^- correlation function from 30-50% central collisions for positive(top) and negative(bottom) kaons for two k_T bins: $0.05 < k_T < 0.35$ GeV/c and $0.35 < k_T < 0.65$ GeV/c. The lines represent the best fits to the data by using the Eq. 3.

The source radii R_{inv} and the λ parameters are obtained from fitting experimental correlation functions with a standard Bowler-Sinyukov form of one dimensional correlation function

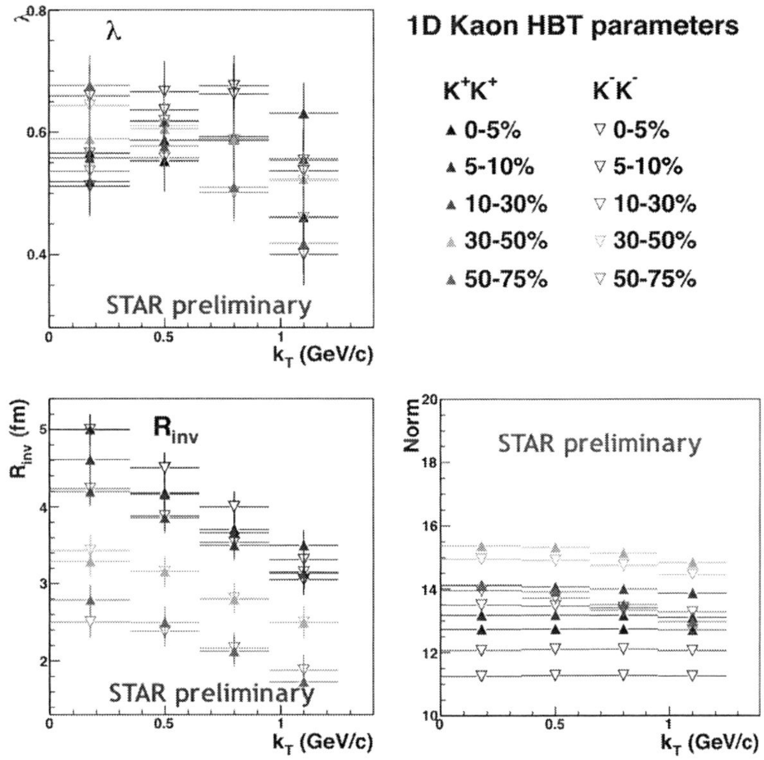

Figure 4. Fit results: λ parameter and source radius R_{inv} as a function of k_T and centrality.

$$CF(q_{inv}) = \left[(1-\lambda) + \lambda K(q_{inv})\, e^{-R_{inv}^2 q_{inv}^2}\right] N, \tag{3}$$

where N is normalization and $K(q_{inv})$ is Coulomb function integrated over source of size R_{inv}. The extracted parameters are shown in Figure 4. As can be seen, the source radii R_{inv} increase with the centrality and decrease with the pair transverse momentum k_T. The errors of fit results are dominated by systematic uncertainties which were estimated by varying the fit ranges. Study of other systematic uncertainties is underway.

6 Comparison of unlike-sign 1D correlation function to Lednicky model

The experimental results of the unlike-sign one-dimensional correlation function can be now compared to the theoretical prediction from Lednicky [7] using a relation:

$$CF(q_{inv}) = \int d^3 r S(r, k^*) \left|\psi_{1,2}(r, k^*)\right|^2, \tag{4}$$

where $S(r, k^*)$ is the source function describing emission of two particles at a relative distance r with the relative momentum k^* in the pair rest frame (PRF). The interaction between the two emitted particles is characterized by their wave function $\psi_{1,2}(r, k^*)$.

Similarly, as assumed for eq. (3), the source is described by one-dimensional Gaussian in the PRF with a parameter R_{inv}. Here the individual R_{inv} are those extracted from fitting like-sign correlation

 DOI: 10.1051/epjconf/201612605008

functions (Figure 4). The used FSI model of Lednicky [7] includes the treatment of $\phi(1020)$ resonance in the final state. The model also introduces a generalized form of smoothness approximation which is needed for correct description of the correlation function in the region of the resonance. Since the theoretical function does not include effects contained in the experimental λ parameter, it is scaled for the comparison according to

$$CF = \left(CF^{theor} - 1\right)\lambda + 1, \tag{5}$$

where λ parameter was obtained from the fit to the like-sign correlation function.

Comparison of unlike-sign one-dimensional correlation functions to model calculations for two collision centralities is shown in Figs. 5 and 6. As can be seen, the model reproduces the overall structure of the observed correlation functions, both at low q_{inv} where the Coulomb and strong interaction in s-wave are present, as well as in the region of $\phi(1020)$ resonance. The agreement in the ϕ region is very good for central collisions, however with decreasing source size the height of the resonance peak is underestimated.

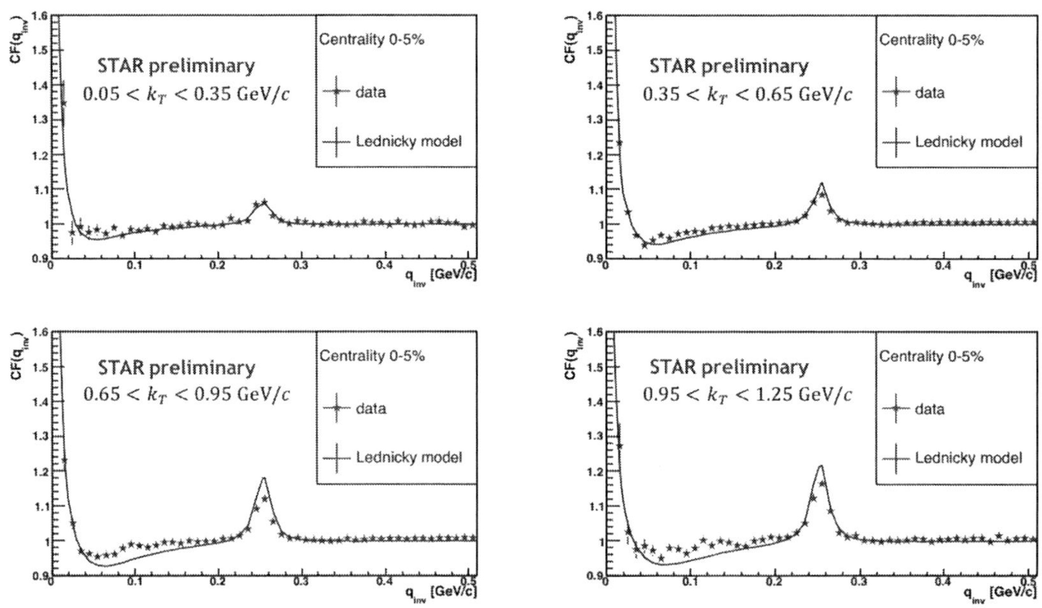

Figure 5. Comparison of experimental K^+K^- correlation functions from Au+Au collisions at $\sqrt{s_{NN}} = 200$ GeV to theoretical calculations for 0-5% centrality for 4 different k_T bins.

7 Summary

In this paper, the preliminary results of a STAR analysis of unlike-sign kaon femtoscopic correlations in Au+Au collisions at $\sqrt{s_{NN}} = 200$ GeV have been presented. The measured K^+K^- correlation function exhibits strong centrality and k_T dependence in the region of $\phi(1020)$ resonance. The obtained correlation function has been compared to a theoretical FSI model with parameters R_{inv}, λ obtained from like-sign correlation functions. The Lednicky FSI model reproduces the correlation function in

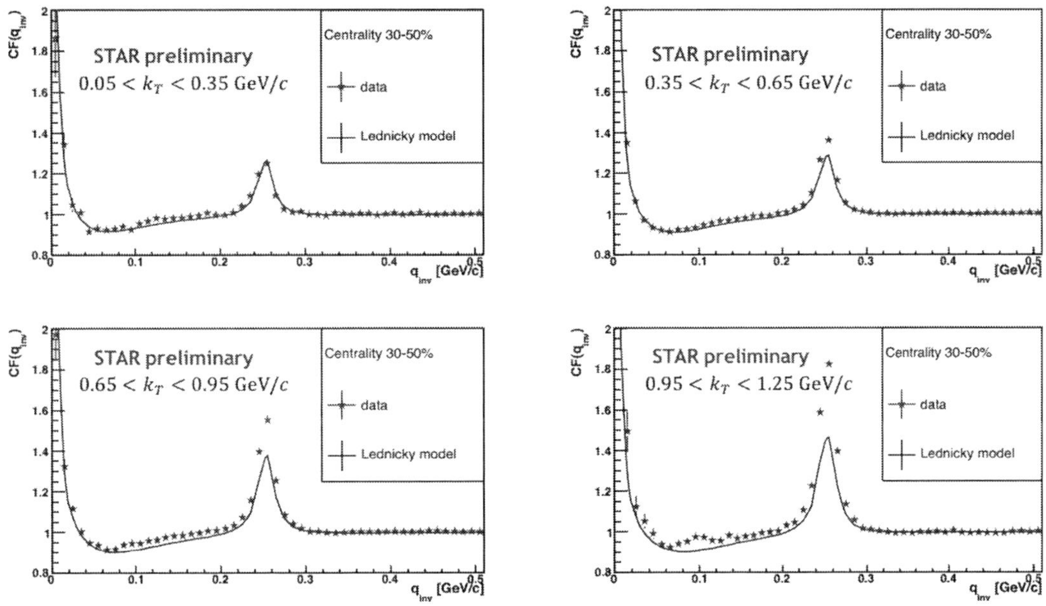

Figure 6. Comparison of experimental K^+K^- correlation functions from Au+Au collisions at $\sqrt{s_{\mathrm{NN}}} = 200$ GeV to theoretical calculations for 30-50% centrality for 4 different k_T bins.

central collisions, but underpredicts the strength of correlation in the $\phi(1020)$ region for peripheral collisions.

Acknowledgments

This work was supported by the grant of the Grant Agency of Czech Republic n.13 - 20841S and by the Grant Agency of the Czech Technical University in Prague, grant No. SGS13/2150HK4/3T/14.

References

[1] G. Goldhaber, S. Goldhaber, W.Y. Lee, A. Pais, Phys. Rev. **120**, 300 (1960)
[2] G.I. Kopylov, Phys. Lett. **B50**, 472 (1974)
[3] B.I. Abelev et al. (STAR), Phys. Rev. **C80**, 024905 (2009)
[4] L. Adamczyk et al. (STAR), Phys. Rev. **C88**, 034906 (2013)
[5] L. Adamczyk et al. (STAR), Nature (2015)
[6] L. Adamczyk et al. (STAR), Phys. Rev. Lett. **114**, 022301 (2015)
[7] R. Lednicky, Phys. Part. Nucl. Lett. **8**, 965 (2011)
[8] K.H. Ackermann et al. (STAR), Nucl. Instrum. Meth. **A499**, 624 (2003)
[9] K.H. Ackermann et al. (STAR), Nucl. Phys. **A661**, 681 (1999)
[10] W.J. Llope (STAR), Nucl. Instrum. Meth. **A661**, S110 (2012)
[11] S.V. Akkelin, Yu.M. Sinyukov, Phys. Lett. **B356**, 525 (1995)

EPJ Web of Conferences **126**, 05009 (2016) DOI: 10.1051/epjconf/201612605009
ICNFP 2015

Fuzzy Geometry of Commutative Spaces and Quantum Dynamics

S.N. Mayburov[1,a]

[1] *Lebedev Institute of Physics, Moscow, Leninski pr. 53*

Abstract. Fuzzy topology and geometry considered as the possible mathematical framework for novel quantum-mechanical formalism. In such formalism the states of massive particle m correspond to the elements of fuzzy manifold called fuzzy points. Due to the manifold weak topology, m space coordinate x acquires principal uncertainty σ_x and described by the positive, normalized density $w(\vec{r}, t)$ in 3-dimensional case. It's shown that the evolution of m state on such 3-dimensional manifold corresponds to Shroedinger dynamics of massive quantum particle.

1 Introduction

It's well known that quantum mechanics (QM) can be consistently described by several alternative formalisms such as, Shroedinger or standard one, algebraic QM, functional integral, etc. [1]. Each of them possesses its own advantages and thus the development of new QM formalisms can bring, in principle, the additional opportunities. In this note we consider the temptative QM formalism based on the specific fuzzy calculus. During the last 50 years the fuzzy set theory and other branches of fuzzy mathematics were applied in a wide range of scientific areas, such as biology, economics and computer science [2, 3]. Meanwhile, from the early days of its formulation and development the significant similarity of that theory and QM formalism was noticed [4–6]. In particular, it was argued that the generic parameter uncertainties which is the cornerstone of fuzzy mathematics are, in fact, similar to QM observable uncertainties [4, 5]. Furthermore, the analysis of quantum particle coordinate measurement evidences that such coordinate can be described as the fuzzy observable [7]. It was shown also that fuzzy phase space is the appropriate tool for the description of noncommutative QM observable measurements [8]. Successful applications of fuzzy methods to particular QM problems demonstrate the close connection between fuzzy mathematics and QM formalism.

Concerning with the general reformulation of QM in fuzzy terms, until now such studies were performed only in the context of fuzzy (multivalued) logics ([9] and refs. therein). In distinction, the approach considered here operates with standard logics and exploits the formalism based on the results of fuzzy topology and geometry [5, 6, 10]. Its application permits to consider the object phase space equipped with fuzzy topology and derive from that the structure of state space which turns out to be equivalent to QM Hilbert space. As the result, it permits to reduce the number of axioms necessary for QM formalism and simplify the description of quantum object evolution. Some intermediate results

[a] e-mail: mayburov@sci.lebedev.ru

on this formalism development were published previously in [11, 12]. In its essence the considered formalism is geometric; the different approaches to QM geometrization studied extensively in the last years, first of all, for the application to quantum gravity theory [13, 14].

2 Geometric Fuzzy Structures

Here we'll consider some examples illustrating the properties of fuzzy structures important for QM formalism construction, the detailed reviews on fuzzy topology and geometry can be found in [6, 10]. Let's start from the case of discrete set, remind that if S is the totally ordered set, then for its elements $\{a_i\}$ the ordering relation between all its element pairs $a_k \leq a_l$ (or vice versa) is fulfilled. However, for partially ordered set (Poset) S^p, some its element pairs can obey to the incomparability relations (IR) between them: $a_j \sim a_k$. If this is the case, then both $a_j \leq a_k$ and $a_k \leq a_j$ propositions are false [17]. Such set can possess some nontrivial properties, to illustrate them, consider poset $S^p = A^p \cup B$, which includes the subset of 'incomparable' elements $A^p = \{a_j\}$, and ordered subset $B = \{b_i\}$. For the simplicity suppose that in B the element's indexes grow correspondingly to their ordering, so that $\forall\, i,\ b_i \leq b_{i+1}$. The relations between an arbitrary pairs a_j, a_i or a_i, b_l can be ordered, as well as incomparable. Let's consider B interval $\{b_l, b_n\}$ and suppose that some A^p element a_j is confined in $\{b_l, b_n\}$, i.e. $b_l \leq a_j$; $a_j \leq b_n$, and simultaneously a_j is incomparable with all internal $\{b_l, b_n\}$ elements: $b_i \sim a_j$; $\forall\, i;\ l+1 \leq i \leq n-1$. In this case a_j is 'smeared' over such interval, which is the rough analogue of a_j coordinate uncertainty relative to B 'coordinate', if to consider the ordered sequence of B elements $\{b_i\}$ as the analogue of coordinate axe. The generalization of poset structure is tolerance space for which the ordering relations can be nontransitive [5].

The next step to fuzzy topology is to detalize the described IR properties introducing the fuzzy relations between S^p elements. For that purpose one can put in correspondence to each a_j, b_i pair of S^p set the nonnegative weight $w_i^j \geq 0$ with norm $\sum_i w_i^j = 1$, so that w_i^j characterizes the rate of closeness (membership) between a_j, b_i [2, 3]. For the example considered above, one can ascribe $w_i^j \neq 0$ to all b_i inside $\{b_l, b_n\}$ interval, $w_i^j = 0$ for other b_i. To illustrate how such structure can appear for the continuum consider the continuous set S^f such that $S^f = A^p \cup X$ where A^p is the same discrete subset, X is the continuous ordered subset. If the flat metrics $\mathcal{M}(x, x')$ is defined on X, then it's equivalent to R^1 real number axe. Then, in this framework the fuzzy relations between elements a_j, x are described by real, nonnegative functions $w^j(x) \geq 0$ with the norm $\int w^j dx = 1$ [3]. In this case $\{a_j\}$ called the fuzzy numbers $\tilde{x}_j$ or in geometric framework the fuzzy points (FPs) [2, 3, 15]. In such approach the ordered point $x_a \in X$ is characterized by $w^a(x) = \delta(x - x_a)$, hence the ordered points and FPs can be regarded formally on the same ground. As the result, the obtained formalism can be considered also as the generalization of 1-dimensional Euclidian geometry for which the positions of some elements $\{a_j\}$ are described by the positive normalized function $w^j(x)$ on R^1; w^j dispersion σ_x characterizes a_j coordinate uncertainty.

In 3-dimensional case as the fundamental set one can consider the set: $C^f = A^p \cup X'$ where A^p defined above, X' is the continuous set. Suppose that on X' 3−dimensional linear space R^3 with flat metrics $\mathcal{M}_{ij}$ is defined. Then FP a_j is described by the nonnegative function $w^j(\vec{r})$ with norm $\int w^j d^3r = 1$. Such C^f is called fuzzy manifold and will be denoted also as $\tilde{R}^3$. Note that in fuzzy mathematics the different FP definitions are exploited, we use here the one given in [3, 15]. Previously, such formalism was called fuzzy (or digital) geometry [6, 15], however, in modern physics such term denotes some noncommutative theories [16], so to stress the difference, we shall exploit for our formalism the term the commutative fuzzy geometry (CFG).

In such formalism $w^j(x)$ *a priori* doesn't have any probabilistic meaning but only the algebraic one, characterizing the properties of fuzzy value $\tilde{x}_j$. To describe the distinction between the fuzzy structure and probabilistic one, the correlation $\kappa_0(x, x')$ defined over w_j support can be introduced; thus if $w_j(x_{1,2}) \neq 0$, then $\forall x_1, x_2$; $\kappa_0(x_1, x_2) = 1$ for FP a_j and $\kappa_0(x_1, x_2) = 0$ for probabilistic a_j distribution. Thus a_j state G on X is described by two functions $G = \{w_j(x), \kappa_0(x, x')\}$. Such description illustrates the difference of such formalism from stochastic geometry, which exploits practically infinite set of parameters. As will be shown below, the similar bilocal correlations can describe some properties of quantum objects.

3 Structure of Fuzzy States

In the considered framework the massive particle of 1-dimensional classical mechanics corresponds to the ordered point $x_a(t) \in X$. By the analogy, in our theory called fuzzy mechanics (FM), in 1-dimensional case the particle m corresponds to FP $a(t)$ and characterized by normalized positive density $w(x, t)$. Beside $w(x)$, m physical state $|g\}$, which called the fuzzy state, can depend on other m degrees of freedom (DFs) which will define its evolution. As the example, consider m average velocity:

$$\bar{v}(t) = \frac{\partial}{\partial t}\int\limits_{-\infty}^{\infty} x w(x, t) dx = \int\limits_{-\infty}^{\infty} x \frac{\partial w}{\partial t} dx \tag{1}$$

It's reasonable to assume that in general $\bar{v}(t)$ can be independent of $w(x, t)$, below we shall look for additional m state DFs in form of real functions $q_{1,...,n}(x, t)$. Let's suppose that in FM m evolution is local, in particular, that:

$$\frac{\partial w}{\partial t}(x, t) = -\Phi(w, q_1, ..., q_n) \tag{2}$$

where Φ is an arbitrary function which depends only on $w(x, t)$, $q_{1,...,n}(x, t)$. From w norm conservation it follows that:

$$\int\limits_{-\infty}^{\infty} \Phi(x, t) dx = -\int\limits_{-\infty}^{\infty} \frac{\partial w}{\partial t}(x, t) dx = -\frac{\partial}{\partial t}\int\limits_{-\infty}^{\infty} w(x, t) dx = 0 \tag{3}$$

If to substitute: $\Phi = \frac{\partial J}{\partial x}$ where $J(x)$ is some differentiable function, then eq. (3) demands:

$$J(\infty, t) - J(-\infty, t) = 0 \tag{4}$$

For normalized density w at $x \to \pm\infty$, $w(x, t) \to 0$, and such assumption is, in fact, trivial. If it's fulfilled, then $J(x)$ can be considered as w flow, and eq. (2) is equivalent to 1-dimensional flow continuity equation [21]:

$$\frac{\partial w}{\partial t} = -\frac{\partial J}{\partial x} \tag{5}$$

$J(x)$ can be decomposed formally as: $J = w(x)v(x)$ where $v(x)$ corresponds to 1-dimensional w flow velocity [21]. In these terms eq.(5) can be written as:

$$\frac{\partial w}{\partial t} = -v\frac{\partial w}{\partial x} - \frac{\partial v}{\partial x} w \tag{6}$$

By the analogy with fluid dynamics, we shall assume that $v(x)$ can be considered as the novel m DF, and there is no other independendent DFs on which m state $|g\}$ can depend. In this case 1-dimensional matrix $g^o = \{w, v\}$ describes the observational $|g\}$ representation, such g^o supposedly contains the information about the expectation values of all m observables. To make the theory most simple it's sensible to look for dynamical $|g\}$ representation η for which the evolution equations will be linear.

This isn't fulfilled for g^o, because the eq. (6) which describes the evolution of its component w is nonlinear. We shall suppose that such η can be also represented as the array of real functions $\{\eta_i(x)\}$, $i = 1, n_a$ which can constitute the complex or quoternionic number or some other algebraic object. Then, the most general η ansatz is $\eta_j(x) = \Upsilon^j_x(w, v)$ where Υ^j_x are some w, v functionals and x is their parameter. For $|g\}$ characterized by two DFs w, v, it's instructive to start the search of minimal $|g\}$ ansatz from complex $\eta(x)$, so $n_a = 2$, not assuming yet that η is normalized as $w(x) = |\eta|^2$.

In FM the evolution of massive particle supposedly can be characterized also by m velocity 'as the whole' $u(t)$ with expectation value $\bar{u}(t)$. Such $u(t)$ is described by the corresponding normalized distribution $w_u(u, t)$ which can be defined in u measurements. Thus, in FM framework, analogously to m coordinate x, such u also can be considered as fuzzy value $\tilde{u}(t)$. Obviously, $\bar{u}(t)$ coincides with $\bar{v}(t)$ of (1) and so it related to m flow velocity $v(x)$:

$$\bar{u} = \int_{-\infty}^{\infty} v(x)w(x)dx \tag{7}$$

In place of u, below it will be convenient to use the variable $p = \mu u$ where μ is the theory constant; for the corresponding distribution $\omega(p)$ it gives $w_u(u) = \mu\omega(\mu u)$. As proposed above, $|g\}$ contains the information on the expectation values of any m observable Q_i, hence any such value should be described by some η functional. In particular, it should be: $\omega(p) = F_p(\eta)$ where F_p is parameter dependent functional. For complex η, it can be shown that F_p related to $\eta(x)$ fourier transform. To prove it and calculate η and ω, let's introduce the auxiliary form: $\varphi(p) = \omega^{\frac{1}{2}} e^{i\beta(p)}$, here $\beta(p)$ is the auxilary real function on which the final ω ansatz wouldn't depend. From the above arguments, $\eta(x)$ is also equal to:

$$\eta(x) = f_x(w, v)e^{i\lambda_x(w,v)} \tag{8}$$

where $f_x(w, v)$, $\lambda_x(w, v)$ are some real functionals. We shall look for ω,β functions such that φ fourier decomposition on X is equal to:

$$\varphi(p) = \int_{-\infty}^{\infty} \eta(x)e^{-ipx}dx = \int_{-\infty}^{\infty} f_x e^{i\lambda_x - ipx}dx \tag{9}$$

w_p is normalized distribution, so the application of Plancherele formulae to that norm gives [19]:

$$\int_{-\infty}^{\infty} \omega(p)dp = \int_{-\infty}^{\infty} \varphi(p)\varphi^*(p)dp = \int_{-\infty}^{\infty} f_x^2(w)dx = 1 \tag{10}$$

To define f_x, one can consider w variation for the equality:

$$\int_{-\infty}^{\infty} [f_x^2(w, v) - w]dx = 0 \tag{11}$$

with additional δw constraint: $\int \delta w dx = 0$. Application of Du Bois-Reymond lemma [18] to eq. (11) together with this constraint gives $f_x^2 = w(x)$, so f_x doesn't depend on $v(x)$. Then $\bar{p}$ can be calculated anew from 2-nd Plancherele formulae :

$$\bar{p} = \int_{-\infty}^{\infty} p\varphi(p)\varphi^*(p)dp = \int_{-\infty}^{\infty} \frac{\partial\lambda_x}{\partial x} f_x^2 dx = \int_{-\infty}^{\infty} \frac{\partial\lambda_x}{\partial x} w(x)dx \tag{12}$$

From the comparison with eq. (7) and $\bar{p} = \mu\bar{u}$ it follows: $\lambda_x = \gamma(x) + \chi(w)$ where γ is the functional:

$$\gamma(x) = \mu \int_{-\infty}^{x} v(\xi) d\xi + c_\gamma \tag{13}$$

here c_γ is an arbitrary real number. $\chi(w)$ is an arbitrary real function which obeys to the condition:

$$\int_{-\infty}^{\infty} \chi(w) \frac{\partial w}{\partial x} dx = 0 \tag{14}$$

and is the analogue of η gauge. The resulting m state in x-representation is equal to:

$$\eta(x) = w^{\frac{1}{2}}(x) e^{i\gamma + i\chi} \tag{15}$$

is the vector (ray) of complex Hilbert space $\mathcal{H}$. $\omega(p)$ and $\beta(p)$ can be calculated from eq. (9) as functions of χ. In particular:

$$\omega(p) = |\int_{-\infty}^{\infty} w^{\frac{1}{2}} e^{i\gamma + i\chi - ipx} dx|^2 \tag{16}$$

is independent of $\beta(p)$, so ω is just w, v functional. $\beta(p)$ is, in fact, the analogue of $\gamma(x)$ for $|g\}$ in p-representation described by $\varphi(p)$.

4 Linear Model of Fuzzy Dynamics

Obtained m state: $\eta = e^{i\chi} g$, where $g(x,t)$ is standard QM wave function, so that $\eta(x,t)$ is its trivial map. Thus one can study first $g(x,t)$ evolution, and then basing on obtained results $\eta(x,t)$ properties will be defined . Evolution equation for g is supposed to be of the first order in time, i.e.:

$$i\frac{\partial g}{\partial t} = \hat{H} g. \tag{17}$$

In general $\hat{H}$ can be the nonlinear operator, for the simplicity we shall consider first the linear case and turn to nonlinear one in the next section. Free m evolution is invariant relative to x space shift to arbitrary x_0 distance performed by the operator $\hat{W}(x_0) = \exp(x_0 \frac{\partial}{\partial x})$. Because of it, the corresponding operator $\hat{H}_0$ should commute with $\hat{W}(x_0)$ for the arbitrary x_0, i.e. $[\hat{H}_0, \frac{\partial}{\partial x}] = 0$. It holds only if $\hat{H}_0$ is differential polinom of the form:

$$\hat{H}_0 = -\sum_{l=1}^{n} b_{2l} \frac{\partial^{2l}}{\partial x^{2l}} \tag{18}$$

where b_{2l} are arbitrary real constants. Suppose that the action of external field on m can be accounted in $\hat{H}$ additively: $\hat{H} = \hat{H}_0 + V(x,t)$ where V is real, nonsingular function. Let's rewrite eq. (17) separating w, γ derivatives:

$$i\frac{\partial g}{\partial t} = (i\frac{\partial w^{\frac{1}{2}}}{\partial t} - w^{\frac{1}{2}} \frac{\partial \gamma}{\partial t}) e^{i\gamma} = e^{i\gamma} \hat{Z} g \tag{19}$$

where $\hat{Z} = e^{-i\gamma} \hat{H}$. Hence:

$$\frac{\partial w^{\frac{1}{2}}}{\partial t} = im(\hat{Z} g) \tag{20}$$

Yet if to substitute $v(x)$ by $\gamma(x)$ in eq. (6) and transform it to $w^{\frac{1}{2}}$ time derivative, then:

$$\frac{\partial w^{\frac{1}{2}}}{\partial t} = -\frac{1}{\mu}\frac{\partial w^{\frac{1}{2}}}{\partial x}\frac{\partial \gamma}{\partial x} - \frac{1}{2\mu}w^{\frac{1}{2}}\frac{\partial^2 \gamma}{\partial x^2} \tag{21}$$

Plainly, to make the formalism consistent, the right parts of equations (20) and (21) should coincide for arbitrary w, γ, otherwise $\hat{H}$ ansatz would be incompatible with w flow continuity described by eqs. (6, 21). Hence $\hat{H}$ ansatz can be obtained from their comparison term by term. In particular, the imaginary part of $\hat{Z}g$ includes the highest γ derivative as the term: $b_{2l}w^{\frac{1}{2}}\frac{\partial^{2l}\gamma}{\partial x^{2l}}$, yet for eq. (21) the highest γ derivative is proportional to $w^{\frac{1}{2}}\frac{\partial^2\gamma}{\partial x^2}$. As the result, it gives: $b_2 = \frac{1}{2\mu}$, for all $l > 1$ it follows that $b_{2l} = 0$, only in this case both expressions for $\frac{\partial w^{\frac{1}{2}}}{\partial t}$ would coincide. Thus g free evolution is described by the only $\hat{H}_0$ term $b_2 = \frac{1}{2\mu}$, so $\hat{H}$ is Schroedinger Hamiltonian for particle with mass μ.

In this framework, the observable p corresponds to the operator $\hat{p} = -i\frac{\partial}{\partial x}$ acting on $g(x)$. Thus, x and p observables are described by the linear self-adjoint operators, which obey to the commutation relation $[\hat{x}, \hat{p}] = i$. By the analogy, we suppose that all m PV observables $\{Q\}$ are the linear, self-adjoint operators on $\mathcal{H}$. The flow velocity $v(x)$ isn't observable, but can be formally defined as the ratio of $J(x), w(x)$ observable expectation values, where w observable is described by the projection operator $\hat{\Pi}(x)$; the operator $\hat{J}(x)$ considered in [20]. As was noticed earlier, the particle evolution in QM in some aspects is similar to the motion of continuous media [21]. This analogy is exploited in hydrodynamical QM model (QFD) [22, 23], its connection with FM will be discussed elsewhere.

Plainly, $\gamma(x)$ corresponds to $|g\}$ quantum phase, so that:

$$k(x, x') = \gamma(x) - \gamma(x')$$

describes the phase correlation between the state components in x, x'. Thus pure FM state can be characterized by the density $w(x)$ and the array of bilocal geometric correlations $\{\kappa_l(x, x')\}$, the first of them: $\kappa_0(x, x')$ was introduced in sect. 2. Until now we've considered only the pure fuzzy states, i. e. the states which aren't the probabilistic mixture of several pure states. Analogously to QM, the mixed states in FM can be defined via the density matrixes, i.e. the positive, trace one operators ρ on $\mathcal{H}$ [1]. In particular, for pure m states:

$$\rho(x, x') = g(x)g^*(x') = [w(x)w(x')]^{\frac{1}{2}}e^{ik(x,x')}$$

is equivalent to $g(x)$, yet such $|g\}$ representation demonstrates in the open the correlation structure of m pure states. Thus, the most consistent FM state ansatz is given by the density matrix ρ. However, the evolution equations for pure states in form of ρ are more complicated then for Dirac vector $g(x)$, and because of it, we shall exploit $g(x)$ throughout our paper.

5 General Fuzzy Dynamics

In the previous section 1-dimensional FM formalism was derived from CFG premises assuming that $|g\}$ evolution is linear and its phase component $\chi(w)$ can be neglected. Now these assumptions will be dropped one by one and general formalism derived. Concerning with nonlinear evolution, the conditions of QM linearity were reconsidered recently by Jordan, and turn out to be essentially weaker than Wigner theorem asserts [24]. In particular, it was proved that if the evolution maps the set of all pure states one to one onto itself, and for arbitrary mixture of orthogonal states $\rho(t) = \sum P_i(t)\rho_i(t)$ all P_i are independent of time, then such evolution is linear. Here $\rho_i(x, x', t) = \psi_i(x, t)\psi_i^*(x', t)$ are the density matrixes of orthogonal pure states ψ_i. Yet for the considered FM formalism first condition is, in fact, generic: no mixed (i.e. probabilistic) state can appear in the evolution of pure state in such geometrodynamical approach. The second condition involves the probabilistic mixture of such orthogonal pure states and also seems to be rather weak assumption.

Now let's return to $\eta(x)$ ansatz of (15), note first that if $\chi(w) \neq 0$ then its appearance demands to introduce at least two additional theory parameters. It can be shown also that $\hat{H}$ linearity demands that $\chi(w) = 0$. Really, it's easy to check that η evolution equation:

$$i\frac{\partial \eta}{\partial t} = \hat{H}\eta \tag{22}$$

for linear $\hat{H}$ ansatz is incompatible with w flow continuity expressed by eq. (6,21) due to the appearance of additional $\hat{H}$ terms which contain χ derivatives. Hence resulting $\eta(x,t)$ coincides with QM wave function, i.e. is Dirac vector in Hilbert space $\mathcal{H}$. Obtained results don't exclude the existence of alternative solutions for $\eta(x)$, it seems, however, that the obtained ansatz is the minimal one.

As follows from the results of sect. 2, 3-dimensional FM, in fact, doesn't demand any principal modification of described formalism. In that case FP a_j is described by the normalized, nonnegative function $w^j(\vec{r})$. The particle m corresponds to FP $a(t)$ characterized by $w(\vec{r},t)$, then analogously to sect. 3, given w evolution depends on local parameters only, it can be expressed as:

$$\frac{\partial w}{\partial t}(\vec{r},t) = -\Phi(\vec{r},t) \tag{23}$$

where Φ is an arbitrary local function. Then from w norm conservation 3-dimensional flow continuity equation follows:

$$\frac{\partial w}{\partial t} = -div\vec{J} \tag{24}$$

One can decompose $\vec{J} = w\vec{v}$ and consider w flow velocity $\vec{v}(\vec{r})$ as independent $|g\}$ parameter. m state supposedly depends on 4 DFs, $w, \vec{v}$, however, analogously to 1-dimensional case, we'll look for minimal m state ansatz $|g\}$ in form of the the complex $w, \vec{v}$ functional $\eta(\vec{r}) = \Upsilon_{\vec{r}}(w,\vec{v})$. For m as the whole, its velocity is supposedly characterized by fuzzy vector $\tilde{\vec{u}}$ which corresponds to distribution $w_u(\vec{u})$, so that:

$$\langle\vec{u}\rangle = \int \vec{u}w_u(\vec{u})d^3u = \int \vec{v}(\vec{r})w(\vec{r})d^3r \tag{25}$$

m kinematical fuzzy momentum defined as: $\tilde{\vec{p}} = \mu\vec{u}$. From that, analogously to eqs. (7 - 15), standard QM ansatz for m state obtained: $g(\vec{r}) = w^{\frac{1}{2}}e^{i\gamma}$, for which g phase $\gamma(\vec{r})$ obeys to the equality $\mu\vec{v} = grad(\gamma)$. m state $g(\vec{r})$ should be defined unambiguously, due to it, this equality imposes constraints on $\vec{v}(\vec{r})$ field [1].

Considering g linear evolution, for free m evolution its operator $\hat{H}_0$ should be the even polinom of the form:

$$\hat{H}_0 = -\sum_{l=1}^{n} b_{2l}\frac{\partial^{2l}}{\partial \vec{r}^{2l}} \tag{26}$$

If the external field action can be described by the addition of real function V to it:

$$\hat{H} = \hat{H}_0 + V(\vec{r},t) \tag{27}$$

then from $\frac{\partial g}{\partial t}$ the term $\frac{\partial w^{\frac{1}{2}}}{\partial t}$ can be extracted and expressed via corresponding w, γ $\vec{r}$-derivatives. From their comparison with corresponding $\hat{H}g$ derivatives the schroedinger equation is obtained for m evolution. The applicability of Jordan theorem to 3-dimensional $\hat{H}$ is obvious, because the derivation of $\hat{H}$ linearity doesn't depend on the dimensiality of coordinate space. The same is true for the proof of uniqueness of $\eta(\vec{r}, t)$ ansatz, i.e. that $\chi(w) = 0$.

Planck constant $\hbar = 1$ in our FM ansatz, but the same value ascribed to it in relativistic unit system in which the velocity of light $c = 1$; in FM framework $\hbar$ only connects x, p scales and doesn't have any other meaning. Note that superposition principle doesn't need to be postulated separately in such formalism. Rather, it follows that the sum of two physical m states $g_{1,2}$ with proper complex coefficients $a_{1,2}$ is also the physical m state. In the described derivation of evolution equation we didn't assume Galilean invariance of FM, rather in our approach it follows from the obtained ansatz, if the reference frame (RF) is regarded as the physical object with mass $\mu \to \infty$ [11, 25].

In our approach the state space is defined by the underlying geometry and corresponding dynamics i.e. is derivable concept. For states of massive particle m it was found to be equivalent to $\mathcal{H}$, but, in principle, it can be different for other systems. The similar features possess the formalism of algebraic QM where the state space is defined by the observable algebra and system dynamics, as the result, the state space can differ from QM Hilbert space [1].

6 Conclusion

As was noticed already, QM can be described by several alternative formalisms, of them the most notorious are algebraic QM and Schroedinger or standard formalism [1]. To discuss the possible advantages of FM formalism it's instructive to compare it with the latter one. From the formal side, standard QM exploits two fundamental structures of different nature: the space-time manifold $R^3 * T$ and function space $\mathcal{H}$ defined on R^3. In distinction, FM formalism involves only one basic structure - the fuzzy manifold $\tilde{R}^3 * T$. FM physical state densities $w(\vec{r})$ are $\tilde{R}^3$ fuzzy points, the equivalence of such states to $\mathcal{H}$ Dirac vectors was proved here. In standard QM the evolution equation or postulated *ad hoc* or derived assuming Galilean invariance of object states [1]. In FM the Schroedinger equation is derived assuming only space-time shift and rotational invariance which are essentially weaker assumptions. The quantum-classical transition in such theory is essentially more simple, it's just the transition from $\tilde{R}^3$ manifold to R^3 one, for which the classical particles correspond to ordered points.

As we know, general relativity is essentially geometric theory, but the attempts to quantize gravity suffer the serious difficulty even at axiomatics level. Hence such geometric QM formalism can help to construct quantum theory of gravity. It can be useful also for development of gauge field theory, which is mainly geometric [26]; in addition, its implications can be important for the analysis of QM foundations. Currently, the main impact of QM geometrization studies is done on the exploit of Hilbert manifolds ([13, 14] and refs. therein), however, the results obtained up to now have quite abstract form, and their applicability to particular problems is questionable. The considered FM formalism possesses simple and logical axiomatics which origin is basically geometrical, so it can become the appropriate part of QM geometrization program.

In conclusion, we have shown that the quantization of elementary systems can be derived directly from axiomatic of set theory and topology together with the natural assumptions about system evolution. It allows to suppose that the quantization phenomenon has its roots in foundations of mathematics [1]. Our approach permits to construct QM formalism starting from geometric concepts and structures only, so in these aspects it's analogous to general relativity construction. In the same time the considered fuzzy manifold describes the possible variant of fundamental pregeometry which is basic component of some quantum gravity theories [16].

References

[1] Jauch J.M.: Foundations of Quantum Mechanics. Reading, Addison-Wesly (1968)
[2] Dubois D., Prade H.: Fuzzy Sets and Systems. Theory and Applications. Academic Press, N-Y (1980)
[3] H.Bandemer, S.Gottwald : Fuzzy sets, Fuzzy Logics, Fuzzy Methods with Apllications. Wiley, N-Y (1995)
[4] C.T. Dodson, Bull. London Math. Soc. 6 (1974) 191.
[5] C. Zeeman, Topology of 3-manifolds, Ed. K. Fort. Prentice-Hall, New Jersey, (1961) 240
[6] T.Poston, Manifold 10 (1971) 25.
[7] S.T.Ali, G. Emch, Journ. Math. Phys. 15 (1974) 176
[8] S.T.Ali, E. Prugovecki, Journ. Math. Phys. 18 (1977) 219
[9] J. Pykaz, Found. Phys. 30(2000) 1503;
Int. J. Theor. Phys., 54 (2015) 4367
[10] A.P. Sostak, J. Math. Sci. 6 (1996) 662
[11] S. Mayburov, J. Phys. A41 (2007) 164071
[12] S. Mayburov, Phys. Part. Nucl. Lett 11 (2014) 1
[13] T. Kibble, Comm. Math. Phys. 62 (1979) 189.
[14] G. Marmo , G.F. Volkert, Phys. Scr. 82 (2010) 038117.
[15] J.J Buckley, E. Eslami Fuzzy Sets Syst., 86 (1997) 179
[16] A.P.Balachandran, S.Kurkcuoglu, S.Vaidia.: Lectures on Fuzzy and Fuzzy SUSY Physics, World Scientific, Singapore (2007)
[17] Schroder B.: Ordered Sets: An Introduction. Boston, Birkhauser (2003)
[18] Gelfand I., Fomin S. : Calculus of Variations. ed. Silverman R. Dover, N-Y (2000)
[19] Rees C.S., Shah S.M., Stanojevic G.V. : Theory and Applications of Fourier Analysis N-Y, Marcel Devker, 1981
[20] Landau L.D.,Lifshitz E.M. Quantum Mechanics. Oxford; N.Y.: Pergamon Press, 1976
[21] Landau L.D.,Lifshitz E.M.: Mechanics of Continuous Media. Oxford; N-Y., Pergamon Press (1976)
[22] E. Madelung, Z. Phys. 40 (1926) 332
[23] S.K. Ghosh, B.M. Deb, Phys. Rep. 92 (1982) 1
[24] T. Jordan, Phys. Rev. A73 (2006) 022101
[25] Y. Aharonov, ,T. Kaufherr Phys. Rev. D30 (1984) 368
[26] Cheng T., Li L. : Gauge Theory of Elementary Particles. Oxford, Claredon (1984)

EPJ Web of Conferences **126**, 05010 (2016) DOI: 10.1051/epjconf/201612605010
ICNFP 2015

On light dilaton extensions of the Standard Model *

Eugenio Megías[1,a], Oriol Pujolàs[2], and Mariano Quirós[2,3,4]

[1] *Max-Planck-Institut für Physik (Werner-Heisenberg-Institut), Föhringer Ring 6, D-80805, Munich, Germany*
[2] *Institut de Física d'Altes Energies (IFAE), The Barcelona Institute of Science and Technology (BIST), Campus UAB, E-08193, Bellaterra (Barcelona), Spain*
[3] *Institució Catalana de Recerca i Estudis Avançats (ICREA), Campus UAB, E-08193, Bellaterra (Barcelona), Spain*
[4] *ICTP South American Institute for Fundamental Research & Instituto de Física Teórica, Universidade Estadual Paulista, São Paulo, Brazil*

Abstract. We discuss the presence of a light dilaton in Conformal Field Theories deformed by a single scalar operator, in the holographic realization consisting of confining Renormalization Group flows. Then, we apply this formalism to study the extension of the Standard Model with a light dilaton in a 5D warped model. We study the spectrum of scalar and vector perturbations, compare the model predictions with Electroweak Precision Tests and find the corresponding bounds for the lightest modes. Finally, we analyze the possibility that the Higgs resonance found at the LHC be a dilaton.

1 Introduction

The discovery of a resonance at the LHC with the properties of a Standard Model-Higgs in 2012 has incited an increasing interest in the study of the nature of the Higgs boson, as this will provide very likely one of the most important indications for the detection of *new physics*. An explanation of the light Higgs mass is quite difficult nowadays without the use of unnatural fine tuning of parameters, an issue that is known as the hierarchy problem, see e.g. Ref. [1] and references therein. So, it would be very interesting to study in what kind of scenarios the Standard Model (SM) can be embedded such that the hierarchy problem is solved. One possibility is that the SM be part of a nearly-conformal sector. These models can be realized by means of strong dynamics/extra-dimensions and the Electroweak scale would arise from the spontaneous breaking of conformal invariance (SBCI). It has been recently appreciated that the SBCI can be naturally associated with the appearance of a light dilaton in the spectrum [2–7]. Within this realization, the dilaton would be the lightest state of the strong/extra-dimensional sector. Going further, it comes out the intriguing possibility that, in fact, the dilaton can play the role of a *Higgs impostor*, so that the 125 GeV resonance would then be the first instance of new physics. The exploration of this possibility will be one of the main motivations of this work. We will do this by considering a class of soft-wall scenario that allows to fully exploit the modelling capabilities of these theories.

*Presented by E. Megías at the 4th International Conference on New Frontiers in Physics (ICNFP 2015), 23-30 August 2015, Kolymbari, Crete, Greece.

[a] e-mail: emegias@mppmu.mpg.de

2 Light dilatons in extra dimensional models

SBCI leads to the existence of a non-zero value for the vacuum expectation value of the dilaton field, which is the Goldstone boson associated with this symmetry. While Conformal Field Theories (CFT) do not naturally exhibit SBCI, it has been recently proposed in Ref. [2] a mechanism that allows for light dilatons in Quantum Field Theories that are close to conformal invariance. This is based on the consideration of certain deformations of CFTs

$$\mathcal{L} = \mathcal{L}^{\text{CFT}} - \lambda O\,, \tag{1}$$

where O is a nearly marginal operator, i.e. $\dim(O) = 4 - \Delta$ with $\Delta \ll 1$. A holographic realization of this mechanism has been checked to work in Refs. [3–6] (see also Ref. [7]). In this Section we describe this mechanism in generic 5D warped models, and then particularize the results to a specific benchmark model.

2.1 Renormalization Group flows and CFT deformations

Let us consider a coupled scalar-gravitational system in 5D defined by the action

$$S = M^3 \int d^4x dy \sqrt{-g}\left(R - \frac{1}{2}(\partial_M \phi)^2 - V(\phi)\right) - M^3 \sum_\alpha \int d^4x dy \sqrt{-g}\, 2\mathcal{V}^\alpha(\phi)\delta(y - y_\alpha)\,, \tag{2}$$

where ϕ is a scalar field, $\mathcal{V}^\alpha$ $(\alpha = 0, 1)$ are UV and IR 4D brane potentials located at $y_0 = y(\phi_0) = 0$ and $y_1 = y(\phi_1)$ respectively, and M is the 5D Planck scale. The deformed CFT of Eq. (1) can be realized by a *domain wall* geometry in which the scalar field develops a profile in the extra dimension y of the form

$$ds^2 = dy^2 + e^{-2A(y)}\eta_{\mu\nu}dx^\mu dx^\nu\,, \qquad \phi = \phi(y)\,. \tag{3}$$

By seeking for solutions of the equations of motion with the near boundary expansion

$$\phi(y) = \lambda\, e^{\Delta y} + \langle O\rangle\, e^{(4-\Delta)y} + \cdots\,, \qquad y \to -\infty\,, \tag{4}$$

one can identify the deformation parameter and the condensate as the leading and sub-leading mode respectively. The holographic Renormalization Group flows can be studied better with the β-function, defined holographically as $\beta(\phi) := -\frac{\partial\phi}{\partial A}$. The equations of motion lead to the following first order differential equation

$$\beta(\phi)\beta'(\phi) = \frac{1}{2}\left(\beta(\phi)^2 - 24\right)\left(\frac{\beta(\phi)}{3} + \frac{V'(\phi)}{V(\phi)}\right)\,. \tag{5}$$

Associated with this equation there is an integration constant which is identified as the condensate $\langle O\rangle$ of the operator along the deformation direction, Eq. (1), and the boundary condition that fixes the physical value of the condensate $\langle O\rangle_{\text{phys}}$ is that the IR end of the flow be the least singular possible, see e.g. [5, 7, 8]. We are interested in *confining flows*, i.e. those which lead to a discrete spectrum in all sectors. In the absence of the IR brane, these flows are characterized by the IR limit

$$\beta(\phi) \to \beta_\infty \qquad \text{with} \qquad \sqrt{6} < -\beta_\infty < 2\sqrt{6}\,, \tag{6}$$

where the lower and upper bounds correspond to the confinement and good IR singularity criterion respectively. A typical profile for such flows is displayed in Fig. 1. It is interesting to see that the flow has three different regimes: i) deformation-dominated regime in the UV, ii) condensate-dominated regime, and iii) confinement region in the IR.

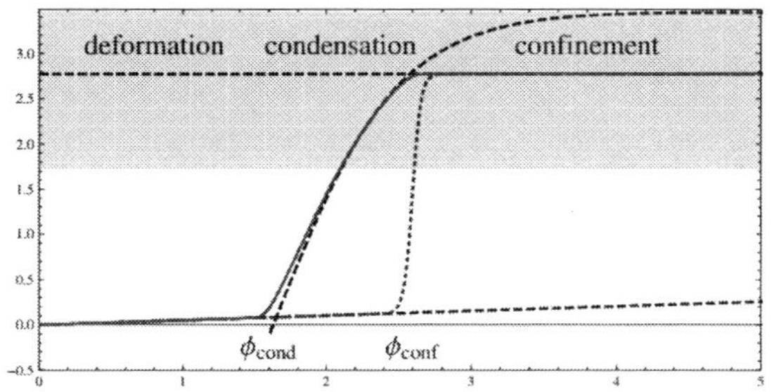

Figure 1. $3\frac{V'(\phi)}{V(\phi)}$ (dotted blue) and $-\beta(\phi)$ (solid red) as a function of ϕ. The three regimes (deformation, condensation and confinement) are clearly distinguishable in $\beta(\phi)$. We have considered Model A of Ref. [5], which is characterized by $\langle O\rangle_{\text{phys}} \neq 0$.

2.2 Spectrum of excitations and type of dilatons

To compute the scalar spectrum and in particular its lightest mode, known as the radion/dilaton, we have to consider a scalar perturbation of the background, and this leads to a contribution in the metric of the form $ds^2 \sim e^{-2(A+F)}$. Then the equation of motion of the excitation modes F_n is written as [9] [1]

$$\ddot{F}_n - 2\dot{A}\dot{F}_n - 4\ddot{A}F_n - 2\frac{\ddot{\phi}}{\dot{\phi}}\dot{F}_n + 4\dot{A}\frac{\ddot{\phi}}{\dot{\phi}}F_n = -e^{2A}m_n^2 F_n\,. \tag{7}$$

This is supplemented with boundary conditions on the branes $\dot{F}_n|_{y_\alpha} = f(\mathcal{V}^\alpha, m_n)$, which are obtained by integrating in a small interval around $y = y_\alpha$. An approximate solution of Eq. (7) assuming light modes, i.e. $m_n^2 \ll \Lambda_{\text{IR}}^2$ where Λ_{IR} is the mass gap in the spectrum, leads to a *mass formula* which allows to distinguish between two kind of light dilatons [5, 10]:

- *Hard dilatons*: those dominated by the value of the β function at the IR brane location. The mass of the dilaton scales like $m_{\text{dil}}^2 \sim \beta_{\text{IR}}^2\Lambda_{\text{IR}}^2$. This is the realization that has been discussed in Refs. [3, 4].
- *Soft dilatons*: those dominated by the value of the β function in the condensation scale. In this case a light dilaton can be realized whenever i) the β function is small at the condensation scale ($\beta_{cond} \ll 1$) and ii) the rise towards confinement ($\beta_{conf} \sim 1$) is fast enough. Then

$$m_{\text{dil}}^2 \sim (\beta_{cond})^\kappa \Lambda_{\text{IR}}^2 \ll \Lambda_{\text{IR}}^2\,, \tag{8}$$

where the precise value of the constant $\kappa > 0$ depends on the particular model [5, 7]. The dilaton turns out to be naturally light for nearly marginal deformations, for which $\beta_{cond} \sim \Delta \ll \beta_{conf}$.

The dilaton corresponds to the fluctuation of the condensate, regardless of whether $\langle O\rangle$ vanishes or not. In the following we will simplify the model by setting $\langle O\rangle = 0$, as the physics is qualitatively similar to the one from models with $\langle O\rangle \neq 0$. This can be done by just considering an analytic β function, as then the integration constant in Eq. (5) is set to vanish. We choose

$$\beta(\phi) = -6ac\left[1 + e^{-a\phi}\right]^{-1}\,, \tag{9}$$

where c and a are parameters which govern the IR value of β and its slope at $\phi = 0$ respectively. We assume that the brane potentials dynamics have fixed (ϕ_0, ϕ_1) to solve the hierarchy problem, i.e. $A(\phi_1) \simeq 35$, where we have normalized the warp factor to $A(\phi_0) = 0$.

We show in Fig. 2 the dilaton mass obtained from a numerical solution of Eq. (7). The realization of the dilaton for $a \lesssim 0.6/c$ and $a \gtrsim 0.6/c$ is hard and soft respectively, and the existence of the peak is a signal of the change of regime between both pictures. We also display in Fig. 2 (B) the spectrum of vector perturbations, which we refer in the following as KK gauge bosons. See Ref. [10] for details.

[1] The dots stand for derivatives with respect to y.

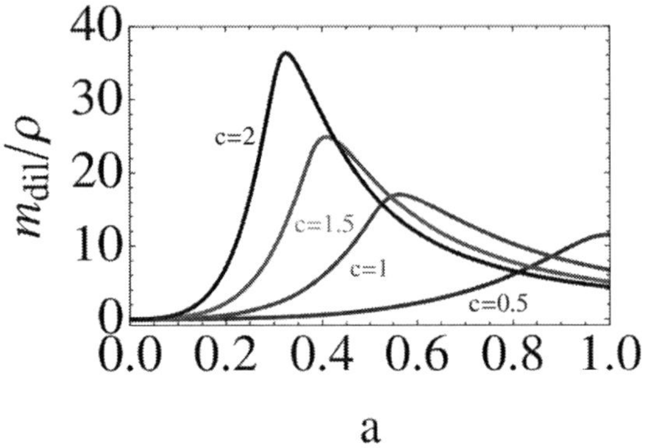

(A)

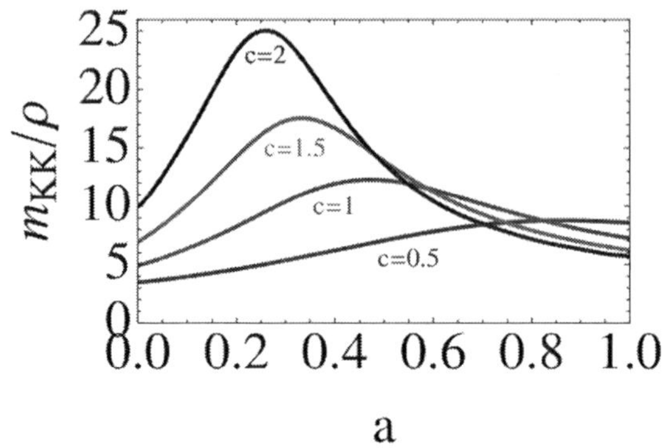

(B)

Figure 2. (A) Mass of the dilaton as a function of the parameter a, normalized to $\rho = k e^{-A(y_1)}$. (B) KK gauge boson masses as a function of the parameter a. In these figures we have plotted the results for $c = 0.5, 1, 1.5$ and 2, and have considered $\phi_1 = 5$.

3 Electroweak breaking

We will now introduce the electroweak sector in the theory. Let us consider the SM propagating in the 5D space described in Section 2. In addition to the 5D $SU(2)_L \times U(1)_Y$ gauge bosons $W^i_M(x, y)$, $B_M(x, y)$ with $i = 1, 2, 3$ and $M = (\mu, 5)$, we define the SM Higgs as

$$H(x, y) = \frac{1}{\sqrt{2}} e^{i\chi(x,y)} \begin{pmatrix} 0 \\ h(y) + \xi(x, y) \end{pmatrix}, \tag{10}$$

where $h(y)$ is the Higgs background. The action of the model is

$$S_5 = \int d^4x dy \sqrt{-g} \left(-\frac{1}{4} \vec{W}^2_{MN} - \frac{1}{4} B^2_{MN} - |D_M H|^2 - V(H) \right), \tag{11}$$

where $V(H)$ is the 5D Higgs potential. Electroweak symmetry breaking is triggered on the IR brane. After the breaking, the lightest modes are separated by a gap from the KK spectrum, and the masses for the W and Z bosons are then approximately given by [2]

$$m^2_V \approx \frac{1}{y_1} \int_0^{y_1} dy M^2_V(y), \qquad V = W, Z, \tag{12}$$

where the y-dependent bulk masses are defined as

$$M_W(y) = \frac{g_5}{2} h(y) e^{-A(y)}, \qquad M_Z(y) = \frac{1}{c_W} M_W(y), \tag{13}$$

and g_5 is the 5D gauge coupling, which is related to the 4D one g by $g_5 = g\sqrt{y_1}$. After solving the equations of motion for the Higgs field $h(y)$, one can obtain from Eqs. (12)-(13) the masses for the gauge bosons. To compare the model predictions with Electroweak Precision Tests (EWPT) a convenient parameterization is using the (S, T, U) variables in [11, 12]. From the fitted values for S and T as [13]:

$$T = 0.05 \pm 0.07, \quad S = 0.00 \pm 0.08, \qquad (90\% \text{ correlation}), \tag{14}$$

DOI: 10.1051/epjconf/201612605010

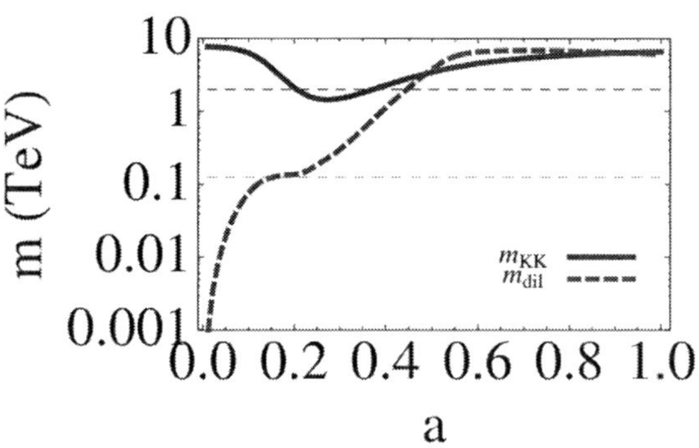

(A)

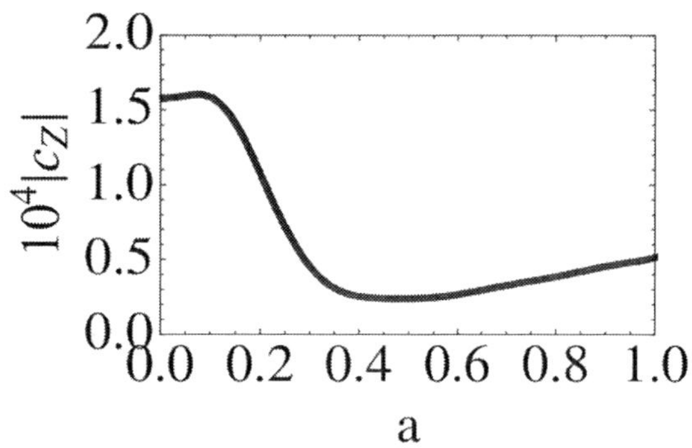

(B)

Figure 3. (A) Bound on KK mass (solid line) as function of a from electroweak observables. The corresponding dilaton mass is in dashed line. We have drawn horizontal dashed lines corresponding to 125 GeV and 2 TeV. (B) Radion coupling to the Z boson as a function of a. The result for W bosons is $|c_W| = (m_W/m_Z)^2 \cdot |c_Z| \approx 0.78\,|c_Z|$. In these figures we have considered $c = 1$ and $\phi_1 = 5$.

one gets the result in Fig. 3 (A). We find a region in a such that $m_{KK} = \mathcal{O}(\text{TeV})$ and $m_{\text{dil}} \lesssim \mathcal{O}(100)$ GeV for $c = 1$. [3] Other values of c lead to similar conclusions.

4 Coupling of the radion/dilaton to Standard Model matter fields

A light dilaton has similar interactions with matter as the Higgs, so an obvious question arises: Is the Higgs-like resonance found at LHC actually a dilaton? There are previous studies in the literature on the coupling of the radion to SM matter fields, and they usually consider that the matter is localized on the IR brane [9, 14]. In this work and motivated by EWPT we are assuming that the matter and Higgs fields are localized in the bulk. If the scalar fluctuation decomposes as $F(x, y) = F(y)\mathcal{R}(x)$, we can compute the coupling of the dilaton to the massive gauge fields (W_μ and Z_μ) normalized as

$$\mathcal{L}_{rad} = -\frac{r(x)}{v}\left\{2c_W\, m_W^2 W_\mu W^\mu + c_Z\, m_Z^2 Z_\mu Z^\mu\right\}, \tag{15}$$

where $r(x)$ is the canonically normalized radion field with kinetic term $\frac{1}{2}\left(\partial_\mu r\right)^2$. The case $c_W = c_Z = 1$ corresponds to the SM Higgs coupling. After expanding Eq. (11) to linear order in the perturbations, and using the massless radion approximation $F = e^{2A}$, one gets the following result for c_V ($V = W, Z$)

$$|c_V| = 2\frac{v\, m_V^2}{\sqrt{6}e^{-A_1} M_{Pl}\,\rho^2}\left(\frac{\int e^{-2A}}{\int e^{2A-2A_1}}\right)^{1/2} ky_1 \int_0^{ky_1} e^{4A-4A_1}\left(\frac{\int_0^y h^2 e^{-2A}}{\int_0^{y_1} h^2 e^{-2A}} - \frac{ky}{ky_1}\right)^2 d(ky)\,, \tag{16}$$

where $M_{Pl} = 2.4 \times 10^{18}$ GeV is the 4D (reduced) Planck mass. We show in Fig. 3 (B) the numerical result of $|c_Z|$ in the same regime of parameters as in Fig. 3 (A). The values are very small so that the present dilaton extension of the SM leads to a LHC phenomenology which deviates from SM predictions by an $\mathcal{O}(10^{-4})$ effect. Unfortunately these tiny effects would be unobservable at the LHC. For the same reason, the possibility of a Higgs impostor is excluded for the present model.

[2] See Ref. [11] for a wide description of the formalism of electroweak breaking by the bulk Higgs.

[3] Extra-dimensional realizations of Higgs impostors require that the KK scale m_{KK} must not exceed a few TeV, because m_{KK} and the electroweak scale $v = 246$ GeV are linked by $m_{KK} \lesssim 4\pi v$.

5 Conclusions and discussion

We have studied a mechanism in holography that allows for a naturally light dilaton in CFTs deformed by a single scalar operator. Two different realizations of a light dilaton have been identified: i) the *hard* realization is induced by the existence of an IR brane, and the dilaton is incarnated by the IR brane location, and ii) the *soft* realization, in which the dilaton is controlled by the condensation threshold of the CFT operator. These results have been confirmed in a 5D warped model which is large enough to exploit the full capabilities of the extra-dimensional models.

The extension of the SM with a light dilaton in the 5D warped model leads to dilaton masses that can naturally be of the order of magnitude of the Higgs mass, and KK vector masses of the order of TeV, all of them compatible with Electroweak Precision Tests. However a first study of the coupling to massive gauge fields suggests that the dilaton couplings to SM particles (affecting e.g. the unitarization of the $V_L V_L$ elastic and inelastic scattering as well as the strength of production of gauge bosons and fermions) are several orders of magnitude smaller than those predicted for the Higgs in the SM in a way that is unobservable at the LHC. For the same reason our results suggest that in the models discussed here the dilaton cannot be a Higgs impostor. In spite of this, some modifications of the warped model that would allow for a sizeable coupling deserve to be studied [10].

Acknowledgements

The work of O.P. and M.Q. is partly supported by the Spanish Consolider-Ingenio 2010 Programme CPAN (CSD2007-00042), by MINECO under Grants CICYT-FEDER-FPA2011-25948 and CICYT-FEDER-FPA2014-55613-P, by the Severo Ochoa Excellence Program of MINECO under the grant SO-2012-0234 and by Secretaria d'Universitats i Recerca del Departament d'Economia i Coneixement de la Generalitat de Catalunya under Grant 2014 SGR 1450. The work of M.Q. is also partly supported by CNPq PVE fellowship project 405559/2013-5. E.M. would like to thank the Institut de Física d'Altes Energies (IFAE), Barcelona, Spain, and the Instituto de Física Teórica IFT UAM/CSIC, Madrid, Spain, for their hospitality during the completion of the final stages of this work. The research of E.M. is supported by the European Union under a Marie Curie Intra-European Fellowship (FP7-PEOPLE-2013-IEF) with project number PIEF-GA-2013-623006.

References

[1] G.F. Giudice, PoS **EPS-HEP2013**, 163 (2013)
[2] R. Contino, A. Pomarol, R. Rattazzi, unpublished work (2010)
[3] B. Bellazzini, C. Csaki, J. Hubisz, J. Serra, J. Terning, Eur. Phys. J. **C74**, 2790 (2014)
[4] F. Coradeschi, P. Lodone, D. Pappadopulo, R. Rattazzi, L. Vitale, JHEP **11**, 057 (2013)
[5] E. Megias, O. Pujolas, JHEP **08**, 081 (2014)
[6] P. Cox, T. Gherghetta, JHEP **02**, 006 (2015)
[7] E. Megias (2015), 1510.01990
[8] I. Papadimitriou, JHEP **05**, 075 (2007)
[9] C. Csaki, M.L. Graesser, G.D. Kribs, Phys. Rev. **D63**, 065002 (2001)
[10] E. Megias, O. Pujolas, M. Quiros, work in progress (2015)
[11] J.A. Cabrer, G. von Gersdorff, M. Quiros, JHEP **05**, 083 (2011)
[12] M.E. Peskin, T. Takeuchi, Phys. Rev. **D46**, 381 (1992)
[13] K.A. Olive et al. (Particle Data Group), Chin. Phys. **C38**, 090001 (2014)
[14] C. Csaki, J. Hubisz, S.J. Lee, Phys. Rev. **D76**, 125015 (2007)

Search for long-lived neutral particles decaying into lepton-jets in 20.3 fb^{-1} proton-proton collisions at $\sqrt{s} = 8$ TeV with the ATLAS detector

Marco Schioppa[1,a] on behalf of the ATLAS Collaboration

[1] *Universita' della Calabria and INFN Cosenza, Rende, Italy*

Abstract. Several models of elementary particle physics beyond the Standard Model, predict the existence of neutral particles that can decay in jets of leptons and light hadrons (lepton-jets). The present contribution collects the results about the lepton-jet search with the ATLAS experiment at the proton-proton LHC collider at $\sqrt{s}$=8 TeV during the entire 2012 data taking (20 fb^{-1}). No excess of events have been observed over the expected background and the exclusion limits for two different models, that predict the Higgs boson to decay in lepton-jets, have been computed. A new lepton-jet search is underway for the new LHC era at ($\sqrt{s}$=13 TeV).

1 Introduction

The Standard Model (SM) of particle physics and its mutual interactions has proven up to now extremely successful, but despite that it still remains incomplete, e.g. it does not explain the nature of dark matter. New long lived light particles are predicted in several extensions of the SM, such as hidden sector scenarios. If the mass of these particles is in the MeV to GeV range, they would decay mainly to leptons and light mesons [1]. One of the most promising model is that in which the hidden sector and the visible one are coupled via the vector portal: a light hidden photon (γ_d) mixes kinetically with the SM photon. If the γ_d is the lightest state of the hidden sector, it decays back to SM particles. Due to the small mass of the γ_d and the size of the kinetic mixing parameter (that control both the BR and lifetime of the γ_d) these particles are produced with a large boost. As a consequence the decay products of the γ_d are collimated jets of pairs of electrons and/or muons and or pions that can be produced far from the IP of the event (displaced lepton-jet). The high resolution and high granularity measurement capability of the ATLAS air-core muon-spectrometer (MS) [2] is ideal for this kind of search. This note refers to the full dataset collected by ATLAS during the 2012 run at $\sqrt{s}$ = 8 TeV, corresponding to an integrate luminosity of 20.3 fb^{-1} as reported in [3]. Search for non-prompt muonic lepton-jet has been performed at ATLAS during the 2011 data taking at $\sqrt{s}$ = 7 TeV as reported in [4].

[a] e-mail: marco.schioppa@cern.ch

2 Lepton-jet definition and selection

The search adopts a generic definition of lepton-jet in order to make the analysis as model-independent as possible. The lepton-jet contains at least one γ_d highly isolated in the inner detector, contained in a narrow angular cone (ΔR) and decaying far from the primary vertex to pairs of electrons/muons/pions. A lepton-jet gun Monte Carlo generator has been developed to optimize search criteria and to produce detection efficiency curves. This MC allows to study the response of the detector to lepton-jets and guide their characterization and the identification of the variables useful for the selection of signal. Only three topologies of lepton-jets are considered: TYPE0, TYPE1 and TYPE2. The first one is a cluster of only muons identified into the MS and no jets in a cone opening ΔR=0.5. It is the signature of a lepton-jet with all γ_d decaying to muon pairs. The second one is a cluster of two muons and jets in a cone opening ΔR=0.5. It the signature of one γ_d decaying to muon pair and the other one in electron or pion pair. The third one are jets with low electromagnetic fraction, narrow width and no muons in a cone opening ΔR=0.5. It is the signature of a lepton-jet with one or two γ_d both decaying to electron/pion pairs. Figures 1 shown a sketch of the three lepton-jet types.
The triggers used to select displaced lepton-jets of TYPE0 and TYPE1 is the unprescaled multi-muon trigger with at least three reconstructed muons in the MS with $p_T \geq 6$ GeV. For the TYPE2 lepton-jets the single jet trigger with lo electromagnetic fraction is used.
The events that pass the trigger filters have to satisfy additional requirements. To separate the signal to the background the following major requirements are made: exactly two reconstructed lepton-jets in the event, both lepton-jet are isolated in the ID (the highest ID $\sum p_T$ in the event must be ≤ 3 GeV), the absolute value of the azimuthal angle $\Delta\phi$ between the two lepton-jets must be ≥ 1, the muons must have no ID track match and the jets must be in timing with the bunch crossing ($-1ns \leq t_{jet} \leq 5ns$).

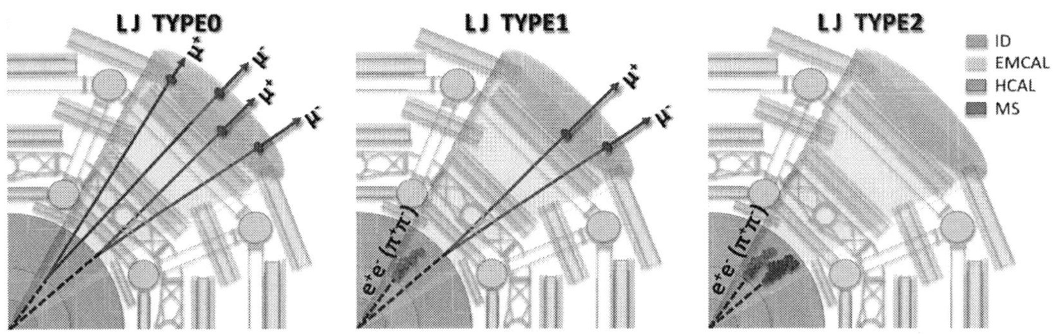

Figure 1. Sketch of the three lepton-jets classification: left TYPE0 lepton-jet (only muons), centre TYPE1 lepton-jet (muons and jets), right TYPE2 lepton-jet (only jets). If the lepton-jet contains only one γ_d it contributes only to TYPE0 and TYPE2.

3 Background

The potential backgrounds in the signal sample include all processes with prompt muons with or without associated jets (e.g. W+jets, Z+jets, ttbar, single-top, etc.), cosmic-rays (e.g. muon bundles in cosmic-ray air showers), beam induced background.
The cosmic-ray events are reduced using jet timing. For muons there is an additional requirement: the perigee to the beam line of the extrapolated MS track has to be close to the PV. For the evaluation of the remain CR contribution the triggers have been used in empty bunches of 2012 data. QCD multi-jet

is reduced using isolation in the ID (less dependent to pile-up) and jet EM fraction/width cuts. The isolation variable is the p_T of the ID tracks associated to the PV in a cone ΔR=0.5 around the lepton-jet line of flight ($p_T^{track} \geq 0.5$ GeV). Figure 2-left shows the ID isolation distribution in the control sample of the 2012 data selected by single-jet triggers. The ID isolation is validated using the muons from Z→ $\mu\mu$ decays. The remain QCD multi-jet background is evaluated using a data-driven matrix method. As uncorrelated variables the $|\Delta\phi|$ and max$\{\sum p_T\}$ are used. The signal region is defined as $|\Delta\phi| \geq 1$rad and max$\{\sum p_T\} \leq 3$ GeV. Figure 2-right shows the event distribution in this plane before the requirements on $|\Delta\phi|$ and max$\{\sum p_T\}$. Indicating with "A" the signal region, the residual background in A can be predicted from the population of the other three regions: $N_A = N_B$ x N_D/N_C.

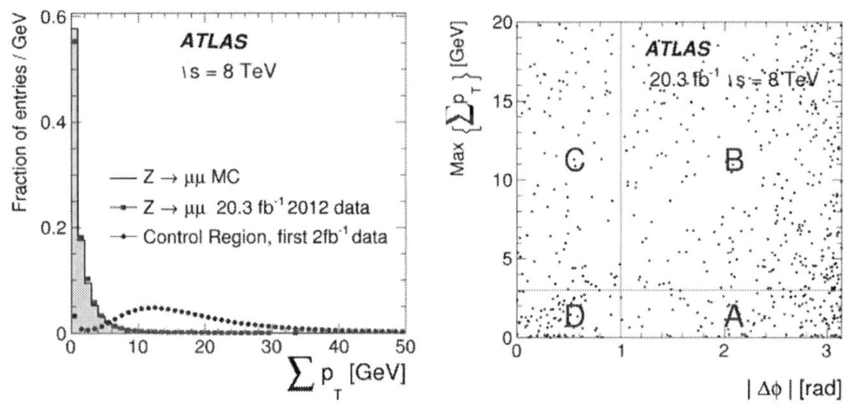

Figure 2. Left plot is the distribution of $\sum p_T$: (filled dot) control sample of the first 2 fb^{-1} of 2012 data, (filled square) $Z \to \mu\mu$ in 2012 data and (solid line) $Z \to \mu\mu$ MC sample. All distributions are normalized to unit area. Right plot is the distribution of lepton-jet events in the ABCD plane before the requirements max$\{\sum p_T\} \leq 3$ GeV and $|\Delta\phi| \geq 1$. Taken from [3].

4 Lepton-jet benchmark models

The performance of the lepton-jets search criteria are evaluated setting limits on two Falkowski-Ruderman-Volansky-Zupan (FRVZ) models [5] which predict non-SM Higgs boson decays to lepton-jets. Figure 3 shows the diagrams for the decay of the Higgs to lepton-jets in the two models. The two diagrams refer to the production of $2\gamma_d$ and $4\gamma_d$ respectively. In both models the decay chain contain two lighter hidden fermions HLSP (Hidden Lightest Stable Particle) that escape to the detection.

5 Results

Table 1 summarizes the data and background results of the search for lepton-jets in the 2012 data sample. Both for all lepton-jet pair events and for the case where the TYPE2-TYPE2 lepton-jets are excluded, the data agree with the background expectation.
The results of the lepton-jets search are used to set upper limits on the Higgs boson decay branching fraction to lepton-jets as a function of the γ_d mean lifetime, according to the FRVZ models. The resulting exclusion limits on the σ x BR, assuming the Higgs boson SM gluon fusion production cross section σ_{SM} = 19.2 pb, are shown in figure 4 as a function of the γ_d mean lifetime (expressed as $c\tau$) for the two benchmark models and for the cases in which all the lepton-jet type events are included

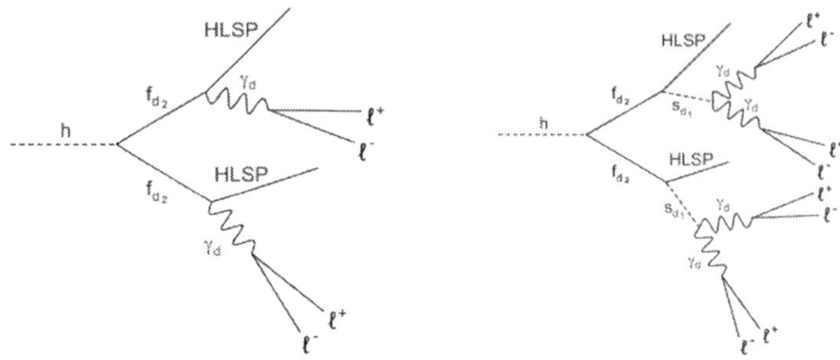

Figure 3. Diagrams of the two FRVZ models used as benchmarks in this search. l^+ l^- corresponds to electron/muon/pion pair decay in the final state.

Table 1. Summary of the lepton-jet (LJ) selection applied to data and background in the full 2012 data sample. The first uncertainty is statistical, while the second is systematic.

	All LJ pair types	TYPE2-TYPE2 LJs excluded
Data	119	29
Cosmic-rays	40 ± 11 ± 9	29 ± 9 ± 29
Multi-jets	70 ± 58 ± 11	12 ± 9 ± 2
Total background	110 ± 59 ± 14	41 ± 12 ± 29

(top plots) and for those in which the TYPE2-TYPE2 events are excluded. The corresponding limits with TYPE2-TYPE2 events excluded are shown in table 2. For the case of a hidden photon which kinetically mixes with the SM photon, these limits can be converted into exclusion limits on the kinetic mixing parameter ϵ. These results are also interpreted in the context of the Vector portal model as exclusion contours in the kinetic mixing parameter ϵ vs γ_d mass plane as shown in figure 5.

Table 2. Ranges of γ_d lifetime $c\tau$ excluded at 95% CL for $H \to 2\gamma_d + X$ and $H \to 4\gamma_d + X$, assuming 10% BR and the Higgs boson SM gluon fusion production cross section. The events TYPE2-TYPE2 are not used.

FRVZ model	Expected $c\tau$ [mm] BR(100%)
Higgs $\to 2\gamma_d$ + X	$14<c\tau<140$
Higgs $\to 4\gamma_d$ + X	$15<c\tau<260$

6 To be continue with Run II data

The opportunity offered by LHC in the era at highest energy ($\sqrt{s}$ = 13 TeV) and high luminosity was caught and a new lepton-jet search is started. An ambitious physics program for this new work is underway. It will include new triggers (e.g. single muon trigger), new TYPEs with EM clusters, the RPC timing for cosmic-rays rejection, the lepton-jet vertex search and the harmonization with prompt-lepton-jet search and extend the analysis to other lepton-jet production mechanisms (e.g. associated production with vector bosons W and Z).

DOI: 10.1051/epjconf/201612605011

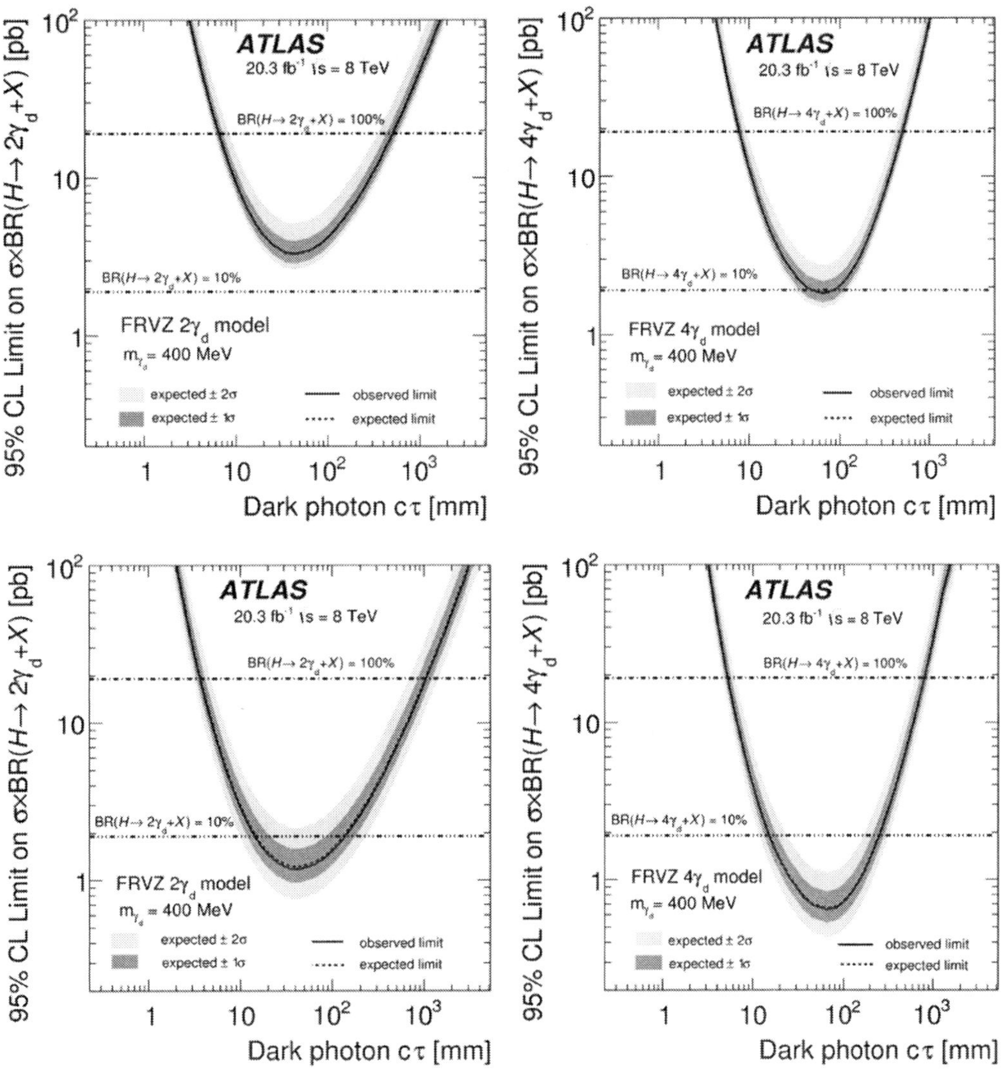

Figure 4. Top plots: 95% upper limits on the σ x BR for the processes $H \rightarrow 2\gamma_d + X$ (left) and $H \rightarrow 4\gamma_d + X$ (right), as a function of the γ_d lifetime ($c\tau$) for the FRVZ benchmark samples. Bottom plots: 95% upper limits excluding the TYPE2-TYPE2 events. The expected limit is shown as the dashed curve and the almost identical solid curve shows the observed limit. The horizontal lines correspond to σ x BR for two values of the BR of the Higgs boson decay to dark photons. Taken from [3].

7 Conclusions

The search for new phenomena beyond the Standard Model is a very active field at the ATLAS experiment. Recent results for lepton-jet search at ATLAS are presented in this note. No excess over the Standard Model expectation has been observed and limits are placed on the two FRVZ benchmark models.

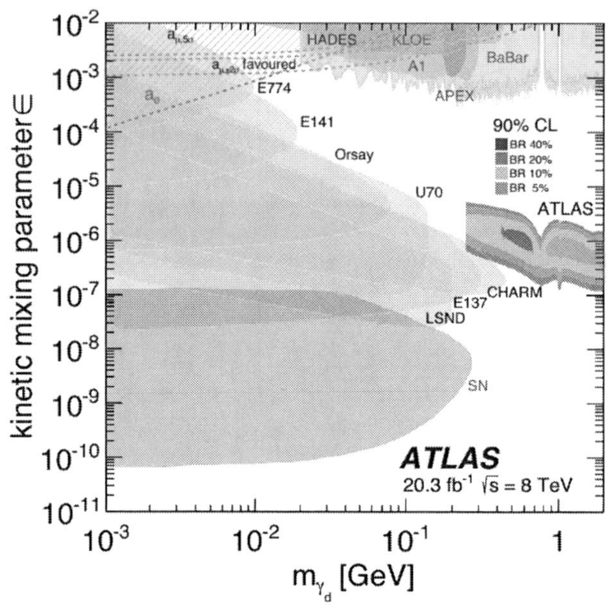

Figure 5. Parameter space exclusion plot for γ_d as a function of its mass and of the kinetic mixing parameter ϵ. The 90% CL exclusion limits are presented assuming the FRVZ model $H \rightarrow 2\gamma_d + X$ with decay branching fraction to γ_d of 5/10/20/40% and the NNLO gluon fusion Higgs production cross section. There are shown the existing 90% CL exclusion regions from other experiments like beam dump, electron and muon anomalous magnetic moment, constraints from astrophysical observations. Taken from [3].

References

[1] M. J. Strassler and K. M. Zurek, Phys. Lett. B **651** 374 (2007)
[2] The ATLAS Collaboration, JINST **3**, S08003 (2008)
[3] The ATLAS Collaboration, JHEP **11**, 088 (2014)
[4] The ATLAS Collaboration, Phys. Lett. B **721**, 32 (2013)
[5] A. Falkowski, J. T. Ruderman, T. Volansky and J. Zupan, JHEP **05** 077 (2010)

Model of Dirac field interacting with material plane within Symanzik's approach

Yu. M. Pismak[1,a] and D. Yu. Shukhobodskaia[1]

[1] *Saint-Petersburg State University*

Abstract. The model for the interaction of a spinor field with a material plane is constructed in the framework of the Symanzik's approach. The characteristics of scattering process of Dirac particles on the plane are calculated. The bounced states localized near the plane are investigated.The model can find application to a wide class of phenomena arising by the interaction of quantum electrodynamics fields with two-dimensional materials.

1 Introduction

Space-time homogeneity and isotropy are typical for usual quantum field theory models of elementary particles [1]. It is a natural assumption in the study of various processes with simplest excitations of quantum vacuum. However, if the properties of the vacuum change essentially as a result of the interaction of quantum fields with macroscopic objects, then such an approach is inapplicable. In this case, quantum macro-effects may appear in dynamics of material bodies which cannot be explained in the framework of classical physics.

Theoretically, this problem was first considered in 1948 by Casimir [2]. He showed that as the result of fluctuations of the quantum vacuum, attraction occurs between two ideally conducting plates of a planar uncharged capacitor. This phenomenon, called the Casimir effect (CE), is observed experimentally [3–6], and the empirical results obtained for materials with high conductivity are, with a high degree of accuracy, in agreement with theoretical ones.

At typical for CE distances of 10 to 1000 nm both classical and quantum properties of the system turn out to be essential. Their combination forms a special nanophysics. Investigation of it is not solely of theoretical interest, understanding its regularities is also important for developing new technological devices, in view of the increasing trend to their miniaturization.

Although theoretical investigations of the CE are the subject of numerous works [6, 7], they are often based on simplified models of a free scalar field theory or free electromagnetic field with fixed boundary conditions, applying only to investigations of some particular aspect of the CE, and ignoring usually specificity of quantum electrodynamics. Such models are not suitable for comprehensively describing a wide range of nanophysical phenomena occurring in a system as a result of the interaction of its quantum degrees of freedom with material bodies of a given shape (classic defect).

[a]e-mail: ypismak@yahoo.com

The results presented in our paper were obtained within the Symanzik's approach [8], in which the interaction of quantized fields with a spatial inhomogeneity is modeled by an additional action functional (defect action), concentrated in the spatial domain where this inhomogeneity - a macroscopic object - is located. An important assumption is also that the standard requirements for quantum field models (locality, renormalizability, symmetry properties) are fulfilled.

Special attention is given to quantum field models with two-dimensional spatial inhomogeneities due to the recent progress in the physics of two-dimensional material. In quantum electrodynamics (QED), the interaction of photons with a two-dimensional defect is entirely determined by the requirements of locality, gauge invariance, and renormalizability and is described by a Chern-Simons action functional with one dimensionless constant characterizing the properties of the material [9]. The Casimir force depends on that constant and thus become nonuniversal. For a planar capacitor it can be not only attractive but also repulsive [9]. The interaction of electromagnetic waves with two-dimensional planar defects in the model with a Chern-Simons potential was investigated in [10] - [12].

In this paper we use Symanzik's approach in the standard QED in (3+1)-dimensional space-time to model the interaction of a two-dimensional material object with the Dirac field. Such a problem for a spinor field was considered in [13]. In the defect action functional, in addition to the photon Chern-Simons action, a purely Dirac contribution must be taken into account in accordance with the basic principles of QED. In this model framework, we investigate scattering of spinor particles on a material plane and also consider bound states localized in its vicinity.

2 Statement of problem

The proposed by Symanzik action functional describing the interaction of the quantum field with material object (defect) has the form:

$$S(\varphi) = S_V(\varphi) + S_{def}(\varphi),$$

where S_V is the action of the original quantum field system and S_{def} is the defect action:

$$S_V(\varphi) = \int L(\varphi(x))d^D x,\ S_{def}(\varphi) = \int_\Gamma L_{def}(\varphi(x))d^{D'} x.$$

Here Γ is a subspace of dimension $D' \leq D$ in a D-dimensional space [8]. The basic principles of QED-gauge invariance, locality, and renormalizability-impose strong constraints on the possible form of the defect action S_{def}. It follows that interaction of the photon field $A_\mu(x)$ with a two-dimensional surface without charges and currents and shape defined by equation $\Phi(x) = 0$, $x = (x_0, x_1, x_2, x_3)$, is described by the Chern-Simons functional

$$S_{def}(A) = \frac{a}{2}\int \varepsilon^{\lambda\mu\nu\rho}\partial_\lambda\Phi(x)A_\mu(x)F_{\nu\rho}(x)\delta(\Phi(x))d^4x,$$

where $F_{\nu\rho}(x) = \partial_\nu A_\rho - \partial_\rho A_\nu$, $\varepsilon^{\lambda\mu\nu\rho}$ is the totally antisymmetric tensor ($\varepsilon^{0123} = 1$), and the parameter a is a dimensionless coupling constant. This expression is the most general form of a gauge-invariant action concentrated on a defect surface being invariant in respect to reparameterization of one and not having any negative-dimension parameters. A fermionic contribution to the defect action in the most general form is written as

$$S_{def}(\bar\psi, \psi) = \sum_{j=1}^{16}\int \alpha_j\bar\psi(x)\Gamma_j\psi(x)\delta(\Phi(x))dx,$$

where Γ_j are the 16 basis Dirac matrices and α_j are dimensionless coupling constants. The full action of the model, which satisfies the basic principles of QED, has the form

$$S(\bar{\psi}, \psi, A) = -\frac{1}{4} F_{\mu\nu} F^{\mu\nu} + \bar{\psi}(i\hat{\partial} - m + ie\hat{A})\psi + S_{def}(A) + S_{def}(\bar{\psi}, \psi).$$

By virtue of the renormalizability requirement, the coupling of fields is described by the standard term $ie\bar{\psi}\hat{A}\psi$ in the QED action.

In this paper we consider the defect, which lie on the material plane $x_3 = 0$. Then in the Dirac part of the action of our model,

$$S(\overline{\psi}, \psi) = \int \overline{\psi}(x)(i\hat{\partial} - m + \Omega(x_3))\psi(x)dx, \tag{1}$$

the coupling of the spinor field to the plane is described by the matrix $\Omega(x_3) = Q\delta(x_3)$. Because $\Omega(x_3)$ and $\delta(x_3)$ have the dimension of mass, the matrix Q is dimensionless. By the symmetry of the system, it must be invariant under transformations in the subgroup of the Poincare group not changing the x_3 coordinate. In the most general case, the matrix Q can be represented as

$$Q = q_1 I + q_2\gamma_5 + q_3\gamma_3 + q_4\gamma_5\gamma_3, \tag{2}$$

where I is the unit $4x4$ matrix and γ_3 and $\gamma_5 = i\gamma_0\gamma_1\gamma_2\gamma_3$ are Dirac matrices, for which we use the representation

$$\gamma_0 = \begin{pmatrix} 1 & 0 & 0 & 0 \\ 0 & 1 & 0 & 0 \\ 0 & 0 & -1 & 0 \\ 0 & 0 & 0 & -1 \end{pmatrix}, \quad \gamma_1 = \begin{pmatrix} 0 & 0 & 0 & 1 \\ 0 & 0 & 1 & 0 \\ 0 & -1 & 0 & 0 \\ -1 & 0 & 0 & 0 \end{pmatrix},$$

$$\gamma_2 = \begin{pmatrix} 0 & 0 & 0 & -i \\ 0 & 0 & i & 0 \\ 0 & i & 0 & 0 \\ -i & 0 & 0 & 0 \end{pmatrix}, \quad \gamma_3 = \begin{pmatrix} 0 & 0 & 1 & 0 \\ 0 & 0 & 0 & -1 \\ -1 & 0 & 0 & 0 \\ 0 & 1 & 0 & 0 \end{pmatrix}.$$

The motion of a spinor particle in the field of the defect $\Omega(x_3)$ is described by the Dirac equation

$$(i\hat{\partial} - m + \Omega(x_3))\psi(x) = 0. \tag{3}$$

It is one of the Euler-Lagrange equations for action (1), obtained by taking the functional derivative of the action with respect to $\overline{\psi}(x)$. Taking the derivative of (1) with respect to $\psi(x)$ yields the second equation:

$$(-i\partial_\mu\overline{\psi}(x))\gamma^\mu + \overline{\psi}(x)(m - \Omega(x_3)) = 0.$$

It can be easily verified that this equation can be obtained for $\overline{\psi}(x) = \psi^*(x)\gamma_0$ by the Dirac conjugation of equation (3), if $\gamma_0\Omega^+(x_3) = \Omega(x_3)\gamma_0$.

We introduce the following notation

$$\bar{x} = (x_0, x_1, x_2), \quad \overline{px} = p_0x_0 - p_1x_1 - p_2x_2.$$

for (2+1)- parts of space-time vectors and represent $\psi(x) = \psi(\bar{x}, x_3)$ as $\psi(x) = \psi_s(x) + \psi_a(x)$, where

$$\psi_s(x) = \frac{1}{2}(\psi(\bar{x}, x_3) + \psi(\bar{x}, -x_3)), \quad \psi_a(x) = \frac{1}{2}(\psi(\bar{x}, x_3) - \psi(\bar{x}, -x_3)).$$

Analysis of the possible singularities of spinor $\psi(x)$ at the point $x_3 = 0$, which was made in the paper [14], leads to the following: for $x_3 \neq 0$, the field $\psi(x)$ satisfies the Dirac equation

$$(i\hat{\partial} - m)\psi(x) = 0, \tag{4}$$

and the relation

$$-i\gamma_3\psi_a(\bar{x}) + Q\psi_s(\bar{x}) = 0. \tag{5}$$

holds at $x_3 = 0$. The general solution of (4) with $x_3 > 0$ can be represented as

$$\psi(x) = \frac{1}{(2\pi)^3}\int e^{i\bar{p}\bar{x}}\psi_+(\bar{p}, x_3)d\bar{p}.$$

Then it follows from (4) that $\psi_+(\bar{p}, x_3)$ satisfies the equation

$$(i\gamma_3\partial_3 + \hat{\bar{p}} + m)\psi_+(\bar{p}, x_3) = 0.$$

Its general solution has the form

$$\psi_+(\bar{p}, x_3) = U(\bar{p}, x_3)\chi_+(\bar{p}),\ U(\bar{p}, x_3) = e^{-i\gamma_3(\hat{\bar{p}}+m)x_3}. \tag{6}$$

where $\chi_+(\bar{p})$ is an arbitrary spinor depending only on $\bar{p}$.

Similarly, for $x_3 < 0$, we obtain

$$\psi(x) = \frac{1}{(2\pi)^3}\int e^{i\bar{p}\bar{x}}\psi_-(\bar{p}, x_3)d\bar{p},\ \psi_-(\bar{p}, x_3) = e^{-i\gamma_3(\hat{\bar{p}}+m)x_3}\chi_-(\bar{p}). \tag{7}$$

From (5), (6), (7) we obtain the relation

$$-i\gamma_3(\chi_+ - \chi_-) + Q(\chi_+ + \chi_-) = 0,$$

which implies that

$$\chi_- = S\chi_+,\ S = (i\gamma_3 + Q)^{-1}(i\gamma_3 - Q). \tag{8}$$

We introduce the notation

$$\kappa(\bar{p}) = \sqrt{\bar{p}^2 - m^2},\ P^{\pm}(\bar{p}) = \frac{1}{2}\left(1 \pm \frac{\gamma_3(\hat{\bar{p}} + m)}{\kappa(\bar{p})}\right).$$

We can then represent $U(\bar{p}, x_3)$ as

$$U(\bar{p}, x_3) = e^{-i\gamma_3(\hat{\bar{p}}+m)x_3} = \cos(\kappa(\bar{p})x_3) - i\sin(\kappa(\bar{p})x_3)\frac{\gamma_3(\hat{\bar{p}} + m)}{\kappa(\bar{p})} = $$
$$= e^{-i\kappa(\bar{p})x_3}P^+(\bar{p}) + e^{i\kappa(\bar{p})x_3}P^-(\bar{p}).$$

The matrices $P^{\pm}(\bar{p})$ are projectors:

$$P^{\pm}(\bar{p})P^{\pm}(\bar{p}) = P^{\pm}(\bar{p}),\ P^+(\bar{p}) + P^-(\bar{p}) = 1,$$
$$P^+(\bar{p})P^-(\bar{p}) = P^-(\bar{p})P^+(\bar{p}) = 0.$$

Each of them has two two-dimensional eigensubspaces L^+ and L^-, which are represented as linear combinations of two mutually orthogonal eigenvectors $v_1^{\pm}$ and $v_2^{\pm}$:

$$L^{\pm}(a_1^{\pm}, a_2^{\pm}) = a_1^{\pm}v_1^{\pm} + a_2^{\pm}v_2^{\pm},$$
$$P^{\pm}(\bar{p})L^{\pm}(a_1^{\pm}, a_2^{\pm}) = L^{\pm}(a_1^{\pm}, a_2^{\pm}),\ P^{\pm}(\bar{p})L^{\mp}(a_1^{\mp}, a_2^{\mp}) = 0.$$

Here, $a_i^{\pm}$, $i = 1, 2$, are complex parameters, and $v_i^{\pm}$, $i = 1, 2$, are spinors, which we choose as

$$v_1^+ = n_1^{-1}\left(-h_1, h_2^-, 1, h_3^-\right),\ v_2^+ = n_2^{-1}\left(h_1, h_2^+, 1, h_3^+\right),$$
$$v_1^- = n_2^{-1}\left(-h_1, h_2^+, 1, -h_3^+\right),\ v_2^- = n_1^{-1}\left(h_1, h_2^-, 1, -h_3^-\right)$$
$$n_1 = 2\sqrt{\frac{p_0 h_2^-}{(p_1 + ip_2)}},\ n_2 = 2\sqrt{\frac{p_0 h_2^+}{(p_1 + ip_2)}},\ h_1 = \frac{\sqrt{p_0^2 - m^2}}{m + p_0},$$
$$h_2^{\pm} = \frac{p_0^2 - m^2 \pm \sqrt{p_0^2 - m^2}\kappa(\bar{p})}{(m + p_0)(p_1 - ip_2)},\ h_3^{\pm} = \frac{\kappa(\bar{p}) \pm \sqrt{p_0^2 - m^2}}{p_1 - ip_2}.$$

These spinors are eigenstates of the helicity operator

$$\sigma(\vec{p}) = \frac{i}{2|\vec{p}|}(\vec{p}\,\vec{s}),\ \vec{s} = (\gamma_2\gamma_3, -\gamma_1\gamma_3, \gamma_1\gamma_2).$$

They satisfy the relations

$$\sigma(\vec{p})|_{p_3=\mp\kappa(\bar{p})} v_1^{\pm} = -\frac{1}{2}v_1^{\pm},\ \sigma(\vec{p})|_{p_3=\mp\kappa(\bar{p})} v_2^{\pm} = \frac{1}{2}v_2^{\pm}$$

We represent the spinors $\chi_{\pm}(\overline{p})$ as

$$\chi_{\pm}(\overline{p}) = b_{\pm}^1(\overline{p})v_1^-(\overline{p}) + b_{\pm}^2(\overline{p})v_2^-(\overline{p}) + c_{\pm}^1(\overline{p})v_1^+(\overline{p}) + c_{\pm}^2(\overline{p})v_2^+(\overline{p}). \tag{9}$$

The equality

$$\chi_{\pm}^*(\overline{p})\gamma_0\gamma_3\chi_{\pm}(\overline{p}) = \overline{\chi}_{\pm}(\overline{p})\gamma_3\chi_{\pm}(\overline{p}) = \frac{\kappa(\overline{p})}{p_0}(|c_{\pm}^1|^2 + |c_{\pm}^2|^2 - |b_{\pm}^1|^2 - |b_{\pm}^2|^2) \tag{10}$$

can be easily verified. Hence, if $p_0 > 0$ and $\kappa(\overline{p}) > 0$, then the amplitudes $c_{\pm}^i$ describe the motion of particles in the direction $x_3 > 0$, and $b_{\pm}^i$ describe the motion in the opposite direction.

Based on Eqs. (3), we investigate the scattering of spinor particles on the defect plane and also the properties of bound states localized in the vicinity of the defect.

3 Scattering of particles on the plane $x_3 = 0$

Let us consider the process where the initial flow of particles I_{in} with the momentum $\vec{p} = p_1, p_2, \kappa(\overline{p})$ moves toward the plane $x_3 = 0$ from the domain $x_3 < 0$. As it was mentioned above, if $p_0 > 0$, then the amplitudes $c_{\pm}^1$ and $c_{\pm}^2$ describe particles moving in the positive direction of the x_3 axis, and $b_{\pm}^1$ and $b_{\pm}^2$ describe particles moving in the negative direction. Hence the properties of the scattering process are determined by the amplitudes c_-^1 and c_-^2. In the domain $x_3 < 0$, there is also the flux of reflected particles specified by the amplitudes b_-^1 and b_-^2. In the domain $x_3 > 0$, there are only particles moving in the positive direction. Their flux is described by the amplitudes c_+^1 and c_+^2. Hence, for the considered process, the reflection and transmission coefficients K_r and K_{tr} have the forms

$$K_r = \frac{|b_-^1|^2 + |b_-^2|^2}{|c_-^1|^2 + |c_-^2|^2},\ K_{tr} = \frac{|c_+^1|^2 + |c_+^2|^2}{|c_-^1|^2 + |c_-^2|^2}.$$

The amplitudes c_-^1, b_-^1 and c_+^1 correspond to particles with negative helicity, and c_-^2, b_-^2, and c_+^2, to those with positive helicity. For the scattering process considered here, the amplitudes c_-^1 and c_-^2 are considered fixed, and b_-^1, b_-^2, c_+^1, and c_+^2 are determined from Eq. (8), which relates χ_+ to χ_-, and from the equations

$$b_+^1 = 0, \;\; b_+^2 = 0. \tag{11}$$

From analysis of Eq. (8), which is necessary for solving our problem, it could be obtained, that choosing the parameters $\varsigma_1, \varsigma_2, \varsigma_3$, and ς_4 to be real, we can represent the matrix S as $S = e^{i\varphi}\mathcal{S}$, where $0 \leq \varphi < 2\pi$ and

$$\begin{gathered} \mathcal{S} = \varsigma_1 I + i\varsigma_2\gamma_5 + i\varsigma_3\gamma_3 + \varsigma_4\gamma_5\gamma_3, \\ \varsigma_1^2 + \varsigma_2^2 - \varsigma_3^2 - \varsigma_4^2 = 1. \end{gathered} \tag{12}$$

From (8), Q can be expressed in terms of $\mathcal{S}$:

$$Q = i\gamma_3(1 - \mathcal{S})(1 + \mathcal{S})^{-1}.$$

Hence, with (12), we express the coefficients q_i, $1 \leq i \leq 4$, in formula (2) in terms of the parameters ς_i:

$$q_1 = -\frac{\varsigma_3}{1+\varsigma_1}, \;\; q_2 = -\frac{i\varsigma_4}{1+\varsigma_1}, \;\; q_3 = 0, \;\; q_4 = -\frac{\varsigma_2}{1+\varsigma_1}. \tag{13}$$

It follows from these relations that q_2 is imaginary and q_1 and q_4 are real. Relations (12) and (13) allow expressing ς_i in terms of q_i:

$$\begin{gathered} \varsigma_1 = -1 + \frac{2}{1 - q_1^2 + q_2^2 + q_4^2}, \;\; \varsigma_2 = -\frac{2q_4}{1 - q_1^2 + q_2^2 + q_4^2}, \\ \varsigma_3 = -\frac{2q_1}{1 - q_1^2 + q_2^2 + q_4^2}, \;\; \varsigma_4 = \frac{2iq_2}{1 - q_1^2 + q_2^2 + q_4^2}. \end{gathered}$$

Substituting the expression for the matrix $\mathcal{S}$ of form (12) and $\chi_\pm$ of form (9) in Eq. (8), we obtain linear equations for the amplitudes $b_\pm^1$, $b_\pm^2$, c_+^1, and c_+^2. Their solution with Eq. (11) taken into account allows finding the fluxes of the reflected and transmitted particles of different polarizations. For compact writing the obtained results we use the following notations

$$\omega = \sqrt{p_0^2 - m^2}, \kappa = \sqrt{p_0^2 - m^2 - p_1^2 - p_2^2}$$

The reflection and transmission amplitudes of considered scattering process read

$$b_-^1 = \frac{\left(c_-^1 m\left(m\varsigma_{33-22} + i\kappa\varsigma_{13+24}\right)\sqrt{\omega^2 - \kappa^2} + c_-^2\kappa\left(\kappa(\omega\varsigma_{14-23} - ip_0\varsigma_{13+24}) + m\left(p_0\varsigma_{22-33} + i\varsigma_{12-34}\omega\right)\right)\right)}{g},$$

$$b_-^2 = \frac{\left(c_-^2 m\left(m\varsigma_{33-22} + i\kappa\varsigma_{13+24}\right)\sqrt{\omega^2 - \kappa^2} + c_-^1\kappa\left(\kappa(\omega\varsigma_{14-23} + ip_0\varsigma_{13+24}) + m\left(p_0\varsigma_{33-22} + i\omega\varsigma_{12-34}\right)\right)\right)}{g},$$

$$c_+^1 = \frac{\kappa\left(c_-^1(\kappa(\omega\varsigma_1 + ip_0\varsigma_2) - im\varsigma_3\omega) + ic_-^2 m\varsigma_2\sqrt{\omega^2 - \kappa^2}\right)}{g},$$

$$c_+^2 = \frac{\kappa\left(c_-^2(\kappa(\omega\varsigma_1 - ip_0\varsigma_2) - im\varsigma_3\omega) + ic_-^1 m\varsigma_2\sqrt{\omega^2 - z^2}\right)}{g},$$

where

$$g = \omega\left(\left(m^2 + \kappa^2\right)\varsigma_2^2 + (\kappa\varsigma_1 - im\varsigma_3)^2\right), \;\; \varsigma_{ik} = \varsigma_i\varsigma_k, \;\; \varsigma_{ik\pm lm} = \varsigma_{ik} \pm \varsigma_{lm}.$$

In the particular case of particle motion orthogonal to the defect plane ($p_1 = 0$, $p_2 = 0, \kappa = \omega$), the amplitudes of reflected and transmitted particles are given by

$$b^1_- = \frac{c^2_- \left(\omega(\omega\varsigma_{14-23} - ip_0\varsigma_{13+24}) + m\left(p_0\varsigma_{22-33} + i\varsigma_{12-34}\omega\right)\right)}{\left((m^2+\omega^2)\varsigma_2^2 + (\omega\varsigma_1 - im\varsigma_3)^2\right)},$$

$$b^2_- = \frac{c^1_- \left(\omega(\omega\varsigma_{14-23} + ip_0\varsigma_{13+24}) + m\left(p_0\varsigma_{33-22} + i\omega\varsigma_{12-34}\right)\right)}{\left((m^2+\omega^2)\varsigma_2^2 + (\omega\varsigma_1 - im\varsigma_3)^2\right)},$$

$$c^1_+ = \frac{c^1_-\omega(\omega\varsigma_1 + ip_0\varsigma_2 - im\varsigma_3)}{\left((m^2+\omega^2)\varsigma_2^2 + (\omega\varsigma_1 - im\varsigma_3)^2\right)}, \quad c^2_+ = \frac{c^2_-\omega(\omega\varsigma_1 - ip_0\varsigma_2 - im\varsigma_3)}{\left((m^2+\omega^2)\varsigma_2^2 + (\omega\varsigma_1 - im\varsigma_3)^2\right)}.$$

4 Bound states

States localized in the vicinity of $x_3 = 0$ arise if $\bar{p}^2 - m^2 < 0$ and $\kappa(\bar{p}) = i|\kappa(\bar{p})|$ is imaginary. For them by $p_0 > m$, the dispersion relation $p_0^2 - p_1^2 - p_2^2 - \xi^{\pm}m^2 = 0$ is satisfied with

$$\xi^{\pm} = 1 - \left(\frac{\varsigma_1\varsigma_3 \pm \sqrt{\varsigma_2^2(1+\varsigma_4^2)}}{\varsigma_1^2 + \varsigma_2^2}\right)^2 .$$

By fixed parameters $\varsigma_i, i = 1\dots4$, there are two possible values for corresponding normalized electric current $\vec{J} = e\bar{\psi}\vec{\gamma}\psi/\bar{\psi}\psi$, which read

$$\vec{J}(x_3,\bar{p}) = \exp\left\{-\sqrt{(1-\xi^{\pm})^2 m}\,|x_3|\right\}\vec{j}(\bar{p}),\ \vec{j}(\bar{p}) = \frac{e\bar{\chi}_+(\bar{p})\vec{\gamma}\chi_+(\bar{p})}{\bar{\chi}_+(\bar{p})\chi_+(\bar{p})} = \frac{e\bar{\chi}_-(\bar{p})\vec{\gamma}\chi_-(\bar{p})}{\bar{\chi}_-(\bar{p})\chi_-(\bar{p})},$$

where e is a charge of the particle. Here, $j_3(\bar{p}) = 0$, and the parallel to the plane $x_3 = 0$ components of $\vec{j}(\bar{p})$ can be written as follows

$$j_1(\bar{p}) = -\frac{ep_1}{m\xi^{\pm}},\quad j_2(\bar{p}) = -\frac{ep_2}{m\xi^{\pm}}.$$

The coefficient $\xi^{\pm}$ at the corresponding values of the parameters ς_i can be positive or zero. For example, $\xi^+ = 0$ at $\varsigma_1 = \varsigma_3$, $\varsigma_2 = 1$, $\varsigma_4 = 0$, and $\xi^- = 0$ at $\varsigma_1 = -\varsigma_3$, $\varsigma_2 = 1$, $\varsigma_4 = 0$. If $\varsigma_2 = 0$, $|\varsigma_1| > \sqrt{1+\varsigma_3^2}$, and $\varsigma_4 = \pm\sqrt{\varsigma_1^2 - 1 - \varsigma_3^2}$, then $\xi^{\pm} > 0$.

The dispersion law $\bar{p}^2 = \xi^{\pm}m^2$ describes free particles with the effective mass $m\sqrt{\xi^{\pm}} < m$ in the $(2+1)$- dimensional space-time with two spatial coordinates and one temporal coordinate if $\xi^{\pm} > 0$. If $\xi^+ = 0$ or $\xi^- = 0$, then the corresponding particles are massless. The motion of such particles explains numerous effects in graphene.

5 Conclusions

We have constructed a model of the interaction of a spinor field with a material plane in the framework of Symanzik's approach. The action functional of the model includes the standard spinor Dirac action in (3+1) dimensional space-time and an additional contribution from the defect, concentrated in the part of space that it occupies (in considered case it was the plane $x_3 = 0$). The action contains three parameters characterizing the properties of the material of the plane. In the model, we calculated the characteristics of a Dirac particle scattering on the defect plane and also studied the properties of states localized near it.

The model and the obtained results can be used to describe theoretically processes of the interaction of electrons, with two-dimensional materials (graphene, thin films, sputter, sharp boundaries of a solid body). Simple modifications of the model allow taking the effects of the action of external electromagnetic fields into account.

Applying Symanzik's approach to QED enables constructing an universal model for a wide class of quantum macroscopic phenomena in two-dimensional materials. These models can prove useful in investigating the Hall effect, plasmon-polariton effects, nanophotonics, properties of topological insulators, thin films, sputters, and sharp boundaries [15].

Acknowledgments. This paper was supported in part by St. Petersburg State University (Grant No. 11.38.660.2013).

References

[1] N. N. Bogoliubov and D. V. Shirkov, *Quantum Fields*, Nauka, Moscow (1993);
C. Itzykson and J.-B. Zuber *Quantum Fields Theory*, McGraw-Hill, New York (1980).

[2] H. B. G Casimir, Proc. K. Ned. Akad. Wet B. **51**, 793-795 (1948).

[3] U. Mohideen and A. Roy, Phys. Rev. Lett. **81**, 4549-4552 (1998); arXiv:physics/9805038v2 (1998);
A. Roy, C.-Y. Lin and U. Mohideen, Phys. Rev. D **60**, 111101(R) (1999); arXiv:quant-ph/9906062v3 (1999).

[4] B. W. Harris, F. Chen and U. Mohideen, Phys. Rev. A **62**, 052109 (2000); arXiv:quant-ph/0005088v1 (2000).

[5] G. Bressi, G. Carugno, R. Onofrio and G. Ruoso, Phys. Rev. Lett. **88**, 041804 (2002); arXiv:quant-ph/ 0203002v1 (2002).

[6] M. Bordag, G. L. Klimchitskaya, U. Mohideen, and V. M. Mostepanenko *Advances in the Casimir Effect*, (Intl. Ser. Monogr. Phys., Vol. 145), Oxford Univ. Press, Oxford (2009); G. L. Klimchitskaya, U. Mohideen and V. M. Mostepanenko, Rev. Mod. Phys. **81**, 1827-1885 (2009); arXiv:0902.4022v1 [cond-mat.other] (2009).

[7] K. A. Milton, J. Phys. A: Math. Gen. **37**, R209-R277 (2004).

[8] K. Symanzik, Nucl.Phys. B **190**, 1-44 (1981). .

[9] V. N. Markov and Yu. M. Pis'mak, J. Phys. A: Math. Gen. **39**, 6525-6532 (2006); arXiv: hep-th/0505218v3 (2005).

[10] V. N. Marachevsky and Yu. M. Pis'mak, Phys. Rev. D **81**, 065005 (2010); arXiv:0907.1985v2 [hep-th] (2009).

[11] D.Yu. Pis'mak and Yu. M. Pis'mak, Theor. Math. Phys. **175**,816-826 (2013); Phys. Part. Nuclei **44**, 450-461 (2013); Theor. Math. Phys. **166**, 1423-1431 (2011).

[12] D.Yu. Pis'mak, Yu. M. Pis'mak and F. J. Wegner, Phys. Rev. E **92**, 013204 (2015); arXiv:1406.1598v1 [hep-th] (2014).

[13] I. V. Fialkovsky, V. N. Markov and Yu. M. Pis'mak, J. Phys. A: Math. Gen. **39**, 6357-6363 (2006); Int. J. Mod. Phys. A **21**, 2601-2616 (2006); arXiv:hep-th/0311236v2 (2003).

[14] D. Yu. Pismak, Yu. M. Pismak, Theor. Math. Phys. **184**, 3, 1329-1341 (2015).

[15] G. Bracco and B. Holst (Eds.), *Surface Science Techniques*, (Springer Ser. Surf. Sci., Vol. 51), Springer, Berlin (2013); M. L. Brongersma and P. G. Kik (Eds.), *Surface Plasmon Nanophotonics*, (Springer Ser. Opt. Sci., Vol. 131), Springer, Berlin (2007); S. A. Maier *Plasmonics: Fundamental and Applications*, Springer, Berlin (2007).

DOI: 10.1051/epjconf/201612605013

Search for space charge effects in the ICARUS T600 LAr-TPC

Marta Torti[1,2,a] on behalf of the ICARUS Collaboration

[1] *Department of Physics, University of Pavia, Italy*
[2] *Istituto Nazionale di Fisica Nucleare (INFN), Section of Pavia, Italy*

Abstract. Space charge in Liquid Argon Time Projection Chamber is due to the accumulation of positive ions, produced by ionizing tracks crossing the detector, which slowly flow toward the cathode. As a consequence, electric field distortions may arise, thus hindering the possibility to produce faithful 3D images of the ionizing events. The presence of space charge becomes relevant for large TPCs operating at surface or at shallow depths, where cosmic ray flux is high. These effects could interest the next phase of the ICARUS T600 detector, which will be deployed at shallow depths as a Far Detector for Short Baseline Neutrino experiment at FNAL dedicated to sterile neutrino searches. In 2001, the first ICARUS T600 module (T300) operated at surface in Pavia (Italy), recording cosmic ray data. In this work, a sample of cosmic muon tracks from the 2001 run was analyzed and results on space charge effects in LAr-TPCs are shown.

1 Introduction

ICARUS T600 is a Liquid Argon Time Projection Chamber (LAr-TPC) that took data from 2010 until 2013 at the Gran Sasso Laboratory (LNGS) in underground conditions [1]. The next phase of the detector foresees its deployment as a Far Detector for Short Baseline Neutrino (SBN) experiment at Fermi National Accelerator Laboratory (FNAL) in USA, dedicated to sterile neutrino searches [2]. This new program will involve three LAr-TPCs, placed at different positions from a neutrino source, the Booster Neutrino Beam (BNB): in addition to the T600 ($\sim$ 600 m), there will be MicroBoone (now running), placed at $\sim$ 470 m, and SBND, at $\sim$ 110 m. The goal of the SBN program is to give a 5σ answer to the LSND anomaly [3] , that can be explained with the presence of a fourth sterile neutrino state.

At FNAL the detector will not operate in underground environment, but it will be at shallow depths, with a high cosmic ray flux. In this condition, space charge effects could be relevant. Due to the accumulation in the drift region of positive ions (Ar^+), produced by ionizing tracks crossing the detector, electric field distortions may arise. In fact, ions survive in the drift region for a very long time because the positive ion mobility in LAr, $\mu_i \sim 1.6 \cdot 10^{-3}\ cm^2 s^{-1} V^{-1}$, is much smaller than the free electrons one, $\mu_e \sim 500\ cm^2 s^{-1} V^{-1}$.

An uniform electric field is fundamental to product faithful 3D images of the ionizing events, in order to have an uniform drift velocity and so the proportionality between drift time and drift coordinate.

[a] e-mail: marta.torti@pv.infn.it

2 The ICARUS T600 detector

ICARUS (Image Cosmic And Rare Underground Signal) T600 detector [4] consists of a large cryostat split into two identical, adjacent half-modules ($3.6 \times 3.9 \times 19.6\ m^3$ each) and filled with about 760 tons of ultra-pure liquid argon: each half-module houses two TPCs (left and right chambers) separated by a common cathode. A uniform electric field (E_D = 500 V/cm) is applied to the LAr bulk.
Charged particles, crossing LAr, produce ionization along their path. Thanks to the low transverse diffusion of charge in LAr, the images of the tracks, produced by ionization electron clouds, are preserved and, drifting along the electric field lines ($v_{drift} \sim 1.6$ mm /μs), are projected onto the anode, as illustrated in figure 1.
The TPC anode is made of three parallel planes of wires, 3 mm apart, facing the 1.5 m drift path (~ 1 ms drift time). Wires are oriented on each plane at a different angle (0°, ± 60°) with respect to the horizontal direction. Therefore, combining the wire/drift coordinates on each plane at a given drift time, a three-dimensional reconstruction of the ionizing event can be obtained, with a resolution of about 1 mm^3. The drift coordinate is obtained by the measurement of the absolute time of the ionizing event, with the scintillation light detection, combined with the electron drift velocity information.
Positive ions are drifted toward the cathode too, but with a velocity ~ 2000 times smaller than the free electrons one, corresponding to a drift time of the order of seconds. If events rate is too high, ions, while reaching the cathode, can interact with ionizing electrons from other events, perturbing their drift path.
A first physical run, with a single T300 module, was performed in 2001 in Pavia (Italy), collecting cosmic ray-induced events. The T600 detector successfully operated from 2010 to 2012 in Gran Sasso underground laboratories (LNGS), where it took data from the CNGS (CERN Neutrino to Gran Sasso) neutrino beam and with cosmic rays, to study neutrino oscillations. After the shut down of the beam, it collected data, focusing on cosmic rays, up to June 2013, when the phase of decommissioning began.

3 Data sample and method

To study possible space charge effects in the T600 detector, a sample of cosmic muon bundles has been considered. These data were taken during the test run in Pavia (Italy), in 2001, where only a T300 module operated at surface. Muon bundles have been used because they are composed of high energy parallel muons crossing entirely the drift region.
The whole sample included 129 muons. Each track was selected with a visual scanning procedure and delta rays were removed. First and last 1.5 cm of each track were not reconstructed, to avoid possible boundary dis-uniformities due to the presence of the wires and cathode planes.
Muon tracks observed to bend, following a parabola, could suggest the presence of possible positive ion accumulation: in fact, due to the dependence of the electron drift velocity on the electric field, the ionization electron arrival time T_{sc} on the anode in presence of space charge effects (apparent track in the figure 2), will be delayed with respect to the time T_u observed in the case of uniform electric field (ionizing track in figure 2).Thus, the bending parameter $\Delta T = T_{sc} - T_u$ is evaluated to search for possible delays.
For each muon track, T_{sc} is measured wire by wire, while T_u is evaluated sketching the muon as a straight line fitting the starting 5 cm and the last 5 cm of the track, where the electric field is fixed. Then, ΔT is evaluated for every point of the remaining part of the track.

4 Results and comparison with simulation

As a first attempt to identify the possible bending of the tracks, for each centimeter of the drift path, the average ΔT from all tracks was calculated; the mean value of the distribution was found to be

DOI: 10.1051/epjconf/201612605013

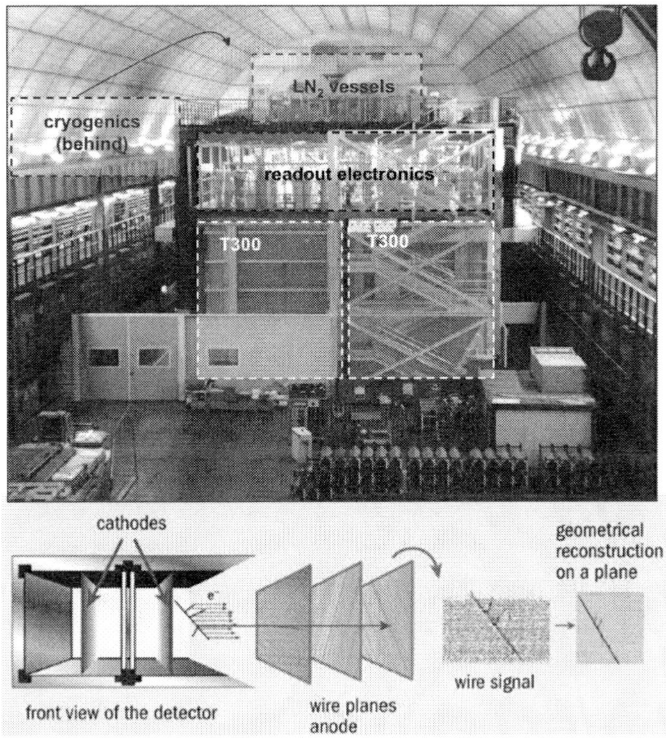

Figure 1. Top: ICARUS T600 detector in the Hall B of LNGS, with a schematic view of different parts. The two T300 modules are enlightened, such as the space occupied by the readout electronics and the cryogenics. **Bottom**: TPC working principle: when a charged particle cross the detector, electrons are drifted through the wire planes, where the charge is collected by the three wire planes. By appropriate voltage biasing, the first two , Induction1 and Induction2 planes, provide signals in a non-destructive way; finally the ionization charge is collected and measured on the last one, the Collection plane. Combining all the information, it is possible to have a 3D reconstruction of the track.

ΔT_{mean} = 0.18 cm, with a RMS of 0.67 cm. Then, ΔT as a function of the drift coordinate was compared with a numerical simulation (figure 3), in which possible electric field distortions caused by space charge were taken into account, due to high cosmic ray flux at surface. Data are in good agreement with the simulation, apart from the first and last sections, due to the ΔT calculation method. The maximum deviation from the linear fit of the tracks was found to be ΔT_{max} = 0.34 cm.
The easiest way to link these results to space charge effects is to analyze cosmic ray data collected underground at LNGS, where the cosmic ray flux is much smaller then at surface and these effects should not be present. A sample of 76 muon tracks was then analyzed as described above. As expected, in LNGS sample, the result is compatible with the absence of these effects: no deviation from the straight track was observed along the entire drift path (figure 4). The mean value of the global ΔT distribution for this sample is ΔT_{mean} = 0.04 cm with a RMS of 0.37 cm.
To better understand if the curvature measured in the Pavia surface sample was due to space charge or to other effects, the analysis was repeated by splitting data between the left (69 muons) and the right (60 muons) TPC chambers of the T300 module. The two chambers show a different distortion behavior (figure 5): in the right chamber the space charge effects were marginal, while in the left one the bending was more accentuated. This can not be explained in terms of space charge: different

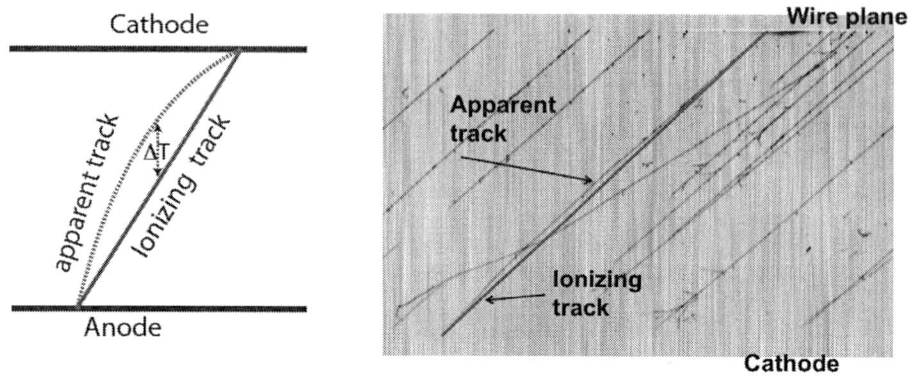

Figure 2. **Left**: schematic view of the bending parameter calculation. **Right**: example of bending parameter calculation on real data.

thermal insulation and LAr re-circulation conditions during the 2001 Pavia run could be responsible of the differences between the chambers.

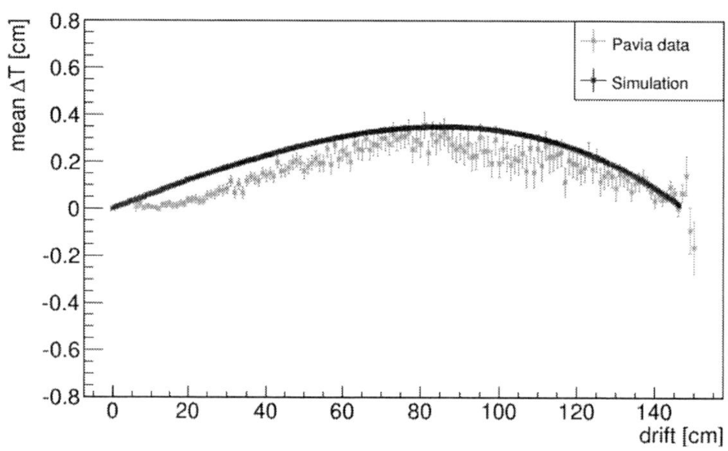

Figure 3. Comparison between Pavia data sample results and simulation of space charge effects due to high cosmic ray flux at surface.

5 Conclusions

Results here presented suggest the presence of small space charge effects inside the ICARUS T300 detector operated at surface in Pavia 2001 run, with a maximum track bending $\Delta T_{max} \sim 0.3$ mm. The analysis of a cosmic muon sample, collected underground at LNGS where this effect should not be present, produced a ΔT flat distribution, confirming that the track distortion observed at surface could be indeed due to space charge effects.
On the other hand, by analyzing the left and the right T300 chambers separately, the bending clearly

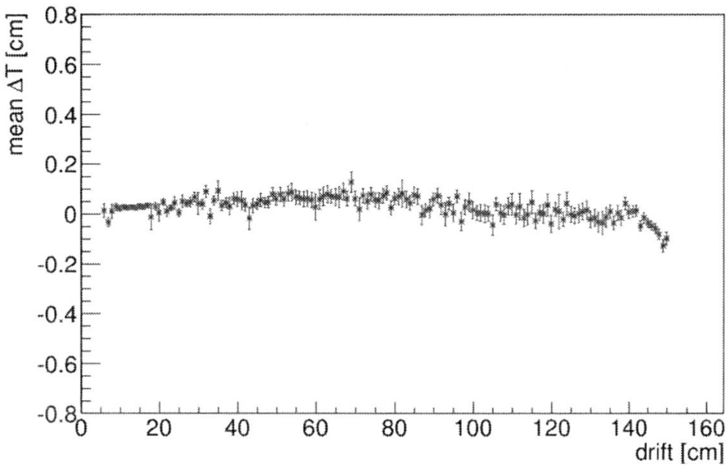

Figure 4. Results for LNGS data sample, collected at underground condition.

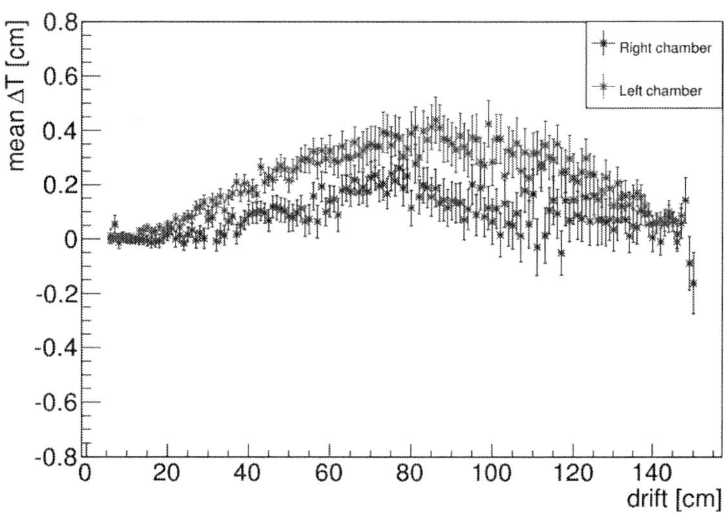

Figure 5. Results for Pavia data sample split between left and right TPC chamber of T300 module.

appears only in the left chamber: this behavior can not be directly ascribed to space charge effects, and more studies are required to fully understand the origin of this small surface muon track bending.

References

[1] C. Rubbia *et al.*, JINST **6** (2011) P07011.

[2] R. Acciari *et al.*, A proposal for a three detector Short Baseline Neutrino oscillation program in the Fermilab Booster Neutrino Beam, arXiv:1503.01520

[3] A. Agiular *et al.* (LSND Collaboration), Phys. Rev. D, **64**, 112007, 2001
[4] S. Amoruso et al.,Nucl. Inst. Meth., **A527** , 329-410, 2004

DOI: 10.1051/epjconf/201612605014

False vacuum as an unstable state

K. Urbanowski[1,a]

[1] *University of Zielona Góra, Institute of Physics, ul. Prof. Z. Szafrana 4a, 65–516 Zielona Góra, Poland.*

Abstract. Calculations performed within the Standard Model suggest that the electroweak vacuum is unstable if $M_H < 126$ GeV, (M_H is the mass of the Higgs particle). LHC discovery of the Higgs boson indicates that $M_H \simeq 125$ GeV. So the vacuum in our Universe may be unstable. We analyze properties of unstable vacuum states from the point of view of the quantum theory. At asymptotically late times the survival probability as a function of time t has an inverse power–like form. We show that at this time region the energy of the false vacuum states tends to the energy of the true vacuum state as $1/t^2$ for $t \to \infty$. This means that the energy density in the unstable vacuum state should have analogous properties and hence the cosmological constant $\Lambda = \Lambda(t)$ too. So Λ in the Universe with the unstable vacuum should have a form of the sum of the "bare" cosmological constant and of the term of a type $1/t^2$: $\Lambda(t) \equiv \Lambda^{bare} + \frac{d}{t^2}$, (where Λ^{bare} is the cosmological constant for the Universe with the true vacuum).

1 Introduction

Intensive studies of properties of the false vacuum and its possible decay began after the publication of pioneer papers by Coleman and his colleagues [1–3]. It was shown, in those papers, that even if the state of the early Universe is too cold to activate a "*thermal*" transition (via thermal fluctuations) from the local minimum of the energy density, which is separated from the absolute minimum by an effective potential barier, to the lowest energy (i.e. *"true vacuum"*) state, a quantum decay from the false vacuum to the true vacuum, may still be possible through a barrier penetration via macroscopic quantum tunneling. Not long ago, the decay of the false vacuum state in a cosmological context has attracted interest, especially in view of its possible relevance in the process of tunneling among the many vacuum states of the string landscape (a set of vacua in the low energy approximation of string theory). In many models the scalar field potential driving inflation has a multiple, low–energy minima or *"false vacuua"*. Then the absolute minimum of the energy density is the *"true vacuum"*.

Krauss and Dent analyzing a false vacuum decay [4, 5] pointed out that in eternal inflation, even though regions of false vacua by assumption should decay exponentially, gravitational effects force space in a region that has not decayed yet to grow exponentially fast. This effect causes that many false vacuum regions can survive up to the times much later than times when the exponential decay law holds. In the mentioned paper by Krauss and Dent the attention was focused on the possible behavior of the unstable false vacuum at very late times, where deviations from the exponential decay law become to be dominat.

[a] e-mail: K.Urbanowski@if.uz.zgora.pl;k.a.urbanowski@gmail.com

DOI: 10.1051/epjconf/201612605014

If to assume that the Standard Model describes correctly processes occurring in the Universe the problem of the vacuum states of the Universe becomes the problem of the electroweak vacuum. In the recent analysis [6] assuming the validity of the Standard Model up to Planckian energies it was shown that a Higgs mass $m_h < 126$ GeV implies that the electroweak vacuum is a metastable state. This means that a discussion of Higgs vacuum stability must be considered in a cosmological framework, especially when analyzing inflationary processes or the process of tunneling among many vacuum states of the string landscape. The discovery of the Higgs–like resonance at 125 — 126 GeV has intensified investigations of the problem of the instability the false vacuum state (see, eg., [6–9]).

The aim of this presentation is to analyze properties of the false vacuum state as an unstable state form the canonical decay times t up to asymptotically late times and to discuss the late time behavior of the energy of the false vacuum states.

2 Unstable states in short

Let $|M\rangle$ be an initial unstable state then the survival probability, $\mathcal{P}(t)$, equals $\mathcal{P}(t) = |a(t)|^2$, where $a(t)$ is the survival amplitude, $a(t) = \langle M|M;t\rangle$, $a(0) = 1$, and, $|M;t\rangle = \exp[-it\mathfrak{H}]\,|M\rangle$, $\mathfrak{H}$ is the total Hamiltonian of the system under considerations. (The units $\hbar = c = 1$ are used in this presentation). The spectrum, $\sigma(\mathfrak{H})$, of $\mathfrak{H}$ is assumed to be bounded from below, $\sigma(\mathfrak{H}) = [E_{min}, \infty)$ and $E_{min} > -\infty$. It is known that the amplitude $a(t)$, and thus the decay law $\mathcal{P}(t)$ of the unstable state $|M\rangle$, are completely determined by the density of the energy distribution function $\omega(\mathcal{E})$ for the system in this state

$$a(t) = \int_{Spec.(\mathfrak{H})} \omega(E)\, e^{-i\,E\,t}\, dE, \tag{1}$$

where $\omega(E) \geq 0$ for $E \geq E_{min}$ and $\omega(E) = 0$ for $E < E_{min}$. From the Paley–Wiener Theorem it follows that in such a case there must be [10] $|a(t)| \;\geq\; A\, \exp[-b\,t^q]$ for $|t| \to \infty$. Here $A > 0$, $b > 0$ and $0 < q < 1$. So the decay law $\mathcal{P}(t)$ of unstable states decaying in the vacuum can not be described by an exponential function of time t at late times, $t \to \infty$, and that for these lengths of time $\mathcal{P}(t)$ tends to zero as $t \to \infty$ slower than any exponential function of t. From models of the decay processes we know that $\mathcal{P}(t) \simeq \exp[-\varGamma_M^0 t]$, (where $\varGamma_M^0$ is the decay rate of the state$|M\rangle$), to a very high accuracy at the canonical decay times t: From t suitably later than the initial instant t_0 up to $t \gg \tau = 1/\varGamma_M^0$, τ is a lifetime), and shorter than $t = T$, where T is the crossover time and it denotes the time t for which the non–exponential deviations of $a(t)$ begin to dominate.

In the case of quasi–stationary (metastable) states sometimes it is convenient to express $a(t)$ in the following form: $a(t) = a_c(t) + a_{non}(t)$, where $a_c(t)$ is the exponential (canonical) part of $a(t)$. In other words:

$$a_c(t) = N\, \exp[-it(E_M^0 - \frac{i}{2}\,\varGamma_M^0)]. \tag{2}$$

Here E_M^0 is the energy of the system in the state $|M\rangle$ measured at the canonical decay times, N is the normalization constant, and $a_{non}(t)$ is the late time non–exponential part of $a(t)$). For times $t \sim \tau$: $|a_c(t)| \gg |a_{non}(t)|$. The crossover time T can be found by solving the following equation: $|a_c(t)|^2 = |a_{non}(t)|^2$. The amplitude $a_{non}(t)$ exhibits inverse power–law behavior at the late time region: $t \gg T$. Indeed, the integral representation (1) of $a(t)$ means that $a(t)$ is the Fourier transform of the energy distribution function $\omega(E)$. Using this fact we can find asymptotic form of $a(t)$ for $t \to \infty$. Results are rigorous. If to assume that $\lim_{E \to E_{min}+}\ \omega(E) \stackrel{\text{def}}{=} \omega_0 > 0$, and that derivatives $\omega^{(k)}(E)$, ($k = 0, 1, 2, \ldots, n$), are continuous in $[E_{min}, \infty)$, (that is if for $E > E_{min}$ all $\omega^{(k)}(E)$ are continuous and all the limits $\lim_{E \to E_{min}+} \omega^{(k)}(E)$ exist) and all these $\omega^{(k)}(E)$ are absolutely integrable functions then [11],

$$a(t) \underset{t \to \infty}{\sim} -\frac{i}{t}\, e^{-i\, E_{min} t} \sum_{k=0}^{n-1} (-1)^k \, (\frac{i}{t})^k \, \omega_0^{(k)} = a_{non}(t), \tag{3}$$

where $\omega_0^{(k)} \stackrel{\text{def}}{=} \lim_{E \to E_{min}+} \omega^{(k)}(E)$. For a more complicated form of the density $\omega(E)$ when

$$\omega(E) = (E - E_{min})^{\lambda} \, \eta(E) \;\in\; L_1(-\infty, \infty), \tag{4}$$

(where $0 \leq \lambda < 1$), and $\eta(E_{min}) \stackrel{\text{def}}{=} \eta_0 > 0$, and $\eta^{(k)}(E) = \frac{d}{dE}\, \eta(E)$, $(k = 0, 1, \ldots, n)$, exist and they are continuous in $[E_{min}, \infty)$, and limits $\lim_{E \to E_{min}+} \eta^{(k)}(E) \stackrel{\text{def}}{=} \eta_0^{(k)}$ exist, and $\lim_{E \to \infty} (E - E_{min})^{\lambda}\, \eta^{(k)}(E) = 0$ for all above mentioned k, then one finds that

$$a(t) \underset{t \to \infty}{\sim} (-1)\, e^{-iE_{min} t} \Big[\Big(-\frac{i}{t}\Big)^{\lambda+1} \Gamma(\lambda + 1)\, \eta_0 \tag{5}$$

$$+ \lambda \Big(-\frac{i}{t}\Big)^{\lambda+2} \Gamma(\lambda + 2)\, \eta_0^{(1)} + \ldots \Big] = a_{non}(t).$$

From (3), (5) it is seen that asymptotically late time behavior of the survival amplitude $a(t)$ depends rather weakly on a specific form of the energy density $\omega(E)$. The same concerns a decay curves $\mathcal{P}(t) = |a(t)|^2$. A typical form of a decay curve, that is the dependence on time t of $\mathcal{P}(t)$ when t varies from $t = t_0 = 0$ up to $t > 30\tau$ is presented in Figs (1), (2). Results presented in these Figures were obtained for the Breit–Wigner energy distribution function, $\omega(E) \equiv \omega_{BW} = \frac{N}{2\pi}\, \Theta(E - E_{min})\, \frac{\Gamma_M^0}{(E - E_M^0)^2 + (\Gamma_M^0/2)^2}$, where $\Theta(E)$ is the unit step function.

3 Instantaneous energy and instantaneous decay rate

From the amplitude $a(t)$ we can extract informations about the decay law $\mathcal{P}(t)$ of the state $|M\rangle$, that is about the decay rate Γ_M^0 of this state, as well as the energy E_M^0 of the system in this state. Indeed if $|M\rangle$ is an unstable (a quasi–stationary) state then $a(t) \cong \exp[-i(E_M^0 - \frac{i}{2}\Gamma_M^0)\, t] \equiv a_c(t)$ for $t \sim \tau$. So, in the case of quasi–stationary states there is

$$E_M^0 - \frac{i}{2}\Gamma_M^0 \equiv i\, \frac{\partial a_c(t)}{\partial t}\, \frac{1}{a_c(t)}, \quad (t \sim \tau). \tag{6}$$

The standard interpretation and understanding of the quantum theory and the related construction of our measuring devices are such that detecting the energy E_M^0 and decay rate Γ_M^0 one is sure that the amplitude $a(t)$ has the canonical form $a_c(t)$ and thus that the relation (6) occurs. Taking the above into account one can define the "effective Hamiltonian", h_M, for the one–dimensional subspace of states $\mathcal{H}_{\|}$ spanned by the normalized vector $|M\rangle$ as follows [11, 12]

$$h_M \stackrel{\text{def}}{=} i\, \frac{\partial a(t)}{\partial t}\, \frac{1}{a(t)} \stackrel{\text{def}}{=} \mathcal{E}_M(t) - \frac{i}{2}\, \gamma_M(t). \tag{7}$$

In general, h_M can depend on time t, $h_M \equiv h_M(t)$. One meets this effective Hamiltonian when one starts with the Schrödinger Equation for the total state space $\mathcal{H}$ and looks for the rigorous evolution equation for the distinguished subspace of states $\mathcal{H}_{\|} \subset \mathcal{H}$ [11, 12]. Thus, one finds the following

DOI: 10.1051/epjconf/201612605014

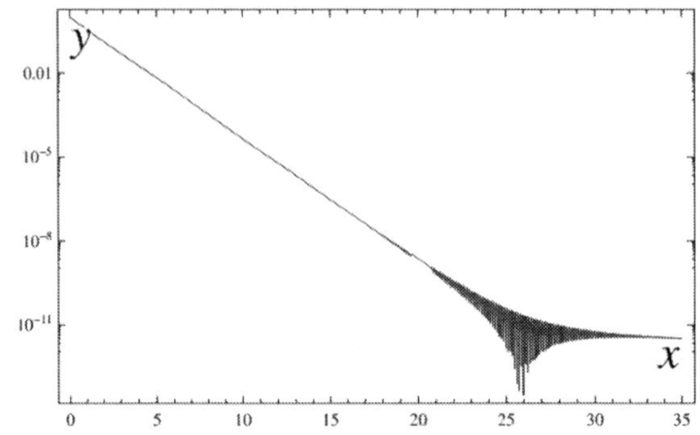

Figure 1. Axes: $y = \mathcal{P}(t)$ — the logarithmic scale, $x = t/\tau$. $\mathcal{P}(t)$ is the survival probability. The time t is measured as a multiple of the lifetime τ. The case $(E_M^0 - E_{min})/\Gamma_M^0 = 50$.

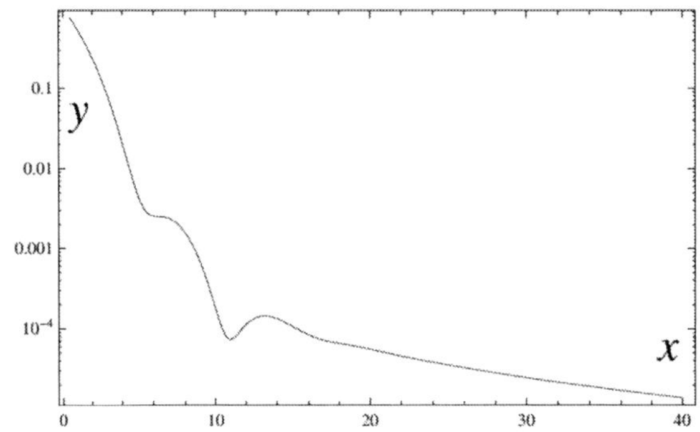

Figure 2. Axes: $y = \mathcal{P}(t)$ — the logarithmic scale, $x = t/\tau$. $\mathcal{P}(t)$ is the survival probability. The time t is measured as a multiple of the lifetime τ. The case $(E_M^0 - E_{min})/\Gamma_M^0 = 1$.

expressions for the energy and the decay rate of the system in the state $|M\rangle$ under considerations, to be more precise for the instantaneous energy $\mathcal{E}_M(t)$ and the instantaneous decay rate, $\gamma_M(t)$ [11],

$$\mathcal{E}_M \equiv \mathcal{E}_M(t) = \Re\,(h_M(t)), \qquad \gamma_M \equiv \gamma_M(t) = -2\,\Im\,(h_M(t)), \tag{8}$$

where $\Re\,(z)$ and $\Im\,(z)$ denote the real and imaginary parts of z respectively.

Starting from the asymptotic expression (5) for $a(t)$ and using (7) after some algebra one finds for times $t \gg T$ that

$$h_M(t)|_{t\to\infty} \simeq E_{min} + (-\frac{i}{t})\,c_1 \;+\; (-\frac{i}{t})^2\,c_2 \;+\; \ldots, \tag{9}$$

where $c_i = c_i^*, \;\; i = 1, 2, \ldots$; (coefficients c_i depend on $\omega(E)$). This last relation means that

$$\mathcal{E}_M(t) \;\simeq\; E_{min} \;-\; \frac{c_2}{t^2}\ldots, \qquad \gamma_M(t) \;\simeq\; 2\,\frac{c_1}{t} \;+\ldots, \quad (\text{for } t \gg T). \tag{10}$$

These properties take place for all unstable states which survived up to times $t \gg T$. Note that from (10) it follows that $\lim_{t\to\infty} \mathcal{E}_M(t) = E_{min}$ and $\lim_{t\to\infty} \gamma_M(t) = 0$.

For the density $\omega(E)$ such that $\omega(E_{min}) > 0$, that is for $a(t)$ having the asymptotic form given by (3) we have

$$c_1 = 1, \quad c_2 = \frac{\omega^{(1)}(E_{min})}{\omega(E_{min})}. \tag{11}$$

For the most general form (4) of the density $\omega(E)$ (i. e. for $a(t)$ having the asymptotic form given by (5)) we have

$$c_1 = \lambda + 1, \quad c_2 = (\lambda + 1)\,\frac{\eta^{(1)}(E_{min})}{\eta(E_{min})}. \tag{12}$$

The energy densities $\omega(E)$ considered in quantum mechanics and in quantum field theory can be described by $\omega(E)$ of the above analyzed forms both.

A typical, general form of

$$\kappa(t) = \frac{\mathcal{E}_M(t) - E_{min}}{E_M^0 - E_{min}}, \tag{13}$$

as a function of time t varying from $t = t_0 = 0$ up to $t > T$ is presented in panels B of Figs (3), (4). These results were obtained for the model considered in the previous Section and correspond with Figs (1), (2) respectively. The crossover time T, that is the time region where fluctuations of $\mathcal{P}(t)$ and $\mathcal{E}_M(t)$ take place depend on the value of the parameter $s_0 = (E_M^0 - E_{min})/\Gamma_M^0$ in the model considered: The smaller s_0 the shorter T.

4 Cosmological applications

Krauss and Dent in their paper [4] mentioned earlier made a hypothesis that some false vacuum regions do survive well up to the time T or later. Let $|M\rangle = |0\rangle^{false}$, be a false, $|0\rangle^{true}$ – a true, vacuum states and E_0^{false} be the energy of a state corresponding to the false vacuum measured at the canonical decay time and E_0^{true} be the energy of true vacuum (i.e. the true ground state of the system). As it is seen from the results presented in previous Section, the problem is that the energy of those false vacuum regions which survived up to T and much later differs from E_0^{false} [13].

So, if one assumes that $E_0^{true} \equiv E_{min}$ and $E_0^{false} = E_M^0$ and takes into account results of the previous Section including those in Figs (3), (4) then one can conclude that the energy of the system in the false vacuum state has the following general properties:

$$\mathcal{E}_0^{false}(t) = E_0^{true} + \Delta E \;\cdot\; \kappa(t), \tag{14}$$

where $\Delta E = E_0^{false} - E_0^{true}$ and $\kappa(t) \simeq 1$ for $t \sim \tau_0^{false} < T$. The quantity $\kappa(t)$ is a fluctuating function of t at $t \sim T$ and $\kappa(t) \propto \frac{1}{t^2}$ for $t \gg T$.

There is at asymptotically late times, $t \gg T$,

$$\mathcal{E}_0^{false}(t) \simeq E_0^{true} - \frac{c_2}{t^2} \ldots \neq E_0^{false}, \tag{15}$$

where $c_2 = c_2^*$ and it can be positive or negative depending on the model considered, and $\gamma_0^{false}(t) \simeq +2\,c_1/t \ldots$ for $t \gg T$. These two last properties of the false vacuum states mean that

$$\mathcal{E}_0^{false}(t) \to E_0^{true} \;\text{ and }\; \gamma_0^{false}(t) \to 0 \;\text{ as }\; t \to \infty. \tag{16}$$

Going from quantum mechanics to quantum field theory one should take into account among others a volume factors so that survival probabilities per unit volume per unit time should be considered.

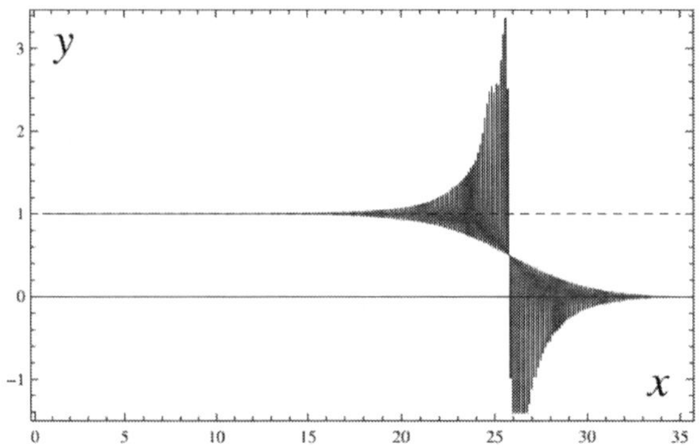

Figure 3. Axes: $y = \kappa(t)$, $x = t/\tau$. The difference of energies $(\mathcal{E}_M(t) - E_{min})$ is measured as a multiple of the difference $(E_M^0 - E_{min})$. The case $(E_M^0 - E_{min})/\Gamma_M^0 = 50$.

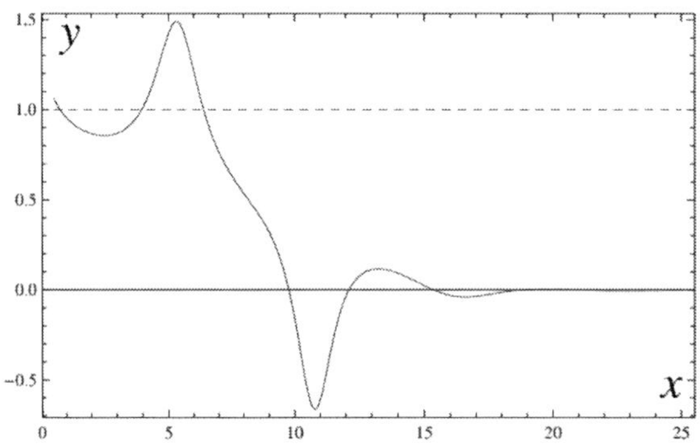

Figure 4. Axes: $y = \kappa(t)$, $x = t/\tau$. The difference of energies $(\mathcal{E}_M(t) - E_{min})$ is measured as a multiple of the difference $(E_M^0 - E_{min})$. The case $(E_M^0 - E_{min})/\Gamma_M^0 = 1$.

The standard false vacuum decay calculations shows that the same volume factors should appear in both early and late time decay rate estimations (see Krauss and Dent [4, 14]). This means that the calculations of cross–over time T can be applied to survival probabilities per unit volume. For the same reasons within the quantum field theory the quantity $\mathcal{E}_M(t)$ can be replaced by the energy per unit volume $\rho_M(t) = \mathcal{E}_M(t)/V$ because these volume factors V appear in the numerator and denominator of the formula (7) for $h_M(t)$. This conclusion seems to hold when considering the energy $\mathcal{E}_0^{false}(t)$ of the system in false vacuum state $|0\rangle^{false}$ because Universe is assumed to be homogeneous and isotropic at suitably large scales. So at such scales to a sufficiently good accuracy we can extract properties of the energy density $\rho_0^{false} = E_M^0/V = E_0^{false}/V$ of the system in the false vacuum state $|0\rangle^{false}$ from properties of the energy $\mathcal{E}_0^{false}(t)$ of the system in this state defining $\rho_0^{false}(t)$ as $\rho_0^{false}(t) = \mathcal{E}_0^{false}(t)/V$. Therefore one can conclude that in the case of a meta–stable (unstable or decaying, false) vacuum the following important property of $\kappa(t)$ holds:

$$\kappa(t) \equiv \frac{\rho_0^{false}(t) - \rho_{bare}}{\rho_0^{false} - \rho_{bare}}, \tag{17}$$

where $\rho_{bare} = E_{min}/V$ is the energy density of the true (bare) vacuum. From the last equation the conclusion follows that

$$\rho_0^{false}(t) - \rho_{bare} = (\rho_0^{false} - \rho_{bare})\,\kappa(t). \tag{18}$$

Hence, because for $t < T$ there is $\kappa(t) = 1$, one conludes that $\rho_0^{false}(t) = \rho_0^{false}$ for $t < T$, whereas for $t \gg T$ there is

$$\rho_0^{false}(t) - \rho_{bare} = (\rho_0^{false} - \rho_{bare})\,\kappa(t) \simeq \pm\, d_2\,\frac{\hbar^2}{t^2}, \quad (t \gg T), \tag{19}$$

where $d_2 = d_2^*$. Analogous relations (with the same $\kappa(t)$) take place for $\Lambda(t) = \frac{8\pi G}{c^2}\,\rho(t)$, or $\Lambda(t) = 8\pi G\,\rho(t)$ in $\hbar = c = 1$ units:

$$\Lambda(t) - \Lambda_{bare} = (\Lambda_0 - \Lambda_{bare})\,\kappa(t), \tag{20}$$

or,

$$\Lambda(t) = \Lambda_{bare} + (\Lambda_0 - \Lambda_{bare})\,\kappa(t). \tag{21}$$

We can expect that Λ_0 equals to the cosmological constant calculated within quantum field theory. From the last equation it is seen that for $t < T$, $\Lambda(t) \simeq \Lambda_0$ for $t < T$ because $\kappa(t < T) \simeq 1$. Now if to assume that Λ_0 corresponds to the value of the cosmological "constant" Λ calculated within the quantum field theory, than swe may expect that $\frac{\Lambda_o}{\Lambda_{bare}} \geq 10^{120}$, (see [15]), which allows one to write down Eq. (21) as follows

$$\Lambda(t) \simeq \Lambda_{bare} + \Lambda_0\,\kappa(t). \tag{22}$$

From the above analysis it follows that for $t \gg T$ there should be (see (19))

$$\Lambda_0\,\kappa(t) \simeq 8\pi G\,\frac{d_2}{t^2} \equiv \pm\frac{\alpha^2}{t^2}, \quad \text{for } (t \gg T). \tag{23}$$

5 Final Remarks

The use of the parametrization following from quantum theoretical treatment of metastable vacuum states can explain why the cosmologies with the time–dependent cosmological constant $\Lambda(t)$ are worth considering and may help to explain the cosmological constant problem [16, 17]. The time dependent Λ of the type $\Lambda(t) = \Lambda_{bare} + \frac{\alpha^2}{t^2}$ was used eg. in [18]. Earlier analogous form of Λ was obtained in [19], where the invariance under scale transformations of the generalized Einstein equations was studied. Such a time dependence of Λ was postulated also in [20] as the result of the analysis of the large numbers hypothesis. The cosmological model with time dependent Λ of the above postulated form was studied also in [21] and in much more recent papers.

The nice feature and maybe even the advantage of the formalism presented in Sect. 4 is that in the case of the universe with metastable (false) vacuum if one takes into account that the decay of this unstable vacuum state is the quantum decay process then it emerges automatically that there have to exist the true ground state of the system that is the true (or bare) vacuum with the minimal energy, $E_{min} > -\infty$, of the system corresponding to him and equivalently, $\rho_{bare} = E_{min}/V$, or Λ_{bare}. What is more, in this case such $\Lambda \equiv \Lambda(t)$ emerges that at suitable late times it has the form described by relations (22), (23). In such a case the function $\kappa(t)$ given by the relation (13) describes time dependence for all times t of the energy density $\rho_M(t)$ or the cosmological "constant" $\Lambda_M(t)$ and it general form is presented in Figs (3), (4). Note that results presented in Sections 2 — 4 are rigorous. The formalism mentioned was applied in [14, 15], where cosmological models with $\Lambda(t) = \Lambda_{bare} \pm \frac{\alpha^2}{t^2}$ were studied: The most promising result is reported in [15] where using the parametrization following from the mentioned quantum theoretical analysis of the decay process of the unstable vacuum state an attempt was made to explain the small today's value of the cosmological constant Λ. This shows that formalism and the approach described in this paper and in [14, 15] is promising and can help to solve the cosmological constant and other cosmological problems and it needs further studies, especially if to take into account the LHC result concerning the mass of the Higgs boson and cosmological consequences of this result.

Acknowledgments: The work was supported in part by the NCN grant No DEC-2013/09/B/ST2/03455.

References

[1] S. Coleman, Phys. Rev. **D 15**, 2929 (1977).
[2] C.G. Callan and S. Coleman, Phys. Rev. **D 16**, 1762 (1977).
[3] S. Coleman and F. de Lucia, Phys. Rev. **D 21**, 3305 (1980).
[4] L. M. Krauss, J. Dent, Phys. Rev. Lett., **100**, 171301 (2008).
[5] S. Winitzki, Phys. Rev. **D 77**, 063508 (2008).
[6] G. Degrassi, *et al.*,JHEP **1208**, 098, (2012).
[7] A. Kobakhidze, A. Spencer–Smith, Phys. Lett. **B 722**, 130,(2013).
[8] J. Elias–Miro, *et al.*, Phys. Lett. **B 709**, 222, (2012).
[9] Wei Chao, *et al.*, Phys. Rev. **D 86**, 113017, (2012).
[10] L. A. Khalfin, Zh. Eksp. Teor. Fiz. **33**, 1371 (1957); [Sov. Phys. JETP **6**, 1053 (1958)].
[11] K. Urbanowski, *Eur. Phys. J. D*, **54**, 25, (2009).
[12] K. Urbanowski, Cent. Eur. J. Phys. **7**, 696 (2009).
[13] K. Urbanowski, *Phys. Rev. Lett.*, **107**, 209001 (2011).
[14] K. Urbanowski, M. Szydłowski, *AIP Conf. Proc.* **1514**, 143 (2013); doi: 10.1063/1.4791743.
[15] M. Szydłowski, Phys. Rev. **D 91**, 123538 (2015).
[16] S. Weinberg, Rev. Mod. Phys. **61**, 1 (1989).
[17] S. M. Carroll, The Cosmological Constant, *Living Rev. Relativity*, 3, (2001), 1; http://www.livingreviews.org/lrr-2001-1.
[18] J. L. Lopez and D. V. Nanopoulos, Mod. Phys. Lett. **A 11**, 1 (1996).
[19] V. Canuto and S. H. Hsieh, Phys. rev. Lett. **39**, 429, (1977).
[20] Y. K. Lau and S. J. Prokhovnik, Aust. J. Phys., **39**, 339, (1986).
[21] M. S. Berman, Phys. Rev. **D 43**, 1075, (1991).

EPJ Web of Conferences **126**, 05015 (2016) DOI: 10.1051/epjconf/201612605015
ICNFP 2015

Effect of entanglement on dual condensates at finite isospin chemical potential from an entangled Polyakov-Nambu-Jona-Lasinio model

Zhao Zhang[1,a] and Qing Miao[1]

[1]*School of Mathematics and Physics, North China Electric Power University, Beijing 102206, China*

Abstract. The dual observables as order parameters for deconfinement transition are investigated at finite isospin chemical potential μ_I in an entangled Polyakov-Nambu-Jona-Lasinio model of QCD. Compared to PNJL, the four quark vertice with entanglement between the Polyakov loop and the chiral and pion condensates are considered in this model. We mainly concentrate on the role of this entanglement on the thermal behavior of the dressed polyakov loop and the dual pion condensate for $\mu_I > m_\pi/2$. We find that the T-dependences of these dual condensates are qualitatively consistent with the PNJL results. Due to the entanglement, the critical temperatures determined by the quark and pion condensates and their corresponding dual partners get close to that extracted from the Polyakov loop.

1 Introduction

The main phenomena in QCD at finite temperature and density are the chiral restoration and deconfining phase transitions. In the chiral limit, the standard order parameter for chiral transition is the quark condensate. However, it is conceptually difficult to define an order parameter for deconfinement in QCD. Usually, the expectation value of the Polyakov loop (PL) is used to indicate the quark deconfining transition. This quantity is a true order parameter for center symmetry in pure Yang-Mills theory. Nevertheless, this symmetry is badly broken by the light dynamical quarks in QCD. These two (effective) order parameters had been extensively studied in lattice QCD at finite T and zero density and it suggests that two pseudo critical temperatures are very close to each other [1, 2].

Recently, some dual observables are introduced as the new order parameters for center symmetry by using the twisted boundary conditions for quarks [3–6]. Especially, it is demonstrated in the formalism of lattice QCD [3, 4] that the dressed Polyakov loop (DPL) interpolates between the chiral condensate and the thin PL. The studies from the functional methods [7–10] and effective models [11–13] also suggest the DPL shows the order parameter-like behavior like the thin PL. In principle, one can construct many dual observables which transform in the same way as the thin PL under the center transformation [5, 6]. They are true order parameters for deconfinement in the heavy quark limit $m \to \infty$.

[a]e-mail: zhaozhang@pku.edu.cn

However, the center symmetry is seriously broken in QCD since the light quark masses are very small. Then, a question naturally arises: to what extent do these quantities still contain the confinement information? Recently, it is demonstrated in the NJL model that the rapid rise of DPL near T_c is totally driven by the chiral transition [14]. The following study based on several variants of NJL with (possible) confining elements suggests that the rapid rise of DPL with T still happens in the chiral transition region and no effect related to the change of the confining properties of the propagator is observed [15]. The authors thus conclude that the DPL obtained in these models is not an appropriate order parameter for deconfinement. In addition, the test of dual observables as order parameters has been extended to finite isospin chemical potential by considering the pion condensation [16]. It is found in PNJL that not only the dressed Polyakov loop, but also the dual pion condensate (DPC) are quite sensitive to the chiral dynamics and pion condensation rather than the thin PL [16]. Thus it may be difficult to distinguish the deconfinement transition from the dual observables , at least in the NJL-type models.

We stress that the PNJL results in [16] may not really true in QCD. In other words, the possible subtle relationship between the center symmetry breaking and chiral transition existing in QCD may not be totally captured by PNJL, even the center symmetry is considered in this model. Note that it derived in Ref.[17] that a particular non-local version of PNJL can re regarded as a low energy limit of QCD. Inspired by this work, the entangled PNJL (EPNJL) model has been proposed [18]. Compared to PNJL, the four-quark interaction is dependent on the Polyakov-loop in EPNJL. So the coupling or entanglement between the PL and quark condensate in EPNJL is somewhat stronger than that in PNJL. The motivation of this paper is to investigate the role of such an enhanced entanglement on the thermal properties of the dual condensates at finite isospin chemical potential for $\mu_I > m_\pi/2$.

2 EPNJL model at finite isospin chemical potential with twisted boundary conditions

2.1 Two flavor EPNJL model and the model parameters

We adopt the following lagrangian of two-flavor EPNJL model

$$\begin{aligned} \mathcal{L} &= \bar{\psi}\left(i\gamma_\mu D^\mu + \gamma_0\hat{\mu} - \hat{m}_0 - i\lambda\gamma_5\tau_1\right)\psi \\ &\quad + g_s(\Phi)\left[\left(\bar{\psi}\psi\right)^2 + \left(\bar{\psi}i\gamma_5\vec{\tau}\psi\right)^2\right] - g_v^s\left(\bar{\psi}\gamma_\mu\psi\right)^2 \\ &\quad - g_v^v\left(\bar{\psi}\vec{\tau}\gamma_\mu\psi\right)^2 - \mathcal{U}\left(\Phi,\bar{\Phi},T\right), \end{aligned} \tag{1}$$

which is an improved version of PNJL used in [16]. The last term in (1) is the pure gauge sector described by the Polyakov loop Φ, which is invariant under the $Z(3)$ transformation. As mentioned, the scalar coupling g_s is explicitly Φ dependent in this model. According to the suggestion in [18], the form of g_s below is employed in this paper:

$$g_s(\Phi) = g_s[1 - \alpha_1\Phi\bar{\Phi} - \alpha_2(\Phi^3 + \bar{\Phi}^3)], \tag{2}$$

which respects the chiral, P, C and the extended $Z(3)$ symmetries. Following [18], both α_1 and α_2 are fixed as 0.2 which can reproduce the available lattice data. The model parameters other than $g_s(\Phi)$ in (1) are all adopted from [16] in this paper. Note that g_s in (2) is also the same as that in [16]. The possible Φ-dependence of $g_v^{v(s)}$ is ignored in our calculation.

2.2 EPNJL with twisted boundary conditions for $\mu_I > m_\pi/2$

It is well known that the charged pion condensation appears for zero and low temperatures for $\mu_I > m_\pi/2$ [19], where μ_I is the isospin chemical potential. The study based on PNJL in [20] suggests that relation between the PL and pion condensate at finite temperature for $\mu_I > m_\pi/2$ is quite similar to that between the PL and quark condensate at zero isospin density. In addition, the test of the dual condensates as order parameters for center symmetry has also been extended to finite isospin chemical potential in [16], where the PNJL model is adopted. Besides the extensively studied DPL, another simple dual observable, namely the dual pion condensate is proposed and studied in [16]. These two dual condensates are defined as

$$\Sigma_\sigma^{(1)} = -\int_0^{2\pi} \frac{d\phi}{2\pi} e^{-i\phi}\sigma(\phi), \tag{3}$$

and

$$\Sigma_\pi^{(1)} = -\int_0^{2\pi} \frac{d\phi}{2\pi} e^{-i\phi}\pi(\phi), \tag{4}$$

respectively. Note that $\sigma(\phi)$ ($\pi(\phi)$) is the generalized quark (pion) condensate for the twisted angle ϕ.

When considering the Φ-dependence of g_s, the two energy gaps in [16] are modified as

$$M = m - 2g_s(\Phi)\sigma(\phi), \quad N = \lambda - 2g_s(\Phi)\pi(\phi). \tag{5}$$

We still use μ' and μ'_I to indicate the shifted quark and isospin chemical potentials when considering the vector interaction

$$\mu' = \mu - 2g_v^s(\rho_u + \rho_d), \quad \mu'_I = \mu_I - 2g_v^v(\rho_u - \rho_d), \tag{6}$$

where $\rho_{u(d)}$ is the u(d) quark density. The thermal potential of EPNJL at the mean field level still takes the form of Eq.(14) in [16].

Under the generalized boundary condition, the modified quark chemical potential μ' in the thermal potential should be replaced by $iT(\phi - \pi)$ [3, 6, 11], which is nothing but an effective imaginary chemical potential. Strictly speaking, the μ' for $\phi \neq \pi$ should also contain the density-related contribution $2g_v^s(\rho_u + \rho_d)$ even the real μ is zero. This is because the imaginary chemical potential also leads to a nonzero baryon number density. It has been shown in [11] that the coupling g_v^s only has significant effect on $\Sigma_\sigma^{(1)}$ for $T > 1.5T_c$ in PNJL. Since we are only interested in the thermal behavior of dual observables near and below T_c, the correction $2g_v^s(\rho_u + \rho_d)$ is ignored in our calculation. Note that the μ'_I is still real and keeps the form as (6). As in [16], we only conduct the calculation for the case with $\mu_I > m_\pi/2$ and zero μ.

3 Results and discussion

3.1 ϕ-dependence of quark and pion condensates

The generalized quark condensate as function of ϕ for $\mu_I = 100\,\text{MeV}$ at different temperatures is shown in Fig. 1. We see that the shapes of $|\sigma(\phi)|$ for temperatures below and above T_c^χ are quite different: for $T = 190\,\text{MeV}$ and $230\,\text{MeV}$ (or $T > T_c^\chi$), the quark condensates are concave lines with $|\sigma(\pi)| < |\sigma(0)|$; but for $T = 150\,\text{MeV}$ and $170\,\text{MeV}$ (or $T < T_c^\chi$), they are

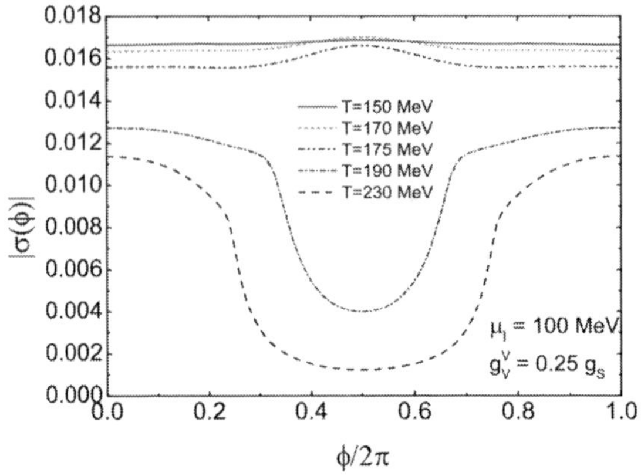

Figure 1. The twisted angle dependence of the quark condensate $\sigma(\phi)$ at $\mu_I = 100$ MeV for different temperatures

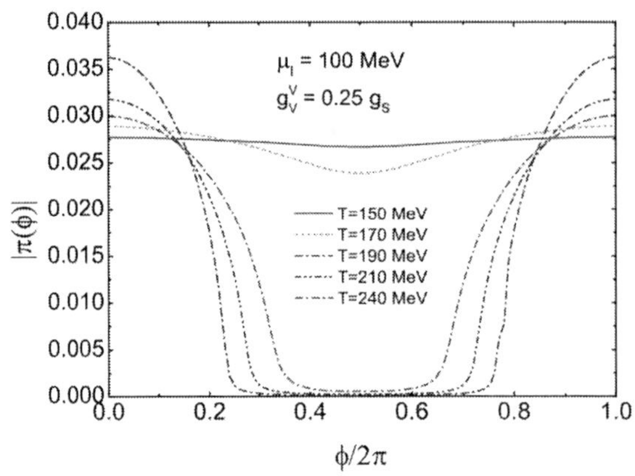

Figure 2. The twisted angle dependence of the pion condensate $\pi(\phi)$ at $\mu_I = 100$ MeV for different temperatures

convex ones with $|\sigma(\pi)| > |\sigma(0)|$. All this is very similar to Fig.1.a in [16] but quite different from what obtained in [3, 8, 11], where only concave curves emerge for vanishing μ and μ_I.

The generalized pion condensate as function of ϕ for $\mu_I = 100\,\mathrm{MeV}$ at different temperatures is shown in Fig. 2. In contrast, Fig. 2 shows that all lines of $|\pi(\phi)|$ at different fixed T are concave curves. We see that in the fermionic-like region (namely the area for $\phi \sim \pi$), $|\pi(\phi)|$ decreases with T but it increases with T in the bosonic-like region (namely the area for ϕ near zero or 2π). In addition, the curve of $|\pi(\phi)|$ becomes more flat with decreasing T. All those is very similar to the ϕ-dependence of the quark condensate obtained for zero μ and μ_I [3, 8, 11], which is qualitatively consistent with the PNJL result, namely Fig.1.b in [16].

3.2 Thermal behaviors of dual condensates for $\mu_I > m_\pi/2$

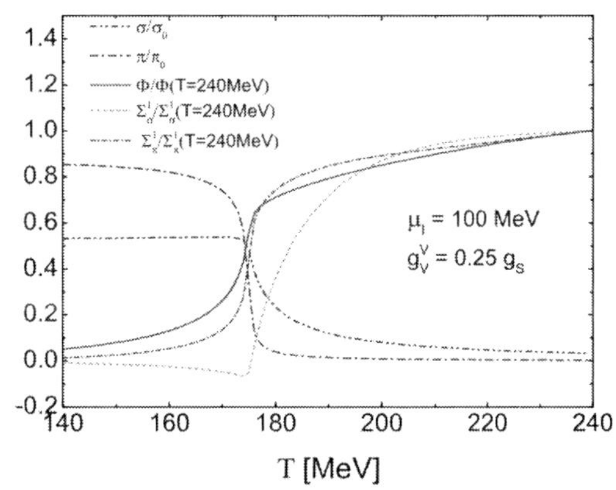

Figure 3. The temperature dependence of the normalized conventional ployakov loop, quark and pion condensates and their corresponding dual parters at $\mu_I = 100$ MeV.

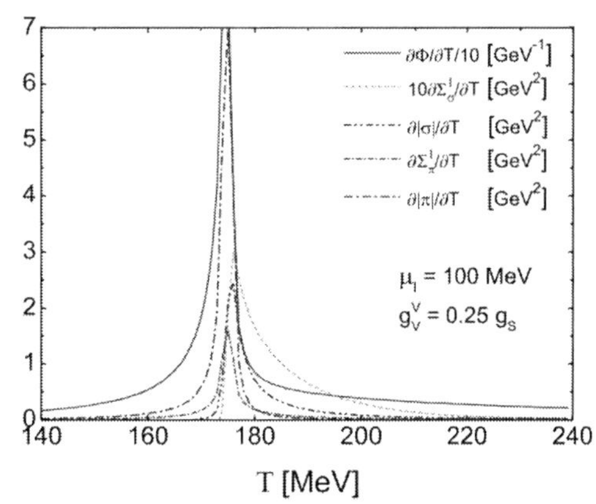

Figure 4. The temperature dependence of the T-derivatives of the conventional ployakov loop, quark and pion condensates and their corresponding dual parters at $\mu_I = 100$ MeV.

Two dual condensates and other three (pseudo-) order parameters as functions of T for $\mu_I = 100\,\text{MeV}$ are shown in Figs. 3. The quark and pion condensates are obtained with physical boundary condition $\phi = \pi$, which are normalized by $\sigma_0 = \sigma(T = 0, \mu = 0, \mu_I = 0)$; the dual condensates and PL are normalized by their corresponding values at $T = 240\,\text{MeV}$. We mainly focus on the thermal behaviors of these quantities near the phase transitions.

Figure. 3 shows that $|\pi(\phi = \pi)|$ (Φ) decreases (increases) monotonically with T [1], but $|\sigma(\phi = \pi)|$ first increases (slowly) up to $T \sim 175$ MeV and then decreases. These results are also qualitatively agreement with that obtained in [16] within the PNJL model.

Consistent with Fig. 2, Fig. 3 indicates that the normalized DPC really behaves like an order parameter for center symmetry: analogous to the DPL obtained in [3, 8, 11], it keeps rather small value in low temperature region and gradually becomes larger with T. Like the thin PL, the DPC increases monotonically with T. However, the normalized DPL in Fig. 3 shows abnormal thermal behavior, which first reduces with T (up to $T \sim 175$MeV) and then raises. Similar to the PNJL result, we see that the DPL is quite sensitive to the T-dependence of the quark condensate rather than the PL near the phase transition region. In addition, the DPL and DPC rise rapidly near the chiral transition and the melting of the pion condensate, respectively. This is also consistent with the PNJL results.

Several susceptibilities defined as the T derivatives of the quantities displayed in Fig. 3 are shown in Fig. 4. As in [14, 16], the peak of a susceptibility is used to locate the critical temperature. Fig. 4 indicates that the PL susceptibility has one peak, which indicates $T_c^P =$ 175 MeV. Different from Fig. 5 in [16], the other susceptibilities also only have one peak, which is very close to T_c^P. We see that the highest peak of $\partial\Sigma_\sigma{}^1/\partial T$ coincides with that of $\partial|\sigma|/\partial T$, and the corresponding critical temperatures $T_c^{d\sigma}$ and T_c^σ are slightly lager than T_c^P. The coincidence of $T_c^{d\sigma}$ and T_c^σ is consistent with [14, 16]. In addition, Fig. 4 also shows that the critical temperatures $T_c^{d\pi}$ and T_c^π (extracting from $\partial\Sigma_\pi{}^1/\partial T$ and $\partial|\pi|/\partial T$, respectively) also coincide, which almost take the same value as T_c^P. So the double peaks of the susceptibilities other than that of the thin PL one observed in PNJL [16] disappear totally due to the enhanced entanglement between the PL and quark and pion condensates in EPNJL model. However, we still notice that there exists a slight difference between T_c^P and $T_c^{d\sigma}$ (or $T_c^{d\pi}$).

4 Conclusion

The investigation of the thermal behavior of dual condensates at finite isospin chemical potential for $\mu_I > m_\pi$ has been performed in an entangled PNJL model. Compared to the PNJL model, the entanglement of the center symmetry breaking and the chiral transition or the evaporation of the pion condensate with T is enhanced in EPNJL. We mainly focus on the role of this strengthened entanglement on the thermal properties of the dual observables by considering the pion condensation.

We find that the twisted angle dependences of the quark and pion condensate obtained in EPNJL are still qualitatively consistent with the PNJL results in [16] and the dual pion condensate also exhibits the order parameter like thermal behavior as the thin PL. Due to the competition between the quark and pion condensates, the abnormal thermal behavior of the dressed PL found in [16] is still observed in EPNJL, which suggests that this dual condensate is also quite sensitive to the chiral transition rather than the deconfinement described by the PL. On the other hand, the critical temperatures extracted from the quark and pion condensates and their dual partners are very close to the one determined by the thin PL. This is quite different from the PNJL results.

Anyway, whether the dual observables carry the deconfinement information is an interesting question. Which result of these two models is more close to QCD can be determined

[1] This is also observed in [20] and other chiral model studies [21]. The reason for such an anomaly is that the quantity $\sqrt{\sigma^2 + \pi^2}$ always decreases with T but $|\pi|$ drops more quickly near the phase transition since λ is zero but m is finite.

by the lattice simulation with physical quark masses, which is free from the sign problem at finite isospin chemical potential.

Acknowledgements

Z.Z. was supported by the NSFC (No.11275069).

References

[1] Y. Aoki, S. Borsanyi, S. Durr, Z. Fodor, S. D. Katz, S. Krieg and K. K. Szabo, JHEP **0906**, 088 (2009).
[2] S. Borsanyi *et al.* [Wuppertal-Budapest Collaboration], JHEP **1009**, 073 (2010).
[3] E. Bilgici, F. Bruckmann, C. Gattringer and C. Hagen, Phys. Rev. D **77**, 094007 (2008).
[4] E. Bilgici, F. Bruckmann, J. Danzer, C. Gattringer, C. Hagen, E. M. Ilgenfritz and A. Maas, Few Body Syst. **47**, 125 (2010).
[5] B. Zhang, F. Bruckmann, C. Gattringer, Z. Fodor and K. K. Szabo, AIP Conf. Proc. **1343**, 170 (2011) [arXiv:1012.2314 [hep-lat]].
[6] J. Braun, L. M. Haas, F. Marhauser and J. M. Pawlowski, Phys. Rev. Lett. **106**, 022002 (2011).
[7] C. S. Fischer, Phys. Rev. Lett. **103**, 052003 (2009).
[8] C. S. Fischer and J. A. Mueller, Phys. Rev. D **80**, 074029 (2009).
[9] C. S. Fischer, J. Luecker and J. A. Mueller, Phys. Lett. B **702**, 438 (2011).
[10] C. S. Fischer, A. Maas and J. A. Muller, Eur. Phys. J. C **68**, 165 (2010).
[11] K. Kashiwa, H. Kouno and M. Yahiro, Phys. Rev. D **80**, 117901 (2009).
[12] R. Gatto and M. Ruggieri, Phys. Rev. D **82**, 054027 (2010).
[13] T. K. Mukherjee, H. Chen and M. Huang, Phys. Rev. D **82**, 034015 (2010).
[14] S. Benic, Phys. Rev. D **88**, 077501 (2013).
[15] F. Marquez, A. Ahmad, M. Buballa and A. Raya, Phys. Lett. B **747**, 529 (2015) [arXiv:1504.06730 [nucl-th]].
[16] Z. Zhang and Q. Miao, arXiv:1507.07224 [hep-ph].
[17] K. I. Kondo, Phys. Rev. D **82**, 065024 (2010) doi:10.1103/PhysRevD.82.065024 [arXiv:1005.0314 [hep-th]].
[18] Y. Sakai, T. Sasaki, H. Kouno and M. Yahiro, Phys. Rev. D **82**, 076003 (2010) doi:10.1103/PhysRevD.82.076003 [arXiv:1006.3648 [hep-ph]].
[19] D. T. Son and M. A. Stephanov. Phys. Rev. Lett. **86** , 592 (2001); Phys. At. Nucl. **64**, 834 (2001).
[20] Z. Zhang and Y. -X. Liu, Phys. Rev. C **75**, 064910 (2007).
[21] L. y. He, M. Jin and P. f. Zhuang, Phys. Rev. D **71**, 116001 (2005).

AUTHOR INDEX

AUTHOR INDEX

AUTHOR INDEX

AUTHOR INDEX

EDP Sciences – Web of Conferences
17, Avenue du Hoggar
Parc d'Activité de Courtabouf
BP 112
F-91944 Les Ulis Cedex A
France

ISBN 978-1-5108-3238-1
9 781510 832381

CURRAN ASSOCIATES INC.
proceedings.com